PROPERTY OF CASB / ENGINEERING
PROPRIÉTÉ du LABORATOIRE d'INGÉNIERIE

Contents

D1710598

Metals Handbook®
Ninth Edition

Volume 17
Nondestructive Evaluation and Quality Control

Prepared under the direction of the
ASM INTERNATIONAL Handbook Committee

Joseph R. Davis, Manager of Handbook Development
Kathleen M. Mills, Manager of Book Production
Steven R. Lampman, Technical Editor
Theodore B. Zorc, Technical Editor
Heather J. Frissell, Editorial Supervisor
George M. Crankovic, Editorial Coordinator
Alice W. Ronke, Assistant Editor
Jeanne Patitsas, Word Processing Specialist
Karen Lynn O'Keefe, Word Processing Specialist

Robert L. Stedfeld, Director of Reference Publications

Editorial Assistance
Lois A. Abel
Wendy L. Jackson
Robert T. Kiepura
Penelope Thomas
Nikki D. Wheaton

PROPERTY OF CASB / ENGINEERING
PROPRIÉTÉ du LABORATOIRE d'INGÉNIERIE

METALS PARK, OH 44073

Copyright © 1989
by
ASM INTERNATIONAL
All rights reserved

No part of this book may be reproduced, stored in a retrieval system, or transmitted, in any form or by any means, electronic, mechanical, photocopying, recording, or otherwise, without the written permission of the copyright owner.

First printing, September 1989

Metals Handbook is a collective effort involving thousands of technical specialists. It brings together in one book a wealth of information from world-wide sources to help scientists, engineers, and technicians solve current and long-range problems.

Great care is taken in the compilation and production of this Volume, but it should be made clear that no warranties, express or implied, are given in connection with the accuracy or completeness of this publication, and no responsibility can be taken for any claims that may arise.

Nothing contained in the Metals Handbook shall be construed as a grant of any right of manufacture, sale, use, or reproduction, in connection with any method, process, apparatus, product, composition, or system, whether or not covered by letters patent, copyright, or trademark, and nothing contained in the Metals Handbook shall be construed as a defense against any alleged infringement of letters patent, copyright, or trademark, or as a defense against liability for such infringement.

Comments, criticisms, and suggestions are invited, and should be forwarded to ASM INTERNATIONAL.

Library of Congress Cataloging-in-Publication Data

ASM INTERNATIONAL

Metals handbook.

Vol. 17: Prepared under the direction of
the ASM INTERNATIONAL Handbook Committee.
Includes bibliographies and indexes.
Contents: v. 1. Properties and selection—
v. 2. Properties and selection—nonferrous alloys
and pure metals—[etc.]—v. 17. Nondestructive
evaluation and quality control.

1. Metals—Handbooks, manuals, etc.
I. ASM Handbook Committee.
II. ASM INTERNATIONAL. Handbook Committee.
TA459.M43 1978 669 78-14934
ISBN 0-87170-007-7 (v. 1)
SAN 204-7586

Printed in the United States of America

Foreword

Volume 17 of *Metals Handbook* is a testament to the growing importance and increased sophistication of methods used to nondestructively test and analyze engineered products and assemblies. For only through a thorough understanding of modern techniques for nondestructive evaluation and statistical analysis can product reliability and quality control be achieved and maintained.

As with its 8th Edition predecessor, the aim of this Volume is to provide detailed technical information that will enable readers to select, use, and interpret nondestructive methods. Coverage, however, has been significantly expanded to encompass advances in established techniques as well as introduce the most recent developments in computed tomography, digital image enhancement, acoustic microscopy, and electromagnetic techniques used for stress analysis. In addition, material on quantitative analysis and statistical methods for design and quality control (subjects covered only briefly in the 8th Edition) has been substantially enlarged to reflect the increasing utility of these disciplines.

Publication of Volume 17 also represents a significant milestone in the history of ASM INTERNATIONAL. This Volume completes the 9th Edition of *Metals Handbook*, the largest single source of information on the technology of metals that has ever been compiled. The magnitude, respect, and success of this unprecedented reference set calls for a special tribute to its many supporters. Over the past 13 years, the ASM Handbook Committee has been tireless in its efforts, ASM members have been unflagging in their support, and the editorial staff devoted and resourceful. Their efforts, combined with the considerable knowledge and technical expertise of literally thousands of authors, contributors, and reviewers, have resulted in reference books which are comprehensive in coverage and which set the highest standards for quality. To all these men and women, we extend our most sincere appreciation and gratitude.

Richard K. Pitler
President,
ASM INTERNATIONAL

Edward L. Langer
Managing Director,
ASM INTERNATIONAL

The Ninth Edition of Metals Handbook
is dedicated to the memory of
TAYLOR LYMAN, A.B. (Eng.), S.M., Ph.D.
(1917-1973)
Editor, Metals Handbook, 1945-1973

Preface

The subject of nondestructive examination and analysis of materials and manufactured parts and assemblies is not new to *Metals Handbook*. In 1976, Volume 11 of the 8th Edition—*Nondestructive Inspection and Quality Control*—provided what was at that time one of the most thorough overviews of this technology ever published. Yet in the relatively short time span since then, tremendous advances and improvements have occurred in the field—so much so that even the terminology has evolved. For example, in the mid-1970s the examination of an object or material that did not render it unfit for use was termed either nondestructive testing (NDT) or nondestructive inspection (NDI). Both are similar in that they involve looking at (or through) an object to determine either a specific characteristic or whether the object contains discontinuities, or flaws.

The refinement of existing methods, the introduction of new methods, and the development of quantitative analysis have led to the emergence of a third term over the past decade, a term representing a more powerful tool. With nondestructive evaluation (NDE), a discontinuity can be classified by its size, shape, type, and location, allowing the investigator to determine whether or not the flaw is acceptable. The title of the present 9th Edition volume was modified to reflect this new technology.

Volume 17 is divided into five major sections. The first contains four articles that describe equipment and techniques used for qualitative part inspection. Methods for both defect recognition (visual inspection and machine vision systems) and dimensional measurements (laser inspection and coordinate measuring machines) are described.

In the second section, 24 articles describe the principles of a wide variety of nondestructive techniques and their application to quality evaluation of metallic, composite, and electronic components. In addition to detailed coverage of more commonly used methods (such as magnetic particle inspection, radiographic inspection, and ultrasonic inspection), newly developed methods (such as computed tomography, acoustic microscopy, and speckle metrology) are introduced. The latest developments in digital image enhancement are also reviewed. Finally, a special six-page color section illustrates the utility of color-enhanced images.

The third section discusses the application of nondestructive methods to specific product types, such as one-piece products (castings, forgings, and powder metallurgy parts) and assemblies that have been welded, soldered, or joined with adhesives. Of particular interest is a series of reference radiographs presented in the article "Weldments, Brazed Assemblies, and Soldered Joints" that show a wide variety of weld discontinuities and how they appear as radiographic images.

The reliability of discontinuity detection by nondestructive methods, referred to as quantitative NDE, is the subject of the fourth section. Following an introduction to this rapidly maturing discipline, four articles present specific guidelines to help the investigator determine the critical discontinuity size that will cause failure, how long a structure containing a discontinuity can be operated safely in service, how a structure can be designed to prevent catastrophic failure, and what inspections must be performed in order to prevent failure.

The final section provides an extensive review of the statistical methods being used increasingly for design and quality control of manufactured products. The concepts of statistical process control, control charts, and design of experiments are presented in sufficient detail to enable the reader to appreciate the importance of statistical analysis and to organize and put into operation a system for ensuring that quality objectives are met on a consistent basis.

This Volume represents the collective efforts of nearly 200 experts who served as authors, contributors of case histories, or reviewers. To all we extend our heartfelt thanks. We would also like to acknowledge the special efforts of Thomas D. Cooper (Wright Research & Development Center, Wright-Patterson Air Force Base) and Vicki E. Panhuise (Allied-Signal Aerospace Company, Garrett Engine Division). Mr. Cooper, a former Chairman of the ASM Handbook Committee, was instrumental in the decision to significantly expand the material on quantitative analysis. Dr. Panhuise organized the content and recruited all authors for the section "Quantitative Nondestructive Evaluation." Such foresight and commitment from Handbook contributors over the years has helped make the 9th Edition of *Metals Handbook*—all 17 volumes and 15,000 pages—the most authoritative reference work on metals ever published.

The Editors

Officers and Trustees of ASM INTERNATIONAL

Richard K. Pitler
President and Trustee
Allegheny Ludlum Corporation
(retired)

Klaus M. Zwilsky
Vice President and Trustee
National Materials Advisory Board
National Academy of Sciences

William G. Wood
Immediate Past President and Trustee
Kolene Corporation

Robert D. Halverstadt
Treasurer
AIMe Associates

Trustees

John V. Andrews
Teledyne Allvac

Edward R. Burrell
Inco Alloys International, Inc.

Stephen M. Copley
University of Southern California

H. Joseph Klein
Haynes International, Inc.

Gunvant N. Maniar
Carpenter Technology Corporation

Larry A. Morris
Falconbridge Limited

William E. Quist
Boeing Commercial Airplanes

Charles Yaker
Howmet Corporation

Daniel S. Zamborsky
Consultant

Edward L. Langer
Managing Director

Members of the ASM Handbook Committee (1988–1989)

Dennis D. Huffman
(Chairman 1986–; Member 1983–)
The Timken Company

Roger J. Austin (1984–)
ABARIS

Roy G. Baggerly (1987–)
Kenworth Truck Company

Robert J. Barnhurst (1988–)
Noranda Research Centre

Peter Beardmore (1986–1989)
Ford Motor Company

Hans Borstell (1988–)
Grumman Aircraft Systems

Gordon Bourland (1988–)
LTV Aerospace and Defense Company

Robert D. Caligiuri (1986–1989)
Failure Analysis Associates

Richard S. Cremisio (1986–1989)
Rescorp International, Inc.

Gerald P. Fritzke (1988–)
Metallurgical Associates

J. Ernesto Indacochea (1987–)
University of Illinois at Chicago

John B. Lambert (1988–)
Fansteel Inc.

James C. Leslie (1988–)
Advanced Composites Products and Technology

Eli Levy (1987–)
The De Havilland Aircraft Company of Canada

Arnold R. Marder (1987–)
Lehigh University

John E. Masters (1988–)
American Cyanamid Company

L.E. Roy Meade (1986–1989)
Lockheed-Georgia Company

Merrill L. Minges (1986–1989)
Air Force Wright Aeronautical Laboratories

David V. Neff (1986–)
Metaullics Systems

Dean E. Orr (1988–)
Orr Metallurgical Consulting Service, Inc.

Ned W. Polan (1987–1989)
Olin Corporation

Paul E. Rempes (1986–1989)
Williams International

E. Scala (1986–1989)
Cortland Cable Company, Inc.

David A. Thomas (1986–1989)
Lehigh University

Kenneth P. Young (1988–)
AMAX Research & Development

Previous Chairmen of the ASM Handbook Committee

R.S. Archer
(1940–1942) (Member, 1937–1942)

L.B. Case
(1931–1933) (Member, 1927–1933)

T.D. Cooper
(1984–1986) (Member, 1981–1986)

E.O. Dixon
(1952–1954) (Member, 1947–1955)

R.L. Dowdell
(1938–1939) (Member, 1935–1939)

J.P. Gill
(1937) (Member, 1934–1937)

J.D. Graham
(1966–1968) (Member, 1961–1970)

J.F. Harper
(1923–1926) (Member, 1923–1926)

C.H. Herty, Jr.
(1934–1936) (Member, 1930–1936)

J.B. Johnson
(1948–1951) (Member, 1944–1951)

L.J. Korb
(1983) (Member, 1978–1983)

R.W.E. Leiter
(1962–1963) (Member, 1955–1958, 1960–1964)

G.V. Luerssen
(1943–1947) (Member, 1942–1947)

G.N. Maniar
(1979–1980) (Member, 1974–1980)

J.L. McCall
(1982) (Member, 1977–1982)

W.J. Merten
(1927–1930) (Member, 1923–1933)

N.E. Promisel
(1955–1961) (Member, 1954–1963)

G.J. Shubat
(1973–1975) (Member, 1966–1975)

W.A. Stadtler
(1969–1972) (Member, 1962–1972)

R. Ward
(1976–1978) (Member, 1972–1978)

M.G.H. Wells
(1981) (Member, 1976–1981)

D.J. Wright
(1964-1965) (Member, 1959-1967)

Policy on Units of Measure

By a resolution of its Board of Trustees, ASM INTERNATIONAL has adopted the practice of publishing data in both metric and customary U.S. units of measure. In preparing this Handbook, the editors have attempted to present data in metric units based primarily on Système International d'Unités (SI), with secondary mention of the corresponding values in customary U.S. units. The decision to use SI as the primary system of units was based on the aforementioned resolution of the Board of Trustees and the widespread use of metric units throughout the world.

For the most part, numerical engineering data in the text and in tables are presented in SI-based units with the customary U.S. equivalents in parentheses (text) or adjoining columns (tables). For example, pressure, stress, and strength are shown both in SI units, which are pascals (Pa) with a suitable prefix, and in customary U.S. units, which are pounds per square inch (psi). To save space, large values of psi have been converted to kips per square inch (ksi), where 1 ksi = 1000 psi. The metric ton (kg $\times$ 10^3) has been shown in megagrams (Mg). Some strictly scientific data are presented in SI units only.

To clarify some illustrations, only one set of units is presented on artwork. References in the accompanying text to data in the illustrations are presented in both SI-based and customary U.S. units. On graphs and charts, grids corresponding to SI-based units appear along the left and bottom edges. Where appropriate, corresponding customary U.S. units appear along the top and right edges.

Data pertaining to a specification published by a specification-writing group may be given in only the units used in that specification or in dual units, depending on the nature of the data. For example, the typical yield strength of aluminum sheet made to a specification written in customary U.S. units would be presented in dual units, but the sheet thickness specified in that specification might be presented only in inches.

Data obtained according to standardized test methods for which the standard recommends a particular system of units are presented in the units of that system. Wherever feasible, equivalent units are also presented. Some statistical data may also be presented in only the original units used in the analysis.

Conversions and rounding have been done in accordance with ASTM Standard E 380, with attention given to the number of significant digits in the original data. For example, an annealing temperature of 1570 °F contains three significant digits. In this case, the equivalent temperature would be given as 855 °C; the exact conversion to 854.44 °C would not be appropriate. For an invariant physical phenomenon that occurs at a precise temperature (such as the melting of pure silver), it would be appropriate to report the temperature as 961.93 °C or 1763.5 °F. In some instances (especially in tables and data compilations), temperature values in °C and °F are alternatives rather than conversions.

The policy on units of measure in this Handbook contains several exceptions to strict conformance to ASTM E 380; in each instance, the exception has been made in an effort to improve the clarity of the Handbook. The most notable exception is the use of g/cm^3 rather than kg/m^3 as the unit of measure for density (mass per unit volume).

SI practice requires that only one virgule (diagonal) appear in units formed by combination of several basic units. Therefore, all of the units preceding the virgule are in the numerator and all units following the virgule are in the denominator of the expression; no parentheses are required to prevent ambiguity.

Authors and Reviewers

D.A. Aldrich
Idaho National Engineering Laboratory
EG&G Idaho, Inc.
Craig E. Anderson
Nuclear Energy Services
Gerald L. Anderson
American Gas and Chemical Company
Glenn Andrews
Ultra Image International
Bruce Apgar
DuPont NDT Systems
R.A. Armistead
Advanced Research and Applications Corporation
Ad Asead
University of Michigan at Dearborn
David Atherton
Queen's University
Yoseph Bar-Cohen
Douglas Aircraft Company
McDonnell Douglas Corporation
R.C. Barry
Lockheed Missiles & Space Company, Inc.
John Bassart
Iowa State University
George Becker
DuPont NDT Systems
R.E. Beissner
Southwest Research Institute
Alan P. Berens
University of Dayton Research Institute
Harold Berger
Industrial Quality, Inc.
Henry Bertoni
Polytechnic University of New York
R.A. Betz
Lockheed Missiles & Space Company, Inc.
Craig C. Biddle
United Technologies Research Center
Kelvin Bishop
Tennessee Valley Authority
Carl Bixby
Zygo Corporation
Dave Blackham
Consultant
Gilbert Blake
Wiss, Janney, Elstner Associates
James Bolen
Northrop Aircraft Division
Jim Borges
Intec Corporation
J.S. Borucki
Ardox Inc.

Richard Bossi
Boeing Aerospace Division
The Boeing Company
Byron Brendan
Battelle Pacific Northwest Laboratories
G.L. Burkhardt
Southwest Research Institute
Paul Burstein
Skiametrics, Inc.
Willard L. Castner
National Aeronautics and Space Administration
Lyndon B. Johnson Space Center
V.S. Cecco
Atomic Energy of Canada, Ltd.
Chalk River Nuclear Laboratories
Francis Chang
General Dynamics Corporation
Tsong-how Chang
University of Wisconsin, Milwaukee
F.P. Chiang
Laboratory for Experimental Mechanics Research
State University of New York at Stony Brook
D.E. Chimenti
Wright Research & Development Center
Wright-Patterson Air Force Base
P. Cielo
National Research Council of Canada
Industrial Materials Research Institute
T.N. Claytor
Los Alamos National Laboratory
J.M. Coffey
CEGB Scientific Services
J.F. Cook
Idaho National Engineering Laboratory
EG&G Idaho, Inc.
Thomas D. Cooper
Wright Research & Development Center
Wright-Patterson Air Force Base
William D. Cowie
United States Air Force
Aeronautical Systems Division
L.D. Cox
General Dynamics Corporation
Robert Cribbs
Folsom Research Inc.
J.P. Crosson
Lucius Pitkin, Inc.
Darrell Cutforth
Argonne National Laboratory
William Dance
LTV Missiles & Electronics Group

Steven Danyluk
University of Illinois
Oliver Darling
Spectrum Marketing, Inc.
E.A. Davidson
Wright Research & Development Center
Wright-Patterson Air Force Base
Vance Deason
EG&G Idaho, Inc.
John DeLong
Philadelphia Electric Company
Michael J. Dennis
NDE Systems & Services
General Electric Company
Richard DeVor
University of Illinois at Urbana-Champaign
Robert L. Ditz
GE Aircraft Engines
General Electric Company
Kevin Dooley
University of Minnesota
Thomas D. Dudderar
AT&T Bell Laboratories
Charles D. Ehrlich
National Institute of Standards & Technology
Ralph Ekstrom
University of Nebraska Lincoln
Robert Erf
United Technologies Research Center
K. Erland
United Technologies Corporation
Pratt & Whitney Group
J.L. Fisher
Southwest Research Institute
Colleen Fitzpatrick
Spectron Development Laboratory
William H. Folland
United Technologies Corporation
Pratt & Whitney Group
Joseph Foster
Texas A&M University
Kenneth Fowler
Panametrics, Inc.
E.M. Franklin
Argonne National Laboratory
Argonne—West
Larry A. Gaylor
Dexter Water Management Systems
David H. Genest
Brown & Sharpe Manufacturing Company
Dennis German
Ford Motor Company

Ron Gerow
Consultant
Scott Giacobbe
GPU Nuclear
Robert S. Gilmore
General Electric Research and Development Center
J.N. Gray
Center for NDE
Iowa State University
T.A. Gray
Center for NDE
Iowa State University
Robert E. Green, Jr.
The Johns Hopkins University
Arnold Greene
Micro/Radiographs Inc.
Robert Grills
Ultra Image International
Donald Hagemaier
Douglas Aircraft Company
McDonnell Douglas Corporation
John E. Halkias
General Dynamics Corporation
Grover L. Hardy
Wright Research & Development Center
Wright-Patterson Air Force Base
Patrick G. Heasler
Battelle Pacific Northwest Laboratories
Charles J. Hellier
Hellier Associates, Inc.
Edmond G. Henneke
Virginia Polytechnic Institute and State University
B.P. Hildebrand
Failure Analysis Associates, Inc.
Howard E. Housermann
ZETEC, Inc.
I.C.H. Hughes
BCIRA International Centre
Phil Hutton
Battelle Pacific Northwest Laboratories
Frank Iddings
Southwest Research Institute
Bruce G. Isaacson
Bio-Imaging Research, Inc.
W.B. James
Hoeganaes Corporation
D.C. Jiles
Iowa State University
Turner Johnson
Brown & Sharpe Manufacturing Company
John Johnston
Krautkramer Branson
William D. Jolly
Southwest Research Institute
M.H. Jones
Los Alamos National Laboratory
Gail Jordan
Howmet Corporation
William T. Kaarlela
General Dynamics Corporation
Robert Kalan
Naval Air Engineering Center
Paul Kearney
Welch Allyn Inc.
William Kennedy
Canadian Welding Bureau
Lawrence W. Kessler
Sonoscan, Inc.
Thomas G. Kincaid
Boston University
Stan Klima
NASA Lewis Research Center
Kensi Krzywosz
Electric Power Research Institute
Nondestructive Evaluation Center
David Kupperman
Argonne National Laboratory
H. Kwun
Southwest Research Institute
J.W. Lincoln
Wright Research & Development Center
Wright-Patterson Air Force Base
Art Lindgren
Magnaflux Corporation
D. Lineback
Measurements Group, Inc.
Charles Little
Sandia National Laboratories
William Lord
Iowa State University
D.E. Lorenzi
Magnaflux Corporation
Charles Loux
GE Aircraft Engines
General Electric Company
A. Lucero
ZETEC, Inc.
Theodore F. Luga
Consultant
William McCroskey
Innovative Imaging Systems, Inc.
Ralph E. McCullough
Texas Instruments, Inc.
William E.J. McKinney
DuPont NDT Systems
Brian MacCracken
United Technologies Corporation
Pratt & Whitney Group
Ajit K. Mal
University of California, Los Angeles
A.R. Marder
Energy Research Center
Lehigh University
Samuel Marinov
Western Atlas International, Inc.
George A. Matzkanin
Texas Research Institute
John D. Meyer
Tech Tran Consultants, Inc.
Morey Melden
Spectrum Marketing, Inc.
Merlin Michael
Rockwell International
Carol Miller
Wright Research & Development Center
Wright-Patterson Air Force Base
Ron Miller
MQS Inspection, Inc.
Richard H. Moore
CMX Systems, Inc.
Thomas J. Moran
Consultant
John J. Munro III
RTS Technology Inc.
N. Nakagawa
Center for NDE
Iowa State University
John Neuman
Laser Technology, Inc.
H.I. Newton
Babcock & Wilcox
G.B. Nightingale
General Electric Company
Mehrdad Nikoonahad
Bio-Imaging Research, Inc.
R.C. O'Brien
Hoeganaes Corporation
Kanji Ono
University of California, Los Angeles
Vicki Panhuise
Allied-Signal Aerospace Company
Garrett Engine Division
James Pellicer
Staveley NDT Technologies, Inc.
Robert W. Pepper
Textron Specialty Materials
C.C. Perry
Consultant
John Petru
Kelly Air Force Base
Richard Peugeot
Peugeot Technologies, Inc.
William Plumstead
Bechtel Corporation
Adrian Pollock
Physical Acoustic Corporation
George R. Quinn
Hellier Associates, Inc.
Jay Raja
Michigan Technological University
Jack D. Reynolds
General Dynamics Corporation
William L. Rollwitz
Southwest Research Institute
A.D. Romig, Jr.
Sandia National Laboratories
Ward D. Rummel
Martin Marietta Astronautics Group
Charles L. Salkowski
National Aeronautics and Space Administration
Lyndon B. Johnson Space Center
Thomas Schmidt
Consultant
Gerald Scott
Martin Marietta Manned Space Systems
D.H. Shaffer
Westinghouse Electric Corporation
Research and Development Center
Charles N. Sherlock
Chicago Bridge & Iron Company

Thomas A. Siewert
National Institute of Standards and Technology

Peter Sigmund
Lindhult & Jones, Inc.

Lawrence W. Smiley
Reliable Castings Corporation

James J. Snyder
Westinghouse Electric Company
Oceanic Division

Doug Steele
GE Aircraft Engines
General Electric Company

John M. St. John
Caterpillar, Inc.

Bobby Stone Jr.
Kelly Air Force Base

George Surma
Sundstrand Aviation Operations

Lyndon J. Swartzendruber
National Institute of Standards and Technology

Richard W. Thams
X-Ray Industries, Inc.

Graham H. Thomas
Sandia National Laboratories

R.B. Thompson
Center for NDE
Iowa State University

Virginia Torrey
Welch Allyn Inc.

James Trolinger
Metro Laser

Michael C. Tsao
Ultra Image International

Glen Wade
University of California, Santa Barbara

James W. Wagner
The Johns Hopkins University

Henry J. Weltman
General Dynamics Corporation

Samuel Wenk
Consultant

Robert D. Whealy
Boeing Commercial Airplane Company

David Willis
Allison Gas Turbine Division
General Motors Corporation

Charles R. Wojciechowski
NDE Systems and Services
General Electric Company

J.M. Wolla
U.S. Naval Research Laboratory

John D. Wood
Lehigh University

Nello Zuech
Vision Systems International

Contents

Inspection Equipment and Techniques

Visual Inspection

VISUAL INSPECTION is a nondestructive testing technique that provides a means of detecting and examining a variety of surface flaws, such as corrosion, contamination, surface finish, and surface discontinuities on joints (for example, welds, seals, solder connections, and adhesive bonds). Visual inspection is also the most widely used method for detecting and examining surface cracks, which are particularly important because of their relationship to structural failure mechanisms. Even when other nondestructive techniques are used to detect surface cracks, visual inspection often provides a useful supplement. For example, when the eddy current examination of process tubing is performed, visual inspection is often performed to verify and more closely examine the surface disturbance.

Given the wide variety of surface flaws that may be detectable by visual examination, the use of visual inspection may encompass different techniques, depending on the product and the type of surface flaw being monitored. This article focuses on some equipment used to aid the process of visual inspection. The techniques and applicability of visual inspection for some products are considered in the Selected References in this article and in the Section "Nondestructive Inspection of Specific Products" in this Volume.

The methods of visual inspection involve a wide variety of equipment, ranging from examination with the naked eye to the use of interference microscopes for measuring the depth of scratches in the finish of finely polished or lapped surfaces. Some of the equipment used to aid visual inspection includes:

- Flexible or rigid borescopes for illuminating and observing internal, closed or otherwise inaccessible areas
- Image sensors for remote sensing or for the development of permanent visual records in the form of photographs, videotapes, or computer-enhanced images
- Magnifying systems for evaluating surface finish, surface shapes (profile and contour gaging), and surface microstructures
- Dye and fluorescent penetrants and magnetic particles for enhancing the observation of surface cracks (and sometimes near-surface conditions in the case of magnetic particle inspection)

This article will review the use of the equipment listed above in visual inspection, except for dye penetrants and magnetic particles, which are discussed in the articles "Liquid Penetrant Inspection" and "Magnetic Particle Inspection," respectively, in this Volume.

Borescopes

A borescope (Fig. 1) is a long, tubular optical device that illuminates and allows the inspection of surfaces inside narrow

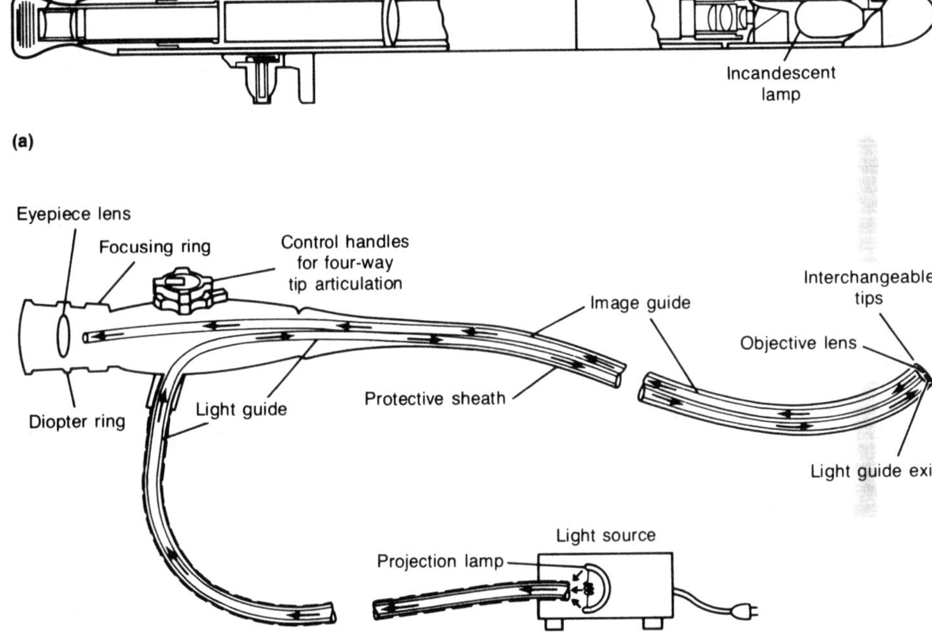

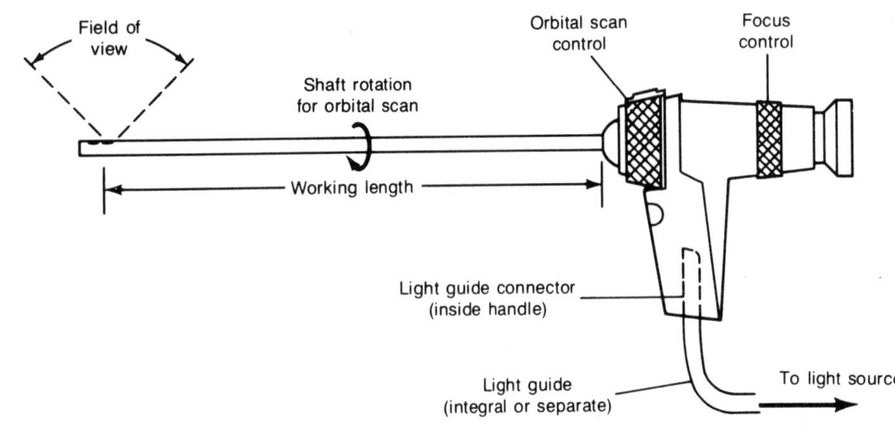

Fig. 1 Three typical designs of borescopes. (a) A rigid borescope with a lamp at the distal end. (b) A flexible fiberscope with a light source. (c) A rigid borescope with a light guide bundle in the shaft

Fig. 2 Typical directions and field of view with rigid borescopes

Fig. 3 Typical chamberscope. Courtesy of Lenox Instrument Company

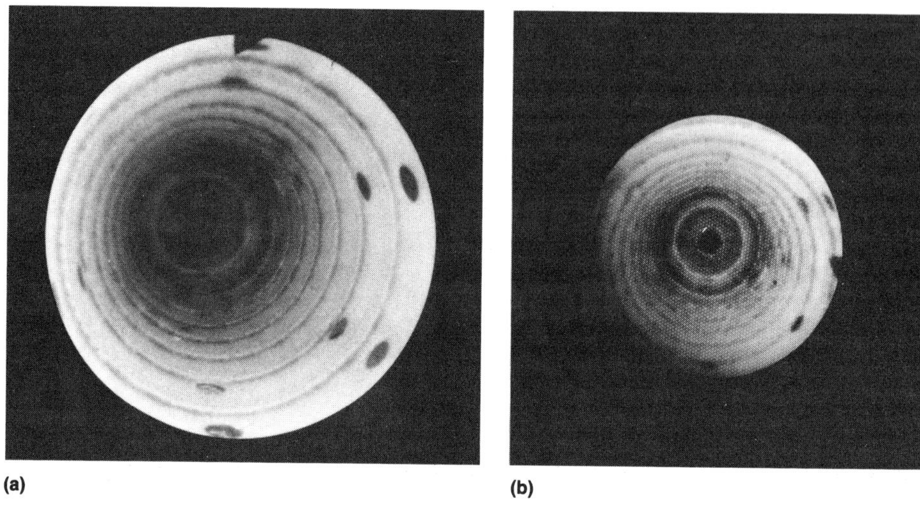

(a) (b)

Fig. 4 Two views down a combustor can with the distal tip in the same position. A fiberscope with smaller diameter fibers and 40% more fibers in the image bundle provides better resolution (a) than a fiberscope with larger fibers (b). Courtesy of Olympus Corporation

tubes or difficult-to-reach chambers. The tube, which can be rigid or flexible with a wide variety of lengths and diameters, provides the necessary optical connection between the viewing end and an objective lens at the distant, or distal, tip of the borescope. This optical connection can be achieved in one of three different ways:

- By using a rigid tube with a series of relay lenses
- By using a tube (normally flexible but also rigid) with a bundle of optical fibers
- By using a tube (normally flexible) with wiring that carries the image signal from a charge-coupled device (CCD) imaging sensor at the distal tip

These three basic tube designs can have either fixed or adjustable focusing of the objective lens at the distal tip. The distal tip also has prisms and mirrors that define the direction and field of view (see Fig. 2). These views vary according to the type and application of borescope. The design of illumination system also varies with the type of borescope. Generally, a fiber optic light guide and a lamp producing white light is used in the illumination system, although ultraviolet light can be used to inspect surfaces treated with liquid fluorescent penetrants. Light-emitting diodes at the distal tip are sometimes used for illumination in videoscopes with working lengths greater than 15 m (50 ft).

Rigid Borescopes

Rigid borescopes are generally limited to applications with a straight-line path between the observer and the area to be observed. The sizes range in lengths from 0.15 to 30 m (0.5 to 100 ft) and in diameters from 0.9 to 70 mm (0.035 to 2.75 in.). Magnification is usually 3 to 4×, but powers up to 50× are available. The illumination system is either an incandescent lamp located at the distal end (Fig. 1a) or a light guide bundle made from optical fibers (Fig. 1c) that conduct light from an external source.

The choice of viewing heads for rigid borescopes (Fig. 2) varies according to the application, as described in the section "Selection" in this article. Rigid borescopes generally have a 55° field of view, although the fields of view can range from 10 to 90°. Typically, the distal tips are not interchangeable, but some models (such as the extendable borescopes) may have interchangeable viewing heads.

Some rigid borescopes have orbital scan (Fig. 1c), which involves the rotation of the optical shaft for scanning purposes. Depending on the borescope model, the amount of rotation can vary from 120 to 370°. Some rigid borescopes also have movable prisms at the tip for scanning.

Rigid borescopes are available in a variety of models having significant variations in the design of the shaft, the distal tip, and the illumination system. Some of these design variations are described below.

Basic Design. The rigid borescope typically has a series of achromatic relay lenses in the optical tube. These lenses preserve the resolution of the image as it travels from the objective lens to the eyepiece. The tube diameter of these borescopes ranges from 4 to 70 mm (0.16 to 2.75 in.). The illumination system can be either a distal lamp or a light guide bundle, and the various features may include orbital scan, various viewing heads, and adjustable focusing of the objective lens.

Miniborescopes. Instead of the conventional relay lenses, miniborescopes have a single image-relaying rod or quartz fiber in the optical tube. The lengths of miniborescopes are 110 and 170 mm (4.3 and 6.7 in.), and the diameters range from 0.9 to 2.7 mm (0.035 to 0.105 in.). High magnification (up to 30×) can be reached at minimal focal lengths, and an adjustable focus is not required, because the scope has an infinite depth of field. The larger sizes have forward, side view, and forward-oblique views. The 0.9 mm (0.035 in.) diam size has only a forward view. Miniborescopes have an integral light guide bundle.

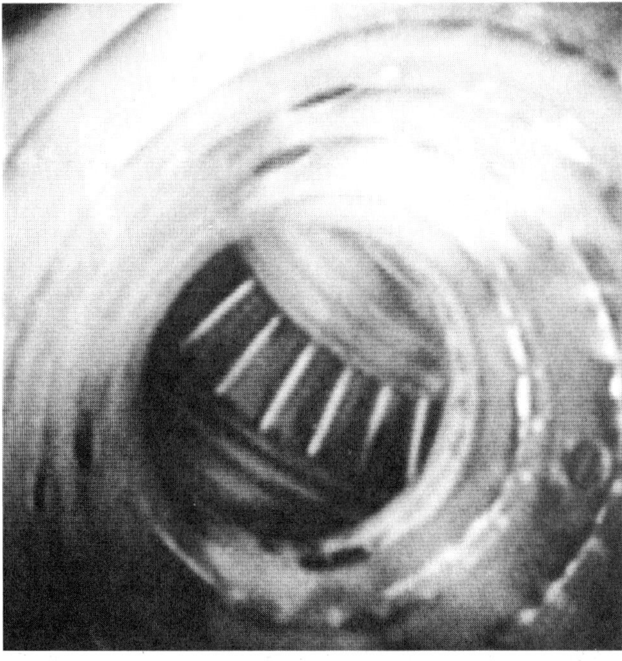

(a)

(b)

Fig. 5 Videoscope images (a) inside engine guide vanes (b) of an engine fuel nozzle. Courtesy of Welch Allyn, Inc.

Hybrid borescopes utilize rod lenses combined with convex lenses to relay the image. The rod lenses have fewer glass-air boundaries; this reduces scattering and allows for a more compact optical guide. Consequently, a larger light guide bundle can be employed with an increase in illumination and an image with a higher degree of contrast.

Hybrid borescopes have lengths up to 990 mm (39 in.), with diameters ranging from 5.5 to 12 mm (0.216 to 0.47 in.). All hybrid borescopes have adjustable focusing of the objective lens and a 370° rotation for orbital scan. The various viewing directions are forward, side, retrospective, and forward-oblique.

Extendable borescopes allow the user to construct a longer borescopic tube by joining extension tubes. Extendable borescopes are available with either a fiber-optic light guide or an incandescent lamp at the distal end. Extendable borescopes with an integral lamp have a maximum length of about 30 m (100 ft). Scopes with a light guide bundle have a shorter maximum length (about 8 m, or 26 ft), but do allow smaller tube diameters (as small as 8 mm, or 0.3 in.). Interchangeable viewing heads are also available. Extendable borescopes do not have adjustable focusing of the objective lens.

Rigid chamberscopes allow more rapid inspection of larger chambers. Chamberscopes (Fig. 3) have variable magnification (zoom), a lamp at the distal tip, and a scanning mirror that allows the user to observe in different directions. The higher illumination and greater magnification of chamberscopes allow the inspection of surfaces as much as 910 mm (36 in.) away from the distal tip of the scope.

Mirror sheaths can convert a direct-viewing borescope into a side-viewing scope. A mirror sheath is designed to fit over the tip of the scope and thus reflect an image from the side of the scope. However, not all applications are suitable for this device. A side, forward-oblique, or retrospective viewing head provides better resolution and a higher degree of image contrast. A mirror sheath also produces an inverse image and may produce unwanted reflections from the shaft.

Scanning. In addition to the orbital scan feature described earlier, some rigid borescopes have the ability to scan longitudinally along the axis of the shaft. A movable prism with a control at the handle accomplishes this scanning. Typically, the prism can shift the direction of view through an arc of 120°.

Flexible Borescopes

Flexible borescopes are used primarily in applications that do not have a straight passageway to the point of observation. The two types of flexible borescopes are flexible fiberscopes and videoscopes with a CCD image sensor at the distal tip.

Flexible Fiberscopes. A typical fiberscope (Fig. 1b) consists of a light guide bundle, an image guide bundle, an objective lens, interchangeable viewing heads, and remote controls for articulation of the distal tip. Fiberscopes are available in diameters from 1.4 to 13 mm (0.055 to 0.5 in.) and in lengths up to 12 m (40 ft). Special quartz fiberscopes are available in lengths up to 90 m (300 ft).

The fibers used in the light guide bundle are generally 30 μm (0.001 in.) in diameter. The second optical bundle, called the image guide, is used to carry the image formed by the objective lens back to the eyepiece. The fibers in the image guide must be precisely aligned so that they are in an identical relative position to each other at their terminations for proper image resolution.

The diameter of the fibers in the image guide is another factor in obtaining good image resolution. With smaller diameter fibers, a brighter image with better resolution can be obtained by packing more fibers in the image guide. With higher resolution, it is then possible to use an objective lens with a wider field of view and also to magnify the image at the eyepiece. This allows better viewing of objects at the periphery of the image (Fig. 4). Image guide fibers range from 6.5 to 17 μm (255 to 670 μin.).

The interchangeable distal tips provide various directions and fields of view on a single fiberscope. However, because the tip can be articulated for scanning purposes, distal tips with either a forward or side viewing direction are usually sufficient. Fields of view are typically 40 to 60°, although they can range from 10 to 120°. Most fiberscopes provide adjustable focusing of the objective lens.

Videoscopes with CCD probes involve the electronic transmission of color or black and white images to a video monitor. The distal end of electronic videoscopes con-

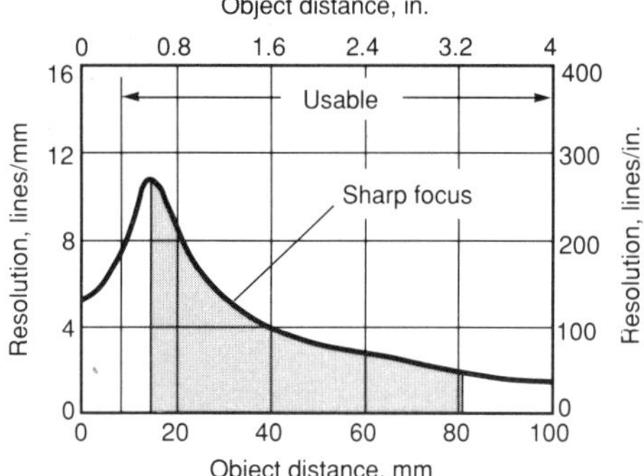

(a)

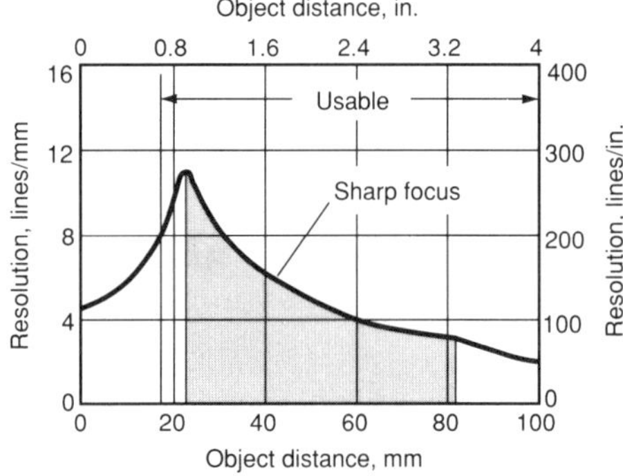

(b)

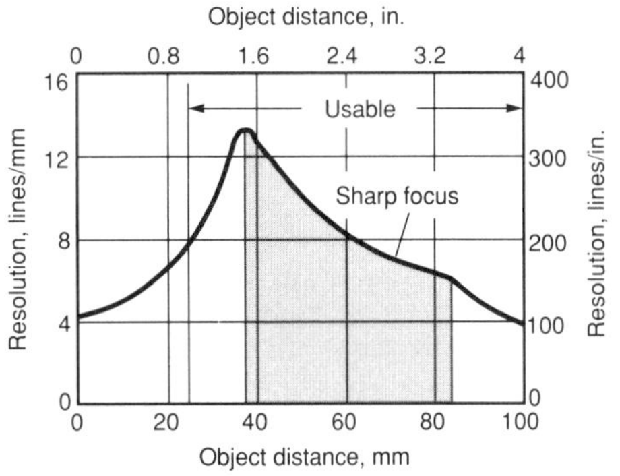

(c)

Fig. 6 Typical resolution of CCD videoscopes with a 90° field of view (a), 60° field of view (b), 30° field of view (c). Source: Welch Allyn, Inc.

tains a CCD chip, which consists of thousands of light-sensitive elements arrayed in a pattern of rows and columns. The objective lens focuses the image of an object on the surface of the CCD chip, where the light is converted to electrons that are stored in each picture element, or pixel, of the CCD device. The image of the object is thus stored in the form of electrons on the CCD device. At this point, a voltage proportional to the number of electrons at each pixel is determined electronically for each pixel

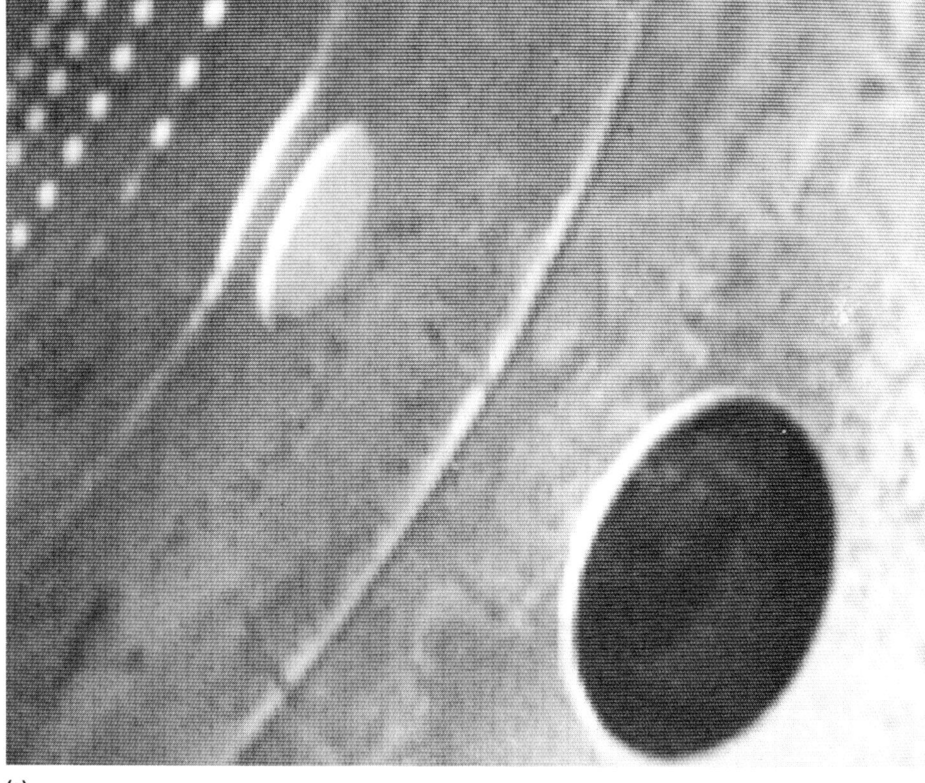

(a)

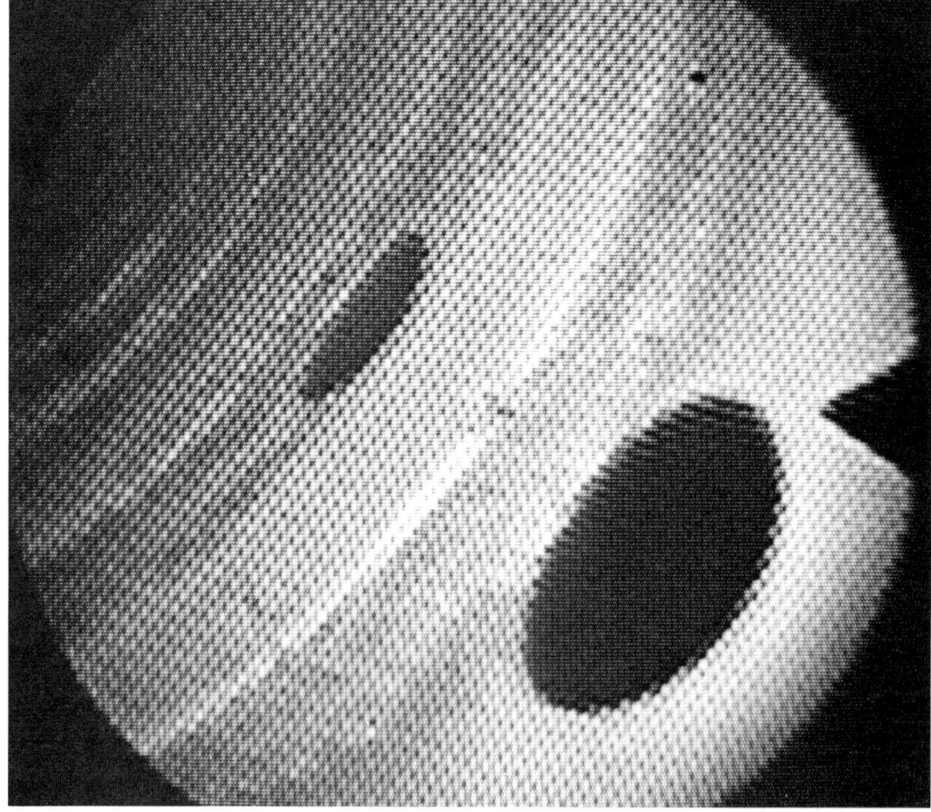

(b)

Fig. 7 Image from a videoscope (a) and a fiberscope (b). In some fiberscope images, voids between individual glass fibers can create a honeycomb pattern that adds graininess to the image. Courtesy of Welch Allyn, Inc.

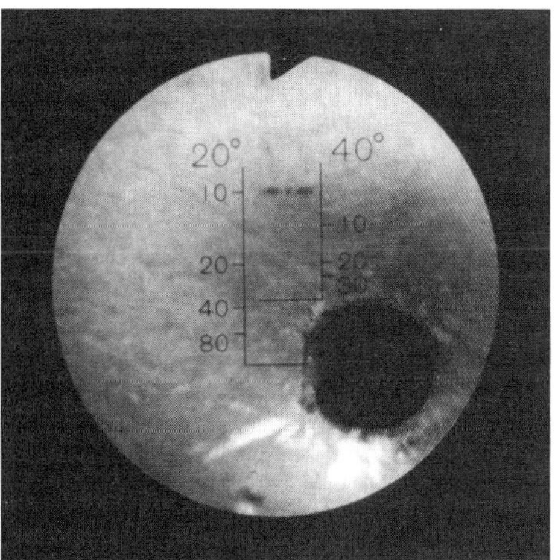

Fig. 8 View through a measuring fiberscope with reticles for 20° and 40° field-of-view lenses. Courtesy of Olympus Corporation

site. This voltage is then amplified, filtered, and sent to the input of a video monitor.

Videoscopes with CCD probes produce images (Fig. 5) with spatial resolutions of the order of those described in Fig. 6. Like rigid borescopes and flexible fiberscopes, the resolution of videoscopes depends on the object-to-lens distance and the fields of view, because these two factors affect the amount of magnification (see the section "Magnification and Field of View" in this article). Generally, videoscopes produce higher resolution than fiberscopes, although fiberscopes with smaller diameter fibers (Fig. 4a) may be competitive with the resolution of videoscopes.

Another advantage of videoscopes is their longer working length. With a given amount of illumination at the distal tip, videoscopes can return an image over a greater length than fiberscopes. Other features of videoscopes include:

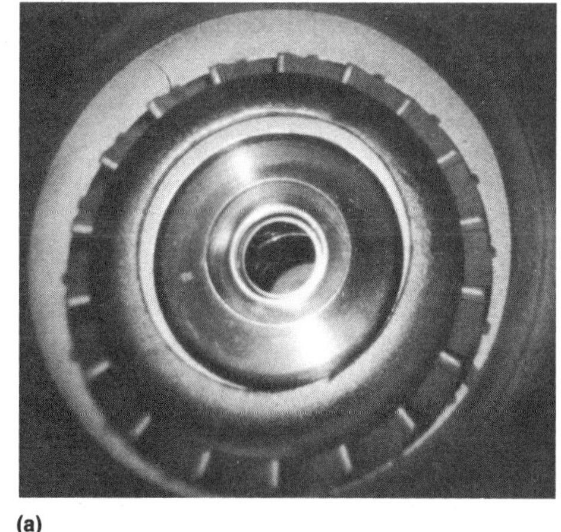

(a)

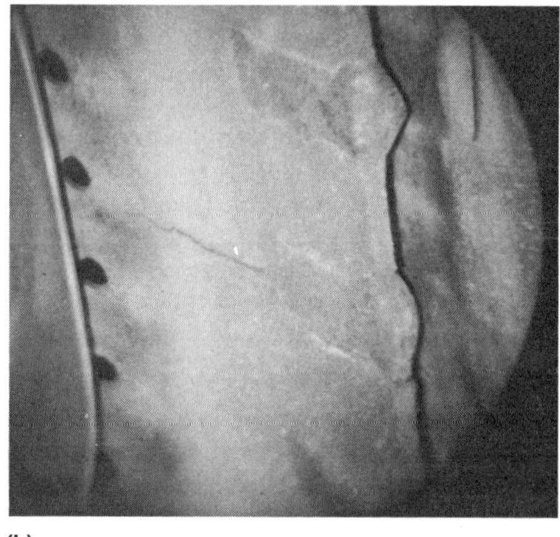

(b)

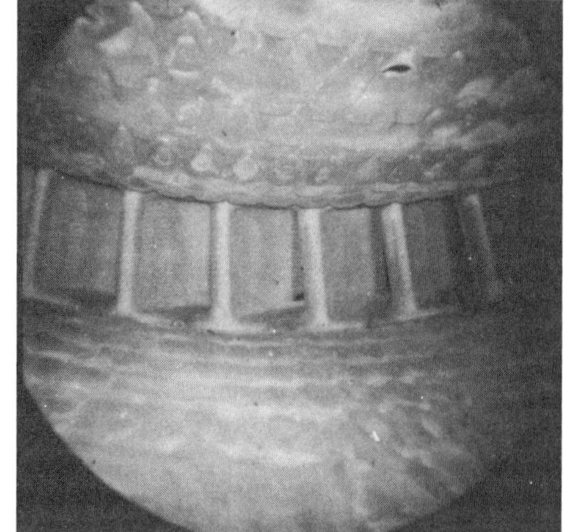

(c)

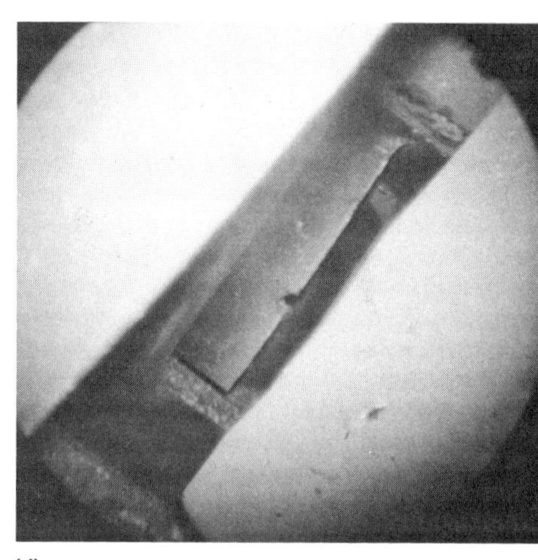

(d)

Fig. 9 Turbine flaws seen through a flexible fiberscope. (a) Crack near a fuel burner nozzle. (b) Crack in an outer combustion liner. (c) Combustion chamber and high pressure nozzle guide vanes. (d) Compressor damage showing blade deformation. Courtesy of Olympus Corporation

- The display can help reduce eye fatigue (but does not allow the capability of direct viewing through an eyepiece)
- There is no honeycomb pattern or irregular picture distortion as with some fiberscopes (Fig. 7)

Fig. 10 In-service defects as seen through a borescope designed for automotive servicing. (a) Carbon on valves. (b) Broken transmission gear tooth. (c) Differential gear wear. Courtesy of Lenox Instrument Company

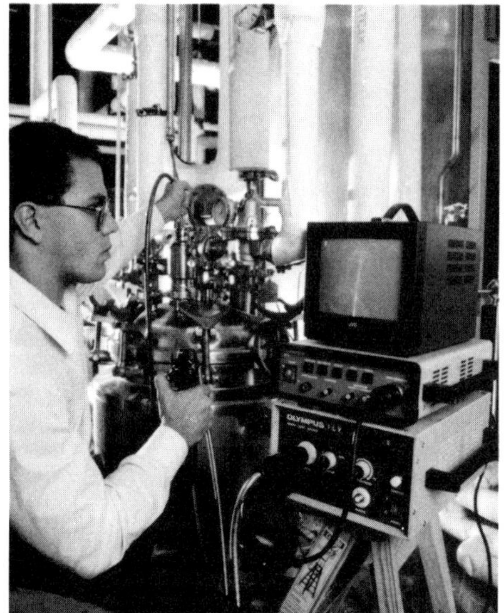

Fig. 11 Operator viewing a weld 21 m (70 ft) inside piping with a videoscope. Courtesy of Olympus Corporation

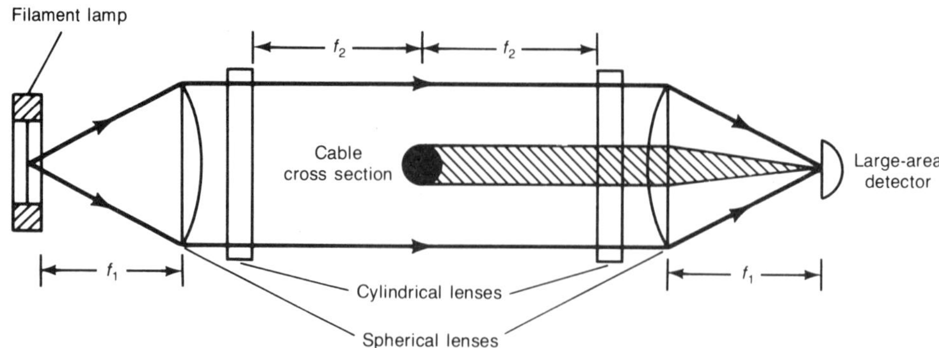

Fig. 12 Schematic of line projection method for monitoring the surface roughness on fast-moving cables

Fig. 13 Setup used in the in-plant trials of the line projection method for monitoring the surface roughness of cables. Courtesy of P. Cielo, National Research Council of Canada

- The electronic form of the image signal allows digital image enhancement and the potential for integration with automatic inspection systems
- The display allows the generation of reticles on the viewing screen for point-to-point measurements

Special Features

Measuring borescopes and fiberscopes contain a movable cursor that allows measurements during viewing (Fig. 8). When the object under measurement is in focus, the movable cursor provides a reference for dimensional measurements in the optical plane of the object. This capability eliminates the need to know the object-to-lens distance when determining magnification factors.

Working channels are used in borescopes and fiberscopes to pass working devices to the distal tip. Working channels are presently used to pass measuring instruments, retrieval devices, and hooks for aiding the insertion of thin, flexible fiberscopes. Working channels are used in flexible fiberscopes with diameters as small as 2.7 mm (0.106 in.). Working channels are also under consideration for the application and removal of dye penetrants and for the passage of wires and sensors in eddy current measurements.

Selection

Flexible and rigid borescopes are available in a wide variety of standard and customized designs, and several factors can influence the selection of a scope for a particular application. These factors include focusing, illumination, magnification, working length, direction of view, and environment.

Focusing and Resolution. If portions of long objects are at different planes, the scope must have sufficient focus adjustment to achieve an adequate depth of field. If the scope has a fixed focal length, the object will be in focus only at a specific lens-to-object distance.

To allow the observation of surface detail at a desired size, the optical system of a borescope must also provide adequate resolution and image contrast. If resolution is adequate but contrast is lacking, detail cannot be observed.

In general, the optical quality of a rigid borescope improves as the size of the lens increases; consequently, a borescope with the largest possible diameter should be used. For fiberscopes, the resolution is dependent on the accuracy of alignment and the diameter of the fibers in the image bundle. Smaller-diameter fibers provide more resolution and edge contrast (Fig. 4), when combined with good geometrical alignment of the fibers. Typical resolutions of videoscopes are given in Fig. 6.

Illumination. The required intensity of the light source is determined by the reflectivity of the surface, the area of surface to be illuminated, and the transmission losses over the length of the scope. At working lengths greater than 6 m (20 ft), rigid borescopes with a lamp at the distal end provide the greatest amount of illumination over the widest area. However, the heat generated by the light source may deform rubber or plastic materials. Fiber-optic illumination in scopes with working lengths less than 6 m (20 ft) is always brighter and is suitable for heat-sensitive applications because filters can remove infrared frequencies. Because

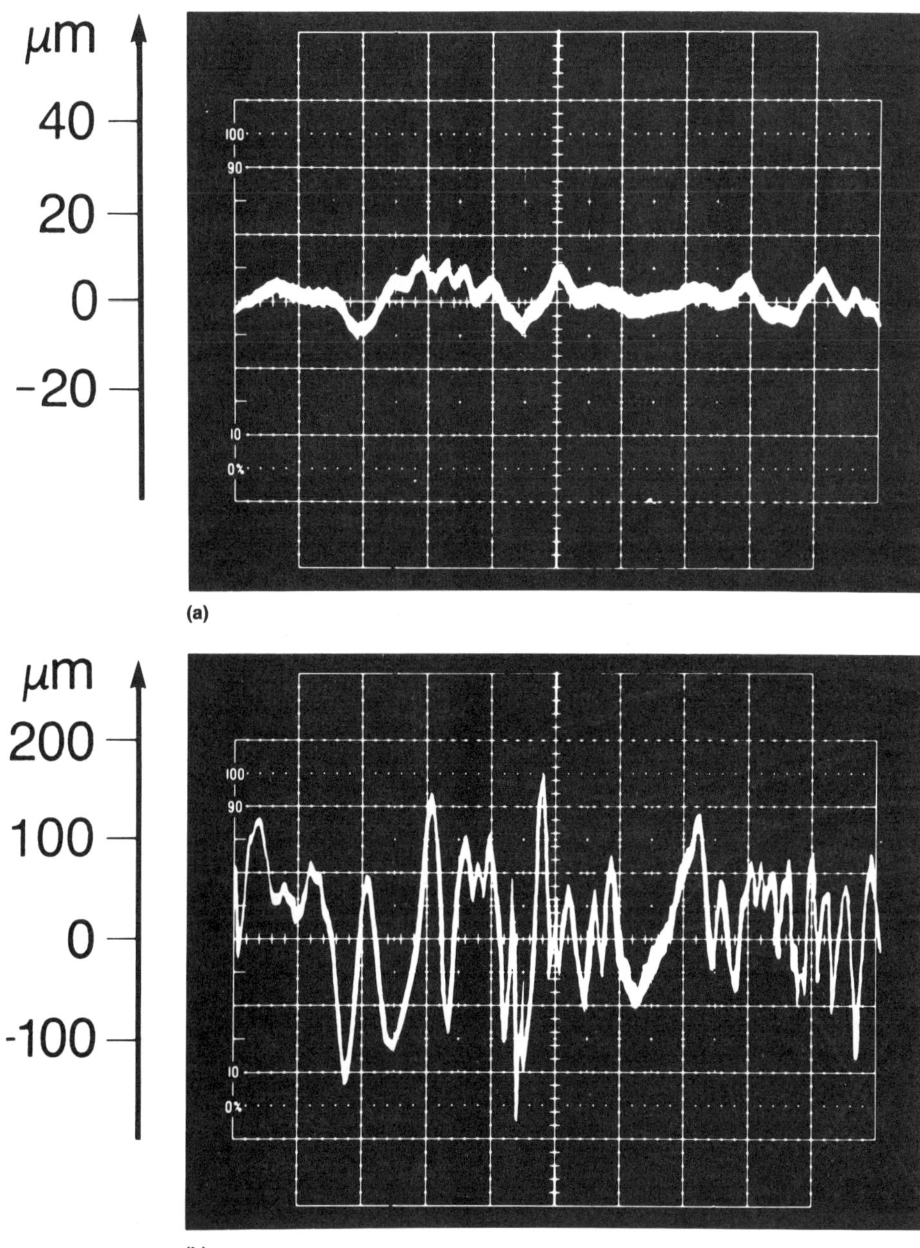

Fig. 14 Examples of signals obtained with the apparatus shown in Fig. 13. (a) Acceptable surface roughness. (b) Unacceptable surface roughness

the amount of illumination depends on the diameter of the light guide bundle, it is desirable to use the largest diameter possible.

Magnification and field of view are interrelated; as magnification is increased, the field of view is reduced. The precise relationship between magnification and field of view is specified by the manufacturer.

The degree of magnification in a particular application is determined by the field of view and the distance from the objective lens to the object. Specifically, the magnification increases when either the field of view or the lens-to-object distance decreases.

Working Length. In addition to the obvious need for a scope of sufficient length, the working length can sometimes dictate the use of a particular type of scope. For example, a rigid borescope with a long working length may be limited by the need for additional supports. In general, videoscopes allow a longer working length than fiberscopes.

Direction of View. The selection of a viewing direction is influenced by the location of the access port in relation to the object to be observed. The following sections describe some criteria for choosing the direction of view shown in Fig. 2. Flexible fiberscopes or videoscopes, because of their articulating tip, are often adequate with either a side or forward viewing tip.

Circumferential or panoramic heads are designed for the inspection of tubing or other cylindrical structures. A centrally located mirror permits right-angle viewing of an area just scanned by the panoramic view.

The forward viewing head permits the inspection of the area directly ahead of the viewing head. It is commonly used when examining facing walls or the bottoms of blind holes and cavities.

Forward-oblique heads bend the viewing direction at an angle to the borescope axis, permitting the inspection of corners at the end of a bored hole. The retrospective viewing head bends the cone of view at a retrospective angle to the borescope axis, providing a view of the area just passed by the advancing borescope. It is especially suited to inspecting the inside neck of cylinders and bottles.

Environment. Flexible and rigid borescopes can be manufactured to withstand a variety of environments. Although most scopes can operate at temperatures from −34 to 66 °C (−30 to 150 °F), specially designed scopes can be used at temperatures to 1925 °C (3500 °F). Scopes can also be manufactured for use in liquid media.

Special scopes are required for use in pressures above ambient and in atmospheres exposed to radiation. Radiation can cause the multicomponent lenses and image bundles to turn brown. When a scope is used in atmospheres exposed to radiation, quartz fiberscopes are generally used. Scopes used in a gaseous environment should be made explosionproof to minimize the potential of an accidental explosion.

Applications

Rigid and flexible borescopes are available in different designs suitable for a variety of applications. For example, when inspecting straight process piping for leaks, rigid borescopes with a 360° radial view are capable of examining inside diameters of 3 to 600 mm (0.118 to 24 in.). Scopes are also used by building inspectors and contractors to see inside walls, ducts, large tanks, or other dark areas.

The principal use of borescopes is in equipment maintenance programs, in which borescopes can reduce or eliminate the need for costly teardowns. Some types of equipment, such as turbines, have access ports that are specifically designed for borescopes. Borescopes provide a means of checking in-service defects in a variety of equipment, such as turbines (Fig. 9), automotive components (Fig. 10), and process piping (Fig. 11).

Borescopes are also extensively used in a variety of manufacturing industries to en-

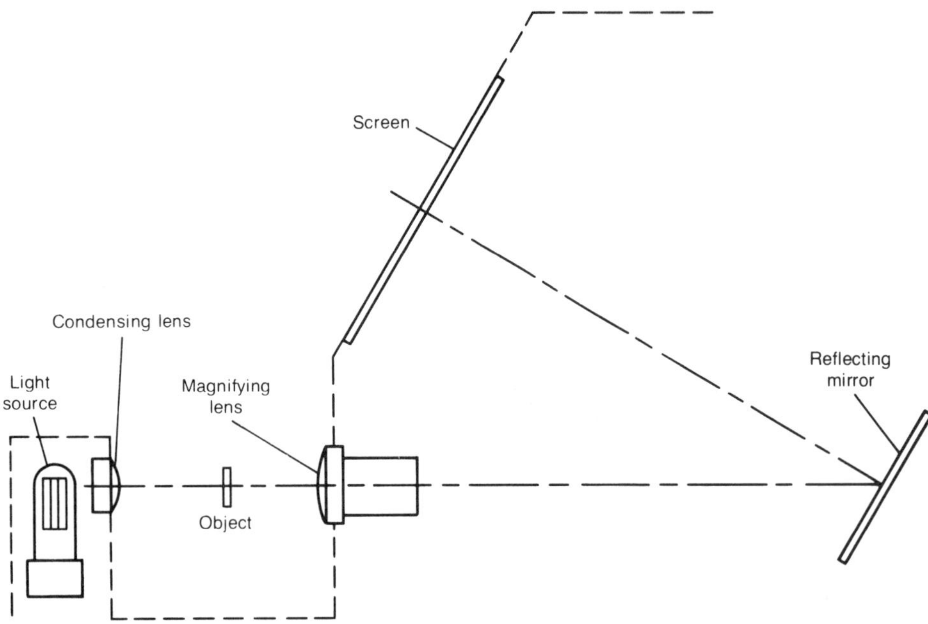

Fig. 15 Schematic of an optical comparator

sure the product quality of difficult-to-reach components. Manufacturers of hydraulic cylinders, for example, use borescopes to examine the interiors of bores for pitting, scoring, and tool marks. Aircraft and aerospace manufacturers also use borescopes to verify the proper placement and fit of seals, bonds, gaskets, and subassemblies in difficult-to-reach regions.

Optical Sensors*

Visible light, which can be detected by the human eye or with optical sensors, has some advantages over inspection methods based on nuclear, microwave, or ultrasound radiation. For example, one of the advantages of visible light is the capability of tightly focusing the probing beam on the inspected surface (Ref 1). High spatial resolution can result from this sharp focusing, which is useful in gaging and profiling applications (Ref 1).

Some different types of image sensors used in visual inspection include:

- Vidicon or plumbicon television tubes
- Secondary electron-coupled (SEC) vidicons
- Image orthicons and image isocons
- Charge-coupled device sensors
- Holographic plates (see the article "Optical Holography" in this Volume)

Television cameras with vidicon tubes are useful at higher light levels (about 0.2 lm/m^2, or 10^{-2} ftc), while orthicons, isocons, and SEC vidicons are useful at lower light levels. The section "Television Cameras" in the article "Radiographic Inspection" in this Volume describes these cameras in more detail.

*Example 1 in this section was adapted with permission from P. Cielo, *Optical Techniques for Industrial Inspection*, Academic Press, 1988.

Charge-coupled devices are suitable for many different information-processing applications, including image sensing in television-camera technology. Charge-coupled devices offer a clear advantage over vacuum-tube image sensors because of the reliability of their solid-state technology, their operation at low voltage and low power dissipation, extensive dynamic range, visible and near-infrared response, and geometric reproducibility of image location. Image enhancement (or visual feedback into robotic systems) typically involve the use of CCDs as the optical sensor or the use of television signals that are converted into digital form.

Optical sensors are also used in inspection applications that do not involve imaging. The articles "Laser Inspection" and "Speckle Metrology" in this Volume describe the use of optical sensors when laser light is the probing tool. In some applications, however, incoherent light sources are very effective in non-imaging inspection applications utilizing optical sensors.

Example 1: Monitoring Surface Roughness on a Fast-Moving Cable. A shadow projection configuration that can be used at high extrusion speeds is shown in Fig. 12. A linear-filament lamp is imaged by two spherical lenses of focal length f_1 on a large-area single detector. Two cylindrical lenses are used to project and recollimate a laminar light beam of uniform intensity, nearly 0.5 mm (0.02 in.) wide across the wire situated near their common focal plane. The portion of the light beam that is not intercepted by the wire is collected on the detector, which has an alternating current output that corresponds to the defect-related wire diameter fluctuations. The wire speed is limited only by the detector response time. With a moderate detector bandwidth of 100 kHz, wire extrusion speeds up to 50 m/s (160 ft/s) can be accepted. Moreover, the uniformity of the nearly collimated projected beam obtained with such a configuration makes the detected signal relatively independent of the random wire excursions in the plane of Fig. 12. It should be mentioned that the adoption of either a single He-Ne laser or an array of fiber-pigtailed diode lasers proved to be inadequate in this case because of speckle noise, high-frequency laser amplitude or mode-to-mode interference fluctuations, and line nonuniformity.

An industrial prototype of such a sensor was tested on the production line at extruding speeds reaching 30 m/s (100 ft/s). Figure 13 shows the location of the sensor just after the extruder die. Random noise introduced by vapor turbulence could be almost completely suppressed by high-pass filtering. Figure 14 shows two examples of signals obtained with a wire of acceptable and unacceptable surface quality. As shown, a roughness amplitude resolution of a few micrometers can be obtained with such a device. Subcritical surface roughness levels can thus be monitored for real time control of the extrusion process.

Magnifying Systems

In addition to the use of microscopes in the metallographic examination of microstructures (see the article "Replication Microscopy Techniques for NDE" in this Volume), magnifying systems are also used in visual reference gaging. When tolerances are too tight to judge by eye alone, optical comparators or toolmakers' microscopes are used to achieve magnifications ranging from 5 to 500×.

A toolmakers' microscope consists of a microscope mounted on a base that carries an adjustable stage, a stage transport mechanism, and supplementary lighting. Micrometer barrels are often incorporated into the stage transport mechanism to permit precisely controlled movements, and digital readouts of stage positioning are becoming increasingly available. Various objective lenses provide magnifications ranging from 10 to 200×.

Optical comparators (Fig. 15) are magnifying devices that project the silhouette of small parts onto a large projection screen. The magnified silhouette is then compared against an optical comparator chart, which is a magnified outline drawing of the workpiece being gaged. Optical comparators are available with magnifications ranging from 5 to 500×.

Parts with recessed contours can also be successfully gaged on optical comparators. This is done with the use of a pantograph. One arm of the pantograph is a stylus that traces the recessed contour of the part, and the other arm carries a follower that is visible in the light path. As the stylus moves, the follower projects a contour on the screen.

ACKNOWLEDGMENT

ASM INTERNATIONAL would like to thank Oliver Darling and Morley Melden of Spectrum Marketing, Inc., for their assistance in preparing the section on borescopes. They provided a draft of a textbook being developed for Olympus Corporation. Thanks are also extended to Virginia Torrey of Welch Allyn, Inc., for the information on videoscopes and to Peter Sigmund of Lindhult and Jones, Inc., for the information on instruments from Lenox, Inc.

REFERENCE

1. P. Cielo, *Optical Techniques for Industrial Inspection*, Academic Press, 1988, p 243

SELECTED REFERENCES

- Robert C. Anderson, *Inspection of Metals: Visual Examination*, Vol 1, American Society for Metals, 1983
- Detecting Susceptibility to Intergranular Attack in Austenitic Stainless Steels, ASTM A 262, *Annual Book of ASTM Standards*, American Society for Testing and Materials
- Detecting Susceptibility to Intergranular Attack in Ferritic Stainless Steels, ASTM A 763, *Annual Book of ASTM Standards*, American Society for Testing and Materials
- Detecting Susceptibility to Intergranular Corrosion in Severely Sensitized Austenitic Stainless Steel, ASTM A 708, *Annual Book of ASTM Standards*, American Society for Testing and Materials
- W.R. DeVries and D.A. Dornfield, *Inspection and Quality Control in Manufacturing Systems*, American Society of Mechanical Engineers, 1982
- C.W. Kennedy and D.E. Andrews, *Inspection and Gaging*, Industrial Press, 1977
- Standard Practice for Evaluating and Specifying Textures and Discontinuities of Steel Castings by Visual Examination, ASTM Standard A 802, American Society for Testing and Materials
- Surface Discontinuities on Bolts, Screws, and Studs, ASTM F 788, *Annual Book of ASTM Standards*, American Society for Testing and Materials
- Visual Evaluation of Color Changes of Opaque Materials, ASTM D 1729, *Annual Book of ASTM Standards*, American Society for Testing and Materials

Laser Inspection

Carl Bixby, Zygo Corporation

THE FIRST LASER was invented in 1960, and many useful applications of laser light have since been developed for metrology and industrial inspection systems. Laser-based inspection systems have proved useful because they represent a fast, accurate means of noncontact gaging, sorting, and classifying parts. Lasers have also made interferometers a more convenient tool for the accurate measurement of length, displacement, and alignment.

Lasers are used in inspection and measuring systems because laser light provides a bright, undirectional, and collimated beam of light with a high degree of temporal (frequency) and spatial coherence. These properties can be useful either singly or together. For example, when lasers are used in interferometry, the brightness, coherence, and collimation of laser light are all important. However, in the scanning, sorting, and triangulation applications described in this article, lasers are used because of the brightness, unidirectionality, and collimated qualities of their light; temporal coherence is not a factor.

The various types of laser-based measurement systems have applications in three main areas:

- Dimensional measurement
- Velocity measurement
- Surface inspection

The use of lasers may be desirable when these applications require high precision, accuracy, or the ability to provide rapid, noncontact gaging of soft, delicate, hot, or moving parts. Photodetectors are generally needed in all the applications, and the light variations or interruptions can be directly converted into electronic form.

Dimensional Measurements

Lasers can be used in several different ways to measure the dimensions and the position of parts. Some of the techniques include:

- Profile gaging of stationary and moving parts with laser scanning equipment
- Profile gaging of stationary parts by shadow projection on photodiode arrays
- Profile gaging of small gaps, and small-diameter parts from diffraction patterns
- Gaging of surfaces that cannot be seen in profile (such as concave surfaces, gear teeth, or the inside diameters of bores) with laser triangulation sensors
- Measuring length, alignments, and displacements with interferometers
- Sorting of parts
- Three-dimensional gaging of surfaces with holograms
- Measuring length from the velocity of moving, continuous parts (see the section "Velocity Measurements" in this article)

These techniques provide high degrees of precision and accuracy as well as the capability for rapid, noncontact measurement.

Scanning Laser Gage. Noncontact sensors are used in a variety of inspection techniques, such as those involving capacitive gages, eddy-current gages, and air gages. Optical gages, however, have advantages because the distance from the sensor to the workpiece can be large and because many objects can be measured simultaneously. Moreover, the light variations are directly converted into electronic signals, with the response time being limited only by the photodetector and its electronics.

Optical sensors for dimensional gaging employ various techniques, such as shadow projection, diffraction phenomena, and scanning light beams. If the workpiece is small or does not exhibit large or erratic movements, diffraction phenomena or shadow projection on a diode array sensor can work well. However, diffraction techniques become impractical if the object has a dimension of more than a few millimeters or if its movement is large. Shadow projection on a diode array sensor may also be limited if the size or movement of the part is too large.

The scanning laser beam technique, on the other hand, is suited to a broad range of product sizes and movements. The concept of using a scanning light beam for noncontact dimensional gaging predates the laser; the highly directional and collimated nature of laser light, however, greatly improves the precision of this method over techniques that use ordinary light. The sensing of outside diameters of cylindrical parts is probably the most common application of a laser scanning gage.

A scanning laser beam gage consists of a transmitter, a receiver, and processor electronics (Fig. 1). A thin band of scanning laser light is projected from the transmitter to the receiver. When an object is placed in a beam, it casts a time-dependent shadow. Signals from the light entering the receiver are used by the microprocessor to extract the dimension represented by the time difference between the shadow edges. The gages can exhibit accuracies as high as ±0.25 μm (±10 μin.) for diameters of 10 to 50 mm (0.5 to 2 in.). For larger parts (diameters of 200 to 450 mm, or 8 to 18 in.), accuracies are less.

There are two general types of scanning laser gages: separable transmitters and receivers designed for in-process applications, and self-contained bench gages designed for off-line applications. The in-process scanning gage consists of a multitasking electronic processor that controls a number of scanners. The separable transmitters and receivers can be configured for different scanning arrangements. Two or more scanners can also be stacked for large parts, or they can be oriented along different axes for dual-axis inspection. High-speed scanners are also available for the detection of small defects, such as lumps, in moving-part applications.

The bench gage is compact and can measure a variety of part sizes quickly and easily (Fig. 2). As soon as measurements are taken, the digital readout displays the gaged dimension and statistical data. It indicates the total number of measurements taken, the standard deviation, and the maximum, minimum, and mean readings of each batch tested.

Applications. A wide variety of scanning laser gages are available to fit specific applications. Measurement capabilities fall within a range of 0.05 to 450 mm (0.002 to 18 in.), with a repeatability of 0.1 μm (5 μin.) for the smaller diameters. Typical applications include centerless grinding, precision machining, extrusions, razor blades, turbine blades, computer disks, wire lines, and plug gages.

Photodiode Array Imaging. Profile imaging closely duplicates a shadowgraph or

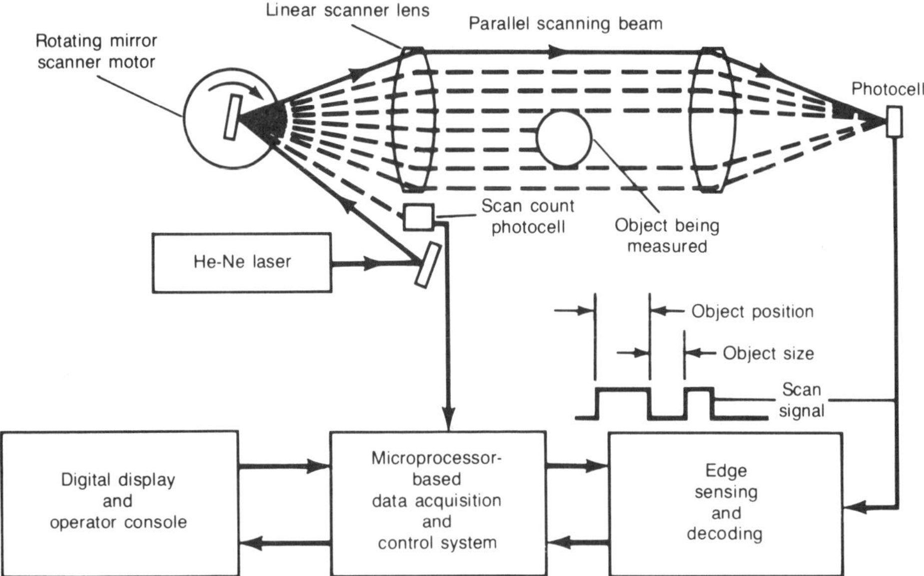

Fig. 1 Schematic of a scanning laser gage

Fig. 2 Self-contained laser bench micrometer

contour projector where the ground-glass screen has been replaced by a solid-state diode array image sensor. The measurement system consists of a laser light source, imaging optics, a photodiode array, and signal-processing electronics (Fig. 3). The object casts a shadow, or profile image, of the part on the photodiode array. A scan of the array determines the edge image location and then the location of the part edges from which the dimension of the part can be determined.

Accuracies of ±0.05 μm (±2 μin.) have been achieved with photodiode arrays. For large-diameter parts, two arrays are used—one for each edge. When large or erratic movement of the part is involved, photodiode arrays are not suitable. However, when part rotation is involved, stroboscopic illumination can freeze the image of the part.

Diffraction Pattern Technique. Diffraction pattern metrology systems can be used on-line and off-line to measure small gaps and small diameters of thin wire, needles, and fiber optics. They can also be used to inspect such defects as burrs on hypodermic needles and threads on bolts.

In a typical system, the parallel coherent light in a laser beam is diffracted by a small part, and the resultant pattern is focused by a lens on a linear diode array. One significant characteristic of the diffraction pattern is that the smaller the part is, the more accurate the measurement becomes. Diffraction is not suitable for diameters larger than a few millimeters.

The distance between the alternating light and dark bands in the diffraction pattern bears a precise mathematical relationship to the wire diameter, the wavelength of the laser beam, and the focal length of the lens. Because the laser beam wavelength and the lens focal length are known constants of the system, the diameter can be calculated directly from the diffraction pattern measurement.

Laser triangulation sensors determine the standoff distance between a surface and a microprocessor-based sensor. Laser triangulation sensors can perform automatic calculations on sheet metal stampings for gap and flushness, hole diameters, and edge locations in a fraction of the time required in the past with manual or ring gage methods.

The principle of single-spot laser triangulation is illustrated in Fig. 4. In this technique, a finely focused laser spot of light is directed at the part surface. As the light strikes the surface, a lens in the sensor images this bright spot onto a solid-state, position-sensitive photodetector. As shown in Fig. 4, the location of the image spot is directly related to the standoff distance from the sensor to the object surface; a change in the standoff distance results in a lateral shift of the spot along the sensor array. The standoff distance is calculated by the sensor processor.

Laser triangulation sensors provide quick measurement of deviations due to changes in the surface. With two sensors, the method can be used to measure part thickness or the inside diameters of bores. However, it may not be possible to probe the entire length of the bore. Laser triangulation sensors can also be used as a replacement for tough-trigger probes on coordinate measuring machines. In this application, the sensor determines surface features and surface locations by utilizing an edge-finding device.

The accuracy of laser triangulation sensors varies, depending on such performance requirements as standoff and range. Typically, as range requirements increase, accuracy tends to decrease; therefore, specialized multiple-sensor units are designed to perform within various specific application tolerances.

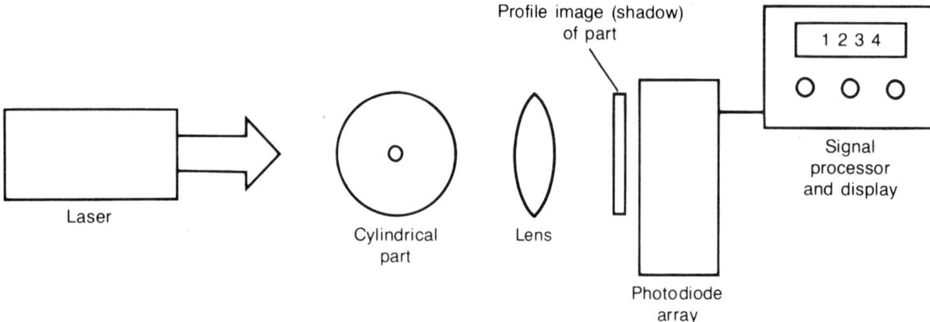

Fig. 3 Schematic of profile imaging. The laser beam passing the edge of a cylindrical part is imaged by a lens onto a photodiode array. A scan of the array determines the edge image location and then the location of the part edges from which the dimension of the part can be determined.

Interferometers provide precise and accurate measurement of relative and absolute length by utilizing the wave properties of light. They are employed in precision metal finishing, microlithography, and the precision alignment of parts.

The basic operational principles of an interferometer are illustrated in Fig. 5. Monochromatic light is directed at a half-silvered mirror acting as a beam splitter that transmits half the beam to a movable mirror and reflects the other half 90° to a fixed mirror. The reflections from the movable and fixed mirrors are recombined at the beam splitter, where wave interference occurs according to the different path lengths of the two beams.

When one of the mirrors is displaced very slowly in a direction parallel to the incident beam without changing its angular alignment, the observer will see the intensity of the recombined beams increasing and decreasing as the light waves from the two paths undergo constructive and destructive interference. The cycle of intensity change from one dark fringe to another represents a half wavelength displacement of movable mirror travel, because the path of light corresponds to two times the displacement of the movable mirror. If the wavelength of the light is known, the displacement of the movable mirror can be determined by counting fringes.

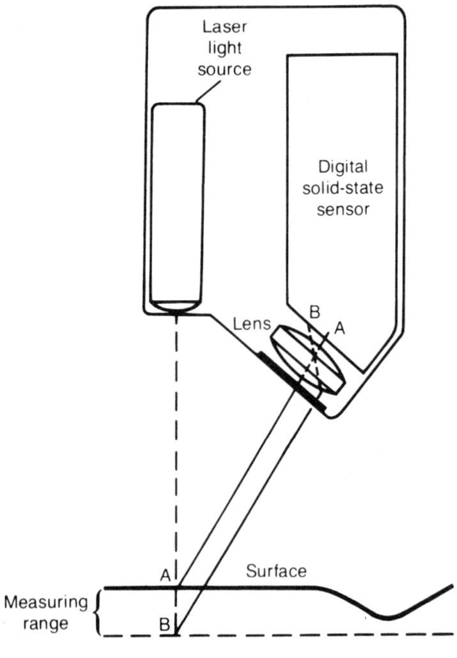

Variations on the basic concepts described above produce interferometers in different forms suited to diverse applications. The two-frequency laser interferometer, for example, provides accurate measurement of displacements (see the following section in this article). Multiple-beam interferometers, such as the Fizeau interferometer, also have useful applications. The Fizeau interferometer has its greatest application in microtopography and is used with a microscope to provide high resolution in three dimensions (see the section "Interference Microscopes" in this article).

Fig. 4 Schematic of laser triangulation method of measurement. As light strikes the surface, a lens images the point of illumination onto a photosensor. Variations in the surface cause the image dot to move laterally along the photosensor.

Displacement and Alignment Measurements. Laser interferometers provide a high-accuracy length standard (better than 0.5 ppm) when measuring linear positioning, straightness in two planes, pitch, and yaw. The most accurate systems consist of a two-frequency laser head, beam directing and splitting optics, measurement optics, receivers, wavelength compensators, and electronics (Fig. 6).

The most important element in the system is a two-frequency laser head that produces one frequency with a P polarization and another frequency with an S polarization. The beam is projected from the laser head to a remote interferometer, where the beam is split at the polarizing beam splitter into its two separate frequencies. The frequency with the P polarization becomes the measurement beam, and the frequency with the S polarization becomes the reference beam (Fig. 7).

The measurement beam is directed through the interferometer to reflect off a moving optical element, which may be a target mirror or retroreflector attached to the item being measured. The reference beam is reflected from a stationary optical element, which is usually a retroreflector. The measurement beam then returns to the interferometer, where it is recombined with the reference beam and directed to the receiver.

Whenever the measurement target mirror or retroreflector moves, the accompanying Doppler effect induces a frequency shift in the returning beam. Because of their orthogonal polarization, the frequencies do not interfere to form fringes until the beam reaches the receiver. Consequently, the receiver can monitor the frequency shift associated with the Doppler effect, which is compared to the reference frequency to yield precise measurement of displacement.

The principal advantage of the two-frequency system is that the distance informa-

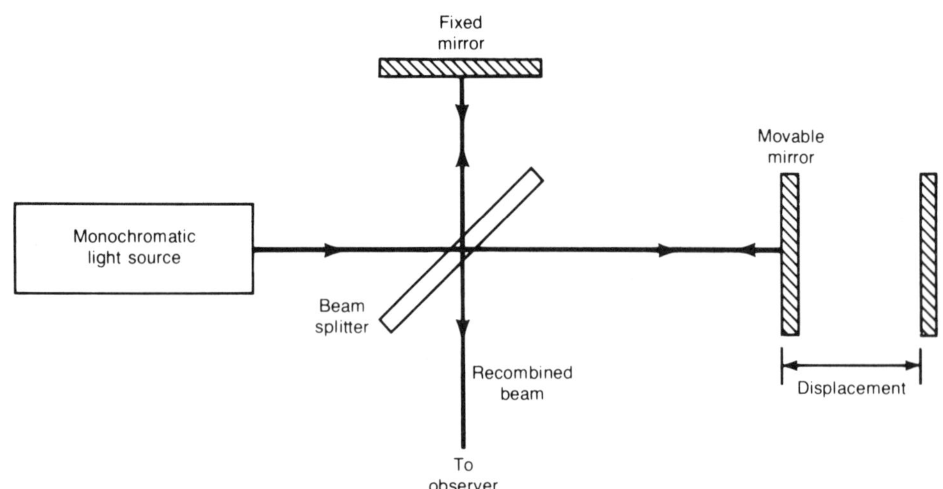

Fig. 5 Schematic of a basic interferometer

Fig. 6 Components of a laser interferometer. The components include a laser head, beam directing and measuring optics, receivers, electronic couplers, and a computer system.

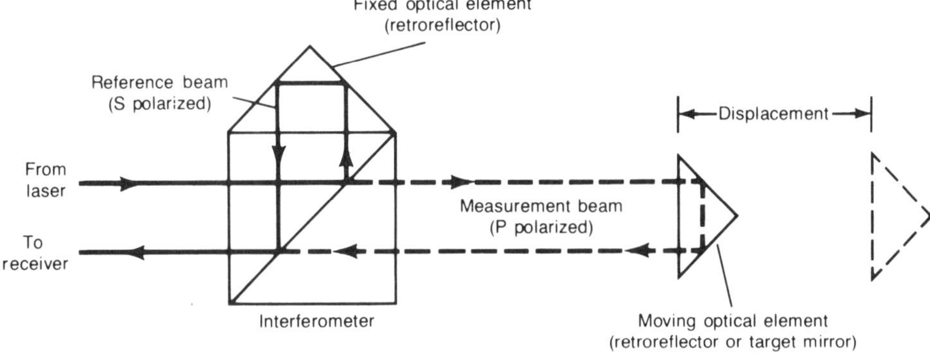

Fig. 7 Schematic of a two-frequency laser interferometer

tion is sensed in terms of frequency. Because a change in frequency is used as the basis for measuring displacement, a change in beam intensity cannot be interpreted as motion. This provides greater measurement stability and far less sensitivity to noise (air turbulence, electrical noise, and light noise). Because motion detection information is embedded in the frequency of the measurement signal, only one photodetector per measurement axis is required; this decreases the sensitivity of optical alignment. Another advantage of the two-frequency interferometer is that the laser head need not be mounted on the machine or instrument being tested.

Typical applications include the calibration of length-measuring standards such as glass scales and the characterization of positional, angular, and straightness errors in precision equipment, such as machine tools, coordinate measuring machines, and X-Y stages. The linear resolution of a two-frequency displacement interferometer is 1 nm (0.05 μin.), the angular resolution is 0.03 arc seconds, and the straightness resolution is 40 nm (1.6 μin.).

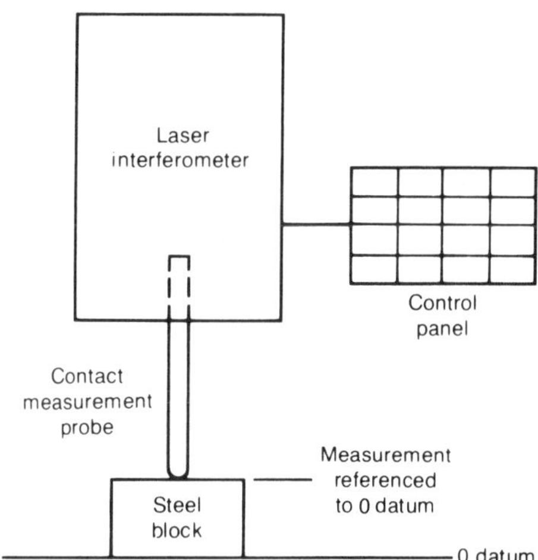

Fig. 8 Schematic of a laser interferometric micrometer

The laser interferometric micrometer uses interferometric technology and a laser beam to perform absolute length measurements to a resolution of 0.01 μm (0.4 μin.) with an accuracy of ±0.08 mm (±0.003 in.). A contact probe interfaces with an internal interferometer that measures changes in distance (Fig. 8). The part to be measured is placed under the probe, and the interference effects are electronically analyzed and displayed in terms of the distance from the probe to the datum (Fig. 8).

Based on user-entered information, the system can automatically compensate for room temperature, humidity, atmospheric pressure, the temperature of the part, and the thermal expansion of the probe. The instrument performs gage comparison, maximum and minimum surface deviation, and total indicator reading measurements. Simple statistical functions include mean and one standard deviation reporting. Actual measurement readings can be compared to user-entered tolerance limits for automatic "go, no-go" testing.

Sorting. Parts can be sorted by dimension, prior to automatic assembly, with an in-process inspection system. A laser beam sorting system can provide accept-or-reject measurements of length, height, diameter, width, thread presence, and count. Each production run of a different part requires a simple setup to accommodate the part to be measured.

In operation, a collimated laser beam is optically processed, focused, and directed onto the part. A photodetector converts the light signals to electrical signals for processing. For length inspection, the laser beam is split into three beams. The center beam is stationary and acts as a reference beam. Both of the other two beams are adjusted independently by micrometer dials to define the distance between an over- or undersize measurement. The parts can then be gravity fed past the laser quantification

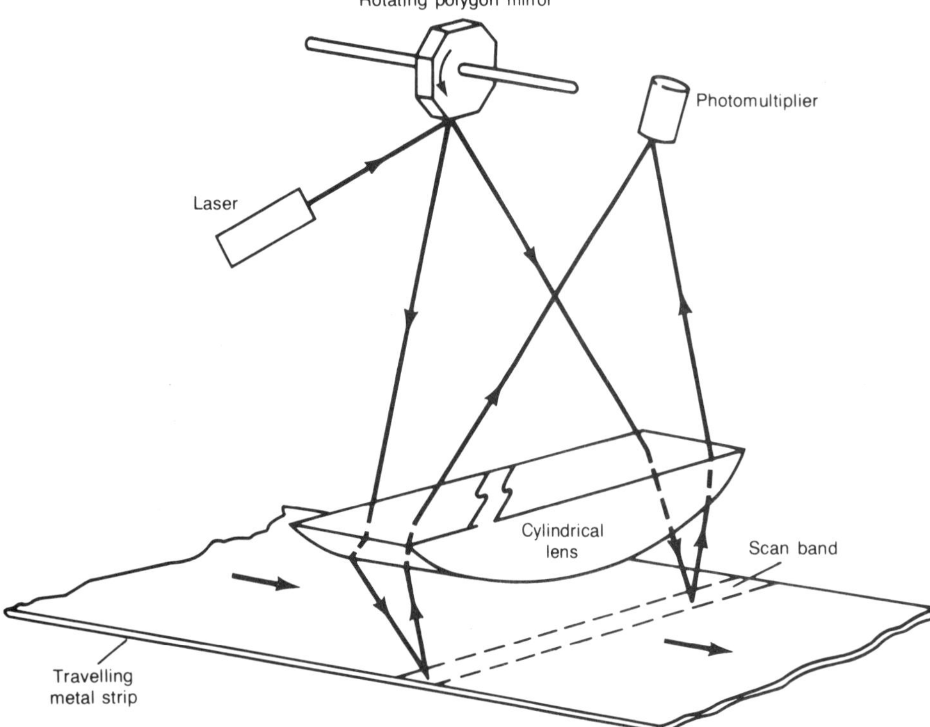

Fig. 9 System for the high-speed scanning of steel sheets for surface defects. In one system, the scanner acquired 10^4 data points per millisecond on a 3 mm^2 (0.005 in.^{2}) sheet.

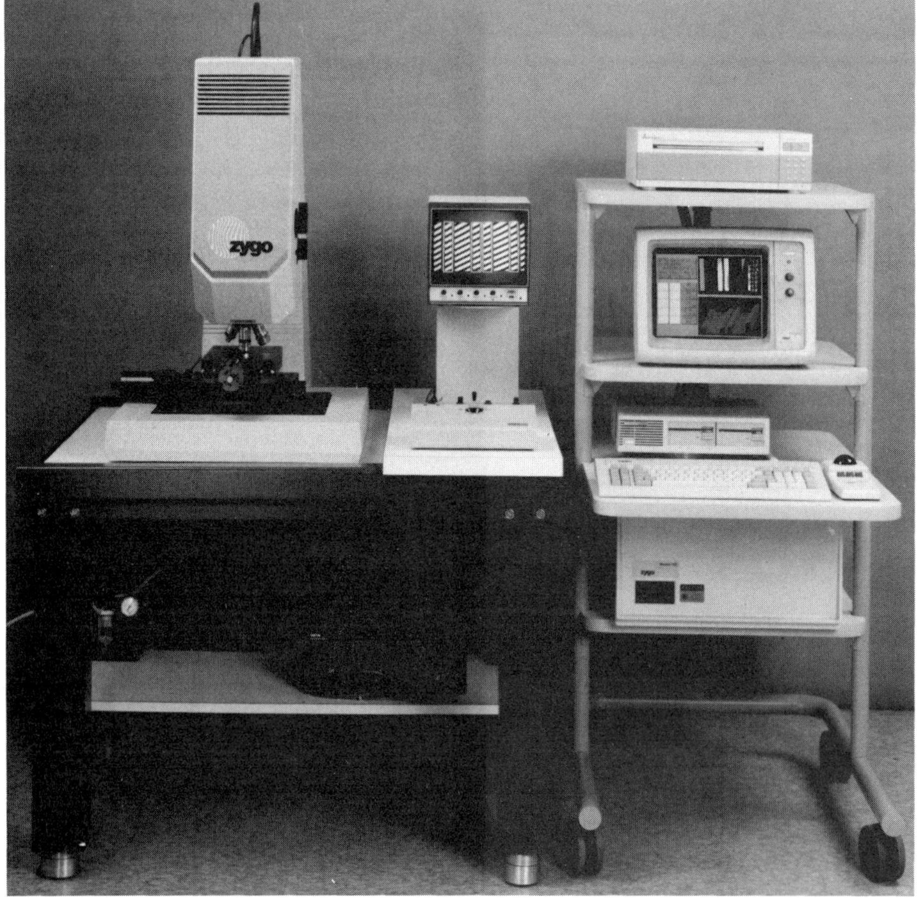

Fig. 11 Laser interference microscope with displays. Fizeau and Mirau interferometers are mounted in the turret.

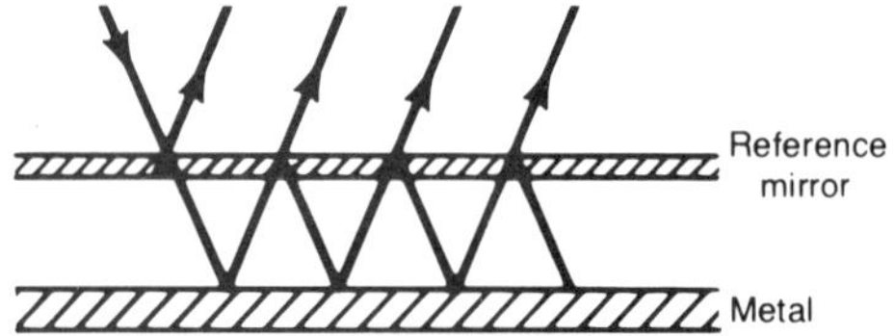

Fig. 10 General principle of a multiple-beam interferometer

system for accept-or-reject measurement. In many applications, parts can be inspected at a rate of 100 to 700 parts per minute. A typical application for a laser-based sorting system is the in-process, accept-or-reject measurement of bolts, nuts, rivets, bearings, tubes, rollers, and stampings.

Holography is an important measurement technique in the three-dimensional contouring of large spatial areas. Holography can determine small deviations (as small as 0.1 μm, or 4 μin.) in surface shape over large areas for all types of surface microstructure. This is accomplished by illuminating both the object and the hologram of its original or desired shape with the original reference wave (see the article "Optical Holography" in this Volume). If the object deviates from its original or desired shape, interference fringes will appear during illumination with the reference wave.

The exactness of the holographic image makes it invaluable for detecting faults in such diverse items as automobile clutch plates, brake drums, gas pipelines, and high-pressure tanks. The holographic image also depicts the vibration pattern of mechanical components and structures such as turbine blades.

Velocity Measurements

Velocity measurements from the Doppler effect on laser light have become a useful tool for measuring gas, liquid, and solid-surface velocities. Many systems are intended for laboratory applications. However, instruments have also been introduced for industrial process control. One application is in the primary metals industries and consists of measuring length in a unidirectional flow process.

The laser Doppler velocity gage is a noncontact instrument that uses laser beams and microprocessors to measure the speed and length of a moving surface. It can measure almost any type of continuously produced material without coming into contact with it, whether it is hot, cold, soft, or delicate. It outputs various types of measurements, such as current speed, average speed, current lengths, and total length.

The instrument consists of a sensor, the controller, and a computer. The sensor emits two laser beams that converge on the surface of the product being measured. The light reflected from the product surface exhibits Doppler shifts because of the movement. The frequency of the beam pointing toward the source of the product is shifted up, and the beam pointing toward the destination is shifted down. The processor measures the frequency shift and uses this information to calculate the speed and length of the product.

Surface Inspection

Surface inspection includes the in-process detection of surface flaws and the measurement of surface defects and roughness. Lasers are used in both of these functions.

One technique of in-process flaw detection is illustrated in Fig. 9. A rotating polygonal mirror scans the laser light across the moving sheet, and a stationary photomultiplier detects the scattered light. If there is a defect on the surface, the intensity of the scattered light increases; if the defect is a crack, the intensity decreases. Similar arrangements can estimate surface roughness by analyzing the envelope of scattered light.

Interference microscopes are used to measure the microtopography of surfaces. The interference microscope divides the light from a single point source into two or more waves. In multiple-beam interference microscopes, this is done by placing a partially transmitting and partially reflecting reference mirror near the surface of the specimen (Fig. 10).

The multiple beams illustrated in Fig. 10 are superimposed after traveling different lengths. This produces interference patterns, which are magnified by the microscope. The interference fringes having a perfectly flat surface appear as straight, parallel lines of equal width and spacing. Height variations cause the fringes to appear curved or jagged, depending on the unit used. With multiple-beam interferometers, height differences as small as $\lambda/200$ can be measured, where λ is the wavelength of the light source.

Lasers can provide a monochromatic light source, which is required in interference microscopes. One such system is shown in Fig. 11. This system involves the use of photodetectors with displays of isometric plots, contour plots, and up to five qualitative parameters, such as surface roughness, camber, crown, radius of curvature, cylindrical sag, and spherical sag.

SELECTED REFERENCES

- D. Belforte, *Industrial Laser Annual Handbook*, Vol 629, Pennwell, 1986
- R. Halmshaw, *Nondestructive Testing*, Edward Arnold, 1987

Coordinate Measuring Machines

David H. Genest, Brown & Sharpe Manufacturing Company

THE COORDINATE MEASURING MACHINE (CMM) fulfills current demands on manufacturing facilities to provide extremely accurate as well as flexible three-dimensional inspection of both in-process and finished parts on the assembly line. Manufacturers are under tremendous financial pressure to increase production and to minimize waste. Part tolerances once quoted in fractional figures are now quoted in thousandths of a millimeter, and manufacturers are under ever-increasing pressure to meet ever more demanding specifications on a regular basis.

The CMM, which first appeared some 25 years ago, has developed rapidly in recent years as the state-of-the-art measuring tool available to manufacturers. The capabilities, accuracy, and versatility of the CMM, as well as the roles it will play in manufacturing, continue to increase and evolve almost daily.

Coordinate measuring machines are an object of intense interest and aggressive development because they offer potentially viable solutions to a number of challenges facing manufacturers:

- The need to integrate quality management more closely into the manufacturing process
- The need to improve productivity and to reduce waste by eliminating the manufacture of out-of-tolerance parts faster
- The realization that the measurement process itself needs to be monitored and verified
- The objective to eliminate fixtures and fixed gages (and their inherent rebuild costs due to evolving products), thus providing increased gaging flexibility
- The potential to incorporate existing technology into new and more efficient hybrid systems

Historically, traditional measuring devices and CMMs have been largely used to collect inspection data on which to make the decision to accept or reject parts. Although CMMs continue to play this role, manufacturers are placing new emphasis on using CMMs to capture data from many sources and bring them together centrally where they can be used to control the manufacturing process more effectively and to prevent defective components from being produced. In addition, CMMs are also being used in entirely new applications—for example, reverse engineering and computer-aided design and manufacture (CAD/CAM) applications as well as innovative approaches to manufacturing, such as the flexible manufacturing systems, manufacturing cells, machining centers, and flexible transfer lines.

Before purchasing a CMM, the user needs to understand and evaluate modern CMMs and the various roles they play in manufacturing operations today and in the future. This article will:

- Define what a CMM is
- Examine various types of machines available
- Outline CMM capabilities
- Examine major CMM components and systems
- Examine various applications in which CMMs can be employed
- Provide guidelines for use in specifying and installing CMMs

Terminology germane to CMMs includes:

- *Ball bar*: A gage consisting of two highly spherical tooling balls of the same diameter connected by a rigid bar
- *Gage*: A mechanical artifact of high precision used either for checking a part or for checking the accuracy of a machine; a measuring device with a proportional range and some form of indicator, either analog or digital
- *Pitch*: The angular motion of a carriage, designed for linear motion, about an axis that is perpendicular to the motion direction and perpendicular to the yaw axis
- *Pixel*: The smallest element into which an image is divided, such as the dots on a television screen
- *Plane*: A surface of a part that is defined by three points
- *Repeatability*: A measure of the ability of an instrument to produce the same indication (or measured value) when sequentially sensing the same quantity under similar conditions
- *Roll*: The angular motion of a carriage, designed for linear motion, about the linear motion axis
- *Yaw*: The angular motion of a carriage, designed for linear motion, about a specified axis perpendicular to the motion direction. In the case of a carriage with horizontal motion, the specified axis should be vertical unless explicitly specified. For a carriage that does not have horizontal motion, the axis must be explicitly specified

CMM Operating Principles

Technically speaking, a CMM is a multiaxial device with two to six axes of travel or reference axes, each of which provides a measurement output of position or displacement. Coordinate measuring machines are primarily characterized by their flexibility, being able to make many measurements without adding or changing tools. As products evolve, the same CMM can generally be used (depending on size and accuracy limitations) simply by altering software instead of altering equipment mechanics or electronics.

Practically speaking, CMMs consist of the machine itself and its probes and moving arms for providing measurement input, a computer for making rapid calculations and comparisons (to blueprint specifications, for example) based on the measurement input, and the computer software that controls the entire system. In addition, the CMM has some means of providing output to the user (printer, plotter CRT, and so on) and/or to other machines in a complete manufacturing system. Coordinate measuring machines linked together in an overall inspection or manufacturing system are referred to as coordinate measuring systems.

When CMMs were first introduced in the late 1950s, they were called universal measuring machines. Today, they are sometimes referred to as flexible inspection systems or flexible gages. The most important feature of the CMM is that it can rapidly and accurately measure objects of widely varying size and geometric configuration—for example, a particular part and the tooling for that part. Coordinate measuring machines can also readily measure the many different features of a part, such as holes, slots, studs, and weldnuts, without needing other tools. Therefore, CMMs can replace

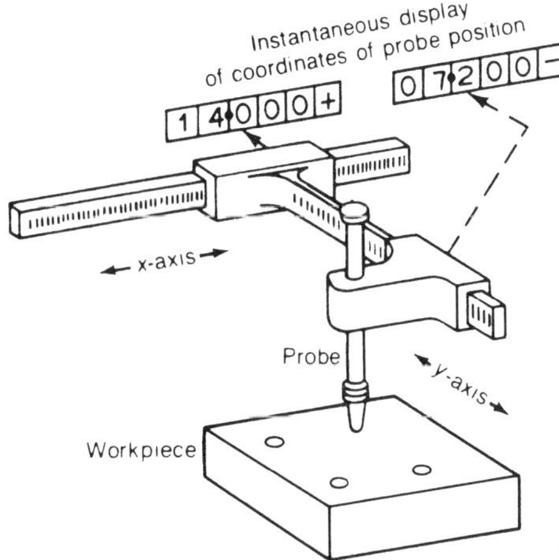

Fig. 1 Elements of a CMM showing typical digital position readout. The probe is positioned by brackets slid along two arms. Coordinate distances from one point to another are measured in effect by counting electronically the lines in gratings ruled along each arm. Any point in each direction can be set to zero, and the count is made in a plus or minus direction from there.

the numerous hand tools used for measurement as well as the open-plate and surface-plate inspection tools and hard gages traditionally used for part measurement and inspection.

Coordinate measuring machines do not always achieve the rates of throughput or levels of accuracy possible with fixed automation-type measuring systems. However, if any changes must be made in a fixed system for any reason—for example, a different measurement of the same part or measurement of a different part—making the change will be costly and time consuming. This is not the case with a CMM. Changes in the measurement or inspection routine of a CMM are made quickly and easily by simply editing the computer program that controls the machine. The greater or more frequent the changes required, the greater the advantage of the CMM over traditional measuring devices. This flexibility, as well as the resulting versatility, is the principal advantage of the CMM.

CMM Measurement Techniques

A CMM takes measurements of an object within its work envelope by moving a sensing device called a probe along the various axes of travel until the probe contacts the object. The precise position of the contact is recorded and made available as a measurement output of position or displacement (Fig. 1). The CMM is used to make numerous contacts, or hits, with the probe; using all axes of travel, until an adequate data base of the surfaces of the object has been constructed. Various features of an object require different quantities of hits to be accurately recorded. For example, a plane, surface, or circular hole can be recorded with a minimum of three hits.

Once repeated hits or readings have been made and stored, they can be used in a variety of ways through the computer and geometric measurement software of the CMM. The data can be used to create a master program, for example, of the precise specifications for a part; they can also be compared (via the software) to stored part specification data or used to inspect production parts for compliance with specifications. A variety of other sophisticated applications are also possible using the same captured measurement data—for example, the reverse engineering of broken parts or the development of part specifications from handmade models.

Coordinate Systems. The CMM registers the various measurements (or hits) it takes of an object by a system of coordinates used to calibrate the axes of travel. There are several coordinate systems in use. The most commonly used system is Cartesian, a three-dimensional, rectangular coordinate system the same as that found on a machine tool. In this system, all axes of travel are square to one another. The system locates a position by assigning it values along the x, y, and z axes of travel.

Another system used is the polar coordinate system. This system locates a point in space by its distance or radius from a fixed origin point and the angle this radius makes with a fixed origin line. It is analogous to the coordinate system used on a radial-arm saw or radial-arm drill.

Types of Measurements. As stated earlier, fundamentally, CMMs measure the size and shape of an object and its contours by gathering raw data through sensors or probes. The data are then combined and organized through computer software programs to form a coherent mathematical representation of the object being measured, after which a variety of inspection reports can be generated. There are three general types of measurements for which CMMs are commonly used, as follows.

Geometric measurement deals with the elements commonly encountered every day—points, lines, planes, circles, cylinders, cones, and spheres. In practical terms, these two-dimensional and three-dimensional elements and their numerous combinations translate into the size and shape of various features of the part being inspected.

A CMM can combine the measurements of these various elements into a coherent view of the part and can evaluate the measurements. It can, for example, gage the straightness of a line, the flatness of a plane surface, the degree of parallelism between two lines or two planes, the concentricity of a circle, the distance separating two features on a part, and so on. Geometric measurement clearly has broad application to many parts and to a variety of industries.

Contour measurement deals with artistic, irregular, or computed shapes, such as automobile fenders or aircraft wings. The measurements taken by a CMM can be easily plotted with an exaggerated display of deviation to simplify evaluation. Although contour measurements are generally not as detailed as geometric measurements, presenting as they do only the profile of an object with its vector deviation from the nominal or perfect shape, they too have broad application.

Specialized surface measurement deals with particular, recurring shapes, such as those found on gear teeth or turbine blades. In general, these shapes are highly complex, containing many contours and forms, and the part must be manufactured very precisely. Tight tolerances are absolutely critical. Because manufacturing accuracy is critical, measurement is also highly critical, and a specialty in measuring these forms has evolved. By its nature, specialized surface measurement is applied to far fewer applications than the other two types.

CMM Capabilities

Coordinate measuring machines have the fundamental ability to collect a variety of different types of very precise measurements and to do so quickly, with high levels of repeatability and great flexibility. In addition, they offer other important capabilities based on computational functions.

Automatic Calculation of Measurement Data. The inclusion of a computer in the CMM allows the automatic calculation of such workpiece features as hole size, boss size, the distance between points, incremental distances, feature angles, and intersections. Prior to this stage of CMM development, an inspector had to write down the measurements he obtained and manually compare them to the blueprint. Not only is such a process subject to error, but it is relatively time consuming. While waiting for the results of the inspection, production decisions are delayed and parts (possibly not being produced to specifications) are being manufactured.

Compensation for Misaligned Parts. Coordinate measuring machines no longer require that the parts being measured be manually aligned to the coordinate system of the machine. The operator cannot casually place the part within the CMM work envelope. Once the location of the appropriate reference surface or line has been determined through a series of hits on the datum features of the part, the machine automatically references that position as its zero-zero starting point, creates an x, y, z part coordinate system, and makes all subsequent measurements relative to that point. In addition, the part does not have to be leveled within the work envelope. Just as the CMM will mathematically compensate if the part is rotationally misaligned, it will also compensate for any tilt in the part.

Multiple Frames of Reference. The CMM can also create and store multiple frames of reference or coordinate systems; this allows features to be measured on all surfaces of an object quickly and efficiently. The CMM automatically switches to the appropriate new alignment system and zero point (origin) for each plane (face) of the part. The CMM can also provide axis and plane rotation automatically.

Probe Calibration. The CMM automatically calibrates for the size and location of the probe tip (contact element) being used. It also automatically calibrates each tip of a multiple-tip probe.

Part Program and Data Storage. The CMM stores the program for a given part so that the program and the machine are ready to perform whenever this part comes up for inspection. The CMM can also store the results of all prior inspections of a given part or parts so that a complete history of its production can be reconstructed. This same capability also provides the groundwork for all statistical process control applications.

Part programs can also be easily edited, rather than completely rewritten, to account for design changes. When a dimension or a feature of a part is changed, only that portion of the program involving the workpiece revision must be edited to conform to part geometry.

Interface and Output. As mentioned earlier, CMMs can be linked together in an overall system or can be integrated with other devices in a complete manufacturing system. The CMM can provide the operator with a series of prompts that tell him what to do next and guide him through the complete measurement routine.

Output is equally flexible. The user can choose the type and format of the report to be generated. Data can be displayed in a wide variety of charts and graphs. Inspection comments can be included in the hard copy report and/or stored in memory for analysis of production runs.

CMM Applications

Coordinate measuring machines are most frequently used in two major roles: quality control and process control. In the area of quality control, CMMs can generally perform traditional final part inspection more accurately, more rapidly, and with greater repeatability than traditional surface-plate methods.

With regard to process control, CMMs are providing new capabilities. Because of the on-line, real-time analytical capability of many CMM software packages, CMMs are increasingly used to monitor and identify evolving trends in production before scrap or out-of-spec parts are fabricated in the first place. Thus, the emphasis has shifted from inspecting parts and subsequently rejecting scrap parts at selected points along the production line to eliminating the manufacture of scrap parts altogether and producing in-tolerance parts 100% of the time.

In addition to these uses, there is a trend toward integrating CMMs into systems for more complete and precise control of production. Some shop-hardened CMMs, also known as process control robots, are being increasingly used in sophisticated flexible manufacturing systems in the role of flexible gages.

Coordinate measuring machines can also be used as part of a CAD/CAM system. The CMM can measure a part, for example, and feed that information to the CAD/CAM program, which can then create an electronic model of the part. Going in the other direction, the model of the desired part in the CAD/CAM system can be used to create the part program automatically.

Types of CMMs

The ANSI/ASME B89 standard formally classifies CMMs into ten different types based on design. All ten types employ three axes of measurement along mutually perpendicular guideways. They differ in the arrangement of the three movable components, the direction in which they move, and which one of them carries the probe, as well as where the workpiece is attached or mounted. However, among the many different designs of CMMs, each with its own strengths, weaknesses, and applications, there are only two fundamental types: vertical and horizontal. They are classified as such by the axis on which the probe is mounted and moves. The ANSI/ASME B89 Performance Standard classifies coordinate measuring machines as:

Vertical	Horizontal
Fixed-table cantilever	Moving ram, horizontal arm
Moving-table cantilever	Moving table, horizontal arm
Moving bridge	Fixed table, horizontal arm
Fixed bridge	
L-shaped bridge	
Column	
Gantry	

In addition, the two types of machine can be characterized to some degree by the levels of accuracy they each achieve (although there is a considerable degree of overlap based on the design of an individual machine), by the size of part they can handle, and by application. The prospective buyer/user of a CMM cannot make an intelligent choice of the type of machine that will best meet his needs, let alone the specific make and model of CMM, until he thoroughly evaluates and plans both the specific intended application of the CMM and the overall manufacturing and quality context in which it will operate. Table 1 provides a general comparison of CMM types, applications, and levels of measurement accuracy a CMM user can expect.

Experience clearly shows that the user who simply buys a CMM and places it in use without considering the impact of the unit on the manufacturing process will almost invariably fail. On the other hand, the user who determines the results desired from the CMM, the resources required to achieve those results, and the effect of the CMM on the overall manufacturing process will likely realize the maximum benefits a CMM can provide.

Evaluation and planning are further discussed in the section "CMM Implementation" in this article.

Vertical CMMs

Vertical CMMs, which have the probe or sensor mounted on the vertical z-axis, have the potential to be the most accurate type. Vertical CMMs in general can be more massive and can be built with fewer moving parts than their horizontal counterparts. They are therefore more rigid and more stable. Their limitation, however, is the size of part they can conveniently handle, because the part to be measured must fit under the structural member from which the probe descends.

In an attempt to overcome this limitation, various designs have been produced. As a result, within the overall category of vertical CMM, there are designs utilizing moving or fixed bridges, cantilevers, gantries, and so on. Each design approach is of course a compromise, because as the size of the work envelope of the machine and the travel distance along the axes increase, so also do the problems of maintaining rigidity and accuracy. A gantry design has proved to be an effective solution to the problem of increasing the capacity of a CMM while maintaining a high level of accuracy. A cantilever design, on the other hand, presents some inherent problems associated with isolating the CMM from floor vibrations and maintaining high precision due to the overhanging unsupported arm.

Cantilever-type CMMs employ three movable components moving along mutually perpendicular guideways. The probe is attached to the first component, which moves vertically (z-direction) relative to the second. The second component moves horizontally (y-direction) relative to the third. The third component is supported at one end only, cantilever fashion, and moves horizontally (x-direction) relative to the machine base. The workpiece is supported on the worktable. A typical machine of this configuration is shown in Fig. 2(a). A modification of the fixed-table cantilever configuration is the moving-table cantilever CMM shown in Fig. 2(b).

The cantilever design provides openness and accessibility from three sides, making it popular for small, manual CMMs. Its cantilevered y-axis places a size limitation on

Table 1 Typical CMM specifications

Application	CMM type	Bearing type	Minimum measurement mm	in.
Laboratory quality(a)				
Laboratory grade	Vertical, moving bridge	Air bearings	<0.003	<0.0001
Clean room	Vertical, moving bridge	Air bearings	<0.003	<0.0001
Production machines				
Open shop	Vertical, moving bridge	Air bearings	<0.013–0.025	<0.0005–0.001
Sheet metal	Horizontal (with fixed x-y axis)	Recirculating bearing packs	<0.025–0.050	<0.001–0.002
Clean room and shop	Vertical, moving bridge; all horizontal	Air/roller bearings	<0.050	<0.002

(a) These CMMs have specific environments into which they must be installed to maintain their rated accuracies.

this configuration. Because of the small y-z assembly, this configuration is lightweight and provides fast measuring speeds in direct computer control (DCC) applications.

Bridge-type CMMs employ three movable components moving along mutually perpendicular guideways. The probe is attached to the first component, which moves vertically (z-direction) relative to the second. The second component moves horizontally (y-direction) relative to the third. The third component is supported on two legs that reach down to opposite sides of the machine base, and it moves horizontally (x-direction) relative to the base. The workpiece is supported on the base.

Moving Bridge. A typical moving-bridge CMM is shown schematically in Fig. 3(a). This configuration accounts for 90% of all CMM sales.

The moving-bridge design overcomes the size limitations inherent in the cantilever design by providing a second leg, which allows for an extended y-axis. The second leg does, however, reduce access to the unit. The limitations of this configuration usually occur because of walking problems associated with retaining drive through just one leg. Higher speeds, achieved by increasing dynamic forces and reducing machine setting time, accentuate the problem. Vertical moving-bridge CMMs can be controlled manually (Fig. 4) and with DCC hardware (Fig. 5).

The fixed-bridge configuration (Fig. 3b) provides a very rigid structure and allows a relatively light moving x-z structure that can achieve fast x-z moves. The moving table in larger machines can become massive, with decreased throughput capability. The influence of part weight on accuracy becomes a consideration for large parts.

L-Shaped Bridge. Another modification of the bridge configuration has two bridge-shaped components (Fig. 3c). One of these bridges is fixed at each end to the machine base. The other bridge, which is an inverted L-shape, moves horizontally (x-direction) on guideways in the fixed bridge and machine base.

The column CMM goes one step further than the fixed bridge in providing a very rigid z-axis configuration, and a two-axis saddle that allows movement in the horizontal (x-y) directions (Fig. 6). High accuracy can be achieved with this design. As in the fixed-bridge configuration, however, part mass and table considerations can restrict measuring volume and speed.

Column CMMs are often referred to as universal measuring machines rather than CMMs by manufacturers. Column units are considered gage-room instruments rather than production-floor machines.

Gantry CMMs employ three movable components moving along mutually perpendicular guideways (Fig. 7). The probe is attached to the probe quill, which moves vertically (z-direction) relative to a crossbeam. The probe quill is mounted in a carriage that moves horizontally (y-direction) along the crossbeam. The crossbeam is supported and moves in the x-direction along two elevated rails, which are supported by columns attached to the floor.

The gantry design has relatively restricted part access unless utilized in very large machines. The machine is physically large with respect to the size of the part. Large axis travels can be obtained, and heavy parts are not a problem, because the weight of the part can be decoupled from the measurement system by proper design of the machine base (foundation in a large machine). This is not as practical in smaller CMMs, and this configuration is most popular for large machines.

The gantry configuration was initially introduced in the early 1960s to inspect large parts, such as airplane fuselages, automobile bodies, ship propellers, and diesel engine blocks. The open design permits the operator to remain close to the part being inspected while minimizing the inertia of the moving machine parts and maintaining structural stiffness.

Horizontal CMMs

Horizontal CMMs, which have the probe mounted on the horizontal y-axis, are generally used in applications in which large parts must be measured—for example, automobile bodies or airplane wings. Horizontal CMMs require no bridge over the part because the part is approached from the side. Therefore, there is substantially less restriction on the sizes of the parts that can be measured.

Most horizontal CMMs, however, do not measure to state-of-the-art levels of accuracy, because of the high cost of achieving such accuracy in machines capable of handling large parts. In general, it is more cost effective to accept accuracy in the 0.050

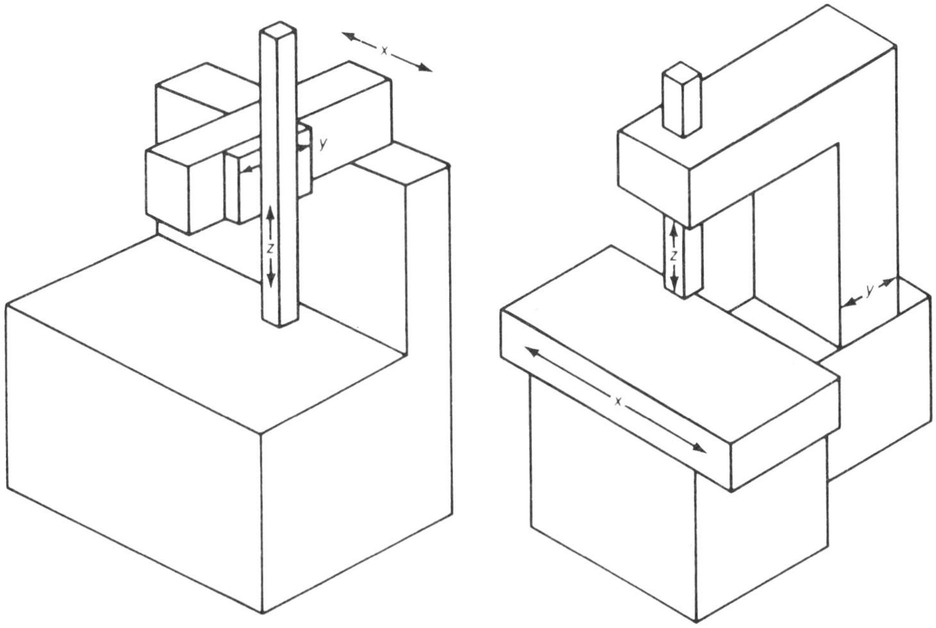

(a) (b)

Fig. 2 Motion of components in cantilever-type CMMs. (a) Fixed table. (b) Moving table

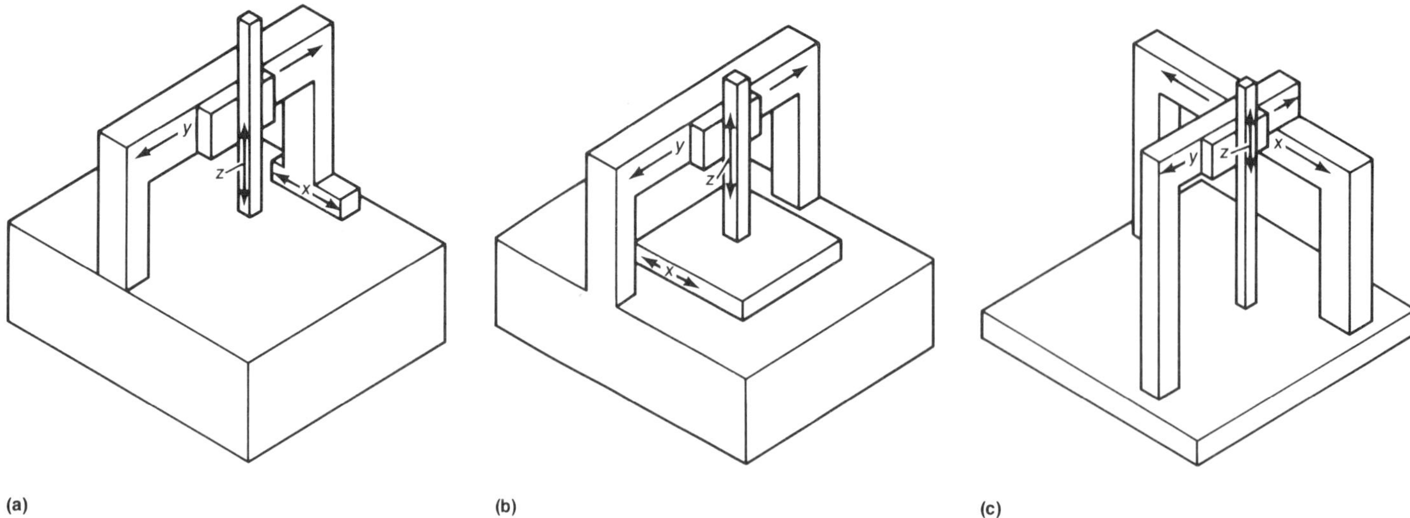

Fig. 3 Motion of components in bridge-type CMMs. (a) Moving bridge. (b) Fixed bridge. (c) L-shaped bridge

mm (0.002 in.) range when gaining the part-handling capabilities of large horizontal machines and to use vertical designs when finer measurement is demanded.

In any CMM design, fewer moving parts and joints will result in higher potential levels of accuracy. This principle has been applied to a class of horizontal CMM called process control robots. In these units, fixed members that move together provide the flexibility and capacity of a horizontal CMM, along with the higher accuracy of a vertical design. The horizontal direction of attack makes these CMMs the logical design for production applications in which horizontal machine tools are used.

In choosing a CMM, the buyer must take into consideration not only the degree of accuracy required but also the location of the unit and the measurements to be taken. In general, when dealing with smaller parts where measurements of very high accuracy are required, the potential user is likely to be best served by a machine located in a clean room environment. Where on-line process control is desired, the appropriate shop-hardened CMM that can be located in the shop itself is the best solution.

Horizontal-Arm CMMs. Several different types of horizontal-arm CMMs are avail-

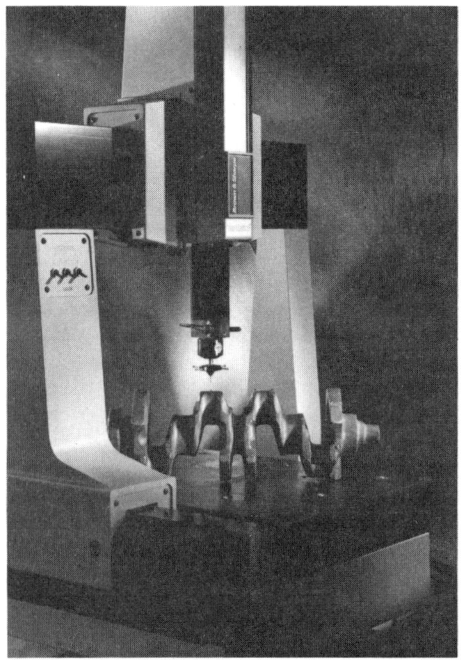

Fig. 4 Vertical moving-bridge CMM inspecting a case-hardened cast steel automotive crankshaft using a touch-trigger probe. This manually operated unit has a granite table.

Fig. 5 Vertical moving-bridge CMM, which incorporates a granite base, inspecting a cast aluminum alloy four-wheel-drive transaxle gear case. The DCC motor drive that controls all of the axial movements is equipped with a remote joystick controller.

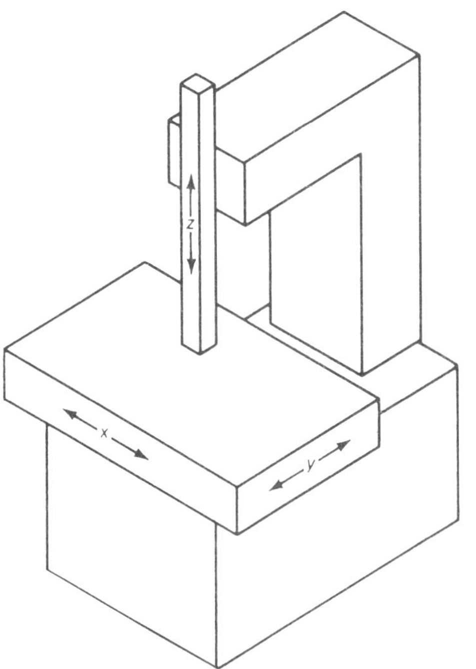

Fig. 6 Schematic of column CMM illustrating movement of probe, column, and table components

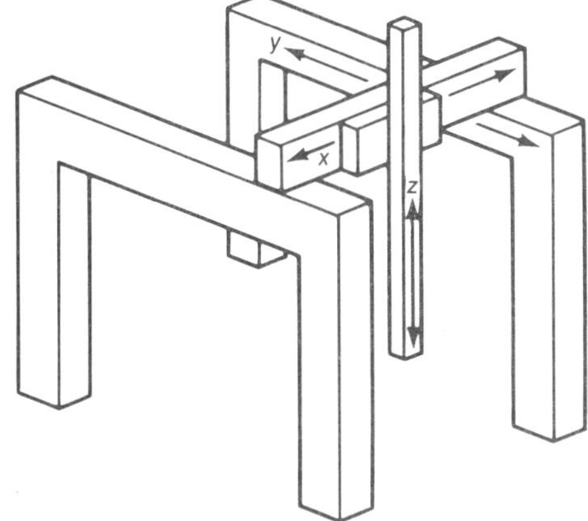

Fig. 7 Schematic of gantry CMM illustrating movement of probe, crossbeam, and elevated rails

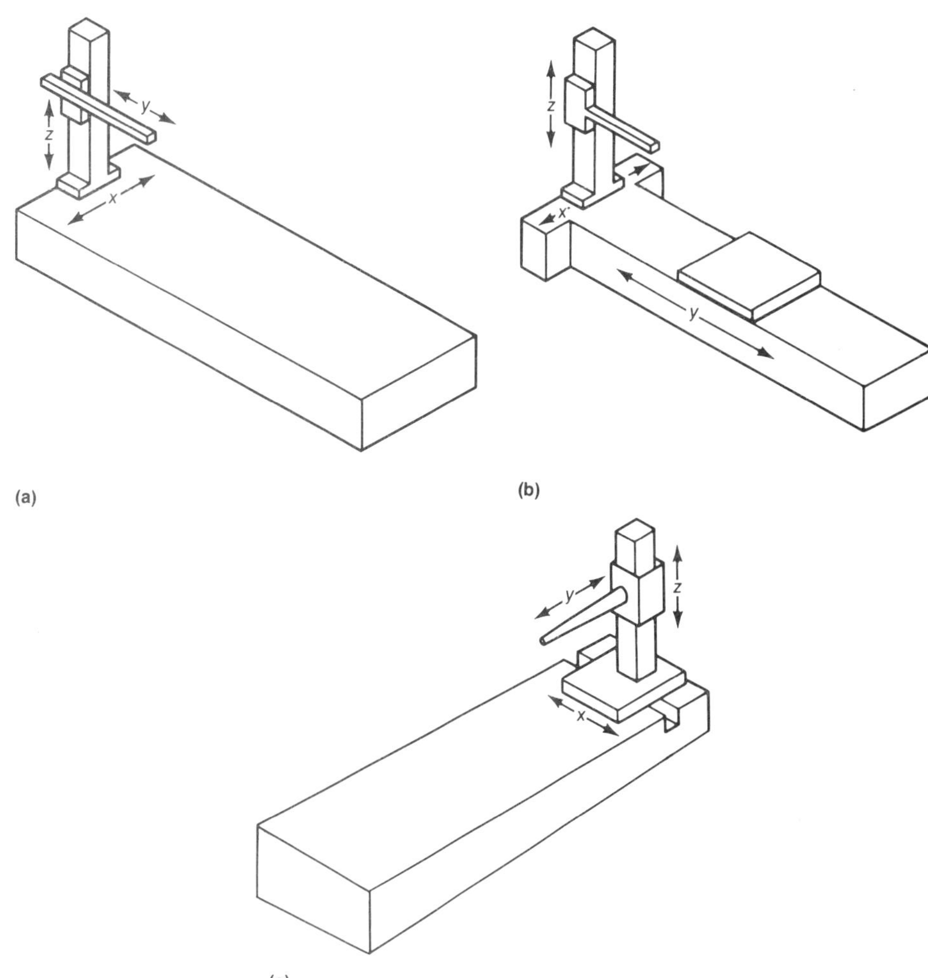

Fig. 8 Schematic illustrating three types of horizontal-arm CMMs. (a) Moving ram. (b) Moving table. (c) Fixed table. Length of table/base is usually two to three times the width. Units with bases capable of accommodating two to three interstate buses situated end-to-end have been built for major automotive manufacturers.

able. As with all CMMs, the horizontal-arm configuration employs three movable components moving along mutually perpendicular guideways.

Horizontal-arm CMMs are used to inspect the dimensional and geometric accuracy of a broad spectrum of machined or fabricated workpieces. Utilizing an electronic probe, these machines check parts in a mode similar to the way they are machined on horizontal machine tools. They are especially suited for measuring large gearcases and engine blocks, for which high-precision bore alignment and geometry measurements are required. Four-axis capability can be obtained by incorporating a rotary table.

Horizontal arms for large machines have a lower profile than vertical arms. For some applications, horizontal access is desirable. For others, it is restrictive and a rotary table is usually required, thus increasing the cost.

Moving-Ram Type. In this design, the probe is attached to the horizontal arm, which moves in a horizontal *y*-direction (Fig. 8a). The ram is encased in a carriage that moves in a vertical (*z*) direction and is supported on a column that moves horizontally (*x*-direction) relative to the base.

Moving-Table Type. In this configuration, the probe is attached to the horizontal arm, which is permanently attached at one end only to a carriage that moves in a vertical (*z*) direction (Fig. 8b) on the column. The arm support and table move horizontally (*x*- and *y*-directions) relative to the machine base. The moving-table horizontal-arm CMM unit shown in Fig. 9 is even more versatile because of the introduction of a rotary moving table.

Fixed-Table Type. In the fixed-table version, the probe is attached to the horizontal arm, which is supported cantilever style at the arm support and moves in a vertical (*z*) direction (Fig. 8c). The arm support moves horizontally (*x*- and *y*-directions) relative to the machine base. Parts to be inspected are mounted on the machine base. The versatility of the fixed-table horizontal-arm design is readily illustrated by its ability to inspect a relatively plane surface such as a door (Fig. 10) as well as an array of cylindrical surfaces encountered in a valve assembly (Fig. 11).

Key CMM Components

As stated earlier, it is virtually impossible for the CMM buyer to select the appropriate unit without carefully evaluating which functions are needed, where the unit will operate, under what conditions, who will operate it, and so on. Every CMM offers trade-offs and its own cost-to-benefits ratio. The CMM that provides state-of-the-art measurement, for example, may not be the best choice if the user cannot provide the correct operating environment. The best CMM is the one that matches the application in terms of design, capacity, accuracy, protection, and so on. Once these factors

Fig. 9 Horizontal-arm moving-table CMM equipped for operation in hostile environments is shown inspecting a cast iron automotive engine block. The unit incorporates a rotary table, a motorized programmable probe, and an environmentally protected automatic probe changer. With its process control robot, this CMM is designed for easy integration and installation in manufacturing cells.

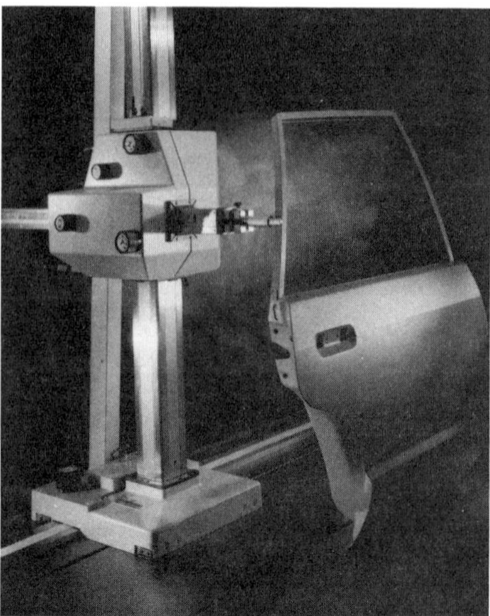

Fig. 10 Horizontal-arm fixed-table CMM using a metal scribing tool to inspect a steel sheet metal automotive door taken from the production line. This manually operated unit has a cast iron table.

have been considered, the CMM buyer should assess the characteristics and capabilities of the following major components of the CMM.

Base/Bed Construction. The first major component to examine is the overall construction of the machine. The CMM must be rigidly built to minimize errors introduced by unintended movement between the machine members and in their movement along the axes of travel. Some designs (such as the cantilever) require a more massive structure to provide the same degree of rigidity than do others (such as the gantry type) that have more inherent rigidness. If the CMM is to operate in the shop, the user should select a design having the same toughness and shielding as a machine tool. If the CMM will be used in a laboratory, a less protected model can be selected.

Traditionally, all measuring machines had cast iron bases or beds, but today granite is also frequently used because of its mass and thermal stability. In the past, it was essential that the bed of the measuring machine be perfectly flat, but with the advent of computer-based error compensation, this characteristic has become far less vital.

Coordinate measuring machines designed for use in the laboratory or clean room are usually built with granite beds. Horizontal-arm CMMs and any CMM intended for use on the factory floor are still built with a cast iron base because of their larger size and the levels of accuracy required in these applications. Steel is also far easier than granite to fabricate and use in integrated systems involving automated materials handling and similar functions.

Key elements of the structure (such as bearing ways and scales) may be of stainless steel, granite, aluminum, or more exotic materials, such as ceramic. One principal factor, however, is that all these critical components must be of the same material. This reduces the effects of expansion and contraction due to temperature changes. The buyer wants a machine that is thermally stable, and the manufacturer can achieve this stability in several ways. A CMM whose key components are of ceramic will be thermally stable because ceramic reacts very slowly to temperature changes. By the same token, a CMM with aluminum components will provide thermal stability because these components react very quickly to temperature changes. In both cases, the user knows with a high degree of certainty the thermal condition of the CMM.

Bearing Construction. Bearings have a direct impact on the accuracy of all CMMs because of their effect on every motion of the machine along its axes. Bearing assemblies can be of the air or roller type or can consist of recirculating bearing packs.

Air Bearings. For CMMs that will be used in a laboratory or clean room, where there is a minimum of dirt or dust, air bearings are the best choice because they move without friction and are therefore the most accurate. Air bearings are not suitable on CMMs located on the shop floor, because the bearings need constant cleaning and require costly air-activated vibration isolation systems, which slow the measuring speed and decrease the accuracy of the CMM. The isolation system also increases the potential for failure because of its complexity and generally fragile design.

Roller Bearings. For many shop applications, such as those for sheet metal grade CMMs, roller bearings are a good choice. Although they provide somewhat lower levels of potential accuracy than air bearings, roller bearings are sturdier and can function in an atmosphere containing some dirt and dust. They are not suitable for CMMs in hostile environments.

Recirculating Bearing Packs. The durability of roller bearings and accuracy approaching that of air bearings are available from recirculating bearing packs such as those typically used on grinding machines. Because these bearing packs are completely sealed, they permit operation in harsh factory environments, and they can handle the weight of larger, more rigid machines. Their inherent stiffness provides volumetric accuracies and repeatabilities equal to those of air bearings at significantly higher accelerations, which shorten measurement cycle times.

Scales and Encoders. The scales of a CMM show where the probe is located on

Fig. 11 Horizontal-arm fixed-table CMM using an edge-finder probe to inspect a precision cast aluminum alloy valve body. This manually operated unit incorporates a cast iron table.

the x, y, and z axes within the work envelope of the machine. Because CMMs measure to such a high degree of resolution, the scales used are machine readable and cannot be read by the naked eye. The CMM encoder reads the scales and inputs this measurement data into the computer for computation. A machine readout amplifies the same measurement data so that they are accessible to the operator by the naked eye. The buyer should make sure that the readout on any CMM is both visible and decipherable.

There are many different kinds of scales—for example, rotary, wire, magnetic, linear metal, and glass. The most accurate scales are glass because this material permits finer etching of measurement lines than steel or other commonly used materials. The resolution of the finest scales is to the submicron level. The buyer should make sure that the scales of any machine intended for shop use are thoroughly sealed and protected from the outside environment.

Sensors and Probes. Sensors and probes are the devices through which CMMs collect their measurement input. A broad variety of probes are currently available, each with its own application, permitting users to obtain very accurate measurements of virtually any type of part feature, contour, surface, and so on.

Hard or fixed probes have been in use since the mid-1970s. They were the original type used on CMMs. The CMM operator manually brings a hard probe into contact with the object being measured and reads (via the machine) the coordinates of the measurement from the machine scales. Hard probes are available in a variety of configurations (ball, tapered plug, and edge) and continue to have broad application and utility, primarily on manual machines.

Of the various types of hard probes, ball probes are used to measure the distance from one point to another. Tapered-plug probes are used for such measurements as the distance between the centerline of two holes. Edge probes are used to locate edge points.

The shortcoming of all hard or fixed probes is that they depend on a subjective element in the performance of the machine operator. Every operator has his own touch, a light or heavy hand in moving and bringing the probe in contact with the part. Just as one operator will use more force than another to close a micrometer on a part and thus obtain a variation in measurement, the CMM operator can unknowingly influence measurements taken with a hard probe.

Touch-trigger or electronic touch-trigger probes are the second major category of probe currently used. Touch-trigger probes operate like an electronic switch that closes when the probe comes in contact with the workpiece measured. The development of touch-trigger probes was a major breakthrough for CMM technology.

Not only do touch-trigger probes remove a major source of operator error, they greatly increase the flexibility of measurements and facilitate direct computer-controlled CMMs. Because of their considerable advantages, touch-trigger probes are replacing hard probes in many applications. Touch-trigger probes are also available in many configurations and types, that is, with extensions, special mounts, multiaxis motorized heads, and so on.

Noncontact trigger probes are used in a similar manner to the contact touch-trigger probes described above. However, with noncontact probes, a beam of light (generally a diode) is used (operating as an optical switch) to probe the workpiece rather than the physical contact of a probe tip. The probe is permanently set to a specific standoff distance at which the laser beam is triggered and measurements are taken. Because the probe never comes in contact with the part, the likelihood of damage to the probe or part is greatly reduced, and the measurement speed is greatly improved with only a slight reduction in accuracy.

Touch Scanners. This type of probe, which produces analog readings instead of the digital measurements produced by the probes mentioned above, is used for taking contour measurements. These probes remain in contact with the workpiece as they move along its surface.

Laser probes primarily provide measurements of contoured surfaces, but do so without coming in contact with the surface. Laser probes project a beam of laser light, called a footprint, onto the surface of the part, the position of which is then read by triangulation through the lens in the probe receptor. This triangulation (much like surveying) in turn provides the actual position and so on, of the feature on the part being measured.

Vision Probes. This type of sensor is the most recent and sophisticated addition to the sensor systems currently in use. Vision is another form of noncontact sensing and is especially useful where very high speed inspection or measurement is required. In this system, the part is not measured directly. Instead, an electronic representation of it is measured and evaluated. The various features of this image of the workpiece (size, shape, location, and so on) are measured in comparison to various electronic models of expected results by counting the pixels of the electronic image. From this, the true nature of the part being inspected can be inferred and reported.

Probe Changers. Probe changers store alternative and/or backup probes, permitting the exchange or replacement of various probes automatically. The measurement routine of a given part may call for the use of different probes or sensors to measure particular features of the part. A deep bore may require the use of a probe with an elongated tip, for example, while other features may require different specialized probes. The type of probe to be used for the measurement of every feature of a part is written into the parts program (the software program that controls that particular CMM operation). By utilizing a probe changer, which attaches to the CMM, the entire measurement routine can be carried out without stopping for probe changes and without operator intervention.

Computer Hardware. The computer is the heart of the CMM system, giving it the capabilities and versatility that are rapidly making CMMs essential to efficient manufacturing. The computer receives that measurement data gathered by the CMM and, taking guidance from the software program, manipulates it into the forms required by the user. The computer also performs the essential function of automatically aligning parts for measurement and automatically compensating for errors within the structure of the machine. In DCC applications, the computer also operates the CMM.

Because of the vital role of the computer, there are several points the CMM buyer must be aware of. Primarily, the buyer will want to be sure that the operating system is designed to run without assistance or information from the user. The CMM user does not want to become involved in computer programming. The user needs a fast, efficient computer that supports a state-of-the-art operating system (and peripherals, that is, printer, plotter, and so on) and has good, widespread service support. The state-of-the-art permits the user to plug in the CMM and, after proper installation, testing, and qualification, insert the software needed (available as off-the-shelf items) and then immediately begin measuring parts.

The user can communicate with and give instructions to the computer (via the software) in a number of convenient ways—for example, with keyboards, function keys, light pens, touch-sensitive video screens, or a mouse. Most of these devices are excellent for a laboratory or office, but in the shop, the simplest methods requiring the fewest parts, such as touch-sensitive screens, tend to work best. No matter what method is used, however, the CMM operator does not need, nor necessarily will he benefit from, specialized computer training.

Computer Software. Although the computer is the heart of the CMM, it is the software that enables the system to fulfill its potential. Not many years ago, obtaining the appropriate application software to run measurement routines necessitated writing a program in Fortran or Basic, but the situation has very nearly completely reversed itself today. Software for the CMM has been refined to the extent that no computer programming knowledge or experience is needed to run even the most sophisticated measurement programs available.

Virtually all CMM software consists of off-the-shelf, menu-driven (that is, the program asks the operator what he wants to do and even prompts the most likely choice) programs that are very user-friendly, with comprehensive, solid help screens available. The result is that the software (combined with the formidable power of the personal computer) for any particular routine is very easy and flexible to use and can be customized to fit individual applications using plain English rather than a specific programming language tied to a complicated operating system.

The range of available software is very wide. For example, programs are available for statistical process analysis and control. Programs for sheet metal applications facilitate the location and measurement of parts containing bolts, weldnuts, and similar features. Other contour programs permit the rapid and accurate alignment of complex, nongeometric shapes without straight edges. There is also a family of software that allows the transfer of data between CMMs and CAD systems, and other programs are available that facilitate the communication of data between CAD systems made by different manufacturers. Development in this area has been so dynamic that standards for data formats have already been developed and accepted by the industry, such as Dimensional Measurement Interface Specification for communication from CMMs to CAD systems, and Initial Graphics Exchange Specification for CAD-to-CAD data exchange. Such integration makes possible the use of CMMs in a variety of new applications, such as for reverse engineering, in which specifications are derived from the measurement of a model or a broken part, and for the development of part programs (measurement routines) directly from CAD data.

Factors Affecting CMM Performance

Because CMMs are highly sophisticated, precision machines, their performance and accuracy are affected by many environmental, design, and operating factors. Most errors in CMMs are systematic (that is, stemming from machine setup and operating conditions) rather than random. Therefore, to obtain a high level of performance, a user must first choose the appropriate machine for the application. The CMM must then be located in a suitable environment, which can range from the laboratory to the shop floor, and it must be installed with all the appropriate safeguards for proper, error-free operation. This section of the article will outline the factors that affect performance and will provide suggestions for ensuring the satisfactory performance of the CMM.

Volumetric Accuracy. Historically, measuring machines were evaluated in terms of linear scale accuracy but as they began to be used to measure more complex parts, it soon became apparent that the linear accuracy of the machine was insufficient to guarantee the level of measurement performance required. It became important, therefore, to evaluate performance on the basis of the accuracy of the CMM at any point within its three-dimensional work envelope, that is, volumetric accuracy.

Until quite recently, there was no recognized standard for gaging and evaluating the volumetric accuracy of a CMM. Under the auspices of the American Society of Mechanical Engineers and the U.S. Bureau of Standards, CMM manufacturers and users developed testing criteria and methods using laser, step gage, and ball bar tests that verify the volumetric and linear accuracy of the CMM. In 1985, those tests were published as part of ANSI National Standard B89.1.12.

There are 21 degrees of freedom or geometric inaccuracy (all of which can be sources of measurement error in a three-axis CMM): roll, pitch, yaw, scale error, repeatability, two straightness errors per axis, plus three squareness errors. To ensure volumetric accuracy, every one of these 21 degrees of freedom must be tested. When testing, the supplier must test the complete CMM system (the mechanics and electronics of the machine plus probe holder, probe, and its measurement software) because it will be used to measure parts to ensure that the results will be accurate predictors of measurement performance.

Historically, errors inherent in machine components were mechanically refined. A cast iron bed was planed, for example, to eliminate irregularities, and granite components were lapped. Today, however, mechanical errors in the CMM are refined with the CMM computer, which automatically compensates for geometric errors with powerful software algorithms in real time as measurements are taken.

Probe Compensation. Every probe must be qualified to locate its position in relation to the x, y, and z axes of the CMM and to compensate for any geometric error introduced into measurement by the difference between the actual diameter of the probe (or footprint) and its diameter as perceived by the CMM in a given position within the work envelope. This compensation must be made for every probe (probe holder, tip, or combination thereof, and so on) in every position required during a measurement routine. When a probe is replaced because of wear, breakage, or substitution, the new probe must again be qualified in every position in which it will be used. The results of this calibration are then entered into the program guiding the CMM, and the inherent error is factored out in the final measurement data.

Environmental factors are critical to the satisfactory performance of any CMM. The user who fails to provide a suitable environment will almost certainly be disappointed with the performance of the machine. Manufacturers will provide potential buyers with environmental guidelines for the operation of the unit, and these guidelines may vary somewhat from manufacturer to manufacturer. In no case will the manufacturer guarantee the performance of the CMM unless the environmental criteria are met. According to the ANSI/ASME B89.1.12 standard for performance evaluation of CMMs, the user has the clear responsibility of providing a suitable performance test environment (on which to base acceptance of the machine from the supplier), either by meeting the guidelines of the supplier or by accepting reduced performance. Among the environmental factors that can alter CMM performance are vibration, airborne particulates, and temperature.

Vibration can greatly affect the measurement accuracy of any CMM. Motion from other machines, compressors and similar equipment, lift trucks, and so on, can be transferred to the CMM through the floor, foundation, or support system on which it stands. This motion, whether continuous vibration, intermittent shock, or combinations of both, will degrade the accuracy and repeatability of measurements taken by the machine because of the relative motion among the probe, the machine parts, and the workpiece. Extremes of motion may even damage the machine itself.

Therefore, the CMM user must use caution and good judgement in selecting a location for the unit. The area must be examined and analyzed for sources of potential vibration and shock, and an appropriate support or isolation system must be used to isolate the machine from the effects of vibration. The manufacturer will supply a statement of acceptable vibration levels as part of the machine specification so that the user can make intelligent judgements about these requirements before the purchase.

Air bearing equipped CMMs, because of their low, damped natural frequency, will require an air-activated vibration isolation system. However, with CMMs that use roller bearing packs, a typical machine tool-like isolated foundation is all that is required.

Airborne Particulates. Although there is no formal standard for acceptable levels of dirt and dust in the operating environment of a CMM, dirt can be very damaging to the performance of any CMM, most especially to any machine that has not been specifically designed and hardened for operation in a hostile environment. Therefore, the CMM user must give careful consideration to dirt and dust in the environment when selecting a CMM, determining its location in the facility, and estimating the resources required to install and operate it.

A CMM equipped with air bearings can operate without adverse effects on its performance in an environment containing dry dust because the bearings themselves will expel air and eject the contaminant. Wet dirt, however, will tend to stick and clog the machine and will probably introduce a source of significant error and possible damage to the machine.

A CMM operating in a clean room or laboratory must be dusted on a daily basis, paying particular attention to the bearing ways, scales, encoder, and so on. Any CMM operating in a harsh or hostile environment must be massive and sturdy to withstand the rigors of its environment, with the vital components mentioned above shielded from dirt, oil, grease, chips, and so on. The performance of any CMM will benefit from a conscientious program of periodic and thorough preventive maintenance.

Temperature exerts a significant influence on the accuracy of the measurements taken by a CMM. The temperature of the various components of the machine and the workpiece should be as uniform and constant as is reasonably possible within the working environment. Although specific operating temperature standards have not been established, various factors that affect operating temperatures, changes in temperature, and CMM accuracy have been identified and must be considered when choosing the location of a CMM unit and establishing the environment in which it will operate.

A CMM should not be exposed to direct sunlight or other powerful sources of radiant energy, such as fluorescent lighting. The rate and speed of air flow around the machine should be limited, as should the amount and rate of temperature change in the surrounding room.

As in the case of vibration and shock, the manufacturer will supply guidelines for an acceptable thermal environment for operation of the CMM. Although adherence to these guidelines will not ensure accuracy, it will shift responsibility for unsatisfactory performance of the machine from the user to the manufacturer.

CMM Implementation

Successful implementation of one or many CMMs depends on thorough, detailed planning and complete integration of the unit into the manufacturing system as a whole. Selection of a CMM should be based on detailed knowledge of the capabilities, strengths, and weaknesses of the manufacturing process. Knowledge of the statistical performance of the process is very helpful and, if available, can be utilized to address many of the key questions that must be answered for successful implementation. However, absolutely essential is a clear recognition of the prevailing philosophy of measurement and a full understanding of the implications of that philosophy.

Quality Plan Objectives. Before any decisions or evaluation can be made regarding CMMs, the buyer will be best served by developing a quality plan that outlines every aspect of the implementation of the CMM, its use, and the role it will play in the manufacturing process. The buyer should consider the following questions and as many more as can be formulated concerning the impact of the CMM on the overall manufacturing process:

- What is desired, and what must be accomplished relative to gaging?
- How will the CMM be integrated into the manufacturing process as a whole?
- What will be the relationship of the CMM to the process as a whole?

A laboratory-grade, vertical CMM can be used off-line, for example, as a master arbiter of part specifications, or a shop-hardened, horizontal CMM can be used on-line for continuing workpiece measurement and real-time statistical process control. Either approach, as well as many others, can be perfectly valid, but any approach must fit smoothly within the overall system. Based on manufacturing capacity and historical records, the buyer must consider such questions as:

- To what degree of accuracy will parts be gaged?
- What parts are critical to manufacturing?
- What are the roles of the critical parts in the final product?
- Which of these parts will be measured?
- How frequently will they be measured?

The answers to such questions will begin to narrow the field of potential CMMs. If a manufacturing process requires accuracy to less than 0.010 mm (0.0004 in.), for example, then production-grade, horizontal machines are eliminated. The buyer knows that the machine must be housed in a clean room environment. The buyer should also be aware that a particular measurement could be far removed from the production line, thus affecting the ability to react quickly to parts that fall below specification.

The buyer must also identify the resources that will be needed for implementation and must realistically evaluate where the resources will be obtained and at what expense. Relevant questions include the following:

- Where will the CMMs be located?
- How much space is required?
- Will the CMMs require a separate room or enclosure, or will they operate on the shop floor?
- Will they require isolation pads or special air bearings?
- Are compressed air, electrical service, and so on, of sufficient quality and readily available?

Environmental concerns must also be carefully analyzed. The following factors must be considered:

- What environment will best suit the CMM?
- Where are the sources of thermal energy that may interfere with CMM operation?
- What sources of vibration are present, and how can the CMM be shielded from them?
- Are dirt, oil, grit, and so on, present in the atmosphere, and how will the CMM be shielded from them?

Ongoing Operations. Answers to the above questions will provide insight and guidance regarding the obstacles that must be overcome for successful implementation as well as the factors that will affect ongoing operations. For example, a hostile environment may lead the user to consider enclosing the CMM in a separate room. However, this will also slow the measuring process compared to a CMM located on the shop floor, because parts will have to soak until they reach thermal equilibrium with the CMM.

In writing the parts programs that guide the measurement routines, the user must determine the information needed and the best way to obtain it. The following questions are relevant:

- What features of the part are critical?
- How many measurements must be taken?
- What probes will be required?

Training of CMM Personnel. Another essential area often overlooked at great expense to the success of CMM utilization is training. This aspect of CMM implementation involves the following considerations:

- What personnel will be utilized to operate CMMs?
- How many will be trained?
- For how long?
- How, when, and where will they be trained?

Experience has shown that personnel with inspection backgrounds tend to make the best CMM operators. If they happen to have an interest in computers, so much the better. It is often more difficult to teach inspection principles to individuals with a computer background than vice versa. Regardless of the personnel selected, it is essential that operator training be given full attention and made an integral part of the implementation plan.

Machine Vision and Robotic Inspection Systems

John D. Meyer, Tech Tran Consultants, Inc.

MACHINE VISION emerged as an important new technique for industrial inspection and quality control in the early 1980s. When properly applied, machine vision can provide accurate and inexpensive 100% inspection of workpieces, thus dramatically increasing product quality. Machine vision is also increasing in use as an in-process gaging tool for controlling the process and correcting trends that could lead to the production of defective parts. Consequently, manufacturers in a variety of industries have investigated this important new technology, regardless of the products being manufactured. As shown in Fig. 1, the automotive and electronics industries combined have purchased nearly 60% of the 2500 machine vision systems installed in 1984.

Machine vision, sometimes referred to as computer vision or intelligent vision, is a means of simulating the image recognition and analysis capabilities of the human eye/brain system with electronic and electromechanical techniques. A machine vision system senses information about an image and analyzes this information to make a useful decision about its content; in much the same way, the eye acts as the body's image sensor, with the brain analyzing this information and taking action based on the analysis.

Therefore, a machine vision system includes both visual-sensing and interpretive capabilities. An image-sensing device, such as a vidicon camera or a charge-coupled device (CCD) image sensor, is nothing more than a visual sensor that receives light through its lens and converts this light into electrical signals. When a data-processing device, such as a microcomputer, is used, these electrical signals can be refined and analyzed to provide an interpretation of the scene that generated the signals. This information can then be used as a basis for taking an appropriate course of action. The entire process of image formation, analysis, and decision making is referred to as machine vision.

This ability to acquire an image, analyze it, and then make an appropriate decision is extremely useful in inspection and quality control applications. It enables machine vision to be used for a variety of functions, including:

- Identification of shapes
- Measurement of distances and ranges
- Gaging of sizes and dimensions
- Determining orientation of parts
- Quantifying motion
- Detecting surface shading

These functional capabilities, in turn, allow users to employ machine vision systems for cost-effective and reliable 100% inspection of workpieces.

The analogy of the human eye/brain system is helpful in understanding machine vision, but the human eye/brain system is extremely complex and operates in ways and at data rates much different from those of commercial machine vision systems. Humans are more flexible and often faster than machine vision systems. On the other hand, machine vision systems provide capabilities not achievable by humans, particularly with respect to consistency and reliability. Table

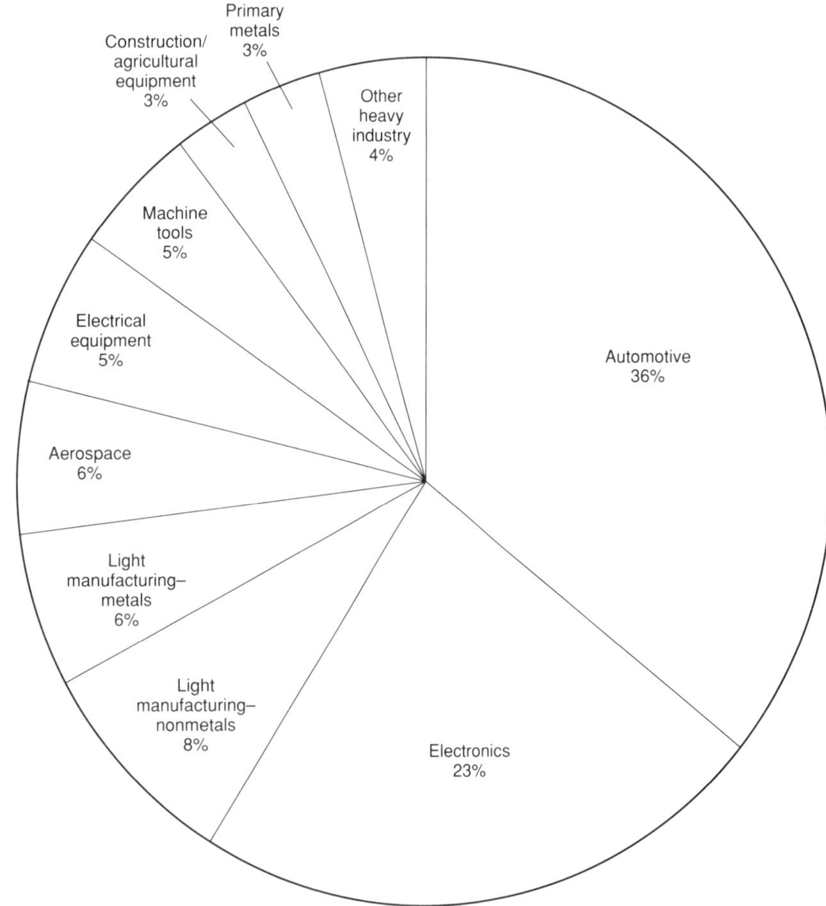

Fig. 1 Breakdown of machine vision systems market by industry in 1984. Based on annual sales of 2500 units. Source: Ref 1 (Tech Tran Consultants, Inc.)

Table 1 Comparison of machine and human vision capabilities

Capabilities	Machine vision	Human vision
Distance	Limited capabilities	Good qualitative capabilities
Orientation	Good for two dimensions	Good qualitative capabilities
Motion	Limited; sensitive to image blurring	Good qualitative capabilities
Edges/regions	High-contrast image required	Highly developed
Image shapes	Good quantitative measurements	Qualitative only
Image organization	Special software needed; limited capability	Highly developed
Surface shading	Limited capability with gray scale	Highly developed
Two-dimensional interpretation	Excellent for well-defined features	Highly developed
Three-dimensional interpretation	Very limited capabilities	Highly developed
Overall	Best for quantitative measurement of structured scene	Best for qualitative interpretation of complex, unstructured scene

Source: Ref 1 (Tech Tran Consultants, Inc.)

Table 2 Evaluation of machine and human vision capabilities

Performance criteria	Machine vision	Human vision
Resolution	Limited by pixel array size	High resolution capability
Processing speed	Fraction of a second per image	Real-time processing
Discrimination	Limited to high-contrast images	Very sensitive discrimination
Accuracy	Accurate for part discrimination based on quantitative differences; accuracy remains consistent at high production volume.	Accurate at distinguishing qualitative differences; may decrease at high volume
Operating cost	High for low volume; lower than human vision at high volume	Lower than machine at low volume
Overall	Best at high production volume	Best at low or moderate production volume

Source: Ref 1 (Tech Tran Consultants, Inc.)

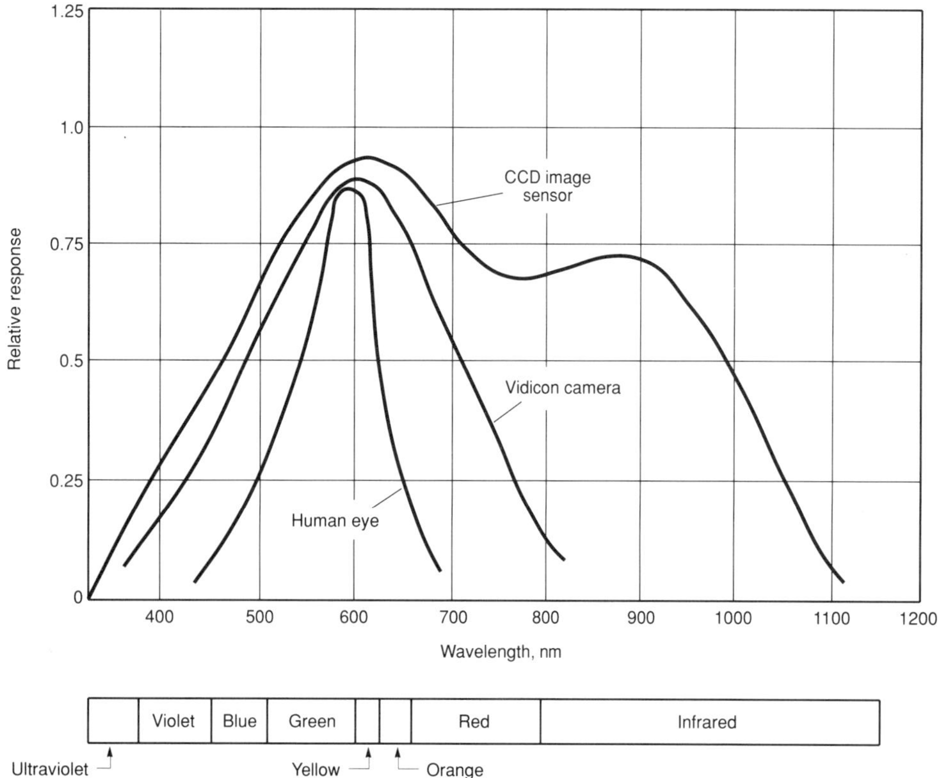

Fig. 2 Spectral response of the human eye, vidicon camera, and CCD image sensor. Source: Ref 2

1 compares human and machine vision capabilities, and Table 2 evaluates the performance of each. In addition, machine vision systems can detect wavelengths in the ultraviolet and infrared ranges, while the human eye is limited to wavelengths in the visible range (Fig. 2).

As indicated in Tables 1 and 2, neither machine vision nor human vision is clearly superior in all applications. Human vision is better for the low-speed, qualitative interpretation of complex, unstructured scenes. An example in which human vision is superior to machine vision is the inspection of automobile body surfaces for paint quality. Human vision can easily and quickly detect major flaws, such as paint sagging, scratches, or unpainted areas, but this same task would be much more difficult and time consuming with machine vision techniques.

Machine vision, on the other hand, is better suited to the high-speed measurement of quantitative attributes in a structured environment. Thus, machine vision is very good at inspecting the masks used in the production of microelectronic devices and at measuring basic dimensions for machined workpieces, for example. Machine vision can not only perform these types of inspection better than humans but can do so reliably, without the fatigue and the errors that confront humans doing these types of repeated inspection tasks.

Machine vision also has several additional important characteristics. First, it is a noncontacting measurement technique. This is particularly important if the workpiece is fragile or distorts during contact or if the workpiece would be contaminated or damaged if it were touched. Second, machine vision can be very accurate. Although accuracy is a function of many variables, including camera resolution, lens quality, field of view, and workpiece size, machine vision systems are often used to make measurements with an accuracy of ± 3 μm (± 120 μin.) or better. Third, machine vision can perform these functions at relatively large standoff distances—up to 1 m (3 ft) or more in some applications. Finally, these capabilities can be provided at relatively low cost. The price of a machine vision system may range from \$5000 to \$500 000, depending on the specific application and the capabilities of the system, but the typical price is less than \$50 000. Collectively, these characteristics of machine vision provide the user with a capability that, for many applications, cannot be matched by human vision or other sensor or inspection technologies.

Machine Vision Process

To understand the capabilities and limitations of machine vision, it is useful to examine how a machine vision system operates. Figure 3(a) shows the key components

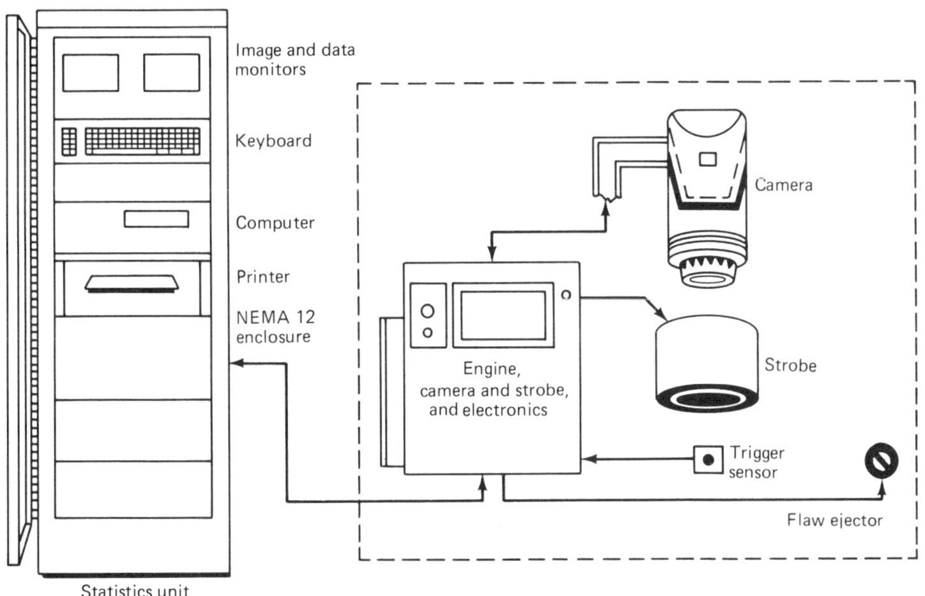

Fig. 3(a) Schematic illustrating the key components of a machine vision system

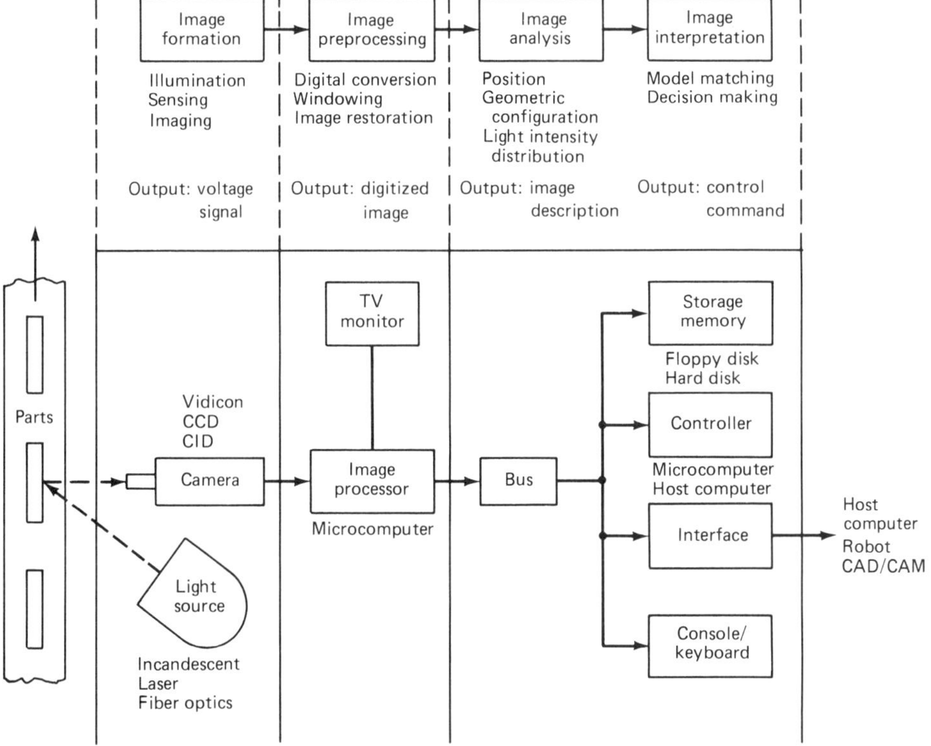

Fig. 3(b) Overview of the machine vision process. CID, charge-injected device. Source: Ref 1 (Tech Tran Consultants, Inc.)

of a machine vision system, and Fig. 3(b) illustrates the process.

The machine vision process consists of four basic steps:

- An image of the scene is formed
- The image is processed to prepare it in a form suitable for computer analysis
- The characteristics of the image are defined and analyzed
- The image is interpreted, conclusions are drawn, and a decision is made or action taken, such as accepting or rejecting a workpiece

Image Formation

The machine vision process begins with the formation of an image, typically of a workpiece being inspected or operated on. Image formation is accomplished by using an appropriate sensor, such as a vidicon camera, to collect information about the light being generated by the workpiece.

The light being generated by the surface of a workpiece is determined by a number of factors, including the orientation of the workpiece, its surface finish, and the type and location of lighting being employed. Typical light sources include incandescent lights, fluorescent tubes, fiber-optic bundles, arc lamps, and strobe lights. Laser beams are also used in some special applications, such as triangulation systems for measuring distances. Polarized or ultraviolet light can also be used to reduce glare or to increase contrast.

Proper Illumination. Correct placement of the light source is extremely important because it has a major effect on the contrast of the image. Several commonly used illumination techniques are illustrated in Fig. 4. When a simple silhouette image is all that is required, backlighting of the workpiece can be used for maximum image contrast. If certain key features on the surface of the workpiece must be inspected, frontlighting would be used. If a three-dimensional feature is being inspected, sidelighting or structured lighting may be required. In addition to proper illumination, fixturing of the workpiece may also be required to orient the part properly and to simplify the rest of the machine vision process.

Once the workpiece or scene has been properly arranged and illuminated, an image sensor is used to generate the electronic signal representing the image. The image sensor collects light from the scene (typically through a lens) and then converts the light into electrical energy by using a photosensitive target. The output is an electrical signal corresponding to the input light.

Most image sensors used in industrial machine vision systems generate signals representing two-dimensional arrays or scans of the entire image, such as those formed by conventional television cameras. Some image sensors, however, generate signals using one-dimensional or linear arrays that must be scanned numerous times in order to view the entire scene.

Vidicon Camera. The most common image sensor in the early machine vision systems was the vidicon camera, which was extensively used in closed-circuit television systems and consumer video recorders. An image is formed by focusing the incoming light through a series of lenses onto the photoconductive faceplate of the vidicon tube. An electron beam within the tube scans the photoconductive surface and produces an analog output voltage proportional to the variations in light intensity for each scan line of the original scene. Normally, the output signal conforms to commercial television standards—525 scan lines inter-

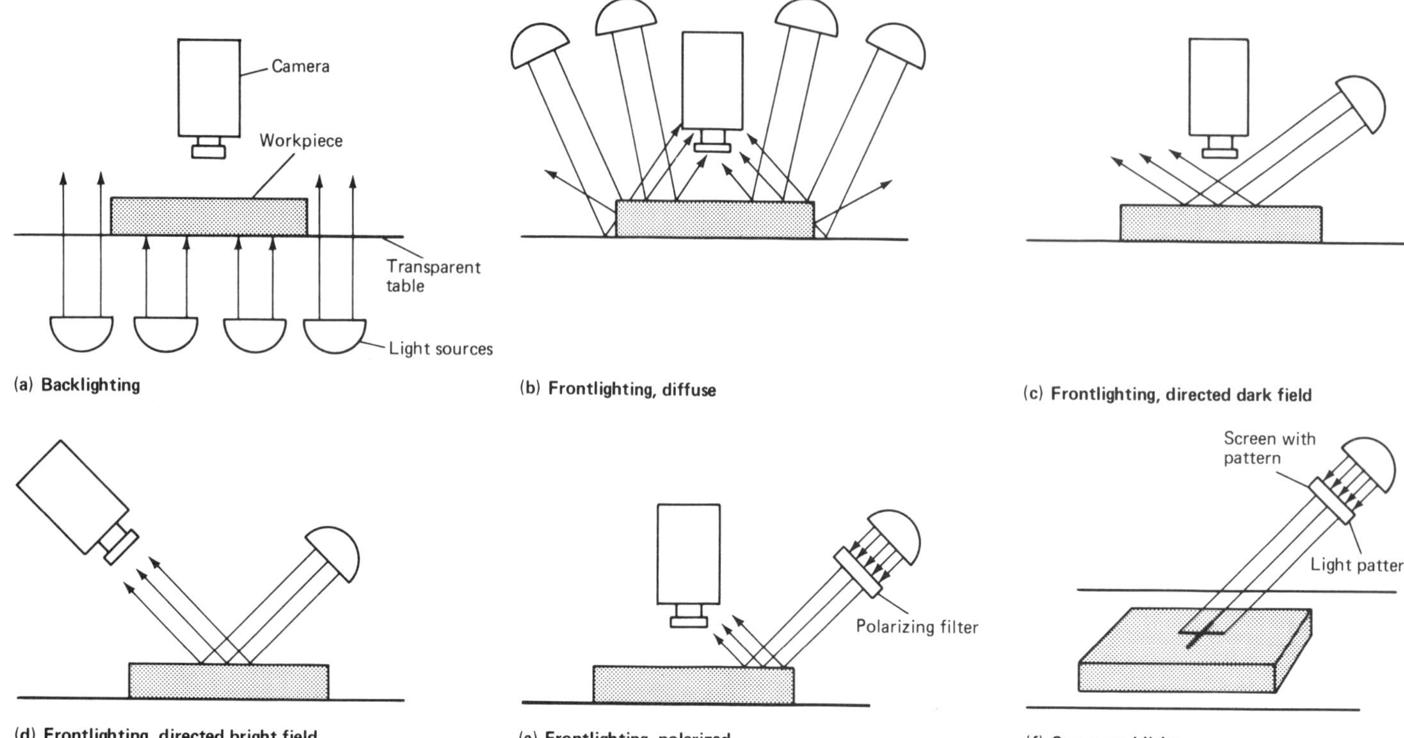

Fig. 4 Schematics of commonly used illumination techniques for machine vision systems. Source: Ref 1 (Tech Tran Consultants, Inc.)

laced into 2 fields of 262.5 lines and repeated 30 times per second.

The vidicon camera has the advantage of providing a great deal of information about a scene at very fast speeds and at relatively low cost. Vidicon cameras do have several disadvantages, however. They tend to distort the image due to their construction and are subject to image burn-in on the photoconductive surface. Vidicon cameras also have limited service lives and are susceptible to damage from shock and vibration.

Solid-State Cameras. Most state-of-the-art machine vision systems use solid-state cameras, which employ charge-coupled (Fig. 5) or charge-injected device image sensors. These sensors are fabricated on silicon chips using integrated circuit technology. They contain matrix or linear arrays of small, accurately spaced photosensitive elements. When light passing through the camera lens strikes the array, each detector converts the light falling on it into a corresponding analog electrical signal. The entire image is thus broken down into an array of individual picture elements known as pixels. The magnitude of the analog voltage for each pixel is directly proportional to the intensity of light in that portion of the image. This voltage represents an average of the light intensity variation of the area of the individual pixel. Charge-coupled and charge-injected device arrays differ primarily in how the voltages are extracted from the sensors.

Typical matrix array solid-state cameras have 256 × 256 detector elements per array, although a number of other configurations are also popular. The output from these solid-state matrix array cameras may or may not be compatible with commercial television standards. Linear array cameras typically have 256 to 1024 or more elements. The use of a linear array necessitates some type of mechanical scanning device (such as a rotating mirror) or workpiece motion (such as a workpiece traveling on a conveyor) to generate a two-dimensional representation of an image.

Selection of a solid-state camera for a particular application will depend on a number of factors, including the resolution required, the lenses employed, and the constraints imposed by lighting cost, and so on. Solid-state cameras offer several important advantages over vidicon cameras. In addition to being smaller than vidicon cameras, solid-state cameras are also more rugged. The photosensitive surfaces in solid-state

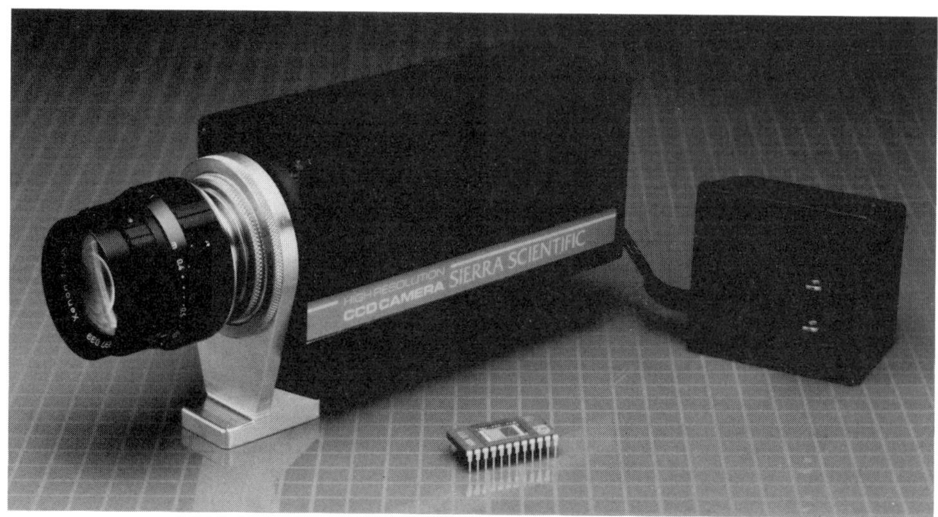

Fig. 5 Typical CCD image sensor. Courtesy of Sierra Scientific Corporation

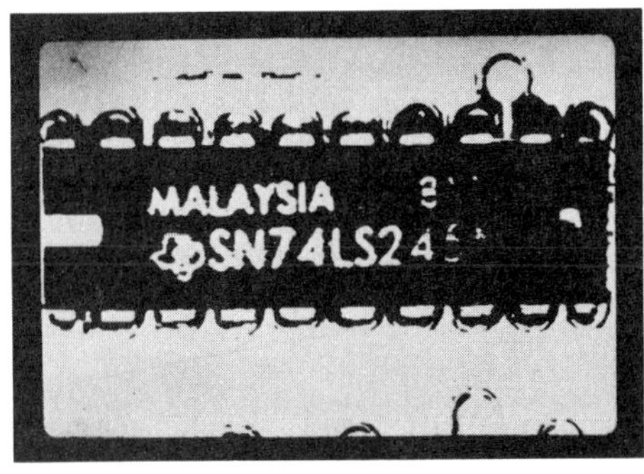

(a)

(b)

(c)

Fig. 6 Gray scale digitization of an IC module on a printed circuit board. (a) Binary. (b) 8-level. (c) 64-level. Courtesy of Cognex Corporation

sensors do not wear out with use as they do in vidicon cameras. Because of the accurate placement of the photo detectors, solid-state cameras also exhibit less image distortion. On the other hand, solid-state cameras are usually more expensive than vidicon cameras, but this cost difference is narrowing.

Although most industrial machine vision systems use image sensors of the types described above, some systems use special-purpose sensors for unique applications. This would include, for example, specialty sensors for weld seam tracking and other sensor types, such as ultrasonic sensors.

Image Preprocessing

The initial sensing operation performed by the camera results in a series of voltage levels that represent light intensities over the area of the image. This preliminary image must then be processed so that it is presented to the microcomputer in a format suitable for analysis. A camera typically forms an image 30 to 60 times per second, or once every 33 to 17 ms. At each time interval, the image is captured, or frozen, for processing by an image processor. The image processor, which is typically a microcomputer, transforms the analog voltage values for the image into corresponding digital values by means of an analog-to-digital converter. The result is an array of digital numbers that represent a light intensity distribution over the image area. This digital pixel array is then stored in memory until it is analyzed and interpreted.

Vision System Classifications. Depending on the number of possible digital values that can be assigned to each pixel, vision systems can be classified as either binary or gray-scale systems.

Binary System. The voltage level for each pixel is assigned a digital value of 0 or 1, depending on whether the magnitude of the signal is less than or greater than some predetermined threshold level. The light intensity for each pixel is therefore considered to be either white or black, depending on how light or dark the image is.

Gray-Scale System. Like the binary system, the gray-scale vision system assigns digital values to pixels, depending on whether or not certain voltage levels are exceeded. The difference is that a binary system allows two possible values to be assigned, while a gray-scale system typically allows up to 256 different values. In addition to white or black, many different shades of gray can be distinguished. This greatly increased refinement capability enables gray-scale systems to compare objects on the basis of such surface characteristics as texture, color, or surface orientation, all of which produce subtle variations in light intensity distributions. Gray-scale systems are less sensitive to the placement of illumination than binary systems, in which threshold values can be affected by lighting.

Binary Versus Gray-Scale Vision Systems. Most commercial vision systems are binary. For simple inspection tasks, silhouette images are adequate (for example, to determine if a part is missing or broken). However, gray-scale systems are used in many applications that require a higher degree of image refinement. Figure 6 shows the effects of gray-scale digitization on the surface of an integrated circuit module.

One of the most fundamental challenges to the widespread use of true gray-scale systems is the greatly increased computer processing requirements relative to those of binary systems. A 256 × 256 pixel image array with up to 256 different values per pixel will require over 65 000 8-bit storage locations for analysis. At a speed of 30 images per second, the data-processing requirement becomes very large, which means that the time required to process large amounts of data can be significant. Ideally, a vision system should be capable of the real-time processing and interpretation of an image, particularly when the system is used for on-line inspection or the guidance and control of equipment such as robots.

Windowing. One approach to reducing the amount of data to be processed, and therefore the time, is a technique known as windowing, which can substantially reduce processing time. This process creates an electronic mask around a small area of an image to be studied. Only the pixels that are not blocked out will be analyzed by the computer. This technique is especially useful for such simple inspection applications as determining whether or not a certain part has been attached to another part. Rather than process the entire image, a window can be created over the area where the attached part is expected to be located. By simply counting the number of pixels of a certain intensity within the window, a quick determination can be made as to whether or not the part is present. A window can be virtually any size, from one pixel up to a major portion of the image.

Image Restoration. Another way in which the image can be prepared in a more suitable form during the preprocessing step is through the techniques of image restoration. Very often an image suffers various forms of degradation, such as blurring of lines or boundaries, poor contrast between image regions, or the presence of background noise. There are several possible causes of image degradation, including:

- Motion of the camera or object during image formation
- Poor illumination or improper placement of illumination
- Variations in sensor response
- Defects or poor contrast on the surface of the subject, such as deformed letters on labels or overlapping parts with similar light intensities

Techniques for improving the quality of an image include constant brightness addition, contrast stretching, and Fourier-domain processing.

Constant brightness addition involves simply adding a constant amount of brightness to each pixel. This improves the contrast in the image.

Contrast stretching increases the relative contrast between high- and low-intensity elements by making light pixels lighter and dark pixels darker.

Fourier-domain processing is a powerful technique based on the principle that changes in brightness in an image can be represented as a series of sine and cosine waves. These waves can be described by specifying amplitudes and frequencies in a series of equations. By breaking the image down into its sinusoidal components, each component image wave can be acted upon separately. Changing the magnitude of certain component waves will produce a sharper image that results in a less blurred image, better defined edges or lines, greater contrast between regions, or reduced background noise.

Additional Data Reduction Processing Techniques. Some machine vision systems perform additional operations as part of the preprocessing function to facilitate image analysis or to reduce memory storage requirements. These operations, which significantly affect the design, performance, and cost of vision systems, differ according to the specific system and are largely dependent on the analysis technique employed in later stages of the process. Two commonly used preprocessing operations—edge detection and run length encoding—are discussed below.

Edge Detection. An edge is a boundary within an image where there is a dramatic change in light intensity between adjacent pixels. These boundaries usually correspond to the real edges on the workpiece being examined by the vision system and are therefore very important for such applications as the inspection of part dimensions. Edges are usually determined by using one of a number of different gradient operators, which mathematically calculate the presence of an edge point by weighting the intensity value of pixels surrounding the point. The resulting edges represent a skeleton of the outline of the parts contained in the original image.

Some vision systems include thinning, gap filling, and curve smoothing to ensure that the detected edges are only one pixel wide, continuous, and appropriately shaped. Rather than storing the entire image in memory, the vision system stores only the edges or some symbolic representation of the edges, thus dramatically reducing the amount of memory required.

Run length encoding is another preprocessing operation used in some vision systems. This operation is similar to edge detection in binary images. In run length encoding, each line of the image is scanned, and transition points from black-to-white or white-to-black are noted, along with the number of pixels between transitions. These data are then stored in memory instead of the original image, and serve as the starting point for the image analysis. One of the earliest and most widely used vision techniques, originally developed by Stanford Research Institute and known as the SRI algorithms, uses run length encoded image data.

Image Analysis

The third general step in the vision-sensing process consists of analyzing the digital image that has been formed so that conclusions can be drawn and decisions made. This is normally performed in the central processing unit of the system. The image is analyzed by describing and measuring the properties of several image features. These features may belong to the image as a whole or to regions of the image. In general, machine vision systems begin the process of image interpretation by analyzing the simplest features and then adding more complicated features until the image is clearly identified. A large number of different techniques are either used or being developed for use in commercial vision systems to analyze the image features describing the position of the object, its geometric configuration, and the distribution of light intensity over its visible surface.

Determining the position of a part with a known orientation and distance from the camera is one of the simpler tasks a machine vision system can perform. For example, consider the case of locating a round washer lying on a table so that it can be grasped by a robot. A stationary camera is used to obtain an image of the washer. The position of the washer is then determined by the vision system through an analysis of the pattern of the black and white pixels in the image. This position information is transmitted to the robot controller, which calculates an appropriate trajectory for the robot arm. However, in many cases, neither the distance between the part and the camera nor the part orientation is known, and the task of the machine vision system is much more difficult.

Object-Camera Distance Determination. The distance, or range, of an object from a vision system camera can be determined by stadimetry, triangulation, and stereo vision.

Stadimetry. Also known as direct imaging, this is a technique for measuring distance based on the apparent size of an object in the field of view of the camera (Fig. 7a). The farther away the object, the smaller will be its apparent image. This approach requires an accurate focusing of the image.

Triangulation is based on the measurement of the baseline of a right triangle formed by the light path to the object, the reflected light path to the camera, and a line from the camera to the light source (Fig. 7b). A typical light source for this technique is an LED or a laser, both of which form a well-defined spot of light (see the section "Laser Triangulation Sensors" in the article "Laser Inspection" in this Volume). Because the angle between the two light paths is preset, the standoff distance is readily calculated. Typical accuracies of 1 μm (40 μin.) can be achieved.

Stereo vision, also known as binocular vision, is a method that uses the principle of parallax to measure distance (Fig. 7c). Parallax is the change in the relative perspective of a scene as the observer (or camera) moves. Human eyesight provides the best example of stereo vision. The right eye views an object as if the object were rotated slightly from the position observed by the left eye. Also, an object in front of another object seems to move relative to the other object when seen from one eye and then from the other. The closer the objects, the greater the parallax. A practical stereo vision system is not yet available, primarily because of the difficulty in matching the two different images that are formed by two different views of the same object.

Object orientation is important in manufacturing operations such as material handling or assembly to determine where a robot may need to position itself relative to a part to grasp the part and then transfer it to another location. Among the methods used for determining object orientation are the equivalent ellipse, the connecting of three points, light intensity distribution, and structured light.

Equivalent Ellipse. For an image of an object in a two-dimensional plane, an ellipse can be calculated that has the same area as the image. The major axis of the ellipse will define the orientation of the object. Another similar measure is the axis that yields the minimum moment of inertia of the object.

Connecting of Three Points. If the relative positions of three noncolinear points in a surface are known, the orientation of the surface in space can be determined by measuring the apparent relative position of the points in the image.

Light Intensity Distribution. A surface will appear darker if it is oriented at an angle other than normal to the light source. Determining orientation based on relative light intensity requires knowledge of the source of illumination as well as the surface characteristics of the object.

Structured light involves the use of a light pattern rather than a diffused light source. The workpiece is illuminated by the structured light, and the way in which the pattern is distorted by the part can be used to determine both the three-dimensional shape and the orientation of the part.

Object Position Defined by Relative Motion. Certain operations, such as tracking or part insertion, may require the vision system to follow the motion of an object. This is a difficult task that requires a series of image frames to be compared for relative changes in position during specified time intervals. Motion in one dimension, as in the case of a moving conveyor of parts, is the least complicated motion to detect. In

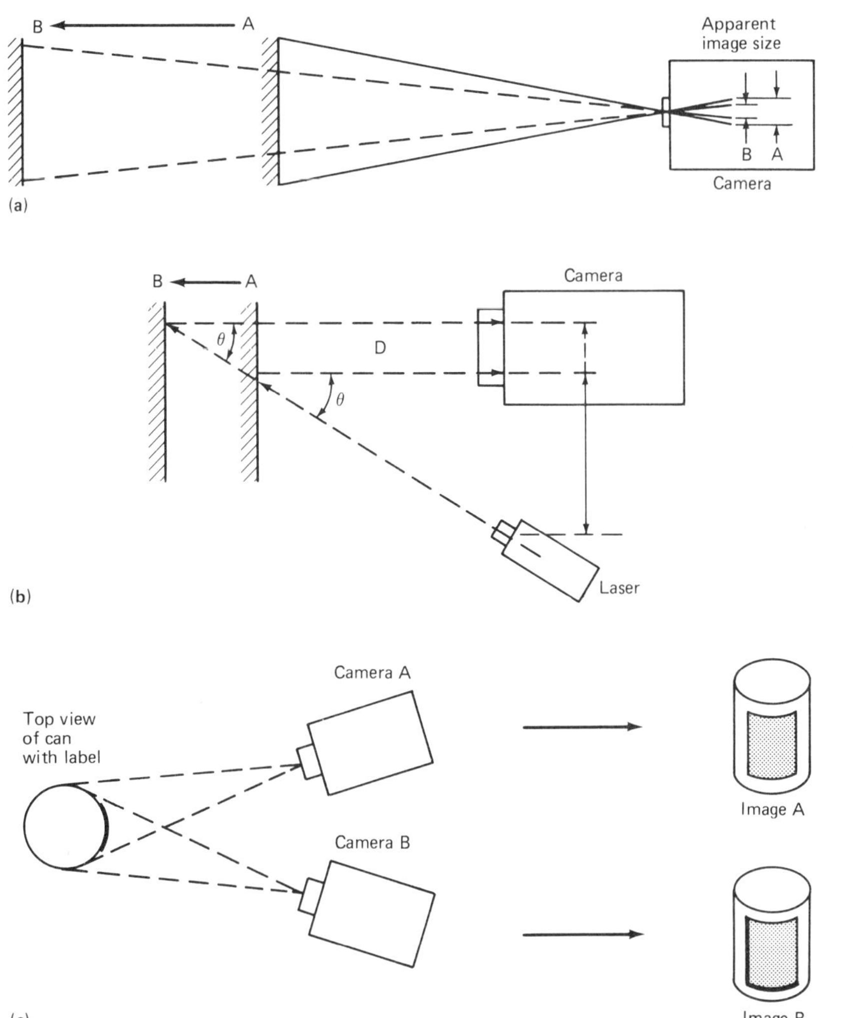

Fig. 7 Techniques for measuring the distance of an object from a vision system camera. (a) Stadimetry. (b) Triangulation. (c) Stereo vision. Source: Ref 1 (Tech Tran Consultants, Inc.)

two dimensions, motion may consist of both a rotational and a translational component. In three dimensions, a total of six motion components (three rotational axes and three translational axes) may need to be defined.

Feature Extraction. One of the more useful approaches to image interpretation is analysis of the fundamental geometric properties of two-dimensional images. Parts tend to have distinct shapes that can be recognized on the basis of elementary features. These distinguishing features are often simple enough to allow identification independent of the orientation of the part. For example, if surface area (number of pixels) is the only feature needed for differentiating the parts, then the orientation of the part is not important. For more complex three-dimensional objects, additional geometric properties may need to be determined, including descriptions of various image segments. The process of defining these elementary properties of the image is often referred to as feature extraction. The first step is to determine boundary locations and to segment the image into distinct regions. Next, certain geometric properties of these regions are determined. Finally, these image regions are organized in a structure describing their relationships.

Light Intensity Variations. One of the most sophisticated and potentially useful approaches to machine vision is the interpretation of an image based on the difference in intensity of light in different regions. Many of the features described above are used in vision systems to create two-dimensional interpretations of images. However, analysis of subtle changes in shadings over the image can add a great deal of information about the three-dimensional nature of the object. The problem is that most machine vision techniques are not capable of dealing with the complex patterns formed by varying conditions of illumination, surface texture and color, and surface orientation.

Another, more fundamental difficulty is that image intensities can change drastically with relatively modest variations in illumination or surface condition. Systems that attempt to match the gray-level values of each pixel to a stored model can easily suffer a deterioration in performance in real-world manufacturing environments. The use of such geometric features such as edges or boundaries is therefore likely to remain the preferred approach. Even better approaches are likely to result from research being performed on various techniques for determining surface shapes from relative intensity levels. One approach, for example, assumes that the light intensity at a given point on the surface of an object can be precisely determined by an equation describing the nature and location of the light source, the orientation of the surface at the point, and the reflectivity of the surface.

Image Interpretation

When the system has completed the process of analyzing image features, some conclusions must be drawn with regard to the findings, such as the verification that a part is or is not present, the identification of an object based on recognition of its image, or the determination that certain parameters of the object fall within acceptable limits. Based on these conclusions, certain decisions can then be made regarding the object or the production process. These conclusions are formed by comparing the results of the analysis with a prestored set of standard criteria. These standard criteria describe the expected characteristics of the image and are developed either through a programmed model of the image or by building an average profile of previously examined objects.

Statistical Approach. In the simplest case of a binary system, the process of comparing an image with standard criteria may simply require that all white and black pixels within a certain area be counted. Once the image is segmented (windowed), all groups of black pixels within each segment that are connected (called blobs) are identified and counted. The same process is followed for groups of white pixels (called holes).

The blobs, holes, and pixels are counted and the total quantity is compared with expected numbers to determine how closely the real image matches the standard image. If the numbers are within a certain percentage of each other, it can be assumed that there is a match.

An example of the statistical approach to image interpretation is the identification of a part on the basis of a known outline, such as the center hole of a washer. As illustrated in Fig. 8, a simple 3 × 3 pixel window can be used to locate the hole of the washer and to distinguish the washer from other distinctly different washers. The dark pixels shown in Fig. 8(a) represent the rough shape of the washer. When the window is centered on the shape, all nine pixels are assigned a value of 0 (white). In Fig. 8(b), a defective washer appears, with the hole skewed to the

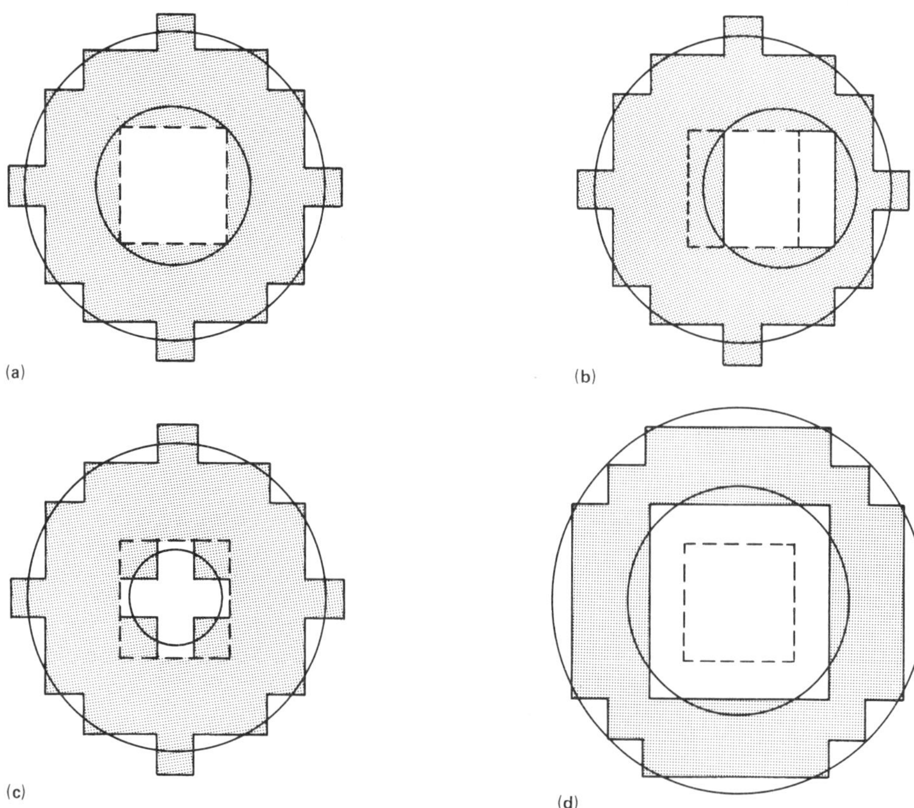

Fig. 8 Examples of the binary interpretation of washers using windowing. (a) Standard washer (9 pixels in window). (b) Washer with off-center hole (6 pixels in window). (c) Washer with small hole (5 white pixels in window). (d) Large washer (9 white pixels in window; need larger window). Source: Ref 1 (Tech Tran Consultants, Inc.)

right. Because the window counts only six white pixels, it can be assumed that the hole is incorrectly formed. In Fig. 8(c), a second washer category is introduced, one with a smaller hole. In this case, only five white pixels are counted, which is enough information to identify the washer as a different type. In Fig. 8(d), a third washer is inspected, one that is larger than the first. The window counts nine white pixels, as in Fig. 8(a). In this case, some ambiguity remains, and so additional information would be required, such as the use of a 5 × 5 window. Another approach is to count all the black pixels rather than the white ones.

Such simple methods are finding useful applications in manufacturing because of the controlled, structured nature of most manufacturing environments. The extent of the analysis required for part recognition depends on both the complexity of the image and the goal of the analysis. In a manufacturing situation, the complexity of the image is often greatly reduced by controlling such factors as illumination, part location, and part orientation. The goal of the analysis is simplified when parts have easily identifiable features, as in the example of the washer.

Much more sophisticated image analysis techniques will be required for machine vision systems to achieve widespread usage. A major reason for using a vision system is to eliminate the need for elaborate jigs and fixtures. A sophisticated vision system can offer a great deal of flexibility by reducing the amount of structure required in presenting parts for inspection, but as structure is reduced, the relative complexity of the image increases (that is, the degree of ambiguity increases). This includes such additional complexities as overlapping parts, randomly aligned parts, and parts that are distinguishable only on the basis of differences in surface features.

Gray-Scale Image Interpretation Versus Algorithms. There are two general ways in which image interpretation capabilities are being improved in vision systems. The first is gray-scale image interpretation, and the second is the use of various algorithms for the complex analysis of image data. The use of gray-scale image analysis greatly increases the quality of the data available for interpreting an image. The use of advanced data analysis algorithms (see Table 3) improves the way in which the data are interpreted. Both of these approaches allow the interpretation of much more complex images than the simple washer inspection example described above. However, even gray-scale image analysis and sophisticated data analysis algorithms do not provide absolute interpretation of images.

Machine vision deals in probabilities, and the goal is to achieve a probability of correct interpretation as close to 100% as possible. In complex situations, human vision is vastly superior to machine systems. However, in many simple manufacturing operations, where inspection is performed over long periods of time, the overall percentage of correct conclusions can be higher for machines than for humans, who are subject to fatigue.

The two most commonly used methods of interpreting images are feature weighting and template matching. These methods are described and compared below.

Feature Weighting. In cases in which several image features must be measured to interpret an image, a simple factor weighting method can be used to consider the relative contribution of each feature to the analysis. For example, the image area alone may not be sufficient to ensure the positive identification of a particular valve stem in a group of valve stems of various sizes. The measurement of height and the determination of the centroid of the image may add some additional information. Each feature would be compared with a standard for a goodness-of-fit measurement. Features that are known to be the most likely indicators of a match would be weighted more than others. A weighted total goodness-of-fit score could then be determined to indicate the likelihood that the object has been correctly identified.

Template Matching. In this method, a mask is electronically generated to match a standard image of an object. When the system inspects other objects in an attempt to recognize them, it aligns the image of each object with that of the standard object. In the case of a perfect match, all pixels would align perfectly. If the objects are not precisely the same, some pixels will fall outside of the standard image. The percentage of pixels in the two images that match is a measure of the goodness-of-fit. A threshold value can then be assigned to test for pass (positive match) or reject (no match) mode. A probability factor, which presents the degree of confidence with which a correct interpretation has been made, is normally calculated, along with the go/no-go conclusions.

Variations on these two approaches are used in most commercially available vision systems. Although conceptually simple, they can yield powerful results in a variety of manufacturing applications requiring the identification of two-dimensional parts with well-defined silhouettes.

With either method, a preliminary session is usually held before the machine is put into use. During this session, several sample known parts are presented to the machine for analysis. The part features are stored and updated as each part is presented, until the machine is familiar with the part. Then, the actual production parts are studied by comparison with this stored model of a standard part.

Table 3 Typical software library of object location and recognition algorithms available from one machine vision system supplier

Tool	Function	Applications
Search	Locates complex objects and features	Fine alignment, inspection, gaging, guidance
Auto-train	Automatically selects alignment targets	Wafer and PCB alignment without operator involvement
Scene angle finder	Measures angle of dominant linear patterns	Coarse object alignment, measuring code angle for reading
Polar coordinate vision	Measures angle; handles circular images	Locating unoriented parts, inspecting and reading circular parts
Inspect	Performs Stanford Research Institute (SRI) feature extraction (blob analysis)	Locating unoriented parts, defect analysis, sorting, inspection
Histograms	Calculates intensity profile	Presence/absence detection, simple inspection
Projection tools	Collapses 2-dimensional images into 1-dimensional images	Simple gaging and object finding
Character recognition	Reads and verifies alphanumeric codes	Part tracking, date/lot code verification
Image processing library	Filters and transforms images	Image enhancement, rotation, background filtering
V compiler	Compiles C language functions incrementally	All
Programming utilities	Handles errors, aids debugging	All
System utilities	Acquires images, outputs results, draws graphics	All
C library	Performs mathematics, creates reports and menus	All

Source: Cognex Corporation

Mathematical Modeling. Although model building, or programming, is generally accomplished by presenting a known sample object to the machine for analysis, it is also possible to create a mathematical model describing the expected image. This is generally applicable for objects that have well-defined shapes, such as rectangles or circles, especially if the descriptive data already exist in an off-line data base for computer-aided design and manufacture (CAD/CAM). For example, the geometry of a rectangular machined part with several circular holes of known diameters and locations can be readily programmed. Because more complex shapes may be difficult to describe mathematically, it may be easier to teach the machine by allowing it to analyze a sample part. Most commercial systems include standard image-processing software for calculating basic image features and comparing with models. However, custom programming for model generation can be designed either by the purchaser or by the vision system supplier. Off-line programming is likely to become increasingly popular as CAD/CAM interface methods improve.

Although the techniques described above apply to many, if not most, of the machine vision systems that are commercially available, there are still other approaches being used by some suppliers, particularly for special-purpose systems for such applications as printed circuit board (PCB) inspection, weld seam tracking, robot guidance and control, and inspection of microelectronic devices and tooling. These special-purpose systems often incorporate unique image analysis and interpretation techniques that exploit the constraints inherent in the applications.

For example, some PCB inspection systems employ image analysis algorithms based on design rules rather than feature weighting or template matching. In the design rule approach, the inspection process is based on known characteristics of a good product. For PCBs, this would include minimum conductor width and spacing between conductors. Also, each conductor should end with a solder pad if the board is correct. If these rules are not complied with, then the product is rejected.

Interfacing

A machine vision system will rarely be used without some form of interaction with other factory equipment, such as CAD/CAM devices, robots, or host computers. This interaction is the final element of the machine vision process, in which conclusions about the image are translated into actions. In some cases, the final action may take the form of cumulative storage of information in a host computer, such as counting the number of parts in various categories for inventory control. In other situations, a final action may be a specific motion, such as the transfer of parts into different conveyors, depending on their characteristics. Vision systems are being increasingly used for control purposes through the combination of vision systems and robots. In this case, the vision system greatly expands the flexibility of the robot.

For most applications, interfacing a machine vision system with other equipment is a straightforward task. Most systems are equipped with a number of input and output ports, including a standard RS232C interface. Connecting a vision system to a robot, however, is much more complicated because of timing constraints, data formats, and the inability of most robot controllers to handle vision system inputs. To overcome this problem, several robot and vision system manufacturers have developed integrated system capabilities.

Machine Vision Applications

Machine vision systems can be considered for use in most manufacturing applications in which human vision is currently required. Human vision is required for applications in which noncontact feedback is used to provide information about a production process or a part. For example, a human welder or machinist uses visual feedback to ensure that the correct relationship is maintained between the tool and the workpiece. Human assemblers visually analyze the position of parts so that other parts can be correctly aligned for insertion or some other form of mating. Quality control inspectors visually check products or parts to ensure that there are no defects, such as missing parts, damage, or incorrect location of various features.

As discussed previously, the primary strength of human vision is its ability to analyze qualitative aspects of an object or a scene. However, humans are not particularly adept at measuring quantitative data. For example, although human vision uses a sophisticated approach for depth perception that allows it to correctly determine the relative distances of objects, it is not able to measure a specific distance to an object other than as a very rough estimate. In addition, human vision can measure dimensions only approximately. Humans must rely on some standard frame of reference for judging an object. A standard retained in the memory does not provide a very good frame of reference from which to make quantitative measurements. It is not absolute, and it will vary from individual to individual. Because humans are also subject to fatigue, the interpretation of a standard may change over time.

Machine vision systems are ideally suited to a number of applications in which their ability to interpret images consistently over long periods of time makes them perform better than humans. Machine vision systems are also beginning to be used in many new and unique applications that simply did not exist previously. This includes, for example, on-line inspections that were not economically feasible before and the use of machine vision to increase manufacturing flexibility and reduce dependence on expensive hard tooling. The net result is both

Table 4 Typical applications of machine vision systems

Area	Applications
Visual inspection	Measurement of length, width, and area
	Measurement of hole diameter and position
	Inspection of part profile and contour
	Crack detection
	On-line inspection of assemblies
	Verification of part features
	Inspection of surface finish
Part identification	Optical character recognition
	Identification of parts for spray painting
	Conveyor belt part sorting
	Bin picking
	Keyboard and display verification
Guidance and control	Vision-assisted robot assembly
	Vision-assisted robot material handling
	Weld seam tracking
	Part orientation and alignment systems
	Determining part position and orientation
	Monitoring high-speed packaging equipment

Source: Ref 1 (Tech Tran Consultants, Inc.)

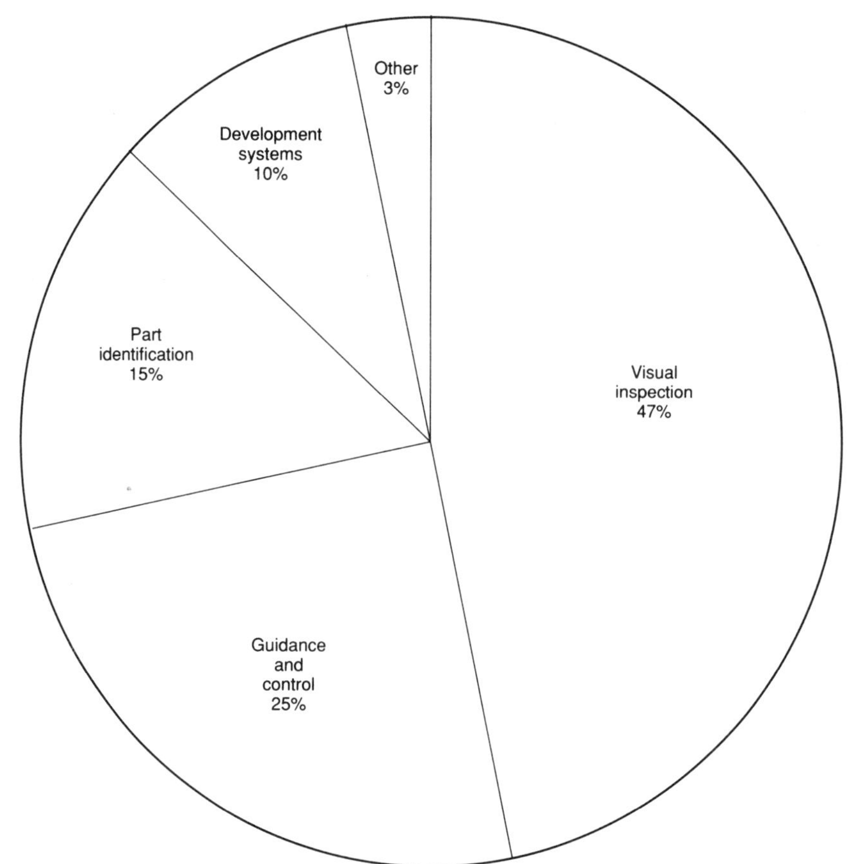

Fig. 9 Breakdown of machine vision systems applications based on a 1984 survey of 2500 installations. Source: Ref 1 (Tech Tran Consultants, Inc.)

improved product quality and lower production costs.

In deciding whether or not machine vision will be effective in a particular application, the user must consider the capabilities of machine vision versus the requirements of the application. Although many applications are suitable for automated vision sensing, there are several complex applications in which the sophisticated recognition capability of human vision is better, such as the inspection of certain complex three-dimensional objects.

In general, machine vision systems are suitable for use in three categories of manufacturing applications:

- Visual inspection of a variety of parts, subassemblies, and finished products to ensure that certain standards are met
- Identification of parts by sorting them into groups
- Guidance and control applications, such as controlling the motion of a robot manipulator

Examples of each of these three areas are listed in Table 4. Figure 9 shows the percentage breakdown of each of these three categories, as well as miscellaneous applications, on the basis of the number of machine vision units purchased in 1984.

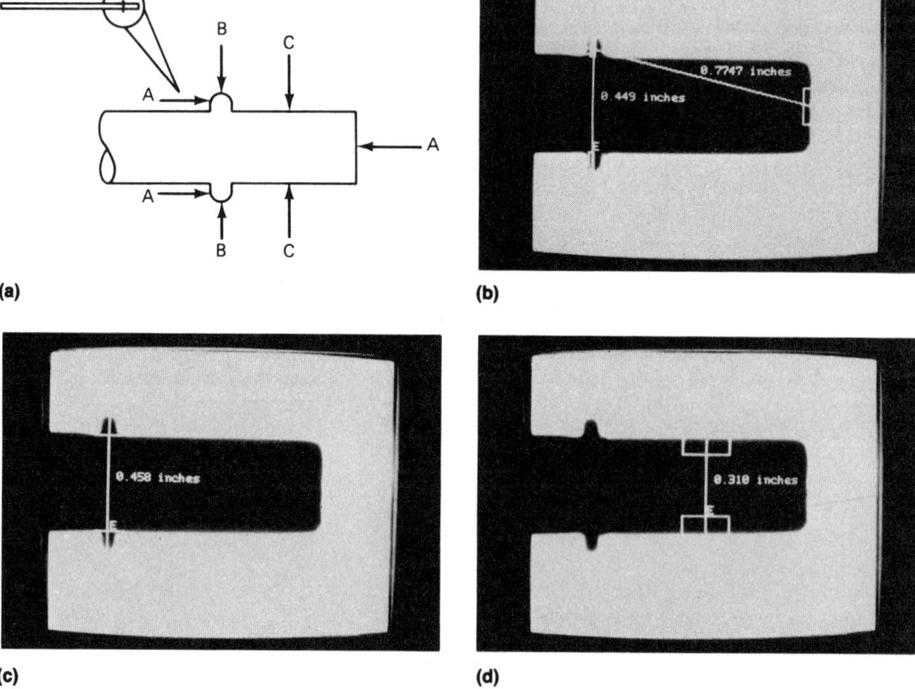

Fig. 10 Machine vision inspection of an automotive fuel sender tube with a cycle time of 2 s per part. (a) Schematic of one end of the fuel tube assembly showing the three critical dimensions in the workpiece. A–A, back of bead to front of tube; B–B, maximum bead diameter; C–C, microfinished end diameter. (b) Screen display of dimension A–A. (c) Monitor image of dimension B–B. (d) Screen display of dimension C–C. Courtesy of Industrial Systems Division, Ball Corporation

Visual Inspection

The ability of an automated vision system to recognize well-defined patterns and to determine if these patterns match those stored in the memory of the system makes it ideal for the inspection of parts, assemblies, containers, and labels. Two types of inspection can be performed by vision systems: quantitative and qualitative. Quantitative inspection is the verification that measurable quantities, such as dimensional measurements or numbers of holes, fall within desired ranges of tolerance. Qualitative inspection is the verification that certain components or properties, such as defects, missing parts, extraneous components, or misaligned parts, are present and located in a certain position. The output from a machine vision inspection task is normally a pass/reject evaluation for the object being inspected, although actual measurement data may also be an output for statistical process control and record-keeping purposes.

Visual inspection represents one of the last manufacturing areas in which automation techniques have been employed. Most visual inspection is performed manually, either by simple observation or by using a measurement tool, such as an optical comparator. One major advantage of using machine vision systems in place of human inspectors is that machine vision systems can perform 100% on-line inspection of parts at high speeds, possibly in working environments that are unpleasant for humans. Human workers are susceptible to fatigue and boredom and require periodic rest periods, while machine vision systems can operate with consistent results over prolonged periods of time. As a result, while human inspectors can anticipate only an 85 to 90% rate of accuracy in many situations, machine vision systems can achieve close to a 100% accuracy rate.

The objects for which vision systems are being used or considered for use in inspection applications include raw materials, machined parts (Fig. 10), assemblies, finished products, containers, labels, and markings. Vision systems have been considered for use with parts or products that have very simple geometries as well as those with very complex geometries, such as aerospace body frames or engine castings.

One technique applicable to the inspection of simple geometries is polar coordinate vision (Fig. 11). This technique can be used to read codes, inspect components, and measure angles in both circular and rectangular-shaped workpieces (Fig. 12). A polar coordinate vision system, can, for example, verify the existence and location of components such as rollers in a bearing assembly in less than ⅛ s as described in Fig. 13.

Polar coordinate vision can be used in a variety of applications in which the features of interest are arranged radially, such as ball bearings held in a circular assembly (Fig. 12 and 13). Polar coordinate vision consists of two major modules. The first module transforms the Cartesian images described in (x,y) coordinates into polar images, that is, it expresses the same information in (r,ϕ) coordinates. When a transformed image is subsequently analyzed by other vision tools, these tools treat the pixel data just as if it were expressed in (x,y) coordinates. The effect of treating a polar image as if it were a Cartesian image can be visualized as cutting a ring and unwrapping it into a bar. The second module measures the angle of the complex scene and searches for this polar model within the polar image to obtain the ϕ position that provides the best model/image match and then reports this ϕ value as the object's angle.

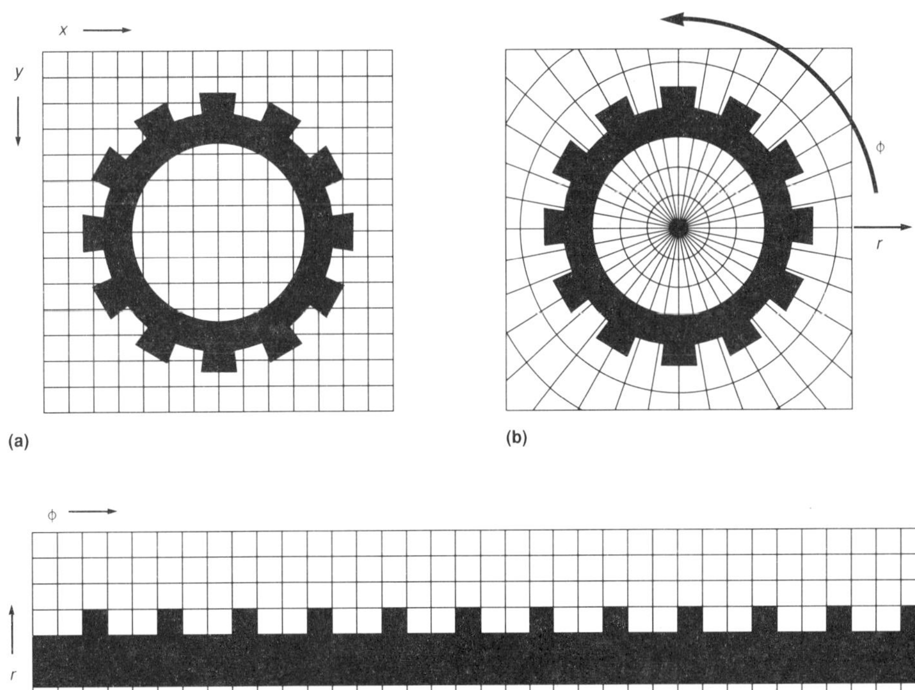

Fig. 11 Steps involved in a polar coordinate transformation. (a) Original system positioned in a Cartesian coordinate system. (b) Same image in a polar coordinate system. (c) A polar transformed image, cut at 0° and unwrapped. Courtesy of Cognex Corporation

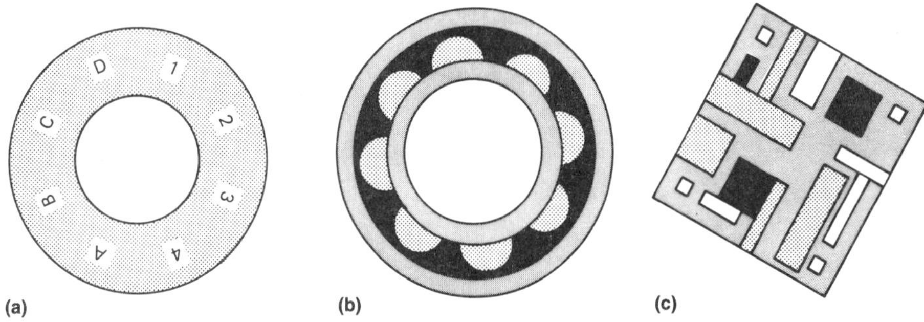

Fig. 12 Typical applications of polar coordinate vision. (a) Reading codes printed around an arc. (b) Inspecting circular objects like ball bearing assemblies. (c) Measuring the angle of complex objects such as semiconductor dies. Courtesy of Cognex Corporation

Machine vision systems are also being used by almost every manufacturing industry inspecting a wide variety of parts and assemblies, including glass containers and textiles, machine parts, microelectronics and PCBs, stampings (Fig. 14) and forgings, fasteners and gears, and pharmaceuticals and food products.

The advantage of using machine vision for low-volume, complex parts is that it is generally both faster and less error prone than human inspection. A complex part may require hundreds of measurements, which must be performed in a logical sequence to avoid missing any steps. The same qualities that make machine vision suitable for high-volume production applications also make it suitable for parts with a large number of features to be inspected.

Inspection tasks are generally more concerned with verifying the presence of or measuring a workpiece than with the actual recognition of objects. Therefore, the most useful vision system capabilities for inspection applications include the ability to seg-

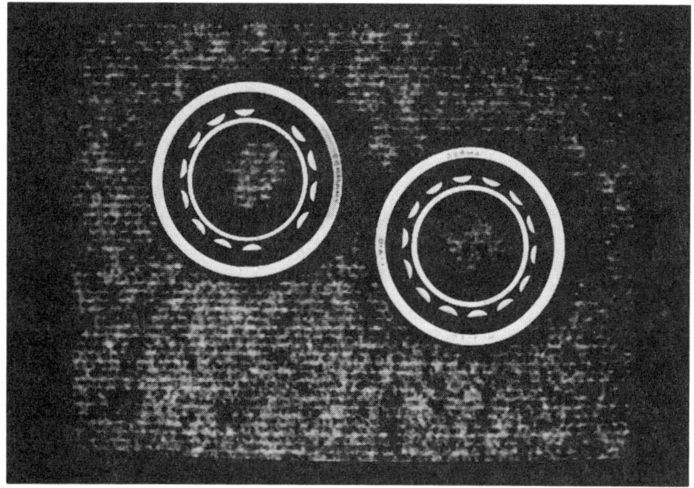

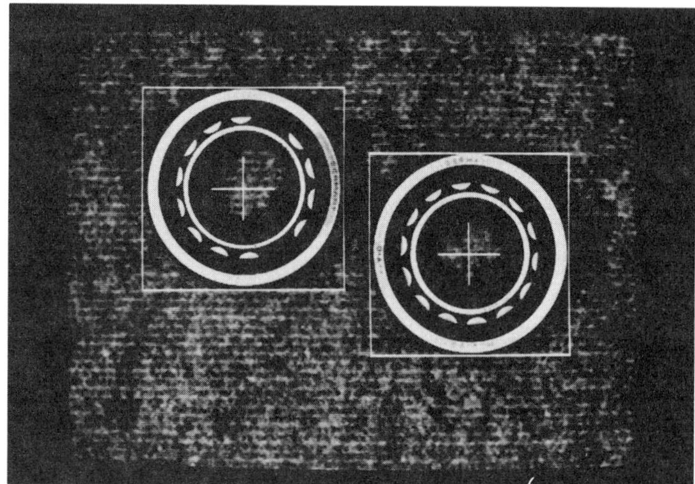

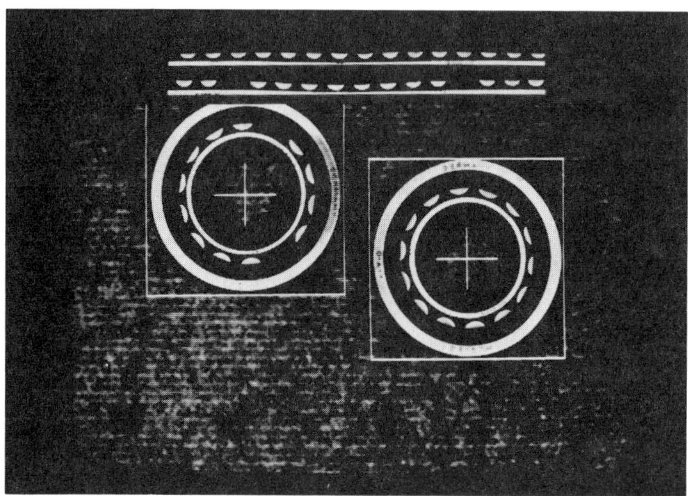

(c)

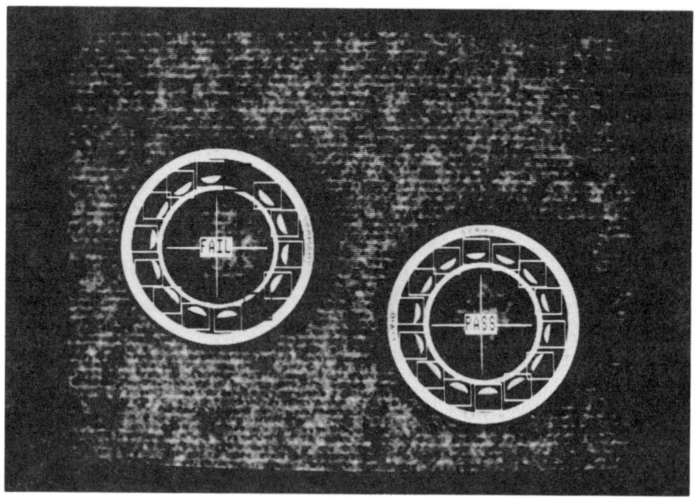

(d)

Fig. 13 Polar coordinate vision used to inspect roller bearing assemblies by verification of roller placement. (a) Actual machine vision monitor image of two bearing assemblies on a conveyor belt. (b) First step, requiring <70 ms, is to locate bearing within the image as indicated by the crosshair and square that show *x-y* position of each assembly. (c) Polar coordinate transformation, requiring <50 ms, of the annulus containing the roller yields the rectangular image shown at the top of the display to aid in the interpretation of the components. (d) System displays complete inspection results by indicating the location of the missing rollers (enclosed in black squares) as well as those rollers correctly positioned (enclosed in white squares) and also specifies which assemblies have passed or failed inspection. Total processing time, which includes part location, image transformation, and roller counting, is <125 ms. Courtesy of Cognex Corporation

ment images by forming edges and the ability to measure the geometric features of these images and segments. Two-dimensional images are generally used to perform these functions. Of the three performance criteria considered here, the most important for visual inspection is the ability to perform the task at high speed.

Visual inspection of all types, including quantitative and qualitative inspection, has been estimated to account for about 10% of the total labor cost of all manufactured durable goods. This excessively high percentage can be significantly reduced when machine vision is employed. Along with lower cost, the quality of the final product can also be increased.

Part Identification

The most fundamental use for a machine vision system is the recognition of an object from its image. Inspection deals with the examination of object characteristics without necessarily requiring that the objects be identified. In part recognition, however, it is necessary to make a positive identification of an object and then to make a decision based on that knowledge. Generally, the decision involves some form of a categorization of the object into one of several groups. This categorization can be in the form of information, such as categories of inventories to be monitored (Fig. 15, 16, and 17), or it can be in a physical form, as in the placement of parts on different conveyor belts, depending on their characteristics.

The process of part identification generally requires strong geometric feature interpretation capabilities because most manufacturing applications allow part differentiation on the basis of differences in silhouette shapes (Fig. 18). As in the inspection applications, processing speed is generally more important than resolution or discrimination ability. Recognition applications often require an interface capability with some form of part-handling equipment, such as an industrial robot.

Guidance and Control

One of the basic limitations of industrial robots in such applications as assembly, machining, welding, or other process-oriented operations is that feedback capabilities are limited. In these applications, parts must be continuously monitored and positioned relative to other parts with a high degree of precision. In some applications, such as bolt tightening, a force feedback capability may be required to determine when the operation should cease. In many other applications, a vision system can be a

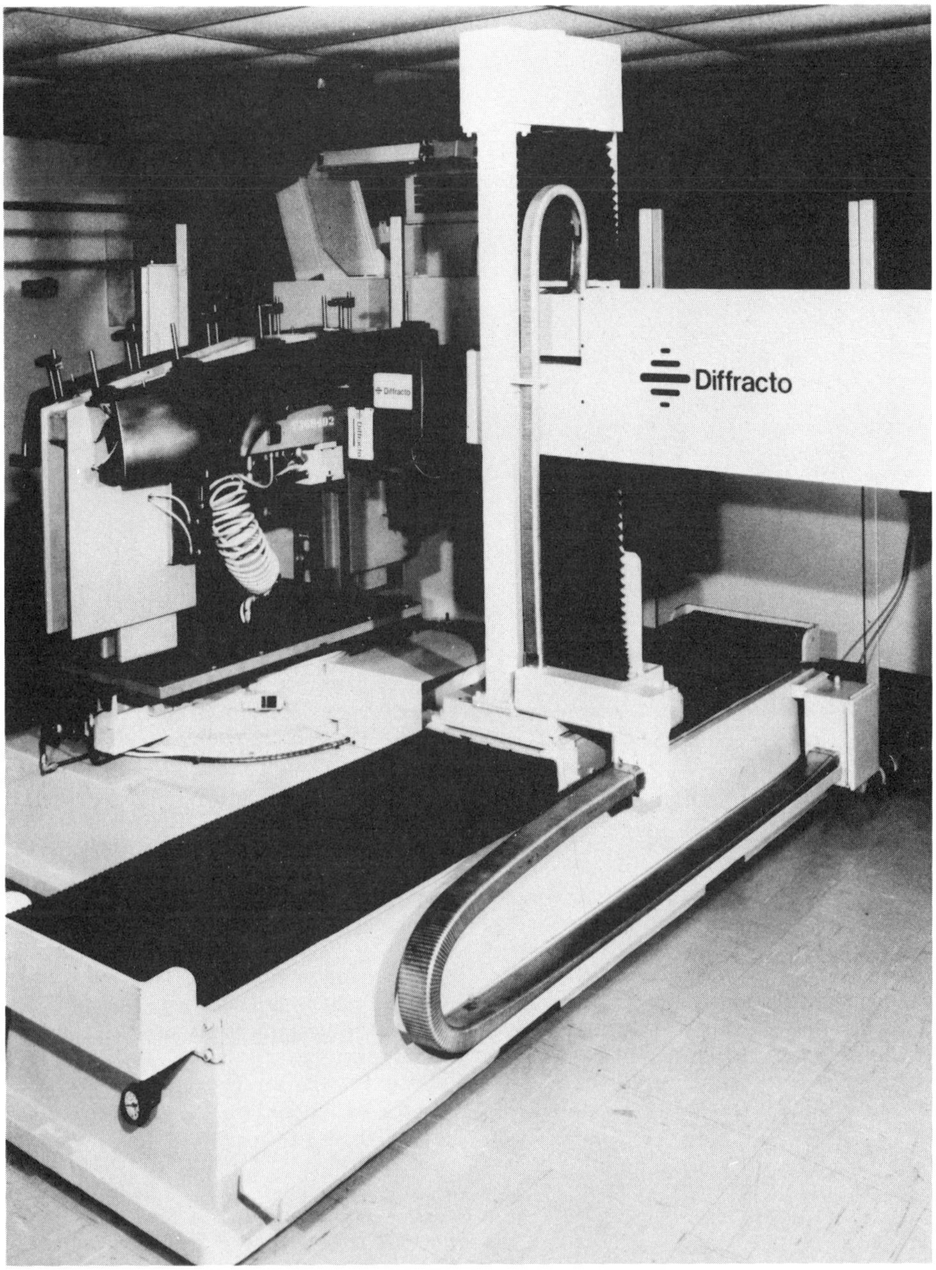

Fig. 14 Robot-based noncontact measuring system shown gaging a sheet metal automotive fender held in a fixture. Fit (gap and flushness) are measured by means of optical triangulation. Courtesy of Diffracto Limited

powerful tool for controlling production operations when combined with other forms of automated equipment.

These guidance and control applications tend to require advanced machine vision system capabilities. Because of the need to determine spatial relationships among objects, the ability to measure distance is often important (Fig. 19), along with the ability to measure the orientation and geometric shape of an object (Fig. 20). In addition, an ideal vision system would allow three-dimensional interpretation of images. Finally, the precise positioning requirements of these tasks means that a high degree of image resolution is desirable.

Additional Applications

As machine vision technology evolves and becomes more widely used within industry, additional applications are beginning to emerge. Many of these applications represent special-purpose equipment that has been designed to satisfy a particular need, while others combine two or more of the previously mentioned functions (visual inspection, part identification, and guidance and control). For example, the system shown in Fig. 21 can inspect as well as sort various fasteners and similar multiple-diameter parts. Still other applications reflect machine vision systems that are incorporated into other equipment.

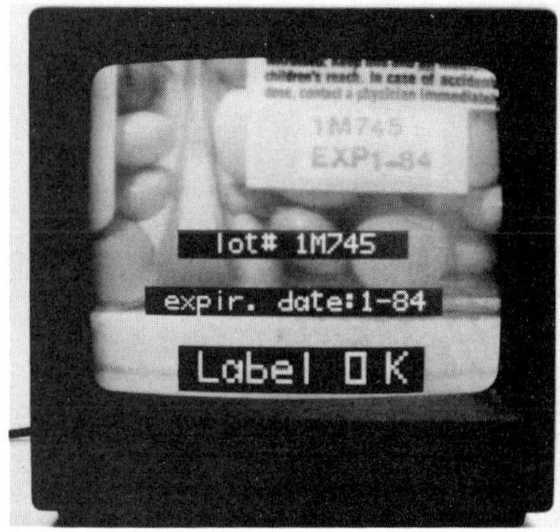

Fig. 15 Monitor screen of a machine vision inspection system verifying the expiration date and lot code on a pharmaceutical bottle. Courtesy of Cognex Corporation

Special-Purpose Systems. In the future, special-purpose systems are likely to represent a large, if not the largest, use of machine vision technology. One of the best examples of this type of system is equipment for inspecting PCBs. A number of companies have either developed or are in the process of developing such systems, which are expected to be widely used by the end of the decade. Similar special-purpose equipment is entering the market for the inspection of thick-film substrates and circuits, surface-mounted devices, and photolithographic artwork.

Embedded Technology. In the embedded technology area, one of the major uses of machine vision is mask alignment for the production of microelectronic devices. Similarly, vision technology is also becoming widely used for controlling other microelectronic fabrication equipment, such as the automation of wire-bonding machines for connecting integrated circuits to their cases. In the future, these two thrusts—special-purpose systems and embedded vision technology—should result in numerous applications unheard of today.

Future Outlook

The potential for using machine vision in manufacturing applications is enormous. Many inspection operations that are now performed manually could be automated by machine vision, resulting in both reduced costs and improved product quality. However, before machine vision can reach its full potential, several basic improvements in the technology must be made.

Limitations of Current Systems

At the present time, there are six key issues that must be addressed by vision system developers. Many organizations are attempting to resolve these issues through

Fig. 16 Machine vision system reading box labels for sortation in a shipping facility. Courtesy of Cognex Corporation

such developments as improved computer hardware or improved software algorithms, but much work remains to be done to develop effective vision systems that are available at a reasonable cost. The following issues represent basic limitations of commercial vision systems:

- Limited 3-D interpretation
- Limited interpretation of surfaces
- Need for structured environment
- Long processing time
- High cost
- Excessive applications engineering

Limited 3-D Interpretation. Most commercial vision systems are two dimensional; that is, they make conclusions about objects from data that are essentially two-dimensional in nature. In many manufacturing situations, an outline of the shape of an object is sufficient to identify it or to determine whether an inspection standard has been achieved. However, in many other operations, such as the inspection of castings, this information is not sufficient. Many more sophisticated operations could be performed with vision systems if the three-dimensional shape of an object could be inferred from an image or a series of images. To accomplish this, vision system suppliers will need to incorporate more sophisticated data interpretation algorithms along with improved system performance (resolution, speed, and discrimination).

Limited Interpretation of Surfaces. Complex surface configurations on objects, such as textures, shadows, and overlapping parts, are difficult for vision systems to interpret. Improved gray-scale image formation capabilities have helped somewhat, but vision systems are extremely limited in their ability to analyze the large amounts of data provided by gray-scale image formation. The ability to accurately interpret light intensity variations over the surface of an object, which is so fundamental to human vision, must be refined if vision systems are to be used for such applications as object recognition or inspection from surface characteristics.

Need for Structured Environment. Although vision systems, being a form of flexible automation, should be able to eliminate the need for elaborate jigs and fixtures, they still require a relatively orderly environment in most current applications. Vision systems have difficulty dealing with overlapping or touching parts; therefore, workpieces must be presented one at a time to the system. Ideally, a vision system should be able to examine parts as humans do—by studying key features no matter how the part is oriented and even if some portions of the parts are obstructed by other overlapping parts.

Long Processing Time. There are constraints on the speed of the manufacturing operation in which a vision system can be used. Only a limited number of real-time (30 images per second) systems have begun to appear on the market. However, most real-time systems are used for simple applications rather than more complex tasks. There is generally a trade-off between the processing time required and the degree of complexity of a processing cycle. An ideal vision system would be capable of performing complex three-dimensional analyses of objects, including surface features, in real time.

High Cost. Although payback periods for vision systems are generally good (1 year or less for some applications), the basic purchase price of many systems is still prohibitively high to promote widespread use of this technology within the manufacturing industry.

Extensive Applications Engineering. It is still nearly impossible to purchase an off-the-shelf vision system and apply it without considerable assistance from a vendor, consultant, or in-house engineering staff. This is partly due to the complexity of

Fig. 17 Machine vision used to read part numbers engraved on semiconductor wafers in order to expedite work tracking of components. Courtesy of Cognex Corporation

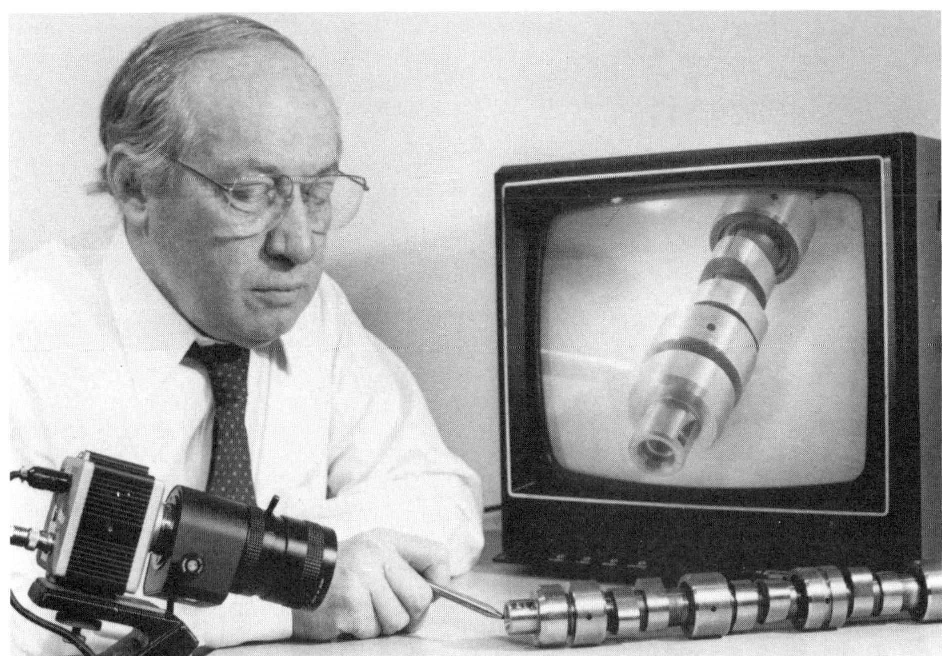

Fig. 18 Machine vision system that controls a multiple-programmed cam grinder being used to identify each specific automotive camshaft to determine machine settings. Courtesy of Industrial Automation Systems, Gould, Inc.

Fig. 19 Adaptive robotic arc welding of automotive bodies using a three-dimensional vision system for seam tracking and measurement of seam width and depth. Both the vision module and the metal-inert gas welding gun are attached to the tool mounting plate located on the robot's wrist. Courtesy of Robotic Vision Systems, Inc.

real-world applications. Other factors include the limitations of current equipment and the lack of trained personnel within user organizations. Application engineering cost and risk and a shortage of trained technical personnel are major barriers to widespread use of industrial vision systems.

Future Developments

Many developmental programs are underway, both in private industry as well as in universities and other research organizations, to develop advanced vision systems that are not subject to the limitations discussed previously.

The solution to these problems is likely to emerge from several important developments expected to occur during the next decade; these developments are discussed in the following sections. However even if no further improvements are made in machine vision systems, the number of systems in use would continue to grow rapidly. Machine vision systems are beginning to be introduced into applications for which they previously would have not even been considered, because of the complexity of the manufacturing process.

Improved Camera Resolution. As solid-state cameras with arrays of 512 × 512 or even 1024 × 1024 pixels are used, image resolution will improve. As a result, the ability of vision systems to sense small features on the surfaces of objects should also improve.

Ability to Sense Color. A few developmental vision systems are already available that sense color. The addition of this capability to commercial vision systems would allow the measurement of one more feature in identifying objects. It would also provide a greater degree of discrimination in analyzing surfaces.

Effective Range Sensing. This is a prerequisite for three-dimensional interpretation and for certain types of robot vision. Based on research such as that being performed on binocular vision, it is likely that a range-sensing capability will become a standard feature of commercial vision systems within a few years.

Ability to Detect Overlap. This capability will improve the ability of vision systems to interpret surfaces and three-dimensional objects. It will also provide a greater degree of flexibility for vision systems. There will no longer be a need to ensure that moving parts on a conveyor are not touching or overlapping, and this will reduce the amount of structure required.

Improved Gray-Scale Algorithms. As vision system hardware becomes capable of forming more complex images, the software algorithms for interpreting these images will improve, including the ability to infer shape from changes in light intensity over an image.

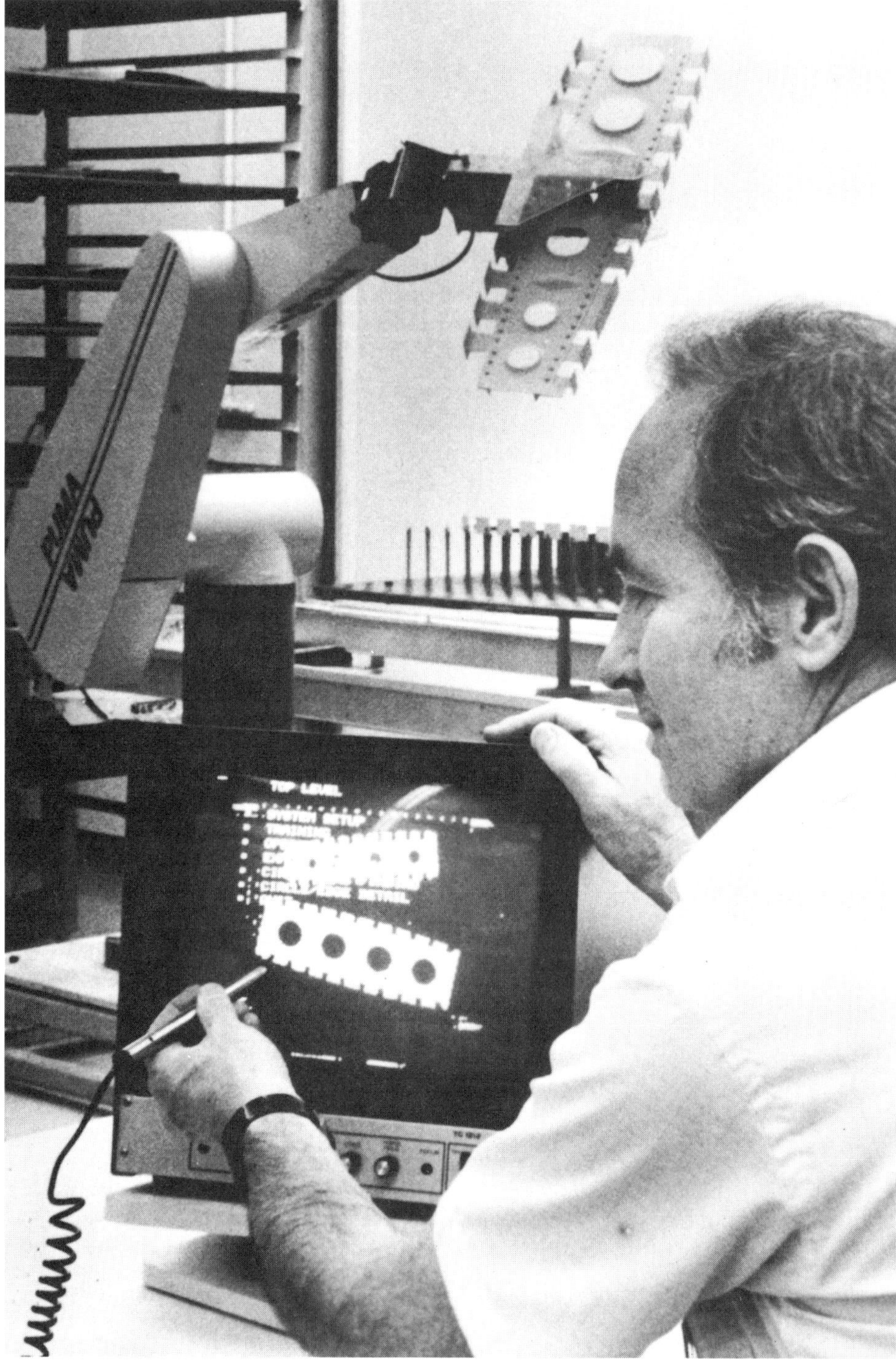

Fig. 20 Robot gripper arm used to maneuver and orient an aluminum aircraft section with the aid of a video camera that helps it see the workpiece and then identifies it from a data base. Courtesy of Lockheed-California Company

Robot Wrist-Mounted Vision System. Based on work being performed at a number of organizations, it is likely that an effective wrist-mounted vision system will be available within the next few years. Mounting the camera on the robot's wrist provides the advantage of greatly reducing the degree of structure required during such operations as robot-controlled welding, assembly, or processing.

Motion-Sensing Capability. There are two elements being developed in this area. First is the ability of a vision system to create and analyze an image of a moving object. This requires the ability to freeze each frame without blurring for analysis by the computer. Second is the more complex problem of determining the direction of motion of an object and even the magnitude of the velocity. This capability will be valuable in such applications as collision avoidance or tracking moving parts.

Parallel Processing of Whole Image. One of the most promising methods of approaching a real-time processing capability is the use of a parallel processing architecture. Several systems currently on the market offer this type of architecture. This approach is likely to be used more extensively in the future.

Standardized Software Algorithms. Although some standard vision system application algorithms are available, most programs for current manufacturing applications are custom designed. It is likely that standard programs will become increasingly available for standard applications. In addition, programming languages will continue to become more user oriented.

Computers Developed Specifically for Vision Systems. Most vision systems today use standard off-the-shelf computers, which tends to limit the data analysis capabilities of the vision system. In the future, especially as sales volumes increase, it is likely that computers will be designed specifically for dedicated use with a vision system. This will reduce processing times and help to reduce system prices. Several systems have been developed with this type of custom computer architecture.

Hard-Wired Vision Systems. To overcome the problem of processing speed, some researchers have suggested the use of hard-wired circuitry rather than microprocessor-based systems. This would significantly speed up image processing times, but may result in less system flexibility and limited capabilities.

Special-Purpose Systems. As discussed previously, there is a growing trend toward special-purpose, rather than general-purpose, vision systems. This permits the system developer to take advantage of prior knowledge concerning the application and to provide only the features and capabilities required, resulting in more cost-effective systems. A number of vendors have already begun to offer special-purpose systems for such applications as weld seam tracking, robot vision, and PCB inspection.

Integration With Other Systems. One of the major problems with the current vision systems is the difficulty in interfacing them with other types of equipment and systems. A number of companies and research organizations are attacking this problem, particularly with respect to special-purpose vision systems.

Optical Computing. It is possible to perform image processing using purely optical techniques, as opposed to the traditional approach of converting an image into an electrical signal and analyzing this symbolic representation of the image. In the optical domain, processing steps such as the

Fig. 21 Noncontact digital computer-vision-based fastener inspection system used to gage and then sort parts. Unit measures 11 parameters for up to 24 different fasteners that are stored in a memory and is capable of sorting up to 180 parts per minute or over 10 000 parts per hour. Courtesy of Diffracto Limited

computation of Fourier transforms take place almost instantaneously. Although optical computing techniques offer considerable promise, it will take a number of years before they become a practical reality.

Custom Microelectronic Devices. As the sales volume for vision systems continues to grow, it will become increasingly feasible to implement portions of the system design in custom microelectronic circuits. This will be particularly true for low-level image-processing functions, such as histogram calculations, convolutions, and edge detectors. Such chips should be available within the next few years.

Innovative Sensor Configurations. A number of researchers are working on unique vision sensors to improve overall performance. This includes novel sensor configurations, such as annular arrangements of detector elements, as well as other camera concepts, such as multiple spectral detectors that sense energy in more than one portion of the electromagnetic spectrum.

Visual Servoing. Several researchers are studying the use of vision systems as an integral feedback component in a motion control system, such as a robot vision system for positioning the manipulator arm. Although vision systems are currently used for robot guidance and control, this is usually accomplished outside the control loop. In visual servoing, on the other hand, the vision system would serve as a position-sensing device or error measurement component on a real-time basis.

REFERENCES

1. "Machine Vision Systems: A Summary and Forecast," 2nd ed., Tech Tran Consultants, Inc., 1985
2. P. Dunbar, Machine Vision, *Byte*, Jan 1986

SELECTED REFERENCES

- I. Aleksander, *Artificial Vision for Robots*, Chapman and Hall, 1983
- D. Ballard and C. Brown, *Computer Vision*, Prentice-Hall, 1982
- J. Brady, *Computer Vision*, North-Holland, 1982
- O. Faugeras, *Fundamentals of Computer Vision*, Cambridge University Press, 1983
- J. Hollingum, *Machine Vision: Eyes of Automation*, Springer-Verlag, 1984
- A. Pugh, *Robot Vision*, Springer-Verlag, 1983

Methods of Nondestructive Evaluation

Guide to Nondestructive Evaluation Techniques

John D. Wood, Lehigh University

NONDESTRUCTIVE EVALUATION (NDE) comprises many terms used to describe various activities within the field. Some of these terms are nondestructive testing (NDT), nondestructive inspection (NDI), and nondestructive examination (which has been called NDE, but should probably be called NDEx). These activities include testing, inspection, and examination, which are similar in that they primarily involve looking at (or through) or measuring something about an object to determine some characteristic of the object or to determine whether the object contains irregularities, discontinuities, or flaws.

The terms irregularity, discontinuity, and flaw can be used interchangeably to mean something that is questionable in the part or assembly, but specifications, codes, and local usage can result in different definitions for these terms. Because these terms all describe what is being sought through testing, inspection, or examination, the term NDE (nondestructive evaluation) has come to include all the activities of NDT, NDI, and NDEx used to find, locate, size, or determine something about the object or flaws and allow the investigator to decide whether or not the object or flaws are acceptable. A flaw that has been evaluated as rejectable is usually termed a defect.

Selection of NDE Methods

The selection of a useful NDE method or a combination of NDE methods first necessitates a clear understanding of the problem to be solved. It is then necessary to single out from the various possibilities those NDE methods that are suitable for further consideration; this is done by reviewing the articles in this Volume and in the technical literature.

Several different ways of comparing the selected NDE methods are presented in this article, but there is no completely acceptable system of comparison, because the results are highly dependent on the application. Therefore, it is recommended that a comparison be developed specifically for each NDE area and application. The final validation of any NDE protocol will depend on acceptance tests conducted using appropriate calibration standards.

Nondestructive evaluation can be conveniently divided into nine distinct areas:

- Flaw detection and evaluation
- Leak detection and evaluation
- Metrology (measurement of dimension) and evaluation
- Location determination and evaluation
- Structure or microstructure characterization
- Estimation of mechanical and physical properties
- Stress (strain) and dynamic response determination
- Signature analysis
- Chemical composition determination

Because two of these areas—signature analysis and chemical composition determination—are usually not considered when NDE applications are discussed and are therefore not covered in this Volume, they will not be discussed further. Information on these subjects can, however, be found in Volume 10 of the 9th Edition of *Metals Handbook*. The remaining seven areas are vastly different and therefore will be covered separately, along with a discussion of the selection of specific NDE methods* for each.

Flaw Detection and Evaluation

Flaw detection is usually considered the most important aspect of NDE. There are many conceivable approaches to selecting NDE methods. One approach is to consider that there are only six primary factors involved in selecting an NDE method(s):

- The reason(s) for performing the NDE
- The type(s) of flaws of interest in the object
- The size and orientation of flaw that is rejectable
- The anticipated location of the flaws of interest in the object
- The size and shape of the object
- The characteristics of the material to be evaluated

The most important question to be answered before an NDE method can be selected is, What is the reason(s) for choosing an NDE procedure? There are a number of possible reasons, such as:

- Determining whether an object is acceptable after each fabrication step; this can be called in-process NDE or in-process inspection
- Determining whether an object is acceptable for final use; this can be called final NDE or final inspection
- Determining whether an existing object already in use is acceptable for continued use; this can be called in-service NDE or in-service inspection

After the reasons for selecting NDE have been established, one must specify which types of flaws are rejectable, the size and orientation of flaws that are rejectable, and the locations of flaws that can cause the object to become rejectable. The type, size, orientation, and location of flaws that will cause a rejection must be determined if possible, using stress analysis and/or fracture mechanics calculations. If definitive calculations are not economically feasible, the type, size, and orientation of flaw that will cause the object to be rejected must be estimated with an appropriate safety factor.

The type, size, orientation, and location of the rejectable flaw are often dictated by a code, standard, or requirement, such as the American Society of Mechanical Engineers Pressure Vessel Code, a Nuclear Regulatory Commission requirement, or the American Welding Society Structural Welding Code. If one of these applies to the object under consideration, the information needed will be available in the appropriate document.

*Throughout this article the term method is used to describe the various nondestructive testing disciplines (for example, ultrasonic testing) within which various test techniques may exist (for example, immersion or contact ultrasonic testing).

Table 1 Volumetric flaw classification and NDE detection methods

Volumetric flaws
Porosity
Inclusions
Slag
Tungsten
Other
Shrinkage
Holes and voids
Corrosion thinning
Corrosion pitting
NDE detection methods
Visual (surface)
Replica (surface)
Liquid penetrant (surface)
Magnetic particle (surface and subsurface)
Eddy current
Microwave
Ultrasonic
Radiography
X-ray computed tomography
Neutron radiography
Thermography
Optical holography
Speckle metrology
Digital image enhancement (surface)

Table 2 Planar flaw classification and NDE detection methods

Planar flaws
Seams
Lamination
Lack of bonding
Forging or rolling lap
Casting cold shut
Heat treatment cracks
Grinding cracks
Plating cracks
Fatigue cracks
Stress-corrosion cracks
Welding cracks
Lack of fusion
Incomplete penetration
Brazing debond
NDE detection methods
Visual
Replication microscopy
Magnetic particle
Magnetic field
Eddy current
Microwave
Electric current perturbation
Magabsorption
Ultrasonic
Acoustic emission
Thermography

Table 3 NDE methods for the detection of surface and interior flaws

Surface
Visual
Replica
Liquid penetrant
Magnetic particle
Magnetic field
Electric current
Magabsorption
Eddy current
Ultrasonic
Acoustic emission
Thermography
Optical holography
Speckle metrology
Acoustic holography
Digital image enhancement
Acoustic microscopy
Interior
Magnetic particle (limited use)
Magnetic field
Electric current perturbation
Magabsorption
Eddy current
Microwave
Ultrasonic
Acoustic emission
Radiography
X-ray computed tomography
Neutron radiography
Thermography (possible)
Optical holography (possible)
Acoustic holography (possible)
Acoustic microscopy (possible)

Volumetric and Planar Flaws. Once the size and orientation of the rejectable flaw have been established, it is necessary to determine which types of flaws are rejectable. In general, there are two types of flaws: volumetric and planar. Volumetric flaws can be described by three dimensions or a volume. Table 1 lists some of the various types of volumetric flaws, along with useful NDE detection methods. Planar flaws are thin in one dimension but larger in the other two dimensions. Table 2 lists some of the various types of planar flaws, along with appropriate NDE detection methods.

Flaw Location, Shape, and Size. In addition to classifying flaws as volumetric or planar, it is necessary to consider the locations of the flaws in the object. Flaws can be conveniently classified as surface flaws or as interior flaws that do not intercept the surface. Table 3 lists NDE methods used to detect surface and interior flaws.

Two additional factors that affect NDE method selection are the shape and size of the object to be evaluated. Tables 4 and 5 compare NDE techniques for varying size (thickness) and shape.

The characteristics of the material that may affect NDE method selection are highly dependent on the specific NDE method under consideration. Table 6 lists a number of NDE methods and the characteristic of critical importance for each.

The specific NDE method can be selected by applying all the previously discussed factors. Because each NDE method has a specific behavior, it is often desirable to select several NDE methods having complementary detection capabilities. For example, ultrasonic and radiographic methods can be used together to ensure the detection of both planar flaws (such as cracks) and volumetric flaws (such as porosity).

Leak Detection and Evaluation

Because many objects must withstand pressure, the nondestructive determination of leakage is extremely important. The NDE area known as leak detection utilizes many techniques, as described in the article "Leak Testing" in this Volume. Each technique has a specific range of applications, and a particular leak detection technique should be selected only after careful consideration of the factors discussed in the aforementioned article.

Metrology and Evaluation

The measurement of dimensions, referred to as metrology, is one of the most widely used NDE activities, although it is often not considered with other conventional NDE activities, such as flaw detection. Although conventional metrology is not specifically discussed in this Volume, modern high-technology metrology is covered in the articles "Laser Inspection," "Coordinate Measuring Machines," and "Machine Vision and Robotic Inspection Systems."

The selection of a metrology system is highly dependent on the specific requirements of a given application. Standard reference works on the topic should be consulted for conventional metrology, and the articles listed above should be studied for selecting new technology. In addition, other NDE methods, such as eddy current, ultrasonic, optical holography, and speckle metrology, often find application in the field of metrology. Selection of these methods for metrology application can be assisted by the information in the articles so titled in this Volume.

Location Determination and Evaluation

An occasional problem is whether an assembled unit (one that contains several parts) actually contains the necessary components. This type of inspection has resulted in an NDE activity that can be termed location determination. The most commonly employed NDE techniques for location determination are x-ray radiography, x-ray computed tomography, and neutron radiography. These techniques and their selection are discussed in separate articles in this Volume.

Structure or Microstructure Characterization

Another interesting area of NDE is microstructural characterization, which can be done *in situ* without damaging the object by using replication microscopy (discussed in the following article in this Volume) or by using conventional optical microscopy techniques with portable equipment, including polishing, etching, and microscopic equip-

Table 4 Comparison of NDE methods based on size of object to be evaluated

The thickness or dimension limitation is only approximate because the exact value depends on the specific physical properties of the material being evaluated.

Surface only but independent of size
Visual
Replica
Digital enhancement
Liquid penetrant
Shallow depth or thin object (thickness ≤1 mm, or 0.04 in.)
Magnetic particle
Magnetic field
Magabsorption
Eddy current
Increased thickness (thickness ≤3 mm, or 0.12 in.)
Microwave
Optical holography
Speckle metrology
Acoustic holography
Acoustic microscopy
Increased thickness (thickness ≤100 mm, or 4 in.)
X-ray computed tomography
Increased thickness (thickness ≤250 mm, or 10 in.)
Neutron radiography(a)
X-ray radiography
Thickest (dimension ≤10 m, or 33 ft)
Ultrasonic

(a) All NDE methods suitable for thick objects can be used on thin objects, except neutron radiography, which is not useful for most thin objects.

Table 5 Comparison of NDE techniques based on shape of object to be evaluated

Simplest shape ↑
Optical holography
Acoustic holography
Acoustic microscopy
Thermography
Microwave
Eddy current
Magnetic particle
Magnetic field
Magabsorption
Neutron radiography
X-ray radiography
Ultrasonic
Liquid penetrant
Digital enhancement
Replica
Visual
X-ray computed tomography
↓ **Most complex shape**

ment. In addition, it is possible to characterize the microstructure through the correlation with some type of NDE information. For example, the transmission of ultrasonic energy has been correlated with the microstructure of gray cast iron.

Microstructure can often be characterized by determining physical or mechanical properties with NDE techniques because there is usually a correlation among microstructure, properties, and NDE response. Characterizing microstructure from NDE responses is a relatively recent area of NDE application, and new developments are occurring frequently.

Estimation of Mechanical and Physical Properties

As discussed previously, the prediction of mechanical and physical properties with NDE techniques is a relatively new application of NDE. Eddy current, ultrasonic, x-ray and neutron radiography, computed tomography, thermography, and acoustic microscopy phenomena are affected by microstructure, which can be related to some mechanical or physical properties. In addition, microwave NDE can be related to the properties of plastic materials. Several technical meetings are held each year to discuss advances in NDE, and these meetings often feature sessions on characterizing microstructure and mechanical and physical properties with NDE techniques. Some of the meetings are listed below:

- Annual Review (every spring), Center for Nondestructive Evaluation, The Johns Hopkins University
- Annual Review of Progress in Quantitative NDE (every summer), The Center for NDE, Iowa State University
- Symposium on Nondestructive Evaluation (every other spring), Nondestructive Testing Information Center, Southwest Research Institute
- Spring and Fall Meetings, American Society for Nondestructive Testing

Table 6 NDE methods and their important material characteristics

Method	Characteristic
Liquid penetrant	Flaw must intercept surface.
Magnetic particle	Material must be magnetic.
Eddy current	Material must be electrically conductive or magnetic.
Microwave	Microwave transmission
Radiography and x-ray computed tomography	Changes in thickness, density, and/or elemental composition
Neutron radiography	Changes in thickness, density, and/or elemental composition
Optical holography	Surface optical properties

Stress/Strain and Dynamic Response Determination

The local strain at a specific location in an object under a specific set of loading conditions can be determined by using strain sensing methods such as photoelastic coatings, brittle coatings, or strain gages. These methods are discussed in the article "Strain Measurement for Stress Analysis" in this Volume. If the stress-strain behavior of the material is known, these strain values can be converted into stress values.

A number of methods have also been developed for measuring residual stresses in materials. These include x-ray diffraction, ultrasonics, and electromagnetics. Surface residual stresses can be measured by x-rays as described in the article "X-Ray Diffraction Residual Stress Techniques" in Volume 10 of the 9th Edition of *Metals Handbook*. Practical application of ultrasonic techniques for characterizing residual stresses have not yet materialized. A number of electromagnetic techniques have, however, been successfully used as described in the articles "Electromagnetic Techniques for Residual Stress Measurements" and "Magabsorption NDE" in this Volume.

Dynamic behavior of an object can be evaluated during real or simulated service by employing strain sensing technology while the object is being dynamically loaded. In addition, accelerometers and acoustic transducers can be used to determine the dynamic response of a structure while it is being loaded. This dynamic response is called a signature and evaluation of this signature is called signature analysis. The nature of this signature can be correlated with many causes, such as machine noise, vibrations, and structural instability (buckling or cracking).

Replication Microscopy Techniques for NDE

A.R. Marder, Energy Research Center, Lehigh University

SURFACE REPLICATION is a well-developed electron microscopy sample preparation technique that can be used to conduct *in situ* measurements of the microstructure of components. The *in situ* determination of microstructural deterioration and damage of materials subjected to various environments is an objective of any nondestructive evaluation (NDE) of structural components. The need to assess the condition of power plant and petrochemical metallic components on a large scale recently led to the application of surface replication to the problem of determining remaining life. The usual method of metallographic investigation, which may involve cutting large pieces from the component so that laboratory preparation and examination can be performed, usually renders the component unfit for service or necessitates a costly repair. As a result, metallographic investigations are avoided, and important microstructural information is not available for evaluating the component for satisfactory performance. Therefore, an *in situ* or field microscopy examination is needed to aid in the proper determination of component life.

The replica technique for the examination of surfaces has been extensively used for studying the structure of polished-and-etched specimens and for electron fractographic examination (see the article "Transmission Electron Microscopy" in Volume 12 of the 9th Edition of *Metals Handbook* for a discussion of replication techniques in fractography). Surface replication was the predominant technique in electron microscopy prior to being supplemented by thin-foil transmission and scanning electron microscopy. Recently, the replication microscopy technique has become an important NDE method for microstructural analysis, and an American Society for Testing and Materials specification has been written for its implementation (Ref 1).

Specimen Preparation

Mechanical Polishing Methods. Components in service usually have a well-developed corrosion or oxidation product or a decarburized layer on the surface that must be removed before replication. Coarse-grinding equipment can be used as long as the proper precautions are taken to prevent the introduction of artifacts into the structure due to overheating or plastic deformation. Sandblasting, wire wheels, flap wheels, and abrasive disks have all been used. After the initial preparation steps are completed, standard mechanical polishing techniques can be used. Field equipment is commercially available to help the metallographer reproduce the preparation steps normally followed in the laboratory. Depending on the material, various silicon carbide abrasive disks of different grit size, together with polishing cloth disks with diamond paste or alumina of varying grit size, can be used to prepare for the etching step. Finally, any appropriate etchant for the material being examined can be applied to develop the microstructure. For the proper identification of such microstructural features as creep cavities, a maximum double or triple etch-polish-etch procedure should be used (Ref 2). The etchants used for the various materials investigated by the replication technique are described in Volume 9 of the 9th Edition of *Metals Handbook* and in Ref 3.

Electrolytic Preparation Technique. Although electrolytic polishing and etching techniques have often been employed as the final mechanical polish step in sample preparation, inherent problems still exist in this process. The electropolishing technique uses an electrolytic reaction to remove material to produce a scratch-free surface. This is done by making the specimen the anode in an electrolytic cell. The cathode is connected to the anode through the electrolyte in the cell. Specimens can be either polished or etched, depending on the applied voltage and current density, as seen in the fundamental electropolishing curve in Fig. 1. However, the pitting region must be avoided so that artifacts are not introduced into the microstructure. It is virtually impossible to prevent pitting without precise control of the polishing variables, and pits can often be mistakenly identified as creep voids.

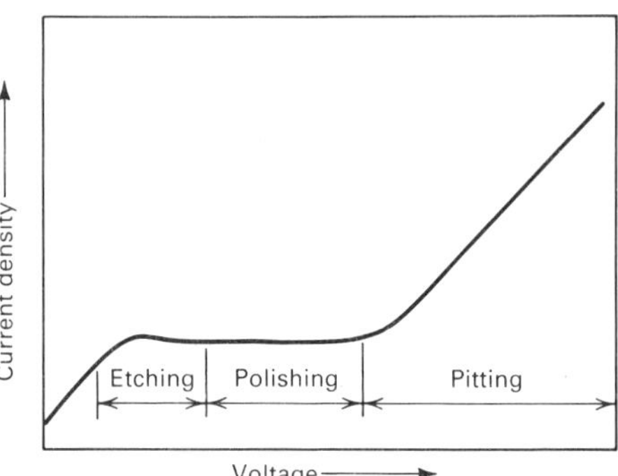

Fig. 1 Current density-voltage curve for electropolishing

Several portable electropolishing units are commercially available. The most important variables (time, bath temperature, electrolyte composition, and the current density-voltage relationship) have been investigated for a selected group of electrolytes (Ref 4). A direct comparison of electropolishing units and the precautions necessary for handling certain electrolytes are given in Ref 5.

It should be noted that there are areas in both fossil and nuclear plants in which neither acid etches nor electropolishing methods and materials are allowed because of the potential for intergranular stress-corrosion cracking. Stainless steel piping in nuclear plants can be replicated to determine defects by manual polishing without etchants. Generator retaining rings have been replicated by manual polishing to resolve NDE indications, because they are extremely sensitive to stress-corrosion cracking and no acids or caustics are allowed to be used (Ref 6).

Replication Techniques

Replication techniques can be classified as either surface replication or extraction replication. Surface replicas provide an image of the surface topography of a speci-

Table 1 Comparison of replica techniques

Type	Advantages	Disadvantages
Surface replicas		
Acetate	Excellent resolution	Coating required
Acrylic........	Direct viewing	Adhesion
Rubber........	Easy removal	Resolution
Extraction replicas		
Direct stripped plastic	Easy preparation	Particle retention
Positive carbon	Excellent particle retention with two-stage etching	Coating required
Direct carbon ..	Excellent resolution	Not applicable to *in situ* studies

men, while extraction replicas lift particles from the specimen. The advantages and disadvantages of some typical replication techniques are given in Table 1.

Surface Replicas. Replication of a surface can involve either direct or indirect methods. In the direct, or single-stage, method, a replica is made of the specimen surface and subsequently examined in the microscope, while in the indirect method, the final replica is taken from an earlier primary replica of the specimen surface. Only the direct method will be considered in this article because it lends itself more favorably to on-site preparation. The most extensively used direct methods involve plastic, carbon, or oxide replica material. All direct methods except plastic methods are destructive and therefore require further preparation of the specimen before making additional replicas.

Plastic replicas lend themselves to in-plant nondestructive examination because of their relative simplicity and short preparation time. Plastic replicas can be examined with the light optical microscope, the scanning electron microscope, and the transmission electron microscope, depending on the resolution required. As illustrated in Fig. 2, the plastic replica technique involves softening a plastic film in a solvent, applying it to the surface, and then allowing it to harden as the solvent evaporates. After careful removal from the surface, the plastic film contains a negative image, or replica, of the microstructure that can be directly examined in the light microscope or, after some preparation, in the electron microscope. Double-faced tape is used to bond the replica to the glass slide in order to obtain large, flat, undistorted replica surfaces.

There are some significant advantages of the replica technique over the use of portable microscopes in the field (Ref 5):

- A permanent record of the specimen is obtained
- Better resolution and higher magnification can be used
- Contamination of the polished surface is minimized
- Time spent in an unpleasant or hazardous environment is minimized
- Scanning electron microscopy can be utilized

Several materials, including acetate, acrylic resin, and rubber, can be used in the surface replica technique (Ref 5). The choice of material depends on the geometry of the component and the microstructural features to be examined.

In the acetate method, an acetate tape is wetted with acetone and applied to the surface; other less volatile solvents, such as methyl acetate, can be used when large areas are replicated. For improved resolution, the back side of the replica can be painted with any fast-drying black paint or ink prior to removal, or for the same effect, evaporated coatings of carbon, aluminum,

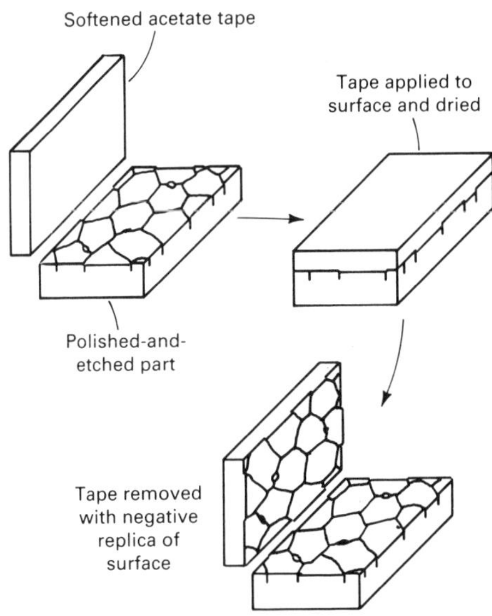

Fig. 2 Schematic of the plastic replica technique

or gold can be applied at a shadow angle of 45° to the front side of the replica after removal.

In the acrylic casting resin method, dams are required because a powder is mixed with a liquid on the surface to be replicated. After hardening, the replica can be examined directly in an optical microscope without further processing. If adhesion is a problem, a composite replica can be made of an initial layer of Parlodian lacquer before the acrylic layer is applied.

In the dental impression rubber method, uncured liquid rubber material (for example, GE RTV60 silicon rubber compound) is poured onto the surface to be replicated and is contained by a dam. After removal, the replica can be examined directly or can be coated for better resolution.

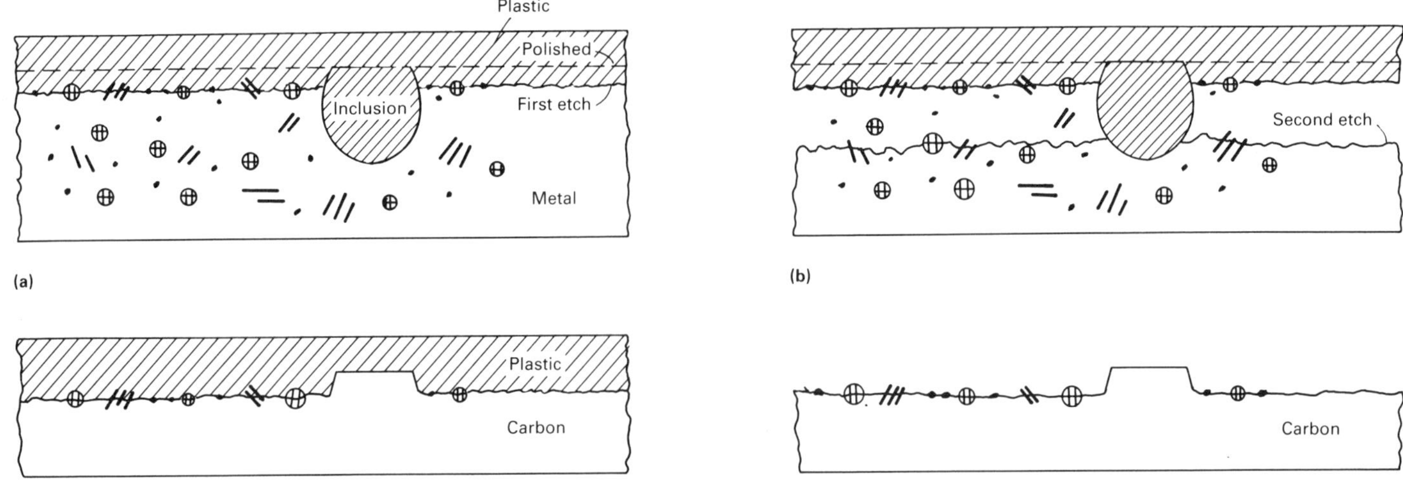

Fig. 3 Positive carbon extraction replication steps. (a) Placement of plastic after the first etch. (b) After the second etch. (c) After the deposition of carbon. (d) The positive replica after the plastic is dissolved

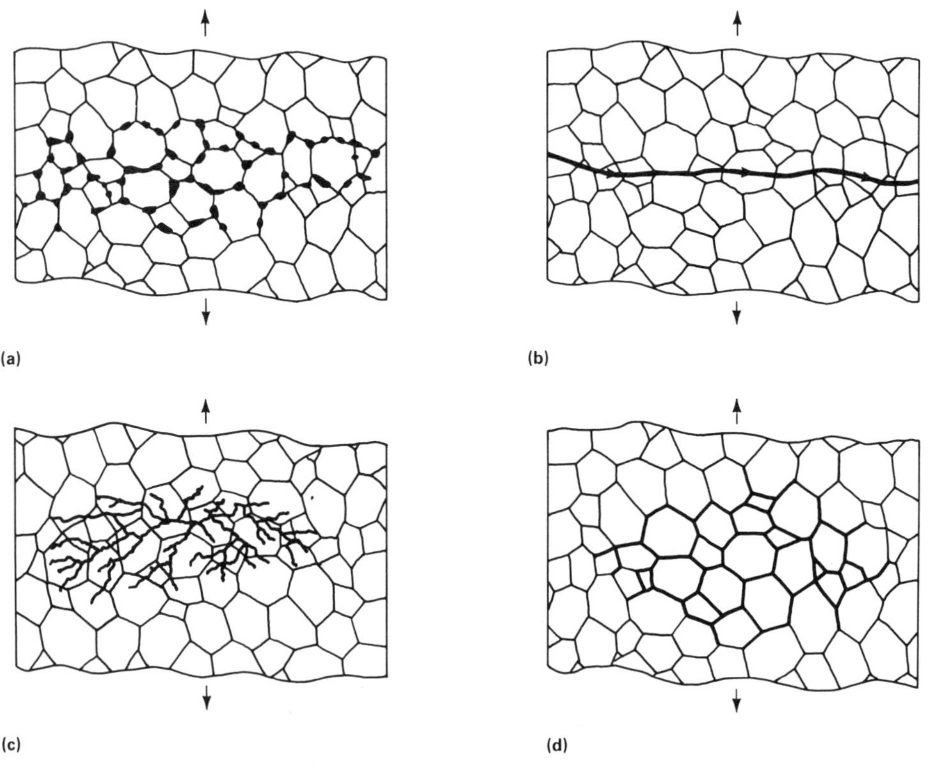

Fig. 4 Propagation of different crack types. (a) Creep. (b) Fatigue. (c) Stress corrosion. (d) Intergranular corrosion

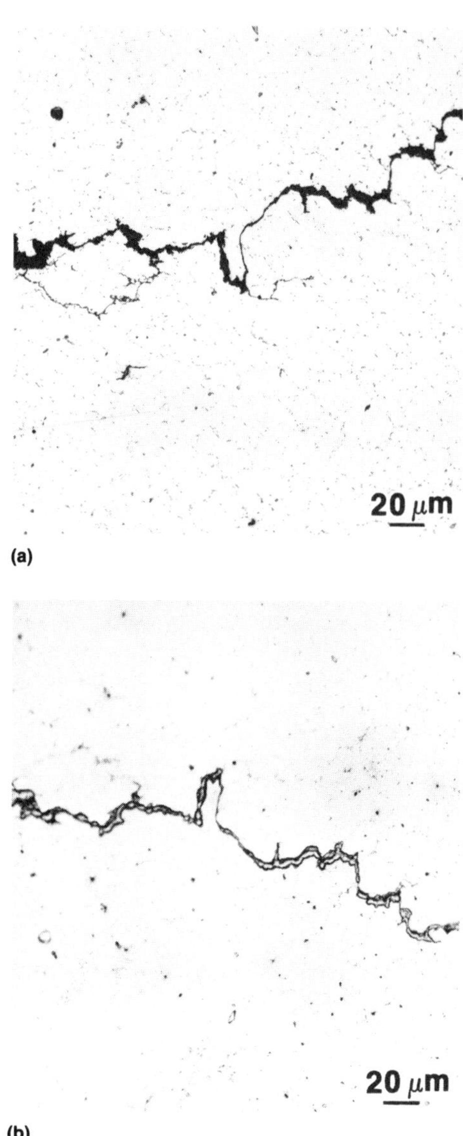

Fig. 5 Surface crack in a boiler tube. Comparison of the (a) actual microstructure and (b) the replica of the crack

Extraction Replicas. Several different extraction replica techniques can be used to characterize small particles that are embedded in a matrix, such as small second-phase particles in a steel (see the article "Analytical Transmission Electron Microscopy" in Volume 10 of the 9th Edition of *Metals Handbook*). More detailed descriptions of the various extraction replica techniques can be found in Ref 7 and 8.

After careful preparation of the surface using normal polishing methods, the first step in producing an extraction replica is to etch the alloy heavily to leave the particles of interest in relief. In the positive carbon extraction replica, as shown in Fig. 3, a piece of solvent-softened polymeric film (cellulose acetate tape) is pressed onto the surface exposed by this first etch (Ref 5). Once the solvent has evaporated, one of two steps can be taken. The tape can be carefully pulled from the specimen to produce a negative of the surface, or the specimen can undergo a second etch to free the particles exposed by the first etch (Fig. 3). In the second etch, the specimen can be etched through the plastic; most plastics are quite permeable to etching solutions, and the specimen etches almost as rapidly as without the plastic film (Ref 9). Carbon is then evaporated in a vacuum onto the plastic replica. The carbon and plastic containing the particles now make up the positive replica. The cellulose acetate is then dissolved, and the positive carbon replica is allowed to dry. It should be noted that for the negative carbon extraction replica technique, vacuum deposition of carbon onto the surface of the specimen is required, and therefore this replica method is not applicable to NDE.

Microstructural Analysis

Crack determination is important to help establish the root cause of a potential failure in a component. After a preliminary evaluation of the crack to assess crack shape and length by using magnetic flux or dye penetrant, the replica method is then used on unetched specimens to assist in the crack evaluation. Figure 4 schematically shows the propagation of different types of cracks in a steel structure (Ref 10). Each crack has its own characteristics, and it is often possible to make a correct determination of crack type. It is important to determine whether the crack is the original defect or has been caused by service conditions or damage. Once the crack type is identified, the proper corrective action, such as eliminating a corrosive environment or reducing stress levels, can be attempted. Figure 5 shows the replication of surface cracks in a boiler tube.

Creep Damage. Creep defects cause the majority of failures in power plant components operating under stress and thermal load, and the replica method is especially suitable for the detection of these defects. Therefore, the replica method has become an especially important tool in the determination of remaining life in such components as boiler tubes, steam piping, and turbine components. The replica method reveals defects due to creep at a much earlier stage than other NDE techniques. Creep defects begin as small holes or cavities at grain boundaries or second phases. With time and stress, these holes or cavities can link up and form cracks that eventually lead to failure of the component (Fig. 6). Creep cracks are usually very localized, and they form in welds, bends, or other highly stressed regions. Determining the remaining life of components normally depends on assessments of regular inspections, as indicated in Table 2. Figure 7 shows a comparison of creep voids in a surface replica and the corresponding bulk microstructure.

Precipitate Analysis. The detection of various deleterious precipitates in components subjected to high temperature and stress can lead to improved life assessment

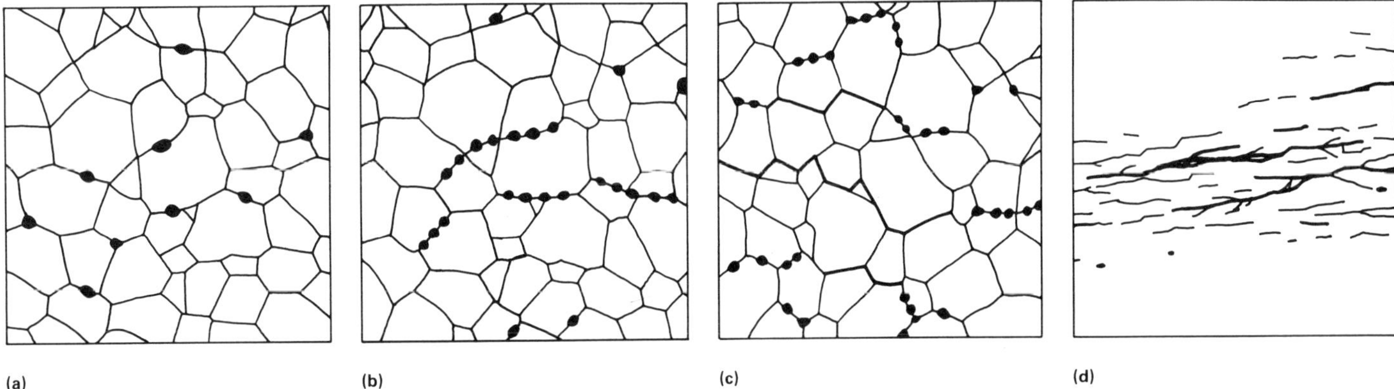

Fig. 6 Schematic of creep crack formation. Small cavities (a) link up over time (b) and form intergranular cracks (c) and eventually macrocracks (d).

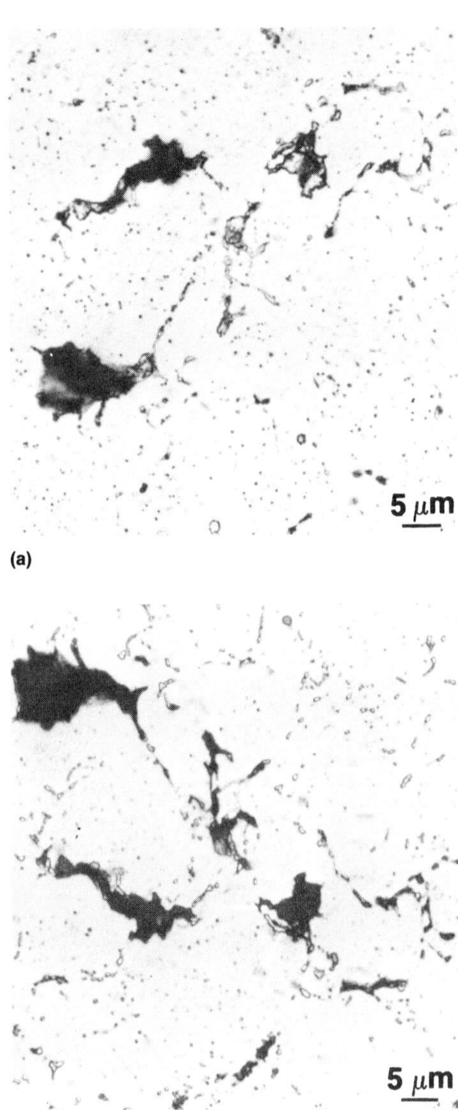

Fig. 7 Comparison of creep voids in (a) a replica and (b) the actual microstructure

analysis of these components. The extraction replication technique is an excellent nondestructive method of detecting these precipitates.

Table 2 Creep damage classification

Class	Nature	Action
1	No creep defects	None
2	A few cavities	Reinspection after 20 000 h of service
3	Coalescent cavities	Reinspection after 15 000 h of service
4	Microscopic creep cracks	Reinspection after 10 000 h of service
5	Macroscopic creep cracks	Management must be informed immediately

Source: Ref 11

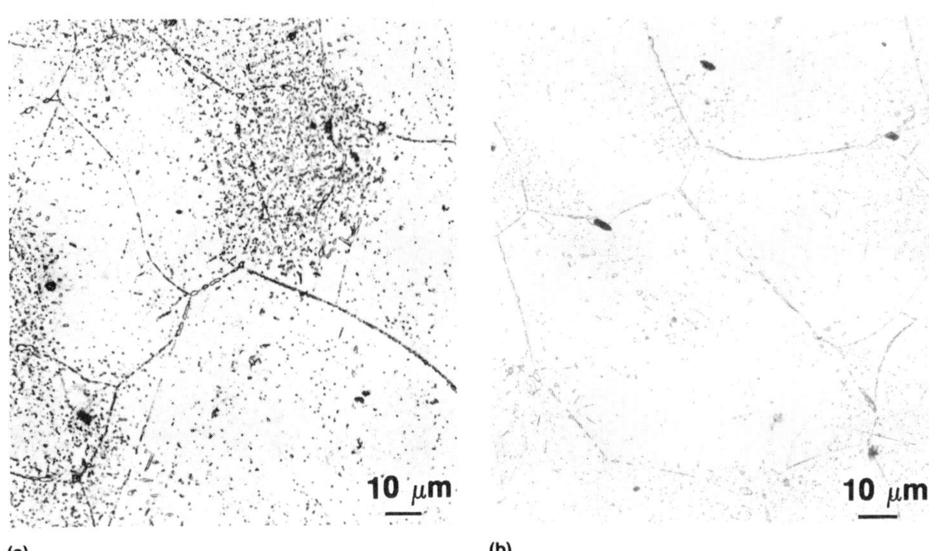

Fig. 8 Comparison of σ-phase formation as seen in (a) a replica and (b) the actual microstructure

Sigma phase is a deleterious FeCr compound that can form in some stainless steels, and its presence can severely limit remaining life. Extraction replicas have been used to determine the amount of σ phase in the microstructure (Ref 12), and the amount of σ phase has been directly related to the creep rate (Ref 13). Figure 8 shows an example of σ phase in an extraction replica.

The composition of carbides, and their stability with time and temperature of exposure, can indicate the remaining life of a component. Extraction replicas have been used to evaluate carbides, and it has been suggested that changes in morphology and chemistry can be used to assist the estimation of effective exposure temperature for use in determining the remaining life of components (Ref 14). Figure 9 shows an example of precipitates extracted from a 200 000-h exposed sample, together with the accompanying chemical analysis.

ACKNOWLEDGMENT

The author would like to acknowledge the contributions of his colleagues A.O. Benscoter, S.D. Holt, and T.S. Hahn in the preparation of this article.

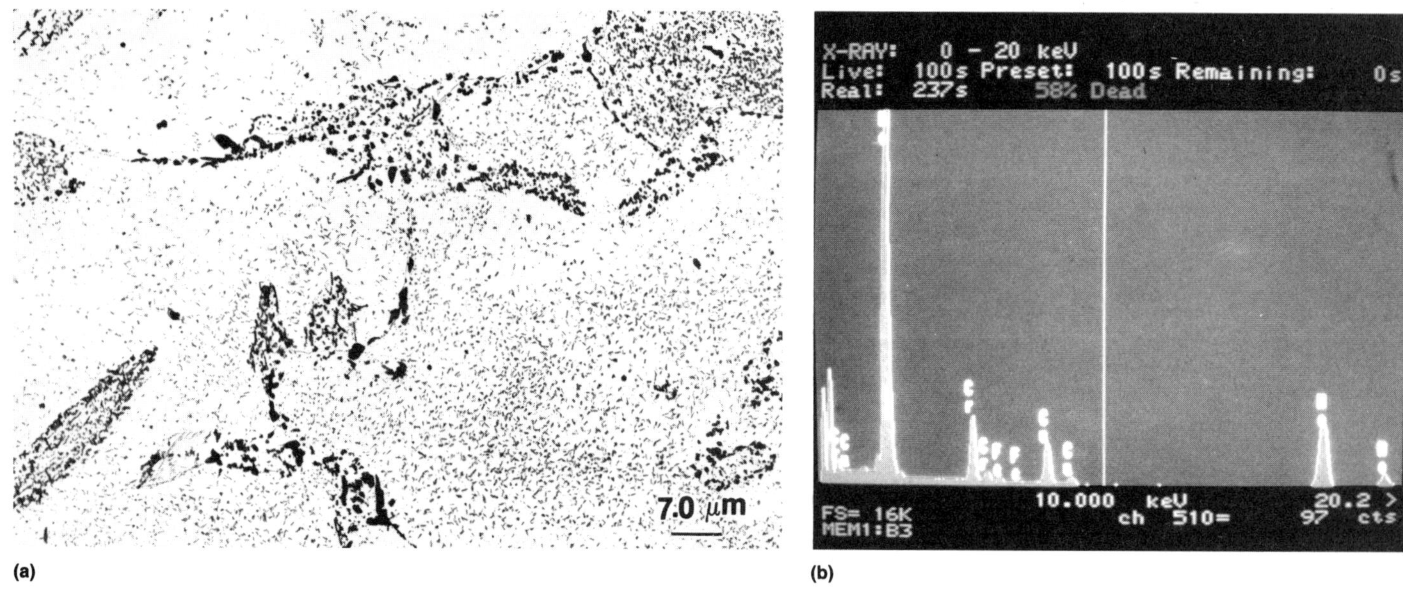

(a) (b)

Fig. 9 Extraction replica of the microstructure (a) and the precipitate microchemical analysis (b) from an extraction replica

REFERENCES

1. "Standard Practice for Production and Evaluation of Field Metallographic Replicas," E 512-87, *Annual Book of ASTM Standards*, American Society for Testing and Materials
2. A.M. Bissel, B.J. Cane, and J.F. DeLong, "Remanent Life Assessment of Seam Welded Pipework," Paper presented at the ASME Pressure Vessel and Piping Conference, American Society of Mechanical Engineers, June 1988
3. G.F. Vander Voort, *Metallography: Principles and Practice*, McGraw-Hill, 1984
4. T.S. Hahn and A.R. Marder, Effect of Electropolishing Variables on the Current Density—Voltage Relationship, *Metallography*, Vol 21, 1988, p 365
5. M. Clark and A. Cervoni, "In Situ Metallographic Examination of Ferrous and Non-Ferrous Components," Canadian Electrical Association, Nov 1985
6. J.F. DeLong, private communication
7. D. Kay, Ed., *Techniques for Electron Microscopy*, Blackwell Scientific Publications, 1965
8. J.W. Edington, *Practical Electron Microscopy in Materials Science*, Van Nostrand Rheinhold, 1976
9. G.N. Maniar and A. Szirmae, in *Manual on Electron Metallography Techniques*, STP 547, American Society for Testing and Materials, 1973
10. P.B. Ludwigsen, Non-Destructive Examination, *Structure*, Sept 1987, p 3
11. B. Neubauer and U. Wedel, NDT: Replication Avoids Unnecessary Replacement of Power Plant Components, *Power Eng.*, May 1984, p 44
12. F. Masuyama, K. Setoguchi, H. Haneda, and F. Nanjo, Findings on Creep-Fatigue Damage in Pressure Parts of Long-Term Service-Exposed Thermal Power Plants, in *Residual Life Assessment Nondestructive Examination and Nuclear Heat Exchanger Materials*, PVP-Vol 98-1, Proceedings of the Pressure Vessels and Piping Conference, American Society of Mechanical Engineers, 1985, p 79
13. T. Fushimi, "Life Evaluation of Long Term Used Boiler Tubes," Paper presented at the Conference on Boiler Tube Failures in Fossil Plants (Atlanta), Electric Power Research Institute, Nov 1987
14. A. Afrouz, M.J. Collins, and R. Pilkington, Microstructural Examination of 1Cr-0.5Mo Steel During Creep, *Met. Technol.*, Vol 10, 1983, p 461

Leak Testing

Revised by Gerald L. Anderson, American Gas and Chemical Company

LEAK TESTING is the branch of nondestructive testing that concerns the escape or entry of liquids or gases from pressurized or into evacuated components or systems intended to hold these liquids. Leaking fluids (liquid or gas) can penetrate from inside a component or assembly to the outside, or vice versa, as a result of a pressure differential between the two regions or as a result of permeation through a somewhat extended barrier. Leak testing encompasses procedures for one or a combination of the following:

- Locating (detecting and pinpointing) leaks
- Determining the rate of leakage from one leak or from a system
- Monitoring for leakage

Leak testing is increasing in importance because of the rising value of, and warranties on, manufactured products and because of the constantly increasing sensitivity of components and systems to external contaminants. Environmental concerns are causing additional emphasis on leak testing and its conduct.

Leak Testing Objectives

Like other forms of nondestructive testing, leak testing has a great impact on the safety and performance of a product. Reliable leak testing decreases costs by reducing the number of reworked products, warranty repairs, and liability claims. The most common reasons for performing a leak test are:

- To prevent the loss of costly materials or energy
- To prevent contamination of the environment
- To ensure component or system reliability

Terminology. The following terms must be understood in their strict definitions within the field of leak testing:

- *Leak*: An actual through-wall discontinuity or passage through which a fluid flows or permeates; a leak is simply a special type of flaw
- *Leakage*: The fluid that has flowed through a leak
- *Leak rate*: The amount of fluid passing through the leak per unit of time under a given set of conditions; properly expressed in units of quantity or mass per unit of time
- *Minimum detectable leak*: The smallest hole or discrete passage that can be detected
- *Minimum detectable leak rate*: The smallest detectable fluid-flow rate

The amount of leakage required for a leak testing instrument to produce a minimum detectable signal can be determined. This amount is generally used to indicate the sensitivity of the instrument. Instrument sensitivity is independent of test conditions, but when an instrument is used in a test, the sensitivity of the test depends on existing conditions of pressure, temperature, and fluid flow.

Measurement of Leakage. A leak is measured by how much leakage it will pass under a given set of conditions. Because leakage will vary with conditions, it is necessary to state both the leak rate and the prevailing conditions to define a leak properly. At a given temperature, the product of the pressure and the volume of a given quantity of gas is proportional to its mass. Therefore, leak rate is often expressed as the product of some measure of pressure and volume per unit of time—for example, torr liters per second (torr · L/s), micron liters per second (μL/s), and atmosphere cubic centimeters per second (atm cm^3/s).

The two most commonly used units of leakage rate with pressure systems are standard cubic centimeters per second (std cm^3/s) and its equivalent, standard atmosphere cubic centimeters per second (atm cm^3/s). The most frequently used unit in vacuum leak testing is torr liters per second. The recent adoption of the Système International d'Unités (SI) system of units has resulted in a new measure of leakage, pascal cubic meters per second (Pa · m^3/s). Another new unit is moles per second (mol/s), which has the advantage that information concerning temperature implied in other units is automatically included in this unit. The term used for leak rate in this article is standard atmosphere cubic centimeter per second, where standard conditions are defined as 1 atm = 101.325 kPa and standard temperature is 273.15 K (0 °C). The conversion factor is std cm^3/s = 4.46 × 10^{-5} mol/s. Another unit of leak rate generally accepted is the gas-flow rate that causes a pressure rise per unit of time of 1 μm Hg/s (40 μin. Hg/s) in a volume of 1 L (0.26 gal.). This is termed the lusec (liter microns per second). The clusec (centiliter microns per second) unit is equal to 0.01 lusec. Boyle's law (PV = K, where P is pressure, V is volume, and K is a constant) enables one to convert these units into more meaningful volumetric terms; these conversions are listed in Table 1, along with conversions for pressures and volumes.

Types of Leaks. There are two basic types of leaks: real leaks and virtual leaks.

A real leak is an essentially localized leak, that is, a discrete passage through which fluid may flow (crudely, a hole). Such a leak may take the form of a tube, a crack, an orifice, or the like. As with flaws, all leaks are not the same. Leaks tend to grow over time, and they tend to operate differently under different conditions of pressure and temperature. A system may also leak because of permeation of a somewhat extended barrier; this type of real leak is called a distributed leak. A gas may flow through a solid having no holes large enough to permit more than a small fraction

Table 1 Conversion factors for quantities related to leak testing

std cm^3/s	Pa · m^3/s, at 0 °C (32 °F)	torr · L/s, at 0 °C (32 °F)
1	0.1	7.6 × 10^{-1}
1 × 10^{-1}	1 × 10^{-2}	7.6 × 10^{-2}
5 × 10^{-2}	5 × 10^{-3}	3.8 × 10^{-2}
1 × 10^{-2}	1 × 10^{-3}	7.6 × 10^{-3}
5 × 10^{-3}	5 × 10^{-4}	3.8 × 10^{-3}
1 × 10^{-3}	1 × 10^{-4}	7.6 × 10^{-4}
5 × 10^{-4}	5 × 10^{-5}	3.8 × 10^{-4}
1 × 10^{-4}	1 × 10^{-5}	7.6 × 10^{-5}
5 × 10^{-5}	5 × 10^{-6}	3.8 × 10^{-5}
1 × 10^{-5}	1 × 10^{-6}	7.6 × 10^{-6}
5 × 10^{-6}	5 × 10^{-7}	3.8 × 10^{-6}
1 × 10^{-6}	1 × 10^{-7}	7.6 × 10^{-7}
5 × 10^{-7}	5 × 10^{-8}	3.8 × 10^{-7}
1 × 10^{-7}	1 × 10^{-8}	7.6 × 10^{-8}
1 × 10^{-8}	1 × 10^{-9}	7.6 × 10^{-9}
1 × 10^{-9}	1 × 10^{-10}	7.6 × 10^{-10}
1 × 10^{-10}	1 × 10^{-11}	7.6 × 10^{-11}
1 × 10^{-11}	1 × 10^{-12}	7.6 × 10^{-12}
1 × 10^{-12}	1 × 10^{-13}	7.6 × 10^{-13}
1 × 10^{-13}	1 × 10^{-14}	7.6 × 10^{-14}

of the gas to flow through any one hole. This process involves diffusion through the solid and may also involve various surface phenomena, such as absorption, dissociation, migration, and desorption of gas molecules.

Virtual leaks involve the gradual desorption of gases from surfaces or escape of gases from nearly sealed components within a vacuum system. It is not uncommon for a vacuum system to have both real leaks and virtual leaks at the same time.

Types of Flow in Leaks

Types of flow in leaks include permeation, molecular flow, transitional flow, viscous flow, laminar flow, turbulent flow, and choked flow. The type of flow corresponding to a specific instance of leakage is a function of pressure differential, type of gas, and size and shape of the leak. Some of these types of flow (which are described in the following sections) also occur at locations such as piping connections and have characteristics similar to leaks.

Permeation is the passage of a fluid into, through, and out of a solid barrier having no holes large enough to permit more than a small fraction of the total leakage to pass through any one hole. This process involves diffusion through a solid and may involve other phenomena, such as absorption, dissociation, migration, and desorption.

Molecular flow occurs when the mean free path of the gas is greater than the longest cross-sectional dimension of the leak. The mean free path is the average distance that a molecule will travel before colliding with another molecule. It is an inverse linear function of the pressure. In molecular flow, the leakage (flow) is proportional to the difference in the pressures. Molecular flow is a frequent occurrence in vacuum testing.

Transitional flow occurs when the mean free path of the gas is approximately equal to the cross-sectional dimension of the leak. The conditions for transitional flow are between those for laminar and molecular flow (see the section "The Equation of the State of an Ideal Gas" in this article).

Viscous flow occurs when the mean free path of the gas is smaller than the cross-sectional dimension of the leak. In viscous flow, the leakage (flow) is proportional to the difference of the squares of the pressures. Viscous flow occurs in high-pressure systems such as those encountered in detector probing applications. Above a critical value of the Reynolds number, N_{Re} (about 2100 for circular-pipe flow), flow becomes unstable, resulting in countless eddies or vortices.

Laminar and Turbulent Flow. Laminar flow occurs when the velocity distribution of the fluid in the cross section of the passage or orifice is parabolic. Laminar flow is one of two classes of viscous flow; the other is turbulent flow. Particles in turbulent flow follow very erratic paths, but particles in laminar flow follow straight lines. The term viscous flow is sometimes incorrectly used to describe laminar flow.

Choked flow, or sonic flow, occurs under certain conditions of configuration and pressure. Assume there exists a passage in the form of an orifice or a venturi, and assume that the pressure upstream is kept constant. If the pressure downstream is gradually lowered, the velocity of the fluid through the throat or orifice will increase until it reaches the speed of sound. The downstream pressure at the time the orifice velocity reaches the speed of sound is called the critical pressure. If the downstream pressure is lowered below this critical pressure, no further increase in orifice velocity can occur, the consequence being that the mass-flow rate has reached its maximum.

Principles of Fluid Dynamics

Leakage generally falls within the discipline of fluid dynamics. However, only the elementary principles of fluid dynamics are needed for most of the practical requirements of leak testing. Mathematical models are very helpful in testing gas systems, and the following discussion is mainly concerned with such systems.

The equation of the state of an ideal gas (that is, a gas made up of perfectly elastic point particles) is:

$$PV = NRT \qquad \text{(Eq 1)}$$

where P is the gas absolute pressure, V is the volume of the vessel containing the gas, T is the absolute temperature of the gas, N is the number of moles of gas, and R is the universal gas constant. The constant, R, is the same for all (ideal) gases; the numerical value of R depends on the system of units in which P, V, and T are measured.

The number of moles of gas, N, is equal to the mass of the gas in grams divided by the gram-molecular weight of the gas. The gram-molecular weight is simply the number of grams numerically equal to the molecular weight of the gas. Regardless of the molecular weight, 1 mol of a gas always contains 6.022×10^{23} (Avogadro's number) molecules.

In leak testing, pressures are measured in terms of atmospheric pressure. By international agreement, the pressure of the standard atmosphere is 1 013 250 dyne/cm^2. This is equivalent to the pressure exerted by a 760 mm (30 in.) column of mercury, at 0 °C (32 °F), under a standard acceleration of gravity of 980.665 cm/s^2 (32.1740 ft/s^2). Another unit of pressure commonly used, especially in vacuum technology, is the torr, which is 1/760 of a standard atmosphere, equivalent to 1 mm (0.04 in.) of mercury. (Conversion factors are given in Table 1.)

The numerical value of the gas constant, R, in two useful sets of units is:

$$R = \frac{0.0820 \text{ atm} \cdot \text{L}}{\text{mol} \cdot \text{K}} = \frac{62.3 \text{ torr} \cdot \text{L}}{\text{mol} \cdot \text{K}} \qquad \text{(Eq 2)}$$

Equations of state, in conjunction with measurements of pressure, volume, and temperature, are used to determine the total quantity of gas in a closed system. The quantity of gas of a given composition can be expressed in terms of the total number of molecules, the total mass of gas, or any quantity proportional to these. For example, suppose a pressure vessel of volume V_1 is pressurized to P_1 (greater than atmospheric pressure) at temperature T_1 and allowed to stand for a period of time. At the end of this period, the pressure is found to have decreased to P_2. If the temperature of the system is unchanged and the gas in the vessel has not changed in volume, the decrease in pressure must be the result of a loss of gas, and a leak must be presumed to be present. However, if temperature T_2 is less than T_1, the decrease in pressure may be caused by a decrease in the gas volume as the result of cooling. If the quantity of gas has not changed, it follows from Eq 1 that P_1, V_1, T_1, V_2, and T_2 must satisfy the relation:

$$\frac{(P_1V_1)}{T_1} = \frac{(P_2V_2)}{T_2} \qquad \text{(Eq 3)}$$

If $V_1 = V_2$, as assumed, then the following relation exists:

$$\frac{P_1}{T_1} = \frac{P_2}{T_2} \qquad \text{(Eq 4)}$$

Thus, if P_2 has decreased in proportion to the decrease in temperature, the system can be presumed not to have leaked at a rate discernible by the pressure gage and thermometer used to monitor the system. If the precision and drift of these instruments are known, a maximum leak rate consistent with their indication of no leakage can be computed.

If the reduction in pressure cannot be accounted for by the reduction in temperature (that is, if Eq 4 is not satisfied), a leak or leaks may be presumed present, and the leak rate can be determined by computing the loss in quantity of gas. Referring to Eq 1, the loss of gas (expressed in number of moles) is given by:

$$\Delta N = N_2 - N_1 = \left[\left(\frac{P_2}{T_2}\right) - \left(\frac{P_1}{T_1}\right)\right]\left(\frac{V}{R}\right) \qquad \text{(Eq 5)}$$

where V is presumed constant.

If desired, ΔN can be converted to mass by multiplying by the mean gram-molecular

Table 2 Mean free path lengths of various atmospheric gases at 20 °C (68 °F) and various vacuum pressures

	Mean free path length at indicated absolute pressure									
	1 μPa (7.5 × 10^{-9} torr)		1 mPa (7.5 × 10^{-6} torr)		1 Pa (7.5 × 10^{-3} torr)		1 kPa (7.5 torr)		100 kPa (750 torr)(a)	
Gas	km	ft × 10^4	m	ft	mm	in.	μm	μin.	nm(b)	Å
Air	6.8	2.2	6.8	22	6.8	0.27	6.8	272	68	680
Argon	7.2	2.4	7.2	24	7.2	0.28	7.2	290	72	720
Carbon dioxide	4.5	1.5	4.5	15	4.5	0.18	4.5	180	45	450
Hydrogen	12.5	4.10	12.5	41.0	12.5	0.492	12.5	500	125	1250
Water	4.2	1.4	4.2	14	4.2	0.16	4.2	170	42	420
Helium	19.6	6.43	19.6	64.3	19.6	0.772	19.6	784	196	1960
Nitrogen	6.7	2.2	6.7	22	6.7	0.26	6.7	268	67	670
Neon	14.0	4.59	14.0	45.9	14.0	0.55	14.0	560	140	1400
Oxygen	7.2	2.4	7.2	24	7.2	0.28	7.2	290	72	720

(a) Approximately atmospheric pressure. (b) 1 nm (nanometer) = 10^{-9} m. Source: Ref 1

weight of the gas. Dividing ΔN by the time span between measurements gives the average leak rate.

For most systems, the volume of the system is assumed to be constant. However, all pressure vessels expand by some amount when they are pressurized; moreover, their volume may vary with thermal expansion or contraction. When very precise estimates of leak rate are required, explicit corrections for these effects must be made.

A given leak may exhibit several types of flow, depending on the pressure gradient, temperature, and fluid composition. Therefore, it is important to identify the type of flow that exists in order to predict the effect of changing these variables. Conversely, with constant temperature, pressure, and fluid composition, the configuration of the leak will determine the type of flow. Because the leak may be a crack, a hole, permeation, or a combination of these, its configuration is often impossible to ascertain, but general empirical guidelines can establish the type of flow for gases. If the leak rate is:

- Less than 10^{-6} atm cm^3/s, the flow is usually molecular
- 10^{-4} to 10^{-6} atm cm^3/s, the flow is usually transitional
- 10^{-2} to 10^{-6} atm cm^3/s, the flow is usually laminar
- Greater than 10^{-2} atm cm^3/s, the flow is usually turbulent

These types of gas flow are more accurately described by the Knudsen number:

$$N_K = \frac{\lambda}{d} \quad \text{(Eq 6)}$$

where N_K is the Knudsen number, λ is the mean free path of the gas, and d is the diameter of the leak. The relationship between N_K and flow regime is:

$N_K < 0.01$ (laminar, or a higher regime)
$0.01 \leq N_K \leq 1.00$ (transitional)
$N_K > 1.00$ (molecular)

Mean Free Path. Values of the mean free path for various gases and pressures are listed in Table 2. In a vacuum system, the mean free path will vary from inches to many feet; when the mean free path is very long, collisions with the chamber surfaces occur more frequently than collisions between molecules. This, in part, is the reason that neither a gas nor leakage diffuses evenly throughout a vacuum system at a rapid rate. The flow of gases in a vacuum is analogous to the flow of current in an electrical system. Every baffle or restriction in the system impedes gas diffusion. Consequently, if there are many impedances between the leak and the leak detector that reflect or absorb the molecules, movement of the gas molecules to the leak detector does not follow theoretical diffusion rates; instead, it is greatly dependent on the configuration of the vacuum system.

The level of vacuum is virtually always described in terms of absolute pressure. However, the mean free path or the concentration of molecules controls such vacuum properties as viscosity, thermal conductance, and dielectric strength. Furthermore, very few vacuum gages actually measure pressure, but instead measure the concentration of molecules. Therefore, in the context of vacuum systems, the term pressure is largely inaccurate, although it remains in use.

Leak Testing of Pressure Systems Without a Tracer Gas

Leak testing methods can be classified according to the pressure and fluid (gas or liquid) in the system. The following sections describe the common fluid-system leak testing methods in the general order shown in Table 3, which also lists methods used in the leak testing of vacuum systems (discussed in the section so titled in this article). Table 4 compares leak testing method sensitivities.

Leak detection by monitoring changes in the pressure of the internal fluid is often used when leak detection equipment is not immediately available. For the most part, detection can be accomplished with instruments that are already installed in the system.

Table 3 Method of leak testing systems at pressure or at vacuum

Gas systems at pressure

Direct sensing
- Acoustic methods
- Bubble testing
- Flow detection

Gas detection
- Smell
- Chemical reaction
- Halogen gas
- Sulfur hexafluoride
- Combustible gas
- Thermal-conductivity gages
- Infrared gas analyzers
- Mass spectrometry
- Radioisotope count
- Ionization gages
- Gas chromatography

Quantity-loss determination
- Weighing
- Gaging differential pressure

Liquid systems at pressure

Unaided visual methods
Aided visual methods
Surface wetting
Weight loss
Water-soluble paper with aluminum foil

Vacuum systems

Manometers
Halogen gas
Mass spectrometry
Ionization gages
Thermal-conductivity gages
Gas chromatography

Table 4 Sensitivity ranges of leak testing methods

Method	Sensitivity range, cm^3/s Pressure	Vacuum
Mass spectrometer	10^{-3} to 10^{-5}	10^{-3} to 10^{-10}
Electron capture	10^{-6} to 10^{-11}	...
Colorimetric developer	1 to 10^{-8}	...
Bubble test—liquid film	10^{-1} to 10^{-5}	10^{-1} to 10^{-5}
Bubble test—immersion	1 to 10^{-6}	...
Hydrostatic test	1 to 10^{-2}	...
Pressure increase	1 to 10^{-4}	1 to 10^{-4}
Pressure decrease/flow	1 to 10^{-3}	...
Liquid tracer	1 to 10^{-4}	1 to 10^{-4}
High voltage	...	1 to 10^{-4}
Halogen (heated anode)	10^{-1} to 10^{-6}	10^{-1} to 10^{-5}
Thermal conductivity (He)	1 to 10^{-5}	...
Gage	...	10^{-1} to 10^{-7}
Radioactive tracer	10^{-13}	...
Infrared	1 to 10^{-5}	...
Acoustic	1 to 10^{-2}	1 to 10^{-2}
Smoke tracer	1 to 10^{-2}	...

Source: Ref 2

Acoustic Methods

The turbulent flow of a pressurized gas through a leak produces sound of both sonic and ultrasonic frequencies (Fig. 1). If the leak is large, it can probably be detected with the ear. This is an economical and fast method of finding gross leaks. Sonic emissions are also detected with such instruments as stethoscopes or microphones,

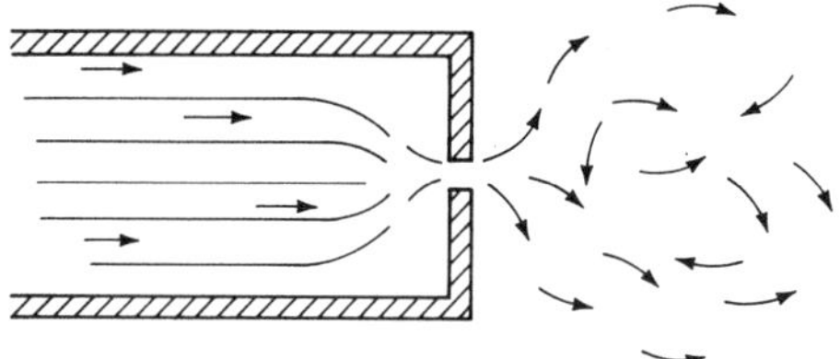

Fig. 1 Turbulence caused by fluid flow through an orifice

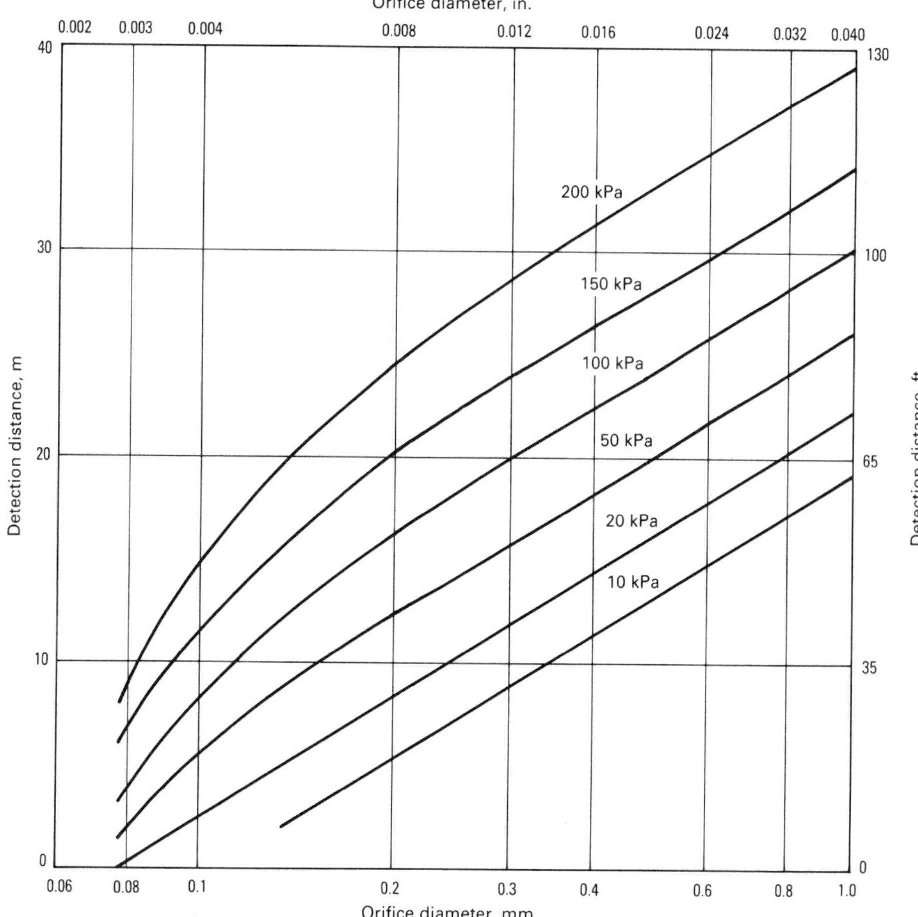

Fig. 2 Relation of orifice diameter to detection distance with an ultrasonic leak detector for various internal air pressures

which have limited ability to locate as well as estimate the approximate size of a leak. Electronic transducers enhance detection sensitivity.

Smaller leaks can be found with ultrasonic probes operating in the range of 35 to 40 kHz, although actual emissions from leaks range to over 100 kHz. Ultrasonic detectors are considerably more sensitive than sonic detectors for detecting gas leaks and from distances of 15 m (50 ft) are capable of detecting air leaking through a 0.25 mm (0.010 in.) diam hole at 35 kPa (5 psi) of pressure. The performance of an ultrasonic leak detector as a function of detection distance, orifice diameter, and internal air pressure is shown in Fig. 2. The sound level produced is an inverse function of the molecular weight of the leaking gas. Therefore, a given flow rate of a gas such as helium will produce more sound energy than the same flow rate of a heavier gas such as nitrogen, air, or carbon dioxide. If background noise is low, ultrasonic detectors can detect turbulent gas leakage of the order of 10^{-2} atm cm^3/s. Ultrasonic leak detectors have also been successfully used with ultrasonic sound generators when the system to be tested could not be pressurized.

Bubble Testing

A simple method for leak testing small vessels pressurized with any gas is to submerge them in a liquid and observe bubbles. If the test vessel is sealed at atmospheric pressure, a pressure differential can be obtained by pumping a partial vacuum over the liquid or by heating the liquid. The sensitivity of this test is increased by reducing the pressure above the liquid, the liquid density, the depth of immersion in the liquid, and the surface tension of the liquid.

Immersion Testing. Oils are a more sensitive medium than ordinary water. Therefore, it is common practice to test electric components in a bath of hot perfluorocarbon. When testing by reducing the pressure above the liquid, several precautions must be observed, particularly if the reduced pressure brings the liquid close to its boiling point. If the liquid begins to boil, a false leak indication will be given. The test vessel must be thoroughly cleaned to increase surface wetting, to prevent bubbles from clinging to its surface, and to prevent contamination of the fluid. If water is used, it must be distilled or deionized and should be handled with minimal sloshing to reduce the absorbed-gas content. A small amount of wetting agent is normally added to water to reduce surface tension. With the addition of the proper wetting agent, water can be even more sensitive than oils. Water-base surfactant solutions can be used successfully to detect leaks as small as 10^{-6} atm cm^3/s.

Immersion testing can be used on any internally pressurized item that would not be damaged by the test liquid. Although this test can be relatively sensitive, as previously stated, it is more commonly used as a preliminary test to detect gross leaks. This method is inexpensive, requires little operator skill for low-sensitivity testing, and enables an operator to locate a leak accurately.

Bubble-forming solutions can be applied to the surface of a pressurized vessel if it is too large or unwieldy to submerge. However, care must be taken to ensure that no bubbles are formed by the process itself. Spraying the bubble solution is not recommended; it should be flowed onto the surface. A sensitivity of about 10^{-5} atm cm^3/s is possible with this method, if care is taken. Sensitivity may drop to about 10^{-3} atm cm^3/s with an untrained worker or to 10^{-2} atm cm^3/s if soap and water is used.

Like immersion testing, the use of bubble-forming solutions is inexpensive and does not require extensive training of the inspector. One disadvantage is that the test does not normally enable the operator to determine the size of a leak accurately.

Flow Detection

Three separate methods are frequently grouped under the heading of flow detection:

- Pressure increase
- Pressure decrease
- Flow

The sensitivity for each of these methods is not the same, nor can any one method always be substituted for the other. If an internally pressurized vessel or system is enclosed within a larger vessel that is then sealed except for a small duct, leakage from the vessel being tested will result in an increase in pressure within the larger enclosing vessel, and gas will flow through the attached duct. If a device sensitive to the movement or flow of gas is installed in the attached duct, it can serve as a leak-rate indicator. Extremely sensitive positive-displacement flowmeters have been developed for this purpose.

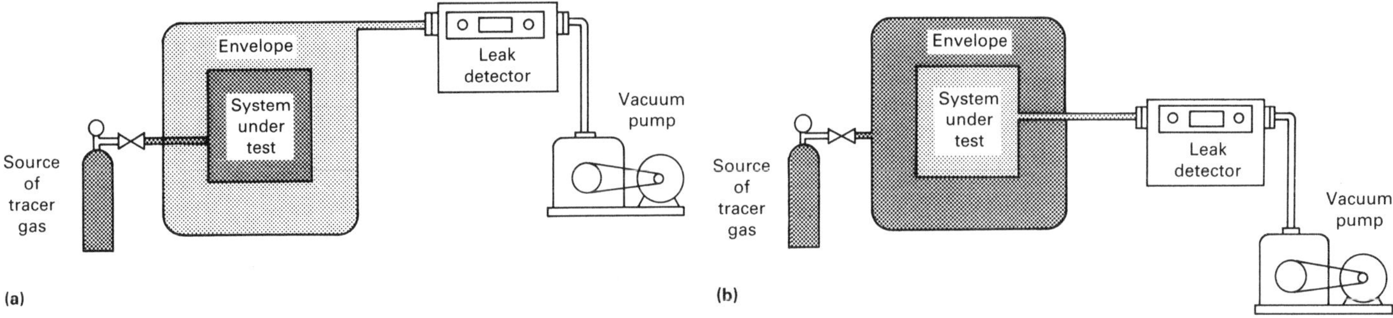

Fig. 3 Modes of leakage measurement used in dynamic leak testing techniques utilizing vacuum pumping. (a) Pressurized system mode for the leak testing of smaller components. (b) Pressurized envelope mode for the leak testing of larger-volume systems

Alternatively, a volumetric-displacement meter, which consists essentially of a cylinder with a movable piston, can be used. When attached to the duct from the enclosing vessel, the piston moves as the pressure of the enclosing vessel rises, effectively increasing the volume of the vessel and returning the internal pressure to the ambient atmospheric pressure. The piston must be very nearly free of frictional drag, and its motion must be accurate horizontally. Sensitive volumetric-displacement meters are equipped with micrometric cathetometers, which can accurately measure extremely small displacements of the piston. Some volumetric-displacement meters will detect leaks of about 10^{-4} atm cm^3/s. This is a very simple method to use when it is inconvenient to measure directly the pressure change of the container being tested.

The simplest method of leak detection and measurement is the bubble tube. If the end of the duct from the outer enclosing vessel empties into a liquid bath, appreciable leakage will generate bubbles. Even very small leak rates can be detected simply by the movement of the liquid meniscus in the tube.

Leak Testing of Pressure Systems Using Specific-Gas Detectors

Many available types of leak detectors will react to either a specific gas or a group of gases that have some specific physical or chemical property in common. Leak-rate measurement techniques involving the use of tracer gases fall into two classifications:

- Static leak testing
- Dynamic leak testing

In static leak testing, the chamber into which tracer gas leaks and accumulates is sealed and is not subjected to pumping to remove the accumulated gases. In dynamic leak testing, the chamber into which tracer gas leaks is pumped continuously or intermittently to draw the leaking tracer gas through the leak detector. The leak-rate measurement procedure consists of first placing tracer gas within (Fig. 3a) or around the whole system being tested (Fig. 3b). A pressure differential across the system boundary is established either by pressurizing one side of the pressure boundary with tracer gas or by evacuating the other side. The concentration of tracer gas on the lower-pressure side of the pressure boundary is measured to determine the leak rate.

Specific-Gas Detection Devices

Some of the more commonly used gas detectors are described below. These range from the simple utilization of the senses of sight and smell, when possible, to complex instrumentation such as mass spectrometers and gas chromatographs.

Odor Detection Via Olfaction. The human sense of smell can and should be used to detect odorous gross leaks. The olfactory nerves are quite sensitive to certain substances, and although not especially useful for leak locations, they can determine the presence of strong odors. However, the olfactory nerves fatigue quite rapidly, and if the leakage is not noted immediately, it will probably not be detected.

Color Change. Chemical reaction testing is based on the detection of gas seepage from inside a vessel by means of sensitive solutions or gas.

The ammonia color change method is probably the best known. In this method, the vessel surface is cleaned and then a calorimetric developer is applied to the surface of the vessel, where it forms an elastic film that is easily removed after the test. The developer is fairly fluid, and when applied with a spray gun it sets rapidly and adheres well to metal surfaces, forming a continuous coating. An air-ammonia mixture (usually varying from 1 to 5% NH_3) is then introduced into the dry vessel. Leakage of gas through the discontinuity causes the indicator to change color. The sensitivity of this method can be controlled by varying the concentration of ammonia, the pressure applied to the air-ammonia mixture, and the time allowed for development.

Indicator tapes are useful for the leak testing of welded joints; the method used for this application is as follows. After the surface of the joint to be tested has been cleaned with a solvent, the indicator tape is fixed on the weld either by a rubber solution applied on the edge of the tape or by a plastic film. Then the inspection gas, which consists of an ammonia-air mixture with 1 to 10% NH_3, is fed into the vessel at excess pressure. If microdiscontinuities exist, the gas leaks out and reacts chemically with the indicator, forming colored spots that are clearly visible on the background of the tape.

The indicator tape method has the following advantages:

- Remote control is possible, ensuring safety for the operators
- Leaks of approximately 10^{-7} atm cm^3/s can be detected
- The color of the tape is not affected by contact with the hands, high humidity, or the passage of time
- Tapes can sometimes be used more than once; if there are spots caused by the action of ammonia, they can be removed by blowing the tape with dry compressed air

Ammonia and Hydrochloric Acid Reaction. Another method involves pressurizing the vessel with ammonia gas, then searching for the leak with an open bottle of hydrochloric acid. A leak will produce a white mist of ammonium chloride precipitate when the ammonia comes into contact with the hydrogen chloride vapor. Good ventilation is necessary because of the noxious characteristics of both hydrogen chloride vapor and ammonia gas.

Ammonia and Sulfur Dioxide Reaction. Another modification of the ammonia chemical indicator test involves the use of ammonia and sulfur dioxide gas to produce a white mist of ammonium sulfide. Sulfur dioxide is not as irritating or corrosive as hydrogen chloride, but it is still a noxious gas and should be used only in well-ventilated areas.

Halide Torch. Still another type of chemical reaction test uses the commercial halide torch, which consists of a gas tank and a brass plate (Fig. 4). Burning gas heats the brass plate; in the presence of halogen gas, the color of the flame changes because of

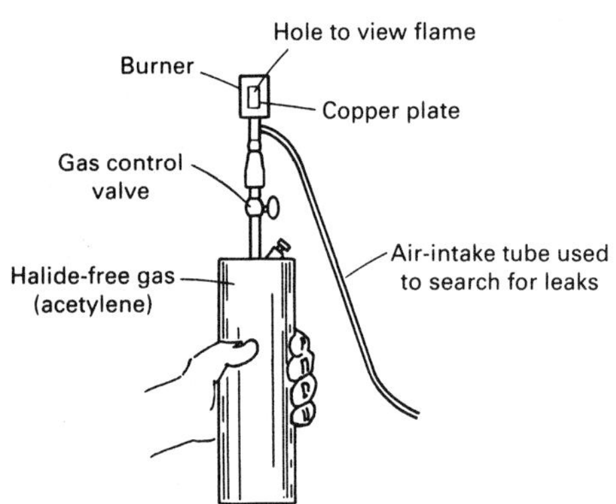

Fig. 4 Halide torch used for leak location

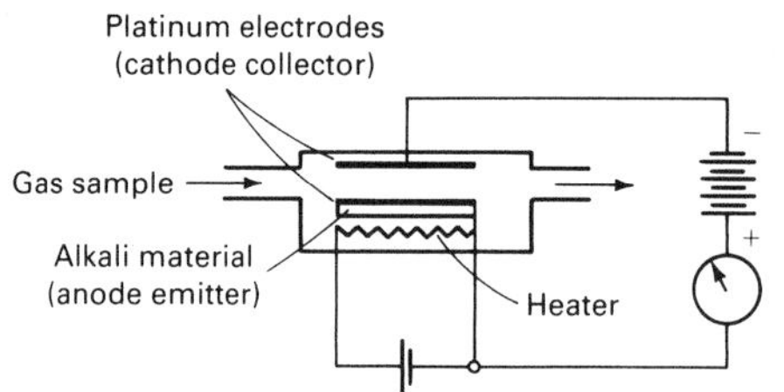

Fig. 5 Schematic of sensing element of a heated anode halogen leak detector used at atmospheric pressure for leak location with detector probe system. Positive-ion current from heated alkali electrode responds to refrigerant gases and other halogenated hydrocarbon tracer gases.

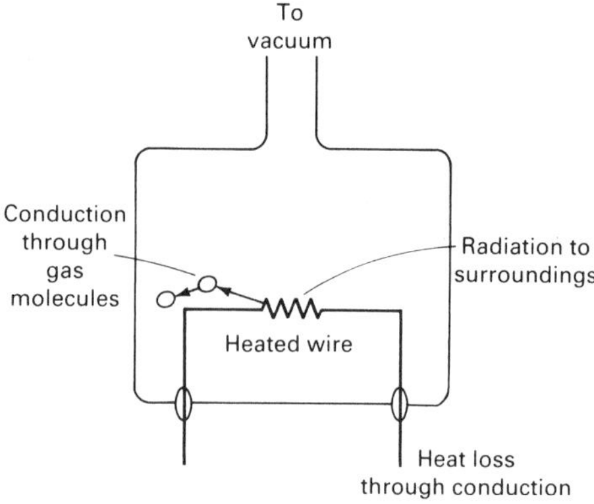

Fig. 6 Schematic of a thermal-conductivity gage using a Pirani-type detector. Thermal losses from the electrically heated resistance wire vary with heat conduction by gas molecules. Heat losses are reduced as gas pressure is lowered.

the formation of copper halide. (The flame is also used to inspirate gas through the probe, which is a length of laboratory tubing.) The halide torch locates leaks as small as 1×10^{-3} atm cm^3/s.

Halogen-Diode Testing. In halogen-diode testing, a leak detector is used that responds to most gases containing chlorine, fluorine, bromine, or iodine. Therefore, one of these halogen-compound gases is used as a tracer gas. When a vessel is pressurized with such a tracer gas, or a mixture of a halogen compound and air or nitrogen, the sniffing probe of the leak detector is used to locate the leak.

The leak detector sensing element operates on the principle of ion emission from a hot plate to a collector (Fig. 5). Positive-ion emission increases with an increase in the amount of halogen-compound gases present. This ion current is amplified to give an electrical leak signal. The sensitivity of halogen detectors operating at atmospheric pressure is about 10^{-6} atm cm^3/s, but this will vary depending on the specific gas that is being used.

Several different types of halogen leak detectors are available. Each includes a control unit and a probe through which air is drawn at about 4.9×10^5 mm^3/min (30 in.^{3}/min). When searching for leaks from an enclosure pressurized with a tracer gas, the probe tip is moved over joints and seams suspected of leaking. The probe tip should lightly touch the surface of the metal as it is moved. Where forced ventilation is required to keep the air free of halogen vapors, it must be stopped during actual testing, or care must be taken to ensure that drafts do not blow the leaking gas away from the test probe. When the probe passes over or close to a leak, the tracer gas is drawn into the probe with the air and through a sensitive element, where it is detected. The leak signal is either audible or visual.

Certain precautions are necessary in this probe exploration. Probing too rapidly may result in missing small leaks. To avoid this risk, the speed at which the probe is moved must be in proportion to the minimum leak tolerance. When testing a vessel for allowable leaks of the order of 0.001 kg (0.04 oz) per year, the probe travel can be 25 to 50 mm/s (1 to 2 in./s), but for smaller leaks the probe speed should be reduced to 13 mm/s (½ in./s).

Sulfur hexafluoride detectors operate on the principle of electron-capture detectors, which are used widely in the field of gas chromatography. The sensing chamber of a sulfur hexafluoride detector consists of a cylindrical cell that has a centrally mounted insulated probe. The inner wall of the cell is coated with a radioactive element (10 gigabecquerel (GBq), or 300 millicuries (mCi), of tritium). Low-energy electrons emitted by the tritium are collected on the central probe by means of a polarizing voltage maintained between the probe and the cell wall. The resulting electric current is amplified and displayed on a conventional meter. Leak-rate sensitivity is 10^{-8} mL/s (3×10^{-10} oz/s), and concentration sensitivity is 1 part sulfur hexafluoride in 10^{10} parts air. The equipment is fully portable.

Pure nitrogen (oxygen free) is passed through the detector and the standing current is set by a zero control to a given position on the meter scale. When an electron-capturing compound such as sulfur hexafluoride enters the cell, the electrons forming the current are captured by the sulfur hexafluoride molecules, resulting in a reduction of the standing current, which is indicated by a meter deflection. This meter reading is proportional to the amount of sulfur hexafluoride entering the cell and is therefore an approximate indication of the size of the leak. The oxygen molecule also possesses the electron-capture characteristic, although to a lesser extent than sulfur hexafluoride; hence the requirement for purging the instrument with oxygen-free nitrogen during testing.

Combustible-gas detectors are often used as monitors or as leak locators where combustible fumes are likely to accumulate, as in basements. These instruments warn of potentially hazardous conditions by their ability to measure combustible-gas mixtures well below the dangerous concentration level. Catalytic combustible gas instruments measure the gas concentration as a percentage of its lower explosive limit. The temperature of a heated catalytic element will rise in the presence of a combustible gas. The minimum sensitivity of a catalytic bead is about 500 ppm, which is a leak rate of approximately 10^{-3} atm cm^3/s. As leak locators, therefore, catalytic elements are not sufficiently sensitive. To locate a combustible gas, a solid-state sensor or flame ionization detector should be used. These sensors can detect a leak of approximately 10^{-5} atm cm^3/s.

Thermal-Conductivity Gages. The thermal conductivity of a gas can be measured by a hot-wire bridge method (Fig. 6). A resistance element, usually a thin wire or filament, is heated electrically and exposed to the gas. The temperature, and therefore the resistance of the wire, depends on the thermal conductivity of the surrounding gas, provided the power input is held constant. Alternatively, the temperature (and resistance) of the wire can be maintained at a constant value, and the required power input measured either directly or indirectly in terms of the applied voltage or current. Many types of thermal-conductivity detectors are commercially available, including differential detectors, simple Pirani gages, hydrogen Pirani gages, differential or trapped Pirani gages, charcoal Pirani gages, and thermocouple gages.

Pirani-Type Detectors. When Pirani-type detectors are used, the search gas should have low density, low viscosity, and low molecular weight. Most important, the gas should have a thermal conductivity markedly different from that of air.

The component or structure to be tested is filled with the tracer gas under positive pressure. This gas will consequently escape through even the most minute leak; hence the need for the desirable physical properties mentioned above. The escaping gas is drawn into the narrow-bore probe of the leak detector by a small suction pump. The sample is then allowed to expand into the sensing head, which contains an electrically heated filament. Simultaneously, a sample of pure air is drawn by the same pump into a second, identical chamber that also contains a heated filament. The two filaments form two arms of a conventional Wheatstone bridge circuit, which is initially balanced by an external variable resistance while both arms are simultaneously exposed to air. As soon as one arm receives a sample containing a trace of the tracer gas, such as helium, heat is extracted at a greater rate because of the substantially greater thermal conductivity of helium than air. This will cause a corresponding change in the resistance of the filament, thus unbalancing the Wheatstone bridge network. This is shown by an appropriate deflection on a center-zero milliammeter; simultaneously, an audio alarm circuit is triggered.

Infrared gas analyzers can detect a gas mixture that has a clear absorption band in the infrared spectrum by comparing it to the absorption characteristics of a pure standard sample of the same gas. The tracer gas used must possess a strong absorption in the infrared region. Nitrous oxide possesses this property markedly. The known characteristic is converted into a measurable response by allowing a heat source to radiate through two absorption tubes that contain the gases under comparison. These tubes are separated by a thin metal diaphragm that, in combination with an adjacent insulated metal plate, forms an electrical condenser. If the system is in balance (that is, if the same gas is in each tube), the heating effect will be equal and no pressure differential on the diaphragm will occur. However, if one tube contains nitrous oxide admixed with air, absorption of heat will occur to a greater extent than in the tube containing pure air. This will cause a higher pressure on one side of the diaphragm, as a result of the increase in temperature, and will cause it to move slightly in relation to the insulated plate. The resulting change in capacity of the condenser is amplified electronically and rendered visible on an output meter.

Infrared absorption is also a very sensitive way to measure small concentrations of hydrocarbons, such as methane. Infrared lasers are becoming more common as monitors for a variety of toxic and combustible gases. For some compounds, infrared laser spectroscopy can detect gas at parts per trillion levels.

Mass Spectrometer Testing. A mass spectrometer is basically a device for sorting charged particles. The sample gas enters the analyzer, where its molecules are bombarded by a stream of electrons emitted by a filament. The bombarded molecules lose an electron and become positively charged ions, which are electrostatically accelerated to a high velocity. Because the analyzer lies in a magnetic field perpendicular to the ion path, the ions travel in distinct, curved paths according to their mass. The radii of these paths are determined by ion mass, the magnitude of initial acceleration, and the strength of the magnetic field. With a constant magnetic field, any group of ions having the same mass can be made to travel the specific radius necessary to strike the ion collector. The positive charge of the ions is imparted to the target, or collector, and the resulting current flow is proportional to the quantity of the ions of that particular mass.

Specialized mass spectrometers are available, such as residual-gas analyzers, partial pressure analyzers, and helium mass spectrometers, which have been tuned to respond only to certain ranges of atomic mass units. In particular, the helium mass spectrometer is constructed so that it does not scan but is tuned to the helium peak. It will detect only helium; all other molecules passing through the detector tube will miss the target or collector because of their differences in mass or momentum from helium.

The theoretical sensitivity of the helium mass spectrometer is about 10^{-12} atm cm^3/s; the sensitivity of the residual-gas analyzer is about one order of magnitude less. General-purpose mass spectrometers have a sensitivity even less than this, depending on the range of atomic mass units that the instrument is designed to measure. Helium mass spectrometers, however, may not detect leaks smaller than approximately 10^{-8} atm cm^3/s in large systems. Because of background, outgassing of sorbed gases, noise, permeation, and other such factors, 10^{-8} to 10^{-9} atm cm^3/s is often the minimum detectable vacuum leak rate for helium mass spectrometers.

Because mass spectrometers must operate in a vacuum, they are ideally suited to the leak testing of vacuum systems. They can also be adapted to testing systems pressurized with a tracer gas by using either a detector probe or enclosure to collect the leakage while monitoring with the mass spectrometer. When used to test a pressure system, the mass spectrometer is much less sensitive and the minimum detectable leak rate is approximately 10^{-4} atm cm^3/s. Special assemblies containing a pump and permeation membrane, which are placed between the detector probe and the instrument manifold inlet, can decrease response time and increase the sensitivity for detector probe tests of pressure systems to 10^{-6} atm cm^3/s. Small units, such as sealed electronic components or nuclear-fuel elements, are often internally pressurized with tracer gas; leakage is detected by placing these units in an evacuated bell jar and monitoring the outgassing with a mass spectrometer (Fig. 7). In general, the basic methods used in mass spectrometer leak testing include detector probe (sniffer) testing, encapsulator testing, tracer probe testing, hood testing, bell jar testing, accumulation testing, pressure vacuum testing, and differential sorption testing.

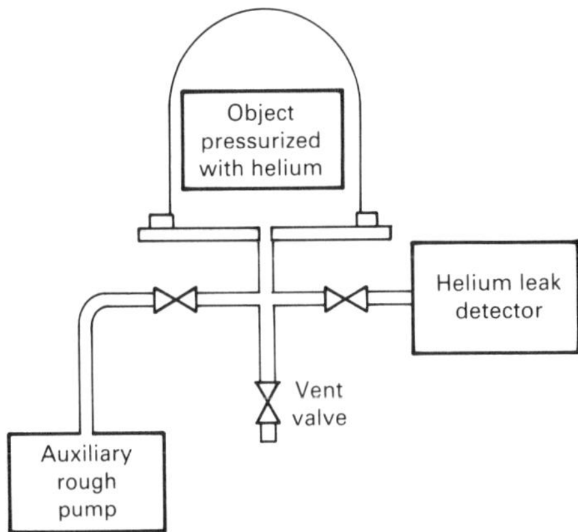

Fig. 7 Schematic of the leak testing of sealed components internally pressurized with helium tracer gas and enclosed in a bell jar

Radioisotope Testing. A number of leak testing techniques have been developed that involve the use of solutions containing radioactive tracers. In these tests, water is the liquid most commonly used, although hydrocarbons are extensively used in the petroleum industry. Soluble salts and compounds containing radioisotopes are added to the liquid. One of the most satisfactory isotopes that can be added to water is the sodium isotope ^{24}Na. The use of this radioisotope is safe because the concentration required for the detection of leaks is less than the drinking tolerance. Furthermore, this radioisotope has a short half-life (15 h) and can be retained in a vessel until its activity has been greatly reduced as a result of radioactive disintegration (decay).

Using this method, the vessel is filled with radioactive sodium bicarbonate solution and pressurized to the level required for a code proof test. After the vessel is held at pressure for a short period of time, it is drained and flushed with fresh water. A gamma-radiation detector is used to determine the location of leaks from the presence of gamma radiation emitted from the radioactive sodium bicarbonate solution that has accumulated around leaks in the vessel.

In another method, small components are backfilled with ^{85}Kr by placing them in a chamber and then pressurizing the chamber with ^{85}Kr (Fig. 8). If a component has a leak, gas will enter it. After the backfilling is complete, the components are air washed, and the leakage is measured with a radiation counter. Under carefully controlled conditions, the radiation counter has a sensitivity of about 10^{-12} atm cm^3/s. The radiation counter is automated and enables rapid, sensitive testing of large quantities of small parts such as microcircuits.

One disadvantage of this method is that organic coatings on the parts may absorb

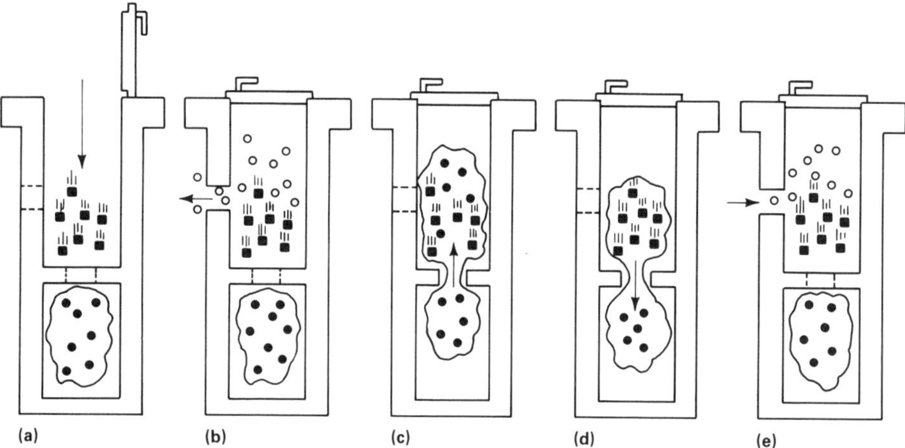

Fig. 8 Schematic of basic steps in the ^{85}Kr tracer-gas bombing of hermetically sealed electronic components. (a) Load parts. (b) Evacuate and remove air. (c) Soak parts in radioactive gas. (d) Recover radioactive gas. (e) Return air

the ^{85}Kr, thus giving a false leak indication. The primary disadvantage of this method is the attendant health hazard. However, some pressurized systems such as pipelines, lend themselves to leak testing by this method.

Ionization Gage Testing. The Phillips vacuum gage is used to measure the pressure of a gas by the intensity of a gas discharge in a magnetic field. However, this form of gas discharge ceases when the gas pressure is reduced to a pressure of approximately 0.13 Pa (10^{-3} torr); therefore, the Phillips gage cannot be used for making measurements at low pressures.

Penning Gage. A vacuum gage based on the same principle as the Phillips gage, but useful at lower pressures, was constructed by replacing the electrode system with a system consisting of two parallel round metal disks (acting as cathode and anode) enclosed in a cylindrical glass jacket. To keep the distance between the pole shoes of the permanent magnet enveloping the glass tube as short as possible, the tube was pinched at that point. This improved instrument is called the Penning gage, and it has a sensitivity about ten times greater than the original Phillips gage. The Penning gage covers a pressure range from 13 to 0.13 mPa (10^{-4} to 10^{-6} torr) and can probably be used even for pressures below 13 μPa (10^{-7} torr).

In any pumping installation, the vacuum gage can also be used to detect and locate leaks. This is done by replacing atmospheric air with tracer gas in each part of the apparatus. When the leak is reached, the tracer gas flows in, and the gage usually registers any change in pressure.

Additional Ionization Detectors. In addition to the Phillips and Penning gages, other ionization detectors include the oxygen-leak detector, the hot-filament ionization gage, the helium-ionization leak detector, the photometric leak detector, and the palladium-barrier detector.

Gas chromatographs are sometimes used for leak detection. In a gas chromatograph, a carrier gas is made to flow through an adsorbent column; then a tracer gas is injected into the carrier gas ahead of the adsorbent column. Next, a gas sample is fractionated into its constituents as a function of the residence time of each constituent in the adsorbent column. The constituents are detected as they leave the adsorbent column, usually by thermal-conductivity instruments, but they may also be detected by infrared analyzers or vapor-density balances. The advantage of the gas chromatograph is that, unlike the mass spectrometer, with proper calibration it will identify the tracer gas as a compound rather than as individual molecules. This can be beneficial if there are several gases in the system; their identification will often aid in locating the source of the leakage. Residual gas analyzers and partial pressure analyzers are used in the same manner.

Specific-Gas Detector Applications

The proper method for using a specific-gas detector is based on the function of the leak detector, the fluid that is leaking, and the type of vessel being tested. Some of the more important considerations in determining the best method of using leak detectors are discussed below.

Probing. Detector probes, which will react to a number of gases, can be used in either of two ways. The first is the scanning mode, and the second is the monitoring of an enclosure placed around the pressurized item (Fig. 9).

Detector Probe Technique. In the detector probe mode, the external surface of the pressurized vessel is scanned either with a portable detector having a short probe attached or with a long probe connected to a stationary leak detector by flexible tubing (Fig. 10). In general, the connection of a long probe to a stationary detector reduces sensitivity because of the slow release of sorbed gases in the probe tubing, which results in a high background reading. Because a correspondingly longer time is required for the gas to flow up the tube to the sensing element, it is difficult to pinpoint the location of the leak. Also undesirable is the long time needed for purging a long-probe system of tracer gas. The purging occurs when a rather large leak is encountered and the probe tubing becomes filled with a high concentration of tracer gas. One minute or more may be required for the leak detector pump to clear the tracer gas from the sensing element. This in turn reduces scanning speed.

There are means for offsetting some of the disadvantages of the long-probe meth-

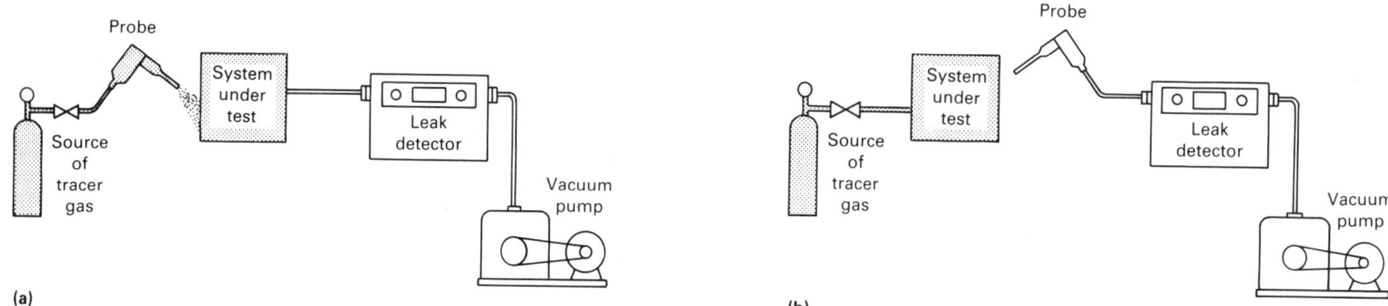

Fig. 9 Tracer-gas detection techniques used to locate leaks with sensitive electronic instruments. (a) Tracer probe. (b) Detector probe

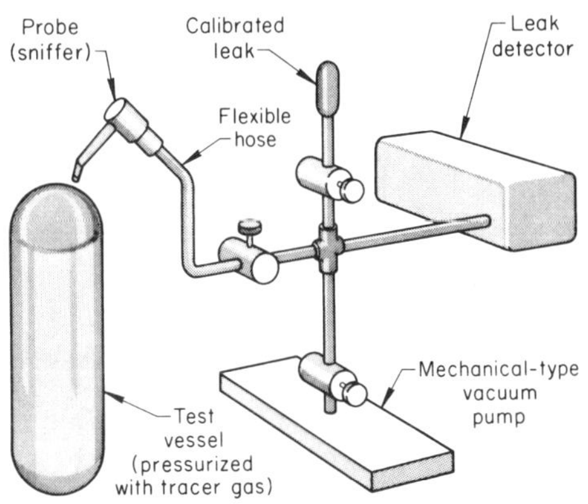

Fig. 10 Schematic showing a long detector probe connected by a flexible hose to a stationary leak detector for scanning the external surface of a pressurized vessel

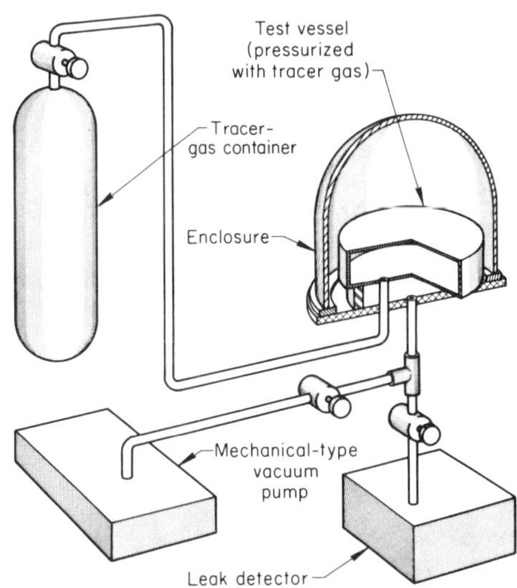

Fig. 11 Schematic of setup for detecting leaks by monitoring an enclosure placed around a vessel pressurized with tracer gas

od. When a helium mass spectrometer is used, carbon dioxide can be injected into the probe line near the tip. The carbon dioxide will act as a carrier fluid for the helium, thus purging the tube walls, and, in conjunction with a high-capacity pump, will carry the helium tracer gas to the leak detector very rapidly. The carbon dioxide is then selectively trapped with liquid nitrogen, leaving the helium. A probe of this type is several times faster than conventional probes and reduces background buildup.

When background buildup and sensitivity losses do occur, accurate measurements of leak rate are difficult to obtain. However, leak-rate measurement can be accomplished by placing a calibrated leak over the probe tip to calibrate the leak-detector-tube-probe system. The leakage can then be measured by placing a suction cup on the probe and applying it to the leak area.

Another, more recent, means of decreasing the disadvantages of the long-probe method is the use of an assembly between the probe and the mass spectrometer manifold port, which contains a permeation membrane and pump. The pump reduces the response time, and the permeation membrane reduces the pressure drop into the mass spectrometer and also causes accumulation of the entering tracer, which increases sensitivity.

Another disadvantage of the long-probe method is that the operator generally cannot observe the reaction of the leak detector. This can be circumvented with earphones, audible signals, and lights, but such devices must be triggered at some threshold level of detector response that may not correspond to the desired leak size. One person can operate the probe while another monitors the leak detector readout, but ideally one person should perform both tasks simultaneously.

Detector Probe Accumulation Technique. The second method of detector probing involves the use of a tracer-gas leak detector to monitor an enclosure that accumulates the leakage (Fig. 11). This method does not provide a means of locating the leak, but it is preferred for leakage measurement and, in many applications, for detecting the existence of a leak.

A large unit can be tested by enclosing it in a plastic bag. The accumulation of tracer gas in the bag is monitored by a leak detector connected to the bag by a short probe or by piping. Leakage measurement consists of inserting a calibrated leak into the enclosure and comparing the test-leak rate with the calibrated-leak rate. One of the problems associated with this method is the construction of a bag that does not leak and has low permeability with respect to tracer gas.

A small unit can be tested by placing it in an enclosure such as a sealed bell jar, which is monitored for the accumulation of tracer gas. Depending on the type of leak testing instrument used, the enclosure may be placed under a vacuum. If relatively large leaks are expected, a supply of tracer gas must be connected to the test unit to ensure that the concentration of its internal tracer gas and the pressure drop across the leak are constant.

Back pressuring is a method of pressure testing that is normally used with small, hermetically sealed electronic components, such as integrated circuits, relays, and transistors. In this method, the test unit is placed in a pressurized container filled with a tracer gas and is kept there for a time to allow tracer gas to flow into the unit through any leaks that may be present.

In using back pressuring, care must be taken to ensure that the leaking components contain tracer gas. Therefore, it is important that the time between back pressuring and leak testing be suitably controlled. For example, in a test specification for transistors, the transistors were subjected to a helium pressure of 690 kPa (100 psi) for 16 h, air washed for 4 min, and then leak tested within 3 h. If more than 3 h had elapsed before leak testing, the transistors were pressurized again. If there is doubt as to how much tracer gas will flow into the components under back pressurization, the following expression can be used for estimation purposes:

$$P_i = P_o\left(\frac{1 - e^{-AT}}{V}\right) \quad \text{(Eq 7)}$$

where P_i is the partial pressure of tracer gas inside the component (in atmospheres), P_o is the pressure of the tracer gas surrounding the component (in atmospheres), A is the conductance of the leak (in cm^3/s) with a pressure differential of 1 atm, V is the internal free volume of the component (in cubic centimeters), and T is the time that pressure P_o is applied (in seconds).

Because leaks are generally expected to be quite small, very sensitive tracer-detector combinations must be used. The tracer gas normally used is helium or ^{85}Kr, and detection is accomplished with a helium mass spectrometer or a nuclear-radiation detector, respectively. Absolute values of the size of the leak are sometimes difficult to determine with back pressuring because of:

- Adsorption and absorption of the tracer gases
- Different detector response times for different leak directions

When testing complex systems, each sealed component can be pressurized with a different tracer gas. Leak detection with a residual-gas analyzer will both detect leakage and provide a determination of which components are leaking.

Leak Testing of Pressure Systems by Quantity Loss

Determining the amount of fluid lost from a vessel is one means of establishing that leakage has occurred. Weighing a vessel to determine fluid loss is practical only if the vessel is of reasonable size and can be weighed on a suitable scale. The sensitivity and accuracy of this method are limited by those of the scale; in general, this method is unsatisfactory unless the weight of the vessel is comparable to or less than the weight of its contents.

Differential-Pressure Method. Gas-quantity loss can be determined by the differential-pressure method. This is done by connecting two identical containers together with a differential-pressure gage between them, as shown in Fig. 12. Differential-pressure gages are sensitive to small pressure differences; therefore, if the two containers are maintained at equal ambient conditions, leaks as small as 10^{-4} atm cm^3/s can be detected in systems operating at pressures

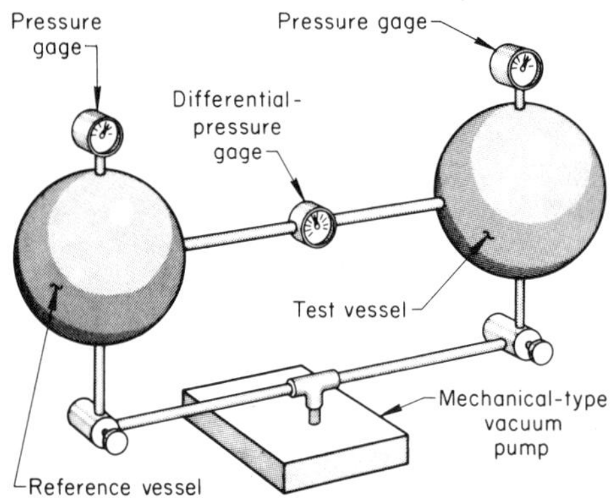

Fig. 12 Schematic showing the use of a differential-pressure gage between a reference vessel and the test vessel to detect leaks in the test vessel

up to 40 MPa (6 ksi). The primary advantages of this method are that temperature effects are largely cancelled and the actual working fluid can be used.

Visual Leak Testing of Pressure Systems

The simplest method of checking a liquid-filled vessel for leaks is visual observation. Like listening and smelling, this is a quick and simple method of testing for gross leaks in liquid-filled containers.

Vision is a variable commodity when considered from the viewpoint of a single individual and is even more variable when considered from the viewpoints of several individuals. This is because variations in the eye, as well as variations in the brain and nervous system, affect both optical resolution and color perception.

Optical aids to vision, such as mirrors, lenses, microscopes, borescopes, fiberoptics, and magnifiers, compensate for some of the limits of the human eye by enlarging small discontinuities. For example, borescopes permit direct visual inspection of the interior of hollow tubes, chambers, and other internal surfaces.

Hydrostatic leak testing requires that a component be completely filled with a liquid, such as water. The normal sensitivity for visual inspection using deionized water as the test fluid is 10^{-2} atm cm^3/s. Therefore, only large discontinuities can be revealed with this method.

The sensitivity of the test can be improved by the addition of:

- A water developer applied to the outside that changes color when contacted by a small leak. This method has replaced the older lime wash method, which required a larger amount of water seepage to be visible
- A concentrate that lowers the surface tension of the water and provides a visible or fluorescent tracer

When conducted properly, the hydrostatic test can achieve the same sensitivity as a

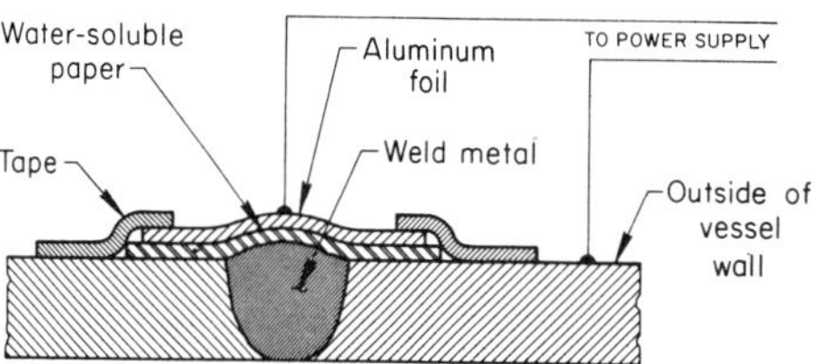

Fig. 13 Cross section of a welded seam in a water-filled vessel showing strips of water-soluble paper and aluminum foil used to detect leaks in the seam

liquid penetrant leak test on large-volume specimens. The penetrant test is two orders of magnitude more sensitive on volumes of typical electronic components.

A recent variation of the hydrostatic test is the stand-pipe test, which is used to detect leaks in underground storage tanks. In this test, the storage tank is overfilled, and the liquid level is monitored by reference to an aboveground liquid column or stand pipe. This method is not very effective and has a sensitivity of between 10^{-1} and 10^{-2} atm cm^3/s.

Detection by Water-Soluble Paper With Aluminum Foil. One method of detecting water leakage employs a strip of aluminum foil laid over a wider strip of water-soluble paper. The strips are then laid over the welded seams of a water-filled vessel, as shown in Fig. 13. If a leak exists, the water-soluble strip will dissolve, indicating the leak location, and the aluminum foil strip will be in electrical contact with the vessel. A corresponding change in resistance indicates that a leak is present.

Liquid Penetrants. The use of a visible liquid penetrant for leak testing is only slightly limited by the configuration of the item to be inspected. Another advantage is that this method can be used on subassemblies prior to completion of the finished vessel. Leak testing with penetrants is equally suited to ferrous, nonferrous, and nonmetallic materials, provided the materials are not adversely affected by the penetrant. Leak testing by this method cannot be substituted for a pressure test when the applied stress and associated proof-test factors are significant.

The liquid penetrant should be applied in a manner that ensures complete coverage of one surface without allowing the penetrant to reach the opposite surface by passing around edges or ends or through holes or passages that are normal features of the design. A sufficient amount of penetrant should be applied to the surface during the period of the test to prevent drying. Immediately after application of the penetrant, the developer should be applied to the opposite surface. The developer must be prevented from contacting the penetrant at edges or ends or through designed holes or passages so that the test results are valid.

If greater sensitivity is desired than is provided by a color-contrast (visible) liquid penetrant test, fluorescent liquid penetrant can be used to increase the sensitivity approximately one-hundredfold. Additional information on the use of liquid penetrant for leak testing is available in the article "Liquid Penetrant Inspection" in this Volume.

Smoke bombs and candles can also be used for detecting leaks. Generally speaking, a volume of smoke sufficient to fill a volume five or six times larger than the area to be tested is required for a smoke test. Medium-size volumes, such as boilers and pressure vessels, can be tested by closing all vents, igniting a smoke candle or smoke bomb, and placing it inside the vessel. All openings should then be closed; almost immediately, escaping smoke will pinpoint any leaks within the sensitivity of this method. When testing boilers and similar equipment with volumes of 2800 m^3 (100 000 ft^3) and more, it is often desirable to apply air pressure.

Smoke candles can provide from 110 m^3 (4000 ft^3) of smoke in 30 s to 3700 m^3 (130 000 ft^3) in 2 to 3 min. The smoke will vary in color from white to gray, depending on density and lighting. It is generated by a chemical reaction, contains no explosive materials, and is nontoxic.

Leak Testing of Vacuum Systems

Vacuum systems are normally tested in the evacuated condition, rather than being internally pressurized. Because some leaks, particularly small ones, are directional, it is desirable to test with the leakage flowing in the operational direction.

Introduction of Tracer Gas Into System

There are two ways in which a tracer gas can be used. One is to tracer probe to both detect and locate the leak by spraying tracer gas over the outer surface of the vessel. The other is to envelop the entire test vessel (or a particular portion of it) in a bath of tracer gas by the use of either rigid or flexible enclosures to both detect and measure leak rate. Calibration is accomplished by injecting a known quantity of tracer gas into the vacuum system or by attaching a calibrated leak to that system.

Helium is one of the most commonly used tracer gases for the following reasons:

- It is basically inert and safe
- It has a small molecular size and thus can flow through very small leaks
- Its diffusion rate is high
- It is easily detected by a mass spectrometer

In testing a vacuum system, even the smallest leaks are important because they have a great influence on the ultimate vacuum that the system can reach. Therefore, the most sensitive test method—helium tracer gas in combination with a helium mass spectrometer—is normally used.

Certain characteristics of high vacuums and high-vacuum systems must be recognized and understood prior to testing such a system for leaks. These characteristics are briefly discussed below.

Tracer-gas diffusion rates are highly variable and are dependent on the type of gas used and the configuration of the system. The diffusion rates of such gases as helium and hydrogen are generally eight or nine times higher than those of the Freons.

Attachment of Mass Spectrometer. Mass spectrometers incorporate pumping systems that can be used to maintain the vacuum if small systems are being tested and the leakage is small. With large systems, it may be necessary to operate system diffusion vacuum pumps backed with forepump mechanical vacuum pumps while the system is being tested. In general, the mass spectrometer should be attached to the foreline of the diffusion pumps, parallel with the forepumps, and not to the test vessel (Fig. 14a). In this manner, the mass spectrometer will receive an amount of tracer gas that is proportional to the pumping rate of the spectrometer and the forepump according to:

$$T_G = \frac{100\ S_1}{(S_2 + S_1)} \quad \text{(Eq 8)}$$

where T_G is the percentage of total tracer-gas leakage flowing to the mass spectrometer, S_1 is the spectrometer pumping speed, and S_2 is the forepump pumping speed. If the mass spectrometer is connected directly to the test vessel, it will receive only a small amount of tracer gas because the spectrometer pumping speed is generally far lower than that of the vacuum chamber diffusion pumps (Fig. 14b). Conversely, when the mass spectrometer is connected to the diffusion pump foreline, the compression of the evacuated gas increases the total gas pressure in the foreline and in turn increases the tracer gas partial pressure in that line, which makes it more detectable than if it were sensed directly from the vessel. If the system throughout is small, however, the system diffusion pumps and roughing pumps can be valved off or backed by the mass-spectrometer pumps, which in turn will maintain the desired vacuum.

System Materials and Cleanliness. Both the vacuum system and the leak testing system should be maintained in a very clean state; otherwise, contaminant outgassing will result, reducing the system vacuum. The proper materials must also be selected to reduce hang-up of the tracer gas; glass or stainless steel is recommended. Grease coatings on seals and the use of rubber should be held to a minimum be-

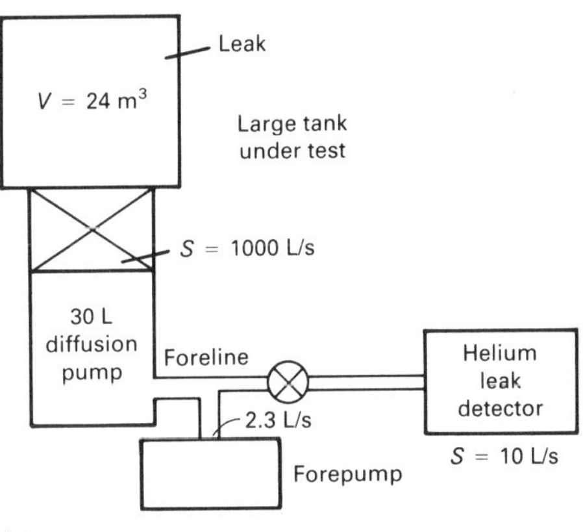

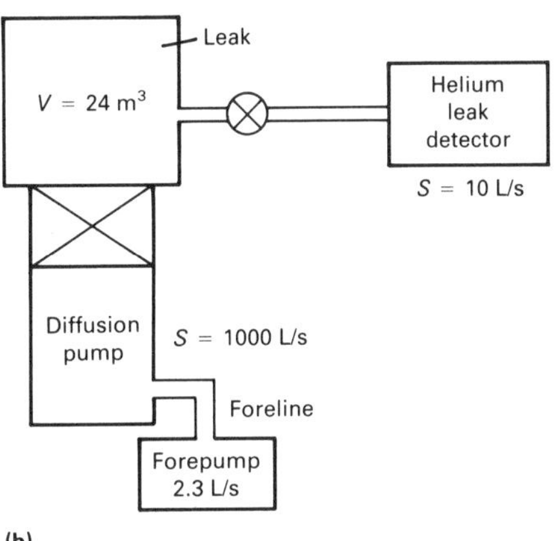

Fig. 14 Schematics of correct (a) and incorrect (b) connections in pumping setups for the vacuum leak testing of large volumes

cause these materials readily absorb helium and then slowly release it, giving a high background count.

Temperature and Tracer-Gas Concentration. For leak testing, the temperature should be held constant, when possible, preferably at the expected operating temperature of the item tested. Also, tracer-gas concentration should be as high as possible to ensure sufficient detector response.

Vacuum Leak Testing Methods

The equipment most commonly used for the leak testing of vacuum systems involves the use of manometers, halogen detectors, mass spectrometers, ionization gages, thermal-conductivity gages, residual gas analyzers, partial pressure analyzers, and gas chromatographs. The methods used with this equipment are listed in Table 3. Methods that can be used with both vacuum and pressure systems are discussed in detail in the section "Leak Testing of Pressure Systems Using Specific-Gas Detectors" in this article.

Manometers. Testing for leaks in systems under vacuum can be done by observing the manometers installed on the system. Although manometers are not as adaptable to locating a leak as regular leak detectors, some indication of leakage can be attained. Because most vacuum gages will react or change their reading in the presence of a tracer gas, spraying tracer gases over individual portions of the vacuum system will produce a gage response and will provide some indication of leak location. The location can also be roughly established by isolating various portions of the system and observing the gage response. The primary evidence of a leak in the system is failure of the system to achieve anticipated pressure levels in a certain pumping time as based on past or calculated performance. However, consideration must be given to other gas loads (virtual leaks) that may exist in the system. Sources of these gas loads are system contaminants, such as water, outgassing of chamber surfaces and test items, and outgassing of entrapped areas such as unvented O-ring grooves.

Pressure-change gages having a pressure-change sensitivity of 0.13 mPa (10^{-6} torr) are available for vacuum applications. Ionization gages will measure vacuums to 1.3 pPa (10^{-14} torr) and can be used to measure changes in pressure. The sensitivity of all pressure-change methods is time dependent (that is, the mass changes per unit of time); therefore, the longer the test duration, the more sensitive the method.

Halogen Detectors. Heated anode halogen detectors can also be used on vacuum systems. With special adapters, they can be used down to 0.13 Pa (10^{-3} torr). Leaks are detected by using an externally applied tracer gas.

Mass spectrometers are well suited for use as leak detectors in vacuum systems because the detector tube of a mass spectrometer must be maintained at a vacuum. Also, mass spectrometers usually have built-in vacuum-pumping systems that can be directly coupled to the system being tested. Helium mass spectrometers are most commonly used. Leaks are detected by spraying or enveloping the outer surface of the system with helium.

Ionization Gages. In addition to detecting leaks in vacuum systems by pressure-change instruments, ionization gages can be used to detect the presence of specific tracer gases. Because each gas traveling through the gage ionizes differently, equal flows of different gases will produce different readings. Leak rates for molecular or laminar flow can be related to ionization-gage readings in response to a tracer gas through:

$$Q_a = \frac{\Delta R S_a}{(\sigma - 1)} \text{ (Molecular)} \quad \text{(Eq 9)}$$

$$Q_a = \frac{\Delta R S_a}{[(\sigma\nu/\mu) - 1]} \text{ (Laminar)} \quad \text{(Eq 10)}$$

where Q_a is the flow of air (in torr liters per second), ΔR is the change in ionization-gage reading (in torr), S_a is the pumping speed (in liters per second), σ is the gage sensitivity factor, $\nu = \eta_a/\eta_t$ (where η_a is the viscosity of air and η_t is viscosity of tracer gas), and $\mu = (M_a/M_t)^{1/2}$ (where M_a is the molecular weight of air and M_t is the molecular weight of tracer gas). Values of σ for ionization gages, and of the gas factors ($\sigma - 1$) and $[(\sigma\nu/\mu) - 1]$, for several gases are as follows:

Gas	σ	Gas factor ($\sigma - 1$)	Gas factor $[(\sigma\nu/\mu) - 1]$
Helium	0.17	−0.83	−0.93
Argon	1.3	0.3	0.85
Carbon dioxide	1.4	0.4	0.4
Water vapor	0.9	−0.1	−0.72
Hydrocarbons	3–10	2–9	1–10
Hydrogen	0.5	−0.5	−0.94

By spraying tracer gas over the outside surface of a vacuum system, leaks can be located by observing the ionization-gage response. The gage reading should decrease in the presence of gages having a negative gas-factor value and should increase for gases having a positive value.

The palladium-barrier gage is a modified ionization gage that has a palladium barrier in front of the ionization chamber. Hydrogen is used as a tracer gas because it is the only gas that will permeate heated palladium into the ionization chamber.

Thermal-Conductivity Gages. The thermal conductivity of a gas is a function of its mean free path; therefore, gages measuring the thermal conductivity of the gas as it passes over a heated filament will indicate vacuum. Several types of these gages, including Pirani gages, are available (see the discussion on thermal-conductivity gages in the section "Leak Testing of Pressure Systems Without a Tracer Gas" in this article).

Gas chromatographs can be used as leak detectors in vacuum systems by condensing in a cold trap a sample of gas drawn from the vacuum system with a vacuum pump. The cold trap is then isolated from the vacuum system with valves, and the frozen gas is released to the chromatograph by warming the trap. The sensitivity of this instrument is limited essentially by the volume of sample drawn into the cold trap, the volume being proportional to the length of time the gas is drawn into the cold trap.

Residual gas analyzers and partial pressure analyzers are similar in operating principle to mass-spectrometer leak detectors. They have the added advantage of being able to scan the molecular peaks of all gases within the sample entering the instrument. This provides the leak-test operator with information as to whether the sample is predominantly leakage or contamination from system surfaces, O-rings, and cleaning agent residues. They have the disadvantage of having no integral pump system. They must be installed behind a pump system on the system being tested in order to obtain any kind of reasonable response time if they are to be used for leak detection and location.

Choosing the Optimum Leak Testing Method

The three major factors that determine the choice of leak testing method are:

- The physical characteristics of the system and the tracer fluid
- The size of the anticipated leak
- The reason for conducting the test (that is, locating or detecting a leak or measuring a leak rate)

System and Tracer-Fluid Characteristics. The physical characteristics of the system play a large role in the selection of leak detection methods. If a tracer fluid, either gas or liquid, is used, it must be nonreactive with the test item or system components. In general, the items or systems should be leak tested with the actual working fluid. This eliminates any errors that might be caused by converting from tracer-fluid leakage to working-fluid leakage. It is also possible that tracer-fluid leakage will occur while working-fluid leakage will not, and vice versa.

If tracer gas is to be used, consideration should be given to the characteristics of the gas. In most cases, gases with high diffusion rates and small molecular size, such as hydrogen and helium, are desirable. On the other hand, when probing the surface of a container filled with tracer gas, it may be desirable to use a persistent gas or a gas with a low diffusion rate. A persistent gas will remain in the leak area longer and may facilitate leak detection and location because of an increase in tracer-gas concentration.

Leak Size. The method and instrument must match or be responsive to the size of the leak. The approximate working ranges of several leak detectors or leak detection methods are given in Fig. 15. If leakage is too great, the leak detector will be swamped. This is normally self-evident because the detector will go to full scale and stay there until it is removed from the leakage source. Some combustible-gas detectors, however, may return to zero in the presence of a high concentration of leakage. For example, if the vapor-to-air ratio of gasoline vapor becomes richer than about one part vapor to nine parts air, the mixture is no longer combustible, and the catalytic detector may indicate zero or some nominal value, despite the fact that high concentrations of gasoline vapor exist.

Small leaks may also produce no response on a given detector, thus eliminating any possibility of finding the leak. It should not be assumed, however, that the leakage is zero. There is universal agreement that the terms zero leakage and no leakage are inaccurate. Specifications or test procedures should define leakage with such terms as minimum detectable leakage or a maximum leak rate.

Leak Location. If leak location is the purpose of the test, methods that include the use of probes or portable detectors are necessary so that the surface of the test vessel can be scanned. In vacuum systems, tracer gas can be sprayed over the vessel surface and its entry point detected by observing the leak detector reaction as the tracer-gas spray is moved along the surface. In pressure systems, bubbles, immersion, liquid penetrants, and chemical indicators provide means of locating leaks through visual observation. Figure 16 outlines various methods of leak detection and leakage rate measurement in terms of methods, sensitivity, and cost.

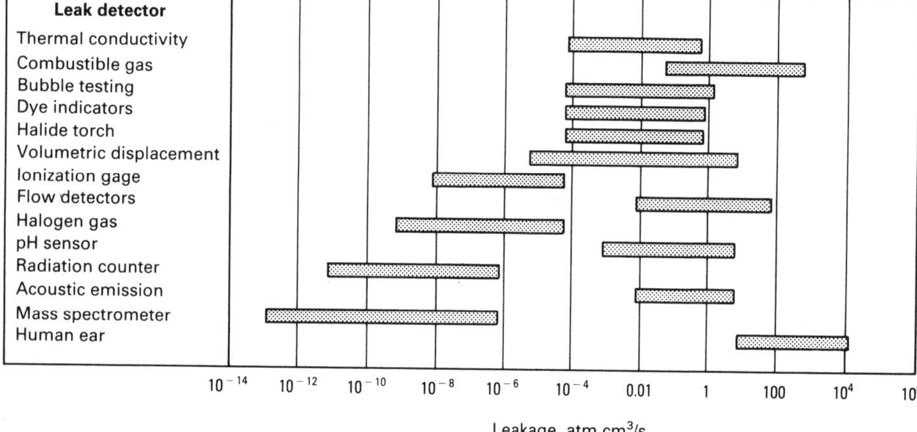

Fig. 15 Approximate working ranges of various leak detectors or leak detection methods

System Response

The response of a leaking system to a leak is important in leak testing; if the response is too slow, it can be very time consuming

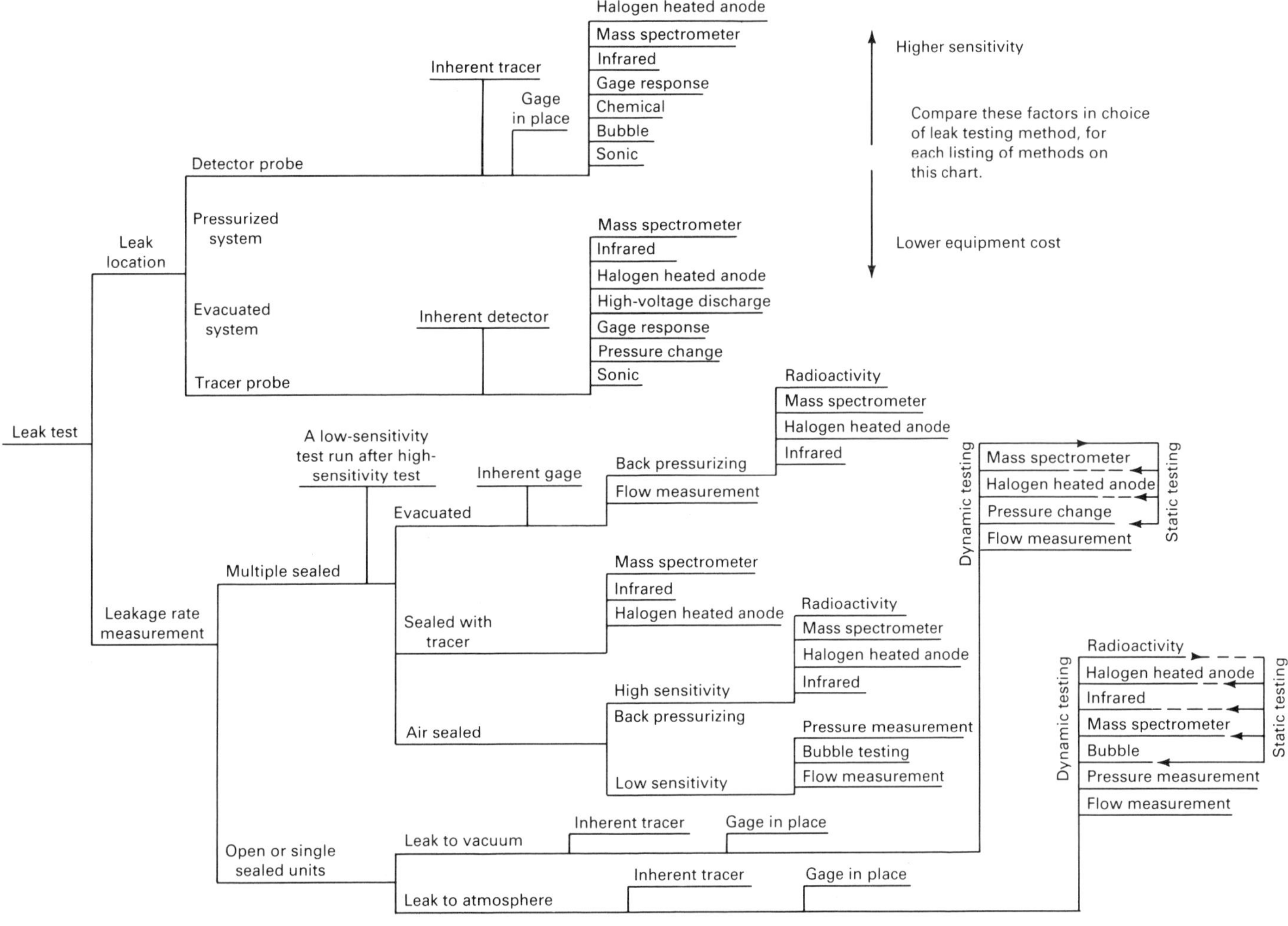

Fig. 16 Decision-tree guide for the selection of leak testing methods. Source: Ref 1

to locate the leak, or the leak can be missed. A number of factors affect system response, including size of the leak, diffusion rate of the tracer gas, absorption and adsorption of the tracer gas, cleanup time of the detector, pumping speed, and volume of the system. Attention must be given to these factors to minimize response time.

Leak Size. Theoretically, leak size does not affect system response time; it only affects the magnitude of the detector readout. However, if some type of leakage accumulation (or mass loss) method is being employed, leak size will affect response time because sufficient leakage must be accumulated (or lost) before the detector will respond.

The smallest leakage to which the detector will respond is termed the minimum detectable leakage and is measured by observing the leak detector reaction to a calibrated or known leakage. When calculating the minimum detectable leakage of a mass spectrometer, it is conventional practice that the noise and random-background count rate be doubled. The sensitivity of the mass-spectrometer detector is defined by:

$$S = \frac{Q}{(D - B_s)} \qquad \text{(Eq 11)}$$

where S is the sensitivity (in atmosphere cubic centimeters per second per scale division), D is the total deflection produced by the calibrated leak (in scale divisions), Q is the calibrated leakage flow (in atmosphere cubic centimeters per second), and B_s is the steady background reading (in scale divisions). In terms of S, the minimum detectable leakage (MDL) can be expressed as:

$$\text{MDL} = 2(N + B_R)(S) \qquad \text{(Eq 12)}$$

where N is the scale deflection produced by instability or noise in system (in scale divisions) and B_R is the random background deflection (in scale divisions).

Tracer-Gas Diffusion Rate. In large, open vessels, gas diffusion is not a major problem. However, in tortuous systems, one investigator found that even after 24 h neither helium nor Freon had evenly diffused throughout the system.

Absorption and adsorption of tracer gas are responsible for hang-up in the system; they can also materially reduce tracer-gas concentration. Permeation also causes difficulties; for example, helium permeates through a 25 mm (1 in.) diam O-ring in about 1 h, thus producing a detectable reading in the system.

Detector cleanup time (the time required for the leak detector to purge itself of tracer gas) can range from a few seconds to several minutes, thus affecting the time between searches for a leak. Cleanup time is usually defined as the time it takes for the meter reading to decay to 37% of its maximum value.

Pumping Speed. It can be shown that a leak detector will exhibit a readout deflection that is 63% of its maximum value at some characteristic time, t, known as the response time. Response time can be calculated as $t = V/S$, where V is the system volume in liters and S is the pumping speed in liters per second. Increasing the pumping speed will decrease the response time, but will introduce a corresponding decrease in the maximum signal. Increasing the pumping speed by a factor of ten will reduce the readout deflection of the detector by a factor of ten, which may cause problems. Calculation of the response time is useful for determining the length of time that a tracer gas must be sprayed over the external surface of a vessel under vacuum in order to detect a leak. For example, if the vessel has a volume of 1000 L and the leak detector has a pumping speed of 5 L/s, then the suspected leak area should be sprayed with search gas for about 200 s (1000/5) for the detector to reach 63% of its maximum value.

System Volume. The larger the volume of the system, the smaller the concentration of leakage. This can be alleviated by closing off part of the system when possible. The enclosures used to trap the leakage should be as small as possible. For example, if the leakage of tracer gas is 10^{-5} atm cm^3/s, if the enclosure volume minus the test unit volume is 10^6 cm^3, and if the detector can detect 1 ppm of search gas, then the tracer gas will take about 28 h to reach detectable concentration.

Measurement of Leak Rate Using Calibrated Leaks

Leak-rate measurement can be accomplished with almost any of the standard leak detecting methods in conjunction with a calibrated or known leak. If the leakage is inward, it can be measured by tracer-gas concentration inside the vessel or system vacuum. If the leakage is outward, it can be collected in an enclosure and the concentration rise within the enclosure monitored; it can also be measured by placing the leak detector probe directly over the leak and isolating the surrounding area with a suction cup. In addition, leaks can be measured by mass change methods.

To calibrate a leak testing instrument to measure leakage rates, its response to a known leakage must be found. Furthermore, the instrument should be calibrated with the smallest leakage to which it will respond, preferably using the fluid that will be utilized during testing, because most leak detectors respond differently to different fluids. Leak detectors for measuring gross changes in mass or flow can be calibrated in a conventional manner by using deadweight testers, nozzles, orifices, and manometers. The more sensitive gas or tracer-gas detectors can be calibrated by injecting known amounts of tracer gas into the system or by using standard or calibrated leaks.

A standard leak can be made by constructing a device with a small orifice or nozzle; by maintaining constant upstream pressure, the flow will be constant. Another method is to maintain the upstream pressure at a level high enough to produce flow of sonic velocity at the nozzle throat, thus making the leakage independent of upstream pressure. Very small calibrated leaks are produced by flow through a capillary tube or by permeation through a membrane (normally glass). The capillary type may or may not have a self-contained gas supply. There is an advantage to capillary-type leaks not having a self-contained gas supply, because a variety of gases can be used with them, although a constant tracer-gas pressure must be supplied. Capillary tubes may be purchased with leak rates ranging from about 10^{-2} to 10^{-7} atm cm^3/s. Specific leak sizes are also available; these calibrated leaks can normally be supplied to match any specified leakage within about 20%. Permeation-type standard leaks are supplied with their own self-contained gas supply, usually helium. Although the flow through the membrane is molecular and therefore pressure dependent, typical helium loss is only 10% in 10 years. These leaks have a flow in the range of 10^{-6} to 10^{-10} atm cm^3/s.

Common Errors in Leak Testing

The most common errors encountered in leak testing are:

- Use of a method that is too sensitive
- Use of only one leak-testing method
- Failure to control the environment in the vicinity of the leak

Because sensitive methods cannot be used in the presence of gross leaks, leak testing should be conducted in two or more stages, beginning with gross testing and progressing to the more sensitive methods. This procedure saves both time and money.

Control of environment in the vicinity of the leak consists of a number of steps, depending on the test method employed. If a tracer gas is being used, the atmosphere must be controlled to prevent contamination by the tracer gas that might be erroneously reported as a leak. If sonic or ultrasonic methods are being used, background noise must be controlled. If small leaks are present, airborne contaminants may cause temporary plugging of the leak. For example, breathing on a molecular leak may cause it to plug due to water vapor in the breath. Particulate matter in the air may plug or partially plug a leak. Table 5 lists the differences in leak tightness requirements for a variety of industrial products.

Table 5 Leak tightness requirements for various products

Leak testing term	Vessel or part (material contained)	Minimum detectable leakage rate, std cm^3/s
Large	Truck (gravel)	1×10^3
	Hourglass (sand)	1×10^2
	Car window (air)	1×10^{-1}
Gross	Truck (oil)	1×10^{-2}
	Bucket (water)	1×10^{-3}
	Storage tank (gasoline)	1×10^{-4}
Small	Pipeline (gas)	1×10^{-5}
	Tanker (liquified natural gas)	1×10^{-6}
Fine	Storage tank (NH_3)	1×10^{-8}
	Heart pacemaker (gas)	1×10^{-11}

REFERENCES

1. Leak Testing, in *Nondestructive Testing Handbook*, R.C. McMaster, Ed., Vol 1, 2nd ed., American Society for Nondestructive Testing, 1982
2. *Encyclopedia of Materials Science and Engineering*, Vol 4, MIT Press, 1986

Liquid Penetrant Inspection

Revised by J.S. Borucki, Ardrox Inc., and Gail Jordan, Howmet Corporation

LIQUID PENETRANT INSPECTION is a nondestructive method of revealing discontinuities that are open to the surfaces of solid and essentially nonporous materials. Indications of a wide spectrum of flaw sizes can be found regardless of the configuration of the workpiece and regardless of flaw orientations. Liquid penetrants seep into various types of minute surface openings by capillary action. Because of this, the process is well suited to the detection of all types of surface cracks, laps, porosity, shrinkage areas, laminations, and similar discontinuities. It is extensively used for the inspection of wrought and cast products of both ferrous and nonferrous metals, powder metallurgy parts, ceramics, plastics, and glass objects.

In practice, the liquid penetrant process is relatively simple to utilize and control. The equipment used in liquid penetrant inspection can vary from an arrangement of simple tanks containing penetrant, emulsifier, and developer to sophisticated computer-controlled automated processing and inspection systems. Establishing procedures and standards for the inspection of specific parts or products is critical for optimum end results.

The liquid penetrant method does not depend on ferromagnetism (as does, for example, magnetic particle inspection), and the arrangement of the discontinuities is not a factor. The penetrant method is effective not only for detecting surface flaws in nonmagnetic metals but also for revealing surface flaws in a variety of other nonmagnetic materials. Liquid penetrant inspection is also used to inspect items made from ferromagnetic steels; generally, its sensitivity is greater than that of magnetic particle inspection.

The major limitation of liquid penetrant inspection is that it can detect only imperfections that are open to the surface; some other method must be used for detecting subsurface flaws. Another factor that can limit the use of liquid penetrants is surface roughness or porosity. Such surfaces produce excessive background and interfere with inspection.

Physical Principles

Liquid penetrant inspection depends mainly on a penetrant's effectively wetting the surface of a solid workpiece or specimen, flowing over that surface to form a continuous and reasonably uniform coating, and then migrating into cavities that are open to the surface. The cavities of interest are usually exceedingly small, often invisible to the unaided eye. The ability of a given liquid to flow over a surface and enter surface cavities depends principally on the following:

- Cleanliness of the surface
- Configuration of the cavity
- Cleanliness of the cavity
- Size of surface opening of the cavity
- Surface tension of the liquid
- Ability of the liquid to wet the surface
- Contact angle of the liquid

The cohesive forces between molecules of a liquid cause surface tension. An example of the influence of surface tension is the tendency of free liquid, such as a droplet of water, to contract into a sphere. In such a droplet, surface tension is counterbalanced by the internal hydrostatic pressure of the liquid. When the liquid comes into contact with a solid surface, the cohesive force responsible for surface tension competes with the adhesive force between the molecules of the liquid and the solid surface. These forces jointly determine the contact angle, θ, between the liquid and the surface (Fig. 1). If θ is less than 90° (Fig. 1a), the liquid is said to wet the surface, or to have good wetting ability; if the angle is equal to or greater than 90° (Fig. 1b and c), the wetting ability is considered poor.

Closely related to wetting ability is the phenomenon of capillary rise or depression (Fig. 2). If the contact angle, θ, between the liquid and the wall of the capillary tube is less than 90° (that is, if the liquid wets the tube wall), the liquid meniscus in the tube is concave, and the liquid rises in the tube (Fig. 2a). If θ is equal to 90°, there is no capillary depression or rise (Fig. 2b). If θ is greater than 90°, the liquid is depressed in the tube and does not wet the tube wall, and the meniscus is convex (Fig. 2c). In capillary rise (Fig. 2a), the meniscus does not pull the liquid up the tube; rather, the hydrostatic pressure immediately under the meniscus is reduced by the distribution of the surface tension in the concave surface, and the liquid is pushed up the capillary tube by the hydraulically transmitted pressure of the atmosphere at the free surface of the liquid outside the capillary tube. Figure 3 clearly shows the forces that cause liquid to rise in a capillary tube.

The height to which the liquid rises is directly proportional to the surface tension of the liquid and to the cosine of the angle of contact, and it is inversely proportional to

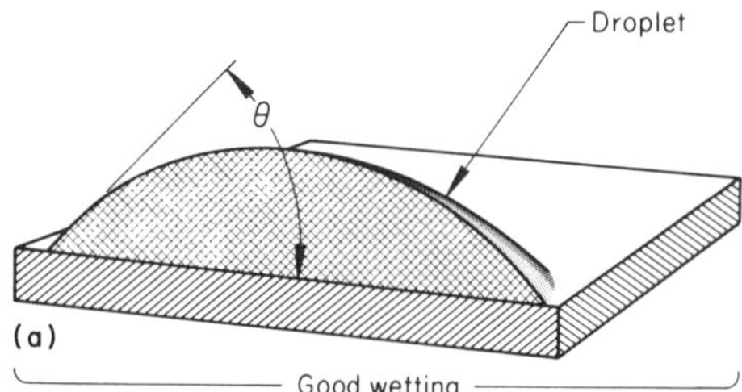

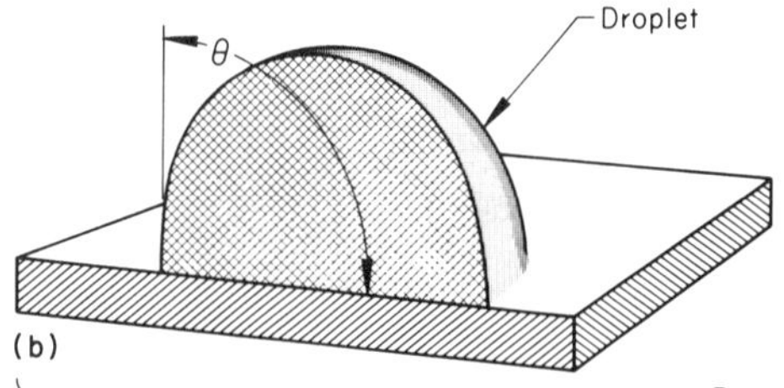

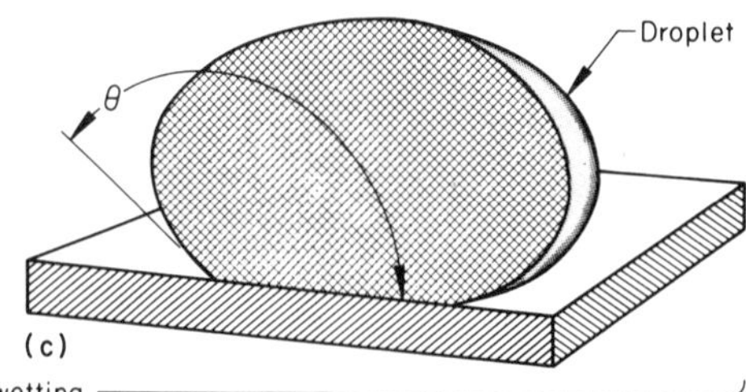

Fig. 1 Wetting characteristics as evaluated by the angle, θ, between a droplet of liquid and a solid surface. Good wetting is obtained when $\theta < 90°$ (a); poor wetting, when $\theta \geq 90°$ (b) and (c).

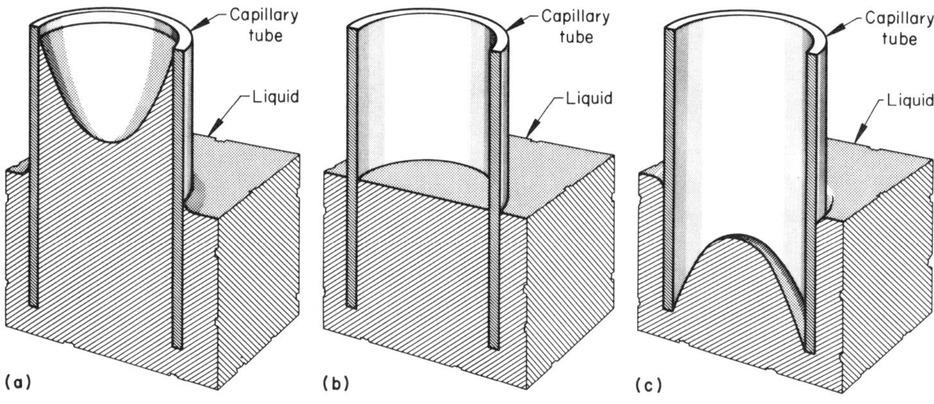

Fig. 2 Rise or depression in small vertical capillary tubes, determined by the contact angle θ, between a liquid and the wall of a capillary tube. (a) $\theta < 90°$: capillary rise. (b) $\theta = 90°$: no capillary depression or rise. (c) $\theta > 90°$: capillary depression

the density of the liquid and to the radius of the capillary tube. If, as in Fig. 4, the capillary tube is closed, a wetting liquid will still rise in the tube; however, there will be extra pressure resulting from the air and vapor compressed in the closed end of the tube, and the capillary rise will not be as great.

These examples of surface wetting and capillary rise illustrate the basic physical principles by which a penetrant enters fine surface discontinuities even though the practical circumstances encountered in the use of liquid penetrants are more complex than these examples may suggest. Cracks, for example, are not capillary tubes, but simulate the basic interaction between a liquid and a solid surface, which is responsible for the migration of penetrants into fine surface cracks.

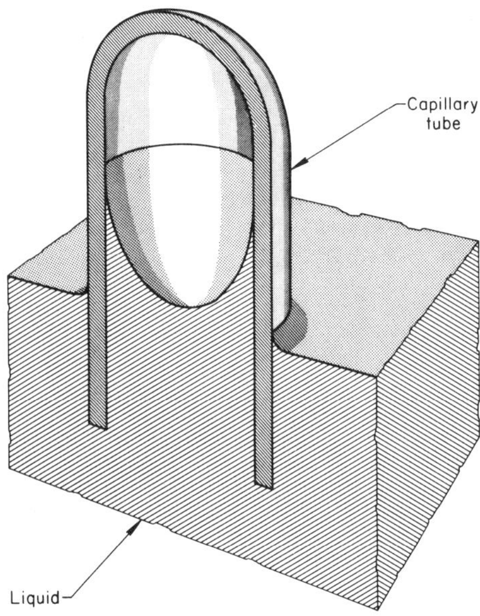

Fig. 4 Schematic showing that the rise and depression of liquids in closed capillary tubes are affected by the compressed air entrapped in the closed end. Compare height of liquid in tube shown in Fig. 3 with that in the tube here.

The viscosity of the liquid is not a factor in the basic equation of capillary rise. Viscosity is related to the rate at which a liquid will flow under some applied unbalanced stress; in itself, viscosity has a negligible effect on penetrating ability. In general, however, very viscous liquids are unsuitable as penetrants because they do not flow rapidly enough over the surface of the workpiece; consequently, they require excessively long periods of time to migrate into fine flaws.

Another necessary property of a penetrant is its capability of dissolving an adequate amount of suitable fluorescent or visible-dye compounds. Finally, the penetrant liquid must be removable with a suitable solvent/remover or emulsifier without precipitating the dye.

Just as it is important that a penetrant enter surface flaws, it is also important that the penetrant be retained in the flaws and emerge from the flaw after the superficial coating is removed from the surface and the developer is applied. It is apparently a paradox that the same interaction between a liquid and a surface that causes the liquid to enter a fine opening is also responsible for its emergence from that opening. The resolution of the paradox is simple: Once the surface has been freed of excess penetrant it becomes accessible to the entrapped liquid, which, under the effect of the adhesive forces between liquid and solid, spreads over the newly cleaned surface until an equilibrium distribution is attained (Fig. 5).

Although in some cases the amount of penetrant in a surface bead at equilibrium is sufficient to be detected visually, sensitivity is vastly increased by the use of a developer, of which there are several types available (forms A, B, C, and D). When applied, a developer forms a fine surface film, that is, a spongelike system of very fine, random capillary paths. If the penetrant contacts the powder, the powder then competes with the freshly cleaned surface of the workpiece for the penetrant as it flows out of the defect. If the developer is properly designed, it readily adsorbs the penetrant from the flaw. The penetrant continues to migrate by capillary action, spreading through the developer until an equilibrium is reached. This migrating action is illustrated schematically in Fig. 6. The visibility of the entrapped

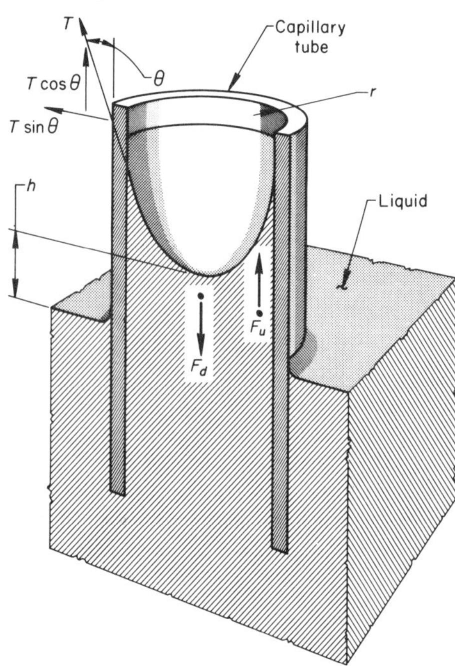

Downward force (F_d) is equal to the weight of the liquid column:

$$F_d = 2\pi r^2 hg\rho$$

Upward force (F_u) is equal to the surface tension times the perimeter of the meniscus:

$$F_u = (T \cos \theta)(2\pi r)$$

Where r is the inside radius of the capillary tube, h is the height of the liquid in the tube, g is acceleration due to gravity, ρ is density of the liquid, T is surface tension, and θ is the contact angle.

Fig. 3 Schematic showing the forces involved in capillary rise: the downward force from weight of the liquid column, and the upward force from surface tension along the meniscus perimeter

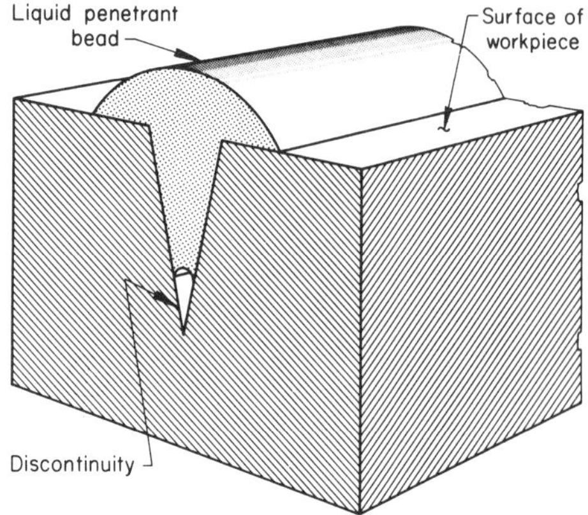

Fig. 5 Bead of liquid penetrant formed when, after excess penetrant has been removed from a workpiece surface, the penetrant remaining in a discontinuity emerges to the surface until an equilibrium is established

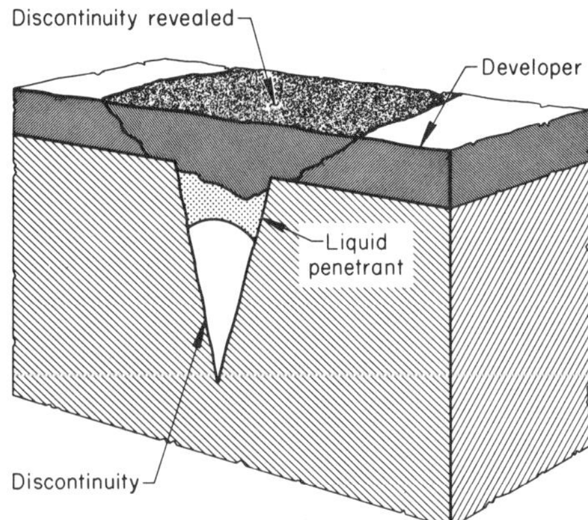

Fig. 6 Sectional view showing result of the migrating action of a liquid penetrant through a developer

penetrant within the flaw is greatly increased by the spreading or enlargement of the indication.

Evolution of the Process

The exact origin of liquid penetrant inspection is not known, but it has been assumed that the method evolved from the observation that the rust on a crack in a steel plate in outdoor storage was somewhat heavier than the rust on the adjacent surfaces as a result of water seeping into the crack and forcing out the oxide it had helped to produce. The obvious conclusion was that a liquid purposely introduced into surface cracks and then brought out again would reveal the locations of those cracks.

The only material that fulfilled the known criteria of low viscosity, good wettability, and ready availability was kerosene. It was found, however, that although wider cracks showed up easily, finer ones were sometimes missed because of the difficulty of detecting, by purely visual means, the small amounts of kerosene exuding from them. The solution was to provide a contrasting surface that would reveal smaller seepages. The properties and availability of whitewash made it the logical choice. This method, known as the kerosene-and-whiting test, was the standard for many years. The sensitivity of the kerosene-and-whiting test could be increased by hitting the object being tested with a hammer during testing. The resulting vibration brought more of the kerosene out of the cracks and onto the whitewash. Although this test was not as sensitive as those derived from it, it was quick, inexpensive, and reasonably accurate. Thus, it provided a vast improvement over ordinary visual examination.

The first step leading to the methods now available was the development of the fluorescent penetrant process by R.C. Switzer. This liquid, used jointly with a powder developer, brought penetrant inspection from a relatively crude procedure to a more scientific operation. With fluorescent penetrant, minute flaws could be readily detected when exposed to ultraviolet light (commonly called black light). This development represented a major breakthrough in the detection of surface flaws.

Switzer's work also included the development of the visible-color contrast method, which allowed for inspection under white light conditions. Although not as sensitive as fluorescent penetrant inspection, it is widely used in industry for noncritical inspection. Through the developments described above, liquid penetrant inspection has become a major nondestructive inspection method.

Penetrant Methods

Because of the vast differences among applications for penetrant inspection, it has been necessary to refine and develop the two types of penetrants (type I, fluorescent, and type II, visible) into four basic methods to accommodate the wide variations in the following principal factors:

- Surface condition of the workpiece being inspected
- Characteristics of the flaws to be detected
- Time and place of inspection
- Size of the workpiece
- Sensitivity required

The four methods are broadly classified as:

- Water washable, method A
- Postemulsifiable lipophilic, method B
- Solvent removable, method C
- Postemulsifiable hydrophilic, method D

These four methods are described below.

Water-washable penetrant (method A) is designed so that the penetrant is directly water washable from the surface of the workpiece; it does not require a separate emulsification step, as does the postemulsifiable penetrant methods. It can be used to process workpieces quickly and efficiently. It is important, however, that the washing operation be carefully controlled because water-washable penetrants are susceptible to overwashing. The essential operations involved in this method are illustrated schematically in Fig. 7.

Postemulsifiable penetrants (methods B and D) are designed to ensure the detection of minute flaws in some materials. These penetrants are not directly water washable. Because of this characteristic, the danger of overwashing the penetrant out of the flaws is reduced. The difference between the water-washable and postemulsifiable method lies in the use of an emulsifier prior to final rinsing. The emulsifier makes the residual surface penetrant soluble in water so that the excess surface penetrant can be removed by the water rinse. Therefore, the emulsification time must be carefully controlled so that the surface penetrant becomes water soluble but the penetrant in the flaws does not. The operations involved in the postemulsifiable method are illustrated schematically in Fig. 8 for the lipophilic system and in Fig. 9 for the hydrophilic system. Despite the additional processing steps involved with the postemulsifiable methods B and D, these methods are the most reliable for detecting minute flaws.

Solvent-removable penetrant (method C) is available for use when it is necessary to inspect only a localized area of a workpiece or to inspect a workpiece at the site rather than on a production inspection basis. Normally, the same type of solvent is used for precleaning and for removing excess penetrant. This penetrant process is convenient and broadens the range of applications of penetrant inspections. However,

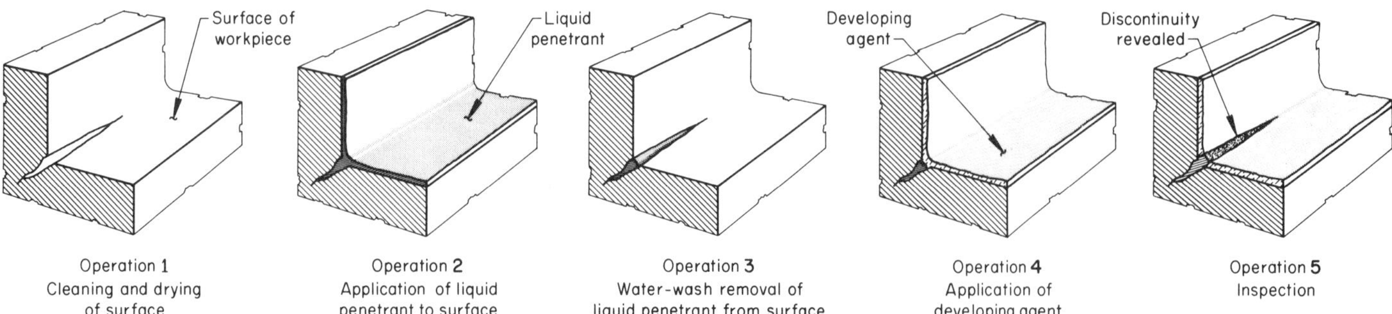

Fig. 7 Five essential operations for liquid penetrant inspection using the water-washable system

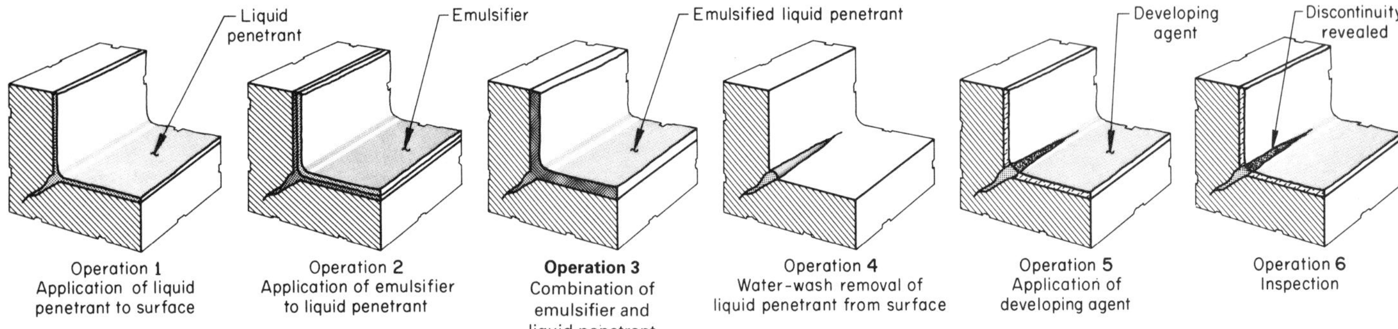

Fig. 8 Operations (in addition to precleaning) for the postemulsifiable, method B, lipophilic liquid penetrant system

because the solvent-removable method is labor intensive, it is not practical for many production applications. When properly conducted and when used in the appropriate applications, the solvent-removable method can be one of the most sensitive penetrant methods available. The operations for this process are illustrated schematically in Fig. 10.

Whichever penetrant method is chosen, the degree and speed of excess penetrant removal depend on such processing conditions as spray nozzle characteristics, water pressure and temperature, duration of the rinse cycle, surface condition of the workpiece, and inherent removal characteristics of the penetrant employed.

Description of the Process

Regardless of the type of penetrant used, that is, fluorescent (type I) or visible (type II), penetrant inspection requires at least five essential steps, as follows.

Surface Preparation. All surfaces to be inspected, whether localized or the entire workpiece, must be thoroughly cleaned and completely dried before being subjected to penetrant inspection. Flaws exposed to the surface must be free from oil, water, or other contaminants if they are to be detected.

Penetration. After the workpiece has been cleaned, penetrant is applied in a suitable manner so as to form a film of the penetrant over the surface. This film should remain on the surface long enough to allow maximum penetration of the penetrant into any surface openings that are present.

Removal of Excess Penetrant. Excess penetrant must be removed from the surface. The removal method is determined by the type of penetrant used. Some penetrants can be simply washed away with water; others require the use of emulsifiers (lipophilic or hydrophilic) or solvent/remover. Uniform removal of excess surface penetrant is necessary for effective inspection, but overremoval must be avoided.

Development. Depending on the form of developing agent to be used, the workpiece is dried either before or directly after application of the developer. The developer forms a film over the surface. It acts as a blotter to assist the natural seepage of the penetrant out of surface openings and to spread it at the edges so as to enhance the penetrant indication.

Inspection. After it is sufficiently developed, the surface is visually examined for indications of penetrant bleedback from surface openings. This examination must be performed in a suitable inspection environment. Visible penetrant inspection is performed in good white light. When fluorescent penetrant is used, inspection is performed in a suitably darkened area using black (ultraviolet) light, which causes the penetrant to fluoresce brilliantly.

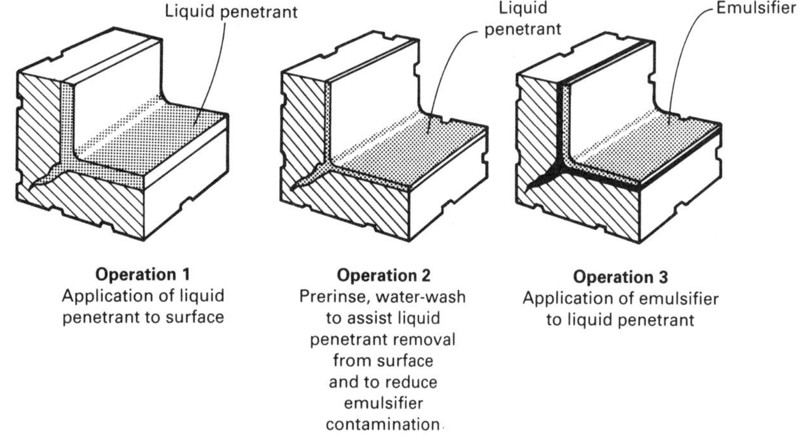

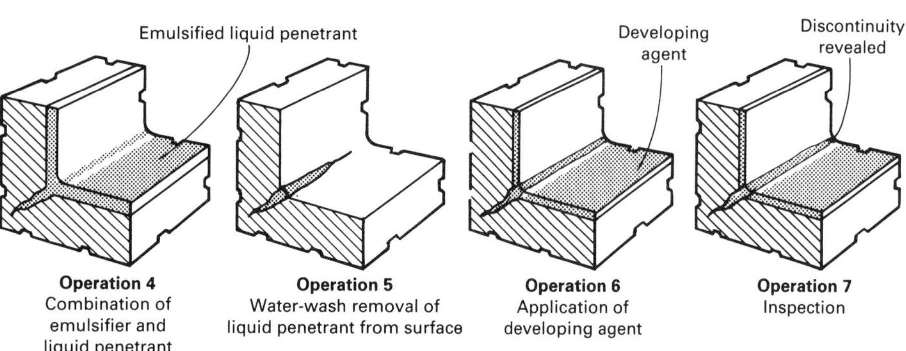

Fig. 9 Operations (in addition to precleaning) for the postemulsifiable, method D, hydrophilic liquid penetrant system

Materials Used in Penetrant Inspection

In addition to the penetrants themselves, liquids such as emulsifiers, solvent/cleaners and removers, and developers are required for conducting liquid penetrant inspection.

Penetrants

There are two basic types of penetrants:

- Fluorescent, type I
- Visible, type II

Each type is available for any one of the four methods (water washable, postemulsifiable lipophilic, and postemulsifiable hydrophilic,

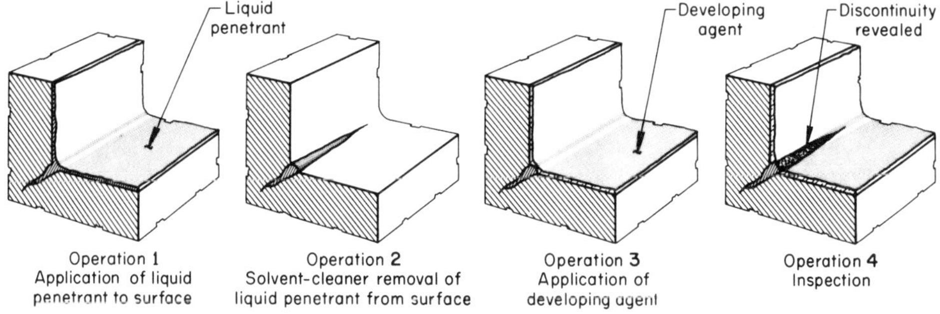

Fig. 10 Operations (in addition to precleaning) for the solvent-removable liquid penetrant system

and solvent removable) mentioned in the section "Penetrant Methods" in this article.

Type I fluorescent penetrant utilizes penetrants that are usually green in color and fluoresce brilliantly under ultraviolet light. The sensitivity of a fluorescent penetrant depends on its ability to form indications that appear as small sources of light in an otherwise dark area. Type I penetrants are available in different sensitivity levels classified as follows:

- *Level ½*: Ultralow
- *Level 1*: Low
- *Level 2*: Medium
- *Level 3*: High
- *Level 4*: Ultrahigh

Type II visible penetrant employs a penetrant that is usually red in color and produces vivid red indications in contrast to the light background of the applied developer under visible light. The visible penetrant indications must be viewed under adequate white light. The sensitivity of visible penetrants is regarded as Level 1 and adequate for many applications.

Penetrant selection and use depend on the criticality of the inspection, the condition of the workpiece surface, the type of processing, and the desired sensitivity (see the section "Selection of Penetrant Method" in this article).

Method A, water-washable penetrants are designed for the removal of excess surface penetrant by water rinsing directly after a suitable penetration (dwell) time. The emulsifier is incorporated into the water-washable penetrant. When this type of penetrant is used, it is extremely important that the removal of excess surface penetrant be properly controlled to prevent overwashing, which can cause the penetrant to be washed out of the flaws.

Methods B and D, lipophilic and hydrophilic postemulsifiable penetrants are insoluble in water and therefore not removable by water rinsing alone. They are designed to be selectively removed from the surface of the workpiece by the use of a separate emulsifier. The emulsifier, properly applied and left for a suitable emulsification time, combines with the excess surface penetrant to form a water-washable surface mixture that can be rinsed from the surface of the workpiece. The penetrant that remains within the flaw is not subject to overwashing. However, proper emulsification time must be established experimentally and maintained to ensure that overemulsification, which results in the loss of flaws, does not occur.

Method C, solvent-removable penetrants are removed by wiping with clean, lint-free material until most traces of the penetrant have been removed. The remaining traces are removed by wiping with clean, lint-free material lightly moistened with solvent. This type of penetrant is primarily used where portability is required and for the inspection of localized areas. To minimize the possibility of removing the penetrant from discontinuities, the use of excessive amounts of solvent must be avoided.

Physical and Chemical Characteristics. Both fluorescent and visible penetrants, whether water washable, postemulsifiable, or solvent removable, must have certain chemical and physical characteristics if they are to perform their intended functions. The principal requirements of penetrants are as follows:

- Chemical stability and uniform physical consistency
- A flash point not lower than 95 °C (200 °F); penetrants that have lower flash points constitute a potential fire hazard
- A high degree of wettability
- Low viscosity to permit better coverage and minimum dragout
- Ability to penetrate discontinuities quickly and completely
- Sufficient brightness and permanence of color
- Chemical inertness with materials being inspected and with containers
- Low toxicity to protect personnel
- Slow drying characteristics
- Ease of removal
- Inoffensive odor
- Low cost
- Resistance to ultraviolet light and heat fade

Emulsifiers

Emulsifiers are liquids used to render excess penetrant on the surface of a workpiece water washable. There are two methods used in the postemulsifiable method: method B, lipophilic, and method D, hydrophilic. Both can act over a range of durations from a few seconds to several minutes, depending on the viscosity, concentration, method of application, and chemical composition of the emulsifier, as well as on the roughness of the workpiece surface. The length of time an emulsifier should remain in contact with the penetrant depends on the type of emulsifier employed and the roughness of the workpiece surface.

Method B, lipophilic emulsifiers are oil based, are used as supplied, and function by diffusion (Fig. 11). The emulsifier diffuses into the penetrant film and renders it spontaneously emulsifiable in water. The rate at which it diffuses into the penetrant establishes the emulsification time. The emulsifier is fast acting, thus making the emulsification operation very critical. The emulsifier continues to act as long as it is in contact with the workpiece; therefore, the rinse operation should take place quickly to avoid overemulsification.

Method D, hydrophilic emulsifiers are water based and are usually supplied as concentrates that are diluted in water to concentrations of 5 to 30% for dip applications and 0.05 to 5% for spray applications. Hydrophilic emulsifiers function by displacing excess penetrant from the surface of the part by detergent action (Fig. 12). The force of the water spray or the air agitation of dip tanks provides a scrubbing action. Hydrophilic emulsifier is slower acting than the lipophilic emulsifier; therefore, it is easier to control the cleaning action. In addition to the emulsifier application, method D also requires a prerinse. Utilizing a coarse water spray, the prerinse helps remove the excess penetrant to minimize contamination of the emulsifier. Of greater significance, only a very thin and uniform layer of penetrant will remain on the surface, thus allowing easy removal of the surface layer with minimum opportunity of removing penetrant from the flaws. This step is required because the penetrant is not miscible with the hydrophilic emulsifier.

The penetrant manufacturer should recommend nominal emulsification times for the specific type of emulsifier in use. Actual emulsification times should be determined experimentally for the particular application. The manufacturer should also recommend the concentrations for hydrophilic emulsifiers.

Solvent Cleaner/Removers

Solvent cleaner/removers differ from emulsifiers in that they remove excess surface penetrant through direct solvent action. There are two basic types of solvent removers: flammable and nonflammable. Flammable cleaners are essentially free of halogens but are potential fire hazards. Nonflammable cleaners are widely used. However, they do contain halogenated sol-

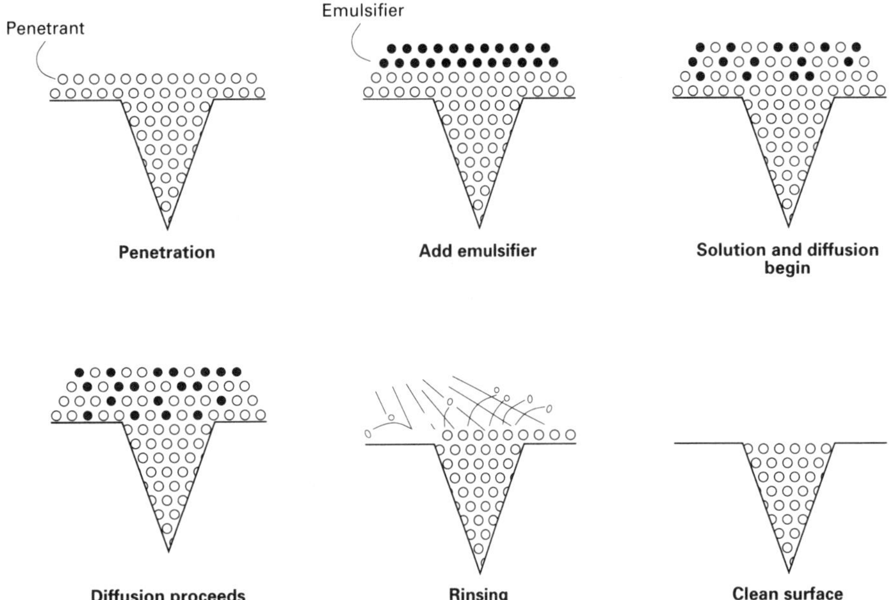

Fig. 11 Elements in the functioning of lipophilic emulsifiers

vents, which may render them unsuitable for some applications.

Excess surface penetrant is removed by wiping, using lint-free cloths slightly moistened with solvent cleaner/remover. It is not recommended that excess surface penetrant be removed by flooding the surface with solvent cleaner/remover, because the solvent will dissolve the penetrant within the defect and indications will not be produced.

Developers

The purpose of a developer is to increase the brightness intensity of fluorescent indications and the visible contrast of visible-penetrant indications. The developer also provides a blotting action, which serves to draw penetrant from within the flaw to the surface, spreading the penetrant and enlarging the appearance of the flaw.

The developer is a critical part of the inspection process; borderline indications that might otherwise be missed can be made visible by the developer. In all applications of liquid penetrant inspection, use of a developer is desirable because it decreases inspection time by hastening the appearance of indications.

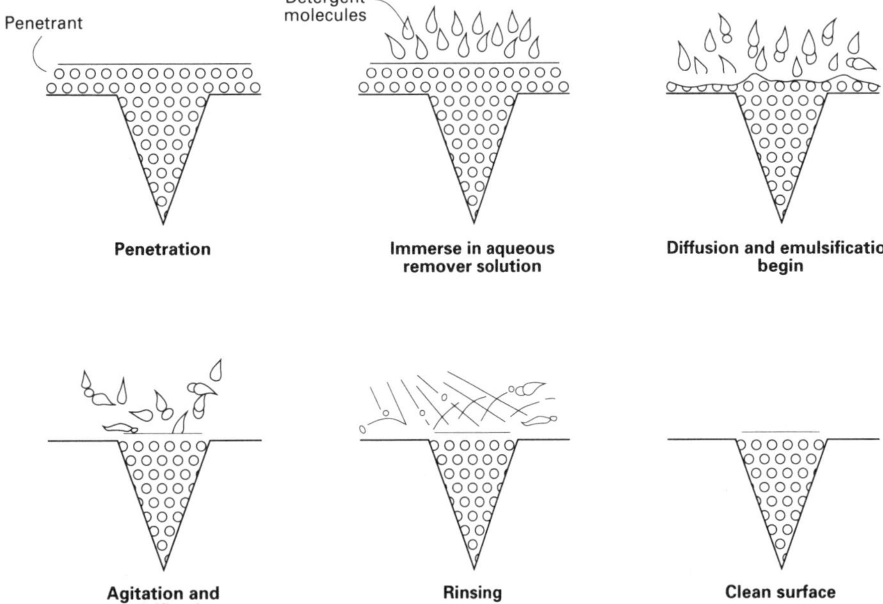

Fig. 12 Elements in the functioning of hydrophilic emulsifiers

Required Properties. To carry out its functions to the fullest possible extent, a developer must have the following properties or characteristics (rarely are all these characteristics present to optimum degrees in any given material or formulation, but all must be considered in selecting a developer):

- The developer must be adsorptive to maximize blotting
- It must have fine grain size and a particle shape that will disperse and expose the penetrant at a flaw to produce strong and sharply defined indications of flaws
- It must be capable of providing a contrast background for indications when color-contrast penetrants are used
- It must be easy to apply
- It must form a thin, uniform coating over a surface
- It must be easily wetted by the penetrant at the flaw (the liquid must be allowed to spread over the particle surfaces)
- It must be nonfluorescent if used with fluorescent penetrants
- It must be easy to remove after inspection
- It must not contain ingredients harmful to parts being inspected or to equipment used in the inspection operation
- It must not contain ingredients harmful or toxic to the operator

Developer Forms. There are four forms of developers in common use:

- Form A, dry powder
- Form B, water soluble
- Form C, water suspendible
- Form D, nonaqueous solvent suspendible

The characteristics of each form are discussed below.

Dry powder developers (form A) are widely used with fluorescent penetrants, but should not be used with visible-dye penetrants because they do not produce a satisfactory contrast coating on the surface of the workpiece. Ideally, dry powder developers should be light and fluffy to allow for ease of application and should cling to dry surfaces in a fine film. The adherence of the powder should not be excessive, as the amount of black light available to energize fluorescent indications will be reduced.

For purposes of storage and handling as well as applications, powders should not be hygroscopic, and they should remain dry. If they pick up moisture when stored in areas of high humidity, they will lose their ability to flow and dust easily, and they may agglomerate, pack, or lump up in containers or in developer chambers.

For reasons of safety, dry powder developers should be handled with care. Like any other dust particle, it can dry the skin and irritate the lining of the air passages, causing irritation. If an operator will be working continuously at a developer station, rubber gloves and respirators may be desirable. Modern equipment often includes an exhaust system on the developer spray booth or on the developer dust chamber that prevents dust from escaping. Powder recovery filters are included in most such installations.

Water-soluble developers (form B) can be used for both fluorescent (type I) or visible (type II) postemulsifiable or solvent-removable penetrants. Water-soluble developers are not recommended for use with water-washable penetrants, because of the potential to wash the penetrant from within the flaw if the developer is not very carefully controlled. Water-soluble developers are supplied as a dry powder concentrate, which is then dispersed in water in recommended proportions, usually from 0.12 to 0.24 kg/L (1 to 2 lb/gal.). The bath concentration is monitored for specific gravity with the appropriate hydrometer. Necessary constituents of the developers include corrosion inhibitors and biocides. The advantages of this form of developer are as follows:

- The prepared bath is completely soluble and therefore does not require any agitation
- The developer is applied prior to drying, thus decreasing the development time
- The dried developer film on the workpiece is completely water soluble and is thus easily and completely removed following inspection by simple water rinsing

Water-suspendible developers (form C) can be used with either fluorescent (type I) or visible (type II) penetrants. With fluorescent penetrant, the dried coating of developer must not fluoresce, nor may it absorb or filter out the black light used for inspection.

Water-suspendible developers are supplied as a dry powder concentrate, which is then dispersed in water in recommended proportions, usually from 0.04 to 0.12 kg/L (⅓ to 1 lb/gal.). The amount of powder in suspension must be carefully maintained. Too much or too little developer on the surface of a workpiece can seriously affect sensitivity. Specific gravity checks should be conducted routinely, using a hydrometer to check the bath concentration. Water-soluble developers contain dispersing agents to help retard settling and caking as well as inhibitors to prevent or retard corrosion of workpieces and equipment, and biocides to extend the working life of the aqueous solutions. In addition, wetting agents are present to ensure even coverage of surfaces and ease of removal after inspection.

Water-suspendible developer is applied before drying; therefore, developing time can be decreased because the heat from the drier helps to bring penetrant back out of surface openings. In addition, with the developer film already in place, the developing action begins at once. Workpieces are ready for inspection in a shorter period of time.

Nonaqueous solvent-suspendible developers (form D) are commonly used for both the fluorescent and the visible penetrant process. This form of developer produces a white coating on the surface of the part. This coating yields the maximum white color contrast with the red visible penetrant indication and extremely brilliant fluorescent indication.

Nonaqueous solvent-suspendible developers are supplied in the ready-to-use condition and contain particles of developer suspended in a mixture of volatile solvents. The solvents are carefully selected for their compatibility with the penetrants. Nonaqueous solvent-suspendible developers also contain surfactants in a dispersant whose functions are to coat the particles and reduce their tendency to clump or agglomerate.

Nonaqueous solvent-suspendible developers are the most sensitive form of developer used with type I fluorescent penetrants because the solvent action contributes to the absorption and adsorption mechanisms. In many cases where tight, small flaws occur, the dry powder (form A), water-soluble (form B), and water-suspendible (form C) developers do not contact the entrapped penetrant. This results in the failure of the developer to create the necessary capillary action and surface tension that serve to pull the penetrant from the flaw. The nonaqueous solvent-suspendible developer enters the flaw and dissolves into the penetrant. This action increases the volume and reduces the viscosity of the penetrant. The manufacturer must carefully select and compound the solvent mixture. There are two types of solvent-base developers: nonflammable (chlorinated solvents) and flammable (nonchlorinated solvents). Both types are widely used. Selection is based on the nature of the application and the type of alloy being inspected.

Selection of Penetrant Method

The size, shape, and weight of workpieces, as well as the number of similar workpieces to be inspected, often influence the selection of a penetrant method.

Sensitivity and Cost. The desired degree of sensitivity and cost are usually the most important factors in selecting the proper penetrant method for a given application. The methods capable of the greatest sensitivity are also the most costly. Many inspection operations require the ultimate in sensitivity, but there are a significant number in which extreme sensitivity is not required and may even produce misleading results.

On a practical basis, the fluorescent penetrant methods are employed in a wider variety of production inspection operations than the visible penetrant methods, which are utilized primarily for localized inspections. As stated earlier, penetrants are classified on the basis of penetrant type:

- *Type I*: Fluorescent
- *Type II*: Visible
- *Method A*: Water washable
- *Method B*: Postemulsifiable-lipophilic
- *Method C*: Solvent removable
- *Method D*: Postemulsifiable-hydrophilic

Penetrants are also classified in terms of sensitivity levels:

- *Level ½*: Ultralow
- *Level 1*: Low
- *Level 2*: Medium
- *Level 3*: High
- *Level 4*: Ultrahigh

Advantages and Limitations of Penetrant Methods. Each penetrant method, whether postemulsifiable (either lipophilic or hydrophilic), solvent removable, or water washable, using fluorescent or visible-dye penetrants, has inherent advantages and limitations.

The postemulsifiable fluorescent penetrant method is the most reliable and sensitive penetrant method. This procedure will locate wide, shallow flaws as well as tight cracks and is ideal for high-production work. On the other hand, emulsification requires an additional operation, which increases cost. Also, this method requires a water supply and facilities for inspection under black light. The postemulsifiable, lipophilic fluorescent penetrant method is less sensitive and less reliable than the hydrophilic method. Its use is therefore declining.

The solvent-removable fluorescent penetrant method employs a procedure similar to that used for the postemulsifiable fluorescent method, except that excess penetrant is removed with a solvent/remover. This method is especially recommended for spot inspection or where water cannot be conveniently used. It is more sensitive than the water-washable system, but the extreme caution and additional time required for solvent removal often preclude its use.

The water-washable fluorescent penetrant method is the fastest of the fluorescent procedures. It is also highly sensitive, reliable, and reasonably economical. It can be used for both small and large workpieces and is effective on most part surfaces. However, it will not reliably reveal open, shallow flaws if overwashed and in some cases, depending on the sensitivity level of the penetrant, will not locate the very tightest cracks. There is also the danger of overwashing by applying water for an excessive period of time or with a pressure sufficient to remove the penetrant from the flaws.

The postemulsifiable visible penetrant method is used whenever sensitivity required is greater than that provided by the water-washable visible penetrant method. However, the additional step of applying emulsifier makes this system more costly than the water-washable visible penetrant dye method that requires water, but otherwise no location limitations are imposed.

The solvent-removable visible penetrant method has a distinct advantage in that all the necessary ingredients are portable; accordingly, it can be used in a practically limitless number of locations, both in the shop and in the field. Because of the problems involved in penetrant removal, however, the method is generally confined to spot inspection or to inspection under circumstances that prohibit the use of other methods because of workpiece size or location.

The water-washable visible penetrant system is the fastest and simplest of all penetrant techniques. It is, however, the least sensitive because the penetrant is likely to be removed from wide, shallow flaws. Therefore, it is most useful in those applications where shallow and relatively wide flaws are not significant. This method is also the least sensitive for locating tight cracks. It requires a water source, but can be performed in almost any location because neither a darkened area nor electricity is required.

Equipment Requirements

The equipment used in the penetrant inspection process varies from spray or aerosol cans to complex, automated, computer-driven processing systems. Some of the more generally used types of equipment are described in the following sections.

Portable Equipment

For occasional inspections, especially in the field, where equipment portability is necessary, minimal kits for either visible or fluorescent penetrant inspection are commercially available. (Generally, portable penetrant applications are limited to localized areas or spot inspections rather than entire part surfaces.)

Such a kit for visible penetrant inspection work includes a precleaner, a penetrant, and a penetrant remover and developer, all in pressurized spray cans. Penetrant removal requires wiping with lint-free cloths or paper towels.

A similar kit is available for fluorescent work; a precleaner, a penetrant, penetrant remover and developer are likewise supplied in pressurized cans. Cleaning is accomplished by wiping with lint-free cloths or paper towels. This kit includes a small, portable black light for conducting the inspection.

Stationary Inspection Equipment

The type of equipment most frequently used in fixed installations consists of a series of modular subunits. Each subunit performs a special task. The number of subunits in a processing line varies with the type of penetrant method used. The subunits are:

- Drain and/or dwell stations
- Penetrant and emulsifier stations
- Pre- and post-wash stations
- Drying station
- Developer station
- Inspection station
- Cleaning stations

The drain or dwell stations are actually roller-top benches that hold the parts during the processing cycle. The usual arrangement is to position a drain or dwell station following each of the dip tanks, the wash station, and the drying oven. The subunits are described in more detail below.

Penetrant Station. The principal requirement of a penetrant station is that it provide a means for coating workpieces with penetrant—either all over, for small workpieces, or over small areas of large workpieces when only local inspection is required. In addition, means should be provided for draining excess penetrant back into the penetrant reservoir, unless the expendable technique is being used. Draining racks usually serve the additional purpose of providing a storage place for parts during the time required for penetration (dwell time).

Small workpieces are easily coated by dipping them into a reservoir of penetrant. This may be done individually or in batches in a wire basket.

The penetrant container should be equipped with an easily removable cover to reduce evaporation when not in use. A drain cock should also be provided to facilitate draining of the tank for cleaning. Containers are usually made of steel, but stainless steel containers should be used with water-base penetrants.

For large workpieces, penetrant is often applied by spraying or flowing. This is done mainly for convenience but also for economy, because the volume of penetrant needed to immerse a large object may be so great as to increase unnecessarily the original cost of installation. A small reservoir of penetrant equipped with a pump, a hose, and a spray or flow nozzle is usually almost as fast a means of coating large objects as the dipping operation. For this type of operation, the penetrant station consists of a suitably ventilated booth with a rotatable grill platform on which the workpiece is set. A drain under the platform returns penetrant runoff to the sump, from which it is pumped back to the spray nozzle. The booth enclosure prevents the overspraying of penetrant on areas outside the penetrant station.

In some applications, it has been found that only a small amount of penetrant is recoverable and reusable, and this has led to the adoption of the expendable technique for some very large workpieces. In this technique, penetrant is sprayed over the workpiece in a penetrant station similar to the one mentioned previously. The penetrant is stored in a separate pressure tank fitted with a hose and a spray nozzle. The spray booth is not equipped with a sump to recover excess penetrant. Instead, the booth is fitted with water spray nozzles and a drain so that it can serve the multiple purpose of draining and washing. A decision to use the expendable technique and related equipment should be based on a careful analysis and consideration of cost, time, rate of production, and handling problems.

Emulsifier Station. The emulsifier liquid is contained in a tank of sufficient size and depth to permit immersion of the workpieces, either individually or in batches. Covers are sometimes provided to reduce evaporation, and drain valves are supplied for cleanout when the bath has become contaminated. Suitable drain racks are also a part of this station and are used to permit excess emulsifier to drain back into the tank.

If large workpieces must be coated with emulsifier, methods must be devised to achieve the fastest possible coverage. Multiple spraying or copious flowing of emulsifier from troughs or perforated pipes can be used on some types of automatic equipment. For the local coating of large workpieces, spraying is often satisfactory, using the expendable technique described for the application of penetrant.

Pre- and Postrinse Stations. The water rinsing (washing) of small workpieces is frequently done by hand, either individually or in batches in wire baskets. The workpieces are held in the wash tank and cleaned with a hand-held spray using water at tap pressure and temperature. The wash trough or sink should be large enough and deep enough so that workpieces can be easily turned to clean all surfaces. Splash shields should separate the rinse station from preceding (penetrant or emulsifier) and succeeding (wet developer) stations. Rinse stations are always equipped with at least one ultraviolet light so that the progress of removal of fluorescent penetrant can be easily followed.

The automatic rinsing of small workpieces is satisfactorily accomplished by means of a rotating table. The basket is placed on the table, and water-spray heads are properly located so as to rinse all surfaces of the workpieces thoroughly.

Specially built automatic washers for rinsing workpieces that are large and of irregular contour are often installed. Spray nozzles must be located to suit the individual application.

The removal of excess penetrant by simply submerging the workpiece in water is generally not recommended. However, in some cases, simple submersion in an air-agitated water bath is satisfactory.

The rinse station is subject to corrosion. All steel should be protected by rustproofing and painting. Most satisfactory, but more costly, is the use of stainless steel equipment.

Drying Station. The recirculating hot-air drier is one of the most important equipment components. The drier must be large enough to easily handle the type and number of workpieces being inspected. Heat input, air flow, and rate of movement of workpieces through the drier, as well as temperature control, are all factors that must be balanced. The drier may be of the cabinet type, or it may be designed so that the workpieces pass through on a conveyor. If conveyor operation is used, the speed must be considered with the required drying cycle.

Electric-resistance elements are frequently used as sources of heat, but gas, hot water, and steam are also used. Heat input is controlled by suitably located thermostats and is determined by workpiece size, composition, and rate of movement.

Integrated equipment invariably includes the recirculating hot-air drier mentioned previously. Makeshift driers are sometimes used—often because nothing better is available. Electric or gas hot-air blowers of commercial design have been used, but because no control of temperature is possible, these are very unsatisfactory and are ordinarily used only on an emergency basis. Infrared lamps are not suitable for drying washed workpieces, because the radiant heat cannot be readily controlled.

Equipment designed to handle workpieces of special size and shape requires a specially designed drier. Each drier is a separate engineering problem involving a special combination of workpiece composition, mass, surface area, speed of movement, and other considerations unique to the circumstances.

Developer Station. The type and location of the developer station depend on whether dry or wet developer is to be used. For dry developer, the developer station is downstream from the drier, but for wet developer it immediately precedes the drier, following the rinse station.

The dry-developer station usually consists of a simple bin containing the powder. Dried workpieces are dipped into the powder, and the excess powder is shaken off. Larger workpieces may not be so easily immersed in the powder, so a scoop is usually provided for throwing powder over the surfaces, after which the excess is shaken off. The developer bin should be equipped with an easily removable cover to protect the developer from dust and dirt when not in use.

Dust control systems are sometimes needed when dry developer is used. Control is accomplished by a suction opening across the back of the bin at the top, which draws off any developer dust that rises out of the bin. The dust-laden air is passed through filter bags, from which the developer dust can be reclaimed for further use (Fig. 13).

Developer powder can also be applied with air pressure. This system requires no bin, but it does require a booth or a cabinet and also makes dust collection mandatory.

Equipment for the automatic application of dry developer consists of a cabinet through which the dried workpieces are passed on a conveyor. The air in the cabinet is laden with dust that is kept agitated by means of a blower. As workpieces pass through, all surfaces are brought into contact with developer powder carried by the air. Air must be exhausted from the cabinet and either recirculated or cleaned by being passed through a dust-collecting filter.

Wet developer, when used, is contained in a tank similar to that used for penetrant or emulsifier. The tank should be deep enough to permit workpieces to be submerged in the developer. There should also be a rack or conveyor on which parts can rest after dipping. This will permit excess developer to run back into the tank.

Suspendible developer baths settle out when not in use; therefore, a paddle for stirring should be provided. Continuous agitation is essential because the settling rate is rapid. Pumps are sometimes incorporated into the developer station for flowing the developer over large workpieces through a hose and nozzle and for keeping the developer agitated.

In automatic units, special methods of applying developer are required. Flow-on methods are frequently used. This technique requires a nozzle arrangement that permits the workpieces to be covered thoroughly and quickly.

Inspection Station. Essentially, the inspection station is simply a worktable on which workpieces can be handled under

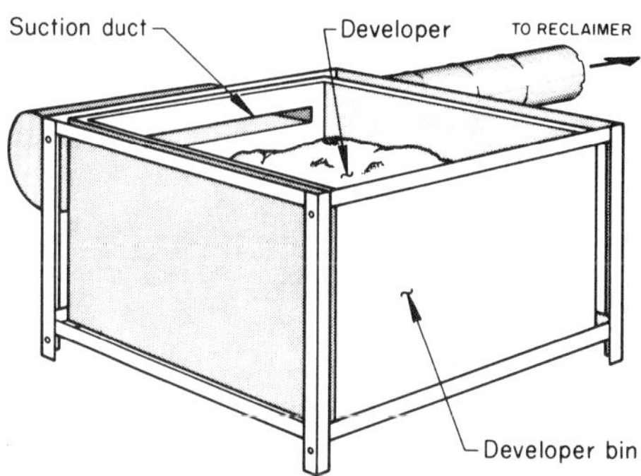

Fig. 13 Dry-developer bin equipped with dust control and reclaimer system

proper lighting. For fluorescent methods, the table is usually surrounded by a curtain or hood to exclude most of the white light from the area. For visible-dry penetrants, a hood is not necessary.

Generally, black (ultraviolet) lights (100 W or greater) are mounted on brackets from which they can be lifted and moved about by hand. Because of the heat given off by black lights, good air circulation is essential in black light booths.

For automatic inspection, workpieces are moved through booths equipped with split curtains, either by hand, monorail, or by conveyor. In some large inspection installations, fully enclosed rooms have been built for black light inspection. Access to the room is provided by a light lock. Inspection rooms must be laid out efficiently to prevent rejected workpieces from reentering the production line. Figures 14 and 15 illustrate typical penetrant inspection components and their layout in an inspection station installation.

Automated Inspection Equipment

For many years, the penetrant inspection of production parts has been a manual op-

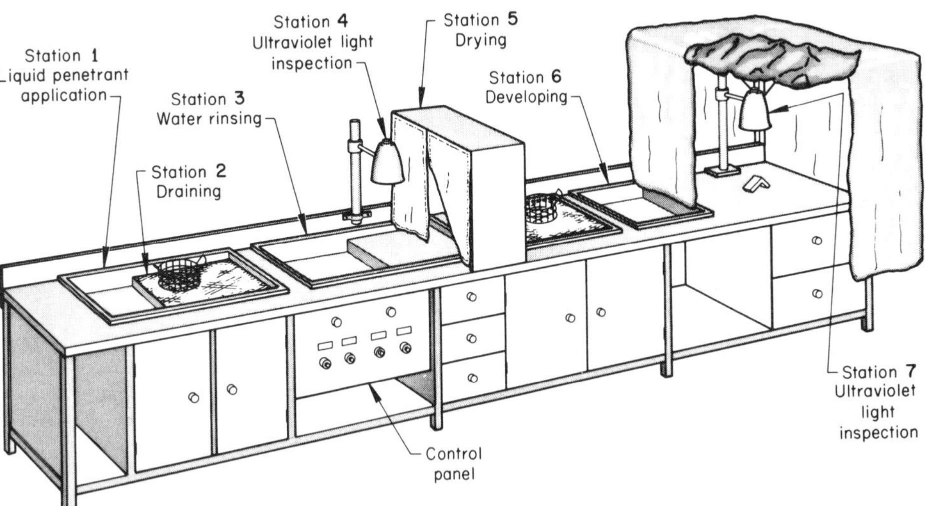

Fig. 14 Typical seven-station package equipment unit for inspecting workpieces by the water-washable fluorescent penetrant system

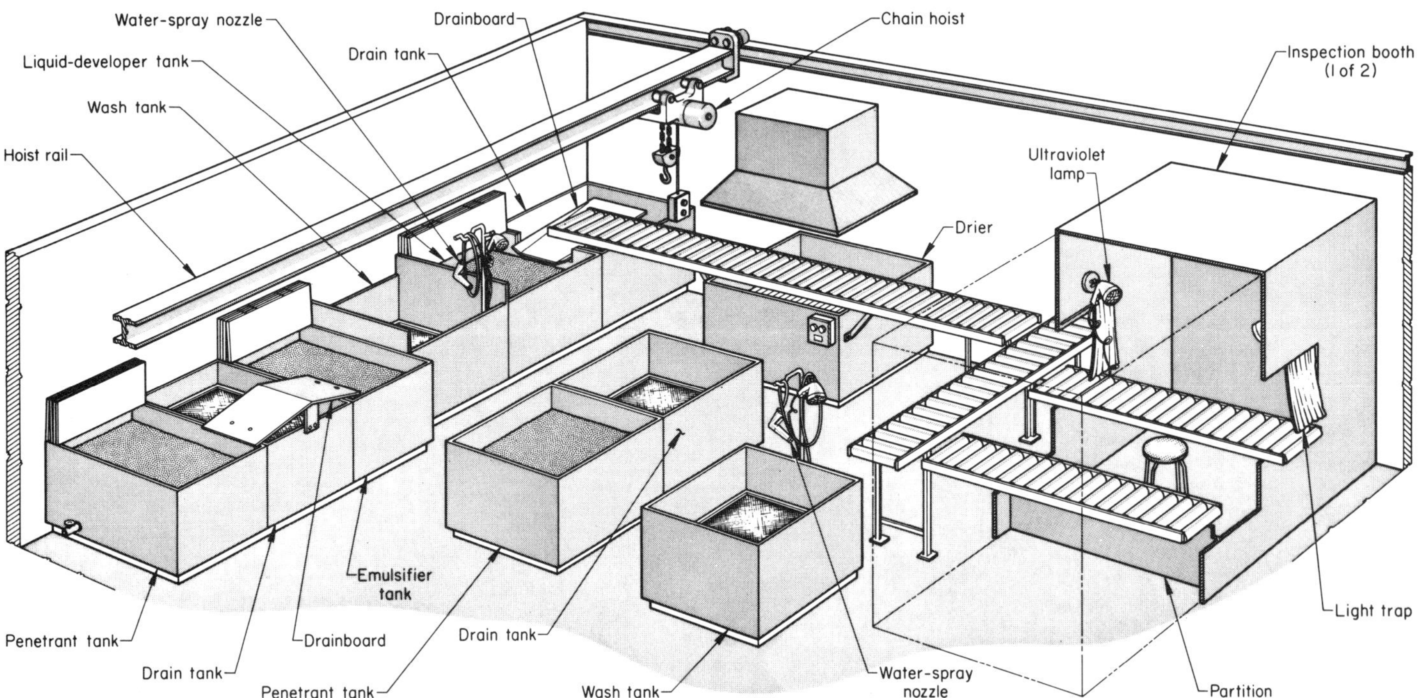

Fig. 15 Arrangement of equipment used in one foundry for the liquid penetrant inspection of a large variety of castings to rigid specifications. Many of the castings require handling by crane or roller conveyor.

eration of moving parts from station to station through the penetrant line. Properly trained and motivated operators will do an excellent job of processing and inspecting parts as well as controlling the process. There are, however, many situations in which manual processing simply cannot keep up with the production rates required or control the process properly.

The use of automated inspection systems, therefore, has become a significant factor in performing penetrant inspections of high-volume production parts. Modern automated penetrant inspection systems provide precise and repeatable process control, improved inspection reliability, increased productivity, and lower inspection costs.

Automated penetrant inspection systems incorporate programmable logic control (PLC) units, which are programmed to control the handling of parts through the system, to control the processing cycle precisely, and to monitor the functions at each processing station. Figures 16 to 18 show typical automated penetrant systems currently in use.

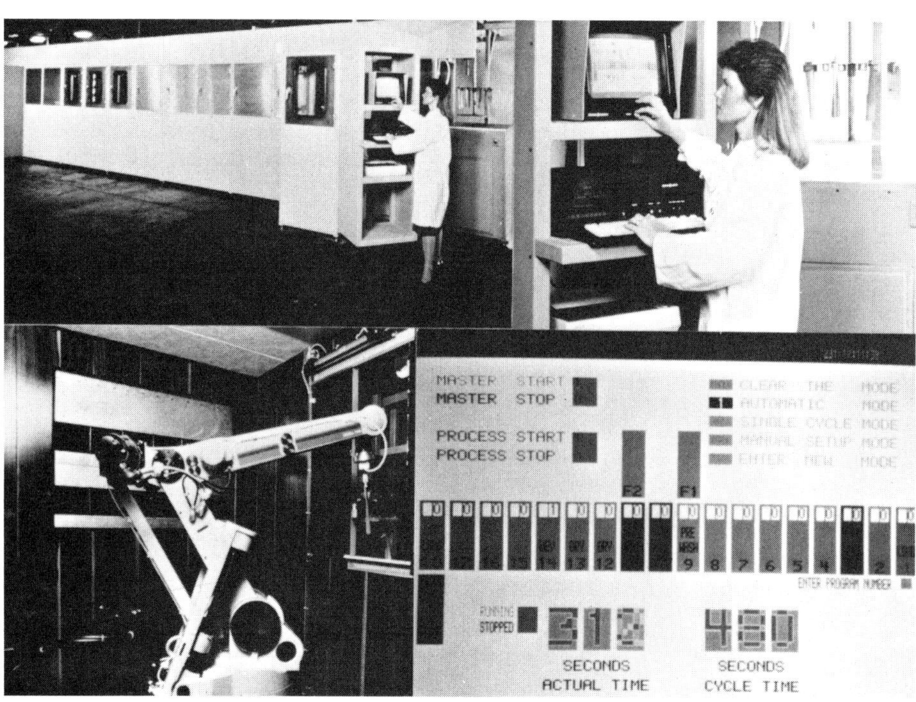

Fig. 16 Automated inspection equipment setup (upper left) with close-up of operator checking PLC panel (upper right), which includes a screen display (lower right). The setup incorporates material handling devices such as the robot shown (lower left) to transfer workpieces from station to station and to apply penetrants and other solutions needed to inspect components.

Precleaning

Regardless of the penetrant chosen, adequate precleaning of workpieces prior to penetrant inspection is absolutely necessary for accurate results. Without adequate removal of surface contamination, relevant indications may be missed because:

- The penetrant does not enter the flaw
- The penetrant loses its ability to identify the flaw because it reacts with something already in it
- The surface immediately surrounding the flaw retains enough penetrant to mask the true appearance of the flaw

Also, nonrelevant (false) indications may be caused by residual materials holding penetrants.

Cleaning methods are generally classified as chemical, mechanical, solvent, or any combination of these.

Fig. 17 Typical automated fluorescent penetrant inspection installation

Chemical cleaning methods include alkaline or acid cleaning, pickling or chemical etching, and molten salt bath cleaning.

Mechanical cleaning methods include tumbling, wet blasting, dry abrasive blasting, wire brushing, and high-pressure water or steam cleaning. Mechanical cleaning methods should be used with care because they often mask flaws by smearing adjacent metal over them or by filling them with abrasive material. This is more likely to happen with soft metals than with hard metals.

Solvent cleaning methods include vapor degreasing, solvent spraying, solvent wiping, and ultrasonic immersion using solvents. Probably the most common method is vapor degreasing. However, ultrasonic immersion is by far the most effective means of ensuring clean parts, but it can be a very expensive capital equipment investment.

Cleaning methods and their common uses are listed in Table 1. A major factor in the selection of a cleaning method is the type of contaminant to be removed and the type of alloy being cleaned. This is usually quite evident, but costly errors can be avoided by accurate identification of the contaminant. Before the decision is made to use a specific method, it is good practice to test the method on known flaws to ensure that it will not mask the flaws.

Equally important in choosing a cleaning method is knowledge of the composition of the workpiece being cleaned. For example, abrasive tumbling can effectively remove burrs from a machined steel casting and leave a surface that is fully inspectable. This method, however, is not suitable for aluminum or magnesium, because it smears these metals and frequently hides flaws. Particular care must be taken in selecting a cleaning method for workpieces fabricated from more than one alloy (brazed assemblies are notable examples); a chemical cleaning method to remove scale, stopoff material, or flux must be chosen carefully to ensure that neither the braze nor the components of the assembly will be attacked.

The surface finish of the workpiece must always be considered. When further processing is scheduled, such as machining or final polishing, or when a surface finish of 3.20 μm (125 μin.) or coarser is allowed, an abrasive cleaning method is frequently a good choice. Generally, chemical cleaning methods have fewer degrading effects on surface finish than mechanical methods (unless the chemical used is strongly corrosive to the material being cleaned). Steam cleaning and solvent cleaning rarely have any effect on surface finish.

Some materials are subject to delayed reactions as a result of improper cleaning. Two notable examples are high-strength steel and titanium. If it is ever necessary to chemically etch a high-strength steel workpiece, it should be baked at an appropriate temperature for a sufficient time to avoid hydrogen embrittlement. This should be done as soon after etching as possible but no later than 1 h. Titanium alloys can be subject to delayed cracking if they retain halogenated compounds and are then exposed to temperatures exceeding 480 °C (900 °F). Consequently, halogenated solvents should not be used for titanium and its alloys if their complete removal cannot be ensured.

Choice of cleaning method may be dictated by Occupational Safety and Health Administration and Environmental Protection Agency health and safety regulations. Quantities of materials that will be used, toxicity, filtering, neutralization and dispos-

Fig. 18 Automated inspection installation for the fluorescent penetrant inspection of large workpieces, such as castings. The installation incorporates a complex roller conveyor system.

Table 1 Applications of various methods of precleaning for liquid penetrant inspection

Method	Use
Mechanical methods	
Abrasive tumbling	Removing light scale, burrs, welding flux, braze stopoff, rust, casting mold, and core material; should not be used on soft metals such as aluminum, magnesium, or titanium
Dry abrasive grit blasting	Removing light or heavy scale, flux, stopoff, rust, casting mold and core material, sprayed coatings, carbon deposits—in general, any friable deposit. Can be fixed or portable
Wet abrasive grit blasting	Same as dry except, where deposits are light, better surface and better control of dimensions are required
Wire brushing	Removing light deposits of scale, flux, and stopoff
High-pressure water and steam	Ordinarily used with an alkaline cleaner or detergent; removing typical machine shop soils such as cutting oils, polishing compounds, grease, chips, and deposits from electrical discharge machining; used when surface finish must be maintained; inexpensive
Ultrasonic cleaning	Ordinarily used with detergent and water or with a solvent; removing adherent shop soil from large quantities of small parts
Chemical methods	
Alkaline cleaning	Removing braze stopoff, rust, scale, oils, greases, polishing material, and carbon deposits; ordinarily used on large articles where hand methods are too laborious; also used on aluminum for gross metal removal
Acid cleaning	Strong solutions for removing heavy scale; mild solutions for light scale; weak (etching) solutions for removing lightly smeared metal
Molten salt bath cleaning	Conditioning and removing heavy scale
Solvent methods	
Vapor degreasing	Removing typical shop soil, oil, and grease; usually employs chlorinated solvents; not suitable for titanium
Solvent wiping	Same as for vapor degreasing except a hand operation; may employ nonchlorinated solvents; used for localized low-volume cleaning

al techniques, and worker protection all are crucial factors.

Penetrant Inspection Processing Parameters

It is extremely important to understand the significance of adhering to the established process parameters for a given application. Failure to control the process parameters will affect the quality of the inspection. For example, excessive overwashing or overemulsification can remove the penetrant from the flaws; minimal washing or underemulsification can result in excessive background, which could mask the flaws and render them undetectable.

Processing time in each station, the equipment used, and other factors can vary widely, depending on workpiece size and shape, production quantities of similar workpieces, and required customer specifications for process parameters.

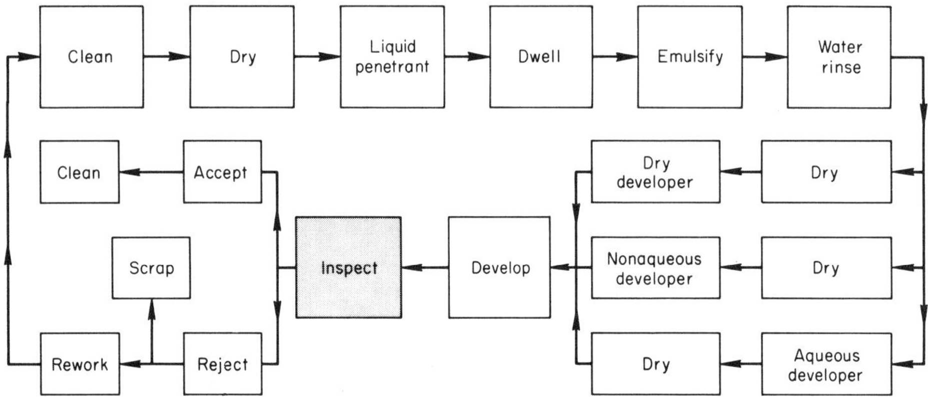

Fig. 19 Processing flow diagram for the postemulsifiable, method B, lipophilic liquid penetrant system

Postemulsifiable Method

The processing cycles for the postemulsifiable processes, method B (lipophilic) and method D (hydrophilic) are illustrated in the processing flow diagrams (Fig. 19 and 20, respectively). The major difference between the two methods, as described below, is the additional prerinse step utilized in method D.

Application of Penetrant. Workpieces should be thoroughly and uniformly coated with penetrant by flowing, brushing, swabbing, dipping, or spraying. Small workpieces requiring complete surface inspection are usually placed in a basket and dipped in the penetrant. Larger workpieces are usually brushed or sprayed. Electrostatic spray application is also very effective and economical. After the workpiece has been coated with a light film of penetrant, it should be positioned so that it can drain and so that excess penetrant cannot collect in pools. Workpieces should not be submerged during the entire penetration dwell time. Heating the workpiece is also not necessary or recommended, because certain disadvantages can occur, such as volatilization of the penetrant, difficulty in washing, and a decrease in fluorescence.

Dwell Time. After the penetrant has been applied to the workpiece surface, it should be allowed to remain long enough for complete penetration into the flaws. Dwell time will vary, depending mainly on the size of the defects sought, cleanliness of the workpiece, and sensitivity and viscosity of the penetrant. In most cases, however, a minimum of 10 min and a maximum of 30 min is adequate for both fluorescent- and visible-penetrant types. A lengthy dwell time could cause the penetrant to begin drying on the surface, resulting in difficult removal. If drying does occur, it is necessary to reapply the penetrant to wet the surface and then begin the removal steps. Recommendations from the penetrant supplier will help establish the time, but experimentation will determine optimum dwell time.

Prerinse. When using method D (hydrophilic), a coarse waterspray prerinse is needed to assist in penetrant removal and to reduce contamination of the emulsifier. A coarse water spray is recommended, using a pressure of 275 to 345 kPa (40 to 50 psi). The prerinse water temperature should be 10 to 40 °C (50 to 100 °F). The prerinse time should be kept to a minimum (that is, 30 to 90 s) because the purpose is to remove excess penetrant so that the emulsifier does not become contaminated quickly.

Emulsifier Application. It is very important that all surfaces of the workpiece be coated with the emulsifier at the same time. Small workpieces are dipped individually or in batches in baskets or on racks, whichever

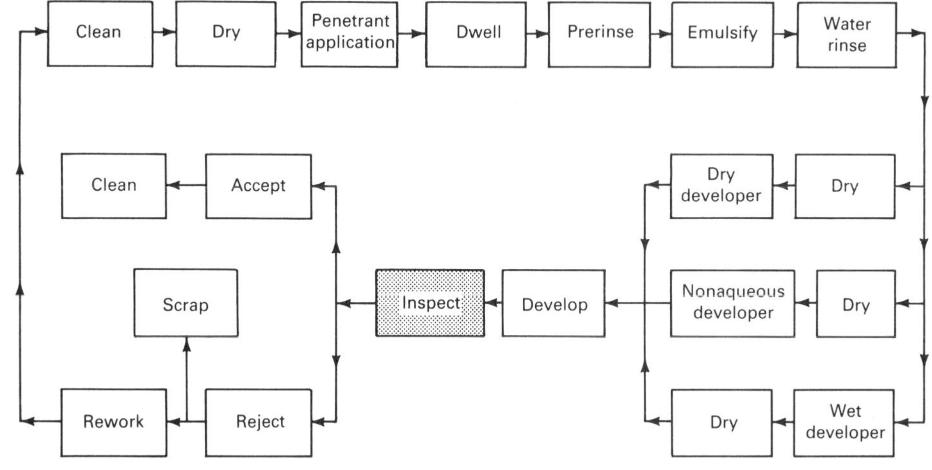

Fig. 20 Processing flow diagram for the postemulsifiable, method D, hydrophilic liquid penetrant system

is the most convenient. For large workpieces, methods must be devised to achieve the fastest possible coverage; two methods often used are spraying or immersing. Localized emulsification of large workpieces can be achieved by spraying. The temperature of the emulsifier is not extremely critical, but a range of 20 to 30 °C (70 to 90 °F) is preferred.

Emulsification Time. The length of time the emulsifier is allowed to remain on the workpiece and in contact with the penetrant is the emulsification time and depends mainly on the type of emulsifier employed, its concentration, and on the surface condition of the workpieces. Recommendations by the manufacturer of the emulsifier can serve as guidelines, but the optimum time for a specific workpiece must be established by experimentation. The surface finish, size, and composition of the workpiece will determine more precisely the choice of emulsifier and emulsification time. Emulsification time ranges from approximately 30 s to 3 min and is directly related to the concentration of the emulsifier. If emulsification time is excessive, penetrant will be removed from the flaws, making detection impossible.

Rinsing. For all methods, removing the penetrant from the workpiece is probably the most important step in obtaining reproducible results. If penetrant removal is performed properly, penetrant will be stripped from the surface and will remain only in the flaws. More variability in individual technique enters into this particular phase of inspection than any other step. Therefore, removal must be performed with the same sequence of operations time after time if results are to be reproducible. This is especially important when inspecting for tight or shallow flaws.

Rinse time should be determined experimentally for specific workpieces; it usually varies from 10 s to 2 min. For spray rinsing, water pressure should be constant. A pressure of about 275 kPa (40 psi) is desirable; too much pressure may remove penetrants from the flaws. A coarse water spray is recommended and can be assisted with air (the combined water and air pressure should not exceed the pressure recommended for water alone). Water temperature should be maintained at a relatively constant level. Most penetrants can be removed effectively with water in a range of 10 to 40 °C (50 to 100 °F).

Drying is best done in a recirculating hot-air drier that is thermostatically controlled. The temperature in the drier is normally between 65 and 95 °C (150 and 200 °F). The temperature of the workpieces should not be permitted to exceed 70 °C (160 °F). Workpieces should not remain in the drier any longer than necessary; drying is normally accomplished within a few minutes. Excessive drying at high temperatures can impair the sensitivity of the inspection. Because drying time will vary, the exact time should be determined experimentally for each type of workpiece.

Developing depends on the form of developer to be used. Various types of developers are discussed below.

Dry-developer powder (form A) is applied after the workpiece has been dried and can be applied in a variety of ways. The most common is dusting or spraying. Electrostatic spray application is also very effective. In some cases, application by immersing the workpiece into the dry powder developer is permissible. For simple applications, especially when only a portion of the surface of a large part is being inspected, applying with a soft brush is often adequate. Excess developer can be removed from the workpiece by a gentle air blast (140 kPa, or 20 psi, maximum) or by shaking or gentle tapping. Whichever means of application is chosen, it is important that the workpiece be completely and evenly covered by a fine film of developer.

Water-soluble developer (form B) is applied just after the final wash and immediately prior to drying by dip, flow-on, or spray techniques. No agitation of the developer bath is required. Removal of the developer coating from the surface of the workpiece is required and easily accomplished because the dried developer coating is water soluble and therefore completely removable by a water rinse.

Water-suspendible developer (form C) is applied just after the final wash and immediately before drying. Dip, flow-on, and spray are common methods of application. Care must be taken to agitate the developer thoroughly so that all particles are in suspension; otherwise, control of the concentration of the applied coating is impossible. Removal of the water-suspendible developer can best be achieved by water spray rinsing. If allowed to remain indefinitely on the workpiece, the developer can become difficult to remove.

Solvent-suspendible nonaqueous developer (form D) is always applied after drying by spraying, either with aerosol containers or by conventional or electrostatic methods. Proper spraying produces a thin, uniform layer that is very sensitive in producing either fluorescent or red visible indications. The volatility of the solvent makes it impractical to use in open tanks. Not only would there be solvent loss, reducing the effectiveness of the developer, but there would also be a hazardous vapor condition. Dipping, pouring, and brushing are not suitable for applying solvent-suspendible developer.

Developing Time. In general, 10 min is the recommended minimum developing time regardless of the developer form used. The developing time begins immediately after application of the developer. Excessive developing time is seldom necessary and usually results in excessive bleeding of indications, which can obscure flaw delineation.

Inspections. After the prescribed development time, the inspection should begin. The inspection area should be properly darkened for fluorescent penetrant inspection. Recommended black light intensity is 1000 to 1600 $\mu W/cm^2$. The intensity of the black light should be verified at regular intervals by the use of a suitable black light meter such as a digital radiometer. The intensity of the black light should be allowed to warm up prior to use—generally for about 10 min. The inspector should allow time for adapting to darkness; a 1-min period is usually adequate. White light intensity should not exceed 20 lx (2 ftc) to ensure the best inspection environment.

Visible-penetrant systems provide vivid red indications that can be seen in visible light. Lighting intensity should be adequate to ensure proper inspection; 320 to 540 lx (30 to 50 ftc) is recommended. Lighting intensity should be verified at regular intervals by the use of a suitable white light

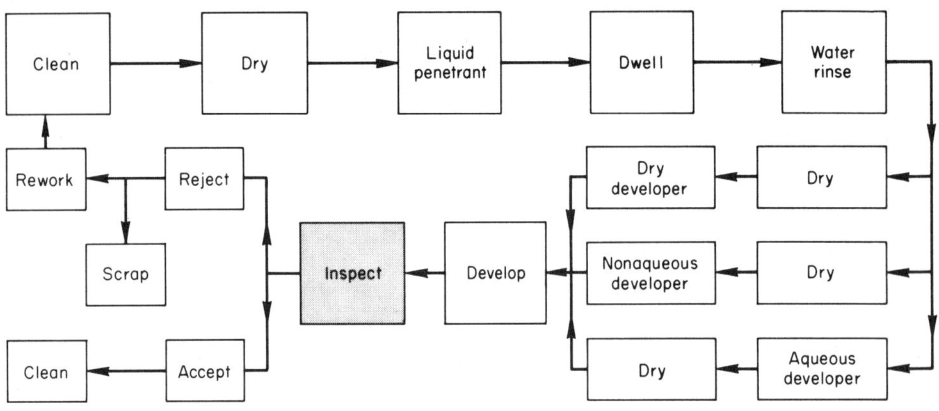

Fig. 21 Processing flow diagram for the water-washable liquid penetrant system

meter such as a digital radiometer. Detailed information on inspection techniques is available in the sections "Inspection and Evaluation" and "Specifications and Standards" in this article.

Water-Washable Method

As indicated by the flow diagram in Fig. 21, the processing cycle for the water-washable method is similar to that for the postemulsifiable method. The difference lies in the penetrant removal step. As discussed in the section "Materials Used in Penetrant Inspection" in this article, the water-washable penetrants have a built-in emulsifier, thus eliminating the need for an emulsification step. One rinse operation is all that is required, and the washing operation should be carefully controlled because water-washable penetrants are susceptible to overwashing.

Rinse time should be determined experimentally for a specific workpiece; it usually varies from 10 s to 2 min. The best practical way of establishing rinse time is to view the workpiece under a black light while rinsing and washing only until the fluorescent background is removed to a satisfactory degree. On some applications, such as castings, an immersion rinse followed by a final spray rinsing is desirable to remove tenacious background fluorescence. This technique, however, must be very carefully controlled to ensure that overwashing does not occur.

For spray rinsing, a nominal water pressure of 140 to 275 kPa (20 to 40 psi) is recommended; too much pressure can result in overwashing, that is, the removal of penetrant from within flaws. Hydro-air spray guns can be used. The air pressure, however, should not exceed 170 kPa (25 psi). The temperature of the water should be controlled to 10 to 40 °C (50 to 100 °F). Drying, developing, and inspection process parameters are the same as the postemulsifiable method process parameters described in the section "Postemulsifiable Method" in this article.

Solvent-Removable Method

The basic sequence of operations for the solvent-removable penetrant system is generally similar to that followed for the other methods. A typical sequence is shown by the flow diagram in Fig. 22. A notable difference is that with the solvent-removable method the excess penetrant is removed by wiping with clean, lint-free material moistened with solvent. It is important to understand that flooding the workpiece to remove excess surface penetrant will also dissolve the penetrant from within the flaws.

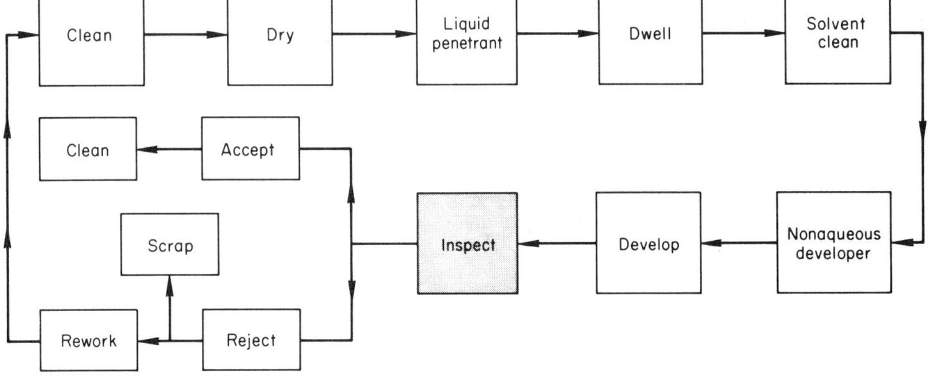

The processing parameters for the use of developer are the same as those described above for the postemulsifiable method. Dry-powder developers, however, are not recommended for use with the visible solvent-removable penetrant method.

Postcleaning

Some residue will remain on workpieces after penetrant inspection is completed. In many cases, this residue has no deleterious effects in subsequent processing or in service. There are, however, instances in which postcleaning is required. Residues can result in the formation of voids during subsequent welding or unwanted stopoff in brazing, in the contamination of surfaces (which can cause trouble in heat treating), or in unfavorable reactions in chemical processing operations.

Drastic chemical or mechanical methods are seldom required for postcleaning. When justified by the volume of work, an emulsion cleaning line is effective and reasonable in cost. In special circumstances, ultrasonic cleaning may be the only satisfactory way of cleaning deep crevices or small holes. However, solvents or detergent-aided steam or water is almost always sufficient. The use of steam with detergent is probably the most effective of all methods. It has a scrubbing action that removes developers, the heat and detergent remove penetrants, it leaves a workpiece hot enough to promote rapid, even drying, and it is harmless to nearly all materials. Vapor degreasing is very effective for removing penetrants, but it is practically worthless for removing developers. It is frequently used in combination with steam cleaning. If this combination is used, the steam cleaning should always be done first because vapor degreasing bakes on developer films.

Where conditions do not warrant or permit permanent cleaning installations, hand wiping with solvents is effective. Dried developer films can be brushed off, and residual penetrants can be rinsed off by solvent spraying or wiped off with a solvent-dampened cloth.

Quality Assurance of Penetrant Inspection Materials

It is important to provide the controls necessary to ensure that the penetrant materials and equipment are operating at an acceptable level of performance. The frequency of the required checks should be based on a facility operating for a full, one-shift operation daily. In general, it is good practice to check the overall system performance on a daily basis. This check

Fig. 22 Processing flow diagram for the solvent-removable liquid penetrant system

should be performed by processing a known defect standard through the line, using appropriate processing parameters and comparing the indications thus obtained to those obtained with fresh, unused penetrant material samples. When the performance of the in-use materials falls below that of the unused materials, the in-use material quality should be checked with the appropriate tests (as described below) and corrected prior to conducting any further penetrant inspection.

Key quality assurance tests to be periodically conducted on in-use penetrants, emulsifiers, and developers are listed in Table 2. Also listed are the intervals at which the light sources and the overall system performance should be checked.

Military Standard 6866 specifies the specific test procedure to use for the tests defined in Table 2. Penetrants applied by spray application from sealed containers are not likely to be exposed to the same working environment as with open dip tanks and are therefore not required to be tested as defined in Table 2 unless contamination is suspected.

Maintenance of Materials

With constant open-tank use, penetrant materials are inherently subject to potential deterioration. Such factors as evaporation losses and contamination from various sources can contribute to deterioration. It is essential, therefore, to monitor the condition of these materials as described in Table 2.

The evaporation of the volatile constituents of penetrants can alter their chemical and performance characteristics, thus resulting in changes in inherent brightness, removability, and sensitivity. Liquid penetrant materials qualified to MIL-I-25135D (and subsequent revisions) have a flash point requirement of a minimum of 95 °C (200 °F) (per Pensky Martens flash point test procedure), assuring the minimization of evaporation losses.

The contamination of water-washable penetrant with water is the most frequent source of difficulty. When present beyond a critical percentage, this contamination will render the penetrant tank useless. For postemulsifiable penetrants, water contamination is not as critical a problem, because water is usually not miscible with postemulsifiable penetrants and will separate from the penetrant, which can then be subsequently removed. Water contamination can be minimized by implementing and following proper processing procedures.

It is important to recognize that acid contamination (carryover from precleaning) will render fluorescent penetrants ineffective. Acid contamination changes the consistency of the penetrant and damages or destroys the fluorescent dye.

Dust, dirt and lint, and similar foreign materials get into the penetrant in the ordinary course of shop usage. These contaminants do no particular harm unless present to the extent that the bath is scummy with floating or suspended foreign material. Reasonable care should be taken to keep the penetrant clean. Workpieces containing adhering sand and dirt from the shop floor should be cleaned before being dipped into the penetrant.

Table 2 Intervals at which solutions, light sources, and system performance should be checked

Test	Minimum test frequency	Requirement
Penetrants		
Fluorescent brightness	Quarterly	Not less than 90% of reference standard
Sensitivity	Monthly	Equal to reference standard
Removability (method A water wash only)	Monthly	Equal to reference standard
Water content (method A water wash penetrant only)	Monthly	Not to exceed 5%
Contamination	Weekly	No noticeable tracers
Emulsifiers		
Removability	Weekly	Equal to reference standard
Water content (method B, lipophilic)	Monthly	Not to exceed 5%
Concentration (method D, hydrophilic)	Weekly	Not greater than 3% above initial concentration
Contamination	Weekly	No noticeable tracers
Developers		
Dry-developer form	Daily	Must be fluffy, not caked
Contamination	Daily	Not more than ten fluorescent specks observed in a 102 mm (4 in.) circle of sample
Aqueous (soluble and suspended) developer		
Wetting/coverage	Daily	Must be uniform/wet and must coat part
Contamination	Daily	Must not show evidence of fluorescence contaminates
Concentration	Weekly	Concentration shall be maintained as specified
Black lights	Daily	Minimum 1000 $\mu W/cm^2$ at 381 mm (15 in.)
White light	Weekly	Minimum 2200 lx (200 ftc)
System performance	Daily	Must equal reference standards

Contamination of the emulsifier must also be considered. Method B, lipophilic emulsifiers inherently become contaminated by penetrant through the normal processing of parts coated with penetrant being dipped into the emulsifier. It is imperative, therefore, that the lipophilic emulsifier have a high tolerance (that is, 10%) for penetrant contamination. Water contamination of the lipophilic emulsifier is always a potential problem due to the nature of the process. Generally, 5% water contamination can be tolerated.

Method D, hydrophilic emulsifiers are not normally subject to appreciable amounts of penetrant contamination, mainly because of the prerinse processing step, which removes most of the excess surface penetrant before emulsification. Because hydrophilic emulsifiers are water based, water contamination is not a problem, except for the fact that the bath concentration must be maintained at the prescribed limits.

In general, emulsifiers that become severely contaminated will not properly emulsify the surface penetrant on the parts. Periodic monitoring is essential.

Developer must also be maintained to ensure proper performance. Contamination of the dry-powder developer with water or moisture in the air can result in caking. Dry developers must remain fluffy and free flowing if they are to perform properly. In addition, contamination from the fluorescent penetrant must not occur. Fluorescent specks in the developer powder could be misinterpreted as an indication. Wet developer (soluble or suspendible) must not become contaminated with penetrant or any contaminant that could affect its ability to wet and evenly cover the workpiece.

Training and Certification of Personnel

The apparent simplicity of the penetrant method is deceptive. Very slight variations in performing the penetrant process and the inspection can invalidate the inspection results by failing to indicate all flaws. Therefore, many companies require that penetrant inspection be conducted only by trained and certified personnel. Minimum requirements for personnel training and certification are described by various military and industry specifications (such as MIL-STD-410 and ASNT SNT-TC-1A). The following are examples of the most commonly followed training programs; however, specific customer training requirements are usually defined within the contract.

Training is minimal for level I penetrant inspection operators (personnel responsible for the processing). However, the penetrant process must be correctly performed to ensure accurate inspection. Operator training consists of the satisfactory completion of a period of on-the-job training, as deter-

Table 3 Common types, locations, and characteristics of flaws or discontinuities revealed by liquid penetrant inspection

Type	Locations	Characteristics
Relevant indications		
Shrinkage cracks	Castings (all metals)—on flat surfaces	Open
Hot tears	Castings (all metals)—at inside corners	Open
Cold shuts	Castings (all metals)—at changes in cross section	Tight, shallow
Folds	Castings (all metals)—anywhere	Tight, shallow
Inclusions	Castings, forgings, sheet, bar—anywhere	Tight, shallow, intermittent
Microshrinkage pores	Castings—anywhere	Spongy
Laps	Forgings, bar—anywhere	Tight, shallow
Forging cracks	Forgings—at inside or outside corners and at changes in cross section	Tight or open
Pipe	Forgings, bar—near geometric center	Irregular shape
Laminations	Sheet—at edges	Tight or open
Center bead cracks	Welds—at center of reinforcement	Tight or open
Cracks in heat-affected zone	Welds—at edge of reinforcement	Tight or open
Crater cracks	Welds—at end of bead	Star-shaped
Porosity	Castings, welds	Spherical
Grinding cracks	Any hard metal—ground surfaces	Tight, shallow, random
Quench cracks	Heat treated steel	Tight to open, oxidized
Stress-corrosion cracks	Any metal	Tight to open; may show corrosion
Fatigue cracks	Any metal	Tight
Nonrelevant indications(a)		
Weld spatter	Arc welds	Spherical or surface
Incomplete penetration	Fillet welds	Open, full weld length
Surface expulsion	Resistance welds	Raised metal at weld edge
Scuff marks	Seam welds	Surface of seam welds
Press-fit interface	Press fits	Outlines press fit
Braze runoff	Brazed parts	Edge of excess braze
Burrs	Machined parts	Bleeds heavily
Nicks, dents, scratches	All parts	Visible without penetrant aids

(a) These may be prohibited flaws, but are usually considered nonrelevant in penetrant testing.

mined by immediate supervision, conducted under the guidance of a certified level I inspector.

Training for level II inspectors (personnel responsible for the inspection and evaluation) is more extensive than that for the level I operators. Training usually consists of 40 h of formal training, followed by several weeks of on-the-job training under the supervision of a designated trainer, usually a certified level II operator.

Certification. Personnel of sufficient background and training in the principles and procedures of penetrant inspection are usually certified by the successful completion of a practical test, which demonstrates their proficiency in penetrant techniques, and a written test, which documents their knowledge of penetrant inspection. Certified personnel are also normally required to pass a periodic eye examination, which includes a color-vision test. Certification can be obtained on-site through a certified level III inspector who may be with an outside source contracted to certify personnel or a company employee who has been certified as a level III inspector by the appropriate agency.

Inspection and Evaluation

After the penetrant process is completed, inspection and evaluation of the workpiece begin. Table 3 lists the more common types of flaws that can be found by penetrant inspection, together with their likely locations and their characteristics.

Inspection Tools. An inspector must have tools that are capable of providing the required accuracy. These tools usually include suitable measuring devices, a flashlight, small quantities of solvent, small quantities of dry developers or aerosol cans of nonaqueous wet developers, pocket magnifiers ranging from 3 to 10×, and a suitable black light for fluorescent penetrants or sufficient white light for visible penetrants. Photographic standards or workpieces that have specific known flaws are sometimes used as inspection aids.

A typical inspection begins with an overall examination to determine that the workpiece has been properly processed and is in satisfactory condition for inspection. Inspection should not begin until the wet developers are completely dry. If developer films are too thick, if penetrant bleedout appears excessive, or if the penetrant background is excessive, the workpiece should be cleaned and reprocessed. When the inspector is satisfied that the workpiece is inspectable, it is examined according to a specified plan to be sure no areas have been missed. An experienced inspector can readily determine which indications are within acceptable limits and which ones are not. The inspector then measures all other indications. If the length or diameter of an indication exceeds allowable limits, it must be evaluated. One of the most common and accurate ways of measuring indications is to lay a flat gage of the maximum acceptable dimension of discontinuity over the indication. If the indication is not completely covered by the gage, it is not acceptable.

Evaluation. Each indication that is not acceptable should be evaluated. It may actually be unacceptable, it may be worse than it appears, it may be false, it may be real, but nonrelevant, or it may actually be acceptable upon closer examination. One common method of evaluation includes the following steps:

- Wipe the area of the indication with a small brush or clean cloth that is dampened with a solvent
- Dust the area with a dry developer or spray it with a light coat of nonaqueous developer
- Remeasure under lighting appropriate for the type of penetrant used

If the discontinuity originally appeared to be of excessive length because of bleeding of penetrant along a scratch, crevice, or machining mark, this will be evident to a trained eye. Finally, to gain maximum assurance that the indication is properly interpreted, it is good practice to wipe the surface again with solvent-dampened cotton and examine the indication area with a magnifying glass and ample white light. This final evaluation may show that the indication is even larger than originally measured, but was not shown in its entirety because the ends were too tight to hold enough penetrant to reach the surface and become visible.

Disposition of Unacceptable Workpieces. A travel ticket will usually accompany each workpiece or lot of workpieces. Provision should be made on this ticket to indicate the future handling of unacceptable material, that is, scrapping, rework, repair, or review board action. There is often room on such tickets for a brief description of the indication. More often, indications are identified directly on the workpiece by circling them with some type of marking that is harmless to the material and not easily removed by accident, but removable when desired.

Reworking an unacceptable flaw is often allowable to some specified limit; indications can be removed by sanding, grinding, chipping, or machining. Repair welding is sometimes needed; in this case, the indication should be removed as in reworking before it is repair welded, or welding may move the flaw to a new location. In addition, it is imperative that all entrapped penetrant be removed prior to repair welding, because entrapped penetrant is likely to initiate a new flaw. Verification that the indication and the entrapped penetrant have been removed is required.

Table 4 Partial listing of standards and specifications for liquid penetrant inspection

Number	Title or explanation of standard or specification
ASTM standards	
ASTM-E-165	Standard Practice for Liquid-Penetrant Inspection Method
ASTM-E-270	Standard Definitions of Terms Relating to Liquid-Penetrant Inspection
ASTM-E-1208	Standard Method for Fluorescent Liquid-Penetrant Examination Using the Lipophilic Post-Emulsification Process
ASTM-E-1209	Standard Method for Fluorescent-Penetrant Examination Using the Water-Washable Process
ASTM-E-1210	Standard Method for Fluorescent-Penetrant Examination Using the Hydrophilic Post-Emulsification Process
ASTM-E-1219	Standard Method for Fluorescent-Penetrant Examination Using the Solvent-Removable Process
ASTM-E-1220	Standard Method for Visible-Penetrant Examination Using the Solvent-Removable Process
ASTM-E-1135	Standard Test Method for Comparing the Brightness of Fluorescent Penetrants
ASTM-D-2512	Compatibility of Materials with Liquid Oxygen (Impact-Sensitivity Threshold Technique)
Test for AMS-SAE specifications	
AMS-2647	Fluorescent Penetrant Inspection—Aircraft and Engine Component Maintenance
ASME specifications	
ASME-SEC V	ASME Boiler and Pressure Vessel Code Section V, Article 6
U.S. military and government specifications	
MIL-STD-6866	Military Standard Inspection, Liquid Penetrant
MIL-STD-410	Nondestructive Testing Personnel Qualifications & Certifications
MIL-I-25135	Inspection Materials, Penetrant
MIL-I-25105	Inspection Unit, Fluorescent Penetrant, Type MA-2
MIL-I-25106	Inspection Unit, Fluorescent Penetrant, Type MA-3
MIL-STD-271 (Ships)	Nondestructive Testing Requirements for Metals

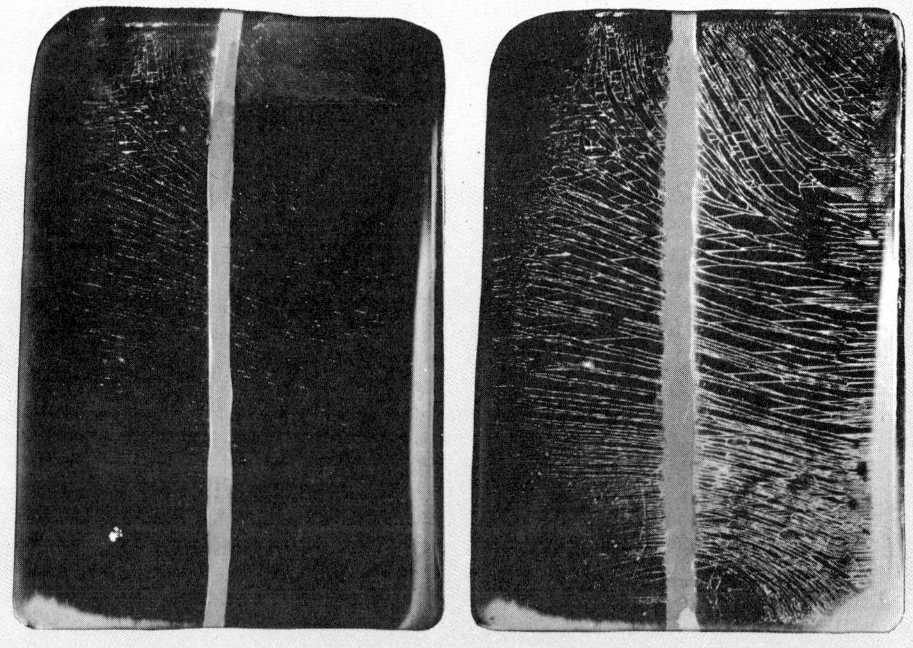

Fig. 23 Comparison of indications on chromium-cracked panels developed with water-washable liquid penetrants of low sensitivity (panel at left) and high sensitivity (panel at right)

Because reworking is usually required, it is good practice to finish it off with moderately fine sanding, followed by chemical etching to remove smeared metal. All traces of the etching fluid should be rinsed off, and the area should be thoroughly dried before reprocessing for reinspection. Reprocessing can be the same as original processing for penetrant inspection, or can be done locally by applying the materials with small brushes or swabs.

False and Nonrelevant Indications. Because penetrant inspection provides only indirect indications or flaws, it cannot always be determined at first glance whether an indication is real, false, or nonrelevant. A real indication is caused by an undesirable flaw, such as a crack. A false indication is an accumulation of penetrant not caused by a discontinuity in the workpiece, such as a drop of penetrant left on the workpiece inadvertently. A nonrelevant indication is an entrapment of penetrant caused by a feature that is acceptable even though it may exceed allowable indication lengths, such as a press-fit interface.

Specifications and Standards

It has not been practical to establish any type of universal standardization, because of the wide variety of components and assemblies subjected to penetrant inspection, the differences in the types of discontinuities common to them, and the differences in the degree of integrity required. Generally, quality standards for the types of discontinuities detected by penetrant inspection are established by one or more of the following methods:

- Adoption of standards that have been successfully used for similar workpieces
- Evaluation of the results of penetrant inspection by destructive examination
- Experimental and theoretical stress analysis

Specifications. Normally, a specification is a document that delineates design or performance requirements. A specification should include the methods of inspection and the requirements based on the inspection or test procedure. With penetrant inspection, this becomes difficult. Too often the wording in quality specifications is ambiguous and meaningless, such as "workpieces shall be free from detrimental defects" or "workpieces having questionable indications shall be held for review by the proper authorities."

Specifications applicable to penetrant inspection are generally divided into two broad categories: those involving materials and equipment, and those concerning methods and standards. There are, however, several standards and specifications that are in common use; some of these are listed in Table 4. Because the equipment used for penetrant inspection covers such a broad scope, that is, ranging from small dip-tank setups to large automated installations, most emphasis in standards and specifications has been placed on the materials used in this inspection process.

Control Systems. In conjunction with the specifications listed in Table 4, several methods and several types of standards are used to check the effectiveness of liquid penetrants. One of the oldest and most frequently used methods involves chromium-cracked panels, which are available in sets containing fine, medium, and coarse cracks. Many other types of inspection standards have been produced—often for specific indications needed for a unique application. A comparison of indications from two water-washable penetrants of different sensitivity that were applied to a chromium-cracked panel containing fine cracks is shown in Fig. 23.

Acceptance and rejection standards for liquid penetrant inspection are usually established for each individual item or group of items by the designer. In most cases, acceptance and rejection standards are based on experience with similar items, the principal factor being the degree of integrity required. At one extreme, for certain noncritical items, the standard may permit some specific types of discontinuities all over the workpiece or in specified areas. Inspection is often applied only on a sampling basis for noncritical items. At the opposite extreme, items are subjected to 100% inspection, and requirements are extremely stringent to the point of defining the limitations on each specific area.

Magnetic Particle Inspection

Revised by Art Lindgren, Magnaflux Corporation

MAGNETIC PARTICLE INSPECTION is a method of locating surface and subsurface discontinuities in ferromagnetic materials. It depends on the fact that when the material or part under test is magnetized, magnetic discontinuities that lie in a direction generally transverse to the direction of the magnetic field will cause a leakage field to be formed at and above the surface of the part. The presence of this leakage field, and therefore the presence of the discontinuity, is detected by the use of finely divided ferromagnetic particles applied over the surface, with some of the particles being gathered and held by the leakage field. This magnetically held collection of particles forms an outline of the discontinuity and generally indicates its location, size, shape, and extent. Magnetic particles are applied over a surface as dry particles, or as wet particles in a liquid carrier such as water or oil.

Ferromagnetic materials include most of the iron, nickel, and cobalt alloys. Many of the precipitation-hardening steels, such as 17-4 PH, 17-7 PH, and 15-4 PH stainless steels, are magnetic after aging. These materials lose their ferromagnetic properties above a characteristic temperature called the Curie point. Although this temperature varies for different materials, the Curie point for most ferromagnetic materials is approximately 760 °C (1400 °F).

Method Advantages and Limitations

Nonferromagnetic materials cannot be inspected by magnetic particle inspection. Such materials include aluminum alloys, magnesium alloys, copper and copper alloys, lead, titanium and titanium alloys, and austenitic stainless steels.

In addition to the conventional magnetic particle inspection methods described in this article, there are several proprietary methods that employ ferromagnetic particles on a magnetized testpiece. Three of these methods—magnetic rubber inspection, magnetic printing, and magnetic painting—are described in the Appendix to this article.

Applications. The principal industrial uses of magnetic particle inspection are final inspection, receiving inspection, in-process inspection and quality control, maintenance and overhaul in the transportation industries, plant and machinery maintenance, and inspection of large components.

Although in-process magnetic particle inspection is used to detect discontinuities and imperfections in materials and parts as early as possible in the sequence of operations, final inspection is needed to ensure that rejectable discontinuities or imperfections detrimental to the use or function of the part have not developed during processing. During receiving inspection, semifinished purchased parts and raw materials are inspected to detect any initially defective material. Magnetic particle inspection is extensively used on incoming rod and bar stock, forging blanks, and rough castings.

The transportation industries (truck, railroad, and aircraft) have planned overhaul schedules at which critical parts are magnetic particle inspected for cracks. Planned inspection programs are also used in keeping plant equipment in operation without breakdowns during service. Because of sudden and severe stress applications, punch-press crankshafts, frames, and flywheels are vulnerable to fatigue failures. A safety requirement in many plants is the inspection of crane hooks; fatigue cracks develop on the work-hardened inside surfaces of crane hooks where concentrated lifting loads are applied. The blading, shaft, and case of steam turbines are examined for incipient failure at planned downtimes.

Advantages. The magnetic particle method is a sensitive means of locating small and shallow surface cracks in ferromagnetic materials. Indications may be produced at cracks that are large enough to be seen with the naked eye, but exceedingly wide cracks will not produce a particle pattern if the surface opening is too wide for the particles to bridge.

Discontinuities that do not actually break through the surface are also indicated in many cases by this method, although certain limitations must be recognized and understood. If a discontinuity is fine, sharp, and close to the surface, such as a long stringer of nonmetallic inclusions, a clear indication can be produced. If the discontinuity lies deeper, the indication will be less distinct. The deeper the discontinuity lies below the surface, the larger it must be to yield a readable indication and the more difficult the discontinuity is to find by this method.

Magnetic particle indications are produced directly on the surface of the part and constitute magnetic pictures of actual discontinuities. There is no electrical circuitry or electronic readout to be calibrated or kept in proper operating condition. Skilled operators can sometimes make a reasonable estimate of crack depth with suitable powders and proper technique. Occasional monitoring of field intensity in the part is needed to ensure adequate field strength.

There is little or no limitation on the size or shape of the part being inspected. Ordinarily, no elaborate precleaning is necessary, and cracks filled with foreign material can be detected.

Limitations. There are certain limitations to magnetic particle inspection the operator must be aware of; for example, thin coatings of paint and other nonmagnetic coverings, such as plating, adversely affect the sensitivity of magnetic particle inspection. Other limitations are:

- The method can be used only on ferromagnetic materials
- For best results, the magnetic field must be in a direction that will intercept the principal plane of the discontinuity; this sometimes requires two or more sequential inspections with different magnetizations
- Demagnetization following inspection is often necessary
- Postcleaning to remove remnants of the magnetic particles clinging to the surface may sometimes be required after testing and demagnetization
- Exceedingly large currents are sometimes needed for very large parts
- Care is necessary to avoid local heating and burning of finished parts or surfaces at the points of electrical contact
- Although magnetic particle indications are easily seen, experience and skill are sometimes needed to judge their significance

Specifications and standards for magnetic particle inspection have been devel-

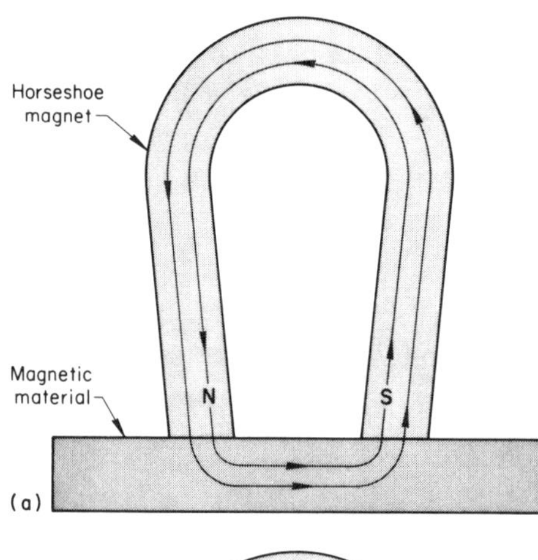

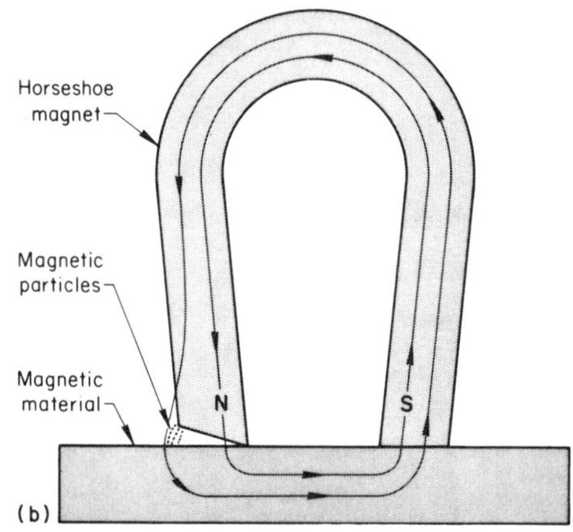

Fig. 1 Schematics of magnetic lines of force. (a) Horseshoe magnet with a bar of magnetic material across poles, forming a closed, ringlike assembly, which will not attract magnetic particles. (b) Ringlike magnet assembly with an air gap, to which magnetic particles are attracted

oped by several technical associations and divisions of the U.S. Department of Defense. Sections III, V, and VIII of the ASME Boiler and Pressure Vessel Code contain specifications for nondestructive inspection of the vessels. Several Aerospace Material Specifications (published by the Society of Automotive Engineers) and standards from the American Society for Testing and Materials cover magnetic particle inspection. Various military standards include specifications for vendors to follow in establishing inspection procedures for military equipment and supplies. American Society for Nondestructive Testing Recommended Practice SNT-TC-1A is a guide to the employer for establishing in-house procedures for training, qualification, and certification of personnel whose jobs require appropriate knowledge of the principles underlying the nondestructive inspection they perform.

Description of Magnetic Fields

Magnetic fields are used in magnetic particle inspection to reveal discontinuities. Ferromagnetism is the property of some metals, chiefly iron and steel, to attract other pieces of ferromagnetic materials. A horseshoe magnet will attract magnetic materials to its ends, or poles. Magnetic lines of force, or flux, flow from the south pole through the magnet to the north pole.

Magnetized Ring. When a magnetic material is placed across the poles of a horseshoe magnet having square ends, forming a closed or ringlike assembly, the lines of force flow from the north pole through the magnetic material to the south pole (Fig. 1a). (Magnetic lines of force flow preferentially through magnetic material rather than through nonmagnetic material or air.) The magnetic lines of force will be enclosed within the ringlike assembly because no external poles exist, and iron filings or magnetic particles dusted over the assembly are not attracted to the magnet even though there are lines of magnetic force flowing through it. A ringlike part magnetized in this manner is said to contain a circular magnetic field that is wholly within the part.

If one end of the magnet is not square and an air gap exists between that end of the magnet and the magnetic material, the poles will still attract magnetic materials. Magnetic particles will cling to the poles and bridge the gap between them, as shown in Fig. 1(b). Any radial crack in a circularly magnetized piece will create a north and a south magnetic pole at the edges of a crack. Magnetic particles will be attracted to the poles created by such a crack, forming an indication of the discontinuity in the piece.

The fields set up at cracks or other physical or magnetic discontinuities in the surface are called leakage fields. The strength of a leakage field determines the number of magnetic particles that will gather to form indications; strong indications are formed at strong fields, weak indications at weak fields. The density of the magnetic field determines its strength and is partly governed by the shape, size, and material of the part being inspected.

Magnetized Bar. A straight piece of magnetized material (bar magnet) has a pole at each end. Magnetic lines of force flow through the bar from the south pole to the north pole. Because the magnetic lines of force within the bar magnet run the length of the bar, it is said to be longitudinally magnetized or to contain a longitudinal field.

If a bar magnet is broken into two pieces, a leakage field with north and south poles is created between the pieces, as shown in Fig. 2(a). This field exists even if the fracture surfaces are brought together (Fig. 2b). If the magnet is cracked but not broken completely in two, a somewhat similar result occurs. A north and a south pole form at opposite edges of the crack, just as though the break were complete (Fig. 2c). This field attracts the iron particles that outline the crack. The strength of these poles will be different from that of the fully broken pieces and will be a function of the depth of the crack and the width of the air gap at the surface.

The direction of the magnetic field in an electromagnetic circuit is controlled by the direction of the flow of magnetizing current through the part to be magnetized. The magnetic lines of force are always at right angles to the direction of current flow. To remember the direction taken by the magnetic lines of force around a conductor, consider that the conductor is grasped with the right hand so that the thumb points in the direction of current flow. The fingers then point in the direction taken by the magnetic lines of force in the magnetic field surrounding the conductor. This is known as the right-hand rule.

Circular Magnetization. Electric current passing through any straight conductor such as a wire or bar creates a circular magnetic

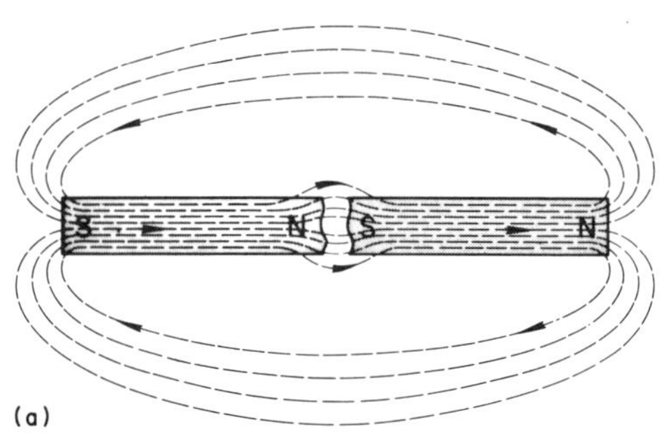

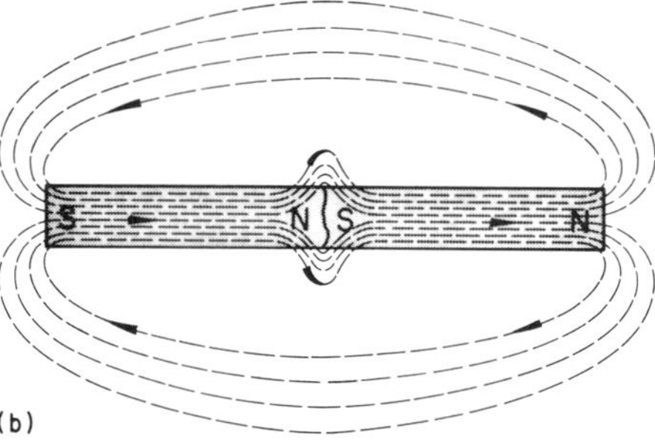

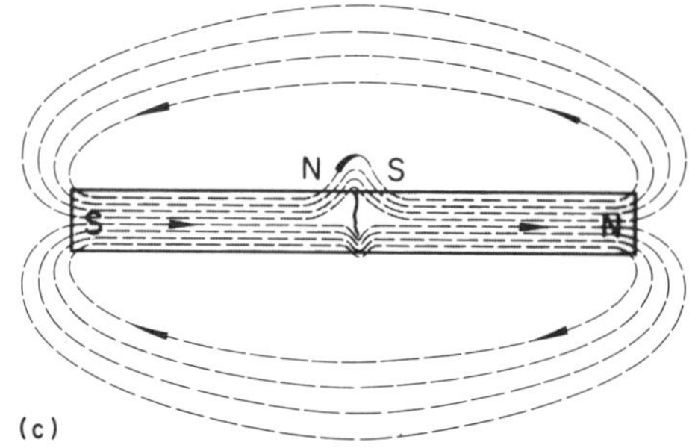

Fig. 2 Leakage fields between two pieces of a broken bar magnet. (a) Magnet pieces apart. (b) Magnet pieces together (which would simulate a flaw). (c) Leakage field at a crack in a bar magnet

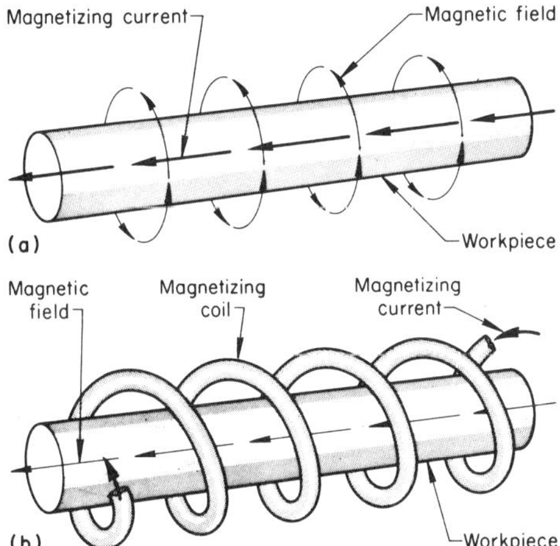

Fig. 3 Magnetized bars showing directions of magnetic field. (a) Circular. (b) Longitudinal

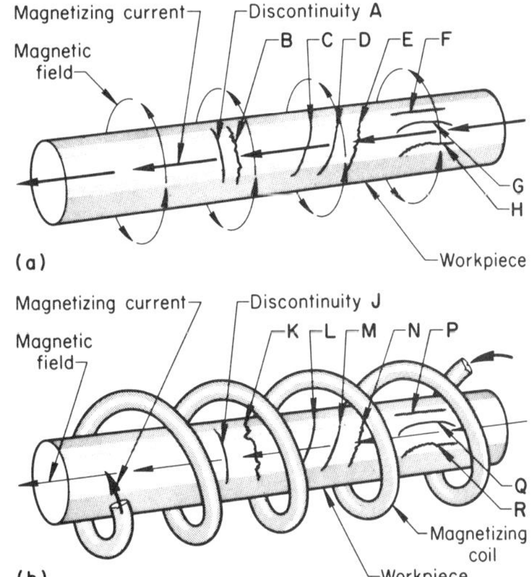

Fig. 4 Effect of direction of magnetic field or flux flow on the detectability of discontinuities with various orientations. (a) Circular magnetization. (b) Longitudinal magnetization. See text for discussion.

field around the conductor. When the conductor of electric current is a ferromagnetic material, the passage of current induces a magnetic field in the conductor as well as in the surrounding space. A part magnetized in this manner is said to have a circular field or to be circularly magnetized, as shown in Fig. 3(a).

Longitudinal Magnetization. Electric current can also be used to create a longitudinal magnetic field in magnetic materials. When electric current is passed through a coil of one or more turns, a magnetic field is established lengthwise or longitudinally, within the coil, as shown in Fig. 3(b). The nature and direction of the field around the conductor that forms the turns of the coil produce longitudinal magnetization.

Effect of Flux Direction. To form an indication, the magnetic field must approach a discontinuity at an angle great enough to cause the magnetic lines of force to leave the part and return after bridging the discontinuity. For best results, an intersection approaching 90° is desirable. For this reason, the direction, size, and shape of the discontinuity are important. The direction of the magnetic field is also important for optimum results, as is the strength of the field in the area of the discontinuity.

Figure 4(a) illustrates a condition in which the current is passed through the part, causing the formation of a circular field around the part. Under normal circumstances, a discontinuity such as A in Fig. 4(a) would give no indication of its presence, because it is regular in shape and lies parallel to the magnetic field. If the discontinuity has an irregular shape but is predominantly parallel to the magnetic field, such as B, there is a good possibility that a weak indication would form. Where the predominant direction of the discontinuity is at a 45° angle to the magnetic field, such as C, D, and E, the conditions are more favorable for detection regardless of the shape of the discontinuity. Discontinuities whose predominant directions, regardless of shape, are at a 90° angle to the magnetic field produce the most pronounced indications (F, G, and H, Fig. 4a).

A longitudinally magnetized bar is shown in Fig. 4(b). Discontinuities L, M, and N, which are at about 45° to the magnetic field, would produce detectable indications as they would with a circular field. Discontinuities J and K would display pronounced indications, and weak indications would be produced at discontinuities P, Q, and R.

Magnetization Methods. In magnetic particle inspection, the magnetic particles can be applied to the part while the magnetizing current is flowing or after the current has ceased, depending largely on the retentivity of the part. The first technique is known as the continuous method; the second, the residual method.

If the magnetism remaining in the part after the current has been turned off for a period of time (residual magnetism) does not provide a leakage field strong enough to produce readable indications when magnetic particles are applied to the surface, the part must be continuously magnetized during application of the particles. Consequently, the residual method can be used only on materials having sufficient retentivity; usually the harder the material, the higher the retentivity. The continuous method can be used for most parts.

Magnetizing Current

Both direct current (dc) and alternating current (ac) are suitable for magnetizing parts for magnetic particle inspection. The strength, direction, and distribution of magnetic fields are greatly affected by the type of current used for magnetization.

The fields produced by direct and alternating current differ in many respects. The important difference with regard to magnetic particle inspection is that the fields produced by direct current generally penetrate the cross section of the part, while the fields produced by alternating current are confined to the metal at or near the surface of the part, a phenomenon known as the skin effect. Therefore, alternating current should not be used in searching for subsurface discontinuities.

Direct Current. The best source of direct current is the rectification of alternating current. Both the single-phase (Fig. 5a) and three-phase types of alternating current (Fig. 5b) are furnished commercially. By using rectifiers, the reversing alternating current can be converted into unidirectional current, and when three-phase alternating current is rectified in this manner (Fig. 5c), the delivered direct current is entirely the equivalent of straight direct current for purposes of magnetic particle inspection. The only difference between rectified three-phase alternating current and straight direct current is a slight ripple in the value of the rectified current, amounting to only about 3% of the maximum current value.

When single-phase alternating current is passed through a simple rectifier, current is permitted to flow in one direction only. The reverse half of each cycle is completely blocked out (Fig. 5d). The result is unidirectional current (called half-wave current) that pulsates; that is, it rises from zero to a maximum and then drops back to zero. During the blocked-out reverse of the cycle, no current flows, then the half-cycle forward pulse is repeated, at a rate of 60 pulses per second. A rectifier for alternating current can also be connected so that the reverse half of the cycle is turned around and fed into the circuit flowing in the same direction as the first half of the cycle (Fig. 5e). This produces pulsating direct current, but with no interval between the pulses. Such current is referred to as single-phase full-wave direct current or full-wave rectified single-phase alternating current.

There is a slight skin effect from the pulsations of the current, but it is not pronounced enough to have a serious impact on the penetrations of the field. The pulsation of the current is useful because it imparts some slight vibration to the magnetic particles, assisting them in arranging themselves to form indications. Half-wave current, used in magnetization with prods and dry magnetic particles, provides the highest sensitivity for discontinuities that are wholly below the surface, such as those in castings and weldments.

Magnetization employing surges of direct current can be used to increase the strength of magnetic fields; for example, a rectifier capable of continuously delivering 400-A

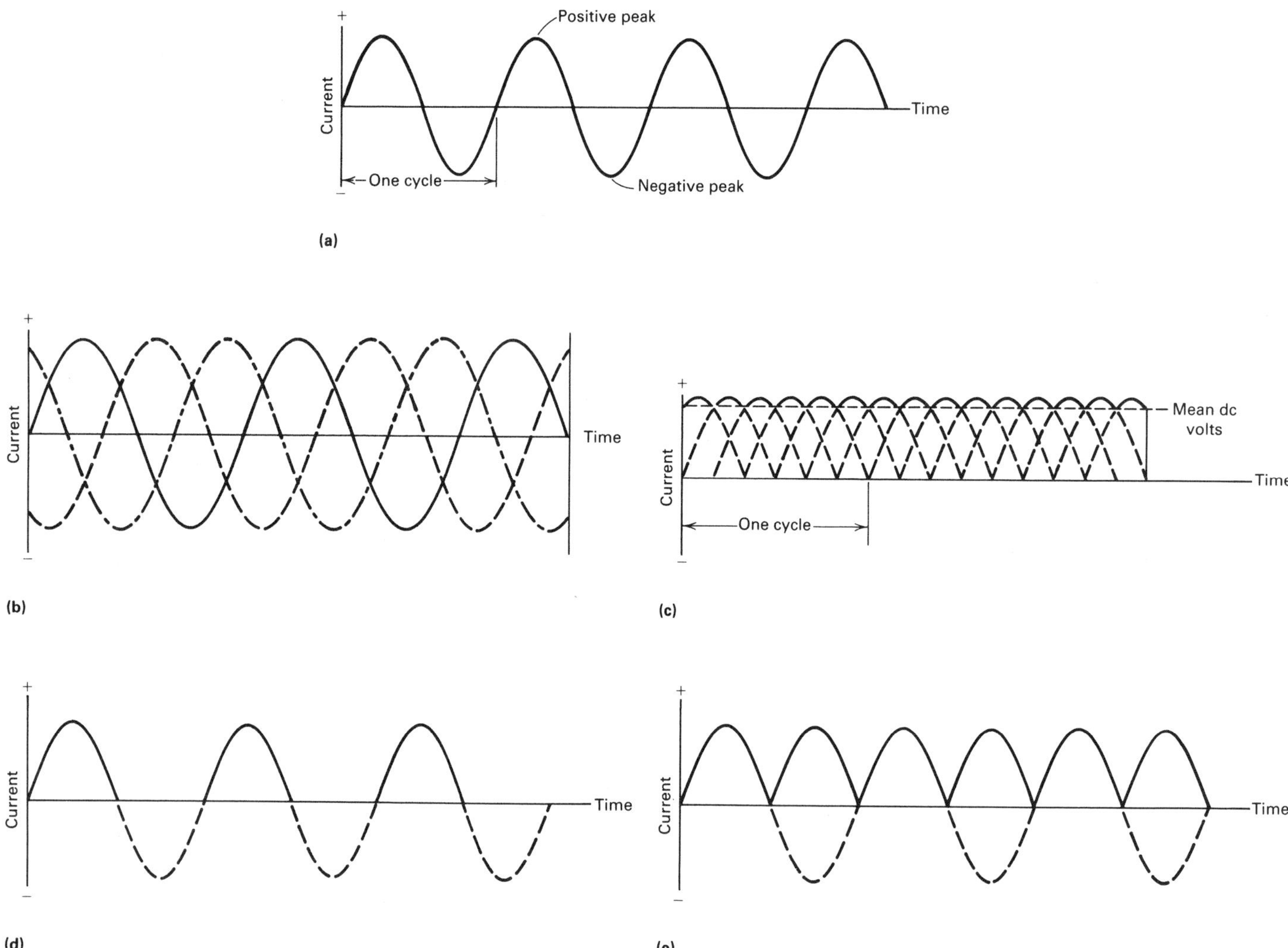

Fig. 5 Alternating current wave forms. (a) Single-phase. (b) Three-phase. (c) Three-phase rectified. (d) Half-wave rectified single-phase. (e) Full-wave rectified single-phase

current can put out much more than 400 A for short intervals. Therefore, it is possible, by suitable current-control and switching devices, to pass a very high current for a short period (less than a second) and then reduce the current, without interrupting it, to a much lower value.

Alternating current, which must be single-phase when used directly for magnetizing purposes, is taken from commercial power lines and usually has a frequency of 50 or 60 Hz. When used for magnetizing, the line voltage is stepped down, by means of transformers, to the low voltages required. At these low voltages, magnetizing currents of several thousand amperes are often used.

One problem encountered when alternating current is used is that the resultant residual magnetism in the part may not be at a level as high as that of the magnetism generated by the peak current of the ac cycle. This is because the level of residual magnetism depends on where in the cycle the current was discontinued.

Power Sources

Early power sources were general-purpose units designed to use either alternating or direct current for magnetization. When direct current was used, it was derived directly from a bank of storage batteries, and a carbon-pile rheostat was used to regulate current level. Subsequent advances in technology have made the storage battery obsolete as a power supply and have given rise to many innovations, especially in the area of current control. Portable, mobile, and stationary equipment is currently available, and selection among these types depends on the nature and location of testing.

Portable equipment is available in lightweight (16 to 40 kg, or 35 to 90 lb) power source units that can be readily taken to the inspection site. Generally, these portable units are designed to use 115-, 230-, or 460-V alternating current and to supply magnetizing-current outputs of 750 to 1500 A in half-wave or alternating current. Machines capable of supplying half-wave current and alternating current and having continuously variable (infinite) current control can be used for magnetic particle inspection in a wide range of applications. Primary application of this equipment is hand-held prod inspection utilizing the half-wave output in conjunction with dry powder. In general, portable equipment is designed to operate with relatively short power supply cables, and the output is very limited when it is necessary to use longer cables.

The major disadvantage of portable equipment is the limited amount of current available. For the detection of deep-lying discontinuities and for coverage of a large area with one prod contact, a machine with higher-amperage output is required. Also, portable equipment cannot supply the full-wave direct current necessary for some inspections and does not have the accessories found on larger mobile equipment.

Mobile units are generally mounted on wheels to facilitate transportation to various inspection sites. Mobile equipment usually supplies half-wave or alternating magnetizing-current outputs. Inspection of parts is accomplished with flexible cables, yokes, prod contacts, contact clamps, and coils. Instruments and controls are mounted on the front of the unit. Magnetizing current is usually controlled by a remote-control switch connected to the unit by an electric cord. Quick-coupling connectors for connecting magnetizing cables are on the front of the unit.

Mobile equipment is usually powered by single-phase, 60-Hz alternating current (230 or 460 V) and has an output range of 1500 to 6000 A. On some units, current control is provided by a power-tap switch, which varies the voltage applied to the primary coil of the power transformer; most of these have either 8 or 30 steps of current control. However, current control on more advanced units is provided either by solid-state phase control of the transformer or by use of a saturable-core reactor to control the transformer. Phase control of the transformer is achieved by silicon-controlled rectifiers or triacs in series with the transformer. A solid-state control circuit is used to rapidly switch the ac supply on and off for controlled fractions of each cycle. A triac provides current control in both directions, while a saturable-core reactor provides current control in one direction only. In a circuit employing a saturable-core reactor to control magnetizing-current output, a silicon-controlled rectifier is used in conjunction with phase control to control a saturable-core reactor that is in series with, and that controls the input to, the power transformer.

Standard instruments and controls on mobile equipment are as follows:

- Ammeters to indicate the magnetizing current flowing through the yokes, prods, or coil as being alternating, half-wave, or direct current
- Switches for controlling the magnetizing or demagnetizing current
- Pilot light to indicate when power to the unit is on
- Current control, either stepped or continuously variable, for controlling the amount of magnetizing and demagnetizing current
- Remote-control cable receptacle that permits turning the magnetizing current on and off at some distance from the unit
- Receptacles to permit the connection of the magnetizing-current cables

Built-in demagnetizers as contained in mobile equipment for magnetic particle inspection are available that demagnetize by one of four methods:

- Low-voltage high-amperage alternating current with a motor-driven power-tap switch, arranged to automatically provide periods of current-on and periods of current-off in succession, with the amount of demagnetizing current reduced with each successive step
- Low-voltage high-amperage alternating current provided by a continuously variable current control that affords complete control of the demagnetization current from full-on to zero
- Current-decay method, in which low-voltage high-amperage alternating current is caused to decay from some maximum value to zero in an automatic and controlled manner. Because the entire cycle can be completed in a few seconds, the current-decay method offers an advantage over some of the more time-consuming methods
- Low-voltage high-amperage direct current, with which demagnetization is accomplished by a motor-driven power-tap switch in conjunction with means for reversing the current direction from positive to negative as the current is systematically reduced in steps of current-on periods followed by current-off periods

Stationary equipment can be obtained as either general-purpose or special-purpose units. The general-purpose unit is primarily intended for use in the wet method and has a built-in tank that contains the bath pump, which continuously agitates the bath and forces the fluid through hoses onto the part being inspected. Pneumatically operated contact heads, together with a rigid-type coil, provide capabilities for both circular and longitudinal magnetization. Self-contained ac or dc power supplies are available in amperage ratings from 2500 to 10 000 A. Maximum opening between contact plates varies from 0.3 to 3.7 m (1 to 12 ft).

Optional features that are available include self-regulating current control, automatic magnetizing circuit, automatic bath applicator, steady rests for heavy parts, and demagnetizing circuitry. Other options are curtains or hoods and an ultraviolet light; these are used during inspection with fluorescent particles.

Stationary power packs serve as sources of high-amperage magnetizing current to be used in conjunction with special fixtures or with cable-wrap or clamp-and-contact techniques. Rated output varies from a customary 4000 to 6000 A to as high as 20 000 A. The higher-amperage units are used for the overall magnetization of large forgings or castings that would otherwise require systematic prod inspection at much lower current levels.

Multidirectional Magnetizing. Some units feature up to three output circuits that are systematically energized in rapid sequence, either electrically or mechanically, for effectively magnetizing a part in several directions in the same time frame. This reveals discontinuities lying in any direction after only a single processing step.

Special-purpose stationary units are designed for handling and inspecting large quantities of similar items. Generally, conveyors, automatic markers, and alarm systems are included in such units to expedite the handling of parts.

Methods of Generating Magnetic Fields

One of the basic requirements of magnetic particle inspection is that the part undergoing inspection be properly magnetized so that the leakage fields created by discontinuities will attract the magnetic particles. Permanent magnets serve some useful purpose in this respect, but magnetization is generally produced by electromagnets or the magnetic field associated with the flow of electric current. Basically, magnetization is derived from the circular magnetic field generated when an electric current flows through a conductor. The direction of this field is dependent on the direction of current flow, which can be determined by applying the right-hand rule (see the section "Description of Magnetic Fields" in this article). General applications, advantages, and limitations of the various methods of magnetizing parts for magnetic particle inspection are summarized in Table 1. Additional information can be found in the article "Magnetic Field Testing" in this Volume.

Yokes

There are two basic types of yokes that are commonly used for magnetizing purposes: permanent-magnet and electromagnetic yokes. Both are hand held and therefore quite mobile.

Permanent-magnet yokes are used for applications where a source of electric power is not available or where arcing is not permissible (as in an explosive atmosphere). The limitations of permanent-magnet yokes include the following:

- Large areas or masses cannot be magnetized with enough strength to produce satisfactory crack indications
- Flux density cannot be varied at will
- If the magnet is very strong, it may be difficult to separate from a part
- Particles may cling to the magnet, possibly obscuring indications

Electromagnetic yokes (Fig. 6) consist of a coil wound around a U-shaped core of soft iron. The legs of the yoke can be either fixed or adjustable. Adjustable legs permit changing the contact spacing and the relative angle of contact to accommodate irregularly-shaped parts. Unlike a permanent-magnet yoke, an electromagnetic yoke can readily be switched on or off. This feature makes it convenient to apply and remove the yoke from the testpiece.

The design of an electromagnetic yoke can be based on the use of either direct or

Table 1 General applications, advantages, and limitations of the various magnetizing methods used in magnetic particle inspection

Application	Advantages	Limitations
Coils (single or multiple loop)		
Medium-size parts whose length predominates, such as a crankshaft or camshaft	All generally longitudinal surfaces are longitudinally magnetized to locate transverse discontinuities.	Part should be centered in coil to maximize length effectively magnetized during a given shot. Length may dictate additional shots as coil is repositioned.
Large castings, forgings, or shafts	Longitudinal field easily attained by wrapping with a flexible cable.	Multiple processing may be required because of part shape.
Miscellaneous small parts	Easy and fast, especially where residual method is applicable. Noncontact with part. Relatively complex parts can usually be processed with same ease as simple cross section.	Length-to-diameter (L/D) ratio is important in determining adequacy of ampere-turns; effective ratio can be altered by utilizing pieces of similar cross-sectional area. Sensitivity diminishes at ends of part because of general leakage field pattern. Quick break of current is desirable to minimize end effect on short parts with low L/D ratios.
Yokes		
Inspection of large surface areas for surface discontinuities	No electrical contact. Highly portable. Can locate discontinuities in any direction, with proper yoke orientation	Time consuming. Yoke must be systematically repositioned to locate discontinuities with random orientation.
Miscellaneous parts requiring inspection of localized areas	No electrical contact. Good sensitivity to surface discontinuities. Highly portable. Wet or dry method can be used. Alternating current yoke can also serve as demagnetizer in some cases.	Yoke must be properly positioned relative to orientation of discontinuity. Relatively good contact must be established between part and poles of yoke; complex part shape may cause difficulty. Poor sensitivity to subsurface discontinuities except in isolated areas
Central conductors		
Miscellaneous short parts having holes through which a conductor can be threaded, such as bearing rings, hollow cylinders, gears, large nuts, large clevises, and pipe couplings	No electrical contact, so that possibility of burning is eliminated. Circumferentially directed magnetic field is generated in all surfaces surrounding the conductor. Ideal for parts for which the residual method is applicable. Lightweight parts can be supported by the central conductor. Multiple turns can be used to reduce the amount of current required.	Size of conductor must be ample to carry required current. Ideally, conductor should be centrally located within hole. Large-diameter parts require several setups with conductor near or against inner surface and rotation of part between setups. Where continuous method is being employed, inspection is required after each setup.
Long tubular parts such as pipe, tubing, hollow shafts	No electrical contact. Both inside (ID) and outside (OD) surfaces can be inspected. Entire length of part is circularly magnetized.	Sensitivity of outer surface to indications may be somewhat diminished relative to inner surface for large-diameter and thick-wall parts.
Large valve bodies and similar parts	Good sensitivity to inner-surface discontinuities	Same as for long tubular parts, above
Direct contact, head shot		
Solid, relatively small parts (cast, forged, or machined) that can be inspected on a horizontal wet-method unit	Fast, easy process. Complete circular field surrounds entire current path. Good sensitivity to surface and near-surface discontinuities. Simple as well as relatively complex parts can usually be easily inspected with one or more shots.	Possibility of burning part exists if proper contact conditions are not met. Long parts should be inspected in sections to facilitate bath application without resorting to an excessively long current shot.
Direct contact, clamps and cables		
Large castings and forgings	Large surface areas can be inspected in a relatively short time.	High amperage requirements (8000–20 000 A) dictate use of special direct current power pack.
Long tubular parts such as tubing, pipe, and hollow shafts	Entire length can be circularly magnetized by contacting end-to-end.	Effective field is limited to outer surface so process cannot be used to inspect inner surface. Part ends must be shaped to permit electrical contact and must be able to carry required current without excessive heating.
Long solid parts such as billets, bars, and shafts	Entire length can be circularly magnetized by contacting end-to-end. Amperage requirements are independent of length. No loss of magnetism at ends	Voltage requirements increase as length increases because of greater impedance of cable and part. Ends of parts must have shape that permits electrical contact and must be capable of carrying required current without excessive heating.
Prod contacts		
Welds, for cracks, inclusions, open roots, or inadequate joint penetration	Circular field can be selectively directed to weld area by prod placement. In conjunction with half-wave current and dry powder, provides excellent sensitivity to subsurface and surface discontinuities. Prods, cables, and power packs can be brought to inspection site.	Only small area can be inspected at one time. Arc burn can result from poor contact. Surface must be dry when dry powder is being used. Prod spacing must be in accordance with magnetizing-current level.
Large castings or forgings	Entire surface area can be inspected in small increments using nominal current values. Circular magnetic field can be concentrated in specific areas likely to contain discontinuities. Prods, cables, and power packs can be brought to the inspection site.	Coverage of large surface areas requires a multiplicity of shots, which can be very time consuming. Arc burn can result from poor contact. Surface must be dry when dry powder is being used.
Induced current		
Ring-shaped parts, for circumferential discontinuities	No electrical contact. All surfaces of part are subjected to a toroidal magnetic field. 100% coverage is obtained in a single magnetization. Can be automated	Laminated core is required through ring to enhance magnetic path. Type of magnetizing current must be compatible with magnetic hardness or softness of metal inspected. Other conductors encircling field must be avoided.
Balls	No electrical contact. Permits 100% coverage for indications of discontinuities in any direction by use of a three-step process with reorientation of ball between steps. Can be automated	For small-diameter balls, use is limited to residual method of magnetization.
Disks and gears	No electrical contact. Good sensitivity at or near periphery or rim. Sensitivity in various areas can be varied by selection of core or pole piece. In conjunction with half-wave current and dry powder, provides excellent sensitivity to discontinuities lying just below the surface	100% coverage may require two-step process. Type of magnetizing current must be compatible with magnetic hardness or softness of metal inspected.

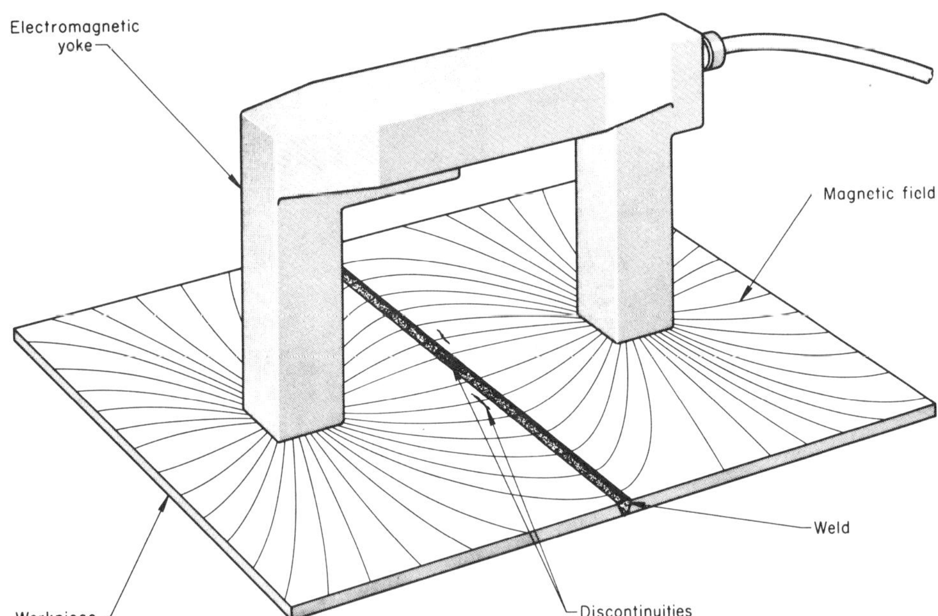

Fig. 6 Electromagnetic yoke showing position and magnetic field for the detection of discontinuities parallel to a weld bead. Discontinuities across a weld bead can be detected by placing the contact surfaces of the yoke next to and on either side of the bead (rotating yoke about 90° from position shown here).

alternating current or both. The flux density of the magnetic field produced by the direct current type can be changed by varying the amount of current in the coil. The direct current type of yoke has greater penetration while the alternating current type concentrates the magnetic field at the surface of the testpiece, providing good sensitivity for the disclosure of surface discontinuities over a relatively broad area. In general, discontinuities to be disclosed should be centrally located in the area between pole pieces and oriented perpendicular to an imaginary line connecting them (Fig. 6). Extraneous leakage fields in the immediate vicinity of the poles cause an excessive buildup of magnetic particles.

In operation, the part completes the magnetic path for the flow of magnetic flux. The yoke is a source of magnetic flux, and the part becomes the preferential path completing the magnetic circuit between the poles. (In Fig. 6, only those portions of the flux lines near the poles are shown.) Yokes that use alternating current for magnetization have numerous applications and can also be used for demagnetization.

Coils

Single-loop and multiple-loop coils (conductors) are used for the longitudinal magnetization of components (Fig. 3b and 4b). The field within the coil has a definite direction, corresponding to the direction of the lines of force running through it. The flux density passing through the interior of the coil is proportional to the product of the current, I, in amperes, and the number of turns in the coil, N. Therefore, the magnetizing force of such a coil can be varied by changing either the current or the number of turns in the coil. For large parts, a coil can be produced by winding several turns of a flexible cable around the part, but care must be taken to ensure that no indications are concealed beneath the cable.

Portable magnetizing coils are available that can be plugged into an electrical outlet. These coils can be used for the in-place inspection of shaftlike parts in railroad shops, aircraft maintenance shops, and shops for automobile, truck, and tractor repair. Transverse cracks in spindles and shafts are easily detected with such coils.

Most coils used for magnetizing are short, especially those wound on fixed frames. The relationship of the length of the part being inspected to the width of the coil must be considered. For a simple part, the effective overall distance that can be inspected is 150 to 230 mm (6 to 9 in.) on either side of the coil. Thus, a part 305 to 460 mm (12 to 18 in.) long can be inspected using a normal coil approximately 25 mm (1 in.) thick. In testing longer parts, either the part must be moved at regular intervals through the coil or the coil must be moved along the part.

The ease with which a part can be longitudinally magnetized in a coil is significantly related to the length-to-diameter (L/D) ratio of the part. This is due to the demagnetizing effect of the magnetic poles set up at the ends of the part. This demagnetizing effect is considerable for L/D ratios of less than 10 to 1 and is very significant for ratios of less than 3 to 1. Where the L/D ratio is extremely unfavorable, pole pieces of similar cross-sectional area can be introduced to increase the length of the part and thus improve the L/D ratio. The magnetization of rings, disks, and other parts with low L/D ratios is discussed and illustrated in the section "Induced Current" in this article.

The number of ampere-turns required to produce sufficient magnetizing force to magnetize a part adequately for inspection is given by:

$$NI = 45\,000\,(L/D) \qquad \text{(Eq 1)}$$

where N is the number of turns in the coil, I is the current in amperes, and L/D is the length-to-diameter ratio of the part. When the part is magnetized at this level by placing it on the bottom of the round magnetizing coil, adjacent to the coil winding, the flux density will be about 110 lines/mm^2 (70 000 lines/in.^{2}). Experimental work has shown that a flux density of 110 lines/mm^2 is more than satisfactory for most applications of coil magnetization and that 54 lines/mm^2 (35 000 lines/in.^{2}) is acceptable for all but the most critical applications.

When it is desirable to magnetize the part by centering it in the coil, Eq 1 becomes:

$$NI = \frac{43\,000r}{\mu_{eff}} \qquad \text{(Eq 2)}$$

where r is the radius of the coil in inches and $\mu_{eff} = (6L/D) - 5$. Equation 2 is applicable to parts that are centered in the coil (coincident with the coil axis) and that have cross sections constituting a low fill factor, that is, with a cross-sectional area less than 10% of the area encircled by the coil.

When using a coil for magnetizing a barlike part, strong polarity at the ends of the part could mask transverse defects. An advantageous field in this area is assured on full wave, three phase, direct current units by special circuitry known as "quick" or "fast" break. A "controlled" break feature on alternating current, half wave, and on single-phase full wave direct current units provides a similar advantageous field.

Central Conductors

For many tubular or ring-shaped parts, it is advantageous to use a separate conductor to carry the magnetizing current rather than the part itself. Such a conductor, commonly referred to as a central conductor, is threaded through the inside of the part (Fig. 7) and is a convenient means of circularly magnetizing a part without the need for making direct contact to the part itself. Central conductors are made of solid and tubular nonmagnetic and ferromagnetic materials that are good conductors of electricity.

The basic rules regarding magnetic fields around a circular conductor carrying direct current are as follows:

- The magnetic field outside a conductor of uniform cross section is uniform along the length of the conductor
- The magnetic field is 90° to the path of the current through the conductor

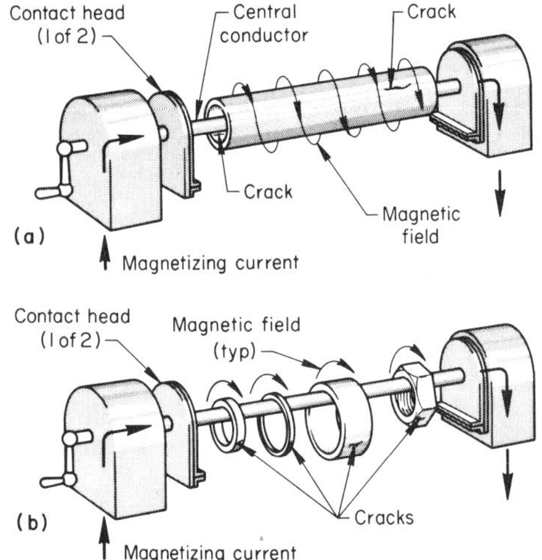

Fig. 7 Use of central conductors for the circular magnetization of long, hollow cylindrical parts (a) and short, hollow cylindrical or ringlike parts (b) for the detection of discontinuities on inside and outside surfaces

- The flux density outside the conductor varies inversely with the radial distance from the center of the conductor

Solid Nonmagnetic Conductor Carrying Direct Current. The distribution of the magnetic field inside a nonmagnetic conductor, such as a copper bar, when carrying direct current is different from the distribution external to the bar. At any point inside the bar, the flux density is the result of only that portion of the current that is flowing in the metal between the point and the center of the bar. Therefore, the flux density increases linearly, from zero at the center of the bar to a maximum value at the surface. Outside the bar, the flux density decreases along a curve, as shown in Fig. 8(a). In calculating flux densities outside the bar, the current can be considered to be concentrated at the center of the bar. If the radius of the bar is R and the flux density, B, at the surface of the bar is equal to the magnetizing force, H, then the flux density at a distance $2R$ from the center of the bar will be $H/2$; at $3R$, $H/3$; and so on.

Solid Ferromagnetic Conductor Carrying Direct Current. If the conductor carrying direct current is a solid bar of steel or other ferromagnetic material, the same distribution of magnetic field exists as in a similar nonmagnetic conductor, but the flux density is much greater. Figure 8(b) shows a conductor of the same diameter as that shown in Fig. 8(a). The flux density at the center is zero, but at the surface it is μH, where μ is the material permeability of the magnetic material. (Permeability is the ease with which a material accepts magnetism.) The actual flux density, therefore, may be many times that in a nonmagnetic bar. Just outside the surface, however, the flux density drops to exactly the same value as that for the nonmagnetic conductor, and the decrease in flux density with increasing distance follows the same curve.

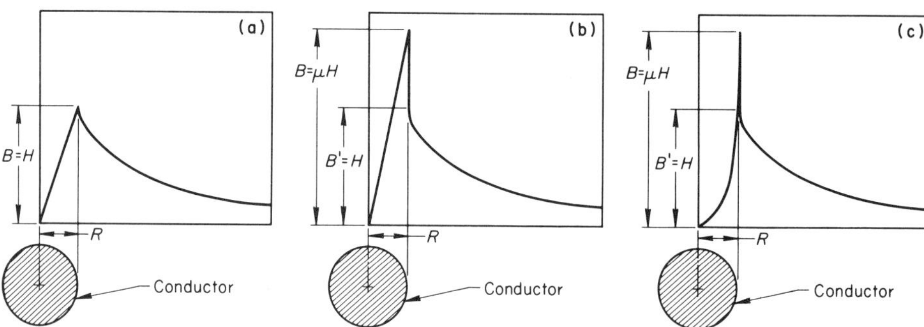

Fig. 8 Flux density in and around solid conductors of the same diameter. (a) Nonmagnetic conductor ($\mu = 1.0$) carrying direct current. (b) Ferromagnetic conductor ($\mu > 1.0$) carrying direct current. (c) Ferromagnetic conductor ($\mu > 1.0$) carrying alternating current. See text for discussion.

Solid Ferromagnetic Conductor Carrying Alternating Current. The distribution of the magnetic field in a solid ferromagnetic conductor carrying alternating current is shown in Fig. 8(c). Outside the conductor, the flux density decreases along the same curve as if direct current produced the magnetizing force; however, while the alternating current is flowing, the field is constantly varying in strength and direction. Inside the conductor, the flux density is zero at the center and increases toward the outside surface—slowly at first, then accelerating to a high maximum at the surface. The flux density at the surface is proportional to the permeability of the conductor material.

Central Conductor Enclosed Within Hollow Ferromagnetic Cylinder. When a central conductor is used to magnetize a hollow cylindrical part made of a ferromagnetic material, the flux density is maximum at the inside surface of the part (Fig. 9). The flux density produced by the current in the central conductor is maximum at the surface of the conductor (H, in Fig. 9) and then decreases along the same curve outside the conductor, as shown in Fig. 8, through the space between the conductor and the inside surface of the part. At this surface, however, the flux density is immediately increased by the permeability factor, μ, of the material of the part and then decreases to the outer surface. Here the flux density again drops to the same decreasing curve it was following inside the part.

This method, then, produces maximum flux density at the inside surface and therefore gives strong indications of discontinuities on that surface. Sometimes these indications may even appear on the outside surface of the part. The flux density in the wall of the cylindrical part is the same whether the central conductor is of magnetic or nonmagnetic material, because it is the field external to the conductor that constitutes the magnetizing force for the part.

If the axis of a central conductor is placed along the axis of a hollow cylindrical part, the magnetic field in the part will be concentric with its cylindrical wall. However, if the central conductor is placed near one point on the inside circumference of the part, the flux density of the field in the cylindrical wall will be much stronger at this point and will be weaker at the diametrically opposite point.

In small hollow cylinders, it is desirable that the conductor be centrally placed so that a uniform field for the detection of discontinuities will exist at all points on the cylindrical surface. In larger-diameter tubes, rings, or pressure vessels, however, the current necessary in the centrally placed conductor to produce fields of adequate strength for proper inspection over the entire circumference becomes excessively large.

An offset central conductor should then be used (Fig. 10). When the conductor passing through the inside of the part is placed against an inside wall of the part, the current levels given in the section "Magnitude of Applied Current" in this article apply except that the diameter will be considered the sum of the diameter of the central conductor and twice the wall thickness. The distance along the part circumference (interior or exterior) that is effectively magnetized will be taken as four times the diameter of the central conductor, as illus-

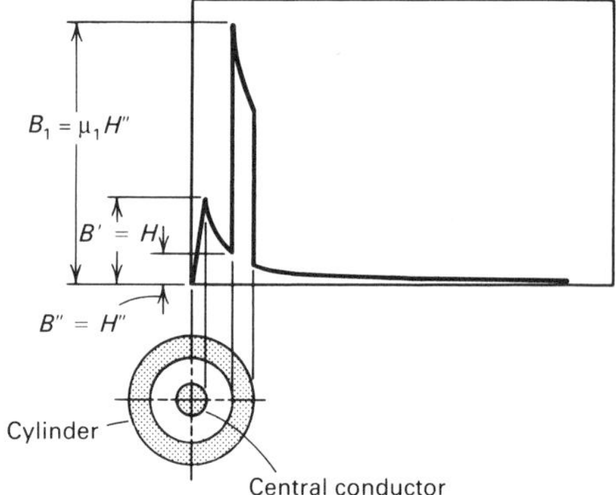

Fig. 9 Flux density in and around a hollow cylinder made of magnetic material with direct current flowing through a nonmagnetic central conductor

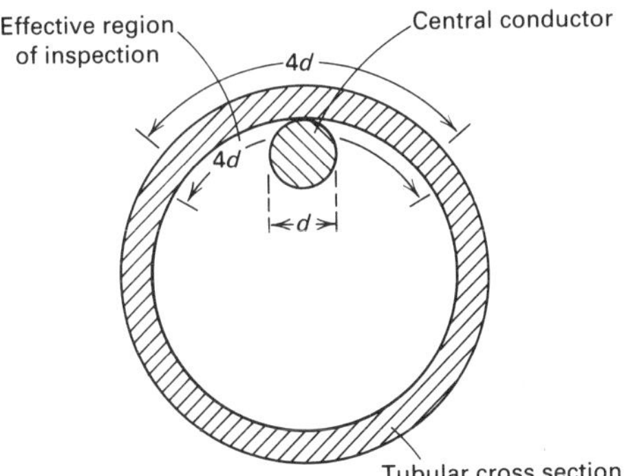

Fig. 10 Schematic showing that the effective region of inspection when using an offset central conductor is equal to four times the diameter of the conductor

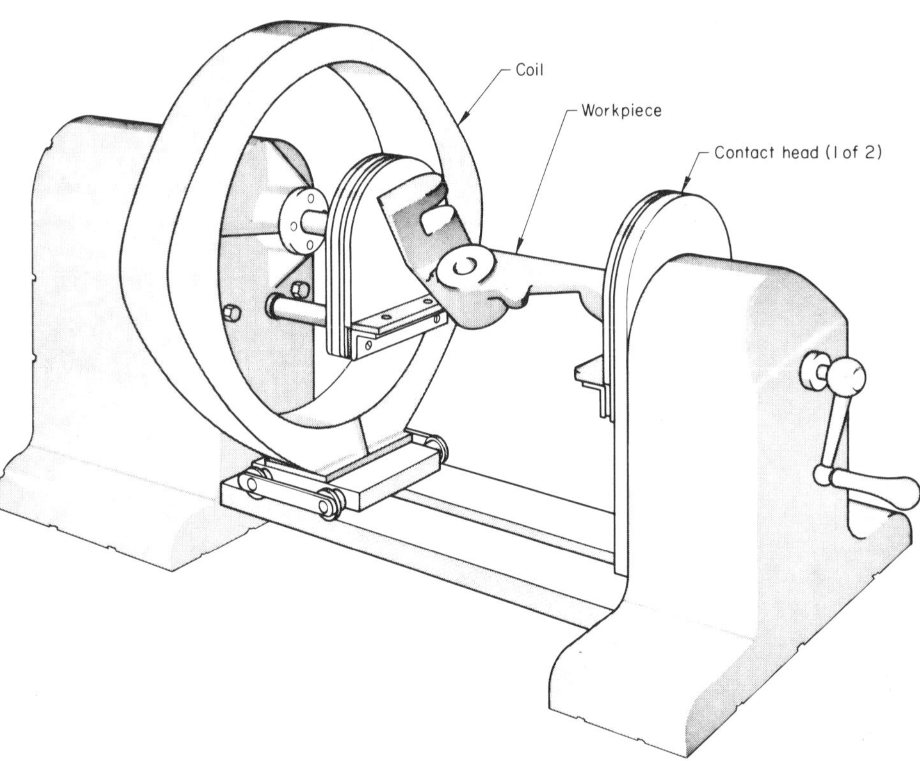

Fig. 11 Bench unit for the circular magnetization of workpieces that are clamped between contact heads (direct-contact, head-shot method). The coil on the unit can be used for longitudinal magnetization.

trated in Fig. 10. The entire circumference will be inspected by rotating the part on the conductor, allowing for approximately a 10% magnetic field overlap.

The diameter of a central conductor is not related to the inside diameter or the wall thickness of the cylindrical part. Conductor size is usually based on its current-carrying capacity and ease of handling. In some applications, conductors larger than that required for current-carrying capacity can be used to facilitate centralizing the conductor within the part.

Residual magnetization is usually employed whenever practicable because the background is minimized and contrast is therefore enhanced. Also, residual magnetization is faster and less critical than continuous magnetization.

The central-conductor type of inspection is sometimes required on components having parallel multiple openings, such as engine blocks. The cylinders can be processed with a single central conductor in the normal manner. However, a multiple central-conductor fixture can be designed that enables the operator to process two or more adjacent cylinders at one time with the same degree of sensitivity as if processed individually. In fact, in the areas between the central conductors, the circular fields reinforce one another to enhance sensitivity.

Direct-Contact Method

For small parts having no openings through the interior, circular magnetic fields are produced by direct contact to the part. This is done by clamping the parts between contact heads (head shot), generally on a bench unit (Fig. 11) that incorporates the source of the current. A similar unit can be used to supply the magnetizing current to a central conductor (Fig. 7).

The contact heads must be constructed so that the surfaces of the part are not damaged—either physically by pressure or structurally by heat from arcing or from high resistance at the points of contact. Heat can be especially damaging to hardened surfaces such as bearing races.

For the complete inspection of a complex part, it may be necessary to attach clamps at several points on the part or to wrap cables around the part to orient fields in the proper directions at all points on the surface. This often necessitates several magnetizations. Multiple magnetizations can be minimized by using the overall magnetization method, multidirectional magnetization, or induced-current magnetization.

Prod Contacts

For the inspection of large and massive parts too bulky to be put into a unit having clamping contact heads, magnetization is often done by using prod contacts (Fig. 12) to pass the current directly through the part or through a local portion of it. Such local contacts do not always produce true circular fields, but they are very convenient and practical for many purposes. Prod contacts are often used in the magnetic particle inspection of large castings and weldments.

Advantages. Prod contacts are widely used and have many advantages. Easy portability makes them convenient to use for the field inspection of large tanks and welded structures. Sensitivity to defects lying wholly below the surface is greater with this method of magnetization than with any other, especially when half-wave current is used in conjunction with dry powder and the continuous method of magnetization.

Limitations. The use of prod contacts involves some disadvantages:

- Suitable magnetic fields exist only between and near the prod contact points. These points are seldom more than 305 mm (12 in.) apart and usually much less; therefore, it is sometimes necessary to relocate the prods so that the entire surface of a part can be inspected
- Interference of the external field that exists between the prods sometimes makes observation of pertinent indications difficult; the strength of the current that can be used is limited by this effect
- Great care must be taken to avoid burning of the part under the contact points. Burning may be caused by dirty contacts, insufficient contact pressure, or excessive currents. The likelihood of such damage is particularly great on steel with a carbon content of 0.3 to 0.4% or more. The heat under the contact points can produce local spots of very hard material that can interfere with later operations, such as machining. Actual cracks are sometimes produced by this heating effect. Contact heating is less likely to be damaging to low-carbon steel such as that used for structural purposes

Induced Current

Induced current provides a convenient method of generating circumferential magnetizing current in ring-shaped parts without making electrical contact. This is accomplished by properly orienting the ring within a magnetizing coil such that it links

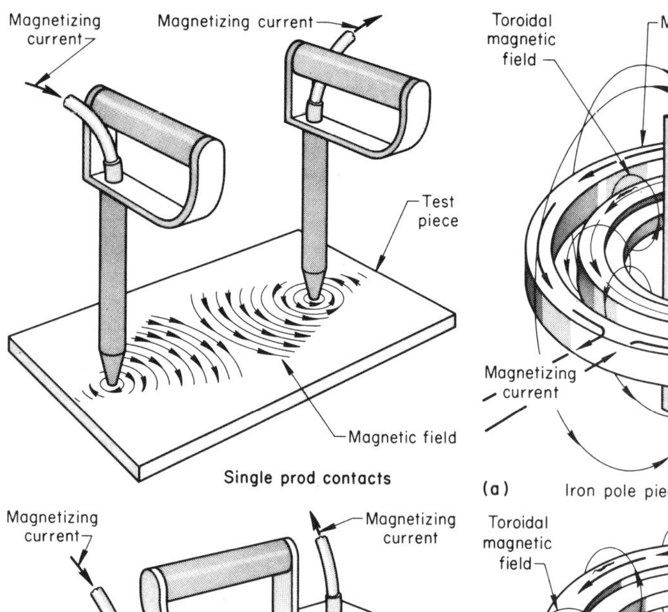

Double prod contacts

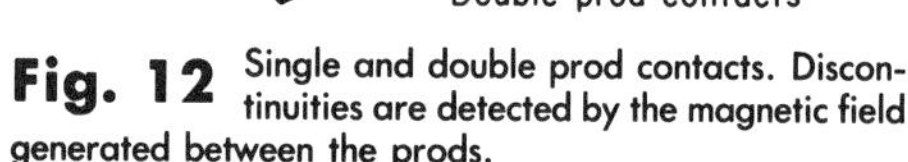

Fig. 12 Single and double prod contacts. Discontinuities are detected by the magnetic field generated between the prods.

or encloses lines of magnetic flux (flux linkage), as shown in Fig. 13(a). As the level of magnetic flux changes (increases or decreases), a current flows around the ring in a direction opposing the change in flux level. The magnitude of this current depends on the total flux linkages, rate of flux linkage changes, and the electrical impedance associated with the current path within the ring. Increasing the flux linkages and the rate of change increases the magnitude of current induced in the ring. The circular field associated with this current takes the form of a toroidal magnetic field that encompasses all surface areas on the ring and that is conducive to the disclosure of circumferential types of discontinuities. This is shown schematically in Fig. 13(b). To enhance the total flux linkages, laminated soft iron pole pieces are usually inserted through the hole in the part as shown in Fig. 13(a).

Direct Versus Alternating Current. The choice of magnetizing current for the induced-current method depends on the magnetic properties of the part to be inspected.

Fig. 13 Induced-current method of magnetizing a ring-shaped part. (a) Ring being magnetized by induced current. Current direction corresponds to decreasing magnetizing current. (b) Resulting induced current and toroidal magnetic field in a ring

In cases in which the residual method is applicable, such as for most bearing races or similar parts having high magnetic retentivity, direct current is used for magnetizing. The rapid interruption of this current, by quick-break circuitry, results in a rapid collapse of the magnetic flux and the generation of a high-amperage, circumferentially directed single pulse of current in the part. Therefore, the part is residually magnetized with a toroidal field, and the subsequent application of magnetic particles will produce indications of circumferentially oriented discontinuities.

Passing an alternating current through a conductor will set up a fluctuating magnetic field as the level of magnetic flux rapidly changes from a maximum value in one direction to an equal value in the opposite direction. This is similar to the current that would flow in a single-shorted-turn secondary of a transformer. The alternating induced current, in conjunction with the continuous method, renders the method applicable for processing magnetically soft, or less retentive, parts.

Applications. The induced-current method, in addition to eliminating the possibility of damaging the part, is capable of magnetizing in one operation parts that would otherwise require more than one head shot. Two examples of this type of part are illustrated in Fig. 14 and 15. These parts cannot be completely processed by one head shot to disclose circumferential defects, because regions at the contact point are not properly magnetized. Therefore, a two-step inspection process would be required for full coverage, with the part rotated approximately 90° prior to the second step. On the other hand, the induced-current method provides full coverage in one processing step. The disk-shaped part shown in Fig. 15 presents an additional problem when the contact method is employed to disclose circumferential defects near the rim. Even when a two-step process is employed, as with the ring shown in Fig. 14, the primary current path through the disk may not develop a circular field of ample magnitude in the rim area. The induced current can be selectively concentrated in the rim area by proper pole piece selection to provide full coverage (rim area) in a single processing step. The pole pieces shown in Fig. 15(b) are hollow and cylindrical, with one on each side of the disk. These pole pieces direct the magnetic flux through the disk such that the rim is the only portion constituting a totally enclosing current path.

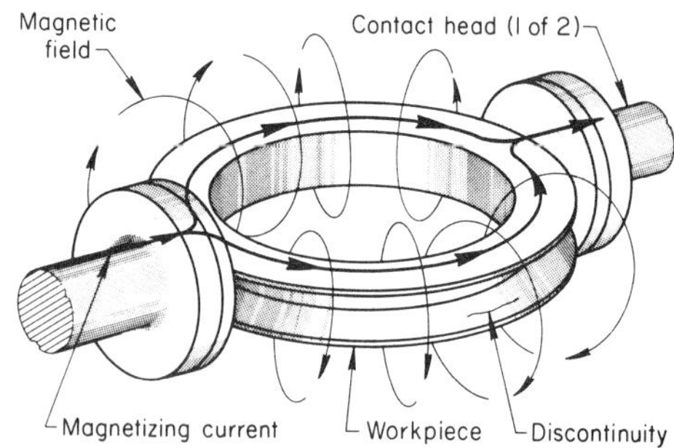

Fig. 14 Current and magnetic-field distribution in a ring being magnetized with a head shot. Because the regions at the contact points are not magnetized, two operations are required for full coverage. With the induced-current method, parts of this shape can be completely magnetized in one operation.

Pole pieces used in conjunction with this method are preferably constructed of laminated ferromagnetic material to minimize the flow of eddy currents within the pole pieces, which detract from the induced (eddy) current developed within the part being processed. Pole pieces can also be made of rods, wire-filled nonconductive tubes, or thick-wall pipe saw cut to break up the eddy-current path. In some cases, even a solid shaft protruding from one side of a gear or disk can be used as one of the pole pieces.

Inspection of Steel Balls. Direct contact is not permitted during the inspection of hardened, finished steel balls for heat treating or grinding cracks, because of the highly polished surface finish. The discontinuities may be oriented in any direction, and 100% inspection of the balls is required. The induced-current method can provide the required in-

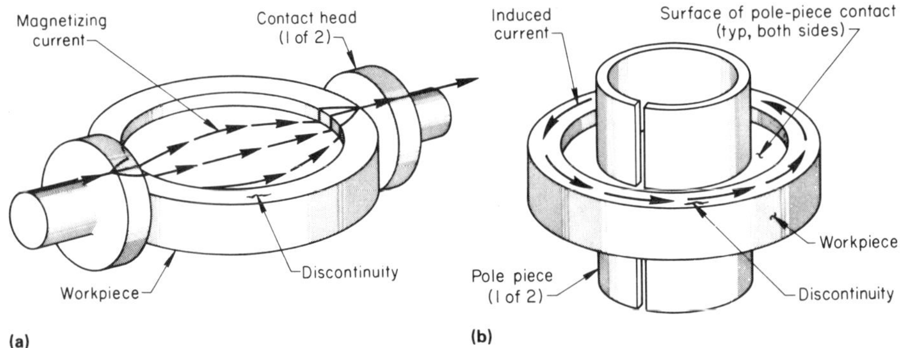

Fig. 15 Current paths in a rimmed disk-shaped part that has been magnetized by (a) head-shot magnetization and (b) induced-current magnetization

spection without damaging the surface finish. The *L/D* ratio of 1:1 for spheres is unfavorable for magnetization with a coil; therefore, laminated pole pieces are used on each side of the balls to provide a more favorable configuration for magnetizing. Because of the highly retentive nature of the material, residual magnetization with direct current and quick-break circuitry is used for magnetizing the balls. The smallness of the heat-treating or grinding cracks and the high surface finish dictate that the inspection medium be a highly oil-suspendible material.

Balls are inspected along the *x*-, *y*-, and *z*-axes in three separate operations. The operation for each axis consists of:

- An induced-current shot
- Bathing the ball with the wet-particle solution
- Inspection while rotating the ball 360°

Rotation and reorientation can be accomplished in a simple manually operated fixture, or the entire operation can be automated.

Permeability of Magnetic Materials

The term permeability is used to refer to the ease with which a magnetic field or flux can be set up in a magnetic circuit. For a given material, it is not a constant value but a ratio. At any given value of magnetizing force, permeability, μ, is B/H, the ratio of flux density, B, to magnetizing force, H. Several permeabilities have been defined, but material permeability, maximum permeability, effective (apparent) permeability, and initial permeability are used with magnetic particle testing.

Material permeability is of interest in magnetic particle inspection with circular magnetization. Material permeability is the ratio of the flux density, B, to the magnetizing force, H, where the flux density and magnetizing force are measured when the flux path is entirely within the material. The magnetizing force and the flux density produced by that force are measured point by point for the entire magnetization curve with a fluxmeter and a prepared specimen of material.

Maximum Permeability. For magnetic particle inspection, the level of magnetization is generally chosen to be just below the knee of a normal magnetization curve for the specific material; the maximum material permeability occurs near this point. For most engineering steels, the maximum material permeability ranges from 0.06 to 0.25 $T/A \cdot m^{-1}$ (500 to 2000 G/Oe) or more. The 500 value is for 400-series stainless steels. Specific permeability values for the various engineering materials are not readily available, but even if they were, they could be misleading. To a large extent, the numerous rules of thumb consider the variations in permeability, so that knowledge of permeability values is not a prerequisite for magnetic particle inspection.

Effective (apparent) permeability is the ratio of the flux density in the part to the magnetizing force, when the magnetizing force is measured at the same point in the absence of the part. Effective permeability is not solely a property of the material, but is largely governed by the shape of the part and is of prime importance for longitudinal magnetization.

Initial permeability is exhibited when both the flux density, B, and the magnetizing force, H, approach zero (Fig. 16a). With increasing magnetizing force, the magnetic field in the part increases along the virgin curve of the hysteresis loop.

Magnetic Hysteresis

Some ferromagnetic materials, when magnetized by being introduced into an external field, do not return to a completely unmagnetized state when removed from that field. In fact, these materials must be subjected to a reversed field of a certain strength to demagnetize them (discounting heating the material to a characteristic temperature, called the Curie point, above which the ferromagnetic ordering of atomic moments is thermally destroyed, or mechanically working the material to reduce the magnetization). If an external field that can be varied in a controlled manner is applied to a completely demagnetized (virgin) specimen and if instrumentation for measuring the magnetic induction within the specimen is at hand, the magnetization curve of the material can be determined. A representative magnetization (hysteresis) curve for a ferromagnetic material is shown in Fig. 16.

As shown in Fig. 16(a), starting at the origin (O) with the specimen in the unmagnetized condition and increasing the magnetizing force in small increments, the flux in the material increases quite rapidly at first,

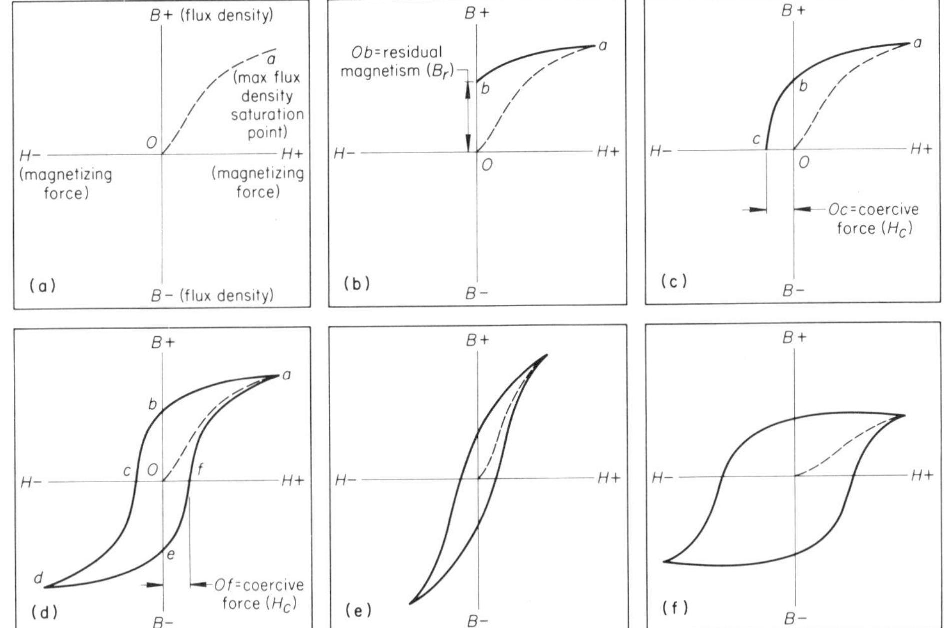

Fig. 16 Representative magnetization (hysteresis) curve for a ferromagnetic material. Dashed line in each of the charts is the virgin hysteresis curve. See text for discussion.

then more slowly until it reaches a point beyond which any increase in the magnetizing force does not increase the flux density. This is shown by the dashed (virgin) curve *Oa*. In this condition, the specimen is said to be magnetically saturated.

When the magnetizing force is gradually reduced to zero, curve *ab* results (Fig. 16b). The amount of magnetism that the steel retains at point *b* is called residual magnetism, B_r.

When the magnetizing current is reversed and gradually increased in value, the flux continues to diminish. The flux does not become zero until point *c* is reached, at which time the magnetizing force is represented by *Oc* (Fig. 16c), which graphically designates the coercive force, H_c, in the material. Ferromagnetic materials retain a certain amount of residual magnetism after being subjected to a magnetizing force. When the magnetic domains of a ferromagnetic material have been oriented by a magnetizing force, some domains remain so oriented until an additional force in the opposite direction causes them to return to their original random orientation. This force is commonly referred to as coercive force.

As the reversed field is increased beyond *c*, point *d* is reached (Fig. 16d). At this point, the specimen is again magnetically saturated. The magnetizing force is now decreased to zero, and the *de* line is formed and retains reversed-polarity residual magnetism, B_r, in the specimen. Further increasing the magnetizing force in the original direction completes the curve *efa*. The cycle is now complete, and the area within the loop *abcdefa* is called the hysteresis curve.

The definite lag throughout the cycle between the magnetization force and the flux is called hysteresis. If the hysteresis loop is slender (Fig. 16e), the indication usually means that the material has low retentivity (low residual field) and is easy to magnetize (has low reluctance). A wide loop (Fig. 16f) indicates that the material has high reluctance and is difficult to magnetize.

Magnetic Particles and Suspending Liquids

Magnetic particles are classified according to the medium used to carry the particles to the part. The medium can be air (dry-particle method) or a liquid (wet-particle method). Magnetic particles can be made of any low-retentivity ferromagnetic material that is finely subdivided. The characteristics of this material, including magnetic properties, size, shape, density, mobility, and degree of visibility and contrast, vary over wide ranges for different applications.

Magnetic Properties. The particles used for magnetic particle inspection should have high magnetic permeability so that they can be readily magnetized by the low-level leakage fields that occur around discontinuities and can be drawn by these fields to the discontinuities themselves to form readable indications. The fields at very fine discontinuities are sometimes extremely weak.

Low coercive force and low retentivity are desirable for magnetic particles. If high in coercive force, wet particles become strongly magnetized and form an objectionable background. In addition, the particles will adhere to any steel in the tank or piping of the unit and cause heavy settling-out losses. Highly retentive wet particles tend to clump together quickly in large aggregates on the test surface. Excessively large clumps of particles have low mobility, and indications are distorted or obscured by the heavy, coarse-grain backgrounds.

Dry particles having coercive force and high retentivity would become magnetized during manufacture or in first use and would therefore become small, strong permanent magnets. Once magnetized, their control by the weak fields at discontinuities would be subdued by their tendency to stick magnetically to the test surface wherever they first touch. This would reduce the mobility of the powder and would form a high level of background, which would reduce contrast and make indications difficult to see.

Effect of Particle Size. Large, heavy particles are not likely to be arrested and held by weak fields when such particles are moving over a part surface, but fine particles will be held by very weak fields. However, extremely fine particles may also adhere to surface areas where there are no discontinuities (especially if the surface is rough) and form confusing backgrounds. Coarse dry particles fall too fast and are likely to bounce off the part surface without being attracted by the weak leakage fields at imperfections. Finer particles can adhere to fingerprints, rough surfaces, and soiled or damp areas, thus obscuring indications.

Magnetic particles for the wet method are applied as a suspension in some liquid medium, and particles much smaller than those for the dry method can be used. When such a suspension is applied over a surface, the liquid drains away, and the film remaining on the surface becomes thinner. Coarse particles would quickly become stranded and immobilized. The stranding of finer particles as a result of the draining away of the liquid occurs much later, giving these particles mobility for a sufficient period of time to be attracted by leakage fields and to accumulate and thus form true indications.

Effect of Particle Shape. Long, slender particles develop stronger polarity than globular particles. Because of the attraction exhibited by opposite poles, these tiny, slender particles, which have pronounced north and south poles, arrange themselves into strings more readily than globular particles. The ability of dry particles to flow freely and to form uniformly dispersed clouds of powder that will spread evenly over a surface is a necessary characteristic for rapid and effective dry-powder testing. Elongated particles tend to mat in the container and to be ejected in uneven clumps, but globular particles flow freely and smoothly under similar conditions. The greatest sensitivity for the formation of strong indications is provided by a blend of elongated and globular shapes.

Wet particles, because they are suspended in a liquid, move more slowly than do dry particles to accumulate in leakage fields. Although the wet particles themselves may be of any shape, they are fine and tend to agglomerate, or clump, into unfavorable shapes. These unfavorable shapes will line up into magnetically held elongated aggregates even when the leakage field is weak. This effect contributes to the relatively high sensitivity of fine wet particles.

Visibility and contrast are promoted by choosing particles with colors that are easy to see against the color of the surface of the part being inspected. The natural color of the metallic powders used in the dry method is silver-gray, but pigments are used to color the particles. The colors of particles for the wet method are limited to the black and red of the iron oxides commonly used as the base for wet particles.

For increased visibility, particles are coated with fluorescent pigment by the manufacturer. The search for indications is conducted in total or partial darkness, using ultraviolet light to activate the fluorescent dyes. Fluorescent magnetic particles are available for both the wet and dry methods, but fluorescent particles are more commonly used with the wet method.

Types of Magnetic Particles. The two primary types of particles for magnetic particle inspection are dry particles and wet particles, and each type is available in various colors and as fluorescent particles. Particle selection is principally influenced by:

- Location of the discontinuity, that is, on the surface or beneath the surface
- Size of the discontinuity, if on the surface
- Which type (wet or dry particles) is easier to apply

Dry particles, when used with direct current for magnetization, are superior for detecting discontinuities lying wholly below the surface. The use of alternating current with dry particles is excellent for revealing surface cracks that are not exceedingly fine, but is of little value for discontinuities even slightly beneath the surface (Fig. 17).

Wet particles are better than dry particles for detecting very fine surface discontinuities regardless of which form of magnetizing current is used. Wet particles are often used with direct current to detect discontinuities

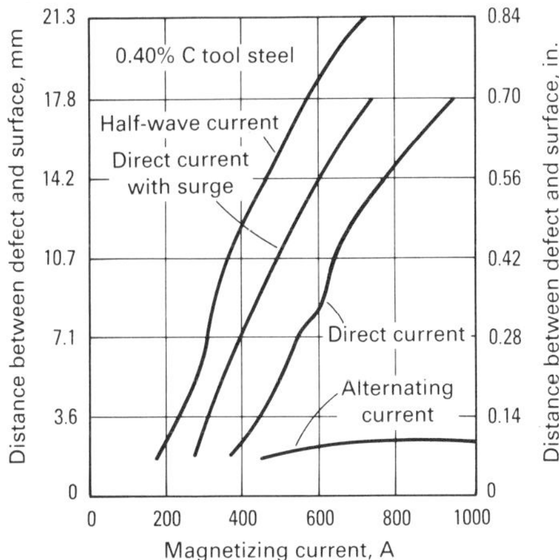

Fig. 17 Comparison of sensitivity of alternating current, direct current, direct current with surge, and half-wave current for locating defects at various distances below a surface by means of the dry-particle continuous magnetizing method. See text for description of test conditions.

Fig. 18 Black magnetic-powder indications of a cold shut in a casting when seen under black light. Courtesy of Magnaflux Corporation

that lie just beneath the surface. The surface of a part can easily be covered with a wet bath because the bath flows over and around surface contours. This is not easily accomplished with dry powders.

Colored particles are used to obtain maximum contrast with the surface of the part being inspected. Black stands out against most light-colored surfaces (Fig. 18), and gray against dark-colored surfaces. Red particles are more visible than black or gray particles against silvery and polished surfaces. Fluorescent particles, viewed under ultraviolet light, provide the highest contrast and visibility.

Dry particles are available with yellow, red, black, and gray pigmented coloring and with fluorescent coatings. The magnetic properties and particle sizes are similar in all colors, making them equally efficient. Dry particles are most sensitive for use on very rough surfaces and for detecting flaws beneath the surface. They are ordinarily used with portable equipment. The reclamation and reuse of dry particles is not recommended.

Air is used to carry the particles to the surface of the part, and care must be taken to apply the particles correctly. Dry powders should be applied in such a way that they reach the magnetized surface in a uniform cloud with a minimum of motion. When this is done, the particles come under the influence of the leakage fields while suspended in air and have three-dimensional mobility. This condition can be best achieved when the magnetized surface is vertical or overhead. When particles are applied to a horizontal or sloping surface, they settle directly to the surface and do not have the same degree of mobility.

Dry powders can be applied with small rubber spray bulbs or specially designed mechanical powder blowers. The air stream of such a blower is of low velocity so that a cloud of powder is applied to the test area. Mechanical blowers can also deliver a light stream of air for the gentle removal of excess powder. Powder that is forcibly applied is not free to be attracted by leakage fields. Neither rolling a magnetized part in powder nor pouring the powder on the part is recommended.

Wet particles are best suited for the detection of fine discontinuities such as fatigue cracks. Wet particles are commonly used in stationary equipment where the bath can remain in use until contaminated. They are also used in field operations with portable equipment, but care must be taken to agitate the bath constantly.

Wet particles are available in red and black colors or as fluorescent particles that fluoresce a blue-green or a bright yellow-green color. The particles are supplied in the form of a paste or other type of concentrate that is suspended in a liquid to produce the coating bath.

The liquid bath may be either water or a light petroleum distillate having specific properties. Both require conditioners to maintain proper dispersion of the particles and to allow the particles the freedom of movement to form indications on the surfaces of parts. These conditioners are usually incorporated in the powder concentrates.

Oil Suspending Liquid. The oil used as a suspending liquid for magnetic particles should be an odorless, well-refined light petroleum distillate of low viscosity having a low sulfur content and a high flash point. The viscosity of the oil should not exceed 3 $\times$ 10^{-6} m^2/s (3 cSt) as tested at 40 °C (100 °F) and must not exceed 5 $\times$ 10^{-6} m^2/s (5 cSt) when tested at the bath temperature. Above 5 $\times$ 10^{-6} m^2/s (5 cSt), the movement of the magnetic particles in the bath is sufficiently retarded to have a definite effect in reducing buildup, and therefore the visibility, of an indication of a small discontinuity. Parts should be precleaned to remove oil and grease because oil from the surface accumulates in the bath and increases its viscosity.

Water Suspending Liquid. The use of water instead of oil for magnetic particle wet-method baths reduces costs and eliminates bath flammability. Figures 19 through 22 illustrate the effectiveness of water-base magnetic particle testing. Water-suspendible particle concentrates include the necessary wetting agents, dispersing agents, rust inhibitors, and antifoam agents.

Water baths ordinarily should not be used where the temperature is below freezing.

Fig. 19 Spindle defects revealed with water-base magnetic particle testing. Courtesy of Circle Chemical Company

However, ethylene glycol can be employed to protect against reasonably low temperatures. Care must be exercised, however, because a high percentage of ethylene glycol can impede particle mobility. For the inspection of large billets, a solution containing three parts water and one part antifreeze has been successfully used.

Because water is a conductor of electricity, units in which water is to be used must be designed to isolate all high-voltage circuits so as to avoid all possibility of operator shock, and the equipment must be thoroughly and positively grounded. Also, electrolysis of parts of the unit can occur if preventive measures are not taken. Units specifically designed to be used with water as a suspensoid are safe for the operator and minimize electrolytic corrosion.

The strength of the bath is a major factor in determining the quality of the indications obtained. The proportion of magnetic particles in the bath must be maintained at a uniform level. If the concentration varies, the strength of the indications will also vary, and the indications may be misinterpreted. Fine indications may be missed entirely with a weak bath. Too heavy a concentration of particles gives a confusing background and excessive adherence of particles at external poles, thus interfering with clean-cut indications of extremely fine discontinuities.

The best method for ensuring optimum bath concentration for any given combination of equipment, bath application, type of part, and discontinuities sought is to test the bath using parts with known discontinuities. Bath strength can be adjusted until satisfactory indications are obtained. This bath concentration can then be adopted as standard for those conditions.

The concentration of the bath can be measured with reasonable accuracy with the settling test. In this test, 100 mL (3.4 oz) of well-agitated bath is placed in a pear-shaped centrifuge tube. The volume of solid material that settles out after a predetermined interval (usually 30 min) is read on the graduated cylindrical part of the tube. Dirt in the bath will also settle and usually shows as a separate layer on top of the oxide. The layer of dirt is usually easily discernible because it is different in color from the magnetic particles.

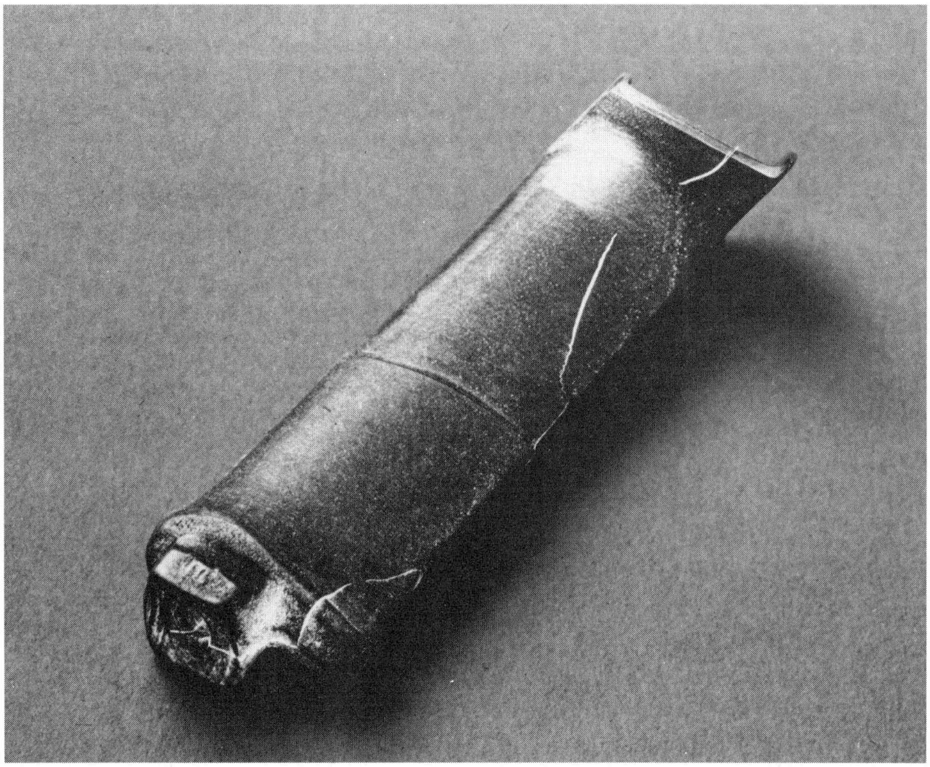

Ultraviolet Light

A mercury-arc lamp is a convenient source of ultraviolet light. This type of lamp emits light whose spectrum has several intensity peaks within a wide band of wavelengths. When used for a specific purpose, emitted light is passed through a suitable filter so that only a relatively narrow band of ultraviolet wavelengths is available. For example, a band in the long-wave ultraviolet spectrum is used for fluorescent liquid penetrant or magnetic particle inspection.

Fluorescence is the characteristic of an element or combination of elements to absorb the energy of light at one frequency and emit light of a different frequency. The fluorescent materials used in liquid penetrant and magnetic particle inspection are combinations of elements chosen to absorb light in the peak energy band of the mercury-arc lamp fitted with a Kopp glass filter. This peak occurs at about 365 nm (3650 Å). The ability of fluorescent materials to emit

Fig. 20 Compressor vane microcracks revealed with water-base magnetic particle testing. Courtesy of Circle Chemical Company

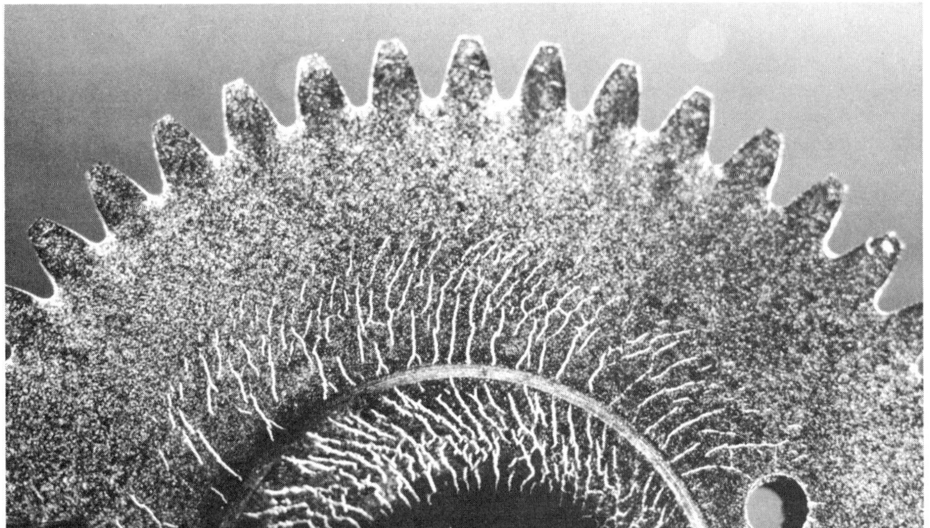

Fig. 21 50 mm (2 in.) diam gear subjected to water-base magnetic particle testing showing cracks 0.25 mm (0.0098 in.) long by 0.1 mm (0.004 in.) wide. Courtesy of Circle Chemical Company

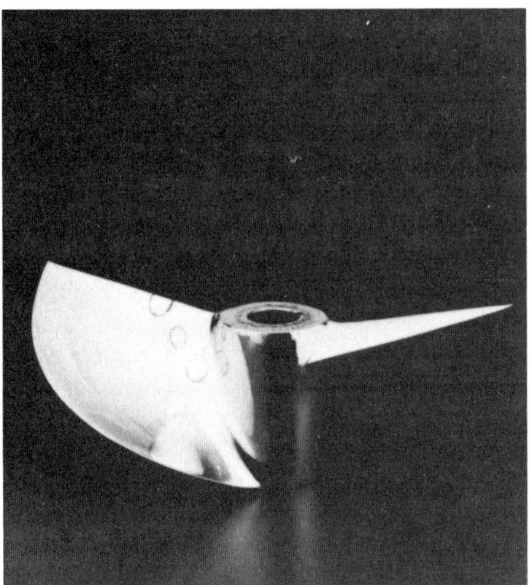

(a)

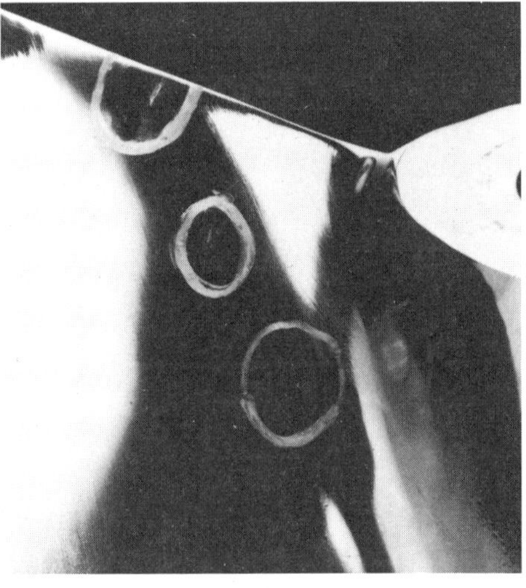

(b)

Fig. 22 Defects in the leading edge and fillet areas of a hydroplane propeller as revealed by water-base magnetic particle testing. (a) Overall view. (b) Close-up of fillet area. Courtesy of Circle Chemical Company

light in the greenish-yellow wavelengths of the visible spectrum depends on the intensity of ultraviolet light at the workpiece surface. In contrast to the harmful ultraviolet light of shorter wavelengths, which damages organs such as the eyes and the skin, the black light of 365 nm (3650 Å) wavelengths poses no such hazards to the operator and provides visible evidence of defects in materials, as shown in Fig. 23 through 28.

Early specifications required 970 lx (90 ftc) of illumination at the workpiece surface measured with a photographic-type light meter, which responds to white light as well as ultraviolet light. Because the ultraviolet output of a mercury-arc lamp decreases with age and with hours of service, the required intensity of the desired wavelength is often absent, although the light meter indicates otherwise. Meters have been developed that measure the overall intensity of long-wave ultraviolet light only in a band between 300 to 400 nm (3000 to 4000 Å) and that are most sensitive near the peak energy band of the mercury-arc lamp used for fluorescent inspection. These meters read the intensity level in microwatts per square centimeter ($\mu W/cm^2$). For aircraft-quality fluorescent inspection, the minimum intensity level of ultraviolet illumination is 1000 $\mu W/cm^2$. High-intensity 125-W ultraviolet bulbs are available that provide up to 5000 $\mu W/cm^2$ at 380 mm (15 in.).

Nomenclature Used in Magnetic Particle Inspection

An indication is an accumulation of magnetic particles on the surface of the part that forms during inspection.

Relevant indications are the result of errors made during or after metal processing. They may or may not be considered defects.

A nonrelevant indication is one that is caused by flux leakage. This type of indication is usually weak and has no relation to a discontinuity that is considered to be a defect. Examples are magnetic writing, change in section due to part design, or a heat affected zone line in welding.

False indications are those in which the particle patterns are held by gravity or surface roughness. No magnetic attraction is involved.

A discontinuity is any interruption in the normal physical configuration or composition of a part. It may not be a defect.

A defect is any discontinuity that interferes with the utility or service of a part.

Interpretation consists of determining the probable cause of an indication, and assigning it a discontinuity name or label.

Evaluation involves determining whether an indication will be detrimental to the service of a part. It is a judgement based on a well-defined accept/reject standard that may be either written or verbal.

Detectable Discontinuities

The usefulness of magnetic particle inspection in the search for discontinuities or imperfections depends on the types of discontinuities the method is capable of finding. Of importance are the size, shape, orientation, and location of the discontinuity with respect to its ability to produce leakage fields.

Surface Discontinuities. The largest and most important category of discontinuity consists of those that are exposed to the surface. Surface cracks or discontinuities are effectively located with magnetic particles. Surface cracks are also more detrimental to the service life of a component than are subsurface discontinuities, and as a result they are more frequently the object of inspection.

Magnetic particle inspection is capable of locating seams, laps, quenching and grinding cracks, and surface ruptures in castings, forgings, and weldments. The method will also detect surface fatigue cracks developed during service. Magnetizing and particle application methods may be critical in certain cases, but in most applications the requirements are relatively easily met because leakage fields are usually strong and highly localized.

For the successful detection of a discontinuity, there must be a field of sufficient strength oriented in a generally favorable direction to produce strong leakage fields. For maximum detectability, the field set up in the part should be at right angles to the length of a suspected discontinuity (Fig. 3 and 4). This is especially true if the discon-

Fig. 23 Casting viewed under black light showing strong magnetic particle indication with minimal background fluorescence. Courtesy of Magnaflux Corporation

tinuity is small and fine. The characteristics of a discontinuity that enhance its detection are:

- Its depth is at right angles to the surface
- Its width at the surface is small, so that the air gap it creates is small
- Its length at the surface is large with respect to its width
- It is comparatively deep in proportion to the width of its surface opening

Many incipient fatigue cracks and fine grinding cracks are less than 0.025 mm (0.001 in.) deep and have surface openings of perhaps one-tenth that or less. Such cracks are readily located using wet-method magnetic particle inspection. The depth of the crack has a pronounced effect on its detectability; the deeper the crack, the stronger the indication for a given level of magnetization. This is because the stronger leakage flux causes greater distortion of the field in the part. However, this effect is not particularly noticeable beyond perhaps 6.4 mm (¼ in.) in depth. If the crack is not close-lipped but wide open at the surface, the reluctance (opposition to the establishment of magnetic flux in a magnetic circuit) of the resulting longer air gap reduces the strength of the leakage field. This, combined with the inability of the particles to bridge the gap, usually results in a weaker indication.

Detectability generally involves a relationship between surface opening and depth. A surface scratch, which may be as wide at the surface as it is deep, usually does not produce a magnetic particle pattern, although it may do so at high levels of magnetization. Because of many variables, it is not possible to establish any exact values for this relationship, but in general a surface discontinuity whose depth is at least five times its opening at the surface will be detectable.

There are also limitations at the other extreme. For example, if the faces of a crack are tightly forced together by compressive stresses, the almost complete absence of an air gap may produce so little leakage field that no particle indication is formed. Shallow cracks produced in grinding or heat treating and subsequently subjected to strong compression by thermal or other stresses usually produce no magnetic particle indications. Sometimes, with careful, maximum-sensitivity techniques, faint indications of such cracks can be produced.

One other type of discontinuity that sometimes approaches the lower limit of detectability is a forging or rolling lap that, although open to the surface, emerges at an acute angle. In this case, the leakage field produced may be quite weak because of the small angle of emergence and the resultant relatively high reluctance of the actual air gap; consequently, very little leakage flux takes the path out through the surface lip of the lap to cross this high reluctance gap. When laps are being sought (usually when newly forged parts are being inspected), high-sensitivity, such as combining dc magnetizing with the wet fluorescent method, is desirable. Figure 29 shows two indications of forging laps in a 1045 steel crane hook.

A seam that was found during the magnetic particle inspection of a forged crane hook is shown in Fig. 30. The seam was present in the material before the hook was forged. The cold shut shown in Fig. 31 was found in the flange of a cast drum after machining. A faint indication was noted in the rough casting, but the size of the cold shut was not known until after machining of the drum.

The magnetic particle inspection of a 460 mm (18 in.) diam internally splined coupling revealed the indications shown in Fig. 32, one of which was along the fusion zone of a repair weld. Routine magnetic particle in-

Fig. 24 Potentially dangerous cracks in a lawn mower blade revealed when fluorescent magnetic particles are exposed to black light. Courtesy of Magnaflux Corporation

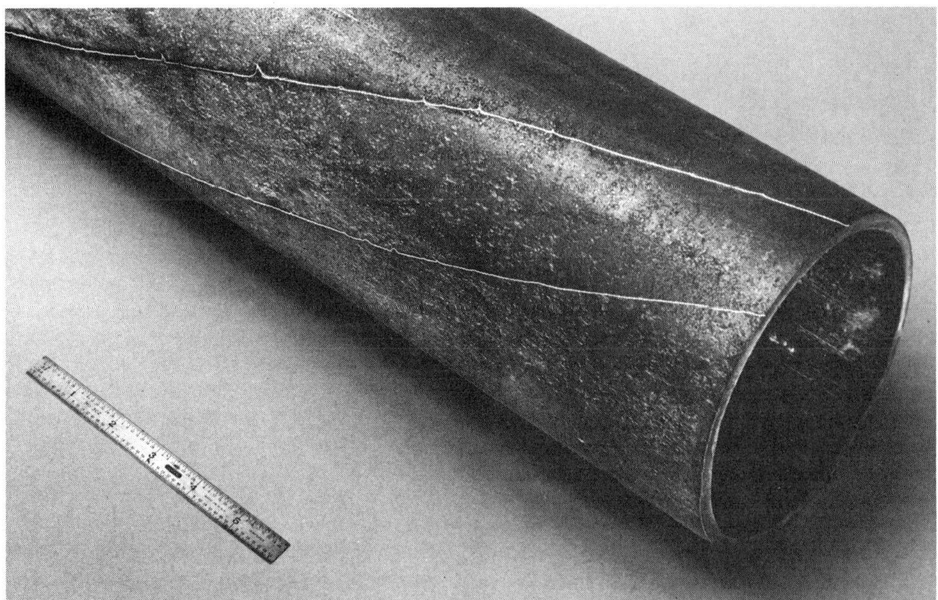

Fig. 25 Fluorescent magnetic particle indications of quenched-and-drawn drill casing tubing that shows rejectable, cracked spiral seams in the tubing when exposed to black light. Courtesy of Magnaflux Corporation

spection of a 1.2 m (4 ft) diam weldment revealed cracks in the weld between the rim and web, as shown in Fig. 33.

Internal Discontinuities. The magnetic particle method is capable of indicating the presence of many discontinuities that do not break the surface. Although radiography and ultrasonic methods are inherently better for locating internal discontinuities, sometimes the shape of the part, the location of the discontinuity, or the cost or availability of the equipment needed makes the magnetic particle method more suitable. The internal discontinuities that can be detected by magnetic particle inspection can be divided into two groups:

- Subsurface discontinuities (those lying just beneath the surface of the part)
- Deep-lying discontinuities

Subsurface discontinuities comprise those voids or nonmetallic inclusions that lie just beneath the surface. Nonmetallic inclusions are present in all steel products to some degree. They occur as scattered individual inclusions, or they may be aligned in long stringers. These discontinuities are usually very small and cannot be detected unless they lie very close to the surface, because they produce highly localized but rather weak fields.

Deep-lying discontinuities in weldments may be caused by inadequate joint penetration, subsurface incomplete fusion, or cracks in weld beads beneath the last weld bead applied. In castings, they result from internal shrinkage cavities, slag inclusions, or gas pockets. The depth to which magnetic particle testing can reach in locating internal discontinuities cannot be established in millimeters, because the size and shape of the discontinuity itself in relation to the size of the part in which it occurs is a controlling factor. Therefore, the deeper the discontinuity lies within a section, the larger it must be to be detected by magnetic particle inspection.

In considering the detectability of a discontinuity lying below the surface, of primary concern are the projected area presented as an obstruction to the lines of force and the sharpness of the distortion of the field produced. It is helpful to think of the magnetic field as flowing through the specimen like a stream of water. A coin, on edge below the surface and at right angles to the surface and flow, would cause a sharp disturbance in the movement of the water, but a straight round stick placed at a similar depth and parallel to the direction of flow would have very little effect. A ball or marble of the same size as the coin would present the same projected area, but would be much less likely to be detected than the coin because the flow lines would be streamlines around the sphere and the disturbance created in the field would be much less sharp.

The orientation of the discontinuity is another factor in detection. The coin-shaped obstruction discussed above was considered to be 90° to the direction of the flux. If the same discontinuity were inclined, either vertically or horizontally, at an angle of only 60 or 70°, there would be a noticeable difference in the amount of leakage field and therefore in the strength of the indication. This difference would result not only because the projected area would be reduced but also because of a streamlining effect, as with the sphere.

The strength and direction of the magnetic field are also important in the detection of deep-lying discontinuities. Direct current yokes are effective if the discontinuity is close to the surface. Prod magnetization using direct current or half-wave current is more effective for discontinuities deeper in the part.

Nonrelevant Indications

Nonrelevant indications are true patterns caused by leakage fields that do not result from the presence of flaws. Nonrelevant indications have several possible causes and therefore require evaluation, but they should not be interpreted as flaws.

Sources of Nonrelevant Indications

Particle patterns that yield nonrelevant indications can be the result of design, fabrication, or other causes and do not imply a condition that reduces the strength or utility of the part. Because nonrelevant indications are true particle buildups, they are difficult to distinguish from buildups caused by flaws. Therefore, the investigator must be aware of design and fabrication conditions that would contribute to or cause nonrelevant indications.

Particle Adherence Due to Excessive Magnetizing Force. One type of nonrelevant indication is that caused by particle adherence at leakage fields around sharp corners, ridges, or other surface irregularities when magnetized longitudinally with too strong a magnetizing force. The use of too strong a current with circular magnetization can produce indications of the flux lines of the external field. Both of the above phenomena (excessive magnetizing force or excessive current) are clearly recognized by experienced operators and can be eliminated by a reduction in the applied magnetizing force.

Mill Scale. Tightly adhering mill scale will cause particle buildup, not only because of mechanical adherence but also because of the difference in magnetic permeability between the steel and the scale. In most cases, this can be detected visually, and additional cleaning followed by retesting will confirm the absence of a true discontinuity.

Configurations that result in a restriction of the magnetic field are a cause of nonrelevant indications. Typical restrictive configurations are internal notches such as splines, threads, grooves for indexing, or keyways.

Abrupt changes in magnetic properties, such as those between weld metal and base metal or between dissimilar base metals, result in nonrelevant indications. Depending on the degree of change in the magnetic property, the particle pattern may consist of loosely adhering particles or may be strong and well defined. Again, it is necessary for the investigator to be aware of such conditions.

Fig. 26 Black light used to reveal magnetic particle indication of a crack in the centering stud of a ball-yoke-type universal-joint section of a drive-shaft assembly. Courtesy of Magnaflux Corporation

Magnetized writing is another form of nonrelevant indication. Magnetic writing is usually associated with parts displaying good residual characteristics in the magnetized state. If such a part is contacted with a sharp edge of another (preferably magnetically soft) part, the residual field is locally reoriented, giving rise to a leakage field and consequently a magnetic particle indication. For example, the point of a common nail can be used to write on a part susceptible to magnetic writing. Magnetic writing is not always easy to interpret, because the particles are loosely held and are fuzzy or intermittent in appearance. If magnetic writing is suspected, it is only necessary to demagnetize the parts and retest. If the indication was magnetic writing, it will not reappear.

Additional Sources. Some other conditions that cause nonrelevant indications are brazed joints, voids in fitted parts, and large grains.

Distinguishing Relevant From Nonrelevant Indications

There are several techniques for differentiating between relevant and nonrelevant indications:

- Where mill scale or surface roughness is the probable cause, close visual inspection of the surface in the area of the discontinuity and use of magnification up to ten diameters
- Study of a sketch or drawing of the part being tested to assist in locating welds, changes in section, or shape constrictions
- Demagnetization and retesting
- Careful analysis of the particle pattern. The particle pattern typical of nonrelevant indications is usually wide, loose, and lightly adhering and is easily removable even during continuous magnetization
- Use of another method of nondestructive inspection, such as ultrasonic testing or radiography, to verify the presence of a subsurface defect

The following two examples illustrate how nonrelevant indications are used in nondestructive testing to verify product quality.

Example 1: Nonrelevant Indications in Electric Motor Rotors. An instance where nonrelevant indications are used to advantage is in the inspection of rotors for squirrel cage electric motors. These rotors are usually fabricated from laminations made of magnetic material. The conductor-bar holes are aligned during assembly of the laminations. The end rings and conductor bars are cast from an aluminum alloy in a single operation. The integrity of the internal cast aluminum alloy conductor bars must be checked to ensure that each is capable of carrying the required electrical current; voids and internal porosity would impair their electrical properties.

These rotors are tested by clamping them between the heads of a horizontal unit and processing them by the continuous method using wet magnetic particles. The end rings distribute the magnetizing current through each of the conductor bars, which constitute parallel paths. All conductor bars that are sound and continuous produce broad, pronounced, subsurface-type indications on the outside surface of the rotor. The absence of such indications is evidence that the conductor bar is discontinuous or that its current-carrying capacity is greatly impaired. This is a direct departure from the customary inspection logic in that negative results, or the absence of indications, are indicative of defects.

Example 2: Nonrelevant Indications Present in Welding A537 Grade 2 Carbon Steel With E8018-C1 Weld Wire. Linear magnetic particle indications have frequently been observed in the heat-affected zone of A537 grade 2 (quenched-and-tempered) materials joined with E8018-C1 (2½% Ni) weld wire. However, when liquid penetrant examinations of these areas are performed, no indications are apparent.

Four types of welds were used in the investigation:

- *Weld A*: a T-joint with fillet welds on both sides (Fig. 34)
- *Weld B*: a butt weld (Fig. 35)
- *Weld C*: a pickup simulating a repair where a fit-up device had been torn off (Fig. 36)
- *Weld D*: a T-joint with fillet welds on both sides, one of which contained a longitudinal crack (Fig. 37)

The fourth sample was made in such a manner that it would contain a linear discontinuity in or near the heat-affected zone.

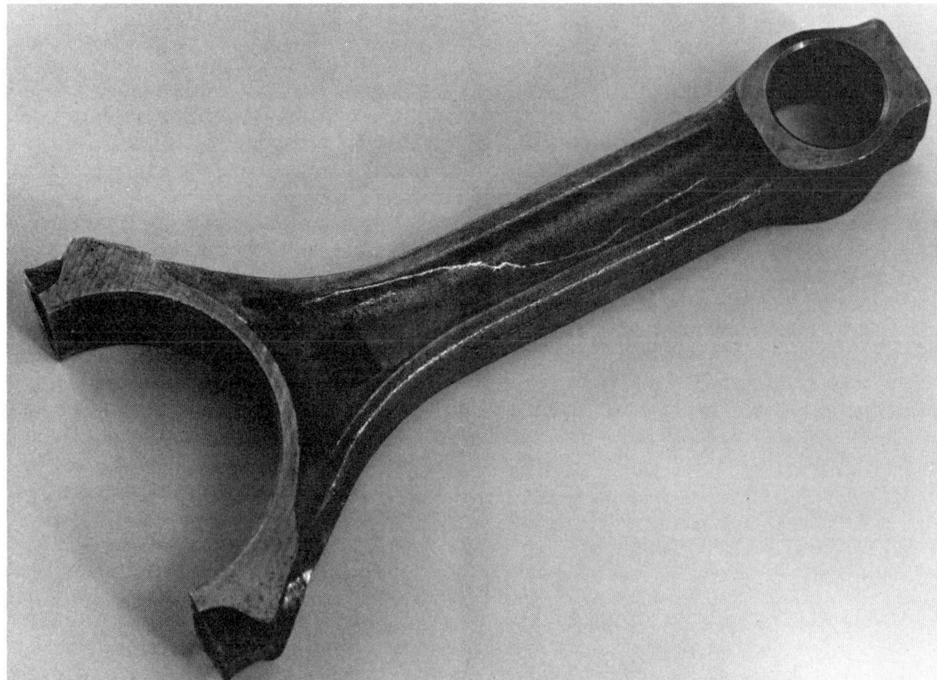

Fig. 27 Black light used to reveal magnetic particle indication of a lap that could lead to potential failure of a forged connecting rod. Courtesy of Magnaflux Corporation

Fig. 28 Black light used to reveal magnetic particle indication of a dangerous crack near a wheel stud. The crack was detected during manufacture. Courtesy of Magnaflux Corporation

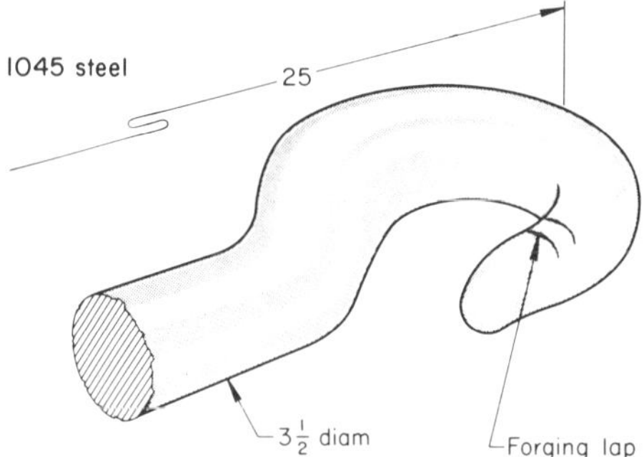

Fig. 29 1045 steel crane hook showing indications of forging laps of the type revealed by magnetic particle inspection. Dimensions given in inches

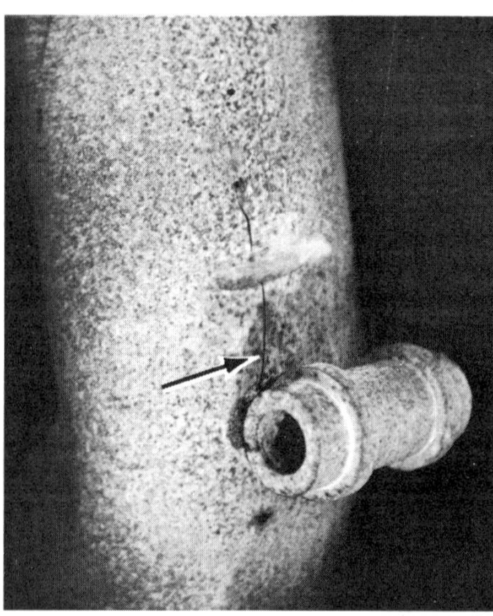

Fig. 30 Magnetic particle indications of a seam (at arrow) in the shank of a forged crane hook

Examination with magnetic particle and liquid penetrant inspection methods yielded the following results:

- *Weld A*: Magnetic particle inspection showed loosely held, slightly fuzzy linear indications at the toes of the fillet, with the bottom edge showing the strongest pattern due to gravity and the configuration. The liquid penetrant examination produced no relevant indications. The etched cross section revealed no discontinuities
- *Weld B*: Magnetic particle inspection showed loosely held, slightly fuzzy linear indications along both edges of the butt weld. The liquid penetrant examination produced no relevant indications. The etched cross section revealed no discontinuities
- *Weld C*: Magnetic particle inspection showed a very loosely held, fuzzy pattern over the entire pickup weld. The liquid penetrant examination produced no relevant indications. The etched cross section revealed only a minor slag inclusion, which had no bearing on the magnetic particle pattern produced
- *Weld D*: Magnetic particle inspection showed a tightly held, sharply defined linear indication along one edge of the fillet with connecting small, transverse linear indications at various locations. The liquid penetrant examination in this case produced the same pattern of indications. The etched cross section revealed a crack completely through the toe of one leg of the fillet

The difference in magnetic properties between this parent material and the weld metal creates magnetic leakage fields in the heat-affected zone or metallic interface of these materials during a magnetic particle inspection. This can result in magnetic par-

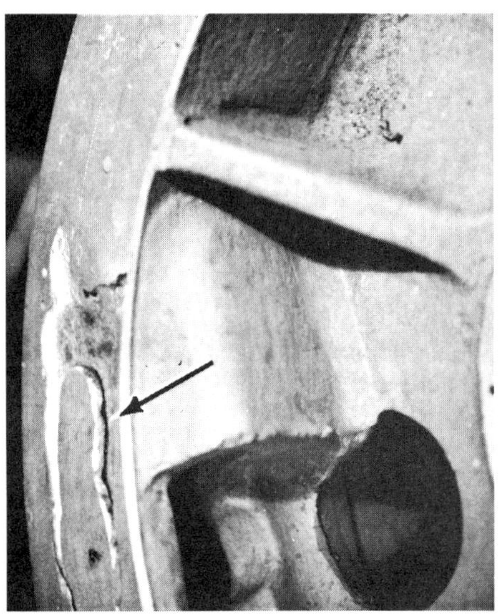

Fig. 31 Cold shut (at arrow) in the flange of a machined cast drum. Magnetic particle inspection revealed faint indications of the cold shut in the rough casting.

Fig. 32 Magnetic particle indications of cracks in a large cast splined coupling. Indication (at arrow) in photo at right is along the fusion zone of a repair weld.

ticle indications in the heat-affected zone along the toe of the fillets, along the edges of butt welds, or over entire shallow pickup welds. The magnetic particle indications produced by metallic interface leakage fields are readily distinguished from indications caused by real discontinuities by their characteristic of being loosely held and slightly fuzzy in appearance and by their location in or at the edge of the heat-affected zone.

General Procedures for Magnetic Particle Inspection

In magnetic particle inspection, there are many variations in procedure that critically affect the results obtained. These variations are necessary because of the many types of discontinuities that are sought and the many types of ferromagnetic materials in which these discontinuities must be detected.

Establishing a set of procedures for the magnetic particle inspection of a specific part requires that the part be carefully analyzed to determine how its size and shape will affect the test results. The magnetic characteristics of the material and the size, shape, location, and direction of the expected discontinuity also affect the possible variations in the procedure. The items that must be considered in establishing a set of procedures for the magnetic particle inspection of a specific part include:

- Type of current
- Type of magnetic particles
- Method of magnetization
- Direction of magnetization
- Magnitude of applied current
- Equipment

Type of Current

The electric current used can be either alternating current or some form of direct current. This choice depends on whether the discontinuities are surface or subsurface and, if subsurface, on the distance below the surface.

Alternating Current. The skin effect of alternating current at 50 or 60 Hz limits its use to the detection of discontinuities that are open to the surface or that are only a few thousandths of an inch below the surface. With alternating current at lower frequencies, the skin effect is less pronounced, resulting in deeper penetration of the lines of force.

The rapid reversal of the magnetic field set up by alternating current imparts mobility to dry particles. Agitation of the powder helps it move to the area of leakage fields and to form stronger indications.

The strength of magnetization, which is determined by the value of the peak current at the top of the sine wave of the cycle, is 1.41 times that of the current indicated on the meter. Alternating current meters indicate more nearly the average current for the cycle than the peak value. Obtaining an equivalent magnetizing effect from straight direct current requires more power and heavier equipment.

Direct current, on the other hand, magnetizes the entire cross section more or less uniformly in a longitudinal direction, and with a straight-line gradient of strength from a maximum at the surface to zero at the center of a bar in the case of circular magnetization. This effect is demonstrated in Fig. 8(a) and (b).

Alternating Current Versus Direct Current. In an experiment designed to compare the effectiveness of 60-Hz alternating current and three types of direct current, 12 holes representing artificial defects were drilled in a 127 mm (5 in.) OD by 32 mm (1¼ in.) ID by 22 mm (⅞ in.) thick ring made of unhardened O1 tool steel (0.40% C). The 12 holes, 1.8 mm (0.07 in.) in diameter and spaced 19 mm (¾ in.) apart, were drilled through the ring parallel to the cylindrical surface at increasing distances from that surface. The centerline distances ranged from 1.8 to 21.3 mm (0.07 to 0.84 in.), in increments of 1.8 mm (0.07 in.). A central conductor, dry magnetic particles, and continuous magnetization were used for this test. The three types of direct current were straight direct current from batteries, three-phase rectified alternating current with surge, and half-wave rectified single-phase 60-Hz alternating current. The threshold values of current necessary to give readable indications of the holes in the ring are plotted in Fig. 17.

Current levels as read on the usual meters were varied from the minimum needed to indicate hole 1 (1.8 mm, or 0.07 in., below the surface) for each type of current, up to a maximum of over 1000 A. To produce an indication at hole 1 using alternating current, about 475 A was required, and at hole

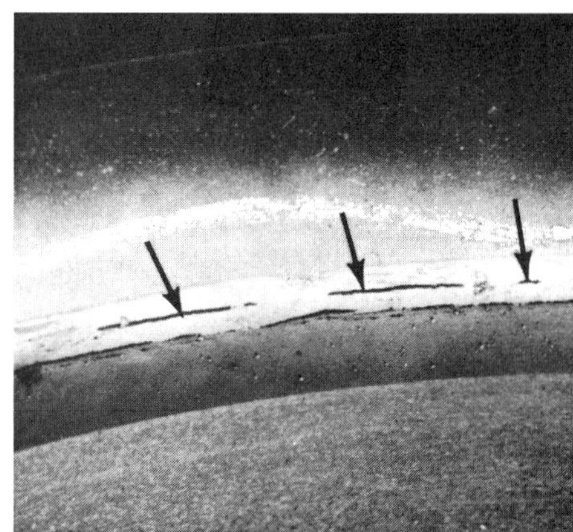

Fig. 33 Indications of cracks (at arrows) in the weld between the web and rim of a 1.2 m (4 ft) diam weldment

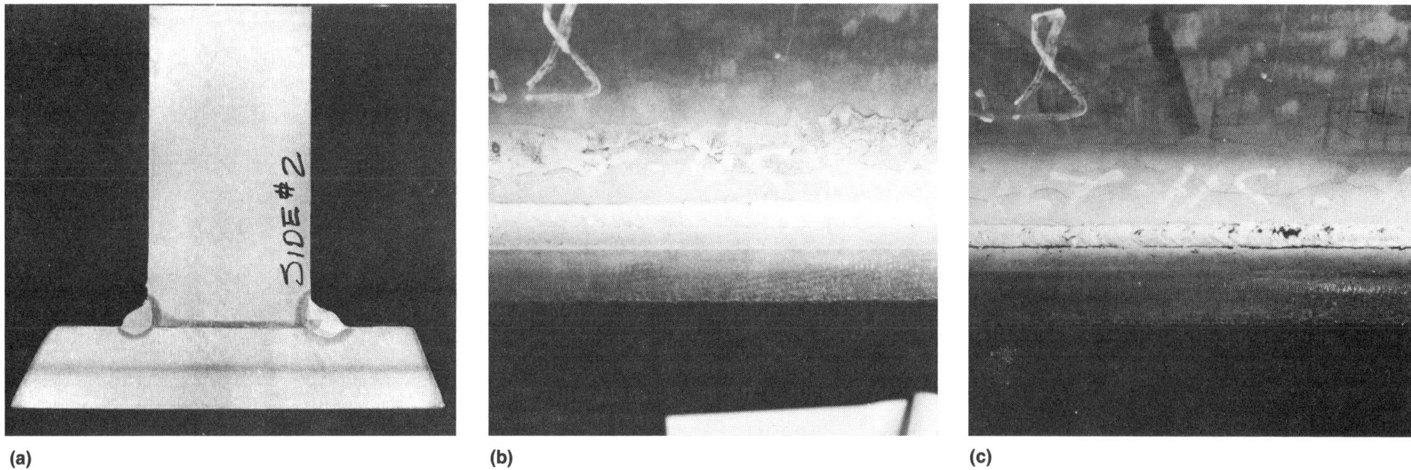

(a) (b) (c)

Fig. 34 T-joint weld of A537 grade 2 (quenched and tempered) material joined with E8018-C1 (2.5% Ni) weld wire showing linear magnetic particle indications. Weld A, shown in cross section (a), was examined along its length using both the (b) liquid penetrant inspection method and the (c) magnetic particle inspection method. See text for discussion. Courtesy of Chicago Bridge & Iron Company

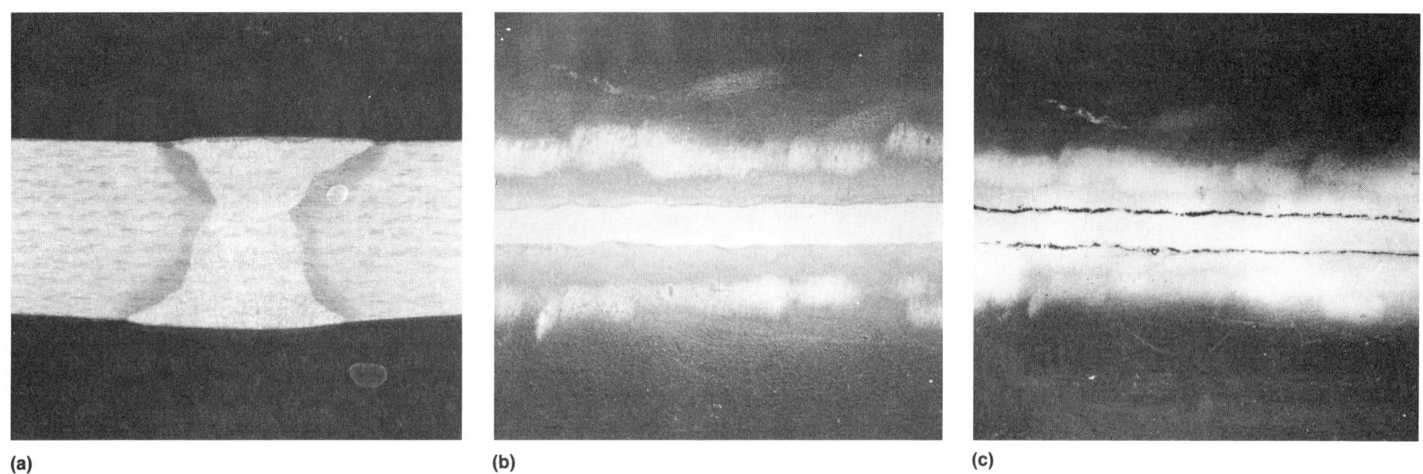

(a) (b) (c)

Fig. 35 Butt weld of A537 grade 2 (quenched and tempered) material joined with E8018-C1 (2.5% Ni) weld wire showing linear magnetic particle indications. Weld B, shown in cross section (a), was checked using both the (b) liquid penetrant inspection method and the (c) magnetic particle inspection method. See text for discussion. Courtesy of Chicago Bridge & Iron Company

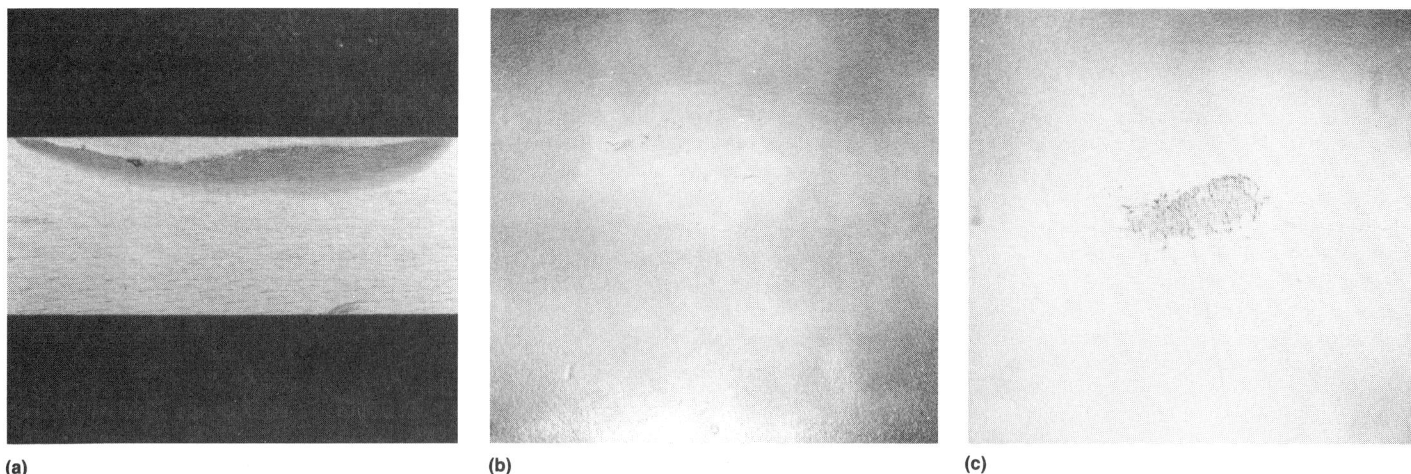

(a) (b) (c)

Fig. 36 A pickup, simulating a repair of A537 grade 2 (quenched and tempered) material joined with E8018-C1 (2.5% Ni) weld wire where a fit-up device had been torn off, showing linear magnetic particle indications. Weld C, shown in cross section (a), was checked using both the (b) liquid penetrant inspection method and the (c) magnetic particle inspection method. See text for discussion. Courtesy of Chicago Bridge & Iron Company

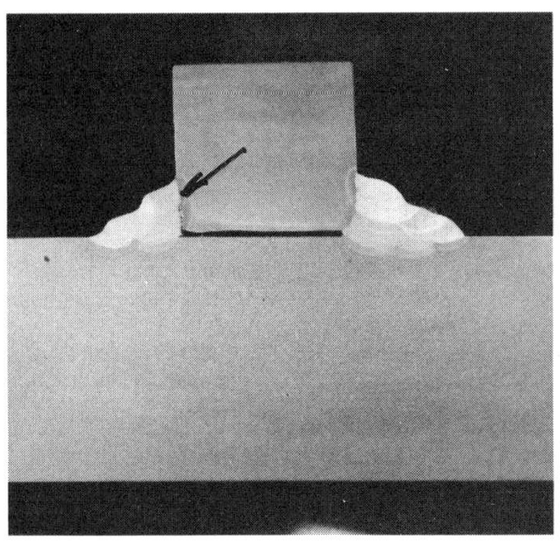

(a)

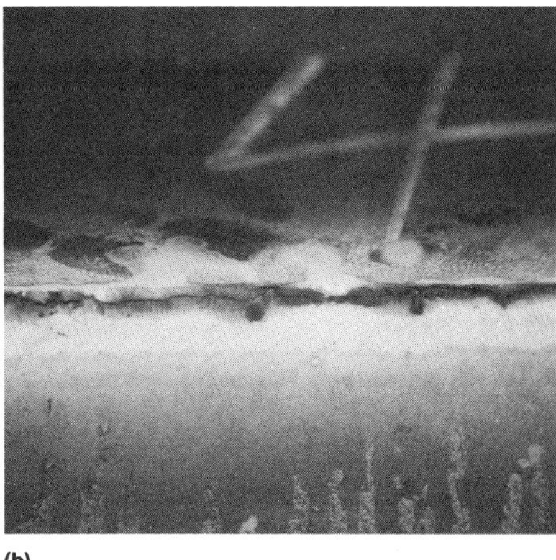

(b)

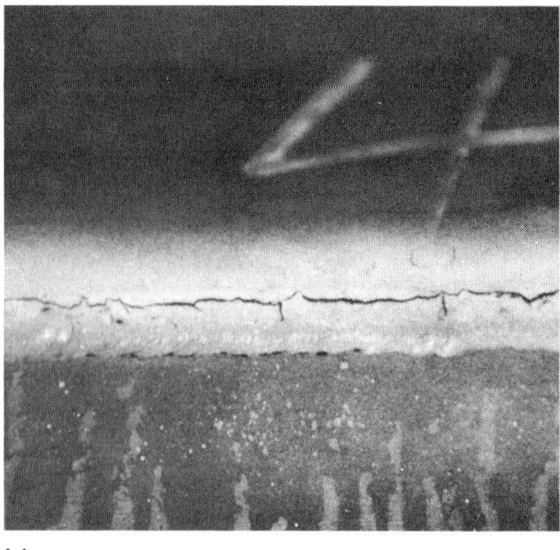

(c)

Fig. 37 T-joint weld of A537 grade 2 (quenched and tempered) material joined with E8018-C1 (2.5% Ni) weld wire showing a longitudinal crack in one of the fillet welds (arrow in cross section) made visible by linear magnetic particle indications. Weld D, shown in cross section (a), was checked using both the (b) liquid penetrant inspection method and the (c) magnetic particle inspection method. See text for discussion. Courtesy of Chicago Bridge & Iron Company

2 (3.56 mm, or 0.14 in., below the surface), over 1000 A. Hole 3 (5.33 mm, or 0.21 in., below the surface) could not be revealed with alternating current at any current level available. Indications at hole 2 were produced using 450-A straight direct current, 320-A direct current preceded by a surge of twice that amperage, and 250-A half-wave current. Indications were produced at hole 12 (21.3 mm, or 0.84 in., below the surface) using 750-A half-wave current, while 975-A straight direct current was required for hole 10 (17.8 mm, or 0.70 in., below the surface).

The current levels needed to produce indications using wet particles were somewhat higher. For example, an indication for hole 1 using direct current and wet particles required approximately 440 A, and for hole 3, approximately 910 A. Over 625 A was required to detect hole 1 using alternating current and wet particles.

The hardness of the testpiece also had an effect on the current level needed to produce indications. At a hardness of 63 HRC, to produce an indication at hole 1, approximately 200 A of half-wave current, 300 A of direct current with surge, and 450 A of direct current were needed. For hole 3, the current levels needed for the three types of current were approximately 1300, 1875, and 2700 A, respectively. Tests similar to the one described above have been performed on ring specimens made of 1020 and 4130 steels (Ref 1).

For the inspection of finished parts such as machined and ground shafts, cams, and gears of precision machinery, direct current is frequently used. Alternating current is used for detecting fine cracks that actually break the surface, but direct current is better for locating very fine nonmetallic stringers lying just beneath the surface.

Method of Magnetization

The method of magnetization refers to whether residual magnetism in the part provides a leakage field strong enough to produce readable indications when particles are applied or if the part must be continuously magnetized while the particles are applied.

Residual Magnetism. The procedure for magnetic particle inspection with residual magnetism, using either wet or dry particles, basically consists of two steps: establishing a magnetic field in the part and subsequently applying the magnetic particles. The method can be used only on parts made of metals having sufficient retentivity. The residual magnetic field must be strong enough to produce leakage fields at discontinuities that in turn produce readable indications. This method is reliable only for detecting surface discontinuities.

Either the dry or the wet method of applying particles can be used with residual magnetization. With the wet method, the magnetized parts can either be immersed in a gently agitated bath of suspended metallic particles or flooded by a curtain spray. The time of immersion of the part in the bath can affect the strength of the indications. By leaving the magnetized part in the bath or under the spray for a considerable time, the leakage fields, even at fine discontinuities, can have time to attract and hold the maximum number of particles. The location of the discontinuity on the part as it is immersed has an effect on particle buildup. Buildup will be greatest on horizontal upper surfaces and will be less on vertical surfaces and horizontal lower surfaces. Parts should be removed from the bath slowly because rapid removal can wash off indications held by weak leakage fields.

Continuous Magnetism. In the continuous method, parts are continuously magnetized while magnetic particles are applied to the surfaces being inspected. In the dry-particle continuous method, care must be taken not to blow away indications held by weak leakage fields. For this reason, the magnetizing current is left on during the removal of excess particles.

In the wet-particle continuous method, the liquid suspension containing the magnetic particles is applied to the part, and the magnetizing current is applied simultaneously with completion of particle application. This prevents washing away of indications held by weak leakage fields. For reliability of results, the wet continuous method requires more attention to timing and greater alertness on the part of the operator than the wet residual method.

The continuous method can be used on any metal that can be magnetized because in this method residual magnetism and retentivity are not as important in producing a leakage field at a discontinuity. This method is mandatory for inspection of low-carbon steels or iron having little or no retentivity. It is frequently used with alternating current on such metals because the ac field produces excellent mobility of dry magnetic particles. Maximum sensitivity for the detection of very fine discontinuities is achieved by immersing the part in a wet-particle bath, passing the magnetizing current through the part for a short time during immersion, and leaving the current on as the part is removed and while the bath drains from the surface.

Direction of Magnetization

The shape and orientation of the suspected discontinuity in relation to the shape and principal axis of the part have a bearing on whether the part should be magnetized in a circular or a longitudinal direction or in both directions. The rule of thumb is that the current must be passed in a direction parallel to the discontinuity. If the principal direction of the discontinuities is unknown, to detect all discontinuities, both circular and longitudinal magnetization must be used; with the prod and yoke methods, the prods or yoke must be repositioned at 90°

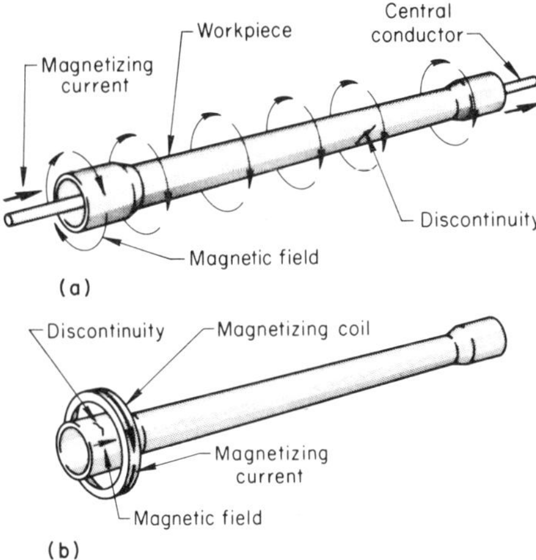

Fig. 38 Magnetic particle inspection of oil well tubing for longitudinal (a) and circumferential (b) discontinuities

from the first magnetizing position. The magnetic particle background held on a part by extraneous leakage fields is minimized when using the circular method of magnetizing because the field generally is self-contained within the part.

Magnitude of Applied Current

The amount of magnetized current or the number of ampere-turns needed for optimum results is governed by the types and minimum dimensions of the discontinuities that must be located or by the types and sizes of discontinuities that can be tolerated.

The amount of current for longitudinal magnetization with a coil is initially determined by Eq 1 and 2. For circular magnetization, when magnetizing by passing current directly through a part, the current should range from 12 to 31 A/mm (300 to 800 A/in.) of the diameter of the part. The diameter is defined as the largest distance between any two points on the outside circumference of the part. Normally, the current used should be 20 A/mm (500 A/in.) or lower, with the higher current values of up to 31 A/mm (800 A/in.) used to inspect for inclusions or to inspect alloys such as precipitation-hardened steels. The prod method of magnetization usually requires 4 to 4.92 A/mm (100 to 125 A/in.) of prod spacing. Prod spacing should not be less than 50 mm (2 in.) nor more than 203 mm (8 in.).

Equipment

Selection of equipment for magnetic particle inspection depends on the size, shape, number, and variety of parts to be tested.

Bench Units. For the production inspection of numerous parts that are relatively small but not necessarily identical in shape, a bench unit with contact heads for circular magnetization, as well as a built-in coil for longitudinal magnetization, is commonly used (Fig. 11).

Portable units using prods, yokes, or hand-wrapped coils may be most convenient for large parts. Half-wave current and dry particles are often used with portable equipment. Wet particles can be used with portable equipment, but the bath is usually not recovered.

Mass Production Machinery. For large lots of identical or closely similar parts, single-purpose magnetization-and-inspection units or fixtures on multiple-purpose units can be used.

Inspection of Hollow Cylindrical Parts

Some hollow cylindrical parts requiring magnetic particle inspection present difficulties in processing because of configuration, extraneous leakage field interference, requirements for noncontact of magnetizing units, the overall time required, or low *L*/*D* ratio. Techniques for inspecting long pieces of seamless tubing (oil well tubing), butt-welded and longitudinally welded carbon steel pipe or tubing, and a cylinder with a closed end are described in the following sections.

Oil well tubing is made of high-strength steel using hot finishing operations and has upset ends for special threading. The discontinuities most expected are longitudinally oriented on the main body of the tube and transversely oriented on the upset ends. For these reasons, the entire length of the tube is circularly magnetized and inspected for longitudinal-type discontinuities. Also, the upset ends are longitudinally magnetized and inspected for transverse-type discontinuities. Tube sections are usually more than 6 m (20 ft) long.

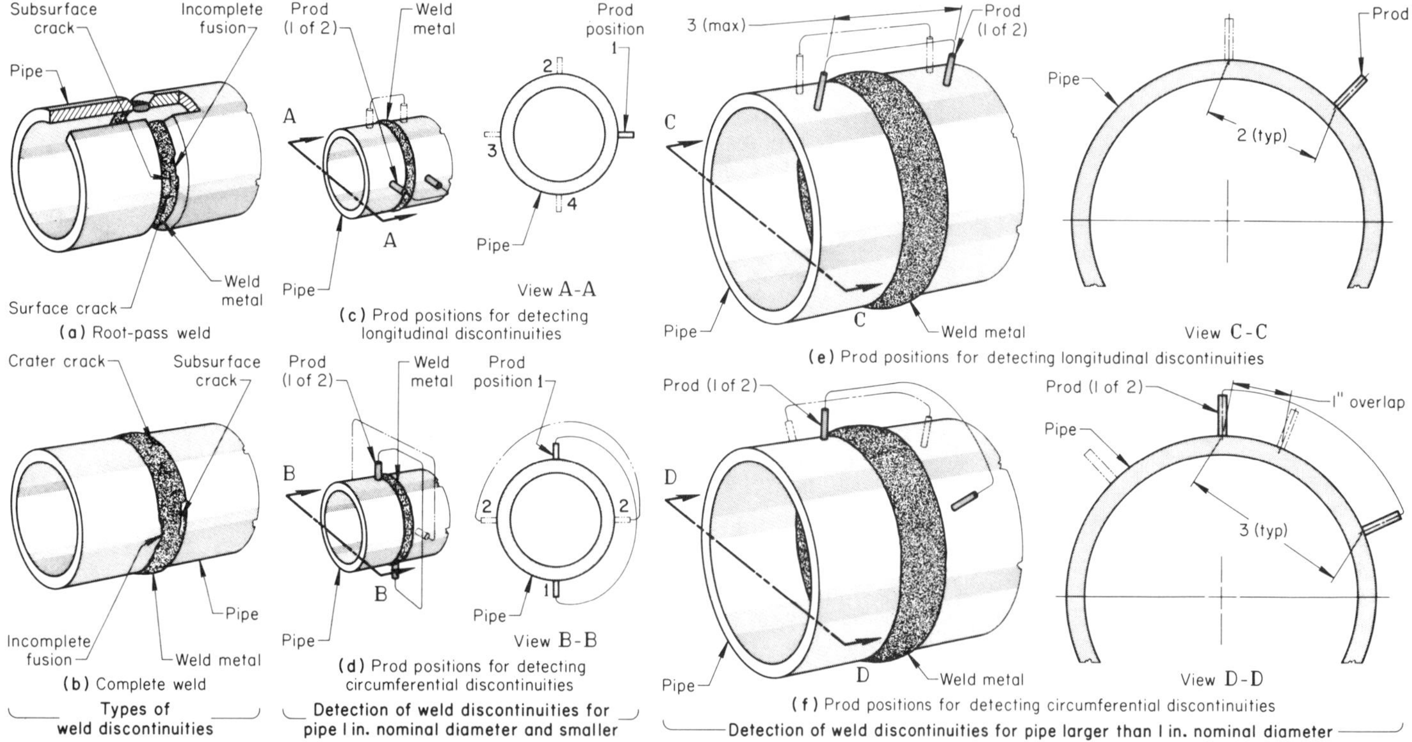

Fig. 39 Magnetic particle inspection for detection of discontinuities in consumable-insert root welds and final welds in carbon steel pipe. Dimensions given in inches

An insulated central conductor is used to introduce circular magnetism instead of passing the current through the material by head contacts. The central conductor facilitates inspection of the inside surface of the upset ends; the direct-contact method would not provide the required field.

The central-conductor technique used for circumferential magnetization of the tube body is illustrated in Fig. 38(a). The magnetic particles are applied to the outside surface of the tube. The residual-magnetism technique is used. The current density for this test is usually 31 to 39 A/mm (800 to 1000 A/in.) of tube diameter.

The encircling-coil technique used to magnetize the upset ends in the longitudinal direction is shown in Fig. 38(b). The residual-magnetism technique is used. Both the inside and outside surfaces are inspected for discontinuities.

Welds in Carbon Steel Pipe. Magnetic particle inspection using the prod technique is a reliable method of detecting discontinuities in consumable-insert root welds and final welds in carbon steel pipe up to 75 mm (3 in.) in nominal diameter. (For larger-diameter pipe, less time-consuming magnetic particle techniques can be used.) The types of discontinuities found in root welds are shown in Fig. 39(a), and those found in final welds are shown in Fig. 39(b).

Placement of the prods is important to ensure reliable inspection of the welds. Circular magnetization, used to check for longitudinal discontinuities, is accomplished by placing the prods at 90° intervals (four prod placements) around the pipe, as shown in Fig. 39(c). For pipes larger than 25 mm (1 in.) in nominal diameter, prods should be spaced around the pipe at approximately 50 mm (2 in.) intervals, as shown in Fig. 39(e). Circumferentially oriented discontinuities are revealed by placing the prods as shown in Fig. 39(d). The prods are placed adjacent to and on opposite sides of the weld bead to ensure flux flow across the weld metal. If the circumferential distance between the prods is greater than 75 mm (3 in.) when positioned as shown in Fig. 39(d), the prods should be positioned as shown in Fig. 39(f). To ensure proper magnetization, the areas inspected should overlap approximately 25 mm (1 in.). Magnetic particles are applied to the weld area while the current is on because of the low retentivity of carbon steel.

A buildup of magnetic particles in the fusion-line crevice of the weld is indicative of either a subsurface discontinuity or a nonrelevant indication because of the abrupt change in material thickness and/or the crevicelike depression between the weld metal and the base metal. However, a true indication, as from incomplete fusion between the weld metal and the base metal, would be a sharply defined particle pattern. This pattern would be difficult if not impossible to remove by blowing with a hand-held powder blower while the magnetizing current is being applied. If the indication at the fusion zone can be blown away with a hand-held powder blower, the indication is nonrelevant. The current used is approximately 3.9 A/mm (100 A/in.) of prod spacing.

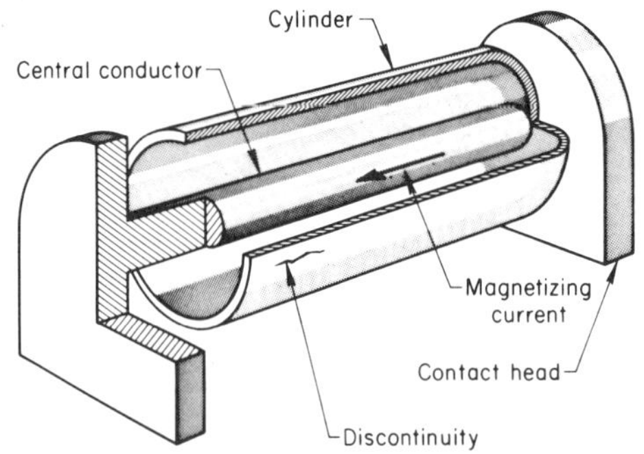

Fig. 40 Use of a central conductor for the magnetic particle testing of a cylinder with one end closed

Hollow cylinders closed at one end, such as drawn shells or forged fluid-power cylinders, can be magnetized circumferentially for the inspection of longitudinal discontinuities using a head shot end-to-end. However, this technique does not provide sensitivity for discontinuities on the inside surface.

As shown in Fig. 40, a central conductor can be used in such a manner that the closed end of the cylinder completes the current path. Also, the open end of the cylinder is accessible for the application of a wet-particle bath to the inside surface, which can then be inspected directly. For thin-wall cylinders, discontinuities on the inside surface produce subsurface-type indications on the outside surface. The central-conductor method for magnetization is advantageous when the inside diameter is too small to permit direct internal viewing.

Inspection of Castings and Forgings

Castings and forgings may be difficult to inspect because of their size and shape. External surfaces can usually be inspected with prods; however, on large parts this can be time consuming, and inspection of interior surfaces may not be adequate.

High-amperage power supplies, in conjunction with flexible cable used with clamps (as contact heads), central conductors, or wrapping, can effectively reduce inspection time because relatively large areas can be inspected with each processing cycle. Figure 41 illustrates the direct-contact, cable-wrap, and central-conductor techniques that can be used to inspect two relatively large parts. The three circuits for each part can be applied on a single-shot basis or, if high-output multidirectional equipment is available, can be combined into a single inspection cycle.

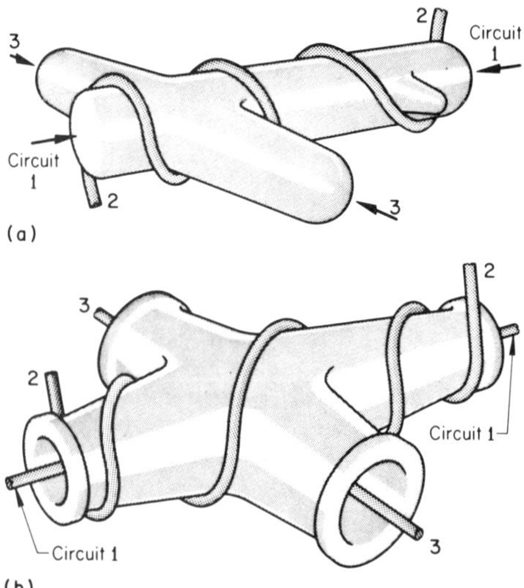

Fig. 41 Methods of using cable for applying magnetizing circuits to large forgings and castings. For the forging in (a), circuits 1 and 3 are head shots and circuit 2 is a cable wrap. For the casting in (b), circuits 1 and 3 are central conductors and circuit 2 is a cable wrap.

Generally, the higher-amperage power supplies are of the dc type. Alternating current and half-wave current supplies are limited to outputs of approximately 5000 to 6000 A because of the reactive impedance component associated with these types of current. Wet particles are generally preferred as the inspection medium in the presence of strong dc fields because wet particles exhibit much more mobility on the surface of a part than dry particles. Wet particles also readily permit full coverage of large surface areas and are easy to apply to internal surfaces.

Crane Hooks. The inspection of crane hooks, as required by the Occupational Safety and Health Act, has focused attention on cracks and other discontinuities in these components. Crane hooks are generally magnetic particle inspected with electromagnetic yokes having flexible legs. Power supplies are 115-V, 60-Hz alternating current and half-wave current. Stress areas in a crane hook are:

- The bight (in tension) on both sides and in the throat (area A, Fig. 42)
- The area below the shank (in compression and tension) on four sides (area B, Fig. 42)
- The shank (in tension), mainly in threads and fillet (area C, Fig. 42)

The steps involved in the magnetic particle inspection of crane hooks are as follows:

1. Remove dirt and oil from hook
2. Magnetize and apply particles to areas A and B in Fig. 42, using a yoke and an ac field parallel to the axis of the hook
3. For hooks out of surface, inspect the shank (area C, Fig. 42), using a yoke and an ac field parallel to the axis of the hook

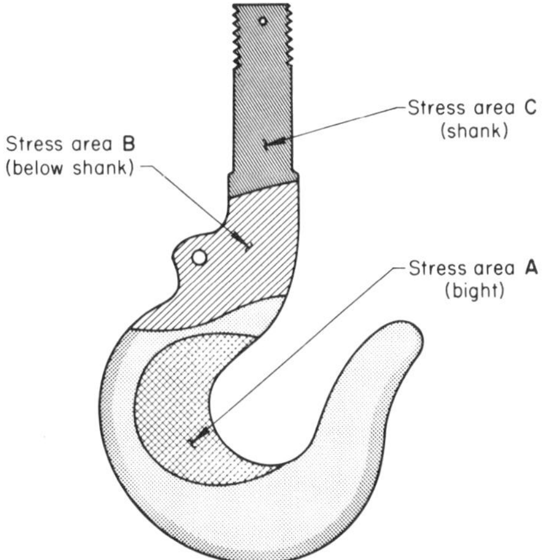

Fig. 42 Forged crane hook showing stress areas subject to inspection

4. For hooks in service, inspect the shank ultrasonically
5. Repeat steps 2 and 3 using a dc field for subsurface indications

The 50 kN (6 tonf) crane hook shown in Fig. 43 was removed from a jib crane after an indication was found during magnetic particle inspection. The discontinuity was found to be a deep forging lap. Sectioning the hook through the lap revealed a discontinuity about 19 mm (¾ in.) deep in a 50 mm (2 in.) square section (section A–A, in Fig. 43). All inspections of the hook were made with the magnetic field parallel to the hook axis, thus inspecting for transverse cracks and other discontinuities. Because of the depth of the lap, the defect was detected even though the field was parallel to its major dimension. A sufficient flux-leakage field occurred at the surface to attract the magnetic particles.

Inspection of a 90 kN (10 tonf) crane hook revealed a forging lap in the area below the shank. The lap was transverse and was located in the fillet below the keeper hole. A fatigue crack had initiated in the lap.

Drive-Pinion Shaft. An annual preventive maintenance inspection was performed on the large forged drive-pinion shaft shown in Fig. 44(a), and detection of a large crack (arrows, Fig. 44b) in the fillet between the shaft and coupling flange prevented a costly breakdown. Three areas were magnetic particle inspected for cracks:

- Along the shaft
- The fillet between the shaft and the coupling flange (cracked area, Fig. 44b)
- At each fillet in the wobbler coupling half (arrows, Fig. 44a)

Inspection was conducted using a portable power source capable of up to 1500-A output in alternating current or half-wave direct current. Double prod contacts and a 4/0 cable were used to introduce the magnetic fields in the shaft. The fillets in the wobbler coupling half were inspected for discontinuities with double prod contacts.

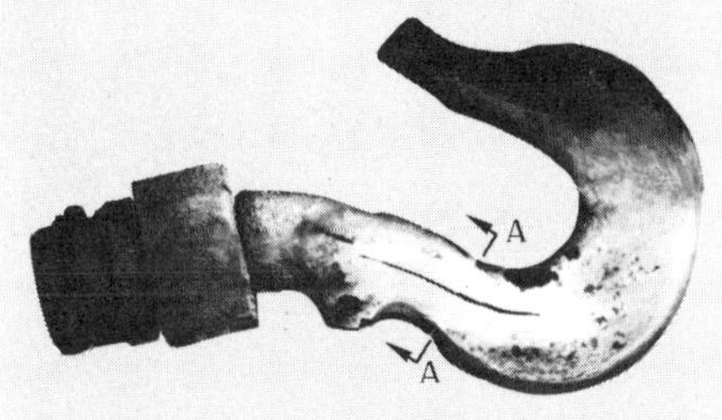

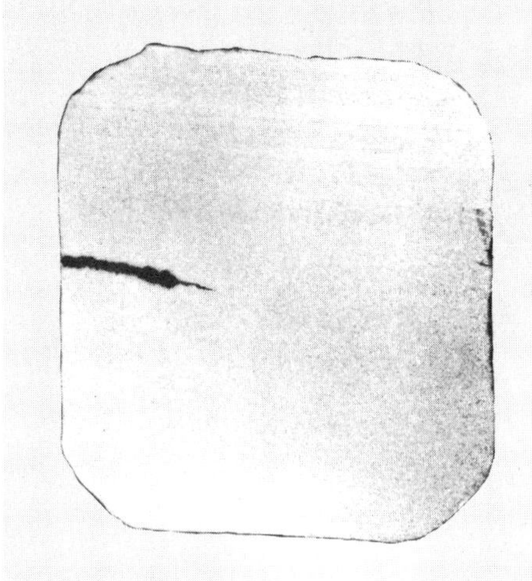

Section A-A

Fig. 43 50 kN (6 tonf) crane hook showing magnetic particle indication of a forging lap and section through hook showing depth of lap

The steps involved in inspecting the shaft were as follows:

1. Clean all areas to be tested
2. Wrap cable around shaft and apply current providing 2900 ampere-turns. Inspect for transverse cracks in the shaft portion
3. Place prods across fillet at coupling flange (Fig. 44b) at spacing of 152 to 203 mm (6 to 8 in.); apply 500-A current, producing a circular field perpendicular to fillet. Inspect for discontinuities parallel to fillet
4. Place prods across fillet at flange on wobbler coupling half and across fillets in shaft portion (arrows, Fig. 44a) at spacing of 152 to 203 mm (6 to 8 in.); apply 500-A current, producing a circular field perpendicular to the fillets. Inspect for discontinuities parallel to the fillets

Disk or Gear on Shaft. A forged or cast disk or gear on a heavy through shaft can be wrapped with a cable to form two opposing coils, as shown for the disk in Fig. 45. The two opposing coils produce radial magnetic fields on each side of the disk. This type of field reveals circumferential discontinuities on the sides of the disk and transverse discontinuities in the shaft. Also, the shaft can be used as a central conductor for the purpose of locating radial discontinuities in the disk and longitudinal discontinuities in the shaft.

Fig. 44 Forged drive-pinion shaft and coupling in which the detection of a crack during preventive maintenance magnetic particle inspection prevented a costly breakdown. (a) Drive-pinion shaft and coupler assembly; arrows show locations of fillets on wobbler coupling half that were inspected. (b) Fillet between shaft and coupling flange showing crack (at arrows) found during inspection

In parts where the shaft extends from one side only, a pole piece can be used to simulate a through shaft. The pole piece should have approximately the same diameter as the shaft.

Y-shaped parts may not be processed on a horizontal wet-particle unit in such a manner that only one shot will be required for complete inspection. This is true even though steps are taken to ensure that the current is divided equally through the two upper legs of the Y. With equally divided current, an area of no magnetization will

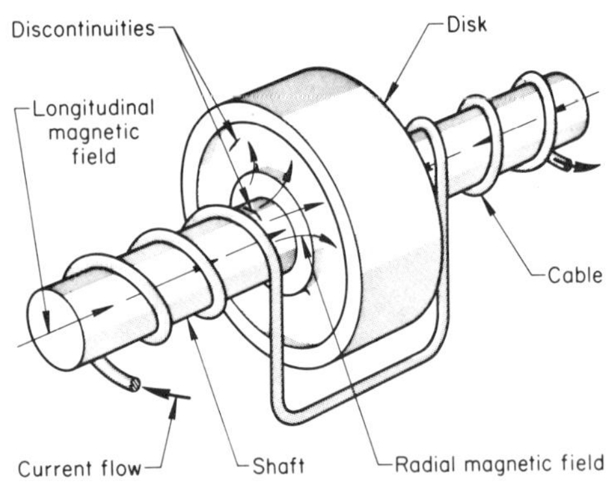

Fig. 45 Disk on a through shaft in which the shaft was cable wrapped to produce a longitudinal magnetic field in the shaft and a radial field in the disk. Using the shaft as a central conductor produced a circular magnetic field in both the shaft and the disk.

exist at the junction of the two legs. A Y-shaped part can be processed in two steps using a conventional horizontal wet-particle unit, or it can be processed in one operation using a modified unit having a double headstock and special current assurance magnetizing circuit.

Inspection of Weldments

Many weld defects are open to the surface and are readily detectable by magnetic particle inspection with prods or yokes. For the detection of subsurface discontinuities, such as slag inclusions, voids, and inadequate joint penetration at the root of the weld, prod magnetization is best, using half-wave current and dry powder. Yokes, using alternating current, direct current, or half-wave current, are suitable for detecting surface discontinuities in weldments.

The positioning of a yoke with respect to the direction of the discontinuity sought is different from the corresponding positioning of prods. Because the field traverses a path between the poles of the yoke, the poles must be placed on opposite sides of the weld bead to locate cracks parallel to the bead (Fig. 6), and adjacent to the bead to locate transverse cracks. Prods are spaced adjacent to the weld bead for parallel cracks and on opposite sides of the bead for transverse cracks.

For applications in which holding the prod contacts by hand is difficult or tiring, prods incorporating magnetic clamps, or leeches, that magnetically hold the prods to the work are available. The prods carrying the magnetizing current are held firmly to the work by an electromagnet. Both prods can be attached by the magnets, or one of the prods can be held magnetically and the other by hand.

There is one type of weld on which the penetrating power of half-wave current results in nonrelevant indications: a T-joint welded from one or both sides for which complete joint penetration is not specified and in which an open root is permissible and almost always present. When half-wave current is used with prods, this open root will probably be indicated on the weld surface. This nonrelevent indication can be eliminated by using alternating current instead of half-wave current.

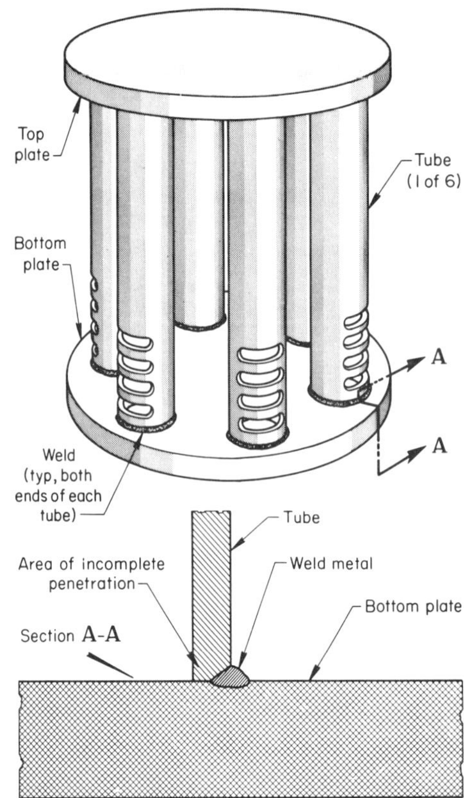

Fig. 46 Assembly in which T-joints between tubes and end plates that were welded (as specified) with partial penetration produced nonrelevant magnetic particle indications when inspected using dc magnetization

A case in which nonrelevant magnetic particle indications were found occurred at the welded T-joints between six tubes and the end plates of a complicated assembly (Fig. 46). The welds were made from the outside of the tubes only. Liquid penetrant and radiographic inspections and metallographic examination disclosed that the integrity of the welds was good. Investigations revealed that the depth of penetration of the field produced by dc magnetization was sufficient to reveal the joint along the inside wall of the cylinder. Inspection by ac magnetization, which had less depth of penetration of the field, eliminated these indications.

The detectability of subsurface discontinuities in butt welds between relatively thin plates can often be improved by positioning a dc yoke on the side opposite the weld bead, as shown in Fig. 47. Magnetic particles are applied along the weld bead. Improvement is achieved because of the absence of extraneous leakage flux that normally emanates from the yoke pole pieces. Techniques developed and used for the magnetic particle inspection of weldments are described in the two examples that follow.

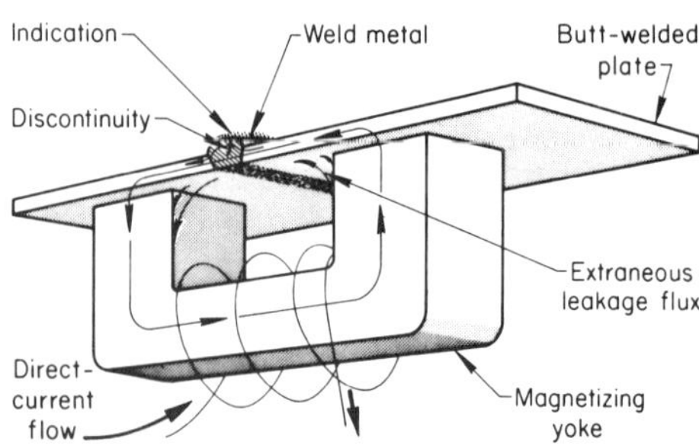

Fig. 47 Method of using a yoke for detecting subsurface discontinuities in butt welds joining thin plates

Example 3: Magnetic Particle Inspection of Shielded Metal-Arc Welds in the Outer Hull of a Deep-Submergence-Vessel Float Structure. The outer hull of the float structure of a deep-submergence vessel was made of 3.2 mm (⅛ in.) thick steel plate and was butt welded using the shielded metal-arc process. For magnetic particle inspection, the weld areas were magnetized with prods, using a half-wave current of 3.9 A/mm (100 A/in.) of prod spacing. Dry, red magnetic particles provided adequate sensitivity and color contrast for inspection.

The crowns on the surfaces of the welds were removed with a flat chisel. Neither grinding of the welds nor the use of pneumatic needle guns was permitted. The surfaces were cleaned of paint, scale, and slag by sand or vapor blasting.

The area to be inspected was shielded from air currents to prevent disturbing the magnetic particle indications. Magnetic particles were applied with a powder dispenser hand-held 305 to 381 mm (12 to 15 in.) from the weld at an angle of 30 to 45° to the surface of the plate along the axis of the weld. The powder dispenser was held stationary once it was positioned for each prod placement. The powder was allowed to float to the weld surface so that a very light dustlike coating of particles settled on the surface being tested. Excessive application of magnetic particles required removal of all particles and reapplication.

Longitudinal discontinuities were detected by placing the prods at a 152 mm (6 in.) spacing parallel to the longitudinal axis of the weld and on the surface of the weld. Successive prod spacings overlapped approximately 25 mm (1 in.) to ensure complete coverage. For the detection of transverse discontinuities, prods were positioned at right angles to the weld axis, 75 to 152 mm (3 to 6 in.) apart on each side of the weld, and at intervals not exceeding one-fourth the prod spacing.

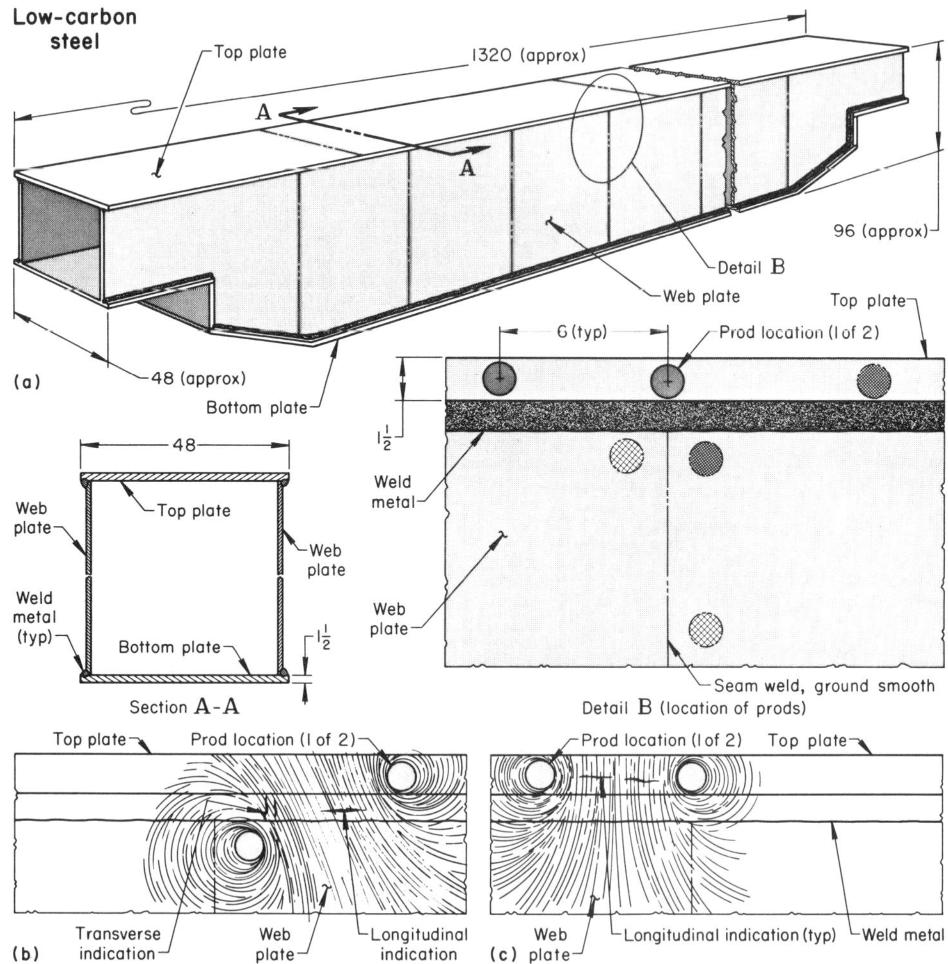

Fig. 48 Magnetic particle inspection of a girder weldment for an overhead crane. (a) Girder weldment, with section showing welds and detail showing prod locations. (b) Magnetograph of magnetic field used to check for discontinuities in the welds. (c) Magnetograph of field used to check for discontinuities in top plate. Note different prod placement in (b) and (c). Dimensions given in inches

Magnetic particle inspection was used instead of radiography. Following are the maximum radiographic acceptance standards from which the magnetic particle procedures were developed:

- A linear or linearly disposed indication 1.6 mm (1/16 in.) or more in length, with its length at least four times greater than its width
- A single slag (subsurface) indication longer than 0.8 mm (1/32 in.), and multiple slag indications closer together than 4.8 mm (3/16 in.) or totaling more than 3.2 mm (1/8 in.) in any 152 mm (6 in.) of weld length
- A cluster of four or more indications of any size spaced within 3.2 mm (1/8 in.) of each other

Example 4: Magnetic Particle Inspection of a Girder Weldment for an Overhead Crane. The girder weldment for an overhead crane (Fig. 48a) was magnetic particle inspected in the presence of the customer's representative. Fabrication of the weldment was complete, and the weldment had been sandblasted in preparation for painting.

The dry-powder, continuous-magnetization method was used with half-wave current. The current used was 3.9 to 5.9 A/mm (100 to 150 A/in.) of prod contact spacing, which was 152 mm (6 in.). Inspected were the fillet welds joining the top and bottom plates to the web plate (section A–A in Fig. 48), and all butt seam welds of the top, bottom, and web plates. The edges of the top and bottom plates were also inspected for discontinuities. All prod-contact points on the girder were ground to ensure good contact; prod locations are shown in detail B in Fig. 48.

During magnetic particle testing of the welds, the prods were alternated on either side of the welds. A magnetograph of the magnetic flux during inspection of a weld joining the top plate to the web plate is shown in Fig. 48(b). Inspection of the edges of the top and bottom plates was performed by placing the prods parallel to the weld with contacts on the edge of the plate; a magnetograph of this test is shown in Fig. 48(c). Longitudinal indications were observed in both the plate and the weld.

Nonrelevant indications that the process was capable of detecting were established prior to inspecting the girder. All the indications detected were verified as either false or nonrelevant, signifying an absence of defects.

Inspection of Billets

A billet is the last semifinished intermediate step between the ingot and the finished shape. Steel billets are rectangular or square and range from 2600 to 32 000 mm^2 (4 to 49 in.^{2}) in cross-sectional area. The magnetic particle inspection of billets requires a large unit equipped to handle billets 50 to 184 mm (2 to 7¼ in.) square and 2.4 to 12 m (8 to 40 ft) long. The amperage setting on the testing unit should be 1200 to 4000 A. The discontinuities shown in Fig. 49 would appear as bright fluorescent indications under ultraviolet light.

Cracks in billets appear as deep vertical breaks or separations in the surface of the steel. Arrowhead cracks (Fig. 49a) occur early in processing, usually as the result of primary mill elongation of an ingot containing a transverse crack. Longitudinal cracks (Fig. 49b) appear as relatively straight lines in the direction of rolling. They are 0.3 m (1 ft) or more in length and usually occur singly or in small numbers.

Seams are longitudinal discontinuities that appear as light lines in the surface of the steel.

Normal seams (Fig. 49c) are similar to longitudinal cracks, but produce lighter indications. Seams are normally closed tight enough that no actual opening can be visually detected without magnetic particle inspection. Seams have a large number of possible origins, some mechanical and some metallurgical.

Brush seams (Fig. 49d) are clusters of short (<102 mm, or 4 in., long) seams that appear as though they had been painted or brushed onto the surface. These defects are usually the result of the removal of metal from the steel surface by scarfing or scaling, exposing ingot blowholes and subsurface porosity. They may range in depth from 0.13 to 7.6 mm (0.005 to 0.300 in.) and may occur either in zones or across the entire surface of the billet.

Laps (Fig. 49e) are longitudinal discontinuities of varying severity that are caused by the formation of ribs or extensions of metal during hot rolling and the subsequent folding over of these protrusions. Laps usually run at acute angles to the surface. They often occur at opposite sides of the billet and frequently span the entire length of the billet.

Scabs (Fig. 49f) appear as extraneous pieces of metal partially welded to the sur-

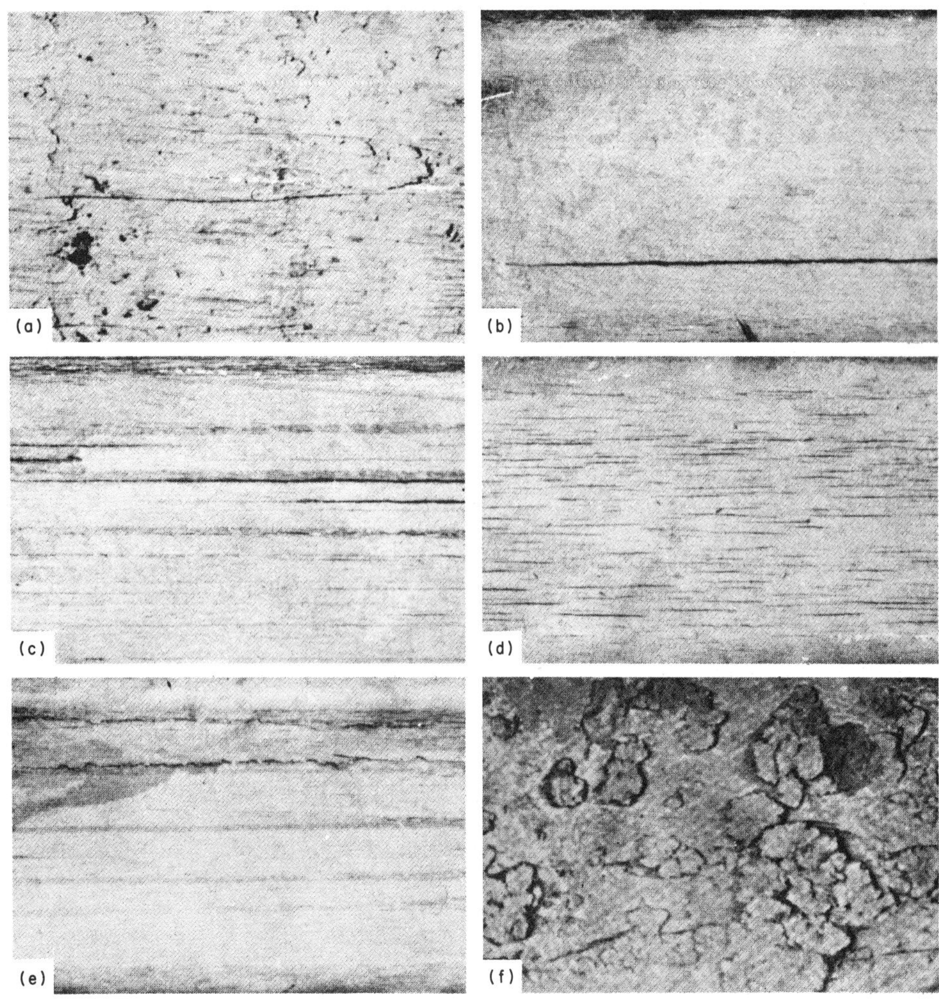

Fig. 49 Discontinuities on the surfaces of steel billets that can be detected by magnetic particle inspection. (a) Arrowhead cracks. (b) Longitudinal cracks. (c) Normal seams. (d) Brush seams. (e) Laps. (f) Scabs. See text for discussion.

face of a steel billet. Two major sources of scabs are the splashing of metal against the mold wall during teeming and the adherence of scarfing wash or fins to blooms after conditioning.

Inspection of Welded Chain Links

Magnetic particle inspection is commonly used for maintenance repair, preventive maintenance, and safety programs. The following procedure for the inspection of welded chain links is a proven method and is suitable for safety programs in which periodic inspection is required. In this procedure, a longitudinal field transverse to the weld will detect discontinuities in the weld and in the link itself. The continuous method, with wet fluorescent particles and ultraviolet light, is used.

The chain is suspended from the hook of a crane or hoist and pulled up through a magnetizing coil. Just below the coil, the wet fluorescent magnetic particles are sprayed onto the chain. Above the coil is the ultraviolet light that renders the indications readable. The chain is inspected in sections using the following steps:

1. Remove all oil or grease from chain links
2. Turn on current in coil
3. Apply the wet fluorescent magnetic particle suspension to each section while the section is in the magnetic field of the coil
4. Turn the current off after the suspension stops flowing and while the section to be inspected is still in the magnetic field of the coil
5. Inspect the section under ultraviolet light for transverse discontinuities in both the weld and the link itself
6. Continue until all sections have passed through the coil in an upward direction and have been inspected

Automated Equipment for Specific Applications

The automation of magnetic particle inspection is often necessary to permit inspection at the required production rate. Loading, processing, conveying, rotating or manipulating, demagnetization, and discharge can all be automated, which results in consistent and effortless processing. The inspector can devote all of his time to actual visual inspection while the machine performs the other functions. Production rates are achieved by having the various functions performed simultaneously at different stations. Where applicable, the cost of automatic equipment can be justified when compared to the number of manual units needed for similar production rates.

The cost and time required for manual inspection can be a factor in automating the process. For example, in billet inspection, everything is automated except the actual visual inspection and marking of imperfections. Manual processing, inspection, and handling of such a product would be extremely costly and would not satisfy the production requirements. The use of automated magnetic particle inspection for bearing rollers, bearing rings, small parts, small castings, large couplings, crankshafts, and steel mill billets is discussed in the following sections.

Bearing rollers are produced at substantial rates, and when magnetic particle inspection is required, it must be conducted at a rate compatible with production. This usually requires automation of the process in such a manner that the various stages are performed simultaneously at individual stations.

The contact method is commonly used for inspecting finished rollers for seams, quench cracks, and grinding cracks. This may involve either the continuous or the residual method, depending on roller material and inspection requirements.

One design of automated equipment for inspecting bearing rollers by the continuous magnetization method incorporates an indexing-turntable arrangement and utilizes two inspectors. The various functions—magnetization, bath application, inspection and demagnetization—are performed simultaneously at the various stations. Each inspector visually inspects the previously processed roller in the fixture, and the roller is removed and placed on the demagnetizing cradle (if acceptable) or disposed of (if defective). The inspector then places a new roller in the fixture for subsequent processing and inspection by the other inspector. Acceptable rollers are automatically stripped from the cradle in tandem, demagnetized, and unloaded into or onto a suitable container.

The production rate of the unit described above is approximately 1000 parts per hour. A similarly designed unit, based on the residual method, utilizes four inspectors and has a capacity of 2000 parts per hour. The residual method permits the use of a magnetizing-current shot of very short duration (approximately 50 ms). This minimiz-

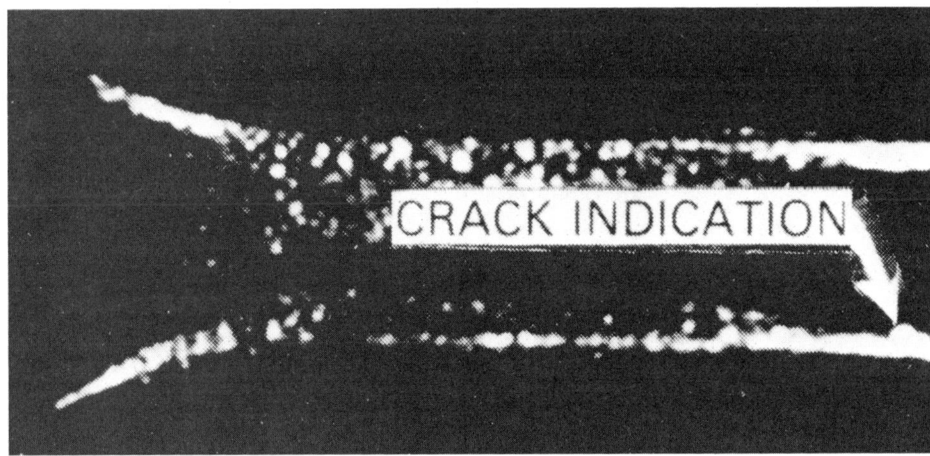

Fig. 50 A small crack indication in a connecting rod that is masked by excessive visual background noise. Courtesy of Y.F. Cheu, General Motors Technical Center

es the heat generated at the points of contact by the flow of magnetizing current.

Where inspection requirements do not allow electrical contact with the rollers due to the fact that any heat generated at the points of contact is objectionable, production units based on the induced current method can be used.

Bearing rings can be completely inspected by a two-step process involving both a central conductor and induced current. The noncontact nature of both magnetizing techniques can be very desirable when highly polished surfaces are present.

This process utilizes automatic equipment for bearing-ring inspection on a high-volume basis. The design is based on a multiple-station, fixturized, indexing conveyor. The fixtures are combination laminated poles and central conductors. In both magnetizing methods, magnetizing current is direct current with quick break, and the wet-particle residual method is employed.

A bearing ring progresses through the unit in the following sequence:

1. Load
2. Magnetize using a central-conductor shot
3. Apply bath
4. Visually inspect for transverse indications
5. Replace on fixture
6. Magnetize using induced current
7. Apply bath
8. Visually inspect for circumferential indications
9. Place on gravity chute for final roll through ac demagnetizing coil and discharge

Those parts that are judged defective at one of the two inspection stations are immediately removed from the system.

Small parts can often be inspected on general-purpose equipment at production rates by separating the processing and inspection functions and providing a convenient means for transporting the parts from one station to the other. A typical arrangement comprises a horizontal wet-particle unit and a conveyorized unit integrated into a complete inspection system. Combinations of this type are available for handling a wide variety of parts. Variations such as automatic magnetization, automatic bath application, and controls for magnetizing a part in several directions at virtually the same time are also available. Special fixturing may or may not be required, depending on production rate and part configuration. Rates of up to 900 parts per hour can be achieved by such arrangements.

Small castings may warrant special consideration because of high-volume inspection techniques as well as manpower requirements. For example, the inspection of disk brake caliper castings could have been accomplished using general-purpose two-step production equipment manned by a loader-operator and four inspectors, but the use of a special-purpose unit for these castings not only reduced manpower requirements from five to three but also reduced the required floor space.

The automated special-purpose unit incorporated an indexing-type conveyor with special fixtures, a half-wave automatic three-circuit processing station, a two-man ultraviolet light inspection booth, and dual demagnetizing coils. The special fixtures properly oriented the casting for processing in addition to conveying the casting through the system. Longitudinal plus circular magnetization was provided by combining coil-wound laminated poles and contact heads. In operation, a casting was loaded onto the fixture, automatically processed, and inspected. Defective castings were discharged into containers; acceptable parts were placed on gravity chutes that carried them through demagnetizing coils into tote boxes.

Large couplings used in conjunction with oil well tubing and casing require circular magnetization for the disclosure of longitudinal defects. In one automated installation, a twin-line unit performs the required processing automatically using central conductors. Special roller-type fixtures facilitate manual rotation during visual inspection. When a defective coupling is encountered, the inspector merely depresses a reject button to direct that coupling into a reject bin when it reaches the unloading station.

Crankshafts. Various types of automatic equipment are available for inspecting crankshafts on a production basis. Design varies with crankshaft size and weight, inspection process and rate, and facilities available for loading and unloading. Most of this equipment incorporates completely automatic processing and handling based on a fixturized indexing conveyor. Some units use power-free fixtures to convey the crankshaft from the loading station to the processing station by gravity rails. This permits loading to be done independently of the processing and inspection cycles and the banking of uninspected crankshafts. The inspector can draw crankshafts from the bank as required. Empty fixtures are automatically returned to the loading area after the unloading operation has been completed.

Connecting Rods (Ref 2). Prior to machining, ferromagnetic parts such as connecting rods usually have rough surfaces with granular texture and sharp edges. The granular surface retains fluorescent magnetic particles that are a source of high visual background noise and degrade the contrast of the crack indication. A small crack indication concealed by high background noise is extremely hard to detect (Fig. 50).

The edges of a part attract magnetic particles that fluoresce brilliantly. The edge indication is bright, long, and thin, similar to a crack indication (Fig. 51). It is consequently very hard to distinguish a crack indication from an edge indication. One may say that edge location is a known parameter: an edge indication can be sorted out because its location is known. However, this would require that parts be precisely positioned with tolerance less than the thickness of a crack indication. This is very difficult to achieve in a high-volume production environment. In addition, because it is possible that a crack may occur on the edge, excluding edge indication based on location would leave cracks undetected on the edges.

Thus, the low contrast and the uncertainty in distinguishing edge indications from crack indications make automation of fluorescent magnetic particle inspection difficult. To alleviate these problems, a new preimage-processing technique was developed to rinse the part immediately after it is processed with magnetic particles and before it is imaged. Figure 52(a) shows a part processed through magnetic particles without rinsing; Fig. 52(b) shows the same part

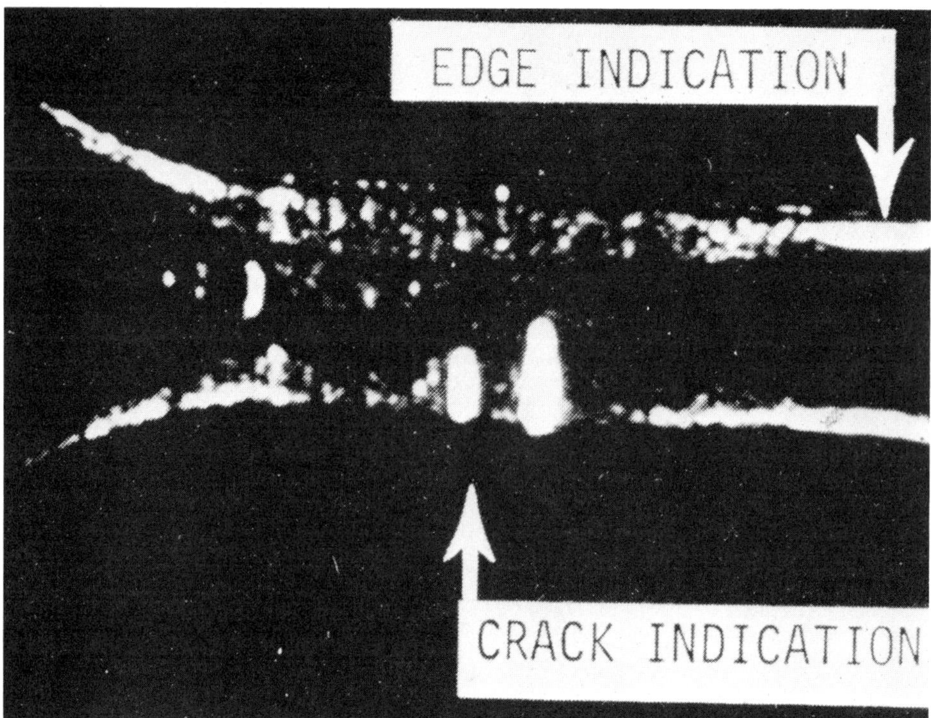

Fig. 51 Crack indication and edge-indication uncertainty in a connecting rod. Courtesy of Y.F. Cheu, General Motors Technical Center

after it is rinsed. Note that rinsing has removed the edge indications while simultaneously enhancing the crack indication.

The intensity of the fluorescent crack indication varies significantly because of the wide range of crack depth, the change of concentration of the magnetic particle solution, and the uneven sensitivity of the sensor. To accommodate these inevitable variations, a local adaptive-thresholding technique was developed and used in the crack-detection algorithm.

The algorithm demarcates an image into regional locales with 16 by 16 pixels in each locale. The average gray level of 16 by 16 pixels is calculated. The locale is then thresholded by twice this average. This locale adaptive-thresholding technique simplifies a gray-level image into a binary image based on relative intensity. The technique thereby minimizes the intensity variation problem and results in a less noisy binary image. Gray-scale digitization and binary image are discussed in the article "Machine Vision and Robotic Inspection Systems" in this Volume.

Figure 53(a) shows a digital image of a connecting rod with 256 levels of gray; the left half of the image is brighter than the right half. If the image is thresholded by a global-constant gray value, the resulting image (a crack indication and bright background noise) is shown in Fig. 53(b). On the other hand, Fig. 53(c) shows a binary image resulting from local adaptive thresholding of Fig. 53(a) as described in the previous paragraph. It is noted that local adaptive thresholding yields a more accurate binary image.

The algorithm to detect a crack from a binary image is based on the length and the aspect ratio (width-to-length ratio) of a crack. If both the length and the aspect ratio meet the specified criteria, a crack is detected. Figure 53(d) shows that a crack is detected and that background noise is ignored.

Example 5: A Ten-Station Rotary Table Integrated With a Machine Vision System to Detect Cracks in Connecting Rods (Ref 2). The basic block diagram of the vision subsystem is shown in Fig. 54. It consists of multiple ultraviolet strobe lamps, an optical filter, a lens, a charge-coupled device (CCD) matrix camera, a frame grabber, and a microcomputer.

The strobe lamp is a xenon flash tube coupled with a band-pass ultraviolet filter to provide near-ultraviolet light. The near-ultraviolet light energizes fluorescent magnetic particles to emit visible light at a wavelength of 525 nm (5250 Å), which serves as a crack indication. The optical filter mounted behind the camera lens is a band-pass green filter to pass the crack indication and to filter out near-ultraviolet and ambient light. Multiple lamps are used to provide uniform and intensified lighting.

The CCD camera is a matrix array with 404 by 256 pixels. The frame grabber is a video digitizer and an image buffer. The digitizer is an 8-bit analog-to-digital (A/D) converter that converts the video into 256 levels of gray at a speed of 30 frames per

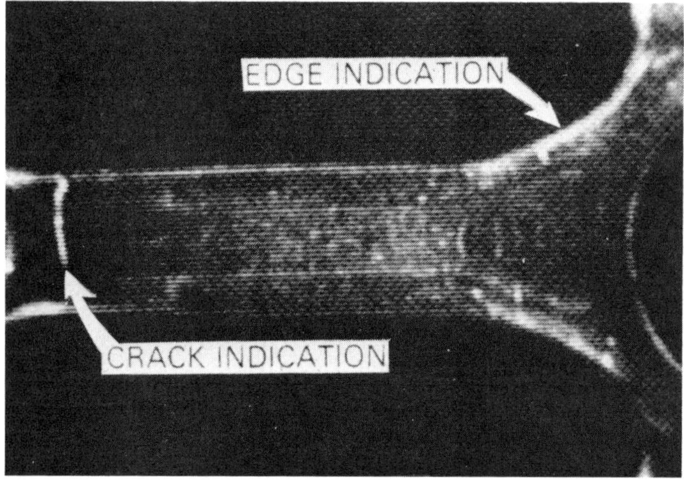

(a)

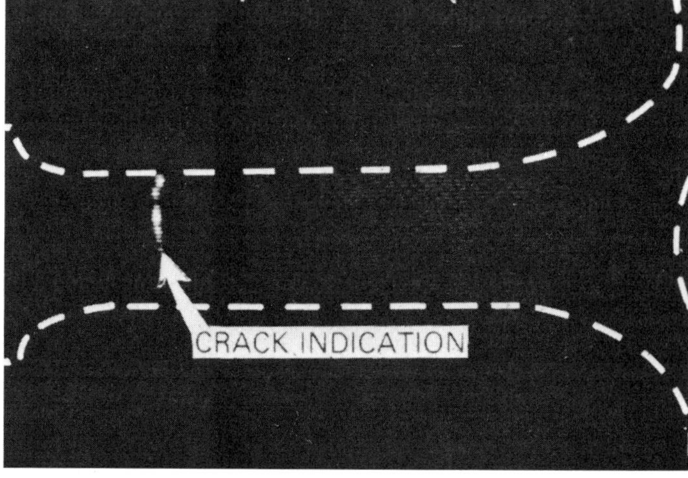

(b)

Fig. 52 A connecting rod processed with magnetic particles and imaged. (a) Before rinsing. (b) After rinsing. Courtesy of Y.F. Cheu, General Motors Technical Center

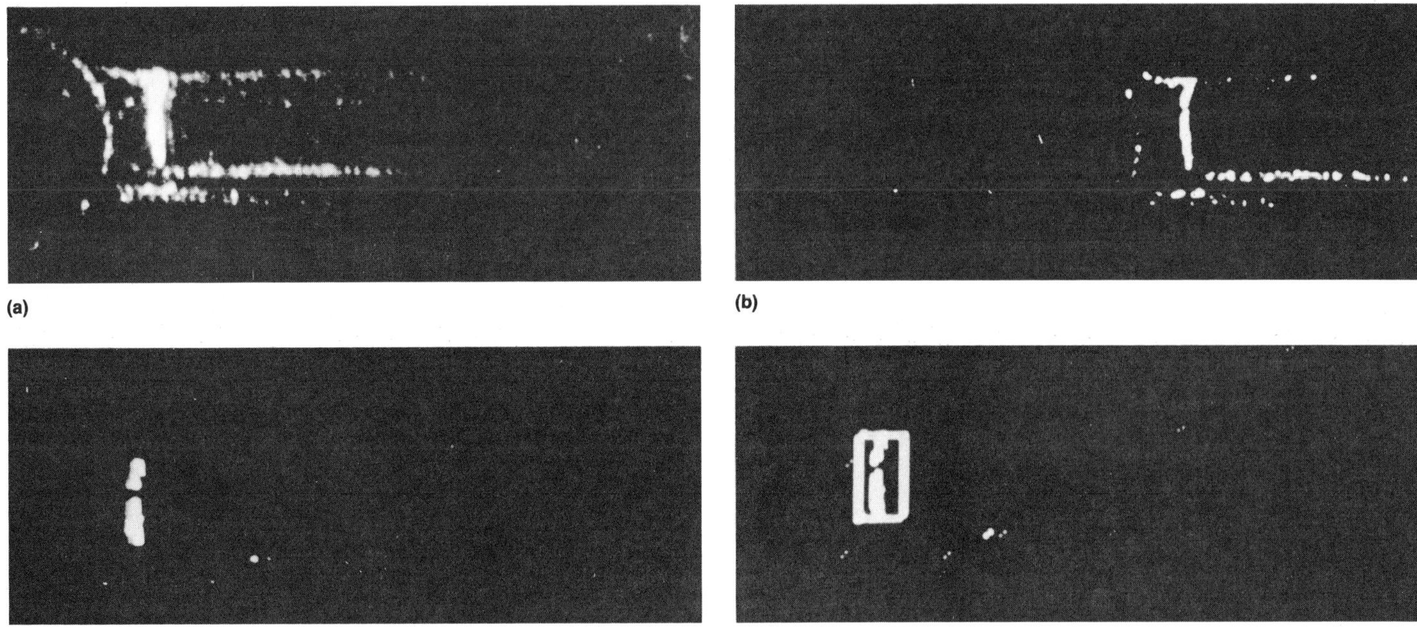

Fig. 53 Digital image processing of a magnetic particle inspected connecting rod. (a) Nonuniform brightness (256 gray-level image). (b) Result of constant global thresholding. (c) Result of local adaptive thresholding. (d) Result of shape discrimination. Courtesy of Y.F. Cheu, General Motors Technical Center

second. The image buffer has a capacity of storing 512 by 512 pixels.

The microcomputer is a 16-bit machine that analyzes the digital image using the digital image processing technique described previously to make an inspection decision. An accept/reject decision resulting from the inspection is fed into a programmable controller. The programmable controller controls the material handling subsystem, which will unload the connecting rod at either the good rod station or the reject rod station.

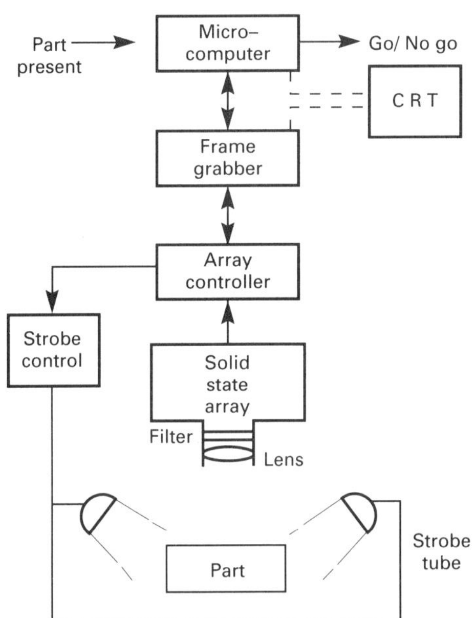

Fig. 54 Block diagram of machine vision subsystem for automated connecting-rod crack detection system. Source: Y.F. Cheu, General Motors Technical Center

The mechanical subsystem is a ten-station rotary index table (Fig. 55). Each station on the table has a gripper. Every 3 s, the table indexes 36°, drops 305 mm (12.0 in.), dwells, and rises 305 mm (12.0 in.). The sequence of operations for inspecting the connecting rods is as follows:

- *Load station*: A magazine loader lifts a connecting rod from the accumulation conveyor into a gripper
- *Magnetizing station*: The rod is magnetized while immersed in the magnetic particle solution
- *Rinse station*: The rod is rinsed with water to remove residual magnetic particles
- *Drain station*: Rinse water from the previous station is allowed to drip away to prevent interference by water droplets with the vision inspection
- *Camera station*: The connecting rod is inspected by the vision subsystem
- *Wrist-pin inspection station*: The wrist-pin hole is checked by an optical sensor to ascertain its opening
- *Rust inhibitor station*: The connecting rod is dipped into a bath of rust inhibitor to prevent rust formation on connecting rods
- *Paint identification station*: Color codes the defect-free rods
- *Unload and demagnetize station*: All defect-free rods are demagnetized and unloaded in a shipping container
- *Unload reject station*: All rejected rods are unloaded onto a table for final inspection and disposition

The sequencing functions of the process are controlled by a programmable controller. To ensure a reliable inspection, the programmable controller also monitors the magnetizing current, the pressure of the rinsing water, and the concentration level of the magnetic particle solution. The machine is capable of inspecting 1200 connecting rods per hour.

Typical examples of the system inspection results are shown in Fig. 56. Figure 56(a) shows a crack of average size, and Fig. 56(b) shows that it is detected.

Steel-mill billets are magnetic particle inspected to locate and mark discontinuities that become elongated by the rolling process. Once located, discontinuities are removed by various means to improve the quality of the subsequently rolled final product.

Usual practice is to circularly magnetize the billet by passing the current end to end and to test by the wet-particle continuous method. The equipment required for performing the handling and processing functions is somewhat different from the normal inspection equipment.

A typical billet inspection unit is capable of handling billets up to 12 m (40 ft) long and weighing as much as 1.8 Mg (4000 lb). Every aspect of handling and processing is automatic. The inspector is required only to view and mark discernible indications. When the inspection has been completed, the inspector depresses a button to initiate processing of the next billet. One design incorporates a ferris wheel to rotate the billets at the inspection station. Two sides of the billet are viewed while it is in the upper position, and the remaining two sides become accessible when the billet is rotated to the lower position. Chain-sling billet

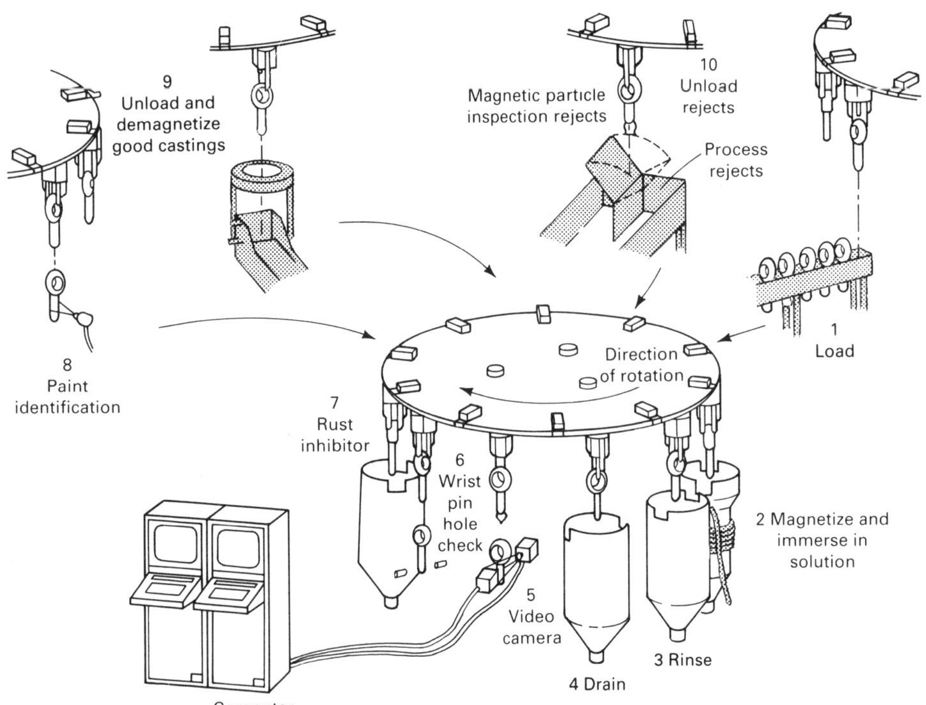

Fig. 55 Schematic of ten-station rotary table material-handling subsystem for automated connecting-rod crack detection system. Source: Y.F. Cheu, General Motors Technical Center

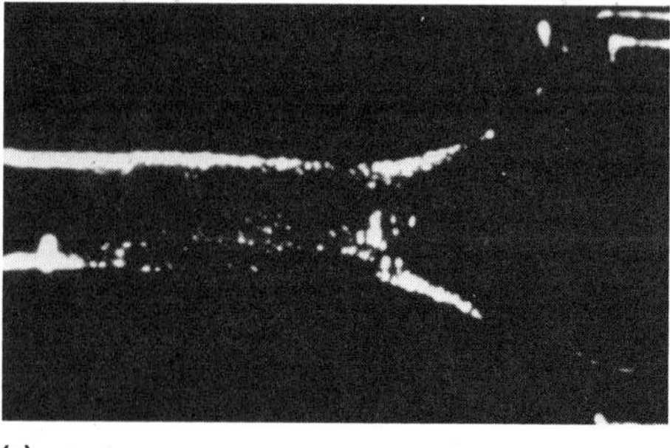

(a)

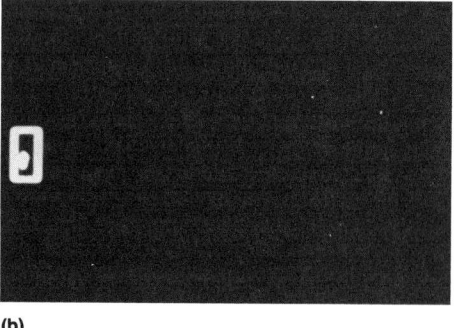

(b)

Fig. 56 An average-sized crack in a connecting rod before (a) and after (b) detection by the automated inspection system. Courtesy of Y.F. Cheu, General Motors Technical Center

turners are employed in other designs to perform billet rotation.

Demagnetization After Inspection

All ferromagnetic materials, after having been magnetized, will retain a residual magnetic field to some degree. This field may be negligible in magnetically soft metals, but in harder metals it may be comparable to the intense fields associated with the special alloys used for permanent magnets.

Although it is time consuming and represents an additional expense, the demagnetization of parts after magnetic particle inspection is necessary in many cases. Demagnetization may be easy or difficult, depending on the type of metal. Metals having high coercive force are the most difficult to demagnetize. High retentivity is not necessarily related directly to high coercive force, so that the strength of the retained magnetic field is not always an accurate indicator of the ease of demagnetizing.

Reasons for Demagnetizing. There are many reasons for demagnetizing a part after magnetic particle inspection (or, for that matter, after magnetization for any other reason). Demagnetization may be necessary for the following reasons:

- The part will be used in an area where a residual magnetic field will interfere with the operation of instruments that are sensitive to magnetic fields or may affect the accuracy of instrumentation incorporated in an assembly that contains the magnetized part
- During subsequent machining, chips may adhere to the surface being machined and adversely affect surface finish, dimensions, and tool life
- During cleaning operations, chips may adhere to the surface and interfere with subsequent operations such as painting or plating
- Abrasive particles may be attracted to magnetized parts such as bearing surfaces, bearing raceways, or gear teeth, resulting in abrasion or galling, or may obstruct oil holes and grooves
- During some electric arc-welding operations, strong residual magnetic fields may deflect the arc away from the point at which it should be applied
- A residual magnetic field in a part may interfere with remagnetization of the part at a field intensity too low to overcome the remanent field in the part

Reasons for Not Demagnetizing. Demagnetization may not be necessary if:

- Parts are made of magnetically soft steel having low retentivity; such parts will usually become demagnetized as soon as they are removed from the magnetizing source
- The parts are subsequently heated above their Curie point and consequently lose their magnetic properties
- The magnetic field is such that it will not affect the function of the part in service
- The part is to be remagnetized for further magnetic particle inspection or for some secondary operation in which a magnetic plate or chuck may be used to hold the part

This last reason may appear to conflict with the last item in the section "Reasons for Demagnetizing." The establishment of a longitudinal field after circular magnetization negates the circular field, because two fields in different directions cannot exist in the same part at the same time. If the magnetizing force is not of sufficient strength to establish the longitudinal field, it should be increased, or other steps should be taken to ensure that the longitudinal field actually has been established. The same is true in changing from longitudinal to circular magnetization. If the two fields (longitudinal and circular) are applied simultaneously, a field will be established that is a vector combination of the two in both strength and direction. However, if the fields are impressed successively, the last field applied, if it is strong enough to establish itself in the part, will destroy the remanent field from the previous magnetization. If the magnetizing force last applied does not equal or exceed the preceding one, the latter may remain as the dominant field.

The limits of demagnetization can be considered to be either the maximum extent to which a part can be demagnetized by available procedures or the level to which the terrestrial field will permit it to become

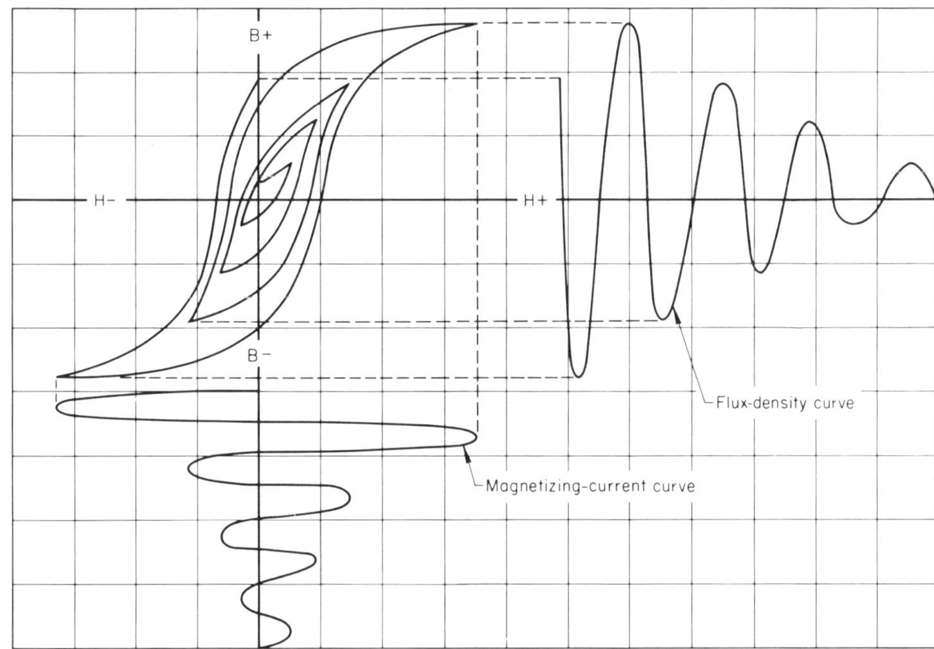

Fig. 57 Current and flux density curves during demagnetization, projected from the hysteresis loop. See text for discussion.

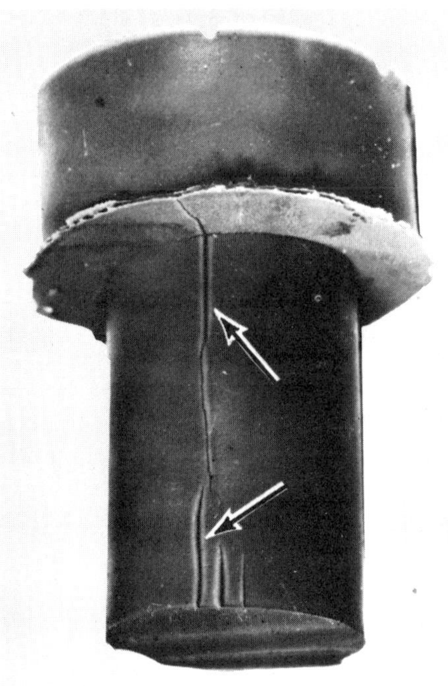

Fig. 58 Indications of discontinuities (arrows) on a magnetic rubber replica removed from a 16 mm (5/8 in.) diam through hole in 24 mm (15/16 in.) thick D-6ac steel plate

demagnetized. These limits can be further modified by the practical degree or limit of demagnetization that is actually desired or necessary.

There are a number of ways of demagnetizing a part, all based on the principle of subjecting the part to a field continually reversing its direction and at the same time gradually decreasing in strength to zero (Fig. 57). The sine wave or curve of a reversing current at the bottom of Fig. 57 is used to generate the hysteresis loops. As the current diminishes in value with each reversal, the loop traces a smaller and smaller path. The curve at the upper right of Fig. 57 represents the flux density in the part as indicated on the diminishing hysteresis loops. Both current and flux density curves are plotted against time, and when the current reaches zero, the field remaining in the part will also have approached zero.

In using this principle, the magnetizing force must be high enough at the start to overcome the coercive force and to reverse the residual field initially in the part. Also, the incremental decrease between successive reductions in current must be small enough so that the reverse magnetizing force will be able, on each cycle, to reverse the field remaining in the part from the last previous reversal.

Demagnetization With Alternating Current. A common method of demagnetizing small to moderate-size parts is by passing them through a coil through which alternating current at line frequency is passing (usually 50 to 60 Hz). Alternatively, the 60-Hz alternating current is passed through a coil with the part inside the coil, and the current is gradually reduced to zero. In the first method, the strength of the reversing field is reduced by axially withdrawing the part from the coil (or the coil from the part) and for some distance beyond the end of the coil (or part) along that axial line. In the second method, gradual decay of the current in the coil accomplishes the same result. Passing a part through an ac coil is usually the faster, preferred method.

Small parts should not be loaded into baskets and the baskets passed through the coil as a unit, because alternating current will not penetrate into such a mass of parts and because only a few parts on the outside edges will be demagnetized (and these possibly only partially demagnetized). Small parts can be demagnetized in multiple lots only if they are placed in a single layer on a tray that holds them apart and in a fixed position with their long axes parallel to the axis of the coil.

Large parts are not effectively demagnetized with 60-Hz alternating current, because of its inability to penetrate. Alternating current with 25-Hz frequency is more effective.

Machines that provide decaying alternating current have a built-in means for automatically reducing the alternating current to zero by the use of step-down switches, variable transformers, or saturable-core reactors. When decaying alternating current is used, the current can be passed directly through the part instead of through a coil. Passing the current through the part is more effective on long, circularly magnetized parts than the coil method, but does not overcome the lack of penetration because of the skin effect, unless frequencies much lower than 60 Hz are used. High field strength ac demagnetizing coils are available with power factor correction, resulting in lower line current.

Demagnetization With Direct Current. Methods of demagnetizing with direct current are essentially identical in principle to the methods just described for alternating current. By using reversing and decreasing direct current, low-frequency reversals are possible, resulting in more complete penetration of even large cross sections.

A commonly used frequency is one reversal per second. It is a successful means of removing circular magnetic fields, especially when the current is passed directly through the part and can be used to demagnetize large parts. When a part in a coil is demagnetized using direct current at one reversal per second, the part remains in the coil for the duration of the entire cycle.

Oscillating circuits are a means of obtaining a reversing decaying current for demagnetizing purposes. By connecting a large capacitance of the correct value across the demagnetizing coil, the coil becomes part of an oscillatory circuit. The coil is energized with direct current; when the source of current is cut off, the resonant resistance-inductance-capacitance circuit oscillates at its own resonant frequency, and the current gradually diminishes to zero.

Yokes, either direct or alternating current, provide a portable means for demagnetizing parts. The space between the poles of the yoke should be such that the parts to be demagnetized will pass between them as

Table 2 Applicability of demagnetizing methods on the basis of part size, metal hardness, and production rate

Method	Part size(a)			Metal hardness(a)			Production rate(a)		
	Small	Medium	Large	Soft	Medium	Hard	Low	Medium	High
Coil, 60-Hz ac	A	A	N	A	A	N	A	A	A
Coil, dc, 30-point reversing step down	N	A	A	A	A	A	A	N	N
Through current, ac, 30-point step down	N	A	A	A	A	A	A	A	N
Through current, ac, reactor decay	N	A	A	A	A	A	A	A	N
Through current, dc, 30-point reversing step down	N	A	A	A	A	A	A	N	N
Yoke, ac	A	(b)	N	A	A	N	A	N	N
Yoke, reversing dc	A	(b)	N	A	A	A	A	N	N

(a) A, applicable; N, not applicable. (b) Used for local areas only

snugly as possible. With alternating current flowing in the coil of the yoke, parts are passed between the poles and withdrawn. Yokes can be used on large parts for local demagnetization by placing the poles on the surface, moving them around the area, and then withdrawing the yoke while it is still energized. Yokes using low-frequency reversing direct current, instead of alternating current, are more effective in penetrating larger cross sections. The applicability of demagnetizing methods, based on part size, metal hardness, and production rate, is given in Table 2.

Appendix: Proprietary Methods of Magnetic Particle Inspection

Several proprietary methods of magnetic particle inspection have been developed for specific applications. Three of these methods, which are described in this section, are magnetic rubber inspection, magnetic printing, and magnetic painting.

Magnetic Rubber Inspection

Henry J. Weltman, Jack D. Reynolds, John E. Halkias, and William T. Kaarlela, General Dynamics Corporation

Magnetic rubber inspection is a nondestructive inspection method for detecting discontinuities on or near the surfaces of parts made of ferromagnetic metals. In this method, finely divided magnetic particles, dispersed in specially formulated room temperature curing rubber, are applied to a test surface, which is subsequently magnetized. The particles are attracted to the flux fields associated with discontinuities. Following cure of the rubber (about 1 h), the solid replica casting is removed from the part and examined, either visually or with a low-power microscope, for concentrations of magnetic particles that are indications of discontinuities on or just below the surface of the testpiece.

Method Advantages and Limitations

Advantages. Magnetic rubber inspection extends and complements the capabilities of other nondestructive inspection methods in certain problem areas. These include:

- Regions with limited visual accessibility
- Coated surfaces
- Regions having difficult-to-inspect shapes and sizes
- Indications requiring magnification for detection or interpretation

The replica castings furnish evidence of machining quality, physical dimensions, and surface conditions. The replicas can also be used to detect and record the initiation and growth of fatigue cracks at selected intervals during a fatigue test. The replicas provide a permanent record of the inspection; however, because the replicas shrink slightly during storage, critical measurements should be made within 72 h of casting. Replicas stored for extended periods may require a light wipe with solvent to remove any secreted fluid.

Limitations. The process is limited to the detection of discontinuities on or near the surfaces of parts made of ferromagnetic metals. It can be used on nonmagnetic metals for surface topography testing only. In this application, surface conditions, tool marks, and physical dimensions will be recorded, but there will be no migration of magnetic particles.

Magnetic rubber inspection is not as fast as other inspection methods, because of the time required to cure the rubber. This is of little disadvantage, however, when a large number of parts are being inspected. By the time all the regions being inspected have been prepared, poured, and magnetized, the first replicas are usually cured and ready for removal and examination.

Procedure

The conventional procedure used in magnetic rubber inspection can be divided into three steps:

- Preinspection preparation of parts
- Catalyzing, pouring, and magnetizing
- Review and interpretation of cured replicas

Preinspection preparation consists of cleaning the part of loose dirt or other contamination. It is often unnecessary to remove paint, plating, or flame-sprayed metal coating, but the removal of such coatings will often intensify any magnetic indications. Coatings thicker than 0.25 mm (0.01 in.) should always be removed. The next step is to prepare a reservoir to hold the liquid rubber on the inspection area. This is accomplished with the use of aluminum foil, aluminum or plastic tubing, and plastic tape and putty to seal the reservoirs against leakage.

Catalyzing and Pouring. The rubber inspection material must be thoroughly mixed before use to ensure a homogeneous dispersion. Black-oxide particles are included in the inspection material. A measured quantity of curing agents is stirred into the rubber, which is then transferred to the prepared reservoir.

Table 3 Flux density and duration of magnetization for various applications of magnetic rubber inspection

Type of area inspected	Flux density		Duration of magnetization, min
	mT	G	
Uncoated holes	5–10	50–100	½
	2.5–5	25–50	1
Coated holes	10–60(a)	100–600	½–1½(a)
Uncoated surfaces	15	150	1
	10	100	3
	5	50	10
	2	20	30
Coated surfaces	5–60	50–600	1–60(a)

(a) Flux density and time depend on the thickness of the coating.

Magnetizing. Continuous or residual magnetism is then induced into the part by using permanent magnets, direct current flowing through the part, or dc yokes, coils, prods, or central conductors. Direct current yokes are preferred for most applications. Because the magnetic particles in the suspension must migrate through the rubber, the duration of magnetism is usually longer than that of the standard magnetic particle method.

The minimum flux density along the surface of the test specimen is 2 mT (20 G); the higher the flux density, the shorter the required duration. Optimum durations of magnetization vary with each inspection task. Some typical examples of flux densities and durations of magnetization are given in Table 3.

As in the standard magnetic particle method, cracks and other discontinuities are displayed more strongly when they lie perpendicular to the magnetic lines of force. Therefore, the magnetizing current should be applied from two directions to increase reliability. This is accomplished by magnetizing in one direction, then moving the magnetizing unit to change the field 90° and remagnetizing on the same replica. Experiments have shown that the second magnetization does not disturb particles drawn to discontinuities during the first magnetization.

Review and Interpretation. Following cure, the replicas are removed from the part and examined for concentration of magnetic particles, which indicates the presence of discontinuities. This examination is best conducted with a low-power microscope (about seven to ten diameters) and a high-intensity light. During this examination, the topography of the replica is noted; tool marks, scratches, or gouges in the testpiece are revealed. Indications on a replica removed from a 16 mm (5/8 in.) diam through hole in 24 mm (15/16 in.) thick D-6ac low-alloy ultrahigh-strength steel plate are shown in Fig. 58.

Alternative Procedure. Another procedure used in magnetic rubber inspection involves placing a thin plastic film between the test surface and the rubber. This can be accomplished by stretching a sheet of polyvinylidene chloride over the test area and painting a thin layer of catalyzed or uncatalyzed rubber over it. The film can then be removed and examined for indications immediately following magnetization, eliminating the need to wait for the rubber to cure.

In addition to providing immediate inspection results, this technique has other advantages:

- No damming is required
- Postinspection cleanup is easier because the rubber never directly comes into contact with the part
- Uncatalyzed rubber can be reused
- Catalyzed rubber can be used if a permanent record is desired

The technique, however, is less sensitive than the conventional magnetic rubber inspection method and is difficult to apply to irregularly shaped surfaces.

Use on Areas of Limited Visual Accessibility

Examples of areas of limited visual accessibility that can be magnetic rubber inspected are holes and the inside surfaces of tubular components. Holes with small diameters, especially if they are threaded, are very difficult to inspect by other nondestructive methods. The deeper the hole and the smaller the diameter, the greater the problem. Liquid penetrant and magnetic particle methods are each highly dependent on the visual accessibility of the part itself; therefore, they are limited in such applications. With the use of magnetic rubber inspection, however, the visibility restriction is removed because replica castings can be taken from the inaccessible areas and examined elsewhere under ideal conditions without any visual limitations.

An application for the inspection of small-diameter holes is illustrated in Fig. 59. The testpiece is a 4.0 mm (5/32 in.) thick D-6ac steel aircraft longeron containing several groups of three nutplate holes (Fig. 59a). Each group consisted of two rivet holes 2.4 mm (3/32 in.) in diameter and a main hole 6.4 mm (1/4 in.) in diameter. Examination of a replica of one group of nutplate holes (Fig. 59b) revealed indications of cracks in one of the rivet holes and in the main hole.

Blind holes present a problem in conventional magnetic particle inspection or in liquid penetrant inspection. If the part is stationary, the inspection fluid will accumulate at the bottom of the hole, preventing inspection of that area. Another problem is directing adequate light into a blind hole for viewing.

Similar visibility problems restrict inspection of the inside surfaces of tubular components. The longer the component and the smaller its diameter, the more difficult it becomes to illuminate the inside surface and to see the area of interest. Magnetic particle, liquid penetrant, and borescope techniques have limited value in this type of application. Grooves, lands, and radical section changes also limit the use of ultrasonic and radiographic methods for the inspection of inside surfaces. The magnetic rubber technique, however, provides replica castings of such surfaces for examination after the replicas have been removed from the components. Some examples of this application include mortar and gun barrels, pipe, tubing, and other hollow shafts.

Use on Coated Surfaces

Coatings such as paint, plating, and flame- or plasma-sprayed metals have always presented difficulties in conventional nondestructive inspection. Liquid penetrants are unsuccessful unless discontinuities in the substrate have also broken the surface of the coating. Even then, it is difficult to determine whether a liquid penetrant indication resulted from cracks in the coating or cracks in the coating plus the substrate. Production ultrasonic techniques have been successfully used to locate discontinuities in coated flat surfaces; however, their ability to detect small cracks less than 2.54 mm (0.100 in.) long by 0.0025 mm (0.0001 in.) wide in bare or coated material is poor to marginal.

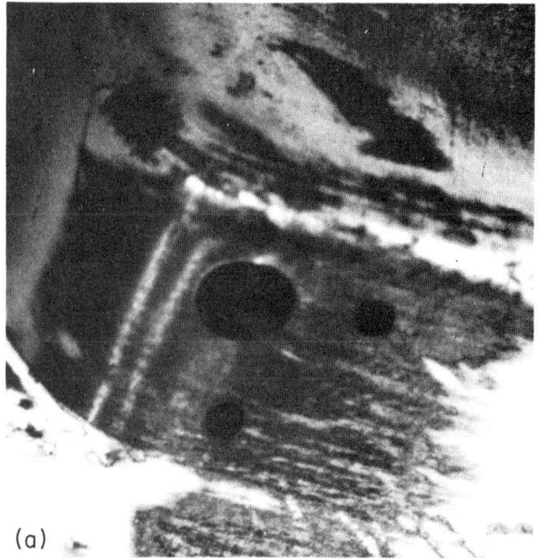
(a)

(b)

Fig. 59 Aircraft longeron (a), of 4.0 mm (5/32 in.) thick D-6ac steel, showing nutplate holes that were magnetic rubber inspected. (b) Cured magnetic rubber replica with indications (arrows) of cracks in the 6.4 mm (1/4 in.) diam main hole and a 2.4 mm (3/32 in.) diam rivet hole

Because most coatings are nonmagnetic, it is possible to use magnetic particle and magnetic rubber techniques to inspect ferromagnetic materials through the coatings. Experience has shown that conventional magnetic particle techniques also become marginal if the coating is 0.10 mm (0.004 in.) thick or greater. However, magnetic rubber inspection has the capability of producing indications through much thicker coatings. Because of the weak leakage field at the surface, the particles used in the conventional magnetic particle method are lightly

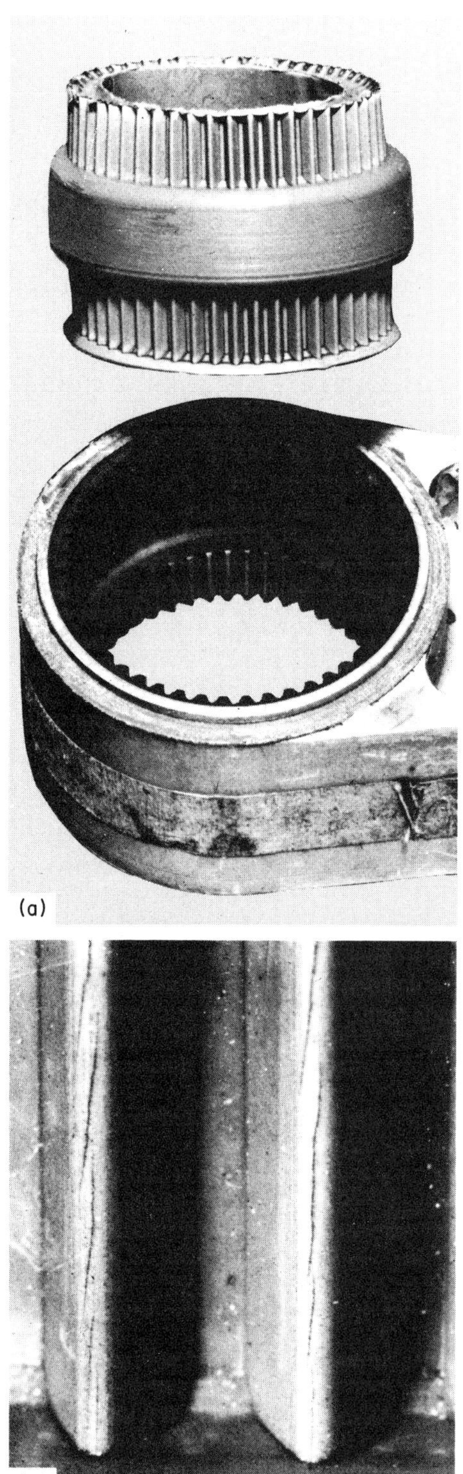

Fig. 60 Magnetic rubber inspection of spline teeth in a 4330 steel bracket for an aircraft-flap actuator. (a) View of bracket with rubber replica removed. (b) Macrograph of replica showing crack indications in roots of teeth

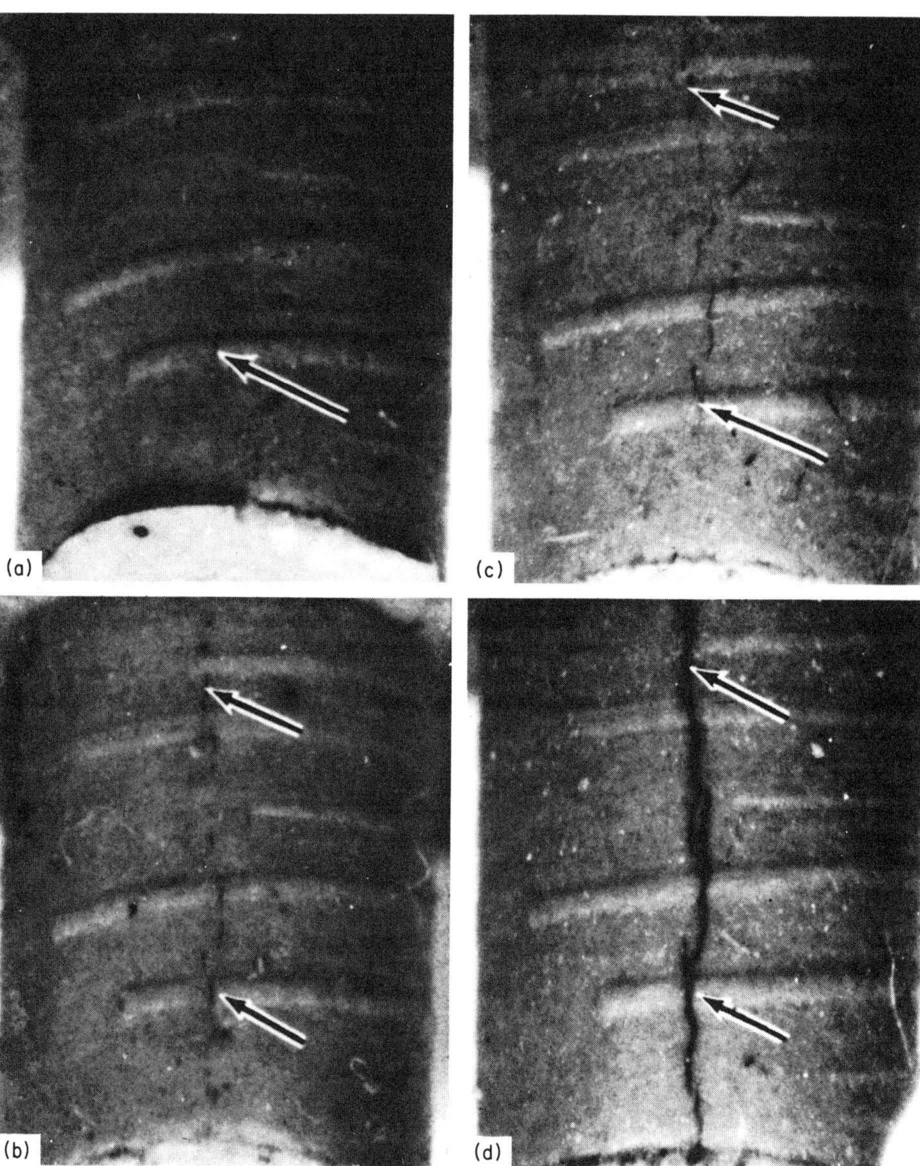

Fig. 61 Magnetic rubber replicas used to monitor crack growth in a hole during fatigue testing of a D-6ac steel aircraft part. Part fractured at 4545 cycles. (a) Initial replica of the hole showing a tool mark (arrow). (b) Replica made after 3500 fatigue cycles. Intensity of indication increased at tool mark (lower arrow), and a new indication was formed (upper arrow). (c) Replica made after 4000 cycles. Indications joined, and growth of crack (arrows) is evident. (d) Replica made after 4500 cycles. Mature fatigue crack (arrows), extending all along hole, is very evident.

attracted to the region of the discontinuity. In the magnetic rubber technique, the reduced particle attraction is compensated for by increasing the time of magnetization, up to several minutes, to ensure sufficient particle accumulation. The attracted particles remain undisturbed until the rubber is cured.

Use on Difficult-to-Inspect Shapes or Sizes

Complex structures exhibiting varying contours, radical section changes, and surface roughness present conditions that make interpretation of data obtained by radiographic, magnetic particle, liquid penetrant, or ultrasonic inspection difficult because of changing film densities, accumulation of excess fluids, and high background levels. As a result, discontinuities in such structures frequently remain undetected. The magnetic rubber process minimizes background levels on the cured replicas with little change in the intensity of any crack indications. Typical items to which magnetic rubber inspection is applicable are multiple gears, internal and external threads, and rifling grooves in gun barrels.

When magnetic particle fluid is applied to a threaded area, some of the liquid is held by surface tension in the thread roots (the most likely area for cracks). This excess fluid masks defect indications, especially when the fluorescent method is used. With the magnetic rubber method, thread root cracks are displayed with little or no interfering background.

Example 6: Magnetic Rubber Inspection of Spline Teeth in an Aircraft-Flap Actuator. The process applied to internal spline teeth is illustrated in Fig. 60(a), which shows an aircraft-flap actuator bracket with the magnetic rubber replica. A macroscopic view of this replica (Fig. 60b) reveals several cracks in the roots of the spline teeth. The bracket was made of 4330 steel. The spline teeth were 16 mm (5/8 in.) long, with 24/48 pitch.

Magnification of Indications

The examination of cast replicas under magnification permits detection of cracks as short as 0.05 mm (0.002 in.). Detection of these small cracks is often important to permit easier rework of the part prior to crack propagation. These cracks are also of interest during fatigue test monitoring.

Discontinuity indications in magnetic particle inspection often result from deep scratches or tool marks on the part surface, and it is difficult to distinguish them from cracks. When the magnetic rubber replicas are viewed under magnification, the topography of the surface is easily seen, and indications from scratches and tool marks can be distinguished from crack indications. This distinction may prevent the unnecessary rejection of parts from service.

Surface Evaluation

Magnetic rubber replicas are reproductions of the test surface and therefore display surface conditions such as roughness, scratches, tool marks, or other machining or service damage. Some surface conditions in holes, such as circumferential tool marks, are usually not harmful. Discontinuous tool marks (from tool chatter) are stress raisers and potential sites of crack initiation and propagation. Axial tool marks, which may result from a fluted reamer, are often not permitted in areas of high stress. Surface studies by magnetic rubber inspection can be applied to areas other than holes, such as rifling lands and grooves.

Use for Fatigue Test Monitoring

Magnetic rubber inspection has been used in the structural laboratory for studies of crack initiation and propagation during fatigue testing. Because each replica casting is a permanent record of the inspection, it is convenient to compare the test results during various increments of a test program.

Example 7: Magnetic Rubber Inspection of Aircraft Structural Part to Monitor Steel Fatigue. An example of the fatigue test monitoring of an aircraft structural part made of D-6ac steel is shown in Fig. 61. The test area was a 4.8 mm (3/16 in.) diam, 5.6 mm (7/32 in.) deep hole. A replica of the hole at the beginning of the test is shown in Fig. 61(a). A few tool marks were noted at this time. After 3500 cycles of fatigue loading, another replica of the hole was made (Fig. 61b). A comparison with the original replica showed some new discontinuity indications growing from the tool marks. After 4000 cycles (Fig. 61c), another replica showed that the indications in the hole were increasing and beginning to join together. This propagation continued until at 4500 cycles the crack extended through the entire hole (Fig. 61d). A few cycles later, the testpiece failed through the hole.

Because the test hole described above was located in an area that was obstructed from view, nondestructive inspection methods requiring viewing the hole would have been very difficult to perform. Moreover, the hole was coated with flame-sprayed aluminum, which would have further limited the applicability of some nondestructive-inspection methods.

Magnetic Printing (U.S. Patent 3,243,876)

Orlando G. Molina, Rockwell International

Magnetic printing employs a magnetizing coil (printer), magnetic particles, and a plastic coating of the surface of the testpiece for the detection of discontinuities and flaws. The process can be used on magnetic materials that have very low magnetic retentivity.

The magnetizing coil, or magnetic printer, consists of a flat coil made of an electrical conductor, and it is connected to a power supply capable of delivering 60-Hz alternating current of high amperage at low voltage. When the coil is energized, a strong, pulsating magnetic field is distributed along the axis of the coil and produces a vibratory effect on the testpiece and the magnetic particles. This vibratory effect causes the magnetic particles to stain or print the plastic coating in regions where magnetic particles have been attracted by changes in magnetic permeability. The magnetic particles are made of ferromagnetic iron oxide (Fe_3O_4) and are similar to those used for conventional magnetic particle inspection.

Magnetically printed patterns are made visible by first spraying the surface of the testpiece with a white plastic coating. The coating provides a contrasting background and a surface on which the particles print. After a print has been obtained and the particles have been removed, the patterns can be fixed by spraying with a clear plastic coating. Because the two coatings are of the same composition, a single film is formed, within which the printed pattern is sandwiched. When dry, the coating can be stripped from the surface of the testpiece and used as a permanent record.

Procedure

The testpiece should be cleaned so that it is free of dirt, moisture, oil, paint, scale, and other materials that can obscure a discontinuity or flaw. The white plastic coating is sprayed onto the test surface in an amount to establish a white background to offset the color of the test surface. The coating should be free of puddles, runs, and signs of orange peel. The coating is dried before application of the magnetizing current and particles.

With the magnetizing current on and properly adjusted, the testpiece and magnetic printer are placed adjacent to each other, and dry printing particles are dusted on the test surface with a powder bulb applicator. The testpiece and printer can be moved relative to each other to obtain uniform printing. When a suitable print has formed, usually after 6 to 12 s, the magnetizing current is turned off. The excess magnetic printing particles can be removed with a gentle air blast or gentle tapping.

If a suitable print has not been obtained, the print can be erased with a damp sponge, and the application of magnetizing current and printing particles repeated. If a permanent record is needed, the printed surface is sprayed with two coats of clear plastic. To assist in removing the coating, a piece of pressure-sensitive clear plastic tape can be applied to the printed surface after the clear plastic coating has dried to the touch. Copies of magnetic printings can be made by conventional photographic contact printing methods, using the magnetic printed record as a negative. A transparency for projection purposes can be made by magnetic printing on the clear coating instead of the white coating. White or clear nitrocellulose lacquer can be used in place of the strippable coatings when a permanent, nonstrippable magnetic print is required.

On some occasions, the magnetic printing particles group together in certain areas of the part surface, reducing the printing capabilities of the particles. When an aluminum alloy plate, such as 2024, is placed beneath the magnetizing coil (with the coil between the aluminum alloy plate and the testpiece), the particles remain in constant dispersion, thus preventing grouping.

A similar inspection method (U.S. Patent 3,826,917) uses a coating, preferably an organic coating, containing fluorescent material and nonfluorescent particles, preferably suspended in a liquid medium. Magnetic-flux lines are established substantially perpendicular to the suspected discontinuities in the surface of the testpiece. The particles agglomerate and form indications on the coating adjacent to the discontinuities. The testpiece is inspected under ultraviolet light to locate and reveal the surface discontinuities. If a permanent record is needed, a clear strippable plastic coating is applied over the magnetic indications of imperfections, and the resulting coating is stripped from the surface.

Applications

Magnetic printing can be used for inspecting ferromagnetic materials of either high or low magnetic retentivity to detect any con-

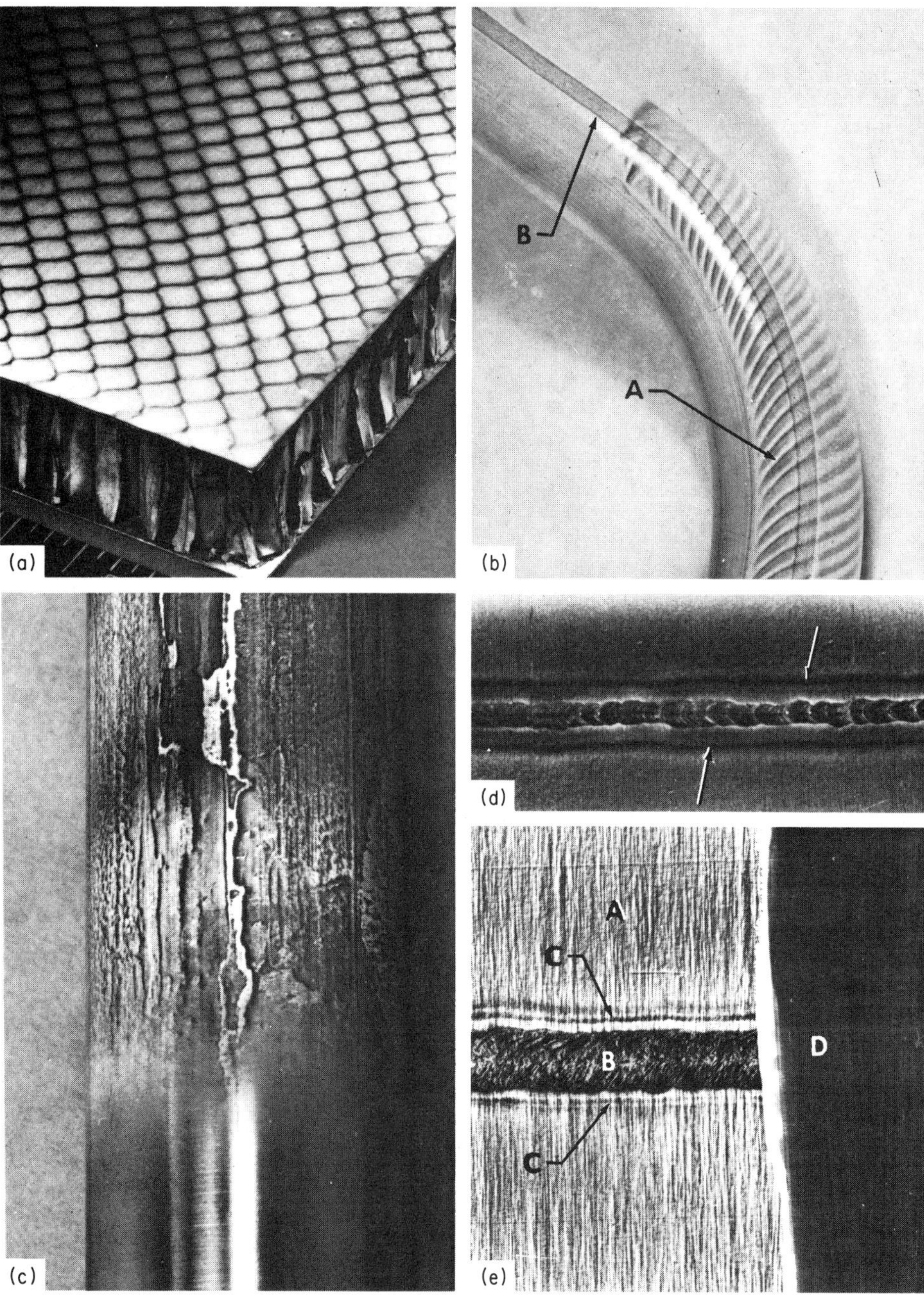

Fig. 62 Typical applications of magnetic printing for the detection of discontinuities and metallurgical flaws. (a) PH 15-7 Mo stainless steel brazed honeycomb panel showing core pattern. (b) AM-350 steel tube showing Lüders lines (at A) and weld bead (at B). (c) AM-350 steel tube; upper region is a magnetic print showing white riverlike areas that are stringers of retained austenite, and lower region is a nonprinted area. (d) Weld bead in PH 15-7 Mo stainless steel sheet showing heat-affected zones (arrows). (e) Print of the machined surface of a PH 15-7 Mo stainless steel weldment showing ferrite stringers in an essentially austenitic matrix (area at A), weld metal (area at B) and the adjacent heat-affected zones (arrows at C's), and surface where the magnetic print had been removed (area at D)

dition that affects magnetic permeability. Some typical applications of magnetic printing are illustrated in Fig. 62, and discussed in the following sections.

Brazed Honeycomb Panels. Figure 62(a) shows a magnetic print of a brazed honeycomb panel made of PH 15-7 Mo stainless steel. Visible in the print is the core pattern otherwise invisible to the eye. Areas of lack of attachment between core cells and facing sheet, puddling of brazing alloy, and the face sheet seam weld have been observed in magnetic prints of honeycomb panels.

Elastic and Plastic Deformation. A response to certain degrees of elastic and plastic deformation in some ferromagnetic materials can be detected by magnetic printing. Indications of localized plastic deformation (Lüders lines) and the seam weld in an AM-350 steel tube are shown in Fig. 62(b). These patterns are obtainable even after the stress has been removed. However, magnetic printing patterns can be obtained when a testpiece is under elastic stress, but they are no longer obtainable when the stress is removed. These phenomena occur because of regions of different magnetic permeability within a given testpiece.

Crack Detection. The magnetic printing method is generally more sensitive in detecting cracks than liquid penetrants. No special orientation of the flux lines is needed to detect cracks at different angles to each other, as is required in conventional magnetic particle inspection. Crack growth in fatigue and tension tests can be monitored by making magnetic prints at intervals during the test. In a tensile test on an AM-350 steel tube specimen, changes were revealed in contained areas of retained austenite. Not only was gradual transformation noticed in the recorded appearance of the metallurgical detail but a distinct indication was recorded in the last print taken before fracture. Some of the magnetic prints showed stress patterns at the ends of cracks.

Metallurgical details not always obtainable by common macroetching methods are usually revealed by magnetic printing. These details include flow lines in extrusions and forgings, as well as stringers of retained austenite. The magnetic printing of an AM-350 steel tube is shown in Fig. 62(c). The white, riverlike areas are stringers of retained austenite; the presence of retained austenite was confirmed by x-ray diffraction and metallographic examination.

Heat-affected zones adjacent to welds can be detected by magnetic printing. The heat-affected zones adjacent to the weld bead in PH 15-7 Mo stainless steel are shown in Fig. 62(d). This weld was subsequently sectioned for magnetic printing and metallographic examination, which verified the presence of heat-affected zones. Figure 62(e) shows ferrite stringers in an essentially austenitic matrix, cast weld metal, heat-affected zones, and an area where the magnetic print had been removed. The print exhibits a three-dimensional effect as if the weld area had not been machined.

Magnetic Painting (U.S. Patent 3,786,346)

D.E. Lorenzi, Magnaflux Corporation

Magnetic painting uses a visually contrasting magnetic particle slurry for flaw detection. A slurry concentrate having a consistency of paint is brush applied to the surface being inspected. Brushing allows for the selective application of the material; the magnetic particles can be spread evenly and thoroughly over the test area of inter-

Fig. 63 Magnetic paint indications of cracks in weld metal

Fig. 64 Wet fluorescent magnetic paint indications of minute grinding cracks in the faces of a small sprocket

est. When the testpiece is subjected to a suitable magnetizing force, flaws appear as contrasting black indications on a light-gray background, as illustrated by the cracks in the weld metal shown in Fig. 63. Wet fluorescent magnetic paint indications of minute grinding cracks in the faces of a small sprocket are shown in Fig. 64. These indications are semipermanent; that is, they remain intact for extended periods of time unless intentionally erased by re-brushing.

Method Advantages

Magnetic paint slurry requires no special lighting aids and is compatible with both the continuous and the residual methods of magnetization. It is nondrying and, depending on the degree of cleanliness required, can be removed with dry rags, paper towels, or prepared cleaning solvents.

Magnetic paint covers dark- and light-colored test surfaces equally well. Consequently, the contrast between indication and background is independent of the test-surface color. In contrast to dry magnetic particles, high wind velocities and wet test surfaces do not constitute adverse inspection conditions with magnetic paint. Magnetic paint can be applied and processed on a testpiece completely immersed in water. The material requires minimal surface preparation of testpieces because it can be applied directly over oily, rusty, plated, or painted surfaces with little or no degradation of performance, provided the coatings are not excessively thick.

Because magnetic paint is a slurry having the consistency of ordinary paint, it can be selectively applied with a brush to any test surface, regardless of its spatial orientation. As a result, there is no material overspray, and any problems associated with airborne magnetic particles and/or liquid are completely eliminated. This becomes a very desirable feature when magnetic particle inspection must be performed on vertical and overhead surfaces.

For applications that require the continuous method of magnetization, the critical sequencing between the application of magnetic particles and the magnetization is eliminated because the magnetic paint is applied before the testpiece is magnetized. In addition, the material can be rebrushed to erase previous results, and the testpiece can be reprocessed without additional slurry.

Performance

Magnetic painting can be used with all standard magnetizing techniques—circular, coil, prods, and yokes, using ac or dc magnetization—and is applicable to the detection of surface as well as subsurface flaws. The material formulation utilizes selective magnetic particles, in flake form, dispersed in a viscous, oily vehicle. The viscosity of the oil-type suspending medium is chosen to restrict substantial lateral mobility of the ferromagnetic flakes while permitting rotary movement of the flakes at the flaw site when acted upon by magnetic leakage fields (Fig. 65).

Magnetic paint appears light gray in color when brush applied to a testpiece. This

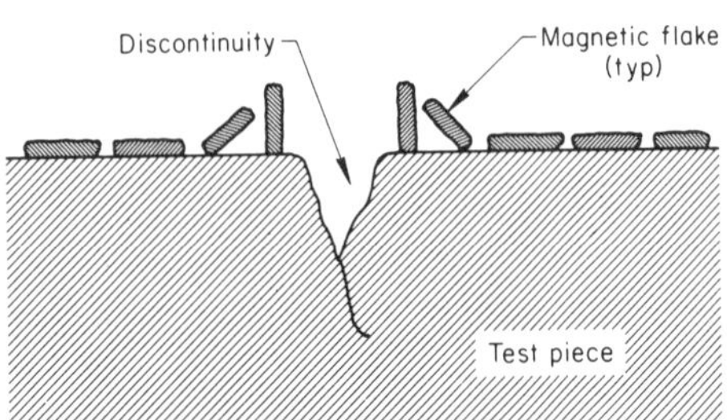

Fig. 65 Schematic of rotation of flakes of magnetic paint at the site of a discontinuity

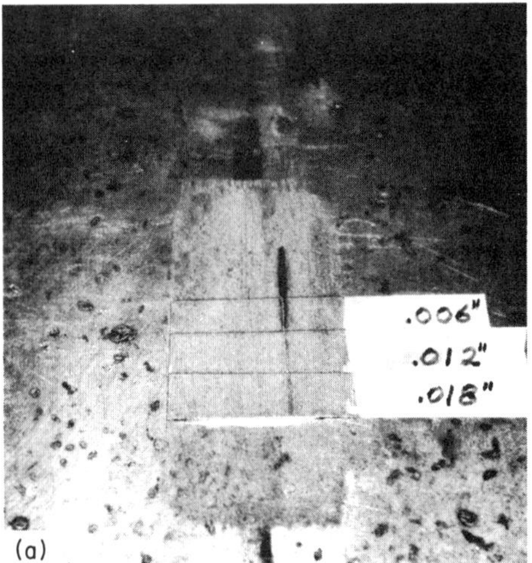

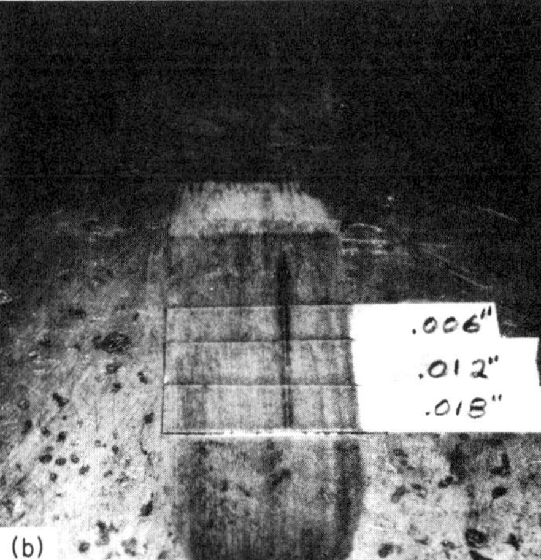

Fig. 66 Effect of oil-to-flake ratio in a magnetic paint slurry mix on the contrast between a flaw indication (dark vertical line near center) and background. (a) 6:1 ratio. (b) 10:1 ratio. Magnetic paint was applied over bare metal (upper band across indication) and over 0.15 mm (0.006 in.), 0.30 mm (0.012 in.), and 0.46 mm (0.018 in.) thicknesses of transparent plastic tape.

indicates that the flakes are oriented with their faces predominantly parallel to the surface and tend to reflect the ambient light. Because the flakes tend to align themselves with a magnetic leakage field, they virtually stand on end when subjected to the leakage field associated with a cracklike discontinuity. These edges, being relatively poor reflectors of light, appear as dark, contrasting lines against the light-gray background. Broad leakage fields result in correspondingly broad dark areas.

The nature of the indication depends, to a significant extent, on the ratio of oil-to-flake used in the slurry mix. The standard mixture provides good contrast between indications and background as well as relatively long permanence. However, the concentration can be diluted, by increasing the oil-to-flake ratio, to achieve greater indicating sensitivity (Fig. 66). Diluting the mixture results in some loss of contrast and indication permanence. Although the material is supplied having an oil-to-flake ratio of the order of 6:1, ratios as high as 20:1 can be used.

Applications

Because magnetic painting is a recent development, field applications have been limited. However, extensive laboratory testing has produced favorable results and suggests that improved testing capabilities can be realized in the following areas of application:

- Inspection of welds in pipelines, tank cars, shipbuilding, pressure vessels, and general structural steel construction
- Field inspection of used drill pipe and tubing
- Overhaul and routine field maintenance on aircraft, trucks, buses, and railroad equipment
- General industrial maintenance inspection of structural parts and equipment components

REFERENCES

1. *Mater. Eval.*, Vol 30 (No. 10), Oct 1972, p 219-228
2. Y.F. Cheu, Automatic Crack Detection With Computer Vision and Pattern Recognition of Magnetic Particle Indications, *Mater. Eval.*, Nov 1984

SELECTED REFERENCES

- "Description and Applications of Magnetic-Rubber Inspection," General Dynamics Corporation
- J.E. Halkias, W.T. Kaarlela, J.D. Reynolds, and H.J. Weltman, MRI—Help for Some Difficult NDT Problems, *Mater. Eval.*, Vol 31 (No. 9), Sept 1973
- "Inspection Process, Magnetic Rubber," MIL-I-83387, Military Specification, U.S. Air Force, Aug 1972
- M. Pevar, New Magnetic Test Includes Stainless Steel, *Prod. Eng.*, Vol 32 (No. 6), 6 Feb 1961, p 41-43

Magnetic Field Testing

R.E. Beissner, Southwest Research Institute

MAGNETIC FIELD TESTING includes some of the older and more widely used methods for the nondestructive evaluation of materials. Historically, such methods have been in use for more than 50 years in the examination of magnetic materials for defects such as cracks, voids, or inclusions of foreign material. More recently, magnetic methods for assessing other material properties, such as grain size, texture, or hardness, have received increasing attention. Because of this diversion of applications, it is natural to divide the field of magnetic materials testing into two parts, one directed toward defect detection and characterization and the other aimed at material properties measurements.

This article is primarily concerned with the first class of applications, namely, the detection, classification, and sizing of material flaws. However, an attempt has also been made to provide at least an introductory description of materials characterization principles, along with a few examples of applications. This is supplemented by references to other review articles.

All magnetic methods of flaw detection rely in some way on the detection and measurement of the magnetic flux leakage field near the surface of the material, which is caused by the presence of the flaw. For this reason, magnetic testing techniques are often described as flux leakage field or magnetic perturbation methods. The magnetic particle inspection method is one such flux leakage method that derives its name from the particular method used to detect the leakage field. Because the magnetic particle method is described in the article "Magnetic Particle Inspection" in this Volume, the techniques discussed in this article will be limited to other forms of leakage field measurement.

Although it is conceivable that leakage field fluctuations associated with metallurgical microstructure might be used in the analysis of material properties, the characterization methods now in use rely on bulk measurements of the hysteretic properties of material magnetization or of some related phenomenon, such as Barkhausen noise. The principles and applications of magnetic characterization presented in this article are not intended to be exhaustive, but rather to serve as illustrations of this type of magnetic testing.

The principles and techniques of leakage field testing and magnetic characterization are described in the two sections that follow. These sections will discuss concepts and methods that are essential to an understanding of the applications described in later sections. The examples of applications presented in the third section will provide a brief overview of the variety of inspection methods that fall under the general heading of magnetic testing.

Principles of Magnetic Leakage Field Testing

Origin of Defect Leakage Fields. The origin of the flaw leakage field is illustrated in Fig. 1. Figure 1(a) shows a uniformly magnetized rod, which consists of a large number of elementary magnets aligned with the direction of magnetization. Inside the material, each magnetic pole is exactly compensated by the presence of an adjacent pole of opposite polarity, and the net result is that interior poles do not contribute to the magnetic field outside the material. At the surfaces, however, magnetic poles are uncompensated and therefore produce a magnetic field in the region surrounding the specimen. This is illustrated in Fig. 1(a) by flux lines connecting uncompensated elementary poles.

If a slot is cut in the rod, as illustrated in Fig. 1(b), the poles on the surface of this slot are now also uncompensated and therefore produce a localized magnetic field near the slot. This additional magnetic field, which is represented by the extra flux lines in Fig. 1(b), is the leakage field associated with the slot.

Figure 1, although adequate for a qualitative understanding of the origin of leakage fields, does not provide an exact quantitative description. The difficulty is the assumption that the magnetization remains uniform when the flaw is introduced. In general, this does not happen, because the presence of the flaw changes the magnetic field in the vicinity of the flaw, and this in turn leads to a change in magnetization near the flaw. With regard to Fig. 1, this means that the strengths and orientations of the elementary dipoles (magnets) actually vary from point to point in the vicinity of the flaw, and this variation also contributes to the flaw leakage field. The end result is that the accurate description of a flaw leakage field poses a difficult mathematical problem that usually requires a special-purpose computer code for its solution.

Experimental Techniques. One of the first considerations in the experimental application of magnetic leakage field methods is the generation of a suitable magnetic field within the material. In some ferromagnetic materials, the residual field (the field that

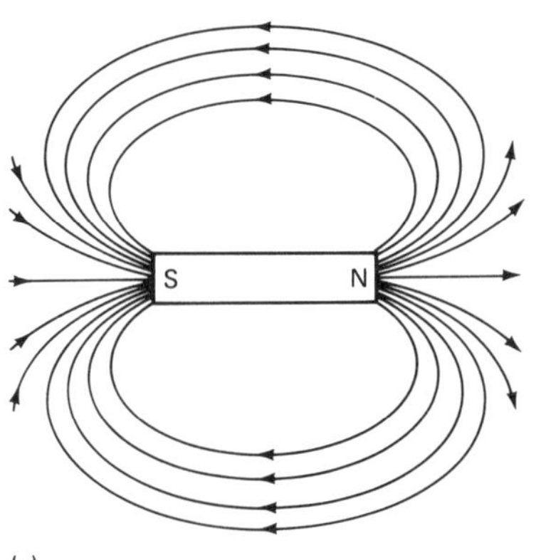

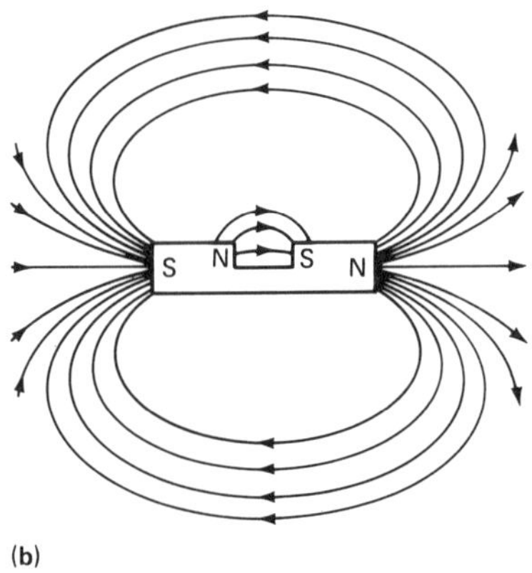

Fig. 1 Origin of defect leakage fields. (a) Magnetic flux lines of a magnet without a defect. (b) Magnetic flux lines of a magnet with a surface defect. Source: Ref 1

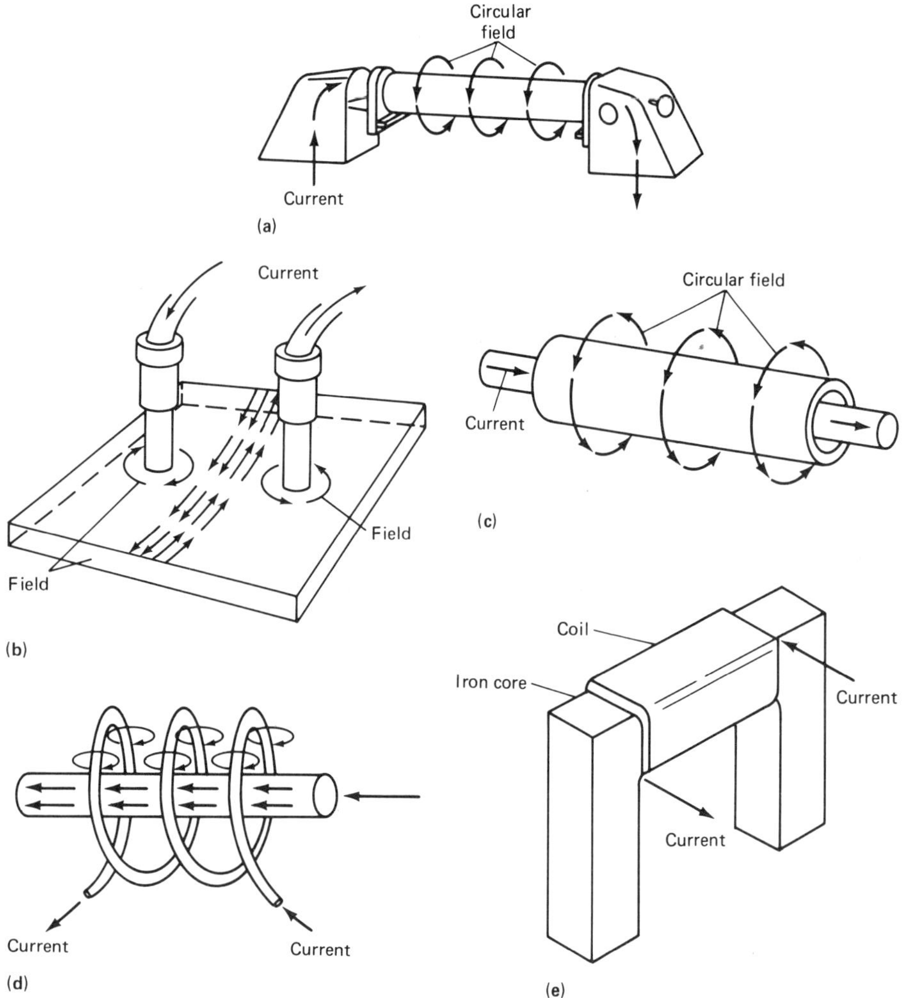

Fig. 2 Methods of magnetization. (1) Head-shot method. (b) Magnetization with prods. (c) Magnetization with a central conductor. (d) Longitudinal magnetization. (e) Yoke magnetization

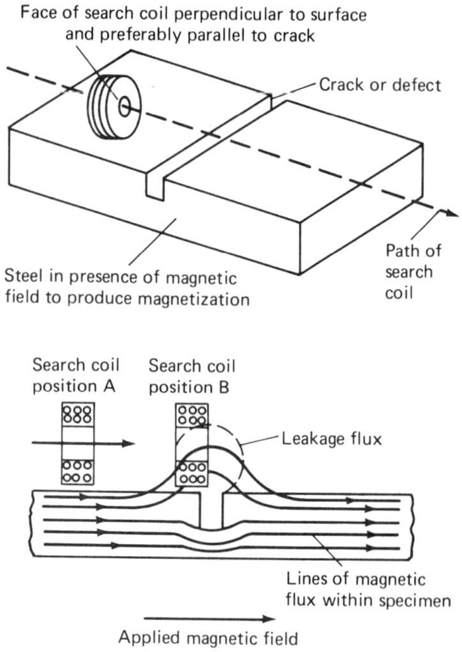

remains after removal of an external magnetizing field) is often adequate for surface flaw detection. In practice, however, residual magnetization is rarely used because use of an applied magnetizing field ensures that the material is in a desired magnetic state (which should be known and well characterized) and because applied fields provide more flexibility (that is, one can produce a high or low flux density in the specimen as desired).

Experience has shown that control of the strength and direction of the magnetization can be useful in improving flaw detectability and in discriminating among different types of flaws (Ref 1-9). In general, the magnitude of the magnetization should be chosen to maximize the flaw leakage field with respect to other field sources that might interfere with flaw detection; the optimum magnetization is usually difficult to determine in advance of a test and is often approached by trial-and-error experimentation. The direction of the field should be perpendicular to the largest flaw dimension to maximize the effect of the flaw on the leakage field.

It is possible to generate a magnetic field in a specimen either directly or indirectly (Ref 10-12). In direct magnetization, current is passed directly through the part. With the indirect approach, magnetization is induced by placing the part in a magnetic field that is generated by an adjacent current conductor or permanent magnet. This can be done, for example, by threading a conductor through a hollow part such as a tube or by passing an electric current through a cable wound around the part. Methods of magnetizing a part both directly and indirectly are illustrated schematically in Fig. 2.

The flaw leakage field can be detected with one of several types of magnetic field sensors. Aside from the use of magnetic particles, the sensors most often used are the inductive coil and the Hall effect device.

The inductive coil sensor is based on Faraday's law of induction, which states that the voltage induced in the coil is proportional to the number of turns in the coil multiplied by the time rate of change of the flux threading the coil (Ref 13). It follows that detection of a magnetostatic field requires that the coil be in motion so that the flux through the coil changes with time.

Fig. 3 Flux leakage measurement using a search coil. Source: Ref 13

The principle is illustrated in Fig. 3, in which the coil is oriented so as to sense the change in flux parallel to the surface of the specimen. If the direction of coil motion is taken as x, then the induced electromotive force, E, in volts is given by:

$$E = 10^{-8} NA \frac{dB}{dx} \frac{dx}{dt}$$

where N is the number of turns in the coil, A is its cross-sectional area, and B is the flux density, in Gauss, parallel to the surface of the part. Thus, the voltage induced in the coil is proportional to the gradient of the flux density along the direction of coil motion multiplied by the coil velocity. Figure 4 shows the flux density typical of the leakage field from a slot, along with the corresponding signal from a search coil oriented as in Fig. 3.

Unlike the inductive coil, which provides a measure of the flux gradient, a Hall effect sensor directly measures the component of the flux itself in the direction perpendicular to the sensitive area of the device (Ref 1). Because the response of a Hall effect sensor does not depend on the motion of the probe, it can be scanned over the surface to be inspected at any rate that is mechanically convenient. In this respect, the Hall device has an advantage over the coil sensor because there is no need to maintain a constant scanning speed during the inspection. On the other hand, Hall effect sensors are more difficult to fabricate, are somewhat delicate compared to inductive coil sensors, and require more complex electronics.

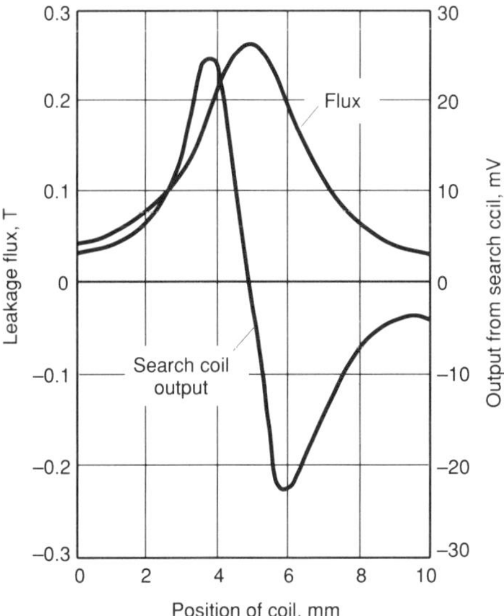

Fig. 4 Leakage flux and search coil signal as a function of position. Source: Ref 13

Other magnetic field sensors that are used less often in leakage field applications include the flux gate magnetometer (Ref 14), magnetoresistive sensors (Ref 15), magnetic resonance sensors (Ref 16), and magnetographic sensors (Ref 17), in which the magnetic field at the surface of a part is registered on a magnetic tape pressed onto the surface.

Analysis of Leakage Field Data. In most applications of the leakage field method, there is a need not only to detect the presence of a flaw but also to estimate its severity. This leads to the problem of flaw characterization, that is, the determination of flaw dimensions from an analysis of leakage field data.

The most widely used method of flaw characterization is based on the assumptions that the leakage field signal amplitude is proportional to the size of the flaw (which usually means its depth into the material) and that the signal amplitude can therefore be taken as a direct measure of flaw severity. In situations where all flaws have approximately the same shape and where calibration experiments show that the signal amplitude is indeed proportional to the size parameter of concern, this empirical method of sizing works quite well (Ref 18).

There are, however, many situations of interest where flaw shapes vary considerably and where signal amplitude is not uniquely related to flaw depth, as is the case for corrosion pits in steel tubing (Ref 19). In addition, different types of flaws, such as cracks and pits, can occur in the same part, in which case it becomes necessary to determine the flaw types present as well as their severity. In such cases, a more careful analysis of the relationship between signal and flaw characteristics is required if serious errors in flaw characterization are to be avoided.

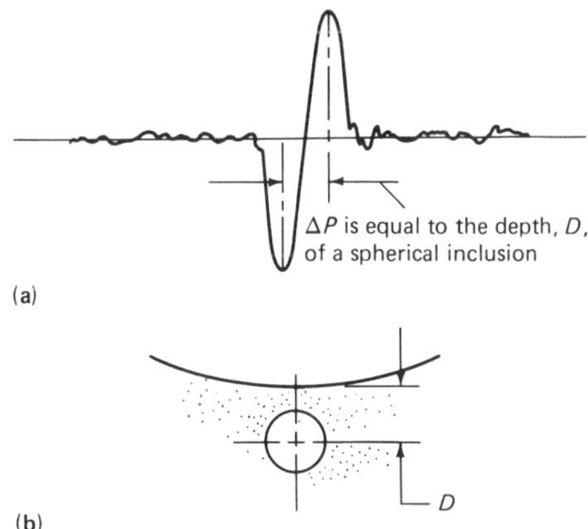

Fig. 5 Dependence of magnetic signal peak separation (a) on the depth of a spherical inclusion (b)

One of the earliest attempts to use a theoretical model in the analysis of leakage field data was based on the analytic solution for the field perturbed by a spherical inclusion (Ref 20, 21). Two conclusions were drawn from this analysis. First, when one measures the leakage flux component normal to the surface of the part, the center of the flaw is located below the scan plane at a distance equal to the peak-to-peak separation distance in the flaw signal (Fig. 5), and second, the peak-to-peak signal amplitude is proportional to the flaw volume. A number of experimental tests of these sizing rules have confirmed the predicted relationships for nonmagnetic inclusions in steel parts (Ref 21).

Further theoretical and experimental data for spheroidal inclusions and surface pits have shown, however, that the simple characterization rules for spherical inclusions do not apply when the flaw shape differs significantly from the ideal sphere. In such cases, the signal amplitude depends on the lateral extent of the flaw and on its volume, and characterization on the basis of leakage field analysis becomes much more complicated (Ref 19, 22).

Finally, there has been at least one attempt to apply finite-element calculations of flaw leakage fields to the development of characterization rules for a more general class of flaws. Hwang and Lord (Ref 23) performed most of their computations for simple flaw shapes, such as rectangular and triangular slots and inclusions, and from the results devised a set of rules for estimating the depth, width, and angle of inclination of a flaw with respect to the surface of the part. One of their applications to a flaw of complex shape is shown in Fig. 6.

The promising results obtained from the finite-element work of Hwang and Lord, as well as the analytically based work on spheroidal flaws, suggest that the estimation of flaw size and shape from leakage field data is feasible. Another numerical method potentially applicable to flux leakage problems is the boundary integral method, which may prove useful in flaw characterization. Unfortunately, much more work must be done on both the theoretical basis and on experimental testing before it will be possible to analyze experimental leakage field data with confidence in terms of flaw characteristics.

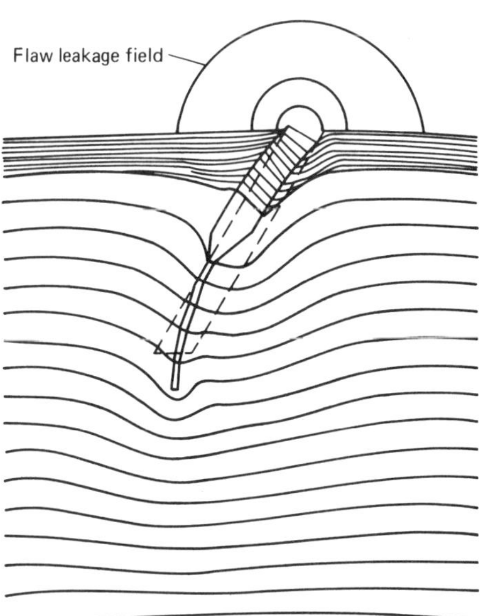

Fig. 6 Characterization of a ferrite-tail type of defect. The dashed line shows the flaw configuration estimated from the leakage field data.

Principles of Magnetic Characterization of Materials

Metallurgical and Magnetic Properties. The use of magnetic measurements to monitor the metallurgical properties of ferromagnetic materials is based on the fact that variables such as crystallographic phase, chemical composition, and microstructure, which determine the physical properties of materials, also affect their magnetic characteristics (Ref 24-26). Some parameters, such as grain size and orientation, dislocation density, and the existence of precipitates, are closely related to measurable characteristics of magnetic hysteresis, that is, to the behavior of the flux density, B, induced in a material as a function of the magnetic field strength, H.

This relationship can be understood in principle from the physical theory of magnetic domains (Ref 27). Magnetization in a particular direction increases as the domains aligned in that direction grow at the expense of domains aligned in other directions. Factors that impede domain growth also impede dislocation motion; hence the connection, at a very fundamental level, between magnetic and mechanical properties.

Other magnetic properties, such as the saturation magnetization, which is the maximum value B can achieve, or the Curie

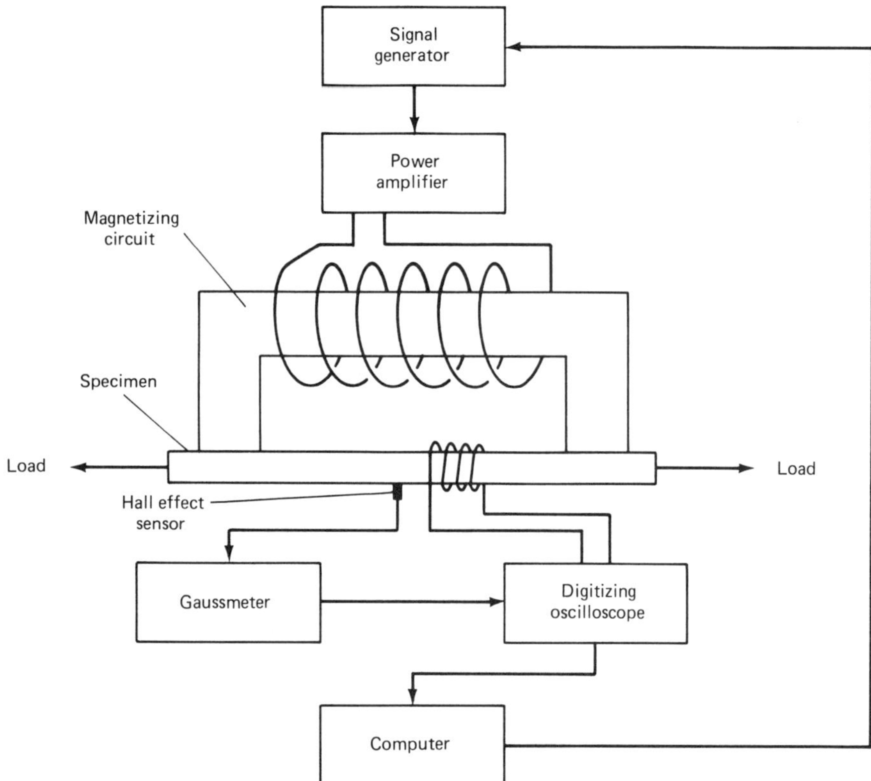

Fig. 7 Experimental arrangement for hysteresis loop measurements. Source: Ref 32

temperature at which there is a transition to a nonmagnetic state, are less dependent on microstructure, but are sensitive to such factors as crystal structure and chemical composition. Interest in the magnetic characterization of materials, principally steels, derives from many such relationships between measurable magnetic parameters and metallurgical properties. These relationships are, however, quite complicated in general, and it is often difficult to determine how or if a particular measurement or combination of measurements can be used to determine a property of interest. Nevertheless, the prospect of nondestructive monitoring and quality control is an attractive one, and for this reason research on magnetic materials characterization continues to be an active field.

It is not the purpose of this article to explore such magnetic methods in depth, but simply to point out that it is an active branch of nondestructive magnetic testing. The more fundamental aspects of the relationship between magnetism and metallurgy are discussed in Ref 26 and 28. Engineering considerations are reviewed in Ref 24. The proceedings of various symposia also contain several papers that provide a good overview of the current status of magnetic materials characterization (Ref 29-31).

Experimental Techniques. A typical setup for measuring the *B-H* characteristic of a rod specimen is shown in Fig. 7. The essential elements are an electromagnet for generating the magnetizing field, a coil wound around the specimen for measuring the time rate of change of the magnetic flux, B, in the material, and a magnetic field sensor, in this case a Hall effect probe, for measuring the magnetic field strength, H, parallel to the surface of the part. The signal generator provides a low-frequency magnetizing field, typically of the order of a few Hertz, and the output of the flux measuring coil is integrated over time to give the flux density in the material. In the arrangement shown in Fig. 7, an additional feature is the provision for applying a tensile load to the specimen for studies of the effects of stress on the hysteresis data. When using a rod specimen such as this, it is important that the length-to-diameter ratio of the specimen be large so as to minimize the effects of stray fields from the ends of the rod on the measurements of B and H.

Another magnetic method that uses a similar arrangement is the measurement of Barkhausen noise (Ref 33, 34). As the magnetic field strength, H, is varied at a very slow rate, discontinuous jumps in the magnetization of the material can be observed during certain portions of the hysteresis cycle. These jumps are associated with the sudden growth of a series of magnetic domains that have been temporarily stopped from further growth by such obstacles as grain boundaries, precipitates, or dislocations. Barkhausen noise is therefore dependent on microstructure and can be used independently of hysteresis measurements, or in conjunction with such measurements, as another method of magnetic testing. The experimental arrangement differs from that shown in Fig. 7 in that a single sensor coil, oriented to measure the flux normal to the surface of the specimen, is used instead of the Hall probe and the flux winding.

The review articles and conference proceedings cited above contain additional detail on experimental technique and a wealth of information on the interpretation of hysteresis and Barkhausen data. However, it should be noted that test methods and data interpretation are often very specific to a particular class of alloy, and techniques that seem to work well for one type of material may be totally inappropriate for another. The analysis of magnetic characterization data is still largely empirical in nature, and controlled testing of a candidate technique with the specific alloy system of interest is advisable.

Applications

Flaw Detection by the Flux Leakage Method. Perhaps the most prevalent use of the flux leakage method is the inspection of ferromagnetic tubular goods, such as gas pipelines, down hole casing, and a variety of other forms of steel piping (Ref 35, 36). In applications in the petroleum industry, the technique is highly developed, but details on inspection devices and methods of data analysis are, for the most part, considered proprietary by the companies that provide inspection services. Still, the techniques currently in use have certain features in common, and these are exemplified by the typical system described below.

The device shown in Fig. 8 is an inspection tool for large-diameter pipelines. Magnetization is provided by a large electromagnet fitted with wire brushes to direct magnetic flux from the electromagnet into the pipe wall. To avoid spurious signals from hard spots in the material, the magnetization circuit is designed for maximum flux density in the pipe wall in an attempt to magnetically saturate the material. Leakage field sensors are mounted between the pole pieces of the magnet in a circle around the axis of the device to provide, as nearly as possible, full coverage of the pipe wall. In most such tools, the sensors are the inductive coil type, oriented to measure the axial component of the leakage field gradient. Data are usually recorded on magnetic tape as the system is propelled down a section of pipe. After the inspection, the recorded signals are compared with those from calibration standards in an attempt to interpret flaw indications in terms of flaw type and size.

In addition to systems for inspecting rotationally symmetric cylindrical parts, flux

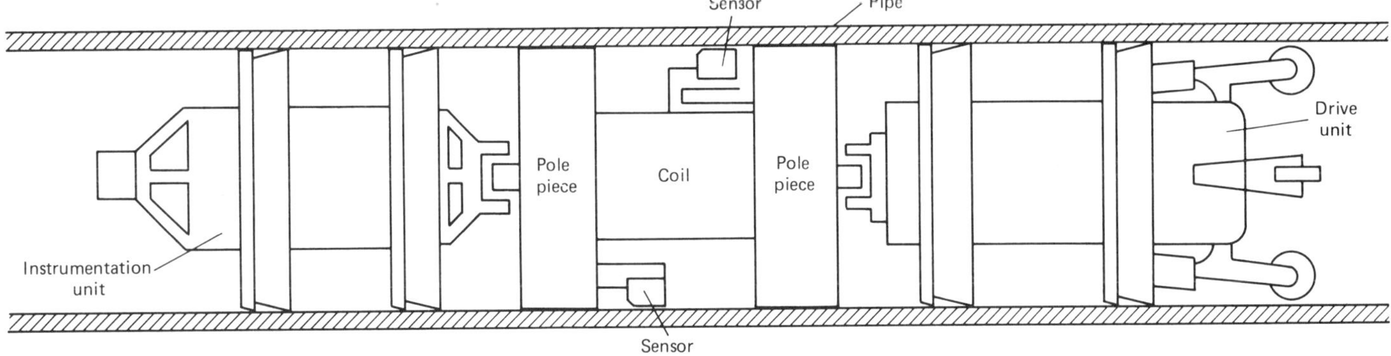

Fig. 8 Typical gas pipeline inspection pig. The tool consists of a drive unit, an instrumentation unit, and a center section with an electromagnetic and flux leakage sensors.

leakage inspection has been applied to very irregular components, such as helicopter rotor blade D-spars (Ref 37), gear teeth (Ref 38), and artillery projectiles (Ref 39). Several of these special-purpose applications have involved only laboratory investigations, but in some cases specialized instrumentation systems have been developed and fabricated for factory use. These systems are uniquely adapted to the particular application involved, and in most cases only one or at most several instrumentation systems have been built. Even in the case of laboratory investigations, special-purpose detection probe and magnetizing arrangements have been developed for specific applications.

One such system for automated thread inspection on drill pipe and collars is described in Ref 40. The device consists of an electromagnet and an array of sensors mounted outside a nonmagnetic cone that threads onto the tool joint. The assembly is driven in a helical path along the threads by a motor/clutch assembly. To minimize the leakage flux signal variations caused by the threads, signals from the sensor array are compared differentially. The system is capable of operating in a high field strength mode for the detection of cracks and corrosion pits and also in a residual field mode for the detection of other forms of damage. At last report, the system was undergoing field tests and was found to offer advantages, in terms of ease of application and defect detection, over the magnetic particle technique normally used for thread inspection.

The flux leakage method is also finding application in the inspection of ropes and cables made of strands of ferromagnetic material. One approach is to induce magnetization in the piece by means of an encircling coil energized by a direct current (dc). With this method, one measures the leakage field associated with broken strands using a Hall effect probe or an auxiliary sensor coil. A complementary method with alternating current (ac), which is actually an eddy current test rather than flux leakage, is to measure the ac impedance variations in an encircling coil caused by irregularities in the cross-sectional area of the specimen. Haynes and Underbakke (Ref 41) describe practical field tests of an instrumentation system that utilizes both the ac and dc methods. They conclude that instrumentation capable of a combination of inspection techniques offers the best possibility of detecting both localized flaws and overall loss of cross section caused by generalized corrosion and wear. They also present detailed information on the practical characteristics of a commercially available device that makes use of both the ac and dc methods.

Another area in which the flux leakage method has been successfully implemented is the inspection of rolling-element antifriction bearings (Ref 42-45). A schematic illustration of the method as applied to an inner bearing race is shown in Fig. 9. In this application, the part is magnetized by an electromagnet, as indicated in Fig. 9(a). The race is then rotated by a spindle, and the surface is scanned with an induction coil sensor. Typically, the race is rotated at a surface speed of about 2.3 m/s (7.5 ft/s), and the active portion of the raceway is inspected by incrementally indexing the sensor across the raceway. Magnetizing fields are applied in the radial and circumferential orientations. It has been shown that radial field inspection works best for surface flaws, while circumferential field inspection shows greater sensitivity to subsurface flaws (Ref 43). Data have been collected on a large number of bearing races to establish the correlation between leakage field signals and inclusion depths and dimensions determined by metallurgical sectioning (Ref 44, 45).

Finally, the flux leakage method has also been adapted to the inspection of steel reinforcement in concrete beams (Ref 46, 47). The basic function of the magnetic field disturbance (MFD) inspection equipment is to provide maps of the magnetic field across the bottom and sides of the beam. An electromagnet on an inspection cart, which is suspended on tracks below the beam, provides a magnetic field that induces magnetization in permeable structures in its vicinity, such as steel rebars, cables, and stirrups. An array of Hall effect sensors distributed across the bottom and sides of the beam measures the field produced by magnetized structures within the beam. If a flaw is present in one of these magnetized structures, it will produce a disturbance of the normal magnetic field pattern associated with the unflawed beam. Thus, the idea behind the MFD system is to search the surface of the beam for field anomalies that

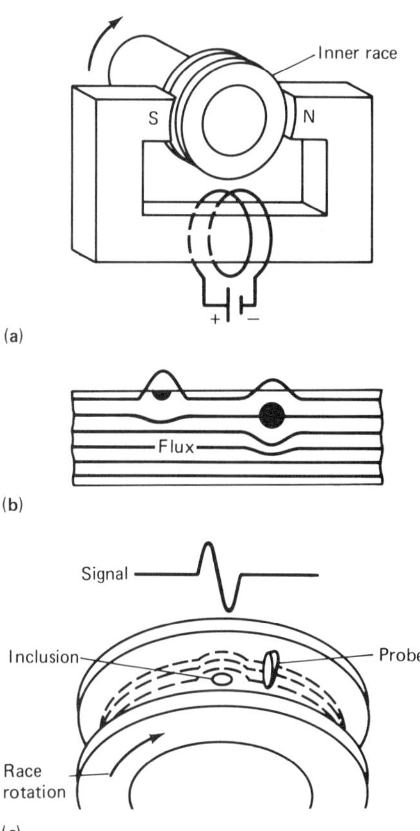

Fig. 9 Flux leakage inspection of a bearing race. (a) Magnetization of inner race. (b) Perturbation in the magnetic flux at the surface of the inner race. (c) Probe scanning the surface

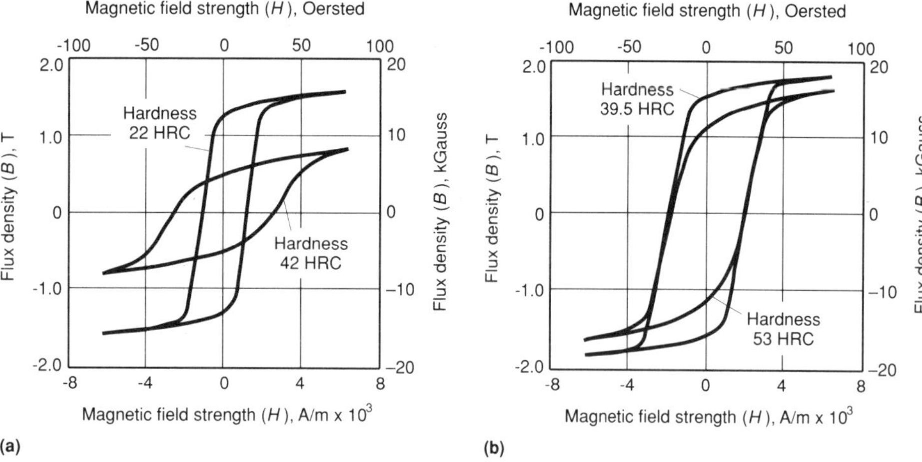

Fig. 10 Effect of mechanical hardness on hysteresis loop data. (a) AISI 410 stainless steel. (b) SAE 4340 steel. Source: Ref 32

indicate the presence of flaws in reinforcing steel within the structure.

A flaw, such as a broken wire in a cable or a fractured rebar, produces a distinctive magnetic field anomaly that depends on the size of the discontinuity and its distance from the sensor. Because the signal shape that results from such an anomaly is known, flaw detection is enhanced by searching magnetic field records for specific signal shapes, that is, those that are characteristic of discontinuities in magnetic materials. In the MFD system, this is accomplished by a computer program that compares signal shapes with typical flaw signal shapes. The program produces a correlation coefficient that serves as a measure of similarity of the observed signal shape to a typical flaw signal shape. Flaw detection is therefore not only enhanced by signal shape discrimination but also automated by computer processing of the magnetic field data.

Laboratory tests have demonstrated the ability of the system to detect fracture in steel rebars and cables in a large prestressed concrete structure (Ref 47). Also planned are field tests of the equipment in the inspection of bridge decks for reinforcement corrosion damage.

Nondestructive Characterization of Materials. Only two examples of magnetic methods for monitoring material properties are given because the examples chosen should suffice to illustrate the types of tests that might be employed. Measurements of magnetic characteristics can, however, provide a wealth of data, and various features of such data can yield information on different material properties. For example, it has been demonstrated that different features of magnetic hysteresis data can be interpreted in terms of heat treatment and microstructure, plastic deformation, residual stress, and mechanical hardness (Ref 25).

An example of the effects of mechanical hardness on hysteresis data is shown in Fig. 10. These data were obtained in the absence of applied tensile stress with the experimental arrangement shown in Fig. 7. Specimens of different hardness were prepared by tempering at different temperatures. The grain size (ASTM No. 7) was the same for all four specimens used in these tests. Other data showed, however, that grain size has little effect on hysteretic behavior for the classes of alloys studied.

The main point illustrated in Fig. 10 is that the mechanically harder specimens of the same alloy are also harder to magnetize; that is, the flux density, B, obtained at a large value of H is smaller for mechanically harder specimens than for softer specimens. For one alloy, AISI 410 stainless steel, the hysteresis loop intersects the $B = 0$ axis at larger values of H for the harder specimen than for the softer specimen; that is, the coercive force is greater for the harder material. However, for the other material, SAE 4340 steel, the coercive force does not change with hardness. This suggests that, for the two alloys considered here, the saturation flux density provides a more reliable measure of hardness than the coercive force.

Mayos *et al.* (Ref 48) used two quite different techniques to measure the depth of surface decarburization of steels. One method was a variation of a standard eddy current test, with the difference from standard practice being that eddy current probe response was measured in the presence of a low-frequency (~0.1 Hz) magnetic field. This arrangement provides a measure of incremental permeability, that is, the magnetic permeability corresponding to changes in the applied field about some quasistatic value. The second method employed was Barkhausen noise analysis.

Depth of decarburization was analyzed by varying the frequency of the excitation field, thus changing the skin depth in the material. Experiments were performed with both artificial samples containing two layers of different carbon content and industrial samples in which carbon concentration varied smoothly with distance from the surface.

It was shown that certain features of both Barkhausen noise and incremental permeability data can be correlated with depth of decarburization. The Barkhausen noise method showed a somewhat stronger sensitivity to depth, but was useful over a smaller range of depths than the incremental permeability method. It can be concluded that both methods are useful, with the optimum choice depending on accuracy requirements and the expected depth of decarburization.

REFERENCES

1. R.E. Beissner, G.A. Matzkanin, and C.M. Teller, "NDE Applications of Magnetic Leakage Field Methods," Report NTIAC-80-1, Southwest Research Institute, 1980
2. G. Dobmann, Magnetic Leakage Flux Techniques in NDT: A State of the Art Survey of the Capabilities for Defect Detection and Sizing, in *Electromagnetic Methods of NDT*, W. Lord, Ed., Gordon and Breach, 1985
3. P. Höller and G. Dobmann, Physical Analysis Methods of Magnetic Flux Leakage, in *Research Techniques in NDT*, Vol IV, R.S. Sharpe, Ed., Academic Press, 1980
4. F. Förster, Magnetic Findings in the Fields of Nondestructive Magnetic Leakage Field Inspection, *NDT Int.*, Vol 19, 1986, p 3
5. F. Förster, Magnetic Leakage Field Method of Nondestructive Testing, *Mater. Eval.*, Vol 43, 1985, p 1154
6. F. Förster, Magnetic Leakage Field Method of Nondestructive Testing (Part 2), *Mater. Eval.*, Vol 43, 1985, p 1398
7. F. Förster, Nondestructive Inspection by the Method of Magnetic Leakage Fields. Theoretical and Experimental Foundations of the Detection of Surface Cracks of Finite and Infinite Depth, *Sov. J. NDT*, Vol 18, 1982, p 841
8. F. Förster, Theoretical and Experimental Developments in Magnetic Stray Flux Techniques for Defect Detection, *Br. J. NDT*, Nov 1975, p 168-171
9. J.R. Barton and F.N. Kusenberger, "Magnetic Perturbation Inspection to Improve Reliability of High Strength Steel Components," Paper 69-DE-58, presented at a conference of the Design Engineering Division of ASME, New York, American Society of Mechanical Engineers, May 1969
10. R.C. McMaster, *Nondestructive Testing Handbook*, Vol II, Section 30, The Ronald Press Company, 1959
11. H.J. Bezer, Magnetic Methods of Non-

destructive Testing, Part 1, *Br. J. NDT*, Sept 1964, p 85-93; Part 2, Dec 1964, p 109-122
12. F.W. Dunn, Magnetic Particle Inspection Fundamentals, *Mater. Eval.*, Dec 1977, p 42-47
13. C.N. Owston, The Magnetic Leakage Field Technique of NDT, *Br. J. NDT*, Vol 16, 1974, p 162
14. F. Förster, Non-Destructive Inspection of Tubing and Round Billets by Means of Leakage Flux Probes, *Br. J. NDT*, Jan 1977, p 26-32
15. A. Michio and T. Yamada, Silicon Magnetodiode, in *Proceedings of the Second Conference on Solid State Devices* (Tokyo), 1970; Supplement to *J. Jpn. Soc. Appl. Phys.*, Vol 40, 1971, p 93-98
16. B. Auld and C.M. Fortunko, "Flaw Detection With Ferromagnetic Resonance Probes," Paper presented at the ARPA/AFML Review of Progress in Quantitative NDE, Scripps Institution of Oceanography, July 1978
17. F. Förster, Development in the Magnetography of Tubes and Tube Welds, *Non-Destr. Test.*, Dec 1975, p 304-308
18. W. Stumm, Tube Testing by Electromagnetic NDE Methods-1, *Non-Destr. Test.*, Oct 1974, p 251-256
19. R.E. Beissner, G.L. Burkhardt, M.D. Kilman, and R.K. Swanson, Magnetic Leakage Field Calculations for Spheroidal Inclusions, in *Proceedings of the Second National Seminar on Nondestructive Evaluation of Ferromagnetic Materials*, Dresser Atlas, 1986
20. G.P. Harnwell, *Principles of Electricity and Magnetism*, 2nd ed., McGraw-Hill, 1949
21. C.G. Gardner and F.N. Kusenberger, Quantitative Nondestructive Evaluation by the Magnetic Field Perturbation Method, in *Prevention of Structural Failure: The Role of Quantitative Nondestructive Evaluation*, T.D. Cooper, P.F. Packman, and B.G.W. Yee, Ed., No. 5 in the Materials/Metalworking Technology Series, American Society for Metals, 1975
22. M.J. Sablik and R.E. Beissner, Theory of Magnetic Leakage Fields From Prolate and Oblate Spheroidal Inclusions, *J. Appl. Phys.*, Vol 53, 1982, p 8437
23. J.H. Hwang and W. Lord, Magnetic Leakage Field Signatures of Material Discontinuities, in *Proceedings of the Tenth Symposium on NDE* (San Antonio, TX), Southwest Research Institute, 1975
24. J.F. Bussière, On Line Measurement of the Microstructure and Mechanical Properties of Steel, *Mater. Eval.*, Vol 44, 1986, p 560
25. D.C. Jiles, Evaluation of the Properties and Treatment of Ferromagnetic Steels Using Magnetic Measurements, in *Proceedings of the Second National Seminar on Nondestructive Evaluation of Ferromagnetic Materials*, Dresser Atlas, 1986
26. R.M. Bozorth, *Ferromagnetism*, Van Nostrand, 1951
27. C. Kittel and J.K. Galt, Ferromagnetic Domain Theory, in *Solid State Physics*, Vol 3, Academic Press, 1970
28. S. Chikazuma and S.H. Charap, *Physics of Magnetism*, John Wiley & Sons, 1964
29. *Proceedings of the 3rd International Symposium on Nondestructive Characterization of Materials*, Springer-Verlag, to be published
30. *Proceedings of the Third National Seminar on Nondestructive Evaluation of Ferromagnetic Materials*, Atlas Wireline Services, 1988
31. D.O. Thompson and D.E. Chimenti, Ed., *Review of Progress in Quantitative NDE*, Vol 7, Plenum Press, 1988
32. H. Kwun and G.L. Burkhardt, Effects of Grain Size, Hardness and Stress on the Magnetic Hysteresis Loops of Ferromagnetic Steels, *J. Appl. Phys.*, Vol 61, 1987, p 1576
33. J.C. McClure and K. Schroeder, The Magnetic Barkhausen Effect, *CRC Crit. Rev. Solid State Sci.*, Vol 6, 1976, p 45
34. G.A. Matzkanin, R.E. Beissner, and C.M. Teller, "The Barkhausen Effect and Its Applications to Nondestructive Evaluation," Report NTIAC-79-2, Southwest Research Institute, 1979
35. P.E. Khalileev and P.A. Grigor'ev, Methods of Testing the Condition of Underground Pipes in Main Pipelines (Review), *Sov. J. NDT*, Vol 10 (No. 4), July-Aug 1974, p 438-459
36. W.M. Rogers, New Methods for In-Place Inspection of Pipelines, in *Proceedings of the 16th Mechanical Working and Steel Processing Conference* (Dolton, IL), Iron and Steel Society of AIME, 1974, p 471-479
37. J.A. Birdwell, F.N. Kusenberger, and J.R. Barton, "Development of Magnetic Perturbation Inspection System (A02GS005-1), for CH-46 Rotor Blades," P.A. No. CA375118, Technical Summary Report for Vertol Division, The Boeing Company, 11 Oct 1968
38. J.R. Barton, "Feasibility Investigation for Sun Gear Inspection," P.A. NA-380695, Summary Report for Vertol Division, The Boeing Company, 25 July 1968
39. R.D. Williams and J.R. Barton, "Magnetic Perturbation Inspection of Artillery Projectiles," AMMRC CTR 77-23, Final Report, Contract DAAG46-76-C-0075, Army Materials and Mechanics Research Center, Sept 1977
40. M.C. Moyer and B.A. Dale, An Automated Thread Inspection Device for the Drill String, in *Proceedings of the First National Seminar on Nondestructive Inspection of Ferromagnetic Materials*, Dresser Atlas, 1984
41. H.H. Haynes and L.D. Underbakke, "Nondestructive Test Equipment for Wire Rope," Report TN-1594, Civil Engineering Laboratory, Naval Construction Battalion Center, 1980
42. J.R. Barton, J. Lankford, Jr., and P.L. Hampton, Advanced Nondestructive Testing Methods for Bearing Inspection, *Trans. SAE*, Vol 81, 1972, p 681
43. F.N. Kusenberger and J.R. Barton, "Development of Diagnostic Test Equipment for Inspection Antifriction Bearings," AMMRL CTR 77-13, Final Report, Contract Nos. DAAG46-74-C-0012 and DAAG46-75-C-0001, U.S. Army Materials and Mechanics Research Center, March 1977
44. J.R. Barton and J. Lankford, "Magnetic Perturbation Inspection of Inner Bearing Races," NASA CR-2055, National Aeronautics and Space Administration, May 1972
45. R.J. Parker, "Correlation of Magnetic Perturbation Inspection Data With Rolling-Element Bearing Fatigue Results," Paper 73-Lub-37, American Society of Mechanical Engineers, Oct 1973
46. F.N. Kusenberger and J.R. Barton, "Detection of Flaws in Reinforcing Steel in Prestressed Concrete Bridge Members," Interim Report on Contract No. DOTFH-11-8999, Federal Highway Administration, 1977
47. R.E. Beissner, C.E. McGinnis, and J.R. Barton, "Laboratory Test of Magnetic Field Disturbance (MFD) System for Detection of Flaws in Reinforcing Steel," Final Report on Contract No. DTFH61-80-C-00002, Federal Highway Administration, 1984
48. M. Mayos, S. Segalini, and M. Putignani, Electromagnetic Nondestructive Evaluation of Surface Decarburization on Steels: Feasibility and Possible Application, in *Review of Progress in Quantitative NDE*, Vol 6, D.O. Thompson and D.E. Chimenti, Ed., Plenum Press, 1987

Electric Current Perturbation NDE

Gary L. Burkhardt and R.E. Beissner, Southwest Research Institute

ELECTRIC CURRENT PERTURBATION (ECP) is an electromagnetic nondestructive evaluation method for detecting and characterizing defects in nonferromagnetic material. Laboratory evaluations have shown that this method can detect very small surface and subsurface cracks in both low- and high-conductivity metals (for example, titanium and aluminum alloys). Results from experiments and an analytical ECP model confirm that linear relationships exist between the ECP signal amplitude and the crack interfacial area and between the signal peak-to-peak separation and the crack length. This article will discuss the principles of the ECP method, typical inspection results, and crack characterization results.

Background

The principle by which ECP detects flaws is illustrated in Fig. 1. An electric current density, $\mathbf{j}_0$, is introduced into the region to be inspected, thus producing an associated magnetic-flux density, $\mathbf{B}_0$ (Fig. 1a). A flaw perturbs the current flow, as shown in Fig. 1(b), and the flux density is changed by an amount $\Delta\mathbf{B}$. Flaw detection is accomplished by sensing this change in flux density, $\Delta\mathbf{B}$, with a magnetic-field sensing device (Fig. 2). The sensor coils are oriented to detect the *x*-axis component of the magnetic flux.

In most applications, the ECP method makes use of an induction coil, much like that of a conventional eddy-current probe, to provide the unperturbed current density, $\mathbf{j}_0$, in the region to be inspected. However, unlike conventional eddy-current methods, the use of a coil to induce an alternating current is not essential. Experiments have in fact used not only ac injection but also dc injection, with results similar to those obtained by induction. When direct current is injected, a Hall effect sensor is used instead of coils.

Another difference between ECP and conventional eddy-current methods lies in the orientation of the sensor. The exciter coil and the sensor coil axes in the ECP method are perpendicular (Fig. 2), which reduces or eliminates direct coupling with

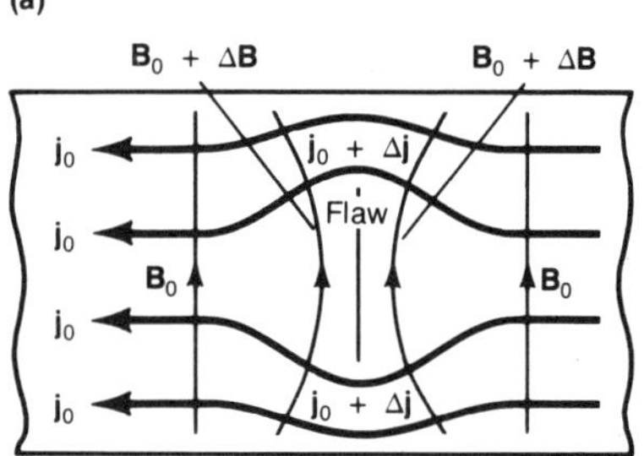

Fig. 1 Electric current density, **j**, and the associated magnetic flux density, **B** (a) with no surface flaw and (b) with a flaw (for example, a crack) perpendicular to the current flow

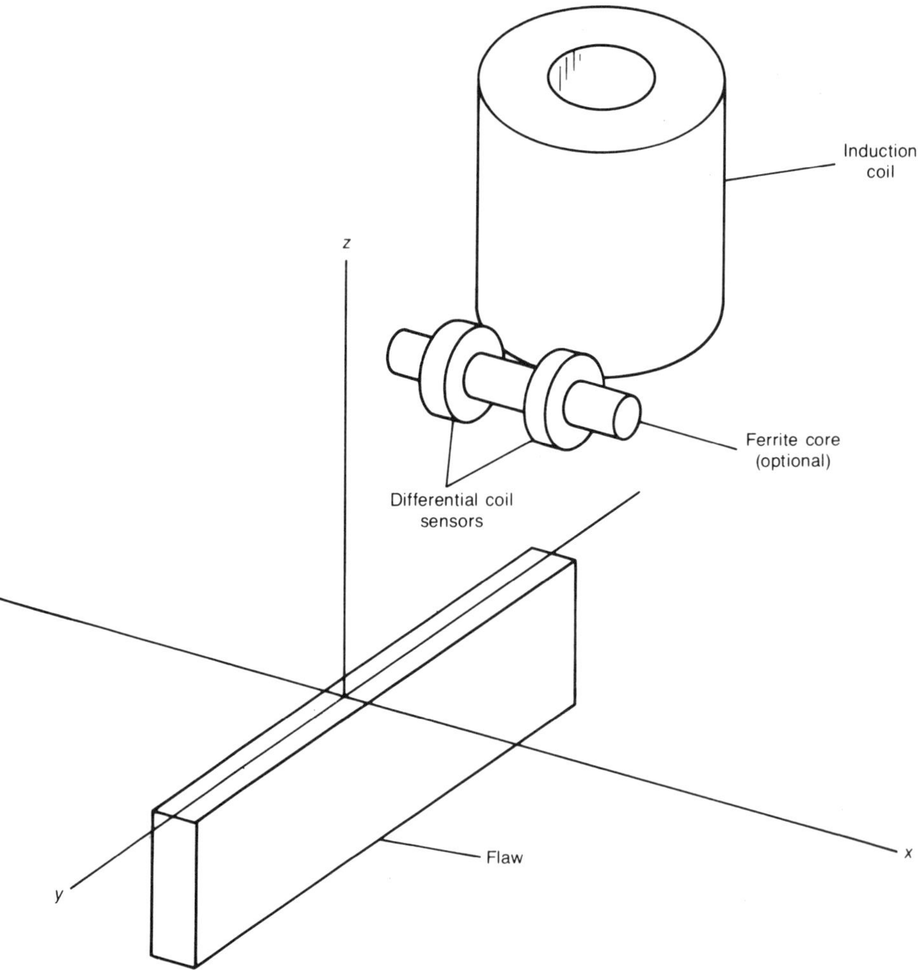

Fig. 2 ECP probe configuration and orientation with respect to a flaw in a part surface. The size difference between the sensor and the induction coil is much greater than shown.

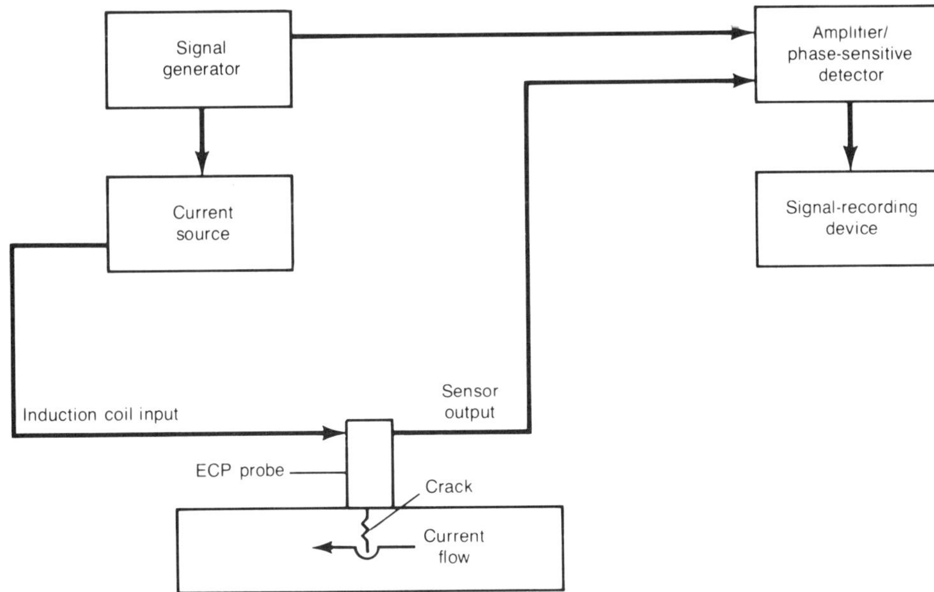

Fig. 3 Block diagram of a typical ECP system

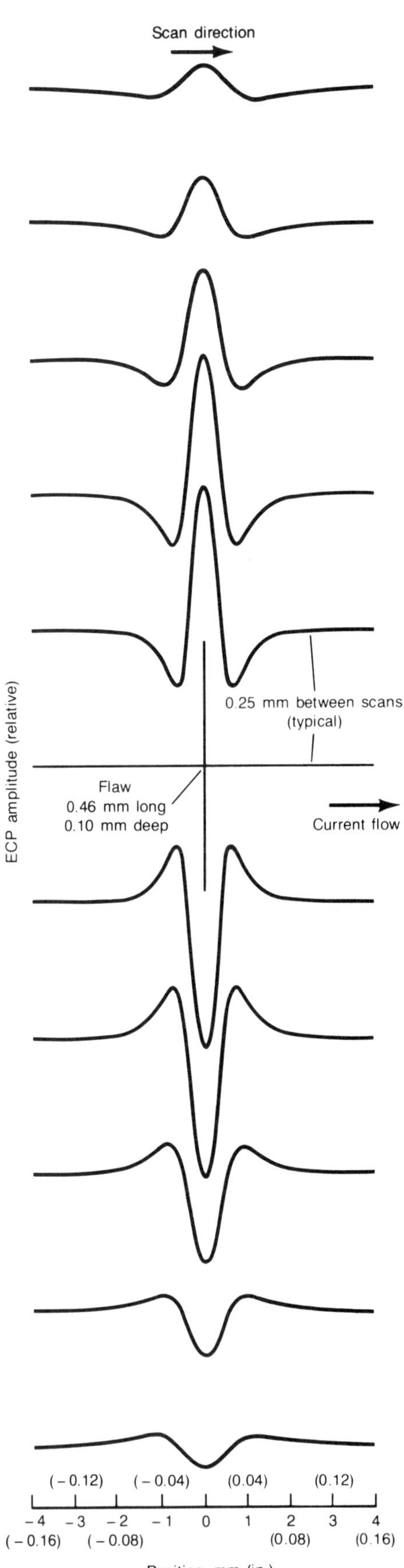

Fig. 4 Calculated ECP signals for scans perpendicular to a 0.46 mm (0.018 in.) long by 0.10 mm (0.004 in.) deep slot

$\mathbf{B}_0$. In addition, the use of differential sensors tends to cancel the effect of remaining coupling. These decoupling features allow the excitation to be much stronger in the ECP method, yielding gains in sensitivity. Probe-to-surface coupling is also minimized, and sensitivity to probe liftoff variations is greatly reduced (Ref 1). Liftoff is a major source of noise with probes using conventional eddy-current coil orientations. The end result is that ECP probes have excellent signal-to-noise characteristics, making them well suited to the detection of small flaws. The penalty for this geometry is directionality: The sensing coils select specific current perturbations and ignore others.

In practice, one noticeable difference between the ECP and conventional eddy-current method is that the frequencies employed with ECP are almost always much lower. The difference in operating frequencies results from the requirement for eddy-current probes to use higher frequencies to reduce liftoff noise by separating the liftoff component of the signal from the flaw component (Ref 2). This is not necessary with ECP because the probe design inherently reduces liftoff noise. Therefore, while many eddy-current probes designed for surface-flaw detection are configured for frequencies of several hundred kilohertz to several megahertz, ECP probes are normally used at frequencies below 100 kHz. Using lower frequencies allows the ECP electronics to be simplified.

A block diagram of a typical ECP system is shown in Fig. 3. A signal generator drives a current source to provide excitation current for the induction coil in the ECP probe. The sensor output is directed to an amplifier and a phase-sensitive detector. Devices such as a digitizing oscilloscope or chart recorder can also be used for recording and displaying ECP signals.

Typical Applications

The following examples describe the evaluation of the ECP method for detecting flaws in a variety of parts. In these examples, the evaluation first involved the machining of small notches in the surfaces of the parts. These notches served as test flaws.

In each of the examples, the induced current, $\mathbf{j}_0$, was perpendicular to the length of the notches. The ECP sensors scanned the surface to determine the ECP response to the test flaws. The scan direction was perpendicular to the flaw in Examples 1 and 5, but was parallel to the flaw in Examples 2 to 4. When the scan direction is perpendicular, a unipolar ECP signal is produced, and when the scan direction is parallel to the flaw, a bipolar ECP signal is produced. The ECP signal shapes for perpendicular and parallel scans are shown in Fig. 4 and 5.

Example 1: Surface Flaws in a Flat Plate. The sensitivity of the ECP method to very small surface flaws in a low-conductivity material was illustrated with flaws in a flat-bar Ti-6Al-4V specimen (Ref 3). The ECP signals obtained by scanning a probe linearly past three electrical discharge machined (EDM) notches are shown in Fig. 6. The 0.47 mm (0.0184 in.) long by 0.17 mm (0.0066 in.) deep notch in this specimen gave a very prominent ECP signal. More important, recognizable signals were also obtained from the two small notches—one 0.20 mm (0.0079 in.) long by 0.06 mm (0.0022 in.) deep and the other 0.22 mm (0.0085 in.) long by 0.05 mm (0.0020 in.) deep. The background signal in the region where no flaws were present was caused by

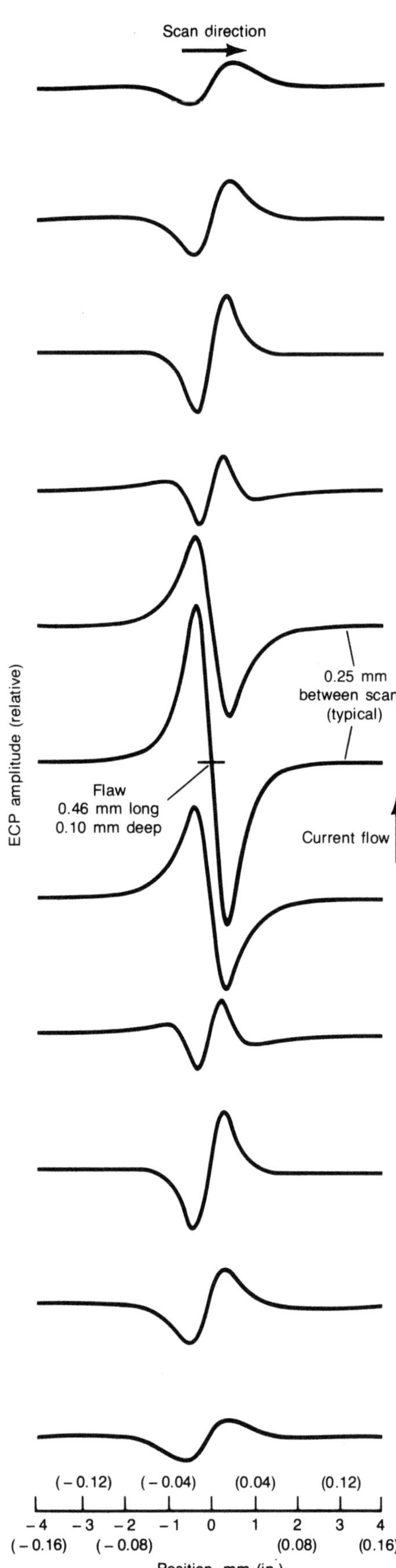

Fig. 5 Calculated ECP signals for scans parallel to a 0.46 mm (0.018 in.) long by 0.10 mm (0.004 in.) deep slot

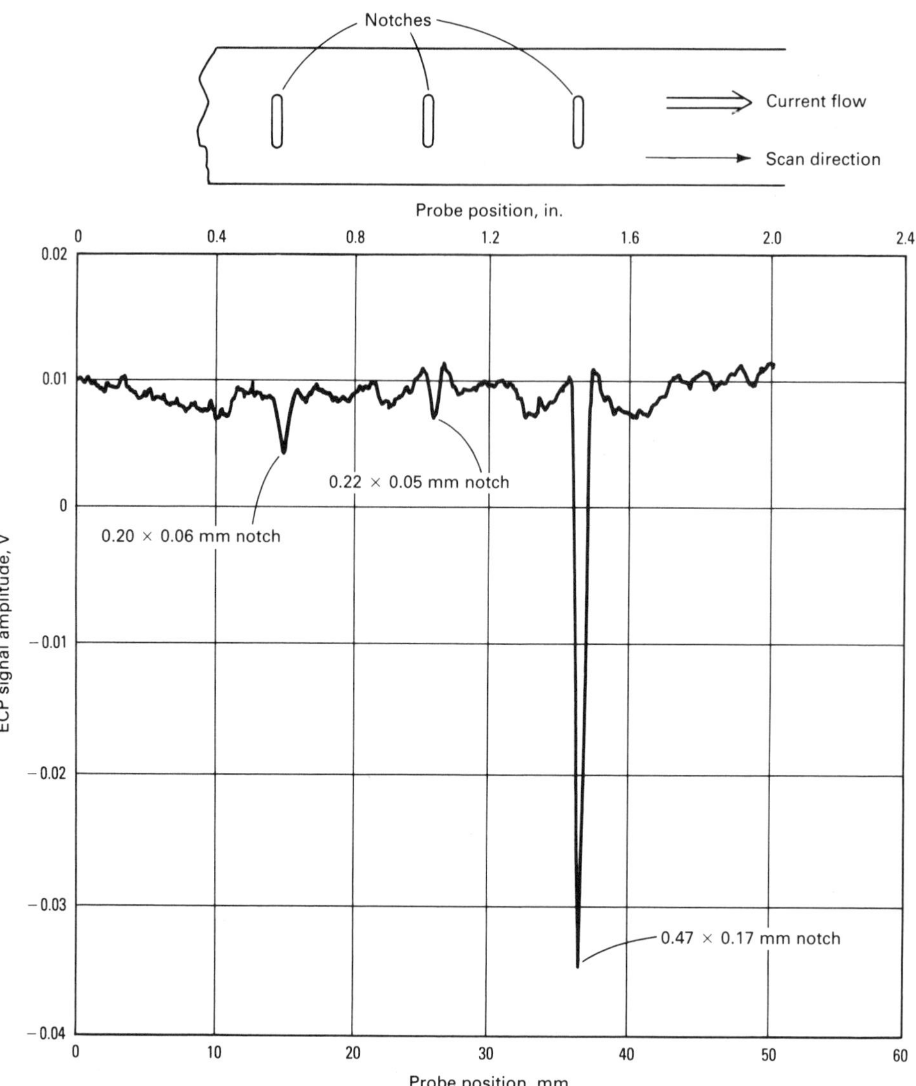

Fig. 6 ECP signals from EDM notches in a Ti-6Al-4V flat-bar specimen. Flaw dimensions are length by depth.

the inhomogeneity of the specimen material. From these data, the inherent sensitivity of the ECP method for detection of very small surface flaws was established.

Example 2: Flaw Detection in Fan-Disk Blade Slots. Evaluations of the ECP method were performed on blade slots in the first-stage fan disk of the F100 gas-turbine engine (Ref 3). Figure 7 shows a photograph of this disk, which was fabricated from Ti-6Al-2Sn-4Zr-6Mo material. The blade-slot configuration is illustrated in Fig. 8. An 8.89 mm (0.350 in.) radius on each side of the blade slot blended into a flat area on the bottom of the blade slot, and the primary area of interest for surface-flaw detection was at the tangency point of this radius. Notches were electrical discharge machined in two blade slots at the tangency point for use as target flaws. The ECP probe scan path was approximately in line with each EDM notch (Fig. 8). Figure 9 shows the ECP signals from two blade slots with surface EDM notches measuring 0.46 mm (0.0182 in.) long by 0.27 mm (0.0105 in.) deep and 0.27 mm (0.0105 in.) long by 0.15 mm (0.0058 in.) deep. Prominent signatures were obtained from both the larger EDM notch (Fig. 9a) and the smaller EDM notch (Fig. 9b); this demonstrated the sensitivity for detecting small flaws in this engine component.

Example 3: Surface Flaws in Thread Roots. The ECP method was evaluated for the detection of flaws in the thread roots of a titanium helicopter rotary wing-head spindle (Ref 4). The threaded end of the spindle is illustrated in Fig. 10. Notches were electrical discharge machined into the thread roots, as shown in Fig. 10. The ECP probe was positioned on the crest of the threads, resulting in a liftoff of 1.22 mm (0.048 in.) from the root. For scanning, the probe was held stationary while the spindle was simultaneously rotated and translated axially so that the probe followed the threads.

The ECP data were obtained on a spindle containing three rectangular notches designated A, B, and C (Fig. 11). These notches

Fig. 7 F100 first-stage fan disk

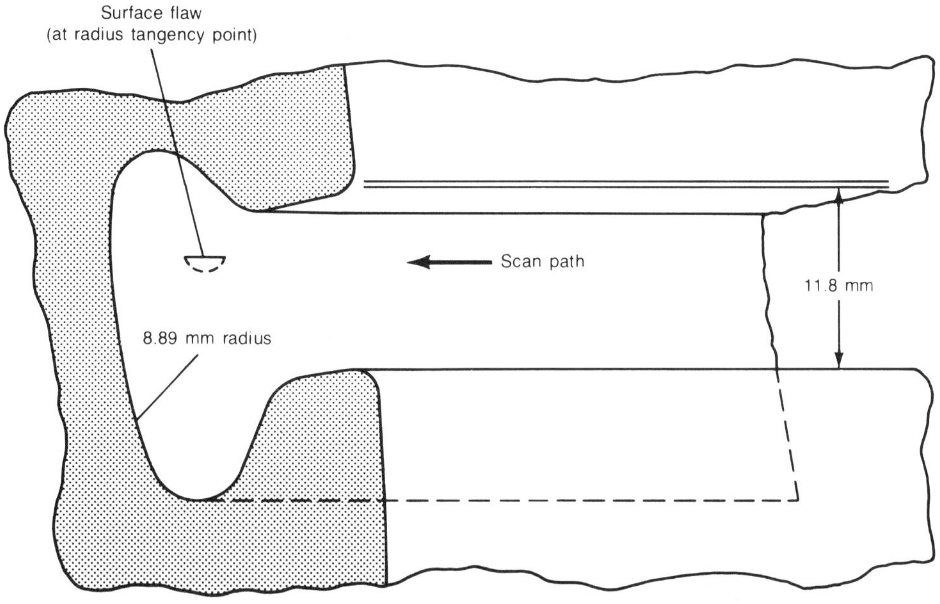

Fig. 8 F100 first-stage fan-disk blade-slot configuration showing a flaw location and ECP probe scan track

were spaced 120° apart around the circumference in the root of one thread. As shown in Fig. 11, prominent ECP signals were obtained from all three defects, including the smallest one, which measured 0.64 mm (0.025 in.) long by 0.43 mm (0.017 in.) deep. These data exhibit several important characteristics. First, the signal background was far above electronic noise and was highly reproducible for repeat scans. The ECP sensitivity was limited only by the signal background obtained from the spindle itself, not by electronic noise. Second, signals were obtained when the probe passed directly over the defects in a thread and when the probe was located at adjacent threads, as indicated by the satellite signals designated A′, B′, and C′ in Fig. 11. Signal polarity reversals were obtained when the probe was positioned away from the flaws in the adjacent threads; the satellite signals can be distinguished from the primary flaw signals by this polarity reversal. These characteristics are typical of the ECP method and demonstrate that it is not adversely affected by the complex specimen geometry imposed by the spindle threads.

Example 4: Flaw Detection Through Spindle Wall Thickness. The helicopter rotary wing-head spindle described in Example 3 was also inspected with an ECP probe located in the spindle bore. In this evaluation, flaws in the thread roots were detected through the 9.32 mm (0.367 in.) spindle wall thickness.

Data obtained from a spindle that contained a flaw 7.75 mm (0.305 in.) long by 2.21 mm (0.087 in.) deep in the root of the first thread are shown in Fig. 12; the flaw is identified as I. A flaw signal was obtained when the probe passed directly over the flaw, and satellite signals (I′) were obtained when the probe was located at the adjacent threads. The flaw-to-sensor distance was large because the inspection was performed through the wall thickness. This resulted in the satellite signal polarity reversal (evident in Example 3) occurring outside the region shown in the plot.

Example 5: Second-Layer Fastener-Hole Cracks. The ECP method was evaluated on a C-5A aircraft spanwise splice-joint specimen to demonstrate the detection of fastener-hole defects in the second layer with the probe positioned on the outer surface of the first layer (Ref 5). A cross section of the two-layer splice joint is illustrated in Fig. 13. The aluminum specimen had top and bottom layer thicknesses of 3.8 mm (0.15 in.) (in the joint region) and a single row of 4.77 mm (0.188 in.) diam tapered titanium fasteners with countersunk heads. The specimen contained 2.5 mm (0.10 in.) deep and 5.1 mm (0.20 in.) long through-thickness radial notches in the second layer.

The ECP probe was positioned on the top layer of the specimen and was scanned linearly along the row of fasteners (Fig. 14),

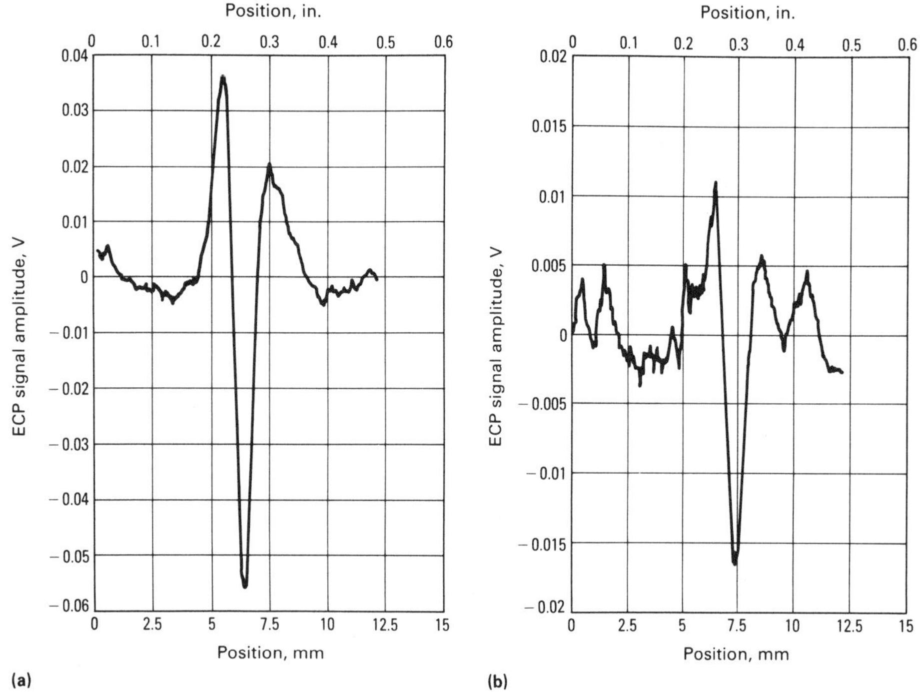

Fig. 9 ECP signals from EDM notches in F100 first-stage fan-disk blade slots. (a) Signal from the 0.46 mm (0.0182 in.) long by 0.27 mm (0.0105 in.) deep notch. (b) Signal from the 0.27 mm (0.0105 in.) long by 0.15 mm (0.0058 in.) deep notch

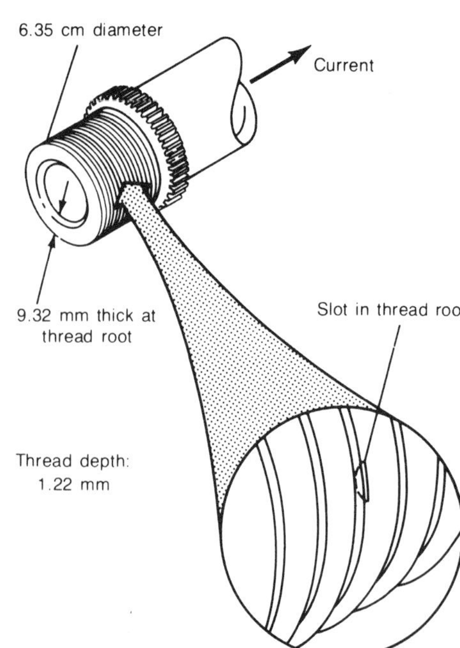

Fig. 10 Threaded end of a helicopter spindle showing a flaw location

with the sensor scan track 13.5 mm (0.53 in.) from the edge of the fastener holes. Strong signatures were obtained from hole No. 11, with a 5.1 mm (0.20 in.) notch oriented toward the scan track side of the fastener hole, and from hole No. 8, with a notch of the same size oriented toward the opposite side of the fastener hole from the scan track. A prominent signal was also obtained from the 2.5 mm (0.10 in.) notch in hole No. 4. Signals from unflawed holes are very small and are not evident at the sensitivity used to obtain the data in Fig. 14. These results show the potential for a rapid, linear scan inspection of fastener holes.

Flaw Characterization

Both experimental data and theoretical modeling results suggest the potential capability of the ECP method for flaw characterization (Ref 6). Signal features related to crack characterization are most apparent when the scan direction is parallel with the crack length (perpendicular to $\mathbf{j}_0$) and directly over it, although these features can be extracted from other scan orientations. When a defect is scanned perpendicular to $\mathbf{j}_0$, a bipolar signal is obtained, as in the scan shown in Fig. 11. The peak-to-peak signal amplitude and the peak-to-peak spacing can be related to the flaw interfacial area and the flaw length, respectively. These relationships will be discussed in this section. The details of theoretical modeling are discussed in Ref 1.

Experimental results were obtained from a single, half-penny-shaped fatigue crack grown with a laboratory fatigue machine in a Ti-6Al-4V rod-type tensile specimen. Fracture of identical specimens containing cracks grown under similar conditions showed that true half-penny-shaped fatigue cracks (2:1 aspect ratio) were obtained.

Theory and experiment were compared for the peak-to-peak signal amplitude as a function of the interfacial area of the crack (the planar area exposed if the specimen were fractured) and for the peak-to-peak separation as a function of crack surface length. These data are shown in Fig. 15. In Fig. 15(a), it can be seen that the theory predicts an approximately linear relationship between signal amplitude and the interfacial area of the crack. Experimental results show very close agreement with theory, indicating that the crack interfacial area for cracks of this type can be estimated by measurement of the ECP signal amplitude.

A plot of the ECP peak-to-peak separation as a function of crack surface length is shown in Fig. 15(b) for both theory and experiment. The theoretical model again

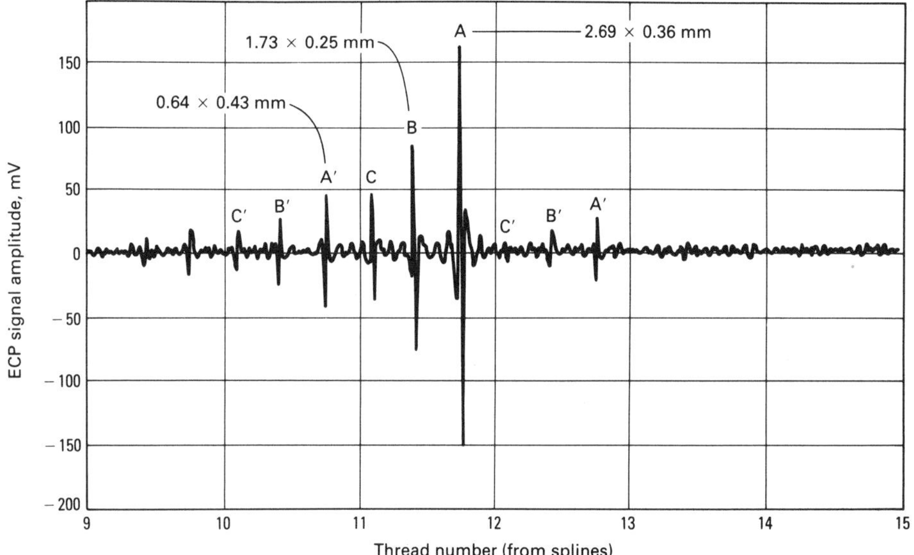

Fig. 11 ECP signals from flaws A, B, and C in a spindle thread root with a probe on the crest of the threads. Signals obtained when the probe was located over the adjacent thread are designated A′, B′, and C′. Flaw dimensions are length by depth.

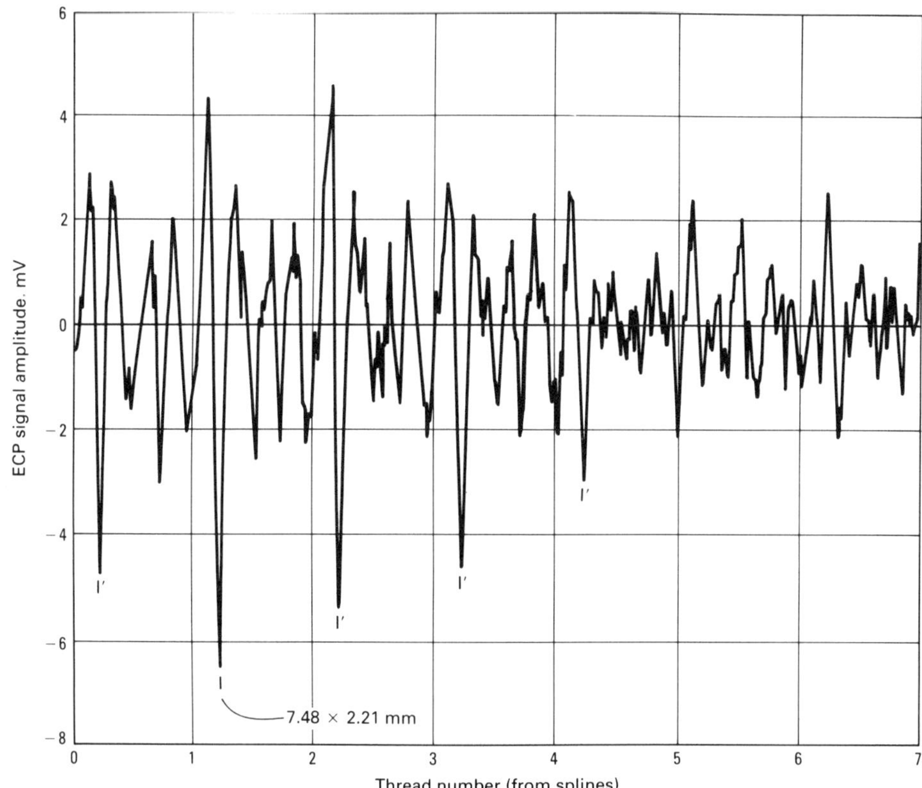

Fig. 12 ECP signals from defect I (7.75 mm, or 0.305 in., long by 2.21 mm, or 0.087 in., deep) in the spindle thread root with a probe located in the spindle bore. The signals obtained when the probe was located at adjacent threads are designated I′.

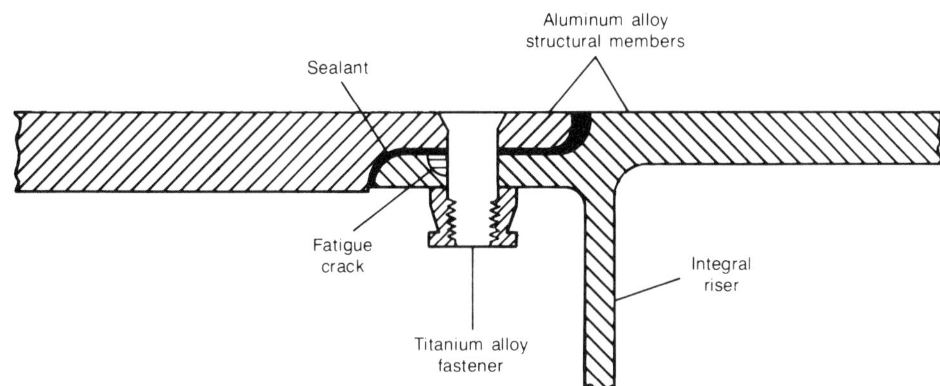

Fig. 13 Schematic of a two-layer structural-fastener configuration

predicts an approximately linear relationship between the peak-to-peak separation and the crack surface length over the range of lengths shown in the plot. Experimental data show a similar relationship, although some departure from linearity is observed. In both cases, a direct relationship is apparent between the peak-to-peak separation and the crack surface length. This shows that the surface length of the defect can be estimated by measuring the peak-to-peak separation of the ECP signal.

These correlations between the signal amplitude and the interfacial area of the crack and between the peak-to-peak separation and the surface length demonstrate that crack parameters can be obtained from signal features, thus establishing the feasibility of crack characterization. The results were obtained with half-penny-shaped defects. However, flaw characterization has also been demonstrated with other aspect ratios (Ref 3).

REFERENCES

1. R.E. Beissner and M.J. Sablik, Theory of Electric Current Perturbation Probe Optimization, *Rev. Prog. Quant. NDE*, Vol 3A, 1984, p 633-641
2. F. Forster and H.L. Libby, Probe Coil-Detection of Cracks in Nonmagnetic Materials, in *Nondestructive Testing Handbook*, Vol 4, *Electromagnetic Testing*, R.C. McMaster, P. McIntire, and M.L. Mester, Ed., 2nd ed., American Society for Nondestructive Testing, 1986, p 185-187
3. R.E. Beissner, G.L. Burkhardt, and F.N. Kusenberger, "Exploratory Development of Advanced Surface Flaw Detection Methods," Final Report, Contract No. F33615-82-6-5020, Report No. AFWAL-TR-84-4121, Wright-Patterson Air Force Base, Sept 1984
4. C.M. Teller and G.L. Burkhardt, Application of the Electric Current Perturbation Method to the Detection of Fatigue Cracks in a Complex Geometry Titanium Part, *Rev. Prog. Quant. NDE*, Vol 2B, 1983, p 1203-1217
5. C.M. Teller and G.L. Burkhardt, NDE of Fastener Hole Cracks by the Electric Current Perturbation Method, *Rev. Prog. Quant. NDE* , Vol 1, 1982, p 399-403
6. C.M. Teller and G.L. Burkhardt, Small Defect Characterization by the Electric Current Perturbation Method, in *Proc. of the Thirteenth Symposium on Nondestructive Evaluation*, Nondestructive Testing Information Analysis Center, April 1981, p 443-453

SELECTED REFERENCES

- R.E. Beissner and G.L. Burkhardt, Application of a Computer Model to Electric Current Perturbation Probe Design, *Rev. Prog. Quant. NDE*, Vol 4, 1985, p 371-378
- R.E. Beissner, C.M. Teller, G.L. Burkhardt, R.T. Smith, and J.R. Barton, Detection and Analysis of Electric-Current Perturbation Caused by Defects, in *Eddy-Current Characterization of Materials and Structures*, STP 722, F. Birnbaum and G. Free, Ed., American Society for Testing and Materials, 1981, p 428-446
- G.L. Burkhardt and R.E. Beissner, Probability of Detection of Flaws in a Gas Turbine Engine Component Using Electric Current Perturbation, *Rev. Prog. Quant. NDE*, Vol 4, 1985, p 333-341
- G.L. Burkhardt, F.N. Kusenberger, and R.E. Beissner, Electric Current Perturbation Inspection of Selected Retirement-For-Cause Turbine Engine Components, *Rev. Prog. Quant. NDE*, Vol 3, 1984, p 1377-1387

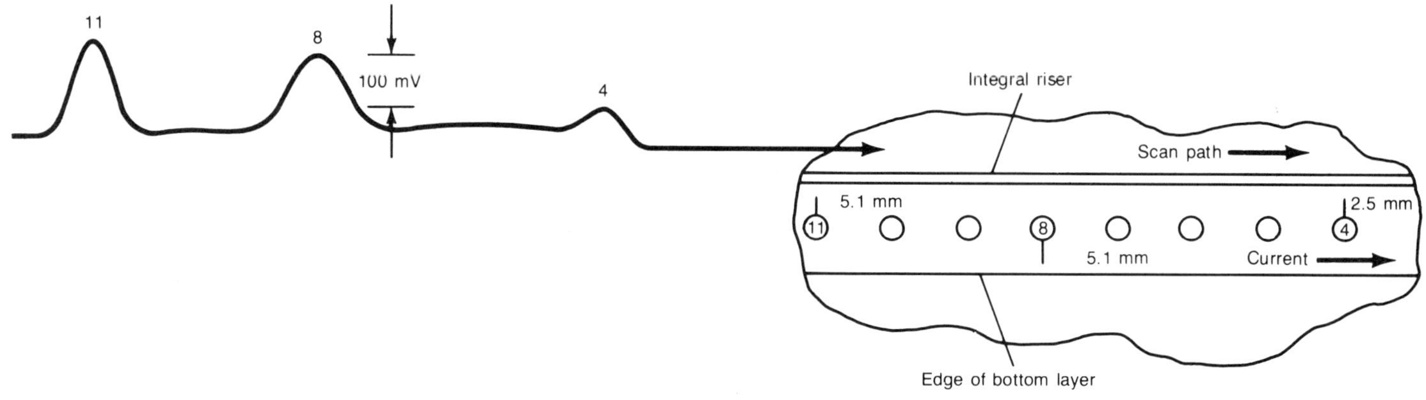

Fig. 14 ECP signals from second-layer flaws in a C-5A wing-splice specimen with the probe on the top layer. Flaw dimensions are the radial length of EDM notches.

- G.L. Burkhardt, F.N. Kusenberger, R.E. Beissner, and C.M. Teller, Electric Current Perturbation Inspection of Complex Geometry Features in Gas Turbine Engine Components, in *Proc. of the Fourteenth Symposium on Nondestructive Evaluation*, Nondestructive Testing Information Analysis Center, April 1983, p 468-474
- G.L. Burkhardt, G.W. Scott, and B.N. Ranganathan, Electric Current Perturbation Method for Inspection of Aluminum Welds, *Rev. Prog. Quant. NDE*, Vol 5, 1986, p 1713-1721
- R. Palanisamy, D.O. Thompson, G.L. Burkhardt, and R.E. Beissner, Eddy Current Detection of Subsurface Cracks in Engine Disk Bolt Holes, *Rev. Prog. Quant. NDE*, Vol 3, 1984, p 643-651

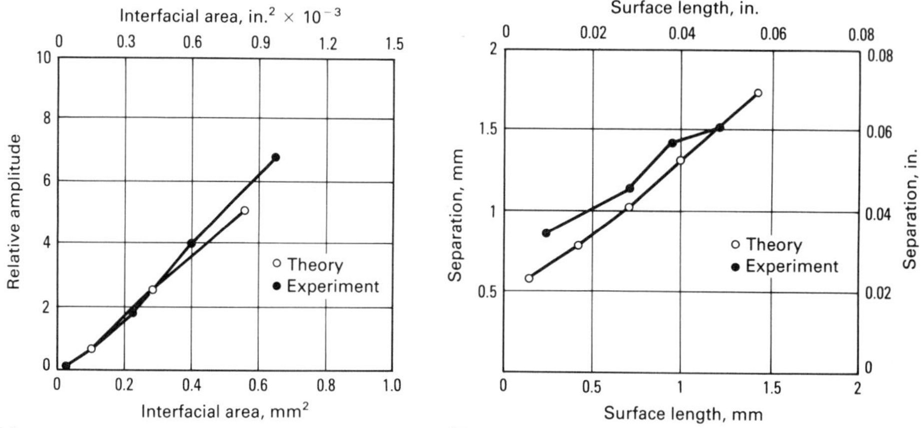

Fig. 15 Flaw characterization of half-penny cracks with the ECP method. (a) Comparisons between the ECP signal amplitude and the crack interfacial area. (b) Comparisons between the ECP signal peak-to-peak separation and the crack length

Magabsorption NDE

William L. Rollwitz, Southwest Research Institute

MAGABSORPTION TECHNIQUES were originated by the author at Southwest Research Institute in June 1957. During electron spin resonance (ESR) measurements on a rocket propellant, it was noticed that an entirely different ESR signal was obtained from the propellant with Fe_3O_4 than from the same propellant without the iron oxide. When the microwave power for the ESR was turned off but the modulation was left on, the different signal disappeared. When the modulation power was turned off, the different signal also disappeared. Therefore, the signal appeared to depend not only on the microwaves but also on the modulation. The frequency of the signal itself was predominantly a second harmonic of the modulation frequency. When the magnetic field was varied from 0 to 0.4 T (4000 G), the ESR-type signal was obscured by the different signal. The shape of the different signal did not change as the field varied from 0 to 0.4 T (4000 G), but the amplitude did. The amplitude decreased almost linearly from 0.4 T (4000 G). The signal had the same shape but at a much lower level as the field increased. The signal was obtained from the sample with iron oxide but not from the sample without the Fe_3O_4. The same type of signal was obtained from a nuclear magnetic resonance (NMR) detector at 30 MHz and from the ESR detector at 9240 MHz. When the modulation was stopped with the NMR and ESR detectors, the signal disappeared.

It was concluded that the shape of this different signal was independent of the frequency used; it was a function of the presence of Fe_3O_4, and it had a fundamental frequency that was twice the modulation frequency. When the signal was observed before the amplitude detection in the NMR detector, it was found to be a 30-MHz carrier amplitude and/or frequency modulated with the different signal whose fundamental frequency was twice the modulation frequency. The phenomenon was not a resonance function of frequency. It was a function of the modulation field and was an amplitude modulation of the radio-frequency (RF) signal used.

A schematic diagram of a system specifically made to detect the different signal is shown in Fig. 1. The sample of Fe_3O_4 in Fig. 1 is in two alternating magnetic fields. The first is an RF field at 30 or 9240 MHz, and the second is a lower-frequency field at 60 Hz. The peak value of the RF field, H_{RF}, is much lower than 8 A/m (0.1 Oe). The peak value of the lower-frequency field is much higher at around 8000 A/m (100 Oe). The RF detection coil is resonated at the radio frequency by the tuning capacitor in the NMR detector. The output of the NMR or ESR amplifier is the special signal amplitude modulated on the RF voltage. The different signal is separated from the RF by an amplitude detector. Because the shape of the different signal was found to be independent of frequency, a much lower frequency was used. For the first detection system specifically made to detect the different signal, a frequency of 0.5 MHz was chosen as shown in Fig. 1. An NMR-type detector was used at 0.5 MHz to detect the different signal.

The resonant circuit of the NMR-type detector can be represented by the diagram in Fig. 2. It is possible to use the techniques used in nuclear magnetic resonance, which relate the very small changes in the energy absorbed by the nuclei from the RF magnetic field as an amplitude modulation of the RF voltage V_s across the RF coil in Fig. 2, where L is the inductance of the RF coil, R_L is the resistance representing the losses in the coil, and C is the capacitance required to tune the circuit to $\omega_o = 2\pi \cdot 0.5 \cdot 10^6$ radians/s. The impedance, Z_o, with no sample in the two fields, is:

$$Z_o = \frac{L}{CR} \quad \text{(Eq 1)}$$

When there is a sample in the two magnetic fields, the impedance is changed to Z_s, which is:

$$Z_s = \frac{L}{CR} \frac{1}{1 + (4\pi FQP/\omega_o H_1^2)} \quad \text{(Eq 2)}$$

where F is the filling factor of the sample in the RF magnetic field, P is the power absorbed by the resonated nuclei sample, Q is the figure of merit of the resonant circuit where $Q = \omega_o L/R$, and H_1 is the peak amplitude of the sinusoidal RF magnetic field. The change of impedance caused by the power absorption by the sample is:

$$\Delta Z = Z_o - Z_s = Z_o \frac{4\pi QFP}{\omega_o H_1^2} \quad \text{(Eq 3)}$$

where

$$\frac{4\pi FQP}{\omega_o H_1^2} \ll 1$$

and

$$1 + \frac{4\pi QFP}{\omega_o H_1^2} \simeq 1$$

Equation 3 shows that the change in the resonant impedance of the detection circuit is directly proportional to the power absorp-

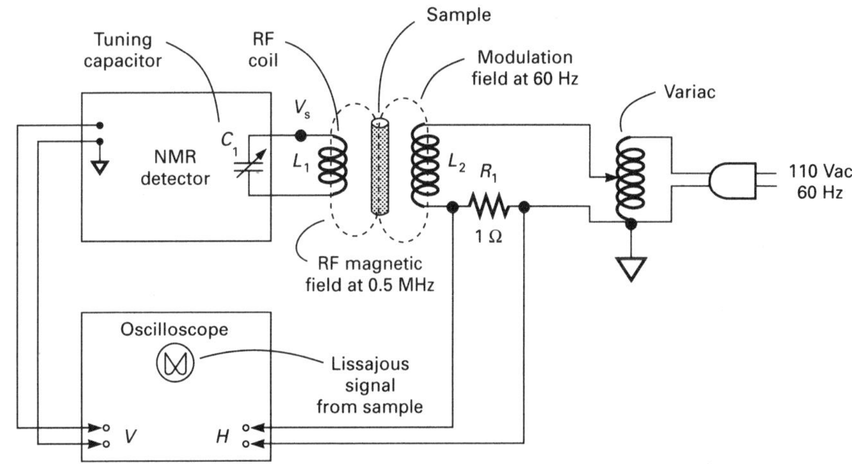

Fig. 1 Schematic of the special test system for the investigation of the detection of the different signal with an NMR-type detector

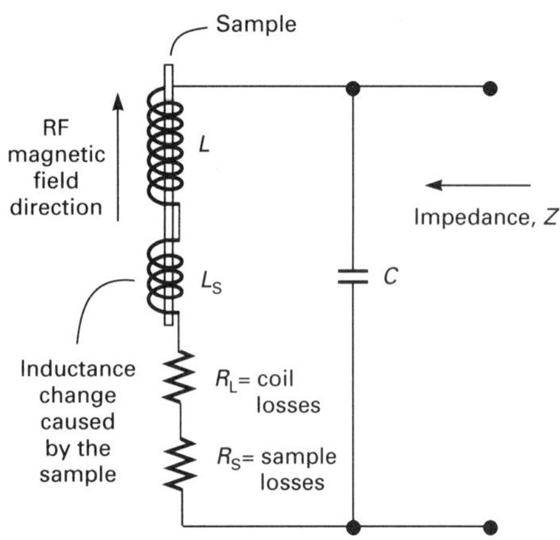

Fig. 2 Resonant circuit showing the sample in the RF magnetic field

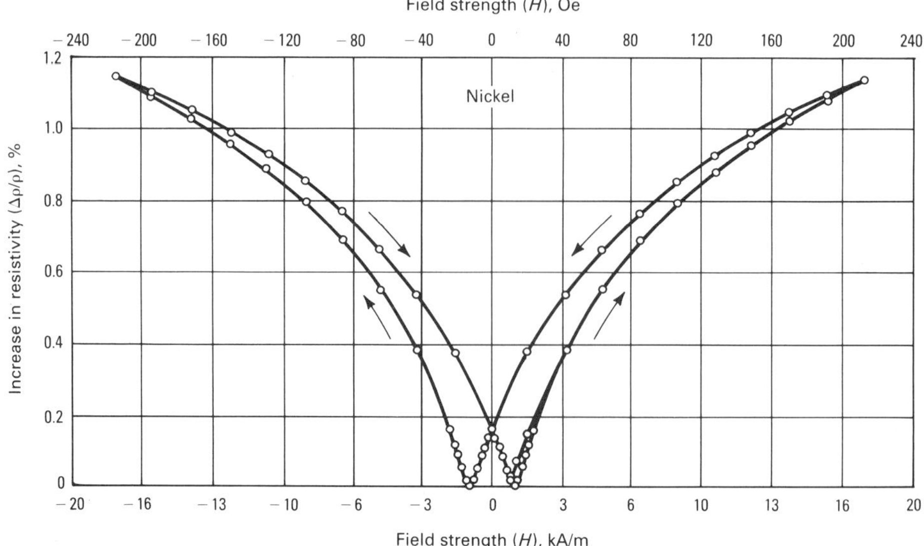

Fig. 3 Hysteresis of the magnetoresistance of nickel as plotted against applied field strength. Source: Ref 2

tion (the rate of energy absorption) by the sample from the RF magnetic field. The nuclear magnetic resonance signal is also directly proportional to the power absorbed by the chosen nuclei from an RF magnetic field. The RF power absorbed by the nuclei is proportional to the complex susceptibility of the sample, which is:

$$\chi = \chi' - \chi'' \quad \text{(Eq 4)}$$

The power absorbed from the RF coil by the sample will therefore have two components—one in phase with the RF magnetic field proportional to χ'' and a second one 90° out of phase with the RF magnetic field proportional to χ'. The first is called the in-phase or absorption component of the NMR signal, and the second is called the quadrature-phase or dispersion component of the signal. The different signal also has in-phase and quadrature-phase components. Therefore, the different signal is directly proportional to the power absorbed by the susceptibility of the sample of iron oxide. The susceptibility has both an in-phase component and a quadrature-phase component.

In October 1957, the author received funding for a program to study the phenomenon in more detail. Just before the project was undertaken, it was found that W.E. Bell of Varian Associates had discovered a similar different signal. He had termed the phenomenon magnetoabsorption and had written a paper with that title in 1956 (Ref 1). Bell attributed the phenomenon to magnetoresistance and called it magnetoabsorption because the apparatus measures the RF absorption in the test specimen as a function of the applied magnetic field.

The magnetoresistance phenomenon is described by Bozorth as the change in the electrical resistivity of iron when it is magnetized (Ref 2). In most magnetic materials, this magnetoresistance is an increase of resistivity with magnetization when the current and magnetization are parallel and a decrease when they are at right angles to each other; the change in resistivity is independent of the sense of field or induction and, in this respect, is like magnetostriction (Ref 2). Reference 2 also gives the curve of $\Delta\rho/\rho$ as a function of the applied field strength; this magnetoresistance curve for nickel is shown in Fig. 3. The magnetoresistance curve has a butterfly shape and hysteresis. The magnetoresistance change is 10 to 15% of the resistivity at saturation, $\Delta\rho_s$. Reference 2 also states that the change in resistance (increase or decrease) is beyond the knee of the magnetization curve.

The effects of both tension and magnetic field on the resistivity of ferromagnetic materials are discussed in Ref 2 with the aid of the domain illustrations in Fig. 4. According to Ref 2, domains having an initially haphazard distribution of orientation (Fig. 4a) are aligned parallel by a strong (saturating) magnetic field (Fig. 4b). Tension applied to an unmagnetized material, with positive magnetostriction, orients the domains along the axis of tension with one-half of the domains antiparallel to the others (Fig. 4c). Because the resistivity depends on magnetization and increases with increasing magnetization, each domain will contribute to the resistivity an amount depending on its orientation. This contribution will be the same for parallel and antiparallel orientations; for the specimen as a whole, $\Delta\rho$ will be the same for Fig. 4(b) and (c) and will be $\Delta\rho_s$. When the magnetostriction is negative, tension orients the domains at right angles to the direction of tension, and the resistivity decreases (Fig. 4d). A magnetic field applied subsequently causes the domains to rotate so as to increase the resistivity (Fig. 4e). When a material having positive magnetostriction is subjected to strong tension, subsequent application of a magnetic field will cause 180° rotations (of the domains) almost exclusively, and these will have no effect on resistivity (going from Fig. 4c to e).

Because of the above discussion from Ref 2, the term magnetoabsorption was adopted by the author for the phenomenon, and the internally sponsored work at Southwest Research Institute was undertaken using the title "Magnetoabsorption" (Ref 3). However, the term magabsorption is now being used because the techniques have developed differently from the magnetoabsorption described by Bell (Ref 1).

One of the magabsorption measurements described in Ref 3 involved measurements on iron carbonyl powders with a bridge-type NMR detector. The in-phase or the loss component and the quadrature or the reactive component of the magabsorption signal were separately detected and recorded. They were displayed as Lissajous figures where the magabsorption signal (in-phase or quadrature-phase) was on the vertical axis of the oscilloscope and a voltage proportional to the 60-Hz modulation was fed to the horizontal axis. The in-phase (resistive) and the quadrature-phase (reactive) components of the magabsorption signals from a 3 μm (120 μin.) sample of carbonyl iron are shown in Fig. 5. The shapes of the quadrature-phase signals are quite different from the in-phase signals in Fig. 5. The resistivity or the in-phase signals have higher harmonic content than the quadrature-phase or reactive signals. Other results obtained during the research summarized in Ref 3 are discussed in the section "Measurements and Applications" in this article.

The above discussion of the effect on the magnetoresistance caused by an applied magnetic field and an applied stress on a piece of ferromagnetic material, coupled with the differences between the magabsorption signals from nickel powder and

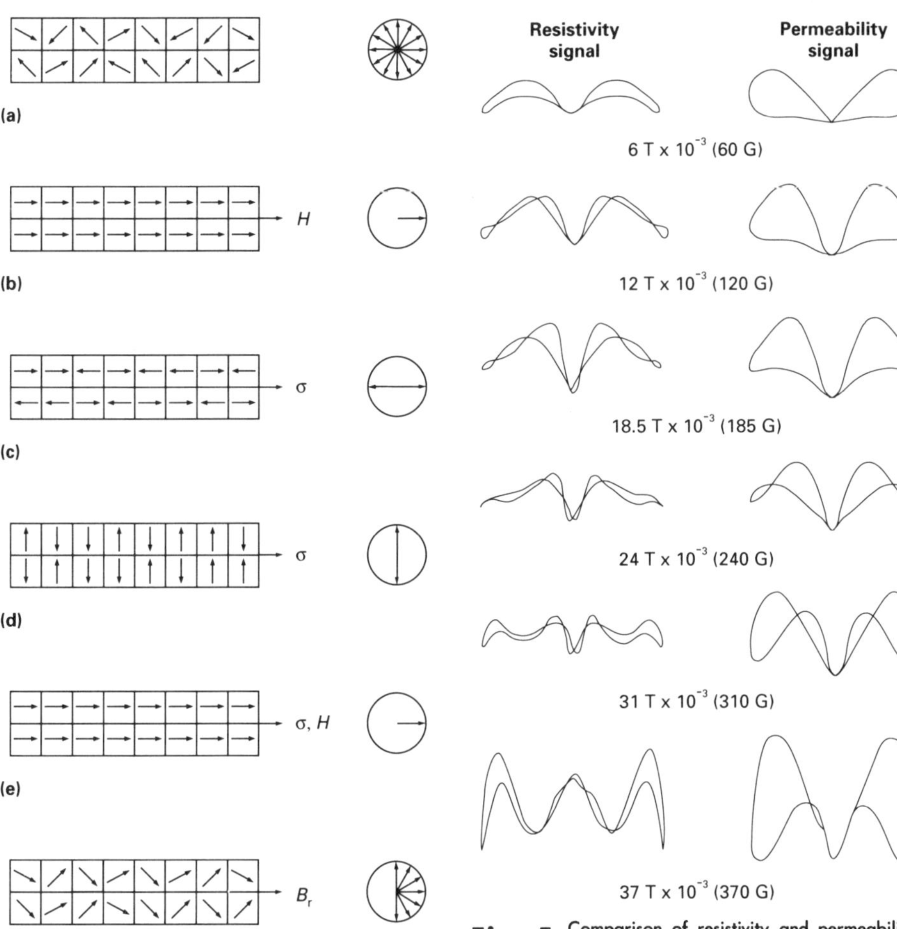

Fig. 4 Domain interpretation of the effects of field and tension on resistance. At right, domain vectors have been moved to a common origin. (a) Unmagnetized. (b) Saturated. (c) Tension applied (positive magnetostriction). (d) Tension applied (negative magnetostriction). (e) Tension and strong field. (f) Residual induction

Fig. 5 Comparison of resistivity and permeability components for a 3 μm (120 μin.) sample of iron carbonyl

nickel plating on a brass screw, led to the belief that magabsorption could be used to measure the stress and material properties of ferromagnetic materials and nickel-plated nonferromagnetic materials. Research in this area was conducted, and it resulted in four final reports (Ref 4-7). This work is briefly described in this article, along with information from Ref 8. Two patents have also been issued for this technology (Ref 9, 10).

Magabsorption Theory

Magabsorption involves the measurement of the change in the energy absorbed from the field of an RF coil, by a ferromagnetic material placed in the field of the coil, as a function of the strength of a magnetic bias field also applied to the materials. The general concept of magabsorption is shown in Fig. 6. The ferromagnetic material is in both the RF magnetic field of strength H_{RF} and the bias magnetic field of strength H_B. The value of H_{RF} is much smaller than H_B.

A thorough understanding of the basic electrical and magnetic properties of ferromagnetic materials is essential to the theoretical treatment of magabsorption. With these properties, a solution can be obtained that has correlated very well with experimental measurements. However, because some of these properties are difficult to describe mathematically, an exact solution is difficult if not impossible. Therefore, simplifying approximations are required to obtain a useful solution.

***B-H* Characteristics.** When a magnetic field is applied to any material, a resulting magnetic flux density, B, is induced. The intensity of the applied magnetic field is called the magnetic field intensity, H. The resulting flux density, B, is directly proportional to the magnetic field intensity, H.

This relationship is normally written in the following form:

$$B = \mu H \quad \text{(Eq 5)}$$

The constant of proportionality, μ, is called the permeability, which may be nonlinear. When a ferromagnetic material is subjected to a magnetic field, the flux density, B, is not a linear function of the magnetic field intensity, H. In other words, the permeability, μ, is not a constant. In addition to this

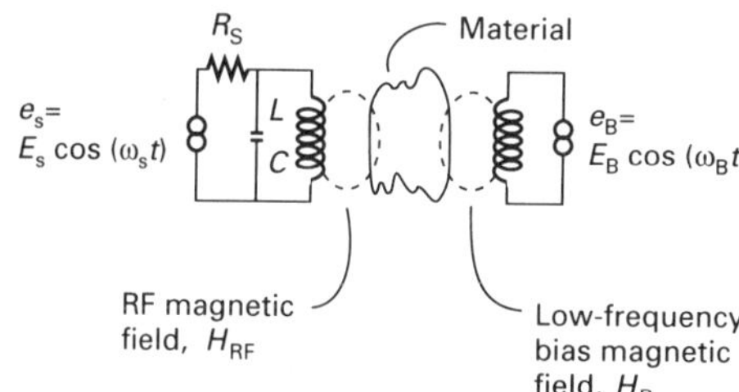

Fig. 6 General concept of magabsorption showing the RF field and the bias field in the material

nonlinearity, the permeability depends on the past magnetic and stress history of the ferromagnetic material. This dependence on past magnetic and stress history is known as hysteresis.

The hysteresis loop is explained by the concept of magnetic domains. Each domain consists of a group of atoms acting as a small magnet. The group of atoms in a small region forms a small magnet because molecular interactions cause the molecular magnetic moments to be aligned parallel to one another. In other words, each domain is spontaneously magnetized to saturation even in the absence of an external magnetic field. Because the directions of magnetization in different domains are not necessarily parallel to one another, the resultant magnetic flux density may be zero. When the specimen is placed in a magnetic field, the resultant magnetic flux density can be changed by two different mechanisms. The first is movement of a domain wall, and the second is directional rotation of the magnetic field of the domain. These two domain processes must stop at a high magnetic field intensity because all the domains become aligned with the applied magnetic field. This limiting process causes the nonlinearity of the *B-H* characteristics. Also, the stability of different domain orientations causes the hysteresis of the *B-H* characteristics. Therefore, the shape of the hysteresis loop and the domain orientation are dependent on the type of material, its stress history, and its magnetic history.

Reversible Permeability. If at some point on the hysteresis loop the magnetic field intensity is reversed from its normal direction, the resulting magnetic flux density will not follow the original hysteresis loop. This deviation from the original hysteresis is explained by the concept of the domain. If the magnetic field intensity is decreased before it reaches the previous limit, the orientation stability of the domain will cause hysteresis.

If the deviation of the magnetic field intensity is a small cyclic field, ΔH, superimposed on a large bias field, H_B, then a smaller inner hysteresis loop will be formed, as shown in Fig. 7. Note that the magnetic biasing field, H_B, is the magnetic field that determines the larger external hysteresis loop. As the size of the small

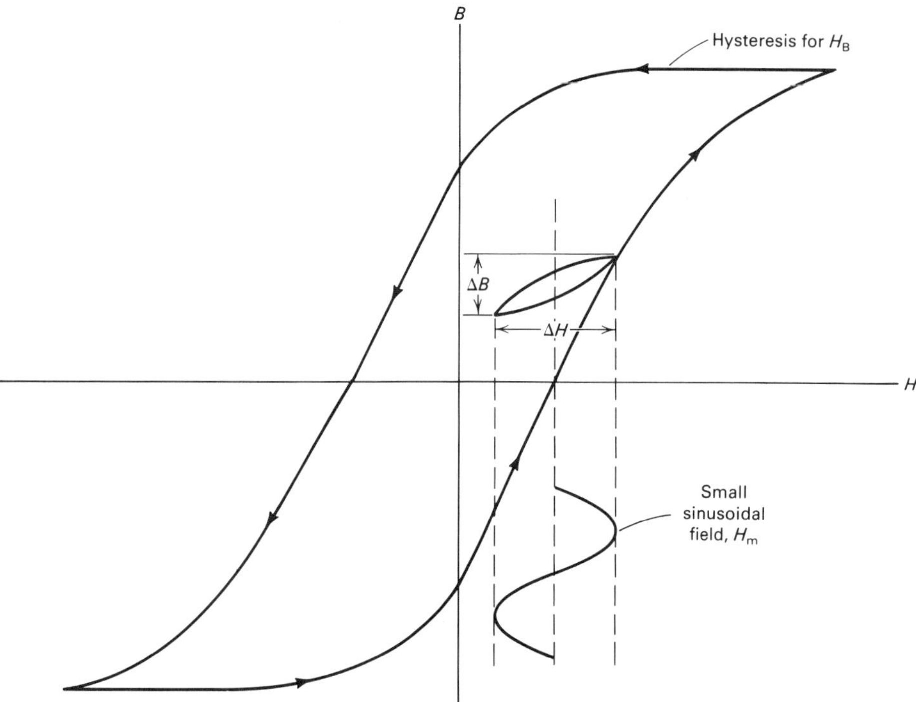

Fig. 7 Superimposed hysteresis loops

cyclic deviation in the magnetic field intensity, ΔH, is decreased, the resulting change in the magnetic flux density, ΔB, is decreased. As ΔH approaches zero, the ratio $\Delta B/\Delta H$ approaches a limiting value $\mu_o\mu_r$:

$$\lim_{\Delta H \to 0} \frac{\Delta B}{\Delta H} = \mu_o\mu_r = \mu \qquad \text{(Eq 6)}$$

where μ_o is the permeability of free space (in henries/meter), μ_r is the relative reversible permeability (unitless), and μ is the reversible permeability (in henries/meter).

The reversible, or alternating current (ac), permeability is the property of great interest in the study of magabsorption. Reversible permeability is dependent on the reversible boundary displacement and reversible rotation of the domains in the ferromagnetic material. Therefore, the reversible permeability is dependent on the type of material and its magnetic and stress history. The relative reversible permeabilities for unannealed nickel, annealed nickel, and iron are shown in Fig. 8. For all the relative reversible permeability curves in Fig. 8, the upper portion of the curve is observed for an increasing bias field, and the lower portion of the curve is observed for a decreasing bias field. Assuming that the ferromagnetic material was originally demagnetized, the plot of μ_r versus H_B is symmetrical about $H_B = 0$.

A block diagram of an apparatus used to measure relative reversible permeability is shown in Fig. 9. A small cyclic deviation, ΔH, in the magnetic field intensity was generated with an exciting coil. The change in the magnetic flux density was detected with two identical pickup coils. The two pickup coils were connected so that the sum of their output voltages would be determined by the derivative of the difference in the magnetic flux in the two coils. If a sample is placed in one of the pickup coils, the output voltage is directly proportional to the derivative of the magnetic flux in the sample. Using an electronic integrator to integrate the output voltage of the two pickup coils, a voltage is obtained that is directly proportional to the magnetic flux in the sample. The output voltage of the integrator was displayed versus the cyclic deviation in the magnetic field intensity on an oscilloscope. The resulting oscilloscope display is the smaller inner hysteresis loop shown in Fig. 7. A biasing magnetic field was applied with the use of larger field coils.

If the cyclic deviation in the magnetic field intensity, ΔH, is held constant in magnitude, the cyclic deviation in the magnetic flux density, ΔB, is directly proportional to the relative reversible permeability, μ_r. The output of the integrator can be converted to dc voltage with an ac-to-dc converter, which can be displayed on an *x-y* recorder versus the biasing magnetic field, H_B.

The Magabsorption Phenomenon. The basic arrangement of magabsorption measurements is shown in Fig. 6. When H_B varies slowly in the material and B is plotted as a function of H, the B/H hysteresis loop of Fig. 10 is obtained. If H_B is applied sinusoidally, the B/H curve will be repeated at the same rate, and B is not only distorted in the shape relative to the sinusoidal mag-

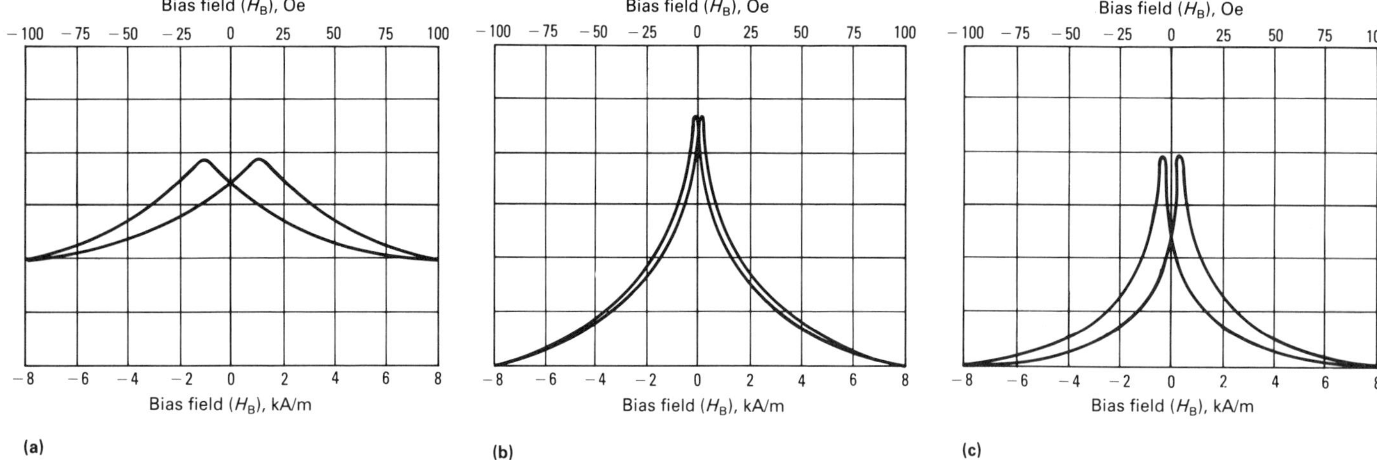

Fig. 8 Relative reversible permeability curves for (a) unannealed nickel wire, (b) annealed nickel wire, and (c) iron thermocouple wire

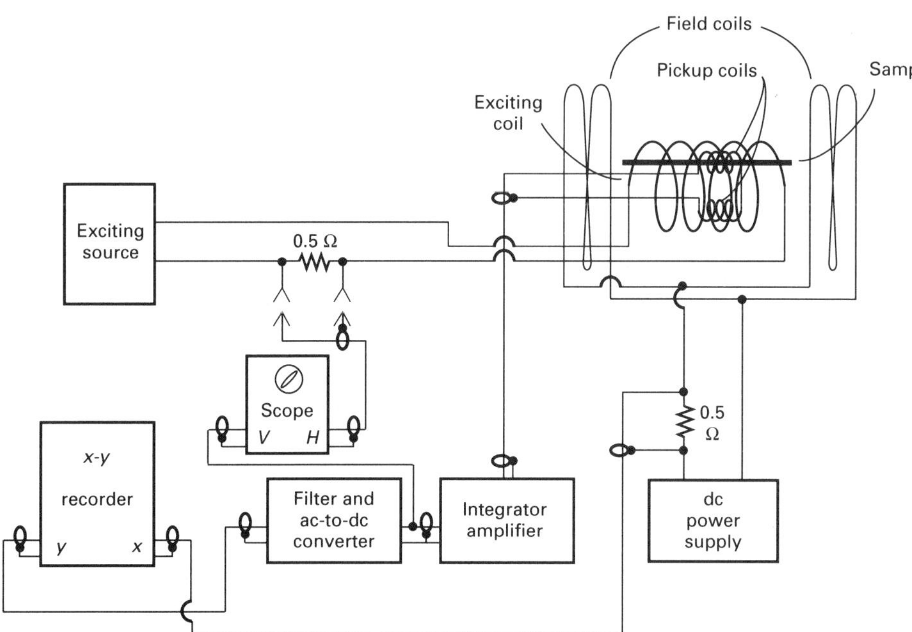

Fig. 9 Block diagram for relative reversible permeability apparatus

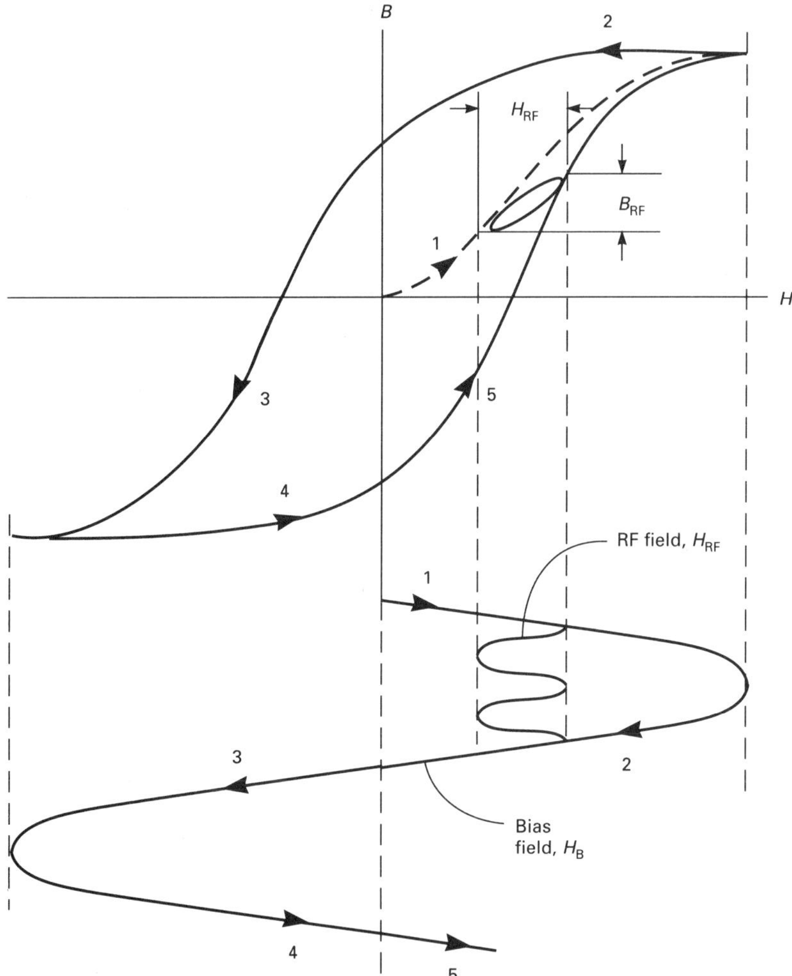

Fig. 10 Hysteresis loops of the material under two magnetomotive forces, H_B and H_{RF}

netomotive force but also shifted in phase. As the frequency of H_B increases, both the distortion in shape and the shift in phase increase. If another magnetomotive force, H_{RF}, of smaller magnitude and higher frequency is added to the material (Fig. 6), there will be a second B_{RF}/H_{RF} loop for the material. This second hysteresis loop is shown at one value of H_B in Fig. 10. As H_B varies through one sine wave, the magnitude and phase of B_{RF} also change. If the RF permeability, $B_{RF}/H_{RF} = \mu_{RF}$, is graphed as a function of H_B, the curve in Fig. 11 is obtained. This RF permeability hysteresis curve is also a function of the frequency and the amplitude of the magnetomotive force, H_B.

The variation in μ_{RF} as a function of H_B is the basis for the magabsorption phenomenon. Because the RF permeability, μ_{RF}, changes as H_B is varied sinusoidally, the energy absorbed by the material from the RF coil also varies as a function of H_B. These variations in the absorption of RF energy by the material can also cause changes in the impedance of the RF coil. For example, if a cylindrical sample is placed in the coil as shown in Fig. 12, the change in coil impedance, ΔZ, is related to the energy absorbed, according to Eq 3.

The variations in coil impedance from magabsorption involve a resistive change, ΔR, and a change in inductance, ΔL. Both of these changes can be related to the RF permeability, μ_{RF}. Because the RF hysteresis loop has a shape other than a straight line (that is, an ellipse), if H_{RF} is H_{RF} exp $i\omega t$, the RF induction is B_{RF} exp $i(\omega t + \theta)$. Therefore, the RF permeability is a complex variable:

$$\mu_{RF} = \mu'_{RF} + \mathbf{i}\mu''_{RF} \quad \text{(Eq 7)}$$

where μ''_{RF} is the loss (ΔR) term and μ'_{RF} is the dispersion (ΔL) term. For a cylindrical sample of radius a and a coil of radius c as shown in Fig. 12, the solution of the Maxwell equations resulted in the following normalized values of ΔR and ΔL for a ferromagnetic material (Ref 4):

$$\Delta R = \frac{\omega_{RF} L_o}{c^2} (-0.18 \times 10^{-7} + 2.23 \times 10^{-4} \cdot a \cdot \sqrt{\mu''_{RF}}) \quad \text{(Eq 8)}$$

$$\Delta L = \frac{\omega_{RF} L_o}{c^2} (0.08 \times 10^{-7} - a^2 + 2.22 \times 10^{-4} \cdot a \cdot \sqrt{\mu'_{RF}}) \quad \text{(Eq 9)}$$

where c is larger than a and μ'_{RF} and μ''_{RF} are functions of the bias field H_B. Because there is no simple equation for μ_{RF} as a function of H_B (Fig. 11), the shape of ΔR and ΔL is most easily obtained graphically, as shown in Fig. 13. The resultant curve, in the top right-hand corner, is either the magabsorption amplitude signal of the loss (ΔR) component or the dispersive (ΔL) component.

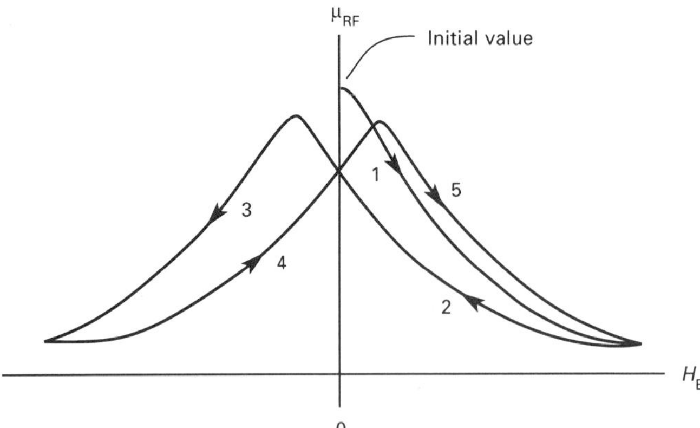

Fig. 11 RF permeability as a function of the alternating magnetomotive force, H_B, at a fixed value of the RF magnetomotive force, H_{RF}. The numbers on the curves correspond to numbers on the curve of H_B in Fig. 10.

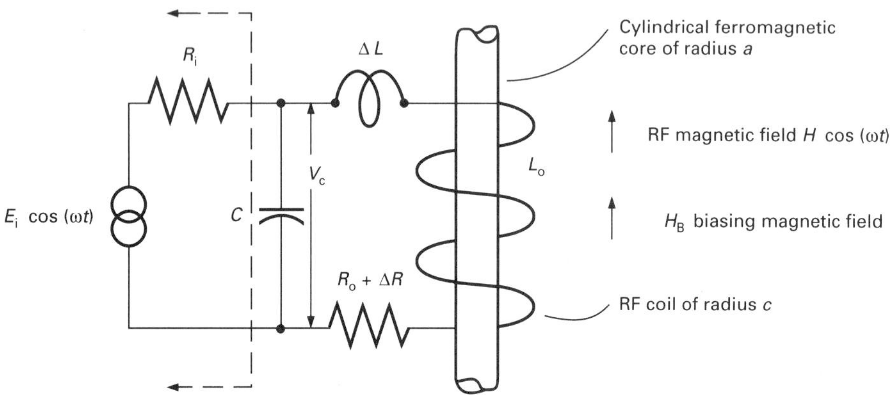

Fig. 12 Basic magabsorption circuit composed of an inductance, L, with the sample core in a biasing field, H_B, and tuned to resonance with capacitance C. The circuit is fed from an RF voltage $e_i = E_i \cos(\omega t)$.

Therefore, the variations in the absorption of RF energy by the material produce a magabsorption signal, which causes the resistance and inductance of the RF coil to change. The basic or fundamental frequency of the magabsorption signal is twice that of the magnetic bias frequency. If H_B has a frequency of 60 Hz, the magabsorption signal has a basic frequency of 120 Hz. However, because the magabsorption signal is not a pure sinusoid (Fig. 14), the entire frequency content of the magabsorption signal has harmonics of the basic frequency. Therefore, the magabsorption signal, v_{MA}, is:

$$v_{MA} = v_o + v_n \sum_{n=0}^{\infty} \cos(2n\omega_B t + \phi_n) \qquad \text{(Eq 10)}$$

where ω_B is the magnetic bias angular frequency and ϕ_n is the phase angle of the harmonic order, n.

A more extensive derivation of the magabsorption signal is discussed in a report of work done for the Air Force Materials Laboratory in Dayton, OH (Ref 4). It was from this report that Eq 8 and 9 were obtained.

Magabsorption Detection

The phenomenon of magabsorption changes the resistance and inductance of the RF coil, which causes the modulation of the RF voltages in the circuit shown in Fig. 6. The RF voltage across the resonant circuit is amplitude modulated by the change in resistance of the coil and is phase or frequency modulated by the change in inductance of the coil. Of course, there is a small additional amplitude modulation caused by the phase modulation. Therefore, if the voltage across the resonant circuit of Fig. 6 is demodulated through both an amplitude detector and a phase detector using a voltage, proportional to the bias current, as a reference signal, the magabsorption amplitude signal is obtained from the amplitude detector, and the magabsorption frequency signal is obtained from a phase or frequency detector.

If the voltage generator e_s and the resistance R_S are those of a self-sustained oscillator, the voltage across the resonant circuit is amplitude modulated by means of the magabsorption amplitude signal, and the frequency of oscillation will be modulated by the magabsorption frequency signal. If the oscillator is operated at a high level of output (A, Fig. 15), the detection sensitivity will be weak for the amplitude modulation. When the oscillator is operated at its marginal point or the point close to where oscillations cease (B, Fig. 15), the slope for the voltage change as a function of resistance change in the coil is very large, and the voltage across the resonant circuit is highly modulated by the magabsorption amplitude signal. Because the resonant circuit alone controls the frequency, the frequency modulation is relatively independent of the oscillation level. It will be true, however, that at the high levels of oscillation the effective Q of the RF coil will be low, and the amplitude modulation of the oscillator output by the magabsorption amplitude signal will be reduced. Under this condition, the amplitude modulation caused by the frequency modulation will be reduced.

The amplitude modulation of the RF signal by the magabsorption signal in Fig. 13 has a fundamental frequency and harmonic frequency components at twice the frequency of the bias field. If ω_B is the bias field frequency, the magabsorption signal frequencies are at $2\omega_B$, $4\omega_B$, $6\omega_B$, and so on. If ω_{RF} is the frequency of the RF field in Fig. 6, the equation for the modulated RF signal, v_s, across the resonant LC circuit in Fig. 6 is:

$$v_s = \sum_{n=0}^{6} \left(1 + \frac{V_B}{V_{RF}} A_n \cos 2n\omega_B t\right) \cdot (V_{RF} \cos \omega_{RF} t) \qquad \text{(Eq 11)}$$

where V_B and V_{RF} are the peak amplitudes of the bias and RF fields and A_n is a multiplier to give the amplitude of the nth harmonic of the magabsorption signal.

With a marginal oscillator as a magabsorption detector, the magabsorption amplitude signal can be obtained with an amplitude modulation detector. The magabsorption frequency signal, however, will be obtained with a frequency discriminator or similar frequency demodulator. Both of these voltages from the demodulators will be similar in shape to the magabsorption curve, and they will be nearly of equal amplitude for the same or similar demodulator constants. Further, each signal can be described by a Fourier series, as indicated in Eq 10.

General Detection Methods. The magabsorption signal as modulation on a carrier can be detected in three ways. First, the modulated carrier can be amplified and the modulation recovered by a diode detector, a coherent detector, or a mixer. This would yield the ΔR component of the magabsorption signal.

In the second method, a receiver or narrow-band amplifier is set to one or more of the sideband frequencies, and the amplitude of that sideband is detected by amplitude demodulation. For example, if the bias frequency, f_B, is 80 Hz and the RF frequency, f_{RF}, is 10 kHz, the magabsorption modulated carrier will have components at 10 160, 10 320, 10 480, 10 640, 9840, 9680, 9520, and 9360 Hz if the magabsorption signal contains only four harmonics. Therefore, the presence of a magabsorption signal

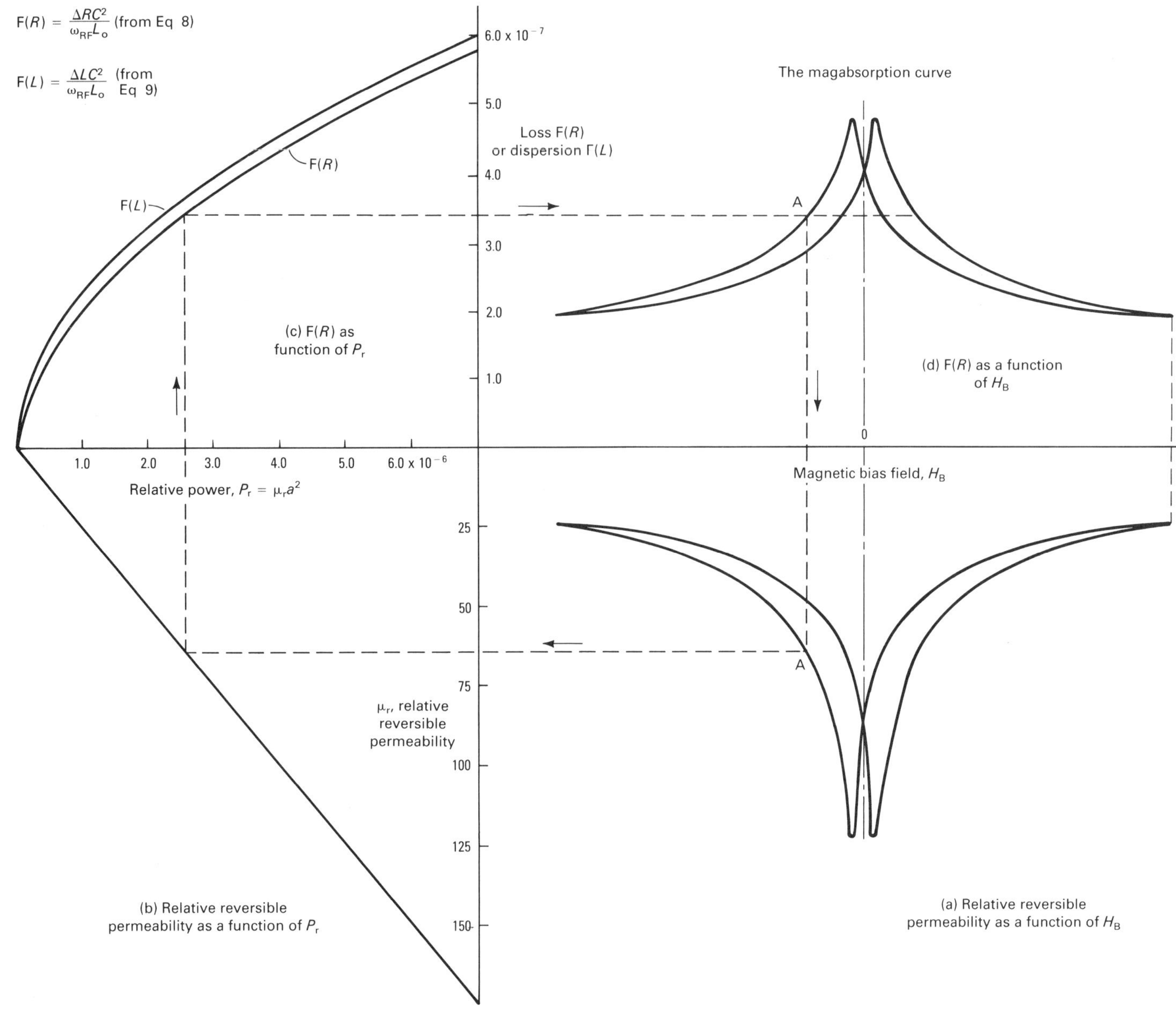

Fig. 13 Graphical derivation of the magabsorption curve from the permeability and F(R) and F(L) curves

could be detected by a narrow-band amplifier tuned to any one of the frequencies given above. Because the 10 160 Hz and 9840 Hz components are the strongest, they will provide the most sensitive detection.

The third method is to amplify the voltage across the resonance circuit and to detect the frequency modulation. This would give the ΔL component of the magabsorption signal. The amplitude demodulation (the first method) gives a mixture of both the ΔR and the ΔL components of the magabsorption signal. In most cases, however, the ΔR component is much larger than the ΔL component because the frequency modulation (ΔL) causes only a small amplitude modulation relative to that caused by ΔR.

The basic magabsorption circuit used with the cylindrical sample of radius a is shown in Fig. 12. The magabsorption phenomenon causes a change in both the resistance, ΔR, and the inductance, ΔL, of the magabsorption detection coil. These variations change the coil impedance, Z_c, by an amount ΔZ such that $Z_c = \Delta Z + Z_o$. The magnitude of the voltage across the parallel resonant circuit at resonance is $V_c = (L/RC)I_c$, where $I_c \simeq (E_i/R_i)\cos(\omega t)$, because R_i is much larger than Z_c. For all magabsorption measurements, a high-Q coil is used. A high-Q coil is defined as one in which $Q = (\omega L/R) > 10$.

For magabsorption measurements on wire, the filling factor, F, which is the ratio of the volume of the wire sample to the volume inside of the RF coil, is less than 0.01. This value of the filling factor keeps the loaded Q of the coil also greater than 10.

The theoretical derivation has shown that $\Delta R \simeq \omega\Delta L$. Taking into account the above assumptions and because $R_i \simeq 10\ L_o/R_oC$ and ΔR is less than 0.1 R_o, the voltage change, ΔV, can be approximated within 1% to be directly proportional to the ΔR from magabsorption. The change in resonant frequency can also be shown to be directly related to ΔL.

The Magabsorption Bridge Detector. The voltage across the resonant RF circuit in Fig. 12 is $V_c + \Delta V$. Although the actual value is approximately 0.1 E_i, there may be problems in amplifying $V_c + \Delta V$ because of the dynamic range of many amplifiers. Therefore, the bridge circuit in Fig. 16 has been used to eliminate V_c and to give an output of only ΔV. The left resonant circuit contains the material sample, while the right

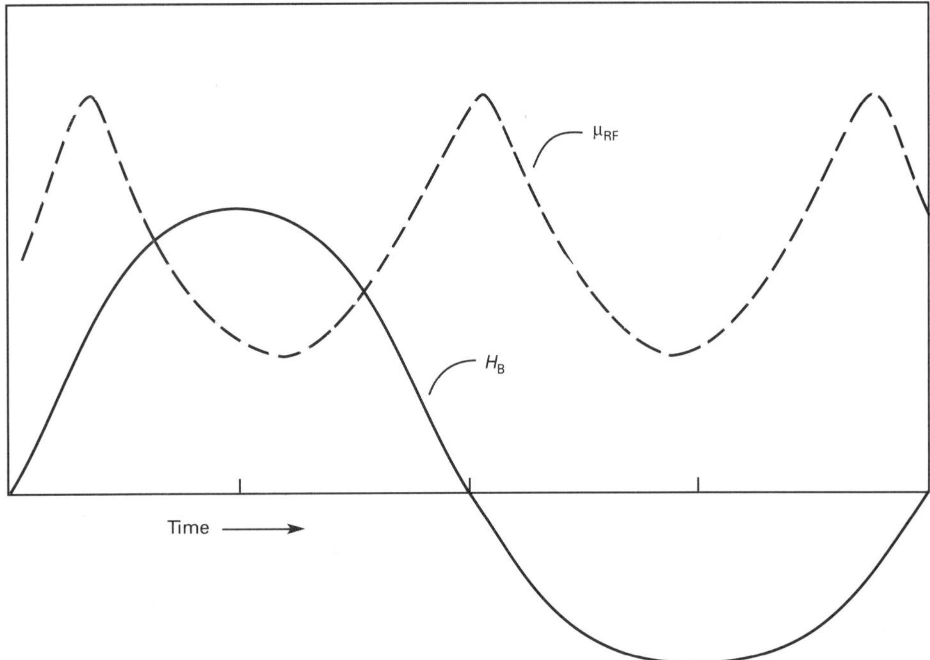

Fig. 14 Time plot of H_B and μ_{RF}. Although the shape of μ_{RF} is distorted relative to a sinusoid, the basic frequency of μ_{RF} is twice that of H_B.

resonant circuit contains no sample. The rectified voltages across each resonant circuit are detected, subtracted, filtered, and supplied to the output. With no magabsorption signal, the output is zero. With a magabsorption signal, the output of the bridge in Fig. 16 is:

$$V_o = \frac{V_c \cdot \Delta R}{11 R_o} \quad \text{(Eq 12)}$$

The magabsorption bridge is used when the variation rate for ΔR is very low (<0.1 Hz). At low variation rates, the output is very stable with changes in time, temperature, and input voltage. The use of the bridge reduces stringent requirements on the amplitude stability and the frequency stability of the source for the input to the bridge at 500 kHz. This bridge was used for all of the earliest magabsorption measurements.

When the bias field is changed or varied, the bridge must be rebalanced for each sample unless a complicated automatic balancing circuit is used. The need to rebalance the bridge can be eliminated if the resonant circuit in Fig. 12 is the resonant circuit of an oscillator.

The Marginal Oscillator Magabsorption Detector. Figure 17 shows the schematic of the circuit when the RF coil is part of an active oscillator circuit. This type of circuit can also be used to detect the ΔR component of the magabsorption signal. This detection will occur only if the oscillator is made to operate as close to Class A or linear conditions as possible. As such, it will be an efficient magabsorption detector.

The detector illustrated in Fig. 17 is called a marginal oscillator because it is operated on the edge of dropping out of oscillation. The presence of a magabsorption sample in the RF coil will change both the series resistance and the inductance of the coil. Instead of ΔL changing only the resonant frequency of the RF coil, ΔL changes both the driving frequency of the oscillator and the resonance frequency of the RF coil by the same amount. Therefore, the resonant frequency of the coil and the frequency fed to the RF coil are the same. They are locked together because the RF resonant circuit controls both. In this way, it is possible to measure the effect of the sample both on the losses from the coil and the inductance of the coil. As stated previously, the losses change the amplitude of the oscillation, while the dispersion changes the oscillation frequency.

For an analysis of the effects of the magabsorption phenomenon on a marginal oscillator, assume that the ratio of the radius of the sample to the radius of the RF coil is such that the effect of the sample is small compared to the magnitudes of the coil inductance and effective resistance. The sample effect will therefore be more like that of a paramagnetic material. The inductance of the coil shown in Fig. 12 or 17 can then be written as:

$$L = L_o(1 + F\chi) \quad \text{(Eq 13)}$$

where L_o is the inductance without the sample, F is the filling factor of the coil-sample system, and χ is the susceptibility of the sample. The susceptibility is usually defined for only diamagnetic or paramagnetic substances by the relation:

$$B = \mu_o H(1 + \chi) \quad \text{(Eq 14)}$$

where $\chi = \chi' - i\chi$; χ'' is the loss term and χ' is the dispersion term.

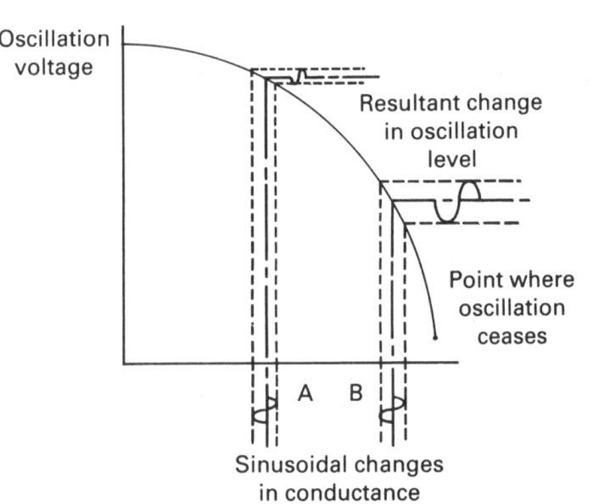

Fig. 15 Oscillation voltage as a function of the resonant circuit conductance. A, operation point for low sensitivity; B, operation point for high sensitivity

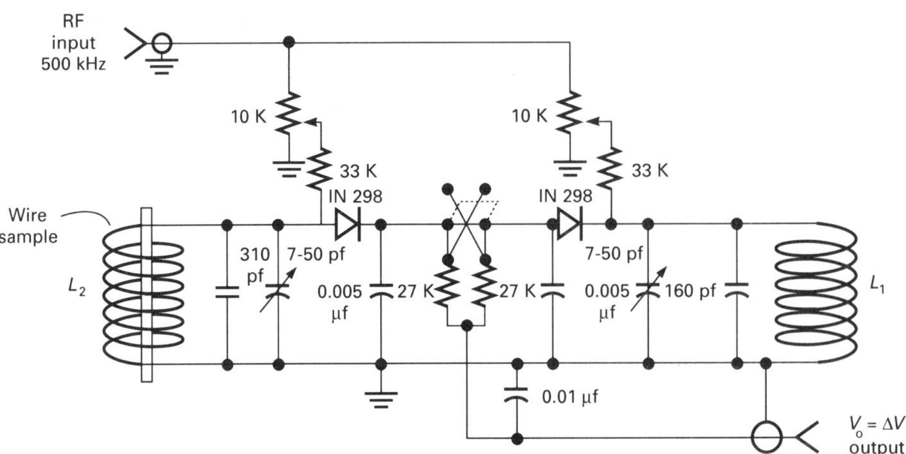

Fig. 16 Magabsorption measurement RF bridge, in which the peak value of two similar simple circuits are subtracted to give the magabsorption signal

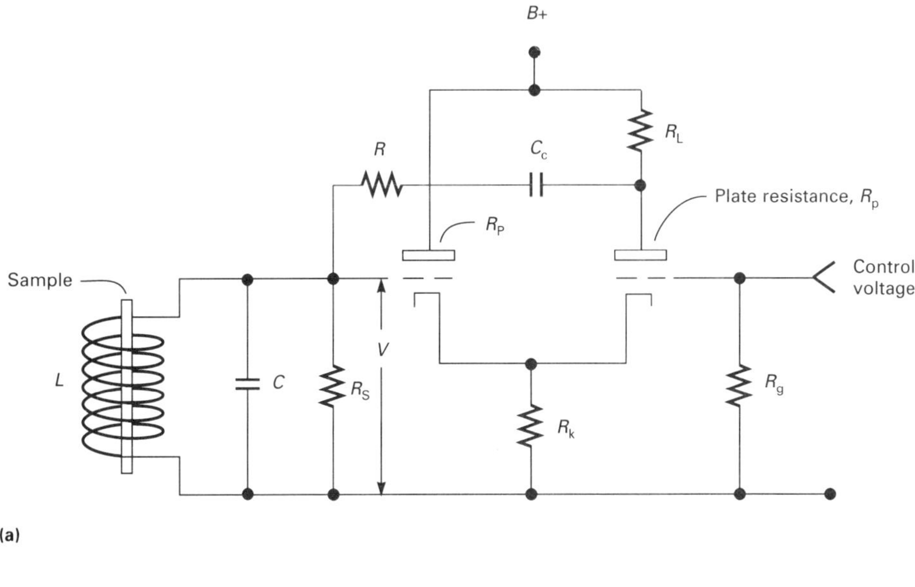

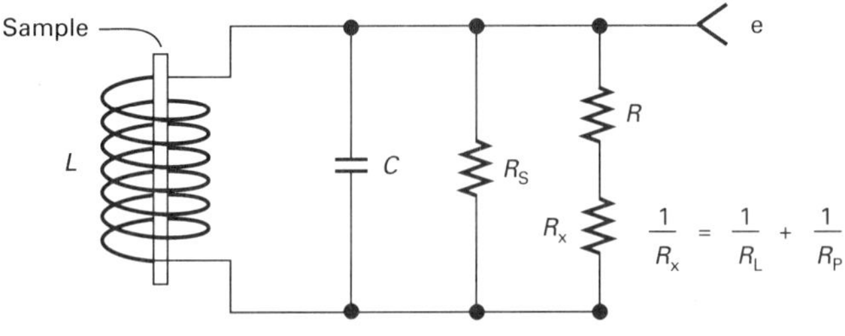

Fig. 17 Marginal oscillator basic circuit (a) and equivalent circuit (b)

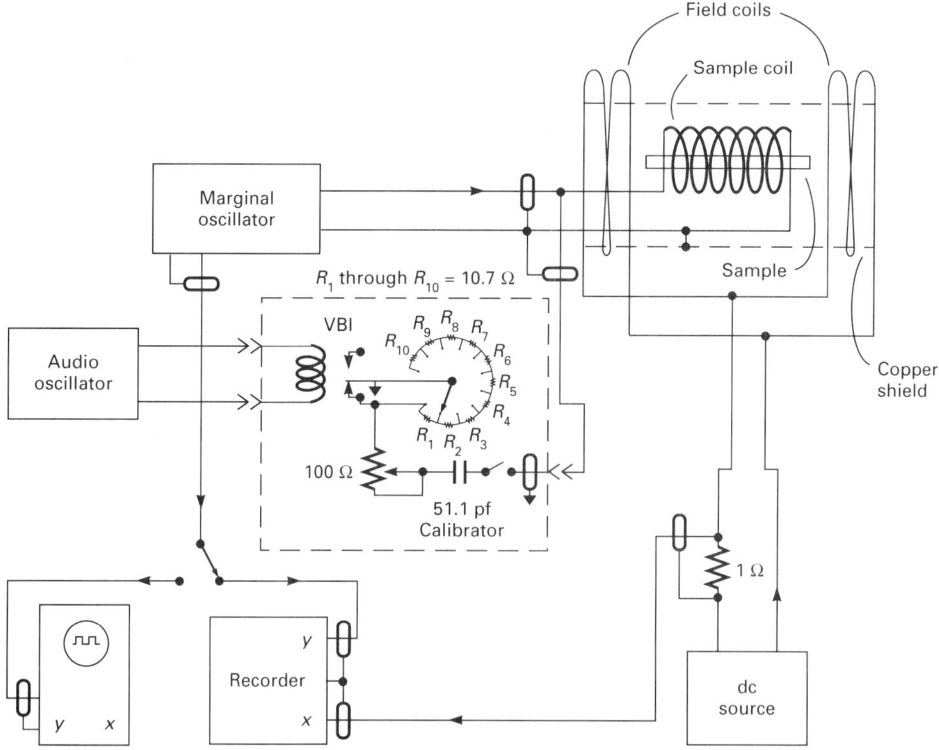

Fig. 18 Block diagram of the system for measuring ΔG versus the voltage output of a marginal oscillator

For paramagnetic materials, the susceptibility is of the order of 10^{-4} to 10^{-6}. For ferromagnetic materials, the susceptibility may be many orders of magnitude larger. However, the susceptibility seen by the RF coil is very small when the filling factor (volume of the core divided by the volume of the coil) for the ferromagnetic core is kept small. Therefore, Eq 14 can be written as:

$$B - \mu_o H(1 + F\chi) \quad \text{(Eq 15)}$$

when the filling factor F is considered. The filling factors used are in the range of 10^{-3} to 10^{-4}. Therefore, the ferromagnetic core material with a large susceptibility and a small filling factor can have the same effect as a paramagnetic material with a small susceptibility and a unity filling factor. With this approach, the frequency shift, Δf, of the oscillator with a sample in the coil is:

$$\Delta f = \frac{F\chi'}{2\pi L_o C} \quad \text{(Eq 16)}$$

and the change in conductance, ΔG, of the resonant circuit is:

$$\Delta G = \frac{F\chi''}{\omega L_o} \quad \text{(Eq 17)}$$

To keep the resonance condition, then, the frequency must be changed by the factor Δf (Eq 16). If this change in frequency is accomplished automatically by making the resonant circuit the frequency-controlling circuit of an oscillator, ΔG can be measured by determining its effect on the impedance of the resonant circuit. The value of Δf can be obtained by measuring the frequency shift of the oscillator. With ΔG and Δf, the values of χ'' and χ' can be calculated and related to the material of the sample inserted in the coil and to the effects of the magnetic bias field.

When the resonant circuit illustrated in Fig. 17 undergoes an impedance change from any variation of χ'', the change in conductance (Eq 17) will result in a change in the voltage across the resonant circuit. If the value of ΔG is much smaller than $1/R_S$ of the circuit in Fig. 17(b) and if the first approximation of linearity is assumed, then the voltage change in the oscillator can be shown to be directly proportional to the conductance change caused by the sample. Moreover, if the conductance change, ΔG, is sinusoidal, a phase change is introduced that will shift the phase of the sidebands relative to the oscillation frequency. Figure 18 shows the block diagram of a system for measuring ΔG versus the voltage change in the oscillator.

It is also interesting to compare a marginal oscillator with the passive resonant circuit of Fig. 12. The effective gain of the oscillator circuit over the passive circuit can range from 5 to 10 with readily attainable values of circuit constants. The oscillator produces a gain over that of the passive system by decreasing the bandwidth or by increasing the effective quality factor. Because the signal is amplified and detected in both cases, the marginal oscillator offers a gain advantage that may improve the signal-to-noise ratio.

Measurements and Applications

The basic method of magabsorption measurement requires a ferromagnetic material that is excited by both an RF magnetic field and another magnetic field with a lower frequency and a much higher field strength. This basic arrangement is shown in Fig. 19 with a stressed wire placed inside the RF coil. Magabsorption measurements can also be performed with magabsorption detection heads (Fig. 20) that are placed on the surface of a specimen.

In the work discussed in Ref 3 to 7, magabsorption measurements were performed on various materials in different applications. Some of the magabsorption signals from a variety of materials are given in Fig. 21. The potential applications of magabsorption measurements, which are described in more detail in the following sections, include:

- Magabsorption measurements on ferromagnetic and ferrimagnetic powders, along with a particle size effect in the magabsorption measurements of the powders that might be used to determine the size or range of particles
- Magabsorption measurements of applied and residual stress in ferromagnetic materials or nonferromagnetic materials having a ferromagnetic coating
- Magabsorption measurements of residual magnetism in ferromagnetic and nonferromagnetic materials (such as some stainless steels) in which the yield point or phase transition temperature has been exceeded

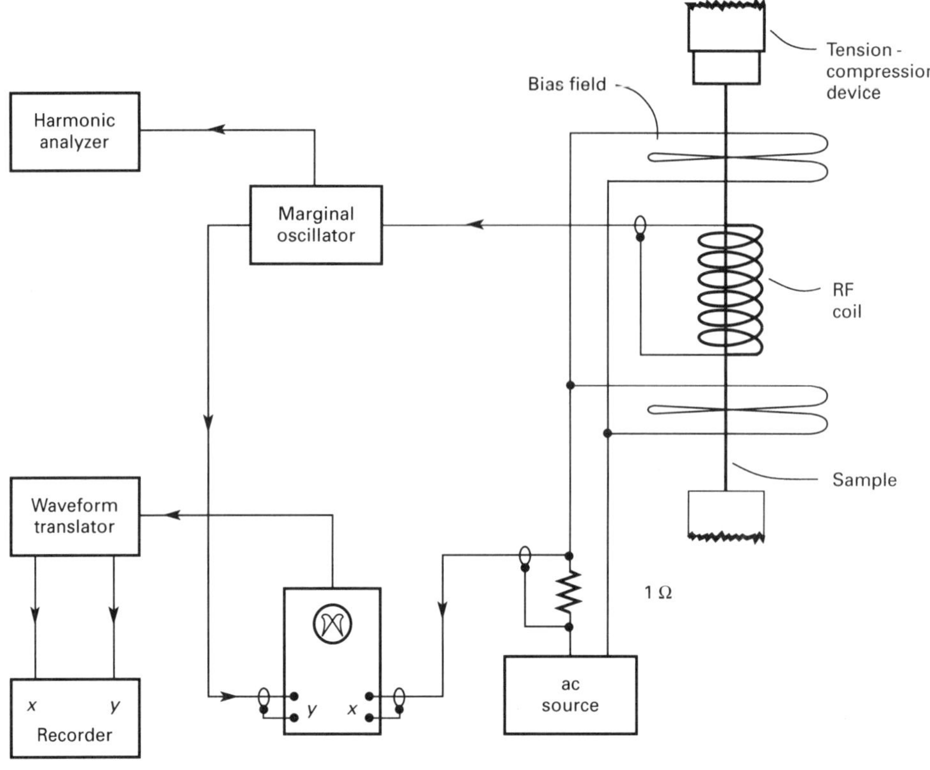

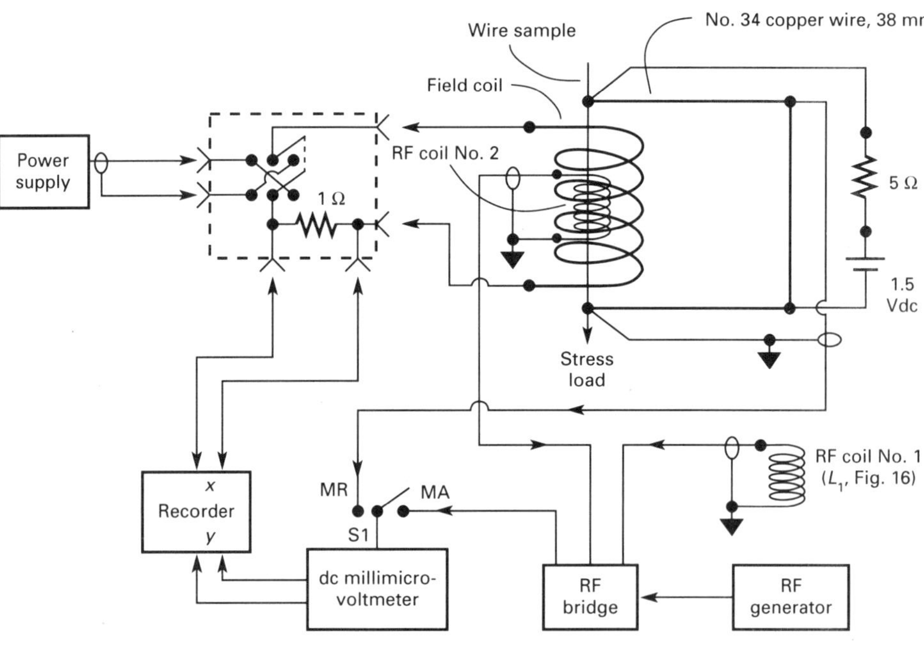

Fig. 19 Two arrangements for measuring magabsorption signals. (a) Block diagram of system with a marginal oscillator for measuring the harmonic content. (b) Block diagram of a system with a bridge detector. Switch S1 connects the circuit for either magnetoresistance (MR) measurements or magabsorption (MA) measurements.

Magabsorption Measurements of Powders

The first magabsorption measurements were made using iron oxide and iron carbonyl (Ref 3). The first measuring instrument was a Q-meter made by Boonton. When a sample is inserted into the coil, the Q of the coil changes from the Q value without a material in the coil. The changes in Q with the addition of samples of iron oxides and carbonyl iron were too small to give a measurable change in Q. However, as the radio frequency was changed from 0.7 to 3.0 MHz, the Q value of the RF coil had a maximum at a different frequency for the four particle diameters used (3, 5, 10, and 20 μm). The 3 μm (120 μin.) sample peaked at 2.6 MHz, the 5 μm (200 μin.) sample at 2.3 MHz, the 10 μm (400 μin.) sample at 1.8 MHz, and the 20 μm (800 μin.) sample at 0.8 MHz.

An NMR detector (a marginal oscillator type) was also used to give the magabsorption signals from the powder samples, and iron carbonyl was the first powdered ferromagnetic material to be measured with a marginal oscillator type magabsorption detector. The magabsorption Lissajous figures for four samples of different particle sizes (5, 8, 10, and 20 μm) are given in Table 1. The fundamental frequency for each magab-

Fig. 20 Closeup view of three detection heads used with magabsorption measurements on a crankshaft throw. Left, head for perpendicular measurements in fillets; middle, head for parallel measurements in fillets; right, head for all measurements in areas having a large radius of curvature

Table 1 Marginal-oscillator signals for iron carbonyl particles

The first four signals were obtained with a peak-to-peak bias field strength, H_B, of 24 × 10^{-3} T (240 G); the last signal, 0.6 × 10^{-3} T (6 G).

Particle size		Oscilloscope Lissajous patterns	Frequency spectrum	Peak-to-peak signal amplitude, mV
μm	μin.			
5	200		120 240	30
8	320		120 240	65
10 (HP)	400		120 240	2500
20	800		120 240	5000
20	800		120 240	38

sorption signal is 120 Hz for a bias frequency of 60 Hz. The first harmonic of the signal is at 240 Hz, also as shown in the "Frequency spectrum" column in Table 1. The vertical amplifier gain control was adjusted for each sample so that the magabsorption signal was of useful amplitude to display the shape of the signal. The magabsorption signal amplitude is very low for the 5 and 8 μm (200 and 320 μin.) samples and is very high for the 10 and 20 μm (400 and 800 μin.) samples when the bias field is 0.024 T (240 G) peak-to-peak. In the bottom row of Table 1, the peak-to-peak value of the bias field is reduced by 40 times to 6 × 10^{-4} T (6 G) peak-to-peak. This bias field reduction increased the ratio of the 240-Hz component relative to the 120-Hz component as shown by a comparison between the data in the second row from the bottom with that from the bottom row. The second row from the bottom uses a peak-to-peak bias field of 0.024 T (240 G), while the bottom row uses only 6 × 10^{-4} T (6 G) peak-to-peak. There is a distinctive change in the shape of the magabsorption signal from when the bias field is reduced from 0.024 to 6 × 10^{-4} T (240 to 6 G).

There is also an increase in the ratio of the 240-Hz component to the 120-Hz component. The ratio of the amplitude of the 240-Hz component to the amplitude of the 120-Hz component is graphed in Fig. 22 for five particle sizes of carbonyl iron powder in a bias field of 0.024 T (240 G) peak-to-peak. Figure 23 shows the root mean square (rms) magnitude of the magabsorption signal plotted as a function of the peak-to-peak amplitude of the bias magnetic field.

During the experiments with powders, it was noticed that there was a difference in the magabsorption signal shape and amplitude for iron oxide powders in different suspension media. The results with iron carbonyl particles indicate that the more tightly the medium holds the particle, or the greater the viscosity of the medium, the larger the magabsorption signal is. It was also noticed that when the iron carbonyl particles were allowed to settle, the signal decreased in amplitude by ten times. When the particles were redistributed, the signal returned to its larger value.

Magabsorption Measurement of Stresses

As discussed in the section "*B-H* Characteristics" in this article and as illustrated in Fig. 4, stress can affect the orientation of magnetic domains. When tension is applied to a saturated ferromagnetic material, with a positive magnetostriction constant, some of the parallel-aligned domains have their direction reversed so that there are domains parallel and antiparallel to the applied field. On the other hand, when tension is applied to a saturated ferromagnetic material such as nickel with a negative magnetostriction constant, some of the parallel-aligned domains have their direction rotated 90° so that there are domains perpendicular to the applied field. Therefore, the peak-to-peak magnitude and the shape of the magabsorption signal depend on the direction of the magnetic bias field relative to the direction of the applied stress and on whether the material has a positive or negative magne-

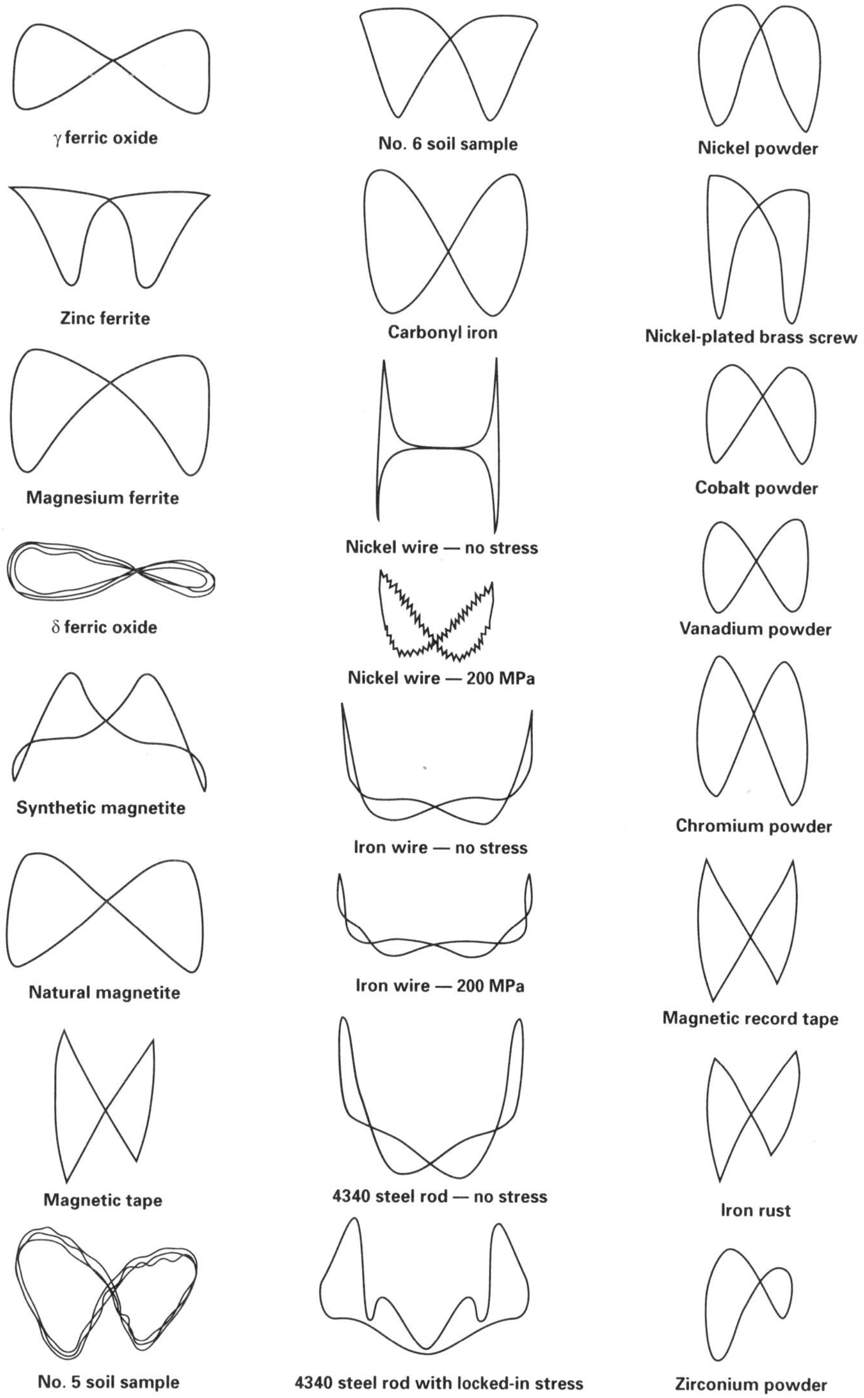

Fig. 21 Magabsorption signals from various materials

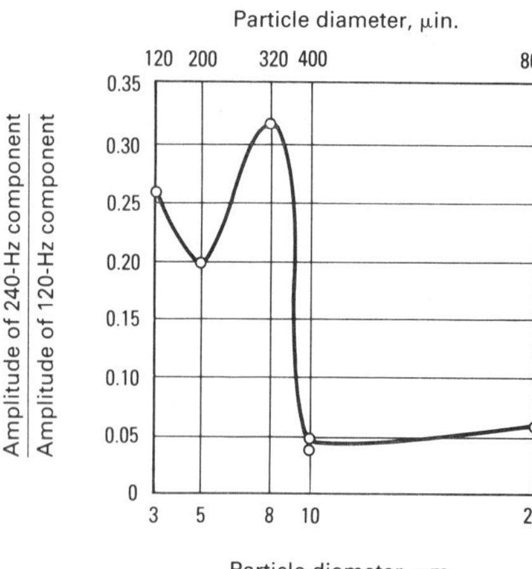

Fig. 22 Ratio of the amplitude of the 240-Hz component to the amplitude of the 120-Hz component for various particle diameters of carbonyl iron with a magnetic field of 24×10^{-3} T (240 G) peak-to-peak

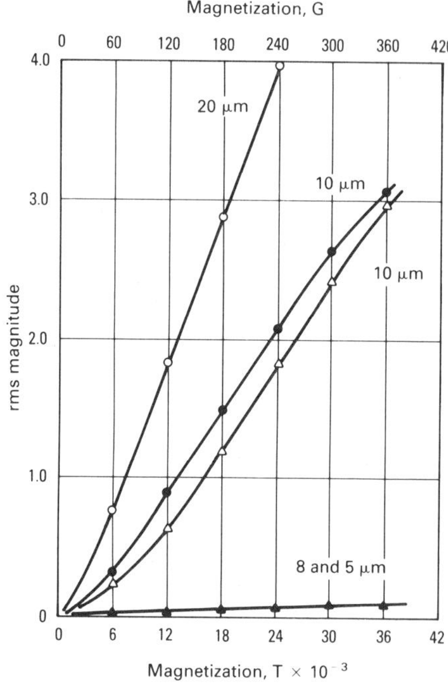

Fig. 23 Root mean square (rms) magnitude of the resistivity signal as a function of magnetization magnitude

tostriction constant. The work on many types of materials has shown that when the magnetic field bias, H_B, is parallel to the direction of the applied stress, the following conditions result:

- For materials with a positive magnetostriction constant, increasing tension increases the signal magnitude, while increasing compression decreases the amplitude (Fig. 24a)
- For materials with a negative magnetostriction constant, increasing tension decreases the signal magnitude, while increasing compression increases the amplitude (Fig. 24b)

These effects of stress on the peak-to-peak amplitude of the magabsorption signal are exactly the opposite for the condition in which the direction of the magnetic bias field is perpendicular to the direction of the stress. The behavior of the magabsorption signal peak-to-peak amplitude with the stress parallel to the direction of the magnetic bias field is shown in Fig. 24. The maximum stress in each case is below the yield point of the material. The solid lines are for the increasing stress in Fig. 24, while the broken lines are for the decreasing stress. In most materials, the magabsorption amplitude for decreasing stress will not follow the curve for increasing stress, because there is a hysteresis.

Applied Stresses in Ferromagnetic Materials. Measurement of the magabsorption signal as a function of stress has been performed on a variety of specimens. Measurements were made on iron and nickel wire and on a variety of bar specimens.

Magabsorption Measurements on Wire. The block diagram of one system used to

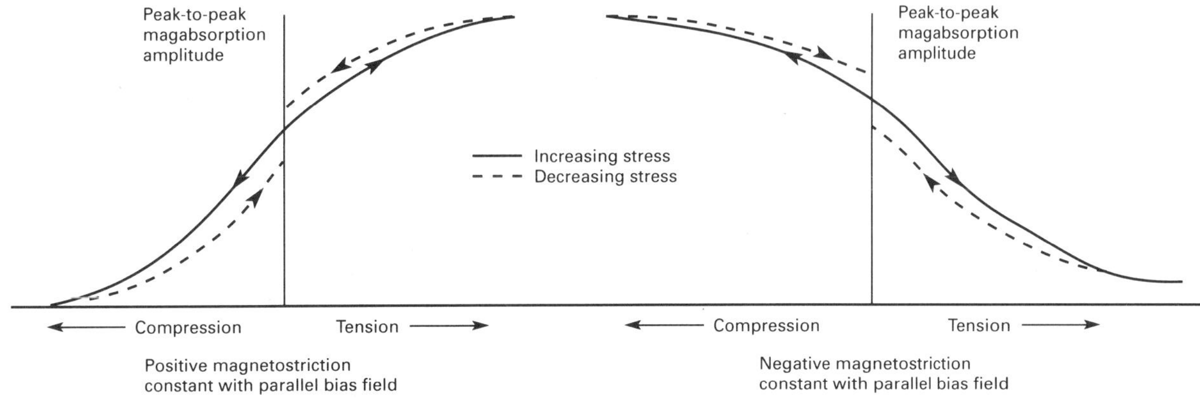

Fig. 24 Amplitude (peak-to-peak) of the magabsorption signals graphed as a function of stress (both tension and compression) for materials with both positive and negative magnetostriction constants. The magnetic bias field is applied parallel to the stress direction.

measure the magabsorption signal in wires is shown in Fig. 19(a). The detection head consists of an RF coil and a Helmholtz pair of coils that supply the bias magnetic field parallel to the axis of the wire. The RF magnetic field is also applied parallel to the axis of the wire. The RF detection coil is fed through a coaxial cable to a marginal oscillator. The output of the marginal oscillator is the magabsorption signal, and it is applied to the *y*-axis input of the oscilloscope. A voltage proportional to the current in the bias field coils is fed to the horizontal input of the oscilloscope.

Another system for measuring the magabsorption signals from wires is a bridge circuit (Fig. 19b). The relative reversible permeability curves for iron wire, unannealed nickel wire, and annealed nickel wire shown in Fig. 8 were taken with an ac bridge. The magabsorption curves will also be similar to the curves of the relative reversible permeability.

Measurements of the magabsorption signal were made on iron and nickel wire as a function of stress. For the positive magnetostriction constant material (iron), the peak-to-peak magnitude of the magabsorption signal from the material increased with tension and decreased with compression when the bias field was parallel to the applied stress. The reverse occurred when the material had a negative magnetostriction constant (nickel). When iron and nickel have residual stress and additional stress is applied, the peak-to-peak magnitude of the magabsorption signal may be less than that for no residual stress.

Magabsorption Measurements on Bar Specimens. Many measurements with magabsorption detector heads (such as the one shown in Fig. 20) have been made on bar specimens with the bias field both perpendicular (Fig. 25a) and parallel (Fig. 25b) to the stress direction.

A number of measurements were made on type 1018 steel bars (Ref 7). Graphs were constructed of the peak-to-peak magabsorption signal amplitude as a function of the stress, both tension and compression, applied as a bending moment. Additional similar measurements were made on bars of 410 and 4340 steel.

Another series of measurements was obtained from a 5046 steel crankshaft throw (Ref 11). The measurement procedure was very similar to that described above except that the detection heads (Fig. 20) were made much smaller and were ground curved to fit the curves of the crankshaft throw.

Stresses in Nonferromagnetic Materials. It has also been shown that when a nonferromagnetic material is coated with a thin layer of ferromagnetic material, there exists a possibility of measuring stress at

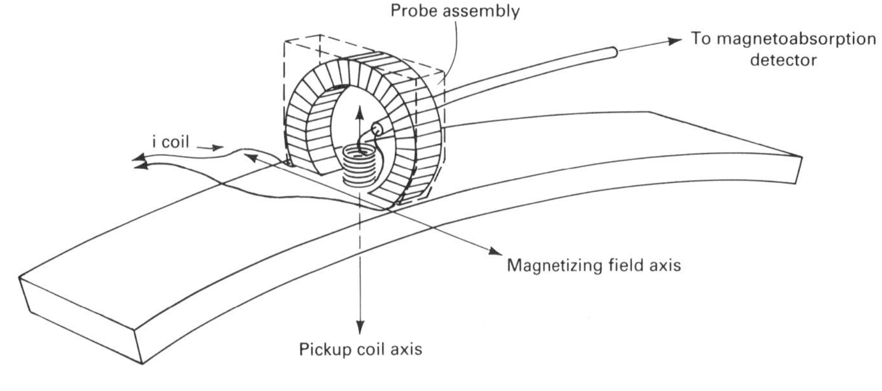

(a)

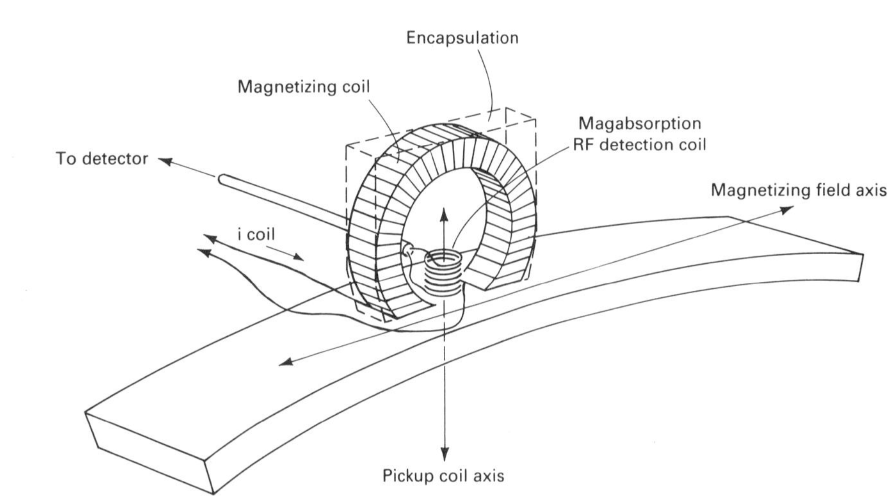

(b)

Fig. 25 Probe specimen geometry for parallel and perpendicular magabsorption measurements on a bent steel bar. (a) Probe in transverse or perpendicular position. (b) Probe in axial or parallel position

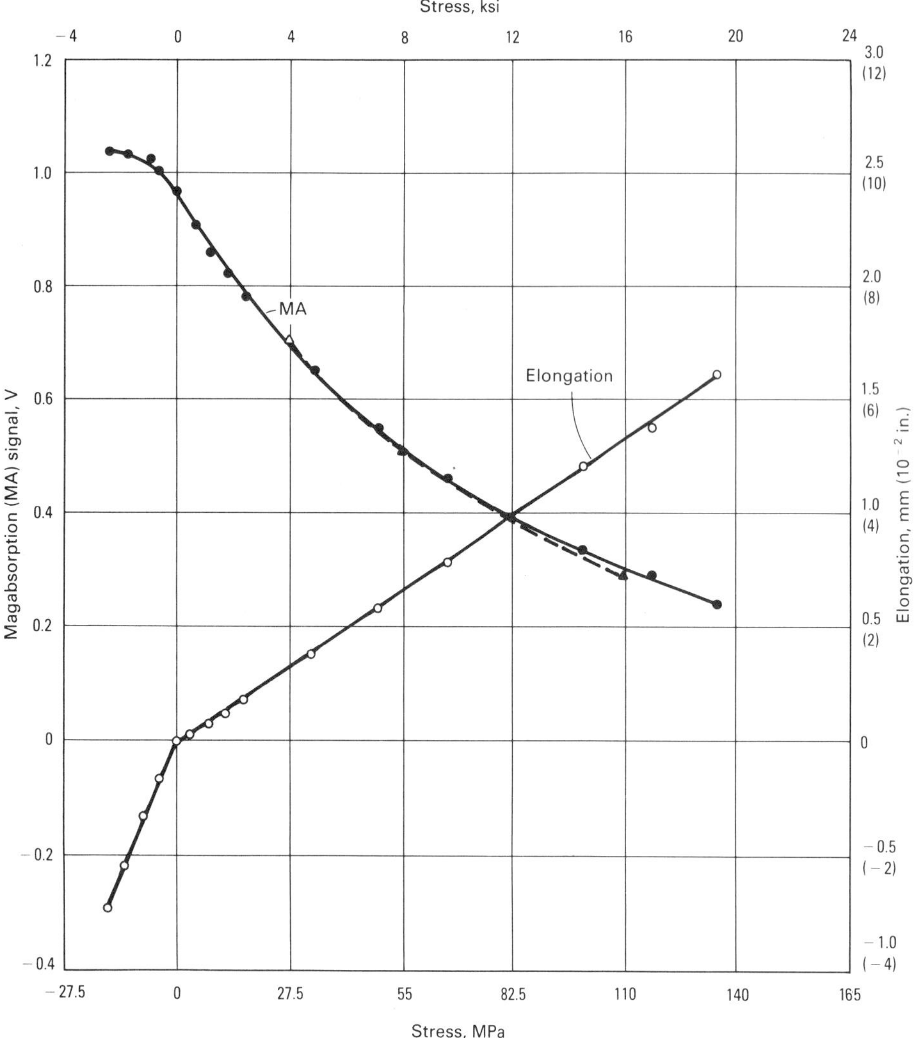

Fig. 26 Variations of the magabsorption signal and strain with stress for a nickel plated 3 mm (1/8 in.) diam aluminum rod

the surface of the nonferromagnetic material from the magabsorption signals of the ferromagnetic coating (Ref 6). This requires good adherence of the coating in order to reduce the distortion of strain transmitted from the base material to the coating. The strain transmitted to the coating may also exhibit additional distortion if the testpiece is not plane-stressed.

If the strains are assumed equal in the plating and in the base material, the stresses are related by:

$$\frac{\tau_1}{E_l} = \frac{\tau_2}{E_2} \qquad \text{(Eq 18)}$$

within the proportional limit where τ_i is the stress and E_i (where $i = 1,2$) is the modulus of elasticity for the two materials. In Eq 18, $i = 1$ refers to the substrate and $i = 2$ refers to the ferromagnetic coating. Thus, if the ratio of the moduli is known and T_2 is the stress in the ferromagnetic coating as measured by the magabsorption technique, then the stress at the surface of the substrate can be determined.

For the experimental work reported in Ref 6, aluminum welding rods were plated with nickel. As mentioned previously, it is important that the coating adheres well to the base material. It has been reported in the literature "that the bond strength of a nickel plating is of the order of the tensile strength of the base material when a phosphoric acid anodizing pretreatment process is employed." Studies of this anodizing process by the Southwest Research Institute have shown no blistering or separation in nickel-plated aluminum samples when subjected to a 180° bend of one thickness radius. Because the nickel plating adheres this well, the strain in the base metal was assumed to be the same as that in the plated material.

A loading device was used to apply the stress (tension) to the nickel-plated aluminum welding rods. Both tension and strain were measured. The marginal oscillator type of magabsorption detector was used to deliver the in-phase magabsorption signal. For each sample, the data were reduced so that strain and peak-to-peak magabsorption magnitude could be plotted as a function of stress. The graph of strain versus stress for a nickel-plated 3 mm (1/8 in.) diam aluminum rod and a graph of the peak-to-peak amplitude of the magabsorption signals from the nickel plating as a function of stress are given in Fig. 26. For this test, the plated aluminum sample is not annealed after plating. The dashed line in Fig. 26 is the empirical (curve-fitted) relationship between the magabsorption amplitude, A, and the applied stress (in psi) is given by the equation $A = 0.96 \exp(-\tau_1/1360)$.

Another plated rod was annealed for 30 min at 360 °C (680 °F). The graphs of strain versus stress and magabsorption as a function of stress are given in Fig. 27. The magabsorption amplitude, A, from the nickel is proportional directly to the applied stress (in psi) to the nickel plated aluminum rod within the range of $2400 \leqq \tau \leqq 10\,000$ psi and can be expressed as $A = -1.31 \times 10^{-4} (\tau - 16\,000)$. In the region of $10\,000 \leq \tau \leq 17\,000$ psi, the curve is fitted by the exponential $A = 3.21 \exp(-\tau/7150)$ as shown by the dashed line in Fig. 27.

These meager data show that with a calibration curve, the stress in nonferromagnetic materials can be measured using the magabsorption signals from a thin nickel plating on the nonferromagnetic material. The magabsorption versus stress graph for the nickel plated aluminum seems to obey nearly the same equation as does the nickel wire. To date, no measurements of the stress in bars of aluminum with small areas of nickel plating have been accomplished as yet, but are planned for the future.

Residual Stresses and Magnetism. Asymmetries in magabsorption signals may be indicative of residual stresses or magnetism. In the case of residual magnetism, the magnetic domains are not entirely haphazard; instead they do some ordering in a particular direction (Fig. 4f). This will produce asymmetries in the magnitude of the magabsorption signal, depending on the orientation of the bias field, H_B, with respect to the orientation of the residual magnetism.

Similarly, residual stresses are indicated by the ratio of magabsorption signals from two orientations (0°, 90°) of the bias coil. When the 0° and 90° amplitudes are equal, the stress is zero whatever the amplitudes are. When the parallel/perpendicular ratio is greater than 1, the stress value is positive and is tensile; when the ratio is less than 1, the stress is negative or compressive. For example, in one investigation, magabsorption measurements were made on steel samples before and after turning, cutting, and shaping operations. For the turning operation, the sample was reduced in diameter with a cutting tool; for the cutting operation, a sample was reduced in thickness by a ram shaper; for the shaping operation, the sample was reduced in thickness by an end mill. With the turning operation, the ratio of the 0° to the 90° magnitude of the magabsorption decreased from 0.91 to 0.89 when a 130 μm (5 mil) cut was made. When a 230 μm (9 mil) cut was made, the magabsorption ratio decreased from 0.89 to 0.65. When a 500 μm (20 mil) cut was made, the magabsorption ratio decreased from 0.65 to 0.60. These changes indicated that the turning operation was placing compressive stress on the testpiece. The testpiece used with the ram shaper was in tension along its length before being reduced in width. A reduction in width of 760 μm (30 mils) by the ram shaper applied perpendicular to the length caused the surface magabsorption signal ratio to indicate compression after the reduction. With the end mill, the reduction in thickness resulted in the stress changing from tensile to compressive.

Example 1: Magabsorption Measurement of Residual Stress in a Crankshaft Throw. Quantitative estimates of residual stress from magabsorption measurements were also performed on a large crankshaft throw (Fig. 28) made of 5046 steel. The estimates first required the development of calibration curves as described below.

The calibration curves were developed from two samples made of the same material as the crankshaft throw (type 5046 steel). The graph of the parallel-versus-perpendicular peak-to-peak values of the magabsorption signals from two of the calibration samples are given in Fig. 29. Two straight lines at angles of 45 and 50° relative to the horizontal axis are also drawn in Fig. 29. The one at 45° is a zero-stress line where the parallel and perpendicular magabsorption signals are the same magnitude. The line at 50° is the calibration line to be used to determine the calibration constant for the estimate of residual stress from the magabsorption measurements. Five stress levels (A, B, C, D, and E in Fig. 29) were applied at the measuring point on each test bar, and the calibration constant was determined as described below.

Previous experiments have indicated that the intersections of the parallel and perpendicular magnitudes for magabsorption signals at one point for the residual stresses seldom occur along the same line as applied stresses. However, it has been indicated that the applied stress lines in Fig. 29 probably can be used to determine the residual stress values in general by following the rule: All points on a radial line from the origin at some angle with respect to the abscissa have the same value of residual stress. The 45° line should be the locus of points of zero stress where the parallel and perpendicular values of the magabsorption curve are equal. With the 45° line as a reference, residual or applied stress can be expressed mathematically as:

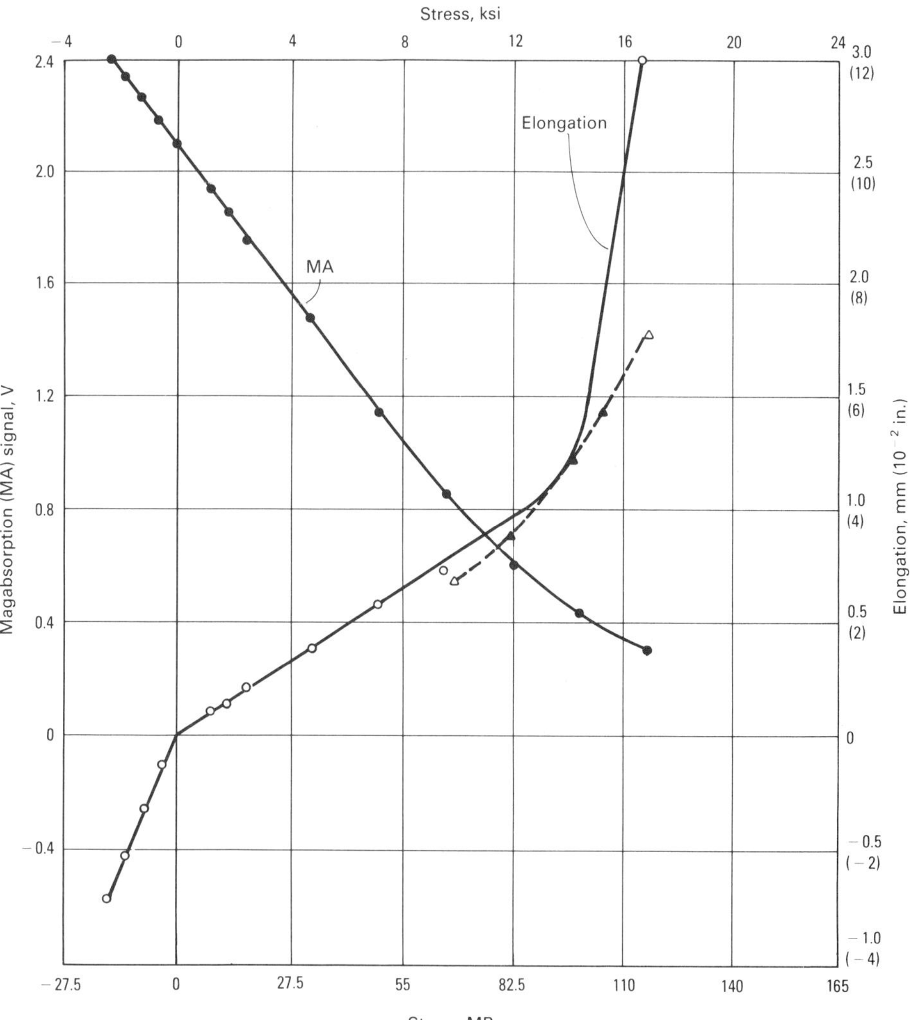

Fig. 27 Variations of the magabsorption signal and stress with strain for an annealed nickel plated 3 mm (1/8 in.) diam aluminum rod

$$\text{Stress (residual or applied)} = K\left[\tan^{-1}\left(\frac{\text{Parallel}}{\text{Perpendicular}}\right) - 45^\circ\right] \quad \text{(Eq 19)}$$

where K is the calibration constant for the material. When the parallel and perpendicular amplitudes are equal, the stress is zero whatever the amplitudes are. When the parallel/perpendicular ratio is greater than one, the stress value is positive and is tension; when the ratio is less than one, the stress is negative or compression.

The calibration constant, K, is obtained from Fig. 29 by the following procedure: (a) draw a calibration line through the origin at an angle for which an applied stress can be assigned to the intersection of the parallel/perpendicular ratio for the applied stress and (b), calculate the value of the constant K by inserting the applied stress and the angle into Eq 19.

Before this procedure could be used with the data in Fig. 29, the usable applied stress curve had to be chosen. One of the applied stress curves from sample No. 7 was chosen because its amplitude was the closest to the residual stress data, and because its curve shape was close to that obtained from the throw. The calibration proceeded as described. The line at 50° was chosen for the first calibration value because it intersected the lowest amplitude (dashed curve) for the data from sample 7 at nearly the value of applied stress level B (closed circle) (55 MPa, or 8 ksi). The 45° line passes through the No. 7 stress curve (dashed) nearly at level C (open triangle), or where an estimated value of 160 MPa (23.5 ksi) had been applied. Therefore, the line at an angle of 50° represents a differential applied stress relative to the 45° line of 160 − 55 = 105 MPa (23.5 − 8 = 15.5 ksi). When several values are taken from several lines both above and below the 45° line, the average value for a line 5° above or below 45° is 100 MPa (15 ksi). Therefore, if both the value of $\tan^{-1}$ 50° and the stress equal to 100 MPa (15 ksi) are inserted into Eq 19, the value of K is found to be 20 MPa/degree (3 ksi/degree). This implies that for every degree

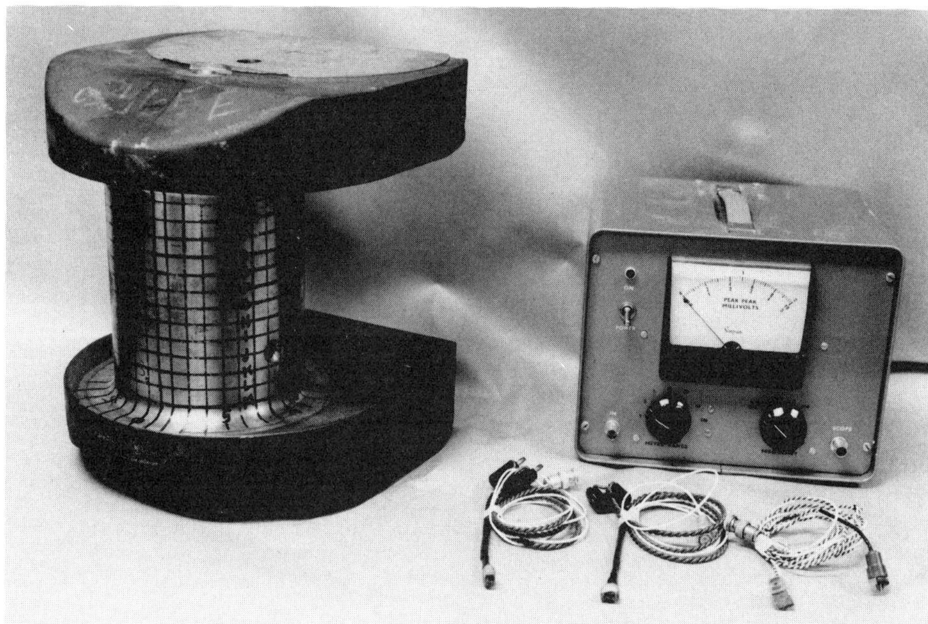

Fig. 28 Magabsorption detector and three detector heads used to perform measurements on the throw of the crankshaft shown on the left. A closeup view of the detector heads is shown in Fig. 20.

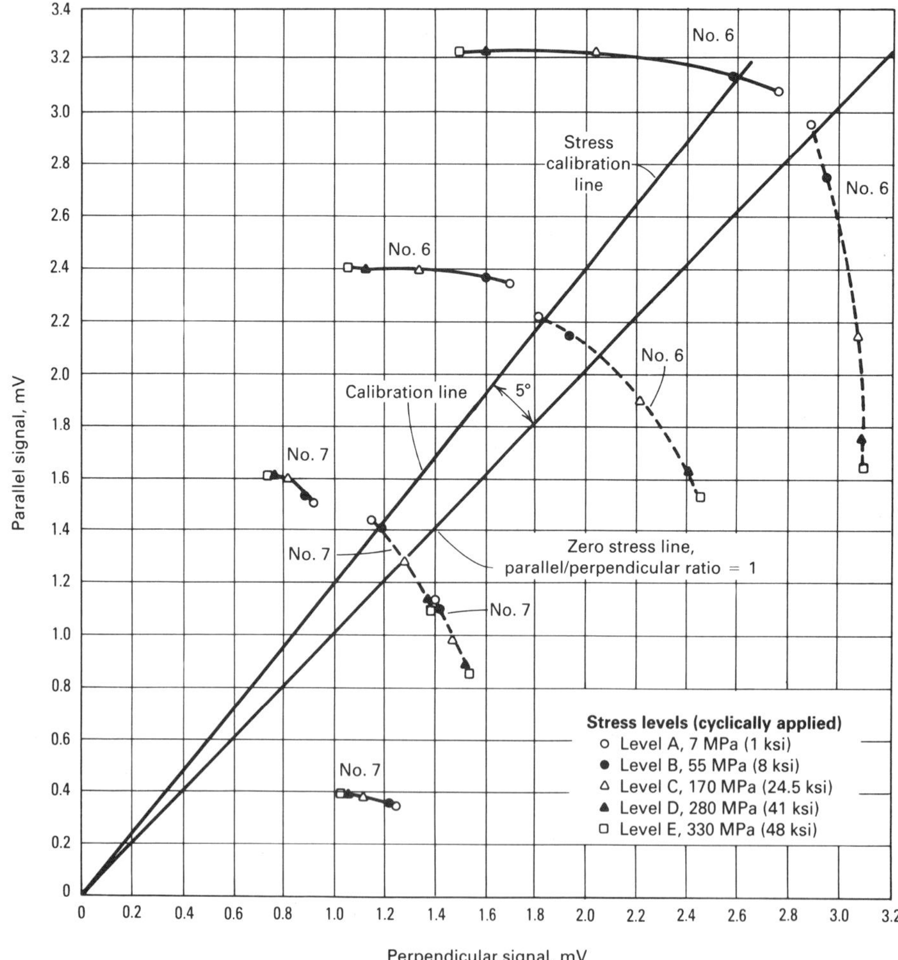

Fig. 29 Graph showing the plot of the parallel/perpendicular ratios for sample 6 (type 5046 steel) and sample 7 (type 5046 steel)

of offset from the 45° line, points along the line at that offset resulting from signal amplitude measurements will have the same value of residual stress.

When the value of K is used, calibration lines can be drawn through zero at useful angles relative to the 45° line. Using these calibration lines and marks, the values of the residual stresses for points on the crankshaft throw were determined. Residual stress values as high as 600 MPa (87 ksi) tension and 90 MPa (13 ksi) compression were determined.

REFERENCES

1. W.E. Bell, *Magnetoabsorption*, Vol 2, Proceedings of the Conference on Magnetism and Magnetic Materials, American Institute of Physics, 1956, p 305
2. R.M. Bozorth, Magnetism and Electrical Properties, in *Ferromagnetism*, D. Van Nostrand, 1951, p 745-768
3. W.L. Rollwitz, "Magnetoabsorption," Final Report, Research Project No. 712-4, Southwest Research Institute, 1958
4. W.L. Rollwitz and A.W. Whitney, "Special Techniques for Measuring Material Properties," Technical Report ASD-TDR-64-123, USAF Contract No. AF-33(657)-10326, Air Force Materials Laboratory, 1964
5. W.L. Rollwitz and J.P. Classen, "Magnetoabsorption Techniques for Measuring Material Properties," Technical Report AFML-TR-65-17, USAF Contract No. AF-33(657)-10326, Air Force Materials Laboratory, 1965
6. W.L. Rollwitz and J.P. Classen, "Magnetoabsorption Techniques for Measuring Material Properties," Technical Report AFML-TR-66-76 (Part I), USAF Contract No. AF-33(657)-10326, Air Force Materials Laboratory, 1966
7. W.L. Rollwitz, "Magnetoabsorption Techniques for Measuring Material Properties. Part II. Measurements of Residual and Applied Stress." Technical Report AFML-TR-66-76 (Part II), USAF Contract No. AF-33(615)-5068, Air Force Materials Laboratory, 1968
8. W.L. Rollwitz, Magnetoabsorption, *Progress in Applied Materials Research*, Vol 6, E.G. Stanford, J.H. Fearon, and W.J. McGonnagle, Ed., Heywood, 1964
9. W.L. Rollwitz, Sensing Apparatus for Use With Magnetoabsorption Apparatus, U.S. Patent 3,612,968, 1971
10. W.L. Rollwitz, J. Arambula, and J. Classen, Method of Determining Stress in Ferromagnetic Members Using Magnetoabsorption, 1974
11. W.L. Rollwitz, "Preliminary Magnetoabsorption Measurements of Stress in a Crankshaft Throw," Summary Report on Project 15-2438, Southwest Research Institute, 1970

Electromagnetic Techniques for Residual Stress Measurements

H. Kwun and G.L. Burkhardt, Southwest Research Institute

RESIDUAL STRESSES in materials can be nondestructively measured by a variety of methods, including x-ray diffraction, ultrasonics, and electromagnetics (Ref 1-3). With the x-ray diffraction technique, the interatomic planar distance is measured, and the corresponding stress is calculated (Ref 4). The penetration depth of x-rays is of the order of only 10 μm (400 μin.) in metals. Therefore, the technique is limited to measurements of surface stresses. Its use has been generally limited to the laboratory because of the lack of field-usable equipment and concern with radiation safety.

With ultrasonic techniques, the velocity of the ultrasonic waves in materials is measured and related to stress (Ref 5). These techniques rely on a small velocity change caused by the presence of stress, which is known as the acoustoelastic effect (Ref 6). In principle, ultrasonic techniques can be used to measure bulk as well as surface stresses. Because of the difficulty in differentiating stress effects from the effect of material texture, practical ultrasonic applications have not yet materialized.

With electromagnetic techniques, one or more of the magnetic properties of a material (such as permeability, magnetostriction, hysteresis, coercive force, or magnetic domain wall motion during magnetization) are sensed and correlated to stress. These techniques rely on the change in magnetic properties of the material caused by stress; this is known as the magnetoelastic effect (Ref 7). These techniques, therefore, apply only to ferromagnetic materials, such as steel.

Of the many electromagnetic stress-measurement techniques, this article deals with three specific ones: Barkhausen noise, nonlinear harmonics, and magnetically induced velocity changes. The principles, instrumentation, stress dependence, and capabilities and limitations of these three techniques are described in the following sections.

Barkhausen Noise

The magnetic flux density in a ferromagnetic material subjected to a time-varying magnetic field does not change in a strictly continuous way, but rather by small, abrupt, discontinuous increments called Barkhausen jumps (after the name of the researcher who first observed this phenomenon), as illustrated in Fig. 1. The jumps are due primarily to discontinuous movements of boundaries between small magnetically saturated regions called magnetic domains in the material (Ref 7-9). An unmagnetized macroscopic specimen consists of a great number of domains with random magnetic direction so that the average bulk magnetization is zero. Under an external magnetic field, the specimen becomes magnetized mainly by the growth of volume of domains oriented close to the direction of the applied field, at the expense of domains unfavorably oriented. The principal mechanism of growth is the movement of the walls between adjacent domains. Because of the magnetoelastic interaction, the direction and magnitude of the mechanical stress strongly influence the distribution of domains and the dynamics of the domain wall motion and therefore the behavior of Barkhausen jumps (Ref 8). This influence, in turn, is used for stress measurements. Because the signal produced by Barkhausen

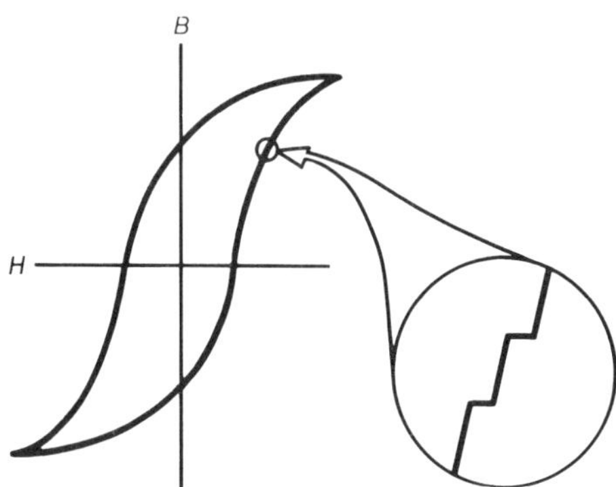

Fig. 1 Hysteresis loop for magnetic material showing discontinuities that produce Barkhausen noise. Source: Ref 2

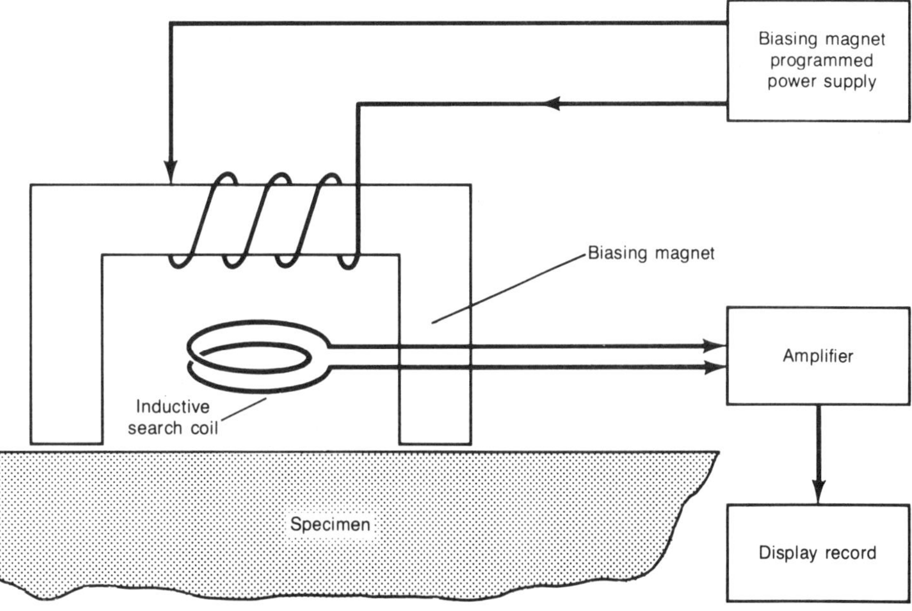

Fig. 2 Arrangement for sensing the Barkhausen effect

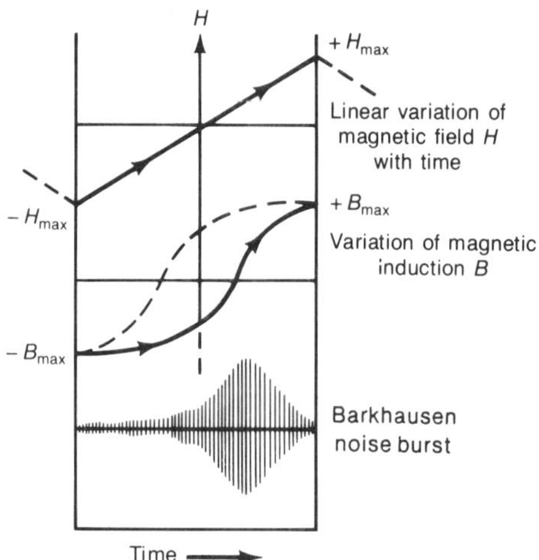

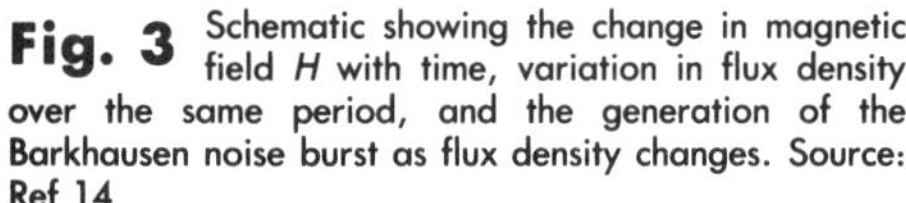

Fig. 3 Schematic showing the change in magnetic field H with time, variation in flux density over the same period, and the generation of the Barkhausen noise burst as flux density changes. Source: Ref 14

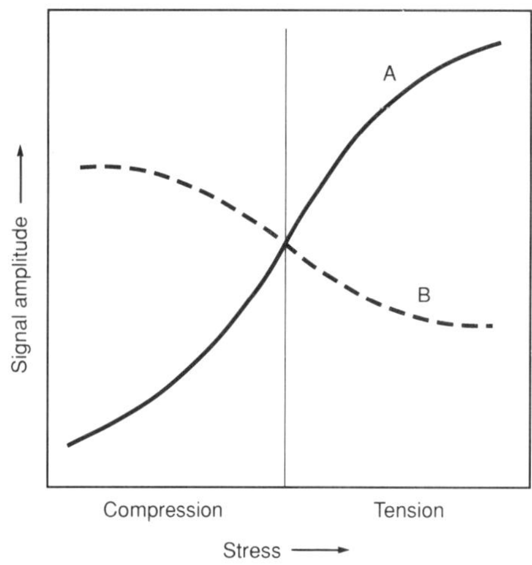

Fig. 4 Typical stress dependence of Barkhausen noise signal amplitude with the applied magnetic field parallel (curve A) and perpendicular (curve B) to the stress direction

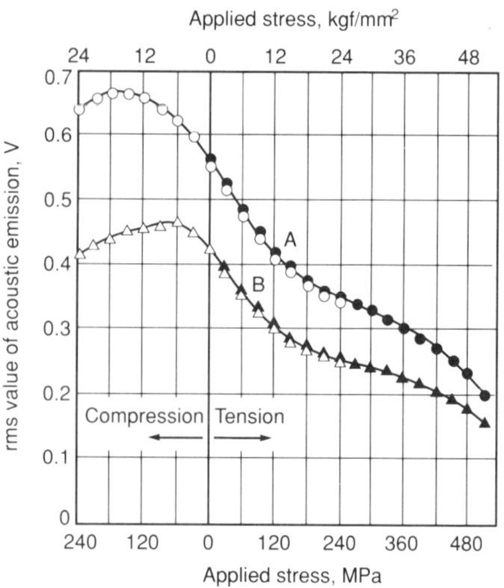

Fig. 5 Dependence of acoustic emission during the magnetization of low-carbon steel on stress. Total gain: 80 dB. Magnetic field strength (rms value): 13 000 A/m (160 Oe) for curve A; 6400 A/m (80 Oe) for curve B. Source: Ref 13

jumps resembles noise, the term Barkhausen noise is often used.

Instrumentation. The arrangement illustrated in Fig. 2 is used for the Barkhausen noise technique (Ref 8). A small C-shaped electromagnet is used to apply a controlled, time-varying magnetic field to the specimen. The abrupt movements of the magnetic domains are typically detected with an inductive coil placed on the specimen. The detected signal is a burst of noiselike pulses, as illustrated in Fig. 3. Certain features of the signal, such as the maximum amplitude or root mean square (rms) amplitude of the Barkhausen noise burst or the applied magnetic field strength at which the maximum amplitude occurs, are used to determine the stress state in the material (Ref 1, 8-12).

In addition to inductive sensing of the magnetic Barkhausen noise, magnetoacoustic Barkhausen activity can also be detected with an acoustic emission sensor (Ref 13). This phenomenon occurs when Barkhausen jumps during the magnetization of a specimen produce mechanical stress pulses in a manner similar to the inductive Barkhausen noise burst shown in Fig. 3. It is caused by microscopic changes in strain due to magnetostriction when discontinuous, irreversible domain wall motion of non-180° domain walls occurs (Ref 14, 15). This acoustic Barkhausen noise is also dependent on the stress state in the material and can therefore be used for stress measurements (Ref 15-19).

Stress Dependence. The magnetic Barkhausen effect is dependent on the stress as well as the relative direction of the applied magnetic field to the stress direction. To illustrate this, Fig. 4 shows a typical stress dependence of the inductively detected Barkhausen noise in a ferrous material. In the case where the magnetic field and the stress are parallel, the Barkhausen amplitude increases with tension and decreases with compression (Ref 8, 10, 20, 21). In the case where the two are perpendicular, the opposite result is obtained. The behavior shown in Fig. 4 holds for materials with a positive magnetostriction coefficient; for materials with a negative magnetostriction, the Barkhausen amplitude exhibits the opposite behavior.

For a given stress, the dependence of the Barkhausen amplitude on the angle between the magnetic field and stress directions is proportional to the strain produced by the stress (Ref 21). Because the Barkhausen noise is dependent on the strain, Barkhausen measurements can be used as an alternative to strain gages (Ref 20, 21).

A typical stress dependence of the acoustic Barkhausen noise is illustrated in Fig. 5, in which the magnetic field is applied parallel to the stress direction. As shown, the amplitude of the acoustic signal decreases with tension. Under compression, it increases slightly and then decreases with an increasing stress level. The acoustic Barkhausen noise, therefore, cannot distinguish tension from compression.

Capabilities and Limitations. Because of the eddy current screening, the inductively detected Barkhausen noise signals reflect the activity occurring very near the surface of the specimen to a depth of approximately 0.1 mm (0.004 in.). Therefore, the Barkhausen noise technique is suitable for measuring near-surface stresses. The effective stress-measurement range is up to about 50% of the yield stress of the material because the change in the Barkhausen noise with stress becomes saturated at these high stress levels.

Barkhausen measurements can usually be made within a few seconds. Continuous measurements at a slow scanning speed (~10 mm/s, or 0.4 in./s) are possible. Preparation of the surface of a part under testing is generally not required. Portable, field-usable Barkhausen instruments are available.

The results of Barkhausen noise measurements are also sensitive to factors not related to stress, such as microstructure, heat treatment, and material variations. Careful instrument calibration and data analysis are essential for reliable stress measurements. As can be seen in Fig. 4, the Barkhausen amplitude at zero stress is approximately isotropic and shows no dependence on the relative orientation of the magnetic field and stress directions. When the specimen is subjected to a stress, the Barkhausen noise exhibits dependence on the magnetic field direction and becomes anisotropic. This stress-induced anisotropy in the Barkhausen noise is effective for differentiating stress from nonstress-related factors whose effects are approximately isotropic. The accuracy of the technique is about ±35 MPa (±5 ksi).

The acoustic Barkhausen noise technique can be used in principle to measure bulk stresses in materials because the acoustic waves travel through materials. However, practical application of this technique is currently hampered by the difficulty in differentiating acoustic Barkhausen noise from other noise produced from surrounding environments.

The inductive Barkhausen noise technique has been used for measuring welding residual stresses (Ref 20, 22), for detecting grinding damage in bearing races (Ref 23), and for measuring compressive hoop stresses in railroad wheels (Ref 24).

Nonlinear Harmonics

Because of the magnetic hysteresis and nonlinear permeability, the magnetic induc-

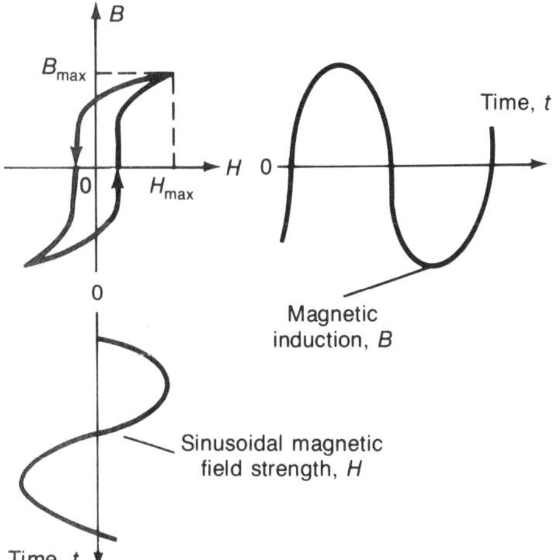

Fig. 6 Distortion of magnetic induction caused by hysteresis and nonlinearity in magnetization curve. The curve for magnetic induction, *B*, is not a pure sinusoid; it has more rounded peaks.

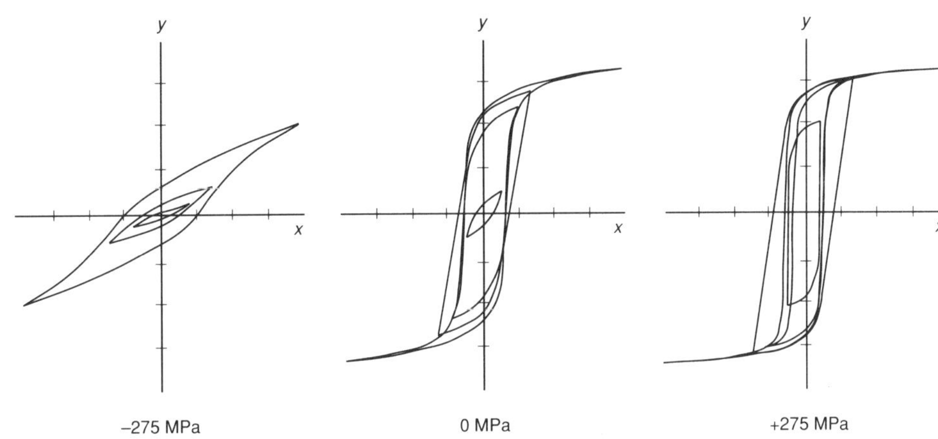

Fig. 7 Hysteresis loops of an AISI 410 stainless steel specimen having ASTM No. 1 grain size and a hardness of 24 HRC under various levels of uniaxial stress. *x*-axis: 1600 A/m (20 Oe) per division; *y*-axis: 0.5 T (5 kG) per division. Source: Ref 25

tion, *B*, of a ferromagnetic material subjected to a sinusoidal, external magnetic field, *H*, is not sinusoidal but distorted, as illustrated in Fig. 6. This distorted waveform of the magnetic induction contains odd harmonic frequencies of the applied magnetic field. Mechanical stresses greatly influence the magnetic hysteresis and permeability of the material (Ref 7). An example of the stress effects on the hysteresis loops is shown in Fig. 7. Accordingly, the harmonic content of the magnetic induction is also sensitive to the stress state in the material. With the nonlinear harmonics technique, these harmonic frequencies are detected, and their amplitudes are related to the state of stress in the material (Ref 26, 27).

Instrumentation. The nonlinear harmonics technique is implemented with the arrangement shown schematically in Fig. 8. The magnetic field is applied to a specimen with an excitation coil, and the resulting magnetic induction is measured with a sensing coil. A sinusoidal current of a given frequency is supplied to the excitation coil with a function generator (or oscillator) and a power amplifier. The induced voltage in the sensing coil is amplified, and the harmonic frequency content of the signal is analyzed. The amplitude of the harmonic frequency, typically the third harmonics, is used to determine the stress.

Stress Dependence. The harmonic amplitudes are dependent on the stress as well as the relative orientation between the stress and the applied magnetic field directions. Like the stress dependence of the Barkhausen noise amplitude illustrated in Fig. 4, the harmonic amplitude for materials with a positive magnetostriction increases with tension when the direction of the stress and the applied field are parallel (Ref 27). When the directions are perpendicular, the opposite result is obtained. As with Barkhausen noise, the nonlinear harmonics depend on strain and can be used to determine stress.

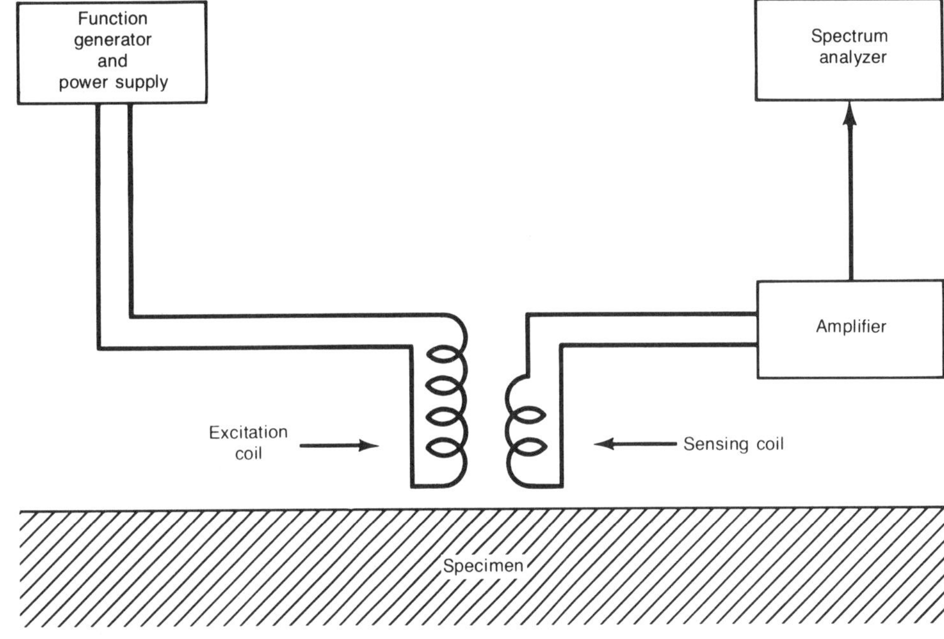

Fig. 8 Block diagram of nonlinear harmonics instrumentation

Capabilities and Limitations. The nonlinear harmonics technique can be used to measure near-surface stresses, with sensing depth approximately equal to the skin depth of the applied magnetic field. Because the skin depth is a function of the frequency of the applied magnetic field, the depth of sensing can be changed by varying the frequency. Therefore, the technique can potentially be used to measure stress variations with depth.

The results of nonlinear harmonic measurements are sensitive to factors not related to stress, such as microstructure, heat treatment, and material variations. The stress-induced anisotropy in the harmonic amplitude has been shown to be effective for differentiating stress from factors not related to stress (Ref 27). When the stress-induced anisotropy is used for stress determination, the accuracy of the technique is about ±35 MPa (±5 ksi). The range of stress to which the technique is effective is up to about 50% of the yield stress of the material, with the response becoming saturated at higher stress levels. With this technique, it would be feasible to measure stress while scanning a part at a high speed (~10 m/s, or 30 ft/s); therefore, this technique has potential for rapidly surveying stress states in pipelines or continuously welded railroad rails (Ref 28).

Magnetically Induced Velocity Changes (MIVC) for Ultrasonic Waves

Because of the magnetoelastic interaction, the elastic moduli of a ferromagnetic

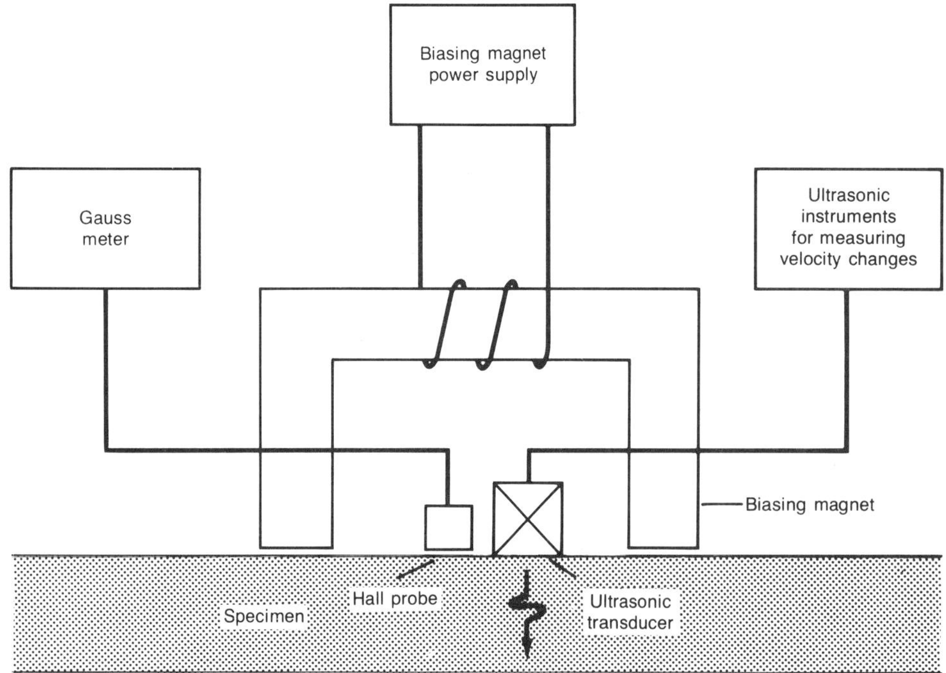

Fig. 9 Block diagram of instrumentation for measuring MIVC for ultrasonic waves

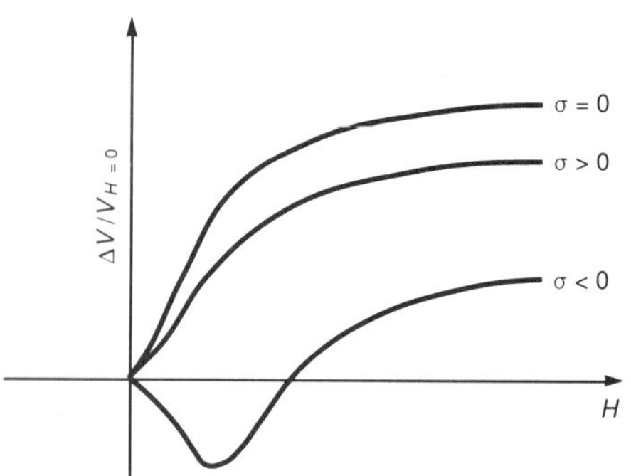

Fig. 10 Schematic showing the change in ultrasonic velocity, ΔV, with magnetic field H under various stress levels, σ

material are dependent on the magnetization of the material. This phenomenon is known as the ΔE effect (Ref 7). Consequently, the velocity of the ultrasonic waves in the material changes when an external magnetic field is applied to the material. This MIVC for ultrasonic waves is characteristically dependent on the stress as well as the angle between the stress direction and the direction of the applied magnetic field (Ref 29-32). This characteristic stress dependence of the MIVC is used for stress determination (Ref 33-36).

Instrumentation. Figure 9 shows a block diagram of instrumentation for measuring MIVC. An electromagnet is used to apply a biasing magnetic field to the specimen. The applied magnetic field is measured with a Hall probe. An ultrasonic transducer is used to transmit ultrasonic waves and to detect signals reflected from the back surface of the specimen. For surface waves, separate transmitting and receiving transducers are used. The shift in the arrival time of the received ultrasonic wave caused by the velocity change due to the applied magnetic field is detected with an ultrasonic instrument. Because MIVC is a small effect (of the order of only 0.01 to 0.1%), the measurements are typically made using the interferometer principle, called the phase comparison technique, in the ultrasonic instrumentation (Ref 33).

Stress Dependence. A typical stress dependence of the MIVC is illustrated in Fig. 10. At zero stress, the MIVC at first generally increases rapidly with the applied magnetic field, *H*, and then gradually levels off toward a saturation value. When the material is subjected to stress, the magnitude of the MIVC decreases, and the shape of the MIVC curve as a function of *H* changes. Under tension, the shape of the MIVC curve remains similar to that at zero stress, but with reduced magnitude in proportion to the stress level. Under compression, the MIVC curve exhibits a minimum, which drastically changes the shape and reduces the magnitude. The magnitude of the minimum and the value of *H* where the minimum occurs increase with stress level.

The detailed stress dependence of MIVC, however, varies with the mode of ultrasonic wave used (longitudinal, shear, or surface) and with the relative orientation between the stress and the magnetic field directions (Ref 29-35). The stress dependence shown in Fig. 10 holds for longitudinal waves in materials with a positive magnetostriction coefficient (Ref 31, 33). With this technique, the stress state in the material, including the magnitude, direction, and sign (tensile or compressive) of the stress, is characterized by analyzing the shape and magnitude of the MIVC curves measured at two or more different magnetic field directions.

Capabilities and Limitations. The MIVC technique can be used to measure bulk and surface stresses by applying both bulk (shear or longitudinal) and surface ultrasonic waves. A measurement can be made within a few seconds. Because the magnitude of MIVC depends on material type, reference or calibration curves must be established for that material type prior to stress measurements. However, this technique is insensitive to variations in the texture and composition of nominally the same material. The accuracy in stress measurements is about ±35 MPa (±5 ksi). This technique has been used to measure residual welding stresses (Ref 34) and residual hoop stresses in railroad wheels (Ref 36).

A relatively large electromagnet is needed to magnetize the part under investigation and may be cumbersome to handle in practical applications. Because of difficulty in magnetizing complex-geometry parts, the application of the technique is limited to simple geometry parts.

REFERENCES

1. M.R. James and O. Buck, Quantitative Nondestructive Measurements of Residual Stresses, *CRC Crit. Rev. Solid State Mater. Sci.*, Vol 9, 1980, p 61
2. C.O. Ruud, "Review and Evaluation of Nondestructive Methods for Residual Stress Measurement," Final Report, NP-1971, Project 1395-5, Electric Power Research Institute, Sept 1981
3. W.B. Young, Ed., *Residual Stress in Design, Process, and Material Selection*, Proceedings of the ASM Conference on Residual Stress in Design, Process, and Materials Selection, Cincinnati, OH, April 1987, ASM INTERNATIONAL, 1987
4. M.R. James and J.B. Cohen, The Measurement of Residual Stresses by X-Ray Diffraction Techniques, in *Treatise on Materials Science and Technology—Experimental Methods*, Vol 19A, H. Herman, Ed., Academic Press, 1980, p 1
5. Y.H. Pao, W. Sachse, and H. Fukuoka, Acoustoelasticity and Ultrasonic Measurement of Residual Stresses, in *Physical Acoustics: Principles and Methods*, Vol XVII, W.P. Mason and R.M. Thurston, Ed., Academic Press, 1984, p 61-143
6. D.S. Hughes and J.L. Kelly, Second-Order Elastic Deformation of Solids, *Phys. Rev.*, Vol 92, 1953, p 1145
7. R.M. Bozorth, *Ferromagnetism*, Van Nostrand, 1951

8. G.A. Matzkanin, R.E. Beissner, and C.M. Teller, "The Barkhausen Effect and Its Applications to Nondestructive Evaluation," State of the Art Report, NTIAC-79-2, Nondestructive Testing Information Analysis Center, Southwest Research Institute, Oct 1979
9. J.C. McClure, Jr., and K. Schroder, The Magnetic Barkhausen Effect, *CRC Crit. Rev. Solid State Sci.*, Vol 6, 1976, p 45
10. R.L. Pasley, Barkhausen Effect—An Indication of Stress, *Mater. Eval.*, Vol 28, 1970, p 157
11. S. Tiitto, On the Influence of Microstructure on Magnetization Transitions in Steel, *Acta Polytech. Scand.*, No. 119, 1977
12. R. Rautioaho, P. Karjalainen, and M. Moilanen, Stress Response of Barkhausen Noise and Coercive Force in 9Ni Steel, *J. Magn. Magn. Mater.*, Vol 68, 1987, p 321
13. H. Kusanagi, H. Kimura, and H. Sasaki, Acoustic Emission Characteristics During Magnetization of Ferromagnetic Materials, *J. Appl. Phys.*, Vol 50, 1979, p 2985
14. D.C. Jiles, Review of Magnetic Methods for Nondestructive Evaluation, *NDT Int.*, Vol 21 (No. 5), 1988, p 311-319
15. K. Ono and M. Shibata, Magnetomechanical Acoustic Emission of Iron and Steels, *Mater. Eval.*, Vol 38, 1980, p 55
16. M. Shibata and K. Ono, Magnetomechanical Acoustic Emission—A New Method for Nondestructive Stress Measurement, *NDT Int.*, Vol 14, 1981, p 227
17. K. Ono, M. Shibata, and M.M. Kwan, Determination of Residual Stress by Magnetomechanical Acoustic, in *Residual Stress for Designers and Metallurgists*, L.J. Van de Walls, Ed., American Society for Metals, 1981
18. G.L. Burkhardt, R.E. Beissner, G.A. Matzhanin, and J.D. King, Acoustic Methods for Obtaining Barkhausen Noise Stress Measurements, *Mater. Eval.*, Vol 40, 1982, p 669
19. K. Ono, "Magnetomechanical Acoustic Emission—A Review," Technical Report TR-86-02, University of California at Los Angeles, Sept 1986
20. G.L. Burkhardt and H. Kwun, "Residual Stress Measurement Using the Barkhausen Noise Method," Paper 45, presented at the 15th Educational Seminar for Energy Industries, Southwest Research Institute, April 1988
21. H. Kwun, Investigation of the Dependence of Barkhausen Noise on Stress and the Angle Between the Stress and Magnetization Directions, *J. Magn. Magn. Mater.*, Vol 49, 1985, p 235
22. L.P. Karjalainen, M. Moilanen, and R. Rautioaho, Evaluating the Residual Stresses in Welding From Barkhausen Noise Measurements, *Materialprüfung*, Vol 22, 1980, p 196
23. J.R. Barton and F.M. Kusenberger, "Residual Stresses in Gas Turbine Engine Components From Barkhausen Noise Analysis," Paper 74-GT-51, presented at the ASME Gas Turbine Conference, Zurich, Switzerland, American Society of Mechanical Engineers, 1974
24. J.R. Barton, W.D. Perry, R.K. Swanson, H.V. Hsu, and S.R. Ditmeyer, Heat-Discolored Wheels: Safe to Reuse?, *Prog. Railroad.*, Vol 28 (No. 3), 1985, p 44
25. H. Kwun and G.L. Burkhardt, Effects of Grain Size, Hardness, and Stress on the Magnetic Hysteresis Loops of Ferromagnetic Steels, *J. Appl. Phys.*, Vol 61, 1987, p 1576
26. N. Davis, Magnetic Flux Analysis Techniques, in *Research Techniques in Nondestructive Testing*, Vol II, R.S. Sharpe, Ed., Academic Press, 1973, p 121
27. H. Kwun and G.L. Burkhardt, Nondestructive Measurement of Stress in Ferromagnetic Steels Using Harmonic Analysis of Induced Voltage, *NDT Int.*, Vol 20, 1987, p 167
28. G.L. Burkhardt and H. Kwun, Application of the Nonlinear Harmonics Method to Continuous Measurement of Stress in Railroad Rail, in *Proceedings of the 1987 Review of Progress in Quantitative Nondestructive Evaluation*, Vol 7B, D.O. Thompson and D.E. Chimenti, Ed., Plenum Press, 1988, p 1413
29. H. Kwun and C.M. Teller, Tensile Stress Dependence of Magnetically Induced Ultrasonic Shear Wave Velocity Change in Polycrystalline A-36 Steel, *Appl. Phys. Lett.*, Vol 41, 1982, p 144
30. H. Kwun and C.M. Teller, Stress Dependence of Magnetically Induced Ultrasonic Shear Wave Velocity Change in Polycrystalline A-36 Steel, *J. Appl. Phys.*, Vol 54, 1983, p 4856
31. H. Kwun, Effects of Stress on Magnetically Induced Velocity Changes for Ultrasonic Longitudinal Waves in Steels, *J. Appl. Phys.*, Vol 57, 1985, p 1555
32. H. Kwun, Effects of Stress on Magnetically Induced Velocity Changes for Surface Waves in Steels, *J. Appl. Phys.*, Vol 58, 1985, p 3921
33. H. Kwun, Measurement of Stress in Steels Using Magnetically Induced Velocity Changes for Ultrasonic Waves, in *Nondestructive Characterization of Materials II*, J.F. Bussiere, J.P. Monchalin, C.O. Ruud, and R.E. Green, Jr., Ed., Plenum Press, 1987, p 633
34. H. Kwun, A Nondestructive Measurement of Residual Bulk Stresses in Welded Steel Specimens by Use of Magnetically Induced Velocity Changes for Ultrasonic Waves, *Mater. Eval.*, Vol 44, 1986, p 1560
35. M. Namkung and J.S. Heyman, Residual Stress Characterization With an Ultrasonic/Magnetic Technique, *Nondestr. Test. Commun.*, Vol 1, 1984, p 175
36. M. Namkung and D. Utrata, Nondestructive Residual Stress Measurements in Railroad Wheels Using the Low-Field Magnetoacoustic Test Method, in *Proceedings of the 1987 Review of Progress in Quantitative Nondestructive Evaluation*, Vol 7B, D.O. Thompson and D.E. Chimenti, Ed., Plenum Press, 1988, p 1429

Eddy Current Inspection

Revised by the ASM Committee on Eddy Current Inspection*

EDDY CURRENT INSPECTION is based on the principles of electromagnetic induction and is used to identify or differentiate among a wide variety of physical, structural, and metallurgical conditions in electrically conductive ferromagnetic and nonferromagnetic metals and metal parts. Eddy current inspection can be used to:

- Measure or identify such conditions and properties as electrical conductivity, magnetic permeability, grain size, heat treatment condition, hardness, and physical dimensions
- Detect seams, laps, cracks, voids, and inclusions
- Sort dissimilar metals and detect differences in their composition, microstructure, and other properties
- Measure the thickness of a nonconductive coating on a conductive metal, or the thickness of a nonmagnetic metal coating on a magnetic metal

Because eddy currents are created using an electromagnetic induction technique, the inspection method does not require direct electrical contact with the part being inspected. The eddy current method is adaptable to high-speed inspection and, because it is nondestructive, can be used to inspect an entire production output if desired. The method is based on indirect measurement, and the correlation between the instrument readings and the structural characteristics and serviceability of the parts being inspected must be carefully and repeatedly established.

Advantages and Limitations of Eddy Current Inspection

Eddy current inspection is extremely versatile, which is both an advantage and a disadvantage. The advantage is that the method can be applied to many inspection problems provided the physical requirements of the material are compatible with the inspection method. In many applications, however, the sensitivity of the method to the many properties and characteristics inherent within a material can be a disadvantage; some variables in a material that are not important in terms of material or part serviceability may cause instrument signals that mask critical variables or are mistakenly interpreted to be caused by critical variables.

Eddy Current Versus Magnetic Inspection Methods. In eddy current inspection, the eddy currents create their own electromagnetic field, which can be sensed either through the effects of the field on the primary exciting coil or by means of an independent sensor. In nonferromagnetic materials, the secondary electromagnetic field is derived exclusively from eddy currents. However, with ferromagnetic materials, additional magnetic effects occur that are usually of sufficient magnitude to overshadow the field effects caused by the induced eddy currents. Although undesirable, these additional magnetic effects result from the magnetic permeability of the material being inspected and can normally be eliminated by magnetizing the material to saturation in a static (direct current) magnetic field. When the permeability effect is not eliminated, the inspection method is more correctly categorized as electromagnetic or magnetoinductive inspection. Methods of inspection that depend mainly on ferromagnetic effects are discussed in the article "Magnetic Particle Inspection" in this Volume.

Development of the Inspection Process

The development of the eddy current method of inspection has involved the use of several scientific and technological advances, including the following:

- Electromagnetic induction
- Theory and application of induction coils
- The solution of boundary-value problems describing the dynamics of the electromagnetic fields within the vicinity of induction coils, and especially the dynamics of the electromagnetic fields, electric current flow, and skin effect in conductors in the vicinity of such coils
- Theoretical prediction of the change in impedance of eddy current inspection coils caused by small flaws
- Improved instrumentation resulting from the development of vacuum tubes, semiconductors, integrated circuits, and microprocessors which led to better measurement techniques and response to subtle changes in the flow of eddy currents in metals
- Metallurgy and metals fabrication
- Improved instrumentation, signal display, and recording

Electromagnetic induction was discovered by Faraday in 1831. He found that when the current in a loop of wire was caused to vary (as by connecting or disconnecting a battery furnishing the current), an electric current was induced in a second, adjacent loop. This is the effect used in eddy current inspection to cause the eddy currents to flow in the material being inspected and it is the effect used to monitor these currents.

In 1864, Maxwell presented his classical dissertation on a dynamic theory of the electromagnetic field, which includes a set of equations bearing his name that describe all large-scale electromagnetic phenomena. These phenomena include the generation and flow of eddy currents in conductors and the associated electromagnetic fields. Thus, all the electromagnetic induction effects that are basic to the eddy current inspection method are described in principle by the equations devised by Maxwell for particular boundary values for practical applications.

In 1879, Hughes, using an eddy current method, detected differences in electrical conductivity, magnetic permeability, and temperature in metal. However, use of the eddy current method developed slowly, probably because such an inspection method was not needed and because further development of the electrical theory was necessary before it could be used for practical applications.

Calculating the flow of induced current in metals was later developed by the solution of Maxwell's equations for specific boundary

*V.S. Cecco, Atomic Energy of Canada Limited, Chalk River Nuclear Laboratories; E.M. Franklin, Argonne National Laboratory, Argonne-West; Howard E. Houserman, ZETEC, Inc.; Thomas G. Kincaid, Boston University; James Pellicer, Staveley NDT Technologies, Inc.; and Donald Hagemaier, Douglas Aircraft Company, McDonnell Douglas Corporation

conditions for symmetrical configurations. These mathematical techniques were important in the electric power generation and transmission industry, in induction heating, and in the eddy current method of inspection.

An eddy current instrument for measuring wall thickness was developed by Kranz in the mid-1920s. An example of early well-documented work that also serves as an introduction to several facets of the eddy current inspection method is that of Farrow, who pioneered in the development of eddy current systems for the inspection of welded steel tubing. He began his work in 1930 and by 1935 had progressed to an inspection system that included a separate primary energizing coil, differential secondary detector coil, and a dc magnetic-saturating solenoid coil. Inspection frequencies used were 500, 1000, and 4000 Hz. Tubing diameters ranged from 6.4 to 85 mm (¼ to 3⅓ in.). The inspection system also included a balancing network, a high-frequency amplifiers, a frequency discriminator-demodulator, a low-frequency pulse amplifier, and a filter. These are the same basic elements that are used in modern systems for eddy current inspection.

Several artificial imperfections in metals were tried for calibrating tests, but by 1935 the small drilled hole had become the reference standard for all production testing. The drilled hole was selected for the standard because:

- It was relatively easy to produce
- It was reproducible
- It could be produced in precisely graduated sizes
- It produced a signal on the eddy current tester that was similar to that produced by a natural imperfection
- It was a short imperfection and resembled hard-to-detect, short natural weld imperfections. Thus, if the tester could detect the small drilled hole, it would also detect most of the natural weld imperfections

Vigners, Dinger, and Gunn described eddy current type flaw detectors for nonmagnetic metals in 1942, and in the early 1940s, Förster and Zuschlag developed eddy current inspection instruments. Numerous versions of eddy current inspection equipment are currently available commercially. Some of this equipment is useful only for exploratory inspection or for inspecting parts of simple shape. However, specially designed equipment is extensively used in the inspection of production quantities of metal sheet, rod, pipe, and tubing.

Principles of Operation

The eddy current method of inspection and the induction heating technique that is used for metal heating, induction hardening, and tempering have several similarities. For example, both are dependent on the principles of electromagnetic induction for inducing eddy currents within a part placed within or adjacent to one or more induction coils. The heating is a result of I^2R losses caused by the flow of eddy currents in the part. Changes in coupling between the induction coils and the part being inspected and changes in the electrical characteristics of the part cause variations in the loading and tuning of the generator.

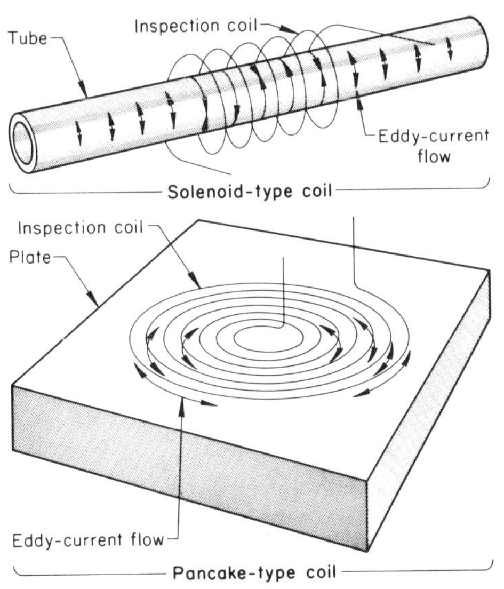

Fig. 1 Two common types of inspection coils and the patterns of eddy current flow generated by the exciting current in the coils. Solenoid-type coil is applied to cylindrical or tubular parts; pancake-type coil, to a flat surface.

The induction heating system is operated at high power levels to produce the desired heating rate. In contrast, the system used in eddy current inspection is usually operated at very low power levels to minimize the heating losses and temperature changes. Also, in the eddy current system, electrical-loading changes caused by variations in the part being inspected, such as those caused by the presence of flaws or dimensional changes, are monitored by electronic circuits. In both eddy current inspection and induction heating, the selection of operating frequency is largely governed by skin effect (see the section "Operating Variables" in this article). This effect causes the eddy currents to be concentrated toward the surfaces adjacent to the coils carrying currents that induce them. Skin effect becomes more pronounced with increase in frequency.

The coils used in eddy current inspection differ from those used in induction heating because of the differences in power level and resolution requirements, which necessitate special inspection coil arrangements to facilitate the monitoring of the electromagnetic field in the vicinity of the part being inspected.

Functions of a Basic System. The part to be inspected is placed within or adjacent to an electric coil in which an alternating current is flowing. As shown in Fig. 1, this alternating current, called the exciting current, causes eddy currents to flow in the part as a result of electromagnetic induction. These currents flow within closed loops in the part, and their magnitude and timing (or phase) depend on:

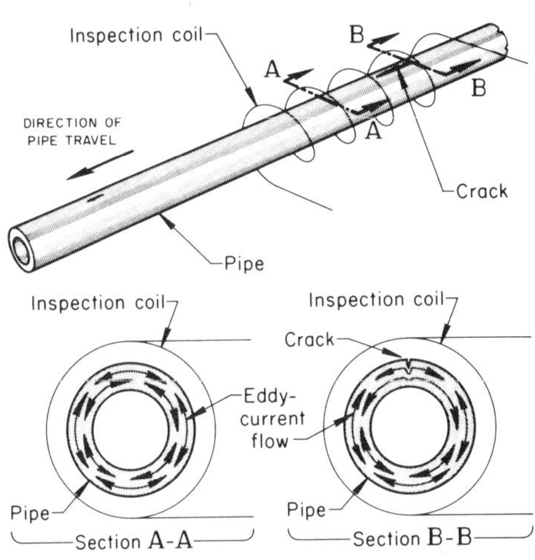

Fig. 2 Effect of a crack on the pattern of eddy current flow in a pipe

- The original or primary field established by the exciting currents
- The electrical properties of the part
- The electromagnetic fields established by currents flowing within the part

The electromagnetic field in the region in the part and surrounding the part depends on both the exciting current from the coil and the eddy currents flowing in the part. The flow of eddy currents in the part depends on:

- The electrical characteristics of the part
- The presence or absence of flaws or other discontinuities in the part
- The total electromagnetic field within the part

The change in flow of eddy currents caused by the presence of a crack in a pipe is shown in Fig. 2. The pipe travels along the length of the inspection coil as shown in Fig. 2. In section A–A in Fig. 2, no crack is present and the eddy current flow is symmetrical. In section B–B in Fig. 2, where a crack is present, the eddy current flow is impeded and changed in direction, causing significant changes in the associated electromagnetic field. From Fig. 2 it is seen that the electromagnetic field surrounding a part depends partly on the properties and characteristics of the part. Finally, the condition of the part can be monitored by observing the effect of the resulting field on the electrical characteristics of the exciting coil, such as its electrical impedance, induced voltage, or induced currents. Alternatively, the effect of the electromagnetic field can be monitored by observing the induced voltage in one or more other coils placed within the field near the part being monitored.

Each and all of these changes can have an effect on the exciting coil or other coil or coils used for sensing the electromagnetic field adjacent to a part. The effects most often used to monitor the condition of the part being inspected are the electrical impedance of the coil or the induced voltage of either the exciting coil or other adjacent coil or coils.

Eddy current systems vary in complexity depending on individual inspection requirements. However, most systems provide for the following functions:

- Excitation of the inspection coil
- Modulation of the inspection coil output signal by the part being inspected
- Processing of the inspection coil signal prior to amplification
- Amplification of the inspection coil signals
- Detection or demodulation of the inspection coil signal, usually accompanied by some analysis or discrimination of signals
- Display of signals on a meter, an oscilloscope, an oscillograph, or a strip chart recorder; or recording of signal data on magnetic tape or other recording media
- Handling of the part being inspected and support of the inspection coil assembly or the manipulation of the coil adjacent to the part being inspected

Elements of a typical inspection system are shown schematically in Fig. 3. The particular elements in Fig. 3 are for a system developed to inspect bar or tubing. The generator supplies excitation current to the inspection coil and a synchronizing signal to the phase shifter, which provides switching signals for the detector. The loading of the inspection coil by the part being inspected modulates the electromagnetic field of the coil. This causes changes in the amplitude and phase of the inspection coil voltage output.

The output of the inspection coil is fed to the amplifier and detected or demodulated by the detector. The demodulated output signal, after some further filtering and analyzing, is then displayed on an oscilloscope or a chart recorder. The displayed signals, having been detected or demodulated, vary at a much slower rate, depending on:

- The speed at which the part is fed through an inspection coil
- The speed with which the inspection coil is caused to scan past the part being inspected

Operating Variables

The principal operating variables encountered in eddy current inspection include coil impedance, electrical conductivity, magnetic permeability, lift-off and fill factors, edge effect, and skin effect. Each of these variables will be discussed in this section.

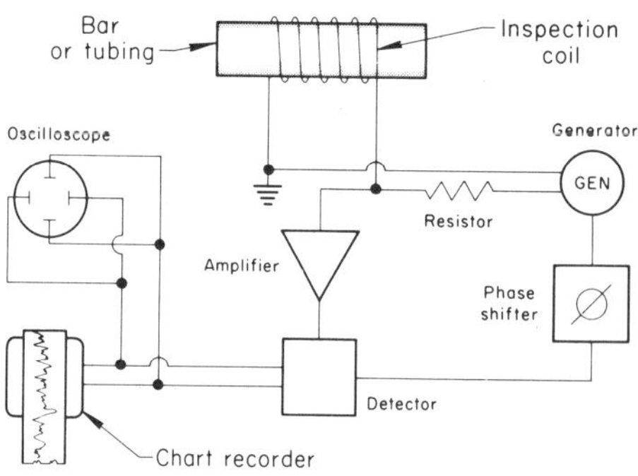

Fig. 3 Principal elements of a typical system for eddy current inspection of bar or tubing. See description in text.

Coil Impedance

When direct current is flowing in a coil, the magnetic field reaches a constant level, and the electrical resistance of the wire is the only limitation to current flow. However, when alternating current is flowing in a coil, two limitations are imposed:

- The ac resistance of the wire, R
- A quantity known as inductive reactance, X_L

The ac resistance of an isolated or empty coil operating at low frequencies or having a small wire diameter is very nearly the same as the dc resistance of the wire of the coil. The ratio of ac resistance to dc resistance increases as either the frequency or the wire diameter increases. In the discussion of eddy current principles, the resistance of the coil wire is often ignored, because it is nearly constant. It varies mainly with wire temperature and the frequency and spatial distribution of the magnetic field threading the coil.

Inductive reactance, X_L, is the combined effect of coil inductance and test frequency and is expressed in ohms. Total resistance to the flow of alternating current in a coil is called impedance, Z, and comprises both ac resistance, R, and inductive reactance, X_L. The impedance can be expressed as $Z = \sqrt{R^2 + X_L^2}$, where $X_L = 2\pi f L_0$, f is the test frequency (in Hertz), and L_0 is the coil inductance (in henrys).

When a metal part is placed adjacent to or within a test coil, the electromagnetic field threading the coil is changed as a result of eddy current flow in the test object. In general, both the ac resistance and the inductive reactance of the coil are affected. The resistance of the loaded coil consists of two components, namely, the ac resistance of the coil wire and the apparent, or coupled, resistance caused by the presence of the test object. Changes in these components reflect conditions within the test object.

Impedance is usually plotted on an impedance-plane diagram. In the diagram, resistance is plotted along one axis and inductive reactance along the other axis. Because

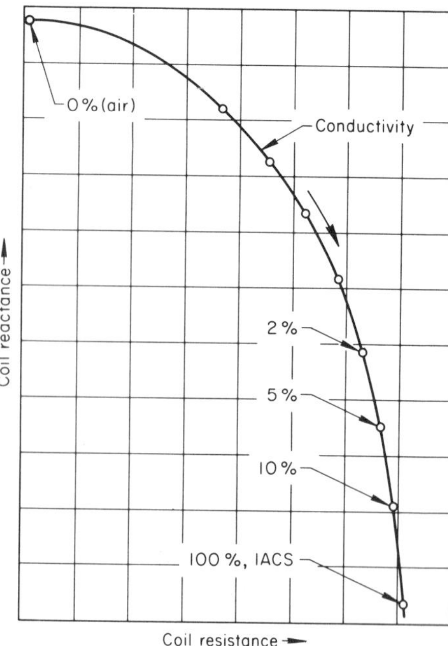

Fig. 4 Typical impedance-plane diagram derived by placing an inspection coil sequentially on a series of thick pieces of metal, each with a different IACS electrical resistance or conductivity rating. The inspection frequency was 100 kHz.

each specific condition in the material being inspected may result in a specific coil impedance, each condition may correspond to a particular point on the impedance-plane diagram. For example, if a coil were placed sequentially on a series of thick pieces of metal, each with a different resistivity, each piece would cause a different coil impedance and would correspond to a different point on a locus in the impedance plane. The curve generated might resemble that shown in Fig. 4, which is based on International Annealed Copper Standard (IACS) conductivity ratings. Other curves would be generated for other material variables, such as section thickness and types of surface flaws.

Impedance Components. Figure 5(a) shows a simplified equivalent circuit of an inspection coil and the part being inspected. The coil is assumed to have inductance, L_0, and negligible resistance. The part being inspected consists of a very thin tube having shunt conductance, G, closely coupled to the coil. When an alternating current is caused to flow into the system under steady-state conditions, some energy is stored in the system and returned to the generator each cycle and some energy is dissipated or lost as heat each cycle. The inductive-reactance component, X_L, of the impedance, Z, of the circuit is proportional to the energy stored per cycle, and the resistance component, R, of the impedance is proportional to the energy dissipated per cycle. The impedance, Z, is equal to the complex ratio of the applied voltage, E, to the current, I, in accordance with Ohm's

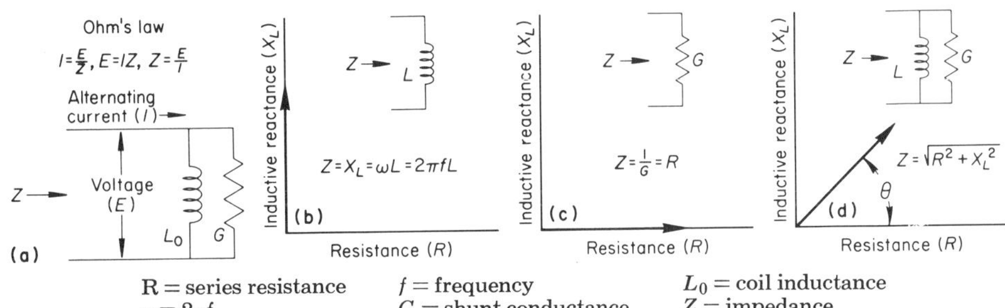

Fig. 5 Simplified equivalent circuit (a) of an eddy current inspection coil and the part being inspected. (b) to (d) Three impedance diagrams for three conditions of the equivalent circuit. See text for explanation.

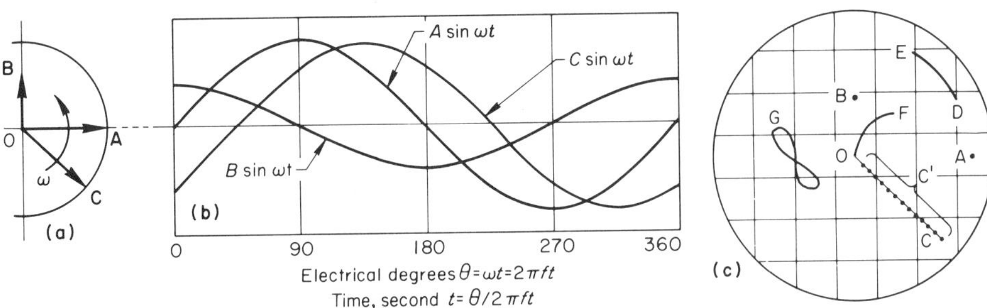

Fig. 6 Phasor representation of sinusoids. See text for explanation.

law. The term complex is used to indicate that, in general, the alternating current and voltage do not have the same phase angle.

Figures 5(b) to (d) show three impedance diagrams for three conditions of the equivalent circuit in Fig. 5(a). When only the coil is present, the circuit impedance is purely reactive; that is, $Z = X_L = \omega L = 2\pi fL$, as shown in Fig. 5(b). When only the conductance of this equivalent circuit is present (a hypothetical condition for an actual combination of inspection coil and part being inspected), the impedance is purely resistive; that is, $Z = 1/G = R$, as shown in Fig. 5(c). When both coil and conductance are connected, the impedance has both reactive and resistive components in the general instance, and the impedance $Z = \sqrt{R^2 + X_L^2}$, as shown in Fig. 5(d). Here, R is the series resistance and X_L is the series reactance. An angle, θ, is associated with the impedance, Z. This angle is a function of the ratio of the two components of the impedance, R and X_L. In Fig. 5(d), this angle, θ, is about 45°.

Points and loci on impedance-plane diagrams can be displayed using phasor representation because of the close relationship between the impedance diagrams and the phasor diagrams. In a given circuit with input impedance Z, applying an impressed fixed current I, will produce a signal voltage E in accordance with Ohm's law ($E = IZ$). This signal voltage can be displayed as a phasor. With I fixed, the signal voltage E is directly proportional to the impedance Z. Thus, the impedance plane can be readily displayed using the phasor technique.

Phasor Representation of Sinusoids. One method often used in signal analysis and in the representation of eddy current inspection signals is the phasor method schematically shown in Fig. 6. In Fig. 6(a) are shown three vectors, **A**, **B**, and **C**, which are rotating counterclockwise with radian velocity $2\pi ft = \omega t$. The equations that describe these vectors are of the form $K \sin (\omega t + \phi)$, where K is a constant equal to **A**, **B**, or **C** and φ is the electrical phase angle. These equations are plotted in Fig. 6(b). The length of the vectors **A**, **B**, and **C** determine the amplitude of the sinusoids generated in Fig. 6(b). The physical angle between the vectors **A** and **B**, or between **A** and **C**, determines the electrical phase angle, φ, between sinusoids. In Fig. 6(b), these angles are +90° and −45°, respectively.

The three vectors, **A**, **B**, and **C**, are considered to be rotating at frequency, f, generating three rather monotonous sinusoids. This system of three vectors rotating synchronously with the frequency of the sinusoids is not very useful, because of its high rate of rotation. However, if rotation is stopped, the amplitudes and phase angles of the three sine waves can be easily seen in a representation called a phasor diagram.

In eddy current inspection equipment, the sine wave signals are often expanded in quadrature components and displayed as phasors on an x-y oscilloscope, shown in Fig. 6(c). Usually, only the tips of the phasors are shown. Thus, **A** and **B** in Fig. 6(c) show the cathode ray beam position representing the two sinusoids of Fig. 6(b). Point C represents a sinusoid $C \sin \omega t$ having the same amplitude as $A \sin \omega t$, but which lags or follows it in phase by an electrical angle equal to 45°. The points indicated as C′ represent sinusoids having the same phase angle as $C \sin \omega t$, but with different amplitudes. The concept of a phasor locus is introduced by varying the amplitude gradually from the maximum at C to zero at the origin O. This results in the beam spot moving from C to O, producing a locus. In contrast, a shift of the phase angle of a sinusoid causes a movement of the phasor tip around the origin O as shown by the arc DE. Here, D represents a sinusoid having the same amplitude as the sinusoid represented by A but leading it by 30°. Increasing this phase angle from 30 to 60° results in the phasor locus DE. When both amplitude and phase changes occur, more complicated loci can be formed as shown at F and G.

Electrical Conductivity

All materials have a characteristic resistance to the flow of electricity. Those with the highest resistivity are classified as insulators, those having an intermediate resistivity are classified as semiconductors, and those having a low resistivity are classified as conductors. The conductors, which include most metals, are of greatest interest in eddy current inspection. The relative conductivity of the common metals and alloys varies over a wide range.

Capacity for conducting current can be measured in terms of either conductivity or resistivity. In eddy current inspection, frequent use is made of measurement based on the International Annealed Copper Standard. In this system, the conductivity of annealed, unalloyed copper is arbitrarily rated at 100%, and the conductivities of other metals and alloys are expressed as a percentage of this standard. Thus, the conductivity of unalloyed aluminum is rated 61% IACS, or 61% that of unalloyed copper. The resistivity and IACS conductivity ratings of several common metals and alloys are given in Table 1.

Many factors influence the conductivity of a metal, notably, temperature, composition, heat treatment and resulting microstructure, grain size, hardness, and residual stresses. Conversely, eddy currents can be used to monitor composition and various metallurgical characteristics, provided their influence on conductivity is sufficient to provide the necessary contrast. For example, it is possible to monitor the heat treatment of age-hardenable aluminum alloys because of the marked effect of hardness on conductivity (Fig. 7).

Magnetic Permeability

Ferromagnetic metals and alloys, including iron, nickel, cobalt, and some of their alloys, act to concentrate the flux of a magnetic field.

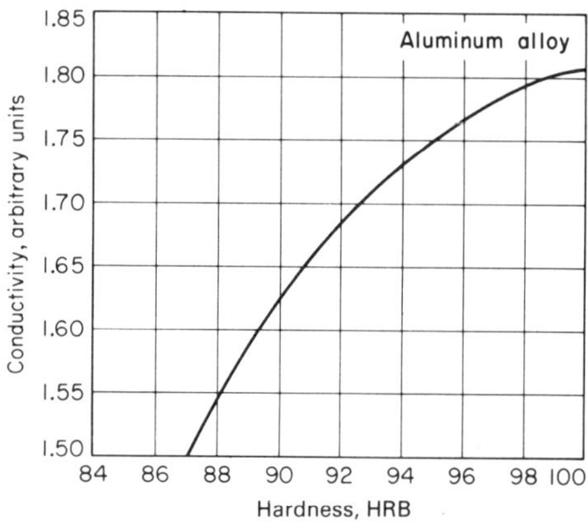

Fig. 7 Relation of hardness and electrical conductivity in an age-hardenable aluminum alloy that permits the eddy current monitoring of heat treatment of the alloy

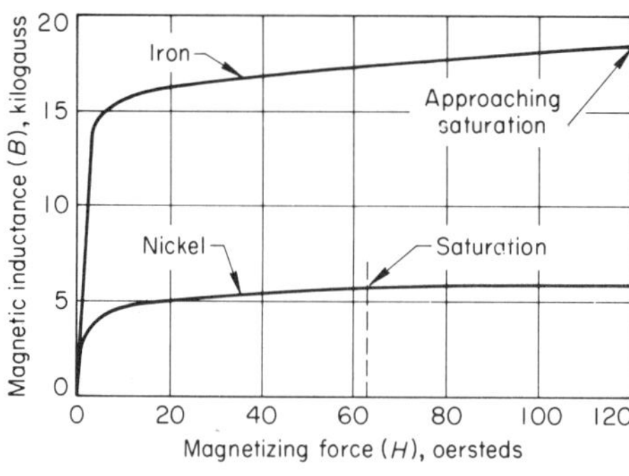

Fig. 8 Magnetization curves for annealed commercially pure iron and nickel

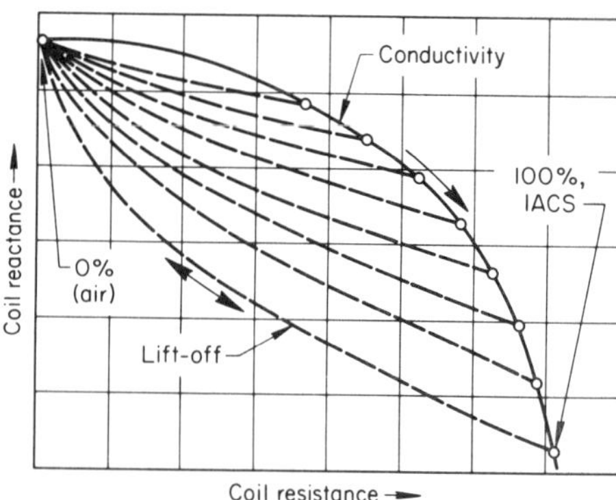

Fig. 9 Impedance-plane diagram showing curves for electrical conductivity and lift-off. Inspection frequency was 100 kHz.

They are strongly attracted to a magnet or an electromagnet, have exceedingly high and variable susceptibilities, and have very high and variable permeabilities.

Magnetic permeability is not necessarily constant for a given material but depends on the strength of the magnetic field acting upon it. For example, consider a sample of steel that has been completely demagnetized and then placed in a solenoid coil. As current in the coil is increased, the magnetic field associated with the current will increase. The magnetic flux within the steel, however, will increase rapidly at first and then level off so that an additionally large increase in the strength of the magnetic field will result in only a small increase in flux within the steel. The steel sample will then have achieved a condition known as magnetic saturation. The curve showing the relation between magnetic field intensity and the magnetic flux within the steel is known as a magnetization curve. Magnetization curves for annealed commercially pure iron and nickel are shown in Fig. 8.

Table 1 Electrical resistivity and conductivity of several common metals and alloys

Metal or alloy	Resistivity, μΩ · mm	Conductivity, % IACS
Silver	16.3	105
Copper, annealed	17.2	100
Gold	24.4	70
Aluminum	28.2	61
Aluminum alloys		
6061-T6	41	42
7075-T6	53	32
2024-T4	52	30
Magnesium	46	37
70-30 brass	62	28
Phosphor bronzes	160	11
Monel	482	3.6
Zirconium	500	3.4
Zircaloy-2	720	2.4
Titanium	548	3.1
Ti-6Al-4V alloy	1720	1.0
Type 304 stainless steel	700	2.5
Inconel 600	980	1.7
Hastelloy X	1150	1.5
Waspaloy	1230	1.4

The magnetic permeability of a material is the ratio between the strength of the magnetic field and the amount of magnetic flux within the material. As shown in Fig. 8, at saturation (where there is no appreciable change in induced flux in the material for a change in field strength) the permeability is nearly constant for small changes in field strength.

Because eddy currents are induced by a varying magnetic field, the magnetic permeability of the material being inspected strongly influences the eddy current response. Consequently, the techniques and conditions used for inspecting magnetic materials differ from those used for inspecting nonmagnetic materials. However, the same factors that may influence electrical conductivity (such as composition, hardness, residual stresses, and flaws) may also influence magnetic permeability. Thus, eddy current inspection can be applied to both magnetic and nonmagnetic materials. Although magnetic conductors also have an electrical conductivity that can vary with changes in material conditions, permeability changes generally have a much greater effect on eddy current response at lower test frequencies than conductivity variations.

The fact that magnetic permeability is constant when a ferromagnetic material is saturated can be used to permit the eddy current inspection of magnetic materials with greatly reduced influence of permeability variations. The part to be inspected is placed in a coil in which direct current is flowing. The magnitude of current used is sufficient to cause magnetic saturation of the part. The inspection (encircling) coil is located within the saturation coil and close to the part being inspected. This technique is generally used when inspecting magnetic materials for discontinuities because small variations in permeability are not of interest and may cause rejection of acceptable material.

Lift-Off Factor

When a probe inspection coil, attached to a suitable inspection instrument, is energized in air, it will give some indication even if there is no conductive material in the vicinity of the coil. The initial indication will begin to change as the coil is moved closer to a conductor. Because the field of the coil is strongest close to the coil, the indicated change on the instrument will continue to increase at a more rapid rate until the coil is directly on the conductor. These changes in indication with changes in spacing between the coil and the conductor, or part being inspected, are called lift-off. The lift-off effect is so pronounced that small variations in spacing can mask many indications resulting from the condition or conditions of primary interest. Consequently, it is usually necessary to maintain a constant relationship between the size and shape of the coil and the size and shape of the part being inspected. The lift-off effect also accounts for the extreme difficulty of performing an inspection that requires scanning a part having a complex shape.

The change of coil impedance with lift-off can be derived from the impedance-plane diagram shown in Fig. 9. When the coil is suspended in air away from the conductor, impedance is at a point at the upper end of the curve at far left in Fig. 9. As the coil approaches the conductor, the impedance moves in the direction indicated by the dashed lines until the coil is in contact with the conductor. When contact occurs, the impedance is at a point corresponding to the impedance of the part being inspected, which in this case represents its conductivity. The fact that the lift-off curves approach the conductivity curve at an angle can be utilized in some instruments to separate lift-off signals from those resulting from variations in conductivity or some other parameter of interest.

Although troublesome in many applications, lift-off can also be useful. For example, with the lift-off effect, eddy current instruments are excellent for measuring the thickness of nonconductive coatings, such as paint and anodized coatings, on metals.

Fill Factor

In an encircling coil, a condition comparable to lift-off is known as fill factor. It is a measure of how well the part being inspected fills the coil. As with lift-off, changes in fill factor resulting from such factors as variations in outside diameter must be controlled because small changes can give large indications. The lift-off curves shown in Fig. 9 are very similar to those for changes in fill factor. For a given lift-off or fill factor, the conductivity curve will shift to a new position, as indicated in Fig. 9. Fill factor can sometimes be used as a rapid method for checking variations in outside diameter measurements in rods and bars.

For an internal, or bobbin-type, coil, the fill factor measures how well the inspection coil fills the inside of the tubing being inspected. Variations in the inside diameter of the part must be controlled because small changes in the diameter can give large indications.

Edge Effect

When an inspection coil approaches the end or edge of a part being inspected, the eddy currents are distorted because they are unable to flow beyond the edge of a part. The distortion of eddy currents results in an indication known as edge effect. Because the magnitude of the effect is very large, it limits inspection near edges. Unlike lift-off, little can be done to eliminate edge effect. A reduction in coil size will lessen the effect somewhat, but there are practical limits that dictate the sizes of coils for given applications. In general, it is not advisable to inspect any closer than 3.2 mm (⅛ in.) from the edge of a part, depending on variables such as coil size and test frequency.

Skin Effect

In addition to the geometric relationship that exists between the inspection coil and the part being inspected, the thickness and shape of the part itself will affect eddy current response. Eddy currents are not uniformly distributed throughout a part being inspected; rather, they are densest at the surface immediately beneath the coil and become progressively less dense with increasing distance below the surface—a phenomenon known as the skin effect. At some distance below the surface of a thick part there will be essentially no currents flowing.

Figure 10 shows how the eddy current varies as a function of depth below the surface. The depth at which the density of the eddy current is reduced to a level about 37% of the density at the surface is defined as the standard depth of penetration. This depth depends on the electrical conductivity and magnetic permeability of the material and on the frequency of the magnetizing current. Depth of penetration decreases with increases in conductivity, permeability, or inspection frequency. The standard depth of penetration can be calculated from:

$$S = 1980 \sqrt{\rho/\mu f} \qquad \text{(Eq 1)}$$

where S is the standard depth of penetration (in inches), ρ is the resistivity (in ohm-centimeters), μ is the magnetic permeability (1 for nonmagnetic materials), and f is the inspection frequency (in hertz). Resistivity, it should be noted, is the reciprocal of conductivity. The standard depth of penetration, as a function of inspection frequency, is shown for several metals at various electrical conductivities in Fig. 11.

The eddy current response obtained will reflect the workpiece material thickness. It is necessary, therefore, to be sure that either the material has a constant thickness or is sufficiently thick so that the eddy currents do not penetrate completely through it. It should be remembered that the eddy currents do not cease at the standard depth of penetration but continue for some distance beyond it. Normally, a part being inspected must have a thickness of at least two or three standard depths before thickness ceases to have a significant effect on eddy current response. By properly calibrating an eddy current instrument, it is possible to measure material thickness because of the varying response with thickness. Changing material thickness follows curves in the impedance plane such as those shown in Fig. 12. As indicated by the curves, measurements of thickness by the eddy current method are more accurate on thin materials (Fig. 12b) than they are on thick materials (Fig. 12a). The opposite is true of thickness measurements made by ultrasonics; thus, the two methods complement each other.

Principal Impedance Concepts

This section considers in detail some of the principal impedance concepts that are fundamental to an understanding and effective application of eddy current inspection.

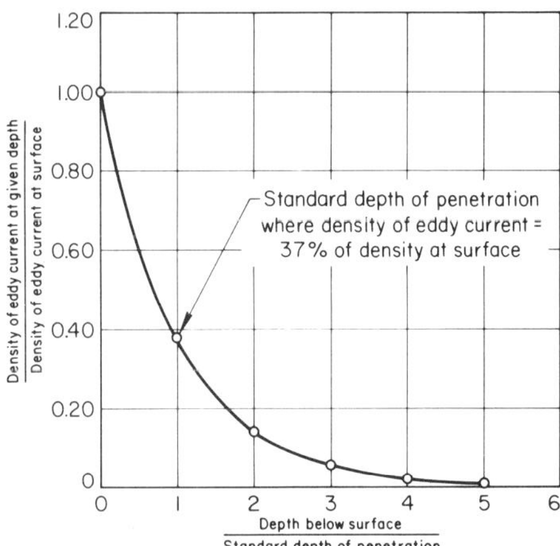

Fig. 10 Variation in density of eddy current as a function of depth below the surface of a conductor—a variation commonly known as skin effect

Impedance of a Long Coil Encircling a Thin-Wall Tube. An impedance diagram for a long coil encircling a thin-wall nonferromagnetic tube, with reactance values plotted as ordinates (horizontal axes) and resistance values plotted as abscissas (vertical axes), is shown in Fig. 13. When a tube being inspected has zero conductance (the empty-coil condition), the impedance point is at A. The coil input impedance is all reactance and is equal to ωL or $2\pi fL$ ohms. The resistance component is zero. The ac resistance of the coil wire is assumed to be constant and is not included in these diagrams. As the conductance of the part being inspected is caused to increase, the impedance, Z, follows the locus ABO, for which an example is shown in Fig. 13. This is a

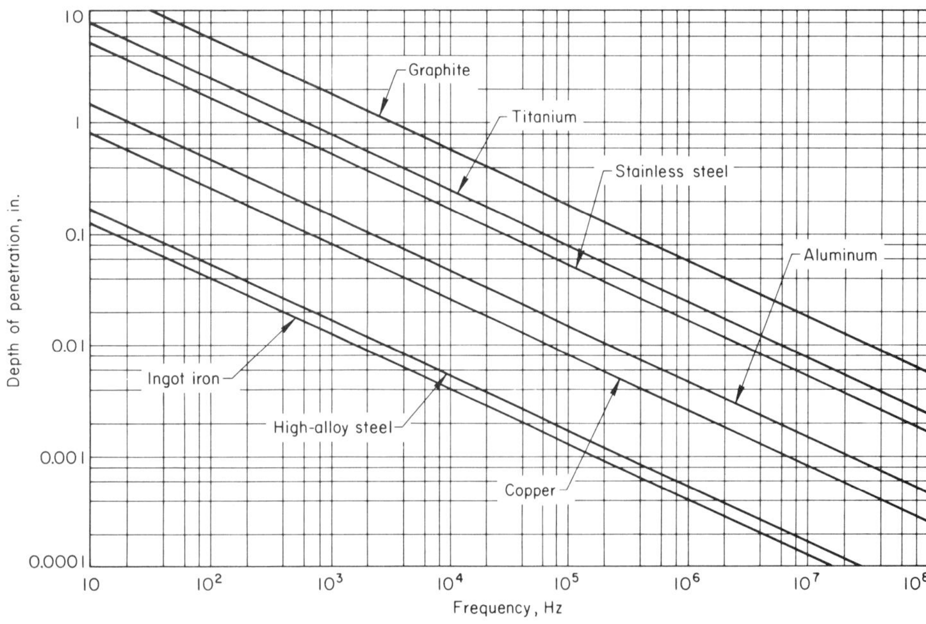

Fig. 11 Standard depths of penetration as a function of frequencies used in eddy current inspection for several metals of various electrical conductivities

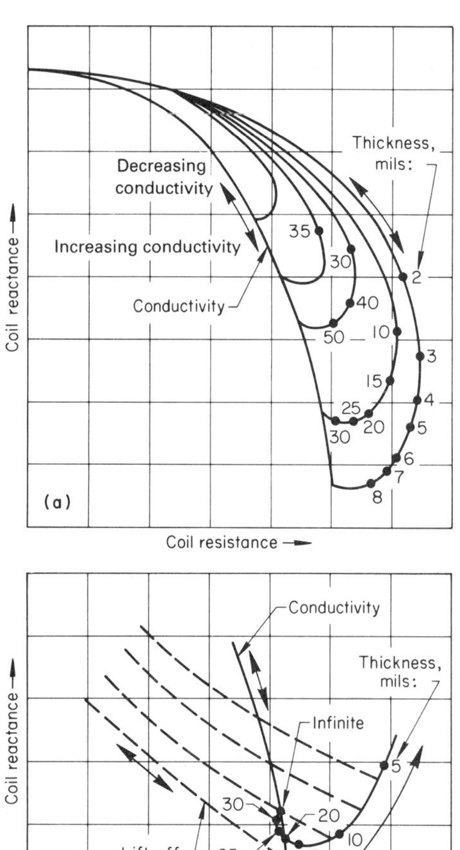

Fig. 12 Typical impedance-plane diagrams for changing material thickness. (a) Diagram for thick material. (b) Diagram for thin material on an expanded scale. Inspection frequency was 100 kHz.

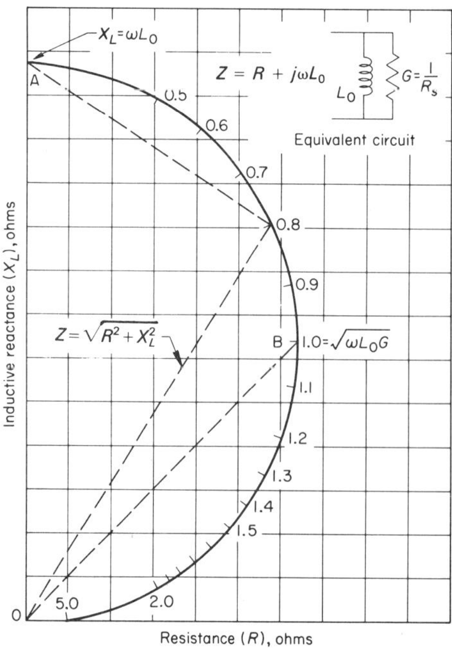

Fig. 13 Impedance diagram for a long coil encircling a thin-wall nonferromagnetic tube, showing also an equivalent circuit. R: series resistance; R_s: effective shunt resistance; ω: $2\pi f$; f: frequency; G: shunt conductance; L_0: coil inductance; Z: impedance; j: $\sqrt{-1}$; $\sqrt{\omega L_0 G}$: dimensionless constant

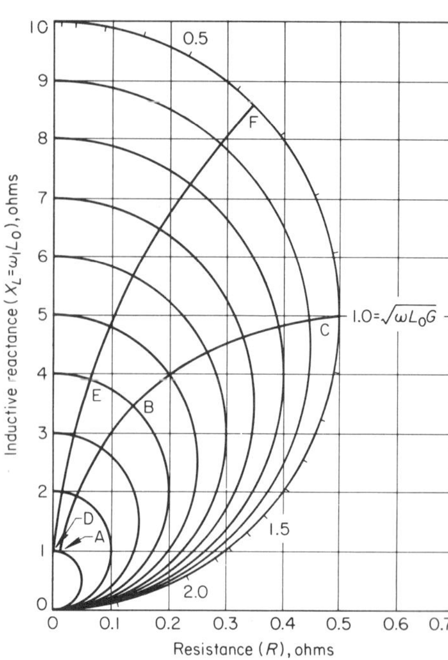

Fig. 14 Impedance diagram for a long coil encircling a thin-wall nonferromagnetic tube showing impedance as a function of frequency

circular arc and occurs as shown in Fig. 13 if the tube wall is very thin compared with the skin depth at the frequency of operation. The impedance locus is marked with reference numbers calculated from the dimensionless constant $\sqrt{\omega L_0 G}$ and placed on the locus at points corresponding to the respective impedance values.

Several characteristics of the eddy current inspection of tubes or bars are shown in Fig. 13. The simplification resulting from the assumption that skin effect is absent alters the detailed loci in important ways, as shown in subsequent diagrams. However, this simplified diagram serves as an introduction to the more detailed diagrams, which include the variations caused by the skin effect. The locus ABO in Fig. 13 shows the effect on the effective coil impedance of changing the conductance of the thin-wall tube; because the tube conductance is proportional to the product of the wall thickness of the tube and the conductivity of the tube material, the impedance loci resulting from variation of thickness coincide with the locus associated with varying tube material conductivity.

Effects of Changing Operating Frequency. One effect of changing operating frequency is to increase the empty-coil reactance in direct proportion to the frequency; thus, the impedance diagram grows in size. However, with the part being inspected in place within the coil, the impedance of the coil for different part conditions and different frequency values changes at different rates as the frequency changes. This is shown in Fig. 14 as a prelude to introducing the concept of impedance normalization. Although frequency is contained in the diagram in Fig. 13, the discussion of that diagram is based on a fixed frequency. In contrast, Fig. 14 shows the impedance of a long coil encircling a thin-wall nonferromagnetic tube as a function of frequency. As in Fig. 13, the shape of the impedance locus is semicircular because of the negligible skin effect, but now there is a separate locus for each frequency considered. Impedance loci are shown for ten different operating frequencies (ω_1 through $10\omega_1$). Each locus represents a condition of maximum coupling between the long solenoid and the encircled tube. This maximum coupling cannot be realized in practice, because the diameter of the tube and of the coil would need to be equal. The coil wire must occupy some space; therefore, it is not possible for the exciting current to flow exactly at the surface.

The ten curves in Fig. 14 show that the impedance of the empty coil, assuming the coil resistance is negligible, increases in direct proportion to increases in operating frequency and that this impedance is reactive. The coil at the operating frequency of ω_1 has a reactance of $\omega_1 L_0$ ohms. At a frequency of $2\omega_1$, the reactance is doubled, and so on, until at $10\omega_1$, the reactance is $10\omega_1 L_0$. In contrast to this linear change of impedance or reactance with frequency, note the nonlinear change of impedance when the coil has a part within it. First, assume that the part being inspected is a thin-wall tube and that its reference number $\sqrt{\omega L_0 G}$ is 0.316 at radian frequency ω_1. This corresponds to point A on the conductance locus of the coil at radian frequency ω_1. Locus ABC shows the change in impedance of this particular combination of coil and tube as the frequency is increased from ω_1 to $10\omega_1$. The impedance variation is far from linear with respect to frequency variation. Locus DEF similarly shows the impedance variation as frequency varies from ω_1 to $10\omega_1$ when the tube reference number $\sqrt{\omega L_0 G} = 0.2$ at radian frequency ω_1.

It is customary to normalize groups of impedance curves, such as those in Fig. 14, by dividing both reactance and resistance values by the impedance or reactance of the empty coil. This transforms all the curves into a single curve, such as the outer or large curve in Fig. 15, which can be used under a wide range of conditions. When using the single curve, the nature of its origin must be recalled in interpreting the real effect of varying frequency. Correct relative changes in impedance are shown on the normalized curve as the frequency is changed in the reference number $\sqrt{\omega L_0 G}$, but the actual growing nature of the impedance plane as frequency is increased is hidden.

Several other characteristics of the impedance diagrams for a long coil encircling a tube or bar are shown for simplified condi-

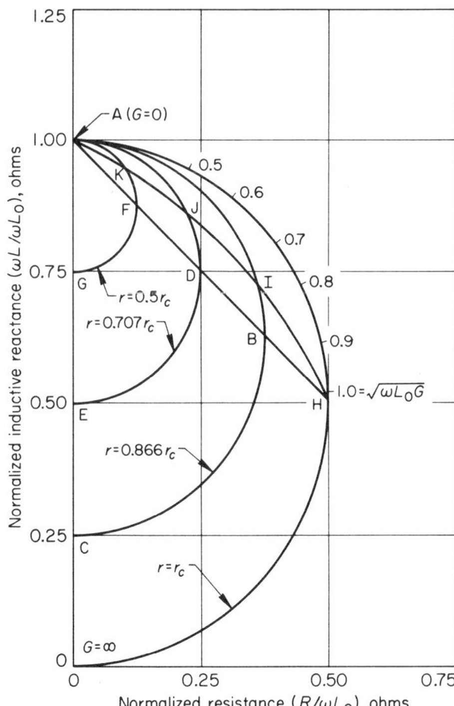

Fig. 15 Effects of variations in tube radius on the impedance of a long coil of fixed diameter encircling a thin-wall nonferromagnetic tube. *G*: conductance; *r*: tube radius; r_c: coil radius

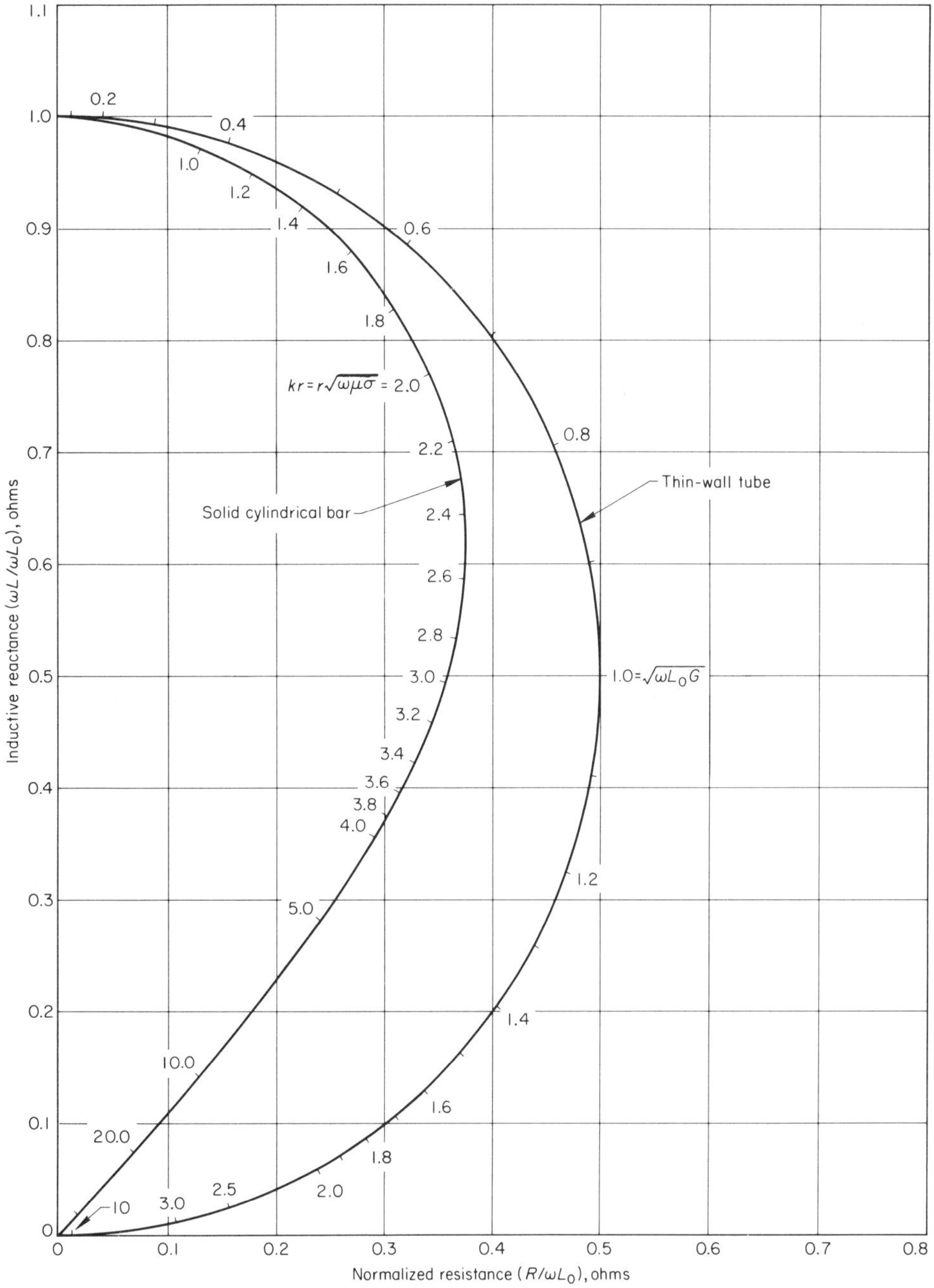

Fig. 16 Normalized impedance diagram for a long coil encircling a solid cylindrical nonferromagnetic bar showing also the locus for a thin-wall tube (which is similar to the loci in Fig. 14 and 15). *k*, electromagnetic wave propagation constant for a conducting material, or $\sqrt{\omega\mu\sigma}$; *r*, radius of conducting cylinder, meters; ω, 2π*f*; *f*, frequency; $\sqrt{\omega L_0 G}$, equivalent of $\sqrt{\omega\mu\sigma}$ for simplified electric circuits; μ, magnetic permeability of bar, or = $4\pi \times 10^{-7}$ H/m if bar is nonmagnetic; σ, electrical conductivity of bar, mho/m; 1.0, coil fill factor

tions in Fig. 15. The tube wall is assumed to be very thin in relation to the skin depth at the frequency of operation. The large semicircular curve represents the locus of impedance resulting from changing tube conductance. Because the tube wall is assumed to be very thin, skin effect is minimal. Maximum coupling exists between the coil and the tube, and because the conductance is equal to the product of conductivity of the tube material and the wall thickness in this simplified example, the conductivity locus and thickness locus are identical. Note, however, that the skin effect must be negligible for this condition to be obtained. The curves of smaller radius (arcs ABC, ADE, and AFG) are for tubes having diameters smaller than the coil diameters. As the tube diameter becomes smaller, the electromagnetic coupling between the coil and tube decreases, and loci such as HBDFA or HIJKA would be generated. The curvature of the loci depends on the rate at which the conductance of the thin-wall tube varies as the radius is decreased. Figure 15, therefore, shows that increases in conductance of the thin-wall tube produce semicircular loci whose radii depend on tube diameter and the amount of coupling (fill factor) between the inspection coil and the tube. The change in conductance may be caused by a change in either the wall thickness or the electrical conductivity of the tube.

Solid Cylindrical Bar. The normalized impedance diagram for a long encircling coil closely coupled to a solid cylindrical nonferromagnetic bar is shown in Fig. 16. The locus for the thin-wall tube in Fig. 16 is similar to that discussed in Fig. 14 and 15. The locus for the solid bar is constructed from an analytical solution of Maxwell's equations for the particular conditions existing for the solid bar. The reference number quantity for the bar is different from that of the thin-wall tube to satisfy the new conditions for the solid bar for which the skin effect is no longer negligible. The new reference number quantity $\sqrt{r^2\omega\mu\sigma}$ or $r\sqrt{\omega\mu\sigma}$ is from the theory developed in the application of Maxwell's equations for a cylindrical conductor. The quantity $\sqrt{\omega\mu\sigma}$ is the electromagnetic wave propagation constant for a conducting material, and the quantity $\sqrt{\omega L_0 G}$ is the equivalent of $\sqrt{\omega\mu\sigma}$ for simplified electric circuits. The quantity $\sqrt{r^2\omega\mu\sigma}$ or $r\sqrt{\omega\mu\sigma}$ is dimensionless and serves as a convenient reference number for use in entering on the impedance diagram.

In Fig. 16, the impedance region between the semicircular locus of the impedance for the thin-wall tube and the locus for the solid cylinder represents impedance values for hollow cylinders or tubes of various wall thicknesses and of materials with different electrical conductivities. In each case, the outer radius of the tube is equal to the radius of the coil—the ideal for maximum coupling. The effect on impedance of changing the outer radius of the tube can be projected from the effects illustrated in Fig. 15, in which a group of electrical-conductivity loci are shown generated by varying the tube radius. The effect on impedance of varying the outer radius or diameter of the solid cylinder is shown in Fig. 17.

The locus resulting from varying the outer diameter of the cylindrical bar does not follow a straight path. The reference number is a function of bar radius, r, and as the radius becomes smaller, the reference number is likewise reduced, producing a curved radius locus such as the locus ABCD in Fig. 17. At lower values of $r\sqrt{\omega\mu\sigma}$, the radius locus intercepts the conductivity locus at slighter angles and nearly parallels the conductivity locus, as shown in locus EFD in Fig. 17. This difference in intercept angle is of importance when it is required to discriminate between conductivity variations and diameter variations. The larger intercept angle permits better discrimination. The factor $(r_a/r_c)^2$, where r_a is the cylindrical bar radius and r_c is the coil radius, is called the fill factor and represents that fraction of the coil area occupied by the bar.

Thickness Loci. Transition from the very thin-wall tube to the solid cylinder produces impedance loci extending generally from the thin-wall tube locus to a final or end point on the solid-cylinder locus determined by the conductivity and frequency, as shown in Fig. 18. These loci curve or spiral in a clockwise direction. When the end points of the thickness loci are points on the solid-cylinder conductivity locus having low values of the reference number $kr = r\sqrt{\omega\mu\sigma}$, the thickness loci are curved but do not rotate around the end point. The spiral effect is more pronounced at the higher values of $kr = r\sqrt{\omega\mu\sigma}$. As in each instance where loci are observed, an opportunity exists to calibrate the instrument readings so that specific points along a locus represent specific values of the variable of interest. In this case, the variable is tube-wall thickness. It is important that other inspection variables such as outer radius of tube and conductivity of tube remain constant if the calibration is to be valid and usable. For example, in Fig. 18, if the tube radius is varied, whole groups of thickness curves can be generated with accompanying changes of position on the phasor diagram and changes in sensitivity.

The phase-discrimination technique is often used to reduce the effect of a particular variable on one output signal channel of the eddy current instrument (Fig. 19). The impedance diagram for a probe or pancake coil in the vicinity of a nonferromagnetic part is shown in Fig. 19(a). Assume that this diagram can be displayed on an oscilloscope. The locus ABCD represents a thickness locus obtained when the probe coil is in position on the surface of the part. The point D is called an end point of the thickness locus and as such is located on a conductivity locus for this particular probe. If C represents the impedance point for nominal or specified thickness, then variations above or below this nominal thickness value will give readings along the thickness loci at perhaps E or F, respectively. If the probe is lifted from the surface for the condition of nominal thickness, locus point C, the locus CGA is generated. This new locus is sometimes called a coil lift-off locus. When thickness is being measured, it is often desired to reduce the effect of small excursions of the probe from the surface. The phase-discrimination technique can be used for this purpose.

The region of the diagram in the vicinity of impedance point C in Fig. 19(a) is shown enlarged in Fig. 19(b). One output channel of the instrument produces beam deflections in the vertical direction, and the other in the horizontal direction. The enlarged diagram in Fig. 19(b) shows that variations in coil lift-off produce signals on both vertical and horizontal channels. Thus, the

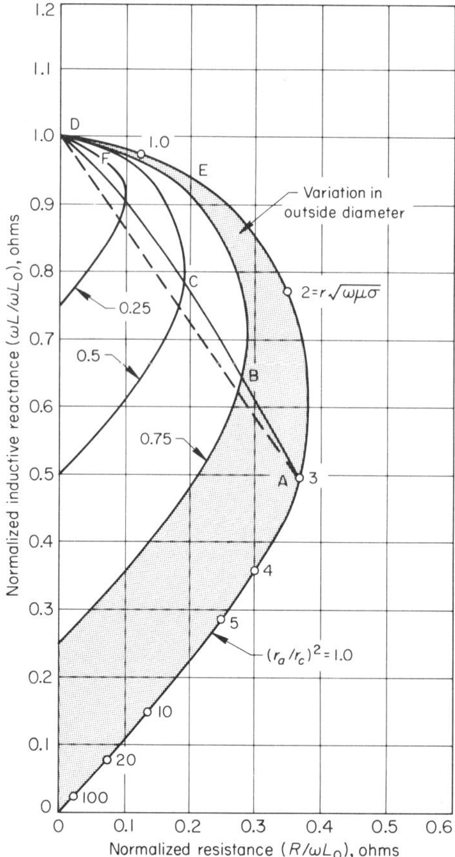

Fig. 17 Effect of variation in bar diameter on the impedance of a long coil encircling a solid cylindrical nonferromagnetic bar

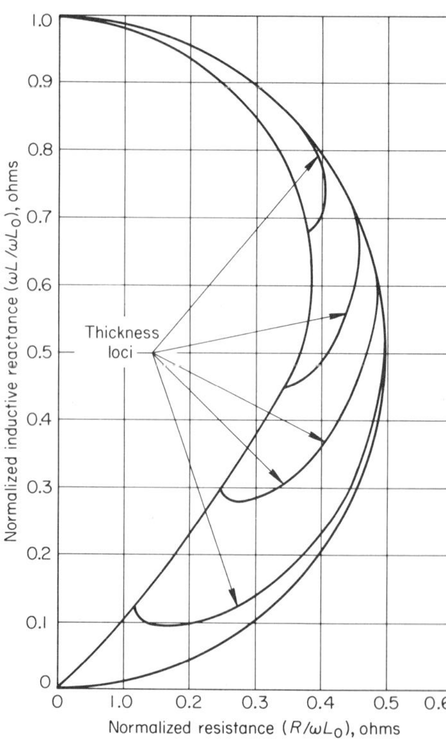

Fig. 18 Impedance diagram showing thickness-loci transition from tube to solid cylinder

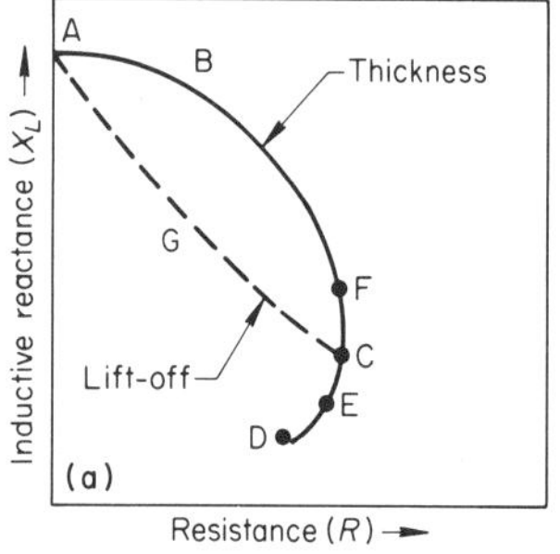

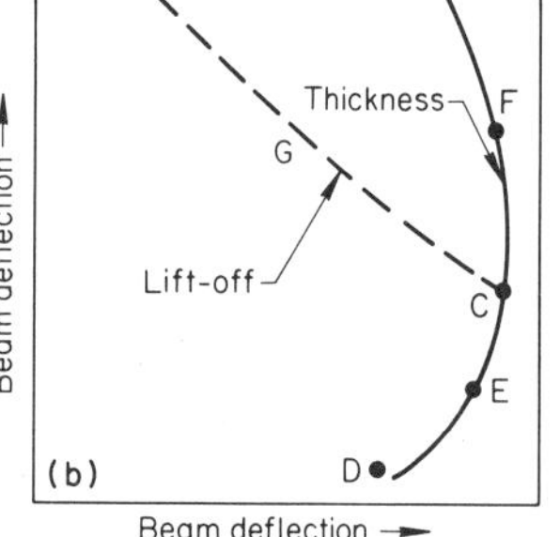

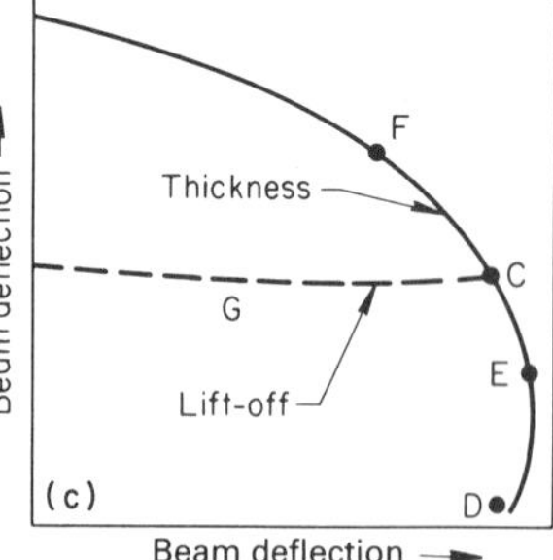

Fig. 19 Phase-discrimination technique for reducing the effect of a particular variable on one output-signal channel during eddy current inspection. (a) Impedance diagram for a probe or pancake coil in the vicinity of a nonferromagnetic part. (b) Enlarged view of the area in the vicinity of impedance point C in diagram (a) as seen on an oscilloscope with the thickness locus in the normal position. (c) Pattern shown in (b) rotated until the coil lift-off locus is horizontal on the oscilloscope. See text for explanation.

lift-off signals will interfere significantly when it is desired to read the effect of thickness variations. Rotation of the phase-shift control, or phase-discrimination control, rotates the pattern or phasor diagram at the output of the instrument. Rotating the pattern in Fig. 19(b) until the lift-off locus in the vicinity of point C is horizontal (Fig. 19c) minimizes the lift-off signal that is effective in the channel producing vertical deflections. Thus, the signal on the vertical channel now varies with thickness variation with little or no effect from small excursions of the probe. Actually, the thickness calibration still changes slightly as a function of probe position, and significant errors can still be introduced by large lift-off excursions of the probe.

Impedance Changes Caused by Small Flaws.* The first widely used formula for the impedance change of an eddy current inspection coil due to a small flaw was given by Burrows (Ref 1). Burrows assumed the flaw to be small enough that the electromagnetic fields were essentially uniform in the vicinity of the flaw. In practice, this means that the flaw is small compared to the standard depth of penetration (the skin depth) and the coil radius. He also assumed that, in general, the flaw was subsurface, and the flaw was small compared to the distance below the surface. For certain flaw geometries, he was also able to apply the formula to surface flaws. Although these restrictions do not fit the description of all flaws of interest, the formula allows calculation of the response to very small flaws at the limit of detectability. This information can be used for estimating ultimate performance and optimizing coil design for detectability.

By restricting the class of materials to nonferromagnetic materials with conductivity σ and the flaws to voids, Burrows's formula can be simplified to the following useful version (Ref 2). If a current I_1 of radian frequency ω in coil 1 causes a uniform magnetic vector potential of magnitude A_1 at the flaw and if a current I_2 in coil 2 causes a uniform potential of magnitude A_2 at the flaw, then the change in the mutual impedance due to the presence of the flaw is:

$$\Delta Z_{12} = \left[(3/2)\sigma\omega^2\left(\frac{A_1}{I_1}\right)\left(\frac{A_2}{I_2}\right)\right] \cdot [\text{Volume} \cdot SOF] \qquad \text{(Eq 2)}$$

where SOF is the shape and orientation factor of the flaw. If there is only one coil, then $A_1 = A_2$, and the mutual impedance becomes the self-impedance of the coil.

In general, the SOF is a complex function of the shape and orientation of the flaw relative to the eddy current fields. However, there are two special cases of interest for which the SOF is relatively simple: the sphere and the circular disk, or penny flaw. For the spherical void, the shape and orientation factor SOF is 1, and the volume is $(4/3)\pi a^3$, where a is the radius of the sphere. For a circular disk oriented with the circle perpendicular to the eddy current field direction, the product of the volume and SOF is $(16/9)a^3$. The same theory can also be applied to surface counterparts of these two flaws: the hemisphere and the half disk, or half penny flaw. In both of these cases, the product of the volume of SOF is half that of their subsurface counterparts, because of the volume being halved.

The other part of Burrows's formula requires determination of the magnitudes of the magnetic vector potentials, A_1 and A_2. Formulas for calculating the magnetic vector potential fields of coils in proximity to conductors are given in Ref 3. These formulas are in the form of definite integrals, which can be evaluated by computer for the geometry of interest.

There have been numerous extensions to the Burrows formulation of the impedance change formula that relax the restriction that the flaw size be small compared to the skin depth and extend the range of flaw shapes for which the SOF can be calculated. However, the solution remains in the basic form given by Burrows, which shows the dependence of the impedance change on the square of the field strength and the effective volume and orientation of the flaw.

*This section was prepared by Thomas Kincaid, Boston University.

Inspection Frequencies

The inspection frequencies used in eddy current inspection range from about 200 Hz to 6 MHz or more. Inspections of nonmagnetic materials are usually performed at a few kilohertz. In general, the lower frequencies, which start at about 1 kHz, are used for inspecting magnetic materials. However, the actual frequency used in any specific eddy current inspection will depend on the thickness of the material being inspected, the desired depth of penetration, the degree of sensitivity or resolution required, and the purpose of the inspection.

Selection of inspection frequency is normally a compromise. For example, penetration should be sufficient to reach any subsurface flaws that must be detected. Although penetration is greater at lower frequencies, it does not follow that as low a frequency as possible should be used. Unfortunately, as the frequency is lowered, the sensitivity to flaws decreases somewhat, and the speed of inspection can be curtailed. Normally, therefore, an inspection frequency as high as possible that is still compatible with the penetration depth required is selected. The choice is relatively simple when detecting surface flaws only, in which case frequencies up to several megahertz can be used. When detecting flaws at some considerable depth below the surface, very low frequencies must be used and sensitivity is sacrificed. Under these conditions, it is not possible to detect small flaws.

In inspecting ferromagnetic materials, relatively low frequencies are normally used because of the low penetration in these materials. Higher frequencies can be used when it is necessary to inspect for surface conditions only. However, even the higher frequencies used in these applications are still considerably lower than those used to inspect nonmagnetic materials for similar conditions.

Selection of operating frequency for the inspection of nonferromagnetic cylindrical bars can be estimated using the chart in Fig. 20. The three main variables on the chart are conductivity, diameter of the part, and operating frequency. A fourth variable, the operating point on the simple impedance curve, is also taken into account. Usually, the desired operating point for cylindrical bars corresponds to a value of $kr = r\sqrt{\omega\mu\sigma}$, which is approximately 4, but which can be in the range 2 to 7. In a typical problem, the two variables of the part, conductivity and radius (or diameter), are known, and it is necessary to find the frequency of operation to determine a particular operating point on the single impedance diagram.

Some typical impedance points and the corresponding kr values are shown in the small impedance diagram at the top left of Fig. 20. A column of values of $r^2\omega\mu\sigma = (kr)^2 = f/f_g$ are also given as a common point between the use of $r\sqrt{\omega\mu\sigma}$ or $f/f_g = r^2\omega\mu\sigma$ as the reference number.

Multifrequency Techniques*

Instrumentation capable of operating at two or more test frequencies has expanded the capabilities of the eddy current method by allowing the user to perform simultaneous tests and to provide for signal combinations using multiparameter techniques. The most widely used application of multifrequency technology has been the inspection of installed heat exchanger tubing. The thin-wall nonferromagnetic tubing used in the production of steam by nuclear power has been extensively tested using these multifrequency techniques since 1977. These steam generators contain thousands of tubes, all of which form a pressure boundary that must be periodically inspected to ensure the safe operating condition of the power plant. In addition to the necessity for inspecting with inside coils, all test probe positioning for tube selection and probe insertion/withdrawal functions must be performed remotely due to the radiation levels in the vicinity of the tube access areas.

*This section was prepared by Howard Houserman, ZETEC, Inc.

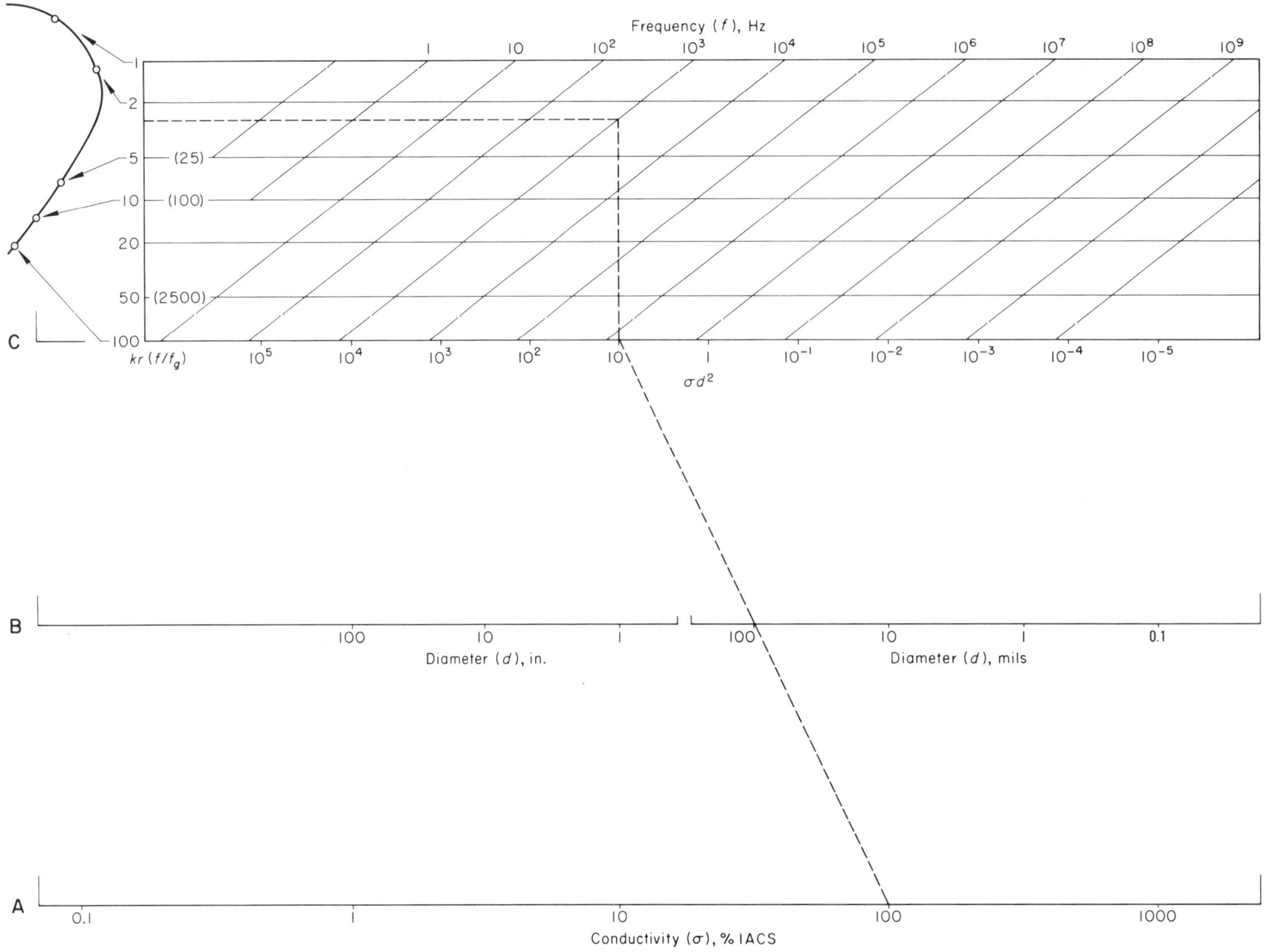

Step 1. Select the value of electrical conductivity (σ) of the bar in per cent IACS (International Annealed Copper Standard) on line A.

Step 2. Select the value of the bar diameter (d) in mils or in inches on either of the scales in line B.

Step 3. Lay a straightedge between these two points, extending the line connecting them to intersect line C.

Step 4. Extend a line vertically from the point on line C found in step 3 until it intersects with a horizontal line corresponding to the desired value of kr.

Step 5. The desired operating frequency is read from the frequency chart (slanted lines), selecting the frequency line that intersects the intersection determined in step 4.

Fig. 20 Chart for selection of frequency for the inspection of nonferromagnetic cylindrical bars

Instrumentation Methods. Test coils are typically excited with multifrequency signals using either continuous or sequential methods.

In the continuous method, excitation currents at each test frequency are simultaneously impressed on the test coil. The test coil outputs are separated by bandpass filters tuned to the individual test frequencies.

The sequential technique relies on switching between test frequencies and is often referred to as a multiplexed system. Detection or demodulation, along with signal display capabilities, is provided for each test frequency. Virtually all the equipment used for this application provides phase and amplitude signal representation through the use of either a cathode ray oscilloscope or a computer screen. The x and y signal components are derived from phase detector circuits electrically separated by 90°. The two signal components are displayed as a single moving point on the screen. Through the use of the storage capabilities of an oscilloscope or other computer methods, the point is traced on the screen to describe what is commonly called a Lissajous signal. Additional instrumentation capabilities can include provisions for both differential and absolute coil arrangements as well as the use of multiple-coil arrays.

Test Frequency and Coil Arrangement Selection. The test frequency for the inspection of installed nonferromagnetic tubing is selected to provide both flaw detection and depth measurement.

Differential Coil Arrangement. With a differential coil arrangement, flaw depth can be related to the change in the angle of the displayed Lissajous signal. Figures 21(a) and 22(a) show the Lissajous signals resulting from artificial flaws in Inconel 600 tubing having a wall thickness of 1.3 mm (0.050 in.) tested at both 400 and 100 kHz. The flaws are 100, 80, 60, 40, and 20% through wall flat-bottom holes originating from the outside diameter of this tube. Although a test frequency of 400 kHz provides a larger range of angles with which to measure flaw depth, testing at 100 kHz can provide for better detection of minor flaws on the tube outside diameter. The capability to test simultaneously at both frequencies can be used to accomplish two goals and to confirm a flawed condition by noting the flaw signal with both tests.

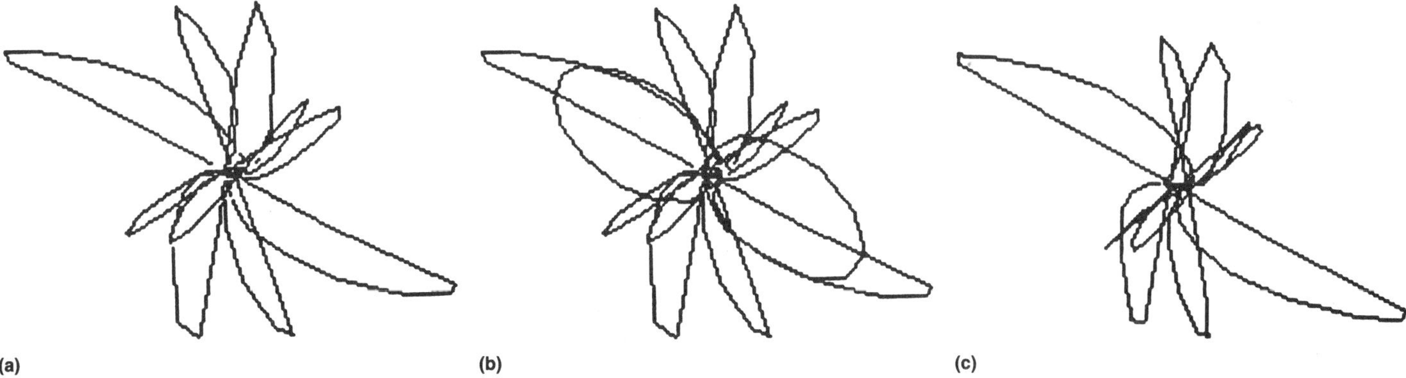

Fig. 21 Lissajous signals resulting from 100, 60, 40, and 20% through wall outside diameter flaws when tested (a) at 400 kHz, (b) at 400 kHz with tube support plate added, and (c) with a digital mixing technique used to eliminate the signal noise that originates from the tube support plate

Absolute Coil Arrangement. Another factor in determining test frequency might be the desire to measure either conductive and magnetic depositions occurring on the tube. A lower test frequency can provide better sensitivity to magnetic deposits that have accumulated on the tube outside diameter. This has been a common practice in an attempt to determine the relationship between deposits on the outer tube wall surface and the presence of flaws originating in the same area. On the other hand, testing at a higher frequency might be desirable to provide increased sensitivity to variations on the inside diameter of the tubing. The latter method, using an absolute coil arrangement, has been implemented to profile the inside diameter of installed tubing that had been incorrectly expanded with mechanical rollers. This test has provided information that enabled an assessment of the induced high-stress areas as well as a basis for selective repair procedures.

The ability to test at multiple frequencies and with both differential and absolute coil arrangements can allow the discrimination and detection of various flaw mechanisms as well as other anomalies of interest in one test scan. This is of great importance where inspection time is critical.

Multiparameter techniques are used to separate test variables by combining the results of testing at more than one frequency. The test variables, or parameters, can include effects such as:

- Lift-off variation caused by probe wobble
- Tube dimension changes resulting from dents, pilgering, and tube expansion processes
- Extraneous signals resulting from tube support plates or depositions (Figures 21(b) and 22(b) illustrate the same flaw signals as shown in Fig. 21(a) and 22(a), with the additional signal resulting from the tube support plate)
- Flaws caused by wastage, cracking, pitting, and so on

By far the most important aspect of the multiparameter method is to provide for the detection and sizing of flawed conditions in the presence of the effects of the other variables.

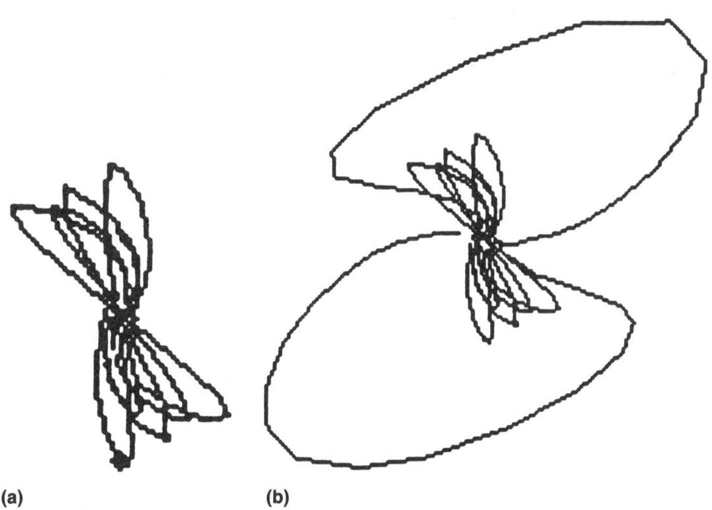

Fig. 22 Lissajous signals resulting from 100, 80, 40, and 20% through wall outside diameter flaws tested at 100 kHz (a) without tube support plate and (b) with tube support plate

One commonly used technique combines the signal from the selected test frequency for flaw detection with a lower frequency to eliminate the effects of the signal resulting from a tube support plate. This mixing process has been accomplished using both analog and digital techniques and combines the output signals from two test frequencies in such a way that the support plate signal is suppressed or eliminated while the signals resulting from flawed conditions remain.

In the analog instrumentation approach, the x and y signal components of the lower-frequency signal are rotated and/or scaled such that the resulting support plate signal closely matches the support plate signal from the test frequency used for flaw detection. The outputs are then combined in a way that subtracts the manipulated lower-frequency signal outputs from the flaw detection frequency output. The result is a mixed signal that shows no response to support plate influences yet can be used to detect and measure flaws.

Digitally, results of equal or better quality can be achieved. Rather than manually manipulating the signal with phase rotators and amplifiers, as is typically done in analog instrumentation, a computer can solve for the best result using mathematical techniques. Figure 21(c) shows the data of Fig. 21(b) resulting from a digital combination or mix used to suppress the support plate response. One approach is to establish a set of simultaneous linear equations prescribing a general signal combination condition. The coefficients of the independent variables are then determined through a least-squares method to provide the signal output desired.

Inspection Coils

The inspection coil is an essential part of every eddy current inspection system. The shape of the inspection coil depends to a considerable extent on the purpose of the inspection and on the shape of the part being inspected. When inspecting for flaws, such as cracks or seams, it is essential that

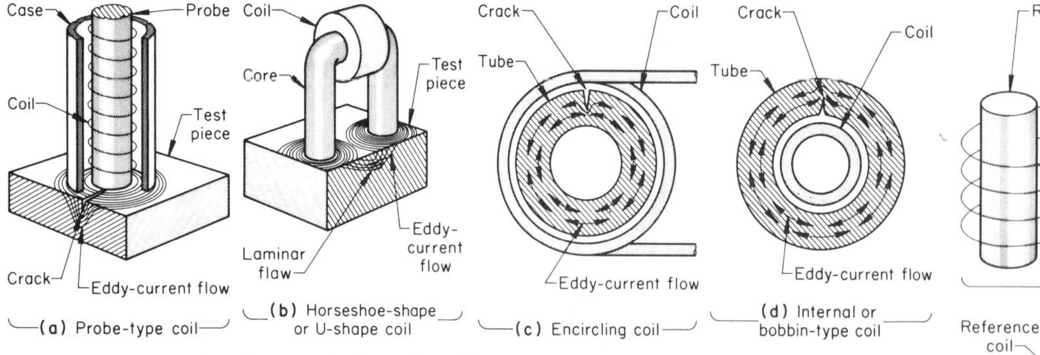

Fig. 23 Types and applications of coils used in eddy current inspection. (a) Probe-type coil applied to a flat plate for detection of a crack. (b) Horseshoe-shaped or U-shaped coil applied to a flat plate for detection of a laminar flaw. (c) Encircling coil applied to a tube. (d) Internal or bobbin-type coil applied to a tube

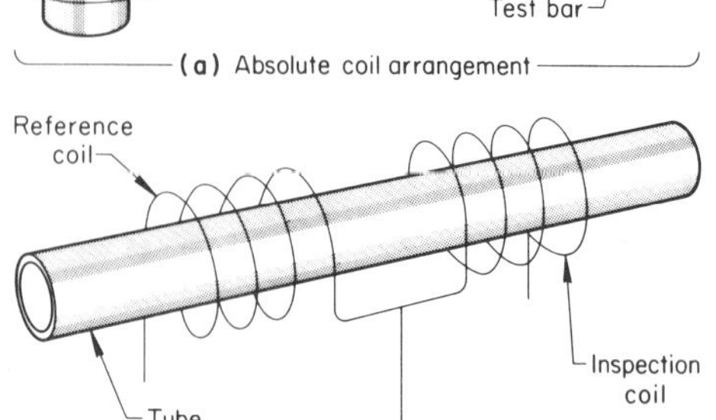

Fig. 24 Absolute and differential arrangements of multiple coils used in eddy current inspection. See text for discussion.

the flow of the eddy currents be as nearly perpendicular to the flaws as possible to obtain a maximum response from the flaws. If the eddy current flow is parallel to flaws, there will be little or no distortion of the currents and therefore very little reaction on the inspection coil.

Probe and Encircling Coils. Of the almost infinite variety of coils employed in eddy current inspection, probe coils and encircling coils are the most commonly used. Normally, in the inspection of a flat surface for cracks at an angle to the surface, a probe-type coil would be used because this type of coil induces currents that flow parallel to the surface and therefore across a crack, as shown in Fig. 23(a). On the other hand, a probe-type coil would not be suitable for detecting a laminar type of flaw. For such a discontinuity, a U-shaped or horseshoe-shaped coil, such as the one shown in Fig. 23(b), would be satisfactory.

To inspect tubing or bar, an encircling coil (Fig. 23c) is generally used because of complementary configuration and because of the testing speeds that can be obtained with this type of coil. However, an encircling coil is sensitive only to discontinuities that are parallel to the axis of the tube or bar. The coil is satisfactory for this particular application because, as a result of the manufacturing process, most discontinuities in tubing and bar are parallel to the major axis. If it is necessary to locate discontinuities that are not parallel to the axis, a probe coil must be used, and either the coil or the part must be rotated during scanning. To detect discontinuities on the inside surface of a tube or when testing installed tubing, an internal or bobbin-type coil (Fig. 23d) can be used. The bobbin-type coil, like the encircling coil, is sensitive to discontinuities that are parallel to the axis of the tube or bar.

Multiple Coils. In many setups for eddy current inspection, two coils are used. The two coils are normally connected to separate legs of an alternating current bridge in a series-opposing arrangement so that when their impedances are the same, there is no output from the pair. Pairs of coils can be used in either an absolute or a differential arrangement (Fig. 24).

Absolute Coil Arrangements. In the absolute arrangement (Fig. 24a), a sample of acceptable material is placed in one coil, and the other coil is used for inspection. Thus, the coils are comparing an unknown against a standard, with the differences between the two (if any) being indicated by a suitable instrument. Arrangements of this type are commonly employed in sorting applications.

Differential Coil Arrangement. In many applications, an absolute coil arrangement is undesirable. For example, in tubing inspection, an absolute arrangement will indicate dimensional variations in both outside diameter and wall thickness even though such variations may be well within allowable limits. To avoid this problem, a differential coil arrangement such as that shown in Fig. 24(b) can be used. Here, the two coils compare one section of the tube with an adjacent section. When the two sections are the same, there is no output from the pair of coils and therefore no indication on the eddy current instrument. Gradual dimensional variations within the tube or gross variations between individual tubes are not indicated, while discontinuities, which normally occur abruptly, are very apparent. In this way, it is possible to have an inspection system that is sensitive to flaws and relatively insensitive to changes that normally are not of interest.

Sizes and Shapes. Inspection coils are available in a variety of sizes and shapes. Selection of a coil for a particular application depends on the type of discontinuity. For example, when an encircling coil is used to inspect tubing or bar for short discontinuities, optimum resolution is obtained with a short coil. Alternatively, a

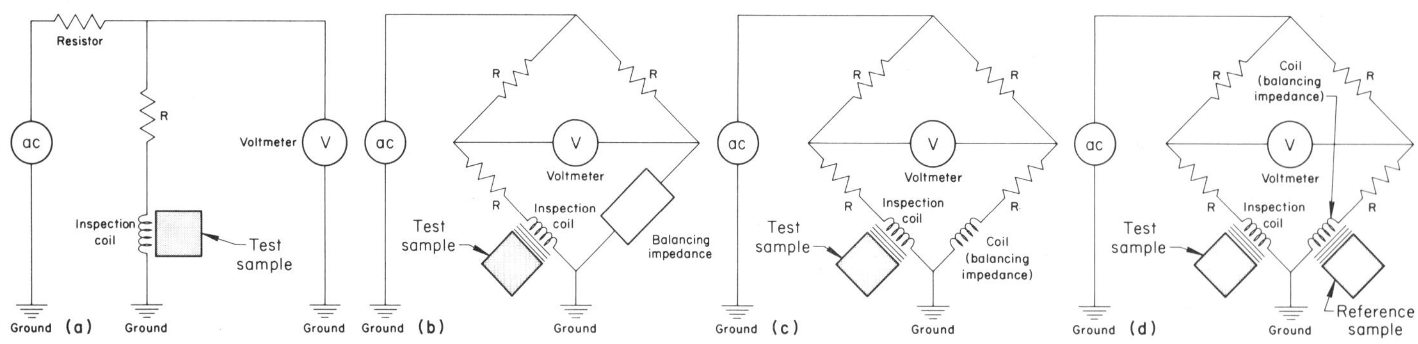

Fig. 25 Four types of eddy current instruments. (a) A simple arrangement, in which voltage across the coil is monitored. (b) Typical impedance bridge. (c) Impedance bridge with dual coils. (d) Impedance bridge with dual coils and a reference sample in the second coil

PROPERTY OF CASB / ENGINEERING
PROPRIÉTÉ du LABORATOIRE d'INGÉNIERIE

short coil has the disadvantage of being sensitive to the position of the part in the coil. Longer coils are not as sensitive to the position of the part, but are not as effective in detecting very small discontinuities. Small-diameter probe coils have greater resolution than larger ones, but are more difficult to manipulate and are more sensitive to lift-off variations.

Eddy Current Instruments

This section discusses the various types of detection and readout instrumentation used in eddy current inspection.

Instrument System Operations

Eddy current instruments can be classified as belonging to one of the following categories:

- Resistor and single-coil system
- Bridge unbalance system
- Induction bridge system
- Through transmission system

Resistor and Single-Coil System. A simple eddy current instrument, in which the voltage across an inspection coil is monitored, is shown in Fig. 25(a). This circuit is adequate for measuring large lift-off variations if accuracy is not extremely important.

Bridge Unbalance System. A circuit designed for greater accuracy is shown in Fig. 25(b). This instrument consists of a signal source, an impedance bridge with dropping resistors, an inspection coil in one leg, and a balancing impedance in the other leg. The differences in voltage between the two legs of the bridge are measured by an ac voltmeter. Alternatively, the balancing impedance in the leg opposite the inspection coil may be a coil identical to the inspection coil, as shown in Fig. 25(c), or it may have a reference sample in the coil, as shown in Fig. 25(d). In the latter, if all the other components in the bridge were identical, a signal would occur only when the inspection coil impedance deviated from that of the reference sample.

There are other methods of achieving bridge balance, such as varying the resistance values of the resistor in the upper leg of the bridge and one in series with the balancing impedance. The most accurate bridges can measure absolute impedance to within 0.01%. However, in eddy current inspection, it is not how an impedance bridge is balanced that is important, but rather how it is unbalanced. Because of the effect of undesired variables, eddy current inspections are seldom performed with the bridge balance.

Figure 26 plots the voltages across an inspection coil, $\mathbf{V}_1$, and a reference coil, $\mathbf{V}_2$, for a bridge such as that shown in Fig. 25(c). For simplification, the loading effects the voltmeter has on the system have been omitted. Voltage inspection values for nine different samples, representing three different levels of magnetic permeability and three different levels of electrical conductivity, are shown. The voltage measured by the voltmeter is the magnitude of the difference between the voltages of the inspection coil and the reference coil. The voltage of the reference coil can be moved to any point on the diagram by varying the components in the bridge. For this particular inspection, the intent was to measure the small permeability changes without being affected by the conductivity changes. If the bridge were balanced at the nominal value so that the $\mathbf{V}_2$ vector terminated on the point $\sigma = \sigma_0$ and $\mu = 1.005$, changes in permeability would result in a large change in the magnitude of the difference voltage, $|\Delta\mathbf{V}|$, but a similarly large change would result if the conductivity varied. However, if the voltage of the reference coil were adjusted equidistant from $\sigma = \sigma_0 - \Delta\sigma$ and $\sigma_0 + \Delta\sigma$, then the same difference voltage will result for both points (for changes in conductivity), but a large change in difference voltage will result with changes in permeability. If the line between $\sigma_0 - \Delta\sigma$ and $\sigma_0 + \Delta\sigma$ is the arc of a circle and if the voltage of the reference coil is adjusted to the center of that circle, then there will be no change in the magnitude of the difference voltage as the conductivity varies between $\sigma_0 - \Delta\sigma$ to $\sigma_0 + \Delta\sigma$, but the voltage will vary with permeability changes. Thus, the effects of the undesired variable—conductivity—can be reduced or eliminated, and the measurement can be made sensitive to the desired variable—permeability. This indicates how bridge unbalance is used to eliminate a single undesired variable.

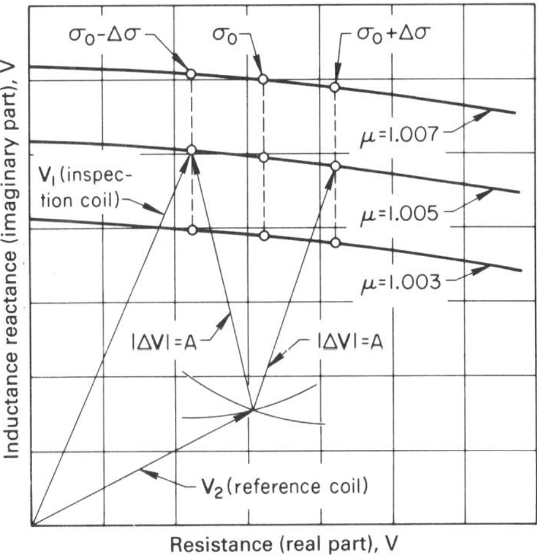

Fig. 26 Complex voltage diagram for an inspection coil and a reference coil. The voltage inspection values are for nine different samples, representing three different levels of magnetic permeability and three different levels of electrical conductivity.

A major limitation of this technique is that there are usually several undesirable variables in any eddy current inspection, and not all of them can be eliminated completely. Normalized coil impedance variations with conductivity and lift-off produce changes in impedance that are not parallel to each other. Regardless of where the reference-voltage point is set, the effects of both variables cannot be eliminated completely. Thus, with a single-frequency inspection, it is preferable to select operating conditions so that undesired variables are all approximately parallel to each other and perpendicular to the variable to be measured. A two-frequency inspection would allow for the measurement of four variables.

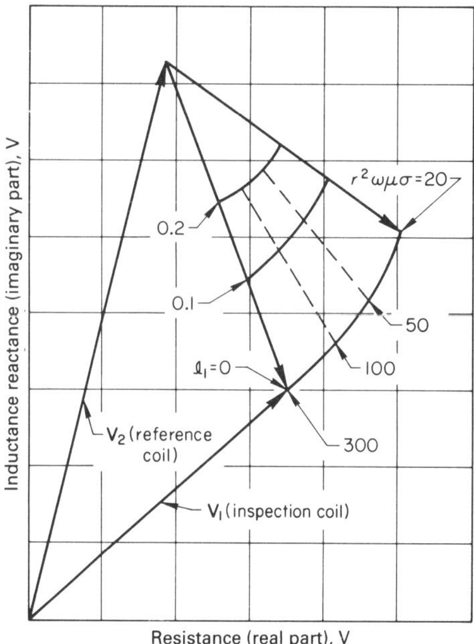

Fig. 27 Complex voltage diagram for inspection coil and reference coil for various electrical conductivities and lift-off

In some applications, it is more desirable to measure the phase rather than the magnitude of the voltage difference. For example, Fig. 27 shows how both the magnitude and phase of the unbalance voltage vary as functions of electrical conductivity and lift-off. Thus, it is possible to measure the phase shift as a function of conductivity and be relatively insensitive to changes in lift-off, the usual method of phase measurement. On the other hand, magnitude can be measured as a function of lift-off and can be relatively insensitive to the conductivity. As a result, the unbalanced bridge can function as a decoding network to separate the two variables.

Temperature stability is also more of a problem when the bridge is unbalanced. When the bridge is balanced and two corresponding components in opposite legs of the bridge drift with exactly the same temperature coefficient, the bridge remains in balance. However, when the bridge is operated in an unbalanced mode, the thermal drift in components fails to cancel by an amount proportional to the amount of unbalance.

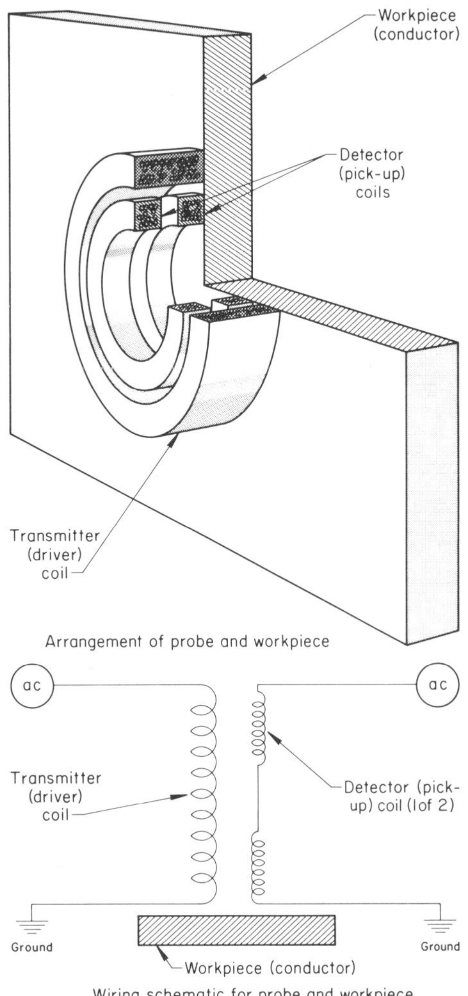

Fig. 28 Reflection (transmit-receive) probe in place at the surface of a workpiece. Schematic shows how power signal is transformer-coupled from a transmitter coil into two detector coils—an inspection coil (at bottom) and a reference coil (at top).

Table 2 Types and capabilities of commercially available eddy current instruments

Type of instrument	Frequency range	Signals measured	Simultaneous frequencies	Features
Resistor and single coil	1 kHz to 5 MHz	Magnitude	1	Direct reading, analog meters
Inspection coil and balance coil, bridge unbalance	1 kHz to 5 MHz	Magnitude, phase x_l, x_r	1	Phase rotation of signals, storage scope, display of impedance planes, continuously variable frequency, *x-y* alarm gates. Portable up to 500 kHz
Inspection coil and variable impedance, bridge unbalance	1 kHz to 2 MHz	Magnitude, phase x_l, x_r	1	Direct digital readout of thickness and electrical conductivity, binary coded decimal output
Induction bridge	100 Hz to 50 MHz	Magnitude, phase	2	Simultaneous measurement of four variables, analog computers, binary coded decimal output, direct digital readout of thickness and lift-off. Portable up to 500 kHz

Another limitation in bridge unbalance procedures is the difficulty in finding the proper unbalance. There are many poor choices for setting the reference voltage, and the few good choices are difficult to identify when only one meter is used to monitor voltage.

Induction Bridge System. Another type of bridge system is an induction bridge, in which the power signal is transformer-coupled into an inspection coil and a reference coil. In addition, the entire inductance-balance system is placed in the probe, as shown in Fig. 28. The probe consists of a large transmitter, or driver, coil, and two small detector, or pickup, coils wound in opposite directions as mirror images to each other. An alternating current is supplied to the large transmitter coil to generate a magnetic field. If the transmitter coil is not in the vicinity of a conductor, the two detector coils detect the same field, and because they are wound in opposition to each other, the net signal is zero. However, if one end of the probe is placed near a metal surface, the field is different at the two ends of the probe, and a net voltage appears across the two coils. The resultant field is the sum of a transmitted signal, which is present all the time, and a reflected signal due to the presence of a conductor (the metal surface). The intensity of the transmitted signal decreases rapidly as the distance between the coil and conductor is increased, and the intensity of the reflected wave does the same. The detector coil nearer the conductor detects this reflected wave, but the other detector coil (the reference coil) does not, because the amplitude of the wave has greatly decreased in the distance from the reflecting metal surface to the rear detector coil.

The magnitude and phase obtained for a system such as this are similar to those in a bridge unbalance system (Fig. 27) with the reference coil in air. However, the effects of temperature variations of the probe can be completely eliminated from the phase-shift measurements.

Through Transmission System. Another system of eddy current measurements is the through transmission system, in which a signal is transmitted from a coil through a metal and detected by a coil on the opposite side of the metal. If the distance between the two coils is fixed and the driving and detecting circuits have high impedances, the signal detected is independent of position of the metal provided it remains between them. This type of measurement completely eliminates lift-off but requires that the two coils be positioned.

Table 2 outlines the types of eddy current instruments commercially available and lists the capabilities. The features listed are intended to cover all instruments of a given type. No one instrument includes all the features listed for a given type of instrument; several manufacturers produce each of the types listed. Each manufacturer will provide more precise details and specifications on instrument type, performance, and coil parameters.

Readout Instrumentation

An important part of an eddy current inspection system is the instrument used for a readout. The readout device can be an integral part of the system, an interchangeable plug-in module, or a solitary unit connected by cable. The readout instrument should be of adequate speed, accuracy, and range to meet the inspection requirements of the system. Frequently, several readout devices are employed in a single inspection system. The more common types of readout, in order of increasing cost and complexity, are discussed in the following sections.

Alarm lights alert the operator that a test parameter limit has been exceeded.

Sound alarms serve the same purpose as alarm lights, but free the attention of the operator so that he can manipulate the probe in manual scanning.

Kick-out relays activate a mechanism that automatically rejects or marks a part when a test parameter has been exceeded.

Analog meters give a continuous reading over an extended range. They are fairly rapid (with a frequency of about 1 Hz), and the scales can be calibrated to read parameters directly. The accuracy of the devices is limited to about 1% of full scale. They can be used to set the limits on alarm lights, sound alarms, and kick-out relays.

Digital meters provide much greater accuracy and range than analog meters. The chance of operator error is much less in reading a digital meter, but fast trends are more difficult to interpret. Although many

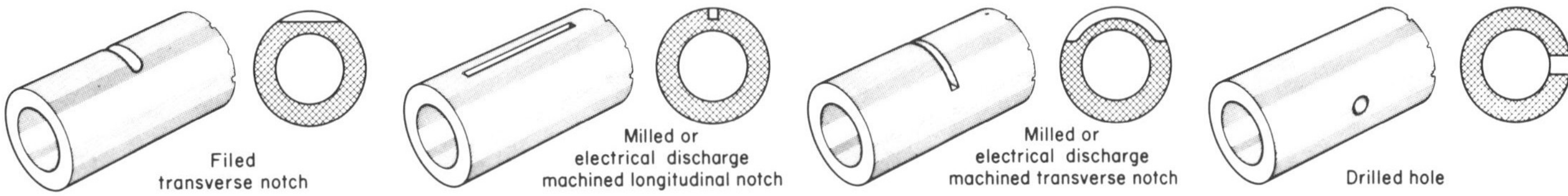

Fig. 29 Several fabricated discontinuities used as reference standards in eddy current inspection

digital meters have binary coded decimal output, they are relatively slow.

***X-y* plotters** can be used to display impedance-plane plots of the eddy current response. They are very helpful in designing and setting up eddy current bridge unbalance inspections and in discriminating against undesirable variables. They are also useful in sorting out the results of inspections. They are fairly accurate and provide a permanent copy.

***X-y* storage oscilloscopes** are very similar to *x-y* plotters but can acquire signals at high speed. However, the signals have to be processed manually, and the screen can quickly become cluttered with signals. In some instruments, high-speed *x-y* gates can be displayed and set on the screen.

Strip chart recorders furnish a fairly accurate (~1% of full scale) recording at reasonably high speed (~200 Hz). However, once on the chart, the data must be read by an operator. Several channels can be recorded simultaneously, and the record is permanent.

Magnetic tape recorders are fairly accurate and capable of recording at very high speed (1 kHz). Moreover, the data can be processed by automated techniques.

Computers. The data from several channels can be fed directly to a high-speed computer, either analog or digital, for online processing. The computer can separate parameters and calculate the variable of interest and significance, catalog the data, print summaries of the result, and store all data on tape for reference in future scans.

Discontinuities Detectable by Eddy Current Inspection

Basically, any discontinuity that appreciably alters the normal flow of eddy currents can be detected by eddy current inspection. With the encircling coil inspection of either solid cylinders or tubes, surface discontinuities having a combination of predominantly longitudinal and radial dimensional components are readily detected. When discontinuities of the same size are located beneath the surface of the part being inspected at progressively greater depths, they become increasingly difficult to detect and can be detected at depths greater than 13 mm (½ in.) only with special equipment designed for this purpose.

On the other hand, laminar discontinuities such as can be found in welded tubes may not alter the flow of the eddy currents enough to be detected unless the discontinuity breaks either the outside or inside surfaces or unless it produces a discontinuity in the weld from upturned fibers caused by extrusion during welding. A similar difficulty could arise in the detection of a thin planar discontinuity that is oriented substantially perpendicular to the axis of the cylinder.

Regardless of the limitations, a majority of objectionable discontinuities can be detected by eddy current inspection at high speed and at low cost. Some of the discontinuities that are readily detected are seams, laps, cracks, slivers, scabs, pits, slugs, open welds, miswelds, misaligned welds, black or gray oxide weld penetrators, pinholes, hook cracks, and surface cracks.

Reference Samples. A basic requirement for eddy current inspection is a reliable and consistent means for setting the sensitivity of the tester to the proper level each time it is used. A standard reference sample must be provided for this purpose. Without this capability, eddy current inspection would be of little value. In selecting a standard reference sample, the usual procedure is to select a sample of product that can be run through the inspection system without producing appreciable indications from the tester. Several samples may have to be run before a suitable one is found; the suitable one then has reference discontinuities fabricated into it.

The type of reference discontinuities that must be used for a particular application are specified (for example, by the American Society for Testing and Materials and the American Petroleum Institute). Some of the major considerations in selecting reference discontinuities are that they:

- Must meet the required specification
- Should be easy to fabricate
- Should be reproducible
- Should be producible in precisely graduated sizes
- Should produce an indication on the eddy current tester that closely resembles those reduced by the natural discontinuities

Several discontinuities that have been used for reference standards are shown in Fig. 29. These include a filed transverse notch, milled or electrical discharge machined longitudinal and transverse notches, and drilled holes. Figure 30 shows the eddy current signals generated when testing a hollow tube with an internal rotating probe and the discontinuities in the positions shown.

Inspection of Tubes*

The techniques used in the eddy current inspection of tubes differ depending on the diameter of the tube. Additional information is available in the article "Remote-Field Eddy Current Inspection" and "Tubular Products" in this Volume.

Tube Outside Diameter Under 75 mm (3 in.)

Tubes up to 75 mm (3 in.) in diameter can be eddy current inspected for discontinuities using an external encircling coil. The diameter limitation is imposed primarily by resolution requirements; that is, as the diameter is increased, the area of a given discontinuity becomes an increasingly smaller percentage of the total inspected area.

The inspection is performed by passing the tube longitudinally through a concentric coil assembly. The coil assembly contains an energizing (primary) coil, a differentially wound detector (secondary) coil, and when inspecting ferromagnetic materials, a magnetic-saturating (direct current) coil. A typical coil assembly and V-roll conveyor for transporting the tube are shown in Fig. 31.

The energizing coil is energized with alternating current at a frequency compatible with the inspection situation (typically 1 kHz for many ferromagnetic products) and induces the eddy currents in the tube. The detector coil monitors the flow of the induced currents and permits detection of current variations, which are indicative of discontinuities. The saturating coil, when used, is energized with direct current at high current levels to produce magnetic saturation in the cylinder. This increases the eddy current penetration and nullifies the effects of magnetic variables that may otherwise degrade the signal-to-noise ratio of the inspection. Because of the orientation of eddy current flow, this type of inspection is best suited to the detection of such discontinuities as pits, slugs, seams, laps, and cracks. The inside diameter of the coil as-

*Examples 1 and 2 were prepared by V.S. Cecco, Atomic Energy of Canada Limited, and Example 3 was prepared by E.M. Franklin, Argonne National Laboratory.

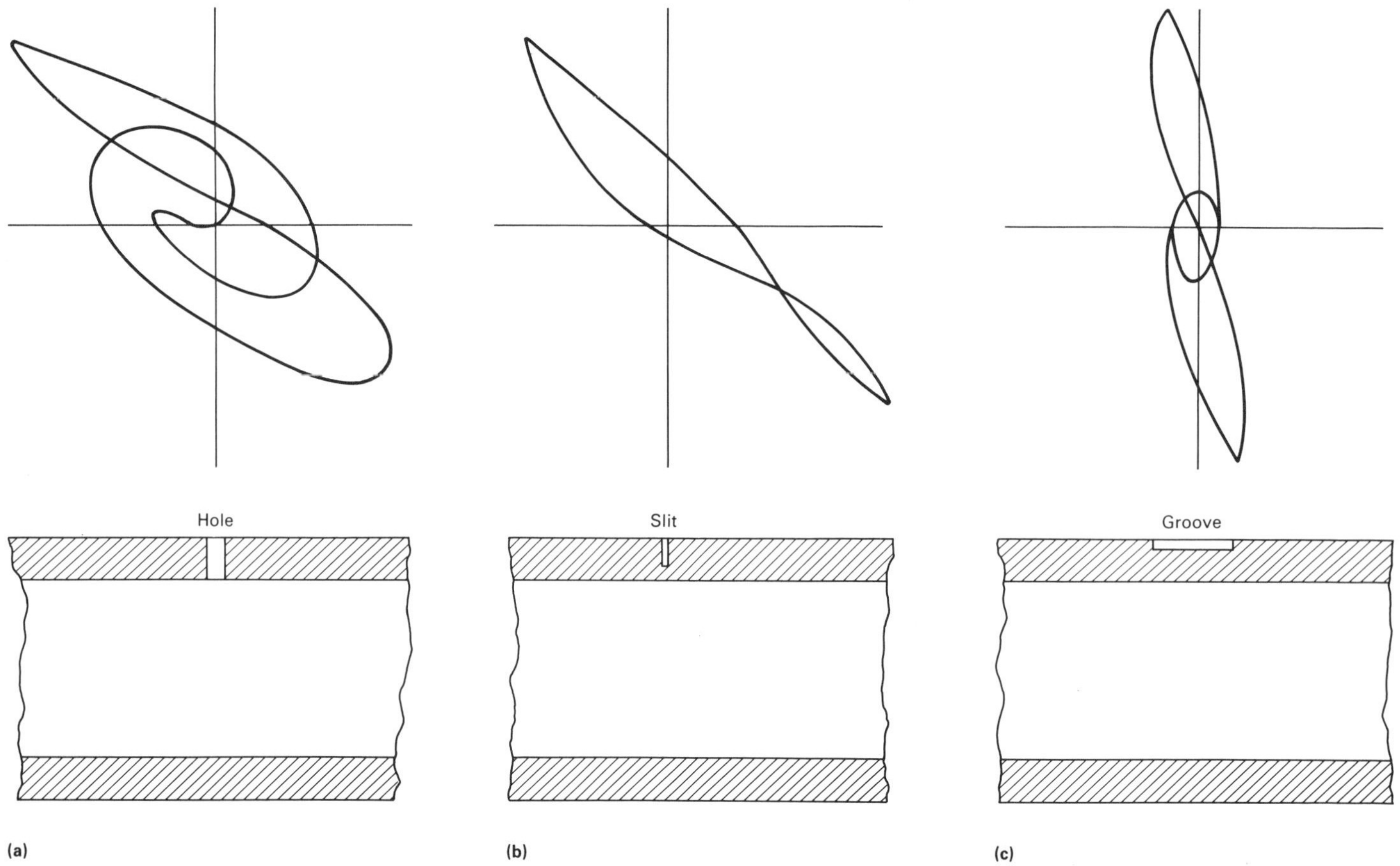

Fig. 30 Eddy current signals obtained with an internal rotating probe operating at 250 kHz for the discontinuities shown. (a) 2 mm (0.08 in.) diam radial hole. (b) OD surface narrow slit. (c) OD surface longitudinal groove. Source: Ref 4

sembly is about 9.5 mm (3/8 in.) larger than the outside diameter of the tube to allow for mechanical or geometric irregularities. The coil is centered on the tube. The cylinder-conveying mechanism must provide smooth, uniform propulsion. Throughput speeds of 30 m/min (100 sfm) are common. Because inspection instrumentation is normally designed for a fixed throughput speed, conveyor speed should be held within about ±10% of nominal.

Encircling coil inspection offers simplicity of application, both electrically (a single inspection coil assembly and circuit for total wall inspection) and mechanically (no scanning mechanisms required). A disadvantage is that the discontinuity is located with respect to only its longitudinal, not its circumferential, position. For best results, each inspection system must be carefully adapted to the product it is to inspect. When this is achieved, a large majority of objectionable discontinuities can be detected at high speed and at low cost using eddy current methods. Eddy current inspection is excellent for detecting discontinuities that are below the surface or within 13 mm (1/2 in.) of the surface. The three examples that follow show typical applications of eddy current inspection on tubes of up to 75 mm (3 in.) outside diameter.

Example 1: Quality Control of Nonferromagnetic Heat Exchanger Tubesheet Rolled Joints. General inspection of heat exchanger tubes can be performed with differential or absolute probes. For dimensional measurements, however, absolute probes are normally used.

The absolute probe (Fig. 32) uses a toroidal reference coil that matches the characteristic impedance of the test coil within 10% over a wide frequency range, making it effective for multifrequency testing. Good

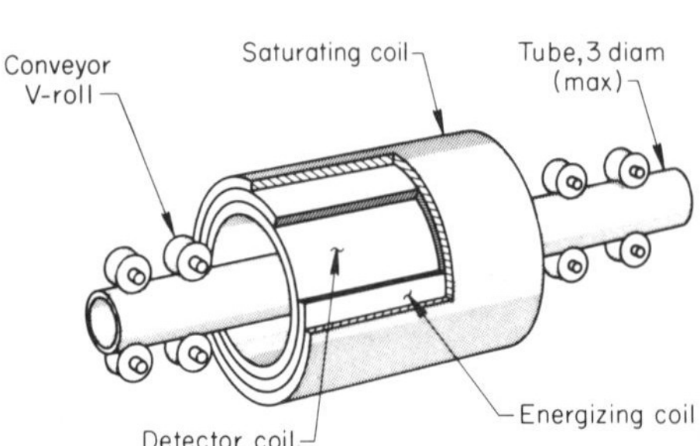

Fig. 31 Setup and encircling coil components for continuous eddy current inspection of ferromagnetic or nonferromagnetic tubes up to about 75 mm (3 in.) in diameter

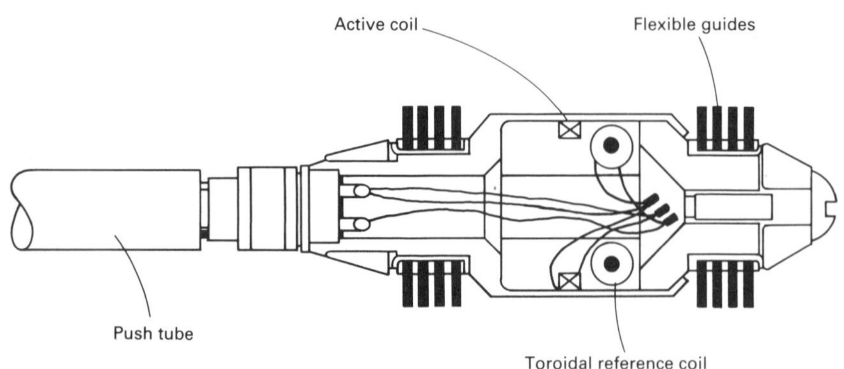

Fig. 32 Absolute bobbin probe with flexible wafter guides used for the eddy current evaluation of heat exchanger tubesheet rolled joints

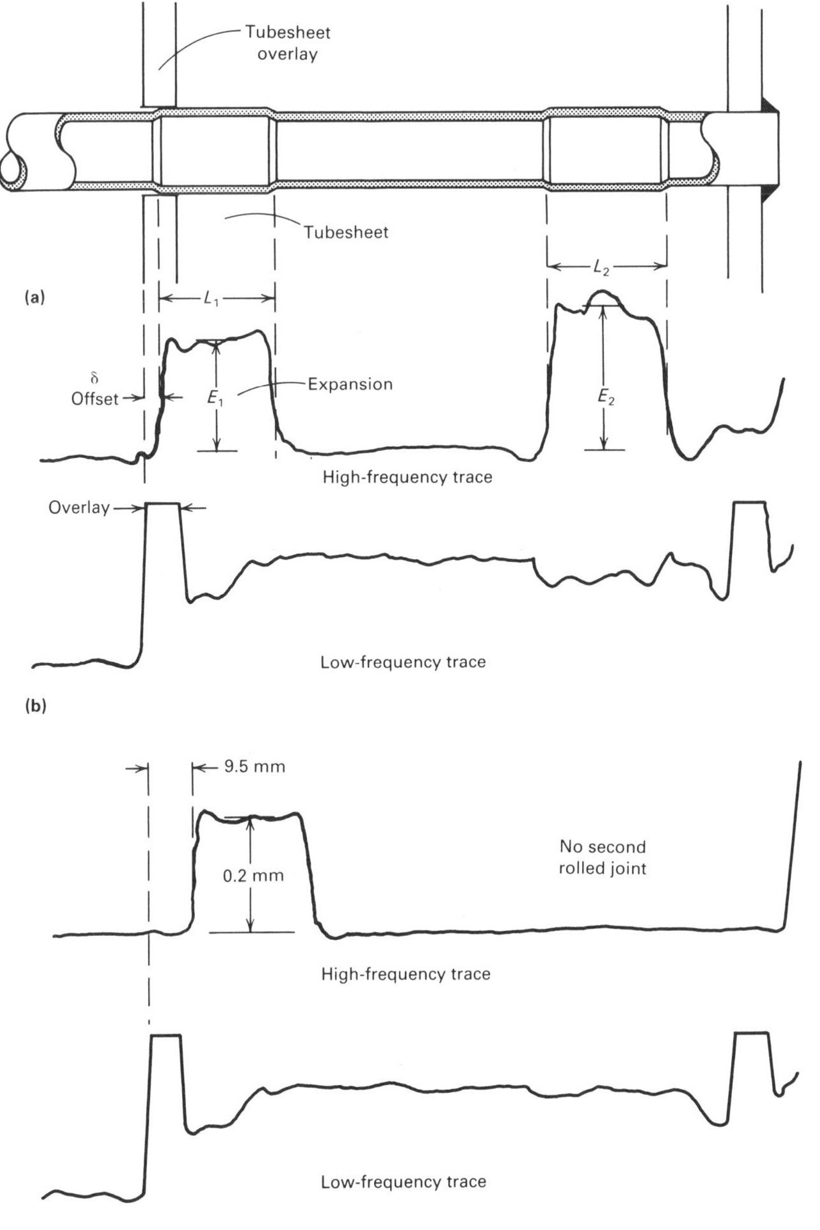

Fig. 33 Dual-frequency inspection of tubesheet showing tube expansion, offset, and overlay regions of rolled areas. (a) Cross-sectional view of tubesheet region. (b) Traces showing typical rolled joints. (c) Traces showing incorrect tubesheet rolled joints. The high-frequency (700 kHz) traces display tube expansion dimensions, while the low-frequency (100 kHz) traces display the tubesheet overlay location.

guidance is essential for absolute probes; flexible wafter guides (Fig. 32) center the probe by peripheral contact, but collapse to cope with diametral variations and deposits.

One application of an absolute probe is the measurement for location of the rolled joint relative to the tubesheet secondary face and the degree of rolling in tubesheet rolled joints. Heat exchangers and steam generators are normally assembled with nonferromagnetic tubes rolled into the tubesheet and then welded at the primary tubesheet face. Rolling is primarily performed to eliminate corrosion-prone crevices. However, if tubes are rolled beyond the tubesheet secondary face, they are prone to cracking. Therefore, the location and, to a lesser extent, the degree of rolling are critical. These dimensions cannot be readily measured directly. A dual-frequency eddy current method, using an absolute probe, can provide such measurements rapidly and reliably. A high test frequency locates tube-expanded regions, and a low frequency simultaneously locates the tubesheet face. As shown in Fig. 33(a), the tubesheet has an Inconel nonferromagnetic overlay, which makes detection easier.

Strip chart traces of a typical tube are illustrated in Fig. 33(b). One channel displays the high-frequency (700 kHz) signal, and the other the low-frequency (100 kHz) signal. The location of the rolled joint relative to the tubesheet secondary face and degree of rolling can be readily determined. Figure 33(c) illustrates the results from an improperly rolled tube, with a large crevice region (9.5 mm, or ⅜ in.) and without the secondary rolled joint. Using this technique, the distance between the rolled joint and tubesheet face can be measured rapidly to an accuracy of ±1 mm (±0.04 in.), and the increase in diameter of the rolled sections can be measured to better than ±10%. No surface preparation of the tubing is required prior to testing.

Example 2: Probes for Inspecting Ferromagnetic Heat Exchanger Tubes. Ferromagnetic tubes present special problems in eddy current testing. Real defect signals are normally indistinguishable from those due to normal permeability variations. Permeability values can range from 20 to several hundred in engineering materials and can vary with composition, cold work, and thermal history. The best remedy for the eddy current testing of magnetic tubes is magnetic saturation in the vicinity of the test coil(s). If greater than 98% saturation can be achieved, the signals from defects and other anomalies display the characteristic phase expected for nonferromagnetic tubes. Unfortunately, not all magnetic tubes can be completely saturated. In these cases, eddy current inspection reduces to a measurement of magnetic perturbation. A thin, internal surface layer adjacent to the probe responds to the distortion of the magnetic flux at defects from the saturation field. This classifies the technique with nondestructive testing methods such as magnetic particle inspection and flux leakage testing (see the articles "Magnetic Particle Inspection" and "Magnetic Field Testing" in this Volume).

The importance of achieving maximum saturation is illustrated in Fig. 34, which shows results from a type 439 stainless steel heat exchanger tube. A 15.9 mm (0.625 in.) OD by 1.2 mm (3/64 in.) thick tube with internal and external calibration defects and shot-peened area was used to compare the performance of various saturation probes. As shown in Fig. 34(a), the external defects ranged from 20 to 100% deep. Figure 34(b) shows the signals obtained with a probe capable of 98% saturation; the eddy current signals from the external calibration holes display the characteristic phase rotation with depth expected for nonmagnetic materials. In contrast, with only 89% saturation, the signals are distorted and indistinguishable from change-in-magnetic-permeability signals (Fig. 34c). From similar tests on other ferromagnetic tubes it has been found that at least 98% saturation is needed (relative permeability < 1.2) for reliable test results. This requires detailed optimization of the saturation magnet design for each ferromagnetic tube material.

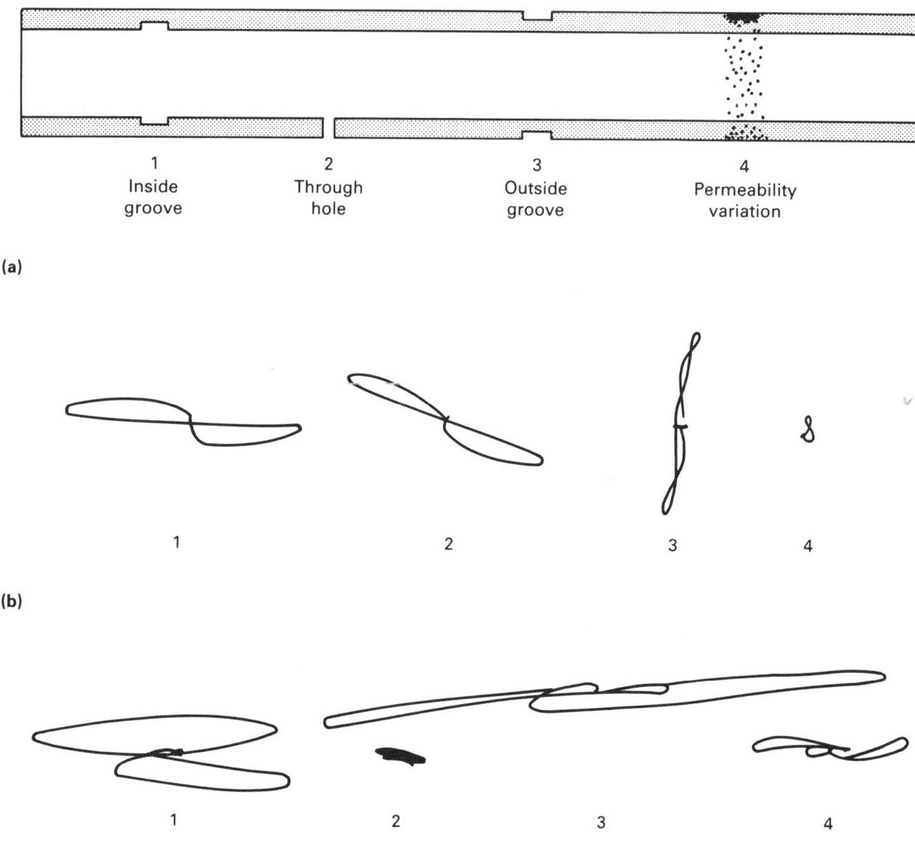

Fig. 34 Signals obtained from type 439 stainless steel calibration tube when using saturation probes. (a) Cross-sectional view of calibration tube showing location of discontinuities. (b) Signals obtained at each discontinuity shown in (a) using a probe with 98% saturation. (c) Signals obtained at each discontinuity shown in (a) using a probe with 89% saturation

Inspection for fretting wear in ferromagnetic tubes presents an even more difficult problem. Probes capable of complete saturation of unsupported tube sections cannot normally saturate the tube under a carbon steel support. This is because the magnetic flux takes the lower-reluctance path through the support rather than along the tube.

To saturate under baffle plates requires a probe with radial magnetization. In this case, the magnetic flux remains almost undisturbed and actually increases slightly under baffle plates. Eddy current signals from a radial saturation probe are illustrated in Fig. 35. The large signal in Fig. 35(b) from the baffle plate is due to the probe sensing an increase in permeability of the Monel tube. As a result, the 40% fretting wear, when under the support, causes only minor distortion to this large signal and is virtually undetectable. The radial saturation probe continues to saturate the Monel tube as it passes under the carbon steel baffle. As illustrated in Fig. 35(c), 40% fretting wear under the baffle plate is readily detectable by the vertical component of the vectorially additive signals.

Permanent-magnet probes can achieve high magnetic saturation and can provide clear, undistorted defect signals. False ferromagnetic indications are eliminated, and defect depth can be determined, eliminating unnecessary tube plugging.

Inspection can be performed on ferromagnetic tubing such as Monel 400, 3Re60, and type 439 stainless steel. Only if thick deposits are present in the bore of the tube must the tube be cleaned prior to testing; no other surface preparation is required. Inspection speeds are comparable to nonferromagnetic testing, typically 0.5 m/s (1.6 ft/s). Inspection costs are similar to those of nonferromagnetic tube inspections, and only conventional instrumentation is needed.

Example 3: Comparison of Skin Depth Test Frequency Versus Foerster Limit Frequency Methods to Obtain Optimum Test Frequency in Eddy Current Inspection of Type 304 Stainless Tubing. In eddy current inspection, test frequency is an essential test parameter. Traditional nonmagnetic testing methods employ the standard depth of penetration (SDP) or skin depth approach for frequency selection (Ref 5, 6). Figure 36 shows a comparison between the SDP and the Förster limit frequency, F_g, approaches to frequency selection for defect testing in stainless tubing filled with conducting material (Ref 7-9). The standard used in the comparison was a type 304 stainless steel tube (6.35 mm, or 0.250 in., OD × 5.54 mm, or 0.218 in., ID) with through holes 3.81 mm (0.150 in.) and 1.11 mm (0.0437 in.) in diameter drilled at opposite ends. At one end, lead slugs were placed inside the tube beneath the through holes.

Figure 36(b) shows the results when the SDP method was used to select frequency. In this case, the lead slugs masked the through holes. For the SDP method, the test frequency was determined as follows:

$$F = \frac{1}{(\mathrm{SDP})^2 \pi \mu \sigma} = 1.2\ \mathrm{MHz} \qquad \text{(Eq 3)}$$

where $\sigma = 0.14 \times 10^7$ mho/m, $\mu = (\mu_{rel} \times \mu_0) = (1.02 \times 4\pi \times 10^{-7}$ H/m), and SDP = 3.81×10^{-4} m (tube wall thickness).

Figure 36(c) shows the results of testing at a frequency determined by the F_g method. In this case, the presence of lead slugs was suppressed. For test situations of this type, the optimum test frequency, F_T, will be in the range of 2 to 20 times F_g or $2 \leqq F_T/F_g \leqq 20$. In this case:

$$F_g = \frac{C}{\mu_{rel} \sigma g_1 g_2} = 0.15\ \mathrm{MHz} \qquad \text{(Eq 4)}$$

where C = 5.066×10^5 (unitless constant of proportionality), $\sigma = 0.14 \times 10^7$ mho/m (conductivity of type 304 stainless steel), $\mu_{rel} = 1.02$ (relative permeability), $g_1 = 5.54 \times 10^{-3}$ m (tube inside diameter), and $g_2 = 6.35 \times 10^{-3}$ m (tube outside diameter).

By testing the multiple frequencies in the range (F_T = 0.3 to 3.0 MHz), an optimum is determined. In this case, 2.1 MHz was found to be the best frequency, using:

$$F_T = 14 F_g \qquad \text{(Eq 5)}$$

Substituting for F_g yields:

$$F_T = 14(0.15\ \mathrm{MHz}) = 2.1\ \mathrm{MHz} \qquad \text{(Eq 6)}$$

In practical applications, the inspector can try both the SDP and F_g approaches for selecting an optimum frequency. A practical application would be the inspection of nuclear fuel rods. Uranium pellets encased in thin-wall metal tubing require defect inspection of the tube wall. Gaps between adjacent fuel pellets yield nonrelevant indications at low values of F_g. Therefore, in this case, the Foerster frequency selection method would produce the best results (Ref 10). It should be noted that the F/F_g equation derived from normalized impedance diagrams does not relate to the depth of penetration and therefore cannot be used to measure skin depth using eddy current testing.

Tube Outside Diameter Over 75 mm (3 in.)

When the diameter of a tube exceeds about 75 mm (3 in.), it is generally no longer practical to inspect with an external encir-

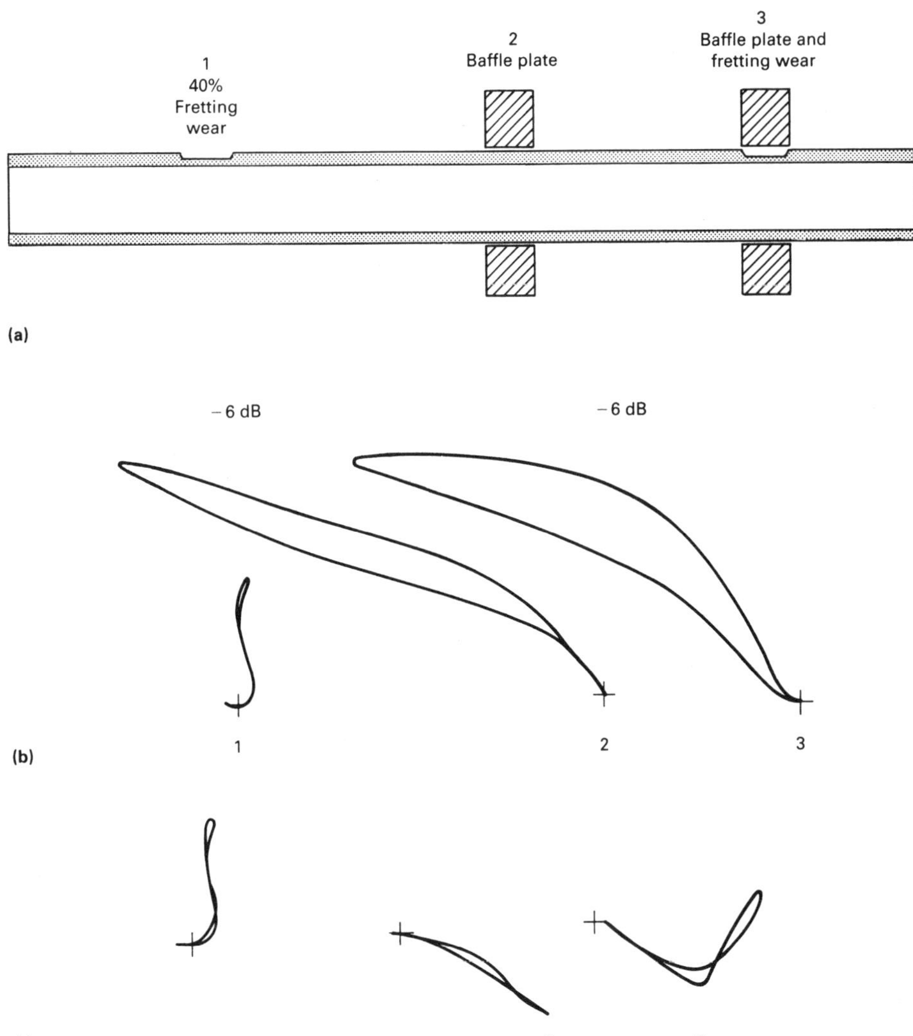

Fig. 35 Signals obtained from Monel 400 calibration tube with simulated fretting wear. (a) Cross-sectional view of calibration tube illustrating locations of fretting wear and carbon steel baffles. (b) Signals obtained using an axial saturation probe at the three locations shown in (a). (c) Signals detected using a radial saturation probe at the three locations shown in (a). The gain of the signals shown in the center and right in (b) has been decreased by a factor of two, as designated by −6 dB above each curve, to reduce the amplitude of the traces in both the *x* and *y* directions. The remaining four traces are drawn to scale.

cling coil for reasons of flaw resolution. A satisfactory technique for larger diameters is the use of multiple probes.

In many respects, a multiple-probe inspection is similar to encircling coil inspection (Fig. 37). An encircling saturating coil is used when inspecting ferromagnetic materials, and an encircling energizing, or primary, coil is employed. Instead of an encircling detector, or secondary, coil, however, the detector is composed of several mechanically and electrically separate probe-coil assemblies. The number of probe assemblies is dependent on the diameter of the tube to be inspected. Each probe assembly consists of several individual probes. The probes are typically about 50 mm (2 in.) long and have a 19 mm (¾ in.) square cross section containing differential windings. The probes are electrically balanced individually and then wired in series. A probe assembly for the inspection of a 24.5 mm (9⅝ in.) diam tube, for example, would contain 14 probes.

As shown in Fig. 37, the probe assemblies are contoured to the curvature of the tube and are designed to ride directly on the surface of the tube on hardened wear shoes embedded in the assemblies. Typically, four probe assemblies are employed. In this case, each assembly inspects slightly more than one-quarter of the circumference of the tube. The assemblies are stagger-mounted so that there is some overlapping of inspected areas. The probe assemblies are each mounted to an arm that brings them into contact with the tube for inspection or retracts them to a protected location when the end of the tube is being inserted into the inspection station.

Test signals from each probe assembly are usually fed to a separate inspection circuit and marking system, although the outputs of the circuits can be combined for operating a common alarm or marker. In terms of area covered per detector, inspection of a 305 mm (12 in.) diam tube in this manner is comparable to inspecting a 75 mm (3 in.) diam tube with an encircling coil.

Another advantage of multiple probes is improved inspection sensitivity, because the detector is always in close proximity to the part being inspected. The use of multiple-probe assemblies also localizes discontinuity position within the sector.

Reflection and Transmission Methods

Discontinuities in nonmagnetic tubing such as that made of copper, brass, or aluminum can be detected by the reflection and transmission methods of eddy current inspection, depending on such variables as the size, location, and orientation of the discontinuity. When the reflection method is employed, both the primary coil, which excites the electromagnetic field, and the secondary coil, which detects the discontinuity, are arranged adjacent to each other on either the outside or inside wall of the tube. When the transmission method is employed, the exciting and receiving coils are placed at opposite walls, either on the outside or inside diameter. With the transmission method, the receiving coil is affected only by those electromagnetic fields that have passed through the entire wall of the tube. Consequently, the transmission method is ideal for indicating tube discontinuities of the same magnitude on inner and outer surfaces with discontinuity signal amplitudes of the same height.

There are two distinct coil designs associated with the reflection and transmission methods:

- The encircling coil, which encircles the part completely
- The rotating probe, which spins around the part in a circular path, with or without making contact with the part

When a cylindrical coil is used inside a tube, it is referred to as an internal coil. Eight possible coil combinations employed in the eddy current reflection and transmission methods are shown in Fig. 38.

The arrangement shown in Fig. 38(a) is used with the reflection method, and both the exciting and receiving coils (encircling coils) are located outside the tube. This arrangement is suitable for detecting outer-surface and transverse discontinuities; it is used for inspecting radioactive fuel cans under water.

The arrangement shown in Fig. 38(b) is also used with the reflection method; both the exciting and receiving coils are inside the tube. The arrangement is suitable for detecting inner- and outer-surface trans-

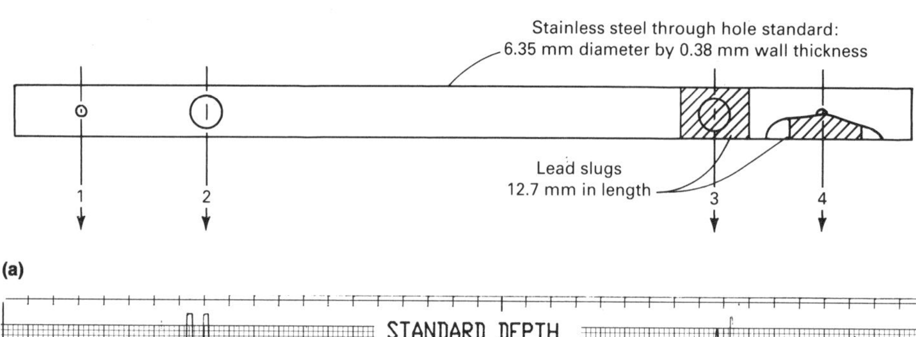

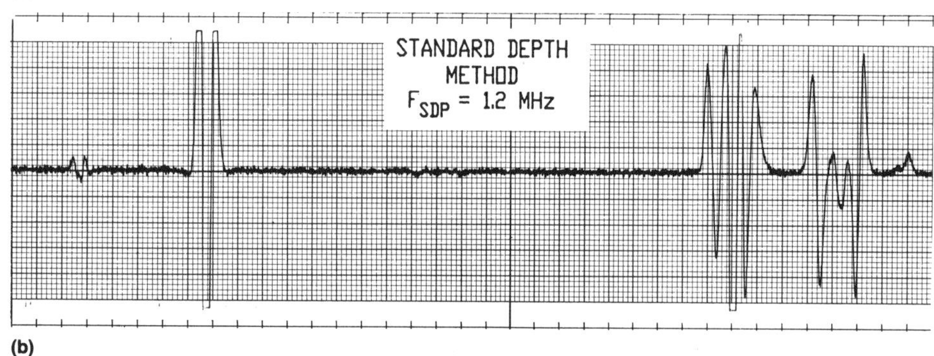

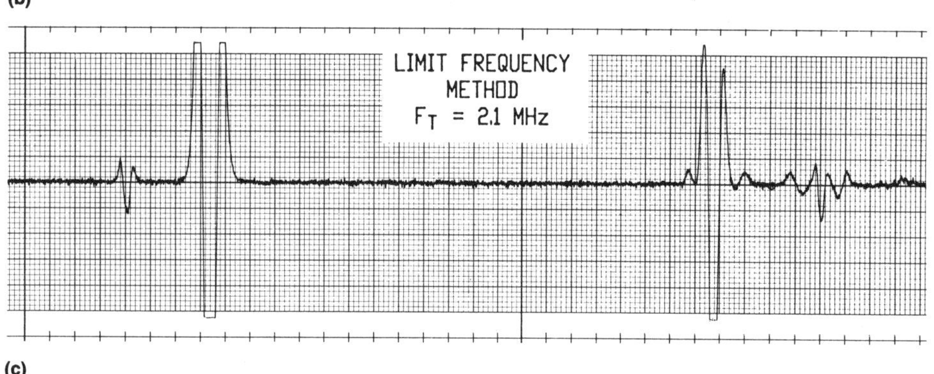

No.	Defect type	Hole diameter through one wall mm	in.
1	Through hole	1.11	0.044
2	Through hole	3.81	0.150
3	Through hole with slug	3.81	0.150
4	Through hole with slug	1.11	0.044

Fig. 36 Comparison of skin depth test frequency, F_{SDP}, versus optimum test frequency, F_T, for a type 304 stainless steel tube. (a) Schematic indicating size and location of discontinuities in tube. (b) Plot of eddy current inspection of tube with 1.2-MHz test frequency. (c) Plot of eddy current inspection of tube with 2.1-MHz optimum test frequency derived from F_g. See text for discussion.

verse discontinuities. It is used for the internal inspection of heat exchanger and reactor tubing.

The arrangement shown in Fig. 38(c) is used with the transmission method and consists of a receiving coil outside the tube and an exciting coil inside the tube. This arrangement provides good sensitivity to both inner- and outer-surface discontinuities and is more sensitive to outer-surface discontinuities than the arrangement shown in Fig. 38(a). It is used in the inspection of six-finned tubes on a continuous basis.

The arrangement shown in Fig. 38(d) is used with the transmission method and consists of an exciting coil outside the tube and a receiving coil inside the tube. The arrangement provides good sensitivity to both outer- and inner-surface discontinuities, but provides increased sensitivity to inner-surface discontinuities. It has been used in conjunction with the arrangement shown in Fig. 38(c) in the inspection of six-finned tubes.

The arrangement shown in Fig. 38(e) is used with the reflection method and consists of an external rotating probe comprising both the exciting and receiving coils. The arrangement has exceptionally high resolving power for detecting discontinuities on or beneath the outer surface of the tube and is capable of detecting the smallest surface blemishes. It is used for inspecting reactor components and fuel elements.

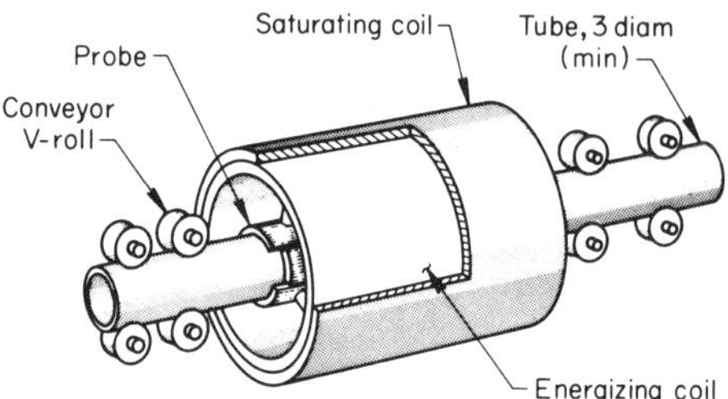

Fig. 37 Multiple-probe setup and encircling coil components for the continuous eddy current inspection of ferromagnetic or nonferromagnetic tubes over 75 mm (3 in.) in diameter

Another rotating probe for use with the reflection method is shown in Fig. 38(f). This probe is similar to that shown in Fig. 38(e), but is located inside the tube and detects inner-surface discontinuities most effectively. The arrangement is used for the inspection of reactor U-shaped heat exchanger tubes for corrosion and cracks.

The arrangements shown in Fig. 38(g) and (h) are combinations used with the transmission method. That shown in Fig. 38(g) is a combination of exciting rotating probe outside the tube and receiving encircling coil inside the tube. It provides high sensitivity in the measurement of wall thickness and eccentricity on both thin-wall and thick-wall tubes. The reverse arrangement is shown in Fig. 38(h); it too is used for measuring tube-wall thickness and eccentricity. Additionally, it is used to detect and to precisely locate surface and subsurface discontinuities.

Inspection of Solid Cylinders

As with tubing, the techniques used in the eddy current inspection of solid cylinders differ depending on the diameter of the cylinder.

Solid Cylinders up to 75 mm (3 in.) in Diameter. Inspection of solid cylinders up to 75 mm (3 in.) in diameter with an external encircling coil is similar to the inspection of tubes in this size range. The limitation regarding resolution of discontinuities applies equally to tubes and solid cylinders.

The inspection is performed by passing the cylinder longitudinally through a concentric coil assembly containing a primary, secondary, and sometimes a saturating coil. When inspecting a solid material using eddy current techniques, it is difficult to detect discontinuities that are located more than 13 mm (½ in.) below the surface. Magnetic field and eddy current densities decrease to zero at the centers of a cylinder. In addition, the magnetic field density decreases exponentially with increasing distance from a short coil. The skin effect adds to the decreases in magnetic field density and eddy current density, but it can be controlled by decreasing the frequency. Eddy current penetration is dependent on the electrical conductivity and magnetic perme-

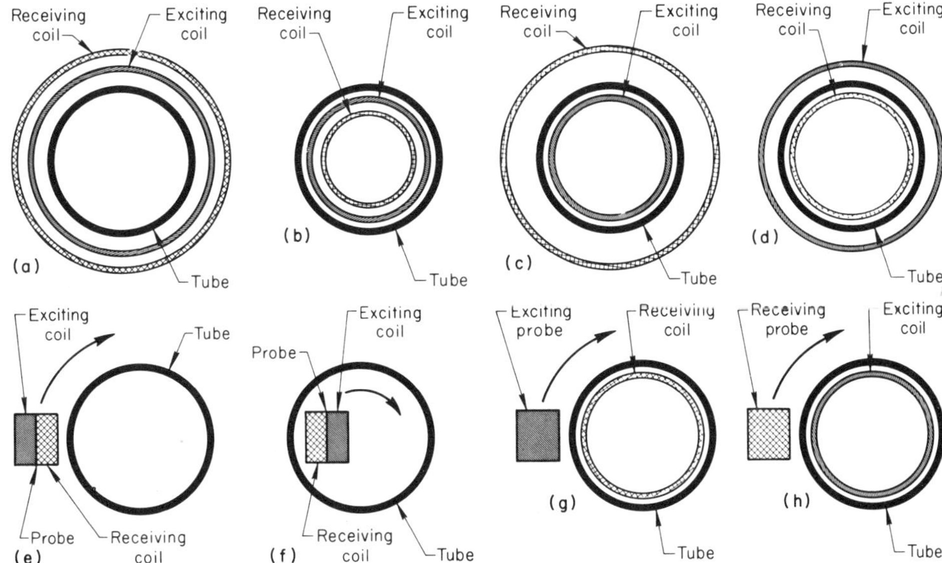

Fig. 38 Eight possible coil combinations employed in the eddy current reflection and transmission methods of inspecting nonmagnetic tubes. See text for description.

ability of the part and on the frequency of the energizing current.

For a given part, the conductivity is fixed. The permeability of ferromagnetic materials can be made to approach unity through use of the saturating coil. This permits greater penetration of the eddy currents. For maximum penetration, a low inspection frequency, such as 400 Hz, is employed. However, as the frequency is lowered, inspection efficiency drops off rapidly, so that some compromise is usually necessary. Also, at low inspection frequencies, the throughput speed of the part is limited because the inspection circuit requires interaction with a certain number of cycles of energizing signal in order to register a discontinuity.

Solid Cylinders Over 75 mm (3 in.) in Diameter. Because of loss of resolution, the inspection of solid cylinders of 75 mm (3 in.) in diameter, using an external encircling coil, generally is not practical. Consequently, the use of multiple probes as described above and shown in Fig. 37 for tubes is applied. The multiple-probe techniques for tubes and solid cylinders are virtually identical.

Machine for Inspection of Tubes on Solid Cylinders

An eddy current inspection machine for detecting surface discontinuities that can be applied to either tubes or solid cylinders having a wide range of diameters is known as the Orbitest machine. Orbitest machines are provided with a rotating drum through which the cylinder to be inspected is conveyed. The drum is so mounted that it is free to float to compensate for any lack of straightness in the part being inspected. Mounted to the drum are two or more search heads, each of which contains one or two search probes. The search probes are caused to engage the cylinder and orbit about it. Signal information is taken from the probes to the detection circuitry by way of slip rings. The longitudinal feeding of the cylinder by conveyor, together with the orbiting of the probes, results in a helical scanning path.

Two paint-marking systems can be utilized at the same time. One orbits with the search heads and marks the precise locations of discontinuities on the cylinder or tube as they are detected; the other is a stationary system that marks with a different color the longitudinal position of deep discontinuities. When a dual marking system is used, the depth thresholds on both marking systems are independently adjustable. In normal operation, one threshold is adjusted to ignore harmless shallow discontinuities and to mark only those that are deep enough to require removal. The second marking system identifies only very deep discontinuities. The deepest discontinuity can be removed first if it is not too deep. If it is too deep, the cylinder or tube can be scrapped before time is wasted in removing shallower discontinuities.

The conveyor speed through the Orbitest machine varies from 12 to 46 m/min (40 to 150 ft/min). Drum rotation speed ranges from 100 to 180 rev/min. Thus, the scanning path can be controlled as required for products of various diameters and for the length of discontinuities to be detected.

Inspection of Round Steel Bars

High-speed automatic eddy current inspection machines have been developed to detect seams, laps, cracks, slivers, and similar surface discontinuities in round steel bars. The machines detect the discontinuities, mark their exact location on the bar, and automatically sort the bars into three cradles in relation to the depth, length, and frequency of the discontinuities. One cradle is for prime bars, a second cradle is for bars that can be salvaged by grinding, and the third cradle is for scrap bars.

Inspection is accomplished by eddy currents that are induced in the bar from a small probe coil, which also serves as the detector coil. When a surface discontinuity is encountered, the eddy currents are forced to flow beneath the discontinuity to complete their path. This increases the length of the path of the eddy currents and thus increases the electrical resistance to the flow of eddy currents in the bar. The change in resistance is proportional to the depth of the discontinuity. The probe coil detects the change in resistance, and this is interpreted by the instrument in terms of the depth of the discontinuity.

The probe coil is about 16 mm (5⁄8 in.) in diameter. It is encapsulated and mounted in a stainless steel housing between two carbide wear shoes. The probe coil is flexibly supported and mechanically biased to ride against the surface of the bar. The bars are rotated and propelled longitudinally. This combination of rotary and longitudinal motion causes the probe (detector) coil to trace on the bar a helical path with a pitch of approximately 75 mm (3 in.). The rolling process by which bars are produced greatly elongates the natural discontinuities so that they are usually several inches in length. The 75 mm (3 in.) pitch of the helical scan has been found to be more than adequate to detect almost all discontinuities.

A nominal helical scanning speed of 30 m/min (100 sfm) has provided good results. Where higher inspection rates are desired, multiple-probe-coil machines are used. Such machines have been built with as many as six probe coils.

The locations of discontinuities are marked exactly on the bar by a small, high-speed rotary milling device mounted downstream of the probe. The milling cutter is brought into momentary contact with the bar precisely one revolution after the discontinuity has passed the probe coil. This provides a shallow mark exactly on the discontinuity. Because the mark is a bright milled spot, it is easy to see and will not smear or rub off.

These machines are completely automatic and, after setup, require no attention other than crane service for loading the feed table and for unloading the cradles after the bars have been inspected. The instrumentation is provided with two separate alarm controls that can be set as required. For example, one alarm might be set to register all discontinuities over 0.25 mm (0.010 in.)

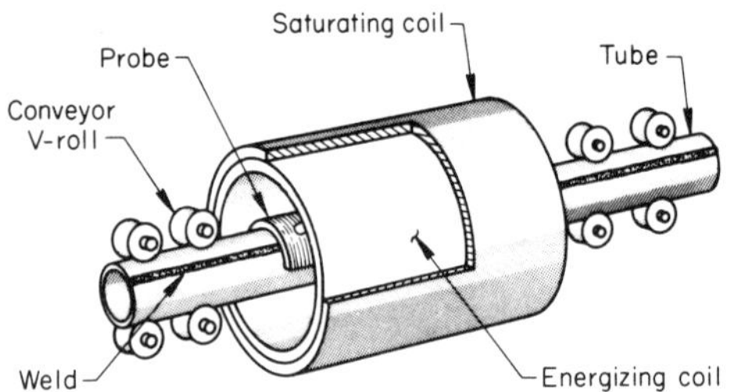

Fig. 39 Setup and coil arrangement for the eddy current inspection of longitudinal welds in ferromagnetic welded tubing

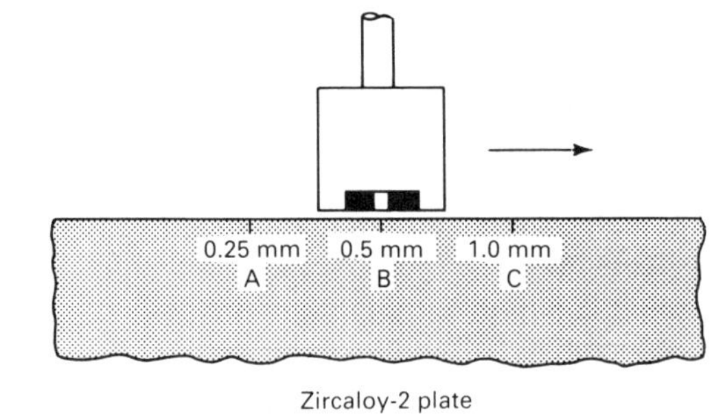

(a)

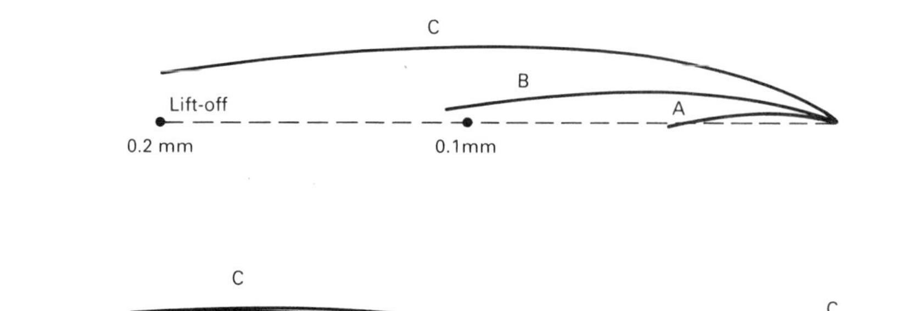

(b)

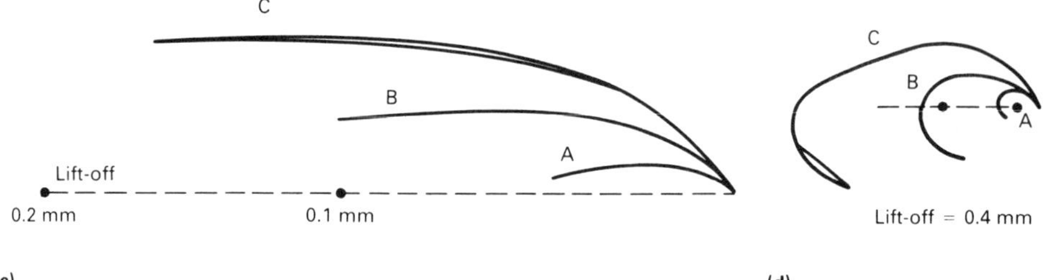

(c) (d)

Fig. 40 Eddy current signals obtained with a conventional probe when inspecting a Zircaloy-2 plate. (a) Location of calibration defects. (b) Impedance display of signals at 200 kHz. (c) Impedance display of signals at 950 kHz. (d) Multifrequency residual signals, a mix of 200 and 950 kHz signals

in depth. The second alarm might be set to register all discontinuities over 1.52 mm (0.060 in.) in depth.

When a bar is being inspected, the outputs from the two alarms are fed into a discontinuity analyzer (computer). This device can be set to respond to a wide variety of conditions, depending on requirements. For example, a bar that produced no response from either alarm would always be rated as prime. However, if it is acceptable for a specified percentage of discontinuities over 0.25 mm (0.010 in.) but less than 1.52 mm (0.060 in.) to be in the material, the discontinuity analyzer can be set to classify as prime whatever length of discontinuity is permissible. Perhaps a few very short discontinuities over 1.52 mm (0.060 in.) can be accepted either as prime or as salvageable. If so, the analyzer is set to meet either of these requirements. The combination of two alarms at different, but adjustable, levels and the versatility of the discontinuity analyzer provide great flexibility in meeting specific material requirements. Inspection efficiency is also improved because the inspection can be adjusted to give a product that meets requirements but is not overgraded. This minimizes costs and scrap losses. These machines have been built to inspect round bars from 9.5 to 114 mm (3⁄8 to 4½ in.) in diameter and from 1.5 to 15 m (5 to 50 ft) in length.

Inspection of Welds in Welded Tubing and Pipe

Longitudinal welds in welded tubing and pipe can be inspected for discontinuities using eddy current techniques with an external encircling primary energizing coil and a probe-type differential detector coil. The probe-type detector coil is located at the longitudinal center in the inner periphery of the primary coil and is arranged so that it inspects the outside surface of the longitudinal weld.

The inspection, as shown in Fig. 39, is performed by passing the tube or pipe longitudinally through the primary energizing coil, causing the probe-type detector coil to traverse the longitudinal weld from end to end. The primary coil is energized with alternating current at a frequency that is suitable for the part being inspected (typically 1 kHz for ferromagnetic products) and induces the eddy currents in the tube or pipe.

For the inspection of ferromagnetic products, a dc magnetic-saturating coil is located concentrically around the primary energizing coil. The dc coil is energized at high current levels to magnetically saturate the tube or pipe. This improves the penetration of the eddy currents and cancels the effect of magnetic variables.

Because of the circumferential orientation of eddy current flow, this type of inspection is effective in detecting most types of longitudinal weld discontinuities, such as open welds, weld cracks, hook cracks, black spots, gray spots, penetrators, and pinholes. Certain types of cold welds having objectionably low mechanical strength may not be detected if the welds are sufficiently bonded to provide a good electrical path for the eddy currents.

With proper wear shoes and a suitable retracting mechanism to lower the probe-type detector coil onto the pipe at the front and retract it at the back, the detector-probe-coil assembly can ride directly on the weld area of the pipe. This provides optimum sensitivity and resolution for the detection of discontinuities. The primary energizing coil may have a clearance of 25 mm (1 in.) or more to provide room for the probe coil and to provide for easy passage of the pipe. It is easy to adjust the energy in the primary energizing coil to compensate for any reasonable amount of primary coil clearance.

It is important that the longitudinal weld be carefully positioned under the probe-type detector coil before the pipe is passed through the tester. It is essential to provide good conveying equipment for the pipe so that, as the pipe is propelled longitudinally, the longitudinal weld will always be located under the detector coil.

There is no limit to the maximum diameter of pipe that can be inspected by this procedure. This eddy current method offers relative simplicity with high sensitivity and resolution. With it, a large majority of objectionable discontinuities can be detected at high speed and low cost. Throughput speeds of 30 m/min (100 sfm) are common. The conveyor speed should be controlled to within about ±10%.

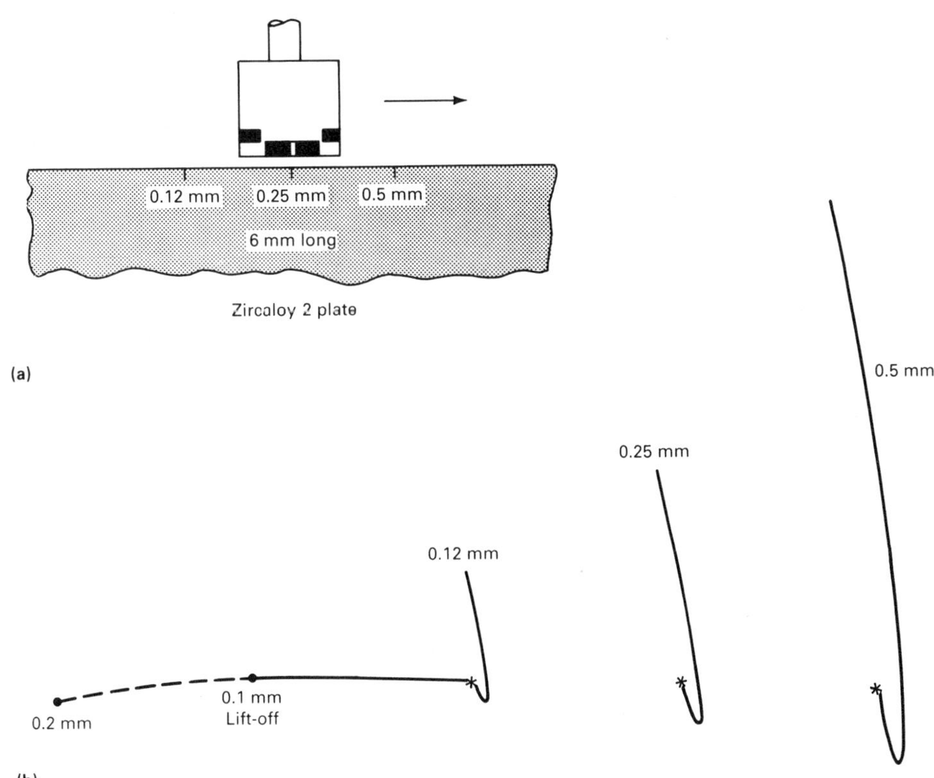

Fig. 41 Eddy current signals obtained with a lift-off discriminating probe when inspecting a Zircaloy-2 plate. (a) Location of calibration defects. (b) Impedance display of signals at 200-kHz test frequency

Inspection of Plates, Skin Sections Panels, and Sheets*

The following four examples illustrate typical applications of eddy current testing to detect flaws in plates, panels, and sheets. Nonferromagnetic materials (Zircaloy-2 and aluminum) are used in all four case histories. Example 7 shows the use of eddy current inspection to gage glue line thickness in adhesive bonding.

Example 4: Eddy Current Inspection to Detect Shallow Surface Defects in Zircaloy-2 Plates. Conventional eddy current testing cannot reliably detect surface defects less than about 0.1 mm (0.004 in.) in depth. This limitation arises because signals from shallow defects are indistinguishable from lift-off noise caused by probe wobble. This problem cannot be overcome with multifrequency testing, because signal phase for shallow defects and lift-off remain nearly identical at all normal test frequencies, as shown in Fig. 40.

An effective method of improving the detection of shallow defects is to obtain clear phase separation between defect and lift-off signals. A lift-off discriminating probe design (Ref 11) consists of two coils that interact to rotate the phase of a defect signal relative to lift-off (Fig. 41a). Phase rotation depends on test frequency, sample resistivity, and coil size and spacing. Through proper choice of probe parameters, defect signals can be rotated 90° from lift-off noise, as shown in Fig. 41(b). This allows positive phase discrimination between lift-off noise and signals from localized defects. Lift-off discriminating surface probes have been used to detect defects as shallow as 0.05 mm (0.002 in.) in zirconium components during in-service inspections. However, these surface probes have one main limitation: Signal phase from defects of varying depth does not change appreciably, so only signal amplitude can be used to estimate defect depth.

It should be noted that no surface preparation of the nonferromagnetic components is required prior to testing. Figure 41 indicates that shallow surface defects can be reliably detected in the presence of probe wobble or lift-off noise.

Example 5: Special Fixture Provides Two-Step Eddy Current Inspection of the Skin Section of the First-Stage Booster of the Saturn Rocket (Ref 12). For continuous inspection, if the configuration of the test article permits, the operation can be automated. Ordinarily, this will require a mechanism that allows the article to move past the test transducer in a repeatable fashion. Depending on the sensitivity required, the mechanism can be:

- A rudimentary arrangement in which the test article is simply moved through the magnetic field of the test transducer
- For more stringent requirements, the test article can be precisely positioned and held fixed while an electromechanical scanning system moves the transducer over it. This arrangement reduces the possibility of the transducer detecting movement or misalignment of the specimen rather than flaws

An example of special fixturing for the inspection of cylindrical-tank-wall skin sections of the Saturn V/S-1C first stage booster is shown in Fig. 42. The contractor, under the direction of Marshall Space Flight Center, assembled this system using off-the-shelf equipment to inspect aluminum skin panels nominally 5.08 mm (0.200 in.) thick. Inspection was accomplished in two steps. First, surface discontinuities whose depths exceed 5% of the part thickness were located with rapid (37 m/min, or 120 sfm) linear scan (Fig. 42a). Second, the length and depth of the discontinuity were measured with the rotating eddy current probe device illustrated in Fig. 42(b).

Example 6: Crack Detection in First Layer or Subsurface Layers of Aluminum-To-Aluminum Structures Joined by Aluminum Rivets. When inspecting large numbers of fastener holes for cracks in the first or second layer of metal structures, it is advisable to use a fixed or adjustable low-frequency sliding eddy current probe (Fig. 43) instead of the slower conventional types. These probes are used by sliding them over the fastener heads and observing the pattern produced on the oscilloscope. The surface may or may not be painted, but it must be cleaned prior to testing. When a crack is present, the dot will move in a different path, producing an indication proportional to its size.

When inspecting for first-layer cracks, the frequency should be chosen to limit the penetration to the crack and to avoid variables in the layers below. For the second (or multiple) layer, the frequency should be low enough to penetrate the total thickness required.

Figure 44 shows a typical presentation obtained with aluminum rivets joining two 1.6 mm (0.063 in.) thick aluminum skins. Similar presentations are obtained for other fastener types and plate thicknesses up to about 12.5 mm (0.5 in.).

A nonconductive straightedge is useful to use as a probe guide and should be aligned to place the probe directly above the fastener heads. Lift-off has been set slightly below horizontal, and the good fastener hole indications slightly above horizontal as the best compromise. With this setup, crack indications give a mainly vertical increase, showing the larger crack as a greater loop.

It is possible to determine crack direction by observing the dot movement on the CRT display on the oscilloscope. If the dot

*Example 4 was prepared by V.S. Cecco, Atomic Energy of Canada Limited, and Examples 6 and 7 were prepared by J. Pellicer, Staveley Instruments.

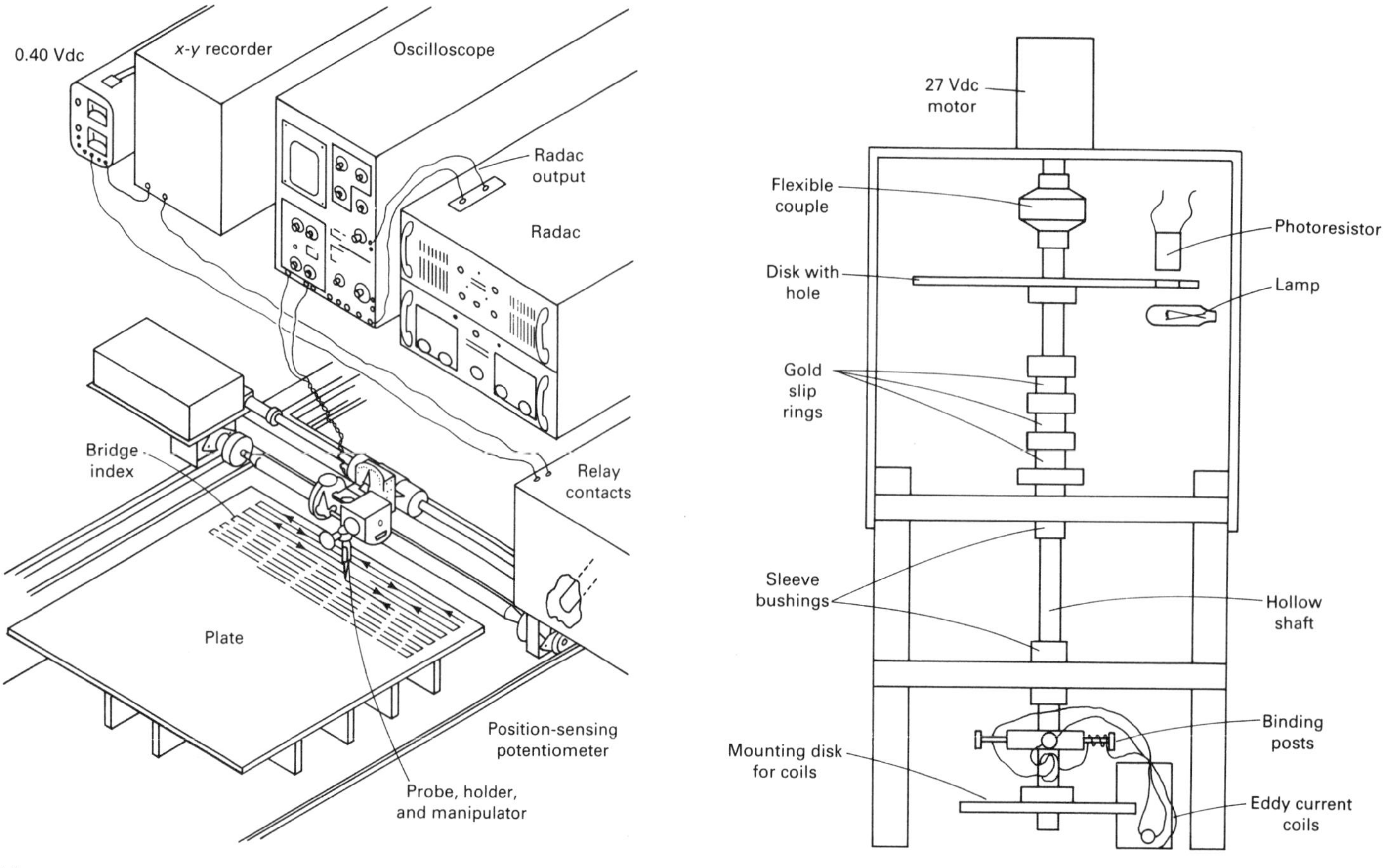

Fig. 42 Eddy current fixturing setup for locating and sizing flaws in Saturn V/S-1C rocket booster tank-wall skin sections using a two-step process. (a) Surface discontinuities with depths ≧5% of skin section thickness are first located with 37 m/min (120 sfm) linear scanner. Radac, rapid digital automatic computing. (b) Rotating eddy current probe device then gages length and depth of discontinuities discovered by linear scanner. Source: Ref 12

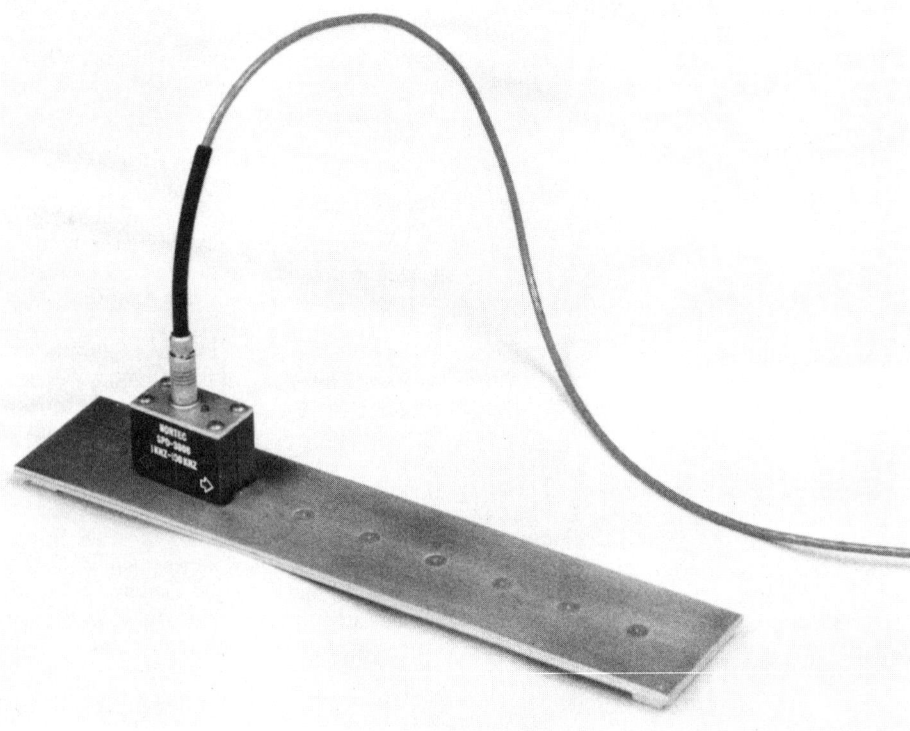

Fig. 43 Low-frequency sliding-type eddy current inspection probe used to detect cracks in the first or subsurface layers of multilayered structures. Courtesy of J. Pellicer, Staveley Instruments

moves first along the good fastener hole indication (that is, clockwise), the crack is located on the opposite side. If the dot moves up at first and then eventually comes down (that is, counterclockwise), the crack is located before the fastener or the near side. A larger crack will eventually move the dot outside the CRT area.

This method allows rapid inspection for fastener hole cracks. The sliding probes are directional units and should be used in the direction of cracking. This is normally easily determined by the stress direction in a given structure. Probes will allow approximately a ±40° angle tolerance.

Example 7: Eddy Current Inspection of Glue-Line Thickness of a Metal-To-Metal Adhesive Joint. The objective was to measure the glue-line thickness between 7075-T6 aluminum sheets having a thickness of 1.6 mm (0.063 in.) and a conductivity of 32 IACS. Although this inspection can be performed with through transmission ultrasonics, a simple eddy current measurement was used, with one side access only (and no couplant needed). The surface may or may not be painted, but should be cleaned prior to testing.

Inspection was accomplished by a simple calibration with the help of samples of the

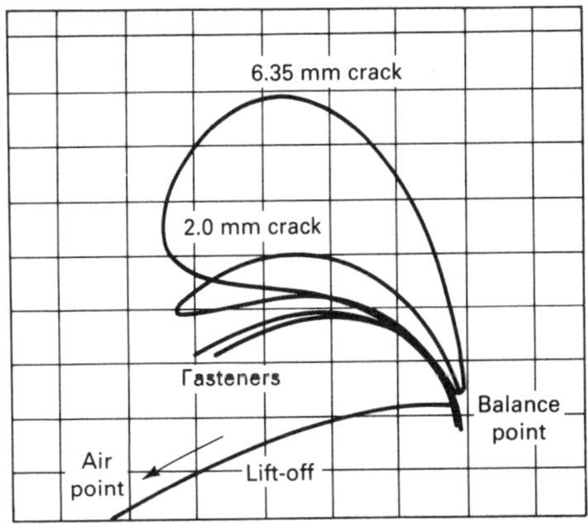

Fig. 44 Typical oscilloscope display obtained using the setup and sliding probe shown in Fig. 43. Instrument test frequency was 2.5 kHz, horizontal gain was 60 dB, and vertical gain was 68 dB with probe drive setting at normal.

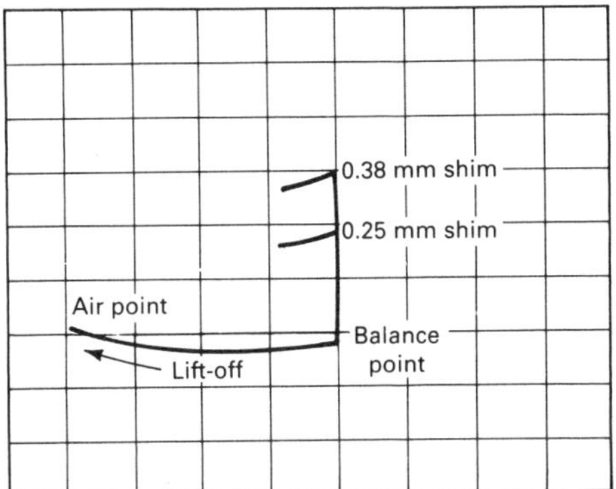

Fig. 45 Typical oscilloscope trace used to measure glue-line thickness in a metal-to-metal adhesive joint. Instrument settings: 3-kHz frequency, 60-dB horizontal gain, 76-dB vertical gain, and maximum probe drive

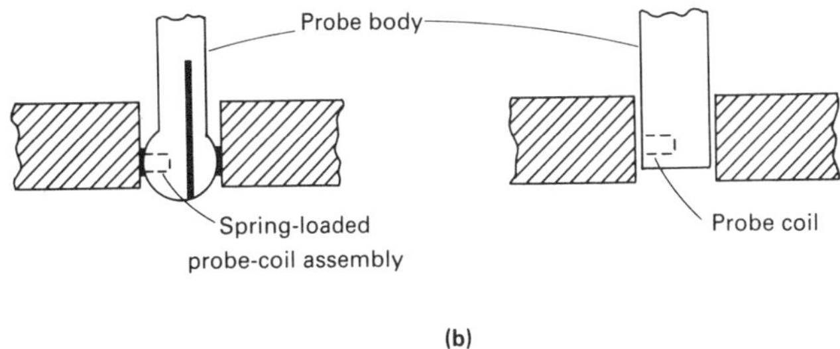

Fig. 46 Types of probe-coil bodies used in eddy current scanners. (a) Contact. (b) Noncontact

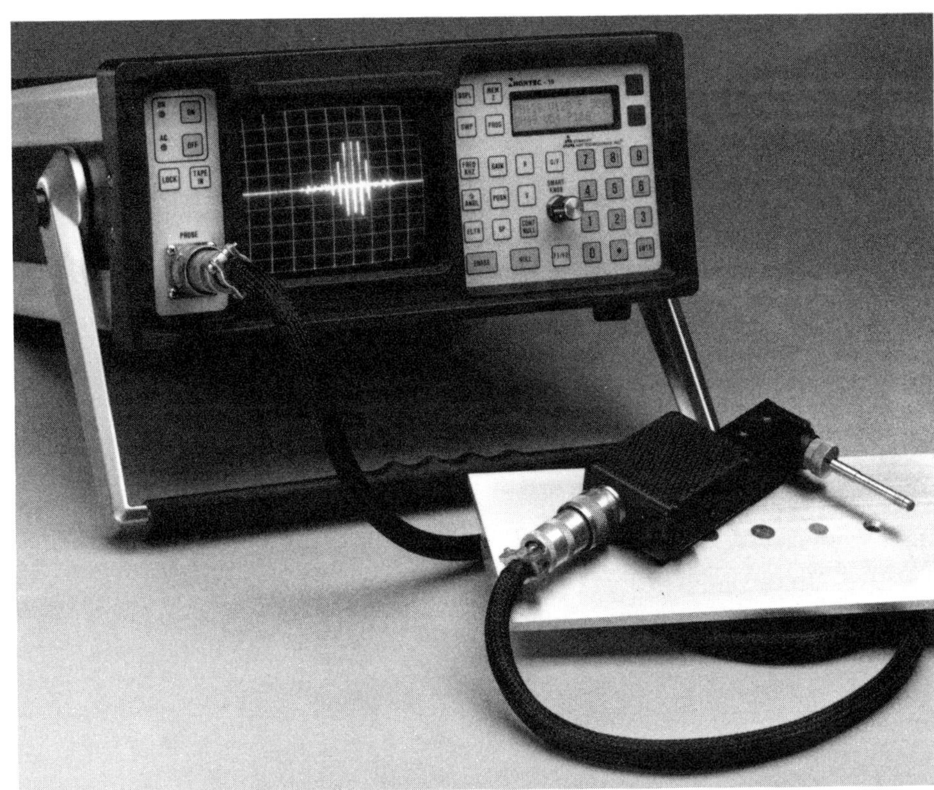

Fig. 47 Portable eddy current instrument that incorporates a hand-held scanner to detect cracks in fastener holes

metal used and nonconductive shims (or paper sheets) up to the desired glue thickness to be measured. Nonconductive shims measuring 0.25 and 0.38 mm (0.010 and 0.015 in.) were used.

A surface probe of the required frequency and an impedance plane (CRT) eddy current instrument were required. The 16 mm (0.63 in.) OD probe selected was usable between 300 Hz and 10 kHz.

Frequency can be adjusted experimentally for best deflection. A good starting point is the frequency of a standard depth of penetration for the first layer only (this can be obtained from a graph or by using the eddy current slide rule). The frequency is then decreased in steps of about 20% to find the best response. For this particular application, the frequency required for one standard depth of penetration was 5 kHz. However, optimum results were obtained at 3 kHz.

With lift-off set horizontal, shim spacing (glue thickness) variations gave a near-vertical movement, as shown in Fig. 45. In addition, the horizontal gain was attenuated to avoid too much movement due to changes in lift-off. The vertical gain was adjusted to give a clear vertical deflection for the shims used.

Conductivity differences between materials prevent a direct correlation between the testpieces and the actual panel to be inspected (unless a sample of the panel itself is available). If necessary, the instrument can be rebalanced on the panel, and the lowest point can be found by conducting a scan. This point is then used as a reference for minimum bond-line thickness. Upward movement of the dot can be read as a thickness increase that was calibrated with the shims previously. Although this eddy current test will not show a lack of bond between parts, it is a simple and reasonably accurate method of measuring bond-line thickness (and therefore bond strength) in nonmagnetic materials.

Inspection of Aircraft Structural and Engine Components*

Eddy current inspection has long been used to examine the quality of aircraft subassemblies. Improvements in this inspection method, especially the miniaturization of the probes themselves, have made it practical to test the intensity of aircraft maintenance procedures for aging commercial and military aircraft without the need to disassemble the aircraft and with the component in place on the plane. This minimizes component removal or disassembly and possibly masks operational and structural defects.

*Example 8 was prepared by J. Pellicer, Staveley Instruments.

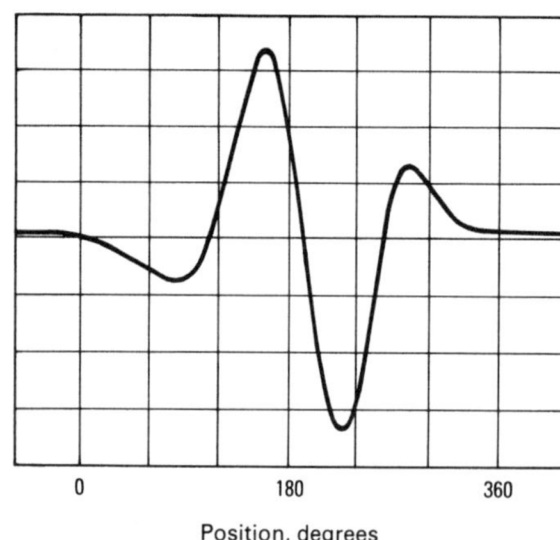

Fig. 48 Screen display using the trigger pulse mode to determine the position of a 1 mm (0.040 in.) long corner crack. The scanner probe indicates that it is located at the 180° position.

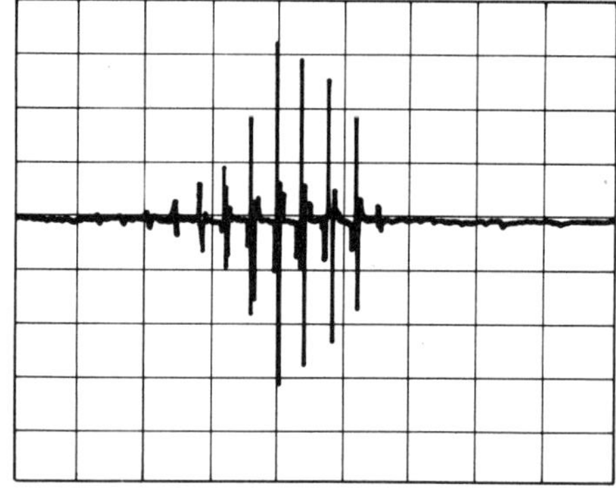

Fig. 49 Screen display using slow sweep mode to determine the crack depth of a 1 mm (0.040 in.) long crack detected by a hand-held scanner probe traversing with a 0.5 mm (0.020 in.) pitch

Inspection of Aircraft Subassemblies

As shown in the three examples that follow, aircraft components such as aluminum structural parts, titanium engine blades, and metallic composite materials can be readily inspected using eddy current inspection techniques.

Example 8: Eddy Current Inspection of Fastener Holes. Originally, inspections for crack detection in fastener holes were performed using unshielded, absolute eddy current probes. Later, shielded differential coils provided an overall improvement in detection capability with a higher signal-to-noise ratio. Manual operation is now giving way to scanners specially designed to ensure good coverage.

Both contact and noncontact probe bodies are used in the scanners. The contact method probe relies on a spring action to ensure proper alignment of the coil assembly and friction against the hole surface (Fig. 46a). Unless ovality is present, a constant contact with the hole surface is maintained, and stable sensitivity is the end result. Nevertheless, in demanding high-gain applications, exact alignment of the scanner assembly is necessary to minimize background surface noise. In the noncontact method, the probe body is designed to be located at a depth of approximately 0.25 mm (0.010 in.) below the hole opening (Fig. 46b).

To maintain constant sensitivity, the probe must be centered in the hole, but the hand-held scanner takes this into account during calibration. Sufficient gain is set to detect the defect regardless of probe position. Scanners with noncontact probe bodies allow higher inspection speeds and typically generate lower background noise than scanners with contact probe bodies.

Some scanners are designed to traverse a helical path in the bore, typically at 0.50 mm (0.020 in.) pitch, to provide good coverage. With this method, rotational speed can be adjusted as required.

Most hand-held scanners (Fig. 47) use noncontact probes and operate at speeds in excess of 1000 rev/min. Because of its high speed of rotation, the probe can be inserted and withdrawn quickly without fear of missing a defect due to lack of coverage.

Use of Trigger Pulse and Slow Sweep Modes to Detect Cracks. In addition to the normal impedance plane display in the CRT of the eddy current instrument, there are two other useful presentations that can be displayed. A trigger pulse can be generated to monitor the hole periphery along the horizontal axis of the CRT (Fig. 48). This signal will display a complete 0 to 360° cycle, along with the crack direction relative to the probe.

Alternatively, a slow sweep signal will display the length of the bore in the horizontal axis to show the depth at which the defect is located (Fig. 49). Crack indications will be obtained with the rotation of the probe, and a profile will emerge that gives the crack length along the bore on the basis of the number of hits detected. In either the trigger pulse or slow sweep mode, the magnitude of the response along the vertical direction displayed on the screen (the crack deflection) is a relative indication of the depth of the crack in the hole. Fastener hole inspection is similar to any other eddy current technique that depends on proper calibration and the use of reference standards having cracks and electric discharge machining notches of known dimensions.

High- and Low-Pass Filters. Scanners operating at constant speed allow for the use of filtering. High-pass filters eliminate slow changes, such as hole ovality or lift-off variations. Low-pass filters reduce high-frequency noise. When combined, high- and low-pass filters considerably enhance the ability of the probe to detect small defects. The use of filters in scanners having manually rotated probes is normally not recommended.

Example 9: Eddy Current Inspection of Titanium Alloy Jet Aircraft Engine Blades for Cracks Resulting From Low-Cycle Fatigue. Titanium alloy fan blades used in jet aircraft engines are subjected to very high stresses during takeoff. Maximum stresses in root sections of the blades may approach or even exceed the yield strength of the blade material, and high residual compressive stresses may cause cracking. Eddy current inspection was applied to the root sections of these fan blades to detect cracking associated with damage resulting from low-cycle fatigue.

The inspection technique utilized a commercial high-frequency (1 to 4 MHz) single-probe, eddy current instrument. The probe coil was wound on a special 1.52 mm (0.060 in.) diam ferrite core to fit the root area of the blade and to improve sensitivity to cracks. The coil was retained in a plastic holder molded to fit the blade root, using a rubbery compound that allowed a small amount of movement. By setting the coil tip slightly beyond the edge of the holder, the flexibility of the bonding compound ensured tip contact with the blade when the probe was positioned on the blade.

The instrument response was recorded on either an *x-y* or a strip chart recorder as the probe was moved across the blade root. The probe motion was represented by the time-base motion of the pen of the *x-y* recorder or by the paper motion of the strip chart recorder.

A blade with an electrical discharge machined crack in the root section was used as a standard. Sensitivity was established by setting the recorder and instrument gain to obtain a given amplitude on the recording for the machined crack in the blade root.

Example 10: Use of Eddy Current Inspection To Determine the Fiber Content of a Metallic Composite Material. A new material, a composite of boron silicon fiber and aluminum, was evaluated for high-strength, lightweight applications. The fiber was fabricated by coating 0.013 mm (0.0005 in.) diam tungsten wire with boron to a diameter of approximately 0.11 mm (0.0045 in.) and then treating the boron with a silicon coating. The fibers were then made into a 0.13 mm (0.005 in.) thick tape by bonding them with aluminum. Structures were made by diffusion bonding layers of the tapes. Optimum properties require that the finished product contain 50% fiber and 50% aluminum.

A commercial single-probe, low-frequency eddy current instrument was used to determine and monitor the fiber content of the finished product by electrical-conductivity measurements. A calibration curve was established by measurements of electrical conductivity on a series of samples containing from 30 to 50% fiber. It was found that conductivity measurements (% IACS) varied by as much as 20 to 30% with

Portion of aluminum cladding corroded away

Contaminants (electrolyte) in moat area formed by fastener and countersink

Paint film

Aluminum skin

Anode

Anode

Intergranular corrosion fissures

Cathode

Cadmium-plated fastener

(a)

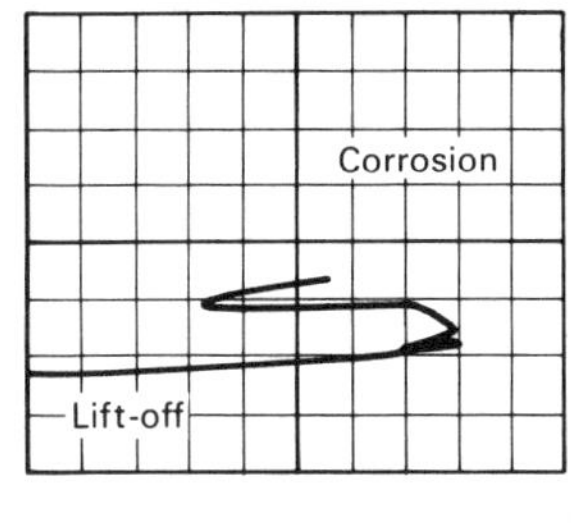

(b)

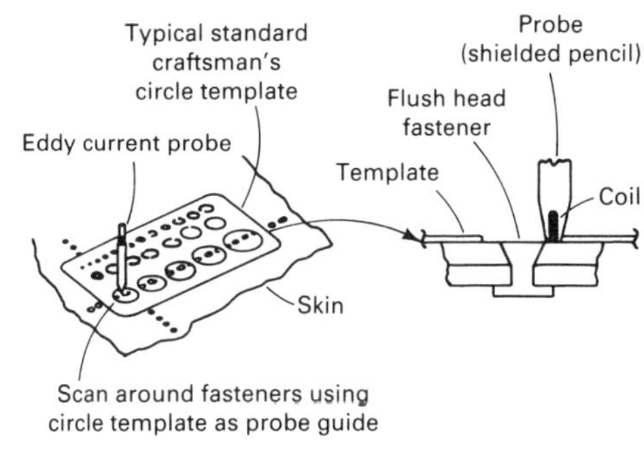

(c)

Fig. 50 Detection of galvanic exfoliation corrosion in aluminum wing skins. (a) Schematic showing source and growth of galvanic exfoliation corrosion. (b) Eddy current impedance responses for exfoliation corrosion around fastener holes in wing skins. (c) Schematic illustrating use of circle template to guide eddy current scanning probe around fasteners. Source: Ref 13

a variation of only 3% in fiber content, thus ensuring close control.

On-Aircraft Eddy Current Inspection (Ref 13)

After an airplane enters service, ongoing inspection and maintenance of its structure are essential to ensure a continuing high level of safety. Experience has shown that the inherent structural integrity of commercial transports has been effectively maintained by operator inspection and maintenance programs specified and approved by the certifying agencies. However, after many years of service, these aircraft will reach an age at which an increase in fatigue cracking and corrosion may be expected. As reports of cracking or corrosion are received, nondestructive inspection (NDI) methods are developed and verified for use by operators to ensure the structural integrity of their aircraft.

Damage tolerance and NDI reliability programs consistently show that eddy current inspection is superior to other nondestructive testing methods for the detection of tight fatigue cracks and corrosion. Therefore, NDI engineers have developed numerous types of eddy current inspection for use in inspecting aircraft at operators maintenance bases.

Existing phase-analysis eddy current instruments enable the inspector to produce impedance-plane responses automatically on the storage oscilloscope. If the test part contains cracks or corrosion, the eddy current response on the oscilloscope immediately establishes their existence and relative severity. These instruments operate from 100 Hz to 6 MHz, allowing the inspector to choose the best frequency for a given material and test. The five examples that follow demonstrate the versatility of eddy current inspection in on-aircraft applications.

Example 11: Galvanic Exfoliation Corrosion of Aluminum Wing Skins. Although contact between two galvanically dissimilar metals, such as steel and aluminum, is known to be a cause of corrosion, the design of aircraft structures occasionally requires that such metals be joined. When this is unavoidable, it is a design requirement that contacting surfaces be electrically insulated with organic paint or sealant or that one of the surfaces be coated with a metallic coating galvanically similar to the other surface. For example, cadmium-plated steel bolts are used on many types of aircraft. The cadmium plating not only protects the steel bolts from corrosion but also provides a surface galvanically similar to aluminum so that the possibility of corrosion is greatly reduced. However, if the cadmium in the plating is depleted or if there is a crevice where moisture can collect between the fastener head and the aluminum skin, pitting and intergranular corrosion may occur (Fig. 50a).

Intergranular corrosion occurs along aluminum grain boundaries, which in sheet and plate are oriented parallel to the surface of the material because of the rolling process. (Intergranular corrosion in its more severe form is exfoliation corrosion.) Exfoliation corrosion is basically the intergranular delamination of thin layers of aluminum parallel to the surface, with white corrosion products between the layers. Where fasteners are involved, the corrosion extends outward from the fastener hole, either from the entire circumference of the hole or in one direction from a segment of the hole. In advanced cases, the surface bulges upward; but in milder cases there may be no telltale bulging, and the corrosion can be detected only using nondestructive testing methods.

To conduct the high-frequency eddy current test, the instrument is calibrated using a corroded sample from a corroded area on the aircraft. Typical eddy current impedance plane response to exfoliation corrosion around installed fasteners is illustrated in Fig. 50(b). The use of high-frequency (100 to 300 kHz) pencil-point probes and a circle template, as illustrated in Fig. 50(c), is necessary to detect very small areas of corrosion. The circle template is centered over each fastener, and a 360° scan is made using the eddy current probe. The corrosion response will appear on the CRT and remain there until electronically erased. This method is quite slow, but has been shown to be accurate in detecting very small areas of corrosion around installed fasteners.

Example 12: Surface Cracks Through Thin Sealant Masking Protruding Fasteners. Sealants are most frequently used around protruding fasteners in the inner wing surface. The sealants are applied to provide a fuel seal (leak prevention) and to prevent corrosion of the fasteners and wing spar caps. Unfortunately, these sealants mask the areas they cover and prevent visual inspection for possible fatigue cracking of the spar caps. Because it is difficult to

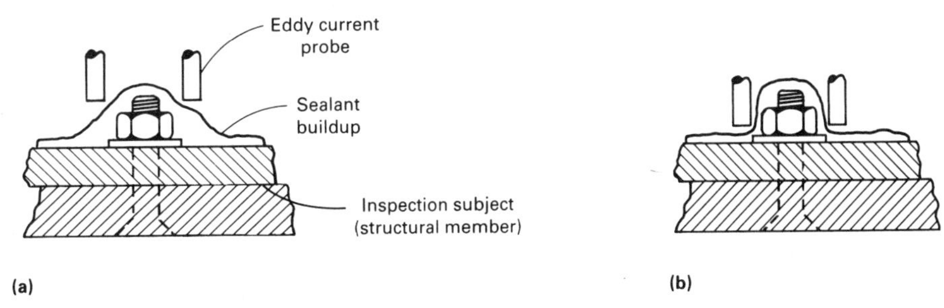

Fig. 51 Sealant trimming requirements for eddy current inspection. (a) Sealant buildup that hinders detection of fatigue cracks in spar caps because probe capability is exceeded. (b) Sealant trimmed to dimension thin enough to allow probe to be positioned properly around fastener. Source: Ref 13

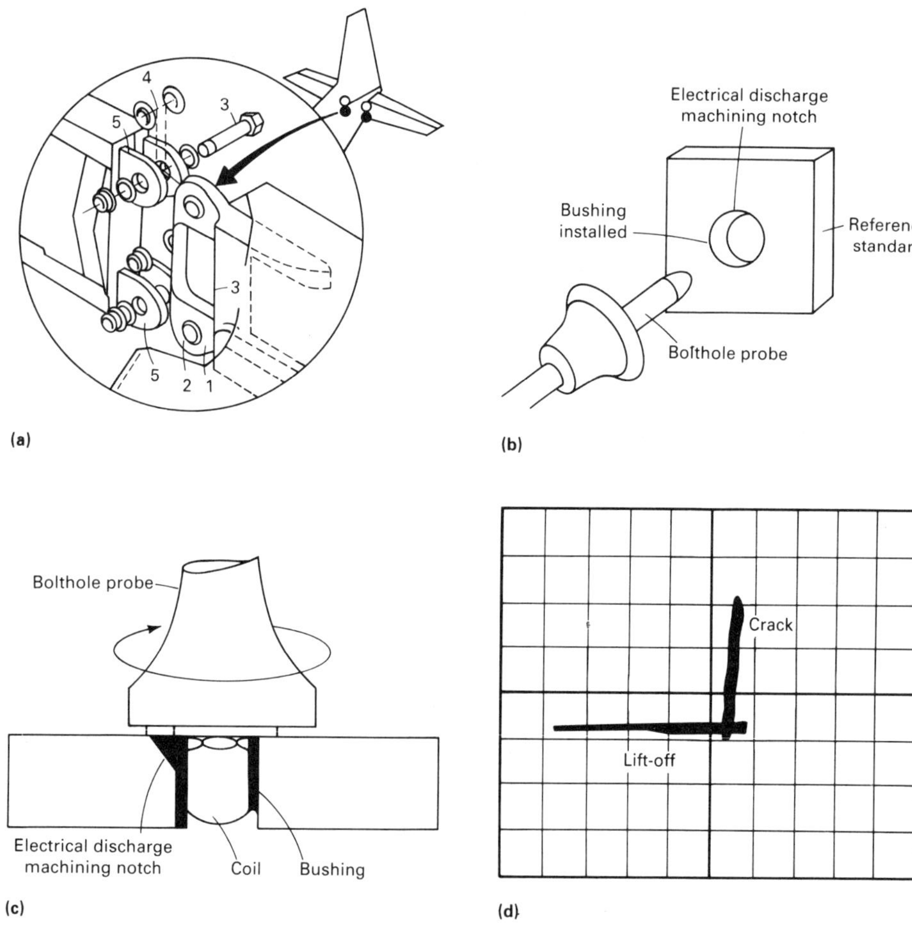

Fig. 52 Eddy current inspection of cracks located under installed bushings. (a) Schematic of typical assembly employing interference-fit bushings in a clevis/lug attachment assembly. (b) Reference standard incorporating an electrical discharge machined corner notch. (c) Probe coil positioned in bolthole and encircled by bushing. (d) CRT display of a crack located under a ferromagnetic bushing. Source: Ref 13

remove and costly to replace the sealants, eddy current inspection through the sealant is a viable crack detection method.

This section describes the steps taken to perform eddy current inspection for surface cracks under sealant up to 1.3 mm (0.050 in.) thick. Similar methods have been developed for eddy current inspection through sealants up to 2.54 mm (0.100 in.) thick.

The equipment required consists of an impedance-plane, flying-dot, multipurpose eddy current instrument. In some cases, the instrument may be too bulky for access to the inner wing surface. If this situation exists, one operator can manipulate the probe using an extended lead connected to the remotely located instrument monitored by a second operator. The probe operator should periodically scan the defect in the reference standard during the inspection to validate the instrument operator's interpretation. The two operators must be in constant communication with each other to ensure defect detection and to prevent inspection in areas where sealant thickness is beyond the range of the equipment calibration. Thick layers of sealant do require trimming to acceptable dimensions, as shown in Fig. 51. The probe used is a reflectance-type model that is operated at 20 kHz. It measures 25.4 mm (1.0 in.) high by 7.6 mm (0.30 in.) in diameter.

Example 13: Cracks Under Nonmagnetic Bushings Used in Clevis/Lug Attachments. There are places in light piston-powered aircraft where bushings are installed by staking or interference fit, and it is difficult and costly to remove them for crack inspection. Bushings are usually used at clevis/lug attachments for assembly union with wing, horizontal stabilizer, and vertical stabilizer box sections with spar caps. A typical joint is illustrated in Fig. 52(a). The bushings in fitting 1 are identified as 2, and the bushings in clevis 5 are identified as 4. The bolt, which is removed, is identified as 3.

A reference standard was made from material of the proper thickness, and the electrical discharge machined corner notch was made at the edge of the appropriate-size hole. The bushing was then installed in the reference standard, as shown in Fig. 52(b). The proper-size bolthole probe was selected and inserted into the bushed hole, and the operating frequency was selected to allow the eddy current to penetrate through the bushing in order to detect the notch (Fig. 52c and d).

After calibration, the bolthole probe was inserted into the appropriate bushed hole in the lug or crevis on the aircraft. The probe was inserted at increments of about 1.59 mm (0.0625 in.) and rotated 360° through each hole to be inspected. The bushing, made of a copper alloy, had a thickness of

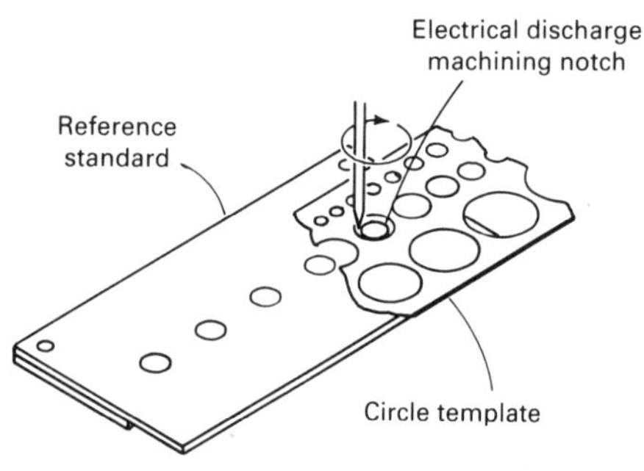

(a)

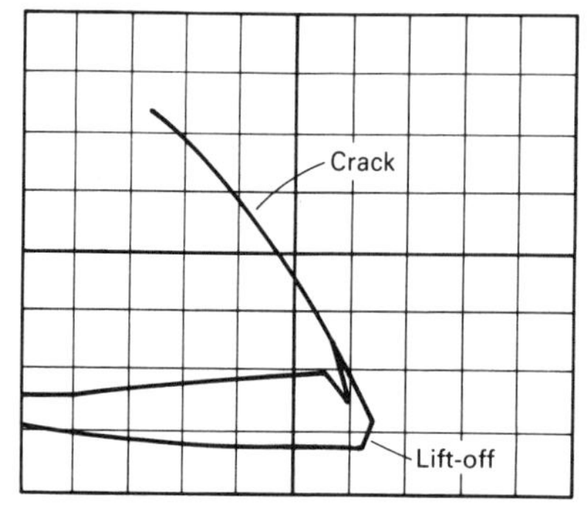

(b)

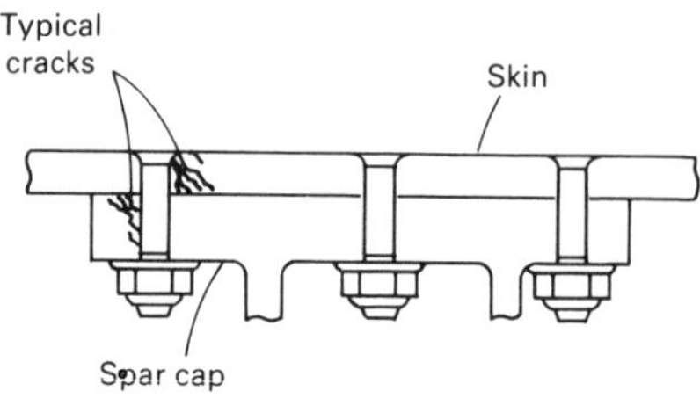

(c)

Fig. 53 High-frequency eddy current inspection of surface and subsurface cracks in aircraft splice joints. (a) Calibration procedure involves introducing an electrical discharge machining notch in the reference standard to scan the fastener periphery using a circle template to guide the probe. (b) CRT trace on an oscilloscope of typical cracks in both skin and spar cap sections shown in (c). Source: Ref 13

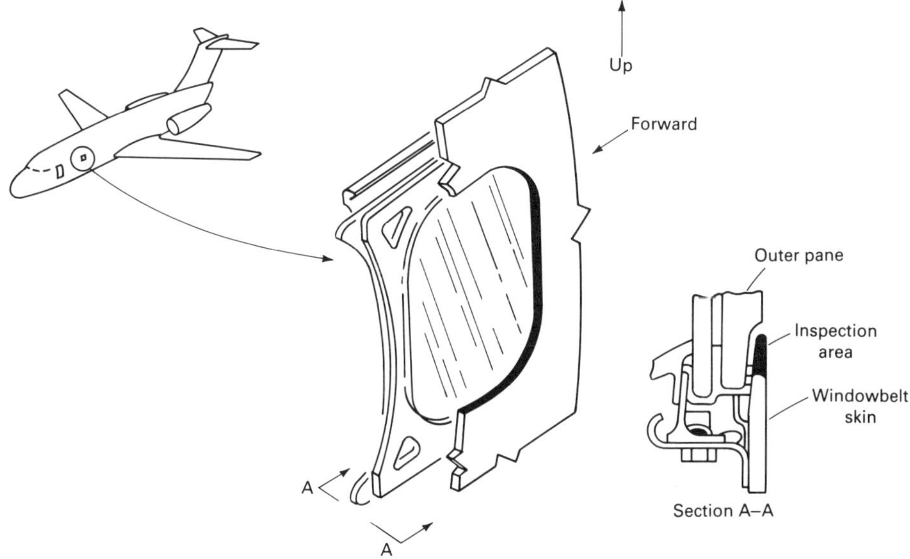

Fig. 54 Location of subsurface corrosion in aircraft windowbelt panels. Source: Ref 13

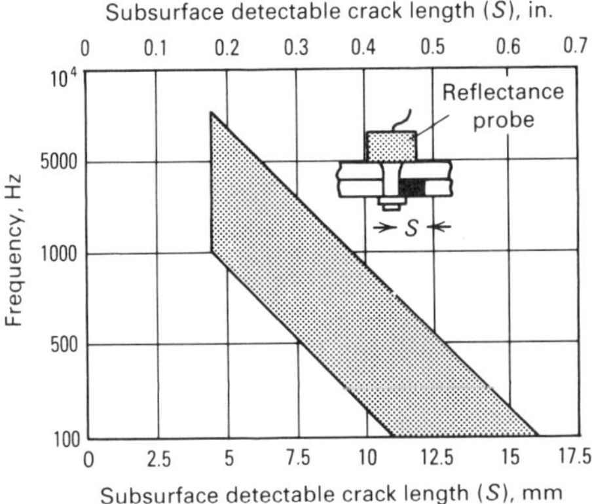

Fig. 56 Plot of operating frequency versus detectable crack length in aluminum structures using reflectance-type (transmit-receive) eddy current probes. Source: Ref 13

about 1.5 mm (0.060 in.) and a conductivity between 25 and 30% IACS, which is easily penetrated at a frequency of 1 to 2 kHz.

Example 14: Detection of Fatigue Cracks in Aircraft Splice Joints. Surface and subsurface fatigue cracks usually occur at areas of high stress concentration, such as splice joints between aircraft components or subassemblies. High-frequency (100 to 300 kHz) eddy current inspection was performed to detect surface cracks with shielded small-diameter probes. A reference standard was made from typical materials, and a small electrical discharge machining notch was placed at the corner of the external surface adjacent to a typical fastener. The high-frequency probe was scanned around the periphery of the fastener using a circle template for a guide, as illustrated in Fig. 53(a).

When subsurface cracks are to be detected, low-frequency eddy current techniques are employed. Basically stated, the thicker the structure to be penetrated, the lower the eddy current operating frequency that is required. However, the detectable flaw size usually becomes larger as the frequency is lowered.

Example 15: Hidden Subsurface Corrosion in Windowbelt Panels. There are various areas of the aircraft where subsurface (hidden) corrosion may occur. If such corrosion is detected, usually during heavy maintenance teardown, a nondestructive testing method can be developed to inspect these areas in the remainder of the fleet. Following is an example of subsurface corrosion detected by low-frequency (<10 kHz) eddy current check of the windowbelt panels. Such inspection is applicable at each window on both sides of the aircraft.

Moisture intrudes past the window seal into the inboard side of the windowbelt panel and causes corrosion thinning of the inner surface (Fig. 54). The eddy current inspection is performed using a phase-sensitive instrument operating at 1 to 2.5 kHz and either a 6.4 or 9.5 mm (0.25 or 0.375 in.) surface probe.

The edge of the inner surface of the windowbelt (where corrosion occurs) tapers from 4.06 to 2.0 mm (0.160 to 0.080 in.) over a distance of 19 mm (0.750 in.). A reference standard simulating various degrees of corrosion thinning (or, in reality, remaining material thickness) is used to calibrate the eddy current instrument (Fig. 55a). The instrument phase is rotated slightly so that probe lift-off response is in the horizontal direction of the CRT. As the probe is scanned across the steps in the standard, the eddy current response is in a vertical direction on the CRT. The amplitude of the response increases as the material thickness decreases (Fig. 55b). As each step in the standard is scanned, the eddy current response may be offset, as shown in Fig. 55(c).

After calibrating the instrument, the inspector scans along the inner edge of the window and monitors the CRT for thinning responses, which are indicative of internal corrosion. When thinning responses are noted, the inspection marks the extent of the corrosion and determines the relative remaining thickness. Results are marked on a plastic overlay or sketch and submitted to the engineering department for disposition. The extent of severe corrosion and whether

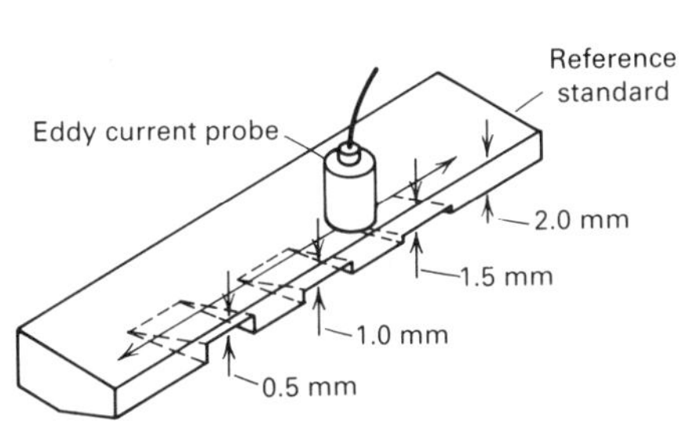

(a)

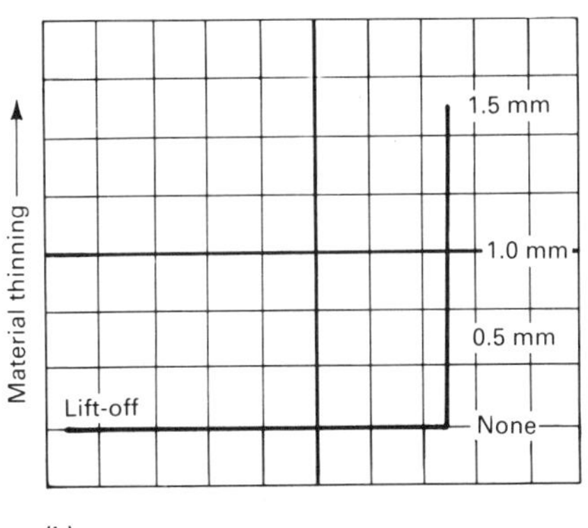

(b)

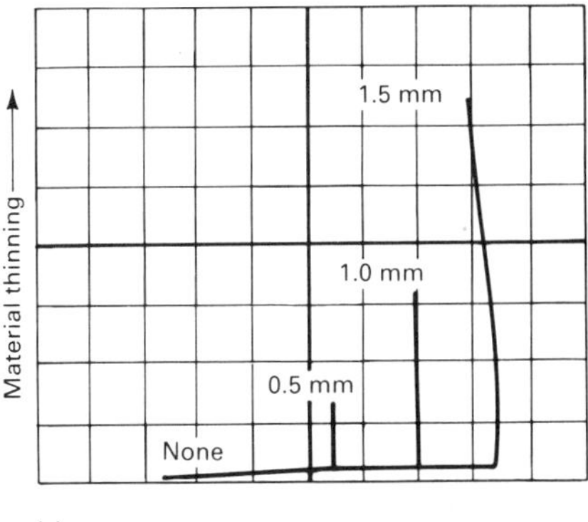

(c)

Fig. 55 Eddy current calibration procedure to detect subsurface corrosion in the aircraft windowbelt panels illustrated in Fig. 51. (a) Reference standard used to simulate varying degrees of corrosion thinning from 0.5 to 2.0 mm (0.020 to 0.08 in.) in 0.5 mm (0.020 in.) increments. (b) Plot of CRT display at 2.25-kHz test frequency. (c) CRT offset display permits resolution of amplitudes at the various material thicknesses. Source: Ref 13

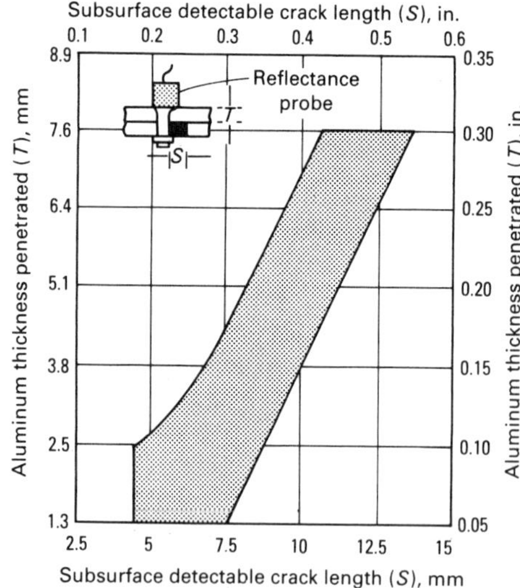

Fig. 57 Plot of detectable crack length versus thickness of overlying aluminum layer for reflectance-type eddy current probes. Source: Ref 13

or not thinning has occurred are determined by removing the internal panels, window, and insulation to expose the corroded areas. The corrosion products are removed, and the thickness is measured using an ultrasonic thickness gage or depth dial indicator.

Effect of Test Frequency on Detectable Flaw Size. Very small surface cracks, extending outward from fastener holes, are detectable using high-frequency, small-diameter eddy current probes. However, the detection of subsurface cracks requires a reduction in operating frequency that also necessitates an increase in the coil (probe) diameter resulting in a larger detectable crack. Because the depth of eddy current penetration is a function of operating frequency, material conductivity, and material magnetic permeability, increased penetration can only be accomplished by lowering the operating frequency. Therefore, the thicker the part to be penetrated, the lower the frequency to be used.

Most of the subsurface crack detection is accomplished with advanced-technology phase-sensitive CRT instruments and reflection (driver/receiver) type eddy current probes. To demonstrate the capability of this technology to detect subsurface cracks in aluminum structures adjacent to fastener holes, Fig. 56 shows a plot of operating frequency versus detectable crack size.

Figure 56 illustrates that the detectable flaw size increases as the frequency is reduced. The simulated subsurface flaws range in length from 4.8 to 12.7 mm (0.1875 to 0.50 in.), and the operating frequency band is from 100 Hz to 10 kHz. In addition, Fig. 57 shows a plot of detectable crack size versus thickness of the aluminum layer penetrated before the eddy currents intercepted the crack in the underlying layer.

The simulated subsurface cracks range in length from 4.8 to 12.7 mm (0.1875 to 0.50 in.). The thickness of the aluminum penetrated before the crack was reached ranged from 1.3 to 7.62 mm (0.050 to 0.300 in.). Although Fig. 57 shows only one overlying layer, the actual specimens contain from one to three layers on top of the layer containing the crack. From Fig. 57, it can be seen that the detectable crack size increases as the overlying layer increases in thickness.

REFERENCES

1. M.L. Burrows, "A Theory of Eddy Current Flaw Detection," University Microfilms, Inc., 1964
2. C.V. Dodd, W.E. Deeds, and W.G. Spoeri, Optimizing Defect Detection in Eddy Current Testing, *Mater. Eval.*, March 1971, p 59-63
3. C.V. Dodd and W.E. Deeds, Analytical Solutions to Eddy-Current Probe-Coil Problems, *J. Appl. Phys.*, Vol 39 (No. 6), May 1968, p 2829-2838
4. R. Halmshaw, *Nondestructive Testing*, Edward Arnold, 1987
5. R.L. Brown, The Eddy Current Slide Rule, in *Proceedings of the 27th National Conference*, American Society for Nondestructive Testing, Oct 1967
6. H.L. Libby, *Introduction to Electromagnetic Nondestructive Test Methods*, John Wiley & Sons, 1971
7. E.M. Franklin, Eddy-Current Inspection—Frequency Selection, *Mater. Eval.*, Vol 40, Sept 1982, p 1008
8. L.C. Wilcox, Jr., Prerequisites for Qualitative Eddy Current Testing, in *Proceedings of the 26th National Conference*, American Society for Nondestructive Testing, Nov 1966
9. F. Foerster, Principles of Eddy Current Testing, *Met. Prog.*, Jan 1959, p 101
10. E.M. Franklin, Eddy-Current Examination of Breeder Reactor Fuel Elements, in *Electromagnetic Testing*, Vol 4, *Nondestructive Testing Handbook*, American Society for Nondestructive Testing, 1986, p 444
11. H.W. Ghent, "A Novel Eddy Current Surface Probe," AECL-7518, Atomic Energy of Canada Limited, Oct 1981
12. "Nondestructive Testing: A Survey," NASA SP-5113, National Aeronautics and Space Administration, 1973
13. D. Hagemaier, B. Bates, and A. Steinberg, "On-Aircraft Eddy Current Inspection," Paper 7680, McDonnell Douglas Corporation, March 1986

Remote-Field Eddy Current Inspection

J.L. Fisher, Southwest Research Institute

REMOTE-FIELD EDDY CURRENT (RFEC) INSPECTION is a nondestructive examination technique suitable for the examination of conducting tubular goods using a probe from the inner surface. Because of the RFEC effect, the technique provides what is, in effect, a through-wall examination using only the interior probe. Although the technique is applicable to any conducting tubular material, it has been primarily applied to ferromagnetics because conventional eddy current testing techniques are not suitable for detecting opposite-wall defects in such material unless the material can be magnetically saturated. In this case, corrosion/erosion wall thinning and pitting as well as cracking are the flaws of interest. One advantage of RFEC inspection for either ferromagnetic or nonferromagnetic material inspection is that the probe can be made more flexible than saturation eddy current or magnetic probes, thus facilitating the examination of tubes with bends or diameter changes. Another advantage of RFEC inspection is that it is approximately equal (within a factor of 2) in sensitivity to axially and circumferentially oriented flaws in ferromagnetic material. The major disadvantage of RFEC inspection is that, when applied to nonferromagnetic material, it is not generally as sensitive or accurate as traditional eddy current testing techniques.

Theory of the Remote-Field Eddy Current Effect

In a tubular geometry, an axis-encircling exciter coil generates eddy currents in the circumferential direction (see the article "Eddy Current Inspection" in this Volume). The electromagnetic skin effect causes the density of eddy currents to decrease with distance into the wall of the conducting tube. However, at typical nondestructive examination frequencies (in which the skin depth is approximately equal to the wall thickness), substantial current density exists at the outer wall. The tubular geometry allows the induced eddy currents to rapidly cancel the magnetic field from the

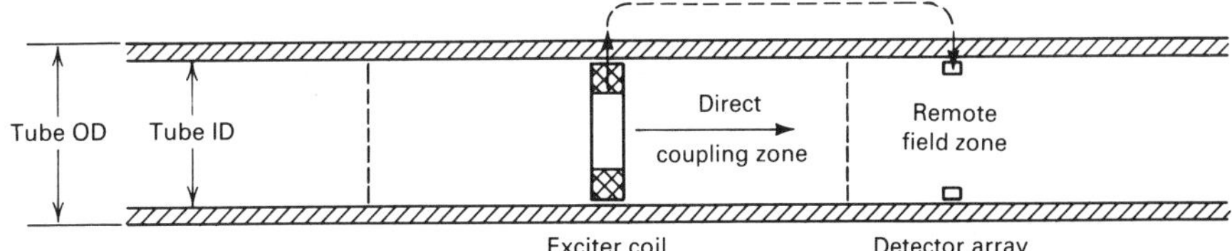

Fig. 1 Schematic showing location of remote-field zone in relation to exciter coil and direct coupling zone

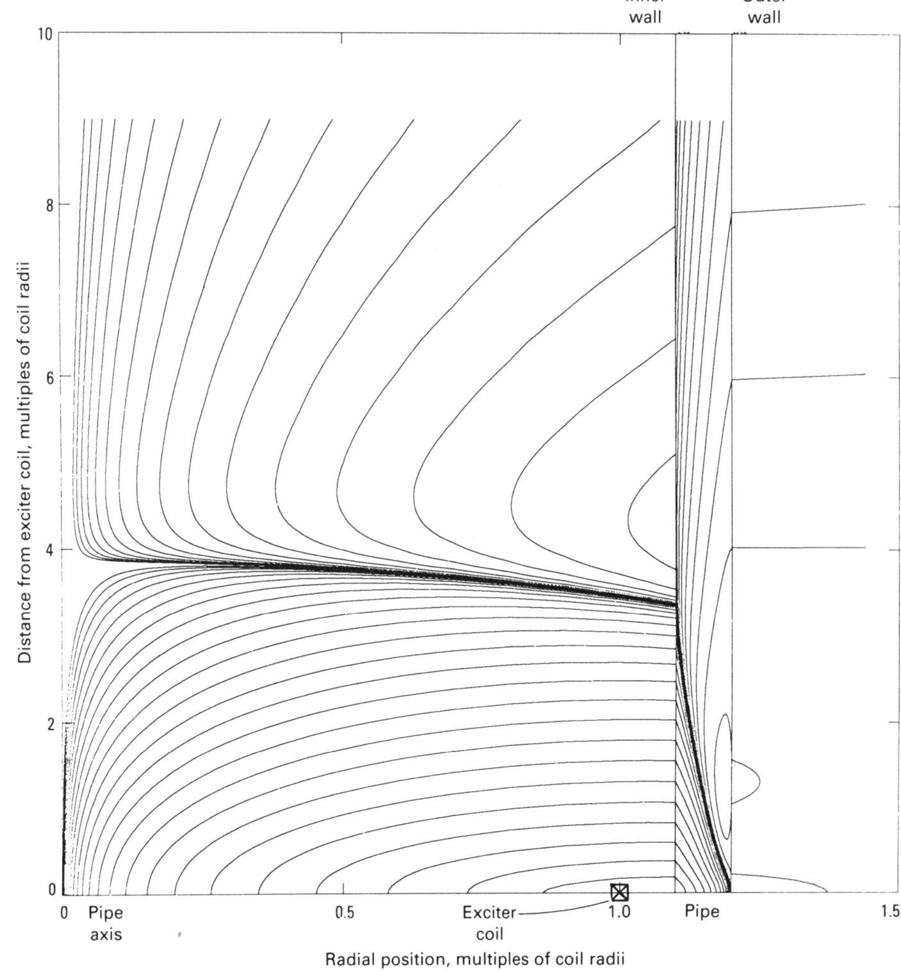

Fig. 2 Instantaneous field lines shown with a log spacing that allows field lines to be seen in all regions. This spacing also emphasizes the difference between the near-field region and the remote-field region in the pipe. The near-field region consists of the more closely spaced lines near the exciter coil in the pipe interior, and the remote-field region is the less dense region further away from the exciter.

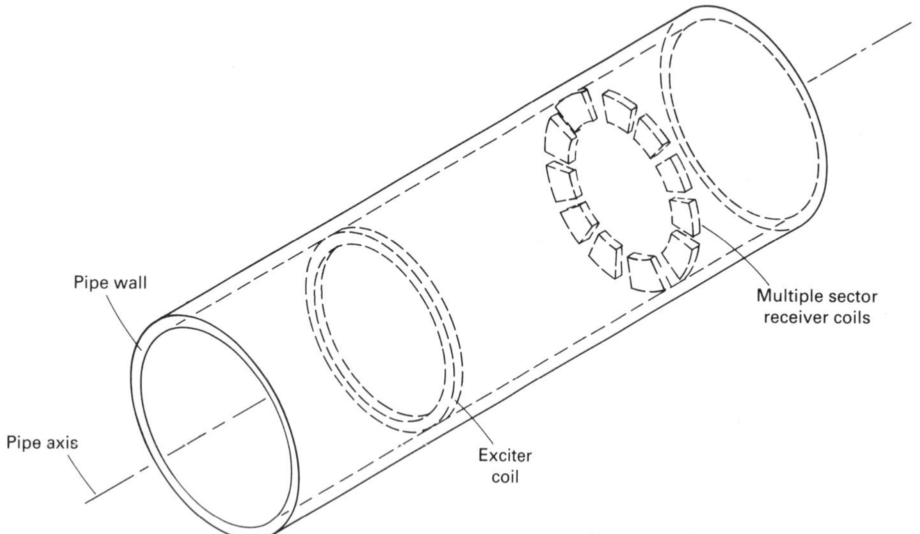

Fig. 3 RFEC configuration with exciter coil and multiple sector receiver coils

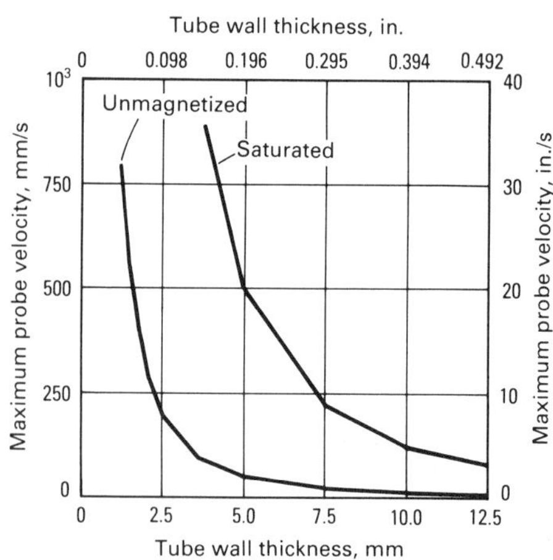

Fig. 4 Relationship between maximum probe speed and tube wall thickness for nominal assumptions of resolution and tube characteristics

exciter coil inside the tube, but does not shield as efficiently the magnetic field from the eddy currents that are generated on the outer surface of the tube. Therefore, two sources of magnetic flux are created in the tube interior; the primary source is from the coil itself, and the secondary source is from eddy currents generated in the pipe wall (Fig. 1). At locations in the interior near the exciter coil, the first source is dominant, but at larger distances, the wall current source dominates. A sensor placed in this second, or remote field, region is thus picking up flux from currents through the pipe wall. The magnitude and phase of the sensed voltage depend on the wall thickness, the magnetic permeability and electrical conductivity of tube material, and the possible presence of discontinuities in the pipe wall. Typical magnetic field lines are shown in Fig. 2.

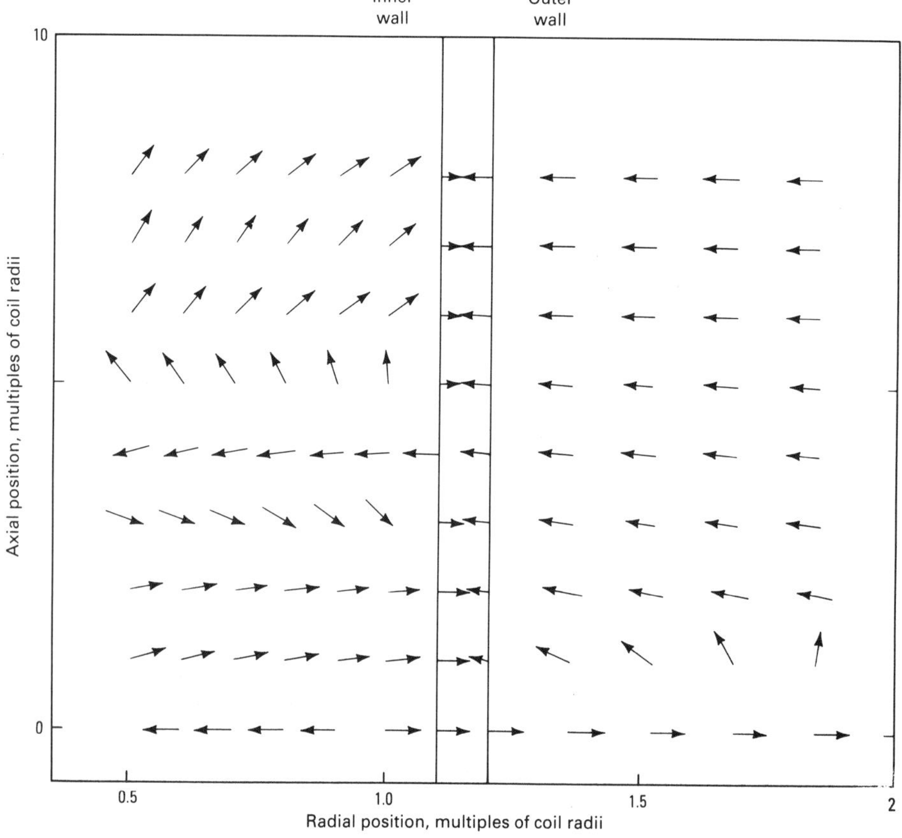

Fig. 5 Poynting vector field showing the direction of energy flow at any point in space. This more directly demonstrates that the direction of energy flow in the remote-field region is from the exterior to the interior of the pipe.

Probe Operation

The RFEC probe consists of an exciter coil and one or more sensing elements. In most reported implementations, the exciter coil encircles the pipe axis. The sensing elements can be coils with axes parallel to the pipe axis, although sensing coils with axes normal to the pipe axis can also be used for the examination of localized defects. In its simplest configuration, a single axis-encircling sensing coil is used. Interest in this technique is increasing, probably because of a discovery by Schmidt (Ref 1). He found that the technique could be made much more sensitive to localized flaws by the use of multiple sector coils spaced around the inner circumference with axes parallel to the tube axis. This modern RFEC configuration is shown in Fig. 3.

The use of separate exciter and sensor elements means that the RFEC probe operates naturally in a driver-pickup mode instead of the impedance-measuring mode of traditional eddy current testing probes. Three conditions must be met to make the probe work:

- The exciter and sensor must be spaced relatively far apart (approximately two or more tube diameters) along the tube axis
- An extremely weak signal at the sensor must be amplified with minimum noise generation or coupling to other signals. Exciter and sensing coils may consist of several hundred turns of wire in order to maximize signal strength

- The correct frequency must be used. The inspection frequency is generally such that the standard depth of penetration (skin depth) is the same order of magnitude as the wall thickness (typically 1 to 3 wall thicknesses)

When these conditions are met, changes in the phase of the sensor signal with respect to the exciter are directly proportional to the sum of the wall thicknesses at the exciter and sensor. Localized changes in wall thickness cause phase and amplitude changes that can be used to detect such defects as cracks, corrosion thinning, and pitting.

Instrumentation

Instrumentation includes a recording device, a signal generator, an amplifier (because the exciter signal is of much greater power than that typically used in eddy current testing), and a detector. The detector can be used to determine exciter/sensor phase lag or can generate an impedance-plane type of output such as that obtained with conventional driver-pickup eddy current testing instruments. Instrumentation developed specifically for use with RFEC probes is commercially available. Conventional eddy current instruments capable of operating in the driver-pickup mode and at low frequencies can also be used. In this latter case, an external amplifier is usually provided at the output of the eddy current instrument to increase the drive voltage. The amplifier can be an audio amplifier designed to drive loudspeakers if the exciter impedance is not too high. Most audio amplifiers are designed to drive a 4- to 8-Ω load.

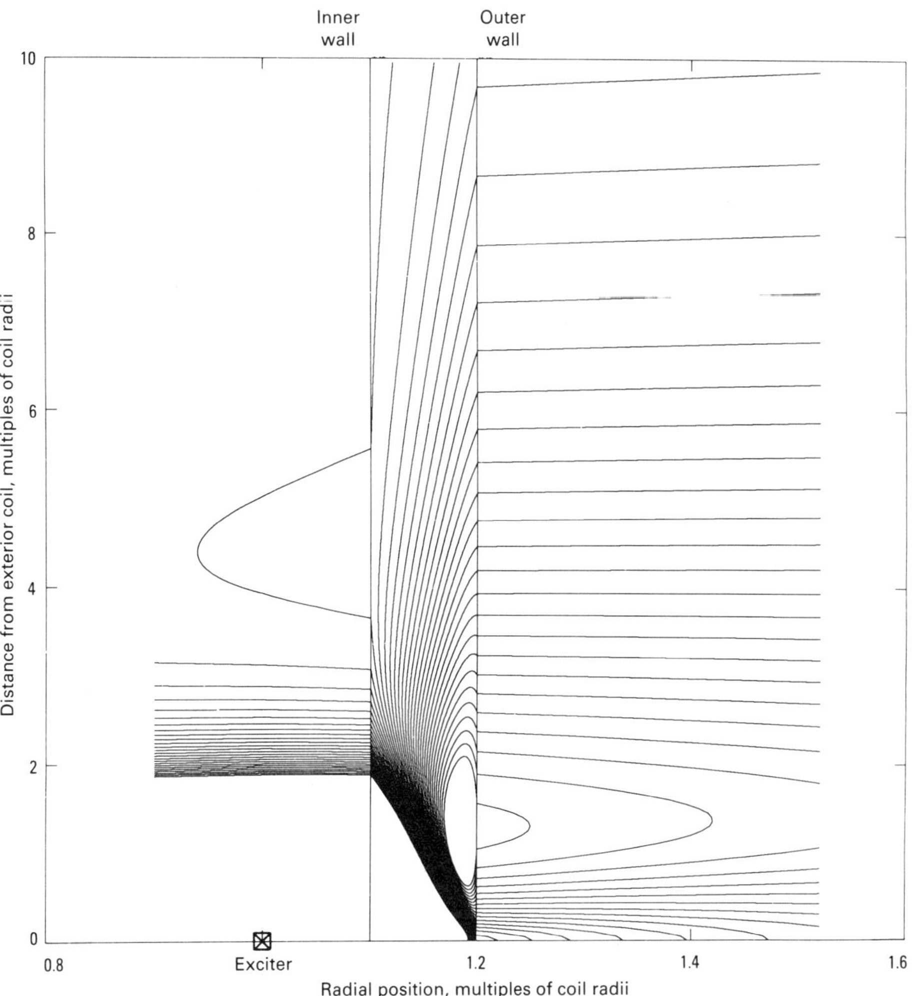

Fig. 6 Magnetic field lines generated by the exciter coil and currents in the pipe wall. The greater line density in the pipe closer to the outside wall in the remote-field region confirms the observation that field energy diffuses into the pipe interior from the exterior. A significant number of the field lines have been suppressed.

Limitations

Operating Frequency. The speed of inspection is limited by the low operating frequency. For example, the inspection of standard 50 mm (2 in.) carbon steel pipe with a wall thickness of 3.6 mm (0.14 in.) requires frequencies as low as 40 Hz. If the phase of this signal is measured (and a phase measurement can be made once per cycle), then only 40 measurements per second are obtained. If a measurement is desired every 2.5 mm (0.1 in.) of probe travel, the maximum probe speed is 102 mm/s (4 in./s), or 6 m/min (20 ft/min). Although this speed may be satisfactory for many applications, the speed must decrease directly in proportion to the spatial resolution required and inversely (approximation is based on simple skin effect model and is generally valid when the skin depth is greater than the wall thickness) with the square of the wall thickness. This limitation is illustrated in Fig. 4 for a range of wall thicknesses.

Effect of Material Permeability. Another limitation is that the magnitude and phase of the sensor signal are affected by changes in the permeability of the material being examined. This is probably the limiting factor in determining the absolute response to wall thickness and the sensitivity to localized damage in ferromagnetic material. This disadvantage can be overcome by applying a large magnetic field to saturate the material, but a bulkier probe that is not easily made flexible would be required.

Effect of External Conductors on Sensor Sensitivity. A different type of limitation is that the sensor is also affected by conducting material placed in contact with the tube exterior. The most common examples of this situation are tube supports and tube sheets. This effect is produced because the sensor is sensitive to signals coming from the pipe exterior. For tube supports, a characteristic pattern occurs that varies when a flaw is present. While allowing flaw detection, this information is probably recorded at a reduced sensitivity. Geometries such as finned tubing weaken the RFEC signal and add additional signal variation to such an extent that the technique is not practical under these conditions.

Difficulty in Distinguishing Flaws. Another limitation is that measuring exciter/sensor phase lag and correlating remaining wall thickness leads to nondiscrimination of outside diameter flaws from inside diameter flaws. Signals indicating similar outside and inside diameter defects are nearly identical. However, conventional eddy current probes can be used to confirm inside-diameter defects.

Current RFEC Research

No-Flaw Models. Most published research regarding RFEC inspection has been concerned with interpreting and modeling the remote-field effect without flaws in order to explain the basic phenomenon and to demonstrate flaw detection results. The no-flaw case has been successfully modeled by several researchers, including Fisher *et al.* (Ref 2), Lord (Ref 3), Atherton and Sullivan (Ref 4), and Palanissimy (Ref 5), using both analytical and finite-element techniques.

This work has shown that in the remote-field region the energy detected by the sensor comes from the pipe exterior and not directly from the exciter. This effect is seen in several different ways. For example, in the Poynting vector plot shown in Fig. 5, the energy flow is away from the pipe axis in the near-field region, but in the remote-field region a large area of flow has energy moving from the pipe wall toward the axis. In the magnetic field-line plot shown in Fig. 6, the magnetic field in the remote-field region is greater near the tube outside diameter than near the inside diameter. This condition is just the opposite from what one would expect and from what exists in the near-field region. The energy diffusion is from the region of high magnetic field concentration to regions of lower field strength.

Flaw models with the RFEC geometry have been generated more rarely. The problem is that realistic flaw models require the use of three-dimensional modeling, something that is difficult to achieve with eddy current testing. One model by Fisher *et al.* (Ref 2) used a boundary-element calculation in conjunction with the two-dimensional, unperturbed-field calculation to predict the response to pitting. The response to outside-diameter and inside-diameter slots has been modeled with two-dimensional finite-element programs (Ref 3).

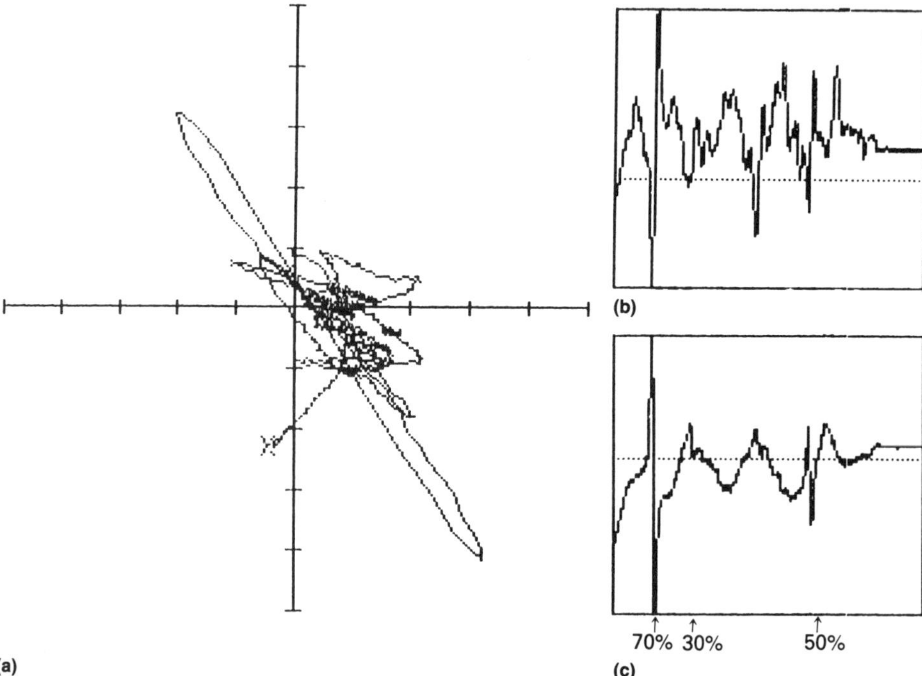

Fig. 7 Signal processing of impedance-plane sensor voltage in RFEC testing. (a) RFEC scan with a B_r probe through a carbon steel tube with outside surface pits that were 30, 50, and 70% of wall thickness depth. Each graduation in *x* and *y* direction is 10 V. (b) Horizontal channel at 0° rotation. (c) Vertical channel at 0° rotation. Signal amplitudes in both (b) and (c) are in arbitrary units. Only the 70% flaw stands out clearly.

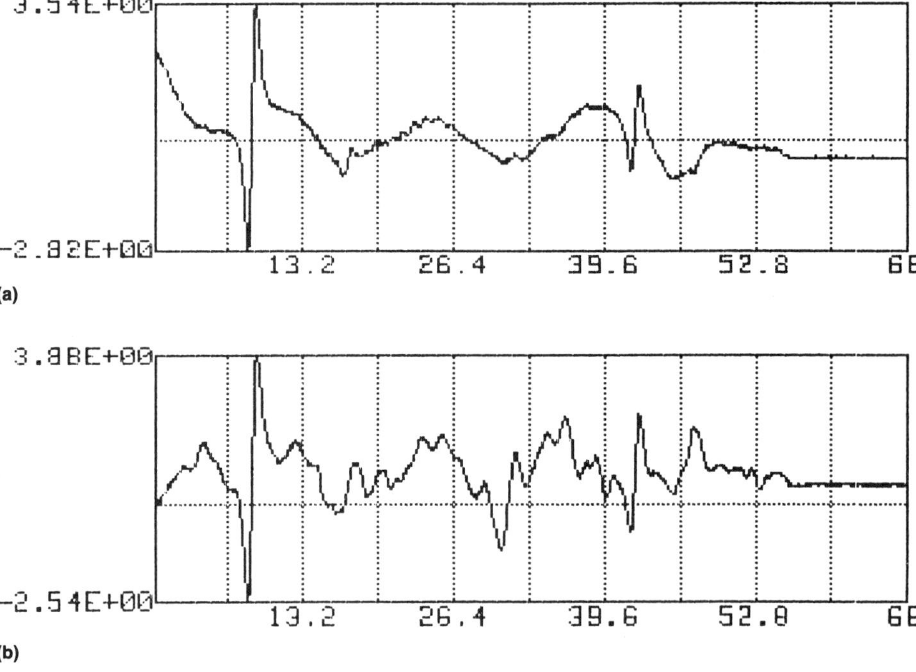

Fig. 8 Horizontal (a) and vertical (b) data of Fig. 7 after 100° rotation. Signal amplitudes are in arbitrary units.

Techniques Used to Increase Flaw Detection Sensitivity

Two general areas—sensor configuration and signal processing—have been identified for improvements in the use of RFEC inspection that would allow it to achieve greater effective flaw sensitivity.

Sensor Configuration. For the detection of localized flaws, such as corrosion pits, the results of the unperturbed and the flaw-response models suggest that a receiver coil oriented to detect magnetic flux in a direction other than axial might provide increased flaw sensitivity. This suggestion was motivated by the fact that the field lines in the remote-field region are approximately parallel to the pipe wall, as shown in Fig. 6. Thus, a sensor designed to pick up axial magnetic flux, B_z, would always respond to the unperturbed (no-flaw) field; a flaw response would be a perturbation to this primary field. If the sensor were oriented to receive radial flux, B_r, then the unperturbed flux would be reduced and the flaw signal correspondingly enhanced. This approach has been successful; a comparison of a B_z sensor and a B_r sensor used to detect simulated corrosion pits showed that the B_r probe is much more sensitive. A B_r sensor would also minimize the transmitter coil signal from a flaw, which is always present when a B_z sensor is used, thus eliminating the double signals from a single source. This configuration appears to be very useful for the detection of localized flaws, but does not appear to have an advantage for the measurement of wall thickness using the unperturbed field.

Signal Processing. The second area of possible improvement in RFEC testing is the use of improved signal-processing techniques. Because it was observed that the exciter/sensor phase delay was directly proportional to wall thickness in ferromagnetic tubes, measurement of sensor phase has been the dominant method of signal analysis (Ref 1). However, it is possible to display both the magnitude and phase of the sensor voltage or, correspondingly, the complex components of the sensor voltage. This latter representation (impedance plane) is identical to that used in modern eddy cur-

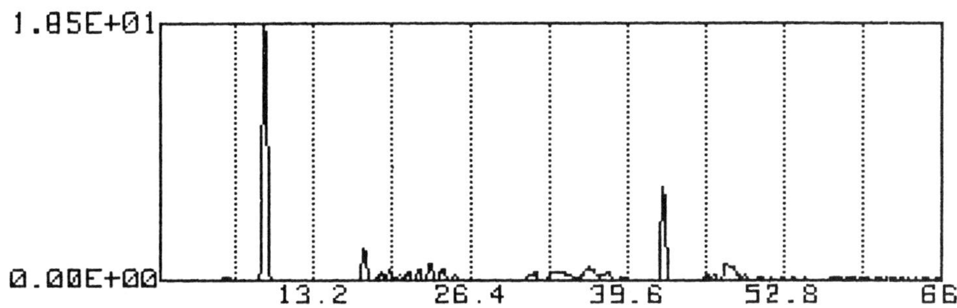

Fig. 9 Data from Fig. 8 processed with the correlation technique. All three flaws are now well defined. Signal amplitude is in arbitrary units.

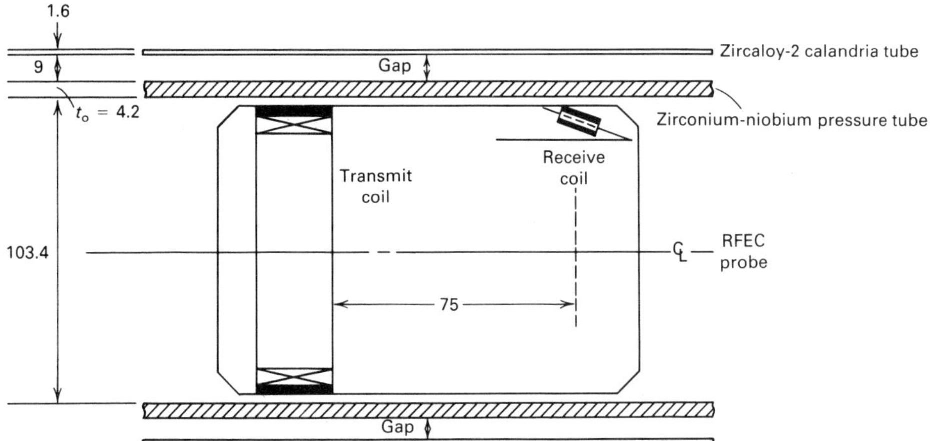

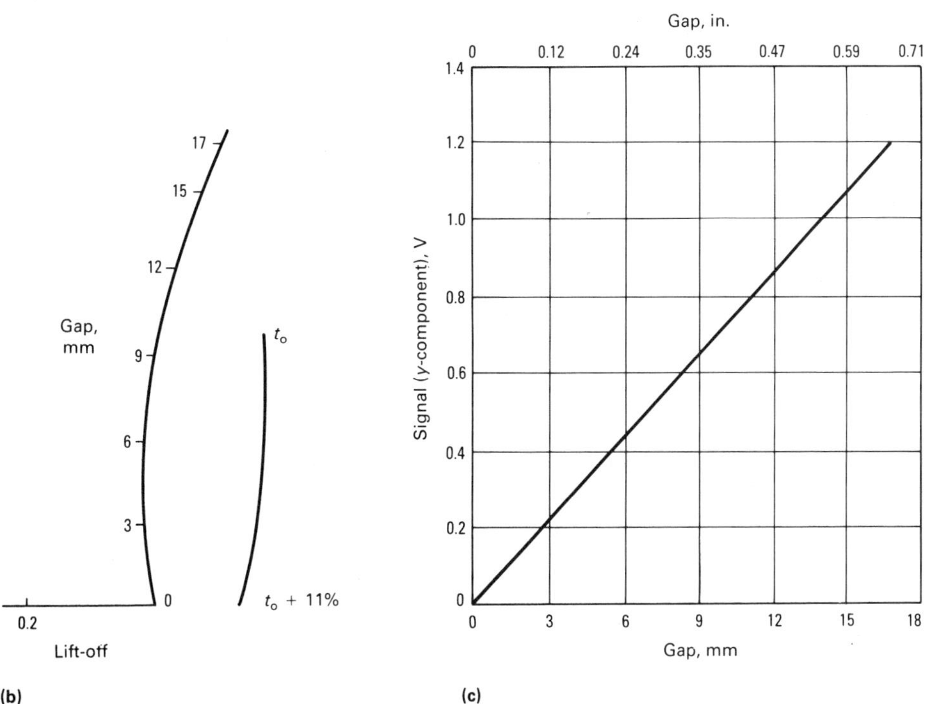

Fig. 10 Gap measurement between two concentric tubes in a nuclear fuel channel with an RFEC probe. (a) Cross-sectional view of probe and test sample. Dimensions given in millimeters. (b) Plot of eddy current signals illustrating effect of gap, wall thickness, and lift-off. Dimensions given in millimeters. (c) Plot of *y*-component of signal versus gap at a frequency of 3 kHz. Source: V.S. Cecco, Atomic Energy of Canada Limited

rent testing instrumentation for probes operated in a driver/pickup mode. Figures 7 and 8 show the results of using this type of display. Figure 7 shows the data from a scan through a carbon steel tube with simulated outside surface pits of 30, 50, and 70% of nominal wall thickness. A B_r probe was used for the experiment. Figure 8 shows the horizontal and vertical channels after the scan data were rotated by 100°. Much of the noise was eliminated in this step.

An additional possible signal-processing step is to use a correlation technique to perform pattern matching. Because the B_r sensor has a characteristic double-sided response to a flaw, flaws can be distinguished from material variations or undesired probe motion by making a test sensitive to this shape. This effect is achieved by convolving the probe signals with a predetermined sample signal that is representative of flaws. The results of one test using this pattern-matching technique are shown in Fig. 9. It is seen that even though the three flaws have a range of depths and diameters, the correlation algorithm using a single sample flaw greatly improves the signal-to-noise ratio.

Other signal-processing techniques that show promise include the use of high-order derivatives of the sensor signal for edge detection and bandpass and median filtering to remove gradual variations and high-frequency noise (Ref 6).

Applications*

Remote-field eddy current testing should be considered for use in a wider range of examinations because its fundamental physical characteristics and limitations are now well understood. The following two examples demonstrate how improvements in RFEC probe design and signal processing have been successfully used to optimize the operation of key components in nuclear and fossil fuel power generation.

Example 1: Gap Measurement Between Two Concentric Tubes in a Nuclear Fuel Channel Using a Remote-Field Eddy Current Probe. The Canadian Deuterium Uranium reactor consists of 6 m (20 ft) long horizontal pressure tubes containing the nuclear fuel bundles. Concentric with these tubes are calandria tubes with an annular gap between them. Axially positioned garter spring spacers separate the calandria and pressure tubes. Because of unequal creep rates, the gap will decrease with time. Recently, it was found that some garter springs were out of position, allowing pressure tubes to make contact with the calandria tubes.

Because of this problem, a project was initiated to develop a tool to move the garter springs back to their design location for

*Example 1 was prepared by V.S. Cecco, Atomic Energy of Canada Limited.

operating reactors. Successful use of this tool would require measurement of the gap during the garter spring unpinching operation. The same probe could also be used to measure the minimum gap along the pressure tube length. Because the gap is gas filled, an ultrasonic testing technique would not have been applicable.

At low test frequencies, an eddy current probe couples to both the pressure tube and the calandria tube (Fig. 10a). The gap component of the signal is obtained by subtracting the pressure tube signal from the total signal. Because the probe is near or in contact with the pressure tube and nominally 13 mm (½ in.) away from the calandria tube, most of the signal comes from the pressure tube. In addition, the calandria tube is much thinner and of higher electrical resistivity than the pressure tube, further decreasing the eddy current coupling. To overcome the problem of low sensitivity with large distances (>10 mm, or 0.4 in.), a remote-field eddy current probe was used. Errors in gap measurement can result from variations in:

- Lift-off
- Pressure tube electrical resistivity
- Ambient temperature
- Pressure tube wall thickness

Multifrequency eddy current methods exist that significantly reduce the errors from the first three of the above-mentioned variations, but not for wall thickness variations. This is because of minimal coupling to the calandria tube and because the signal from the change in gap is similar (in phase) to the signal from a change in wall thickness (at all test frequencies).

The inducing magnetic field from an eddy current probe must pass through the pressure tube wall to sense the calandria tube. The upper test frequency is limited by the high attenuation through the pressure tube wall and the lower test frequency by the low coupling to the pressure tube and calandria tube. In addition, at low test frequency, there is poor signal discrimination because of variations in lift-off, electrical resistivity, wall thickness, and gap. In practice, 90° phase separation between lift-off and gap signals gives optimum signal-to-noise and signal discrimination. For this application, the optimum test frequency was found to be between 3 and 4 kHz.

Typical eddy current signals from a change in pressure tube to calandria tube gap, lift-off, and wall thickness are shown in Fig. 10(b). The output signal for the complete range in expected gap is linear, as shown in Fig. 10(c). To eliminate errors introduced from wall thickness variations, the probe body includes an ultrasonic normal beam transducer located between the eddy current transmit and receive coils. This combined eddy current and ultrasonic testing probe assembly with a linear output provides an accuracy of ±1 mm (±0.04 in.) over the complete gap range of 0 to 18 mm (0 to 0.7 in.).

Example 2: RFEC Inspection of Carbon Steel Mud Drum Tubing in Fossil Fuel Boilers. Examination of the carbon steel tubing used in many power boiler steam generators is necessary to ensure their safe, continuous operation. These tubes have a history of failure due to attack from acids formed from wet coal ash, as well as erosion from gas flow at high temperatures. Failure of these tubes from thinning or severe pitting often requires shutdown of the system and results in damage to adjacent tubes or to the entire boiler. The mud drum region of the steam generator is one such area. Here the water travels up to the superheater section, and the flame from the burners is directed right at the tubes just leaving the mud drum. The tubes in this region are tapered and rolled into the mud drum (Fig. 11), then they curve sharply upward into the superheater section.

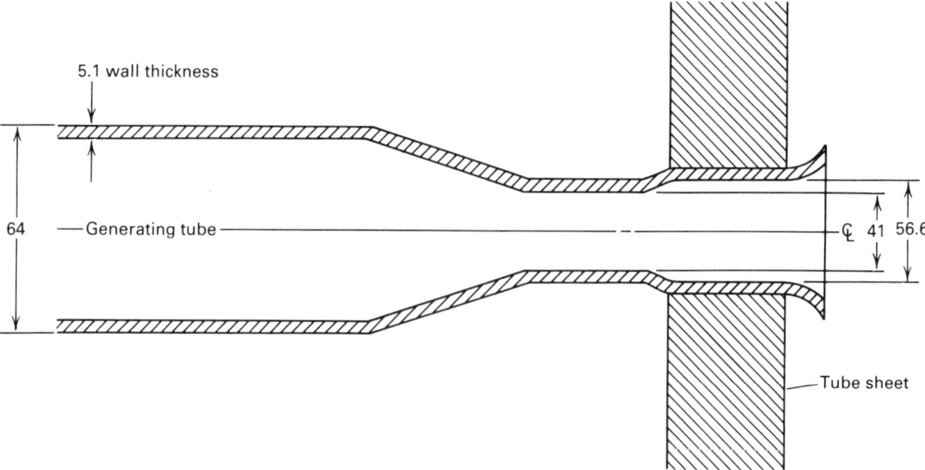

Fig. 11 Geometry and dimensions of 64 mm (2.5 in.) OD carbon steel generating tubes at a mud drum. The tube is rolled into a tube sheet to provide a seal as shown. Dimensions given in millimeters

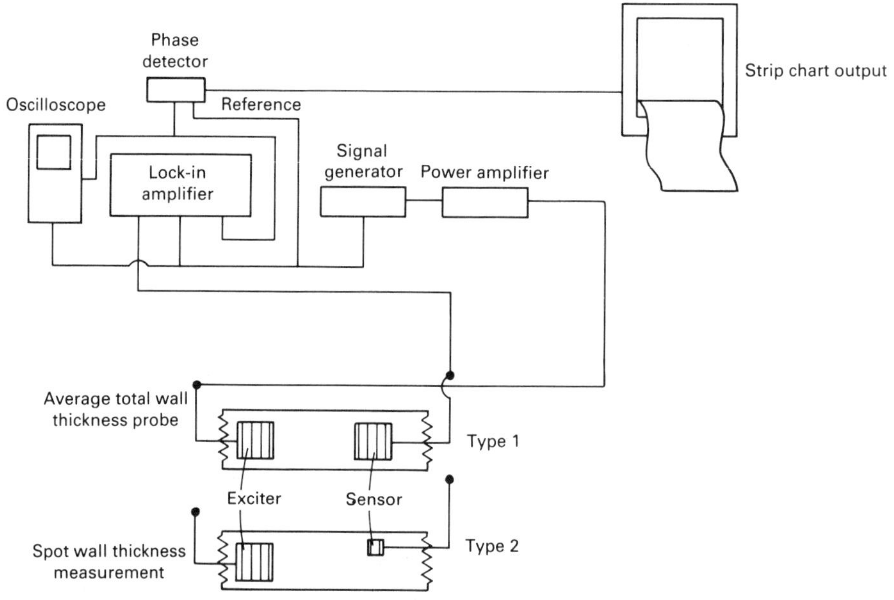

Fig. 12 Breadboard instrumentation necessary to excite and receive the 45-Hz signal. Analysis was based on the phase difference between the reference and received signals.

The tubes are constructed of carbon steel typically 64 mm (2.5 in.) in outside diameter, and 5.1 mm (0.2 in.) in wall thickness. A conventional eddy current test would be ineffective, because the eddy currents could not penetrate the tube wall and still influence coil impedance in a predictable manner. Ultrasonic testing could be done, but would be slow if 100% wall coverage were required. The complex geometry of the tube, the material, and the 41 mm (1.6 in.) access opening necessitated that a different technique be applied to the problem.

Because the RFEC technique behaves as if it were a double through transmission effect, changes in the fill factor are not as significant as they would be in a conventional eddy current examination. This allowed the transmitter and receiver to be designed to enter through the 41 mm (1.6 in.) opening. Nylon brushes recentered the RFEC probe in the 56.6 mm (2.23 in.) inside diameter of the tube. To maintain flexibility of the assembly and still be able to position the inspection head, universal joints were used between elements.

The instrumentation used in this application was a combination of laboratory standard and custom-designed components (Fig. 12). The Type 1 probe was first used to locate any suspect areas, then the Type 2 probe was used to differentiate between general and local thinning of the tube wall. This technique was applied to the examination of a power boiler, and the results were compared to outside diameter ultrasonic readings where possible. Agreement was within 10%.

REFERENCES

1. T.R. Schmidt, The Remote-Field Eddy Current Inspection Technique, *Mater. Eval.*, Vol 42, Feb 1984
2. J.L. Fisher, S.T. Cain, and R.E. Beissner, Remote Field Eddy Current Model, in *Proceedings of the 16th Symposium on Nondestructive Evaluation* (San Antonio, TX), Nondestructive Testing Information Analysis Center, 1987
3. W. Lord, Y.S. Sun, and S.S. Udpa, Physics of the Remote Field Eddy Current Effect, in *Reviews of Progress in Quantitative NDE*, Plenum Press, 1987
4. D.L. Atherton and S. Sullivan, The Remote-Field Through-Wall Electromagnetic Technique for Pressure Tubes, *Mater. Eval.*, Vol 44, Dec 1986
5. S. Palanissimy, in *Reviews of Progress in Quantitative NDE*, Plenum Press, 1987
6. R.J. Kilgore and S. Ramchandran, NDT Solution: Remote-Field Eddy Current Testing of Small Diameter Carbon Steel Tubes, *Mater. Eval.*, Vol 47, Jan 1989

Microwave Inspection

William L. Rollwitz, Southwest Research Institute

MICROWAVES (or radar waves) are a form of electromagnetic radiation located in the electromagnetic spectrum at the frequencies listed in Table 1. Major subintervals of the microwave frequency band are designated by various letters; these are listed in Table 2. The microwave frequency region is between 300 MHz and 325 GHz. This frequency range corresponds to wavelengths in free space between 1000 cm and 1 mm (40 and 0.04 in.).

Although the general nature of microwaves has been known since the time of Maxwell, not until World War II did microwave generators and receivers useful for the inspection of material become available. One of the first important uses of microwaves was for radar. Their first use in nondestructive evaluation (NDE) was for components such as waveguides, attenuators, cavities, antennas, and antenna covers (radomes). The interaction of microwave electromagnetic energy with a material involves the effect of the material on the electric and magnetic fields that constitute the electromagnetic wave, that is, the interaction of the electric and magnetic fields with the conductivity, permittivity, and permeability of the material. Microwaves behave much like light waves in that they travel in straight lines until they are reflected, refracted, diffracted, or scattered. Because microwaves have wavelengths that are 10^4 to 10^5 times longer than those of light waves, microwaves penetrate deeply into materials, with the depth of penetration dependent on the conductivity, permittivity, and permeability of the materials. Microwaves are also reflected from any internal boundaries and interact with the molecules that constitute the material. For example, it was found that the best source for the thickness and voids in radomes was the microwaves generated within the radomes. Both continuous and pulsed incident waves were used in these tests, and either reflected or transmitted waves were measured.

Table 1 Divisions of radiation, frequencies, wavelengths, and photon energies of the electromagnetic spectrum

Division of radiation	Frequency, Hz	Wavelength, m	Photon energy J	Photon energy eV
Radio waves (FM and TV)	3×10^8	1	1.6×10^{-25}	10^{-6}
Microwaves	3×10^9	10^{-1}	1.6×10^{-24}	10^{-5}
	3×10^{10}	10^{-2}	1.6×10^{-23}	10^{-4}
	3×10^{11}	10^{-3}	1.6×10^{-22}	10^{-3}
Infrared	3×10^{12}	10^{-4}	1.6×10^{-21}	10^{-2}
	3×10^{13}	10^{-5}	1.6×10^{-20}	10^{-1}
Visible light	3×10^{14}	10^{-6}	1.6×10^{-19}	1
Ultraviolet light	3×10^{15}	10^{-7}	1.6×10^{-18}	10
	3×10^{16}	10^{-8}	1.6×10^{-17}	10^2
X-ray and γ-ray radiation	3×10^{17}	10^{-9}	1.6×10^{-16}	10^3
	3×10^{18}	10^{-10}	1.6×10^{-15}	10^4
	3×10^{19}	10^{-11}	1.6×10^{-14}	10^5
	3×10^{20}	10^{-12}	1.6×10^{-13}	10^6
	3×10^{21}	10^{-13}	1.6×10^{-12}	10^7
Cosmic ray radiation	3×10^{22}	10^{-14}	1.6×10^{-11}	10^8

Microwave Inspection Applications

The use of microwaves for evaluating material properties and discontinuities in materials other than radomes began with the evaluation of the concentration of moisture in dielectric materials. Microwaves of an appropriate wavelength were found to be strongly absorbed and scattered by water molecules. When the dry host material is essentially transparent to the microwaves, the moisture measurement is readily made.

Next, the thickness of thin metallic coatings on nonmetallic substrates and of dielectric slabs was measured. In this case, incident and reflected waves were allowed to combine to form a standing wave. Measurements were then made on the standing wave because it provided a scale sensitive to the material thickness.

The measurement of thickness was followed by the determination of voids, delaminations, macroporosity, inclusions, and other flaws in plastic or ceramic materials. Microwave techniques were also used to detect flaws in bonded honeycomb structures and in fiber-wound and laminar composite materials. For most measurements, the reflected wave was found to be most useful, and the use of frequency modulation provided the necessary depth sensitivity. Success in these measurements also indicated that microwave techniques could give information related to changes in chemical or molecular structure that affect the dielectric constant and dissipation of energy at microwave frequencies. Some of the properties measured include polymerization, oxidation, esterification, distillation, and vulcanization.

Advantages. In comparison with ultrasonic inspection and x-ray radiographic inspection, the advantages of inspection with microwaves are as follows:

- Broadband frequency response of the coupling antennas
- Efficient coupling through air from the antennas to the material
- No material contamination problem caused by the coupling
- Microwaves readily propagate through air, so successive reflections are not obscured by the first one

Table 2 Microwave frequency bands

Band designator	Frequency range, GHz
UHF	0.30–1
P	0.23–1
L	1–2
S	2–4
C	4–8
X	8–12.5
K_u	12.5–18
K	18–26.5
K_a	26.5–40
Q	33–50
U	40–60
V	50–75
E	60–90
W	75–110
F	90–140
D	110–170
G	140–220
Y	170–260
J	220–325

Source: Ref 1

- Information concerning the amplitude and phase of propagating microwaves is readily obtainable
- No physical contact is required between the measuring device and the material being measured; therefore, the surface can be surveyed rapidly without contact
- The surface can be scanned in strips merely by moving the surface or by scanning the surface with antennas
- No changes are caused in the material; therefore, the measurement is nondestructive
- The complete microwave system can be made from solid-state components so that it will be small, rugged, and reliable
- Microwaves can be used for locating and sizing cracks in materials if the following considerations are followed. First, the skin depth at microwave frequencies is very small (a few micrometers), and the crack is detected most sensitively when the crack breaks through the surface. Second, when the crack is not through the surface, the position of the crack is indicated by a detection of the high stresses in the surface right about the subsurface crack. Finally, microwave crack detection is very sensitive to crack opening and to the frequency used. Higher frequencies are needed for the smaller cracks. If the frequency is increased sufficiently, the incident wave can propagate into the crack, and the response is then sensitive to crack depth

Limitations. The use of microwaves is in some cases limited by their inability to penetrate deeply into conductors or metals. This means that nonmetallic materials inside a metallic container cannot be easily inspected through the container. Another limitation of the lower-frequency microwaves is their comparatively low power for resolving localized flaws. If a receiving antenna of practical size is used, a flaw whose effective dimension is significantly smaller than the wavelength of the microwaves used cannot be completely resolved (that is, distinguished as a separate, distinct flaw). The shortest wavelengths for which practical present-day microwave apparatus exists are of the order of 1 mm (0.04 in.). However, the development of microwave sources with wavelengths of 0.1 mm (0.004 in.) are proceeding rapidly. Consequently, microwave inspection for the detection of very small flaws is not suited for applications in which flaws are equal to or smaller than 0.1 mm (0.004 in.). Subsurface cracks can be detected by measuring the surface stress, which should be much higher in the surface above the subsurface crack.

Physical Principles of Microwaves

In free space, an electromagnetic wave is transverse; that is, the oscillating electric and magnetic fields that constitute it are transverse to the direction of travel of the wave. The relative directions of these two fields and the direction of propagation of the wave are shown schematically in Fig. 1. As the wave travels along the z-axis, the electric and magnetic field intensities at an arbitrary fixed location in space vary in magnitude. A particularly simple form of a propagating electromagnetic wave is the linearly polarized, sinusoidally varying, plane electromagnetic wave illustrated in Fig. 2. The magnitude of the velocity, v, at which a wave front travels along the z-axis is given by the relation $v = f\lambda$, where f is frequency and λ is wavelength. In free space, this velocity is the speed of light, which has the value 2.998×10^8 m/s and is usually designated by the letter c.

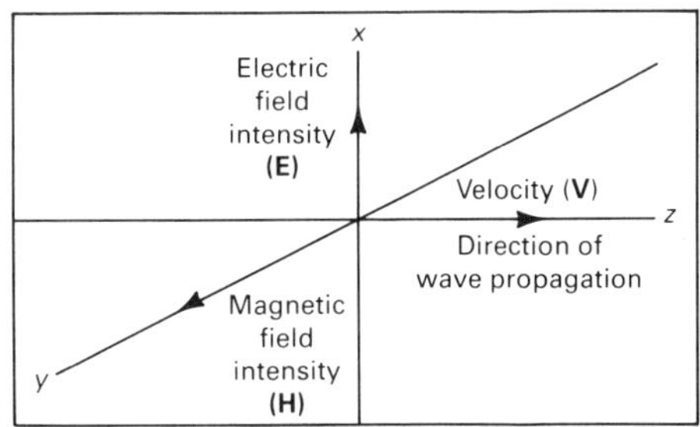

Fig. 1 Relative directions of the electric field intensity (**E**), the magnetic field intensity (**H**), and the direction of propagation (z) for a linearly polarized, plane electromagnetic wave

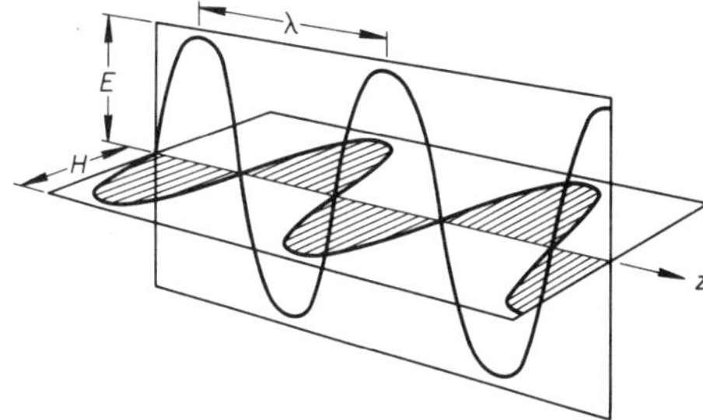

Fig. 2 Diagram of a linearly polarized, sinusoidally varying, plane electromagnetic wave propagating in empty space. λ, wavelength; z, direction of wave propagation; E, amplitude of electric field; H, amplitude of magnetic field

In microwave inspection, a homogeneous material medium can be characterized in terms of a magnetic permeability, μ; a dielectric coefficient, ϵ; and an electrical conductivity, σ. In general, these quantities are themselves functions of the frequency, f. Moreover, μ and ϵ must usually be treated as complex quantities, rather than as purely real ones, to account for certain dissipative effects. However, a wide variety of applications occur in which μ and ϵ can be regarded as mainly real and constant in value. The magnetic permeability, μ, usually differs only slightly from its value in vacuum, while the dielectric coefficient, ϵ, usually varies between 1 and 100 times its value in vacuum. The electrical conductivity, σ, ranges in value from practically zero (10^{-16} $\Omega \cdot$ mm) for good insulators to approximately 10^7 $\Omega \cdot$ mm for good conductors such as copper.

For an electromagnetic wave incident upon a material, a part of the incident wave is transmitted through the surface and into the material, and a part of it is reflected. The sum of the reflected energy and refracted energy (transmitted into the material) equals the incident energy. If the reflected wave is subtracted in both amplitude and phase from the incident wave, the transmitted wave can be determined. When the reflected wave is compared, in both amplitude and phase, with the incident wave, information about the surface impedance of the material can be obtained.

Plane electromagnetic waves propagating through a conductive medium diminish in amplitude as they propagate, falling to 37% of their amplitude at a reference position in distance, referred to as the skin depth, measured along the direction of propagation. The skin depth, δ, in a good conductor ($\sigma \gg \omega\epsilon$, where ω is the angular frequency) is given by the relation $\delta = (2/\mu\sigma\omega)^{1/2}$. The velocity, v, of an electromagnetic wave propagating in a nonconductor is given by the relation $v = 1/(\mu\epsilon)^{1/2}$. This velocity can be expressed relative to the velocity of electromagnetic waves in vacuum, the ratio being the index of refraction, n, where $n = c/v = (\mu\epsilon/\mu_0\epsilon_0)^{1/2}$ with μ_0 and ϵ_0 being the magnetic permeability and dielectric coefficient of free space. The phase velocity of a plane harmonic electromagnetic wave in a conductive medium can be given as $v = \omega\sigma = (2\omega/\mu\delta)^{1/2}$, where δ is the skin depth. Therefore, the velocity, v, depends strongly on the frequency, ω, even if magnetic permeability, μ, and electrical conductivity, σ, do not themselves depend on ω. As a result, a conductive medium is said to be highly dispersive, because a wave packet, which comprises sinusoidal components of many different frequencies, disperses (spreads) as it propagates. In conductive media, the magnetic component of an electromagnetic wave does not propagate in phase with the electric component. Assuming $|\epsilon| \ll |\sigma|$, the surface impedance of a material is (Ref 2):

$$Z_s = \sqrt{\frac{\mu}{\epsilon - j(\sigma/\omega)}} \approx \sqrt{\frac{j\omega\mu}{\sigma}} \qquad \text{(Eq 1)}$$

where μ is the complex permeability, σ is the conductivity, ϵ is the complex permittivity, ω is the angular frequency, and $j = \sqrt{-1}$.

The skin depth, δ, of an electromagnetic wave propagating in a weakly conductive medium ($\sigma \ll \omega\epsilon$) is given approximately by the relation $\delta \simeq 2/\sigma \sqrt{\epsilon/\mu}$. In this case, the wavelength is approximately given by:

$$\lambda \simeq \lambda_0\left(\frac{1}{n}\right)\left[1 - \frac{1}{8}(\sigma/\epsilon\omega)^2\right] \qquad \text{(Eq 2)}$$

where λ_0 is the wavelength of the wave in vacuum, n is the index of refraction of the medium. Equation 1 is sufficiently accurate for most materials having electrical conductivity low enough for practical microwave inspection involving transmission. The criterion for its validity is that the nonattenuative wavelength, λ_0/n, be short compared to the skin depth, δ.

Reflection and Refraction. The laws of reflection and refraction of microwaves at interfaces between mediums of differing, electromagnetic properties are essentially the same as those for the reflection and refraction of visible light. The angle of refraction is determined by Snell's law, $n_2 \sin \phi = n_1 \sin \theta$, where n_1 and n_2 are, respectively, the indexes of refraction of the two media; θ is the angle of incidence; and φ is the angle of refraction. The medium in which the incident ray occurs is arbitrarily designated by the subscript 1, and the other medium by the subscript 2.

For linearly polarized plane waves incident perpendicularly on an interface separating two dielectric mediums, the amplitudes of the reflected and transmitted waves, respectively, are given by:

$$E_{\text{max, reflected}} = [(n_2 - n_1)/(n_1 + n_2)]\, E_{\text{max, incident}} \quad \text{(Eq 3a)}$$

$$E_{\text{max, transmitted}} = [2n_1/(n_1 + n_2)]\, E_{\text{max, incident}} \quad \text{(Eq 3b)}$$

Analogous relations hold for the amplitudes of the magnetic field. For angles of incidence other than zero, the corresponding relations are more complicated and will not be quoted here. The amplitudes of reflected and refracted waves vary as a function of the angle of incidence for a typical choice of the ratio n_1/n_2, as shown in Fig. 3. The shapes of the curves vary somewhat with the dielectric constant (Ref 3).

The curves in Fig. 3 show the amplitude reflection factor as a function of the angle of incidence when the polarization factor (direction of the electric field vector) is either parallel or perpendicular to the plane of the interface. The curves in Fig. 3(a) represent the conditions in which the microwaves are entering the material, and the curves in Fig. 3(b) are for waves leaving the material. In Fig. 3(a), the reflection factor increases steadily to unity at 90° for the perpendicular polarization factor. The reflection factor for the parallel polarization decreases to zero at the Brewster angle and then increases to total reflection as a function of increasing angle of incidence. In Fig. 3(b), the same occurrence takes place except that the unity reflection factor for both polarization factors occurs at the critical angle of incidence rather than at 90°. The critical angle for reflection, from Snell's law, is equal to arc sin $(1/\epsilon^{1/2})$, where ε is the dielectric constant of the material. The Brewster angle for the case of reflection is equal to the arc tan $(1/\epsilon^{1/2})$. The scattering cross section oscillates with decreasing magnitude as the ratio of the

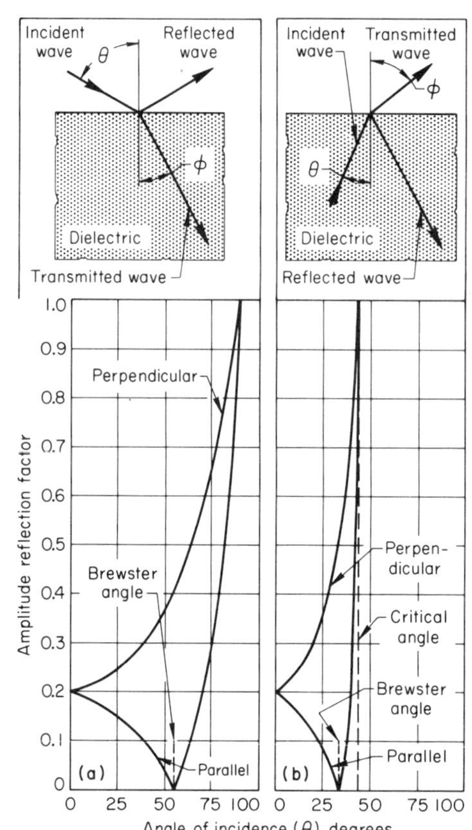

Fig. 3 Representative reflection of a linearly polarized plane electromagnetic wave at a dielectric interface with electric field parallel or perpendicular to the plane of incidence. (a) Wave entering dielectric. (b) Wave leaving dielectric

circumference to wavelength varies from 0.3 to 10. Ratios from 0.3 to 10 are in the resonance region. The maximum scattering occurs when the circumference equals the wavelength. In the optical region, the scatter cross section equals the physical cross section, and $\sigma/\pi r^2 \approx 1$, where r is the radius of the sphere. Thus, scattering is proportional to physical cross section (πr^2).

Absorption and Dispersion of Microwaves. Microwaves are also affected during their propagation through a homogeneous nonmetallic material, primarily by the interaction of the electric field with the dielectric (molecular) properties of the nonmetallic material. The storage and dissipation of the electric field energy by the polarization and conduction behavior of the material are the basic factors under consideration.

Polarization and conduction are the cumulative results of molecular-charge-carrier movement in the nonmetallic material. Polarization involves the action of bound charges in the form of permanent or induced dipoles. Conduction refers to whatever small amount of free charge carriers (electrons) is present. As a microwave passes through the material, the dipoles oscillate because of the cyclic nature of the force on them from the electric field. The dipole oscillation alternately stores and dissipates the electric field energy. Conduction currents only dissipate the energy, that is, convert it to heat.

The dielectric properties of a nonmetallic material are normally expressed in terms of the dielectric constant (permittivity) and loss factor. The dielectric constant is related to the amount of electric field energy that the dipoles in a material temporarily store and release during each half cycle of the electric field change. Materials with a high dielectric constant have a great storage capacity. This reduces the electric field strength, the velocity, and the wavelength. The loss factor, or loss tangent, expresses the dissipation of energy caused by both conduction and dipole oscillation losses. In other words, the dielectric constant measures the energy storage, while the loss factor measures the dissipation of electromagnetic energy by nonmetallics.

Standing Waves. Interference conditions usually prevail when microwaves are used for nondestructive inspection because of the wavelengths and velocities involved, the coherent nature of the microwaves used, the high transparency of most nonmetallic materials, and the fact that the thicknesses of the nonmetallic materials are within several wavelengths. The familiar standing wave is the usual wave pattern resulting during nondestructive inspection. A standing wave (Fig. 4) is produced when two waves of the same frequency are propagating in opposite directions and interfere with each other. The result is the formation of a total field whose maximum and minimum points remain in a fixed or standing position. Both component waves are still traveling, and only the resultant wave pattern is fixed. A simple way to form standing waves is to transmit a coherent wave normal to a surface. The incident and reflected waves interfere and cause a standing wave. The standing wave wavelength and peak amplitude change along the standing wave pattern and are related to the velocity and attenuation experienced by the wave in a given medium. The technique by which standing waves are formed with microwave radiation is used to make accurate measurements of thickness where conventional caliper methods are especially difficult. The technique involved is discussed in the section "Thickness Gaging" in this article.

Scattering of Microwaves. Microwaves reflect from inhomogeneities by a process known as scattering. The scattering is generally used to describe wave interaction with small particles or inhomogeneities. The term reflection is generally used to describe wave interaction with surfaces that are large compared to wavelength. When the surface is not smooth on a scale commensurate with the wavelength of the microwaves used, the reflected wave is not a simple single wave, but is essentially a com-

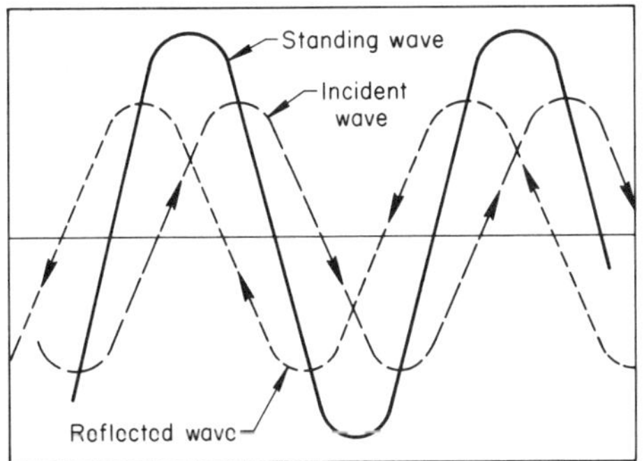

Fig. 4 Pattern of a standing wave formed by interference of an incident wave and a reflected wave

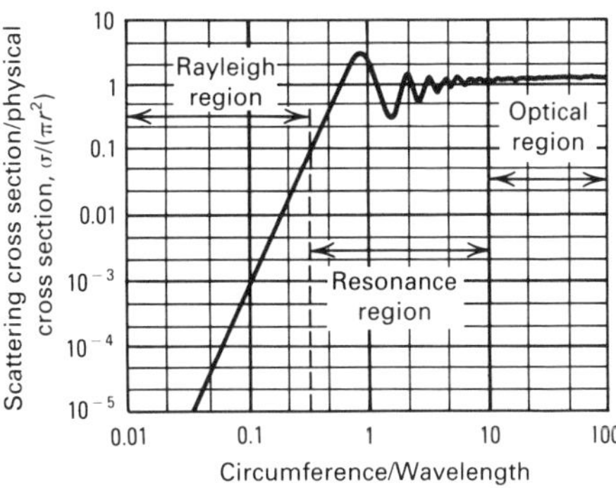

Fig. 5 Microwave scattering by metal spheres of various sizes

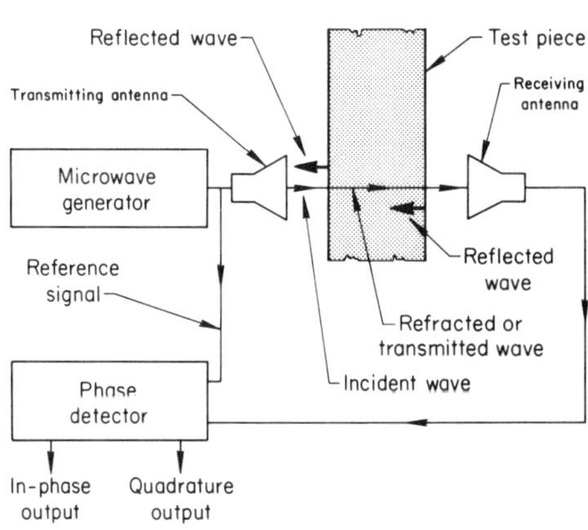

Fig. 6 Diagram of the basic components of the transmission technique used for microwave inspection

posite of many such waves of various relative amplitudes, phases, and directions of propagation. This effect is greatest when the wavelength is comparable to the dimensions of the irregularities. Under these circumstances, the radiation is said to be scattered. The scattering behavior of metal spheres having differing ratios of circumference to the wavelength of incident microwave radiation is shown in Fig. 5. The scattering cross section is proportional to the amplitude of the wave scattered in the direction back toward the source of the incident wave.

The scattering is small for values of the ratio of the circumference to the wavelength less than 0.3, and scattering varies as the fourth power of this ratio. Ratios below 0.3 lie in what is known as the Rayleigh region.

Mathematically, the dielectric constant and loss factor are expressed in combined forms as a complex permittivity:

$$\epsilon^* = \epsilon' - j\epsilon'' \qquad \text{(Eq 4)}$$

where ϵ^* is the complex permittivity or complex dielectric constant, ϵ' is the permittivity or dielectric constant, j is the phasor operator $[(-1)^{1/2}]$, and ϵ'' is the loss factor. Equation 4 shows that the dielectric constant is 90° out of phase with respect to the loss factor. The loss tangent is equal to the ratio of ϵ''/ϵ'. Wave velocity and attenuation are related to dielectric properties and therefore serve as a means of measurement.

The dielectric properties of a nonmetallic material are frequently affected by other material properties of industrial importance. The degree of correlation depends on the frequency of the electromagnetic wave, and sensitive measurements can often be made with microwaves.

Special Techniques of Microwave Inspection

The following general approaches have been used in the development of microwave nondestructive inspection:

- Fixed-frequency continuous-wave transmission
- Swept-frequency continuous-wave transmission
- Pulse-modulated transmission
- Fixed-frequency continuous-wave reflection
- Swept-frequency continuous-wave reflection
- Pulse-modulated reflection
- Fixed-frequency standing waves
- Fixed-frequency reflection scattering
- Microwave holography
- Microwave surface impedance
- Microwave detection of stress corrosion

Each of these techniques uses one or more of the several processes by which materials can interact with microwaves, namely, reflection, refraction, scattering, absorption, and dispersion. These techniques will be briefly described from the standpoint of instrumentation and are grouped to four areas:

- Transmission
- Reflection
- Standing wave
- Scattering

Transmission Techniques

The basic components of the transmission technique are shown schematically in Fig. 6. A microwave generator feeds both a transmitting antenna and a phase-sensitive detector (or comparator). The transmitting antenna produces the electromagnetic wave that is incident on one face of the material to be inspected. At the surface, the incident wave is split into a reflected wave and a transmitted or refracted wave. The transmitted wave goes through the material into the receiving antenna. All of the transmitted wave will not pass through the second face of the material because some of it will be reflected at the second surface. The microwave signal from the receiving antenna can be compared in amplitude and phase with the reference signal taken directly from the microwave generator.

The reference signal can be taken to be of the form $V_{ref} = V_0 \cos \omega t$. The received signal, V_{rec}, is then of the form:

$$V_{rec} = V' \cos(\omega t + \phi) = (V' \cos \phi) \cos \omega t - (V' \sin \phi) \sin \omega t \qquad \text{(Eq 5)}$$

Because it is the coefficient of the term that varies in phase with the reference signal, the quantity $V' \cos \phi$ is referred to as the in-phase component, and the quantity $V' \sin \phi$ is termed the quadrature component. Standard electronic phase-sensitive detectors are available that can detect each of these components separately. The transmission technique can have three variations:

- Fixed-frequency continuous wave
- Variable-frequency continuous wave
- Pulse-modulated wave

Fixed-Frequency Continuous-Wave Transmission. In this technique, the frequency of the microwave generator is constant. It is used either when the band of frequencies required for the desired interaction is very narrow or when the band of frequencies is so broad that the changes of material properties with frequency are very small and therefore not especially frequency sensitive. The fixed-frequency continuous-wave transmission technique is the only one of the transmission techniques in which both components (in-phase and quadrature-phase) can be detected with little mutual interference. When separation (the ability to separate the components) is important, this technique is generally used.

Swept-Frequency Continuous-Wave Transmission. Some microwave interactions are frequency sensitive in that their resonant frequency shifts with changes in material properties. For others, the response as a function of frequency over substantial bandwidth must be used. The swept-frequency continuous-wave transmission technique provides for a transmission measurement over a selected range of frequencies. The fixed-frequency microwave generator shown in Fig. 6 is replaced with a swept-frequency generator whose frequency is programmed to vary automatically. With currently available generators, a frequency band of one octave or more can be electronically swept (from 1 to 2 GHz, for example). Broad band amplifiers of low noise and high gain also make it possible to detect transmitted signals through materials having very high attenuation. Multioctave generators from 100 kHz to 4 GHz or 10 MHz to 40 GHz are available. Vector network analyzers provide broadband amplitude and phase.

Pulse-Modulated Transmission. Although phase measurements can be made on the transmitted wave, they are only relative to the reference wave. There is no simple method for tagging a particular sine wave crest relative to another to measure transmission time. Therefore, when a measurement of the time of transmission is required, the pulse-modulated technique is used. To produce the pulse modulation, the microwave generator is gated on and off. The phase-sensitive detector in the receiver is usually replaced by a peak-value detector. Thus, the receiver output consists of pulses that are delayed a finite time relative to the transmitted pulse. An oscilloscope with an accurate horizontal-sweep rate can be used to display these pulses. Swept-frequency measurements give group delay information. The time domain features of vector network analyzers can also be useful.

Reflection Techniques

The reflection techniques are of two types: single antenna and dual antenna. The single-antenna system, in which incident and reflected waves are both transmitted down the waveguide between the microwave generator and the antenna, is shown schematically in Fig. 7(a). The phase detector is set so that it compares the phase of the reflected wave with that of the incident wave. This gives two output signals that are respectively proportional to the in-phase and quadrature components in the reflected wave. Such a system works well only for normal or near-normal incidence.

The dual-antenna reflection system (Fig. 7b) operates at any angle of incidence for which there is appreciable reflection. A comparison of Fig. 7(b) with Fig. 6 shows that the dual-antenna reflection equipment is essentially the same as that used for transmission measurements. In the transmission technique, the reflected wave is not used. In the reflection technique, the transmitted wave is ordinarily not used except as a reference.

The boundary conditions must be complied with at the surface of the material. Therefore, the reflected wave, in principle, has the same information about the bulk microwave properties of the material as the refracted wave. However, the wave reflected from the first surface does not contain any information about the inhomogeneous properties of the material within the sample being tested. There are further reflections from any internal discontinuities or boundaries, which ultimately add to the surface-reflected wave when refracted at the surface. In this manner, properties beneath the surface are sensed. If the component being inspected is a plate that is backed with a layer of conductive material, the wave reflected from this metal face traverses the material twice, and it too adds to the surface-reflected wave to provide information about the interior of the material.

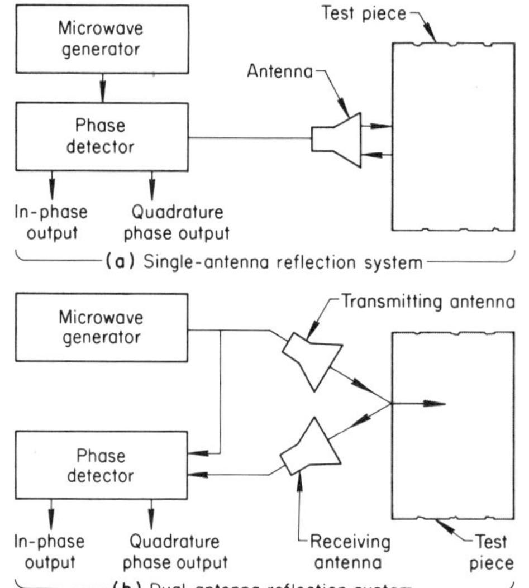

Fig. 7 Diagram of the single-antenna and dual-antenna reflection systems used for microwave inspection

Fixed-Frequency Continuous-Wave Reflection. The simplest microwave reflectometer is based on the fixed-frequency technique. The microwave signal is incident on the material from an antenna; the reflected signal is picked up by the same antenna. Both the in-phase and quadrature-phase components of the reflected wave can be determined. In practice, most such techniques have used only the amplitude of the reflected signal. The dual-antenna reflection technique (Fig. 7b) can also be used at a fixed frequency. The fixed-frequency continuous-wave technique has two limitations. First, the depth of a flaw cannot be determined, and second, the frequency response of the material cannot be determined. For these reasons, a swept-frequency technique may be more useful.

Swept-Frequency Continuous-Wave Reflection. When the interaction between a material and microwaves is frequency sensitive, a display of reflection as a function of frequency may be valuable. Because phase-sensitive detection over a wide range of frequencies is difficult, only the amplitude of the reflected signal is usually used as the output in swept-frequency techniques. However, phase-sensitive detection over a wide range of frequencies has recently been simplified with the use of vector network analyzers.

Depth can be measured with a swept-frequency technique if the reflected signal is mixed with the incident signal in a nonlinear element that produces the difference signal. The difference frequency, then, is a measure of how much farther the reflected signal has traveled than the incident signal. Thus, not only can the presence of an internal reflector be determined but also its depth. Depth can also be measured on a vector network analyzer utilizing time domain techniques.

Another application of this technique employs a slow sweep of frequency to identify specific layers of several closely spaced layers of material. The reflection at even multiples of one-fourth wavelength is larger than at odd multiples of one-fourth wavelength. The reflected signal identifies specific frequencies for which the layer spacing is at even or odd integral multiples of the quarter-wavelength. This same effect is used, for example, to reduce reflection from lenses by coating them with layers of dielectric.

Pulse-Modulated Reflection. The depth of a reflection, in principle, can also be determined by pulse modulating the incident wave. When the reflected time-delayed pulse is compared in time with the incident pulse and when the velocity of the wave in the material is known, the depth to the site of the reflection can be determined. In both frequency and time domain modulation, the nature of the reflector is determined by the strength of the reflected signal. The limitation of pulse modulation is that the pulses required are very narrow if shallow depths are to be determined. For this reason, the use of frequency modulation has been developed and used.

Standing Wave Techniques

A standing wave is obtained from the constructive interference of two waves of the same frequency traveling in opposite directions. The result is a standing wave in space. If a small antenna is placed at a fixed point in space, a voltage of constant amplitude and frequency is detected. Moving the antenna to another location would give a voltage of a different constant amplitude with the same frequency. The graph of the amplitude of the voltage as a function of position (distance) along a pure standing wave is shown in Fig. 8. One antenna is needed to produce the incident wave, which can interact with the reflected wave to produce the standing wave. Another antenna or probe is needed to make measurements along the standing wave. Thus, the dual-antenna system shown in Fig. 7(b) could be used to both make and measure microwave standing waves. The receiving antenna must not interfere with the incident wave. A single antenna fed through a circulator can also be used to separately transmit the incident wave and the reflected wave.

Scattering Techniques

The dual-antenna system diagrammed in Fig. 7(b) must be used for scattering measurements because the angles of the dif-

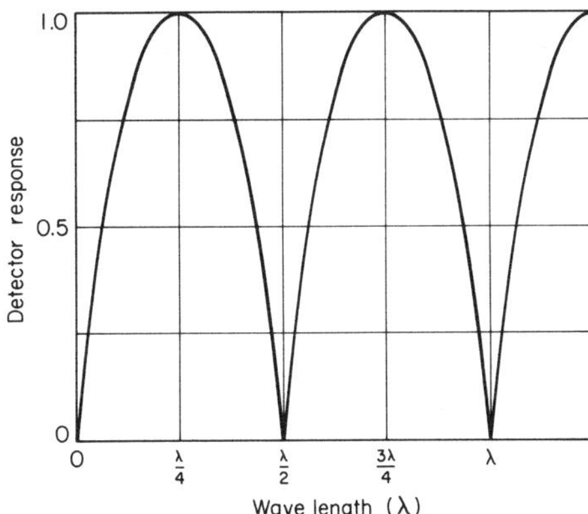

Fig. 8 Relation of electromagnetic wave amplitude (detector response) to the distance along a standing wave

fused or reradiated waves can be over a solid angle. To measure all of the scattered radiation, the entire sphere around the irradiated object or material should be scanned and the detected signal graphed as a function of position.

Microwave Holography

When two electromagnetic waves (Fig. 2) occupy the same space, the resulting electromagnetic wave is the vector sum of the two. If the two waves have the same source, they will have the same frequency and will add or subtract as vectors according to their relative phase angles. These additions and subtractions will cause an interference pattern, which is a plane of interference composed of places where they add and places where they subtract. Wave front reconstruction is discussed in Ref 4 to 8.

When the electromagnetic waves are in the infrared through ultraviolet range of frequencies and when a photographic film or plate is used at the plane of interference, a hologram is produced on the film. This hologram is a picture of the interference pattern. When laser light passes through the hologram, the result is a three-dimensional projection of the object from which the hologram was made. The image projected in this manner is sometimes called the hologram.

For electromagnetic waves of microwave frequencies (300 MHz to 300 GHz), the interference detector, instead of being a photographic film, is a microwave receiving antenna feeding a receiver in which the peak amplitude of the received signal is detected. This gives a voltage proportional to the magnitude of the vector sum of the wave reflected from the object and a reference signal derived directly from the source of the incident microwave irradiating the object. The microwave holographic system is illustrated in Fig. 9. When the antenna probe scans the plan of interference, the interference pattern could be drawn out with the voltage from the detector. When the voltage is amplified to feed a lamp or an oscilloscope, the interference pattern can be visually observed or photographed. The photograph is then a microwave hologram from which a three-dimensional visual image of the object can be projected using light. Microwave holography is discussed in more detail in the section "Microwave Holography Practice" in this article. Additional information is available in the articles "Optical Holography" and "Acoustical Holography" in this Volume.

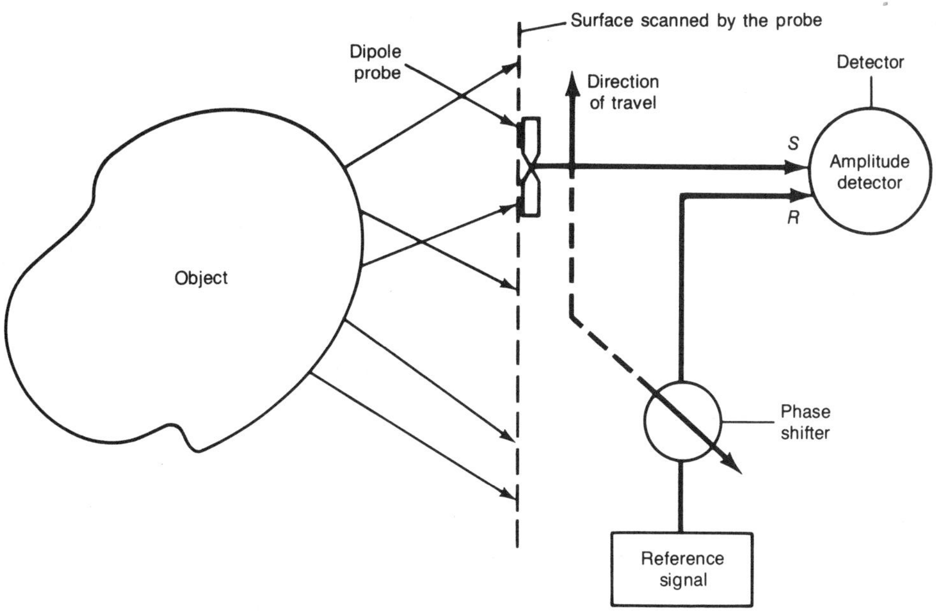

Fig. 9 Microwave holography with locally produced nonradiated reference wave

Instrumentation

The high frequencies of microwaves do not allow the use of standard electronic circuits (which possess discrete inductance, capacitance, and resistance). At microwave frequencies, these quantities become distributed along the propagation path. Therefore, microwave circuit design and description are handled from a wave standpoint, rather than from the conduction of electrons through wires and components.

Hollow metal tubes (typically, rectangular in cross section) called waveguides are normally used to convey microwaves between two parts of a circuit. Sometimes the microwaves are conveyed, as a wave, along coaxial or parallel conductors (Ref 3).

Microwave energy is conveyed through a waveguide in electromagnetic patterns called modes. The patterns consist of repetitive distributions of the electrical and magnetic fields along the axis of the waveguide. A classic transverse electromagnetic (TEM) wave (Fig. 2) cannot propagate in a waveguide. Modes are identified as either transverse electric (TE) or transverse magnetic (TM), referring to whether the electric or the magnetic field is perpendicular to the waveguide axis, or direction of propagation. High-order modes produce complicated patterns, while lower-order modes produce relatively simple patterns. Practically all microwave nondestructive inspection (NDI) circuits use the lowest-order transverse electric mode ($TE_{1,0}$) (Ref 3).

The waveguide pattern for the $TE_{1,0}$ mode is shown in Fig. 10. The electric field is represented by the solid lines that are vertical in the waveguide. The intensity of the electric field varies sinusoidally along and across the waveguide, with a peak intensity in the middle of the waveguide and zero intensity at either sidewall (end view, Fig. 10). A uniform (constant) electric field exists in the B (height) direction, and a sinusoidal variation occurs every one-half wavelength along the length of the waveguide (front view, Fig. 10). A top view of the waveguide shows that the magnetic field is in the form of loops, spaced at intervals of one-half wavelength.

Each mode has a cutoff frequency below which it cannot be propagated down a given

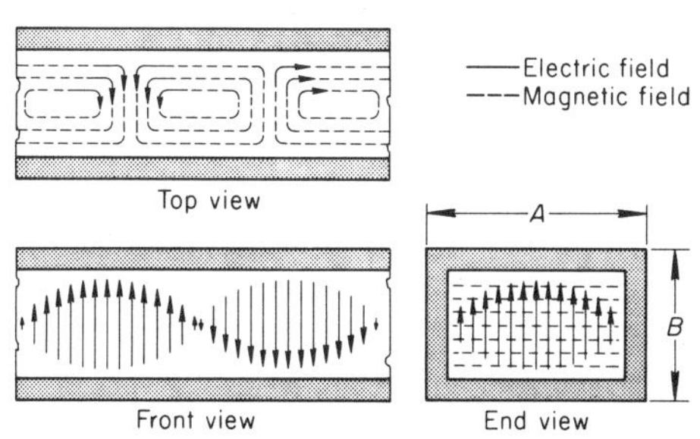

Fig. 10 Diagrams showing the lowest-order transverse electric mode ($TE_{1,0}$) in a waveguide

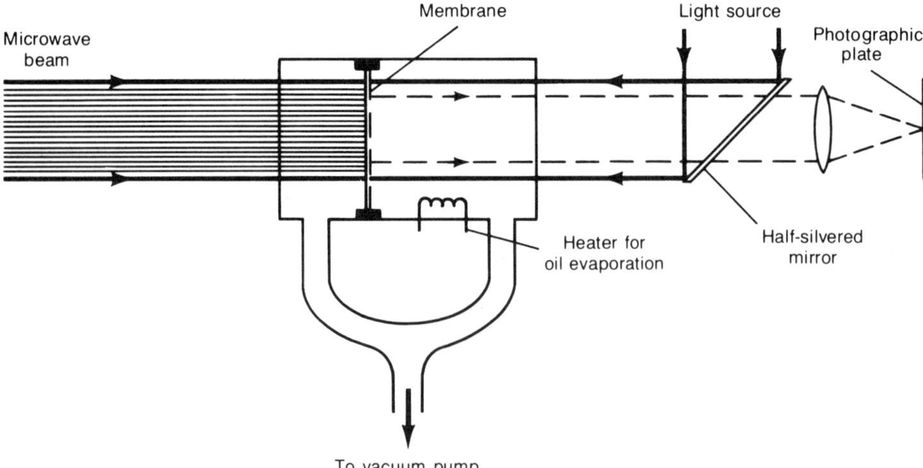

Fig. 11 Experimental setup used for microwave thermography

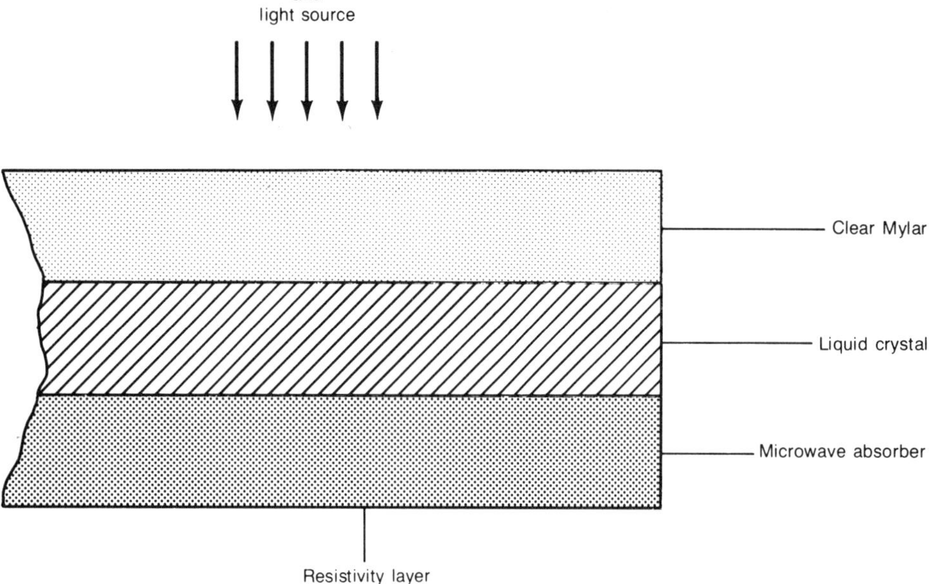

Fig. 12 Schematic of a microwave liquid crystal display. A resistivity layer provides electromagnetic energy-to-heat conversion, a liquid crystal layer provides pattern definition, and a Mylar section provides support and protection from liquid crystal contamination.

size of waveguide. For the commonly used $TE_{1,0}$ mode, the cutoff frequency is equal to the free-space velocity divided by twice the A dimension. This means that the A (width) dimension of the waveguide must be at least one-half the free-space wavelength for the $TE_{1,0}$ mode to propagate. Therefore, large waveguides imply low frequencies, and small ones imply high frequencies.

In the past, microwaves were mainly generated by special vacuum tubes called reflex klystrons. Klystrons generate microwave radiation by the synchronized acceleration and deceleration of electrons into bunches through a resonant cavity. The microwave radiation, frequently emitted through a mica window, is virtually monochromatic and phase coherent. Currently, microwaves are also generated over the range of 300 MHz to 300 GHz by solid-state or semiconducting devices. The advantage of these sources is that their outputs are single frequencies that are coherent. The frequency can also be varied over a wide range.

Microwave energy can be detected by devices such as special semiconductor diodes, bolometers (barretters), or thermistors. Phase-sensitive detectors can also be used. The diode, the most common detector used for nondestructive inspection, generates an electromotive force proportional to the microwave power level impinging on it. Another type of detector operates on the principle of converting the microwave power to heat (barretters and thermistors). For many years, the evapograph has been used to record an infrared image on a thermosensitive layer (Ref 9). The infrared radiation causes localized heating on the heat-sensitive layer so that the infrared image is recorded. In 1967, this same technique was reportedly being used to observe and record microwave electromagnetic wave interference patterns (Ref 10).

The experimental setup used is illustrated in Fig. 11. The microwave image is produced on the evaporated-oil membrane. The membrane is composed of a thin plastic sheet with two coatings. The coating that receives the microwave energy is a thin, vacuum-deposited layer of a low-conductivity metal such as bismuth or lead. This first layer absorbs a fraction (10 to 50%) of the incident microwave power pattern. This locally produced heat profile is transmitted through the plate sheet to the second layer on the other side of the plastic membrane. The second layer is a thin film of hexadecane ($C_{16}H_{34}$, with melting point of 20 °C, or 68 °F) or similar oil, which was deposited from a vapor and is held in the equilibrium state between condensation and reevaporation. The local heat caused by the incident microwave power causes reevaporation of the oil film. When the oil film is illuminated with light, as shown in Fig. 11, the oily film presents a uniform interference color. Under the microwave heating, the oil coating presents a microwave interference pattern seen as a colored visual image on the oil film. This colored visual image is also a microwave hologram. Good holograms have been obtained with a microwave energy of 80 mW/cm^2 (520 mW/in.^{2}) in a few minutes (Ref 9).

A newer microwave liquid crystal display (MLCD) has made it possible to obtain real-time two-dimensional color display of microwaves having frequencies up to and above 300 GHz (Ref 10). Figure 12 illustrates the basic operating mechanism of the MLCD. The display consists of an absorbing layer with a typical resistivity of 5 kΩ/cm^2 and a liquid crystal layer, both bonded to a flexible Mylar plastic sheet. The liquid crystal and resistive layers are 0.025 mm (0.001 in.) thick, while the Mylar support is 0.075 mm (0.003 in.) thick. Absorbed power as low as 1 mW/cm^2 (6.5 mW/in.^{2}) will form display patterns. The absorbing layer converts electromagnetic energy into heat, and the liquid crystal layer selectively reflects light according to its temperature. The above mechanism produces two-dimensional color displays, electromagnetic field patterns, and interference field patterns with incident microwave radiation as low as 1 mW/cm^2 (6.5 mW/in.^{2}).

Metal horns are usually used to radiate or pick up microwave beams. The directionality of the radiation pattern is a function of its aperture size in terms of number of wavelengths across the aperture. Horns with aperture dimensions of several wave-

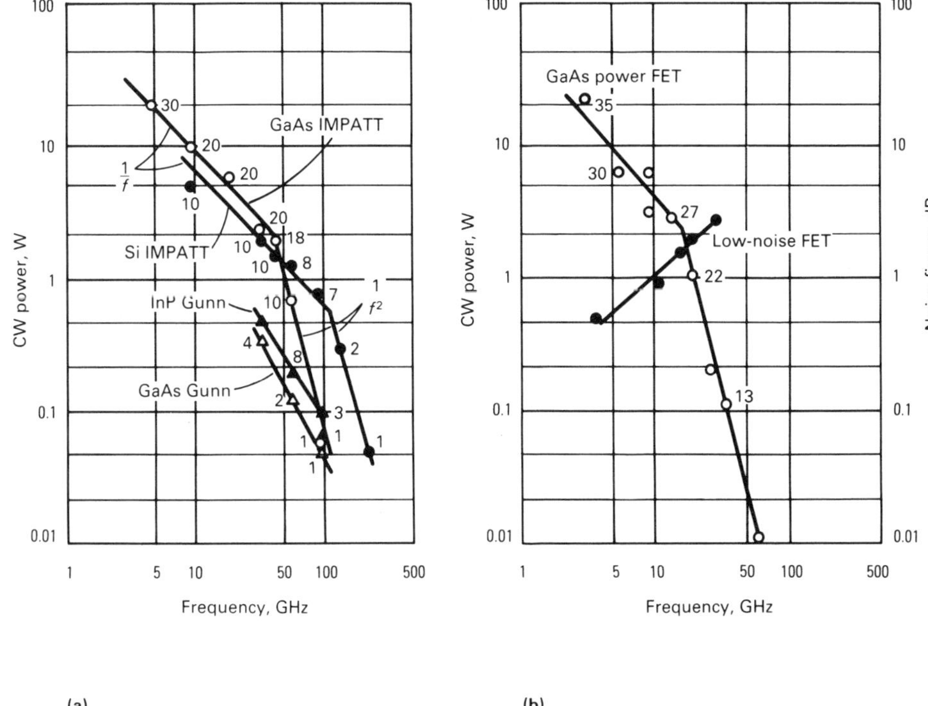

Fig. 13 Plots of continuous wave power versus frequency to obtain efficiency (in %) of various solid-state devices. (a) Efficiency data for InP and GaAs Gunn diodes; and GaAs and Si IMPATT diodes. Data obtained from Hughes, MA/COM, Raytheon, TRW, Varian. (b) Efficiency data for power and low noise FETs. Data obtained from Avantek, Hughes, MSC, Raytheon, TI. Source: Ref 11

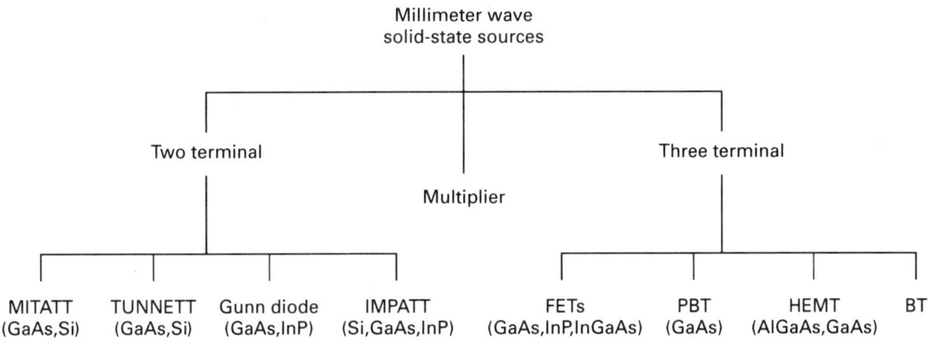

Fig. 14 Solid-state microwave sources by categories. HEMT, high electron-mobility transistor

lengths produce fairly good directionality. Most close-up NDI applications utilize even smaller horns, regardless of their larger beam spread.

Most of the microwave equipment used in the past for nondestructive inspection operated at frequencies in the vicinity of 10 GHz (X band). The free-space wavelength at 10 GHz is 30 mm (1.18 in.). A few applications have used frequencies as low as 1.0 GHz and as high as 100 GHz. The waveguides used for nondestructive inspection at 10 GHz are typically 25 mm (1 in.) wide and 13 mm (½ in.) high, and horn sizes range from 25 × 25 mm (1 × 1 in.) in aperture to 127 mm (5 in.) or more. The currently available solid-state sources from 1 to 300 GHz should make microwave NDE readily accomplished in this range.

One of the developing areas of great interest to those wanting to use microwave frequencies for NDE is that in solid-state and vacuum-tube sources and transmitters. The information in Fig. 13(a) and (b) gives the status of solid-state devices as of October 1983. No more recent information in summary form was found up to October 1988. The solid-state sources available in 1983 are listed by categories in Fig. 14. In Fig. 13(a) and (b), the numbers are the efficiencies (in percent) of power output divided by the input dc power. Efficiency values as high as 30% are obtained with an output of 20 W at 5 GHz using a GaAs, impact avalanche transit time (IMPATT) diode. The IMPATT diode has a negative resistance region, which comes from the effects of phase shift introduced by the solid-state sources, that closely follows the $1/f$ and the $1/f^2$ slopes shown in Fig. 13(a) and (b). The transition from the $1/f$ to the $1/f^2$ slope falls between 50 and 60 GHz. The transition for silicon impact avalanche transit time (Si IMPATT) diodes is between 100 and 120 GHz. As shown in Fig. 13(b), in the 40 to 60 GHz region, the GaAs IMPATTs show higher power and efficiency, while the Si IMPATTs are produced with higher reliability and yield (Ref 11).

The Gunn diode oscillators have better noise performance than the IMPATTs. They are used as local oscillators for receivers and as primary source where continuous-wave (CW) powers of up to 100 mW are required. Indium phosphide Gunn devices show higher power and efficiency than gallium arsenide Gunn devices.

The performance in power and low noise for GaAs field-effect transistors (FETs) is shown in Fig. 13(b). Again, the number by the data points is the efficiency associated with that power. Higher powers can and will be achieved by using optimized in-package matching circuitry.

The noise figure of the low-noise GaAs-FET devices ranges from 14 to 7 dB over the frequency range of 3.5 to 60 GHz. Peak powers 3 to 6 dB higher have been obtained from power GaAs FETs, while 5 to 10 dB increases in power are possible with IMPATTs. The power combining of IMPATT devices was also demonstrated successfully by 1983. Peak powers of several hundred watts were obtained in the X band (8 to 12.5 GHz, as shown in Table 2). Broadband CW combiners in the U band (46 to 56 GHz) have been reported that deliver several watts of power (Ref 11).

The state-of-the-art performance for various microwave vacuum tubes is shown in Fig. 15(a) and (b). Again, the numbers by the data points are the efficiency in percent. The parenthetical letter represents the company that supplied the data and developed the device. The devices represented in Fig. 15(a) are the traveling wave tube (TWT), the helix TWT, the backward wave oscillator (BWO) tube, the gyrotron, the orotron, the extended interaction amplifier (EIA), and the extended interaction oscillator (EIO). The devices in Fig. 15(b) include the crossed-field amplifier (CFA), TWT, klystron, and gyrotron. Power levels from these devices cover the range of 1 to 3 kW over the frequency range of 5 to 270 GHz. The devices in Fig. 15(b) can supply almost 5 MW peak power at 1 GHz and 1.5 kW peak power at 300 GHz (Ref 11).

Impressive power and voltage gain values are also available. Over 50 dB of gain is available in a 93 to 95 GHz TWT at an average output level of 50 W (Ref 11). The most impressive power achievements have been made with the gyrotrons (Ref 11). Many researchers are examining the cyclo-

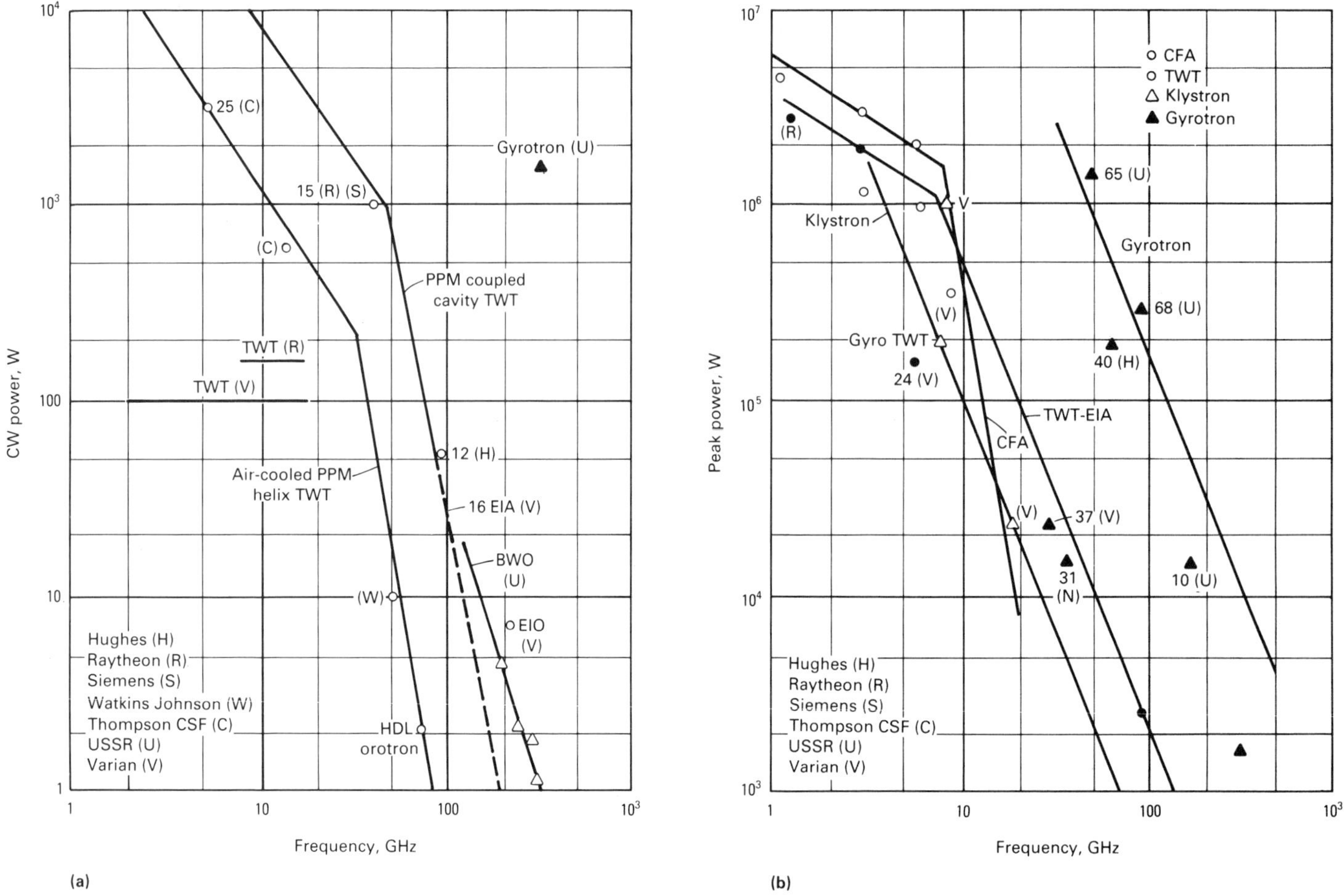

Fig. 15 Plots of power versus frequency to obtain efficiency (in %) of various microwave vacuum tubes. (a) Continuous-wave power. (b) Peak power. Source: Ref 12

tron resonance principle to build amplifiers (Ref 11). Recent developments in the helix plane circuit promise to provide ultralow-cost, high-average-power TWTs (Ref 11). Increased activity from 1983 through 1987 by tube designers has resulted in TWTs exceeding 1 kW continuous wave and 100 W continuous wave at 100 GHz (Ref 12). Gyrotrons in 1987 could generate close to 200 kW of continuous wave at 140 GHz (Ref 12). Although these large powers are not usually used in NDE, developments such as these benefit the whole field of microwave NDE. The sizes of these vacuum-tube amplifiers prevent their use for most NDE applications. All of the continuing developments in solid-state sources and transmitters are usable for microwave NDE.

In the early development of microwaves, systems consisted of a large number of devices performing specific single functions. The system engineer became a master of optimizing interfaces and writing device specifications with adequate margin to overcome interactions. This led to reduced overall system performance. The need for improvement led to the supercomponent concept. The microwave supercomponent was thought to be the answer to the problem of the complex and costly device interfaces. The supercomponent was defined as a stand-alone device that contains multiple functions of a generic nature (sources, amplifiers, mixers, and so on) and support functions such as switches, isolators, filters, attenuators, couplers, control circuits, and supply circuits, all of which are tightly packaged and combined to meet a single subsystem specification. One of the earliest microwave supercomponents was the combination of a TWT with a solid-state power supply in mid-1960. This combination eliminated the need to adjust this interface after the device left the factory (Ref 13).

The concept of supercomponents is best utilized when the system designer works actively with the manufacturer of the supercomponent. The designer should keep in mind that he may be dealing with a supercomponent engineer who is more skilled in individual devices than in system work. Therefore, simply writing a specification and submitting it will often not produce the desired results. Good communication is necessary between the system designer and the supercomponent engineer to ensure that both understand each other's needs so that they can jointly arrive at the best combination of performance and cost. Most companies that are truly dedicated to supplying supercomponents are continuously training engineers who have the necessary broad background to work successfully at producing supercomponents.

The supercomponent or subsystem concept has and is maturing rapidly. It has progressed beyond a risky, expensive strategy to one that offers solid ground for improvement of microwave NDE equipment in terms of size, weight, cost performance, reliability, and maintainability. For some NDE applications, the use of supercomponents may mean the difference between successful and mediocre operation. Other information on supercomponents can be found in Ref 14 and 15.

Microwave instrumentation for nondestructive inspection can be set up for reflectometry through transmission and scattering techniques. Usually, the single-frequency design is used, but swept-frequency systems have been constructed for certain applications.

Some applications require the use of through transmission or wave-scattering systems involving a separate transmitter and receiver. As with ultrasonics, through transmission is performed by placing the transmitter on one side of the material and the receiver on the opposite side (see the

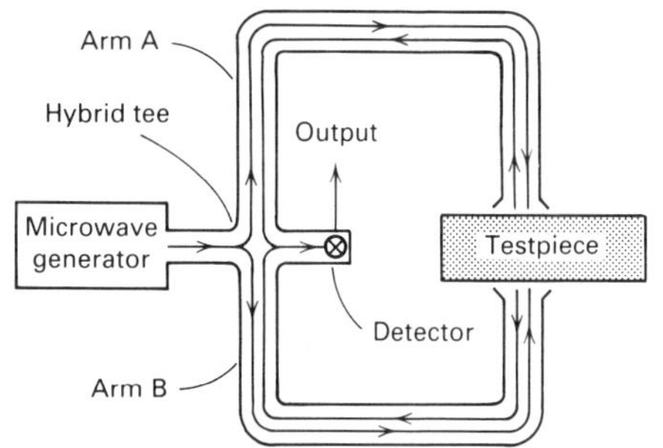

Fig. 16 Diagram of a reflectometer for determining metal thickness using microwave techniques. Arms A and B differ in length by an integral number of half-wavelengths for detector output null.

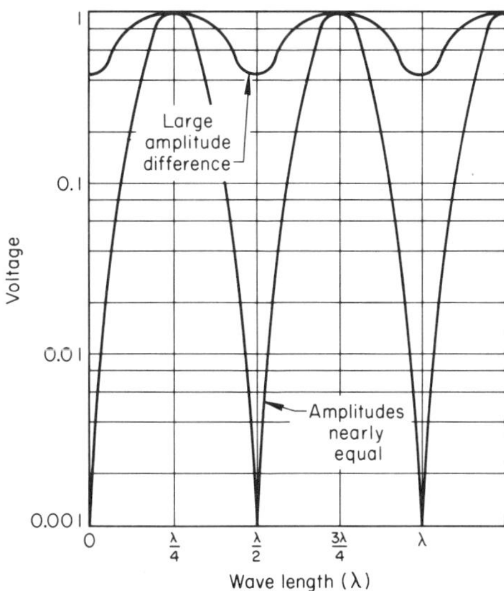

Fig. 17 Response of a detector moved along the path of interfering waves for large amplitude difference and amplitudes nearly equal

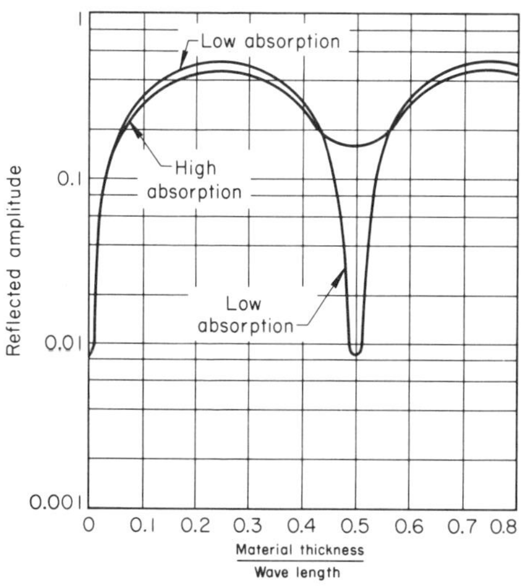

Fig. 18 Relation of the amplitude of reflected microwaves and the ratio of material thickness to wavelength. Source: Ref 3

article "Ultrasonic Inspection" in this Volume). Scattering setups orient the receiver horn at some oblique angle to the transmitted beam. The receiver is sometimes placed against a side surface (for example, the side of a block) so that its direction is oriented 90° with respect to the polarization of the transmitted beam. If measurable scattering sites exist in the material, they can frequently be detected by scanning the obliquely oriented transmitter and receiver.

Microwave flaw detectors, based on swept-frequency heterodyning principles, can be used to measure flaw depth. This type of flaw detector transmits a signal whose frequency sweeps from a maximum to a minimum value at a specified repetition rate. A discontinuity in the material under test reflects a portion of this energy back to the reflectometer. The reflected energy interferes with the transmitted energy by a process known as heterodyning. The heterodyning action produces a beat frequency equal to the instantaneous difference between the frequencies of the transmitted and delayed received waves. This difference in frequency, caused by timing differences between the waves, is proportional to the distance to the flaw.

Swept-frequency approaches are also used for microwave thickness gaging and density determinations. The frequency that yields maximum power reflection is related to either density or thickness.

The pulse-echo system, as used in ultrasonic and many radar designs, cannot be applied to microwave flaw detection or thickness gaging, because the wave velocities are too great and the distances too short. The necessary electronic resolving powers, well into the picosecond (10^{-12} s) range, are not available for accurately measuring flaw depth. This is one major reason why frequency-modulated and standing wave designs are used. Another potential problem with pulse-echo resolving methods is that the frequency of such short-duration pulses would be too high for desirable propagation behavior in nonmetallic materials.

Microwave instruments for nondestructive inspection based on Doppler techniques are used for industrial in-motion applications, particularly in Europe. The Doppler principle is based on a frequency shift in a wave that is reflected from a moving target. The amount of frequency shift is proportional to the velocity of the target.

Thickness Gaging

Thickness measurements can be made with microwave techniques on both metallic and nonmetallic materials. For metals, two reflected waves are used, one from each side of the spectrum, as shown in Fig. 16. The measurement is made using the standing wave technique. When the wave is incident on a metal (electrically conductive), most of the wave is reflected; only a small amount is transmitted (refracted). The transmitted wave is highly attenuated in the metal within the first skin depth. For nonmetallic materials (electrically nonconductive), the reflected wave is much smaller than the incident wave, so that any standing wave that does develop does not have a large amplitude.

As shown in Fig. 17, the distance between the minima of the standing wave is equal to one-half wavelength; that is, when a detector antenna moves a distance of one-half wavelength along a standing wave, the amplitude changes. Therefore, with a reflectometer (shown schematically in Fig. 16), the thickness can be measured by forming a standing wave in the detector arm of a hybrid tee, a standard microwave device. The wave from the generator is split by the hybrid tee into two waves of equal amplitude. Each of these waves travels down separate waveguides, is reflected from the metal surface, and travels back through the waveguide to the hybrid tee. The hybrid tee then recombines the two reflected waves, directing them into the detector. One arm of the system has been changed in length to allow the sample to be inserted. When the difference between the respective round-trip distances traveled by the two waves is one-half wavelength, the two waves interfere destructively, and a minimal voltage is read on the detector. The frequency of the generator can be adjusted so that a minimal voltage is obtained for any given thickness of specimen. The specimen thickness can then be calculated because the wavelength is known and the distance to one side of the sample is measured or calibrated.

The reflectometer shown in Fig. 16 can also be adapted to measure the thickness of an electrically nonconductive material if one side is coated with a thin film of a conductive material. The difference between the two arm lengths would then be a distance equivalent to twice the thickness of the nonconductive material. If the velocity of the wave in the material and the frequency are known, the thickness can be computed.

If the dielectric material whose thickness is being measured also has absorption, the reflected wave is reduced in amplitude such that a smaller-amplitude standing wave is obtained. The smaller-amplitude standing wave decreased the accuracy of the measurements of the position of the minimum. A comparison of the standing waves for low and high absorption is given in Fig. 18.

The range over which the simple thickness meter shown in Fig. 16 can be used unambiguously is approximately one-fourth of the wavelength of the microwaves in the material being measured. If the thickness varies more than one-fourth wavelength, ambiguous results will be obtained. The nominal thickness of most materials can usually be estimated or is known within

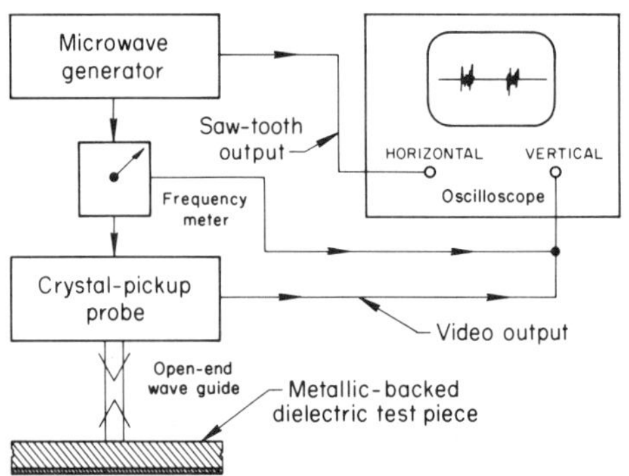

Fig. 19 Diagram of a single-side microwave thickness gage. See text for discussion.

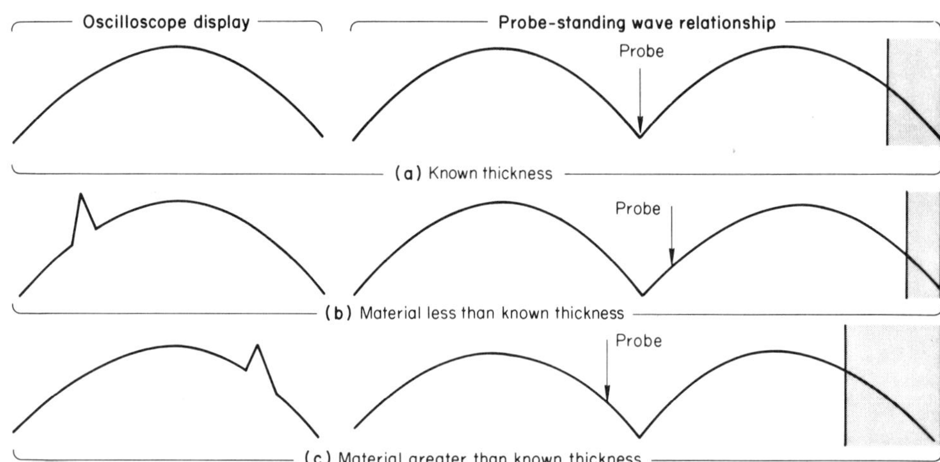

Fig. 20 Scope display and standing wave relationship for a single-sided microwave thickness gage. See also Fig. 19.

manufacturing tolerances to within plus or minus one-eighth wavelength, so that the ambiguity can be resolved.

A diagram of another simple microwave thickness gage is shown in Fig. 19. This single-side gage has a direct waveguide to the testpiece, which must be backed by a thin metallic coating. The standing wave that is set up in the waveguide is detected by a probe in the waveguide itself. If the frequency were fixed and the probe were movable, it would trace out a curve similar to those shown in Fig. 18. In practice, however, the probe is fixed and the frequency is swept. A material of known thickness is inserted, and the center frequency and frequency sweep width for the microwave generator are adjusted until a single one-half wavelength standing wave is observed on the oscilloscope. The sawtooth output from the microwave generator is proportional to the instantaneous frequency shift, so that horizontal position on the oscilloscope display is proportional to frequency.

Also observed on the oscilloscope is the output signal from the frequency meter, which occurs as a sharp peak when the frequency of the oscillator passes through the frequency to which the meter is set. This absorption peak appears superimposed on the standing wave as a pulse (Fig. 20). When the equipment is properly set, the peak is not seen except at the null of the standing wave, as shown in Fig. 20(a). When the material of known thickness is replaced by one that is thinner, the peak appears on the standing wave, as shown in Fig. 20(b). When the material is greater than the known thickness, the display is as shown in Fig. 20(c). By observing the movement of the peak with respect to the standing wave, the material can be calibrated in thickness relative to the known thickness.

The use of a microwave thickness gage in a variety of production operations has demonstrated its suitability for nondestructive inspection and quality control of dielectric components. The data for the graph in Fig. 21(a) was obtained from polyethylene coating on common aluminum foil wrapping material. Measurements were made on

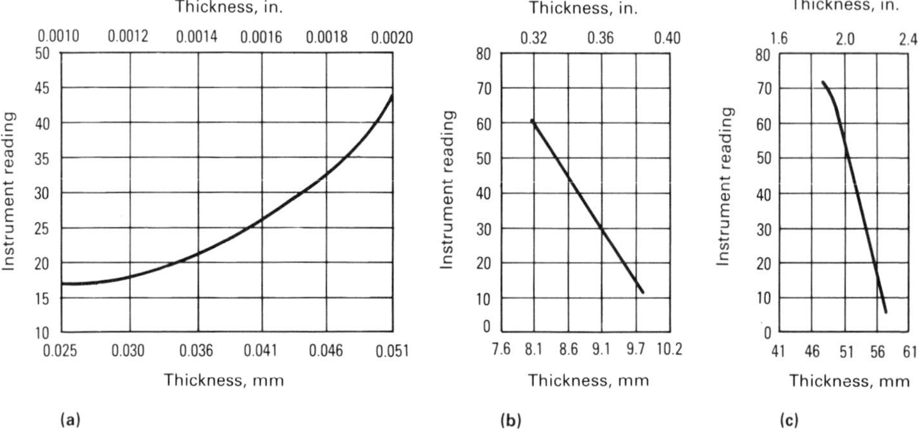

Fig. 21 Readings of an amplitude-sensitive reflectometer as a function of film thickness for (a) polyethylene, (b) fiberglass plate, and (c) polyurethane

coatings of thicknesses of 25, 38, 44, and 51 μm (0.0010, 0.0015, 0.00175, and 0.0020 in.). The data for Fig. 21(b) were obtained from a fiberglass-reinforcement resin plate used for fabricating boat hulls. The data for Fig. 21(c) were obtained from aluminum-backed thick polyurethane foam used as a thermal insulation blanket for space vehicles. All the data in Fig. 21 were taken with a simple amplitude-sensitive reflectometer operating at 9.4 GHz. An antenna (a 25 mm, or 1 in., square horn) was used to obtain the data in Fig. 21(b) and (c). The data in Fig. 21(a) were obtained with an antenna known as a short-field probe (Ref 16).

Detection of Discontinuities

Discontinuities such as cracks, voids, delaminations, separations, and inclusions predominantly reflect or scatter electromagnetic waves. Wherever these types of flaws occur, there is a more or less sharp boundary between two materials having markedly different velocities for electromagnetic waves. At these boundaries, which are usually thin compared to the wavelength of electromagnetic radiation being used, the electromagnetic wave is reflected, refracted, or scattered. The reflected or scattered radiation has appreciable amplitude only if the minimum dimension of the discontinuity is larger than about one-half the wavelength of the incident radiation in the material being tested.

Porosity and regions of defective material such as departures from nominal composition do not produce strong reflection or scattering. They do influence the attenuation and the velocity of the electromagnetic wave. When there is absorption, the transmitted wave has an exponential decay with respect to the distance traveled.

Continuous-wave reflectometers are those in which the amplitude of the standing wave is measured by a detector in a combination send-receive, slotted microwave coaxial transmission line. They can be used to detect discontinuities in such components as glass-filament solid-rocket motor chambers using a frequency of 12.4 to 18.0 GHz (Ref 17). Reflections from internal discontinuities change the amplitude of the standing wave measured by the detector and give a

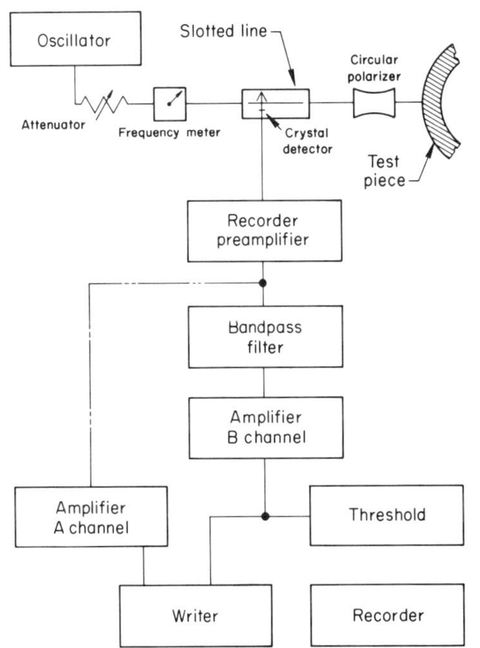

Fig. 22 Diagram of a continuous-wave microwave system used to detect discontinuities

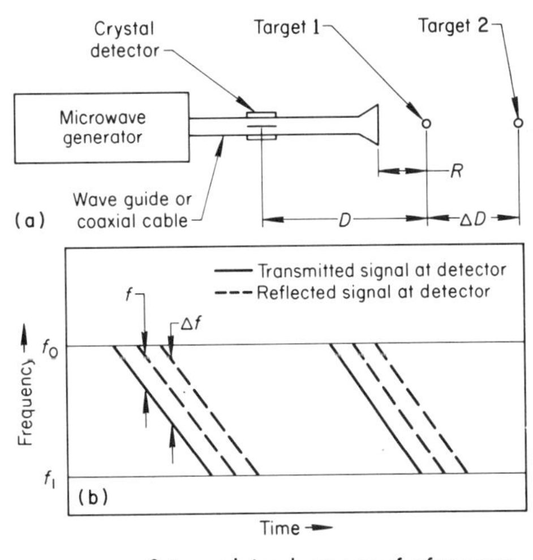

Fig. 23 Setup and signal response of a frequency-modulated microwave reflectometer. (a) Schematic showing components of a frequency-modulated microwave reflectometer. (b) Paths of transmitted and reflected signals at detector in relation to frequency and time

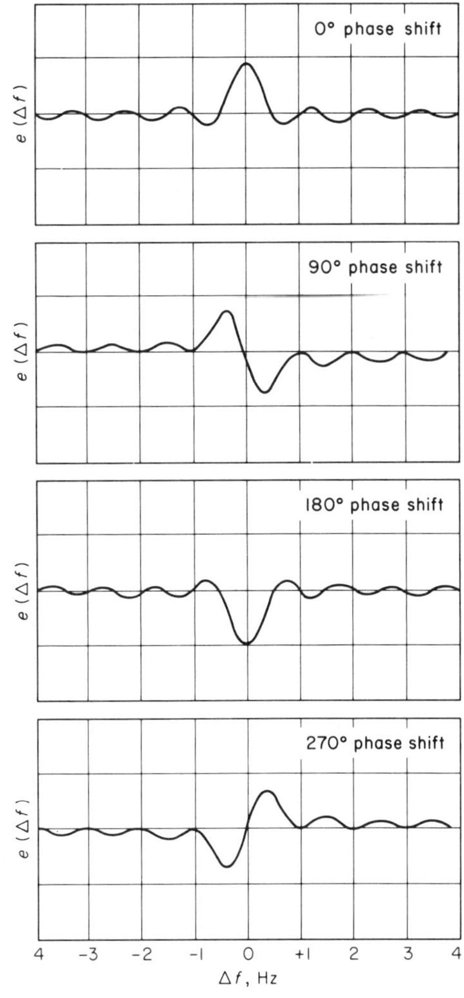

Fig. 24 Theoretical indications of reflections, $e(\Delta f)$, produced by a frequency-modulated microwave reflectometer during phase shifts of 0°, 90°, 180°, and 270° as functions of frequency shift of the modulation Δf

change in the output. When the material is scanned and the reflectometer signal is recorded by intensity modulating a pen recorder, the discontinuities can be observed as light or dark areas. This type of C-scan, when connected with level-sensing equipment, can be used to record only variations above or below a certain level, thus showing only the discontinuities. A diagram of a continuous-wave microwave system that incorporates a slotted coaxial transmission line is shown in Fig. 22.

The detection of voids depends on a change in the wave reflected into the antenna resulting from the added reflection from the discontinuity caused by the void. The system shown in Fig. 22 can be modified for these measurements by substituting a directional coupler for the slotted line. This coupler allows the output signal to be proportional only to the amplitude of the reflected signal. The reflected signal from components such as a rocket-motor chamber consists of a large reflection from the outer surface of the case plus smaller ones from the glass insulation and the insulation/propellant interfaces. At any void or unbonded area, an additional reflection occurs that increases the overall reflection into the antenna. A standing wave can be set up by the reflection in the material so that, at some frequencies, complete cancellation occurs for reflections that are one-fourth wavelength apart. Because of this possibility, frequencies should be carefully chosen. Several frequencies are used to ensure that no cancellation will occur.

In the frequency-modulated reflectometer, a linear sweep is made of the entire frequency range of two generators: 1 to 2 GHz and 12.4 to 18 GHz (roughly the L band and the K_u band, respectively, per Table 2). A separate system is used for each band; each band is swept at a rate of 40 sweeps per second. Figure 23(a) shows the basic components of the frequency-modulated reflectometer (Ref 18). A graph of the signal at the detector as a function of time is shown in Fig. 23(b). The transmitted signal from the antenna is reflected back into the antenna and the detector from targets 1 and 2. The reflected signals also vary in frequency, as do the transmitted signals, but at the detector they are displaced in time, as shown by the dashed lines in Fig. 23(b). Therefore, at any instant of time, there are three signals at the detector: the first from the transmitted wave, the second from target 1, and the third from target 2.

Each of these signals is of a different frequency, corresponding to the different propagation times involved. The frequency difference, Δf, is equal to the rate of sweep (in hertz) times the delay (in seconds). The phase of the reflected signal, ϕ, relative to the reference (transmitted) signal is $40\pi f/c$. Both the frequency and the phase of the reflected signals are proportional to the distance to the respective reflectors. Small changes in the position of the reflector cause relatively large changes in phase, which in turn cause large changes in the shape of the signal obtained at the crystal detector. As the reflector moves through a distance equal to phase shifts of 0°, 90°, 180°, and 270°, the output of the detector, $e(\Delta f)$, is a function of the frequency difference and changes in shape as shown by the curves in Fig. 24.

The arrangement of equipment for a frequency-modulated microwave reflectometer using L-band frequencies is shown schematically in Fig. 25. In this setup, separate antennas are used for sending and receiving. The detector senses both the transmitted signal and the reflected signals. These are mixed in the detector to produce a difference frequency equal to the reflected wave, f, minus the reference frequency, f_0, generated by the swept oscillator. A separate beat frequency is developed for each reflected signal. For two discrete reflectors, the output signal of the crystal detector consists of two frequencies, one for each target.

To determine the number of reflected objects and their locations, the signal from the detector passes through a cross-correlation spectrum analyzer. The readout of the signal processor is a plot of the amplitude of the detector signal as a function of frequency. To accomplish this, the signal is cut into many segments of equal duration. If the spacing of these segments corresponds to one of the frequency components of the signal, an output results. The spacing of these segments is varied by the long, saw-

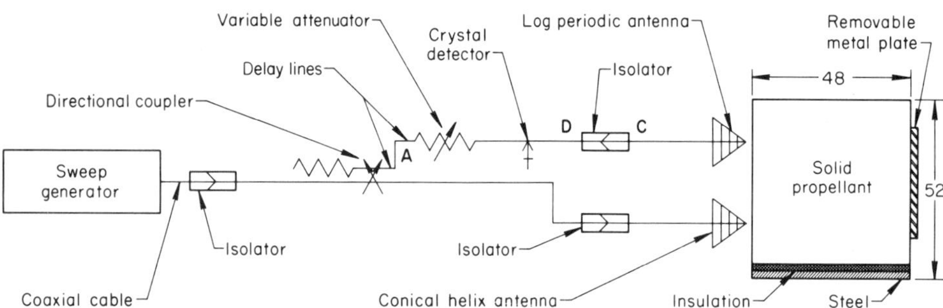

Fig. 25 Schematic of components of a frequency-modulated microwave reflectometer using L-band frequencies. Dimensions given in inches. Source: Ref 18

Table 3 Frequency bands of frequency-modulated reflectometers

Frequency, GHz	Type	Resolution in air mm	Resolution in air in.	Resolution in plastics(a) mm	Resolution in plastics(a) in.
0.5–1.0	Coaxial cable	300	12	150	5.9
1.0–2.0	Coaxial cable	150	5.9	75	3.0
2.0–4.0	Coaxial cable	75	3.0	37	1.5
4.0–8.0	Coaxial cable	37	1.5	18	0.71
8.0–12.4	Waveguide	34	1.3	17	0.67
12.4–18.0	Waveguide	28	1.1	14	0.55
18.0–26.0	Waveguide	19	0.75	10	0.39
26.0–40.0	Waveguide	11	0.43	5.5	0.22

(a) This assumes a refractive index of 2.

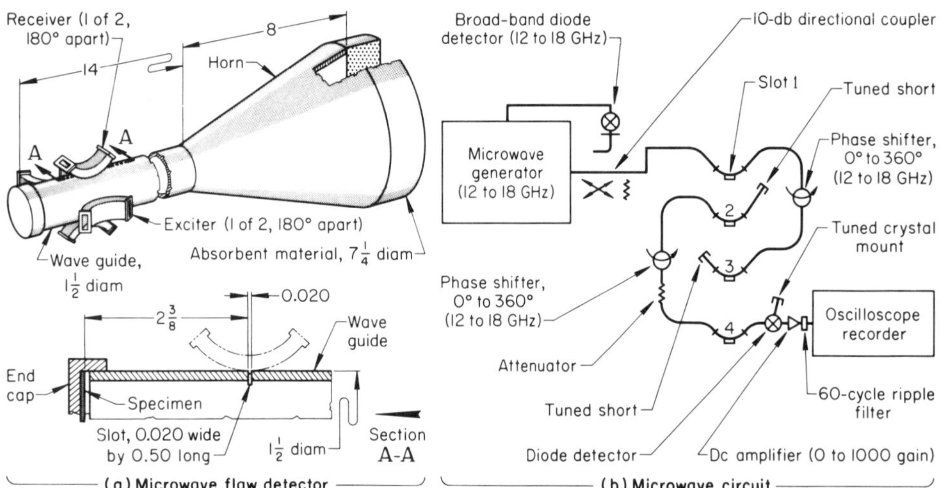

Fig. 26 "Nonresonant" standing-wave system that can be used for detecting cracks in metal parts. Dimensions given in inches. Source: Ref 19

tooth voltage, A, so that all the possible frequencies in the signal are swept. Therefore, the output contains a voltage for each of the frequency components of the signal. Each component is displayed as a dot on an oscilloscope. The vertical displacement of this dot is proportional to the amplitude reached by the integrator for that particular frequency in the integration time. The K-band equipment is similar to that shown in Fig. 25 for the L band except that a waveguide is used rather than a coaxial cable.

Frequency-modulated reflectometers typically operate in the range of frequency bands given in Table 3. Sources are now available to frequencies beyond 200 GHz. These would produce resolution to 1 mm (0.04 in.). Application of these higher frequencies to inspection and measurement has been limited by the relatively high price of the microwave sources.

Applications have been used in penetrating rock, soil, and thick solid propellant samples in the low bands; concrete in the medium bands; and plastics and composite materials such as filament-wound glass, Kevlar, and honeycomb in the higher bands.

Microwave Detection of Surface Cracks in Metals

When an electromagnetic wave is incident on a metallic surface that has slits or cracks, the metallic surface reradiates (reflects) a signal because of induced current. Under the proper conditions, the reflected wave combines with the incident wave to produce a standing wave. The reflection from a surface without a crack is different from that surface with a crack and depends on the direction of the incident wave polarization relative to the crack. When the crack is long and narrow and the electric field is perpendicular to the length of the crack, the reflected wave (and therefore any standing wave) is affected by the presence of the crack. The amount of change is used to determine the size and depth of the crack.

Standing Wave System. The most sensitive means of detecting the small crack-related changes in the standing wave is to use the standing wave in a resonant cavity (Ref 19). Such a resonant system is also sensitive to variables other than surface cracks. However, a nonresonant standing wave system can be used. One such system is shown in Fig. 26. The detection head with the test specimen forming one end of the system is shown in Fig. 26(a). The excitation is fed into two slots, while the receiver is fed from two other slots. A diagram of the microwave circuit is shown in Fig. 26(b). The test specimen is mounted so that it can be rotated.

In the system described above and shown in Fig. 26, the test surface acts as a short for one end of the waveguide section. In order that the only waves present in the waveguide section are those incident on and reflected from the test specimen, the opposite end of the waveguide section is made nonreflective by using a matching horn and absorbing material. The exciter slots are so arranged that the cylindrical transverse electric ($TE_{1,1}$) standing wave is set up by the reflections from a flat (crack-free) specimen.

The goal of the design is to have none of the transverse magnetic ($TM_{1,1}$ or $TM_{2,1}$) modes excited when the test specimen is crack free. The presence of a crack in the surface of the test specimen disturbs the flow of current in the end required by the transverse electric mode so that the higher-order transverse magnetic modes will be excited. Because the higher modes are specifically caused by the currents due to the crack in the end plate, when the end plate is rotated these modes will also rotate past the receiver slots. With the $TM_{1,1}$ mode, a receiver output shows two peaks for a 360° rotation; with the $TM_{2,1}$ mode, four peaks are obtained from a full rotation. A crack at the exact center of the test surface does not produce a signal.

A suitable operating frequency was found by sweeping the frequency while the specimen, with a groove 20.3 mm (0.8 in.) long and 0.76 mm (0.030 in.) deep, was being rotated. The correct value is indicated by a zero-dc level and a signal response to rota-

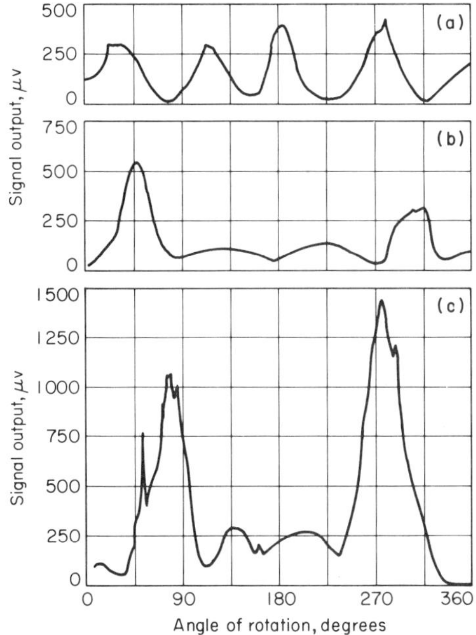

Fig. 27 Signal outputs from the standing wave crack detection system in Fig. 26 for side-notched specimens. (a) Crack-free specimen. (b) Fatigue crack in one specimen. (c) Fatigue crack in two specimens. The rotating specimen holder was designed so that two specimens could be inspected at the same time.

tion of the groove that established the reference signal level. The microwave circuits were then tuned to this frequency. For the groove that was 20.3 mm (0.8 in.) long and 0.25 mm (0.010 in.) deep, a frequency of 15.965 GHz was used.

The influence of burrs was studied with a burr 0.0089 mm (0.00035 in.) high on a groove 0.08 mm (0.003 in.) deep. The average signal with the burr had a peak-to-peak value of 15 μV, while the average signal without the burr was 6 μV.

Variations of signal outputs from fatigue cracks are shown in Fig. 27. The fatigue cracks were much smaller than the notches used for calibration, being approximately 3.2 mm (0.125 in.) long and 1.25 μm (50 μin.) wide. The fourfold variation shown in Fig. 27(a) was caused by the asymmetry of the interfaces between the sample holder and the sample. The presence of one crack gave the results shown in Fig. 27(b). When a second crack was added, the results in Fig. 27(c) show that the two outside variations were increased in value. The peaks shown in Fig. 27(c) were probably caused by a mixture of the two TM modes rather than higher transverse magnetic modes.

Chemical Composition of Dielectric Materials

In many cases, chemical changes in dielectric materials, such as plastics, can be detected or even continuously monitored by microwaves, and the results of chemical changes can be measured. The criterion is that the chemical change must affect the dielectric properties (dielectric constant and/or loss tangent) of the material for electromagnetic frequencies in the microwave region. Some of the chemical or molecular-level applications that have been successfully investigated include polymerization, degree of cure, oxidation, esterification, distillation, vulcanization, evaporation, and titration (end points).

Microwave techniques have been used to measure specific gravity, homogeneity during blending, and vibration (or displacement). Several investigators have measured the glass-to-resin ratio of some aerospace composites (Ref 17). Studies such as these can be made with either a through transmission system or a reflectometer because the changes in chemical composition cause a change in the velocity of propagation. The changes in propagation velocity change the amount and angle of the reflected and transmitted energy.

Moisture Analysis Using Microwaves

The free (unbound) moisture content of many dielectric materials can be accurately measured with microwave techniques. Microwaves are strongly absorbed and scattered by water molecules because water exhibits a broadband rotational relaxation in the microwave region. Because many completely dry host materials are transparent in the same frequency range, a moisture-measuring technique is possible. This technique has found widespread use on both a continuous-process and laboratory-sample basis, especially for plastic and ceramic materials. Moisture measurement of polyethylene powder using microwaves has been done by through-transmission techniques at a frequency of 9.4 GHz (Ref 16). Measurements similar to those for the plastic powders can also be obtained for solid shapes, slurries, and liquids. The technique (as described here) is not applicable for gases.

Because water is a polar molecule, its dielectric properties are a function of temperature. Therefore, if substantial material temperature changes occur, compensation is necessary in the instrument calibration. For example, a decrease in temperature of 12 °C (22 °F) will normally cause an error of ±0.2% in the moisture readout. The effect is more pronounced in materials having a high moisture content. Fortunately, the product temperature of most plastic-process lines is kept reasonably constant at a given location. The temperature effect is even less of a problem with plastics because they are relatively dry at the point of measurement.

Microwave Measurement of Material Anisotropy

The directionally dependent properties of some materials can be measured by the use of linearly polarized microwaves. Measurements of anisotropy, on both the molecular scale and macroscopic scale, are made by rotating the sensing head relative to the material under test and observing the signal as a function of polar angle. Certain linearly oriented polymers can be used as examples of molecular-level measurements; wood and fiberglass-resin structures are examples of the macroscopic level. The direction of glass fibers can be readily determined. The direction of the fibers in paper can also be detected with linearly polarized microwaves and antenna rotation.

Fiber-matrix composites have also been measured for directionality. One instance was a multiple-ply composite having a fiberglass-resin matrix and unidirectional boron filaments. The boron filaments, approximately 0.13 mm (0.005 in.) in diameter, contained a tungsten-fiber center (0.013 mm, or 0.0005 in., in diameter) onto which the boron had been deposited. Being electrically conductive, the overall composite acted on the microwaves very much like a diffraction grating. Alignment shifts of only a few degrees were detected easily by monitoring either the reflected or the transmitted microwave component. When the boron filaments were perpendicular with respect to the polarization of the microwave beam (electric field vector), a maximum amount of energy was transmitted (or a minimum amount of energy reflected). When the boron filaments and the polarization were parallel, the reverse was true. By continuously rotating the sample, a sinusoidal pattern of filament direction versus signal amplitude was obtained.

Stress-Corrosion Microwave Measurements

Materials such as aluminum, magnesium, and titanium are subject to stress-corrosion cracking or fatigue when they are under stress in a corrosion-producing environment. Cracks and fatigue can appear at stress values much less than the normal yield in a corroded material. Early fatigue may occur at less than rated load in a structure that has corroded materials. For reliable structures that may be constructed from corroded materials or from materials corroded after construction, it is necessary to have a method for detecting corrosion-prone and corroded areas of materials.

Several mechanisms have been suggested to explain stress-corrosion cracking and fatigue (Ref 20). Three of them are electromechanical, mechanical, and surface energy. Similarities among these mechanisms suggest an approach to the detection of corroded or corrosion-prone areas of the above materials.

All of these theories suggest that corrosion and the subsequent cracking begin at crystal lattice imperfections. The imperfec-

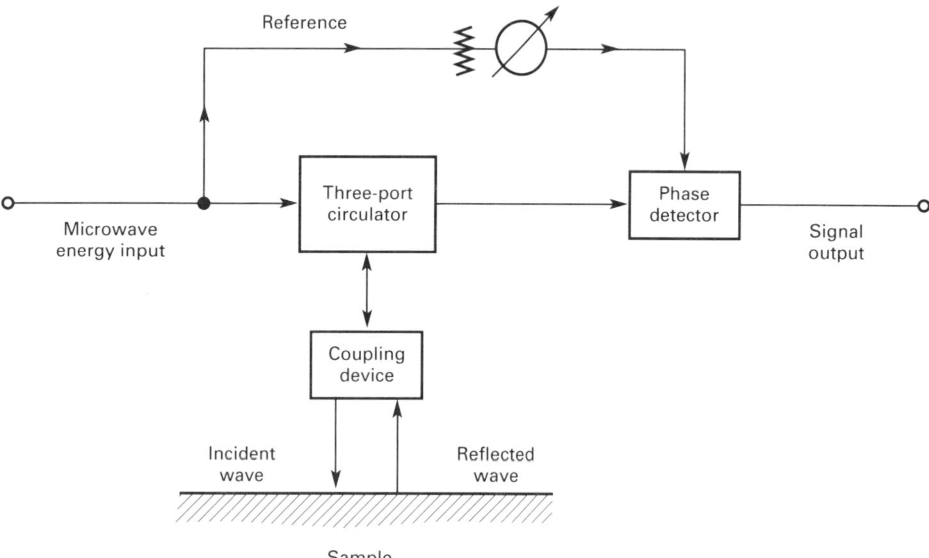

Fig. 28 Basic microwave surface impedance measurement technique

tions considered are dislocations, vacancies, and impurities. It was assumed that if the existence of these imperfections could be detected by a nondestructive method, then that method could be used to evaluate the stress-corrosion susceptibility of materials and structures so as to eliminate corrosion-prone or corroded materials.

The correlation between electrical resistivity and dislocation density has been discussed by many authors (Ref 21-24). It has also been found that the concentration of impurities varies the electrical resistivity of a material. It is further mentioned that there is a difference in the surface resistivity between dislocations caused by vacancies and impurities. Surface resistivity can be measured with high-frequency electromagnetic waves because they penetrate to a depth of only 10 to 0.1 μm (400 to 4 μin.) or less into the surface and because the wave reflected from the surface has its phase changed from that of the incident wave by the action of the surface impedance, which includes the resistance of the material. Any protective coating such as paint or oxides on the surface of the metal will allow most of the energy in the electromagnetic wave to pass through them and to be reflected from the conductive surface, if the frequency used is in the microwave region or below. However, the higher the frequency, the greater will be the magnitude of the change in the surface impedance caused by imperfections. Therefore, although frequencies from microwaves through the ultraviolet can be used, less interferences from surface coatings coupled with a relatively high sensitivity will be found in the microwave region from 1 to 100 GHz.

Therefore, a measurement of the microwave surface impedance by means of a comparison of the reflected wave with the incident wave should yield a measure of the imperfection density. A microwave system for measuring the surface impedance of a metal can be adapted and calibrated for the determination of the corrosion-prone areas in materials. Measurements of resistivity have been made on many metals and alloys, including aluminum, showing the resistivity to be dependent on the dislocation density (Ref 21-24). Also, a difference in resistivity of dislocations related to vacancies and impurities has been measured (Ref 21-24). The imperfection densities have been determined with an electron microscope, and good agreement has been obtained between the experimental and theoretical values for the resistivity of dislocations (Ref 21-24). For aluminum, the average dislocation resistance was found to be $7 \times 10^{-19}\ D \cdot \Omega \cdot \mathrm{cm}^3$, where D is the dislocation density of $D \cdot \mathrm{cm}^{-2}$ (Ref 21-24). The vacancy resistivity was also found to be $1.4 \times 10^{-6}\ \Omega \cdot \mathrm{cm}$ per atomic percent vacancies (Ref 21-24). Investigators have made measurements on surface properties of metals, such as surface impedance, skin depth, and conductivity (Ref 25). Most of these investigations have been conducted to determine the electrical properties of metals rather than to measure the density of their imperfections. One investigation was made using microwaves to study the development of early surface damage during the fatigue of high-strength aluminum alloys (Ref 25). The microwave measurements were related to the fatigue period for the test parameters rather than to the material parameters, such as the density of imperfections.

A vector network analyzer can be used to obtain magnitude and phase information of the reflected wave. The interval processing capabilities of present-day vector network analyzers provide the data in a variety of formats.

The basic concepts of the microwave method of measuring surface impedance are shown in Fig. 28. The output from a microwave generator is fed through a circulator and through a coupling device such as an antenna to provide an electromagnetic wave incident upon the surface whose impedance is to be determined.

A small part of the incident wave is transmitted into the surface of the material, and the remainder is reflected. Therefore, the reflected wave contains the information about the impedance of the surface. If the reflected wave is compared in both amplitude and phase with a reference signal taken directly from the microwave generator, the output of this amplitude-sensitive phase detector will be information about the surface impedance of the material. The surface impedance of a material having a complex permeability μ, conductivity σ, and permittivity ϵ is, at the angular frequency ω, given in Eq 1 (Ref 2). Because all three terms can be complex at microwave frequencies, the impedance is complex, having both real and imaginary parts.

A highly sensitive but slowly scanning method for measuring surface impedance is shown in Fig. 29. The microwave energy delivered to the system is generated by the source. Frequency control is provided by a dc voltage, giving a convenient method of sweeping the frequency. The precise frequency is then determined by a frequency counter or wavemeter, and a calibrated attenuator is used to control the amount of power passing down the waveguide to the cavity. Isolation between the system and the source can be obtained by the use of a ferrite isolator, which allows power to pass through the isolator in one direction only. A ferrite three-port circulator is an ideal device for this type of measurement. It is a wide absorption cavity and provides isolation between the detector and the klystron.

The cavity shown in the simplified circuit in Fig. 29 is an absorption-type cavity. Its distinguishing feature is that only one port is provided for coupling energy into and out of the cavity. Another type of cavity that has obtained prominence is the transmission cavity, in which there are two coupling ports. Energy is transmitted through the transmission cavity. When using the absorption cavity, one end or a side, depending on whether the cavity is circular or rectangular, is replaced by a sample to be measured. In this manner, the difference in skin depth or depth of penetration of the electromagnetic energy of the sample and a standard sample can be measured in terms of a change in quality factor. The frequency can be swept through the center frequency of the cavity and the detected output recorded on a chart recorder or other recording instrument. A curve identical to a parallel resonant-circuit frequency-response

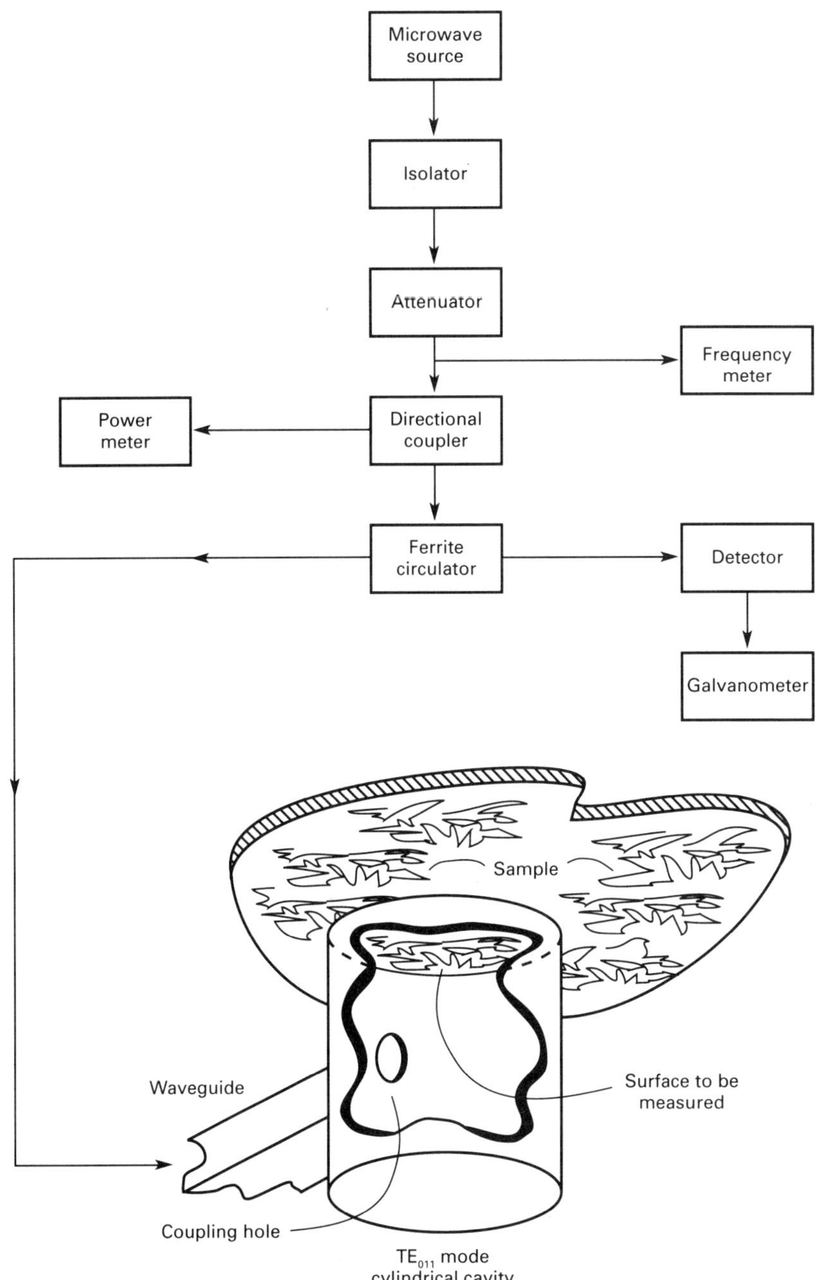

Fig. 29 Typical microwave measurement circuit

curve will be measured directly. The quality factor Q is defined as (Ref 25):

$$Q = \frac{f_0}{\Delta f} \quad \text{(Eq 6)}$$

where f_0 is the resonant frequency and Δf is the frequency difference between half-power frequencies or bandwidth.

A relation for the skin depth of an unknown sample in terms of the skin depth of a standard sample and the difference in their quality factors for a circular cavity has been derived (Ref 1):

$$S_x = S_a + \frac{2V'}{S'_a}\left(\frac{1}{Q_x} - \frac{1}{Q_a}\right) \quad \text{(Eq 7)}$$

where S_x is the skin depth of the unknown sample, S_a is the skin depth of the standard sample, Q_x is the quality factor of the unknown sample, Q_a is the quality factor of the known standard, and $2V'/S'_a$ is the constant depending on cavity and frequency.

The skin depth, S_x, for a good conductor has been derived in terms of the frequency, permeability, and conductivity (Ref 25):

$$S_x = \left(\frac{2}{\omega\mu\sigma}\right)^{1/2} \quad \text{(Eq 8)}$$

where ω is the angular frequency, μ is the permeability, and σ is the conductivity (=1/resistivity).

Because resistivity, ρ, and conductivity are reciprocal, the preceding has shown that the resistivity, and therefore a measure of the imperfection density, can be obtained from microwave measurements. It should be stressed that only the resistivity at or near the surface can be measured for a good conductor and that this is the region where stress-corrosion fatigue begins.

As shown in Fig. 28 and 29, the phase detection required to obtain the amplitude and phase of the surface impedance is accomplished at microwave frequencies. This will improve the signal-to-noise ratio, but the two output voltages, one proportional to the magnitude and one proportional to the phase, are expected to be of very low level and essentially direct current. Additional gain at direct current will be required, which can cause drift in the output. A much preferred situation would be one in which ac amplification is used.

To use ac amplification, the signal must be either amplitude, frequency, or phase modulated. Such modulation can be provided in two ways:

- Modulate the microwave source
- Modulate the effect being measured (in this case, the surface impedance)

Modulation of the microwave source is readily accomplished. If the modulation frequency is made high enough, the signal-to-noise ratio will be improved because the noise output of the phase detector diodes is lower at higher modulation frequencies than it is at direct current. Additional signal-to-noise improvement will be obtained if the demodulated output is again phase detected with the original modulation as the reference.

Modulation of the surface impedance can be accomplished with the electroreflectance effect in nonferromagnetic materials (Ref 26). When this is done, the modulation will be proportional only to the surface impedance. Therefore, the component of the reflected signal that is proportional to the surface impedance can be readily separated from the other effects. Additional signal-to-noise improvement can be obtained when phase detection is used at the modulation frequency.

The electroreflectance effect is understood to be caused by a change in the energy bands in the surface of the material, which in turn is caused by an electric field also in the surface (Ref 26). In metals, it is very difficult to use externally applied high electric fields to accomplish this modulation. However, when an electrolyte is used, it is possible to obtain large fields at the surface of the sample with small applied voltages because a dipole layer is formed in the electrolyte at the interface between the electrolyte and the sample.

In experiments in which light has been used to measure the reflectance, the metal sample was placed in a fused-quartz cell (Ref 26). A 35-Hz modulating voltage of only 2-V peak was applied between the sample and a platinum electrode, which are immersed in the $KCl\text{-}H_2O$ electrolyte. The reflected light was detected by photomultiplier tubes. The output of the photomultiplier tubes was a 35-Hz voltage proportional to the change in reflectance. After this 35-Hz signal was phase detected, using the original modulation as a reference, a dc voltage was obtained that is also proportional to the change in reflectance caused by the modulating electric field. No information was found on the phase angle of the 35-Hz signal relative to the modulation.

Experiments need to be conducted to show that the changes in the reflectance are caused by changes in the optical and microwave constants of the materials and that the electrolyte may act as a sensitive probe for the reflectivity of the metal. Therefore, by moving the probe while the whole surface is illuminated by microwave electromagnetic waves, the whole surface can be scanned. The larger problem is the manner in which the electroreflectance can be used in practical systems.

To modulate the surface impedance of a ferromagnetic material, the technique of magabsorption can be used (see the article "Magabsorption NDE" in this Volume). For magabsorption, a magnetomotive force is applied to the surface of the material by means of a low-frequency, many-turn coil. The microwave signal, reflected from the surface of the material inside of the magabsorption coil, will be amplitude modulated at a fundamental frequency twice the modulation frequency. The Lissajous figure obtained by applying the detected reflected microwave signal (the magabsorption signal) can be used to determine the surface impedance of the material. Changes will indicate the areas that are stress-corrosion prone or already corroded or fatigued.

Microwave Eddy Current Testing

Microwave crack detection can be thought of as detection by microwave eddy currents (Ref 27). The use of microwave frequencies produces some distinct differences between microwave and conventional eddy current testing (Ref 27) (see the articles "Eddy Current Inspection" and "Remote-Field Eddy Current Inspection" in this Volume). Three of the major differences are as follows:

- Because the skin depth at microwave frequencies is very small (typically a few micrometers), a crack must break through the surface of the metal in order to be detected
- Radiating and nonradiating probes can be used at microwave frequencies
- If the crack is open, energy storage within the crack begins to dominate the crack response as the frequency is increased. This means that a microwave crack detection system is very sensitive to crack opening. Eventually, if the frequency is increased enough, a wave can propagate within the crack, and the crack response also becomes very sensitive to crack depth

In eddy current testing at lower frequencies there are two distinct approaches. The first is the single-coil system in which the induced eddy currents change the impedance of the coil by a factor of ΔZ. The second approach is the introduction of a second sensor (an induction coil or a Hall-effect device) that senses both the applied field and the secondary field arising from induced eddy currents. Numerous variations of these two basic schemes have been explored at low frequencies and at microwave frequencies. The microwave version of the low-frequency single-coil system will be considered first.

One method that has been applied to the analysis of microwave eddy current sensors for calculating the change in the probe impedance, ΔZ_c, caused by a crack is based on the Lorentz reciprocity theorem (Ref 28). Basically, the reciprocity theorem relates the change in impedance produced by the flaw to an integral of certain magnetic fields on the surface of a volume that encloses the flaw (Ref 27). The basic equation used for analysis of microwave eddy currents is (Ref 27):

$$\Delta Z_c = \frac{1}{I \cdot I'} \iint_{S_F} [(\mathbf{n} \times \mathbf{E}') \cdot \mathbf{H} - (\mathbf{n} \times \mathbf{E}) \cdot \mathbf{H}'] dS \qquad \text{(Eq 9)}$$

where I is the probe current without a flaw, I' is the probe current with a flaw, S_F is the surface including the flaw, $\mathbf{H}$ is the magnetic field without a flaw, $\mathbf{H}'$ is the magnetic field with a flaw, $\mathbf{E}$ is the electric field without a flaw, $\mathbf{E}'$ is the electric field with a flaw, and $\mathbf{n}$ is the unit vector normal to the surface S_F. Equation 9 shows that the change in probe impedance is not influenced by the electromagnetic fields that are perpendicular to the surface S_F (Ref 27).

Equation 9 can be simplified by making realistic assumptions. First, the radiation fields can be neglected as long as the probe and the flow are much smaller than a wavelength. It should be assumed that the flow is rectangular with depth a and width Δu and that $a \ll \Delta u$.

If the ratio of flaw depth to skin depth is much larger than unity, as is typical at microwave frequencies, then:

$$\Delta Z_c = \left(\frac{H_0}{I}\right) [jka(\pi a/2 + \Delta u)\eta_0 - Z_s \cdot \Delta u] \qquad \text{(Eq 10)}$$

where k is 2π divided by the free-space wavelength ($ka \ll 1$), H_0 is the complex amplitude of the magnetic field tangent to the metal surface, η_0 is the intrinsic impedance of free space or 120π, and Z_s is the surface impedance. From Eq 10, if Δu is small and H_0/I is real, then ΔZ_c is essentially inductive.

If the frequency is high enough so that the flaw depth divided by the skin depth is much greater than unity, the change in impedance is:

$$\Delta Z_c = \left(\frac{H_0}{I}\right)^2 [2Z_s \left(a - \frac{\Delta u}{2}\right) + jka\eta_0 \cdot \Delta u - 1.56/\sigma] \qquad \text{(Eq 11)}$$

where σ is the conductivity of the surface. Equation 11 shows that if the slot width is small and H_0/I is real, then ΔZ_c will have both resistive and inductive components. If the slot width is not small, then ΔZ_c becomes predominantly inductive at high frequencies.

Research on microwave crack detection has been ongoing since the late 1960s (Ref 19, 29, 30, and 31). The block diagrams of three microwave crack detection systems are given in Fig. 30 to 32.

An earlier microwave crack detection scheme is shown in block diagram form in Fig. 30. It is a basic reflectometer with two additional features. The first one is a microwave bridge to null out the background signal (that is, the surface reflection). The second is a rotating waveguide joint that provides polarization of the incident wave. This effectively tags the signal scattered by the crack because the crack is a polarizing filter that scatters maximum when the incident electric field is perpendicular to the length of the crack (Ref 29). This is a type of mode conversion in which the crack converts a fraction of the incident wave (a mode) into an orthogonally polarized wave (an independent mode).

Another system of mode conversion is used in the microwave crack detection system given in Fig. 31. The diameter of the circular waveguide connecting the mode coupler to the horn antenna is chosen so that either the TM_{01} or the TE_{01} circular waveguide mode is excited and radiated. The horn is placed close to the sample so that the incident fields at the sample surface are essentially those of the waveguide mode. A crack in the surface converts some of the incident energy to the TE_{11} mode, which is separated from the other modes with a tapered waveguide and a Faraday-rotor modulator.

The block diagram of the third microwave eddy current detection system is given in Fig. 32. It also uses mode conversion but without polarization modulation (Ref 31). A linearly polarized incident wave is partially converted to orthogonal polarization when the incident polarization is at an angle of 0 to 90° degrees relative to the slot (crack)

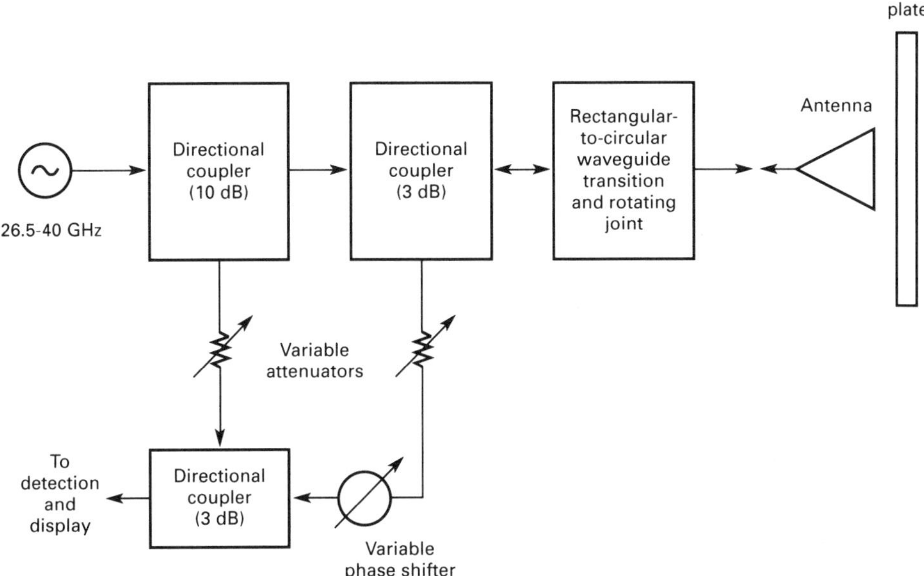

Fig. 30 Microwave crack-detection system. Source: Ref 29

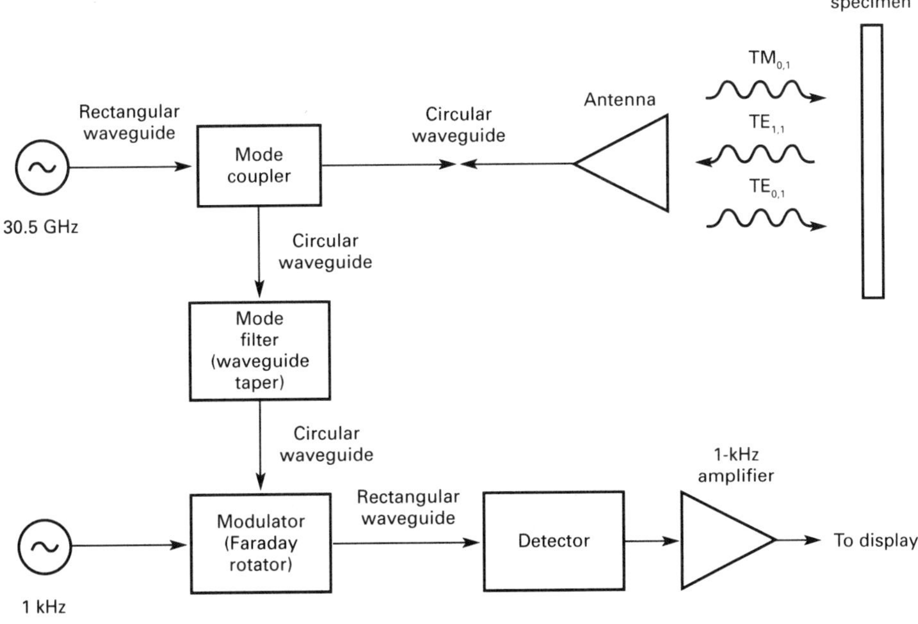

Fig. 31 Mode-converting crack-detection system. Source: Ref 30

(Ref 31). The homodyne detection system provides outputs that are in-phase (I) and quadrature-phase (Q) and that can be displayed in the same way as the conventional (low-frequency) eddy current system.

The antenna used in the system was a lens-focused horn with a beam width at its focal point of about 3.5 mm (0.14 in.) at the operating frequency of 100 GHz. This is an example of a radiating probe that searches for flaws in the far field.

The results from the three systems reported indicate that the crack is detectable provided that the crack is open, the crack has a high Q, and the crack length is larger than one-half wavelength (Ref 31). In practice, one would vary the frequency until the crack signal is normal to the lift-off circle. Knowing the frequency and having a model for the electromagnetic propagation within the crack, the depth could be determined. This feature is self-calibrating because the phase of the crack signal is measured relative to the lift-off circle. (The lift-off signal roughly follows a circular path on the oscilloscope. This path is called the lift-off circle. The diameter of the lift-off circle depends on the system gain.)

Microwave eddy current systems have also been built using planar transmission lines rather than waveguides so that they are of the near-field type. This probe has a higher sensitivity than the others in Fig. 30 to 32. The probes use a pair of adjacent coupled striplines, as shown in Fig. 33. The coupled striplines can support two orthogonal TEM modes, the even (Fig. 33a) and the odd (Fig. 33b). The direction of propagation is perpendicular to the plane of Fig. 33. For the even mode, the two strips are at the same potential relative to ground. For the odd mode, the potentials with respect to ground have potentials that are of the same magnitude with respect to ground but of opposite polarity. The strips carry opposite but equal currents.

In the even mode, there will be no conversion to the odd mode as long as symmetry is maintained. However, when a surface crack in one ground plane perturbs the current under one strip but not the other, some power will be converted to the odd mode. Such a pair of coupled-strip transmission lines can be used as a crack detector if the surface to be inspected is one of the ground planes, if the strips are excited in the even mode, and only if the odd mode is detected. This type of probe depends on mode conversion to achieve a high sensitivity, as have the ones discussed previously.

A cross section of the probe used to inspect 175 mm shells is shown in Fig. 34. Two important features of the stripline probe are as follows:

- The striplines can be as long as desired
- The stripline probe can be made to conform to a curved surface if thin, flexible dielectric substrate is used

In Fig. 34, the shell body forms one ground plane of the strip transmission line as is necessary for detecting open cracks.

The system used with the coupled stripline to detect cracks is shown in Fig. 35. By adjusting the variable attenuator and phase shifter connected to the output of the probe, it was found possible to achieve 50 to 75 dB of dynamic range (Ref 27). Quadrature detection and subsequent squaring and adding produced a phase-insensitive output. The frequency used was near 10 GHz. When the system illustrated in Fig. 35 was connected to the crack detector shown in Fig. 34, electric discharge machining slots as small as 2.54 mm (0.100 in.) long, 0.038 mm (0.0015 in.) wide, and 0.13 mm (0.005 in.) deep could be detected. It was, however, difficult to maintain high isolation between the even and odd modes (bridge balance) as the probe scanned the surface. This is usually because of slight lift-off of the detector from the shell. Perhaps lift-off effects could have been reduced if the in-phase and quadrature-phase signals were used rather than only the amplitude.

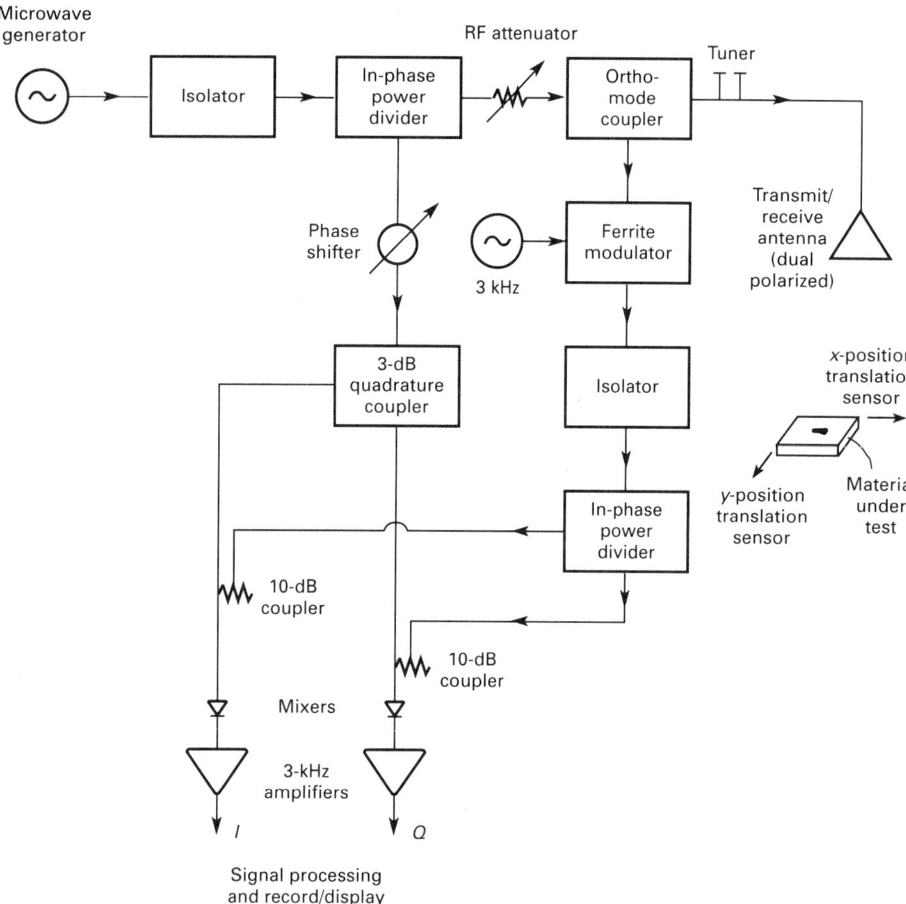

Fig. 32 Microwave system for measuring cross-polarized backscatter using homodyne detection. Source: Ref 31

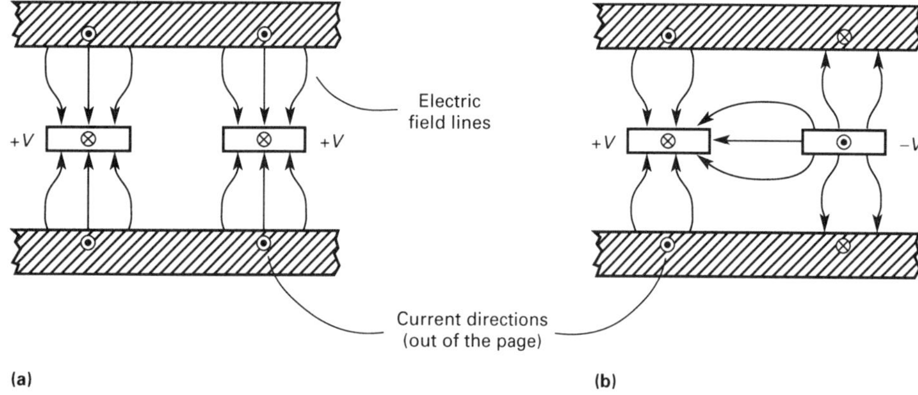

Fig. 33 Two orthogonal modes for a pair of coupled striplines (support dielectric not shown). (a) Even mode. (b) Odd mode. Source: Ref 32

Ferromagnetic Resonance Eddy Current Probes

An entirely different eddy current detection head (Ref 34-39) is illustrated in Fig. 36. Instead of electromagnetic resonance, this probe makes use of ferromagnetic resonance (FMR) (Ref 40). Ferromagnetic materials exhibit a natural dispersion in their permeability. Such an effect has been studied in ferrites that typically have two resonances—one in the range of 3 to 500 MHz and a second at 1200 MHz without an external applied magnetic field (Ref 40). Evidence shows that the higher resonances are due to electron spin (Ref 40). These spin resonances occur naturally because there is an internal anisotropy magnetic field. This internal magnetic field tends to keep the electron magnetization vector aligned with a preferred crystal axis (Ref 40). Most of these resonances are very broad.

Studies of the magnetic resonance of the garnets have given some narrow ferromagnetic resonance lines. Yttrium iron garnet (YIG) gives a very narrow ferromagnetic resonance line (Ref 40). The line with a polycrystalline YIG shows internal field values of 33, 14, and 4.0 $kA \cdot m^{-1}$ (420, 180, and 50 Oe) (Ref 40). The properties are modified by substituting for part of the yttrium, chromium, and rare-earth elements such as gadolinium and dysprosium. An exceptionally narrow line is obtained when the substituted YIG is in the form of polished, single-crystal spheres. The best result was a width of 41 $A \cdot m^{-1}$ (0.52 Oe) (Ref 40). The value of the internal field was 0.175 T (1750 G). The *Q*-factor for single crystals in the form of 0.8 mm (0.03 in.) diam spheres, with a line width of 185 $A \cdot m^{-1}$ (2.3 Oe), is around 3000 in the frequency range of 9.5 to 67 GHz. External magnetic fields are required to cover this range of frequencies.

The resonator illustrated in Fig. 36 is a single crystal YIG sphere less than 1 mm (0.04 in.) in diameter. The resonant frequency of the sphere is determined by the strength and direction of an applied field, $\mathbf{H}_{DC}$, shown in Fig. 36. The internal magnetic field in the YIG comes from the precessing magnetic field of the electrons in the YIG sphere because of their interaction with the applied magnetic field $\mathbf{H}_{DC}$. The radian precession frequency, ω_0 is determined by:

$$\omega_0 = \gamma \mathbf{H}_{DC} \quad \text{(Eq 12)}$$

where $\gamma = 2.8 \times 10^6$ rad/Oe · s. For a resonant frequency range of 500 to 2000 MHz, the value of $\mathbf{H}_{DC}$ is varied from 14.2 to 56.8 $A \cdot m^{-1}$ (178 to 714 Oe). Magnetic field values in this range are readily supplied by small samarium-cobalt and other types of modern magnets.

Unlike lower-frequency eddy current probes, the coil shown in Fig. 36 is a single turn whose diameter is 1.5 times the diameter of the YIG sphere or from 0.25 to 0.75 mm (0.01 to 0.03 in.). The diameter of the wire used is 0.13 mm (0.005 in.).

The ferromagnetic resonance is a condition of the ferromagnetic material. When the YIG sphere is so small, boundary conditions must be taken into account. These lead to an infinite number of resonance modes, called magnetostatic modes, each with a characteristic distribution of the magnetization within the sphere and a characteristic frequency that depends on $\mathbf{H}_{DC}$. When the microwave eddy current probe in Fig. 36 is fed as indicated in Fig. 37, the characteristics of a material placed within the radio-frequency (RF) field ($\mathbf{H}_{RF}$) of the coil, as shown in Fig. 36, will cause a

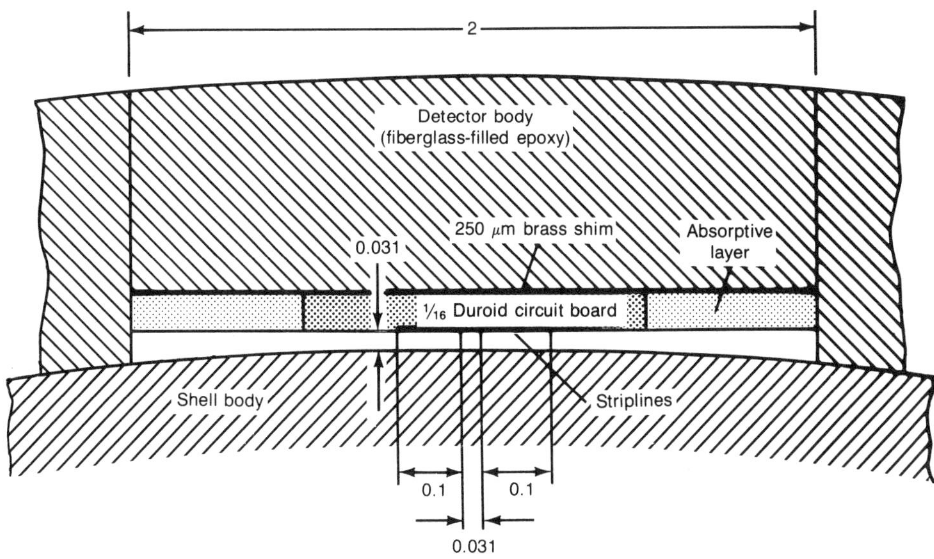

Fig. 34 Cross section of curved stripline crack detector. Dimensions given in inches. Source: Ref 33

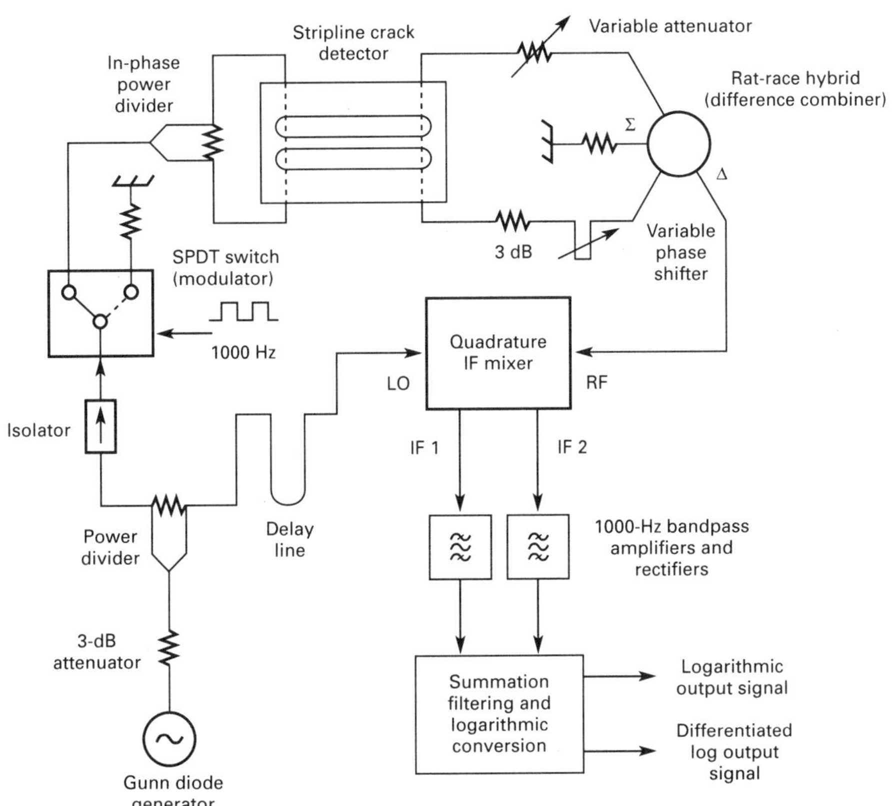

Fig. 35 System for coupled-stripline surface crack detector. LO, local oscillator; IF, intermediate frequency. Source: Ref 33

change in the losses and the mode coupling in the YIG. The material close to the probe causes changes in both the amplitude and frequency of the modes. This in turn will change the energy reflected from the single-turn loop inductor.

The dominant mode is one in which the magnetization is uniformly distributed throughout the sphere and is precessing uniformly. The precession frequency is given by Eq 12 and can be represented by an electron dipole moment of strength ($\mathbf{m}_{RF}$ V_s), where V_s is the volume of the sphere. The dipole moment, because of the cross product between $\mathbf{H}_{DC}$ and $\mathbf{m}_{RF}$, is precessing about $\mathbf{H}_{DC}$ at a frequency ω, according to the right-hand rule, in a plane perpendicular to $\mathbf{H}_{DC}$, as shown in Fig. 36. The precession is at the frequency of the drive current, I, applied to the coupling coil. The dipole moment $\mathbf{m}_{RF}$ obtains its maximum amplitude when the precession frequency ω_0 (Eq 14) equals the precession frequency ω. The time-varying magnetic fields are created outside of the sphere by this rotating dipole moment. These outside time-varying magnetic fields interact with a flaw in the material and produce a flaw-detecting signal in the same way that the time-alternating magnetic fields do from low-frequency (conventional) eddy current probes.

In practice, the YIG sphere is mounted in a depression on a piece of microwave circuit board with a loop etched on the board. Wire leads connect the end of the loop to a semirigid coaxial transmission line. A small samarium-cobalt magnet in a brass holder supplies the bias field. The placement of the magnet can be adjusted to change the field orientation and the resonant frequency. This eddy current probe, because of its small size, gives excellent spatial resolution, discrimination against edge effects, and accessibility to restricted corners.

The high sensitivity and spatial resolution of the FMR eddy current probe have been experimentally verified (Ref 34, 39). In one case, the FMR probe was found to be 40 times more sensitive to an open slot than a commercial 100-kHz probe. The correlation between the flaw signature, detected by the FMR probe, and the width of the opening of a stressed fatigue crack has also been demonstrated.

Another group of researchers has applied the FMR probe to aluminum, type 316 stainless steel, Armco iron, and XC-38 (Ref 41). They used the YIG and gallium-doped yttrium iron garnet (GaYIG). The main difference between this garnet and the previous garnet is that the saturation magnetization is fixed at 0.075 T (750 G) for YIG, while that of the GaYIG varies with the amount of gallium. The resonator principle that was used is given in Fig. 38. The characterization of the spectrum line is done by measuring the reflection coefficient of the YIG resonator in Fig. 38. A microwave signal is applied onto the garnet by means of the copper loop. In the absence of resonance, the wave is reflected. As the resonance is approached, the incident wave is absorbed by the garnet, and the reflected wave is highly attenuated. The reflection coefficient is characterized either by a scaler analysis or by a vector analysis.

In air (uncoupled to a sample), the resonant frequency is proportional to the applied bias field $\mathbf{H}_D$ in Fig. 38. The proportionality constant is 2.8 MHz/Oe. The resonance lines for YIG are located above 3 GHz. For GaYIG, the resonance frequency can go down to less than 1 GHz.

The garnets used have diameters of 375 or 500 μm (0.015 or 0.020 in.). The only garnets used were the YIG, which required a bias magnetic field between 80 and 160

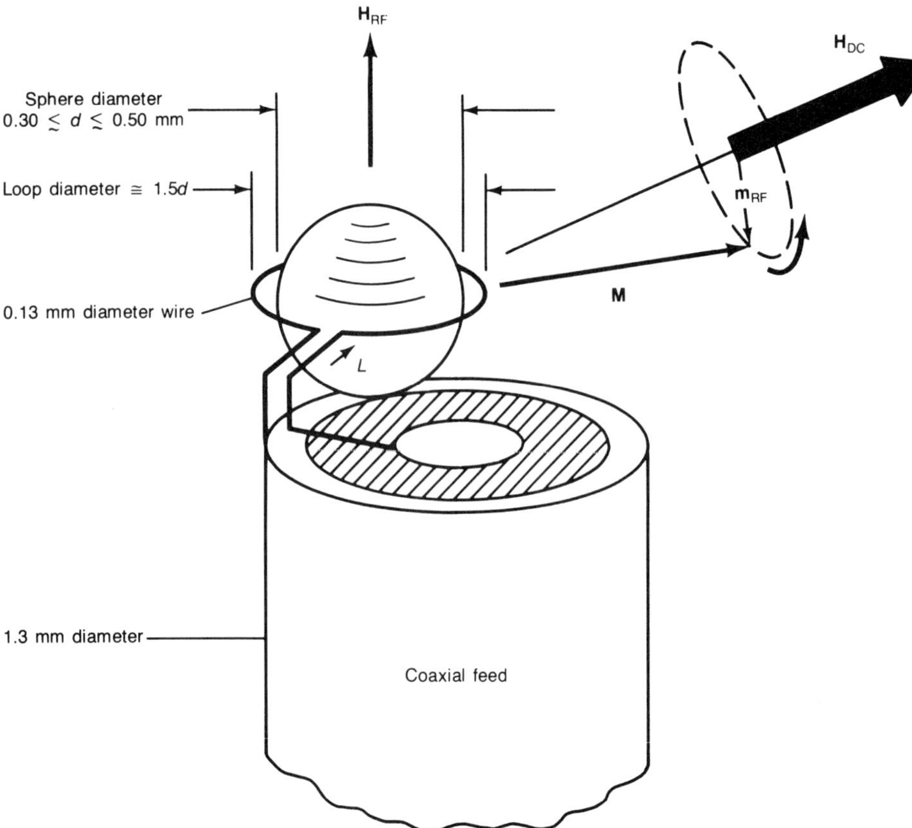

Fig. 36 Typical dimensions and feed arrangement for a ferromagnetic resonance eddy-current probe. **M**, internal magnetization. Source: Ref 34

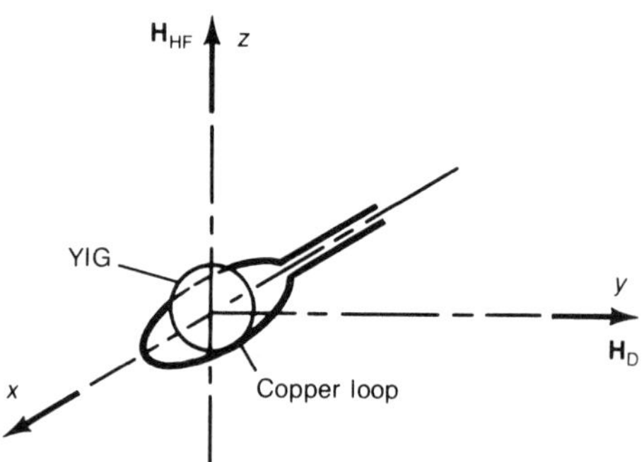

Fig. 38 Schematic of YIG resonator principle. $\mathbf{H_D}$, continuous magnetic field; $\mathbf{H_{HF}}$, microwave magnetic field. Source: Ref 41

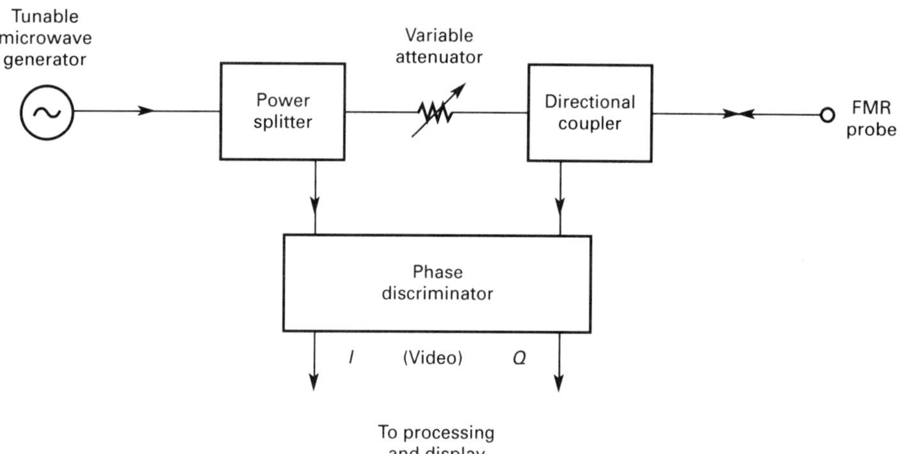

Fig. 37 Microwave system for measuring the reflection from an FMR probe. Source: Ref 38

kA · m^{-1} (1000 and 2000 Oe). The microwave loop is made on a printed circuit. Experience with microwave oscillators has shown that this loop should have a diameter of about 1.5 times that of the garnet and that the copper width should be 0.1 mm (0.004 in.).

In Ref 41, three different probes (A, B1, B2) were used. These three detection heads are illustrated in Fig. 39. These three probes are configured for use as follows:

- Probe A is for use with a nonmagnetic metal where the magnetic field is parallel to the surface of the metal and the fringe field of probe A will be in the surface of the nonmagnetic metal
- Probe B1 is for use with magnetic material where the FMR probe is between the metal and the magnet, providing the bias field for the YIG, and the RF field is perpendicular to the magnetic metal surface
- Probe B2 is for use with magnetic material where the FMR probe is between the magnet and the metal and the RF field from the YIG-coil system is parallel to the metal surface

Head B2 imposes a minimum distance of the order of approximately 1 mm (0.04 in.) between the metal and the garnet because of the position of the loop. This drawback can be overcome by using a half-loop. In head B1, there are numerous spurious modes, which are almost as intense as the main mode. However, this head appears to be of particular value in the separation of the distance effect from the flaw effect.

The measurement principle is described as the characterization of the resonance line and of its alteration in the presence of a flaw. The block diagram of the electronics used with the FMR probes is shown in Fig. 40. Figure 40(a) is for the scalar analysis, and Fig. 40(b) is for the vector analysis. For the scalar analysis, a wobbulator (German for signal generator, specifically a frequency-modulated RF source) sends a wave through the coupler and through the circulator to the probe. The signal from the probe goes through the circulator and through a detector to the vertical axis of the oscilloscope. The signal used to vary the frequency is sent to the horizontal axis of the oscilloscope so that the oscilloscope presents the resonance of the probe as a function of frequency. An automatic control device makes it possible to determine the frequency, f, of the peak. The scanning of the metal takes place in a single direction, and the curves $f(x)$ are plotted. The flaw is characterized as a frequency shift.

For the vector analysis, the block diagram shown in Fig. 40(b) is used. The reflected wave is sent to a vector analyzer, which makes it possible to display the reflection coefficient in the phase plane. The influence of the flaw is characterized by the displacement of the resonance loop in the phase plane. The synchronizer allows phase locking on the wobbulator to provide greater frequency stability when required.

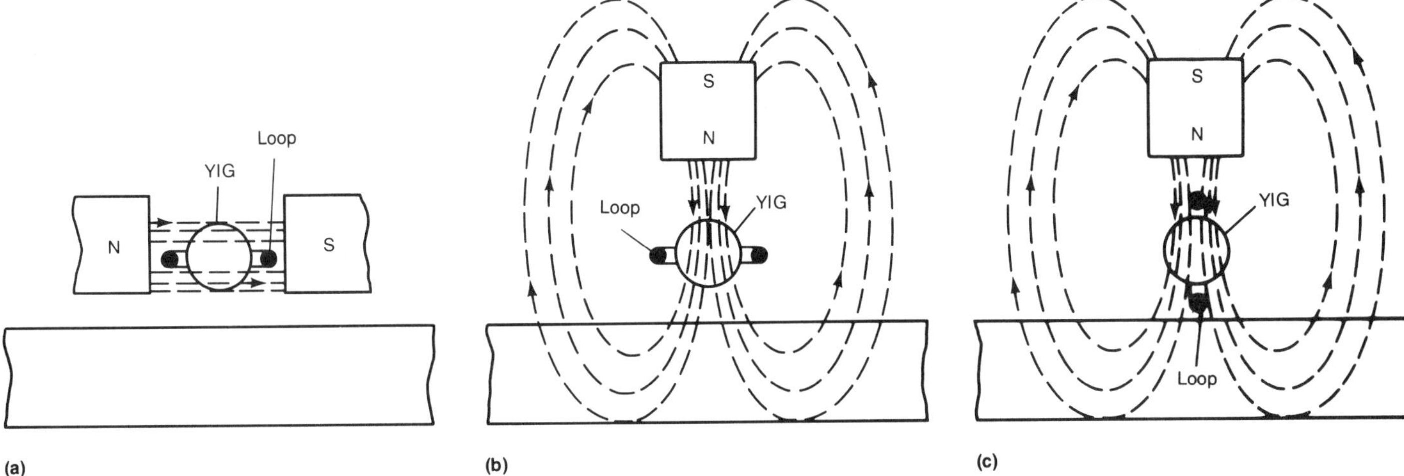

Fig. 39 Three microwave eddy current detection head configurations that use a YIG coil. (a) A, for nonmagnetic metal. (b) B1, for magnetic metal, RF field perpendicular to magnetic metal surface. (c) B2, for magnetic metal, RF field parallel to magnetic metal surface. Source: Ref 41

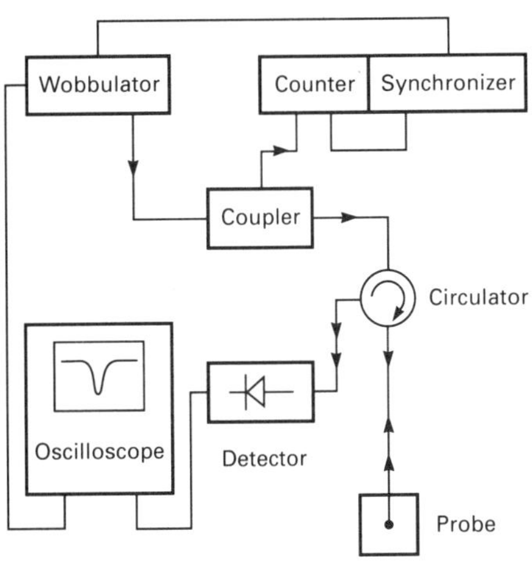

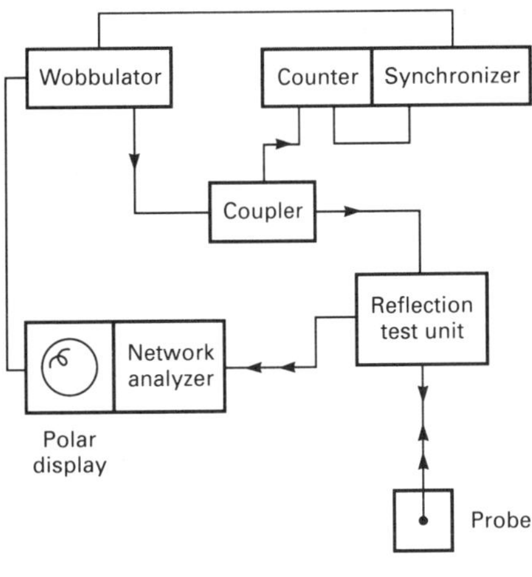

Fig. 40 Schematic diagram for (a) scalar analysis and (b) vector analysis. Source: Ref 41

Three metals were studied: aluminum, type 316 stainless steel, and Armco iron. The samples are surface ground to an arithmetic average roughness, R_a, of 0.1 μm (4 μin.).

The flaws used are straight artificial grooves, cutting the samples across their entire length. The parameter investigated is the frequency of the resonance peak. The results for aluminum and stainless steel are comparable. The differences between scalar analysis of the aluminum and the Armco iron were significant.

The lift-off effect is basically different for the aluminum and the Armco iron but the same for the aluminum and the stainless steel. For a lift-off, Δh, of more than 300 μm (0.012 in.), the frequency remains stable for aluminum, while from 300 to 100 μm (0.012 to 0.004 in.), the frequency increases from 4014 to 4024 MHz. For Armco iron, the frequency increases almost exponentially from 3800 MHz at 600 μm (0.024 in.) lift-off to 4300 MHz at 100 μm (0.004 in.). The frequency variation in iron is due to the variation of the space between the magnetic and the metal. In the nonmagnetic metals, the frequency variation is due to the eddy currents in the metal.

In the configurations illustrated in Fig. 39 and 40, the sensitivity of the frequency measurement is 100 kHz, mostly limited by instabilities. The graph of frequency shift caused by a 100 × 100 μm (0.004 × 0.004 in.) flaw in each of the three metals with a lift-off of 100 μm (0.004 in.) is smallest for aluminum, medium for the stainless steel, and largest for the Armco iron. The values of frequency shift are 4 MHz for aluminum, 12 MHz for stainless steel, and 27 MHz for Armco iron. Flaws of 10 × 5 μm (400 × 200 μin.) were lost in the background for all three materials.

The experiments confirmed that FMR probes have better sensitivity to flaws for ferromagnetic materials than for nonmagnetic metals. On nonmagnetic metals, the frequency shift is due to eddy currents, while in ferromagnetic metals, the frequency shift is caused by a modification of the bias field by the flaw. Because a frequency shift of 0.1 MHz can be measured, a field variation of 0.05% can be detected.

Eddy Currents for Holes and Areas of Small Radii

To make microwave eddy current measurements in holes and in areas of small radii of curvature, a slow-wave structure has been used at Southwest Research Institute. This slow-wave structure is similar to a helical antenna or a helix TWT except that the last turn of the helix is shorted. This shorting forms a shorted spiral transmission line without a shield. In the test unit, the length and diameter will be adjusted so that when this helix is placed within the hole, it forms a helical resonator when the last turn is not shorted or a spiral delay line when the helix is shorted. When the helical antenna is used with the system shown in Fig. 32, the amplitude and phase signals will measure the surface impedance, which can be related to its corrosion susceptibility. Multiple helical probes could be constructed so that a number of hole sizes could be measured.

For an area with a small radius of curvature, the slow-wave helix could be made flexible so as to conform to the definite configuration of the part, or it could be scanned over the surface. By calibration, then, the surface impedance could be related to the corrosion susceptibility of the surface under inspection.

In any case, the microwave signal is carried either by a wire of the helix so that it has sensitivity both inside and outside of the helix or by a hollow pipe of circular or rectangular cross section that can be made sensitive for surface impedance measurements through slots or holes in the pipe. Therefore, a multitude of configurations are possible to meet the needs for surface impedance measurements on the different shapes of aircraft components.

Microwave Holography Practice

A hologram is a recording of an interference pattern formed by superimposing a reference electromagnetic wave on the object-scattered electromagnetic wave (Ref 42). Because the interference depends on the relative amplitudes and phases of the incident and scattered waves, highly coherent waves (same frequency and same phase) must be used to keep the phase relationship constant during the generation of the microwave hologram.

The reference wave is called $\mathbf{E}_1$, and it is assumed to be a plane wave (same frequency, same phase). The object-scattered wave is $\mathbf{E}_2$, and any point on the interference pattern in the *x-y* plane will be described by an intensity $(I(x,y))$ variation:

$$I(x,y) = (\mathbf{E}_1 + \mathbf{E}_2)(\mathbf{E}_1 + \mathbf{E}_2)^* = |\mathbf{E}_1|^2 + |\mathbf{E}_2|^2 + \mathbf{E}_1\mathbf{E}_2^* + \mathbf{E}_1^*\mathbf{E}_2 \quad \text{(Eq 13)}$$

where the asterisk indicates the complex conjugate. In the photographic record, the last two terms, $\mathbf{E}_1\mathbf{E}_2^*$ and $\mathbf{E}_1^*\mathbf{E}_2$, are the only contributors because they are the relative signal-bearing terms. The $|\mathbf{E}_1|^2$ and $|\mathbf{E}_2|^2$ terms are the zero-order diffracted terms upon reconstruction. When the interference pattern is illuminated with a third plane wave, $\mathbf{E}_3$, a modulated wave, $\mathbf{E}_4$, is generated, which is given by:

$$\mathbf{E}_4 = \mathbf{E}_3\mathbf{E}_1\mathbf{E}_2^* + \mathbf{E}_3\mathbf{E}_1^*\mathbf{E}_2 \quad \text{(Eq 14)}$$

Because both $\mathbf{E}_1$ and $\mathbf{E}_3$ are plane waves, $\mathbf{E}_1 = \mathbf{E}_1^*$ and the product terms $\mathbf{E}_3\mathbf{E}_1$ and $\mathbf{E}_3\mathbf{E}_1^*$ are constants. Equation 15 states that the object-scattered wave $\mathbf{E}_2$ and its conjugate $\mathbf{E}_2^*$ are reconstructed so that virtual and real object images are reconstructed.

Zone Plates. A hologram can be looked upon as a microwave interference pattern consisting of many coherently superimposed zone plates. Then, holography, whether at microwave or light frequencies, becomes a relatively simple diffraction process that is fully understandable without concern for phase, modulation, coding, or some of the other concepts often mentioned as means to assist the understanding of holograms. Rogers was the first to note the similarity of holograms and zone plates (Ref 43). Later, he used holograms to generate microwave radar zones plates. A zone plate is defined as a diffracting screen so constructed that it obstructs alternate Fresnel zones of a wave front.

Figure 41 illustrates a zone plate with an opaque disk at its center element. The open spaces permit the passage of energy that will add at the focal point, *f*. The opaque rings prevent the passage of energy that would interfere at the focal point. The opaque or blocking zones can be replaced

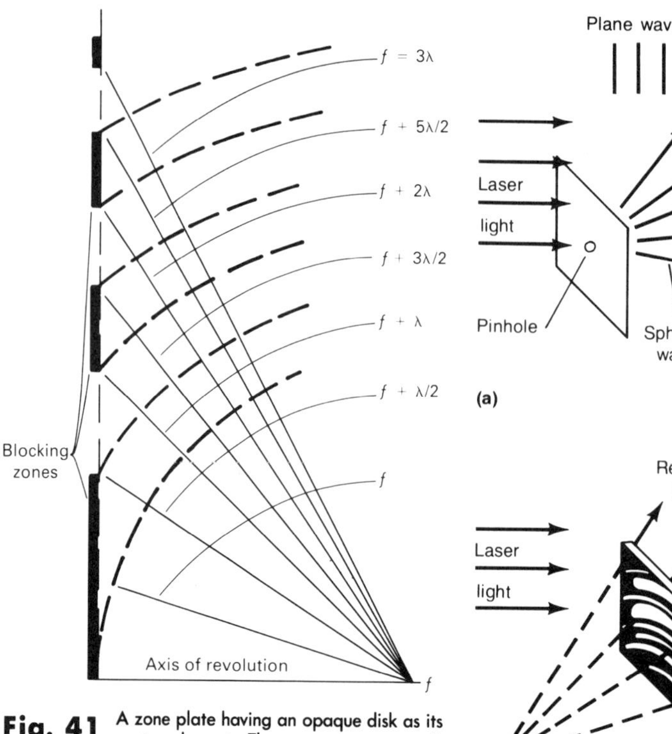

Fig. 41 A zone plate having an opaque disk as its center element. The open spaces permit the passage of energy that will add at the focal point, and the opaque rings prevent the passage of energy that would interfere at that point. Source: Ref 43

Fig. 42 Schematic of the Gabor holographic. Construction (a) and reconstruction (b) when recording a point source of light (pinhole in an opaque screen). The interference pattern generated corresponds to the pattern of a two-dimensional zone plate. Source: Ref 43

with open spaces, and the zone plate will function the same as the one shown in Fig. 41. The zone plate also generates a set of diverging waves, which is very important to the hologram. It is the diverging waves that give the three-dimensional view to the object obtained from the hologram.

The design procedure for zone plates is indicated in Fig. 41. Circles are drawn with centers at the desired focal point and with radii differing by one-half the design wavelength. The radii of these circles intercept the plane of the zone plate. The first circle has a radius of *f*. The radius of the second circle is $(f + \lambda/2)$. Where the second circle intersects the zone plate, the first transparent ring starts. The end of the first transparent ring is where the third circle of radius $(f + 3\lambda/2)$ intersects the zone plate. The third circle is also the start of the first opaque or blocking zone. This process continues with the radius of each successive circle being increased by $\lambda/2$ over the radius of the previous circle. The intersection of the first circle of radius $(f + \lambda/2)$ with the zone plate marks the end of the first or central blocking zone.

Optical Holography. The way a hologram of a point source of light might be made is shown in Fig. 42. In the developed photographic zone plate shown in Fig. 42(a), the open spaces pass only the energy that will add constructively at a desired focal point. The opaque rings prevent the passage of energy that would interfere destructively at the desired focal length.

Where laser light illuminates the hologram in the reconstruction process, shown in Fig. 42(b), diffraction causes converging waves to create a real image of the spot of light at the focal point *F*. Diffraction at the zone plate also generates diverging waves that create, for a viewer, the illusion that there is a point of light also located at the conjugate focal point F_c in Fig. 42(b). The diverging light is indistinguishable from the light from the pin hole, and the viewer imagines he sees the second source of light located in the space behind the illuminated zone plate (hologram). Detailed information is available in the article "Optical Holography" in this Volume.

Microwave Holography. Microwaves in microwave holography are analogous to laser light in optical holography. A microwave hologram can be defined as the photographically recorded interference pattern between a set of coherent (at the same frequency and phase) microwaves of interest and a coherent reference wave generated by the same source. This method is still used today.

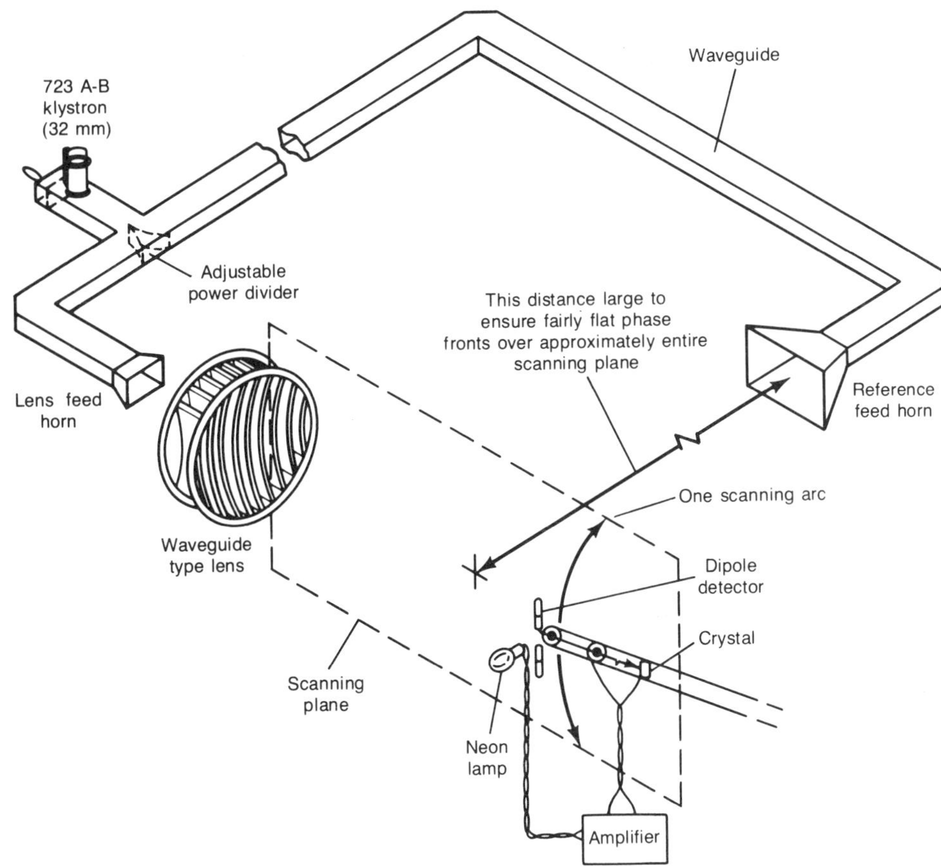

Fig. 43 Procedure for making a microwave hologram by photographic scanning. Source: Ref 44, 45

The equipment for making a microwave hologram is illustrated in Fig. 43. The interference patterns were made visible by photographic scanning. In making the pattern, two wave sets are required. The first is the incident wave from the feed horn that is passed through a type of waveguide lens, and the second is the reference wave. These two wave sets cause an interference pattern over the scanning plane where the incident wave and the reference wave are combined. A small dipole probe antenna, with a small neon lamp attached, scanned the interference plate. The signal picked up by the small dipole is peak detected, and the peak-detected signal is amplified and applied to the neon lamp, causing its brightness to vary. A camera set at time exposure recorded the neon lamp brightness as a function of the position of the dipole detector. The resulting photograph is the microwave interference pattern (a microwave hologram).

Two observations need to be made. First, photographic plates sensitive to microwaves have been studied for many years without success. Second, microwave interference patterns can be made by two methods. The first method is photographs of the interference pattern recorded on a microwave liquid crystal display. The second method is through the use of the thermograph. Both of these techniques are discussed in the section "Instrumentation" in this article.

A third observation must also be made. In Fig. 43, the hologram is made photographically by converting the interference pattern, point-by-point, from a microwave intensity to light intensity and then photographing the light intensity point-by-point. Also in Fig. 43, the reference is an electromagnetic wave that is added to the electromagnetic waves reflected or refracted by the object. There is another way to achieve the same result. At microwave frequencies over the complete band of interest (300 to 300 000 MHz) it is possible to produce an interference pattern without producing a radiated reference electromagnetic plane wave or a reference spherical wave. This can be done by adding, before detection, a locally produced reference signal to the signal detected from the object, as shown in Fig. 9.

To simulate a plane wave, using a reference wave, **R**, added to the object signal, **S**, the phase of the locally produced signal must vary linearly with the displacement of the probe in Fig. 9 in its direction of travel (Fig. 9). This linear variation in phase is most readily achieved by linking the motion of the probe to the rotations of the phase shifter (Ref 46).

The simulation of a spherical wave is more complicated. Within the Fresnel approximation, the phase can be made to vary quadratically (in-phase and quadrature-phase) with position, and this will suffice (Ref 46). This method of using a local reference signal, **R**, with a programmable phase leads to more possibilities (Ref 46). For example, it is possible to simulate a slow wave in which the dielectric constant is larger than that of free space so that the wave travels slower than in free space. Because the phase is rotated in synchronism with the probe motion, the phase shifter can be rotated at any desired rate to simulate an arbitrarily large wave vector. This makes it possible to shift the spectrum of the object, even when it occupies a wide band, enough to avoid overlap with the image spectrum. This separates real and virtual images in the reconstruction.

Another possibility is the simulation of a noiselike reference wave in which the phase is made to vary in some arbitrary manner, such as $\phi(x,y)$, as a function of position. In this case, for the purpose of reconstruction, the function $\phi(x,y)$ must be stored for use during reconstruction (Ref 47). Further, the phase can be varied so that a simulation of the use of a diffuse reference wave is obtained. The same diffuse reference wave must be used for reconstruction. The advantages of using a programmable reference wave are (Ref 47):

- Easier instrumentation even in the case of a plane wave reference
- More flexibility
- The possibility of readily simulating complex modulation schemes

A further possibility is shown in Fig. 44(a), in which a product detector is used instead of the intensity-measuring device or a square-law detector, which is composed of a multiplier and a low-pass filter. It is also called a correlator and is widely used in interferometry and radio astronomy.

If **R** (reference) and **S** (signal) are the two inputs to the product detector and if **S** is from the object and **R** is from the locally produced reference, the output of the product detector is Re **R*S** (the asterisk means a complex conjugate and Re means the real part of the product of **R*** and **S**). The output recorded as a function of the probe position ($f(x,y)$, Fig. 44a) is called a product hologram. It can be transcribed to a photographic plate and used exactly as the intensity hologram in the reconstruction. The only difference is that the product hologram does not contain the unwanted terms $|\mathbf{R}|^2$ and $|\mathbf{S}|^2$ because these have been filtered out. In some cases of intensity holography, the spectrum of $|\mathbf{S}|^2$ could overlap **R*S** and **RS***, resulting in distortions. Because the product hologram is not always positive, it must be transcribed with its sign.

At microwave frequencies, it is relatively easy to record both phase and amplitude of the field instead of just the intensity. This

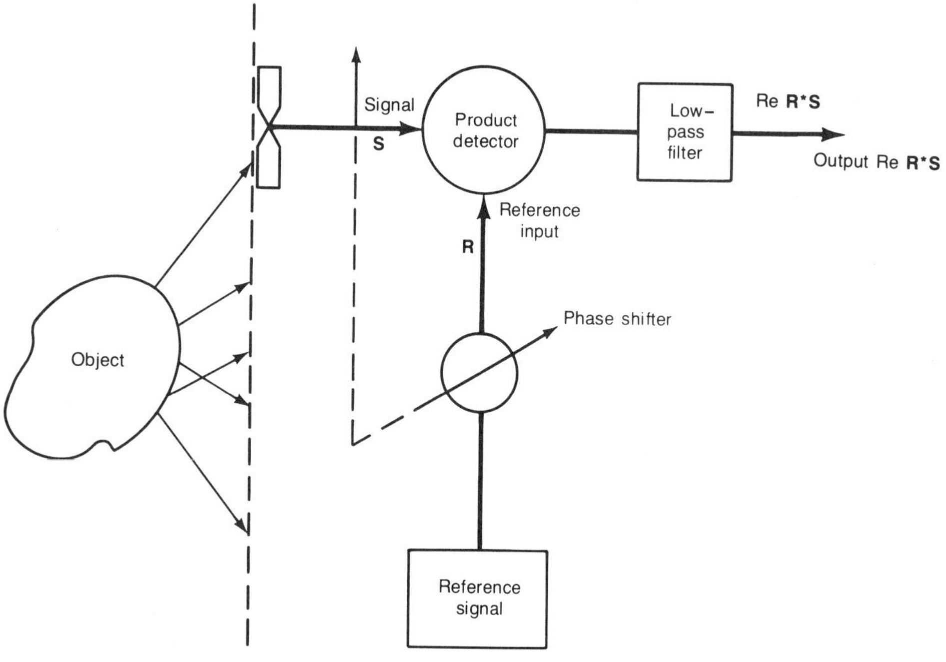

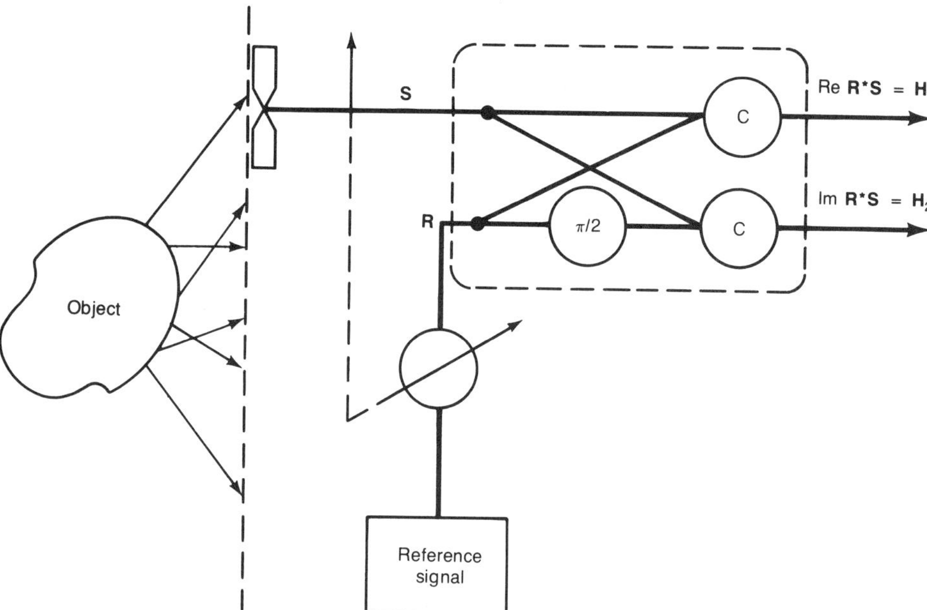

Fig. 44 Recording of (a) microwave product hologram and (b) microwave complex hologram $\mathbf{H}_1 + j\mathbf{H}_2$. C, product detector. Source: Ref 47

can be done with two product detectors. One product detector correlates **R** and **S** and the other j**R** and **S**, where j**R** is **R** shifted by 90°. The two outputs are the real and imaginary parts of **R***S. Recorded as two functions, $\mathbf{H}_1(x,y)$ and $\mathbf{H}_2(x,y)$, they form the complex hologram.

For the complex hologram, the shifting of the phase of the reference is dispensed with. The function of phase shift was only to make it possible to recover the phase after recording only one intensity pattern. This phase recovery can be understood by a comparison with the complex hologram. If the phase of the reference wave, **R**, varies very rapidly compared to the signal, **S**, that is, if the signal **S** does not change appreciably as **R** goes through a complete phase rotation, two intensity measurements taken at points where the phase of **R** differs by 90° will be equivalent to a measurement of **R*S**. The complex hologram consists of two maps, $\mathbf{H}_1(x,y)$ = Re **R*S** and $\mathbf{H}_2(x,y)$ = Im **R*S**, as functions of position in the plane of the record. If it is desired to reproduce **S** at some point, then multiplying signal **R** by $\mathbf{H}_1$ and j**R** by $\mathbf{H}_2$ and adding the results will accomplish it. If $k|\mathbf{R}|^2$ is taken as unity, then the output is precisely **S**.

It is interesting that in contrast to the intensity hologram, only one image, the virtual image, is produced. If the real image is desired, all that needs to be done is to change the +90° phase shift to −90°.

Microwave and millimeter wave holography has several advantages over optical holography (Ref 47):

- More favorable propagation through the atmosphere so that images of reflecting objects can be made through fog, haze, and other optically opaque dielectric barriers
- The availability of well-established and more flexible detection and electronic processing methods so that image separation and enhancement can be achieved
- The information content is considerably lower than optical holograms, which permits real-time operation

Microwave holography has been used to aid airport security through the detection and identification of concealed weapons and to provide close-range visualization of airport runways in inclimate weather.

Example 1: Use of Microwave Holography to Detect Concealed Weapons Aboard Aircraft Baggage. A holographic imaging experiment was conducted to produce a visible image of guns at 70 GHz. The simplified block diagram of the system used is shown in Fig. 45(a). The weapon was irradiated with 70 GHz (λ = 4.3 mm, or 0.17 in.) in a CW mode. The output of the klystron used was 750 mW, and it was fed to a parabolic antenna that produced uniform irradiance of 0.2 mW/cm² (1.3 mW/in.²). The phase distribution of the wave field scattered by the weapon was mapped by a spiral scanning pyramidal horn that had a 10 × 7.7 mm (0.39 × 0.30 in.) receiving aperture. The output of the receiver phase detection system is a value proportional to the phase of the field of the detected object. The display on an oscilloscope was photographed to produce a phasigram from which the image was retrieved by the method shown in Fig. 45(b). The reconstructed image of a gun showed that its general outline was that of the gun used. The edges of the gun were emphasized, as expected. The linear resolution at a range of 1 m (3.3 ft) from the antennas is about 10 mm (0.4 in.).

Example 2: Use of Microwave Holography to Provide Close-Range Visualization of Airport Runways. The microwave holographic system was used for the close-range detection of a landing strip that was marked at regular intervals with retroflectors (Ref 47). The microwave system was constructed to visualize holographically the distribution of the retroreflectors on the runway. The landing aircraft would illuminate the marked runway with two micro-

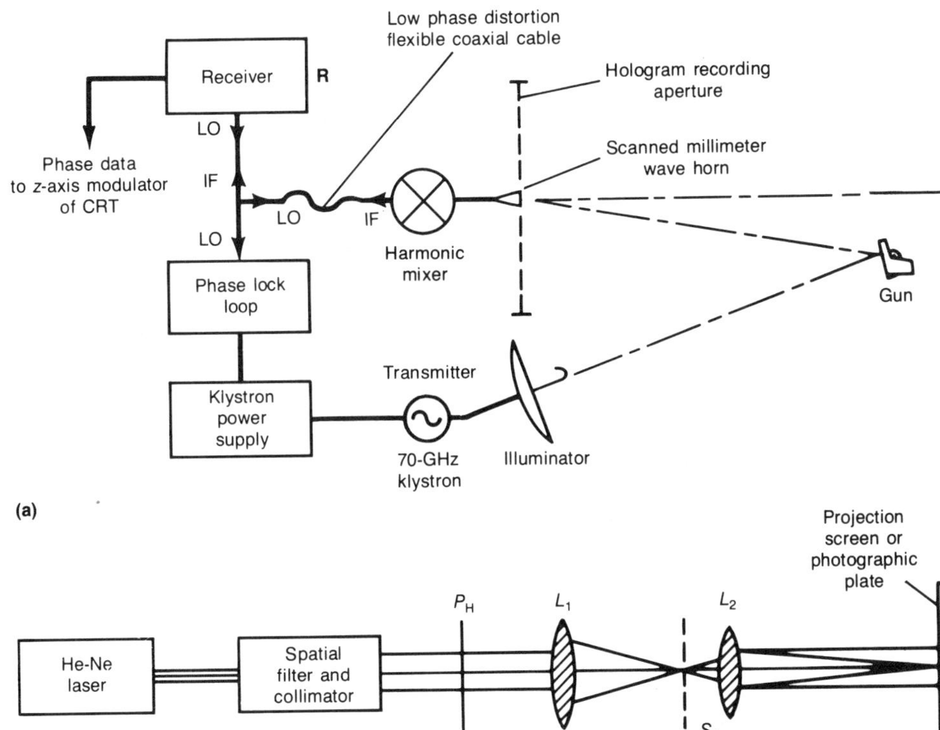

Fig. 45 Recording and reconstruction configurations for the detection of concealed weapons. (a) Simplified block diagram of recording system. (b) Image retrieval arrangement. IF, intermediate frequency; LO, local oscillator. Source: Ref 47

wave beams. Because the one intended for imaging lies in the millimeter range, its wavelength is short enough so that it can easily resolve the target (the runway). The longer-wavelength radiation is used to realize a target-derived reference beam. This gives the advantages of reducing the required number of receiver elements in the airborne hologram recording array and relaxes the accuracy with which the position of the receiver elements in the array must be known. The imaging wavelength of 3 mm (0.12 in.) and a retroreflector separation of 6 m (20 ft) made the hologram recording array length, to resolve two adjacent retroreflectors, 1.5 m (4.9 ft) at a maximum operational range of 3 km (9800 ft) (Ref 47).

Incoherent-to-Coherent Image Converters. Three incoherent-to-coherent image converters are listed in Table 4. Although these devices were primarily developed for large-screen images, they are also potentially useful in real-time analog imaging reconstruction from incoherent light pattern displays of hologram intensity or phase information. The basic common property of these devices is their ability to transform an intensity pattern into a spatial amplitude or phase modulation of a visible wave front; this is accomplished by the reflection or transmission of an interrogating beam through them.

For example, the γ-Ruticon can be used to reconstruct millimeter or microwave holograms using the arrangement shown in Fig. 46. The γ-Ruticon consists of a layered structure made of a conductive transparent substrate (S), a thin photoconductive layer (C), an elastomer (E), and a thin flexible reflective gold electrode (G). A bias voltage, V_B, is applied between the conducting layers, as shown in the detail in Fig. 46. The device is capable of converting the hologram intensity or phase distribution data displayed on the CRT face plate into a demagnified relief pattern of the flexible gold surface. This surface behaves as a phase-modulated hologram from which the image can be retrieved with the aid of a helium-neon laser beam (Ref 47). The device has the advantage of being:

- Repeatedly usable
- Compatible with the CRT display
- Capable of a fast enough response time that allows TV frame rates and therefore real-time operation

Both the photoconductor-liquid crystal sandwich and the ferroelectric photoconductor image camera (FERPIC), a storage and display device, can also be used in real-time image retrieval from microwave holograms by incorporating them in schemes such as that shown in Fig. 46.

Microwaves have been used for flaw detection in some articles, such as solid propellant grains, because of the difficulties involved in their inspection (Ref 54). The resolution of defects has not been particularly successful, especially for spherical voids. Holography and optical data processing are used for microwave inspection because they permit synthetic aperture techniques that integrate wide-angle signals from spherical reflectors and scatterers. This system avoids the use of a spheric correcting lens in CW holograms and in a swept-frequency, side-looking radar, which is useful at very short range.

Microwaves require a very large hologram to record much detail. The use of two very different wavelengths for recording and readout, as described previously, causes some severe problems. As a result, the expected advantages of microwave holography are not realized in terms of a realistic image of the object.

Side-looking radar is another approach. It is more attractive and feasible because there is a method of obtaining depth resolution at close range. The method uses a swept frequency rather than a pulse echo for depth resolution.

An optical coherent signal processor for A-scan data is illustrated in Fig. 47. It uses a collimator, which is a transform lens to produce the transform plane. The image lens produces the output, which is photographically recorded. The results are a series of in-line spots on the film. The first spot is for the front surface. Crack simulations were made and tested at 25, 50, 75, and 102 mm (1, 2, 3, and 4 in.). Each crack was clearly and cleanly defined.

An A-scan can be readily converted into a B-scan (Ref 54). The horizontal sweep for the oscilloscope, shown in Fig. 48, is replaced by a sweep synchronizer, and a cylinder lens replaces the spherical transform lens in the optical processor. This

Table 4 Characteristics of incoherent-to-coherent image converters

Device	Threshold of sensitivity, mJ/cm^2	Contrast ratio	Resolution, lines/mm	Read in time, s	Erasure time, s	Operational mode
FERPIC (Ref 48)	10 mW/cm^2	40:1	50	1 PVK 0.01 CdS	Same as read in	Storage
γ-Ruticon (Ref 49)	3×10^{-2}	...	80	TV frame rates	0.01 (80%)	Real time
Photoconductor liquid crystal sandwich (Ref 50, 51)	5×10^{-3}	30:1	10–20 (Ref 50, 51)	0.001 (Ref 52) 0.0001 (Ref 51)	$\gtrsim 3 \times 10^{-3}$	Real time or storage

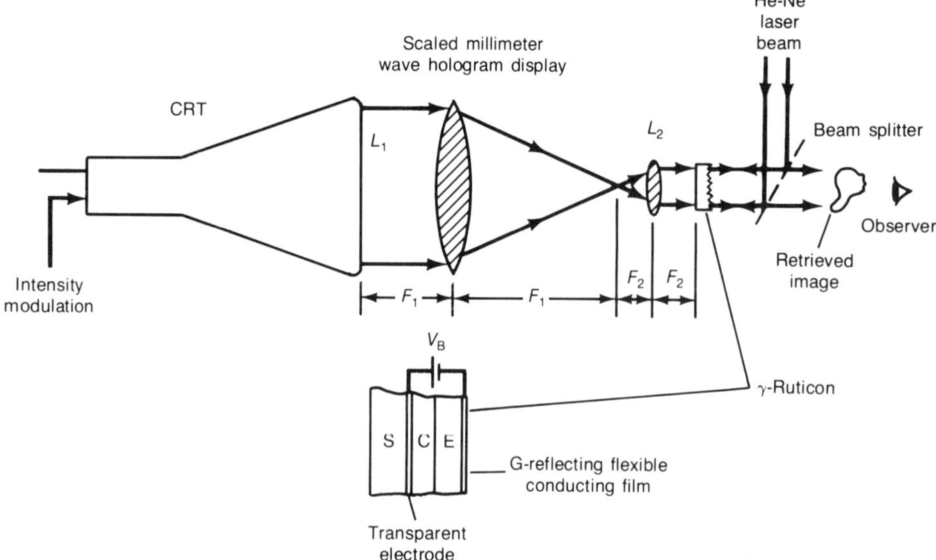

Fig. 46 Real-time millimeter wave hologram reconstruction utilizing a γ-Ruticon. See text for discussion. Source: Ref 47

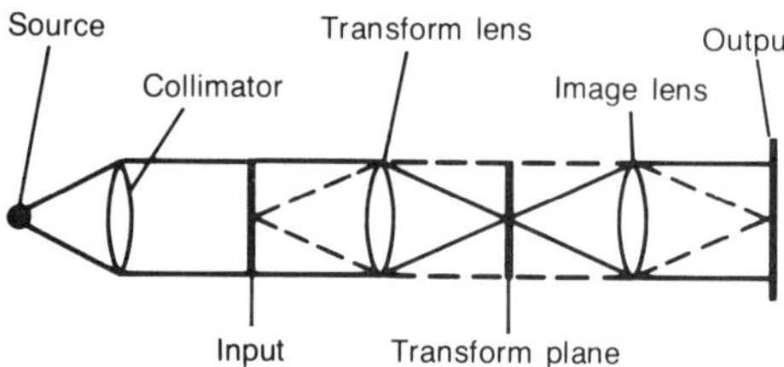

Fig. 47 Coherent processor for optical processing of A-scan data. Source: Ref 47

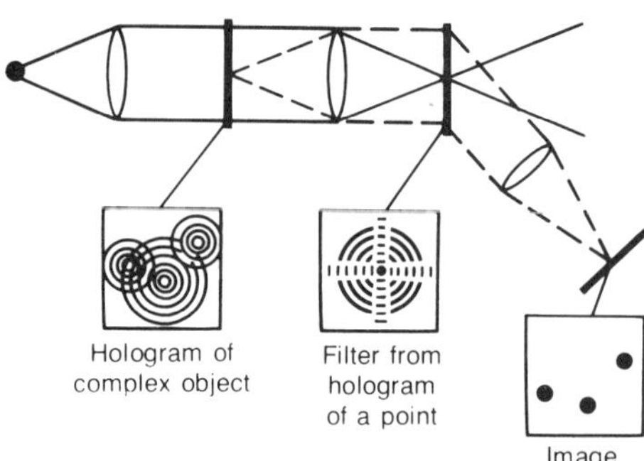

Fig. 50 Forming an image from a CW hologram using a Vander Lugt filter. Source: Ref 54

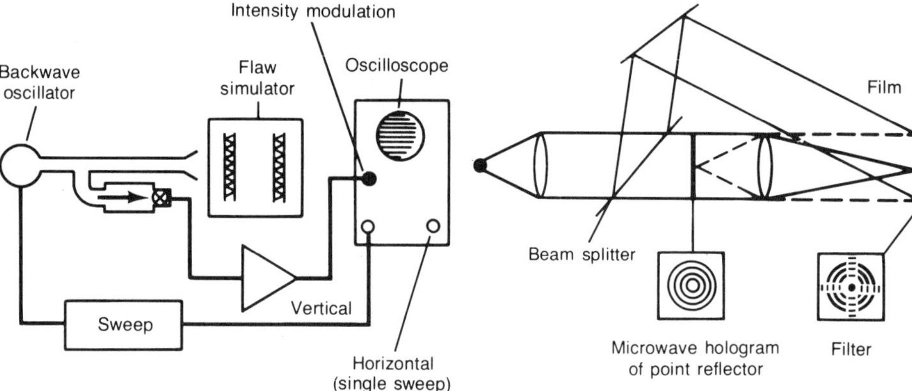

Fig. 48 Optical recording of A-scan data. Source: Ref 47

Fig. 49 Making of a Vander Lugt filter for image formation with CW holograms. Source: Ref 54

arrangement, along with three simulated delaminations, superficially resembles a side-looking radar system, but it does not have side-looking radar presentation, and it is further restricted to plane delaminations. Cylindrical flaws were placed in the simulator, but their round shape was not resolved.

To convert optically processed B-scans to swept-frequency, side-looking radar scans, all that is needed is to perform the optical processing in two dimensions (Ref 54). The scan of the antenna in this case synthesizes a hologram similar to that formed by an incident plane wave and the spherical waves reflected from the object. The incident wave is properly synthesized and serves as the reference beam of the hologram. The object beam travels twice the path from the antenna to the object. This is also true of airborne side-looking radars. The only difference is the way in which depth resolution is obtained. The hologram is composed of one-dimensional interference patterns produced along the scan direction. Their location in the other dimension is determined by the distance from the antenna to the object. The spacing of the interference fringes is also a function of distance, and this causes a tilt to the image plane. In airborne radars, a conical lens is used to make a correction so that all of the object points are imaged in the focal plane.

Fortunately, there is another approach to image formation from this type of hologram. The first step is to produce a Vander Lugt filter. The hologram of a point is made and used to make the Vander Lugt filter (Ref 54). On the hologram, an object point is represented by an interference point. If the pattern is recognizable, the object point that produced it can be identified. The Vander Lugt filter is made with the processor illustrated in Fig. 49. When the filter made in this manner is placed in Fig. 50, the correctly reconstructed image is formed. The filter is used with the processor shown in Fig. 49, which was used to make the filter but without the reference beam inserted by the beam splitter in Fig. 49. The forming of an image from a CW hologram using the Vander Lugt filter is accomplished with the equipment shown in Fig. 50. The processor shown in Fig. 50 will mark a spot on the output frame for each piece of the hologram contained in the input.

REFERENCES

1. *Electromagnetic Testing*, Vol 4, 2nd ed., *Nondestructive Testing Handbook*, American Society for Nondestructive Testing, 1986
2. H.E. Bussey, Standards and Measurements of Microwave Surface Impedance, Skin Depth, Conductivity, and Q, *IRE Trans. Instrum.*, Vol 1-9, Sept 1960, p 171-175
3. A. Harvey, *Microwave Engineering*, Academic Press, 1963
4. R.P. Dooley, X-Band Holography, *Proc. IEEE*, Vol 53 (No. 11), Nov 1965, p 1733-1735
5. W.E. Kock, A Photographic Method for Displaying Sound Wave and Microwave Space Patterns, *Bell Syst. Tech. J.*, Vol 30, July 1951, p 564-587
6. E.N. Leith and J. Upatnieks, Photography by Laser, *Sci. Am.*, Vol 212 (No. 6), June 1965, p 24
7. G.W. Stroke, *An Introduction to Current Optics and Holography*, Academic Press, 1966

8. G.A. Deschamps, Some Remarks on Radiofrequency Holography, *Proc. IEEE*, Vol 55 (No. 4), April 1967, p 570
9. G.W. McDaniel and D.Z. Robinson, Thermal Imaging by Means of the Evapograph, *Appl. Opt.*, Vol 1, May 1962, p 311
10. P.H. Kock and H. Oertel, Microwave Thermography 6, *Proc. IEEE*, Vol 55 (No. 3), March 1967, p 416
11. H. Heislmair *et al.*, State of the Art of Solid-State and Tube Transmitters, *Microwaves and R.F.*, April 1983, p 119
12. H. Bierman, Microwave Tubes Reach New Power and Efficiency Levels, *Microwave J.*, Feb 1987, p 26-42
13. K.D. Gilbert and J.B. Sorci, Microwave Supercomponents Fulfill Expectations, *Microwave J.*, Nov 1983, p 67
14. E.C. Niehenke, Advanced Systems Need Supercomponents, *Microwave J.*, Nov 1983, p 24
15. W. Tsai, R. Gray, and A. Graziano, The Design of Supercomponents: High Density MIC Modules, *Microwave J.*, Nov 1983, p 81
16. R.J. Botsco, Nondestructive Testing of Plastic With Microwaves, Parts 1 and 2, *Plast. Des. Process.*, Nov and Dec 1968
17. M.W. Standart, A.D. Lucian, T.E. Eckert, and B.L. Lamb, "Development of Microwave NDT Inspection Techniques for Large Solid Propellant Rocket Motors," NAS7-544, Final Report 1117, Aerojet-General Corporation, June 1969
18. A.D. Lucian and R.W. Cribbs, The Development of Microwave NDT Technology for the Inspection of Nonmetallic Materials and Composites, in *Proceedings of the Sixth Symposium on Nondestructive Evaluation of Aerospace and Weapons Systems Components and Materials*, Western Periodicals Co., 1967, p 199-232
19. L. Feinstein and R.J. Hruby, Surface-Crack Detection by Microwave Methods, in *Proceedings of the Sixth Symposium on Nondestructive Evaluation of Aerospace and Weapons Systems Components and Materials*, Western Periodicals Co., 1967, p 92-106
20. F.H. Haynie, D.A. Vaughan, P.D. Frost, and W.K. Boyd, "A Fundamental Investigation of the Nature of Stress-Corrosion Cracking in Aluminum Alloys," AFML-TR-65-258, Air Force Materials Laboratory, Oct 1965
21. R.M.J. Cotterill, An Experimental Determination of the Electrical Resistivity of Dislocations in Aluminum, *Philos. Mag.*, Vol 8, Nov 1963, p 1937-1944
22. D. Nobili and L. Passari, Electrical Resistivity in Quenched Aluminum—Alumina Alloys, *J. Nucl. Mater.*, Vol 16, 1965, p 344-346
23. Z.S. Basinski, J.S. Dugdale, and A. Howie, The Electrical Resistivity of Dislocations, *Philos. Mag.*, Vol 8, 1963, p 1989
24. L.M. Clarebrough, M.E. Hargreaves, and M.H. Loretts, Stored Energy and Electrical Resistivity in Deformed Metals, *Philos. Mag.*, Vol 6, 1961, p 807
25. E.C. Jordan, *Electromagnetic Waves and Radiating Systems*, Prentice-Hall, 1950, p 237
26. J. Feinleib, Electroreflectance in Metals, *Phys. Rev. Lett.*, Vol 16 (No. 26), June 1966, p 1200-1202
27. A.J. Bahr, *Microwave Nondestructive Testing Methods*, Vol 1, *Nondestructive Testing and Tracts*, W.J. McGonnagle, Ed., Gordon & Breach, 1982, p 49-72
28. E.E. Collin, *Field Theory of Guided Waves*, McGraw-Hill, 1960, p 39-40
29. L. Feinstein and R.J. Hruby, Paper 68-321, presented at the AIAA/ASME 95th Structures, Structural Dynamics and Materials Conference, American Institute of Aeronautics and Astronautics/American Society of Mechanical Engineers, April 1968
30. R.J. Hruby and L. Feinstein, A Novel Nondestructive, Nonconducting Method of Measuring the Depth of Thin Slits and Cracks in Metals, *Rev. Sci. Instrum.*, Vol 41, May 1970, p 679-683
31. A.J. Bahr, Microwave Eddy-Current Techniques for Quantitative Non-Destructive Evaluation, in *Eddy-Current Characterization of Materials and Structures*, STP 722, G. Birnbaum and G. Freed, Ed., American Society for Testing and Materials, 1981, p 311-331
32. L.A. Robinson and U.H. Gysel, "Microwave Coupled Stripline Surface Crack Detector," Final Report, Contract DAAG46-72-C-0019, SRI Project 1490, Stanford Research Institute, Aug 1972
33. U.H. Gysel and L. Feinstein, "Design and Fabrication of Stripline Microwave Surface-Crack Detector for Projectiles," Final Report, Contract DAAG46-73-C-0257, SRI Project 2821, Stanford Research Institute, Sept 1974
34. B.A. Auld, F. Muennemann, and D.K. Winslow, "Observation of Fatigue-Crack Closure Effects with the Ferromagnetic-Resonance Probes," G.L. Report 3233, E.L. Ginzton Laboratory, Stanford University, March 1981
35. B.A. Auld, *et al.*, Surface Flaw Detection With Ferromagnetic Resonance Probes, in *Proceedings of DARPA/AFML Review of Progress in Quantitative NDE* (La Jolla, CA), Defense Advanced Research Projects Agency, July 1980
36. B.A. Auld, New Methods of Detection and Characterization of Surface Flaws, in *Proceedings of DARPA/AFML Review of Progress in Quantitative NDE* (Cornell University), Defense Advanced Research Projects Agency, June 1977
37. B.A. Auld, *et al.*, Surface Flaw Detection With Ferromagnetic Resonance Probes, in *Proceedings of DARPA/AFML Review of Progress in Quantitative NDE* (La Jolla, CA), Defense Advanced Research Projects Agency, July 1978
38. B.A. Auld, *et al.*, Surface Flaw Detection With Ferromagnetic Resonance Probes, in *Proceedings of DARPA/AFML Review of Progress in Quantitative NDE* (La Jolla, CA), Defense Advanced Research Projects Agency, July 1979
39. B.A. Auld and D.K. Winslow, Microwave Eddy Current Experiments With Ferromagnetic Probes, in *Eddy Current Characterization of Material and Structures*, STP 722, G. Birnbaum and G. Free, Ed., American Society for Testing and Materials, 1981, p 348-366
40. A.F. Harvey, Properties and Applications of Gyromagnetic Media, in *Microwave Engineering*, Academic Press, 1963, p 352-358
41. S. Segaline, *et al.*, Application of Ferromagnetic Resonance Probes to the Characterization of Flaws in Metals, *J. Nondestr. Eval.*, Vol 4 (No. 2), 1984, p 51-58
42. W.E. Kock, Microwave Holography, in *Holographic Nondestructive Testing*, R.K. Erf, Ed., Academic Press, 1974, p 373-403
43. G.L. Rogers, Gabor Diffraction Microscopy: The Hologram as a Generalized Zone-Plate, *Nature*, Vol 166 (No. 4214), 1950, p 237
44. W.E. Kock and F.K. Harvey, Sound Wave and Microwave Space Patterns, *Bell Syst. Tech. J.*, Vol 20, July 1951, p 564
45. W.E. Kock, Hologram Television, *Proc. IEEE*, Vol 54 (No. 2), Feb 1966, p 331
46. G.A. Deschamps, Some Remarks on Radio-Frequency Holography, *Proc. IEEE*, Vol 55 (No. 4), April 1967, p 570
47. N.H. Farhat, Microwave Holography and Its Applications in Modern Aviation, in *Proceedings of the Engineering Applications of Holography* (Los Angeles), Defense Advanced Research Projects Agency, 1972, p 295-314
48. J.R. Maldonado and A.H. Meitzler, Strain-Biased Ferroelectric Photoconductor Image Storage and Display Devices, *Proc. IEEE*, Vol 59 (No. 3), March 1971
49. N.K. Sheridon, "A New Optical Recording Device," Paper presented at the 1970 IEEE International Electron Devices meeting, Washington, Institute of Electrical Engineers, Oct 1970
50. G. Assouline, *et al.*, Liquid Crystal and Photoconductor Image Converter,

Proc. IEEE, Vol 59, Sept 1971, p 1355-1357

51. T.D. Beard, "Photoconductor Light-Gated Crystal Used for Optical Data," Paper presented at the 1971 Annual Fall Meeting, Ottawa, Canada, Optical Society of America, Oct 1971
52. R.B. MacAnally, Liquid Crystal Displays for Matched Filtering, *Appl. Phys. Lett.*, Vol 18 (No. 2), 15 Jan 1971
53. G.H. Heilmeier, Liquid Crystal Display Devices, *Sci. Am.*, Vol 222 (No. 4), April 1970, p 100
54. R.W. Cribbs and B.L. Lamb, Resolution of Defects by Microwave Holography, in *Proceedings of the Symposium on Engineering Applications of Holography*, Defense Advanced Research Projects Agency, 1972, p 315-323

Ultrasonic Inspection

Revised by Yoseph Bar-Cohen, Douglas Aircraft Company, McDonnell Douglas Corporation; Ajit K. Mal, University of California, Los Angeles; and the ASM Committee on Ultrasonic Inspection*

ULTRASONIC INSPECTION is a nondestructive method in which beams of high-frequency sound waves are introduced into materials for the detection of surface and subsurface flaws in the material. The sound waves travel through the material with some attendant loss of energy (attenuation) and are reflected at interfaces. The reflected beam is displayed and then analyzed to define the presence and location of flaws or discontinuities.

The degree of reflection depends largely on the physical state of the materials forming the interface and to a lesser extent on the specific physical properties of the material. For example, sound waves are almost completely reflected at metal/gas interfaces. Partial reflection occurs at metal/liquid or metal/solid interfaces, with the specific percentage of reflected energy depending mainly on the ratios of certain properties of the material on opposing sides of the interface.

Cracks, laminations, shrinkage cavities, bursts, flakes, pores, disbonds, and other discontinuities that produce reflective interfaces can be easily detected. Inclusions and other inhomogeneities can also be detected by causing partial reflection or scattering of the ultrasonic waves or by producing some other detectable effect on the ultrasonic waves.

Most ultrasonic inspection instruments detect flaws by monitoring one or more of the following:

- Reflection of sound from interfaces consisting of material boundaries or discontinuities within the metal itself
- Time of transit of a sound wave through the testpiece from the entrance point at the transducer to the exit point at the transducer
- Attenuation of sound waves by absorption and scattering within the testpiece
- Features in the spectral response for either a transmitted or a reflected signal

Most ultrasonic inspection is done at frequencies between 0.1 and 25 MHz—well above the range of human hearing, which is about 20 Hz to 20 kHz. Ultrasonic waves are mechanical vibrations; the amplitudes of vibrations in metal parts being ultrasonically inspected impose stresses well below the elastic limit, thus preventing permanent effects on the parts. Many of the characteristics described in this article for ultrasonic waves, especially in the section "General Characteristics of Ultrasonic Waves," also apply to audible sound waves and to wave motion in general.

Ultrasonic inspection is one of the most widely used methods of nondestructive inspection. Its primary application in the inspection of metals is the detection and characterization of internal flaws; it is also used to detect surface flaws, to define bond characteristics, to measure the thickness and extent of corrosion, and (much less frequently) to determine physical properties, structure, grain size, and elastic constants.

Basic Equipment

Most ultrasonic inspection systems include the following basic equipment:

- An electronic signal generator that produces bursts of alternating voltage (a negative spike or a square wave) when electronically triggered
- A transducer (probe or search unit) that emits a beam of ultrasonic waves when bursts of alternating voltage are applied to it
- A couplant to transfer energy in the beam of ultrasonic waves to the testpiece
- A couplant to transfer the output of ultrasonic waves (acoustic energy) from the testpiece to the transducer
- A transducer (can be the same as the transducer initiating the sound or it can be a separate one) to accept and convert the output of ultrasonic waves from the testpiece to corresponding bursts of alternating voltage. In most systems, a single transducer alternately acts as sender and receiver
- An electronic device to amplify and, if necessary, demodulate or otherwise modify the signals from the transducer
- A display or indicating device to characterize or record the output from the testpiece. The display device may be a CRT, sometimes referred to as an oscilloscope; a chart or strip recorder; a marker, indicator, or alarm device; or a computer printout
- An electronic clock, or timer, to control the operation of the various components of the system, to serve as a primary reference point, and to provide coordination for the entire system

Advantages and Disadvantages

The principal advantages of ultrasonic inspection as compared to other methods for nondestructive inspection of metal parts are:

- Superior penetrating power, which allows the detection of flaws deep in the part. Ultrasonic inspection is done routinely to thicknesses of a few meters on many types of parts and to thicknesses of about 6 m (20 ft) in the axial inspection of parts such as long steel shafts or rotor forgings
- High sensitivity, permitting the detection of extremely small flaws
- Greater accuracy than other nondestructive methods in determining the position of internal flaws, estimating their size, and characterizing their orientation, shape, and nature
- Only one surface needs to be accessible
- Operation is electronic, which provides almost instantaneous indications of flaws. This makes the method suitable for immediate interpretation, automation, rapid scanning, in-line production monitoring, and process control. With most systems, a permanent record of inspection results can be made for future reference
- Volumetric scanning ability, enabling the inspection of a volume of metal extending

*Charles J. Hellier, *Chairman*, Hellier Associates, Inc.; William Plumstead, Bechtel Corporation; Kenneth Fowler, Panametrics, Inc.; Robert Grills, Glenn Andrews, and Mike C. Tsao, Ultra Image International; James J. Snyder, Westinghouse Electric Corporation, Oceanic Division; and J.F. Cook and D.A. Aldrich, Idaho National Engineering Laboratory, EG&G Idaho, Inc.; Robert W. Pepper, Textron Specialty Materials

from front surface to back surface of a part
- Nonhazardous to operations or to nearby personnel and has no effect on equipment and materials in the vicinity
- Portability
- Provides an output that can be processed digitally by a computer to characterize defects and to determine material properties

The disadvantages of ultrasonic inspection include the following:

- Manual operation requires careful attention by experienced technicians
- Extensive technical knowledge is required for the development of inspection procedures
- Parts that are rough, irregular in shape, very small or thin, or not homogeneous are difficult to inspect
- Discontinuities that are present in a shallow layer immediately beneath the surface may not be detectable
- Couplants are needed to provide effective transfer of ultrasonic wave energy between transducers and parts being inspected
- Reference standards are needed, both for calibrating the equipment and for characterizing flaws

Applicability

The ultrasonic inspection of metals is principally conducted for the detection of discontinuities. This method can be used to detect internal flaws in most engineering metals and alloys. Bonds produced by welding, brazing, soldering, and adhesive bonding can also be ultrasonically inspected. In-line techniques have been developed for monitoring and classifying material as acceptable, salvageable, or scrap and for process control. Both line-powered and battery-operated commercial equipment is available, permitting inspection in shop, laboratory, warehouse, or field.

Ultrasonic inspection is used for quality control and materials inspection in all major industries. This includes electrical and electronic component manufacturing; production of metallic and composite materials; and fabrication of structures such as airframes, piping and pressure vessels, ships, bridges, motor vehicles, machinery, and jet engines. In-service ultrasonic inspection for preventive maintenance is used for detecting the impending failure of railroad-rolling-stock axles, press columns, earthmoving equipment, mill rolls, mining equipment, nuclear systems, and other machines and components.

Some of the major types of equipment that are ultrasonically inspected for the presence of flaws are:

- *Mill components*: Rolls, shafts, drives, and press columns
- *Power equipment*: Turbine forgings, generator rotors, pressure piping, weldments, pressure vessels, nuclear fuel elements, and other reactor components
- *Jet engine parts*: Turbine and compressor forgings, and gear blanks
- *Aircraft components*: Forging stock, frame sections, and honeycomb sandwich assemblies
- *Machinery materials*: Die blocks, tool steels, and drill pipe
- *Railroad parts*: Axles, wheels, track, and welded rail
- *Automotive parts*: Forgings, ductile castings, and brazed and/or welded components

The flaws to be detected include voids, cracks, inclusions, pipe, laminations, debonding, bursts, and flakes. They may be inherent in the raw material, may result from fabrication and heat treatment, or may occur in service from fatigue, impact, abrasion, corrosion, or other causes.

Government agencies and standards-making organizations have issued inspection procedures, acceptance standards, and related documentation. These documents are mainly concerned with the detection of flaws in specific manufactured products, but they also can serve as the basis for characterizing flaws in many other applications.

Ultrasonic inspection can also be used to measure the thickness of metal sections. Thickness measurements are made on refinery and chemical-processing equipment, shop plate, steel castings, submarine hulls, aircraft sections, and pressure vessels. A variety of ultrasonic techniques are available for thickness measurements; several use instruments with digital readout. Structural material ranging in thickness from a few thousandths of an inch to several feet can be measured with accuracies of better than 1%. Ultrasonic inspection methods are particularly well suited to the assessment of loss of thickness from corrosion inside closed systems, such as chemical-processing equipment. Such measurements can often be made without shutting down the process. Special ultrasonic techniques and equipment have been used on such diverse problems as the rate of growth of fatigue cracks, detection of borehole eccentricity, measurement of elastic moduli, study of press fits, determination of nodularity in cast iron, and metallurgical research on phenomena such as structure, hardening, and inclusion count in various metals.

For the successful application of ultrasonic inspection, the inspection system must be suitable for the type of inspection being done, and the operator must be sufficiently trained and experienced. If either of these prerequisites is not met, there is a high potential for gross error in inspection results. For example, with inappropriate equipment or with a poorly trained operator, discontinuities having little or no bearing on product performance may be deemed serious, or damaging discontinuities may go undetected or be deemed unimportant.

The term flaw as used in this article means a detectable lack of continuity or an imperfection in a physical or dimensional attribute of a part. The fact that a part contains one or more flaws does not necessarily imply that the part is nonconforming to specification nor unfit for use. It is important that standards be established so that decisions to accept or reject parts are based on the probable effect that a given flaw will have on service life or product safety. Once such standards are established, ultrasonic inspection can be used to characterize flaws in terms of a real effect rather than some arbitrary basis that may impose useless or redundant quality requirements.

General Characteristics of Ultrasonic Waves

Ultrasonic waves are mechanical waves (in contrast to, for example, light or x-rays, which are electromagnetic waves) that consist of oscillations or vibrations of the atomic or molecular particles of a substance about the equilibrium positions of these particles. Ultrasonic waves behave essentially the same as audible sound waves. They can propagate in an elastic medium, which can be solid, liquid, or gaseous, but not in a vacuum.

In many respects, a beam of ultrasound is similar to a beam of light; both are waves and obey a general wave equation. Each travels at a characteristic velocity in a given homogeneous medium—a velocity that depends on the properties of the medium, not on the properties of the wave. Like beams of light, ultrasonic beams are reflected from surfaces, refracted when they cross a boundary between two substances that have different characteristic sound velocities, and diffracted at edges or around obstacles. Scattering by rough surfaces or particles reduces the energy of an ultrasonic beam, comparable to the manner in which scattering reduces the intensity of a light beam.

Analogy With Waves in Water. The general characteristics of sonic or ultrasonic waves are conveniently illustrated by analogy with the behavior of waves produced in a body of water when a stone is dropped into it. Casual observation might lead to the erroneous conclusion that the resulting outward radial travel of alternate crests and troughs represents the movement of water away from the point of impact. The fact that water is not thus transported is readily deduced from the observation that a small object floating on the water does not move away from the point of impact, but instead merely bobs up and down. The waves travel outward only in the sense that the crests and troughs (which can be compared to the compressions and rarefactions of sonic waves in an elastic medium) and the energy associated with the waves propagate radially outward. The water particles remain in place and oscillate up and down from their normal positions of rest.

Continuing the analogy, the distance between two successive crests or troughs is the wavelength, λ. The fall from a crest to a trough and subsequent rise to the next crest (which is accomplished within this distance) is a cycle. The number of cycles in a specific unit of time is the frequency, f, of the waves. The height of a crest or the depth of a trough in relation to the surface at equilibrium is the amplitude of the waves.

The velocity of a wave and the rates at which the amplitude and energy of a wave decrease as it propagates are constants that are characteristic of the medium in which the wave is propagating. Stones of equal size and mass striking oil and water with equal force will generate waves that travel at different velocities. Stones impacting a given medium with greater energy will generate waves having greater amplitude and energy but the same wave velocity.

The above attributes apply similarly to sound waves, both audible and ultrasonic, propagating in an elastic medium. The particles of the elastic medium move, but they do not migrate from their initial spacial orbits; only the energy travels through the medium. The amplitude and energy of sound waves in the elastic medium depend on the amount of energy supplied. The velocity and attenuation (loss of amplitude and energy) of the sound waves depend on the properties of the medium in which they are propagating.

Wave Propagation. Ultrasonic waves (and other sound waves) propagate to some extent in any elastic material. When the atomic or molecular particles of an elastic material are displaced from their equilibrium positions by any applied force, internal stress acts to restore the particles to their original positions. Because of the interatomic forces between adjacent particles of material, a displacement at one point induces displacements at neighboring points and so on, thus propagating a stress-strain wave. The actual displacement of matter that occurs in ultrasonic waves is extremely small. The amplitude, vibration mode, and velocity of the waves differ in solids, liquids, and gases because of the large differences in the mean distance between particles in these forms of matter. These differences influence the forces of attraction between particles and the elastic behavior of the materials.

The concepts of wavelength, cycle, frequency, amplitude, velocity, and attenuation described in the preceding section "Analogy With Waves in Water" in this article apply in general to ultrasonic waves and other sound waves. The relation of velocity to frequency and wavelength is given by:

$$V = f\lambda \qquad \text{(Eq 1)}$$

where V is velocity (in meters per second), f is frequency (in hertz), and λ is wavelength (in meters per cycle). Other consistent units of measure can be used for the variables in Eq 1, where convenient.

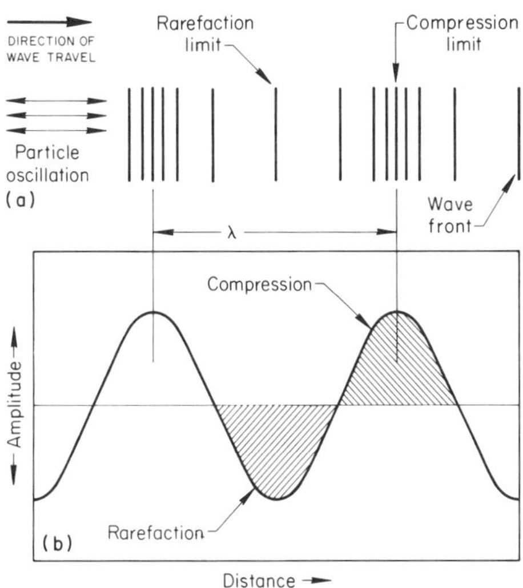

Fig. 1 Schematic of longitudinal ultrasonic waves. (a) Particle oscillation and resultant rarefaction and compression. (b) Amplitude of particle displacement versus distance of wave travel. The wavelength, λ, is the distance corresponding to one complete cycle.

On the basis of the mode of particle displacement, ultrasonic waves are classified as longitudinal waves, transverse waves, surface waves, and Lamb waves. These four types of waves are described in the following sections.

Longitudinal waves, sometimes called compression waves, are the type of ultrasonic waves most widely used in the inspection of materials. These waves travel through materials as a series of alternate compressions and rarefactions in which the particles transmitting the wave vibrate back and forth in the direction of travel of the waves.

Longitudinal ultrasonic waves and the corresponding particle oscillation and resultant rarefaction and compression are shown schematically in Fig. 1(a); a plot of amplitude of particle displacement versus distance of wave travel, together with the resultant rarefaction trough and compression crest, is shown in Fig. 1(b). The distance from one crest to the next (which equals the distance for one complete cycle of rarefaction and compression) is the wavelength, λ. The vertical axis in Fig. 1(b) could represent pressure instead of particle displacement. The horizontal axis could represent time instead of travel distance because the speed of sound is constant in a given material and because this relation is used in the measurements made in ultrasonic inspection.

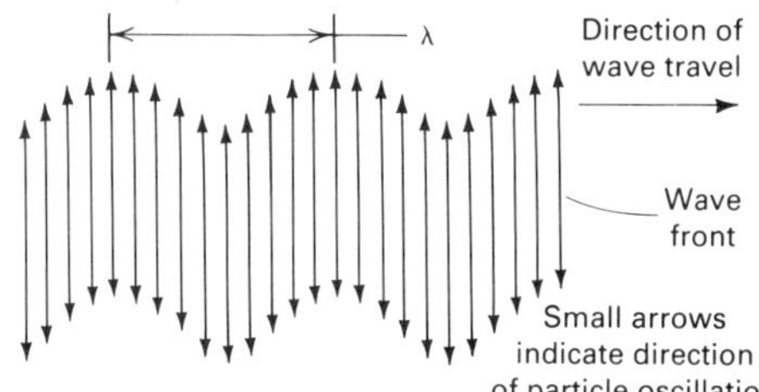

Fig. 2 Schematic of transverse (shear) waves. The wavelength, λ, is the distance corresponding to one complete cycle.

Longitudinal ultrasonic waves are readily propagated in liquids and gases as well as in elastic solids. The mean free paths of the molecules of liquids and gases at a pressure of 1 atm are so short that longitudinal waves can be propagated simply by the elastic collision of one molecule with the next. The velocity of longitudinal ultrasonic waves is about 6000 m/s (20 000 ft/s) in steel, 1500 m/s (5000 ft/s) in water, and 330 m/s (1080 ft/s) in air.

Transverse waves (shear waves) are also extensively used in the ultrasonic inspection of materials. Transverse waves are visualized readily in terms of vibrations of a rope that is shaken rhythmically, in which each particle, rather than vibrating parallel to the direction of wave motion as in the longitudinal wave, vibrates up and down in a plane perpendicular to the direction of propagation. A transverse wave is illustrated schematically in Fig. 2, which shows particle oscillation, wave front, direction of wave travel, and the wavelength, λ, corresponding to one cycle.

Unlike longitudinal waves, transverse waves cannot be supported by the elastic collision of adjacent molecular or atomic particles. For the propagation of transverse waves, it is necessary that each particle exhibit a strong force of attraction to its neighbors so that as a particle moves back and forth it pulls its neighbor with it, thus causing the sound to move through the material with the velocity associated with transverse waves, which is about 50% of the longitudinal wave velocity for the same material.

Air and water will not support transverse waves. In gases, the forces of attraction between molecules are so small that shear waves cannot be transmitted. The same is true of a liquid, unless it is particularly viscous or is present as a very thin layer.

Surface waves (Rayleigh waves) are another type of ultrasonic wave used in the inspection of materials. These waves travel along the flat or curved surface of relatively thick solid parts. For the propagation of waves of this type, the waves must be

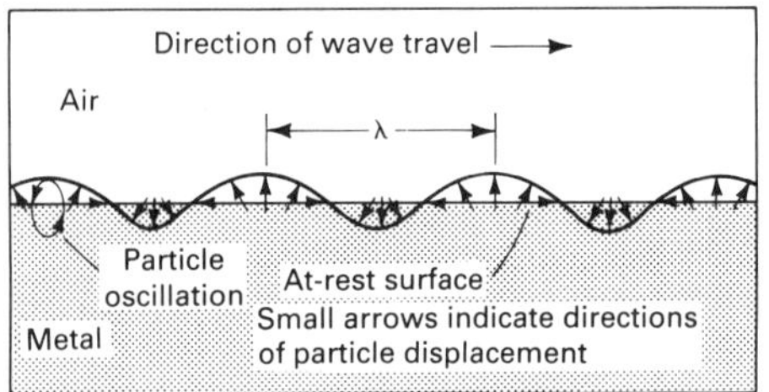

Fig. 3 Diagram of surface (Rayleigh) waves propagating at the surface of a metal along a metal/air interface. The wavelength, λ, is the distance corresponding to one complete cycle.

traveling along an interface bounded on one side by the strong elastic forces of a solid and on the other side by the practically negligible elastic forces between gas molecules. Surface waves leak energy into liquid couplants and do not exist for any significant distance along the surface of a solid immersed in a liquid, unless the liquid covers the solid surface only as a very thin film.

Surface waves are subject to attenuation in a given material, as are longitudinal or transverse waves. They have a velocity approximately 90% of the transverse wave velocity in the same material. The region within which these waves propagate with effective energy is not much thicker than about one wavelength beneath the surface of the metal. At this depth, wave energy is about 4% of the wave energy at the surface, and the amplitude of oscillation decreases sharply to a negligible value at greater depths.

Surface waves follow contoured surfaces. For example, surface waves traveling on the top surface of a metal block are reflected from a sharp edge, but if the edge is rounded off, the waves continue down the side face and are reflected at the lower edge, returning to the sending point. Surface waves will travel completely around a cube if all edges of the cube are rounded off. Surface waves can be used to inspect parts that have complex contours.

In surface waves, particle oscillation generally follows an elliptical orbit, as shown schematically in Fig. 3. The major axis of the ellipse is perpendicular to the surface along which the waves are traveling. The minor axis is parallel to the direction of propagation. surface waves can exist in complex forms that are variations of the simplified waveform illustrated in Fig. 3.

Lamb waves, also known as plate waves, are another type of ultrasonic wave used in the nondestructive inspection of materials. Lamb waves are propagated in plates (made of composites or metals) only a few wavelengths thick. A Lamb wave consists of a complex vibration that occurs throughout the thickness of the material. The propagation characteristics of Lamb waves depend on the density, elastic properties, and structure of the material as well as the thickness of the testpiece and the frequency. Their behavior in general resembles that observed in the transmission of electromagnetic waves through waveguides.

There are two basic forms of Lamb waves:

- Symmetrical, or dilatational
- Asymmetrical, or bending

The form is determined by whether the particle motion is symmetrical or asymmetrical with respect to the neutral axis of the testpiece. Each form is further subdivided into several modes having different velocities, which can be controlled by the angle at which the waves enter the testpiece. Theoretically, there are an infinite number of specific velocities at which Lamb waves can travel in a given material. Within a given plate, the specific velocities for Lamb waves are complex functions of plate thickness and frequency. The specific velocities of Lamb waves are discussed in Ref 1 and 2.

In symmetrical (dilatational) Lamb waves, there is a compressional (longitudinal) particle displacement along the neutral axis of the plate and an elliptical particle displacement on each surface (Fig. 4a). In asymmetrical (bending) Lamb waves, there is a shear (transverse) particle displacement along the neutral axis of the plate and an elliptical particle displacement on each surface (Fig. 4b). The ratio of the major to minor axes of the ellipse is a function of the material in which the wave is being propagated.

Major Variables in Ultrasonic Inspection

The major variables that must be considered in ultrasonic inspection include both the characteristics of the ultrasonic waves used and the characteristics of the parts being inspected. Equipment type and capability interact with these variables; often, different types of equipment must be selected to accomplish different inspection objectives.

The frequency of the ultrasonic waves used affects inspection capability in several ways. Generally, a compromise must be made between favorable and adverse effects to achieve an optimum balance and to overcome the limitations imposed by equipment and test material.

Sensitivity, or the ability of an ultrasonic inspection system to detect a very small discontinuity, is generally increased by using relatively high frequencies (short wavelengths). Resolution, or the ability of the system to give simultaneous, separate indications from discontinuities that are close together both in depth below the front surface of the testpiece and in lateral position, is directly proportional to frequency bandwidth and inversely related to pulse length. Resolution generally improves with an increase of frequency.

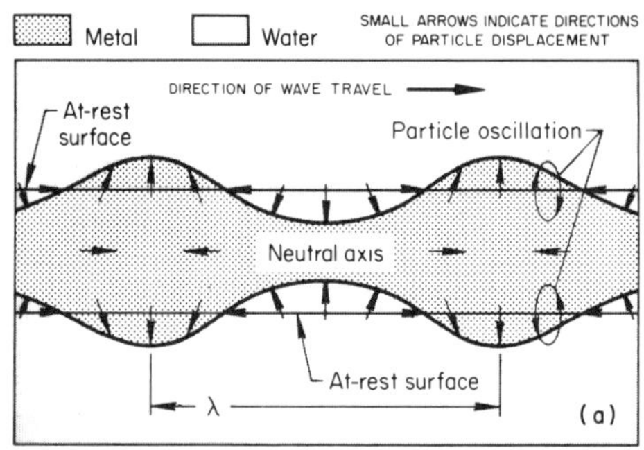

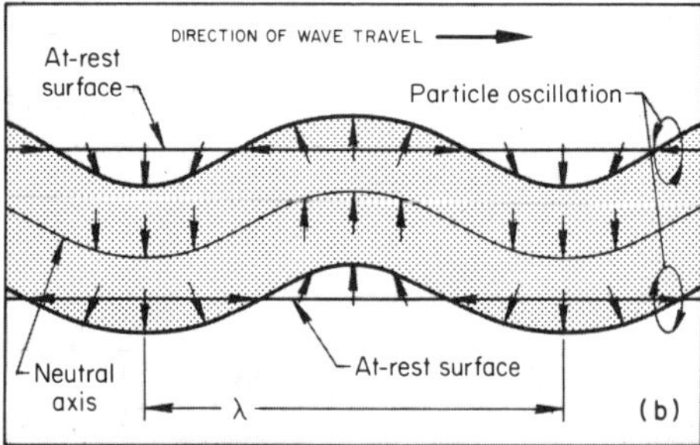

Fig. 4 Diagram of the basic patterns of (a) symmetrical (dilatational) and (b) asymmetrical (bending) Lamb waves. The wavelength, λ, is the distance corresponding to one complete cycle.

Penetration, or the maximum depth (range) in a material from which useful indications can be detected, is reduced by the use of high frequencies. This effect is most pronounced in the inspection of metal that has coarse grain structure or minute inhomogeneities, because of the resultant scattering of the ultrasonic waves; it is of little consequence in the inspection of fine-grain, homogeneous metal.

Beam spread, or the divergence of an ultrasonic beam from the central axis of the beam, is also affected by frequency. As frequency decreases, the shape of an ultrasonic beam increasingly departs from the ideal of zero beam spread. This characteristic is so pronounced as to be observed at almost all frequencies used in inspection. Other factors, such as the transducer (search unit) diameter and the use of focusing equipment, also affect beam spread; these are discussed in greater detail in the sections "Beam Spreading" and "Acoustic Lenses" in this article. Sensitivity, resolution, penetration, and beam spread are largely determined by the selection of the transducer and are only slightly modified by changes in other test variables.

Acoustic Impedance. When ultrasonic waves traveling through one medium impinge on the boundary of a second medium, a portion of the incident acoustic energy is reflected back from the boundary while the remaining energy is transmitted into the second medium. The characteristic that determines the amount of reflection is the acoustic impedance of the two materials on either side of the boundary. If the impedances of the two materials are equal, there

Table 1 Acoustic properties of several metals and nonmetals

Material	Density (ρ), g/cm³	Sonic velocities, 10^5 cm/s V_l(a)	V_t(b)	V_s(c)	Acoustic impedance (Z_l), 10^6 g/cm² · s(d)
Ferrous metals					
Carbon steel, annealed	7.85	5.94	3.24	3.0	4.66
Alloy steel					
Annealed	7.86	5.95	3.26	3.0	4.68
Hardened	7.8	5.90	3.23	. . .	4.6
Cast iron	6.95–7.35	3.5–5.6	2.2–3.2	. . .	2.5–4.0
52100 steel					
Annealed	7.83	5.99	3.27	. . .	4.69
Hardened	7.8	5.89	3.20	. . .	4.6
D6 tool steel					
Annealed	7.7	6.14	3.31	. . .	4.7
Hardened	7.7	6.01	3.22	. . .	4.6
Stainless steels					
Type 302	7.9	5.66	3.12	3.12	4.47
Type 304L	7.9	5.64	3.07	. . .	4.46
Type 347	7.91	5.74	3.10	2.8	4.54
Type 410	7.67	5.39	2.99	2.16	4.13
Type 430	7.7	6.01	3.36	. . .	4.63
Nonferrous metals					
Aluminum 1100-O	2.71	6.35	3.10	2.90	1.72
Aluminum alloy 2117-T4	2.80	6.25	3.10	2.79	1.75
Beryllium	1.85	12.80	8.71	7.87	2.37
Copper 110	8.9	4.70	2.26	1.93	4.18
Copper alloys					
260 (cartridge brass, 70%)	8.53	3.83	2.05	1.86	3.27
464 to 467 (naval brass)	8.41	4.43	2.12	1.95	3.73
510 (phosphor bronze, 5% A)	8.86	3.53	2.23	2.01	3.12
752 (nickel silver 65-18)	8.75	4.62	2.32	1.69	4.04
Lead					
Pure	11.34	2.16	0.70	0.64	2.45
Hard (94Pb-6Sb)	10.88	2.16	0.81	0.73	2.35
Magnesium alloy M1A	1.76	5.74	3.10	2.87	1.01
Mercury, liquid	13.55	1.45	. . .	. . .	1.95
Molybdenum	10.2	6.25	3.35	3.11	6.38
Nickel					
Pure	8.8	5.63	2.96	2.64	4.95
Inconel	8.5	5.82	3.02	2.79	4.95
Inconel X-750	8.3	5.94	3.12	. . .	4.93
Monel	8.83	5.35	2.72	2.46	4.72
Titanium, commercially pure	4.5	6.10	3.12	2.79	2.75
Tungsten	19.25	5.18	2.87	2.65	9.98
Nonmetals					
Air(e)	0.00129	0.331	. . .	. . .	0.00004
Ethylene glycol	1.11	1.66	. . .	. . .	0.18
Glass					
Plate	2.5	5.77	3.43	3.14	1.44
Pyrex	2.23	5.57	3.44	3.13	1.24
Glycerin	1.26	1.92	. . .	. . .	0.24
Oil					
Machine (SAE 20)	0.87	1.74	. . .	. . .	0.150
Transformer	0.92	1.38	. . .	. . .	0.127
Paraffin wax	0.9	2.2	. . .	. . .	0.2
Plastics					
Methylmethacrylate (Lucite, Plexiglas)	1.18	2.67	1.12	1.13	0.32
Polyamide (nylon)	1.0–1.2	1.8–2.2	. . .	. . .	0.18–0.27
Polytetrafluoroethylene (Teflon)	2.2	1.35	. . .	. . .	0.30
Quartz, natural	2.65	5.73	. . .	. . .	1.52
Rubber, vulcanized	1.1–1.6	2.3	. . .	. . .	0.25–0.37
Tungsten carbide	10–15	6.66	3.98	. . .	6.7–9.9
Water					
Liquid(f)	1.0	1.49	. . .	. . .	0.149
Ice(g)	0.9	3.98	1.99	. . .	0.36

(a) Longitudinal (compression) waves. (b) Transverse (shear) waves. (c) Surface waves. (d) For longitudinal waves $Z_l = \rho V_l$. (e) At standard temperature and pressure. (f) At 4 °C (39 °F). (g) At 0 °C (32 °F)

will be no reflection; if the impedances differ greatly (as between a metal and air, for example), there will be virtually complete reflection. This characteristic is used in the ultrasonic inspection of metals to calculate the amounts of energy reflected and transmitted at impedance discontinuities and to aid in the selection of suitable materials for the effective transfer of acoustic energy between components in ultrasonic inspection systems.

The acoustic impedance for a longitudinal wave, Z_l, given in grams per square centimeter-second, is defined as the product of material density, ρ, given in grams per cubic centimeter, and longitudinal wave velocity, V_l, given in centimeters per second:

$$Z_l = \rho V_l \qquad \text{(Eq 2)}$$

The acoustic properties of several metals and nonmetals are listed in Table 1. The acoustic properties of metals and alloys are influenced by variations in structure and metallurgical condition. Therefore, for a given testpiece the properties may differ somewhat from the values listed in Table 1.

The percentage of incident energy reflected from the interface between two materials depends on the ratio of acoustic impedances (Z_2/Z_1) and the angle of incidence. When the angle of incidence is 0° (normal incidence), the reflection coefficient, R, which is the ratio of reflected beam intensity, I_r, to incident beam intensity, I_i, is given by:

$$R = I_r/I_i = [(Z_2 - Z_1)/(Z_2 + Z_1)]^2 = [(r - 1)/(r + 1)]^2 \qquad \text{(Eq 3)}$$

where Z_1 is the acoustic impedance of medium 1, Z_2 is the acoustic impedance of medium 2, and r equals Z_2/Z_1 and is the impedance ratio, or mismatch factor. With T designating the transmission coefficient, $R + T = 100\%$, because all the energy is either reflected or transmitted, and T is simply obtained from this relation.

The transmission coefficient, T, can also be calculated as the ratio of the intensity of the transmitted beam, I_t, to that of the incident beam, I_i, from:

$$T = I_t/I_i = 4Z_2Z_1/(Z_2 + Z_1)^2 = 4r/(r + 1)^2 \qquad \text{(Eq 4)}$$

When a longitudinal ultrasonic wave in water (medium 1) is incident at right angles to the surface of an aluminum alloy 1100 testpiece (medium 2), the percentages of acoustic energy reflected and transmitted are calculated as shown below (the calculations are based on data from Table 1):

Impedance ratio (r) = $Z_2/Z_1 = 1.72/0.149 = 11.54$

Reflection coefficient (R) = $[(r - 1)/(r + 1)]^2 = (10.54/12.54)^2 = 0.71 = 71\%$

Transmission coefficient (T) = $1 - R = 0.29 = 29\%$

The same values are obtained for R and T when medium 1 is the aluminum alloy and medium 2 is water. For any pair of materials, reversing the order of the materials does not change the values of R and T.

Angle of Incidence. Only when an ultrasonic wave is incident at right angles on an interface between two materials (normal incidence; that is, angle of incidence = 0°) do transmission and reflection occur at the interface without any change in beam direction. At any other angle of incidence, the phenomena of mode conversion (a change in the nature of the wave motion) and

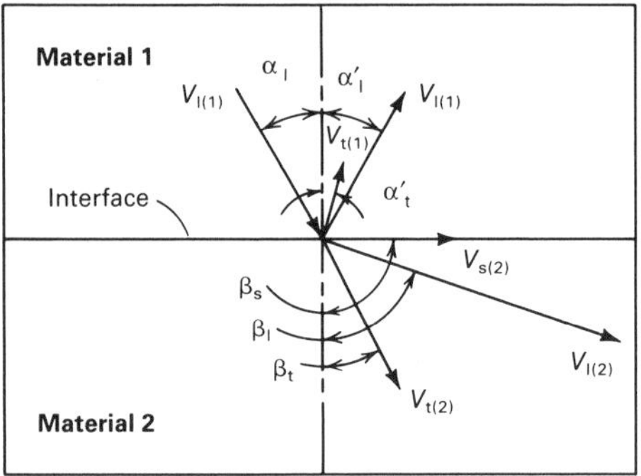

Fig. 5 Diagram showing relationship (by vectors) of all possible reflected and refracted waves to an incident longitudinal wave of velocity $V_{l(1)}$ impinging on an interface at angle α_l relative to normal to the interface. See text for explanation of vectors.

refraction (a change in direction of wave propagation) must be considered. These phenomena may affect the entire beam or only a portion of the beam, and the sum total of the changes that occur at the interface depends on the angle of incidence and the velocity of the ultrasonic waves leaving the point of impingement on the interface. All possible ultrasonic waves leaving this point are shown for an incident longitudinal ultrasonic wave in Fig. 5. Not all the waves shown in Fig. 5 will be produced in any specific instance of oblique impingement of an ultrasonic wave on the interface between two materials. The waves that propagate in a given instance depend on the ability of a waveform to exist in a given material, the angle of incidence of the initial beam, and the velocities of the waveforms in both materials.

The general law that describes wave behavior at an interface is known as Snell's law. Although originally derived for light waves, Snell's law applies to acoustic waves (including ultrasound) and to many other types of waves. According to Snell's law, the ratio of the sine of the angle of incidence to the sine of the angle of reflection or refraction equals the ratio of the corresponding wave velocities. Snell's law applies even if mode conversion takes place. Mathematically, Snell's law can be expressed as:

$$\sin \alpha / \sin \beta = V_1/V_2 \qquad \text{(Eq 5)}$$

where α is the angle of incidence, β is the angle of reflection or refraction, and V_1 and V_2 are the respective velocities of the incident and reflected or refracted waves. Both α and β are measured from a line normal to the interface.

Equation 6 is the general relationship applying to reflection and refraction, taking into account all possible effects of mode conversion for an incident longitudinal ultrasonic wave, as shown in Fig. 5:

$$\sin \alpha_l/V_{l(1)} = \sin \alpha_l'/V_{l(1)} = \sin \alpha_t'/V_{t(1)}$$
$$\sin \beta_l/V_{l(2)} = \sin \beta_t/V_{t(2)} \qquad \text{(Eq 6)}$$

where:

α_l is the angle of incidence for incident longitudinal wave in material 1,

α'_l is the angle of reflection for reflected longitudinal wave in material 1 = α_l,

α'_t is the angle of reflection for reflected transverse wave in material 1,

β_l is the angle of refraction for refracted longitudinal wave in material 2,

β_t is the angle of refraction for refracted transverse wave in material 2,

$V_{l(1)}$ is the velocity of incident longitudinal wave in material 1 = velocity of reflected longitudinal wave in material 1,

$V_{t(1)}$ is the velocity of reflected transverse wave in material 1,

$V_{l(2)}$ is the velocity of refracted longitudinal wave in material 2, and

$V_{t(2)}$ is the velocity of refracted transverse wave in material 2.

For quantities that are shown in Fig. 5 but do not appear in Eq 6,

β_s is the angle of refraction for refracted surface (Rayleigh) wave in material 2 = 90°, and $V_{s(2)}$ is the velocity of refracted surface (Rayleigh) wave in material 2. Equation 6 can apply to similar relationships for an incident transverse (instead of longitudinal) wave by substituting the term $\sin \alpha_t/V_{t(1)}$ for the first term, $\sin \alpha_l/V_{l(1)}$. Correspondingly, in Fig. 5, the incident longitudinal wave at angle α_l (with velocity $V_{l(1)}$ in material 1) would be replaced by an incident transverse angle α_t equal to α_t' (with velocity $V_{t(1)}$).

Critical Angles. If the angle of incidence (α_l, Fig. 5) is small, sound waves propagating in a given medium may undergo mode conversion at a boundary, resulting in the simultaneous propagation of longitudinal and transverse (shear) waves in a second medium. If the angle is increased, the direction of the refracted longitudinal wave will approach the plane of the boundary ($\beta_l \rightarrow 90°$). At some specific value of α_l, β_l will exactly equal 90°, above which the refracted longitudinal wave will no longer propagate in the material, leaving only a refracted (mode-converted) shear wave to propagate in the second medium. This value of α_l is known as the first critical angle. If α_l is increased beyond the first critical angle, the direction of the refracted shear wave will approach the plane of the boundary ($\beta_t \rightarrow 90°$). At a second specific value of α_l, β_t will exactly equal 90°, above which the refracted transverse wave will no longer propagate in the material. This second value of α_l is called the second critical angle.

Critical angles are of special importance in ultrasonic inspection. Values of α_l between the first and second critical angles are required for most angle-beam inspections. Surface wave inspection is accomplished by adjusting the incident angle of a contact-type search unit so that it is a few tenths of a degree greater than the second critical angle. At this value, the refracted shear wave in the bulk material is replaced by a Rayleigh wave traveling along the surface of the testpiece. As mentioned earlier in this article, Rayleigh waves can be effectively sustained only when the medium on one side of the interface (in this case, the surface of the testpiece) is a gas. Consequently, surface wave inspection is primarily used with contact methods.

In ordinary angle-beam inspection, it is usually desirable to have only a shear wave propagating in the test material. Because longitudinal waves and shear waves propagate at different speeds, echo signals will be received at different times, depending on which type of wave produced the echo. When both types are present in the test material, confusing echo patterns may be shown on the display device, which can lead to erroneous interpretations of testpiece quality. Frequently, it is desirable to produce shear waves in a material at an angle of 45° to the surface. In most materials, incident angles for mode conversion to a 45° shear wave lie between the first and second critical angles. Typical values of α_l for all three of these—first critical angle, second critical angle, and incident angle for mode conversion to 45° shear waves—are listed in Table 2 for various metals.

Table 2 Critical angles for immersion and contact testing, and incident angle for 45° shear wave transmission, in various metals

Metal	First critical angle, degrees(a), for: Immersion testing(b)	First critical angle, degrees(a), for: Contact testing(c)	Second critical angle, degrees(a), for: Immersion testing(b)	Second critical angle, degrees(a), for: Contact testing(c)	45° shear wave incident angle, degrees(a), for: Immersion testing(b)	45° shear wave incident angle, degrees(a), for: Contact testing(c)
Steel	14.5	26.5	27.5	55	19	35.5
Cast iron	15–25	28–50	...	...	...	...
Type 302 stainless steel	15	28	29	59	19.5	37
Type 410 stainless steel	11.5	21	30	63	20.5	39
Aluminum alloy 2117-T4	13.5	25	29	59.5	20	37.5
Beryllium	6.5	12	10	18	7	12.5
Copper alloy 260 (cartridge brass, 70%)	23	44	46.5	...	31	67
Inconel	11	20	30	62	20.5	38.5
Magnesium alloy M1A	15	27.5	29	59.5	20	37.5
Monel	16.5	30	33	79	23	44
Titanium	14	26	29	59	20	37

(a) Measured from a direction normal to surface of test material. (b) In water at 4 °C (39 °F). (c) Using angle block (wedge) made of acrylic plastic

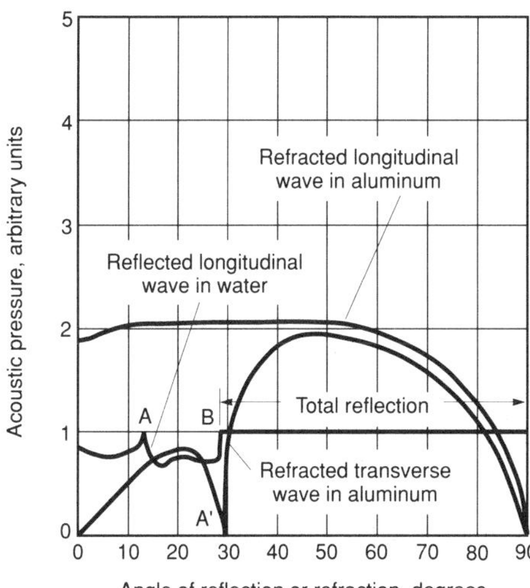

Fig. 6 Variation of acoustic pressure with angle of reflection or refraction during immersion ultrasonic inspection of aluminum. The acoustic pressure of the incident wave equals 1.0 arbitrary unit. Points A and A′ correspond to the first critical angle, and point B to the second critical angle, for this system.

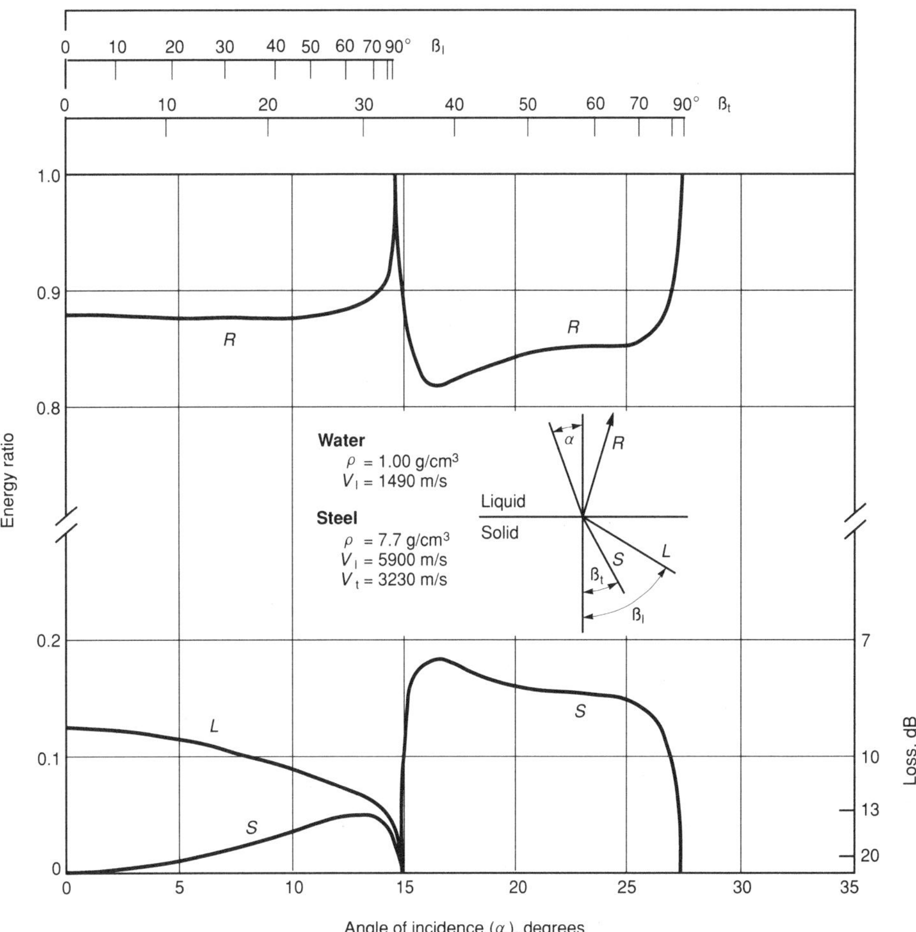

Fig. 7 Partition of acoustic energy at a water/steel interface. The reflection coefficient, *R*, is equal to 1 − (*L* + *S*), where *L* is the transmission coefficient of the longitudinal wave and *S* is the transmission coefficient of the transverse (or shear) wave.

Beam Intensity. The intensity of an ultrasonic beam is related to the amplitude of particle vibrations. Acoustic pressure (sound pressure) is the term most often used to denote the amplitude of alternating stresses exerted on a material by a propagating ultrasonic wave. Acoustic pressure is directly proportional to the product of acoustic impedance and amplitude of particle motion. The acoustic pressure exerted by a given particle varies in the same direction and with the same frequency as the position of that particle changes with time. Acoustic pressure is the most important property of an ultrasonic wave, and its square determines the amount of energy (acoustic power) in the wave. It should be noted that acoustic pressure is not the intensity of the ultrasonic beam. Intensity, which is the energy transmitted through a unit cross-sectional area of the beam, is proportional to the square of acoustic pressure.

Although transducer elements sense acoustic pressure, ultrasonic systems do not measure acoustic pressure directly. However, receiver-amplifier circuits of most ultrasonic instruments are designed to produce an output voltage proportional to the square of the input voltage from the transducer. Therefore, the signal amplitude of sound that is displayed on an oscilloscope or other readout device is a value proportional to the true intensity of the reflected sound.

The law of reflection and refraction described in Eq 5 or 6 gives information regarding only the direction of propagation of reflected and refracted waves and says nothing about the acoustic pressure in reflected or refracted waves. When ultrasonic waves are reflected or refracted, the energy in the incident wave is partitioned among the various reflected and refracted waves. The relationship among acoustic energies in the resultant waves is complex and depends both on the angle of incidence and on the acoustic properties of the matter on opposite sides of the interface.

Figure 6 shows the variation of acoustic pressure (not energy) with angle of reflection or refraction (α'_l, β_l, or β_t, Fig. 5) that results when an incident longitudinal wave in water having an acoustic pressure of 1.0 arbitrary unit impinges on the surface of an aluminum testpiece. At normal incidence ($\alpha_l = \alpha'_l = \beta_l = 0°$), acoustic energy is partitioned between a reflected longitudinal wave in water and a refracted (transmitted) longitudinal wave in aluminum. Because of different acoustic impedances, this partition induces acoustic pressures of about 0.8 arbitrary unit in the reflected wave in water and about 1.9 units in the transmitted wave in aluminum. Although it may seem anomalous that the transmitted wave has a higher acoustic pressure than the incident wave, it must be recognized that it is acoustic energy, not acoustic pressure, that is partitioned and conserved. Figure 7 illustrates the partition of acoustic energy at a water/steel interface.

In Fig. 6, as the incident angle, α_l, is increased, there is a slight drop in the acoustic pressure of the reflected wave, a corresponding slight rise in the acoustic pressure of the refracted longitudinal wave, and a sharper rise in the acoustic pressure of the refracted transverse wave. At the first critical angle for the water/aluminum interface (α_l = 13.6°, β_l = 90°, and β_t = 29.2°), the acoustic pressure of the longitudinal waves reaches a peak, and the refracted waves go rapidly to zero (point A′, Fig. 6). Between the first and second critical angles, the acoustic pressure in the reflected longitudinal wave in water varies as shown between points A and B in Fig. 6. The refracted longitudinal wave in aluminum meanwhile has disappeared. Beyond the second critical angle (α_l = 28.8°), the transverse wave in aluminum disappears, and there is total reflection at the interface with no partition of energy and no variation in acoustic pressure, as shown to right of point B in Fig. 6.

Curves similar to those in Fig. 6 can be constructed for the reverse instance of incident longitudinal waves in aluminum impinging on an aluminum/water interface, for incident transverse waves in aluminum, and for other combinations of wave types and materials. Details of this procedure are available in Ref 1. These curves are important because they indicate the angles of incidence at which energy transfer across the boundary is most effective. For example, at an aluminum/water interface, peak transmission of acoustic pressure for a returning transverse wave echo occurs in the sector from about 16 to 22° in the water relative to a line normal to the interface. Consequently, 35 to 51° angle beams in aluminum are the most efficient in transmitting detectable echoes across the front surface during immersion inspection and can therefore resolve smaller discontinuities than beams directed at other angles in the aluminum.

Attenuation of Ultrasonic Beams

The intensity of an ultrasonic beam that is sensed by a receiving transducer is considerably less than the intensity of the initial transmission. The factors that are primarily responsible for the loss in beam intensity can be classified as transmission losses, interference effects, and beam spreading.

Transmission losses include absorption, scattering, and acoustic impedance effects at interfaces. Interference effects include diffraction and other effects that create wave fringes, phase shift, or frequency shift. Beam spreading involves mainly a transition from plane waves to either spherical or cylindrical waves, depending on the shape of the transducer-element face. The wave physics that completely describe these three effects are discussed in Ref 1 and 2.

Acoustic impedance effects (see the section "Acoustic Impedance" in this article) can be used to calculate the amount of sound that reflects during the ultrasonic inspection of a testpiece immersed in water. For example, when an ultrasonic wave impinges at normal incidence ($\alpha_1 = 0°$) to the surface of the flaw-free section of aluminum alloy 1100 plate during straight-beam inspection, the amount of sound that returns to the search unit (known as the back reflection) has only 6% of its original intensity. This reduction in intensity occurs because of energy partition when waves are only partly reflected at the aluminum/water interfaces. (Additional losses would occur because of absorption and scattering of the ultrasonic waves, as discussed in the sections "Absorption" and "Scattering" in this article.)

Similarly, an energy loss can be calculated for a discontinuity that constitutes an ideal reflecting surface, such as a lamination that is normal to the beam path and that interposes a metal/air interface larger than the sound beam. For example, in the straight-beam inspection of an aluminum alloy 1100 plate containing a lamination, the final returning beam, after partial reflection at the front surface of the plate and total reflection from the lamination, would have a maximum intensity 8% of that of the incident beam. By comparison, only 6% was found for the returning beam from the plate that did not contain a lamination. Similar calculations of the energy losses caused by impedance effects at metal/water interfaces for the ultrasonic immersion inspection of several of the metals listed in Table 1 yield the following back reflection intensities, which are expressed as a percentage of the intensity of the incident beam:

Material	Back reflection intensity, % of incident beam intensity
Magnesium alloy M1A	11.0
Titanium	3.0
Type 302 stainless steel	1.4
Carbon steel	1.3
Inconel	0.7
Tungsten	0.3

The loss in intensity of returning ultrasonic beams is one basis for characterizing flaws in metal testpieces. As indicated above, acoustic impedance losses can severely diminish the intensity of an ultrasonic beam. Because a small fraction of the area of a sound beam is reflected from small discontinuities, it is obvious that ultrasonic instruments must be extremely sensitive to small variations in intensity if small discontinuities are to be detected. The sound intensity of contact techniques is usually greater than that of immersion techniques; that is, smaller discontinuities will result in higher amplitude signals. Two factors are mainly responsible for this difference, as follows.

First, the back surface of the testpiece is a metal/air interface, which can be considered a total reflector. Compared to a metal/water interface, this results in an approximately 30% increase in back reflection intensity at the receiving search unit for an aluminum testpiece coupled to the search unit through a layer of water.

Second, if a couplant whose acoustic impedance more nearly matches that of the testpiece is substituted for the water, more energy is transmitted across the interface for both the incident and returning beams. For most applications, any couplant with an acoustic impedance higher than that of water is preferred. Several of these are listed in the nonmetals group in Table 1. In addition to the liquid couplants listed in Table 1, several semisolid or solid couplants (including wallpaper paste, certain greases, and some adhesives) have higher acoustic impedances than water.

The absorption of ultrasonic energy occurs mainly by the conversion of mechanical energy into heat. Elastic motion within a substance as a sound wave propagates through it alternately heats the substance during compression and cools it during rarefaction. Because heat flows so much more slowly than an ultrasonic wave, thermal losses are incurred, and this progressively reduces energy in the propagating wave. A related thermal loss occurs in polycrystalline materials; a thermoelastic loss arises from heat flow away from grains that have received more compression or expansion in the course of wave motion than did adjacent grains. For most polycrystalline materials, this effect is most pronounced at the low end of the ultrasonic frequency spectrum.

Vibrational stress in ferromagnetic and ferroelectric materials generated by the passage of an acoustic wave can cause motion of domain walls or rotation of domain directions. These effects may cause domains to be strengthened in directions parallel, antiparallel, or perpendicular to the direction of stress. Energy losses in ferromagnetic and ferroelectric materials may also be caused by a microhysteresis effect, in which domain wall motion or domain rotation lags behind the vibrational stress to produce a hysteresis loop.

In addition to the types of losses discussed above, other types exist that have not been accounted for quantitatively. For example, it has been suggested that some losses are caused by elastic-hysteresis effects due to cyclic displacements of dislocations in grains or grain boundaries of metals.

Absorption can be thought of as a braking action on the motion of oscillating particles. This braking action is more pronounced when oscillations are more rapid, that is, at high frequencies. For most materials, absorption losses increase directly with frequency.

Scattering of an ultrasonic wave occurs because most materials are not truly homogeneous. Crystal discontinuities, such as grain boundaries, twin boundaries, and minute nonmetallic inclusions, tend to deflect small amounts of ultrasonic energy out of the main ultrasonic beam. In addition, especially in mixed microstructures or anisotropic materials, mode conversion at crystallite boundaries tends to occur because of slight differences in acoustic velocity and acoustic impedance across the boundaries.

Scattering is highly dependent on the relation of crystallite size (mainly grain size) to ultrasonic wavelength. When grain size is less than 0.01 times the wavelength, scatter is negligible. Scattering effects vary approximately with the third power of grain size, and when the grain size is 0.1 times the wavelength or larger, excessive scattering may make it impossible to conduct valid ultrasonic inspections.

In some cases, determination of the degree of scattering can be used as a basis for acceptance or rejection of parts. Some cast irons can be inspected for the size and distribution of graphite flakes, as described in the section "Determination of Microstructural Differences" in this article. Similarly, the size and distribution of microscopic voids in some powder metallurgy parts, or of strengtheners in some fiber-reinforced or dispersion-strengthened materials, can be evaluated by measuring attenuation (scattering) of an ultrasonic beam.

Diffraction. A sound beam propagating in a homogeneous medium is coherent; that is, all particles that lie along any given plane parallel to the wave front vibrate in identical patterns. When a wave front passes the edge of a reflecting surface, the front bends around the edge in a manner similar to that in which light bends around the edge of an opaque object. When the reflector is very small compared to the sound beam, as is usual for a pore or an inclusion, wave bending (forward scattering) around the edges of the reflector produces an interference pattern in a zone immediately behind the reflector because of phase differences among different portions of the forward-scattered beam. The interference pattern consists of alternate regions of maximum and minimum intensity that correspond to regions where interfering scattered waves are respectively in phase and out of phase.

Diffraction phenomena must be taken into account during the development of ultrasonic inspection procedures. Unfortunately, only qualitative guidelines can be provided. Entry-surface roughness, type of machined surface, and machining direction influence inspection procedures. In addition, the roughness of a flaw surface affects its echo pattern and must be considered.

A sound beam striking a smooth interface is reflected and refracted; but the sound field maintains phase coherence, and beam behavior can be analytically predicted. A rough interface, however, modifies boundary conditions, and some of the beam energy is diffracted. Beyond the interface, a coherent wave must re-form through phase reinforcement and cancellation; the wave then continues to propagate as a modified wave.

The influence on the beam depends on the roughness, size, and contour of the modifying interface. For example, a plane wave striking a diaphragm containing a single hole one wavelength in diameter will propagate as a spherical wave from a point (Huygens) source. The wave from a larger hole will re-form in accordance with the number of wavelengths in the diameter. In ultrasonic inspection, a 2.5 μm (100 μin.) surface finish may have little influence at one inspection frequency and search-unit diameter, but may completely mask subsurface discontinuities at other inspection frequencies or search-unit diameters.

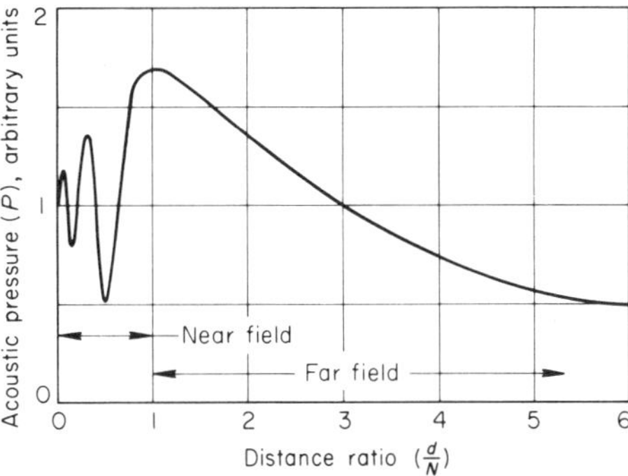

Fig. 8 Variation of acoustic pressure with distance ratio for a circular search unit. Distance ratio is the distance from the crystal face, *d*, divided by the length of the near field, *N*.

Near-Field and Far-Field Effects. The face of an ultrasonic-transducer crystal does not vibrate uniformly under the influence of an impressed electrical voltage. Rather, the crystal face vibrates in a complex manner that can be most easily described as a mosaic of tiny, individual crystals, each vibrating in the same direction but slightly out of phase with its neighbors. Each element in the mosaic acts like a point (Huygens) source and radiates a spherical wave outward from the plane of the crystal face. Near the face of the crystal, the composite sound beam propagates chiefly as a plane wave, although spherical waves emanating from the periphery of the crystal face produce short-range ultrasonic beams referred to as side lobes. Because of interference effects, as these spherical waves encounter one another in the region near the crystal face, a spatial pattern of acoustic pressure maximums and minimums is set up in the composite sound beam. The region in which these maximums and minimums occur is known as the near field (Fresnel field) of the sound beam.

Along the central axis of the composite sound beam, the series of acoustic pressure maximums and minimums becomes broader and more widely spaced as the distance from the crystal face, d, increases. Where d becomes equal to N (with N denoting the length of the near field), the acoustic pressure reaches a final maximum and decreases approximately exponentially with increasing distance, as shown in Fig. 8. The length of the near field is determined by the size of the radiating crystal and the wavelength, λ, of the ultrasonic wave. For a circular radiator of diameter D, the length of the near field can be calculated from:

$$N = \frac{(D^2 - \lambda^2)}{4\lambda} \qquad \text{(Eq 7)}$$

When the wavelength is small with respect to the crystal diameter, the near-field length can be approximated by:

$$N = \frac{D^2}{4\lambda} = \frac{A}{\pi\lambda} \qquad \text{(Eq 8)}$$

where A is the area of the crystal face.

At distances greater than N, known as the far field of the ultrasonic beam, there are no interference effects. At distances from N to about $3N$ from the face of a circular radiator, there is a gradual transition to a spherical wave front. At distances of more than about $3N$, the ultrasonic beam from a rectangular radiator more closely resembles a cylindrical wave, with the wave front being curved about an axis parallel to the long dimension of the rectangle.

Near-field and far-field effects also occur when ultrasonic waves are reflected from interfaces. The reasons are similar to those for near-field and far-field effects for transducer crystals; that is, reflecting interfaces do not vibrate uniformly in response to the acoustic pressure of an impinging sound wave. Near-field lengths for circular reflecting interfaces can be calculated from Eq 7 and 8. Table 3 lists near-field lengths corresponding to several combinations of radiator diameter and ultrasonic frequency. The values in Table 3 were calculated from Eq 7 for circular radiators in a material having a sonic velocity of 6 km/s (4 miles/s) and closely approximate actual lengths of near fields for longitudinal waves in steel, aluminum alloys, and certain other materials. Values for radiators with diameters of 25, 13, and 10 mm (1, ½, and ⅜ in.) correspond to typical search-unit sizes, and values for radiators with diameters of 3 and 1.5 mm (⅛ and 0.060 in.) correspond to typical hole sizes in standard reference blocks.

Table 3 Near-field lengths for circular radiators in a material having a sonic velocity of 6 km/s (4 miles/s)

Frequency, MHz	Wavelength		Near-field length for radiator with diameter of: 25 mm (1 in.)		13 mm (½ in.)		9.5 mm (⅜ in.)		3.2 mm (⅛ in.)		1.5 mm (0.060 in.)	
	mm	in.	cm	in.	cm	in.	cm	in.	cm	in.	cm	in.
1.0	6.0	0.24	2.5	1.0	0.52	0.20	0.23	0.09	...	...	...	...
2.0	3.0	0.12	5.3	2.1	1.3	0.50	0.68	0.27	0.009	0.0035	...	...
5.0	1.2	0.04	13.4	5.3	3.3	1.3	1.9	0.75	0.18	0.07	0.02	0.008
10.0	0.6	0.02	27	11	6.7	2.6	3.8	1.5	0.40	0.16	0.08	0.03
15.0	0.4	0.015	40	16	10	4.0	5.7	2.2	0.62	0.24	0.14	0.055
25.0	0.24	0.009	67	26	17	6.7	9.4	3.7	1.04	0.41	0.24	0.095

Table 4 Approximate attenuation coefficients and useful depths of inspection for various metallic and nonmetallic materials

Using 2-MHz longitudinal waves at room temperature

Attenuation coefficient, dB/mm (dB/in.)	Useful depth of inspection, m (ft)	Type of material inspected
Low: 0.001–0.01 (0.025–0.25)	1–10 (3–30)	Cast metals: aluminum(a), magnesium(a). Wrought metals: steel, aluminum, magnesium, nickel, titanium, tungsten, uranium
Medium: 0.01–0.1 (0.25–2.5)	0.1–1 (0.3–3)	Cast metals(b): steel(c), high-strength cast iron, aluminum(d), magnesium(d). Wrought metals(b): copper, lead, zinc. Nonmetals: sintered carbides(b), some plastics(e), some rubbers(e)
High: >0.1 (>2.5)	0–0.1 (0–0.3)	Cast metals(b): steel(d), low-strength cast iron, copper, zinc. Nonmetals(e): porous ceramics, filled plastics, some rubbers

(a) Pure or slightly alloyed. (b) Attenuation mostly by scattering. (c) Plain carbon or slightly alloyed. (d) Highly alloyed. (e) Attenuation mostly by absorption. (f) Excessive attenuation may preclude inspection.

Beam Spreading. In the far field of an ultrasonic beam, the wave front expands with distance from a radiator. The angle of divergence from the central axis of the beam from a circular radiator is determined from ultrasonic wavelength and radiator size as follows:

$$\gamma = 2 \sin^{-1}\left(0.5 \frac{\lambda}{D}\right) \quad \text{(Eq 9)}$$

where γ is the angle of divergence in degrees, λ is the ultrasonic wavelength, and D is the diameter of a circular radiator. Equation 9 is valid only for small values of λ/D, that is, only when the beam angle is small.

When the radiator is not circular, the angle of divergence cannot be assessed accurately by applying Eq 9. For noncircular search units, beam spreading is most accurately found experimentally.

Beam diameter also depends on the diameter of the radiator and the ultrasonic wavelength. The theoretical equation for −6 dB pulse-echo beam diameter is:

$$-6 \text{ dB beam diameter} = 1.032 \frac{D}{4} \cdot S \quad \text{(Eq 10)}$$

where S is the focusing factor and is $\leqq 1$. Focusing the transducer ($S < 1$) produces a smaller beam. For a flat (that is, nonfocused) transducer ($S = 1$), the beam has a diameter of 0.25 D at the near-field distance N, where N depends on the ultrasonic wavelength as defined in Eq 7.

The overall attenuation of an ultrasonic wave in the far field can be expressed as:

$$P = P_0 \exp(-\alpha L) \quad \text{(Eq 11)}$$

where P_0 and P are the acoustic pressures at the beginning and end, respectively, of a section of material having a length L and an attenuation coefficient α. Attenuation coefficients are most often expressed in nepers per centimeter or decibels per millimeter. Both nepers and decibels are units based on logarithms—nepers on natural logarithms (base e) and decibels on common logarithms (base 10). Numerically, the value of α in decibels per millimeter (dB/mm) is equal to 0.868 the value in nepers per centimeter.

A table of exact attenuation coefficients for various materials, if such data could be determined, would be of doubtful value. Ultrasonic inspection is a process subject to wide variation in responses, and these variations are highly dependent on structure and properties in each individual testpiece. Attenuation determines mainly the depth to which ultrasonic inspection can be performed as well as the signal amplitude from reflectors with a testpiece. Table 4 lists the types of materials and approximate maximum inspection depth corresponding to low, medium, and high attenuation coefficients. Inspection depth is also influenced by the decibel gain built into the receiver-amplifier of an ultrasonic instrument and by the ability of the instrument to discriminate between low-amplitude echoes and electronic noise at high gain settings.

Basic Inspection Methods

The two major methods of ultrasonic inspection are the transmission method and the pulse-echo method. The primary difference between these two methods is that the transmission method involves only the measurement of signal attenuation, while the pulse-echo method can be used to measure both transit time and signal attenuation.

The pulse-echo method, which is the most widely used ultrasonic method, involves the detection of echoes produced when an ultrasonic pulse is reflected from a discontinuity or an interface of a testpiece. This method is used in flaw location and thickness measurements. Flaw depth is determined from the time-of-flight between the initial pulse and the echo produced by a flaw. Flaw depth might also be determined by the relative transit time between the echo produced by a flaw and the echo from the back surface. Flaw sizes are estimated by comparing the signal amplitudes of reflected sound from an interface (either within the testpiece or at the back surface) with the amplitude of sound reflected from a reference reflector of known size or from the back surface of a testpiece having no flaws.

The transmission method, which may include either reflection or through transmission, involves only the measurement of signal attenuation. This method is also used in flaw detection. In the pulse-echo method, it is necessary that an internal flaw reflect at least part of the sound energy onto a receiving transducer. However, echoes from flaws are not essential to their detection. Merely the fact that the amplitude of the back reflection from a testpiece is lower than that from an identical workpiece known to be free of flaws implies that the testpiece contains one or more flaws. The technique of detecting the presence of flaws by sound attenuation is used in transmission methods as well as in the pulse-echo method. The main disadvantage of attenuation methods is that flaw depth cannot be measured.

The principles of each of these two inspection methods are discussed in the following sections, along with corresponding forms of data presentation, interpretation of data, and effects of operating variables. Subsequent sections describe various components and systems for ultrasonic inspection, reference standards, and inspection procedures and applications. In addition, the article "Boilers and Pressure Vessels" in this Volume contains information on advanced ultrasonic techniques.

The application of ultrasonic techniques also involves other methods, such as acoustical holography, acoustical microscopy, the frequency modulation technique, spectral analysis, and sound conduction. The first two of these methods are discussed in the articles "Acoustical Holography" and "Acoustic Microscopy" in this Volume. The other three methods are briefly summarized below.

The frequency modulation (FM) method, which was the precursor of the pulse-echo method, is another flaw detection technique. In the FM method, the ultrasonic pulses are transmitted in wave packets whose frequency varies linearly with time. The frequency variation is repeated in successive wave packets so that a plot of frequency versus time has a sawtooth pattern. There is a time delay between successive packets. Returning echoes are displayed on the readout device only if they have certain characteristics as determined by the electronic circuitry in the instrument. Although not as widely used as the pulse-echo method, the FM method has a lower signal-to-noise ratio and therefore somewhat greater resolving power.

Spectral analysis, which can be used in the through transmission or pulse-echo methods, involves determination of the frequency spectrum of an ultrasonic wave after it has propagated through a testpiece. The frequency spectrum can be determined either by transmitting a pulse and using a fast Fourier transform to obtain the frequency spectrum of the received signal or by sweeping the transmission frequency in real time and acquiring the response at each frequency. The increasing use of the pulse method is attributed to improvements in the speed of digital fast Fourier transform devices.

Spectral analysis is used in transducer evaluations and may be useful in defect characterization. However, because the spectral signatures of defects are influenced by several other factors (such as the spectrum of the input pulse, coupling details, and signal attenuation), defect characterization primarily involves the qualitative interpretation of echoes in the time domain (see the section "Interpretation of Pulse-Echo Data" in this article).

Spectral analysis can also be used to measure the thickness of thin-wall specimens. A short pulse of ultrasound is a form of coherent radiation; in a thin-wall specimen that produces front and back wall echoes, the two reflected pulses show phase differences and can interfere coherently. If the pulse contains a wide band of frequencies, interference maxima and minima can occur at particular frequencies, and these can be related to the specimen thickness.

Sound conduction is utilized in flaw detection by monitoring the intensity of arbitrary waveforms at a given point on the testpiece. These waveforms transmit ultrasonic energy, which is fed into the testpiece at some other point without the existence of a well-defined beam path between the two points. This method is of relatively minor importance and is not discussed in this article.

Pulse-Echo Methods

In pulse-echo inspection, short bursts of ultrasonic energy (pulses) are introduced into a testpiece at regular intervals of time. If the pulses encounter a reflecting surface, some or all of the energy is reflected. The proportion of energy that is reflected is highly dependent on the size of the reflecting surface in relation to the size of the incident ultrasonic beam. The direction of the reflected beam (echo) depends on the orientation of the reflecting surface with respect to the incident beam. Reflected energy is monitored; both the amount of energy reflected in a specific direction and the time delay between transmission of the initial pulse and receipt of the echo are measured.

Principles of Pulse-Echo Methods

Most pulse-echo systems consist of:

- An electronic clock
- An electronic signal generator, or pulser
- A sending transducer
- A receiving transducer
- An echo-signal amplifier
- A display device

In the most widely used version of pulse-echo systems, a single transducer acts alternately as a sending and receiving transducer. The clock and signal generator are usually combined in a single electronic unit. Frequently, circuits that amplify and demodulate echo signals from the transducer are housed in the same unit. Specific characteristics of transducers and other equipment are discussed in subsequent sections of this article.

A pulse-echo system with a single transducer operates as follows. At regular intervals, the electronic clock triggers the signal generator, which imposes a short interval of high-frequency alternating voltage or a unipolar (negative) spike on the transducer. Simultaneously, the clock activates a time-measuring circuit connected to the display device. The operator can preselect a constant interval between pulses by means of a pulse-repetition rate control on the instrument; pulses are usually repeated 60 to 2000 times per second. In most commercially available flaw detectors, the pulse-repetition rate is controlled automatically except for some larger systems. Also, most systems are broadband when they transmit, but may be tuned or filtered for reception. The operator can also preselect the output frequency of the signal generator. For best results, the frequency (and sometimes the pulse-repetition rate) should be tuned to achieve the maximum response of the transducer (resonance in the vibrating element) and maximum signal-to-noise ratio (lowest amount of electronic noise) in the electronic equipment.

The transducer then converts the pulse of voltage into a pulse of mechanical vibration having essentially the same frequency as the imposed alternating voltage. The mechanical vibration (ultrasound) is introduced into a testpiece through a couplant and travels by wave motion through the testpiece at the velocity of sound, which depends on the material. When the pulse of ultrasound encounters a reflecting surface that is perpendicular to the direction of travel, ultrasonic energy is reflected and returns to the transducer. The returning pulse travels along the same path and at the same speed as the transmitted pulse, but in the opposite direction. Upon reaching the transducer through the couplant, the returning pulse causes the transducer element to vibrate, which induces an alternating electrical voltage across the transducer. The induced voltage is instantaneously amplified (and sometimes demodulated), then fed into the display device. This process of alternately sending and receiving pulses of ultrasonic energy is repeated for each successive pulse, with the display device recording any echoes each time.

Theoretically, the maximum depth of inspection is controlled by the pulse-repetition rate. For example, if a 10 MHz pulse is transmitted at a pulse-repetition rate of 500 pulses per second, a longitudinal wave pulse can travel almost 12 m (40 ft) in steel or aluminum before the next pulse is triggered. This means one pulse can travel to a depth of 6 m (20 ft) and return before the next pulse is initiated.

Practically, however, inspection can be performed only to a depth that is considerably less than the theoretical maximum. Sound attenuation in a testpiece can limit the path length. The practical limit varies with the type and condition of the test material, test frequency, and system sensitivity. Furthermore, it is highly desirable for all ultrasonic vibrations (including successively re-reflected echoes of the first reflected pulse) to die out in the testpiece before the next initial pulse is introduced. As a rule, the pulse-repetition rate should be set so that one pulse can traverse the testpiece enough times to dissipate the sonic energy to a nondisplayable level before the next pulse is triggered. Both sound attenuation and pulse reverberation are of little consequence except when inspecting large parts (for example, in the axial inspection of long shafts).

Pulse-echo inspection can be accomplished with longitudinal, shear, surface, or Lamb waves. Straight-beam or angle-beam techniques can be used, depending on testpiece shape and inspection objectives. Data can be analyzed in terms of type, size, location, and orientation of flaws, or any combination of these factors. It should be noted, however, that some forms of data presentation are inherently unable to pinpoint the location of flaws unless the flaws are favorably oriented with respect to the transmitted sonic beam. Similarly, type, location, and orientation of flaws often influence the procedures and techniques used to estimate flaw size.

Sometimes it is advantageous to use separate sending and receiving transducers for pulse-echo inspection. (Separate transducers are always used for through transmission inspection.) Depending mainly on geometric considerations, as discussed later in this article, these separate transducers can be housed in a single search unit or in two separate search units. The term pitch-catch is often used in connection with separate sending and receiving transducers, regardless of whether reflection methods or transmission methods are involved.

Presentation of Pulse-Echo Data

Information from pulse-echo inspection can be displayed in different forms. The basic data formats include:

- *A-scans*: This format provides a quantitative display of signal amplitudes and time-of-flight data obtained at a single point on the surface of the testpiece. The A-scan display, which is the most widely

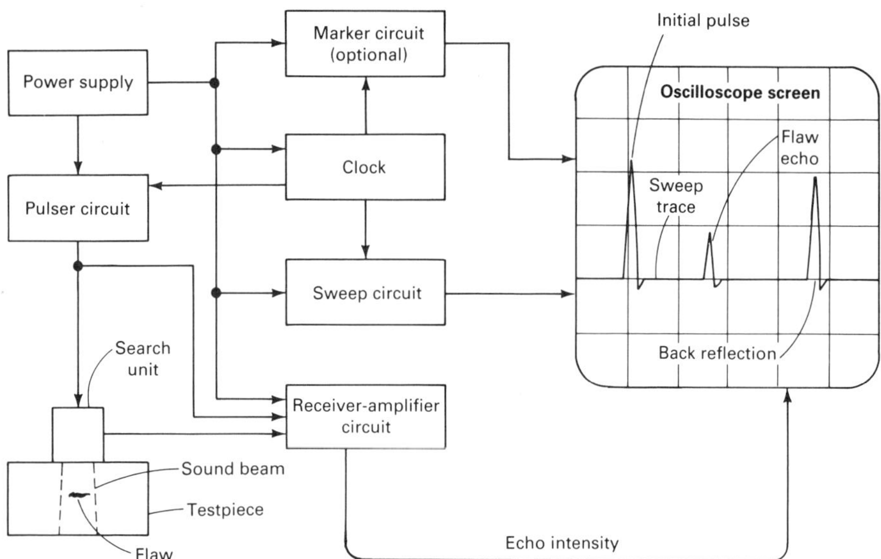

Fig. 9 Typical block diagram of an analog A-scan setup, including video-mode display, for basic pulse-echo ultrasonic inspection

used format, can be used to analyze the type, size, and location (chiefly depth) of flaws

- *B-scans*: This format provides a quantitative display of time-of-flight data obtained along a line of the testpiece. The B-scan display shows the relative depth of reflectors and is used mainly to determine size (length in one direction), location (both position and depth), and to a certain degree the shape and orientation of large flaws
- *C-scans*: This format provides a semiquantitative or quantitative display of signal amplitudes obtained over an area of the testpiece surface. This information can be used to map out the position of flaws on a plan view of the testpiece. A C-scan format also records time-of-flight data, which can be converted and displayed by image-processing equipment to provide an indication of flaw depth

A-scan and B-scan data are usually presented on an oscilloscope screen; C-scan data are recorded by an *x-y* plotter or displayed on a computer monitor. With computerized data acquisition and image processing, the display formats can be combined or processed into more complex displays.

A-scan display is basically a plot of amplitude versus time, in which a horizontal baseline on an oscilloscope screen indicates elapsed time while the vertical deflections (called indications or signals) represent echoes (Fig. 9). Flaw size can be estimated by comparing the amplitude of a discontinuity signal with that of a signal from a discontinuity of known size and shape; the discontinuity signal also must be corrected for distance losses.

Flaw location (depth) is determined from the position of the flaw echo on the oscilloscope screen. With a calibrated time base (the horizontal sweep of the oscilloscope), flaw location can be measured from the position of its echo on the horizontal scale calibrated to represent sound travel within the test object. The zero point on this scale represents the entry surface of the testpiece.

Display Modes. A-scan data can be displayed in either of two modes—radio frequency (RF) mode, in which the individual cycles comprising each pulse are visible in the trace; or video mode, in which only a rectified voltage corresponding to the envelope of the RF wave packet is displayed. The video mode is usually suitable for ordinary ultrasonic inspection, but certain applications demand use of the RF mode for optimum characterization of flaws.

System Setup. A typical A-scan setup that illustrates the essential elements in a basic system for pulse-echo inspection is shown in Fig. 9. These elements include:

- Power supply, which may run on alternating current or batteries
- Electronic clock, or timing circuit, to trigger pulser and display circuits
- Pulser circuit, or rate generator, to control frequency, amplitude, and pulse-repetition rate of the voltage pulses that excite the search unit
- Receiver-amplifier circuit to convert output signals from the search unit into a form suitable for oscilloscope display
- Sweep circuit to control (a) time delay between search-unit excitation and start of oscilloscope trace and (b) rate at which oscilloscope trace travels horizontally across the screen
- Oscilloscope screen, including separate controls for trace brightness, trace focus, and illuminated measuring grid

The search unit and the coaxial cable, although not strictly part of the electronic circuitry, must be matched to the electronics. Otherwise, the response of the transducer element to excitation voltages and the output voltage corresponding to echo vibrations can exhibit excessive ringing or an apparently low sensitivity.

Signal Display. The oscilloscope screen in Fig. 9 illustrates a typical video-mode A-scan display for a straight-beam test (as defined earlier in this section). The trace exhibits a large signal corresponding to the initial pulse, shown at left on the screen, and a somewhat smaller signal corresponding to the back reflection, at right on the screen. Between these two signals are indications of echoes from any interfaces within the testpiece; one small signal corresponding to the flaw shown in the testpiece, also illustrated in Fig. 9, appears between the initial pulse and the back reflection on the screen. The depth of the flaw can be quickly estimated by visual comparison of its position on the main trace relative to the positions of the initial pulse and back reflection. Its depth can be more accurately measured by counting the number of vertical reference lines from either the initial pulse or the back reflection of the flaw signal location on the screen in Fig. 9.

Applications. The A-scan display is not limited to the detection and characterization of flaws; it can also be used for measuring thickness, sound velocities in materials of known thickness, attenuation characteristics of specific materials, and beam spread of ultrasonic beams. Commercial instruments are usually adequate for these purposes, as well as for detecting the small cracks, porosity, and inclusions that are within the limits of resolution for the particular instrument and inspection technique. In addition to conventional single-transducer pulse-echo inspection, A-scan display can be used with transmission or reflection techniques that involve separate sending and receiving transducers.

B-scan display is a plot of time versus distance, in which one orthogonal axis on the display corresponds to elapsed time, while the other axis represents the position of the transducer along a line on the surface of the testpiece relative to the position of the transducer at the start of the inspection. Echo intensity is not measured directly as it is in A-scan inspection, but is often indicated semiquantitatively by the relative brightness of echo indications on an oscilloscope screen. A B-scan display can be likened to an imaginary cross section through the testpiece where both front and back surfaces are shown in profile. Indications from reflecting interfaces within the testpiece are also shown in profile, and the position, orientation, and depth of such interfaces along the imaginary cutting plane are revealed.

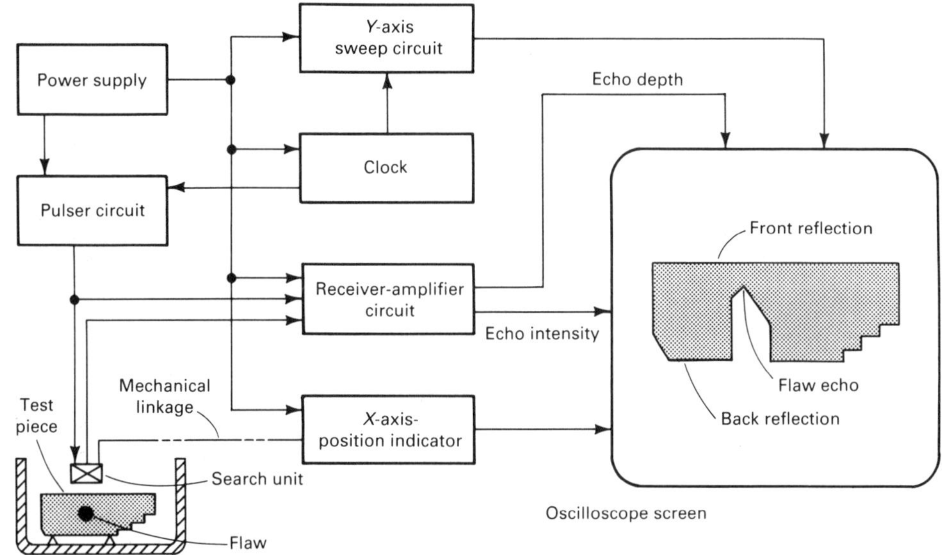

Fig. 10 Typical B-scan setup, including video-mode display, for basic pulse-echo ultrasonic inspection

System Setup. A typical B-scan system is shown in Fig. 10. The system functions are identical to the A-scan system except for the following differences.

First, the display is generated on an oscilloscope screen that is composed of a long-persistence phosphor, that is, a phosphor that continues to fluoresce long after the means of excitation ceases to fall on the fluorescing area of the screen. This characteristic of the oscilloscope in a B-scan system allows the imaginary cross section to be viewed as a whole without having to resort to permanent imaging methods, such as photographs. (Photographic equipment, facsimile recorders, or *x-y* plotters can be used to record B-scan data, especially when a permanent record is desired for later reference.)

Second, the oscilloscope input for one axis of the display is provided by an electromechanical device that generates an electrical voltage or digital signals proportional to the position of the transducer relative to a reference point on the surface of the testpiece. Most B-scans are generated by scanning the search unit in a straight line across the surface of the testpiece at a uniform rate. One axis of the display, usually the horizontal axis, represents the distance traveled along this line.

Third, echoes are indicated by bright spots on the screen rather than by deflections of the time trace. The position of a bright spot along the axis orthogonal to the search-unit position axis, usually measured top to bottom on the screen, indicates the depth of the echo within the testpiece.

Finally, to ensure that echoes are recorded as bright spots, the echo-intensity signal from the receiver-amplifier is connected to the trace-brightness control on the oscilloscope. In some systems, the brightnesses corresponding to different values of echo intensity may exhibit enough contrast to enable semiquantitative appraisal of echo intensity, which is related to flaw size and shape.

Signal Display. The oscilloscope screen in Fig. 10 illustrates the type of video-mode display that is generated by B-scan equipment. On this screen, the internal flaw in the testpiece shown at left in Fig. 10 is shown only as a profile view of its top reflecting surface. Portions of the testpiece that are behind this large reflecting surface are in shadow. The flaw length in the direction of search-unit travel is recorded, but the width (in a direction mutually perpendicular to the sound beam and the direction of search-unit travel) is not recorded except as it affects echo intensity and therefore echo-image brightness. Because the sound beam is slightly conical rather than truly cylindrical, flaws near the back surface of the testpiece appear longer than those near the front surface.

Applications. The chief value of B-scan presentations is their ability to reveal the distribution of flaws in a part on a cross section of that part. Although B-scan techniques have been more widely used in medical applications than in industrial applications, B-scans can be used for the rapid screening of parts and for the selection of certain parts, or portions of certain parts, for more thorough inspection with A-scan techniques. Optimum results from B-scan techniques are generally obtained with small transducers and high frequencies.

C-scan display records echoes from the internal portions of testpieces as a function of the position of each reflecting interface within an area. Flaws are shown on a readout, superimposed on a plan view of the testpiece, and both flaw size (flaw area) and position within the plan view are recorded. Flaw depth normally is not recorded, although it can be measured semiquantitatively by restricting the range of depths within the testpiece that is covered in a given scan. With an increasing number of C-scan systems designed with on-board computers, other options in image processing and enhancement have become widely used in the presentation of flaw depth and the characterization of flaws. An example of a computer-processed C-scan image is shown in Fig. 11, in which a graphite-epoxy sample with impact damage was examined using time-of-flight data. The depth of damage is displayed with a color scale in the original photograph.

System Setup. In a basic C-scan system, shown schematically in Fig. 12, the search unit is moved over the surface of the testpiece in a search pattern. The search pattern may take many forms; for example, a series of closely spaced parallel lines, a fine rastor pattern, or a spiral pattern (polar scan). Mechanical linkage connects the search unit to *x*-axis and *y*-axis position indicators, which in turn feed position data to the *x-y* plotter or facsimile device. Echo-recording systems vary; some produce a shaded-line scan with echo intensity recorded as a variation in line shading, while others indicate flaws by an absence of shading so that each flaw shows up as a blank space on the display (Fig. 12).

Gating. An electronic depth gate is another essential element in C-scan systems. A depth gate is an electronic circuit that allows only those echo signals that are received within a limited range of delay times following the initial pulse or interface echo to be admitted to the receiver-amplifier circuit. Usually, the depth gate is set so that front reflections and back reflections are just barely excluded from the display. Thus, only echoes from within the testpiece are recorded, except for echoes from thin layers adjacent to both surfaces of the testpiece. Depth gates are adjustable. By setting a depth gate for a narrow range of delay times, echo signals from a thin slice of the testpiece parallel to the scanned surface can be recorded, with signals from other portions being excluded from the display.

Some C-scan systems, particularly automatic units, incorporate additional electronic gating circuits for marking, alarming, or charting. These gates can record or indicate information such as flaw depth or loss of back reflection, while the main display records an overall picture of flaw distribution.

Interpretation of Pulse-Echo Data

The interpretation of pulse-echo data is relatively straightforward for B-scan and C-scan presentations. The B-scan always records the front reflection, while internal echoes or loss of back reflection, or both, are interpreted as flaw indications. Flaw

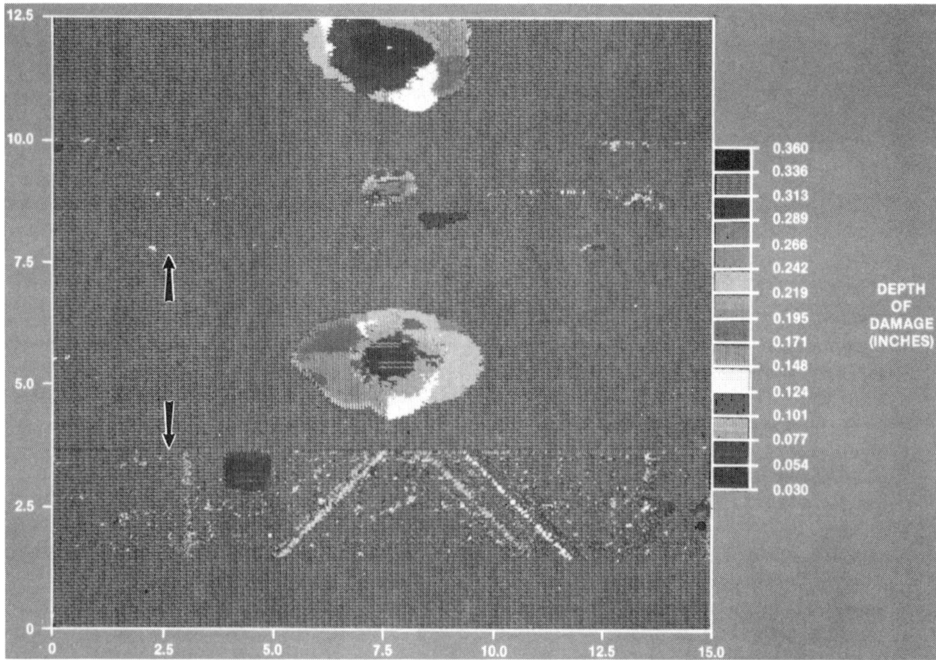

Fig. 11 Time-of-flight C-scan image of impact damage in graphite-epoxy laminate supported by two beams (arrows)

depth is measured as the distance from the front reflection to a flaw echo, with the latter representing the front surface of the flaw. The length of a flaw can be measured as a proportion of the scan length or can be estimated visually in relation to total scan length or to the size of a known feature of the testpiece. The position of a flaw can be determined by measuring its position along the scan with respect to either a predetermined reference point or a known feature of the testpiece. C-scan presentations are interpreted mainly by comparing the x and y coordinates of any flaw indication with the x and y coordinates of either a predetermined reference point or a known feature of the testpiece. The size of a flaw is estimated as a percentage of the scanned area. If a known feature is the size or position reference for the interpretation of either B-scan or C-scan data, it is presumed that this feature produces an appropriate echo image on the display.

In contrast to normal B-scan and C-scan displays, A-scan displays are sometimes quite complex. They may contain electronic noise, spurious echoes, or extra echoes resulting from scattering or mode conversion of the transmitted or interrogating pulse, all of which must be disregarded in order to focus attention on any flaw echoes that may be present. Furthermore, flaw echoes may exhibit widely varying shapes and amplitudes. Accurate interpretation of an A-scan display depends on the ability of the operator to:

- Recognize the type of flaw based on echo shape or echo-intensity effects
- Determine flaw location by accurately measuring echo position on the time trace
- Estimate flaw size, mainly from echo amplitudes with or without simultaneously manipulating the search unit
- Assess the quality of the testpiece by evaluating the A-scan data in terms of appropriate specifications or reference standards

Basic A-scan displays are of the type shown in Fig. 13 for the immersion inspection of a plate containing a flaw. The test material was 25 mm (1 in.) thick aluminum alloy 1100 plate containing a purely reflecting planar flaw. The flaw depth was 45% of plate thickness (11.25 mm, or 0.44 in.), exactly parallel to the plate surfaces, and had an area equal to one-third the cross section of the sound beam. Straight-beam immersion testing was done in a water-filled tank. There were negligible attenuation losses within the test plate, only transmission losses across front and back surfaces.

Figures 13(a), (b), and (c), respectively, illustrate the inspection setup, the complete video-mode A-scan display, and the normal video-mode display as seen on the oscilloscope screen. The normal display (Fig. 13c) represents only a portion of the complete display (Fig. 13b). The normal display is obtained by adjusting two of the oscilloscope controls (horizontal position and horizontal sweep) to display only the portion of the trace corresponding to the transit time (time of flight) required for a single pulse of ultrasound to traverse the testpiece from front surface to back surface and return. Also, the gain in the receiver-amplifier is adjusted so that the height of the first back reflection equals some arbitrary vertical distance on the screen, usually a convenient number of grid lines.

As illustrated in Fig. 13(b), there is a tendency for echoes to reverberate, that is, to bounce back and forth between reflecting surfaces. Each time an echo is reflected from the front surface, a portion of the sound wave energy escapes through the boundary to impinge on the transducer and produce an indication on the display. In Fig. 13(b), the indications labeled 1 through 6 are reverberations of the back reflection, those labeled A through K are reverberations of the primary flaw echo, and those labeled X through Z are reverberations of a subordinate flaw echo induced by re-reflection of the first back reflection.

Only a few types of flaws will produce the types of indications described above. Most flaws are not exactly parallel to the surface of the testpiece, not truly planar but have

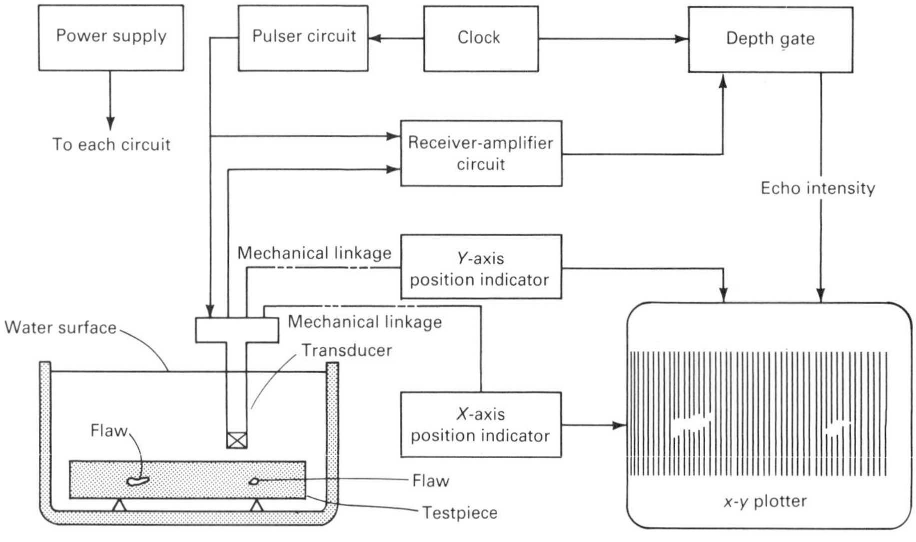

Power supply
To each circuit
Pulser circuit
Clock
Depth gate
Receiver-amplifier circuit
Echo intensity
Mechanical linkage
Y-axis position indicator
Mechanical linkage
Water surface
Transducer
Flaw
X-axis position indicator
Flaw
Testpiece
x-y plotter

Fig. 12 Typical C-scan setup, including display, for basic pulse-echo ultrasonic immersion inspection

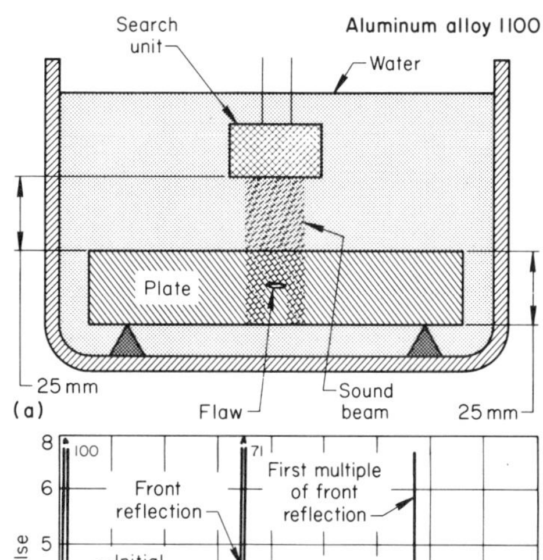

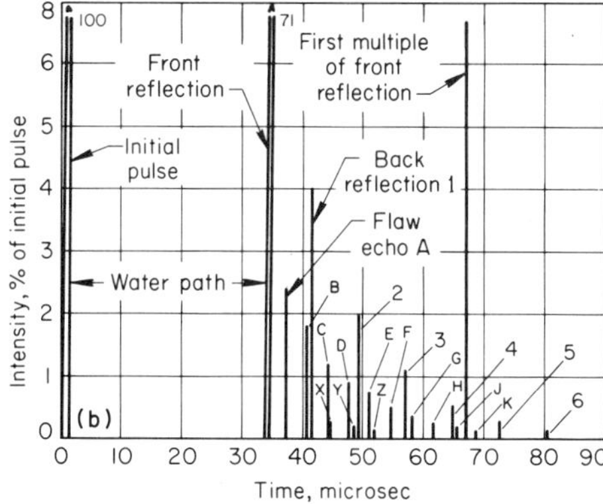

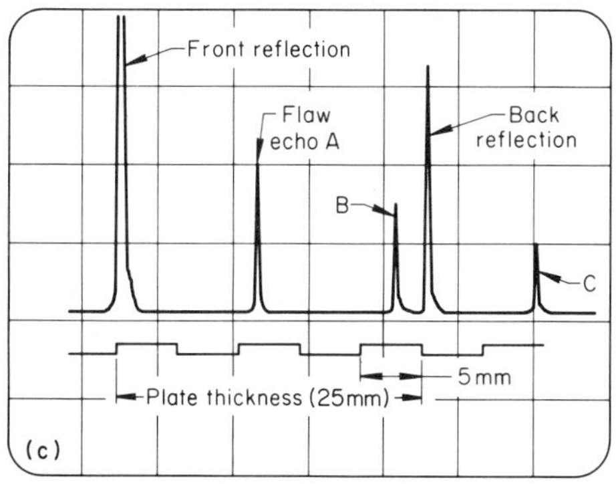

Fig. 13 Schematic of straight-beam immersion inspection of a 25 mm (1 in.) thick aluminum alloy 1100 plate containing a planar discontinuity showing (a) inspection setup, (b) complete video-mode A-scan display, and (c) normal oscilloscope display

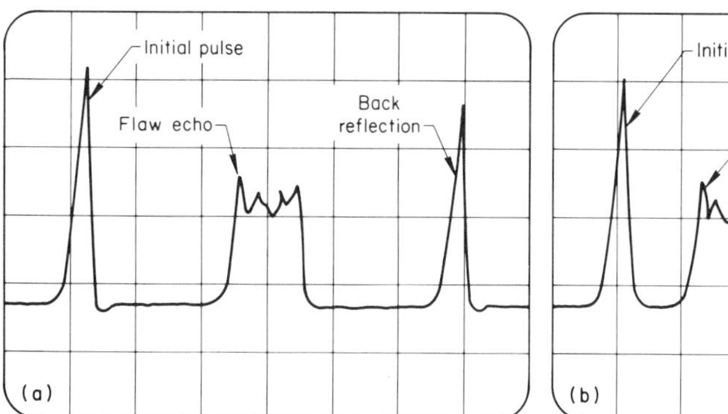

Fig. 14 A-scan displays of broadened-echo indications from curved rough or scattering interfaces showing (a) indications with back reflection and (b) indications without back reflection. See text for discussion.

rough or curved interfaces, not ideal reflectors, and of unknown size. These factors, together with the specific sound-attenuating characteristics of the bulk material, affect the size and shape of the echo signals. The following sections describe how specific material conditions produce and modify A-scan indications.

Echo shape is primarily affected by the shape, orientation, and sound-reflecting characteristics of an interface. Metal/air interfaces produce sharp indications if the interfaces are relatively smooth and essentially parallel to the front surface. If an interface is curved (such as the surface of a large pore) or rough (such as a crack, seam, or lamination) or if it is not ideally reflecting (such as the surface of a metallic inclusion or a slag inclusion), the interface will produce a broadened echo indication, as shown in Fig. 14. If the interface is smaller in area than the cross section of the ultrasonic beam or if ultrasonic waves are transmitted through the interface, a back-surface echo (back reflection) will appear to the right of the flaw echo on the oscilloscope screen, as shown in Fig. 14(a). However, if the flaw is larger than the ultrasonic beam or if the back surface is not normal to the direction of wave travel, no back reflection will appear on the screen, as shown in Fig. 14(b). Often, the amplitude of a broad indication will decrease with increasing depth, as in Fig. 14(b), especially when the echo is from a crack, seam, or lamination rather than an inclusion. Sometimes, especially if the echo is from a spherical flaw or from an interface that is not at right angles to the sound beam, the echo amplitude will increase with depth.

Echo amplitude, which is a measure of the intensity of a reflected sound beam, is a direct function of the area of the reflecting interface for flat parallel reflectors. If the interface is round or curved or is not perpendicular to the sound beam, echo amplitude will be reduced. The effects of roughness, shape, and orientation of the interface on echo amplitude must be understood because these factors introduce errors in estimates of flaw size.

Flaw size is most often estimated by comparing the amplitude of an echo from an interface of unknown size with the amplitude of echoes from flat-bottom holes of different diameter in two or more reference blocks. To compensate for any sound attenuation within the testpiece, these guidelines should be followed:

- Reference holes should be about the same depth from the front (entry) surface of the reference block as the flaw is from the front surface of the testpiece
- Reference blocks should be made of material with acoustic properties similar to those of the testpiece
- The sound beam should be larger than the flaw. (This can best be determined by moving the search unit back and forth on the surface of the part being inspected relative to a position centered over the flaw and observing the effect on both flaw echo and back reflection. If the search unit can be moved slightly without affecting the height of either the flaw echo or back reflection, it can be assumed that the sound beam is sufficiently larger than the flaw)
- Control settings on the instrument and physical arrangement of search unit, couplant, and specimen are the same regardless of whether the specimen is a testpiece or a reference block

In practice, a calibration curve is constructed using reference blocks, as described in the section "Determination of Area-Amplitude and Distance-Amplitude Curves" in this article. Flaw size is then determined by reading the hole size corresponding to the amplitude of the flaw echo directly from the calibration curve. Flaw size determined in this manner is only an estimate of minimum size and should not be assumed equal to the actual flaw size. The amount of sound energy reflected back to the search unit will be less than that from a flat-bottom hole of equal size if an interface has a surface rougher than the bottom surfaces of the reference holes, is oriented at an angle other than 90° to the sound beam, is curved, or transmits some of the sound energy rather than acting as an ideal reflector. Therefore, to produce equal echo heights, actual flaws having any of these characteristics must be larger than the minimum size determined from the calibration curve. This is why flaw sizes are frequently reported as being no smaller than *x*, where *x* is the flaw size that has been estimated from the calibration curve.

It may seem logical to estimate flaw size by comparing the amplitude of a flaw echo to the amplitude of the back reflection. Although an assumption that the ratio of flaw-echo height to back reflection amplitude is equal to the ratio of flaw area to sound beam cross section has been used in the past, this assumption should be considered to be completely unreliable, even

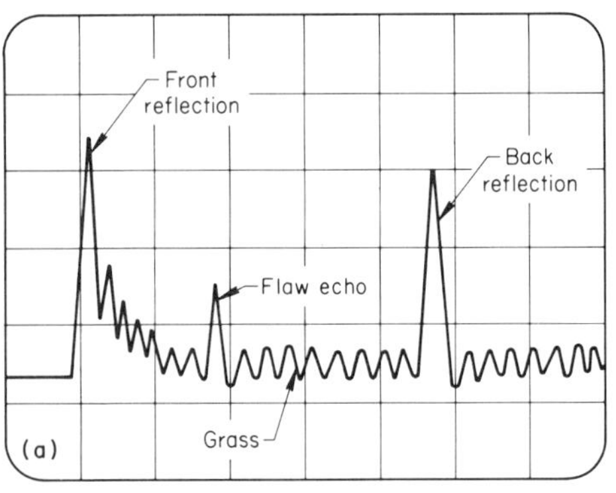

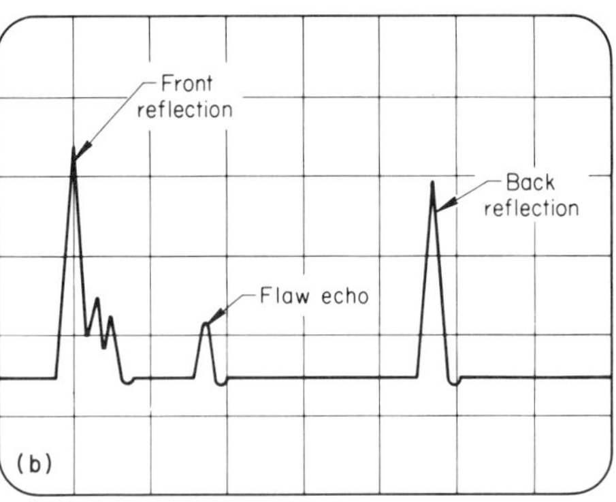

Fig. 15 A-scan displays showing (a) appearance of electronic noise as waviness (grass) and (b) grass filtered out by use of a reject circuit with some attendant loss of echo-signal amplitude

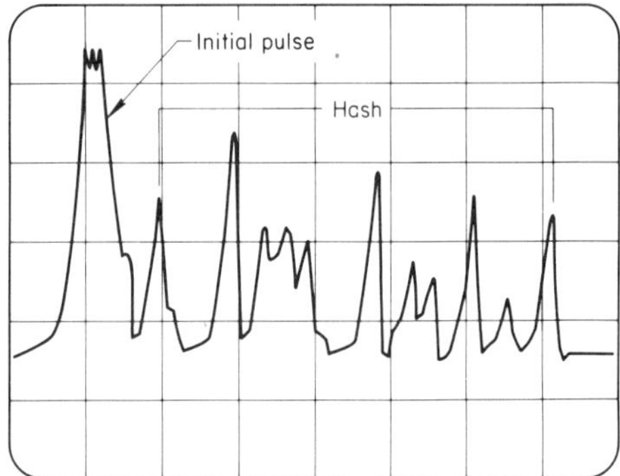

Fig. 16 A-scan display showing coarse-grain indications (hash) that interfere with detection of discontinuities

when distance-amplitude corrections are applied.

Loss of Back Reflection. If a flaw is larger than a few percent of the cross section of a sound beam, the amplitude of the back reflection is less than that of a similar region of the testpiece (or of another testpiece) that is free of flaws. Because sound travels essentially in straight lines, the reflecting interfaces within the testpiece (flaws) cast sound shadows on the back surface, in a manner similar to that in which opaque objects introduced into a beam of light cast shadows on a screen. Sound shadows reduce the amount of energy reflected from the back surface by reducing the effective area of the sound beam. The back reflection is not reduced in direct proportion to the percentage of the original sound beam intercepted by the flaw; the exact proportion varies widely. This effect is termed loss of back reflection, regardless of whether the back-surface signal echo is lost completely or merely reduced in amplitude.

A flaw indication is produced when an internal interface reflects sound onto the receiving transducer. A loss of back reflection can occur even if no flaw indication appears on the A-scan display. If the sound is reflected to the side, where the reflection cannot be picked up by the transducer, there is still a loss of back reflection because of the shadow effect. This provides an additional means of detecting the presence of flaws. Although no direct indication shows on the oscilloscope screen, the size of a flaw can be estimated from the percentage lost from the height of the back reflection indication. This estimate is generally less accurate than an estimate made from an actual flaw indication. There is no assurance that only one flaw produces a given loss of back reflection; other factors, such as excessive roughness of the back surface or internal microporosity, can also reduce the amplitude of the back reflection.

One means of distinguishing whether a certain loss of back reflection is due to the presence of identifiable flaws is to move the search unit back and forth about a mean position over the suspected flaw. If the back reflection rises and falls as the search unit is moved, the presence of specific identifiable flaws can be presumed. Angle-beam techniques or other nondestructive inspection methods can then be used for positive identification of the flaw. However, if the back reflection remains relatively steady as the search unit is moved but the amplitude of the indication is measurably lower than the expected or standard value, the material presumably contains many small flaws distributed over a relatively broad region. This material condition may or may not be amenable to further study using other ultrasonic techniques or other nondestructive methods.

Spurious indications from reflections or indications of sources other than discontinuities are always a possibility. Reflections from edges and corners, extra reflections due to mode conversion, and multiple reflections from a single interface often look like flaw indications. Sometimes, these false, or nonrelevant, indications can be detected by correlation of the apparent flaw location with some physical feature of the testpiece. On other occasions, only the experience of the operator and thorough preliminary analysis of probable flaw types and locations can separate nonrelevant indications due to echoes from actual flaws. As a rule, any indication that remains consistent in amplitude and appearance as the search unit is moved back and forth on the surface of the testpiece should be suspected of being a nonrelevant indication if it can be correlated with a known reflective or geometric boundary. Nonrelevant indications are more likely to occur in certain types of inspection—for example, in longitudinal wave inspection from one end of a long shaft, inspection of complex-shape testpieces, inspection of parts where mixed longitudinal and shear waves may be present, and various applications of shear wave or surface wave techniques.

There are certain other types of indications that may interfere with the interpretation of A-scan data. All electronic circuits generate a certain amount of noise consisting of high-frequency harmonics of the main-signal frequency. Electronic noise is generally of low amplitude and is troublesome only when the main signal is also of low amplitude. In ultrasonic inspection, electronic noise can appear on an A-scan display as a general background, or waviness (called grass), in the main trace at all depths (Fig. 15a). This waviness, or grass, is more pronounced at the higher gain settings. Many instruments are equipped with reject circuits that filter out grass, although usually with some attendant loss of echo-signal amplitude, as shown in Fig. 15(b). When reject circuits are used, they should be adjusted so that grass is reduced only enough not to be a hindrance. If too much rejection is used, small-amplitude echoes will be suppressed along with the grass, and there will be a loss in sensitivity of the inspection technique and the linearity of the instrument will be affected.

A second type of interference occurs when coarse-grain materials are inspected. Reflections from the grain boundaries of coarse-grain materials can produce spurious indications throughout the test depth (Fig. 16). This type of interference, called hash, is most often encountered in coarse-grain steels; it is less troublesome with fine-grain steels or nonferrous metals. Sometimes, hash can be suppressed by adjusting the frequency and pulse length of the ultrasonic waves so that the sound beam is less sensitive to grain-boundary interfaces.

Angle-Beam Techniques

In angle-beam techniques, the incident sound pulse enters the testpiece at an oblique angle instead of at a right angle. In contrast with straight-beam tests, this approach eliminates echoes from the front and back surfaces and only displays reflections from discontinuities that are normal to the incident beam. Only rarely will a back surface be oriented properly to give a back reflection indication.

Figure 17 shows the arrangement of an angle-beam technique with a contact trans-

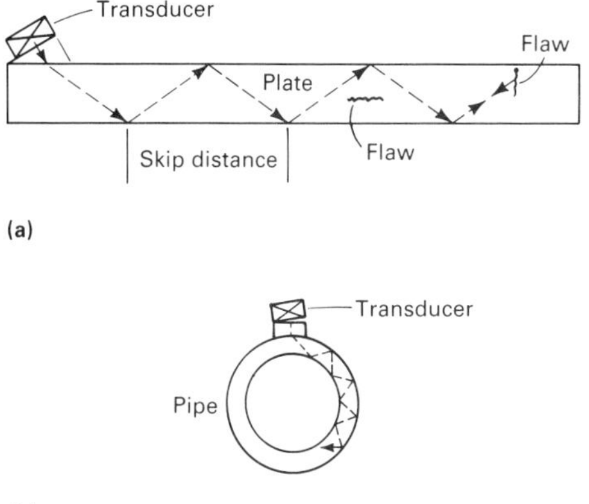

Fig. 17 Angle-beam testing with a contact transducer on a plate (a) and pipe (b)

ducer on a pipe and a plate. The sound beam enters the test material at an angle and propagates by successive zigzag reflections from the specimen boundaries until it is interrupted by a discontinuity or boundary where the beam reverses direction and is reflected back to the transducer. According to the angle selected, the wave modes produced in the test material may be mixed longitudinal and shear, shear only, or surface modes. Usually, angle-beam testing is accomplished with shear waves, although refracted longitudinal waves or surface waves can be used in some applications.

Angle-beam techniques are used for testing welds, pipe or tubing, sheet and plate material, and specimens of irregular shape (such as welds) where straight beams are unable to contact all of the surface. Angle-beam techniques are also useful in flaw location when there is a loss of back reflection. In flaw location, the time base (horizontal scale) on the oscilloscope must be carefully calibrated because in angle-beam testing there is no back reflection echo to provide a reference for depth estimates of the flaw. Usually, an extended time base is used so that flaws are located with one or two skip distances from the search unit (see Fig. 17 for the definition of skip distance).

Figure 17(a) shows how a shear wave from an angle-beam transducer progresses through a flat testpiece—by reflecting from the surfaces at points called "nodes". The linear distance between two successive nodes on the same surface is called the "skip distance" and is important in defining the path over which the transducer should be moved for reliable and efficient scanning. The skip distance can easily be measured by using a separate receiving transducer to detect the nodes or by using an angle-beam test block, or it can be calculated. Once the skip distance is known, the region over which the transducer should be moved to scan can be determined.

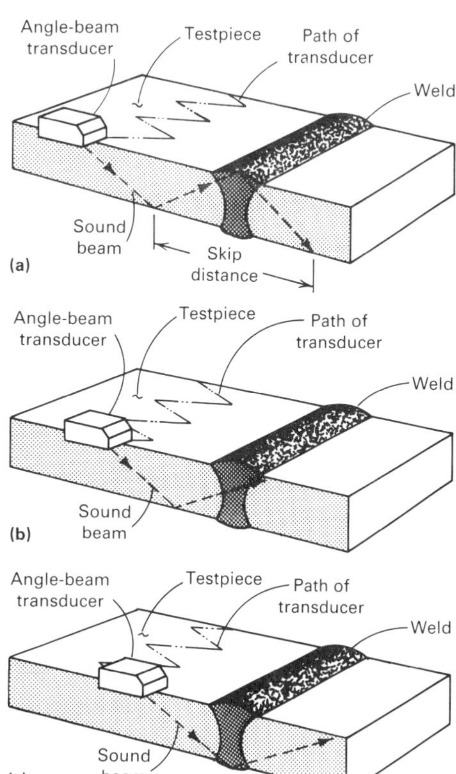

Fig. 18 Three positions of the contact type of transducer along the zigzag scanning path used during the manual angle-beam ultrasonic inspection of welded joints. The movement of the sound beam path across the weld is shown on a section taken along the centerline of the transducer as it is moved from the far left position in the scanning path (a), through an intermediate position (b), to the far right position (c).

Moving the search unit back and forth between one-half skip distance and one skip distance from an area of interest can be used not only for the purpose of defining the location, depth, and size of a flaw but also for the general purpose of initially detecting flaws. Figure 18 illustrates this back-and-forth movement as a way of scanning a weld for flaws.

Sometimes, moving the search unit in an arc about the position of a suspected flaw or swiveling the search unit about a fixed position can be equally useful (Fig. 19a). As shown in Fig. 19(b), traversing the search unit in an arc about the location of a gas hole produces little or no change in the echo; the indication on the oscilloscope screen remains constant in both amplitude and position on the trace as the search unit is moved. On the other hand, if the search unit were to be swiveled on the same spot, the indication would abruptly disappear after the search unit had been swiveled only a few degrees.

If the flaw is a slag inclusion (Fig. 19a), swiveling the search unit on the same spot causes the echo indication to vary randomly; some peaks rise and others fall, and the position of the signal shifts in either direction on the time trace, as indicated by arrows on the oscilloscope screen display in Fig. 19(c). Traversing the search unit in an arc would cause the signal to vary randomly in amplitude and to broaden slightly rather than shift in position.

If the flaw is a crack (Fig. 19a), swiveling the search unit in either direction away from the direction of maximum echo amplitude causes the peak to fall rapidly, accompanied by a slight shift to the right on the time trace, as indicated by arrows on the display in Fig. 19(d). Traversing the search unit in an arc would cause the echo signal to broaden slightly and fall rapidly with no change in position.

Surface Wave Technique. A special adaptation of the angle-beam technique results in the propagation of a surface wave, as discussed in the section "Angle-Beam Units" in this article. Surface waves are mainly used for the detection of shallow surface cracks and other similar flaws occurring at or just below the surface of the testpiece. This technique is most effective when flaws are most likely to extend to the surface or to be located in the dead zone for other techniques. Display appearance is

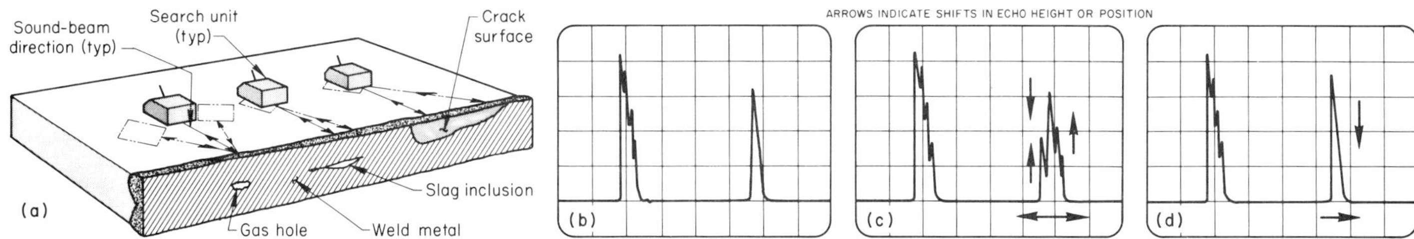

Fig. 19 Angle-beam inspection of a weldment showing effect of search-unit movements on oscilloscope screen display patterns from three different types of flaws in welds. (a) Positions of search units on the testpiece. (b) Display pattern obtained from a gas hole as the result of traversing the search unit in an arc about the location of the flaw. (c) Display pattern obtained from a slag inclusion as the result of swiveling the search unit on a fixed point. (d) Display pattern obtained from a crack, using the same swiveling search-unit movement as in (c).

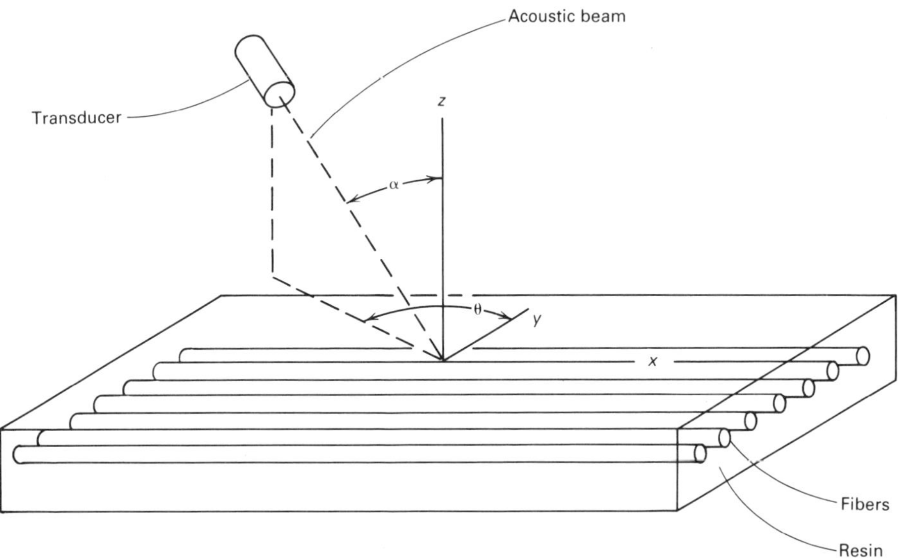

Fig. 20 Schematic of polar backscattering setup. α, angle of incidence; θ, polar angle (the angle between the y-axis and the projection of the beam path on the x-y plane)

similar to that for ordinary angle-beam testing; only flaw indications are displayed on an extended-time sweep trace.

Polar backscattering is basically an angle-beam technique, in which a single transducer has an oblique incidence with the front and back surface of a testpiece. Like other angle-beam techniques, this approach eliminates the detection of reflections from the front and back of the testpiece and only accounts for scattering from discontinuities that are normal to the ultrasonic pulse. In polar backscattering, however, the primary objective is to measure the amplitude of scattering as a function of transducer orientation. Polar backscattering is a useful nondestructive evaluation method for composite materials because their defects mostly have angle-dependent characteristics, which can be revealed by this method (Ref 3).

The test setup for polar backscattering is shown schematically in Fig. 20. Generally, fibers or discontinuities backscatter when the polar angle is normal to their surfaces. Figure 21 shows the response from a cross-ply SiC/Ti laminate tested at an incidence angle of 16.5°. The maxima in backscattering are observed each time the ultrasonic beam is normal to a fiber axis. The finite width of the peaks on the angular spectrum is determined by the transducer and the fiber directivity. The test can be performed in the frequency range of 1 to 25 MHz, using broadband pulses.

Polar backscattering can be used to detect matrix cracking because matrix cracks generate much higher backscattering than the backscattering of the fibers (about 30 dB for graphite-epoxy). By setting a gate with a cutoff level above the backscattering amplitude of the fibers, matrix cracking can be easily detected, as indicated in Fig. 22. This concept can also be applied to the detection of transverse cracking in ARALL laminates, which consist of aramid-aluminum layers (Ref 4).

Porosity is another type of defect that can be detected by polar backscattering. Generally, porosity accumulates between the layers of a composite laminate. Because the layers are randomly spread and have no preferred orientation, they generate backscattering at all polar angles. This behavior of porosity is shown in Fig. 23, in which the backscattering responses are shown for two laminates. As can be seen, porosity introduces an increase in scattering for angles that are not normal to the fiber axis as compared to the response from a defect-free laminate. This behavior of the fibers generates a spatial window in the backscattering through which defects of different scattering directivity can be characterized.

Backscattering can also be used to detect corrosion in various metals. This is feasible because corrosion disrupts the surface of the tested metal (Ref 4). Pitting corrosion, as well as scale, can be very easy to detect when testing aluminum plates at a 16° angle of incidence.

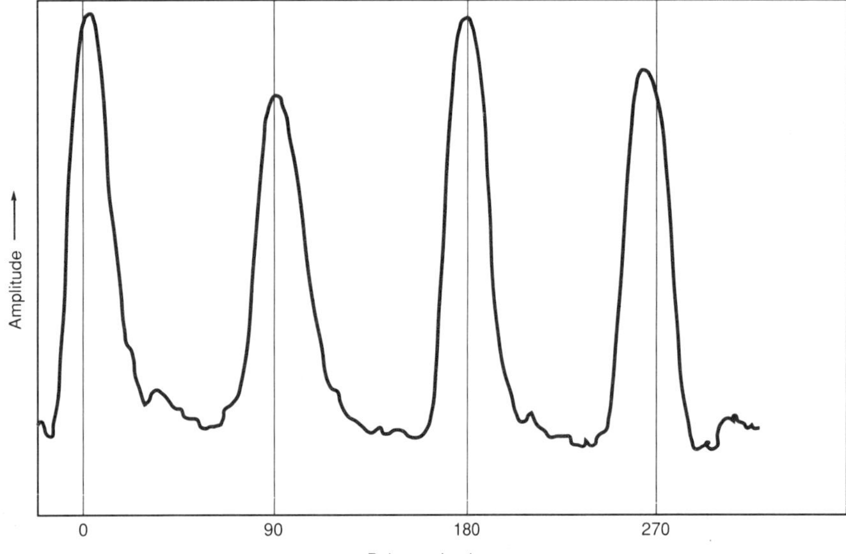

Fig. 21 Polar backscattering response from a SiC/Ti cross-ply laminate. The angle of incidence was 16.5°.

Transmission Methods*

Regardless of whether transmission ultrasonic testing is done with direct beams or reflected beams, flaws are detected by comparing the intensity of ultrasound transmitted through the testpiece with the intensity transmitted through a reference standard made of the same material. Transmission testing requires two search units—one to transmit the ultrasonic waves and one to receive them. Immersion techniques or water-column (bubbler or squirter) techniques (see the section "Water-Column Designs" in this article) are most effective because they provide efficient and relatively uniform coupling between the search units and the testpiece.

Good coupling is critical to transmission methods because variations in sound transmission through the couplants have corresponding effects on measured intensity. These variations in measured intensity introduce errors into the test results and frequently lead to invalid tests. For example, if go/no-go testing is being done and the criterion for rejection is a 10% loss in transmitted intensity, variations of 10% or more in coupling efficiency can cause the rejection of flaw-free testpieces. In addition to good

*Examples 1 and 2 in this section were provided by Robert W. Pepper, Textron Specialty Materials.

(a)

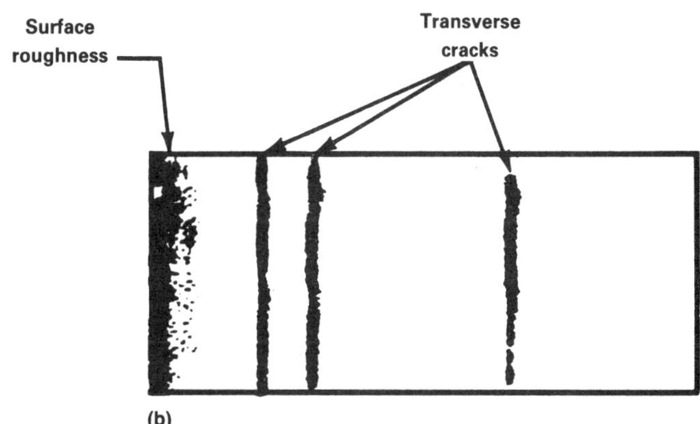

(b)

Fig. 22 C-scan image of polar backscattering from transverse cracks in a graphite-epoxy laminate. (a) Fatigued sample. (b) Statically loaded sample

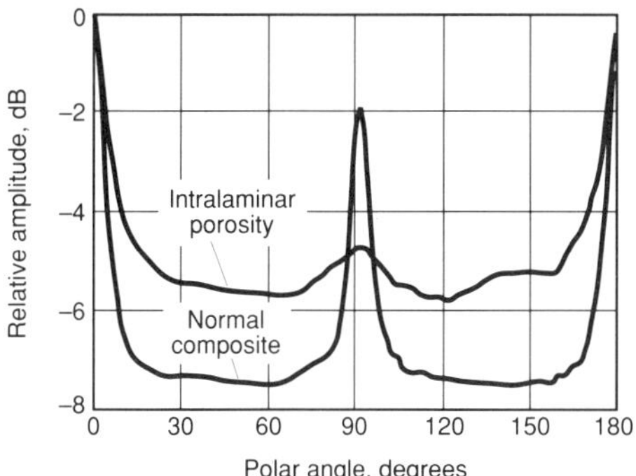

Fig. 23 Effect of porosity on the polar backscattering from a graphite-epoxy laminate

coupling, accurate positioning of the search units with respect to each other is critical. Once proper alignment of the search units is established, they should be rigidly held in position so that no variations in measured sound intensity can result from relative movement between them. Scanning is then accomplished by moving testpieces past the search units.

Displays of transmission test data can be either oscilloscope traces, strip chart recordings, or meter readings. Oscilloscopes are used to record data mainly when pulsed sound beams are used for testing; strip charts and meters are more appropriate for continuous beams. With all three types of display, alarms or automatic sorting devices can be used to give audible warning or to shunt defective workpieces out of the normal flow of production.

Pitch-catch testing can be done either with direct beams (through transmission testing) or with reflected beams using two transducers. The transducers may be housed in separate search units—one sending and the other receiving—or they may be combined in a single search unit. In both instances, pulses of ultrasonic energy pass through the material, and the intensities of the pulses are measured at the point of emergence. An oscilloscope display is triggered simultaneously with the initial pulse, and the transmitted-pulse indication appears on the screen to the right of the initial-pulse indication in a manner quite similar to the back reflection indication in pulse-echo testing. A major advantage of pitch-catch testing is that disturbances and spurious indications can be separated from the transmitted pulse by their corresponding transit times. Only the intensity of the transmitted pulse is monitored; all other sound waves reaching the receiver are ignored. An electronic gate can be set to operate an alarm or a sorting device when the monitored intensity drops below a preset value.

When reflected pulses are used, the technique is almost identical to the loss of back reflection technique that is often used in ordinary pulse-echo testing. In reflected-beam transmission testing, however, no attempt is made to evaluate any signal other than the main reflected pulse; echo signals that would be carefully interpreted in pulse-echo testing are not considered in transmission testing.

Continuous-beam testing does not require a pulser circuit or an oscilloscope. The initial intensity is not monitored, just the transmitted intensity. In this type of testing, considerable interference from standing waves occurs when the ultrasound of a single frequency is introduced into a part. Usually, only a small amount of the direct beam is absorbed by the receiver; the remainder is reflected back and forth repeatedly within the testpiece, soon filling the entire volume of material with a spatial field of standing waves. These standing waves create interference patterns of nodes and antinodes that alter the intensity of the direct beam. Small differences in dimensions or sound-transmission properties between two testpieces of the same design can result in large differences in the measured sound beam intensity because of a shift in the spatial distribution of standing waves.

Standing waves are avoided when pulsed ultrasonic beams are used. They cannot be eliminated when continuous beams are used; however, the effect can be made relatively constant by "wobbling" the test frequency, that is, by rapidly modulating the test frequency about the fundamental test frequency. Either periodic or aperiodic modulation can be used with equivalent results as long as the range of modulation is wide enough. The required range can be estimated from:

$$\frac{f_1}{f_2} = \frac{4n}{(4n + 1)} \quad \text{(Eq 12)}$$

where f_1 and f_2 are the results of the frequency range and n is the number of wavelength modes in the shortest direct-beam path from transmitter to receiver. The value of n is determined from:

$$n = \frac{fx}{V} \quad \text{(Eq 13)}$$

where f is the fundamental frequency, x is the path length of the direct beam, and V is the velocity of the particular wave form involved.

The frequency modulation range $(f_2 - f_1)$ varies with the length of the direct-beam path, as shown in Fig. 24 for a sound wave velocity of 6 km/s (4 miles/s). When applied to the straight-beam, longitudinal wave inspection of steel plate, which has a sound wave velocity of about 6 km/s (4 miles/s), the horizontal scale in Fig. 24 can be read directly as plate thickness. The frequency modulation range $(f_2 - f_1)$ is independent of fundamental frequency, f, but when it is considered as a percentage of the fundamental frequency, higher test frequencies require a lower percentage of modulation to average out standing-wave effects. This is important, because some search units lose sensitivity when the operating frequency differs from the design frequency by more than a few percent. Consequently, higher frequencies are required for the transmission testing of thin testpieces than for thick testpieces. In some cases, equipment limitations may make it impossible to use transmission methods on thin testpieces.

Applications. The main application of transmission methods is the inspection of plate for cracks and laminations that have relatively large dimensions compared to the size of the search units. The following two examples illustrate the variation of ultrason-

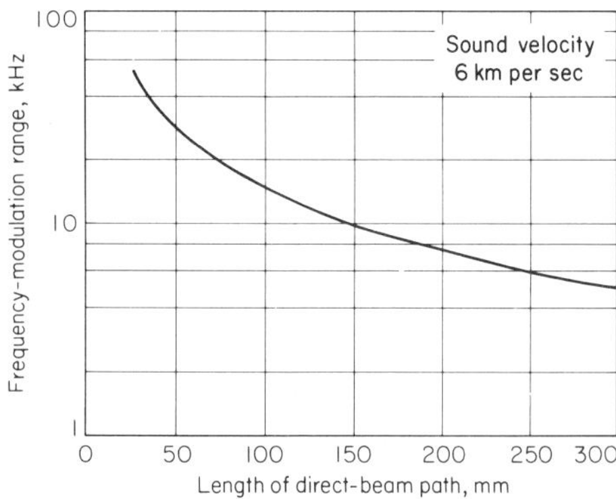

Fig. 24 Effect of direct-beam path length on frequency modulation range needed to avoid standing waves in the continuous-beam transmission testing of a material in which sound velocity is 6 km/s (4 miles/s)

ic transmission in the nondestructive evaluation of metal-matrix composite panels.

Example 1: Ultrasonic Inspection of Titanium-Matrix Composite Panels. Three titanium-matrix composite panels were made available for nondestructive characterization by ultrasonic inspection, velocity measurements, and film radiography. Only one panel showed significant anomalies, as revealed by the ultrasonic C-scan (Fig. 25). No significant variation was seen in the velocity or x-ray inspection data.

The panel was then sectioned parallel to the fiber lay-up, and tensile bar specimens were removed from good (zone B, Fig. 25) and poor (zone A) sound-transmission regions. The results of tensile testing from good and bad C-scan zones showed no correlation. This should be expected because the fiber strength dominates and because the matrix contribution is minimal even with porosity or laminar-matrix defects.

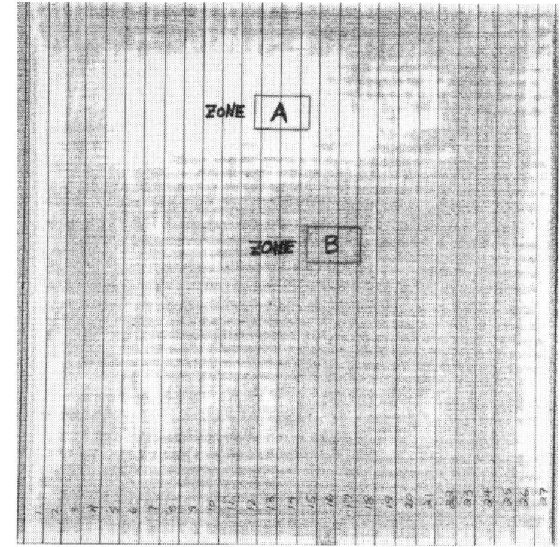

Fig. 25 C-scan of ultrasonic signal amplitudes after transmission through a titanium-matrix composite panel (six plies with 0° fiber orientation). Zone A indicates a region of poor sound transmission; Zone B is a region of good sound transmission. Courtesy of Textron Specialty Materials

The broken tensile specimens were then polished and photomicrographs taken from the good and bad C-scan zones. Figure 26(a) shows the region of poor ultrasonic transmission (zone A, Fig. 25). Inadequate consolidation, porosity sites, bunched fibers, and large grain sizes are visible throughout this zone. Specimens sectioned through the region of good ultrasonic transmission (zone B) exhibited no porosity (Fig. 26b).

Example 2: Ultrasonic Inspection of Surface Cracks and Delamination in a Metal-Matrix Composite Panel. Figure 27(a) shows an ultrasonic C-scan of a titanium-matrix composite with surface cracks. This scan was performed with a 50-MHz focused transducer at 10× magnification. Penetration of the sound was minimal because of the high frequency being used, although there is some indication of poor consolidation (delamination) to the right of the centered crack.

Figure 27(b) shows an ultrasonic C-scan of the same defect area produced with a 10-MHz focused transducer. The C-scan at this frequency provided better resolution of the delaminated area, although it was found to be inadequate for determining fiber integrity and surface cracking conditions.

Lamb Wave Testing. For the high-speed testing of a plate, strip, or wire, where the thickness is of the order of a few wavelengths, there is considerable benefit in using Lamb waves (Ref 5). Lamb waves (or plate waves) are elastic waves that propagate in plates of finite thickness as guided waves and are associated with particle motion in a plane normal to the surface. As mentioned in the section "Lamb Waves" in this article, Lamb waves can be symmetrical or asymmetrical, and they can be of different modes or mixed modes, with the velocity depending on the mode.

The present understanding of Lamb waves is based on theoretical considerations and empirical observations; the precise particle motion involved has not been established unequivocally. However, knowledge about the phenomena associated with Lamb waves has been developed, so that there is a sound basis for their use in nondestructive inspection.

Lamb waves are generated by the oblique incidence of an ultrasonic wave with a properly selected transmission frequency. In addition to the common piezoelectric transducer, Lamb waves can be generated and detected with an electromagnetic-acoustic

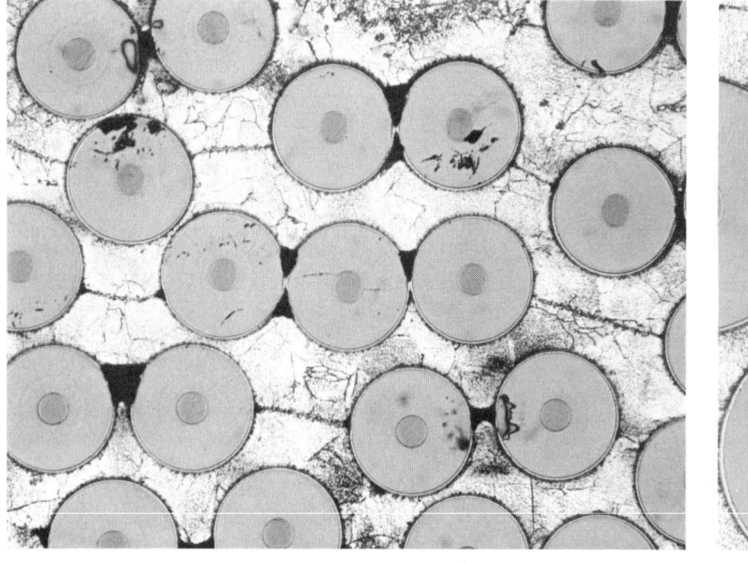
(a)

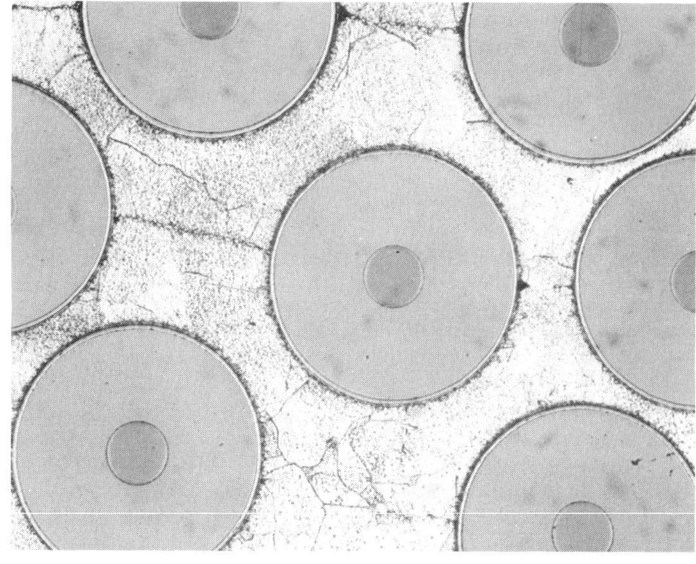
(b)

Fig. 26 Photomicrographs of specimens taken from the good and bad C-scan zones shown in Fig. 25. (a) Specimen from zone A. 115×. (b) Specimen from zone B. 230×

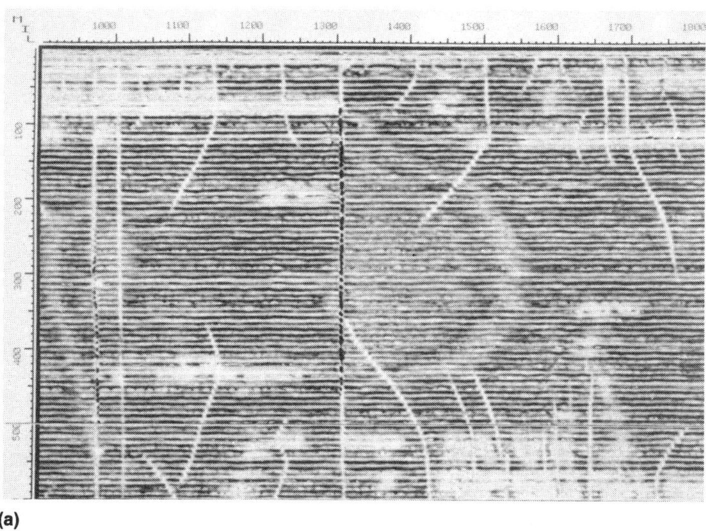
(a)

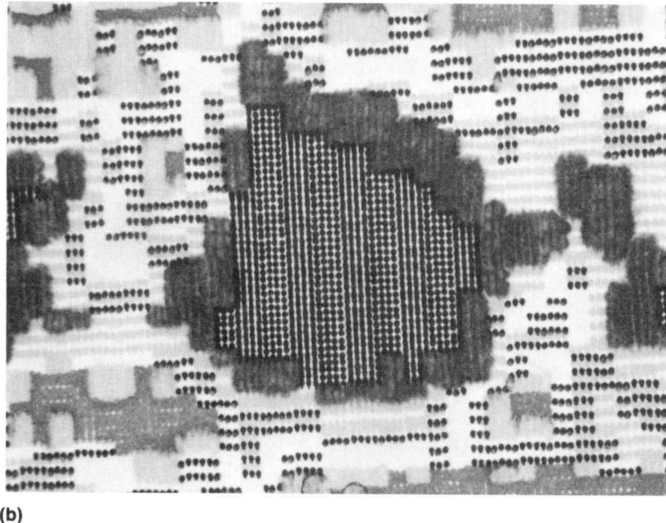
(b)

Fig. 27 Ultrasonic C-scans of a titanium-matrix composite panel with surface cracks and delamination. (a) C-scan using a 50-MHz focused transducer at a magnification of 10×. (b) C-scan of same defect using a 10-MHz focused transducer

(EMA) probe with the appropriate coil configuration (see the section "EMA Transducers" in this article).

The most widely applied technique is to monitor the transmission along a plate between two probes with a fixed spacing. For example, the quality of a spot or seam weld between two sheets can be monitored by the transmission along one sheet and through the spot weld (Fig. 28). This can be done as on-line monitoring of the welding operation, but temperature effects at the weld nugget and mode conversions at the liquid/metal interface complicate the interpretation of the results. In wire specimens, the waves are usually known as rod waves, and there is an advantage in generating them by magnetostriction, with a coil over the end, so that no surface contact is needed. Spirally rotating surface waves can also be generated in a wire with a pair of probes angled to the wire axis.

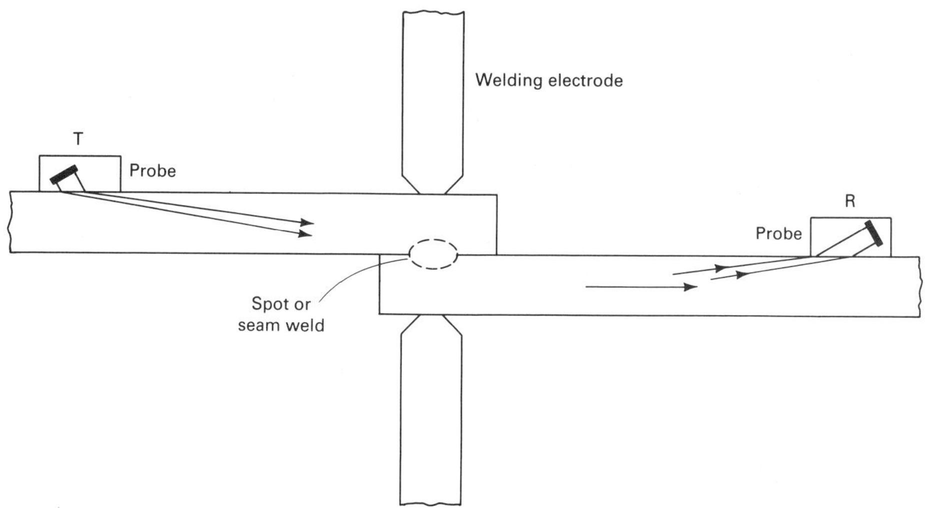

Fig. 28 Use of Lamb waves to inspect a spot-welded joint. Source: Ref 5

Leaky Lamb Wave Testing. In recent years, substantial progress has been made in the understanding of the Lamb wave. This has resulted from the development of sophisticated theoretical and experimental techniques and the use of powerful computing tools. It is now possible to predict and measure the response of Lamb waves even in anisotropic layered materials, such as fiber-reinforced composites. One approach in this effort involves the use of the leaky Lamb wave (LLW) phenomenon. Recent research indicates that this phenomenon can be used to detect and characterize defects that would remain undetected in conventional ultrasonic techniques (Ref 6).

In LLW testing, Lamb waves are induced in a plate through a mode conversion by obliquely insonifying the plate at a properly selected transmission frequency. If a laminate is immersed in a fluid, the Lamb waves leak energy into the fluid at an angle that is established by Snell's law and the Lamb wave phase velocity. When an LLW is induced, the reflected field is distorted. The specular component of the wave and the leaky wave interfere, with a phase cancellation occurring and a null being generated between them. A schematic of LLW field behavior is shown in Fig. 29. For nondestructive evaluation, one can monitor changes in the frequency of given LLW modes or changes in amplitude at the null zone.

A theory has been developed and corroborated regarding the behavior of LLWs in composites (Ref 7). A matrix method solved the problem of wave propagation in multilayered anisotropic media subjected to time-harmonic (single-frequency) or transient (pulse) disturbances. The method was applied to obtain a formal solution of the response of layered composite plates. The solution led to stable numerical schemes for the evaluation of the displacement and stress fields within the laminate. This analysis of LLW behavior was applied to both multiorientation laminates and bonded structures with different interface conditions.

Leaky Lamb wave tests are performed with a pitch-catch setup and flat broadband transducers; the receiver is placed at the null zone of the LLW field. The LLW field can be tested at various angles of incidence in the frequency range of 0.1 to 15.0 MHz using either tone bursts or pulses. For tone burst, signals with a duration sufficiently long to establish a steady-state condition are used. The signals are either displayed as a function of time for a single frequency or as a function of frequency in a sweep mode. For tests with short-duration pulses, the transducers are first adjusted to place the receiver at the null zone, using the tone-burst setup. Then the transducer is substi-

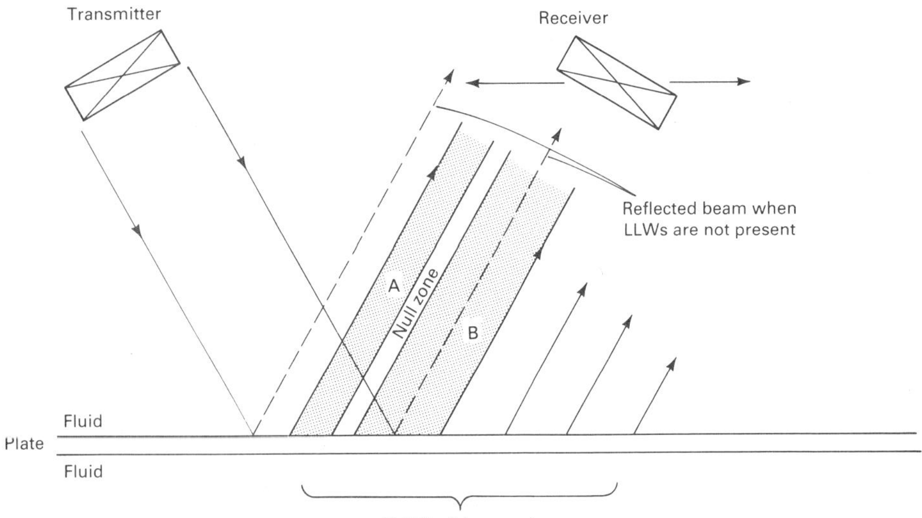

Fig. 29 Schematic of LLW phenomenon. Zone A represents the specular component, while zone B represents the LLW component. The presence of LLWs shifts the observation of reflected energy into zone A.

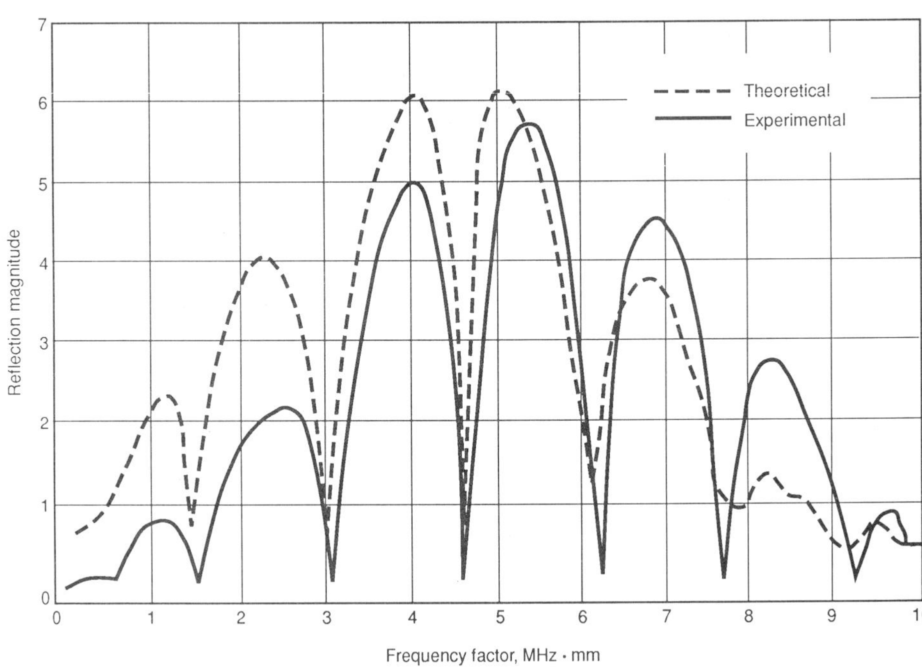

Fig. 30 Experimental and theoretical LLW spectral response of a unidirectional graphite-epoxy laminate obtained with a 15° angle of incidence. The *x*-axis is usually expressed in frequency times thickness to eliminate the effect of total thickness of the plate from the data.

tuted with a pulser-receiver, and the signals are tested from the A-scan display. The two aspects of LLW phenomena that are commonly measured are described below.

Reflection Field. At specific angles of incidence, the spectral response, namely, the reflected amplitude as a function of frequency, is examined. An example of such spectra for a unidirectional graphite-epoxy laminate tested along the fibers is shown in Fig. 30. The reflected field can also be analyzed in the time domain using pulses and commercial pulser-receivers, with the receiving transducer placed at the null zone (Ref 6). In this case, two parameters can be used to examine the composite laminate: the amplitude and the time-of-flight.

Using amplitude measurements while C-scanning a laminate can reveal material property variations and many types of defects. These include delaminations, ply-gap, porosity, and resin/fiber ratio changes. Further details are obtained when using time-of-flight measurements because the depths of the discontinuities are also presented. An example of a time-of-flight LLW C-scan is shown in Fig. 31, in which different colors were assigned to the various time-of-flight ranges to indicate the depth of the defects in the sample. The dark lines along the C-scan image are parallel to the fiber orientation and are a result of the migration of porosity (microballoons of 0.04 mm, or 0.0016 in., diameter) from the center of the sample outward during the laminate cure.

Dispersion curves provide a plot of the phase velocity of Lamb waves as a function of frequency. The *x*-axis is usually expressed in frequency times thickness; therefore, the effect of the total thickness of the plate is eliminated from the data. The modes are determined from the reflected spectra by finding the frequencies at which the minima occur. By an inversion process, one can employ the theory to determine the elastic properties of composite laminates from experimental dispersion curves.

Electronic Equipment

Although the electronic equipment used for ultrasonic inspection can vary greatly in detail among equipment manufacturers, all general-purpose units consist of a power supply, a pulser circuit, a search unit, a receiver-amplifier circuit, an oscilloscope, and an electronic clock. Many systems also include electronic equipment for signal conditioning, gating, automatic interpretation, and integration with a mechanical or electronic scanning system. Moreover, advances in microprocessor technology have extended the data acquisition and signal-processing capabilities of ultrasonic inspection systems.

Power Supply. Circuits that supply current for all functions of the instrument constitute the power supply, which is usually energized by conventional 115-V or 230-V alternating current. There are, however, many types and sizes of portable instruments for which the power is supplied by batteries contained in the unit.

Pulser Circuit. When electronically triggered, the pulser circuit generates a burst of alternating voltage. The principal frequency of this burst, its duration, the profile of the envelope of the burst, and the burst repetition rate may be either fixed or adjustable, depending on the flexibility of the unit.

Search Units. The transducer is the basic part of any search unit. A sending transducer is one to which the voltage burst is applied, and it mechanically vibrates in response to the applied voltage. When appropriately coupled to an elastic medium, the transducer thus serves to launch ultrasonic waves into the material being inspected.

A receiving transducer converts the ultrasonic waves that impinge on it into a corresponding alternating voltage. In the pitch-catch mode, the transmitting and receiving transducers are separate units; in the pulse-echo mode, a single transducer alternately serves both functions. The various types of search units are discussed later in this article.

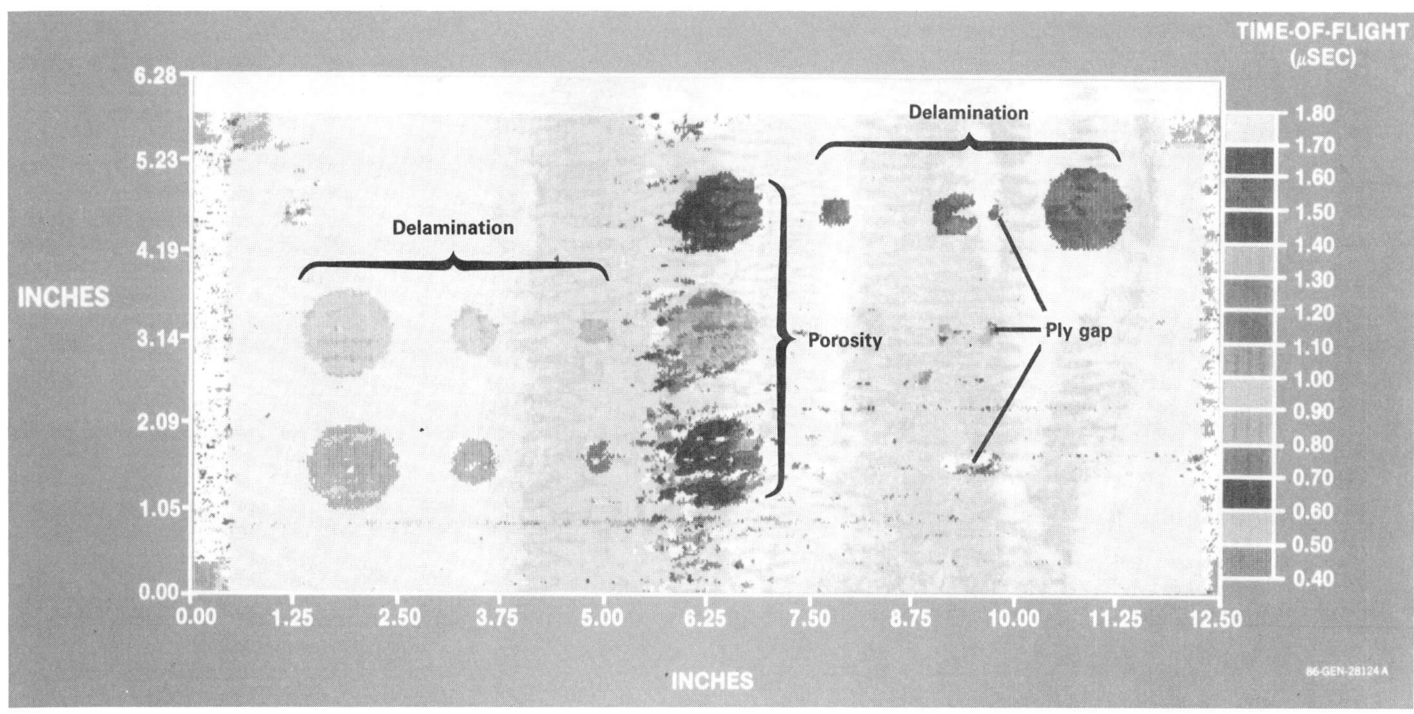

Fig. 31 Pulsed LLW C-scan showing time-of-flight variations in a graphite-epoxy laminate tested at 45° with the fibers

Receiver-amplifier circuits electronically amplify return signals from the receiving transducer and often demodulate or otherwise modify the signals into a form suitable for display. The output from the receiver-amplifier circuit is a signal directly related to the intensity of the ultrasonic wave impinging on the receiving transducer. This output is fed into an oscilloscope or other display device.

Oscilloscope. Data received are usually displayed on an oscilloscope in either video mode or radio frequency mode. In video-mode display, only peak intensities are visible on the trace; in the RF mode, it is possible to observe the waveform of signal voltages. Some instruments have a selector switch so that the operator can choose the display mode, but others are designed for single-mode operation only.

Clock. The electronic clock, or timer, serves as a source of logic pulses, reference voltage, and reference waveform. The clock coordinates operation of the entire electronic system.

Signal-conditioning and gating circuits are included in many commercial ultrasonic instruments. One common example of a signal-conditioning feature is a circuit that electronically compensates for the signal-amplitude loss caused by attenuation of the ultrasonic pulse in the testpiece. Electronic gates, which monitor returning signals for pulses of selected amplitudes that occur within selected time-delay ranges, provide automatic interpretation. The set point of a gate corresponds to a flaw of a certain size that is located within a prescribed depth range. Gates are often used to trigger alarms or to operate automatic systems that sort testpieces or identify rejectable pieces.

Image- and Data-Processing Equipment. As a result of the development of microprocessors and modern electronics, many ultrasonic inspection systems possess substantially improved capabilities in terms of signal processing and data acquisition. This development allows better flaw detection and evaluation (especially in composites) by improving the acquisition of transient ultrasonic waveforms and by enhancing the display and analysis of ultrasonic data. The development of microprocessor technology has also been useful in portable C-scan systems with hand-held

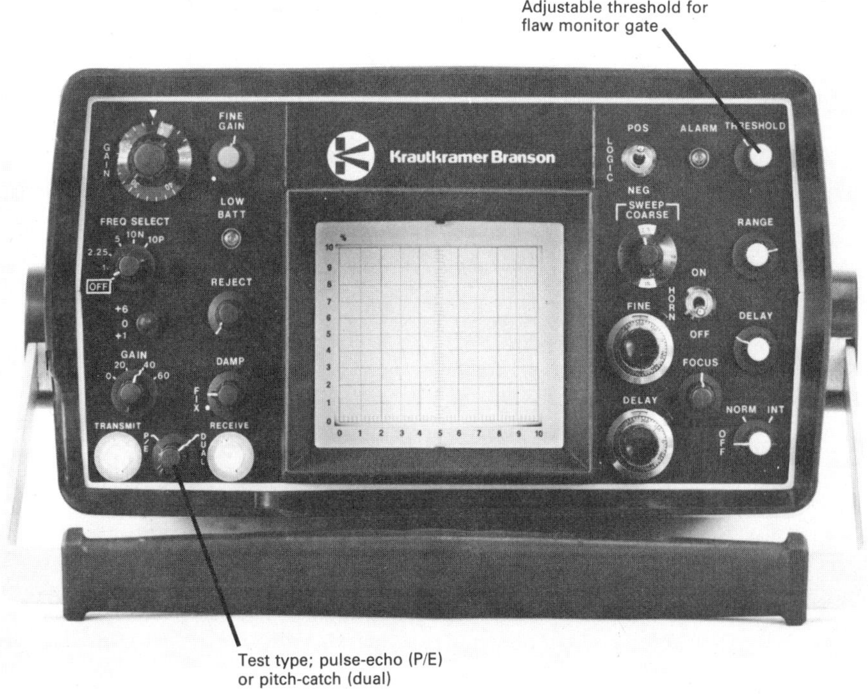

Fig. 32 Typical pulse-echo ultrasonic instrument

transducers (see the section "Scanning Equipment" in this article).

In an imaging system, the computer acquires the position of the transducer and the ultrasonic data from the point. Combining the two, an image is produced on the computer monitor, either in color or in gray-scale shades; each point is represented by a block of a predetermined size. To expedite the inspection, it is common to use large blocks. Areas that require special attention are then inspected with a higher resolution by using smaller blocks. Built-in software in the system allows users to analyze available information under precise, controlled conditions and enables simulation of a top view regardless of transducer angle.

Control Systems

Even though the nomenclature used by different instrument manufacturers may vary, certain controls are required for the basic functions of any ultrasonic instrument. These functions include power supply, clock, pulser, receiver-amplifier, and display. In most cases, the entire electronic assembly, including the controls, is contained in one instrument. A typical pulse-echo instrument is shown in Fig. 32.

The power supply is usually controlled by switches and fuses. Time delays can be incorporated into the system to protect circuit elements during warm-up. The pulses of ultrasonic energy transmitted into the testpiece are adjusted by controls for pulse-repetition rate, pulse length, and pulse tuning. A selector for a range of operating frequencies is usually labeled "frequency," with the available frequencies given in megahertz.

For single-transducer inspection, transmitting and receiving circuits are connected to one jack, which is connected to a single transducer. For double-transducer inspection, such as through transmission or pitch-catch inspection, a T (transmit) jack is provided to permit connecting one transducer for use as a transmitter, and an R (receive) jack is provided for the use of another transducer for receiving only. A selector switch (test switch) for through (pitch-catch) or normal (pulse-echo) transmission is provided for control of the T and R jacks.

Gain controls for the receiver-amplifier circuit usually consist of fine- and coarse-sensitivity selectors or one control marked "sensitivity." For a clean video display, with low-level electronic noise eliminated, a reject control can be provided.

The display (oscilloscope) controls are usually screwdriver-adjusted, with the exception of the scale illumination and power on/off. After initial setup and calibration, the screwdriver-adjusted controls seldom require additional adjustment. The controls and their functions for the display unit usually consist of the following:

- Controls for vertical position of the display on the oscilloscope screen
- Controls for horizontal position of display on the oscilloscope screen
- Controls for brightness of display
- Control for adjusting focus of trace on the oscilloscope screen
- Controls to correct for distortion or astigmatism that may be introduced as the electron beam sweeps across the oscilloscope screen
- A control that varies the level of illumination for a measuring grid usually incorporated in the transparent faceplate covering the oscilloscope screen
- Timing controls, which usually consist of sweep delay and sweep-rate controls, to provide coarse and fine adjustments to suit the material and thickness of the testpiece. The sweep-delay control is also used to position the sound entry point on the left side of the display screen, with a back reflection or multiples of back reflections visible on the right side of the screen
- On/off switch

In addition to those listed above, there are other controls that may or may not be provided, depending on the specific type of instrument. These controls include the following:

- A marker circuit, which provides regularly spaced secondary indications (often in the form of a square wave) on or below the sweep line to serve the same purpose as scribe marks on a ruler. This circuit is activated or left out of the display by a marker switch for on/off selection. Usually there will also be a marker-calibration or marker-adjustment control to permit selection of marker-circuit frequency. The higher the frequency, the closer the spacing of square waves, and the more accurate the measurements. Marker circuits are controlled by timing signals triggered by the electronic clock. Most modern ultrasonic instruments do not have marker circuits
- A circuit to electronically compensate for a drop in the amplitude of signals reflected from flaws located deep in the testpiece. This circuit may be known as distance-amplitude correction, sensitivity-time control, time-corrected gain, or time-varied gain
- Damping controls that can be used to shorten the pulse duration and thus adjust the length of the wave packet emanating from the transducer. Resolution is improved by higher values of damping
- High-voltage or low-voltage driving current, which is selected for the transducer with a transducer-voltage switch
- Gated alarm units, which enable the use of automatic alarms when flaws are detected. This is accomplished by setting up controllable time spans on the display that correspond to specific zones within the testpiece. Signals appearing within the gates may automatically operate visual or audible alarms. These signals may also be passed on to display devices or strip-chart recorders or to external control devices. Gated alarm units usually have three controls: the gate-start or delay control, which adjusts the location of the leading edge of the gate on the oscilloscope trace; the gate-length control, which adjusts the length of the gate or the location of the gate trailing edge; and the alarm-level or sensitivity control, which establishes the minimum echo height necessary to activate an alarm circuit. A positive/negative logic switch determines whether the alarm is triggered above or below the threshold level

Transducer Elements

The generation and detection of ultrasonic waves for inspection are accomplished by means of a transducer element acting through a couplant. The transducer element is contained within a device most often referred to as a search unit (or sometimes as a probe). Piezoelectric elements are the most commonly used transducer in ultrasonic inspection, although EMA transducers and magnetostriction transducers are also used.

Piezoelectric Transducers

Piezoelectricity is pressure-induced electricity; this property is characteristic of certain naturally occurring crystalline compounds and some man-made materials. As the name piezoelectric implies, an electrical charge is developed by the crystal when pressure is applied to it. Conversely, when an electrical field is applied, the crystal mechanically deforms (changes shape). Piezoelectric crystals exhibit various deformation modes; thickness expansion is the principal mode used in transducers for ultrasonic inspection.

The most common types of piezoelectric materials used for ultrasonic search units are quartz, lithium sulfate, and polarized ceramics such as barium titanate, lead zirconate titanate, and lead metaniobate. Characteristics and applications of these materials are summarized in Table 5.

Quartz crystals were initially the only piezoelectric elements used in commercial ultrasonic transducers. Properties of the transducers depended largely on the direction along which the crystals were cut to make the active transducer elements. Principal advantages of quartz-crystal transducer elements are electrical and thermal stability, insolubility in most liquids, high mechanical strength, wear resistance, excellent uniformity, and resistance to aging. A limitation of quartz is its comparatively low electromechanical conversion efficien-

Table 5 Characteristics and applications of transducer (piezoelectric) elements

Piezoelectric element	Characteristics of piezoelectric elements(a): Efficiency: Transmit	Receive	Coupling: To water	To metal	Tolerance to elevated temperature	Damping ability	Undesired modes (inherent noise)	Suitability of element in(a): Contact inspection: Straight-beam	Angle-beam	Immersion inspection
Quartz	P	G	G	F	G	F	G	G	F	G
Lithium sulfate	F	E	E	P	P	E	E	P	F	E
Barium titanate	G	P	G	G	P	P	P	G	G	F
Lead zirconate titanate	E	F	F	E	E	F	P	E	E	F
Lead metaniobate	G	F	G	E	E	E	G	E	E	G

(a) E, excellent; G, good; F, fair; P, poor

cy, which results in low loop gain for the system.

Lithium Sulfate. The principal advantages of lithium sulfate transducer elements are ease of obtaining optimum acoustic damping for best resolution, optimum receiving characteristics, intermediate conversion efficiency, and negligible mode interaction. The main disadvantages of lithium sulfate elements are fragility and a maximum service temperature of about 75 °C (165 °F).

Polarized ceramics generally have high electromechanical conversion efficiency, which results in high loop gain and good search-unit sensitivity. Lead zirconate titanate is mechanically rugged, has a good tolerance to moderately elevated temperature, and does not lose polarization with age. It does have a high piezoelectric response in the radial mode, which sometimes limits its usefulness.

Barium titanate is also mechanically rugged and has a high radial-mode response. However, its efficiency changes with temperature, and it tends to depolarize with age, which makes barium titanate less suitable for some applications than lead zirconate titanate.

Lead metaniobate exhibits low mechanical damping and good tolerance to temperature. Its principal limitation is a high dielectric constant, which results in a transducer element with a high electrical capacitance.

Selection of a piezoelectric transducer for a given application is done on the basis of size (active area) of the piezoelectric element, characteristic frequency, frequency bandwidth, and type (construction) of search unit. Descriptions of various types of search units with piezoelectric elements are given in the section "Search Units" in this article. Different piezoelectric materials exhibit different electrical-impedance characteristics. In many cases, tuning coils or impedance-matching transformers are installed in the search-unit housing to render a better impedance match to certain types of electronic instrumentation. It is important to match impedances when selecting a search unit for a particular instrument.

Both the amount of sound energy transmitted into the material being inspected (radiated power) and beam divergence are directly related to the size (active area) of the transducer element. Thus, it is sometimes advisable to use a larger search unit to obtain greater depth of penetration or greater sound beam area.

Each transducer has a characteristic resonant frequency at which ultrasonic waves are most effectively generated and received. This resonant frequency is determined mainly by the material and thickness of the active element. Any transducer responds efficiently at frequencies in a band centered on the resonant frequency. The extent of this band, known as bandwidth, is determined chiefly by the damping characteristics of the backing material that is in contact with the rear face of the piezoelectric element.

Narrow-bandwidth transducers exhibit good penetrating capability and sensitivity, but relatively poor resolution. (Sensitivity is the ability to detect small flaws; resolution is the ability to separate echoes from two or more reflectors that are close together in depth.) Broad-bandwidth transducers exhibit greater resolution, but lower sensitivity and penetrating capability, than narrow-bandwidth transducers.

Operating frequency, bandwidth, and active-element size must all be selected on the basis of inspection objectives. For example, high penetrating power may be most important in the axial examination of long shafts. It may be best to select a large-diameter, narrow-bandwidth, low-frequency transducer for this application, even though such a transducer will have both low sensitivity (because of low frequency and large size) and low resolution (because of narrow bandwidth).

When resolution is important, such as in the inspection for near-surface discontinuities, use of a broad-bandwidth transducer is essential. Penetrating capability probably would not be very important, so the relatively low penetrating power accompanying the broad bandwidth would not be a disadvantage. If necessary, high sensitivity could be achieved by using a small, high-frequency, broad-bandwidth transducer; an increase in both sensitivity and penetrating power would require the use of a large, high-frequency transducer, which would emit a more directive ultrasonic beam. Resolution can also be improved by using a very short pulse length, an immersion technique, or delay-tip or dual-element contact-type search units.

Array Transducers. In recent years, there has been a growing need to increase the speed of ultrasonic inspections. The fastest means of scanning is the use of an array of transducers that are scanned electronically by triggering each of the transducers sequentially. Such transducers consist of several crystals placed in a certain pattern and triggered one at a time, either manually or by a multiplexer. Array transducers can either transmit normal to their axis or can have an angle beam. To perform beam steering, sound is generated from the various crystals with a predetermined phase difference. The degree of difference determines the beam angle.

EMA Transducers

Electromagnetic-acoustic (EMA) phenomena can be used to generate ultrasonic waves directly into the surface of an electrically conductive specimen without the need for an external vibrating transducer and coupling. Similar probes can also be used for detection, so that a complete noncontact transducer can be constructed. The

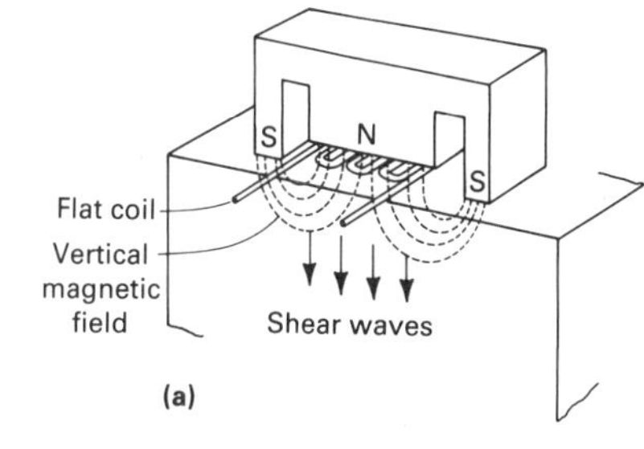

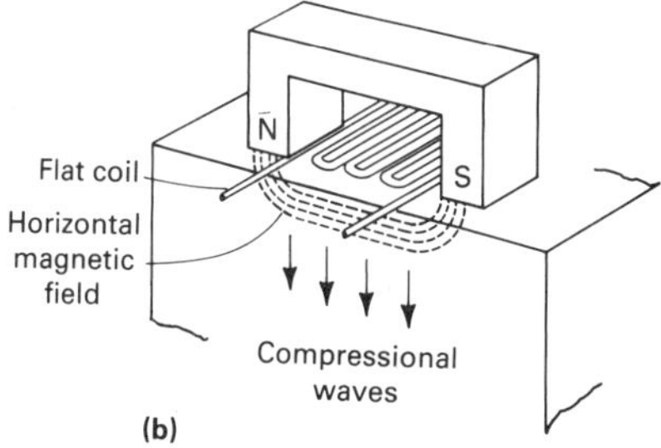

Fig. 33 Schematic of EMA transducer. (a) Arrangement for the production of shear waves. (b) Arrangement for the production of compressional waves

method is therefore particularly suitable for use on high-temperature specimens, rough surfaces, and moving specimens. The received ultrasonic signal strength in EMA systems is 40 to 50 dB lower than a conventional barium titanate probe, but input powers can be increased.

The principle of EMA transducers is illustrated in Fig. 33. A permanent magnet or an electromagnet produces a steady magnetic field, while a coil of wire carries an RF current. The radio frequency induces eddy currents in the surface of the specimen, which interact with the magnetic field to produce Lorentz forces that cause the specimen surface to vibrate in sympathy with the applied radio frequency. When receiving ultrasonic energy, the vibrating specimen can be regarded as a moving conductor or a magnetic field, which generates currents in the coil. The clearance between the transducer and the metal surface affects the magnetic field strength and the strength of the eddy currents generated, and the ultrasonic intensity falls off rapidly with increasing gap. Working at 2 MHz, a gap of 1.0 to 1.5 mm (0.04 to 0.06 in.) has been found to be practicable provided it is kept reasonably constant (Ref 5).

Magnetostriction Transducers

Although magnetostriction transducers are seldom used in the ultrasonic inspection of metals, magnetostriction has advantages in the Lamb wave testing of wire specimens (see the section "Lamb Wave Testing" in this article). Magnetostrictive materials change their form under the influence of a magnetic field, and the most useful magnetostrictive material in practice is nickel. A nickel rod placed in a coil carrying a current experiences a change in length as a function of the current through the coil. A stack of plates of magnetostrictive material with a coil through them can produce an ultrasonic beam at right angles to the plate stack, and the frequency depends on the thickness. Transducers of this type are useful for very low ultrasonic frequencies (<200 kHz).

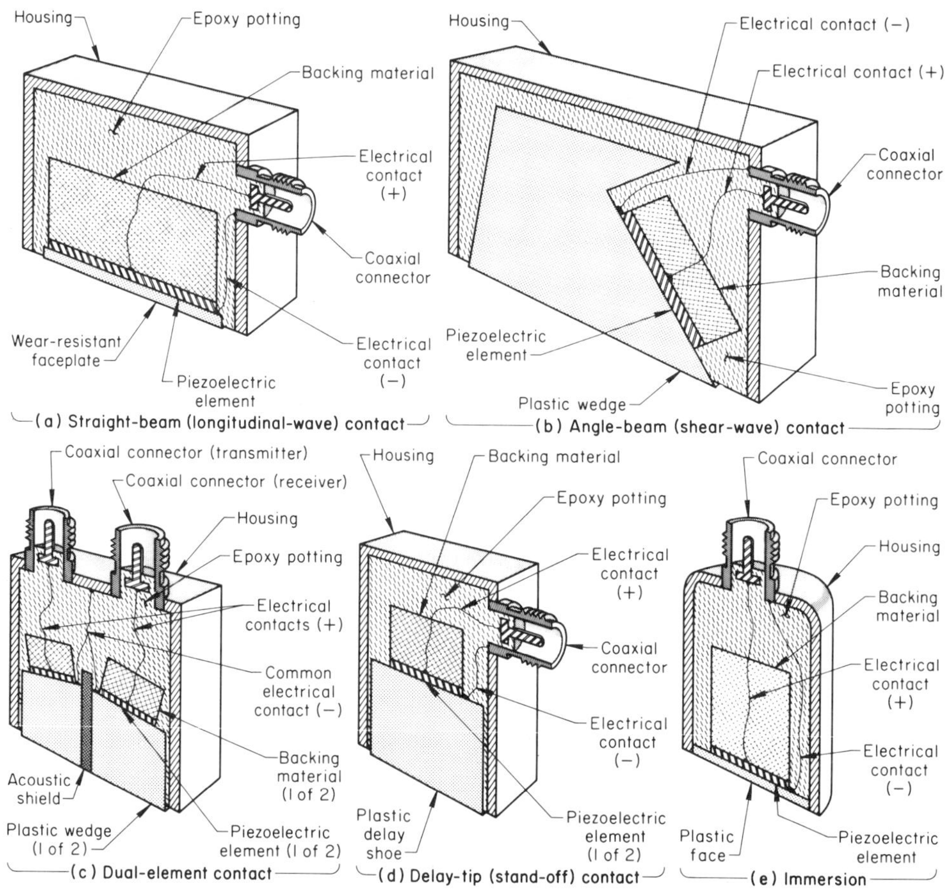

Fig. 34 Sectional views of five typical piezoelectric search units in ultrasonic inspection

Couplants

Air is a poor transmitter of sound waves at megahertz frequencies, and the impedance mismatch between air and most solids is great enough that even a very thin layer of air will severely retard the transmission of sound waves from the transducer to the testpiece. To perform satisfactory contact inspection with piezoelectric transducers, it is necessary to eliminate air between the transducer and the testpiece by the use of a couplant.

Couplants normally used for contact inspection include water, oils, glycerin, petroleum greases, silicone grease, wallpaper paste, and various commercial pastelike substances. Certain soft rubbers that transmit sound waves may be used where adequate coupling can be achieved by applying hand pressure to the search unit.

The following should be considered in selecting a couplant:

- Surface finish of testpiece
- Temperature of test surface
- Possibility of chemical reactions between test surface and couplant
- Cleaning requirements (some couplants are difficult to remove)

Water is a suitable couplant for use on a relatively smooth surface; however, a wetting agent should be added. It is sometimes appropriate to add glycerine to increase viscosity; however, glycerine tends to induce corrosion in aluminum and therefore is not recommended in aerospace applications.

Heavy oil or grease should be used on hot or vertical surfaces or on rough surfaces where irregularities need to be filled.

Wallpaper paste is especially useful on rough surfaces when good coupling is needed to minimize background noise and yield an adequate signal-to-noise ratio.

Water is not a good couplant to use with carbon steel testpieces, because it can promote surface corrosion. Oils, greases, and proprietary pastes of a noncorrosive nature can be used.

Heavy oil, grease, or wallpaper paste may not be good choices when water will suffice, because these substances are more difficult to remove. Wallpaper paste, like some proprietary couplants, will harden and may flake off if allowed to stand exposed to air. When dry and hard, wallpaper paste can be easily removed by blasting or wire brushing. Oil or grease often must be removed with solvents.

Couplants used in contact inspection should be applied as a uniform, thin coating to obtain uniform and consistent inspection results. The necessity for a couplant is one of the drawbacks of ultrasonic inspection and may be a limitation, such as with high-temperature surfaces. When the size and shape of the part being inspected permit, immersion inspection is often done. This practice satisfies the requirement for uniform coupling.

Search Units

Search units with piezoelectric transducers are available in many types and shapes. Variations in search-unit construction include transducer-element material; transducer-element thickness, surface area, and shape; and type of backing material and degree of loading. A reference that is fre-

Table 6 Primary applications of four basic types of ultrasonic search units with piezoelectric transducers

Straight-beam contact-type units

Manufacturing-induced flaws
- Billets: inclusions, stringers, pipe
- Forgings: inclusions, cracks, segregations, seams, flakes, pipe
- Rolled products: laminations, inclusions, tears, seams, cracks
- Castings: slag, porosity, cold shuts, tears, shrinkage cracks, inclusions

Service-induced flaws
- Fatigue cracks, corrosion, erosion, stress-corrosion cracks

Angle-beam contact-type units

Manufacturing-induced flaws
- Forgings: cracks, seams, laps
- Rolled products: tears, seams, cracks, cupping
- Welds: slag inclusions, porosity, incomplete fusion, incomplete penetration, dropthrough, suckback, cracks in filler metal and base metal
- Tubing and pipe: circumferential and longitudinal cracks

Service-induced flaws
- Fatigue cracks, stress-corrosion cracks

Dual-element contact-type units

Manufacturing-induced flaws
- Plate and sheet: thickness measurements, lamination detection
- Tubing and pipe: measurement of wall thickness

Service-induced flaws
- Wall thinning, corrosion, erosion, stress-corrosion cracks

Immersion-type units

Manufacturing-induced flaws
- Billets: inclusions, stringers, pipe
- Forgings: inclusions, cracks, segregations, seams, flakes, pipe
- Rolled products: laminations, inclusions, tears, seams, cracks
- Welds: inclusions, porosity, incomplete fusion, incomplete penetration, drop through, cracks, base-metal laminations
- Adhesive-bonded, soldered or brazed products: lack of bond
- Composites: voids, resin rich, resin poor, lack of filaments
- Tubing and pipe: circumferential and longitudinal cracks

Service-induced flaws
- Corrosion, fatigue cracks

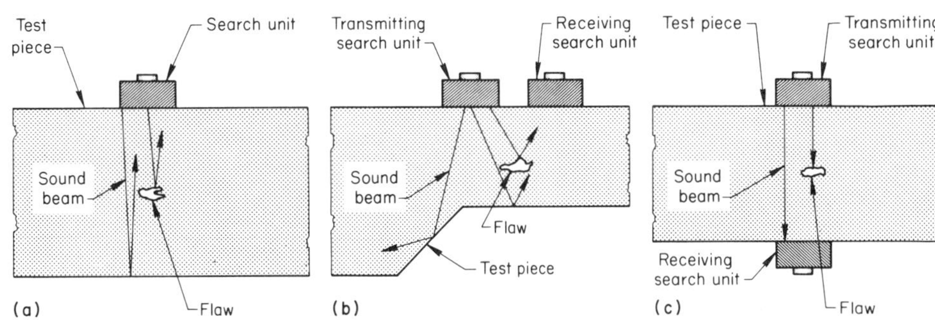

Fig. 35 Applications of the straight-beam (longitudinal wave) contact-type search unit illustrated in Fig. 34(a) showing reflection techniques with (a) single search unit and (b) two search units, and (c) through transmission technique with two search units

quently used in determining search-unit characteristics is American Society for Testing and Materials (ASTM) specification E 1065.

The four basic types of search units are the straight-beam contact type, the angle-beam contact type, the dual-element contact type, and the immersion type; Table 6 lists their primary areas of application. Sectional views of these search units, together with a special type (delay-tip contact-type search unit), are shown in Fig. 34. These units, as well as several other special types of search units, are discussed in this section.

Contact-Type Units

Although contact-type search units can sometimes be adapted to automatic scanning, they are usually hand held and manually scanned in direct contact with the surface of a testpiece. A thin layer of an appropriate couplant is almost always required for obtaining transmission of sound energy across the interface between the search unit and the entry surface.

Straight-Beam Units. Figure 34(a) illustrates a straight-beam (longitudinal wave) contact-type search unit. In service, this unit is hand held and manually scanned in direct contact with the surface of the testpiece. The contact face is subject to abrasion and in most cases is required to couple efficiently to metals having high acoustic impedances. This type of search unit projects a beam of ultrasonic vibrations perpendicular to the entry surface. It can be used for either the reflection (echo) method or the through transmission method.

Echo inspection can be performed with either one or two search units. With the technique using a single search unit, the search unit acts alternately as transmitter and receiver. It projects a beam of longitudinal waves into the material being inspected and receives echoes reflected back to it from the opposite surface and from flaws in the path of the beam, as shown in Fig. 35(a).

The technique with two search units is used when the testpiece is of irregular shape and reflecting interfaces or back surfaces are not parallel with the entry surface. One search unit is the transmitter, and the second search unit is the receiver. The transmitting search unit projects a beam of vibrations into the material; the vibrations travel through the material and are reflected back to the receiving search unit from flaws or from the opposite surface, as shown in Fig. 35(b).

In the through transmission technique, two search units are used—one as a transmitter and the second as a receiver. The transmitting search unit projects a beam of vibrations into the testpiece; these vibrations travel through the material to the opposite surface, where they are picked up by the receiving search unit (Fig. 35c). Flaws in the path of the beam cause a reduction in the amount of energy (sound beam intensity) passing through to the receiving search unit. For optimum results when using two search units, it may be desirable to use different transmitter and receiver materials, depending on the electrical characteristics of the ultrasonic instrument used. For example, a polarized-ceramic search unit can be used as the transmitter, and a lithium sulfate search unit as the receiver.

Angle-Beam Units. Figure 34(b) illustrates the construction of an angle-beam contact-type search unit. A plastic wedge between the piezoelectric element and the contact surface establishes a fixed angle of incidence for the search unit. The plastic wedge must be designed to reduce or eliminate internal reflections within the wedge that could result in undesired false echoes.

Angle-beam search units are used for the inspection of sheet or plate, pipe welds or tubing, and testpieces having shapes that prevent access for straight beam. Angle-beam search units can be used to produce shear waves or combined shear and longitudinal waves, depending on the wedge angle and testpiece material. There is a single value of wedge angle that will produce the desired beam direction and wave type in any given testpiece. A search unit having the appropriate wedge angle is selected for each specific application.

The surface wave search unit is an angle-beam unit insofar as it uses a wedge to position the crystal at an angle to the surface of the testpiece. It generates surface waves by mode conversion as described in the section "Critical Angles" in this article. The wedge angle is chosen so that the shear wave refraction angle is 90° and the wave resulting from mode conversion travels along the surface.

Dual-Element Units. Figure 34(c) shows a dual-element contact-type search unit. Dual-element units provide a method of increasing the directivity and resolution capabilities (especially near-surface resolution) in contact inspection. By splitting the transmitting and receiving functions, two-transducer, send and receive inspection can be done with a single search unit. The dual-element design allows the receiving function to be electrically and acoustically

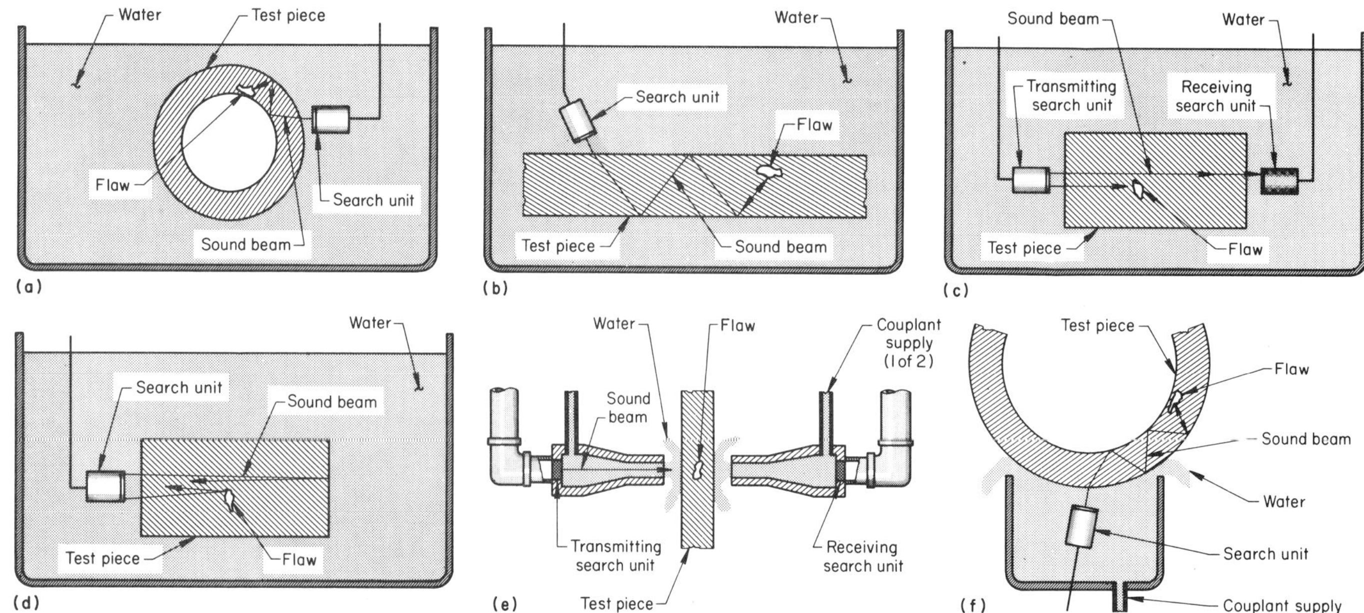

Fig. 36 Schematic views of several immersion techniques using search units of the basic construction shown in Fig. 34(e). (a) Angle-beam pulse-echo inspection of pipe or tube. (b) Angle-beam pulse-echo inspection of sheet or plate. (c) Through transmission inspection. (d) Straight-beam pulse-echo inspection. (e) Through transmission inspection using squirter-type search units. (f) Angle-beam pulse-echo inspection of pipe or tubing using a bubbler-type unit

isolated from effects of the excitation pulse by the use of an acoustical barrier separating the transmitter element from the receiver. The receiving transducer is always in a quiescent state and can respond to the signal reflected from a flaw close to the testpiece surface.

Delay-Tip Units. Figure 34(d) illustrates the construction of a delay-tip (standoff) contact-type search unit. Its primary application is in thickness measurement and for thin materials that require a high degree of resolution.

The delay shoe allows the indication from the front surface of the testpiece to be delayed by the transmission time through the delay shoe. This separates the front-surface signal from the large excitation pulse, thus eliminating much of the dead zone encountered in contact inspection with a search unit that does not have a delay shoe. Reducing the extent of the dead zone allows the piezoelectric crystal to respond to front and back reflections that occur close together in time. This provides improved accuracy in the thickness measurement of thin plate and sheet.

Paintbrush Transducers. The inspection of large areas with small single-element transducers is a long and tedious process. To overcome this and to increase the rate of inspection, the paintbrush transducer was developed. It is so named because it has a wide beam pattern that, when scanned, covers a relatively wide swath in the manner of a paintbrush. Paintbrush transducers are usually constructed of a mosaic or series of matched crystal elements. The primary requirement of a paintbrush transducer is that the intensity of the beam pattern not vary greatly over the entire length of the transducer. At best, paintbrush transducers are designed to be survey devices; their primary function is to reduce inspection time while still giving full coverage. Standard search units are usually employed to pinpoint the location and size of a flaw that has been detected by a paintbrush transducer.

Immersion-Type Units

The advantages of immersion inspection include speed of inspection, ability to control and direct sound beams, and adaptability for automated scanning. Angulation is used in immersion inspection to identify more accurately the orientation of flaws below the surface of the testpiece. If the direction of the sound beam, its point of entry, and its angle of incidence are known, the direction and angle of refraction within the testpiece can be calculated from Eq 5. Only a single search unit is required for conventional immersion inspection, regardless of incident angle. The construction of a conventional immersion-type search unit is shown in Fig. 34(e); several inspection techniques are shown in Fig. 36.

There are three broadly classified scanning methods that utilize immersion-type search units:

- Conventional immersion methods in which both the search unit and the testpiece are immersed in liquid (Fig. 36a to d)
- Squirter and bubbler methods in which the sound is transmitted in a column of flowing water (Fig. 36e and f)
- Scanning with a wheel-type search unit (Fig. 37), which is generally classified as an immersion method because the transducer itself is immersed

Basic Immersion Inspection. In conventional immersion inspection, both the search unit and the testpiece are immersed in water. The sound beam is directed into the testpiece using either a straight-beam (longitudinal wave) technique or one of the various angle-beam techniques, such as shear, combined longitudinal and shear, or Lamb wave. Immersion-type search units are basically straight-beam units and can be used for either straight-beam or angle-beam inspection through control and direction of the sound beam (Fig. 36a to d).

In straight-beam immersion inspection, the water path (distance from the face of the search unit to the front surface of the testpiece) is generally adjusted to require a longer transit time than the depth of scan (front surface to back surface) so that the first multiple of the front reflection will appear farther along the oscilloscope trace than the first back reflection (Fig. 13b). This is done to clear the displayed trace of signals that may be misinterpreted. Water path adjustment is particularly important when gates are used for automatic signaling and recording. Longitudinal wave velocity in water is approximately one-fourth the velocity in aluminum or steel; therefore, on the time base of the oscilloscope, 25 mm (1 in.) of water path is approximately equal to 100 mm (4 in.) of steel or aluminum. Therefore, a rule of thumb is to make the water path equal to one-fourth the testpiece thickness plus 6 mm (¼ in.).

Water-Column Designs. In many cases, the shape or size of a testpiece does not lend itself to conventional immersion inspection in a tank. The squirter scanning method, which operates on the immersion principle, is routinely applied to the high-speed scanning of plate, sheet, strip, cylindrical forms, and other regularly shaped testpieces (Fig. 36e). In the squirter method, the sound beam is projected into the material through a column of water that flows through a nozzle on the search unit. The sound beam can be directed either perpendicular or at an angle to the surface of the testpiece. Squirter methods can also be adapted for through transmission inspection. For this type of inspection, two squirter-type search units are used, as shown in Fig. 36(e). The bubble method (Fig. 36f) is a minor modification of the squirter method that gives a less directional flow of couplant.

Wheel-type search units (Fig. 37) operate on the immersion principle in that a sound beam is projected through a liquid path into the testpiece. An immersion-type search unit, mounted on a fixed axle inside a liquid-filled rubber tire, is held in one position relative to the surface of the testpiece while the tire rotates freely. The wheel-type search unit can be mounted on a stationary fixture and the testpiece moved past it, or it can be mounted on a mobile fixture that moves over a stationary testpiece, as shown in Fig. 37(a). The position and angle of the transducer element are determined by the inspection method and technique to be used and are adjusted by varying the position of either the immersion unit inside the tire or the mounting yoke of the entire unit.

For straight-beam inspection, the ultrasonic beam is projected straight into the testpiece perpendicular to the surface of the testpiece (Fig. 37b). Applications of the straight-beam technique include the inspection of plate for laminations and of billet stock for primary and secondary pipe.

In angle-beam inspection, ultrasound is projected into the material at an angle to the surface of the testpiece (Fig. 37c). The most commonly used angle of sound beam propagation is 45°; however, other angles can be used where required. The beam can be projected forward, in the direction of wheel rotation, or can be projected to the side, 90° to the direction of wheel rotation. The flexibility of the tire permits angles other than those set by the manufacturer to be obtained by varying the mounting-yoke axis with respect to the surface of the testpiece.

Special wheel-type search units can be made with beam direction or other features tailored to the specific application. One such unit is the cross-eyed Lamb, which utilizes two transducer elements that are mounted at angles to the axle so that the two sound beams cross each other in a forward direction, as shown in Fig. 37(d). This unit is used in the Lamb wave inspection of narrow, thin strip for edge nicks and laminations. By selecting incident angles and beam directions that are appropriate for the thickness, width, and material of the strip being inspected, Lamb waves are set up by the combined effect of the two ultrasonic beams.

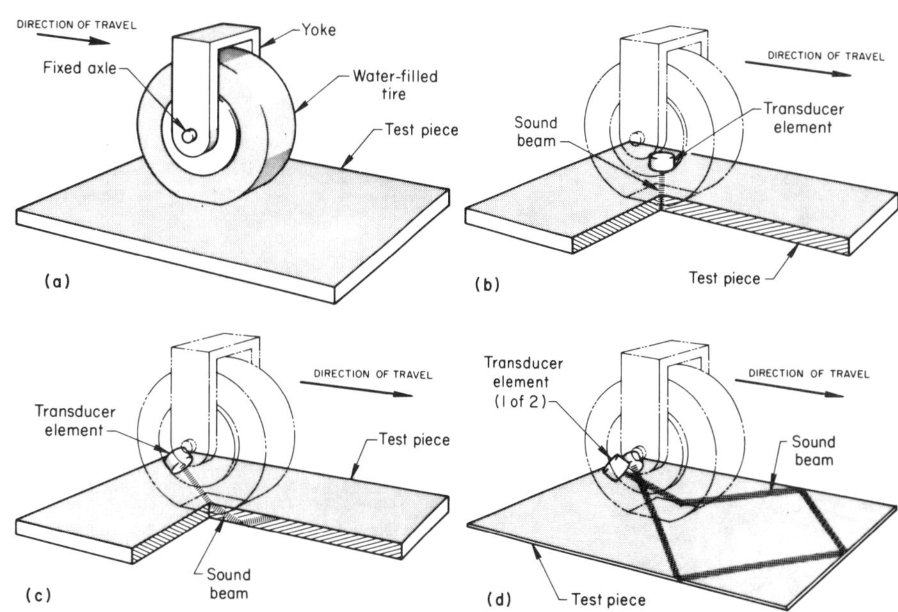

Fig. 37 Several techniques and applications for wheel-type search units. (a) Typical setup for a wheel-type search unit. (b) Straight-beam inspection with beam entering the testpiece perpendicular to the surface. (c) Angle-beam inspection with beam entering the testpiece at 45° to the surface. Beam can also be directed forward or to the side at 90° to the direction of wheel rotation. (d) Use of two transducers to cross and angle the beams to the sides and forward cross-eyed Lamb unit

Focused Units

Sound can be focused by acoustic lenses in a manner similar to that in which light is focused by optic lenses. Most acoustic lenses are designed to concentrate sound energy, which increases beam intensity in the zone between the lens and the focal point. When an acoustic lens is placed in front of the search unit, the effect resembles that of a magnifying glass; that is, a smaller area is viewed but details in that area appear larger. The combination of a search unit and an acoustic lens is known as a focused search unit or focused transducer; for optimum sound transmission, the lens of a focused search unit is usually bonded to the transducer face. Focused search units can be immersion or contact types.

Acoustic lenses are designed similarly to optic lenses. Acoustic lenses can be made of various materials; several of the more common lens materials are methyl methacrylate, polystyrene, epoxy resin, aluminum, and magnesium. The important properties of materials for acoustic lenses are:

- Large index of refraction in water
- Acoustic impedance close to that of water or the piezoelectric element
- Low internal sound attenuation
- Ease of fabrication

Acoustic lenses for contour correction are usually designed on the premise that the entire sound beam must enter the testpiece normal to the surface of the testpiece. For example, in the straight-beam inspection of tubing, a narrow diverging beam is preferred for internal inspection and a narrow converging beam for external inspection. In either case, with a flat-face search unit there is a wide front-surface echo caused by the inherent change in the length of the water path across the width of the sound beam, which results in a distorted pattern of multiple back reflections (Fig. 38a). A cylindrically focused search unit completely eliminates this effect (Fig. 38b).

The shapes of acoustic lenses vary over a broad range; two types are shown in Fig. 39—a cylindrical (line-focus) search unit in Fig. 39(a) and a spherical (spot-focus) search unit in Fig. 39(b). The sound beam from a cylindrical search unit illuminates a rectangular area that can be described in terms of beam length and width. Cylindrically focused search units are mainly used for the inspection of thin-wall tubing and round bars. Such search units are especially sensitive to fine surface or subsurface cracks within the walls of tubing. The sound beam from a spherical search unit illuminates a small circular spot. Spherical transducers exhibit the greatest sensitivity and resolution of all the transducer types; but the area covered is small, and the useful depth range is correspondingly small.

Focusing can also be achieved by shaping the transducer element. The front surface of a quartz crystal can be ground to a cylindrical or spherical radius. Barium titanate can be formed into a curved shape before it is polarized. A small piezoelectric element can be mounted on a curved backing member to achieve the same result.

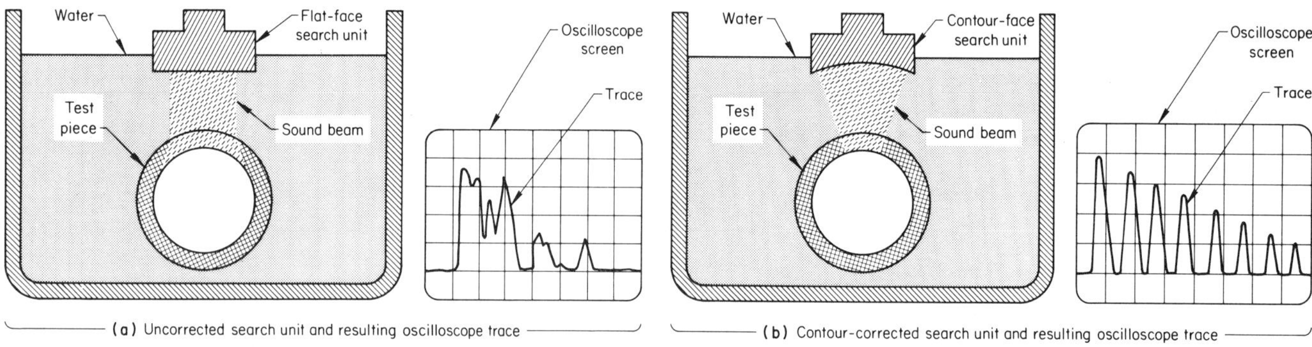

Fig. 38 Comparison of oscilloscope traces resulting from straight-beam immersion inspection by (a) an uncorrected (flat-face) search unit and (b) a contour-corrected (contour-face or cylindrically focused) search unit

Focal Length. Focused transducers are described by their focal length, that is, short, medium, long, or extralong. Short focal lengths are best for the inspection of regions of the testpiece that are close to the front surface. The medium, long, and extralong focal lengths are for increasingly deeper regions. Frequently, focused transducers are specially designed for a specific application. The longer the focal length of the transducer, the deeper into the testpiece the point of high sensitivity will be.

The focal length of a lens in water has little relation to its focal depth in metal, and changing the length of the water path in immersion inspection produces little change in focal depth in a testpiece. The large differences in sonic velocity between water and metals cause sound to bend at a sharp angle when entering a metal surface at any oblique angle. Therefore, the metal surface acts as a second lens that is much more powerful than the acoustic lens at the transducer, as shown in Fig. 40. This effect moves the focal spot very close to the front surface, as compared to the focal point of the same sound beam in water. This effect also causes the transducer to act as a notably directional and distance-sensitive receiver, sharpens the beam, and increases sensitivity to small reflectors in the focal zone. Thus, flaws that produce very low amplitude echoes can be examined in greater detail than is possible with standard search units.

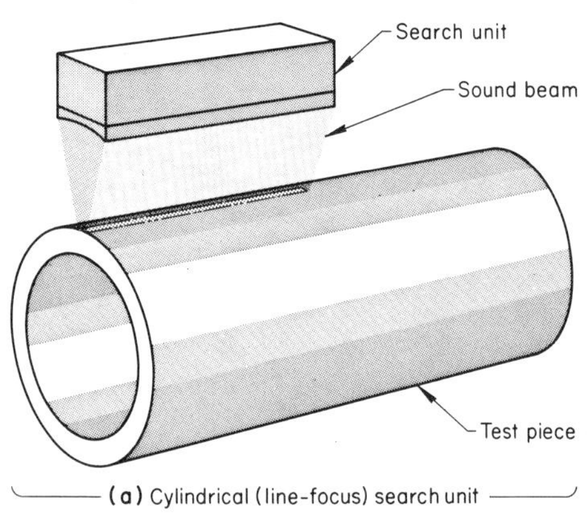

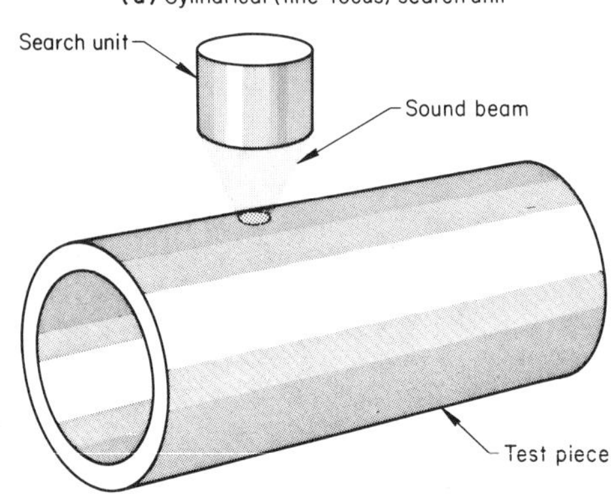

Fig. 39 Two types of focused search units, and the sound beam shapes produced

Useful Range. The most useful portion of a sound beam starts at the point of maximum sound intensity (end of the near field) and extends for a considerable distance beyond this point. Focusing the sound beam moves the maximum-intensity point toward the transducer and shortens the usable range beyond.

The useful range of focused transducers extends from about 0.25 to approximately 250 mm (0.010 to 10 in.) below the front surface. In materials 0.25 (0.010 in.) or less in thickness, resonance or antiresonance techniques can be used. These techniques are based on changes in the duration of ringing of the echo, or the number of multiples of the back-surface echo.

The advantages of focused search units are listed below; these advantages apply mainly to the useful thickness range of 0.25 to 250 mm (0.010 to 10 in.) below the front surface:

- High sensitivity to small flaws
- High resolving power
- Low effects of surface roughness
- Low effects of front-surface contour
- Low metal noise (background)

The echo-masking effects of surface roughness and metal noise can be reduced by concentrating the energy into a smaller beam. The side lobe energy produced by a flat transducer is reflected away by a smooth surface. When the surface is rough, some of the side lobe energy returns to the transducer and widens the front reflection, causing loss of resolving power and increasing the length of the dead zone. The use of a focusing lens on a transducer will reduce the side lobe energy, thus reducing the effects of rough surfaces. The limitation of focused search units is the small region in the test part in the area of sound focusing that can be effectively interrogated.

Noise. Material noise (background) consists of low-amplitude, random signals from numerous small reflectors irregularly distributed throughout the testpiece. Some of the causes of metal noise are grain boundaries, microporosity, and segregations. The larger the sound beam area, the greater the number of small reflectors encountered by the beam. If echoes from several of these small reflectors have the same transit time, they may produce an indication amplitude that exceeds the acceptance level. Focused beams reduce background by reducing the volume of metal inspected, which reduces the probability that the sound beam will encounter several small reflectors at the same depth. Echoes from discontinuities of unacceptable size are not affected and will rise well above the remaining background.

Focused search units allow the highest possible resolving power to be achieved with standard ultrasonic equipment. When a focused search unit is used, the front surface of the testpiece is not in the focal

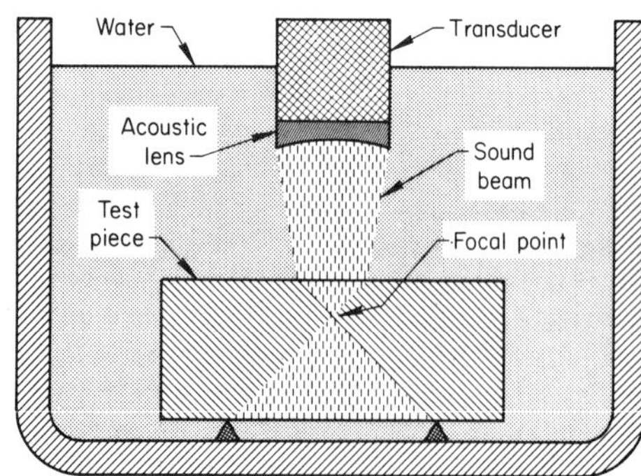

Fig. 40 Effect of a metal surface on the convergence of a focused sound beam. See text for discussion.

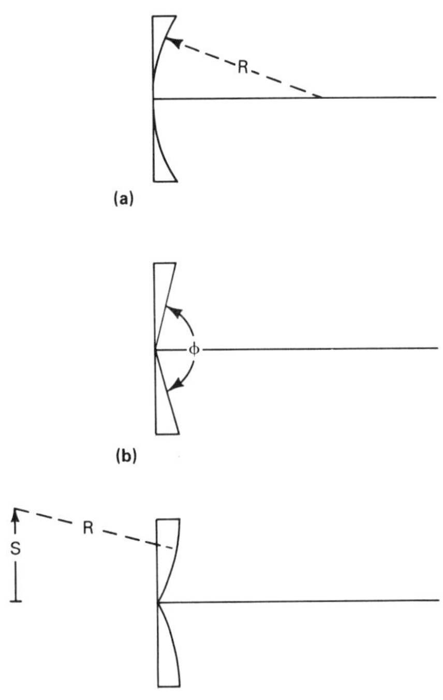

Fig. 41 Aperture geometries for transducer focusing. (a) Spherical. (b) Conical. (c) Toroidal

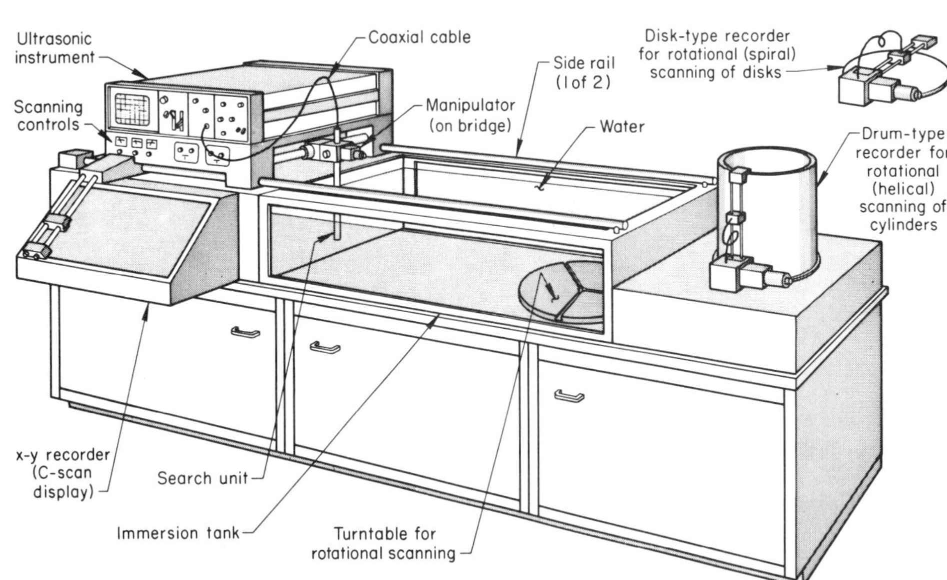

Fig. 42 Principal components of a universal unit for the immersion scanning of testpieces of various shapes and sizes. See text for discussion.

zone, and concentration of sound beam energy makes the echo amplitude from any flaw very large. The resolving power of any system can be greatly improved by using a focused transducer designed specifically for the application and a defined region within the testpiece. Special configurations consist of spherical, conical, and toroidal apertures (Fig. 41) with improvements in beam width and depth of field.

Scanning Equipment

Scanning of testpieces is accomplished by the use of a search unit connected to the ultrasonic instrument by a suitable coaxial cable. During scanning, the search unit is continuously moved over the surface of the testpiece, often in a predetermined search pattern.

Scanning can be done without immersion (contact inspection), provided there is a suitable couplant covering the entire area to be scanned. For example, computerized C-scan systems can be built as a portable system with a hand-held transducer. Although portable C-scan systems are still in their infancy, their impact on in-service inspection is increasing. Some applications of portable ultrasonic imaging systems are discussed in Ref 8. For testpieces of appropriate size and shape, immersion scanning is widely used. Tanks of various designs and shapes have been built for the immersion of specific testpieces. In addition, standard units (often referred to as universal units or installations) and computerized motion control are available. Computerized systems are capable of following contours at high speed and thus overcoming one of the major limitations of ultrasonics, namely, the need to maintain normal incidence of the ultrasonic beam to ensure the highest defect detection sensitivity.

Most universal units consist of a bridge and manipulator that support and position a pulse-echo inspection system over a large water tank, as shown in Fig. 42. Drive mechanisms move the bridge along the side rails of the tank while traversing mechanisms move the manipulator from side to side along the bridge. Most of these units are automated, although some early units were manually operated.

Adjustable brackets and lazy-susan turntables can be provided on the tank bottom for support of testpieces. Clean, deaerated water containing a wetting agent usually covers a testpiece by a foot or more, providing excellent coupling between search unit and testpiece. For operator comfort, water temperature is usually kept at about 21 °C (70 °F), often by automatic control.

The immersion-scanning unit shown in Fig. 42 represents a high degree of versatility in its capability to accommodate testpieces of various sizes and shapes and to scan using various search patterns. The entire instrument cabinet and search unit can be moved longitudinally on side rails along the length of the tank. Also, the search unit is suspended from a manipulator mounted on a bridge, thus enabling it to be moved from side to side. These movements of the search unit are most appropriate for scanning rectangular flat testpieces.

The universal unit illustrated in Fig. 42 can also be used for scanning testpieces of various circular shapes. For example, a round disklike workpiece can be placed on the turntable (shown at right end of tank) and rotated while the search unit is moved radially outward, thus producing a spiral scanning pattern. The spiral pattern is displayed on a disk-type recorder. A cylindrical testpiece can be scanned in a helical pattern, with a drum-type recorder being used to display the data. Accessory equipment is also available for scanning long cylinders in a horizontal position.

Inspection Standards

The standardization of ultrasonic inspection allows the same test procedure to be conducted at various times and locations, and by both customer and supplier, with reasonable assurance that consistent results will be obtained. Standardization also provides a basis for estimating the sizes of any flaws that are found.

An ultrasonic inspection system includes several controls that can be adjusted to display as much information as is needed on the oscilloscope screen or other display device. If the pulse length and sensitivity controls (or their equivalents) are adjusted to a high setting, numerous indications may appear on an A-scan display between the front-surface and back-surface indications. On the other hand, at a low setting, the trace between front-surface and back-surface indications may show no indications even if flaws of prohibited size are present. Inspection or reference standards are used as a guide for adjusting instrument controls to reveal the presence of flaws that may be considered harmful to the end use of the product and for determining which indications come from flaws that are insignificant, so that needless reworking or scrapping of satisfactory parts is avoided.

The inspection or reference standards for pulse-echo testing include test blocks containing natural flaws, test blocks containing artificial flaws, and the technique of evaluating the percentage of back reflection. Inspection standards for thickness testing can be plates of various known thicknesses or can be stepped or tapered wedges.

Test blocks containing natural flaws are metal sections similar to those parts being inspected. Sections known to contain natural flaws can be selected for test blocks.

Test blocks containing natural flaws have only limited use as standards, for two principal reasons:

- It is difficult to obtain several test blocks that give identical responses. Natural flaws vary in shape, surface characteristics, and orientation, and echoes from natural flaws vary accordingly
- It is often impossible to determine the exact nature of a natural flaw existing in the test block without destructive sectioning

Test blocks containing artificial flaws consist of metal sections containing notches, slots, or drilled holes. These test blocks are more widely accepted as standards than are test blocks that contain natural flaws.

Test blocks containing drilled holes are widely used for longitudinal wave, straight-beam inspection. The hole in the block can be positioned so that ultrasonic energy from the search unit is reflected either from the side of the hole or from the bottom of the hole. The flat-bottom hole is used most because the flat bottom of the hole offers an optimum reflecting surface that is reproducible. A conical-bottom hole, such as is obtained with conventional drills, is undesirable, because a large portion of the reflected energy may never reach the search unit. Differences of 50% or more can easily be encountered between the energy reflected back to the search unit from flat-bottom holes and from conical-bottom holes of the same diameter. The difference is a function of both transducer frequency and distance from search unit to hole bottom. Figure 43(a) shows a typical design for a test block that contains a flat-bottom hole. In using such a block, high-frequency sound is directed from the surface called "entry surface" toward the bottom of the hole, and the reflection from it is used either as a standard to compare with signal responses from flaws or as a reference value for setting the controls of the ultrasonic instrument.

In the inspection of sheet, strip, welds, tubing, and pipe, angle-beam inspection can be used. This type of inspection generally requires a reference standard in the form of a block that has a notch (or more than one notch) machined into the block. The sides of the notch can be straight and at right angles to the surface of the test block, or they can be at an angle (for example, a 60° included angle). The width, length, and depth of the notch are usually defined by the applicable specification. The depth is usually expressed as a percentage of testpiece thickness. Figure 43(b) shows a typical design for a test block for the angle-beam inspection of plate.

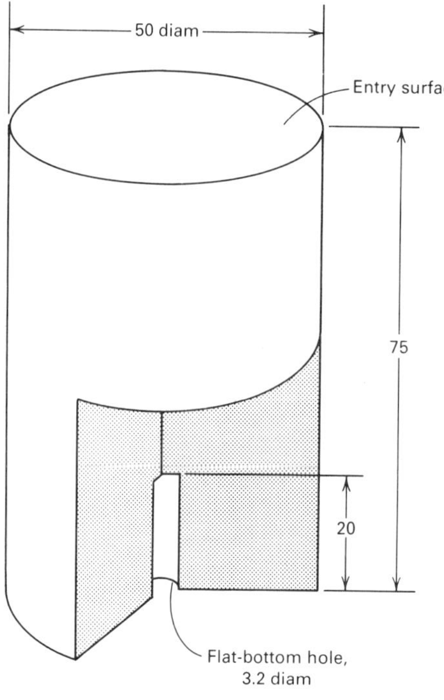

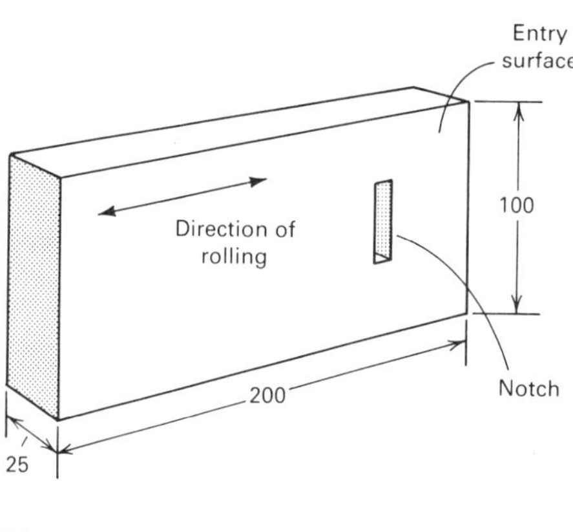

Fig. 43 Typical designs of test blocks that are used as reference standards in the detection of flaws by ultrasonic inspection. (a) Test block for straight-beam inspection. (b) Test block for angle-beam inspection. Dimensions given in millimeters

In some cases, it may be necessary to make one of the parts under inspection into a test block by adding artificial discontinuities, such as flat-bottom holes or notches. These artificial discontinuities can sometimes be placed so that they will be removed by subsequent machining.

When used to determine the operating characteristics of the instruments and transducers or to establish reproducible test conditions, a test block is called a calibration block. When used to compare the height or location of the echo from a discontinuity in a part to that from an artificial discontinuity in the test block, the block is called a reference block. The same physical block, or blocks, can be used for both purposes. Blocks whose dimensions have been established and sanctioned by any of the various groups concerned with materials-testing standards are called standard reference blocks.

Reference blocks establish a standard of comparison so that echo intensities can be evaluated in terms of flaw size. Numerous factors that affect the ultrasonic test can make exact quantitative determination of flaw size extremely difficult, if not impossible. One factor is the nature of the reflecting surface. Although a flat-bottom hole in a reference block has been chosen because it offers an optimum reflecting surface and is reproducible, natural flaws can be of diverse shape and offer nonuniform reflecting surfaces. The origin of a flaw and the amount and type of working that the product has received will influence the shape of the flaw. For example, a pore in an ingot might be spherical and therefore scatter most of the sound away from the search unit, reflecting back only a small amount to produce a flaw echo. However, when worked by forging or rolling, a pore usually becomes elongated and flat and therefore reflects more sound back to the search unit.

On the screen, the height of the echo indication from a hole varies with the distance of the hole from the front (entry) surface in a predictable manner based on near-field and far-field effects, depending on the test frequency and search-unit size, as long as the grain size of the material is not large. Where grain size is large, this normal variation can be altered. Figure 44 shows the differences in ultrasonic transmissibility that can be encountered in reference blocks of a material with two different grain sizes. It can be seen in Fig. 44 that for the austenitic stainless steel inspected, increasing the grain size affected the curve of indication height versus distance from the entry surface so that the normal increase in height with distance in the near field did not occur. This was caused by rapid attenuation of ultrasound in the large-grain stainless steel. In some cases where the grain size is quite large, it may not even be possible to obtain a back reflection at normal test frequencies.

In the inspection of aluminum, a single set of reference blocks can be used for most parts regardless of alloy or wrought mill product. This is considered acceptable practice because ultrasonic transmissibility is about the same for all aluminum alloy compositions. For ferrous alloys, however, ultrasonic transmissibility can vary considerably with composition. Consequently, a single set of reference blocks cannot be

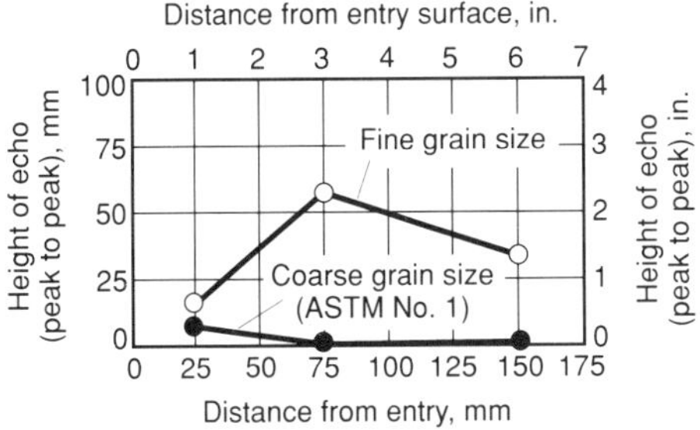

Fig. 44 Distance-amplitude curves for type 304 stainless steel showing effect of grain size on ultrasonic transmissibility. Tests were conducted at 5 MHz with a 30 mm (1⅛ in.) diam search unit on three reference blocks having various metal distances to 20 mm (¾ in.) deep, 3 mm (⅛ in.) diam, flat-bottom holes.

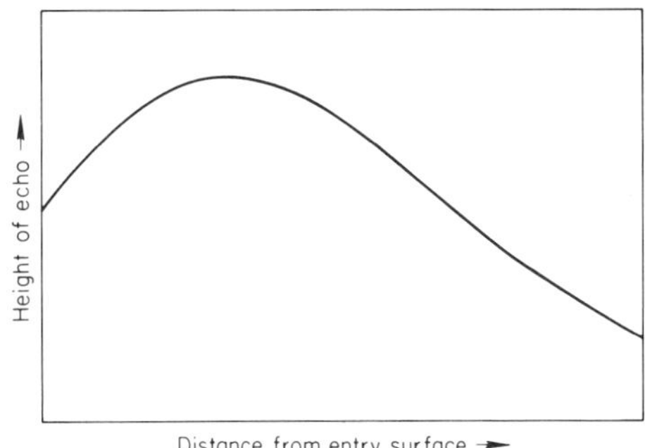

Fig. 45 Effect of distance (depth) of flaws of equal size from the entry surface of a testpiece on the echo height of ultrasonic indications

used when inspecting various products made of carbon steels, stainless steels, tool steels, low-alloy steels, and high-temperature alloys. For example, if a reference block prepared from fine-grain steel were used to set the level of test sensitivity and the material being inspected were coarse grained, flaws could be quite large before they would yield an indication equal to that obtained from the bottom of the hole in the reference block. Conversely, if a reference block prepared from coarse-grain steel were used when inspecting fine-grain steel, the instrument could be so sensitive that minor discontinuities would appear to be major flaws. Thermal treatment can also have an appreciable effect on the ultrasonic transmissibility of steel. For this reason, the stage in the fabrication process at which the ultrasonic inspection is performed may be important. In some cases, it may determine whether or not a satisfactory ultrasonic inspection can be performed.

Reference blocks are generally used to adjust the controls of the ultrasonic instrument to a level that will detect flaws having amplitudes above a certain predetermined level. It is usual practice to set instrument controls to display a certain amplitude of indication from the bottom of the hole or from the notch in a reference block and to record and evaluate for accept/reject indications exceeding that amplitude, depending on the codes or applicable standards. The diameter of the reference hole and the number of flaw indications permitted are generally related to performance requirements for the part and are specified in the applicable codes or specifications.

The following should be considered when setting controls for inspection:

- The larger the section or testpiece, the greater the likelihood of encountering flaws of a particular size
- Flaws of a damaging size may be permitted if found to be in an area that will be subsequently removed by machining or that is not critical
- It is generally recognized that the size of the flaw whose echo exceeds the rejection level usually is not the same as the diameter of the reference hole. In a reference block, sound is reflected from a nearly perfect flat surface represented by the bottom of the hole. In contrast, natural flaws are usually neither flat nor perfectly reflecting
- The material being inspected may conduct sound differently from the material of the reference block. Normally, a reference block will be made from material of the same general type as that being inspected
- The depth of a flaw from the entry surface will influence the height of its echo that is displayed on the oscilloscope screen. Figure 45 shows the manner in which echo height normally varies with flaw depth. Test blocks of several lengths are used to establish a reference curve for the distance-amplitude correction of inspection data

Percentage of Back Reflection. As an alternative to reference blocks, an internal standard can be used. In this technique, the search unit is placed over an indication-free area of the part being inspected, the instrument controls are adjusted to obtain a predetermined height of the first back reflection, and the part is evaluated on the basis of the presence or absence of indications that equal or exceed a certain percentage of this predetermined amplitude. This technique, known as the percentage of back reflection technique, is most useful when lot-to-lot variations in ultrasonic transmissibility are large or unpredictable—a condition often encountered in the inspection of steels.

The size of a flaw that produces a rejectable indication will depend on grain size, depth of the flaw below the entry surface, and test frequency. When acceptance or rejection is based on indications that equal or exceed a specified percentage of the back reflection, rejectable indications may be caused by smaller flaws in coarse-grain steel than in fine-grain steel. This effect becomes less pronounced, or is reversed, as the transducer frequency and corresponding sensitivity necessary to obtain a predetermined height of back reflection are lowered. Flaw evaluation may be difficult when the testpiece grain size is large or mixed.

Generally, metallurgical structure such as grain size has an effect on ultrasonic transmissibility for all metals. The significantly large effect shown in Fig. 44 for type 304 stainless steel is also encountered in other materials. The magnitude of the effect is frequency dependent, that is, the higher the test frequency, the greater the attenuation of ultrasound. In any event, when the grain size approaches ASTM No. 1, the effect is significant regardless of alloy composition or test frequency.

Other techniques are used besides the reference block technique and the percentage of back reflection technique. For example, in the inspection of stainless steel plate, a procedure can be used in which the search unit is moved over the rolled surface, and the display on the screen is observed to determine whether or not an area of defined size is encountered where complete loss of back reflection occurs with or without the presence of a discrete flaw indication. If such an area is encountered, the plate is rejected unless the loss of back reflection can be attributed to surface condition or large grain size.

Another technique that does not rely on test blocks is to thoroughly inspect one or more randomly chosen sample parts for natural flaws by the method to be used on production material. The size and location of any flaws that are detected by ultrasonic testing are confirmed by sectioning the part. The combined results of ultrasonic and destructive studies are used to develop standards for instrument calibration and to define the acceptance level for production material.

Thickness Blocks. Stepped or tapered test blocks are used to calibrate ultrasonic equipment for thickness measurement. These blocks are carefully ground from material similar to that being inspected, and the exact thickness at various positions is marked on the block. Either type of block can be used as a reference standard for resonance inspection; the stepped block can also be used for transit-time inspection.

Standard Reference Blocks

Many of the standards and specifications for ultrasonic inspection require the use of standard reference blocks, which can be prepared from various alloys, may contain holes, slots, or notches of several sizes, and may be of different sizes or shapes. The characteristics of an ultrasonic beam in a testpiece are affected by the following variables, which should be considered when selecting standard reference blocks:

- Nature of the testpiece
- Alloy type

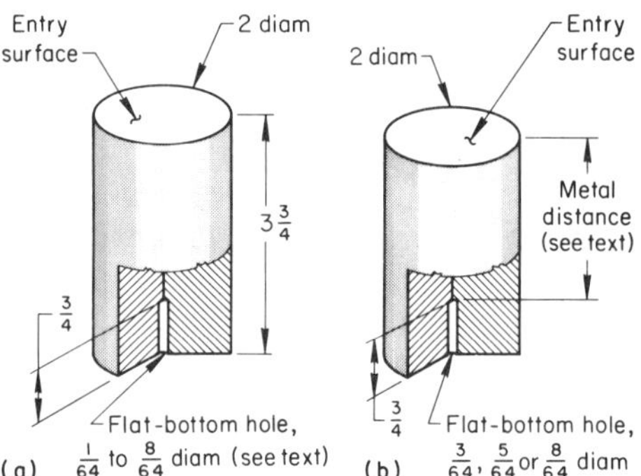

Fig. 46 Standard reference blocks for use in straight-beam ultrasonic inspection. (a) Area-amplitude block. (b) Distance-amplitude block. Dimensions given in inches

- Grain size
- Effects of thermal or mechanical processing
- Distance-amplitude effects (attenuation)
- Flaw size
- Direction of the ultrasonic beam

Information regarding standard reference blocks is available in ASTM E 127 (see also the "Selected References" at the end of this article). Three types of standard blocks are ordinarily used for calibration or reference: area-amplitude blocks, distance-amplitude blocks, and blocks of the type sanctioned by the International Institute of Welding (IIW). These blocks must be prepared from material having the same or similar alloy content, heat treatment, and amount of hot or cold working as the material to be inspected to ensure equal sonic velocity, attenuation, and acoustic impedance in both the reference standard and the testpiece. If blocks of identical material are not available, the difference between the material in the testpiece and the material used in the standard reference blocks must be determined experimentally.

Area-amplitude blocks provide artificial flaws of different sizes at the same depth. Eight blocks made from the same 50 mm (2 in.) diam round stock, each 95 mm (3¾ in.) high, constitute a set of area-amplitude blocks. The block material must have the same acoustic properties as the testpiece material. Each block has a 20 mm (¾ in.) deep flat-bottom hole drilled in the center of the bottom surface (Fig. 46); the hole diameters vary from 0.4 to 3.2 mm (1⁄64 to 8⁄64 in.). The blocks are numbered to correspond with the diameter of the holes; that is, block No. 1 has a 0.4 mm (1⁄64 in.) diam hole, No. 2 has a 0.8 mm (2⁄64 in.) diam hole, and so on, up to No. 8, which has a 3.2 mm (8⁄64 in.) diam hole. Similar area-amplitude blocks made from 49 mm (1 15⁄16 in.) square stock are sometimes known as Series-A blocks.

The amplitude of the echo from a flat-bottom hole in the far field of a straight-beam search unit is proportional to the area of the bottom of the hole. Therefore, these blocks can be used to check the linearity of a pulse-echo inspection system and to relate signal amplitude to the area (or, in other words, the size) of the flaw. Because a flat-bottom hole is an ideal reflector and most natural flaws are less than ideal in reflective properties, an area-amplitude block defines a lower limit for the size of a flaw that yields a given height of indication on the oscilloscope screen; that is, if the amplitude of the indication from a flaw in a testpiece is six scale units high and this is also the amplitude of the indication from a 2.0 mm (5⁄64 in.) diam flat-bottom hole at the same depth as the flaw, it is certain that the flaw is no smaller in area than the 2.0 mm (5⁄64 in.) diam hole. Unfortunately, with ultrasonic inspection methods, it is not possible to determine how much larger than the reference hole the flaw actually is.

Distance-amplitude blocks provide artificial flaws of a given size at various depths. From ultrasonic wave theory, it is known that the decrease in echo amplitude from a flat-bottom hole using a circular search unit is inversely proportional to the square of the distance to the hole bottom. Distance-amplitude blocks (also known as Series-B or Hitt blocks) can be used to check actual variations of amplitude with distance for straight-beam inspection in a given material. They also serve as a reference for setting or standardizing the sensitivity (gain) of the inspection system so that readable indications will be displayed on the oscilloscope screen for flaws of a given size and larger, but the screen will not be flooded with indications of smaller discontinuities that are of no interest. On instruments so equipped, these blocks are used to set the sensitivity-time control or distance-amplitude correction so that a flaw of a given size will produce an indication on the oscilloscope screen that is of a predetermined height regardless of distance from the entry surface.

There are 19 blocks in a Series-B set. All are 50 mm (2 in.) diam blocks of the same material as that being inspected, and all have a 20 mm (¾ in.) deep flat-bottom hole drilled in the center of the bottom surface (Fig. 46). The hole diameter is the same in all the blocks of a set; sets can be made with hole diameters of 1.2, 2.0, and 3.2 mm (3⁄64, 5⁄64, and 8⁄64 in.). The blocks vary in length to provide metal distances of 1.6 to 145 mm (1⁄16 to 5¾ in.) from the top (entry) surface to the hole bottom. Metal distances are:

- 1.6 mm (1⁄16 in.)
- 3.2 mm through 25 mm (⅛ in. through 1 in.) in increments of 3.2 mm (⅛ in.)
- 32 mm through 150 mm (1¼ in. through 5¾ in.) in increments of 13 mm (½ in.)

Each Series-B block is identified by a code number consisting of a digit, a dash, and four more digits. The first digit is the diameter of the hole in one sixty-fourths of an inch. The four other digits are the metal distance from the top (entry) surface to the hole bottom in one hundredths of an inch. For example, a block marked 3-0075 has a 1.2 mm (3⁄64 in.) diam hole and a 20 mm (¾ in.) metal distance.

ASTM blocks can be combined into various sets of area-amplitude and distance-amplitude blocks. The ASTM basic set consists of ten 50 mm (2 in.) diam blocks, each with a 20 mm (¾ in.) deep flat-bottom hole drilled in the center of the bottom surface. One block has a 1.2 mm (3⁄64 in.) diam hole at a 75 mm (3 in.) metal distance. Seven blocks have 20 mm (5⁄64 in.) diam holes at metal distances of 3.2, 6.4, 13, 20, 40, 75, and 150 mm (⅛, ¼, ½, ¾, 1½, 3 and 6 in.). The remaining blocks have 3.2 mm (8⁄64 in.) diam holes at 75 and 150 mm (3 and 6 in.) metal distances. The three blocks with a 75 mm (3 in.) metal distance and hole diameter of 1.2, 2.0, and 3.2 mm (3⁄64, 5⁄64, and 8⁄64 in.) form an area-amplitude set, and the set with the 2.0 mm (5⁄64 in.) diam holes provides a distance-amplitude set. In addition to the basic set, ASTM lists five more area-amplitude standard reference blocks and 80 more distance-amplitude blocks.

Each ASTM block is identified by a code number, using the same system as that used for the Series-B set. The dimensions of all ASTM blocks are given in ASTM E 127, which also presents the recommended practice for fabricating and checking aluminum alloy standard reference blocks. The recommended practice for fabricating and control of steel standard reference blocks is found in ASTM E 428.

IIW blocks are mainly used to calibrate instruments prior to contact inspection using an angle-beam search unit; these blocks are also useful for:

- Checking the performance of both angle-beam and straight-beam search units
- Determining the sound beam exit point of an angle-beam search unit
- Determining refracted angle produced
- Calibrating sound path distance
- Evaluating instrument performance

The material from which a block is prepared is specified by the IIW as killed, open hearth or electric furnace, low-carbon steel in the normalized condition and with a grain size of McQuaid-Ehn No. 8. All IIW standard reference blocks are of the same size and shape; official IIW blocks are dimensioned in the metric system of units. One of the standard English-unit designs is shown in Fig. 47(a).

The miniature angle-beam block is based on the same design concepts as the IIW block, but is smaller and lighter. The miniature angle-beam block is primarily used in the field for checking the characteristics of angle-beam search units and for calibrating the time base of ultrasonic instruments. One of several alternative En-

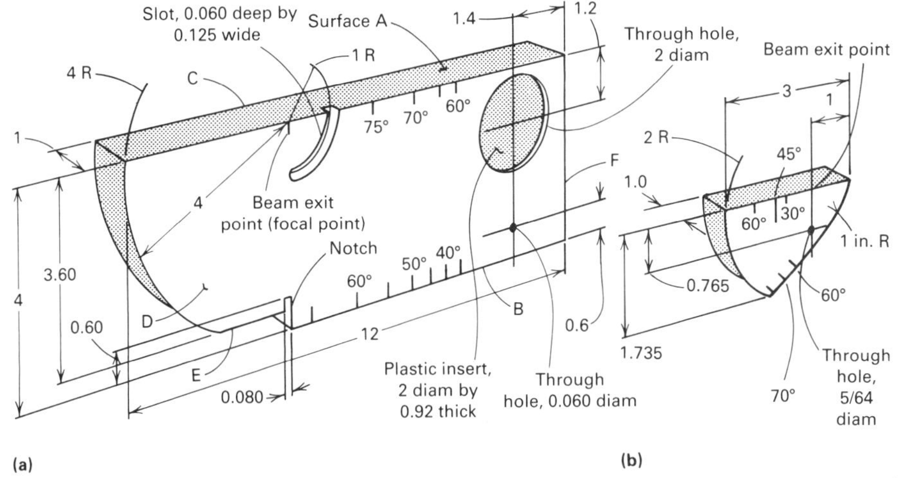

Fig. 47 Two standard reference blocks used in ultrasonic inspection. (a) IIW, type 1, block. (b) Miniature angle-beam block. Dimensions given in inches

glish-unit designs is shown in Fig. 44(b). With the miniature block, beam angle and index point can be checked for an angle-beam search unit, and metal-distance calibration can be made for either angle-beam or straight-beam search units.

The ASTM reference plate (Fig. 48) is another type of standard reference block that can be used for both straight-beam and angle-beam inspection. This block has limited usefulness, however, because:

- The thinness of the plate leads to many false reflections when the block is used for straight-beam calibration or reference purposes
- When the block is used for angle-beam inspection, there is little correlation between hole depth and echo amplitude

The use of the ASTM reference plate for inspecting steel is described in ASTM A 503.

Determination of Area-Amplitude and Distance-Amplitude Curves

Determination of the area-amplitude and distance-amplitude responses of a straight-beam pulse-echo ultrasonic inspection system for flaw detection requires a series of test blocks containing flat-bottom holes. Any of the standard reference blocks of this type can be used. Listed below are typical procedures using ASTM blocks.

Area-amplitude curves are determined by the procedure in the list that follows:

1. Select a No. 5 block
2. Position the search unit on the top (entry) surface of the block when using the contact technique, or over the block and at a distance to give a water path of 75 mm (3 in.) when using the immersion technique. If the immersion technique is used, adjust the water path to within ±0.8 mm (±1⁄32 in.) using a gage, and position the ultrasonic beam perpendicular to the top surface of the block by tilting (angulating) the search unit slightly until the front-reflection indication on the oscilloscope screen is maximum
3. Move the search unit horizontally until the hole-bottom indication is maximum
4. Adjust the gain control on the ultrasonic instrument so that the height of the hole-bottom indication is about one-third the maximum indication height obtainable on the screen
5. Replace the No. 5 block with the other area-amplitude blocks in the set, each in turn
6. With each replacement block, repeat steps 2 and 3. Keep the water path constant for all immersion tests
7. Read the heights of the hole-bottom indications, without changing the gain-control setting that was established with the No. 5 block
8. Plot the heights of each hole-bottom indication against the square of the block number for that indication, as shown in Fig. 49. The square of the block number is proportional to the area of the hole bottom
9. Draw a straight line through the indication height for the No. 5 block and the origin. Significant deviations from this line of ideal linearity are noted (shaded region, Fig. 49) and used as corrections to the indications from actual flaws

Distance-amplitude curves are determined by the following procedure:

1. Select a No. 5-0300 block
2. Position the search unit as in step 2 for area-amplitude calibration
3. Move the search unit as in step 3 for area-amplitude calibration
4. Adjust the gain control to give a hole-bottom indication height about one-fourth of the maximum obtainable
5. Replace block No. 5-0300 with the other distance-amplitude blocks in the set, each in turn
6. With each replacement block, repeat steps 2 and 3 as for area-amplitude calibration

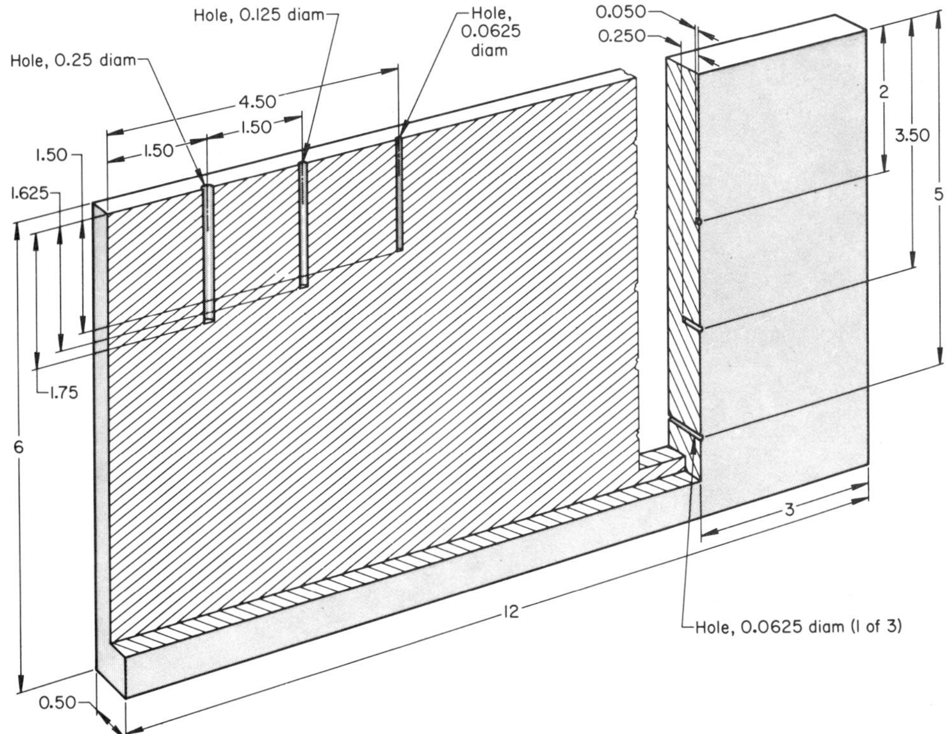

Fig. 48 ASTM standard reference plate for use in angle-beam ultrasonic inspection. Dimensions given in inches

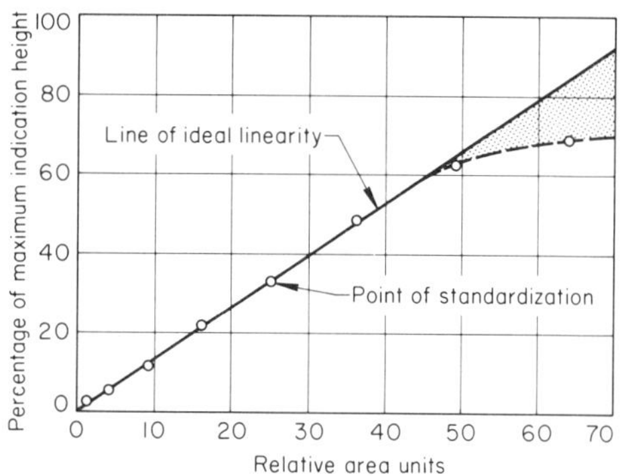

Fig. 49 Typical area-amplitude response curve from ultrasonic test blocks having flat-bottom holes

7. Read the indication height as for area-amplitude calibration
8. Plot the heights of each hole-bottom indication against the metal distance for that indication. The resulting curve is similar to that in Fig. 45, with indication height increasing as metal distance in the near-field region is increased, and decreasing with metal distance in the far-field region

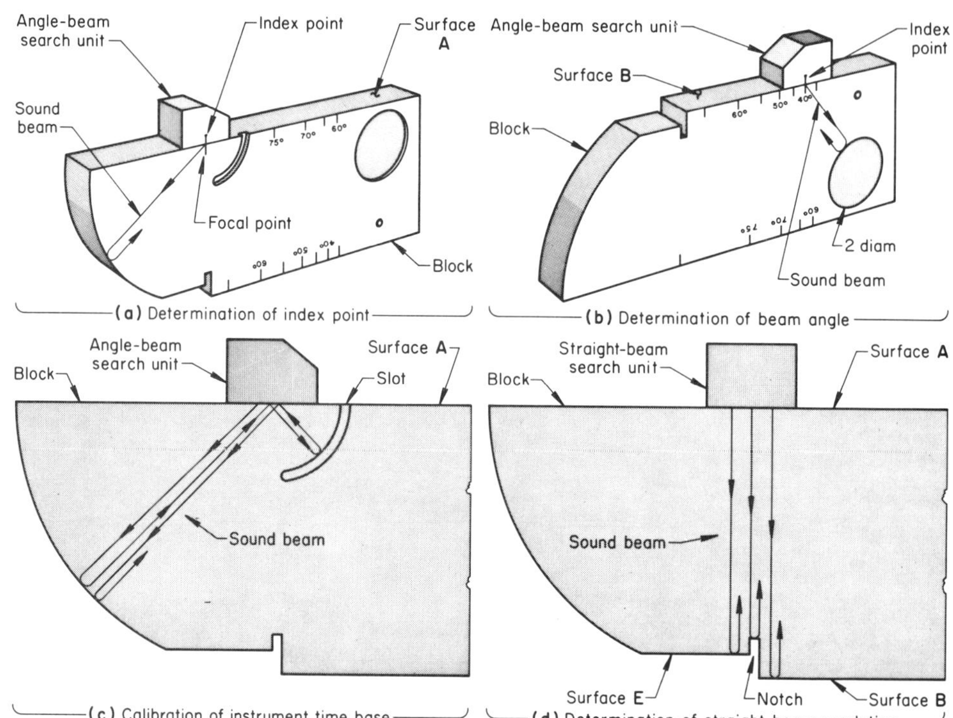

Fig. 50 Positioning of angle-beam and straight-beam search units for calibration using the IIW standard reference block shown in Fig. 47(a). See text for explanation. Dimensions given in inches

Calibration

In many cases, the determination of area-amplitude and distance-amplitude curves constitutes the calibration of a straight-beam pulse-echo system for the detection of flaws. Calibration of an angle-beam pulse-echo ultrasonic test system for flaw detection can be accomplished using an IIW standard calibration block such as the one shown in Fig. 47(a). The large (100 mm, or 4 in., radius) curved surface at one end of the block is used to determine the index point of angle-beam search units (the point where the beam leaves the unit). The mark labeled "beam exit point" in Fig. 47(a) is at the center of this 100 mm (4 in.) radius. Regardless of the angle of the search unit, a maximum echo is received from the curved surface when the index point of the search unit is at the "beam exit point." To determine the index point of an angle-beam search unit, the search unit is placed on surface A in the position shown in Fig. 50(a) and is moved along the surface of the block until the echo is at maximum amplitude. The point on the search unit that is directly over the beam exit point of the block can then be marked as the index point of the search unit.

Beam Angle. Once the index point of the search unit is marked, the 50 mm (2 in.) diam hole is used to determine the angle of the beam in the low-carbon steel from which the block is prepared. The search unit is placed on surface A or surface B and is aimed toward the 50 mm (2 in.) diam hole. Then, the search unit is moved along the surface until a maximum-amplitude echo is received. At this position, the index point on the search unit indicates the beam angle, which is read from one of the degree scales marked along the sides of the block at the edges of surfaces A and B (Fig. 50b).

Beam spread can be determined by moving the search unit in one direction (toward either higher or lower beam angles) from the point of maximum-amplitude echo until the echo disappears and noting the beam angle at the index point. The search unit is then moved in the opposite direction, past the point of maximum-amplitude echo to the point where the echo again disappears. The beam spread is the difference between the angles indicated by the index point at these two extreme positions.

Time Base. The 25 mm (1 in.) radius curved slot (Fig. 47a) is used in conjunction with the 100 mm (4 in.) radius surface for calibrating the time base of the ultrasonic instrument being used with the angle-beam search unit. A direct reflection from the 100 mm (4 in.) radius surface represents a metal distance of 200 mm (8 in.). Similarly, a beam traveling from the beam exit point to the 100 mm (4 in.) radius, then reflecting in turn to surface A, then to the 25 mm (1 in.) radius, back to surface A, once again to the 100 mm (4 in.) radius, and finally back to the search unit, as shown in Fig. 50(c), will give an echo indication at 460 mm (18 in.) metal distance. The time base and marker spacing can be adjusted until the echoes corresponding to 200, 250, 450, 500, 700 mm (8, 10, 18, 20, 28 in.) and so on are aligned with appropriate grid lines on the screen or with a convenient number of marker signals. Upon completion of these adjustments, the instrument time base is calibrated in terms of metal distance in steel. Metal distance can be converted to depth from the entry surface by dividing by 2 and multiplying by the cosine of the beam angle when using a full "vee" path.

The calibration for steel can be converted to a calibration for another material by multiplying metal distance by the ratio of sound velocity in the other material to the corresponding sound velocity in steel (Table 1); transverse wave velocities are ordinarily used because most angle-beam inspection is done with this type of ultrasonic wave. Depth in another material is determined from metal distance as described in the preceding paragraph, but values of metal distance and beam angle for the new material are used instead of the values for steel.

Linearity. Calibration in terms of metal distance or reflector depth assumes a linear oscilloscope sweep for the instrument, which can be checked using a straight-beam search unit. The search unit is placed on either surface C or D to obtain multiple echoes from the 25 mm (1 in.) thickness. These echo indications will be aligned with evenly spaced grid lines or scale marks if the time base is linear. Linearity within ±1% (or less) of the full-scale value of thickness is usually obtainable.

Resolution. A straight-beam search unit, as well as the instrument, can be checked for back-surface resolution by placing the search unit on surface A and reflecting the beam from the bottom of the 2 mm (0.080

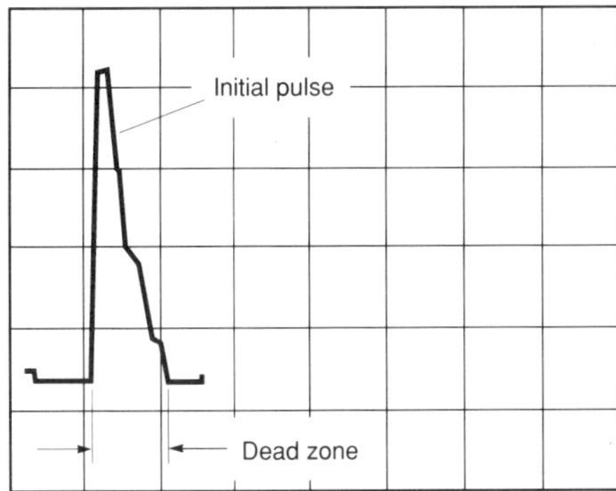

Fig. 51 Schematic showing how dead zone is measured against vertical reference lines. The vertical reference lines represent a unit of vertical thickness.

in.) wide notch and from surfaces B and E on either side of it, as shown in Fig. 50(d). With proper resolution, the indications from these three surfaces should be clearly separated and not overlapped so as to appear as one broad, jagged indication. Because resolution is affected by test conditions and by characteristics of the search unit and instrument amplifier, this degree of resolution sometimes may not be obtainable.

The dead zone is the depth below the entry surface that cannot be inspected because the initial pulse interferes with echo signals. An indication of the length of the dead zone of a straight-beam search unit can be obtained by placing the search unit on surface A or F in line with the 50 mm (2 in.) diam hole (Fig. 47a). When the search unit is placed on surface A, a discernible echo from the 50 mm (2 in.) diam hole indicates a dead zone of less than 5 mm (0.2 in.). Similarly, when the search unit is placed on surface F, a discernible echo from the 50 mm (2 in.) diam hole indicates a dead zone of less than 10 mm (0.4 in.). Alternatively, the length of the dead zone can be measured by calibrating the time base of the instrument, then measuring the width of the initial-pulse indication at the base of the signal, as illustrated schematically in Fig. 51.

Sensitivity. The relative sensitivity of an angle-beam search unit in combination with a given instrument can be defined by placing the unit on either surface A or B and reflecting the beam from the side of the 1.5 mm (0.060 in.) diam hole (Fig. 47a). The position of the search unit is adjusted until the echo from the hole is maximum, then the gain of the instrument is adjusted to give the desired indication height.

When no back reflection is expected, the sensitivity of a straight-beam system is defined by placing the search unit on either surface B or F in line with the 1.5 mm (0.060 in.) diam hole. The position of the search unit is adjusted until the echo from the side of the hole is maximum, then the gain of the instrument is adjusted to give the desired indication height.

When a back reflection is expected, a plastic insert can be used in the 50 mm (2 in.) diam hole to gage the sensitivity of a straight-beam system. The plastic material and insert thickness are specified to have the absorption characteristics of 50 mm (2 in.) of steel. With this calibration, the search unit is placed on the side of the insert facing surface C, and the number of echoes and the height of the last echo indication are noted.

Range Setting. The range for a search-unit and instrument system for straight-beam inspection can be set for various distances by use of the IIW block. From surface F to the 2.0 mm (0.080 in.) wide notch is 200 mm (8 in.), from surface A to surface B is 100 mm (4 in.), from surface E to surface A is 91.4 mm (3.60 in.), and from surface C to surface D is 25 mm (1 in.).

Miniature-Block Application. The miniature angle-beam block shown in Fig. 47(b) is used in a somewhat similar manner as the larger IIW block. Both the 25 and 50 mm (1 and 2 in.) radius surfaces provide ways for checking the location of the index mark of a search unit and for calibrating the time base of the instrument in terms of metal distance. The small hole provides a reflector for checking beam angle and for setting the instrument gain.

Reference Plate Application. The use of the ASTM reference plate (Fig. 48) for calibrating ultrasonic equipment is described in ASTM A 503.

Detection of Flaws

Ultrasonic inspection has been successfully used to detect flaws in cast and wrought metal parts and in welded, brazed, and bonded joints during research and development, production, and service. Contact inspection is more widely used than immersion inspection, not only because it involves equipment that is portable (allowing field inspection) but also because it is versatile and applicable to a wide range of situations. In contact inspection, however, a substance such as oil, grease, or paste is spread on the surface of the part to act as a couplant; this substance is sometimes difficult to remove after inspection. By contrast, the water used as a couplant in immersion inspection provides good acoustic coupling (even to irregular, rough surfaces such as might occur on unfinished sand castings), but presents no removal problem. Immersion inspection is also especially adaptable to mechanized, production line applications.

The following sections discuss the application of both contact and immersion types of ultrasonic inspection for the detection of flaws in various types of products. In many cases, the inspection procedures used for these products are covered in standards and specifications prepared by various governmental agencies, technical and trade societies, and manufacturers. These standards and specifications usually contain the following three basic elements:

- A description of the inspection method, including details on such variables as ultrasonic frequency, type and diameter of search unit, and type of couplant
- Type and use of calibration and reference standards
- A criterion for rejection, usually expressed as maximum allowable flaw-indication amplitude

Inspection of Castings

Both contact inspection and immersion inspection are used to detect in castings such flaws as porosity, tears and cracks, shrinkage, voids, and inclusions. Figure 52 illustrates typical ultrasonic indications from four types of flaws found in castings. Although immersion inspection is preferred for castings having rough and irregular surfaces, any one of the inspection techniques previously described can be used. The choice of technique in a specific instance will depend mainly on casting size and shape.

Because many castings are coarse grained, low-frequency ultrasound may have to be used. Castings of some materials are so coarse grained that extensive scattering makes ultrasonic inspection impractical. These materials include some brasses, stainless steels, titanium alloys, and cast irons. A more extensive discussion is provided in the article "Castings" in this Volume.

Inspection of Primary-Mill Products

Ultrasonic inspection is the most reliable method of detecting internal flaws in forging, rolling, or extrusion billets, in rolling blooms or slabs, and in bar stock. It is capable of detecting relatively small flaws deep within a testpiece. By contrast, eddy current and magnetic particle inspection become less sensitive to internal flaws as the depth of the flaw below the surface increases; detection of internal flaws with these methods is limited to relatively shallow surface layers. Radiography is the only other volumetric method available for inspecting these primary-mill products, but it is slower, more costly, and generally less reliable than ultrasonic inspection for these specifications.

Ultrasonic inspection is used to detect and locate porosity, pipe, internal ruptures, flakes, and nonmetallic inclusions in billets, blooms, slabs, and bar stock of various sizes up to 1.2 m (4 ft) thick in such metals as aluminum, magnesium, titanium, zirconium, carbon steel, stainless steel, high-tem-

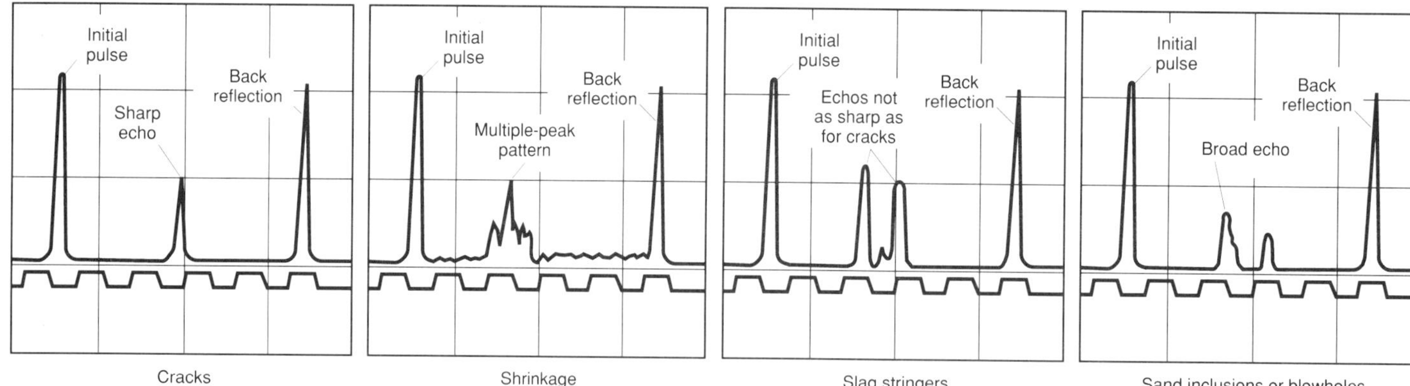

Fig. 52 Ultrasonic indications from four types of flaws found in castings

perature alloys, and uranium. These products are usually inspected with a straight-beam contact-type search unit, which is often hand held; immersion inspection is also used. One technique for inspecting these product forms is to transmit a sound beam the length of the product; if a strong back reflection is obtained, it is unlikely that there is serious pipe present. However, complete inspection requires testing with the beam transverse to the longitudinal axis. Because complete inspection is relatively slow and tedious, it is sometimes performed only on top cuts, where pipe is most likely to occur. For a higher degree of quality assurance, mechanized contact inspection of an entire lot can be done. Figure 53 shows a typical mechanized setup that uses a wheel-type search unit for the pulse-echo inspection of steel billets. A similar setup uses a water-column search unit instead of the wheel units. Setups for through transmission inspection use a pair of wheel or water-column units.

Inspection of Forgings

Forgings can be ultrasonically inspected for internal flaws such as pipe, internal ruptures, flakes, and nonmetallic inclusions. Inspection is usually done with frequencies of 1 to 5 MHz, with the beam normal to both the surface of the forging and the direction of maximum working; this orientation is best for detecting most flaws in forgings. Angle-beam inspection employing shear waves is sometimes used for rings or hollow forgings.

Contact inspection is performed on most forgings that have fairly uniform dimensions. Because of the difficulty in complete contact inspection of irregularly shaped forgings, immersion inspection may be preferred. Alternatively, contact inspection can be performed at an earlier stage of production before the shape becomes too irregular; however a rough machining of the forging surface is recommended.

Contact inspection is also used for inspecting forgings in service. For example, railroad freight-car-axle journals can be inspected using the special hand-held search unit shown in Fig. 54, which was developed for this purpose. The unit fits on the end of the axle being inspected. Surrounding a centering pin are six straight-beam transducers. The widening sound beams from these transducers give complete coverage of the journal portion of the axle.

Inspection of Flat-Rolled Products

Rolled strip, sheet, and plate can be ultrasonically inspected by using either straight-beam or angle-beam pulse-echo techniques for contact or immersion inspection; contact inspection is more widely used. With straight-beam inspection from the top surface of a testpiece, planar discontinuities to which flat-rolled products are susceptible are readily detected, and their locations and limits are easily and accurately determined. However, straight beams cannot be used on strip or sheet too thin to allow resolution of the first back reflection from the initial pulse. Therefore, the inspection of thin products is usually performed with a Lamb wave angle-beam technique.

Straight-beam top inspection of rolled plate is used to detect laminations, excessive stringers of nonmetallic inclusions, and similar flaws. Laminations are particularly detrimental when pieces cut from the plate are to be subsequently welded to form large structural assemblies. Laminations usually occur centered in the thickness of the plate and are usually centered in the as-rolled width of the plate. Laminations do not extend to the surface except at sheared or flame-cut edges and may be difficult to detect visually unless the lamination is gross. For these reasons, ultrasonic inspection is the only reliable way to inspect a plate for laminations.

The straight-beam top inspection of steel plate is usually performed at a beam frequency of 2.25 MHz with search units 20 to 30 mm (¾ to 1⅛ in.) in diameter. Higher or lower frequencies may be necessitated by grain size, microstructure, or thickness of

Fig. 53 Mechanized setup for the pulse-echo ultrasonic inspection of steel billets using a 250 mm (10 in.) diam wheel-type search unit and a longitudinal wave straight beam at 0° angle of incidence

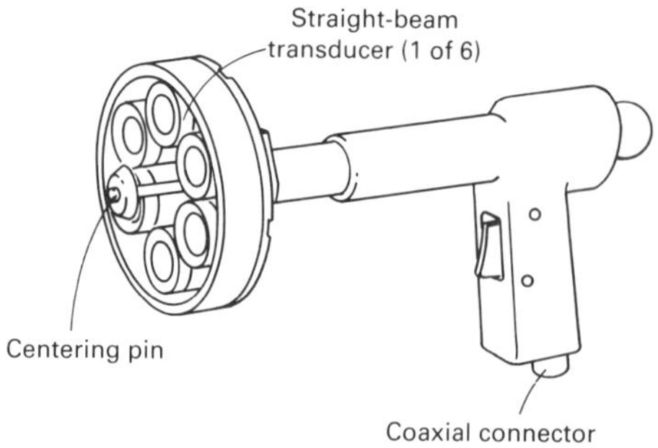

Fig. 54 Hand-held straight-beam contact-type search unit for the in-service ultrasonic inspection of railroad freight-car-axle journals

the material being inspected. The surface of the plate should be clean and free of loose scale that could hinder acoustic transmission. Couplants used for contact inspection include oil, glycerin, or water containing a wetting agent and a rust inhibitor. These are usually applied in a thin layer with a brush.

The contact top inspection of aluminum and other nonferrous plate is usually done with a technique similar to that used for ferrous plate. Frequencies of 1 to 5 MHz, search units 13 to 40 mm (½ to 1½ in.) in diameter, and oil couplants are most often used for nonferrous plate.

Prior to inspecting production material, the controls of the ultrasonic instrument are adjusted to obtain a level of sensitivity that will ensure detection of less severe flaws than would be cause for rejection. How this is established is determined by the specific calibration procedure that applies. Those used in plate inspection include test blocks, percentage of back reflection, and loss of back reflection. Test blocks are usually made of a section of plate that is the same thickness, type, and condition (for example, annealed, pickled, or ground) as the plate to be inspected. The test block will generally contain one or more drilled holes, which are usually flat bottomed. If more than one reference hole is contained in the test block, they will normally be at different depths. The most commonly used depth is midthickness. Instrument controls are adjusted to obtain from the hole a readable echo on the oscilloscope screen. Rejection of material is generally based on the presence of indications exceeding that from the reference hole in the test block. If there are several holes at different depths in the test block, they are used to correct for the effect of metal distance. The depth at which a flaw lies below the entry surface affects the response from the flaw. The test block hole with a depth closest to the depth of the flaw in the plate is used to estimate flaw size. Holes ranging from 3 to 13 mm (⅛ to ½ in.) in diameter can be used for reference, depending on desired sensitivity and specified maximum allowable flaw size.

When the percentage of back reflection technique is employed, the search unit is placed over a sound (indication-free) region of the plate, and the instrument controls are adjusted to obtain a readable indication of the first back reflection. For accuracy in determining flaw size, the first back reflection indication displayed on the screen of the ultrasonic instrument is usually kept at less than full screen height and in the linear portion of oscilloscope response. The plate is then scanned by moving the search unit in a uniform manner over the surface of the plate. The plate to be used for nuclear power purposes generally requires a 100% coverage with a small percentage of overlap (for example, 10% of the search-unit diameter) between successive scans. Some specifications call for inspection along grid lines. Rejection is based on the presence of indications that exceed a predetermined percentage of the reference back reflection indication. The percentage of back reflection technique can be used if test blocks are not available or where extreme variation in ultrasonic transmissibility from one lot of material to another exists and it is not feasible to make a test block from each.

The loss of back reflection technique is the technique most often used for the inspection of plate for planar discontinuities, especially laminations. This technique is similar to the percentage of back reflection technique except that rejection is based on complete or partial loss of the back reflection indication. The system is calibrated by placing the search unit over a sound (lamination-free) region of the plate to be inspected. The pattern on the oscilloscope screen is then a series of multiple back reflection indications from the bottom surface of the plate, as shown for the top inspection of 28.4 mm (1.12 in.) thick steel plate in Fig. 55(a). The instrument controls are adjusted to produce a first back reflection height of about 75 to 100% of full screen height. General scanning of the plate is done at this sensitivity level. Minor sensitivity adjustments can be made to accommodate surface roughness, provided no laminations or other flaws are encountered.

As the search unit passes over the edge of a lamination during scanning, additional indications appear on the oscilloscope screen approximately midway between the initial pulse and the first back reflection and between the multiple back reflection indications. As the search unit is moved farther over the laminated region, indications of the laminations become stronger until finally the original pattern is replaced by a series of multiple lamination indications spaced at approximately half the thickness of the plate (Fig. 55b). If the lamination is gross, the number of multiple indications will decrease.

When a flaw indication is observed during general scanning, the gain (sensitivity) of the instrument is adjusted to produce a first back reflection in an adjoining indication-free area of the plate that is 80 ±5% of full screen height. Then the indication is scanned in all directions to determine the extent of the flawed region. All flaw indications are evaluated on the basis of maximum obtainable response.

The size of a lamination can be determined by moving the search unit back and forth across the flawed region and, when lamination indications drop to 40% of full screen height, marking the plate at a location corresponding to the center of the search unit. The specification will usually describe acceptance in terms of the maximum size of the region over which complete loss of back reflection can occur. Partial loss of back reflection generally must exceed a certain percentage of the normal back reflection, and this loss usually must be accompanied by the presence of lamination indications. Specifications may require a search unit of minimum diameter for the inspection (for example, a minimum diameter of 25 mm, or 1 in., is often specified for steel). The loss of back reflection technique described above has been used on steel plate from 15 to 75 mm (⅝ to 3 in.) thick. With modifications, this technique can be applied to both thinner and thicker material. Further information on the straight-beam top inspection of steel plates is contained in ASTM A 435 and A 578.

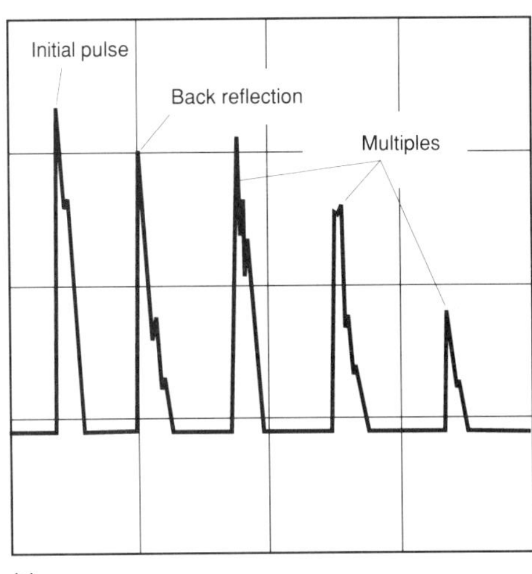

(a)

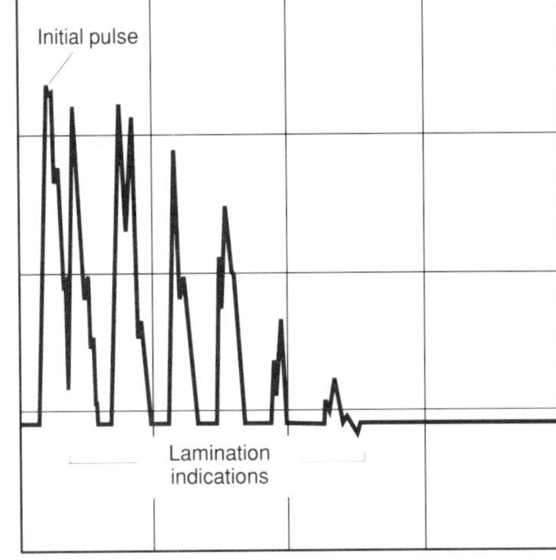

(b)

Fig. 55 Displays of multiple patterns produced during ultrasonic top inspection of steel plate using straight-beam pulse-echo techniques. (a) CRT pattern of back reflection indications from a lamination-free region of a 25 mm (1 in.) thick steel plate. (b) Pattern of indications from a lamination at one-half the plate thickness in an imperfect region of the plate

Straight-beam edge inspection can be used when large quantities of steel plate with widths of 1.2 to 2.1 m (4 to 7 ft) and thicknesses of 20 to 30 mm (¾ to 1⅛ in.) are to be inspected. Preliminary screening by inspecting each plate from the edge can reduce the amount of top inspection described above by as much as 80%. Edge inspection can be done while the plate is stacked for storage and requires no handling of individual plates for the inspection. This inspection is rapid and will reveal those plates that are suspect and should be top inspected. Plates that pass the edge inspection need not necessarily be further inspected unless required by specifications.

In edge inspection, an ultrasonic beam is directed from one edge only through the plate perpendicular to the direction of rolling. This technique directs the sound beam at the narrowest dimension of laminations, yet when laminations or other flaws are present there is a partial loss of normal back reflection (often to less than 50% of the back reflection from lamination-free plate). The search unit usually recommended for inspecting steel plate is 22 mm (⅞ in.) in diameter and has a plastic wear face. On thin steel plates, where making contact with the edge of the plate is more difficult, a 13 mm (½ in.) diam search unit can be used. Poor edge condition frequently makes it difficult to obtain good coupling. In most cases, grease will provide sufficient coupling. If the edge is too concave to be filled in with grease, a plastic shoe may help. If none of these techniques brings the back reflection to an acceptable level, the plate should be set aside for top inspection. Even if it takes extra time to make good edge contact, this technique is still more rapid than top inspection.

The screen pattern produced by edge inspection shows a cluster of indications near the search unit that resulted from the high power needed to penetrate the plates; the back reflection will appear as a group of multiple signals that result from partial mode conversion from longitudinal waves to shear waves, as shown in Fig. 56(a) for the edge inspection of 1.5 m (5 ft) wide steel plate. When the back reflection falls below 50% of full screen height, the plate is marked for top inspection. When a lamination is especially severe, it will produce echoes at a point between the back reflection and initial-pulse indications, as shown in Fig. 56(b). On a steel plate 2.4 m (8 ft) long, four checks at equal intervals along one edge are usually sufficient.

Angle-beam inspection can be used to inspect rolled plates. With angle-beam inspection, it is possible to inspect the entire width of the plate merely by moving the search unit along one edge of the plate. However, some specifications require that for total coverage of the plate the search unit be placed on the top surface, directed toward an edge, and moved away from the edge until the opposite edge is reached. Similar scans are repeated with overlapping (for example, 10% of the search-unit diameter) for successive scans. This procedure is continued until the entire plate has been covered. Some specifications call for a second angle-beam inspection in a direction 90° from the first.

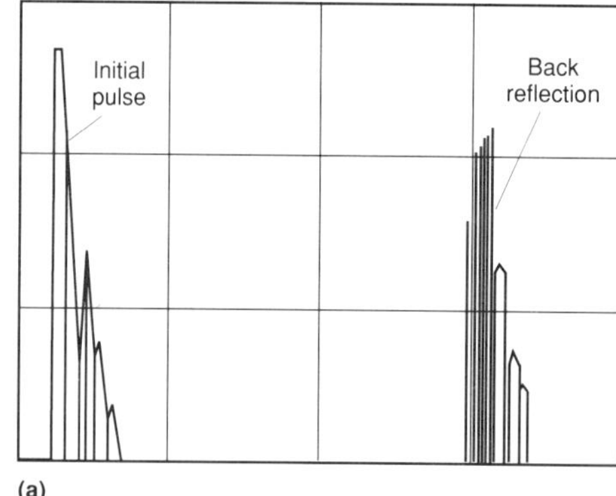

(a)

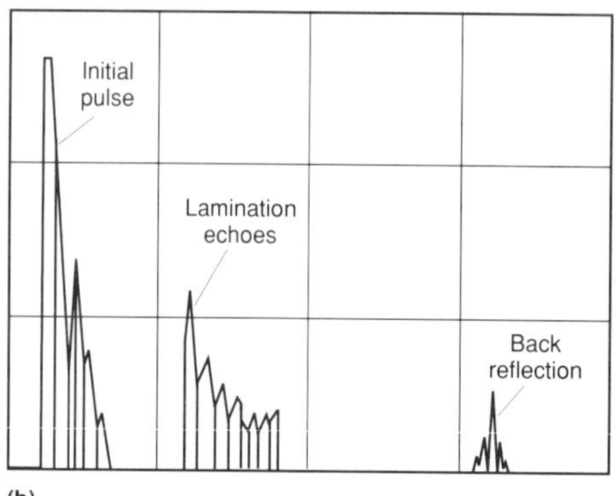

(b)

Fig. 56 Displays of multiple patterns produced during the ultrasonic edge inspection of steel plate using straight-beam pulse-echo techniques. (a) Pattern from a lamination-free region in a 1500 mm (60 in.) wide steel plate. (b) Pattern of indications from a lamination in the plate. The presence of a lamination has reduced the back reflection indication to less than 50% of full screen height.

The search unit used for the angle-beam inspection of plate is usually one that provides 45 to 60° shear waves in the material being inspected. As with straight-beam inspection, the plate surface should be clean and free of loose scale, and a suitable couplant should be used. Prior to angle-beam inspection of plate, the instrument controls are adjusted to obtain a readable response on the oscilloscope screen from a square notch or V-notch in a test plate. The test plate is generally similar to the plate to be inspected with respect to type, thickness, surface finish, and microstructure. The notch is usually 3 or 5% of the plate thickness in depth, 25 mm (1 in.) long or longer, and approximately twice as wide as it is deep.

The search unit is oriented to a position on the test surface so as to receive the maximum reflected energy from the notch; instrument sensitivity is adjusted to display the notch indication at 50 to 75% of full screen height. Often, a distance-amplitude curve is established by repositioning the search unit at several distances from the notch (for example, at ½, 1, and 1½ skip distances, depending on thickness of the plate) and the various responses noted. This curve is the 100% reference line for reporting flaw-echo amplitudes. When evaluating a flaw to determine whether or not the plate is rejectable, it is important to check at a metal distance comparable to that used in initial calibration using the notch in the test plate.

Angle-beam inspection can detect laps and inclusions and possibly laminations not parallel to the plate surface. However, it is difficult to detect laminations parallel to the plate surface. It is hard to distinguish between a signal from the top of the plate, for example, and a lamination of this type near the top. Therefore, it may be desirable to make a rapid sampling of the plate condition using straight-beam longitudinal waves. If laminations of the type just mentioned are found to be absent, angle-beam shear waves can be used to inspect the entire plate. Of course, if it is necessary to check the plate for cracks on or just beneath the surface, surface waves may be used.

Because straight-beam inspection cannot be used on strip or sheet too thin to allow the back reflection to be resolved from the initial pulse, inspection of these products is accomplished by angle-beam (Lamb wave) inspection. The procedure is much the same as for the angle-beam inspection of plate. The search unit is usually mounted in a fixed position at one edge of the strip or sheet. The sound beam is directed across the width to the far edge while the strip or sheet moves past the search unit. Coupling is maintained by continuously feeding a suitable liquid (alcohol can be used for steel products) ahead of the search unit. The location of a flaw is determined by noting the position of the flaw echo between the initial pulse and edge reflection echo on the oscilloscope screen. An alarm gate is often used to signal the presence of a flaw, thus eliminating the need for constant observation of the screen by an operator.

The mechanized inspection of flat-rolled products has been accomplished by several methods; for example, wheel-type search units have been used for mechanized angle-beam inspection, and water-column search units have been used in transmitting and receiving parts to make printed plan views of the responses from a plate as it travels between the pairs. The immersion inspection of plate is highly adaptable to mechanization; some setups are designed for simultaneous thickness gaging and flaw detection.

Inspection of Extrusions and Rolled Shapes

Flaws in extrusions and in rolled shapes are usually longitudinally oriented, that is,

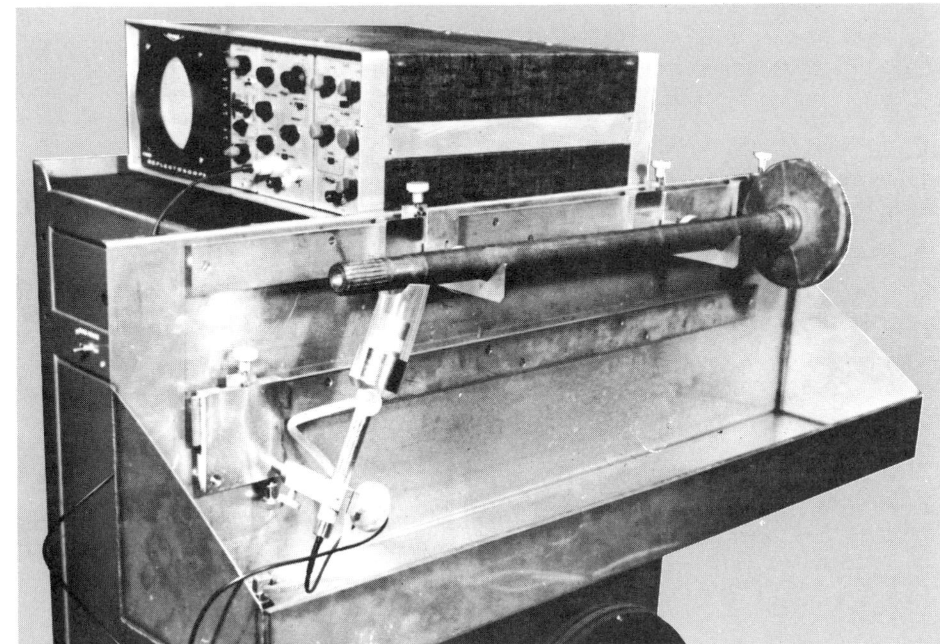

(a) Portable inspection unit

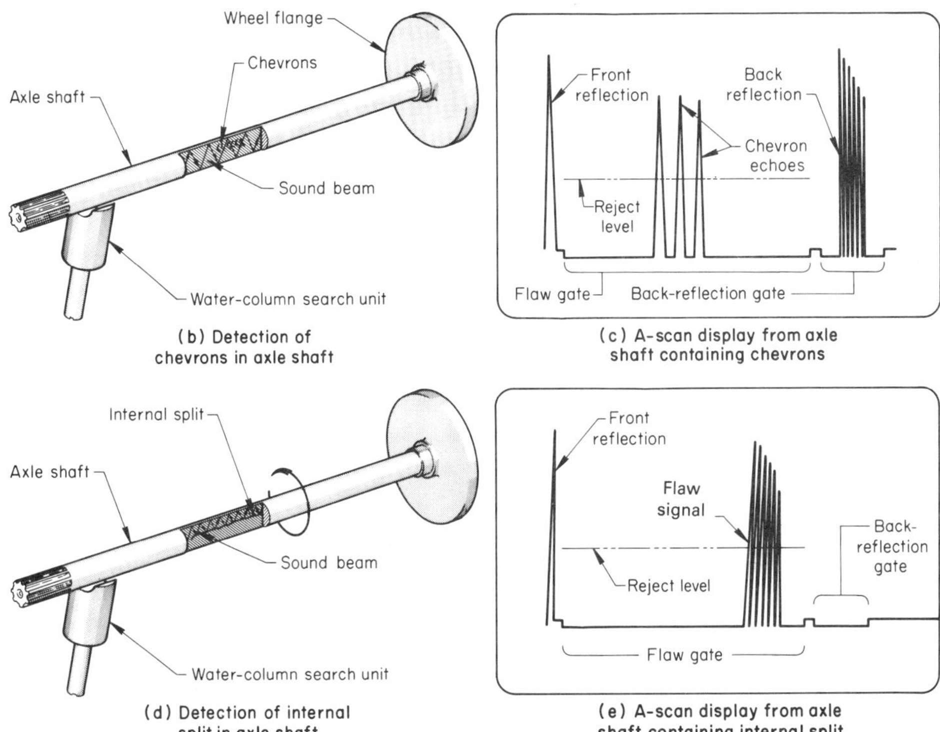

(b) Detection of chevrons in axle shaft

(c) A-scan display from axle shaft containing chevrons

(d) Detection of internal split in axle shaft

(e) A-scan display from axle shaft containing internal split

Fig. 57 Ultrasonic inspection of a cold extruded steel automobile-axle shaft for chevrons and internal splits using a portable inspection unit

parallel to the direction of working. Both contact and immersion inspection are used to inspect for these flaws, usually employing their longitudinal-beam or angle-beam techniques. In some cases, surface waves are used to detect surface cracks or similar flaws.

Aluminum extrusions may be immersion inspected with C-scan water-tank equipment using the straight-beam pulse-echo technique. The usual types of flaws encountered in this test are laps and seams, but linear porosity may also be found in some products. If a second inspection of a region where flaws are suspected is desired, it is often done by straight-beam contact inspection using water or oil as the couplant.

Cold-extruded steel parts are subject to internal bursts called chevrons. This type of flaw is easily detected either by straight-beam (longitudinal wave) inspection with the search unit contacting the end of the extrusion or by angle-beam (shear wave) inspection with the search unit contacting the side of the extrusion.

In one procedure used to inspect cold extruded automobile-axle shafts for chevrons, a variable-angle water-column search unit was used. The axle shaft being inspected was held horizontally by a fixture in the portable inspection unit shown in Fig. 57(a). The variable-angle water-column search unit that was used produced a sound beam at 1.6 to 2.25 MHz and was adjusted so that the beam entered the shaft at 45° to the shaft axis. This allowed the beam to travel the length of the shaft (Fig. 57b) and be reflected from the wheel flange. A dual-gated instrument was employed to allow a fail-safe system and to ensure proper inspection of each shaft.

As the shaft was transferred to the inspection fixture, it closed a limit switch that started the testing sequence. A 5-s time delay after the water column was activated allowed the exclusion of air bubbles and ensured effective coupling of the search unit to the shaft. The instrument then "looked" for the back reflection from the flange in the back reflection gate. If no signal was present, the instrument indicated a no-test condition, and the operator repositioned the shaft and started the sequence a second time. When a back reflection signal was present in the back reflection gate and no signal above a preset amplitude was present in the flaw gate, the instrument indicated an acceptable part. When a signal above the preset amplitude occurred in the flaw gate, as shown in Fig. 57(c), the instrument indicated the presence of damaging chevrons, and the part was automatically marked with paint for rejection.

Another type of internal flaw, a longitudinal split of the core material, was occasionally introduced into axle shaft by an induction-hardening process. The inspection procedure used to detect this type of flaw was the same as for chevrons except that the shaft was rotated during the inspection portion of the testing sequence (Fig. 57d). As the shaft was rotated and the plane of the split assumed a horizontal orientation, the length of the sound beam path was so shortened that the signal moved over into the flaw gate, where it appeared as a flaw indication (Fig. 57e).

Steel bars and pipe are usually contact inspected for internal flaws, using either the straight-beam or the angle-beam technique, or both techniques simultaneously. One method of inspecting round bar is to scan along the bar with a straight-beam search unit having a curved plastic shoe to achieve good contact with the bar. Because the surface of the bar is curved, the sound beam enters the bar at various angles. This may produce both longitudinal and shear waves

Fig. 58 Mechanized equipment for simultaneous ultrasonic and eddy current inspection of round bar or pipe. 1, support frame; 2, drive trolley; 3, carriage; 4, saddle; 5, cable track; 6, manual controls; 7, sensor for signaling end of bar or pipe; 8, eddy current probe; 9, straight-beam ultrasonic search unit; 10, 45° angle-beam ultrasonic search unit; 11, test bar

in the bar and provides for inspection of the entire cross section of the bar.

Another method of inspecting round bar can also be used on pipe. In this method, the bar or pipe is rotated to provide inspection of the entire cross section. Mechanized equipment that uses this method (Fig. 58) is designed to provide inspection simultaneously by both ultrasonic and eddy current techniques. This equipment can inspect bars and pipes from 75 to 300 mm (3 to 12 in.) in diameter and from 3 to 17 m (10 to 56 ft) in length.

The equipment consists of a free-standing support frame to which is mounted a motor-driven trolley (on a monorail track that runs the entire length of the support structure, about 21 m, or 70 ft). The trolley, in turn, supports and propels a carriage-and-saddle assembly. The saddle, suspended beneath the carriage, carries three flaw sensors—two ultrasonic search units and one eddy current probe—and is self-aligning when in contact with a pipe or bar during testing. Also mounted on the carriage are a manual control station and an air cylinder that lowers and raises the saddle. A flexible cable track follows the movement of the carriage and saddle assembly. Instrumentation is enclosed in two separate cabinets, and a control console houses all the controls and relays associated with the system. Operation of the equipment is described below.

The bar or pipe to be inspected is placed on a rotator (not shown in Fig. 58) that turns at an adjustable speed. The operator then initiates the test cycle, and the control system activates an air cylinder that lowers the saddle into contact with the rotating part and starts the flow of the water couplant. The trolley and carriage then move the saddle along the rotating test bar or pipe at a speed (also adjustable) that ensures complete inspection of the part.

Ultrasonic inspection is accomplished by two independent water-column search units employing the pulse-echo technique; one unit uses a straight beam and the other an angle beam. Eddy current inspection is performed by a special parallel-rods proximity probe covering approximately a 75 mm (3 in.) scan path. The eddy current instrument has two adjustable inspecting alarm thresholds to provide a capability for inspecting the ends of a part at higher sensitivity than the middle. A chart recorder is also provided to record the magnitude of any flaw signal. If a flaw is detected by any of these sensors, the control system lowers an associated marking device to produce a permanent mark (unique for each sensor) in the region of the flaw. When the end of the bar or pipe is reached, a sensor signals the control system, and the saddle is raised and returned, at higher speed, to the start position.

Steel rails for railroads can be contact inspected in service with special equipment that detects cracks—either transverse cracks emanating from the edges of the railhead or from hydrogen flakes in the center of the railhead, or cracks emanating from fishplate boltholes. Most larger rail-testing equipment employs the pulse-echo method. Often this equipment is mounted on special railroad cars. Most inspection with this equipment is done by angle-beam techniques, but straight-beam inspection is also used. With some equipment, coupling is through an oil film; in other equipment, wheel-type search units are used.

Seamless tubes and pipes can be contact tested on the mechanized equipment shown in Fig. 58, but immersion inspection is used more often. Usually, the tube or pipe is rotated and driven past the immersed search units, which are positioned to produce an angle beam in the wall of the testpiece. The immersion tank has glands on either side through which the part passes during inspection. Further information on the ultrasonic inspection of seamless tubes and pipe is available in the articles "Tubular Products" and "Boilers and Pressure Vessels" in this Volume.

Inspection of Welded Joints

Welded joints can be ultrasonically inspected using either the straight-beam or the angle-beam technique. Angle-beam inspection is used most often, one reason being that the transducer does not have to be placed on the weld surface, but is typically placed on the relatively smooth base metal surface adjacent to the weld. With angle-beam inspection, the wedge angle is usually selected to produce shear waves in the part being inspected at an optimum angle to detect serious flaws.

The types of flaws usually encountered in welds are porosity, entrapped slag, incomplete penetration, incomplete fusion, and cracks. Serious flaws, such as cracks and incomplete fusion, usually extend longitudinally along the weld and give especially clear signals when the sound beam strikes them at right angles. Spherical porosity will produce a small-amplitude echo, even when the sound beam strikes at an angle to the joint. Slag inclusions may produce stepped indications, which are maximum at right angles to the orientation of the slag. A large inclusion may produce multiple signals as different portions of the inclusion are scanned. The ultrasonic inspection of welds is discussed in the articles "Weldments, Brazed Assemblies, and Soldered Joints" and "Tubular Products" in this Volume.

Inspection of Bonded Joints

If the shape of a joint is favorable, ultrasonic inspection can be used to determine the soundness of joints bonded either adhesively or by any of the various metallurgical methods, including brazing and soldering. Both pulse-echo and resonance techniques have been used to evaluate bond quality in brazed joints.

A babbitted sleeve bearing is a typical part having a metallurgical bond that is ultrasonically inspected for flaws. The bond between babbitt and backing shell is inspected with a straight-beam pulse-echo technique, using a contact-type search unit applied to the outside of the steel shell. A small-diameter search unit is used to ensure adequate contact with the shell through the couplant. Before inspection, the outside of the steel shell and the inside of the cast babbitt liner are machined to a maximum surface roughness of 3.20 μm (125 μin.) (but the liner is not machined to final thickness).

During inspection, the oscilloscope screen normally shows three indications: the initial pulse, a small echo from the bond line (due to differences in acoustical impedance of steel and babbitt), and the back reflection from the inside surface of the liner. Regions where the bond line indication is minimum are assumed to have an acceptable bond. Where the bond line signal increases, the bond is questionable. Where there is no back reflection at all from the inside surface of the liner (babbitt), there is no bond.

Inspection of other types of bonded joints is often done in a manner similar to that described above for babbitted bearings. An extensive discussion of the ultrasonic inspection of various types of adhesive-bonded joints (including two-component lap joints, three component sandwich structures, and multiple-component laminated structures) is available in the article "Adhesive-Bonded Joints" in this Volume. Additional information can be found in Ref 9 for the immersion inspection of lap joints in Ref 10 for inspecting multiple-component laminated structures using resonance techniques. Reference 11 describes the ultrasonic inspection of a type of laminated structure of which one of the components of the structure was a composite material; the inspection was able to detect flaws in the composite material as well as unbonded regions between the components. Reference 12 deals with the inspection of bonded joints by LLW testing.

Crack Monitoring

Laboratory and in-service monitoring of the initiation and propagation of cracks that are relatively slow growing (such as fatigue cracks, stress-rupture cracks, and stress-corrosion cracks) has been accomplished with ultrasonic techniques. An example of the ultrasonic detection of stress-rupture cracks resulting from creep in reformer-furnace headers is given in the article "Boilers and Pressure Vessels" in this Volume. A relatively new and improved approach for monitoring the growth of cracks is done with ultrasonic imaging techniques.

Monitoring of fatigue cracks in parts during laboratory tests and while in service in the field has been extensively done using ultrasonic techniques. Reference 13 describes the use of surface waves to detect the initiation of cracks in cylindrical compression-fatigue testpieces having a circumferential notch. The surface waves, which were produced by four angle-beam search units on the circumference of each testpiece, were able to follow the contour of the notch and detect the cracks at the notch root. Monitoring the crack-growth rate was accomplished by periodically removing the cracked testpiece from the stressing rig and measuring the crack size by straight-beam, pulse-echo immersion inspection. It was found necessary to break open some of the cracked testpieces (using impact at low temperature) and visually measure the crack to establish an accurate calibration curve of indication height versus crack size.

The use of pulse-echo techniques for monitoring fatigue cracks in pressure vessels in laboratory tests is described in Ref 14. These techniques use several overlapping angle-beam (shear wave) search units, which are glued in place to ensure reproducible results as fatigue testing proceeded.

The in-service monitoring of fatigue cracking of machine components is often accomplished without removing the component from its assembly. For example, 150 mm (6 in.) diam, 8100 mm (320 in.) long shafts used in pressure rolls in papermaking machinery developed fatigue cracks in their 500 mm (20 in.) long threaded end sections after long and severe service. These cracks were detected and measured at 3-month intervals, using a contact-type straight-beam search unit placed on the end of each shaft, without removing the shaft from the machine. When the cracks were found to cover over 25% of the cross section of a shaft, the shaft was removed and replaced. In another case, fatigue cracking in a weld joining components of the shell of a ball mill 4.3 m (14 ft) in diameter by 9.1 m (30 ft) long was monitored using contact-type angle-beam search units. The testing was done at 3-month intervals until a crack was detected; then it was monitored more frequently. When a crack reached a length of 150 mm (6 in.), milling was halted and the crack repaired.

Dimension-Measurement Applications

Ultrasonic inspection methods can be used for measurement of metal thickness. These same methods can also be used to monitor the deterioration of a surface and subsequent thinning of a part due to wear or corrosion and to determine the position of a solid object or liquid material in a closed metallic cavity.

Thickness measurements are made using pulse-echo techniques. Resonance techniques were also used in the past, but have become obsolete. The results can be read on an oscilloscope screen or on a meter, or they can be printed out. Also, the same data signals can be fed through gates to operate sorting or marking devices or to sound alarms. Resonance thickness testing was most often applied to process control inspection where opposite sides of the testpieces are smooth and parallel, such as in the inspection of hollow extrusions, drawn tubes, tube bends, flat sheet and plate, or electroplated parts.

The maximum frequency that can be used for the test determines the minimum thickness that can be measured. The maximum thickness that can be measured depends on such test conditions as couplant characteristics, test frequency, and instrument design and on material type, metallurgical condition, and surface roughness.

Pulse-echo thickness gages with a digital readout (Fig. 59) are widely used for thickness measurement. Pulse-echo testing can measure such great thickness that it can determine the length of a steel reinforcing rod in a concrete structure, provided one end of the rod is accessible for contact by the search unit. Although pulse-echo testing is capable of measuring considerable thicknesses, near-field effects make the use of

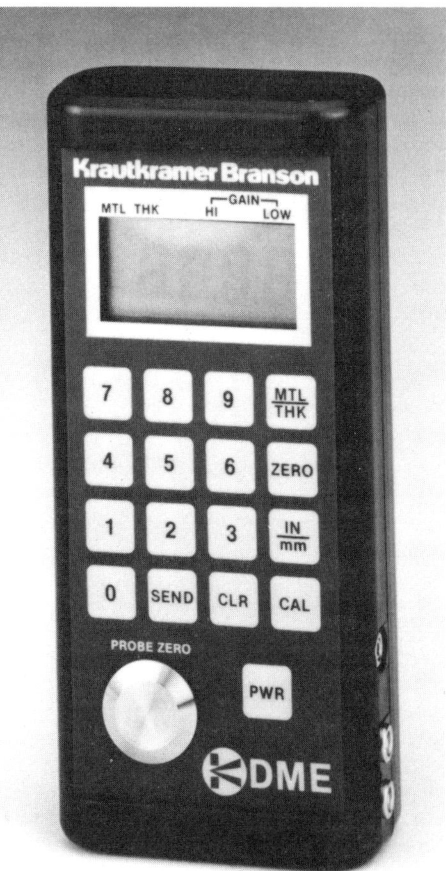

Fig. 59 Hand-held ultrasonic thickness gage

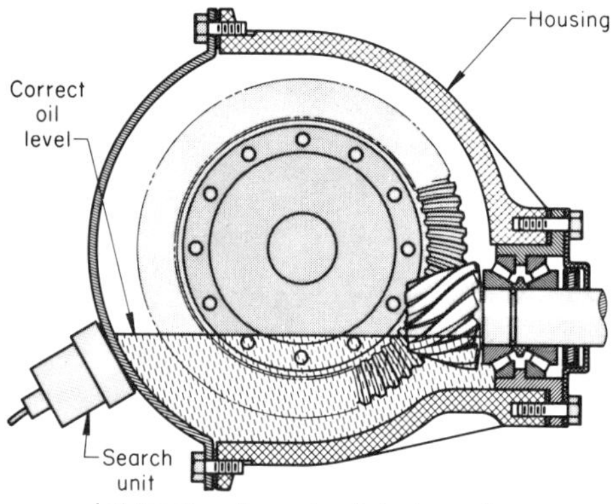

(a) Position of search unit for inspection

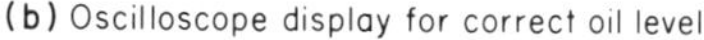

(b) Oscilloscope display for correct oil level

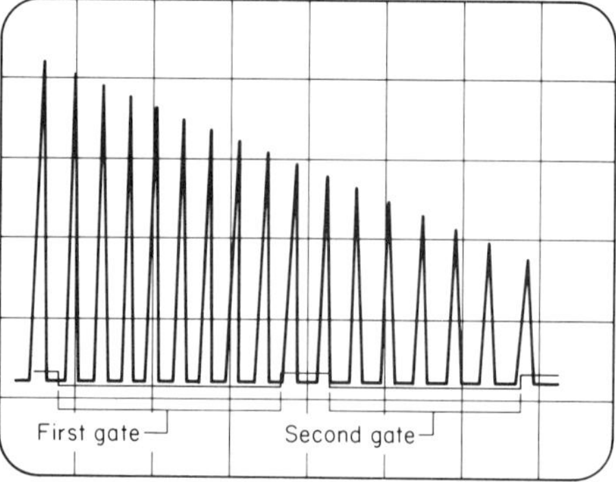

(c) Oscilloscope display for incorrect oil level

Fig. 60 Method of determining correct oil level in an automobile differential housing by use of an ultrasonic pulse-echo system. See text.

pulse-echo testing ineffective on very thin materials.

Position measurements of solid parts or liquid materials in closed metallic cavities are usually made with pulse-echo-type equipment. One technique is to look for changes in back reflection intensity as the position of the search unit is changed. In one variation of this technique, the oil level in differential housings was checked to see if the automated equipment used to put the oil in the housing on an assembly line had malfunctioned. The test developed for this application utilized a dual-gated pulse-echo system that employed a 1.6-MHz immersion-type search unit with a thin, oil-filled rubber gland over its face. The search unit was automatically placed against the outside surface of the housing just below the proper oil level, as shown in Fig. 60(a). With oil at the correct level, sufficient beam energy was transmitted across the boundary between the housing wall and the oil to attenuate the reflected beam so that multiple back reflections were all contained in the first gate (Fig. 60b). The lack of oil at the correct level allowed the multiple back reflections to spill over into the second gate (Fig. 60c). Thus, the test was a fail-safe test that signaled "no test" (no signal in the first gate), "go" (signals in the first gate only), and "no go" (signals in both gates).

In another position measurement system, a set of two contact-type 4-MHz search units was utilized in a through transmission pitch-catch arrangement to determine the movement of a piston in a hydraulic oil accumulator as both precharge nitrogen-gas pressure and standby oil pressure varied (Fig. 61). The two search units were placed 180° apart on the outside surface of the accumulator wall at a position on the oil side of the piston, as shown in Fig. 61. When a high-energy pulse was sent from the transmitting unit, the beam was able to travel straight through the oil, and a strong signal was picked up by the receiving unit. However, as the search units were moved toward the piston (see locations drawn in phantom in Fig. 61), the sloping sides of the recess in the piston bottom deflected the beam so that very little signal was detected by the receiving unit.

Determination of Microstructural Differences

Ultrasonic methods can be used to determine microstructural differences in metals. For this, contact testing with the pulse-echo technique is used. The testing can be either the measurement of ultrasonic attenuation or the measurement of bulk sound velocity.

The attenuation method is based on the decay of multiple echoes from testpiece surfaces. Once a standard is established, other testpieces can be compared to it by comparing the decay of these echoes to an exponential curve. This test is especially suited for the microstructural control of production parts, in which all that is necessary is to determine whether or not the parts conform to a standard.

An example of the use of ultrasonic attenuation in the determination of differences in microstructure is the control of graphite-flake size in gray iron castings, which in turn controls tensile strength. In one application, a water-column search unit that produced a pulsed beam with a frequency of 2.25 MHz was used to test each casting across an area of the casting wall having uniform thickness and parallel front and back surfaces. A test program had been first carried out to determine the maximum size of graphite flakes that could be permitted in

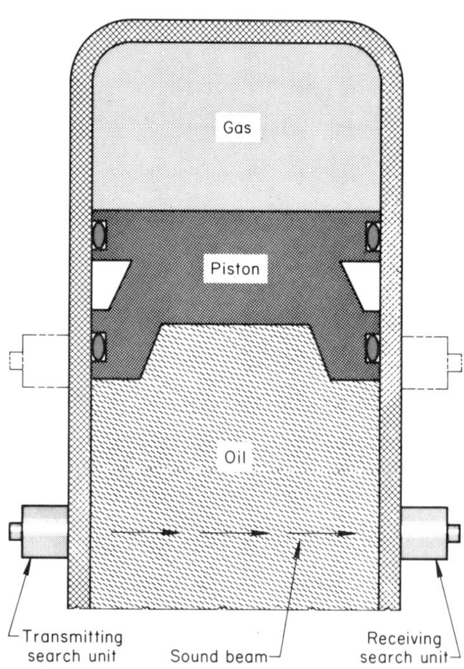

Fig. 61 Setup for determining the position of a piston in a hydraulic oil accumulator by use of two contact search units utilizing a through transmission arrangement

the casting and still maintain a minimum tensile strength of 200 MPa (30 ksi). Then, ultrasonic tests were made on sample castings to determine to what intensity level the second back reflection was lowered by the attenuation effects of graphite flakes larger than permitted. Next, a gate was set on the ultrasonic instrument in the region of the second back reflection, and an alarm was set to signal whenever the intensity of this reflection was below the allowable level. The testing equipment was then integrated into an automatic loading conveyor, where the castings were 100% inspected and passed or rejected before any machining operation.

Velocity Measurements. When considering the compressional and shear wave velocities given in Table 1, there may be small deviations for crystalline materials because of elastic anisotropy. This is important and particularly evident in copper, brass, and austenitic steels. The following example illustrates the variation of sound velocity with changes in the microstructure of leaded free-cutting brass.

Example 3: Measuring the Bulk Sound Velocity in Leaded Free-Cutting Brass.* Leaded free-cutting brass has a matrix microstructure of α phase with a dispersion of lead particles. Beta-phase stringers may also be present. The lead distribution varies in different heat lots from a total random distribution to confinement at grain boundaries.

*This example was provided by James J. Snyder, Westinghouse Electric Corporation, Oceanic Division.

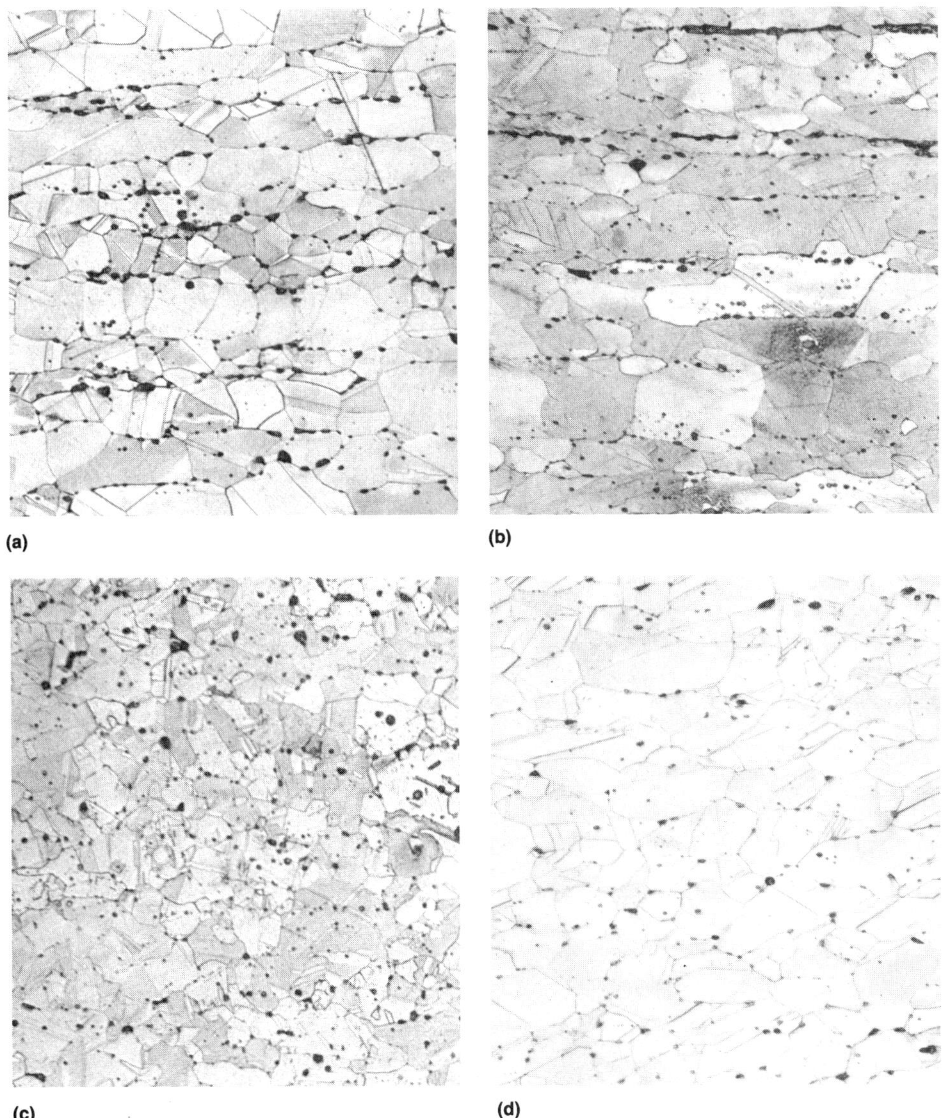

Fig. 62 Photomicrographs of samples tested in Example 3. (a) Sample A. (b) Sample B. (c) Sample C. (d) Sample D. Etched with NH_4OH plus H_2O_2. All 200×

This example shows a relationship correlating the sound velocity characteristic to the type of lead distribution in the brass. Two grades of leaded brass are considered: UNS C35300 and C36000. Both alloys are similar in chemical composition, mechanical properties, and sound velocity characteristics.

Four samples with flat parallel surfaces were prepared from 21 mm (0.83 in.) diam bar stock. These cylindrical samples were 17 mm (0.67 in.) in thickness (height). A similar-size velocity standard made from C36000 leaded brass was available. This standard test block had been measured for ultrasound transmission by the National Bureau of Standards (NBS).

The determination of sound velocity involves the measurement of sample thickness and the measurement of transit times with an A-scan display of ultrasonic echoes. The thickness of the samples and the velocity standard block are measured to an accuracy of ±0.02 mm (±0.001 in.), and the A-scan display provides the relative time or distance between the interfacial and back echoes for the samples and the velocity standard. If the time sweep on the A-scan display is not accurately calibrated, the velocity standard provides the reference for comparing the relative transit times of the A-scan data. The bulk sound velocity of the samples can then be determined by comparing the A-scan spacing (transit times) of echoes in the sample to the echo spacing in the velocity standard. Further details on the measurement of ultrasonic velocity are available in ASTM E 494-75.

With this general procedure, the back-echo distances on the A-scan oscilloscope display were measured using the electronic gate function control. The frequency used was 10 MHz, with a 75 mm (3 in.) focus transducer. The back-echo spacing on the A-scan display was recorded, and bulk sound velocities calculated. The sound velocity values are listed below:

Identity	Material	Condition	Velocity m/s	Velocity ft/s
NBS	C36000	Half hard	4149	13 610
A	C36000	Half hard	3839	12 600
B	C35300	Half hard	4136	13 570
C	C35300	Annealed	4454	14 610
D	C35300	Half hard	4381	14 370

Photomicrographs of samples A, B, C, and D in Fig. 62 show the matrix microstructures and lead distributions. In Fig. 62(a) and (b), the dark lead particles are distributed primarily at grain boundaries. These two samples also have lower bulk sound velocity transmission characteristics. Samples C and D have a significantly more random lead distribution within the α-phase grains and higher sound velocity transmission values.

Corrosion Monitoring

Ultrasonic inspection can be used for the *in situ* monitoring of corrosion by measuring the thickness of vessel walls with ultrasonic thickness gages. The advantage of this method is that internal corrosion of a vessel can be monitored without penetration.

There are, however, some disadvantages. Serious problems may exist in equipment that has a metallurgically bonded internal lining, because it is not obvious from which surface the returning signal will originate. A poor surface finish, paint, or a vessel at high or low temperature may also complicate the use of contact piezoelectric transducers (although this difficulty might be addressed by noncontact *in situ* inspection with an EMA transducer).

Despite these drawbacks, ultrasonic thickness measurements are widely used to determine corrosion rates. To obtain a corrosion rate, a series of thickness measurements is made over an interval of time, and the metal loss per unit time is determined from the measurement samples. Hand-held ultrasonic thickness gages are suitable for these measurements and are relatively easy to use. However, depending on the type of transducer used, the ultrasonic thickness method can overestimate metal thicknesses when the remaining thickness is under approximately 1.3 mm (0.05 in.).

Another corrosion inspection method consists of monitoring back-surface roughness with ultrasonic techniques. The following example describes an application of this method in the monitoring of nuclear waste containers.

Example 4: Ultrasonic Examination of Nuclear Waste Containers to Detect Corrosion.* Before low-level transuranic

*This example was provided by J.F. Cook and D.A. Aldrich, Idaho National Engineering Laboratory.

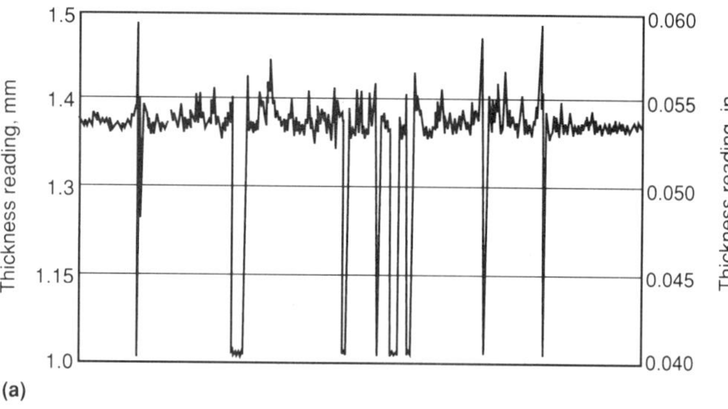

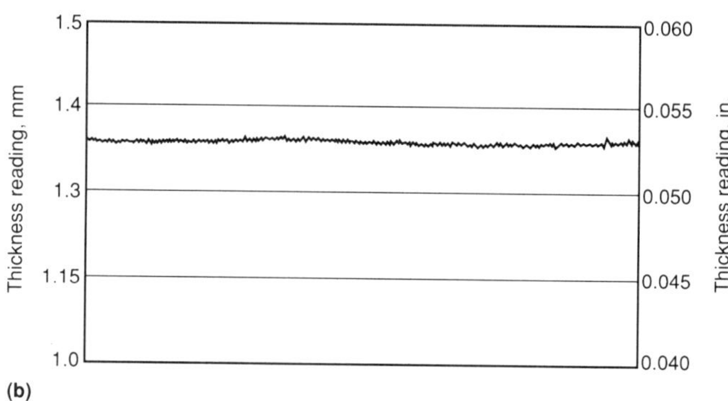

Fig. 63 Variations in ultrasonic indications from (a) corroded and (b) uncorroded areas

waste containers are transferred from interim storage (up to 20 years) to long-term storage, the containers are inspected for corrosion. Previously, the most commonly used procedure was visual examination of waste drum external surfaces and a few trial contact ultrasonic thickness measurements. However, visual examination detects only external damage, and sampling with contact ultrasonic thickness readings does not detect subtle changes resulting from minor internal corrosion. Also, the large number of drums to be examined means that manual examination is not compatible with the goal of keeping personnel radiation exposure as low as reasonably achievable.

An automated ultrasonic inspection was developed for monitoring the corrosion in 210 L (55 gal.) nuclear waste containers (carbon steel drums with a nominal wall thickness of 1.4 mm, or 0.055 in.). An initial evaluation was made of both ultrasonic and eddy current methods for assessing container integrity. Potential problems with obtaining the required depth of penetration with eddy currents and the permeability variation in drums resulted in selection of the ultrasonic method.

The ultrasonic system consisted of an eight-channel multiplexed ultrasonic instrument, a drum rotation fixture, bubbler-type search-unit holders, CRT display, strip chart recorder, analog and digital recording, and digital data processing (Ref 15). Eight locations were scanned during drum rotation, and data were taken every 1 mm (0.04 in.) around the circumference as the drum rotated. The search units (10-MHz focused transducers) were designed to tolerate an angle change of ±5° and had a maximum beam size of 1.3 mm (0.050 in.) at the focal point to provide for measurements over rough surfaces. The system was able to measure the thickness of steel in the range of 0.5 to 3 mm (0.020 to 0.120 in.).

Figure 63 shows indications from corroded and uncorroded areas. This illustrates the ability of the system to detect back-surface roughness due to corrosion buildup. Correlation with metallography shows that the system can detect oxide buildup of 250 μm (10 mils) and more, as well as deterioration as small as 25 μm (1.0 mil).

Stress Measurements

With ultrasonic techniques, the velocity of ultrasonic waves in materials can be measured and related to stress (Ref 16). These techniques rely on the small velocity changes caused by the presence of stress, which is known as an acoustoelastic effect. The technique is difficult to apply because of the very small changes in velocity with changes in stress and because of the difficulty in distinguishing stress effects from material variations (such as texture; see Ref 17). However, with the increased ability to time the arrival of ultrasonic pulses accurately (±1 ns), the technique has become feasible for a few practical applications, such as the measurement of axial loads in steel bolts and the measurement of residual stress (Ref 5).

The real limitation of this technique is that in many materials the ultrasonic pulse becomes distorted, which can reduce the accuracy of the measurement. One way to avoid this problem is to measure the phase difference between two-tone bursts by changing the frequency to keep the phase difference constant (Ref 5). Small specimens are used in a water bath, and the pulses received from the front and back surfaces overlap.

The presence of stress also rotates the plane of polarization of polarized shear waves, and there is some correlation between the angle of rotation and the magnitude of the stress. Measurement of this rotation can be used to measure the internal stress averaged over the volume of material traversed by the ultrasonic beam.

REFERENCES

1. A.J. Krautkramer and H. Krautkramer, *Ultrasonic Testing of Materials*, 1st ed, Springer-Verlag, 1969
2. D. Ensminger, *Ultrasonics*, Marcel Dekker, 1973
3. Y. Bar-Cohen, NDE of Fiber Reinforced Composites—A Review, *Mater. Eval.*, Vol 44 (No. 4), 1986, p 446-454
4. Y. Bar-Cohen, Nondestructive Characterization of Defects Using Ultrasonic Backscattering, in *Ultrasonic International 87*, Conference Proceedings, Butterworth, 1987, p 345-352
5. R. Halmshaw, *Nondestructive Testing*, Edward Arnold, 1987, p 198, 143, 211
6. Y. Bar-Cohen and A.K. Mal, Leaky Lamb Waves Phenomena in Composites Using Pulses, in *Review of Progress in Quantitative NDE*, Vol 8, D.P. Thompson and D.E. Chimenti, Ed., Plenum Press, 1989
7. A.K. Mal and Y. Bar-Cohen, Ultrasonic Characterization of Composite Laminates, in *Wave Propagation in Structural Composites*, Proceedings of the Joint ASME and SES meeting, AMD-Vol 90, A.K. Mal and T.C.T. Ting, Ed., American Society of Mechanical Engineers, 1988, p 1-16
8. R.H. Grills and M.C. Tsao, Nondestructive Testing With Portable Ultrasonic Imaging System, in *Corrosion Monitoring in Industrial Plants Using Nondestructive Testing and Electromechanical Methods*, STP 908, American Society for Testing and Materials, 1987, p 89-101
9. E.A. Lloyd, Non-Destructive Testing of Bonded Joints—A Case for Testing Laminated Structures by Wide-Band Ultrasound, *Non-Destr. Test.*, Dec 1974, p 331-334
10. P.J. Highmore, Non-Destructive Testing of Bonded Joints—The Depth Location of Non-Bonds in Multi-Layered Media, *Non-Destr. Test.*, Dec 1974, p 327-330
11. W.E. Garland, P.O. Ritter, and J.K. Fee, Applications of Ultrasonic Inspection to Composite Materials, in *Proceedings of the 7th Symposium on Nondestructive Evaluation of Components and Materials in Aerospace and Nuclear Applications*, Western Periodicals Company, 1969, p 226-238

12. Y. Bar-Cohen and A.K. Mal, Ultrasonic NDE of Adhesive Bonding, in *Ultrasonic Testing*, Vol 8, ASNT Handbook, American Society for Nondestructive Testing, to be published
13. C.E. Lautzenheiser, A.G. Pickett, A.R. Whiting, and A.W. Wilson, "Ultrasonic Studies of Fatigue Crack Initiation and Propagation," Paper presented at the Spring Convention, Los Angeles, American Society for Nondestructive Testing, Feb 1965
14. C.E. Lautzenheiser, A.R. Whiting, and R.E. Wylie, Crack Evaluation and Growth During Low-Cycle Plastic Fatigue—Nondestructive Techniques for Detection, *Mater. Eval.*, May 1966, p 241-248
15. J.F. Cook, D.A. Aldrich, B.C. Anderson, and V.S. Scown, Development of an Automated Ultrasonic Inspection System for Verifying Waste Container Integrity, *Mater. Eval.*, Dec 1984
16. Y.H. Pao, W. Sachse, and H. Fukuoka, Acoustoelasticity and Ultrasonic Measurement of Residual Stresses, in *Physical Acoustics: Principles and Methods*, Vol XVII, W.P. Mason and R.N. Thurston, Ed., Academic Press, 1984, p 61-143
17. A.V. Clark, J.C. Moulder, R.B. Mignogna, and P.P. DelSanto, "Comparison of Several Ultrasonic Techniques for Absolute Stress Determination in the Presence of Texture," National Bureau of Standards, 1987

SELECTED REFERENCES

- L. Adler and D. Fitting, *Ultrasonic Spectral Analysis for Nondestructive Evaluation*, Plenum, 1981
- G.V. Blessing, "An Assessment of Ultrasonic Reference Block Calibration Methodology," NBSIR 83-2710, National Bureau of Standards, 1983
- C.E. Burley, Calibration Blocks for Ultrasonic Testing, in *Nondestructive Testing Standards—A Review, STP 642*, H. Berger, Ed., American Society for Testing and Materials, 1977
- S. Golan, "A Comparison of American and European Ultrasonic Testing Standards," NBSIR 79-1790, National Bureau of Standards, 1979
- R.E. Green, *Ultrasonic Investigation of Mechanical Properties*, Academic Press, 1973
- G.C. Knollman and R.C. Yee, Ultrasonic-Image Evaluation of Microstructural Damage Accumulation in Materials, *Exp. Mech.*, Vol 28 (No. 2), June 1988, p 110-116
- J. Krautkramer and H. Krautkramer, *Ultrasonic Testing of Materials*, 3rd ed., Springer-Verlag, 1983
- T.M. Mansour, Ultrasonic Inspection of Spot Welds in Thin-Gage Steel, *Mater. Eval.*, Vol 46 (No. 5), April 1988, p 650-658
- M.G. Silk, *Ultrasonic Transducers for NDE*, Adam Hilger, 1984
- M. Stringfellow and B. Hawker, Scanner Improves Ultrasonic Inspection of Austenitic Material, *Atom*, No. 381, July 1988, p 16-21
- A. Vary, *Material Analysis by Ultrasonics*, Noyes Data Corporation, 1987
- A. Vary, Concepts for Interrelating Ultrasonic Attenuation, Microstructure, and Fracture Toughness in Polycrystalline Solids, *Mater. Eval.*, Vol 46 (No. 5), April 1988, p 642-649
- S.Y. Zhang, J.Z. Shen, and C.F. Ying, The Reflection of the Lamb Wave by a Free Plate Edge: Visualization and Theory, *Mater. Eval.*, Vol 46 (No. 5), April 1988, p 638-641

Acoustic Emission Inspection

Adrian A. Pollock, Physical Acoustics Corporation

ACOUSTIC EMISSIONS are stress waves produced by sudden movement in stressed materials. The classic sources of acoustic emissions are defect-related deformation processes such as crack growth and plastic deformation. The process of generation and detection is illustrated in Fig. 1. Sudden movement at the source produces a stress wave, which radiates out into the structure and excites a sensitive piezoelectric transducer. As the stress in the material is raised, many of these emissions are generated. The signals from one or more sensors are amplified and measured to produce data for display and interpretation.

The source of the acoustic emission energy is the elastic stress field in the material. Without stress, there is no emission. Therefore, an acoustic emission (AE) inspection is usually carried out during a controlled loading of the structure. This can be a proof load before service, a controlled variation of load while the structure is in service, a fatigue test, a creep test, or a complex loading program. Often, a structure is going to be loaded anyway, and AE inspection is used because it gives valuable additional information about the performance of the structure under load. Other times, AE inspection is selected for reasons of economy or safety, and a special loading procedure is arranged to meet the needs of the AE test.

Relationship to Other Test Methods

Acoustic emission differs from most other nondestructive testing (NDT) methods in two key respects. First, the signal has its origin in the material itself, not in an external source. Second, acoustic emission detects movement, while most other methods detect existing geometrical discontinuities. The consequences of these fundamental differences are summarized in Table 1.

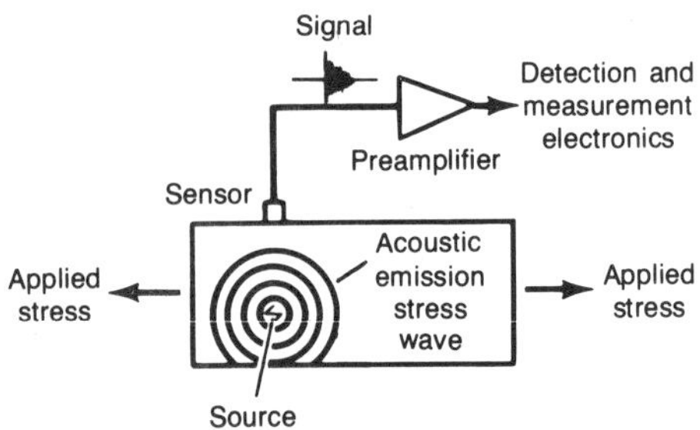

Fig. 1 Basic principle of the acoustic emission method

Often in NDT there is no one method that can provide the whole solution; for cost effectiveness, technical adequacy, or both, it is best to use a combination of methods. Because acoustic emission has features that distinguish it so sharply from other methods, it is particularly useful when used in combination with them.

A major benefit of AE inspection is that it allows the whole volume of the structure to be inspected nonintrusively in a single loading operation. It is not necessary to scan the structure looking for local defects; it is only necessary to connect a suitable number of fixed sensors, which are typically placed 1 to 6 m (4 to 20 ft) apart. This leads to major savings in testing large structures, for which other methods require removal of insulation, decontamination for entry to vessel interiors, or scanning of very large areas.

Typically, the global AE inspection is used to identify areas with structural problems, and other NDT methods are then used to identify more precisely the nature of the emitting defects. Depending on the case, acceptance or rejection can be based on AE inspection alone, other methods alone, or both together.

Table 1 Characteristics of acoustic emission inspection compared with other inspection methods

Acoustic emission	Other methods
Detects movement of defects	Detect geometric form of defects
Requires stress	Do not require stress
Each loading is unique	Inspection is directly repeatable
More material-sensitive	Less material-sensitive
Less geometry-sensitive	More geometry-sensitive
Less intrusive on plant/process	More intrusive on plant/process
Requires access only at sensors	Require access to whole area of inspection
Tests whole structure at once	Scan local regions in sequence
Main problems: noise related	Main problems: geometry related

Range of Applicability

Acoustic emission is a natural phenomenon occurring in the widest range of materials, structures, and processes. The largest-scale acoustic emissions are seismic events, while the smallest-scale processes that have been observed with AE inspection are the movements of small numbers of dislocations in stressed metals. In between, there is a wide range of laboratory studies and industrial testing.

In the laboratory, AE inspection is a powerful aid to materials testing and the study of deformation and fracture. It gives an immediate indication of the response and behavior of a material under stress, intimately connected with strength, damage, and failure. Because the AE response of a material depends on its microstructure and deformation mode, materials differ widely in their AE response. Brittleness and heterogeneity are two major factors conducive to high emissivity. Ductile deformation mechanisms, such as microvoid coalescence in soft steels, are associated with low emissivity.

In production testing, AE inspection is used for checking and controlling welds (Ref 1), brazed joints (Ref 2), thermocompression bonding (Ref 3), and forming operations such as shaft straightening (Ref 4) and punch press operations. In general, AE inspection can be considered whenever the process stresses the material and produces permanent deformation.

In structural testing, AE inspection is used on pressure vessels (Ref 5), storage tanks (Ref 5), pipelines and piping (Ref 6), aircraft and space vehicles (Ref 7), electric utility plants (Ref 7), bridges (Ref 7), railroad tank cars (Ref 8), bucket trucks (Ref 7), and a range of other equipment items. Acoustic emission tests are performed on both new and in-service equipment. Typical uses include the detection of cracks, corrosion, weld defects, and material embrittlement.

Procedures for AE structural testing have been published by The American Society of Mechanical Engineers (ASME), the Ameri-

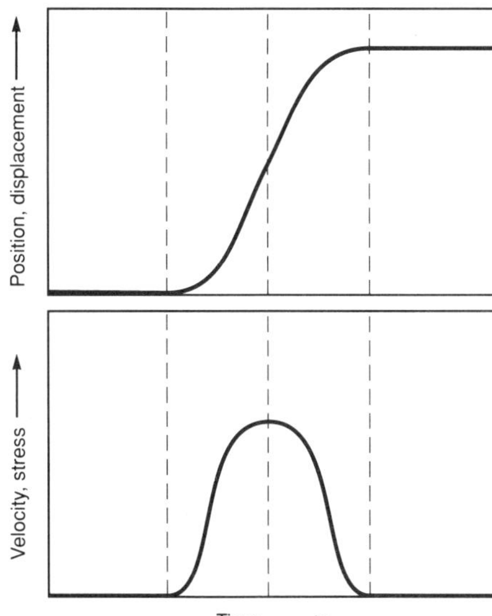

Fig. 2 Primitive AE wave released at a source. The primitive wave is essentially a stress pulse corresponding to a permanent displacement of the material.

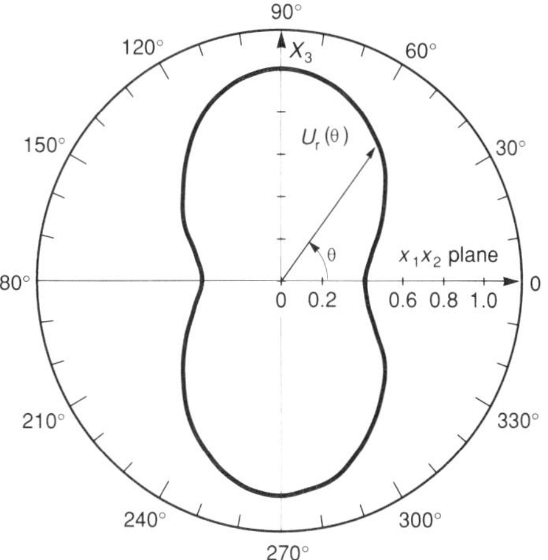

Fig. 3 Angular dependence of acoustic emission radiated from a growing microcrack. Most of the energy is directed in the 90 and 270° directions, perpendicular to the crack surfaces.

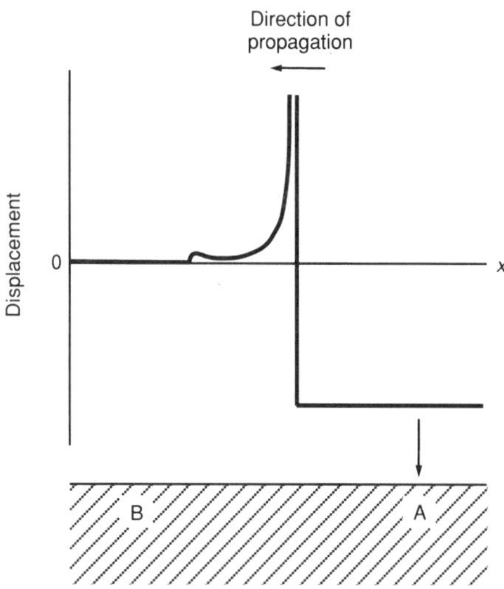

Fig. 4 Displacement waveform produced by an abrupt application of a downward force at point A. The waveform can be viewed as either a "snapshot" of surface displacement (in which case x = position) or as a graph of the displacement at point B as a function of time (x = time).

can Society for Testing and Materials (ASTM), and other organizations. Successful structural testing comes about when the capabilities and benefits of AE inspection are correctly identified in the context of overall inspection needs and when the correct techniques and instruments are used in developing and performing the test procedure (Ref 9).

Acoustic emission equipment is highly sensitive to any kind of movement in its operating frequency range (typically 20 to 1200 kHz). The equipment can detect not only crack growth and material deformation but also such processes as solidification, friction, impact, flow, and phase transformations. Therefore, AE techniques are also valuable for:

- In-process weld monitoring (Ref 10)
- Detecting tool touch and tool wear during automatic machining (Ref 10)
- Detecting wear and loss of lubrication in rotating equipment (Ref 10), and tribological studies (Ref 11)
- Detecting loose parts and loose particles (Ref 12)
- Detecting and monitoring leaks, cavitation, and flow (Ref 12, 13)
- Monitoring chemical reactions, including corrosion processes (Ref 14), liquid-solid transformations, and phase transformations (Ref 14)

When these same processes of impact, friction, flow, and so on, occur during a typical AE inspection for cracks or corrosion, they constitute a source of unwanted noise. Many techniques have been developed for eliminating or discriminating against these and other noise sources. Noise has always been a potential barrier to AE applicability. This barrier is constantly being explored and pushed outward, bringing previously impractical projects into the realm of feasibility.

Acoustic Emission Waves and Propagation

The primitive wave released at the AE source is illustrated in Fig. 2. The displacement waveform is basically a steplike function corresponding to the permanent change associated with the source process. The corresponding velocity and stress waveforms are basically pulselike. The width and height of the primitive pulse depend on the dynamics of the source process. Source processes such as microscopic crack jumps and precipitate fractures are often completed in a few microseconds or fractions of a microsecond, so the primitive pulse has a correspondingly short duration. The amplitude and energy of the primitive pulse vary over an enormous range from submicroscopic dislocation movements to gross crack jumps. The primitive wave radiates from the source in all directions, often having a strong directionality depending on the nature of the source process, as shown in Fig. 3. Rapid movement is necessary if a significant amount of the elastic energy liberated during deformation is to appear as an acoustic emission.

The form of the primitive wave is profoundly changed during propagation through the medium, and the signal emerging from the sensor has little resemblance to the original pulse. This transformation of the AE waveform is important both to the researcher interested in source function analysis and to the practical NDT inspector interested in testing structures. The researcher who wants to determine the original source waveform uses broadband sensors and performs a detailed analysis of the early part of the received signal. This is an important but very demanding line of inquiry. It may take an hour of computer time to process a single waveform. Most materials-oriented researchers, along with NDT inspectors, are interested in broader statistical features of the AE activity and do not need to know the precise details of each source event. They use narrowband sensors and electronic equipment that measures only a few features of the received waveform but is able to process hundreds of signals per second. The salient wave propagation factors are different for these two lines of work, as discussed below.

Factors in Source Function Analysis. The relationship between the source pulse and the resulting movement at the point of detection has been intensively studied during the last 10 to 15 years. Research groups at the NDT Centre at Harwell, UK (Ref 15), the National Bureau of Standards (Ref 16), Cornell University (Ref 17), and the University of Tokyo (Ref 18) have led the attack on this surprisingly difficult problem. A long-term goal of these studies has been to learn how to calculate a description of the source event from observation of the output of a distant sensor.

The difficulty of the problem is indicated in Fig. 4, which shows the vertical component of the surface movement at point B, resulting from the abrupt application of a vertical force at point A on a semi-infinite body. Even with this simple geometry and source function, the resulting waveform is quite complicated. In the case of a plate, it is more complicated yet, because the second surface also plays its role in the elastodynamics of the wave propagation process. In plates, the motion at the point of detection depends strongly on the ratio of source distance to plate thickness.

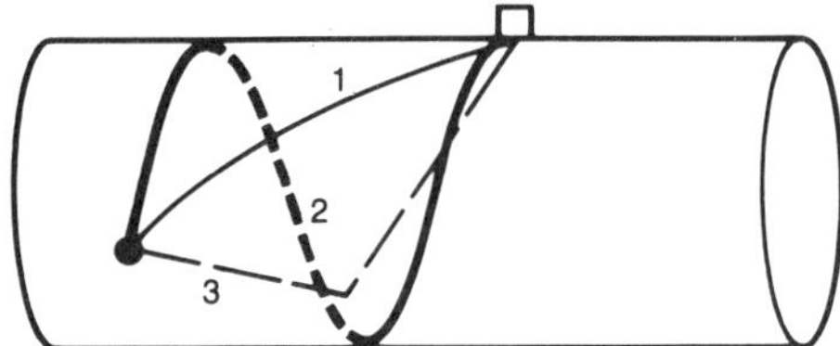

Fig. 5 Three possible paths from source to sensor in a water-filled pipe. 1, direct path; 2, spiral path; 3, waterborne path

In addition, the source function is not instantaneous, it will in fact be a force dipole and/or double couple rather than a point force, its orientation is in general unknown, and horizontal as well as vertical components of motion need to be considered. With these complications, it has understandably taken many years of effort to develop mathematical theory, computational tools, and experimental techniques equal to the task of calculating the source function.

In recent years, the leading laboratories have attained the ability to quantify crack growth increments, orientations, and time characteristics in some of the simpler specimen geometries ordinarily encountered (Ref 19). High-fidelity sensors must be used, and the analysis involves only the first part of the AE waveform, which is recorded in full detail with a high-performance transient recorder. This effort in the characterization of the source pulse has been one of the leading areas of AE research, and it can be expected eventually to yield returns in applied NDT.

Factors in Source Location and Typical AE Measurements. Whereas source function analysis utilizes only the first part of the AE waveform, mainstream AE technology accepts the waveform in its entirety. The later part of this waveform is made up of many components reaching the sensor by a variety of paths. Figure 5 illustrates this principle, but shows only a few of the indefinitely large number of possible paths by which the wave can reach the sensor. Typically, the highest peak in the waveform is produced, not by the first component, but by the constructive interference of several of the later components. The AE wave bounces around the testpiece, repeatedly exciting the sensor until it finally decays away. This decay process may take 100 μs in a highly damped, nonmetallic material or tens of milliseconds in a lightly damped, metallic material—much longer than the source event, which is usually finished in a few microseconds or less.

It is important to understand that the shape of the received waveform is fundamentally the result of these wave propagation processes. Other important aspects of wave propagation in typical AE testing are attenuation and wave velocity. Attenuation is the loss of signal amplitude due to geometric factors and material damping as the wave travels through the material (Ref 20). Attenuation governs detectability at a distance and is therefore an important factor in choosing sensor positions and spacing. Acoustic emission procedures typically call for attenuation measurements to be made before a test and specify permissible sensor spacing based on these measurements.

Wave velocity is an additional factor to be considered when AE technology is used for source location. Source location is an important technique that is widely used both in laboratory studies and in structural testing. It is particularly significant in testing large structures, for which AE inspection is used to identify active regions for conclusive follow-up inspection with other NDT methods. Large cost savings have been realized through this combination of global AE inspection and focused inspection by other methods.

There are several strategies for source location. Zone location places the source within a broad area. Point location places the source precisely, by calculating from the relative arrival times of the AE wave at several sensors. Wave velocity is involved in these calculations. The attainable accuracy is governed by wave propagation processes and depends on such factors as geometry, plate thickness, and contained fluids. In effect, these factors render the wave velocity uncertain and this leads to errors in source location. In favorable cases, the attainable accuracy is better than 1% of the sensor spacing; in unfavorable cases, worse than 10%. The wave propagation effects underlying these variations are reviewed in Ref 20.

Acoustic Emission Sensors and Preamplifiers*

The key element in an AE resonant sensor is a piezoelectric crystal (transducer) that converts movement into an electrical voltage. The crystal is housed in a suitable enclosure with a wear plate and a connector, as shown in Fig. 6. The sensor is excited by the stress waves impinging on its face, and it delivers an electrical signal to a nearby preamplifier and then to the main signal-processing equipment. The preamplifier can be miniaturized and housed inside the sensor enclosure, facilitating setup and reducing vulnerability to electromagnetic noise.

Sensor Response. One of the most sought-after properties in an AE sensor is high sensitivity. Although high-fidelity, flat frequency response sensors are available, most practical AE testing employs resonant-type sensors that are more sensitive, as well as less costly, than the flat frequency response type. These sensors have one or more preferred frequencies of oscillation, governed by crystal size and shape. These preferred frequencies actually dominate the waveform and spectrum of the observed signal in typical AE testing.

*Example 1 was provided by Phil Hutton, Battelle Northwest Laboratory.

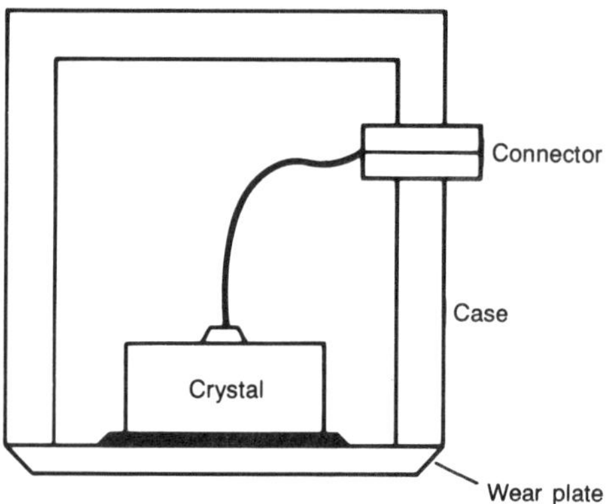

Fig. 6 Typical construction of an AE resonant sensor

The sensitivity calibration of AE sensors was the subject of a substantial developmental program at the National Bureau of Standards (NBS) through the late 1970s. This program has led to the routine availability of NBS-traceable plots showing the absolute sensitivity of AE sensors in volts per unit velocity as a function of frequency (Ref 21).

Acoustic Emission Waveform Transformation. In addition to the wave propagation factors discussed earlier, the transformation of a single AE waveform is further compounded by the sensor response. When a resonant sensor is excited by a broadband transient pulse, it rings like a bell at its own natural frequencies of oscillation. Therefore, the electrical signal at the sensor output is the product of this ringing, thus compounding the effects of multiple paths and multiple wave modes by which the wave travels from source to sensor. A typical AE signal from a piezoelectric sensor is shown in Fig. 7; the radical difference between this observed signal and the simple waveform at the source (Fig. 2) cannot be overemphasized.

Frequency Response. By selecting a resonant sensor from the wide range available, one can effectively choose the monitoring frequency. This is a useful feature that allows the inspector to make a suitable trade-off between the desired detection range and the prevailing noise environment. In practice, the vast majority of AE testing is well performed with sensors that are resonant at about 150 kHz.

Preamplifier Response. The signals generated by the sensor are amplified to provide a higher, more usable voltage. This is accomplished with a preamplifier, which is placed close to (or even inside) the sensor so as to minimize pickup of electromagnetic interference. The preamplifier has a wide

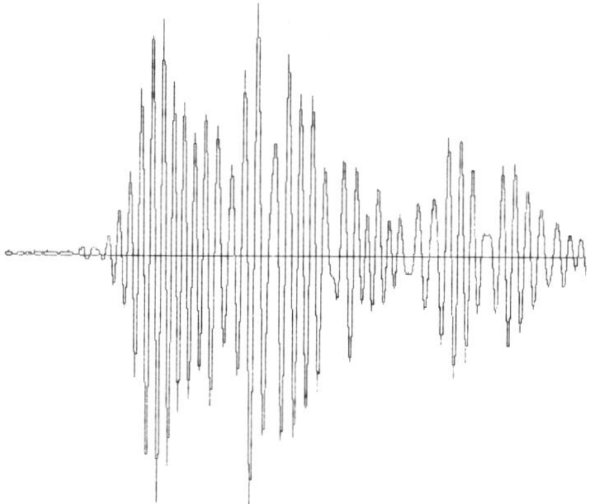

Fig. 7 Typical AE burst-type waveform recorded from a flat-bottom storage tank with a transient recorder

dynamic range and can drive the signal over a long length of cable so that the main instrumentation can be placed hundreds of meters from the testpiece if necessary.

The preamplifier typically provides a gain of 100 (40 dB) and includes a high-pass or bandpass filter to eliminate the mechanical and acoustical background noise that prevails at low frequencies. The most common bandpass is 100 to 300 kHz, encompassing the 150 kHz resonant frequency of the most commonly used sensor. Other operating frequencies can be used, but there are limitations. At lower frequencies, there are increasing problems with mechanical background noise. At higher frequencies, the wave attenuates (damps out) more rapidly, and the detection range of the sensor will be smaller. Choice of operating frequency is therefore a trade-off between noise and detection range. Lower frequencies are used on pipelines, where detection range is at a premium, and in geological work because rocks and soils are highly attenuating. Higher frequencies are used to test steam lines in electricity generating stations, where background noise is unusually high.

Attainable Sensitivity. Preamplifiers inevitably generate electronic noise, and it is this noise that sets the ultimate limit to the smallest movement detectable with AE equipment. The smallest signal that can be detected is about 10 μV at the transducer output, corresponding to a surface displacement of about 25 pm (1×10^{-6} μin.) for a typical high-sensitivity sensor. This sensitivity is more than enough for most practical NDT applications.

Installation. Typically, the sensor is coupled to the testpiece with a fluid couplant and is secured with tape, an adhesive bond, or a magnetic hold-down device. In some applications, however, the AE sensor may be mounted on a waveguide, as in Example 1.

After the sensor is installed and connected to the monitoring equipment, system performance is checked by "lead break" before monitoring begins. This involves the breaking of a lead pencil near the sensor to verify the response from an acoustic signal. Properly performed, the lead break delivers a remarkably reproducible signal that closely matches the "point impulse loading" source discussed above (see the section "Factors in Source Function Analysis" in this article).

Example 1: Acoustic Waveguide Sensors Used in Monitoring the Cooling of Molten Vitrified Nuclear Waste. Acoustic emission monitoring was used to help correlate cracking in vitrified high-level waste with cooling procedures. There was a need for a method capable of performing in an environment consisting of approximately 900 °C (1650 °F) temperatures and 500 Gy/h (50 000 rad/h) gamma radiation for the continuous monitoring of vitrified waste in canisters during cooling to detect glass cracking. Waveguide sensors about 4.6 m (15 ft) long were used; one end was submerged in the glass, and a sensing crystal and preamplifier were positioned on the other end. The signal from the sensor was passed through coaxial cables to the outside of the hot cell, where it was received by an AE monitor system for analysis. At the end of the testing, the AE sensors had been in the environment for 120 days, and the accumulated dose of gamma radiation had reached 14×10^5 Gy (14×10^7 rad). The sensors were still functioning properly.

Instrumentation Principles

During an AE test, the sensors on the testpiece produce any number of transient signals. A signal from a single, discrete deformation event is known as a burst-type signal. This type of signal has a fast rise time and a slower decay, as illustrated in Fig. 7. Burst-type signals vary widely in shape, size, and rate of occurrence, depending on the structure and the test conditions. If there is a high rate of occurrence, the individual burst-type signals combine to form a continuous emission. In some cases, AE inspection relies on the detection of continuous emission (see the sections "Mechanisms of AE Sources" and "Leak Testing" in this article).

The instrumentation of an AE inspection provides the necessary detection of continuous emissions or detectable burst-type emissions. Typically, AE instrumentation must fulfill several other requirements:

- The instrumentation must provide some measure of the total quantity of detected emission for correlation with time and/or load and for assessment of the condition of the testpiece
- The system usually needs to provide some statistical information on the detected AE signals for more detailed diagnosis of source mechanisms or for assessing the significance of the detected signals
- Many systems can locate the source of detectable burst-type emissions by comparing the arrival times of the wave at different sensors. This is an important capability of great value in testing both large and small structures
- The systems should provide a means for discriminating between signals of interest and noise signals from background noise sources such as friction, impact, and electromagnetic interference

Instruments vary widely in form, function, and price. Some are designed to function automatically in automated production environments. Others are designed to perform comprehensive data acquisition and extensive analysis at the hands of skilled researchers. Still others are designed for use by technicians and NDT inspectors performing routine tests defined by ASME codes or ASTM standards.

Signal Detection and Emission Counts. After sensing and preamplification, the signal is transmitted to the main instrument, where it is further amplified and filtered. Next is the critical step of detecting the signal. This is accomplished with a comparator circuit, which generates a digital output pulse whenever the AE signal exceeds a fixed threshold voltage. The relationship between signal, threshold, and threshold-crossing pulses is shown in Fig. 8. The threshold level is usually set by the operator; this is a key variable that determines test sensitivity. Depending on instrument design, sensitivity may also be controlled by adjusting the amplifier gain.

One of the oldest and simplest ways to quantify AE activity is to count the threshold-crossing pulses generated by the comparator (Fig. 8). These acoustic emission counts are plotted as a function of time or load, either as an accumulating total or in the form of a count rate histogram. The all-hardware AE systems of the early 1970s could draw these count and count rate displays on *x-y* recorders as the test proceeded, and much of the early AE literature presents results in this form. Figure 9, a typical plot of this type, shows cumulative counts as a function of applied load during a rising-load test on a precracked specimen of high-strength steel. The vertical scale is 10 000 counts full-scale. The vertical steps on the first parts of the plot are individual AE events. The larger events score several hundred counts each. By 35 kN (8000 lbf), 10 000 counts have been accumulated. The pen resets to the bottom of the graph, and resumes plotting. As the load rises, the AE rate increases, and the individual events are no longer discernible on the plot. As the specimen approaches failure, there are multiple resets of the pen corresponding to the

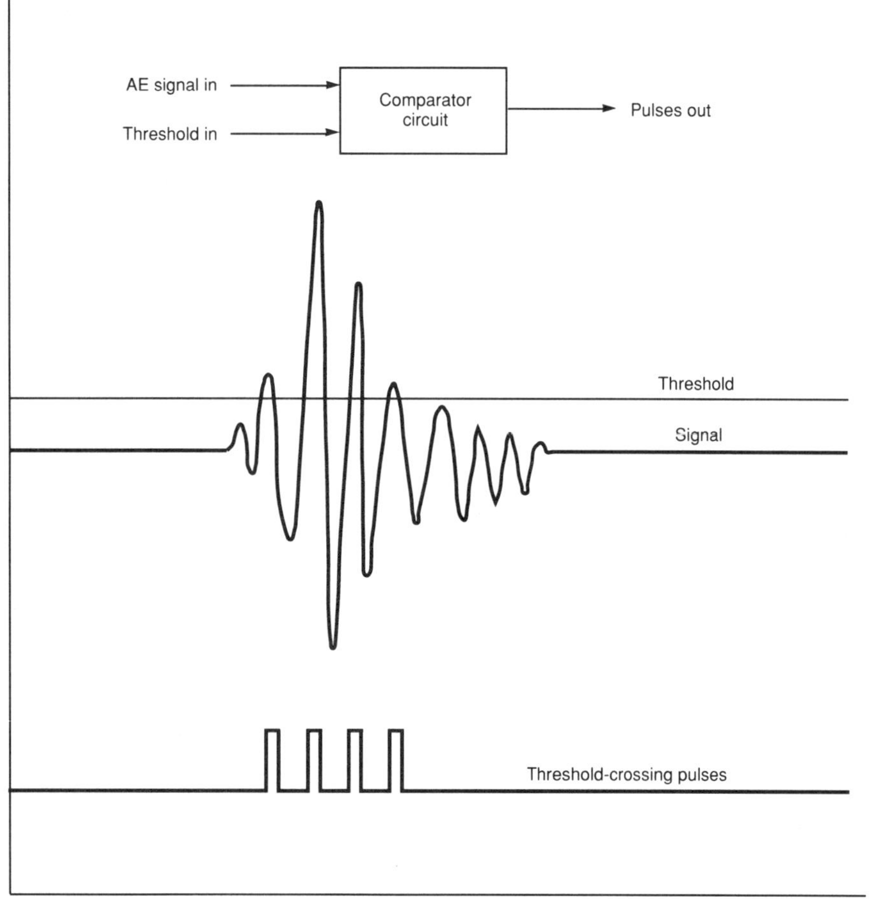

Fig. 8 Principle of AE signal detection and threshold-crossing counts

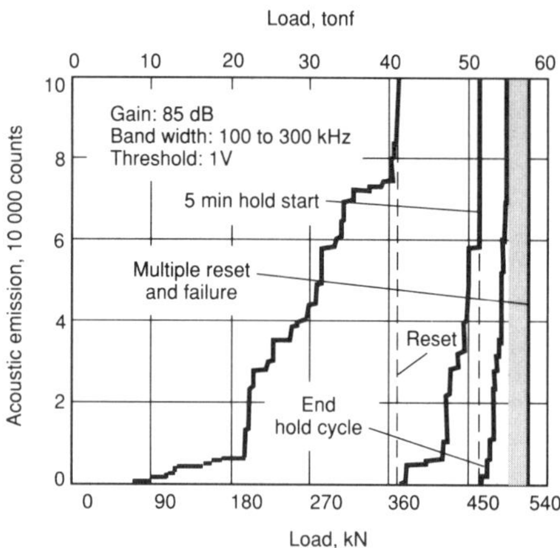

Fig. 9 Acoustic emission from a welded three-point bend specimen of 12% Ni maraging steel. Steps in the curve are discrete, burst-type emissions caused by plastic zone growth and later, crack front movement.

generation of hundreds of thousands of AE counts.

Hit-Driven AE Systems. All-hardware systems reached an apex of development in the late 1970s, but they were eventually superseded by computer-based systems. The development of AE technology coincided with the development of computers, and computers were probably used earlier for AE inspection than for any other NDT method. Computers were first used for AE multichannel source location systems around 1970. Although source location was the first task (and a very advanced one), computers soon came into use for the more general purposes of AE data storage, analysis, and display. At the same time, personnel involved in AE inspection became interested in other signal features of burst-type emissions beyond the threshold-crossing counts (see the section "Signal Measurement Parameters" in this article).

These trends led to a new principle of AE instrumentation that has dominated the technology ever since. This principle involves the measurement of key parameters of each hit, that is, each AE signal that crosses the threshold. A digital description of each hit is generated by the front-end hardware and is passed in sequence with other hit descriptions through a computer system, which provides data storage, a variety of graphical displays, and replay for posttest analysis.

A generic block diagram is shown in Fig. 10, and a typical modern system is shown in Fig. 11. The larger, multichannel systems divide the data-processing tasks among many microprocessors. In the system shown in Fig. 11, for example, a separate microprocessor serves each pair of signal measurement channels. The highest priority for this microprocessor is to read the results of each signal measurement as soon as the measurement process is completed, so that the measurement circuitry can be reset for the next event. The front-end microprocessor can rapidly store several hundred hit descriptions in its buffer, pending further processing. With this parallel processing architecture, added channels will automatically bring added data processing power. With the front-end buffers supplemented by other, even larger buffers in the later stages of the microcomputer network, the system has the versatility to absorb sudden surges of AE activity and to handle widely varying data rates in an optimum manner (Ref 22).

Signal Measurement Parameters. The five most widely used signal measurement parameters are counts (Fig. 8), amplitude, duration, rise time, and the measured area under the rectified signal envelope (MARSE) (Fig. 12). Some tests make do with fewer parameters, and some tests use others, such as true energy, counts-to-peak, average frequency, or spectral moment. However, the five principal parameters have become well standardized and accepted through the market processes of the last 10 years.

Along with these signal parameters, the hit description passed to the computer typically includes important external variables, such as the time of detection, the current value of the applied load, the cycle count (if it is a cyclic fatigue test), and the current level of continuous background noise. The length of the total hit description is usually between 20 and 40 bytes.

Amplitude, *A*, is the highest peak voltage attained by an AE waveform. This is a very important parameter because it directly determines the detectability of the AE event. Acoustic emission amplitudes are directly related to the magnitude of the source event, and they vary over an extremely wide range from microvolts to volts. Of all the conventionally measured parameters, amplitude is the one best suited to developing statistical information in the form of distribution functions (Ref 23). The amplitudes of acoustic emissions are customarily expressed on a decibel (logarithmic) scale, in which 1 μV at the transducer is defined as 0dBae, 10 μV is 20dBae, 100 μV is 40dBae, and so on.

Counts, *N*, are the threshold-crossing pulses (sometimes called ringdown counts) discussed above. This is one of the oldest and easiest ways of quantifying the AE signal. Counts depend on the magnitude of the source event, but they also depend strongly on the acoustic properties and re-

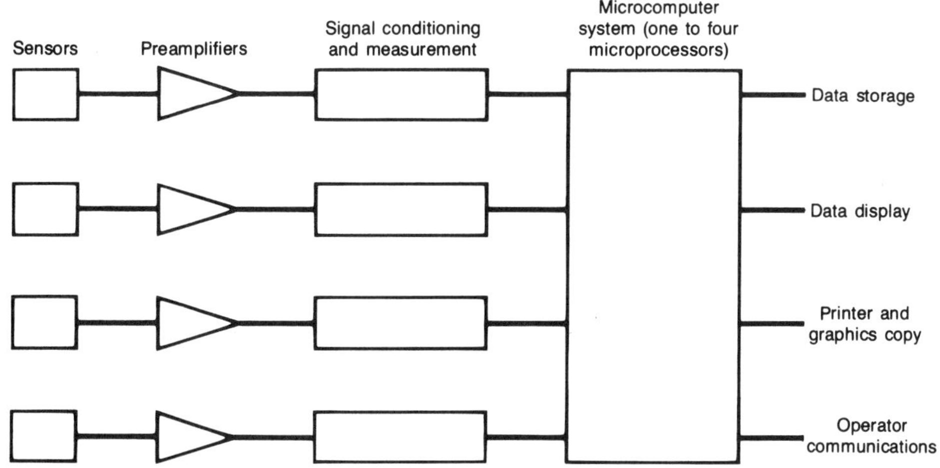

Fig. 10 Generic block diagram of a four-channel acoustic emission system

verberant nature of the specimen and the sensor.

MARSE, sometimes known as energy counts, *E*, is the measured area under the rectified signal envelope. As a measure of the AE signal magnitude, this quantity has gained acceptance and is replacing counts for many purposes, even though the required circuitry is relatively complex. MARSE is preferred over counts because it is sensitive to amplitude as well as duration, and it is less dependent on threshold setting and operating frequency. Total AE activity must often be measured by summing the magnitudes of all the detected events; of all the measured parameters, MARSE is the one best suited to this purpose.

Duration, *D*, is the elapsed time from the first threshold crossing to the last. Directly measured in microseconds, this parameter depends on source magnitude, structural acoustics, and reverberation in much the same way as counts. It is valuable for recognizing certain long-duration source processes such as delamination in composite materials (Ref 24), and it can be useful for noise filtering and other types of signal qualification.

Rise time, *R*, is the elapsed time from the first threshold crossing to the signal peak. Governed by wave propagation processes between source and sensor, this parameter can be used for several types of signal qualification and noise rejection.

Multichannel Considerations. Measurement of the signal proceeds simultaneously on every channel that detects (is hit by) the AE wave. Acoustic emission systems are available in sizes from 1 channel to over 100 channels, depending on the size and complexity of the structure to be tested. Typical laboratory systems have 2 to 6 channels, while most structural tests are accomplished with 12 to 32 channels.

An individual AE event may hit just one channel or it may hit many channels, depending on the strength of the event, the wave attenuation in the structure, and the sensor spacing. Therefore, an early task for the multichannel system is to determine whether a group of closely spaced hits on different channels is from the same source event. Depending on the system design, this can be accomplished either in hardware or in software. The second, third, and later hits from a source event can be either retained for the purposes of source location or discarded to keep the data clean and simple. After this task of event/hit identification has been performed, the system can deal in event descriptions as well as hit descriptions. The event description usually includes channel identification and relative timing information for all the channels involved, along with the signal characteristics of the first hit and perhaps the other hits as well.

The stream of hit (or event) descriptions is passed through a central processor that coordinates the tasks of data storage, display, and operator communications. In larger systems, these tasks can be divided among several processors. In many systems, the entire stream of hit descriptions is stored to disk; this provides unlimited posttest analysis capability. Full data storage is a vital aspect of applied AE technology. It reduces dependence on the on-site operator for ultimate test results, allowing him to concentrate on the vital task of correct data collection (Ref 11).

Data Displays. A software-based, hit-driven AE system can produce many types of graphic displays. The operator is not limited to what can be observed during the test, because the results can be refined, filtered, and redisplayed in any manner during the posttest analysis.

Broadly, AE data displays can be classed as:

- History plots that show the course of the test from start to finish
- Distribution functions that show statistical properties of the emission
- Channel plots showing the distribution of detected emissions by channel
- Location displays that show the position of the AE source
- Point plots showing the correlation between different AE parameters
- Diagnostic plots showing the severity of AE indications from different parts of the structure

Some of these generic display types are illustrated in Fig. 13.

Figures 13(a) and (b) show history plots of AE data versus time in cumulative and

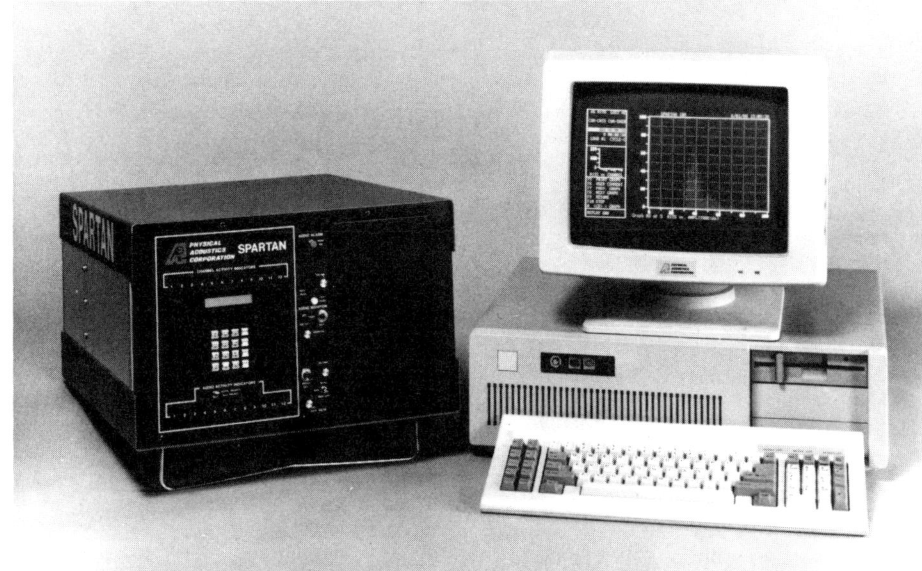

Fig. 11 Typical general-purpose AE instrument—a 12-channel data acquisition system with computer for data display, storage, and analysis. Courtesy of Physical Acoustics Corporation

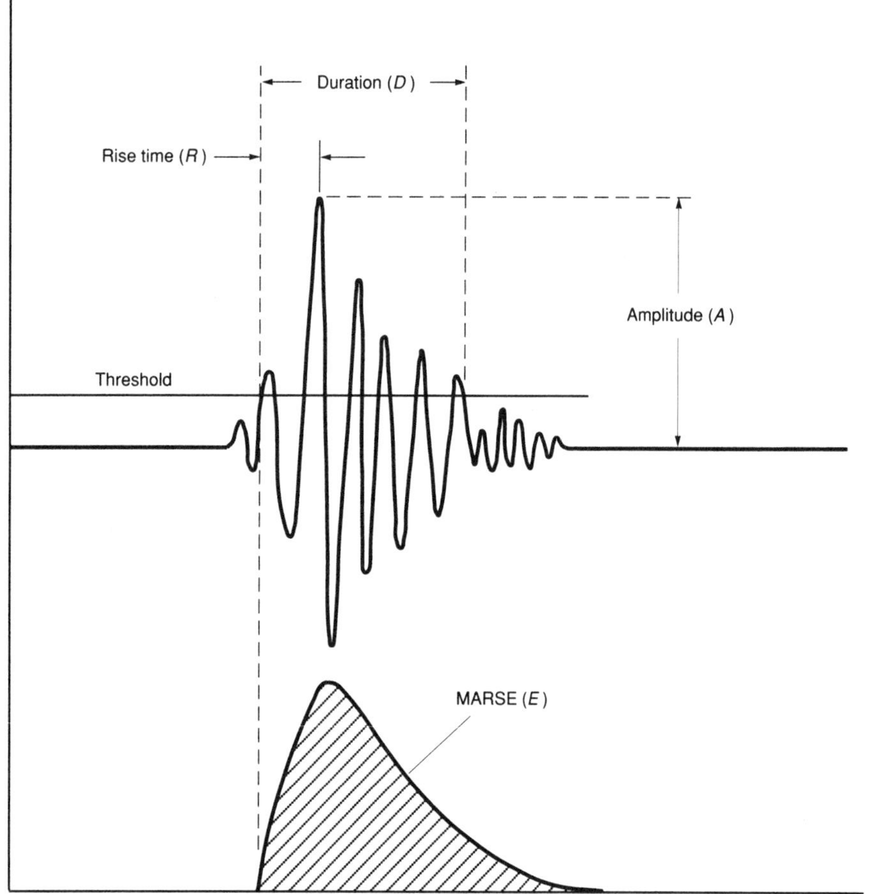

Fig. 12 Commonly measured parameters of a burst-type acoustic emission signal

rate form, respectively. A cumulative plot is the more convenient format for reading off a total emission quantity, while a rate plot highlights the changes in activity that occur during the test.

Figure 13(c) is a history plot of AE data versus load. This is the most fundamental plot because it directly relates cause to effect. This type of plot is especially useful for separating good parts from bad; bad parts characteristically begin to emit at lower loads and give more emission than good parts at all load levels. This basic plot of AE data versus load is also the best way to display the Kaiser and Felicity effects, as shown in Fig. 14.

Figures 13(d) and (e) show the cumulative and differential forms, respectively, of the amplitude distribution function. The x-axis shows amplitude, and the y-axis shows how many hits had that amplitude (differential form) or exceeded it (cumulative form). The differential amplitude distribution (Fig. 13e) is useful for distinguishing between deformation mechanisms and for observing changes in AE intensity as the test proceeds. The cumulative form (Fig. 13d) is more useful for quantitative modeling and for assessing how the detectability of AE will be affected by changes in test sensitivity. The amplitude distribution is a standard AE display, and the underlying theory is well developed (Ref 23). Distribution functions using the other signal measurement parameters are also employed for special purposes.

Figure 13(f) is a planar source location display. This display is basically a map of the structure, with the computed location of each emission event shown as a single point in the appropriate position. Sensor locations are shown as large dots, providing a reference frame. The eye is drawn to clusters of located events, which correspond to the most active sources, typically structurally significant defects.

Figure 13(g) is a point plot of counts (or duration) versus amplitude. Each hit is shown as one point on the display, and its position shows information about the size and shape of the waveform. This type of display is used for data quality evaluation, specifically for identifying some commonly encountered types of unwanted noise (Ref 25). Acoustic emission signals from impulsive sources typically form a diagonal band running across this display. Noise signals from electromagnetic interference fall below the main band (circled area, lower right, Fig. 13g) because they are not prolonged by acoustic reverberation. Noise signals from friction and leaks fall above the main band (circled area, upper left, Fig. 13g) because the source process is extended in time, not a short impulse. This is only one of the many point plots that have proved useful in practical AE testing.

The typical software-based system can generate many displays simultaneously in memory while the test is running, presenting them to the operator upon demand. In addition to these graphic displays, the system may present tabulated data and/or listings of the individual event or hit descriptions.

Special-Purpose AE Systems. The software-based, hit-driven AE system has the architecture of choice for application development and general-purpose laboratory and structural testing, but not all AE systems require this kind of computational power and versatility of display. Once the needs of the test have been defined, simpler equipment is often appropriate for routine application.

Production testing can often be done with a basic, all-hardware instrument that simply measures counts or energy and trips an alarm when the emission exceeds a predetermined quantity. Automatic self-checking for good sensor contact can be incorporated into the function of such an instrument.

Resistance weld monitoring and feedback control is accomplished with all-hardware systems that have special gates, timers, and interfaces to synchronize the AE monitoring with the operation of the weld controller. Other types of weld monitoring instruments incorporate pattern-recognition algorithms for automatically recognizing and classifying specific kinds of weld defects.

Leak testing is a major and relatively simple application of AE instrumentation (see the section "Structural Test Applications" in this article). Leak testing can be performed with instruments that measure only the root mean square (rms) voltage of the continuous emission from a leak. Sometimes, detectability is enhanced by the occurrence of burst-type signals from particle impact or structural degradation of the local material. Small size is a major advantage when the instrument has to be carried around an industrial complex or power generating plant.

Specific Applications. Instrument manufacturers have also developed special instruments for specific, well-established applications, such as bucket truck testing and tank car testing. These instruments are based on the applicable codes or standard procedures for performing the test. Simplification of hardware and software leads to a lower-cost instrument. Customized software provides more positive guidance and fewer operator choices, so that a lower level of skill can be used on-site and the test can be performed reliably and economically.

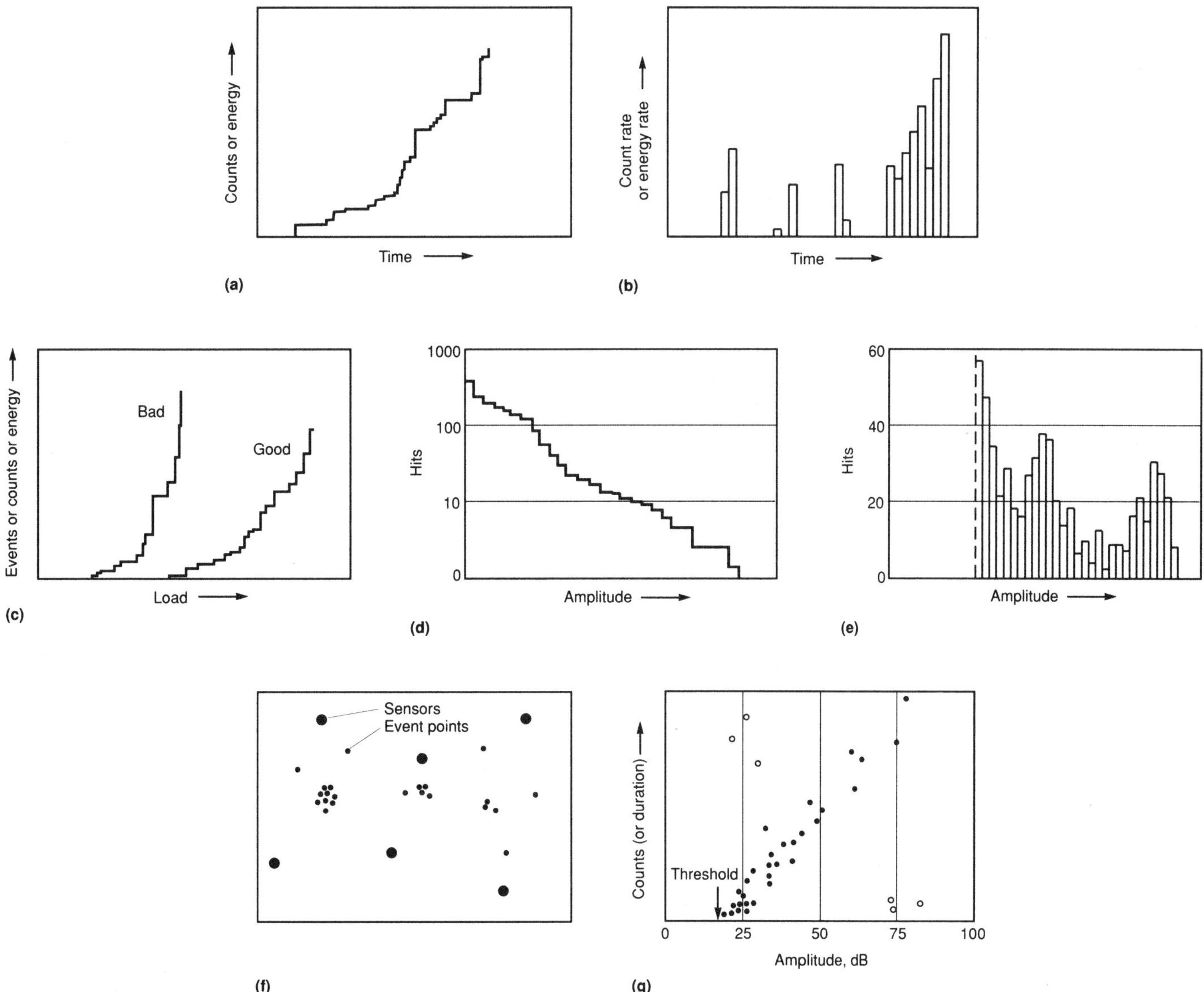

Fig. 13 Typical AE data displays. (a) History plot of the cumulative count or energy. (b) History plot of the count rate or energy rate. (c) History plot of AE data versus load. (d) Cumulative amplitude distribution showing the number of hits that exceeded an amplitude. (e) Differential amplitude distribution showing the number of hits of a particular amplitude. (f) Planar source location display. (g) Point plot of counts (or duration) versus amplitude

Noise

Precautions against interfering noise are an integral part of AE technology. Enormous progress has been made since the early days when students worked at night, using specially constructed loading machines in underground laboratories to avoid disruption of their experiments by street traffic and people moving nearby. With current technology, many tests can be performed without special measures, and a wide range of techniques have been developed to make AE inspection applicable in extremely noisy environments.

A basic starting point is the selection of an appropriate frequency range for AE monitoring. The acoustic noise background is highest at low frequencies. The 100 to 300 kHz range has proved suitable for perhaps 90% of all AE testing. In noisy environments (an electric power plant, for example), higher frequencies, such as 500 kHz, have been necessary to reduce the noise detected from fluid flow. Because higher frequencies bring reduced detection range, there is an inherent trade-off between detection range and noise elimination.

Acoustic noise sources include fluid flow in pumps and valves, friction processes such as the movement of structures on their supports, and impact processes such as rain and wind-blown cables striking the structure. Electrical and electromagnetic noise sources include ground loops, power switching circuits, radio and navigation transmitters, and electrical storms.

Noise problems can be addressed in many ways. First, it may be possible to stop the noise at the source. Second, it may be possible to eliminate an acoustic source by applying impedance-mismatch barriers or damping materials at strategic points on the structure. Electrical noise problems, which are often the result of poor grounding and shielding practices, can be eliminated by proper technique or by using differential sensors or sensors with built-in preamplifiers. If these measures are inadequate, the problem must be dealt with by hardware or software in the AE instrument itself.

Sensitivity adjustments, including floating-threshold techniques, can be very effective as long as they do not also cause the loss of essential AE data. Methods for selective acceptance and recording of data based on time, load, or spatial origin are well developed. Beyond this, because noise sources often give characteristically different waveforms, they can often be separated from true acoustic emissions by computer inspection of the measured signal characteristics (Ref 25). This can be accomplished

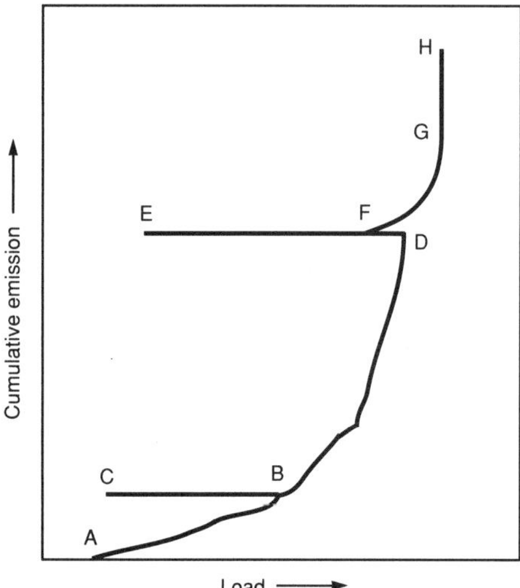

Fig. 14 Basic AE history plot showing Kaiser effect (BCB), Felicity effect (DEF), and emission during hold (GH)

immediately after measurement (front-end filtering), during the display process (graphical filtering), or after the test by playing the data through a posttest filtering program or advanced waveform analysis package.

Through the development and application of these techniques, AE inspection has been brought into service in increasingly demanding environments, and this trend is expected to continue. Examples of difficult applications in which noise elimination was key to the successful use of AE inspection include the on-line monitoring of welding (Ref 1, 26) and the detection of fatigue crack growth in flying aircraft (Ref 7).

Load Control and Repeated Loadings

Because acoustic emission is produced by stress-induced deformation of the material, it is highly dependent on the stress history of the structure. Emission/stress/time relationships also depend on the material and on the type of deformation producing the emission. Some materials respond almost instantly to applied stress, emitting and then quickly stabilizing. Other materials take some time to settle down after a load is applied; this is readily observed in materials that show viscoelastic properties, such as resin-matrix composites. In other cases, a constant load may produce ongoing damage, and the structure may never stabilize. An example of this is hydrogen-induced cracking, which may proceed under constant load to failure, with continual emission.

Acoustic emission testing is often carried out under conditions of rising load. The first load application will typically produce much more emission than subsequent loadings. In fact, for instantaneously plastic materials, subsequent loadings should produce no emissions at all until the previous maximum load is exceeded. This behavior was first reported by Kaiser in 1950 (Ref 27) and has been a leading influence in the development of AE test methodology. Dunegan (Ref 28) showed that for materials that obey the Kaiser effect, emission on a repeat loading will indicate that structural damage occurred between the first loading and the repeat. This became the conceptual basis of much of the AE testing of the 1970s, when the first AE field test organizations undertook periodic inspection of pressure vessels and other structures.

Recent test strategies pay much attention to emission that occurs at loads below the previous maximum and to emission that continues when the load is held at a constant level. The evidence is that structurally significant defects will tend to exhibit these behaviors, while emission related to stabilization of the structure, such as the relief of residual stress, will tend not to recur when the structure is loaded again.

Figure 14 is a generic illustration of these contrasting behaviors. Emission is observed upon initial loading from A to B, but not upon unloading (B to C). Upon reapplying the load, there is no emission (line is horizontal) until B is reached again; this is the Kaiser effect. The load is increased to D, with more emission, and another unload-reload cycle is applied. This time, because of the higher stress levels, significant defects begin to emit at point F, below the previous maximum load. This behavior is known as the Felicity effect. It can be quantified with the Felicity Ratio (FR):

$$\text{FR} = \frac{\text{Load at which emission begins again}}{\text{Previous maximum load}}$$

Technically, the Kaiser effect can be construed as a Felicity Ratio of 1.0 or greater. Systematic decreases in the Felicity Ratio as material approaches failure have been well documented for fiber-reinforced plastics (Ref 29) and a Felicity Ratio less than 0.95 is cause for rejection of an FRP tank or pressure vessel tested by AE inspection according to ASME Article 11 (Ref 30). Under ASME Article 12 (Ref 31) for the AE testing of metal pressure vessels, it is in some cases admissible to ignore AE data from the first loading of a vessel and to consider only AE data from a second loading. The basis for this is that much emission on the first loading comes from local yielding (structurally insignificant), while only the significant defects will emit on the second loading (Felicity Ratio < 1).

Figure 14 also illustrates the graphical appearance of emission continuing during a load-hold period (G to H). The Felicity effect and the occurrence of emission during load holds may share a common underlying explanation; both are associated with the unstable nature of structurally significant defects. Emission during load holds has been known since the early years of AE inspection (Ref 28) and was incorporated in FRP evaluation criteria in the mid 1970s. In the late 1980s, emission during hold has been made the entire basis of Monsanto's successful procedure for the AE testing of railroad tank cars (Ref 8). In this interesting development, data analysis is greatly simplified because the background noise sources present during rising load are much less obtrusive during the load-hold periods.

Careful attention must be paid to the loading schedule if AE testing is to be successful. Procedures for an AE test typically specify the loads that must be applied (relative to the working load or design load) and the upper and lower limits on the loading rate. Fiber-reinforced plastic tanks and vessels must be conditioned by a period at reduced load before the AE test is conducted (Ref 30). An AE test can be invalidated if the structure is inadvertently loaded beforehand to the AE test pressure. For success in dealing with these points, there must be good communication and coordination between the personnel loading the structure and those collecting the AE data.

Acoustic Emission in Materials Studies*

Acoustic emission is a remarkable tool for studying material deformation because the information it provides is both detailed and immediate. With its sensitivity to microstructure and its intimate connection with failure processes, AE inspection can give unique insights into the response of material to applied stress. Acoustic emission analysis is most useful when used in conjunction with other diagnostic techniques, such as stress-strain measurements, microscopy, crack-opening-displacement measurements and potential drop (for crack growth), or ultrasonic damping measurements (for dislocation studies). Acoustic emission complements these techniques and offers additional information on the dynamics of the underlying deformation processes, their interplay, and the transitions from one type of deformation to another.

Many materials studies involve the development of a test approach for eventual field application. Such work can be valuable, but it is subject to the difficulty of simulating defect emissivities and other field conditions in the laboratory. Laboratory tests are often done with simple uniaxial stresses applied parallel to the rolling direction, while materials in industrial service are often subjected to complex biaxial or triaxial stress fields. In such cases, the acoustic emissions from the laboratory tests will not be a good model of the acoustic emissions from materials in industrial service.

*Example 3 was provided by J.M. Wolla, U.S. Naval Research Laboratory.

Mechanisms of AE Sources. Needless to say, acoustic emissions are not generated by the reversible, homogeneous alteration of interatomic spacings that constitutes elastic deformation. Acoustic emissions are only generated when some abrupt and permanent change takes place somewhere in the material. Mechanisms that produce acoustic emissions in metals include the movement and multiplication of dislocations; slip; twinning; fracture and debonding of precipitates, inclusions, and surface layers; some corrosion processes; microcrack formation and growth; small and large crack jumps; and frictional processes during crack closure and opening.

The amount of AE energy released depends primarily on the size and speed of the local deformation process. The formation and movement of a single dislocation does produce an AE stress wave, but it is not a large enough process to be detected in isolation. However, when millions of dislocations are forming and moving at the same time during yielding of a tensile specimen, the individual stress waves overlap and superimpose to give a detectable result. The result is a continuous excitation of the specimen and sensor that is detectable as soon as the voltage it produces becomes comparable with the background noise. The higher the strain rate and the larger the specimen, the larger this signal becomes. This so-called continuous emission is different from burst-type emission in that the individual source events are not discernible. Continuous emission is best measured with rms or energy rate measuring circuitry.

Continuous emission from the plastic deformation of steels, aluminum alloys, and many other metals has been extensively studied, and there have been many detailed findings relating acoustic emissions to dislocation activity and precipitates, microstructure, and materials properties (Ref 32). Such studies can yield valuable insights for alloy and material development. Most studies have focused primarily on continuous emission during and after yield; burst-type emissions sometimes observed in the nominally elastic region are less well explained.

The following example illustrates the dependence of continuous emission on microstructure. The fracture of small-scale precipitates (in this case, pearlite lamellae) generates continuous emission, which can be related to the microstructure that results from heat treating.

Example 2: Relation of Acoustic Emissions With the Optimum Heat Treating of Ferritic/Pearlitic Steel. Figure 15 illustrates the dependence of continuous acoustic emissions on the microstructure of a deep-drawing ferritic/pearlitic steel subjected to a spheroidizing heat treatment to improve its formability. Data arc shown from representative underannealed, optimally annealed, and overannealed conditions. Figure 15 shows AE energy rate as a function of time from dog-bone tensile specimens pulled to failure in a screw-driven test machine. All the graphs display peaks around the yield region, a common feature in the high-sensitivity tensile testing of unflawed specimens. Figure 15 also shows a second, shallow peak at higher strain levels.

The interesting result is that the optimally annealed specimen shows a much smaller peak (gives much less emission) than the other two specimens. The explanation is found by carefully relating AE behavior to microstructural deformation processes. It is known that dislocations can pile up against pearlite lamellae during plastic deformation, eventually causing the lamellae to fracture. This fracture of pearlite lamellae is believed to be the cause of the first peaks in Fig. 15.

With the test material in the underannealed condition, microscopy reveals the presence of many untransformed pearlite lamellae that can intercept the moving dislocations, so the peak is high. With the test material in the optimally annealed condition, microscopy shows that virtually all the lamellae have been transformed to spheroids. These have a smaller cross-sectional area and present less of an obstacle to the moving dislocations, so deformation can proceed without breaking pearlite. Ductility is enhanced, and there is very little emission from this optimally annealed material.

With the test material in the overannealed condition, microscopy shows that additional carbon has come out of solution, growing the spheroids and forming doglegs at the grain boundaries. These larger particles interfere more strongly with dislocation motion and produce larger emissions when fractured, so the emission peak is strong again. It is an interesting result that the optimum material condition is the condition of lowest emissivity, suggesting that AE inspection could be used for inspection and quality control of this material as well as for research.

Acoustic emission from crack growth is of the greatest interest for practical NDT applications of the AE phenomenon. By virtue of the stress concentrations in their vicinity, cracks and other defects will emit during rising load, while unflawed material elsewhere is still silent. Acoustic emissions from crack initiation and growth have been extensively reported in the literature. Many of these reports deal with specialized forms of crack growth, such as fatigue, stress-corrosion cracking, and hydrogen embrittlement (Ref 33).

It is useful to distinguish between AE signals from the plastic zone at the crack tip and AE signals from movement of the crack front itself. Growth of the plastic zone typically produces many emissions of rather low amplitude. These emissions are typically ascribed to the fracture of precipitates and inclusions (for example, manganese sulfide stringers in steels), and the triaxial nature of the stress field is implicated in the emissivity of these sources.

Acoustic emissions from crack front movement depend critically on the nature of the crack growth process. Microscopically rapid mechanisms such as brittle intergranular fracture and transgranular cleavage are readily detectable, even when the crack front is only advancing one grain at a time at subcritical stress levels. Slow, continuous crack growth mechanisms such as microvoid coalescence (ductile tearing) and active path corrosion are not detectable in themselves, but if general yield has not occurred, they may be detectable through associated plastic zone growth. Quantitative theory, which explains why some processes are detectable and others are not, was developed by Wodley and Scruby (Ref 33). The possibility of silent crack growth in ductile materials caused much consternation when it was first recognized in laboratory conditions, but it has not been a deterrent in real-life NDT, in which emission from defects is characteristically enhanced by environmental embrittlement, emissive corrosion products, crack face friction, or emissive nonmetallic materials entrained in the defect during the fabrication process.

Many fracture mechanics models have been developed to relate acoustic emissions to crack growth parameters. An important early approach was to relate acoustic emissions to the plastic zone size with the hope of estimating directly the stress intensity factor at defects found in the field (Ref 34, 35). Other models relate acoustic emissions to crack tip movement in situations of cyclic fatigue (Ref 33) or stress-corrosion cracking (Ref 36) for various materials. These models are commonly framed as power-law relationships, with the acoustic emission described by conventional parameters such as threshold-crossing counts, N. In the more recent but difficult technique of source function analysis discussed in the section "Wave Propagation Effects" in this article, individual crack growth increments can be quantified in absolute terms by computer-intensive analysis of the early portion of the AE waveform.

Nonmetallic layers on metal surfaces also exhibit acoustic emissions for potential NDT applications. Examples of acoustic emissions from nonmetallic layers include:

- The acoustic emission from high-temperature oxidation (Ref 37)
- The extensive study of acoustic emissions from room-temperature corrosion processes (Ref 38, 39)
- The use of acoustic emissions to optimize the performance of ceramic coatings used in high-temperature components (Ref 40)

Metal-Matrix Composites. The following example illustrates one application in-

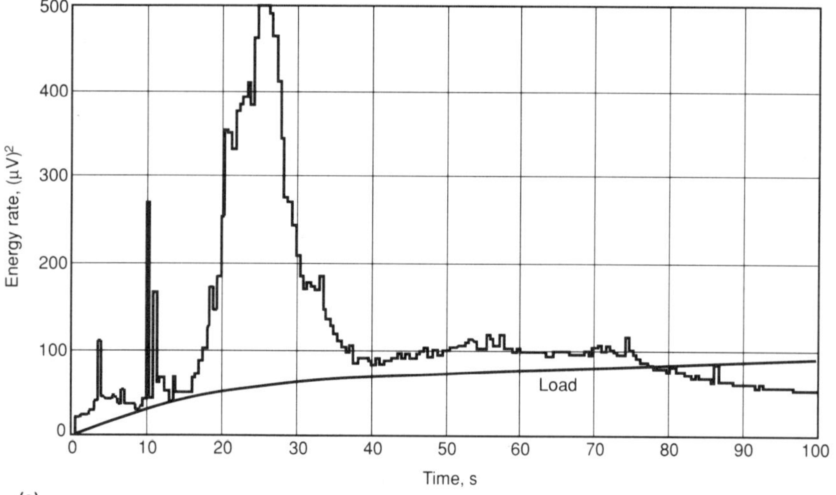

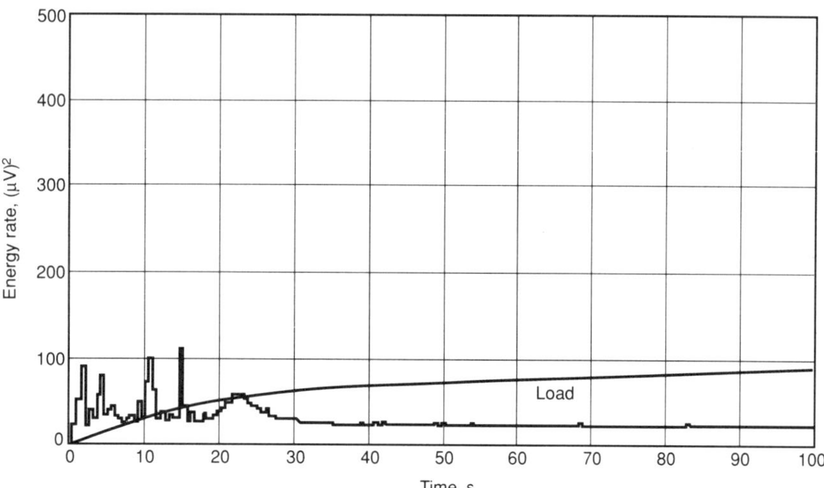

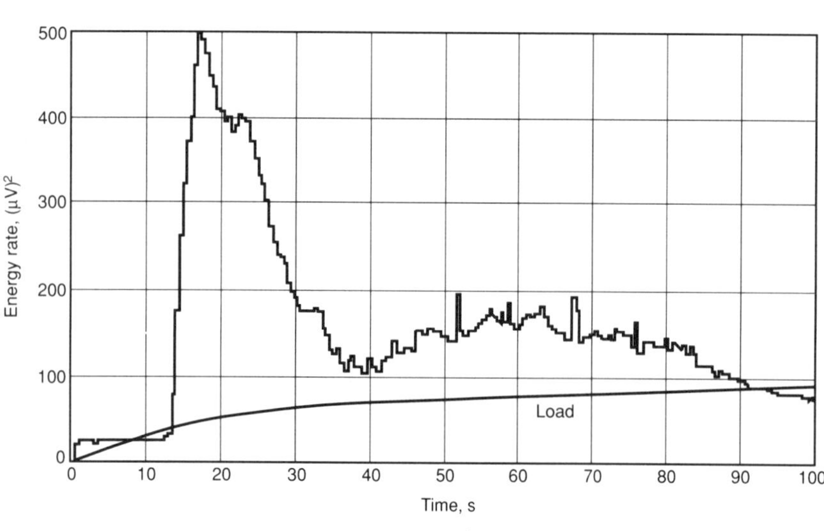

Fig. 15 Acoustic emission energy rate (expressed as the mean square of sensor voltage) and load versus time. Optimum formability corresponds to the lowest emissivity in a deep-drawing steel subjected to spheroidization heat treatment. (a) Underannealed: 80% pearlite, 20% spheroids. (b) Optimally annealed: 100% spheroids. (c) Overannealed: 30% elongated spheroids and doglegs

volving the testing of a metal-matrix composite.

Example 3: Acoustic Emissions From the Microcracking of the Brittle Reaction Zones of Two Metal-Matrix Composites. During the tensile testing of two metal-matrix composites, microcracking of the brittle interphase between the fibers and the matrix produced distinguishable peaks in AE count rates well before ultimate failure from ductile failure of the matrix. This may suggest a potential application of AE inspection for the real-time monitoring of structures made of these or similar metal-matrix composites to provide indications of structural problems before critical damage occurs.

The materials tested were metal-matrix composites that consisted of a titanium (Ti-6Al-4V) matrix reinforced with continuous, large-diameter silicon carbide (SiC, ≈ 0.142 mm, or 0.0056 in., in diameter) or boron carbide coated boron ($B(B_4C)$, ≈ 0.145 mm, or 0.0057 in., in diameter) fibers (with fiber volume fractions of 0.205 and 0.224, respectively). Standard straight-edge tensile test coupons were used. Specimens were cut with the fibers either parallel or perpendicular to the load axis (longitudinal or transverse tension specimens, respectively). Steel end tabs were used, and all surfaces were sanded and cleaned.

Specimens were tested to failure in a servohydraulic testing machine operated at constant crosshead displacement. For each test, a single AE transducer was coupled to the midpoint of the specimen (within the gage section) with vacuum grease, and the acoustic count rate was measured as a function of the longitudinal displacement (strain). After each test, the fracture surface was examined with optical and scanning electron microscopes to determine the fracture processes that occurred.

The values given in Table 2 for rupture or failure strains of the fibers and the brittle reaction compounds formed at the fiber/matrix interface during the hot pressing process were used to find the correlations between fracture processes and AE count rates, with differences in the AE count rates between the two materials related to the differences in their brittle components. As shown in Fig. 16(a), for longitudinal tension specimens of $B(B_4C)$/Ti-6Al-4V, there was a distinguishable rise in the AE count rate near the rupture strain of titanium diboride and a peak near the rupture strain for boron carbide. The final peak resulted from ultimate fiber failure. For transverse tension specimens, Fig. 16(b) and (c) show large peaks in AE count rate near the rupture strains of the major brittle components (titanium diboride in $B(B_4C)$/Ti-6Al-4V and titanium carbide in SiC/Ti-6Al-4V). There were also minor peaks near the rupture strains of the other brittle components. The larger brittle reaction zone in $B(B_4C)$/Ti-

Table 2 Brittle phase mechanical properties

Metal-matrix composite	Brittle compound	Failure strain, %
$B(B_4C)$/Ti-6Al-4V	Titanium diboride	0.25
	Boron carbide	0.57
	Boron	0.80
SiC/Ti-6Al-4V	Titanium carbide	0.28
	Titanium silicide	0.66
	Silicon carbide	0.91

6Al-4V relative to SiC/Ti-6Al-4V results in the larger area for the AE count rate plot. The ultimate failure in the transverse specimens consisted largely of ductile matrix failure with lower AE count rates (relative to the microcracking).

Use of AE Inspection in Production Quality Control

In a small but important class of applications, AE inspection is applied during a manufacturing process to check the quality of the product or one of its components before final assembly and/or delivery. Of the production testing applications discussed in the section "Range of Applicability" in this article, common application of AE inspection in production quality control is the monitoring of welding and shaft straightening processes. Other efforts have been directed toward the inspection of integrated circuits. In the early 1970s, for example, an entire satellite launch mission failed because of a loose particle inside the cavity of a single integrated circuit. As a result, integrated circuits for critical applications are now routinely tested by particle impact noise detection technology, an inexpensive derivative of AE testing (Ref 12). During the manufacturing process, other types of flaws in integrated circuits can also be effectively controlled with AE inspection. Acoustic emissions from bonding processes and from ceramic substrate cracking were investigated by Western Electric researchers during the 1970s and were used as accept/reject criteria for parts on automated assembly lines (Ref 3).

The AE monitoring of welding processes has been part of the technology since its early days. Slag-free, more-automated weld techniques such as resistance welding, laser and electron beam welding, and gas tungsten arc and gas metal arc welding are the easiest to monitor. In the case of resistance welding, AE monitoring is carefully synchronized to the weld cycle, and the various phases of the weld process are treated separately. Emission during solidification and cooling is correlated with nugget size and therefore with weld strength, while high-amplitude signals from expulsions can be used to switch off the weld current at the optimum time to avoid overwelding and to save power and electrode life. In the case of gas tungsten arc and gas metal arc welding, real-time computer algorithms have been developed to recognize the characteristic AE signatures of particular types of defects and to report these defects while the weld is being made. These procedures are effective even in the presence of substantial background noise. Gas tungsten arc welded injector tubes for the space shuttle are among the welded components routinely monitored by AE inspection in the production environment.

Shaft straightening is another production process that lends itself to quality control by AE monitoring. Forged shafts are routinely straightened in special machinery that detects any imperfections in alignment and applies suitable bending forces to correct them. The quality of the product is threatened by microcracking of the hardened surface of the shaft during the bending process. Acoustic emission inspection detects this very effectively and is incorporated into the machinery to warn personnel and to halt the process when potentially damaging microcracking occurs (Ref 4).

In welding and shaft straightening, the stresses that activate acoustic emissions are already present in the normal production processes (in welding, they are thermal stresses). In other cases, the stress is applied for the express purpose of AE testing. This is akin to the loadings routinely applied for the AE inspection of new and in-service pressure vessels and other large structures. Examples include the production testing of brazed joints (Ref 2), and the proof testing of welds in steel ammunition-belt links described in the following example.

Example 4: Acoustic Emission Inspection of Projection Welds in 1050 Steel Ammunition-Belt Links. The ammunition-belt link shown in Fig. 17(a) was made of 0.81 mm (0.032 in.) thick 1050 steel strip. The steel was preformed into link halves that were joined by two projection welds on each side where the sections overlapped. Although the welding schedules were carefully controlled to produce good resistance welds, there was a significant potential for producing some faulty welds in the mass-produced links. In a good weld, a weld nugget is formed that is usually stronger than the base metal in tension; that is, the base metal will tear before the weld will break. In a poor weld, the joint interface is literally just stuck together, and a moderate force, particularly one imposed by impact, will cause the joint to fracture at the interface. A preliminary feasibility investigation showed that poor welds produced more acoustic emission under load than good welds, even though the load was insufficient to break a poor weld.

Proof-Testing Equipment. A mechanical link tester (Fig. 17b) was designed to apply both a shear load and a bending load to the ammunition-belt link at the welded joints. This simulated the service load that would be imposed on the link.

Initially, piezoelectric sensors were attached to each link before testing to monitor acoustic emissions. This was the simplest and most direct method of confirming feasibility. Because attaching sensors directly to the link was not feasible for production testing, piezoelectric sensors were embedded in the spreader arms of the link-test fixtures in an area adjacent to the welded joints in the link.

A spreader force of 270 N (60 lbf) on the link provided a link-spreader-arm interface pressure of about 35 MPa (5 ksi), which

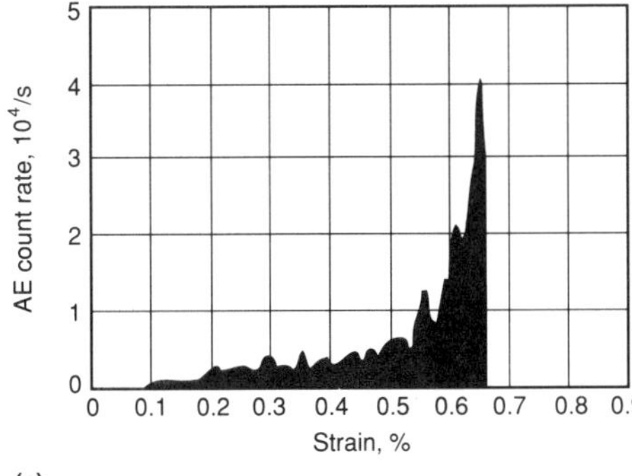

(a)

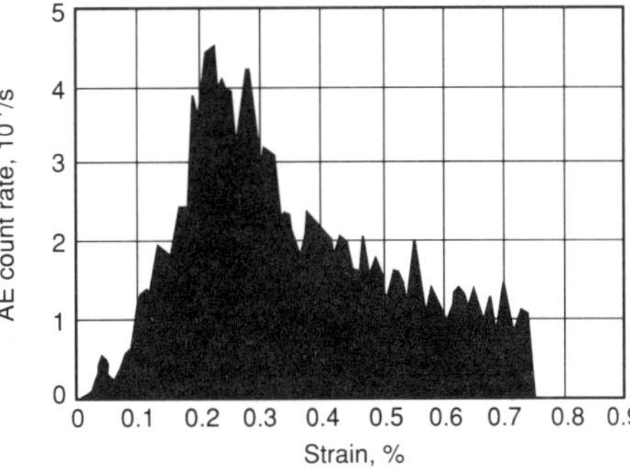

(b)

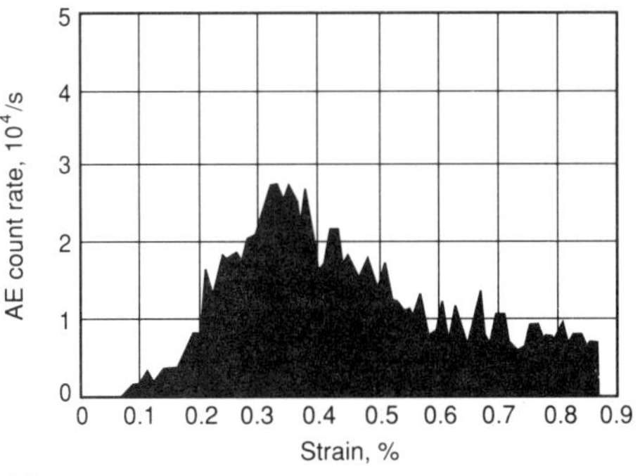

(c)

Fig. 16 Area plot of AE count rate versus strain. (a) Longitudinal tension test of a $B(B_4C)$/Ti-6Al-4V specimen. (b) Transverse tension test of a $B(B_4C)$/Ti-6Al-4V specimen. (c) Transverse tension test of a SiC/Ti-6Al-4V specimen

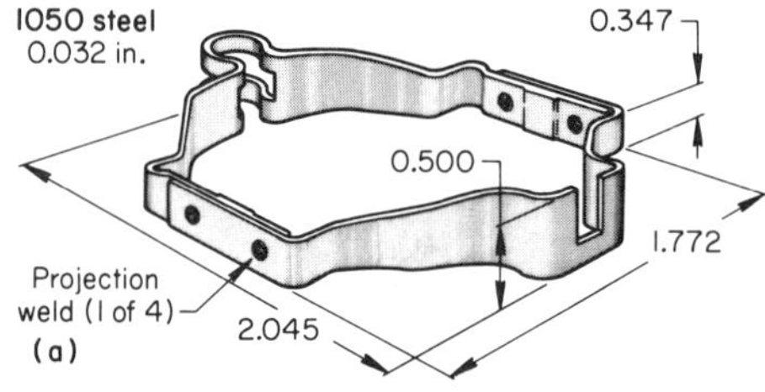

Fig. 17 Ammunition-belt link, of 1050 steel, joined by four projection welds that were inspected by AE monitoring during proof testing in the fixture shown. Dimensions given in inches. Source: Ref 41

provided good coupling of acoustic information across the interface. The sliding action of the spreader mechanism produced a wide-frequency noise range that could not be electronically filtered without also filtering the acoustic emission. This problem was overcome by gating out the noise from moving parts of the link-stressing mechanism and monitoring for acoustic emission during static stressing of the link after the spreader arms had reached full displacement. A microswitch was installed in the fixture to turn on the AE monitoring system in proper relation to operation of the spreader arm.

Acceptance Levels. The form of signal energy analysis that produced the best results consisted of electronically integrating for the area under the half wave rectified envelope of the emission signal in terms of volts amplitude and time duration. The analyzer used for the production application produced a dc voltage proportional to the total AE energy measured. The system sensitivity was adjusted so that an energy analog output voltage of 10 V represented the division point between a good and a bad projection weld in a link. If the welded joint generated enough acoustic emission to produce a 10-V energy output, the link was rejected. If the value was less than 10 V, the link was accepted. The selective ejection function of the mechanical tester was designed to eject the links into the accept or reject container based on an electronic switching function that was controlled by the output voltage of the emission analyzer.

The monitoring system was calibrated by introducing an artificial signal into sensors in the spreader arms, where it was detected and processed by the monitoring system. A 10-V, 10-μs pulse was fed into these sensors from a pulse generator. The resulting signal was reproducible, was a reasonable simulation of the real data, and was simple to generate. Monitoring by acoustic emission was the only available nondestructive method that could perform the necessary 100% inspection of these projection resistance welds.

Structural Test Applications

Acoustic emission inspection has been successfully applied in the structural testing of aircraft, spacecraft, bridges, bucket trucks, buildings, dams, military vehicles, mines, piping systems, pipelines, pressure vessels, railroad tank cars, rotating machinery, storage tanks, and other structures. The typical goal of an AE structural test is to find defects and to assess or ensure structural integrity. Acoustic emission inspection has been described as condition monitoring of static plant (Ref 42), which is parallel to the vibration monitoring techniques that are effectively used to monitor the condition of rotating plant. Both of these methods are useful for predicting failures (Ref 43) and reducing maintenance costs.

Key to structural testing with AE inspection are the stress concentrations that cause defects and other areas of weakness to emit while the rest of the structure is silent. Acoustic emission inspection thus highlights the regions that threaten the integrity of the structure. As a whole-structure test using fixed sensors, AE inspection is normally complemented by other NDT methods that are used to follow up the AE findings and to assist in determining the type, severity, and acceptability of the AE sources.

A major advantage of AE inspection is that it does not require access to the whole examination area. Removal of external insulation or internal process fluids, typically a major expense associated with other NDT methods, is not required for the AE test. In fact, this procedure can be avoided altogether if the AE test indicates that the structure is in good condition.

For AE inspection to function reliably as a whole-structure test, the structure must be loaded in such a way as to stimulate emission from all structurally significant defects. Continuous monitoring in service is a possible test approach that has been applied, for example, to aircraft (Ref 7) and nuclear reactors. This approach guarantees appropriate stressing, but it is difficult because a small amount of emission from defect growth must be separated from a large amount of noise over a long time period. More commonly, the AE test is conducted over the course of a few minutes or hours, during which the structure is stimulated by applying a controlled stress (Ref 44). In most cases, this is satisfactorily accomplished by going somewhat above the normally apparent service loads (for example, 110% of the working pressure for an in-service pressure vessel, 200% of the rated load for an aerial manlift device). However, there are cases in which this approach will not work. For example, if defects are being induced in service by thermal stresses rather than mechanical stresses, an applied mechanical loading may not give a good match to the stress field that is causing the defects to grow in service. The defects, effectively unstressed, may not emit. To overcome this problem, inspectors testing steam lines in electric power plants have conducted AE monitoring during thermal overloads and cool-downs and have reported better success from this type of stressing.

In performing a successful AE test, careful attention must be paid to the type, magnitude, and rate of the applied stimulation (loading). Previously applied stresses will have a very strong influence on the emission that will be observed, as discussed in the section "Load Control and Repeated Loadings" in this article. Precautions must be taken to avoid inadvertent loadings of the structure. Many tests have been spoiled when site personnel, eager to ensure that the pressure system is leaktight, have taken the vessel up to the test pressure before the arrival of the AE inspectors. Accurate load measurement and the ability to hold load at a constant value are other requirements that may demand special attention from site personnel.

Load history is less important for leak testing because leak testing relies primarily on the detection of turbulent flow through the leak orifice. Major structural test applications of acoustic leak testing include flat-bottom storage tanks and nuclear reactor components. In the case of nuclear reactor components, millions of dollars have been saved through the selected use of AE instrumented inspection technology as an alternative to hydrotesting (Ref 45).

Data evaluation procedures depend on the context and content of the test. In one-of-a-kind and developmental testing, the skill and experience of the investigator are of prime importance. This fact inhibited the widespread use of AE inspection until standard test procedures started to become available in the late 1970s. The development of standard test procedures made it possible for AE tests to be efficiently conducted by regular NDT inspectors (given the proper training), while the more innovative investigators moved on to the development of new application areas. Some of the well-developed and standardized structural test applications of AE inspection will be briefly summarized below.

Bucket Trucks. The AE inspection of aerial manlift devices (bucket trucks) was pioneered by the author for Georgia Power Company in 1976 and was carried forward into routine practice by independent testing laboratories and several electric utilities in the years that followed. The ASTM F-18 Committee on Electrical Protective Equipment for Workers published the Standard Practice on the subject in 1985 (Ref 46).

First intended for use on the fiberglass boom sections of insulated bucket trucks, the method was soon extended to cover the pedestal, pins, and other metal components. An estimated 70 000 to 100 000 AE tests have been conducted up to 1988. Bucket trucks can develop problems through accidents, overloads, and fatigue in service. A thorough, regular inspection and test program can identify potential problems before they cause injuries or downtime (Ref 47).

Acoustic emission inspection is a major part of the structural integrity evaluation that complements functional tests of the bucket truck. Of all inspection methods, it is the most effective for detecting problems in fiberglass components, while for metal parts and 100% structural coverage, it serves as a cost-saving screening test that directs the inspector's attention to problem areas. The AE test is preceded by a visual inspection, and any AE indications are normally followed up with magnetic particle, dye penetrant, or ultrasonic inspection.

The AE test typically requires 12 to 16 sensors. System performance is checked by lead break before monitoring begins (see the section "Installation" in this article). Monitoring begins with a noise-check period, followed by two loadings to a predetermined proof load. Emissions during rising load, load hold and load release are separately recorded.

Data evaluation procedures are difficult to spell out because of the wide variety of possible noise and AE sources and the wide range of bucket truck constructions. The experienced inspector uses his knowledge of the truck as he evaluates high- and low-amplitude emission on the different channels during the different stages of the test. Aware of possible noise sources, he looks for indications that may lead to confirmation of damaged fiberglass, cracked metal components, or maintenance problems such as lack of lubrication in visually inaccessible areas.

Using an AE instrument specially designed for economical bucket truck testing, an experienced test crew can perform five to ten AE tests in a single day. When the required visual, operational, ultrasonic, magnetic particle, and dye penetrant testing operations are also performed, typically two to three trucks can be inspected per day.

Jumbo Tube Trailers. Acoustic emission inspection for testing jumbo tube trailers was developed by Blackburn and legitimized in lieu of hydrotest by a Department of Transportation exemption granted in 1983 (Ref 48). These tube trailers carry large volumes of industrial gases on the public highways, typically at a pressure of 18 200 kPa (2640 psi). Fatigue cracks can grow in service, but the hydrotest will not detect these cracks unless they actually cause rupture of the tube. The AE test will detect the cracks while they are still subcritical, during a 10% overpressure applied during the normal filling operation. The AE test is therefore more meaningful than the hydrotest. The AE test is also less expensive and avoids disassembly of the trailer and contamination of the internals with water.

The trailer typically holds 12 tubes, which are all tested at the same time. The AE test requires just two sensors on each 10 m (34 ft) long tube; wave propagation and attenuation characteristics are very favorable in this structure, and AE sources can be accurately located. If ten or more valid events are located within a 200 mm (8 in.) axial distance on the cylindrical portion of the tube, ultrasonic inspection is carried out and the tube is accepted or rejected based on the ultrasonic evaluation of flaw depth. The accept/reject criteria are based on a conservative fracture mechanics analysis of in-service fatigue crack growth. Between March 1983 and March 1988 about 1700 jumbo tubes were AE tested, and the method has been further extended to other shippers of compressed gas and other tube types.

Fiberglass Tanks, Pressure Vessels, and Piping. In the 1970s, the chemical industry was experiencing worldwide failures of fiberglass storage tanks and pressure vessels. Causes included inadequate design and fabrication, mishandling during transportation, and misuse of this relatively unfamiliar material. The situation was aggravated by the lack of a viable inspection method.

An AE test methodology was pioneered by Monsanto, resulting in elimination of the tank failure problem (Fig. 18). The method came into widespread acceptance through the formation and activities of the Committee for Acoustic Emission in Reinforced Plastics (CARP), which is currently affiliated with the American Society for Nondestructive Testing. The written procedure developed by CARP and first published by the Society for the Plastics Industry in 1982 was the basis for the introduction of AE inspection into the ASME Boiler and Pressure Vessel Code in 1983 (Ref 30, 49). An estimated 5000 tests have been carried out using this procedure as of 1988. The work of CARP was also extended to cover methodology for testing fiberglass piping (Ref 50).

The AE test typically requires 8 to 30 sensors, depending on the size of the vessel or tank. High-frequency (typically 150 kHz) channels are used to cover regions of known high stress, such as knuckles, nozzles, and manways; low-frequency (typically 30 kHz) channels having a larger detection range are used to complete the coverage of less critical regions. In the case of a storage vessel, the test is typically conducted during filling with process fluid after an appropriate conditioning period with the contents at a reduced level. In the case of a pressure vessel, appropriate overpressure is applied. The loading is conducted in several stages, with load-hold emission, Felicity Ratio, and other accept/reject criteria evaluated at each stage. System performance checks and background noise checks are part of the test procedure.

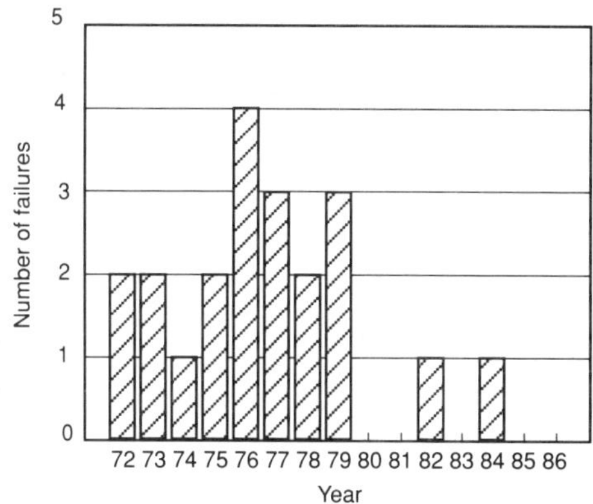

Fig. 18 Failures of FRP tanks. An FRP tank failure problem was eliminated by 100% AE inspection starting in 1979. The isolated failures in 1982 and 1984 occurred after these tanks had failed the AE test, and damage-preventive measures were taken.

Metal Pressure Vessels and Storage Tanks. In the 1970s, many research and engineering organizations and test companies were active in the AE testing of metal pressure vessels. A 1979 survey estimated that about 600 pressure vessels had been tested by AE inspection on a production basis up to that time, mostly in the petroleum, chemical, and nuclear industries (Ref 51). (Although tests on pipes, heat exchanger tubing, and miscellaneous components were much more numerous, pressure vessels have always been a focal point for AE testing.) Much of this testing was done with *ad hoc*, undocumented procedures that relied heavily on the individual experience of the teams performing the work. The main emphasis was usually placed on source location, technically a most attractive feature of AE testing. Located sources would be graded according to their AE activity and/or intensity, and the vessel owner would be advised regarding which areas to inspect with other NDT methods. Many structural defects were successfully identified with these methods.

A significant maturing of the technology took place when Fowler at Monsanto engaged on a systematic program of methods

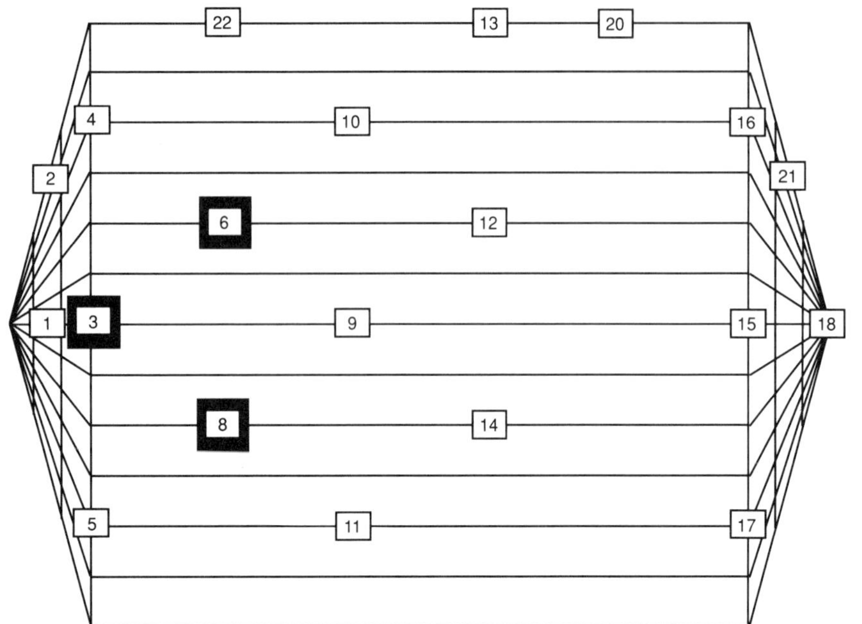

Fig. 19 Display of MONPAC test results for an ethylene storage vessel. In this case, no significant emissions were detected, and there were only minor indications from zones 3, 6, and 8. The need to take the vessel off-line for other forms of inspection was avoided.

development, using the results of follow-up inspection to refine and improve the AE data analysis procedures. Starting in 1979, this program included destructive tests on decommissioned vessels, field tests on many hundreds of in-service vessels and tanks, and development of analytical procedures for recognizing and eliminating extraneous noise (Ref 25). The program de-emphasized the calculation of source location (which requires several sensors to be hit) because in practice many AE events hit only one sensor. To use all the AE hits, zone location was employed instead of point location. The outcome of this program was a comprehensive test procedure backed by detailed case histories and available to licensees under the trademark MONPAC. This procedure has been applied to approximately 2000 metal vessels and storage tanks as of 1988 (Ref 8).

A typical MONPAC test result is shown in Fig. 19. Here a 30-year old ethylene storage bullet was AE tested on-line, raising the pressure by turning off a compressor. Results are presented in the form of an unfolded map of the vessel, with the evaluation of appropriate zones on the map being displayed in color codes (Fig. 19 only shows the color as a dark region). The zone evaluations indicated "no significant emissions," so a hazardous and possibly deleterious internal inspection was avoided (Ref 42).

Other MONPAC tests have found external and internal corrosion, stress-corrosion cracking, weld cracking, lack of fusion, lack of penetration, and embrittled material. Plant shutdowns have been shortened through early diagnosis of major problems, or have been eliminated through positive demonstration of structural integrity. Savings achieved through this method run to tens of millions of dollars.

An AE methodology for metal vessel testing has also been introduced into the ASME Boiler and Pressure Vessel Code as of the December 1988 Addendum (Ref 31). The article states requirements for written test procedure, personnel qualification, equipment, system calibration, preexamination measurements, background noise check, and vessel pressurization. An illustrative loading schedule and sensor layouts are included. Evaluation criteria are to be supplied, class by class, by the referencing code section; they will be based on emission during load hold, count rate, number of hits, large amplitude hits, MARSE, and activity trends. This entry into the ASME Code represents a major milestone in the establishment and maturity of AE technology.

REFERENCES

1. *Acoustic Emission Testing*, Vol 5, 2nd ed., *Nondestructive Testing Handbook*, R.K. Miller and P. McIntire, Ed., American Society for Nondestructive Testing, 1987, p 275-310
2. T.F. Drouillard and T.G. Glenn, Production Acoustic Emission Testing of Braze Joint, *J. Acoust. Emiss.*, Vol 1 (No. 2), 1985, p 81-85
3. S.J. Vahaviolos, Real Time Detection of Microcracks in Brittle Materials Using Stress Wave Emission (SWE), *IEEE Trans.*, Vol PHP-10 (No. 3), Sept 1974, p 152-159
4. K.H. Pärtzel, Acoustic Emission for Crack Inspection During Fully Automatic and Manual Straightening of Transmission Shafts, in *Proceedings of the Acoustic Emission Symposium* (Bad Nauheim), J. Eisenblätter, Ed., Deutsche Gesellschaft für Metallkunde, 1988, p 157-164
5. *Acoustic Emission Testing*, Vol 5, 2nd ed., *Nondestructive Testing Handbook*, R.K. Miller and P. McIntire, Ed., American Society for Nondestructive Testing, 1987, p 167-186, 187-193
6. D.L. Parry, Industrial Application of Acoustic Emission Analysis Technology, in *Monitoring Structural Integrity by Acoustic Emission*, STP 571, American Society for Testing and Materials, 1975, p 150-183
7. *Acoustic Emission Testing*, Vol 5, 2nd ed., *Nondestructive Testing Handbook*, R.K. Miller and P. McIntire, Ed., American Society for Nondestructive Testing, 1987, p 421-424, 434-443, 226-259, 333-339, 267-271
8. T.J. Fowler, Recent Developments in Acoustic Emission Testing of Chemical Process Equipment, in *Progress in Acoustic Emission IV*, Proceedings of the Ninth International Acoustic Emission Symposium, The Japanese Society for Non-Destructive Inspection, 1988, p 391-404
9. P.C. Cole, Acoustic Emission, in *The Capabilities and Limitations of NDT*, Part 7, The British Institute of Non-Destructive Testing, 1988
10. *Acoustic Emission Testing*, Vol 5, 2nd ed., *Nondestructive Testing Handbook*, R.K. Miller and P. McIntire, Ed., American Society for Nondestructive Testing, 1987, p 278, 472-484, 213-219
11. S.L. McBride, Acoustic Emission Measurements on Rubbing Surfaces, in *Proceedings of the World Meeting on Acoustic Emission* (Charlotte, NC), Acoustic Emission Group, March 1989
12. *Acoustic Emission Testing*, Vol 5, 2nd ed., *Nondestructive Testing Handbook*, R.K. Miller and P. McIntire, Ed., American Society for Nondestructive Testing, 1987, p 402-408, 194-202
13. A.A. Pollock and S.-Y.S. Hsu, Leak Detection Using Acoustic Emission, *J. Acoust. Emiss.*, Vol 1 (No. 4), 1982, p 237-243
14. *Acoustic Emission Testing*, Vol 5, 2nd ed., *Nondestructive Testing Handbook*, R.K. Miller and P. McIntire, Ed., American Society for Nondestructive Testing, 1987, p 58-61, 84-88
15. C.B. Scruby, Quantitative Acoustic Emission Techniques, in *Research Techniques in Nondestructive Testing*, Vol VIII, R.S. Sharpe, Ed., Academic Press, 1985, p 141-210
16. *Acoustic Emission Testing*, Vol 5, 2nd ed., *Nondestructive Testing Handbook*, R.K. Miller and P. McIntire, Ed.,

American Society for Nondestructive Testing, 1987, p 64-90

17. Y.H. Pao, Theory of Acoustic Emission, in *Elastic Waves and Non-Destructive Testing of Materials*, AMD-20, American Society of Mechanical Engineers, 1978, p 107-128
18. M. Enoki, T. Kishi, and S. Kohara, Determination of Microcracking Moment Tensor of Quasi-Cleavage Facet by AE Source Characterization, in *Progress in Acoustic Emission III*, Proceedings of the Eighth International Acoustic Emission Symposium, The Japanese Society for Non-Destructive Inspection, 1983, p 763-770
19. S. Yuyama, T. Imanaka, and M. Ohtsu, Quantitative Evaluation of Microfracture Due to Disbonding by Waveform Analysis of Acoustic Emission, *J. Acoust. Soc. Am.*, Vol 83 (No. 3), 1988, p 976-983; Vol 82 (No. 2), 1987, p 506-512
20. A.A. Pollock, Classical Wave Theory in Practical AE Testing, in *Progress in Acoustic Emission III*, Proceedings of the Eighth International Acoustic Emission Symposium, The Japanese Society for Non-Destructive Inspection, 1986, p 708-721
21. "Standard Method for Primary Calibration of Acoustic Emission Sensors," E 1106-86, *Annual Book of ASTM Standards*, American Society for Testing and Materials
22. S.J. Vahaviolos, 3rd Generation AE Instrumentation Techniques for High Fidelity and Speed of Data Acquisition, in *Progress in Acoustic Emission III*, Proceedings of the Eighth International Acoustic Emission Symposium, The Japanese Society for Non-Destructive Inspection, 1986, p 102-116
23. A.A. Pollock, Acoustic Emission Amplitude Distributions, in *International Advances in Nondestructive Testing*, Vol 7, Gordon & Breach, 1981, p 215-239
24. M.R. Gorman and T.H. Rytting, Long Duration AE Events in Filament Wound Graphite/Epoxy in the 100-300KHz Band Pass Region, in *First International Symposium on Acoustic Emission From Reinforced Composites*, The Society of the Plastics Industry, 1983
25. T.J. Fowler, Experience With Acoustic Emission Monitoring of Chemical Process Industry Vessels, in *Progress in Acoustic Emission III*, Proceedings of the Eighth International Acoustic Emission Symposium, The Japanese Society of Non-Destructive Inspection, 1986, p 150-162
26. *Acoustic Emission Testing*, Vol 5, 2nd ed., *Nondestructive Testing Handbook*, R.K. Miller and P. McIntire, Ed., American Society for Nondestructive Testing, 1987, p 340
27. J. Kaiser, Erkenntnisse und Folgerungen aus der Messung von Geräuschen bei Zugbeanspruchung von Metallischen Werkstoffen, *Arch. Eisenhüttenwes.*, Vol 24 (No. 1-2), 1953, p 43-45
28. H.L. Dunegan and A.S. Tetelman, Acoustic Emission, *Res. Dev.*, Vol 22 (No. 5), 1971, p 20-24
29. *Acoustic Emission Testing*, Vol 5, 2nd ed., *Nondestructive Testing Handbook*, R.K. Miller and P. McIntire, Ed., American Society for Nondestructive Testing, 1987, p 426
30. "Acoustic Emission Examination of Fiber-Reinforced Plastic Vessels," Boiler and Pressure Vessel Code, Article 11, Subsection A, Section V, American Society of Mechanical Engineers, 1983
31. "Acoustic Emission Examination of Metallic Vessels During Pressure Testing," Boiler and Pressure Vessel Code, Article 12, Subsection A, Section V, American Society of Mechanical Engineers, Dec 1988, Addendum
32. C.R. Heiple and S.H. Carpenter, Acoustic Emission Produced by Deformation of Metals and Alloys—a Review, *J. Acoust. Emiss.*, Vol 6 (No. 3), 1987, p 177-204; Vol 6 (No. 4), 1987, p 215-237
33. *Acoustic Emission Testing*, Vol 5, 2nd ed., *Nondestructive Testing Handbook*, R.K. Miller and P. McIntire, Ed., American Society for Nondestructive Testing, 1987, p 49-61, 55-57, 78
34. H.L. Dunegan, D.O. Harris, and C.A. Tatro, Fracture Analysis by Use of Acoustic Emission, *Eng. Fract. Mech.*, Vol 1 (No. 1), 1968, p 105-122
35. I.G. Palmer and P.T. Heald, The Application of Acoustic Emission Measurements to Fracture Mechanics, *Mater. Sci. Eng.*, Vol 11 (No. 4), 1973, p 181-184
36. H.H. Chaskelis, W.H. Callen, and J.M. Krafft, "Acoustic Emission From Aqueous Stress Corrosion Cracking in Various Tempers of 4340 Steel," NRL Memorandum Report 2608, Naval Research Laboratory, 1973
37. A. Ashary, G.H. Meier, and F.S. Pettit, Acoustic Emission Study of Oxide Cracking During Alloy Oxidation, in *High Temperature Protective Coating*, The Metallurgical Society, 1982, p 105
38. A.A. Pollock, Acoustic Emission Capabilities and Applications in Monitoring Corrosion, in *Corrosion Monitoring in Industrial Plants Using Nondestructive Testing and Electrochemical Methods*, STP 908, G.C. Moran and P. Labine, Ed., American Society for Testing and Materials, 1986, p 30-42
39. S.H. Yuyama, Fundamental Aspects of Acoustic Emission Applications to the Problems Caused by Corrosion, in *Corrosion Monitoring in Industrial Plants Using Nondestructive Testing and Electrochemical Methods*, STP 908, G.C. Moran and P. Labine, Ed., American Society for Testing and Materials, 1986, p 43-47
40. P. Pantucek and U. Struth, Behaviour of Thermal Barrier and of Corrosion Protective Coating Systems Under Combined Thermal and Mechanical Loads (Mechanical Compatibility Problems and Potential Solutions), in *Ceramic Coatings for Heat Engines*, I. Kuernes, W.J.G. Bunk, and J.G. Wurm, Ed., Advanced Materials Research and Development for Transport, MRS-Europe, Symposium IX (Nov 1985), Les Editions de Physique Vol IX, Les Ulis Cedex, France, 1986, p 117–138
41. P.H. Hutton, Acceptance Testing Welded Ammunition Belt Links Using Acoustic Emission, in *Monitoring Structural Integrity by Acoustic Emission*, STP 571, American Society for Testing and Materials, 1974, p 107-121
42. P.T. Cole, 1987 Acoustic Emission Technology and Economics Applied to Pressure Vessels and Storage Tanks, in *Proceedings of the Fourth European Conference on Nondestructive Testing* (London), Pergamon Press, 1987, p 2892
43. J.M. Carlyle, R.S. Evans, and T.P. Sherlock, "Acoustic Emission Characterization of a Hot Reheat Line Rupture," Paper presented at the NDE Symposium at the ASME Piping and Pressure Vessel Conference (Honolulu), American Society of Mechanical Engineers, June 1989
44. "Standard Practice for Acoustic Emission Monitoring of Structures During Controlled Stimulation," E 569-85, *Annual Book of ASTM Standards*, American Society for Testing and Materials
45. D.P. Weakland and D.P. Grabski, Consider Instrumented Inspection of Safety-Related Nuclear Systems, *Power*, Vol 131 (No. 3), March 1987, p 61-63
46. "Standard Test Method for Acoustic Emission for Insulated Aerial Personnel Devices," 914-85, *Annual Book of ASTM Standards*, American Society for Testing and Materials
47. K. Moore and C.A. Larson, Aerial Equipment Requires Thorough, Regular Inspection, *Transmiss. Distrib.*, Jan 1984, p 23-27
48. P.R. Blackburn and M.D. Rana, Acoustic Emission Testing and Structural Evaluation of Seamless Steel Tubes in Compressed Gas Service, *J. Pressure Vessel Technol.* (Trans. ASME), Vol 108, May 1986, p 234-240
49. "Recommended Practice for Acoustic Emission Testing of Fiberglass Reinforced Resin (RP) Tanks/Vessels," The Society of the Plastics Industry, 1987

50. Recommended Practice for Acoustic Emission Testing of Fiberglass Reinforced Plastics Piping Systems, in *First International Symposium on Acoustic Emission From Reinforced Plastics*, The Society of the Plastics Industry, July 1983 (see also ASTM Standard Practice E 1118-86)
51. J.C. Spanner, Acoustic Emission: Who Needs It—and Why?, in *Advances in Acoustic Emission*, H.L. Dunegan and W.F. Hartman, Ed., Proceedings of the International Conference on Acoustic Emission (Anaheim, CA), Dunhart Publishers, 1979

Radiographic Inspection

Revised by the ASM Committee on Radiographic Inspection*
Chairman: Arnold Greene, Micro/Radiographs Inc.

RADIOLOGY is the general term given to material inspection methods that are based on the differential absorption of penetrating radiation—either electromagnetic radiation of very short wavelength or particulate radiation—by the part or testpiece (object) being inspected. Because of differences in density and variations in thickness of the part or differences in absorption characteristics caused by variations in composition, different portions of a testpiece absorb different amounts of penetrating radiation. These variations in the absorption of the penetrating radiation can be monitored by detecting the unabsorbed radiation that passes through the testpiece.

The term radiography often refers to the specific radiological method that produces a permanent image on film (conventional radiography) or paper (paper radiography or xeroradiography). In a broad sense, however, radiography can also refer to other radiological techniques that can produce two-dimensional, plane-view images from the unabsorbed radiation. Recently, the American Society of Testing and Materials (ASTM) defined radioscopy as the term to describe the applications when film or paper is not used and defined radiology as the general term covering both techniques. However, the term radioscopy has not received wide acceptance yet, and this article considers the following two techniques as radiographic inspection (with x-rays or γ-rays):

- *Film or paper radiography*: A two-dimensional latent image from the projected radiation is produced on a sheet of film or paper that has been exposed to the unabsorbed radiation passing through the testpiece. This technique requires subsequent development of the exposed film or paper so that the latent image becomes visible for viewing
- *Real-time radiography (also known as radioscopy)*: A two-dimensional image can be immediately displayed on a viewing screen or television monitor. This technique does not involve the creation of a latent image; instead, the unabsorbed radiation is converted into an optical or electronic signal, which can be viewed immediately or can be processed in near real time with electronic and video equipment

The principal advantage of real-time radiography over film radiography is the opportunity to manipulate the testpiece during radiographic inspection. This capability allows the inspection of internal mechanisms and enhances the detection of cracks and planar defects by manipulating the part to achieve the proper orientation for flaw detection. Moreover, part manipulation in real-time radiography simplifies three-dimensional (stereo) dynamic imaging and the determination of flaw location and size. In film radiography, however, the position of a flaw within the volume of a testpiece cannot be determined exactly with a single radiograph; depth parallel to the radiation beam is not recorded. Consequently, other film techniques, such as stereoradiography, triangulation, or simply making two or more film exposures (with the radiation beam being directed at the testpiece from a different angle for each exposure), must be used to locate flaws more exactly within the testpiece volume.

Although real-time radiography enhances the detection and location of flaws by allowing the manipulation of the testpiece during inspection, another important radiological technique with enhanced flaw detection and location capabilities is computed tomography. Unlike film and real-time radiography, computed tomography (CT) involves the generation of cross-sectional views instead of a planar projection. The CT image is comparable to that obtained by making a radiograph of a physically sectioned thin planar slab from an object. This cross-sectional image is not obscured by overlying and underlying structures and is highly sensitive to small differences in relative density. Moreover, CT images are easier to interpret than radiographs (see the article "Industrial Computed Tomography" in this Volume).

All of the terms and techniques in the preceding discussion refer to radiological inspection with penetrating electromagnetic radiation in the form of x-rays or γ-rays. Other forms of radiation include subatomic particles that are generated during nuclear decay. The most commonly known subatomic particles are α particles, β particles, and neutrons, all of which are emitted from the nuclei of various atoms during radioactive decay. Beta particles and neutrons are sufficiently penetrating to be useful for radiography, but neutrons are more widely used. More information on neutron radiography is available in the article "Neutron Radiography" in this Volume.

Uses of Radiography

Radiography is used to detect the features of a component or assembly that exhibit a difference in thickness or physical density as compared to surrounding material. Large differences are more easily detected than small ones. In general, radiography can detect only those features that have an appreciable thickness in a direction parallel to the radiation beam. This means that the ability of the process to detect planar discontinuities such as cracks depends on proper orientation of the testpiece during inspection. Discontinuities such as voids and inclusions, which have measurable thickness in all directions, can be detected as long as they are not too small in relation to section thickness. In general, features that exhibit a 1% or more difference in absorption compared to the surrounding material can be detected.

Although neither is limited to the detection of internal flaws, radiography and ultrasonics are the two generally used nondestructive inspection methods that can satisfactorily detect flaws that are completely internal and located well below the surface of the part. Neither method is lim-

*Merlin Michael, Rockwell International; John J. Munro III, RTS Technology, Inc.; R.A. Betz and R.C. Barry, Lockheed Missiles & Space Co., Inc.; G.B. Nightingale, General Electric Company; Thomas A. Siewert, National Institute of Standards and Technology; Craig E. Anderson, Nuclear Energy Services; Theodore F. Luga, Consultant; William H. Folland, United Technologies, Pratt & Whitney; George Surma, Sundstrand Aviation Operations; Ralph McCullough, Texas Instruments Inc.; Richard W. Thams, X-Ray Industries, Inc.; Bruce Apgar, George Becker, and William E.J. McKinney, DuPont NDT Systems; Samuel A. Wenk, Consultant

Table 1 Comparison of suitability of three radiographic methods for the inspection of light and heavy metals

Inspection application	Suitability for light metals(a)			Suitability for heavy metals(a)		
	Film with x-rays	Real-time radiography	Film with γ-rays	Film with x-rays	Real-time radiography	Film with γ-rays
General						
Surface cracks(b)	F(c)	F	F(c)	F(c)	F	F(c)
Internal cracks	F(c)	F	F(c)	F(c)	F	F(c)
Voids	G	G	G	G	G	G
Thickness	F	G	F	F	G	F
Metallurgical variations	F	F	F	F	F	F
Sheet and plate						
Thickness	G(d)	G	G(d)	G(d)	G	G(d)
Laminations	U	U	U	U	U	U
Voids	G	G	G	G	G	G
Bars and tubes						
Seams	P	P	P	P	P	P
Pipe	G	G	G	G	G	F
Cupping	G	G	G	G	G	F
Inclusions	F	F	F	F	F	F
Castings						
Cold shuts	G	G	G	G	G	G
Surface cracks	F(c)	F	F(c)	F(c)	F	F(c)
Internal shrinkage	G	G	G	G	G	G
Voids, pores	G	G	G	G	G	G
Core shift	G	G	G	G	G	G
Forgings						
Laps	P(c)	P(c)	P(c)	P	P	U
Inclusions	F	F	F	F	F	U
Internal bursts	G	G	G	F	F	G
Internal flakes	P(c)	P	U	P(c)	P	U
Cracks and tears	F(c)	F	F(c)	F(c)	F	F(c)
Welds						
Shrinkage cracks	G(c)	G	G(c)	G(c)	G	G(c)
Slag inclusions	G	G	G	G	G	G
Incomplete fusion	G	G	G	G	G	G
Pores	G	G	G	G	G	G
Incomplete penetration	G	G	G	G	G	G
Processing						
Heat-treat cracks	U	U	U	P	U	U
Grinding cracks	U	U	U	U	U	U
Service						
Fatigue and heat cracks	F(c)	F	P(c)	P	P	P
Stress corrosion	F	F	P	F	F	P
Blistering	P	P	P	P	P	P
Thinning	F	F	F	F	F	F
Corrosion pits	F	F	P	G	G	P

(a) G, good; F, fair; P, poor; U, unsatisfactory. (b) Includes only visible cracks. Minute surface cracks are undetectable by radiographic inspection methods. (c) Radiation beam must be parallel to the cracks, laps, or flakes. In real-time radiography, the testpiece can be manipulated for proper orientation. (d) When calibrated using special thickness gages

ited to the detection of specific types of internal flaws, but radiography is more effective when the flaws are not planar, while ultrasonics is more effective when the flaws are planar. In comparison to other generally used nondestructive methods (for example, magnetic particle, liquid penetrant, and eddy current inspection), radiography has three main advantages:

- The ability to detect internal flaws
- The ability to detect significant variations in composition
- Permanent recording of raw inspection data

Applicability. Radiographic inspection is extensively used on castings and weldments, particularly where there is a critical need to ensure freedom from internal flaws. For example, radiography is often specified for the inspection of thick-wall castings and weldments for steam-power equipment (boiler and turbine components and assemblies) and other high-pressure systems. Radiography can also be used on forgings and mechanical assemblies, although with mechanical assemblies radiography is usually limited to inspection for condition and proper placement of components or for proper liquid-fill level in sealed systems. Certain special devices are more satisfactorily inspected by radiography than by other methods. For example, radiography is well suited to the inspection of semiconductor devices for voids in the element mount area, in the case seal area, and in plastic molding compounds used to encapsulate some devices. Radiography is also used to inspect for cracks, broken wires, foreign material, and misplaced and misaligned elements. High-resolution real-time imaging with microfocus x-ray sources has made it possible to use radiography as a failure analysis tool for semiconductors and other electronic components. Real-time imaging allows analysis from a variety of angles, while microfocus adds the capability of detecting flaws as small as 0.025 mm (0.001 in.) in the major dimension. New uses of radiography have also occurred with the inspection of composites. In such applications, the sensitivity is maximized with the use of high-intensity, low-energy radiation.

The sensitivity of x-ray and γ-ray radiography to various types of flaws depends on many factors, including type of material, type of flaw, and product form. (Type of material in this context is usually expressed in terms of atomic number—for example, metals having low atomic numbers are classified as light metals, and those having high atomic numbers are heavy metals.) Table 1 indicates the general degree of suitability of the three main radiographic methods for the detection of discontinuities in various product forms and applications.

Radiography can be used to inspect most types of solid material, with the possible exception of materials of very high or very low density. (Neutron radiography, however, can often be used in such cases, as discussed in the article "Neutron Radiography" in this Volume.) Both ferrous and nonferrous alloys can be radiographed, as can nonmetallic materials and composites.

There is wide latitude both in the material thickness that can be inspected and in the techniques that can be used. Numerous special techniques and special devices have been developed for the application of radiography to specific inspection problems, including even the inspection of radioactive materials. Most of these specialized applications are not discussed in this article, but several can be found in articles in this Volume that deal with the use of radiography in the inspection of specific product forms.

In some cases, radiography cannot be used even though it appears suitable from Table 1, because the part is accessible from one side only. Radiography typically involves the transmission of radiation through the testpiece, in which case both sides of the part must be accessible. However, radiographic and radiometric inspection can also be performed with Compton scattering, in which the scattered photons are used for imaging. With Compton scattering, inspection can be performed when only one side is accessible. Another method of inspecting a region having one inaccessible side is to use probes with a microfocus x-ray tube (see the section "Microfocus X-Ray Tubes" in this article).

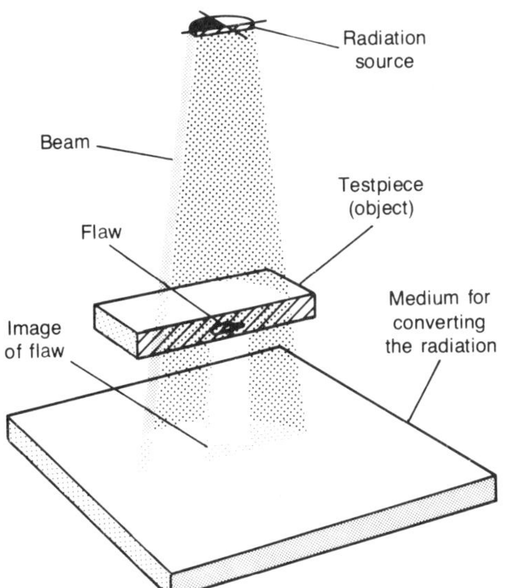

Fig. 1 Schematic of the basic elements of a radiographic system showing the method of sensing the image of an internal flaw in a plate of uniform thickness

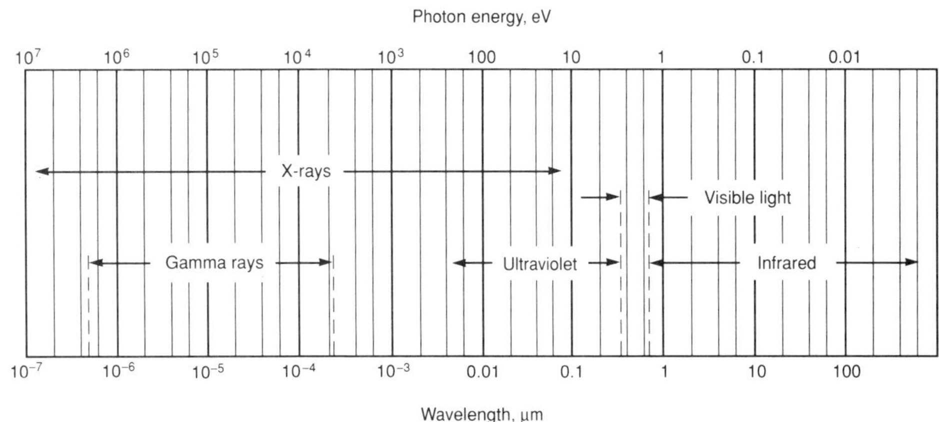

Fig. 2 Schematic of the portion of the electromagnetic spectrum that includes x-rays, γ-rays, ultraviolet and visible light, and infrared radiation showing their relationship with wavelength and photon energy

Limitations. Compared to other nondestructive methods of inspection, radiography is expensive. Relatively large capital costs and space allocations are required for a radiographic laboratory, although costs can be reduced when portable x-ray or γ-ray sources are used in film radiography. Capital costs can be relatively low with portable units, and space is required only for film processing and interpretation. Operating costs can be high; sometimes as much as 60% of the total inspection time is spent in setting up for radiography. With real-time radiography, operating costs are usually much lower, because setup times are shorter and there are no extra costs for processing or interpretation of film.

The field inspection of thick sections can be a time-consuming process because the effective radiation output of portable sources may require long exposure times of the radiographic film. Radioactive (γ-ray) sources are limited in their output primarily because high-activity sources require heavy shielding for the protection of personnel. This limits field usage to sources of lower activity that can be transported. The output of portable x-ray sources may also limit the field inspection of thick sections, particularly if a portable x-ray tube is used. Portable x-ray tubes emit relatively low-energy (300 keV) radiation and are limited in the radiation output. Both of these characteristics of portable x-ray tubes combine to limit their application to the inspection of sections having the absorption equivalent of 75 mm (3 in.) of steel. Instead of portable x-ray tubes, portable linear accelerators and betatrons provide high-energy (>1 MeV) x-rays for the radiographic field inspection of thicker sections.

Certain types of flaws are difficult to detect by radiography. Cracks cannot be detected unless they are essentially parallel to the radiation beam. Tight cracks in thick sections may not be detected at all, even when properly oriented. Minute discontinuities such as inclusions in wrought material, flakes, microporosity, and microfissures may not be detected unless they are sufficiently segregated to yield a detectable gross effect. Laminations are nearly impossible to detect with radiography; because of their unfavorable orientation, delaminations do not yield differences in absorption that enable laminated areas to be distinguished from delaminated areas.

It is well known that large doses of x-rays or γ-rays can kill human cells, and in massive doses can cause severe disability or death. Protection of personnel—not only those engaged in radiographic work but also those in the vicinity of radiographic inspection—is of major importance. Safety requirements impose both economic and operational constraints on the use of radiography for inspection.

Principles of Radiography

Three basic elements of radiography include a radiation source, the testpiece or object being evaluated, and a sensing material. These elements are shown schematically in Fig. 1. The testpiece in Fig. 1 is a plate of uniform thickness containing an internal flaw that has absorption characteristics different from those of the surrounding material. Radiation from the source is absorbed by the testpiece as the radiation passes through it; the flaw and surrounding material absorb different amounts of radiation. Thus, the intensity of radiation that impinges on the sensing material in the area beneath the flaw is different from the amount that impinges on adjacent areas. This produces an image, or shadow, of the flaw on the sensing material. This section briefly reviews the general characteristics and safety principles associated with radiography.

Radiation Sources

Two types of electromagnetic radiation are used in radiographic inspection: x-rays and γ-rays. X-rays and γ-rays differ from other types of electromagnetic radiation (such as visible light, microwaves, and radio waves) only in their wavelengths, although there is not always a distinct transition from one type of electromagnetic radiation to another (Fig. 2). Only x-rays and γ-rays, because of their relatively short wavelengths (high energies), have the capability of penetrating opaque materials to reveal internal flaws.

X-rays and γ-rays are physically indistinguishable; they differ only in the manner in which they are produced. X-rays result from the interaction between a rapidly moving stream of electrons and atoms in a solid target material, while γ-rays are emitted during the radioactive decay of unstable atomic nuclei.

The amount of exposure from x-rays or γ-rays is measured in roentgens (R), where 1 R is the amount of radiation exposure that produces one electrostatic unit (3.33564×10^{-10} C) of charge from 1.293 mg (45.61×10^{-6} oz) of air. The intensity of an x-ray or γ-ray radiation is measured in roentgens per unit time.

Although the intensity of x-ray or γ-ray radiation is measured in the same units, the strengths of x-ray and γ-ray sources are usually given in different units. The strength of an x-ray source is typically given in roentgens per minute at one meter (RMM) from the source or in some other suitable combination of time or distance units (such as roentgens per hour at one meter, or RHM). The strength of a γ-ray source is usually given in terms of the radioactive decay rate, which has the traditional unit of a curie (1 Ci = 37×10^9 disintegrations per second). The corresponding unit in the Système International d'Unités (SI) system is a gigabecquerel (1 GBq = 1×10^9 disintegrations per second).

The spectrum of radiation is often expressed in terms of photon energy rather than as a wavelength. Photon energy is measured in electron volts, with 1 eV being the energy imparted to an electron by an accelerating potential of 1 V. Figure 2 shows the radiation spectrum in terms of both wavelength and photon energy.

Production of X-Rays. When x-rays are produced from the collision of fast-moving electrons with a target material, two types of x-rays are generated. The first type of x-ray is generated when the electrons are rapidly decelerated during collisions with atoms in the target material. These x-rays have a broad spectrum of many wavelengths (or energies) and are referred to as continuous x-rays or by the German word bremsstrahlung, which means braking radiation. The second type of x-ray occurs when the collision of an electron with an atom of the target material causes a transition of an orbital electron in the atom, thus leaving the atom in an excited state. When the orbital electrons in the excited atom rearrange themselves, x-rays are emitted that have specific wavelengths (or energies) characteristic of the particular electron rearrangements taking place. These characteristic x-rays usually have much higher intensities than the background of bremsstrahlung having the same wavelengths.

Production of γ-Rays. Gamma rays are generated during the radioactive decay of both naturally occurring and artificially produced unstable isotopes. In all respects other than their origin, γ-rays and x-rays are identical. Unlike the broad-spectrum radiation produced by x-ray sources, γ-ray sources emit one or more discrete wavelengths of radiation, each having its own characteristic photon energy (or wavelength).

Many of the elements in the periodic table have either naturally occurring radioactive isotopes or isotopes that can be made radioactive by irradiation with a stream of neutrons in the core of the nuclear reactor. However, only certain isotopes are extensively used for radiography (see the section "Gamma Ray Sources" in this article).

Image Conversion

The most important process in radiography is the conversion of radiation into a form suitable for observation or further signal processing. This conversion is accomplished with either a recording medium (usually film) or a real-time imaging medium (such as fluorescent screens or scintillation crystals). The imaging process can also be assisted with the use of intensifying or filtration screens, which intensify the conversion process or filter out scattered radiation.

Recording media provide a permanent image that is related to the variations in the intensity of the unabsorbed radiation and the time of exposure. With a recording medium such as film, for example, an invisible latent image is formed in the areas exposed to radiation. These exposed areas become dark when the film is processed (that is, developed, rinsed, fixed, washed, and dried), with the degree of darkening (the photographic density) depending on the amount of exposure to radiation. The film is then placed on an illuminated screen so that the image formed by the variations in photographic density can be examined and interpreted.

Real-time imaging media provide an immediate indication of the intensity of radiation passing through a testpiece. With fluorescent screens, for example, visible light is emitted with a brightness that is proportional to the intensity of the incident x-ray or γ-ray radiation. This emitted light can be observed directly, amplified, and/or converted into a video signal for presentation on a television monitor and subsequent recording. The various types of imaging systems are described in the section "Real-Time Radiography" in this article.

Intensifying and filtration screens are used to improve image contrast, particularly when the radiation intensity is low or when the radiation energy is high. The screens are useful at higher energies because the sensitivity of films and fluorescent screens decreases as the energy of the penetrating radiation increases. The various types of screens are discussed in the section "Image Conversion Media" in this article.

Image Quality and Radiographic Sensitivity

The quality of radiographs is affected by many variables, and image quality is measured with image-quality indicators known as penetrameters. These devices are thin specimens made of the same material as the testpiece; they are described in more detail in the section "Identification Markers and Penetrameters (Image-Quality Indicators)" in this article. When placed on the testpiece during radiographic inspection, the penetrameters measure image contrast and, to a limited extent, resolution. Detail resolution is not directly measured with penetrameters, because flaw detection depends on the nature of the flaw, its shape, and orientation to the radiation beam.

Image quality is governed by image contrast and resolution, which are also sometimes referred to as radiographic contrast and radiographic definition. These two factors are interrelated in a complex way and are affected by several factors described in the sections "Radiographic Contrast" and "Radiographic Definition" in this article. In real-time systems, image contrast and resolution are also described in terms of the detective quantum efficiency (DQE) (also known as the quantum detection efficiency, or QDE) and the modulation transfer function (MTF), respectively. These terms, which are also used in computed tomography, are defined in the text and in Appendix 2 of the article "Industrial Computed Tomography" in this Volume.

Radiographic sensitivity, which should be distinguished from image quality, generally refers to the size of the smallest detail that can be seen on a radiograph or to the ease with which the images of small details can be detected. Although radiographic sensitivity is often synonymous with image quality in applications requiring the detection of small details, a distinction should be made between radiographic sensitivity and radiographic quality. Radiographic sensitivity refers more to detail resolvability, which should be distinguished from spatial resolution and contrast resolution. For example, if the density of an object is very different from the density of the surrounding region, the flaw might be resolved because of the large contrast, even if the flaw is smaller than the spatial resolution of the system. On the other hand, when the contrast is small, the area must be large to achieve resolvability.

Radiographic contrast refers to the amount of contrast observed on a radiograph, and it is affected by subject contrast and the contrast sensitivity of the image-detecting system. Radiographic contrast can also be affected by the unsharpness of the detected image. Figure 3, for example, shows how contrast is affected when the sharp edge of a step is blurred because of unsharpness. If the unsharpness is much smaller than d in Fig. 3, then the contrast is not reduced and the edges in the image are easily defined. However, if d is much smaller than the unsharpness, then the contrast is reduced. If the unsharpness is too large, the image cannot be resolved.

Subject contrast is the ratio of radiation intensities transmitted by various portions of a testpiece. It depends on the thickness, shape, and composition of the testpiece; the intensity of the scattered radiation; and the spectrum of the incident radiation. It does not depend on the detector, the source strength, or the source-to-detector distance. However, it may depend on the testpiece-to-detector distance because the intensity of scattered radiation (from within the testpiece) impinging upon a detector is almost eliminated when projective enlargement approaches a magnification factor of three (see the section "Enlargement" in this article).

In general, a radiograph should be made with the lowest-energy radiation that will transmit adequate radiation intensities to

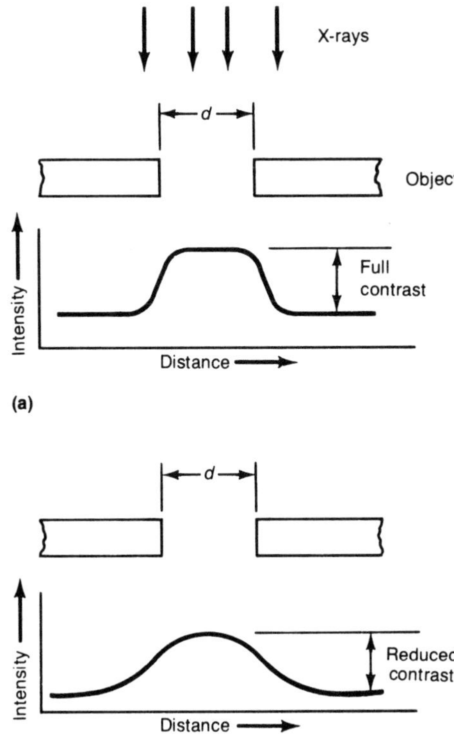

Fig. 3 Effect of geometric unsharpness on image contrast. (a) Flaw size, *d*, is larger than the unsharpness, then full contrast occurs. (b) Flaw size, *d*, is smaller than the unsharpness, then contrast is reduced.

the detector, because long wavelengths tend to improve contrast. However, radiation energies that are too low produce excessive amounts of scattered radiation that washes out fine detail. On the other hand, energies that are too high, although they reduce scattered radiation, may produce images having contrast that is too low to resolve small flaws. For each situation, there is an optimum radiation energy that produces the best combination of radiographic contrast and definition, or the greatest sensitivity.

The amount of subject contrast can be modeled as follows. If there is a change in thickness, Δx, of a testpiece with a uniform attenuation coefficient, μ, then the subject contrast (without scattering) is equal to $\mu\Delta x$. Similarly, if there is a change in the attenuation coefficient, $\Delta\mu$, over a distance X, then the subject contrast (without scatter) is equal to $X\Delta\mu$. If there is scattering, the subject contrast is reduced such that:

$$\text{Subject contrast} = \frac{\mu\Delta x \text{ (or } X\Delta\mu)}{1 + I_S/I_D} \qquad \text{(Eq 1)}$$

where I_S is the intensity of the scattered radiation and I_D is the intensity of the direct (or primary) radiation passing through the testpiece. The scattering may occur from within the testpiece (Fig. 4a) or from the surroundings (Fig. 4b and c). The various mechanisms of scattering and attenuation are described in the section "Attenuation of Electromagnetic Radiation" in this article.

Contrast sensitivity refers to the ability of responding to and displaying small variations in subject contrast. Contrast sensitivity depends on the characteristics of the image detector and on the level of radiation being detected (or on the amount of exposure for films). The relationship between the contrast sensitivity and the level of radiation intensity (or film exposure) can be illustrated by considering two extremes. At low levels of radiation intensity, the contrast sensitivity of the detectors is reduced by a smaller signal-to-noise ratio, while at high levels, the detectors become saturated. Consequently, contrast sensitivity is a function of dynamic range.

In film radiography, the contrast sensitivity is:

$$\text{Contrast sensitivity \%} = \frac{2.3\Delta D^{100}}{G_D} \qquad \text{(Eq 2)}$$

where ΔD is the smallest change in photographic density that can be observed when the film is placed on an illuminated screen. The factor G_D is called the film gradient or film contrast. The film gradient is the inherent ability of a film to record a difference in the intensity of the transmitted radiation as a difference in photographic density. It depends on film type, development procedure, and film density. For all practical purposes, it is independent of the quality and distribution of the transmitted radiation (see the section "Film Radiography" in this article).

Contrast sensitivity in real-time systems is determined by the number of bits (if the image is digitized) and the signal-to-noise ratio (which is affected by the intensity of the radiation and the efficiency of the detector). The factors affecting the contrast sensitivity and signal-to-noise ratio of real-time systems are discussed in the section "Modern Image Intensifiers" in this article. The best contrast sensitivity of digitized images from fluorescent screens is about 8 bits (or 256 gray levels).

Another way of specifying the contrast sensitivity of fluorescent screens is with a gamma factor, which is defined by the fractional unit change in screen brightness, $\Delta B/B$, for a given fractional change in the radiation intensity, $\Delta I/I$. Most fluorescent screens have a gamma factor of about one, which is not a limiting factor. At low levels of intensity, however, the contrast is reduced because of quantum mottle (see the section "Screen Unsharpness" in this article) just as unsharpness reduced the contrast in the example illustrated in Fig. 3.

Dynamic range, or latitude, describes the ability of the imaging system to produce a suitable signal over a range of radiation intensities. The dynamic range is given as the ratio of the largest signal that can be measured to the precision of the smallest signal that can be discriminated. A large dynamic range allows the system to maintain contrast sensitivity over a wide range of radiation intensities or testpiece thicknesses. Film radiography has a dynamic range of up to 1000:1, while digital radiography with discrete detectors can achieve 100 000:1. The latitude, or dynamic range, of film techniques is discussed in the section "Exposure Factors" in this article.

Radiographic definition refers to the sharpness of a radiograph. In Fig. 5(a), for example, the radiograph exhibits poor radiographic definition even when the contrast is high. Better radiographic definition is achieved in Fig. 5(b), even though the contrast is lower.

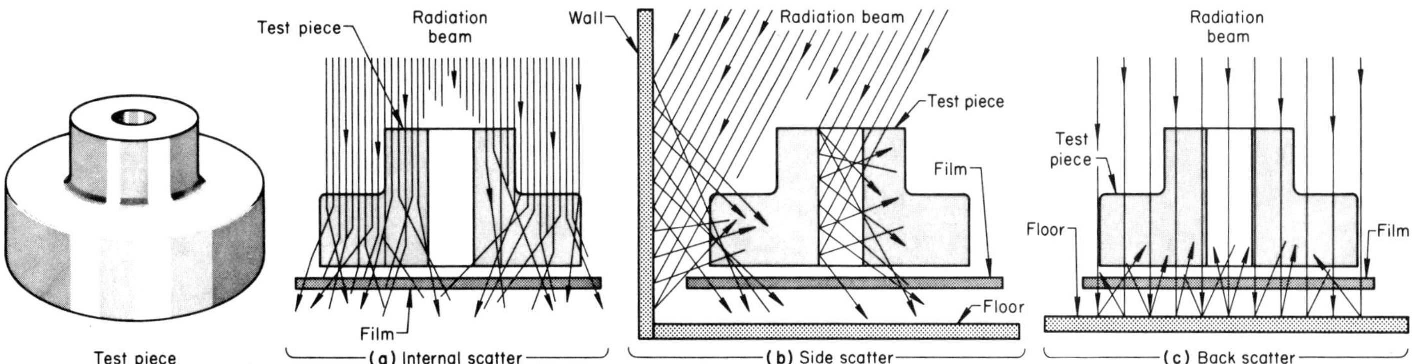

Fig. 4 Schematic of the three types of scattered radiation encountered when using x-rays and γ-rays

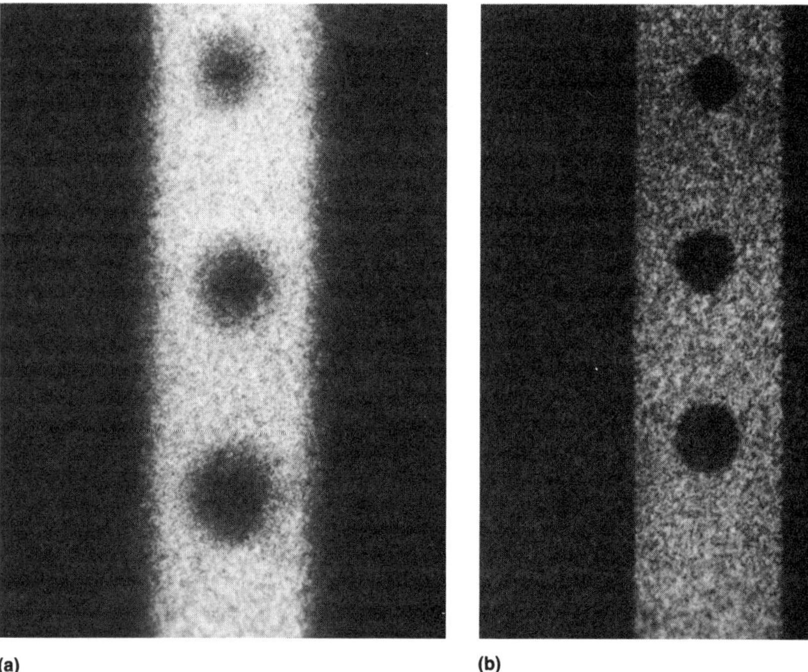

Fig. 5 Radiographs with poor and improved radiographic definition. (a) Advantage of a higher radiographic contrast is offset by poor definition. (b) Despite lower contrast, better detail is obtained with improved definition. Source: Ref 1

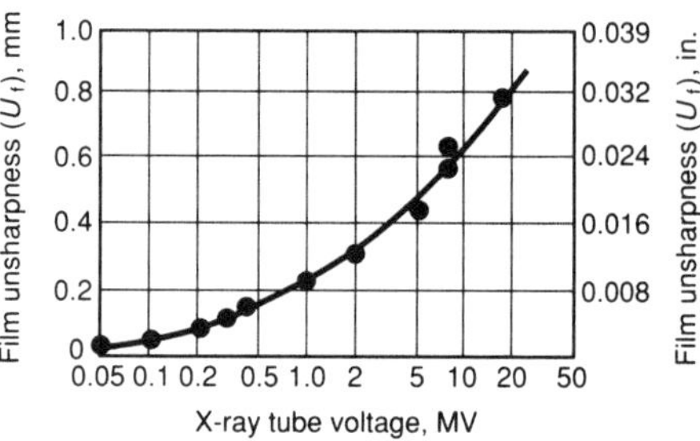

Fig. 6 Experimental curve of film unsharpness, U_f, against x-ray energy (filtered sources). Source: Ref 2

The degree of radiographic definition is usually specified in line pairs per millimeter or by the minimum distance by which two features can be physically distinguished. The factors affecting radiographic definition are geometric unsharpness and the unsharpness (or spatial resolution) of the imaging system. The net effect of these two unsharpnesses is not directly additive, because the combination of the two unsharpnesses is complex. For most practical radiography, the total unsharpness, U_T, is:

$$U_T^2 = U_g^2 + U_d^2 \qquad \text{(Eq 3)}$$

where U_g is the geometric unsharpness and U_d is the unsharpness or spatial resolution of the image-detecting system. Another important type of unsharpness occurs when in-motion radiography is performed. This type of unsharpness, termed motion unsharpness, is discussed in the section "In-Motion Radiography" in this article. Geometric unsharpness is usually the largest contributor to maximum unsharpness in film radiography, while the unsharpness of fluorescent screens in real-time radiography often overshadows the effect of geometric unsharpness.

Distortion and Geometric Unsharpness. Because a radiograph is a two-dimensional representation of a three-dimensional object, the radiographic images of most testpieces are somewhat distorted in size and shape compared to actual testpiece size and shape. The mechanisms of geometric unsharpness and distortion are described in the section "Principles of Shadow Formation" in this article. The severity of unsharpness and distortion depends primarily on the source size (focal-spot size for x-ray sources), source-to-object and source-to-detector distances, and position and orientation of the testpiece with respect to source and detector.

Film unsharpness specifies the spatial resolution of the film and must not be confused with film graininess. Film unsharpness and film graininess must be treated as two separate parameters.

Film unsharpness depends not only on the film grain size and the development process but also on the energy of the radiation and the thickness and material of the screens. When a quantum of radiation sensitizes a silver halide crystal in a film, adjacent crystals may also be sensitized if the quantum has sufficient energy to release electrons during the absorption by the first crystal. Consequently, a small volume of crystals would be sensitized, thus producing a small disk instead of a point image. Figure 6 illustrates an experimental curve of film unsharpness versus radiation energy. The screens used will also affect the unsharpness with the production of electrons.

Screen Unsharpness. In addition to the spatial resolution established by the video or image-processing equipment, the spatial resolution of real-time systems is also affected by screen unsharpness. Screen unsharpness depends not only on the grain size and thickness of the screen but also on the detection efficiency and the energy and intensity of the radiation. Statistical fluctuations (quantum mottle) of screen brightness becomes more significant at low levels of screen brightness, and this can occur when either the radiation intensity or the detection efficiency is lowered. Nevertheless, screen unsharpness due to statistical fluctuation can be eliminated with the image-processing operation of frame summation, which makes an image representing the amount of radiation exposure during the frame-summing interval.

Projective Magnification With Microfocus X-Ray Sources. Even though the imaging systems have limits in their spatial resolution, detail of the testpiece can be enlarged by projective magnification (see the section "Enlargement" in this article). This process of image magnification minimizes the limits of graininess and unsharpness caused by the imaging system. With microfocus x-ray sources, a resolution greater than 20 line pairs per millimeter (or a spatial resolution of 0.002 in.) can be achieved.

Detail Perceptibility of Images. Studies have shown that the perceptibility of image details depends principally on the product: $MTF^2 \cdot DQE$; where MTF is the modulation transfer function and DQE is the detective quantum efficiency of the imaging system. These terms, as previously mentioned in this section, are described in the article "Industrial Computed Tomography" in this Volume.

Radiation Safety

The principles of radiation safety have evolved from experience in health physics, which is the study of the biological effects of ionizing radiation. There are two main aspects of safety: monitoring radiation dosage and protecting personnel. This section summarizes the major factors involved in both. More detailed information on radiation safety is available in the Selected References at the end of this article.

Radiation Units. Ordinarily, x-ray and γ-ray radiation is measured in terms of its ionizing effect on a given quantity of atoms. The roentgen, as described in the section "Radiation Sources" in this article, is a unit derived on this basis for x-rays and γ-rays. However, even though the amount of x-ray and γ-ray exposure can be quantified in the roentgen unit, the effects of radiation on the human body also depend on the amount of energy absorbed and the type of radiation. Consequently, the roentgen equivalent man (rem) unit and the SI-based sievert (Sv) unit have been developed for the purposes of radiation safety. As described below, these two units include safety factors assigned to the absorbed energy dose of a particular type of radiation (where the absorbed energy dose is measured either in the radiation absorbed dose, or rad, unit or the SI-based gray unit).

Rem and Rad Units. The specification of radiation in rems is equal to the product of the rad and a factor known as the relative biological effectiveness (rbe). The rbe (as established by the U.S. Government) is 1.0 for x-rays, γ-rays, or β particles; 5.0 for thermal neutrons; 10 for fast neutrons; and 20 for α particles. One rad represents the absorption of 100 ergs of energy per gram of irradiated material; it applies to any form of penetrating radiation.

Sievert and Gray Units. A sievert is the SI-based version of the rem unit. The same rbe factors are used, but the sievert unit is derived from the absorbed dose specified in the SI-based gray (Gy) unit. Because 1 Gy represents the absorption of one joule of energy per kilogram of material, then 1 Gy = 100 rads and 1 Sv = 100 rem.

Units for X-Rays and γ-Rays. Because 1 R (which applies only to the ionizing effect of x-rays or γ-rays) corresponds to the absorption of only about 83 ergs per gram of air, the unit of 1 R can be considered equal to 0.01 Gy (1 rad) for radiation-safety measurements. With this approach, then, the measurement of x-ray and γ-ray dose in roentgens can be considered equivalent to direct measurement of dose in rems because the rbe is equal to one. This considerably simplifies the process of monitoring dose levels.

Maximum Permissible Dose. Current practice in health physics specifies a threshold value of accumulated dose above which an individual should not be exposed. The value of this threshold, or maximum permissible dose, increases at the rate of 0.05 Sv/yr (5 rem/yr) for all persons over the age of 18. This so-called banking concept is based on experience and assumes that no person will be exposed to penetrating radiation before the age of 18. The total amount of radiation absorbed by any individual should never exceed his continually increasing accumulated permissible dose; furthermore, no individual should be exposed to more than 0.12 Sv (12 rem) in any given year, regardless of other factors. For administrative purposes, maximum permissible dose rate is evaluated for each individual on a quarterly or weekly basis. The applicable dose rates are 0.0125 Sv (1¼ rems) per calendar quarter, or about 1 millisievert (1 mSv, or 100 mrem) per week.

Radiation Protection. The two main factors in radiation protection are controlling the level of radiation exposure and licensing the facility. The main criterion in controlling radiation exposure is the philosophy outlined in Regulatory Guide 8.10 of the United States Nuclear Regulatory Commission, which specifies the objective of reducing radiation exposures to a level as low as reasonably achievable. The licensing process involves the issuance of permission for operating a radiographic facility; permission may be required from federal, state, and local governments.

The federal licensing program is concerned with those companies that use radioactive isotopes as sources. However, in some localities, state and local agencies exercise similar regulatory prerogatives. To become licensed under any of these programs, a facility or operator must show that certain minimum requirements have been met for the protection of both operating personnel and the general public from excessive levels of radiation. Although local regulations may vary in the degree and type of protection afforded, certain general principles apply to all.

First, the amount of radiation that is allowed to escape from the area over which the licensee has direct and exclusive control is limited to an amount that is safe for continuous exposure. In most cases, total dose values of 0.02 mSv (2 mrem) in 1 h, 1 mSv (100 mrem) in 7 consecutive days, and 5 mSv (500 mrem) in a calendar year can be considered safe. These values correspond to those adopted by the federal government for radioactive sources and are also applicable under most local regulations. For purposes of control, unrestricted areas within a given facility are often afforded the same status as areas outside the facility. An unrestricted area is usually defined as an area where employees are not required to wear personal radiation-monitoring devices and where there is unrestricted access by either employees or the public.

Second, main work areas are shielded in accordance with applicable requirements. Some facilities must be inspected by an expert in radiation safety in order to qualify for licensing; it is usually recommended that the owner consult an expert when planning a new facility or changes in an older facility to ensure that all local requirements are met. Appropriate shielding usually consists of lead or thick concrete on all sides of the main work area. The room in which the actual exposures are made is the most heavily shielded and may be lined on all sides with steel or lead. Particular attention is given to potential paths of leakage such as access doors and passthroughs and to seemingly unimportant paths of leakage such as the nails and screws that attach lead sheets to walls and doors.

Third, portable radiation sources are used in strict accordance with all regulations. In most cases, safe operation is ensured by a combination of:

- Movable shielding (usually lead)
- Restrictions on the intensity and direction of radiation emitted from the source during exposure
- Exclusion of all personnel from the immediate area during exposure

The best protection is afforded by distance because radiation intensity decreases in proportion to the square of the distance from the source. As long as personnel stay far enough away from the source while an exposure is being made, portable sources can be used with adequate safety.

Finally, access to radiographic work areas, including field sites and radioactive-source storage areas, must at all times be under the control of competent and properly trained radiographers. Radiographers must be responsible for admitting only approved personnel into restricted work areas and must ensure that each individual admitted to a restricted work area carries appropriate devices for monitoring absorbed radiation doses. In addition to keeping records of absorbed dose for all monitored personnel, radiographers must maintain accurate and complete records of radiation levels in both restricted and adjacent unrestricted areas.

Radiation Monitoring. Every safety program must be controlled to ensure that both the facility itself and all personnel subject to radiation exposure are monitored. Facility monitoring is generally accomplished by periodically taking readings of radiation leakage during the operation of each source under maximum-exposure conditions. Calibrated instruments can be used to measure radiation dose rates at various points within the restricted area and at various points around the perimeter of the restricted area. This information, in conjunction with knowledge of normal occupancy, is used to evaluate the effectiveness of shielding and to determine maximum duty cycles for x-ray equipment.

To guard against the inadvertent leakage of large amounts of radiation from a shielded work area, interlocks and alarms are often required. Basically, an interlock disconnects power to the x-ray source if an access door is opened. Alarms are connected to a separate power source, and they activate visible and/or audible signals whenever the radiation intensity exceeds a preset value, usually 0.02 mSv/h (2 mrem/h).

All personnel within the restricted area must be monitored to ensure that no one absorbs excessive amounts of radiation. Devices such as pocket dosimeters and film badges are the usual means of monitoring, and often both are worn. Pocket dosimeters should be direct reading. One disadvantage of the remote-reading type is that it must be brought to a charger unit to be read. Both types are sensitive to mechanical shock, which could reduce the internal charge, thus implying that excessive radiation has been absorbed. If this should happen, the film badge should be developed immediately and the absorbed dose evaluated.

Under normal conditions, pocket dosimeters are read daily, and the readings are noted in a permanent record. At less frequent intervals, but at least monthly, film badges are developed and are evaluated by comparison with a set of reference films. Because of various factors, the values of absorbed radiation indicated by dosimeters and film badges may differ. This is particularly true at low dose rates and with high radiation energies (1 MeV or more). However, the accuracy of both dosimeters and film badges increases with increasing dose, and at dose rates near the maximum allowable dose of 0.0125 Sv (1¼ rem) per calendar quarter, the readings usually check within ±20%.

Access Control. Permanent facilities are usually separated from unrestricted areas by shielded walls. Sometimes, particularly in on-site radiographic inspection, access barriers may consist only of ropes. In such cases, the entire perimeter around the work area should be under continual visual surveillance by radiographic personnel. Signs that carry the international symbol for radiation must be posted around any high-radiation area. This helps to inform bystanders of the potential hazard, but should never be assumed to prevent unauthorized entry into the danger zone. In fact, no interlock, no radiation alarm, and no other safety device should be considered a substitute for constant vigilance on the part of radiographic personnel.

X-Ray Tubes

X-ray tubes are electronic devices that convert electrical energy into x-rays. Typically, an x-ray tube consists of a cathode structure containing a filament and an anode structure containing a target—all within an evacuated chamber or envelope (Fig. 7). A low-voltage power supply, usually controlled by a rheostat, generates the electric current that heats the filament to incandescence. This incandescence of the filament produces an electron cloud, which is directed to the anode by a focusing system and accelerated to the anode by the high voltage applied between the cathode and the anode. Depending on the size of the focal spot achieved, x-ray tubes are sometimes classified into three groups:

- Conventional x-ray tubes with focal-spot sizes between 2 by 2 mm (0.08 by 0.08 in.) and 5 by 5 mm (0.2 by 0.2 in.)
- Minifocus tubes with focal-spot sizes in the range of 0.2 mm (0.008 in.) and 0.8 mm (0.03 in.)
- Microfocus tubes with focal-spot sizes in the range of 0.005 mm (0.0002 in.) and 0.05 mm (0.002 in.)

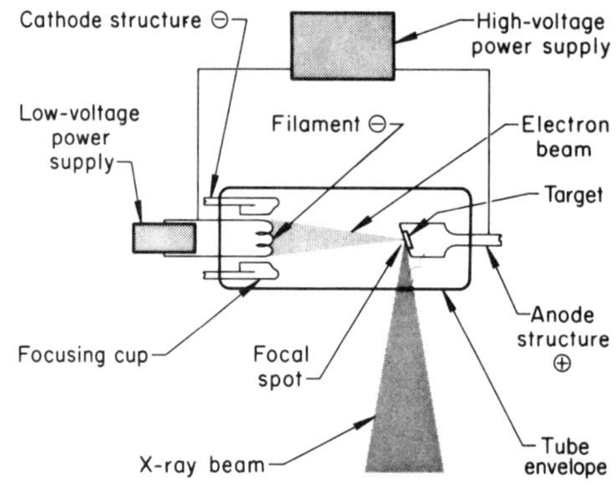

Fig. 7 Schematic of the principal components of an x-ray unit

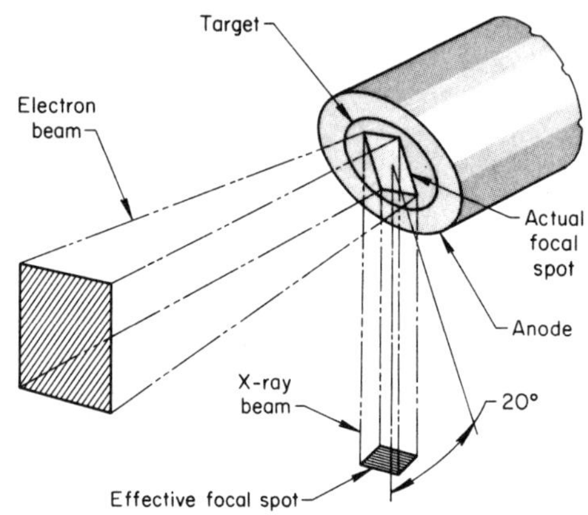

Fig. 8 Schematic of the actual and effective focal spots of an anode that is inclined at 20° to the centerline of the x-ray beam

The design of conventional and microfocus x-ray tubes is discussed in the following section.

When the accelerated electrons impinge on the target immediately beneath the focal spot, the electrons are slowed and absorbed, and both bremsstrahlung and characteristic x-rays are produced. Most of the energy in the impinging electron beam is transformed into heat, which must be dissipated. Severe restrictions are imposed on the design and selection of materials for the anode and target to ensure that structural damage from overheating does not prematurely destroy the target. Anode heating also limits the size of the focal spot. Because smaller focal spots produce sharper radiographic images, the design of the anode and target represents a compromise between maximum radiographic definition and maximum target life. In many x-ray tubes, a long, narrow, actual focal spot is projected as a roughly square effective focal spot by inclining the anode face at a small angle (usually about 20°) to the centerline of the x-ray beam, as shown in Fig. 8.

Tube Design and Materials

The cathode structure in a conventional x-ray tube incorporates a filament and a focusing cup, which surrounds the filament. The focusing cup, usually made of pure iron or pure nickel, functions as an electrostatic lens whose purpose is to direct the electron beam toward the anode. The filament, usually a coil of tungsten wire, is heated to incandescence by an electric current produced by a relatively low voltage, similar to the operation of an ordinary incandescent light bulb. At incandescence, the filament emits electrons, which are accelerated across the evacuated space between the cathode and the anode. The driving force for acceleration is a high electrical potential (voltage) between anode and cathode, which is applied during exposure.

The anode usually consists of a button of the target material embedded in a mass of copper that absorbs much of the heat generated by electron collisions with the target. Tungsten is the preferred material for traditional x-ray tubes used in radiography because its high atomic number makes it an efficient emitter of x-rays and because its high melting point enables it to withstand the high temperatures of operation. Gold and platinum are also used in x-ray tubes for radiography, but targets made of these metals must be more effectively cooled than targets made of tungsten. Other materials are used, particularly at low energies, to take advantage of their characteristic radiation. Most high-intensity x-ray tubes have forced liquid cooling to dissipate the large amounts of anode heat generated during operation.

Tube envelopes are constructed of glass, ceramic materials or metals, or combinations of these materials. Tube envelopes must have good structural strength at high temperatures to withstand the combined effect of forces imposed by atmospheric pressure on the evacuated chamber and radiated heat from the anode. The shape of the envelope varies with the cathode-anode arrangement and with the maximum rated voltage of the tube. Electrical connections for the anode and cathode are fused into the walls of the envelope. Generally, these are made of metals or alloys having thermal-expansion properties that match those of the envelope material.

X-ray tubes are inserted into metallic housings that contain an insulating medium such as transformer oil or an insulating gas. The main purpose of the insulated housing is to provide protection from high-voltage electrical shock. Housings usually contain quick disconnects for electrical cables from the high-voltage power supply or transformer. On self-contained units, most of which are portable, both the x-ray tube and the high-voltage transformer are contained in a single housing, and no high-voltage cables are used.

Microfocus X-Ray Tubes. Developments in vacuum technology and manufacturing processes have led to the design and man-

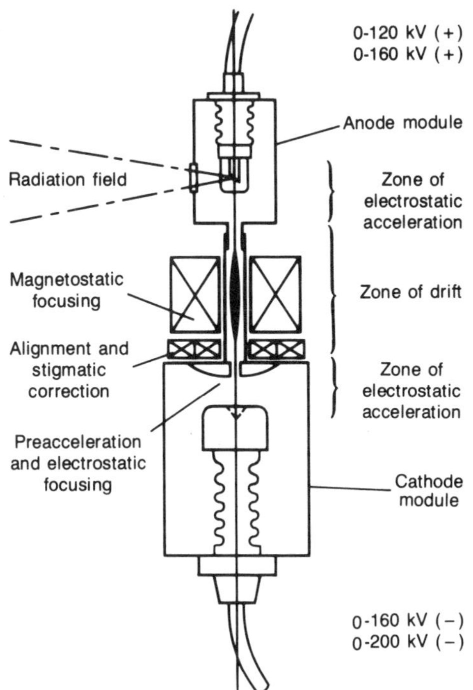

Fig. 9 Schematic of a microfocus x-ray tube

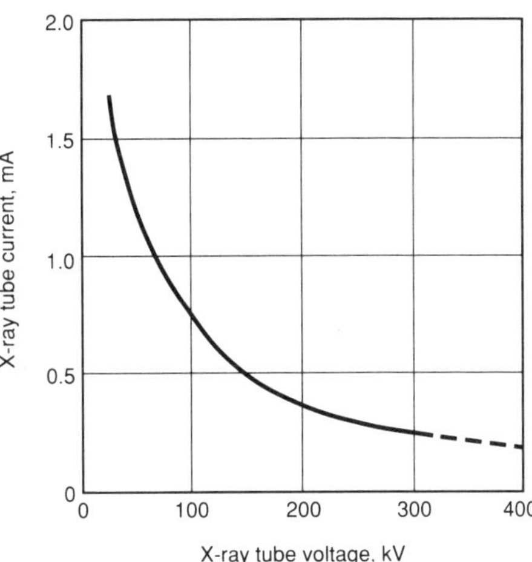

Fig. 10 Maximum ratings for no burn-in with 200-kV cathode/120-kV anode for electron beam widths of ≈10 μm (0.4 mil)

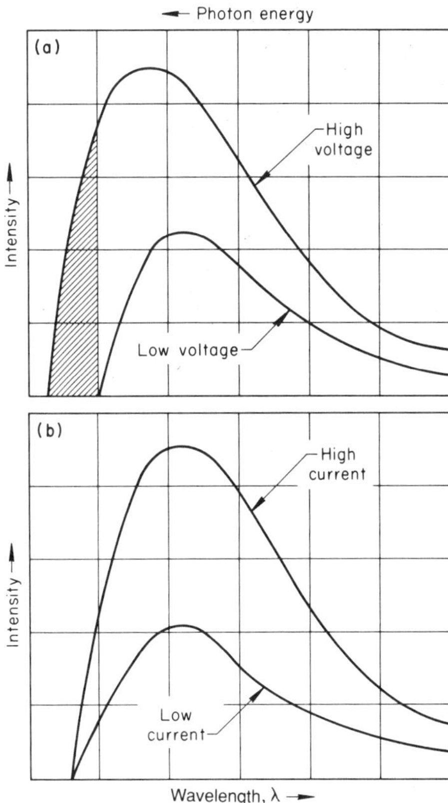

Fig. 11 Effect of (a) tube voltage and (b) tube current on the variation of intensity with wavelength for the bremsstrahlung spectrum of an x-ray tube. See text for discussion.

ufacture of microfocus x-ray systems. Some of these systems incorporate designs that allow the opening of the tube head and the replacement of component parts, such as targets and filaments (Fig. 9). A vacuum is generated by the use of a roughing pump and turbomolecular pumps that rapidly evacuate and maintain the system to levels as low as 0.1×10^{-3} Pa (10^{-6} torr). Electrostatic or magnetic focusing and *x-y* deflection are used to provide very small focal spots and to guide the beam to various locations on the target. Focal-spot sizes are adjustable from 0.005 to 0.2 mm (0.0002 to 0.008 in.). If the target becomes pitted in one area, the beam can be deflected to a new area, extending the target life. When the target, filament, or other interior component is no longer useful in producing the desired focal-spot size or x-ray output, the tube can be opened and the component replaced at minimal cost. The tube can then be evacuated to operating levels in a few minutes or hours, depending on the length of time the tube is open to the atmosphere and the amount of contamination present.

These systems are available with voltages varying from 10 to 360 kV at beam currents from 0.01 to 2 mA. To avoid excessive pitting of the target, the beam current is varied according to the desired focal-spot size and/or kilovolt level (Fig. 10).

Microfocus x-ray systems having focal spots that approach a point source are useful in obtaining very high resolution images. A radiographic definition of 20 line pairs per millimeter (or a spatial resolution of 0.002 in.), using real-time radiography has been achieved with microfocus x-ray sources. This high degree of radiographic definition is accomplished by image enlargement, which allows the imaging of small details (see the section "Enlargement" in this article).

Microfocus x-ray systems have found considerable use in the inspection of integrated circuits and other miniature electronic components. Microfocus x-ray systems with specially designed anodes as small as 13 mm (0.5 in.) in diameter and several inches long also enable an x-ray source to be placed inside otherwise inaccessible areas, such as aircraft structures and piping. The imaging medium is placed on the exterior, and this allows for the single-wall inspection of otherwise uninspectable critical components. Because of the small focal spot, the source can be close to the test area with minimal geometric unsharpness (see the section "Principles of Shadow Formation" in this article for the factors that influence geometric unsharpness).

X-Ray Tube Operating Characteristics

There are three important electrical characteristics of x-ray tubes:

- The filament current, which controls the filament temperature and in turn the quantity of electrons that are emitted
- The tube voltage, or anode-to-cathode potential, which controls the energy of impinging electrons and therefore the energy or penetrating power, of the x-ray beam
- The tube current, which is directly related to filament temperature and is usually referred to as the milliamperage of the tube

The strength, or radiation output, of the beam is approximately proportional to milliamperage, which is used as one of the variables in exposure calculations. This radiation output, or R-output, is usually expressed in roentgens per minute (or hour) at 1 m (as mentioned in the section "Radiation Sources" in this article).

X-Ray Spectrum. The output of a radiographic x-ray tube is not a single-wavelength beam, but rather a spectrum of wavelengths somewhat analogous to white light. The lower limit of wavelengths, λ_{min}, in nanometers, at which there is an abrupt ending of the spectrum, is inversely proportional to tube voltage, V. This corresponds to an upper limit on photon energy, E_{max}, which is proportional to the tube voltage, V:

$$E_{max} = \mathrm{a}V \quad \text{(Eq 4)}$$

where a = 1 eV/volt.

Figure 11 illustrates the effect of variations in tube voltage and tube current on photon energy and the intensity (number of photons). As shown in Fig. 11(a), increasing the tube voltage increases the intensity of radiation and adds higher-energy photons to the spectrum (crosshatched area, Fig. 11a). On the other hand, as shown in Fig. 11(b), increasing the tube current increases the intensity of radiation but does not affect the energy distribution.

The energy of the x-rays is important because higher-energy radiation has greater penetrating capability; that is, it can pene-

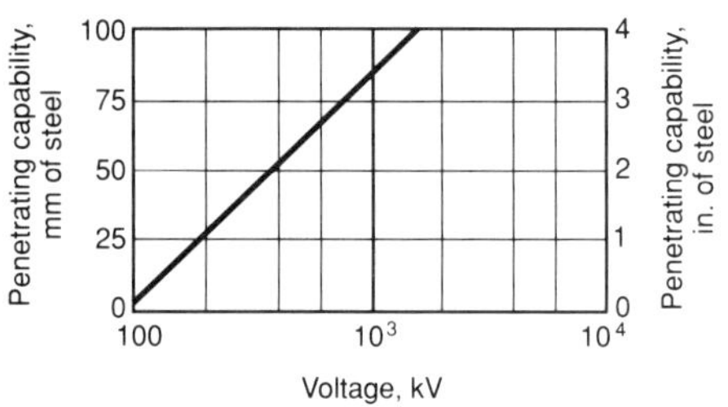

Fig. 12 Effect of tube voltage on the penetrating capability of the resulting x-ray beam

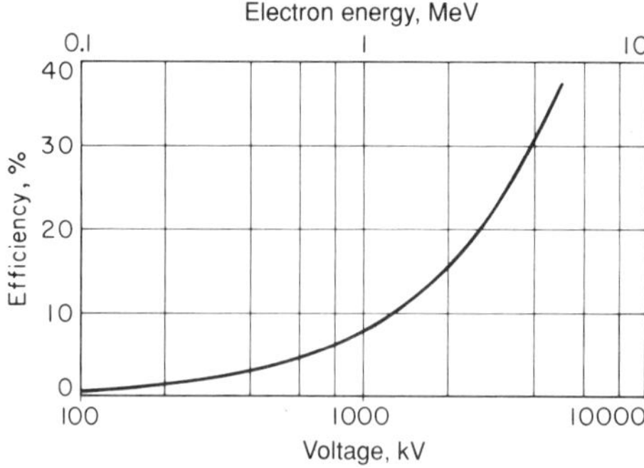

Fig. 13 Effect of tube voltage or electron energy on the efficiency of energy conversion in the target of an x-ray source

trate through greater thickness of a given material or can penetrate denser materials than can lower-energy radiation. This effect is shown in Fig. 12, which relates the penetrating capability to tube voltage. An applied voltage of about 200 kV can penetrate steel up to 25 mm (1 in.) thick, while almost 2000 kV is needed when the thickness to be penetrated is 100 mm (4 in.).

As previously stated, most of the energy in the electron stream is converted into heat rather than into x-rays. The efficiency of conversion, expressed in terms of the percentage of electron energy that is converted into x-rays, varies with electron energy (or tube voltage), as shown in Fig. 13. At low electron energies (tube voltages of about 100 to 200 kV), conversion is only about 1%; about 99% of the energy must be dissipated from the anode as heat. However, at high electron energies (above 1 MeV), the process is much more efficient, varying from about 7% conversion efficiency at 1 MeV to 37% at 5 MeV. Therefore, in addition to producing radiation that is more penetrating, high-energy sources produce greater intensity of radiation for a given amount of electrical energy consumed.

The R-output of an x-ray tube varies with tube voltage (accelerating potential), tube current (number of electrons impinging on the target per unit time), and physical features of the individual equipment. Because of the last factor, the R-output of an individual source also varies with position in the radiation beam, position usually being expressed as the angle relative to the central axis of the beam.

Effect of Tube Voltage. Both the mean photon energy and the R-output of an x-ray source are altered by changes in tube voltage. The effect of tube voltage on the variation of intensity (R-output) is shown in Fig. 11(a). The overall R-output varies approximately as the square root of tube voltage. The combined effect of greater photon energy and increased R-output produces, for film radiography, a decrease in exposure time of about 50% for a 10% increase in tube voltage. The effect is similar with other permanent-image recording media, as in paper radiography and xeroradiography.

Effect of Tube Current. The spectral distribution of wavelengths is not altered by changes in tube current; only the R-output (intensity) varies (Fig. 11b). Because tube current is a direct measure of the number of electrons impinging on the target per unit of time, and therefore the number of photons emitted per unit of time at each value of photon energy, R-output varies directly with tube current. This leads to the so-called reciprocity law for radiographic exposure, expressed as:

$$it = \text{constant} \qquad \text{(Eq 5)}$$

where i is the tube current (usually expressed in milliamperes) and t is the exposure time.

The reciprocity law (Eq 5) is valid for any recording medium whose response depends solely on the amount of radiation impinging on the testpiece, regardless of the rate of impingement (radiation intensity). For example, the reciprocity law is valid for most film and paper radiography and most xeroradiography but not for fluoroscopic screens or radiometric detectors, both of which respond to radiation intensity rather than to total amount of radiation. However, radiographic films, when used with fluorescent screens, exhibit reciprocity law failure because the response of film emulsions varies with the rate of impingement of photons of visible light (screen brightness) and with total exposure. If tube current is decreased and exposure time is increased according to the reciprocity law, a fluorescent screen will emit the same amount of light (same number of photons) but at a lower level of brightness over a longer period of time. The lower brightness level will result in lower film density compared to an equivalent exposure made with a higher tube current.

Deviations from reciprocity law are usually small for minor changes in tube current, causing little difficulty in resolving testpiece features or interpreting radiographs made on screen-type film. However, when the tube current is changed by a factor of four or more, there may be a 20% or more deviation from reciprocity law, and it will be necessary to compensate for the effect of screen brightness on film density.

In most cases, when tube voltage is maintained constant, exposures made in accordance with the reciprocity law should produce identical film densities. This assumes that tube voltage does not vary with tube current. Heavy-duty equipment designed for stationary installations contains electrical circuitry (current stabilizers and voltage compensators) that tend to maintain R-output in accordance with the reciprocity law. However, equipment intended for on-site use is often designed to minimize weight, size, and cost and does not contain such complex circuitry. Consequently, the R-output and x-ray spectrum of many portable or transportable machines vary with tube current. These types of machines may exhibit apparent reciprocity law failure at both ends of the useful range of tube currents. For example, at currents approaching the maximum rated current, the tube voltage tends to be depressed because of the heavy electrical load on the transformer. This produces lower values of both R-output and mean photon energy than would normally be expected. Conversely, at very low tube currents, the tube voltage may actually exceed the calibrated value because the transformer is more efficient. This produces higher values of both R-output and mean photon energy than are normally expected. These deviations from expected values are not always indicated on the monitoring instruments attached to x-ray machines.

Because of deviations from reciprocity law caused by equipment characteristics, it is often desirable to prepare exposure charts (usually graphs) for each x-ray unit. Such charts or graphs should be based on film exposure data at several values of tube voltage within the useful kilovoltage range of the unit. Use of these charts for determining exposure times, especially at abnormally high or low tube currents, will help to avoid unsatisfactory image quality.

Effect of Electrical Waveform. In all x-ray tubes that are powered by transformer circuits, the waveform of the tube voltage affects R-output. X-ray tubes are direct current (dc) devices, yet almost all power supplies are driven by alternating current (ac). If ac power is supplied directly to the x-ray tube, the tube itself will provide half-wave rectification. Many low-power x-ray machines use self-rectifying x-ray tubes. As illustrated in Fig. 14(a), half-wave rectification results in a sinusoidal variation of instantaneous tube voltage with time, except that no tube current flows (and there is no effective tube voltage) during the portion of each cycle when the anode is electrically negative, and consequently there is no emission of x-rays during that portion of the cycle.

In a full-wave-rectified circuit (Fig. 14b), the instantaneous tube voltage varies sinusoidally from zero to peak voltage (kVp) and back to zero, then repeats the sinusoi-

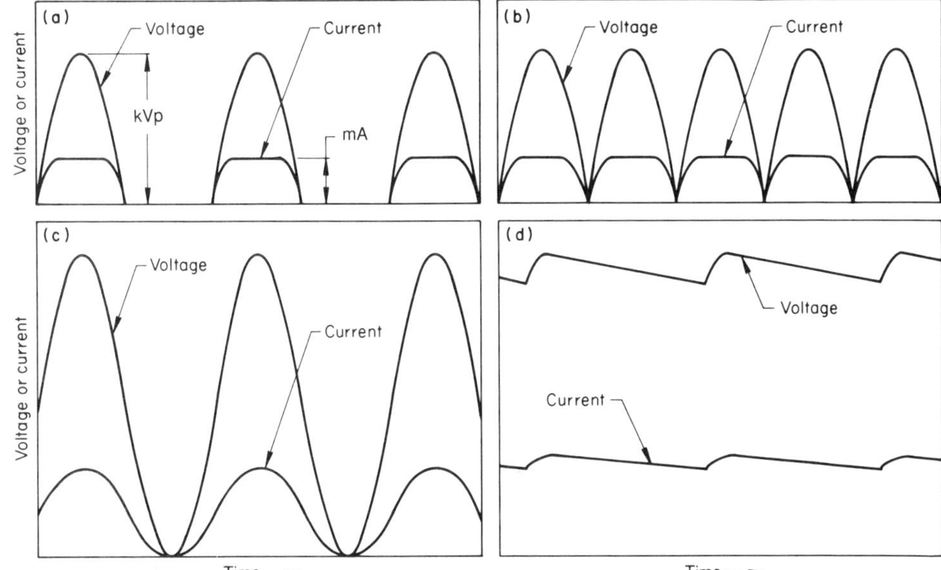

Fig. 14 Waveforms for accelerating potential (tube voltage) and tube current for four generally used types of x-ray-tube circuits. (a) Half-wave rectified circuit and (b) full-wave rectified circuit, in which tube voltage is equal to the peak voltage of the electrical transformer. (c) Villard-type circuit and (d) constant-potential circuit, in which tube voltage is twice the peak voltage of the transformer

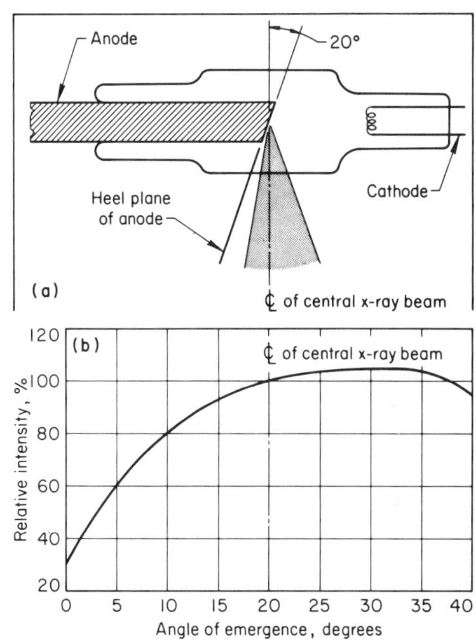

Fig. 15 Heel effect in a typical conventional x-ray tube having an anode face inclined 70° to the tube axis (20° anode heel angle). (a) Diagram showing relation of useful beam (crosshatched region) to tube components. (b) Graph of beam intensity (expressed as a percentage of the intensity of the central beam) as a function of angle of emergence relative to the anode heel plane

dal pattern during the second half of each cycle. This effectively doubles the time during which x-rays are emitted on each cycle, thus doubling the R-output compared to half-wave-rectified equipment.

In Villard-circuit equipment, the electrical input to the x-ray tube varies sinusoidally, but as Fig. 14(c) shows, the zero potential occurs at the minimum instantaneous tube voltage rather than at midrange. Consequently, the accelerating potential across the tube has a peak value (kVp) that is twice the peak voltage of the transformer; for example, 2-MeV electrons can be produced with transformer rated at 1 MVp. X-rays produced by Villard-circuit equipment are harder (that is, of higher mean photon energy) than x-rays produced by rectified equipment having a transformer that operates at the same peak voltage. Villard circuits exhibit a variation in instantaneous tube current that parallels the variation in instantaneous tube voltage.

A modified Villard circuit, also known as a Greinacher circuit or a constant-potential circuit, produces an accelerating potential and instantaneous tube current that are nearly constant, varying only slightly with time, as shown in Fig. 14(d). The chief value of constant-potential equipment lies in the fact that its R-output is about 15% higher than for a full-wave-rectified unit of equivalent tube voltage. This means that a nominal 15% reduction in exposure times can be achieved by using constant-potential instead of rectified equipment.

Heel Effect. X-ray tubes exhibit a detrimental feature known as the heel effect. When the direction in which x-rays are emitted from the target approaches the anode heel plane, the intensity of radiation at a given distance from the focal spot is less than the intensity of the central beam because of self-absorption by the target. Figure 15(a) shows the heel effect schematically, and Fig. 15(b) is a graph of the variation of intensity as a function of angle of emergence relative to the anode heel plane. The direction of the central beam is at 20°—for this particular tube design, an angle equal to the anode heel angle.

Radiographs of large-area testpieces that are made at relatively short source-to-detector distances will exhibit less photographic density (film) or less brightness (real-time) in the region where the incident radiation is less intense because of the heel effect. This can lead to errors in interpretation unless the heel effect is recognized. The consequences of heel effect can be minimized by using a source-to-detector distance that ensures that the desired area of coverage is entirely within the useful beam emanating from the x-ray tube or by making multiple exposures with a reduced area of coverage for each exposure.

The heel effect also influences image sharpness when there is some geometric enlargement. The projected size of the focal spot is reduced near the "heel," which reduces the geometric unsharpness of the radiographic image nearest the heel. Conversely, the radiographic image is less sharp on the cathode side of the radiation beam.

Stem Radiation. In an x-ray tube, electrons that produce the useful x-rays are known as primary electrons. The kinetic energy of the primary electrons is converted partly to radiation and partly to heat when these electrons collide with the target. However, some of the primary electrons collide with other components inside the tube envelope, giving rise to slower-moving electrons known as secondary electrons or stray electrons. Some of these stray electrons bounce around inside the tube and eventually strike the anode. This produces x-rays, called stem radiation, that are of relatively long wavelength because of the low kinetic energy of the stray electrons that produced them. Stem radiation, which is about 1/200 as intense as the useful beam, is produced all over the anode surface, not only at the focal spot. Because stem radiation is projected from a comparatively large area instead of emanating only from the focal spot, it invariably causes image unsharpness.

Some x-ray tubes are designed to eliminate this phenomenon by the use of hollow-end anodes called hooded anodes (Fig. 16). Hooded anodes reduce stem radiation both by shielding the target from some of the stray electrons and by absorbing much of the radiation produced by stray electrons that enter through the open end of the anode cavity.

Tube Rating. X-ray tubes produce a great amount of heat. At 100-kV tube voltage, only about 1% of the electrical energy is converted to x-rays; the remaining 99% is converted to heat. Heat removal constitutes the most serious limitation on x-ray tube design. The size of the focal spot and the design of the anode are the main factors that determine the rating of a particular x-ray

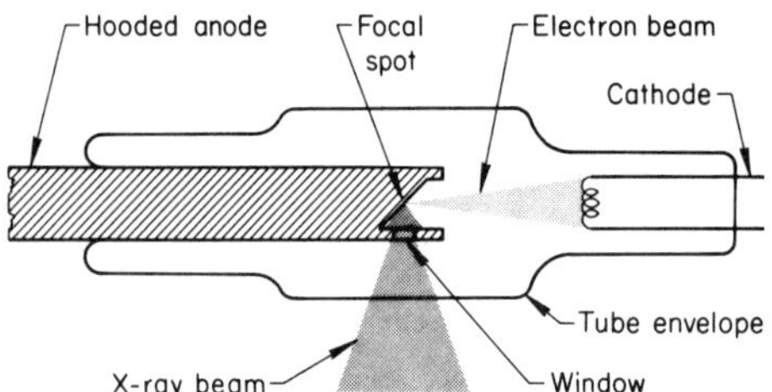

Fig. 16 Schematic of an x-ray tube having a hooded anode to minimize stem radiation

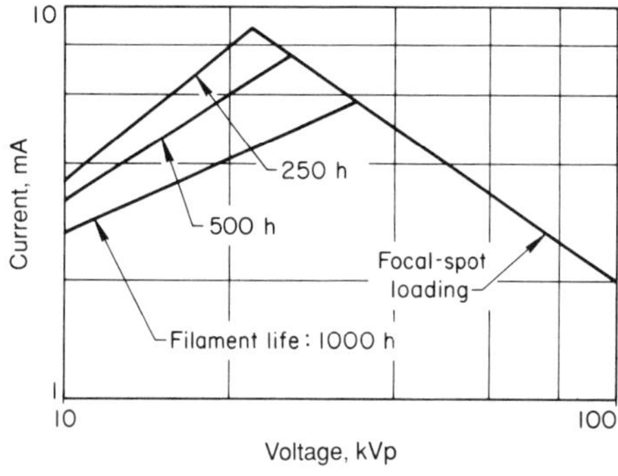

Fig. 17 Tube-rating chart for a typical beryllium-window tube with a 0.3 mm (0.012 in.) focal spot. Tube rating is for continuous operation with a full-wave-rectified single-phase power supply. Below 22.5 kVp, tube current is limited by desired filament life; above 34.5 kVp, tube current is limited by focal-spot loading. Filament voltage is 1.7 to 2.6 V, and filament current is 3.2 to 4.0 A.

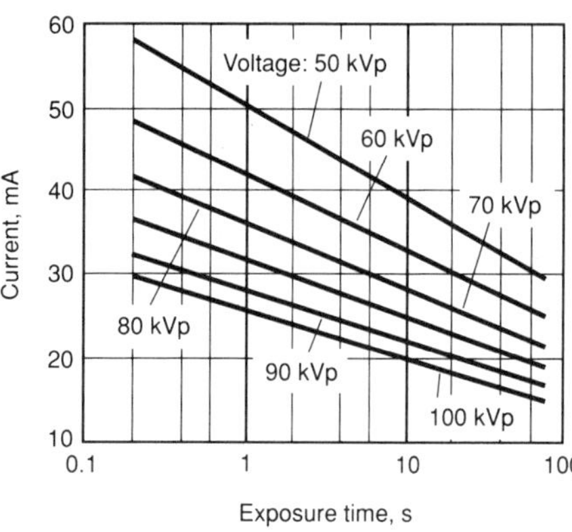

(a)

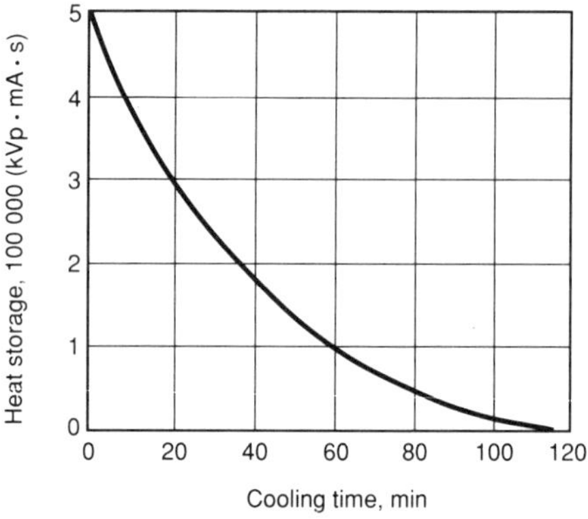

(b)

Fig. 18 Tube-rating chart (a) and cooling chart (b) for a typical industrial x-ray tube having a 1.5 mm (0.06 in.) focal spot. Tube rating is for intermittent operation with a full-wave-rectified single-phase power supply; a minimum 5-s waiting time between exposures is specified. Cooling chart is for an oil-filled housing without an air circulator. When housing is fitted with an optional air circulator (fan), cooling times are reduced to one-half the values determined from the chart.

tube. A tube rating is a limiting or maximum allowable combination of tube voltage (kilovoltage) and tube current (milliamperage).

Most industrial tubes are rated for continuous service at the maximum tube current that will give reasonable target life at maximum tube voltage. Therefore, if a tube has a continuous rating of 10 mA at 150-kV constant potential, the rating is 1500 W. Any product of kilovoltage and milliamperage equal to 1500 will not exceed the heat limit on the anode. The tube should be able to operate continuously at 20 mA and 75 kV, 30 mA and 50 kV, and so on.

However, another limitation on operating characteristics is that, to sustain the same tube current, the filament must be heated to a higher temperature at low tube voltages than at higher tube voltages. Therefore, the tube current at low voltage must be limited to achieve satisfactory service life. Figure 17 shows a tube-rating chart for continuous operation in which the total length of time for each exposure is not a factor in determining operating conditions. On the left side of the chart, at low tube voltages, operating conditions are determined by filament life; on the right side of the chart, at high tube voltages, anode heating (focal-spot loading) limits operating conditions. Any combination of tube voltage and tube current below the curve can be used without impairing the service life of either anode or filament. In practice, unless there are special considerations, values approaching the continuous-duty rating are ordinarily used to minimize exposure time.

Some tube manufacturers use an alternative form of rating (intermittent rating), in which the time involved in making a particular film (or paper) exposure enters into the rating. As illustrated in Fig. 18(a), for each combination of tube current and tube voltage there is a maximum exposure time that must not be exceeded. Longer exposure times can cause the target to melt. Normally, there is a minimum waiting time between exposures for tubes rated for intermittent operation. For the tube rating shown in Fig. 18(a), the waiting time is specified as 5 s when maximum exposure times are used. Some manufacturers have developed rating systems in which the minimum waiting time can be calculated.

Tubes that are rated for intermittent operation can accumulate only a certain amount of heat, then they must be allowed to cool before additional exposures are made. Figure 18(b) is the cooling chart used in conjunction with the tube-rating chart in Fig. 18(a). According to the cooling chart, the tube can store 500 000 units of heat (one heat unit equals 1 mA · 1 kVp · 1 s). Therefore, if a series of exposures is to be made at 70 kVp, 30 mA, and 5 s, each exposure will expend energy equal to 10 500 heat units. The number of successive film exposures that can be made, with a 5-s wait between exposures, is 47 (500 000 heat units divided by 10 500 heat units per exposure). After the 47th exposure, the tube must be allowed to cool according to the cooling curve in Fig. 18(b). The tube can be allowed to cool for any length of time, and then another exposure can be made as long as the additional exposure does not increase the stored heat above 500 000 heat units. After the tube has reached its capacity for heat storage, waiting times between successive exposures must be in accordance with the cooling curve. In this case, as shown in Fig. 18(b), 2 h would be required (or 1 h with an optional air circulator) for the tube to cool completely.

Inherent Filtration. In the radiography (film or real-time) of thin or lightweight materials, which requires low-energy radiation, filtration by the glass walls of the x-ray tube becomes a problem. Ninety-five percent of a 30-kV x-ray beam is absorbed by the glass walls of an ordinary x-ray tube. Consequently, in a tube used to radiograph thin or lightweight materials, a beryllium window is fused into the glass wall in the path of the x-ray beam. Beryllium is one of the lightest of metals and is more transparent to x-rays than any other metal. The beryllium-window tube has a minimum of inherent filtration and allows most of the very low energy x-rays to escape from the tube, as shown by the comparison with an ordinary oil-insulated glass tube in Fig. 19. The results are quite noticeable with both film and real-time radiography, particularly in contrast improvement. This effect is sometimes noticeable even beyond 150 kV. Very costly 250- or 300-kV tubes with beryllium windows are sometimes used when thin, light materials as well as thick, dense materials must be inspected with the same apparatus. In beryllium-window tubes designed for very low voltage work, the window can be as thin as 0.25 mm (0.010 in.). Consequently, to avoid rupture due to external pressure on the evacuated tube, thin beryllium windows are also of small diameter; this restricts the size of the x-ray beam and the corresponding field diameter.

In general, the higher the tube voltage, the larger the tube must be, and consequently the thicker the glass walls must be to support the internal elements and to withstand external atmospheric pressure. Furthermore, as the tube voltage increases, there is more likely to be high-voltage arc-

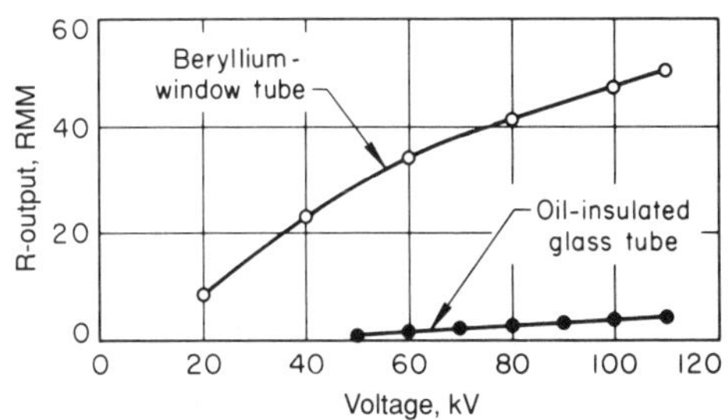

Fig. 19 Comparison of the R-output of a beryllium-window tube with that of an ordinary oil-insulated glass x-ray tube

ing from the tube to various parts of the housing. High-voltage tubes are generally surrounded by oil, which not only insulates the tube from the housing but also is often the primary means of extracting heat from the anode. In some tube assemblies, the oil surrounding the tube may be 50 mm (2 in.) thick or more. Both the oil surrounding the tube and the window in the housing that allows the useful beam of radiation to escape act as filters, particularly affecting the long-wavelength portion of the emitted radiation. Above about 200 kV, inherent filtration by components of the tube assembly is less important, primarily because inherent filtration by the testpiece would not allow the long-wavelength portion of the spectrum to reach the film anyway.

High-Energy X-Ray Sources

Above about 400 kV, the conventional design of an x-ray tube and its high-voltage iron-core transformer becomes cumbersome and large. Although x-ray machines with iron-core transformers have been built for 600 kV (maximum), there are no commercial versions operating above 500 kV. For higher-energy x-rays, other designs are used. Some of the machine designs for the production of high-energy x-rays include:

- Linear accelerators
- Betatrons
- Van de Graaff generators
- X-ray tubes with a resonant transformer

Linear accelerators, which produce high-velocity electrons by means of radio-frequency energy coupled to a waveguide, have extended industrial radiography to about 25-MeV photon energy. Betatrons, which accelerate electrons by magnetic induction, are used to produce 20- to 30-MeV x-rays. Portable linear accelerators and betatrons are also used in field inspection. The electron energy of portable units is of the order of 1.5 MeV for portable linear accelerators and 2 to 6 MeV for portable betatrons. Van de Graaff generators and x-ray tubes with resonant transformers are less widely used in industrial radiography. The Van de Graaff generator is an electrostatic device that operates from 500 kV to about 6 MV. X-ray tubes with resonant transformers were developed in the 1940s, and some units are still in operation. The output of these units is limited to about 4000 keV (4 MeV) in maximum photon energy that can be produced. Above these energies, the efficiency of resonant transformers is somewhat impaired. The radiation spectrum is broad, containing a large amount of radiation produced at much lower energies than the maximum. In addition, the accelerating potential fluctuates, which makes focusing of the electron beam difficult and creates a focal spot that is much larger than desired for optimum radiographic definition.

Table 2 Penetration ranges of x-ray tubes and high-energy sources

Maximum accelerating potential	Penetration range in steel, mm	in.
X-ray tubes		
150 kV	Up to 15	Up to 5/8
250 kV	Up to 40	Up to 1½
400 kV	Up to 65	Up to 2½
1000 kV (1 MV)	5–90	¼–3½
High-energy sources		
2.0 MeV	5–250	¼–10
4.5 MeV	25–300	1–12
7.5 MeV	60–460	2¼–18
20.0 MeV	75–610	3–24

In terms of penetrating capability, expressed as the range of steel thickness that can be satisfactorily inspected, Table 2 compares high-energy sources with conventional x-ray tubes. The maximum values in Table 2 represent the thicknesses of steel that can be routinely inspected using exposures of several minutes' duration and with medium-speed film. Thicker sections can be inspected at each value of potential by using faster films and long exposure times, but for routine work the use of higher-energy x-rays is more practical. Sections thinner than the minimum thicknesses given in Table 2 can be penetrated, but radiographic contrast is not optimum.

R-Output of High-Energy Sources. The R-output of a pulsed x-ray generator such as a linear accelerator or betatron depends on several factors, including pulse length and frequency of pulse repetition, although the output of x-ray generators of similar type and design varies approximately as $V^{2.7}$. For high-energy sources, electron beam current generally is not a satisfactory measure of R-output, and comparisons of beam current with the tube current of conventional x-ray tubes have little meaning. In fact, even the measurement of R-output is complicated and difficult, especially for energies exceeding 3 MeV. It is often more realistic to estimate the output of high-energy sources from film exposure data.

In terms of effective output, linear accelerators have the highest beam strength; machines with estimated outputs as high as 25 000 RMM have been built. Betatrons, although capable of producing higher-energy radiation than linear accelerators, are limited in beam strength to about 500 RMM. By contrast, 200 RMM is about the maximum output of conventional rectified, Villard-circuit or constant-potential equipment, of resonant-transformer-based machines, or of machines powered by an electrostatic Van de Graaff generator. Portable betatrons and linear accelerators have an R-output in the range of 1 to 3 RMM.

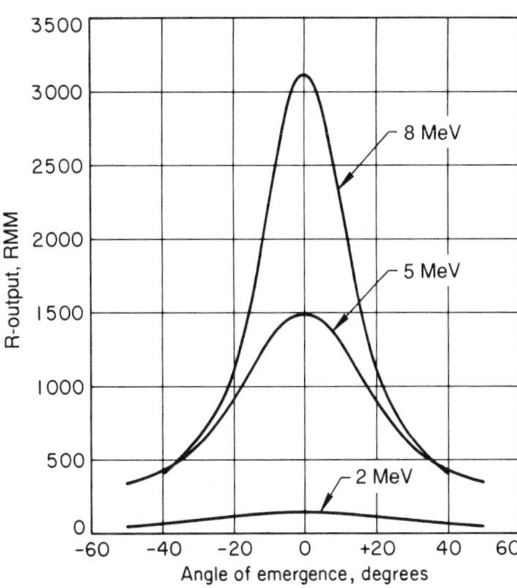

Fig. 20 Variation of R-output with angle of emergence relative to the central beam of a high-energy source. Curves were calculated for x-rays emitted at three levels of electron energy from back side of tungsten target, assuming electron impingement normal to target surface.

With most high-energy sources (in which the x-ray beam is emitted from the back side of the target in approximately the same direction of travel as the impinging stream of electrons), the combined effects of self-absorption and electron scattering within the target itself produce a variation in intensity with angle relative to the axis of the central beam that parallels the heel effect in conventional x-ray tubes. The magnitude of the effect varies with electron energy and is extremely pronounced at energies exceeding 2 MeV. The variation of R-output with angle of emergence, shown in Fig. 20 for three values of electron energy, is of considerable practical importance because this effect can severely limit the area of coverage, especially for short source-to-detector distances. The calculated angles between the central beam and the focus of points of half intensity (that is, where the intensity is one-half that of the central beam), as well as the corresponding field diameters at 1 m (3.3 ft), are given in Table 3 for several values of electron energy, assuming normal incidence of the electrons on the target.

As Table 3 indicates, the useful field produced by a parallel stream of electrons with energy above 10 MeV is inconveniently small for most applications. Fortunately,

Table 3 Half intensity and field diameter at various electron energy values

Electron energy, MeV	Angle to half intensity, degrees	Field diameter at 1 m (3.3 ft) mm	in.
5	11	380	15.0
10	7	244	9.6
18	5	174	6.9
31	3.4	119	4.7
50	1.9	66	2.6

Table 4 Characteristics of γ-ray sources used in industrial radiography

γ-ray source	Half-life	Photon energy, MeV	Radiation output, RHM/Ci(a)	Penetrating power, mm (in.) of steel
Thulium-170	128 days	0.054 and 0.084(b)	0.003	13 (½)
Iridium-192	74 days	12 rays from 0.21–0.61	0.48	75 (3)
Cesium-137	33 years	0.66	0.32	75 (3)
Cobalt-60	5.3 years	1.17 and 1.33	1.3	230 (9)

(a) Output for typical unshielded, encapsulated sources; RHM/Ci, roentgens per hour at 1 m per curie. (b) Against strong background of higher-MeV radiation. (c) Derived primarily from radioactive decay of daughter products

many high-energy sources employ some form of device to focus the electron beam on the target so that a converging beam rather than a parallel beam of electrons strikes the target. Originally intended to achieve a reduced focal-spot size, beam convergence has the secondary effect of increasing the angle to half intensity, thus increasing the effective field at any focal distance. For most high-energy sources, the effect of electron beam convergence is to at least double the field diameter.

In addition to electron beam convergence, almost all high-energy sources are equipped with a field compensator, which further increases the effective field size. A field compensator is a metallic filter that is thicker in the center than it is at the edges. By progressively absorbing less radiation with increasing angle of emergence relative to the central beam, a field compensator can almost double the effective field size. The intensity of the entire beam of radiation is reduced by a field compensator, but even a reduction in central beam intensity of about 35% does not seriously increase film exposure times for most applications. In fact, the increase in effective field size usually more than compensates for the loss of intensity.

There is one type of application in which the inherent variation in intensity with angle of emergence is beneficial. A solid sphere or solid cylinder varies continuously in thickness across a diametral plane. When incident radiation from a uniform-intensity source penetrates a solid sphere or cylinder parallel to a diametral plane, there is a progressive variation in transmitted intensity, because of thickness variations, which leads to underexposure of the central area of the radiograph and overexposure along the edges, often resulting in unsatisfactory definition. With a nonuniform radiation field, however, inherent variations in penetrated thickness are at least partly counterbalanced by variations in incident intensity across the radiation field, which results in a much better resolution of testpiece detail.

Gamma-Ray Sources

Gamma rays are high-energy electromagnetic waves of relatively short wavelength that are emitted during the radioactive decay of both naturally occurring and artificially produced unstable isotopes. In all respects other than their origin, γ-rays and x-rays are identical. Unlike the broad-spectrum radiation produced by an x-ray tube, γ-ray sources emit one or more discrete wavelengths of radiation, each having its own characteristic photon energy.

The two most common radioactive isotopes used in radiography are iridium-192 and cobalt-60. Ytterbium-169 has also gained a measure of acceptability in the radiography of thin materials and small tubes, especially boiler tubes in power plants. Cesium-137, although not used in radiography, has been useful in radiometry and computed tomography.

Radiation Characteristics of γ-Ray Sources. Because gamma radiation is produced by the radioactive decay of unstable atomic nuclei, there is a continuous reduction in the intensity of emitted radiation with time as more and more unstable nuclei transform to stable nuclei. This reduction follows a logarithmic law, and each radioactive isotope has a characteristic half-life, or amount of time that it takes for the intensity of emitted radiation to be reduced by one-half. The term half-life should not be misinterpreted as meaning that the intensity of emitted radiation will be zero at the end of the second half-life. Rather, the intensity remaining at the end of a second half-life will be one-half the intensity at the end of the first half-life, or one-fourth of the initial intensity. The intensity at the end of the third half-life will be one-eighth of the initial intensity, and so on.

As described in the section "Radiation Sources" in this article, the strength of γ-ray sources is the source activity given in the units of gigabecquerel or curies. If the source activity is known at any given time, a table of radioactive-decay factors for the specific radioactive isotope (or the corresponding logarithmic decay formula) can be used to calculate source activity at any subsequent time. Then, radiation intensity, which is usually measured in roentgens per hour, can be found by multiplying source activity by the specific γ-ray constant given in roentgens per hour per gigabecquerel (or curie). Although radiation output is constant for a given isotope, large-size sources emit slightly less radiation per gigabecquerel (curie) than small-size sources; that is, large-size sources have somewhat lower effective output. The difference between calculated radiation output and effective radiation output is the result of self-absorption within the radioactive mass, in which γ-rays resulting from decay of atoms in the center of the mass are partly absorbed by the mass itself before they can escape. Most source manufacturers measure source activity in units of effective gigabecquerel (or curies), which is the net of self-absorption.

Specific activity, commonly expressed as gigabecquerels (curies) per gram or gigabecquerels (curies) per cubic centimeter, is a characteristic of radioactive sources that expresses their degree of concentration. A source that has a high specific activity will be smaller than another source of the same isotope and activity that has a lower specific activity. Generally, sources of high specific activity are more desirable because they have lower self-absorption and provide less geometric unsharpness in the radiographs they produce than sources of low specific activity.

The more important characteristics of γ-ray sources are summarized in Table 4. Source activity and specific activity are not listed, because these characteristics vary according to the physical size of the source, the material and design of its encapsulation, and the degree of concentration at the time the source was originally produced.

Physical Characteristics of γ-Ray Sources. Radioactive sources, which are usually metallic but may be salts or gases adsorbed on a block of charcoal, are encapsulated in a protective covering, usually a thin stainless steel sheath or a thicker sheath of aluminum. Encapsulation prevents abrasion, spillage, or leakage of the radioactive material; reduces the likelihood of loss or accidental mishandling; and provides a convenient means by which wires, rods, or other handling devices can be attached to the source.

Gamma-ray sources are housed in protective containers made of lead, depleted uranium, or other dense materials that absorb the γ-rays, thus providing protection from exposure to the radiation. Two basic types of containers are generally used. One type incorporates an inner rotating cylinder that contains the isotope and is rotated toward a conical opening in the outer cylinder to expose the source and allow the escape of radiation. This type is often referred to as a radioisotope camera. The second type incorporates a remote-controlled mechanical or pneumatic positioner that moves the encapsulated radioactive source out of the container and into a predetermined position where it remains until exposure is completed; the source is then returned to the protective container, again by remote control. This type is more versatile than the first type and is much more extensively used. Remote control of the positioner allows the operator to remain at a safe distance from the source while manipulating the capsule out of and into the protective container.

Attenuation of Electromagnetic Radiation

X-rays and γ-rays interact with any substance, even gases such as air, as the rays pass through the substance. It is this interaction that enables parts to be inspected by differential attenuation of radiation and that enables differences in the intensity of radiation to be detected and recorded. Both these effects are essential to the radiographic process. The attenuation characteristics of materials vary with the type, intensity, and energy of the radiation and with the density and atomic structure of the material.

The attenuation of electromagnetic radiation is a complex process. Because of their electromagnetic properties, x-rays and γ-rays are affected by the electric fields surrounding atoms and their nuclei. It is chiefly interaction with these electric fields that causes the intensity of electromagnetic radiation to be reduced as it passes through any material.

The intensity of radiation varies exponentially with the thickness of homogeneous material through which it passes. This behavior is expressed as:

$$I = I_0 \exp(-\mu t) \qquad \text{(Eq 6)}$$

where I is the intensity of the emergent radiation, I_0 is the initial intensity, t is the thickness of homogeneous material, and μ is a characteristic of the material known as the linear absorption coefficient. The coefficient μ is constant for a given situation, but varies with the material and with the photon energy of the radiation. The units of μ are reciprocal length (for example, cm^{-1}).

The absorption coefficient of a material is sometimes expressed as a mass absorption coefficient (μ/ρ), where ρ is the density of the material. Alternatively, the absorption coefficient can be expressed as an effective absorbing area of a single atom; this property, called the atomic absorption coefficient (μ_a) or cross section, is equal to the linear absorption coefficient divided by the number of atoms per unit volume. The cross section, usually expressed in barns (1 barn $= 10^{-24}$ cm^2), indicates the probability that a photon of radiation will interact with a given atom.

Atomic Attenuation Processes

Theoretically, there are four possible interactions between a photon (quantum) of electromagnetic radiation and material. Also, there are three possible results that an interaction can have on the photon. Thus, there are 12 possible combinations of interaction and result. However, only four of these have a high enough probability of occurrence to be important in the attenuation of x-rays or γ-rays. These four combinations are photoelectric effect, Rayleigh scattering, Compton scattering, and pair production.

The photoelectric effect is an interaction with orbital electrons in which a photon of electromagnetic radiation is consumed in breaking the bond between an orbital electron and its atom. Energy in excess of the bond strength imparts kinetic energy to the electron.

The photoelectric effect generally decreases with increasing photon energy, E, as $E^{-3.5}$, except that, at energies corresponding to the binding energies of electrons to the various orbital shells in the atom, there are abrupt increases in absorption. These abrupt increases are called absorption edges and are given letter designations corresponding to the electron shells with which they are associated. At photon energies exceeding an absorption edge, the photoelectric effect again diminishes with increasing energy.

For elements of low atomic number, the photoelectric effect is negligible at photon energies exceeding about 100 keV. However, the photoelectric effect varies with the fourth to fifth power of atomic number; thus, for elements of high atomic number, the effect accounts for an appreciable portion of total absorption at photon energies up to about 2 MeV.

Rayleigh scattering, also known as coherent scattering, is a form of direct interaction between an incident photon and orbital electrons of an atom in which the photon is deflected without any change in either the kinetic energy of the photon or the internal energy of the atom. In addition, no electrons are released from the atom. The angle between the path of the deflected photon and that of the incident radiation varies inversely with photon energy, being high for low-energy photons and low for high-energy photons. There is a characteristic photon energy, which varies with atomic number, above which Rayleigh scattering is entirely in the forward direction and no attenuation of the incident beam can be detected.

Rayleigh scattering is most important for elements of high atomic number and low photon energies. However, Rayleigh scattering never accounts for more than about 20% of the total attenuation.

Compton scattering is a form of direct interaction between an incident photon and an orbital electron in which the electron is ejected from the atom and only a portion of the kinetic energy of the photon is consumed. The photon is scattered incoherently, emerging in a direction that is different from the direction of incident radiation and emerging with reduced energy and a correspondingly lower wavelength. The relationship of the intensity of the scattered beam to the intensity of the incident beam, scattering angle, and photon energy in the incident beam is complex, yet is amenable to theoretical evaluation. Compton scattering varies directly with the atomic number of the scattering element and varies approximately inversely with the photon energy in the energy range of major interest.

Pair production is an absorption process that creates two 0.5-MeV photons of scattered radiation for each photon of high-energy incident radiation consumed; a small amount of scattered radiation of lower energy also accompanies pair production. Pair production is more important for heavier elements; the effect varies with atomic number, Z, approximately as $Z(Z + 1)$. The effect also varies approximately logarithmically with photon energy.

In pair production, a photon of incident electromagnetic radiation is consumed in creating an electron-positron pair that is then ejected from an atom. This effect is possible only at photon energies exceeding 1.02 MeV because, according to the theory of relativity, 0.51 MeV is consumed in the creation of the mass of each particle, electron, or positron. Any energy of the incident photon exceeding 1.02 MeV imparts kinetic energy to the pair of particles.

The positron created by pair production is destroyed by interaction with another electron after a very short life. This destruction produces electromagnetic radiation, mainly in the form of two photons that travel in opposite directions, each photon having an energy of about 0.5 MeV. Most of the electrons created by pair production are absorbed by the material, producing bremsstrahlung of energy below 0.5 MeV.

Total absorption is the sum of the absorption or scattering effects of the four processes. The atomic absorption coefficient can be expressed as:

$$\mu_a = \sigma_{pe} + \sigma_R + \sigma_C + \sigma_{pr} \qquad \text{(Eq 7)}$$

where σ_{pe} is the attenuation due to the photoelectric effect, σ_R is that due to Rayleigh scattering, σ_C is that due to Compton scattering, and σ_{pr} is that due to pair production.

From calculated atomic absorption coefficients, the mass absorption coefficient can be determined by multiplying μ_a by

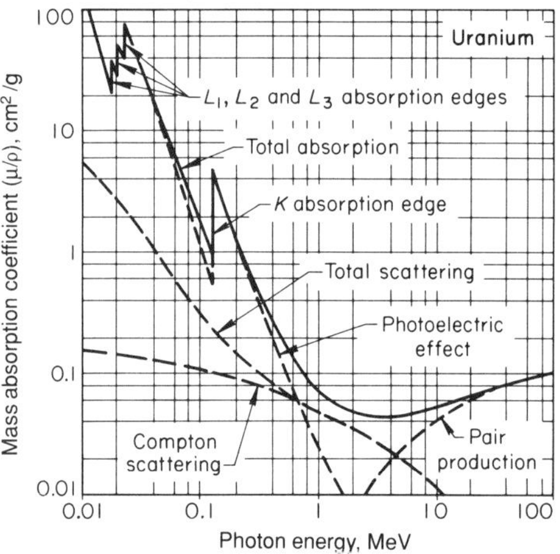

Fig. 21 Calculated mass absorption coefficient for uranium as a function of photon energy (solid line) and contributions of various atomic processes (dashed lines). Rayleigh scattering is the difference between total scattering and Compton scattering.

Avogadro's number, N, and dividing by the atomic weight of the element, A:

$$\frac{\mu}{\rho} = \frac{\mu_a N}{A} \quad \text{(Eq 8)}$$

Figure 21 shows the variation of the mass absorption coefficient for uranium as a function of photon energy and indicates the amount of total absorption that is attributable to the various atomic processes over the entire range of 10 keV to 100 MeV. Tables of cross sections and absorption coefficients for 40 of the more usually encountered elements at various photon energies from 0.01 to 30 MeV are presented in Ref 1. Absorption coefficients can be evaluated for alloys, compounds, and mixtures by calculating a weighted-average mass absorption coefficient in proportion to the relative abundance of the constituent elements by weight.

Effective Absorption of X-Rays

The preceding discussion of mass absorption and atomic absorption characteristics is based on the assumption of so-called narrow-beam geometry, in which any photon that is scattered, no matter how small the angle, is considered absorbed. Experimentally, narrow-beam geometry is approximated by interposing a strong absorber such as lead between the material being evaluated and the radiation detector, then allowing the beam of attenuated radiation to pass through a small hole in the absorber to impinge on the detector. Only total absorption can be determined experimentally; however, it has been shown that the sum of calculated absorption coefficients for the various atomic processes correlates closely with experimentally determined narrow-beam absorption coefficients for a wide variety of materials and photon energies.

Broad-Beam Absorption. The evaluation of absorption coefficients under narrow-beam geometry is mainly a convenience that simplifies theoretical calculations. Whenever there is a measurable width to the radiation beam, narrow-beam conditions no longer exist, and absorption must be evaluated under so-called broad-beam geometry. Under broad-beam geometry, photons that are scattered forward at small angles are not lost, but increase the measured intensity of the attenuated beam. Therefore, the broad-beam absorption coefficient of any material is less than the narrow-beam absorption coefficient.

There are many degrees of broad-beam geometry, depending on the portion of the attenuated-beam intensity that is comprised of scattered radiation. However, broad-beam absorption characteristics are usually determined in such a manner that variations in the areas of either the radiation beam or the testpiece have no effect on the intensity of the transmitted radiation. Therefore, broad-beam absorption coefficients are useful because the ordinary radiographic arrangement (in which the film or screen is placed in close proximity to the testpiece) effectively duplicates this condition.

Effect of Radiation Spectrum. The theoretical assessment of absorption characteristics is based on an assumption of monoenergetic radiation (radiation having a single wavelength). As Fig. 11 shows, the actual output of an x-ray tube is not monoenergetic; for example, 200-kV x-rays have an upper limit of 200 keV of photon energy and are comprised of a broad spectrum of x-rays having lower photon energies. Even if an incident beam were monoenergetic, it would soon contain x-rays of lower photon energies because of Compton scattering. Therefore, in addition to the effect of broad-beam geometry, the effective absorption coefficients of materials are further modified by the multienergetic character of impinging radiation. The correction of calculated absorption coefficients for the known wavelength distribution of a beam of x-rays is time consuming but not impossible.

Empirical values of absorption coefficients based on broad-beam experiments are the values that are most useful in exposure calculations, because these values are corrected for both broad-beam geometry and the variation of intensity with photon energy in the incident radiation. Published values of calculated absorption coefficient are based on narrow-beam geometry. If these values are used to estimate radiographic exposures, the estimates may be inaccurate, especially for low-to-medium energy radiation from an x-ray tube and for relatively thick testpieces.

Scattering Factor. The ratio of the intensity of scattered radiation, I_S, to the intensity of direct radiation, I_D, impinging on the detector is known as the scattering factor and is important in determining image quality. Scattered radiation does not reveal characteristics of the testpiece; it only reduces radiographic contrast, obscuring detail and reducing radiographic sensitivity. The scattering factor is a function of photon energy and the thickness, density, and material type (μ/ρ) of the testpiece. Figure 22 illustrates an example of experimental values for the build-up factor ($1 + I_S/I_D$) with steel.

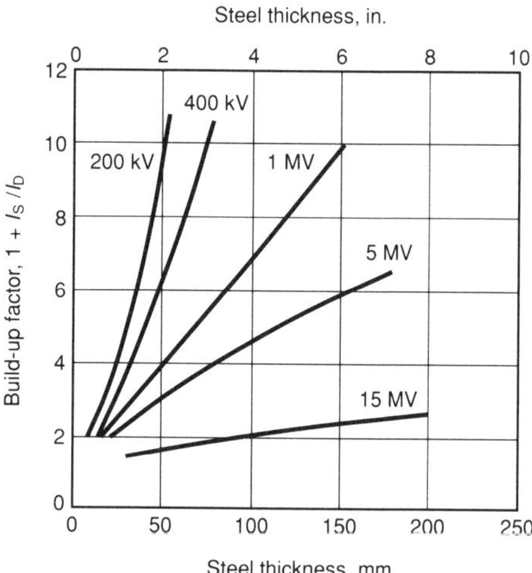

Fig. 22 Experimental values of build-up factor for different thicknesses of steel and different x-ray energies. Source: Ref 2

Because radiation of low photon energy is scattered more than radiation of high photon energy, it is desirable to use relatively high-energy radiation for optimum radiographic subject contrast. This can be accomplished by either raising the voltage of the x-ray tube, which moves the entire bremsstrahlung spectrum to higher photon energies, or by interposing a filter between the source and the testpiece, which effectively raises photon energy by selectively absorbing or scattering the long wavelength portion of the spectrum. The optimum choice of filter material and thickness varies with different combinations of testpiece material and thickness, film or detector type, and bremsstrahlung spectrum as determined by the type of x-ray tube and tube voltage used. Also, the optimum combinations of filter material and thickness, and tube current and exposure time, must be evaluated on the basis of a compromise between image quality and costs. Generally, filters made of metals having high atomic numbers (lead, for example) produce a cleaner image because of a reduction in the amount of scattered radiation. However, such filters also cause a large reduction in the intensity of the radiation beam, which leads to greater expense because of the need for longer exposure times with films or more sensitive detector systems in real-time radiography. Also, because long wavelength radiation is absorbed preferentially, there is a reduction in the ability to image very small flaws.

Table 5 Approximate radiographic absorption equivalence for various metals

Materials	X-rays, keV					X-rays, MeV			γ-rays			
	50	100	150	220	450	1	2	4 to 25	Ir-192	Cs-137	Co-60	Ra
Magnesium	0.6	0.6	0.05	0.08	...	...	...	...	...	...	...	...
Aluminum	1.0	1.0	0.12	0.18	...	...	...	...	0.35	0.35	0.35	0.40
Aluminum alloy 2024	2.2	1.6	0.16	0.22	...	...	...	...	0.35	0.35	0.35	...
Titanium	...	...	0.45	0.35	...	...	...	...	...	...	...	...
Steel	...	12.0	1.0	1.0	1.0	1.0	1.0	1.0	1.0	1.0	1.0	1.0
18-8 stainless steel	...	12.0	1.0	1.0	1.0	1.0	1.0	1.0	1.0	1.0	1.0	1.0
Copper	...	18.0	1.6	1.4	1.4	...	...	1.3	1.1	1.1	1.1	1.1
Zinc	...	...	1.4	1.3	1.3	...	...	1.2	1.1	1.0	1.0	1.0
Brass(a)	...	...	1.4	1.3	1.3	1.2	1.2	1.2	1.1	1.1	1.1	1.1
Inconel alloys	...	16.0	1.4	1.3	1.3	1.3	1.3	1.3	1.3	1.3	1.3	1.3
Zirconium	...	...	2.3	2.0	...	1.0	...	...	...	...	...	...
Lead	...	...	14.0	12.0	...	5.0	2.5	3.0	4.0	3.2	2.3	2.0
Uranium	...	...	...	25.0	...	...	...	3.9	12.6	5.6	3.4	...

(a) Containing no tin or lead; absorption equivalence is greater than these values when either element is present.

Raising the voltage of the x-ray tube, which is sometimes restricted by equipment limitations, does not necessarily involve a compromise between image quality and costs. In fact, high-voltage x-ray equipment is frequently capable of producing higher-quality radiographs at reduced cost. Image quality is improved primarily by reducing the proportion of the image background that is directly attributable to scattered radiation. For example, 80% or more of the photographic density of a given film may be attributed to non-image-forming scattered radiation when low-voltage (soft) x-rays are employed. On the other hand, for a 100 mm (4 in.) thick steel testpiece, only about 40% of the photographic density is caused by scattered radiation when the incident beam is 1.3-MeV radiation and only about 15% with 20-MeV x-rays.

Radiographic Equivalence. The absorption of x-rays and γ-rays by various materials becomes less dependent on composition as radiation energy increases. For example, at 150 keV, 25 mm (1 in.) of lead is equivalent to 350 mm (14 in.) of steel, but at 1000 keV, 25 mm (1 in.) of lead is equivalent to only 125 mm (5 in.) of steel. Approximate radiographic absorption equivalence factors for several metals are given in Table 5. When film exposure charts are available only for certain common materials (such as steel or aluminum), exposure times for other materials can be estimated by determining the film exposure time for equal thickness of a common material from the chart, then multiplying by the radiographic equivalence factor for the actual testpiece material (see the section "Exposure Factors" in this article for detailed information on radiographic exposure).

Principles of Shadow Formation

The image formed on a radiograph is similar to the shadow cast on a screen by an opaque object placed in a beam of light. Although the radiation used in radiography penetrates an opaque object whereas light does not, the geometric laws of shadow formation are basically the same. X-rays, γ-rays, and light all travel in straight lines. Straight-line propagation is the chief characteristic of radiation that enables the formation of a sharply discernible shadow. The geometric relationships among source, object, and screen determine the three main characteristics of the shadow—the degrees of enlargement, distortion, and unsharpness (Fig. 23). This theoretical unsharpness is added to the scattering from the atomic attenuation processes (photoelectric effect, Rayleigh and Compton scattering, and pair production, as mentioned in the previous section) to obtain the total unsharpness and blurring of the image.

Enlargement. The shadow of the object (testpiece) is always farther from the source than the object itself. Therefore, as illustrated for a point source in Fig. 23(a), the dimensions of the shadow are always greater than the corresponding dimensions of the object. Mathematically, the size of the image or degree of enlargement can be calculated from the relationship:

$$M = \frac{S_i}{S_o} = \frac{L_i}{L_o} \quad \text{(Eq 9)}$$

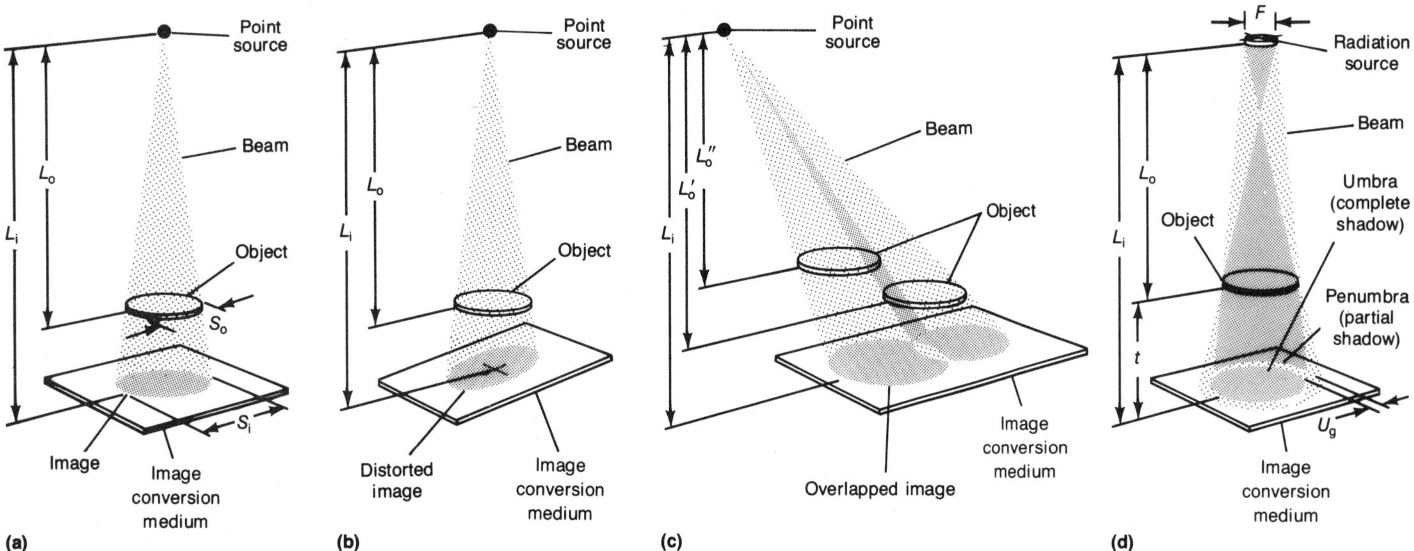

Fig. 23 Schematic of the effect of geometric relationships on image formation with point sources and a radiation source of finite size. (a) Image enlargement, (b) image distortion, and (c) image overlap for point sources of radiation. (d) Degree of image unsharpness from a radiation source of finite size. L_o, source-to-object distance; L_i, source-to-image distance; S_o, size of object; S_i, size of image; U_g, geometric unsharpness; F, size of focal spot; t, object-to-image distance

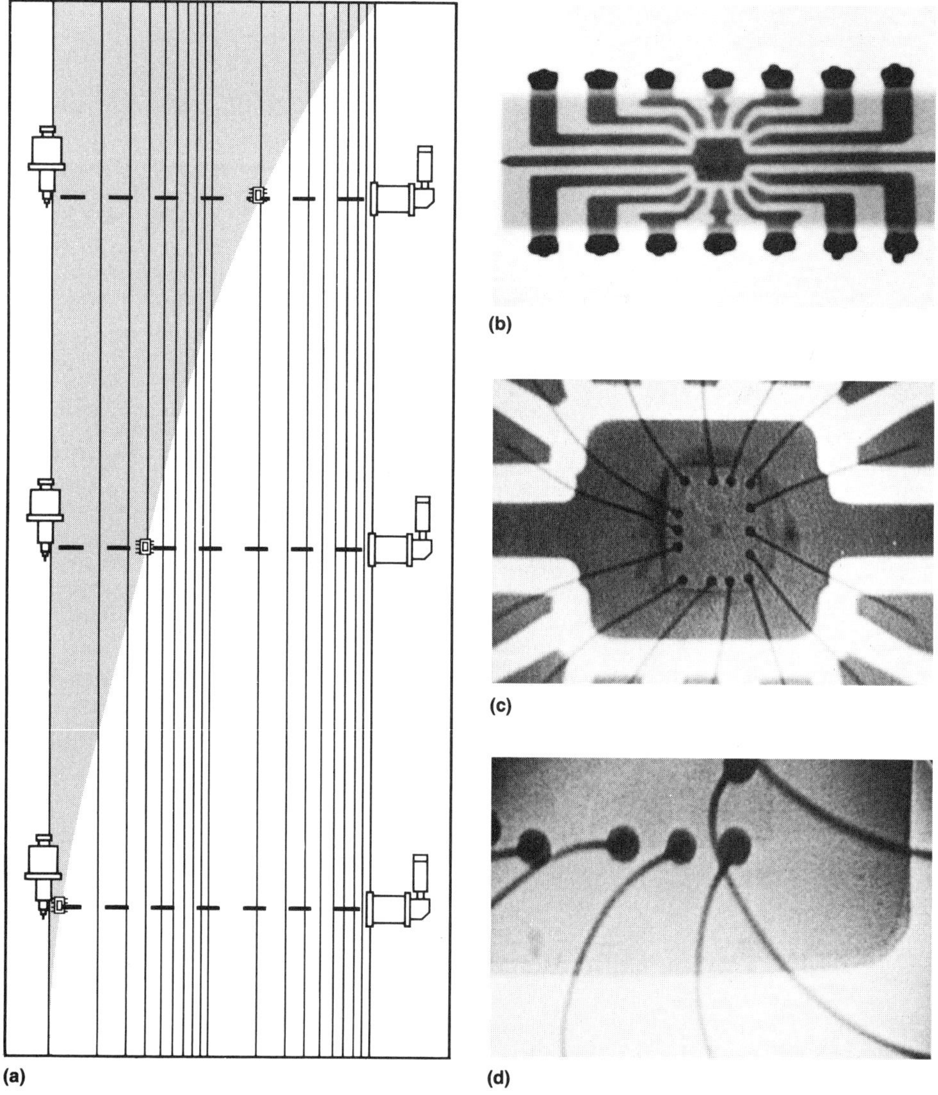

Fig. 24 Geometric magnification of radiographic images with a microfocus x-ray tube. (a) Greater geometric magnification and unsharpness occur when the testpiece is moved away from the detector and toward the radiation source. With a microfocus x-ray tube, however, the amount of geometric unsharpness (Eq 11) is minimized, as shown by an integrated circuit magnified at 1.25× (b), 20× (c), and 40× (d). Courtesy of Rockwell International

where M is the degree of enlargement (magnification), S_i is the size of the image, S_o is the size of the object, L_i is the source-to-film or source-to-detector distance, and L_o is the source-to-object distance.

Variations in the position of a given object relative to the source and recording surface affect image size. For example, when the source-to-image distance, L_i, is decreased without changing the distance of the object from the image, the size of the image, S_i, is increased. Alternatively, when the source-to-object distance, L_o, is increased without changing the distance of the recording surface from the source (L_i constant), the size of the image is decreased.

Film Radiography. The effect of enlargement is normally of little consequence in film and paper radiography, mainly because the recording medium is placed close behind the testpiece to minimize geometric unsharpness. Even with this arrangement, however, images of portions of the testpiece farthest from the recording plane will be larger than corresponding portions of the testpiece itself. This effect is greatest for short source-to-image distances.

Microradiography and Real-Time Radiography. There are certain circumstances when enlargement is advantageous. In real-time radiography, for example, the significance of screen unsharpness can be minimized by positioning a testpiece closer to the source. Details that are otherwise invisible in a radiograph can also be enlarged enough to become visible. This technique has been used for microradiography, in which considerable enlargement renders minute details visible. Because of geometric unsharpness, however, only sources with extremely small focal spots can be used to produce enlarged radiographs. For example, focal spots approximately 5 μm (0.2 mil) in diameter are used for microradiography (Fig. 24). One shortcoming of this technique may be a low level of radiation intensity, which could limit applications to the inspection of low-density materials. Nevertheless, real-time microradiography has found application in the inspection of high-quality castings and high-integrity weldments.

Reduction of Scattering. Another advantage of placing the testpiece closer to the source is a reduction of secondary (scattered) radiation. When the magnification factor, M, in Eq 9 is greater than 3, the build-up factor, $1 + I_S/I_D$, approaches 1 (Ref 2).

Distortion. As long as the plane of a two-dimensional object and the plane of the imaging surface are parallel to each other, the image will be undistorted regardless of the angle at which the beam of radiation impinges on the object. In addition, the degree of enlargement at different points in a given image is constant because the ratio L_i/L_o is invariant. However, as shown in Fig. 23(b), if the plane of the object and the plane of the recording surface are not parallel, the image will be distorted because the degree of enlargement is different between rays passing through any two points on the surface of the object. This is the main reason radiographic images almost always present a distorted picture of testpiece features; the shape of most testpieces is such that one or more of its features is not parallel to the imaging surface. The degree of distortion is directly related to the degree of nonparallelism; small amounts of nonparallelism produce small degrees of distortion, and large amounts produce large degrees of distortion.

Although there is no distortion of images of features that are parallel to the imaging surface, spatial relationships among testpiece features are distorted, with the amount of distortion being directly related to the cosine of the angle of the beam to the imaging surface. For example, two circular features that are parallel to the imaging surface but at different object-to-detector distances may produce two separate circular images, or one noncircular image because two images overlap, as shown in Fig. 23(c), depending on the direction of the radiation beam. In most actual radiographs, regions where images of different features overlap will appear as regions of greater or lesser contrast than adjacent regions where the images do not overlap, depending on whether the features absorb less radiation or more radiation than portions of the testpiece surrounding the features. For example, overlapping images of voids or cavities will appear darker on film or brighter on fluorescent screens in regions of overlap because less radiation is absorbed along paths intersecting overlapping voids than along paths that intersect only one void.

Geometric Unsharpness. In reality, any radiation source is too large to be approximated by a point. Conventional x-ray tubes have focal spots between 2 by 2 mm (0.08 by 0.08 in.) and 5 by 5 mm (0.2 by 0.2 in.) in size, while microfocus x-ray systems have focal-spot sizes as small as 5 μm (0.2 mil). Even high-energy sources have focal spots of appreciable size, although seldom exceeding 2 mm (0.08 in.) in diameter. Gamma-ray sources vary widely in size, depending on source strength and specific activity, but are seldom less than about 2.5 mm (0.1 in.) in diameter.

Radiographic definition varies according to the geometric relationships among source size, source-to-object distance, and object-to-image distance. When radiation from a source of any finite size produces a shadow, that portion of the image that is in shadow for radiation emanating from all points on the surface of the source is a region of complete shadow known as the umbra. Portions of the image that are in shadow for radiation emanating from any portion of the source, but not in shadow for radiation from some portion, are regions of partial shadow known as the penumbra. The degree of geometric unsharpness is equal to the width of the penumbra. Mathematically, the geometric unsharpness, U_g, is determined from the laws of similar triangles, as illustrated in Fig. 23(d), and can be expressed as:

$$U_g = \frac{F \cdot t}{L_o} \qquad \text{(Eq 10)}$$

where F is the size of the focal spot or γ-ray source, t is the object-to-image distance, and L_o is the source-to-object distance. In most cases, t is considered to be the difference between the source-to-image distance, L_i, and the source-to-object distance, L_o; therefore, Eq 10 can be alternatively expressed as:

$$U_g = \frac{F(L_i - L_o)}{L_o} = F(M - 1) \qquad \text{(Eq 11)}$$

The size of the penumbra, or the geometric unsharpness, can be reduced by lengthening the source-to-object distance, by reducing the size of the focal spot, or by reducing the object-to-image distance. In practice, the source size is determined by the characteristics of a given x-ray tube or by the physical dimensions of a radioactive pellet, and the object-to-image distance is minimized by placing the surface of the imaging medium as close as possible to the testpiece. Therefore, the only variable over which the radiographer has any appreciable amount of control is source-to-object distance. However, as discussed in the following section in this article, this distance is a significant variable affecting the intensity of the penetrating radiation. Consequently, the radiographer often must compromise

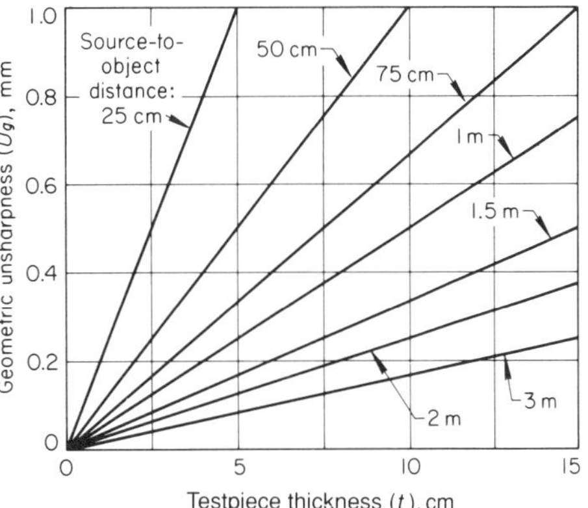

Fig. 25 Relation of geometric unsharpness to testpiece thickness for various source-to-object distances when the source is 5 mm (0.2 in.) in diameter

between maximum definition (minimum unsharpness) and the cost associated with the lower radiation intensity that occurs when the source-to-object distance is increased. Higher costs in film radiography occur with a lower intensity of radiation because exposure time must be increased. In real-time radiography, costs may be less of a factor, but increases in detector efficiency or image-processing time might be required.

Geometric unsharpness is one of several unsharpness factors and is usually the largest contributor to unsharpness in film radiography. When the distance between the actual surface of the imaging medium and the adjacent (facing) surface of the testpiece is neglected (which is an appropriate assumption in film radiography because this distance is usually quite small in relation to testpiece thickness), the geometric unsharpness can be calculated for any source size and can be expressed as a series of straight-line plots relating geometric unsharpness, U_g, to testpiece thickness, t, for various values of the source-to-object distance, L_o. A typical series is shown in Fig. 25 for a 5 mm (0.2 in.) diam source. It is often helpful to prepare graphs such as the one in Fig. 25 for each source size used.

In applications where the maximum unsharpness must be kept below a specific known value to resolve certain types and sizes of flaws, the radiographer can determine the minimum source-to-object distance for a given part from Eq 10. At this value of L_o, the types and sizes of flaws to which the specified value of U_g applies will be resolved if the flaws are in the plane of the testpiece farthest from the imaging surface. Flaws in planes closer to the imaging surface will be even more clearly resolved.

Shadow Intensity and the Inverse-Square Law. The intensity of radiation that penetrates a testpiece and produces a satisfactory image is governed by the energy and spectral quality of the incident radiation, the strength of the radiation, and the source-to-detector distance. In practice, the energy of the incident radiation, which depends mainly on the tube voltage of an x-ray machine or on the radioactive isotope in a γ-ray source, is chosen to be sufficiently penetrating for the type of material and thickness to be inspected. With this factor fixed, the remaining interrelated factors—the source strength (determined by tube current in milliamperes for x-ray sources or by source activity in becquerels, or curies, for γ-ray sources) and the source-to-detector distance—determine the intensity of radiation impinging on the detector.

The intensity of the shadow image determines the amount of exposure time for a given (fast or slow) film or the detector efficiency required in real-time radiography. Similar to visible light, x-rays or γ-rays diverge when emitted from the radiation source and cover an increasingly larger area, with lessened intensity, as the distance from the source increases. The intensity, or amount of radiation falling on a unit area per unit time, varies inversely with the square of the distance from the source. This can be expressed mathematically as:

$$IL^2 = \text{constant} \qquad \text{(Eq 12)}$$

where I is the intensity of radiation at a given distance, L, from a source of given strength. More often, the inverse-square law is expressed as a ratio:

$$\frac{I_1}{I_2} = \frac{L_2^2}{L_1^2} \qquad \text{(Eq 13)}$$

where the subscripts 1 and 2 refer to different points along a line radiating from the source. Because of this inherent characteristic of radiation, if the radiation has a certain intensity at 1 m (3.3 ft) from the source, it will have four times that intensity at 0.5 m (1.65 ft) but only one-fourth that intensity at 2 m (6.6 ft) and only one-ninth that intensity at 3 m (9.9 ft).

Scattered Radiation. When a beam of x-rays or γ-rays strikes a testpiece, secondary radiation is scattered in all directions. This causes a haze over all or part of the image, thus reducing contrast and visibility of detail. Methods for reducing scattered radiation are discussed in the section "Control of Scattered Radiation" in this article.

Internal scatter occurs within the testpiece being radiographed (Fig. 4a). Internal scatter is fairly uniform throughout a testpiece of uniform thickness, but adversely affects definition by diffusing or obscuring the outline of the image.

Side scatter consists of secondary radiation generated by walls, internal holes in the testpiece (as shown in Fig. 4b), or objects near the testpiece. This radiation may pass through the testpiece and may obscure the image of the testpiece.

Back scatter consists of rays that are generated from the floor or table (Fig. 4c) or from any other object that is located on the opposite side of the imaging surface from the testpiece and source. Back scatter increases the background noise and therefore reduces contrast and definition in the recorded image.

Image Conversion Media

After penetrating the testpiece, the x-rays or γ-rays pass through a medium on the imaging surface. This medium may be a recording medium (such as film) or a medium that responds to the intensity of the radiation (such as fluorescent screen or a scintillation crystal in a discrete detector). These two types of media provide the images for subsequent observation.

Radiography can also utilize radiographic screens, which are pressed into intimate contact with the imaging medium. Radiographic screens include:

- Metallic screens placed over film, paper, or the screens used in real-time systems
- Fluorescent screens placed over film or photographic paper
- Fluorometallic screens placed over film

These screens are used to improve radiographic contrast by intensifying the conversion of radiation and by filtering the lower-energy radiation produced by scattering.

Recording Media

By definition, a radiographic recording medium provides a permanent visible image of the shadow produced by the penetrating radiation. Permanent images are recorded on x-ray film, radiographic paper, or electrostatically sensitive paper such as that used in the xeroradiographic process.

X-ray film is constructed of a thin, transparent plastic support called a film base, which is usually coated on both sides (but occasionally on one side only) with an emulsion consisting mainly of grains of silver salts dispersed in gelatin (Fig. 26). These salts are very sensitive to electromagnetic radiation, especially x-rays, γ-rays, and visible light. The film base is usually tinted blue and is about 0.18 mm (0.007 in.) thick. An adhesive undercoat fastens the emulsion to the film base. A very thin but tough coating of gelatin called a protective overcoat covers the emulsion to protect it against minor abrasion. The total thickness of the x-ray film is approximately 0.23 mm (0.009 in.), including film base, two emulsions, two adhesive undercoats, and two protective overcoats.

The presence of emulsions on both sides of the base effectively increases the speed of the x-ray film because penetrating radiation affects each emulsion almost equally. In addition, thin emulsions allow development, fixing, and washing of the exposed film to be accomplished effectively and in a reasonably short time. Where two emulsions are used, two images are produced—

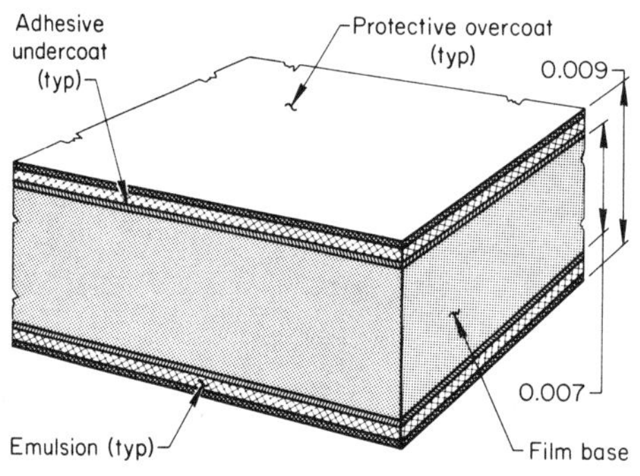

Fig. 26 Schematic cross section of a typical x-ray film. Dimensions given in inches

one in the front emulsion and one in the back emulsion. Viewing these images with the unaided eye causes no problems, because the images are separated by only a few hundredths of a millimeter of film base. However, if magnification is used, parallax can cause the two images to be seen slightly separated, which sometimes hampers interpretation. When appropriate, x-ray film with only one emulsion (single-coated x-ray film) can be used to avoid parallax.

When electromagnetic radiation and certain forms of particulate radiation (namely, electrons) react with the sensitive emulsion of x-ray films, they produce a latent image in the emulsion, which cannot be detected visually or by ordinary physical measurements. To produce a visible image, the exposed film must be chemically processed. When the film is treated with a developer, selective chemical reaction converts the exposed silver halide grains in the latent image into black metallic silver. The metallic silver remains suspended in the gelatin and is responsible for blackening or shades of gray in the developed (visible) image. After development, the film is treated with a chemical called a fixer, which converts unexposed silver halide grains into a water-soluble compound. The reaction products of development and fixing are washed off, and the film is dried. Detailed information on the processing of radiographic film is available in the Appendix to this article.

Although x-ray film is sensitive to light, the characteristics of its emulsion are different from those of emulsions used in photography. Industrial x-ray film is of two general types: that used for direct exposure (called direct-exposure, no-screen, or nonscreen film) and that used with fluorescent screens (often called screen-type film).

Most industrial x-ray film is of the direct-exposure type and is available in various combinations of film speed, gradient, and graininess. The choice of film for a given application depends on the type of radiography to be performed. Film types and selection are discussed in the section "Film Radiography" in this article.

Screen-type films, which are marketed primarily for medical radiography, are only occasionally used for industrial applications—for example, when a low-power x-ray machine is used and exposure time is excessive with direct-exposure x-ray film. Screen-type films are more sensitive to light than to x-radiation and are particularly sensitive to the wavelengths emitted by the fluorescent screen with which they are used. Although blue-sensitive emulsion types are the ones most often used for industrial radiography, other emulsions sensitive to ultraviolet and to green screen light are available.

Radiographic Paper. Ordinary photographic paper can be used to record x-ray images, although its characteristics are not always satisfactory. Photographic paper has a low speed, and the resulting image is low in contrast. However, photographic paper used with fluorescent screens can be effective for some applications.

Specially designed radiographic paper (also known as paper-base film or opaque-base film) is essentially opaque and has only a single emulsion. The emulsion is made of gelatin and silver salts, as in x-ray film, but it also contains developer chemicals. After exposure, the developing action is triggered by an alkaline solution called an activator. Instead of being fixed after development of the radiographic paper, chemicals remaining in the emulsion are made chemically inactive by stabilization processing, which is performed very rapidly in a small automatic processor. The images are similar to those on film radiographs and will not deteriorate for 2 to 3 months. If it is desired to store the radiographs for a longer period, they can be fixed in ordinary x-ray fixer, washed, and dried.

Although paper radiographs can be exposed directly to x-radiation, it is preferred to make the exposure with fluorescent screens that produce ultraviolet light. Fluorescent-screen exposures are preferred because exposure time is short and radiographic contrast is high. With direct exposure, the radiographic contrast is low and the corresponding exposure time is long.

The image is viewed in reflected light like a photograph; however, the greatest amount of detail can be seen when the image is viewed in specular light from a mirrorlike reflector (like a spotlight) directed at about 30° to the surface and to one side of the radiograph. The density (blackening) of paper radiographs, which is known as reflection density, is quite different from the transmission density of film, but the differences do not cause problems in interpretation.

Radiographic paper can exhibit adequate sensitivity, which in many respects matches or exceeds that of fast direct-exposure x-ray films. Radiographic paper does not match the sensitivities of slow x-ray films, but because of their speed, convenience, and low cost, radiographic papers are being used both for the radiography of materials that do not require critical examination and for in-process control.

A new (3M) dry silver film has recently been introduced that is developed with heat in about 12 s. It also is exposed with a fluorescent screen and achieves resolutions up to 7 line pairs per millimeter (or a resolution of 0.0055 in.). Viewing is by reflected or transmitted light.

Xeroradiography (dry radiography) is a form of imaging that uses electrostatic principles for the formation of a radiographic image. In film radiography, a latent image is formed in the emulsion of a film; in xeroradiography, the latent image is formed on a plate coated with selenium. Before use, the plate is given an even charge of static electricity over the entire surface. As soon as the plate is charged, it becomes sensitive to light as well as to x-radiation and must be protected from light by a rigid holder similar to a film cassette. In practice, the holder is used for radiography as though it contained film. X-radiation will differentially discharge the plate according to the amount of radiation received by different areas. This forms an electrostatic latent image of the testpiece on the plate.

The exposed plate is developed by subjecting the plate, in the absence of light, to a cloud of very fine plastic powder, each particle in the cloud being charged opposite to the electrostatic charges remaining on the plate. The charged plastic powder is attracted to the residual charges on the plate. The visible radiographic image at this state is not permanent, because the powder is simply held in place by electrostatic charges. However, the image can be made permanent by placing a piece of specially treated paper over the plate and transferring the powder to the paper, which is then heated to fix the powder in place.

Selenium-coated plates can be easily damaged by fingerprints, dirt, and abrasion. For this reason, automated equipment is used for charging and for development and image transfer to paper.

The image produced by xeroradiography can be either positive or negative, depending on the color of the plastic powder (toner) and the color of the paper upon which the image is deposited. If a negative xeroradiographic image is desired, dark-colored paper with light-colored toner can be used; a positive image would require light-colored paper and dark-colored toner.

Because of an electrostatic edge effect, xeroradiographic images exhibit excellent detail; even small indications in the image are sharply outlined. Optimum results from xeroradiography are more likely when the user develops special techniques rather than employing the radiographic techniques normally used with film.

Lead Screens

It is the combination of filtration and intensification that makes lead screens the most widely used in industrial radiography. Lead absorbs radiation to a greater extent than most other materials, with the amount of absorption depending largely on the penetrating quality of the radiation (photon energy or wavelength). High-energy (short-wavelength) radiation passes through lead much more readily than low-energy radiation; in other words, low-energy radiation is more readily absorbed by a lead screen than high-energy radiation. Because scattered radiation from a testpiece is always of a lower energy than the incident beam passing through a testpiece, a lead screen will absorb a relatively high percentage of unwanted scattered radiation, but will absorb a somewhat lower percentage of the image-forming radiation. This effect is known as filtration, and lead screens are sometimes referred to as lead-filter screens.

Filtration of Secondary Radiation. As discussed in the earlier section "Principles of Shadow Formation" in this article, secondary radiation can arise by internal scatter from within the testpiece and by back scatter from objects behind the film, detector, or screen. Because of the need to filter out both internal scatter and back scatter, two screens are normally used in film radiography. The screen that faces the top of the film is referred to as the front screen, and the screen behind the film toward the table or floor is referred to as the back screen. Both screens absorb scattered radiation.

In practice, the front screen is sometimes the thinner of the two because the image-forming radiation must always pass through this screen. The usual thickness of the front lead screen is 0.13 or 0.25 mm (0.005 to 0.010 in.), but may be greater or less depending on the type of material being inspected and the photon energy of the incident beam. The main function of the beam screen is to absorb unwanted back scatter; back screens can be of any thickness that performs this function adequately, although usually they are of the same thickness as the front screen.

In the radiography of thin-gage or low-density materials, in which low photon energies are used, care must be exercised to ensure that the front screen does not excessively filter the image-forming radiation. Excessive filtration affects subject contrast and tends to reduce radiographic sensitivity. In such cases, lead screens less than 0.13 mm (0.005 in.) thick should be used; lead screens 0.05 or 0.025 mm (0.002 or 0.001 in.) thick or less are available. For thicker testpieces or denser material, where higher photon energies are used, the front screen can be thicker than 0.13 mm (0.005 in.), ranging as high as 1 mm (0.040 in.) when betatrons or high-MeV linear accelerators are used. The back screen can be 0.5 to 10 mm (0.020 to 0.400 in.) thick in these cases.

Intensification. When lead is excited by x-ray or γ-ray radiation, it produces electrons. The number of electrons emitted is directly proportional to the photon energy of the radiation that passes through the testpiece and reaches the screens.

In film radiography, the emitted electrons expose additional silver halide crystals in the film emulsion. After development, film densities are greater than they would have been without the intensifying action of emitted electrons. Intensification not only increases overall photographic density, thus requiring shorter exposure times for producing a given density, but also enhances radiographic contrast, thus improving the ability to resolve small flaws.

In real-time radiography with a fluorescent-screen system, a similar effect occurs. The phosphor material used in fluorescent screens is generally more sensitive to electrons than to primary x-rays. With high-energy x-rays, the electrons generated from a suitable lead screen (or other heavy metal, such as tantalum or tungsten) can be used to enhance the imaging process. These screens are often useful in real-time radiography with MeV radiation.

Because lead screens have properties of both filtration and intensification, there is a combination of thickness of the subject material and photon energy of the incident radiation at which intensification just balances filtration and there is no net advantage. With steel testpieces in film radiography, this null point is generally considered to occur at a combination of 6 mm (¼ in.) testpiece thickness and 140-keV x-rays when a 0.13 mm (0.005 in.) thick front screen and a 0.25 mm (0.010 in.) thick back screen are used. At lower tube voltages or with thinner testpieces, the filtration effect is dominant, resulting in longer exposure times. At high voltages or with thicker testpieces, the intensification factor becomes dominant, and exposure times can be reduced to as much as one-third of the time for direct exposure (no screens), at a tube voltage of 200 to 300 kV. With cobalt-60 radiation and steel testpieces, the exposure time using lead screens is about one-third that for direct exposure.

With light metals such as aluminum, the null point occurs at a greater thickness than for steel. With metals of greater atomic number than steel, it occurs in thinner sections. For both lighter and heavier metals, the null point will occur at a radiation energy different from 140-keV x-rays. Although electrons are produced throughout the volume of lead on both screens when excited by x-rays or γ-rays, electrons produced by radiation having photon energies below 1 MeV are largely low-energy electrons. Such electrons are readily absorbed by the volume of lead in the screen. Only those electrons produced at the surface adjacent to the film escape to intensify the latent image on the film. Therefore, the closer the film emulsion is to the surface of the lead, the more effectively the electrons interact with the emulsion. This is why lead screens should always be in intimate contact with the film.

Low-energy electrons have little penetrating capability. They will affect the emulsion closest to the screen, but will not penetrate the film base to affect the emulsion on the other side. Although electrons have little penetrating capability, they will penetrate interleaving paper, so this should always be removed from the film to avoid a paper-pattern image on the radiograph. Similarly, dust, dirt, lint, and other foreign material between the film and screen must be avoided to prevent extraneous images (artifacts) on the radiograph.

Precautions. Lead sheet of usual screen thicknesses is easily bent. For this reason, lead screens are often backed by cardboard or other material to facilitate handling. Even so, the care and handling of lead screens is important. Deep scratches or dents must be avoided because they can appear as artifacts on the radiographic image. Wrinkles or folds in the screens can also be detected on the radiograph. Chemical spills, dust, and dirt must be carefully removed from the screen before use. Oxidation of the screen, which occurs with age and appears as a gray coating, does not seem to affect the utility of the screen. Some screens are coated with a special material to prevent oxidation. Although coated screens exhibit a reduced capability for intensification, they are easier to keep clean and free of tiny scratches. The use of spray lacquers or acrylics to protect the surface of the screens should be avoided; such coatings often have a highly detrimental effect on the radiographic image.

The composition of the lead used for screens is important. Pure lead is soft and may rub off on the film to produce lead smudge on the radiograph. Lead screens made from 94Pb-6Sb alloy are most commonly used because they are harder and more resistant to scratching. However, care must be taken that there is no segregation of antimony (which appears as shiny or different-colored streaks on the screen), because antimony segregation produces low-density streaks on the radiograph.

Screens must be flat and free of roll marks or chatter. Variations in screen thickness result in areas of poor contact with the film, which can produce fuzzy areas in the radiograph.

Prolonged contact with lead screens can produce an effect called lead-screen fog. Consequently, films should never be left in contact with lead screens longer than is reasonably necessary. This is particularly important under conditions of high temperature and humidity, or within 24 h after cleaning screens with very fine steel wool or other abrasive.

As a general rule, lead screens should be used whenever they can improve the resolution of detail, even though the exposure time may be longer. A single back screen can be used for intensification purposes, but in the absence of a front screen, forward scatter may be present in the radiograph. Sometimes, the single back screen technique is useful in the radiography of very thin materials at low photon energies. If scatter is a problem, other means of control (such as a copper filter located at the tube) are quite effective. Lead screens should always be used in radiography using high photon energies (above 300-keV x-rays or with most γ-ray sources) to avoid extraneous paper patterns or other effects due to irradiation of the film holder.

Certain types of film packaging incorporate thin lead foil (0.025 or 0.050 mm, or 0.001 or 0.002 in. in thickness) on both sides of the film, with the film-foil composite contained in a lighttight envelope. These prepackaged composites are convenient because the envelopes containing screens and film can be used as a film holder or cassette. In addition, loading in the darkroom is eliminated, and there is less likelihood of foreign particles being present to create artifacts on the radiograph.

Lead Oxide Screens

A variation of lead screens is lead oxide screens, which are made by evenly coating a paper base with lead oxide. The result is an extremely flexible screen, equivalent to about 0.013 mm (0.0005 in.) of lead foil, that is lightweight and free of antimony segregation. Lead oxide screens are available only in lighttight envelopes containing screens on both sides of the film.

The principle of lead oxide screens is essentially the same as for lead screens. The main differences are a lesser degree of filtration than lead screens and a greater intensification below 140 keV as well as slightly lesser intensification at 300 keV than with 0.13 mm (0.005 in.) thick lead screens. Although lead oxide screens are particularly advantageous for the radiography of light alloys or thin material, they can also be used in the radiography of heavier material.

Screens of Metals Other Than Lead

Foils of many of the heavier metals can be as effective as lead for radiographic screens, but usually are not as practical, either because the foils are not as flexible or because of cost.

Gold screens perform as well or better than lead screens, but gold costs considerably more and tends to work harden when bent, which eventually cracks the screen.

Tantalum screens exhibit slightly lower filtration but higher intensification than lead. However, tantalum foil is stiff and springy, which does not allow it to be shaped to fit around a testpiece. Tantalum foil can be used in solid cassettes, can be polished, and is quite resistant to minor abrasion, but is expensive to use.

Depleted-uranium screens exhibit greater filtration than lead screens. However, uranium is brittle, is somewhat difficult to obtain in thin sheets, and is also expensive to use.

Copper screens have been used, especially with cobalt-60 radiography. Copper has a lower degree of filtration than lead and a lower intensification factor, but copper provides greater radiographic sensitivity.

Composite screens of lead, copper, and aluminum (and sometimes other metal foils) to control the radiation reaching the film have occasionally been used with supervoltage x-ray machines for the inspection of thick steel testpieces.

Fluorescent Intensifying Screens

The efficiency of film and paper radiography can be improved by fluorescent intensifying screens, which emit radiation in the ultraviolet, blue, or green portion of the electromagnetic spectrum. These screens are called fluorescent intensifying screens because they fluoresce, or produce light, when excited by x-rays or γ-rays. Certain compounds, such as calcium tungstate or barium lead sulfate, often containing trace elements of some other chemical or phosphor, have the property of emitting light immediately upon excitation by short-wavelength radiation. Crystals of these chemicals are finely powdered, mixed with a binder, and coated on some mildly flexible support such as cardboard or plastic to make a fluorescent screen. A thin, tough, transparent overcoat is applied to the sensitized surface of the screen to prevent damage to the crystals during use.

Screen Speeds. Fluorescent intensifying screens, widely used in medical radiography, are available in a variety of speeds. Perhaps the most common of these are blue-emitting screens, which can be characterized in speed as very slow, slow, medium, medium high, high, and super. In industrial radiography, the screen most often used (with the appropriate blue-sensitive screen-type film) is the medium-speed fluorescent screen.

By the addition of proprietary rare-earth phosphors to the fluorescent crystals, green-emitting screens are produced that, when used with the appropriate green-sensitive film, have relatively fast screen speeds characterized as high speed, superspeed, and beyond superspeed.

With the addition of barium lead sulfate and a proprietary trace element, the crystals can be made to fluoresce in the ultraviolet range. Used with the appropriate ultraviolet-sensitive film or radiographic paper, these screens have speeds that are equivalent to the slow and medium-high-speed blue-emitting screens.

For the greatest radiographic effect, fluorescent screens should be used with a film that is sensitive to the particular wavelengths of light emitted by the screen. In general, though not always true, the slower the screen-film combination, the better the radiographic definition. However, even slow-speed combinations can reduce exposure times by 20 to 98% compared to the fastest direct-exposure film, depending on photon energy, type of screen, type of film, and testpiece material.

Disadvantages. The main reason for using fluorescent intensifying screens in industrial radiography is to reduce exposure times. However, in comparison with lead screens, fluorescent screens have the following disadvantages. First, the screen light reaches the film as cones of light from each fluorescing crystal. Because this tends to blur the image, radiographic definition using lead screens is superior. As in real-time radiography, inherent unsharpness in a fluorescent screen often overshadows any effect of geometric unsharpness.

Second, the purely statistical variations in the number of x-ray quanta (the total amount of x-radiation) absorbed from one small area to another on the screen result in uneven brightness, which in turn is recorded on the film as screen mottle (also known as quantum mottle). Screen mottle is particularly apparent at photon energies in the range 150 to 300 keV. There is essentially no screen mottle associated with lead screens.

Finally, lead screens filter out scattered radiation in addition to intensifying the radiation beam, but fluorescent screens have little or no filtering capability and intensify scattered radiation in the same proportion as transmitted radiation. However, they can be placed between lead screens, using the lead to protect against secondary radiation and back scatter.

Precautions in Use. The combination of a screen and the film that is sensitive to the wavelength of the screen light results in minimum exposures. However, fluorescent screens can be used with the direct-exposure films ordinarily used in industrial radiography. With direct-exposure film, the intensification factor may range from 2 to 25 times, but screen mottle and noticeable light spreading will be less because of the lower speed of the screen-film combination.

Fluorescent intensifying screens are often permanently mounted inside rigid film holders (cassettes). Cassettes of this type are designed to provide even pressure between screen and film to maintain intimate contact during exposure. Lack of contact results in unsharpness in the image and should be carefully avoided. Fluorescent screens may not be permanently mounted and can be used in regular film holders. Under these conditions, however, maintenance of good contact between fluorescent screen and film is more difficult.

The fluorescent crystals on the surface of the screens can be easily damaged by handling, and the screen cannot be repaired. Because fluorescent screens are expensive, great care must be taken to avoid scratches, abrasion, or chemical spills, which can damage the screen and affect the radiographic image. It is even more important to avoid fingerprints, stains, grease, dirt, and dust on the surface of the screens. Because these all absorb screen light, their image will be transmitted to the radiograph. If the screen becomes dirty it may be cleaned, but the manufacturer's cleaning recommendations should be followed to avoid damage to the protective overcoat.

Because of the high radiation intensity used in industrial radiography, great care must be exercised not to expose any portion of a fluorescent screen to the full intensity of the primary beam. When exposed to high-intensity radiation, a screen may continue to fluoresce after being removed from the beam; this semipermanent effect, known as screen lag, will produce false images on subsequent radiographs.

Fluorescent intensifying screens are most often used for the radiography of thick testpieces when the x-ray machine has limited penetrating capability. With medium-speed, blue-sensitive fluorescent intensifying screens (with appropriate screen-type film), it is possible to radiograph 75 mm (3 in.) of steel with 250-keV x-rays in a reasonable exposure time. Fluorescent screens are rarely used with γ-rays or with photon energies exceeding 1 MeV because of screen mottle and low intensification factors. In all cases where fluorescent screens are used, reciprocity law failure is likely to occur; reciprocity law failure (discussed in the section "X-Ray Tubes" in this article) is not a problem in direct or lead-screen exposures.

Fluorometallic Screens

Fluorometallic screens consist basically of lead screens placed on either side of a fluorescent-screen/film combination. The theory behind combining lead and fluorescent screens is that the lead preferentially absorbs scattered radiation (which fluorescent screens will not do) and that the fluorescent screen next to the film intensifies the radiation beam to shorten the exposure.

Commercial screens incorporating lead and fluorescent materials laminated together require significantly shorter exposure times than lead screens, and the image quality exceeds that of fluorescent screens. When direct-exposure film is used with fluorometallic screens, the exposure time can be reduced to as little as one-seventh of that for lead screens with almost the same radiographic sensitivity. When screen-type film is used, exposure time can be reduced to as little as one-ninth of that for other methods. The speed factor varies with type of film, type of fluorometallic screen, and photon energy of radiation. With high-voltage x-rays, and with cobalt-60 γ-rays, the speed factor is about two to seven.

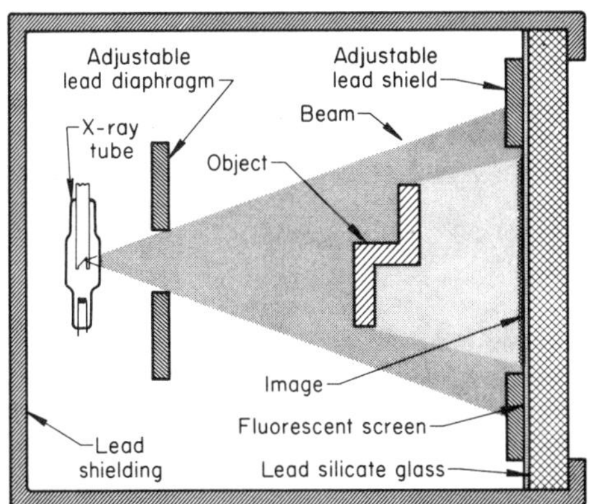

Fig. 27 Diagram of the components and principles of operation of a fluoroscope

Real-Time Radiography

Various types of image conversion techniques allow the viewing of radiographic images while the testpiece is being irradiated and moved with respect to the radiation source and the radiation detector. These radioscopic techniques can be classified as real-time radiography (also known as real-time radioscopy) and near real-time radiography (or near real-time radioscopy). The distinction between these two classifications is that the formation of near real-time images occurs after a time delay, and this requires limitation of the test object motion. An example of near real-time radiographic imaging involves the use of discrete detectors (primarily linear arrays) that scan the area being irradiated. The outputs are then processed digitally to form images in near real time. This technique is often referred to as digital radiography.

Background. The predecessor of the modern methods of real-time radiography is fluoroscopy. This technique, which is now largely obsolete, involves the projection of radiographic images on a fluorescent screen (Fig. 27). The screen consists of fluorescent crystals, which emit light in proportion to the intensity of the impinging radiation. The radiographic image can then be viewed, with appropriate measures taken to protect the viewer from radiation.

The main problem with fluoroscopes is the low level of light output from the fluo-

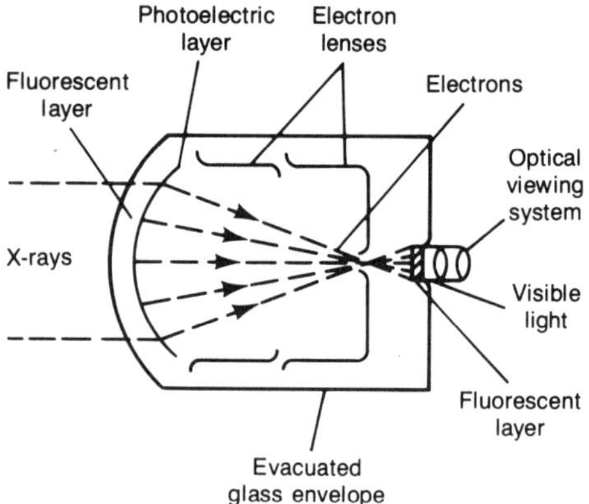

Fig. 28 Schematic of an image-intensifier tube

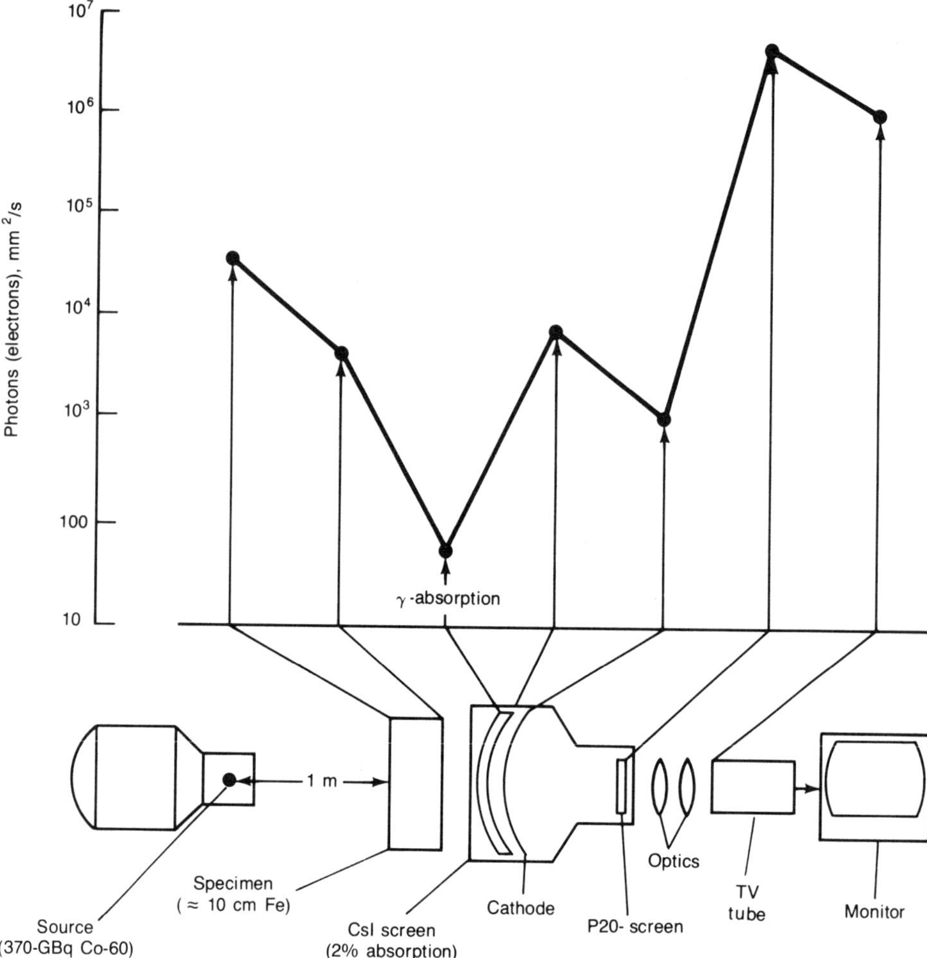

Fig. 29 Schematic of a typical radioscopic system using an x-ray image intensifier

rescent screen. This requires the suppression of background light and about 30 min for the viewer's eyes to become acclimated. Moreover, radiation safety dictates viewing through leaded glass or indirectly by mirrors. Because of these limitations, other methods have been developed to improve safety and to amplify the images from fluorescent screens.

One of the early systems (1950s) involved the development of image-intensifier tubes. Image-intensifier tubes (Fig. 28) are glass-enclosed vacuum devices that convert a low-intensity x-ray image or a low-brightness fluorescent-screen image into a high-brightness visible-light image. Image intensification is achieved by a combination of electronic amplification and image minification. The image brightness at the output window of an image-intensifier tube is about 0.3×10^3 cd/m² (10^{-1} lambert) as compared to about 0.3 cd/m² (10^{-4} lambert) for conventional fluoroscopic screen.

These early image intensifiers were originally developed for medical purposes and were limited to applications with low-energy radiation (because of low detection efficiencies at high energies). Consequently, industrial radiography with these devices was restricted to aluminum, plastics, or thin sections of steel.

By the mid-1970s, other technological developments led to further improvements in real-time radiography. These advances included high-energy x-ray sensitivity for image intensifiers, improved screen materials, digital video processing for image enhancement, and high-definition imaging with microfocus x-ray generators.

Modern Image Intensifiers. Although the early image intensifiers were suitable for medical applications and the inspection of light materials and thin sections of steel, the image quality was not sufficient for general use in radiography. Therefore, image intensifiers had to be redesigned for industrial material testing (Ref 3). The modern image intensifier is a very practical imaging device for radiographic inspection with radiation energies up to 10 MeV. With the image intensifier, a 2% difference in absorption can be routinely achieved in production inspection applications. The typical dynamic range of an image intensifier before image processing is about 2000:1.

A schematic of a modern x-ray image intensifier is shown in Fig. 29, along with a graph indicating the level of signal strength as the radiation passes through a sample and impinges on the entrance screen of the image-intensifier tube. These screens are usually made of a scintillating material, such as cesium iodide (CsI), and fixed to a photocathode. The energy of the incident radiation quanta generates electrons that produce light in the CsI entrance screen (approximately 150 photons per absorbed gamma quantum). To avoid degradation of the image quality by lateral dispersion of the light in the conversion screen, the CsI scintillating material, with a cubic crystalline structure, is grown under controlled conditions, resulting in small, needle-shaped elements. This structure causes the scintillating screen to act as a fiber-optical faceplate. Light generated in one crystal needle does not spread laterally, but is confined to the needle in a direction parallel to the incident radiation. Therefore, the thickness of the conversion screen does not cause appreciable deterioration in the spatial resolution of the system.

The light from the scintillating screen then impinges on a photocathode in contact with the entrance screen. The photocathode emits photoelectrons. The electron image produced at the cathode is reduced by a factor of ten and is intensified by means of an accelerating voltage. The final phosphor screen presents a relatively bright image (approximately 5 million photons per second per square millimeter), caused by the impinging electrons. The image then passes through an optical system, which directs the image to a television camera tube, such as a vidicon or plumbicon tube. Vidicons have a dynamic range of 70:1; plumbicons, 200:1.

Contrast Sensitivity. Figure 29 shows the importance of detection efficiency with regard to the contrast sensitivity of the radioscopic system. The example is for a 370 GBq (10 Ci) cobalt-60 source, but a similar analysis can be performed for an x-ray source. A cobalt-60 source with an activity of 3700 GBq (100 Ci) generates 60 000 photons per second per square millimeter at 1 m with energies of 1.1 and 1.3 MeV. After passing through a 100 mm (4 in.) thick steel specimen, the radiation is attenuated in intensity by a factor of 15. Therefore, only 4000 photons per second per square millimeter are available for imaging. The detection efficiency of the CsI entrance screen in a typical modern image-intensifier tube is approximately 2%. This reduces the photon intensity to 80 per second per square millimeter. With this intensity, the laws of statistics (see the section "Quantum Noise (Mottle) and the Detective Quantum Efficiency" in the article "Industrial Computed Tomography" in this Volume) predict a noise-to-signal ratio of approximately 11%, which limits the attainable contrast resolution.

From Fig. 29, it is apparent that the critical factor for determining the contrast resolution is the relatively low detection efficiency of the entrance screen. To overcome this limitation, one must either increase the detection efficiency of the conversion screen or collect the photons over a longer period by summing television frames to improve the photon statistics in each picture element. The detection efficiency of the entrance screen can be improved by reducing the energy of the radiation or by making the entrance screen thicker. The minimum energy of the radiation is generally dictated by the need to have sufficient energy to penetrate the subject and to provide sufficient intensity for imaging at the conversion screen. Therefore, it is generally impossible to reduce the energy to improve conversion efficiency. Although a thicker entrance screen would improve detection efficiency, it would reduce the attainable spatial resolution. In general, improving the detection efficiency and improving the spatial resolution are competitive processes. The most conventional means of enhancing contrast resolution is through the summation of television frames with a digital image-processing system.

The spatial resolution of a real-time system is principally defined by the size and thickness of the detectors (or grain size of fluorescent crystals), the raster scan of the television system, and, for quantum statistical reasons, the intensity of the radiation. The dependencies are plotted in Fig. 30 for a typical image intensifier. Curve A shows a spatial resolution of about four line pairs per millimeter (or a resolution of 0.01 in.) at 10 R/h without television monitoring. Curve C shows the typical resolution of an image intensifier with television monitoring.

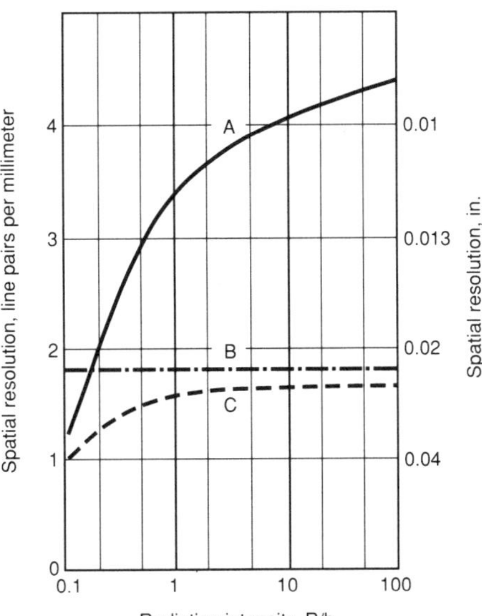

Fig. 30 Spatial resolution dependence of an image-intensifier system as a function of radiation intensity at the entrance screen. A, image-converter resolution; B, 625 TV lines limit; C, combined resolution of image converter and TV line

Real-Time Radiography With Fluorescent Screens. Because of the development of low-level television camera tubes and low-noise video circuitry, the dim images on a fluorescent screen can be monitored with video systems. The contrast sensitivity and the spatial resolution of fluorescent-screen systems are comparable to those of image intensifiers, but the use of fluorescent screens is limited to lower radiation energies (below about 320 keV without intensifying screens and about 1 MeV with intensifying screens). Nevertheless, fluorescent screens can provide an unlimited field of view, while image intensifiers have a field of view limited to about 300 mm (12 in.). Example 1 in this article describes an application in which a fluorescent screen (with isocon camera) was preferred over an image intensifier. The dynamic range of systems with fluorescent screens can vary from 20:1 for raw images to 1000:1 with digital processing and a large number of frames averaged.

Television Cameras. The types of low-level television cameras available include image orthicons, image isocons, and secondary electron-coupled vidicons. Radiographic inspection is also performed with x-ray sensitive television cameras, which require no extra conversion or optics. X-ray sensitive television cameras are limited in image size and sensitivity and are primarily used in the inspection of low-density materials and electronic components.

Image orthicons and image isocons are return beam tubes with internal electron multipliers. The orthicon is useful to light levels of about 1.076×10^{-3} lm/m^2 (10^{-4} ftc), and the isocon is useful to 1.076×10^{-4} lm/m^2 (10^{-5} ftc). The isocon provides the best noise performance and resolution of all tubes for low-light levels and static scenes, but degrades for moving scenes and is highly complex. Both orthicons and isocons exhibit little target overloading damage. Both the image orthicon and isocon camera tubes must be carefully adjusted for optimum performance; the isocon is the most difficult to adjust and demands a skilled technician. Both are temperature sensitive and should be operated under stabilized conditions. The dynamic range for the isocon is about 1000:1.

Secondary electron-coupled vidicons employ an internal initial stage of target amplification, using a secondary target. They are similar in performance to isocons, are less complex, but differ in lag and require special protection circuitry to prevent target destruction from overloading. Their primary use is in low-light level, low-contrast applications down to 1.076×10^{-4} lm/m^2 (10^{-5} ftc).

Fluorescent screens used in real-time radiography are similar in construction and properties to fluorescent intensifying screens for film radiography. The commonly used fluorescent screens consist of a plastic substrate coated with a powdered luminescent material and an epoxy binder. The spatial resolution of the screen depends on the grain size of the luminescent material and the range and distribution of the conversion electrons and light photons. The x-ray absorption properties of the screen depend on the atomic weight and density of the luminescent material. The light conversion efficiency, the color of the emitted light, and the speed of the screen are additional properties that are dependent on the characteristics of the luminescent material.

A wide variety of fluorescent screens are available. Screens using zinc sulfide as the fluorescent material have a high light output and a short afterglow, but the contrast and sharpness of this type of screen are poor. Zinc sulfide screens also use cadmium, which has a higher atomic number and a higher density than zinc. Other phosphor materials used include calcium tungstate and phosphors with rare-earth materials. Until the 1970s, many applications utilized the popular calcium tungstate screens. With the availability of rare-earth materials in high states of purity and at reasonable costs, however, one can design new phosphors with improved x-ray absorptions, greater density, and higher x-ray to light conversion efficiencies. Screens have been made of such materials as gadolinium oxysulfide doped with terbium, LaOBr:Tm, BaFCl:Eu, or $YTaO_2$:Nb. These screens are discussed more fully in Ref 4. It should be noted that many rare-earth screens are not as effective as calcium tungstate in absorbing very high energy x-rays.

Because the intensity of the image on a fluorescent screen is rather low, very sensitive TV cameras are required for remote viewing. Different types of low light level cameras are available that have one- or two-stage light amplifiers or extremely sensitive camera tubes. When designing a fluoroscopic system, it is important to recognize that high light amplification from a double-stage amplifier generally yields poor spatial resolution and contributes additional noise to the image. In many cases, depending on the size of the fluorescent screen, the camera can also be the limiting factor in image quality. Therefore, both the entrance screen and the camera must be optimized for the particular inspection application.

Digital Radiography. A third method of radiographic imaging involves the formation of an image by scanning a linear array of discrete detectors along the object being irradiated. This method directly digitizes the radiometric output of the detectors and generates images in near real time. Direct digitization (as opposed to digitizing the output of a TV camera or image intensifier) enhances the signal-to-noise ratio and can result in a dynamic range up to 100 000:1. The large dynamic range of digital radiography allows the inspection of parts having a wide range of thicknesses and densities. Discrete detector arrangements also allow the reduction of secondary radiation from scattering by using a fan-beam detector arrangement like that of computed tomography (CT) systems. In fact, industrial CT systems are used to obtain digital radiographs (see the article "Industrial Computed Tomography" in this Volume).

The detectors used in digital radiography include scintillator photodetectors, phosphor photodetectors, photomultiplier tubes, and gas ionization detectors. Scintillator and phosphor photodetectors are compact and rugged, and they are used in flying-spot and fan-beam detector arrangements. Photomultiplier tubes are fragile and bulky, but do provide the capability of photon counting when signal levels are low. Gas ionization detectors have low detection efficiencies but better long-term stability than scintillator- and phosphor-photodetector arrays.

A typical phosphor-photodetector array for the radiographic inspection of welds consists of 1024 pixel elements with 25 μm (1 mil) spacing, covering 25 mm (1 in.) in length perpendicular weld seam. The linear photodiode array is covered with a fiberoptic faceplate and can be cooled in order to reduce noise. For the conversion of the x-rays to visible light, fluorescent screens are coupled to the array by means of the fiber optics. A linear collimator parallel to the array is arranged in front of the screen. The resolution perpendicular to the array is defined by the width of this slit and the speed of the manipulator. A second, single-element detector is provided to detect instabilities of the x-ray beam. Using 100-kV radiation, a spatial resolution of 0.1 mm (0.004 in.) can be achieved with a scanning speed of 1 to 10 mm/s (0.04 to 0.4 in./s).

The data from the detector system are digitized and then stored in a fast dual-ported memory. This permits quasi-simultaneous access to the data during acquisition. Before the image is stored in the frame buffer and displayed on the monitor, simple preprocessing can be done, such as intensity correction of the x-ray tube by the data of the second detector and correction of the sensitivity for different array elements. If further image processing or automatic defect evaluation is required, the system can be equipped with fast image-processing hardware. All standard devices for digital storage can be utilized.

Image Processing. Because real-time systems generally do not provide the same level of image quality and contrast as radiographic film, image processing is often used to enhance the images from image intensifiers, fluorescent screens, and detector arrays. With image processing, the video images from real-time systems can compete with the image quality of film radiography. Moreover, image processing also increases the dynamic range of real-time systems beyond that of film (which typically has a dynamic range of about 1000:1).

Images can be processed in two ways: as an analog video signal and as a digitized signal. An example of analog processing is to shade the image after the signal leaves the camera. Shading compensates for irregularities in brightness across the video image due to thickness or density variations in the testpiece. This increases the dynamic range (or latitude) of the system, which allows the inspection of parts having larger variations in thickness and density.

After analog image processing, the images can be digitized for further image enhancement. This digitization of the signal may involve some detector requirements. In all real-time radiological applications, the images have to be obtained at low dose rates (around 20 μR/s). In digital x-ray imaging, however, the most often used dose is around 1 mR, in order to reduce the signal fluctuations that would result from a weak x-ray flux. This means that only a short exposure time (of the order of a few milliseconds) and a frequency of several images can be used if kinetic (motion) blurring is to be avoided. The resulting requirements for the x-ray detector are therefore:

- The capability of operating properly in a pulsed mode, which calls for a fast temporal response
- An excellent linearity, to allow the use of the simplest and most efficient form of signal processing
- A wide operating dynamic range in terms of dose output (around 2000:1)

Once the image from the video camera has been digitized in the image processor, a variety of processing techniques can be implemented. The image-processing techniques may range from the relatively simple operation of frame integration to more complex operations such as automatic defect evaluation. The following discussions briefly describe three image-processing operations. More detailed information on image processing is available in the article "Digital Image Enhancement" in this Volume.

Frame integration (or summing) is a simple technique that can improve the signal-to-noise ratio and decrease (or eliminate) the screen mottle associated with low levels of radiation intensity. This technique adds the digital images from a number of frames and displays the result on a video monitor. Frame integration can be done in real time, that is, at standard television frequency of 30 frames per second. The number of television frames that can be integrated depends on the number of frame buffers available for this procedure. Typical summation times range from a few seconds for applications using x-rays with high intensities up to 1 min or more for applications using radiation from low-activity radionuclide sources. However, these times are small compared to the time required for exposing and developing x-ray film.

Image subtraction can be used to improve sensitivity through the subtraction of one integrated image from a second integrated image. The advantage of the subtraction technique is that all areas of the scene having the same brightness cancel each other to produce an extremely flat or uniformly gray image. With such an image available, the contrast and brightness can then be electronically enhanced to reveal details not visible in either a real-time or a simple integrated image. Example 1 in this article describes one method called pairs subtraction.

Automated defect evaluation systems have been developed for the radiographic inspection of welds (Ref 5). The automatic detection of defects with such systems is based on a mathematical algorithm that evaluates a set of parameters for each indication and compares this set to a catalog of known defects. For any particular problem, the inspection parameters must be selected to provide optimum and reproducible results. To compare different radiographs and to extract features such as the depth of a defect, it is necessary to calibrate the gray levels against material thickness and to calibrate the linear dimensions. Once established, the optimized conditions must be consistently used for all inspections.

Figure 31 shows a flow chart of an automatic image evaluation procedure, and Fig. 32 shows the images from different steps in the procedure. The first step is to remove irrelevant details from the digitized image. This can be done by subtracting a background image. This background image is generated by creating an unsharp image with a low-pass filter. Although radiographs usually have a center brightness that differs from that at the edges, this gradient is eliminated by subtracting the background

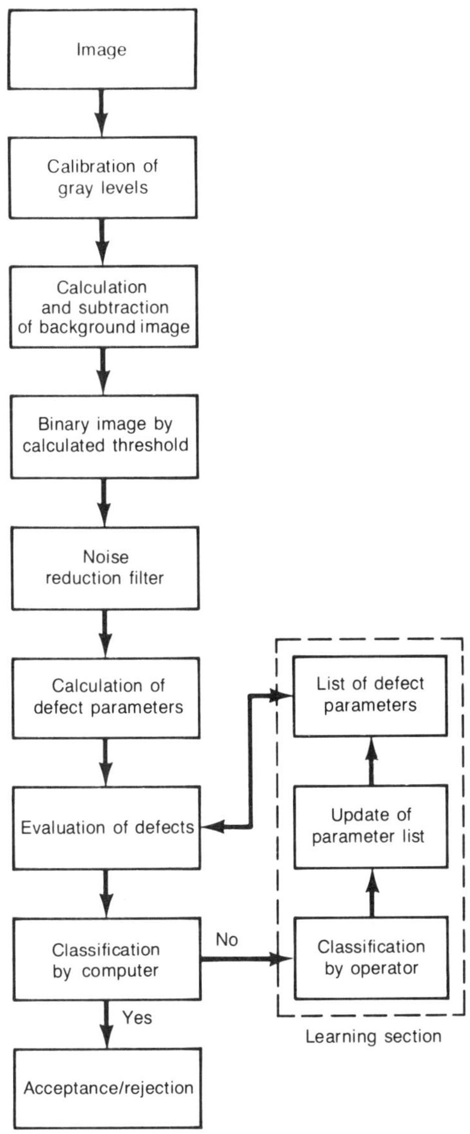

Fig. 31 Flow chart of an automatic image evaluation procedure

image from the original image. At this state (Fig. 32b), it is possible to set a threshold and to identify as possible defects all pixels whose gray levels exceed this threshold.

In the next step, these potential defects are examined. First, all very small indications are examined, such as single pixels or groups of only a few pixels, which result from image noise. Larger details, which are obviously separated by one or two missing pixels, are connected (Fig. 32c). This algorithm prevents the computer from spending time analyzing noise that exceeds the threshold.

For each potential defect, a number of parameters are evaluated. These parameters include position, orientation, and a set of descriptors for size and shape. The values of these parameters are compared with the mean value and standard deviation for each parameter in the catalog data already stored. Where some consistency is found between measured and stored data, the program identifies the defects (Fig. 32d) and calculates the probability that the indication is a particular type of defect. Where there is insufficient consistency between the measured and stored data, the program calls for the assistance of an interpreter and enters a learn mode. Once the interpreter has evaluated the defect, the program stores the values for each parameter of the defect, along with the evaluation in the catalog. This information is now considered in future automated evaluations. Relevant defects cause the rejection of the workpiece.

Applications. Real-time radiography is often applied to objects on assembly lines for rapid inspection. Digital radiography with discrete detectors, in particular, is being more widely used because it provides an unlimited field of view and the largest dynamic range of all the radiographic methods. Because a larger dynamic range allows the inspection of parts having a wider range of thicknesses, digital radiography has found application in the inspection of castings (see the article "Castings" in this Volume).

Radiography is also an accepted method of detecting internal discontinuities in welds. Image intensifiers are the most widely used system for the real-time radiographic inspection of welds, although digital radiography with discrete detectors is also used. Projective magnification with microfocus x-ray sources is also useful in weld and diffusion bond inspection.

Advantages and Limitations. Real-time radiography provides immediate information without the delay and expense of film development. Real-time systems also allow the manipulation of the testpiece during inspection and the improvement of resolution and dynamic range by digital image enhancement. Real-time radiography with discrete detectors (that is, digital radiography) also provides good scatter rejection.

Nevertheless, film radiography is a simpler technique that represents less of a capital investment. The sensitivity and resolution of real-time systems are usually not as good as those of film, although real-time radiography can approach the sensitivity and resolution of film radiography when good inspection techniques are used. Good technique is often more important than the details of the imaging system itself.

Example 1: Design of a Rapid, High-Sensitivity Real-Time Radiographic System. A system was needed for the real-time radiographic inspection of jet-engine turbine blades. These were critical nickel alloy castings that required inspection to at least a 2-1T (1.4%) penetrameter sensitivity level through thicknesses that ranged from about 0.5 to more than 13 mm (0.02 to 0.5 in.). The volume of parts produced necessitated inspection at a rate of 60 parts per hour or more. For consistency, all aspects of the inspection were to be computer controlled. In addition, it was necessary to visually compare the actual parts with their radiographic images to separate castings with excess metal due to mold flaws that might be acceptable from castings with high-density inclusions that might be rejectable.

Image Processing. A preliminary evaluation of the parts indicated that the 2-1T penetrameter sensitivity requirement could be met with suitable image-processing techniques. Both analog and digital processing were used for these complex parts.

The analog processing technique used was an electronic adjustment called shading. Shading is performed in the camera control unit after the signal leaves the camera and before it is sent to the image processor for digitizing and further processing. Shading compensates for variations in brightness across the video image due to thickness variations in the part. In effect, proper shading adjustment increases the latitude (dynamic range) of the image. In the case of the airfoils to be inspected with the real-time radiographic system, shading permitted blade thicknesses ranging from 0.5 to 4 mm (0.02 to 0.16 in.) to be adequately imaged in a single view.

Once the image from the video camera had been digitized in the image processor, a great many processing techniques were implemented. The simplest was to integrate the image. This technique added the digital images from a number of frames and displayed the result on the video monitor. Integration improved the signal-to-noise ratio over that of a single frame from the video camera. For these castings, 128 frames of integration produced good results.

To achieve the required sensitivity, an image subtraction technique was also used. This was an extremely useful processing technique in which one integrated image was subtracted from a second integrated image. There are a number of variations of this technique. It was experimentally determined that the best technique for this application was the one termed "pairs subtraction." To produce a "pairs subtraction," the part was displaced slightly between the two integrated images that were subsequently subtracted from each other.

X-Ray Source. Because of the range of thicknesses to be inspected, two x-ray sources were selected for the system—one rated at 160 kV for the thinner sections and the other at 320 kV for the thicker sections. The use of 0.4 mm (0.015 in.) focal spots for the 160-kV unit and 0.8 mm (0.03 in.) focal spots for the 320-kV unit provided sufficient resolution to detect 0.25 mm (0.01 in.) defects even at the relatively short source-to-screen distances available in the final real-time radiographic system design. Because it was required that the blade be exposed to only one source of radiation at a

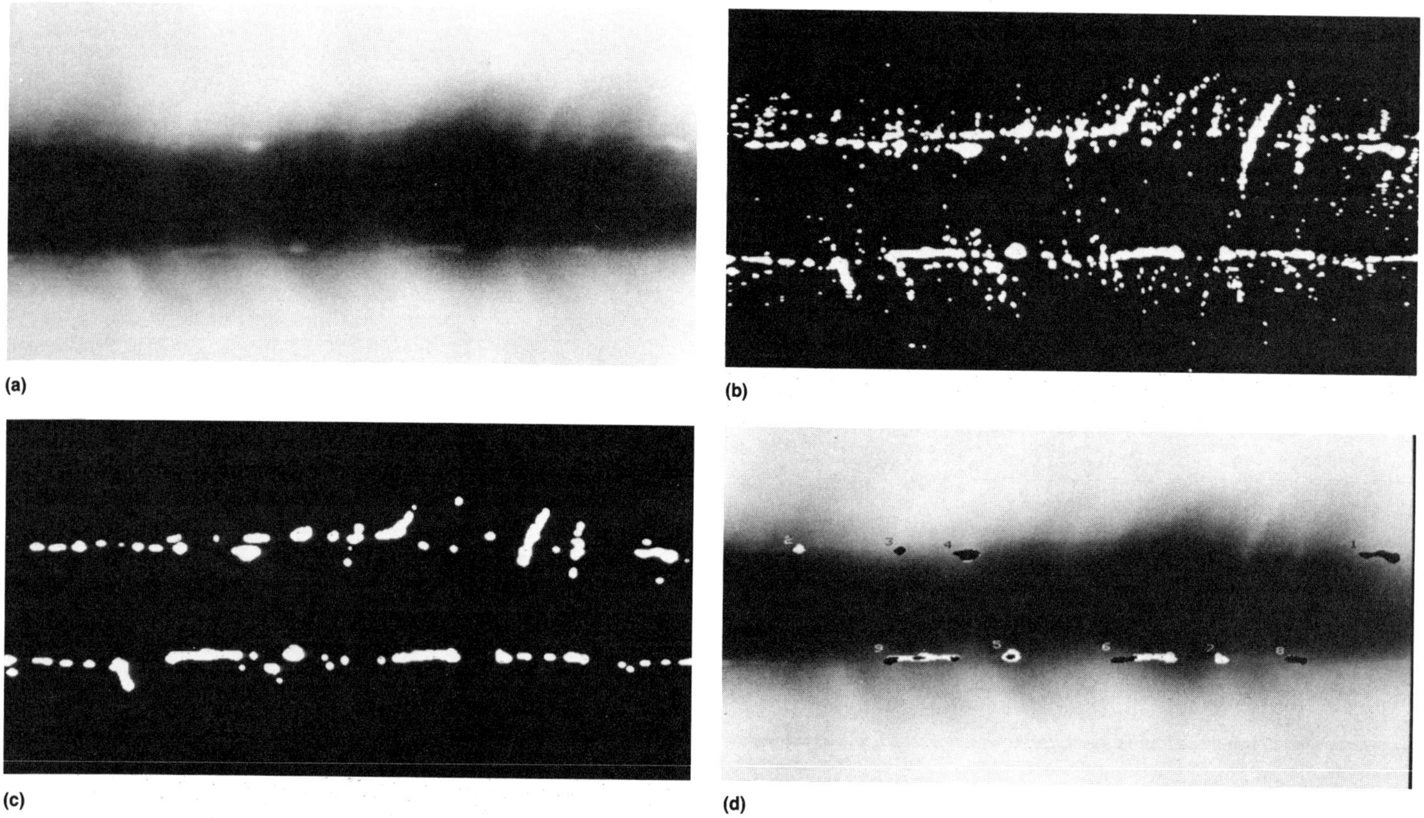

(a)

(b)

(c)

(d)

Fig. 32 Images developed by an image-processing procedure that identifies and evaluates the images of weld defects in real-time radiographs. (a) Original image. (b) Binary image. (c) Binary image after noise reduction. (d) Original image with defects identified. Courtesy of RTS Technology, Inc.

time and that the x-ray tubes be run continuously, two computer-controlled, high-speed lead shutters were designed to block the unwanted radiation.

Imaging System. The imaging camera, which was housed in its own shielded enclosure, included a 100 × 100 mm (4 × 4 in.) rare-earth fluorescent screen for converting x-rays to visible light. Light from the screen was directed to the objective lens of an image isocon camera by a 45° front surface mirror (Fig. 33). An isocon camera was selected over an image intensifier for two reasons. First, an isocon camera is much less susceptible to blooming due to bypass radiation. This permitted the use of relatively loose masking around the part. Second, computer-selectable shading channels were available with the isocon camera; therefore, shading could be preadjusted for different views. The computer program could then set the required shading channel during an inspection. The camera system was designed to include remote controls to facilitate focusing of the camera and adjustment of the field of view from the full-screen size down to a 25-mm (1-in.) square to increase the optical magnification if necessary.

Part Handling. To minimize part-handling time, a very accurate, high-speed, commercially available robot with four axes of movement (*X*, *Y*, *Z*, and rotation) was selected. The robot was equipped with a computer and sophisticated operating programs. After the robot was taught a few fixed points, it determined its own path from one position to another. This was especially convenient because a large number of parts were to be placed into the real-time radiographic system in a tray. The robot only needed to be told the number of rows and columns in the tray and taught

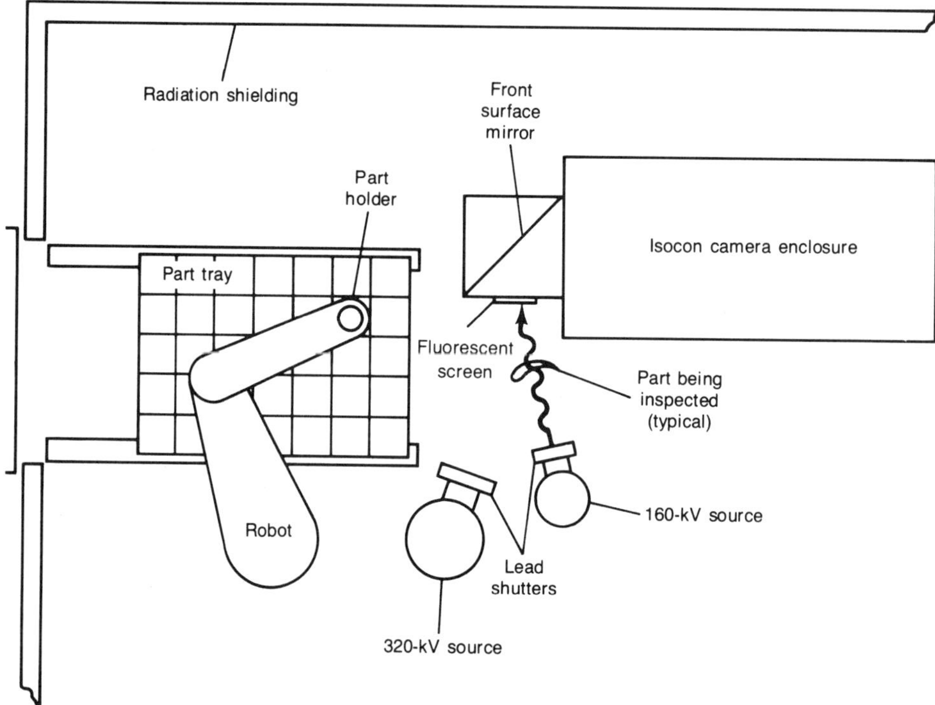

Fig. 33 Schematic of the equipment described in Example 1

three corner pickup locations. From this information, it could calculate the exact pickup points for all other locations in the tray.

The part containers required considerable development before consistent robot pickup was achieved. In the final design, parts were dropped into molded plastic holders with no special care, and they were positioned well enough for the robot to pick them up with a high degree of reliability.

System Layout and Operating Console. To permit installation in a factory, the x-ray sources, imaging camera, and robot were located in an exempt, radiation-shielded enclosure. Trays of parts to be inspected were inserted through a sliding access door from an adjacent environmental enclosure.

An environmental enclosure was included for the operator's station to provide a relatively clean environment in the manufacturing area. The inspection operator's station was designed to include a desktop console with a video monitor on which either real-time or processed images could be viewed. The console also included a computer terminal and keyboard; a joystick control was used for performing various image processor functions and for manual control of the robot. The x-ray controls, the robot controller, the isocon camera controls, and the computer and image processor were located near the console.

A console for reviewing images while examining parts visually was also provided in the environmental enclosure. In addition to the video monitor, image processor, and computer terminal, a voice recognition system was added so that commands and information could be input to the computer without using the keyboard.

System Software. Three separate programs were developed for the real-time radiographic system: Inspect, Review, and Supervisor. The computer selected included a multiuser operating system to allow the Inspect and Review programs to be run concurrently at different terminals. All the programs were menu driven; the operator simply entered information or exercised options when prompted to do so by the menu on the computer terminal.

The Inspect program was designed for day to day inspection operation. In this program, a preestablished scan plan controlled all the parameters required for an inspection. As long as the operator continued accepting views, the program moved through the plan for each part in turn until all parts in the tray had been completed.

In running the scan plan, the computer interfaced with the x-ray machines to control the kilovolt and milliamp settings, the shutters, and x-ray on and off. It also interfaced with the isocon camera control unit to turn the camera beam on and off and to call the required shading channel, and it interfaced with the robot controller to direct the part movements.

The software was designed so that the scan plan could be interrupted at any time. One option in the interrupt mode gave the operator complete manual control of all system functions through the terminal keyboard. The ability to call an image quality indicator (IQI) to check for proper radiographic sensitivity was also designed into the interrupt mode. When the IQI was called for, the part was returned to its tray position, and the IQI was picked up and moved into the x-ray beam. An image of the IQI was generated using the same x-ray parameters and image processing as employed for the part view being inspected at the time.

The Review program was designed to allow stored images to be recalled when the parts were available for visual review. In addition to accessing images, this program allowed various enhancement techniques to be performed on the image. All program functions could be exercised using the voice recognition capability so that the operator's hands were free to handle parts. Once a final disposition for a part was entered, the views for that part were automatically deleted from memory.

The Supervisor program was written to include a number of program functions that were to be protected from normal operator access. These included changing the lists of authorized users and their passwords, access to an operations log, and creation or modification of scan plans.

System Performance. The sensitivity of the system was demonstrated with plaque penetrameters, a wire penetrameter, and an actual part with small holes. The plaque penetrameter sensitivity of the system is shown in Fig. 34. Figure 34(a) shows a pairs-subtracted image of a 0.025 mm (0.001 in.) thick penetrameter through 1.25 mm (0.05 in.) of material, and Fig. 34(b) shows a 0.125 mm (0.005 in.) penetrameter through 6.4 mm (0.25 in.) of material. All three holes can be seen near the centers of the circles as adjacent white and black areas that result from the pairs subtraction technique.

Figure 35 shows a pairs-subtracted image of an actual part in which a series of holes was drilled using the electrical discharge machining process. The holes have a nominal diameter of 0.25 mm (0.01 in.) and a depth that is 2% of the material thickness at the location where they were drilled. The material thickness in this airfoil view ranged from 1 to approximately 4 mm (0.04 to 0.16 in.).

Film Radiography

Film selection and exposure time are two major factors in film radiography. The selection of radiographic film for a particular application is generally a compromise between the desired quality of the radiograph and the cost of exposure time. The quality of the radiograph depends mainly on film density, gradient, graininess, and fog, which are functions of film type and development procedure. Exposure time depends mainly on film speed and on radiation intensity at the film surface. The factors affecting film selection and exposure time are discussed in more detail in this section.

Characteristics of X-Ray Film

Three general characteristics of film—speed, gradient, and graininess—are primarily responsible for the performance of the film during exposure and processing and for the quality of the resulting image. Film speed, gradient, and graininess are interrelated; that is, the faster the film, the larger the graininess and the lower the gradient—and vice versa. Film speed and gradient are derived from the characteristic curve for a film emulsion, which is a plot of film density versus the exposure required for producing that density in the processed film. Graininess is an inherent property of the emulsion, but can be influenced somewhat by the conditions of exposure and development.

There are other characteristics of x-ray film that can be used to produce special effects. However, the following discussion is mainly confined to film speed, gradient, graininess, and the factors that affect them.

Density. The quantitative measure of the blackening of a photographic emulsion is called density. Density is usually measured directly with an instrument called a densitometer.

There are two kinds of density:

- Transmission density, which is associated with transparent-base radiographic film
- Reflection density, which is associated with opaque-base imaging material such as radiographic paper

In the following discussion, the unqualified term density will refer to transmission density.

Transmission density is defined by:

$$D = \log \left(\frac{I_0}{I_t}\right) \qquad \text{(Eq 14)}$$

where D is the density, I_0 is the intensity of light incident on the film, and I_t is the intensity of light transmitted through the film. Reflection density is defined by:

$$D_r = \log \left(\frac{I_0}{I_r}\right) \qquad \text{(Eq 15)}$$

where D_r is the reflection density, I_0 is the light intensity incident on the radiographic image, and I_r is the light intensity reflected from the image. Although Eq 14 and 15 are similar, the densities are quite different.

The ratio of incident intensity, I_0, to transmitted intensity, I_t, in Eq 14 is called opacity. The inverse of this ratio, I_t/I_0, is called transmittance and indicates the frac-

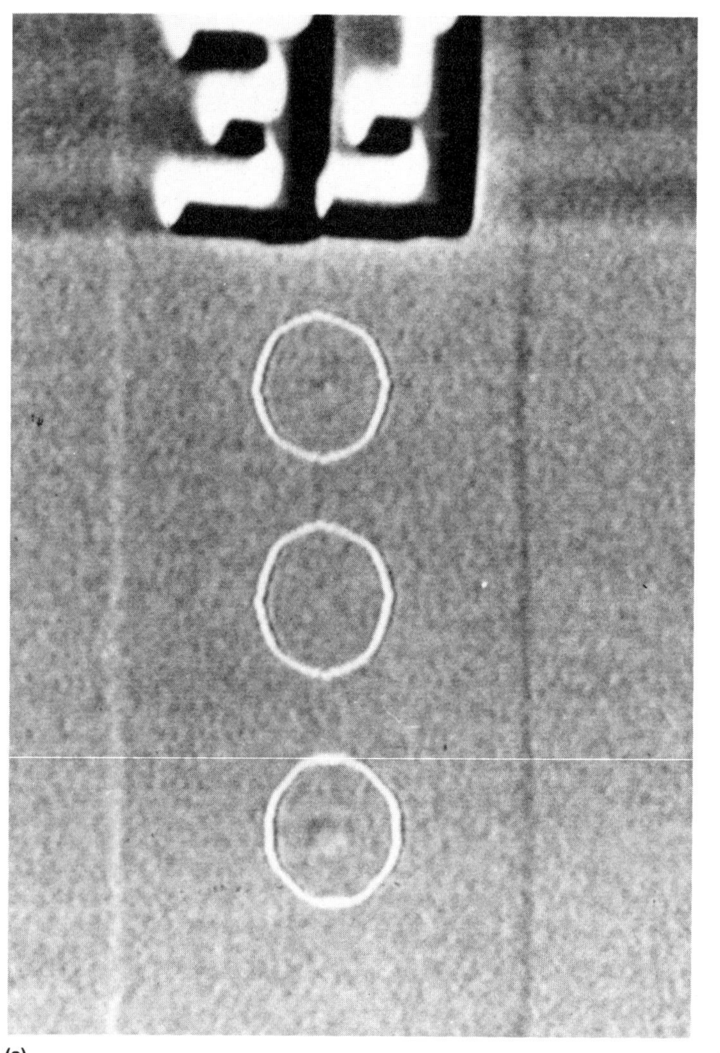

(a)

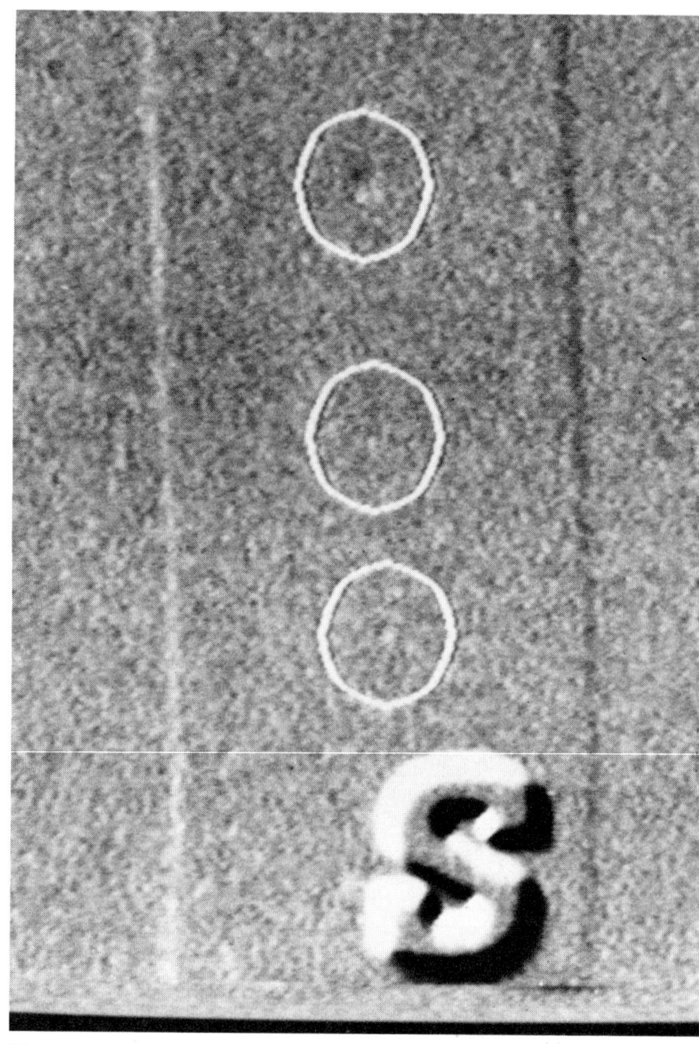

(b)

Fig. 34 Examples of plaque penetrameter indications. (a) 0.025 mm (0.001 in.) thick penetrameter on a testpiece with a thickness of 1.3 mm (0.050 in.). (b) 0.125 mm (0.005 in.) penetrameter on a testpiece with a thickness of 6.4 mm (0.25 in.)

tion of incident light transmitted through the film. The relationship of density to opacity and transmittance is as tabulated below:

Density, log (I_0/I_t)	Opacity, I_0/I_t	Transmittance, I_t/I_0
0	1	1.00
0.3	2	0.50
0.6	4	0.25
1.0	10	0.10
2.0	100	0.01
3.0	1 000	0.001
4.0	10 000	0.0001

If the light transmitted through the film is half of the incident light (transmittance = 0.5), the density is only 0.3, and for a density of 1.0, only one-tenth of the incident light intensity is transmitted. The fact that only 1/100 of the incident light is transmitted at a density of 2.0 is the reason that it is necessary to use high-intensity illuminators for viewing industrial radiographs at densities exceeding 2.0.

Exposure is the intensity of radiation multiplied by the time during which it acts, that is, the amount of energy that reaches a particular area of the film and that is responsible for producing a particular density on the developed film. Exposure can be specified either in absolute units (such as ergs per square centimeter or roentgens) or in relative units (where one particular exposure is used as a reference and all others are specified in terms of that reference). Unless equipment and time are available for making the precise radiation measurements that are required for defining exposure in absolute units, relative exposure is much more convenient than absolute and is equally as useful.

In discussing film properties, only absolute exposure or relative exposure as defined above is appropriate. Relative exposure is used in the present discussion.

Characteristic Curves. The relation between the exposure applied to a given type of radiographic film and the resulting density is expressed in a curve known as the characteristic curve of that particular type of film. Other names for this curve are the H and D curve, D log E curve, or sensitometric curve. Such curves are determined by applying a series of known exposures to the film and, after processing the film according to a standard procedure, reading the resulting densities. The curve is generated by plotting density against the logarithm of relative exposure. Figure 36(a) shows the characteristic curves of three commercial films exposed to x-radiation between lead screens.

Relative exposure is used partly because there are no convenient units suitable to all kilovoltages and scattering conditions and partly because it is easy to determine the logarithm of relative exposure. Using the logarithm of relative exposure instead of only relative exposure has several other advantages; for example, the otherwise long scale is compressed, and ratios of intensities or exposures (which are determined by simply subtracting logarithms) are usually more significant in radiography than actual exposures or intensities.

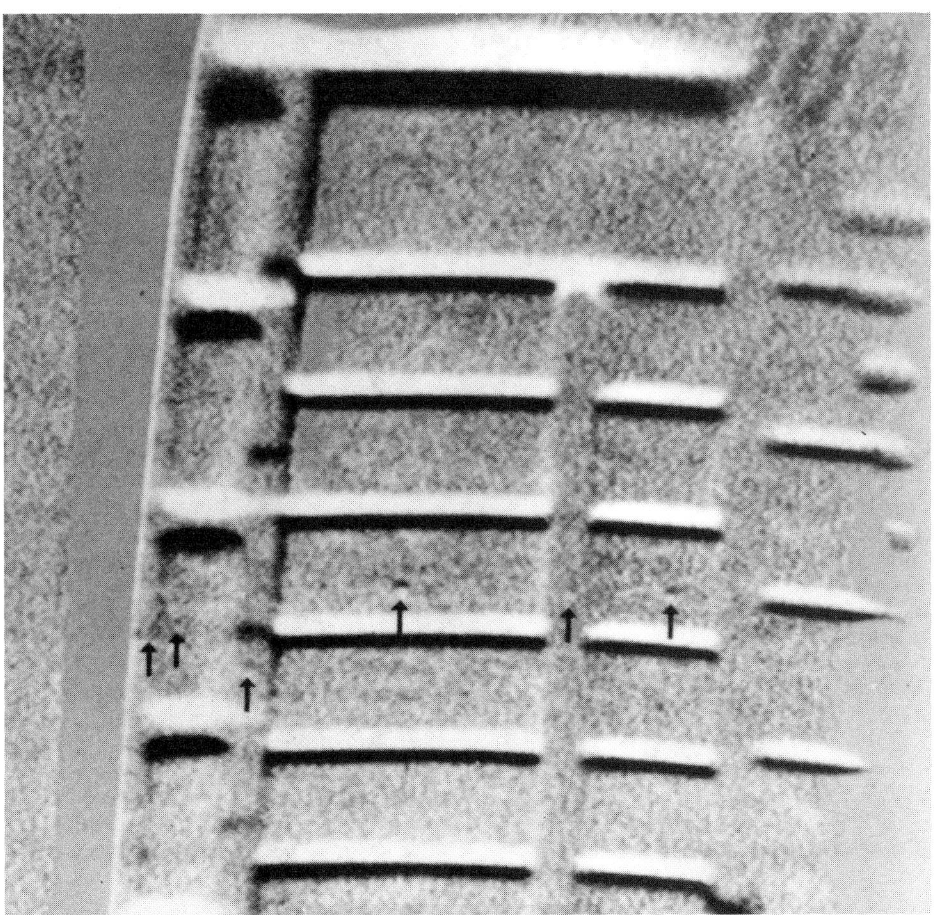

Fig. 35 Radiographic image of holes with a nominal diameter of 0.25 mm (0.01 in.) and a depth of 2% of wall thickness. Holes are indicated with arrows. Courtesy of Lockheed Missiles & Space Company, Inc.

Characteristic curves are very useful in determining the speed and gradient of the film as well as indicating the type of film. For example, in Fig. 36(a), characteristic curves for films A and B are J-shaped, which is typical of industrial x-ray film of types 1, 2, and 3. The curve for film C begins to flatten at a density between 3.0 and 3.5, giving it the S-shape typical of type 4 x-ray films used with fluorescent screens.

Film speed is inversely related to the time required for a given intensity of radiation to produce a particular density on the film; the shorter the exposure, the faster the film. In absolute units, film speed is inversely proportional to the total energy (roentgens) of a particular radiation spectrum (wavelength distribution at a given kilovoltage) that produces a given density on the film. For most practical applications, it is convenient and effective to deal with relative speeds. In using relative speeds, film speeds are expressed in terms of the speed of one particular film whose relative speed is arbitrarily assigned a value, for example, 100. Thus, if film A requires half the exposure of film B, the slower film (film B) is chosen as the standard and assigned an arbitrary speed of 100. Film A, which is twice as fast, would have a relative speed of 200.

To avoid making absolute measurements to determine film speed, it is convenient to refer to a group of film curves such as those in Fig. 36(a). Curves positioned to the left of the chart require less exposure for a given density; those to the right, more exposure. Thus, for a density of 2.0, film C in Fig. 36(a) is faster than film A, and both are faster than film B. Their relative speed is calculated by determining the differences in log relative exposure and converting to the antilog. Film B is the slowest film and requires a log relative exposure of 2.5 for a density of 2.0. Film B is chosen as the standard and is assigned a speed of 100. Film A requires a log relative exposure of 1.9 for the same density. Subtracting 1.9 from 2.5 gives a difference of 0.6, for which the antilog is about 4, so the relative speed of film A is four times that of film B, or has a relative speed of 400. Similarly, at a density of 2.0, film C requires a log relative exposure of 1.6. Subtracting 1.6 from 2.5 gives a log difference of 0.9, for which the antilog is about 8. Thus, film C is eight times faster than film B, or has a relative speed of 800 at a density of 2.0.

Another advantage of using groups of characteristic curves is that the visual assessment of relative speeds can be easily made. For example, although films A and B have similarly shaped curves running almost parallel, the curve for film C is radically different. As calculated above, film C has a relative speed of 800 at a density of 2.0; however, at greater densities, the relative speed of film C is lower than 800, and at lesser densities (down to a density of 1.0), the relative speed of film C is greater than 800. Therefore, whenever relative speeds are used, the density at which they were determined must be given.

Film gradient, also called film contrast, is a measure of the slope of the characteristic curve. If the difference is great, the gradient (contrast) is said to be high. If the difference is slight, the gradient is said to be low. The contrast seen on a radiograph is known as radiographic contrast and is composed of two factors:

- Subject contrast, which is the result of variations in the amount of radiation absorbed by the testpiece and which causes variations in radiation intensity impinging on the film
- Film gradient (or film contrast), which is a measure of the response of the emulsion to the intensity of impinging radiation and is a characteristic of a given film

Film gradient is determined from the characteristic curve by finding the slope of the curve at a given density. The slope of the characteristic curve changes continuously over its entire length, as indicated in Fig. 36(a). The steeper the slope of the curve over a range of relative exposures, the greater the difference in density and therefore the greater the resolution of detail; thus, high gradient is important for good radiography.

In Fig. 36(b) tangents to a characteristic curve for a typical x-ray film have been drawn at two points, and the two corresponding gradients (*a*/*b* and *c*/*d*) have been evaluated. Note that the gradient varies from 0.8 in the shallow portion of the curve (called the toe) to 5 in the steeper portion of the curve.

Two regions of a testpiece that are slightly different in thicknesses will transmit slightly different intensities of radiation to the film. On the characteristic curve, this would represent a small difference in the log relative exposure. For example, assume that at a given kilovoltage the thinner section transmits 20% more radiation than the thicker section. The difference in log relative exposure is 0.08 and is independent of tube current, exposure time, or source-to-film distance. In the toe of the characteristic curve where the gradient is only 0.8, the density difference is only 0.06, as indicated in Fig. 36(c). However, in the steeper portion of the curve where the gradient is 5, the density difference is 0.40. This effect of film gradient is the main reason why it is best to use high exposures, obtaining the

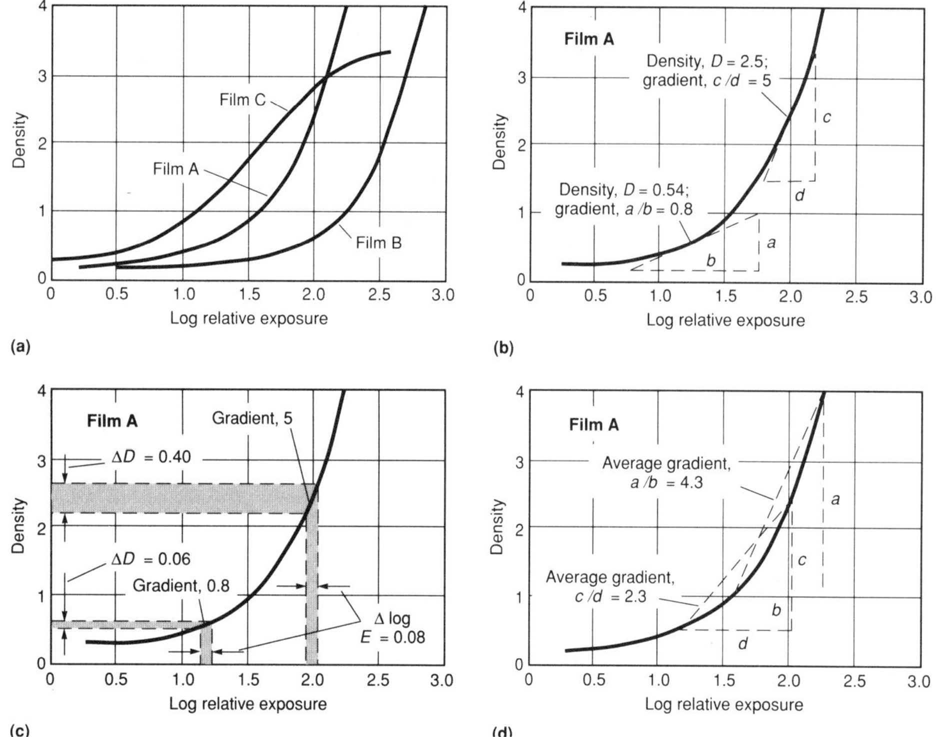

Fig. 36 Characteristic curves for x-ray film that determine type of film and film gradient, speed, and density. (a) Typical curves for three industrial x-ray films exposed to x-radiation between lead screens. (b) Evaluation of gradients at two points on the curve for film A in (a). (c) Density differences (ΔD) corresponding to a 20% difference in relative exposure (Δ log E = 0.08) determined for the two values of gradients evaluated in (b). (d) Average gradients for film A determined over two density ranges

highest density that can be viewed on a given illuminator for the greatest resolution of minor differences in transmitted radiation.

It is often more convenient to express gradient as an average over a given useful range of densities than for a single density. This is a simple calculation in which the difference between the two densities is divided by the difference between the log relative exposures for these densities; this ratio, known as the average gradient, is shown in Fig. 36(d).

Graininess. The silver halide grains that are contained in the emulsion of x-ray film are minute and can be seen only with a high-power microscope, such as an electron microscope. Even though the emulsion on each side of the film is only about 0.013 mm (0.0005 in.) thick, the grains are piled on top of each other in countless numbers. When the exposed-and-processed radiograph is viewed, these small individual silver grains appear grouped together in relatively large masses. This clumping, which is visible to the unaided eye or at low magnification, produces the visual impression called graininess.

All films exhibit some degree of graininess. In general, slower films have less graininess than faster films. This general relation is reflected in Table 6, which is an old classification scheme specified by the American Society for Testing and Materials (ASTM) in the earlier 1984 edition of ASTM E 94. Although this classification scheme is somewhat arbitrary, it is still sometimes specified in codes and specifications. Type 1 films have the least graininess, type 3 films exhibit the most graininess, and type 2 films are intermediate.

The energy of the radiation also affects the graininess of films. As the penetrating capability of radiation increases (photon energy increases and wavelength decreases), the graininess of all films increases, but the rate of increase may be different for different films. As a result, it is possible that of two film types, one will show more graininess at long wavelengths while the other shows more graininess at short wavelengths. The graininess of the images that are produced at high photon energies makes the inherently fine-grain films such as types 1 and 2 particularly useful at photon energies of 1 MeV or more.

Another source of graininess occurs in radiography using fluorescent intensifying screens. The graininess seen in fluorescent-screen radiographs is primarily caused by the graininess that is inherent in the screen. This effect overshadows the effect of film graininess, especially with films that are particularly sensitive to screen fluorescence. The graininess of fluorescent-screen radiographs increases significantly with increased kilovoltage.

Table 6 General characteristics of the four types of radiographic film specified in the earlier (1984) edition of ASTM E 94

These groupings are given only for qualitative comparisons. For a more detailed discussion on film classification, see the section "Film Types" in this article.

Film type	Film characteristic: Speed	Gradient	Graininess
1	Low	Very high	Very fine
2	Medium	High	Fine
3	High	Medium	Coarse
4(a)	Very high(b)	Very high(b)	(c)
	Medium(d)	Medium(d)	Medium(d)

(a) Normally used with fluorescent screens. (b) When used with fluorescent screens. (c) Graininess is mainly a characteristic of the fluorescent screens. (d) When used for direct exposure or with lead screens

Lead screens have little effect on the graininess of the image; only slight increases in graininess at equal densities have been noted when lead-screen radiographs are compared with direct radiographs made at the same photon energy.

Spectral Sensitivity. The shape of the characteristic curve of a given x-ray film is for all practical purposes unaffected by the wavelength distribution in the x-ray or γ-ray beam used for the exposure. However, the sensitivity of the film in terms of roentgens required to produce a given density is strongly affected by radiation energy (beam spectrum of a given kilovoltage or given γ-ray source).

Figure 37 shows the exposure required for producing a density of 1.0 on radiographic film developed in an x-ray developer for 5 min at 20 °C (68 °F). The exposures were made directly, without screens. The spectral-sensitivity curves for all x-ray films have approximately the same general features as the curves shown in Fig. 37. The details, however, differ from film to film; for example, the ratio of maximum-to-minimum sensitivity of the film to roentgens of exposure varies. The details of the curves also differ with the degree that long wavelengths have been filtered from the beam (Fig. 37).

The spectral sensitivity of a given film, or the difference of spectral sensitivity between films, is usually considered in the preparation of exposure charts and tables of relative speeds. For example, in tables of relative speeds, the speed of a particular film usually varies with tube voltage. Also, in the preparation of exposure calculators for γ-rays, where roentgens necessary to expose a given film to produce a given density are extensively used, spectral sensitivity is important. As a result, manufacturers of radiographic film often furnish tables of spectral sensitivity for the films they manufacture, indicating the roentgens required to produce several net densities at

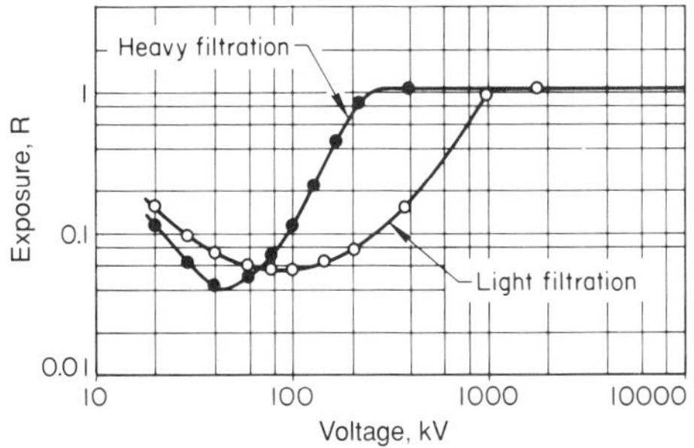

Fig. 37 Spectral-sensitivity curves for a radiographic film showing exposure required to produce a density of 1.0

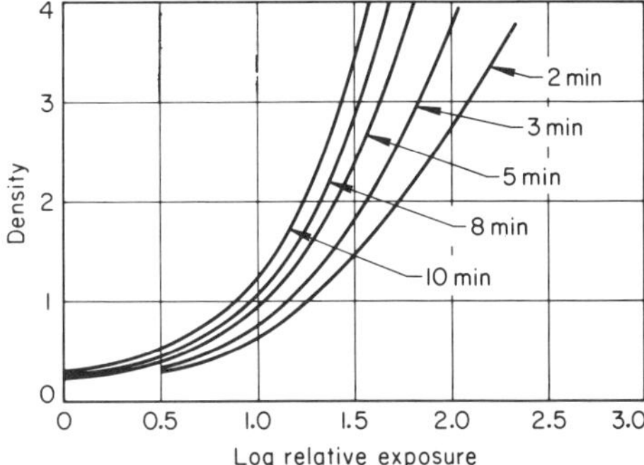

Fig. 38 Effect of various developing times on the characteristic curve of a radiographic film developed at 20 °C (68 °F) in a manual process

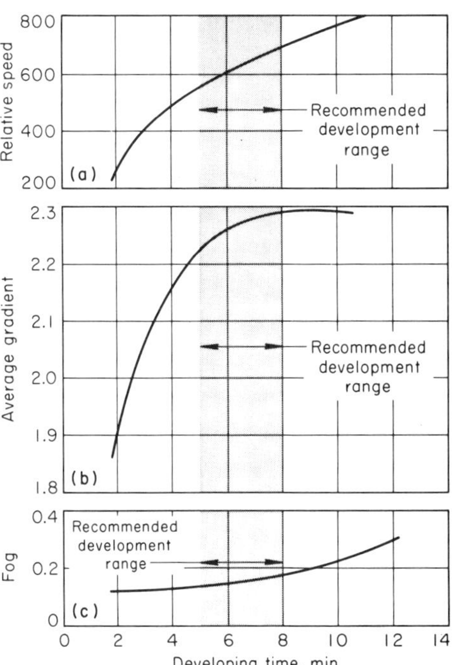

Fig. 39 Effect of developing time on (a) relative speed, (b) average gradient, and (c) fog for a radiographic film. These graphs were derived from the same data as Fig. 38.

various x-ray tube voltages and γ-ray photon energies.

Effect of Development on Film Characteristics. Although the shape of the characteristic curve is not affected by variations in photon energy, it is affected by variations in the degree of development. The degree of development depends on the type of developer used (including its degree of activity), temperature, and developing time. These factors are discussed in the Appendix to this article. Within limits, an increased degree of development increases the speed and gradient of a given film. If the limits are exceeded (that is, the film is overdeveloped), film speed based on density (density above fog) no longer increases and may even decrease. Also, the fog (gray cloudiness) resulting from excessive action of developer chemicals on unexposed grains will increase, and the gradient may decrease.

Figure 38 shows the characteristic curve of a radiographic film as affected by various times of manual development at 20 °C (68 °F) in an x-ray developer mixed from powder. As the developing time increases, the characteristic curve progressively becomes steeper (gradient increases) and positioned more to the left on the chart (speed increases). It is from such curves that values of relative speed and average gradient can be determined and developing recommendations can be derived.

Figure 39 shows the values of relative speed, average gradient, and fog plotted against developing time for the same film and processing conditions detailed in Fig. 38. These curves indicate that at an 8-min developing time the average gradient has reached its peak and further development would decrease film gradient. The speed, gradient, and fog curves of Fig. 39 are characteristic of this type of film, and similar curves are different for other types of films.

The degree of development also affects graininess. For example, if development time is increased to produce higher film speed, the graininess is also increased. On the other hand, a developer or developing technique that gives an appreciable reduction in graininess will also result in an appreciable loss of film speed. Adjustments in developing techniques made to compensate for changes in temperature or in activity of a developer have little effect on graininess.

The variation of film gradient, speed, fog, and graininess with development time and temperature for a given radiographic film and developer solution makes it imperative that standardized procedures for development of radiographic film be meticulously followed. Unless this is done, it is impossible to obtain consistent resolution of variations in radiation intensity. Inconsistency in the resolution of variations in radiation intensity can seriously impair the ability to detect flaws in metals.

Film Types and Selection

The selection of radiographic film for a particular application is generally a compromise between the desired quality of the radiograph and the cost of exposure time. This compromise occurs because slower films generally provide a higher film gradient and a lower level of graininess and fog.

Film Types. The classification of radiographic film is complicated, as evidenced by changes in ASTM standard practice E 94. The 1988 edition of ASTM E 94 references a new document (ASTM E 746-87), which describes a standard test method for determining the relative image-quality response of industrial radiographic film. Careful study of this document is required to arrive at a conclusive classification index suitable for the given radiographic film requirements of a facility.

Earlier editions (1984 and prior) of ASTM E 94 contained a table listing the characteristics of industrial films grouped into four types. The general characteristics of these four types are summarized in Table 6. This relatively simple classification method is referenced by many codes and specifications, which may state only that a type 1 or 2 film can be used for their specification requirements. However, because of this relatively arbitrary method of classification, many film manufacturers may be reluctant to assign type numbers to a given film. Moreover, the characteristics of radiographic films can vary within a type classification of Table 6 because of inherent variations among films produced by different manufacturers under different brand names and because of variations in film processing that affect both film speed and radiographic density. These variations make it essential that film processing be standardized and that characteristic curves for each brand of film be obtained from the film manufacturer for use in developing exposure charts.

Because the variables that govern the classification of film are no longer detailed in ASTM E 94-88, it is largely the responsibility of the film manufacturer to determine the particular type numbers associated with his brand names. Some manufacturers indicate the type number together with the brand name on the film package. If there is doubt regarding the type number of a given brand, it is advisable to consult the manufacturer. Most manufacturers offer a brand of film characterized as very low speed, ultrahigh gradient, and extremely fine grain.

Film selection for radiography is a compromise between the economics of exposure (film speed and latitude) and the quality desired in the radiograph. In general, fine-grain, high-gradient films produce the highest-quality radiographs. However, because of the low speed typically associated with these films, high-intensity radiation or long exposure times are needed. Other fac-

Table 7 Guide to the selection of radiographic film for steel, aluminum, bronze, and magnesium in various thicknesses

Thickness mm	(in.)	Type of film(a) for use with these x-ray tube voltages, or radioactive isotopes: 50–80 kV	80–120 kV	120–150 kV	150–250 kV	Ir-192	250–400 kV	1 MeV	Co-60	2 MeV	Ra	6–31 MeV
Steel												
0–6	(0–¼)	3	3	2	1	...	...	...	...	...	...	...
6–13	(¼–½)	4	3	2	2	...	1	...	...	...	...	...
13–25	(½–1)	...	4	3	2	2	2	1	...	1	2	...
25–50	(1–2)	...	...	...	3	2	2	1	2	1	2	1
50–100	(2–4)	...	...	...	4	3	4	2	2	2	3	1
100–200	(4–8)	...	...	...	...	...	4	3	3	2	3	2
>200	(>8)	...	...	...	...	...	...	...	...	3	...	2
Aluminum												
0–6	(0–¼)	1	1	...	...	...	...	...	...	...	...	...
6–13	(¼–½)	2	1	1	1	...	...	...	...	...	...	...
13–25	(½–1)	2	1	1	1	...	1	...	...	...	...	...
25–50	(1–2)	3	2	2	1	1	1	...	...	...	...	...
50–100	(2–4)	4	3	2	2	1	2	...	...	...	...	...
100–200	(4–8)	...	4	3	3	2	3	...	...	...	...	...
>200	(>8)	...	...	...	...	4	...	...	...	...	...	...
Bronze												
0–6	(0–¼)	4	3	2	1	1	1	1	...	...	...	...
6–13	(¼–½)	...	3	2	2	2	1	1	...	1	...	...
13–25	(½–1)	...	4	4	3	2	2	1	2	1	2	...
25–50	(1–2)	...	...	4	4	3	3	1	2	1	2	1
50–100	(2–4)	...	...	...	...	3	4	2	3	2	3	1
100–200	(4–8)	...	...	...	...	...	...	3	3	2	...	2
>200	(>8)	...	...	...	...	...	...	...	...	3	...	2
Magnesium												
0–6	(0–¼)	1	1	...	...	...	...	...	...	...	...	...
6–13	(¼–½)	1	1	1	...	...	...	...	...	...	...	...
13–25	(½–1)	2	1	1	...	1	...	...	...	...	...	...
25–50	(1–2)	2	1	1	1	1	...	...	...	...	...	...
50–100	(2–4)	3	2	2	1	2	...	...	...	...	...	...
100–200	(4–8)	...	3	2	2	3	...	...	...	...	...	...
>200	(>8)	...	...	...	4	...	...	...	...	...	...	...

(a) These recommendations represent a usually acceptable level of radiographic quality and are based on the qualitative classification of films defined in Table 6. Optimum radiographic quality will be promoted by use of the lowest-number film type that economic and technical considerations will allow. The recommendations for type 4 film are based on the use of fluorescent screens.

tors affecting radiographic quality and film selection are the type and thickness of the testpiece and the photon energy of the incident radiation.

Although the classification of film is more complex than the types given in Table 6, a general guide is that better radiographic quality will be promoted by the lowest type number in Table 6 that economic and technical considerations will allow. In this regard, Table 7 suggests a general comparison of film characteristics for achieving a reasonable level of radiographic quality for various metals and radiation-source energies. It should be noted, however, that the film types are only a qualitative ranking of the general film characteristics given in Table 6. Many radiographic films, particularly those designed for automatic processing, cannot be adequately classified according to the system in Table 6. This compounds the problem of selecting film for a particular application.

Film latitude, which is the range of testpiece thickness that can be recorded with a single exposure, also influences film selection (see the following section in this article for a discussion of latitude). High-gradient films generally have narrow latitude, that is, only a narrow range of testpiece thickness can be imaged with optimum density for interpretation. If the testpiece is of nonuniform thickness, more than one exposure may have to be made (using different x-ray spectra or different exposure times) for complete inspection of the piece. The number of exposures, as well as the exposure times, can often be reduced by using a faster film of lower gradient but wider latitude, although there is usually an accompanying reduction in ability to image small flaws.

Exposure Factors

The exposure time in film radiography depends mainly on film speed, the intensity of radiation at the film surface, the characteristics of any screens used, and the desired level of photographic density. In practice, the energy of the radiation is first chosen to be sufficiently penetrating for the type of material and thickness to be inspected. The film type and the desired photographic density are then selected according to the sensitivity requirements (Eq 2) for recording the expected variations in the intensity of the transmitted radiation. Once these factors are fixed, then the source strength, the source-to-film distance, and the characteristics of any screens used determine the exposure time.

With a given type of film and screen, the exposure time to produce the desired photographic density obeys the reciprocity law for equivalent exposures (Eq 5). The reciprocity law can be modified to include the inverse-square law (Eq 13), which models the relation between radiation intensity at the film and the source-to-film distance. Therefore, because the intensity is inversely proportional to the square of distance from the source according to Eq 13, the reciprocity law for equivalent exposures with an x-ray tube (Eq 5) can be rewritten as:

$$\frac{i_1 t_1}{L_1^2} = \frac{i_2 t_2}{L_2^2} \qquad \text{(Eq 16)}$$

where i is the tube current, t is the exposure time, L is the source-to-film distance, and the subscripts refer to two different combinations that produce images with the desired photographic density. The parallel expression that applies to exposures made with a γ-ray source is:

$$\frac{a_1 t_1}{L_1^2} = \frac{a_2 t_2}{L_2^2} \qquad \text{(Eq 17)}$$

where a is the source strength in gigabecquerel (curies).

For practical applications using the same quality of x-rays (same kilovoltage) or the same radioactive source, Eq 16 or 17 is applied in the following manner. If a satisfactory radiograph of a given testpiece can be obtained at a 1 m (3.3 ft) source-to-film distance, 20-mA tube current, and 10-s exposure time, an equivalent radiograph can be obtained in 6.4 s with 20-mA tube current or in 8 s with 16-mA tube current when the distance is reduced to 0.8 m (2.6 ft). However, if the source-to-film distance is increased to 2 m (6.6 ft), it will take 40 s at 20 mA or 32 s at 25 mA or 16 s at 50 mA to produce equivalent film density.

Similarly, if a satisfactory radiograph can be obtained with a 440 GBq (12 Ci) cobalt-60 source at 1 m (3.3 ft) source-to-film distance in 20 min, a radiograph of equivalent density will require a 45-min exposure time with the same 440 GBq (12 Ci) source if the distance is increased to 1.5 m (4.9 ft).

Exposure Charts for X-Ray Radiography. A starting point must be determined for the calculations described in the discussions above. The starting point is ordinarily derived from exposure charts of the type shown in Fig. 40. Equipment manufacturers usually publish exposure charts for each type of x-ray generator that they manufacture. These published charts, however, are only approximations; each particular unit and each installation is unique. Radiographic density is affected by such factors as radiation spectrum, film processing, setup technique, amount and type of filtration, screens, and scattered radiation.

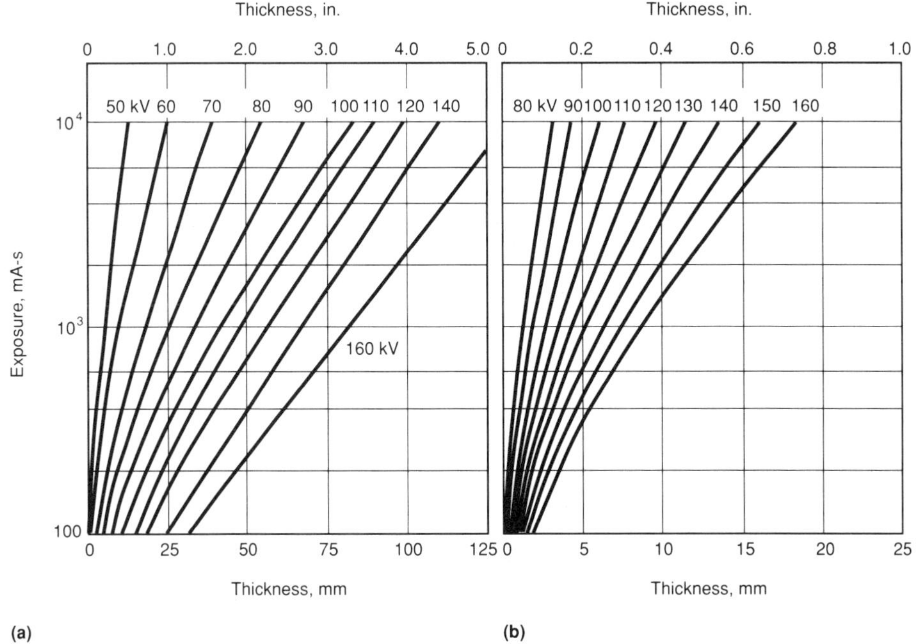

Fig. 40 Typical radiographic exposure charts for (a) aluminum and (b) steel for a film density of 2.0 without screens that relate exposure to thickness of testpieces for several values of tube voltage. Charts for aluminum and steel were prepared specifically for an Andrex 160-kV directional x-ray machine, using a source-to-film distance of 910 mm (36 in.) and Industrex AA film (Eastman Kodak) developed in a manual process for 7 min in PIX developer (Picker).

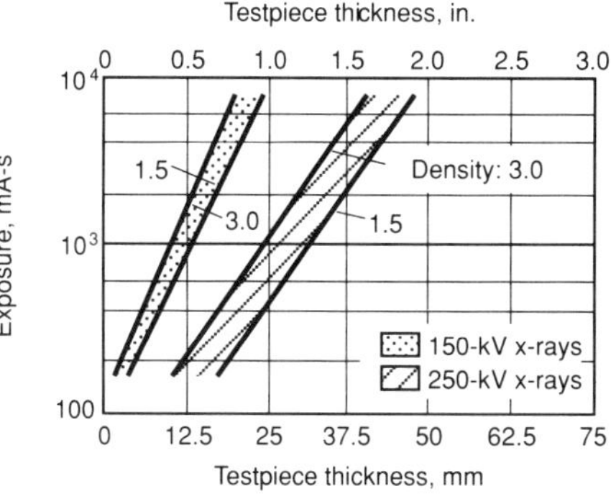

(a)

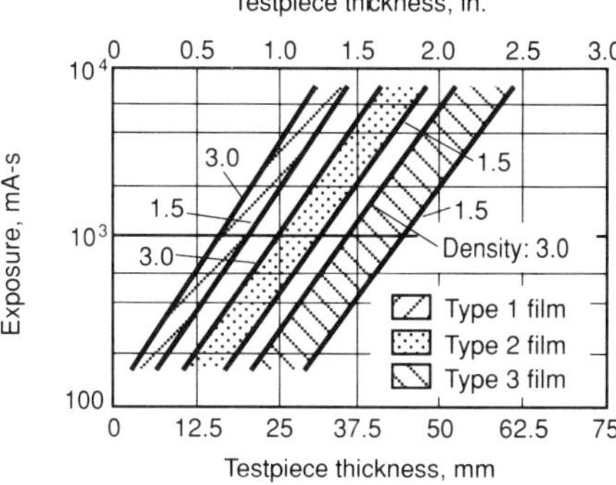

(b)

Fig. 41 Latitude curves for the radiographic inspection of steel at film densities ranging from 1.5 to 3.0. (a) Effect of tube voltage on latitude of a type 2 radiographic film. (b) Effect of film type on latitude for radiography using 250-kV x-rays. Curves were prepared from data obtained using a 250-kV x-ray machine, 910-mm (36-in.) source-to-film distance, and 0.25 mm (0.010 in.) thick lead screens. Films were processed using a standardized technique.

Although published exposure charts are acceptable guides for equipment selection, more accurate charts that are prepared under normal operating conditions are recommended for each x-ray machine. A simple method for preparing accurate exposure charts is as follows:

1. Make a series of radiographs of a calibrated multiple-thickness step wedge, using several different values of exposure at each of several different tube voltage settings
2. Process the exposed films together under conditions identical to those that will be used for routine application
3. From the several densities corresponding to different thicknesses, determine which density (and thickness) corresponds exactly with the density desired for routine application. This step must be done with a densitometer because no other method is accurate. If the desired density does not appear on the radiograph, the thickness corresponding to the desired density can be found by interpolation
4. Using the thickness determined in step 3 and the tube voltage (kilovoltage) and exposure (milliamp-second or milliamp-min) corresponding to that piece of film, plot the relation of thickness to exposure on semilogarithmic paper with exposure on the logarithmic scale
5. Draw lines of constant tube voltage through the corresponding points on the graph

The resulting chart will be similar to those in Fig. 40, but will be accurate for the x-ray machine, type of film, and film development technique that were used. Different charts should be prepared for different film types, both with and without screens.

Charts prepared as described above will be strictly accurate only for testpieces of uniform thickness. Some adjustment in exposure or source-to-film distance will have to be made for the more usual circumstances involving testpieces of nonuniform thickness.

Latitude charts can sometimes be used to help determine the correct exposure when the thickness of a testpiece varies within the area of coverage. Latitude is a range of metal thickness that produces a specific range of density in the processed film for a given combination of radiation spectrum, source-to-film distance, and exposure. The bands in Fig. 41 are latitude curves for the inspection of steel testpieces using x-rays from a 250-kV x-ray machine. As shown in Fig. 41(a), increasing the tube voltage not only decreases the exposure needed to radiograph a section of given nominal thickness but also increases the range of testpiece thickness that is satisfactorily recorded on the radiograph. For example, at a nominal steel thickness of 15 mm (⅝ in.), about 2200 mA-s exposure is needed to produce a satisfactory image on type 2 film with 150-kV x-rays; the image will have satisfactory density for a range of thickness from about 14 to 18 mm (0.57 to 0.70 in.). On the other hand, when 250-kV x-rays are used, an equivalent image can be produced on type 2 film with only about 220 mA-s exposure, and thicknesses from about 13 to 20 mm (0.50 to 0.78 in.) can be recorded with satisfactory density.

Alternatively, latitude charts for different types of film can be used to determine the ranges of thickness covered by each type at a given exposure (Fig. 41b). The difference in latitude for different types of film is one basis for the use of multiple-film techniques. For example, if a steel testpiece is radiographed using 250-kV x-rays and a film holder loaded with one sheet each of types 1, 2, and 3 film, then a range of testpiece thickness of 13 to 40 mm (0.5 to 1.6 in.) can be recorded with a single exposure of 600 mA-s. Each of the films is viewed and interpreted separately for the range of thickness corresponding to its optimum density range. For thicknesses corresponding to gaps between optimum density ranges for the different films, enough contrast and detail usually exist in one or both of the films on either side of the gap to yield a satisfactory image.

When filters are used or when other variations in technique are introduced, further adjustments in exposure or source-to-film distance will have to be made. In fact, if large numbers of a given part are to be inspected, it is usually worthwhile to make a series of radiographs of a representative testpiece using different tube voltages and different exposures. From the results of these experiments a standard setup and exposure can be established.

When there is only one opportunity to inspect a part for which there is no established standard technique or when circumstances will not permit experimentation, one of several alternative techniques can be used. For example, replicate radiographs can be made using different values of one or more of the following: tube voltage, tube current, exposure time, source-to-film distance, or film speed. Alternatively, film holders can be loaded with two or more sheets of film, either of the same type or of different types; the resulting radiographs can be viewed both as double-film and single-film images to obtain wider latitude. The latter technique, using two or more sheets of film, is especially useful when there are large variations in testpiece thickness. Sometimes, it can be advantageous to adopt the technique for routine applications. Another method involves making duplicate radiographs with and without a filter or with and without lead screens. In such cases, exposure time and tube current will have to be varied to compensate for either beam attenuation with a filter or the combined effects of filtration and intensification with lead screens.

Exposure charts apply only to the material of which the step wedge was composed. Most often, this is a standard material such as aluminum or steel. However, exposure charts for a standard material can be used to determine exposure factors for other materials by applying radiographic equivalence factors such as those listed in Table 5. First, the exposure is derived from the exposure chart as if the part were actually made out of the standard material. The exposure so derived is then multiplied by the radiographic equivalence factor from Table 5. For example, if a 13 mm (½ in.) thick part made of titanium is to be radiographed using 150-kV x-rays without screens, the exposure for a 13 mm (½ in.) thick part made of steel can be determined from the exposure chart in Fig. 40(b). That exposure would be 4000 mA-s for a source-to-film distance of 910 mm (36 in.). From Table 5, the radiographic equivalence of titanium for 150-kV x-rays is 0.45. Thus, for 150-kV x-rays and a source-to-film distance of 910 mm (36 in.), the exposure for the titanium part would be 0.45 times 4000 mA-s, or 1800 mA-s. A reasonable exposure for this part would be 2 min (120 s) at 15-mA tube current.

In addition to exposure charts, nomograms or specially constructed slide rules are often used for calculating radiographic exposures. These devices can be constructed using the same type of information as that used in constructing exposure charts. The main advantage of these devices is speed in making a calculation, which can reduce setup time and produce economic benefits.

Gamma-ray exposure charts are constructed in a manner similar to the exposure charts used in determining x-ray exposures. Instead of expressing the exposure in milliampere-seconds, γ-ray exposures are expressed in curie-hours or curie-minutes. To use a γ-ray exposure chart, the source strength must be known. Source strength decreases exponentially with time, and each radioactive isotope had a characteristic half-life. This behavior can be used to determine the strength of a radioactive source at any one time, provided its strength at one time is known. (Normally, source manufacturers provide the source strength as of a given date along with each new source). A graph of source strength versus time is constructed as described in the following paragraph.

On the logarithmic scale of semilogarithmic graph paper, plot the known source strength. (For convenience, the date when the determination of source strength was made should be noted.) Divide the linear scale into convenient units of time (with the known source strength corresponding to zero time), extending the scale at least one half-life. At the time corresponding to one half-life, plot a point corresponding to half of the known source strength that was plotted at zero time. Draw a straight line between the two points, extending the line as far beyond the second point as desired. This line represents the source strength at any time; if desired, the linear scale may now be renumbered using dates, so that at any time the source strength corresponding to a given date can be seen at a glance.

Figure 42 shows two types of γ-ray exposure charts for cobalt-60 radiation and testpieces of steel, gray iron, or ductile iron. In Fig. 42(a), exposure in curie-minutes to produce an average photographic density of 2.0 is read directly for each combination of testpiece thickness and source-to-film distance. Exposure time is determined by dividing exposure determined from the chart by the source strength determined as outlined in the preceding paragraph. In Fig. 42(b), an exposure factor is read directly for each combination of testpiece thickness and desired photographic density. The exposure time is calculated from:

$$t = E \times \frac{L_i^2}{S} \quad \text{(Eq 18)}$$

where t is the exposure time in minutes, E is the exposure factor, L_i is the source-to-film distance (in inches), and S is the source strength (in curies). For example, if it is desired to radiograph a 3 in. thick steel testpiece on Industrex AA film with a photographic density of 3.0 and if a 15-Ci cobalt-60 source is to be used at a source-to-film distance of 12 in., then the corresponding exposure factor is 0.58 (dashed line, Fig. 42), and from Eq 18, the exposure time is $t = 0.58 \times 12^2/15 = 5.6$ min.

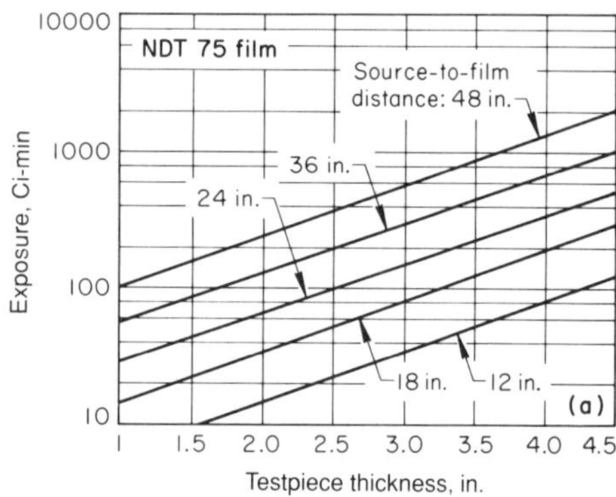

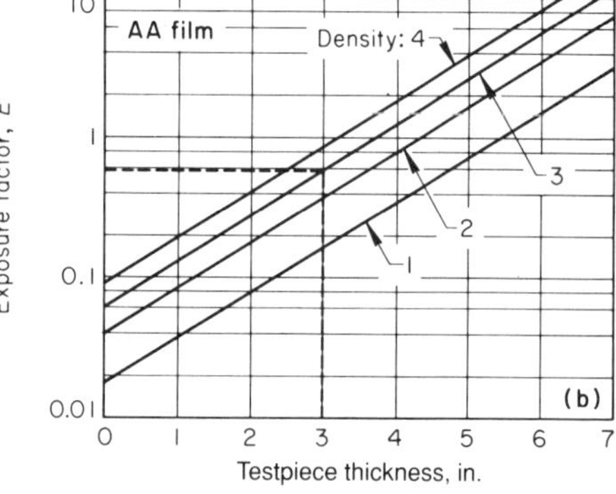

Fig. 42 Two types of exposure chart for computing γ-ray exposures that apply to Co-60 radioisotopes and steel, gray iron, or ductile iron testpieces. (a) Exposure in curie-minutes to produce a photographic density of 2.0 on NDT 75 film (DuPont) as a function of testpiece thickness for various source-to-film distances. (b) Exposure factor, E, for Industrex AA film (Eastman Kodak) as a function of testpiece thickness; four lines of constant photographic density are shown.

As with x-ray exposures, latitude charts such as the one shown in Fig. 43 can be used to determine the range of thickness that can be recorded with a given exposure. By comparing Fig. 43 with Fig. 41, it can be seen that the latitude of type 2 film with cobalt-60 radiation is almost four times the latitude of type 2 film with 250-kV x-rays. Also, as with x-ray exposures, nomograms or slide rules can be constructed to simplify calculation of γ-ray exposures.

Selection of View

The view selected for the radiography of a testpiece is a major factor that controls the ability to detect certain types of flaws. In some circumstances, although flaws are detected, the selected view presents an unsatisfactory or distorted picture of the relationship of the flaws to testpiece shape. For example, a crack in the fillet of a T-shape section of a casting or in a weld in a T-shape weldment is most likely to be revealed when

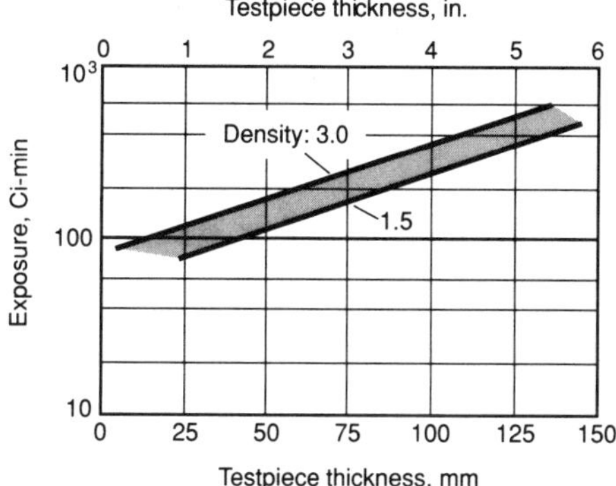

Fig. 43 Latitude curve for radiographic inspection of steel using Co-60 γ-rays to produce a film density range of 1.5 to 3.0. Curve is for type 2 film exposed at a source-to-film distance of 810 mm (32 in.) and with 0.25 mm (0.010 in.) thick lead screens both front and back.

the radiation is directed along the bisector of the angle between the legs of the section, because most cracks in such a location run perpendicular to the surface midway between the legs of the T. When the radiation is directed at an equal angle to both legs, it is most likely to be parallel to any crack that is present, which is the most favorable orientation for revealing cracks. With this orientation, however, there is no distinct line of demarcation between the sections of different thickness, and the portions of the testpiece that have the greatest object-to-detector distance will be recorded with a considerable degree of unsharpness and distortion.

Radiographic Shadows. As discussed in the section on "Principles of Shadow Formation" in this article, radiographic images can be compared to the shadows formed when a beam of light is interrupted by an opaque object. The difference between shadows and radiographic images is that in the umbra of a shadow there is a complete absence of radiation from the light source, while in the umbral region of a radiograph the amount of radiation that falls on the image conversion medium depends mainly on the absorption characteristics of the testpiece. Ordinarily, only the umbral region of a radiograph is of interest because it is only in this region that internal features of the testpiece are revealed.

When a section of the testpiece contains a discontinuity that is essentially planar, such as a crack or cold shut, the planar discontinuity will be revealed only when the radiation is parallel or nearly parallel to that discontinuity (Fig. 44a). Such a discontinuity can be revealed as a difference in absorption compared to the surrounding material only when the radiation impinges on the edge of the discontinuity, as shown at left in the actual setup in Fig. 44(a). In all other orientations, such as that shown at right in the actual setup in Fig. 44(a), the effective thickness of the testpiece is the

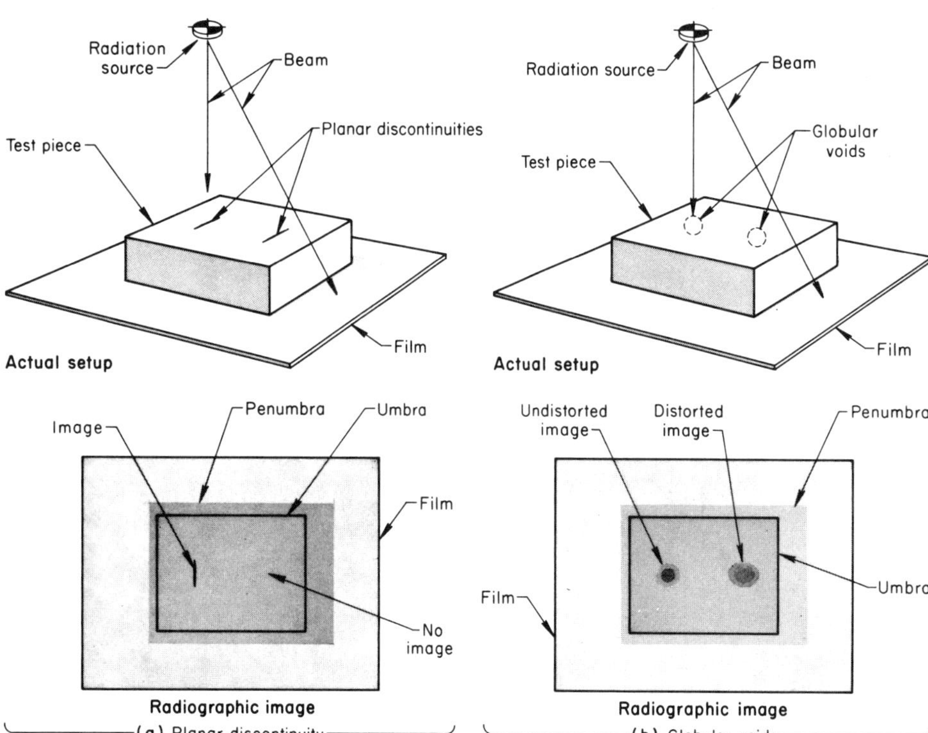

Fig. 44 Effect of direction of radiation beam on the appearance of (a) planar discontinuities (such as cracks or cold shuts) and (b) globular voids. See text for discussion.

same, whether in the region of the planar discontinuity or not, and there is no difference in absorption.

When a section of the testpiece contains a flaw that is spherical in shape, such as a globular void or inclusion, the flaw will be revealed regardless of viewing direction. However, as shown in Fig. 44(b), the shape of a void will be undistorted only when it is aligned with the portion of the radiation beam that is perpendicular to the plane of the recording medium. Distortion is less of a problem for long source-to-detector distances and small testpieces than for short source-to-detector distances or large testpieces.

The size of almost all spherical flaws is difficult to assess. The thickness of a spherical flaw is greatest along a line through the center of the flaw and decreases to nil at the outer edge. Correspondingly, the thickness of the surrounding material is at least at the center of the spherical flaw. Thus, for example, the amount of radiation passing through the testpiece is greatest in the shadow of the center of a void. Progressively outward from the center, the image of the void will grow gradually less dense as the effective thickness of the void grows smaller, and finally the image of the void will fade into the image of the surrounding material. Images of spherical flaws are more distinct and true in size when the flaws account for a substantial portion of the thickness of the section or when the features have absorption characteristics that are markedly different from those of the matrix. Also, the image will be larger than the flaw itself, less distinct, and more likely to be distorted the farther the flaw is above the plane of the image conversion medium.

Figure 45 shows the effect of viewing direction on the image of a solid block of material. When the center of the block is aligned with the central beam and one side of the block rests on the image conversion medium (in this case, film as shown in setup 1 in Fig. 45), the central region of the image presents an undistorted view of the lower surface of the block. Surrounding the undistorted region is a region representing an enlarged view of the upper portion of the block, and surrounding this region is the penumbra. When the block is tilted onto one edge but still aligned with the central beam (setup 2, in Fig. 45), the entire image is distorted and is surrounded by a distorted penumbra. When the block rests on the film but the radiation beam is inclined to both the block and the film (setup 3, Fig. 45), only the image of the lower surface of the block remains undistorted. The remainder of the image is enlarged or otherwise distorted; the greater the object-to-detector distance, the greater the enlargement or distortion. The penumbra is enlarged or distorted similarly to the way the image is affected.

It is important to understand that almost all radiographic images are somewhat distorted in size or shape or both, as compared to the actual features they represent. When

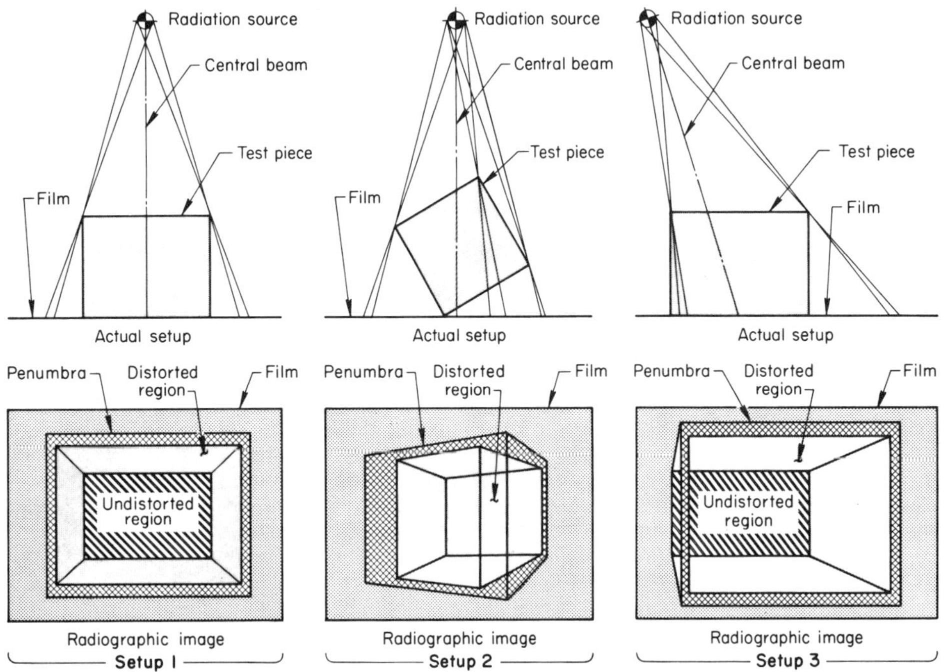

Fig. 45 Effect of viewing direction on the radiographic image of a solid block of material. See text for discussion.

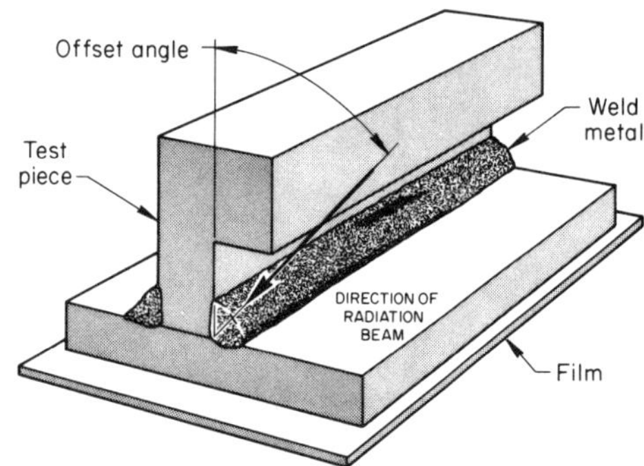

Fig. 46 Use of offset radiation beam to avoid superimposing image of one section of a complex-shape part on another when inspecting difficult-to-reach regions

viewing or interpreting radiographs, the individual features in the image should be recognized as projected shadows of the actual features in the testpiece.

Inspection of Simple Shapes

It is usually best to direct the radiation at right angles to a surface, along a path that presents minimum thickness to the radiation. This not only increases the radiation intensity reaching the detector but, more important, also ensures that any internal feature (except for a planar feature that is not parallel to the radiation beam) will generate the greatest subject contrast. Obviously, for a flaw of a given size, the difference between the amount of radiation transmitted through the material in the area of the flaw and the amount transmitted through an adjacent area will be greater if the flaw occupies a greater portion of the testpiece thickness.

When the presence of planar discontinuities (cracks) is suspected, radiation must be directed essentially parallel to the expected crack plane, regardless of testpiece thickness in that direction. In film radiography, any increased inspection cost resulting from increased exposure time is well justified because only a view parallel to the crack plane will reveal a crack.

Flat plates have the simplest shape and are usually inspected by radiography as an integral part of a more complex assembly or component. The most favorable direction for viewing a flat plate is one in which the radiation impinges perpendicular to the surface of the plate and penetrates the shortest dimension. The image is produced with minimum distortion and with maximum sensitivity and resolution of internal features.

When inspecting a difficult-to-reach region of a complex-shape part, it may be necessary to avoid superimposing the image of one section on another, by selecting a viewing direction other than 90° for a flat-plate section, as shown in Fig. 46 with film as the conversion medium. Nevertheless, except when there is a good reason to do otherwise, it is always desirable to use a viewing direction as close to 90° as possible. Except in the most unusual circumstances, views parallel or nearly parallel to the plate surface should be avoided. These views give severely reduced resolution of detail, a high degree of image distortion, and an excessively low level of radiation intensity transmitted through the testpiece.

When large areas are being radiographed, they should be inspected as a series of radiographs, each one overlapping the area of coverage of all adjacent radiographs. Use of a relatively short source-to-detector distance and multiple, overlapping radiographs is frequently more satisfactory than a single large radiograph. The long source-to-detector distance necessary to avoid the type of distortion illustrated in Fig. 44 and 45 reduces the intensity of transmitted radiation and may require frame summing in real-time radiography or longer exposures in film radiography. In film radiography, the longer exposures of the single-exposure technique may make the single-exposure technique uneconomical as compared to a multiple-exposure technique.

Curved plates are most satisfactorily inspected using views similar to those for flat plates. For optimum resolution of detail, the image conversion plane should be shaped to conform with that of the back surface of the curved plate. If the curved plate has its convex side toward the incident radiation, it is usually advantageous to minimize distortion by making multiple radiographs with reduced individual areas of coverage. If the curved plate has its concave side toward the incident radiation, a distortion-free image can be achieved by placing the source at the center of the radius of curvature. In this latter case, there is no angular limit to the area that can be inspected in a single radiograph, except for any limitations imposed by the directionality of the radiation field. Frequently, this technique is used for the on-site radiography of large-diameter pipes. The source is placed on the centerline of the pipe, and simultaneous exposures are made on one or more films wrapped around the outside of the pipe. The entire circumference is inspected with one shot because the source emits equal amounts of radiation in all directions.

This panoramic technique can also be used for the simultaneous inspection or exposure of several small parts. In a panoramic exposure, testpieces are placed in a circle around a source that emits equal radiation intensity in all directions. Separate films are placed behind each testpiece, and all are exposed at one time.

Solid cylinders can be inspected either by a longitudinal view, which is generally satisfactory only for short large-diameter cylinders, or by a transverse view, which is most satisfactory for relatively small-diameter cylinders. If the cylinder is neither short nor small in diameter, either view poses problems to the radiographer. However, in such cases, the transverse view is usually satisfactory.

The thickness of a cylindrical testpiece varies across a diametral plane in a similar manner as for a sphere, being thickest in the center and progressively thinning out to nil

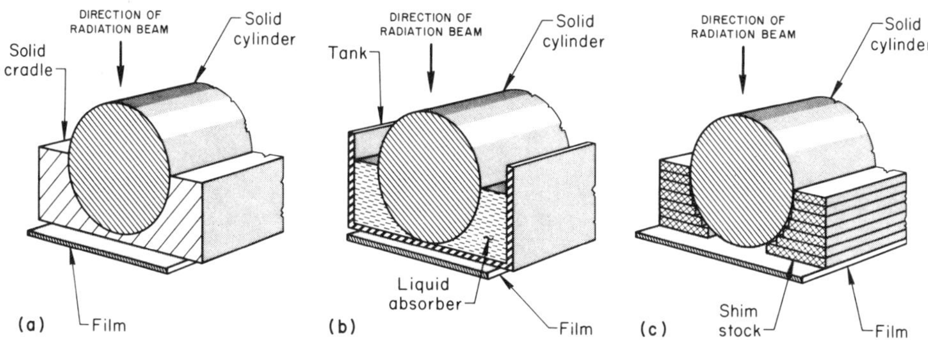

Fig. 47 Three alternative methods of equalizing radiographic density to avoid overexposure of the image of thinner portions of a solid cylinder. The solid cradle (a), liquid absorber (b), and shim stock (c) have radiographic absorption characteristics equivalent to those of the cylinder.

thickness at the edges. Edge definition is relatively good for light-metal cylinders less than about 50 mm (2 in.) in diameter and for heavy-metal cylinders less than about 25 mm (1 in.) in diameter. When the diameter of the cylinder exceeds these values, multiple radiographs using different tube voltages are usually needed to increase the latitude (or dynamic range) of the inspection. In real-time radiography, image processing can also increase the latitude for inspecting the inner and outer portions of the cylinder. In film radiography, double-film techniques can also correct for the inevitable overexposure of the outer portions of the image. An alternative technique is to make a radiograph at the proper exposure for all but the outer portions of the image, then rotate the cylinder 90° and make a second radiograph at the same exposure.

Sometimes, section-equalizing techniques are helpful in film (or paper) radiography. In a section-equalizing technique, the outer edges of the cylinder, where the section is thinnest, are built up to present a greater radiographic density to the x-rays. Close-fitting solid cradles, liquid absorbers, or many layers of shim stock—all having radiographic absorption characteristics equivalent to those of the cylinder—are alternative means of equalizing radiographic density (Fig. 47). Regardless of the form of the absorber, it is placed in contact with the cylinder around one-half of the circumference, as shown in Fig. 47. The effect of a section-equalizing technique is to reduce the subject contrast by about one-half. This is usually sufficient to ensure that the entire image of the cylinder will have a photographic density within the range that can be interpreted correctly.

Because the probable orientation of planar discontinuities is difficult to define for cylinders, it is good practice to make more than one radiograph, rotating the cylinder equal amounts between them. Two radiographs, 90° apart, would be considered a minimum number.

In general, hollow cylinders (or tubular sections) present fewer problems to the radiographer than solid cylinders. For example, in hollow cylinders or tubular sections, there is much less subject contrast because of the nonabsorbing cavity in the center. The inspection of hollow cylinders is discussed in the section "Inspection of Tubular Sections" in this article.

Inspection of Complex Shapes

The inspection of complex shapes most often requires multiple exposures, usually with different viewing directions. The selection of view for each exposure depends primarily on the shape of the section of the testpiece to be inspected with that exposure and the probable orientation of suspected flaws. In principle, the view should be chosen with the objectives of minimizing geometric unsharpness and of reducing the image to a simple shape for easy interpretation. A small variation in testpiece thickness is desirable because this will highlight subject contrast due to flaws. However, the use of high-energy sources, multiple radiographs with different radiation energy levels, or double-film techniques with a single view can circumvent most problems arising from excessive material-thickness variations over a given area of coverage. The following example illustrates the use of three different views to obtain complete coverage of a cast stainless steel impeller.

Example 2: Three-View, Nine-Shot Inspection of a Large Cast CF-8M Stainless Steel Impeller. A large impeller cast from CF-8M stainless steel (Fig. 48) was specified to be radiographically inspected in accordance with ASTM E 94, E 1030, and E 142, using the standard radiographs of ASTM E 186 and E 446 as a reference for acceptance. The impeller was to be inspected in the as-cast, center-bored condition; specified image quality with plaque penetrameters was 2-2T.

Three different views were chosen to provide complete coverage of the impeller. These are shown in Fig. 48 as view A, selected to inspect the hub (a hollow cylinder) with a single shot; view B, using two shots with each providing coverage of half of the 18 volutes in the impeller; and view C,

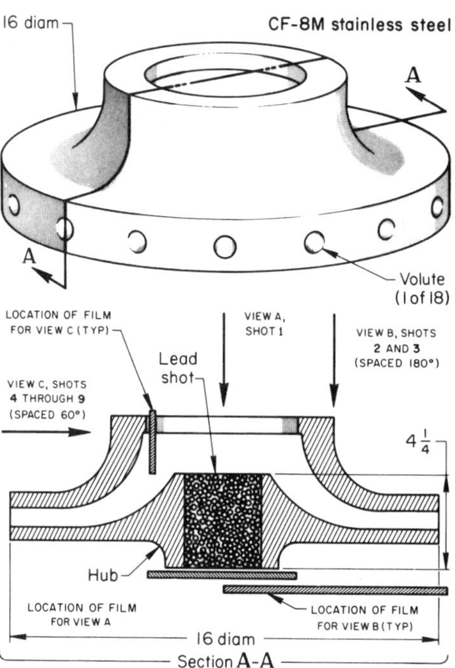

Fig. 48 Cast CF-8M stainless steel impeller that required three different viewing directions and nine separate shots to provide complete coverage in radiographic inspection. Dimensions given in inches

using six shots with each providing coverage of the thicker section of three volutes in a direction that is perpendicular to view B.

View A presented certain problems, namely, the high absorption of the 115 mm (4½ in.) thick hub, which required high-intensity radiation of relatively high energy for adequate penetration, and the bore area along the central axis, which allowed incident radiation to impinge directly on the film and favored undercut because of side scattering along the cylindrical wall of the bore. The hub could not be radiographed in a radial plane because of excessive interference by the surrounding volutes, so the bore was filled with lead shot to absorb the direct radiation in this region and to eliminate undercut.

All radiographs were made with a source-to-film distance of 1.5 m (5 ft), and 1-MV x-rays were used from a source having an 8 mm (0.3 in.) focal-spot size. Shot 1 (view A) was made with 3.0 mA tube current and 4.7-min exposure time on a 180 × 430 mm (7 × 17 in.) double film of NDT 75 (DuPont). Shots 2 and 3 (view B) were made with 1.0-mA tube current and 2.2-min exposure time on three 350 × 430 mm (14 × 17 in.) films of different types—one sheet each of NDT 45 (DuPont), Industrex M, and Industrex AA (Eastman Kodak). Shots 4 to 9 (view C) were made with 1.0-mA tube current and 2.5-min exposure time on 125 × 175 mm (5 × 7 in.) double films of Industrex M (Eastman Kodak). Lead screens, 0.13 mm (0.005 in.) thick, were used both front and back.

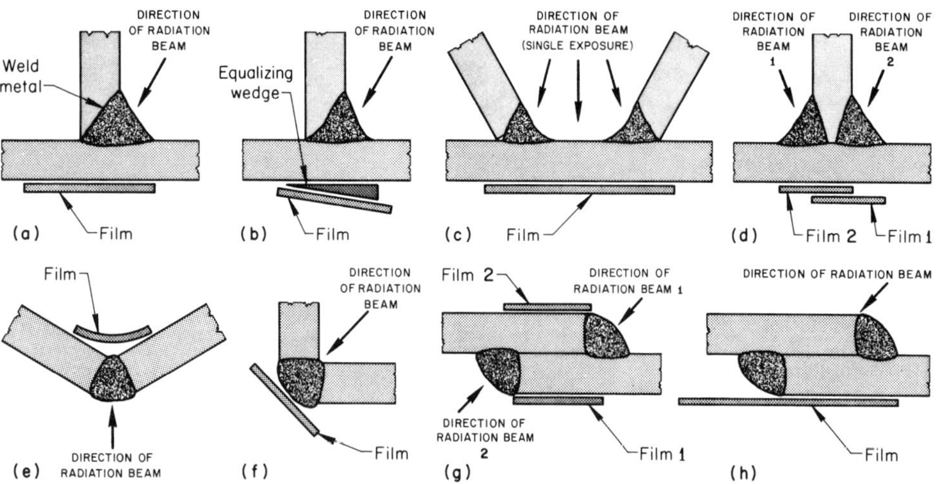

Fig. 49 Preferred viewing directions for the radiographic inspection of several types of fillet-welded joints. (a) Single-fillet T-joint. (b) Single-fillet T-joint with equalizing wedge. (c) Two adjacent single-fillet joints radiographed simultaneously. (d) Double-fillet T-joint. (e) Corner joint with film positioned at the inside surface. (f) Corner joint with film positioned at the outside surface. (g) and (h) Alternative views for double-welded lap joint

For control of image quality, standard ASTM penetrameters were used. The penetrameter for view A was 2 mm (0.08 in.) thick and was placed on a 110 mm (4¼ in.) thick block adjacent to the impeller. Four penetrameters were used for view B: a 0.25 mm (0.010 in.) thick penetrameter on a 13 mm (½ in.) thick block, a 0.38 mm (0.015 in.) thick penetrameter on a 20 mm (¾ in.) thick block, a 0.5 mm (0.020 in.) thick penetrameter on a 25 mm (1 in.) thick block, and a 1.15 mm (0.045 in.) thick penetrameter on a 57 mm (2¼ in.) thick block. For view C, a 0.38 mm (0.015 in.) thick penetrameter was placed directly on the part. All penetrameters and blocks were made of austenitic stainless steel.

This procedure gave both the required complete coverage and the specified 2-2T image quality. Castings were routinely inspected to a minimum quality level corresponding to severity level 1 (in ASTM E 186) in the hub area and to severity level 2 to 3 (in ASTM E 446) in the shroud area.

Inspection of Weldments

The radiographic inspection of weldments is usually intended to cover only welds and heat-affected zones. The quality of the base metal beyond the heat-affected zones is seldom of interest at this stage of inspection. Therefore, the viewing direction is usually selected to reveal clearly the features of welds and heat-affected zones, regardless of the overall shape of the weldment. See the article "Weldments, Brazed Assemblies, and Soldered Joints" in this Volume for further information on the NDE of welds.

Arc Welds. Generally, arc-welded joints are radiographed using a viewing direction normal to the surface of the weld. For butt joints, this view directs the central beam perpendicular to the surfaces of the adjacent plates, but for other types of joints such as T-joints and lap joints, the central beam is usually inclined to both adjacent legs. The maximum area of coverage for a single radiograph is normally recommended to be no more than would result in a 6 to 10% difference in penetrated thickness between the center of the area and the extremities. Mathematically, this can be expressed for flat plates of uniform thickness as:

$$\frac{x}{2L} \leq 0.35 \text{ to } 0.46 \qquad \text{(Eq 19)}$$

where x is the field size, L is the source-to detector distance, and the values 0.35 and 0.46 are derived from the laws of similar triangles for 6 and 10% increase in penetrated thickness, respectively. The numerical range in Eq 19 is equivalent to a radiation-beam cone angle of about 39 to 49°. This range is not intended to limit the thickness range inspected in a single radiograph but rather to ensure the detection of transverse cracklike discontinuities by keeping the image relatively sharp and undistorted.

Views normal to the surface of the weld are not always satisfactory. Flaws in a weld such as dense inclusions due to the deposition of metal from a tungsten electrode, excessive segregation in light alloys, porosity, and shrinkage can be revealed regardless of view. However, cracks or nonfusion along the edge of the fusion zone can be best revealed when the viewing direction is parallel to the face of the joint. The detection of centerline cracks or incomplete penetration requires a normal view. Only square-groove butt joints, U-groove joints, and J-groove joints can be satisfactorily inspected with a normal view; other types of joints, including V-groove joints, bevel joints, flare joints, and fillets, require two views, one parallel to each original prepared edge, for correct inspection for cracks or regions of incomplete fusion. Certain flaws cannot be detected at all because they lie along planes that are more or less parallel to the weld surface. The shape of most weldments prohibits selection of a viewing direction parallel to the weld surface, which is the most favorable view for detecting underbead cracking or delamination. (Ultrasonic inspection is a much more suitable method for detecting either underbead cracking or delamination.)

The most favorable view for fillet welds is a direction that roughly bisects the angle between the legs of the section. Preferred arrangements for several types of fillet welds are shown in Fig. 49. Single-fillet T-joints can be inspected with the simple arrangement shown in Fig. 49(a) or with an arrangement incorporating an equalizing wedge (Fig. 49b) to reduce the inherent subject contrast caused by variations in penetrated thickness across the weld zone. The arrangement in Fig. 49(b) is more relevant for film radiography than for real-time radiography because image processing can be used in real-time radiography. In film radiography, the high subject contrast can result in overexposure of the thin section on one side of the weld or in underexposure of both the weld and thick sections. Either underexposure or overexposure reduces sensitivity and may not resolve images of flaws having low subject contrast. An equalizing wedge is most effective when the size of the fillet weld exceeds the thickness of the thin section. The technique is mainly used on straight seams; special-shape wedges would be required for other seam shapes.

Occasionally, the shape of the weldment is such that two or more seams can be inspected with a single exposure in film radiography (Fig. 49c). The savings in time and material often justify selection of a single view whose area of coverage encompasses two or more seams. However, if the shape of the seam is such that a single view would give severely reduced sensitivity, multiple views should be specified on the radiographic standard shooting sketch.

T-joints that are welded on both sides usually cannot be inspected satisfactorily with a single radiographic view or exposure. Two radiographs, one for each fillet (Fig. 49d), are needed. Corner welds can be inspected from either side, depending on which side provides access (Fig. 49e and f). Lap joints can be inspected using multiple views or by a single view, as shown in Fig. 49(g) and (h). The main factor that determines which technique is most appropriate is the amount of separation between welds in a double-welded joint. When multiple views are used, the central beam of radiation should be aligned with the weld for each view, as in Fig. 49(g). When a single view is used, the central beam should be aligned with a spot midway between the welds, as in Fig. 49(h). Only a single radiograph is needed for each weld segment in a single-welded lap joint.

The techniques discussed above, particularly those for the inspection of fillet welds, are also appropriate for the inspection of junction areas of castings, forgings, and formed parts. Circumferential welds, which are one of the more usual types of arc welds inspected by radiography, can be inspected with techniques discussed in the section "Inspection of Tubular Sections" in this article.

Resistance welds, mainly spot welds and seam welds, are used to join relatively thin sheets. Most often, a resistance-welded joint can be considered a type of lap joint, with the weld positioned inboard of the overlapping edge. Usually, the only outward sign of weld position is a slight depression on the surface of the workpiece resulting from electrode pressure applied during welding.

In the radiography of spot welds and seam welds, the central beam is ordinarily directed normal to the surface of one of the sheets and is centered on the depression that indicates the location of the weld. Radiography can detect discontinuities in both weld metal and base metal, including cracks, inclusions and porosity in the weld, cracking and deformation in the base metal, segregation in the fusion zone of resistance welds in light alloys, and expulsion of molten metal between the faying surfaces. Normally, deficiencies such as underbead cracking and incomplete fusion are not revealed by radiography, although incomplete fusion can sometimes be inferred from the presence of other discontinuities. The area of coverage should be limited in a manner similar to that suggested for arc welds so that maximum sensitivity to cracklike discontinuities can be maintained.

A view other than the normal view is sometimes dictated by the shape of the weldment, especially when portions of the weldment may cast shadows that would detract from the clarity of the weld-zone image. Even in such cases, it is best to select a view as near to normal as possible so as to minimize distortion. In contrast to arc welds, in which an oblique view adversely affects both the intensity of the transmitted radiation and the sensitivity, only sensitivity is affected to any great degree by using an oblique view for the inspection of resistance welds. Most resistance-welded assemblies are thin enough that the section can be penetrated by relatively low-energy radiation even with oblique views. The sensitivity, expressed as minimum detectable discontinuity size, varies with the cosecant of the angle between the viewing direction and the sheet surface. This relationship indicates that the smallest value of minimum detectable discontinuity size occurs with a 90° viewing direction, and viewing directions between 90° and 65° give no more than about a 10% increase in this value. The main disadvantage of oblique views is their inability to resolve the images of transverse cracks (through-thickness cracks) in either the weld or adjacent heat-affected zones.

Inspection of Tubular Sections

Although film radiography is not well suited to the inspection of continuously produced tubular products such as tubing and pipe, the method is suited to the inspection of tubular sections in a wide variety of products and assemblies. Real-time radiography, on the other hand, is more suited to the inspection of continuously produced tubular products, but not tubular sections in large, complex assemblies.

Radiography is the method most often specified for the inspection of welded joints between tubes or pipes or between a tube and a pressure vessel when the weldment is manufactured to ASME Boiler and Pressure Vessel Code specifications. Other hollow sections such as cylindrical bosses on castings and forgings, brazed tube-and-socket joints, large-diameter transmission pipe (line pipe), and longitudinally welded square or circular structural tubing, are frequently inspected by radiography. There are three major inspection techniques for tubular sections:

- The double-wall, double-image technique
- The double-wall, single-image technique
- The single-wall, single-image technique

The double-wall, double-image technique is mainly applicable to sections with an outside diameter of no more than 90 mm (3½ in.). This technique produces a radiograph in which the images of both walls of a tubular section are superimposed on one another. The beam of radiation is directed toward one side of the section, and the radiation conversion medium surface is placed on the opposide side, usually tangent to the section.

As shown in Fig. 50(a), two radiographs 90° apart are required to provide complete coverage when the ratio of outside diameter to inside diameter is 1.4 or less. In exposure 1 in Fig. 50(a), the area between the phantom lines is recorded as a through-thickness image, and the area outside the lines is recorded essentially as an image in the plane of the tube wall that exhibits too much subject contrast for meaningful interpretation. In exposure 2, the area outside the lines in exposure 1 is recorded as a through-thickness image, and there is a certain amount of overlap between the through-thickness images of exposures 1 and 2, which is shown as darker shading in exposure 2 in Fig. 50(a).

When the ratio of outside diameter to inside diameter is greater than 1.4—that is, when radiographing a thick-wall tube—the number of views required to provide complete coverage can be determined by multiplying that ratio by 1.7 and rounding off to the next higher integer. For example, to examine a 50 mm (2 in.) OD cylinder with a 25 mm (1 in.) diam axial hole, a total of 1.7 × 2 = 3.4, or 4 views, will provide complete coverage. The circumferential displacement between each view is found by dividing 180° by the number of shots. In this case, 180°/4, or 45°, rotation of the cylinder with respect to the viewing direction is required between exposures, as shown in Fig. 50(b) The area within the phantom lines for each exposure in Fig. 50(b) indicates the region of a 50 mm (2 in.) OD cylinder with a 25 mm (1 in.) diam axial hole that is recorded as a through-thickness image; heavy shading indicates regions recorded as a through-thickness image on preceding views. If only three shots are used for a cylinder with a 2:1 ratio of outside to inside diameters, there is only a minimal amount of overlap between shots, inadequate to ensure that all detectable discontinuities in the regions of overlap will be recorded.

When an odd number of views are required for complete coverage, the angular spacing between shots can be determined by dividing 360° by the number of views as an alternative to dividing 180° by the number. This alternative cannot be used when the number of views is even, because half of the resulting radiographs would be mirror images of the remaining radiographs and sections of the outside circumference would not receive adequate coverage.

The double-wall, double-image technique can be applied to the inspection of diagonally opposite corner welds in a section of rectangular structural tubing constructed by welding four pieces of metal strip together to form a box section. Only two views are required for the inspection of all four longitudinal welds, as shown in Fig. 50(c). As with circular sections, hollow box sections can be inspected by the double-wall, double-image technique only when they are small; that is, when section height in direction or selected view is 100 mm (4 in.) or less.

A variation of the double-wall, double-image technique, sometimes called the corona or offset technique, is often used for the inspection of circumferential butt welds in small-diameter tubing and pipe. In the corona technique, the central beam is directed at an acute angle to the run of the tube (Fig. 50d) so that the weld is projected on the film as an ellipse rather than a straight band. The offset angle of the radiation beam shown in Fig. 50(d) must be large enough that the image of the upper section of the weld zone does not overlap the image of the lower portion, but not so large as to introduce an unnecessary degree of distortion. Also, the larger the offset angle, the greater the probability that the technique will fail to detect incomplete fusion at the

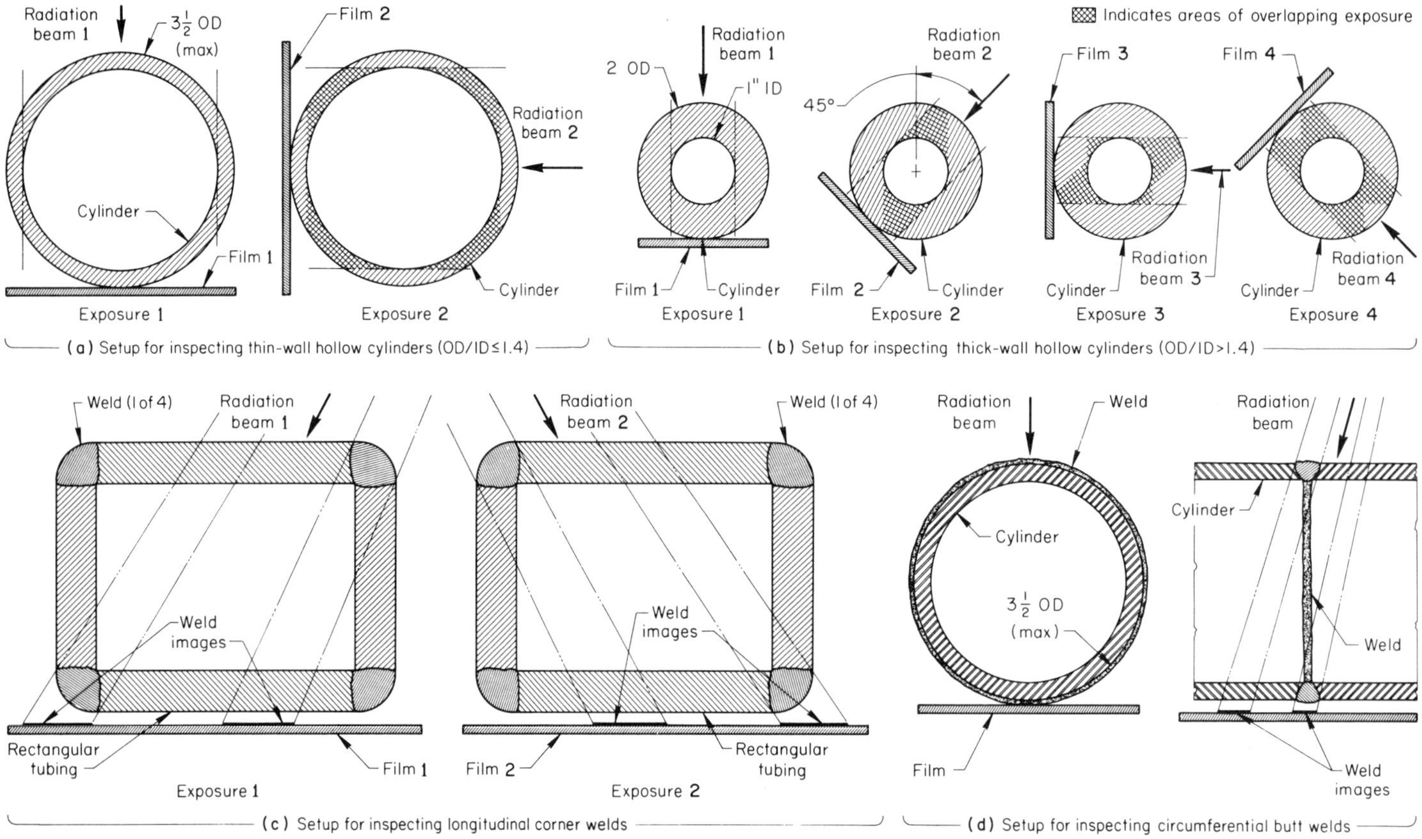

Fig. 50 Diagrams illustrating the relation of radiation beam, testpiece, and film for the double-wall, double-image technique of inspecting hollow cylinders or welds in tubular sections. Dimensions given in inches

root of the plain butt weld. Incomplete fusion at the root can have the radiographic appearance of a root void, but if it exists in a tightly butted joint or if the root gap is filled by capillary action of the molten weld metal, incomplete fusion has the appearance of a cracklike discontinuity lying in the plane of the joint. The incident radiation must be parallel to the plane of the joint so as to detect incomplete root fusion in a tight joint.

The correct number of views and the circumferential location of corresponding views can be determined for the corona technique in the same manner as for the basic double-wall, double-image technique. Sometimes, as when a weld must be inspected for both radial cracks and internal voids, it is beneficial to use both the basic technique and the corona technique—the former to detect cracks and the latter to locate voids.

In all variations of the double-wall, double-image technique, the image of the section of the cylinder wall that is closest to the radiation source will exhibit the greatest amount of enlargement, the greatest degree of shape distortion, and the greatest degree of unsharpness. Selection of proper source-to-detector distance should be based on obtaining the specified image quality for that section of the cylinder. Also, penetrameters should be located where they will evaluate the image of the section closest to the source.

The double-wall, single-image technique is mainly applicable to hollow cylinders and tubular sections exceeding 90 mm (3½ in.) in outside diameter. This technique produces a radiographic image of only the section of the wall that is closest to the imaging plane, although the radiation penetrates both walls. The source is positioned relatively close to the section, so that blurring caused by geometric unsharpness in the image of the cylinder wall closest to the source makes the image completely indistinguishable. Only the image of the wall section closest to the detector is sharply defined.

In film (or paper) radiography, exposures are calculated on the basis of double the wall thickness of the hollow section, as they are for the double-wall, double-image technique. The area of coverage is limited by geometric unsharpness and distortion at the extremities of the resolved image for hollow cylinders that are less than about 380 mm (15 in.) in outside diameter. For larger cylinders, film size is the usual limiting factor. Usually, at least five separate exposures equally spaced around the circumference are required for complete coverage. There must be enough overlap between adjacent exposures to ensure that all of the outside circumference is clearly recorded.

The radiation is usually directed normal to the surface of the cylinder. However, in some cases, it may be advantageous to use viewing direction similar to that of a corona exposure. For example, if a circumferential weld has a high and irregular crown, it may be desirable to have the radiation that impinges on the area of the weld being inspected pass through the tube wall adjacent to the weld so that the incident radiation is attenuated by a more uniform and thinner section, as shown in Fig. 51.

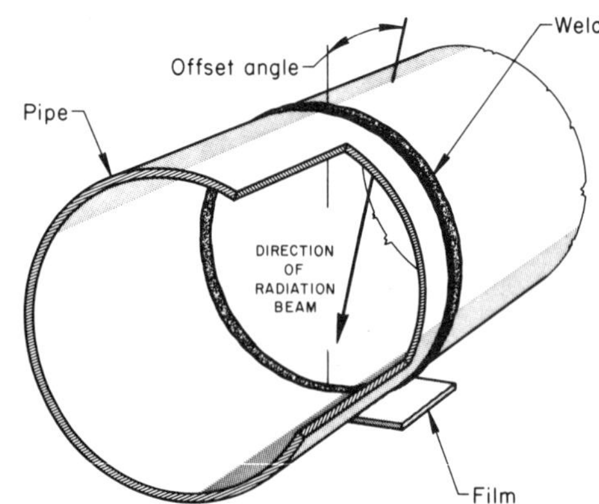

Fig. 51 Schematic of the double-wall, single-image inspection technique applied to a circumferential butt weld in a large-diameter pipe showing relation of radiation beam, weld, and film

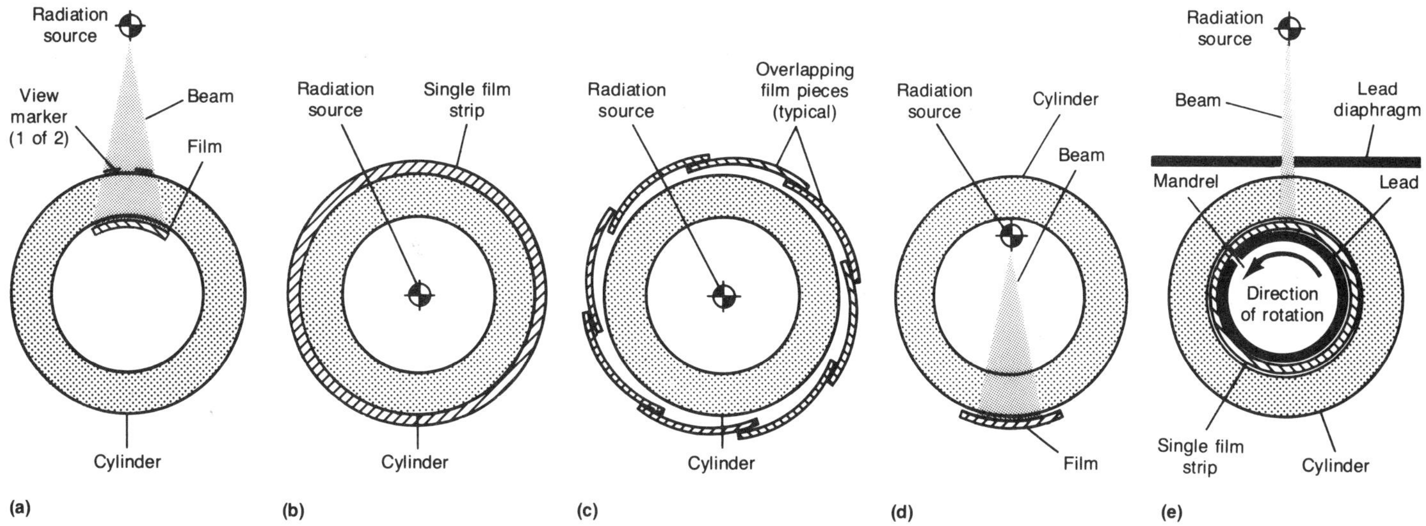

Fig. 52 Schematic of techniques employing different arrangements of radiation source, cylinder, and film for the single-wall, single-image radiography of cylindrical sections using external and internal radiation sources. See text for discussion.

The single-wall, single-image technique is an alternative to either of the double-wall techniques and can be used only when the interior of a section is accessible. With this technique, the radiation source can be placed outside the cylinder and the radiation detector inside the cavity (Fig. 52a), or the detector can be placed outside the cylinder and the radiation source inside (Fig. 52b). In both setups, only a single wall is radiographed. A significant advantage of the single-wall technique is that it is more sensitive than double-wall techniques because any flaw that may be present occupies a greater percentage of the penetrated thickness.

When using the technique shown in Fig. 52(a), the cylinder is rotated to generate a series of radiographs. In film radiography, the amount of rotation must be controlled so that successive radiographs contain regions of overlap. The correct amount of overlap can be proved by placing lead view markers on the surface that is remote from the film; if the image of a view marker appears in both adjacent radiographs, complete coverage is ensured.

In the technique shown in Fig. 52(b), the radiation source is placed inside the cylinder, and one continuous piece of film is wrapped around the outside of the cylinder. Alternatively, several overlapping pieces can be used, as shown in Fig. 52(c). The radiation source, which is usually either a rod-anode type of x-ray tube having 360° radiation emission or a γ-ray source, is placed on the central axis of the hollow cylinder. The entire circumference is inspected with one exposure, regardless of whether a single strip of film or several overlapping pieces are used.

Special adaptations of these techniques can be used. For example, when maximum sensitivity is required for the technique in Fig. 52(b), the radiation source can be moved to a position close to the inside surface of the cylinder diametrically opposite a single piece of film, as illustrated in Fig. 52(d). This adaptation can almost double the source-to-film distance, thus reducing geometric unsharpness by almost 50%. Although there is a gain in sensitivity, there is a loss in efficiency because the entire circumference can no longer be inspected by a single exposure. Several exposures must be made, with the radiation source and film repositioned to several points equally spaced around the circumference to obtain complete coverage.

Another special adaptation is shown in Fig. 52(e). In this arrangement, a continuous piece of film is placed in contact with the inside surface of the cylinder. The film is held in place by a close-fitting, lead-covered mandrel. Incident radiation is directed through a slit in a lead diaphragm to impinge on a narrow section of the circumference of the cylinder. The cylinder, film, and mandrel are rotated slowly past the slit one or more times, so that the total time that a given point on the cylinder is in front of the slit equals the required exposure time. Radiographs produced in this manner are remarkably clear and distinct compared to other single-wall or double-wall techniques. Because the radiation beam impinges on only a narrow region of the cylinder and because the film is in contact with the inner surface, both geometric unsharpness and variations in transmitted intensity caused by variations in penetrated thickness are minimized. As a result, the subject contrast and definition of internal flaws are maximum. The technique shown in Fig. 52(e) is a form of in-motion radiography.

In-Motion Radiography

In-motion radiography is useful when large areas must be inspected. Generally, real-time radiography is more suitable for in-motion radiography because a detector can be used to scan the test area during motion. Nevertheless, in-motion film radiography can be used to inspect longitudinal welds in very large pipe, such as line pipe. In one method, a portable x-ray unit is attached to a traveling carriage inside the pipe and is positioned so that a narrowly collimated x-ray beam impinges on the weld, as shown in Fig. 53(a). The carriage moves slowly along the pipe, successively exposing a series of films placed along the outside surface over the weld. The correct rate of travel is equal to the width of the beam in the direction of relative motion divided by the required exposure time.

An alternative setup for in-motion film radiography, in which a rod-anode x-ray tube is substituted for the portable x-ray unit, is shown in Fig. 53(b). The principle of operation is the same; only the x-ray equipment and method of travel are different. This setup is particularly advantageous when two or more welded joints are to be radiographed because the joints can be radiographed simultaneously. With a rod-anode tube, radiation is emitted 360° around the axis of the tube. As long as the center of the target is maintained on the axis of the pipe, the disk of radiation that emerges between the two lead disks that act as a collimator will produce equal nominal film densities on all radiographs. Even with the portable x-ray unit illustrated in Fig. 53(a), the source-to-film distance must be kept constant because if the distance is permitted to vary there will be a gradual change in nominal film density from radiograph to radiograph as the radiation source moves along the pipe.

The principal advantage of in-motion radiography is that less time is needed for setup and operation. The inspection of long

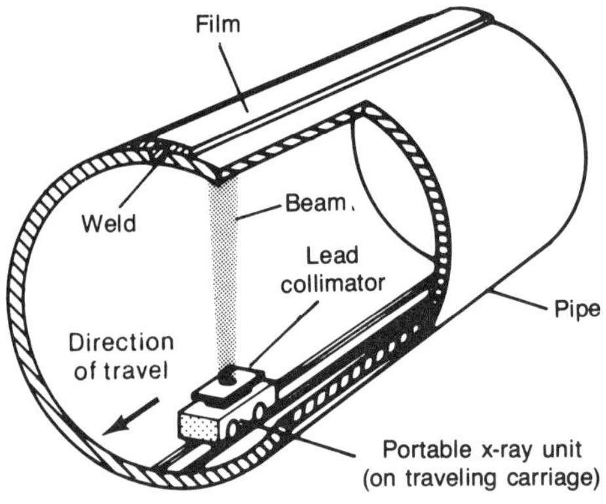

(a)

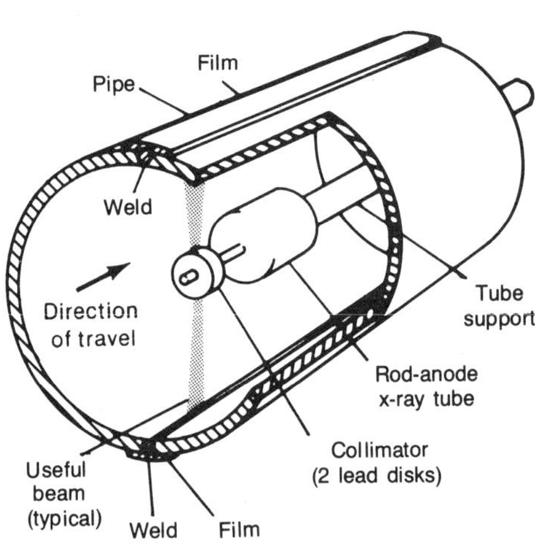

(b)

Fig. 53 Arrangements for the in-motion single-wall, single-image film radiography of longitudinal welds in line pipe, using (a) a conventional portable x-ray unit and (b) a rod-anode x-ray tube

welds by conventional methods requires frequent repositioning of the testpiece with respect to the source and film; consequently, there is a considerable amount of time spent in setting up for successive exposures. With in-motion radiography, a much smaller proportion of total inspection time is required for setup. Therefore, in-motion radiography is seldom, if ever, accomplished with a radioactive source.

Techniques similar to those described above for radiographing longitudinal welds in large-diameter pipe can also be adapted for other product forms when radiographic inspection is to be applied to long testpieces. Also, it may be desirable to adapt in-motion radiography for the inspection of relatively small parts, especially when the diverging beam used in normal radiography would cause unacceptable distortion at the edges of the radiograph.

In general, a real-time radiographic technique is preferred for in-motion radiography. However, if only film techniques are available, then in-motion radiography would be considered whenever more than four to six shots are required for complete coverage of a long straight or circumferential weld in a relatively thin section. It is essential that the source move at constant speed in film radiography because any variations in speed will produce parallel bands of overexposure and underexposure in the radiograph, oriented at right angles to the direction of relative motion.

Motion Unsharpness. The real-time and film techniques used for in-motion radiography cause a specific type of unsharpness called motion unsharpness. Any amount of relative motion of source or testpiece with respect to the image conversion plane results in blurring of the image; the greater the amount of motion, the greater the blurring. In conventional still radiography, it is considered good practice to eliminate all sources of motion, even vibrations, so that motion unsharpness does not detract from radiographic quality. The amount of unsharpness caused by relative motion of the source with respect to the testpiece and detector in in-motion radiography can be evaluated from characteristics of the setup, much as geometric unsharpness is evaluated for conventional techniques. Motion unsharpness , U_m, is described by:

$$U_m = \frac{wt}{L_o} \quad \text{(Eq 20)}$$

where w is the width of the radiation beam in the direction of motion, t is the penetrated thickness, and L_o is the source-to-object distance.

Motion unsharpness is generally greater for in-motion radiography than geometric unsharpness for an equivalent conventional technique. This limits the use of in-motion techniques to thin testpieces or to situations in which a long source-to-detector distance can be used. It is essential that there be no relative movement between testpiece and detector, so that only the movement of the source contributes to motion unsharpness.

Identification Markers and Penetrameters (Image-Quality Indicators)

It is important to relate a particular radiographic image to the direction of the radiation beam used to generate that image. Furthermore, it is important that any permanent record of inspection be traceable to the particular testpiece that was inspected or to the production lot represented by that testpiece. Identification markers are used for these purposes.

In addition to identification, the image should contain some means of evaluating the radiographic procedure in terms of its sensitivity to test conditions. This sensitivity, known as image quality, bears directly on the ability of the radiographic process to record images of small flaws and is usually determined by the use of penetrameters, which are also called image-quality indicators.

Identification Markers

Identification markers are made of lead or lead alloy and are usually in the form of a coded series of letters and numbers. The markers are placed on the testpiece or on the film adjacent to the testpiece during setup. When the testpiece is radiographed, a distinct, clear image of the radiographically dense identification markers is produced at the same time. Identification markers must be located so that their projected shadows do not coincide with the shadows of any regions being inspected in the testpiece. Because markers are radiographically dense, their images will obscure any coinciding image of the testpiece.

Both view identification and testpiece identification almost always appear in coded form. View identification is usually a simple code (such as A, B, C, or 1, 2, 3) that relates some inherent feature of the testpiece or some specific location on the testpiece to the view used; the term sequence numbering, which is used in some specifications, refers to view identification. Often, the location of view markers is handwritten in chalk or crayon directly on the testpiece so that the radiographic image can be correlated with the testpiece itself during interpretation and evaluation of the radiograph. Sometimes, the locations of markers are steel-stamped on the surface of the part being radiographed and become a permanent reference. It is advisable always to mark the location of view markers because it may be necessary to recheck the setup procedure against specifications. Although the specific location of view markers is largely a matter of personal preference, several suggested arrangements are discussed in this section. It is required only that the markers be placed so that their images on the radiograph are legible and do not obscure any area being inspected.

View markers need not be different if identical viewing directions are used for radiographing a series of testpieces of the same type and size. Actually, it is much simpler to begin at the letter "A" or the number "1" for view identification on each individual testpiece. Most radiographers adopt standard methods of identifying views and indicate the type of view markers to be used on the specification sheet for the radiographic procedure, which is commonly called a radiographic standard shooting sketch.

The code for identification of the testpiece is usually more complex than the code used for identification of viewing direction. Each radiographic laboratory must adopt a system of identification that is suited to its specific requirements. As a minimum requirement, the identification code

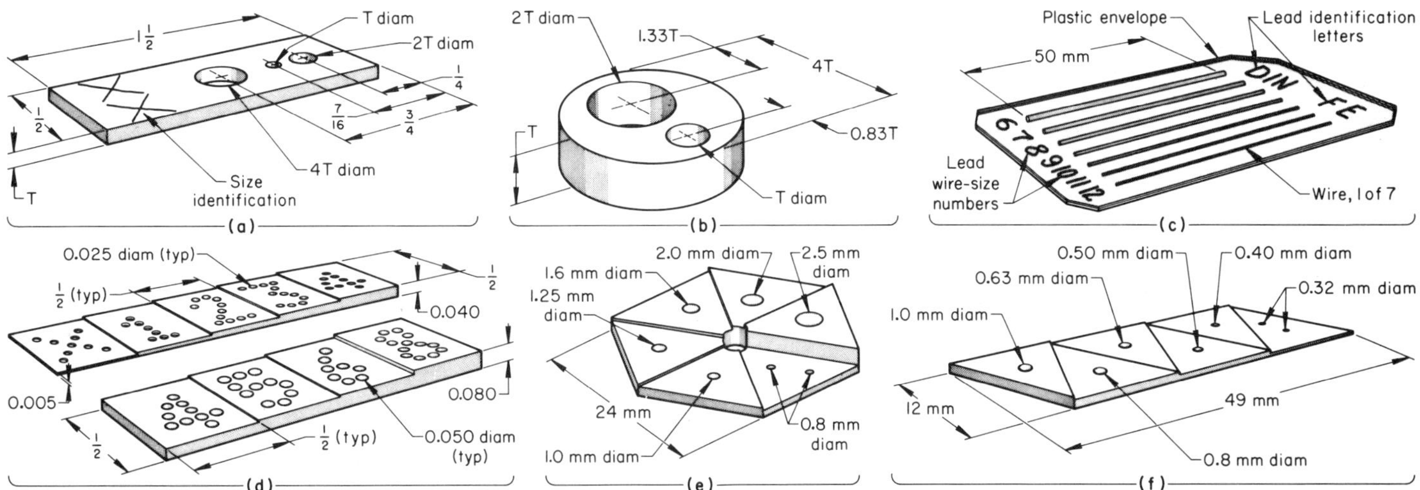

Fig. 54 Designs of several widely used penetrameters (image-quality indicators). (a) Rectangular plaque-type penetrameter (ASTM-ASME standard) for plaque thicknesses of 0.13 to 1.3 mm (0.005 to 0.050 in.). (b) Circular plaque-type penetrameter (ASTM-ASME standard) for plaque thicknesses of 4.6 mm (0.180 in.) or more. (c) Typical wire-type penetrameter (Deutsche Industrie Norm standard DIN 54109). (d) Square-step step wedge penetrameter used by British Welding Research Association (BWRA standard). (e) Hexagonal and (f) linear triangular-step step wedge penetrameters used by the French Navy (AFNOR standard). Dimensions given in inches, except where otherwise indicated

must enable each radiograph to be traced to a particular testpiece or section of a testpiece. Identification codes can be based on part number, lot number, inspection date, customer code, or manufacturing code; or they can be merely a series of consecutive multiple-digit numbers, with the pertinent data concerning testpiece identification recorded in a logbook opposite the corresponding testpiece number.

Penetrameters

Penetrameters, or image-quality indicators, are of known size and shape and have the same attenuation characteristics as the material in the testpiece. They are placed on the testpiece or on a block of identical material during setup and are radiographed at the same time as the testpiece. Penetrameters are preferably located in regions of maximum testpiece thickness and greatest testpiece-to-detector distance and near the outer edge of the central beam of radiation. Because of this location, the degree to which features of the penetrameter are visible in the developed image is a measure of the quality of that image. The image of the penetrameter that appears on the finished radiograph is evaluated during interpretation to ensure that the desired sensitivity, definition, and contrast have been achieved in the developed image.

Penetrameters of different designs have been developed by various standards-making organizations; several of the standard designs most widely used are shown in Fig. 54. Regardless of the design, all penetrameters have the following in common:

- Material used for penetrameters is the same as that of the testpiece or has the same absorption characteristics. Suitable penetrameter materials for various metallic testpieces are grouped in ASTM E 1025 and E 142 as shown in Table 8
- In use, the penetrameter is normally placed directly on the surface of the testpiece that faces the source. Alternatively, when the testpiece is small or when its shape is unfavorable, the penetrameter is placed on a block or shim of the same nominal composition and thickness as the testpiece, with the upper surface of the block at the same distance from the recording plane as the upper surface of the testpiece. In pipe radiography, when permitted by specification, the penetrameter can be placed on the surface of the pipe that faces the film. This film-side placement of the penetrameter is usually read to a tighter sensitivity requirement (for example, 2-1T instead of 2-2T)
- The location of the penetrameter should be such that its projected image does not coincide with any area of interest in the image of the testpiece
- The image of the penetrameter is viewed at the same time as the image of the testpiece; the two images appear simultaneously on the recording medium. The image of the penetrameter is evaluated separately from the image of the testpiece and is used solely as a direct check on the quality of the recorded image
- Image-quality levels are usually expressed as the size of the smallest penetrameter feature (such as hole size or wire size) that is clearly visible in the processed image
- A penetrameter is never used as a size standard against which flaw sizes are compared

Applicable codes, specifications, or purchase agreements usually determine the type of penetrameter to be used. Even when the specification does not require the use of a penetrameter, it is advisable to use a penetrameter to ensure that appropriate image quality has been achieved. Some of the standard penetrameters, including those illustrated in Fig. 54, are described below.

Plaque-type penetrameters consist of strips of materials of uniform thickness with holes drilled through them. There are two general types of plaque penetrameters specified by ASTM and the American Society of Mechanical Engineers (ASME): rectangular plaque penetrameters (Fig. 54a) and circular plaque penetrameters (Fig. 54b).

The rectangular plaque design shown in Fig. 54(a) is specified by ASTM and ASME for plaque thicknesses of 0.13 to 1.3 mm (0.005 to 0.050 in.). The holes in rectangular plaque penetrameters are T, 2T, and 4T in diameter, where T is the thickness of the plaque. A notch system (Fig. 55) is also used to identify the ASTM grade of the penetrameter. For example, the penetrameter shown in Fig. 54(a) would be a grade 1 penetrameter for testpiece materials in group 1 of Table 8.

The circular plaque design is larger than the rectangular plaque design and is specified for plaque thicknesses of 1.5 to 4 mm (0.060 to 0.160 in.). Figure 54(b) shows the circular design specified by ASTM and ASME for plaque-type penetrameters with thicknesses of 4.6 mm (0.180 in.) or more.

Various degrees of image quality can be measured by using plaque-type penetrameters of different thicknesses. Sensitivity is usually expressed in terms of penetrameter thickness (as a percentage of testpiece), and resolution is determined by the smallest hole size visible in the radiograph. For example, an image-quality level of 2-2T indicates that the thickness of the penetram-

Table 8 ASTM penetrameter material grades for the material groups in ASTM E 1025 and E 142

Material groups of testpieces	Suitable penetrameter grade and material
Group Mg: magnesium	Penetrameter grade 000: made of all magnesium, or magnesium being the predominant constituent
Group Al: aluminum	Penetrameter grade 00: made of all aluminum, or aluminum being the predominant constituent
Group Ti: titanium	Penetrameter grade 0: made of all titanium, or titanium being the predominant constituent
Group 1: all carbon steels, all low-alloy steels, all stainless steels, manganese-nickel-aluminum bronze	Penetrameter grade 1: carbon steel or type 304 stainless steel
Group 2: all aluminum bronzes, all nickel-aluminum bronzes, Haynes alloy IN-100	Penetrameter grade 2: may be aluminum bronze (Alloy No. 623 of ASTM Specification B 150) or equivalent, or nickel-aluminum bronze (Alloy No. 630 of Specification B 150) or equivalent
Group 3: nickel-chromium-iron alloys and 18% Ni maraging steel. Some specific alloys include: Haynes alloy No. 713C, Hastelloy alloy D, G.E. alloy SEL, Haynes Stellite alloy No. 21, GMR-235 alloy, Haynes alloy No. 93, Inconel X, Inconel 718, Haynes Stellite alloy No. 6 S 816	Penetrameter grade 3: nickel-chromium-iron alloy UNS N06600 or equivalent
Group 4: nickel, copper, all the nickel-copper series or copper-nickel series of alloys, all the brasses (copper-zinc alloys) exclusive of leaded brasses. There is no restriction on using grade 4 penetrameters for the leaded brasses, because they are more attenuating of radiation, the degree depending on the lead content. This would be equivalent to using a lower grade penetrameter, which is permissible.	Penetrameter grade 4: may be 70Ni-30Cu alloy (Monel) (Class A or B of ASTM Specification B 164) or equivalent, or 70Cu-30Ni alloy (Alloy G of ASTM Specification B 161) or equivalent
Group 5: tin bronzes including gun metal and valve bronze, but excluding any leaded tin bronzes of higher lead content than valve bronze. There is no restriction on using grade 5 penetrameters for bronzes of higher lead content, because they become more attenuating of radiation with increasing lead content. This would be equivalent to using a lower grade penetrameter, which is permissible.	Penetrameter grade 5: tin bronze (Alloy D of ASTM Specification B 139) or equivalent

eter equals 2% of section thickness and the 2T hole is visible. If image quality of 1-1T were required, a radiograph would be acceptable if the outline of a 1% penetrameter were distinguishable. Alternatively, image quality can be expressed as a percentage only. In the ASTM or ASME systems, the equivalent sensitivity in percent is based on visibility of the 2T hole. Table 9 lists equivalent sensitivities for various standard image-quality levels.

Wire-type penetrameters are widely used in Europe, and a standard design is used in the United Kingdom, Germany, the Netherlands, and Scandinavia and by the International Organization for Standardization (ISO) and the International Institute of Welding (IIW). In the United States, the penetrameter design specified in ASTM E 747 is widely used. The ISO design of wire-type penetrameters has a group (typically seven) of straight 30 mm (1.2 in.) wires made of the same material as the testpiece. The diameters of the seven wires are sized in a geometric progression from a range of 21 wire sizes with a numbered geometric progression of diameters ranging from wire number 1 (0.032 mm, or 0.00126 in., in diameter) to wire number 21 (3.200 mm, or 0.126 in., in diameter). This ISO standard is similar to the standard of Deutsche Industrie Norm (DIN 54109), which consists of sixteen wire sizes of three metals—steel, aluminum, and copper. However, the wire numbers in the DIN standard are the reverse of the ISO standard; in the DIN standard, wire diameters decrease in geometric progression from wire number 1 (which has a 3.20 mm, or 0.126 in., diameter) to wire number 16 (which has a 0.10 mm, or 0.004 in., diameter).

The wire sizes on a wire penetrameter consist of different groupings. In the DIN system, for example, a wire penetrameter may have one of three groupings of seven wire sizes:

- One group contains wire numbers 1 through 7, which correspond to wire diameters of 3.20 through 0.80 mm (0.126 through 0.031 in.)
- The second group (Fig. 54c) contains wire numbers 6 through 12, which correspond to wire diameters of 1.00 through 0.25 mm (0.039 through 0.010 in.)
- The third group contains wire numbers 10 through 16, which correspond to wire diameters of 0.40 through 0.10 mm (0.016 through 0.004 in.)

Regardless of whether the wire type is designed to ISO, DIN, or ASTM specifications, image quality is denoted by the wire number of the thinnest wire distinguishable on the radiograph. In contrast to the plaque

Grade 000 penetrameter for materials Group Mg

Grade 00 penetrameter for materials Group Al

Grade 0 penetrameter for materials Group Ti

Grade 1 penetrameter for materials Group 1

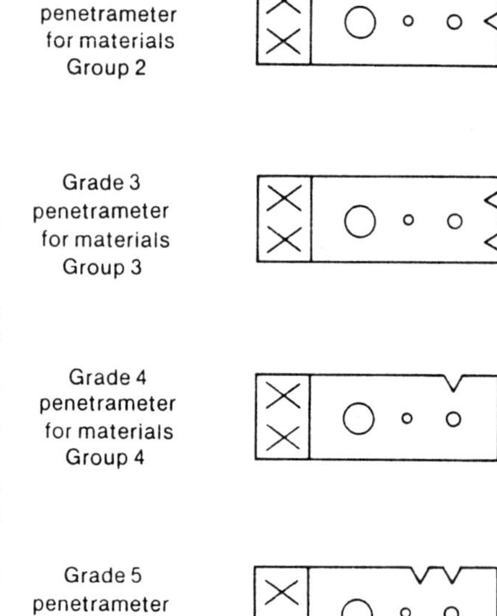

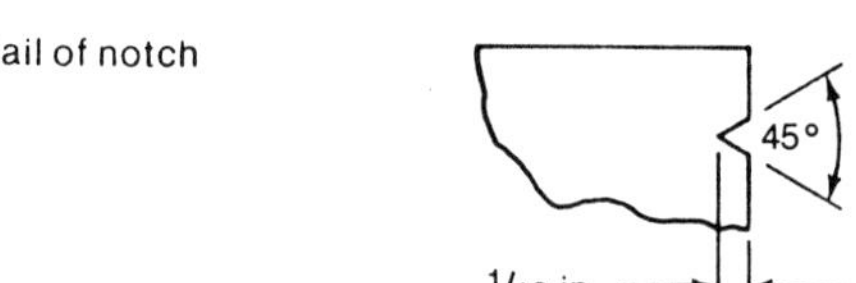

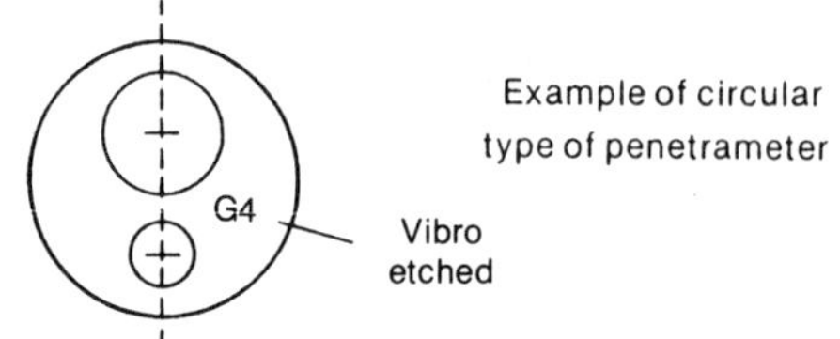

Fig. 55 Identification system of ASTM penetrameter material composition grades

Table 9 Equivalent sensitivities of various standard ASTM or ASME sensitivity levels

Equivalent sensitivity is a percentage equivalent for penetrameter thickness in which 2T is the smallest distinguishable hole size. For example, 1-1T is equivalent to 0.7-2T.

Image-quality level	Penetrameter thickness, % of testpiece thickness	Smallest visible hole size	Equivalent sensitivity, %
1-1T	1	1T	0.7
1-2T	1	2T	1.0
2-1T	2	1T	1.4
2-2T	2	2T	2.0
2-4T	2	4T	2.8
4-2T	4	2T	4.0

Table 10 DIN specification for minimum image quality and equivalent-sensitivity range for each range of testpiece thickness

Minimum image quality is expressed as wire number (BZ) of thinnest wire distinguishable in radiograph.

Testpiece thickness, mm	Testpiece thickness, in.	High-sensitivity level (category 1): Wire No., BZ	High-sensitivity level (category 1): Equivalent sensitivity, %	Normal-sensitivity level (category 2): Wire No., BZ	Normal-sensitivity level (category 2): Equivalent sensitivity, %
0–6	>0–0.25	16	1.7 min	14	2.7 min
6–8	0.25–0.30	15	2.0–1.6	13	3.3–2.5
8–10	0.30–0.40	14	2.0–1.6	12	3.1–2.5
10–16	0.40–0.60	13	2.0–1.3	11	3.2–2.0
16–25	0.60–1.00	12	1.6–1.0	10	2.5–1.6
25–32	1.00–1.25	11	1.3–1.0	9	2.0–1.6
32–40	1.25–1.60	10	1.3–1.0	8	2.0–1.6
40–50	1.60–2.00	9	1.3–1.0	7	2.0–1.6
50–80	2.00–3.15	8	1.3–0.8	6	2.0–1.3
80–200	3.15–8.00	7	1.0–0.4	...	...
80–150	3.15–6.00	...	...	5	1.6–0.8
150–170	6.00–6.70	...	...	4	1.1–0.9
170–180	6.70–7.00	...	...	3	1.2–1.1
180–190	7.00–7.50	...	...	2	1.4–1.3
190–200	7.50–8.00	...	...	1	1.7–1.6

system, however, the wire system does not provide constant sensitivity, because the sensitivity varies with testpiece thickness. Therefore, the equivalent sensitivities of wire penetrameter indications are defined for a range of testpiece thickness (as indicated in Table 10 for the DIN system). Table 11 lists the wire sizes equivalent to a 2-2T sensitivity for a variety of testpiece thicknesses.

Step wedge penetrameters usually have either an arithmetic or a geometric progression of step thicknesses. A plain step wedge penetrameter is useful only for determining the ability of a radiograph to resolve variations in testpiece thickness; it cannot be used to evaluate the effect of imaging unsharpness, which is often the chief factor that determines image quality. However, if a plain step wedge is modified by drilling holes in each step, it becomes sensitive to imaging unsharpness. This type of design is used by the British Welding Research Association (BWRA) and the French Navy (AFNOR). In the BWRA design (Fig. 54d), the holes in a given step wedge are all of the same size—0.635 mm (0.025 in.) diameter for the step wedge ranging from 0.13 to 1.0 mm (0.005 to 0.040 in.) in step thickness and 1.3 mm (0.050 in.) diameter for the step wedge ranging from 1.0 to 2.0 mm (0.040 to 0.080 in.) in step thickness. AFNOR step wedges (Fig. 54e and f) have holes that are the same in diameter as the step thickness. The BWRA standard specifies a uniform increment between step thicknesses; the AFNOR standard specifies a constant ratio between successive thicknesses (similar to that of successive wire diameters in the DIN standard). The BWRA standard incorporates only two penetrameters, both of which are shown in Fig. 54(d). The AFNOR standard specifies penetrameters having four different series of step thicknesses, as follows:

Series	Step thicknesses, mm
1	0.125, 0.16, 0.20, 0.25, 0.32, 0.40
2	0.32, 0.40, 0.50, 0.63, 0.80, 1.00
3	0.80, 1.00, 1.25, 1.60, 2.00, 2.50
4	2.00, 2.50, 3.20, 4.00, 5.00, 6.30

The four series cover the entire range of testpiece thicknesses from about 4 to 300 mm (1.6 to 12 in.). Series 3 is illustrated as a hexagonal penetrameter in Fig. 54(e), and series 2 is illustrated as a linear penetrameter in Fig. 54(f).

Both the BWRA and AFNOR penetrameters are sensitive to image definition and to contrast. Definition is judged on the visibility of holes in the specified step. Sometimes, the image quality defined by a step wedge penetrameter is somewhat ambiguous because the image can be evaluated both on the visibility of steps and on the visibility of holes. When contrast is good and definition is poor, more steps than holes can be seen on the radiograph. However, when the image quality is judged as intended—that is, on the visibility of individual holes for AFNOR penetrameters and on the visibility of the symbol (not necessarily the visibility of individual holes) for BWRA penetrameters—the step wedge penetrameters are quite sensitive to variations in radiographic technique. One minor limitation of AFNOR penetrameters is that the hole size in the thinnest steps are comparable to the size of graininess visible in the radiograph. Sometimes it is not easy to be certain that a hole is visible, but the use of two holes in the thinnest steps partly overcomes this limitation.

Penetrameters for electronic components use spherical particles and wire sizes, typical of the wire used in such devices, for determinations of resolution or sensitivity. The wires are arranged in a three-dimensional grid with close-toleranced spacing for the purpose of providing a measure of distortion. The image of the grid is measured on the radiograph, and the distortion is calculated by:

$$D = \frac{(S_m - S_A)\,100}{S_A} \qquad \text{(Eq 21)}$$

where D is the percent distortion, S_m is the wire spacing as measured on the radiograph, and S_A is the actual wire spacing.

The ASTM standard E 801 describes a set of eight penetrameters having different cover thicknesses and wire sizes. The cover densities and wire sizes are typical of the case materials and internal connecting wires of electronic components. Two of these penetrameters are typically used for each exposure—usually the number having the density closest to that of the component being radiographed and the next-higher-number penetrameter. Because electronic components are typically exposed in groups, the penetrameters are placed at opposing corners of the group edges. This ensures that the worst-case parameter values will be indicated.

Table 11 Wire sizes equivalent to 2-2T hole-type levels

Minimum specimen thickness, mm	Minimum specimen thickness, in.	Wire diameter, mm	Wire diameter, in.
6.35	0.25	0.1(a)	0.004(a)
9.5	0.375	0.13(a)	0.005(a)
13	0.500	0.16	0.0063
16	0.625	0.2	0.008
19	0.750	0.25	0.010
22	0.875	0.33	0.013
25	1.00	0.4	0.016
32	1.25	0.5	0.020
38	1.5	0.64	0.025
44	1.75	0.81	0.032
50	2	1.0	0.040
65	2.5	1.3	0.050
75	3	1.6	0.063
90	3.5	2.0	0.080
100	4	2.5	0.100
125	5	3.2	0.126
150	6	4.0	0.160

(a) Wire diameters for use with specimens less than 13 mm (½ in.) in thickness do not represent the true 2-2T level. They follow the same relationship as the hole type. Source: ASTM E 747

Placement of Identification Markers and Penetrameters

The location of identification markers with respect to the testpiece is important only to the extent that shadows cast by the identification markers should not obscure shadows cast by the testpiece itself. This is accomplished most easily by attaching the lead letters or numbers to the film holder in a region outside the area being inspected, usually along the edges of the holder. When it is important to ensure that identification markers do not obscure the image of some well-defined region of the testpiece, such as a weld, it may be desirable to attach the identification markers to the testpiece adjacent to that region.

When several views of the same testpiece are to be shot, it is good practice to attach an identification (view) marker to the testpiece at each end of the area to be inspected in each view. These markers should be left in place until after the adjacent exposures have been shot. Each view marker should be visible in two adjacent radiographs; if it is not, incomplete coverage has been obtained. In some codes and specifications, this practice (known as sequence numbering) is required.

It is desirable to mark the testpiece with chalk, crayon, or a metal stamp to indicate the exact location of identification markers. This can avoid possible difficulties either in identifying defective testpieces or in correlating radiographs with testpieces.

The placement of penetrameters is important because incorrect placement with respect to the testpiece can result in an incorrect assessment of image quality. On simple shapes, especially flat plates and similar shapes of uniform thickness, it is seldom necessary to be concerned about factors other than placing the penetrameter where it will properly represent maximum unsharpness, will not obscure any region being inspected, and will be located in the outer cone of the radiation beam.

When the shape of a testpiece is complex or when there is a large variation in the thickness of the testpiece, placement of penetrameters can be critical. Several suggested means of achieving proper placement of penetrameters are shown in Fig. 56 for welds between plates of different thickness and for circumferential welds in pipe. When no level testpiece surface is available for placement of the penetrameter, penetrameter blocks placed beside the testpiece are the only reasonable alternative. It is sometimes advantageous to use a stepped wedge as a penetrameter block, with the penetrameter on each step. For example, in the technique used for three-view inspection of the large cast stainless steel impeller discussed in Example 2 in this article, four penetrameters ranging in thickness from 0.25 to 1.15 mm (0.010 to 0.045 in.) were used for six of the exposures. The four penetrameters, each placed on a different block between 13 and 57 mm (½ and 2¼ in.) thick, were needed for assurance that the specified level of image quality was achieved over the entire range of impeller thicknesses.

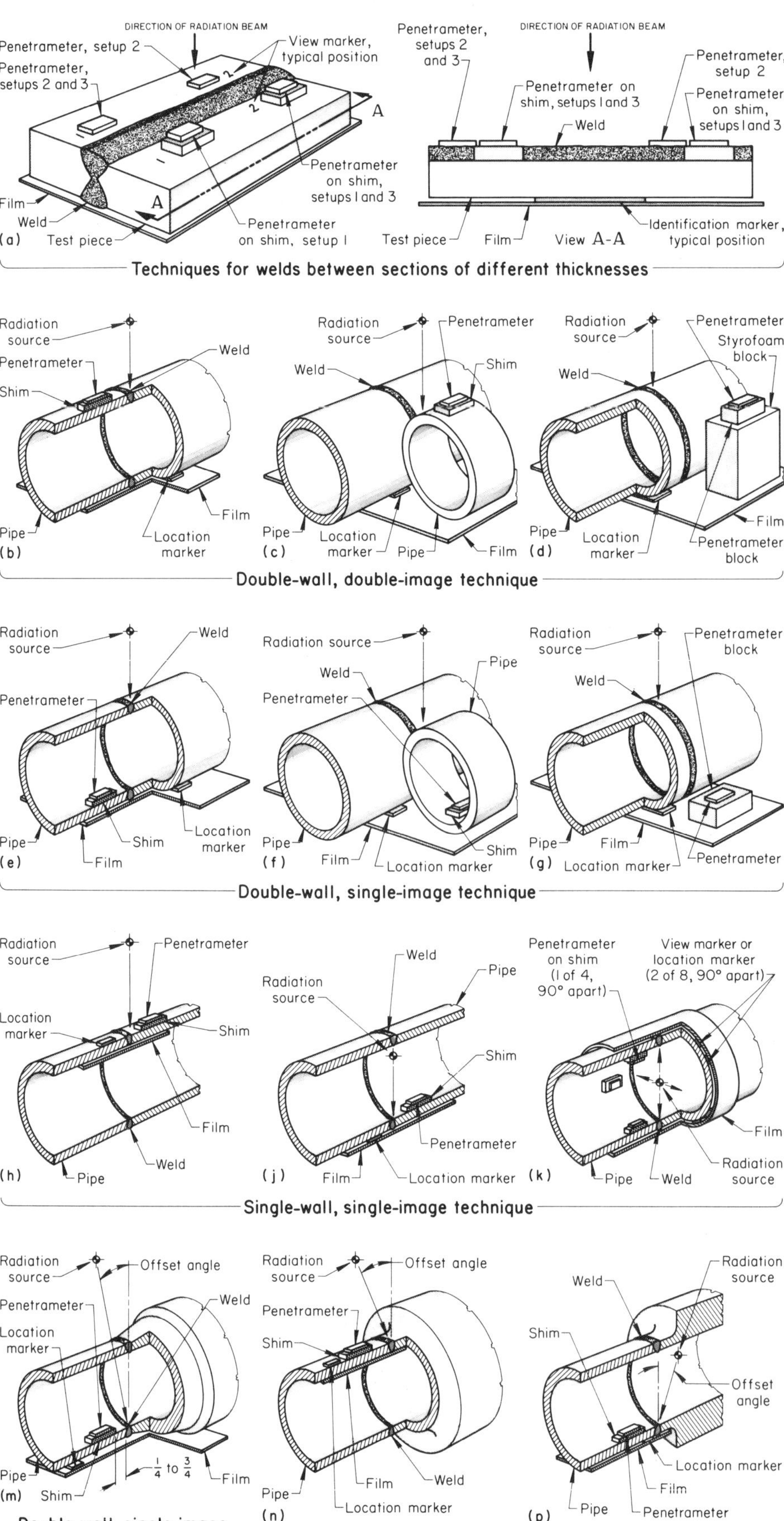

Fig. 56 Correct placement of view markers, location markers, and penetrameters for radiographic inspection. Dimensions given in inches

Even though the following discussion and Fig. 56 illustrate the placement of markers and penetrameters on weldments, similar locations for markers and penetrameters can be used on testpieces that do not contain welds. In all arrangements, penetrameters should be placed in the outer cone of the radiation beam.

Radiography of Plates. Figure 56(a) illustrates three alternative arrangements of penetrameters and identification markers for the radiography of a weld joining one plate to another plate of different thickness. In all three arrangements, the identification markers and penetrameters are placed parallel to the weld. View markers and penetrameters are usually placed 3 to 20 mm (1/8 to 3/4 in.) from the edge of the weld zone, but no more than 40 mm (1½ in.). Testpiece identification markers, however, can be placed farther away if necessary to ensure that their image is outside the image of the weld zone in the processed radiograph. Identification markers are usually placed on the film but view markers should be placed on the surface of the testpiece closest to the radiation source so that correct overlap between adjacent exposures can be verified. (If the view markers were located on the film side, a portion of the testpiece directly above the view markers could be missed even though the images of the markers appeared in adjacent radiographs.)

In Fig. 56(a), the preferred setup (setup 1) has two penetrameters located on the thinner plate. In the alternative setups (setups 2 and 3), two penetrameters are located on the thicker plate (setup 2) or one penetrameter on each plate (setup 3). Shims made of an alloy that has the same absorption characteristics as the weld metal are used under the penetrameters in each instance to compensate for any difference between the thickness of the weld zone, including reinforcement, and the thickness of the plate on which the penetrameter is located. Any shim used should be larger than the penetrameter placed on it, so that the image of the penetrameter can be clearly seen within the umbral image of the shim. Also, the direction of radiation with respect to shim and penetrameter location should be considered, especially with thick shims or penetrameter blocks, to ensure that the shim properly represents the effective penetrated thickness of the testpiece.

Some codes and specifications require that the image of a penetrameter be used to evaluate the quality of only that portion of the radiographic image of the testpiece that has similar photographic density. Strict limits can be placed on the allowable density difference between penetrameter image and testpiece image. For this reason, it may be necessary to use two or more penetrameters to evaluate image quality in different regions on the radiograph. When plaque-type penetrameters are used, plaques of different thickness are used for different regions, depending on testpiece thickness in each region.

Radiography of Cylinders. Figures 56(b) to (d) illustrate alternative locations for markers and penetrameters for the double-wall, double-image radiography of hollow cylinders or welded pipe. These alternatives can be used for either normal or offset (corona) views. When the penetrameter is placed on the cylinder itself, as shown in Fig. 56(b), or on a short section of pipe having the same diameter and wall thickness as the pipe being inspected, as shown in Fig. 56(c), any shim that is used under the penetrameter should be only thick enough to compensate for weld reinforcement; that is, twice the nominal reinforcement for a normal view, but equal to the nominal reinforcement for an offset (corona) view. If the testpiece is a plain cylinder or if a circumferential butt weld is flush with the surface, no shim is needed. When a penetrameter block is used to provide equivalent penetrated thickness under the penetrameter, the block thickness should equal twice the nominal wall thickness of the cylinder plus twice the nominal weld reinforcement. Also, the penetrameter block should be set on a block of Styrofoam or similar nonabsorbing material so that the upper surface of the penetrameter block is aligned with the upper surface of the pipe, as shown in Fig. 56(d). When a short section of pipe is used under the penetrameter (Fig. 56c), the radiation source should be centered between the pipe being inspected and the short pipe section; otherwise, the radiation source should be centered above the pipe being inspected. To ensure that the penetrameter image is within the umbral region of the image of a shim or penetrameter block, the penetrameter should be aligned with the edge of the shim or block closest to the central beam of radiation.

Figures 56(e) to (g) illustrate alternative setups for the double-wall, single-image radiography of hollow cylinders or welded pipe. These alternatives are suitable for both normal and offset (corona) views. As for a double-wall, double-image technique, the penetrameter can be placed on the pipe itself—on a short section of pipe being inspected or on a penetrameter block. The setup illustrated in Fig. 56(e) can be used when there is access to the inside of the pipe for placement of the penetrameter. When a shim is used under the penetrameter (Fig. 56e and f), it should be equal to the height of nominal weld reinforcement, regardless of the view that is used. When there is no reinforcement, no shim is needed. If the penetrameter is placed on a penetrameter block, as in Fig. 56(g), the block should be equal to twice the nominal wall thickness plus the nominal height of weld reinforcement, not plus twice the nominal reinforcement as with the double-wall, double-image technique. Radiation-source location with respect to the testpiece and the location of the penetrameter on the block or shim are the same as for a double-wall, double-image technique.

In any setup for single-wall, single-image radiography, the penetrameters can be placed only on the testpiece because the film is always on one side of the wall and the source on the other side. Figures 56(h) to (k) illustrate alternative arrangements for single-wall, single-image radiography. Shims, when used, need only compensate for any weld reinforcement. When the radiation source is external, as in Fig. 56(h), location markers should be placed on the outside surface for assurance that the correct overlap between adjacent exposures has been achieved. There should be a minimum of one penetrameter and one set of view and location markers per film, except that there should be three or more penetrameters and sets of markers (spaced equally around the circumference of the pipe) when a 360° simultaneous exposure is made on a single strip of film, as shown in Fig. 56(k). A minimum of three penetrameters is needed for assurance that the radiation source was actually located on the central axis of the cylinder and that equal intensity of radiation was incident on the entire circumference. When a 360° simultaneous exposure is made on overlapping pieces of film, not only should penetrameters be placed so that one appears on each piece of film, but also view markers and location markers should be placed so that they coincide with the regions of overlap between adjacent pieces of film.

Radiography of Flanges. Although single-image techniques (especially the single-wall, single-image technique) are ordinarily used with a normal (vertical) viewing direction, there are applications in which an offset view is advantageous. Three setups for the single-image radiography of flanged pipe using offset views are illustrated in Fig. 56(m), (n), and (p). The principles of location-marker and penetrameter placement are similar to those previously discussed for normal views; the only difference is that extra precautions must be taken to ensure that the projected images of markers or penetrameters do not fall on the image of any region being inspected.

Control of Scattered Radiation

Although secondary radiation can never be completely eliminated, numerous means are available to reduce its effect. The various methods, which are discussed below in terms of x-rays, include:

- Use of lead screens
- Protection against back scatter and scatter from external objects

- Use of masks, diaphragms, collimators, and filtration

Most of the same principles for reducing the effect of secondary x-rays apply also to γ-ray radiography. However, differences in application arise because of the highly penetrating characteristics of gamma radiation. For example, a mask for use with 200-kV x-rays could be light enough in weight for convenient handling, yet a mask for use with cobalt-60 radiation would be much thicker, heavier, and more cumbersome. In any event, with either x-rays or γ-rays, the means for reducing the effects of secondary radiation must be selected with consideration of cost and convenience as well as the effectiveness.

Lead screens placed in contact with the front and back emulsions of the film diminish the effect of scattered radiation from all sources by absorbing the long-wavelength rays. They are the least expensive, most convenient, and most universally applicable means of combating the effects of secondary radiation. Lead screens lessen the amount of secondary radiation reaching the film or detector, regardless of whether the screens increase or decrease the intensity of detected radiation. (The intensifying effect of lead screens is discussed in the section "Image Conversion Media" in this article.)

Sometimes, the use of lead screens requires increased exposure time (or image processing in the case of real-time monitoring). If high radiographic quality is desired, lead screens should not be abandoned merely because the photon energy is so low that they exhibit no intensifying action. However, at a sufficiently low photon energy, depending on the testpiece, the absorption of transmitted image-forming radiation by the front screen will degrade image quality. Under these conditions, a front screen should not be used, but a back screen will reduce back-scattered radiation without affecting the image-forming radiation and should be used. In general, lead screens should be used whenever they improve the quality of the radiographic image.

Protection Against Back Scatter. Severe back scatter can produce an image of the back of the cassette or film holder on the film, superimposed on the image of the testpiece. To prevent back scatter from reaching the film, it is customary to place a sheet of lead in back of the cassette or film holder. The thickness needed depends on radiation quality; for example, 3 mm (1/8 in.) of lead for 250-kV x-rays and 6 mm (1/4 in.) of lead for 1-MeV x-rays or for Ir-192 or Co-60 γ-rays. At 100 kV and lower, the lead that is frequently incorporated into the back of the cassette or film holder usually provides sufficient protection from back scatter. Radiographic tables or stands can also be covered with lead to reduce back scatter. Because providing protection against back scatter can usually be done simply and conveniently, it is better to overprotect than to underprotect.

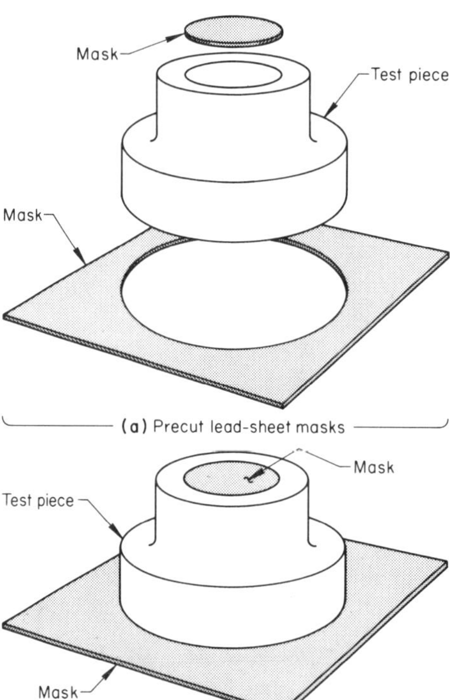

Fig. 57 Use of lead-sheet masks on a testpiece for reducing secondary radiation

For assurance of adequate protection from back-scattered radiation, a characteristic lead symbol (such as a 3 mm, or 1/8 in., thick letter "B") can be attached to the back of the cassette or film holder and a radiograph made in the normal manner. If a low-density image of the symbol appears on the radiograph, it is an indication that protection against back-scattered radiation is insufficient, and additional precautions must be taken. In the event that the image of the symbol is darker than the surrounding image, the intensification effect of lead is the probable cause of the dark image of the symbol. This effect is very rarely observed, and then only when there is little or no filtration, such as in direct or fluorescent-screen exposures or when very thin lead screens are used.

Masks and Diaphragms. Secondary radiation originating in sources outside the testpiece is most serious for testpieces that have high absorption for x-rays (most metals) because secondary radiation from external sources may be large compared with the image-forming radiation that reaches the film through the testpiece. Often, the most satisfactory method of reducing this secondary radiation is by the use of cutout lead masks or some other form of lead-sheet mask mounted over or around the testpiece (Fig. 57).

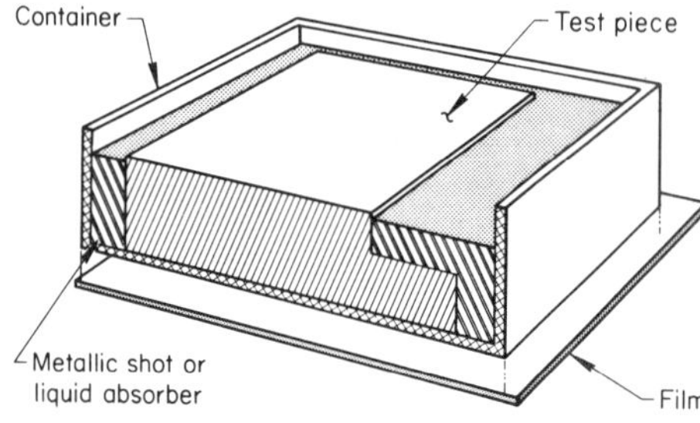

Fig. 58 Setup for radiography using either metallic shot or a liquid absorber as a mask to control secondary radiation

Copper or steel shot having a diameter of 0.25 mm (0.01 in.) or less is an effective and convenient mask. Metallic shot is also very effective for filling cavities in irregular-shape testpieces such as castings, where the normal exposure for thick areas would result in overexposure for thinner areas. Masking can also be accomplished by using barium clay, lead putty, or liquid absorbers such as a saturated solution of lead acetate plus lead nitrate. This solution is made by dissolving approximately 1.6 kg (3½ lb) of lead acetate in 4 L (1 gal.) of hot water and adding approximately 1.4 kg (3 lb) of lead nitrate. [*CAUTION: Care should be exercised at all times when using liquid absorbers, because of their highly poisonous or lethal nature.*]

When metallic shot or a liquid absorber is used as a mask, the testpiece is placed in a container made of aluminum or thin sheet steel, and the metallic shot or liquid absorber is poured in around the testpiece (Fig. 58). A form of masking called blocking, which consists of placing lead blocks at the edges of the testpiece or placing lead plugs in internal holes, also prevents side scatter from reaching the film.

Lead diaphragms limit the area covered by the x-ray beam. Diaphragms are particularly useful when the desired cross section of the beam is a circle, square, or rectangle. Figure 59 shows the combined use of metallic shot, a lead mask, and a lead diaphragm to control scattered radiation.

Collimators. Side scatter from walls, equipment, and other structures in the x-ray room can be greatly reduced by improving the directionality of the x-ray beam. Directionality can be improved by the use of a collimator, which is often a thick lead diaphragm with a small hole through the middle. A collimator absorbs most of the diverging radiation that surrounds the central beam, thus eliminating most of the rays that could be scattered from nearby surfaces. Although considered good practice, removing all unnecessary equipment and other material from the x-ray room is sometimes impossible or impractical. In such cases, a collimator placed at the exit port of the radiation source can substantially reduce, if not eliminate, unwanted side scatter.

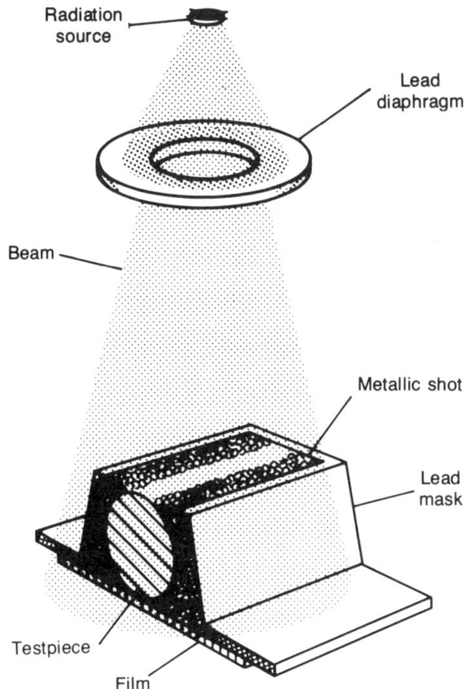

Fig. 59 Use of a combination of metallic shot, a lead mask, and a lead diaphragm to control scattered radiation

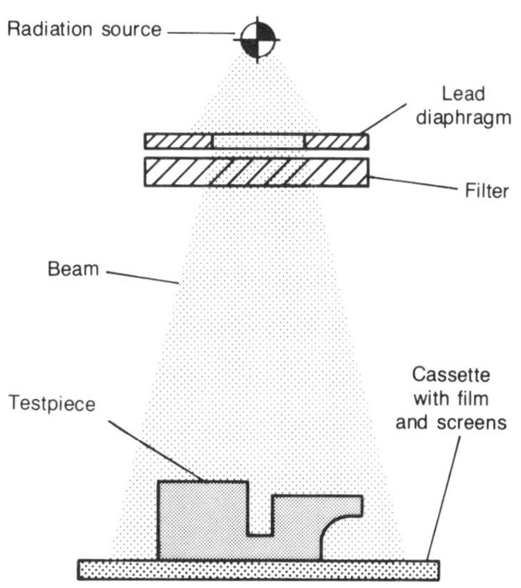

Fig. 60 Use of a lead diaphragm to limit the included angle of the x-ray beam, and use of a filter to reduce subject contrast and to eliminate much of the secondary radiation that causes undercutting

Filtration. In addition to the filtering effect of lead screens, secondary x-rays can be filtered by using thin copper or lead sheets between the testpiece and the cassette or film holder. Filtration is never used in gamma radiography, because of the essentially monochromatic nature of the beam.

When the testpiece has very thin sections adjacent to thick sections or when the direct beam can strike the detector after passing around the testpiece, undercutting may be encountered. If undercutting occurs, additional filtration (that is, more than can be achieved with conventional lead screens) is necessary. Additional filtration is accomplished by placing a filter at or near the x-ray tube, as shown in Fig. 60. This may adequately eliminate overexposure in thin regions of the testpiece and also along the perimeter of the testpiece. Such a filter is particularly useful for reducing undercutting when a lead mask around the testpiece is impractical or when the testpiece may be damaged by masking with liquid absorbers or metallic shot. Filtration of the incident radiation beam reduces undercut by selectively attenuating the long-wavelength portion of the x-ray spectrum. Long wavelengths do not contribute significantly to the detection of flaws but only produce secondary radiation that reduces radiographic contrast and definition.

The choice of a filter material should be made on the basis of availability and ease of handling. For the same filtering effect, the thickness of filter required is less for those materials that have lower absorption coefficients. Often, copper or brass is the most useful because filters of these materials will be lightweight enough to handle but not so thin that they are easily bent or broken.

Definite rules for filter thicknesses are difficult to formulate because the amount of filtration required depends not only on the materials and thickness range of the testpiece but also on the homogeneity of the testpiece and on the amount of undercutting that is to be eliminated. In the radiography of aluminum, a filter of copper about 4% as thick as the thickest area of the testpiece is usually satisfactory. With a steel testpiece, a copper filter ordinarily should be about 20%, or a lead filter about 3%, as thick as the thickest area of the testpiece for optimum filtration. These values are maximum values, and depending on circumstances, useful radiographs can often be made with far less filtration.

Radiation Diffraction. A special form of scattering due to x-ray diffraction is occasionally encountered in radiographic inspection. Diffraction of radiation is observed most often in the radiography of thin testpieces having a grain size large enough to be an appreciable fraction of the part thickness. Castings made of austenitic corrosion-resistant and heat-resistant stainless steel or of Inconel and other nickel-base alloys are the products most likely to exhibit diffraction in radiographs.

The radiographic appearance of this type of scattering can be confused with the mottled appearance sometimes produced by porosity, segregation, or spongy shrinkage. Diffraction patterns can be distinguished from these conditions in the testpiece by making successive radiographs with the testpiece rotated between exposures 1 to 5° about an axis perpendicular to the beam. A mottled pattern due to porosity or segregation will be only slightly changed, but a pattern due to diffraction effects will show a marked change. The radiographs of some testpieces will show mottling from both diffraction and porosity, and careful interpretation of the radiographs is needed to differentiate between them.

Mottling due to diffraction can be reduced, and sometimes eliminated, by raising x-ray tube voltage and by using lead screens. Filters will usually aid in the control of diffraction. Raising the tube voltage and filtration are often of positive value even though radiographic contrast and sensitivity are reduced.

Sometimes, diffraction cannot be reduced. In such cases, two radiographs made as described above can be used to identify diffraction.

Scattering at High Photon Energies. Lead screens should always be used when the radiation energy exceeds 1 MeV. Use of the usual 0.13 mm (0.005 in.) thick front screen and 0.25 mm (0.010 in.) thick back screen is both satisfactory and convenient. Some users find 0.13 mm (0.005 in.) thick front and back screens adequate when filters are used both front and back of the cassette or film holder. Other users consider 0.25 mm (0.010 in.) thick front and back screens of value because of greater selective absorption of scattered radiation from the testpiece.

Filtration of the incident x-ray beam offers no improvement in radiographic quality. However, filters at the film improve radiographs for the inspection of uniform sections.

Lead filters are most convenient for energies above 1 MeV. Care should be taken to minimize mechanical damage to the filter because filter defects could be confused with characteristics of the material being inspected.

It is important to block off all radiation except the effective beam with heavy shielding at the anode. This is usually recognized by manufacturers of high-voltage x-ray equipment. For example, in some linear accelerators, depleted-uranium collimators confine the beams to a 22° included angle. Unless a high-energy x-ray beam is well collimated, radiation striking the walls of the x-ray room will generate secondary radiation and thus seriously degrade the quality of the radiograph. This will be especially noticeable if the testpiece is thick or has projecting parts that are not immediately adjacent to the filter.

Interpretation of Radiographs

Proper identification of both the radiograph and the testpiece, clarity of the penetrameter, suitability of radiographic techniques, adequacy of coverage, and the

Table 12 Probable causes and corrective action for various types of deficient image quality or artifacts on processed radiographic film

Quality or artifact	Probable cause	Corrective action
Density too high	Overexposure	View with higher-intensity light. Check exposure (time and radiation intensity); if as specified, reduce exposure 30% or more.
	Overdevelopment	Reduce development time or developer temperature.
	Fog	See "Fog" below.
Density too low	Underexposure	Check exposure (time and radiation intensity); if as specified, increase exposure 40% or more.
	Underdevelopment	Increase development time or developer temperature. Replace weak (depleted) developer.
	Material between screen and film	Remove material.
Contrast too high	High subject contrast	Increase tube voltage.
	High film contrast	Use a film with lower contrast characteristics.
Contrast too low	Low subject contrast	Reduce tube voltage.
	Low film contrast	Use a film with higher contrast characteristics.
	Underdevelopment	Increase development time or developer temperature. Replace weak (depleted) developer.
Poor definition	Testpiece-to-film distance too long	If possible, decrease testpiece-to-film distance; if not, increase source-to-film distance.
	Source-to-film distance too short	Increase source-to-film distance.
	Focal spot (or γ-ray source) too large	Use smaller source or increase source-to-film distance.
	Screens and film not in close contact	Ensure intimate contact between screens and film.
	Film graininess too coarse	Use finer-grain film.
Fog	Light leaks in darkroom	With darkroom unlighted, turn on all lights in adjoining rooms; seal any light leaks.
	Exposure to safelight	Reduce safelight wattage. Use proper safelight filters.
	Stored film inadequately protected from radiation	Attach strip of lead to loaded film holder and place in film-storage area. Develop test film after 2 to 3 weeks; if image of strip is evident, improve radiation shielding in storage area.
	Film exposed to heat, humidity, or gases	Store film in a cool, dry place not subject to gases or vapors.
	Overdevelopment	Reduce development time or developer temperature.
Fog (continued)	Developer contaminated	Replace developer.
	Exposure during processing	Do not inspect film during processing until fixing is completed.
Finely mottled fog	Stale film	Use fresh film.
Fog on edge or corner	Defective cassette	Discard cassette.
Yellow stain	Depleted developer	Replace developer solution.
	Failure to use stop bath or to rinse	Use stop bath, or rinse thoroughly between developing and fixing.
	Depleted fixer	Replace fixer solution.
Dark circular marks	Film splashed with developer prior to immersion	Immerse film in developer with care.
Dark spots or marblelike areas	Insufficient fixing	Use fresh fixer solution and proper fixing time.
Dark branched lines and spots	Static discharge	Unwrap film carefully. Do not rub films together. Avoid clothing productive of static electricity.
Dark fingerprints	Touching undeveloped film with chemically contaminated fingers	Wash hands thoroughly and dry, or use clean, dry rubber gloves.
Light fingerprints	Touching undeveloped film with oily or greasy fingers	Wash hands thoroughly and dry, or use clean, dry rubber gloves.
Dark spots or streaks	Developer contaminated with metallic salts	Replace developer solution.
Crescent-shaped light areas	Faulty film handling	Keep film flat during handling. Use only clean, dry film hangers.
Light circular patches	Air bubbles on film during development	Agitate immediately upon immersion of film in developer.
Circular or dropshaped light patches	Water or fixer splashed on film before development	Avoid splashing film with water or fixer solution.
Light spots or areas	Dust or lint between screens and film	Keep screens clean.
Sharply outlined light or dark areas	Nonuniform development	Agitate film during development.
Reticulation (leather-grain appearance)	Temperature gradients in processing solutions	Maintain all solutions at uniform, constant temperature.
Frilling (loosening of emulsion from film base)	Fixer solution too warm	Maintain correct temperature of the fixing solution.
	Fixer solution depleted	Replace fixer solution.

techniques of image processing in real-time systems or the precision and uniformity of film processing in film radiography all offset the image that is being interpreted. Film radiographs and real-time radiographs are interpreted similarly with respect to the recognition of flaws and testpiece features. The primary difference is that film radiographs are negative images, while real-time radiographs are generally positive images (which may also be enhanced with image processing).

A qualified interpreter must:

- Define the quality of the radiographic image, which includes a critical analysis of the radiographic procedure and the image-developing procedure
- Analyze the image to determine the nature and extent of any abnormal condition in the testpiece
- Evaluate the testpiece by comparing interpreted information with standards or specifications
- Report inspection results accurately, clearly, and within proper administration channels

Proper identification of both the radiograph and testpiece is an absolute necessity for correlation of the radiograph with the corresponding testpiece. Identification includes both identification of the testpiece and identification of the view or area of coverage.

Poor-quality film radiographs are usually reshot. However, reshooting radiographs increases inspection costs, not only because the original setup must be duplicated and a new exposure made but also because the testpiece must be retrieved and taken to the radiographic laboratory. With on-site radiography, which involves transporting radiographic equipment to the site and returning the exposed films to the laboratory for processing, especially high costs may be involved when poor-quality radiographs must be reshot. Table 12 lists some of the usual causes of poor quality in a radiographic image and indicates the usual corrective action required to eliminate each cause.

Personnel engaged in the interpretation of radiographs should possess certain qualifications. Some qualification recommendations are included in personnel standards published by the American Society for Nondestructive Testing (Ref 6), several governmental agencies, and many private manufacturers. Usually, a minimum level of visual acuity, minimum standards of education and training, and demonstrated proficiency are required of all interpreters of radiographs.

Viewing of radiographs should be carried out in an area with subdued lighting to minimize distracting reflections from the viewing surface. Audible distractions, which interfere with concentration, can best be avoided by locating the work area away from the main production floor or other high-noise area.

Radiographic film images are viewed on an illuminated screen. The viewing apparatus should have an opal-glass or plastic screen large enough to accommodate the largest film to be interpreted. The screen should be illuminated from behind with light of sufficient intensity to reveal variations in photographic density up to a nominal film density of at least 3.0. There may be a need for a smaller, more intensely illuminated viewer for evaluating small areas of film having densities up to 4.5 or more. Viewing screens of high-intensity illuminators should be cooled by blowers or other suitable apparatus to prevent excessive heat from damaging films and to extend lamp life.

When interpreting paper radiographs or xeroradiographs, specular light as from a spotlight or high-intensity lamp should be directed onto the radiograph from the side at an angle of about 30°. Background lighting should be heavily subdued.

A densitometer can be provided for accurate evaluation of small variations in photographic density or for quantitative evaluation of radiographic and processing techniques. A transmission densitometer is used with films, and a reflection densitometer is used with paper radiographs.

Radiographic Acceptance Standards. Usually, a series of radiographs that exhibit various types and sizes of flaws should be selected for acceptance standards. Parts that contain similar flaws should be performance tested to determine the least acceptable condition. The radiograph of the least acceptable part then becomes the minimum acceptance standard for similar parts. Often, the acceptance standard is defined as a length or area of the image that may contain no more than a specified number of flaws of a given size and type. Certain types of flaws, such as cracks or incomplete fusion, may be prohibited regardless of size. The interpreter must determine the degree of imperfection, as related to the minimum acceptance standard, and then decide whether minimum soundness requirements have been met.

Table 13 Reference radiographs in ASTM standards

ASTM standard	Subject of radiographs in standard
E 155	Aluminum and magnesium castings
E 186	Heavy-wall (50–115 mm, or 2–4½ in.) steel castings
E 192	Investment steel castings for aerospace applications
E 242	Appearance of radiographic images as certain parameters are changed
E 272	High-strength copper-base and nickel-copper alloy castings
E 280	Heavy-wall (115–300 mm, or 4½–12 in.) steel castings
E 310	Tin bronze castings
E 390	Steel fusion welds
E 431	Semiconductors and related devices
E 446	Steel castings up to 50 mm (2 in.) in thickness
E 505	Aluminum and magnesium die castings
E 689	Ductile cast irons
E 802	Gray iron castings up to 115 mm (4½ in.) in thickness

Obviously, no single standard can be applied universally to radiographic inspection. However, flaws that are frequently encountered have been reproduced in sets of reference radiographs such as those published by ASTM (Table 13). Reference radiographs depict various types of flaws that may occur in castings or weldments and are graded according to flaw size and severity.

Codes or specifications for radiographic inspection, particularly those that have been standardized by an industry through a trade association or a professional society or those that have been adopted by a governmental agency or a prime contractor, may refer to published reference radiographs. In such cases, the code or specification should designate one or more reference radiographs in a specific set as the minimum standard for acceptance. Although it is considered most desirable to have an acceptance standard that is based on actual service data, standardized codes or specifications usually define rigid acceptance criteria that do not allow for variations in specific design features of similar products.

Elements of Radiograph Interpretation

A penetrameter, when clearly defined in a radiograph, signifies that the image has a certain level of sensitivity for detecting voids in the testpiece. Because the image of a penetrameter usually represents a condition of maximum enlargement and maximum unsharpness, it should never be used as a standard against which the size of a discontinuity is measured.

A radiographic standard shooting sketch (RSSS) or other appropriate information about radiographic technique should be provided so that the interpreter can determine if the technique was adequate prior to interpretation. In particular, the technique should be analyzed for its ability to detect the flaws that are most likely to be present. For example, if planar flaws such as cracks, laps, or seams are likely to have a certain orientation, the radiation beam should be directed parallel to the probable plane of the flaw. As a minimum, the RSSS should contain entries listing the following:

- X-ray source energy (tube voltage) or type of radioisotope
- Tube current in milliamperes, or radioisotope-source strength in curies
- Focal-spot size of x-ray source or physical size of radioisotope source
- Exposure time (film or paper)
- Type of film (brand identification)
- Source-to-detector distance
- Type and thickness of screens
- Description of filters, masks, and diaphragms
- Viewing direction and method of marking
- Description and location of penetrameters
- Description (sketch) and major dimensions of testpiece

If radiography is being performed to satisfy requirements of a code, the minimum information appearing in the RSSS can be specified by the code.

Radiographic coverage, which refers to the percentage of area or volume of a testpiece that appears in a radiograph or series of radiographs, must be evaluated to ensure that all regions of the testpiece to be inspected have been radiographed with adequate clarity. It is not always necessary to have complete coverage; sometimes only a section of the testpiece is inspected by radiography. Often, reference to the RSSS can help in determining whether there is adequate radiographic coverage. Coverage can also be specified on engineering drawings, in oral or written instructions, or in written customer specifications.

In film radiography, the handling and processing of films account for a significant portion of the total inspection procedure. Improper handling and processing, especially darkroom techniques, are the greatest causes of poor quality in radiographic images. Normally, the interpreter should inspect each radiograph to ensure that it is free of artifacts. It is especially important that there be no unidentified artifacts that could be cause for future controversy when the radiograph is specified to be of archival quality. Radiographs made for process-control purposes may be accepted if they contain artifacts that do not impair diagnostic value, although even this is considered bad practice.

Image Analysis. There are three basic types of flaws that can be detected by radiography: voids, inclusions, and dimensional irregularities. Although external dimensional irregularities are ordinarily checked by other inspection methods, radiography can detect gross deviations from the correct shape and more important, can detect certain types of internal deviations, such as core shift in castings. Voids and inclusions may exist in almost any form, ranging from a planar feature (cracks, tears, and cold shuts) to a spherical feature (shrinkage, oxide inclusions, porosity, and entrapped slag).

Five characteristics of a discontinuity (type, specific shape, size, orientation, and location) define its overall effect on a part. Type and location of a flaw govern relative importance. Specific shape mainly influences stress concentration. Flaw size and orientation determine the effect on nominal stress. Each of the five characteristics should be identified for each flaw. This will provide a meaningful basis for comparison with standards. Sometimes, however, a standardized code or specification may not recognize that certain flaw characteristics can affect product performance; for example, in some codes, flaw shape and location have no bearing on acceptance.

The specific radiographic appearance of several different types of flaws is discussed and illustrated in the following sections. Radiographic standards such as those published by ASTM (Table 13) illustrate the radiographic appearance of many types and sizes of the more commonly encountered flaws, both in steel and in light alloys.

The specific shape of a flaw, especially one having sharp terminal points, could be cause for rejection because of the high degree of stress concentration associated with sharp features. Rounded features induce a lower degree of stress concentration and are usually more acceptable, depending on size and location.

Many flaws not only concentrate stress but also reduce the load-carrying capacity of a component by reducing the cross-sectional area. Obviously, large flaws are more damaging in this regard than small flaws. Some parts are designed with a large factor of safety and can tolerate the presence of many small flaws. However, if small flaws are aligned or clustered, there can be too great a reduction in safety factor in a local region. Often, specifications permit certain types of flaws as long as prescribed limits of size and concentration are not exceeded. Here, the interpreter must be guided by the tolerance specified by the design engineer.

Specimen Evaluation. A flaw detected by radiography is meaningless unless compared to a known condition. As discussed previously, a limit of acceptability that defines the radiographic appearance of the least acceptable condition should be established. Therefore, specimen evaluation consists of comparing the interpreted image with the least acceptable condition in terms of the type, size, quality, and severity level of any flaws that are found. The result of this comparison is a judgment to accept or reject the part.

The fact that one or more flaws are found in a part does not necessarily mean that the part is unserviceable; this can be determined only by destructive testing or by comparison of the part with another part that is known to be unserviceable. Even when a code or specification requires rejection if a flaw of a particular type or size is found, it does not necessarily follow that the part can neither be used "as is" nor repaired to a serviceable condition. In some cases, the ultimate disposition of a part is decided only after further examination. Such decisions can be made by management or by a committee known as a Material Review Board, which may include a representative of the customer.

Reports. The results of specimen evaluation, especially the decision to accept or reject a part, must be reported immediately. This is usually done on a standard report form through a well-defined administrative channel. In most cases, delays in forwarding inspection reports can cause costly delays in subsequent operations.

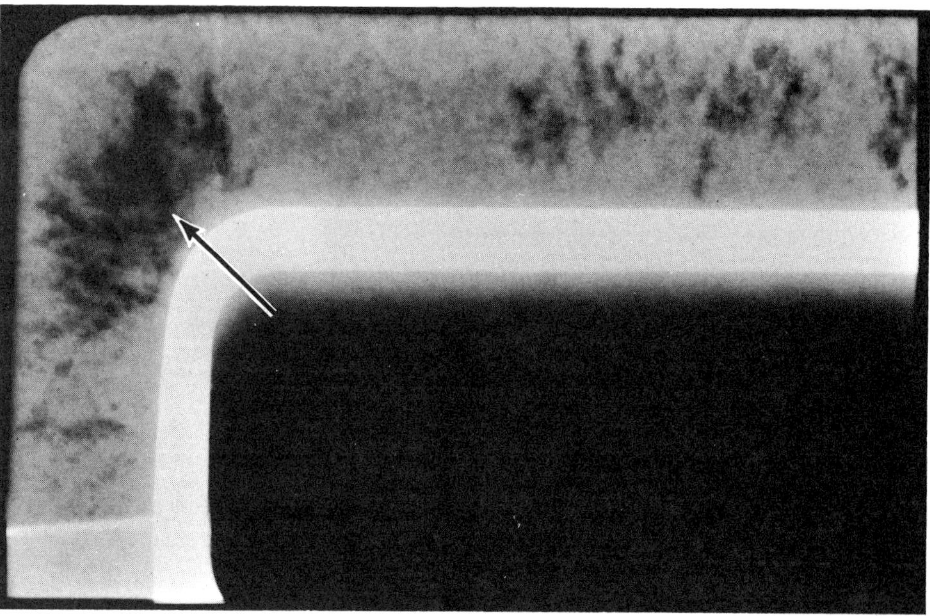

Fig. 61 Radiographic appearance of gross shrinkage porosity (arrow) in an aluminum alloy 319 manifold casting. Radiograph was made at 85 kV with 1-min exposure.

Radiographic Appearance of Specific Types of Flaws

The radiographic appearance of many of the more usual types of flaws found in castings and weldments is described in this section. The descriptions apply specifically to images on film radiographs, although paper radiographs and xeroradiographs will exhibit similar images. Real-time images of the same types of flaws will be reversed in tone (dark tones in a radiograph will be light in a fluoroscopic image and vice versa) but otherwise will be similar to the images described here.

Some of these types of flaws are not unique to castings and weldments. For example, cracks can be found in any product form. Surface inspection methods such as liquid penetrant or magnetic particle inspection are more appropriate than radiography for detecting most surface cracks, yet some forms of metal separation (forging bursts, for example) are entirely internal and cannot be found by surface methods.

Flaws in Castings

It is possible in most cases to identify radiographic images of the usual types of flaws in castings. The main types of foundry flaws that can be identified radiographically are described in the paragraphs that follow. Specific examples can be found in reference radiographs, such as those published by ASTM (Table 13).

Microshrinkage appears as dark feathery streaks or dark irregular patches, corresponding to grain-boundary shrinkage. This condition is most often found in magnesium alloy castings.

Shrinkage porosity (spongy shrinkage) appears as a localized honeycomb or mottled pattern (Fig. 61). Spongy shrinkage may be the result of improper pouring temperature or alloy composition.

Gas porosity appears as round or elongated smooth, dark spots. It occurs individ-

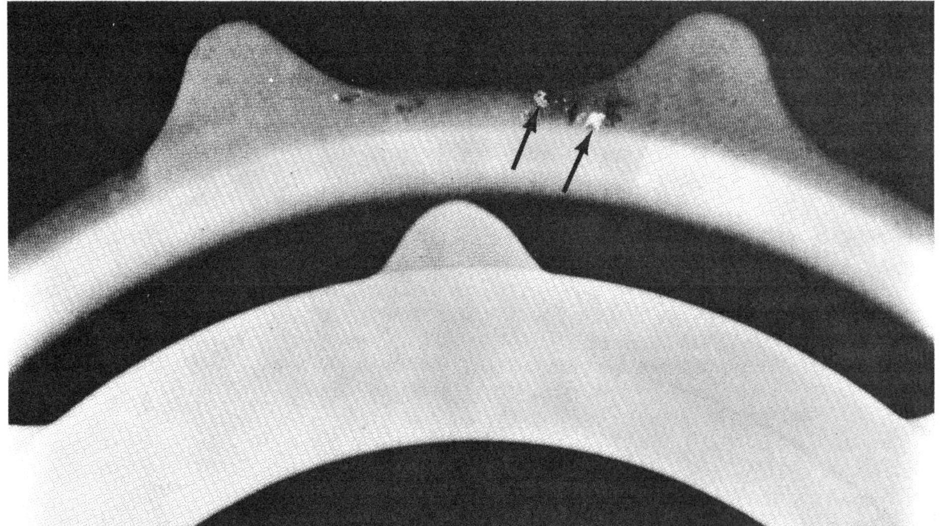

Fig. 62 Radiographic appearance of dross inclusions (arrows) in the outer flange of a cast aluminum alloy 355 housing body. Radiograph made at 75 kV, 1-min exposure.

Fig. 63 Uneven wall thickness in an internal passage of a casting caused by core shift (top right). This radiograph, of an aluminum alloy casting about 3 to 6 mm (1/8 to 1/4 in.) thick, was made at 65 kV with an exposure time of 1 min.

ually or in clusters or may be distributed randomly throughout the casting. This condition is caused by gas released during solidification or by the evaporation of moisture of volatile material from the mold surface.

Dispersed Discontinuities. Although the flaws usually encountered in light-alloy castings are similar to those in ferrous castings, a group of irregularities called dispersed discontinuities may be present in the former. These dispersed discontinuities, prevalent in aluminum and magnesium alloy castings, consist of tiny voids scattered throughout part or all of the casting. Gas porosity and shrinkage porosity in aluminum alloys are examples of dispersed discontinuities. On radiographs of sections more than 13 mm (½ in.) thick, it is difficult to distinguish images corresponding to the individual voids. Instead, dispersed discontinuities may appear on film deceptively as mottling, dark streaks, or irregular patches that are only slightly darker than the surrounding regions.

Tears appear as ragged dark lines of variable width having no definite line of continuity. Tears may exist in groups, starting at a surface, or they may be internal. Tears usually result from normal contraction of the casting during or immediately after solidification.

Cold cracks generally appear as single, straight, sharp dark lines and are usually continuous throughout their lengths. Cold cracks are produced by internal stresses caused by thermal gradients and may occur upon cooling from elevated temperatures during flame cutting, grinding, or quenching operations.

Cold shuts appear as distinct dark lines of variable length and smooth outline. Cold shuts are formed when two bodies of molten metal flowing from different directions contact each other but fail to unite. Cold shuts may be produced by interrupted pouring, slow pouring, or pouring the metal at too low a temperature.

Misruns appear as prominent dark areas of variable dimensions with a definite smooth outline. Misruns are produced by failure of the molten metal to completely fill a section of casting mold, leaving the region void.

Inclusions of foreign material in the molten metal may be poured into the mold. They appear as small lighter or darker areas in a radiograph, depending on the absorption properties of the included material as compared to those of the alloy. Sand inclusions appears as gray or light spots of uneven granular texture and have indistinct outlines. Inclusions lighter than the parent metal appear as isolated irregular or elongated variations of film blackening. Occasionally, an inclusion will have absorption characteristics equivalent to those of the matrix and will go undetected, although normally an inclusion that exhibits a radiographic contrast of about 1.4 to 2.3% can be seen. A contrast of 1.4 to 2.3% corresponds to about 0.005 to 0.01 density difference between adjacent areas on the film. Dross inclusions in the outer flange of a casting are shown in Fig. 62.

Unfused chaplets usually appear in outline conforming to the shape of the chaplet. The outline is caused by a lack of bond between the chaplet and the cast metal.

Core shift can be detected when the view makes it impossible to measure deviation from a specified wall thickness (Fig. 63). Core shift may be caused by jarring the mold, insecure anchorage of cores, or omission of chaplets.

Centerline shrinkage is localized along the central plane of a wall section, irrespective of the position occupied by the section in the mold. Such shrinkage is composed of a network of numerous filamentary veinlets leading in the direction of the nearest riser and can sometimes be mistaken for tears.

Shrinkage cavities occur when insufficient feeding of a section results in a continuous cavity within the section. Shrinkage cavities appear on the radiograph as dark areas that are indistinctly outlined and have irregular dimensions.

Segregation is the separation of constituents in an alloy into regions of different chemical compositions. This condition, seen mostly in aluminum alloys, appears as lighter areas on the film that produce a somewhat mottled appearance.

Surface irregularities may produce an image corresponding to any deviation from normal surface profile. It is possible to confuse these with internal flaws unless the casting is visually inspected at the time of interpretation.

Flaws in Weldments

In welding, heat must be carefully controlled to produce fusion and adequate penetration. Too much heat can cause porosity, cracks, and undercutting; too little heat can cause inadequate joint penetration and incomplete fusion.

Most weld flaws consist of abrupt changes in homogeneity and can be readily detected by radiographic inspection. Stresses that are induced in the metal by welding but that are not accompanied by a physical separation of material will not be detected by radiography. Also, cracks not aligned with the x-ray beam may be missed. Conditions that can be detected in welds by radiography are described below. Several

Fig. 64 Radiograph showing a large crack in a multiple-pass butt weld in 57 mm (2¼ in.) thick steel plate. The crack mainly follows the edge of the weld, but both ends turn in toward the center. The weld joined a 57 mm (2¼ in.) thick plate to a 70 mm (2¾ in.) thick plate that had been tapered to 57 mm (2¼ in.) at the edge of the weld groove. Radiograph was made with 1-MeV x-rays on Industrex AA film.

reference radiographs of flaws in weldments are provided in the article "Weldments, Brazed Assemblies, and Soldered Joints" in this Volume.

Undercutting appears as a dark line of varying width along the edge of the fusion zone. A fine dark line in this darker area could indicate a crack and should be further investigated.

Incomplete fusion appears as an elongated dark line. It sometimes appears very similar to a crack or an inclusion and could even be interpreted as such. Incomplete fusion occurs between weld and base metal and between successive beads in multiple-pass welds. Incomplete penetration along one or both sides of the weld zone has an appearance similar to that of incomplete fusion.

Cracks are frequently missed if they are very small (such as check cracks in the heat-affected zone) or are not aligned with the radiation beam. When present in a radiograph, cracks often appear as fine dark lines of considerable length but without great width. Even some fine crater cracks are readily detected. In weldments, cracks may be transverse or longitudinal and may be either in the fusion zone or in the heat-affected zone of the base metal. Figure 64 illustrates a large crack in a steel weldment.

Porosity (gas holes) consists commonly of spherical voids that are readily recognizable as dark spots, the radiographic contrast varying directly with diameter. These voids may be randomly dispersed, in clusters, or may even be aligned along the centerline of the fusion zone. Occasionally, the porosity may take the form of tubes (worm holes) aligned along the direction of the weld solidification front. These appear as lines with a width of several millimeters.

Slag inclusions are usually irregularly shaped and appear to have some width. Inclusions are most frequently found at the edge of the weld, as illustrated in Fig. 65. In location, elongated slag deposits are often found between the first root pass and subsequent passes or along the weld-joint interface, while spherical slag inclusions can be distributed anywhere in the weld. The density of a slag inclusion is nearly uniform throughout and has less contrast than porosity, because an inclusion can be considered as a pore with absorbing material. Tungsten inclusions or a very high density (barium-containing) slag will appear as white spots.

Incomplete root penetration appears as a dark straight line through the center of the weld. The width of the indication is determined by the root gap and amount of weld penetration.

Flaws in Semiconductors

Voids in semiconductors occur in several different sites, depending on the type of construction. In hermetic integrated circuits and low-power transistors, voiding typically occurs at the die (semiconductor element) and header (case) interface. The voids in this area will have the same approximate density as the areas of the header that are undisturbed by the die mount. In hermetic power transistors, voids may occur at the die mount substrate to header interface and at the die to die mount substrate interface. It is not possible to differentiate between interfaces after the device is sealed. Therefore, two radiographs are required—one after substrate attachment and a second after die attachment. Orientation is critical because the radiographs must be compared to determine the total area of voiding. In plastic-encapsulated semiconductors, voids in the encapsulating material are discernible by density differences, just as a void in other materials or in a weld.

Extraneous material is frequently missed because of its small size and very thin cross section. Conductive material as small as 0.025 mm (0.001 in.) in its major dimension may cause failure of a semiconductor device. In some cases, multiple views may be required to determine whether or not expulsed die mount material is attached to the device case. Because of the very small geometries used, multiple views may also be required to determine if there is adequate clearance between internal connecting wires.

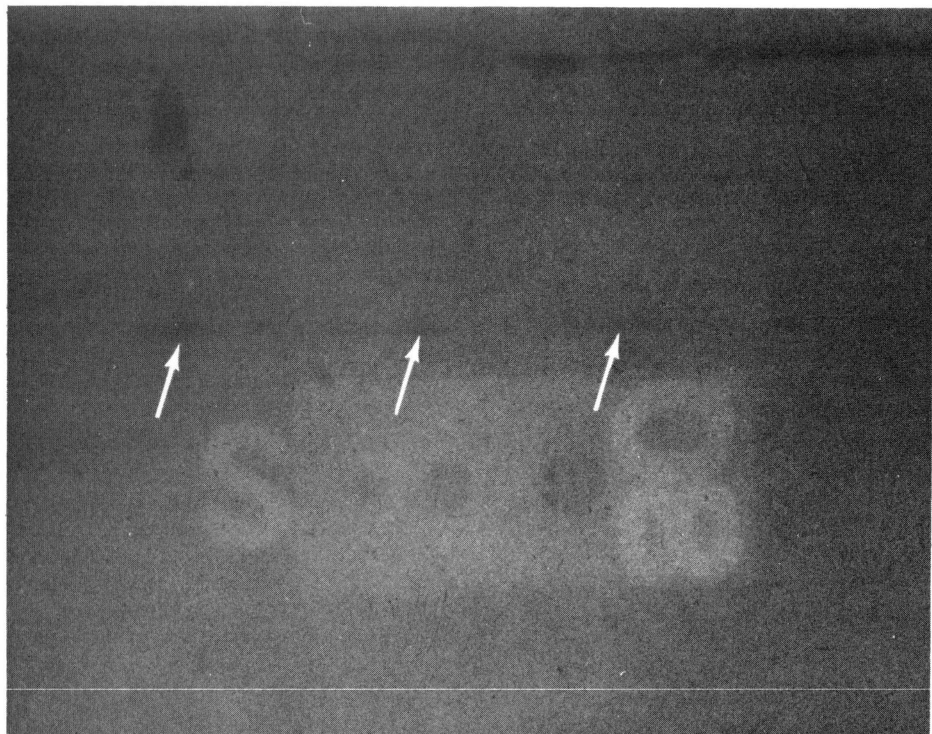

Fig. 65 Radiograph showing a crack (dark line at top) and entrapped slag inclusions (dark spots at arrows) on opposite sides of a multiple-pass butt weld joining two 180 mm (7 in.) thick steel plates. Radiograph was made with 1-MeV x-rays on Industrex AA film.

Electronic devices can also be inspected after they are placed on the circuit board, but the board and its other components will reduce contrast and interfere with the resolution of fine features. The advantage of inspecting after the devices are placed on the board is that connections from the device to the board can be inspected also. Because the solders used to form these joints contain elements of high atomic number, they form high contrast images. Radiographic inspection can detect flaws such as solder balls (undesired solder that has been expelled from the joint), bridging, misregistered devices, and joints without solder. Laminographic systems are available that can emphasize various planes within the circuit board (see the article "Industrial Computed Tomography" in this Volume).

Appendix: Processing of Radiographic Film

In the processing of radiographic film, an invisible latent image produced on the film by exposure to x-rays, γ-rays, or light is made visible and permanent. Film processing is an exacting and important part of the radiographic procedure. Poor processing can be just as detrimental to the quality of a radiograph as poor exposure practice.

Two methods of processing can be employed: manual processing, which is carried out by hand in trays or deep tanks, and automatic processing, which is accomplished in automated equipment. Guidelines for the control and maintenance of manual and automatic radiographic film processing equipment are specified in ASTM E 999.

Manual Film Processing

The manual processing of radiographic film is carried out in a processing room (darkroom) under subdued light of a particular color to which the film is relatively insensitive. The film is first immersed in a developer solution that causes areas of the film that have been exposed to radiation to become dark; the amount of darkening for a given degree of development depends on the degree of exposure. After development, the film is rinsed, preferably in an acid stop bath that arrests development. Next, the film is placed in a fixing solution that dissolves the undarkened portion of the film and hardens the emulsion. The film is subsequently washed to remove the fixing chemicals and soluble salts, then dried.

Although trays and other containers for photographic processing have been used, the usual method of processing industrial radiographic film by hand is the rack-and-tank method. In this method, the processing solutions and wash water are contained in tanks (Fig. 66) deep enough for the film to be hung vertically on developing hangers, or racks. The advantages to this method are:

- Processing solutions have free access to both sides of the film
- Both emulsion surfaces are uniformly processed to the same degree
- The all-important factor of temperature can be controlled by regulating the temperature of the water bath in which the tanks are immersed
- The equipment does not require much space
- There is a savings in time compared to tray processing

Handling of Film. The processing room and all equipment and accessories must always be kept scrupulously clean and used only for the purpose of handling and processing film. Spilled solutions should be wiped up immediately to avoid extraneous spots on the radiographs. Floating thermometers, film hangers, and stirring rods should be thoroughly rinsed in clean water after each use to avoid contamination of chemicals or streaking of film.

Film and radiographs should always be handled with dry hands. Abrasion, static electricity, water, or chemical spots will result in extraneous marks (artifacts) on the radiographs. Medicated hand creams should be avoided; rubber gloves should be used.

Development Procedure. Prepared developers are ordinarily used to ensure a carefully compounded chemical that gives uniform results. Commercial x-ray developers are of two types: automatic and manual. Both are comparable in performance and effective life, but the liquids are easier to mix. Developing time for industrial x-ray films depends mainly on type of developer. Normal developing time for all films is 8 min in a given developer at 20 °C (68 °F). More exact tables can be obtained from the manufacturers of developers.

When exposed film is immersed in the developer solution, the chemicals penetrate the emulsion and begin to act on the sensitized (exposed) grains in the emulsion, reducing the grains to metallic silver. The longer the development time, the more metallic silver is formed and the blacker (more dense) the image on the film becomes.

The rate of development is heavily dependent on the temperature of the solution; the higher the temperature, the faster the development. Conversely, if the developer temperature is low, the reaction is slow, and if the film were developed for 5 min at 16 °C (60 °F) instead of the normal 20 °C (68 °F), the resulting radiograph would be underdeveloped. Within certain, rather narrow, temperature limits, the rate of development can be compensated for by increasing or decreasing developing time. Exceeding these temperature limits usually gives unpredictable results.

The concept of time-temperature development should be used in all radiographic work to avoid inconsistent results. In this concept, the temperature of the developer is always kept within a small range. The developing time is adjusted to temperature so that the degree of development remains essentially constant. If this procedure is not followed, the results of even the most accurate radiographic technique will be nullified.

Inspection of the film at various intervals during development under safelight conditions (called sight development) should be avoided. It is extremely difficult to judge from the appearance of a developed but unfixed radiograph what its appearance will be in the dry, finished state, particularly with regard to contrast. Sight development can also lead to a high degree of fog caused by exposure to safelights during development.

A major advantage of standardized time-temperature development is that the processing procedure is essentially constant,

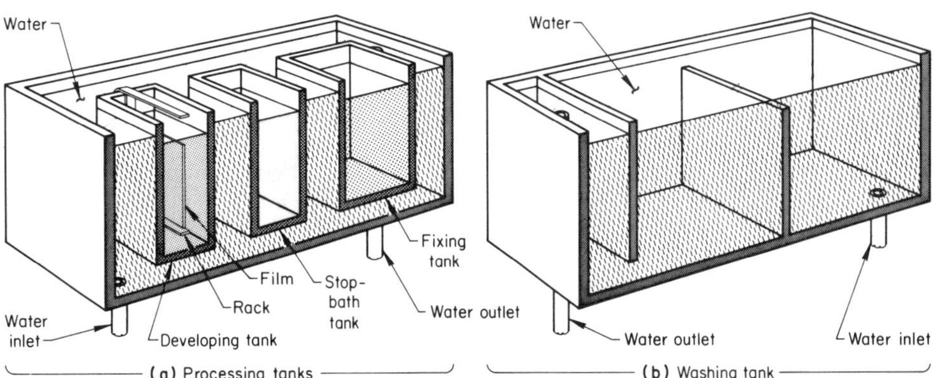

Fig. 66 Typical unit for the manual development of radiographic film by a rack-and-tank method. (a) Processing tanks containing developer, stop bath, and fixer. (b) Cascade (countercurrent) washing tank

and an accurate evaluation of exposure time can be made. This alone can avoid many of the errors that can otherwise occur during exposure. Increased developing time will produce greater graininess in the radiographic image, increased film speed, and in many cases increased radiographic contrast. Although increased contrast or film speed is often desirable, maximum recommended development times should not be exceeded.

Control of Temperature and Time. Because the temperature of the processing solutions has such a large influence on their chemical activity, careful control of temperature—particularly of the developer—is extremely important. A major rule in processing is to check the developer temperature before films are immersed in the developer so that the timer can be accurately set for the correct processing time.

Ideally, the developer should be at 20 °C (68 °F). At temperatures below 16 °C (60 °F), developer action is significantly retarded and is likely to result in underdevelopment. At temperatures exceeding 24 °C (75 °F), the radiograph may become fogged, and the emulsion may be loosened from the base, causing permanent damage to the radiograph.

Where the water temperature in the master tanks surrounding the solution tanks may be below 20 °C (68 °F), hot and cold water connections to a mixing valve supplying the master tank should be used. In warm environments, refrigerated or cooled water may be necessary. Under no circumstances should ice be placed directly into the solution tanks for cooling purposes, because melting ice will dilute, and may contaminate, the solutions. If necessary, ice can be placed in the water bath in the master tanks for control of the solution temperature.

Control of time should be done by setting a processing timer at the time the film is immersed in the developer. The film should be moved to the rinse step as soon as the timer alarm sounds.

Agitation During Development. A good radiograph is uniformly developed over the entire film area. Agitating the film during the course of development is the main factor that eliminates streaking on the radiograph.

When a film is immersed vertically in the developer and is allowed to develop without movement, there is a tendency for certain areas of the film to affect the areas directly below them. The reaction products of development have a higher specific gravity than the developer. As these products diffuse out of the emulsion, they flow downward over the surface of the film and affect the development of the areas over which they pass. As a result, uneven development of affected areas forms streaks, as shown in Fig. 67. This is sometimes referred to as bromide drag. The greater the film density

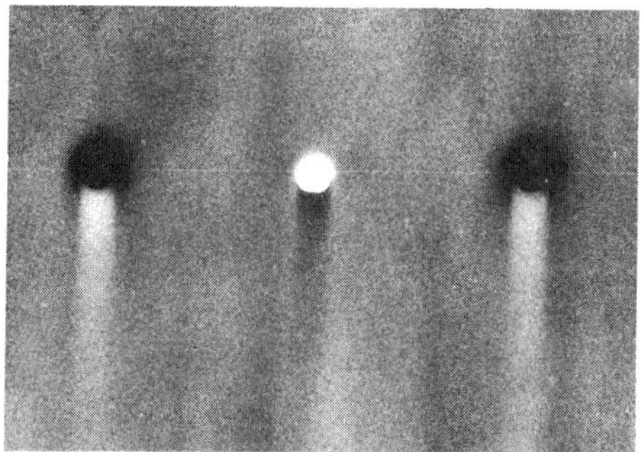

Fig. 67 Streaking, or bromide drag, that can result when a film is immersed in developer solution without agitation

from which reaction products emanate, the greater the effect on adjacent portions of the film.

When the film is agitated, spent developer at the surface of the film is renewed, preventing uneven development. Immediately after hanger and film are immersed in developer, hangers should be tapped sharply two or three times to dislodge any air bells clinging to the emulsion. Although lateral movement of the hanger provides perhaps the best agitation, the size and shape of the solution tanks usually limit the extent of lateral movement, thus making this type of agitation ineffective. Vertical movement works well; it consists of lifting the hanger completely out of the solution, then immediately replacing it in the tank two or three times at intervals of about 1 min throughout the developing time.

Agitation with stirrers or circulating pumps is not recommended. This type of agitation often produces a liquid flow pattern that causes more uneven development than no agitation at all. Nitrogen-burst techniques, although rarely used, can provide adequate agitation.

Activity of Developer Solution. The developing power of the solution decreases when film after film is developed, partly because the developing agent is consumed in converting exposed silver bromide to metallic silver and partly because of the retarding action of accumulated reaction products. The magnitude of this decrease depends on the number of films processed and on their average size and density. Even when the developer is not used, its activity will slowly decrease because of oxidation of the developing agent. The effect of oxidation is often apparent after as little as 1 month of inactivity. Little can be done to control the effects of oxidation except using a lid on the developer tank. Although this is only partly effective in preventing oxidation, it is always good practice to cover developer tanks when not in use in order to prevent contamination.

Replenishment of Developer. If the reduction of developing action is the result of the processing of many films, it is possible to compensate for decreased chemical activity by using a replenishment technique. When done correctly, replenishment can maintain uniform development for a long period of time. Replenishment cannot be used to counteract oxidation or contamination of the developer solution.

Most manufacturers of x-ray developers provide for replenishment either by supplying a separate chemical or by using the developer mixed to a different concentration from that of the original developer solution. The correct quantity of replenisher needed for maintaining consistent properties of the developer solution depends on the size and average density of the radiographs being processed. For example, a dense image over the entire radiograph will use up more developing agents and exhaust the developer to a greater degree than if the film were developed to a lower density. The quantity of replenisher required will depend on the type of subject being radiographed; the following is provided only as a guide:

Density of radiograph (% of background exposed)	Replenisher required(a)	
	g	oz
Low (<15%)	57	2
Medium (~40%)	64	2¼
High (~75%)	70	2½
Extremely high (>90%)	78	2¾

(a) Approximate quantity for each 350 × 430 mm (14 × 17 in.) sheet of film processed

Usually, replenishers are so compounded that drainage of films back into the developer tank should be avoided. The developer being drained is essentially spent; therefore, the developer solution in the tank is more rapidly contaminated with reaction products when this spent solution is drained back.

A systematic procedure should be used so that a fixed quantity of developer is removed from the tank for each film that is developed. A 350 × 430 mm (14 × 17 in.) film mounted on its hanger will normally carry with it 78 to 85 g (2¾ to 3 oz) of developing solution as it is removed from the tank. Because this is approximately the amount of replenisher needed for low-density and medium-density films, it is only necessary to replenish the developer tank to a given liquid level. However, for high-density and extremely high-density radiographs, it would be necessary to remove and discard some of the original developer each time replenisher is added.

If replenisher is added frequently and in small quantities, fluctuations in film density due to changes in chemical activity of the developer will tend to even out. However, if replenisher is added infrequently, a fluctuation in film density will become apparent, which may lead to considerable difficulty in consistently obtaining the required image quality in successively processed radiographs. If replenishment is controlled by maintaining a specific level in the development tank, replenisher should be added when the level of the solution drops by 6 mm (¼ in.).

Arresting Development. After development is completed, the action of the developer absorbed in the emulsion must be arrested by an acetic acid stop bath or at least by prolonged rinsing in clean running water. If these steps are omitted, the developing action continues for a short time in the fixer. This can produce uneven density or streaking on the radiograph and will reduce the life of the acidic fixer solution because of neutralization by the alkaline developer solution.

If the smell of ammonia is detected anywhere—developer, stop bath, or fixer—the solution has become contaminated and must be changed. When development is complete, films should be removed from the developer without draining back.

Films should remain in the stop bath for 30 s with moderate agitation before being transferred to the fixer solution. The stop bath acts as a replenisher for the fixer, so the films can be transferred directly to the fixer without draining back into the stop-bath solution.

If a stop bath is not used, films should be rinsed in running water for at least 2 min. If the flow of water in the rinse tank is only moderate, the film should be agitated so that the rinse will effectively avoid streaks on the radiograph.

Fixing. The purpose of the fixer solution is to remove all of the undeveloped silver salts in the emulsion, leaving the developed silver as a permanent image. Another important function of the fixer is to harden the gelatin of the emulsion so that the film will be able to withstand drying in warm air and will not be tacky to the touch when it is dry and ready for viewing. Portions of the film that have received lower amounts of radiation appear cream colored when the stop-bath procedure has been completed. In the fixing bath, this cream color gradually disappears until it is no longer visible. The interval of time required for this change to take place is called the clearing time. This is the time required to dissolve the undeveloped silver halide out of the emulsion. An equal amount of time is required to allow the dissolved silver salts to diffuse out of the emulsion and the gelatin to harden. The total fixing time, therefore, should be at least twice the clearing time but should not exceed 15 min to avoid loss of density on the film. Films should be agitated vigorously every 2 min during fixation to ensure uniform action of the fixing chemicals.

In performing its function, the fixer solution accumulates soluble salts, which gradually inhibits its chemical activity. As a result of this, and possibly also because of dilution with water, the rate of fixation decreases and hardening action is impaired in proportion to the number of films processed. The usefulness of fixer solution is ended when it loses its acidity or when fixing requires an unusually long interval. At this point, the fixer solution must be discarded. Fixing films in exhausted fixer frequently causes colored stains, known as dichroic fog, to appear on the radiograph. Fixer replenisher is usually available from the manufacturer of the fixer. It is advisable not to substitute brands (or even types) of replenishers for a given brand of fixer; the replenisher may not be compatible with the fixer.

Washing. After radiographic films have been fixed, they should be washed in running water, ensuring that the emulsion area of the film receives frequent changes of water. Proper washing also requires that the hanger bar and top film clips be covered completely by running water.

Effective washing of the film depends on a sufficient flow of water to rapidly carry off the fixer and to allow adequate time for fixer chemicals to diffuse out of the film. In general, the hourly flow of water in the washing tank should be from one to two times the volume of the tank. Under these conditions, and at water temperatures between 16 and 21 °C (60 and 70 °F), films require about 30 min of washing.

The washing tank should be large enough to wash films as quickly as possible. Too small a tank encourages insufficient washing, which may lead to discoloration or fading of the image later when radiographs are in storage.

Drying. When films are removed from washing tanks, water droplets cling to the surface of the emulsion. If the films are then dried rapidly, the areas under the droplets dry more slowly than surrounding areas. Such uneven drying causes distortion of the gelatin, changing the density of the image. Uneven drying often results in visible spots on the finished radiograph that interfere with accurate interpretation.

Water spots usually can be prevented by immersing washed films for 1 to 2 min in a wetting-agent solution and allowing them to drain for a few minutes before placing them in a drier. This procedure permits surplus water to drain off of the film more evenly, significantly reducing drying time and the likelihood that the finished radiograph will exhibit water spots.

It is important to use wetting agents that are compatible with x-ray film emulsions. Some commercial wetting agents are not compatible with film and therefore should not be used.

If only a few radiographs are processed daily, racks that allow the film to be air dried under room conditions of temperature and humidity can be used, although this method of drying requires considerable time. To avoid spots on the radiographs, the racks that hold the film hangers should be positioned so that films do not touch each other, and water should not be splashed on drying films. Radiographs dry best in constantly changing warm dry air.

The fastest way to dry films is in commercial film driers. These driers incorporate fans and heaters, and some driers use chemical desiccants to remove water from the air.

Regardless of the method of drying, a radiograph is considered dry when the film is dry under the hanger clips. When this stage is reached, the hangers are removed from the drier or rack, and the film is carefully removed from the hanger for interpretation.

In manual processing, when the film is clipped to the hanger, pins in the hanger clip penetrate the film and leave sharp projections in the corners of the film. It is usually desirable to cut off the corners containing these clip marks to prevent scratching of other radiographs during handling and reading.

Automatic Film Processing

Although expertly done manual processing is difficult to surpass, it is equally difficult to maintain because of human factors. Automatic processing, which accurately controls temperature, time, agitation, and replenishment, delivers a dry radiograph in a short time. In addition to the advantage of shorter processing time, automatic processing can ensure that variability of time, temperature, and activity of the solutions is eliminated.

Many automatic processors incorporate a roller-transport mechanism that carries the film itself through the entire process cycle without the need for hangers. Figure 68 illustrates a typical automatic processor that incorporates a roller-transport mechanism.

Automatic processing is a carefully controlled system in which the film, processing chemicals and their replenishment, temperature of solutions, travel speed of the roller-transport mechanism, and drying conditions all work together for consistent development of latent images. The advantages of roller-type automatic processors are:

- *Rapid processing of radiographs*: Depending on many variables, particularly the kind of radiographic film and the type of processing chemicals, the elapsed time from exposed film to finished radiograph can be as short as 4½ min and usually not more than 15 min. This represents a savings of at least 45 min compared to manual processing
- *Uniformity of radiographs*: Most automatic processors carefully control the time and temperature of processing. Combined with accurate automatic re-

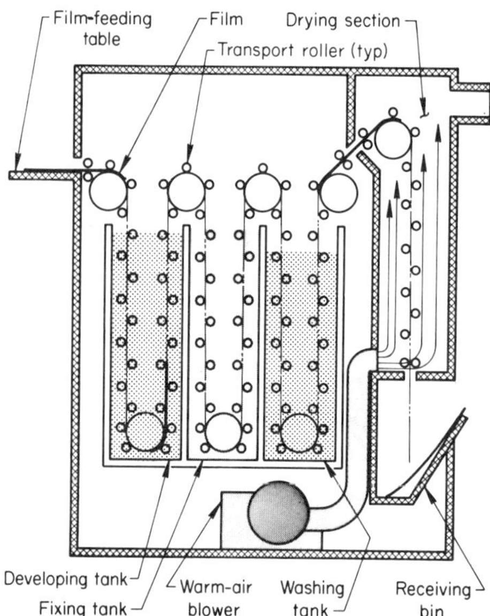

Fig. 68 Cross section of an automatic film processor showing roller-transport mechanism and locations of components

plenishment of solutions and constant agitation of the film as it travels through the processor, accurate time-temperature control produces day-to-day consistency and freedom from processing artifacts seldom achieved in manual processing. Because processing variables are virtually eliminated, optimum exposure techniques can be established

- *Conservation of space*: Most models of automatic processors require no more than 1 m^2 (10 ft^2) of floor space. Some models can even be installed on a workbench. This means that hand tanks and drying facilities can be eliminated from the processing room. All that is needed is a bench for loading and unloading of film, film-storage facilities, and a small open area in front of the processor feed tray. Most processors are installed so that the film is fed from the processing room (darkroom) and emerges as a complete radiograph in the adjacent room

Automatic processors contain a developing tank, a fixing tank, a washing tank, and a drying section (Fig. 68). The stop-bath and wetting-agent tanks are eliminated because automatic processors have squeegee rollers at the exit of each tank that reduce retention of residual solution by the film to a minimum, and minor solution contamination of the fixer is corrected by replenishment. The rollers at the exit of the fixer reduce the amount of residual fixer to only that absorbed in the emulsion as the film enters the wash cycle. Therefore, the running wash water is virtually free of fixer chemicals. Finally, the exit rollers of the wash tank squeeze wash water from the film, so that the film enters the drier in almost a damp-dry condition.

Processing chemicals must be specially formulated for automatic processors. The developer must operate at higher temperatures than for manual processing and usually contains a hardening agent to condition the emulsion so that the film can be moved by the roller transport system without slippage. The fixer is also specially formulated to fix the emulsion in a relatively short time at higher temperatures than for manual processing and to condition the film for proper drying. It should be noted that developers and fixers formulated for manual processing usually are not suitable for automatic processing.

Replenishment of Developer and Fixer. In an automatic processor, replenishment of developer and fixer is automatically controlled. An adjustable, positive-metering device controls both developer and fixer; sometimes separate metering devices are used for developer and fixer. Usually, the replenisher pump is activated when the leading edge of the film enters a film sensor in the processor and continues to pump until the trailing edge of the film passes these rollers. Thus, the amount of replenishment is controlled by the length of film passing through the entrance rollers.

Replenishment rates for developer and fixer are normally supplied by film manufacturers. Obviously, these are only guidelines because if radiographs are routinely of higher or lower densities than the average on which the manufacturer's recommendations are based, the replenishment rate may have to be adjusted upward and downward for optimum results.

The procedure for checking replenishment rates and frequency of replenishment is given in the operator's manual for the processor. The accuracy of replenishment is important. Too little or too much replenishment can adversely affect film densities, lead to transport difficulties, reduce processing uniformity, and shorten the useful life of the processing solutions.

Film-Feeding Procedures. Because replenisher pumps are controlled by the length of film fed into the processor, it is obvious that feeding single, narrow-width films will cause excessive replenishment. Therefore, whenever possible, narrower films should be fed side by side. Films should be fed into the processor parallel to the side of the feed tray. Multiple films should have a space between them to avoid overlapping and should be started together into the processor to avoid excessive replenishment.

Three rolls of the roller-transport system must always be in contact with the film to maintain proper travel through the processor. Thus, there is a lower limit to the length of film to be processed. Although this depends on roller diameter and spacing, the usual lower limit is about 125 mm (5 in.).

In general, roll films having widths from 16 to 430 mm (0.6 to 17 in.) can be processed in most automatic processors. The processing of roll films requires a somewhat different procedure than for sheet film. Because roll film is wound on spools, it frequently has an inherent curl that can cause the film to wander out of the roller system. To avoid wandering, a sheet of leader film can be attached to the leading edge of the roll. Ideally, the leader should be unprocessed radiographic film, preferably wider than the roll being processed and at least 250 mm (10 in.) long. The leader may be attached to the roll by means of pressure-sensitive polyester tape about 25 mm (1 in.) wide. Suitable types of tape must be composed of materials that are not soluble in the solutions. To avoid transport problems, care must be taken that adhesive from the tape does not come in contact with the rollers.

It is important to feed narrow widths of roll film parallel to the edge of the feed tray. If quantities of such film are normally processed, it is usually advisable to provide a guide in the feed tray to make sure each film is parallel to the others and to the sides of the tray. If this is not done, there is a possibility of the films overlapping somewhere in the transport systems.

If only one long roll of narrow film is to be processed, the replenisher pumps will keep running and the result may be excessive replenishment. To avoid this, replenisher pumps should be turned off for a portion of the feed time.

Preventive Maintenance. Most of the downtime and other problems related to the operation of automatic processors stem from the lack of maintenance. Service problems can be minimized by well-established maintenance of the processor and by good housekeeping. Each processor manufacturer recommends daily and weekly cleanup procedures that take only a few minutes to perform. These procedures, usually available in the form of check lists, are necessary for reliable operation of the processor and for production of radiographs of optimum quality.

Precautions. Automatic processing is a system in which the film, chemical, and processing equipment all have to work together for optimum processing. For example, if radiographs leave the processor wet, it could well be an indication of a problem with one of the chemicals and not the result of incorrect drying temperature.

Some films can be successfully processed more rapidly than others. However, changing the speed of the processor to a value other than that for which the processor was designed can unbalance the system and require adjustments in film characteristics, replenishment rates, temperatures, and perhaps other conditions in order to restore optimum processing quality. Care must be taken that the storage life of the radiograph is not impaired by changes in the processor.

Table 14 Probable causes and corrective action for various types of deficient image quality or artifacts from automatic film processing

See Table 12 for factors not specifically associated with automatic film processing.

Quality or artifact	Probable cause	Problem and corrective action
Density too high	Overdevelopment	*Developer temperature too high:* Follow temperature recommendations for developer and processor used. Check temperature of incoming water. Check accuracy of thermometers used. *Improperly mixed chemicals:* Follow instructions for preparations of solutions.
Density too low	Underdevelopment	*Improper replenishment of developer:* Check for clogged strainers or pinched tubing in developer replenishment system. *Developer temperature too low:* Follow temperature recommendations for developer and processor used. Check temperature of incoming water. Check accuracy of thermometers used.
	Contamination	*Fixer in developer tanks:* Use extreme care when installing or removing fixer rack in processor. Always use splash guard when fixer rack is being removed or replaced. Do not exchange racks between fixer and developer compartments.
Contrast too low	Underdevelopment	See *Underdevelopment* above.
	Contamination	See *Contamination* above.
Fog	Overdevelopment	*Developer temperature too high:* Follow temperature recommendations for developer and processor used. Check temperature of incoming water. Check accuracy of thermometers used.
Poor drying	Processing	*Underreplenishment of solutions (particularly fixer):* Check for clogged strainers or pinched tubing in replenishment system. *Inadequate washing:* Check flow of wash water.
	Drying	*Dryer temperature too low:* Follow recommendations for film type and processor involved.
Streaks	Associated with tempo of work	*Long interval between feeding of films:* "Delay streaks" (uneven streaks in direction of film travel) caused by interval of 15 min or more in feeding of successive films, which results in drying of solutions on processor rollers exposed to air. Wipe down exposed rollers with damp cloth. Process unexposed 14 × 17 film before processing radiograph. (Some processors are equipped with "rewet rollers" which prevent delay streaks.)
	Associated with development	*Clogged developer recirculation system:* Change filter cartridge regularly. Check recirculation pump. (This defect is associated with a rapid rise in developer temperature.)
Streaks (continued)	Associated with dryer	*Dirty tubes in dryer:* (Causes regular streaks, visible by reflected light only.) Clean dryer tubes. *High dryer temperature:* (Causes irregular streaks or blotches, visible by reflected light only.) Reduce temperature to recommended value.
Guide marks (regularly spaced scratches or high density lines)	Improperly adjusted guides in processor	Check clearances between guide devices and adjacent rollers or other components
Roller abrasions (random fine lines in the direction of film travel)	Stopped or hesitating rollers	Be sure all rollers are in their proper positions, and that end play is sufficient for rollers to turn freely.
Random scratches and spots	Dirt on feed tray	Clean processor feed tray frequently with soft cloth. If the atmosphere is dirty, the processing room should be fed with filtered air and kept at a pressure above that outside.
Irregular deposit (often light in color, generally elongated in direction of film travel)	Caused by dirt or precipitate in water supplied to washing section.	If condition is temporary, clean wash rack and replace wash water in processor; drain wash tank when shutting processor down. If condition persists, use filters in incoming water lines. (Some dirt deposits can be removed from the dry radiograph by gentle rubbing with dry cotton or a soft cloth.)
Dark lines and spots	Pressure marks caused by build-up of foreign matter on rollers or by improper roller clearances, usually in developer section.	Clean rollers thoroughly and maintain proper clearances.
	"Black comets" with tails extending in direction of film travel caused by rust or other iron particles dropping on film, usually at entrance assembly.	Clean all entrance assembly components. Apply a light coat of grease to microswitch springs and terminals. If air contains iron-bearing dust, filter air supply to processing room.
	"Pi lines." (So called because they occur 3.14 times the diameter of a roller away from the leading edge of a film.)	Most common in newly-installed or freshly cleaned processors. Tends to disappear with use of processor. (Some processors are equipped with buffer rollers at the exit of the wash rack, which remove the deposit before the radiograph enters the dryer.)

Table 14 lists some typical problems with automatic film processing.

Tests for Removal of Fixer

If radiographic films are not properly washed, fixer chemicals (largely thiosulfate salts) remain in the emulsion and affect the storage life of the radiographs. Radiographs intended for storage of 3 to 10 years are usually referred to as having commercial quality. Those to be kept for 20 years or more are known as having archival quality. Archival quality is of considerable importance in complying with certain codes, standards, and specifications. If residual thiosulfate in the radiographs exceeds a certain maximum allowable level, the radiograph is likely to become useless because of fading or a change in color during long-term storage.

The American National Standards Institute has three important documents relating to this problem: ANSI PH 4.8 (1985), ANSI PH 1.41 (1984), and ANSI PH 1.66 (1985).

The first document (ANSI PH 4.8) describes two methods of determining residual thiosulfate in radiographs. Both of these are laboratory procedures for evaluating unexposed but processed (clear) areas of a radiograph. The methylene blue test must be performed within two weeks of processing the film, but the silver densitometric test can be performed at any time after processing. The second document (ANSI PH 1.41) specifies, among other things, the maximum level of thiosulfate in grams per unit area for archival storage. The third document (ANSI PH 1.66) gives other helpful recommendations for storage of radiographs. Procedures given in all three ANSI documents are accepted as valid procedures by most codes, standards, and specifications.

There are other tests that are easier to perform than those described in ANSI PH 4.8, but they indicate more than the residual level of thiosulfate and therefore are considered only estimates rather than accurate determinations. In one test, a solution is made of 710 mL (24 oz) of water, 120 mL (4 oz) of acetic acid (28%), and 7 mL (¼ oz) of silver nitrate and diluted with 950 mL (32 oz) of water prior to use. When a drop of the diluted solution is placed on a clear area of a radiograph, it turns brown. The amount of residual fixer chemicals is estimated by matching the brown spot on the film with one of the patches on a standard test strip. Several manufacturers produce estimating kits, but for accurate determination of residual thiosulfate it is necessary to perform either the methylene blue test or the silver densitometric test described in ANSI PH 4.8.

Microfilming of Radiographs

The microfilming of radiographs has recently been implemented on a commercial basis for the storage of industrial radiographs, and it has been in use for some time in the medical field. The microreduction of radiographs has been successfully performed by the commercial nuclear power plant industry as a solution to problems associated with radiographic film deterioration, large-volume archival storage, and general records management.

Because of improper initial processing or failure to store and handle processed film using sound practices, radiographs are subject to deterioration or degradation. Specific examples of improper processing include:

- Inadequate replenishment of chemicals (developer and fixer in particular)
- Failure to maintain processing temperatures within manufacturer's specified ranges
- Inadequate agitation
- Rushing processing times
- Failure to maintain water quality

The consequences can be failure to remove thiosulfate, resulting in a green, yellow, amber, or brown tint or image destruction, and failure to remove uncombined silver resulting in general darkening of the image.

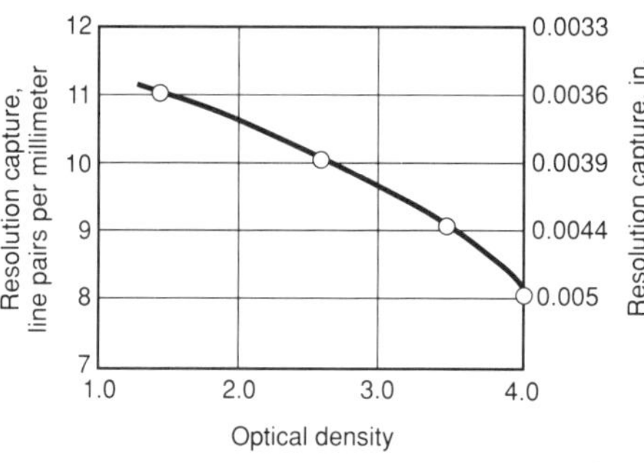

Fig. 69 Spatial resolution versus optical density for a microfilm (15:1 reduction) of a radiograph with a spatial resolution in the range of 8 line pairs per millimeter (or 0.005 in. resolution)

Examples of poor storage and handling practices include:

- Handling radiographs without the use of gloves to protect the film from acids, chemicals, and perspiration on the hands
- Placing rubber bands or paper clips on the film
- Storing hard copy information produced using a diazo process with film
- Stacking film, especially film not interleaved, causing it to adhere
- Filing wet film
- Placing inventory or acceptance punches in the area of interest
- Failure to store the radiographs in low-humidity/room-temperature conditions
- Failure to store radiographs in acid-free interleaving paper

The microfilming of radiographs has been used to transfer the deteriorated or degraded image to a stable, archival medium. Microfilming is performed using a planetary camera system with illuminators to photograph the radiographic image on 35 mm roll format. Spatial resolutions in excess of 120 line pairs per millimeter (or <0.0003 in.) on the resulting image have been produced. With a reduction factor of 15:1, information on the original radiograph in the range of 8 line pairs per millimeter (or a resolution of 0.005 in.) has been reproduced within a density of 4.0 (Fig. 69), and spatial resolutions as high as 11 lines per millimeter (0.0035 in.) have been produced at lower densities.

Because of the wide range of optical densities associated with industrial radiographs and the inability of high-resolution microfilms to reproduce wide latitudes of density, the microfilming of radiographs may involve one or more exposures of the same radiograph to capture all of the information present. Sensitometry checks of the process result in density capture ranges (accurate contrast reproduction) as high as 1.5 density points per exposure.

Special viewing equipment is necessary for use with the microfilm. Equipment is available that will allow accurate dimensional measurements from the microfilm, density determination, binocular viewing, and adjustable light intensities, as with a typical radiograph viewer.

REFERENCES

1. *Radiography & Radiation Testing*, Vol 3, 2nd ed., *Nondestructive Testing Handbook*, American Society for Nondestructive Testing, 1985
2. R. Halmshaw, *Nondestructive Testing*, Edward Arnold Publishers, 1987
3. R. Link, W. Nuding, and K. Sauerwein, *Br. J. Non-Destr. Test.*, Vol 26, 1984, p 291-295
4. L.H. Brixner, *Mater. Chem. Phys.*, Vol 16, 1987, p 253-281
5. W. Daum, P. Rose, H. Heidt, and J.H. Builtjes, *Br. J. Non-Destr. Test.*, Vol 29, 1987, p 79-82
6. "Nondestructive Testing Personnel Qualification and Certification, Supplement A, Radiographic Testing Method," ASNT-TC-1A, American Society for Nondestructive Testing

SELECTED REFERENCES

- H. Cember, *Introduction to Health Physics*, 2nd ed., Pergamon Press, 1983
- R.V. Ely, *Microfocal Radiography*, Academic Press, 1980
- D.A. Garrett and D.A. Bracher, Ed., *Real Time Radiologic Imaging*, STP 716, American Society for Testing and Materials, 1980
- "General Safety Standard for Installations Using Non-Medical X-Ray and Sealed Gamma-Ray Sources, Energies up to 10 MeV," ANSI N43.3 (Handbook 114), American National Standards Institute, New York
- R. Halmshaw, *Industrial Radiology: Theory and Practice*, Applied Science, 1982
- *Handbook on Radiographic Apparatus and Techniques*, 2nd ed., International Institute of Welding, London, 1973
- "Industrial Radiographic Terminology," E 52, Annual Book of ASTM Standards, American Society for Testing and Materials
- *Industrial Radiography*, Agfa-Gevaert Handbook (revised edition), Mortsel, 1986
- "Instrumentation and Monitoring Methods for Radiation Protection," Report 57, National Council on Radiation Protection and Measurements, 1978
- G.F. Knoll, *Radiation Detection and Measurement*, John Wiley & Sons, 1979
- "Metallography and Nondestructive Testing," Volume 03.03 of Section 3, Metals Test Methods and Analytical Procedures, *Annual Book of ASTM Standards*, American Society for Testing and Materials
- *Methods for Radiographic Examination of Fusion-Welded Butt-Joints in Steel*, BS 2600, Parts 1, 2, British Standards Institute, London, 1983
- *Methods for Radiographic Examination*

of Fusion-Welded Circumferential Butt-Joints in Steel Pipes, BS 2910, British Standards Institute, London, 1985

- *Methods for Radiography and Classification for Steel Welds*, JIS/Z3104, Japanese Standards Association, Tokyo, 1968
- K.Z. Morgan and J.E. Turner, *Principles of Radiation Protection*, John Wiley & Sons, 1973
- "Nondestructive Testing Personnel Qualification and Certification, Supplement A, Radiographic Testing Method," ASNT-TC-1A, American Society for Nondestructive Testing
- "Nondestructive Testing: Radiography," Report DDC-TAS-71-54-1, Defense Documentation Center, AD-733860, 1971
- "Performance Specifications for Direct Reading and Indirect Reading Pocket Dosimeters for X and Gamma Radiation," ANSI-N13.5, American National Standards Institute, 1972
- *Personnel Dosimetry Systems for External Radiation Exposures*, Technical Report Series No. 109, International Atomic Energy Agency, 1970
- "Radiation Protection Design Guidelines for 0.1-100 MeV Particle Accelerator Facilities," Report 51, National Council on Radiation Protection and Measurements, 1977
- "Radiation Protection Instrumentation Test and Calibration," ANSI-N323, American National Standards Institute, 1978
- *Radiography & Radiation Testing*, Vol 3, 2nd ed., *Nondestructive Testing Handbook*, American Society for Nondestructive Testing, 1985
- "Recommendations on Limits for Exposure to Ionizing Radiation," Report 91 (supersedes Report 39), National Council on Radiation Protection and Measurements, Bethesda, Maryland
- *Recommended Practice for the Radiographic Examination of Fusion Butt-Welds in Steel Plates from 0.5-50 mm Thick*, IIS/IIW-432-73, International Institute of Welding, London, 1973
- *Specification for Image Quality Indicators for Radiography and Recommendations for Their Use*, BS 3971, British Standards Institute, London, 1980
- "Standards for Protection Against Radiation," Code of Federal Regulations, Title 10 (Energy) Part 20, U.S. Government Printing Office
- *Testing of Welds of Metallic Materials by X- or Gamma-Rays*, DIN 54 111, Deutches Institut für Normung, Berlin, 1973

Industrial Computed Tomography

Michael J. Dennis, General Electric Company, NDE Systems and Services

COMPUTED TOMOGRAPHY (CT), in a general sense, is an imaging technique that generates an image of a thin, cross-sectional slice of a testpiece (Fig. 1a). The CT imaging technique differs from other imaging methods in that the energy beam and the detector array in CT systems lie in the same plane as the surface being imaged. This is unlike typical imaging techniques, in which the energy beam path is perpendicular to the surface being imaged. Moreover, because the plane of a CT image is parallel with the energy beam and detector scan path, CT systems require a computing procedure to calculate, locate, and display the point-by-point relative attenuation of the energy beam passing through the structures within the thin, cross-sectional slice of the testpiece.

Computed tomography has been demonstrated with a number of different types of energy beams, such as ultrasound, electrons, protons, α particles, lasers, and microwaves. In industrial nondestructive evaluations (NDE), however, only x-ray computed tomography is considered to have widespread value. X-ray computed tomography collects and reconstructs the data of x-ray transmission through a two-dimensional slice of an object to form a cross-sectional image without interference from overlying and underlying areas of the object. The CT image represents the point-by-point linear attenuation coefficients in the slice, which depend on the physical density of the material, the effective atomic number of the material, and the x-ray beam energy. The CT image is unobscured by other regions of the testpiece and is highly sensitive to small density differences (<1%) between structures. Computed tomography systems can also produce digital radiography (DR) images, and the DR and CT images can be further processed or analyzed within the computer. A series of CT images can be used to characterize the object volume, with the data reformatted to display alternate planes through the component or to present three-dimensional surfaces of structures within the object.

Historical Background

Although the mathematical principle of computed tomography was developed early in this century by Radon (Ref 1), application of the technology occurred much later. Techniques were developed in the 1950s for radioastronomy (Ref 2), and experimental work progressed through the 1960s, primarily using nuclear tracer and electron microscopy data (Ref 3-5). In the late 1960s and early 1970s, G. Hounsfield at EMI, Ltd., in England developed the first commercial x-ray CT system, also known as a computerized assisted tomography (CAT) scanner (Ref 6). Commercial CT scanners also found increased acceptance with the availability of inexpensive computing power, and industrial testing was considered an appropriate application in this early development of commercial CT systems. These early efforts, however, were directed primarily toward applications for medical diagnostic imaging. There was some limited use of computed tomography in the 1970s for industrial and research applications; this work was primarily done on existing medical systems.

Since the end of the 1970s, dedicated efforts have been applied to the use of

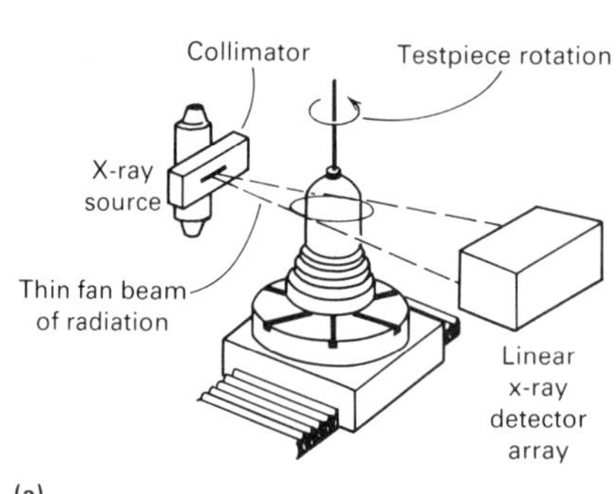

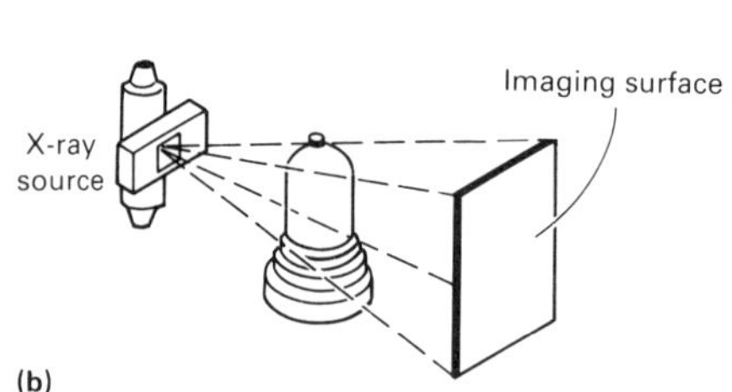

Fig. 1 Comparison of computed tomography (a) and radiography (b). A high-quality digital radiograph (b) of a solid rocket motor igniter shows a serious flaw in a carefully oriented tangential shot. A CT image (a) at the height of the flaw shows the flaw in more detail and in a form an inexperienced viewer can readily recognize. Courtesy of J.H. Stanley, ARACOR

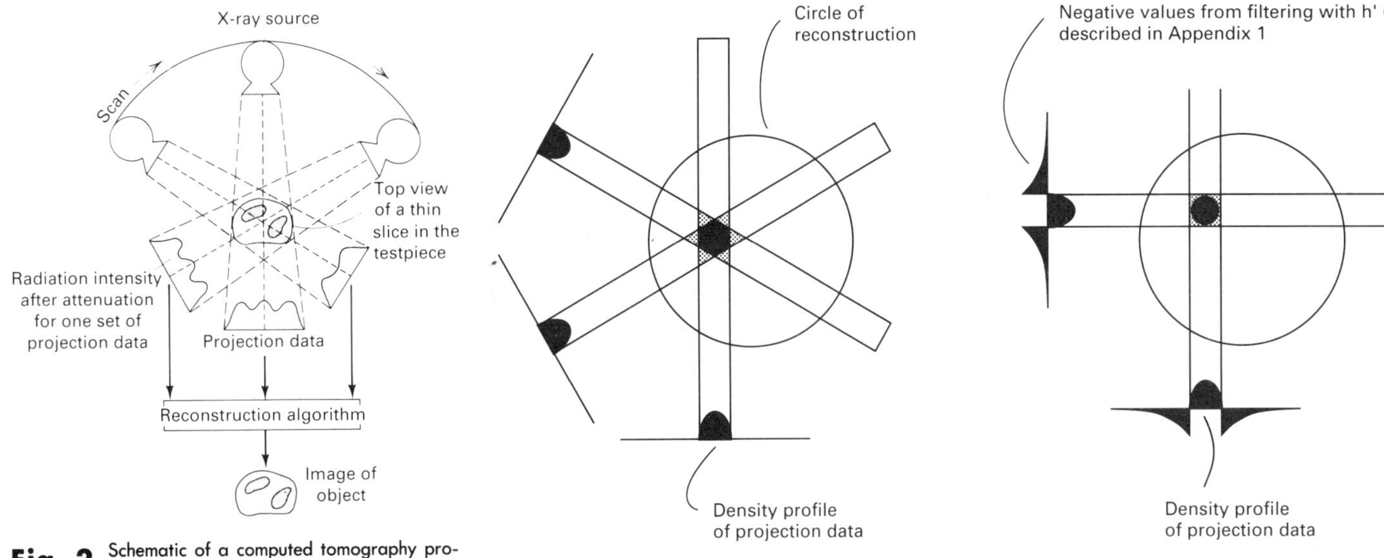

Fig. 2 Schematic of a computed tomography process

Fig. 3 Schematic of CT image reconstruction by (a) simple backprojection and (b) filtered backprojection. (a) Simple backprojection produces a blurred and broadened image. (b) Filtering can subtract the smear from other projections and reduce the broad blurring when backprojection is performed from many angles.

computed tomography in industry. The petroleum industry was one of the early major users of medical-type scanners to analyze oil drilling core samples and to assist in analyzing secondary oil recovery methods in rock samples. Medical CT systems in hospitals and industry are also used to nondestructively test industrial components, primarily carbon composite materials and light metal alloy structures (Ref 7). Diagnostic medical scanners, however, are designed for a specific task. They provide high-quality images of the human body, but are not suitable for large, dense objects.

Industrial CT systems are now manufactured to address specific inspection objectives that extend beyond the capabilities of medical computed tomography. Key differences may include the ability to handle dense materials and larger objects, the use of higher-energy x-ray sources, the use of systems with higher resolution, and the ability to operate within the manufacturing environment. The U.S. military services have actively driven the development of dedicated industrial CT systems. In the late 1970s and early 1980s, several programs were initiated with specific inspection objectives. Early programs included systems for inspecting large rocket motors, and systems for inspecting small precision castings in aircraft engines. Since these early systems, advancements have been made to extend the range of capabilities and applications.

Basic Principles of Computed Tomography

The basic technique of computed tomography is illustrated in Fig. 2. A thin slice of the testpiece is interrogated with a thin beam of radiation, which is attenuated as it passes through the testpiece. The fraction of the x-ray beam that is attenuated is directly related to the density and thickness of material through which the beam has traveled and to the composition of the material and the energy of the x-ray beam. Computed tomography utilizes this information, from many different angles, to determine cross-sectional configuration with the aid of a computerized reconstruction algorithm. This reconstruction algorithm quantitatively determines the point-by-point mapping of the relative radiation attenuation coefficients from the set of one-dimensional radiation measurements.

The CT scanning system contains a radiation source and radiation detector along with a precision manipulator to scan a cross-sectional slice from different angles. The x-ray detector is usually a linear array, that is, a series of individual x-ray sensors arranged in a line. The x-ray source is collimated to form a thin fan beam that is wide enough to expose all of the detector elements (Fig. 1a). The narrow beam thickness defines the thickness of the cross-sectional slice to be measured. The data acquisition system reads the signal from each individual detector, converts these measurements to numeric values, and transfers the data to a computer to be processed. To obtain the full set of transmission data required to produce the CT image, the object, source, or detector moves while a sequence of measurements is made. This motion may involve rotation of the test object relative to the source and the detector array (Fig. 1a) or a combination of rotation and translation. The various types of scanning geometries are discussed in the section "CT System Design and Equipment" in this article.

The CT image reconstruction algorithm generates a two-dimensional image from the set of one-dimensional radiation measurements taken at different scanning angles. The reconstruction algorithms used to generate the CT image fall into two groups: transform techniques and iterative techniques. The transform techniques are generally based on the theorem of Radon, which states that any two-dimensional distribution can be reconstructed from the infinite set of its line integrals through the distribution. (A line integral in CT is the sum of linear attenuation coefficients for one ray of radiation transmitted across the length of a cross-sectional slice.) The iterative techniques are generally algebraic methods that reconstruct the two-dimensional image by performing iterative corrections on an initial guess of what the image might look like. Iterative methods are rarely used, except when only limited data are available (as when a weak γ-ray source would require an inordinate amount of time in obtaining measurements from many scanning angles).

Some of the techniques used in CT image reconstruction are discussed in Appendix 1 in this article. The method most commonly used is the filtered (or convolution) backprojection technique. This technique filters (or convolutes) the projection data and then backprojects the filtered data into the two-dimensional image matrix. The backprojection process is the mathematical operation of mapping the one-dimensional projection data back into a two-dimensional grid, and it is equivalent to smearing the filtered projection data through the corresponding object space (Fig. 3b). The backprojected image values are calculated by taking each point in the image matrix and summing all filtered projection values that pass through that point. The sharpening filter is applied to the measured projection data to compensate for the blurring of the image that is a

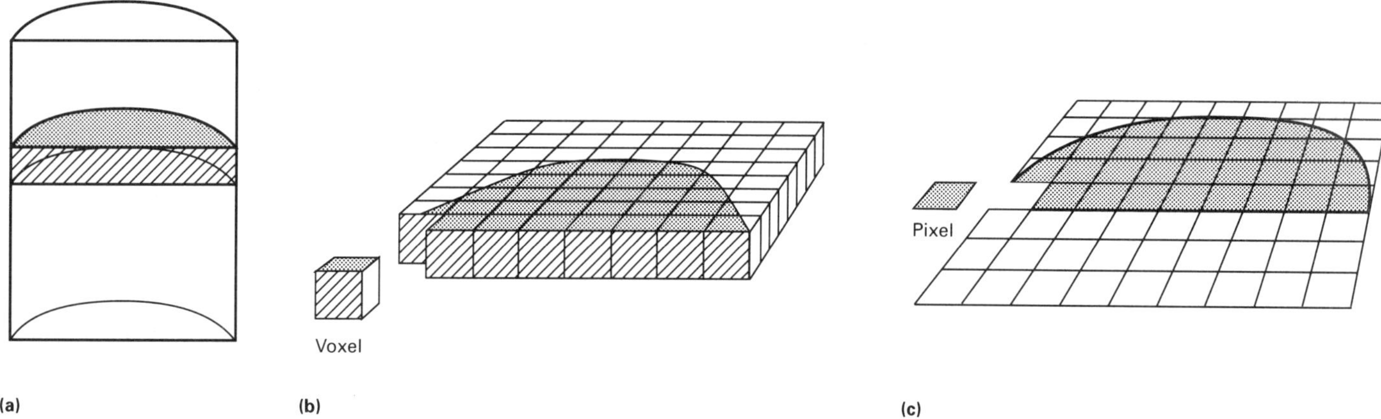

Fig. 4 Components of a CT image. (a) The x-ray transmission measurements and resultant CT image correspond to a defined slice through the object. (b) The cross-sectional slice can be considered to contain a matrix of volume elements, or voxels. (c) The reconstructed CT image consists of a matrix of picture elements, or pixels, each having a numeric value proportional to the measured x-ray attenuation characteristics of the material in the corresponding voxel.

consequence of the backprojection process (Fig. 3a). Additional information on the basic concepts of reconstruction processing are presented in Appendix 1, along with references to more detailed descriptions of the mathematical reconstruction processes. A glossary of terms and definitions is provided in Appendix 2.

The CT Image. The CT reconstruction process yields a two-dimensional array of numbers corresponding to the cross section of the object. Each of these numbers is a pixel or picture element, in the cross-sectional image (Fig. 4c). The reconstructed image is an array of many pixel values per image, and the arrays have sizes such as 128 × 128, 256 × 256, 512 × 512, or 1024 × 1024 pixels per array. Because the scan transmission measurements are made with an x-ray beam of some thickness (Fig. 4a), pixel value in the two-dimensional image corresponds to a volume of the material in the object referred to as a voxel, or volume element (Fig. 4b).

The pixel values are usually an integer value that is proportional to the average linear x-ray attenuation coefficient of the material in the corresponding voxel. The linear attenuation coefficient is approximately proportional to the physical density of the material and is a function of the effective atomic number of the material and the spectral distribution of the x-ray beam. Because these pixel values are more or less proportional to the density of the material, they are sometimes referred to as density or x-ray density values. This is a useful concept in interpreting images, but it is necessary to remember that other factors affect this value when objects contain multiple materials or when images obtained at different energies are compared.

The scaling of the linear attenuation coefficient into an integer pixel value, as well as the range of coefficients covered by the CT number scale, is somewhat arbitrary among systems. Medical systems utilize an offset scale that is normalized to water. These systems assign air a CT value of −1000, water is at 0, and a material twice as attenuative as water has a CT value of +1000. Industrial CT systems normally set the value for air at 0, but no standard scaling of the CT values is applicable to all the various materials and x-ray energies used.

To permit interpretation of the information in the reconstructed array of numbers, this information is displayed visually as an image. Human perception is limited, however, and cannot distinguish the many thousands of density levels that may be present in a single image. For distinguishing small density differences over a wide range of densities, the display system uses an interactive CT number windowing capability. This display windowing presents a limited range of CT numbers over the full gray-scale range of the visual display and is a powerful tool in analyzing subtle features in the image data. Color can also be used to enhance the displayed image where ranges of CT numbers are presented as a particular color. The color scales can be set up as a rainbow of colors, as hues of a particular color, or as a hot body spectrum mimicking the colors emitted as an object is heated.

Capabilities

X-ray computed tomography is a radiological procedure and has many of the same considerations as film and real-time radiography. The system must be designed to permit sufficient penetration of the x-rays through the object. Radiation shielding is required as part of the system or installation. Immersion is not required, and air gaps, multiple materials, complex shapes, and surface irregularities do not present a significant difficulty. The displayed parameter corresponds to the linear attenuation coefficient of the materials within the test specimen, and structures differing by a fraction of a percent can be visualized. In a CT image, the structures of interest are not obscured by other structures, and the sensitivity and ability to detect a feature are less dependent on the presence and overall thickness of other structures.

Because the CT image corresponds to a thin section through the object, features can be localized in three dimensions. This can be highly valuable in determining whether a flaw is within a critical area and in providing specific information on the repairability of a component and how the repair should be implemented. This spatial information also permits one to make certain dimensional measurements from the image that may otherwise be difficult or impossible to obtain nondestructively. The ability to provide spatially specific structure and density information also opens up the opportunity to obtain three-dimensional data representations of physical components for computer-aided design documentation and computer-aided engineering analysis.

Computed tomography also has its limitations. Producing high-quality images requires a stable radiation source, a precise mechanical manipulator, a sensitive and highly linear x-ray detector system, considerable computing power (usually with high-speed array processors), and a high-quality image display. Consequently, the capital costs for industrial CT systems tend to be well above those of most routine inspection systems.

The fact that the CT image corresponds to a thin section through the testpiece provides beneficial information and sensitivity. This can produce throughput difficulties, however, if information is required over a large volume. Many thin CT slices are required for fully characterizing an entire component. Typically, full-volume CT scanning is limited to certain low-volume, high-value testpieces or to the engineering evaluation of prototype or test specimens. In higher-volume production, CT images are often taken at specific critical positions for flaw detection or dimensional analysis. The digital radiography mode of these systems or some other conventional inspection

Table 1 Comparison of ultrasound, radiography, and x-ray computed tomography

Performance characteristic	Ultrasound	Radiography	Computed tomography
Parameter imaged	Acoustic impedance differences, attenuation, velocity	Attenuation	Attenuation
Handling	Immersion, squirter	Flexible	Scanner size limitation
Surface roughness	Poor	Fair	Good
Complex structures	Poor	Fair	Good
Limitations	Needs acoustic coupling; not suitable for acoustically noisy materials	Penetration one side	Penetration over a 360° scan; limited volume imaged per scan
System costs	Medium	Medium	High
Flaw detection			
Voids	Good	Good	Good
Inclusions	Good	Fair	Good
Porosity and microshrink	Good	Fair	Good
Density variations	Fair	Fair	Good
Cracks	Good (not aligned)	Fair (separated and aligned)	Fair (separated)
Debonds	Good	Fair (separated and aligned)	Fair (separated)

technique is often used to identify suspicious indications, and CT imaging is used to characterize the structure or flaw to provide more information for the accept/reject/repair decision on the component.

Another limitation of industrial computed tomography does not pertain to its ability to provide specific information but to the relative newness of the technology and the ability to obtain and use CT information. Critical inspections are normally defined as part of a component design, and prior designs did not consider the potential of computed tomography. Inspection requirements can be changed, but must be done on a component-by-component basis. This has limited much of the use of computed tomography for existing components to in-process inspection for process control and minimizing scrap costs or to providing improved information for material-review decisions.

Some of the current component designs require CT inspection to ensure reliability, to reduce the assumed maximum contained defect used in the design, or to design structures that do not provide good access for ultrasonic or routine radiographic inspection. This ability to characterize the interior structures of complex shapes and assemblies reduces the constraints for inspectability and therefore provides design engineers with the freedom to design.

Detection Capabilities. The ability to present a density (or linear attenuation coefficient) map across a slice through a specimen allows the visualization of many types of structures and flaws. The ability of computed tomography to display specific features will vary with the capabilities of the specific system and the operating parameters used to obtain the image. Assuming adequate capabilities for a particular scanning situation, some generalizations can be made regarding the suitability of computed tomography for various flaw types:

- Voids and inclusions are readily detectable high-contrast targets. Targets considerably smaller than the resolution of the system can be seen, but will have less contrast from the background material. High-structure noise, as in composites, can further limit the detectability of small voids or inclusions
- Porosity and microshrink reduce the density of the material and are usually visible if distributed over a multipixel area. The percentage of porosity can be quantified with properly controlled procedures
- Relative density measurements for a material or for materials with the same effective atomic number can be made. Absolute density can be obtained with properly controlled procedures. Alternative techniques, such as dual energy scanning, may improve capabilities with multiple materials
- Separated or displaced cracks are detectable. As with voids, the separation may be well below the resolution of the system, but contrast will drop as the separation decreases. Detectability may be affected by structure noise or if the crack is adjacent to a high-contrast boundary
- Disbonds and delaminations are detectable if separated. They are similar to cracks, with the additional complication of adjacent structures with the same orientation
- Residual core material in castings is readily seen in bulk. Thin surface layers can be difficult to detect
- Machining defects, such as overdrills and scarfing grooves, are readily seen. Computed tomography can assist in quantifying the extent of machining defects

Table 2 Comparison of performance characteristics for film radiography, real-time radiography, and x-ray computed tomography

Digital radiographic imaging performance with discrete element detector arrays is comparable to CT performance values.

Performance characteristic	Film radiography	Real-time radiography(a)	Computed tomography
Spatial resolution(b)	>5 line pairs/mm	~2.5 line pairs/mm	0.2–4.5 line pairs/mm
Absorption efficiencies, %			
Absorption efficiency (80 keV)	5	20	99
Absorption efficiency (420 keV)	2	8	95
Absorption efficiency (2 MeV)	0.5	2	80
Sources of noise	Scatter, poor photon statistics	Scatter, poor photon statistics	Minimal scatter
Dynamic range	200–1000	500–2000	Up to 1×10^6
Digital image processing	Poor; requires film scanner	Moderate to good; typically 8-bit data	Excellent; typically 16-bit data
Dimensioning capability	Moderate; affected by structure visibility and variable radiographic magnification	Moderate to poor; affected by structure visibility, resolution, variable radiographic magnification, and optical distortions	Excellent; affected by resolution, enhanced by low contrast detectability

(a) General characteristics of real-time radiography with fluorescent screen-TV camera system or an image intensifier. (b) Can be improved with microfocus x-ray source and geometric magnification.

Comparison to Other NDE Methods

X-ray computed tomography provides one method of detecting and locating the internal flaws of a testpiece. Other common NDE methods of internal flaw detection and location include radiography and ultrasound, which are compared with x-ray computed tomography in Table 1. Less common methods in industrial testing are nuclear tracer imaging, nuclear magnetic resonance imaging, and x-ray backscatter imaging. Acoustic emission, holography, and other techniques provide indirect or bulk information on the internal structure without specifically defining the structure.

Computed Tomography and Radiography. X-ray computed tomography is a radiological technique, and has many of the same benefits and limitations as film and real-time radiography. The primary difference is the nature of the radiological image. Radiography compresses the structural information from a three-dimensional volume into a two-dimensional image. This is useful in that it allows a relatively large volume to be interrogated and represented in a single image. This compression, however, limits the information and reduces the sensitivity to small variations. Radiographic images can be difficult to interpret because of shadows from overlying and underlying structures superimposed on the features of interest. Further, a single radiographic image does not provide sufficient information to localize a feature in three dimensions. Some of the performance characteristics of radiography and computed tomography are compared in Table 2.

Nuclear Tracer Imaging. Both nuclear imaging and magnetic resonance imaging are widely used for medical diagnostic imaging, but are infrequently used for industrial structural imaging. Nuclear tracer imaging yields an image of the distribution of a radioactive substance within an object. It has the potential for imaging flow paths through a structure or for the absorption of the tracer in open, porous areas of the object. Tracer studies obviously require the capability to handle unsealed radioactive materials, and the images are affected by the attenuating structure of the object between the tracer and the detector. Nuclear tracer imaging comprises two different advanced CT techniques that are similar to x-ray computed tomography. These are single photon emission computed tomography (SPECT) and positron emission tomography (PET) scanning, which produce cross-sectional displays of the tracer concentration (Ref 8).

Nuclear magnetic resonance (NMR) imaging, or magnetic resonance imaging, is similar to x-ray computed tomography in the type of images produced and in some of the methods for processing the NMR data. For NMR imaging, the object is placed in a strong magnetic field. The magnetic dipole of certain nuclei, such as hydrogen, will tend to line up with the magnetic field. These nuclei will precess, or rotate, at a particular frequency that is dependent on the nuclide and the magnetic field strength. Adding energy in the form of radiowaves at this specific resonance frequency will cause the nucleus to be pushed out of alignment with the magnetic field. The nucleus, after some short period of time, will then fall back into alignment with the magnetic field and emit a radiowave in the process. The emitted radiowaves are monitored, the time constants measured, and the magnetic fields varied in order to produce cross-sectional images of the particular nuclide being monitored.

Nuclear magnetic resonance imaging has been highly successful for the diagnostic medical imaging of the water-based human body, but has severe limitations for most industrial imaging requirements. Nuclear magnetic resonance is sensitive only to nuclides with a nuclear dipole moment, that is, nuclei with an odd number of protons or neutrons. The nuclide of interest is generally required to be in a liquid or gelatinous state because the time constants for solids are too short. In addition, the object cannot contain ferromagnetic materials, which would disrupt the uniform magnetic field, nor can it contain highly conductive materials that would interfere with the transmission of the radiowaves. This technique has appeared useful for analyzing fluids in rock samples for the petroleum industry (Ref 9, 10). Some limited work has been demonstrated for industrial testing in which a liquid tracer, such as alcohol, was used to image the porosity within carbon composite structures (Ref 11, 12).

Backscatter or Compton imaging is a technique in which the object is irradiated by a narrow x-ray beam, and the Compton scattering of this radiation by the object is monitored (Ref 13). The amount of scattered radiation produced is directly related to the electron density of the material, which corresponds well to the physical density of the material.

The proportion of the x-ray beam scattered in a small volume of the object tends to be relatively small, and the x-rays are scattered in all directions. A portion of this scattered radiation is measured by detectors typically placed near the entry point of the x-ray beam.

In general, radiography or computed tomography is preferred for internal inspection where it is practical to use. The major advantage offered by backscatter imaging is that it can be implemented from one side of the object. Other x-ray imaging methods require the source on one side of the object and the film or detector on the opposite side in order to measure x-ray transmission. Backscatter imaging provides a means of obtaining near-subsurface information, particularly if the object is very massive, does not permit x-ray transmission through it, or does not permit access to the opposite side (Ref 14, 15). Like computed tomography, backscatter imaging can also generate images of planar sections through the testpiece.

Industrial CT Applications

Computed tomography provides a unique means of obtaining very specific information on the interior structure of components. Computed tomography technology is very versatile and is not restricted by the shape or composition of the object being inspected. Also, CT systems can be configured to inspect objects of greatly varying sizes, weights, and densities. Generally, systems designed to accommodate large objects do not provide spatial resolution as high as systems designed for the inspection of smaller objects.

The major limitations of computed tomography are the relatively high equipment costs and the limited throughput for evaluating large volumes. Although CT systems improve sensitivity by generating images of a thin cross section, this procedure also produces throughput difficulties if information is required over a large volume. Many thin CT slices are required for the full characterization of an entire component. Typically, full-volume CT scanning is limited to certain low-volume, high-value components or to the engineering evaluation of prototype or test specimens. In higher-volume production, CT images are often taken at specific critical positions for flaw detection or dimensional analysis. The digital radiography mode of CT systems (see the section "Special Features" in this article) or some other conventional inspection technique is often used to identify suspicious indications. The CT imaging mode is then used to characterize the structure or flaw to provide more information for the accept/reject/repair decision on the component.

The petroleum industry uses computed tomography for geological analysis and for fluid dynamics research studies (Ref 16-22). Rock core samples are obtained from drilling sites to evaluate the oil production capabilities of a well. Computed tomography can be used to help evaluate these core samples with regard to heterogeneity, fracturing, or contamination by the drilling process. Small plugs are often taken from these core samples for extensive analysis of porosity, permeability, oil content, and other parameters. The CT data can assist in locating appropriate areas for these plug samples and in avoiding nonrepresentative samples and erroneous results.

Computed tomography is also used in dynamic flow studies of oil well core samples. The improved understanding of oil flow through the rock and the use of water, carbon dioxide, surfactants, and other chemicals to push the oil to a production well can have significant economic benefits. The test core samples are placed in sealed containers with the appropriate plumbing, and the flow experiments can be conducted at high pressures and elevated temperatures to simulate reservoir conditions.

The flow rates are typically slow, and a study can last many hours. A series of CT slices is obtained at various times in the study to characterize the fluid position. To distinguish the various fluids, contrast media or highly attenuating dopants, such as iodine, can be added to one of the fluids. The resulting data are processed to determine the percentage of saturation or volume concentration of the various fluids and gas-

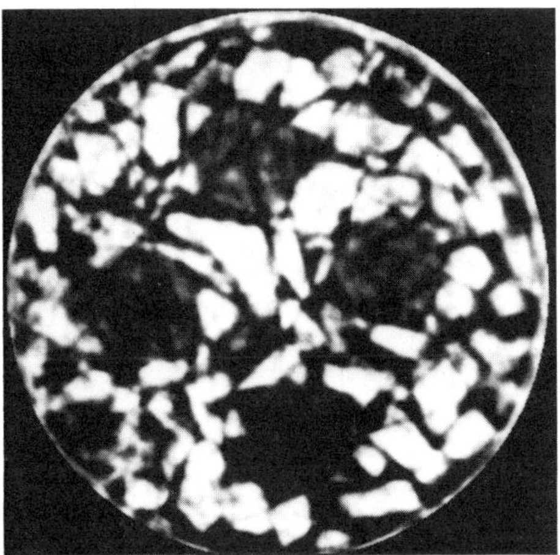

Fig. 5 CT image of a rock sample with oil and water. The lighter circular areas at the top and right are areas containing water, while the darker regions at the bottom and left are areas with oil. Courtesy of B.G. Isaacson, Bio-Imaging Research, Inc.

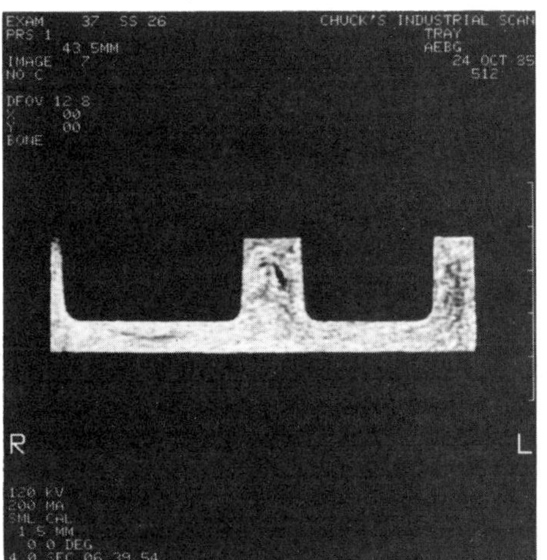

Fig. 6 CT image of a molded chopped-fiber carbon composite showing a resin flow (left) and a void in the center rib

es present (Fig. 5). Dual energy imaging is sometimes used to assist in quantifying these fluid concentrations. Dual energy imaging uses CT scans of a section acquired at different x-ray energies to evaluate the effective atomic number of the materials present.

Another research example in the energy industry was a U.S. Department of Energy funded study of the process of burning coal (Ref 23). In this study, a controlled furnace was placed in a medical CT system. The thermal fronts and relative density changes were imaged by computed tomography while the effluent gases and other parameters were monitored.

Composite structures, by their nature are complex materials that can pose challenges in design, fabrication, and inspection. Because of the high strength-to-weight ratio of these materials, composite usage is increasing, especially in the weight-sensitive aerospace industry. Computed tomography imaging has found considerable use in assisting the engineering development, problem solving, and production of composite components (particularly carbon composite components). The ability to track the integrity of individual components can greatly simplify process analysis relative to a sampled destructive study.

Computed tomography has been used to evaluate the fabrication process for certain composite components, including molded chopped-fiber components, as shown in Fig. 6. Resin flow patterns (Fig. 6) can be seen, and an individual component can be evaluated at different stages of curing to determine the point at which processing defects are created in the component. Resin flows in composite materials are generally visible, assuming a difference in the linear attenuation coefficient between the fiber and resin material. Low-density resin flows with widths near the image resolution may appear similar to a crack with subresolution separation. The use of dual energy techniques may help in distinguishing between similar indications. Fiber orientation in composite materials is sometimes seen, especially with chopped-fiber molded components. The ability to discriminate individual fiber bundles is highly dependent on fiber bundle size and system resolution. Computed tomography can detect indications of waviness in composite layers and porpoising (out-of-plane waviness). Radiographic tracer fibers appear with very high contrast.

Operational tests and tests to failure can also benefit from detailed data obtained at various stages of the test. One example of this type of use is the evaluation of rocket motor nozzles before and after firing tests (Ref 24). In this case, the degree and depth of charring were evaluated and compared with theoretical models. Computed tomography has been used in a number of situations to assist in the evaluation of valuable prototype components.

Some of the highest interest in the use of new techniques occurs with engineering problem solving. An example is the failure of the space shuttle launched communication satellites to obtain proper orbit in February 1984. In this case, computed tomography demonstrated interlaminar density variations (Fig. 7b) in some of the rocket motor exit cones of the type used with the satellites. These variations had not been detected by routine inspection of the components. Computed tomography provided a mechanism to evaluate the existing inventories, permitting judgments to be made on the flight-worthiness of the remaining hardware (Ref 25).

Computed tomography is also used in the production inspection of certain composite hardware. This is particularly true in the case of composite rocket motor hardware because of the experiences discussed above and the active role governmental agencies have played in driving this capability. The increased use of composite materials in critical structures is increasing the role of computed tomography in the production inspection of composite components.

Rocket Motors. In addition to individual rocket motor composite components, CT systems are used to scan full rocket motors. Some of these systems use linear accelerator x-ray sources up to 25 MV and are designed to inspect solid propellant rockets in excess of 3 m (9.8 ft) in diameter. These systems can be used to evaluate the assembly of the various motor components and to evaluate the integrity and fit of the casing, liners, and propellant (Ref 26, 27).

Precision Castings and Forgings. Computed tomography systems are used for the production inspection of small, complex

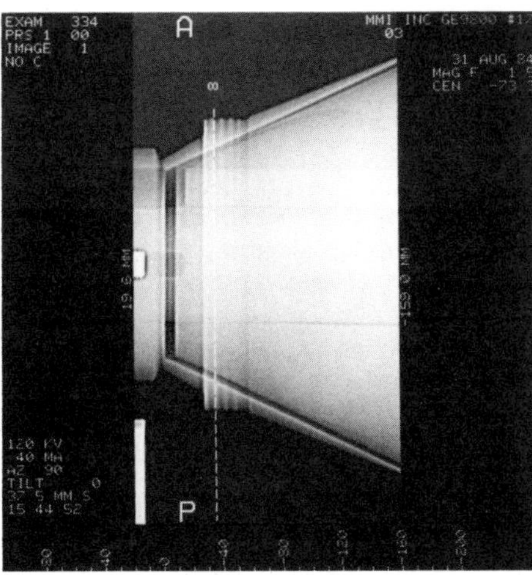

(a)

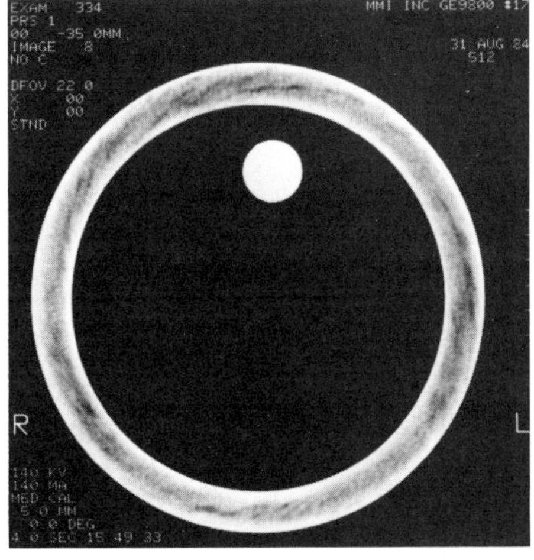

(b)

Fig. 7 Digital radiograph (a) of rocket exit cone showing the position for CT scanning through the thread portion of the component. (b) CT image through threaded area of rocket exit cone. CT revealed the extent of interply density variations.

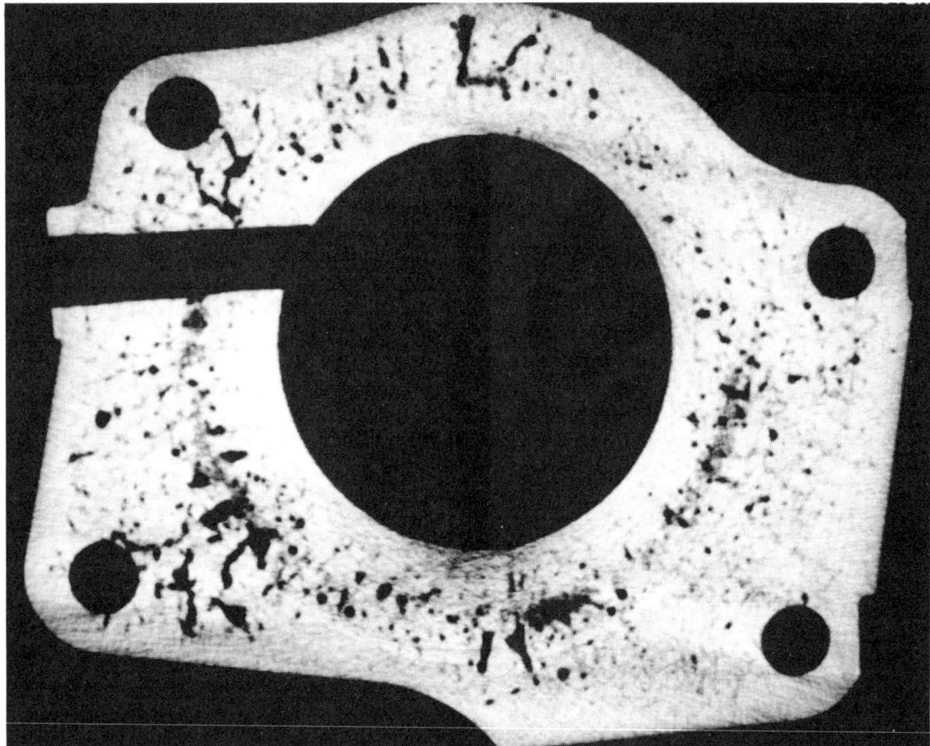

Fig. 8 CT image of cast housing showing severe shrinkage and porosity

precision castings and forgings, especially turbine blades and vanes used in aircraft engines and liquid propellant rocket motor pumps. The ability to localize material flaws and passage dimensions has also permitted the reworking of complex castings and fabrications. In the aircraft engine industry, computed tomography is considered an enabling technology that will allow large, complex structural castings to replace fabricated components, reducing both component weight and manufacturing costs. Computed tomography is also used to analyze the wall thickness of certain used blades to evaluate if sufficient material remains for refurbishing the component. This has permitted the overhaul and reuse of components that would otherwise be questionable.

Flaw Detection. Because of throughput considerations for production, casting flaws are generally detected in the digital radiography mode (see the section "Special Features" in this article for a discussion of the digital radiography mode). Computed tomography is used to detect flaws in critical areas and to further evaluate flaw indications detected by digital radiography or other methods (Ref 27-29). The specific data provided by computed tomography for the evaluation of flaw indications allow for a more informed accept/reject decision. Porosity and microshrink reduce the density of the material and are usually visible if distributed over a multipixel area (Fig. 8). The percentage of porosity can be quantified with properly controlled procedures. Other detectable conditions include casting bridges or fins, residual core material in the casting (Fig. 9a), and machining defects (Fig. 9b).

Dimensional Measurements. Because of the density sensitivity of the CT data and the averaging of the density values within a measured voxel, dimensional measurements can be made with better precision than the resolution of the image. The use of computed tomography for wall thickness gaging has advantages over other gaging techniques in that the measurements are not operator dependent, and precise information can be obtained from regions with internal walls and sharp curvatures.

Figure 10 shows a typical CT image of a turbine blade with the measured wall thickness information displayed. In this case, the wall thicknesses were measured with automated software, which makes many measurements along each wall segment (between ribs), marks the location of the thinnest portion, and posts the measured data in a table. The quantitative data can also be passed to a statistical software package for process control analysis. Applications are being pursued for using this dimensional information for subsequent machining operations, such as adjusting power levels for laser drilling. The ability to provide spatially specific structure and density information also opens up the opportunity to obtain three-dimensional data representations (Fig. 11) of physical components for computer-aided design documentation and computer-aided engineering analysis.

Engineering Ceramics and Powder Metallurgy Products. The sensitivity of computed tomography to density differences of a fraction of 1% can be applied to a number of material characterization problems. Materials that vary in density or composition because of the fabrication process may benefit from this capability. Injection-molded powder metal components have been scanned with the potential of better density characterization and process control in the green state to produce an improved sintered product. The computed tomography of advanced ceramic materials is also increasing, especially for high-stress and complex-shaped components (Ref 12, 30).

Assembled Structures. In general, the advantages of computed tomography over alternative inspection methods increase with the complexity of the structure. The ability of computed tomography to image the components in an assembly does not reduce the need for component inspection prior to assembly. This ability can be a valuable tool, however, when quality questions arise regarding the assembled product. Computed tomography can also be used to verify proper assembly or help evaluate damage or distortion in components caused by the fabrication process.

Examples and potential applications for computed tomography cover a wide range. They include the evaluation of composite spars in helicopter rotor blades, detonators and arming devices, thermal batteries, and a variety of electronic assemblies. The major limitation is the capacity of the system and economic considerations, especially for large production volume items. Computed tomography can provide worthwhile capabilities even for low-cost, high-volume components by assisting engineering development and problem solving.

CT System Design and Equipment

Computed tomography systems are relatively complex compared to some of the other NDE imaging systems. The production of high-quality CT images requires a stable radiation source, a precise mechanical manipulator, a sensitive x-ray detector system with high accuracy and a wide dynamic range, considerable computing power (usually with high-speed array processors), and radiation shielding for personnel safety. Consequently, the capital costs for industrial computed tomography systems tend to be well above those of most routine inspection systems.

The basic functional block diagram of an x-ray CT system is illustrated in Fig. 12. The major components include:

- A precision mechanical manipulator for scanning the testpiece with a radiation source and detector array. Several differ-

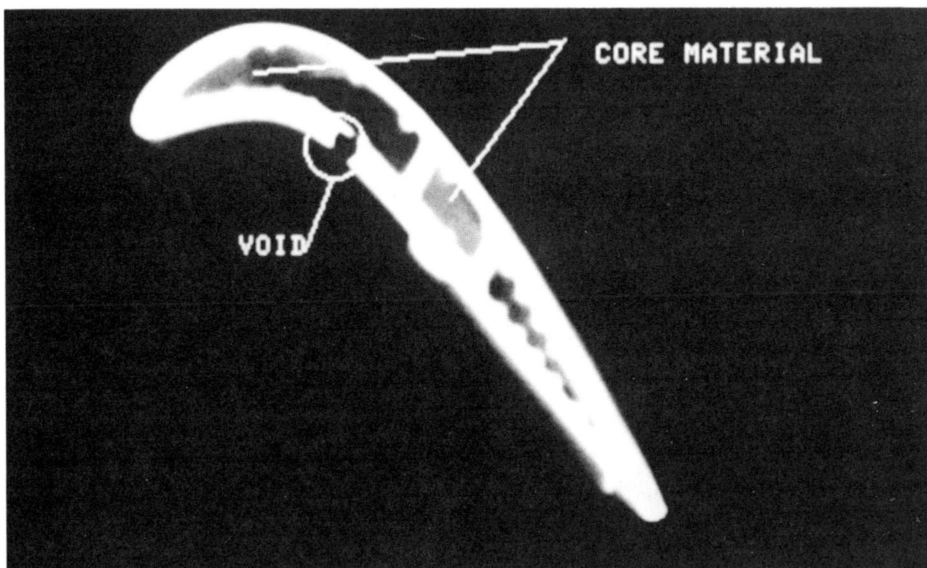

(a)

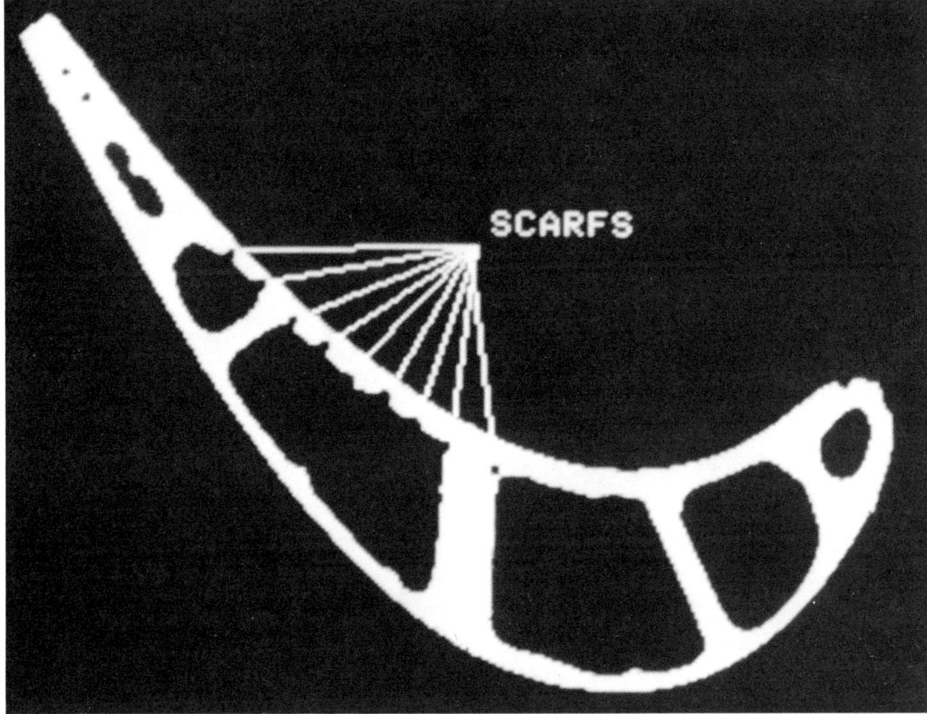

(b)

Fig. 9 CT images of a turbine blade. (a) Residual core material from the casting process. (b) Scarfing caused by near-tangential drilling of nearby walls

ent scanning geometries are used to acquire the ordered set of transmission data (see the section "CT Scanning Geometries" in this article). The mechanical subsystem may also include automated parts handling for loading and unloading the inspection station (Fig. 13)

- Radiation shielding for the safe use of the radiation source. The high-energy radiation sources used in industrial CT systems are typically shielded by placing the radiation source detectors in a room with shielded walls. Self-contained, shielded cabinet systems (Fig. 13) are often used for industrial systems with x-ray sources operating below 500 kV
- A radiation source, such as an isotopic gamma source, an x-ray system, or a linear accelerator. The radiation sources used in industrial CT systems are similar to the x-ray and γ-ray sources used in industrial radiography. These radiation sources are discussed in the article "Radiographic Inspection" in this Volume
- Detector arrays for measuring the transmitted radiation. Industrial CT systems primarily use either gas ionization detectors or scintillation detectors
- A data acquisition system for converting the measured signal into a digital number and transmitting the data to the computer system
- A computer with appropriate software for controlling the data acquisition and reconstructing the cross-sectional image from the measured data
- A display system for displaying the cross-sectional image

In addition to these individual parts of a CT system, the overall capabilities of the total system are also important. The key capabilities of a CT system are:

- The ability to handle the range of components to be inspected
- The throughput or rate at which the components can be tested
- The ability to provide sufficient image quality for the specified inspection task by considering the selectable operational parameters of CT systems that affect image quality

CT Scanning Geometries

The CT scanning geometry is the approach used to acquire the necessary transmission data. In general, many closely spaced transmission measurements from a number of angles are needed. Four types of CT scanning geometries are shown in Fig. 14. Much of the historical progression of the data acquisition techniques has been driven by the need for improvements in data acquisition speed for medical imaging. In medical CT systems, the radiation source and detector are moved, not the patient. Industrial CT systems, however, often manipulate the component; this can simplify the mechanical mechanism and help maintain precise source-detector alignment and positioning. In either case, the relative positions of the object to the measured ray paths are equivalent. Some of the nomenclature derived from medical computed tomography does not readily accommodate some of the manipulator configurations that are practical in industrial imaging.

Single-Detector Translate-Rotate Systems. The first commercial medical CT scanner developed by EMI, Ltd., uses a single detector to measure all of the data for a cross section. The x-ray source and detector are mounted on parallel tracks on a rotating gantry (Fig. 14a). The source and detector linearly traverse past the specimen and make a series of x-ray transmission measurements (240 measurements for a 160 × 160 image array). These measurements correspond to the transmitted intensity through the object along a series of parallel rays. The source-detector mechanics are rotated by 1°, and another linear traverse is

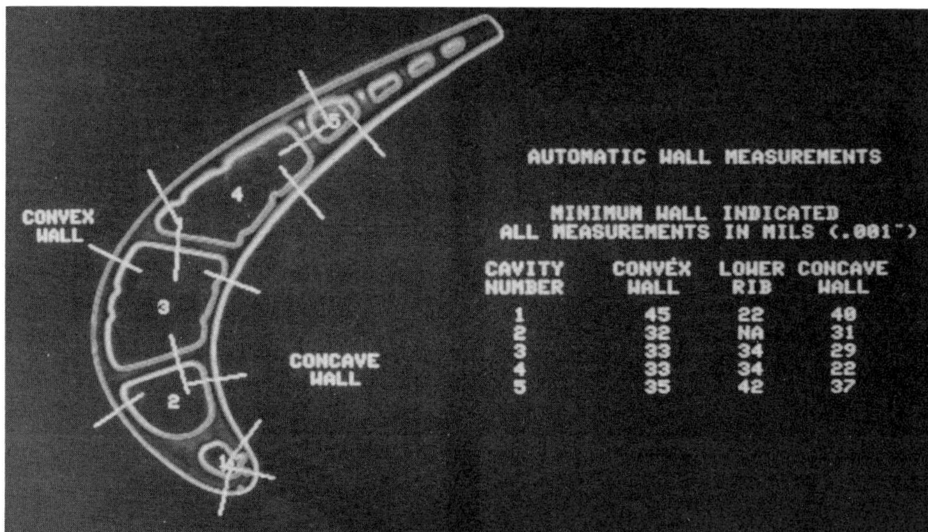

CONVEX WALL
CONCAVE WALL
AUTOMATIC WALL MEASUREMENTS
MINIMUM WALL INDICATED
ALL MEASUREMENTS IN MILS (.001")
CAVITY NUMBER | CONVEX WALL | LOWER RIB | CONCAVE WALL
1 | 45 | 22 | 40
2 | 32 | NA | 31
3 | 33 | 34 | 29
4 | 33 | 34 | 22
5 | 35 | 42 | 37

Fig. 10 Cross-sectional CT image through a turbine blade with wall thickness measurement locations and values determined with automated software

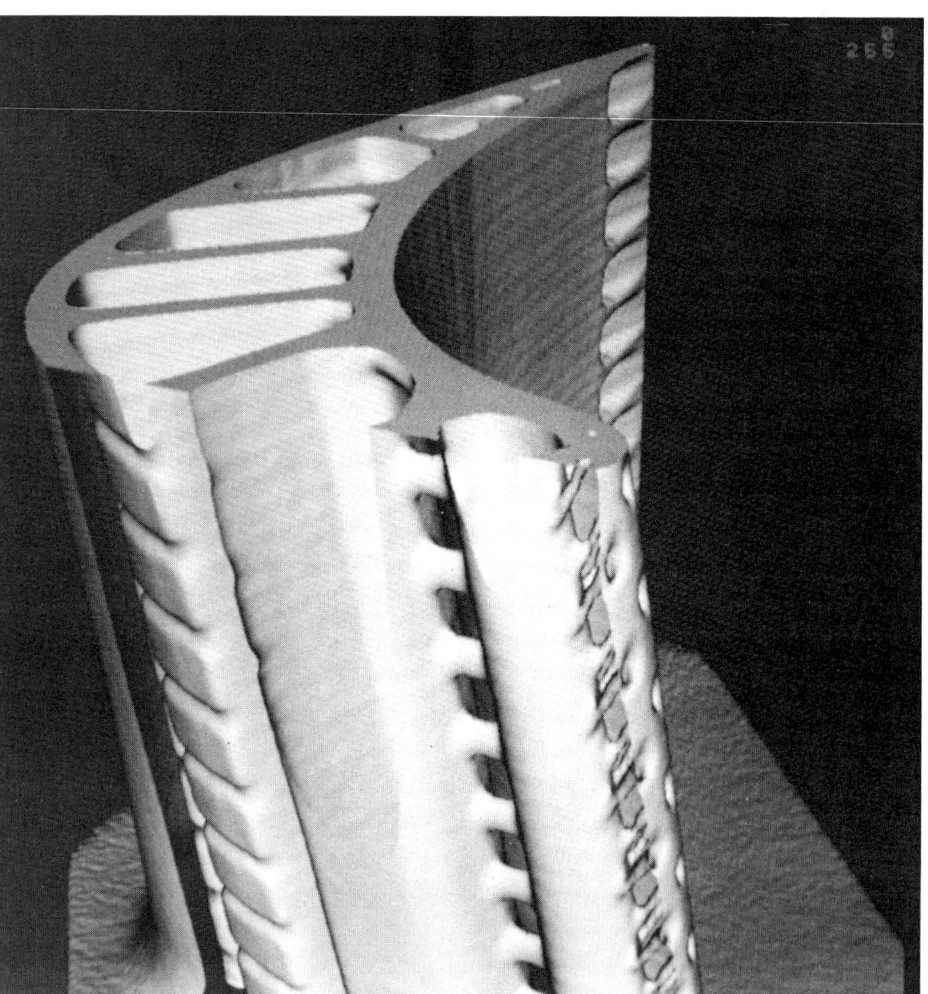

Fig. 11 Three-dimensional surface image of a turbine blade generated from CT data

made. This provides another set of parallel rays, but at a different angle through the object. This process is repeated until data are obtained over a full 180° rotation of the source-detector system.

The original EMI scanner uses two detectors to collect data for two adjacent slices simultaneously; it is relatively slow, with nearly a 5-min scan time to produce a low-resolution image. Despite its limitations, this system was a significant breakthrough for diagnostic medicine.

Single-detector translate-rotate CT systems are sometimes referred to as first-generation systems. The use of the generation nomenclature was originated by medical manufacturers to emphasize the newness of their designs and is sometimes used as a matter of convenience to describe the operation of a system. Current commercial systems do not use the single-detector approach, because of its limited throughput. The simplicity of this approach, however, makes it suitable for applications in basic research.

Multidetector Translate-Rotate Systems. To improve the data acquisition speed, multiple detectors can be used to make a number of simultaneous transmission measurements. The second-generation or multidetector translate-rotate systems use a series of coarsely spaced detectors to acquire x-ray transmission profiles from a number of angles simultaneously (Fig. 14b). The data measurements are still acquired during the translation motion, but the system makes a much larger rotational increment. The distance translated needs to be somewhat longer than that for a single detector system in order to have all source-detector rays pass over the entire object width, but fewer translations are needed to collect a full set of data. The mechanical system of a second-generation CT scanning geometry for industrial applications may range from a design for small components (Fig. 15) to a design for large castings (Fig. 16).

Rotate-Only Systems. Further improvement in data acquisition speed can be achieved by using a broad fan beam of radiation spanning the entire width of the object, eliminating the translation motion entirely. The fast scanning speed of rotate-only systems is important in medical imaging to minimize patient motion during the scan and for high patient throughput. Nearly all current medical scanners are rotate-only systems, with some of these systems making over 1 million transmission measurements for a complete scan in less than 2 s.

There are two basic configurations of rotate-only systems. Third-generation or rotate-rotate systems use many (sometimes over 1000) closely spaced detector elements that are fixed relative to the x-ray beam (Fig. 14c). The source-detector combination may rotate together around the specimen, as in medical scanners (hence the term rotate-rotate). In industrial systems, the object can rotate in order to obtain data from all angles. The reconstruction process for these systems is slightly different than for the other configurations because a transmission profile or view corresponds with a set of rays in the shape of a fan. The use of the term rotate-only in this article will refer to third-generation systems.

Fourth-generation or stationary-detector rotate-only systems have a stationary ring of detectors encircling the specimen, with the x-ray source rotating around the specimen (Fig. 14d). The x-ray fan beam exposes only a portion of the detectors at any moment in time. The data are usually regrouped into fan beam sets, with the detector at the apex of the fan, or into an equivalent parallel ray set. This configuration is generally less flexible, cannot utilize the ability to manipulate the object, and is normally not found in systems designed for industrial use. Fourth-generation geometries require less stringent detector element matching by a factor of 100 or so and are less susceptible to normalization and circular artifacts.

Operational Differences and Variations. Each of the CT scanning geometries has its advantages and limitations. The different approaches each have certain data acquisition parameters that tend to be fixed and others that are readily adjustable.

One of the major factors in determining image resolution is the spacing between the measured transmitted rays. (Other factors include source size, source-object and object-detector distances, detector aperture, and the reconstruction algorithm.) The resolution at the center of the image field and in the radial direction at the periphery is highly dependent on the ray spacing in a transmission profile. This ray spacing can be easily adjusted in translate-rotate systems by adjusting the linear distance moved between measurements during translation.

On rotate-only (third-generation) systems, this ray spacing is fixed by the spacing of the detector elements. Consequently, these systems use a densely packed detector array, with element spacings well below 1 mm (0.040 in.). The ray spacing through the object can be improved by moving the object closer to the x-ray source or by acquiring interleaving data over multiple rotations with the detector array shifting a fraction of the detector spacing.

The circumferential resolution at the periphery of the image is dependent on the angular separation between views and the diameter of the image field. The ray spacing along the circumference or the number of angular views acquired is a readily adjusted parameter on the rotate-only systems.

This angular separation is fixed by the detector array spacing on multidetector translate-rotate systems. Some variation can be achieved by changing the source-detector spacing, or interleaved angular measurements can be obtained with small angular increments.

Multidetector translate-rotate systems can adjust for object size by changing the translation distance to the object diameter plus the width of the fan beam at the object. This tends to be inefficient for a wide fan beam system with small objects, but gives considerable flexibility for large objects.

Rotate-only systems have a field-of-view defined principally by the detector array size and the source-object and object-detector distances (object magnification). These systems can be used for larger fields-of-view by shifting the detector or object between rotations to cover a wider field. However, this rotate-shift-rotate approach still acquires

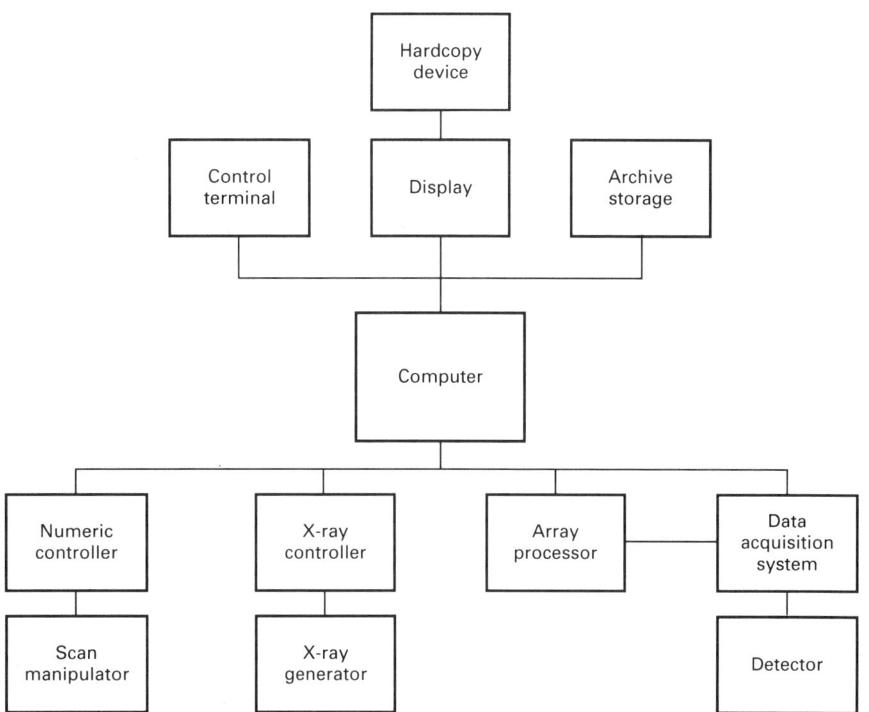

Fig. 12 Functional block diagram of a typical CT system

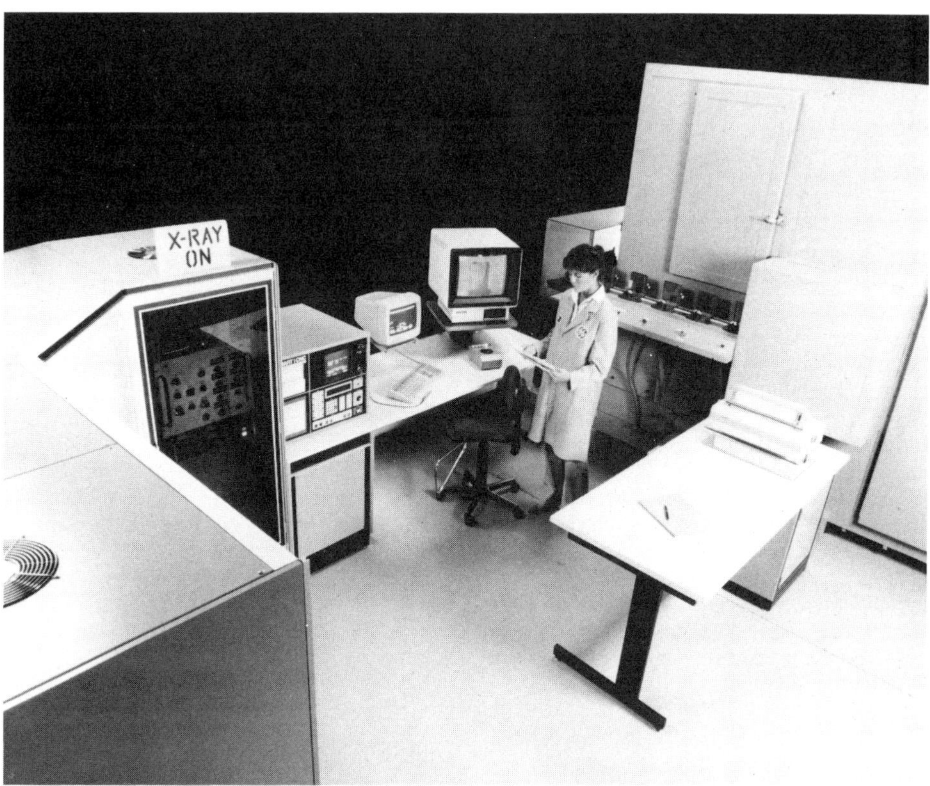

Fig. 13 Automated industrial computed tomography system for the production inspection of small components. Parts-loading conveyor and cabinetized x-ray system are included in this setup.

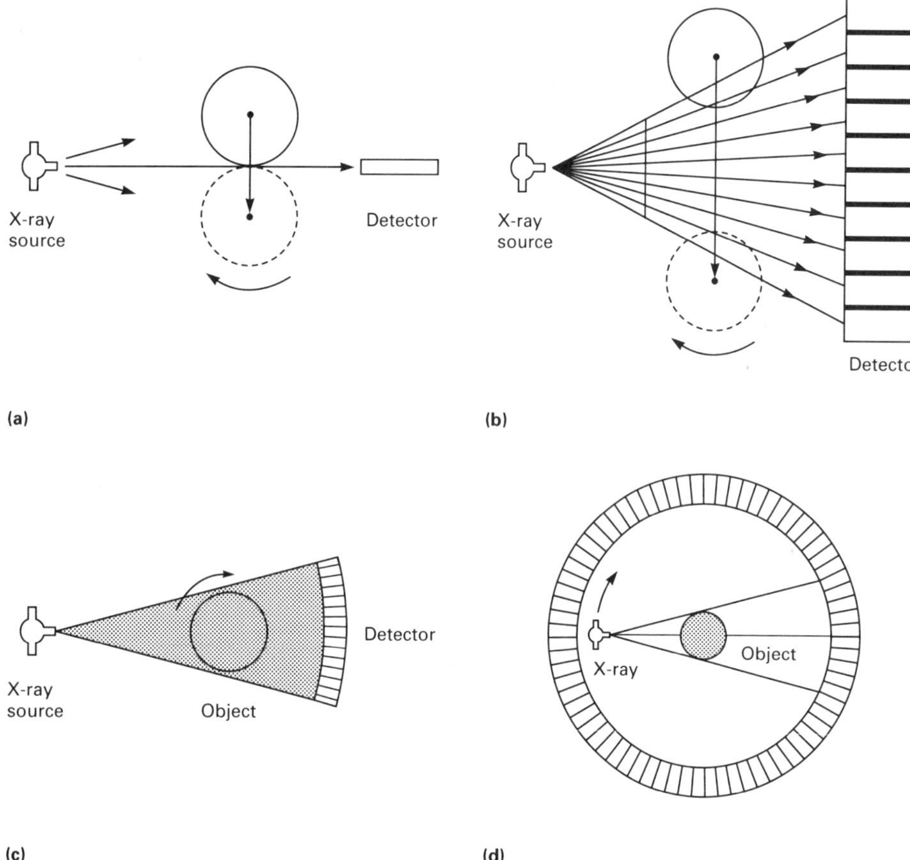

Fig. 14 Basic CT scanning geometries. (a) Single-detector translate-rotate, first-generation system. (b) Multidetector translate-rotate, second-generation system. (c) Rotate-only (rotate-rotate), third generation system. (d) Stationary-detector rotate-only, fourth-generation system. In medical systems, the source and detector are manipulated instead of the object.

fan beam data during the rotation motion, versus the parallel ray data collected during translation on a translate-rotate system.

Another difference is in acquiring data for DR mode imaging. The DR images are commonly used on CT systems to locate the areas in a component where the CT slices should be obtained or can be used for radiographic inspection of the component.

The high-resolution detector of the rotate-only systems normally requires a single z-axis translation to produce a high-quality DR image. For even higher resolution, interleaved data can be obtained by repeating the rotational scan with a shift of a fraction of the detector spacing. For large objects, a scan-shift-scan approach can be used.

Translate-rotate systems are typically less efficient at acquiring digital radiographic data if the wider detector spacing requires multiple scans with shifts to provide adequate interleaving data. If the detector field-of-view does not cover the full width of the object, the object can be translated and the sequence repeated.

Considerable flexibility can exist for various requirements. The mechanical packaging and implementation of the particular data acquisition system are the limiting factors with regard to the range of capabilities. The different acquisition approaches, however, can determine the efficiency of a system in acquiring appropriate data as well as the subsequent imaging throughput.

Radiation Sources

Computed tomography reconstructions can be performed with any set of measured data along lines through an object. This can include the use of ultrasound, neutrons, charged particle beams, nuclear tracer distributions (PET and SPECT imaging), or emitted radio-frequency emissions from nuclei (NMR imaging). In most practical industrial applications, however, x-ray or γ-ray radiation is used in computed tomography.

The three types of radiation sources typically used in industrial computed tomography are x-ray tubes, γ-ray sources, or a high-energy x-ray source such as a linear accelerator. Each of these sources has advantages and disadvantages. The two types of x-ray sources can provide a high-intensity beam, but they also produce a broad spectrum that complicates the quantitative analysis of the projection data. Gamma-ray sources have a narrow energy spectrum but tend to be of low intensity. Other factors include the size of the source and the energy of the beam.

The ideal radiation source would provide a high-intensity beam of x-ray photons at a single energy (or optimum spectral spread) emanating from a very small area with an energy capable of transmitting a reasonable fraction of the x-rays through the object.

Gamma-ray sources are used in some industrial computed tomography systems. A major advantage of γ-ray sources is that the high-energy photons produced by a radioactive source are all at specific energies, while x-ray sources produce photons over a wide range of energies. Changes in the average energy transmitted through various thicknesses of material can cause inconsistencies in the measured data, resulting in errors in the reconstructed image.

The most significant disadvantage of γ-ray sources is the limited intensity or number of gamma photons produced per second. The intensity can be increased by using more radioactive material, but this requires a larger radioactive source, which adversely affects the spatial resolution of the system. Also, because the energy of a γ-ray source is dependent on the radioactive material, the effective energy cannot be readily changed for different imaging requirements.

X-Ray Sources. The number of high-energy photons included in the CT scan measurements is the primary factor affecting the statistical noise in the reconstructed CT image. To collect the maximum number of photons in the least amount of time, x-ray sources are generally used rather than radioactive γ-ray sources. For x-ray sources up to 500 kV in operating voltage, x-ray tubes are used. When higher-energy x-rays are needed to penetrate thick or dense testpieces, linear accelerators are often used.

The key characteristics of an x-ray source include the operating voltage range, the effective size of the focal spot, and the operating power level. These characteristics are important in both radiography and computed tomography (see the article "Radiographic Inspection" in this Volume). In computed tomography, however, the stability of the x-ray source is especially important. Voltage variations are particularly disruptive in that they change the effective energy of the x-ray beam and can cause image artifacts. Current variations are less of a problem because x-ray intensity can be monitored by reference detectors.

X-ray collimators are radiation shields with open apertures that shape the x-ray beam striking the object and the detector. For CT systems, the radiation field is typically a thin fan beam wide enough to cover the linear detector array. A collimator (Fig. 1a) is located between the x-ray source and the object to shape the beam; normally, a

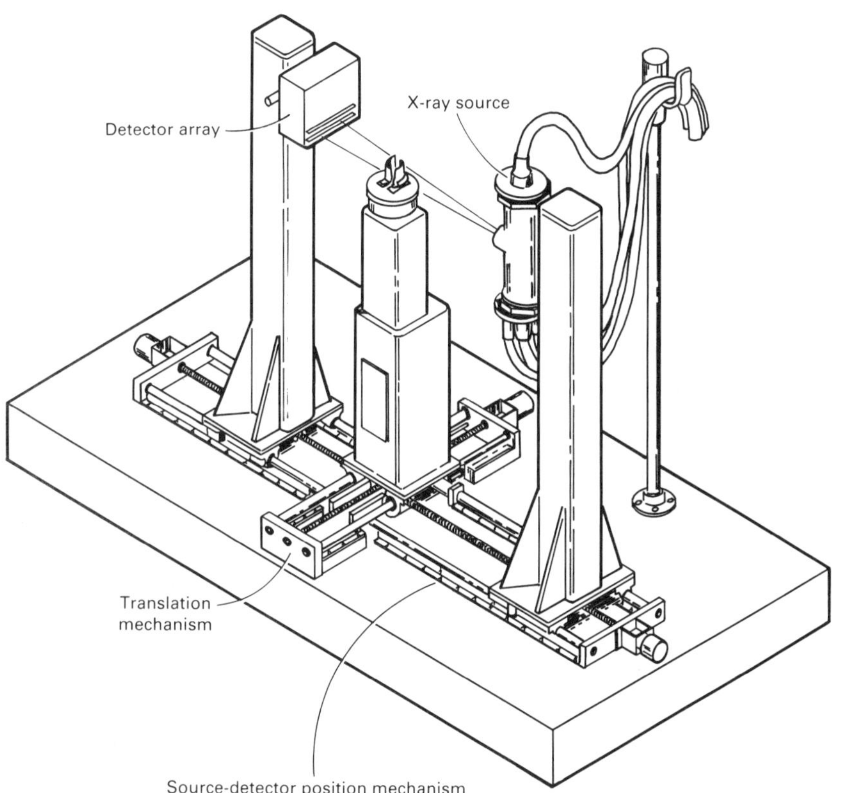

Fig. 15 CT and DR scanner designed for the nondestructive testing and dimensional analysis of a variety of parts up to 300 mm (12 in.) in diameter and 600 mm (24 in.) in length

second collimator is also placed between the object and the detector array to further define the object volume being sampled. One or both of the x-ray collimators may have an adjustable slot spacing to permit operator selection of the slice thickness.

One of the benefits of using a thin fan beam of radiation for CT and DR imaging is that most of the radiation scattered by the object will miss the detector array and not be measured. This improves the quality of the measured data over that obtained by large-field radiography.

Systems that use coarsely spaced detectors, such as multidetector translate-rotate systems, may also have detector aperture width collimators. These collimators reduce the effective size of the detector element, thus improving the transmission data resolution.

X-Ray Detectors

Detector Characteristics. The ability to measure the transmitted x-ray intensity efficiently and precisely is critical to x-ray CT imaging. The features of the detector that are important to imaging performance include efficiency, size, linearity, stability, response time, dynamic range, and energy range of effectiveness.

Detector efficiency is a quantitative measure of the effectiveness of the detector in intercepting, capturing, and converting the energy of the x-ray photons into a measurable signal. This efficiency is a factor in the image quality for a given x-ray source output, or the exposure time required to collect a sufficient amount of radiation. The three components of overall detector efficiency are described below.

The geometrical efficiency is the fraction of the transmitted beam passing through the measured slice volume that is incident on the active detectors. It is equal to the active detector element width in the plane of the slice divided by the center-to-center spacing between detector elements. The collection efficiency is the fraction of the energy incident on an active detector area that is absorbed in the detector. It is dependent on the atomic number and density of the detector material and on the size and depth of the detector. Lack of absorption occurs because of x-ray photons passing through the detector without interacting or because of interactions in which some of the energy is lost due to scatter or characteristic x-ray emissions from the detector material. Conversion efficiency is the fraction of absorbed energy that is converted into a measurable signal.

Detector size consists of both the detector element width and height. The detector height determines the maximum slice thickness that can be measured. Increasing the slice thickness increases the collected x-ray intensity, thus reducing the image noise, but decreases *z*-axis (interplane) resolution and may increase partial volume blurring and partial volume artifacts. The measured slice thickness is adjusted by the slice thickness collimators to be less than the detector height.

The detector element width is a factor in determining the resolution of the CT image. On some CT systems, particularly the multidetector translate-rotate systems, the detector aperture width can also be adjustable with a collimator. This offers higher potential resolution with a lower geometrical efficiency.

Detector Consistency. Detector linearity is the ability to produce a signal that is proportional to the incident x-ray intensity over a wide range of intensities. Detector stability concerns the ability to produce a consistent response to a signal without drifting over time. Channel-to-channel uniformity relates to the consistency between the detector elements in their signal response, noise, aperture size, and other characteristics. These parameters are important for CT detector selection to produce the consistent data set required for image reconstruction. For reproducible characteristics, some degree of inconsistency may be correctable by computer processing of the measured data.

The detector response time is the effective time for the signal to settle and not be significantly influenced by prior incident intensities. The response time is a critical factor in determining how rapidly independent samples can be collected and the quality of the data. Some CT systems interrogate each detector at rates up to 10 000 times per second.

Dynamic range is the range of intensities over which accurate measurement can be made. It is usually specified as a ratio of the maximum signal output to the minimum output. The minimum signal output is limited by the electronic noise of the detector and its electronics. The level of electronic noise determines how finely the signal can be effectively sampled for processing by the computer.

Gas ionization detectors consist of an ionization chamber with a positive and negative electrode (Fig. 17). When radiation interacts with the gas inside the detector, the effect is to knock orbital electrons out of atoms, thus ionizing these atoms. The freed electrons in this gas are pulled to the positive electrode, and the positively charged atoms (anions) are pulled to the negative electrode. This principle is used in basic radiation survey meters, which use an ionization chamber to measure the amount of ionization occurring within a volume of air or gas.

Gas ionization detectors have been shown to be particularly effective in rotate-only CT scanners. Rotate-only systems require a closely spaced detector array with good detector uniformity to provide the finely sampled projection data for image reconstruction. These systems may contain well over 1000 individual detector elements in a contiguous array.

Fig. 16 Industrial computed tomography system x-ray cell for the inspection of large fabricated components and castings with multiaxis parts manipulator

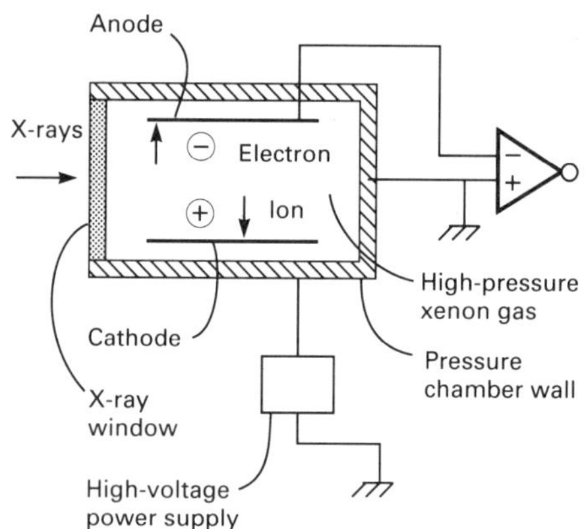

Fig. 17 Typical configuration of a xenon gas ionization detector

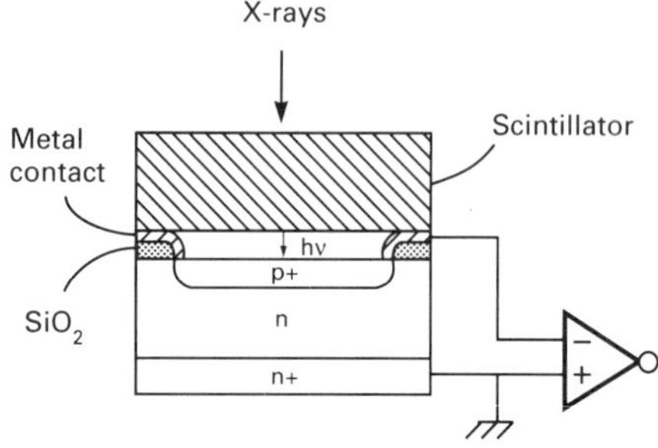

Fig. 18 Configuration for a scintillation crystal-photodiode detector

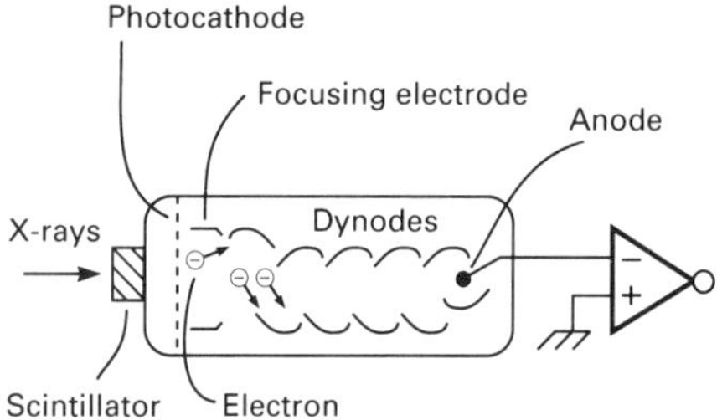

Fig. 19 Configuration for a scintillation crystal-photomultiplier tube detector

Ion chamber detectors define the individual array elements by the placement of the collecting electrodes of each detector element. Medical CT systems arrange the collecting electrodes between the detector elements, with a typical detector spacing of 1.2 mm (0.05 in.). Higher-resolution industrial detectors place the discrete electrodes above and below the x-ray fan beam, with typical detector spacings of 0.25 mm (0.01 in.). The spread of the resolution response function for tightly spaced detector systems, such as high-resolution ionization detectors, may be greater than the physical detector spacing because a small percentage of the intensity incident on a detector element may scatter into adjacent detector elements, contributing to crosstalk between detectors.

Because of the high packing density, with little or no gap between active detector elements, the geometrical efficiency of gas ionization detector arrays is very good. Because a gas is used to stop the x-ray beam, the collection efficiency is typically lower than a high-density solid detector material of the same dimensions. The conversion efficiency of ionization detectors is very good in that few of the ions produced by the radiation are lost due to recombination before reaching the collecting electrode.

To improve the efficiency of the detector at stopping and measuring the ionizing radiation, a high atomic number gas at high pressure is used. The entire detector array is contained in a single pressure vessel that provides a common gas mixture and pressure for all detector elements and aids in producing a uniform detector response. Xenon, an inert gas, is used because of its high atomic number ($Z = 54$). High pressures are used that can increase the physical density of the gas up to 1.5 g/cm^3 (0.9 oz/in.^{3}).

Xenon ionization detectors provide high linearity over a wide range of x-ray intensities and have a reasonably fast response time associated with the travel time of the ions in the electric field of the chamber. Of particular benefit for third-generation rotate-only scanners is that xenon detectors are highly stable and do not exhibit radiation damage effects or other long-time-constant exposure effects that may distort the measured data.

Scintillation detectors, which have been used for many years to measure radionuclide emissions, are used in both translate-rotate and rotate-only CT systems. The two basic designs are shown in Fig. 18 and 19. The most preferred and widely used design in CT systems is the scintillation crystal-photodiode detector system (Fig. 18).

Scintillators. As with the gas ionization detectors, the x-ray radiation causes elec-

Table 3 Scintillator materials

Material	Decay constant, μs	Afterglow, % at 3 ms	Index of refraction	Density, g/cm^3	Conversion efficiency, % (versus NaI)
NaI(Tl)	0.23	0.5–5	1.85	3.67	100
CsI(Tl)	1.0	0.5–5	1.8	4.51	45
BGO	0.3	0.005	2.15	7.13	8
$CaWO_4$	0.5–20	1–5	1.92	6.12	50
$CdWO_4$	0.5–20	0.0005	2.2	7.90	65

trons in the atoms in the absorbing material to be excited or ejected from the atom. These electrons rapidly recombine with the atoms and return to their ground-state configuration, emitting a flash of visible light photons in the process. The term scintilla means spark or flash. Scintillation detectors typically use relatively large transparent crystals. The scintillator is encased in a light-reflective coating and is optically coupled to a sensitive light detector, such as a photomultiplier tube (PMT) (Fig. 19) or photodiode (Fig. 18).

Sodium iodide doped with thallium [NaI(Tl)] coupled to a photomultiplier tube is a standard scintillation detector used for nuclear radiation measurements. This type of detector was used on the original EMI medical scanner. However, new scintillating materials have resulted in improved performance for certain detector characteristics. The scintillation materials of choice are cadmium tungstate ($CdWO_4$) and bismuth germanate (BGO). The properties of the material are important to the efficiency and speed of the detector and its ability to provide accurate data. Table 3 lists some of these materials along with some of their key properties.

The relatively high density and effective atomic number are key attributes of many scintillator materials, providing a highly effective ability to stop and absorb the x-ray radiation. The scintillation conversion efficiency, or fraction of x-ray energy converted to light, is about 15% for NaI(Tl). Table 3 lists the conversion efficiencies relative to sodium iodide with a standard light detector. The efficiency of the scintillator in producing light is important in that some of this light is lost by absorption in the scintillator, reflective walls, and optical coupling material. A high index of refraction increases the difficulty of optically coupling the scintillator.

The principal decay constant (time for the signal to decay to 37% of the maximum) is also listed in Table 3. Other lower-intensity but longer-duration decay constants may also exist. Long afterglow decay constants (from 1 to 100 s) in sodium iodide are one of its significant deficiencies for high-speed data acquisition. In addition to afterglow, the scintillator materials may also exhibit temporary radiation damage, such as reduced optical transparency, which also has an associated time constant for repair. The effect of prior radiation exposure on the measured signal is especially important when a wide range of intensities are to be measured by the detector. Careful selection and characterization of the scintillation detector material are necessary for high-speed CT data acquisition.

Photomultiplier Tube. The light detector is another important element in scintillation detector systems. Early systems utilized PMT optical sensors (Fig. 19). A photomultiplier tube is a vacuum tube with a photocathode that produces free electrons when struck by light. These electrons are accelerated and strike a positive electrode (dynode) that releases additional free electrons. A series of dynodes is used to obtain very high levels of amplification with low background noise. For use on CT systems with high x-ray intensities, the photomultiplier tube is used to produce a signal current proportional to the rate at which x-ray energy is absorbed in the scintillator.

Scintillation crystal-photodiode array detectors have important advantages over photomultiplier tubes in CT applications. Although photomultiplier tubes are relatively efficient, they are subject to drifts in gain and are relatively bulky. Crystals with photodiodes are more stable and permit the use of small, tightly packed detector arrays. Therefore, photodiodes are commonly used as the optical sensor for CT scintillation detectors. Photodiodes produce a small current when light strikes the semiconductor junction. This current is then amplified by a low-noise current-to-voltage converter to produce the measured signal.

Very high resolution detector systems have been fabricated by coupling very thin slabs of scintillation material to high-resolution photodiode arrays. These photodiode arrays may have 40 or more individual photodiodes per millimeter on a single chip. Detectors with very small, tightly packed scintillation crystals may be particularly appropriate for the materials evaluation of small samples at low x-ray energies.

Quantum Noise (Mottle) and the Detective Quantum Efficiency. The beam intensity being measured consists of a finite number of photons. Even with an ideal detector, repeated measurement will contain a certain degree of random variation, or noise. The standard deviation of the number of photons detected per measurement is $\sqrt{N}$, where N is the average number of photons detected per measurement. Therefore, the signal-to-noise ratio (SNR) can be given by:

$$\text{SNR} = \frac{N}{\sqrt{N}} = \sqrt{N} \quad \text{(Eq 1)}$$

Just as fast films that require less exposure tend to produce grainy images, the fewer the number of photons detected, the lower the signal-to-noise ratio. Because this type of noise is due to the measurement of a finite number of discrete particles, or quanta, it is referred to as quantum noise.

Because practical radiation detectors are imperfect devices, not all x-ray photons are absorbed in the detector. In gas ionization detectors, each photon absorbed produces a finite number of measured electrons with their own quantum statistics. In scintillation detectors, a finite number of light photons reach the light sensor, and a finite number of electrons flow in the sensor. Each of these steps has its own quantum statistics (that is, quantum noise). The detector electronics, particularly the initial amplifiers, introduce additional noise into the measured signal, which adds to the quantum noise.

The overall efficiency of the detector, including the noise added to the signal in the detection process, can be characterized by comparing its performance to that of an ideal detector. The detective quantum efficiency (DQE) of a detector is the ratio of the lower x-ray beam intensity needed with an ideal detector to the intensity incident on the actual detector for the same signal-to-noise ratio. If the only increase in noise versus an ideal detector is due to the absorption efficiency, that is, the fraction of photons absorbed in the detector, the DQE would equal the absorption efficiency.

Other losses of the signal occur in the detection process, and other sources of noise can be introduced into the signal to further reduce the signal-to-noise ratio for actual measurements. The result in the quality of the measured signal is effectively the same as detecting fewer photons. The DQE can be determined by comparing the measured signal-to-noise ratio to that calculated for an ideal detector as follows:

$$\text{DQE} = \frac{(\text{SNR}_{\text{detected}})^2}{(\text{SNR}_{\text{incident}})^2} \quad \text{(Eq 2)}$$

The designers of CT detector systems strive to maximize overall detector efficiency while meeting the size, speed, stability, and other operational requirements of a practical CT system.

Data Acquisition System

The data acquisition system is the electronic interface between the detector system and the computer. It provides further amplification of the detector signals, multiplexes the signals from a series of detector channels, and converts these analog voltage or current signals to a binary number that is transmitted to the computer for processing.

Some of the factors key to this electronics package include low noise, high stability, ability to calibrate offset and gain variations, linearity, sensitivity, dynamic range, and sampling rate. The electronics of the data acquisition system should be well matched to the detector system to minimize degradations to the data accuracy and scanning performance.

The dynamic range of the detector system is characterized by the ratio of the maximum signal to the noise. The dynamic range of the data acquisition system is given by the range of numbers that can be transmitted to the computer, often specified by the number of bits per measurement. It is not uncommon for CT systems to have a dynamic range of 20 bits (2^{20}), or 1×10^6 to 1.

Although it is difficult to build practical analog-to-digital converters with a 20-bit dynamic range, this electronics problem has been solved in several different ways. One way is to use an autoranging or floating-point converter. In this case, a linear converter is preceded by an adjustable-gain amplifier. One of several ranges can be selected by changing the gain of the amplifier, usually by a factor of two. If an autoranging amplifier having four range settings corresponding to gains of 1, 2, 4, and 8 precedes a 16-bit linear converter, the overall dynamic range is 20 bits, while the sensitivity varies from 1 to 2^{16} for high intensities to 1 in 2^{20} for low intensities relative to the maximum measurable intensity.

Computer System

Computer systems vary considerably in architecture, with computational performance steadily increasing. The availability of practical minicomputer systems was the enabling technology for the development of computed tomography in the early 1970s. Computed tomography imaging has benefited from the computational improvements by permitting increasing amounts of transmission data to be measured to produce larger, more detailed image matrices with less processing time. High-speed computer systems are also enabling increasingly sophisticated data processing for image enhancement, alternate data presentations, and automated analysis.

Evaluating computer system performance parameters is difficult, especially with the use of distributed processing. It is common to have several processing systems controlled by a central processor or operating relatively independently in a network configuration. This distributed processing may include programmable controllers and numerical controllers for automated parts handling, manipulator control, and safety monitoring; specialized array processors for image reconstruction; image processor for image enhancement and analysis; and display processors for presentation.

Because computed tomography uses large amounts of data (millions of transmission measurements to produce an image containing up to several million image values) and because several billion mathematical operations are required for image reconstruction, high-speed array processors are used for image reconstruction. Array processors are very efficient devices for performing standard vector or array operations. The CT reconstruction technique commonly used also requires a backprojection operation in which the projection values are mapped back into the two-dimensional image matrix. Consequently, most CT systems contain specialized processors designed to perform this operation efficiently in addition to standard array operations.

Data handling and archiving can also become significant considerations because of the size of the data files. Optical disks can be beneficial for compact archiving as an alternative to magnetic tape. High-speed data networking can also provide improved utility in transferring data to alternate workstations for further review and processing.

Image Quality

Flaw sensitivity and image quality are the two basic criteria for evaluating the ability of any imaging system to define sufficiently certain structures for a range of test object situations. Although image quality is a requirement in achieving adequate flaw sensitivity, the evaluation of flaw sensitivity can be ambiguous in relation to inspection task requirements. The analysis for the sensitivity of defined features in specific components can be quite complex and may involve probability-of-detection or relative-operating-curve analysis. These analyses are also dependent on the perception and skill of the interpreter and on the decision criteria, which may emphasize maximizing true positive or minimizing false negative indication detection.

To evaluate and monitor the image quality and performance of an imaging system, a number of simpler methods can be used to characterize the resolution, sensitivity, and accuracy of a system. The three basic factors affecting image quality and system performance are:

- Spatial resolution
- Image contrast
- Image artifacts

These three factors are discussed in the following sections.

Spatial Resolution

Spatial resolution is a measure of the ability of an imaging system to identify and distinguish small details. As in measuring the resolution of a radiographic system, an image of a regularly repeating high-contrast pattern is typically evaluated. The equivalent to a radiography line pair gage for computed tomography is a bar pattern resolution test phantom.

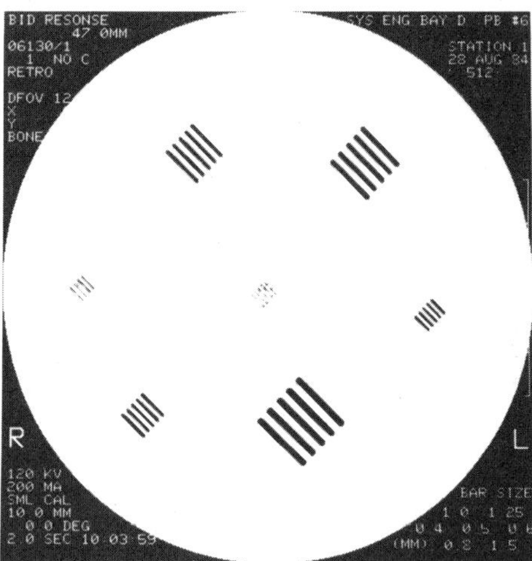

Fig. 20 CT image on a medical CT system of a resolution bar phantom. The center pattern has 0.5 mm (0.020 in.) bars on 1.0 mm (0.040 in.) spacings.

Resolution Test Patterns. A CT resolution bar phantom consists of a stack of alternating high- and low-density layers of decreasing thickness. This phantom is scanned such that the layers are perpendicular to the slice plane. The resultant CT image displays a bar pattern, and the minimum resolvable bar size (layer thickness) is determined (Fig. 20). This measure can be reported as the bar size or as line pairs per millimeter.

Alternatively, resolution hole patterns can be used. This type of phantom is a uniform block of material (usually cylindrical) containing a series of drill holes. The holes are often arranged in rows, with the hole spacings twice the hole diameter. Hole patterns can also be placed in test specimens to demonstrate resolution in the materials to be scanned. As with the bar pattern, the minimum resolvable hole size is determined as a measure of the resolution.

Some differences may exist in the measured results for a system with the different test patterns described. Bar patterns will generally indicate a higher-resolution performance than holes, probably because of the improved perceptibility of the larger structures.

If the system resolution is approaching the display resolution for an image, the relative position of the test structures to the voxels may affect the results. This can be avoided by orienting the test pattern at a 45° angle to the pixel row pattern of the displayed image.

Point Spread Function (PSF). The image produced by a system is a blurred reproduction of the actual object distribution. A more complex way of characterizing resolution and ability to reproduce the object structure is by measuring the degree of image blurring or the PSF of the system. The PSF is the image response to a very small or pointlike feature. This measurement can be made on a CT system by imaging a fine wire oriented perpendicular to the slice plane (additional information on the point spread function is available in the section "Convolution and Filters" in Appendix 1 in this article).

If the pixel spacing is close to the resolution of the system, the extent of the PSF may extend for only a few pixels and be poorly characterized. The PSF can also be determined from measurements of the line spread function (LSF) or edge response function (Ref 31). The LSF is a normalized plot of the image across a line, such as a very thin sheet of metal oriented perpendicular to the slice plane. The edge response function is the plot of the image across a low-density to high-density boundary.

It is helpful in making LSF and edge response function measurements to orient the line or edge slightly out of alignment with the columns of displayed pixels. By determining the relative position of data from several rows, one can obtain a finer sampling of these functions. These functions can be partially characterized as a single parameter by reporting the full width at half the maximum value of the spread function.

Modulation Transfer Function (MTF). Fourier theory states that a signal or an object can be described by a series of sinusoidal functions. The MTF is a plot of the ability of the imaging system to transfer a sinusoidal signal versus the frequency of that signal. The MTF for an imaging system is equivalent to evaluating the frequency response of an audio system, except that the MTF is in terms of spatial frequency (cycles per unit distance).

The bar resolution test measurements provide an approximation to the MTF values. Thick bars show maximum contrast between the high- and low-density materials, corresponding to the difference in linear attenuation coefficient of the materials. As the bars become thinner, the overlap of the blurring at the edges tends to reduce the apparent CT number of the high-density material and raise the values for the low-density material. The contrast or difference in CT number diminishes and can be reported as a fraction of the large bar contrast.

The MTF is similar to a plot of contrast versus cycles per millimeter, except that it is for an idealized test object with a sinusoidal density variation. Because fabricating such a sinusoidal distribution is impractical, the MTF is normally determined by taking the normalized Fourier transform of a measured PSF. In reporting specific values from the MTF function, the spatial frequency (line pairs per millimeter) at which the contrast transfer falls below a specified value (such as 50, 20, or 10%) is sometimes given.

Factors Affecting Resolution. One of the major factors in determining resolution is the spacing between the measured transmitted rays. The resolution at the center of the image field and in the radial direction at the periphery is highly dependent on the ray spacing in a transmission profile. This ray spacing can be easily adjusted in translate-rotate systems by adjusting the linear distance moved between measurements during translation.

Other factors that affect resolution include source size, degree of geometric magnification, detector aperture, and the reconstruction algorithm. The resulting resolution of the reconstructed image may also be limited by the display system.

The measured spatial information is dependent on the sample spacing of the data and on the degree of blurring that occurs in these data. Practical measurements are made with radiation sources of a finite size and a detector with a definable aperture width. A single transmission measurement is therefore a type of average over a ray of some width. This ray profile is dependent on the size and shape of the focal spot, the width of the detector aperture, and the relative position between the source and the detector. An approximation to the effective beam width, known as the effective aperture size, is given by (Ref 32):

$$a_{\text{eff}} = \frac{a^2}{M^2} + \frac{s^2}{[M/(M-1)]^2} \quad \text{(Eq 3)}$$

where a is the detector aperture and s is the width of the x-ray focal spot. The variable, M, is the magnification factor given by:

$$M = \frac{\text{SDD}}{\text{SOD}} \quad \text{(Eq 4)}$$

where SDD is the source-detector distance and SOD is the source-object distance. If the magnification factor is close to 1, as in contact film radiography, the detector resolution is the critical factor. As the magnification increases, a greater burden is placed on having a small x-ray source in order to produce a high-resolution image.

If the sampled data are at a finer spacing than the effective aperture, deconvolution processing of the measured data can reduce the effective aperture somewhat, but the result will be an increase in noise in the data (see the section "Convolutions and Filters" in Appendix 1 of this article). Sample spacings through the object below one-half the effective aperture, however, have little benefit because of the lack of appropriately detailed (high-frequency) information in the measurements (Ref 31).

The sample spacing can also be a limiting factor to the resolution of the system. The Nyquist sampling theorem states that sampling frequency must be at least twice the maximum spatial frequency of the structure being measured. This becomes clear if one thinks of making sampled measurements across the bar resolution pattern. To define the one line-pair per millimeter bars, two samples per millimeter are required—one for the high-density bar and the other for the low-density space between bars.

Given an adequately small effective aperture, the sample spacing within a view is the limiting resolution factor in the radial dimension from the center of the image. However, the circumferential resolution, perpendicular to the radius, is primarily associated with the angular separation between views. This can present particular difficulties near the periphery of the object, where the angular separation of the data corresponds to an increasingly large linear distance along the circumference. The larger the diameter of the object, the larger the number of angular views required to maintain a given circumferential resolution at the edges.

The measured transmission values are processed by a reconstruction algorithm to obtain the image, with the filtered backprojection technique being the common method used. The filter function can be modified or the image postprocessed to yield a certain degree of smoothing in the image in order to minimize the image noise. This smoothing is obviously an additional factor affecting the system resolution.

Also part of the reconstruction process is the backprojection of the projection data into the image matrix. The rays corresponding to the measured data generally do not pass through the center of each affected pixel; therefore, interpolation between measured values is required. The method of interpolation can also affect the image resolution and noise.

The reconstructed image is a two-dimensional array of data with a defined pixel spacing in terms of object dimensions. The Nyquist sampling theorem also applies in this case, and details smaller than the pixel spacing cannot be accurately defined. The pixel spacing can be a significant limitation where a large object is viewed in a single image. The quality of the display system or recording media can also further degrade the resolution contained in the resultant image.

Many factors can affect the final resolution of the images. Appropriate system design and scan parameter selection are required for optimizing the image quality for a given task.

Image Contrast

As with other imaging systems, the amount of contrast on a CT image is governed by object contrast, the contrast resolution of the detector-readout system, and the size of a discontinuity with respect to the spatial resolution of the system. The following sections describe these factors affecting image contrast, along with the use of contrast-detail-dose diagrams for gaging these interrelationships in a particular scanning situation.

Object contrast refers to the contrast generated by variations in the attenuation of the radiation propagating through the testpiece. Object contrast is a function of the relative differences in the linear attenuation coefficients and is affected by the x-ray energy and the composition and density of the materials.

In order for a feature to be detected or identified, it must have a displayed density noticeably different from that of the surrounding material. Computed tomography images are a scaled display of the linear attenuation coefficients of the materials within the specimen. The relative contrast between a feature and the surrounding background material is the normalized difference between the two linear attenuation coefficients, that is:

$$\text{Contrast (\%)} = \frac{|\mu - \mu_b|}{\mu_{ref}} \cdot 100\% \qquad \text{(Eq 5)}$$

where μ and μ_b are the linear attenuation coefficient of the feature of interest and the background material, respectively. The reference coefficient, μ_{ref}, is normally that of the background material, but the maximum coefficient of the two materials can be used, especially if the background material is air or has a very low relative value.

The linear attenuation coefficient values are a function of the physical density of the material, the composition or effective atomic number of the material, and the effective energy of the x-ray beam. The effect of the atomic number on the linear attenuation coefficient is especially important at lower energies (sub-MeV) and for high atomic number materials; in these cases, photoelectric absorption interactions are a significant factor in the overall attenuation. This effect of atomic number can yield a much larger relative contrast at the low energies relative to high-energy scanning; however, sufficient beam energy is required for penetrating the full thickness of the object and collecting an adequate photon signal.

At very high energies, where Compton scattering interactions overwhelmingly dominate, atomic number has relatively little effect. Under these conditions, the linear attenuation coefficient tends to be proportional to the electron density of the material, which corresponds quite closely to the physical density of the material. For situations in which the composition of the material is uniform, the linear attenuation coefficient is directly proportional to the physical density of the material.

Size and Partial Volume Effect. In addition to the true object contrast, other factors also influence the measured and displayed contrast between structures. When the voxel or sampled volume corresponding to a pixel contains two or more separate structures, the resulting pixel value is a volume average of the structures. This is one of the causes for the rounded profile or intermediate CT numbers found along boundaries. The edge pixels correspond to a combination of the high- and low-density materials.

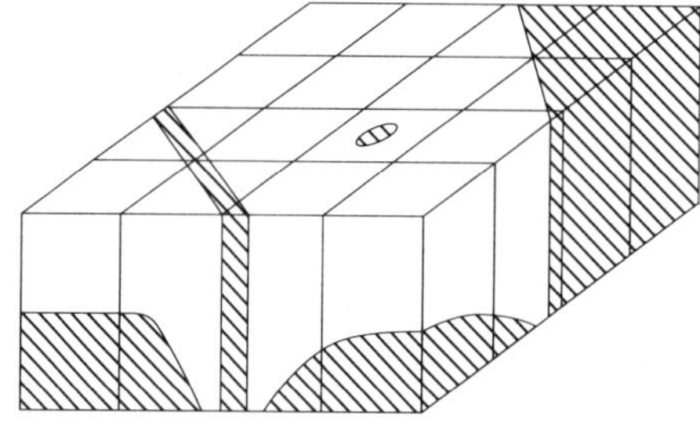

Fig. 21 Schematic of structures partially filling a sampled voxel. The structures yield a measured CT value that correspond to a volume average of the two materials.

With small features, such as subresolution inclusions or cracks, only part of the voxel is occupied by this structure (Fig. 21). In these cases, the feature may be detectable, but the measured or apparent contrast is less than the difference between the two materials. The smaller the feature, the less the imaged contrast. For situations in which the CT number values are known for the two materials, this partial volume effect can be used to estimate the size of the feature or to measure structural dimensions with subresolution accuracy.

The resolution characteristics of the imaging system and the degree of blurring in the image also reduce the apparent contrast difference for small structures. Where the ideal image of a small structure would be a single pixel with a CT number corresponding to the average linear attenuation coefficient within the voxel, the image blurring spreads this measured signal over a large area. This can be thought of as an extension of the partial volume effect, in which the volume corresponding to a pixel is the three-dimensional sensitivity distribution of object locations affecting the pixel value.

Noise and Contrast Sensitivity. The contrast sensitivity of the imaging system is defined by the signal-to-noise ratio and is affected by unwanted random variations in the measured data and the reconstructed image. Image noise is a primary limiting factor in distinguishing low-contrast structures and in detecting high-contrast subresolution features that appear in the image as low-contrast structures. The random uncertainty of the CT numbers causes subtle features to be lost in the texture of the noise.

A primary contributor to image noise is the quantum noise (or mottle) in the measured data due to measurement of a finite number of x-ray photons (Eq 1). Additional sources of noise are the data acquisition system electronics and roundoff errors in the computer processing.

Measurement of Noise. The typical measurement of noise is the standard deviation of the CT number values over an area of many pixels. The CT number standard deviation is commonly measured near the center of the cylindrical test object and may vary with radial position. Because the results are dependent on the quantity of x-rays transmitted through the object, the size and material of the test object should be specified. The noise is often reported as a percentage of the CT number difference between the test object and air.

A more complete method of characterizing the noise is to determine the noise power spectrum of the system over a series of images. The noise power spectrum defines the noise content in the image versus spatial frequency. As a consequence of the filtering operation of the reconstruction process, CT images predominantly contain high-frequency noise.

Factors Affecting Noise. Because x-ray photon statistics (that is, quantum noise) is a major contributor to the noise, factors that increase the number of photons detected tend to reduce the noise (Eq 1). The factors affecting the number of detected photons are:

- The intensity of the radiation source
- The source-to-detector distance
- The attenuation of radiation passing through the testpiece (which in turn is a function of the thickness and material of the testpiece and the energy of the incident radiation)
- The overall detection efficiency of the detector

The overall efficiency with which the useful x-ray beam is captured and measured depends on the geometrical, absorption, and conversion efficiencies of the detector. The geometrical efficiency increases with the number of detectors and the detector aperture width. The absorption and conversion efficiencies depend on the detector design, including the size and material composition of the detector. The number of photons detected is also directly proportional to the scanned slice thickness.

In addition to the transmitted primary photons, which constitute the x-ray signal, a certain amount of the detected radiation is scattered radiation. This scattered radiation reduces contrast by creating a field of background noise. Reduction of scatter, by tight collimation or increasing the object-to-detector distance, reduces this contribution to noise.

The reconstruction process also affects the resultant image noise. For ideal noiseless data, the shape of the reconstruction filter is a high-pass ramp function, which amplifies the high-frequency noise. Because the signal content at these high frequencies may be low due to other factors (such as the effective detector aperture), image quality can be improved for certain imaging tasks by windowing this ramp function and by reducing the amplification of the high-frequency signal and noise. This is effectively the same as smoothing the final image. A CT system may have selectable filters or algorithms that provide variable degrees of smoothing.

The high-frequency content of the noise caused by the reconstruction filter is negatively correlated. This means that if one pixel has a large positive noise fluctuation, surrounding pixels are more likely to have a negative fluctuation. Because the image noise is negatively correlated, filter windowing or smoothing can be particularly advantageous in improving the detectability of low-contrast structures (Ref 33). For situations in which the resolution is limited by the reconstruction process and the effect of the effective aperture is small, the number of detected photons required to maintain a given signal-to-noise ratio is inversely proportional to the resolution cubed (Ref 34). Under these conditions, improving the resolution by a factor of two has the same effect on the signal-to-noise ratio as increasing the number of photons detected by a factor of eight.

Other limitations in system implementation can also contribute to the noise. The high-sensitivity detector and data acquisition system electronics adds some electronic noise to the signal. The reconstruction process consists of a large number of computational operations, and the roundoff errors in these calculations can also contribute to image noise. In addition, noise included in reference measurements used to normalize the scan projection data adds to the noise of the reconstructed image.

Contrast-Detail-Dose (CDD) Diagram. The perception of a feature is dependent on the contrast between the feature and the background, the size of the feature, the noise in the image, and the display window settings used to view the image data. One method of gaging the range of structures that may be resolved is by determining the CDD characteristics for a particular scanning situation (Ref 35, 36).

The CDD diagram is a plot of the imaged percent contrast versus the diameter of resolvable cylindrical test structures. Figure 22 shows a CDD diagram for two similar systems with different detector spacings. Figure 22 also shows the effect of changing the object dose, or number of photons collected, on the ability to resolve features.

For high-contrast structures, the ability to resolve the features has little dependence on the dose or image noise, but is dependent on the resolution characteristics of the system. System A, which resolves smaller-diameter features, has a smaller detector aperture and sample spacing than system B (Fig. 22).

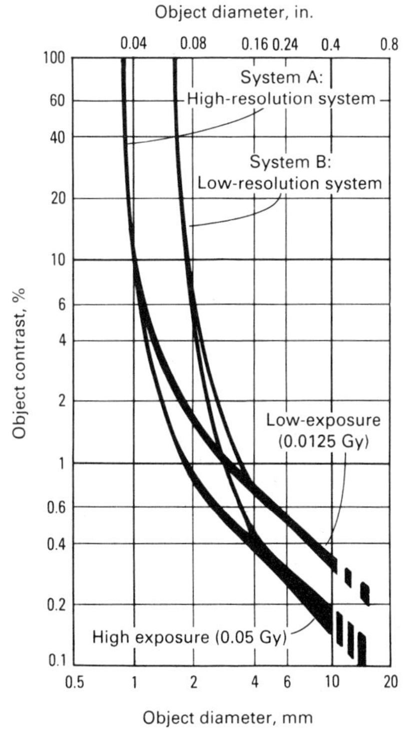

Fig. 22 CDD diagram for two similar CT systems. The resolvability is limited by system resolution for high-contrast objects and by noise for low-contrast structures.

As the contrast of the features decreases and the feature contrast approaches the noise variation in the image, larger structures are required in order to distinguish them from the surrounding noise (Ref 37, 38). With the low-contrast structures, the importance of the overall resolution capability of the system is reduced, and the level of noise in the image becomes more critical. The two systems in Fig. 22 have similar performance for low-contrast features, with the object dose and the level of noise in the image becoming the dominant factor.

System resolution can be measured for different contrast structures with a low-contrast detectability test phantom such as the one shown in Fig. 23. This image demonstrates the ability of computed tomography to display very low contrast, large-area structures. The measured results of these tests depend on factors that affect the structure contrast versus the noise, including the size and the material of the test phantom.

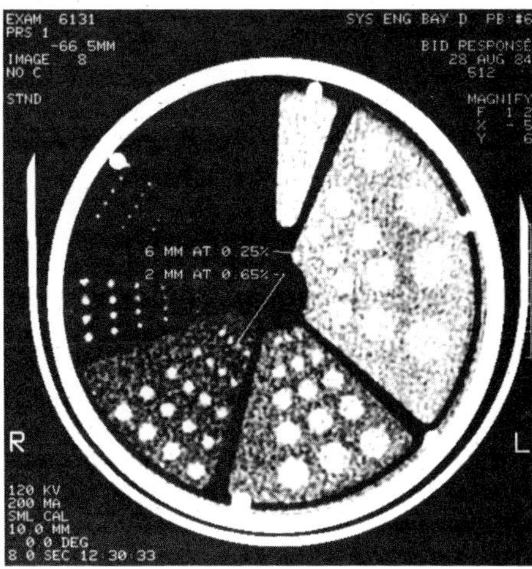

Fig. 23 CT image of low-contrast detectability test phantom

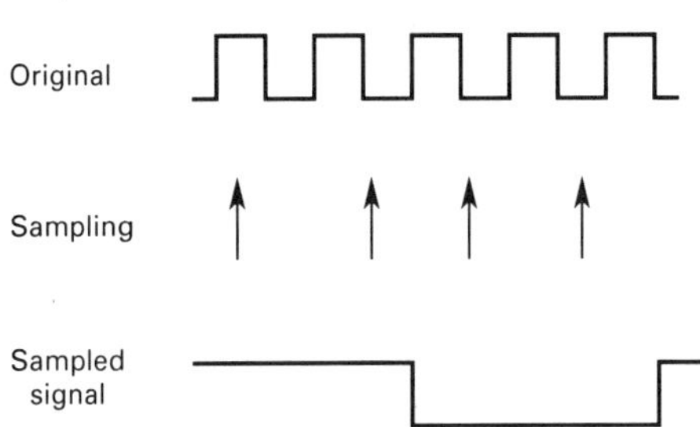

Fig. 24 Illustration of undersampling of a distribution containing high-frequency information causing the false appearance of low-frequency information or aliasing

Image Artifacts

Artifacts are image features that do not correspond to physical structures in the object. All imaging techniques are subject to certain types of artifacts. Computed tomography imaging can be especially susceptible to artifacts because of its sensitivity to small object differences and because each image point is calculated from a large number of measurements (Ref 39).

The types of artifacts that may be produced and the factors causing them must be understood in order to prevent their interpretation as physical structures and to know how scan operating parameters can be adjusted to reduce or eliminate certain image artifacts. In general, the two major causes of CT image artifacts are the finite amount of data for generating the reconstructed image and the systemic errors in the CT process and hardware. Some of the factors that can cause the generation of artifacts include inaccuracies in the geometry, beam hardening, aliasing, and partial penetration (Ref 40-42).

Aliasing and Gibbs Phenomena. Because of the finite number of measurements, certain restrictions are placed on the resolvable feature size. A series of adjacent x-ray transmission measurements allows the characterization of object structures that are larger than this sample spacing. Coarse sampling of detailed features can create falsely characterized data. The effect of this undersampling of the data is termed aliasing, in which high-frequency structures mimic and are recorded as low-frequency data. This effect can cause image artifacts that appear as a series of coarsely spaced waves or herringbone-type patterns over sections of the image (Fig. 24).

Computed tomography systems do not measure the transmission data for infinitely thin rays. The detector has an aperture of some finite thickness, and very fine structures within this broad ray tend to be smoothed out or eliminated from the signal before being measured. This effective smoothing of the data prior to sampling acts to limit the spatial frequency content prior to sampling and minimizes the potential for aliasing artifacts.

Aliasing can also occur as a result of the reconstruction process. If one attempts to reconstruct an image in which the pixel spacing is much greater than the finely sampled projection data, the high-frequency information in the projection data can cause aliasing artifacts in the image. This can be eliminated through appropriate selection of the reconstruction filter function to filter out or eliminate spatial frequencies higher than that contained in the reconstructed image matrix.

The Gibbs phenomenon is also a type of artifact associated with the limited spatial frequency response of the system. It is a consequence of a sharp frequency cutoff in the reconstruction filter function. This causes overshooting or oscillations along high-contrast boundaries. The range and spacing of these oscillations decrease as the cutoff frequency increases. This type of artifact can be minimized by windowing the reconstruction filter function such that it has a smooth roll-off, rather than a sharp cutoff, at the high frequencies.

View aliasing is a form of aliasing resulting from an insufficient number of angular views. This artifact is typically a radial streaking pattern that is apparent toward the outer edges of the image or emanates from a high-contrast structure within the cross-sectional slice. The number of angular views required is dependent on the diameter of the object, the circumferential resolution of the image, and the effective aperture width.

Partial Volume Artifacts. The partial volume effect is the volume averaging of the linear attenuation coefficients of multiple structures within a voxel. Because of the exponential nature of the x-ray attenuation, the actual averaging process can cause nonlinearities or inconsistencies in the measured projection data that produce streaking artifacts in the image.

In Eq 8 to 10 in Appendix 1, the attenuation through a series of materials is shown as the exponential of the sum of individual material attenuation properties. This relationship is used to obtain the line integral of the linear attenuation coefficients of the object along a ray, called a projection value. With the finite size of the measured x-ray beam, if two materials are positioned such that part of the measured ray passes through one material and the other part of the ray passes through the other material, the attenuation equation is the sum of exponential terms rather than the exponential of a sum.

Situations in which only part of a measured ray passes through a structure include rays tangential to high-contrast boundaries and object structures that may penetrate only partially into the measured slice thickness (Fig. 21). If the two materials have considerably different attenuation coefficient values, this inconsistency in the measured data can become significant and cause streaks emanating along edges, from high-contrast features, or between high-contrast features. Artifacts can be particularly prevalent between multiple high-contrast features that only partially intersect the slice thickness.

Partial volume artifacts can be reduced by decreasing the effective ray size with thinner slice thickness and narrower effective aperture. Smoothing or windowing the reconstruction filter also reduces the sensitivity to these artifacts. Another approach is to reduce the contrast of the boundaries by immersing or padding the object with a material having intermediate or similar attenuation values.

Beam Hardening. X-ray sources produce radiation with a range of photon energies up to the maximum energy of the electron beam producing the radiation. The lower-energy, or soft, photons tend to be less penetrating and are attenuated to a greater degree by an object than the higher-energy photons. Consequently, the effective energy of a beam passing through a thick object section is higher than that of a beam traversing a thin section. This preferential transmission of the higher-energy photons and the resulting increase in effective energy is referred to as beam hardening.

Changes in the effective energy of the x-ray beam due to the degree of attenuation in the object cause inconsistencies in the measured data (Ref 43). X-rays that pass through the center of a cylindrical object will have a higher effective energy than those traversing the periphery. This leads to a lower measured linear attenuation coefficient and lower CT number values in the center of the reconstructed image. This CT number shading artifact is referred to as cupping, corresponding to the shape of a plot of CT numbers across the object.

The effect of beam hardening can be reduced with several techniques. The original EMI scanner used a constant-length water bath that yielded a relatively uniform degree of attenuation from the center to the periphery. The addition of x-ray beam filtration reduces the soft x-rays and reduces the degree of beam hardening. Compensating or bow tie shaped x-ray filters are sometimes used in medical CT systems to increase the beam filtration toward the periphery and to have a smaller attenuation variation across a cylindrical-shaped object. Normalizing the data with a cylindrical object of similar size and material as the test object also reduces the effect of the beam hardening.

Beam hardening is often compensated for in the processing software. If the material being scanned is known, the measured transmission value can be empirically corrected. Difficulty occurs, however, when the object consists of several materials with widely varying effective atomic numbers. The extent of the beam hardening depends on the relative degree the attenuation is from the highly energy dependent and atomic number dependent photoelectric absorption versus the less energy dependent and atomic number independent Compton scattering process. Different materials have differing degrees of attenuation of the low-energy versus high-energy photons for a given overall level of attenuation.

Other processing techniques for minimizing beam hardening artifacts are sometimes used. If the object is composed of two specific materials that are readily identifiable in the image, an iterative beam hardening correction can be implemented (Ref 44). This approach identifies the distribution of the second material in the image and implements a correction of the projection data for an improved second reconstruction of the data. Dual energy techniques can also be used. These techniques use data obtained at multiple energies to determine the relative photoelectric and Compton attenuations and can produce images that are fully corrected for beam hardening.

Scatter Radiation. Detected scatter radiation produces a false detected signal that does not correspond to the transmitted intensity along the measured ray. The amount of scatter radiation detected is much lower than that encountered with large-area radiographs because of the thin fan beam normally used in computed tomography. The sensitivity of computed tomography, however, makes even the low levels of scatter a potential problem.

The scatter contribution across the detector array tends to be a slowly varying additive signal. The effect on the measured data is most significant for highly attenuated rays, in which the scatter signal is relatively large compared to the primary signal. The additional scattered photons measured make the materials along the measured ray appear less attenuating, which is the same effect caused by beam hardening.

The types of artifacts caused by scatter radiation are similar to and often associated with beam hardening. In addition to the cupping-type artifact, beam hardening and scatter can also cause broad, low CT number bands between high-density structures. The rays that pass through both of the high-density structures are highly attenuated, and the increase in effective energy and the scatter signal makes the materials along these rays appear to be less attenuating than is determined from other view angles.

Because of the similarities of the effect, the basic beam hardening correction often provides some degree of compensation for scatter. The amount of scatter radiation detected can be modeled, and specific scatter correction processing can be implemented. Fundamental dual energy processing, however, does not correct for the detected scatter radiation.

Control of Scattered Radiation. System design can also minimize the detected scatter radiation. A tightly collimated, thin x-ray beam minimizes the amount of scatter radiation produced. Increasing the distance between the object and the detector decreases the fraction of the scatter radiation that is detected. The use of focused collimators or directionally sensitive detectors also reduces the amount of detected scatter radiation. Because the scatter radiation varies slowly with distance, special reference detectors outside of the primary radiation beam can be used to measure the level of the scatter signal.

Other Systemic Errors. A number of other systemic errors in data measurement can produce inconsistent data and can result in artifacts or degraded image quality. These inconsistencies may be in the measured signal or may be the result of poor characterization of the spatial position of the measured ray paths.

In medical CT imaging, patient motion is a major concern. If the object being scanned moves or changes configuration during data acquisition, the measured data set does not correspond to a specific object. In addition to image blurring, streaking from boundaries and high-density structures occurs. This is less of a problem with rigid structures in industrial imaging. However, objects undergoing dynamic changes relative to the scan time, as in a fluid flow study, or poorly fixtured components that wobble or vibrate during data acquisition can degrade the image quality.

Data are acquired along many ray paths, and the precise combination of this data produces the reconstructed image. Imprecise manipulator motions, geometric misalignment, or insufficiently characterized positioning can also cause blurring and image artifacts. An example is a series of tangential streaks forming a star pattern on translate-rotate systems that can be caused by misalignments between traverses.

Inconsistencies in the data measurements may be due to variations in x-ray source output, the detector, or the detector electronics. The x-ray tube output is normally measured by reference detectors, and the data are scaled appropriately. This is effective for tube current changes that do not alter the effective energy of the x-ray beam. Changes in the tube voltage, however, produce changes in both x-ray intensity and effective energy and are more difficult to correct. A highly stable x-ray generator with feedback control of the voltage is normally required.

Measurement errors may also be the result of nonlinear detector performance over its entire dynamic range, detector overranging, or differences between detector elements in a multidetector array. Detector overranging for a limited number of measurements typically produces well-defined streaks through the image. Third-generation, rotate-only CT systems in particular require a highly consistent detector array because uncharacterized variations between detectors cause concentric ring artifacts. Oscillating changes in the detected signal can also occur because of mechanical vibration of the detector or components within the detector or because of the pickup of electronic interference, especially at the line voltage frequency.

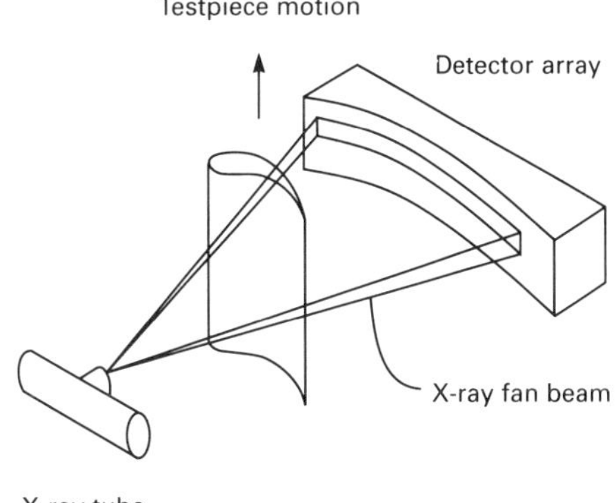

Fig. 25 Digital radiography mode. The component moves perpendicular to the fan beam, and the radiographic data are acquired line-by-line.

Special Features

The components that constitute a CT system often allow for additional flexibility and capabilities beyond providing the cross-sectional CT images. The x-ray source, detector, and manipulation system provide the ability to acquire conventional radiographic images. The acquisition of digital images along with the computer system facilitates the use of image processing and automated analysis. In addition, the availability of the cross-sectional data permits three-dimensional data processing, image generation, and analysis.

Digital Radiography. One of the limitations of CT inspection is that the CT image provides detailed information only over the limited volume of the cross-sectional slice. Full inspection of the entire volume of a component with computed tomography requires many slices, limiting the inspection throughput of the system. Therefore, CT equipment is often used in a DR mode during production operations, with the CT imaging mode used for specific critical areas or to obtain more detailed information on indications found in the DR image. Digital radiography capabilities and throughput can be significant operational considerations for the overall system usage.

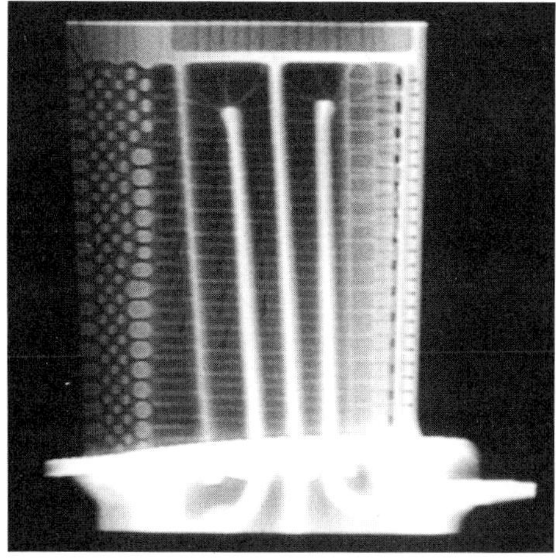

Fig. 26 Digital radiograph of an aircraft engine turbine blade (nickel alloy precision casting) from an industrial CT system

Computed tomography systems generally provide a DR imaging mode, producing a two-dimensional radiographic image of the overall testpiece. The high-resolution detector of the rotate-only systems normally requires a single z-axis translation (Fig. 25) to produce a high-quality DR image. For even higher resolution, interleaved data can be obtained by repeating with a shift of a fraction of the detector spacing. For large objects, a scan-shift-scan approach can be used. The translate-rotate systems are typically less efficient at acquiring DR data because the wider detector spacing requires multiple scans, with shifts to provide adequate interleaving of data. If the width of the radiographic field does not cover the full width of the object, the object can be translated and the sequence repeated.

The capabilities and use of these DR images are generally the same as discussed for radiographic inspection. The method of data acquisition is different on the CT systems; the data are acquired as a sequence of lines or line segments. The use of a thin fan beam with a slit-scan data acquisition is a very effective method of reducing the relative amount of measured scatter radiation. This can significantly improve the overall image contrast (Fig. 26). In addition, the data acquisition requirements for CT systems provide for a high sensitivity and very wide dynamic range detector system.

Image Processing and Analysis. Because the DR and CT images are stored as a matrix of numbers, the computer can be used as a tool to obtain specific image information, to enhance images, or to assist in analyzing image data.

Display Features. A number of display features are used to obtain or present specific image information. A region-of-interest program defines a portion of the image and provides information on the pixel values within the region, such as minimum, maximum, and average CT number values; the standard deviation of the CT numbers; or the area or number of pixels contained in the region. A histogram plot can display the relative frequency of occurrence of CT number values.

Cursor functions can be used to point to specific features, to annotate the images, and to determine coordinate locations, distances between points, or angles between lines. A plot or profile can be generated from the CT numbers along a defined line.

Image Processing. The image itself can be processed to enhance boundaries and details or to smooth the image to reduce noise. Linear operators or filters used to sharpen the image, however, will increase the image noise, and smoothing filters that reduce noise also blur or decrease the sharpness of the image. These processing steps, however, can assist in improving the visibility of specific types of structures. Nonlinear processing techniques can be used to enhance or smooth the image while minimizing the detrimental effects of the processing. Nonlinear techniques include median filtering and smoothing limited to statistically similar pixel values; both techniques reduce the image noise while maintaining sharper boundary edges. These nonlinear techniques, however, typically require more computation and can be difficult to implement in fast array processors.

Image Analysis. Digital images can be analyzed for specific features or to measure definable parameters contained in the image data. Where the image-processing capabilities are generic for any image, automated image analysis software is written to analyze specific features in specific components. The software can verify the presence of specific necessary components or can search for defined indications. Automated analysis is generally computational intensive, and identifying a broad range of indications is a highly complex task. Consequently, the automated analysis of flaw indications has generally been limited to a few, narrowly scoped analysis tasks.

The automated measurement of specific design parameters of components can be more readily defined and implemented. The cross-sectional presentation of the component structure in the CT image allows for the measurement of critical dimensions, wall thicknesses, cord lengths, and curvatures. The automated measurement of critical wall thicknesses on complex precision castings is a standard feature on CT systems used in the manufacture of aircraft engine turbine blades (Fig. 10).

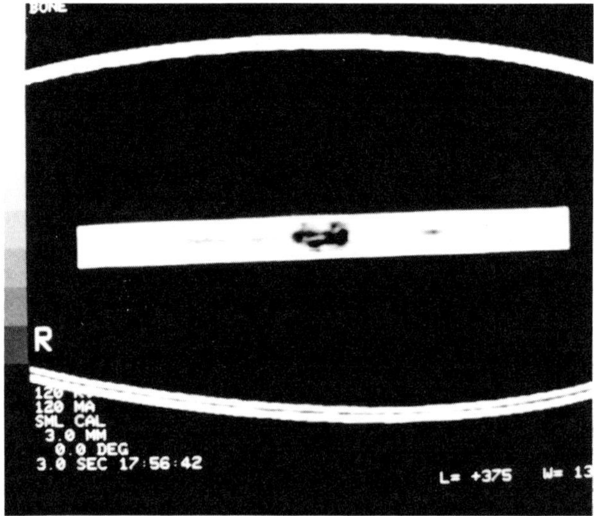

(a)

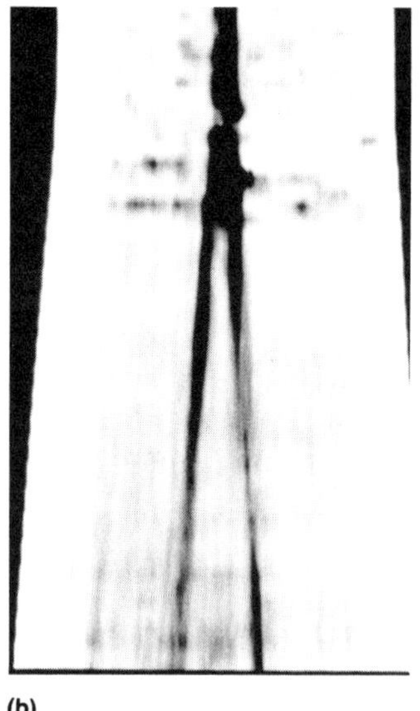

(b)

Fig. 27 CT image (a) across a sample helicopter tail rotor blade showing outer fiberglass airfoil and center composite spar. (b) Planar reformation through the composite spar from a series of CT slices. The dark vertical lines are normal cloth layup boundaries, while the mottling at the top is due to interplanar waviness of the cloth layup.

Planar and Three-Dimensional Image Reformation. Having a stack of adjacent cross-sectional CT slices characterizes the density distribution within the scanned volume in three dimensions. With this set of data in the computer, alternate planes through the object can be defined, and the CT image data corresponding to these planes can be assembled. These CT planar reformations allow the presentation of CT images in planes other than the planes in which the data were originally acquired, including planes that CT data acquisition would not be feasible because of component size and shape (Fig. 27). The CT image data can also be presented along other nonplanar surfaces. The CT data that correspond to a conical surface on a rocket exit cone or along other structures can be presented. The data can also be reformatted to alternate coordinates, such as presenting the data for a tubular section with the horizontal axis of the display corresponding to an angular position and with the radial distance along the vertical axis of the display.

Structures within a scanned object can also be specifically characterized. Three-dimensional surface imaging identifies the surface of a structure in a stack of CT images, defines surface tiles corresponding to this surface, and produces a computer-assisted design perspective display of the structure, as shown in Fig. 11 (Ref 45). Because the CT images also display interior surfaces, the computer can be used to slice open these surface models to display interior features. These capabilities hold the potential for improving the component design cycle by documenting the configuration of physical prototypes and operational components.

Computed tomography can also be used to define actual components, including their internal density distribution, as a direct input for finite-element analysis (FEA). Automated meshing techniques are being analyzed that could significantly reduce the time required to generate FEA models. In addition, this direct FEA modeling of actual components could assist in the nondestructive analysis and characterization of components that are to be failure tested and may allow improved calibration of the engineering models from the test data.

Dual Energy Imaging. In the range of x-ray energies used in industrial computed tomography, attenuation of the x-ray photons predominantly occurs by either photoelectric absorption or Compton scattering. The probability of attenuation due to photoelectric absorption decreases more rapidly than the probability of attenuation due to Compton scattering. Consequently, photoelectric absorption predominates at low energies, and Compton scattering is the primary interaction for high-energy photons. In addition, the probability for attenuation by photoelectic absorption is highly dependent on the atomic number of the absorber, while Compton scattering is relatively independent of atomic number.

These differences between photoelectric and Compton attenuation cause difficulties in correcting x-ray transmission data for beam hardening in structures having multiple materials. This difference in the attenuation process can be used, however, to obtain additional information on the composition of the scanned object (Ref 46). If the object is scanned at two separate energies, the data can be processed to determine a pair of images corresponding to the photoelectric and Compton attenuation differences. The pair of images can be a photoelectric and Compton image or can correspond to physical density and effective atomic number. This pair of basic images can be combined to form a CT image without beam hardening variations.

The data for the basis images are a result of the differences in the high- and low-energy scans and are highly sensitive to random variations. As a consequence, these basis images tend to have a very high noise level. The image noise level for the combined pseudo-monoenergetic image, however, is equivalent to or lower than the noise in either of the original single energy images. Accurate implementation of dual energy processing requires consistent data between the high- and low-energy scans and characterization of the differences, such as that due to the detection of Compton scatter. Dual energy imaging can also be implemented with the DR data to determine the basis data corresponding to the sum along the measured ray paths.

Partial Angle Imaging. All of the CT imaging techniques discussed have considered the ability to obtain transmission data from all angles in the plane of the cross-sectional slice. This is highly suitable for objects that can be readily contained within a tight cylindrical workspace, but it can be impractical for large planar structures. Other components, such as large ring and tubular structures, can be conventionally scanned given a large enough workspace, but it is advantageous to minimize the source-to-detector distance and to image through a single wall in order to have reasonable x-ray intensities and inspection throughput.

Methods have been investigated for medical imaging for reconstruction from partial data sets. Industrial imaging has the advantage, however, in being able to apply *a priori* information from the component design or from measured external contours to compensate for the missing data (Ref 47).

Another related approach is to use methods based on the focal-plane tomography or laminography techniques developed in the early 1920s. Focal-plane tomography involves moving the x-ray source and film relative to the object such that features in the object are blurred, except for the features in the focal plane. This method does not necessarily eliminate all of the structures outside of the focal plane as in computed tomography, however, the data are obtained without circling the object.

The capabilities of the conventional focal-plane tomography approach can be enhanced by using digital processing techniques. With a series of digital radiographs, the images can be shifted and combined within the computer to yield a series of focal planes at different levels in the object from one set of data. In addition, image filtering can be applied, similar to the filtering in the filtered-backprojection CT reconstruction, to enhance the features in the focal plane and to improve the elimination of out of plane structures.

Appendix 1: CT Reconstruction Techniques

Computed tomography requires the reconstruction of a two-dimensional image from the set of one-dimensional radiation measurements. It is useful to know some of the basic concepts of this process in order to understand the information presented in the CT image and the factors that can affect image quality.

The reconstruction process consists of two basic steps:

- The conversion of the measured radiation intensities into projection data that correspond to the sum or projection of x-ray densities along a ray path
- The processing of the projection data with a reconstruction algorithm to determine the point-by-point distribution of the x-ray densities in the two-dimensional image of the cross-sectional slice

The development of projection data is common to all CT reconstruction techniques, while the reconstruction algorithm can be approached by one of several mathematical methods (Ref 48-50).

Of the various reconstruction algorithms developed for computed tomography, there are two basic methods: transform methods and iterative methods. Transform methods are based on analytical inversion formulas of the projection data values given by Eq 9. Transform methods are fast compared to iterative methods and produce good-quality images. The two main types of transform methods are the filtered-backprojection algorithm and the direct Fourier algorithm. The filtered-backprojection technique is the most commonly used method in industrial CT systems.

Projection Data

The first step in the reconstruction of a CT image is the calculation of projection data. The measured transmitted intensity data are normalized by the expected unattenuated intensity (intensity without the object). A logarithm is taken of these relative intensity measurements to obtain the projection data. The projection data values correspond to the integral or sum of the linear attenuation coefficient values along the line of the transmitted radiation. The reconstruction process seeks to determine the distribution of linear attenuation coefficients (or x-ray densities) in the object that would produce the measured set of transmission values.

The projection data values for a narrow monoenergetic beam of x-ray radiation can be theoretically modelled by first considering Lambert's law of absorption:

$$I = I_0 \exp(-\mu s) \qquad \text{(Eq 6)}$$

where I is the intensity of the beam transmitted through the absorber, I_0 is the initial intensity of the beam, s is the thickness of the absorber, and μ is the linear attenuation coefficient of the absorber material. The linear attenuation coefficient corresponds to the fraction of a radiation beam per unit thickness that a thin absorber will attenuate (absorb and scatter). This coefficient is dependent on the atomic number of the materials and on the energy of the x-ray beam and is proportional to the density of the absorber.

If instead of a single homogenous absorber there were a series of absorbers, each of thickness s, the overall transmitted intensity would be:

$$I = I_0 \exp[(\mu_1 + \mu_2 + \mu_3 \ldots + \mu_i)s] \qquad \text{(Eq 7)}$$

where μ_i is the linear attenuation coefficient of the ith absorber.

Considering a two-dimensional section through an object of interest, the linear attenuation coefficients of the material distribution in this section can be represented by the function, $\mu(x, y)$, where x and y are the Cartesian coordinates specifying the location of points in the section. Using the integral equivalent to Eq 7, the intensity of radiation transmitted along a particular path is given by:

$$I = I_0 \exp\left[-\int_{\text{source}}^{\text{detector}} \mu(x,y)\, ds\right] \qquad \text{(Eq 8)}$$

where ds is the differential of the path length along the ray. The objective of the reconstruction program in a CT system is to determine the distribution of $\mu(x, y)$ from a series of intensity measurements through the section. Dividing both sides of Eq 8 by I_0 and taking the natural logarithm of both sides of the equation yields the projection value, p, along the ray:

$$P = \ln\left(\frac{I_0}{I}\right) = \int_{\text{source}}^{\text{detector}} \mu(x,y)\, ds \qquad \text{(Eq 9)}$$

Equation 9, known as the Radon transformation of $\mu(x, y)$, is a fundamental equation of the CT process. It states that taking the logarithm of the ratio of the unattenuated intensity to the transmitted intensity yields the line integral along the path of the radiation through the two-dimensional distribution of linear attenuation coefficients. The

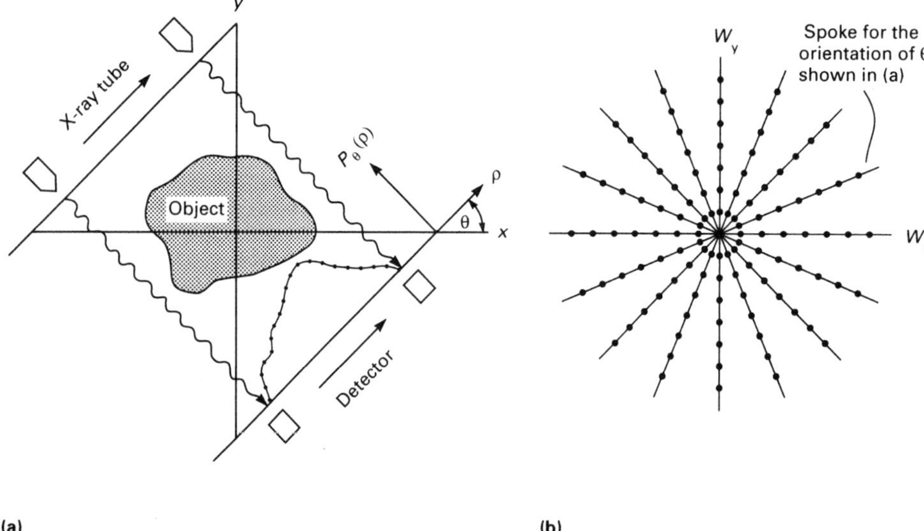

Fig. 28 Data points in (a) direct space and (b) frequency space. The data points obtained in the orientation shown in (a) correspond to the data points on one spoke in the two-dimensional Fourier transform space (b).

inversion of Eq 9 was solved in 1917, when Radon demonstrated in principle that $\mu(x, y)$ could be determined analytically from an infinite set of these line integrals. Similarly, given a sufficient number of projection values (or line integrals) in tomographic imaging, the cross-sectional distribution, $\mu(x, y)$, can be estimated from a finite set of projection values.

The projection value as determined by Eq 9 is based on several assumptions that are not necessarily true in making practical measurements. If an x-ray source is used, the x-ray photon energies or spectra range from very low energies up to energies corresponding to the operating voltage of the x-ray tube. The energy of a beam can be characterized by an effective or average energy. The effective energy, as well as the associated linear attenuation coefficients of the material, will vary with the amount of material of filtration through which the beam passes. The increase in effective energy caused by the preferential absorption of the less penetrating, lower-energy photons when passing through increasing thicknesses of material is referred to as beam hardening. Beam hardening can cause nonlinearities in the measured projection values relative to Eq 9 and can cause shading artifacts in the image. With knowledge of the type of material in the object being imaged, corrections can be made to minimize this effect.

Other variations may also occur in measuring the transmitted intensity values and determining the projection values. These include the measurement of scatter radiation, the stability of the x-ray source, or any intensity or time-dependent nonlinearities of the detector. To the extent that these systemic variations can be characterized or monitored, they can be corrected in the data processing. Another variance that cannot be eliminated is the statistical noise of the measurement due to the detection of a finite number of photons.

Direct Fourier Reconstruction Technique

The direct Fourier reconstruction technique utilizes the Fourier transformation of projection data. A Fourier transform is a mathematical operation that converts the object distribution defined in spatial coordinates into an equivalent sinusoidal amplitude and phase distribution in spatial frequency coordinates. The one-dimensional Fourier transformation of a set of projection data at a particular angle, θ, is described mathematically by:

$$P(\rho, \theta) = \int_{-\infty}^{\infty} p(r,\theta) \exp[-2\pi\rho r]\, dr \qquad \text{(Eq 10)}$$

where r is the spatial position along the set of projection data and ρ is the corresponding spatial frequency variable.

Fourier reconstruction techniques (and the filtered-backprojection method) are based on a mathematical relationship known as the central projection theorem or central slice theorem. This theorem states that the Fourier transform of a one-dimensional projection through a two-dimensional distribution is mathematically equivalent to the values along a radial line through the two-dimensional Fourier transform of the original distribution. For example, given one set of projection data measurements (Eq 9) through an object at a particular angle, taking the one-dimensional Fourier transform (Eq 10) of this data profile provides data values along one spoke in frequency space (Fig. 28). Repeating this process for a number of angles defines the two-dimensional Fourier transform of the object distribution in polar coordinates (Fig. 28b). Taking the inverse two-dimensional Fourier transform of this polar data array yields the object distribution in spatial coordinates.

Although the direct Fourier technique is potentially the fastest method, the technique has not yet achieved the image quality of the filtered-backprojection method, because of interpolation problems. Typical computer methods and display systems are based on rectangular grids rather than polar distributions, and several variations of the direct Fourier reconstruction technique exist for interpolating the data from polar coordinates to a Cartesian grid (usually interpolating the data in the spatial frequency domain). The interpolation techniques can be complex, and the quality of the images from measured data have generally been poorer than some of the other reconstruction methods. Interesting results have been obtained in industrial multiplanar microtomography research, but this method is not typically used in commercial systems (Ref 51).

Filtered-Backprojection Technique

The filtered-backprojection technique is the most commonly used CT reconstruction algorithm. Before discussing this method, a brief discussion of filtering and the simple backprojection method is provided.

Convolutions and Filters. A convolution is a mathematical operation in which one function is smeared by another function. Mathematically it can be defined as:

$$g(x) = f(x) * h(x) \qquad \text{(Eq 11)}$$

$$g(x) = \int_{-\infty}^{\infty} f(x)\, h(x-u)\, du \qquad \text{(Eq 12)}$$

for one dimension, or

$$g(x,y) = f(x,y) * h(x,y) \qquad \text{(Eq 13)}$$

$$g(x,y) = \int_{-\infty}^{\infty}\int_{-\infty}^{\infty} f(x,y)\, h(x-u,y-v)\, du\, dv \qquad \text{(Eq 14)}$$

for two dimensions where * is the symbol for the convolution operation. If the system is a digital system with a discrete number of samples, the corresponding equation to Eq 12 is:

$$g_i = \sum_{k=-\infty}^{\infty} f_i\, h_{i+k} \qquad \text{(Eq 15)}$$

and two-dimensional equivalent to Eq 13 is:

$$g_{i,j} = \sum_{k=-\infty}^{\infty} \sum_{l=-\infty}^{\infty} f_{i,j}\, h_{i+k,j+l} \qquad \text{(Eq 16)}$$

Convolutions are a common physical operation in data acquisition and imaging systems whether or not the term convolution is used. All imaging systems have limitations in their ability to reproduce an

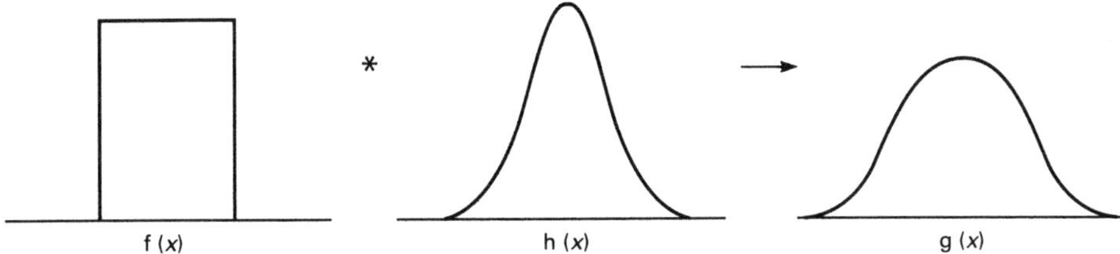

Fig. 29 Convolution operation (*) in which a distribution f(x) is blurred by a function h(x) to form the blurred distribution g(x). The function h(x) is analogous to the PSF of an image system or a smoothing convolution filter in image processing.

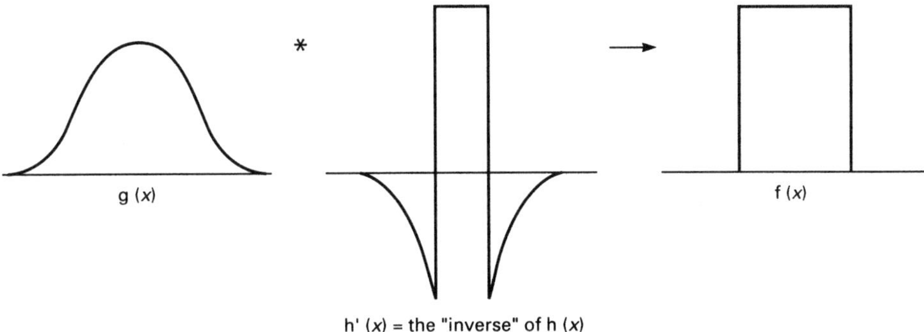

Fig. 30 Convolution operation with the sharpening filter h′(x). This sharpening of g(x) [or a restoration of f(x) with the "inverse" of h(x)] is analogous to image process filtering or to the filtering of the projection data in the filtered-backprojection reconstruction technique.

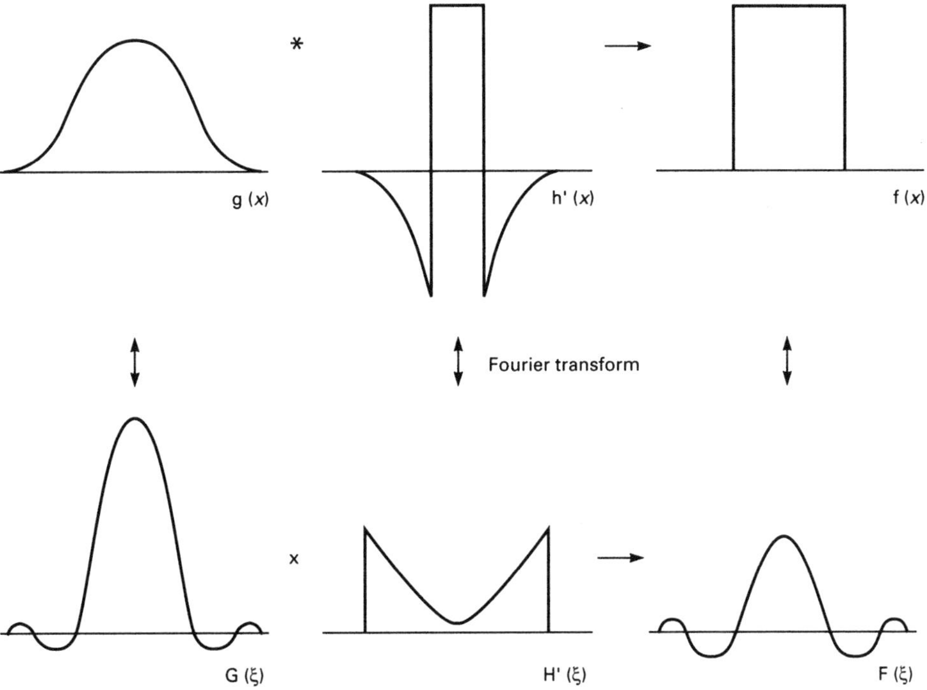

Fig. 31 The process of filtering according to the convolution theorem. Filtering operations can be performed as a convolution (*) of the spatial functions, (top) or as a multiplication (×) of the Fourier transform of these functions (bottom). Either technique can be used in the filtered-backprojection reconstruction process.

object faithfully. One method of quantifying the quality of an image is to define its point spread function or blurring function. A measured image is the convolution of the object function with this blurring function (Fig. 29). This is represented in Eq 13, in which g(x, y) is the resultant image formed by convoluting the object function f(x, y) with the blurring function h(x, y). If the object function was an infinitely small target or a point, then f(x, y) is a delta function with a value of zero everywhere but at one location, and the image g(x, y) will be equal to the PSF, h(x, y). This is how the point spread function is defined and measured.

Convolutions can also be used in image processing to smooth (Fig. 29) or sharpen (Fig. 30) an image. Smoothing convolution filters are typically square (averaging) or bell shaped, while sharpening convolution filters often have a positive value in the center, with adjacent values being negative.

Reference has already been made to Fourier transforms, as in Eq 10, and to their ability to transform spatial data into corresponding spatial frequency data. Filtering operations, such as image smoothing or sharpening, can be readily performed on the data in the spatial frequency domain. According to the convolution theorem, convolution operations in the spatial domain correspond to simple function multiplications in the spatial frequency domain (Fig. 31), that is:

$$g(x) = f(x) * h(x) \qquad \text{(Eq 17)}$$

is equivalent to

$$G(\xi) = F(\xi)\,H(\xi) \qquad \text{(Eq 18)}$$

where G(ξ), F(ξ), and H(ξ) are the Fourier transformed functions of g(x), f(x), and h(x), where ξ is the spatial frequency conjugate of x. The functional multiplication in Eq 18 is simply the multiplication of the values of F(ξ) and H(ξ) at every value of ξ. The two-dimensional convolution of Eq 13 has its counterpart to Eq 18 where the functions G, F, and H are the two-dimensional Fourier transforms of g(x, y), f(x, y), and h(x, y). Note that the spatial frequency variable, ξ, has units of 1/distance.

Because the blurring or convolution process is represented in the spatial frequency domain as a simple functional multiplication, the blurred image conceptually can be easily restored to a closer representation of the original object distribution. If the Fourier transform of the image, G(ξ), is filtered by multiplying by the function H′(ξ) where H′(ξ) = 1/H(ξ), the result is the original object frequency distribution, F(ξ). In practice, this restoration is limited by the frequency limit of H(ξ) leading to division by zero, and by the excessive enhancement of noise along with the signal at frequencies with small H(ξ) values. This restoration process or "deconvolution" of the blurring function can likewise be performed as a convolution in the spatial domain as illustrated in Fig. 30.

Backprojection is the mathematical operation of mapping the one-dimensional projection data back into a two-dimensional grid. This is done intuitively by radiographers in interpreting x-ray films. If a high-density inclusion or structure is visible in two or more x-ray films taken at different angles, the radiographer can mentally backproject along the corresponding ray paths to determine the intersection of the rays and the location of the structure in space.

Mathematically, this is done by taking each point on the two-dimensional image grid and summing the corresponding projection value from each angle from which projection data were acquired. This backprojection process yields a maximum density at the location of the structure where the lines from the rays passing through this structure cross (Fig. 3a). The resultant image is not an accurate representation of the structure, however, in that these lines form a star artifact (Fig. 3a) extending in all directions from the location of the high-density structure. For a very large number of projection angles, the density of this structure is smeared over the entire image and decreases in amplitude with $1/r$, where r is the distance from the structure. This simple backprojected image, f_b, can be represented by the convolution of the true image, f, with the blurring function, $1/r$, or:

$$f_b(r,\theta) = f(r,\theta) * (1/r) \quad \text{(Eq 19)}$$

With ideal data, this blurring function can be removed by a two-dimensional deconvolution or filtering of the blurred image. The appropriate filtering function can be determined by using the convolution theorem to transform Eq 19 into its frequency domain equivalent of:

$$F_b(\rho,\theta) = F(\rho,\theta)\,(1/\rho) \quad \text{(Eq 20)}$$

where the function $(1/\rho)$ is the Fourier transform of $1/r$ in polar coordinates. Dividing both sides by $(1/\rho)$ yields:

$$F(\rho,\theta) = \rho\, F_b(\rho,\theta) \quad \text{(Eq 21)}$$

Therefore, the corrected image, f, can be obtained by determining a simple backprojection, f_b, taking its Fourier transform, filtering, and then taking the inverse Fourier transform to get back into spatial coordinates. Alternatively, the filter function, ρ, can be transformed into spatial coordinates, and the backprojected image can be corrected by a two-dimensional convolution operation. Equations 19 to 21 become somewhat more complex when evaluated in rectangular coordinates rather than as represented in polar coordinates.

This approach tends to produce poor results with actual data. Filtering out this blurring function from the projection data prior to backprojection, however, is quite effective and is the basis for the most commonly used reconstruction method, the filtered-backprojection reconstruction technique.

Filtered-Backprojection Reconstruction. According to the central slice theorem discussed in the section "Direct Fourier Reconstruction Technique" in this Appendix, the Fourier transform of the one-dimensional projection data through a two-dimensional distribution is equivalent to the radial values of the two-dimensional Fourier transform of the distribution. Consequently, the filtering operation performed in Eq 21 can be performed on the projection data prior to backprojection. This is the conceptual basis for the filtered-backprojection reconstruction technique (Ref 52, 53), and it is illustrated schematically in Fig. 3(b).

As with other filtering operations, this correction can be implemented as a convolution in the spatial domain or as a functional multiplication in frequency domain. Fourier filtered backprojection is performed by taking the measured projection data, Fourier transforming it into the frequency domain, multiplying by the ramp-shaped ρ filter (which enhances the high spatial frequencies relative to the low frequencies), taking the inverse Fourier transform of this corrected frequency data, and then backprojecting the filtered projection data onto the two-dimensional grid.

If the filtering is performed in the spatial domain by convoluting the measured projection data with a spatial filter equivalent to the inverse Fourier transform of ρ, the process is often referred to as the convolution-backprojection reconstruction method. The frequency filter has the shape of a ramp, enhancing the high frequency (or sharp detail information) of the projection data. Therefore, the convolution function, or kernel, has the expected shape for a sharpening filter. The positive central value is surrounded by negative values that diminish in magnitude with distance from the center. Because of the similar results obtained by Fourier filtering and convolution filtering and the uncertainty with which the filtering approach is often implemented on specific systems, the terms filtered-backprojection reconstruction and convolution-backprojection reconstruction are sometimes used interchangeably.

The filtered-backprojection reconstruction technique is the method that is commonly used in both medical and industrial tomography systems. The method is more tolerant of measured data imperfections than some of the other techniques. Filtered-backprojection reconstruction provides relatively fast reconstruction times and permits the processing of data after each view is acquired. With appropriate computational hardware, the displayed image can be available almost immediately after all the data have been acquired.

Variations in the filter functions used result from the physical limitations encountered with actual data (particularly its finite quantity) and the boundary assumptions made in the calculations. Additional windowing filters are sometimes combined with the theoretically derived filter to smooth or sharpen the image with no additional processing time.

Another variation used in special situations consists of combinations of the reconstruction techniques, such as performing a filtered-backprojection reconstruction followed by iterative processing. This may be beneficial in cases where the data set is limited and additional known information on the object's shape and content can be applied through the iterative steps to provide an improved image.

Iterative Reconstruction Techniques

Another broad category of methods is the iterative reconstruction algorithms. With this approach, an initial guess is made of the density distribution of the object. This initial guess may result from knowledge of the nominal design of the object or may assume a homogenous cylinder of some density. The computer then calculates the projection data values that would be measured for this assumed object. Each calculated value is compared to the corresponding measured projection data value, and the difference between these values is used to adjust the assumed density values along this ray path. This correction to the assumed distribution is applied successively for each measured ray. An iteration is completed when the image has been corrected along all measured rays, and the process begins again with the first ray. With each iteration, the approximated distribution (or reconstructed image) improves its correspondence to the object distribution.

There are numerous variations of iterative processing that may be additive or multiplicative, weighted or unweighted, restricted or unrestricted, and may specify the order in which the projection data are processed. Some of the more common techniques are the iterative least squares technique, the simultaneous iterative reconstruction technique, and the algebraic reconstruction technique (Ref 54-56).

Iterative reconstruction techniques are rarely used. They require all data to be collected before even the first iteration can be completed, and they are very process intensive. Iterative techniques may be beneficial for selected situations where the data set is limited or distorted. Iterative techniques can be effective with incomplete sets of projection data or with irregular data collection configurations. Known information about the object, such as the design, material densities, or external contours, can be used, along with optimization criteria, to aid in determining the solution. This can allow the reconstruction process to be less sensitive to missing or inaccurate projection data. In addition, reconstruction-dependent corrections can be incorporated into these techniques, such as spectral correction for multiple materials or attenuation corrections in emission computed tomography (SPECT) of radionuclide distributions.

Appendix 2: Computed Tomography Glossary

Afterglow. A detrimental property of a scintillator in which the emission of light continues for a period of time after the exciting radiation has been discontinued.

Algorithm. A procedure for solving a mathematical problem by a series of operations. In computed tomography, the mathematical process used to convert the transmission measurements into a cross-sectional image. Also known as reconstruction algorithm.

Aliasing. An artifact in discretely measured data caused by insufficient sampling of high-frequency data, with this high-frequency information falsely recorded at a lower frequency. See *Nyquist sampling frequency* .

Aperture, collimator. The opening of the collimator enabling variations in the shape of the measured x-ray beam.

Archival storage. Long-term storage of information. In computed tomography, this can be accomplished by storing the information on digital magnetic tape, on optical disk, or as images on x-ray film.

Array, data. An ordered set or matrix of numbers.

Array, detector. An ordered arrangement of individual detector elements that operates as a unit, such as a linear detector array.

Array processor. A special-purpose computer processor used to perform special functions at high speeds.

Artifact. An error or false structure in an image that has no counterpart in the original object.

Attenuation of x-rays. Absorption and scattering of x-rays as they pass through an object.

Backprojection. The process of mapping one-dimensional projection data back into a two-dimensional image array. A process used in certain CT reconstruction algorithms.

Beam hardening. The increase in effective energy of a polyenergetic (for example, x-ray) beam with increasing attenuation of the beam. Beam hardening is due to the preferential attenuation of the lower-energy, or soft, radiation.

Capping. Artifact characterized by dark shading or lowering of pixel values toward edges; opposite of cupping. Sometimes caused by an overcompensating beam hardening correction to the data.

CAT. See *computed tomography* .

Collimator. The x-ray system component that confines the x-ray beam to the required shape. An additional collimator can be located in front of the x-ray detector to further define the portion of the x-ray beam to be measured.

Computed tomography (CT). The collection of transmission data through an object and the subsequent reconstruction of an image corresponding to a cross section through this object. Also known as computerized axial tomography, computer-assisted tomography, or CAT scanning.

Contrast-detail-dose diagram. A plot of the minimum percent contrast needed to resolve a feature versus the feature size. The ability to visualize low-contrast structures tends to be limited by image noise, while small high-contrast structures are resolution limited.

Contrast scale. The change of CT number that occurs per given change in linear attenuation coefficient. Contrast scale is dependent on the x-ray beam energy and the scaling parameters in the reconstruction process.

Contrast sensitivity. The ability to differentiate material density (or linear attenuation coefficient) differences with respect to the surrounding material.

Convolution. A mathematical process used in certain reconstruction algorithms. An operation between two functions in which one function is blurred or smeared by another function.

Coronal plane. A medical term for a plane that divides the body into a front and back section. A *y-z* planar presentation, which may be mathematically constructed from a series of cross-sectional slices. See *sagittal plane* and *reformation, planar*.

Crosstalk, detector. The unwanted pickup of the signal associated with neighboring detector elements. Crosstalk may be caused by radiation scatter or electronic interference.

CT numbers. The numbers used to designate the x-ray attenuation in each picture element of the CT image.

Cupping. Artifact characterized by dark shading in the center of the field of view. May be caused by beam hardening.

Data acquisition system (DAS). The components of a CT system used to collect and digitize the detected x-ray signal.

Density resolution. The measure of the smallest density difference of an image structure that can be distinguished.

Detectability, low contrast. The minimum detectable object size for a specified percent contrast between the object and its surroundings as measured with a test phantom.

Detective quantum efficiency (DQE). The fraction of the beam intensity needed to produce the same signal-to-noise ratio as a particular detector if it were replaced with an ideal detector. Also known as the quantum detection efficiency.

Detector aperture. The physical dimensions of the active area of the detector, including any restriction by the detector collimation. Used in particular for the detector dimension in the plane of the CT scan.

Detector, x-ray. Sensor array used to measure the x-ray intensity. Typical detectors are high-pressure gas ionization, scintillator-photodiode detector arrays, and scintillator-photomultiplier tubes.

Digital radiography (DR). Radiographic imaging technology in which a two-dimensional set of x-ray transmission measurements is acquired and converted to an array of numbers for display with a computer. Computed tomography systems normally have DR imaging capabilities. Also known as digital radioscopy.

Display resolution. Number of picture elements (pixels) per unit distance in the object.

Dual energy imaging. The process of taking two identical scans (DR or CT) at two different x-ray energies and processing the data to produce alternate images that may be insensitive to beam hardening artifacts or may be particularly sensitive to physical density, or the effective atomic number of the materials contained in the object.

Dynamic range. The range of operation of a device between its upper and lower limit. This range can be given as a ratio (example 100:1) of the upper to lower limits, the number of measurable steps between the upper and lower limits, the number of bits (needed to record this number of measurable steps), or the minimum and maximum measurable values.

Edge enhancement. A mathematical manipulation in which rapid density changes are enhanced or sharpened by means of differentiation or high-pass filters. This operation will also increase image noise.

Effective atomic number. For an element, the number of protons in the nucleus of an atom. For mixtures, an effective atomic number can be calculated to represent a single element that would have attenuation properties identical to those of the mixture.

Effective x-ray energy. The monoenergetic beam energy that is attenuated to the same extent by a thin absorber as is the given polyenergetic x-ray beam.

Field-of-view (FOV). The maximum diameter of an object that can be imaged. Also known as scan FOV. The object dimension corresponding to the full width of a displayed image (display FOV).

Focal spot. The area of the x-ray tube from which the x-rays originate. The effective focal-spot size is the apparent size of this area when viewed from the detector.

Fourier transformation. A mathematical process for changing the description of a function by giving its value in terms of its sinusoidal spatial (or temporal) frequency components instead of its spatial coordinates (or vice versa).

Full width at half maximum (FWHM). A parameter that can be used to describe a distribution such as the point spread function.

Geometries, CT. The geometrical configuration and mechanical motion to acquire the x-ray transmission data. Typically, parallel beam data are acquired from translate-rotate data acquisition, and fan beam data are obtained from rotate-only movement of the object (or of the source and detector about the object).

Histogram. A plot of frequency of occurrence versus the measured parameter. In a CT image, the plot of the number of pixels with a particular CT number value versus CT number.

Histogram equalization. An image display process in which an equal number of image pixels is displayed with each displayable gray shade or color.

Hysteresis. Dependence of a measured signal on the previous measurement history.

Intensity. The energy per unit area per unit time incident on a surface.

Ionization. The process in which neutral atoms become charged by gaining or losing an electron.

Iterative reconstruction. A method of reconstruction that produces an improving series of estimated images that is repeated until a suitable quality level is obtained.

Kernel. A function that is convoluted with a data function. The kernel in the convolution-backprojection reconstruction algorithm can be modified to enhance sharpness or reduce noise. Also known as convolution kernel.

Kiloelectron volt (keV). A unit of energy usually associated with individual particles. The energy gained by an electron when accelerated across 1000 V.

Linear attenuation coefficient. The fraction of an x-ray beam per unit thickness that a thin object will absorb or scatter (attenuate). A property proportional to the physical density and dependent on the atomic number of the material and the energy of the x-ray beam.

Linearity, detector. A measure of the consistency in detector sensitivity versus the radiation intensity level.

Matrix. An array of numbers arranged in two dimensions (rows and columns).

Megaelectron volt (MeV). A unit of energy usually associated with a particle. The energy gained by an electron accelerated across 1 000 000 V.

Modulation transfer function (MTF). A measure of the spatial resolution of an imaging system that involves the plot of image contrast (system response) versus the spatial frequency (line pairs per millimeter) of the contrast variations on the object being imaged. A plot of these two variables gives a curve representing the frequency response of a system. The MTF can also be determined from the Fourier transform of the point spread function.

Noise. Salt-and-pepper appearance of an image caused by variations in the measured data. Image noise can be affected by choice of reconstruction algorithm and by image processing, such as sharpening and smoothing operations.

Noise, quantum (or photon). Noise due to statistical variations in the number of x-ray photons detected. An increase in the number of photons measured decreases the relative quantum noise.

Noise, structural. Unwanted detail of an object from overlying structures or due to granularity or material structure.

Noise, structured. Nonrandom noise, such as due to reconstruction algorithm filtering or systemic errors in the measurements.

Nyquist sampling frequency. The ability to characterize a signal from a series of discrete samples requires that the signal be sampled at a minimum of twice the frequency of the highest frequency contained in the signal. The undersampling of a signal causes the high-frequency signals to mimic or appear as lower-frequency signals, an effect termed *aliasing*.

Partial volume artifact. Streaking or shadowing artifacts caused by high-density structures partially intercepting the finite-sized measured x-ray beam. The effect is due to the summation of multiple ray paths in an inhomogenous mixture.

Partial volume effect. The effect of measuring a density lower than the true structure density due to the fact that only a part of the structure is within the measured voxel, that is, within the full slice width or resolution element.

Phantom. A test object used to measure the response of the CT system. Phantoms can be used to calibrate the system or to measure performance parameters.

Photodiode. A semiconductor device that permits current flow proportional to the intensity of light striking the diode junction.

Photomultiplier tube (PMT). A light-sensitive vacuum tube with multiple electrodes providing a highly amplified electrical output.

Photon. A quantum, or particle, of electromagnetic radiation (such as light or x-rays).

Pixel. Shortened term for picture element. A pixel is the smallest displayable element of an image; a single value of the image matrix. A pixel corresponds to the measurement of a volume element (voxel) in the object.

Point spread function (PSF). The image response of a system to a very small, high-amplitude object; the image blurring function.

Polychromatic. Photons with a mixture of colors, or energies. Same as polyenergetic.

Polychromatic artifacts. Image distortion such as image shading or low-density bands between high-density structures caused by the polychromatic x-ray beam. When an x-ray beam travels a longer path or through a high-absorption region, the low-energy components of the polychromatic beam are preferentially absorbed, resulting in a higher effective beam energy and creating artifacts in the image.

Polychromatic x-ray spectra. An x-ray beam that consists of a range of x-ray energies. The maximum amplitude and effective energy of an x-ray beam are always less than the corresponding voltage applied to the tube. The maximum energy of an x-ray beam corresponds to the peak x-ray tube voltage.

Projection. A set of contiguous measured line integrals (projection data points) through an object. May be parallel ray or divergent (fan) ray projection data set. Also called a view.

Projection data. The logarithm of the normalized transmitted intensity data. The line integral of the linear attenuation coefficient through an object.

Quantum detection efficiency. See *detective quantum efficiency*.

Radionuclide. A specific radioactive material.

Reconstruction. The process by which raw digitized detector measurements are transformed into a cross-sectional CT image.

Reformation, planar. A displayed image comparable to a measured CT image, but along a planar section that cuts across a series of measured CT slices. The computer selects the data corresponding to the selected plane from a stack of CT slices through a volume of the object.

Reformation, 3-D surface. A perspective display of a structure; an image that appears as a photograph of the structure.

Resolution, high contrast. A measure of the ability of an imaging system to present multiple high-contrast structures and fine detail; measurements made with line pair gage or bar resolution phantom. See *spatial resolution*.

Resolution, low contrast. The minimum detectable spacing between specified objects for a specified percent contrast between the objects and their background as measured with a test phantom.

Sagittal plane. A medical term for a plane that divides the body into left and right sides. A *y-z* planar presentation, which can be mathematically constructed from a series of cross-sectional slices. See *coronal plane* and *reformation, planar*.

Scan. The operation of acquiring the data for a CT or DR image. Sometimes used in reference to the CT or DR image.

Scattering of x-rays. One of the two ways in which the x-ray beam interacts with matter and the transmitted intensity is diminished in passing through object. The x-rays are scattered in a direction different from the original beam direction. Such scattered x-rays falling on the detector do not add to the radiological image. Because they reduce the contrast in that image, steps are usually taken to eliminate the scattered radiation. Also known as Compton scatter.

Scintillator. A material that produces a rapid flash of visible light when an x-ray photon is absorbed. The scintillator is coupled with a light-sensitive detector to measure the x-ray beam intensity.

Sharpening. Image-processing technique that enhances edge contrast and increases image noise. The opposite of *smoothing*.

Slice. The cross-sectional plane through an object that is scanned to produce the CT image. See also *tomographic plane*.

Slice thickness. The height or *z*-axis dimension of the measured x-ray beam normally measured at the center of the object. The slice thickness along with the image resolution defines the size of the measured volume corresponding to a pixel CT value.

Slit-scan radiography. Method of producing an x-ray image in which a thin x-ray beam produces the image one line at a time. A method used to reduce measured scatter radiation. The DR mode of most CT systems uses a slit-scan technique with a linear detector array.

Smoothing. Image-processing technique that averages the image data, reducing edge sharpness and decreasing image noise. Opposite of *sharpening*.

Spatial resolution. A measure of the ability of an imaging system to represent fine detail; the measure of the smallest separation between individually distinguishable structures. See *resolution, high contrast, point spread function*, and *modulation transfer function*.

Spectrum, x-ray. The plot of the intensity or number of x-ray photons versus energy (or wavelength).

Test phantom. See *phantom*.

Tomographic plane. A section of the part imaged by the tomographic process. Although in CT the tomographic plane or slice is displayed as a two-dimensional image, the measurements are of the materials within a defined slice thickness associated with the plane.

Tomography. From the Greek "to write a slice or section." The process of imaging a particular plane or slice through an object.

View. A set of measured parallel or fan beam data for a particular angle through the center of the object. Also known as a profile or *projection*.

Voxel. Shortened term for volume element. The volume within the object that corresponds to a single pixel element in the image. The box-shaped volume defined by the area of the pixel and the height of the slice thickness.

Window, display. The range of CT values in the image that are displayed. The display window can normally be adjusted interactively by the operator to view different density ranges in the reconstructed image data.

REFERENCES

1. J. Radon, Uber die bestimmung von funktionen durch ihre Integralwerte langs gewisser Mannigfaltigkeiten. Saechsische Akademie der Wissenschaften, Leipzig, *Berichte uber di Verhandlungen*, Vol 69, 1917, p 262-277
2. R.N. Bracewell, Strip Integration in Radio Astronomy, *Aust. J. Phys.*, Vol 9, 1956, p 198-217
3. A.M. Cormack, Representation of a Function by Its Line Integrals, With Some Radiological Applications, *J. Appl. Phys.*, Vol 34, 1963, p 2722-2727
4. D.E. Kuhl and R.Q. Edwards, Image Separation Radioisotope Scanning, *Radiology*, Vol 80, 1963, p 653-662
5. D.J. DeRosier and A. Klug, Reconstruction of Three-Dimensional Structures From Electron Micrographs, *Nature*, Vol 217, 1968, p 130-134
6. G. Hounsfield, Computerized Transverse Axial Scanning Tomography: 1. Description of System, *Br. J. Radiol.*, Vol 46, 1973, p 1016
7. C. Johns and J. Gillmore, CAT Scans for Industry, *Quality*, Feb 1989, p 26-28
8. M. Ter-Pergossian, M.M. Phelps, and G.L. Brownell, Ed., *Reconstruction Tomography in Diagnostic Radiology and Nuclear Medicine*, University Park Press, 1977
9. H.J. Vinegar, X-Ray CT and NMR Imaging of Rocks, *J. Petro. Tech.*, March 1986, p 257-259
10. W.P. Rothwell and H.J. Vinegar, Petrophysical Application of NMR Imaging, *Appl. Opt.*, Vol 24 (No. 3), Dec 1985
11. R. Shuck and S. Jones, Magnetic Resonance Imaging From Medicine to Industry, *Adv. Imag.*, Vol 4 (No. 2), 1989, p 20
12. W.A. Ellingson, J.L. Ackerman, B.D. Sawicka, S. Groenemeyer, and R.J. Kriz, Applications of Nuclear Magnetic Resonance Imaging and X-Ray Computed Tomography for Advanced Ceramics Materials, in *Proceedings of the Advanced Ceramic Materials and Components Conference* (Pittsburgh, PA), April 1987
13. P.G. Lale, The Examination of Internal Tissues, Using Gamma Ray Scatter With a Possible Extension to Megavoltage Radiography, *Phys. Med. Biol.*, Vol 4, 1959
14. H. Strecker, Scatter Imaging of Aluminum Castings Using an X-Ray Beam and a Pinhole Camera, *Mater. Eval.*, Vol 40 (No. 10), 1982, p 1050-1056
15. R.H. Bossi, K.D. Friddell, and J.M. Nelson, Backscatter X-Ray Imaging, *Mater. Eval.*, Vol 46, 1988, p 1462-1467
16. S.Y. Wang, S. Agral, and C.C. Gryte, Computer-Assisted Tomography for the Observation of Oil Displacement in Porous Media, *J. Soc. Pet. Eng.*, Vol 24, 1984, p 53
17. S.Y. Wang *et al.*, Reconstruction of Oil Saturation Distribution Histories During Immiscible Liquid-Liquid Displacement by Computer Assisted Tomography, *AIChE J.*, Vol 30 (No. 4), 1984, p 642-646
18. S.L. Wellington and H.J. Vinegar, "CT Studies of Surfactant-Induced CO_2 Mobility Control," Paper 14393, presented at the 1985 Annual Technical Conference and Exhibition, Las Vegas, Society of Petroleum Engineers, Sept 1985
19. H.J. Vinegar and S.L. Wellington, Tomographic Imaging of Three-Phase Flow Experiments, *Rev. Sci. Instrum.*, Vol 58 (No. 1), 1987, p 96-107
20. S.L. Wellington and H.J. Vinegar, X-Ray Computerized Tomography, *J. Petro. Tech.*, Vol 39 (No. 8), 1987, p 885-898
21. E.M. Withjack, "Computed Tomography for Rock-Property Determination and Fluid-Flow Visualization," Paper 16951, presented at the 1987 SPE Annual Technical Conference and Exhibition, Dallas, Society of Petroleum Engineers, Sept 1987
22. E.M. Withjack, "Computed Tomography Studies of 3-D Miscible Displacement Behavior in a Laboratory Five-Spot Model," Paper 18096, presented at the 1988 SPE Annual Technical Conference and Exhibition, Houston, Society of Petroleum Engineers, 1988
23. D.H. Maylotte, P.G. Kosky, C.L. Spiro, and E.J. Lamby, "Computed Tomography of Coals," U.S. DOE Contract DE-AC21-82MC19210, 1983
24. A.R. Lowrey, K.D. Friddell, and D.W. Cruikshank, "Nondestructive Evaluation of Aerospace Composites Using Medical Computed Tomography (CT) Scanners," Paper presented at the ASNT Spring Conference, Washington, D.C., American Society for Nondestructive Testing, March 1985
25. R.G. Buckley and K.J. Michaels, "Computed Tomography: A Powerful Tool in Solid Motor Exit Cone Evaluation," Paper presented at the ASNT Spring Conference, Washington, D.C., American Society for Nondestructive

Testing, March 1985
26. B.J. Elson, Computerized X-Ray to Verify MX Motors, *Aviat. Week Space Technol.*, Vol 149, 16 April 1984
27. R.A. Armistead, CT: Quantitative 3D Inspection, *Adv. Mater. Process.*, March 1988, p 42-48
28. P.D. Tonner and G. Tosello, Computed Tomography Scanning for Location and Sizing of Cavities in Valve Castings, *Mater. Eval.*, Vol 44, 1986, p 203
29. B.D. Sawicka and R.L. Tapping, CAT Scanning of Hydrogen Induced Cracks in Steel, *Nucl. Instr., Methods*, 1987
30. T. Taylor, W.A. Ellingson, and W.D. Koenigsberg, Evaluation of Engineering Ceramics by Gamma-Ray Computed Tomography, *Ceram. Eng. Sci. Proc.*, Vol 7, 1986, p 772-783
31. S.M. Blumenfeld and G. Glover, Spatial Resolution in Computed Tomography, in *Radiology of the Skull and Brain*, Vol 5, *Technical Aspects of Computed Tomography*, T.H. Newton and D.G. Potts, Ed., C.V. Mosby Company, 1981
32. M.W. Yester and G.T. Barnes, Geometrical Limitations of Computed Tomography (CT) Scanner Resolution, *Appl. Opt. Instr. Med. VI, Proc. SPIE*, Vol 127, 1977, p 296-303
33. P.M. Joseph, Image Noise and Smoothing in Computed Tomography (CT) Scanners, *Opt. Eng.*, Vol 17, 1978, p 396-399
34. H.H. Barrett and W. Swindell, *Radiological Imaging: The Theory of Image Formation, Detection, and Processing*, Academic Press, 1981
35. G. Cohen and F.A. DiBianca, The Use of Contrast-Detail-Dose Evaluation of Image Quality in Computed Tomography Scanners, *J. Comput. Asst. Tomogr.*, Vol 3, 1979, p 189
36. Standard Guide for Computed Tomography (CT) Imaging, American Society for Testing and Materials, 1989
37. K.M. Hanson, Detectability in the Presence of Computed Tomography Reconstruction Noise, *Appl. Opt. Instr. Med. VI, Proc. SPIE*, Vol 127, 1977, p 304
38. E. Chew, G.H. Weiss, R.H. Brooks, and G. DiChiro, Effect of CT Noise on the Detectability of Test Objects, *Am. J. Roentgenol.*, Vol 131, 1978, p 681
39. P.M. Joseph, Artifacts in Computed Tomography, *Phys. Med. Biol.*, Vol 23, 1978, p 1176-1182
40. R.H. Brooks, *et al.*, Aliasing: A Source of Streaks in Computed Tomograms, *J. Comput. Asst. Tomgr.*, Vol 3 (No. 4), 1979, p 511-518
41. D.A. Chesler, *et al.*, Noise Due to Photon Counting Statistics in Computed X-Ray Tomography, *J. Comput. Asst. Tomgr.*, Vol 1 (No. 1), 1977, p 64-74
42. K.M. Hanson, Detectability in Computed Tomographic Images, *Med. Phys.*, Vol 6 (No. 5), 1979
43. W.D. McDavid, R.G. Waggener, W.H. Payne, and M.J. Dennis, Spectral Effects on Three-Dimensional Reconstruction From X-Rays, *Med. Phys.*, Vol 2 (No. 6), 1975, p 321-324
44. P.M. Joseph and R.D. Spital, A Method for Correcting Bone Induced Artifacts in Computed Tomography Scanners, *J. Comput. Asst. Tomogr.*, Vol 2 (No. 3), 1978, p 100-108
45. H.E. Cline, W.E. Lorensen, S. Ludke, C.R. Crawford, and B.C. Teeter, Two Algorithms for the Three-Dimensional Reconstruction of Tomograms, *Med. Phys.*, Vol 15 (No. 3), 1988, p 320-327
46. R.E. Alvarez and A. Macovski, Energy Selective Reconstructions in X-Ray Computerized Tomography, *Phys. Med. Biol.*, Vol 21, 1976, p 733-744
47. K.C. Tam, Limited-Angle Image Reconstruction in Non-Destructive Evaluation, in *Signal Processing and Pattern Recognition in Nondestructive Evaluation of Materials*, C.H. Chen, Ed., NATO ASI Series Vol F44, Springer-Verlag, 1988
48. R. Brooks and G. DiChiro, Principles of Computer Assisted Tomography (CAT) in Radiographic and Radioisotopic Imaging, *Phys. Med. Biol.*, Vol 21, 1976, p 689-732
49. G.T. Herman, *Image Reconstruction From Projections: Implementation and Applications*, Springer-Verlag, 1979
50. G.T. Herman, *Image Reconstruction From Projections: The Fundamentals of Computerized Tomography*, Academic Press, 1980
51. B.P. Flannery, H.W. Deckman, W.G. Roberge, and K.L. D'Amico, Three-Dimensional X-Ray Microtomography, *Science*, Vol 237, 18 Sept 1987, p 1439-1444
52. G.N. Ramachandran and A.V. Lakshminarayanan, Three-Dimensional Reconstruction From Radiographs and Electron Micrographs: III. Description and Application of the Convolution Method, *Indian J. Pure Appl. Phys.*, Vol 9, 1971, p 997
53. L.A. Shepp and B.F. Logan, The Fourier Reconstruction of a Head Section, *Trans. IEEE*, Vol NS-21, 1974, p 21-43
54. R. Gordon, R. Bender, and G.T. Herman, Algebraic Reconstruction Techniques (ART) for Three-Dimensional Electron Microscopy and X-Ray Photography, *J. Theor. Biol.*, Vol 29, 1970, p 471-481
55. P. Gilbert, Iterative Methods for the Three-Dimensional Reconstruction of an Object From Projections, *J. Theor. Biol.*, Vol 36, 1972, p 105-117
56. G.T. Herman and A. Lent, Iterative Reconstruction Algorithms, *Comput. Biol. Med.*, Vol 6, 1976, p 273-294

Neutron Radiography

Harold Berger, Industrial Quality, Inc.

NEUTRON RADIOGRAPHY is a form of nondestructive inspection that uses a specific type of particulate radiation, called neutrons, to form a radiographic image of a testpiece. The geometric principles of shadow formation, the variation of attenuation with testpiece thickness, and many other factors that govern the exposure and processing of a neutron radiograph are similar to those for radiography using x-rays or γ-rays. These topics are extensively covered in the article "Radiographic Inspection" in this Volume and will not be discussed here.

This article will deal mainly with the characteristics that differentiate neutron radiography from x-ray or γ-ray radiography, as discussed in Ref 1 to 9. Neutron radiography will be described in terms of its advantages for improved contrast on low atomic number materials, the discrimination between isotopes, or the inspection of radioactive specimens.

Principles of Neutron Radiography

Neutron radiography is similar to conventional radiography in that both techniques employ radiation beam intensity modulation by an object to image macroscopic object details. X-rays or γ-rays are replaced by neutrons as the penetrating radiation in a through-transmission inspection. The absorption characteristics of matter for x-rays and neutrons differ drastically; the two techniques in general serve to complement one another.

Neutrons are subatomic particles that are characterized by relatively large mass and a neutral electric charge. The attenuation of neutrons differs from the attenuation of x-rays in that the processes of attenuation are nuclear rather than ones that depend on interaction with the electron shells surrounding the nucleus.

Neutrons are produced by nuclear reactors, accelerators, and certain radioactive isotopes, all of which emit neutrons of relatively high energy (fast neutrons). Because most neutron radiography is performed with neutrons of lower energy (thermal neutrons), the sources are usually surrounded by a moderator, which is a material that reduces the kinetic energy of the neutrons.

Neutron Versus Conventional Radiography. Neutron radiography is not accomplished by direct imaging on film, because neutrons do not expose x-ray emulsions efficiently. In one form of neutron radiography, the beam of neutrons impinges on a conversion screen or detector made of a material such as dysprosium or indium, which absorbs the neutrons and becomes radioactive, decaying with a short half-life. In this method, the conversion screen alone is exposed in the neutron beam, then immediately placed in contact with film to expose it by autoradiography. In another common form of imaging, a conversion screen that immediately emits secondary radiation is used with film directly in the neutron beam.

Neutron radiography differs from conventional radiography in that the attenuation of neutrons as they pass through the testpiece is more related to the specific isotope present than to density or atomic number. X-rays are attenuated more by elements of high atomic number than by elements of low atomic number, and this effect varies relatively smoothly with atomic number. Thus, x-rays are generally attenuated more by materials of high density than by materials of low density. For thermal neutrons, attenuation generally tends to decrease with increasing atomic number, although the trend is not a smooth relationship. In addition, certain light elements (hydrogen, lithium, and boron), certain medium-to-heavy elements (especially cadmium, samarium, europium, gadolinium, and dysprosium), and certain specific isotopes have an exceptionally high capability of attenuating thermal neutrons (Fig. 1). This means that neutron radiography can detect these highly attenuating elements or isotopes when they are present in a structure of lower attenuation.

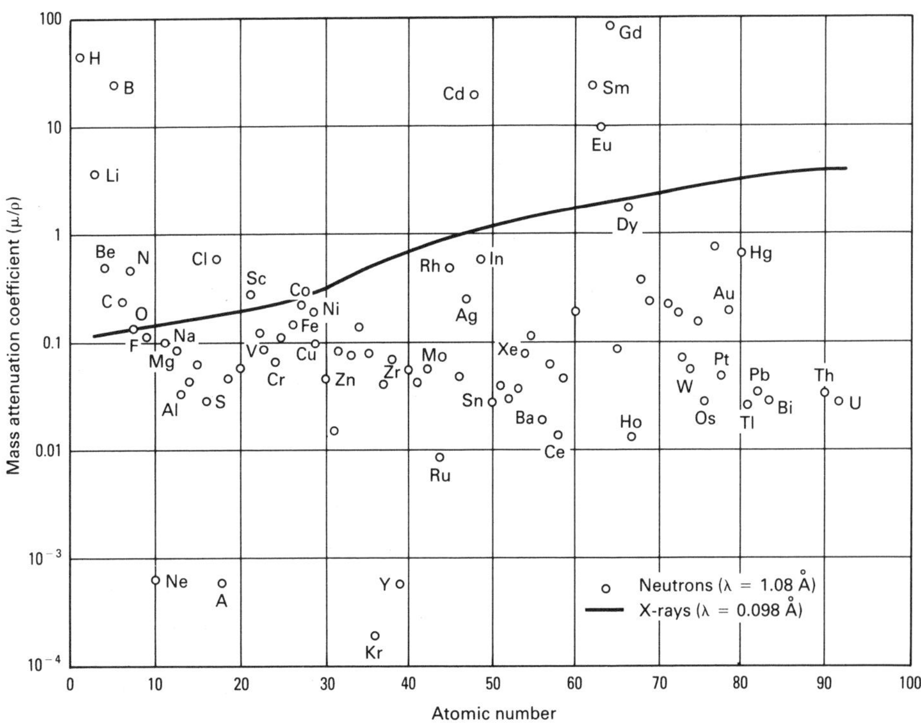

Fig. 1 Mass attenuation coefficients for the elements as a function of atomic number for thermal (4.0×10^{-21} J, or 0.025 eV) neutrons and x-rays (energy ~ 125 kV). The mass attenuation coefficient is the ratio of the linear attenuation coefficient, μ, to the density, ρ, of the absorbing material. Source: Ref 6

Table 1 Characteristics of neutron radiography at various neutron-energy ranges

Type of neutrons	Energy range, J (eV)	Characteristics
Cold	$<1.6 \times 10^{-21}$ (<0.01)	High-absorption cross sections decrease transparency of most materials, but also increase efficiency of detection. An advantage is reduced scatter at energies below the Bragg cutoff, where neutrons can no longer undergo Bragg reflection.
Thermal	1.6×10^{-21} to 8.0×10^{-20} (0.01 to 0.5)	Good discrimination between materials, and ready availability of sources
Epithermal	8.0×10^{-20} to 1.6×10^{-15} (0.5 to 10^4)	Excellent discrimination for particular materials by working at energy of resonance. Greater transmission and less scatter in samples containing materials such as hydrogen and enriched reactor fuels
Fast	1.6×10^{-15} to 3.2×10^{-12} (10^4 to 2.0×10^7)	Good point sources are available. At low-energy end of spectrum, fast-neutron radiography may be able to perform many inspections done with thermal neutrons, but with a panoramic technique. Good penetration capability because of low-absorption cross sections in all materials. Poor material discrimination

Thermal (slow) neutrons permit the radiographic visualization of low atomic number elements even when they are present in assemblies with high atomic number elements such as iron, lead, or uranium. Although the presence of the heavy metals would make detection of the light elements virtually impossible with x-rays, the attenuation characteristics of the elements for slow neutrons are different, which makes detection of light elements feasible. Practical applications of neutron radiography include the inspection of metal-jacketed explosives, rubber O-ring assemblies, investment cast turbine blades to detect residual ceramic core, and the detection of corrosion in metallic assemblies.

Using neutrons, it is possible to detect radiographically certain isotopes—for example, certain isotopes of hydrogen, cadmium, or uranium. Some neutron image detection methods are insensitive to background γ-rays or x-rays and can be used to inspect radioactive materials such as reactor fuel elements. In the nuclear field, these capabilities have been used to image highly radioactive materials and to show radiographic differences between different isotopes in reactor fuel and control materials. The characteristics of neutron radiography complement those of conventional radiography; one radiation provides a capability lacking or difficult for the other.

Neutron Sources

The excellent discrimination capabilities of neutrons generally refer to neutrons of low energy, that is, thermal neutrons. The characteristics of neutron radiography corresponding to various ranges of neutron energy are summarized in Table 1. Although any of these energy ranges can be used for radiography, this article emphasizes the thermal-neutron range, which is the most widely used for inspection.

In thermal-neutron radiography, an object (testpiece) is placed in a thermal-neutron beam in front of an image detector. The neutron beam may be obtained from a nuclear reactor, a radioactive source, or an accelerator. Several characteristics of these sources are summarized in Table 2. For thermal-neutron radiography, fast neutrons emitted by these sources must first be moderated and then collimated (Fig. 2). The radiographic intensities listed in Table 2 typically do not exceed 10^{-5} times the total fast-neutron yield of the source. Part of this loss is incurred in moderating the neutrons, and the remainder in bringing a collimated beam out of a large-volume moderator.

Table 2 Properties and characteristics of thermal-neutron sources

Type of source	Typical radiographic intensity, $n/cm^2 \cdot s$	Spatial resolution	Exposure time	Characteristics
Radioisotope	10^1 to 10^4	Poor to medium	Long	Stable operation, low-to-medium investment cost, possibly movable
Accelerator	10^3 to 10^6	Medium	Average	On-off operation, medium cost, possibly movable
Subcritical assembly	10^4 to 10^6	Good	Average	Stable operation, medium-to-high investment cost, movement difficult
Nuclear reactor	10^5 to 10^8	Excellent	Short	Medium-to-high investment cost, movement difficult

Collimation is necessary for thermal-neutron radiography because there are no useful point sources of low-energy neutrons. Good collimation in thermal-neutron radiography is comparable to small focal-spot size in conventional radiography; the images of thick objects will be sharper with good collimation. On the other hand, it should be noted that available neutron intensity decreases with increasing collimation.

Nuclear Reactors. Many types of reactors have been used for thermal-neutron radiography. The high neutron flux generally available provides high-quality radiographs and short exposure times. Although truck-mounted reactors are technically feasible, a reactor normally must be considered a fixed-site installation, and testpieces must be taken to the reactor for inspection. Investment costs are generally high, but small medium-cost reactors can provide good results. When costs are compared on the basis of available neutron flux (typically, 10^{12} $n/cm^2 \cdot s$ flux is often available at collimator entrance, and 10^6 to 10^7 $n/cm^2 \cdot s$ flux is available at the film plane), reactor sources can be less costly than other sources.

Accelerators. The accelerators most often used for thermal-neutron radiography are:

- The low-voltage type employing the reaction ${}^3_1H + {}^2_1H \rightarrow {}^4_2He + {}^1_0n$, a (d,T) generator, where n, d, and T represent the neutron, deuteron (the nucleus of a deuterium atom, D or ${}^2_1H$, that consists of one neutron and one proton), and tritium (${}^3_1H$), respectively
- High-energy x-ray machines, in which (x,n) reactions are used, where x represents x-ray radiation
- Van de Graaff accelerators
- More recently, high-energy linear accelerators and cyclotrons to generate neutrons by charged-particle reactions on beryllium or lithium targets

Low-Voltage Accelerators. A (d,T) generator provides fast-neutron yields in the range of 10^{10} to 10^{12} n/s. Target lives in sealed neutron tubes are reasonable (100 to 1000 h, depending on yield), and the sealed-tube system presents a source similar to that of certain types of x-ray machines.

High-Energy X-Ray Machines. An (x,n) neutron source is a high-energy x-ray source such as a linear accelerator that can be converted for the production of neutrons by adding a suitable secondary target—for example, beryllium. X-rays having energies above an energy threshold level cause the secondary target to emit neutrons; in beryllium, the threshold x-ray energy for neutron production is 2.67×10^{-13} J (1.66 MeV). Useful neutron radiography has been performed with an 8.8×10^{-13} J (5.5 MeV)

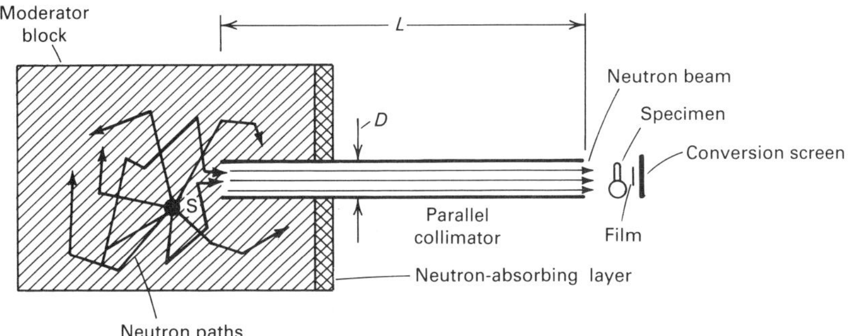

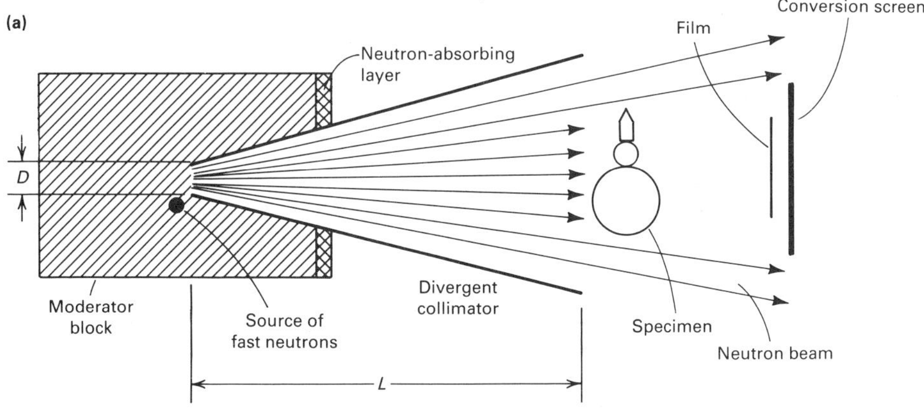

Fig. 2 Thermalization and collimation of beam in neutron radiography. Neutron collimators can be of the parallel-wall (a) or divergent (b) type. The transformation of fast neutrons to slow neutrons is achieved by moderator materials such as paraffin, water, graphite, heavy water, or beryllium. Boron is a typically used neutron-absorbing layer. The *L*/*D* ratio, where *L* is the total length from the inlet aperture to the detector (conversion screen) and *D* is the effective dimension of the inlet of the collimator, is a significant geometric factor that determines the angular divergence of the beam and the neutron intensity at the inspection plane.

linear accelerator having an x-ray output of 0.17 C/kg · min (650 R/min) at 1 m (3 ft). Changeover time from neutron emission to x-ray emission for this source was only 1 h. Beam intensities for neutron radiography with this source were about 5×10^4 n/cm^2 · s, with reasonable beam collimation.

Van de Graaff Accelerators. Much higher beam intensities have been obtained by the acceleration of deuterons onto a beryllium target in a 3.2×10^{-13} J (2.0 MeV) Van de Graaff generator. An intensity of 1.2×10^6 n/cm^2 · s was achieved (with medium collimation), and it is estimated that an acceleration voltage of 4.8×10^{-13} J (3.0 MeV) would improve beam intensity by a factor of approximately six.

The principle of the Van de Graaff machine is illustrated in Fig. 3. A rotating belt transports the charge from a supply to a high-voltage terminal. An ion source within the terminal is fed deuterium gas from a reservoir frequently located within the terminal. A radio-frequency system ionizes the gas, and positive ions are extracted into the accelerator tube. The terminal voltage of about 3 MV is distributed by a resistor chain over about 80 gaps forming the accelerator tube, all of which is enclosed in a pressure vessel filled with insulating gas (N_2 and CO_2 at 2.0 MPa, or 290 psi).

The particle beam is extracted along flight tubes. In a typical neutron reaction, the beam bombards a water-cooled beryllium target in the center of the water moderator tank, which also serves as a partial shield. The higher-energy accelerators indicated above can provide neutron yields of 10^{13} n/s and moderated, well-collimated beam intensities of the order of 10^6 n/cm^2 · s.

A few 4.8×10^{-13} J (3.0 MeV) Van de Graaff generators have recently been placed in service for thermal-neutron radiography. In one such Van de Graaff system designed for neutron radiography, deuterons (4.8×10^{-13} J, or 3 MeV; 280 μA) are accelerated onto a disk-shaped, water-cooled beryllium metal target. Neutrons in the range of 3.2 to 9.6×10^{-13} J (2 to 6 MeV) are emitted preferentially in the forward direction and are moderated in water. The 4π (solid angle) yield of 5×10^{11} n/s produces a peak thermal neutron flux of 2×10^9 n/cm^2 · s. At a collimator ratio of 36:1, the typical exposure time for high-quality film (3×10^9 n/cm^2) is about 2 h.

The accelerator tank for the 4.8×10^{-13} J (3 MeV) machine measures 5.2 m (17 ft) in length and 1.5 m (5 ft) in diameter. The weight is 6100 kg (13 500 lb). The dimensions of the water tank are approximately 1 m (3 ft) on each side. Neutron beams can be extracted through three horizontal beam collimators. Unlike reactors, subcritical multipliers, or (d, T) accelerators, the Van de Graaff accelerators utilize no radioactive source material and sometimes require less stringent license processes.

Other acceleration machines or reactions can be used for thermal-neutron radiography. However, those described above have been most widely used.

Radioactive Sources. There are many possible radioactive sources. The characteristics of several radioisotopes that are commonly used are summarized in Table 3.

Radioisotopes offer the best prospect for a portable neutron-radiographic facility, but it should be recognized that the thermal-neutron intensity is only about 10^{-5} of the total fast-neutron yield from the source. Consequently, neutron radiography using a radioisotope as a neutron source normally requires long exposure times and fast films. For example, a typical 3.7×10^{11} Bq (10 Ci) source would provide a total fast-neutron yield of the order of 10^7 n/s. The radiographic intensity would be about 10^2 n/cm^2 · s, and a typical exposure time using a fast film/converter-screen combination would be about 1 h. Californium-252, usually purchased in the form shown in Fig. 4, has been the most frequently used radioactive source for neutron radiography.

Subcritical Assembly. Another type of source that has received some attention is a subcritical assembly. This type of source is similar to a reactor, except that the neutron flux is less and the design is such that criticality cannot be achieved. A subcritical assembly offers some of the same neutron multiplication features as a reactor. It is

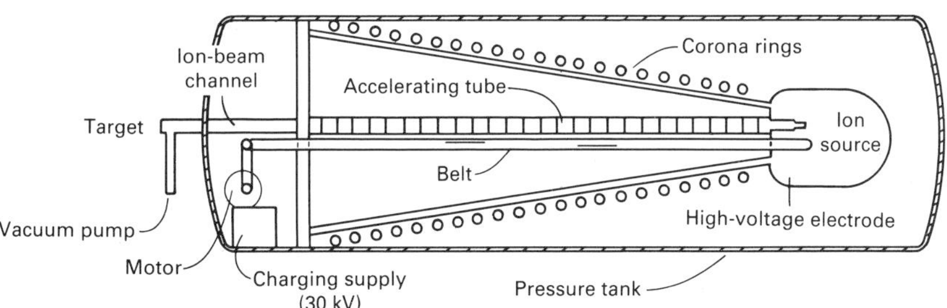

Fig. 3 Cross section showing Van de Graaff principle as it is applied to neutron radiography. Source: Ref 6

Table 3 Properties and characteristics of several radioisotopes used for thermal-neutron radiography

Radioisotope	Neutron reaction	Half-life	Characteristics
^{124}Sb-Be	(γ, n)	60 days	Short half-life, high γ-ray background, low neutron energy easily thermalized, low cost, high neutron yield
^{210}Po-Be	(α, n)	138 days	Short half-life, low γ-ray background, low cost
^{241}Am-Be	(α, n)	458 years	Long half-life, easily shielded γ-ray background, high cost
^{241}Am-^{242}Cm-Be	(α, n)	163 days	Short half-life, high neutron yield, medium cost
^{244}Cm-Be	(α, n)	18.1 years	Long half-life, low γ-ray background, potential low cost
^{252}Cf	Spontaneous fission	2.65 years	Long half-life, small size, low neutron energy easily thermalized, high neutron yield

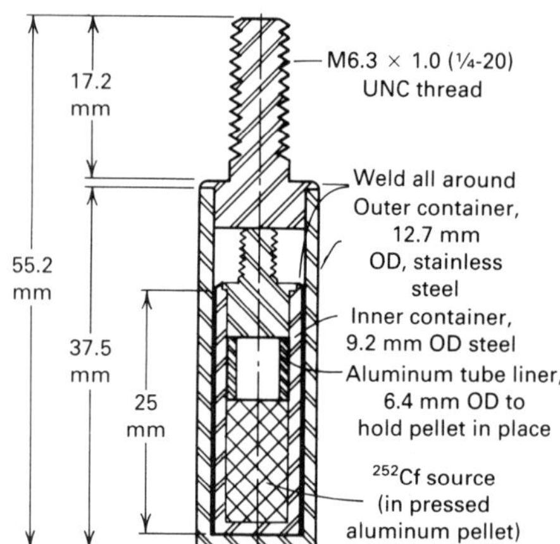

Fig. 4 Cross section of doubly encapsulated ^{252}Cf source. Source: Ref 6

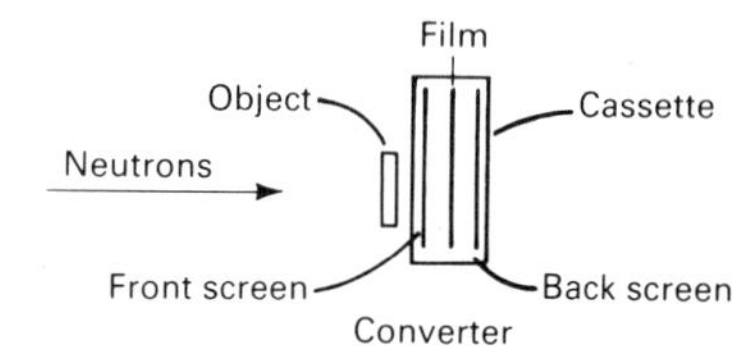

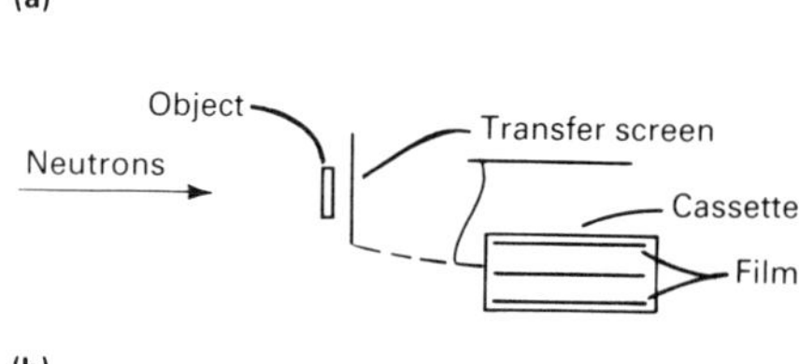

Fig. 5 Schematics of neutron radiography with film using the direct-exposure method (a) and the transfer method (b). The cassette is a light-tight device for holding film or conversion screens and film in close contact during exposure.

somewhat easier to operate, and safety precautions are less stringent, because it is not capable of producing a self-sustaining neutron chain reaction.

Attenuation of Neutron Beams

Unlike electrons and electromagnetic radiation, which interact with orbital electrons surrounding an atomic nucleus, neutrons interact only with atomic nuclei. Usually, neutrons are deflected by interaction with the nuclei, but occasionally a neutron is absorbed into a nucleus. When a neutron collides with the nucleus of an atom and is merely deflected, the neutron imparts some of its kinetic energy to the atom. Both the neutron and the atom move off in different directions from the original direction of motion of the neutron. This process, known as scattering, reduces the kinetic energy of the neutron and the probability that the neutron will pass through the object (testpiece) in a direction that will permit it to be detected by a device placed behind the object.

True absorption of neutrons occurs when they are captured by nuclei. The capture of a neutron transforms the nucleus to the next-higher isotope of the target nucleus and sometimes produces an unstable nucleus that then undergoes radioactive decay. The probability that a collision between a neutron and a nucleus will result in capture is known as the capture cross section and is expressed as an effective area per atom. (The capture cross section is usually measured in barns, 1 barn equaling 10^{-24} cm^2 or 1.6×10^{-25} in.^{2}.) The capture cross section varies with neutron energy, atomic number, and mass number. For thermal neutrons (energy of about 4.0×10^{-21} J, or 0.025 eV), the average capture cross section varies randomly with atomic number, being high for certain elements and relatively low for other elements. The cross section actually varies by isotope rather than element. However, radiographers usually consider an average cross section for an element. For intermediate neutrons (energies of 8.0×10^{-20} to 1.6×10^{-15} J, or 0.5 eV to 10 keV) and for fast neutrons (energies exceeding 1.6×10^{-15} J, or 10 keV), the capture cross section is normally smaller than that for thermal neutrons, and there is much less variation with atomic number. For fast neutrons, most elements are similarly absorbing, and scattering is the dominant process of attenuation.

In relation to other types of penetrating radiation, many materials interact less with neutrons. Therefore, neutrons can sometimes be used to inspect greater thicknesses than can be conveniently inspected with electromagnetic radiation. The combined effect of scattering and capture can be expressed as a mass-absorption coefficient; this coefficient is used to determine the exposure factor for the neutron radiography of a given object (testpiece). For a given material, attenuation varies exponentially with thickness, and the basic law of radiation absorption (discussed in the article "Radiographic Inspection" in this Volume) applies to neutron attenuation as well as to the attenuation of electromagnetic radiation.

Neutron Detection Methods

Detection methods for neutron radiography generally use photographic or x-ray films. In the so-called direct-exposure method, film is exposed directly to the neutron beam, with a conversion screen or intensifying screen providing the secondary radiation that actually exposes the film (Fig. 5a). Alternatively, film can be used to record an autoradiographic image from a radioactive image-carrying screen in a technique called the transfer method (Fig. 5b).

Direct-Exposure Method. Conversion screens of thin gadolinium foil or a scintillator have been most widely used in the direct-exposure method. When bombarded with a beam of neutrons, some of the gadolinium atoms absorb some of the neutrons and then promptly emit γ-rays. The γ-rays in turn produce internal conversion electrons that actually expose the film; these are directly related in intensity to the intensity of the neutron beam. Scintillators, on the other hand, are fluorescent materials often made of zinc sulfide crystals that also contain a specific isotope, such as $^6_3$Li or $^{10}_5$B. In a neutron beam, these isotopes react with neutrons as follows:

$$^6_3\text{Li} + ^1_0\text{n} \rightarrow ^3_1\text{H} + \alpha$$

$$^{10}_5\text{B} + ^1_0\text{n} \rightarrow ^7_3\text{Li} + \alpha$$

The α particles emitted as a result of these reactions cause the zinc sulfide to fluoresce, which in turn exposes the film. Gadolinium oxysulfide, a scintillator originally developed for conventional radiography, is now widely used for neutron radiography.

Scintillators provide useful images with total exposures as low as 5×10^5 n/cm^2. The high speed and favorable relative neutron/gamma response of scintillators make them attractive for use with nonreactor neutron sources. For high-intensity sources, gadolinium screens are widely used. Gadolinium screens provide greater uniformity and im-

Fig. 6 Positive print of a thermal-neutron radiograph of five irradiated nuclear fuel elements, taken to determine if dimensional changes occurred during irradiation. Radiograph was made using a dysprosium transfer-screen method. Dark squares in middle element are voids.

age sharpness (high-contrast resolution of 10 μm, or 400 μin., has been reported), but an exposure about 30 or more times that of a scintillator is required, even with fast films. Excessive background radiation should be kept to a minimum because it can have a detrimental effect on image quality

In the transfer method, a thin sheet of metal called a transfer screen, which is usually made of indium or dysprosium, is exposed to the neutron beam transmitted through the specimen. Neutron capture by the isotope $^{115}_{49}In$ or $^{164}_{66}Dy$ induces radioactivity, indium having a half-life of 54 min and dysprosium a half-life of 2.35 h. The intensity of radioactive emission from each area of the transfer screen is directly related to the intensity of the portion of the transmitted neutron beam that induced radioactivity in that area. The radiograph to be interpreted is made by placing the radioactive transfer screen in contact with a sheet of film. The β particle and γ-ray emissions from the transfer screen expose the film, with film density in various portions of the developed image being proportionally related to the intensity of radioactive emission.

The transfer method is especially valuable for inspecting a radioactive specimen. Although the radiation emitted by the specimen (especially γ-rays) causes heavy film fogging during conventional radiography or direct-exposure neutron radiography, the same radiation will not induce radioactivity in a transfer screen. Therefore, a clear image of the specimen can be obtained even when there is a high level of background radiation.

In comparing the two primary detection methods, the direct-exposure method offers high speed, unlimited image-integration time, and the best spatial resolution. The transfer method offers insensitivity to the γ-rays emitted by the specimen and greater contrast because of lower amounts of scattered and secondary radiation.

Real-time imaging, in which light from a scintillator is observed by a television camera, can also be used for neutron radiography. Because of low brightness, most real-time neutron radiographic images are enhanced by an image-intensifier tube, which may be separate or integral with the scintillator screen. This method can be used for such applications as the study of fluid flow in a closed system or the study of metal flow in a mold during casting. The lubricants moving in an operating engine have been observed with the real-time neutron imaging method.

Applications

Various applications concerning the inspection of ordnance, explosive, aerospace, and nuclear components are discussed in Ref 1 to 9. The presence, absence, or correct placement of explosives, adhesives, O-rings, plastic components, and similar materials can be verified. Nuclear fuel and control materials can be inspected to determine the distribution of isotopes and to detect foreign or imperfect material. Ceramic residual core in investment cast turbine blades can be detected. Observations of corrosion in metal assemblies are possible because of the excellent neutron sensitivity to the hydrogenous corrosion product. Hydride deposition in metals and diffusion of boron in heat treated boron-fiber composites can be observed. The following examples illustrate the application of neutron radiography to the inspection of radioactive materials and several assemblies of metallic and nonmetallic components.

Example 1: Thermal-Neutron Radiography Used to Determine Size of Highly Radioactive Nuclear Fuel Elements. Highly radioactive nuclear fuel elements required size measurements to determine the extent of dimensional changes that may have occurred during irradiation. Generally, inspection is done in a hot cell, but because hot-cell inspection is a relatively long, tedious, and costly procedure, neutron radiography was selected.

The fuel elements to be inspected consisted of 6.4 mm (¼ in.) diam cylindrical pellets of UO_2-PuO_2; the plutonium content was 20%, and the uranium was enriched in ^{235}U. The pellets had been irradiated to 10% burnup, which resulted in a level of radioactivity of 3×10^{-2} C/kg · h (10 KR/h) at 0.3 m (1 ft).

Five elements were selected for inspection. A neutron radiograph was taken by activating 0.25 mm (0.010 in.) thick dysprosium foil with a transmitted beam of thermal neutrons. An autoradiograph of the activated-dysprosium transfer screen on a medium-speed x-ray film yielded the result shown in the positive print in Fig. 6.

Both ^{235}U and plutonium have high attenuation coefficients for thermal neutrons. The high contrast of the fuel pellets made it possible to measure pellet diameter directly from the neutron radiographs. These measurements were both repeatable and statistically significant within 0.013 mm (0.0005 in.). Later, radiographic measurements were compared with physical measurements made in a hot cell. The two sets of values corresponded within 0.038 mm (0.0015 in.).

Example 2: Indium-Resonance Technique for Determining Internal Details of Highly Radioactive Nuclear Fuel Elements. The five nuclear fuel elements inspected for dimensional changes in Example 1 were further inspected for internal details. This was necessary because the thermal-neutron inspection procedure did not reveal any internal details; it only shadowed the pellets, as shown by the positive print in Fig. 6.

To inspect for internal details, an indium-resonance technique, which utilizes epithermal neutrons, was used. In this technique, a collimated neutron beam was filtered by 0.5 mm (0.02 in.) of cadmium to remove most of the thermal neutrons. Filtering produced a neutron beam with a nominal average energy somewhat above thermal. The epithermal-neutron beam passed through the fuel elements and activated a sheet of indium foil. Neutrons with an energy of about 2.34×10^{-19} J (1.46 eV), which is the resonance-absorption energy for indium, were primar-

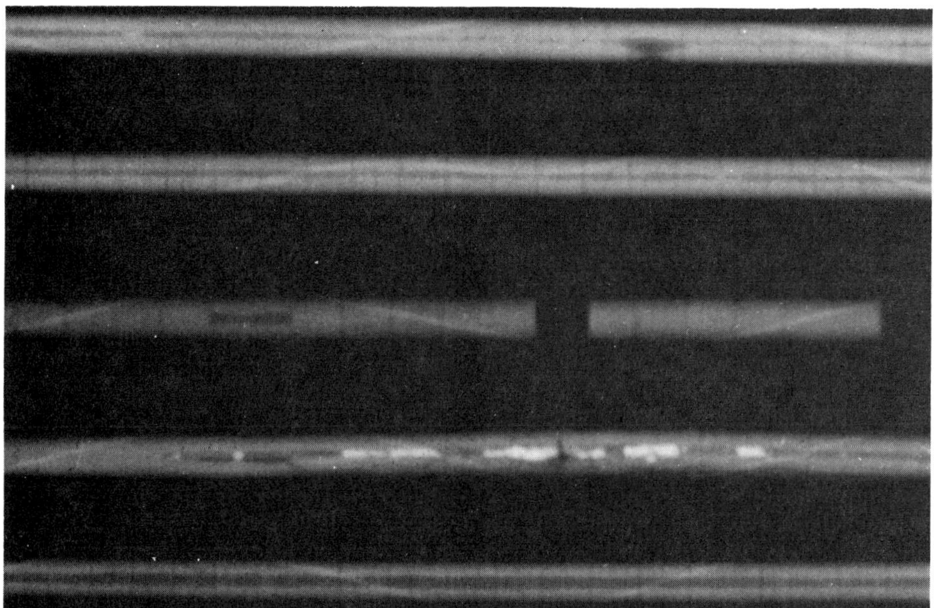

Fig. 7 Positive print of a neutron radiograph of the same five nuclear fuel elements shown in Fig. 6. Radiograph was made with epithermal neutrons and an indium-resonance technique, and it reveals internal details not shown in the thermal-neutron radiograph in Fig. 6.

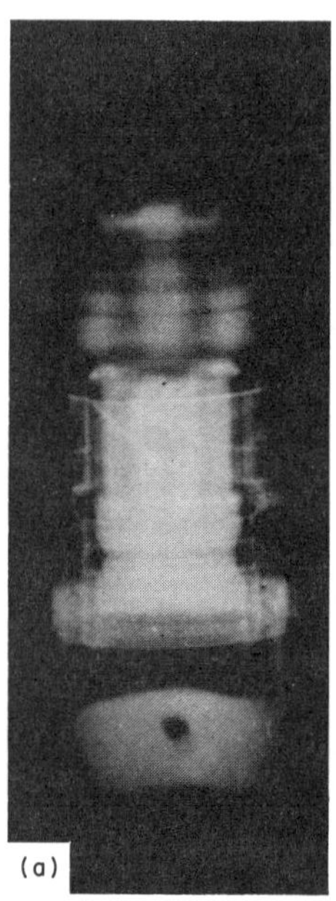

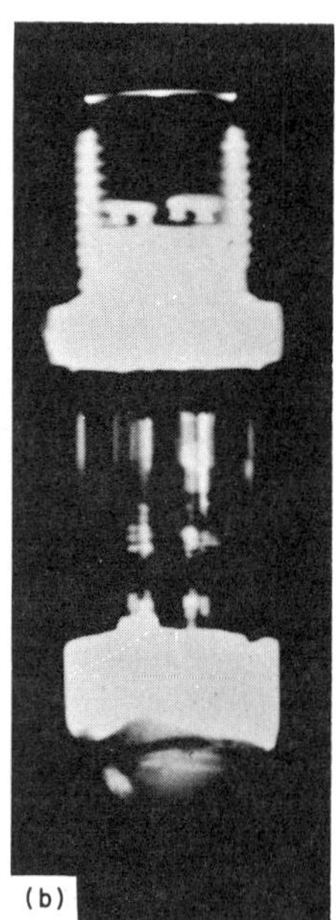

Fig. 8 Comparison of positive prints of a thermal-neutron radiograph (a) and a conventional radiograph (b) of a 50 mm (2 in.) long explosive device. Neutron radiograph reveals details of paper, explosive compound, and plastic components not revealed by x-rays.

ily involved in activation. The positive print of a radiograph made with epithermal neutrons shown in Fig. 7 reveals considerable internal details, in contrast to the lack of internal details in Fig. 6.

With epithermal neutrons, there was less attenuation by the fuel elements than with thermal neutrons. Therefore, internal details that were not revealed by thermal-neutron radiography—such as cracking or chipping of fuel pellets, and dimensional features of the central void in the fuel pellets (including changes in size and accumulation of fission products)—were revealed with the indium-resonance technique.

Example 3: Use of Conventional and Neutron Radiography to Inspect an Explosive Device for Correct Assembly. Small explosive devices assembled from both metallic and nonmetallic components required inspection to ensure correct assembly. The explosive and the components made of paper, plastic, or other low atomic number materials, which are less transparent to thermal neutrons than to x-rays, could be readily observed with thermal-neutron radiography. Metallic components were inspected by conventional x-ray radiography.

A positive print of a thermal-neutron, direct-exposure radiograph of a 50 mm (2 in.) long explosive device is shown in Fig. 8(a). The radiograph was made on Industrex R film (Eastman Kodak), using a gadolinium-foil screen. Total exposure was 3×10^9 n/cm^2, which was achieved with an exposure time of 4 to 5 min. At the top in Fig. 8(a), just inside the stainless steel cap, can be seen a line image that corresponds to a moisture absorbent made of chemically treated paper. Below the paper is a mottled image, which is the explosive charge. Below the explosive charge are plastic components and, at the very bottom, epoxy.

A conventional radiograph of the same device is shown in Fig. 8(b). The metallic components, which were poorly delineated in the thermal-neutron radiograph, are more clearly seen in Fig. 8(b). Together, the two radiographs verified that both metallic and nonmetallic components were correctly assembled.

Example 4: Use of Neutron Radiography to Detect Corrosion in Aircraft Components. Aluminum honeycomb components are extensively used for aircraft construction. The aluminum material is subject to corrosion if exposed to water or humid environments. Thermal-neutron radiography is an excellent method of detecting hidden corrosion in these assemblies. The corrosion products are typically hydroxides or water-containing oxides; these corrosion products contain hydrogen, a material that strongly attenuates thermal neutrons. The aluminum metal, on the other hand, is essentially transparent to the neutrons. Therefore, a thermal-neutron radiograph of a corroded aluminum honeycomb assembly shows the corrosion product and other attenuating components such as adhesives and sealants. Figure 9 depicts a thermal-neutron radiograph of an aluminum honeycomb assembly showing the beginnings of corrosion. The white line image across the middle of the radiograph represents the adhesive coupling together two core sections. The faint white smears in the upper half of the image and the double dot in the lower left area are images of corrosion as disclosed by the thermal-neutron radiograph. Developmental work has shown that thermal-neutron imaging techniques are capable of detecting the corrosion product buildup represented by an aluminum metal loss of 25 μm (1000 μin.). The neutron method, therefore, is a very sensitive technique for the detection of corrosion.

Example 5: Use of Neutron Radiography to Detect Corrosion in Adhesive-Bonded Aluminum Honeycomb Structures. Aluminum corrosion of aircraft surfaces has plagued both military and civilian aircraft. Identification of this corrosion has been difficult, at best, usually being detected after the corrosion has caused the part to fail. Of the nondestructive testing methods used to detect aluminum corrosion, thermal neutron radiography has proved the most sensitive method to date.

The detection of aluminum corrosion is based on the attenuation properties of hydrogen associated with the corrosion products rather than aluminum and aluminum oxide with their low attenuation coefficients. Depending on the environment, the corrosion products include aluminum trihydrates, monohydrates, and various other aluminum salts. Because the linear attenuation coefficient for aluminum is similar to that of water and about 28 times greater than that for aluminum, a 0.13 mm (0.005 in.) corrosion layer should be detectable under optimum conditions.

The sensitivity standard plate for aluminum corrosion fabricated by the Aeronauti-

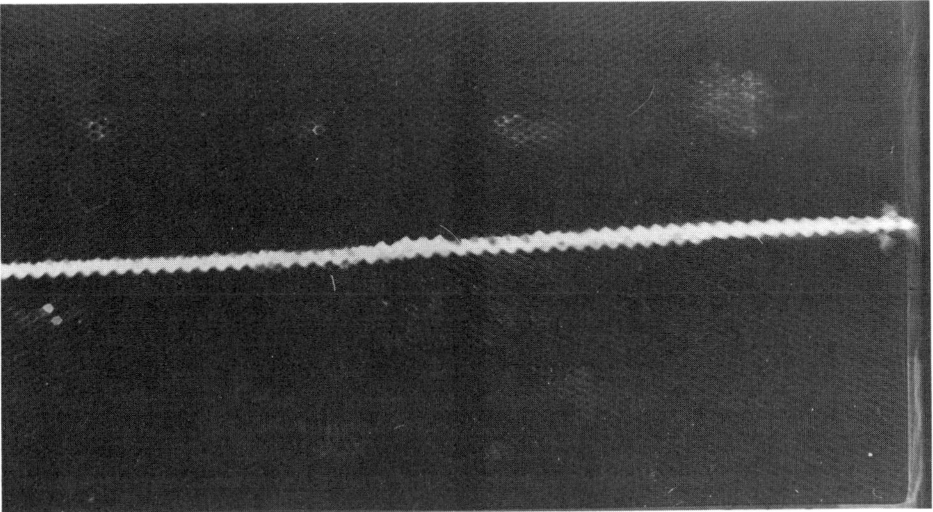

Fig. 9 Thermal-neutron radiograph of aluminum honeycomb aircraft component showing early evidence of hydrogen corrosion. See text for discussion. Courtesy of D. Froom, U.S. Air Force, McClellan Air Force Base

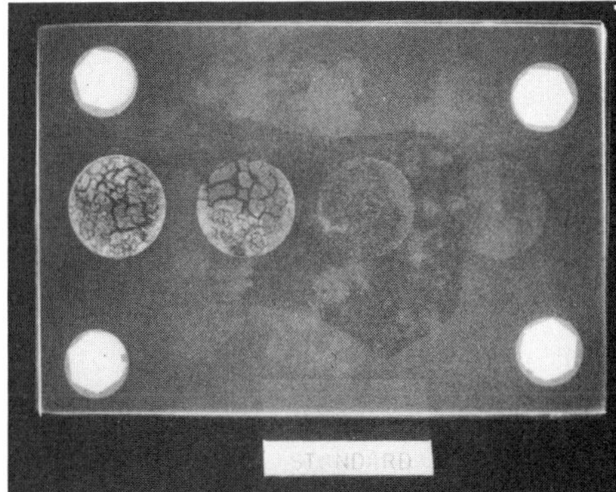

Fig. 10 Standard plate for aluminum corrosion detection contains 0.13 to 0.61 mm (0.005 to 0.024 in.) thick corrosion products. Courtesy of R. Tsukimura, Aerotest Operations Inc.

cal Research Laboratories (Australia) contains corrosion products varying from 0.13 to 0.61 mm (0.005 to 0.024 in.) thick (Fig. 10).

Aluminum corrosion of honeycomb structures is complicated by the bonding adhesives that may appear similar in a neutron radiograph (Fig. 11a) (see the article "Adhesive-Bonded Joints" in this Volume). Tilting of the honeycomb structure will alleviate this problem by allowing adhesive found along the bond lines to be distinguished from the randomly distributed corrosion products (Fig. 11b).

Example 6: Use of Neutron Radiography to Verify Welding of Dissimilar Materials (Titanium and Niobium). Exotic metal welded joints are a product of the extremely cold environment of space and man's desire to explore the vast emptiness of space. For space vehicles, attitude control rockets provide the fine touch for proper vehicle alignment.

For one application, a titanium-niobium welded joint was required between the lightweight propellant tank and the nozzle section. Attempts to verify weld integrity using conventional radiography were not productive.

Thermal neutron radiography provided the image required to ensure quality welds. This defect standard weld shows the porosity at the seam and the similar thermal neutron attenuation for both titanium (Ti) and niobium (formerly known as columbium, Cb) (Fig. 12a). For comparison, the x-ray radiograph image is also shown (Fig. 12b).

Example 7: Use of Neutron Radiography to Detect Core Material Still Remaining in the Interior Cooling Passages of Air-Cooled Turbine Blades. Investment casting of turbine blades using the lost wax process results in relatively clean castings. As the demand for higher-powered turbine engines has increased, the interior cooling passages for air-cooled turbine blades have become more and more complex. Concurrently, the removal of the core material has become increasingly more difficult. Incomplete removal of the core results in restricted flow through the cooling passages and possible failure of the overheated blade.

Previously, visual inspection was the nondestructive inspection method of choice for residual core detection. However, current designs preclude the use of borescopes and other visual means for the interior passages. X-radiography has proved rather ineffective in detecting residual core material. Thermal neutron radiography is the nondestructive testing method of choice, especially when gadolinium oxide (Gd_2O_3) is used to dope the core material (1 to 3% by weight) (Fig. 13) prior to casting the blade.

When concerns about the possible detrimental effects of Gd_2O_3 during the casting process prevents its use in the core material, a procedure was developed to tag the residual core material after the core removal process. The castings are dipped in a gadolinium solution [$Gd(NO_3)_2$ in solution] to impregnate any residual core, which is then imaged and subsequently detected by neutron radiography.

The blades shown in Fig. 14 have been tagged. The neutron radiograph shows any residual core material greater than 0.38 mm

(a)

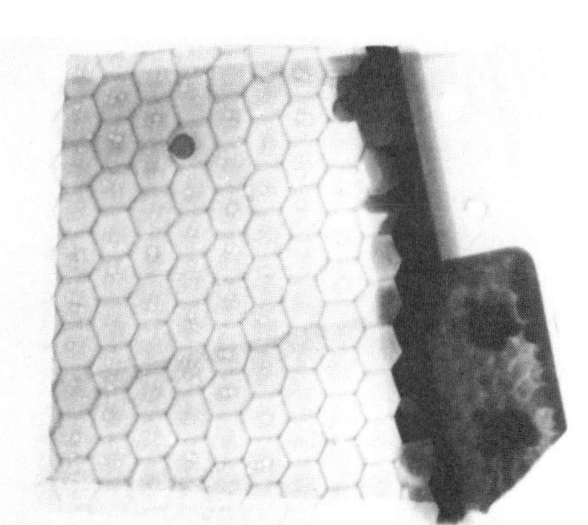

(b)

Fig. 11 Effect of bonding adhesives on the quality of neutron radiographs obtained when checking for aluminum corrosion in honeycomb structures. Radiograph taken (a) normal to specimen surface and (b) tilted at any angle other than 90° to specimen surface. Courtesy of R. Tsukimura, Aerotest Operations Inc.

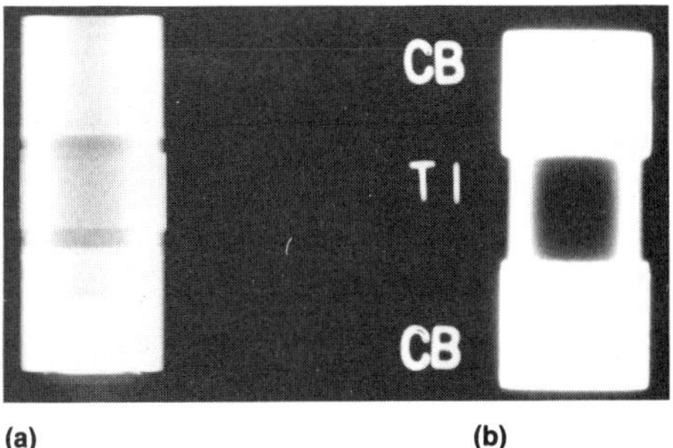

(a) (b)

Fig. 12 Comparison of thermal neutron (a) and x-ray (b) radiographs of a titanium-niobium welded joint. Courtesy of R. Tsukimura, Aerotest Operations Inc.

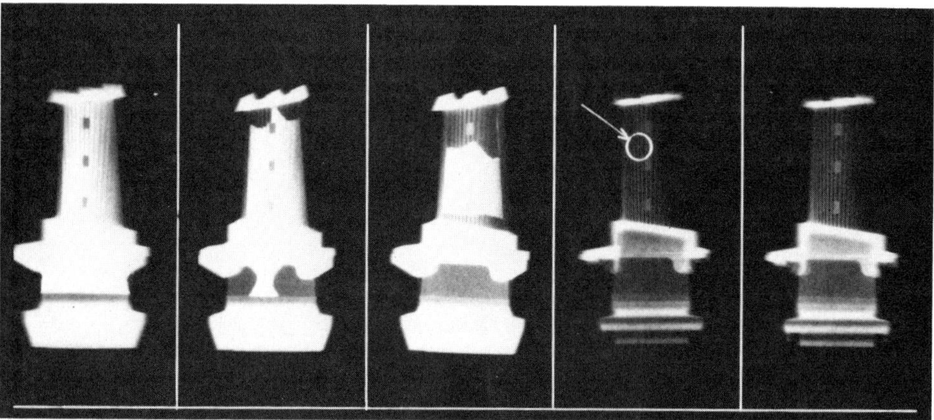

Fig. 13 Residual core material in a gas-cooled aircraft-engine turbine blade as detected by thermal neutron radiography. The excess core material, tagged with 1.5% Gd_2O_3, is shown circled in the second photo from the right. Courtesy of R. Tsukimura, Aerotest Operations Inc.

(0.015 in.) in diameter. Figure 15 is a schematic of typical core fragments in investment cast turbine blades detected by thermal neutron radiography. Image clarity of gadolinium tagged or doped cores is much greater than that of normal cores.

Example 8: Use of Neutron Radiography to Verify Position of Explosive Charges and Seating of O-Ring Seals in Explosive Bolt Assemblies. There are many critical applications of explosive release devices in aircraft, space, and missile systems. Nondestructive testing is an important step in the quality control portion of the production cycle for these units. Thermal neutron radiography has proved an indispensable tool in the nondestructive testing arsenal, particularly for thick-walled, metal devices, such as explosive bolts (Fig. 16).

The inner details of explosive bolts can be imaged only by thermal neutron radiography methods (Fig. 17). This particular type of bolt from a missile system is activated from the bottom by actuating the firing pin onto the primer. The short section of mild detonating cord carries the energy to the output charge, which fractures the bolt and allows the bolt to be severed.

In addition to the explosive charges, the internal O-ring seals, including the concentric pair around the firing pin and for the body are readily visible. For safety's sake, determining the presence of the shear pin can also be accomplished through the use of thermal neutron radiography.

Example 9: Application of Neutron Radiography to Determine Potting Fill Levels in Encapsulated Electronic Filters. Electronic filters are an integral component of all space and satellite systems. Because the cost of these satellites is very high and the cost to repair them even more prohibitive, high reliability filters are necessary.

A common mode of filter failure is that caused by inadequate potting of the internal components and the subsequent physical breakdown of the filter during periods of high vibration, such as that encountered during vehicle launch. Thermal neutron radiography is the method of choice for determining potting fill levels in encapsulated filters.

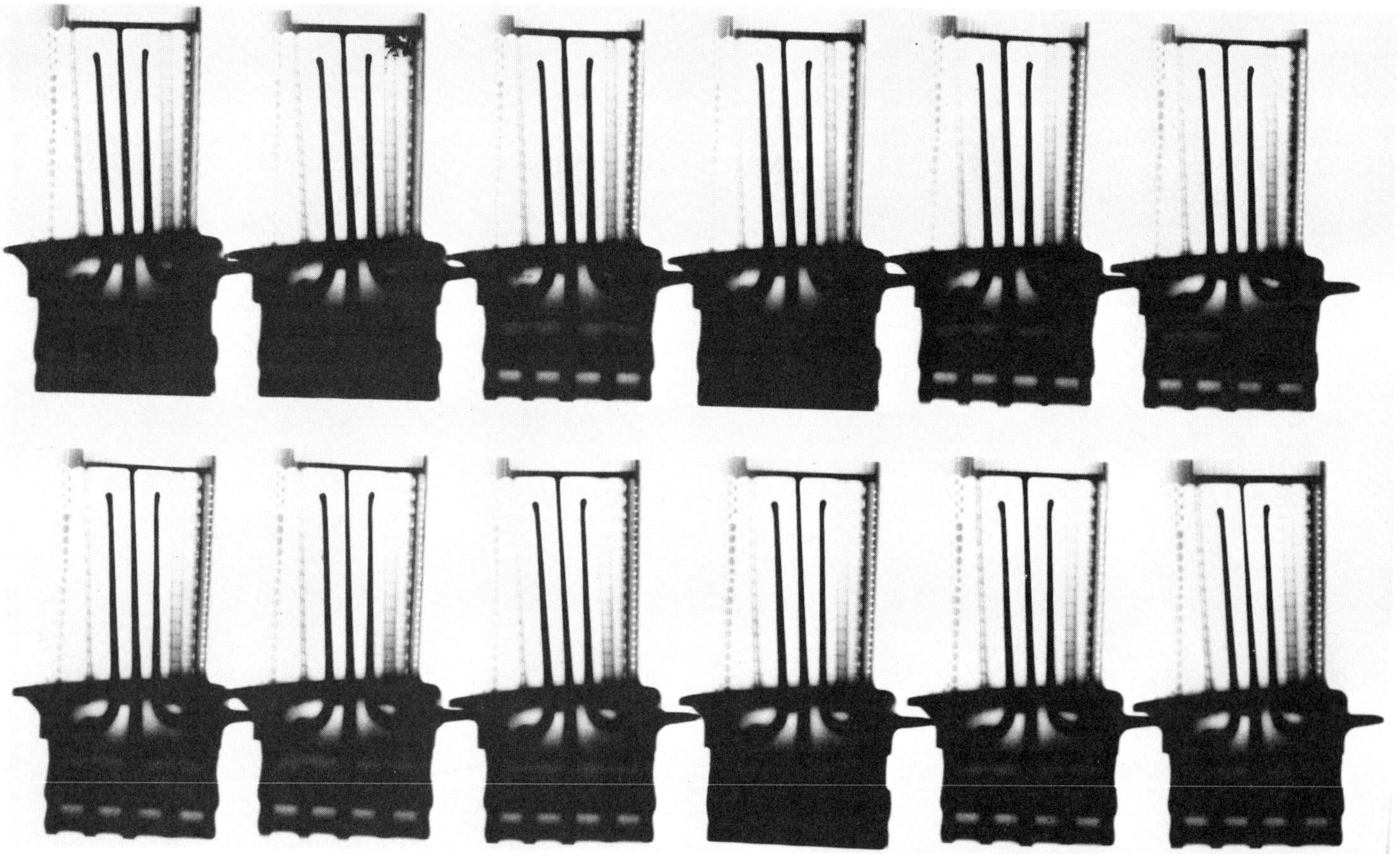

Fig. 14 Thermal neutron radiograph of 12 turbine blades tagged with Gd_2O_3 solution. One of the 12 blades (located in the top row and second from the left) contains residual core material in its upper right-hand corner cooling passage. Courtesy of R. Tsukimura, Aerotest Operations Inc.

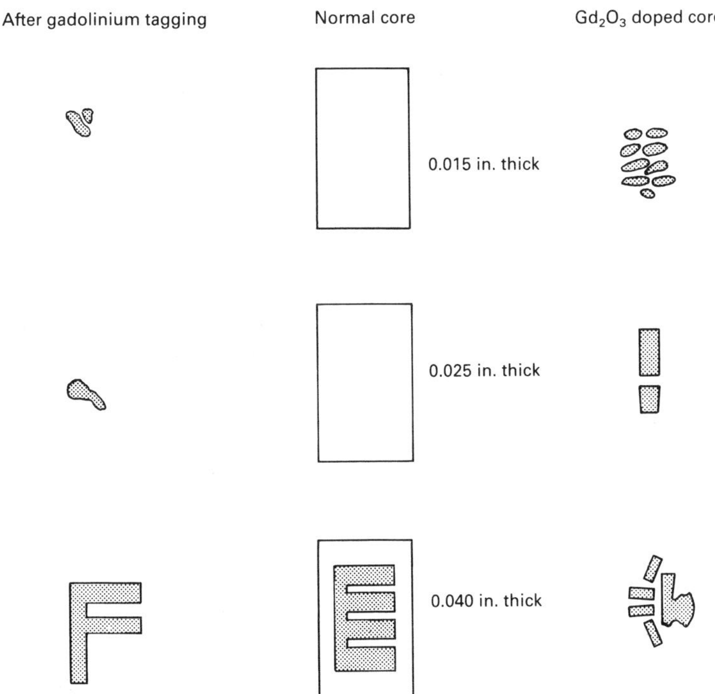

Fig. 15 Schematic of turbine blade core standards: gadolinium [$Gd(NO_3)_3$ in solution] tagged core, normal core (no gadolinium tagging or doping), and Gd_2O_3 doped core. Typical core fragments of various thicknesses are shown. Source: R. Tsukimura, Aerotest Operations Inc.

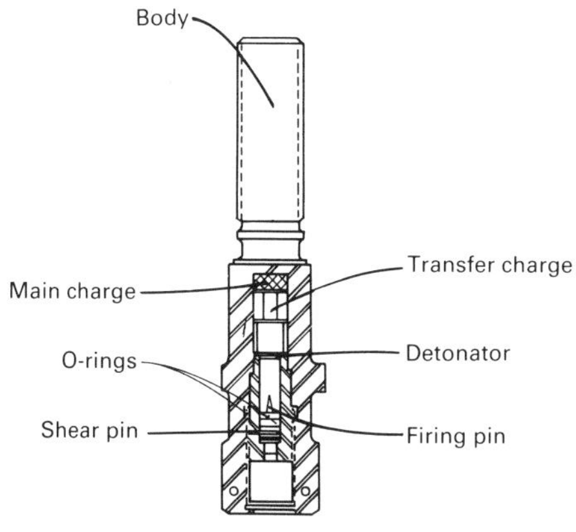

Fig. 16 Schematic showing location of critical components that comprise an explosive bolt. Source: R. Tsukimura, Aerotest Operations Inc.

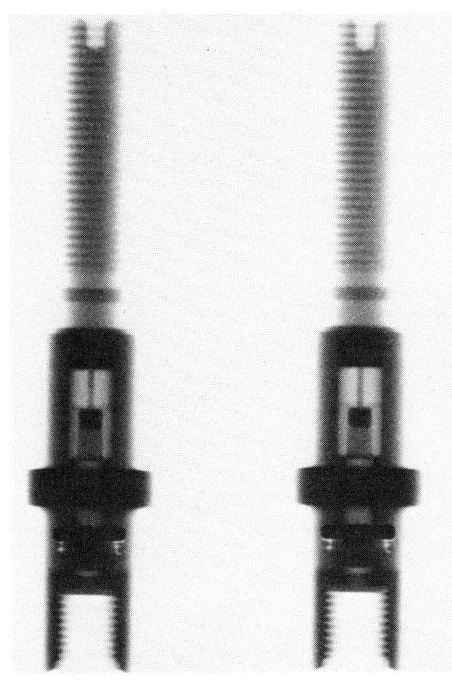

Fig. 17 Thermal neutron radiograph showing two sample bolts identical to the workpiece shown schematically in Fig. 16. Courtesy of R. Tsukimura, Aerotest Operations Inc.

The potting material attenuates the thermal neutrons and appears as the light density area. Voids in the potting material, the fill level, and the distribution can readily be detected with neutron radiography.

REFERENCES

1. H. Berger, Ed., *Practical Applications of Neutron Radiography and Gaging*, STP 586, American Society for Testing and Materials, 1976
2. Neutron Radiography Issue, *At. Energy Rev.*, Vol 15 (No. 2), 1977, p 123-364
3. N.D. Tyufyakov and A.S. Shtan, *Principles of Neutron Radiography*, Amerind Publishing, 1979 (translated from the Russian)
4. P. Von der Hardt and H. Rottger, Ed., *Neutron Radiography Handbook*, D. Reidel Publishing, 1981
5. J.P. Barton and P. Von der Hardt, Ed., *Neutron Radiography*, D. Reidel Publishing, 1983
6. L.E. Bryant and P. McIntire, Ed., Radiography and Radiation Testing, in *Nondestructive Testing Handbook*, Vol 3, American Society for Nondestructive Testing, 1985
7. "Standard Practices for Thermal Neutron Radiography of Materials," E 748, Annual Book of ASTM Standards, American Society for Testing and Materials
8. J.P. Barton, G. Farny, J.L. Person, and H. Rottger, Ed., *Neutron Radiography*, D. Reidel Publishing, 1987
9. H. Berger, Some Recent Developments in X-Ray and Neutron Imaging Methods, in *Nondestructive Testing*, Vol 1, J.M. Farley and R.W. Nichols, Ed., Pergamon Press, 1988, p 155-162
10. A. Ridal and N.E. Ryan, in *Neutron Radiography, Proceedings of the Second World Conference* (Paris, June 1986), J.L. Barton *et al.*, Ed., D. Reidel Publishing, 1987, p 463-470

Thermal Inspection

Grover Hardy, Wright Research and Development Center, Wright-Patterson Air Force Base,
and James Bolen, Northrop Aircraft Division

THERMAL INSPECTION comprises all methods in which heat-sensing devices are used to measure temperature variations in components, structures, systems, or physical processes. Thermal methods can be useful in the detection of subsurface flaws or voids, provided the depth of the flaw is not large compared to its diameter. Thermal inspection becomes less effective in the detection of subsurface flaws as the thickness of an object increases, because the possible depth of the defects increases.

Thermal inspection is applicable to complex shapes or assemblies of similar or dissimilar materials and can be used in the one-sided inspection of objects. Moreover, because of the availability of infrared sensing systems, thermal inspection can also provide rapid, noncontact scanning of surfaces, components, or assemblies.

Thermal inspection does not include those methods that use thermal excitation of a test object and a nonthermal sensing device for inspection. For example, thermally induced strain in holography or the technique of thermal excitation with ultrasonic or acoustic methods does not constitute thermal inspection.

Principles of Thermal Inspection

The basic principle of thermal inspection involves the measurement or mapping of surface temperatures when heat flows from, to, or through a test object. Temperature differentials on a surface, or changes in surface temperature with time, are related to heat flow patterns and can be used to detect flaws or to determine the heat transfer characteristics of a test body. For example, during the operation of a heating system, a hot spot detected at a joint in a heating duct may be caused by a hot air leak. Another example would be a hot spot generated when an adhesive-bonded panel is uniformly heated on one side. A localized debonding between the surface being heated and the substructure would hinder heat flow to the substructure and thus cause a region of higher temperature when compared to the rest of the surface. Generally, the larger the imperfection and the closer it is to the surface, the greater the temperature differential.

Heat Transfer Mechanisms. Heat will flow from hot to cold within an object by conduction and between an object and its surroundings by conduction, convection, and radiation. Within a solid or liquid, conduction results from the random vibrations of individual atoms or molecules that is transferred via the atomic bonding to neighboring atoms or molecules. In a gas, the same process occurs but is somewhat impeded by the greater distance between the atoms or molecules and the lack of bonds, thus requiring collisions to transfer the energy. When a gas or liquid flows over a solid, heat is transferred by convection. This occurs from the collisions between the atoms or molecules of the gas or liquid with the surface (conduction) as well as the transport of the gas or liquid to and from the surface. Convection depends upon the velocity of the gas or liquid, and cooling by convection increases as the velocity of the gas or liquid increases.

Radiation is the remaining mechanism for heat transfer. Although conduction and convection are generally the primary heat transfer mechanisms in a test object, the nature of thermally induced radiation can be important, particularly when temperature measurements are made with radiation sensors.

Electromagnetic radiation is emitted from a heated body when electrons within the body change to a lower energy state. Both the intensity and the wavelength of the radiation depend on the temperature of the surface atoms or molecules. For a blackbody, the radiation wavelength and spectral emission power vary as a function of temperature, as shown in Fig. 1. At 300 K (27 °C), the temperature of a warm day, the dominant wavelength is 10 μm (400 μin.), which is in the infrared region (Fig. 2). A surface would have to be much hotter for the dominant wavelength to fall in the visible region below 0.7 μm (30 μin.). Examples of this are red-hot steel in a forge, a light bulb filament, and the sun (indicated as solar radiation in Fig. 1 at a temperature of 5800 K). However, most subjects in thermal inspection methods will be at temperatures near room temperature and will emit in the infrared region.

Material Heat Transfer Characteristics. The heat transfer mechanisms of conduction, convection, and radiation are affected by material heat transfer characteristics, and thermal inspection depends on local variations in the material heat transfer characteristics. The material characteristics that affect conduction and convection are as follows:

- Specific heat, c, is the amount of heat a mass of material will absorb for a given temperature interval
- Density, ρ, is the mass per unit volume of the material
- Thermal conductivity, k, is the amount of heat that flows in a given direction when there is a temperature difference across the material in that direction
- Thermal diffusivity, α, is the speed at which the heat flows away from a region of higher temperature to the surrounding material
- Convection heat transfer coefficient, h, is a measure of how efficiently heat is exchanged between a surface and a flowing gas or liquid
- Temperature, T, is a measure of the heat energy (local thermal agitation) contained at each point in the test object

Thermal inspection depends on differences in these material characteristics to establish a measurable, and usually localized, temperature differential. For example, when a test body with variations in density and specific heat is heated or cooled from a state of uniform temperature, the change in temperature will occur more slowly in the regions with a higher density and/or specific heat. This difference in the rate of change of temperature within the body produces temperature differentials, which may be measurable.

The important material characteristic in radiation heat transfer is the emissivity, ϵ, of a test surface. The emissivity indicates the efficiency of a surface as a radiator (or absorber) of electromagnetic radiation. Blackbodies, the most efficient radiators and absorbers of electromagnetic radiation,

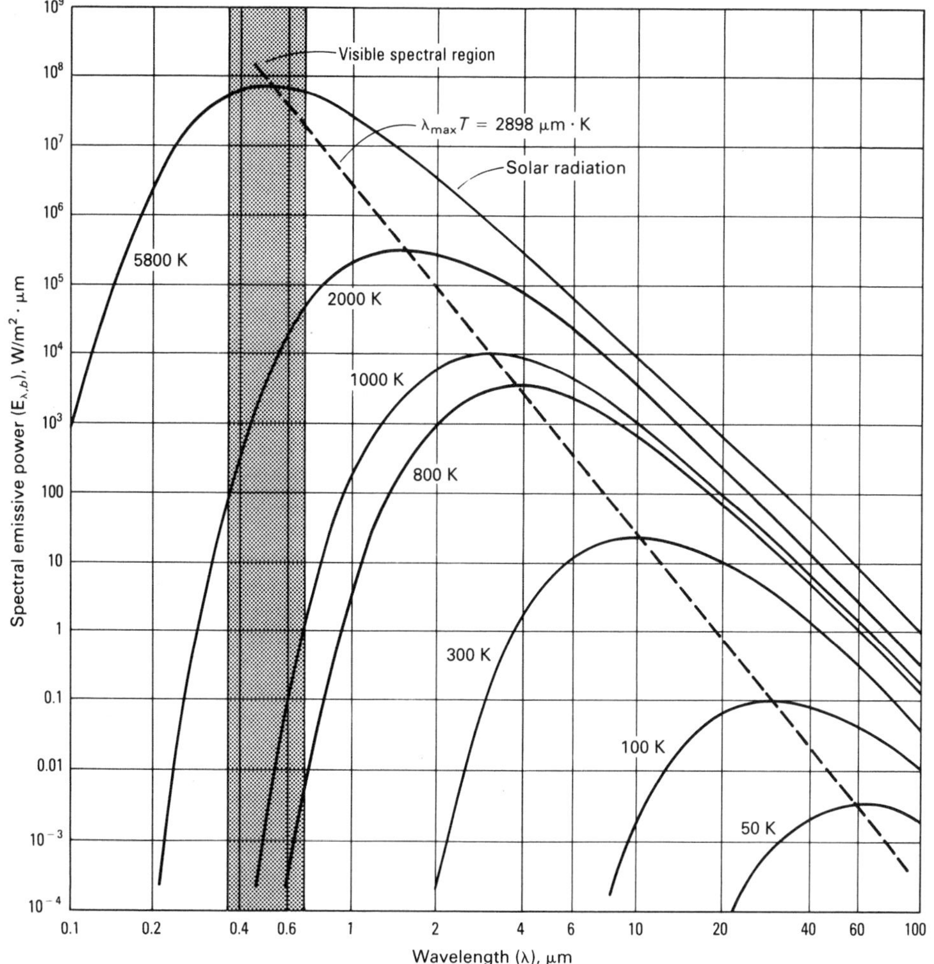

Fig. 1 Spectral blackbody emissive power

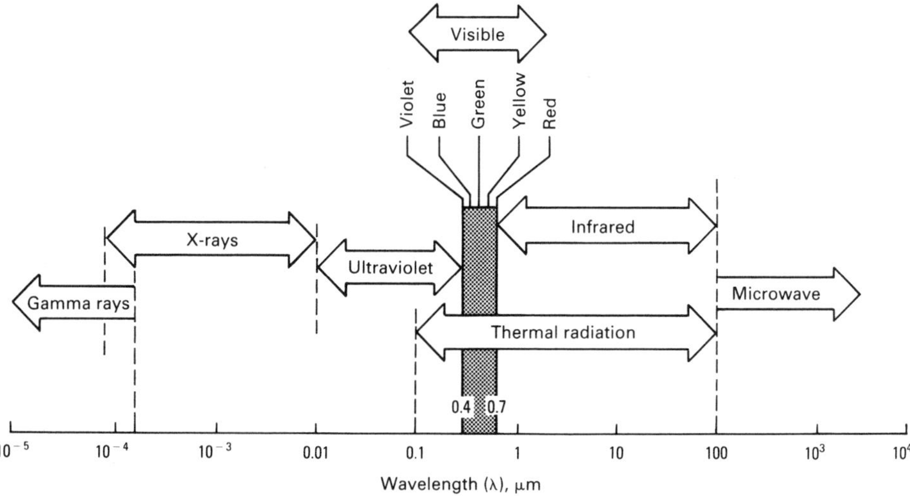

Fig. 2 Spectrum of electromagnetic radiation

have an emissivity of 1.0. All other bodies have an emissivity less than 1.0.

Emissivity is a function of several variables, including color and surface roughness. Like other material heat transfer characteristics, variations in emissivity are important in thermal inspection. This is particularly true when surface temperatures are measured with infrared sensors. Variations in emissivity change the power of radiation emitted at a given temperature and thus affect infrared temperature measurements.

Surface Preparation

The surface condition of the test object is important for thermal inspection. Inspection results can be influenced by variations in surface roughness, cleanliness, foreign material (such as decals), and the uniformity and condition of paint or other surface coatings. A good practice is to clean the surface, remove or strip poorly adhering coatings (if present), and then apply a uniform coating of readily removable flat-black paint. This will allow uniform heat transfer into (or from) the subject and will also produce a reasonably uniform emissivity.

Establishing Heat Flow

In thermal inspection, the test object can be classified as either thermally active or thermally passive. Thermally active test objects generate the necessary heat flow during their operation, while thermally passive test objects require an external heat source or heat sink.

Thermally Active Test Objects. Some test objects can be inspected without the application or removal of heat because they are involved in a process that either generates or removes heat. When a defect results in an abnormal temperature distribution on the surface, no external heating or cooling is required. When the heat transfer process is transient, the timing of the inspection is important. An example of this would be a fluid-contaminated honeycomb panel on an aircraft that has just landed after a long flight at high altitudes. Although the entire aircraft would be warming up from the cooler temperatures experienced at high altitude, the contaminated regions would not warm as rapidly as the uncontaminated areas and therefore could be detected as cool spots in the structure if the inspection were performed immediately after landing. However, if the inspection were delayed, the entire structure would reach an equilibrium temperature, and the contaminated regions would no longer be detectable.

When the heat transfer process is in a steady-state condition, timing no longer becomes critical. An example is an electronic circuit board. Defective electronic components, in which the defect changes the electrical resistance of the component, will be either hotter or cooler than the same component properly operating on another circuit board. Another example would be a blocked tube in a heat exchanger. Temperatures along the tube would be different from temperatures along adjacent, unblocked tubes.

Thermal Excitation of Passive Test Objects. Thermally passive test objects require an external heat source or heat sink to establish the flow of heat to or from the test object. Generally, infrared or thermal measurement techniques become more sensitive as the average temperature of the subject increases. Consequently, the most common form of excitation is heating. However, in cases where additional heating could cause damage, cooling is used to create the required heat flow.

Precautions. One primary concern in heating or cooling the test object is that the thermal changes must not be intrusive. The rate of heating or cooling must be below the point of producing damaging thermal stresses. For example, a chilled glass placed in hot water may crack because the exterior is expanding rapidly while the interior is still cool and because the induced mechanical stress is sufficient to cause fracture. The degree of heating of metals must be controlled so as not to affect the heat treatment, and excessive heat can degrade the material properties of adhesives and matrices in reinforced resin composites.

Point heating can be generated with either a laser or a spherically focused infrared or visible light source to apply heat to a very localized area. For a very thin, homogeneous subject, the heat will flow outward in a circle, and the temperature change will be inversely proportional to the square of the distance from the center of the circle or the irradiated point. For a very thick, homogeneous subject, the heat can also flow in the thickness direction; consequently, the temperature change will be inversely proportional to the cube of the distance to the point of irradiation. Most subjects are somewhere between these two extremes. Because of the rapid change in temperature across the subject, the use of point heating requires rapid-response, continuous time-based measurements, which may be enhanced with pulsed heating techniques (Ref 1-3).

Line heating is similar to point heating and is usually accomplished with a linear heat source such as a quartz-tube lamp or other heat source with a reflector that concentrates the heat into a line on the surface of the test object. For a thin panel, the heat will flow away from the line on both sides, and the temperature change will be inversely proportional to the distance from the line. For a very thick, homogeneous subject, the temperature change will be inversely proportional to the square of the distance to the line of irradiation. Because the temperature changes rapidly with distance, line heating should be monitored with rapid-response, continuously monitored, time-based equipment (tracking the time to reach a given temperature by scanning the detector outward from the line source). Another technique involves scanning the linear heat source across the surface.

Area heating typically involves uniform heating on one surface of the test object. A bank of quartz heaters, a hot plate with a high thermal conductivity material such as copper as a face sheet, or other similar devices can be used as area heaters. If heat is applied to the same side of the object that is subsequently monitored for temperature differences, the heat source is usually switched off to prevent interference with the detector from the source. If the opposite surface is heated, the heat source usually will not interfere with the temperature sensor and can be left on to provide a steady-state condition. The heat source can also be switched off to produce a transient measurement condition.

Vibration-Induced Heating. Exciting a specimen on a shaker table or with a high-power speaker will cause the specimen to respond with its natural modes of vibration. These vibrations will induce localized areas of stress and strain in the object. An anomalous area will respond differently to these induced stresses and, as a result, may be hotter or cooler than the surrounding areas if the anomaly is not at a vibration node (Ref 4-6).

Mechanically Induced Heating. Mechanically loading the specimen will cause heating in those areas that are plastically deformed. Because deformation of the specimen is normally destructive (except in those cases where deformation is part of the intended processing), mechanically induced heating is probably best suited to research studies. Examples are investigations of fatigue and crack growth, analysis of the response of a component to loading, and optimization of deformation processes such as forging (Ref 7-9).

Electrically Induced Heating. Passing current through a specimen, or inductive heating, can be applied only to materials that are electrically conductive. The heat produced is a product of the square of the electrical current and the resistance. If the anomaly of interest locally changes the electrical resistance, the local current flow and the local temperature will also change. One shortcoming of this method is that electrically conductive materials often have good thermal conductive properties, thus making the local changes in temperature short in duration (Ref 10).

Area cooling can be used when heating may be damaging or impractical. Partial immersion of a test object in a cold fluid such as ice water or liquid nitrogen may provide sufficient heat flow to reveal anomalous areas when monitoring the surface temperature as the specimen cools or returns to ambient. As with area heating, a steady-state temperature profile obtained by continued cooling of one surface of the specimen may be useful. For example, filling a cooler with ice and water would allow the exterior detection of cold spots caused by faults in the insulation or a defective lid seal.

Total immersion of the test object in a cold fluid can also be considered a cooling method. In this case, the transient surface temperatures are monitored while the test object returns to ambient temperature.

Cold-thermal-wave excitation involves preheating the test object, followed by cooling with an air jet (see Example 2). The cold-thermal-wave approach is attractive for the thermal testing of metallic structures for a number of reasons. It is an inexpensive technique for fast thermal stimulation of large areas; preheating can be performed with low-rate elements such as hot-wire heaters, although in several on-line applications the part may already be above ambient temperature; and finally, the use of a cold thermal source eliminates self-emission noise problems.

Thermal Inspection Equipment

The temperature sensors used in thermal inspection can be separated into two categories: noncontact temperature sensors and contact temperature sensors. Other equipment includes recording instruments and calibration sources.

Noncontact Temperature Sensors

Noncontact temperature sensors depend on the thermally generated electromagnetic radiation from the surface of the test object. At moderate temperatures, this energy is predominately in the infrared region. Therefore, noncontact measurements in thermal inspection primarily involve the use of infrared sensors.

Infrared imaging equipment is available with a wide range of capabilities. The simplest systems are responsive to the near-infrared portion of the optical spectrum. These include night-vision devices and vidicon systems with silicon or lead sulfide sensors (Ref 11). Silicon sensors provide sensitivity for temperatures above 425 °C (800 °F), while lead sulfide sensors respond to temperatures above 200 °C (400 °F).

Hand-held scanners are portable imaging systems capable of responding in the far-infrared portion of the optical spectrum (wavelengths of 8 to 12 μm). This range is emitted by objects at or near room temperature. In general, hand-held scanners have poor imaging qualities and are not suitable for the accurate measurement of local temperature differences. However, they can be useful for detecting hot spots, such as overheated components, thermal runaway in an electronic circuit, or unextinguished fires (Ref 11).

High-resolution infrared imaging systems are required for most part inspection applications. These systems use either pyroelectric vidicon cameras with image-processing circuitry or cryogenically cooled mechanical scanners to provide good-quality image resolution (150 pixels, or picture elements, per scan line) (Ref 11) and temperature sensitivity to 0.1 °C (0.2 °F) (Ref 12). One system is claimed to have a temperature resolution of 0.001 °C (0.002 °F) (Ref 12). In addition to good image resolution and temperature sensitivity, response times of the order of 0.1 s or less facilitate the detection of transient temperature changes or differentials. These imaging systems will use either a gray scale or a color scale correlated to temperature ranges to depict the temperature distribution within the image.

Thermal wave interferometer systems combine modulated laser excitation with rapid phase and amplitude sensing that can be scanned across a surface to produce an image (Ref 1, 12). One application for this type of system is the inspection of plasma-sprayed coatings. The system senses the interaction between the thermal waves of the laser and the thermal variations from coating defects and thickness variations.

Radiometers and pyrometers are devices for measuring radiation, or spot or line temperatures, respectively, without the spatial resolution needed for an imaging system. Radiometers, because they usually have slow response times, are most useful for monitoring constant or slowly varying temperatures. Pyrometers are used as noncontacting thermometers for temperatures from 0 to 3000 °C (32 to 5400 °F). Newer instruments can superimpose a line trace of the temperature on the visible-light image of the surface or scene being viewed (Ref 13). Radiometers and pyrometers are usually rugged, low-cost devices that can be used in an industrial environment for the long-term monitoring of processes.

Contact Temperature Sensors

Contact temperature sensors include material coatings and thermoelectric devices. Material coatings are relatively low in cost and simple to apply, but they may have the disadvantage of providing qualitative temperature measurements (the exception is coatings with liquid crystals, which can be calibrated to show relatively small changes in temperature). Another disadvantage of coatings is that they may change the thermal characteristics of the surface.

Cholesteric liquid crystals are greaselike substances that can be blended to produce compounds having color transition ranges at temperatures from −20 to 250 °C (−5 to 480 °F) (Ref 13). Liquid crystals can be selected to respond in a temperature range for a particular test and can have a color response for temperature differentials of 1 to 50 °C (2 to 90 °F) (Ref 13). When illuminated with white light while in their color response range, liquid crystals will scatter the light into its component colors, producing an iridescent color that changes with the angle at which the crystals are viewed. Outside this color response range, liquid crystals are generally colorless. The response time for the color change varies from 30 to 100 ms (Ref 13). This is more than adequate to allow liquid crystals to show transient changes in temperature (Ref 14). The spatial resolution obtainable can be as small as 0.02 mm (0.0008 in.) (Ref 14). In addition, because the color change is generally reversible, anomalies can be evaluated by repeating the test as many times as needed.

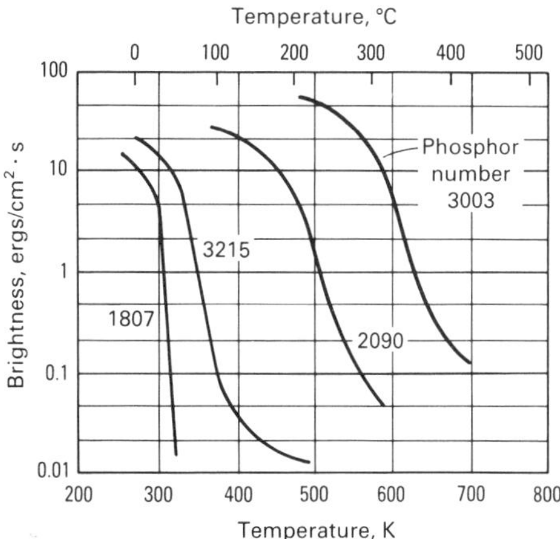

Fig. 3 Relative brightness of four thermally quenched phosphors (U.S. Radium Radelin phosphor numbers) as a function of temperature

Techniques for applying liquid crystals are relatively straightforward once the proper blend of compounds is selected. Because liquid crystals function by reflecting light, they are more readily seen when used against a dark background. Therefore, if the specimen is not already dark, covering the surface with a removable, flat-black coating is strongly recommended before application. The crystals can then be applied by pouring, painting, spraying, or dipping. Care must be taken that the specimen or the coating is not attacked by the solvent base used with the liquid crystals. The applied film of liquid crystals must be of uniform thickness to prevent color irregularities caused by thickness differences rather than temperature differences. A good film thickness is about 0.02 mm (0.0008 in.) (Ref 13). Successive layers used to build up the film thickness should not be allowed to dry between coats. A coating of proper thickness will have a uniform, low-gloss appearance when viewed with oblique illumination.

Thermally quenched phosphors are organic compounds that emit visible light when excited by ultraviolet light. The brightness of a phosphor is inversely proportional with temperature over a range from room temperature to about 400 °C (750 °F), as indicated in Fig. 3. Some phosphors exhibit a change in brightness of as much as 25%/°C (14%/°F). An individual phosphor should be selected to cover the temperature range used for a particular inspection. The coating is applied by painting a well-agitated mixture of the phosphor onto the surface to a thickness of about 0.12 mm (0.0047 in.).

Other Coatings. Heat-sensitive paints, thermochromic compounds, heat-sensitive papers, and meltable frosts and waxlike substances can also be used to indicate surface temperatures. These coatings are useful for determining when a surface has exceeded a certain temperature. A few of the coatings, such as the photochromic paints and thermochromic compounds, have reversible changes that can be used to evaluate indications by retesting. Each of the coatings can be applied directly to the surface. After some experimentation, the coatings could be used for specialized thermal inspections.

Thermoelectric devices are widely used for measuring temperature. Typical thermoelectric devices are thermocouples, thermopiles, and thermistors.

Thermocouples consist of a pair of junctions of two different metals. As the temperature of one of the junctions is raised, an electromotive force (voltage) relative to the other (reference) junction is produced that is proportional to the temperature difference between the two junctions. Thermocouples are usually used in a bridge circuit, with the reference junction maintained at a known and stable temperature. Thermocouples can be placed in contact with the surface of the subject or can be used near the surface to measure the air temperature.

Thermopiles are multiple thermocouples used electrically in series to increase the output voltage. Although thermopiles have a greater output (resulting in greater sensitivity) than individual thermocouples, they also have a slower response time because of the increased mass. Thermopiles are used as a sensing element in radiometers.

Thermistors are electrical semiconductors that use changes in electrical resistance to measure temperature. Thermistors are usually used in a bridge circuit, with one of the thermistors maintained at a known and stable temperature.

Recording Equipment

When high-resolution imaging systems are used, recording equipment, such as a videotape recorder, is extremely useful for analyzing transient effects or for reviewing techniques. Videotape recordings showing the time-varying response of a location on the specimen can be used to estimate the size and depth of the anomalies (Ref 2, 5, 15). Such recorded data can be processed with digital data-processing techniques to enhance the detection of temperature differences and to suppress spurious noise signals (Ref 4) (see the section "Digital Image Enhancement" in this article). Photographic techniques can also be used to record the thermal images of specimens.

When coatings are used, the recording equipment usually consists of color photographs of liquid crystals, heat-sensitive paints, thermochromic materials, and wax sticks. Black-and-white photographs are adequate for most heat-sensitive papers, melting paint materials, and thermally quenched phosphors. For time-based measurements with liquid crystals, a video recorder can provide an excellent recording of color changes as the test progresses. The recording equipment for thermoelectric devices is usually time-based chart recorders or digital recorders.

Temperature Calibration Sources

Temperature calibration sources are needed for all devices used to measure temperatures. Some systems have built-in or internal calibration sources. External sources can vary from very simple devices, such as a container of ice and water or boiling water, to thermocouple-controlled heated plates that can be adjusted to the desired temperature.

Inspection Methods

Steady-state methods are used to detect anomalies where temperatures change very little with time. Many of the thermally active objects or processes can be observed under steady-state conditions. For example, a steam line that regularly carries steam can be inspected for insulation defects by looking for hot spots. These types of anomalies usually produce large differences in temperature, and the resultant images or contact coating indications are easily interpreted.

Steady-state methods are more challenging with the thermal excitation of passive objects. Uniform heating and cooling are essential, and the heating and cooling rates must be adjusted to allow the temperature differences caused by the defects to be maintained. A honeycomb panel can be inspected for liquid intrusion or disbonds by heating one side and cooling the opposite side. The cooled side is viewed for the inspection, and either infrared imaging or contact coatings can be used. Areas of liquid intrusion will produce warmer temperatures than the surroundings, while disbonds will produce cooler temperatures.

In general, anomalies must be large, must be close to the surface, or must create large temperature differences to be detectable with steady-state methods. With active heating or cooling, steady-state measurements will be more effective when made on the surface where the heat transfer is by convection.

Time-based methods are used to detect anomalies where temperatures change during the inspection. Temperature differences may develop and then disappear as a subject changes from one overall temperature to another (Ref 2, 15). The change may be actively or passively produced. For example, a hot forging exiting a forging press or die will cool to room temperature and will produce a temperature difference at a forging lap for only a short time. An aircraft structure with hidden corrosion can be actively heated from one side and then allowed to cool. Areas with trapped water will cool more slowly and create a temporary warm spot, while thinned or corroded areas will cool slightly faster than the rest of the structure and produce a temporary cool spot.

In general, time-based methods can provide the maximum detection sensitivity and can permit inspection from one side. For infrared or photothermal imaging methods utilizing active heating, inspection can be performed after the heat source is turned off, thus eliminating the interference in the image from the heat source itself. Multiple time-based measurements are required for quantitative interpretation of thermal images (Ref 5). For example, a video recording will capture transient temperature differences produced with surface coating methods on a continuous basis and can be analyzed frame by frame to evaluate the indications as they appear and disappear.

Image interpretation has proved to be the most difficult part of many thermal inspection applications. Strong indications with large temperature differences are the easiest to interpret. Strong indications will usually provide a truer image of the anomaly than weak indications. An anomaly close to the surface will produce a stronger indication than an identical anomaly far from the surface, and the resultant indication will more accurately portray the size and shape of the source.

Thermal images reflect the heat flow in the structure. If the structure has an area with a subsurface support or cavity, the heat will flow faster or slower, respectively, into these areas. The edges of the structure will experience more convective and radiative heat transfer than the remainder. Support of the structure will also affect the image because the support is usually another source of heat transfer.

Thermal images can also be influenced by other factors. The heating source, as well as nonuniformities in the emissivity of the specimen caused by variations in surface roughness and color, can produce indications. Consequently, the first step in image interpretation is to look at the strength of the image and closely examine the surface and structure of the specimen as well as the uniformity of the heating source. When no obvious surface or structural correlation exists between the specimen and the image, an internal anomaly should be suspected. If available or possible, digital image enhancement, such as filtering, image averaging, or reference image subtraction, should be used to eliminate nonrelevant indications. An indication that persists is probably a valid indication of an anomaly.

Digital image enhancement can be used to improve the quality of thermal images. Spatial filtering can be used to smooth the temperature data by eliminating the high-frequency noise caused by heating devices and external sources. One technique involves mathematically replacing each pixel value with the average value of its four nearest neighbors (Ref 4). Signal-averaging techniques can also be used to reduce measurement noise if the image does not vary too rapidly with time. Averaging individual pixel values from 100 successive image frames will eliminate 90% of the noise and will increase sensitivity (Ref 4).

Space-domain and time-domain subtraction functions can also be used to enhance flaw detection. Space-domain subtraction functions can eliminate temperature differentials from repetitive sources of noise (such as local variations in surface emissivity), patterns resulting from a nonuniform (or line or spot) heat source, and local variations in heat loss unrelated to anomalies (such as convection patterns or conduction into supporting structures). If self-referencing is not possible, a defect-free reference specimen with similar characteristics can be used to provide a reference image in a space-domain subtraction algorithm (Ref 15). For example, an image taken when a component is new or when a process is functioning properly can be subtracted from an image taken at a later date to indicate damage, flaw propagation, or changes in the process (Ref 4).

When the thermal noise is not repetitive (as when surface emissivity is unpredictable or when reflections occur from an unknown background source), time-domain subtraction functions can enhance flaw detection. A graphic example of time-domain image subtraction is shown in Fig. 4. An irregular pattern of reflective noise from an infrared heat source was eliminated by subtracting two images obtained 3 s and 5 s after initial heating to reveal a defect in a graphite epoxy sheet (Ref 15).

Quantitative Methods

In general, thermographic methods do not lend themselves to the rapid derivation of quantitative data on anomalies. One exception is the case of thin coatings, in which anomaly size and shape correspond very closely to the size and shape of infrared images or contact coating indications from optimized inspections. Similarly, quantitative estimates of very thin coating thicknesses are possible with proper time-based techniques (Ref 1).

Quantification of anomaly size following the detection by thermal methods is usually more readily accomplished by application of a second nondestructive evaluation (NDE) method. For example, determining the size and depth of a delamination detect-

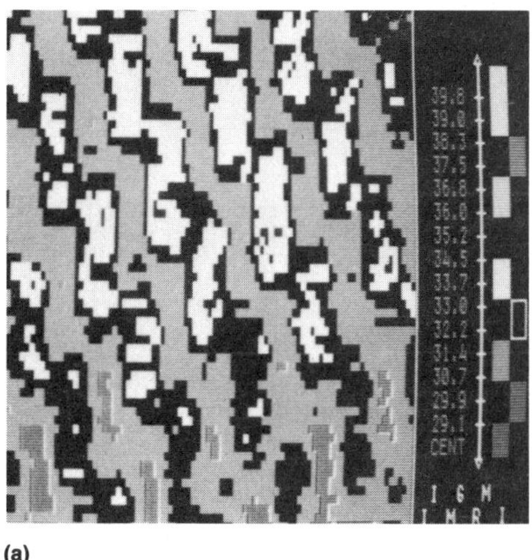

(a)

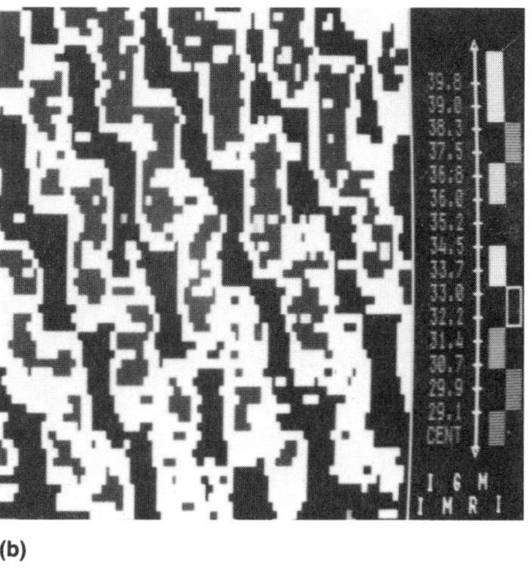

(b)

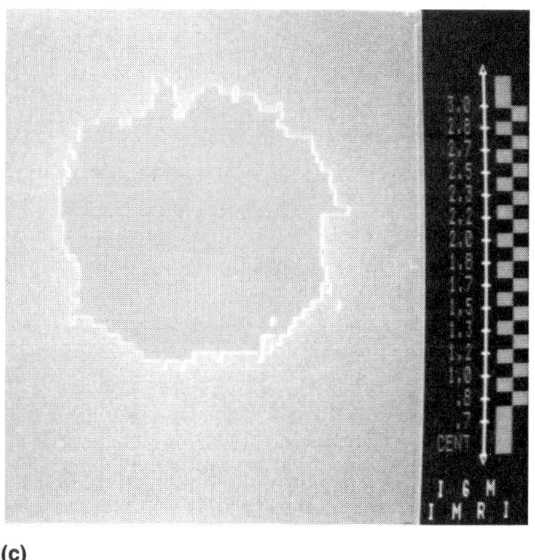

(c)

Fig. 4 Example of time-domain image processing. (a) Image over a defect (10 × 10 mm and 0.5 mm deep, or 0.4 × 0.4 in. and 0.2 in. deep) obtained 3 s after initial heating. (b) Image over the same defect obtained 5 s after initial heating. (c) Image of defect after subtracting image (a) from image (b). Courtesy of P. Cielo, National Research Council of Canada

ed in a graphite/epoxy laminate by thermal methods would be best achieved by ultrasonic techniques. A crack detected in a rocket propellant tank would be best quantified by other methods, such as multiple x-ray films or x-ray computer-aided tomography.

There are some cases in which quantification from thermal indications is always desirable. For example, in a process producing hot specimens, rapid feedback may be necessary because time is too short to wait for the specimens to cool for a subsequent inspection. Another example would be a coating process that needs a rapid thickness measurement to ensure proper thickness.

Most quantitative methods are still under development. Nearly all the quantification methods must be tailored to specific applications. Although some methods can be approached from a theoretical basis, most require an empirical development effort involving reference standards, multiple time-based measurements, and interpretive methods or rules (Ref 5, 13).

Reference standards are as necessary for thermal methods as they are for other NDE methods. Standards ensure consistent performance of the thermal activation methods and the temperature-sensing materials or equipment. Reference standards are essential for quantitative techniques.

A reference standard should closely represent the thermal and surface characteristics of the components or items inspected. For example, if the subject is a solder joint, the reference should be a solder joint with the same contact area, heat sink, wire sizes, and other characteristics. The standard must also provide an anomaly-free condition and a second unacceptable condition associated with a size or the presence of an anomaly. A standard for quantitative methods requires a progression of anomaly sizes and locations. The standard needs to cover the potential accept/reject conditions that can exist in the components to be inspected. This typically necessitates a test program to establish the anomaly sizes and locations that are unacceptable for proper operation of the components to be inspected (Ref 16).

Multiple time-based measurements are needed to establish correlations between a response to thermal activation and anomaly size and location (Ref 5, 16). For pulsed or modulated thermal excitation, the time delay of the temperature response can be recorded as a relative phase change, which may be related to a parameter such as coating thickness (Ref 1) or delamination depth (Ref 3). Alternatively, variations in isotherm position with time can be monitored as a function of anomaly size and depth to establish known time-based response characteristics. Figure 5 shows how the effusivity (which is a function of the pulsed heat input and the time-dependent temperature change) directly over a 20 × 20 mm (0.8 × 0.8 in.) void in a carbon-epoxy laminate varies with time for void depths of 0.3, 1.12, and 2.25 mm (0.012, 0.044, and 0.088 in.).

Quantitative interpretation rules are needed before the time-based data can be used in an evaluation. A simple example is the phase change response as a function of coating thickness of yttria-stabilized zirconium on a nimonic substrate, as shown in Fig. 6. For an acceptable coating thickness of 0.45 to 0.6 mm (0.018 to 0.023 in.), the data show that a 0.5 Hz modulation of the thermal excitation will supply the necessary phase sensitivity and that acceptable coating thickness will be indicated by phase angles ranging from 5 to 13°. This information then provides the accept/reject criteria for tests conducted under these conditions.

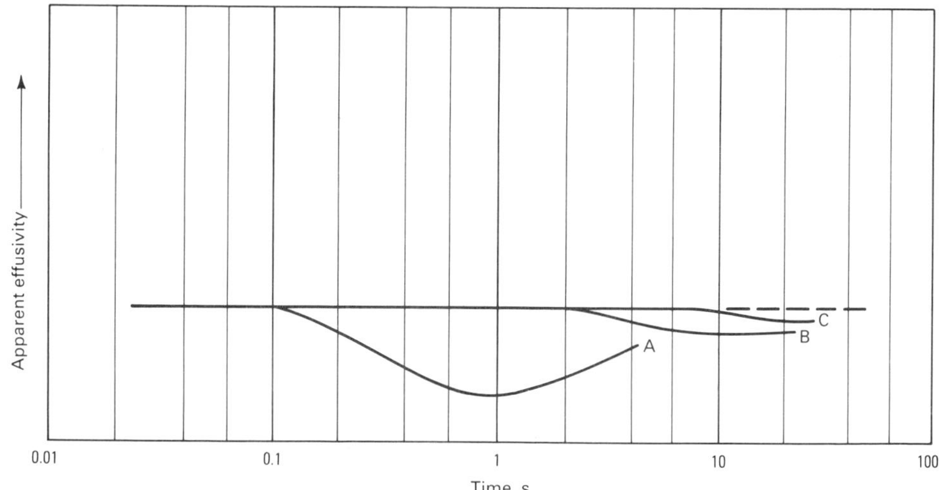

Fig. 5 Experimental apparent-effusivity curves, after correction, of the carbon-epoxy laminate sample in the regions with embedded defects of 20 × 20 mm (0.8 × 0.8 in.) area at three depths. A, 0.3 mm (0.012 in.); B, 1.12 mm (0.044 in.); C, 2.25 mm (0.088 in.). Source: Ref 2

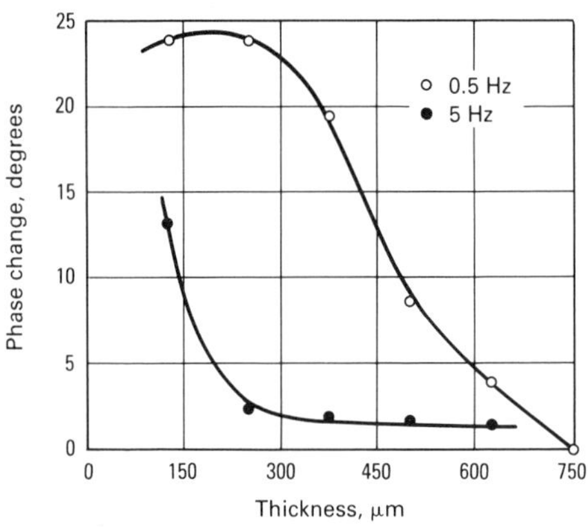

Fig. 6 Variations in the phase of temperature changes versus the coating thickness of a yttria-stabilized zirconia coating on a nimonic substrate. The phase of the temperature changes are relative to the modulation (0.5 and 5 Hz) of the thermal excitation. Source: Ref 1

For a three-variable situation in which void thickness, width and depth beneath a surface are determined, at least three time-based measurements are needed (Ref 5). The resulting calculations relating defect geometry and location are complicated and can be performed only with computer assistance.

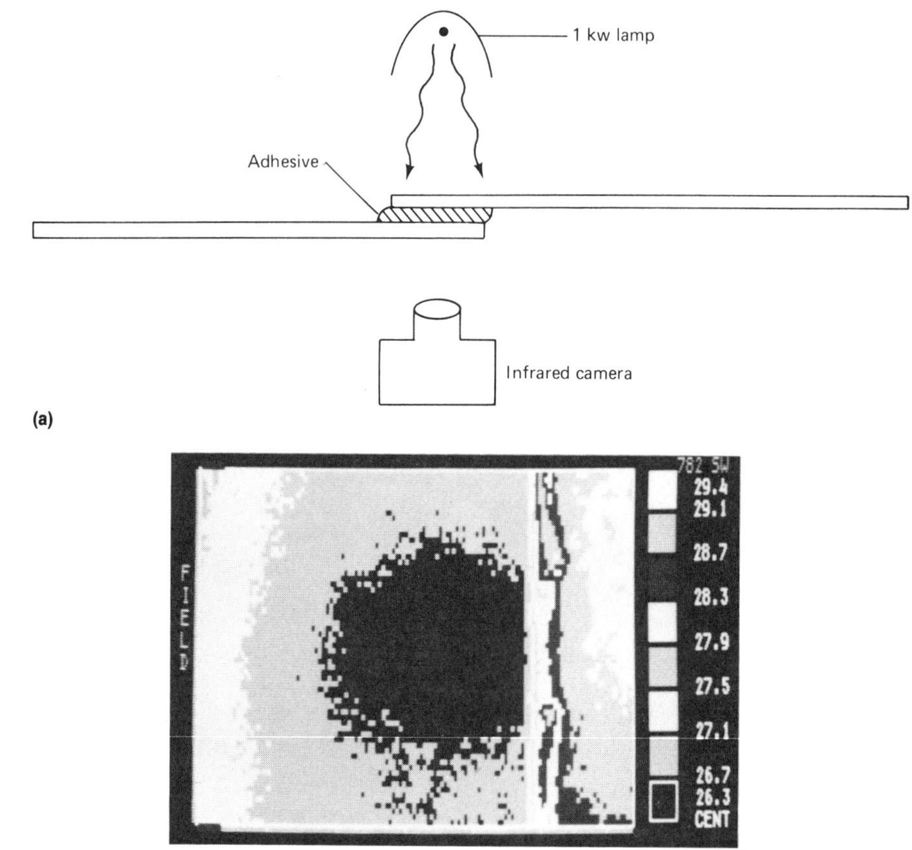

Fig. 7 Thermographic inspection of adhesive-bonded aluminum sheets. (a) Experimental configuration. (b) Thermal image of a 15 × 15 mm (0.6 × 0.6 in.) disbonding in an image area of 40 × 50 mm (1.6 × 2.0 in.). Courtesy of P. Cielo, National Research Council of Canada

Applications

The applications discussed below provide examples of how thermal methods have been used. For ease of discussion, the applications have been divided into the following categories:

- Hot and cold equipment
- Process control
- Liquid intrusion
- Disbonds, delaminations, and voids
- Electronic devices
- Research

Hot and Cold Equipment. Many types of equipment that conduct or generate heat during operation are likely candidates for thermographic methods. Heating ducts, steam lines (Ref 13), radiators (Ref 17), heat exchangers (Ref 15), exhaust systems/chimney stacks (Ref 13), and refrigeration systems are heat transfer devices for which thermal inspection techniques can be used during periodic inspections for leaks, clogged passages, and missing or defective insulation. Furnaces, ovens, salt baths, autoclaves, reaction stacks, and hot manufacturing equipment (such as presses and rolling mills) may also require periodic inspection for unnecessary heat losses. Thermographic techniques are also used to inspect cryogenic tanks.

Other equipment, such as bearings, slides, brakes, transmitting antennas, and electrical equipment, generates heat during operation. Localized hot areas are usually a symptom of a mechanical or electrical malfunction, and early detection provides the opportunity to replace the defective components during regularly scheduled maintenance or before more serious damage occurs.

Process Controls. Thermal inspection methods are appropriate for certain processes in which the products are above room temperature as they exit a process. Examples are heat-set and heat-shaped plastics, hot-worked metal components, hot coating processes, and weld components. Components can be monitored as they exit the process. Abnormal temperatures would indicate an out-of-control process, and corrections can be made to prevent the production of large numbers of defective parts. An example of an actively heated process control method is the technique for monitoring phosphate coating thickness using infrared absorption characteristics (Ref 18).

Liquid Intrusion. Water or fuel intrusion in honeycomb structures is a significant problem for aircraft maintenance. Inspections can be performed immediately after flight to detect such liquids as the structure warms to ambient temperature. Water intrusion in roofs can also be detected in the evening as ambient temperature drops. This method has been reported to be successful in detecting leaks and retained moisture in insulation (Ref 19). The level of fluids in a sealed tank can also be determined thermographically. This is accomplished by either heating or cooling the tank and then noting the location of a sharp differential in resulting tank temperature.

Disbonds, Delaminations, and Voids. Modern fabrication techniques rely heavily on the use of bonding for structures and protective coatings. Thermal techniques are good candidates for the detection of disbonds, delaminations, and voids in thin laminates, honeycomb to thin face sheets, and protective coatings (Ref 15). As the specimen thickness increases, thermal inspection becomes less effective because the possible depth of the defect may be greater.

Thermal inspection may also be difficult with bonded metal structures, such as adhesive-bonded aluminum. The radiative heating of aluminum is relatively inefficient because of the low surface absorptivity, and the thermal-emission signal is low because of the low emissivity. For such reasons, as well as to avoid reflective noise, the aluminum parts must normally be black-painted prior to thermal inspection. Moreover, the high thermal diffusivity of aluminum requires a high thermal-power injection to

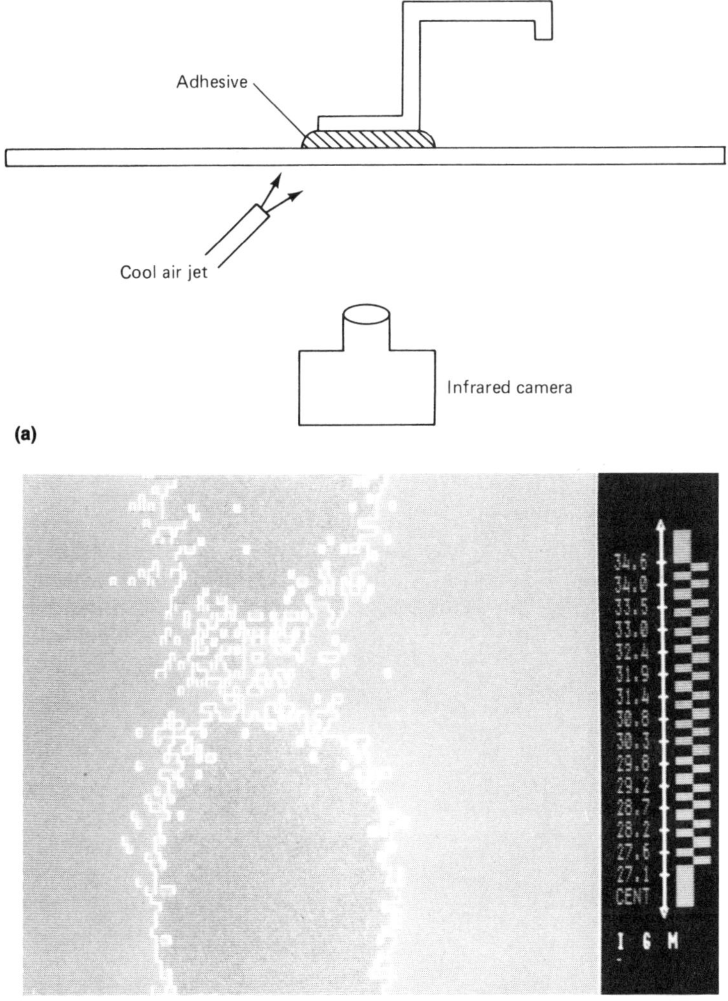

Fig. 8 Cold-thermal-wave technique for inspecting a preheated, adhesive-bonded aluminum structure. (a) Experimental configuration. (b) Thermal image of a 10 × 10 mm (0.4 × 0.4 in.) disbonding. Courtesy of P. Cielo, National Research Council of Canada

produce a visible thermal pattern before thermal uniformization is reached within the structure.

Nevertheless, the possibility of evaluating adhesive-bonded aluminum structures at the rapid pace afforded by thermal inspection is attractive, particularly for online applications (Ref 20). The following two examples describe the thermal inspection of adhesive-bonded aluminum after black-painting.

Example 1: Thermal Inspection of Adhesive-Bonded Aluminum Sheets (Ref 15). Figure 7 shows the detection by a transmission configuration of a 1.5 × 1.5 cm (0.6 × 0.6 in.) lack-of-adhesive defect on a lap joint with nearly 1 mm (0.04 in.) thick skin. The hotter portions on the left and right of the thermal image correspond to the single-sheet material, which was heated at a faster rate than the adhesive joint. The heating rate must be sufficiently fast to avoid significant thermal propagation from the hot single-sheet to the adhesive point. Properly shaped shading masks on the lamp side can be used to avoid single-sheet overheating when the sample geometry is repetitive. A line heating and sample displacement configuration of the type described in Ref 21 can also be used for the fast-response testing of longitudinal joints.

Example 2: Cold-Thermal-Wave Technique for the Thermal Inspection of an Adhesive-Bonded Aluminum Structure (Ref 15). Rapid back heating procedures are difficult to apply when analyzing complex structures of the type shown in Fig. 8. In this case, a cold-thermal-wave approach was applied by which the whole structure was previously warmed up uniformly by nearly 10 °C (20 °F) above ambient, and an ambient air jet was used to rapidly cool the inspected face. The thermal image shows a relatively warmer central area corresponding to the adhesive-bonded strip (oriented vertically in the thermograph). The cooler area in the center of the strip corresponds to a lack-of-adhesive unbond.

Electronic devices typically fail by local overheating, corrosion, or poor solder joints. Infrared microscopes have been used to inspect microelectronics for overheating, and laser-generated heat has been used to inspect circuit-board solder joints. More conventional thermal inspection methods have been used to inspect solder joints during cool-down after fabrication or during circuit operation. Solar cells have also been screened for defects with thermal techniques.

Research. Several investigators have used the heat generated by permanent deformation to track the onset of failure in tensile overloading and crack propagation in fatigue testing (Ref 7-9). Samples of components can be loaded to failure under thermographic monitoring to pinpoint the origin and propagation of deformation (Ref 7).

REFERENCES

1. D.P. Almond, P.M. Patel, and H. Reiter, The Testing of Plasma Sprayed Coatings by Thermal Wave Interferometry, *Mater. Eval.*, Vol 45, April 1987, p 471-475
2. D.L. Balageus, A.A. Déom, and D.M. Boscher, Characterization and Nondestructive Testing of Carbon-Epoxy Composites by a Pulsed Photothermal Method, *Mater. Eval.*, Vol 45, April 1987, p 461-465
3. J.C. Murphy and G.C. Watsel, Photothermal Methods of Optical Characterization of Materials, *Mater. Eval.*, Vol 44, Sept 1986, p 1224-1230
4. P. Potet, P. Jeanin, and C. Bathias, The Use of Digital Image Processing in Vibrothermographic Detection of Impact Damage in Composite Materials, *Mater. Eval.*, Vol 45, April 1987, p 466-470
5. T.V. Baughn and D.B. Johnson, A Method for Quantitative Characterization of Flaws in Sheets by Use of Thermal Response Data, *Mater. Eval.*, Vol 44, June 1986, p 850-858
6. E.G. Henneke II and T.S. Jones, Detection Damage in Composite Materials by Vibrothermography, in *Nondestructive Evaluation and Flaw Criticality of Composite Materials*, STP 696, R.B. Pipes, Ed., American Society for Testing and Materials, 1979
7. B.I. Sandor, D.T. Lohr, and K.C. Schmid, Nondestructive Testing Using Differential Infrared Thermography, *Mater. Eval.*, Vol 45, April 1987, p 372-395
8. Y. Huang, J. Xu, and C.H. Shih, Applications of Infrared Techniques to Research on Tensile Tests, *Mater. Eval.*, Vol 38, Dec 1980, p 76-78
9. Y. Huang, S.X. Li, S.E. Lin, and C.H. Shih, Using the Method of Infrared Sensing for Monitoring Fatigue Process of Metals, *Mater. Eval.*, Vol 42, July 1984, p 1020-1024

10. D.R. Green and J.A. Hassberger, Infrared Electro-Thermal Examination of Stainless Steel, *Mater. Eval.*, Vol 35, March 1977, p 39-43
11. J. Newitt, Application of Specific Thermal Imaging, *Mater. Eval.*, Vol 45, May 1987, p 500-504
12. Infrared Thermal Testing, Product Showcase, *Mater. Eval.*, Vol 45, April 1987, p 403-414
13. H. Kaplan and R. Friedman, Two New Portable Infrared Instruments for Plant Inspection, *Mater. Eval.*, Vol 39, Feb 1980, p 175-179
14. J.H. Williams, Jr., B.R. Felenchak, and R.J. Nagem, Quantitative Geometric Characterization of Two Dimensional Flaws Via Liquid Crystal Thermography, *Mater. Eval.*, Vol 41, Feb 1983, p 190-201, 218
15. P. Cielo *et al.*, Thermographic Nondestructive Evaluation of Industrial Materials and Structures, *Mater. Eval.*, Vol 45, April 1987, p 452-460
16. J.H. Williams, Jr., and R.J. Nagem, A Liquid Crystal Kit for Structural Integrity Assessment of Fiberglass Watercraft, *Mater. Eval.*, Vol 41, Feb 1983, p 202-210
17. E.P. Papadakis, H.L. Chesney, and R.G. Hurley, Quality Assurance of Aluminum Radiators by Infrared Thermography, *Mater. Eval.*, Vol 42, March 1984, p 333-336
18. T.M. Mansour, Nondestructive Thickness Measurement of Phosphate Coatings by Infrared Absorption, *Mater. Eval.*, Vol 41, March 1983, p 302-308
19. E. Feit, Infrared Thermography Saves Shelburne Middle School Time and Money, *Mater. Eval.*, Vol 45, April 1987, p 400-401
20. P. Cielo, R. Lewak, and D.L. Balageas, Thermal Sensing for Industrial Quality Control, in *Thermosense VIII*, Vol 581, Society of Photo Interpretive Engineers, 1985, p 47-54
21. H. Tretout and J.Y. Marin, Transient Thermal Technique for IR NDT of Composite Materials, in *IR Technology and Applications*, Vol 590, Society of Photo Interpretive Engineers, 1985, p 277-292

SELECTED REFERENCES

- R.E. Engelhardt and W.A. Hewgley, Thermal and Infrared Testing, in *Nondestructive Testing: A Survey*, SP-5113, National Aeronautics and Space Administration, 1973, p 118-140
- F.P. Incropera and D.P. DeWitt, *Fundamentals of Heat and Mass Transfer*, 2nd ed., John Wiley & Sons, 1985
- W.N. Reynolds and G.M. Wells, Video-Compatible Thermography, *Br. J. Non-Destr. Test.*, Vol 26 (No. 1), 1984, p 40-44
- A.J. Rogovsky, Ultrasonic and Thermographic Methods for NDE of Composite Tubular Parts, *Mater. Eval.*, Vol 43, April 1985, p 547-555

Optical Holography

Revised by James W. Wagner, The Johns Hopkins University

HOLOGRAPHY is a process for creating a whole image—that is, a three-dimensional image—of a diffusely reflecting object having some arbitrary shape. More precisely, it is a means of recording and subsequently reconstructing wave fronts that have been reflected from or transmitted through an object of interest. Because the entire wave front and not just a two-dimensional image is reconstructed, an image of the original object can be viewed with full depth of field, location, and parallax.

The process involves two steps. In the first step, both the amplitude and phase of any type of coherent wave motion emanating from the object are recorded by encoding this information in a suitable medium. This recording is called a hologram. At a later time, the wave motion is reconstructed from the hologram by a coherent beam in a process that results in the regeneration (reconstruction) of an image having the true shape of the object. The utility of holography for the nondestructive inspection of materials, components, and structures lies in the fact that this reconstructed image can then be used as a kind of three-dimensional template against which any deviations in the shape or dimensions of the object can be observed and measured.

In principle, holography can be performed with any wavelike radiation encompassed in the entire electromagnetic spectrum, any particulate radiation (such as neutrons and electrons) that possesses wave-equivalent properties, and nonelectromagnetic wave radiation (such as sound waves). The two methods currently available for practical nondestructive inspection are:

- Optical holography, using visible light waves, which is discussed in this article
- Acoustical holography, using ultrasonic waves, which is discussed in the article "Acoustical Holography" in this Volume

An optical holographic system can be designed to do only one task or various tasks. A system can be purchased as a unit or assembled in-house from components, or holography can be purchased as an outside service. The selection of a system requires weighing the advantages and disadvantages of each type. The factors governing the selection of holographic systems are discussed in the Appendix to this article.

Uses of Optical Holographic Interferometry

Optical holographic interferometry has been successfully used both in research and testing applications as a noncontacting tool for displacement, strain and vibration studies, depth-contour mappings, and transient/dynamic phenomena analyses. Specific applications of optical holography in nondestructive evaluation include:

- Detection of debonds within honeycomb-core sandwich structures
- Detection of unbonded regions within pneumatic tires and other laminates
- Detection of cracks in hydraulic fittings
- Qualitative evaluation of turbine blades

These uses and others are described in detail in later sections of this article. The advantages and disadvantages of using optical holographic interferometry for nondestructive inspection are described below.

The advantages of using optical holographic interferometry for nondestructive inspection include the following:

- It can be applied to any type of solid material—ferromagnetic or nonferromagnetic; metallic, nonmetallic, or composite; electrically or thermally conductive or nonconductive; and optically transparent or opaque
- It can be applied to test objects of almost any size and shape, provided a suitable mechanism exists for stressing or otherwise exciting the object
- Pulsed-laser techniques allow the inspection of test objects in unstable or hostile environments
- It has an inherent sensitivity to the displacement or deformation of at least one-half an optical wavelength, or about 125 nm (1250 Å). This permits the use of low levels of stress during inspection. Further, special analysis techniques can provide improved sensitivity of almost 1000-fold
- It does not rely on data acquisition by either point-by-point determinations or scanning processes; instead, three-dimensional images of interference fringe fields are obtained of the entire surface (front, back, and internal, if desired) or a large fraction thereof
- It allows flexibility of readout. For example, in flaw detection applications, the images can be examined purely qualitatively for localized fringe anomalies within the overall fringe field. If, on the other hand, the application involves the strain analysis of an object subjected to a specific type of stress, the image can be analyzed to yield a highly quantitative point-by-point map of the resulting surface displacements
- It permits comparison of the responses of an object to two different levels of stress or other excitation. The frame of reference for this differential measurement is usually, but not always, the unstressed or natural state of the test object. This differential type of measurement contrasts with absolute types, which are made without a frame of reference, and with comparative measurements, in which a similar, but different, object is used as a frame of reference
- The interferograms can be reconstructed at any later time to produce three-dimensional replicas of the previously recorded test results

The disadvantages of using optical holographic interferometry for nondestructive inspection include the following:

- Although no physical contact with the test object is required to effect the interaction of the coherent light with either the test object or the photographic plate, it is often necessary to provide fixturing not only for the test object but also for the stressing source. The success of a holographic inspection procedure depends largely on the adequacy of design and the practical performance of both the fixturing and the stress-imparting mechanisms
- It is limited to test objects with wall or component thicknesses that will offer sufficiently large displacements without requiring stressing forces that will cause rigid-body movement or damage to the object. For sandwich structures, the thickness of the skin is the limiting factor; the maximum skin thickness that can be

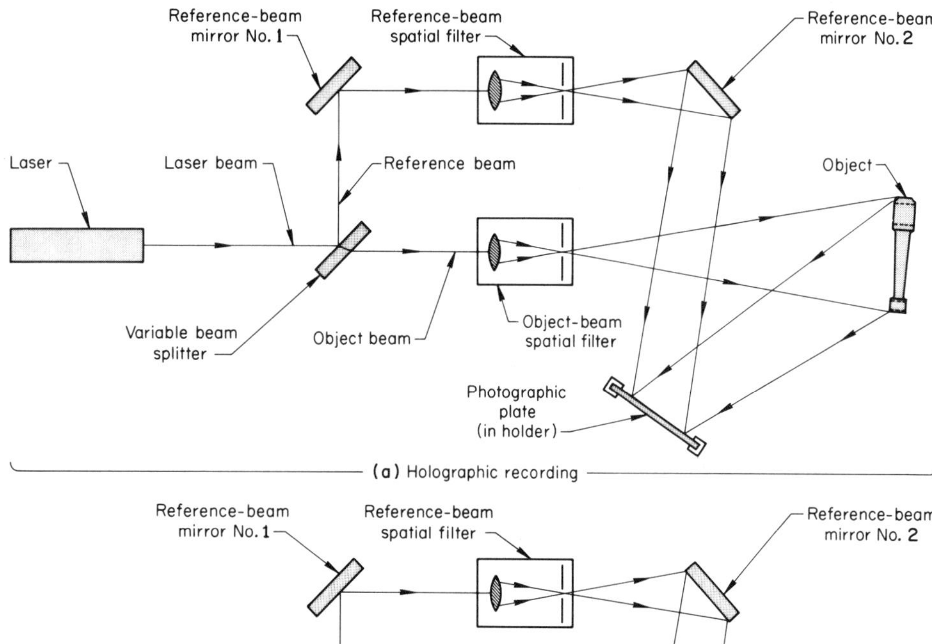

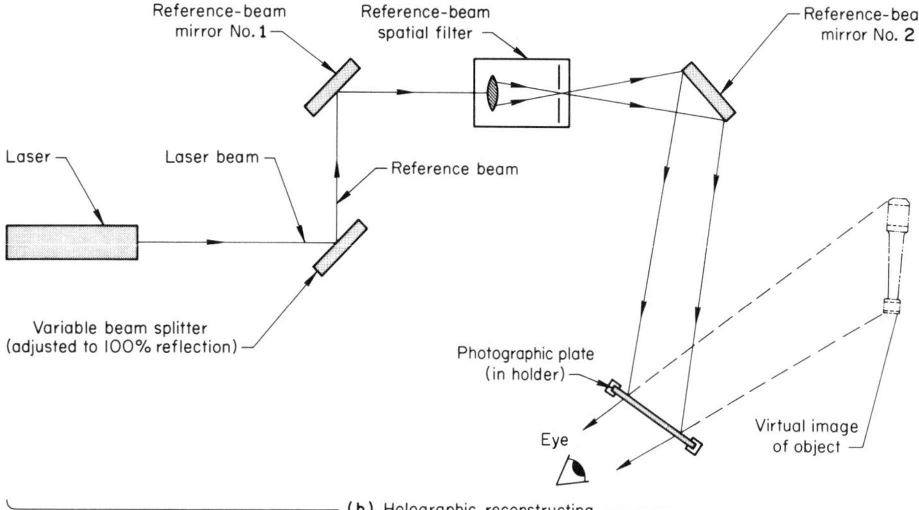

Fig. 1 Schematics of the basic optical systems used in continuous-wave holography. (a) Holocamera used to record hologram of an object on a photographic plate. (b) Optical system for reconstructing a virtual image of the object from the hologram on the photographic plate

tested is a function of the stressing method

- Holographic methods are best suited to diffusely reflecting surfaces with high reflectivity. Removable coatings are often sprayed onto strongly absorbing materials and specularly reflecting surfaces
- Although holographic interferometry is capable of accurately locating a flaw within the surface area of the test object being inspected, the cross-sectional size of the flaw can often be only approximately determined, and information concerning the depth of the flaw, when obtainable at all, is qualitative in nature. Furthermore, a direct correspondence does not always exist between the shape of a flaw-indicative fringe anomaly and the actual shape of the flaw and between the size of the fringe anomaly and that of the flaw. For a given set of test conditions in which only the type of the applied stress is varied, the fringe anomalies indicating the presence of a flaw will vary. In many cases, flaw detection by means of holographic inspection is impeded by the appearance of spurious fringes associated with rigid-body motion and/or rigid-body displacements
- Where visual interpretation of interferograms is to be performed, holographic interferometry may be limited in its dynamic range for some applications. However, electronic methods for fringe interpretation have been used to provide increased sensitivity and dynamic range
- Test results sometimes cannot be analyzed because of the localization of the interference fringes in space rather than on the surface of the test object (in real-time holographic interferometry) or on the reconstructed image of the object (in double-exposure holographic interferometry)
- With the exception of holographic-contouring applications, holographic interferometry is currently limited to differential tests, in which the object is compared to itself after it has been subjected to changes in applied stress. Comparative tests, in which a given object can be compared to a standard object, are not feasible with holographic interferometry because of inherent random variations in test object surfaces
- Personnel performing holographic inspection must be properly trained. The greater the sophistication of the equipment, the greater the required operating skill. Because lasers are a necessary part of holographic equipment, operating personnel must adhere strictly to existing safety codes, particularly with respect to eye safety

Holographic Recording

When visible light waves are employed in holography, the hologram is recorded with an optical system often referred to as a holocamera (Fig. 1a). A monochromatic laser beam of phase-coherent light is divided into two beams by a variable beam splitter. One beam—the object beam—is expanded and filtered by a lens/pinhole spatial filter that reduces the effects of dirt and dust in the beam path. The expanded divergent beam is then directed to illuminate the object uniformly. A portion of the laser light reflected from the object is intercepted by a high-resolution photographic plate, as shown in Fig. 1(a). The second beam—the reference beam—originating from the beam splitter diverges from a second spatial filter and is steered onto the photographic plate directly without reflecting from the object. With either the object beam or reference beam absent, a uniformly exposed photographic plate will result. However, with both coherent beams falling on the plate simultaneously, an interference pattern is generated as a result of the coherent interaction of the two beams, and this pattern is recorded by the photographic emulsion. In fact, this complex interference pattern recorded on the photographic plate contains all the information necessary to reconstruct an extremely high fidelity, three-dimensional image of the object.

Fine-Grain Photographic Emulsions Required for Fringe Resolution. The details of the interference pattern recorded on the film are extremely fine, requiring that high-resolution emulsions be used in very stable recording systems. The fixed wavelength of the light used and the varying magnitude of the offset angle (the angle between the object beam and reference beam at any point on the photographic plate) determine the orientation and spacing of the fringes comprising the recorded interference pattern. The fringe spacing in turn influences the resolution requirements of the recording medium. These requirements are generally of the order of 500 to 3000 lines per millimeter. Thus, the photographic emulsion must have sufficient resolution and contrast to record these interference fringes clearly.

Ordinary photographic emulsions are usually unsatisfactory, because they can resolve at most only about 200 lines per millimeter; exceptionally fine grain emulsions are required.

Optical Path Length Variations. The difference between the path lengths of the reference beam and the object beam must be less than the temporal coherence length of the laser beam and should be as close to zero as possible to obtain fringes of the highest contrast. The holocamera must be isolated from vibration to the extent that there is less than one-fourth to one-eighth wavelength variation in the object beam and reference beam path lengths during an exposure. Such variations can be caused by the relative motion of optical elements in the beam paths, by room vibration, or by temperature changes. Air turbulence can also cause optical path changes that contribute to reduced interference fringe contrast.

Reference Beam Versus Object Beam Intensity Variations. In the absence of optical path length variations, the contrast of the interference fringes is directly related to the amplitude of the reflected light from the object and the amplitude of the reference beam. Thus, when recording a hologram, the reference and object beam intensities are adjusted with a variable beam splitter and/or appropriate attenuating filters to obtain optimum contrast. Although optimum fringe contrast would be obtained for equal object and reference intensities at the film plate, the variation in reflected intensities from point to point on a diffuse object, coupled with concerns for recording linearity, dictates that the reference beam intensity be two to ten times brighter than the object beam intensity at the film plate. A more detailed discussion of the effects of this and other recording parameters is given in the section "Effects of Test Variables," (specifically the discussion "Exposure Parameters") in this article.

Continuous-Wave (CW) Lasers Versus Pulsed Lasers. Two generic classes of laser sources are available for holography: continuous-wave lasers and pulsed lasers. Continuous-wave lasers are characterized by the continuous (steady-state) emission of coherent light at relatively low power. Pulsed lasers are characterized by the pulsed emission of coherent light at relatively high power over sometimes extremely short intervals of time.

Depending on the laser power, a few seconds of exposure of a holographic plate with a CW laser can yield the same results as a single-pulse exposure with a pulsed laser with regard to achieving an exposure level suitable for reconstructing a holographic image. Therefore, a CW laser source can be effectively used to record holograms of objects that are stationary throughout the duration of the exposure. A pulsed-laser source must be used to record holograms of objects undergoing rapid changes by freezing the motion (that is, by recording information over a very small interval of time). Regardless of which class of laser source is used for recording the hologram, a CW laser is always used for reconstruction.

Holographic Reconstruction

In the reconstruction process (Fig. 1b), the complex interference pattern recorded on the hologram is used as a diffraction grating. When the grating is illuminated with the reference beam only, three angularly separated beams emerge: a zero-order, or undeflected, beam and two first-order diffracted beams. The diffracted beams reconstruct real and virtual images of the object to complete the holographic process. The real image is pseudoscopic, or depth inverted, in appearance. Therefore, the virtual image (also referred to as the true, primary, or nonpseudoscopic image) is the one that is of primary interest in most practical applications of holography. In Fig. 1(b), only the first-order diffracted beam that yields the virtual image has been shown; the other two beams were omitted for reasons of clarity. If the original object is three dimensional, the virtual image is a genuine three-dimensional replica of the object, possessing both parallax and depth of focus. However, if the configuration of the optical system or the wavelength of light used during reconstruction differs from that used during recording, then distortion, aberration, and changes in magnification can occur. (The holographic recording and reconstruction systems can be designed to minimize these effects.)

The light intensity in the reconstructed image depends on the diffraction efficiency of the hologram. This efficiency is a function of several recording parameters, the most significant of which is the type of recording medium (film) used. Under ideal conditions, gelatin or polymeric films, which modify only the phase of the reconstructing light without absorbing it, can provide nearly 100% holographic diffraction efficiency. Commercially available holocameras using thermoplastic phase holograms, which can be developed by a thermal process without removing the film from its holder, provide diffraction efficiencies up to about 20%. High-resolution black-and-white film plates, which are used to form absorption holograms, have theoretical efficiencies up to 6.25%. For a typical hologram of this type, however, the intensity is usually less than 3% of the incident reconstructing light intensity.

Once reconstructed, the light in the image beam can be used just as one would use the light from an illuminated object viewed through a window. Pictures of the reconstructed image can be recorded photographically or electronically.

Interferometric Techniques of Inspection

When optical holography is applied to the inspection of parts, the generation of a three-dimensional image of the object is of little value. Furthermore, for opaque materials, optical holography is strictly limited to surface observations. Therefore, if optical holography is to be implemented for nondestructive evaluation and inspection, supplementary means must be used to stress or otherwise excite test objects so as to produce surface manifestations of the feature of interest. It is for the measurement of such manifestations that the techniques of optical holography have been further developed to form the subfield of holography termed optical holographic interferometry.

As in conventional interferometry, holographic interferometric measurements can be made with great accuracy (to within a fraction of the wavelength of the light being used). Although conventional interferometry is usually restricted to the examination of objects possessing highly polished surfaces and simple shapes, holographic interferometry can be used to examine objects of arbitrary shape and surface condition. Because holographic interferometry produces a three-dimensional fringe-field image, which can be examined from many different perspectives (limited only by the size of the hologram), a single holographic interferogram is equivalent to a series of conventional two-dimensional interferograms.

As with conventional interferometry, holographic interferometry is the comparison of a test wave front with a known master wave front. Deviations of the test wave front from the master result in the appearance of interferometric fringes. In contrast to conventional interferometry, in which the master wave front is generated by appropriate imaging optics and mirrors, holographic interferometry systems use a holographically recorded image of the test object itself as a master against which the test object is compared after exposure to, for example, an applied stress. Depending on the inspection technique used, the interferometric comparison of the test object with its reference image can be made continuously (real time) or at selected instants in time (multiple exposure). This latter technique takes advantage of the fact that more than one hologram can be made on the same recording medium. Periodic vibratory motion can be examined with time-average holographic interferometry, while high-speed displacements or deformations can be studied with pulse methods.

The following sections will provide brief and qualitative descriptions of the basic holographic interferometry methods. A discussion of the quantitative information that can be obtained with these methods is presented later.

Real-Time Interferometry

In the real-time technique of holographic interferometry, a CW laser is used to make a hologram of the object in its natural or undisturbed state or in some other useful reference state, and the holographic reconstruction is compared to the object itself as it is deformed in real time. With the hologram being used as an observation window, it is possible to observe the resulting interference pattern produced by the interaction of the recorded object beam (also referred to as the stored, or frozen, beam) with the modified, real-time object beam (sometimes referred to as the live beam). With this method, it is possible to observe interference fringes resulting from differential, stroboscopic, and time-averaged motion of the object surface. Stroboscopic visualization can be obtained by chopping the laser source at a frequency and phase linked to the object excitation source. For time-varying fringe fields, time averaging is performed by the observer's eye, permitting the identification of nodes and antinodes of displacement when the object is excited at some vibratory resonance.

Multiple-Exposure Interferometry

The double-exposure, or differential, technique of holographic interferometry involves the use of either a continuous-wave or a pulsed laser to record two holograms—one of the object in its natural or undisturbed state (or in some other useful reference state) and the other of the object in some changed or disturbed state. Both holographic recordings are made on the same piece of film so that in the reconstruction process, the two recorded object beams interfere with each other to produce a fringe field in which the contour and spacing describe the changes that occurred between the two exposures. Because both the master and the test images are holographically recorded, high-speed disturbances such as stress wave propagation or ballistic impact can be examined using the double-exposure technique and double-pulsed laser sources. Laser pulse durations of several nanoseconds can be used to freeze surface motion that cannot be readily observed in real time.

One important variation of the double-exposure recording process incorporates two angularly separated reference beams. One of the beams is used for each exposure. When the resulting hologram is reconstructed by either beam alone, the image in the primary diffraction order contains no interference fringes, because only a single corresponding object beam is reconstructed. With both reference beams illuminating the hologram, the reconstructed image appears as one that might be obtained from a conventional double-exposure hologram with a single reference beam. The advantage of the dual-reference double-exposure technique derives from the ability to independently access the two reconstructed images to be interfered. By changing the frequency or phase of one reconstructing beam relative to the other, the observed fringes can be made to move in such a manner as to facilitate the quantitative interpretation of the interferometric information with measurement accuracies approaching 1/1000 of a fringe.

The stroboscopic technique of holographic interferometry involves the multiple recording of certain predetermined positions of an object in periodic motion, usually at the vibrational antinodes, the positive and negative maximum amplitudes of vibration. A pulsed laser or a continuously chopped CW laser synchronized with the source of vibrational excitation is ordinarily used to make these recordings. In the reconstruction process, fringes of high contrast are produced that enable the measurement of large amplitudes of vibration. By appropriately timing the laser pulses during holographic recording, the phase relationships of the antinodes can also be obtained.

Time-Average Interferometry

The continuous-exposure technique (time-average technique) of holographic interferometry permits mapping of the displacements of the surface of an object in periodic motion. A CW laser is ordinarily used to make this recording. Because the holographic exposure time is usually much greater than the period of object motion, the hologram is a record of the time-averaged irradiance distribution at the hologram plane. In the fringe field resulting upon reconstruction, bright fringes correspond to nodal regions (regions of zero or very small amplitudes of vibration). Fringe order increases as one approaches antinodal regions (regions of large amplitudes of vibration). By using appropriate fringe analysis techniques, the amplitudes of vibration for each point on the surface of the object can be determined. Because the fringe contrast decreases as the amplitude of motion increases, the displacement amplitudes analyzed in this way should not be too great.

Holographic Contouring

In addition to being a powerful tool for studying the response of an object of complex shape to a variety of applied stresses, holographic interferometry also provides a convenient means of evaluating the shape itself. With the appropriate techniques, it is possible to produce interferometric fringes that are a function of surface contour rather than surface deformation or displacement. Principally, two double-exposure recording methods are employed for this purpose: dual wavelength and dual refractive index methods.

The dual wavelength method for holographic contouring requires that the two holographic exposures be made at slightly different laser wavelengths. A CW gas laser such as an argon ion laser is usually used for recording purposes because the laser cavity can be easily tuned to any of several operating wavelengths. Upon reconstruction of the hologram with one or the other of the two recording wavelengths, a fringe pattern, directly related to the surface relief of the object, is observed superimposed on the image.

The dual refractive index method requires that the object to be contoured be viewed through a flat window in a test cell or chamber containing the object. Between the holographic double exposures, the refractive index of the medium surrounding the object in the chamber is changed. The contour fringe rate is determined by the change in refractive index. When a high fringe rate is desired, liquids such as mixtures of water and alcohol can be used. For lower contour intervals, the pressure of air within the test chamber can be regulated to control the refractive index.

Methods of Stressing for Interferometry

The nondestructive inspection of materials or structures by optical holography is potentially applicable to any problem in which stressing the part will produce a change in its shape that is indicative of the property or flaw being sought. Because holographic interferometry is most sensitive to displacements out of the plane of the specimen surface, stressing methods must be sought that will cause changes in the out-of-plane direction, which are detected by the recording of a holographic interferogram. The interferogram will consist of uniformly spaced dark and light bands (fringes), with a flaw being characterized as a distinguishable anomaly, such as a circular set of closed fringes. The part can be subjected to any of several types of stress, including acoustic, thermal, pressure (or vacuum), and mechanical. The type of stress used depends on:

- The physical characteristics of the part
- The type of flaw under inspection
- The accessibility of the part (whether the part can be tested by itself or must be tested as an integral part of a more complex system)

Acoustic Stressing. By combining optical holography with acoustic-stressing techniques, it is possible to record interferometrically the physical properties of relatively large surfaces as they are deformed by acoustic wave propagation and therefore to detect flaws. The acoustic stressing is done at sonic and moderately ultrasonic frequencies (usually less than 100 kHz). Excitation at frequencies in this range can be an important advantage of this approach over conventional ultrasonic inspection at megahertz frequencies because the effects of grain scattering, sonic attenuation, surface roughness, and shape complexity are all considerably minimized (see the article "Ultrasonic Inspection" in this Volume). This increases detection probability, permits the inspection of much larger areas at one time, and generally eases the transducer-coupling problem.

Application of acoustic-stressing systems for inspection is readily accomplished by electrically driving a piezoelectric transducer that has been bonded to the surface of the test object or to a fixture in which the object can be solidly mounted. The transducer vibrations are transmitted through the bond and cause flexural displacements of the component. In this way, with a proper choice of the driving frequency, a resonant-plate mode can be established in the test object. Additional transducers can be used as sensors to aid in establishing plate resonances. This technique of excitation has the advantage of producing a relatively broad range of frequencies—from a few hundred hertz to several hundred kilohertz.

In situations where larger-amplitude vibrations are required, a transducer can be mechanically coupled to a single point on the test object through a solid exponential horn (acoustic transformer). The horn consists of a cylinder whose radius decreases exponentially from one end to the other, with a piezoelectric transducer mounted on the larger end. The smaller end is pressed against the test object, and the transducer is driven at resonance to couple a large amount of acoustic energy into the test object. A typical setup is shown in Fig. 2. Such drive systems are generally limited to operation within narrow bands of frequency that are centered about the design frequency and its resonances. With this type of drive at 50 kHz, a peak displacement of about 10 μm (400 μin.) is obtained at the horn output. Horn-coupling excitation eliminates the bonding process required in the simple piezoelectric technique described above and is the most practical technique of acoustic stressing at high frequencies. This single-point method of excitation also permits the establishment of resonance throughout the entire test object, thus permitting inspection of the full surface at one time.

Standing, traveling, and surface waves can be utilized for inspection. Each is discussed below.

Standing Acoustic Waves. With standing waves, surfaces over debonds and voids, for example, can readily be made to vibrate, thus producing an easily distinguishable change in the mode pattern. Figure 3 shows two examples of areas of debonds detected with standing acoustic waves. Resonances within the skin covering the debonded area may require the application of increasingly

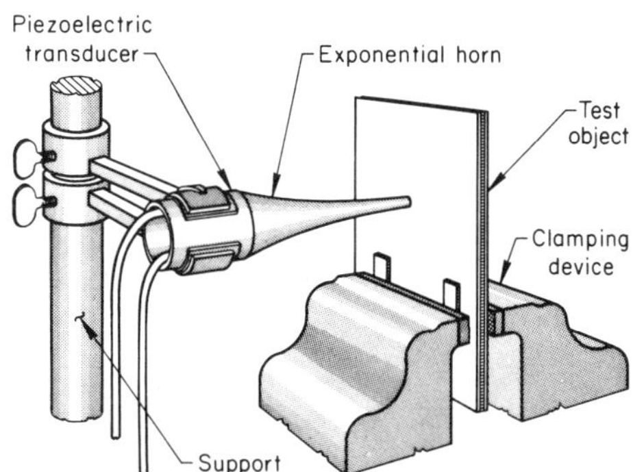

Fig. 2 Typical setup for the acoustic stressing of a test object using a piezoelectrically driven exponential horn

higher frequencies as the size of the debonded region to be detected becomes smaller. Broadband acoustic excitation over a band of frequencies is often used to excite motion in debonded regions that may vary in size.

Traveling Acoustic Waves. With traveling waves, acoustic energy will be scattered or absorbed at locations where the test object has a flaw. This scattered energy can be observed as an anomaly in the surface displacement in a holographic reconstruction of the test object. It is important to note, however, that the dynamic characteristic of traveling waves requires pulsed-laser holographic procedures, while standing waves are amenable to both CW laser and pulsed-laser holography.

Surface waves (Rayleigh waves) offer another acoustic approach, but require some sophistication in their generation to achieve sufficient amplitude for easy holographic observation. They are, however, easily attenuated by changes in structure and therefore offer potential for the detection of such flaws as surface microcracks in metal components. Electromagnetic drive systems can also be used to excite the test object and are generally quite easy to effect. For inspecting nonferromagnetic materials, some preliminary preparation, such as coating with magnetic paint, is required.

Thermal Stressing. The use of thermal stressing for successful holographic nondestructive inspection relies on changes in the surface deformation caused by differences in thermal expansion that are a direct result of a flaw or feature of interest. For example, when there is a lack of intimate contact at a debond between two components made of materials with different thermal conductivities, the lower rate of heat removal from the debonded region can affect the surface expansion in such a way as to create an anomaly in the interferometric fringe pattern of the hologram. In general, thermal stressing causes time-varying changes in surface displacement and therefore is best used in conjunction with real-time CW laser or pulsed-laser holographic techniques.

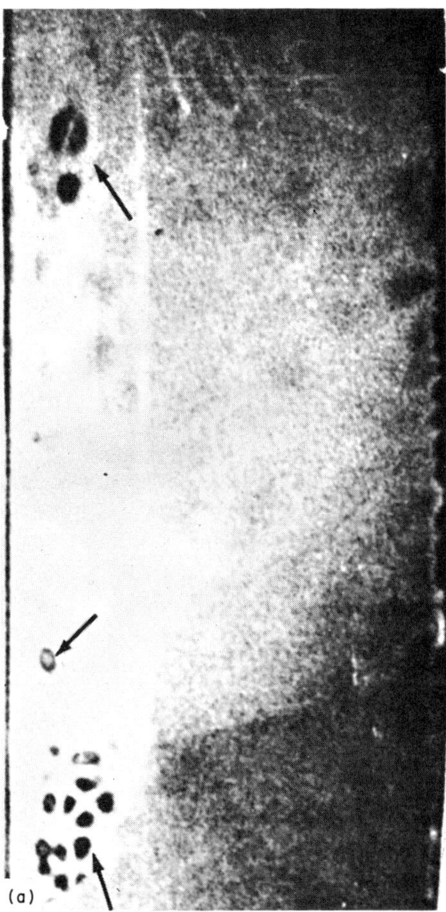

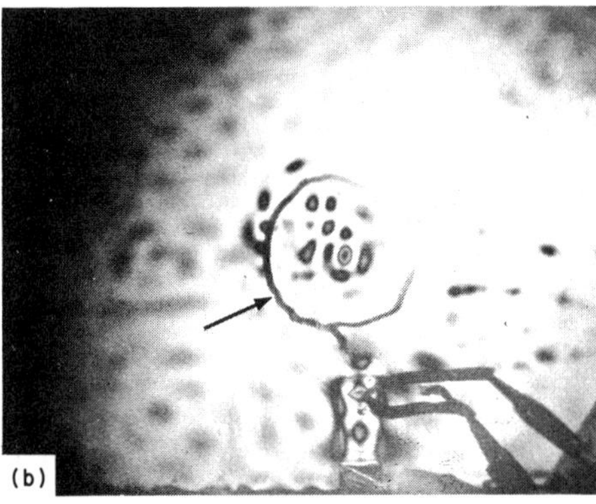

Fig. 3 Optical holographic interferograms showing areas of debonds (arrows) that were detected with standing acoustic waves. (a) Debonds between a compressor fan blade and a metal strip protecting the leading edge. (b) Debonds between the metal face sheet and the honeycomb-cellular core of a sandwich structure

Many methods are available for applying a thermal stress to a component under test. The method used and its applicability depend on the particular application.

Radiant Heat Sources. Radiation from a heat lamp or a photoflood lamp directed onto the front or back surface of the component is one of the simplest methods.

Electrical-resistance heating tape applied to a specific area of the test object is a method for preferential thermal stressing, as is a heat gun.

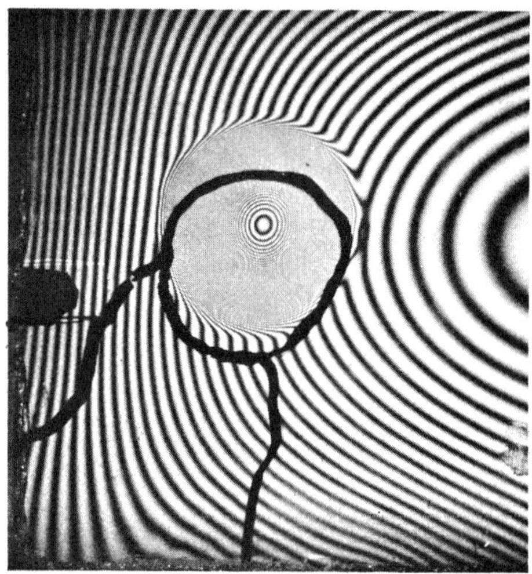

Fig. 4 Optical holographic interferogram of a pressure-stressed honeycomb-core sandwich panel. The sandwich panel was stressed by the atmospheric pressure of the air sealed in the core cavities when the air surrounding the panel was evacuated. An area of debond between the core and the near faceplate (of aluminum) is indicated by the fringe pattern in the vicinity of the ink marks on the faceplate. The background fringes were caused by general movement of the faceplate during evacuation.

Resistance heating of the object itself can be performed by passing electrical current through it.

Hot-air heating or liquid-nitrogen cooling can also be utilized as a thermal stressing method. For hollow or channeled components, the heated or cooled gases can be passed through the test object where accessible.

In addition to steady-state and transient heating methods, periodic heating of the object and synchronized strobing of the real-time holographic imaging system have been effective in locating surface-breaking cracks and similar defects.

Pressure or Vacuum Stressing. One of the most effective methods of stressing hollow components for holographic inspection is pressurization or evacuation. This method is most easily implemented in situations where access is available to the internal cavity. However, it is not limited to this type of test object. For any sealed-off structure that is manufactured at atmospheric pressure, external evacuation can be utilized to create a pressure differential between the cavity and the environment. An example of this latter approach is presented in Fig. 4, which shows an area of debond between the core and an aluminum faceplate of a honeycomb-core sandwich panel.

In general, very mild pressure stressing (0 to 70 kPa, or 0 to 10 psi) is sufficient to produce a surface change capable of detection by the holographic interferometric technique. However, in some cases (for example, the inspection of diffusion bonds and electron beam and resistance welds), large pressure differentials of the order of 700 kPa (100 psi) are required to create enough surface deformation for holographic interferometric detection. In these cases, the possibility of inspection by pressure stressing of hollow components when internal access is not available is precluded because external evacuation would not provide a sufficient pressure differential.

Mechanical Stressing. Another method for deforming a test object during holographic interferometric inspection consists of subjecting it to simple mechanical force. There are numerous ways of applying mechanical force, but the technique chosen will depend on the shape of the test object and the type of flaw being sought. One method that has found relatively wide acceptance for the vibration analysis of components is the use of mechanical shakers.

Inspection Procedures

As previously mentioned in the section "Holographic Recording" in this article, optical holography utilizes both continuous-wave lasers and pulsed lasers.

Continuous-Wave Techniques

Generally, in developing a procedure for the holographic inspection of a specific component or class of components, the common CW technique of laser holography is initially used to:

- Study engineering aspects of the problem
- Evaluate various stressing methods
- Develop an effective test procedure

Subsequently, specialized adaptations of the CW technique can be designed to eliminate or minimize the potentially deleterious effects of environmental vibration and the associated need for massive vibration-isolated tables. Indeed, for the inspection or testing of manufactured parts from an assembly line, it may be necessary to provide sufficient vibration isolation from the factory environment. In such situations, with adequate vibration isolation, the straightforward, widely used and well-developed conventional holographic procedures can be implemented.

However, in many cases, inspection cannot be done with a CW technique. This is especially true when:

- The most effective stressing procedures involve noncyclic dynamic loading
- The inspection must be performed without using a stable (holographic) table
- The inspection area cannot be darkened to facilitate recording of the hologram
- Specialized test facilities must be used

In such situations, holography using a pulsed-laser beam offers one practical approach to nondestructive inspection.

Pulsed-Laser Techniques

Continuous-wave lasers are used in optical holography at widely varying power levels and wavelengths. However, when high-speed transient stressing is used or when the test environment is unstable or hostile, a pulsed-laser technique has been most successful for making holograms. The most common pulsed lasers used for holographic applications are flash lamp pumped solid-state lasers such as ruby and frequency-doubled neodymium-doped yttrium aluminum garnet (Nd:YAG).

The process of making holographic interferograms with a pulsed laser can be more complicated than it is with CW lasers. Because laser pulse duration determines the exposure times, which may be of the order of several nanoseconds, precise synchronization of the laser with the excitation source is usually required. An additional complication arises if a stable environment is not used for the pulsed system. In this case, if a single exposure of a specimen is made, followed a few seconds later by a second exposure of the same object without applying any excitation, the resulting hologram will be found to be covered with a number of uninterpretable fringes caused by atmospheric changes and rigid-body specimen displacements that occurred between exposures. If this procedure is used to record an interferogram in the same manner as with the CW technique, the fringes resulting from the presence of a flaw may not even be visible. One solution to this problem is to pulse the second exposure a sufficiently short time after the first so that no significant rigid-body motion can occur between exposures. Elaborate switching systems have been created to perform this function. A typical pulse sequence might be a first exposure of 20 ns, a delay of 25 μs, and then a second exposure of 20 ns.

With pulsed-laser techniques, the nature of the excitation required for the detection of flaws can be quite different from that used with CW techniques. One method is to impact the test object and thus trigger the pulse sequence. Fringes tend to form around or in the vicinity of flaws because of a difference in the mechanical properties and therefore in the response of the surface. These effects are entirely too brief to record by CW techniques. The same type of effects will be recorded by pulsed-laser techniques if the component is thermally excited before the pulse sequence in such a way that the specimen is still deforming thermally during exposure.

Because the requirement of environmental stability is greatly reduced when pulsed-laser techniques are used, the laser and the necessary ancillary optical components can be installed in an existing testing facility to perform holographic interferometry. The laser head and its power supply can be placed in any convenient, nearby location. A beam splitter and several mirrors are used to generate two approximately equal optical beam paths to illuminate the test object and

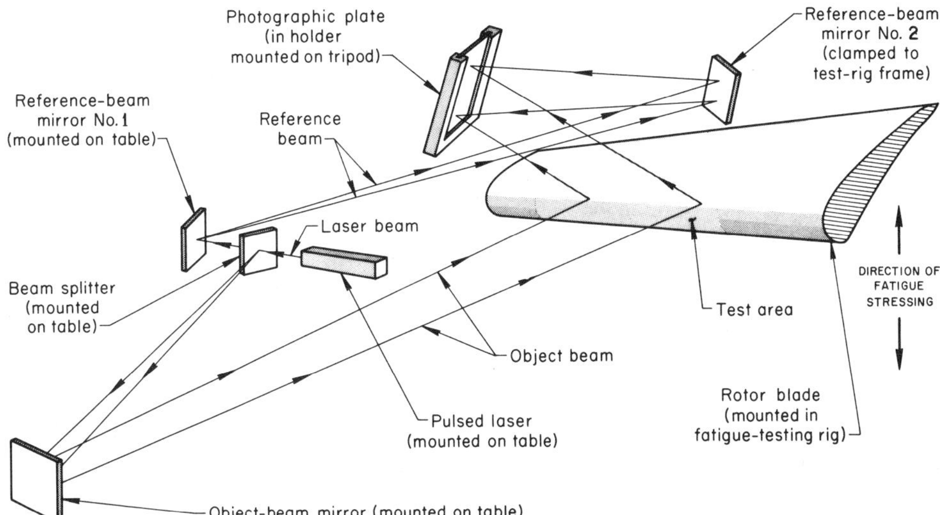

Fig. 5 Schematic of a portable setup for producing a pulsed-laser holographic interferogram of an area on a sandwich-structure helicopter-rotor blade being subjected to fatigue stressing

to provide a reference beam, and the photographic plate is clamped in position to view the test area of interest. A portable pulsed-laser setup is shown in Fig. 5.

Pulsed-laser systems can be subdivided on the basis of the operational mode used to generate the holographic interferogram, that is, Q-switched and conventional pulse mode. The conventional pulsing of a flash lamp pumped solid-state laser occurs simultaneously with the excitation of the flash lamps, so that the laser output may last for up to 2 ms. By placing a Q-switching element in the laser cavity, short pulses (tens of nanoseconds) can be produced during a single flash lamp cycle.

Q-Switched Pulse Mode. Double exposures using two Q-switched pulses from separate flash lamp cycles, perhaps operationally the easiest to effect, permit the recording of an essentially static test object in two different stress states. The distinction between this and the conventional CW holographic interferometric procedure lies in the extremely short duration of each individual exposure, which can often obviate the need for sophisticated vibration-isolation tables. Some stability must be retained, however, because gross movement of the object would simply result in the formation upon reconstruction of two images (similar to a double-exposed photograph) without the characteristic fringe pattern associated with holographic interferometric inspection. In practice, a holographic exposure of the test object is recorded, a stressing force is applied, and a second exposure is recorded, with the time interval between exposures (several seconds to many minutes) being dependent on the mechanics of the test.

Double pulsing of the Q-switch during a single flash lamp cycle is more widely used and perhaps the best known pulsed holographic interferometry technique. The method uses a double Q-switching action to generate both output pulses during a single optical pumping cycle of the laser. As such, the exposure-time interval is variable from microseconds to 1 or 2 ms; the lower limit is a function of the laser characteristics and Q-switching electronics, and the upper limit is controlled more by the characteristics of the flash lamps used to optically pump the laser. Consequently, this mode of operation greatly reduces restrictions placed on the holographic process by virtue of the physical environment and is directly applicable to many dynamic events. Generating two matching pulses, however, becomes more difficult as the pulse-separation time exceeds 200 μs because of the dynamic thermal conditions in the laser cavity. The result is images with contour fringes that modulate displacement fringes, thus obscuring the information sought. In addition, the stressing force appropriate to manifesting the flaw of interest must be capable of rapid application—a feature that may limit the effectiveness of the method in some instances.

Conventional Pulse Mode. In a conventional pulse (non-Q-switched) mode, a long-duration (1 to 2 ms) pulse is used for recording interferograms. This approach offers several advantages. First, the pulse duration is short enough to isolate objects that are moving because of environmental vibrations with frequencies of 100 Hz and less, while still permitting time-average studies of intentionally induced vibratory excitation of a component at frequencies of 1000 Hz and higher. Second, the total energy in the laser pulse is increased by a factor of ten over a Q-switched pulse for the same flash lamp power, permitting larger surface coverage. The most important aspect of these features is the vibration-analysis capability, for it permits application of the pulsed technique in those situations where cyclic acoustic excitation is the preferred method of stressing the testpiece.

Effects of Test Variables

In performing nondestructive inspection by optical holographic techniques, several important test variables should be considered. Some variables, such as exposure time and reference-to-object-beam intensity ratios, are controllable by the operator, while others are not (for example, the physical test environment). Although there are considerable interrelationships among test variables, for the purposes of this discussion they have been divided into the following eight categories:

- Object size
- Exposure parameters
- Surface finish
- Surface preparation
- Surface condition
- Whole-body motion
- Ambient illumination
- Environmental vibration

Object Size

There are of course theoretical limits on the size of object that can be inspected by optical holography, but generally there are practical considerations that set the upper bounds.

Laser Coherence Length Limitations. The theoretical limitation is imposed by the finite coherence length of the laser determined by the natural broadening of the laser spectral line. Because the instantaneous frequency of the light leaving the laser cavity varies over some narrow range with time, light exiting the laser at one instant may not be coherent with that emitted at some other instant. Therefore, the coherence length is the distance that light leaving the laser can travel before there is a shift in instantaneous laser frequency sufficient to produce light that is no longer coherent with that which left the laser initially. As a result, to produce successful holographic recordings, the path length difference between the reference and object beams must not differ by more than that coherence length.

With essentially flat objects mounted parallel with, and close to, the holographic plate, it is necessary to consider the difference in path length between the point on the object closest to the film plate (essentially the normal between the two) and the point on the object farthest from the plate. Generally, if this difference exceeds the coherence length, thus prohibiting complete holographic coverage, it is possible to move the holographic plate farther away from the object, thus decreasing the length differential between these two points on the object. In addition, if the object itself also has considerable depth along a path normal to the holographic plate, conventional holographic procedures will limit the portion recorded to a distance equal to a maximum of one-half the coherence length of the laser, depending on recording geometry.

However, because lasers with coherence lengths of the order of several meters are readily available and because inspection by optical holography generally involves normal or near-normal observation of the surface of the object, other practical limitations will almost always prevail in establishing maximum object size and/or surface coverage.

Vibration-Isolation Table Capacity Limitations. When a CW laser is used, the object and holographic components must be rigidly mounted to a common fixture, which is itself isolated from the environment. This is generally accomplished with vibration-isolation tables, which in turn place a practical upper limit on the size of the object that can be recorded; commercial tables are available that are large enough to permit the holography of objects with several square feet of surface area and weighing several hundred pounds. When using a pulsed laser, because vibration isolation is less important, the holographic plate can be placed at as great a distance from the object as is desired or is physically possible. In this situation, the practical limits of size are then set by the available energy output of the laser; commercial systems with outputs of up to 3 to 5 J/pulse (0.8 to 1.4 mWh/pulse) are currently available, permitting coverage of areas in excess of 9.3 m^2 (100 ft^2) in a single hologram.

Exposure Parameters

When using a CW laser and proper vibration isolation and control over the ambient light, the exposure time is a function of the film sensitivity, the size of the object, and its reflective properties. A proper ratio must be maintained between the reference beam intensity and object beam intensity, which is most easily controlled with adjustable beam splitters. (Neutral-density filters, and polarizing filters if the laser has a polarized output, can also be used in the reference beam path for preferential control of the beam ratio.) It is generally desirable to use a reference-beam intensity somewhat higher than the object-beam intensity, giving a ratio as measured at the holographic plate perhaps as high as 2:1 for real-time, 3:1 for double-exposure, and 4:1 for time-average techniques. (Measurement at the holographic plate automatically accounts for the reflective properties of the object.) However, the optimum exposure time and beam ratio for a particular component are best determined by trial and error; exposure times of seconds to minutes are not uncommon, and beam ratios as high as 10:1 or 20:1 are sometimes used. The important consideration in judging the results of varying the exposure time and the beam ratio is the fringe contrast in the reconstructed image, with the goal being high-contrast fringes.

When using a pulsed laser, the exposure time is fixed by the laser characteristics. Although some control of exposure time can be exercised when operating in the conventional pulse mode (1 to 2 ms) by controlling the laser pumping variables such as voltage and capacitor storage, the more commonly used Q-switch systems have fixed exposure times in the 10 to 50 ns region.

Surface Finish

Investigations of surface finish have demonstrated that it imposes no severe limitations on the holographic process as applied to the nondestructive inspection of metal parts (Ref 1). Surfaces varying from 0.1 μm (4 μin.) (very highly polished) to 25 μm (1000 μin.) root mean square roughness (quite rough) are easily recorded with no significant effect on interferometric fringe location. However, optimum contrast is not always obtained, particularly with polished surfaces, which can act as specular reflectors.

Specular reflection of the light source onto the hologram reduces image visibility and causes image smearing. In addition, specular reflections from low-roughness metal surfaces are brightest in a direction that is both a specular direction and a direction perpendicular to the finishing direction. Either specular directions or finishing directions may limit the area that can be recorded.

Surface Preparation

Because surface roughness imposes no severe limitations on holographic inspection, no preparation of the object surface is required; this is one of the most important advantages of this type of inspection. However, where permissible, surface preparation can be used to facilitate the inspection process by:

- Allowing greater area coverage and therefore reduced inspection time
- Eliminating the bothersome specular effect noted above and therefore simplifying data readout
- Producing a more uniformly bright reconstruction, which improves the overall fringe contrast and eases interpretation of the reconstructed image
- Permitting greater reflectivity from nonmetallic materials such as polymers and composites

To increase the intensity of the reflected light (increasing the area coverage and often improving the fringe contrast) or simply to minimize specular effects, any reasonably diffuse, flat, water-base white paint is generally satisfactory and easy to remove. Still greater increases in reflected light intensity (permitting even larger area coverage or reduced exposure time when using a CW laser) can be achieved with metallic paints or reflective paint of the type often used on highway signs. Regardless of the paint used, it must be allowed to dry completely. If the paint structure changes between exposures, no fringes will be produced.

Surface Condition

Because optical holographic inspection compares the surface of the test object in two states of stress, it is important that the condition of the surface remain as constant as possible between exposures. Dust, dirt, grease, oil, corrosion, wear, pitting, and loss of paint might be erroneously identified as displacement caused by a flaw if proper care is not exercised during the inspection procedure.

Whole-Body Motion

Quite often, the stressing force applied to manifest a flaw as a change in the surface shape also causes a general overall movement of the object under test. This movement is sometimes referred to as rigid-body or whole-body motion. The movement may or may not prevent proper interpretation of the reconstructed holographic interferogram, but it can be disconcerting. In certain situations (when performing real-time CW holography), fringe-control techniques can be used to compensate for rigid-body motion.

Fringe-Control Mirror. One approach to fringe control requires the incorporation of an additional mirror, having two axes of rotation, as the last element in the optical train of the object-illumination beam. If the object now rotates slightly during application of the stressing force, a linear phase change will be introduced into the optical beam reflected to the real-time hologram. The optical interference between this beam and the holographically reconstructed beam will introduce a fringe pattern along the object when it is viewed through the real-time hologram. By tilting the object beam in an appropriate manner, using the additional fringe-control mirror placed into the system, adequate compensation for the phase change introduced by the object rotation can often be accomplished. Therefore, most of the unwanted fringes can be eliminated, and the anomalous fringe pattern around a flaw can be better revealed (Fig. 6).

The appropriate setting of the fringe-control mirror can best be determined by slightly rotating the mirror about each axis of rotation while simultaneously viewing the object through the hologram. By noting the direction of rotation that tends to broaden and separate the fringes, the appropriate rotational direction can be found. In practice, several such adjustments about each

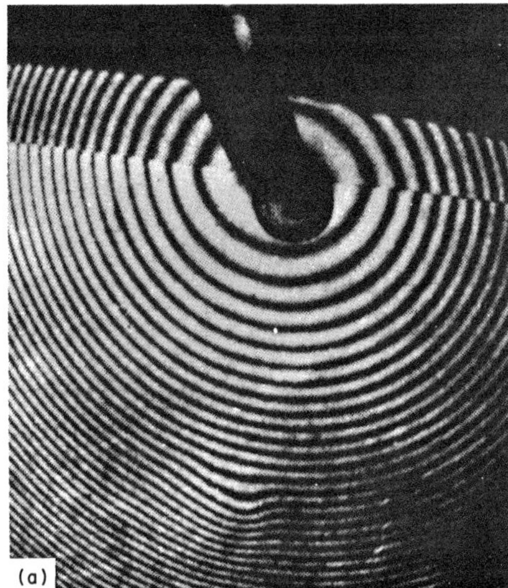

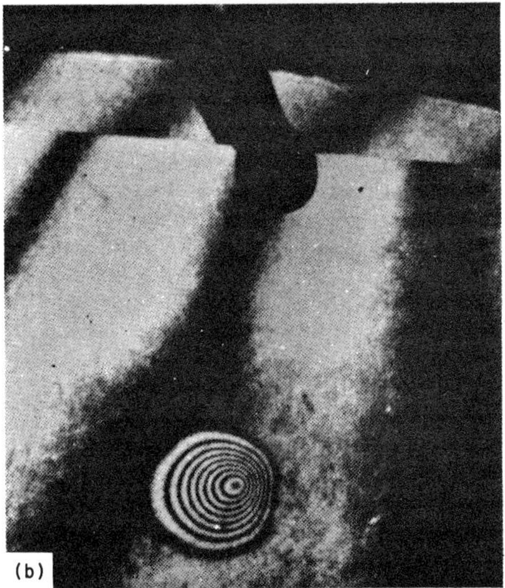

Fig. 6 Optical holographic interferograms of a clamped sandwich panel, illustrating use of fringe-control technique to eliminate unwanted fringes. (a) Interferogram showing unwanted fringes caused by stresses from a clamp (at top center). (b) Interferogram of the same clamped panel as in (a), but recorded with the use of a fringe-control mirror to eliminate the clamping-stress fringes, thus revealing a fringe pattern (at bottom center) caused by a flaw

axis are usually required to produce the broadest spacing of fringes on the object.

Phase-Change Compensating Lens. Another method of fringe control incorporates a lens, which can be translated along its optical axis, in the object beam path to compensate for undesirable phase changes caused by object movement (Ref 2). Furthermore, this method can be adapted to double-exposure techniques by using a different reference beam for each exposure and both reference beams in the reconstruction step.

Ambient Illumination

Holography is most easily performed in a darkened area. In the presence of any ambient illumination that cannot be controlled, pulsed-laser techniques are usually the most desirable. With such techniques, the holographic plate can be shuttered in synchronization with the laser pulse (or pulses, if used in the double-pulse mode) to hold fogging of the film to an acceptable level.

Focal-Plane Shutters. Specially modified focal-plane shutters to provide a full opening for a few milliseconds are quite satisfactory. This opening duration is long enough to encompass both pulses when used in the double-pulse mode and is short enough to permit holography in almost any factory environment. Successful tests have been performed with as high as 2.2 klx (200 ftc) of reflected ambient light, as measured at the film plane, using shutter durations of the order of 75 ms (Ref 1).

Interference Filters. If necessary, the level of ambient radiation reaching the film can be further reduced by using optical filters matching the spectral content of the laser. Generally, a fairly broadband red filter is more than sufficient for ruby lasers. Highly tuned interference filters, which are available to match all common laser sources, can be used to essentially eliminate any ambient light problem. However, care must be exercised in the application of these filters because their transmission characteristics are highly dependent on incidence angle, thus greatly restricting the permissible optical configuration. Also, they are expensive, especially in larger sizes. However, they are useful for very specific configurations and applications, especially in the holography of self-luminous objects.

Air Turbulence and Environmental Vibration

The holographic recording process is extremely sensitive to variations in optical path length that may occur during film exposure. Unwanted variations in optical paths may be caused by turbulence in the air path through which beams are propagating or by sources of rumble and vibration in the environment.

Air turbulence problems are most severe when relatively long holographic exposures are called for and can be dealt with directly by shielding the holographic apparatus or enclosing the beam paths in a series of interconnecting tubes. These effects can also be minimized by using optical fibers to deliver the light beams prior to beam expansion. Electronic feedback is sometimes used to control the position of a mirror to compensate for fluctuations in refractive index caused by air turbulence.

Environmental Vibration Problems. In most cases, air turbulence effects are less troublesome than the effects of environmental vibration. Pulsed-laser systems have been effectively used for practical testing outside of a vibration-controlled optical laboratory. Pulsed lasers are extremely effective in these situations and therefore allow the use of nondestructive inspection by optical holography in most factory environments. By maintaining the pulse-separation interval during the recording of a double-exposure holographic interferogram at or below a few hundred microseconds, any environmental vibration effects below 1000 Hz are obviated. Although the higher-frequency vibration of objects would hamper successful testing, such vibrations are generally of sufficiently low amplitude in normal industrial environments to be of no consequence.

If the use of pulsed-laser systems is undesirable or if the nature of the test (such as low-frequency vibration analysis) necessitates CW techniques, various methods are available for vibration compensation. These techniques include:

- Speckle reference beam holography (Ref 3)
- Focused-image holography (Ref 4)
- Optical processing of the reference beam (Ref 5)
- Attachment of the reference beam mirror (Ref 6) or the photographic plate itself to the object (Ref 7)

All of these techniques can generally be classed as local reference beam generation methods, in which the motion of the object is automatically compensated for by a comparable change in the reference-beam path.

Electronic-feedback servosystem methods to provide reference beam compensation by both phase modulation (Ref 8) and frequency modulation (Ref 9) are also available. Finally, the use of scanned-beam holography offers an approach to compensation for object movement (Ref 10). In general, these methods require a reasonably good working knowledge of holography because of their more complex nature and would be further troubled by any existing ambient illumination. These object motion compensation methods, as well as fringe-control techniques, are discussed in Ref 11.

Readout Methods

When using a reconstructed interferometric image for holographic inspection, something more than a simple visual display is generally required. A permanent record of the image, such as a photograph, is most commonly used. If further data reduction is to be done on the interferogram, some form of electronic image is more convenient than a photographic image. Photographic and electronic methods of readout of holographic interferograms are discussed below.

Photographic Readout

Photographic readout with a small-format camera, such as a 35 mm single lens reflex camera, is adequate for many inspection tasks. A reflex-type camera is recommended because the picture can be composed. A wide variety of film is available for 35 mm cameras. The film can be obtained economically in bulk form and loaded into cartridges as needed. Slightly more expensive, but more convenient, are preloaded cartridges of 20 or 36 exposures.

Interchangeable Lenses. It is convenient, but not essential, for the camera to have interchangeable lenses. Most needs can be met with a standard 55 mm lens—the faster the better. For photographing remote images, a telephoto lens (for example, one with a 135 mm focal length) is often desirable. Extension tubes 25 and 50 mm lengths are an inexpensive but useful way to obtain closer working distances from the standard 55 mm lens. Several wide-angle to medium-telephoto macrozoom lenses are available. Generally, in making holographic reconstructions, the light levels are fairly low, so lenses with small *f*-numbers are recommended for shorter exposures and smaller speckle size.

The better the quality of the photographic negative, the better the final print. For the best negatives possible, a large-format camera, such as a 100 by 125 mm (4 by 5 in.) view camera, should be used. Such a camera also provides the option of using Polaroid film, which is often quite convenient. A standard 135 mm lens will serve for most purposes. A 90 mm wide-angle lens and a 250 mm telephoto lens will accommodate unusual situations.

Photographic Film Specifications. The choice of photographic film is not critical. Some inadequacies in the original reconstruction can be partially compensated for in the darkroom. For example, the recording of high-contrast, high-order fringes in a time-average interferogram requires a dynamic range not obtainable with existing film material. To compensate for the overbright, zero-order fringe and to enhance the weaker, higher-order fringes, a superproportional reducer such as a 20% solution of ammonium persulfate can be used. The reducer takes silver from those portions of the negative where there is the most, thus enhancing the weakly recorded higher-order fringes. Special high-contrast copy films, such as Kodak 410, are available and are excellent for making reconstructions that are to be used for quantitative data reduction. For maximum resolution in the reconstruction, moderate-speed or slow-speed films are best. In general, the faster the film, the grainier the image. When a high *f*-number camera is used, however, graininess of the image will be determined by speckle size, not film noise. For qualitative purposes or for a record only, almost any film can be used. For convenience, high-speed films, such as Polaroid type 57 (ASA 3000), are favored.

It is not possible to record everything that can be observed in the holographic image. Lateral motion often prevents a fringe pattern from localizing on the surface of the object. As a consequence, when a large-aperture optical system having a very short depth of field is focused on the object, the fringes may vanish. They lie outside the depth of field of the lens and are blurred out. To increase the depth of field, the lens aperture can be decreased, but this also increases the size of the speckles in the resulting reconstruction.

Electronic Readout

Photographs serve well for archival purposes. However, if there is to be any further manipulation of the data, position and density information in the photographic negative or in the image must be converted to numerical values. This can be accomplished by using a standard television-type video system and one of several flexible, multiple-purpose systems for electronic image processing. Analog image processors permit several types of image enhancement, including edge enhancement, boundary detection, area determination, particle counts, and color coding of displays by density level.

The digital image processing of holographic images permits direct and automatic interpretation of fringe patterns, resulting in psuedo-three-dimensional displays of object contour or deformation. Computer algorithms are available for the conversion of fringes from a single interferogram image. In addition, the holographic recording and readout system can be altered so that several images of the interferogram can be processed to produce displacement information, providing in some cases nearly a 1000-fold improvement in sensitivity over conventional holographic interferometry. Among these techniques are phase stepping and heterodyne holographic interferometry. The principles behind their operation are explained in the section "Interpretation of Inspection Results" in this article.

For inspection functions, many of the standard video methods are applicable. Because the diffraction of a typical hologram is low (<3% of the incident light), it is advisable to use a television-type video system with a camera capable of operating at low light levels, although a standard television camera is often satisfactory. Cameras capable of operating at low light levels are discussed in the section "Television Cameras" of the article "Radiographic Inspection" in this Volume.

The resolution available with electronic readout varies from that afforded by charge coupled device array cameras and standard 525-line video cameras to 1000- and 2000-line high-resolution video cameras. For even higher resolution, a special scanner-type readout system, such as an image dissector, flying spot, or laser scanner, is required. However, because these special scanning systems generally work from photographic film inputs, their high resolution is gained at the expense of not being able to take advantage of the real-time nature of electronic readout.

Data Storage and Retrieval. When using electronic readout, videotapes can be made for archival purposes. For rapid response and more accurate timing, video storage disks and scan convertors can be used. In building a library of patterns, any of the electronic storage media mentioned above can be used to advantage in conjunction with photography. For example, a video system can be used to isolate and temporarily store a pattern of interest, which can then be photographed from the monitor for permanent or individual storage. Photographs are much less expensive than any form of electronic storage.

Interpretation of Inspection Results

The holographic process can be considered merely an optical high-fidelity recording process; the reconstructed image of an object is indiscernible from the object seen with the unaided eye. Whatever the source, the observer sees an intensity distribution of light that produces, using the optical system of the eye, an image of the object.

When a holographic interferogram is viewed, a system of fringes is seen overlaid on the object (Fig. 7). Different object motions produce different fringe patterns. In general, the more complex the object motion, the more complex the fringe pattern. The fringe pattern in a holographic interferogram of a diffusely reflecting object can be used to extract quantitative information about the object. The procedures for interpreting the patterns depend on the type of hologram and are presented for the most common cases later in this section.

It is important to remember that optical holographic interferometry measures directly only surface deformation or displacement. The information in the fringe pattern and the holographic image are directly related to the fine structure of the surface. Should the fine structure of the surface be altered between the "before" and "after" exposures in a double-exposure interferogram, which is the most common form of holographic interferogram, the loss of contrast in the resulting fringe pattern will reflect that change. The general change in the shape of the test object is portrayed by the fringe pattern itself. It is only this aspect of the interpretation of the fringes that is considered in the following discussion.

Qualitative Results

There are many cases in which quantitative results from fringe analysis are neither necessary nor desirable. For example, qualitative information is sufficient in the inspection of laminated or sandwich struc-

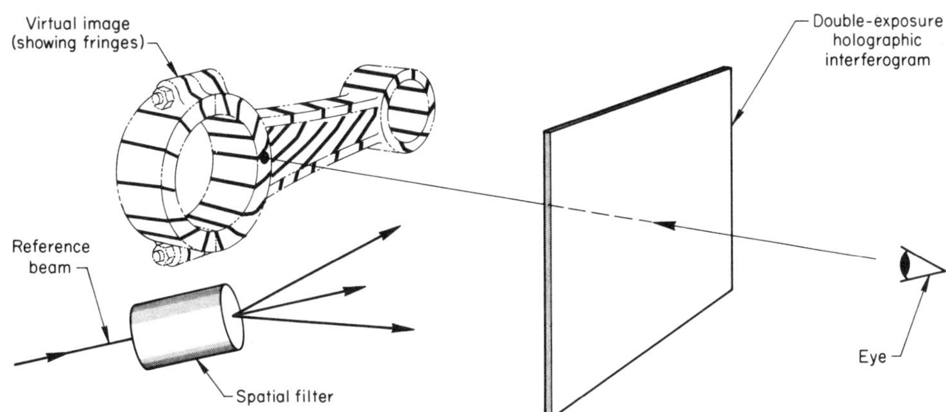

Fig. 7 Schematic of the reconstruction process of a virtual image of an object showing fringes from a double-exposure holographic interferogram

tures for debonded regions. It is of no particular significance how much the debonded region has puckered. What is important is the extent of the debonded region and whether or not it is a debilitating flaw in the structure. The question of what size a debonded region must be in order to qualify as a debilitating flaw is outside the realm of holography. The inspector's experience and the intelligent use of the inspection device are what matter in this area of interpretation. The question of pattern recognition and what constitutes a pathological or bad specimen arises. Perhaps the safest way to proceed is to compile a library of anomalous fringe patterns from known bad or pathological specimens. Representative types of fringe anomalies include:

- Independent, localized fringe systems
- Abrupt changes in fringe continuity
- Abrupt changes in fringe density
- Significant changes in fringe shape

Depending on the amount of data or samples to be treated, the library of patterns can be further classified into types of anomalies as experience dictates.

Out-of-Plane Displacements Versus In-Plane Displacements. In general, out-of-plane displacements lead to fringe patterns that localize close to the surface of the object, while in-plane displacements lead to fringe patterns that do not localize on the object. If the object and the fringe pattern are to be recorded in the same photograph, they must both be within the depth of field of the lens of the recording camera. The concept of homologous rays is useful in understanding fringe localization (Ref 12).

The Characteristic Function

To understand what the fringes in a holographic interferogram represent, the concept of the characteristic function is useful (Ref 13). The distribution of the object intensity and the system of fringes observed in the interferogram can be considered to be the product of the square of the characteristic function, M, and the image intensity distribution, $I_0(\mathbf{x})$, that is:

$$\langle I(\mathbf{x})\rangle = M^2 I_0(\mathbf{x}) \quad \text{(Eq 1)}$$

Wherever the characteristic function goes to zero, fringes are seen on the object. In this case, M^2 might well be referred to as a fringe function. Only when the character of the object motion is well known can the characteristic function be determined. Once the characteristic function is known, quantitative information can be extracted from the interferogram. For step-function motion, such as that used to produce double-exposure interferograms, the characteristic function, M, is a cosine function of position and object displacement.

Out-of-Plane Displacements

If it is known that the object of a double-exposure interferogram was in some motionless state during an initial exposure and that it was in a different motionless state during the second exposure, then one is able directly to deduce the characteristic function. In this situation, the characteristic function is cos $(\mathbf{K} \cdot \mathbf{d}/2)$, where $\mathbf{K}$ represents the sensitivity vector and $\mathbf{d}$ is the vector connecting the old position, P, of a point on the surface of the object with the new position, P', as shown in Fig. 8. The sensitivity vector is defined as $\mathbf{k}_2 - \mathbf{k}_1$, where $\mathbf{k}_2$ and $\mathbf{k}_1$ are the propagation vectors of the light scattered from the point in the directions of unit vectors $\mathbf{n}_2$ (that is, toward the interferogram and observer) and $\mathbf{n}_1$ (which is on the line that emanates from the focal point of the object beam spatial filter and passes through the point on the surface of the object), respectively. Each value of $\mathbf{k}$ can be determined from the formula $\mathbf{k} = (2\pi/\lambda)\mathbf{n}$, where λ is the wavelength of the object beam.

Dark fringes result whenever cos $(\mathbf{K} \cdot \mathbf{d}/2) = 0$; that is, whenever:

$$(\mathbf{K} \cdot \mathbf{d}/2) = \left[\frac{(2n-1)}{2}\right]\pi,\ n = 1,2,\ldots \quad \text{(Eq 2)}$$

Writing in scalar form and solving for the component of $\mathbf{d}$ parallel to $\mathbf{K}$, the following expression for out-of-plane displacements is obtained:

$$|\mathbf{d}| \cos\alpha = \left[\frac{(2n-1)}{2}\right]\left[\frac{\lambda}{2\sin\theta}\right] \quad \text{(Eq 3)}$$

That is, each fringe except the first one represents a displacement of $\lambda/(2 \sin\theta)$ from the previous one. The first fringe represents only half that displacement. The relations among $\mathbf{K}$, $\mathbf{k}_1$, $\mathbf{k}_2$, $\mathbf{d}$, α, and θ are illustrated in Fig. 8.

The description given above is applicable in those cases when the fringes localize on the object and a definite order number, n, is assignable to each fringe. This holds in cases of out-of-plane displacement. This is the most sensitive way to apply holographic interferometry. Out-of-plane displacements from 15 down to 1.25 μm (600 to 50 μin.) can be measured in this way. By assuming continuity and choosing the best fit for the data, displacements can be measured with an accuracy of $\lambda/25$.

In-Plane Displacements

In-plane displacements can also be measured, but generally with lower precision than out-of-plane displacements. The fringe

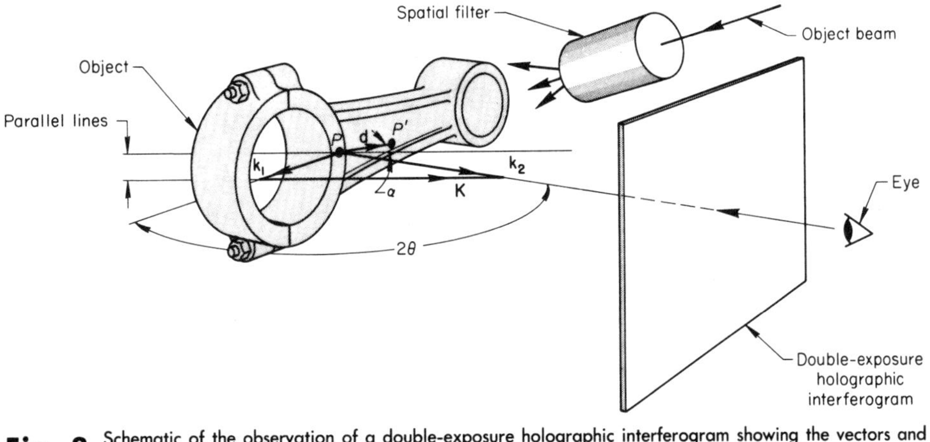

Fig. 8 Schematic of the observation of a double-exposure holographic interferogram showing the vectors and angles relating to out-of-plane displacement of a point on the surface of the object from position P to position P'. See text for explanation of the symbols.

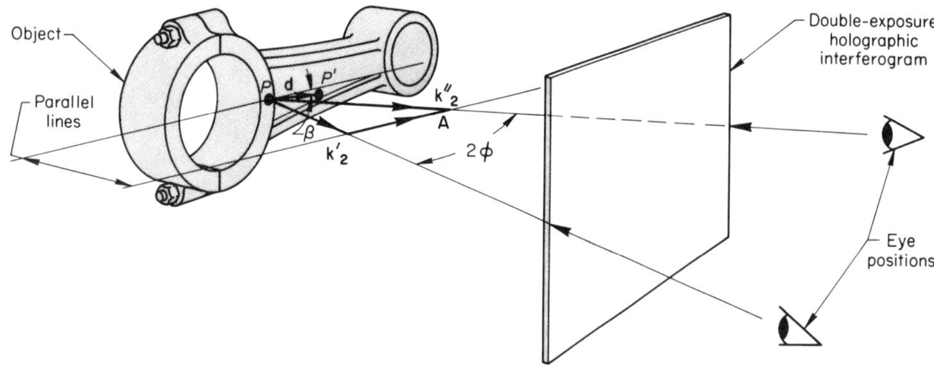

Fig. 9 Schematic of the observation of a double-exposure holographic interferogram showing the vectors and angles relating to in-plane displacement of a point on the surface of the object from position P to position P'. See text for explanation of symbols.

pattern for an in-plane displacement will exhibit parallax when viewed through the interferogram. As the observer's eye moves, the fringe pattern moves in either the same or the opposite direction with respect to a given point on the object surface. Each dark fringe that crosses a given point during eye movement corresponds to a value of zero for the characteristic function.

Scanning the interferogram from one side to the other, as shown in Fig. 9, defines a new vector, $\mathbf{A} = \mathbf{k}_2'' - \mathbf{k}_2'$. The displacement parallel to **A** can be determined from the number of fringes that are observed to cross point P. Because zero values of the characteristic function occur at multiples of π, the expression for in-plane displacements that is equivalent to Eq 2 is:

$$(\mathbf{A} \cdot \mathbf{d}/2) = n\pi \qquad \text{(Eq 4)}$$

Writing in scalar form and solving for the component of **d** parallel to **A**, the following expression for in-plane displacements is obtained:

$$|\mathbf{d}| \cos \beta = \frac{(n\lambda)}{(2 \sin \phi)} \qquad \text{(Eq 5)}$$

That is, the component of **d** lateral to the line of sight is proportional to the number of fringes crossing point P and is inversely proportional to the sine of the half-angle between the two views. The angle between **A** and **d** is called β.

It is possible to measure in-plane displacements from 5 up to 150 μm (200 μin. to 0.006 in.) using double-exposure holographic interferometry (Ref 14). An error of ±3 μm (±118 μin.) can be expected on the large-displacement end of the range because of the limiting size of the interferogram (the limiting magnitude of **A**).

For an object surface that undergoes a general displacement involving both in-plane and out-of-plane motion, the fringe pattern can be viewed from several (at least three) directions to obtain the projection of the surface displacement of each image point along at least three different sensitivity vectors, $\mathbf{k}_{1,2,3}$. The three sensitivity vectors can then be considered to be a set of nonorthogonal coordinate axes that can be rotated and projected onto a conventional coordinate system to obtain general object point displacements. Although the description of this process is a bit confusing, the mathematical and computer algorithms used to perform these manipulations are quite straightforward.

Time-Average Interferograms

The characteristic function for time-average holographic interferograms differs from that of the double-exposure case. If it is known that the object is undergoing strictly sinusoidal motion during the time of exposure of the holographic interferograms, then the characteristic function is J_0 $(\mathbf{K} \cdot \mathbf{d})$, where J_0 is the zero-order Bessel function of the first kind and **d** is the vector displacement. This function behaves similarly to a cosine function with regard to its zero values; however, it is not strictly periodic with zeroes existing at regular intervals. The first and second zeroes occur when the argument is 2.4048 and 5.5201. After that, the zero values can be approximated by those given by the asymptotic limit for large argument (large x); that is:

$$J_0(x) \rightarrow \sqrt{2/\pi x} \cos [x - (\pi/4)] \qquad \text{(Eq 6)}$$

For example, for the third fringe, the error is 15 parts in about 8600. None of the measurements to determine the values of θ or λ (Fig. 8) is likely to be this accurate, so use of the zero values for the asymptotic limit is generally well justified. Writing Eq 6 in scalar form and solving for the component of **d** parallel to **K** yields the following: For the first fringe:

$$|\mathbf{d}| \cos \alpha = 2.4048\ \lambda/(4\pi \sin \theta) \qquad \text{(Eq 7)}$$

For the second fringe:

$$|\mathbf{d}| \cos \alpha = 5.5201\ \lambda/(4\pi \sin \theta) \qquad \text{(Eq 8)}$$

For succeeding fringes, with $n \geqq 3$:

$$|\mathbf{d}| \cos \alpha = (n - 1/4)\ [\lambda/(4 \sin \theta)] \qquad \text{(Eq 9)}$$

To a good approximation, the first fringe represents a displacement of about $3\lambda/(16 \sin \theta)$, with succeeding fringes representing steps of $\lambda/(4 \sin \theta)$.

In addition to differences in the location of zeroes, the characteristic (Bessel) function for this case dramatically decreases in amplitude with increasing fringe order. Because of the decreasing brightness of the fringes and the limited dynamic range of the reconstruction film, it is difficult to record much more than seven fringes in photographically produced reconstructions. Even when a superproportional reducer is used on the reconstruction negative to increase the visibility of the higher-order fringes, it is difficult to work to much more than 30 fringes. In addition, it is difficult to work with a slope on the object in excess of about 0.6% with either double-exposure or time-average holograms, because of the high frequency of the fringes produced.

High-Resolution Interpretation Methods

As mentioned previously, consideration has been given only to the relationship between holographic interference fringes and object surface motion. In fact, the appearance and apparent location of fringes in a reconstructed image depend not only on displacement but also on object surface reflectivity and fringe brightness or contrast. Therefore, the intensity at each point in a reconstructed holographic interferogram is a function of these three variables and not simply surface displacement. Using high-resolution methods such as phase stepping (Ref 15, 16) and heterodyning (Ref 17), one can compute directly all three variables at each point in the image to a degree of accuracy up to 1000 times better than can be achieved by simple fringe counting. In this way, displacements as small as 0.25 nm (2.5 Å) can be detected in principle.

To perform either of these interpretation methods, independent control of the two interfering images must be available during reconstruction. This is a natural consequence of real-time holographic interferometry because the reference and object beams can be altered independently. In double-exposure methods, however, a dual-reference arrangement as described previously must be used to permit independent control.

Phase-Stepping Methods. For phase stepping, several video images are recorded of the fringe pattern with a small phase difference introduced between the reconstructed images prior to recording each video image. The phase shift can be performed in several ways, but perhaps the most common method is to use a mirror mounted on an electromechanical translation device such as a piezoelectric element. If the phase shift imposed prior to each video recording is known, then only three images need be recorded. Because the intensity at each point on the image is known to be a function

Table 1 Wavelengths and temporal coherence lengths of the six types of laser beams in common use for holography

Type of laser beam	Wavelengths nm	Å	Electromagnetic spectrum	Temporal coherence length Minimum mm	in.	Typical mm	in.
Helium-neon	633	6330	Orange-red	152	6	457	18
Helium-cadmium	422	4220	Deep blue	75	3	305	12
	325	3250	Ultraviolet	75	3	305	12
Argon	514	5140	Green	25	1	914	36(a)
	488	4880	Blue	25	1	914	36(a)
	(b)	(b)	(b)	25	1	914	36(a)
Krypton	647	6470	Red	25	1	914	36
	(c)	(c)	(c)	25	1	914	36
Ruby	694	6940	Deep red	0.8	0.030	914	36
Nd:YAG	1064	10640	Infrared	(d)		(d)	
	532(e)	5320(e)	Green(e)	(d)		(d)	

(a) Typical temporal coherence length achieved when laser incorporates an etalon (selective filtering device) to make it suitable for holography. (b) Six other visible lines. (c) Nine other visible lines. (d) Varies with cavity design. (e) Frequency-doubling crystals

of the three variables described above, intensity information from the three images can be used to solve a series of three equations in three unknowns. In addition to providing automated interpretation of fringe patterns, phase stepping affords an increase in displacement sensitivity by as much as 100-fold (1/100 of a fringe) relative to fringe-counting methods. In practice, most investigators report a sensitivity boost of about 30.

Heterodyning Method. Still higher holographic sensitivity can be obtained with heterodyne holographic interferometry. As with phase stepping, independent control of the interfering images must be provided either by real-time analysis or dual-reference methods. Instead of introducing a phase shift between several recorded images, a fixed frequency shift is introduced in one reconstructing beam relative to the other. Typically, acousto-optic phase shifters are used to produce a net frequency shift of the order of 100 kHz. As a result, fringes once visible in the reconstructed interferogram are now blurred because of their apparent translation across the image field at a 100 kHz rate. A single-point optical detector placed in the image plane can detect this fringe motion and will produce a sinusoidal output signal as fringes pass by the detection spot. By comparing the phase of this sinusoidal signal to that obtained from some other point on the image, the difference in displacement or contour can be electronically measured. An entire displacement map can be obtained by scanning the optical detector over the entire image. Because scanning is required, the speed of heterodyne holographic interferometry is relatively slow. Sensitivities approaching 1/1000 of a fringe have been obtained, however.

Holographic Components

The basic components of a holocamera are the:

- Light source (laser)
- Exposure controls
- Beam splitter
- Beam expanders (spatial filters)
- Mirrors
- Photographic plate or film holder
- Lenses
- Mounts for the equipment
- Tables to support the holographic system

Components and complete holographic systems are commercially available (Ref 18, 19).

Laser Sources

The characteristics of six types of lasers commonly used for holography are listed in Table 1. Helium-neon, argon, and ruby lasers are the most common. Helium-cadmium and krypton lasers, although not used as frequently, can fulfill special requirements for CW applications. Frequency-doubled Nd:YAG lasers are finding increasing popularity for pulsed holographic applications.

Helium-neon lasers are the most popular laser source when low powers are sufficient. Excitation of the gas is achieved through glow discharge. These lasers have excellent stability and service life with relatively low cost. Another type of laser is usually considered only when a helium-neon laser will not perform as required. A 20-mW helium-neon laser in a stable system can conveniently record holograms of objects 0.9 m (3 ft) in diameter. (Within the limits of coherence length and exposure time, as discussed previously, even larger objects could be recorded.) Such a laser consumes 125 W of 110-V electrical power and operates in excess of 5000 h without maintenance. A 5-mW laser records objects 460 mm (18 in.) in diameter and operates for more than 10 000 h without maintenance.

Helium-cadmium lasers are closely related to helium-neon lasers, with the following differences:

- Tube life is poor by comparison (approximately 1000 to 2000 h)
- The principal visible wavelength—422 nm (4220 Å) is 30% shorter (Table 1), which provides increased sensitivity and allows the use of recording mediums sensitive to blue light
- They have an output in the ultraviolet (325 nm, or 3250 Å), which is half the wavelength of helium-neon lasers and produces doubly sensitive displacement measurements
- There is more danger to the eyes at the shorter wavelengths produced by helium-cadmium lasers

Argon and krypton ion lasers can be the least expensive holographic sources on the basis of light output per dollar. Laser outputs of 1 W with 9 m (30 feet) of coherence length are available. Low-power argon lasers, however, are more expensive than helium-neon lasers. The use of an argon laser should be considered over a helium-neon laser in the following situations:

- When stability or dynamic conditions necessitate short exposures requiring high light power
- When the recording of large objects requires higher power to record good holograms
- When the recording medium requires high-power blue or green light
- When the holographic system needs the higher sensitivity provided by the shorter wavelengths of the argon laser

Argon lasers in the 1- to 4-W output range, equipped with an etalon (a selective filter required to achieve long coherence length), are excellent holographic light sources; however, a heium-neon or a helium-cadmium laser may be preferred for the following reasons:

- Argon lasers consume thousands of watts of electrical power and require water cooling
- Gas excitation is by electric arc, which generates high electrical and thermal loads on components and makes reliability and stability lower than with a helium-neon laser
- The output power is well above that which causes damage to the eyes, especially at the shorter wavelengths produced by argon lasers. Most available data indicate that helium-neon lasers are incapable of causing the damage that could be caused by an argon laser. Therefore, safety requirements for argon lasers must be more stringent

Krypton lasers are essentially the same as argon lasers except that the tubes are filled with krypton gas instead of argon gas. The output wavelengths are longer (Table 1) and the power is lower than for an argon laser. A 2-W argon laser produces 0.8 W in its most powerful line (514 nm, or 5140 Å), while the same laser device filled with krypton produces 0.5 W at 647 nm (6470 Å) and 1.3 W total. Argon and krypton gases can be combined in the tube to give custom outputs over a wide range of wavelengths.

Ruby lasers use rods of ruby instead of a gas-filled glass tube as a lasing medium. Excitation of the medium is by optical pumping using xenon flash lamps adjacent to the ruby rod. Ruby requires such high energy inputs to lase that the waste heat cannot be removed fast enough to sustain continuous output. For this reason, ruby lasers are always operated in a pulsed mode, and the output is usually measured in joules of energy per pulse (1 J of energy released per second is 1 W of power). Peak output powers of ruby lasers exceed 10 MW, requiring extensive safety precautions.

The development of pulsed ruby lasers for holography has progressed with the need to record holograms of moving (or highly unstable) objects. Ruby lasers have been extensively used to record the shock waves of aerospace models in wind tunnels, for example. Most holographic interferometry done with a ruby laser uses a double-pulse technique. The tasks that require a ruby laser are those that cannot be done with a helium-neon or an argon laser. Ruby lasers can routinely generate 1-J, 30-ns pulses of holographic-quality and relatively long coherence length light, which is sufficient for illuminating objects up to 1.5 m (5 ft) in diameter and 1.8 m (6 ft) deep.

The problem with ruby lasers lies in generating the two matching pulses required to record a suitable interferogram. Most lasers can be either pulsed once during each of two consecutive flash lamp cycles or Q-switch pulsed twice in the same flash lamp pulse to record differential-velocity interferograms. The pulse-separation time in the one flash lamp pulse mode extends to 1 ms. Generating two matching pulses becomes more difficult as the pulse-separation time exceeds 200 ms because of the dynamic thermal conditions in the laser cavity. The result is images with contour fringes that modulate displacement fringes, thus obscuring the information sought.

The operation of a ruby laser, when changing pulse-separation time or energy, requires the possible adjustment of flash lamp voltages, flash lamp timing with respect to the Q-switch timing, Q-switch voltages, and system temperatures. These conditions change as the laser system ages. Setting up the system requires many test firings to achieve stable performance. In short, operation of the laser requires high operator skill. In addition, the high performance of these systems requires care in keeping the optical components clean; buildup of dirt can burn the coatings on expensive optical components. The periodic replacement of flash lamps and other highly stressed electrical and optical components is to be expected. A helium-neon or a krypton laser is usually needed to reconstruct a ruby-recorded hologram for data retrieval. Differences between recording and reconstruction wavelengths lead to aberrations and changes in magnification in the reconstructed images.

Nd:YAG Lasers. Pulsed Nd:YAG lasers are constructed similarly to ruby lasers. Instead of a ruby rod, however, a neodymium-doped yttrium aluminum garnet rod is substituted as the lasing medium. The Nd:YAG laser is more efficient than the ruby system, but it operates in the near infrared at a wavelength of 1.064 μm (41.89 μin.). Frequency-doubling crystals with efficiencies of approximately 50% are used to produce light at a more useful green wavelength of 532 nm (5320 Å). All of the pulsed modes of operation available with the ruby system are also available with Nd:YAG system. The reconstruction of pulsed holograms can be performed with an argon ion laser at 514 nm (5140 Å). Owing to somewhat better thermal properties, continuous-wave Nd:YAG lasers are available with power capabilities well over 50 W (multimode), but their application in holography is still quite limited.

Exposure Controls

Most holographic systems control light by means of a mechanical or electrical shutter attached to the laser or separately mounted next to the laser. More sophisticated systems have photodetectors in the optical system and associated electronics that integrate the light intensity and close the shutter when the photographic plate or film has been properly exposed. Holographic systems that require strobing capabilities use acousto-optic modulators that can modulate the laser beam at rates up to at least 10 MHz and with 85% efficiency. It should be noted that a strobed system with a 5% duty cycle will have an effective brightness of 5% of normal (a 20-mW laser is effectively a 1-mW laser).

Beam Splitters

A piece of flat glass is usually a sufficient beam splitter for a production holographic system designed for recording only. If the system is to be used for recording and reconstruction or for real-time analysis, there are two approaches:

- The less expensive system uses a beam splitter that splits 20 to 30% of the light into the reference beam; a variable attenuator or a filter wheel is used to adjust the reference beam to the proper intensity
- The more expensive approach is to use a variable beam splitter, which consists of a wheel that varies the split from 95-to-5% to 5-to-95% as the wheel is rotated

Beam Expanders and Spatial Filters

Beam Expanders. Expansion of the narrow laser beam is required to illuminate the test object as well as the holographic film. A short focal length converging lens is often used for this purpose, ultimately causing the beam to diverge for distances greater than the focal length of the lens. For high-power pulsed-laser sources, a diverging lens must be used because the field strengths may become so intense at the focus of a converging lens that dielectric breakdown of the air may occur.

Spatial Filters. An unfiltered expanded laser beam usually displays diffraction rings and dark spots arising from extraneous particles on the beam-handling optical components. These rings and spots detract from the visual quality of the image and may even obscure the displacement-fringe pattern. For most CW holographic systems, laser powers are sufficiently low that spatial filters can be used to clean up the laser beam. Spatial filters basically consist of a lens with a short focal length and an appropriate pinhole filter. By placing a pinhole of the proper size at the focal point of the lens, only the laser light unscattered by dust and imperfections on the surfaces of the optical components can pass through the pinhole. The result is a uniform, diverging light field.

Pinhole Size and Alignment Specifications. A good spatial filter uses a high-quality microscope objective lens; a round, uniform pinhole in a foil of stainless steel or nickel; and a mount that allows the quick and stable positioning of the lens and pinhole. A complete analysis of the best pinhole size includes the factors of beam diameter, wavelength, and objective power. If the pinhole is too small, light transmission will suffer, and alignment will be very sensitive. As the pinhole size is increased, alignment is easier to achieve and maintain. As the pinhole size becomes too large, it begins to allow off-center, scattered light to pass through, with the result that the diverged beam will contain diffraction rings and other nonuniformities associated with dust and dirt. The pinhole will then begin to transmit information to construct the diffraction field of the particle. This does not prevent the recording of holograms; it only generates unwanted variations in light intensity. As a general rule, the magnification power of the objective multiplied by the pinhole diameter (in microns) should equal 200 to 300.

The position of the pinhole should be adjusted at a laser power level below 50 mW. A misaligned pinhole at a high power level can be burned by the intense point of light, rendering the pinhole useless. High-magnification spatial filters require the most care. With proper alignment, standard pinholes will function without degradation when the laser output power in watts multiplied by the objective magnification does not exceed 20.

Mirrors

Most holographic mirrors are front-surface coated. Second-surface-coated mirrors are generally unsatisfactory, because of losses at the front surface and the fact that the small reflection that occurs at the front surface generates unwanted fringes in the light field, which interfere with interpretation. Mirrors for holography do not need to be ultraflat. Inexpensive front-surface-coated mirrors are readily available in sizes up to 610 mm (24 in.). Metal-coated mirrors are usually the least expensive, but they can cause a 15% loss of reflected light. Dielectric-coated mirrors have greater than 99.5% reflectivity, but are much more expensive than metal-coated mirrors and are sensitive to the reflection angle. All mirrors, like all other optical components, require some care with regard to cleanliness. The cleaning of some mirrors is so critical that in many applications it is best to use inexpensive metal-coated mirrors, which can be periodically replaced. The manufacturer should be consulted in each instance as to the proper procedure for cleaning each particular type of mirror. Because mirrors reflect light rather than transmit it, they are a particularly sensitive component in a holographic system. They must be rigidly mounted and should be no larger than necessary.

Photographic Plate and Film Holders

Photographic plate and film holders perform the following two functions:

- They hold the plate (or film) stable during holographic recording
- They permit precise repositioning of the plate (or film) for real-time analysis

The first function is not difficult to achieve, but the second function is. If real-time analysis is not required, glass plates or films will work in almost any holder or transport mechanism. Real-time work requires special considerations. The problems inherent in real-time work can be handled by the use of replaceable plate holders, in-place liquid plate processors, and nonliquid plate processing.

Replaceable Plate Holders. With replaceable plate holders, the photographic plate is placed in the holder, exposed, and removed for processing. After processing, the plate is put back in the holder; the plate must be as close as possible to its original position in the holder to permit real-time analysis. Some plate holders have micrometer adjustments to dial out residual fringes. As a production method, the use of replaceable plate holders is very slow.

In-place plate processing is accomplished by using a liquid-gate plate holder (termed a real-time plate holder), which has a built-in liquid tank with appropriate viewing windows. The plate is immersed in the liquid in the tank (usually water) and allowed to soak for 15 to 30 s. Upon exposure, the tank is drained of the immersing liquid, and the plate is developed in place by pumping in the proper sequence of developing chemicals. After the plate is developed, the developing chemicals are replaced with the original immersing liquid, and the hologram is viewed through the gate of liquid. This procedure not only permits processing of the plate without disturbing its position but also eliminates the problem of emulsion swelling and shrinking, which causes residual fringes in many real-time setups. Plate development can take less than 30 s; total processing time is 1 min or less. Commercial systems are available that cycle the appropriate liquids through the cell as well as provide film advance for holographic films in a continuous-roll format.

Another holographic camera system permitting in-place development uses a thermoplastic recording medium that is developed by the application of heat. Such systems are available from at least two commercial suppliers. One system permits erasure and reexposure of the thermoplastic film plate with cycle times of just under 1 min. The plates can be reexposed at least 300 times. These systems and the high-speed liquid-gate processing systems mentioned above eliminate many of the inconveniences associated with holographic film handling and processing.

Nonliquid Plate Processing. Other in-place processing systems have been devised. Nonliquid plate processing using gases for self-development holds much promise for holographic recording. Photopolymers are promising as production recording media because they can generate a hologram quickly and inexpensively. For one photopolymer film, the photopolymer is exposed at a much higher energy level than is a silver emulsion (2 to 5 mJ/cm^2 versus 20 μJ/cm^2 or less for silver emulsions). After exposure, the hologram is ready to use. However, to prevent further photoreaction during viewing, the hologram is fixed by a flash of ultraviolet light.

Lenses

Lenses are required in some holographic systems. If the function of the lens is to diverge or converge a light beam, almost any quality of lens will suffice. However, if precise, repeatable control is desired, the lenses may need to be diffraction limited. Analysis of a proposed holographic system is sometimes best done by trial and error or by use of the best possible components, rather than by attempting a complicated mathematical computation.

Lenses can be antireflection coated if needed or desired. For example, lenses used in pulsed ruby systems for diverging a raw beam should be fused-silica negative lenses with a high-power antireflection coating. Some low-power ruby laser systems, however, have operated satisfactorily with uncoated lenses. Guidelines for using lenses with ruby lasers are available from the laser manufacturer.

Mounts

Mounts for the holographic components should be carefully chosen. Mounts that require adjustments should be kept to as few as possible. All fixed mounts should be bolted or welded in place. Some attention should be given to mount material; aluminum, for example, is generally a good material, but because of its high coefficient of thermal expansion, an alternative material might be more suitable in a given application. Mounts that are rugged and rigidly built should be selected. Holographic components should be positioned as close to the supporting structure as is practical.

Holographic Tables

Holographic components must be mounted with sufficient rigidity and isolation from ambient vibration to maintain their dimensional relationships within a few millionths of an inch during recording and real-time analysis. As discussed previously, the use of pulsed lasers to generate double-exposure holograms requires very little vibration isolation so long as the separation between exposures is short. When vibration isolation is required, the designer must exercise care in the design of the structure used to support the holographic system in order to isolate the structure from outside excitation. This design involves the three following considerations:

- Building the structure with sufficient rigidity to reduce the deflection of components to within holographic limits
- Building the structure with sufficient damping capacity to absorb excitation energy and to prevent excessive resonant-vibration amplitudes
- Building the structure with sufficient mass to increase inertia and therefore decrease response from outside driving forces

Small, low-cost holographic systems are usually supported by one of a variety of vibration-isolation tables, which float on three or four rubber air bladders or small inner tubes. The holographic components are screwed, clamped, magnetically held, or simply set in place on the table. As the size of the holographic system (and therefore the size of the table) increases, more care is needed to maintain stability. There are three basic types of large holographic tables:

- Honeycomb tables
- Slabs
- Weldments

Honeycomb tables, made of honeycomb-core sandwiched panels, are extensively used for holography. They can range in thickness from 50 mm to 0.9 m (2 in. to 3

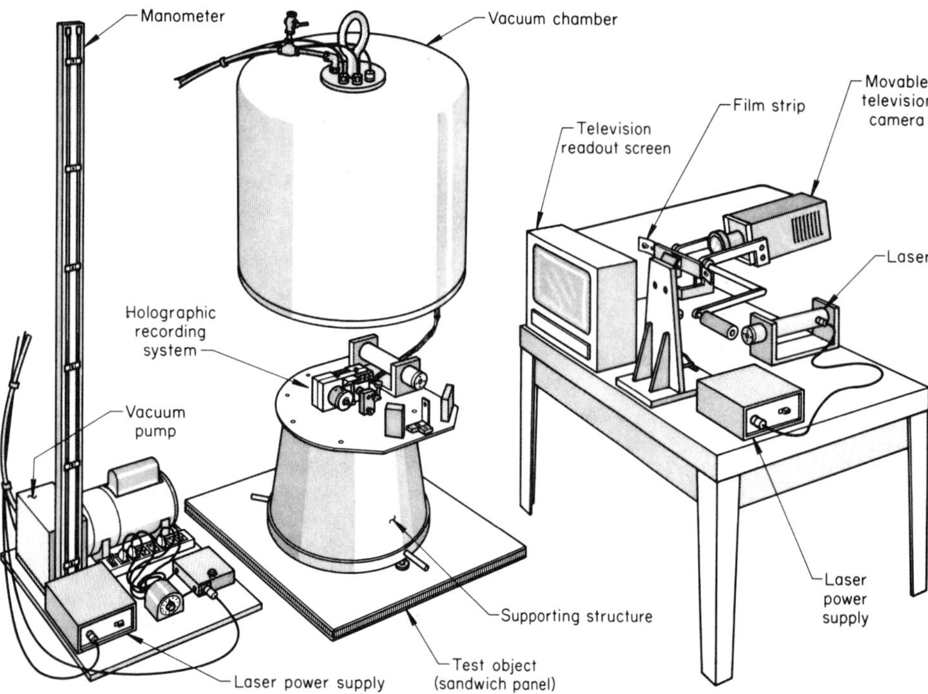

(a) Equipment for vacuum stressing and holographic recording (b) Equipment for holographic reconstruction

Fig. 10 Portable holographic analyzer for the inspection of sandwich panels with use of vacuum stressing. See description in text.

ft). The outer skin is usually a ferromagnetic stainless steel, but for increased temperature stability, an outer skin of Invar can be used. Honeycomb tabletops weigh less than 10% as much as a steel table of equivalent rigidity. They can be fabricated from vibration-damping materials to make them acoustically dead. The tables are usually floated on three or four air mounts. Air mounts (generally forming a leg for the table) are large air cylinders with rolling-diaphragm pistons that contact the table. A servovalve inputs or exhausts air at the cylinder to maintain constant height of the leg and keeps the table level as components are moved about. Air mounts provide excellent isolation by virtue of their low resonant frequencies, typically 1 to 2 Hz. The holographic components can be set on the honeycomb table, attached with magnetic clamps, or screwed down utilizing an array of drilled-and-tapped holes in the upper-skin centers.

Most solid tabletops used in holography are flat within 0.025 mm (0.001 in.) or less, while honeycomb tables (1.2 × 2.4 m, or 4 × 8 ft) are flat within 0.10 to 0.25 mm (0.004 to 0.010 in.). This difference does not hamper the performance of most holographic systems.

Slabs are the least costly type of support for a large holographic system. They are usually made from steel or granite and floated on a vibration-isolation system. Many low-cost supports have been laboratory constructed by floating a surplus granite or steel surface plate on an array of tire tubes. It is usually difficult to dampen vibrations that reach the surface of a slab. For this reason, the performance of a slab degrades when the test objects are large and/or the ambient noise level is high (particularly from air-conditioning systems that emit low-frequency noise). This problem can be minimized by selecting a material (such as gray iron) that has naturally high damping capacity rather than a material that has a ring.

Most components that are attached to three-point mounts need not be rigidly attached to the slab, but other components (and particularly the object being vibrated or otherwise stressed) need to be rigidly mounted to ensure stability. To facilitate the mounting of components, the slab top may require tapped holes, T-slots, or a coating of tacky wax, or it may need to be ferromagnetic.

Weldments are heavily braced frames or plates generally designed as part of a portable or otherwise special system used to analyze very large or unusual test objects. Weldments are generally used where slabs or honeycomb tables are not suitable, although a slab or honeycomb-core sandwich panel may be part of the structure for mounting the components.

Types of Holographic Systems

There are basically two types of holographic systems: stationary and portable. Both will be discussed in this section.

Stationary Holographic Systems

A holographic system is considered stationary when it is of such size, weight, or design that it can be utilized only by bringing the test object and required stressing fixtures to the system for analysis. Stationary systems are usually dependent on building services, requiring compressed air for the vibration-isolation system, electric power for the laser and other electronic components, and running water for processing the holograms and cooling the laser (for example, as required for an argon laser). Most stationary systems operate in a room with light and air control to achieve high stability and the low light levels required for recording, processing, and viewing holograms. As the size of the table increases, the stability requirements become more difficult to satisfy. As a result, the cost of a stationary system generally increases approximately exponentially with object size.

Stationary systems are used in the following cases:

- Production line inspection of small objects
- Where required flexibility in the type and size of test objects is needed for developmental work
- Inspection of a large or awkward structure that cannot be holographed by a portable system

Portable Holographic Systems

A portable holographic system can be moved to the test object and operated with minimal setup time. A portable system built for the Apollo lunar exploration program was designed to record holograms of lunar soil. It was battery powered, weighed 7.89 kg (17.4 lb), and occupied less than 0.017 m^3 (0.6 ft^3) of area. A portable holographic system used for developmental work, particularly in wind tunnels, has been transported throughout the United States by semitrailer truck. The components of the system are mounted in cabinets or in frames on wheels, and upon arrival at the test site, the system is unloaded and set up in several hours. A portable system used to inspect sandwich-structure helicopter-rotor blades is shown schematically in Fig. 5; a portable system used to inspect sandwich panels is shown in Fig. 10.

Reflection Holographic Systems. In a simplified type of portable system, a tripod-mounted laser and spatial filter project light directly through a holographic plate fastened to the test object. A reflection hologram is formed by interference between the light traveling through the plate and the light reflected back to the plate from the object. In this system, the reference beam and the object beam strike the emulsion from opposite sides of the plate, resulting in the reconstruction of a virtual image produced by reflection. This configuration differs from the holographic systems described earlier, in which the two recording beams strike the plate from the same side and, during reconstruction, the virtual image is produced by transmission.

An important consideration in designing reflection holographic systems is that reflection holograms are more sensitive than transmission holograms to photographic emulsion shrinkage, which may take place during the development and drying processes. This shrinkage causes the image formed during reconstruction to be produced at a slightly shorter wavelength (a hologram recorded with red laser light will reconstruct best in yellow or green light). Unless the reconstructing light matches this shorter wavelength, the image will be quite faint. Therefore, white light, which contains all the required wavelengths for efficient image production, is often used as the reconstructing reference beam instead of laser light.

Because, as described above, the film plate also serves as a beam splitter and can be mounted to the test object itself, reflection holographic systems can be quite insensitive to object vibration. The major critical stability requirement is the relationship between the object and the holographic plate fastened to it. If the emulsion shrinkage is fairly uniform and the holographic plate is in close proximity to the test object, the image formed will be bright and clear. The plate must be dried carefully, however, to avoid variations in emulsion thickness, which would cause variation in the color of the image. Because the diffraction of light changes with color, variations in color will cause smearing of the image and loss of resolution. The greater the distance from the object to the plate, the greater the smearing.

Portable systems can be used in the following cases:

- Field inspection
- When the size or configuration of a test object is such that it is more practical to attach the holographic system to the object than vice versa
- When the test object is in an environment or in a structure required as part of the experiment

Portable holographic systems are usually designed to inspect a specific part or a range of small parts. The questions that establish criteria for designing a portable system are the following:

- How is the system to be powered?
- How is stability between the system and the test object to be maintained both during and between exposures?
- How can critical adjustments be made or eliminated?
- How is the photographic plate to be handled and processed?
- How can the holographic components and the stressing fixture be designed into a workable system within the definition of portable?

Typical Holographic Testing Applications

Among the applications for which holographic testing is utilized are the following:

- Inspection of sandwich structures for debonds
- Inspection of laminates for unbonded regions
- Inspection of metal parts for cracks
- Inspection of hydraulic fittings
- Measuring of small crack displacements
- Vibration analysis of turbine and propeller blades
- Holographic contouring
- Characterization of composite materials

Inspection of Sandwich Structures for Debonds

A sandwich structure usually consists of face sheets separated by a lightweight core. The face sheets are designed to carry in-plane loads; the core is designed to stiffen the face sheets and prevent them from buckling and to carry normal loads in compression or shear. The core can be a solid material, such as balsa wood, or a cellular material, such as foam plastic or honeycomb construction. The face sheets and core are usually held together by an adhesive or braze material.

The extensive use of sandwich structures in widely varying applications has created some unusual inspection problems. The main area of interest is the quality of the attachment of the face sheets to the core; large areas of structures must be inspected inexpensively for unbonded or unbrazed areas (debonds) and for poorly bonded or brazed regions. The inspection for poorly bonded or brazed regions has not yet been satisfactorily accomplished on a production basis, but is currently the subject of various research programs. A secondary area of interest is the edge-closure assembly, which surrounds the sandwich structure. The configuration of the closure varies depending on whether the sandwich is brazed or bonded and on the types of materials involved. Sandwich structures cause inspection problems because of the number of parts being joined together and the abrupt changes in thickness or solidity of the assembly. An additional complication arises in designing a holographic inspection system for sandwich structures in that structures range in area from several square inches to several square feet and their contours vary from simple flat panels to complex curved shapes, such as helicopter rotor blades.

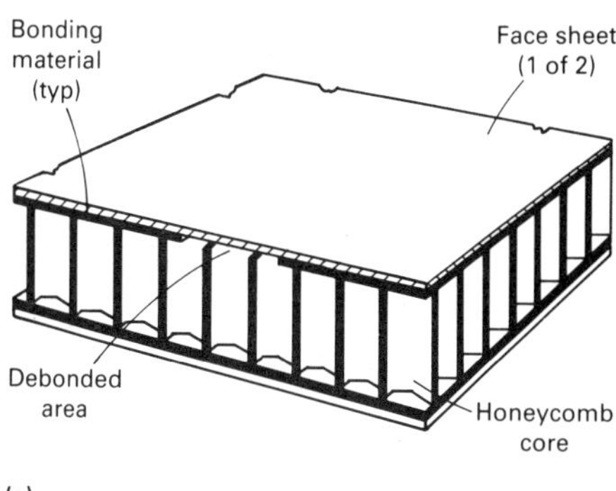

(a)

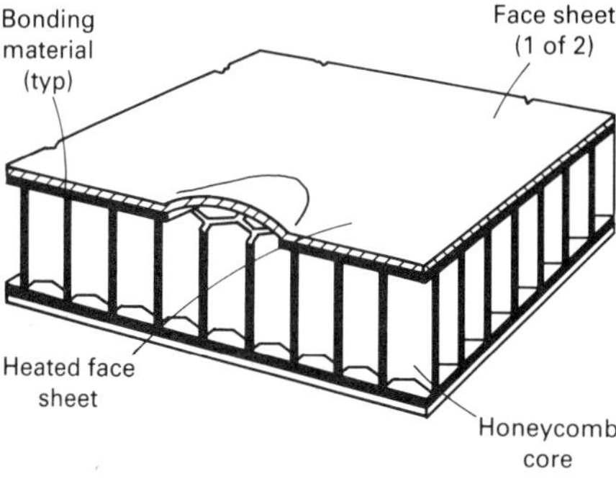

(b)

Fig. 11 Honeycomb-core panel illustrating the detection of debond by thermal stressing. (a) Section through the region of debond. (b) Same section as in (a) showing bulge in face sheet over the region of debond, caused by gentle heating of the face sheet

Both sides and all edges of the structures must be inspected.

When inspecting sandwich structures, it is necessary to determine which stressing technique or combination of techniques will best detect the types of flaws likely to be present. If several techniques are chosen, it must be decided whether to apply them simultaneously or sequentially. The two stressing techniques that have been found to work well for the routine inspection of sandwich structures are thermal stressing (Fig. 11) and vacuum stressing (Fig. 4), although stress for inspection can be provided by acoustical loading (Fig. 3), fatigue loading (Fig. 5), or impact loading.

Techniques for inspecting sandwich structures are well established and documented (Ref 20). Most inspection of sandwich structures is being done, and probably will be done, using CW techniques, with pulsed-laser techniques being used only for special applications.

Example 1: Detection of a Debond in a Honeycomb-Core Panel With Thermal-Stressing Holographic Techniques. As an example of holographic inspection using thermal stressing, assume that a honeycomb-core sandwich panel, such as that illustrated in Fig. 11(a), contains a debond. If the face sheet over the debond is gently heated, the region over the debond will

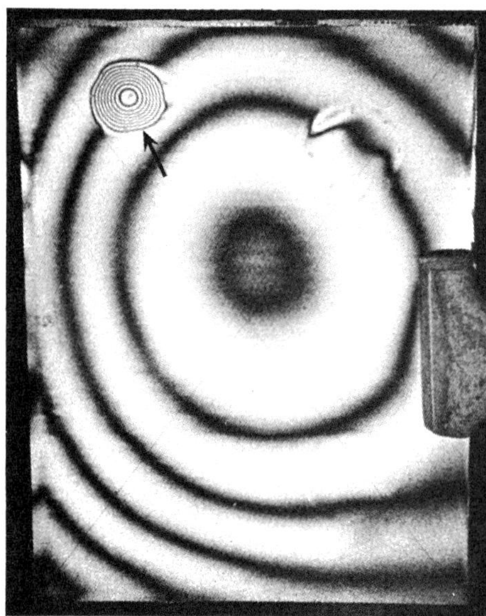

Fig. 12 Double-exposure time-lapse interferogram of a thermally stressed honeycomb-core sandwich panel showing a fringe pattern (arrow) contouring a region of the front face sheet over a debond. The panel had metal face sheets 500 μm (0.02 in.) thick and was heated about 2.8 °C (5 °F) between exposures, which were made with a pulsed ruby laser. The fringes indicate a maximum displacement over the debond of 3 μm (120 μin.). The background fringes were caused by general movement of the face sheet due to heating.

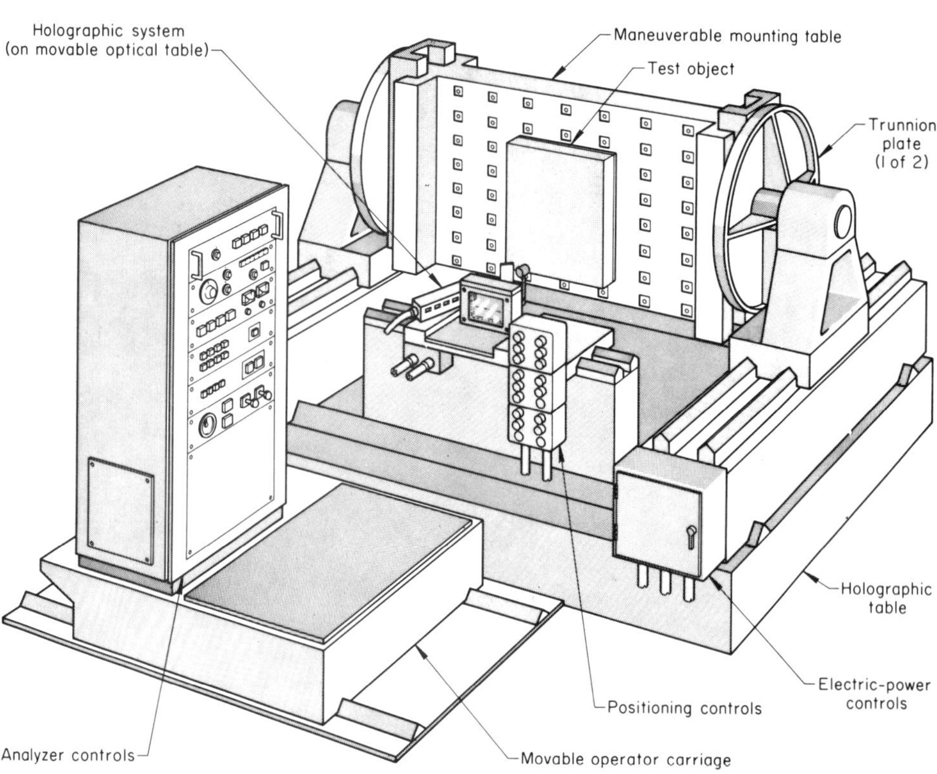

Fig. 13 Commercial holographic analyzer used for the inspection of sandwich structures. See text for description.

become hotter faster because the heat in that region is not conducted away to the core. The result of this differential-temperature field is a slight bulge in the heated face sheet (Fig. 11b). Using either real-time or double-exposure time-lapse holographic techniques, an image will be formed in which the region of the bulge is contoured by a set of fringes representing lines of constant displacement between the two images.

A typical set of fringes caused by a debond in a honeycomb-core sandwich panel is illustrated by the interferogram in Fig. 12, which demonstrates the sensitivity of inspection by thermal stressing. This interferogram was obtained using the double-exposure technique and a pulsed ruby laser, which has a wavelength of 694 nm (6940 Å) (Table 1). Using Eq 3, it can be seen that each fringe, except the first, represents an out-of-plane displacement of 0.694/(2 sin 90°), or about 0.33 μm (13 μin.). The region above the debond in the panel, which had a face sheet 500 μm (0.02 in.) thick and was heated only about 2.8 °C (5 °F), has a maximum displacement of about 3 μm (120 μin.), or only about 0.6% of the face sheet thickness.

Example 2: Holographic Detection of a Debond in a Sandwich Structure Using the Vacuum-Stressing Method. Inspection by vacuum stressing has also been found to be effective for detecting debonds in sandwich structures. Figure 10 shows the essential components of a portable holographic analyzer for the inspection of sandwich panels. This analyzer consists of two separate sets of equipment: one set for vacuum stressing and holographic recording (Fig. 10a) and one set for holographic reconstruction (Fig. 10b). The holographic recording system is mounted on the top of a hollow supporting structure that rests on the test panel. The recording system contains a 3-mW CW helium-neon laser and, through the use of suitable optical components, encompasses a 460 mm (18 in.) diam circular field of view of the portion of the panel surface beneath the supporting structure.

During an inspection, the vacuum chamber, which is made of fiberglass, is lowered over the recording system until contact is made with the surface of the panel. A first exposure is made of the panel in its unstressed state. A second exposure is made when the internal pressure has been reduced by approximately 7 kPa (1 psi). The pressure of the ambient air in the core voids pushes the face sheets out at unbonded regions. The double-exposure hologram is recorded as a circular field 8 mm (0.3 in.) in diameter on a 16 mm (⅝ in.) film strip. The film is then advanced, the vacuum is released, the system is moved to the next location on the panel, and the sequence is repeated. The total time required for constructing a double-exposure hologram is usually about 1 min or less. By using a film strip 2030 mm (80 in.) long, it is possible to record a total of more than 200 double-exposure holograms.

A typical commercial holographic analyzer for the inspection of sandwich structures is illustrated in Fig. 13. It consists of a 3.7 × 2.4 m (12 × 8 ft) table supported on air bearings. On this table is a part-holding mounting plate, which is supported by two 1220 mm (48 in.) diam trunnion plates. The mounting plate can be rotated and translated to view either flat panels or curved shapes. The holographic system usually uses a 50-mW helium-neon laser, but can also use higher-powered argon lasers. The part to be inspected is held in place on the mounting plate by a series of vacuum cups or clamps. Thermal, vibrational, pressure or vacuum stressing can be applied. A part measuring up to 1.5 × 1.8 m (5 × 6 ft) can be inspected with this analyzer. Other analyzers are available that are designed to handle either smaller parts or larger parts (up to 1.8 × 6.1 m, or 6 × 20 ft).

Approximately 0.19 or 0.28 m^2 (2 or 3 ft^2) of a part surface can be viewed in a single hologram with the analyzer illustrated in Fig. 13. The exposure and processing of the hologram are controlled automatically by the analyzer. With the 50-mW helium-neon laser, approximately 2.3 m^2 (25 ft^2) of surface can be inspected per hour. Either double-exposure or real-time interferograms can be made with this analyzer. Using the real-time technique, the inspected areas can be recorded by taking still photographs through the hologram or preferably by re-

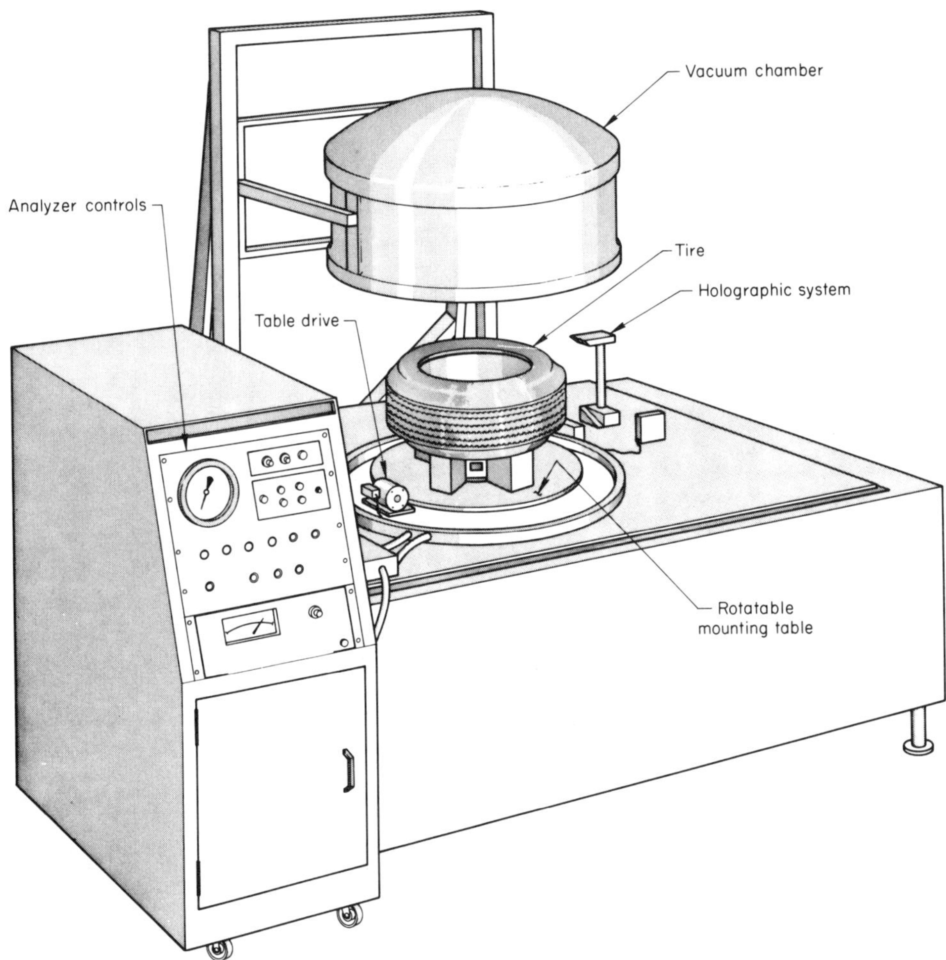

Fig. 14 Commercial holographic analyzer used for the inspection of pneumatic tires. See text for a description of the inspection procedure.

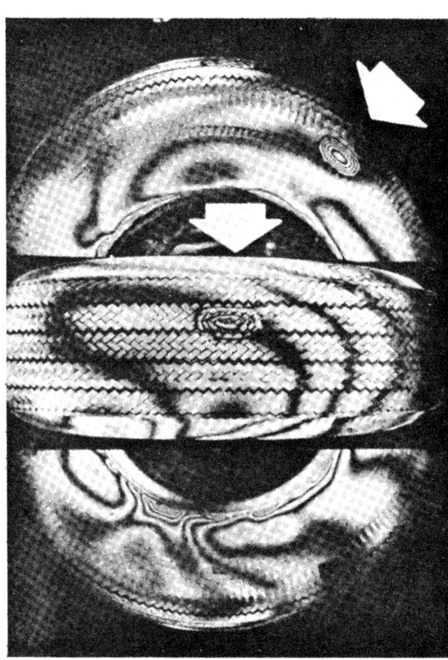

Fig. 15 Double-exposure time-lapse interferogram of a defective pneumatic tire that was stressed by inflation to 345 kPa (50 psi) between exposures. The contoured fringe patterns (arrows) indicate regions of the sidewall and tread where there is no bond between layers of the cured tire. General movement of the tire during inflation caused the background fringes.

cording the transient patterns on videotape. Still photographs give an indication of obvious flaws, but far smaller flaws can be detected by an experienced inspector viewing a sweeping fringe pattern. A library of videotapes showing various flaws can be used as a training and qualifying aid. As in other inspection techniques, standards and qualifying procedures must be established. Application and further development of automated fringe interpretation methods could also be used to advantage for this type of inspection.

Inspection of Laminates for Unbonded Regions

Standard optical holographic systems are readily used to identify the presence of flaws in laminated composites and structures. Generally, double-exposure techniques are the most convenient, although both real-time techniques and time-average techniques (for oscillating systems) can also be used. Because the most common flaw in laminates is a lack of bonding, the critical problem is to select the most suitable stressing technique that will reveal the presence of the unbonded region by inducing some differential movement between the bonded and unbonded regions.

Example 3: Holographic Tire Analyzer for Detecting Flaws With Vacuum-Stressing Technique. An example of a laminated composite is the pneumatic tire, which is constructed of layers of rubber and fabric. (Some of the fabrics used in modern tires contain steel wires.) The commercial holographic tire analyzer illustrated in Fig. 14 is capable of inspecting tires ranging in size up to a maximum outside diameter of 1145 mm (45 in.) at a rate of more than 12 tires per hour when auxiliary semiautomatic options are employed. The inspection procedure uses the vacuum-stressing technique described previously; a double-exposure interferogram is made of each quadrant of the inner walls of the tire undergoing inspection. Flaw anomalies are manifested by minute changes in the inner-wall topography occurring as a result of the pressure differential existing between the unvented tire flaws and the evacuated chamber.

Example 4: Holographic Tire Analyzer for Detecting Defects Using 345 kPa (50 psi) Inflation Pressure. A different type of holographic tire analyzer employed to obtain the interferogram shown in Fig. 15 used inflation to 345 kPa (50 psi) rather than vacuum stressing. With this analyzer, both of the sidewalls and the tread portion of the tire can be inspected simultaneously for all unbonded regions.

Additional Techniques for Stressing Laminates. Other laminates can be stressed by heating (as with printed circuit boards), by ultrasonic excitation, and by the addition and removal of a mechanical load.

Example 5: Holographic Inspection of Coprene Rubber-Stainless Steel Laminate Using Mechanical Loading Technique. An example of the use of the mechanical-loading method is a simple rubber-steel laminate that was stressed in cantilever bending while clamped along one edge. Holographic exposures made before and after stressing reveal a field of closely spaced fringes (Fig. 16) due to anelastic effects that prevent full recovery. Local perturbations in this field due to an unbonded region between the rubber sheet and the metal substrate can be readily identified as an anomaly in the fringe pattern.

Inspection of Metal Parts for Cracks

Optical holographic inspection has not proved to be an effective method for finding even large cracks in metal parts. The primary reason for this is the difficulty of stressing the test object in such a manner as to create a difference in displacement that is

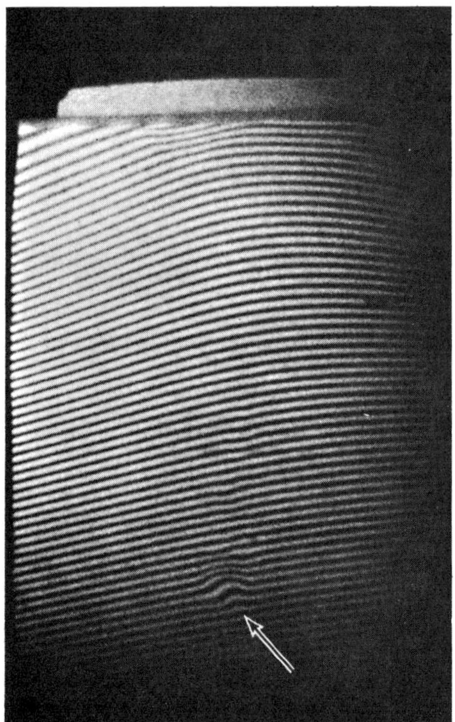

Fig. 16 Double-exposure time-lapse interferogram of a defective rubber-steel laminate that was mechanically stressed between exposures showing a local perturbation in the fringe pattern (arrow) over an unbonded region. The laminate was 125 mm (4.92 in.) long by 90 mm (3.5 in.) wide and was composed of a 0.94 mm (0.037 in.) thick sheet of coprene rubber bonded to a 1.93 mm (0.075 in.) thick sheet of stainless steel. The rubber face was painted with a bright white reflecting paint to aid in holographic inspection.

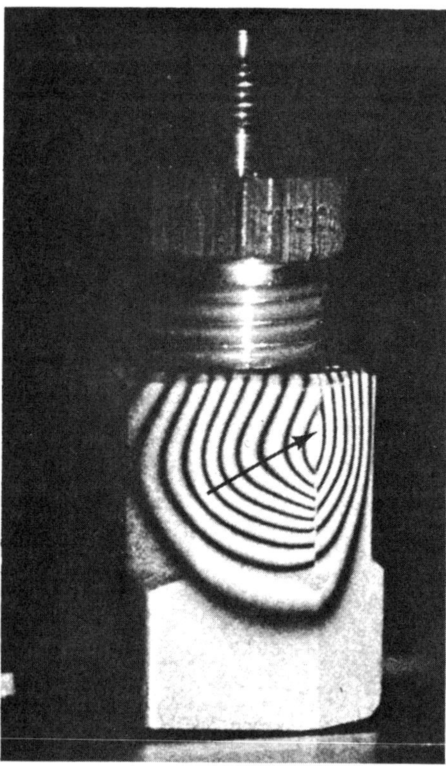

Fig. 17 Double-exposure interferogram of a cracked hydraulic fitting. The fitting was stressed between exposures by having a mating fitting screwed into it. The holocamera used a 15-mW helium-neon laser. The discontinuity in the fringe pattern (arrow) was caused by a small crack in the fitting.

easily detectable optically. Speckle-pattern techniques can be useful for the detection of in-plane displacements associated with surface cracks.

In general, only those cracks greater in length than the thickness of the part are detectable by optical holographic inspection. Several stressing methods have been found to be useful, such as mechanical stressing by means of loading fixtures or interference fasteners as well as thermal stressing by means of heat lamps or cold liquids or solids. Either real-time techniques or double-exposure techniques can be used. Also, high-resolution techniques, such as phase stepping and heterodyne holographic interferometry, have been used (see the section "Interpretation of Inspection Results" in this article). Time-average time-lapse techniques are generally not successful for detecting the effects of cracks on vibratory patterns.

Example 6: Holographic Inspection of Hydraulic Fittings to Detect Cracks. An instance of the successful detection of cracks by optical holography was the inspection of small hydraulic fittings. Radiographic studies using x-rays, eddy current testing, and other forms of nondestructive inspection did not provide reliable detection of these small cracks or sufficient data on crack growth characteristics.

To solve the inspection problem, conventional double-exposure time-lapse holography was employed using a 15-mW helium-neon laser and portable optical components. The plate holder was a conventional static type and held 100 × 125 mm (4 × 5 in.) glass plates. The entire optical system was mounted on a commercial 1.2 × 2.4 m (4 × 8 ft) vibration-isolation (holographic) table.

Inspection consisted of first making a holographic exposure of a fitting held statically in a vise-type clamping fixture. After the first exposure, a mating fitting was screwed into the test fitting to approximate normal in-use loading, and a second exposure was then made on the same holographic plate. The resulting interferogram is shown in Fig. 17. Each interference fringe in this reconstruction represents a displacement between exposures of approximately one-half the wavelength of the helium-neon light (λ = 633 nm, or 6330 Å), or about 0.33 μm (13 μin.). The discontinuity in the fringe pattern indicates that relative motion, or slippage, occurred along the front vertical edge of the fitting during loading. Inspection at still higher loads revealed a small crack along that edge.

The use of optical holographic inspection in this application permitted relatively inexperienced personnel to pinpoint small cracks in several hydraulic fittings and to study the propagation of these cracks under varying load conditions. This brief study program required approximately 6 h of engineering time and no special fixturing or tooling.

Example 7: Small Crack Displacement Measurements Using Heterodyne Holographic Interferometry. As mentioned above, the small out-of-plane displacements associated with most cracks make it difficult to apply conventional (homodyne) holographic interferometry techniques. Still, small displacements do occur and can be visualized holographically using high-resolution techniques. For example, Fig. 18 shows the results of a heterodyne analysis of a region of a holographic interferogram near a surface-breaking crack. Total displacements of only 5 to 6 nm (50 to 60 Å) were observed in this case with a background noise floor of about 0.6 nm (6 Å).

A dual-reference, double-exposure recording technique was used where a bending stress was applied to the cracked specimen between the holographic double exposures. The resulting hologram was reconstructed with a 100-kHz frequency difference imposed between the two reconstructing beams so that the intensity of the image varied sinusoidally at a 100-kHz rate.

Although each point on the image varies in intensity at 100 kHz, the relative phase of these oscillations varies from point to point on the image, depending on the amount of displacement recorded between holographic exposures. Therefore, when a small detector is scanned over the image, the phase of its output signal can be compared with some reference phase from another (fixed) point on the object, as shown in Fig. 19. Because a phase difference of 360° corresponds to a single interferometric fringe, the resulting map of phase difference is directly related to surface displacement. Electronic phase measurement accuracy to 0.36° corresponds to 1/1000 of one fringe. Uncertainties resulting from the effects of speckle and the environment do not permit meaningful measurements below this level.

Vibration Analysis of Turbine and Propeller Blades

Vibration analysis using routine optical holographic techniques can significantly contribute to the inspection and evaluation of turbine and propeller blades in both the design and manufacturing stages (Ref 21-23). A recommended approach to turbine blade evaluation is the simultaneous holographic recording of both sides of the blade as it is excited into vibration with shaker tables (at frequencies generally limited to less than 50 kHz), air-horn vibrators, elec-

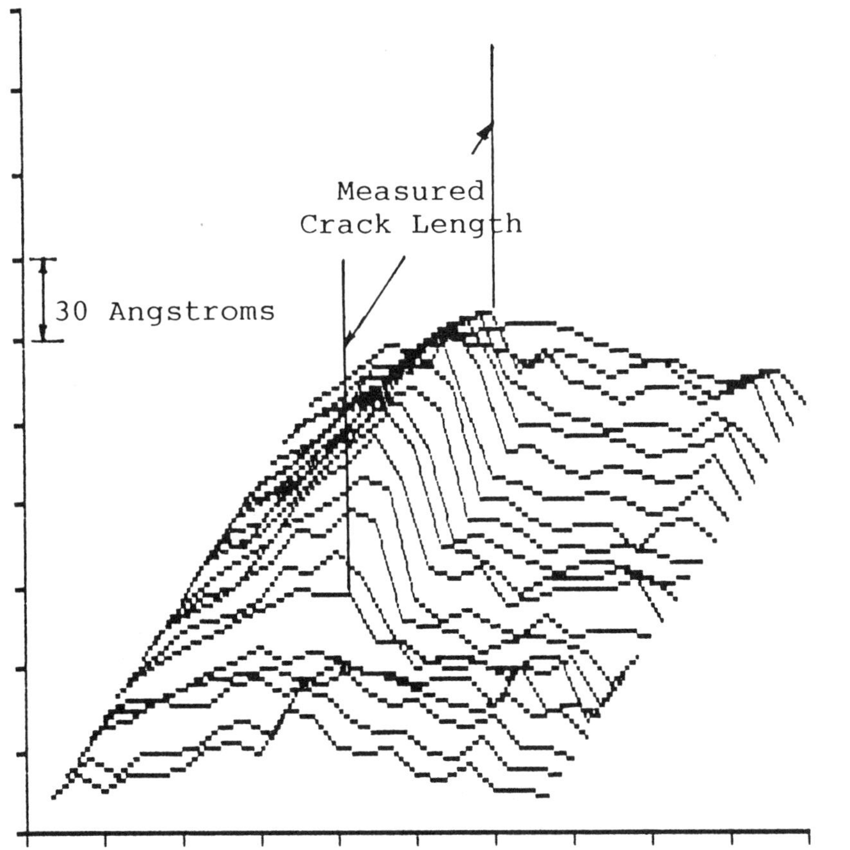

Fig. 18 Results of heterodyne holographic interferometry showing minute displacements adjacent to a surface-breaking crack in a nickel-base superalloy

tromagnetic drive systems (where blade materials permit), or piezoelectric transducers (bonded or clamped to the blade). By the use of simultaneous recording, information over and above the straightforward recording of vibrational mode patterns can be obtained. Differences in the mode patterns for the two sides of the blade will suggest an absence of structural integrity at those points, while differences in vibrational amplitude (for example, in hollow blades) can offer a relative measure of blade wall thickness (that is, a check on cooling passage alignment).

Holographic techniques are applicable to all types of blades, both solid and hollow and of almost any size and shape, as well as entire turbine wheels or propeller assemblies. However, some experimental problems may be encountered in the dual-sided simultaneous recording of large parts, thus necessitating a two-step procedure. (Caution must then be exercised to ensure that the driving frequency and amplitude are identical for both of the holograms.)

For blades of reasonable size (up to perhaps 100 to 125 mm, or 4 to 5 in., chord by 460 to 610 mm, or 18 to 24 in., length), the simultaneous recording of both sides of the blade can be accomplished by a standard holographic system through the use of mirrors (taking care to maintain the path lengths of the reference beam and the object beam within the coherence length of the laser). An alternative method of simultaneous recording is to use a dual holographic system. A reasonably compact (1.2 × 1.2 m, or 4 × 4 ft) setup of a dual system can be assembled by first splitting the incoming laser beam into two beams (one for each hologram), each of which is subsequently split again into a reference beam and an object beam by a symmetrical arrangement of the various optical components, as shown in Fig. 20.

An example of the type of data that can be obtained by simultaneous recording is illustrated in Fig. 21, which shows interferograms of both sides of a hollow blade whose pressure wall is about 250 to 375 μm (10 to 15 mil) thinner than its suction wall. Although the two interferograms in Fig. 21 were not recorded simultaneously, identical ultrasonic frequencies and amplitudes were used to excite the blade during recording. This produced fringe patterns in solid regions (the leading and trailing edges) that are essentially the same on the front and back surfaces. However, over the hollow region of the blade, the vibrational amplitude on the pressure side is measurably larger (each fringe represents approximately 0.30 μm, or 12 μin., of out-of-plane displacement) than that on the suction side. In addition, the vibrational amplitude increases from the root to the tip of the blade, indicating a decreasing wall thickness for both the pressure and the suction sides. Pressure stressing is an alternative to vibrational excitation for the holographic inspection of hollow turbine blades for detecting weakened structural characteristics.

To establish the most favorable driving force and frequency for a particular blade, a somewhat extensive series of tests is recommended (perhaps including sectioning of the blades) to correlate the results and to establish standards of acceptance. Although such preliminary testing may be costly, for large production runs this inspection procedure could be valuable for accurate wall thickness gaging in thin-wall structures, in which standard ultrasonic pulse-echo techniques are the most difficult to effect.

The ability to observe one entire surface of a large test object at one time, rather than in a series of limited views, is one of the most important benefits of using holography for nondestructive inspection. One entire surface of an 810 mm (32 in.) diam jet engine fan assembly can be recorded in a single hologram (Fig. 22), considerably facilitating the performance of a vibrational-mode analysis that alternatively would require transducers placed over the entire assembly.

Holographic Contouring

When evaluating the shape of an opaque object, interference fringes are generated on the object that represent depth or elevation contours. These fringes are usually generated most readily by optically interacting two images of the object that have been slightly displaced from one another, although one method simply requires projecting onto the object a set of fringe surfaces whose normals are roughly parallel to the line of sight of the viewer. The three principal techniques for holographic contouring are discussed below.

Multiple-Source Contouring. In the multiple-source technique (which is not really holographic contouring, but rather holographic interferometry), an optical system is used that incorporates a rotatable mirror for steering the object beam (Fig. 23). A fixed contour map of the object can be generated with this system by making two holographic exposures on the photographic plate that differ only by a slight rotation of the object beam steering mirror, which shifts the virtual image of the object-illumination source.

Alternatively, a single holographic exposure can be made with the optical system illustrated in Fig. 23, and after processing and repositioning in the plate holder, a real-time contour map of the object can be generated by a similar slight rotation of the object beam steering mirror. In either case, when the system is suitably arranged for proper viewing, the altitude contours on the

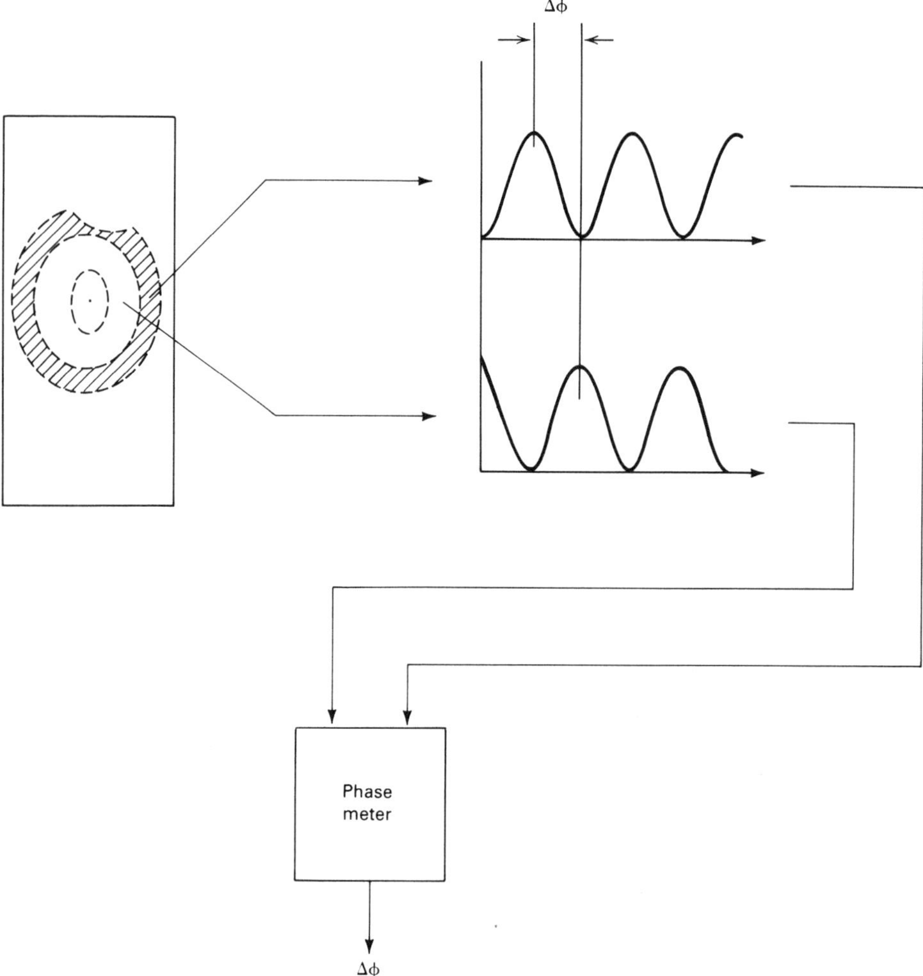

Fig. 19 Dual-detector readout for heterodyne holographic interferometry

Fig. 21 Continuous-exposure interferograms of both sides of a turbine blade that were recorded while the blade was being vibrated. The interferograms were recorded at identical driving frequencies and amplitudes, such as might be obtained with the compact dual holographic system shown in Fig. 20. The holocameras used helium-neon lasers.

map of the surface have a separation distance, Δh, given by:

$$\Delta h = \frac{\lambda}{(2 \sin \alpha)} \quad \text{(Eq 10)}$$

where λ is the wavelength of the laser light source and α is the angle of mirror rotation. Suitable arrangement of the system requires that the line of sight through the hologram

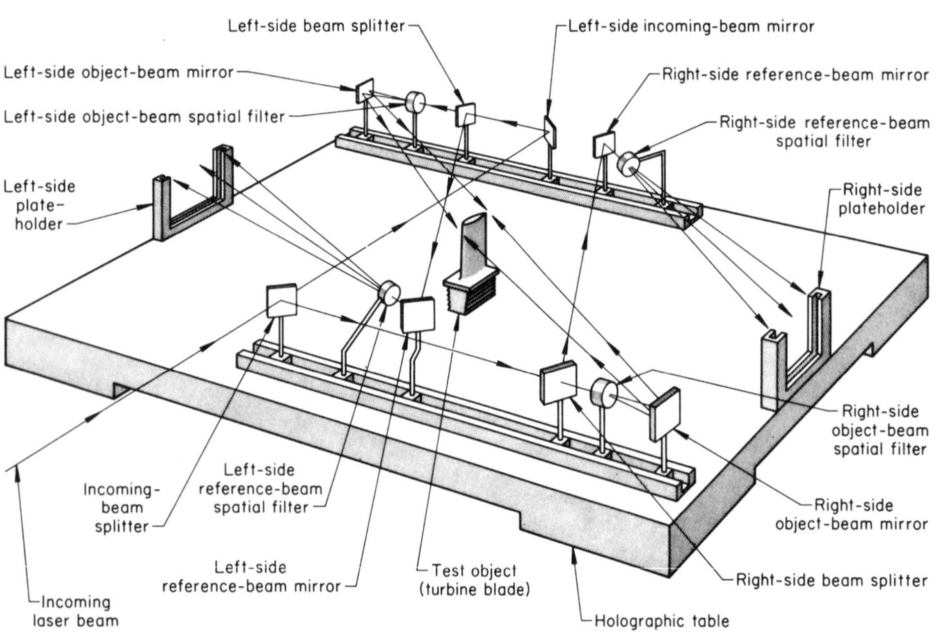

Fig. 20 Schematic of a compact dual holographic system for the simultaneous recording of both sides of a turbine blade

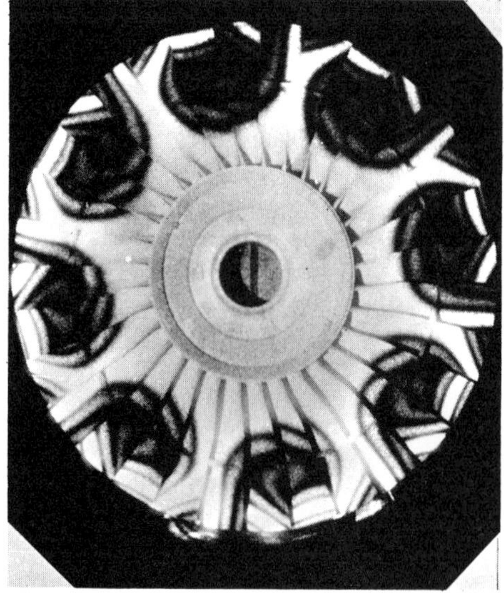

Fig. 22 Continuous-exposure interferogram of one side of an 813 mm (32 in.) diam jet engine fan assembly that was excited at a frequency of 670 Hz showing a 4-nodal-diam mode of vibration. The holocamera used a helium-neon laser.

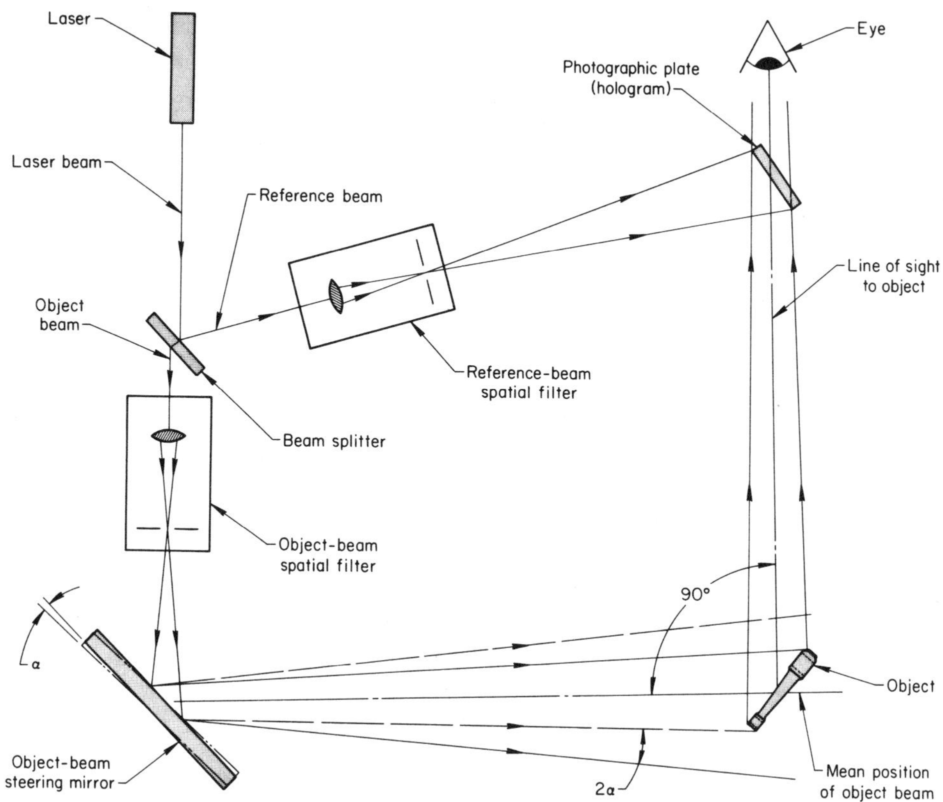

Fig. 23 Schematic of an optical system for multiple-source holographic contouring. Source: Ref 24

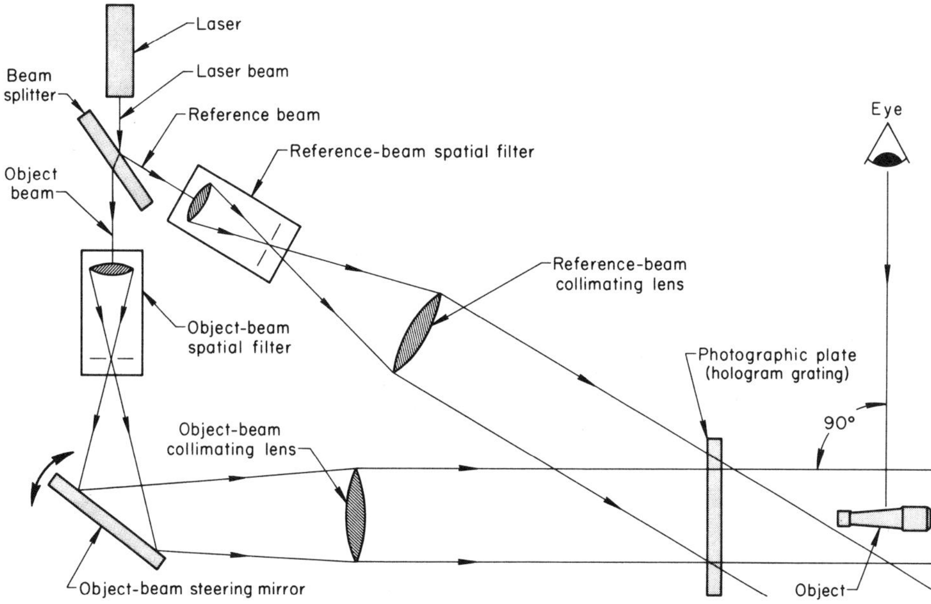

Fig. 24 Schematic of a real-time optical system for holographic contouring using the multiple-source technique. See text for discussion.

to the object be in the same plane as the angle of rotation of the object beam and be perpendicular to the mean position of the reflected object beam, as shown in Fig. 23. The multiple-source technique has the advantage of providing for an almost unlimited range of contour separations. However, because of the orthogonal illumination and observation directions, shadowing effects are a handicap, and no reentrant surfaces can be contoured by this technique.

Real-Time Contouring. An optical system for the real-time generation of a contour map by the multiple-source technique is shown in Fig. 24. This system uses collimated reference and object beams to generate hologram gratings by means of two successive exposures at different settings of the object beam steering mirror. Subsequent illumination of a hologram grating by only the collimated reference beam produces a simultaneous reconstruction of both the original and the rotated object beams, which propagate and interfere beyond the hologram grating. The spacing of the resulting vertical-standing-wave interference planes is given by Eq 10 and can be readily calibrated by direct measurement of the fringe field on the hologram itself or by exposing a sheet of film oriented at right angles to the propagation direction of the reconstructed object beams and measuring the fringes recorded there. Finally, suitable orientation of the object anywhere in the reconstructed object beams yields a real-time contour map when viewed at right angles to the object beams.

Example 8: Real-Time Contouring of Electroplated Copper Foil Using a Multiple-Source Holographic System. The multiple-source system shown in Fig. 24 was used in a study of the tensile properties of electrodeposited metal foils. In the course of this study, the shapes of permanently bulged specimens (Fig. 25a) of copper foil 0.0889 mm (0.00350 in.) thick were measured. The bulges, which were 29 mm (1.14 in.) in diameter, ranged from a few millimeters to half a centimeter in height; therefore, a range of sensitivities was needed.

Because of shadowing, only one side or half of the specimen could be contoured at a time, but multiple recordings with the specimen revolved to expose all sides were made and are shown combined in Fig. 25(b). For this contouring, the fringe separation was 0.229 mm (0.00902 in.); other gratings were made for separations up to 5 mm (0.20 in.) and down to 0.1 mm (0.004 in.). Because no high-precision repositioning is required and the reconstructed fringe field is permanent, this represents a convenient and practical means of real-time contouring. However, if a single exposure is recorded in the hologram and is accurately repositioned to provide an initially clear field with appropriately balanced beams, simple adjustments of the object beam steering mirror provide the capability of a real-time variation of the sensitivity (fringe separation) of the contour map when needed.

Multiple-Wavelength Contouring. Constant-range contours measured from an origin at the center of the hologram can be generated with the multiple-wavelength technique (Ref 24). Because a change in wavelength produces changes in position as well as in magnification and the phase of the object images, the multiple-wavelength technique is more complicated than the multiple-source technique. One configuration has been demonstrated in which both the illumination and viewing directions lie along a line through the center of the holo-

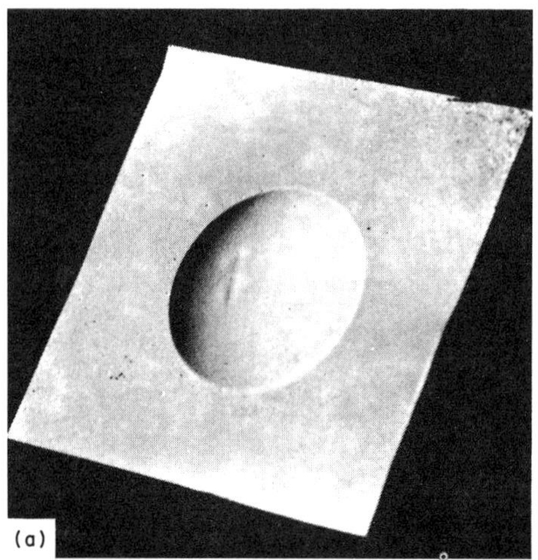
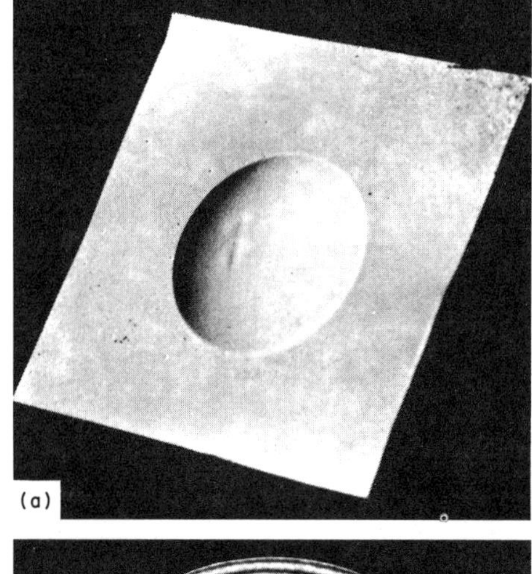

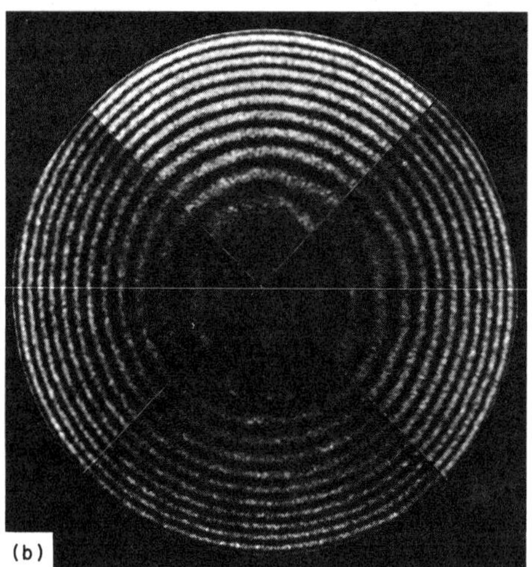

Fig. 25 Example of real-time contouring using the multiple-source holographic system illustrated in Fig. 24. (a) Photograph of bulged specimen of electrodeposited copper foil 0.0889 mm (0.00350 in.). The bulge was 29 mm (1.14 in.) in diameter. (b) Composite of two contour maps. The specimen was rotated between recordings to expose both sides of the bulge. The fringe separation in this map is 0.229 mm (0.00902 in.).

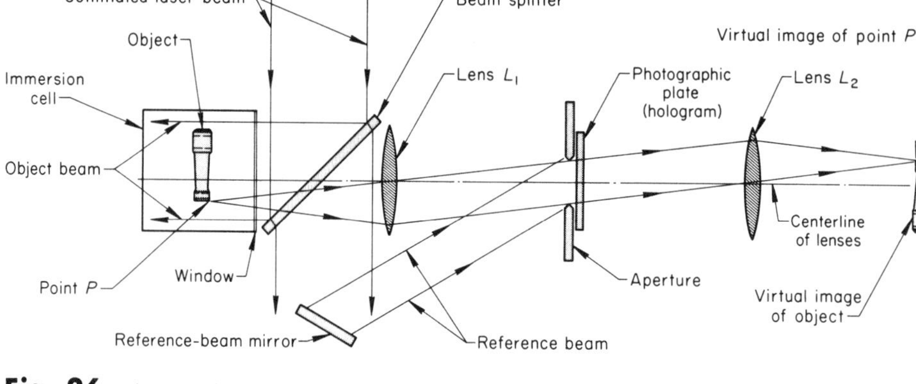

Fig. 26 Schematic of the optical system for multiple-index holographic contouring

gram and the object (Ref 25). This configuration yields fringes with separations, Δr, given by:

$$\Delta r = \frac{\lambda_1 \lambda_2}{(2|\lambda_1 - \lambda_2|)} \quad \text{(Eq 11)}$$

where λ_1 and λ_2 are the wavelengths of the two laser light sources. Because λ_1 and λ_2 are not arbitrarily variable, only discrete sensitivities are available, all of which are quite high. However, because of the more normal illumination conditions, the problems of shadowing associated with the multiple-source technique are completely eliminated.

Multiple-Index Contouring. In this technique, the gaseous or liquid medium surrounding an object is changed between successive exposures of the hologram (for double-exposure interferometry) (Ref 25, 26). This changes the optical-path length by slightly altering the index of refraction of the medium and is most readily accomplished by the use of an immersion cell with windows through which the object can be illuminated and viewed simultaneously. In the optical system for the multiple-index technique (Fig. 26), lenses L_1 and L_2 act as a telescope with its viewing direction normal to the cell window. The aperture between the lenses is used to limit aberrations. The hologram can be constructed at any plane to the right of the beam splitter (for example, at the aperture, as shown in Fig. 26). When this system is used, the contour separation distance, Δh, is given by:

$$\Delta h = \frac{\lambda}{2(|n_1 - n_2|)} \quad \text{(Eq 12)}$$

where λ is the wavelength of the laser light source and n_1 and n_2 are the values of the index of refraction of the cell media used for the two exposures.

Example 9: Multiple-Index Holographic Contouring Used to Examine Wear of Articulating Surfaces of an Artificial Knee Implant. Figure 27 shows the results of a multiple-index holographic-contouring technique applied to the lower (tibial) com-

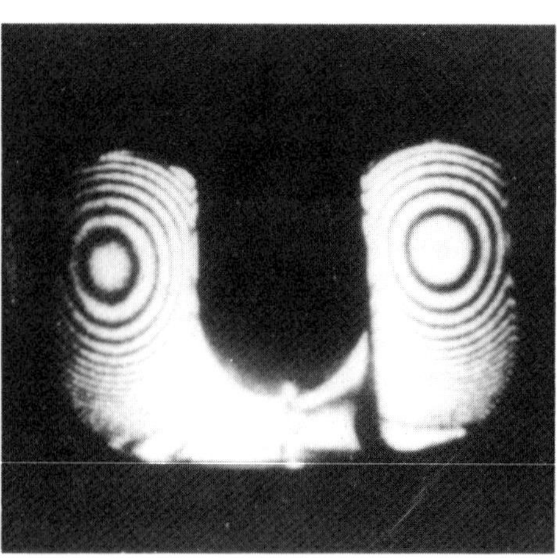

Fig. 27 Dual refractive index contour hologram of an artificial knee implant component

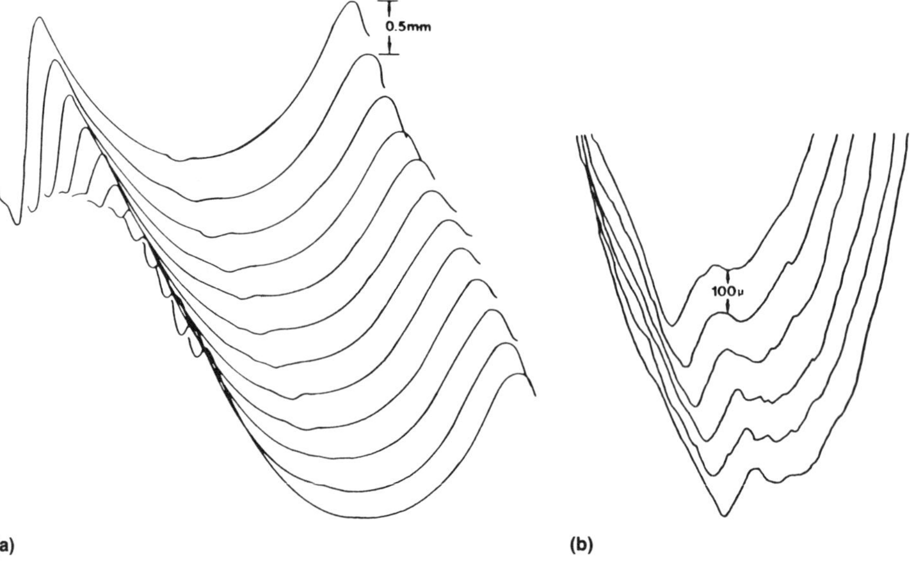

Fig. 28 Results of heterodyne holographic analysis. (a) Surface gouge in a portion of an artificial knee implant. (b) Higher-resolution scanning that reveals details of the gouge

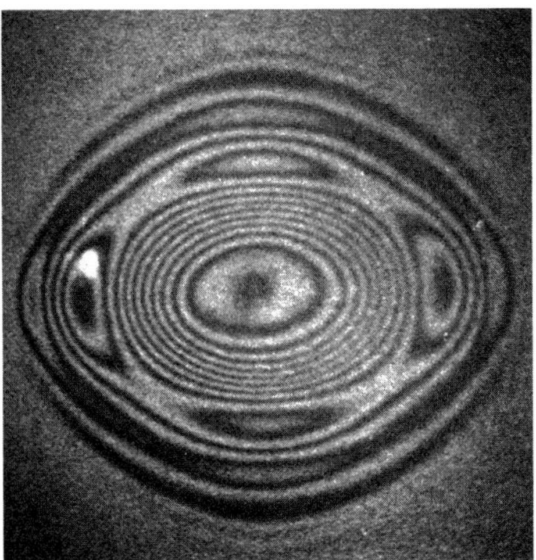

Fig. 29 Double-exposure pulsed hologram showing displacements associated with large-amplitude acoustic waves traveling in a centrally excited composite sheet

ponent of an artificial knee implant (Ref 27, 28). The material is a high-density polymer, and the intent of the holographic contouring was to examine the articulating surfaces of these two sockets for wear resulting from *in vitro* testing. Because wear tracks are not visible from the interferometric images, high-sensitivity heterodyne holographic techniques were again applied. Figure 28 shows the results of the heterodyne analysis over a region of one of the knee sockets. From these scans a gouge in the material is clearly observed; the details are revealed upon further amplification of the phase-difference signal (Fig. 28b).

Multiple-Index Method Versus Multiple-Wavelength Method. The multiple-index technique offers results equivalent to those of the multiple-wavelength technique. (Note that if the effective wavelengths in Eq 11 are defined as $\lambda_1 = \lambda/n_1$ and $\lambda_2 = \lambda/n_2$, Eq 11 becomes Eq 12.) The two techniques have the same types of restrictions and advantages with regard to sensitivity, but the multiple-index optical system is easier to arrange. Both techniques can be performed in either the double-exposure or the real-time mode.

Characterization of Composite Materials

The high-speed pulsed holographic interferometry of transient acoustic waves is used to help determine material properties and to identify certain types of defects in graphite-reinforced epoxy laminate sheets (Ref 29).

Example 10: Holographic Interferometry of Composite Sheet Material. Large-amplitude acoustic waves were generated in the sheet materials by direct laser pulse excitation for thin sheets or by laser detonation of a very small chemical explosive in thicker ones. Double-pulsed holographic exposures were made, with the first exposure at the instant of excitation and the second exposure occurring after several microseconds of delay. The resulting interferograms are shown in Fig. 29.

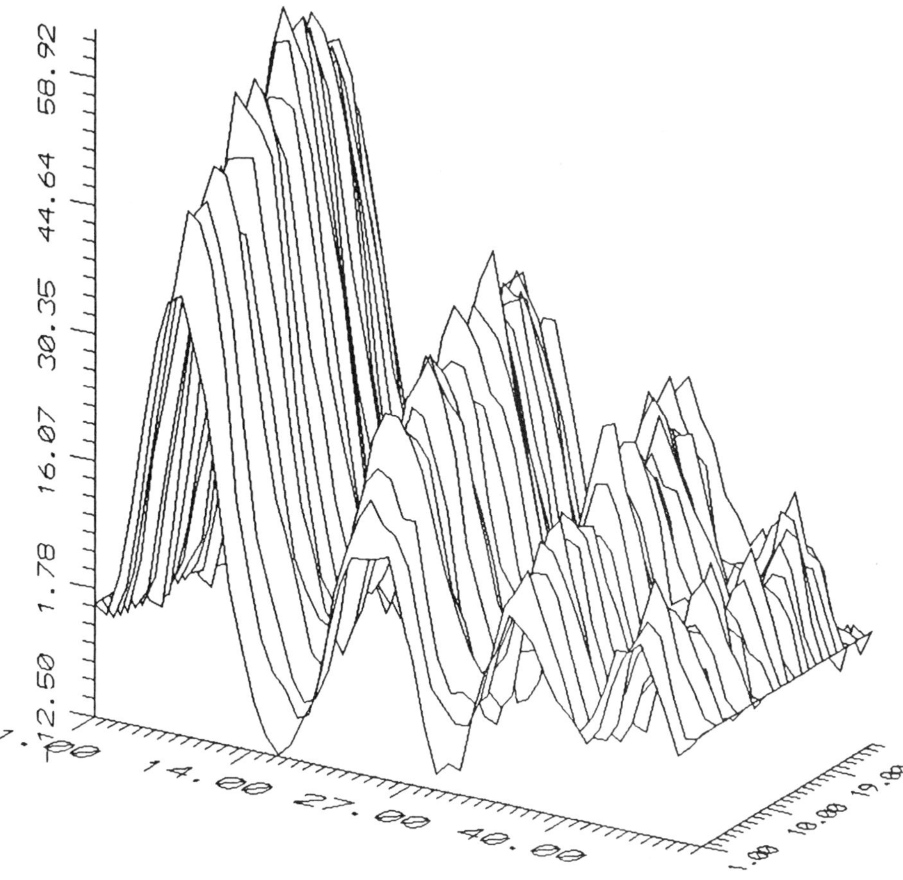

Fig. 30 Surface displacements resulting from Lamb wave propagation in an aluminum sheet. These subfringe displacements were extracted from the holographic interferogram using phase-step (quasi-heterodyne) holographic interferometry.

In this case, the composite sheet was six plies thick, with the fibers running in only one direction in each ply. The ply orientation stacking sequence was 0_2-90_4-0_2.

From the holographic reconstruction, one observes flexural waves (asymmetric Lamb waves) emanating from the point of excitation. They are clearly least attenuated in the directions of the reinforcing fibers. From their velocities, one can compute the flexural stiffness and estimate the effective Young's modulus as a function of direction in the material. Again, in cases where features smaller than a single fringe must be resolved or when it is not practical to use high excitation forces, phase stepping or heterodyne techniques can be applied (Ref 30).

Figure 30 shows the displacements associated with flexural wave propagation in a sheet where the peak displacement amplitude is about one-third of an interferometric fringe. These displacement data were extracted using phase-stepping methods on a hologram in which not even a single complete fringe was observed. With phase stepping, sensitivities to about 1/30 of a fringe are readily obtained with ultimate sensitivities to about 1/100 achievable under near ideal conditions. (Heterodyne sensitivities approach 1/1000 of a fringe.)

Appendix: Selection of Holographic Systems

In general, the considerations that apply to the selection of a holographic system are similar to the considerations that apply to the selection of any other major piece of capital equipment. The selection will be based primarily on the ability of the equipment to perform the tasks that are of chief concern. Beyond that, selection will be governed by such factors as the space and support facilities required, personnel and training, system versatility, and cost. This section outlines the advantages and disadvantages of purchasing holography as an outside service and of operating an in-house system.

Contract (Purchased) Holography

An initial approach for a company with little or no holographic equipment or technical skill might be to purchase the holographic work on a contract basis from a company that performs this work as a service. This would probably be the most economical approach for a company that has only a few parts to inspect or wishes to run only a single development program employing holographic analysis.

The advantages of purchasing holographic work are as follows:

- No capital investment and minimal in-house technical skill are required
- The holographic work can be purchased as required, even on a daily or a per-piece basis
- Some companies will conduct holographic developmental work at low cost if favorable results will promote the sale of equipment
- Purchasing holographic work is a good way to test an application prior to investing in equipment and personnel training
- This method allows the buyer to utilize holography with a very short lead time

The disadvantages of purchasing holographic work are the following:

- Test objects must be shipped to and from the service company. Often, the travel of some in-house personnel may also be required for program success
- Contracted holographic work costs may be high, depending on the equipment and personnel required. This charge must be compared with the cost of suitable alternatives
- Personnel of the service company must be educated concerning the holographic techniques and results as applied to the application in question

In-House Holographic Systems

If the amount of work or other factors make it desirable to have in-house capability, a suitable holographic system can be installed. This can be a commercially available system, such as a tire analyzer or a sandwich-panel analyzer, designed to perform only one task; a commercially available system, such as a portable holocamera or a holographic interference camera, designed for a variety of tasks; or a system built in-house from separately purchased or fabricated components.

Assembled Systems

In-house units can be purchased piecemeal or as one complete system.

The advantages of in-house assembly of a system from components are as follows:

- The system can be built to fit specific or universal needs
- In-house engineering talent can be utilized to reduce capital expenditures by 40% or more
- Such a system can utilize readily available components
- Building a system is an excellent learning experience for the novice desiring to make future use of holographic techniques
- The system can be built and modified to match the expanding skills of the operator

The disadvantages of in-house assembly are the following:

- Such systems usually require a long lead time to become operational
- Designing and in-house assembly of a system require either developed or purchased technical skill that can be justified only when it is readily available or when many possible applications exist

Purchased Systems

Purchasing a commercial system has the advantage that the vendor can assume complete responsibility for a system to meet the specifications of the buyer. Details of the components do not need clarification in the bid specification. A specification, for example, need simply state that the system shall be capable of finding a flaw in a given part, of a given length, with a specified confidence level.

The purchased system can be either a complete system designed especially for one task or a system designed for a variety of tasks.

Complete System Designed for One Specific Task. The advantages of purchasing a complete system designed especially for one task are as follows:

- Minimum in-house holographic technical skill is required
- A system that has been previously built or designed can be delivered in a short time

The disadvantages are the following:

- Special systems are usually expensive because of the labor required for design, construction, and testing. Costs per unit drop rapidly, however, if more than one unit is built
- It may be difficult to modify the system for other tasks, although many holographic components can be removed for use in another system
- A system that must be designed and built for a special task will require a long lead time

Multifunctional Complete System Designed for Numerous Tasks. Although a special system may be the only means for testing a part of unusual design or size, it is often advantageous to purchase a system designed for a variety of jobs. The advantages of this type of system are as follows:

- Such systems can be less expensive than special systems because they are in volume production, which reduces design costs per unit
- Such systems have flexibility in change-over to inspect different kinds of parts. This usually requires little more than changing the stressing fixture
- The buyer is free to make his own setups
- Delivery time is short; some systems can be purchased off the shelf
- Many vendors will train in-house personnel at no additional cost

The disadvantages are the following:

- The system will be limited by the size or weight of the test object it can handle
- The system may not be as stable as a specialized system because of the flexibility required in the optical components and mounts

REFERENCES

1. R.K. Erf *et al.*, "Nondestructive Holographic Techniques for Structures Inspection," AFML-TR-72-204, Air Force Materials Laboratory, Oct 1972 (AD-757 510)
2. L. Kersch, Advanced Concepts of Holographic Nondestructive Testing, *Mater. Eval.*, Vol 29, 1971, p 125
3. J.P. Waters, Object Motion Compensation by Speckle Reference Beam Holography, *Appl. Opt.*, Vol 11, 1972, p 630
4. L. Rosen, Focused-Image Holography, *Appl. Phys. Lett.*, Vol 9, 1969, p 1421
5. H.J. Caulfield *et al.*, Local Reference Beam Generation in Holography, *Proc. IEEE*, Vol 55, 1967, p 1758
6. V.J. Corcoran *et al.*, Generation of a Hologram From a Moving Target, *Appl. Opt.*, Vol 5, 1966, p 668
7. D.B. Neumann *et al.*, Object Motion Compensation Using Reflection Holography, *J. Opt. Soc. Am.*, Vol 62, 1972, p 1373
8. D.B. Neumann *et al.*, Improvement of Recorded Holographic Fringes by Feedback Control, *Appl. Opt.*, Vol 6, 1967, p 1097
9. H.W. Rose *et al.*, Stabilization of Holographic Fringes by FM Feedback, *Appl. Opt.*, Vol 7, 1968, p a87
10. J.C. Palais, Scanned Beam Holography, *Appl. Opt.*, Vol 9, 1970, p 709
11. R.K. Erf, Ed., *Holographic Nondestructive Testing*, Academic Press, 1974
12. J.Ch. Vienot, J. Bulabois, and J. Pasteur, in *Applications of Holography*, Proceedings of the International Symposium on Holography, Laboratory of General Physics and Optics, 1970
13. J.E. Sollid and J.D. Corbin, Velocity

Measurements Made Holographically of Diffusely Reflecting Objects, in *Proceedings of the Society of Photo-Optical Instrumentation Engineers*, Vol 29, 1972, p a125
14. J.E. Sollid, A Comparison of Out-of-Plane Deformation and In-Plane Translation Measurements Made With Holographic Interferometry, in *Proceedings of the Society of Photo-Optical Instrumentation Engineers*, Vol 25, 1971, p 171
15. P. Hariharan, Quasi-Heterodyne Hologram Interferometry, *Opt. Eng.*, Vol 24 (No. 4), 1985, p 632-638
16. W. Juptner *et al.*, Automatic Evaluation of Holographic Interferograms by Reference Beam Shifting, *Proc. SPIE*, Vol 398, p 22-29
17. R. Dandliker and R. Thalmann, Heterodyne and Quasi-Heterodyne Holographic Interferometry, *Opt. Eng.*, Vol 24 (No. 5), 1985, p 824-831
18. J.D. Trolinger *et al.*, Putting Holographic Inspection Techniques to Work, *Lasers Applic.*, Oct 1982, p 51-56
19. *The Optical Industry and Systems Purchasing Directory*, 34th ed., Laurin Publishing Company, 1988
20. R.C. Grubinskas, "State of the Art Survey on Holography and Microwaves in Nondestructive Testing," MS 72-9, Army Materials and Mechanics Research Center, Sept 1972, p a40-46
21. R. Aprahamian *et al.*, "An Analytical and Experimental Study of Stresses in Turbine Blades Using Holographic Interferometry," Final Report AM 71-5 under NASC Contract N00019-70-C-0590, July 1971
22. J. Waters *et al.*, "Investigation of Applying Interferometric Holography to Turbine Blade Stress Analysis," Final Report J990798-13 under NASC Contract N00019-69-C-0271, Feb 1970 (available as AD 702 420)
23. *Proceedings of the Symposium on Engineering Applications of Holography*, Society of Photo-Optical Instrumentation Engineers, 1972
24. B.P. Hildebrand and K.A. Haines, Multiple-Wavelength and Multiple-Source Holography Applied to Contour Generation, *J. Opt. Soc. Am.*, Vol 57 (No. 2), 1967, p a155-162
25. J.S. Zelenka and J.R. Varner, Multiple Index Holographic Contouring, *Appl. Opt.*, Vol 8 (No. 7), 1969, p 1431-1434
26. T. Tsuruta, N. Shiotake, J. Tsujuichi, and K. Matsuda, Holographic Generation of Contour Map of Diffusely Reflecting Surface by Using Immersion Method, *Jpn. J. Appl. Phys.*, Vol 6, 1967, p 661-662
27. J.W. Wagner, High Resolution Holographic Techniques for Visualization of Surface Acoustic Waves, *Mater. Eval.*, Vol 44 (No. 10), 1986, p 1238-1242
28. J.W. Wagner, Examples of Holographic Versus State-of-the-Art in the Medical Device Industry, in *Holographic Nondestructive Testing (NDT)—Status and Comparison With Conventional Methods: Critical Review of Technology*, Vol 604, Conference Proceedings, Los Angeles, CA, Society of Photo-Optical Instrumentation Engineers, 1986
29. M.J. Ehrlich and J.W. Wagner, Anisotropy Measurements and Determination of Ply Orientation in Composite Materials Using Holographic Mapping of Large Amplitude Acoustic Waves, in *Proceedings of the Third International Symposium on Nondestructive Characterization of Materials* (Saarbrucken, West Germany), Oct 1988
30. J.W. Wagner and D.J. Gardner, Heterodyne Holographic Contouring for Wear Measurement of Orthopedic Implants, in *Proceedings of the International Congress on Applications of Lasers and Electro-Optics* (Los Angeles, CA), Laser Institute of America, 1984

SELECTED REFERENCES

- R.J. Collier, C.B. Burckhardt, and L.H. Lin, *Optical Holography*, Academic Press, 1971
- R.K. Erf, Ed., *Holographic Nondestructive Testing*, Academic Press, 1974
- C.M. Vest, *Holographic Interferometry*, John Wiley & Sons, 1979

Speckle Metrology

F.P. Chiang, Laboratory for Experimental Mechanics Research,
State University of New York at Stony Brook

MEASUREMENT TECHNIQUES or devices traditionally employ either a regular geometric pattern or a constant as a base transducer. The change of the regularity of the pattern or the response of the base is converted into the measured quantity. Rulers, pressure and strain gages, moiré and interferometric methods, and so on, are all representative of these devices or techniques. Since the 1970s, techniques have been developed that employ random patterns as transducers. Measurement is made not by knowing the deviation from regularity but by determining the correlation between two random signals. This has opened a new field of activity in which new measurement instruments are being developed. In the discussion that follows, the basic principle of the random speckle technique will be described, together with its application to the metrological measurement of metals. Speckle metrology is also known as laser speckle photography (LSP).

Laser Speckle Patterns

Speckle patterns occur naturally in everyday life. From the metrological point of view, they did not attract attention until the advent of the laser. When a coherent laser beam illuminates an optically rough surface (that is, not a mirrorlike surface), the reflected wavelets from each point of the surface mutually interface to form a very complicated pattern. The process is analogous to the ripple pattern formed on the surface of a pond that has been disturbed by raindrops. The only difference is that water speckles form on the surface of the pond, but laser speckles fill the entire space covered by the reflected wavelets. A typical laser speckle pattern is shown in Fig. 1.

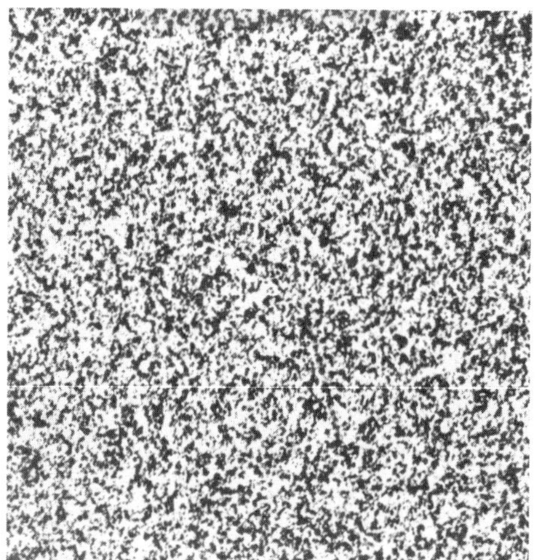

Fig. 1 Typical laser speckle pattern

Recording and Delineation of Speckle Displacement

The speckle methods that will be described in the following sections originated from a physical phenomenon first observed by Burch and Tokarsky (Ref 1). They found that when two random patterns are superimposed with an in-plane displacement between them and illuminated with a coherent beam of light, the far field diffraction spectrum is that of a single speckle pattern modulated by a series of Young's fringes whose orientation and spacing are identical to those formed by two point sources located at the ends of the displacement vector experienced by the speckles.

When applying the laser speckle technique, a coherent laser beam is used to illuminate a metal surface (or any optically rough surface). Depending on the information sought, one selects a certain section of the resulting volumetric speckle pattern to be photographed by a camera. When the surface is deformed, the speckle pattern moves accordingly. The spatial movement of the speckle is quite complicated and is governed by a set of three equations (Ref 2).

Conceptually, laser speckles can be viewed as tiny particles attached to the surface by rigid but massless wires. They shift as the surface shifts and tilt in much the same way as light rays are tilted by a mirror. The displaced speckles are photographed again on the same film through double exposure, or a separate recording is made and then superimposed on the first one. The two superimposed speckle patterns are called a specklegram. The speckle displacement can be delineated by sending a narrow laser beam at the points of interest and receiving their corresponding far field diffraction spectrum on a screen. The experimental arrangement is shown in Fig. 2(a). The observer sees a circular diffraction halo (assuming the aperture of the camera lens is circular) modulated by a series of uniformly spaced straight fringes, that is, the Young's fringes. The intensity distribution of the diffraction spectrum can be expressed as follows (Ref 3):

$$I(\mathbf{u}) = 4 \cos k \left(\frac{\mathbf{u} \cdot \mathbf{d}}{2L}\right) I_h(\mathbf{u}) \qquad \text{(Eq 1)}$$

where $\mathbf{u}$ is the position vector at the receiving plane, $k = 2\pi/\lambda$ with λ being the wavelength of light, $\mathbf{d}$ is the displacement vector at the point of probing, L is the distance between the specklegram and the receiving screen, and I_h is the so-called halo function, which is the diffraction spectrum of a single speckle pattern.

Equation 1 can be converted into (Ref 3):

$$|\mathbf{d}| = \frac{\lambda L}{S} \qquad \text{(Eq 2)}$$

where $|\mathbf{d}|$ is the magnitude of the displacement vector and S is the fringe spacing.

These fringes represent the displacement vector of a small cluster of speckles illuminated by the laser beam (Fig. 2b). The displacement of the speckles can be converted into the displacement of the specimen, and depending on the experimental arrangement, different information can be obtained.

Measurement of In-Plane Deformation

For this application, the recording camera is directly focused on the flat surface of the specimen. (The method can also be applied to measuring interior in-plane displacement if the material is transparent to the radiation used, as discussed in Ref 4.) Under either a plane-strain or (generalized) plane-stress condition, the speckle movement is largely confined to the plane of the surface. Therefore, the observer again photographs the speckle pattern formed on the surface after the specimen has been subjected to deformation. The resulting specklegram can be probed point by point to yield the displacement information, as

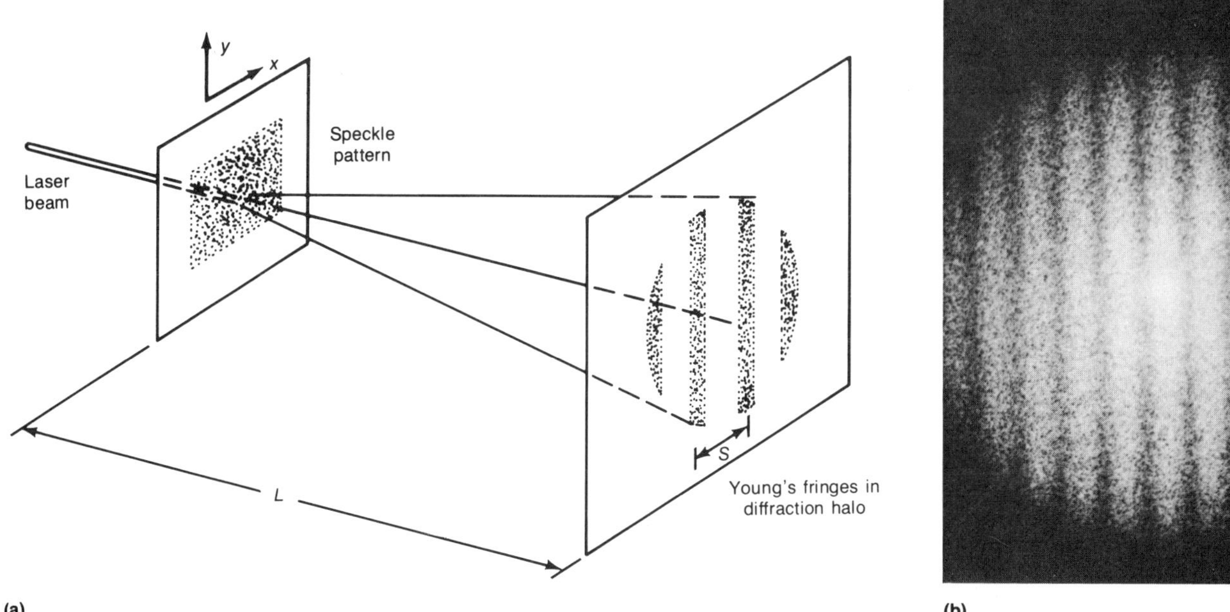

Fig. 2 Pointwise Fourier processing that results in Young's fringes representing the displacement vector at the point of probing. (a) Schematic of the optical setup. (b) Resulting Young's fringe in a diffraction halo

described in the previous section in this article.

Alternatively, the observer can obtain full-field displacement information by using the following optical spatial filtering scheme. The specklegram is placed inside a convergent laser beam, as shown in Fig. 3(a). The diffraction spectrum at the focal plane is also governed by Eq 1. At the focal plane, a mask with a small hole situated at **u** is used to block all light except that passing through the small aperture. A second lens is used to receive the filtered light and forms an image of the specimen surface, which is now modulated by a series of fringes governed by (Ref 3):

$$|\mathbf{u}| \cos \theta = \frac{nL}{|\mathbf{u}|}, \; n = 0, \pm 1, \pm 2, \ldots \qquad \text{(Eq 3)}$$

where θ is the angle between the displacement vector **d** and the aperture vector **u**. Therefore, the fringes are contours of displacement component resolved along the direction **u**. Displacement contours along any direction can be obtained by simply varying the direction of the aperture, and the sensitivity of the contour can be changed continuously by sliding the aperture along the direction of the position vector of the aperture. An example of this application is also shown in Fig. 3(b), in which the fringes are contours of opening (that is, vertical) displacement of a cracked aluminum specimen.

Another example of this type of application is illustrated in Fig. 4, which shows the thermal strain patterns of a heated aluminum plate. The plate is heated with a torch at the lower left-hand corner until it is red hot. It is then left to cool by natural convection. A thermocouple is embedded nearby to monitor the temperature. The

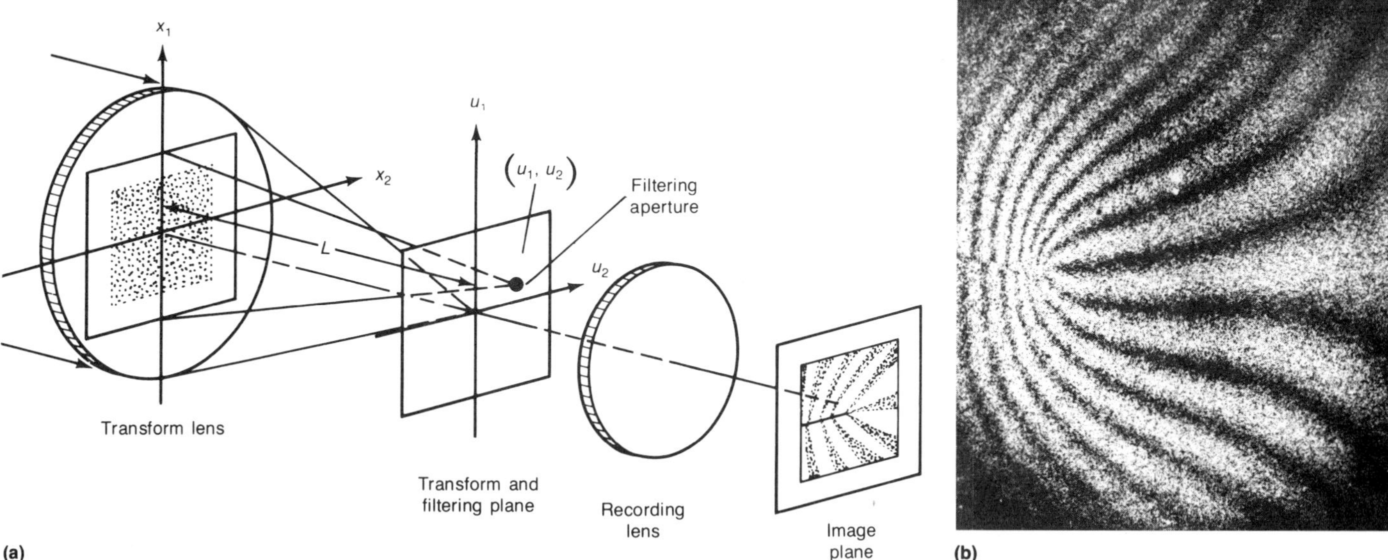

Fig. 3 Optical spatial filtering used to generate full-field displacement contour (isothetic) fringes. (a) Components of optical spatial filter setup. (b) Fringe pattern obtained for an aluminum specimen containing a crack. Each fringe represents 0.005 mm (0.0002 in.) in vertical displacement.

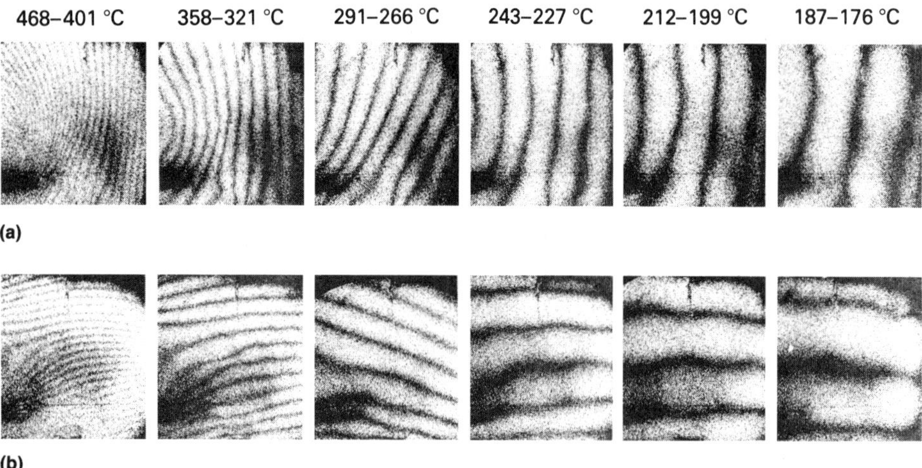

Fig. 4 Thermal strain determination using laser speckle method. (a) u field. (b) v field. Fringes show the residual strain field between two temperature levels of an aluminum plate torch heated to red hot and allowed to cool by natural convection. Fields u and v are the contour fringe displacements in the x- and y-directions, respectively.

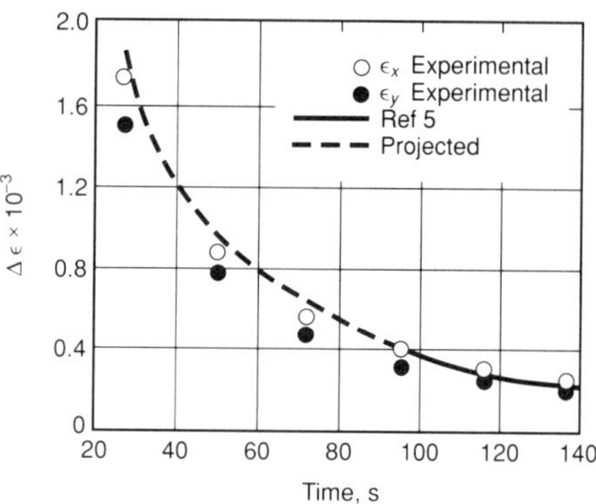

Fig. 5 Comparison between theoretical and experimental thermal strain relaxation at a point as a function of time

patterns are the result of differential thermal strain from the temperatures indicated in Fig. 4. Figure 4(a) shows the displacement of the contour fringes along the x-direction; Fig. 4(b), those along the y-direction. These fringe patterns are governed by:

$$u = \frac{n\lambda L}{u_x}, \ n = 0, \pm 1, \pm 2, \ldots \quad \text{(Eq 4)}$$

$$v = \frac{n\lambda L}{u_y}, \ n = 0, \pm 1, \pm 2, \ldots \quad \text{(Eq 5)}$$

where u and v are the displacement components along the x- and y-directions, respectively, and u_x and u_y are the distances of the filtering aperture along the x- and y-directions, respectively. As the plate cools, the temperature tends toward equilibrium, as evidenced by the uniformity of the fringe patterns at lower temperatures. Thermal strains measured by the speckle method compared quite well with the existing data, as shown in Fig. 5.

Measurement of Out-of-Plane Deformation

The laser speckle method can also be used to measure the out-of-plane deformation of a specimen with a slightly different experimental arrangement. Instead of photographing the speckles on or very close to the surface, one can photograph the speckles at a certain finite distance (a few centimeters, for example) from the surface. When the surface experiences out-of-plane deformation, the spatial speckles are tilted, resulting in speckle displacement at the defocused plane. By recording these speckles before and after deformation, the resulting specklegram is processed in the same way as described in the preceding section in this article. The resulting fringes are governed by (Ref 3):

$$\frac{\partial w}{\partial x} = \frac{n\lambda L}{2Au_x}, \ n = 0, \pm 1, \pm 2, \ldots \quad \text{(Eq 6)}$$

$$\frac{\partial w}{\partial y} = \frac{n\lambda L}{2Au_y}, \ n = 0, \pm 1, \pm 2, \ldots \quad \text{(Eq 7)}$$

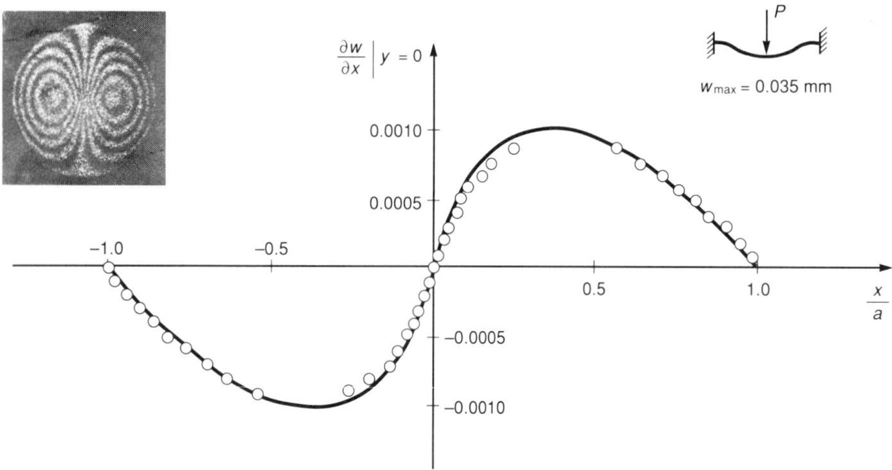

Fig. 6 Slope fringes of a clamped circular plate under concentrated load

where $\partial w/\partial x$ and $\partial w/\partial y$ are the slopes of the surface along the x- and y-directions, respectively, and A is the defocused distance. An example of this application is shown in Fig. 6, in which the fringes are the slope fringes $\partial w/\partial x$ of a clamped thin circular plate transversely loaded at the center. A comparison between the experimental result and the theoretical result is also shown in Fig. 6.

White Light and Electron Microscopy Speckle Methods

Laser speckle is a natural result of coherent radiation, but speckle pattern can also be artificially created. For example, the aerosol particles of black paint sprayed onto a white surface (or vice versa) create a pattern quite similar to laser speckles.

White Light Method. Another efficient way of creating artificial speckle is to cover the surface with a layer of retroreflective paint. When illuminated by incoherent white light (or any visible light), the individual glass beads embedded in the paint converge the light rays and send them back in approximately the same direction. When imaged and recorded on film, the beads appear as sharply defined, random speckles. These patterns can be used for metrological measurements as well. Any speckle pattern can be used if it yields a distinct pattern and can be recorded for future reference. A few examples are given in Ref 6. These are classified as white light speckles because they are observable under the illumination of incoherent white light. The identical procedures used for laser speckles can be adapted to record and process white light speckles. The only exception is that because white light speckles are located on the surface of the specimen, they cannot be tilted in the same way as laser speckles. However, through the use of defocused photography, depth information can nevertheless be obtained (Ref 6).

The sensitivity of any speckle method is controlled by the sizes of the smallest speckles. The laws of physics dictate that one cannot observe, and therefore record, an object smaller than the wavelength of the radiation that is used to illuminate the object. Therefore, using light in the visible spectrum implies that the smallest observable speckle is about 0.5 μm (20 μin.) in diameter. This is essentially the resolution limit (that is, the smallest measurable displacement) of the speckle method. To increase the sensitivity of the technique, it is necessary to resort to radiation sources having shorter wavelengths.

Electron Microscopy Speckle Method. One option is to use the electron microscope speckle method (Ref 7). Through vacuum deposition or some other type of chemical processing, speckles only a small fraction of a micron in size can be created. These speckles are observable only through an electron microscope. Figure 7 shows two such patterns as observed by scanning electron microscopy and transmission electron microscopy. They were recorded on photographic film and then processed using a laser beam. A typical Young's fringe pattern from a double-exposure electron microscopy specklegram is also shown in Fig. 7. With this approach, the sensitivity of the speckle method can be increased by two orders of magnitude.

Determination of Surface Roughness, Plastic Strain, and Fatigue

When a narrow laser beam impinges on a metal surface, the reflected speckle pattern carries the roughness information of the surface as well. Because plastic strain increases the surface roughness (until the saturation stage is reached), the pattern can also be used to monitor plastic strain. Furthermore, if this technique is used to monitor plastic strain at the tip of a notch, for example, the process of metal fatigue resulting in crack initiation and crack propagation can be ascertained.

Figure 8 shows a typical relationship between surface roughness and plastic strain for an 1100-00 aluminum specimen. As long as the total effective plastic strain is the same, the same surface roughness results irrespective of the path along which the plastic strain is attained.

The variation of light intensity of a laser beam reflected from a rough surface can be expressed:

$$\frac{d\langle I\rangle}{\langle I\rangle}|V = 0 \approx -2g\left(\frac{d\sigma}{\sigma}\right) \quad \text{(Eq 8)}$$

along the direction of mirror reflection and:

$$\frac{d\langle I\rangle}{\langle I\rangle}|V \neq 0 \approx 2\left(\frac{d\sigma}{\sigma}\right) - \frac{1}{2}V^2T^2\left(\frac{dT}{T}\right) \quad \text{(Eq 9)}$$

along all other directions. In Eq 8 and 9, $\langle I\rangle$ is the average intensity, V is the direction of diffracted light, σ is the standard deviation of the surface height function, T is the correlation length of the surface height function, and:

$$g = \left[\frac{2\pi}{\lambda}(1 + \cos\theta)\sigma\right]^2 \quad \text{(Eq 10)}$$

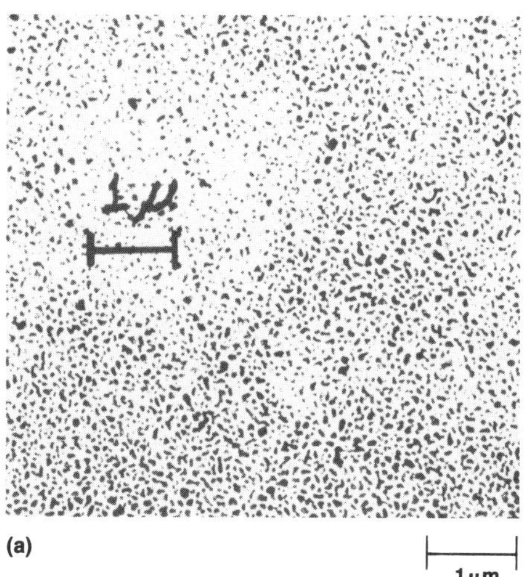

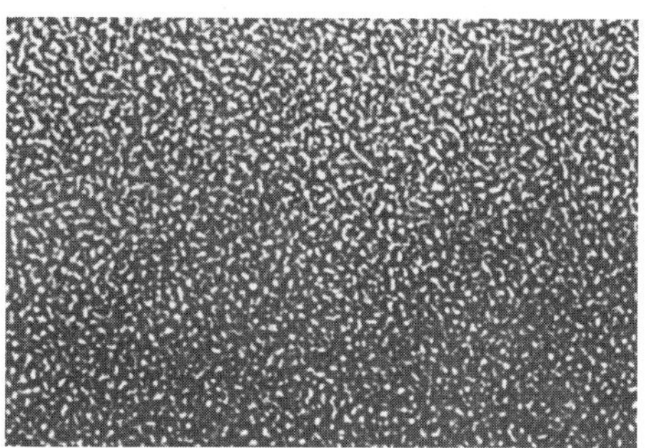

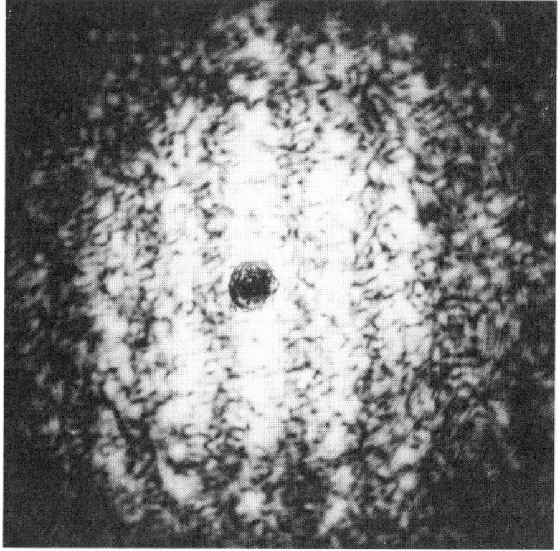

Fig. 7 Electron microscopy speckle method for microdeformation measurement. (a) Speckles formed by vacuum depositing gold particles on graphite as recorded by scanning electron microscopy. (b) Extremely small speckles revealed by transmission electron microscopy. (c) Young's fringes representing displacement of the order of 0.001 nm (0.01 Å)

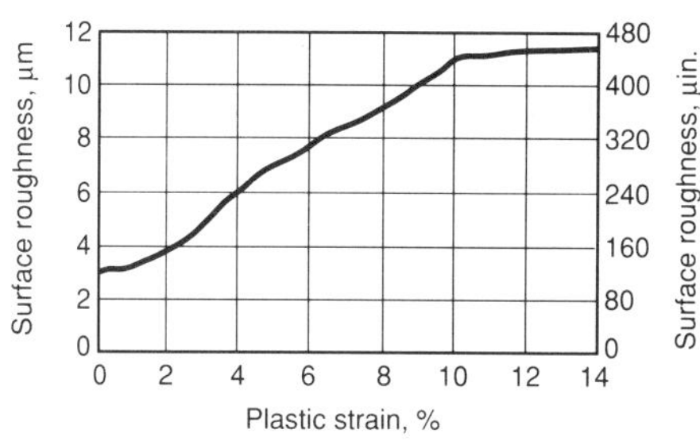

Fig. 8 Plot of surface roughness versus plastic strain for a polished 1100-00 aluminum specimen

where λ is the wavelength of the incident light and θ is the incident angle. Typical diffraction patterns from a metal surface at two levels of plastic strain are shown in Fig. 9 and 10.

Both the standard deviation, σ, and the correlation length, T, of the surface height function appear in Eq 9. Therefore, it is more advantageous to use this part of the deflected light for the characterization of surface roughness. This fact is also qualitatively evident from the two diffraction patterns shown in Fig. 9 and 10.

There are many ways to extract information from such a diffraction pattern. One approach is to window out a portion of the higher spatial frequency domain of the diffraction spectrum, digitize it into gray levels (see the article "Machine Vision and Robotic Inspection Systems" in this Volume), and then calculate the statistical contrast value of the pattern using:

$$CON = \Sigma(i - j)^2P(i, j) \quad \text{(Eq 11)}$$

where $P(i, j)$ is the probability density of a pair of gray levels occurring at two points separated by a certain distance.

The *CON* value is used to calibrate the measured strains of an aluminum specimen (1100-H14) in tension. The *CON* value increases monotonically with respect to total strain until about 1.2% and plastic strain until about 0.9%. Beyond these points, the *CON* values start to decrease with increased strain. However, in the discussion that follows, it is shown that the *CON* value is nevertheless quite effective in predicting the onset of fatigue crack initiation and propagation.

The *CON* value method is applied to monitoring plastic strains at nine points surrounding the tip region of a saw-cut notch in an aluminum specimen under cyclic bending stress. The geometry of the specimen and the configuration of the points are shown in Fig. 11 and 12, respectively.

The specimen is mounted in a fatigue machine capable of rendering cyclic bending load. At predetermined cycles, the machine is stopped, and a laser beam is used to illuminate the nine selected points sequentially. The resulting diffraction spectra are received on a ground glass, and a window of

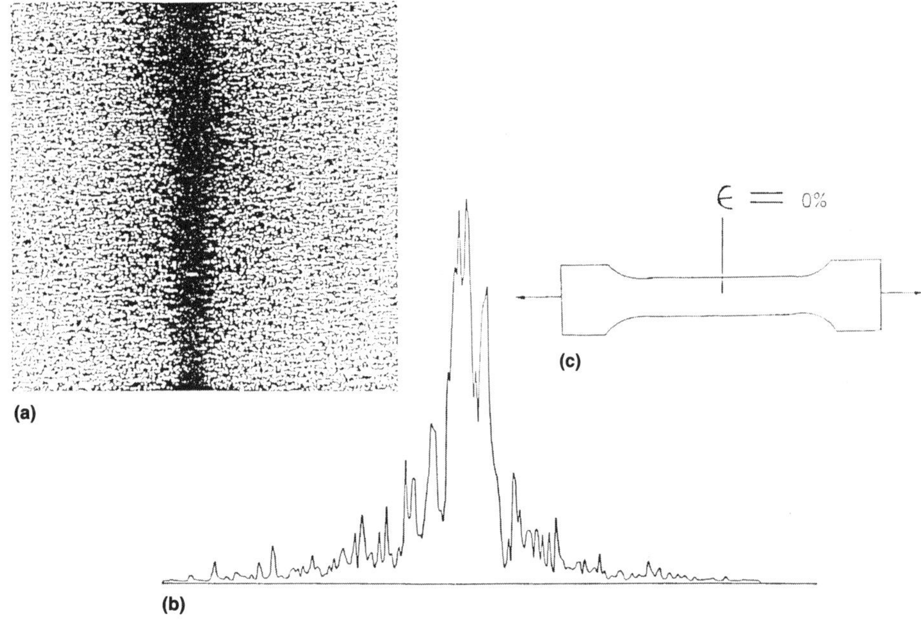

Fig. 9 Speckle metrology data for an aluminum 1100 specimen in the original (as-received) state (ϵ = 0%). (a) Speckle spectrum. (b) Intensity profile of speckle spectrum. (c) Schematic of sampling procedure for testing plastic strain

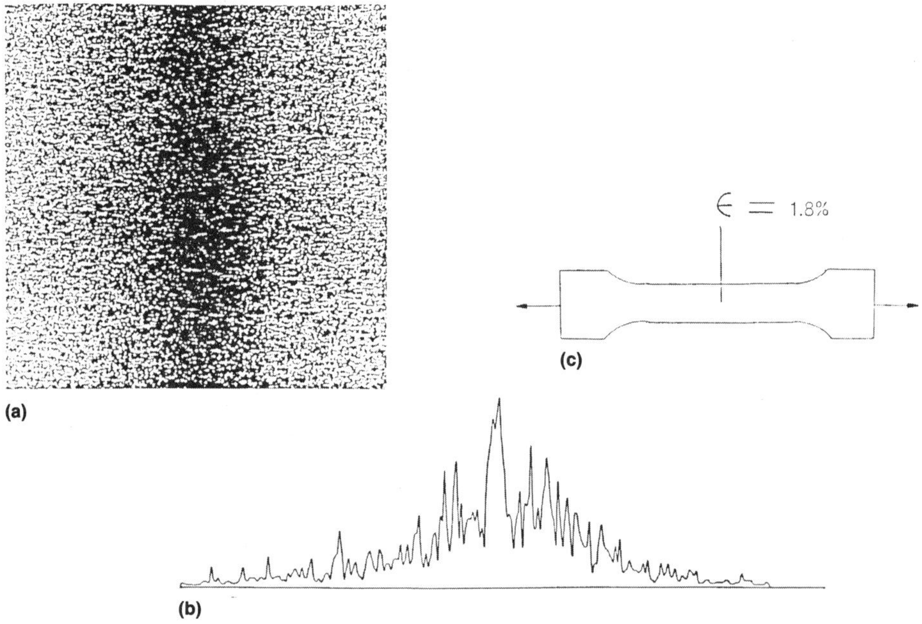

Fig. 10 Speckle metrology data for an 1100 aluminum specimen subjected to 1.8% plastic strain (ϵ = 1.8%). (a) Speckle spectrum. (b) Intensity profile of speckle spectrum. (c) Schematic of sampling procedure for testing plastic strain

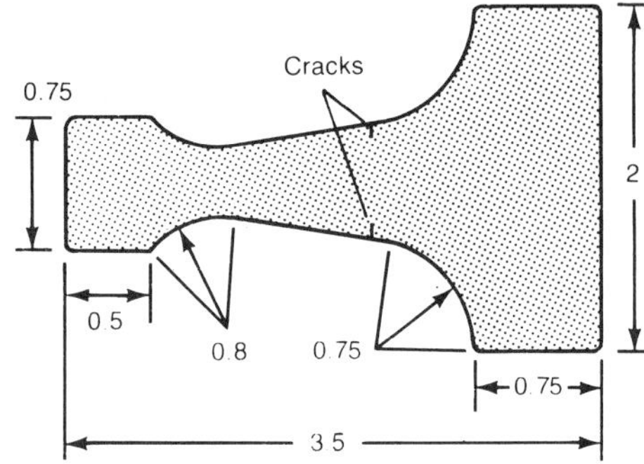

Fig. 11 Schematic showing cracks in a fatigue specimen made of 2.3 mm (0.09 in.) thick 1100-00 aluminum. Dimensions given in inches

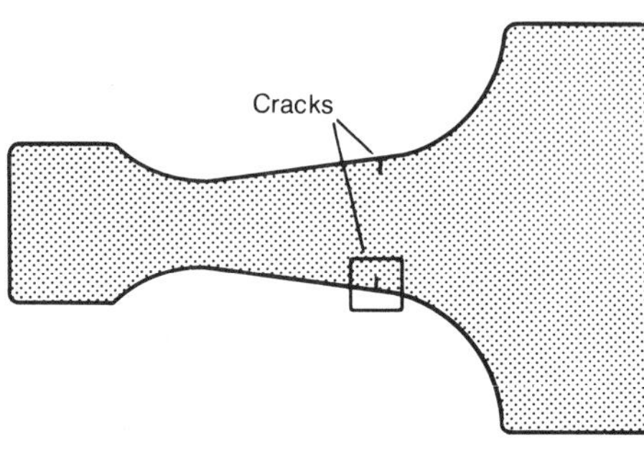

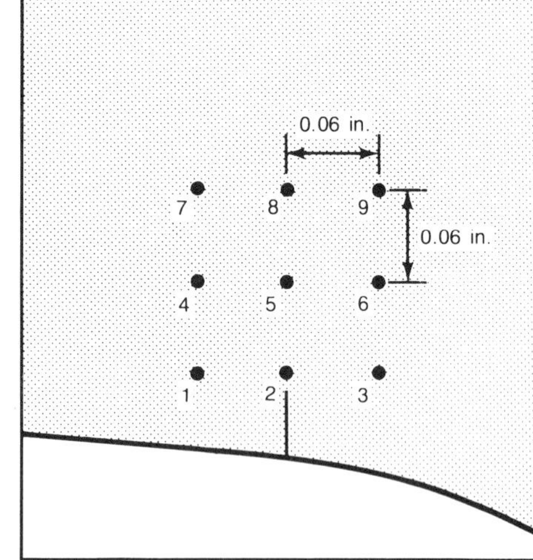

Fig. 12 Schematic of aluminum specimen showing cracks (a) and the nine test points arranged around the tip of the bottom crack (b)

their high-frequency region is digitized. A *CON* value is then calculated for each of these digitized images. The results obtained are plotted in Fig. 13.

It is interesting to note that along points 2, 5, and 8, which pinpoint the location of the eventual crack propagation path, the *CON* value increases dramatically at certain cycles of loading. On the other hand, the *CON* value stays almost stationary at the six remaining points, which are only 1.5 mm (0.06 in.) away from the path of crack propagation. This experiment was performed on an aluminum specimen with an initial surface roughness of 0.7 μm (28 μin.). Similar results were obtained for specimens of other initial roughnesses.

Future Outlook

The speckle method is a relatively new entry in the arsenal of metrological techniques, and it is still evolving. Current research is emphasizing the use of automation in speckle metrology (Ref 8). Researchers are focusing their efforts on two promising techniques. One is the automatic processing of the Young's or full-field fringes, and the other consists of digitizing the speckles and processing them directly. The latter is very appealing because it would eliminate the use of photographic film. However, with the current technology, the solid-state detector does not yet have resolution compa-

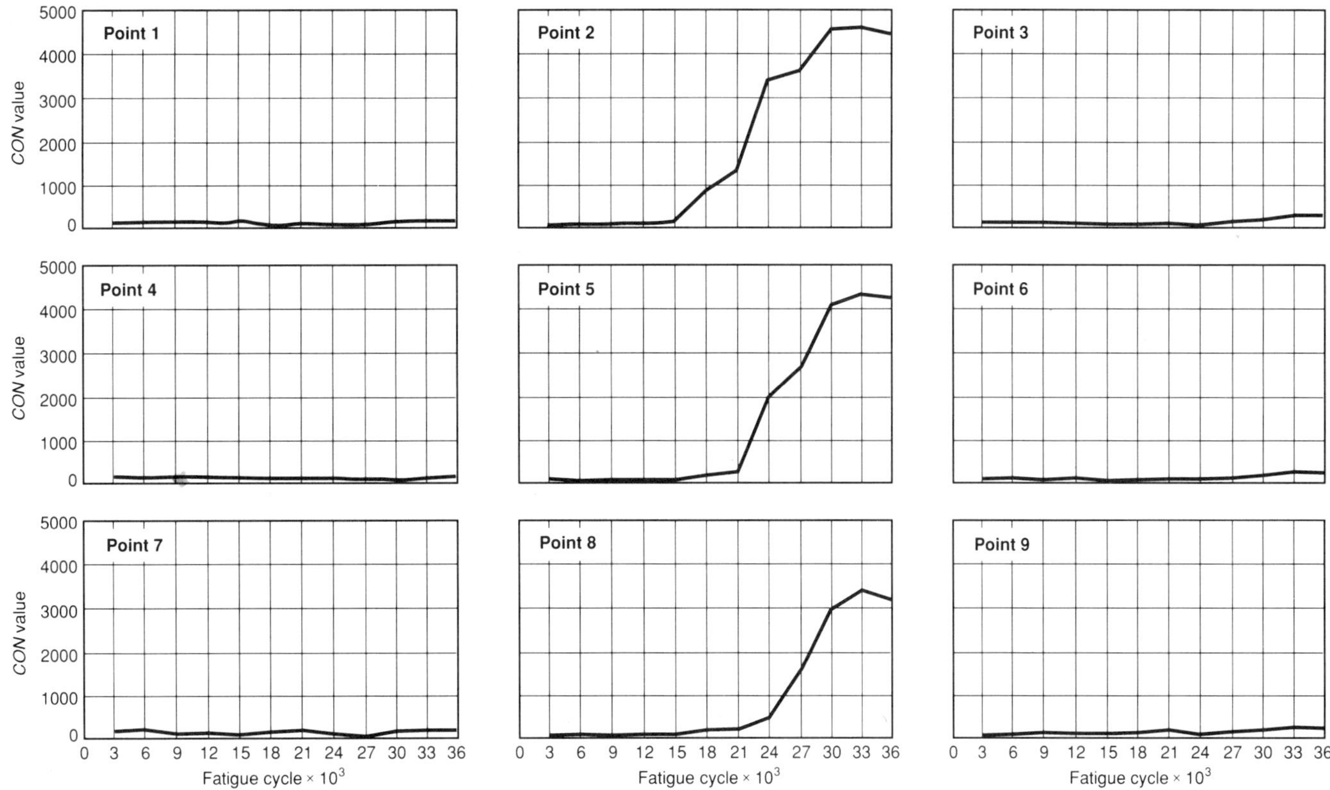

Fig. 13 Results of using *CON* value of a laser speckle pattern to monitor the onset of fatigue crack initiation and the path of propagation for the sample described in Fig. 12

rable to that of a photographic emulsion. Other research efforts have been geared toward utilizing a combination of speckle technique and holographic interferometry (Ref 8).

REFERENCES

1. J.M. Burch and J.M.J. Tokarski, Production of Multiple Beam Fringes From Photographic Scatterers, *Optica Acta*, Vol 15 (No. 2), 1968
2. D.W. Li and F.P. Chiang, Laws of Laser Speckle Movement, *Opt. Eng.*, Vol 25 (No. 5), May 1986, p 667-670
3. F.P. Chiang, A Family of 2-D and 3-D Experimental Stress Analysis Techniques Using Laser Speckles, *Solid Mech. Arch.*, Vol 3 (No. 1), 1978, p 1-32
4. F.P. Chiang, A New Three-Dimensional Strain Analysis Technique by Scattered Light Speckle Interferometry, in *The Engineering Uses of Coherent Optics*, E.R. Robertson, Ed., Cambridge University Press, 1976, p 249-262
5. F.P. Chiang, J. Adachi, and R. Anastasi, Thermal Strain Measurement by One-Beam Laser Speckle Method, *Appl. Opt.*, Vol 19 (No. 16), 1980, p 2701-2704
6. A. Asundi and F.P. Chiang, Theory and Application of White Light Speckle Methods, *Opt. Eng.*, Vol 21 (No. 4), 1982, p 570-580
7. F.P. Chiang, Speckle Method With Electron Microscopes, in *Proceedings of the Annual Spring Meeting*, Society for Experimental Mechanics, 1982, p 24-28
8. F.P. Chiang, Some Recent Advances in Speckle Techniques for Photomechanics and Optical Metrology, in *Optical Testing and Metrology*, Vol 661, Conference Proceedings, Society of Photo-Optical Instrumentation Engineers, 1986, p 249-261

Acoustical Holography

Revised by B.P. Hildebrand, Failure Analysis Associates, Inc.

ACOUSTICAL HOLOGRAPHY is the extension of holography into the ultrasonic domain. The principles of acoustical holography are the same as those of optical holography (see the article so titled in this Volume) because the laws of interference and diffraction apply to all forms of radiation obeying the wave equation. Differences arise only because the methods for recording and reconstructing the hologram must accommodate the form of radiation used. This need to accommodate the form of radiation restricts the practical range of sound wave frequency that can be used in acoustical holography, as discussed later in this article.

At present, only two types of basic systems for acoustical holography are available:

- The liquid-surface type
- The scanning type

These utilize two different detection methods, and these methods in turn dictate the application of the systems to nondestructive inspection. Neither of these two types of systems relies on the interferometric techniques of optical holographic inspection, in which information on flaws at or near the surface of a test object is obtained from the pattern formed by interference between two nearly identical holographic images that are created while the object is differentially stressed. Instead, systems for acoustical holography obtain information on internal flaws directly from the image of the interior of the object. The principles of flaw detection by ultrasonic waves, as well as the equipment used, are discussed in the article "Ultrasonic Inspection" in this Volume. Ultrasonic holography is discussed in detail in Ref 1.

Liquid-Surface Acoustical Holography

The basic system for liquid-surface acoustical holography is similar to the basic system for optical holography except for the method of readout.

Acoustical System. In liquid-surface systems (Fig. 1), two separate ultrasonic transducers supply the object beam and reference beam, which are both usually pulsed. The two transducers and the test object are immersed in a water-filled tank. The test object is positioned in the object plane of an acoustic (ultrasonic) lens, which is also immersed in the tank.

The practical limits for the object beam transducer in a commercial system are a wave frequency of 1 to 10 MHz, a pulse length of 50 to 300 μs, and a pulse-repetition rate of 60 to 100 per second. This transducer is placed so that its beam passes through the test object. (Occasionally, the object beam transducer is glued to the bottom of the test object, but more often there is a liquid path between them.) As the object beam passes

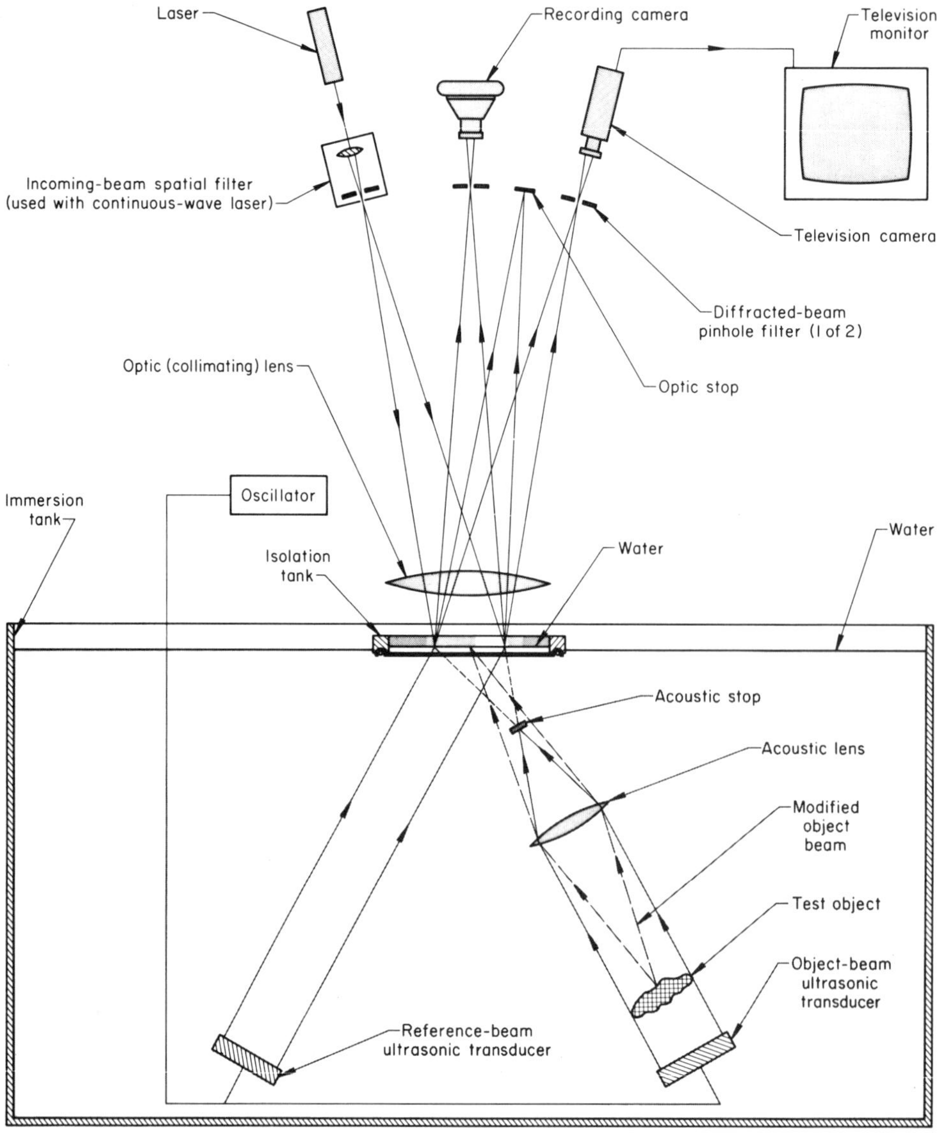

Fig. 1 Schematic of the basic acoustical and optical systems used in nondestructive inspection by liquid-surface acoustical holography. See text for description.

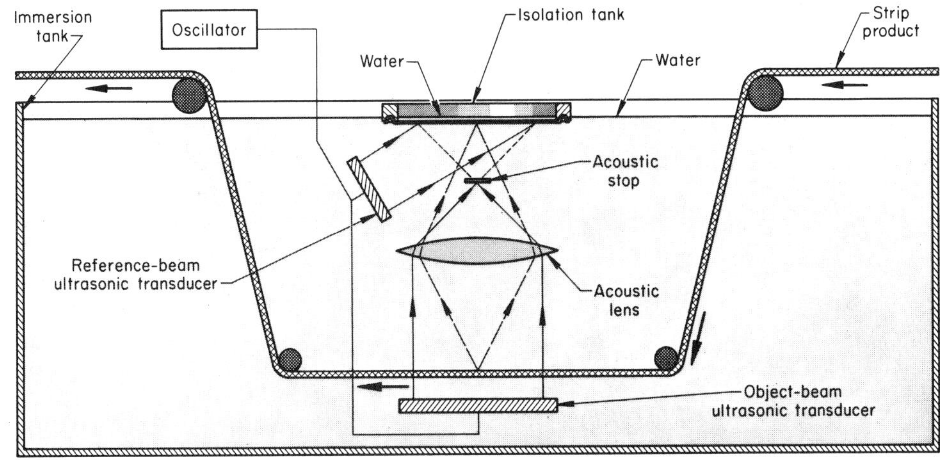

Fig. 2 Schematic of the acoustical portion of a liquid-surface acoustical holography system that is arranged to inspect strip products. The optical portion of this system is identical to that in Fig. 1.

through the test object, it is modified by the object. The modification is generally in both amplitude and phase. The object beam then passes through an acoustic lens, which focuses the image of the test object at the liquid surface. This image contains a wave front nearly identical to that emanating from the test object. (A small disk of a material having very high acoustic impedance can be placed at the focal point of the acoustic lens to prevent any unmodified portion of the object beam from reaching the surface of the liquid.)

The reference beam transducer is connected to the same oscillator as the object beam transducer so that it emits a second wave front coherent with the wave front from the object beam transducer. The reference beam transducer is aimed at the same region of the liquid surface as the object beam, where the wave fronts interfere. The pressure of the beams distorts the liquid surface in this region, levitating portions of it to form a static ripple pattern that acts as a relief hologram. Pulsed ultrasonic beams are used more often than continuous beams. With pulsed beams, the time of arrival of the reference beam at the liquid surface can be varied to compensate for any variations in transit time of the object beam as it traverses the test object and for any significant differences in the path lengths of the beams; this ensures that both beams will arrive at the liquid surface at precisely the same instant to form the ripple pattern.

A water-filled isolation tank is situated inside the immersion tank and serves to prevent ambient vibrations from interfering with the formation of the hologram. A second reason that pulsed ultrasonic beams are used more often than continuous beams is to avoid the creation of standing waves in the immersion tank that would make isolation more difficult.

Although liquid-surface acoustical holography is ideally suited to the use of the ultrasonic object beam in the transmission mode, theoretically there is no reason that the beam cannot be used in the reflection mode to form the image. Practically, however, the high level of object beam intensity and the high quality of object beam transducer required for reflection-mode operation prevent the liquid-surface system from being effective in this mode.

Optical System. In contrast to optical holography, in which the holographic pattern is photographed and the resulting hologram is used to diffract light, in acoustical holography the ripple pattern on the liquid surface is used directly as a grating to diffract light. Therefore, this is a real-time system. The image is reconstructed from the hologram by reflecting a beam of coherent light from the ripple pattern in the isolation tank (Fig. 1). The beam is generated by a laser and passes through an optical system before illuminating the ripple pattern. When a pulsed laser is used, the timing of its pulse is coordinated with the timing of the ultrasonic pulse so that the liquid surface in the isolation tank is illuminated only when there is a stable ripple pattern present. The zero order of diffracted light carries no useful information, so it is blocked by an optically opaque stop. The first two orders of diffracted light carry the image; after the beams are passed through spatial filters, one might be used for photography and the other for television pickup or other purposes. Methods of readout are discussed in greater detail in a later section of this article.

Object Size and Shape. The size of test object that can be completely imaged at any instant is limited by the size of the object beam transducer and lens (typically to a maximum of 152 mm, or 6 in., in diameter). For large test objects, a large object beam transducer and lens must be used, or the test object must be translated through the object beam. Translation in one direction is used for narrow objects; wide objects require translation in two directions. A large, thick object requires translation along the lens axis, as well as along the two other axes, in order to inspect the entire volume of the object. This is because at any instant only a thin segment of the object is in the object plane of the lens and therefore is imaged.

Despite the physical limitations imposed by liquid-surface acoustical holography, there are many inspection problems that can be accommodated. The fact that the system is in real time allows large test objects to be moved through the object plane fast enough to permit high inspection rates. Strip products are especially suitable for inspection by liquid-surface acoustical holography because they are essentially two dimensional and can easily be translated through the object plane. This type of arrangement is shown in Fig. 2. Small parts can be moved through the object plane in a similar manner by the addition of a conveyor belt.

The system is usually pulsed at a rate compatible with television frame rates. Typically, 60 holograms per second are produced, which yields flicker-free television images. Because pulse lengths are generally 50 to 300 μs, strip velocities of 50 to 300 mm/s (2 to 12 in./s) are possible without blurring. With this arrangement, a photograph of one of the holograms of the moving object can be taken at any time by setting the shutter speed of the camera at 1/60 s.

Sensitivity and Resolution. The sensitivity of liquid-surface systems is fixed somewhat by the necessity to produce a ripple pattern on the liquid surface. For a typical system, the minimum intensity of the ultrasonic beams that can be detected is about 10^{-3} to 10^{-6} W/cm^2.

The detail that can be resolved by liquid-surface acoustical holography is determined by the focal length and linear aperture of the ultrasonic lens and the wavelength of the ultrasonic beam through the use of the formulas:

$$\Delta x = 1.22\lambda\ (f/A) \qquad \text{(Eq 1)}$$

$$\Delta z = 2\lambda\ (f/A)^2 \qquad \text{(Eq 2)}$$

where Δx is the resolution in the object plane (that is, lateral resolution), Δz is the resolution perpendicular to the object plane (that is, depth resolution), λ is the ultrasonic wavelength in the immersion liquid, f is the focal length of the ultrasonic lens in the immersion liquid, A is the linear aperture (diameter) of the ultrasonic lens, and f/A is the f-stop number of the ultrasonic lens. For an $f/3$ lens and a frequency of 5 MHz, flaws in steel or aluminum should be resolved to about 1 mm (0.04 in.) transversely and about 5 mm (0.20 in.) in depth; that is, small flaws separated by 1 mm (0.04 in.) transversely or by 5 mm (0.20 in.) in depth can be individually seen, and the size of large flaws can be measured to an accuracy of 1 mm

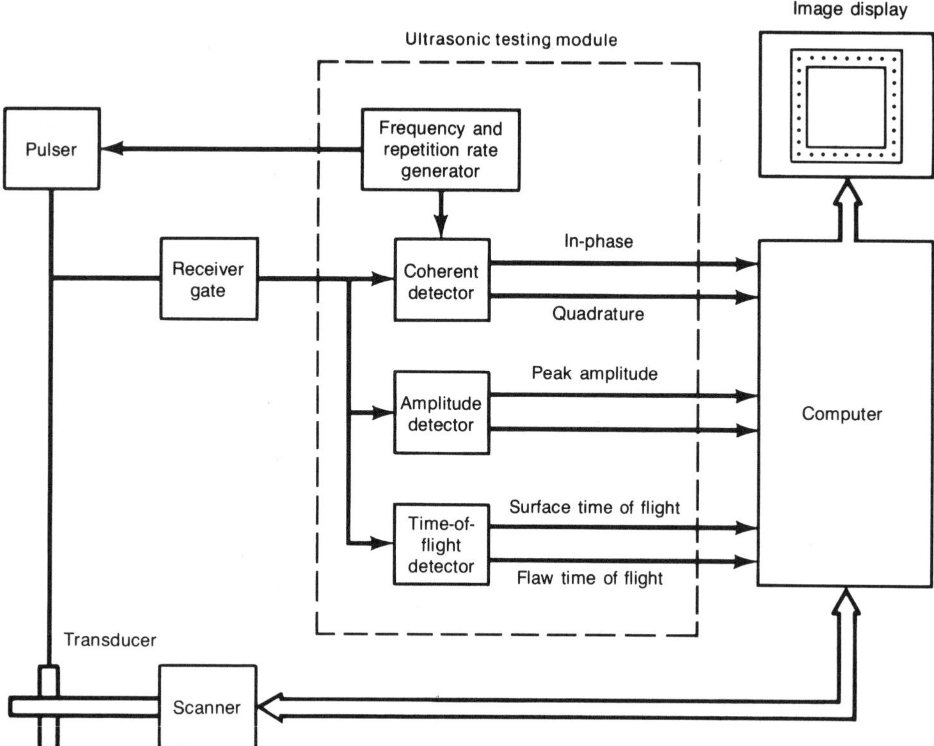

Fig. 3 Block diagram of a commercial ultrasonic and light detection system with capability for A-scan, B-scan, and C-scan imaging and holography

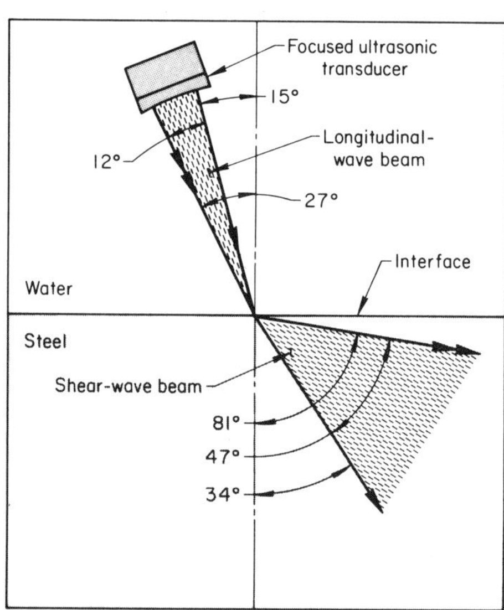

Fig. 4 Diagram of refraction of a longitudinal wave ultrasonic beam focused on a water/steel interface at incident angles that produced a shear wave beam in the steel

(0.04 in.) transversely and 5 mm (0.20 in.) in depth.

The values of *f*-stop number for the ultrasonic lenses used in liquid-surface systems are typically 2 to 3. Although the resolution of a system increases with the frequency of the ultrasound used, absorption and scattering of the ultrasound also increase, so that for most industrial testing the frequency range is limited to between 3 and 7 MHz. Sometimes 1 MHz is used, as for rubber products.

Scanning Acoustical Holography

The basic system for scanning acoustical holography is shown in Fig. 3. No reference beam transducer is required in this system, because electronic phase detection is used to produce the hologram; that is, the required interaction (mixing) between the piezoelectrically detected object beam signal and the simulated reference beam signal occurs in the electronic domain. A pulser circuit (consisting of a continuous-wave oscillator and a pulse gate that is triggered by an electronic clock) and a power amplifier feed a single focused ultrasonic transducer that is scanned over an area above the test object while alternately transmitting and receiving the ultrasonic signals. The transducer and the test object are immersed in a water-filled tank or coupled by a water column.

The signal is pulsed so that time gating can be used to reject undesired surface echoes. The pulse length can be set to any desired value from a few periods of the wave frequency to an upper limit of 50% of the time between successive pulses. Long pulse lengths are used to examine regions lying deep within the metal, while short pulse lengths are required for the regions near the surface so that transmitted energy is not mixed with reflected energy. The practical limits of the transducer in a commercial system are a wave frequency of 1 to 10 MHz, a pulse length of 5 to 20 μs, and a pulse-repetition rate of 500 to 1000 per second. The frequency band of a given transducer is relatively narrow (±5% of the mean frequency).

The transducer is positioned so that its focal point is at the surface of the test object. In this way, the pulse enters the object and travels through it as a spherical wave front diverging from a point on the surface. Because of refraction at the water/metal interface, the solid angle formed by the conical beam of energy in the metal will be greater than that formed by the focused conical beam from the transducer. This is illustrated in Fig. 4 by a longitudinal wave ultrasonic beam focused at the surface of a steel test object and angled to produce shear waves in the steel. In this situation, the range of angles of incidence must lie between 15 and 27° to produce only shear waves in the steel, and this range of angles will produce a 47° cone of shear-wave energy in the steel. (Refraction at water/metal interfaces is discussed in the article "Ultrasonic Inspection" in this Volume.)

Either longitudinal or shear waves can be used in the scanning system illustrated in Fig. 3. For the longitudinal wave mode, the transducer is oriented parallel to the object surface; for the shear wave mode, the transducer is oriented at the appropriate angle of incidence. With either mode, the wave travels through the object to a flaw, where some of the energy is reflected back to the transducer. This reflected signal is passed through a transmit-receive switch to a gate that is triggered by the electronic clock through a time-delay circuit adjusted to compensate for the water-path transit time. This gate is adjusted in delay and width so that it will pass the reflections from the particular thin segment of the object under examination during the scan and will reject all others.

The reflected signal is split and sent to two balanced mixers, which are also fed by signals derived from the oscillator. These signals are phase shifted by 90° from each other. The outputs of the two mixers are therefore equivalent to the sine and cosine Fourier coefficients of the reflected signal. These coefficients are complete descriptors of the phase and amplitude of the signal reflected by the flaw and thus constitute a hologram.

The outputs of the balanced mixers are sampled, converted to digital numbers by an analog-to-digital converter, and stored in the memory of a personal computer. Thus, the result of a two-dimensional scan of a planar surface is a matrix of complex numbers recorded on the computer disk.

Reconstruction. The results of the detection procedure described above constitute a measurement of the reflected wave front in the scan plane. Because this wave front resulted from the propagation of the reflected sound from the flaw, it must have propagated to the scan plane. Therefore, it should be possible to obtain an image of the flaw by propagating the wave front back to its source. This is done by computer in the following sequence:

- Compute the two-dimensional Fourier transform of the detected wave front as represented by the matrix of Fourier coefficients
- Multiply the transformed data by a complex propagator, the value of which is determined by the distance to the desired image plane
- Compute the inverse Fourier transform of the back-propagated data to return to image coordinates
- Display the magnitude of the resultant matrix of complex numbers representing the image

All these steps can be accomplished in 20 s on a personal computer interfaced with an array processor. The image is displayed on a high-resolution black and white or color monitor.

Because there is no conversion of wavelengths, as there was in the original optical reconstruction method, there is no image distortion. Therefore, the relationship between the size of the image and the size of the object is simply the magnification factor of the display.

Object Size. Although the scanning system is too slow to be useful for the total inspection of large objects, taking about 10 min for a 152 mm (6 in.) square hologram, the real-time advantage held by the liquid-surface system will disappear when piezoelectric arrays and associated electronics become available, because electronic scanning can then be used.

Sensitivity and Resolution. Because the scanning system uses a piezoelectric element as a detector, it is more sensitive than the liquid-surface system by a factor of about 10^6. For a typical scanning system, the minimum intensity of ultrasonic beam that can be detected is about 10^{-12} W/cm^2. Mainly because of its much greater sensitivity, scanning acoustical holography appears to hold more long-term promise for the nondestructive inspection of metal products than does the liquid-surface system. It also provides an image of the flaw for subsequent determination of both size and depth within the resolving capacity of the system, which is obtainable from Eq 1 and 2 using the *f*-stop number of the focused transducer. Transducers having an *f*-stop number as low as 2 are available for operation in the longitudinal wave mode.

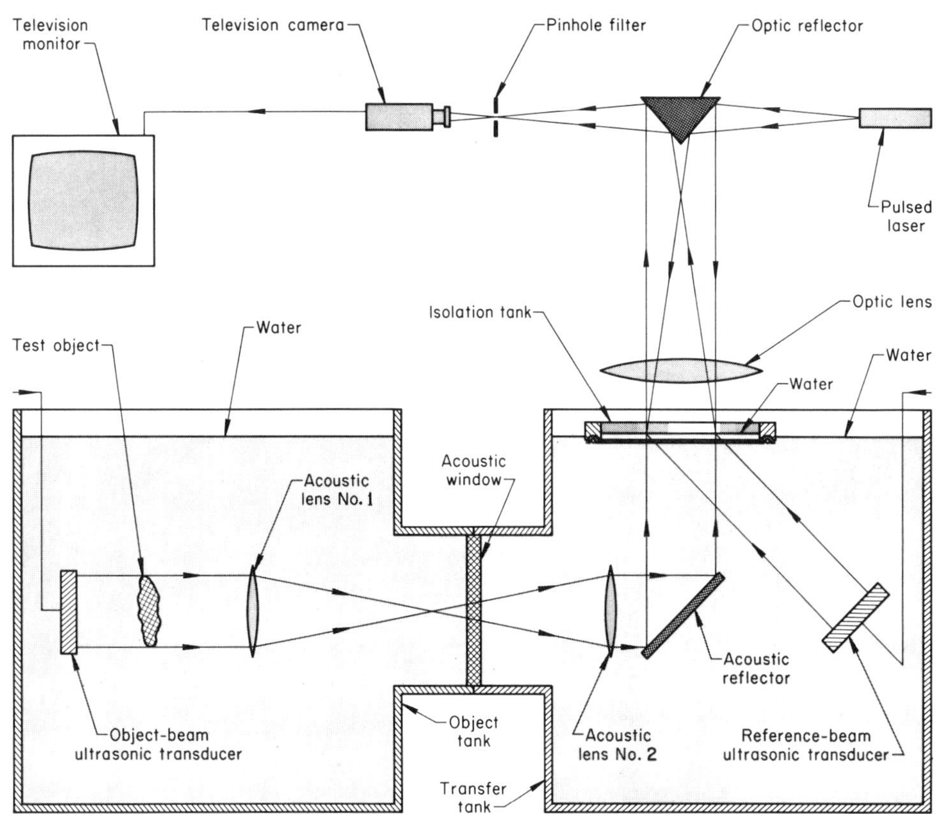

Fig. 5 Schematic of the acoustical and optical systems in the equipment for liquid-surface acoustical holography shown in Fig. 6

Commercial Liquid-Surface Equipment

A schematic of one type of commercially available system for liquid-surface acoustical holography is presented in Fig. 5. The standard size of the effective object field in this system, which is dependent on the diameters of the two transducers, is 114 mm (4½ in.) in diameter; test objects larger than this can be inspected by moving them through the object field.

This system is well stabilized against vibration through the use of an object tank, a transfer tank, and an isolation tank. The ultrasonic beam is relayed by acoustic lenses from the object tank to the transfer tank through an acoustic window that connects them. The beam is then relayed by an acoustic reflector to the isolation tank, where the holographic ripple pattern is formed. Although the acoustic window permits transfer of the ultrasonic beam, it forms an effective barrier to the transfer of gross movements from the object tank to the transfer tank, where they could disturb the isolation tank and therefore the ripple pattern.

The pulsed object beam transducer, which is typically 127 mm (5 in.) in diameter, emits a wave train several hundred wavelengths long (100 to 300 μs in duration) 60 times per second. Therefore, the acoustic framing rate is 60 Hz. Similar wave trains are emitted by the reference beam transducer and are delayed in time to arrive at the liquid surface in coincidence with the object beam. The delay time, controlled manually by the operator, depends on the position of the object beam transducer and the thickness of the metal object being examined. The reconstructed image is continuously displayed on the closed-circuit television monitor or on a ground-glass screen.

This system is essentially stationary because the imaging tank must be leveled and the optical system correspondingly aligned and because the system is quite large and heavy. Typical dimensions of the equipment are indicated in Fig. 6. It weighs approximately 450 kg (1000 lb).

Equipment with a configuration other than that shown in Fig. 5 and 6 has been designed for particular applications. Figure 7 illustrates an alternative configuration used in a piece of liquid-surface equipment designed for the examination of metal strip. This strip moves continuously through the immersion tank by means of entrance and exit water seals. The dimensions of this equipment are shown in Fig. 8. Normally, this equipment operates at 3 to 5 MHz, but frequencies of 1 to 9 MHz can be used. The equipment is designed so that neither mechanical vibration nor electrical static is detrimental to its proper functioning.

Commercial Scanning Equipment

Commercially available equipment, such as that described in Ref 2, for inspection by

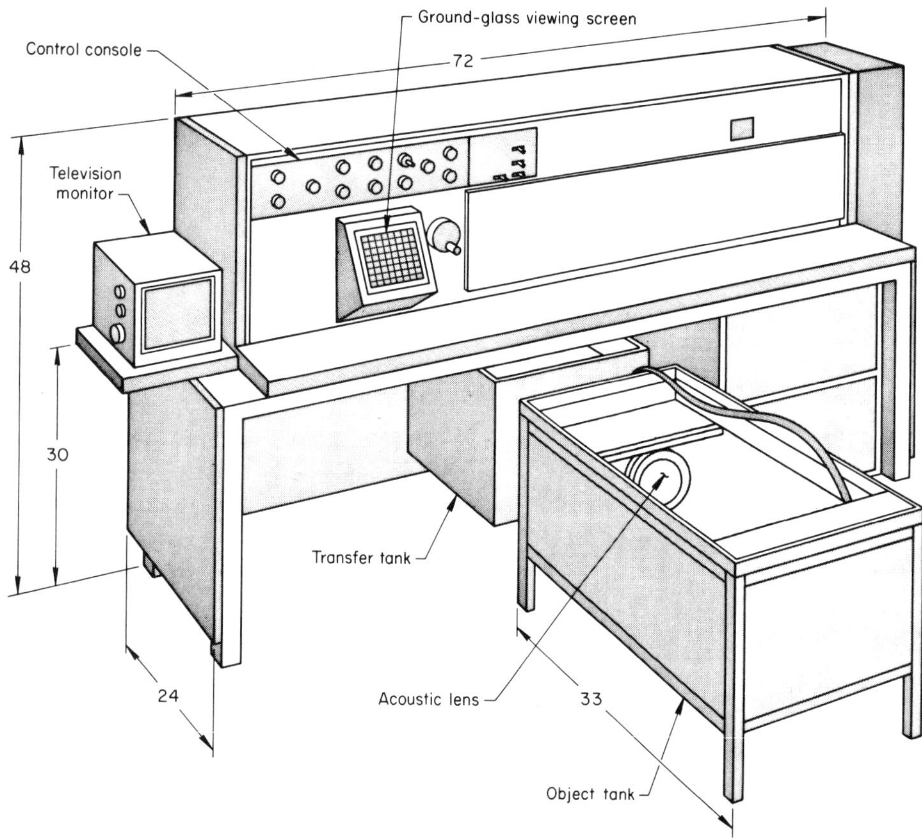

Fig. 6 One type of commercially available equipment for the inspection of small metal objects by liquid-surface acoustical holography. Dimensions given in inches

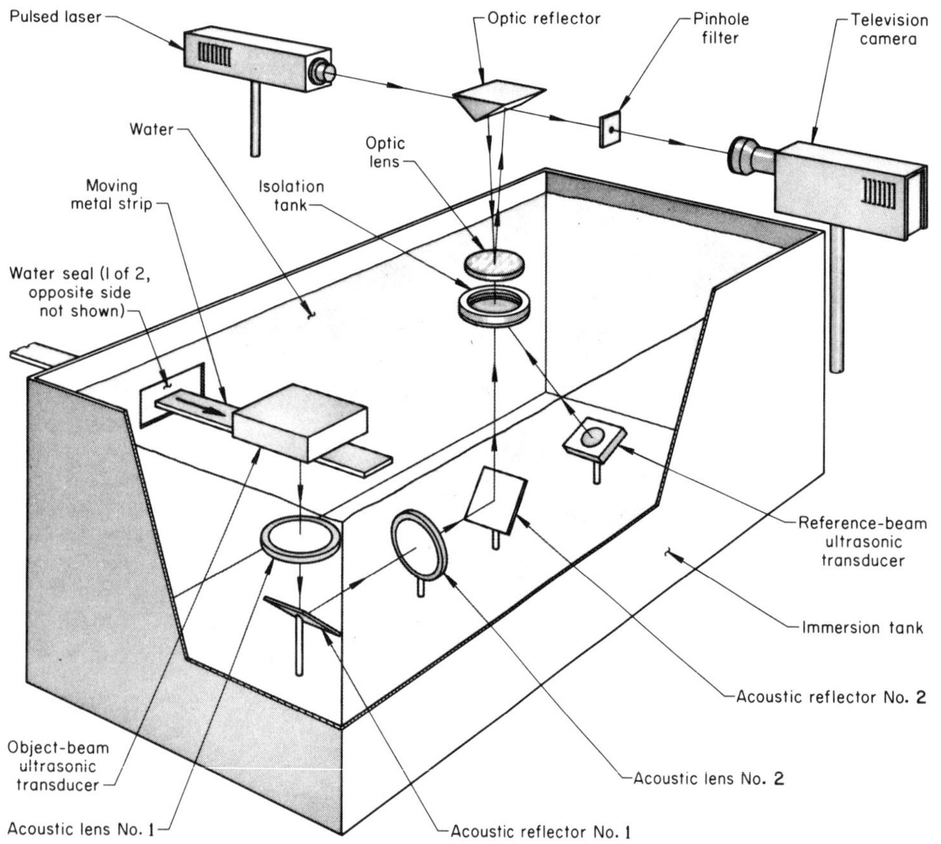

Fig. 7 Schematic of the acoustical and optical systems in the equipment shown in Fig. 6 for the inspection of a continuously moving strip of metal by liquid-surface acoustical holography

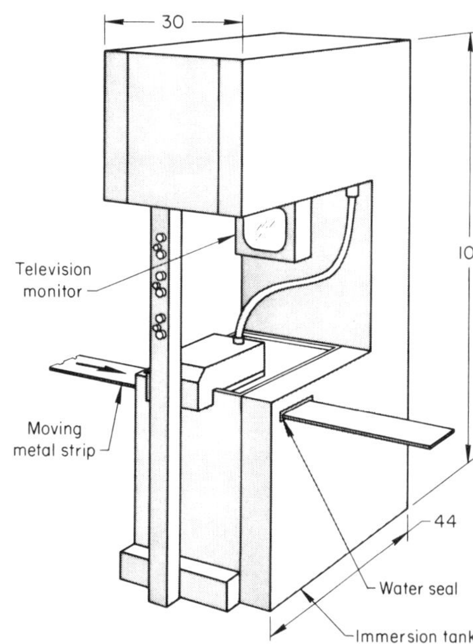

Fig. 8 One type of commercially available equipment for the inspection of a continuously moving strip of metal by liquid-surface acoustical holography. See Fig. 7 for a diagram of the acoustical and optical systems employed. Dimensions given in inches

scanning acoustical holography is often packaged in several units for portability. Figure 9 shows some of the components in one type of commercial system. Each of these units is readily transportable by one or two persons.

The attachment fixture is designed to hold the scanner in position even on surfaces with complex contours. The fixture permits preadjustment of the focal plane of the transducer. After the attachment fixture is in place and aligned, the scanner is mounted on it. The scanner is digitally controlled and is capable of a raster scan pattern in either the x or the y direction. There are six settings of index spacing that can be selected for successive scans: 0.15, 0.30, 0.46, 0.60, 0.76, and 0.91 mm (0.006, 0.012, 0.018, 0.024, 0.030, and 0.036 in.). The largest area that can be recorded in a single hologram with this scanning system is about 305 mm (12 in.) square. However, by systematically repositioning the scanner, much larger areas can be examined.

A series of 25 mm (1 in.) diam transducers with 102 mm (4 in.) focal lengths is used in conjunction with this scanner. Each transducer has a different characteristic operating frequency, so that inspections can be made using any appropriate frequency within the range of 0.5 to 10 MHz. The transducer and bubbler can be oriented for either straight-beam or angle-beam testing. A signal gate in the reflection-signal processor unit can be adjusted to reject reflections from all depths in the object except that under examination. Both the gate width and the gating delay are readily adjustable.

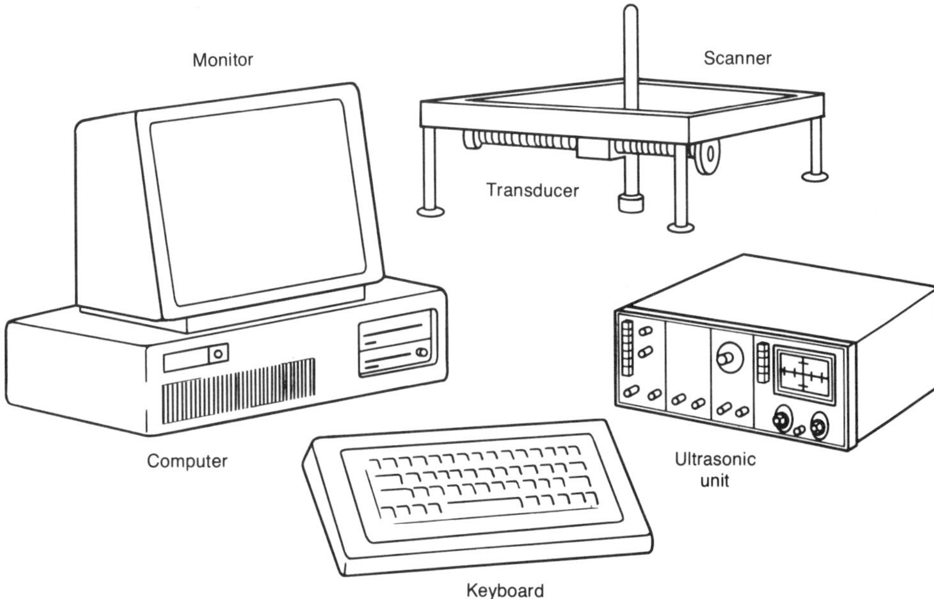

Fig. 9 Key components of a commercially available holography system: personal computer, ultrasonic unit, and scanner

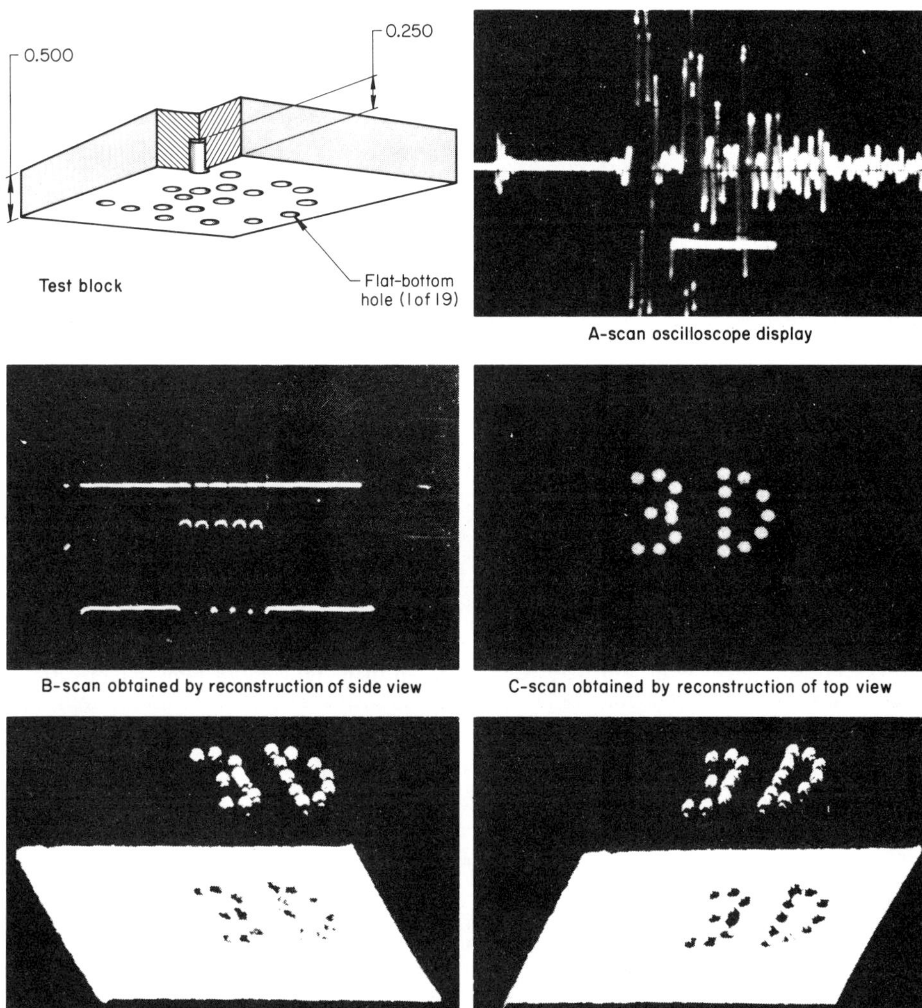

Fig. 10 Test block with flaws simulated by pattern of flat-bottom holes and displays obtained by scanning the block with the equipment for scanning acoustical holography. Dimensions given in inches

Some models of this equipment also provide switch selectability among A-scan, B-scan, and C-scan modes of display, as well as a combination B-scan and C-scan mode of display that presents an angled view of the flaw image. This angle can be varied to enhance visualization of a flaw (see the article "Ultrasonic Inspection" in this Volume for a description of A-scan, B-scan, and C-scan ultrasonic display modes). The three display modes are illustrated in Fig. 10 for a test block with flat-bottom holes that simulate flaws. The A-scan display was obtained on an oscilloscope screen. The amplitude of the reflected signal is displayed in the vertical direction; time of signal flight, which is proportional to depth in the test block, is displayed horizontally. The B-scan display was obtained by a reconstruction of the hologram produced by orienting the transducer for angle-beam testing, which provided essentially a profile of the region being examined (in this case, the row of five holes in the straight portion of the "D"). The C-scan display was obtained with the transducer oriented for straight-beam testing. The pseudo-three-dimensional image was obtained by the reconstruction of B-scan and C-scan holograms combined.

Comparison of Liquid-Surface and Scanning Systems

The outstanding feature of the liquid-surface system of acoustical holography is that it provides a real-time image, while the image provided by the scanning system requires reconstruction. The real-time feature makes the liquid-surface system suitable for the rapid inspection of large amounts of material on a continuous basis. In contrast, inspection with the scanning system is relatively slow. Photographic or videotape records for later study can be made with either system.

The prominent feature of the scanning system is its ability to determine accurately the position and dimensions of flaws lying deep in opaque test objects, especially when only one side of the object is accessible. (In contrast to the scanning system, the liquid-surface system is usually operated in the transmission mode, which requires access to both sides of the test object.) Although both systems offer about the same resolution, the sensitivity of the scanning system is greater by a factor of about 10^6. The excellent flaw-measuring ability of the scanning system can be used to characterize the flaws detected previously by faster inspection methods, such as scanning with a conventional ultrasonic search unit.

Another important feature of the scanning system is that the commercial equipment for this system is usually transportable, while liquid-surface equipment is usually stationary. In addition, scanning transducers can be coupled to very large test objects by water columns, while inspection by the liquid-surface system usually requires that the test objects be small enough to be placed in a water-filled tank and completely immersed.

A useful advantage of the scanning system is its capability for selectively producing either longitudinal or shear waves in the volume of metal under examination through adjustment of the angle of incidence of the incoming ultrasonic beam. The liquid-surface system is limited to straight-beam testing, which produces only longitudinal waves.

The scanning system can also be used for accurate surface contouring. Even surfaces inside vessels can be contoured, without appreciable loss in surface detail, by the use of a scanning system placed outside the vessel.

Readout Methods

The readout portion of an acoustical holography inspection system provides the means by which the acoustically derived images generated during the reconstruction process are rendered visible for observation and analysis. A variety of methods are available to accomplish this end. One of the most rudimentary but effective methods consists of projecting the image onto the surface of a ground-glass screen for direct viewing. For optimal viewing of the screen, however, the test results must be observed in a darkened area or through a hood that has been placed over the screen to shield out ambient light.

If a permanent record of the test results is required, a loaded film holder can be substituted for the ground-glass screen to record the projected image photographically. This elementary photographic approach requires control of the film-exposure time interval. Exposure-time control can be effected by:

- Simply turning the laser on and off
- Opening and closing a diaphragm or electro-optical shutter placed directly in front of the laser or light source
- Blocking and passing the laser beam with a hand-held piece of cardboard or any other optically opaque substance
- Removing and inserting a rectangular light shield located immediately in front of the film plane

An alternative and more convenient approach is simply to substitute a conventional photographic camera for the film holder. Polaroid films, because of their very short processing times (15 s per individual exposure for black-and-white films) and their ease of usage, have been extensively used for holographic recording. This type of film is available as individual 4 × 5 in. sheet films, 3 × 4 in. film packs, and 3 × 4 in. rolls.

Television Readout. The use of a closed-circuit television system for image readout has the following useful advantages:

- When making real-time observations, the necessity for direct viewing through laser-illuminated screens, which can be tiresome to the eye, is obviated
- A magnification of the image, which is limited only by the size of the monitor screen, is realized without the usual attendant loss in intensity (attributed to the inverse-square law)
- Permanent records of the monitor displays can be made photographically or on videotape
- Special devices can be inserted in the electronic circuits for image enhancement

Time-invariant (steady-state) monitor displays can be photographically recorded with any conventional still camera. If the display varies with time—which is possible with acoustical holography—or if the recording of a sequence of motionless images is desired, the imaging data can be effortlessly recorded on videotape. A movie camera can also be used for this purpose, provided it is accurately synchronized with the monitor display. In both cases, an audio track can be added to describe the nature of the recording. From a recording standpoint, a videotape recorder offers the following advantages:

- The recording medium does not have to be chemically processed, as movie film does
- Playback of the imaging data can immediately follow the recording phase
- The recording medium is erasable and reusable
- Recordings can be made in full color as well as in black and white
- Many videotape recorders are equipped with a variable-frame-rate feature that enables the motion to be slowed to any speed or stopped

Electronic image enhancement can take several forms. By means of a display technique referred to as level slicing, points of uniform intensity in the input image can be displayed as contours. These contours are used in pairs to form the boundaries of bands corresponding to the extremities of an intensity window of finite but variable width. On a black-and-white monitor, the bands are discernible as different shades of gray; on a color monitor, as distinct differences in hue.

Other display formats include isometric and profile displays. Not only can regions within the same image be investigated, but comparisons can be made between two independently selected images to determine the existence of differences in image content. The image signals can also be digitized to permit a better quantitative analysis of the imaging data. Once digitized, the data can be fed into a computer for storage, recall, and subsequent numerical analysis. The nature of the inspection problem, the capabilities of the acoustical holographic approach, and the financial resources available are prominent factors in determining the degree of analytical sophistication to which the data will be subjected.

Calibration

In conventional ultrasonic systems, the amplitude of the flaw signal is compared to the amplitude of the signal from a standard discontinuity in a test block made of metal similar to that of the test object (see the article "Ultrasonic Inspection" in this Volume). This technique, however, is restricted because signal amplitude depends on the orientation and roughness of the flaw, not simply on its size. In contrast, an acoustical holographic system provides a series of two-dimensional images at different depths. As long as sufficient ultrasonic energy is received to produce an image, the size (and shape) of the flaw can be determined directly from the image. This direct determination provides a better estimate of size than that obtained from conventional ultrasonic systems, which determine size primarily from the total integrated sound energy reflected by the flaw and returned to the transducer. In fact, the flaw image produced by acoustical holography is the most precise characterization obtainable by any nondestructive inspection system used singly. Although the image produced by acoustical holography is not compared to standard discontinuities in test blocks in the same way that comparisons are made in conventional ultrasonic systems, test blocks are required to verify the sensitivity and resolution of the system.

Test Blocks. Figure 11(a) shows an aluminum alloy 1100-F test block that has been used for verification of the sensitivity and lateral resolution of a scanning acoustical holography system (Ref 3). This block contains 15 flat-bottom holes, 3.18 mm (0.125 in.) in diameter and 78 mm (3.1 in.) deep, drilled in three rows from the bottom surface. In the first row, the holes are separated by 1 mm (0.04 in.); in the second row, by 2 mm (0.08 in.); and in the third row, by 4 mm (0.16 in.).

Using a focused ultrasonic transducer with a 25 mm (1 in.) diameter and a 102 mm (4 in.) focal length, two holograms were constructed of the region of the block containing the holes. When the frequency used to construct the hologram was 3.12 MHz (theoretical lateral resolution in the object plane of 2 mm, or 0.08 in., in aluminum), the holes in the left row merged while those in the middle row remained completely separate (Fig. 11b). This shows that the actual

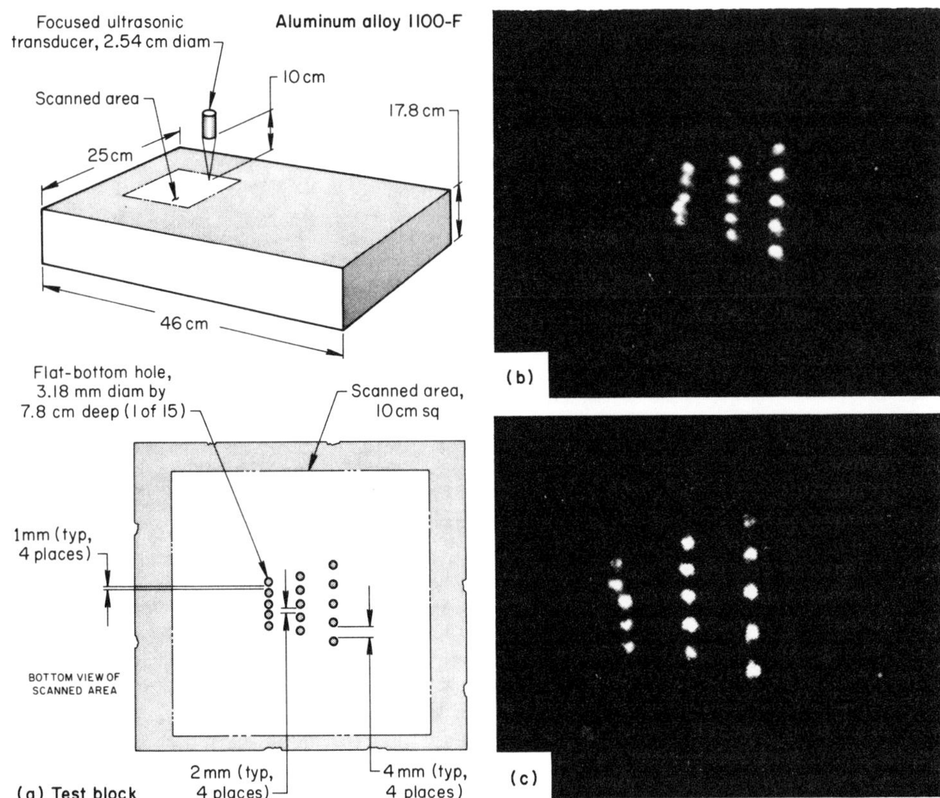

Fig. 11 Test block (a) made to measure resolution in the object plane and to locate the position of the focused ultrasonic transducer during inspection by scanning acoustical holography. (b) and (c) Reconstructed images of scanned area taken at 3.12 and 5.1 MHz, respectively

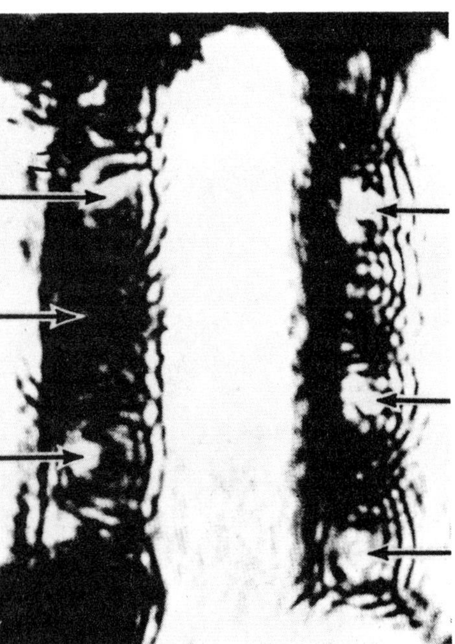

Fig. 12 Reconstructed image of spot-welded test strips that were inspected by liquid-surface acoustical holography at 3 MHz. Arrows indicate spot-welded areas.

lateral resolution was slightly better than 2 mm (0.08 in.). When the frequency was increased to 5.1 MHz (theoretical lateral resolution of 1 mm, or 0.04 in.), the actual lateral resolution also increased to more than 1 mm (0.04 in.) (the holes in the left row in Fig. 11(c) are completely separate).

Interpretation of Results

The optical image that is reconstructed from an acoustical hologram is a two-dimensional presentation of a portion of a three-dimensional object. This portion consists of a subsurface layer of finite thickness oriented perpendicularly to the direction of propagation of the ultrasonic beam. The layer thickness, in turn, is a function of Δz, the depth-resolution capability of the acoustical holographic inspection system. When viewing this two-dimensional presentation, it is important to keep in mind which area is being viewed and which viewing direction (mode of ultrasonic wave propagation) was used to produce it.

When viewing ultrasonic images, the viewer should not make interpretations based on experience with optic images. The long ultrasonic waves limit the resolution so that the images produced are poor compared to those produced by visible light waves. In addition, the coherence of the ultrasonic energy produces an effect analogous to the results obtained when viewing an object illuminated by coherent light, that is, a speckled image having highlights.

One further difficulty in the interpretation of images obtained using acoustical holographic techniques, in addition to not seeing conventional three-dimensional images, is that natural flaws are not of regular geometric configuration. The simple-shaped calibration holes in test blocks are useful in determining the sensitivity and resolution of the system, but experience with natural flaws is necessary for a correct interpretation of the reconstructed image. For this reason, it is best to obtain holograms of the flaw from two directions, but accessibility and orientation of the flaw are often restrictive. When experience is gained with different types of flaws, such as cracks, pores, and incomplete weld fusion, it is often found that they exhibit characteristic images from which comparisons can be made. Like other methods of nondestructive inspection, acoustical holography is not universal in application; often several methods must be used to reinforce its findings.

Applications

Liquid-surface and scanning systems of acoustical holography have been used commercially to inspect various types and sizes of welds and to continuously inspect sleeve-bearing stock. In addition, scanning acoustical holography has been used commercially for contouring. These applications are discussed in the following sections.

Inspection of Welds in Thin Materials. The liquid-surface equipment shown in Fig. 7 and 8 has been used to inspect welds in materials that are a few thousandths of an inch to several inches thick. For example, this equipment was used at 3 MHz to inspect spot welds between two layers of aluminum 4.11 to 4.17 mm (0.162 to 0.164 in.) thick. The reconstructed images of two welded test strips, which were used in checking machine settings for a subassembly, are shown in Fig. 12. The good welds can easily be distinguished from the poor or less satisfactory welds. The clearly defined bright spots in Fig. 12 are good welds; the smaller, less clearly defined spots are less satisfactory welds; and the completely dark spot (center arrow on left side of Fig. 12) is a poor weld.

Another example of a weld in thin material inspected by liquid-surface acoustical holography is the weld in 1.6 mm (1/16 in.) thick titanium plate shown in Fig. 13(a). The reconstruction of the image, which was produced at 5 MHz, is shown in Fig. 13(b). The dark regions of the weld are porous; the bright regions are sound.

Inspection of Welds in Thick Materials. Most pressure vessels and pressure vessel components are thick, massive structures that contain many feet of multiple-pass welds. These welds require extensive non-

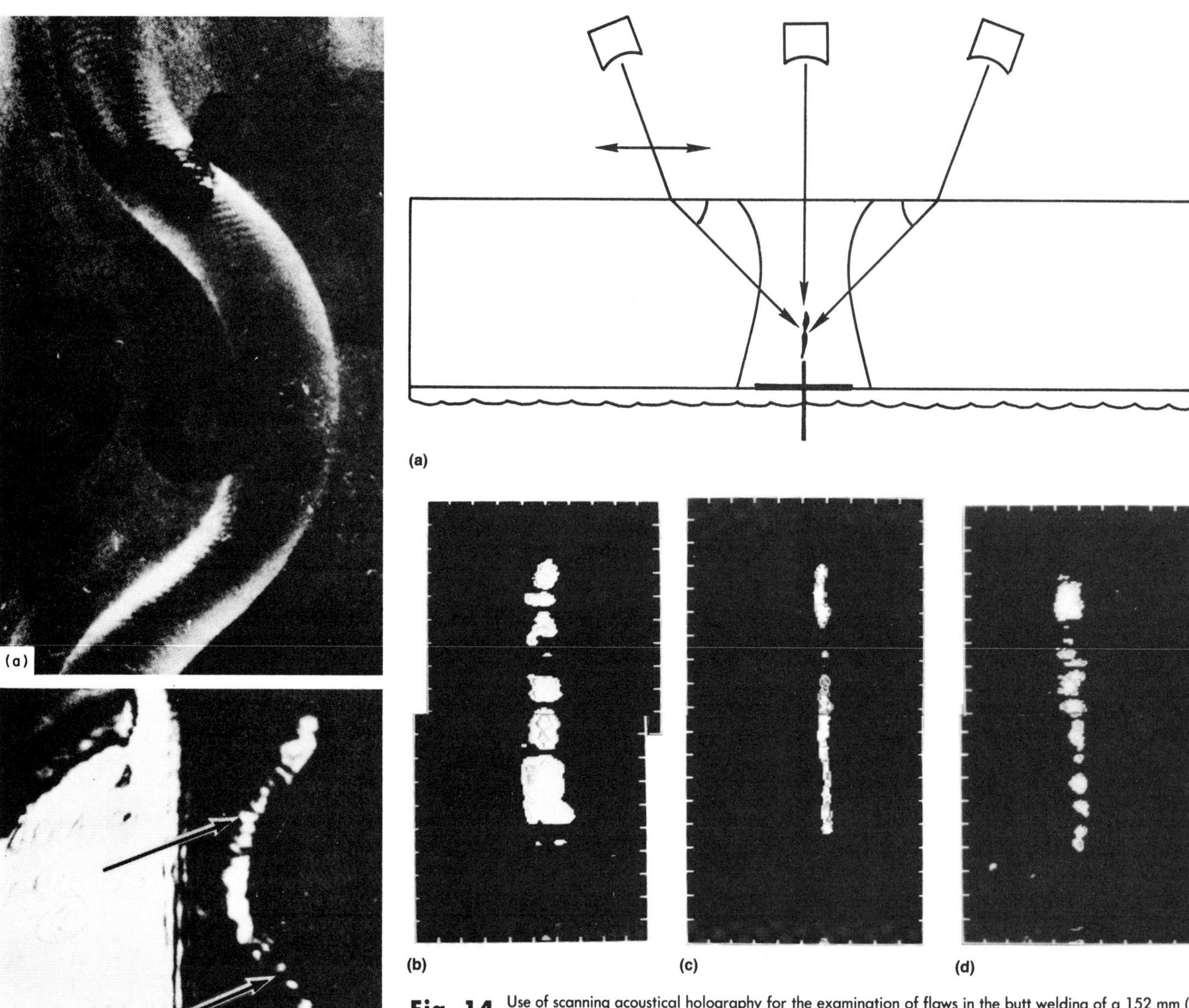

Fig. 13 Sine-shaped weld (a) in 1.6 mm ($^1/_{16}$ in.) thick titanium plate. (b) Reconstructed image of weld, produced at 5 MHz by liquid-surface acoustical holography. Arrows indicate regions containing porosity.

Fig. 14 Use of scanning acoustical holography for the examination of flaws in the butt welding of a 152 mm (6 in.) thick steel plate. (a) Transducer locations above plate surface. (b) Flaw detected with +45° shear sound wave. (c) Flaw detected with 0° longitudinal sound wave. (d) Flaw detected with −45° shear sound wave

destructive inspection to ensure consistent high integrity of the vessel. Scanning acoustical holography, using equipment similar to that shown in Fig. 9, has been used for the inspection of these thick welds.

An example of the use of this equipment is the evaluation of a flaw in the weld of a 152 mm (6 in.) thick steel plate. To obtain sufficient information to dimension the flaw accurately, three holograms were made, as shown in Fig. 14(a). These were made by examining the flaw with ±45° shear and 0° longitudinal sound waves. The resulting images are shown in Fig. 14(b) to (d). The data were taken at 2 MHz with a 25 mm (1 in.) diam transducer focused at the surface of the plate. A 102 mm (4 in.) focal length produced a cone of sound of approximately 60° for the longitudinal and 40° for the shear holograms.

These three images, when properly thresholded, can be used to calculate the position and size of the flaw in three dimensions. This particular flaw was found to be 18° from vertical, with a length of 121 mm (4.76 in.) and a through wall dimension of 22.4 mm (0.88 in.). Destructive examination showed the flaw to be 114 mm (4.48 in.) in length and 13.5 mm (0.53 in.) in through wall dimension. A complete analysis of test results with ultrasonic holography is presented in Ref 4, indicating a root mean square error in estimating flaw height of ±5.33 mm (±0.21 in.) and length of ±5.59 mm (±0.22 in.).

Inspection of Sleeve-Bearing Stock. Sleeve-bearing stock for automobile engines, consisting of 0.25 mm (0.010 in.) thick leaded aluminum alloy bonded to a 2.54 mm (0.100 in.) thick steel base, has been continuously inspected by acoustical holography for unbounded regions between the two layers, using the liquid-surface equipment illustrated in Fig. 7 and 8. The stock consists of strips from 102 to 203 mm (4 to 8 in.) wide and several hundred feet long. During manufacture, the strip moves through the bonding line at a rate of approximately 152 mm/s (6 in./s).

Formerly, the strip was inspected visually for lack of bonding, as evidenced by the

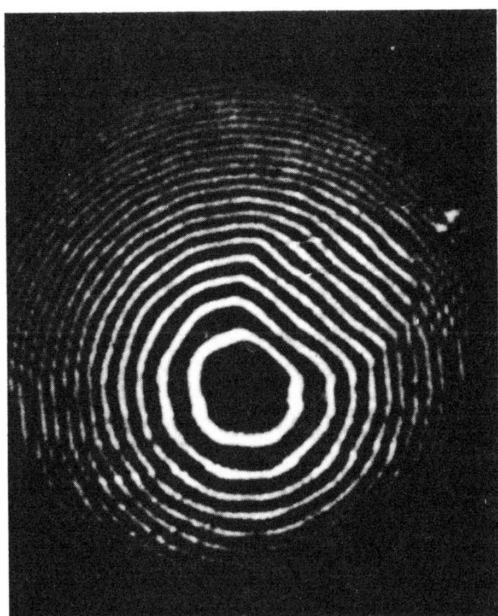

Fig. 15 Contour image of a spherical surface with one flat circular region (above center at right). Image was made at a frequency of 8 MHz using scanning acoustical holography. Each full fringe represents a change in elevation of 95 μm (0.00375 in.).

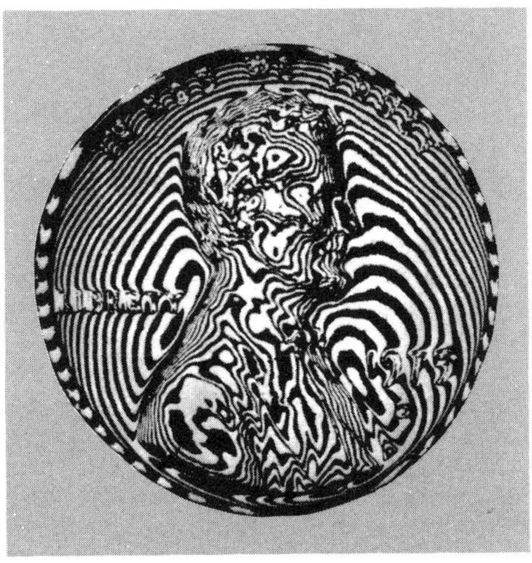

Fig. 16 Contour image of a Lincoln U.S. penny, made at 50 MHz using scanning acoustical holography with the coin slightly tilted with respect to the scanning plane

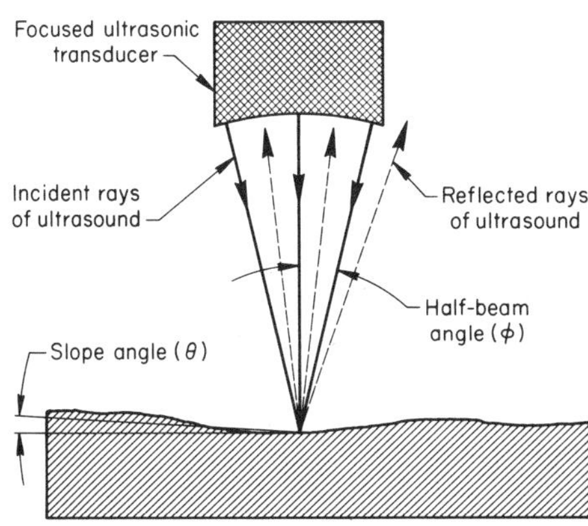

Fig. 17 Schematic of the reflection of rays of ultrasound from a surface being contoured by scanning acoustical holography

presence of blistered areas on the aluminum side. However, because not all the regions of poor bonding formed blisters, some regions were not physically detectable. For this reason, destructive tests using a chisel were sometimes required. With acoustical holographic equipment, unbonded regions as small as 1.6 mm (1/16 in.) in diameter are readily visible on the television monitor.

The contouring of surfaces is readily carried out with commercially available systems for scanning acoustical holography, such as those shown in Fig. 3 and 9. Contour accuracies are limited by the mechanical accuracy of the scanning equipment, but can be better than 1/20 of the wavelength of ultrasound in water at 5 MHz, or about 15 μm (600 μin.). Two examples of contour images formed by acoustical holography are shown in Fig. 15 and 16.

When a scanning acoustical holography system is used in the contouring mode, the transducer is focused on the surface to be contoured, with the central ray approximately perpendicular to the surface. The arrangement is the same as that used in forming acoustical holograms of regions lying deep within a solid material, except that a focused-image hologram is formed; that is, the scan plane is made to lie coincident with the surface to be contoured. Gating is used to eliminate the reflections from the back (far) surface of the object so that they will not interfere with the interpretation of the contours formed over the image of the front (near) surface.

Inaccessible surfaces (far surfaces) of opaque materials can also be readily contoured by moving the focal point of the transducer to the plane of the back (far) surface of the object to be contoured. In this case, the reflections from the front (near) surface are gated out.

There is no distinction between the hologram and the image in this case, because the contour lines are the hologram fringes. As in all ultrasonic inspection, a coupling medium must be used between the transducer and the surface to be examined. The object to be contoured can be immersed in water along with the scanning transducer, but this is not necessary. More often, the transducer is mounted in a water-filled cylinder that has a sliding seal in contact with the surface. The surface need not be perfectly smooth. The relatively minor loss of liquid couplant in scanning over a rough surface is replenished constantly by means of a water pump and reservoir.

A technique that has been used to achieve high resolution in optical (nonholographic) interferometry is multiple-beam interferometry (Ref 5). With this technique, contours of very smooth surfaces have been determined optically to an accuracy of better than 0.5 nm (5 Å); that is, better than λ/1000, where λ is the wavelength of the light. A similar technique can be applied to acoustical contouring by setting the reflection-signal gate to accept only the *n*th reflection, where the integral value of *n* depends on the smoothness of the surface and the amount of energy reflected at the surface. In ordinary contouring by acoustical holography, $n = 1$ and each fringe in the image represents a half-wavelength change in elevation. At 5 MHz in water, a half wavelength is 150 μm (0.006 in.). However, if the reflection-signal gate is set to accept the tenth reflection, each fringe will represent 1/20 wavelength, or about 15 μm (600 μin.).

Figure 17 illustrates the reflection of an ultrasonic beam from a sloping surface. As the slope angle, θ, of the surface increases in magnitude, less and less of the reflected beam is intercepted by the transducer. This limits the surface slope that can be properly contoured. This limit depends partly on the half-beam angle, ϕ, of the transducer, which in currently available systems is about 7°. A conservative rule for estimating the maximum slope angle that will be properly contoured is:

$$\theta_{max} \approx \phi/2n \quad \text{(Eq 3)}$$

Cylindrical and spherical surfaces can, in theory, be contoured to show deviations from the ideal surface by causing the focal point of the transducer to follow a scan plane shaped to the ideal surface, with the central ray kept normal to the surface.

REFERENCES

1. B.P. Hildebrand and B.B. Brenden, *An Introduction to Acoustical Holography*, Plenum Press, 1972
2. B.P. Hildebrand, T.J. Davis, A.J. Boland, and R.L. Silta, A Portable Digital Ultrasonic Holography System for Imaging Flaws in Heavy Section Materials, *IEEE Trans.*, SV-31, No. 4, 1984, p 287-294
3. B.P. Hildebrand and H.D. Collins, Evaluation of Acoustical Holography for the Inspection of Pressure Vessel Sections, *Mater. Res. Stand.*, Vol 12 (No. 12), Dec 1972, p 23-32
4. D.E. MacDonald and E.K. Kietzman, "Comparative Evaluation of Acoustic Holography Systems," Report NP-5130, Electric Power Research Institute, 1987
5. S. Tolansky, *Multiple-Beam Interferometry of Surfaces and Films*, Clarendon Press, 1948

Strain Measurement for Stress Analysis

L.D. Lineback, Measurements Group, Inc.

EXPERIMENTAL STRESS ANALYSIS TECHNIQUES, in one form or another, can be applied at all stages in the life of load-bearing parts, members, and structures. They are used in new-product engineering, from preliminary design evaluation to production design testing; in proof and overload testing; in failure analysis, from design defect detection to redesign validation; and in materials and structural research.

Because stress is not directly measurable, the experimental analysis of engineering stresses must be based on strains, which can be readily measured by a number of techniques. With measured values of strain and a knowledge of the mechanical properties of the material from which the structure is fabricated, stresses can be calculated from the appropriate stress-strain relationship of the material.

The usual objective of stress analysis is to locate and measure the most significant strains in a part, member, or structure. These include not only the high-level strains located at stress concentrations where the product will most likely fail under load but also the lowest ones in overdesigned regions where material can be safely removed to save on costs and improve performance. Except for the simplest of designs, location of these significant strains will normally require a full-field measurement technique such as photoelasticity or brittle coatings.

Once the areas of significant strains are identified, the stress analyst generally makes one or more measurements of strain at a point to accurately determine the magnitude and direction of the principal strains. Depending on the nature of the load, type of material, and mode of failure, it may also be necessary to measure the maximum shear strain. For making these point measurements of strain, both electrical resistance strain gages and photoelastic coatings are widely used.

In addition to the kinds of strains to be measured, it is important to consider a number of other practical criteria when selecting the appropriate strain measurement technique. These generally include:

- Location of test site (laboratory or field)
- Type of loading (static, dynamic, or a combination)
- Anticipated strain range
- Required resolution and accuracy of measurement
- Operating environment (temperature, medium, and so on)
- Duration of test (from microseconds to decades)

For specialized measurements, other parameters may also require consideration.

Resistance Strain Gage Method

The bonded resistance strain gage is the most powerful and widely used strain measurement technique in the field of experimental stress analysis. Basically, the strain gage (Fig. 1) is an extremely thin, small, strain-sensitive electrical resistor bonded to a flexible backing material that, when adhesively bonded to the structure or test part, transforms surface strains into changes of electrical resistance. The resulting changes in resistance are then read out directly as strain, load, pressure, torsion, and so forth, on the appropriate instrument.

Early bondable strain gages were manufactured by forming small-diameter (0.025 mm, or 0.001 in.) constantan wire into a suitable grid geometry and bonding the grid in place on a paper carrier. The active gage lengths were relatively large, typically 6 mm (0.25 in.) or longer. Today, most precision strain gages are made by etching grids into thin metal foil held in place by a compatible backing. The active gage lengths vary from 0.2 to 100 mm (0.008 to 4.000 in.). Typical gage thickness for foil gages, including backing and grid, is 0.05 mm (0.002 in.) or less.

Gage Performance. Resistance strain gages are designed to measure strain in the axial direction of the grid (Fig. 1) and to be as insensitive as possible to strains in the transverse direction. Because strain gage grids tend to yield an integrated average of the strains under them, very short gages are

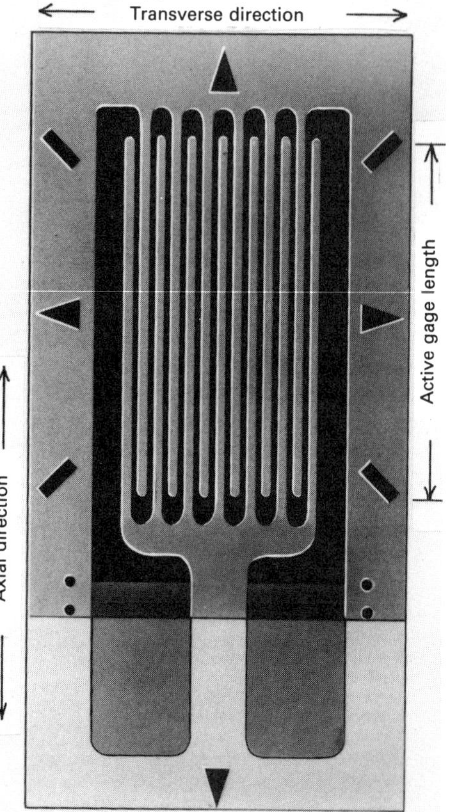

Fig. 1 Modern foil strain gage with a polyimide backing and encapsulation. About 6× actual size

generally used when the strain gradient is steep. Measuring strain over as small an area as possible tends to minimize any undesirable strain integration. Longer gage lengths are intended for applications in which the mean strain over a considerable length is more representative. For example, gages with longer grids are commonly used on filament-wound composite structures to average out the strains in these nonhomogeneous materials consisting of filaments and matrix. Very long gages are used in making measurements on the surface of concrete structures having large-diameter aggregate in the cement matrix.

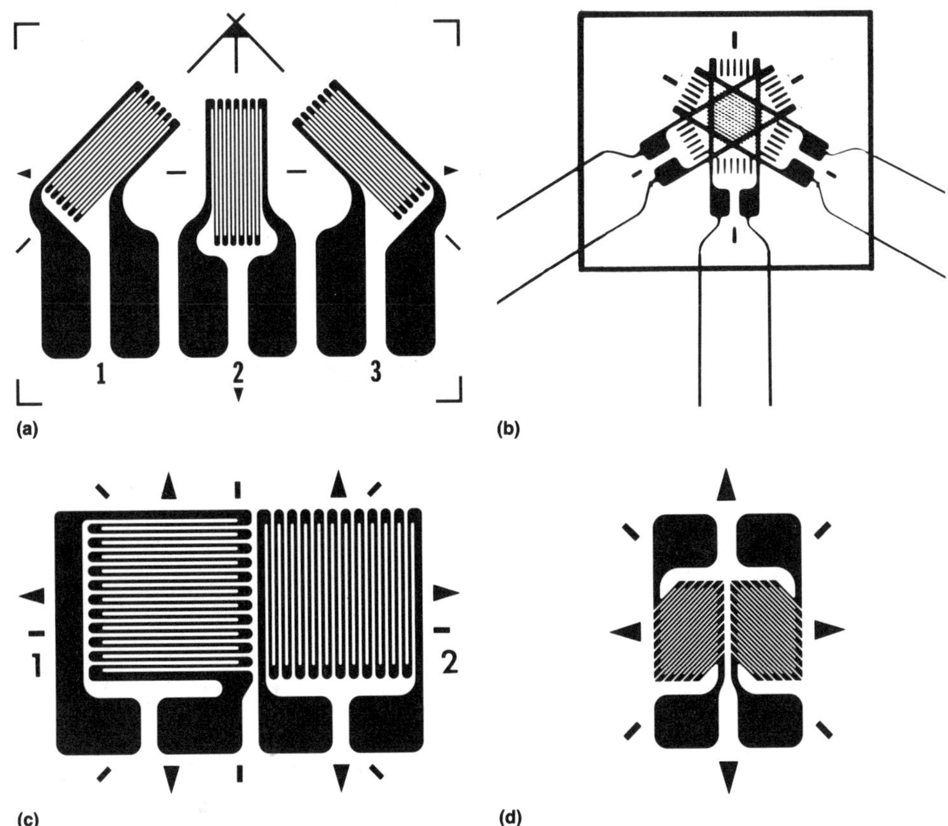

Fig. 2 Typical grid geometries. (a) Three-element rosette with planar construction. (b) Three-element rosette with stacked construction. (c) Two-element T rosette. (d) Two-element pattern for measuring shear strain

Gage sensitivity to strain is defined as the gage factor, GF, which is the ratio of unit change of resistance ($\Delta R/R$) to unit strain ($\Delta L/L$):

$$\frac{\Delta R/R}{\Delta L/L} = \text{GF} \qquad \text{(Eq 1)}$$

The gage factor is a function of the gage design, as well as the alloy used to make the gage grid, its thermomechanical history, and, to a lesser extent, the measurement temperature. Gage factors at room temperature for most typical strain gage alloys range from two to four.

Grid Geometries. Because the grid of a strain gage is designed to measure normal strains in the axial direction of the grid, various gage geometries are used during strain measurements. Because gages with single grids measure strains in a single direction, their use should be restricted to well-defined uniaxial stress states, that is, pure tension or compression. When a point measurement is required in a biaxial stress field and the directions of the principal strains are unknown or uncertain, a three-element strain gage rosette (Fig. 2a and b) should be used. With grids typically oriented at 0-45-90° or 0-60-120°, these gages enable three separate strain measurements to be made about a point in order to solve for the three unknowns (the two principal strain magnitudes and their direction). When the principal axes are known, only two independent strain measurements are required to determine the magnitudes. For these measurements, two-element "T" rosettes (Fig. 2c) are used. Rosettes are available in both stacked (grids on top of one another) and planar (grids on a single surface) geometries. Active grid lengths typically vary from 0.4 to 13 mm (0.015 to 0.500 in.) for rosettes.

Shear strains cannot be directly measured with strain gages. However, the magnitude of shear strain at a point is always equal to the difference in two orthogonal normal strains at that point with a direction that is intermediate (45°) to the orthogonal normal strains.

When the direction of the maximum shear strain is known (pure torsion, for example), the magnitude can be determined by measuring the two associated orthogonal normal strains. To ensure a proper alignment of sensing grids, gages with special shear patterns are available (Fig. 2d). For these gages, two grids are oriented at right angles to one another and at 45° angles to the longitudinal axis of the gage. When properly connected to a strain indicator, the indicated strain will be equal to the shear strain along the gage axis.

Gage Materials. The gage user has a choice of foil and backing materials to suit a wide range of performance and operating conditions. The most widely selected strain-sensing alloy for use on most materials is a copper-nickel alloy known as constantan. With a gage factor (defined in the section "Gage Performance" in this article) of about 2.1 for most grid geometries, constantan can be thermomechanically processed to minimize the effects of thermally induced resistance changes (thermal output) that occur when gages experience a change in temperature while strains are being measured. The measurement range of constantan gages is generally limited to strains of 5% or less, except for gages made from fully annealed material, which have a limit of 20%.

For elevated-temperature measurements, that is, up to 290 °C (550 °F) long-term or 400 °C (750 °F) short-term, a nickel-chromium alloy similar to Karma (Ni-20Cr-3Fe-3Al) is used for strain gage grids. With a gage factor similar to that of constantan, this alloy exhibits a higher fatigue life. The grids of gages having the greatest resistance to fatigue are made of iron-nickel-chromium-molybdenum alloy (Iso-elastic). This alloy has a high gage factor (about 3.5), but cannot be treated to minimize thermal effects. As a result, Iso-elastic gages are normally used only in dynamic applications. The strain measurement range of both Karma and Iso-elastic gages is limited to strains of 2% or less. Iso-elastic gages become very nonlinear at strains above 0.5%.

The strain gage backing serves to hold the sensing grid in place. The backing (along with the adhesive) also electrically isolates the grid from the test specimen and transfers the strains from the surface of the test part to the sensing grid. For stress analysis work, gage backings are commonly made of a tough, flexible polyimide. For extreme temperatures, that is, below −45 °C (−50 °F) or above 120 °C (250 °F), a more stable glass-reinforced epoxy phenolic is commonly used.

Gage Installation. The problem of where to install strain gages requires either a fundamental understanding of the mechanics involved or the use of photoelastic or brittle coatings to locate high and low stress regions. Photoelastic and brittle coatings will also indicate the directions of the maximum and minimum principal strains.

After the appropriate location is determined, careful installation of the gage is required in order to achieve accurate and reliable strain measurements. To accomplish this requires strict adherence to the recommended installation procedures supplied by the gage manufacturer, including the use of the proper accessory tools and supplies.

The adhesive, which must transfer the strain from the specimen surface to the gage backing, is of paramount importance in gage installation. Specially qualified cyanoacrylate adhesives are popular because of their ease of application and short, room-tem-

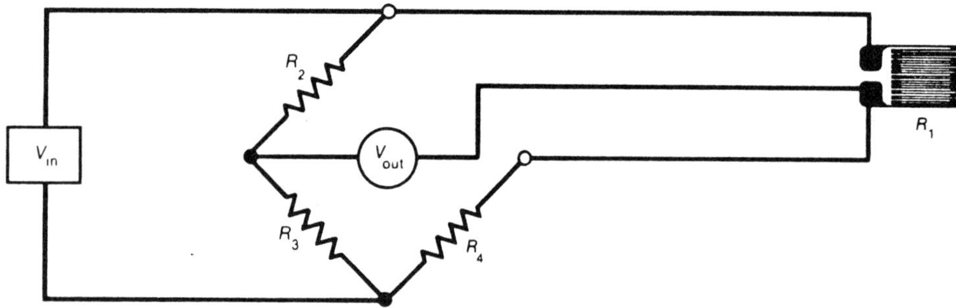

Fig. 3 Wheatstone bridge configuration with quarter-bridge, three-wire circuit

perature curing cycle. However, these adhesives are degraded by time, humidity, elevated temperature, and moisture absorption. Under these conditions, an epoxy-base strain gage adhesive would be a better selection. In addition to having higher elongation capabilities, epoxies have longer cure times. Some also require elevated-temperature cures.

Care must be taken not only in selecting the best adhesive but also in choosing the correct leadwires, solders, and protective coating for the test environment. While most tests are performed in a laboratory-like environment at room temperature, by using the proper techniques and materials for bonding, wiring, and protecting strain gages, it is possible to make good strain measurements under a wide variety of difficult environmental conditions over an extended period of time. Technical data on this subject are available in a variety of forms (training programs, videotapes, publications, applications engineering assistance) from strain gage manufacturers.

Instrumentation. Because the resistance changes per unit of microstrain (μm/m) are so small (0.00024 ohms/μm/m for a 120-ohm gage with a gage factor of 2.000), it is usually not practical to measure the resistance directly. As a result, electrical circuits are commonly employed to convert the resistance changes into a voltage signal, which can be measured with recording instruments.

Modern strain gage instruments generally employ a Wheatstone bridge (Fig. 3) as the primary sensing circuit. A stable, high-gain dc amplifier is then used to amplify the small bridge output signal to a level suitable for driving some form of display or output device. In addition to these two basic components, a typical strain indicator (Fig. 4) includes the bridge power supply and built-in bridge completion resistors, along with balance and gain controls, provision for shunt calibration, and various convenience features.

For a given strain, the bridge output can be increased by increasing the bridge voltage. While it is usually desirable to obtain as high an output as possible, there are limitations set by the self-heating effects, which vary with the current through the gage. The power densities that can be tolerated by the gage are strongly influenced by the specimen, which serves as a heat sink. For most practical purposes, bridge voltages are normally in the range of 1 to 15 V, with most measurements being made at voltages in the range of 2 to 5 V.

An important feature to be considered when dynamic measurements are involved is the frequency response or bandwidth of the analog output from the amplifier. Bandwidth specifications are normally referred to the −0.5 dB (5% loss of output) or the −3 dB (30% loss of output) point. Errors due to the frequency response of the instrumentation must be assessed by dynamic calibration with a frequency generator.

The output from the amplifier is normally in the form of an analog voltage (up to 10 Vdc), which can be fed into an appropriate indicating or recording device. For static measurements, the analog output will typically be fed to an analog-to-digital converter, with the digital signal being passed to a digital display (often built into the instrument) or to a computer. For dynamic measurements, the analog voltages are normally fed into an oscilloscope, *x-y* recorder, magnetic tape recorder, or recording oscillograph.

Errors. The stability and repeatability of present-day gages and instrumentation enable resolutions of strain as small as 1 μm/m. In terms of accuracy, however, measurement errors depend on a number of factors, including, primarily, gage factor uncertainties, thermal output, signal attenuation in installations with long leadwires, Wheatstone bridge nonlinearity, gage alignment errors, incomplete transmission of strain through the adhesive, sensitivity of the gage to transverse strains, and reception of noise from outside the bridge. When these sources of error are properly handled, strain measurements having inaccuracies that total 1% or less can readily be made. Inaccuracies may be higher at elevated temperatures.

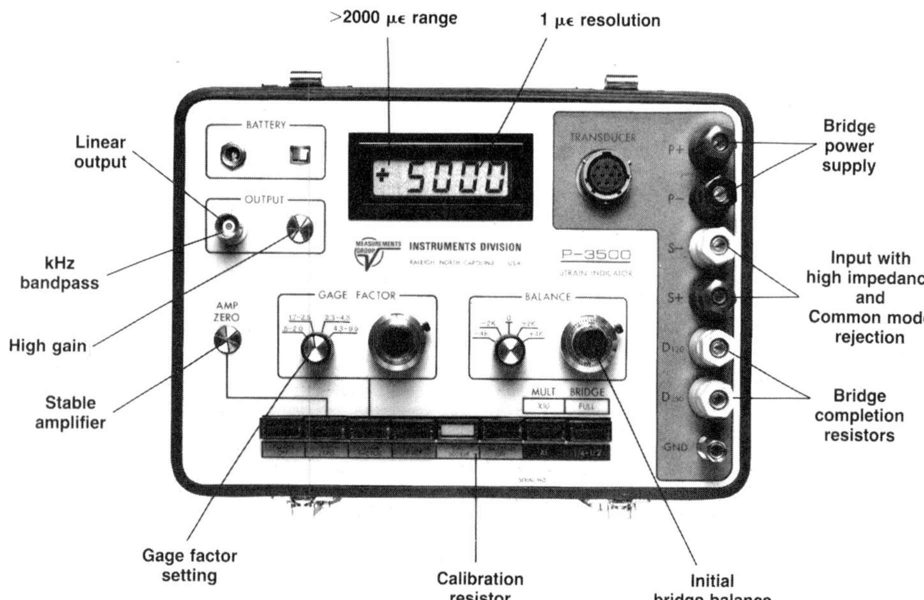

Fig. 4 Features of a high-quality strain indicator

Photoelastic Coating Method

Photoelasticity is a visual full-field technique for measuring strain. When a photoelastically active material is strained and viewed under polarized light, distinct color bands (fringes) are observed (see Fig. 24a in the section "Use of Color for NDE" in this Volume). Aside from its colorful and aesthetically pleasing character, the photoelastic fringe pattern provides valuable information for the stress analyst.

Photoelasticity embraces three broad techniques: photoelastic coating, two-dimensional model analysis, and three-dimensional model analysis. Model analysis is useful at the early design stage of a product, before actual parts are made. However, the

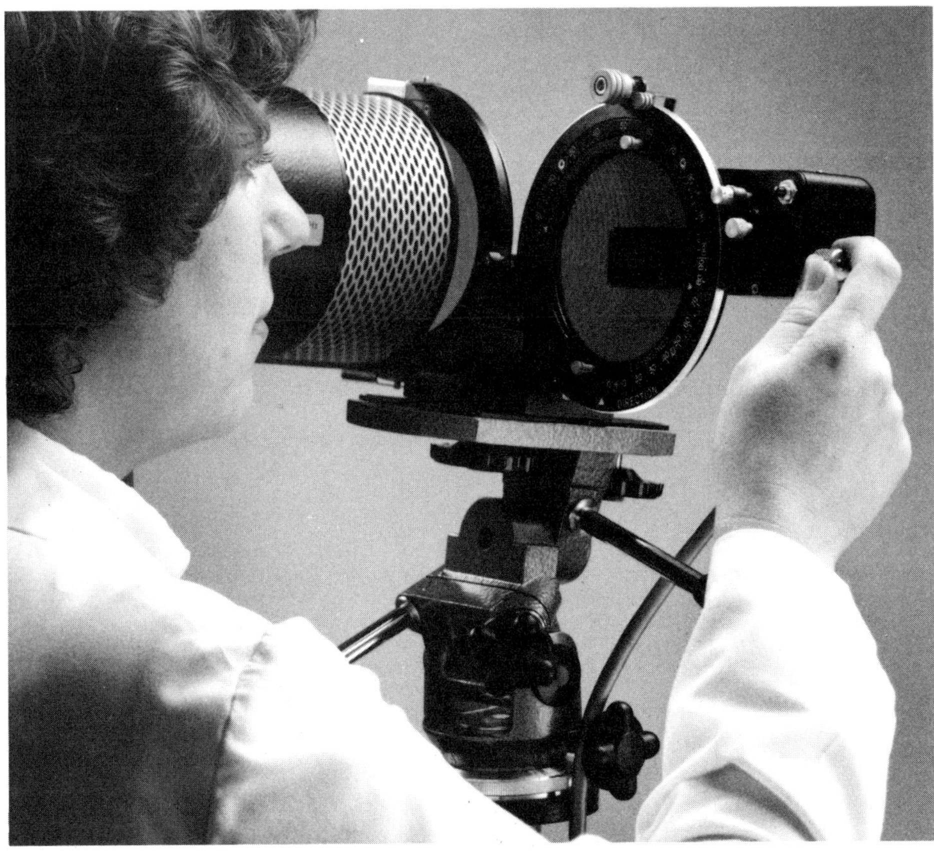

Fig. 5 Reflection polariscope with a null-balance compensator being adjusted by the operator

Fig. 6 Pump housing being coated with photoelastic material

most widely used of the three techniques is the photoelastic coating method, which involves the bonding of a thin photoelastic layer to the surface of a part. When the part is loaded, the strains on its surface are transmitted to the coating. The principal-strain differences in the coating can then be determined with a reflection polariscope (Fig. 5).

The photoelastic coating method has many advantages over other methods of experimental stress analysis. It enables the investigator to measure surface strains in parts and structures of virtually any size, shape, or material operating under actual service conditions at or near room temperature. Additional advantages include:

- The strain field is visually displayed over the entire coated area
- The gage length is effectively zero, permitting accurate measurement of stress gradients
- Yielding is easily detected because a portion of the strain pattern remains visible after the loads are removed
- Assembly strains can be measured
- The combined effects of various load conditions can be observed and measured
- Dynamic analysis can be performed on the part or structure
- Residual stresses can be determined
- Understressed areas can be easily identified for elimination of excess material

The photoelastic coating material, usually a polymer, can be applied either by bonding prefabricated sheets to the surfaces of flat specimens or by the contour forming of photoelastic material during polymerization to complex curved surfaces (Fig. 6).

Interpretation of the Photoelastic Pattern. When a photoelastic material is subjected to applied loads, changes in optical properties take place that are proportional to the strains developed; the material becomes birefringent.

Polarized light passing through a stressed material splits into two rays, each vibrating along a principal stress axis and traveling at a different speed. A relative phase shift between the two rays produces the color interference pattern, which is observed and analyzed with the polariscope. With this instrument, the plane of vibration can quickly be established to permit measurement of the principal stress direction at every point in the field. The phase shift is easily measured by optical compensation to determine the strain magnitude. In the case of a uniaxial stress condition, the stress is proportional to the product of the measured phase shift and the fringe value of the coating (microstrain per fringe). In a biaxial stress condition, the product is equal to the difference in principal strains, or the shear strain.

The photoelastic color sequence observed with increasing stress is black, yel-

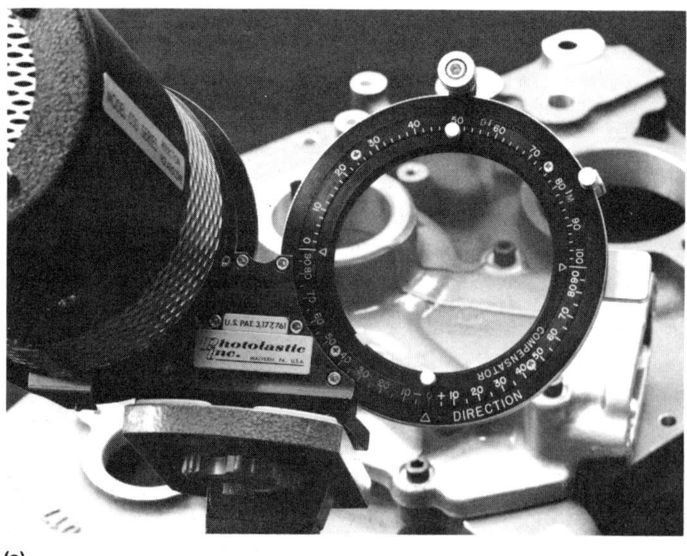

(a)

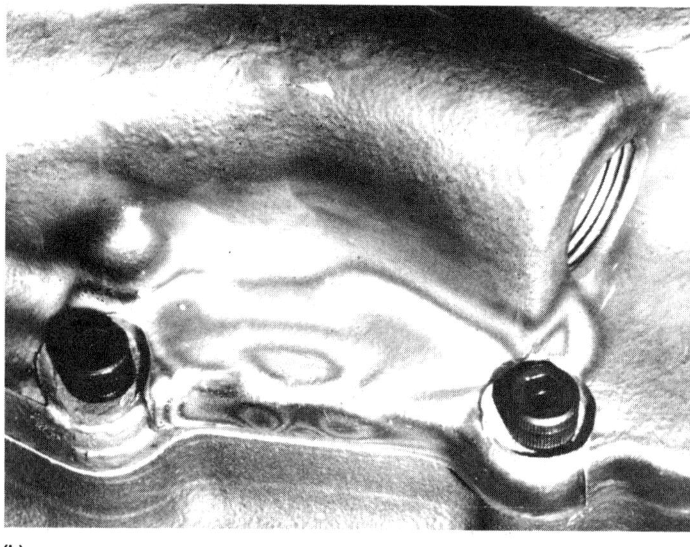

(b)

Fig. 7 Full-field capability of photoelastic coatings. (a) Measuring the effects of stress concentration near a corner of a complex part. (b) Regions near the bolts that may contain overlooked assembly stresses

low, red, blue-green, yellow, red, green, yellow, red, and green (see Fig. 24b in the section "Use of Color for NDE" in this Volume). A constant-color band is a contour of constant difference in principal strains. A band separating the red and green colors is defined as a fringe. The higher the fringe order, the higher the strain; the closer the bands are together, the steeper the stress gradient, indicative of a stress concentration. A uniform color area represents a constant stress area. In brief, the overall stress distribution can be easily and quickly studied by recognizing fringes, their order, and their location with respect to one another.

Fig. 8 Cracks in a brittle coating normal to the maximum tensile strains

At any point on the band of colors, a positive identification of the fringe order is easily made by the null-balance method with a compensator. The compensator is a device that produces a calibrated phase shift equal in magnitude but opposite in sign to that occurring in the observed color pattern on the coating. When superimposed, the photoelastic effect is canceled and a black (zero) fringe is evident. This method of measurement is accurate to one-fiftieth of a fringe, or about 10 to 20 microstrain for most coatings, when coupled with a modern photoelastic data-acquisition system.

At a free boundary, where there is only one nonzero principal stress, tangent to the boundary, the maximum principal stress can be determined directly. However, because the photoelastic method of strain measurement indicates the difference in principal strains, it is sometimes necessary for the stress analyst to separate the principal strains at points other than at a free boundary. Such a situation is illustrated in Fig. 7. Separation requires that a second measurement be made at each point of interest. The most direct approach is to apply a special strain gage directly to the coating. This gage measures the sum of any two orthogonal strains at that point (an invariant). Adding the sum and difference of principal strains together and dividing the result by 2 yields the maximum principal strain at that point.

A wide variety of coating materials and adhesives are available. Because the optical signal is a function of both the optical properties of the material and its thickness, selection of the proper coating should be made on the basis of anticipated strain levels.

Additional considerations in coating selection include reinforcement effect, test temperature, and maximum elongation. The modulus of elasticity of coating materials ranges from 4 to 3000 MPa (0.6 to 450 ksi). Maximum allowable elongations are from 3 to 50%, or more. Testing temperatures are normally limited to the vicinity of room temperature, but the range can be extended up to about 200 °C (400 °F) under certain circumstances. Special equipment is available for casting contourable sheets from liquid materials.

Reflection polariscopes contain all the optical components necessary for photoelastic coating analysis, including a suitable light source. Stroboscopic light sources are available for making dynamic measurements. Other common options include a digital indicator, null-balance compensator (Fig. 5) camera, and telemicroscope.

Brittle-Coating Method

For this method of strain measurement, a specially compounded, strain-sensitive lacquer is sprayed onto the surface or part to be examined. After it has dried and the part has been loaded, the lacquer develops a series of cracks normal to the direction of the maximum tensile strain on the surface. The crack pattern (Fig. 8) provides an overall picture of the tensile strain distribution, with an indication of the strain magnitude and direction. Brittle coatings, therefore, are useful for obtaining a quick indication of likely locations of critical stresses or stress concentrations. They are not used for detailed measurements, but for a fast, economical means of determining exact locations and orientations for subsequent strain gage installations.

The brittle-coating test is particularly suitable for inaccessible areas. For example, in vehicle testing, a part to be investigated can be coated, assembled in the vehicle, and then disassembled for examination upon completion of a road test. The resultant crack pattern helps to ensure that any subsequent strain gage investigation is centered at the point of significant strains. The knowledge that measurements are taken in the correct place can reduce the test time in subsequent strain gage studies.

SELECTED REFERENCES

- J.W. Dally and W.F. Riley, *Experimental Stress Analysis*, 2nd ed., McGraw-Hill, 1978
- A.S. Kobayashi, Ed., *Handbook on Experimental Mechanics*, Prentice-Hall, 1987
- J.M. Whitney, I.M. Daniel, and R.B. Pipes, *Experimental Mechanics of Fiber Reinforced Composite Materials*, 2nd ed., Society for Experimental Mechanics, Inc., 1984
- A.L. Window and G.S. Holister, Ed., *Strain Gage Technology*, Applied Science, 1982
- F. Zandman, S. Redner, and J.W. Dally, *Photoelastic Coatings*, The Iowa State University Press, 1977

Digital Image Enhancement

T.N. Claytor and M.H. Jones, Los Alamos National Laboratory

IMAGE PROCESSING AND IMAGE ENHANCEMENT have become staple methods for analyzing x-ray, tomographic, nuclear magnetic resonance (NMR), ultrasonic, thermographic, and other nondestructive evaluation (NDE) data. With image enhancement, the interpretation of complex data becomes easier for the operator. Fewer errors are made, more subtle features can be detected, and quantitative measurements are facilitated. The disadvantages of image processing are that it usually requires much more time than, for example, the visual reading of film, and it requires the operators to have basic computer skills and a knowledge of image-processing tools. This article presents an overview of image enhancement techniques with application to the needs of NDE.

Important image-processing functions for NDE applications include:

- Input device control
- Palette adjustment
- Contrast enhancement
- Scaling
- Point, and line density measurement
- A variety of filtering techniques
- Image addition and subtraction
- Trend removal (field flattening)
- Edge enhancement

In addition, there are technique-dependent processing functions, such as tomographic reconstruction or time and frequency domain processing for NMR and ultrasonic data. Numerous general image processing textbooks, handbooks, and literature overviews that address techniques used at large, in medical practice, and in industrial NDE are available (Ref 1-10).

NDE Digital Image Enhancement Systems

The basic elements of an image-processing station are shown in Fig. 1. The station consists of:

- A computer
- Imaging software
- Mass storage
- Hard copy
- High-resolution display
- Videotape or optical disk input and output
- Other input devices
- Network interface for communication to other workstations

Data can be input from a variety of devices, it can be displayed and manipulated on a screen, the data can be archived, hardcopies can be made, and information can be exchanged with other computers. Although it is relatively easy to obtain a subset of the system shown in Fig. 1, it is difficult to find compatible hardware and software for all functions. As has been the case for the last decade, hardware and software for image-processing applications are rapidly evolving into more powerful and inexpensive products.

The selection of a computer and software for processing is dependent mainly on the use and throughput required. Many acquisition systems, such as real-time radiographic, infrared, and ultrasonic systems, have (optional) custom image-processing capability. A high-speed computer workstation is usually desirable for multiuser use, for the use of many computationally intensive algorithms, or for the development of algorithms.

More than 50 computers have been ranked according to time required to solve a standard set of linear equations (Ref 11). From this ranking, as well as additional independent tests, the approximate time required for various image-processing functions is given in Table 1 if the code is written in C and the images are 8-bit 1024^2. These times are very approximate because:

- The efficiency of the compiler may vary by as much as a factor of two
- Algorithms vary in coding efficiency
- Many image-processing functions are performed on integer numbers rather than in floating point form (the speed at which integer arithmetic is computed may vary greatly from the floating point value depending on the machine)

To achieve the fastest processing speeds, the computer can be augmented with an

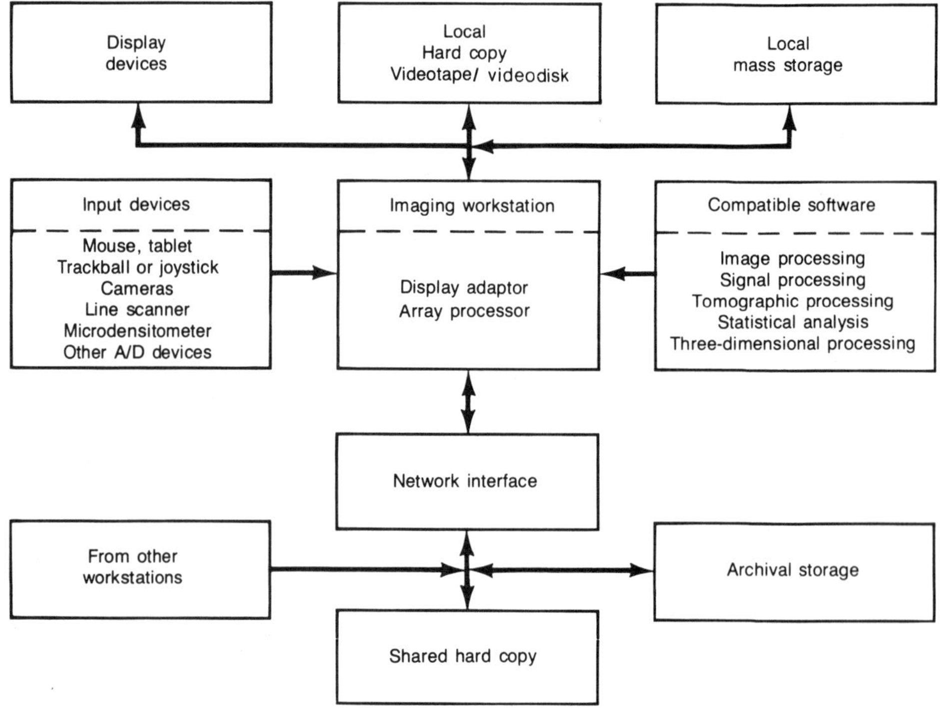

Fig. 1 Typical workstation configuration for use with NDE applications. A/D, analog-to-digital

Table 1 The 8-bit 1024^2 image-processing capabilities of numerous computers

Relative performance	Computer	Time per image operation, s(d): Linear histogram equalization	Image addition	3 × 3 low-pass filter	Rotate	3 × 3 Sobel filter	Fast Fourier transform
Class 1	PC-AT 386 20 MHz/wcp(a); Sun 3/260; Microvax II	15	32	56	110	135	352
Class 2	Sun 3/260 FPA(b); DEC VAX 8550/8800; HP 9000 840/850; MIPS Computer Systems, Inc. M800 w/R 2010 FP(c)	3.0	6.4	11.2	21.8	27	70
Class 3	HP 9000 835; MIPS Computer Systems, Inc. M120-5	1.1	2.3	4.0	7.8	9.6	25

(a) Wcp with coprocessor (80836). (b) FPA, floating point accelerator. (c) FP, floating point. (d) Typical performance

array processor or other hardware that performs a specific function rapidly. For example, 3 × 3 filtering can be performed on a 1024^2 by 8-bit image in 0.1 s, or fast Fourier transform (FFT) can be executed on a 1024^2 by 8-bit image in less than 10 s with the aid of an array processor. As computer processing power increases, more complicated processing functions will become practical for routine use. Digitally enhanced images are also discussed in the articles "Industrial Computed Tomography," "Thermal Inspection," and "Acoustic Microscopy" in this Volume.

Image Capture and Acquisition System

An important part of image enhancement is the system that captures or acquires the data. In many cases, an image will be recorded on film to be digitized, or the data will come directly from an inherently digital system, such as real-time radiography, thermographic, or ultrasonic imaging system. The quality and dynamic range of the image will influence subsequent processing. It is usually desirable to have interscene dynamic ranges greater than 6 bits (64 levels). The eye has a dynamic range of about 10^{10} (photopic and scotopic combined), an interscene dynamic range of about 150, and a brightness discrimination of 2% (50 levels) (Ref 3). The schematic of a typical display system is shown in Fig. 2. Each pixel can represent a gray level or color, and if the pixel is composed of 8 bits, 256 gray shades or colors can be displayed. Because the eye has an approximate dynamic range of 150, a 256-level display is adequate for gray-level viewing. For computational purposes, for the display of color, and for the ability to fully digitize film, it is sometimes desirable to have a 12 (or more) bit pixel, which can be compressed into 8 bits for display. The display size indicated in Fig. 2 is $N \times N$; for most systems, N should be at least 512, with 1024 preferred.

Table 2 lists several types of digitizers that are currently available. Although charge-coupled device (CCD) cameras and light tables are the standard for the efficient input of medium-resolution optical data, there are other alternatives better suited to specific applications. The vidicons offer the highest frame rates in formats of 1024^2 and more, the line scanners have excellent resolution at moderate prices, and the microdensitometers are the most precise with regard to photometric and geometric quantities. Most microdensitometers will digitize to 12 bits, which limits their dynamic range to about 3.6 in film density. As noted in Table 2, the standard scanning microdensitometer (row one) will take longer (up to several hours) to scan film if the film density is over 2. The laser scanners are faster (depending on the make) because of the more intense laser light source and the fact that one axis is scanned with the light beam rather than mechanically.

Linear CCD or photodiode arrays are good choices for film digitization and can have even better dynamic ranges than those listed in Table 2. When these devices are cooled, each 7° C (13 °F) reduction in temperature reduces the root mean square noise by a factor of two. The charge-coupled device and charge injection device are capable of good dynamic range and fair resolution, but have certain artifacts (Ref 12). Again, if the devices are cooled, the noise floor is reduced, and they can be integrated for long periods to enhance the dynamic range and sensitivity. Cooled CCDs are available that rival the low light sensitivity of silicon-intensified targets (SITs), but not the frame speed.

In general, the frame rate, dynamic range, and resolution are all interrelated. The interscene dynamic range is listed as the maximum achievable for the microdensitometers and as the dynamic range that can be achieved at the given frame rate for the other devices. The faceplate illumination given for the tubes and the solid-state detectors assumes mid-level illumination (halfway between saturation and preamplifier noise) (Ref 13). This may vary among tubes and generic types by a factor of three to four (Ref 14). The interscene dynamic range will also vary greatly, but can be maximized by the proper selection of a tube such that a dynamic range of 200 can be achieved at 33 ms/field with a high resolution (>1000 lines).

The quoted resolution for the tubes is at a modulation transfer function (MFT) of 5%, which means that the contrast between a black line and a white line is only 5% at the stated resolution. This, of course, is measured at optimum illumination and at the center of the tube image field. Under other conditions, the resolution will be less. For comparison, a 1024^2 CCD camera may have an MTF of 5% at 750 lines. The lags quoted for the tubes are typical for the particular type at 3 TV fields or 50 ms. Tube cameras are

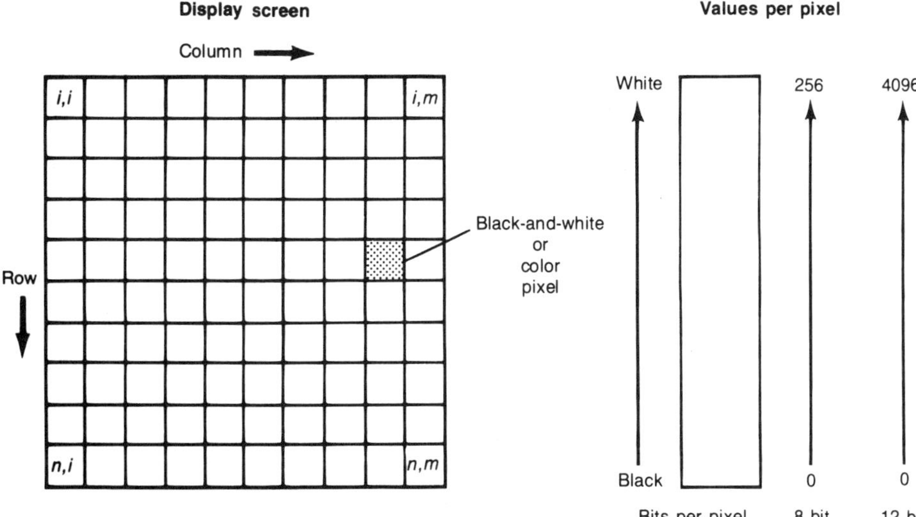

Fig. 2 Concept of image display (a) and representation (b) in computer memory

Table 2 Typical characteristics of image digitization devices

Type		Image quality	Pixel size or pixel number resolution	Acquisition time per 1024^2 by 8-bit image	Interscene dynamic range	Image area, mm × mm (in. × in.)	Face plate illumination lx	Face plate illumination ftc
Point source and detector	Microdensitometer	Excellent	5–75 μm (200 μin.–0.0030 in.)(a)	≥15 min	≤4096	500 × 500 (20 × 20)	...	...
	Laser scanner	Excellent	35 or 75 μm (0.0014 or 0.0030 in.)(a)	≥1 min	≤4096	430 × 350 (17 × 14)	...	...
Linear detector	Linear CCD array	Good	≤4096	≥500 ms	>1500	...	...	...
	Photodiode array	Good	≤1024	≥5 s	1000	...	...	...
Solid-state cameras	CCD	Fair	1320 × 1035	≥140 ms	500	...	1.5	0.14
	CID	Fair	776 × 512	33 ms(b)	200	...	0.1	0.009
Tube type cameras (vidicon family)	Vidicon (Sb_2S_3)	Fair	>1500	33 ms, lag = 20%	80–200	...	5	0.5
	Newvicon (ZnSe)	Fair	800	33 ms, lag = 10–20%	50–200	...	0.5	0.05
	Pasecon/Chalnicon (CdSe)	Fair	1600	33 ms, lag = 5–10%	30–60	...	1	0.09
	Saticon (Se+Te+As)	Fair	1200	33 ms, lag = 3%	50–160	...	6	0.6
	Plumbicon (PbO)	Fair	1200	33 ms, lag = 4%	80–160	...	2.5	0.23
	SIT (Silicon)	Fair	750	33 ms, lag = 7%	60	...	0.01	0.0009

(a) Aperture. (b) 776 × 512

primarily used in radiation environments or in specialized applications, such as high-resolution real-time radiography, in which the frame rate is higher than that achievable by current CCD designs. Charge-coupled device camera design is rapidly evolving for high-resolution scientific use and can be expected to improve with regard to real-time frame rates (Ref 12, 15).

It should be noted that a 355 × 432 mm (14 × 17 in.) film digitized at a resolution of 50 μm (0.002 in.) with 12-bit accuracy will consume a 92-Mbyte file. Just writing or reading this file to or from a hard disk could take up to 8 min (optical disks take even longer). Even with high-density optical disks and data compression, the digital storing of high-resolution radiographs represents a formidable problem.

Image Processing

Prior to image processing, it may be necessary to perform some type of preprocessing on the data. Most often, an image file will need to be converted into the workstation standard from some other type of format; for example, the files may be 8-, 16-, or 24-bit uncompressed format with a variable-length header and need to be converted to 8-bit format with no header. Typically, a problem arises when a file must be read that is in an unknown size, a compressed format, or has a limited color or nonstandard palette.

Other preprocessing functions operate on the raw data in some manner before inputting it for image processing. For example, the system should have the capability to average, filter, and acquire full frames and control the scanning parameters on the digitizer so that the noise level can be minimized. In other cases, more complicated operations are called for, such as tomographic reconstruction, synthetic aperture operations, or an FFT. Shown in Fig. 1(a) and (b) in the section "Use of Color for NDE" in this Volume are tomographs taken before and after oversampling by a factor of four and Wiener filtering of the input data set to correct for the point spread function (Ref 16). The preprocessing operation greatly increases the smoothness of the lines and reduces the artifacts.

Image Enhancement

There are three major scientific applications for image processing:

- Enhancement of an image to facilitate viewing
- Manipulation and restoration of an image
- Measurement and separation of features

These functions are listed in Table 3. Many of the functions have a dual purpose and, as will be shown, are usually combined to form other functions.

Table 3 Image-processing software algorithms useful for NDE applications

Image enhancement	Image operations	Information extraction
Contrast stretching	Scaling	Image statistics
Histogram equalization	Translation	Point, line, angle perimeter, area, measurement
Contouring	Rotating	FFT transformation, one and two dimensional
Thresholding	Registration	Correlation
Composite image building	Warping	Edge detection
Palette operations	Combining	Deblurring
Color model selection	Filtering	Motion restoration
True color representation	Thickening, thinning	Pattern recognition
	Noise cleaning	
	Trend removal	

Contrast Stretching and Histogram Equalization. The concept of contrast stretching is shown in Fig. 3(a). The basic operation is given by the pixel value transform:

$$s = T(r)$$
$$s = T(r) = ar + b \qquad \text{Linear}$$
$$T(r) = a \log(r) + b \qquad \text{Logarithmic} \qquad \text{(Eq 1)}$$

where r is the original pixel value and s is the transformed pixel value.

The linear stretch and offset method is the most commonly applied, while transforms of the nonlinear type can be used to convert film density to integrated dose or to linearize the film transfer function to account for base density. Although the manual contrast stretch is often used, a more automated contrast stretch is available with an operation known as histogram equalization. The general form is:

$$P_r(k) = \frac{n_k}{n}$$

$$s_k = s_{max} \sum_{j=0}^{k} P_r(j) \qquad \text{Linear}$$

$$s_k = -\frac{1}{a} \ln \left[1 - \sum_{j=0}^{k} P_r(j)\right] \qquad \text{Exponential}$$

$$s_k = s_{max} \left(\sum_{j=0}^{k} P_r(j)\right)^3 \qquad \text{Cube} \qquad \text{(Eq 2)}$$

A simple linear equalization example is shown in Fig. 3(b). Occasionally, it may be desirable to use the cubic or logarithmic equalization. The cubic emphasizes the lower values, making the image darker, while the exponential makes the image much brighter. Other types of transfer functions are often used to remove nonlinear response in the imaging or camera system. For isolated defects, a manual stretch usually gives good results. On textured materials, however, histogram equalization often works well. An example of a linear histogram equalization of a low-contrast ultra-

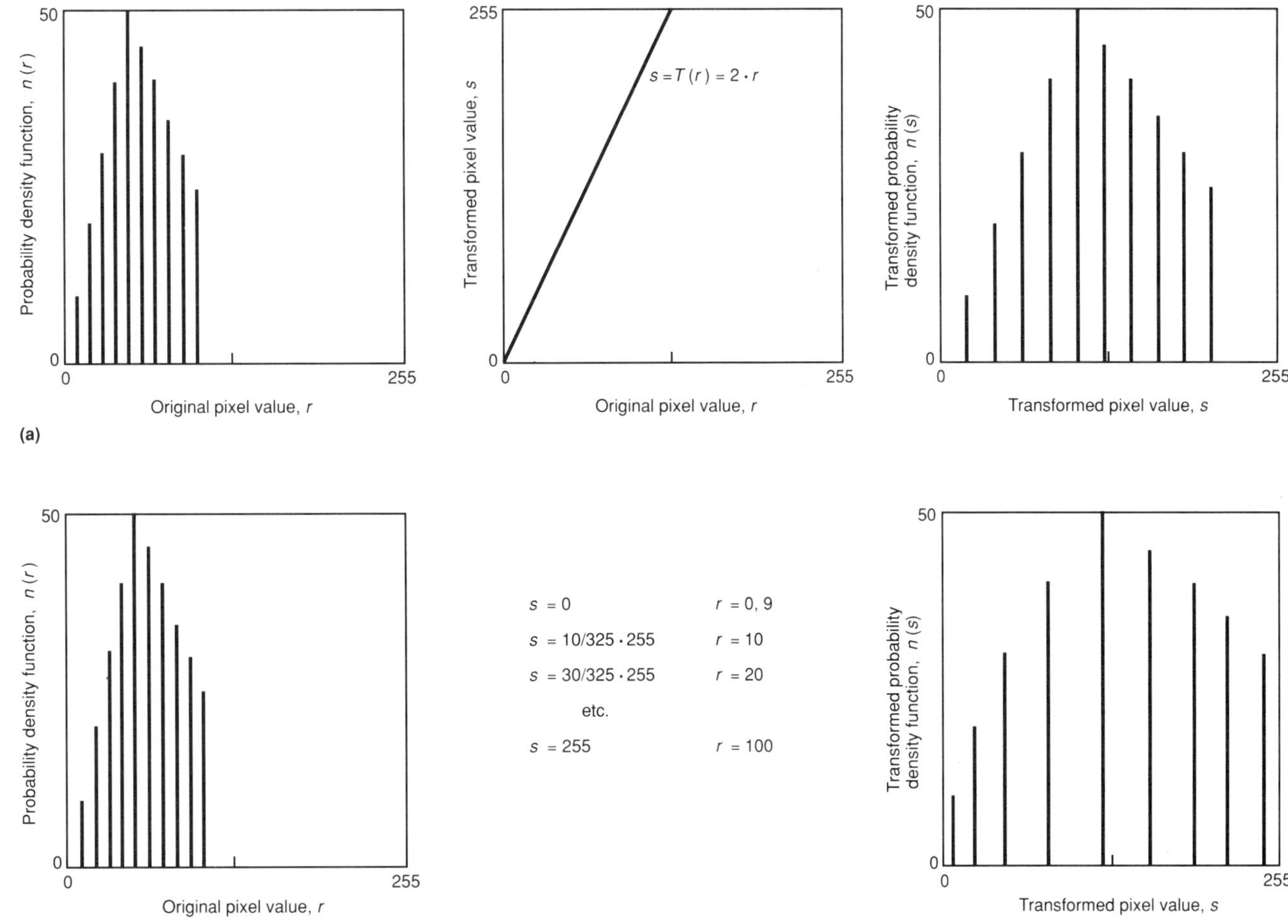

Fig. 3 Concept of histogram stretching and equalization. (a) Histogram is stretched with a linear transformation. (b) Histogram is equalized such that the probability density is constant.

sonic image of a bond line in an explosively bonded steel-to-aluminum plate is shown in Fig. 2 in the section "Use of Color for NDE" in this Volume.

Contouring and Thresholding. Contouring and thresholding are often combined with palette operations to show gradations in intensity or to highlight defects. Color should be used to detect gradients in intensity because the eye is sensitive to many more colors (approximately 10^4 shades) than gray levels. Thresholding can be done with a binary picture (black and white), a color scale, or a combination of gray and color. The concept of thresholding has particular application in the generation of a mask for future image-processing operations or the removal of the background. Contouring is used to change only those values between two pixel levels to a certain value. It can also be used in outlining high-definition features before further processing.

Composite Image Building. The ability to display composite images is useful in comparing the difference between images produced by different modalities, such as a radiograph and an ultrasonic or NMR image. Because image quality suffers greatly when displayed at resolutions of less than 512^2 it is imperative to have at least a 1024^2 display or to use two monitors if detailed comparisons are to be made. Figure 4 illustrates how x-ray radiographs and neutron radiographs can be displayed together and combined to enhance voids in a composite material. As shown in Fig. 4(c), the images have simply been added to enhance the detection of voids. The absorption of neutrons in the epoxy is higher than in the alumina and the reverse is true for the absorption of x-rays. However, the voids do not absorb in either case, and the contrast is therefore improved if the images are added.

Palette Control. Color can be displayed as a true color map or as a pseudocolor representation. In a pseudocolor representation, the digital data values are mapped to any color that can be produced by the combination of the red, green, and blue (RGB) values specified by a look-up table (LUT). A typical 8-bit pseudocolor system is shown in Fig. 5(a). With only an 8-bit-deep pixel as shown in Fig. 2, a type of true color is possible; however, each color is represented only by eight or fewer color levels (Fig. 5b). With 8-bits per color (as in Fig. 5c), the colors can be shaded continuously and are suitable for three-dimensional displays in which depth cueing is indicated by decreasing the luminance. Eight-bit pixels with 24-bit LUTs can produce good color scenes by adjusting the LUT to include only those colors that appear in a scene. However, the palette is then image specific.

Several types of pseudocolor and linear gray-scale palettes are shown in Fig. 4 in the section "Use of Color for NDE" in this Volume. The spectrum palette (or inverse spectrum) is preferred for range of color, but is not suitable for displaying rapid changes in pixel level. Other palettes, such as the complementary palettes shown in Fig. 4(d) and (e), can be used with good results on figures that have gradual changes in gray level. A typical threshold palette is shown in Fig. 4(b). All values below 192 will appear as a gray scale, while those equal to or above 192 will appear as red. The gray scale is still used to display lower-intensity data so that the operator can still detect unusual, yet not rejectable, anomalies. In contrast stretching, when further work on

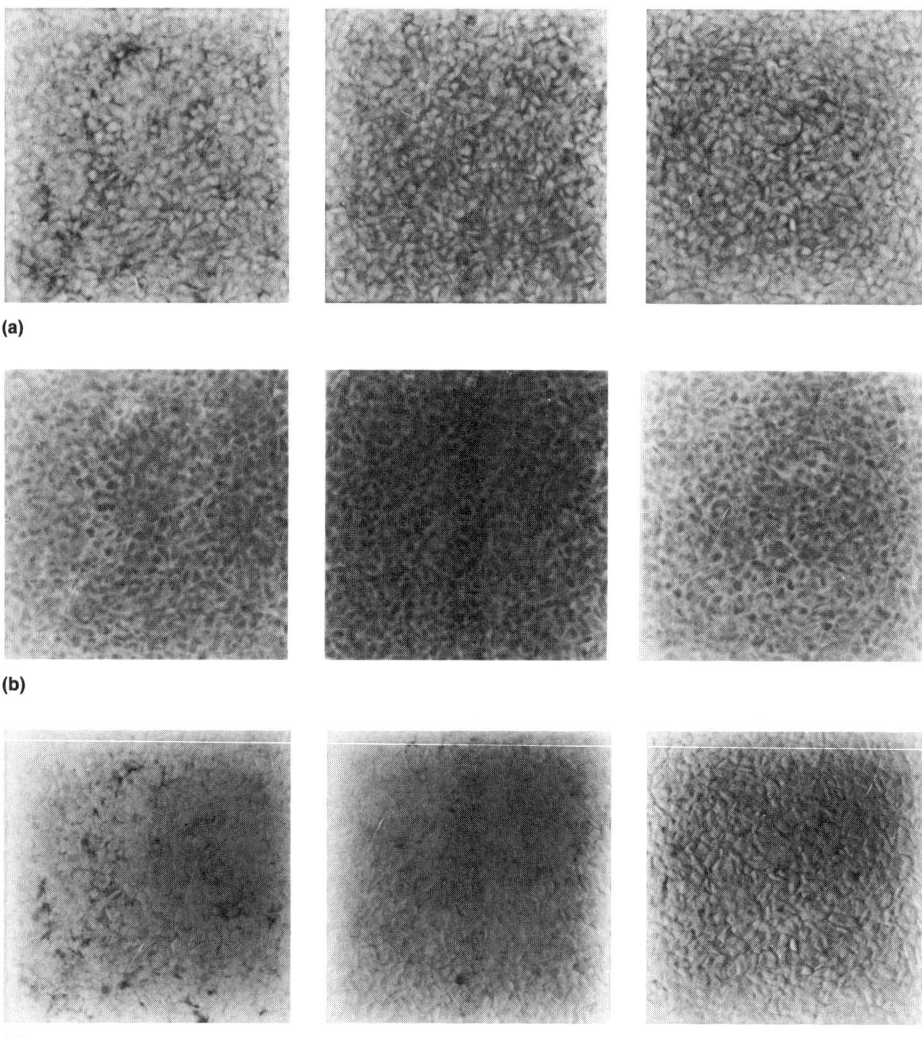

Fig. 4 Voids detected in three 60 × 60 × 10 mm (2.4 × 2.4 × 1 in.) alumina-filled epoxy tiles using three different imaging techniques. (a) Conventional radiography. (b) Neutron radiography. (c) Combined x-ray and neutron radiographic image. Building a composite image of both modalities indicates that there are voids in the sample. The addition of the two images enhances the voids because of the differential absorption in the matrix between the x-ray and neutron images.

the image is anticipated, the palette (that is, LUT values) is altered, leaving the pixel values intact.

Color Models and the Use of Color. There are two main color models: the red, green, blue (RGB) model and the hue, saturation, luminance (HSL) model. High-resolution imaging systems use the RGB model, while the National Television Systems Committee has adopted the HSL model for broadcast television. The colors in the RGB model are simply an additive mixture of the Commission Internationale de l'Éclairage monochromatic primary colors (red = 700 nm, or 7000 Å; green = 546.1 nm, or 5461 Å; and blue = 435.8 nm, or 4358 Å). When equal intensities of these colors are combined, a nearly white light results. In the HSL model, the primaries are transformed such that the hue value represents the color (0 to 360°; red = 0, 360; green = 120; and blue = 240), the saturation is the color intensity (values 0 to 1), and the luminance is the amount of brightness (0 to 1). If the saturation is 0 and the luminance is 1, the color will be white. The advantage of the HSL model is that a single hue can be manipulated in intensity more easily than with the RGB model.

Although both color and black-and-white images contain the same information, there are features that are much more obvious in color plots than in corresponding black-and-white plots, and vice versa. In particular, a black-and-white representation is appropriate when there is a high spatial frequency inherent in the image; in such a case, the mind-eye combination interprets this as a texture. On the other hand, color is preferred when there are a few isolated objects or low spatial frequency. This makes it easier for the image interpreter to discern small changes in image density through the use of contrast enhancement. Black-and-white images are shown in Fig. 6 and 7, and the color versions are shown in Fig. 6(c) and (d) in the section "Use of Color for NDE" in this Volume.

Image Operations

Listed in the second column of Table 3 are some of the more important image operations that are routinely used to manipulate images.

Geometric Processes. The scaling, rotation, translation, warping, and registration of an image are classified as geometric processes.

The scaling of an image is an important function because it is often used before images are combined. Scaling is used to magnify or shrink the images permanently to fit printer sizes or simply to view areas closely.

Rotation and translation are important mainly if two images need to be registered closely and combined or compared. The operations are described below. The scaling operation can involve interpolation, or it can merely involve a duplication of pixels. In the case of integer scaling (×2, ×3), the pixels are replicated in the x and y by the integer. For noninteger scaling (such as magnification by 1.5), every other pixel is replicated. Image interpolation and noninteger scaling should be avoided unless the image is subsequently filtered or unless interpolation and scaling are the last steps in a processing chain before printing. The printer will often perform interpolation and dithering (adding a small random number to the value) to obtain an image with a smoother appearance:

$$\begin{vmatrix} u \\ v \end{vmatrix} = \begin{vmatrix} \cos\theta & -\sin\theta \\ \sin\theta & \cos\theta \end{vmatrix} \begin{vmatrix} x \\ y \end{vmatrix} \quad \text{Rotation}$$

$$\begin{matrix} ab \\ cd \end{matrix} \rightarrow \begin{matrix} aabb \\ aabb \\ ccdd \\ ccdd \end{matrix} \quad \text{Scale} \times 2 \qquad \text{(Eq 3)}$$

Warping is a nonlinear scaling process. The equation for a sixth-order warp is:

$$u = a_1 + a_2x + a_3y + a_4xy + a_5x^2 + a_6y^2$$
$$v = b_1 + b_2x + b_3y + b_4xy + b_5x^2 + b_6y^2 \quad \text{(Eq 4)}$$

This transform can be of use if an object is at an angle to the viewing plane and it is necessary to measure a feature of interest or to correct the image for the viewing angle.

Image registration can be performed by trial and error with the rotation, translation, and scaling functions. The goodness of fit can be calculated with the correlation function or by subtraction. Registration is usually a prior step to image combination.

Image Combination. Image combination operations generally include the following: AND, NAND, OR, NOR, XOR, difference, subtract, add, multiply, divide, and average.

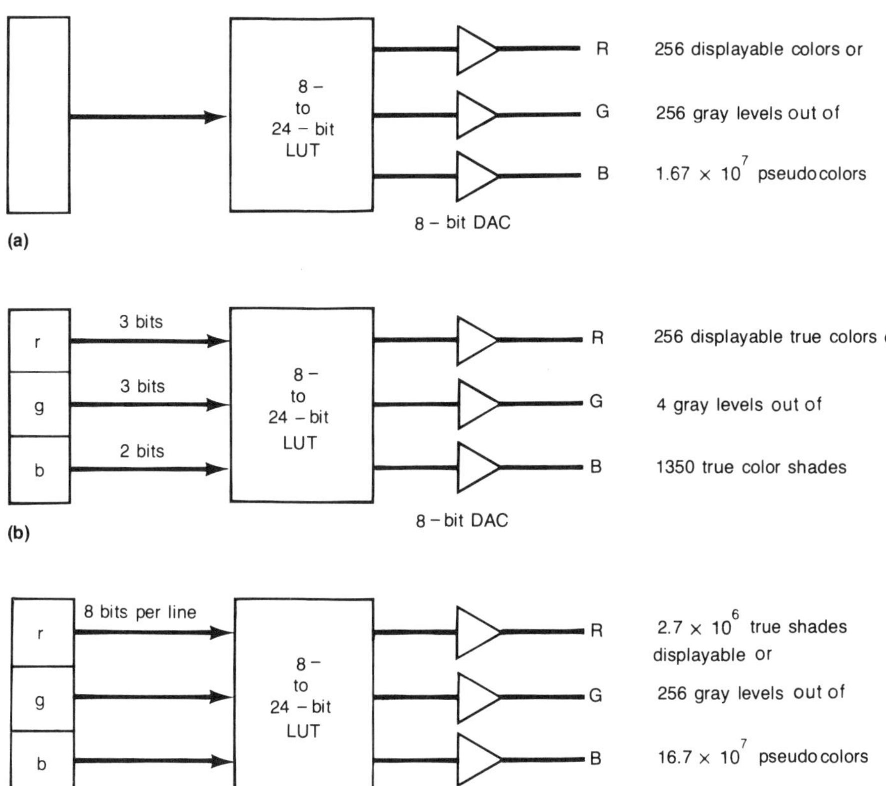

Fig. 5 Concept of a look-up table for three types of color displays. (a) 8-bit pseudocolor representation. (b) 8-bit true color map. (c) 24-bit true color map. The LUT is controlled by the palette, which specifies what color composition (R,G,B) will be assigned to one of 256 levels, as in (a). The digital-to-analog converter (DAC) converts the digital signal to RGB for input to the monitor. Shown in (b) is a way to display true color with an 8-bit pixel. Better true color can be obtained by using a 24-bit pixel, with each byte (8 bits) assigned to a specific primary color, as in (c).

All these operations require two operands. One will be the data in the primary image, the other can be either a constant or data in a secondary image. The image combination operations are used to mask certain areas of images, to outline features, to eliminate backgrounds, to search for commonality, and/or to combine images. For example, a video of a part can be taken, edge enhanced, boosted, ANDed with 128 added to 127 and scaled and registered with a radiograph or other image, and then added to the other image.

Filtering. There are many types of spatial filters. The most useful are low pass, high pass, edge, noise, and morphological, as listed in Table 4.

Many filters are classified as convolution filters. The convolution filter uses the discrete form of the two-dimensional convolution integral to compute new pixel values. In Eq 5, the image matrix is defined as $f(x,y)$, and the filter is a small 3×3 or (in general, $m \times n$) matrix called a kernel, K. In practice, the kernel is limited to a size of less than 20×20 because the FFT filtering procedure becomes faster than direct convolution for large kernels. The discrete form of the convolution integral is:

$$g(x,y) = \sum_{m=0}^{M-1} \sum_{n=0}^{N-1} f(m,n)\, K_{(x-m,\ y-n)} \qquad \text{(Eq 5)}$$

Because kernels are often symmetric about a 180° rotation about the center pixel, algorithms that use the correlation integral will give identical results to the convolution operation. The discrete form of the correlation integral is:

$$g(x,y) = \sum_{m=0}^{M-1} \sum_{n=0}^{N-1} f(m,n)\, K_{(x+m,\ y+n)} \qquad \text{(Eq 6)}$$

In this representation, one can visualize the kernel sliding over the image, with each pixel under the kernel being multiplied by the kernel value and then all values being summed to produce the new center pixel. It is customary to perform the convolution such that the initial and final image sizes are identical, with the outer row and column of the image not convolved with the kernel. Some of the most useful convolution kernels are shown in Fig. 8. Those kernels that add to 0, such as the Laplacian and the horizontal and vertical edge kernels, will remove most of the image except the edge components. The low-pass and noise reduction filters will blur the image and remove point noise, while the high-pass filters tend to sharpen the images and accentuate noise.

Other filters that are useful in image processing are the median, erosion, dilation, and unsharp mask filters. The median filter selects the median value in an image neighborhood (usually 3×3) and replaces the center pixel with the median value. This filter can be very effective in removing isolated pixel noise without blurring the image as severely as the low-pass filter. The effect of a low-pass and a median filter is shown on a noisy ultrasonic image of a hot-rolled plate in Fig. 3(a), (b), and (c) in the section "Use of Color for NDE" in this Volume. The erosion and dilation filters work by replacing the center pixel value in a neighborhood with the smallest or largest index in the neighborhood, respectively. They can be used to thicken or to thin boundaries. The unsharp mask filter algorithm is:

$$g(x,y) = 2 \cdot f(x,y) - LPF(f)_{xy} \qquad \text{(Eq 7)}$$

where LPF is the low-pass filter. The new image is the difference between the original image multiplied by a factor and the low-pass filtered image. This operation tends to enhance the contrast and to sharpen the edges slightly.

The Roberts and Sobel filters are similar types of edge detectors. The Roberts filter algorithm is:

$$g(x,y) = \{[f(x,y) - f(x+1,y+1)]^2 + [(f(x,y+1) - f(x+1,y)]^2\}^{1/2} \qquad \text{(Eq 8a)}$$

This algorithm operates on a 2×2 neighborhood and will enhance high frequencies, while the Sobel filter operates with a 3×3 kernel and will produce good (but thick) outlines of images. The Sobel filter algorithm is:

$$\begin{aligned} g(x,y) = (\{[(f(x+1,y-1) + 2 \cdot f(x+1,y) \\ + f(x+1,y+1)] - [(f(x-1,y-1) \\ + 2 \cdot f(x-1,y) + f(x-1,y+1)]\}^2 \\ + \{[f(x-1,y-1) + 2 \cdot f(x,y-1) \\ + f(x+1,y-1)] - [f(x-1,y+1) \\ + 2 \cdot f(x,y+1) + f(x+1,y+1)]\}^2)^{1/2} \end{aligned} \qquad \text{(Eq 8b)}$$

A summary of the effect of various filters and edge detectors is shown in Fig. 9.

There is a class of filters that restore image quality by considering *a priori* knowledge of the noise or the degradation mechanism. An example is the inverse filter. In this case, the image has been degraded and is reestimated by the use of the inverted degrading transfer function and a noise model to minimize the least square error. This type of filter can be useful if a transfer function model is derived for the imaging chain.

For noise cleaning, the filters listed in Table 4 and the median filter can be used. Fast Fourier transform techniques are particularly effective for the removal of periodic noise, especially filtering in the frequency domain.

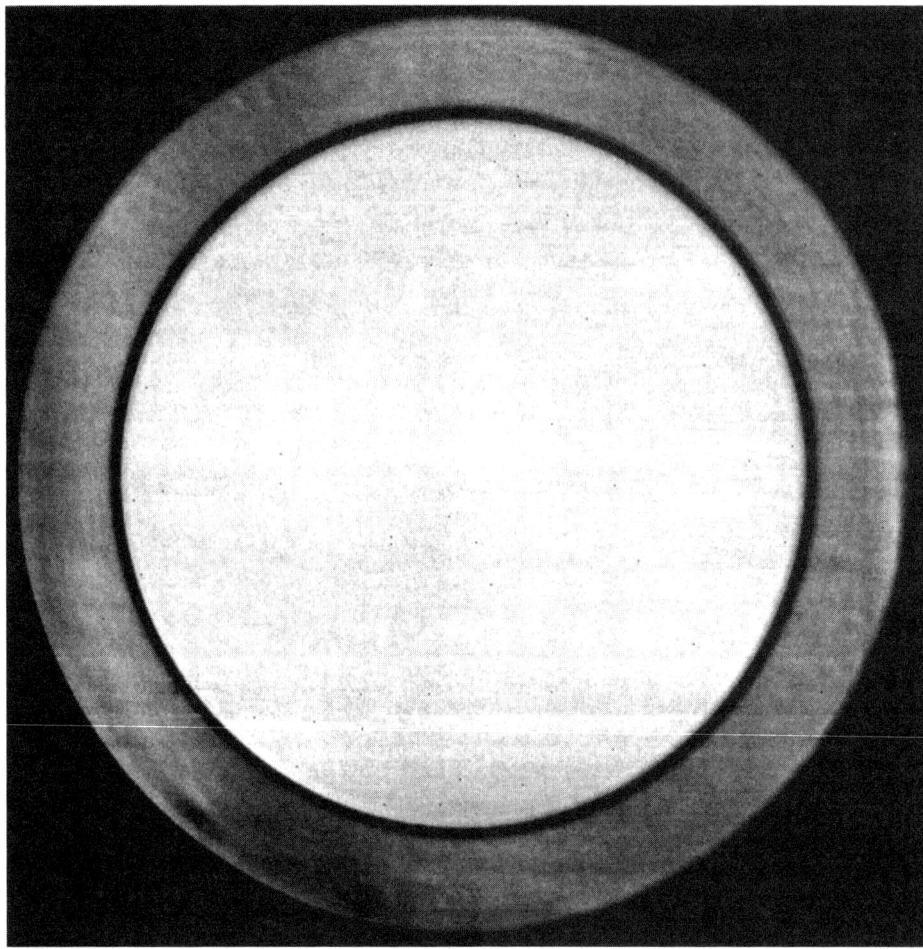

Fig. 6 Ultrasonic image of a hot isostatic pressed tungsten plate prepared from powder (99% dense). The black-and-white image shows small changes in density that are not easily discernible. The density changes are enhanced by the use of color, as shown in Fig. 6(c) of the section "Use of Color for NDE" in this Volume. The ring is caused by a transition from one thickness to another (the outer thickness was about 50% of the inner circle).

Trend removal, or field flattening, is an important processing function to perform before contrast enhancement. In most cases, an x-ray film will not have a uniform density even if the part is uniform. In addition, variations in density may occur when the part is digitized with a camera. When the technique of contrast enhancement is attempted, parts of the image will saturate, and defects will be difficult to detect, as in the ceramic disk shown in Fig. 6(a) of the section "Use of Color for NDE" in this Volume. Listed below are techniques used to reduce the effect:

1. High-pass filter using an FFT
2. Low-pass filter using a large kernel or FFT, and subtracted from the original image
3. Polynominal fitted to image line and fitted curve subtracted from original

Most of these algorithms produce artifacts and are not entirely satisfactory, especially if there is a high spatial frequency superimposed on a low-frequency trend. Figure 6(b) of the section "Use of Color for NDE" in this Volume shows the results of using technique 2 (similar to unsharp masking) on a noisy image with a top-to-bottom trend.

Information Extraction

Listed in the third column of Table 3 are image-processing functions that extract quantitative information concerning images.

Image Statistics and Measurement. The most useful image statistic needed for subsequent processing is the histogram or the probability density function. This will define the dynamic range of the image and determine subsequent processing. Also of use is the line profile and frequency analysis of the line or small area of interest. The line profile is of use when quantitative analysis needs to be performed on an image—for example, measurement of the MTF of the imaging system or the variation in pixel value across a particular defect (such as a void). The line profile can also be useful in determining the amount of trend present in an image.

Simple measurement functions that are often used are:

- Determination of the pixel value at a particular point (such as on a defect)
- Conversion of the pixel value to film density or part density
- The distance from one pixel to another in terms of pixel units or engineering units
- The angle of one line with respect to another

It is very desirable to be able to measure both the perimeter and the area of a defect. In many cases, such as a part with many voids, it may be desirable to obtain the total area of all defects above (or below) a threshold and their radius in terms of pixel values or engineering units and then to produce a histogram of the defect radii. This type of quantitative analysis is useful for material evaluation as a function of processing variables.

Fast Fourier Transform. The FFT algorithm is central to many filtering and information extraction schemes. The two-dimensional FFT is analogous to the one-dimensional FFT in that the function extracts frequency-dependent information from the waveform or image. A discrete form of the two-dimensional FFT algorithm is:

$$F(x,y) = \sum_{m=0}^{M-1} \sum_{n=0}^{N-1} f(m,n) \exp \left[-j2\pi \left(\frac{mx}{M} + \frac{ny}{N} \right) \right] \quad \text{(Eq 9)}$$

Specific methods for calculating Eq 9 are given in Ref 6 and 17.

The most direct uses of the FFT and inverse FFT are as filters to eliminate periodic noise from an image. The noise may be a periodic pattern, as in the case of the filament-wound vessel shown in Fig. 10(a), or it could be 60-cycle noise. The general filter procedure is shown in Fig. 10. The image is simply transformed, edited, and inverse transformed. In Fig. 10(a), the original radiograph shows a section of a filament-wound vessel with a cut in the windings. After the transform, the signal from the filament-wound structure is seen to radiate from the center at 45° (Fig. 10b). The filament-wound structure is removed from the image by removing the frequency components of the structure as shown in Fig. 10(c). As shown in Fig. 10(c) the zero frequency component (center pixel) is left intact to restore the average gray-level value. The image is then inverse transformed to recreate the original image minus the 45° filament pattern. Figure 10(d) shows the pattern after the inverse transform; the cut section is readily apparent.

A correlation function can be used directly to detect objects or features of an image that are of interest or to align images for combination. If two images that are

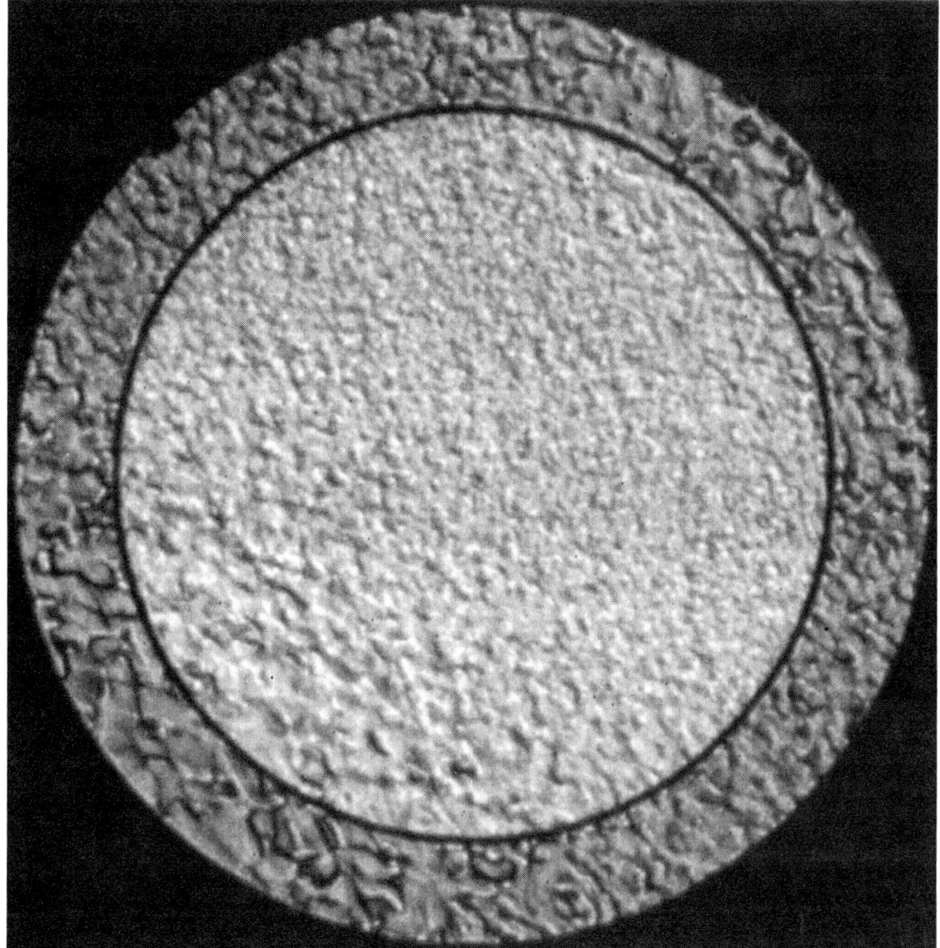

Fig. 7 Tungsten plate similar to the one shown in Fig. 6, except that this part was fabricated from a plate that was rolled rather than hot isostatic pressed. The black-and-white image shows a texture imparted to the material due to the inclusion of 2% porosity. The color version of the plate, Fig. 6(d) in the section "Use of Color for NDE" in this Volume, shows how difficult the texture is to interpret in a color representation using a spectrum palette.

$$\begin{bmatrix} -1 & -1 & -1 \\ -1 & 9 & -1 \\ -1 & -1 & -1 \end{bmatrix} = 1 \quad \begin{bmatrix} 0 & -1 & 0 \\ -1 & 5 & -1 \\ 0 & -1 & 0 \end{bmatrix} = 1 \quad \frac{1}{8}\begin{bmatrix} -1 & -1 & -1 \\ -1 & 16 & -1 \\ -1 & -1 & -1 \end{bmatrix} = 1$$

High pass

$$\frac{1}{9}\begin{bmatrix} 1 & 1 & 1 \\ 1 & 1 & 1 \\ 1 & 1 & 1 \end{bmatrix} = 1 \quad \begin{bmatrix} -1 & -1 & -1 \\ 1 & 1 & 1 \\ 0 & 0 & 0 \end{bmatrix} = 0 \quad \begin{bmatrix} -1 & 1 & 0 \\ -1 & 1 & 0 \\ -1 & 1 & 0 \end{bmatrix} = 0$$

Low pass — Horizontal edge — Vertical edge

$$\begin{bmatrix} -1 & -1 & -1 \\ -1 & 8 & -1 \\ -1 & -1 & -1 \end{bmatrix} = 0 \quad \begin{bmatrix} 1 & 1 & 1 \\ -1 & -2 & 1 \\ -1 & -1 & 1 \end{bmatrix} = 0$$

Laplacian — Compass (NE)

$$\frac{1}{9}\begin{bmatrix} 1 & 1 & 1 \\ 1 & 1 & 1 \\ 1 & 1 & 1 \end{bmatrix} = 1 \quad \frac{1}{10}\begin{bmatrix} 1 & 1 & 1 \\ 1 & 2 & 1 \\ 1 & 1 & 1 \end{bmatrix} = 1$$

Noise removal

Fig. 8 Various convolution kernels and their use

generated from different modalities are to be combined and a registration mark is not present, the two images can be correlated over a region of interest and moved until the correlation coefficient peaks. This is best accomplished in images having edges or texture, such as those shown in Fig. 4 and 7.

The edge detection filters described in the section "Image Operations" in this article are often used to detect or outline images; in particular, the Sobel filter works well on high-definition or noisy edges. In radiographs having noisy, fuzzy, and low-contrast edges, the Laplacian, Roberts, Sobel, and simple convolution filters will usually not produce good results. Edge followers, Hough transforms, and other *ad hoc* algorithms that depend on knowledge of the rough shape of the curve (template matching) to be outlined work satisfactorily in most cases. These algorithms, although of some use, are usually not part of a processing package because of their specialized applications.

Table 4 Commonly used image filters

Convolution filters	Other filters
High pass	Median
Low pass	Erosion
Horizontal edge	Dilation
Vertical edge	Unsharp mask
Laplacian	Roberts
Compass	Sobel or Prewitt
	Wiener

Advanced Functions. General formulations of image restoration functions such as motion restoration, deblurring, maximum entropy, and pattern recognition are not commonly found in general image-processing software packages, probably because of the computation required and the limited demand for these applications. These algorithms are discussed in Ref 3 (motion restoration), Ref 1 and 3 (deblurring), Ref 9 (maximum entropy), and Ref 1 and 10 (pattern recognition). The algorithms are of use in the restoration of flash radiographs or high-speed video as well as the correction of point-spread function in tomography or ultrasonics. Maximum entropy methods may be of use in the enhancement of NMR signals and other low signal-to-noise data, while pattern recognition can be of use in automating the detection of flawed, incomplete, or misassembled parts.

Image Display

A wide variety of devices and techniques are available for the display and output of the processed data.

CRT Displays. The CRT display is the standard means of output for all imaging applications because the range of colors, brightness, and dynamic range cannot be matched by any other display technology. Black-and-white monitors can have resolutions of up to 2000 × 1500 pixels, bandwidths of the order of 200 MHz, and a brightness of 515 cd/m^2 (150 ft-L) at a dynamic range in excess of 256. Color monitors may have 1280 × 1024 pixels and bandwidths of 250 MHz. High-resolution color monitors have a brightness of 85 cd/m^2 (25 ft-L) or more, with a dynamic range of 256 when viewed in dim light.

Recent advances in tube design have darkened the shadow mask by a factor of four to allow increased room illumination with no degradation in dynamic range. The only other major advance in CRT technology is the recent introduction of flat-screen tubes with resolutions up to 640 × 480 pixels.

The standards for video output are the RS-170, which specifies 525 lines of interlaced video at a frame rate of 30 Hz,

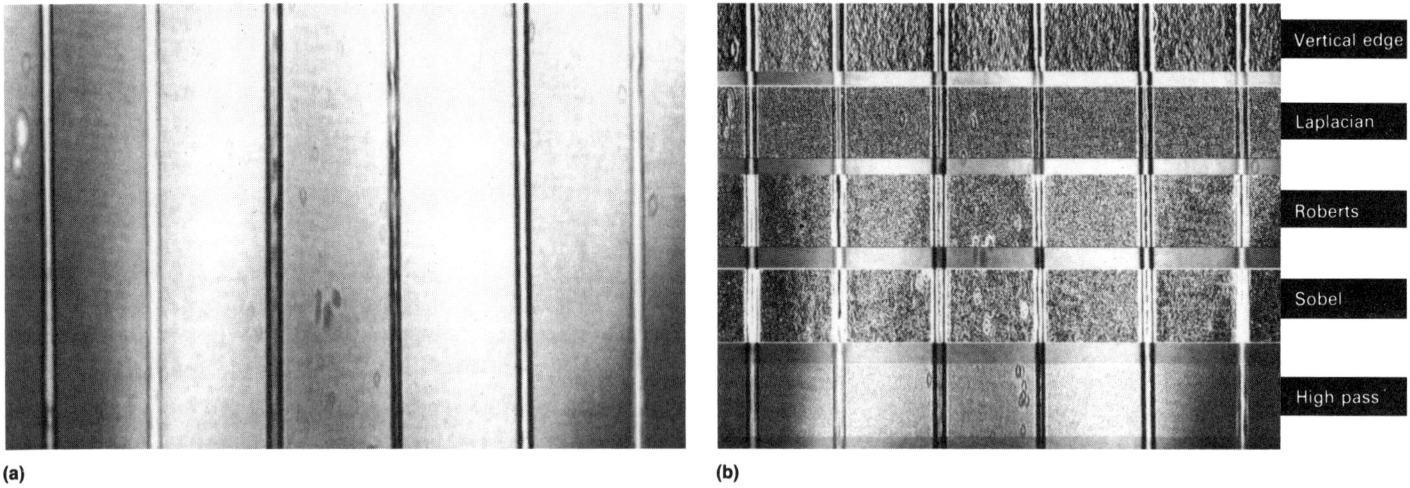

Fig. 9 Ultrasonic image of an adhesive bond (250 μm, or 0.01 in., thick) between two opaque plastic parts. (a) Unfiltered. (b) Filtered with numerous filter types (vertical edge, high pass, Laplacian, Roberts, and Sobel)

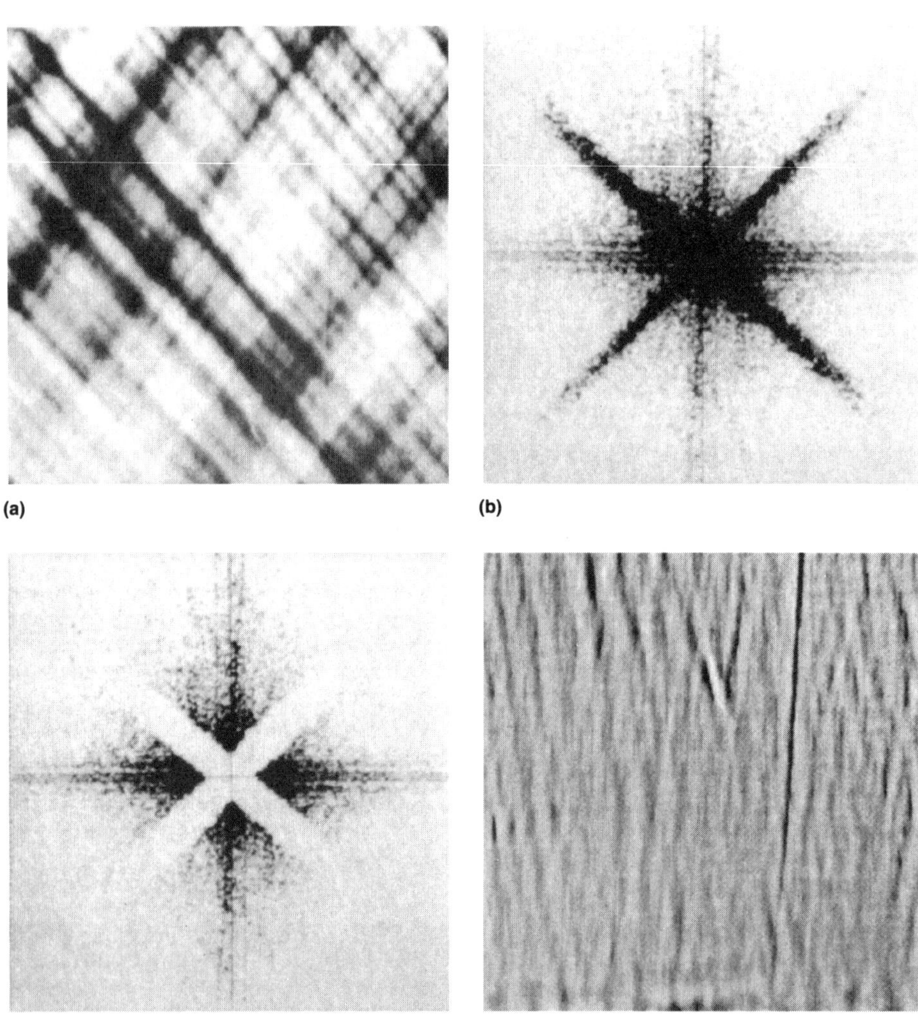

Fig. 10 General procedure used to filter an image with the FFT. (a) Original image. (b) Image transformed to the frequency domain. (c) Image edited. (d) Image inverse transformed into the filtered image

and RS-343-A, which specifies 1023 interlaced lines also at a 30-Hz frame rate. Most modern monitors have special circuitry that enables the monitor to sync to various standards with either the sync superimposed on the green signal or as a separate input. Most high-resolution (1280 × 1024) imaging systems use a 483 mm (19 in.) tube operating in a noninterlaced mode at 60 to 64 Hz for the primary display.

Table 5 Typical characteristics of color and black-and-white hard copy output devices

Device	Resolution, dpi	Image quality Paper	Film
Color			
Camera	...	Excellent	Very good
Thermal transfer	300	Very good	Very good
Ink-jet	180	Good	Fair
Black and white			
Camera	...	Excellent	Very good
Laser film	300	Very good	Very good
Thermal paper	300	Good	...
Thermal transfer	300	Fair	Fair
Laser printer	300	Poor	Poor
Ink-jet	180	Poor	Very poor

Printers. Various color and black-and-white printers are available for hard copy. A simple method of hard copy for color and black-and-white is to photograph the CRT screen or a small (80 mm, or $3\frac{5}{32}$ in.) very flat screen directly with an 8 × 11 in. or smaller Polaroid-type camera. The disadvantages of this method are the cost of the instant film in the large format and the small size of the image in the 4 × 5 in. format. However, the color quality of the image is excellent.

The two other types of color printers for small-system color output are ink-jet printers and thermal-transfer printers. The ink-jet printers are very inexpensive and produce vivid colors at 180 dpi on paper, but have poor contrast on transparency material. The thermal copiers are more expensive but produce high-resolution copies (300 dpi) by transferring color from a Mylar sheet onto paper or film with very good color and color density. The Polaroid cameras can reproduce all the colors seen on the screen, while the ink-jet and thermal-transfer print-

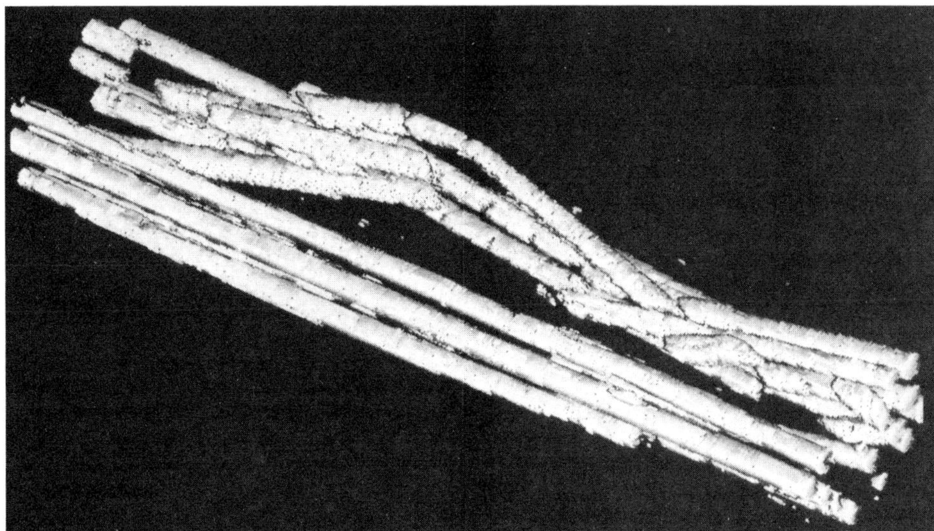

Fig. 11 Three-dimensional perspective plot representation of a bundle of failed, bent fuel rods. The representation was reconstructed from three-dimensional data sets obtained from a series of 250 individual 128 × 128 pixel tomographic slices of the rod bundle.

ers will produce at least 162 colors at 150 or 90 dpi, with 2 × 2 pixel enlargement.

Table 5 lists the possible color and black-and-white hard copy output devices suitable for printing images. Impact printers and plotters are not listed, because of the limited number of colors and poor resolution. Also not listed in Table 5 are the color printers based on the Xerox process.

There are many more black-and-white printers available than color; however, the high-quality printers are restricted to the camera type and the laser film printer. The laser film printer writes with a modulated laser directly on a special dry silver paper or film that is then developed with heat. These printers are capable of producing 64 shades of gray on paper and 128 shades on film at 300 dpi with very dense blacks. The other devices listed in Table 5 have poor resolution or low contrast. The other black-and-white printers produce poor plots because they are binary printers (a pixel may be black or white) at 300 dpi.

To achieve a gray scale, the printer must use a 2 × 2, 3 × 3 or other super pixel size, and this decreases the effective resolution. If a 2 × 2 pixel is used, the printer can print only five shades of gray at 150 dpi; if 3 × 3 pixels are used (100 dpi), ten shades are possible. The printers can actually produce many more shades by programming the device driver for a super pixel dithering. In this case, the program looks at a large super pixel of 5 × 5, determines whether or not the pixel should be smaller, and adjusts the dot density with a dither so that, effectively, about 26 or more shades of gray are perceived with a resolution of about 100 dpi.

Videotape and Videodisk. A common use of videotape is to input TV frames from a dynamic test to the image-processing system. For example, a high-speed video can be made of a pressure vessel as it bursts. Later, the sequential frames are captured and enhanced to determine the shape of the initiation crack or the speed of crack growth. Conversely, radiographs can be made periodically during the slow evolution of a part under the influence of some external variable (for example, void coalescence in ceramics as a function of temperature). The radiographs are then digitized and interpolated to form a time-compressed video.

Videodisks can be made in a write once read many format with relatively inexpensive machines. Videodisks offer better resolution and more convenient frame access than videotape. Another alternative for the display of dynamic data is to use the hard disk in the imaging workstation. Hard disk transfer rates (300 kbytes/s) allow small (200 × 200) movies to be shown on a workstation screen for tens of seconds (depends on disk size) at seven frames per second. Ultimately, it will be possible to expand compressed data quickly enough so that it can be read off a hard disk or a removable digital optical disk in near real time for significant durations.

The processing of tape can be tedious because of the amount of data required for a few seconds of video and also because of the limited resolution. However, the techniques of video data display are a very powerful enhancement tool because of the attraction of the eye to changing features.

Pseudo Three-Dimensional Images. All discussion of output data display has been in terms of a flat two-dimensional representation. This is the display of preference for processing because the algorithms are available or easily coded. However, depending on the data and the method of acquisition, other types of displays may be more appropriate. Shown in Fig. 10(a) and (b) in the section "Use of Color for NDE" in this Volume are two ways to present two-dimensional data in three dimensions. In Fig. 10(a), a two-dimensional ultrasonic microscopy image of a circuit board is shown with two defective metallic conductor pads (Ref 18). The image lines are drawn, as in the terrain-mapping method, with high values elevated with respect to low values and with a hidden line algorithm. Another way to present two-dimensional image data is shown in Fig. 10(b). In this case, the two-dimensional ultrasonic data were taken from a C-scan of a filament-wound sphere with Teflon shims imbedded in the matrix (Ref 19). Presenting the two-dimensional data in this manner results in a realistic display and facilitates interpretation.

Tomographic, ultrasonic, and NMR data sets may consist of many two-dimensional images that can be stacked to yield a picture of a true three-dimensional object. There are two ways to represent this three-dimensional data. In the first method, the data are modeled as a geometric solid or surface; in the second method, the data are displayed as raw data, or voxels (volume elements). The first method is used in solids-modeling workstations for design engineering. It can also be used to image NDE data, especially if the number of modeled elements is very large (10^5 or more) or has many surfaces. Figure 11 shows a three-dimensional reconstruction of a bundle of bent fuel rods after a reactor accident test. The image was constructed of 250 individual 128 × 128 pixel tomograms. A type of surface modeling that treats surfaces as polygons was used to model the rods. An inspection of the original tomograms shows that it is impossible to quickly grasp the shape of the bent rods from simple two-dimensional plots (Ref 20). Another three-dimensional reconstruction of this failed assembly can be found in Fig. 9 in the section "Use of Color for NDE" in this Volume.

Figure 8 in the section "Use of Color for NDE" shows a tomogram of a turbine blade that was displayed in terms of voxel imaging (also known as volume rendering), rather than modeling the surfaces by polygons or other shapes. Volume elements are created from the raw data, and these voxels are then projected to a view surface to produce an image. The raw data are not actually displayed. Figure 8 also demonstrates how the surfaces can be made transparent and the lighting model adjusted for a realistic effect. Graphics accelerators are available so that the images shown in Fig. 8 can be rotated, lighting sources adjusted, and textures added in almost real time (see Fig. 7 of the section "Use of Color for NDE" in this Volume). Three-dimensional image processing is in its infancy, but many medical and NDE applications are envisioned (Ref 21).

ACKNOWLEDGMENT

The authors would like to thank their colleagues at several national laboratories, universities, and industrial research departments for discussions concerning uses of enhancement and for contributions of images to the article.

REFERENCES

1. W.K. Pratt, *Digital Image Processing*, John Wiley & Sons, 1978
2. M.H. Jacoby, Image Data Analysis, in *Radiography & Radiation Testing*, Vol 3, 2nd ed., *Nondestructive Testing Handbook*, American Society for Nondestructive Testing, 1985
3. R.C. Gonzalez and P. Wintz, *Digital Image Processing*, Addison-Wesley, 1977
4. G.A. Baxes, *Digital Image Processing: A Practical Primer*, Prentice-Hall, 1984
5. W.B. Green, *Digital Image Processing: A Systems Approach*, Van Nostrand Reinhold, 1983
6. H.K. Huang, *Element of Digital Radiology: A Professional Handbook and Guide*, Prentice-Hall, 1987
7. A. Rosenfeld and A.A. Kak, *Digital Picture Processing*, Academic Press, 1982
8. K.R. Castleman, *Digital Image Processing*, Prentice-Hall, 1979
9. B.R. Frieden, Image Enhancement and Restoration, in *Picture Processing and Digital Filtering*, Vol 6, *Topics in Applied Physics*, T.S. Huang, Ed., Springer-Verlag, 1979, p 177
10. D.H. Janney and R.P. Kruger, Digital Image Analysis Applied to Industrial Nondestructive Evaluation and Automated Parts Assembly, *Int. Adv. Nondestr. Test.*, Vol 6, 1979, p 39-93
11. J.J. Dongarra, "Performance of Various Computers Using Standard Linear Equations Software in a Fortran Environment," Technical Memorandum 23, Argonne National Laboratory, 1989
12. J.R. Janesick, T. Elliott, S. Collins, M.M. Blouke, and J. Freeman, Scientific Charge Coupled Devices, *Opt. Eng.*, Vol 26 (No. 8), 1987, p 692-714
13. I.P. Csorba, *Image Tubes*, Howard W. Sams & Co., 1985
14. G.I. Yates, S.A. Jaramillo, V.H. Holmes, and J.P. Black, "Characterization of New FPS Vidicons for Scientific Imaging Applications," LA-11035-MS, US-37, Los Alamos National Laboratory, 1988
15. L.E. Rovich, *Imaging Processes and Materials*, Van Nostrand Reinhold, 1989
16. K. Thompson, private communication, Sandia National Laboratories, 1988
17. C.S. Burrus and T.W. Parks, *DFT/FFT and Convolution Algorithms: Theory and Implementation*, John Wiley & Sons, 1985
18. J. Gieske, private communication, Sandia National Laboratories, 1988
19. W.D. Brosey, Ultrasonic Analysis of Spherical Composite Test Specimens, *Compos. Sci. Technol.*, Vol 24, 1985, p 161-178; private communication, Sandia National Laboratories, 1988
20. C. Little, private communication, Sandia National Laboratories, 1988
21. A.R. Smith, *Geometry and Imaging: Two Distinct Kinds of Graphics*, to be published 1989; private communication

Acoustic Microscopy

Lawrence W. Kessler, Sonoscan, Inc.

ACOUSTIC MICROSCOPY is the general term applied to high-resolution, high-frequency ultrasonic inspection techniques that produce images of features beneath the surface of a sample. Because ultrasonic energy requires continuity of materials to propagate, internal defects such as voids, inclusions, delaminations, and cracks interfere with the transmission and/or reflection of ultrasound signals. Compared to conventional ultrasound imaging techniques, which operate in the 1 to 10 MHz frequency range, acoustic microscopes operate up to and beyond 1 GHz, where the wavelength is very short and the resolution correspondingly high. In the early stages of acoustic microscopy development, it was envisioned that the highest frequencies would dominate the applications. However, because of the high-attenuation properties of materials, the lower frequency range of 10 to 100 MHz is extensively used. Acoustic microscopy is recognized as a valuable tool for nondestructive inspection and materials characterization. Acoustic microscopy comprises three different methods:

- Scanning laser acoustic microscopy (SLAM), which was first discussed in the literature in 1970 (Ref 1)
- C-mode scanning acoustic microscopy (C-SAM), which is the improved version of the C-scan instrumentation (Ref 2)
- Scanning acoustic microscopy (SAM), which was first discussed in the literature in 1974 (Ref 3)

Each of these methods has a specific range of utility, and most often the methods are noncompetitive with regard to applications. That is, only one method will be best suited to a particular inspection problem.

Acoustic microscopes are practical tools that have emerged from the laboratory to find useful applications within industry. They can be applied to a broad range of problems that previously had no solutions, and they have been especially useful in solving problems with new high-technology materials and components not previously available. The three acoustic microscope types can be as different from each other as are microradiography and electron microscopy. This discussion should provide the potential user with an awareness of the techniques and their distinctions in order to maximize the opportunities for the successful use of acoustic microscopy.

Fundamentals of Acoustic Microscopy Methods

As a general comparison between the methods, the scanning laser acoustic microscope is primarily a transmission mode instrument that creates true real-time images of a sample throughout its entire thickness (reflection mode is sometimes employed). In operation, ultrasound is introduced to the bottom surface of the sample by a piezoelectric transducer, and the transmitted wave is detected on the top side by a rapidly scanning laser beam.

The other two types of microscopes are primarily reflection mode instruments that use a transducer with an acoustic lens to focus the wave at or below the sample surface. The transducer is mechanically translated (scanned) across the sample in a raster fashion to create the image. The C-mode scanning acoustic microscope can image several millimeters or more into most samples and is ideal for analyzing at a specific depth. Because of a very large top surface reflection from the sample, this type of microscope is not effective in the zone immediately below the surface unless the Rayleigh-wave mode to scan near-surface regions is used along with wide-aperture transducers. The scanning acoustic microscope uses this Rayleigh wave mode and is designed for very high resolution images of the surface and near-surface regions of a sample. Penetration depth is intrinsically limited, however, to only one wavelength of sound because of the geometry of the lens. For example, at 1 GHz, the penetration limit is about 1 μm (40 μin.). The C-mode scanning acoustic microscope is designed for moderate penetration into a sample, and transmission mode imaging is sometimes employed. This instrument uses a pulse-echo transducer, and the specific depth of view can be electronically gated. More detailed discussions of each acoustic microscopy technique follow.

SLAM Operating Principles

A collimated plane wave of continuous-wave (CW) ultrasound at frequencies up to several hundred megahertz is produced by a piezoelectric transducer located beneath the sample, as illustrated in Fig. 1. Because this ultrasound cannot travel through air (making it an excellent tool for crack, void, and disbond detection), a fluid couplant is used to bring the ultrasound to the sample. Distilled water, spectrophotometric-grade alcohol, or other more inert fluids can be

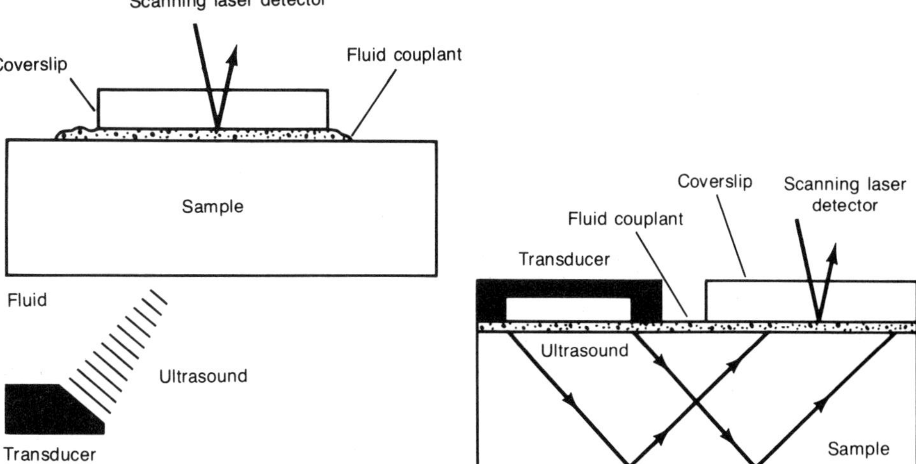

Fig. 1 Simplified view of the methods for producing through transmission and noncollinear reflection mode acoustic images with the scanning laser acoustic microscope

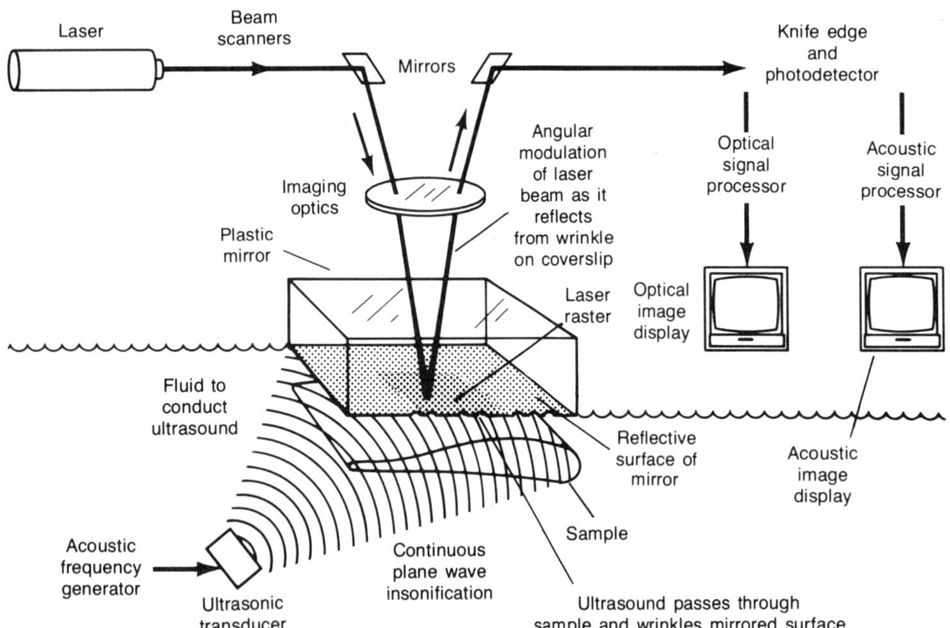

Fig. 2 Schematic showing principal components of a scanning laser acoustic microscope. The unit employs a plane wave piezoelectric transducer to generate the ultrasound and a focused laser beam as a point source detector of the ultrasonic signal. Acoustic images are produced at a rate of 30 images per second.

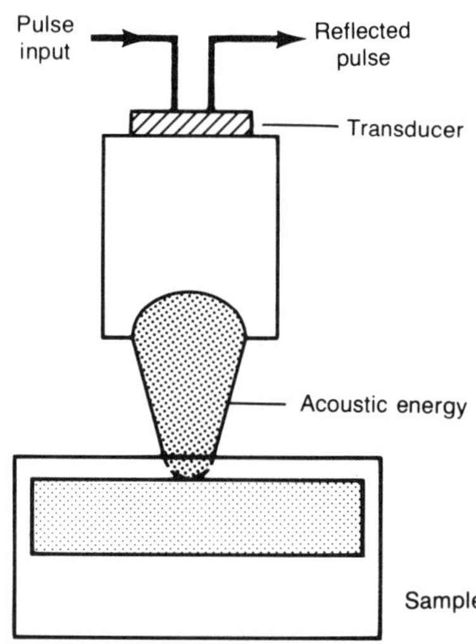

Fig. 3 Schematic of the C-mode scanning acoustic microscope. This instrument incorporates a reflection, pulse-echo technique that employs a focused transducer lens to generate and receive the ultrasound signals beneath the surface of the sample.

used, depending on user concerns for sample contamination. When the ultrasound travels through the sample, the wave is affected by the homogeneity of the material. Wherever there are anomalies, the ultrasound is differentially attenuated, and the resulting image reveals characteristic light and dark features, which correspond to the localized acoustic properties of the sample. Multiple views can be made to determine the specific depth of a defect, as is performed by stereoscopy (Ref 4).

A laser beam is used as an ultrasound detector by means of sensing the infinitesimal displacements (rippling) at the surface of the part created by the ultrasound. In typical samples that do not have polished, optically reflective surfaces, a mirrored plastic block, or coverslip, is placed in close proximity to the surface and is acoustically coupled with fluid. The laser is focused onto the bottom surface of the coverslip, which has an acoustic pattern that corresponds to the sample surface. By rapid sweeping of the laser beam, the scanning laser acoustic microscope images are produced in real time (that is, 30 pictures per second) and are displayed on a high-resolution video monitor. In contrast to other less accurate uses of the term real time in industry today, the scanning laser acoustic microscope can be used to observe events as they occur—for example, a crack propagating under an applied stress. The images produced by SLAM are shadowgraph mode images of structure throughout the thickness of the sample. This provides the distinct advantage of simultaneous viewing of the entire thickness of the sample, as in x-ray radiography. In situations where it is necessary to focus on one specific plane, holographic reconstruction of the SLAM data can be employed (Ref 5, 6).

Figure 2 shows a system block diagram for the scanning laser acoustic microscope. In addition to an acoustic image on a CRT, an optical image is produced by means of the direct scanned laser illumination of the sample surface. For this mode, the reflective coating on the coverslip is made semitransparent. The optical image serves as a reference view of the sample for the operator to consult for landmark information, artifacts, and positioning of the sample to known areas. The SLAM acoustic images also provide useful and easily interpreted quantitative data about the sample. For example, the brightness of the image corresponds to the acoustic transmission level. By removing the sample and restoring the image brightness level with a calibrated electrical attenuator placed between the transducer and its electrical driver, precise insertion loss data can be obtained (Ref 7, 8). With the acoustic interference mode, the velocity of sound can be measured in each area of the sample (Ref 9, 10). When these data are used to determine regionally localized acoustic attenuation loss, modulus of elasticity, and so on, the elastic microstructure can be well characterized.

The simplest geometries for SLAM imaging are flat plates or disks. However, with proper fixturing, complex shapes and large samples can also be accommodated. For example, tiny hybrid electronic components, large (254 × 254 mm, or 10 × 10 in.) metal plates, aircraft turbine blades and ceramic engine cylinder liner tubes have been routinely examined by SLAM.

C-SAM Operating Principles

The C-mode scanning acoustic microscope is primarily a pulse-echo (reflection-type) microscope that generates images by mechanically scanning a transducer in a raster pattern over the sample. A focused spot of ultrasound is generated by an acoustic lens assembly at frequencies typically ranging from 10 to 100 MHz. A schematic diagram is shown in Fig. 3.

The ultrasound is brought to the sample by a coupling medium, usually water or an inert fluid. The angle of the rays from the lens is generally kept small so that the incident ultrasound does not exceed the critical angle of refraction between the fluid coupling and the solid sample. Note that the focal distance into the sample is shortened considerably by the liquid-solid refraction. The transducer alternately acts as sender and receiver, being electronically switched between the transmit and receive modes. A very short acoustic pulse enters the sample, and return echoes are produced at the sample surface and at specific interfaces within the part. The return times are a function of the distance from the interface to the transducer. An oscilloscope display of the echo pattern, known as an A-scan, clearly shows these levels and their time-distance relationships from the sample surface, as illustrated in Fig. 4. This provides a basis for investigating anomalies at specific levels within a part. An electronic gate selects information from a specific level to be imaged while it excludes all other echoes. The gated echo brightness modulates a CRT that is synchronized with the transducer position.

Compared to older conventional C-scan instruments, which produce a black/white

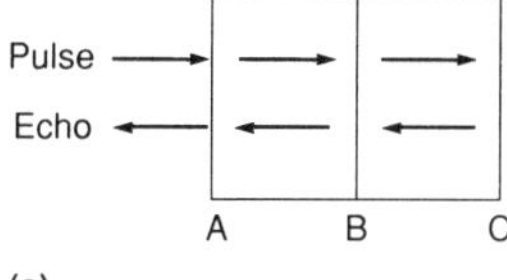

(a)

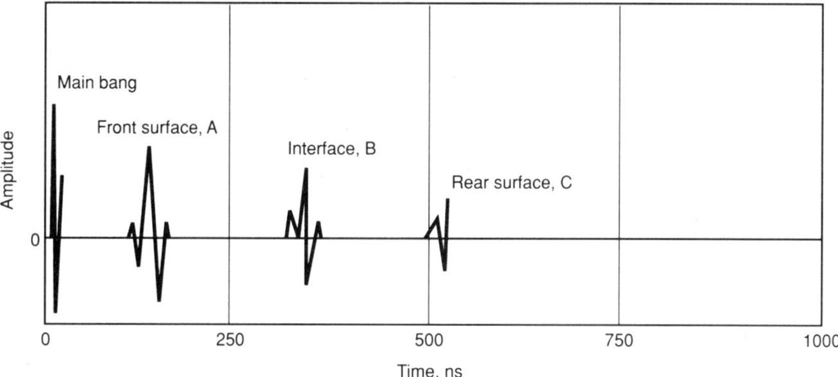

(b)

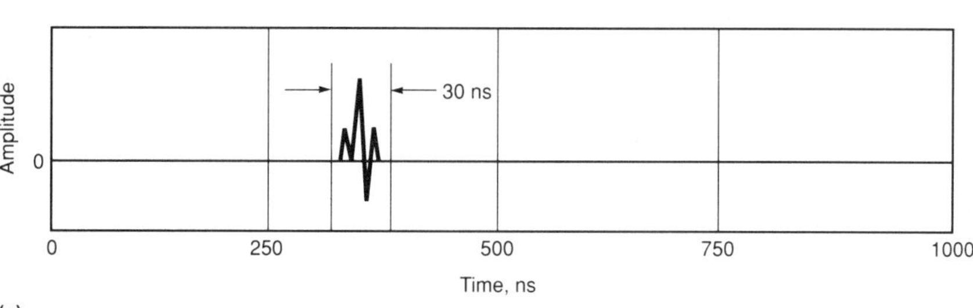

(c)

Fig. 4 Signal/source interfaces and A-scan displays for typical sample. (a) Simplified diagram showing pulses of ultrasound being reflected at three interfaces: front surface, A; material interface, B; and rear interface, C. Main bang is the background electrical noise produced by transducer excitation. (b) Plot of amplitude versus time as transducer is switched between transmit and receive modes at each interface during a time span of under 1 μs. (c) Signal at interface B gated for a duration of 30 ns

output on thermal paper when a signal exceeds an operator-selected threshold, the output of the C-mode scanning acoustic microscope is displayed in full gray scale, in which the gray level is proportional to the amplitude of the interface signal (gray-scale digitization is discussed in the article "Machine Vision and Robotic Inspection Systems" in this Volume). The gray scale can be converted into false color, and as shown in Fig. 5, the images can also be color coded with echo polarity information (Ref 11). That is, positive echoes, which arise from reflection from a higher-impedance interface, are displayed in a gray scale having one color scheme, while negative echoes, from reflections off of lower-impedance interfaces are displayed in a different color scheme. This allows quantitative determination of the nature of the interface within the sample. For example, the echo amplitude from a plastic-ceramic boundary is very similar to that from a plastic-airgap boundary, except that the echoes are 180° out of phase. Thus, to determine whether or not an epoxy is bonded to a ceramic, echo amplitude analysis alone is not sufficient. The color-coded enhanced C-mode scanning acoustic microscope is further differentiated from conventional C-scan equipment by the speed of the scan. Here, the transducer is positioned by a very fast mechanical scanner that produces images in tens of seconds instead of tens of minutes for typical scan areas to cover the size of an integrated circuit.

With regard to the depth zone within a sample that is accessible by C-SAM techniques, it is well known that the large echo from a liquid/solid interface (the top surface of the sample) masks the small echoes that may occur near the surface within the solid material. This characteristic is known as the dead zone, and its size is usually of the order of a few wavelengths of sound or more. Far below the surface, the maximum depth of penetration is determined by the attenuation losses in the sample and by the geometric refraction of the acoustic rays, which shorten the lens focus by the solid material. Therefore, depending on the depth of interest within a sample, a proper transducer and lens must be used for optimum results.

SAM Operating Principles

The scanning acoustic microscope is primarily a reflection-type microscope that generates very high resolution images of surface and near-surface features of a sample by mechanically scanning a transducer in a raster pattern over the sample (Fig. 6 and 7). In the normal mode, an image is generated from echo amplitude data over an x,y scanned field-of-view. As with SLAM, a transmission interference mode can be configured for velocity of sound measurements. In contrast to C-SAM, a more highly focused spot of ultrasound is generated by a very wide angle acoustic lens assembly at frequencies typically ranging from 100 to 2000 MHz. The angle of the sound rays is well beyond the critical cutoff angle, so that there is essentially no wave propagation into the material. There is a Rayleigh (surface) wave at the interface and an evanescent wave that reaches to about one wavelength depth below the surface. As in the other techniques, the ultrasound is brought to the sample by a coupling medium, usually water or an inert fluid. The transducer alternately acts as sender and receiver, being electronically switched between the transmit and receive modes. However, instead of a short pulse of acoustic energy, a long pulse of gated radio-frequency (RF) energy is used. No range gating is possible, as in C-SAM, because of the basic design concept of the SAM system. The returned acoustic signal level is determined by the elastic properties of the material at the near-surface zone. The returned signal level modulates a CRT, which is synchronized with the transducer position. In this way, images are produced in a raster scan on the CRT. As with C-SAM, complete images are produced in about 10 s.

With the SAM technique operating at very high frequencies, it is possible to achieve resolution approaching that of a conventional optical microscope. This technique is employed in much the same way as an optical microscope, with the important exception that the information obtained relates to the elastic properties of the material. Even higher resolution than an optical microscope can be obtained by lowering the temperature of operation to near 0 K and using liquid helium as a coupling fluid. The wavelength in the liquid helium is very short compared to that of water, and submicron resolution can be obtained (Ref 12).

The SAM technique has also been found to be useful for characterizing the elastic properties of a sample over a microscopic-size area, which is determined by the focal-spot size of the transducer. In this method, the reflected signal level is plotted as a function of the distance between the sample and the lens. Because of the leaky surface waves generated by mode conversion at the liquid/solid interface as the sample is defo-

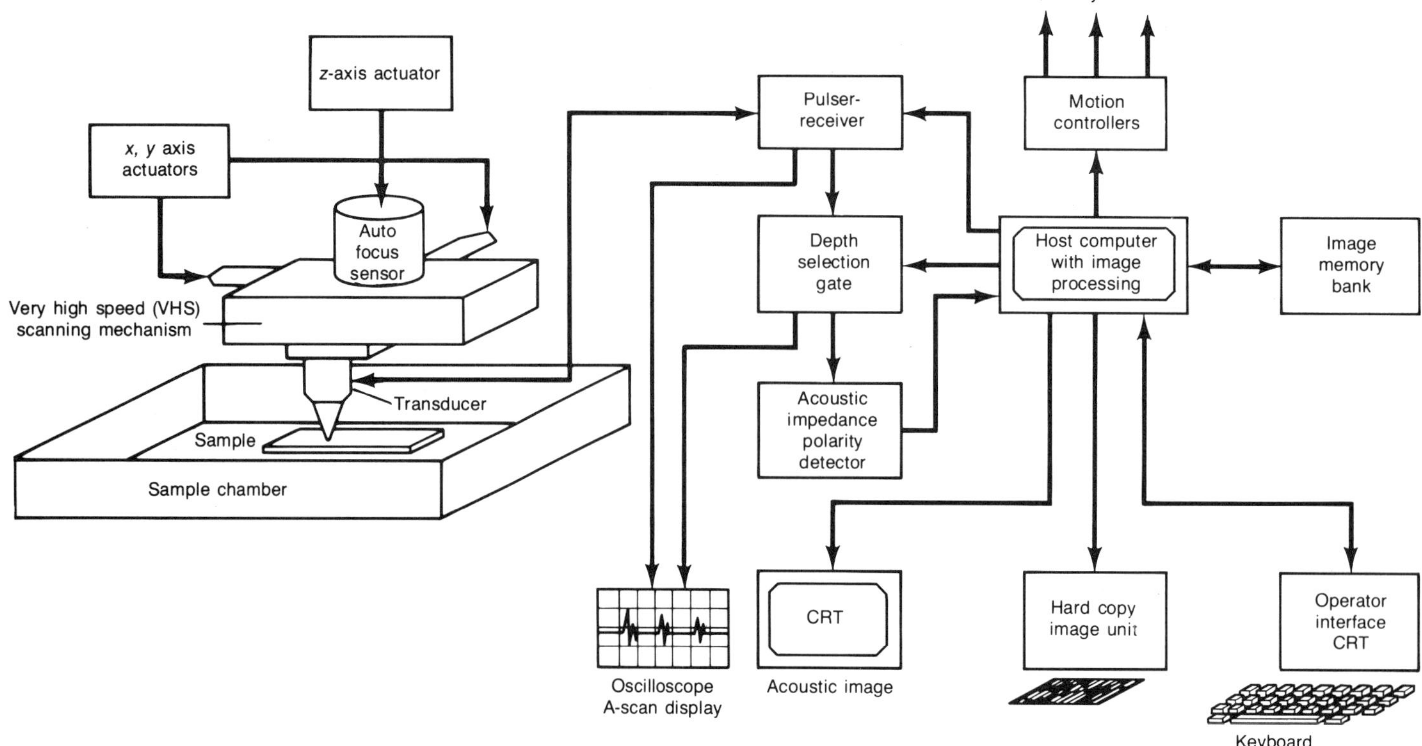

Fig. 5 Schematic and block diagram of the C-mode scanning acoustic microscope. The instrument employs a very high speed mechanical scanner and an acoustic impedance polarity detector to produce high-resolution C-scan images.

cused toward the lens, an interference signal is produced between the mode converted waves and the direct interface reflection. The curve obtained is known as the *V(z)* curve; by analyzing the periodicity of the curve, the surface wave velocity can be determined. Furthermore, defocusing the lens enhances the contrast of surface features that do not otherwise appear in the acoustic image (Ref 13, 14). In addition, by using a cylindrical lens instead of a spherical lens, the anisotropy of materials can be uniquely characterized (Ref 15).

Comparison of Methods to Optimize Use of Techniques

Figure 8 illustrates the zones of application for all three types of acoustic microscopy techniques. The differences are substantial with regard to the potential for visualizing features within a sample, and they should be carefully weighed before a particular method is selected. Table 1 lists some of the major advantages and benefits of the techniques. However, generalizations are sometimes difficult to make. Proposed or laboratory-demonstrated solutions to some of the limitations of each technique may already exist. Table 1 was prepared on the basis of instrumentation and techniques that are commercially available and therefore should be used only as a general guide.

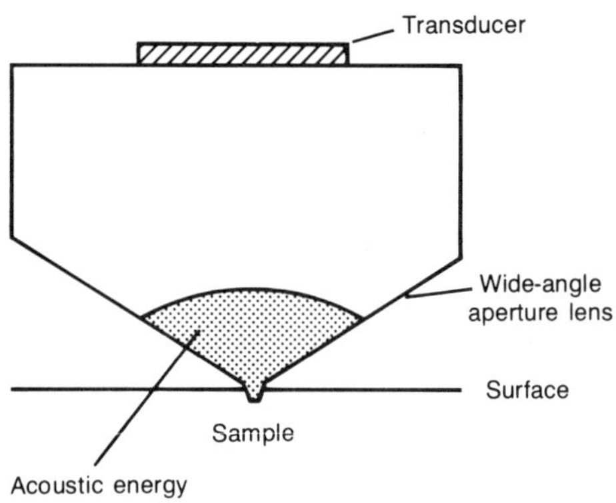

Fig. 6 Schematic of the scanning acoustic microscope lens used for interrogating the surface zone of a sample

Acoustic Microscopy Applications*

It is difficult to document concisely the very broad range of applications for acoustic microscopy, but a few generalizations can be made that follow the examples of conventional ultrasonic nondestructive testing (see the articles "Ultrasonic Inspection" and "Adhesive-Bonded Joints" in this Volume). Acoustic microscopy is compatible with most metals, ceramics, glasses, polymers, and composites (made from combinations of the above materials). The compatibility of a material is ultimately limited by ultrasound attenuation caused by scattering, absorption, or internal reflection. In metals, the grain structure causes scattering losses; and in ceramics, the porosity may cause losses. The magnitude of these effects generally increases with ultrasound frequency, and the dependence is monotonic.

Figure 9 shows a general guide to acoustic microscopy applications with respect to imaging. The quantitative aspects have not yet found widespread industrial acceptance, although they are extremely important and form the basis for materials characterization. The techniques of SLAM, C-SAM, and SAM all produce quantitative data in addition to images. Acoustic microscopy methods are compared below with a typical C-scan ultrasound method in terms of frequency employed:

Method	Frequency range, MHz
C-scan ultrasound	1–10
Acoustic microscopy	
C-SAM	10–100
SLAM	10–500
SAM	100–2000

The most popular application of SLAM and C-SAM is the nondestructive evaluation of bonding, delamination, and cracks in materials. These instruments are often used for process and quality control, although a significant percentage of the devices are placed in analytical and failure analysis laboratories. The most popular application of SAM utilizes its very high magnification mode and is employed as a counterpart to conventional optical and electron microsco-

*Examples 1 and 3 were prepared by G.H. Thomas, Sandia National Laboratories, Livermore, CA. Example 4 was prepared by R.S. Gilmore, General Electric Research & Development Center.

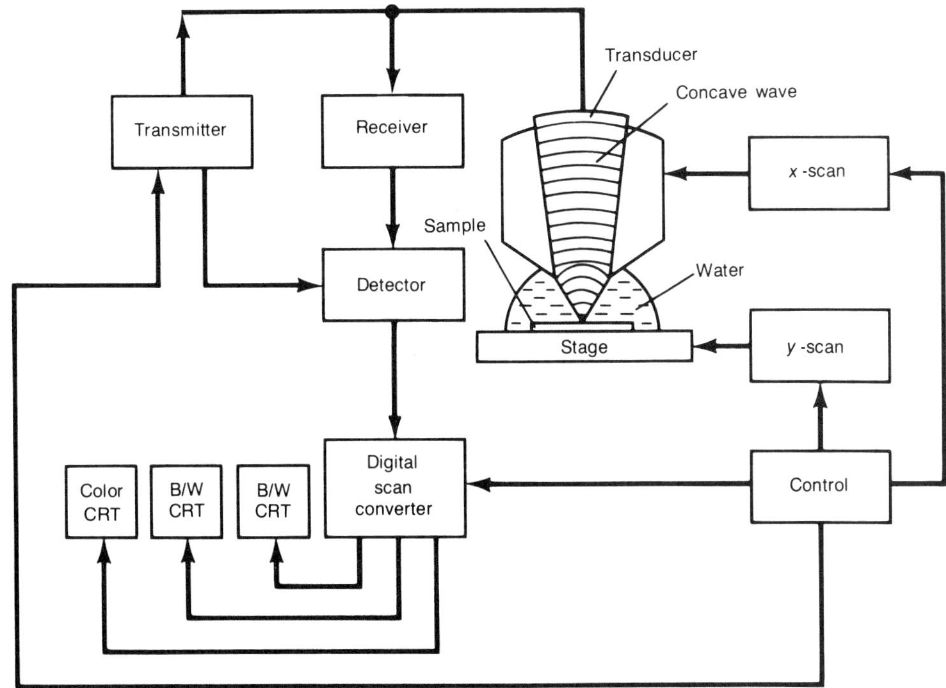

Fig. 7 Block diagram of a reflective-type SAM system that uses mechanical scanning of a highly focused transducer to investigate the surface zone of a sample at high magnification

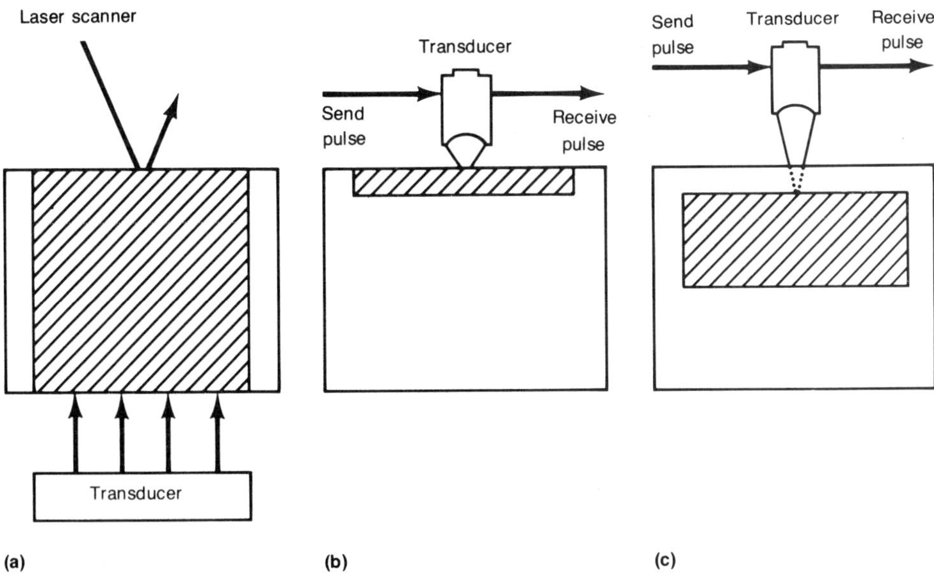

Fig. 8 Simplified comparison of three acoustic microscopy techniques, particularly their zones of application (crosshatched area) within a sample. (a) SLAM. (b) SAM. (c) C-SAM

py, that is, to see fine detail at and near surfaces. The scanning acoustic microscope, like the other acoustic microscopy methods, produces image contrast, which is a function of the elastic properties of a material, where other nonacoustic techniques may not.

Composite Materials

Composite materials represent an exciting challenge for the materials scientist and for nondestructive testing. With the combination of materials having different properties and with manufactured anisotropy, acoustic microscopy can clearly define the relevant property distribution of materials at the microscopic level. The examples shown in this section are polymer materials reinforced with fibers. Metal-matrix composites and ceramic-matrix composites can be similarly studied.

In general, the combinations of materials having different mechanical properties result in interfaces that cause scattering of ultrasound and differential attenuation within the field-of-view. This can be important in determining the population density shifts of fibers within a sample.

SLAM Images. Acoustic microscopy can be used to differentiate fibers that are bonded well to the matrix from fibers that are separated from the matrix. Excessive stress, for example, would cause such a separation. As an example of this, Fig. 10 shows two nominally identical tensile-test bars made of carbon fiber reinforced plastic (CFRP). One of the bars has been stressed to a level that produced a 1.5% strain. The other sample has not been stressed. Figures 11(a) and (b) compare SLAM images of these samples at 100 MHz. The lighter image, Fig. 11(a), shows texturing that follows the fiber direction. Figure 11(b) shows the much darker, attenuating characteristics of the stressed sample. The excess attenuation is due to the separation of the fiber from the matrix.

Figure 11 in the section "Use of Color for NDE" in this Volume shows a 30-MHz transmission SLAM image of a Kevlar fiber-reinforced plastic that has been cut perpendicular to the fibers, thus producing extensive delamination (red) and cracks along the axis of the sample. Defects anywhere throughout the thickness will block the ultrasound transmission.

Figure 12 in this article shows a curved CFRP component. This sample is more difficult to image than the tensile-test bars because it has a complex shape; in addition, it has fibers oriented in a variety of different directions for directional strength.

In Fig. 13, a 30-MHz SLAM transmission image, a transition is seen between different fiber orientations in the sample; Fig. 14 in the section "Use of Color for NDE" in this Volume shows a large anomaly that is located in the middle of the curved portion. Scanning laser acoustic microscopy can be used to image complex-shape samples even though the curvature may cause some restriction of the field-of-view due to critical angle effects and lenslike action by the sample.

C-SAM Image. Figure 14 shows a 50-MHz reflection image in the C-SAM image of a CFRP tensile-test bar similar to that shown in Fig. 10. The acoustic lens is focused near the surface of the sample, and the electronic gate was set to receive a portion of the backscattered signal. Because the sample is constructed with fibers distributed throughout the volume, acoustic energy is scattered from all depths, and distinct, time-separated echoes were not generated. When the transducer was focused deep into the sample, there was less definition of the fibers in the acoustic image, similar to what is seen in the SLAM image.

Ceramic Materials

Ceramic materials are used in a variety of applications. Some are used in the electronics industry as substrates for delicate hybrid circuits. In structural applications, ceramics are used where high temperature and light

Table 1 Comparison of industrial acoustic microscopy techniques

Parameter	SLAM	C-SAM	SAM
General description	Utilizes CW, plane wave ultrasonic illumination of sample and scanning focused laser beam detection of ultrasound; simultaneous optical images and acoustic images are produced. SLAM produces images in real time, which is the fastest of all acoustic microscopy techniques.	High-resolution focused-beam C-scan(a). Utilizes pulse-echo mode and has full gray-scale image output. Images are produced by mechanically scanning the transducer over the sample area.	Utilizes very highly focused acoustic lenses to image exterior surfaces by means of surface acoustic wave generation. Images are produced by mechanically scanning the transducer over the sample area. SAM produces the highest resolution of all acoustic microscopy techniques.
Primary use	Nondestructive testing and materials characterization	Nondestructive testing and materials characterization	Materials characterization and research
Frequency range	10–500 MHz	10–100 MHz	100 MHz–2 GHz
Resolution	Resolution limited to the wavelength of ultrasound within the sample material	Resolution limited to the wavelength of ultrasound within the sample material, multiplied by a factor, typically 2–10, due to the lens design	Resolution limited to the wavelength of ultrasound within the coupling fluid at the surface of the sample
Imaging mode	Through transmission (primarily); off-axis reflection; orthoscopic view of sample	Reflection (primarily); orthoscopic view of sample	Reflection (only); orthoscopic view of sample
Image information	Image produced is an acoustic shadowgraph representing the entire thickness of the sample.	Image is produced at any selected depth within the sample, but is affected by wave propagation through all the levels prior to the focus location.	Image is produced only at the sample surface, and the depth of the information extends into the material a distance of one wavelength of sound.
Ultrasound mode	Continuous-wave and frequency-modulated ultrasound waves	Short pulses of ultrasound, less than a few cycles of RF in duration	Gated RF (or tone burst) containing many cycles (10–100) of RF at frequency selected
Ultrasound signal timing	Not applicable due to CW	Echoes are spread out in time, and one is selected and electronically gated for desired image depth within sample.	Gated at the surface of the sample where virtually all the ultrasound is reflected. There are no subsurface echoes to select.
Transducer type	Plane wave transducer for sample illumination and focused laser beam for detection	Focused transducer used for transmit and receive mode. The angle of the rays is less than the critical angles between the coupling fluid and the sample. This is known as a low-numerical-aperture transducer.	Focused transducer in which the geometric angle of the rays is much greater than the critical angles in order to excite surface modes. This is known as a high-numerical-aperture transducer.
Image outputs	Amplitude mode: Records level of ultrasound transmission at each x,y coordinate of the scan	C-mode: Records echo amplitudes at each x,y coordinate of the scan; image output on CRT	Normal mode: Records reflected energy from surface at each x,y position. Image output on CRT. $V(z)$, or acoustic material signature graph, records the change in reflected signal level at any x,y coordinate as the z position is varied (Ref 12–14). This will characterize the material by means of its surface acoustic wave velocity.
	Interference mode: Records velocity of sound variations within sample over the field-of-view at each frequency	A-scan: Oscilloscope display of the echo pattern as a function of time (distance into sample) at each x,y coordinate. This can be used to measure the depth of a feature or the velocity of sound in the material. The A-scan is used to determine the setting of the echo gate, which is critical to the composition and interpretation of the images.	Interference mode: Records velocity of sound variations within sample over the field-of-view at each frequency
	Frequency scan mode: Similar to amplitude mode except that the insonification frequency is swept over a range to eliminate speckle and other artifacts of coherent imaging		
	Optical mode: Laser scanned optical image produced in synchrony with the acoustic images		
Depth of penetration	Penetration is limited by acoustic attenuation characteristic of sample.	In addition to attenuation by the sample, penetration is limited by focal length of lens and geometric refraction of the rays, which causes shortening of the focus position below the surface. There is also a dead zone just under the surface due to the large-amplitude front-surface echo, which masks smaller signals occurring immediately thereafter. This can be rectified with a high-numerical-aperture transducer.	Limited to a distance of one wavelength of sound below the surface. There is essentially no wave propagation into the sample.
Imaging speed	True real-time imaging: 30 frames/s; fastest of all acoustic microscopes	10 s to 30 min per frame; varies greatly among manufacturers	10 to 20 s/frame
New developments	Holographic reconstruction of each plane through the depth of the sample (Ref 5, 6)	Acoustic impedance polarity detector to characterize the physical properties of the echo producing interfaces (Ref 11)	Low-temperature liquid helium stages for extremely high resolution images (Ref 15)

(a) C-scan produces ultrasonic images by mechanically translating a pulsed transducer in an x,y plane above a sample while recording the echo amplitude within a preset electronic time gate. The transducer may be planar or focused. The frequencies of operation are typically 1–10 MHz, and the data are usually displayed as a binary brightness level on a hard copy output unit, such as thermal paper. A threshold level selected by the operator determines the transition between what amplitude of echo is displayed as a bright or dark image print.

	C-Scan		C-SAM	SLAM		SAM	
Frequency	1 MHz	10 MHz	30 MHz	100 MHz	200 MHz	500 MHz	1 GHz
Wavelength	1.5 mm			15 μm			1.5 μm

- –Medical ultrasound–
- –Conventional ultrasonic NDT–
- –Course grain metals: defect detection–
- –Cracks in plastic encapsulated IC devices–
- –Composite materials–
- –Cracks in ceramic IC packages–
- –Spot welds–
- –Heat sealing of food pouches–
- –Ceramic chip capacitors: delaminations and cracks–
- –Hermetic seal reliability–
- –Polymer-foil package lamination–
- –Integrated circut die attach–
- –Thick film adhesion and porosity–
- –Laser spot welds–
- –Fine ceramics: defect detection–
- –Fine grain metals: defect detection–
- –Seam welds on tin cans–
- –Lead bonds on hybrid circuts–
- –Ceramic substrate porosity and cracks–
- –Cracks in silicon wafers–
- –Thin film adhesion–
- –Grain structure determination–
- –Fine line inspection on silicon–

Fig. 9 Comparison of acoustic microscopy applications with C-scan applications, based on transducer frequency and wavelength

weight are important. Silicon nitride, silicon carbide, and zirconia are receiving much attention for future engine applications. However, because of the inherent brittleness of ceramic materials, small defects are very critical to the structural integrity of the materials. The successful use of these materials necessitates careful nondestructive screening of the samples.

Aluminum Oxide. Figure 15 shows an optical picture of an aluminum oxide panel with laser-machined holes. This sample is an electronics-grade material that is 99.5% pure and has very low porosity. When the ceramic powder is first compressed in the green state, opportunities arise for segregated low-density areas to occur. After the sintering operation, these areas are usually found to be very porous. If one of these areas happens to coincide with the site of laser machining, the stresses that occur can cause the material to crack. Upon visual examination, it may be very difficult to detect fine cracks, even if they come to the surface. Dye penetrants can be used to increase the visual contrast. Unfortunately, in many applications, dyes are considered to be contaminants and therefore cannot be used for nondestructive testing. Because of the discontinuity in material property, a crack will reflect acoustic waves and produce high-contrast images.

Fig. 10 Photograph of two CFRP tensile-test bars—one unstressed and one stressed. The stressed bar was subjected to a force that yielded a 1.5% strain.

Figure 16 shows a 100-MHz SLAM acoustic micrograph of a ceramic sample with a crack that can be seen as a dark line originating from one of the holes. Surrounding the area of the crack are several small, dark patches, which arise from localized increases in porosity. In the acoustic image, the pores may not be visible individually if they are smaller than the wavelength of sound. In this case, the pores are only a few microns in size, and the wavelength is about 25 μm (0.001 in.). The porous areas are detectable by virtue of excess ultrasound scattering, which causes the differential attenuation. Correlative analyses have shown that pores in the 1 μm (40 μin.) size range cause detectable attenuation increases (Fig. 16). However, the presence of a single 1 μm (40 μin.) pore would be difficult to detect unless the frequency of the ultrasound was increased to 1 GHz or more.

It is significant to note that attenuation changes in ceramics correlate well with variations in strength (Ref 16). Figure 15 in the section "Use of Color for NDE" in this Volume shows a digitally enhanced pseudocolor image of a ceramic sample that is useful for quantitative analysis of the gray scale and for identifying precisely the acoustic signal levels without relying on operator interpretation of the gray scale on the CRT screen.

Example 1: Use of C-SAM to Detect Internal Porosity Defects in an Alumina Ceramic Disk. Acoustic microscopy is a powerful tool for nondestructively evaluating ceramic materials and displaying inter-

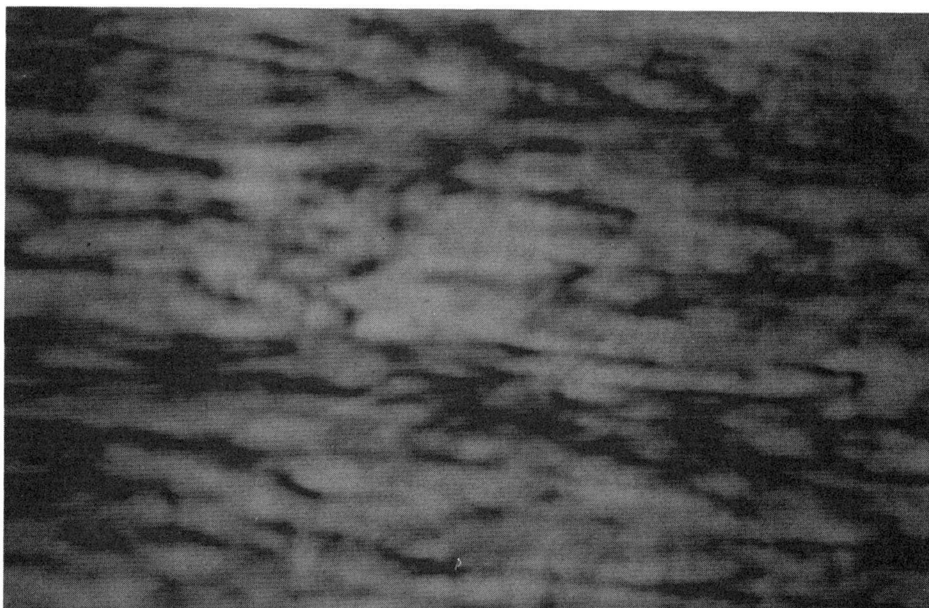

(a)

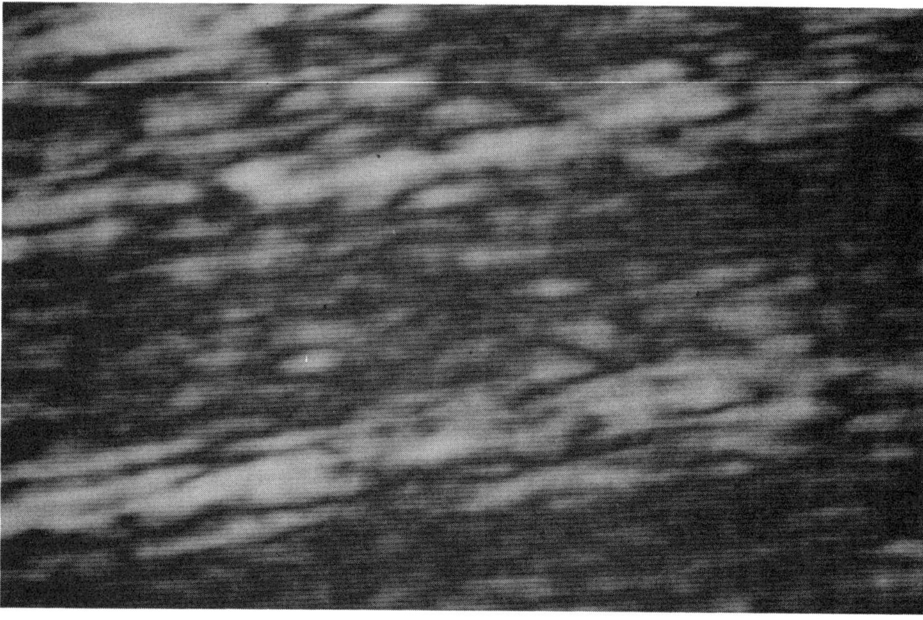

(b)

Fig. 11 100-MHz SLAM acoustic micrographs of the tensile bar samples shown in Fig. 10. (a) In the unstressed bar, the texture corresponds to fiber population shifts and small areas of disbond. (b) In the stressed bar that was pulled to a 1.5% strain level, there is no visible evidence of fracture. Field of view: both 3 × 2.25 mm. However, the increased acoustic attenuation (dark zones) present in (b) indicates stress damage to the material.

nal defects and density gradients such as porosity. An acoustic microscope scanning at 50 MHz with an f/1 lens on the transducer was used to evaluate a 6.4 mm (0.25 in.) thick aluminum oxide ceramic disk. No surface preparation of the sample was required prior to scanning.

The acoustic microscope, with a 25 μm (0.001 in.) acoustic resolution, scanned a 12.7 × 12.7 mm (0.50 × 0.50 in.) area of the ceramic disk. A pulse-echo technique (C-SAM) displayed internal reflections. The black areas (arrows, Fig. 17), which were detected as having the highest-amplitude reflection, indicate areas of porosity of the order of 10 to 20 μm (400 to 800 μin.) in diameter.

Example 2: Analysis of Alumina Ceramic Disks Supplied by a Variety of Manufacturers and Subjected to SLAM, C-SAM, and SAM Evaluation. Disks of alumina were obtained from various manufacturers as part of a study on the effects of sintering temperature changes on mechanical strength and ultrasonic properties. The disks were approximately 25 mm (1 in.) diameter and 6 mm (15/64 in.) thick. Figure 18 shows a through-transmission 10-MHz SLAM image of an alumina disk. Higher SLAM frequencies could not be used, because of excessive attenuation. This is unusual for ceramic materials unless they are very porous or otherwise defective. Most of the area of this sample is nontransmissive (dark), which indicates a large internal defect. A subsequent cross section of this part, shown in Fig. 19, reveals a large crack that correlated to the indication in the SLAM image. The crack is not parallel to the surface; this was not known prior to cross sectioning.

Figures 16(a) and (b) in the section "Use of Color for NDE" in this Volume are 15-MHz reflection C-SAM images of the same part focused to different depths. The radical changes in echo pattern at different depth locations in the disk result in widely different images. This is due to the nonparallel nature of the crack. In these figures, the darkest features correspond to the greatest echo amplitude. The C-mode scanning acoustic microscope clearly shows anomalies in this part, and by sequential scanning at a variety of depths, an overall diagnosis of the part can be assembled.

Figure 20 shows a SAM image produced at 180 MHz in which the lens was focused slightly below the surface (20 μm, or 800 μin., in water) to highlight subsurface information. The predominant features in Fig. 20 are surface scratches, which may also be associated with near-surface cracks. The grain structure and porosity of this sample are finer than can be visualized at 180 MHz.

Another alumina sample produced under different conditions was found to be much more transparent acoustically. Figure 21 shows a 30-MHz SLAM image of a typical area of this disk. The higher frequencies are associated with higher magnifications and therefore smaller fields-of-view. The ceramic appears fairly homogeneous at this level of magnification. Figure 22, a 100-MHz SLAM image of the sample in Fig. 21, shows that some textural inhomogeneities are beginning to appear. The texture may be due to nonuniform segregation of porosity in the sample and throughout its thickness. Figure 23, a 50-MHz C-SAM image of the disk, shows large pores (white) about 1 mm (0.04 in.) below the surface. Figure 24 shows a 180-MHz SAM image made under conditions identical to Fig. 20. The fine texture at the surface of this sample is due to porosity. The differences between the two samples are clearly evident acoustically.

Metals

In typical optical microscopy and metallography, it is necessary to polish and etch a sample to reveal the microstructural pat-

Fig. 12 Complex-shaped CFRP component. The fibers are arranged to impart directional strength to certain critical areas of the sample.

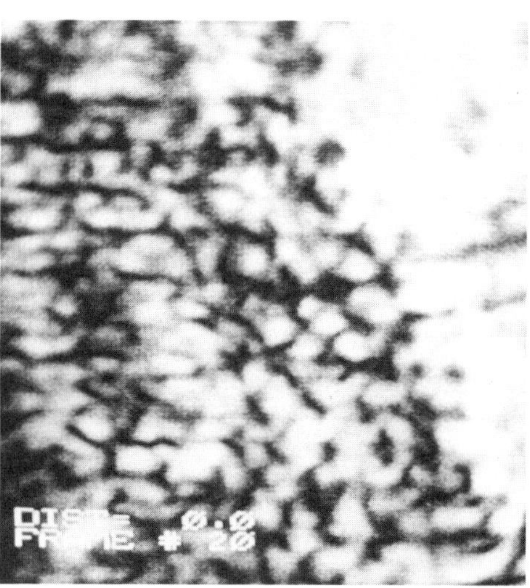

Fig. 13 30-MHz SLAM acoustic micrograph of the sample shown in Fig. 12. The left portion of the micrograph shows a more complex fiber network than the area on the right. A color image of the curved CFRP component is shown in Fig. 14 in the section "Use of Color for NDE" in this Volume. Field of view: 14 × 10.5 mm

tern. With SAM, this may not be necessary, as evidenced by the above. As one further illustration of SAM, Fig. 25 shows a 400-MHz image, focused 23 μm (920 μin.) below the surface, of a manganese-zinc ferrite material. The different phases of this important material can be visualized without the use of etchants. For certain metallic samples, that is, those with smooth, polished surfaces, SAM can be used for nondestructive testing as well as for metallographic analysis.

Figure 26 shows the metallographic structure of low-carbon steel; the specimen was polished but not etched. Although the images are similar to optical displays, the acoustic microscope is sensitive to acoustic properties and will generate images of surface and subsurface structure. The low-carbon steel was scanned at 1.3 GHz over an area measuring 100 × 100 μm (0.004 × 0.004 in.). A high-resolution scanning acoustic microscope was used to generate this image from variations in the acoustic properties of the steel sample. Details of the grain structure, grain boundaries, and impurities in the specimen are visible in Fig. 26. The darker areas indicate contaminants trapped in the metal.

Microelectronic Components

In the field of microelectronics, reliability is affected by tiny defects that may be harbored within a component or assembly but that are not detectable by electrical testing. The defects may grow with time or with thermal cycling, and when they reach a critical size, electrical performance may be affected; that is, a connection may break, or an electrical characteristic may change. Integrated circuits, ceramic capacitors, and

Fig. 14 50-MHz C-SAM reflection mode micrograph of a CFRP test sample. The ultrasound was focused near the top surface of the sample. Field of view: 19 × 14 mm

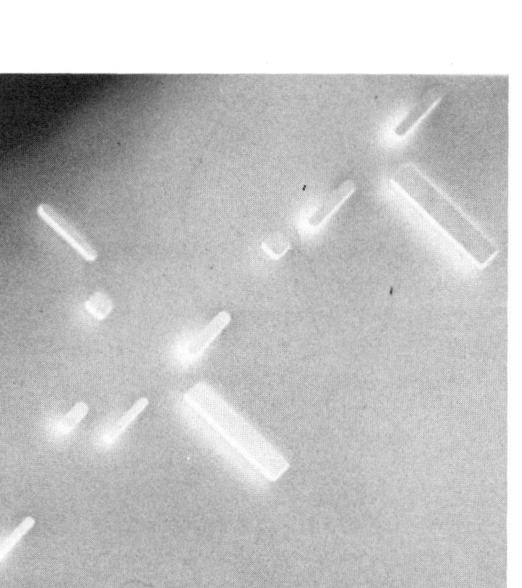

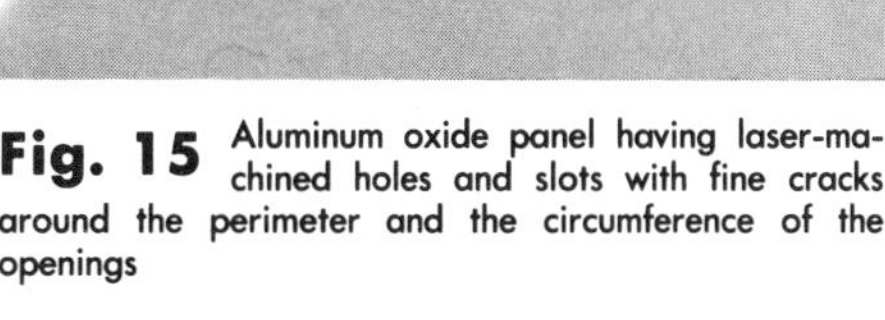

Fig. 15 Aluminum oxide panel having laser-machined holes and slots with fine cracks around the perimeter and the circumference of the openings

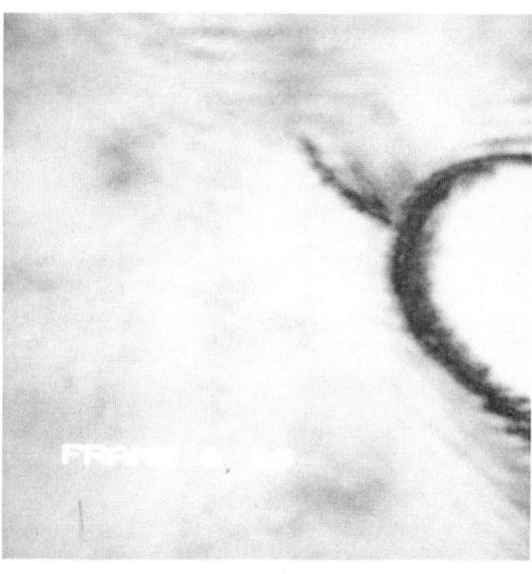

Fig. 16 100-MHz SLAM acoustic micrograph of an alumina panel section that contains a hole showing a crack with very high contrast. A digitally enhanced image of this alumina panel is also shown in Fig. 15 in the section "Use of Color for NDE" in this Volume. Field of view: 3 × 2.25 mm

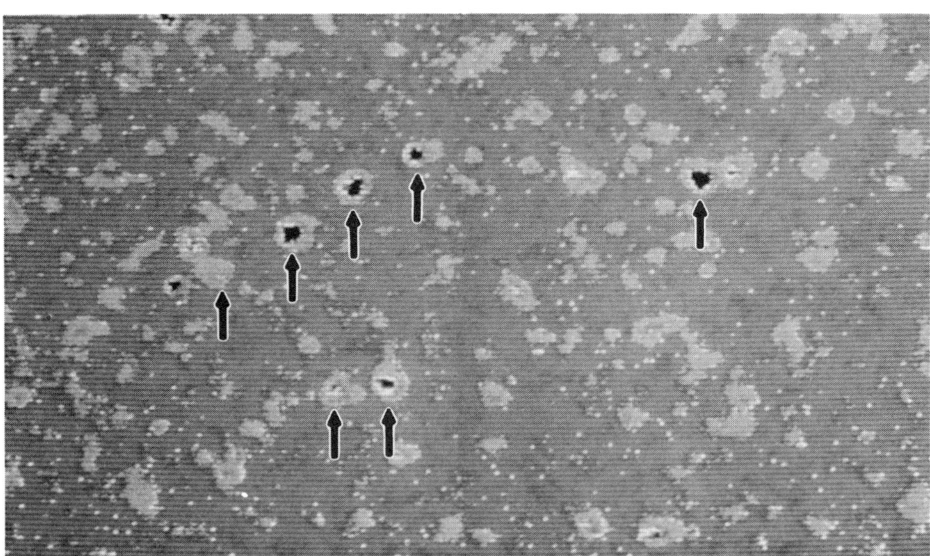

Fig. 17 A portion of the 160 mm^2 (0.25 in.^{2}) area of an alumina ceramic disk scanned by an acoustic microscope showing the presence of porosity in the disk (arrows). Courtesy of G.H. Thomas, Sandia National Laboratories

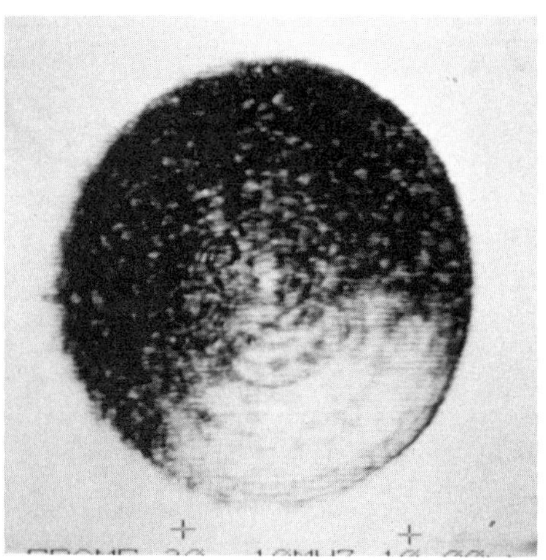

Fig. 18 Low-frequency 10-MHz SLAM image of a 25 mm (1 in.) diam alumina test disk. The disk is very attenuating to ultrasound because of internal defects that cover about 75% of the area (dark zones). Field of view: 35 × 26 mm

thyristors that contain defects of this type will be examined in this section.

Integrated Circuits (ICs). A typical silicon IC chip (die) generates heat during its normal operation. The heat must be dissipated to stabilize the electrical behavior of the semiconductor. Therefore, the die is bonded onto a thermally conductive substrate, such as copper, aluminum oxide, or beryllium oxide. A high-reliability IC must also be hermetically sealed from ionic contamination and moisture in the atmosphere; therefore, a sealed ceramic or ceramic/metal package is used to house the die and to serve as an interconnection vehicle. Figure 27 shows a typical ceramic IC package containing a die bonded to a gold-plated surface. The die bond is also referred to as a die-attach.

To complete the component, tiny interconnection wires are attached from the die to the corresponding metallization sites (bond pads) inside the ceramic package. A metal cover (lid) is then soldered to the top of the package. Apart from defects in the silicon chip itself, the life of the component depends on the quality of the die-attach, the integrity of the interconnection wire bonds, and the lid seal, all of which require good bonding.

Conventional Radiography Limitations. Because of its wide acceptance, conventional or x-ray radiography has usually been the first method of investigation for virtually all internal examinations (see the article "Radiographic Inspection" in this Volume). However, it has finally been realized that x-ray techniques indicate only the presence, absence, and density of materials through which the beam passes. On the positive side, if a poor bond is caused by the absence of a material, for example, solder, there will be less x-ray absorption and therefore a correct indication of a defect. However, if all the materials are present and in their proper proportions, there can be no differential x-ray absorption and no defect identification.

For example, suppose that the die is uniformly coated with solder but that there is no adhesion of the solder to the substrate due to contamination of the surface or improper metallurgy. Suppose also that the solder bonds well to the die and to the metallic layer but that for some reason the metallic layer becomes detached from the substrate. Both of these conditions are common problems that cannot be detected by x-ray but, because of the thin interface air gaps, cause complete reflection of ultrasound from, and no transmission of ultrasound across, the interface (Ref 17). The U.S. military standards for the inspection of circuits now include acoustic microscopy inspection (Ref 18).

Another common situation involves a die-attach material, such as epoxy, that is not

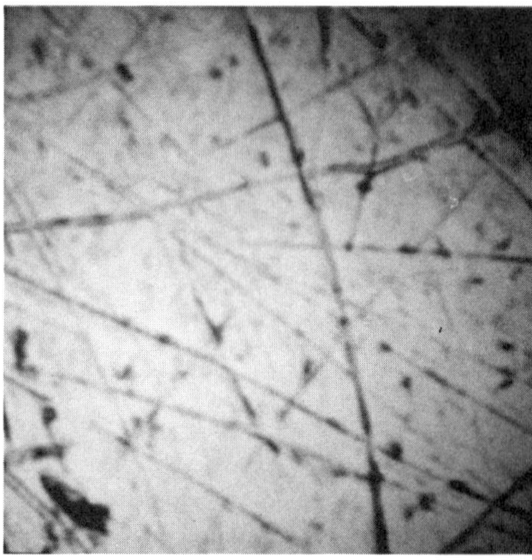

Fig. 20 SAM image at 180 MHz of the sample shown in Fig. 18 and 19 revealing features of the ceramic at and near the surface. Color images of this disk are shown in Fig. 16(a) and (b) in the section "Use of Color for NDE" in this Volume. Field of view: 1 × 1 mm

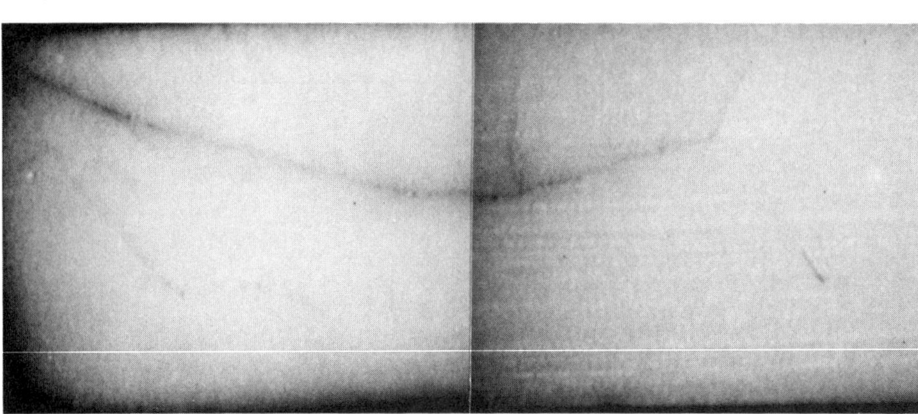

Fig. 19 Cross sectioning of the alumina test disk shown in Fig. 18 reveals a large crack that correlates with the area of low acoustic transmission. This confirms the presence of an internal defect.

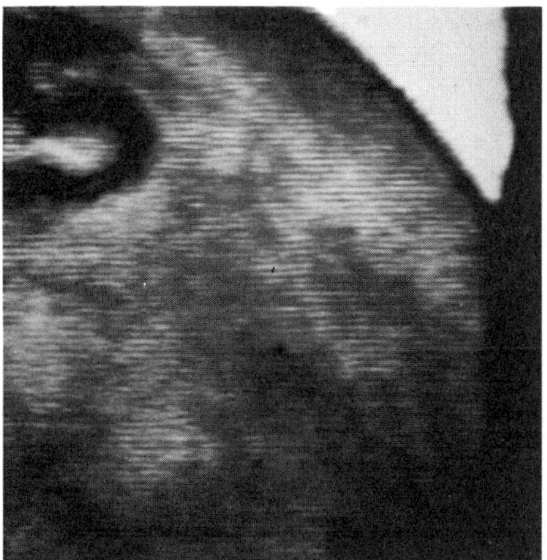

Fig. 21 SLAM images at 30 MHz of an alumina test disk similar in size to that shown in Fig. 18. This sample was quite transparent to the ultrasound, as evidenced by the bright, relatively uniform appearance of the acoustic image. Field of view: 14 × 10.5 mm

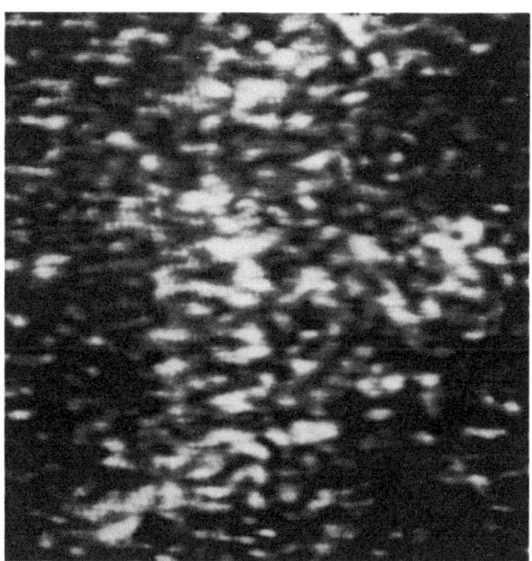

Fig. 22 SLAM 100-MHz transmission acoustic image of the sample shown in Fig. 21. At this higher frequency, textured variations in the sample are evident that may be due to micro (subresolution) porosity segregations, which cause differential absorption. Field of view: 3.5 × 2.6 mm

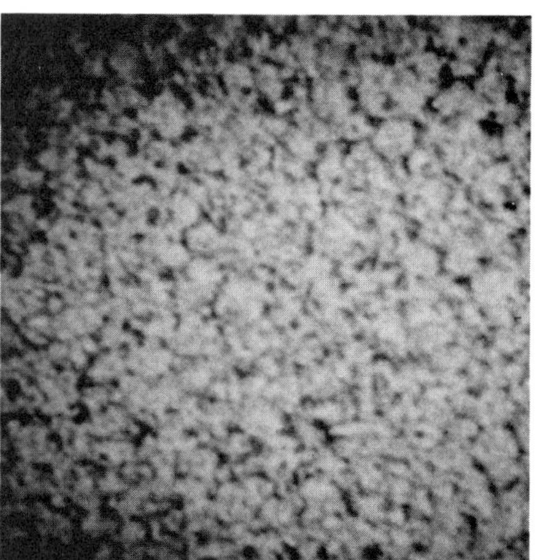

Fig. 24 SAM surface mode image at 180 MHz made under conditions identical to Fig. 20. In this image, the fine texture corresponds to porosity of the sample. Note the sharp contrast between the microstructure of this sample and that of Fig. 20. Field of view: 1 × 1 mm

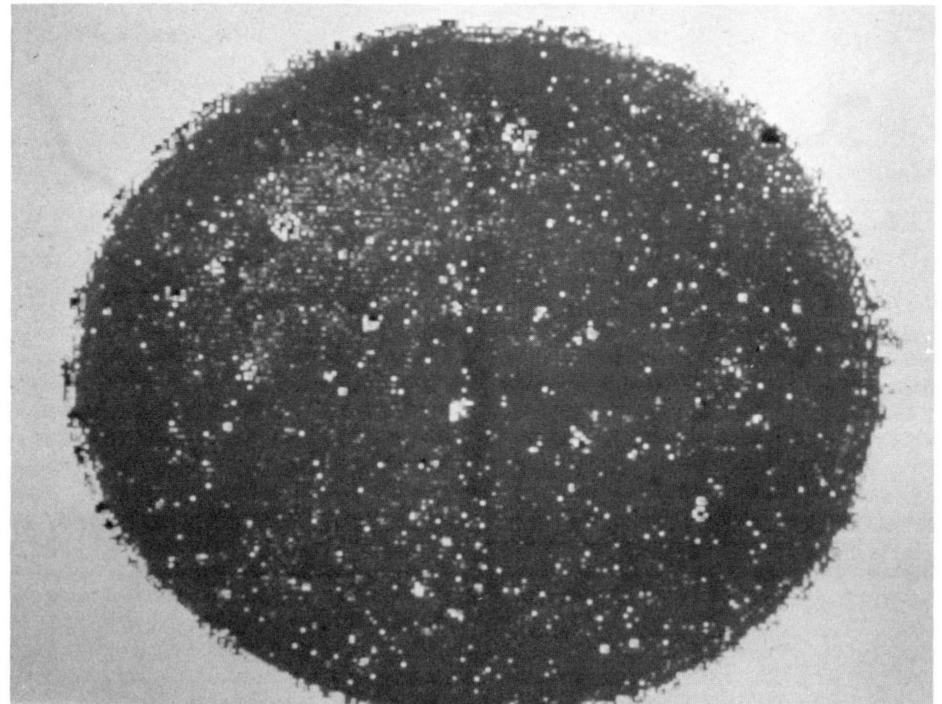

Fig. 23 C-SAM reflection mode image at 50 MHz made by setting the gate and focus to about 1 mm (0.04 in.) below the surface. The white circular spots correspond to individual pores located at this depth. Field of view: 30 × 30 mm

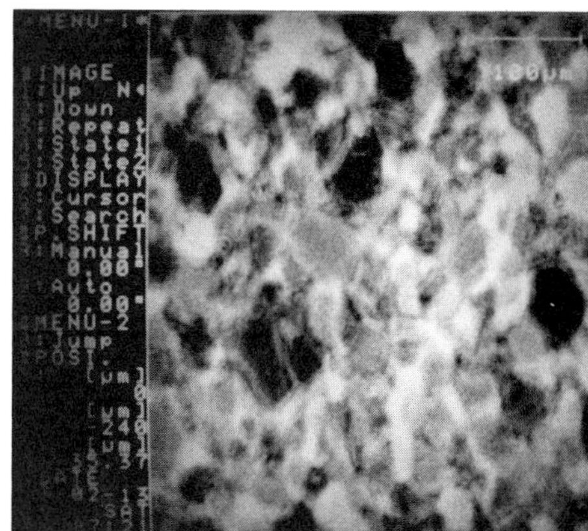

Fig. 25 SAM surface mode image at 400 MHz of a manganese-zinc ferrite sample that was polished metallurgically but not chemically etched. The elastic property differences between the various phases of this material are responsible for the contrast shown in this image. Courtesy of Honda Electronics Company, Ltd.

very absorptive to x-rays, as is the typical lead tin solder alloy. If the die is bonded to a ceramic substrate that is also not very absorptive, low-energy x-rays can be used to detect voids (but not delaminations) in the die-attach material. However, if the die is attached to a metal substrate that absorbs x-rays, much higher energy is required for penetration. In this case, the die-attach material absorption becomes a negligible fraction of the total absorption signal, and neither disbonds nor voids can be detected.

Advantages of Acoustic Microscopy. The die-attach can be inspected by SLAM techniques in those cases in which ultrasound can be coupled to the top and bottom surfaces of the IC, that is, when there is no lid enclosing the package. When the lid is on, the air gap within the die cavity precludes the transmission of ultrasound, and a reflection method, such as C-SAM must be used. This is not true for plastic packages, for which SLAM and C-SAM can be used.

Figure 28 illustrates the SLAM inspection of an IC. Figure 29 shows a transmission SLAM image of a die-attach. The dark zones represent little or no ultrasound transmission. A white line generated by the SLAM image analysis system outlines the die region.

Figure 30 shows the SLAM data after digital processing to highlight the areas of bond (white) and disbond (black). Because of its real-time imaging speed, the scanning laser acoustic microscope would be useful for quality control inspection of the die-attach before a ceramic IC is completely assembled.

Figure 31 illustrates the inspection of an IC within a closed, hermetic package by means of C-SAM. In this case, the ultrasound accesses the die-attach interface through the ceramic substrate. Access

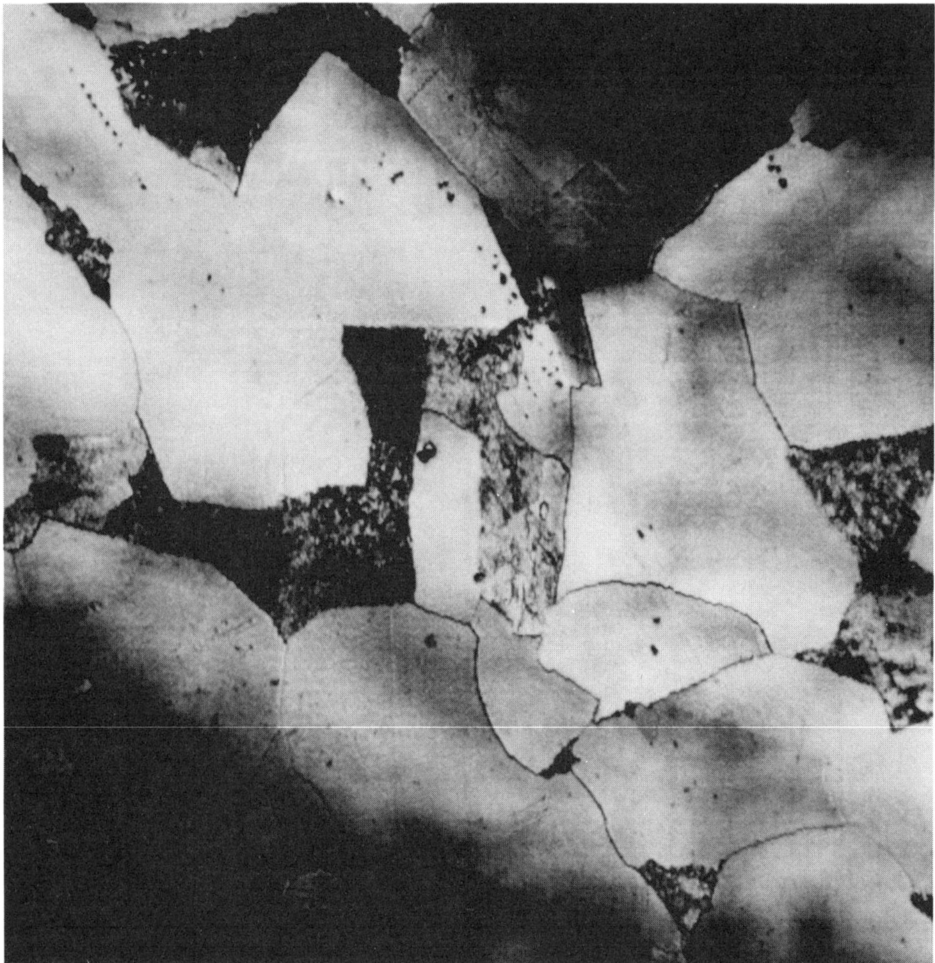

Fig. 26 SAM image at 1.3 GHz of a polished but unetched manganeze-zinc low-carbon steel. Darker regions indicate the presence of typical contaminants trapped in the metal. Courtesy of G.H. Thomas, Sandia National Laboratories

Fig. 27 Typical ceramic-packaged IC showing the silicon die, which is bonded to the metallized surface

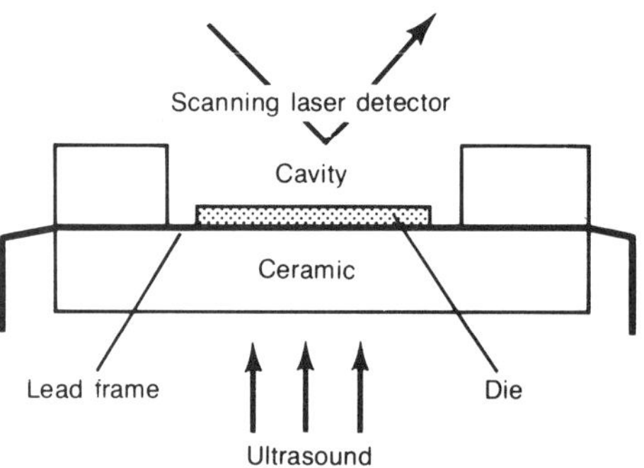

Fig. 28 Schematic illustrating use of the SLAM through transmission technique to evaluate the die-attach bond between the silicon die and the ceramic package

through the lid is precluded by the air gap. Because of the differences in acoustic impedance, Z, between the ceramic and silicon materials, an echo will be returned from the die-attach interface in the case of a good bond or a bad bond, although the amplitudes of the echoes may be different. The acoustic impedances of a few typical materials are listed below:

Material	Acoustic impedance, 10^6 rayl
Air, vacuum	0
Water	1.5
Plastic	2.0–3.5
Glass	15
Aluminum	17
Silicon	20
Beryllia	32
Copper	42
Alumina	21–45
Tungsten	104

More complete information on acoustic impedances can be found in Ref 19. If the amplitude of an echo relative to the amplitude of the incident acoustic pulse is denoted by R, then the value of R can be easily determined by the formula:

$$R = \frac{Z_2 - Z_1}{Z_2 + Z_1} \quad \text{(Eq 1)}$$

where Z_1 is the acoustic impedance of the material through which the ultrasound is traveling before reaching the interface and Z_2 is the acoustic impedance of the material encountered on the other side of the interface. Using Eq 1 and considering the interface materials to be alumina ($Z_1 = 35$) and silicon ($Z_2 = 20$) a good bond will have an R value of 0.27. In the case of a complete disbond, $Z_2 = 0$ (for air) and the R value is 1.0. In interpreting C-SAM images, because the ratio of signal levels between good and bad will be 3.7, care must be exercised to calibrate the system on a known sample.

Figure 13 in the section "Use of Color for NDE" in this Volume shows a reflection mode C-SAM image produced at 50 MHz of the sample shown in Fig. 29. To assist the low-contrast differentiation of typical black-and-white gray-scale images, the C-SAM data have been digitally processed and pre-

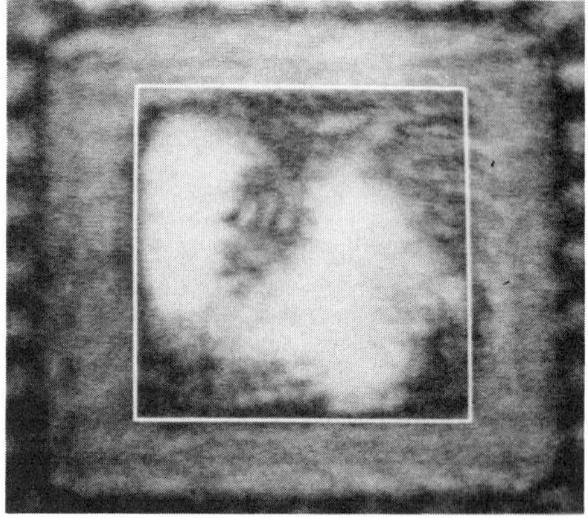

Fig. 29 30-MHz transmission SLAM image of a die-attach. The dark areas of the image correspond to little or no acoustic wave transmission due to the lack of bonding between the die and the ceramic. The white rectangle outlining the die is generated by an acoustic image analysis computer, which is used to determine the percent area of disbond. A 50-MHz reflection mode C-SAM color image of this integrated circuit is shown in Fig. 13 in the section "Use of Color for NDE" in this Volume. Field of view: 14 × 10.5 mm

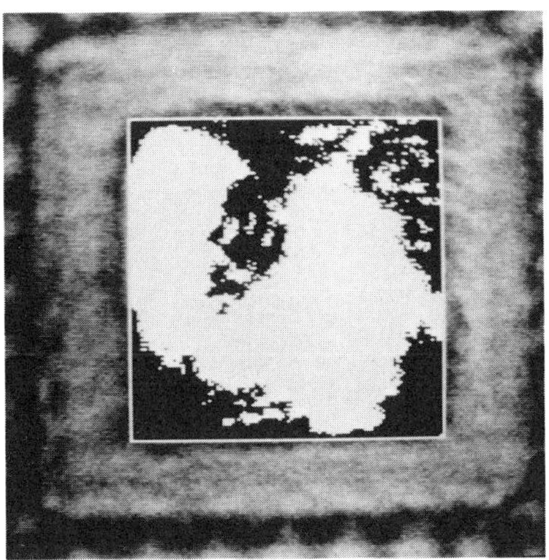

Fig. 30 Computer-analyzed SLAM acoustic image of Fig. 29 in which the disbonds are clearly displayed as black. A 50-MHz reflection mode C-SAM color image of this IC is shown in Fig. 13 in the section "Use of Color for NDE" in this Volume.

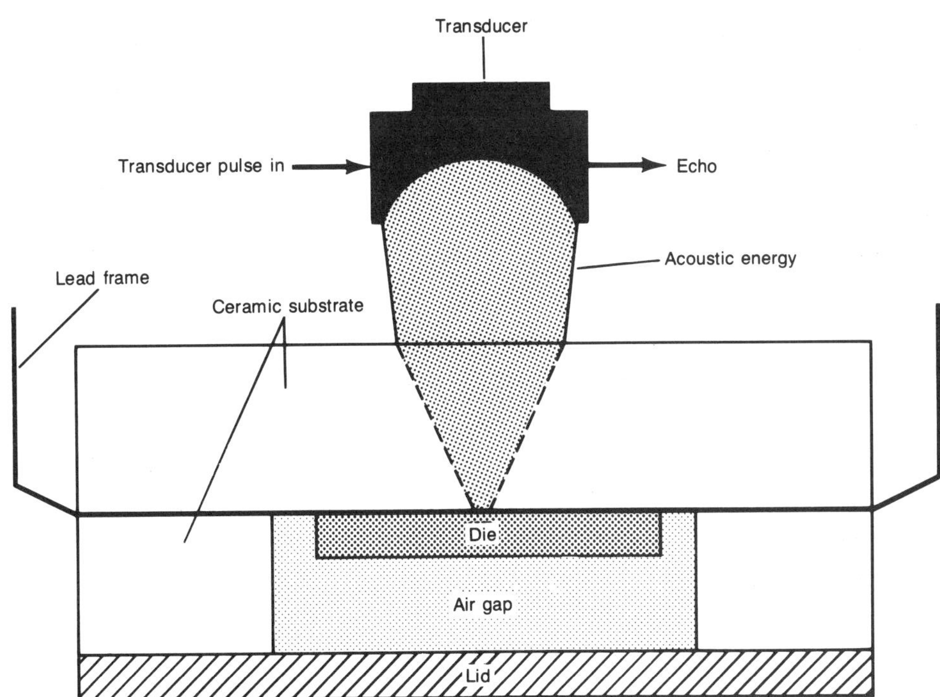

Fig. 31 Schematic illustrating use of the C-SAM reflection technique to evaluate the die-attach bond between the silicon die and the ceramic package of a ceramic dual-in-line package IC. With this technique, the ultrasound access to the bond layer is obtained by applying the pulse to the ceramic package, rather than through the lid, to avoid the air gap over the die.

sented as a pseudo-color-enhanced image in which red indicates disbond. The time gate was selected for the die-attach interface; therefore, unbonded regions of the die do not appear in the image, because from the ceramic side, the ultrasound cannot differentiate the air gap surrounding the die from the air gap under the disbonded areas of the die. In the case of unknown materials or in the case of not having an appropriate calibration sample, it may be more difficult to differentiate the bond versus the nobond condition because of nearly equal echo levels. Therefore, it is usually suggested to remove the lid from one sample of the lot and perform SLAM imaging. With SLAM, the amplitude of signal transmission across a two-material interface, T, is governed by:

$$T = 1 - R \qquad \text{(Eq 2)}$$

In the case of a good bond, $T = 0.73$, and in the case of a bad bond $T = 0$. The ratio of signal levels is limited only by the dynamic range of the instrument and can be typically 1000:1. To clarify this further, the ability of any electronic instrument to measure a zero signal level is determined by the signal-to-noise ratio of the instrument, which can typically exceed 60 dB.

Another important issue arises in ICs when the die is encapsulated in plastic molding compound instead of a ceramic package. In this case, both SLAM and C-SAM can be employed. When SLAM is employed, disbonds will appear nontransmitting and therefore dark (or whichever color is assigned). However, with C-SAM, more care must be taken in echo-level discrimination. The acoustic impedance of plastic is low ($Z = 3$) relative to ceramic; however, the echo magnitude from a plastic/air interface will still be 1.0. The echo from a plastic/silicon interface, however, will be 0.75, which is very close to 1.0, thus making the interpretation of echo level possibly difficult. Fortunately, however, the polarity of the echoes is different for air and silicon interfaces. In the case of $Z_2 > Z_1$ (plastic/silicon), the echo is positive and in the case of $Z_2 < Z_1$ (plastic/air), the echo is negative, according to Eq 2.

In the color-coded enhanced C-mode scanning acoustic microscope, the images can be displayed as echo-magnitude images or as echo-magnitude and polarity images (Ref 11). This is illustrated in Fig. 32 and in Fig. 12, which can be found in the section "Use of Color for NDE" in this Volume; both of these figures show a plastic-encapsulated IC that has disbonded areas between the plastic molding compound and the metal lead frame. As shown in Fig. 12 in the section "Use of Color for NDE," the color differentiation between positive echoes (blue) and negative echoes (disbond) (red) is essential for correctly diagnosing the integrity of this device. The echo signal levels from the plastic/lead frame interface are similar to the levels from the plastic/air gap interface; therefore, the good versus bad condition cannot be analyzed in the black-and-white presentation of Fig. 32. A through transmission SLAM image, such as that shown in Fig. 33, is clearer with regard to differentiating between bond (bright) and disbond (dark). A more thorough discussion of the acoustic microscopy analysis of plastic-encapsulated ICs can be found in Ref 20.

At this point it should be mentioned that the interpretation of echoes is not always simple. Equations 1 and 2 are restricted to an interface of two thick materials. If there is a thin layer, echoes from the front and back surfaces of the thin layer will merge, and the resulting pulse shape will become distorted. More detailed analysis of the echo shape will be needed to determine the nature of the interface (Ref 21). Complete information on IC packaging is available in Ref 22.

Integrated Circuit Wire Bonding. Another problem with ICs is the connection of the tiny wires to the silicon chip. High-speed, densely spaced ICs may contain well over 300 leads. Traditional techniques of wire bonding use 0.018 mm (0.0007 in.) diam gold wires that are individually point-by-point stitch bonded from pads on the chip to pads on the package. Newer methods involve gang bonding, that is, the simultaneous bonding of a network frame of leads mechanically fixed in position by a polyimide film (tape). These are created by photo etching a pattern in a solid sheet of 0.025 mm (0.001 in.) thick copper that is bonded to the film. A sample produced by this method is shown in Fig. 34. This process, known as tape automated bonding (TAB), may soon become a widely used method for assembling ICs of all types.

In order for the process to become acceptable, however, the bond integrity of each lead must be ensured. With conventional stitch wire bonding, each wire can

Fig. 32 C-SAM reflection mode image at 15 MHz of a plastic-encapsulated IC showing a suspicious area of the lead frame. In this image, the brightness of the image (towards white) represents the magnitude of the echoes from the interface. There is no regard for the polarity of the echoes. The color version of this image, shown in Fig. 12 in the section "Use of Color for NDE" in this Volume, is a dramatic improvement over the black-and-white image shown here and clearly differentiates the bonded regions from the unbonded regions of the lead frame.

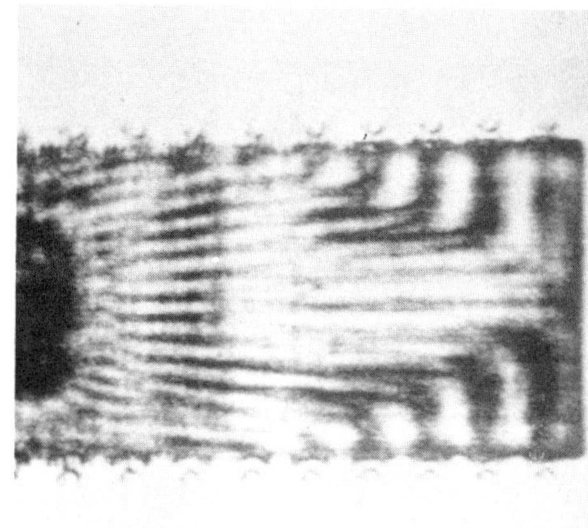

Fig. 33 SLAM through transmission image produced at 10 MHz of sample shown in Fig. 32 of this article and Fig. 12 in the section "Use of Color for NDE" in this Volume. In this image, the disbonded zones are presented as dark. Careful analysis of Fig. 33 indicates that the disbonded areas in black-and-white are larger than those indicated in the color micrograph. This is explained by the fact that the scanning laser acoustic microscope shows the disbonds on either side of the lead frame, that is, two interfaces, while the C-SAM image shows only one interface. When the IC was turned over and the opposide side of the lead frame examined, the SLAM and C-SAM information agreed. Field of view: 35 × 26 mm

be stressed up to a few grams-force in a device called a nondestructive mode pull tester. Unfortunately, the high population density of TAB leads precludes pull testing, except in the case where the device is destroyed to measure the quality of the lead bonds. Acoustic microscopy, particularly SLAM, was determined to be a very reliable nondestructive test method (Ref 23, 24). The notion behind developing the test was that if the areas of bonding of the leads could be measured and if the areas of the bonds correlated with the mechanical strength of the bonds upon destruction, then the test method could be employed nondestructively on subsequent samples (Ref 25).

The good correlations described in Ref 23 and 24 are summarized and illustrated in the discussion that follows. Figures 35 and 36 show 200-MHz SLAM images of lead bonds on two IC chips. Some of the leads are bonded well, as indicated by the clear, bright areas, and other leads are obviously disbonded, completely or partially. Figure 37 shows a graph in which the area of bond, relative to a 100% maximum, is plotted for leads 1 through 68 around the perimeter of a chip. Pull strength relative to a maximum value of 100% is also given for each lead. The pull tests were performed to the point of failure of each lead, and the peak force was recorded. This acoustic microscopy procedure is under consideration as a standard test method for the military.

Example 3: Use of SAM to Evaluate an IC on a Silicon Wafer. Scanning acoustic microscopy offers a technique for generating high-resolution images of material structure and defects in ICs located on silicon crystal substrates. When the IC is scanned with a scanning acoustic microscope having a 2-GHz transducer, the image produced measures approximately 62 × 62 μm (0.0025 × 0.0025 in.). This image displays the metallization on an IC. A high-resolu-

Fig. 34 IC bonded to a frame of tiny leads that have been simultaneously bonded by a process known as tape automated bonding. This process is being used to perform interconnections on densely packed ICs having over 300 leads per chip.

Fig. 35 SLAM acoustic image at 200 MHz of TAB-bonded leads on an IC. The bright areas at the tips of each lead indicate good-quality bonds. Field of view: 1.75 × 1.3 mm

Fig. 36 SLAM acoustic image at 200 MHz of TAB-bonded leads that are not of as uniform quality as those shown in Fig. 35. The poor bonds are mechanically weak and are more likely to suffer from long-term reliability problems. Field of view: 1.75 × 1.3 mm

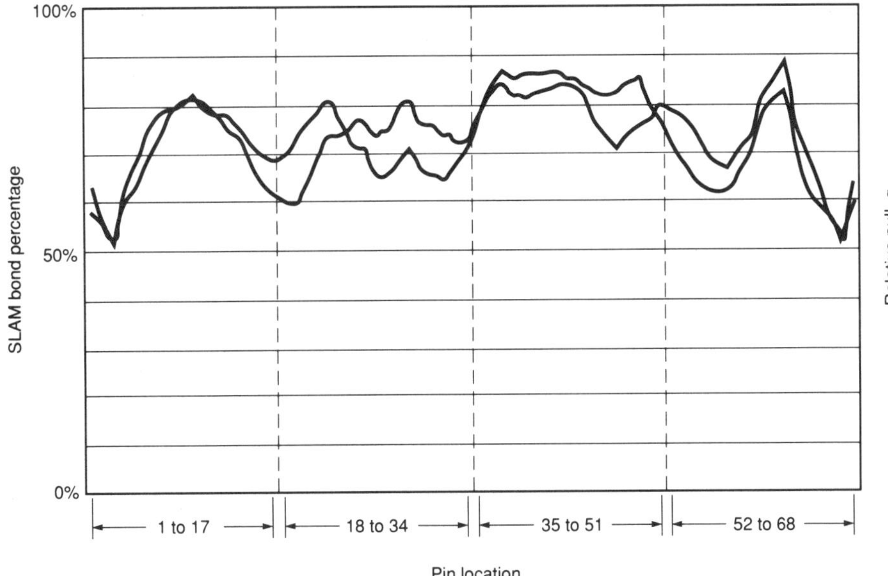

Fig. 37 Plot of bond area percentage versus pin location to verify that SLAM bond strength analysis agrees with destructive pull tests. Graph shows the similarity between the relative strength of TAB-bonded leads, as determined by destructive pull tests, and the areas of bond, as determined by 200-MHz SLAM, on the same sample plotted as a function of location around the perimeter of an IC chip. The locations are denoted by pin numbers 1 through 68. Pressure readings for inner lead bonding are averages obtained for numerous samples that were tested.

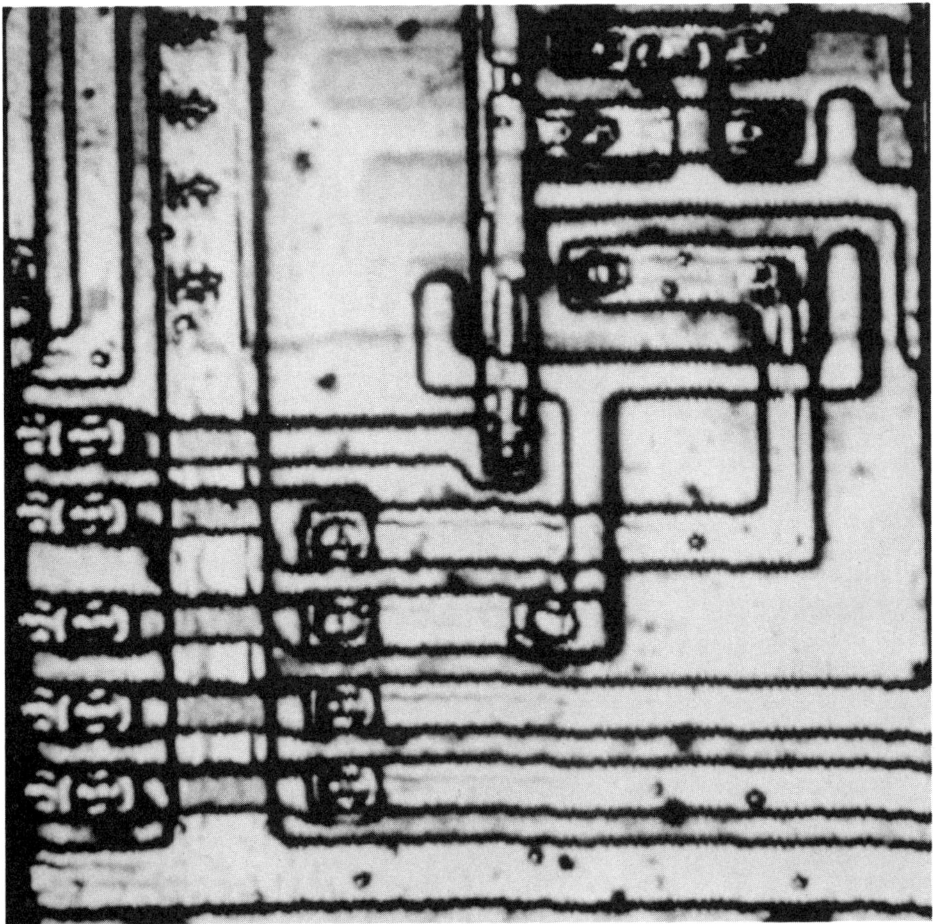

Fig. 38 Pulse-echo SAM micrograph of 4 to 6 μm (160 to 240 μin.) wide aluminum lines on the silicon substrate of an IC. Dark spots are silicon nodule defects, the source of which is silicon bubbling up from the substrate of the wafer. Courtesy of G.H. Thomas, Sandia National Laboratories, Livermore, CA

tion, pulse-echo scan shows 4 to 6 μm (160 to 240 μin.) wide aluminum lines deposited on a silicon substrate (Fig. 38). The small dark spots randomly scattered throughout the metallization are silicon nodules, which bubble up from the substrate, creating small defects in the conducting material. Depending on the focal depth, such structures as defects in the metallization, delaminations or disbonds between conductors, metal line thickness, and buried *p*-type material can be measured or imaged. The shaded fingers midway on the right side in Fig. 38 indicate doped material below the surface. No preparation of the sample prior to testing was necessary.

Ceramic capacitors are passive electronic components used in virtually every electronic circuit to store electrical charge. They are manufactured to be physically small, yet they must have very high capacitance values, which can be obtained with specially formulated, high dielectric constant materials. Figure 39 shows various miniature multilayer capacitors. The basic construction of a capacitor involves two parallel plate electrodes separated by a small distance that is occupied by the dielectric. Capacitance values are inversely proportional to the dielectric thickness and are proportional to the total area of the electrodes.

To increase the area in a small space, multiple layers are used and are electrically connected together in parallel. The problems associated with capacitors that cannot be detected electrically are internal cracks in the ceramic, large voids in the dielectric layers, delaminations between the layers, and high porosity of the dielectric ceramic material. This is illustrated in Fig. 40. Most ceramic capacitors have barium titanate in the dielectric formulation and are therefore x-ray opaque.

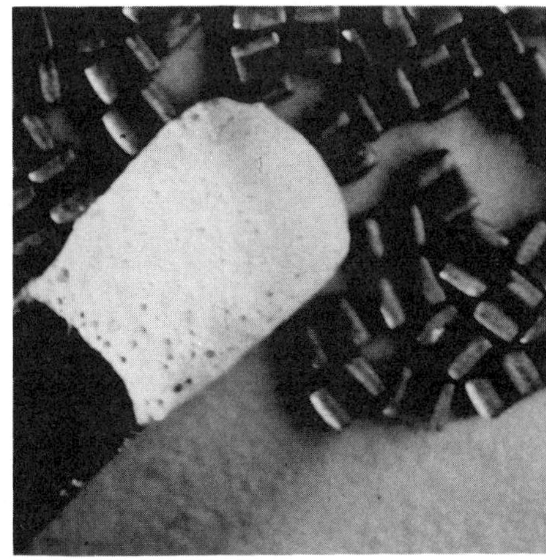

Fig. 39 Miniature ceramic capacitors constructed of multiple layers of high dielectric constant ceramic alternated with layers of conductive metallization. Relative size of single capacitor is evident when compared to the head of a match.

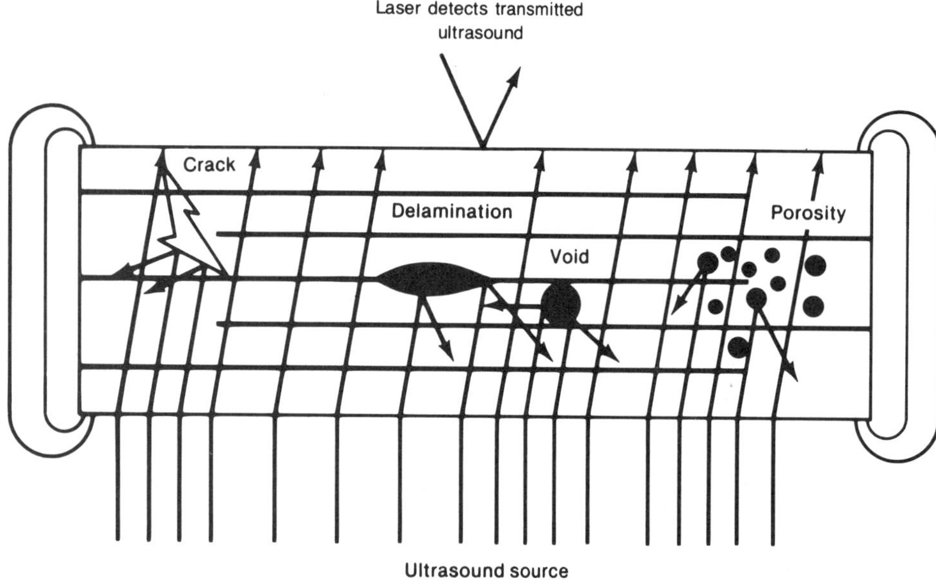

Fig. 40 Diagram illustrating the layered construction used in multilayer ceramic capacitors and the locations of potential defects in these electronic components

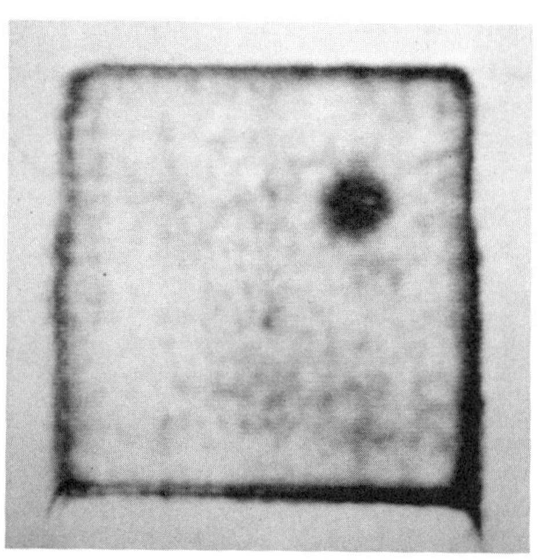

Fig. 42 SLAM image at 30 MHz of a plastic-encapsulated ceramic capacitor having a significant defect that can cause long-term reliability problems. This defect is not evident by electrical testing. Field of view: 14 × 10.5 mm

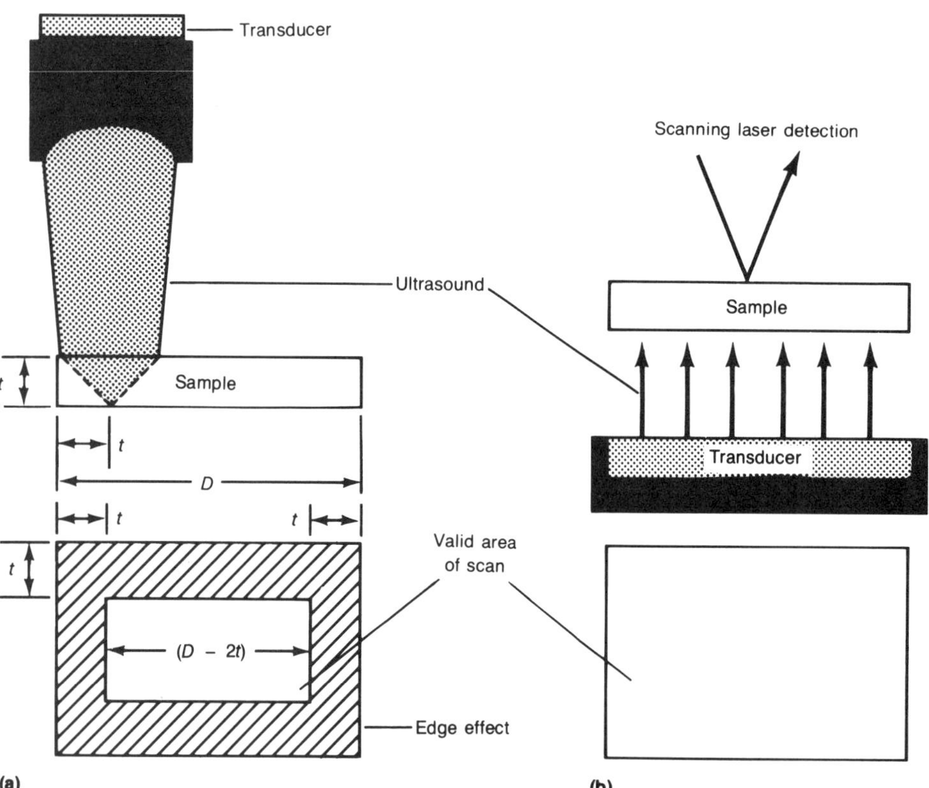

Fig. 41 Edge effect in (a) C-SAM and (b) SLAM. In the case of C-SAM, when the transducer is too close to the left edge of a sample having thickness t, the acoustic beam becomes cut off, and the echo signal does not return from the bottom surface (side view, upper figure). Assuming a ray angle of 45°, the edge of the sample cannot be inspected to within a distance t, thus limiting the area of scan of a sample (top view, lower figure). If the sample has a lateral dimension $D \leqq 2t$, the focused reflection mode cannot be employed in a valid manner. By way of comparison, in the case of SLAM, because collimated energy is used, the edge effect is minor and limited only by diffraction of the illumination wave at the edges (also shown in side view, upper figure, and top view, lower figure).

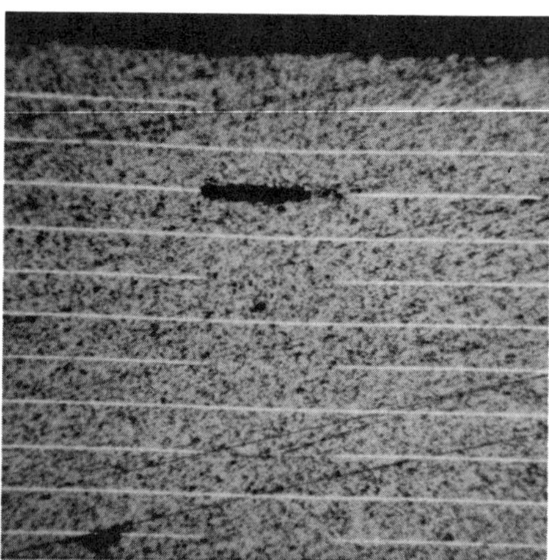

Fig. 43 Cross-sectional analysis of the capacitor shown in Fig. 42 revealing a delamination in the location indicated by SLAM. The cut edge of the capacitor was polished and placed under a conventional microscope.

Ultrasound C-scan was suggested as a method for detecting flaws in ceramic capacitors (Ref 26). However, it was later found to be useful only for the large components because of edge effects that are characteristic of all C-scan and C-SAM images. Scanning laser acoustic microscope techniques are not necessarily limited by the edge effects, as illustrated in Fig. 41.

Figure 42 shows a 30-MHz SLAM image of a ceramic capacitor that has a relatively uniform appearance in transmission. The dark circular spot is a delamination that was confirmed by destructive cross sectioning of the component, as shown in Fig. 43.

The analysis of ceramic capacitors by SLAM has become a standard method for the nondestructive evaluation and screening of ceramic capacitors for internal defects (Ref 27). Investigations have been performed to correlate the information obtained nondestructively with that obtained by conventional destructive physical analysis. Quantitative data have been acquired and associated with the physical properties of the materials. Users of the technique have reported significant improvements in reliability as a result of the SLAM screening of parts.

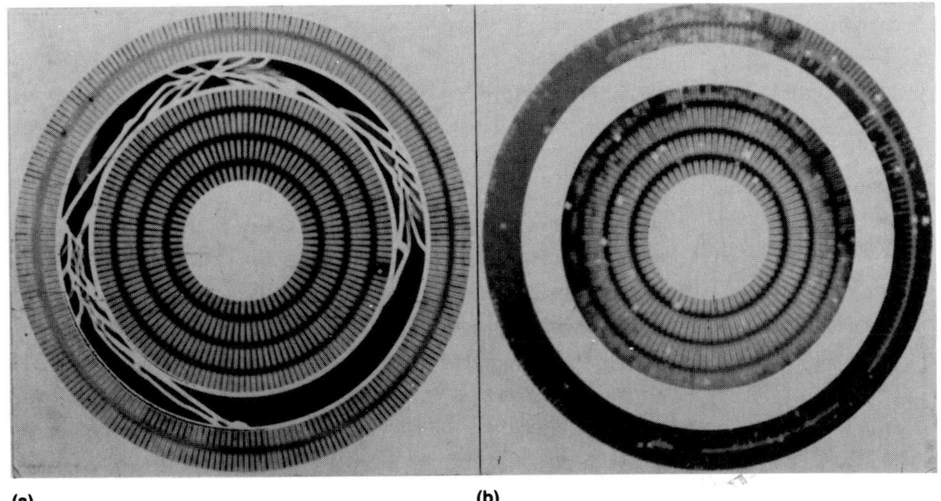

Fig. 44 Thyristor gate arrays imaged through 0.51 mm (0.020 in.) of overlying [100] cut silicon. The annular rings combined with the underlying silicon wafer make up a 5-kV, 2-kA gate-turnoff power device. (a) This device shows well-attached and well-formed finger arrays. (b) This device shows poorly formed and poorly attached arrays. The cracked spacer ring is not critical to the operation of (a) but was included to display cracks as they would appear if the array annuli had cracked. Courtesy of R.S. Gilmore, General Electric Research and Development Center

Silicon Thyristors. The silicon thyristor, which is basically three close coupled *p-n* junctions in the form of a *p-n-p-n* structure, is an important switching device. The thyristor has two stable states: the on state with high current and low voltage and the off state with low current and high voltage. Because of its two stable states and the low power dissipation in these two states, the thyristor has unique applications ranging from speed control in home appliances to switching and power inversion in high-voltage transmission lines. The following example illustrates the use of acoustic microscopy to evaluate whether or not a thyristor has been correctly assembled and will operate according to design specifications.

Example 4: Use of Acoustic Microscopy to Evaluate Gate-Turnoff Silicon-Controlled Thyristor Integrity. The performance of 77 mm (3.0 in.) diam, 5-kV, 2-kA silicon power thyristors is dependent on good electrical attachment between the gate array and the underlying silicon device. Gate-turnoff (GTO) thyristors are fabricated by the solid-state attachment of a conducting pattern between one large [100] cut silicon wafer and two smaller concentric annular wafer segments also having [100] cut orientations. Both the underlying wafer and the annular segments are 0.51 mm (0.020 in.) thick. The large underlying wafer is 77 mm (3.0 in.) in diameter, and the larger-diameter annulus is approximately 51 mm (2.0 in.) in diameter. The successful fabrication of this silicon-controlled power device requires that the conducting patterns be of good quality, that the wafers be intact (no cracks), and that the bonds be well attached.

Acoustic microscopy was chosen to inspect the conducting patterns at the interface between the overlying annuli and the underlying silicon wafer. The surfaces presented to the interrogating ultrasonic beam were lapped and coated with approximately 1.0 μm (40 μin.) of gold for good electrical contact.

The images shown in Fig. 44 were made by focusing a 50-MHz beam on the interface between the annuli and the underlying wafer. This produced a resolution spot size of approximately 100 μm (0.004 in.). The beam was then scanned, and a 1024 × 1024 pixel image was acquired by recording and displaying the echo from the described interface.

Figure 44 shows two silicon-controlled GTO thyristors that contain most of the desirable and undesirable features that can be produced in the fabrication of a silicon power device. The device in Fig. 44(a) shows well-developed finger arrays except for the two outer fingers at the approximate three o'clock position in the inner ring. A silicon spacer ring used to keep the two annuli centered was badly cracked, but this ring will be removed in a subsequent fabrication operation. The spacer ring is included in this image to show how cracks would appear in the silicon. The device in Fig. 44(b) shows very poorly formed finger arrays, and the resulting device would probably be inoperable if it were assembled and tested. This nondestructive evaluation of GTO thyristors demonstrated that acoustic microscopy is a valuable tool for the process and quality control of silicon power devices.

REFERENCES

1. A. Korpel, L.W. Kessler, and P.R. Palermo, Acoustic Microscope Operating at 100 MHz, *Nature*, Vol 232 (No. 5306), 1970, p 110-111
2. Product Bulletin, Sonoscan, Inc.
3. R.A. Lemons and C.F. Quate, Acoustic Microscope—Scanning Version, *Appl. Phys. Lett.*, Vol 24 (No. 4), 1974, p 163-165
4. L.W. Kessler and D.E. Yuhas, Acoustic Microscopy—1979, *Proc. of IEEE*, Vol 67 (No. 4), April 1979, p 526-536
5. Z.C. Lin, H. Lee, G. Wade, M.G. Oravecz, and L.W. Kessler, Holographic Image Reconstruction in Scanning Laser Acoustic Microscopy, *Trans. IEEE*, Vol UFFC-34 (No. 3), May 1987, p 293-300
6. B.Y. Yu, M.G. Oravecz, and L.W. Kessler, "Multimedia Holographic Image Reconstruction in a Scanning Laser Acoustic Microscope," Paper presented at the 16th International Symposium on Acoustical Imaging (Chicago, IL), Sonoscan, Inc., June 1987; L.W. Kessler, Ed., *Acoustical Imaging*, Vol 16, Plenum Press, 1988, p 535-542
7. L.W. Kessler, VHF Ultrasonic Attenuation in Mammalian Tissue, *Acoust. Soc. Am.*, Vol 53 (No. 6), 1973, p 1759-1760
8. M.G. Oravecz, Quantitative Scanning Laser Acoustic Microscopy: Attenuation, *J. Phys. (Orsay)*, Vol 46, Conf. C10, Supplement 12, Dec 1985, p 751-754
9. S.A. Goss and W.D. O'Brien, Jr., Direct Ultrasonic Velocity Measurements of Mammalian Collagen Threads, *Acoust. Soc. Am.*, Vol 65 (No. 2), 1979, p 507-511
10. M.G. Oravecz and S. Lees, Acoustic Spectral Interferometry: A New Method for Sonic Velocity Determination, in *Acoustical Imaging*, Vol 13, M. Kaveh, R.K. Mueller, and T.F. Greenleaf, Ed., Plenum Press, 1984, p 397-408
11. F.J. Cichanski, Method and System for Dual Phase Scanning Acoustic Microscopy, Patent Pending
12. J.S. Foster and D. Rugar, Low-Temperature Acoustic Microscopy, *Trans. IEEE*, Vol SU-32, 1985, p 139-151
13. K.K. Liang, G.S. Kino, and B.T. Khuri-Yakub, Material Characterization by the Inversion of V(z), *Trans. IEEE*, Vol SU-32, 1985, p 213-224
14. R.D. Weglein, Acoustic Micro-Metrology, *Trans. IEEE*, Vol SU-32, 1985, p 225-234
15. J. Kushibiki and N. Chubachi, Material Characterization by Line-Focus-Beam Acoustic Microscope, *Trans. IEEE*, Vol SU-32, 1985, p 189-212
16. M. Oishi, K. Noguchi, T. Masaki, and M. Mizushina, Defect Characterization of Tetragonal Zirconia Polycrystals by a Scanning Laser Acoustic Microscope, in *Proceedings of the International Meeting on Advanced Materials* (Tokyo), Materials Research Society MRS, 1988
17. L.W. Kessler, J.E. Semmens, and F. Agramonte, Nondestructive Die Attach Bond Evaluation Comparing Scanning

Laser Acoustic Microscopy (SLAM) and X-Radiography, in *Proceedings of the 35th Electronic Components Conference* (Washington, D.C.), Institute of Electrical and Electronics Engineers, 1985, p 250-258

18. MIL-STD-883C, Method 2030, U.S. Government Printing Office
19. A.R. Selfridge, Approximate Material Properties in Isotropic Materials, *Trans. IEEE*, Vol SU-32 (No. 3), 1985, p 381-394
20. J.E. Semmens and L.W. Kessler, Nondestructive Evaluation of Thermally Shocked Plastic Integrated Circuit Packages Using Acoustic Microscopy, in *Proceedings of the International Symposium on Testing and Failure Analysis*, ASM INTERNATIONAL, 1988, p 211-215
21. S. Lees, Ultrasonic Measurement of Thin Layers, *Trans. IEEE*, Vol SU-18 (No. 2), 1971, p 81-86
22. *Packaging*, Vol 1, *Electronic Materials Handbook*, ASM INTERNATIONAL, 1989
23. J.E. Semmens and L.W. Kessler, Nondestructive Evaluation of TAB Bonds by Acoustic Microscopy, in *Proceedings of the ISHM Conference* (Seattle, WA), International Society for Hybrid Microelectronics, 1988, p 455-463
24. "Nondestructive Evaluation of Metallurgical Tape Bonds," Report 88-7114, Rome Air Development Center, Griffiss Air Force Base (summary available from Sonoscan, Inc.), 1988
25. L.W. Kessler, Acoustic Microscopy: Nondestructive Tool for Bond Evaluation on TAB Interconnections, in *Proceedings of the ISHM Symposium* (Dallas, TX), International Society for Hybrid Microelectronics, 1984, p 79-84
26. G.R. Love, Nondestructive Testing of Monolithic Ceramic Capacitors, in *Proceedings of the 1973 International Microelectronics Symposium*, 1A3-1A8, International Society for Hybrid Microelectronics, 1973
27. L.W. Kessler and J.E. Semmens, Nondestructive SLAM Analysis of Ceramic Capacitors: An Overview of Experiences From 1980-1989, in *Proceedings of the Capacitor and Resistor Technology Symposium* (Orlando, FL), Components Technology Institute

Use of Color for NDE

MAJOR DEVELOPMENTS in digital electronics have led to the commercial availability of a wide range of specialized devices for image processing and enhancement. With digital image enhancement systems, data interpretation becomes easier, difficult-to-detect features are more readily discernible, and quantitative analysis is possible. Among the numerous image-processing operations available (see the article "Digital Image Enhancement" in this Volume), color-enhanced images produce some of the more valuable, if not spectacular, results.

This collection of color images is intended to assist the reader in recognizing the advantages and/or limitations of pseudocolor in NDE. This collection both complements and supplements articles in this Handbook that describe in detail the principles and applications of methods that utilize color for discontinuity characterization. Numerous comparisons are made between figures appearing in color in this section and their black-and-white counterparts found elsewhere in this Volume. The types of color images appearing in this section include:

Inspection method	Fig. No.
Ultrasonic inspection	2, 3, 6(c), 6(d), 10, 21, 23
Acoustic microscopy	5, 11–16
Thermal analysis	17–20, 22
Computed tomography	1, 7–9
Digital radiography	6(a), 6(b)
Stress analysis(a)	24

(a) The photoelastic fringe patterns shown in Fig. 24 are the result of stress/strain analysis. See the article "Strain Measurement for Stress Analysis" for details.

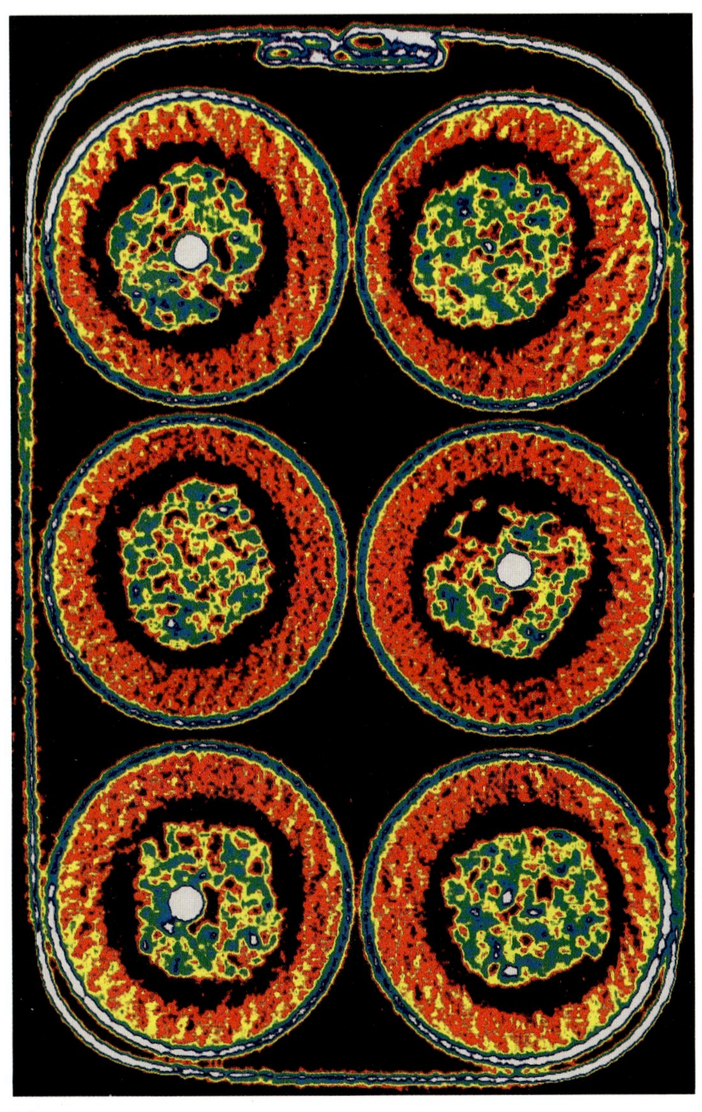

(a)

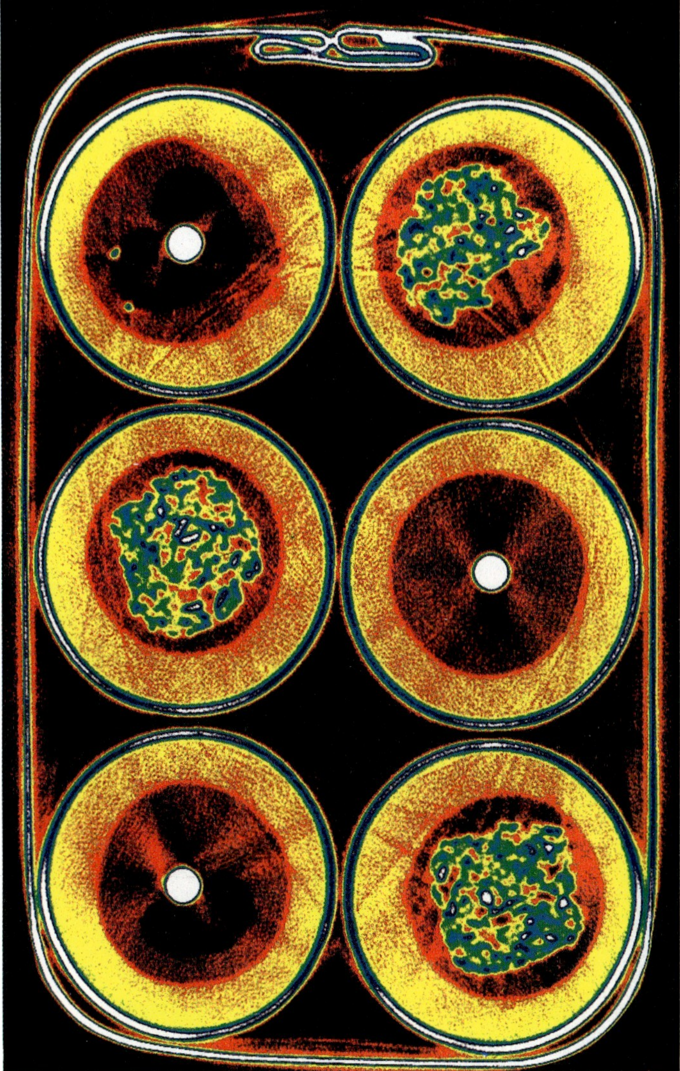

(b)

Fig. 1 Computed tomographic reconstruction of a 9-V battery. (a) Image reconstructed from 500 views showing evidence of undersampling. (b) Image reconstructed from 1000 views, the data filtered with a Wiener filter before backprojection to correct for the point spread of the beam. (T.N. Claytor, Los Alamos National Laboratory)

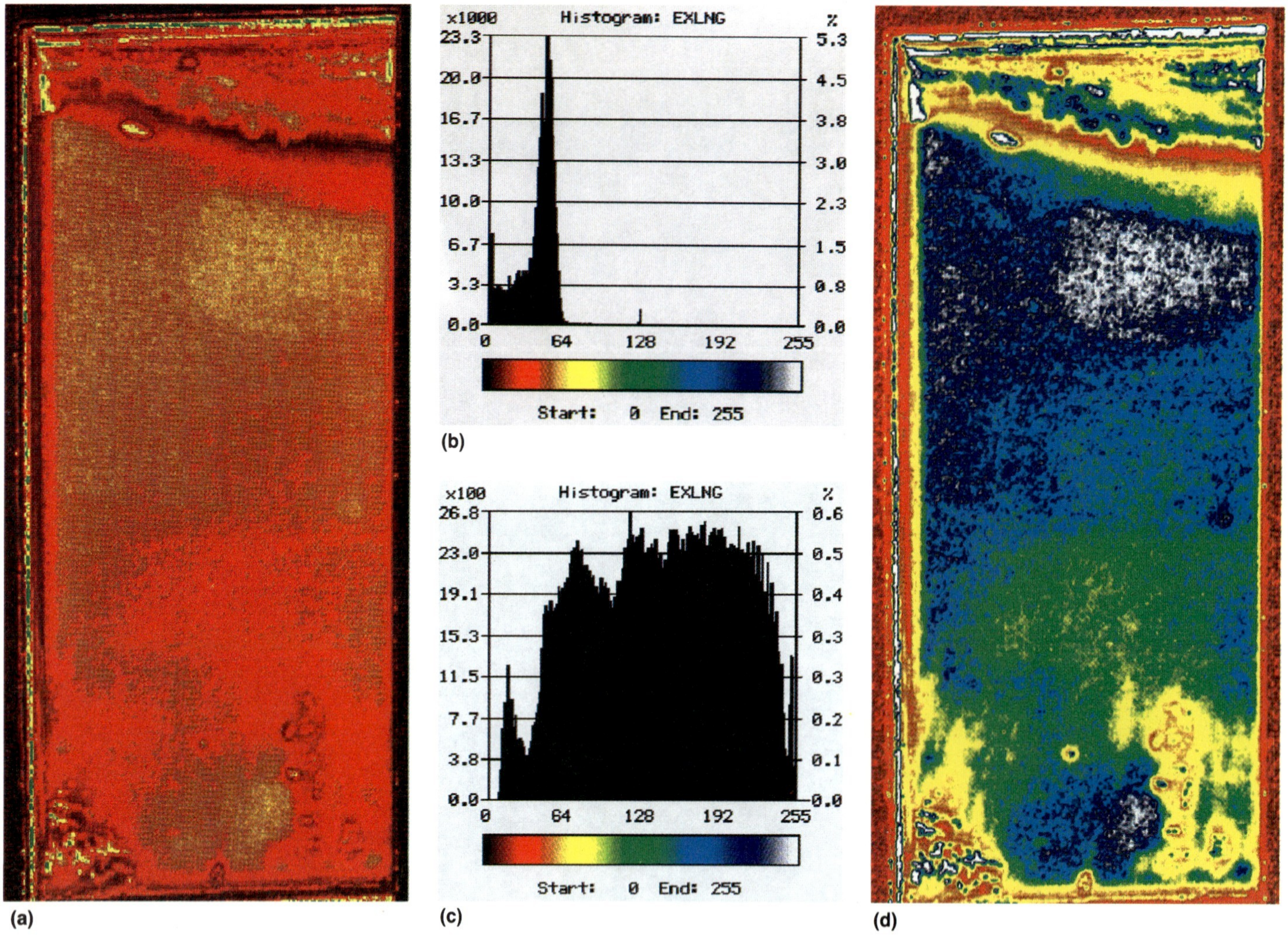

x1000 Histogram: EXLNG %

23.3 20.0 16.7 13.3 10.0 6.7 3.3 0.0

5.3 4.5 3.8 3.0 2.3 1.5 0.8 0.0

0 64 128 192 255

Start: 0 End: 255

(b)

x100 Histogram: EXLNG %

26.8 23.0 19.1 15.3 11.5 7.7 3.8 0.0

0.6 0.5 0.4 0.3 0.3 0.2 0.1 0.0

0 64 128 192 255

Start: 0 End: 255

(a) (c) (d)

Fig. 2 Linear histogram equalization (image enhancement technique) of a low-contrast ultrasonic image of a bond line in an explosively bonded steel-to-aluminum plate. The detonation wave travels from top to bottom, with higher values of intensity indicating better bonding. (a) and (b) Digitized image with 64 levels and corresponding histogram. (c) and (d) After the histogram equalization process, the colors are more widely separated so that the subtler features of the bonding can be seen. (T.N. Claytor, Los Alamos National Laboratory)

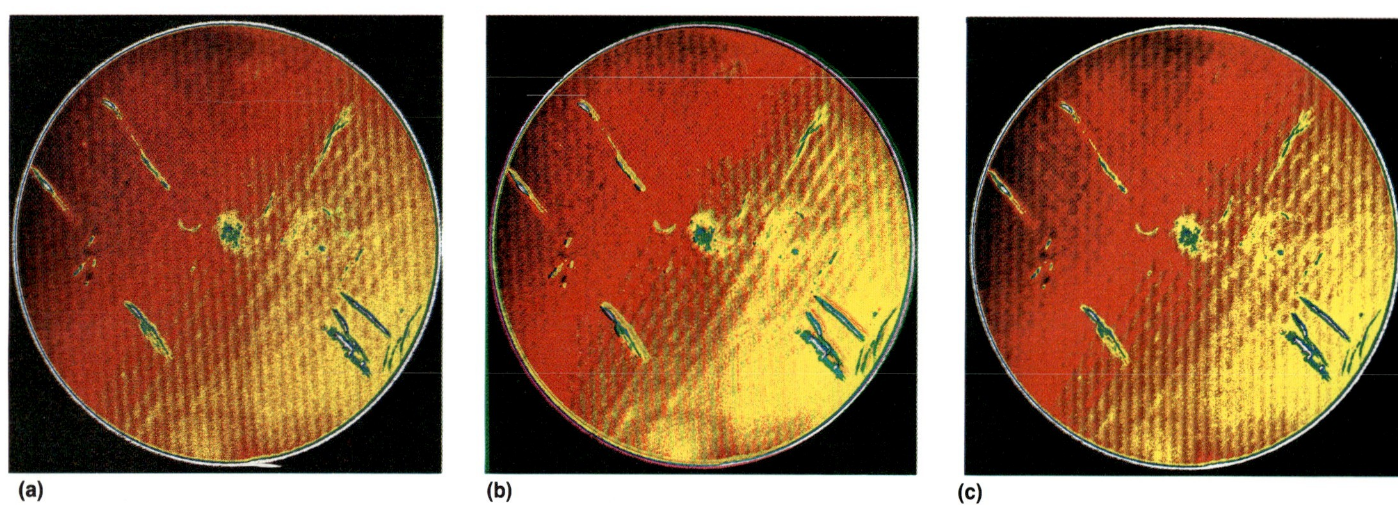

(a) (b) (c)

Fig. 3 Effects of a low-pass filter and a median filter on a slightly noisy ultrasonic image of a cross-rolled tungsten plate with internal rolling tears. (a) The original image has point noise and horizontal streaks. (b) After low-pass filtering, the point noise has been eliminated, but the streaks are still visible. (c) After median filtering, the point noise is eliminated, the streaks are less prominent, and the sharpness of the original image is preserved. (T.N. Claytor, Los Alamos National Laboratory)

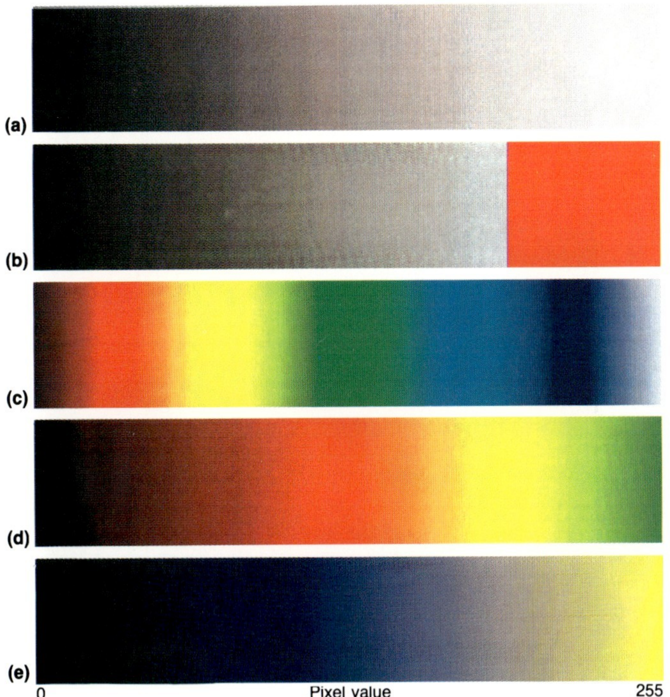

Fig. 4 Palettes useful for black-and-white and color presentation of data. (a) Standard linear gray scale. (b) Gray scale with a red threshold for identifying out-of-bound conditions. (c) The most colorful palette spectrum. (d) and (e) Complementary palettes. (T.N. Claytor, Los Alamos National Laboratory)

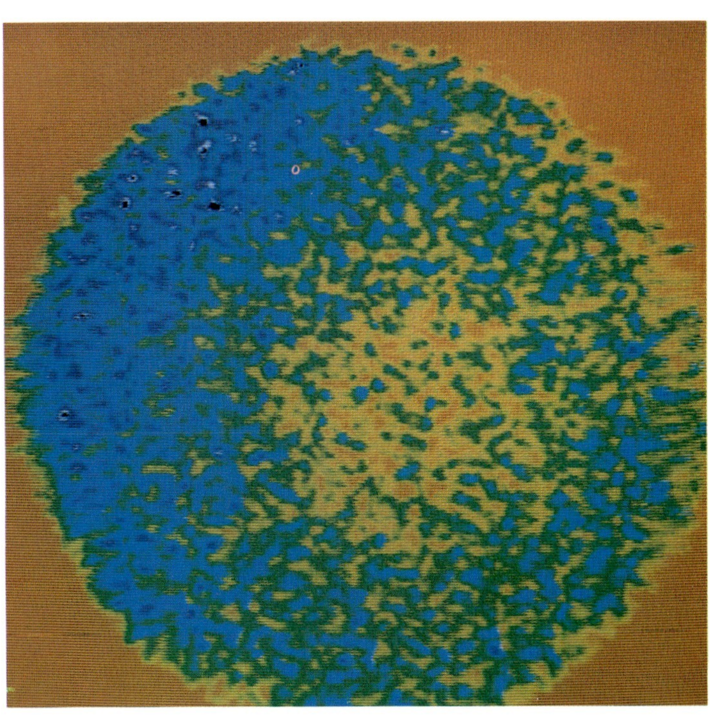

Fig. 5 50-MHz acoustic microscope image of a stainless steel-to-stainless steel diffusion bond with a silver interlayer. The blue, white, and black areas indicate poorer bonding. The pattern displayed in this image matched the ductile/brittle pattern on the fracture surface. (G.H. Thomas, Sandia National Laboratories)

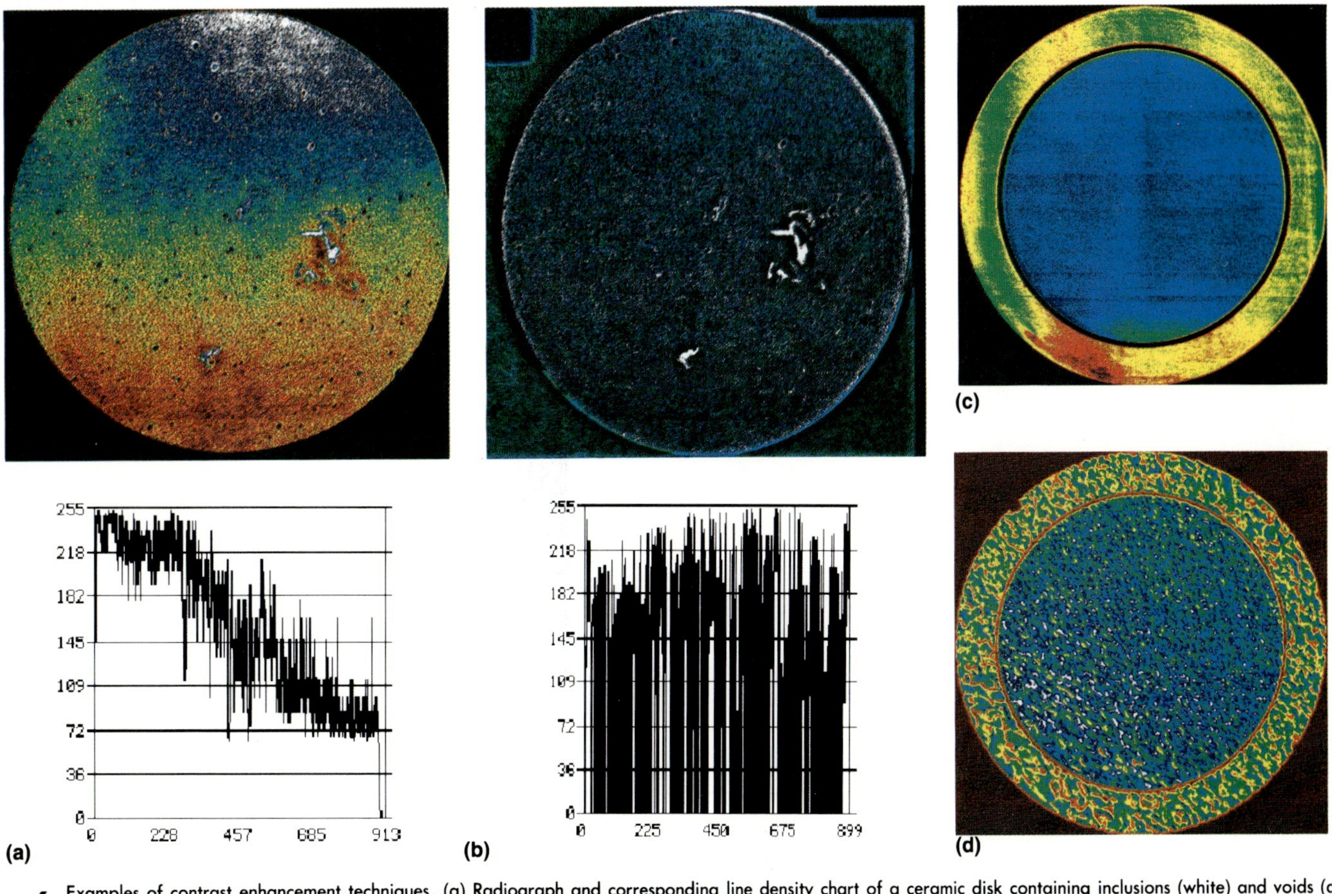

Fig. 6 Examples of contrast enhancement techniques. (a) Radiograph and corresponding line density chart of a ceramic disk containing inclusions (white) and voids (dark areas). (b) After a field-flattening contrast enhancement procedure, the large voids and inclusions are more easily detected at the expense of smaller voids. (c) Ultrasonic image of a HIPed tungsten plate showing density variations enhanced by the use of color. (d) Ultrasonic image of a hot-rolled tungsten plate showing texture variations. Compare Fig. 6(c) and (d) with Fig. 6 and 7 in the article "Digital Image Enhancement." (T.N. Claytor, Los Alamos National Laboratory)

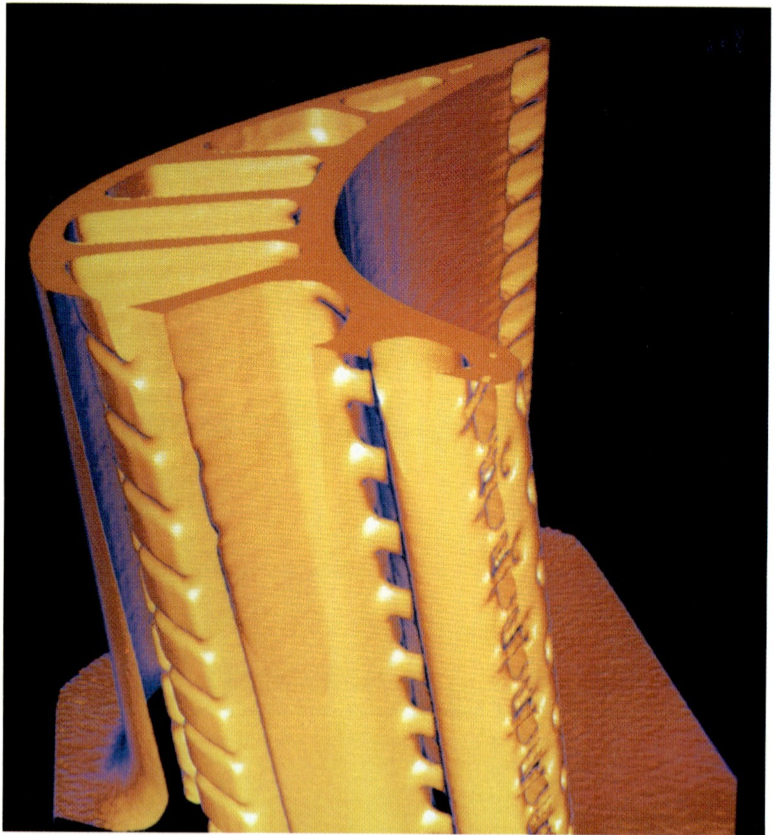

Fig. 7 Digital replica of a three-dimensional surface image of a turbine blade generated from computed tomography data (M.J. Dennis, General Electric NDE Systems and Services)

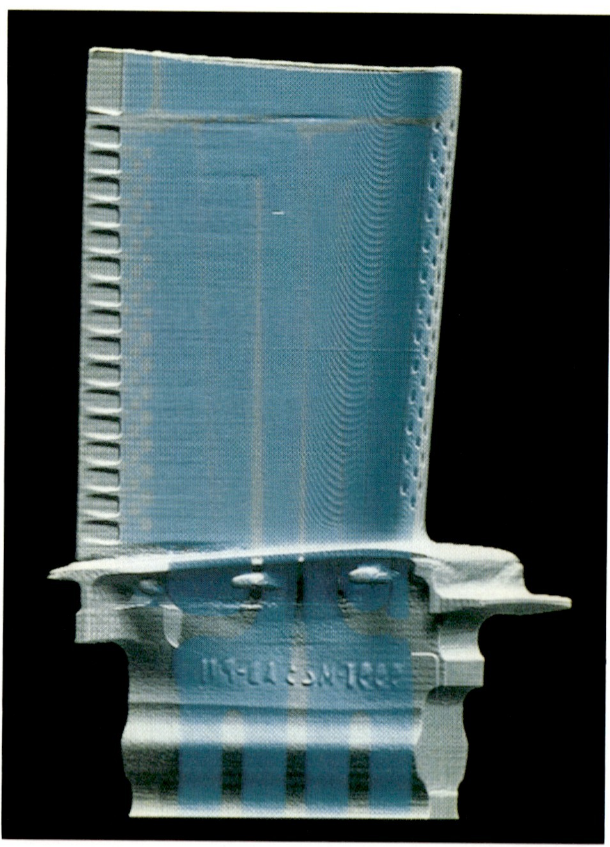

Fig. 8 Voxel reconstruction of a turbine blade derived from computed tomography slices showing internal cooling air passage surfaces outlined in blue (T.N. Claytor, Los Alamos National Laboratory)

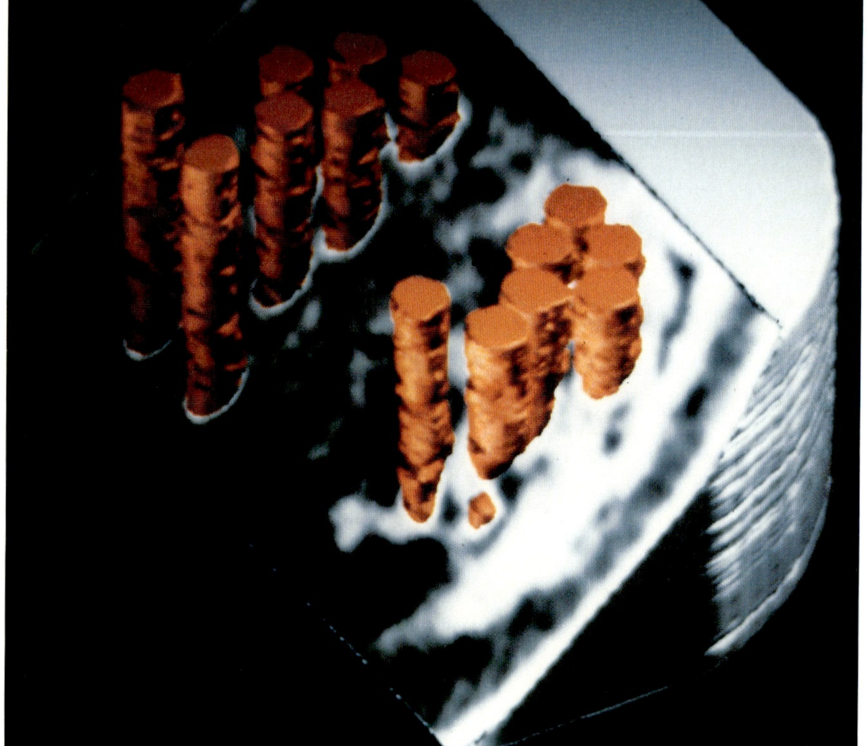

Fig. 9 Three-dimensional tomographic end view of a failed nuclear fuel rod bundle showing white insulation surrounding rods cut away (with threshold operation) at an angle to expose relative positions of orange rods (T.N. Claytor, Los Alamos National Laboratory)

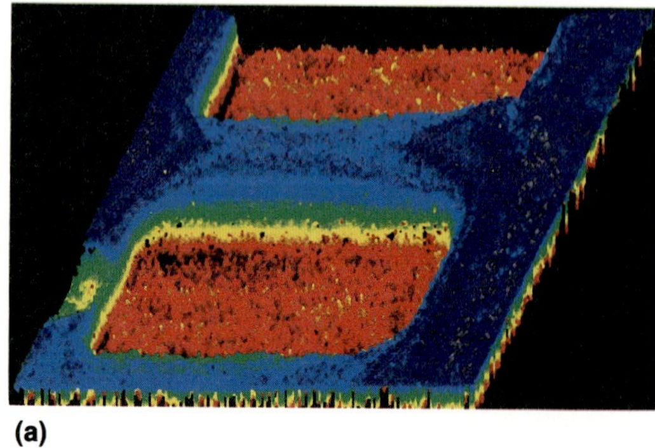

(a)

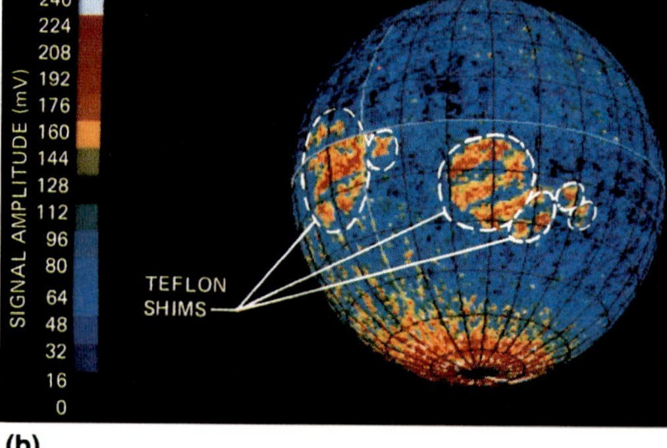

(b)

Fig. 10 Representations of two-dimensional data in a three-dimensional format. (a) Ultrasonic image of defects in circuit board pads. (b) C-scan of a filament-wound vessel with defects. (T.N. Claytor, Los Alamos National Laboratory)

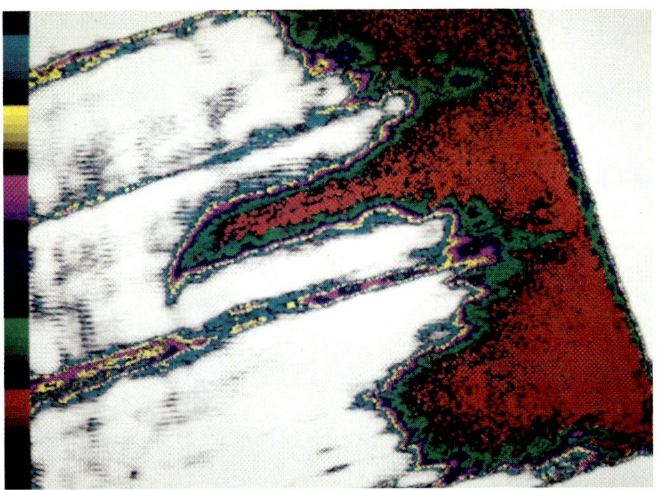

Fig. 11 30-MHz SLAM acoustic image of a Kevlar-reinforced composite test coupon with delamination and cracks. The damage was induced by cutting the bar perpendicular to fibers. The extent of the damage is shown by the ingression of color, with red/green indicating the most extensive damage. The colors are associated with levels of acoustic transmission. A high level (white) indicates a homogeneous material, while a lower level (red/green) indicates discontinuities. (L.W. Kessler, Sonoscan, Inc.)

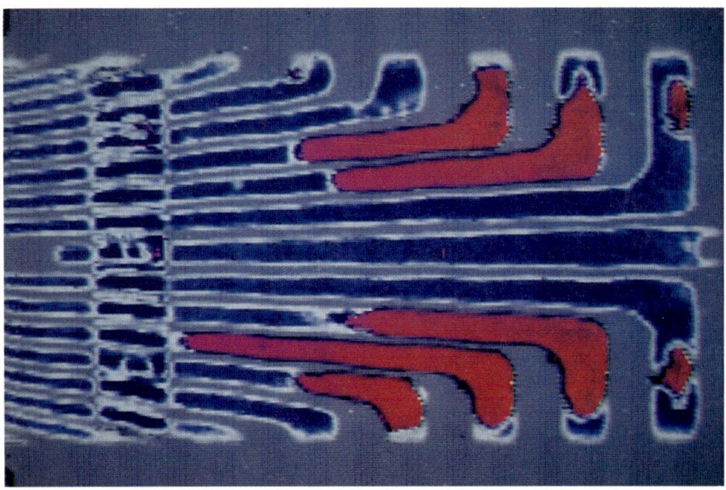

Fig. 12 15-MHz C-SAM reflection mode image of a plastic-encapsulated integrated circuit showing defects in the lead frame. In this image, the echo polarity information is presented by color scale changes in which disbonds are red (negative polarity) and good bonds (positive echoes) are blue. Without this polarity information, it is difficult to make a correct interpretation of the bonding condition in this sample (see Fig. 32 in the article "Acoustic Microscopy"). (L.W. Kessler, Sonoscan, Inc.)

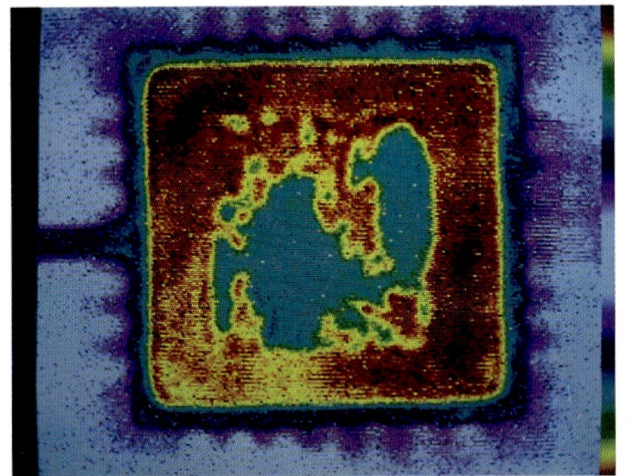

Fig. 13 50-MHz C-SAM reflection mode image of the same die-attach sample shown in Fig. 29 and 30 in the article "Acoustic Microscopy." In this image, areas of disbond are indicated by the color red. (L.W. Kessler, Sonoscan, Inc.)

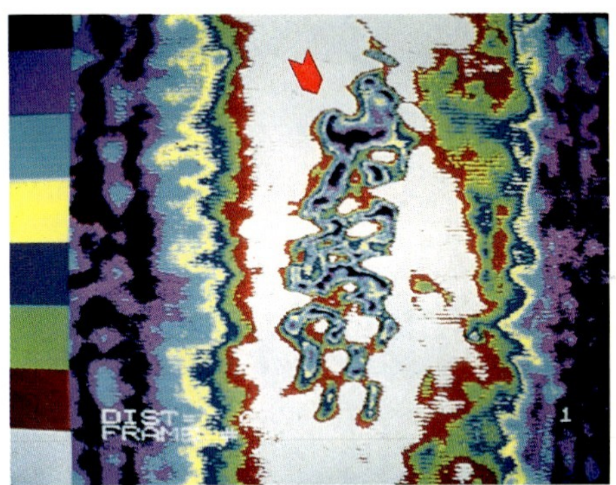

Fig. 14 30-MHz SLAM acoustic micrograph of the carbon fiber reinforced polymer component shown in Fig. 13 of "Acoustic Microscopy." Long, irregular defect (arrow) corresponds to blistering of one of the laminates. (L.W. Kessler, Sonoscan, Inc.)

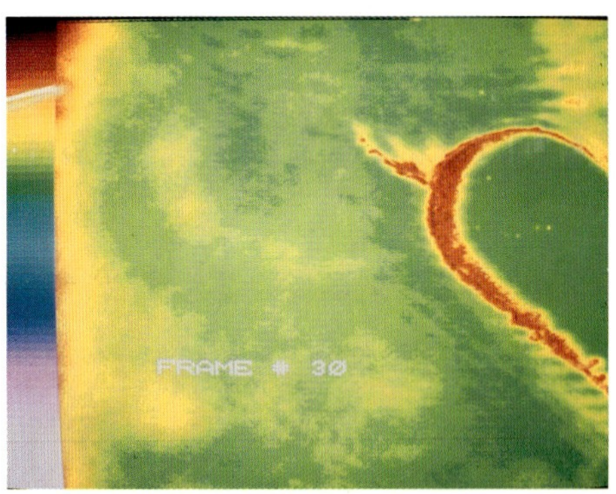

Fig. 15 100-MHz SLAM micrograph of the alumina panel shown in Fig. 16 in the article "Acoustic Microscopy." Color is used to delineate a crack with a specific acoustic brightness level. (L.W. Kessler, Sonoscan, Inc.)

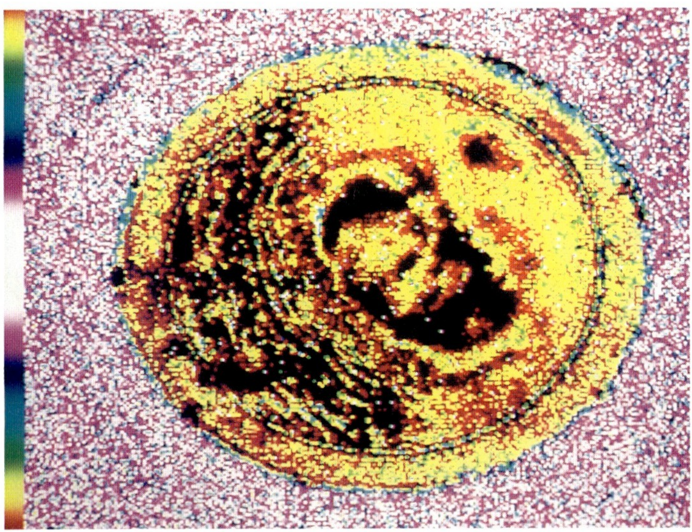

(a)

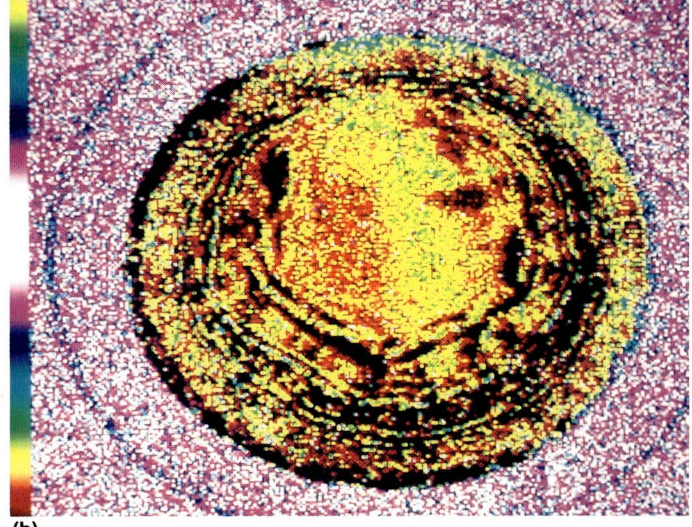

(b)

Fig. 16 15-MHz reflection mode C-SAM images of internal defects in the alumina test disk shown in Fig. 18 in the article "Acoustic Microscopy." In (a), the darker features correspond to higher-magnitude echo levels than the yellow. In (b), the focus and gate levels were changed slightly. Because the defects are not parallel to the surface of the part, echoes will not return at different times; therefore, a fixed-gate position will show echoes only at the depth selected. (L.W. Kessler, Sonoscan, Inc.)

Fig. 17 Thermal images of graphite-epoxy sheets with 20 × 20 mm implanted Teflon defects at 0.3 mm depth (a) and 2.25 mm depth (b). Image processing (c) improved the detectability of the 2.25 mm depth defect. (P. Cielo, National Research Council of Canada)

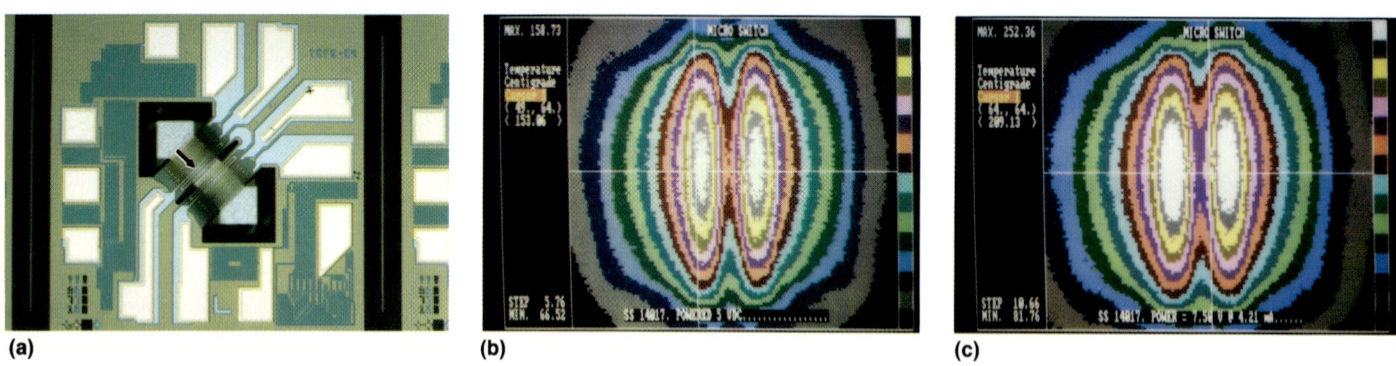

Fig. 18 Thermal analysis of a silicon mass airflow sensor. (a) Heater resistor (arrow) under analysis, 50×. (b) and (c) Thermal maps indicating a 100 °C rise in temperature of an exposed heat resistor chip under no-flow conditions and normal voltage (b) and under 150% voltage (c). (R.W. Gehman, Micro Switch)

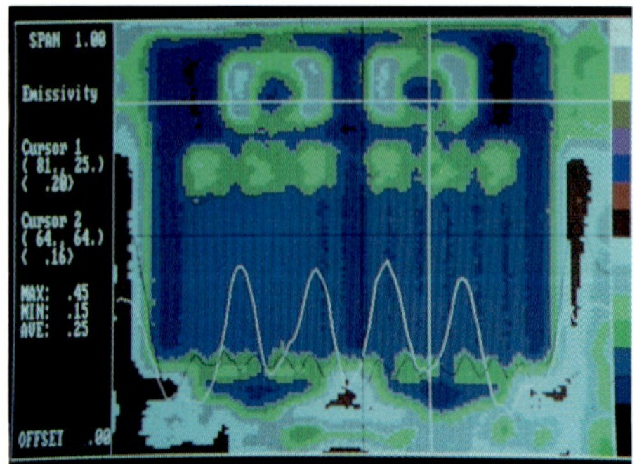

Fig. 19 Emissivity display of a GaAs FET semiconductor chip. Color scale at right represents 16 emissivity levels between lowest (black) and highest (white) values. (J.O. Brown, EDO Corp.)

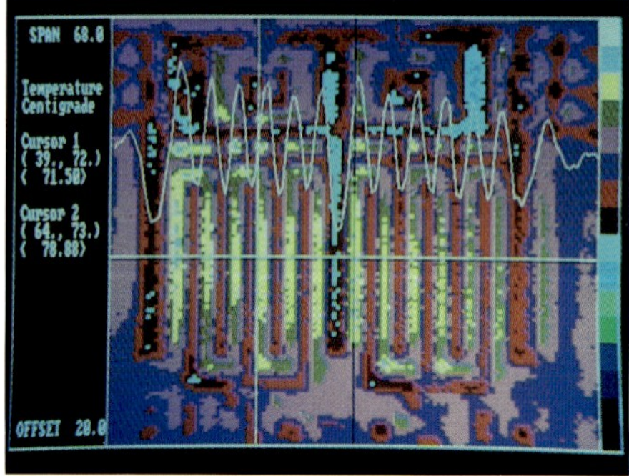

Fig. 20 Temperature display of a GaAs FET device showing a substrate bonding flaw as an oval-shaped anomaly (yellow) across center. Area displayed is 0.6 mm across. (J.O. Brown, EDO Corp.)

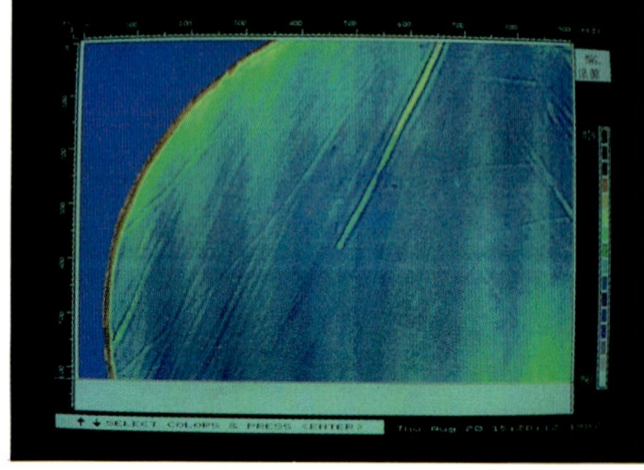

Fig. 21 50-MHz surface wave ultrasonic image of scratch and lapping marks in Si_3N_4. The smallest visible scratches are 25 to 50 μm wide. (T. Nelligan, Panametrics Inc.)

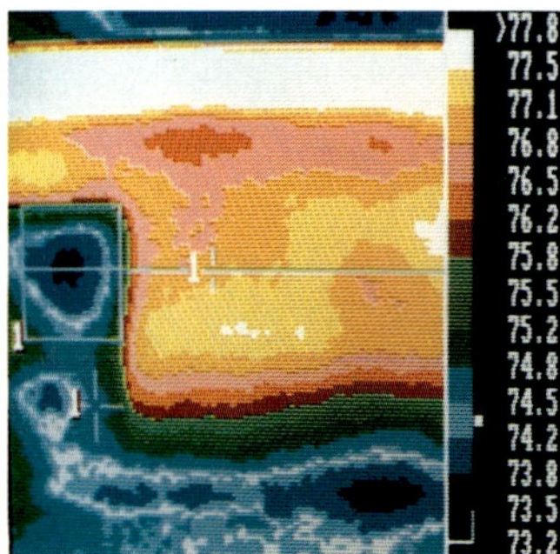

Fig. 22 Thermal imaging of two simulated defects (Teflon inserts) in a graphite-epoxy laminate (D. Mauro, Techmarketing Inc./AGEMA Infrared Systems)

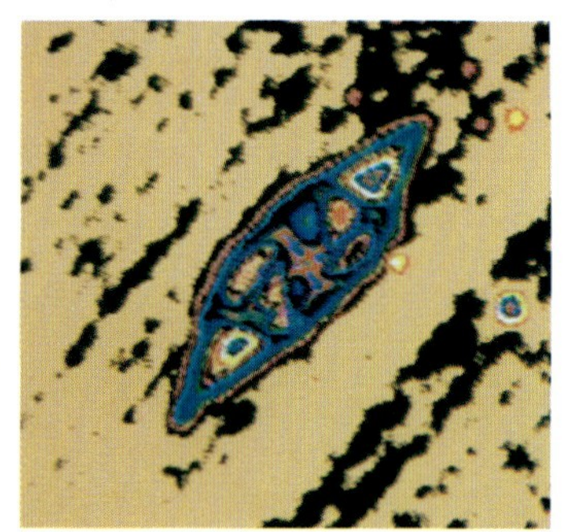

Fig. 23 Ultrasonic image of an impacted region in a 1.4 mm thick graphite-reinforced polymer sheet. The damaged region is ~6 mm long. (M.C. Bhardwaj, Ultron Laboratories, Inc.)

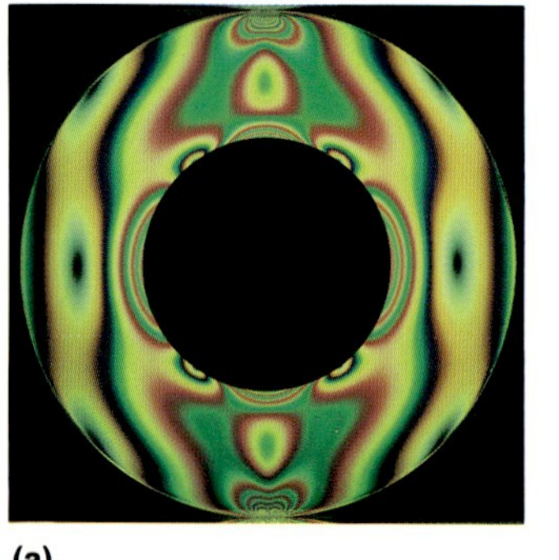

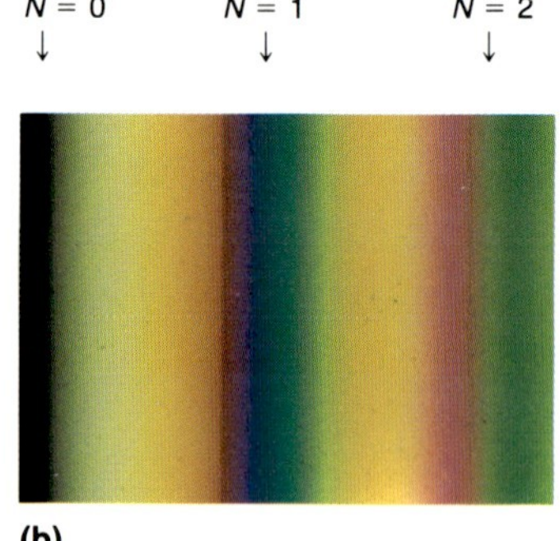

Fig. 24 Photoelastic fringe patterns produced when a photoelastic material is strained and viewed under polarized light (a) and the photoelastic color sequence viewed with a circular polariscope (b). *N*, fringe order phase shift. (L.D. Lineback, Measurements Group, Inc.)

Nondestructive Inspection of Specific Products

Forgings

IN FORGINGS of both ferrous and nonferrous metals, the flaws that occur most often are caused by conditions that exist in the ingot, by subsequent hot working of the ingot or the billet, and by hot or cold working during forging. The nondestructive inspection (NDI) methods most commonly used to detect these flaws include visual, magnetic particle, liquid penetrant, ultrasonic, eddy current, and radiographic inspection. This article discusses the applications of these methods to forgings. Information on the equipment and techniques used in these inspection methods is available in the articles so titled in this Volume.

Flaws Originating in the Ingot

Many large open-die forgings are forged directly from ingots. Most closed-die forgings and upset forgings are produced from billets, rolled bar stock, or preforms. Many, though by no means all, of the imperfections found in forgings can be attributed to conditions that existed in the ingot, sometimes even when the ingot has undergone primary reduction prior to the forging operation. Some, but again by no means all, of the service problems that occur with forgings can be traced to imperfections originating in the ingot.

Chemical Segregation. The elements in a cast alloy are seldom distributed uniformly. Even unalloyed metals contain random amounts of various types of impurities in the form of tramp elements or dissolved gases; these impurities are also seldom distributed uniformly. Therefore, the composition of the metal or alloy will vary from location to location. Deviation from the mean composition at a particular location in a forging is termed segregation. In general, segregation is the result of solute rejection at the solidification interface during casting. For example, the gradation of composition with respect to the individual alloying elements exists from cores of dendrites to interdendritic regions. Segregation therefore produces a material having a range of compositions that do not have identical properties.

Forging can partially correct the results of segregation by recrystallizing or breaking up the grain structure to promote a more homogeneous substructure. However, the effects of a badly segregated ingot cannot be totally eliminated by forging; rather, the segregated regions tend to be altered by the working operation, as shown in Fig. 1.

In metals, the presence of localized regions that deviate from the nominal composition can affect corrosion resistance, forging and joining (welding) characteristics, mechanical properties, fracture toughness, and fatigue resistance. In heat-treatable alloys, variations in composition can produce unexpected responses to heat treatments, which result in hard or soft spots, quench cracks, or other flaws. The degree of degradation depends on the alloy and on process variables. Most metallurgical processes are based on an assumption that the metal being processed is of a nominal and reasonably uniform composition.

Ingot Pipe and Centerline Shrinkage. A common imperfection in ingots is the shrinkage cavity, commonly known as pipe, often found in the upper portion of the ingot. Shrinkage occurs during freezing of the metal, and eventually there is insufficient liquid metal near the top end to feed the ingot. As a result, a cavity forms, usually approximating the shape of a cylinder or cone—hence the term pipe. Piping is illustrated in Fig. 2. In addition to the primary pipe near the top of the ingot, secondary regions of piping and centerline shrinkage may extend deeper into an ingot (Fig. 3).

Fig. 1 Microstructural bonding due to chemical segregation and mechanical working

Primary piping is generally an economic concern, but if it extends sufficiently deep into the ingot body and goes undetected, it can eventually result in a defective forging. Detection of the pipe can be obscured in some cases if bridging has occurred.

Piping can be minimized by pouring ingots with the big end up, by providing risers in the ingot top, and by applying sufficient hot-top material (insulating refractories or exothermic materials) immediately after pouring. These techniques extend the time that the metal in the top regions of the ingot remains liquid, thus minimizing the shrinkage cavity produced in this portion of the ingot.

On the other hand, secondary piping and centerline shrinkage can be very detrimental because they are harder to detect in the mill and may subsequently produce centerline defects in bar and wrought products. Such a material condition may indeed provide the flaw or stress concentrator for a forging burst in some later processing operation or for a future product failure.

High Hydrogen Content. A major source of hydrogen in certain metals and alloys is the reaction of water vapor with the liquid metal at high temperatures. The water vapor may originate from the charge materials, slag ingredients and alloy additions, refractory linings, ingot molds, or even the atmosphere itself if steps are not taken to prevent such contamination. The resulting hydrogen goes into solution at elevated temperatures; but as the metal solidifies after pouring, the solubility of hydrogen decreas-

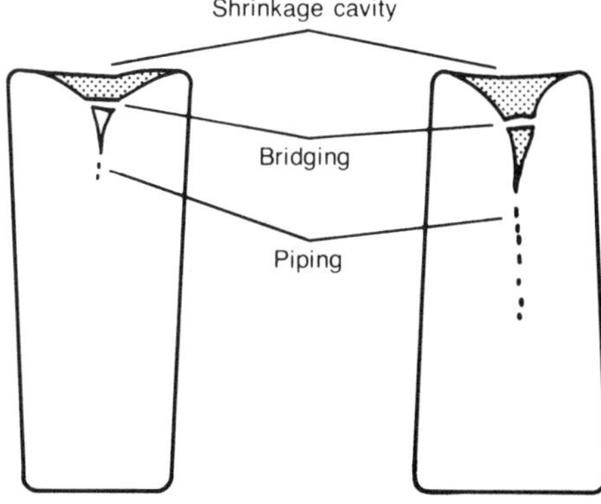

Fig. 2 Schematic showing piping in top-poured ingots

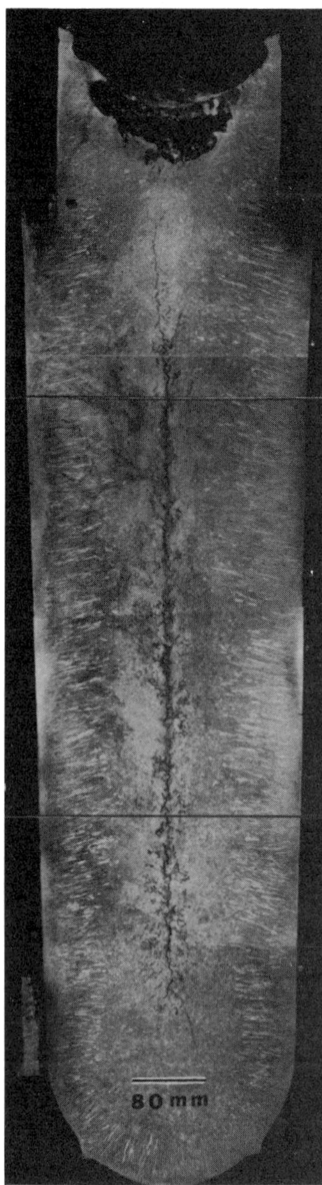

Fig. 3 Longitudinal section through an ingot showing extensive centerline shrinkage

es, and it becomes entrapped in the metal lattice.

Hydrogen concentration in excess of about 5 ppm has been associated with flaking, especially in heavy sections and high-carbon steels. Hydrogen flakes (Fig. 4) are small cracks produced by hydrogen that has diffused to grain boundaries and other preferred sites, for example, inclusion/matrix interfaces. However, hydrogen concentrations in excess of only 1 ppm have been related to the degradation of mechanical properties in high-strength steels, especially ductility, impact behavior, and fracture toughness.

Metals can also possess a high hydrogen content without the presence of flakes or voids. In this case, the hydrogen may cause embrittlement of the material along selective paths, which can drastically reduce the resistance of a forged part to crack propagation resulting from impact loading, fatigue, or stress corrosion.

In cases where hydrogen-related defects can serve as the initiation site for cracking and thus increase the likelihood of future failures, it is advisable to use a thermal treatment that can alleviate this condition. For example, slow cooling immediately following a hot-working operation or a separate annealing cycle will relieve residual stresses in addition to allowing hydrogen to diffuse to a more uniform distribution throughout the lattice and, more important, to diffuse out of the material.

Nonmetallic inclusions, which originate in the ingot, are likely to be carried over to the forgings, even though several intermediate hot-working operations may be involved. Also, additional inclusions may develop in the billet or in subsequent forging stages.

Most nonmetallic inclusions originate during solidification from the initial melting operation. If no further consumable-remelting cycles follow, as in air-melted or vacuum-induction products (with no remelting cycle to follow), the size, frequency, and distribution of the nonmetallic inclusions will not be altered or reduced in size or frequency during further processing. If a subsequent vacuum-remelting operation is used, the inclusions will be lessened in size and frequency and will become more random in nature. If an electroslag-remelting cycle is used, a more random distribution of inclusions will result.

Two kinds of nonmetallic inclusions are generally distinguished in metals:

- Those that are entrapped in the metal inadvertently and originate almost exclusively from particles of matter that are occluded in the metal while it is molten or being cast
- Those that separate from the metal because of a change in temperature or composition

Inclusions of the latter type are produced by separation from the metal when it is in either the liquid or the solid state. Oxides, sulfides, nitrides, or other nonmetallic compounds form droplets or particles when these compounds are produced in such amounts that their solubility in the matrix is exceeded.

Air-melted alloys commonly contain inclusions mainly of these chemical characteristics. Vacuum- or electroslag-remelted alloys more commonly contain conglomerates of any of these types, frequently combined with carbon or the hardening element or elements that precipitate during stabilization and aging cycles to form inclusions such as titanium carbonitrides or carbides. Homogenizing cycles are normally used for the ingot prior to conversion or at an early stage of conversion.

Because these compounds are products of reactions within the metal, they are normal constituents of the metal, and conventional melting practices cannot completely eliminate such inclusions. However, it is desirable to keep the type and amount of inclusions to a minimum so that the metal is relatively free from those inclusions that cause the most problems.

Of the numerous types of flaws found in forgings, nonmetallic inclusions appear to contribute significantly to service failures, particularly in high-integrity forgings such as those used in aerospace applications. In many applications, the presence of these inclusions decreases the ability of a metal to withstand high static loads, impact forces, cyclical or fatigue loading, and sometimes corrosion and stress corrosion. Nonmetallic inclusions can easily become stress concentrators because of their discontinuous nature and incompatibility with the surrounding composition. This combination may very well yield flaws of critical size that, under appropriate loading conditions, result in complete fracture of the forged part.

Unmelted electrodes and shelf are two other types of ingot flaws that can impair forgeability. Unmelted electrodes (Fig. 5a) are caused by chunks of electrodes being eroded away during consumable melting and dropping down into the molten material as a solid. Shelf (Fig. 5b) is a condition resulting from uneven solidification or cooling rates at the ingot surfaces.

The consumable-melting operation has occasionally been continued to a point where a portion of the stinger rod is melted into the ingot, which may be undesirable because the composition of the stinger rod may differ from that of the alloy being melted. To prevent this occurrence, one practice is to weld a wire to the stinger rod, bend the wire down in tension, and weld the other end of the wire to the surface of the electrode a few inches below the junction of the stinger rod and the electrode. When the electrode has been consumed to where the wire is attached to it, the wire is released and springs out against the side of the crucible, thus serving as an alarm to stop the melting. A disadvantage of this practice is that the wire may become detached and contaminate the melt.

Flaws Caused by Processing of the Ingot or Billet

Flaws that occur during the preliminary reduction of ingots or billets prior to final forging include internal bursts and various surface flaws, such as laps, seams, slivers, rolled-in scale, ferrite fingers, fins, overfills, and underfills. Some flaws of these types may also occur during final forging (see next section).

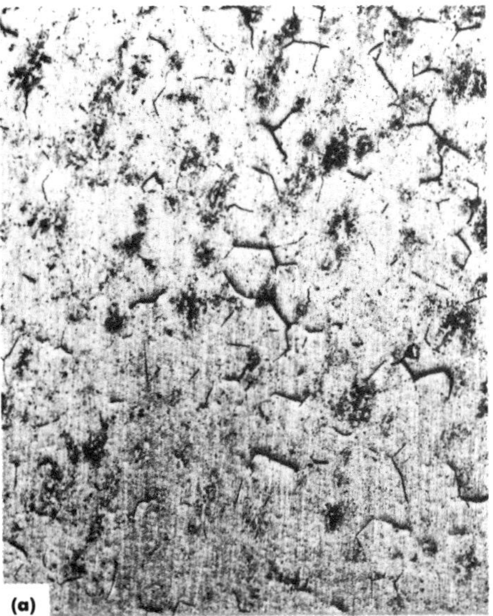

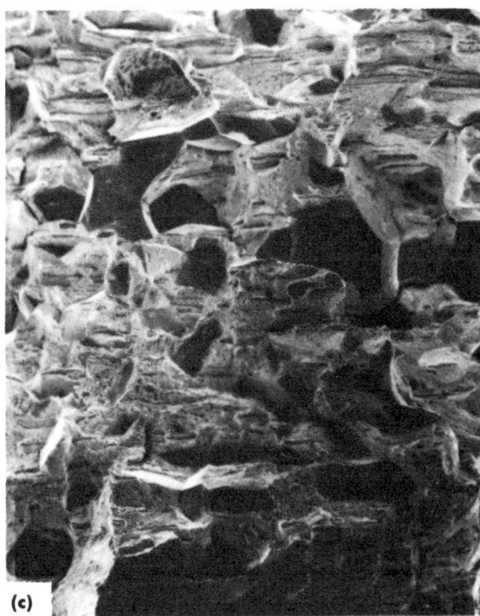

Fig. 4 Hydrogen flaking in an alloy steel bar. (a) Polished cross section showing cracks due to flaking. (b) Fracture surface containing hydrogen flakes. Note the reflective, faceted nature of the fracture. (c) SEM micrograph showing the intergranular appearance of the flakes in this material

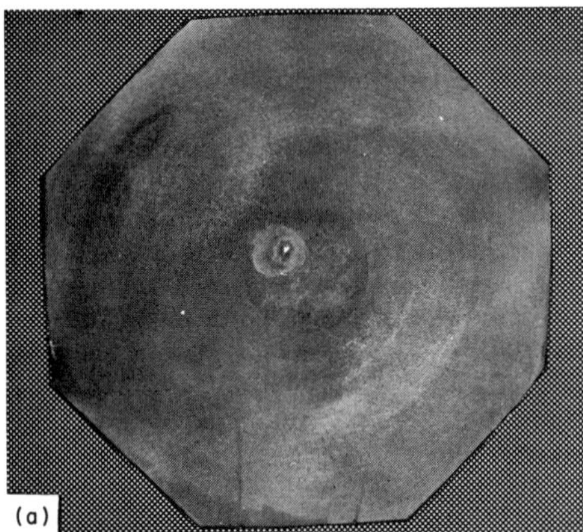

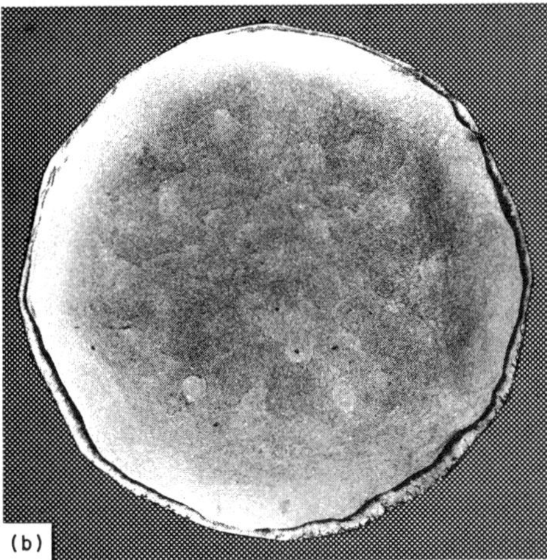

Fig. 5 Sections through two heat-resistant alloy ingots showing flaws that can impair forgeability. (a) Piece of unmelted consumable electrode (white spot near center). (b) Shelf (black line along edge) resulting from uneven solidification of the ingot

Bursts. Where the work metal is weak, possibly from pipe, porosity, segregation, or inclusions, the tensile stresses can be sufficiently high to tear the material apart internally, particularly if the hot-working temperature is too high. Such internal tears are known as forging bursts or ruptures. Similarly, if the metal contains low-melting phases resulting from segregation, these phases may cause bursts during hot working of the ingot or billet. Bursts can also occur during the forging operation.

Figure 6 shows a large forging burst that occurred during the forging of an electroslag-remelted ingot. The cause was traced to a weak solidification plane near the bottom of the ingot combined with higher forging temperatures than normal.

Laps are surface irregularities that appear as linear defects and are caused by the folding over of hot metal at the surface. These folds are forged into the surface, but are not metallurgically bonded (welded) because of the oxide present between the surfaces. Therefore, a discontinuity with a sharp notch is created.

A seam is a surface defect that also appears as a longitudinal indication and is a result of a crack, a heavy cluster of nonmetallic inclusions, or a deep lap (a lap that intersects the surface at a large angle). A seam can also result from a defect in the ingot surface, such as a hole, that becomes oxidized and is prevented from healing during working. In this case, the hole simply stretches out during forging or rolling, producing a linear cracklike seam in the workpiece surface.

Other Surface Defects. Slivers are loose or torn pieces of steel rolled into the surface. Rolled-in scale is scale formed during rolling. Ferrite fingers are surface cracks that have been welded shut but still contain the oxides and decarburization. Fins and overfills are protrusions formed by incorrect reduction during hot working. Underfills are the result of incomplete working of the section during reduction.

Flaws Caused by the Forging Operation

Flaws produced during the forging operation (assuming a flaw-free billet or bar) are the result of improper setup or control. Proper control of heating for forging is necessary to prevent excessive scale, decarburization, overheating, or burning. Excessive scale, in addition to causing excessive metal loss, can result in forgings with pitted surfaces. The pitted surfaces are caused by the scale being hammered into the surface and may result in unacceptable forgings.

Severe overheating causes burning, which is the melting of the lower melting point constituents. This melting action severely reduces the mechanical properties of the metal, and the damage is irreparable. Detection and sorting of forgings that have been burned during heating can be extremely difficult.

In many cases, the flaws that occur during forging are the same as, or at least similar to, those that may occur during hot working of the ingots or billets; these are described in the previous section.

Internal flaws in forgings often appear as cracks or tears, and they may result either from forging with too light a hammer or from continuing forging after the metal has cooled down below a safe forging temperature. Bursts, as described above, may also occur during the forging operation.

A number of surface flaws can be produced by the forging operation. These flaws are often caused by the movement of metal over or upon another surface without actual

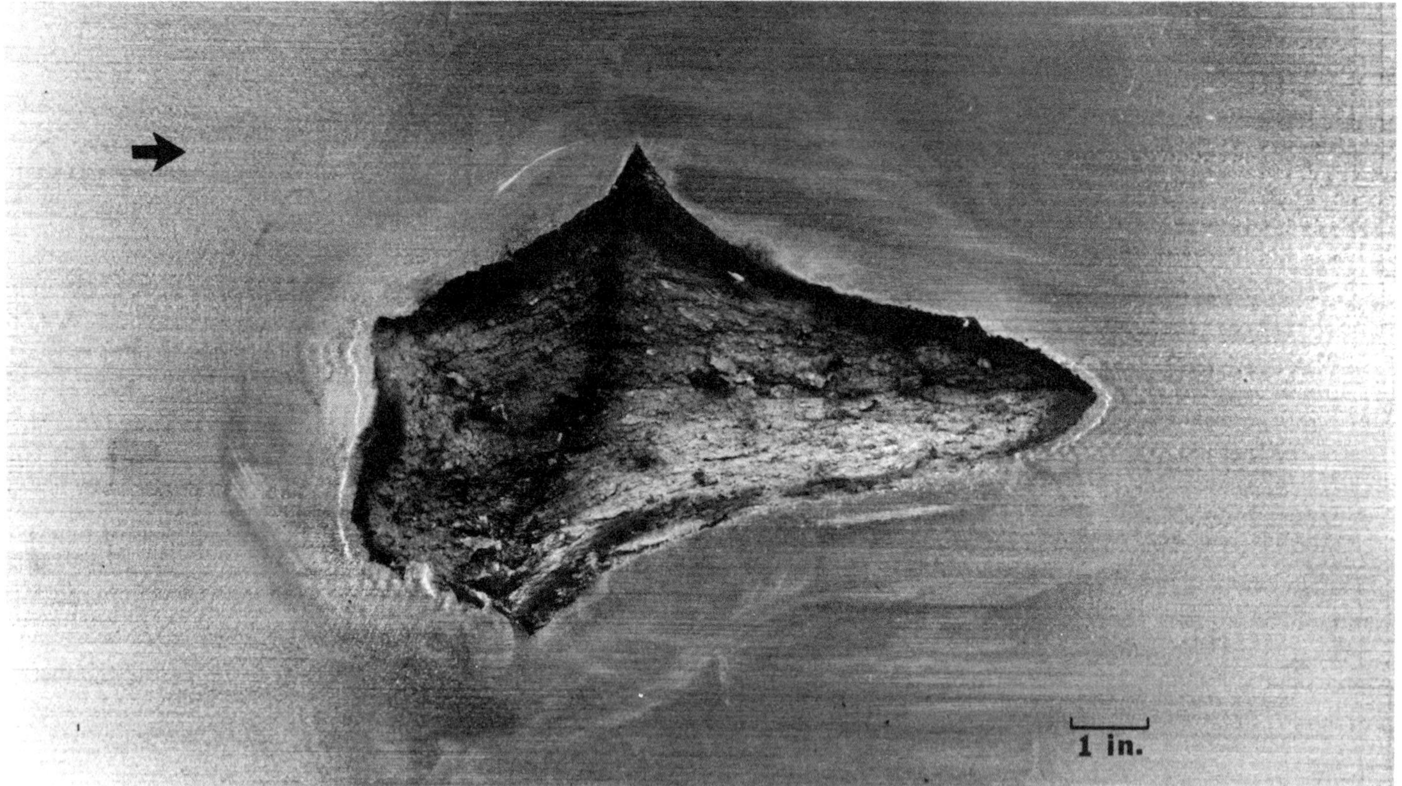

Fig. 6 Cross section of a forged bar showing a forging burst. The burst is located approximately at the centerline of the workpiece. Arrow indicates the direction of working.

welding or fusing of the surfaces; such flaws may be laps or folds (described previously).

Cold shuts often occur in closed-die forgings. They are junctures of two adjoining surfaces caused by incomplete metal fill and incomplete fusion of the surfaces.

Surface flaws weaken forgings and can usually be eliminated by correct die design, proper heating, and correct sequencing and positioning of the workpieces in the dies.

Shear cracks often occur in steel forgings; they are diagonal cracks occurring on the trimmed edges and are caused by shear stresses. Proper design and condition of trimming dies to remove forging flash are required for the prevention of shear cracks.

Other flaws in steel forgings that can be produced by improper die design or maintenance are internal cracks and splits. If the material is moved abnormally during forging, these flaws may be formed without any evidence on the surface of the forging.

Selection of Inspection Method

The principal factors that influence the selection of an NDI method for forgings include degree of required integrity of the forging, metal composition, size and shape of the forging, and cost. There are sometimes other influential factors, such as the type of forging method used.

For high-integrity forgings, it is often required that more than one inspection method be employed because some inspection methods are capable of locating only surface flaws; therefore, one or more additional methods are required for locating internal flaws. For example, many forgings for aerospace applications are inspected with liquid penetrants (or with magnetic particles, depending on the metal composition) for locating surface flaws, then by ultrasonics for detecting internal flaws.

Certain characteristics or conditions unique to forgings can create service problems, yet these conditions are not easily detected by nondestructive inspection. Exposed end grain, which can lead to poor corrosion resistance or to susceptibility to stress-corrosion cracking, is the most prevalent of the undesirable conditions. The strength of the forging can be adversely affected if the grains flow in an undesirable direction (as in grain reversal) or if grain flow is confined to only a portion of the section being forged rather than being well distributed. Both exposed end grain and poor grain flow can be most effectively corrected by redesigning the forging or forging blank, particularly with regard to flash. It is virtually impossible to detect exposed end grain nondestructively; only rarely can poor grain flow be detected nondestructively. Both conditions are much more readily analyzed by sectioning and macroetching sample forgings in preproduction stages of product development.

When certain steels or nonferrous alloys are forged at too high a temperature or sometimes when a part cools too slowly after forging, there is a potential for grain size in the finished forging to be excessively large. Such a condition is difficult to detect nondestructively, except with ultrasonics, and then only when the grains are very large. Even with very large grains, ultrasonic inspection cannot determine grain size quantitatively, nor can it detect large grains reliably. Only the possibility that large grains are present can be inferred from excessive attenuation of the ultrasonic beam.

Effect of Type of Forging

Many of the types of flaws that can occur in forgings do so without particular regard to the type of forging; that is, open-die, closed-die, upset, or rolled. However, there are many cases in which a specific type of flaw is more likely to occur in one type of forging than in another. Additional information on the types of forging processes discussed below is available in Volume 14 of the 9th Edition of *Metals Handbook*.

Open-Die Forgings. Most forgings produced in open dies are relatively large; therefore, their size is likely to impose some restrictions not only on the inspection method used but also on the system within a given inspection method. For large open-die forgings, NDI methods (other than visual) are generally limited to magnetic particle or liquid penetrant inspection (for surface discontinuities) or to ultrasonic inspection (for internal flaws). In general, the flaws likely to be found in open-die forgings are similar to those that may occur in other hot-worked shapes—with the exception of forging laps and cold shuts, which usually occur only in closed-die forgings.

Closed-Die and Upset Forgings. The discontinuities in closed-die forgings that can be detected by liquid penetrant inspection or magnetic particle inspection (if the forging is ferromagnetic) are the following:

- Forging laps, which can be caused by incorrect die design, use of incorrect size of forging stock, excessive local conditioning of forging stock for removal of surface flaws, and excessively sharp corners in the forging stock
- Seams, due to incomplete removal of seams from the forging stock
- Surface cracks, caused by incorrect forging temperature, nonductile metallic or nonmetallic segregates in the forging stock, or surface contamination from the furnace atmosphere or other contaminants in the furnace or on the forging (such as high-sulfur fuels for heating nickel alloys or leaded crayons used for marking parts before heating)
- Quench cracks
- Cold-straightening cracks

The likelihood that any of the above discontinuities will appear in a closed-die forging produced in a press or by upsetting is more prevalent than for hammer forgings, because press and upset forgings permit no opportunity to monitor the workpiece being forged during the forging operation. During hammer forging, the top surface of the forging is visible between hammer blows. The bottom surface may also be visible at times—particularly for large forgings that are raised intermittently during forging for descaling and lubricating the bottom surface or for forgings of temperature-sensitive alloys that are raised off the bottom die to permit heat recovery to the bottom surface.

Any seam in the forging stock or incipient laps or cracks will probably develop into significant forging laps or cracks if not detected during formation. Consequently, in hammer forging, a large percentage of such potentially scrap forgings can be removed from the production run and can either be salvaged by removing the discontinuity prior to finishing or scrapped at that point to avoid wasted forging time. Also, multiple-cavity hammer forgings permit inspection of the parts and blending out of minor laps or superficial cracks before finish forging.

Discontinuities that are not detectable by either magnetic particle or liquid penetrant inspection are noted in the following list. Most of these can be detected by ultrasonic inspection:

- Flakes, due to the absorption of hydrogen
- Pipe, due to center shrinkage in the ingot and subsequent insufficient reduction of forging stock
- Subsurface nonmetallic segregation
- Subsurface cracking, which may occur in certain alloys, particularly during the forging of irregular sections
- Weak centers in forging stock, caused by insufficient reduction from the ingot
- Subsurface cracks caused by forging material having comparatively cold centers, and generally occurring in large forging billets heated for insufficient time
- Rewelded forging laps, formed and rewelded during forging. With subsequent hammer blows, the lap forms, the scale is knocked or blown off, and the lapped metal rewelds, forming a healed lap with transverse grain flow and, possibly, entrapped scale

The presence of a rewelded forging lap in a suspected area can be checked by removing the surface metal below the decarburized layer and polishing this surface and swabbing with cold ammonium persulfate, thus revealing the decarburization at both sides of the lap (if present). The condition can be eliminated with corrections in blocker-die design.

Ring-Rolled Forgings. Discontinuities in forgings produced by ring rolling may be either inherited from the ingot or mechanically induced in forging operations, much the same as for forgings produced completely in hammers or presses. Inherited discontinuities are common to all products produced from bar or billet and can usually be traced back to the composition, cleanliness, or condition of the ingot. Although these discontinuities are not found only in ring-rolled forgings, they probably account for the vast majority of known discontinuities in ring-rolled forgings. Typical discontinuities are inclusions, porosity, hot-top remnants, and segregation.

Ultrasonic inspection is a reliable method of detecting the presence of inherited discontinuities. It is always advisable to inspect the material before it is ring rolled. Extremely large billets (1200 mm, or 48 in., in diameter and larger) may have surface conditions that cause problems relative to sound entry. Large billets may exhibit structural conditions, depending on the amount of reduction from the ingot, that are too large for a complete ultrasonic inspection. Smaller, well-worked billets can be examined at 2.25 MHz. The larger billets require the use of 1.0-MHz crystals, and even then ultrasonic penetration is not always possible.

Final proof that the forging is free of inherited discontinuities is accomplished through ultrasonic inspection of the completed forging. Depending on the final machined shape, certain ring-rolled forgings may require a preliminary ultrasonic inspection before final machining. When an extreme change in contour prevents a complete ultrasonic inspection of the final shape, inspection can be performed on portions of the forged ring.

Externally induced mechanical discontinuities that have been found in ring-rolled forgings include surface-related laps, cracks, and exfoliations. Normally, these discontinuities can be detected visually either during the manufacturing process or in the machined condition after rolling. However, ultrasonic inspection can be valuable for determining the depth of surface-related discontinuities and for detecting them even when they have been obscured by subsequent working or metal movement. Magnetic particle inspection is also used to detect surface-related, externally induced mechanical discontinuities in ferromagnetic ring-rolled forgings. Liquid penetrant inspection has often been successfully used to detect surface flaws in nonferromagnetic rings.

Mechanically induced internal discontinuities (known as strain-induced porosity) can occur in certain materials. Some nickel or titanium alloys have an inherent susceptibility to these types of discontinuities in portions of the ring stretched at critical temperatures. This may result from improper stock distribution, improper rolling techniques, or improper tooling. In extreme cases, in which the induced porosity is excessive, the rupturing may progress to one or more external surfaces. Either form of internal mechanically induced discontinuity may initiate in an area where inherited discontinuities are present.

Effect of Forging Material

Some types of forging flaws are unique to specific work metals, and may influence the choice of inspection method.

Steel Forgings. The most common surface flaws in steel forgings are seams, laps, and slivers. Other surface flaws include rolled-in scale, ferrite fingers, fins, overfills, and underfills. The most common internal flaws found in steel forgings are pipe, segregation, nonmetallic inclusions, and stringers.

Either magnetic particle or liquid penetrant inspection can be used for steel forgings, although magnetic particle inspection is usually preferred. Only liquid-penetrant inspection can be used for some stainless steel or nonferrous forgings. The selection of an inspection method depends on the size and shape of the forging and on whether the forging can be moved to the inspection station or the inspection equipment can be moved to the forging. For either inspection method, systems are available for inspecting forgings of almost unlimited size and weight (see articles on the specific inspection methods in this Volume). In most cases, magnetic particle inspection is less ex-

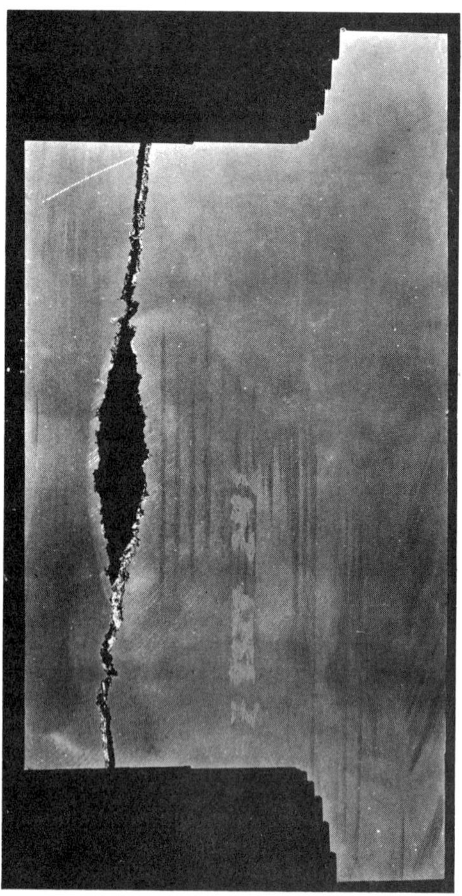

Fig. 7 Section through a heat-resistant alloy forging showing a central discontinuity that resulted from insufficient homogenization during conversion. Step machining was used to reveal the location of the rupture; original diameter is at right.

pensive and faster than liquid penetrant inspection.

Heat-Resistant Alloy Forgings. Most of the flaws found in forgings of heat-resistant alloys can be categorized as those related to scrap selection, melting, or primary conversion to bar or billet or those that occur during forging or heat treatment. Tramp elements such as lead or zinc have been present in the makeup of scrap charges at levels that have caused hot shortness and a degradation of hot-tensile ductility occurring near 370 °C (700 °F) of air-melted alloys.

No NDI method would reliably evaluate the presence or absence of possible tramp-element contamination, and composition checks or hot-tensile checks could be considered to be too random for complete assurance of the presence or absence of contamination. The positive corrective action is the use of a vacuum-remelted product.

Melt-related discontinuities, such as inclusions, pipe, unhealed center conditions, flakes, or voids, are the types of discontinuities that most frequently exist in heat-resistant alloy forgings. Ultrasonic inspection can detect and isolate these conditions when they exist. Segregated structures, unmelted electrodes, or portions of stringer rods are types of discontinuities that may sometimes be found in heat-resistant alloys.

Flakes (internal cracks) can be produced each time the material is heated and cooled to room temperature. The random orientation does not always present a properly oriented reflector for ultrasonic inspection, but in most cases flaking can be detected ultrasonically with a high degree of reliability.

Unmelted pieces of electrode or shelf conditions appear infrequently in vacuum-melted alloys. Either of these conditions can seriously degrade forgeability. Macroetching or another appropriate type of surface inspection of the machined forging or the billet is the most effective method of detecting unmelted electrodes or shelf conditions.

Seams are common to rolling practices, and are readily detected by visual, magnetic particle, or liquid penetrant inspection. Grinding cracks are caused by severe grinding, which promotes network-type cracking on the surface of the material being conditioned. The network-type cracking may be present immediately after grinding or may not occur until subsequent heating for further forging. Seams and grinding cracks will cause severe surface rupturing during forging. Center bursts occur during conversion to bar or billet if reduction rates are too severe or temperatures are incorrect; they are readily detected by ultrasonic inspection.

Ingot pipe, unhealed center conditions, or voids are melt-related discontinuities, but their occurrence in forgings is often a function of reduction ratio. The conversion practice must impart sufficient homogenization or healing to produce a product with sound center conditions. An example of an unsound condition that did not heal is shown in Fig. 7. Macroetching and ultrasonic inspection methods are the most widely used for identifying regions of unsoundness.

The nickel-base heat-resistant alloys are highly susceptible to surface contamination during heating for forging. Fuel oils containing sulfur will induce a grain-boundary attack, which will cause subsequent rupturing during forging. Paint or marking crayons with high levels of similar contaminants will cause similar areas of grain-boundary contamination. Surface contamination is not normally detected by NDI methods prior to heating and processing, but if present at a level high enough to cause contamination, rupturing during forging will occur that can be detected by visual inspection. If the contamination occurs after final forging, with no subsequent metal deformation, the contaminated areas will be apparent as areas of intergranular attack. Macroetching, followed by liquid penetrant inspection, should be used.

Advanced forging processes, such as isothermal and hot-die forging, and the increasing use of computer modeling have greatly reduced problems associated with heat-resistant alloy forgings. These developments are outlined in Volume 14 of the 9th Edition of *Metals Handbook*.

Nickel Alloy Forgings. The discontinuities that occur in nickel alloy forgings are generally of the same type as those found in heat-resistant alloy forgings; namely, cracks (external and internal), tears, seams, laps, coarse-grain wrinkles, inclusions, and pipe. Although all metals may be subject to thermal cracking during forging, the age-hardenable nickel alloys are more vulnerable than most other metals, thus requiring close temperature control during forging to avoid large temperature gradients.

Internal discontinuities in nickel alloy forgings can be located by ultrasonic inspection. Liquid penetrants are most often used to inspect for surface flaws; magnetic particles can be used if the alloy is sufficiently magnetic.

Aluminum Alloy Forgings. Common surface discontinuities in aluminum alloy forgings are laps, folds, chops, cracks, flow-throughs, and suck-ins (Fig. 8). The generation of these discontinuities is associated with the forging operation, processing practices, or design. Cracks can also result from seams in the forging stock.

The internal discontinuities that occur in aluminum alloy forgings are ruptures, cracks, inclusions, segregation, and occasionally porosity (Fig. 8). Ruptures and cracks are associated with temperature control during preheating or forging or with excessive reduction during a single forging operation. Cracks can also occur in stock that has been excessively reduced in one operation. Inclusions, segregation, and porosity result from forging stock that contains these types of discontinuities.

The inspection of aluminum alloy forgings takes two forms: in-process inspection and final inspection. In-process inspection, using such techniques as statistical process control and/or statistical quality control, is used to determine if the product being manufactured meets critical characteristics and if the forging processes are under control. Final inspection, including mechanical property testing, is used to verify if the completed forging product conforms with all drawing and specification criteria. Typical final inspection procedures used for aluminum alloy forgings include dimensional checks, heat treatment verification, and nondestructive evaluation.

Dimensional Inspection. All final forgings are subjected to dimensional verification. For open-die forgings, final dimensional inspection may include verification of all required dimensions on each forging or the use of statistical sampling plans for groups or lots of forgings. For closed-die forgings,

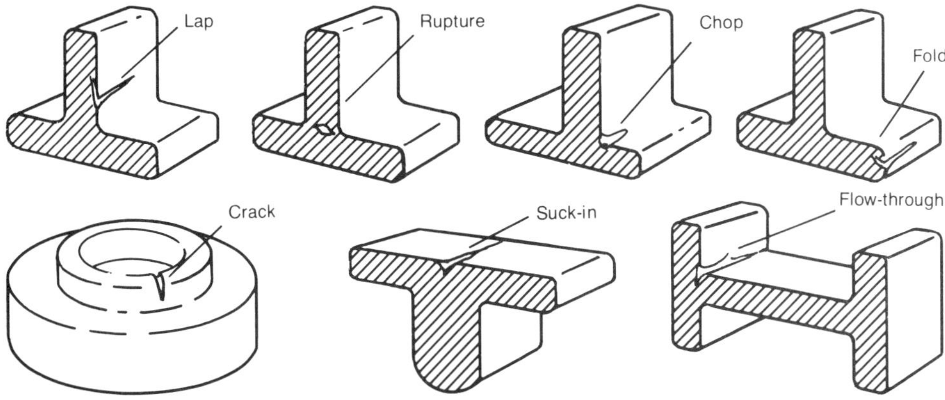

Fig. 8(a) Typical discontinuities found in aluminum alloy forgings. See text for discussion.

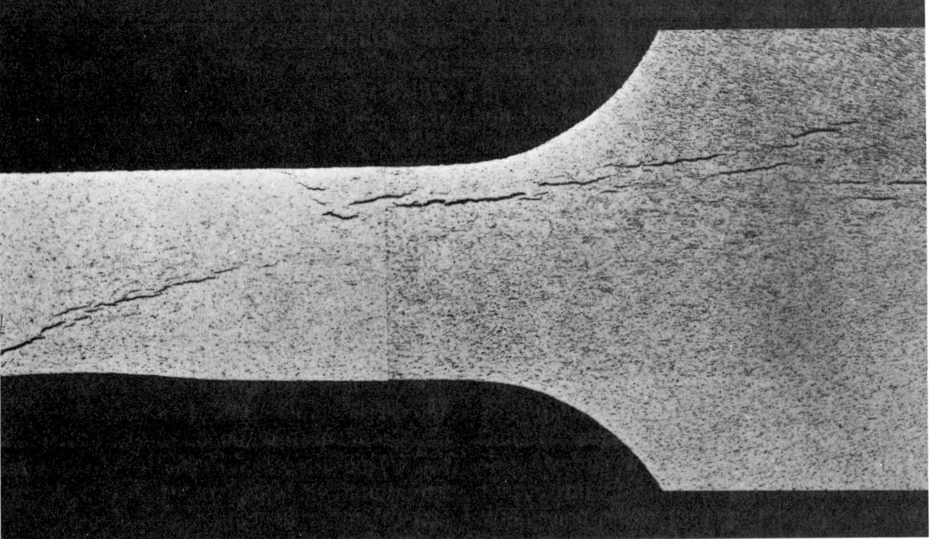

Fig. 8(b) Band of shrinkage cavities and internal cracks in an alloy 7075-T6 forging. The cracks developed from the cavities, which were produced during solidification of the ingot and which remained during forging because of inadequate cropping. Etched with Keller's reagent. 9×

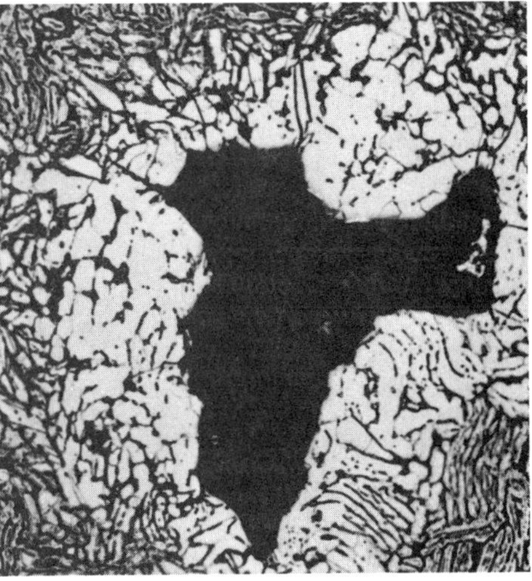

Fig. 9 Ti-8Al-Mo-1V, as forged. Ingot void (black), surrounded by a layer of oxygen-stabilized α (light). The remaining structure consists of elongated α grains in a dark matrix of transformed β. Etched with Kroll's reagent (ASTM 192). 25×

conformance of the die cavities to the drawing requirements, a critical element in dimensional control, is accomplished prior to placing the dies in service by using layout inspection of plaster or plastic casts of the cavities. With the availability of computer-aided design (CAD) data bases on forgings, such layout inspections can be accomplished more expediently with computer-aided manufacturing (CAM) driven equipment, such as coordinate measuring machines or other automated inspection techniques. With verification of die cavity dimensions prior to use, final part dimensional inspection may be limited to verifying the critical dimension controlled by the process (such as die closure) and monitoring the changes in the die cavity. Further, with high-definition and precision aluminum forgings, CAD data bases and automated inspection equipment, such as coordinate measuring machines and two-dimensional fiber optics, can be used in many cases for actual part dimensional verification.

Heat Treatment Verification. Proper heat treatment of aluminum alloy forgings is verified by hardness measurements and, in the case of 7*xxx*-T7*xxx* alloys, by eddy current inspection. In addition to these inspections, mechanical property tests are conducted on forgings to verify conformance to specifications. Mechanical property tests vary from the destruction of forgings to tests of extensions and/or prolongations forged integrally with the parts.

Nondestructive Inspection. Aluminum alloy forgings are frequently subjected to nondestructive inspection to verify surface or internal quality. The surface finish of aluminum forgings after forging and caustic cleaning is generally good. A root mean square (rms) surface finish of 3.2 μm (125 μin.) or better is considered normal for forged and etched aluminum alloys; under closely controlled production conditions, surfaces smoother than 3.2 μm (125 μin.) rms can be obtained. Selection of NDI requirements depends on the final application of the forging. When required, satisfactory surface quality is verified by liquid penetrant, eddy current, and other techniques. Aluminum alloy forgings used in aerospace applications are frequently inspected for internal quality using ultrasonic inspection techniques.

Magnesium alloy forgings are subject to the same types of surface and internal discontinuities as aluminum alloy forgings. In addition, surface cracks are common in magnesium alloy forgings and are usually caused by insufficient control of the forging temperature.

Visual inspection and liquid penetrant inspection are used to detect surface discontinuities. Ultrasonic inspection is used to locate internal discontinuities.

Titanium Alloy Forgings. Discontinuities that are most likely to occur in titanium alloy forgings are usually carried over in the bar or billet. Typical discontinuities in titanium alloy forgings are α-stabilized voids, macrostructural defects, unsealed center conditions, clean voids, and forging imperfections.

Alpha-stabilized voids are among the most common discontinuities found in forgings of titanium alloys. Investigation and research have determined that voids surrounded by oxygen-stabilized α grains may be present in the ingot (Fig. 9). Because of the size of these voids and the coarse-grain nature of the ingot, they cannot be detected until the ingot has been suitably reduced in cross section and refined in structure. When the structure has been refined, the voids can be detected by ultrasonic inspection. Also, when the section is reduced sufficiently, radiographic inspection can be effectively used.

Alpha voids do not readily deform during forging, nor do they align with the flow pattern, as do typical inclusions in carbon

Fig. 10 Ti-6Al-4V α-β processed billet illustrating the macroscopic appearance of a high-aluminum defect. 1.25×

or alloy steel. In most cases, α voids appear to be somewhat globular. Extremely small voids do not present an especially ideal target or reflector for ultrasonic energy. Attempts to correlate size with amplitude of indication obtained during ultrasonic inspections have not been completely reliable. For critical-application forgings, the material is most often inspected twice—once in the bar or billet form before forging and again after forging. Because forging further refines structure and reorients possible discontinuities in relation to the sound-entry surface, the forging operation probably enhances the possibility of detecting these discontinuities.

Macrodefects. Three principal defects are commonly found in macrosections of ingot, forged billet, or other semifinished product forms. These include high-aluminum defects (Type II defects), high-interstitial defects (Type I defects or low-density interstitial defects), and β flecks. High-aluminum defects are areas containing an abnormally high amount of aluminum. These are soft areas in the material (Fig. 10) and are also referred to as α segregation. Defects referred to as β segregation are sometimes associated with α segregation. These are areas in which aluminum is depleted. The high-interstitial defects (Fig. 11) are normally high in oxygen and/or nitrogen, which stabilize the α phase. These defects are hard and brittle; they are normally associated with porosity, as illustrated in Fig. 9.

Beta flecks are regions enriched in a β-stabilizing element due to segregation during ingot solidification. Figure 12 shows the macroscopic appearance of β flecks in a Ti-6Al-6V-2Sn forging billet.

Unsealed center conditions are associated with insufficient ingot reduction. These are more prevalent in the larger stock sizes (>230 mm, or 9 in., in diameter) and are normally removed by adequate croppage at the mill. Clean voids describe a condition that can be associated with unsatisfactory sealing of porosity elsewhere in the ingot or through center porosity formed during ingot reduction.

Nondestructive Inspection. Ultrasonic inspection is the most definitive and practical method of inspecting titanium alloy forgings. Inspection techniques are normally tailored to the rejection level indicated in the specifications and to the physical condition of the material being inspected. Surface conditions usually must be ideal, grain size must be fine, and structural conditions must be controlled. Most airframe or similar static parts are inspected with equipment settings based on a No. 3 flat-bottom-hole standard. For the examination of critical rotating forgings for aircraft gas-turbine engines, it is not uncommon to inspect to the equivalent of a No. 1 flat-bottom-hole standard. Experience with these highly critical forgings, which in service rotate at high speed in the presence of extreme temperature and pressure, has indicated that small α voids can initiate cracks and have caused catastrophic failures.

For satisfactory ultrasonic inspection of forgings to these stringent requirements, special techniques and equipment are usually required. Specially designed ultrasonic electronic equipment is used with focused or otherwise unique transducers. Also required are an immersion tank with rotating devices, automatic small incremental indexing devices, and automatic alarms for signal level. Special reference blocks are required, along with the usual flat-bottom-hole reference blocks. The correct indexing increment must be established, the linear alignment of the ultrasonic unit must be verified, and calibration checks must be made. All information must be recorded and retained for future reference.

Visual Inspection

Despite the many sophisticated inspection methods available, unaided visual inspection is still important and is often the sole method of inspecting forgings used for common hardware items. Under proper lighting conditions, the trained eye can detect several types of surface imperfections, including certain laps, folds, and seams. Visual inspection is often used first, then questionable forgings are further examined by macroetching and inspection with macrophotography or some type of nondestructive method.

The only equipment necessary for visual inspection is a bench on which to place the forging and suitable cranes or hoists for forg-

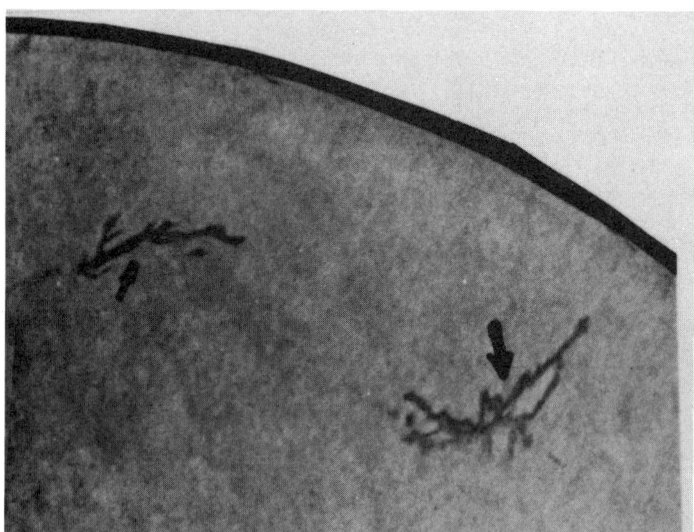

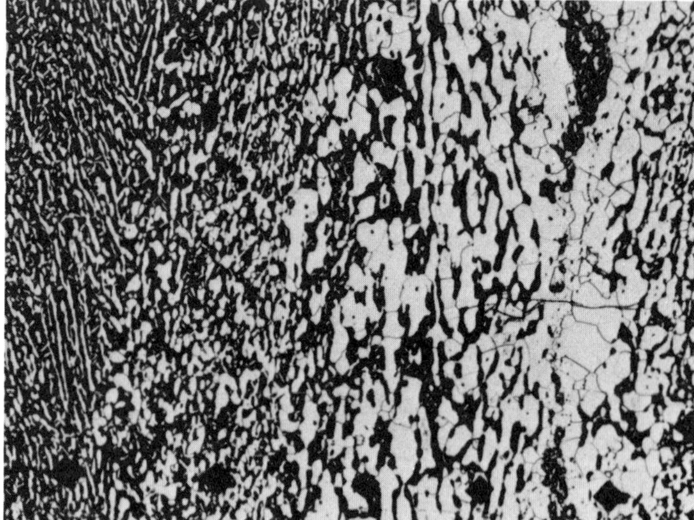

Fig. 11 Macrodefects in titanium billets. Left: Ti-6Al-4V α-β processed billet illustrating macroscopic appearance of a high-interstitial defect. Actual size. Right: at 100×. The high oxygen content results in a region of coarser and more brittle oxygen-stabilized α than observed in the bulk material.

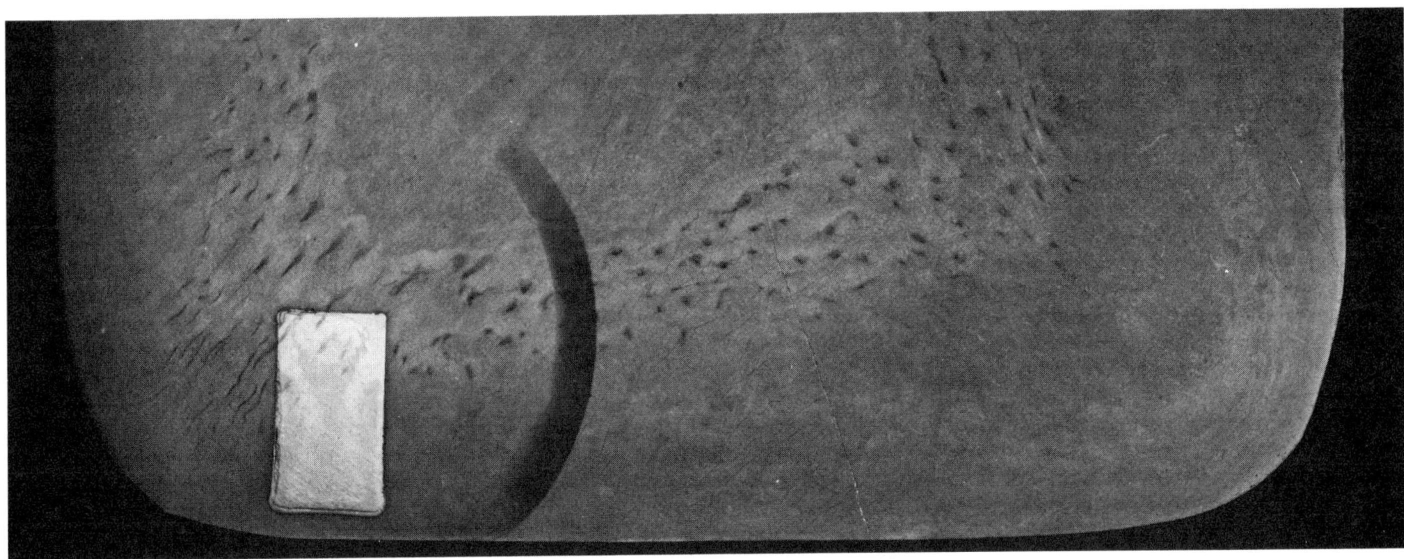

Fig. 12 Ti-6Al-6V-2Sn α-β forged billet illustrating macroscopic appearance of β flecks that appear as dark spots. Etched with 8 mL HF, 10 mL HF, 82 mL H_2O, then 18 g/L (2.4 oz/gal.) of NH_4HF_2 in H_2O. Less than 1×

ings that are too heavy to lift by hand. Good and well-controlled lighting conditions are essential. Optical aids such as magnifying glasses that can magnify up to about ten diameters are often used to increase the effectiveness of visual inspection.

Magnetic Particle Inspection

Magnetic particle inspection is useful for detecting surface imperfections as well as certain subsurface imperfections that are within approximately 3 mm (1/8 in.) of the surfaces in forgings of steel, some grades of stainless steel, and other ferromagnetic metals. Magnetic particle inspection can be used with fluorescent particles and ultraviolet light. Detailed information is available in the article "Magnetic Particle Inspection" in this Volume.

The advantages of magnetic particle inspection include the following:

- Almost instant results can be obtained in locating surface and certain subsurface imperfections
- Equipment can be transported to the forging, or the forging can be transported to the inspection station, as dictated by the size and shape of forging
- Preparation of the forging is minimal, mainly involving the removal of surface contaminants that would prevent magnetization or inhibit particle mobility
- Routine inspection work can be effectively done by relatively unskilled labor properly trained in interpretation
- For forgings that are simple in configuration, and when justified by the quantity, magnetic particle inspection can be automated
- For some forgings, electronic sensing can be used, thus reducing the chances of human error and increasing inspection reliability
- Many forgings have sufficient retentivity to permit the use of multidirectional magnetization, thus permitting the inspection of indications in all orientations with a single preparation. Retentivity must be checked for the particular forging before a decision is made to use multidirectional magnetization
- The cost of magnetic particle inspection is generally lower than that for several other inspection methods in terms of investment in equipment, inspection materials, and inspection time

The limitations of the magnetic particle inspection of forgings are generally the same as for inspecting other workpieces and include the following:

- The method is applicable only to forgings made from ferromagnetic metals
- Because magnetic particle inspection is basically an aided visual inspection, under most circumstances, its effectiveness is subject to the visual acuity and judgment of the inspector
- Magnetic particle inspection is generally limited to detecting imperfections that are within about 3 mm (1/8 in.) of the surface of the forging
- Because the forging must be thoroughly magnetized, magnetic particle inspection is likely to be ineffective unless scale, grease, or other contaminants are removed from the forging. Such surface contaminants inhibit the mobility of the particles necessary to delineate the indications
- Following inspection, the forging usually must be demagnetized, depending mainly on the retentivity of the particular metal, subsequent shop operations, and end use

Detection of Surface Discontinuities. Magnetic particle inspection and liquid penetrant inspection are both widely used for detecting discontinuities in steel forgings, although the former is the more widely used. As described above and in the articles that deal with the specific inspection methods in this Volume, one advantage of using the magnetic particle technique is its ability to detect certain subsurface discontinuities that are not open to the surface. Subsurface discontinuities cannot be located with liquid penetrants. Also, some surface discontinuities may be so packed with scale that liquid penetrant techniques are marginal or infeasible. Therefore, in most cases, magnetic particle inspection is preferred to liquid penetrant inspection. Continuous magnetization is usually prescribed for inspecting steel forgings, because at the stage in which the forgings are inspected they are in an annealed or semiannealed condition and consequently have poor retentivity of magnetism.

Two inspection methods are available: dry powder and wet. Selection between the dry and the wet methods may sometimes be purely arbitrary, although it is usually based on the available equipment and the size of the forgings being inspected. The dry-powder method is used to a greater extent for large forgings. Similarly, selection between fluorescent and nonfluorescent particles may often be arbitrary, although the size of the forging can be a major factor, because if the fluorescent method is used the forging must usually be of such size and shape that it can be inspected under ultraviolet light, with white light substantially eliminated.

Many specific procedures have been established for in-plant use. The dry-powder and wet techniques adopted in one plant for the inspection of ferromagnetic metal forgings are described below.

Dry-Powder Technique. The contact method of magnetization was selected to inspect the ferromagnetic materials. Prods are used to pass direct current or rectified alternating current through the workpiece. The magnetic particles are nontoxic, finely divided ferromagnetic material of high permeability and low retentivity, free from rust, grease, dirt, or other materials that may interfere with the proper functioning of the magnetic particles. The particles must also exhibit good visual contrast with the forging being inspected.

Inspection is by the continuous-current method; that is, the magnetizing current remains on during the period of time that the magnetic particles are being applied and also while the excess particles are being removed. Prods are spaced 150 to 200 mm (6 to 8 in.) apart, except where restricted by configuration. The magnetic field is induced in two directions, 90° apart. The current used is 4 to 5 A/mm (100 to 125 A/in.) of prod spacing and is kept on for a minimum of ⅕ s.

Dry magnetic particles are applied uniformly to the surface, using a light dustlike technique. Excess particles are removed by a dry-air current of sufficient force to remove the excess particles without disturbing particles that show indications. The nozzle is held obliquely about 35 to 50 mm (1½ to 2 in.) above the test area. Nozzle size and air pressure result in a pressure (measured by a manometer) of 25 to 40 mm (1.0 to 1.5 in.) of water at an axial distance of 25 mm (1 in.) from the nozzle and 7.5 to 15 mm (0.3 to 0.6 in.) of water at 50 mm (2 in.) from the nozzle.

A 100 mm (4 in.) grid pattern over the entire forging surface is normally used for evaluation (Fig. 13). The prods are placed 200 mm (8 in.) apart (Fig. 13), except where restricted by the shape of the forging, when using this grid pattern.

Prods are placed on the surface to be tested in the proper position, as shown by position 1 in Fig. 13, and the current is turned on (4 to 5 A/mm, or 100 to 125 A/in., of prod spacing). The powder is applied, the excess particles are removed, and the current is turned off. Inspection is conducted during application of the powder and after removal of the excess particles.

The next step is to reposition the prods 90°, as indicated by position 2 in Fig. 13; the above procedure is then repeated. When the shape of the forging does not permit a full 90° rotation with the established prod spacing, the prod spacing can be changed, provided it is not less than 50 mm (2 in.) nor more than 200 mm (8 in.) between prods.

Wet Technique. The magnetic particles selected are nonfluorescent and suspended in a liquid vehicle. The magnetizing equipment is capable of inducing a magnetic flux of suitable intensity in the desired direction by both the circular and the longitudinal

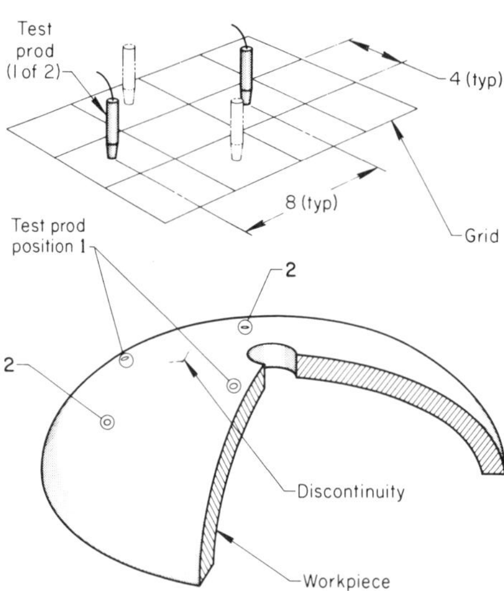

Fig. 13 Grid pattern and prod positions used in one plant for the magnetic particle inspection of forgings using prods and the dry-powder technique

methods. Direct current from generators, storage batteries, or rectifiers is used to induce the magnetic flux. Circular magnetization is obtained by passing the current through the forging being examined or through a central conductor to induce the magnetic flux. Longitudinal magnetization is obtained by using a solenoid, coil, or magnet to induce the magnetic flux. The magnetic particles are nontoxic and exhibit good visual contrast. The viscosity of the suspension vehicle for the particles must be a maximum of 5×10^{-6} m^2/s (5.0 centistokes) at any bath temperature used. The magnetic particles are limited to 28 to 40 g (1.0 to 1.4 oz) of solid per gallon of liquid vehicle.

The liquid used as a vehicle for the magnetic particles may be a petroleum distillate such as kerosene. Tap water with suitable rust inhibitors and wetting and antifoaming agents can be substituted for the petroleum distillate. The water should contain about 0.3% antifoam agent, 3.9% rust inhibitor, and 12.8% wetting agent.

Inspection is carried out by the continuous method. For this method, the magnetizing circuit is closed just before applying the suspension or just before removing the forging from the suspension. The circuit remains closed for approximately ⅕ to ½ s.

For circular magnetization, an ammeter is used to verify the presence of adequate field strength. For verifying adequate field strength in longitudinal magnetization, a field indicator is useful. Typical current levels utilized to provide an adequate field strength are 4 to 12 A/mm (100 to 300 A/in.) of diameter of the surface being examined, although current levels of up to 30 A/mm (750 A/in.) of diameter have been used.

The magnetizing force for longitudinal magnetization is 2000 to 4000 ampere-turns per 25 mm (1 in.) of diameter of the surface being examined. If both the inside and outside diameters of cylindrical parts are to be inspected, the larger diameter is used in establishing the current. If it is impractical to attain currents of the calculated magnitude, a magnetic field indicator is used to verify the adequacy of the magnetic field.

Suspensions must be tested daily or when they appear to have become discolored by oil or contaminated by lint. Common practice is to test the suspension at the beginning of each operating shift. The suspension test is conducted as follows:

- Let the pump motor run for several minutes to agitate a normal mixture of particles and vehicle
- Flow the bath mixture through the hose and nozzle for a few minutes to clear the hose
- Fill the centrifuge tube to the 100-mL line
- Place the centrifuge tube and stand in a location free from vibration
- Let the tube stand for 30 min for particles to settle out
- After 30 min, readings for settled particles should be 1.7 to 2.4 mL. If the reading is higher, add vehicle; if lower, add particle powder to the suspension

Example 1: Magnetic Particle Inspection of 1541 Steel Connecting Rods by the Multidirectional System. Connecting rods for piston engines were manufactured from 1541 steel by forging in closed dies. Because connecting rods used in piston engines are subject to considerable stress, it is necessary to ensure the integrity of these parts, especially in heavy-duty applications, such as high-performance automobiles and trucks. Accordingly, fluorescent magnetic particle inspection of the forged connecting rod was done immediately after unwanted scale and dirt were removed by shotblasting. Bidirectional (longitudinal and transverse) magnetic fields were generated rapidly and sequentially by the use of a multidirectional magnetizing machine, as shown in Fig. 14. The use of this machine allowed for a single inspection to detect all discontinuities regardless of their orientation.

The process of manufacturing connecting rods usually results in only longitudinally oriented flaws, but, transversely oriented flaws occasionally do occur. Hot tears along the trim line and metal folds or laps in the web area are of primary interest in the inspection. Worn tooling, overheating, and underheating of the forging blanks, as well as improper trimming, are the primary causes of these flaws.

Inspection Procedure. The inspection of these connecting rods utilized a multidirectional system, and the inspection procedure was as follows. Initially, the connecting rod was pneumatically clamped in the contact heads (Fig. 14) and then automatically

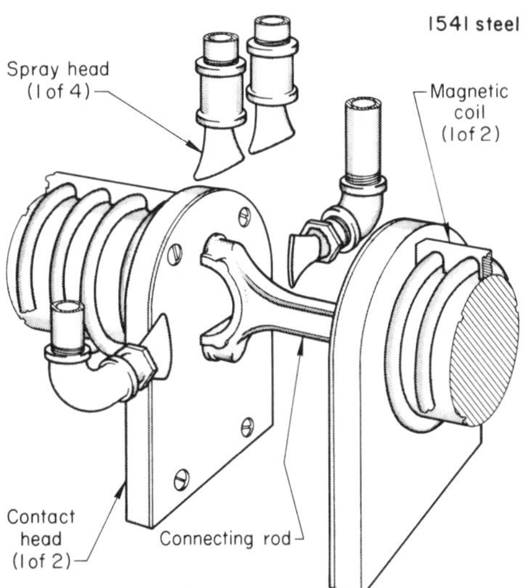

Fig. 14 Setup for the inspection of forged 1541 steel automotive connecting rods by the magnetic particle method using a multidirectional magnetizing system

flooded with a suspension of fluorescent magnetic particles. The magnetizing fields were induced while the suspension was draining from the part. Current settings for circular magnetization were 1000 A; for longitudinal magnetization, 4500 ampere-turns equivalent. The magnetizing current automatically switched back and forth, thus revealing any imperfections regardless of their orientation. Inspection of the forging was done under an ultraviolet light of no less than 2700 lx (250 ftc) at the surface of the forging. Following completion of inspection, the forging was demagnetized so that metal chips did not adhere to the surface during subsequent machining operations. The principal disadvantage of this inspection procedure was that the production rate was limited because the procedure was performed manually.

Example 2: Use of Electromagnetic, Ultrasonic, and Magnetic Particle Methods for the Inspection of Forged 1046 Steel Steering Knuckles. Automobile steering knuckles similar to the forged 1046 steel knuckle shown in Fig. 15 are regarded as extremely high-integrity parts. At one plant, to provide the required degree of quality control, 100% of all steering knuckles were nondestructively inspected using three different techniques—electromagnetic comparison for hardness and ultrasonics for internal soundness at one location, followed by in-line fluorescent magnetic particle inspection.

Inspection Procedures. Figure 15 shows the orientation of the steering knuckle during inspection with the electromagnetic test coil and ultrasonic transducer. Electromagnetic comparator coils were used to evaluate hardness in the bearing area; the ultrasonic 60° shear wave method was used to detect internal discontinuities in the spindle portion of the steering knuckle. The spindle of the knuckle was immersed in a dielectric oil to provide couplant for the ultrasonic inspection; a 5.0-MHz transducer was used. For electromagnetic inspection, a 60-Hz frequency was used for the hardness detection comparator. Electromagnetic and ultrasonic inspections were combined both for efficiency and compatibility of the two methods.

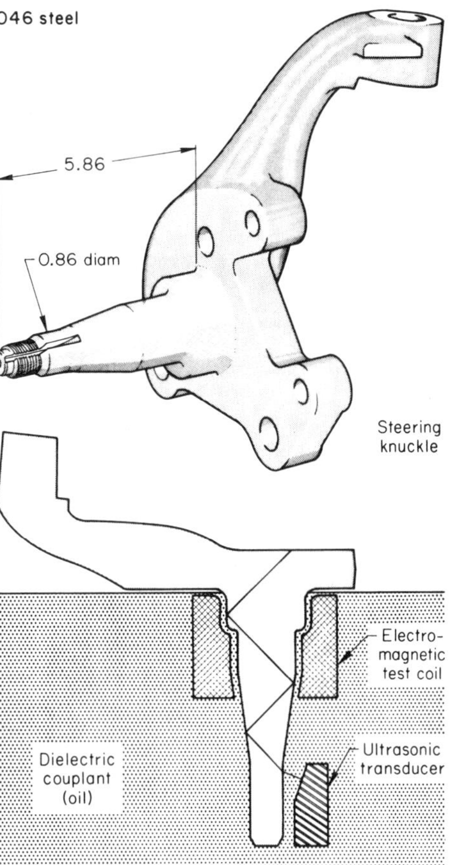

Fig. 15 Steering knuckle, forged from 1046 steel, that was inspected by a combination of electromagnetic, ultrasonic, and magnetic particle methods. The setup for electromagnetic and ultrasonic inspection is shown. Dimensions given in inches

All steering knuckles were fluorescent magnetic particle inspected following the electromagnetic-ultrasonic inspection. This inspection was accomplished by magnetizing the steering knuckle between contact plates, flooding with the particle suspension, and viewing under an ultraviolet light having an intensity of 2700 lx (250 ftc) minimum. The steering knuckles were then transported by conveyor to a magnetizing coil for magnetizing in the other direction so that circumferentially oriented indications would be shown under a subsequent second ultraviolet light inspection. These nondestructive tests resulted in the detection of both internal and external flaws as well as the detection of variations in hardness.

Liquid Penetrant Inspection

Liquid penetrant inspection is a versatile NDI process and is widely used for locating surface imperfections in all types of forgings, either ferrous or nonferrous, although it is more frequently used on nonferrous forgings. There is no limitation on the size or shape of a forging that can be liquid penetrant inspected.

Any of the three basic liquid penetrant systems (water-washable, postemulsifiable, and solvent-removable) can be used to inspect forgings. The product or product form is not a principal factor in the selection of a system. The fundamentals of the three systems, as well as the advantages and limitations of each for inspecting various products (including forgings), are presented in the article "Liquid Penetrant Inspection" in this Volume.

Advantages. Among the advantages of liquid penetrant inspection of forgings are the following:

- There are no limitations on metal composition or heat-treated condition
- There are no limitations imposed on the size or shape of the forging that can be inspected
- Liquid penetrant inspection can be done with relatively simple equipment
- Training requirements for inspectors are minimal
- Inspection can be performed at any stage of manufacture
- Liquid penetrant materials can be taken to the forgings or the forgings taken to the inspection station, depending on the size and shape of the forgings

The limitations of the liquid penetrant inspection of forgings are basically the same as those for the inspection of other workpieces. The characteristics of the surface of a forging sometimes impose specific limitations. The most important general limitations are:

- Liquid penetrant inspection is restricted to detecting discontinuities that are open to the surface
- Liquid penetrant inspection is basically a visual aid; therefore, results depend greatly on the visual acuity and judgment of the inspector
- Satisfactory inspection results require that the surface of the forging be thoroughly cleaned before inspection. The presence of surface scale can cause inaccurate readouts. If the surface of the forging is excessively scaled, it should be pickled or grit-blasted, preferably pickled. The forgings should also be cleaned to remove surface contaminants, such as grease and oil
- Liquid penetrant inspection is slower than magnetic particle inspection

Liquid Penetrant Detection of Flaws in Steel Forgings

Factors affecting the selection of a special penetrant system for inspecting steel forgings include available equipment; size,

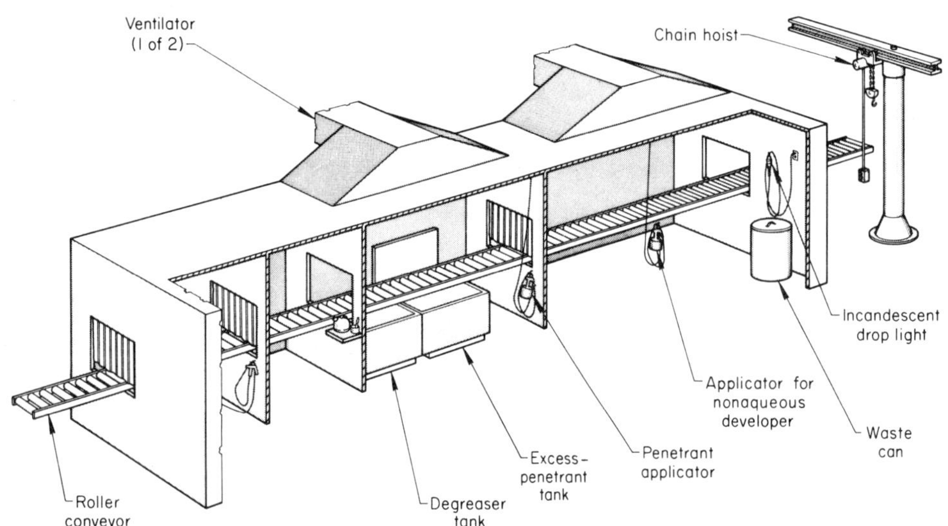

Fig. 16 Equipment layout for the liquid penetrant inspection of type 316 stainless steel forged valve bodies and bonnet forgings by the solvent-removable visible-dye system

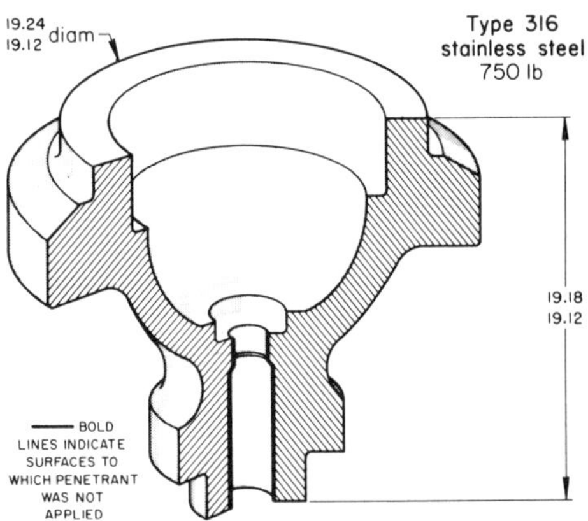

Fig. 17 Bonnet forging of type 316 stainless steel that was liquid penetrant inspected for surface flaws by the solvent-removable visible-dye system. Dimensions given in inches

shape, and surface conditions of the forgings; degree of sensitivity required; whether or not the entire forging requires inspection; and cost. Regardless of which system is used, the degree of success achieved depends greatly on the surface conditions of the forging. Rough, scaly surfaces are likely to result in either false indications or obscuring of meaningful flaws. Pickled surfaces are preferred. Abrasive blasting is usually satisfactory for cleaning forging surfaces, although overblasting must be avoided or some flaws may be tightly closed and prevent the penetrant from entering. The postemulsifiable and solvent-removable liquid penetrant systems are most often used to inspect steel forgings.

The postemulsifiable system is generally preferred to the water-washable system for forgings because of its greater sensitivity. Either the fluorescent-penetrant or the visible-dye technique can be used. Selection depends largely on whether ultraviolet-light inspection can be used. For forgings of a size and shape that can be immersed in tanks and inspected in a booth, the fluorescent technique is usually preferred.

The solvent-removable system is especially well adapted to applications in which only a portion of the forging requires inspection. Equipment for this system can be minimal and completely portable or may involve more elaborate systems used on a production basis, as described in the following examples.

Example 3: Solvent-Removable Liquid Penetrant Inspection of Type 316 Stainless Steel Forged Valve Bodies. Figure 16 illustrates the equipment layout used for the liquid penetrant inspection of 150 and 200 mm (6 and 8 in.) type 316 stainless steel forged valve bodies by the solvent-removable visible-dye system. This same equipment layout was also used for many other sizes and types of forgings requiring liquid penetrant inspection, such as the bonnet forgings described in the next example. The roller conveyor shown in Fig. 16 provided easy handling of relatively large forgings and, at the same time, allowed for the variations in time cycles that were compatible with the various sizes and shapes of forgings.

Inspection Procedure. The processing cycle that proved satisfactory for inspection of the forged valve bodies was as follows:

- Preclean by gritblasting, followed by a light etch in a 30% HNO_3, 10% HF, and 60% H_2O (by volume) acid solution. Forgings were also ground and polished as required so that the surface roughness was 6 μm (250 μin.) maximum
- Vapor degrease (wiping with 1,1,1-trichloroethane is an acceptable alternative)
- Apply solvent-removable visible-dye penetrant with a spray gun at a pressure of approximately 140 kPa (20 psig). Workpiece temperature should be kept within the temperature range of 10 to 40 °C (50 to 100 °F)
- Allow dwell time (penetration) of 15 to 20 min
- Rough wipe with lint-free cloth or absorbent paper
- Finish wipe with cloth or absorbent paper moistened with solvent remover (same group as the penetrant)
- Air dry for 5 to 10 min
- Apply nonaqueous developer, using a spray gun or pressurized spray cans, and allow drying time of 7 to 30 min prior to inspection
- Inspect in a booth using adequate light intensity, although at a level that will avoid producing a glare from the white surface

Acceptable workpieces were then cleaned by flushing with tap water, wiping with 1,1,1-trichloroethane, and air drying. They were then stamped and moved to the next processing operation.

Forgings that failed to meet acceptance standards were marked by encircling the flaw with a crayonlike marker that did not contaminate the surface. The flaws were then ground and polished, and the forging was reinspected in accordance with the above procedure.

Example 4: Solvent-Removable Liquid Penetrant Inspection of Type 316 Stainless Steel Bonnet Forgings. The solvent-removable visible-dye system of liquid penetrant inspection and the equipment shown in Fig. 16 were used to detect surface imperfections in 340 kg (750 lb) type 316 stainless steel bonnet forgings (Fig. 17).

Precleaning. Forged bonnets were delivered to the penetrant inspection operation in the machined condition. Machined workpieces were lightly etched in a 30% HNO_3, 10% HF, and 60% H_2O (by volume) acid solution and thoroughly water rinsed (spray or immersion) to remove all traces of acid. Workpieces were then swabbed with a 5% ammonia solution and subsequently rerinsed in water (spray or immersion). They were then allowed to dry at room temperature. Immediately preceding application of the penetrant, the etched forgings were precleaned with 1,1,1-trichloroethane and allowed to dry at room temperature for at least 5 min.

Inspection Procedure. The forgings were inspected in accordance with the following procedure:

- Apply solvent-removable visible-dye penetrant (group 1, MIL-I-25135) to the surfaces of the workpieces with a spray gun attached to the shop air line (brushing is an acceptable alternative). The penetrant need not be applied to the inside bore surfaces (as indicated in Fig. 17), because these surfaces are to be machined and subsequently inspected. Temperature of the workpiece for this operation as well as for subsequent operations is maintained within 15 to 50 °C (60 to 120 °F)

- Allow dwell time of 15 to 25 min
- Remove excess penetrant, using a solvent remover of the same family as the penetrant, by wiping with lint-free cloths or absorbent paper
- Final wipe with lint-free cloth or absorbent paper dampened slightly with solvent remover
- Dry at room temperature for 5 to 10 min
- Apply nonaqueous developer with a spray gun, pressurized spray can, or airbrush. Agitate developer periodically during application. Apply with short dusting strokes so that a uniform but thin layer is applied
- Perform inspection no sooner than 7 min and no later than 30 min following developer application
- Inspect, using adequate white light, but at a level that will avoid glare
- Clean inspected forgings by wiping with clean cloths moistened with 1,1,1-trichloroethane

Indications were evaluated in accordance with specifications listed as follows:

- *Linear indications*: None longer than 4.2 mm (1/6 in.)
- *Rounded indications*: None greater than 4.8 mm (3/16 in.) in diameter
- *Linearly disposed rounded indications*: Four or more in line separated by 1.6 mm (1/16 in.) or less, measured edge to edge, are evaluated as linear indications
- *Multiple indications*: No more than ten within any area of 40 cm^2 (6 $in.^2$) and whose major dimension is no more than 150 mm (6 in.)

Repair and Reinspection. Forgings containing flaws in excess of the above acceptance standards were repaired by grinding or polishing, provided these operations did not impair dimensional requirements. Forgings that were repaired by grinding or polishing were subject to reinspection using the technique and acceptance standards outlined above.

Liquid Penetrant Detection of Flaws in Heat-Resistant Alloy Forgings

Because most heat-resistant alloy forgings are nonmagnetic, the use of magnetic particle inspection for detecting surface flaws cannot be considered. Liquid penetrants are extensively used for inspecting the surfaces of high-integrity forgings. Critical forgings such as these require close quality control surveillance. Following production penetrant inspection using group VI fluorescent penetrant, it is desirable to conduct quality control overchecks on samples selected from previously inspected batches. These overchecks are performed using highly sensitive, high-resolution penetrants, as described in the following examples.

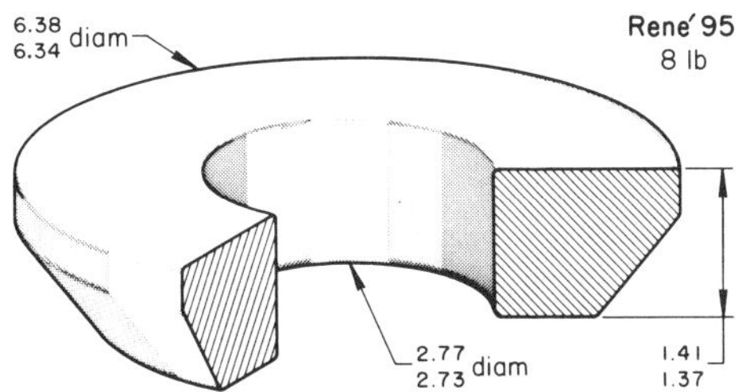

Fig. 18 Forged René 95 cooling plate for a jet engine that was first inspected by the postemulsifiable fluorescent penetrant system and then reinspected by a high-resolution penetrant after being accepted. Dimensions given in inches

Example 5: Routing Liquid Penetrant Inspection of Jet-Engine Cooling Plates Forged From René 95. Cooling plates (Fig. 18) used in the rear of several jet engines were forged from René 95, which is a nickel-base heat-resistant alloy.

Inspection Procedures. Forgings of this type were inspected in accordance with type 1, method B (MIL-I-6866), which is a postemulsifiable fluorescent penetrant inspection using group VI (MIL-I-25135) penetrants, emulsifiers, and dry developers. The following procedure was used:

- Clean machined forgings with solvent (1,1,1-trichloroethane) by immersion, swabbing, or wiping
- Lightly etch each forging with an acid mixture of 80% HCl, 13% HNO_3, and 7% HF (by volume)
- Rinse (immerse or spray) in tap water to remove all traces of acid
- Air dry prior to penetrant application. Inspect visually for adequacy of etch and to ascertain that forgings are dry
- Using a basket-type container, immerse several forgings at a time into group VI (MIL-I-25135) fluorescent penetrant. Allow to soak in penetrant or rest on drain table for at least 30 min
- Immerse container in emulsifier (group VI, MIL-I-25135), or hose on emulsifier, taking care to completely cover each forging completely. Allow emulsification time of 2½ min maximum
- Wash off excessive emulsified penetrant, using water at 40 °C (100 °F) maximum. Forgings can be immersed in agitated water wash or can be cleaned by a water spray located at the wash station
- Verify adequacy of removal of excess penetrant by use of ultraviolet light located at the wash station
- Dry in oven at 60 to 80 °C (140 to 180 °F) for 7 min maximum
- Visually inspect forging for adequacy of drying cycle
- Apply dry developer over entire surface of forging, using compressed air with a dry developer-applicator hose
- Allow development time of 10 to 15 min

Following development, the forgings were inspected in a darkened booth using a portable ultraviolet light (intensity: 1060 $\mu W/cm^2$, or 6840 $\mu W/in.^2$). Unacceptable indications were recorded; acceptance standards were as follows:

- No linear indications are permitted
- Scattered nonlinear indications are permitted if they are not visible at one diameter in white light (1075 lx, or 100 ftc, at surface subject to inspection), if they are separated by at least 3 mm (1/8 in.), and if there are no more than ten in any 25 mm (1 in.) diam area

All rejected forgings were scrapped or salvaged. Forgings that were salvaged were subject to reinspection in accordance with the above acceptance standards. Following inspection, the forgings were cleaned by solvent or water (spray or immersion) to remove retained developer. They were then dried at room temperature; a warm air blast was used to speed up the drying process.

Example 6: Quality Assurance Liquid Penetrant Inspection of Jet-Engine Cooling Plates Forged From René 95. Because of the critical application for the forged jet-engine cooling plate described in Example 5 and illustrated in Fig. 18, and because of the type of flaws expected (very tight forging cracks at the periphery and excessive porosity), sample forgings were periodically selected at random from lots that had been accepted after being inspected by the procedure described in Example 5. The sample forgings were subjected to liquid penetrant inspection using a commercially available high-resolution fluorescent penetrant to evaluate the adequacy of each cycle used in the production penetrant inspection described in Example 5. In particular, the high-resolution penetrant was used to determine the effectiveness of salvage grinding and polishing operations utilized to remove unacceptable indications detected in the original inspection. It is important to recognize that the high-resolution penetrants are not normally adaptable to production liquid penetrant inspection. The following procedure was employed when conducting the quality assurance inspection:

- Thoroughly clean the surface of the forging, using high-resolution precleaner. When necessary, before precleaning, lightly etch with an 80% HCl, 13% HNO_3, and 7% HF (by volume) acid solution; rinse; and dry. Allow precleaner to dry for 4 to 5 min
- Apply high-resolution fluorescent penetrant over entire surface of the forging, using a fine brush or the dauber attached to the cap of the can containing the penetrant. Allow penetrant to dry completely (approximate drying time, 1 min)

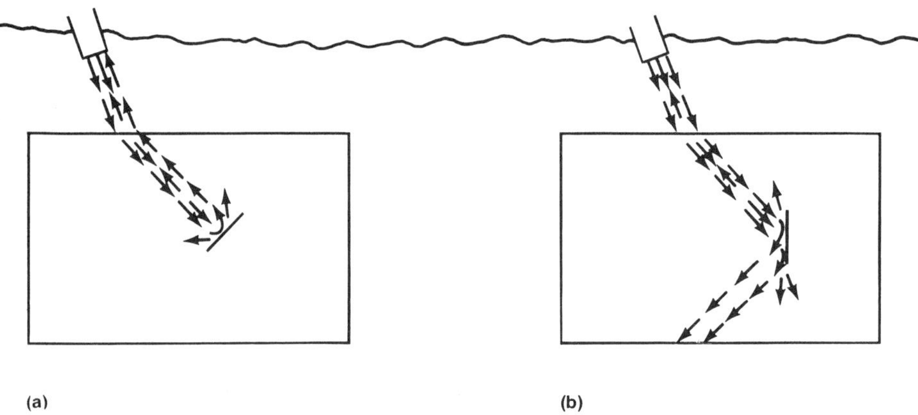

Fig. 19 Comparisons of discontinuities at normal orientation versus radial orientation. (a) Discontinuity normal to sound energy, almost total reflection of energy back toward transducer. (b) Discontinuity at 0° with respect to 45° sound energy path, almost no energy reflected back toward transducer

- Soak a piece of cheesecloth or equivalent material with high-resolution penetrant remover and remove the dried penetrant from the surface of the forging, using a circular motion. The penetrant remover will liquefy excess surface penetrant. Remove all traces of excess penetrant by scrubbing with a clean, lint-free cloth under a flood of tap water at room temperature. Finally, wipe surfaces with a clean, lint-free cloth that has been dampened with tap water at room temperature
- Allow surface to dry thoroughly. Drying may be aided by a blast of clean air or by means of a fan
- Apply nonaqueous developer from agitated spray can, in light, uniform dusting strokes. Hold can approximately 150 to 250 mm (6 to 10 in.) from the surface of the forging
- Allow a minimum of 10 min to elapse before inspection and evaluation of indications. Evaluate indications under ultraviolet light (intensity: 1060 $\mu W/cm^2$, or 6840 $\mu W/in.^2$) in a darkened inspection booth. Acceptance standards are the same as those listed in Example 5
- Postclean, using either the high-resolution precleaner or a water wash, and then dry at room temperature

Flaws in excess of acceptance standards detected with the high-resolution penetrants determined the disposition of the lot or modifications to the production penetrant cycles and appropriate action taken.

Ultrasonic Inspection

Ultrasonic inspection is used to detect both large and small internal flaws in forgings. Detailed information on the fundamentals of this method can be found in the article "Ultrasonic Inspection" in this Volume.

Forgings, by their nature, are amenable to ultrasonic inspection. Both longitudinal or shear wave (straight or angle beam) techniques are utilized. The size, orientation, location, and distribution of flaws influence the selection of technique and the inspection results. Consider, for example, Fig. 19, which shows the influence of flaw orientation on signal response. There are, however, some definite limitations. All ultrasonic systems currently in use generate sound electrically and transmit the energy through a transducer to the forging. Because the relationship of sound transmitted to sound received is a factor in the inspectability of a forging, particular attention must be given to the surface condition of the forging. Although techniques and couplants can enhance the energy transmission from the transducer to the forging, as-forged surfaces impair the effectiveness of ultrasonic inspection. Near-surface flaws are most difficult to detect, and a dead zone* at the entry surface often interferes. Because of the difficulty involved in detecting surface flaws by ultrasonic inspection, another method, such as magnetic particle or liquid penetrant inspection, is often used in conjunction with ultrasonic inspection to inspect high-integrity forgings thoroughly.

Complex shapes are difficult to inspect ultrasonically because of the problems associated with sound-entry angle. Most ultrasonic inspection of forgings uses techniques that send waves into the forging perpendicular to the surface. Radii, fillets, and similar configurations must receive special treatment if all areas of the forging must be inspected. The special treatment involves the use of a standoff that has an end contoured to fit the inspection surface or the use of a small-diameter or focused transducer.

Application. In certain cases, where the end use of a forging is considered critical, ultrasonic inspection is used to inspect the wrought material before it is worked. Surface or internal flaws that are not detected before a billet is forged may not be detected in the final forging and will therefore be present in the finished part. Ultrasonic inspection is often used as part of a completely diagnostic inspection of a forging from newly designed dies, where use of the finished part does not warrant inspection of every part.

Quality control measures often include the ultrasonic inspection of random samples from a particular forging. This provides the necessary assurance that the process is under control and that variables affecting internal quality have not been inadvertently introduced.

Ultrasonic inspection is often used in the further evaluation of flaws detected by other nondestructive methods. This reduces the possibility that a particular forging will be unsuitable for its intended service.

Ultrasonic inspection can be used on every forging to validate its integrity for extremely rigorous requirements. This applies in particular to forgings for nuclear and aerospace applications, where rigid standards of acceptance have been established. Standards and criteria have been set up to detect material inclusions, internal voids, laminations, and other conditions. In addition, the inspection of every forging by ultrasonics has been effective in detecting excessive grain size and other structural conditions.

Ultrasonic inspection is often used to qualify a particular lot of forgings that has been subjected to certain variations in approved processing procedures. A notable instance is the use of ultrasonics to determine the presence of thermal flakes or in locating quench cracks.

Basic Procedures for Ultrasonic Inspection

Many specific procedures have been developed for the inspection of forgings by ultrasonics. The basic procedures described here are based on those developed for use in a specific plant.

Forgings of regular (symmetrical) shape are ultrasonically inspected at final inspection (prior to shipment). If possible, the entire forging is ultrasonically inspected.

Irregularly shaped forgings, which have a configuration that precludes complete inspection prior to shipment, are ultrasonically examined at the latest stage of processing that will permit examination of the entire forging (preliminary inspection). These complex shapes are then reexamined to the extent practical before shipping (final inspection). To ensure complete inspection coverage of the forging, the transducer is indexed with at least 15% overlap with each

*The term dead zone refers to the depth below the entry surface where any echoes from flaws cannot be detected because they are masked by the trailing edge of the transmitted pulse. The dead zone depends on sound velocity, test frequency, and pulse length and is different for different combinations of test metal, transducer, and test instrument.

pass. The rate of scanning does not exceed 150 mm/s (6 in./s).

The surface roughness of forgings from which ultrasonic examination is conducted should be no greater than 6 μm (250 μin.), unless otherwise specified. The forging surfaces should also be free of extraneous material, such as loose scale, paint, dirt, or other contaminants.

Equipment. An ultrasonic, pulsed, reflection-type instrument is used to inspect forgings. The instrument is capable of operating at three or more frequencies; for example, 1.0, 2.25, and 5.0 MHz. The instrument provides a linear presentation (within ±5%) to approximately 75% of the screen height or a minimum of 40 mm (1½ in.). Signal attenuators and calibrated gain control, if used, are accurate over the range used to ±10% of the nominal attenuation ratio. Forgings are examined by the longitudinal wave technique at 2.25 MHz unless adequate penetration cannot be obtained, in which case the inspection is conducted at 1.0 MHz.

Longitudinal Wave Ultrasonic Inspection

Where possible, all forgings are scanned in two directions perpendicular to each other. Disk (pancake) forgings are examined axially from at least one end face and radially from the circumference, whenever practical. Ring forgings are examined radially from at least the outside surface and axially from at least one end face.

Cylindrical forgings are examined radially from the circumference. The axial examination of cylindrical forgings can be conducted when practical, using the back-reflection method of longitudinal wave inspection.

Back-Reflection Method. Longitudinal wave inspection by back reflection is often performed with a 2.25-MHz transducer. A higher or lower frequency can be used, when necessary, to obtain a meaningful inspection result.

The sensitivity for longitudinal wave calibration is established by adjusting the instrument controls to obtain a 40 mm (1½ in.) back-reflection height from the opposite side of the forging over an area free of indications (this level corresponds to approximately 75% of the useful screen height of the instrument). Calibration must be reestablished for each change of thickness greater than 10%. The calibration level and minimum-back-reflection requirements as outlined in this section apply to areas with parallel surfaces.

Routine scanning is done at the calibration gain level. When specified, the instrument gain level is increased to at least two times the calibration gain level. Flaw amplitudes and back-reflection reduction are evaluated with the gain setting at the calibration level of 40 mm (1½ in.) back-reflection height in an indication-free area as close as practical to the area of indications.

All areas of the forging are examined with sufficient gain to provide a minimum back-reflection amplitude of 40 mm (1½ in.), except for areas where reduction of the back reflection is associated with ultrasonic indications.

Recording of indications detected by back-reflection method is done as follows:

- Any indication having an amplitude that equals or exceeds 10% of the calibration back-reflection level is reported in increments of 10
- Any traveling indication equal to or exceeding 5% of the calibration back reflection is recorded. Traveling indications are defined as indications whose leading edge moves a distance equivalent to 25 mm (1 in.) or more of metal depth with movement of the transducer over the surface of the forging. Any cluster of indications equal to or exceeding 5% of the calibration back reflection is recorded. A cluster of indications is defined as five or more indications located in a volume representing a 50 mm (2 in.) or smaller cube in the forging
- Any continuous indication equal to or exceeding 5% of the calibration back reflection is recorded. A continuous indication is defined as an indication that can be held on the same plane over a square area whose sides are greater than twice the diameter of the search unit
- Any reduction in back reflection (associated with a discontinuity indication) exceeding 20% of the calibration back reflection, measured in increments of 10%, is recorded

Distance-Amplitude-Correction (DAC) Curve Method. The following refers to the primary test direction, which is the radial direction for hollow and shaft-type forgings and the axial direction for upset disk-type forgings. Test sensitivity is established with reference flat-bottom-hole standards machined into either the forging itself or separate reference blocks. If separate reference blocks are used for instrument calibration, they should be of material having an attenuation coefficient within ±25% of forged production material. A DAC curve is established for longitudinal wave inspection of the forging in the primary test direction through the use of reference blocks of various thicknesses and hole sizes, as explained in the article "Ultrasonic Inspection" in this Volume. The size of the flat-bottom hole and the metal distance from entry surface to the hole bottom depend on the maximum machined thickness in the primary test direction at the time of initial ultrasonic inspection.

The following method is used to prove that the calibration blocks have attenuation characteristics within 25% of the production forging. An acoustic-decay pattern (the initial three back reflections) is determined for the calibration-block material while that material is still part of a larger section, so that there is no side-effect cancellation or reinforcement on the exponential decay pattern. An acoustic-decay pattern is also determined for the production forging at the midlength, using the same type of instrument and transducer as used for determining the acoustic-decay pattern of the calibration-block material. The amplitude of the first back reflection should be the same for both tests and within the vertical linearity of the instrument. The calibration-block material, which is fabricated in accordance with either ASTM E 428 or E 127, is considered compatible with the production forging if the sum of the amplitudes of three successive back reflections from the production forging is within 25% of the sum of the amplitudes of three successive back reflections from the calibration-block material or if the slope of the plotted back-reflection amplitudes of the calibration-block material is within 25% of the slope of the plotted back-reflection amplitudes of the production forging at the same sound-travel distance.

Recording of indications detected by the DAC-curve method is determined by the types and magnitudes of indications. Those recorded are any indication whose amplitude equals or exceeds 100% of the established DAC curve; traveling, clustered or continuous indications (as defined above) that have amplitudes equal to or greater than 50% of the DAC curve; and any reduction in back reflection (associated with a discontinuity indication) exceeding 20% of the calibration back reflection, measured in increments of 10%.

Calibration: Primary Test Direction. At the time of initial inspection, the areas of greatest and least acoustical penetrability are determined by scanning selected passes, which are representative of the forging cross section, at a fixed instrument gain level. The amplitude of the first back reflection is monitored during these scans to determine the areas of greatest and least penetrability.

The amplitude of the reflection from the flat-bottom hole in the longest calibration block is set at 13 mm (0.5 in.) sweep to peak, and the other points of the DAC curve are obtained at this gain setting. This curve can be marked on the face of the screen or plotted on graph paper. This is the fixed DAC curve that will be used to evaluate the indications. Any portion of the DAC curve above the vertical linearity limits of the instrument is brought on screen with the attenuator for evaluation of the indications in this area.

After establishing the fixed DAC curve, the amplitude of the back reflection over the area of greatest acoustical penetrability is determined at this instrument gain level. The transducer is then coupled to the forging over the area of least penetrability, and the gain setting of the instrument is increased to bring the back reflection to the same amplitude as the amplitude obtained over the area of greatest acoustical penetrability (a minimum of 40 mm, or 1½ in., sweep to peak). For scanning, the instrument gain level is increased a minimum of twice the final gain setting.

Any indication, detected while scanning, whose amplitude equals or exceeds the reporting level is evaluated. This evaluation is conducted by readjusting the back reflection in a discontinuity-free area of the forging as close as practical to the area of indication to equal the same height of back reflection obtained in the area of greatest acoustic penetrability when the DAC curve was established. The search unit is repositioned over the area of the indication, and the evaluation to the DAC curve is made.

Back-reflection reduction is evaluated by adjusting the first back reflection to the maximum back reflection within the vertical linearity limits of the instrument in a discontinuity-free area of the forging as close as practical to the area of the indication to be evaluated. The search unit is repositioned over the area containing the indication, and the amplitude is recorded.

For contoured forgings at final inspection, ultrasonic inspection in the primary test direction is performed using a DAC curve established from the same reference blocks used for preliminary inspection. The last point of the curve is 13 mm (0.5 in.) sweep to peak, and the other points of the DAC curve are obtained at that gain setting. This is the fixed DAC curve that will be used to evaluate any indications. Compensation for the areas of greatest and least penetrability does not apply at final inspection with a contoured forging.

Calibration: Secondary Test Direction. The secondary test direction is approximately at 90° to the primary direction. The transducer is coupled to the face of the forging, and the amplitude of the first back reflection is set within the vertical linearity limit of the instrument (minimum 40 mm, or 1½ in., sweep to peak). The DAC curve is established at this gain setting, using the same blocks that were used for the longitudinal wave inspection in the primary test direction. The last point of the DAC curve is extended parallel to the sweep line to midlength plus 25 mm (1 in.). Any indications will be evaluated against this DAC curve. When the configuration of the forging at final inspection due to contour machining prevents ultrasonic examination in the axial direction to midlength plus 25 mm (1 in.), the DAC curve will be extended parallel to the sweep line to the deepest location, wherever possible, for evaluation of the indications.

Shear Wave Ultrasonic Inspection

Shear wave inspection is performed from the circumference of rings and hollow forgings that have axial length greater than 50 mm (2 in.) and a ratio of outside to inside diameters of less than 2:1. A 1.0-MHz, 45° angle-beam search unit is used when shear wave inspecting forgings having a ratio of outside to inside diameters of less than 1.4:1. Forgings with a ratio of outside to inside diameters of 1.4:1 or greater, but less than 2:1, will be shear wave tested at a beam angle of less than 45°. Shear wave inspection is performed by scanning in both circumferential directions (clockwise and counterclockwise) along the periphery of the forging.

Inside diameter and outside diameter calibration notches are cut axially in the surface of the forging or a similar testpiece (preferably, excess metal or test metal). The sides of the notches should be smooth and parallel to the axis of the forging.

The calibration notches are 25 mm (1 in.) long, V-shaped or rectangular, with a width not exceeding twice the depth. These notches should have a depth of 3% of the maximum section thickness (based on thickness at initial ultrasonic inspection) or 6 mm (¼ in.), whichever is smaller. If at final inspection it is necessary to place the calibration notches in the forging, these will have to be weld repaired.

The sensitivity for angle-beam inspection of the forging is established by adjusting the instrument controls to obtain a minimum 13 mm (0.5 in.) sweep-to-peak indication from the outside diameter calibration notch. The response from the inside diameter calibration notch is then obtained at this sensitivity level. The peaks of the indications from the inside diameter and outside diameter notches are connected to establish a reference line. Gain is increased two times for scanning. For contoured forgings at final inspection, the inspection is conducted to the extent that contour permits a meaningful examination.

Recalibration

Any realignment of the transducer with respect to the material or any change in the search unit, couplant, or instrument settings from that used for calibration necessitates recalibration. A calibration check is performed at least once every 8-h shift. If a 15% or greater decrease in the sensitivity level is observed during a calibration check, then the required calibration is reestablished, and all material examined in the preceding calibration period is reexamined.

Ultrasonic Inspection of Specific Forgings*

The procedure used for the ultrasonic inspection of a given forging depends greatly on the size and shape of the forging, and sometimes more importantly on the degree of inspection that is required. For example, a dome-shaped forging such as the low-alloy steel forging shown in Fig. 20 has a symmetrical shape that readily permits inspection by either the immersion or the contact ultrasonic method with no great problems other than the handling equipment needed.

*Example 11 was prepared by Carol Miller, Wright Research & Development Center, Wright-Patterson Air Force Base. The research was conducted by Automation Industries, Inc., under Air Force contract.

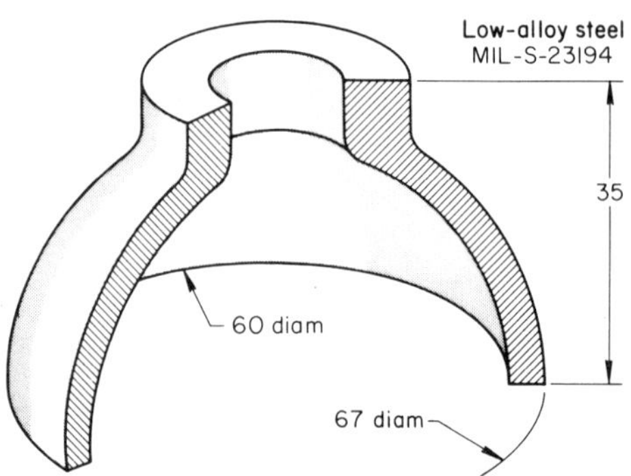

Fig. 20 Symmetrical, dome-shaped forging that was ultrasonically inspected by the longitudinal wave technique on the inside and outside of the dome and on the ends, and by the circumferential shear technique on the outside of the dome and the neck portion. Dimensions given in inches

The forging illustrated in Fig. 20 was inspected by using the longitudinal wave technique on the inside diameter, outside diameter, and ends of the forging, plus the circumferential shear technique on the outside diameter. In addition, circumferential shear was conducted on the outside diameter of the neck portion of the forging.

The landing-gear forging illustrated in Fig. 21 has a more intricate shape and is large, which makes it more difficult to inspect. For forgings such as this, the best procedure is thorough inspection of the billets prior to forging. A common practice is to ultrasonically inspect the billet material to an agreed-on specification. This minimizes the amount of inspection required after forging. A forging such as the one shown in Fig. 21 would be ultrasonically inspected by immersion, longitudinal wave, hand-scan techniques in all areas where the shape permits. The following examples describe procedures used for the ultrasonic inspection of a variety of steel and aluminum alloy forgings.

Example 7: Ultrasonic Inspection of Forged Medium-Carbon Steel Axle Shafts for the Detection of Chevrons and Burning. Automotive rear-axle shafts (Fig. 22) are made from medium-carbon steels by first forging in closed dies to form the hub end and then by cold extrusion to form the shaft section. Cold-extruded rear-axle drive shafts are prone to develop serious internal flaws known as chevrons or cup-cone ruptures. Chevrons are internal ruptures that may or may not extend to the surface of the extruded stock (most often they do not extend to the surface). There may be any

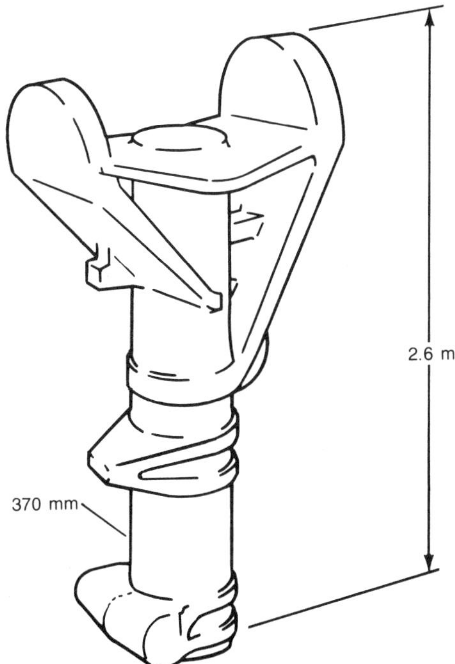

Fig. 21 Landing-gear forging typical of complex forgings that are preferably ultrasonically inspected in billet form because the intricate forged shape prevents thorough ultrasonic inspection

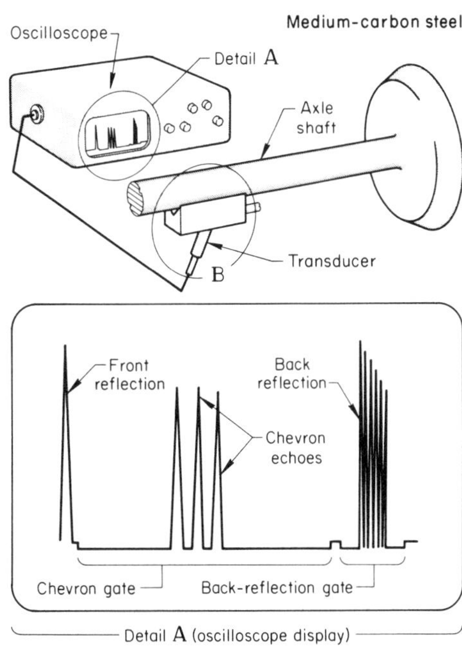

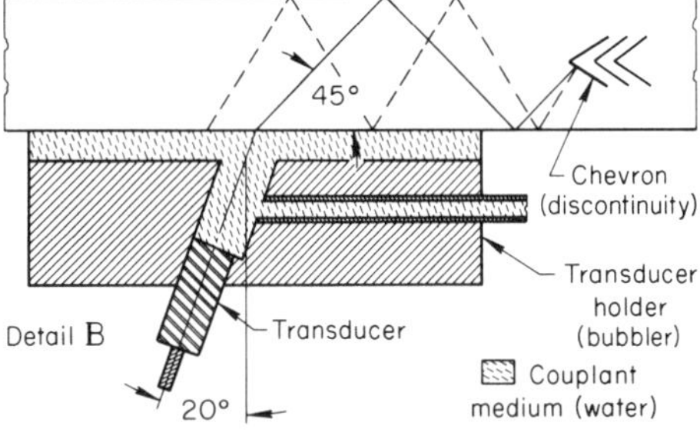

Fig. 22 Setup and operating essentials for the ultrasonic inspection of the shaft portion of a forged medium-carbon steel axle shaft

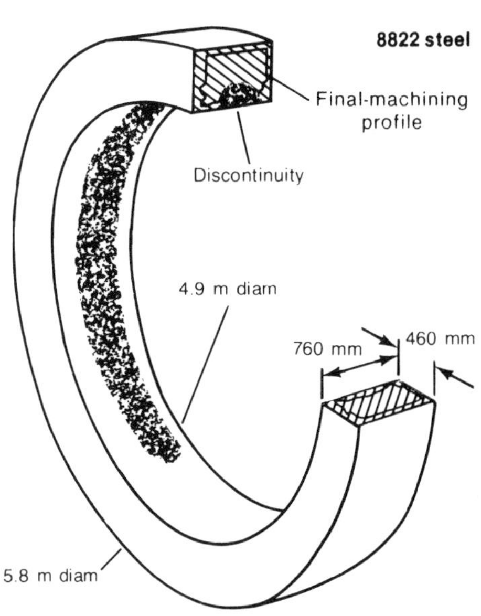

Fig. 23 Large rolled ring that was ultrasonically inspected as a 1.7 m (68 in.) diam billet and as a rolled ring

number of chevrons produced during the extrusion of a single axle shaft. Therefore, it is necessary to inspect the entire length of each shaft.

Inspection Procedure. To perform a 100% ultrasonic inspection of these axle shafts in a plant where production was 2 to 3 million units annually, an automated system was used. The test station was supplied by a conveyor.

Figure 22 illustrates the position of the axle shaft when the ultrasonic test was performed. A transducer holder, with a fixed angle fabricated of hardened tool steel with no adjustment, coupled the ultrasound from the transducer to the shaft end. A detailed view of the transducer holder and V-block is shown in detail B in Fig. 22. Transducers with a frequency of 1.6 to 2.25 MHz were used to perform the ultrasonic inspection. The transducer incident angle was approximately 20°, which yielded a 45° shear wave generated within the shaft. This permitted the detection of internal flaws. The oscilloscope display in detail A in Fig. 22 illustrates the interaction of an ultrasonic beam with a chevron. Two electronic gates were utilized to monitor two separate areas of the shaft, the area from where the sound enters the shaft to approximately 150 mm (6 in.) from the end of the flange. This was a positive gate, commonly called the chevron channel. The second gate was positioned to monitor the back-reflection signal and was a negative gate.

The acceptance of any axle shaft required a strong back reflection with no signal in the chevron channel gate. In addition to detecting chevrons, this inspection procedure detected the results of any burning that had occurred during heating. A burned forging attenuated the back reflection and caused the forging to be rejected.

Example 8: Ultrasonic Inspection of a 1.7 m (68 in.) Diam 8822 Steel Billet and the Ring Into Which It Was Rolled. The large steel ring shown in Fig. 23 was produced for a nuclear application from a 1.7 m (68 in.) diam billet of 8822 steel. The large billet was first forged into a doughnut preform in a large press, then formed into the ring by ring rolling.

A ring of this type requires a high degree of integrity; therefore, inspections were performed at various stages of manufacture. Routine ultrasonic inspection of the billet revealed a gross center condition in approximately 67% of the length. Inspection was performed from the outside diameter and end faces to the maximum extent possible. On billets of this size, however, the surface condition and the lack of refinement of the internal structure have an adverse effect on the ability to perform a thorough inspection by ultrasonics.

When the billet was preformed into the doughnut shape, the portion that had been the center of the billet was removed. The doughnut was then ring rolled into a large ring having a wall 460 mm (18 in.) thick (Fig. 23). It was hoped that the area indicated as defective in the billet stage would have been removed, but because it was uncertain whether all of the defective area had been removed, the ring was machined on the outside diameter and on one end face to a surface finish of 6 μm (250 μin.), thus permitting a more sensitive inspection.

Primary consideration was given to performing a straight-beam ultrasonic inspection from the outside surface in the radial direction. During ring rolling, included foreign materials become aligned in the circumferential direction to provide a lamellar-type discontinuity. This type of discontinuity is most readily detected from the perpendicular direction or, in this case, either from the outside or inside surfaces of the ring. Because the original discontinuities in the billet were in the center and would remain in a similar position during forging, it was anticipated that if remnants of the center condition were still present, they would be near or adjacent to the inside or bore of the ring. Therefore, a radial ultrasonic inspection from the outside using the straight-beam technique was considered to be the best method of detection.

Inspection Procedure. A straight-beam inspection was instituted as indicated, and calibration was established using the back-surface-reflection method to determine whether adequate ultrasonic penetration was available. A 25 mm (1 in.) diam, 2.25-MHz transducer was used in conjunction with the ultrasonic unit.

The outside surface was scanned in overlapping paths to ensure complete coverage. During the inspection, areas of indications were noted at approximately midheight and adjacent to the bore area; one area approximately 7 m (24 ft) long in a circumferential direction was encountered during the inspection; several smaller spot-type areas were also noted.

Evaluation of indications on the basis of applicable American Society of Mechanical Engineers (ASME) Boiler and Pressure Vessel Codes revealed that the indications would be acceptable. However, for closer evaluation, a straight-beam inspection from the face and a circumferential angle-beam inspection, in two 180° opposing directions, were performed. Straight-beam inspection from the face did not reveal indications at a normal gain or sensitivity level; they could be detected only at an extremely high sensitivity level. Angle-beam inspections showed very small low-level (approximately 10% maximum) occasional indications in the area previously noted.

Although not required by the ASME code, an axial angle-beam inspection from the outside was performed, mainly in the area of indications. Indications were readily detectable, but were not considered serious enough to reject the forging. On the basis of results of the ultrasonic inspections, the ring was considered acceptable within the required specifications. Magnetic particle inspection revealed a few small indications in the areas tested. Because severity did not appear to be indicative of the ultrasonic results, the area was conditioned by grinding and polishing to obtain an additional inspection at a greater depth from the inside surface. Subsequent magnetic particle inspection indicated a much more severe condition, which was not acceptable.

Conclusions. Metallographic investigation showed that the indications could be classified as areas of chemical segregation and nonmetallic inclusions (predominantly sulfides). These were evidently present in the center of the billet and remained in the final ring. Therefore, the ring was considered unsatisfactory for the application. Removal of the questionable metal and repair by welding was considered, but the cost would have been excessive. Replacement of the defective ring from an acceptable billet was the most economical solution.

Example 9: Ultrasonic Inspection of a 4340 Steel Shaft Forging. The pulse-echo technique of ultrasonic inspection was selected for locating flakes, bursts, and pipe in large 4340 steel shaft forgings such as the one shown in Fig. 24.

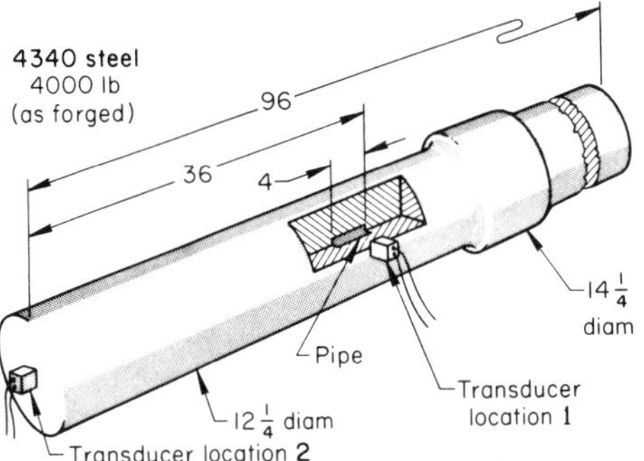

Fig. 24 Transducer locations for radial and longitudinal ultrasonic inspection of a large steel shaft forging for detection of flakes, bursts, and pipe. Dimensions given in inches

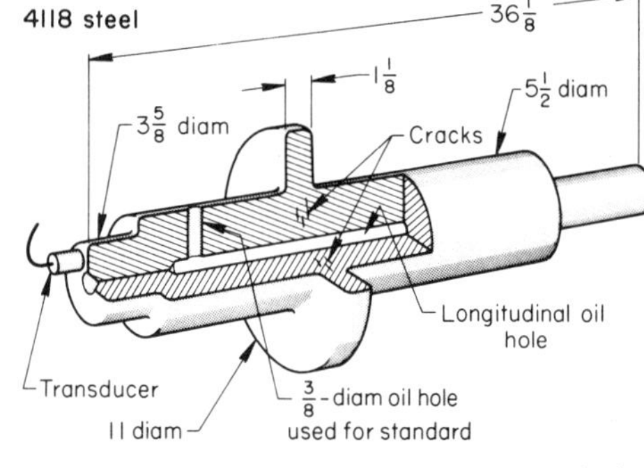

Fig. 25 Upset shaft forging, of 4118 steel, that was ultrasonically inspected by the pulse-echo longitudinal wave technique. Dimensions given in inches

Equipment Specified. The ultrasonic instrument specified was a pulsed, reflection-type unit capable of testing at 1 to 5 MHz. It was also required that the instrument provide linear presentation (within ±5%) for at least 75% of the screen height (sweep line to top of screen). It was further specified that the electronic apparatus contain a signal attenuator (accurate over its useful range to ±10% of the nominal attenuation ratio), which allows measurement of signals beyond the linear range of the instrument. Transducers 25 to 28 mm (1 to 1⅛ in.) in diameter or 25 mm (1 in.) square were used.

Couplants having good wetting characteristics, such as SAE No. 20 or No. 30 motor oil, glycerin, pine oil, water, or a nonionic water-soluble polymer, were specified. (The above couplants, however, may not be comparable to one another, and the same couplant must be used for calibration and recording of results.) Test blocks containing flat-bottom holes were used for calibrating equipment in accordance with the above apparatus requirements. Such test blocks were used to establish recording levels for longitudinal testing.

Surface Preparation. It was required that all surfaces be machined so that surface roughness did not exceed 6 μm (250 μin.) and that surfaces contacted by transducers should be free from all extraneous material, such as scale, dirt, and paint, before the couplant was applied.

Inspection Procedure. Ultrasonic inspection was performed after heat treatment, but prior to drilling holes, cutting keyways, or otherwise machining to a contour. The entire forging was inspected using a test frequency of 2.25 MHz. Search patterns that incorporated 15% overlap between passes were scanned at a speed of less than 150 mm/s (6 in./s). Scanning from two mutually perpendicular directions (radial and longitudinal) provided complete coverage. Referring to Fig. 24, when indications were found with the transducer oriented for radial scanning (transducer location 1), the operator moved the transducer to location 2, on the end of the forging, to determine the size and axial location of the flaw. In the forging shown in Fig. 24, pipe that was 100 mm (4 in.) long by 25 mm (1 in.) thick was found in the center of the cross section and 915 mm (36 in.) from one end of the forging.

Acceptance Standards. Forgings showing a single flaw indication greater than that from a 13 mm (½ in.) diam flat-bottom hole were not acceptable. Forgings containing clustered or traveling indications greater than 50% of that from a 13 mm (½ in.) diam flat-bottom hole were not acceptable. Forgings containing a complete loss of back reflection not attributable to size, shape, or grain size were not acceptable; loss of back reflection not accompanied by an indication from a flaw was further investigated by the use of other search units. Complete loss of back reflection was assumed when the back reflection fell to less than 5% of the reference back reflection.

Example 10: Ultrasonic Inspection of an Upset-Forged 4118 Steel Shaft. Field failures were encountered in the upset portion of forged 4118 steel shafts (Fig. 25). The failures were traced to internal cracks in the upset flange area that were initiated from gross nonmetallics. These nonmetallics, which were originally located near the center of the cross section, were pushed outward into an almost radial plane in the upsetting of the flange. When the shaft was heat treated, these planes of inclusions developed internal cracks.

The location of these cracks, their orientation, and the size of the shaft all indicated that ultrasonics would be the best method for detection. Two approaches were feasible. The first was to send a longitudinal beam in from the end of the shaft, and the second was to use a shear wave from the outside diameter of the shaft adjacent to the flange. The first technique was selected because it permitted testing of shafts already assembled.

A 22 mm (⅞ in.) diam, 2.25-MHz transducer was used with a plastic wear face and an oil couplant. The shaft had a radially drilled oil hole 9 mm (⅜ in.) in diameter (note location in Fig. 25). This radial hole was used as a landmark and standard to set the sensitivity. The signal from this hole was set to give full-screen height from a known flaw-free shaft. The end of the shaft was scanned in a zone extending from the center out 40 mm (1½ in.) radially. The area around the longitudinal oil hole within the upset section was the area where the flaws occurred (Fig. 25).

When the shaft was tested from the short end (measured from the upset), a portion of the test area was in the shadow of the radial oil hole. In addition, because of the distance to the radial hole, the second multiple of the radial-hole signal occurred at the same point on the screen as did the flaws. For this reason and because of variation in flaw orientation, it was desirable to test the shaft from the long end as well as from the short end. In the assembled machines, the only access was to the long end of the shaft. The same standard was used from the long end of the shaft, but an instrument having higher sensitivity was required to produce a full-screen signal from the radial hole because of the added length of beam path.

The rejection level was set at 20% of full screen and was based on the size of flaws observed when the shafts were cut up. The flange portion was first cut out of the shaft and retested with a smaller-diameter transducer to better define the extent of the flaw. The section was further cut until one edge of the flaw was exposed. When possible, this reduced section was broken open through the flaw. The fracture face provided the best view of the extent and shape of the flaws. If the inclusion flaws were larger than 20 mm (¾ in.) in diameter, they were considered rejectable even if they had not cracked in heat treatment. An attempt was made to test the shafts in the as-forged condition, but this approach was not reliable because some inclusions developed cracks during heat treatment.

Similar but slightly larger shafts were found to have the same types of flaws. These shafts were tested with the same technique as described above and initially showed a very high rate of rejection. No flaws were found when the rejected shafts were sectioned, and when the shafts were retested, no flaws were found. After some investigation, it was realized that the flaw signals were false. When a portion of the beam struck the oily surface of the longitudinal oil hole, a mode-converted wave was reflected back to the search unit. The time of travel of the mode-converted false signal coincided with the time of travel of the true flaw signals, causing rejection of the shafts. The problem was solved by removing the oil film from the longitudinal oil hole.

Example 11: Ultrasonic Inspection of B-58 Nose Landing Gear Trunnion. The development of this inspection focused on the nose landing gear trunnions (Fig. 26). Initiation sites for stress-corrosion cracking were found to originate on the trunnion-bore inside surfaces at the forging parting plane (Fig. 27). The trunnion is made from forged aluminum alloy 7075-T6. After examination of previous failures, it was shown that these cracks may propagate to a severe condition before the trunnion fails. A manual shear wave ultrasonic inspection was developed that concentrated on these initiation sites.

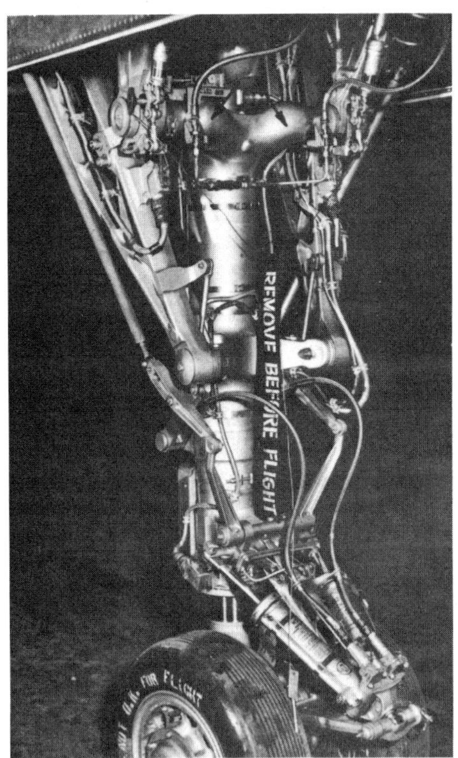

Fig. 26 Location of nose landing gear trunnion (arrows) on a B-58 aircraft. Courtesy of Tom Dusz, The University of Dayton Research Institute

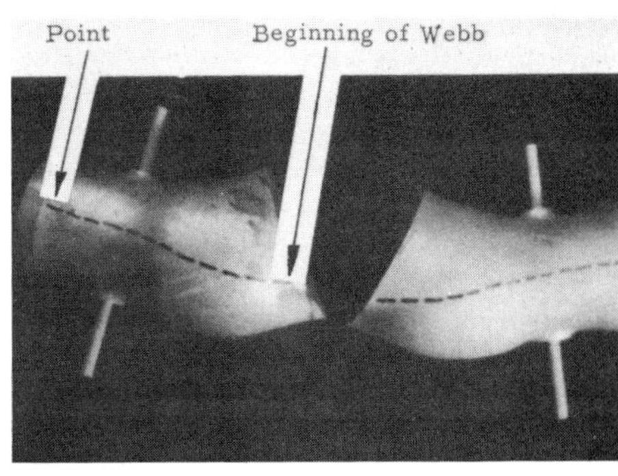

(a)

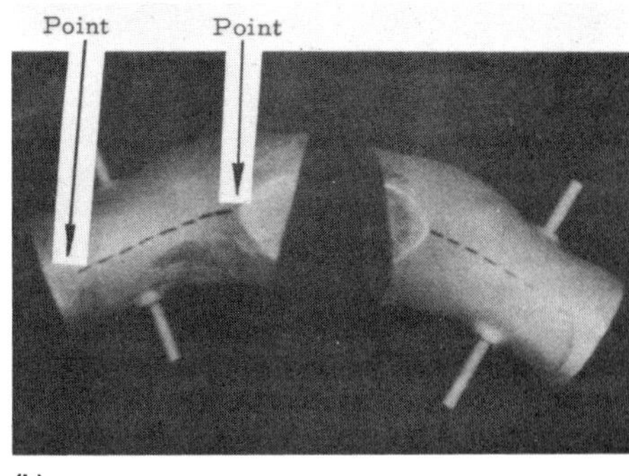

(b)

Fig. 27 Location of parting line on trunnion. (a) Top view of parting line. (b) Bottom view of parting line. Courtesy of Tom Dusz, The University of Dayton Research Institute

Inspection Procedure. The equipment needed for this inspection includes a flaw detection instrument capable of a test frequency of 5 MHz; a 5-MHz search unit with a special lens contoured to 45 mm (1.75 in.) cylindrical radius; a reference standard containing artificial notches 0.25, 0.50, and 1.25 mm (0.010, 0.020, and 0.050 in.) deep; and a clean oil couplant. The reference standard is manufactured from the outer bore section of an actual trunnion. The sensitivity cali-

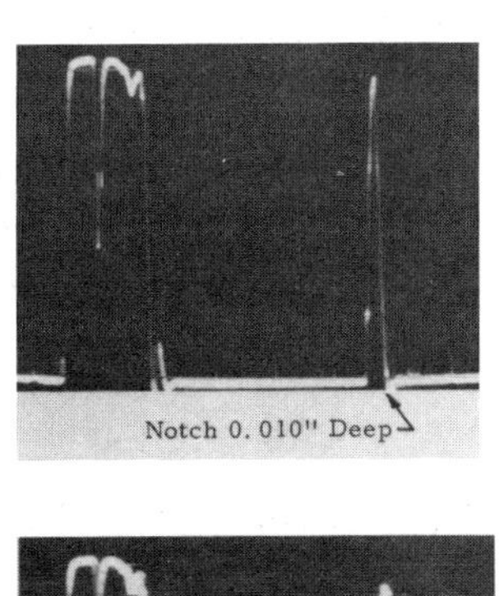

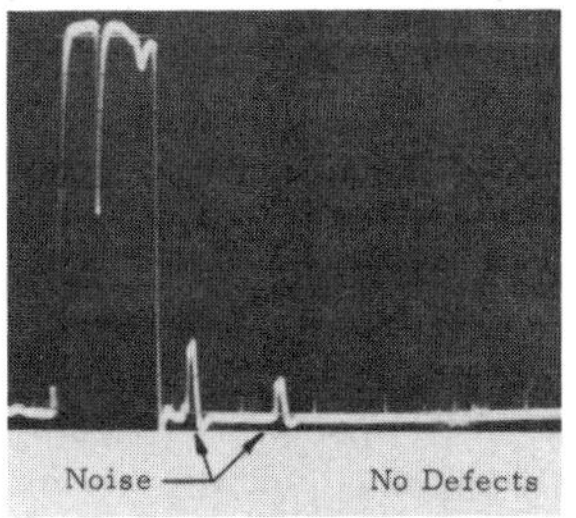

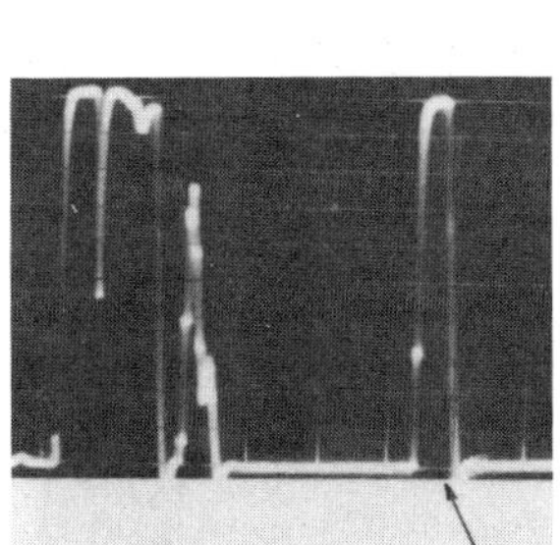

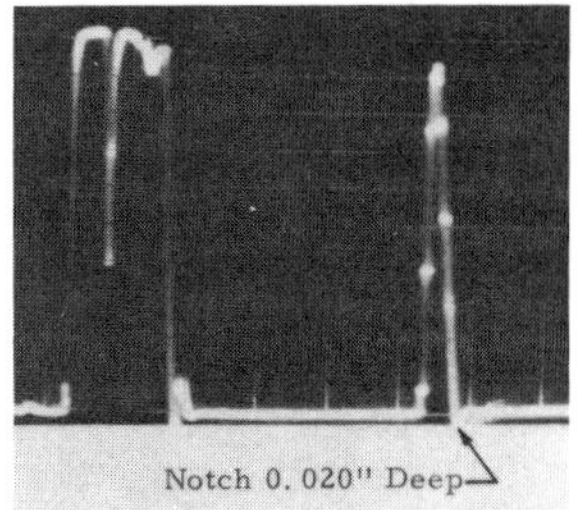

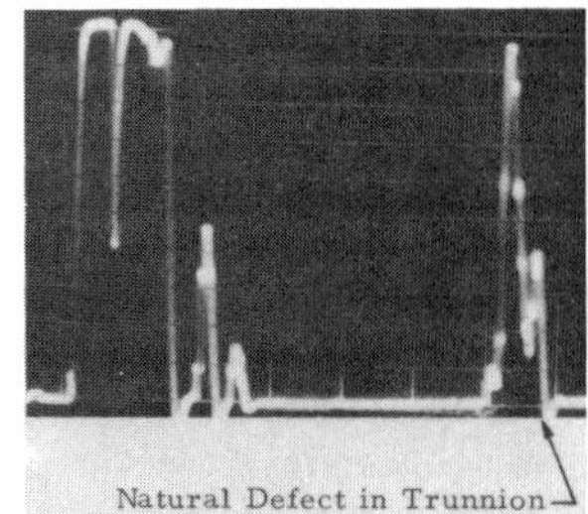

Fig. 28 Typical signals that may appear during inspection. Dimensions given in inches. Courtesy of Tom Dusz, The University of Dayton Research Institute

bration is performed by applying sufficient couplant to the reference standard and by positioning the search unit over the 1.25 mm (0.050 in.) deep notch. A full-scale amplitude signal is obtained. The standard is scanned, and the signals are observed from the 0.25 and 0.50 mm (0.010 and 0.020 in.) deep notches. The instrument sweep delay is adjusted until the first reflected signals from the artificial defects reach their peak amplitude at the center of the screen. The sensitivity, pulse length, and reject controls are used to obtain a 95% amplitude signal from the 0.25 mm (0.010 in.) deep notch. The part is cleaned with an acceptable solvent to remove dirt, grease, tar, loose paint, and other debris. The forging parting plane and two planes parallel and 40 mm (1.5 in.) on either side of the forging parting plane are marked with ink that will not come off the part during the inspection. The areas of concern include the entire area between the parting line and the parallel lines, the bolt hole to the web on top, and the ground area on the bottom.

A defect signal first appears near the center of the screen and moves to the left as the search unit is scanned closer to the parting line. When the search unit is scanned from the parting line back toward the parallel line, the defect signal moves from the left of the screen (near the front surface noise) to the center of the screen. Figure 28 shows some typical amplitude signals that may appear on the screen. Any defect indications showing a 25% and greater amplitude signal are considered for further disassembly evaluation and/or replacement.

Conclusion. The development of this testing technique enabled inspectors to locate very gross or severe conditions before trunnion failure occurred. This greatly reduced the amount of on-aircraft failures.

Example 12: Ultrasonic Inspection by the Immersion Technique of an Aluminum Alloy Forging Having No Parallel Surfaces. The aluminum alloy forging shown in Fig. 29(a) was ultrasonically inspected by the immersion (water) technique using a hand-held search unit. Details of the inspection conditions are given in the table accompanying Fig. 29.

Reference standards were used for calibrating the oscilloscope that corresponded to full- and half-metal thickness of the thickest section of the forging and for a metal thickness of 6 mm (¼ in.). Equivalent flat-bottom-hole sizes required for the inspection class were used in the calibration standards.

The inspection was performed to determine whether internal discontinuities of any type were present. For the type of forging shown in Fig. 29(a), because of the continuously changing metal thicknesses and surface contours, immersion testing using a hand-held search unit was the most suitable method of ultrasonic inspection. This inspection procedure is called scrubbing. The search unit consisted of a transducer fitted with a hollow plastic tube, called a stand-off tube, to maintain a constant water distance of 75 mm (3 in.) from the entry surface (see transducer assembly, Fig. 29b). The cable and scrubber housing were attached to the scrubber cap. This portion of the unit was wrapped with watertight rubber tape because the unit is partially or totally submerged during use. The stand-off tube, while submerged, was completely filled with water, with no air gaps or bubbles present.

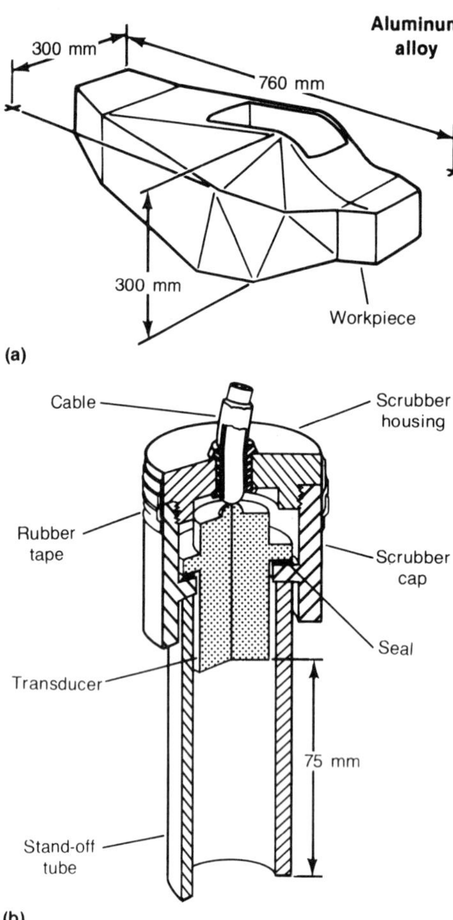

Inspection technique	Immersion, hand-held unit
Mode of sound	Longitudinal wave
Frequency, MHz	5
Cable length, m (ft)	6 (20)
Water distance, mm (in.)	75 (3)
Stand-off tube, mm (in.)	25 (1) ID, 30 (1¼) OD

Fig. 29 Aluminum alloy forging (a) having no parallel sides that was ultrasonically inspected by the immersion technique, using the transducer assembly and stand-off tube shown in (b) and the conditions listed in the accompanying table

During inspection of the forging, the operator had to be aware of the metal thickness under examination at any particular moment. As a result of nonparallel surfaces, the screen presentation usually did not have a directly reflected first back reflection, because the back surface was not normal to the ultrasound beam path. There was a series of lesser-amplitude back reflections resulting from the sound wave bouncing around within the forging and ultimately being picked up by the search unit. These back reflections moved back and forth for short distances along the trace line. The operator had a good knowledge of the changing metal thicknesses and was able to determine the presence of a discontinuity relative to where an indication appeared on the trace line. Substantiation of a discontinuity was accomplished by inspecting the area from the opposite surface and establishing its depth relative to total metal thickness.

Eddy Current and Electromagnetic Inspection

In the nondestructive inspection of forgings, the eddy current method is commonly used to detect flaws, while the electromagnetic method is used to detect differences in microstructure, chemical composition, or hardness. Electromagnetic inspection, which is restricted to ferromagnetic materials, is sometimes categorized as a modification of the eddy current method because both techniques are based on electromagnetic principles.

In concept, eddy current equipment functions by the introduction of relatively high-frequency alternating currents into the surface areas of conductive materials. The response of the material to the induced field is then measured by a mechanism sensitive to the induced field. Detailed information is available in the article "Eddy Current Inspection" in this Volume.

Detection of Flaws

The detection of flaws in forgings by eddy current inspection is almost always done with a system consisting of a single probe that is connected to an instrument generically known as a defectometer. This system is balanced with the probe in air and is further balanced to a null value on sound material of the same composition, heat treatment, and surface condition as the forging to be inspected. Areas of the surface of the forging where flaws are suspected are scanned with the probe, which searches for an unbalance due to the flaw. Generally, the scanning is done in two directions approximately at right angles to each other.

The advantages of eddy current inspection for flaw detection include the following:

- The unbalance level can be adjusted and calibrated with notches of known depth in the same material in the same condition, which can give a reasonable estimate of the depth of the flaw. This estimate can be helpful in reaching a decision as to the serviceability of the forging
- In the event that flaws are oriented in one direction only, as with seams or rolling

laps in the original stock, the technique can be automated

- Threshold gates, which are automatic signal-monitoring networks, are available to automatically signal flaws of a sufficient magnitude to be judged defective. These signals can in turn be used to mark or otherwise identify flawed forgings or locations on the forging having flaws
- The automated tester can be operated by unskilled personnel once the system has been calibrated. Solid-state electronics and their adaptation to eddy current equipment permit very stable instrumentation with no need for constant adjustment

The disadvantages of eddy current inspection for detection of flaws include the following:

- The correlation between unbalance signal and flaw depth is often not linear, for a variety of reasons. Accordingly, this method is frequently used as a go/no-go device at a depth of flaw, plus or minus a band of uncertainty. The band of uncertainty must be determined by experimental methods
- In ferrous materials, variations in the decarburization levels can render the method invalid
- This method is not suited to the detection of deep subsurface flaws

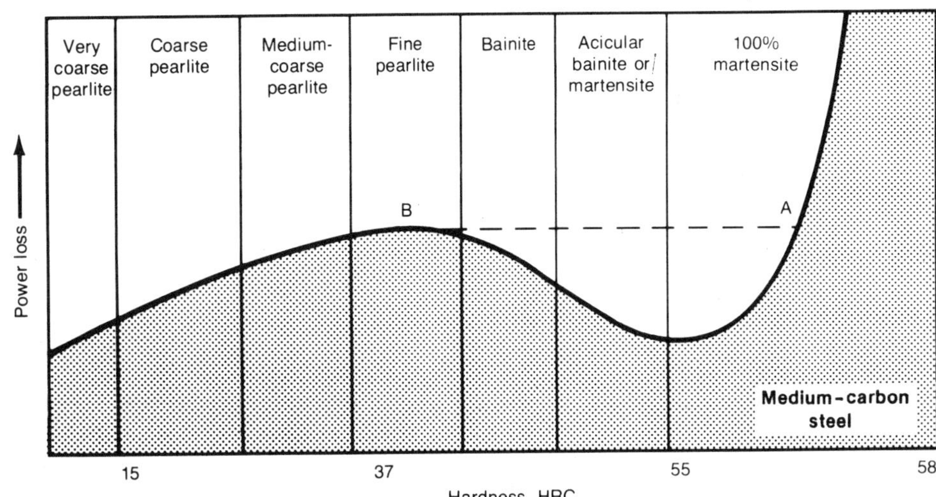

Fig. 30 Relation of an electromagnetic S-curve to variations in hardness and microstructure of a medium-carbon steel forging

Detection of Differences in Microstructure

Differences in microstructure, which usually register as differences in hardness, can be detected by electromagnetic inspection using either encircling coils or spot probes as pickups. Regardless of whether a probe or a coil is used, the instrumentation must be set up and balanced in accordance with the manufacturer's directions. This is done without forgings in or near the pickup, with forgings in the pickup, or sometimes both with and without forgings.

Once the test setup has been established, it is necessary to have good-quality forgings of known electromagnetic response available to ensure that the instrumentation has not varied. These forgings (which are sometimes referred to as masters) must be available to personnel who are trained in the setup and maintenance of the equipment.

The advantages of electromagnetic inspection for detecting differences in microstructure include the following:

- The equipment can be electronically gated based on the response of the instrumentation to the properties of the forging
- Electromagnetic inspection can be readily automated for properties throughout a forging (usually an encircling coil) or for properties at a specific location (using a spot probe)
- Once it has been properly set up, the operation can be effectively run by unskilled personnel

The disadvantages of electromagnetic inspection for detecting differences in microstructure include the following:

- A given response can indicate more than one condition in the forging; for this reason, testing technique must be developed very carefully. For example, as illustrated in Fig. 30 by an electromagnetic S-curve for a medium-carbon steel forging, when two variables (hardness and microstructure) exist, the curve shows essentially the same height in power loss for 37 HRC as for 56.5 HRC (dashed line B-A, in Fig. 30) even though the microstructure is widely different
- Development of techniques can be done only by trained personnel; even then, a great deal of experimentation is usually required to develop procedures that will yield accurate results

Radiographic Inspection

Radiography (γ-ray or x-ray) is not extensively used for the inspection of forgings for two reasons. First, the types of discontinuities most commonly located by radiography (gas porosity, shrinkage porosity, and shrinkage cavities) are not usually found in forgings. Second, for the types of internal discontinuities that are commonly found in forgings (inclusions, pipe, bursts, or flakes), ultrasonic inspection is more effective, more adaptable, and more economical.

Radiographic techniques can sometimes be helpful in the further investigation of known internal discontinuities in forgings when the presence of these discontinuities has been determined earlier by ultrasonic inspection. In sections that are not too thick to penetrate with available radiographic equipment, the size, orientation, and possibly the type of discontinuities can be evaluated by radiography.

Castings*

INSPECTION PROCEDURES for castings are established at the foundry to ensure conformance with customer drawings and documents, which are frequently based on various government, technical society, or commercial specifications. For a foundry to ensure casting quality, inspection procedures must be efficiently directed toward the prevention of imperfections, the detection of unsatisfactory trends, and the conservation of material—all of which ultimately lead to reduction in costs. Inspectors should be able to assess on sight the probable strong and weak points of a casting and know where weaknesses and faults would most likely be found.

The inspection of castings normally involves checking for shape and dimensions, coupled with aided and unaided visual inspection for external discontinuities and surface quality. Chemical analyses and tests for mechanical properties are supplemented by various forms of nondestructive inspection, including leak testing and proof loading, all of which are used to evaluate the soundness of the casting. These inspections add to the cost of the product; therefore, the initial consideration must be to determine the amount of inspection needed to maintain adequate control over quality. In some cases, this may require full inspection of each individual casting, but in other cases sampling procedures may be sufficient.

Inspection Categories

Methods for Determining Surface Quality. Cracks and other imperfections at the surface of a casting can be detected by a number of inspection techniques, including visual inspection, chemical etching, liquid penetrant inspection, eddy current inspection, and magnetic particle inspection (which can also reveal discontinuities situated immediately below the surface). All these inspection methods require clean and relatively smooth surfaces for effective results.

Methods for Detecting Internal Discontinuities. The principal nondestructive methods used for detecting internal discontinuities in castings are radiography, ultrasonic inspection, and eddy current inspection. Of these methods, radiography is the most highly developed technique for detailed inspection; it can provide a pictorial representation of the form and extent of many types of internal discontinuities. Ultrasonic inspection, which is less universally applicable, can give qualitative indications of many discontinuities. It is especially useful in the inspection of castings of fairly simple design, for which the signal pattern can be most reliably interpreted. Ultrasonic inspection can also be used to determine the shape of graphite particles in cast iron. Eddy current and other closely related electromagnetic methods are used to sort castings for variations in composition, surface hardness, and structure.

Infrared thermography (thermal inspection) has also occasionally been proposed as a method for detecting subsurface defects. However, its successful uses have generally been restricted to the detection of larger defects because of the relatively slow rates at which heat can be put into a component and because of the relatively low sensitivity of infrared detectors. Increased use of thermal inspection may occur with the introduction of pulsed video thermography, in which a very short burst of intense heat is directed at the component. The presence of near-surface defects influences the rate at which heat is dissipated from the surface, and temperature variations are detected with a high-resolution infrared camera recorded onto videotape and presented as an image on a TV monitor. The method was developed for the detection of small defects in composites and in aerospace turbine engine blades, but some initial results obtained with cast iron test plates have proved promising (Ref 1).

Methods for Dimensional Inspection. A number of techniques are used to determine the dimensional accuracy of castings. These include manual checks with micrometers, manual and automatic gages, coordinate measuring machines, and three-dimensional automatic inspection stations (machine vision systems). This article will discuss the use of coordinate-measuring machines. Additional information on methods for dimensional inspection can be found in the Section ''Inspection Equipment and Techniques'' in this Volume.

Casting Defects

Foundrymen refer to the deviations in less-than-perfect castings as discontinuities, but these imperfections are more commonly termed casting defects. Some casting defects may have no influence on the function or the service life of cast components, but will give an unsatisfactory appearance or will make further processing, such as machining, more costly. Many such defects can be easily corrected by shotblast cleaning or grinding. Other defects that may be more difficult to remove can be acceptable in some locations. The casting designer must understand the differences and must write specifications that meet the true design needs.

Classification of Casting Defects. Foundrymen have traditionally used rather unique names, such as rattail, scab, buckle, snotter, and shut, to describe various casting imperfections (such terms are defined in the ''Glossary of Terms'' in Volume 15 of the 9th Edition of *Metals Handbook*). Unfortunately, foundrymen may use different nomenclature to describe the same defect. The International Committee of Foundry Technical Associations has standardized the nomenclature, starting with the identification of seven basic categories of casting defects:

- Metallic projections
- Cavities
- Discontinuities
- Defects
- Incomplete casting
- Incorrect dimension
- Inclusions or structural anomalies

In this scheme, the term discontinuity has the specific meaning of a planar separation of the metal, that is, a crack.

Table 1 presents some of the common defects in each category. In general, defects that can serve as stress raisers or crack promoters are the most serious. These include preexisting cracks, internal voids, and nonmetallic inclusions.

Common Inspection Procedures

The inspection of castings is most often limited to visual and dimensional inspections, weight testing, and hardness testing. However, for castings that are to be used in critical applications, such as automotive or

*By the ASM Committee on Nondestructive Inspection of Castings. Lawrence E. Smiley, Reliable Castings Corporation; Bruce G. Isaacson, Bio-Imaging Research Inc.; R.A. Armistead, Advanced Research and Applications Corporation; I.C.H. Hughes, BCIRA International Centre for Cast Metals Technology (Great Britain); John Johnston, Krautkramer Branson; Carol Miller, Wright Research & Development Center, Wright-Patterson Air Force Base

Table 1 International classification of common casting defects

No.	Description	Common name	Sketch
Metallic Projections			
A 100: Metallic projections in the form of fins or flash			
A 110: Metallic projections in the form of fins (or flash) without change in principal casting dimensions			
A 111	Thin fins (or flash) at the parting line or at core prints	Joint flash or fins	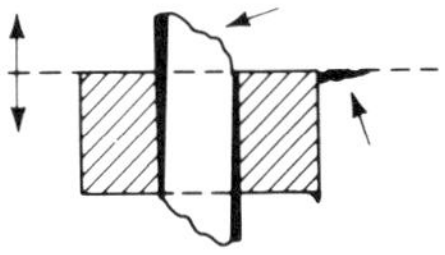
A 112	Projections in the form of veins on the casting surface	Veining or finning	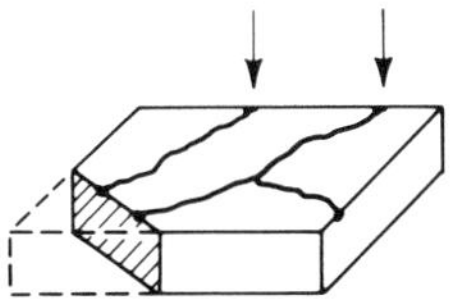
A 113	Network of projections on the surface of die castings	Heat-checked die	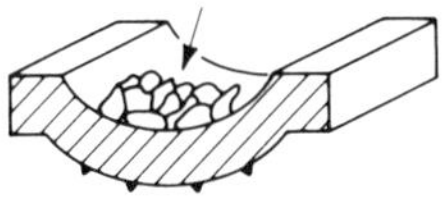
A 114(a)	Thin projection parallel to a casting surface, in re-entrant angles	Fillet scab	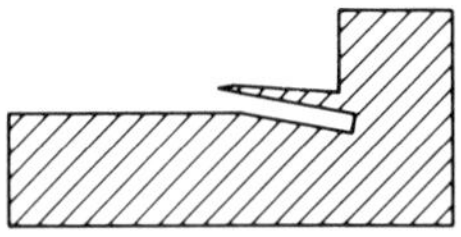
A 115	Thin metallic projection located at a re-entrant angle and dividing the angle in two parts	Fillet vein	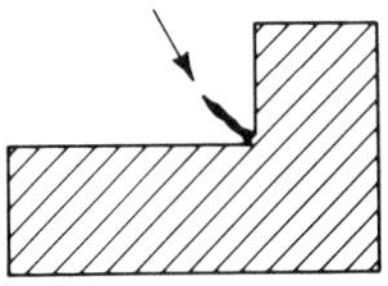
A 120: Metallic projections in the form of fins with changes in principal casting dimensions			
A 123(a)	Formation of fins in planes related to direction of mold assembly (precision casting with waste pattern); principal casting dimensions change	Cracked or broken mold	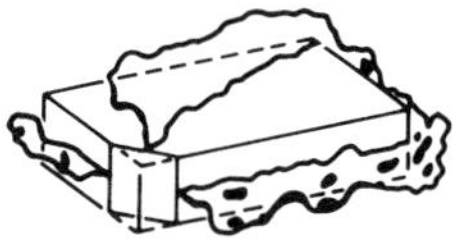
A 200: Massive projections			
A 210: Swells			
A 212(a)	Excess metal in the vicinity of the gate or beneath the sprue	Erosion, cut, or wash	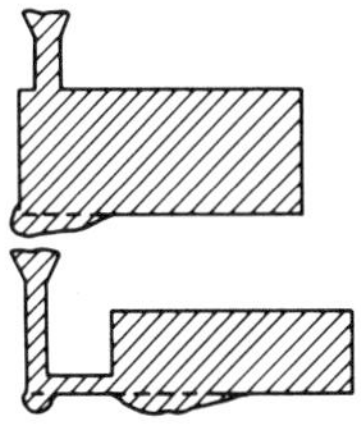
A 213(a)	Metal projections in the form of elongated areas in the direction of mold assembly	Crush	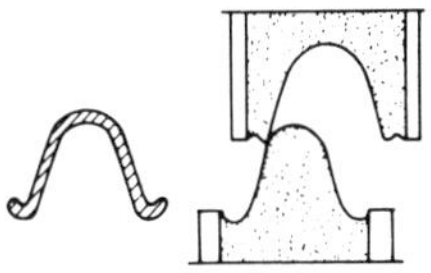
A 220: Projections with rough surfaces			
A 221(a)	Projections with rough surfaces on the cope surface of the casting	Mold drop or sticker	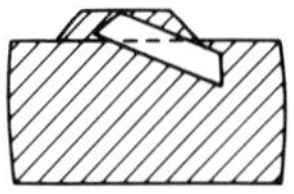
A 222(a)	Projections with rough surfaces on the drag surface of the casting (massive projections)	Raised core or mold element cutoff	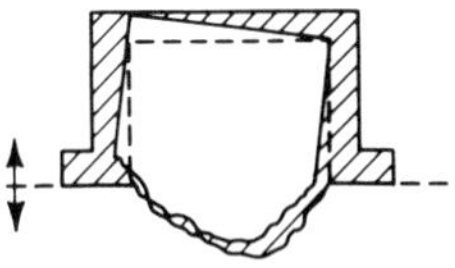
A 223(a)	Projections with rough surfaces on the drag surface of the casting (in dispersed areas)	Raised sand	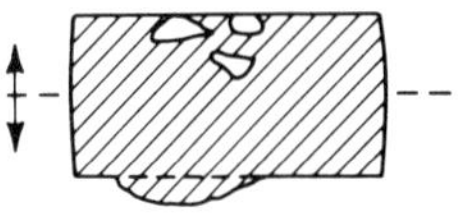
A 224(a)	Projections with rough surfaces on other parts of the casting	Mold drop	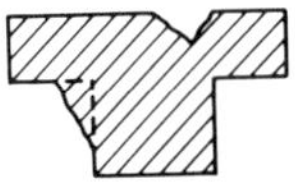
A 225(a)	Projections with rough surfaces over extensive areas of the casting	Corner scab	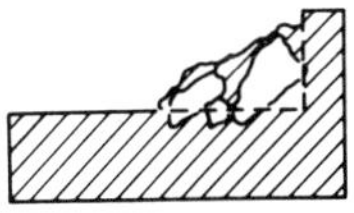
A 226(a)	Projections with rough surfaces in an area formed by a core	Broken or crushed core	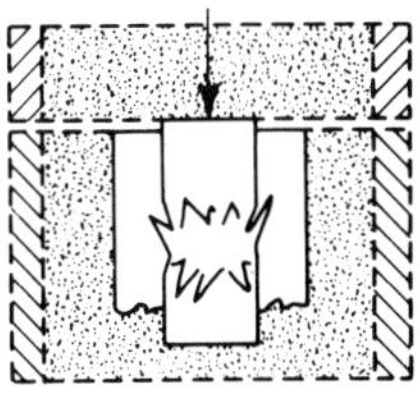

(continued)

(a) Defects that under some circumstances could contribute, either directly or indirectly, to casting failures. Adapted from *International Atlas of Casting Defects*, American Foundrymen's Society, Des Plaines, IL

Table 1 (continued)

No.	Description	Common name	Sketch
	Cavities		
B 100:	**Cavities with generally rounded, smooth walls perceptible to the naked eye (blowholes, pinholes)**		
B 110:	**Class B 100 cavities internal to the casting, not extending to the surface, discernible only by special methods, machining, or fracture of the casting**		
B 111(a)	. . . Internal, rounded cavities, usually smooth-walled, of varied size, isolated or grouped irregularly in all areas of the casting	Blowholes, pinholes	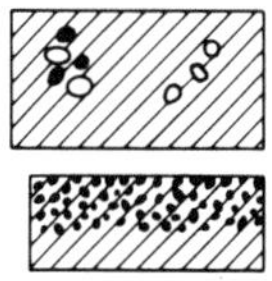
B 112(a)	. . . As above, but limited to the vicinity of metallic pieces placed in the mold (chills, inserts, chaplets, etc.)	Blowholes, adjacent to inserts, chills, chaplets, etc.	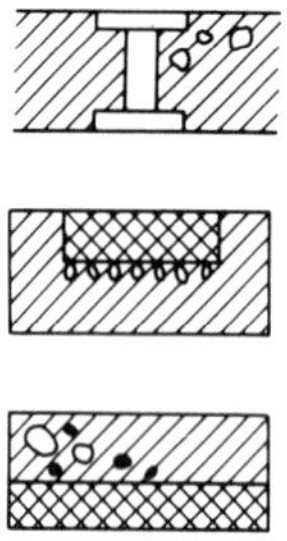
B 113(a)	. . . Like B 111, but accompanied by slag inclusions (G 122)	Slag blowholes	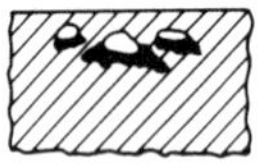
B 120:	**Class B 100 cavities located at or near the casting surface, largely exposed or at least connected with the exterior**		
B 121(a)	. . . Exposed cavities of various sizes, isolated or grouped, usually at or near the surface, with shiny walls	Surface or subsurface blowholes	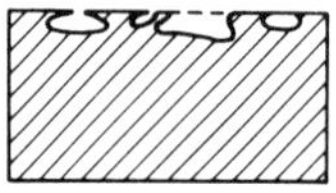
B 122(a)	. . . Exposed cavities, in re-entrant angles of the casting, often extending deeply within	Corner blowholes, draws	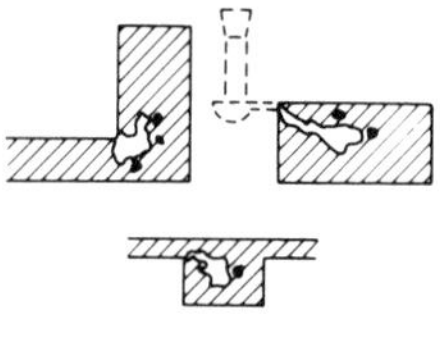
B 123	. . . . . Fine porosity (cavities) at the casting surface, appearing over more or less extended areas	Surface pinholes	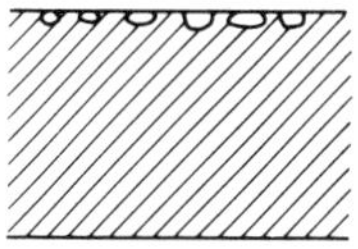
B 124(a)	. . . Small, narrow cavities in the form of cracks, appearing on the faces or along edges, generally only after machining	Dispersed shrinkage	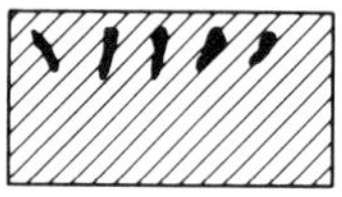
B 200:	**Cavities with generally rough walls, shrinkage**		
B 210:	**Open cavity of Class B 200, sometimes penetrating deeply into the casting**		
B 211(a)	. . . Open, funnel-shaped cavity; wall usually covered with dendrites	Open or external shrinkage	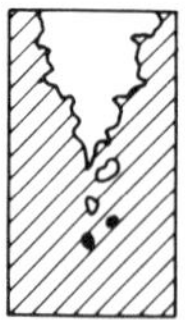
B 212(a)	. . . Open, sharp-edged cavity in fillets of thick castings or at gate locations	Corner or fillet shrinkage	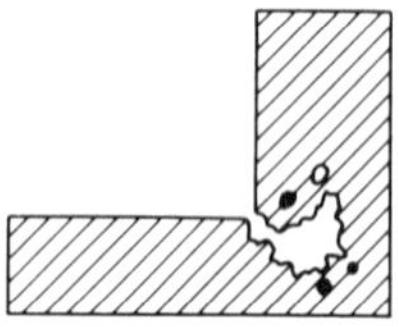
B 213(a)	. . . Open cavity extending from a core	Core shrinkage	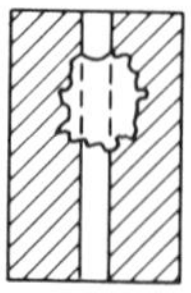
B 220:	**Class B 200 cavity located completely internal to the casting**		
B 221(a)	. . . Internal, irregularly shaped cavity; wall often dendritic	Internal or blind shrinkage	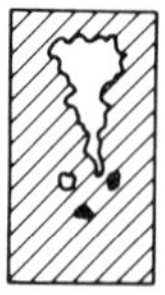
B 222(a)	. . . Internal cavity or porous area along central axis	Centerline or axial shrinkage	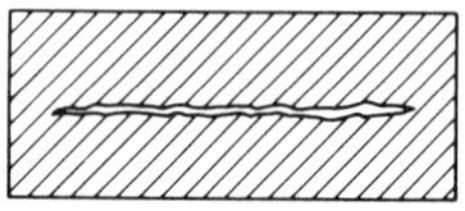
B 300:	**Porous structures caused by numerous small cavities**		
B 310:	**Cavities according to B 300, scarcely perceptible to the naked eye**		
B 311(a)	. . . Dispersed, spongy dendritic shrinkage within walls of casting; barely perceptible to the naked eye	Macro- or micro-shrinkage, shrinkage porosity, leakers	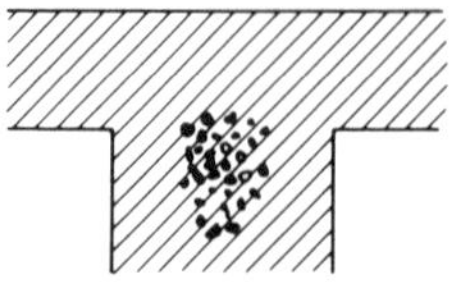

(continued)

(a) Defects that under some circumstances could contribute, either directly or indirectly, to casting failures. Adapted from *International Atlas of Casting Defects*, American Foundrymen's Society, Des Plaines, IL

Table 1 (continued)

No.	Description	Common name	Sketch
	Discontinuities		
C 100:	**Discontinuities, generally at intersections, caused by mechanical effects (rupture)**		
C 110:	**Normal cracking**		
C 111(a)	. . . Normal fracture appearance, sometimes with adjacent indentation marks	Breakage (cold)	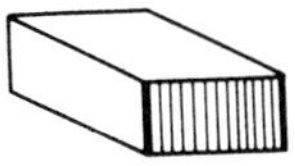
C 120:	**Cracking with oxidation**		
C 121(a)	. . . Fracture surface oxidized completely around edges	Hot cracking	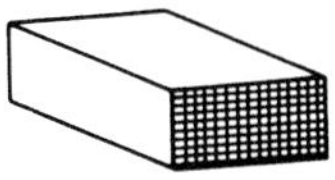
C 200:	**Discontinuities caused by internal tension and restraints to contraction (cracks and tears)**		
C 210:	**Cold cracking or tearing**		
C 211(a)	. . . Discontinuities with squared edges in areas susceptible to tensile stresses during cooling; surface not oxidized	Cold tearing	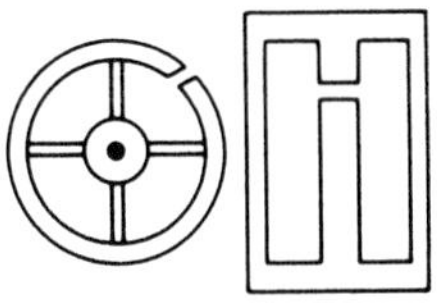
C 220:	**Hot cracking and tearing**		
C 221(a)	. . . Irregularly shaped discontinuities in areas susceptible to tension; oxidized fracture surface showing dendritic pattern	Hot tearing	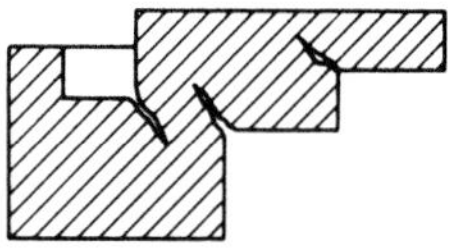
C 222(a)	. . . Rupture after complete solidification, either during cooling or heat treatment	Quench cracking	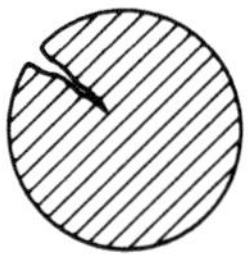
C 300:	**Discontinuities caused by lack of fusion (cold shuts); edges generally rounded, indicating poor contact between various metal streams during filling of the mold**		
C 310:	**Lack of complete fusion in the last portion of the casting to fill**		
C 311(a)	. . . Complete or partial separation of casting wall, often in a vertical plane	Cold shut or cold lap	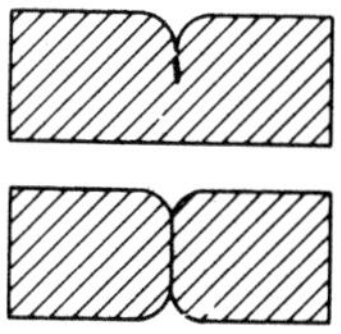
C 320:	**Lack of fusion between two parts of casting**		
C 321(a)	. . . Separation of the casting in a horizontal plane	Interrupted pour	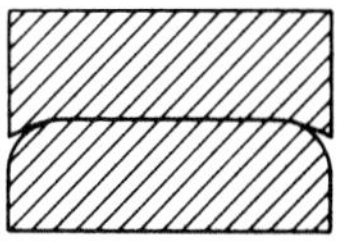
C 330:	**Lack of fusion around chaplets, internal chills, and inserts**		
C 331(a)	. . . Local discontinuity in vicinity of metallic insert	Chaplet or insert cold shut, unfused chaplet	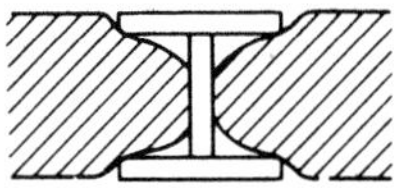
C 400:	**Discontinuities caused by metallurgical defects**		
C 410:	**Separation along grain boundaries**		
C 411(a)	. . . Separation along grain boundaries of primary crystallization	Conchoidal or "rock candy" fracture	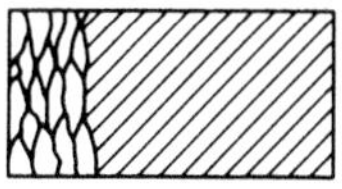
C 412(a)	. . . Network of cracks over entire cross section	Intergranular corrosion	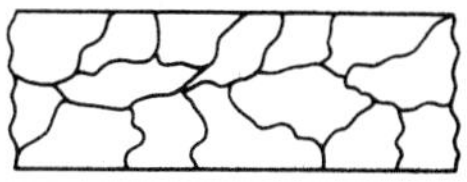
	Defective Surface		
D 100:	**Casting surface irregularities**		
D 110:	**Fold markings on the skin of the casting**		
D 111	. . . . . Fold markings over rather large areas of the casting	Surface folds, gas runs	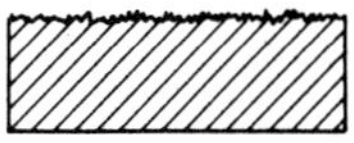

(continued)

(a) Defects that under some circumstances could contribute, either directly or indirectly, to casting failures. Adapted from *International Atlas of Casting Defects*, American Foundrymen's Society, Des Plaines, IL

Table 1 (continued)

No.	Description	Common name	Sketch
D 112	Surface shows a network of jagged folds or wrinkles (ductile iron)	Cope defect, elephant skin, laps	
D 113	Wavy fold markings without discontinuities; edges of folds at same level, casting surface is smooth	Seams or scars	
D 114	Casting surface markings showing direction of liquid metal flow (light alloys)	Flow marks	
D 120: Surface roughness			
D 121	Depth of surface roughness is approximately that of the dimensions of the sand grains	Rough casting surface	
D 122	Depth of surface roughness is greater than that of the sand grain dimensions	Severe roughness, high pressure molding defect	
D 130: Grooves on the casting surface			
D 131	Grooves of various lengths, often branched, with smooth bottoms and edges	Buckle	
D 132	Grooves up to 5.1 mm (0.2 in.) in depth, one edge forming a fold which more or less completely covers the groove	Rat tail	
D 133	Irregularly distributed depressions of various dimensions extending over the casting surface, usually along the path of metal flow (cast steel)	Flow marks, crow's feet	
D 134	Casting surface entirely pitted or pock-marked	Orange peel, metal mold reaction, alligator skin	
D 135	Grooves and roughness in the vicinity of re-entrant angles on die castings	Soldering, die erosion	
D 140: Depressions in the casting surface			
D 141	Casting surface depressions in the vicinity of a hot spot	Sink marks, draw or suck-in	
D 142	Small, superficial cavities in the form of droplets of shallow spots, generally gray-green in color	Slag inclusions	
D 200: Serious surface defects			
D 210: Deep indentation of the casting surface			
D 211	Deep indentation, often over large area of drag half of casting	Push-up, clamp-off	
D 220: Adherence of sand, more or less vitrified			
D 221	Sand layer strongly adhering to the casting surface	Burn on	
D 222	Very adherent layer of partially fused sand	Burn in	

(continued)

(a) Defects that under some circumstances could contribute, either directly or indirectly, to casting failures. Adapted from *International Atlas of Casting Defects*, American Foundrymen's Society, Des Plaines, IL

Table 1 (continued)

No.	Description	Common name	Sketch
D 223	Conglomeration of strongly adhering sand and metal at the hottest points of the casting (re-entrant angles and cores)	Metal penetration	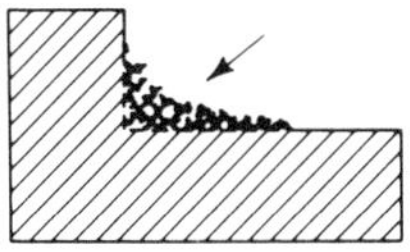
D 224	Fragment of mold material embedded in casting surface	Dip coat spall, scab	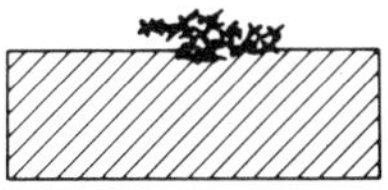
D 230:	**Plate-like metallic projections with rough surfaces, usually parallel to casting surface**		
D 231(a)	Plate-like metallic projections with rough surfaces parallel to casting surface; removable by burr or chisel	Scabs, expansion scabs	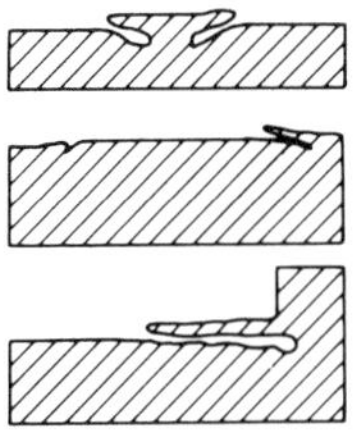
D 232(a)	As above, but impossible to eliminate except by machining or grinding	Cope spall, boil scab, erosion scab	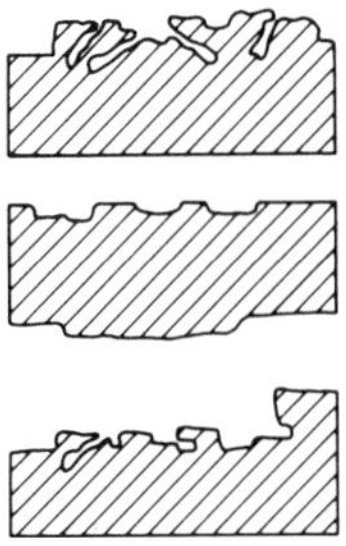
D 233(a)	Flat, metallic projections on the casting where mold or core washes or dressings are used	Blacking scab, wash scab	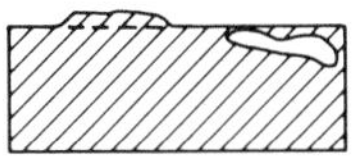
D 240:	**Oxides adhering after heat treatment (annealing, tempering, malleablizing) by decarburization**		
D 241	Adherence of oxide after annealing	Oxide scale	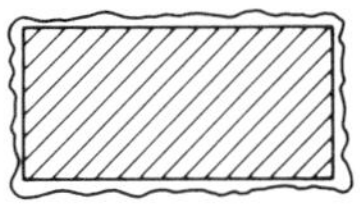
D 242	Adherence of ore after malleablizing (white heart malleable)	Adherent packing material	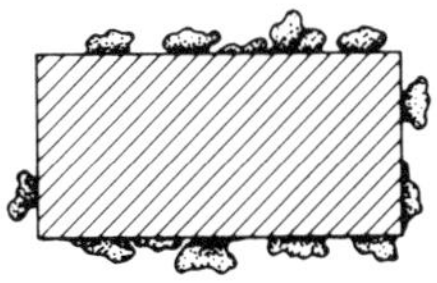

No.	Description	Common name	Sketch
D 243	Scaling after anneal	Scaling	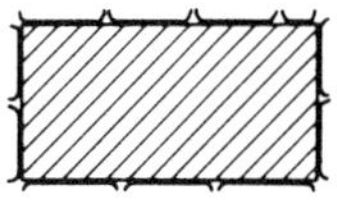

Incomplete Casting

No.	Description	Common name	Sketch
E 100:	**Missing portion of casting (no fracture)**		
E 110:	**Superficial variations from pattern shape**		
E 111	Casting is essentially complete except for more or less rounded edges and corners	Misrun	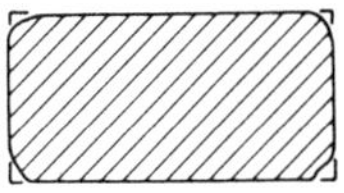
E 112	Deformed edges or contours due to poor mold repair or careless application of wash coatings	Defective coating (tear-dropping) or poor mold repair	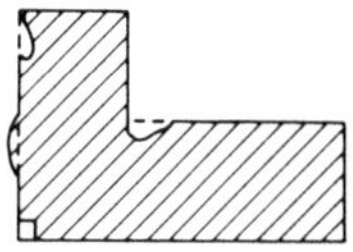
E 120:	**Serious variations from pattern shape**		
E 121	Casting incomplete due to premature solidification	Misrun	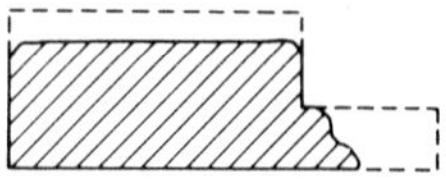
E 122	Casting incomplete due to insufficient metal poured	Poured short	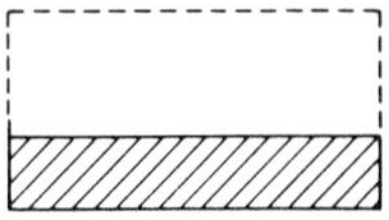
E 123	Casting incomplete due to loss of metal from mold after pouring	Runout	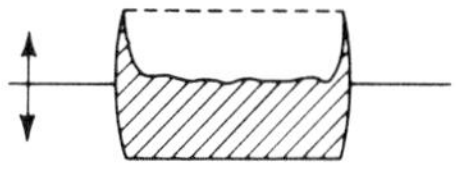
E 124	Significant lack of material due to excessive shot-blasting	Excessive cleaning	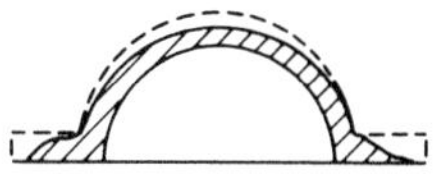

(continued)

(a) Defects that under some circumstances could contribute, either directly or indirectly, to casting failures. Adapted from *International Atlas of Casting Defects*, American Foundrymen's Society, Des Plaines, IL

Table 1 (continued)

No.	Description	Common name	Sketch
E 125	Casting partially melted or seriously deformed during annealing	Fusion or melting during heat treatment	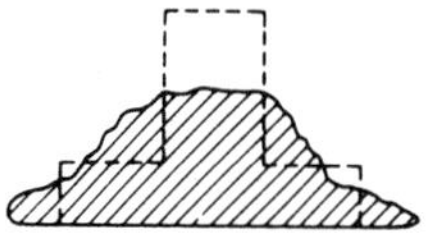
E 200: Missing portion of casting (with fracture)			
E 210: Fractured casting			
E 211	Casting broken, large piece missing; fractured surface not oxidized	Fractured casting	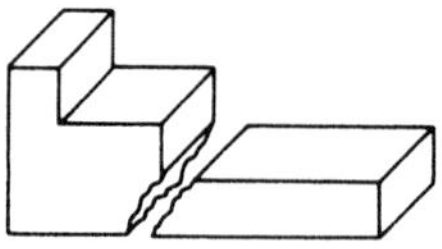
E 220: Piece broken from casting			
E 221	Fracture dimensions correspond to those of gates, vents, etc.	Broken casting (at gate, riser, or vent)	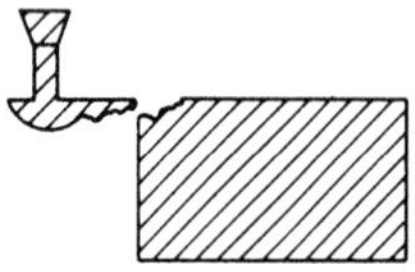
E 230: Fractured casting with oxidized fracture			
E 231	Fracture appearance indicates exposure to oxidation while hot	Early shakeout	
Incorrect Dimensions or Shape			
F 100: Incorrect dimensions; correct shape			
F 110: All casting dimensions incorrect			
F 111	All casting dimensions incorrect in the same proportions	Improper shrinkage allowance	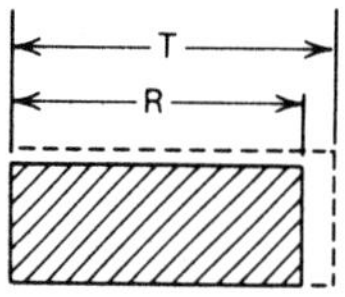 T, R
F 120: Certain casting dimensions incorrect			
F 121	Distance too great between extended projections	Hindered contraction	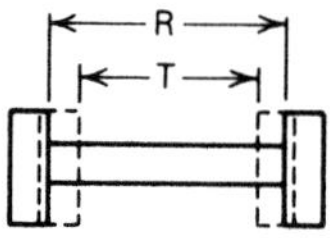 R, T
F 122	Certain dimensions inexact	Irregular contraction	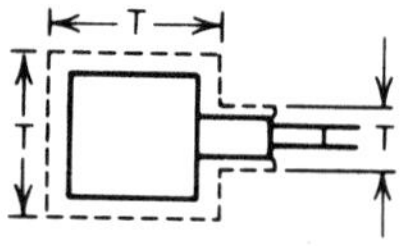 T, T, T
F 123	Dimensions too great in the direction of rapping of pattern	Excess rapping of pattern	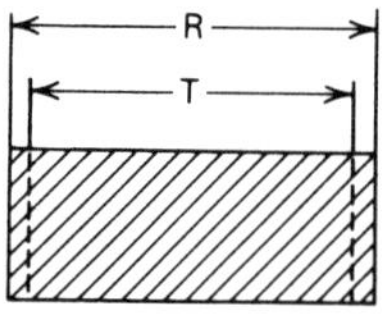 R, T
F 124	Dimensions too great in direction perpendicular to parting line	Mold expansion during baking	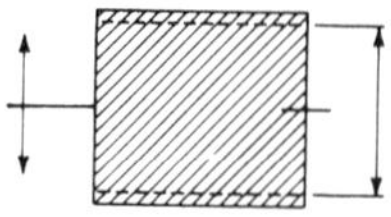
F 125	Excessive metal thickness at irregular locations on casting exterior	Soft or insufficient ramming, mold-wall movement	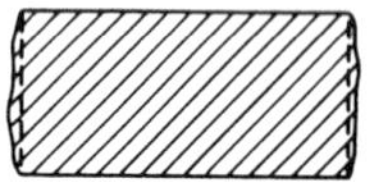
F 126	Thin casting walls over general area, especially on horizontal surfaces	Distorted casting	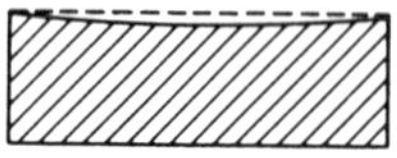
F 200: Casting shape incorrect overall or in certain locations			
F 210: Pattern incorrect			
F 211	Casting does not conform to the drawing shape in some or many respects; same is true of pattern	Pattern error	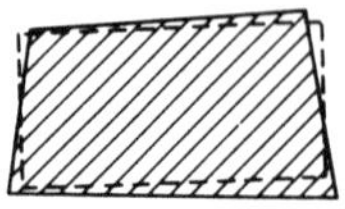
F 212	Casting shape is different from drawing in a particular area; pattern is correct	Pattern mounting error	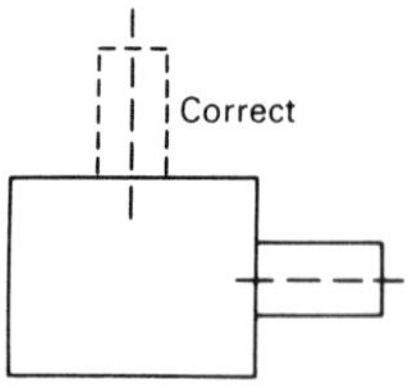 Correct

(continued)

(a) Defects that under some circumstances could contribute, either directly or indirectly, to casting failures. Adapted from *International Atlas of Casting Defects*, American Foundrymen's Society, Des Plaines, IL

Table 1 (continued)

No.	Description	Common name	Sketch
F 220: Shift or Mismatch			
F 221	Casting appears to have been subjected to a shearling action in the plane of the parting line	Shift	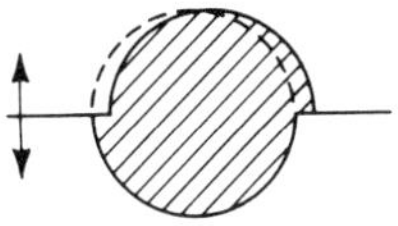
F 222	Variation in shape of an internal casting cavity along the parting line of the core	Shifted core	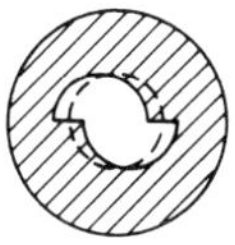
F 223	Irregular projections on vertical surfaces, generally on one side only in the vicinity of the parting line	Ramoff, ramaway	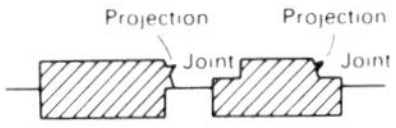 Projection Joint Projection Joint
F 230: Deformations from correct shape			
F 231	Deformation with respect to drawing proportional for casting, mold, and pattern	Deformed pattern	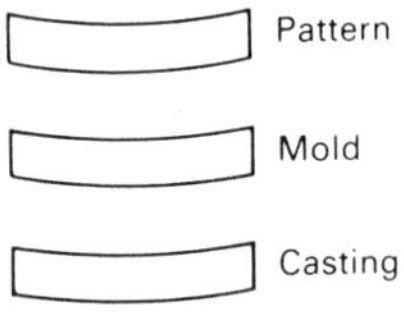 Pattern Mold Casting
F 232	Deformation with respect to drawing proportional for casting and mold; pattern conforms to drawing	Deformed mold, mold creep, springback	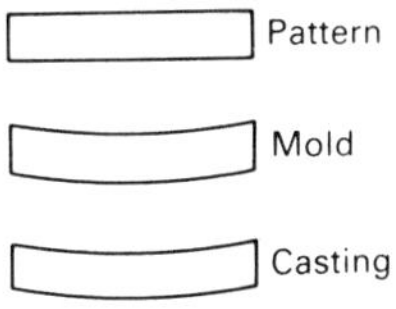 Pattern Mold Casting
F 233	Casting deformed with respect to drawing; pattern and mold conform to drawing	Casting distortion	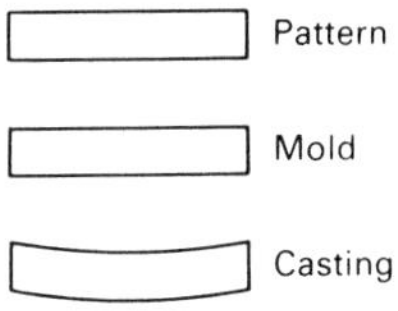 Pattern Mold Casting
F 234	Casting deformed with respect to drawing after storage, annealing, machining	Warped casting	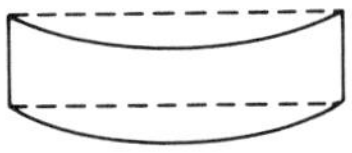
Inclusions or Structural Anomalies			
G 100: Inclusions			
G 110: Metallic inclusions			
G 111(a)	Metallic inclusions whose appearance, chemical analysis or structural examination show to be caused by an element foreign to the alloy	Metallic inclusions	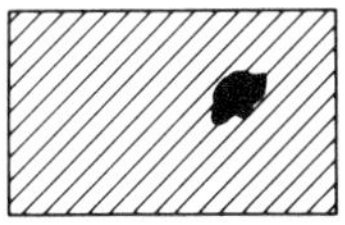
G 112(a)	Metallic inclusions of the same chemical composition as the base metal; generally spherical and often coated with oxide	Cold shot	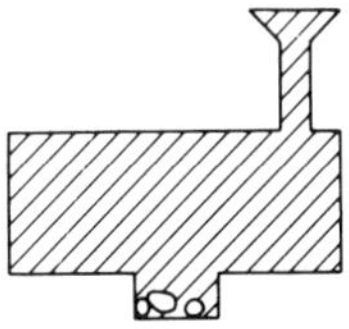
G 113	Spherical metallic inclusions inside blowholes or other cavities or in surface depressions (see A 311). Composition approximates that of the alloy cast but nearer to that of a eutectic	Internal sweating, phosphide sweat	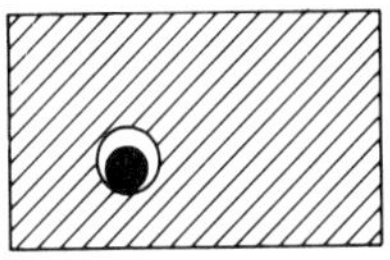
G 120: Nonmetallic inclusions; slag, dross, flux			
G 121(a)	Nonmetallic inclusions whose appearance or analysis shows they arise from melting slags, products of metal treatment or fluxes	Slag, dross or flux inclusions, ceroxides	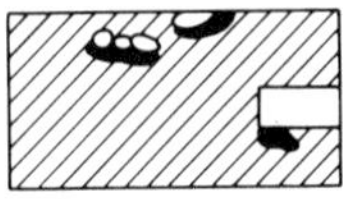
G 122(a)	Nonmetallic inclusions generally impregnated with gas and accompanied by blowholes (B 113)	Slag blowhole defect	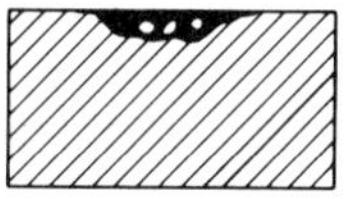
G 130: Nonmetallic inclusions; mold or core materials			
G 131(a)	Sand inclusions, generally very close to the surface of the casting	Sand inclusions	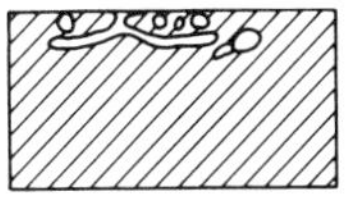
G 132(a)	Inclusions of mold blacking or dressing, generally very close to the casting surface	Blacking or refractory coating inclusions	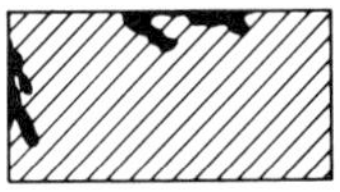

(continued)

(a) Defects that under some circumstances could contribute, either directly or indirectly, to casting failures. Adapted from *International Atlas of Casting Defects*, American Foundrymen's Society, Des Plaines, IL

Table 1 (continued)

No.	Description	Common name	Sketch	No.	Description	Common name	Sketch
G 140:	**Nonmetallic inclusions; oxides and reaction products**						
G 141	Clearly defined, irregular black spots on the fractured surface of ductile cast iron	Black spots		G 143(a)	Folded films of graphitic luster in the wall of the casting	Lustrous carbon films, or kish tracks	
G 142(a)	Inclusions in the form of oxide skins, most often causing a localized seam	Oxide inclusion or skins, seams		G 144	Hard inclusions in permanent molded and die cast aluminum alloys	Hard spots	

(a) Defects that under some circumstances could contribute, either directly or indirectly, to casting failures. Adapted from *International Atlas of Casting Defects*, American Foundrymen's Society, Des Plaines, IL

aerospace components, additional methods of nondestructive inspection are used to determine and to control casting quality.

Visual inspection of each casting ensures that none of its features has been omitted or malformed by molding errors, short running, or mistakes in cleaning. Most surface defects and roughness can be observed at this stage.

Initial sample castings from new pattern equipment should be carefully inspected for obvious defects. Liquid penetrant inspection can be used to detect surface defects. Such casting imperfections as shrinks, cracks, blows, or dross usually indicate the need for adjustment in the gating or foundry techniques. If the casting appears to be satisfactory upon visual inspection, internal quality can be checked by radiographic and ultrasonic inspection.

The first visual inspection operation on the production casting is usually performed immediately after shakeout or knockout of the casting, ensuring that major visible imperfections are detected as quickly as possible. This information, promptly relayed to the foundry, permits early corrective action to be taken with a minimum of scrap loss. The size and complexity of some sand castings require that the gates and risers be removed to permit proper inspection of the casting. Many castings that contain numerous internal cores or have close dimensional tolerances require a rapid but fairly accurate check of critical wall dimensions. In some cases, an indicating-type caliper gage is suitable for this work, and special types are available for casting shapes that do not lend themselves to the standard types. Ultrasonic inspection is also used to determine wall thickness in such components as cored turbine blades made by investment casting. New developments in visual inspection procedures for examining component appearance are mainly based on vision systems that use electronic cameras coupled to computer-assisted image-processing systems (Ref 1).

With the development of high-sensitivity cameras having exposure times of 1/1000 s, components can be inspected on moving belts. Flexibility for examining three-dimensional components can be achieved with an array of cameras multiplexed to a common image processor or with a computer-controlled camera scanning system. Such systems have been successfully applied to the inspection of printed circuit boards in the electronics industry and engineering subassemblies in automobile manufacture. These tests usually operate on a go/no-go basis; either the assembly is complete with connections correctly made or it is not correct. This is a far easier task than evaluating casting quality.

Studies that have been carried out to assess the possibility of extending such methods to iron castings have not given encouraging results. Contrast between defective and nondefective areas is low, illumination is critical, and consistent standards of inspection are difficult to maintain because of differences in reflectivity of the casting surfaces depending on whether or not they have been recently shotblasted. Even the simple task of identifying castings to determine their type is best carried out by examining their backlighted silhouette, and this provides no advantage in examining their quality.

Dimensional Inspection. Consistency of dimensions is an inescapable requirement of premium-quality castings supplied as near-net shape components on which subsequent high-speed machining operations are to be carried out (Ref 1). Customers will not accept increased machining costs due to inconsistencies in dimensions nor will they tolerate damage to flexible machining systems or transfer times resulting from poor control and inspection in foundries. Variations in dimensions represent one of the most common complaints with regard to the machinability of iron castings.

Prevention is within the control of foundries. Differences in pattern size when using multipattern plates can be virtually eliminated by the use of computer-aided design and manufacturing methods and computer numerical control machines in patternmaking (see the article "Patterns and Patternmaking" in Volume 15 of the 9th Edition of *Metals Handbook*). Better process control and methoding can eliminate variations in dimensions due to changes in metal composition or feeding methods. Variations in mold rigidity, caused by inadequate compaction with green sand, or the use of cold sand or insufficient curing times with cold-setting systems, which cause casting dimensions to fall outside the preset tolerance limits, can be greatly reduced by good molding and coremaking practices (see the articles "Sand Molding" and "Coremaking" in Volume 15 of the 9th Edition of *Metals Handbook*). Because the dimensions and weight of iron castings are directly related to their soundness and are dependent on mold rigidity, the measurement of size or weight provides a simple test for checking casting integrity and for monitoring the consistency of the moldmaking process (see the section "Weight Testing" in this article).

Casting dimensions are usually checked with dial gages, vernier calipers, micrometers, or vertical height gages, which may be hand held or incorporated into acceptance fixtures. Wall thickness measurements can be made with small hand-held ultrasonic thickness gages. Under ideal conditions, the accuracy of these instruments is claimed to be ±0.01 mm (±0.0004 in.), but this is rarely achieved in practice because the surfaces are not parallel and are not machined. Instruments are available that display variations in thickness from some preset standard and provide a digital readout and a permanent record of results for statistical analysis.

Developments lie in the use of measuring systems employing capacitance, electrical contact, or linear displacement transducers. Such systems are capable of high accuracy and the output can be linked directly to microcomputers for data recording and statistical analysis to meet the requirements of statistical process control. The use of computer-aided dimensional control and statistical process control is discussed later in this article.

Laser methods of measurement using beam displacement or time-lapse techniques are available for use in machine shops where accurate measurement is required for control of automatic machining processes. At present, they are generally not well suited for measuring castings, because of their high cost and because it is difficult to make precise measurements on components having a complex shape with curved or as-cast surfaces. As these laser methods become more widely used in other industries, lower-cost systems will become available, and these might be developed for foundry use (see the article "Laser Inspection" in this Volume).

Weight Testing. Many intricately cored castings are extremely difficult to measure accurately, particularly the internal sections. It is important to ensure that these sections are correct in thickness for three main reasons:

- There should be no additional weight that would make the finished product heavier than permissible
- Sections must not be thinner than designed so as not to decrease the strength of the casting
- If hollow cavities have been reduced in area by increasing the metal thickness of the sections, any flow of liquid or gases is reduced

A ready means of testing for these discrepancies is by accurately weighing each casting or by measuring the displacement caused by immersing the casting in a liquid-filled measuring jar or vessel. In certain cases in which extreme accuracy is demanded, a tolerance of only ±1% of a given weight may be allowed.

Hardness testing is often used to verify the effectiveness of the heat treatment applied to actual castings. Its general correlation with the tensile strength of many ferrous alloys allows a rough prediction of tensile strength to be made.

The Brinell hardness test is most frequently used for casting alloys. A combination of large-diameter ball (5 or 10 mm) and heavy load (500 to 3000 kgf) is preferred for the most effective representation because a deep impression minimizes the influence of the immediate surface layer and of the relatively coarse microstructure. The Brinell hardness test is unsuitable for use at high hardness levels (above ~600 HB), because distortion of the ball indenter can affect the shape of the indentation.

Either the Rockwell or the Vickers (136° diamond pyramid) hardness test is used for alloys of extreme hardness or for high-quality and precision castings in which the large Brinell indentation cannot be tolerated. Because of the very small indentations produced in Rockwell and Vickers tests, which use loads of 150 kg or less, results must be based on the average of a number of determinations. Portable hardness testers or ultrasonic microhardness testers can be used on large castings that cannot be placed on the platform of a bench-type machine. More detailed information on hardness testing is available in Volume 8 of the 9th Edition of *Metals Handbook*.

The hardness of ferrous castings can be determined from the sonic velocity of the metal if all other test conditions remain constant. This has been demonstrated on chilled rolls in determining the average hardness of the core.

Computer-Aided Dimensional Inspection

The use of computer equipment in foundry inspection operations is finding more acceptance as the power and utility of available hardware and software increase. The computerization of operations can reduce the man-hours required for inspection tasks, can increase accuracy, and can allow the analysis of data in ways that are not possible or practical with manual operations. Perhaps the best example of this, given the currently available equipment, is the application of computer technology to the dimensional inspection of castings.

Importance of Dimensional Inspection. One of the most critical determinants of casting quality in the eyes of the casting buyer is dimensional accuracy. Parts that are within dimensional tolerances, given the absence of other casting defects, can be machined, assembled, and used for their intended functions with minimal testing and inspection costs. Major casting buyers are therefore demanding statistical evidence that dimensional tolerances are being maintained. In addition, the statistical analysis of in-house processes has been demonstrated to be effective in keeping those processes under control, thus reducing scrap and rework costs.

The application of computer equipment to the collection and analysis of dimensional inspection data can increase the amount of inspection that can be performed and decrease the time required to record and analyze the results. This furnishes control information for making adjustments to tooling on the foundry floor and statistical information for reporting to customers on the dimensional accuracy of parts. Additional information on dimensional accuracy in castings can be found in the article "Dimensional Tolerances and Allowances" in Volume 15 of the 9th Edition of *Metals Handbook*.

Typical Equipment. A typical installation for the dimensional inspection of castings consists of an electronic coordinate measuring machine, a microcomputer interfaced to the coordinate measuring machine controller with a data transfer cable, and a software system for the microcomputer (Fig. 1). The software system should be capable of controlling the functions and memory storage of the coordinate measuring machine as well as recalling and analyzing the data it collects. The software serves as the main control element for the dimensional inspection and statistical reporting of results. Such software can be purchased or, if the expertise is available, developed in-house for highly specialized requirements.

Coordinate measuring machines typically record dimensions along three axes from data points specified by the user. Depending on the sophistication of the controller, such functions as center and diameter finds for circular features and the electronic rotation of measurement planes can be performed. Complex geometric constructions, such as the intersection points of lines and planes and out-of-roundness measurements, are typically off-loaded for calculation into the microcomputer. The contact probe of the coordinate measuring machine can be manipulated manually, or in the case of direct computer controlled machines, the probe can be driven by servomotors to perform the part measurement with little operator intervention. More detailed information on these machines can be found in the article "Coordinate Measuring Machines" in this Volume.

The Measurement Process. Figure 2 illustrates the general procedure that is followed in applying semiautomatic dimensional inspection to a given part. The first step is to identify the critical part dimensions that are to be measured and tracked. Nominal dimensions and tolerances are usually taken from the customer's specifications and blueprints. Dimensions that are useful in controlling the foundry process can also be selected. A data base file, including a description and tolerance limits for each dimension to be checked, is then created using the microcomputer software system.

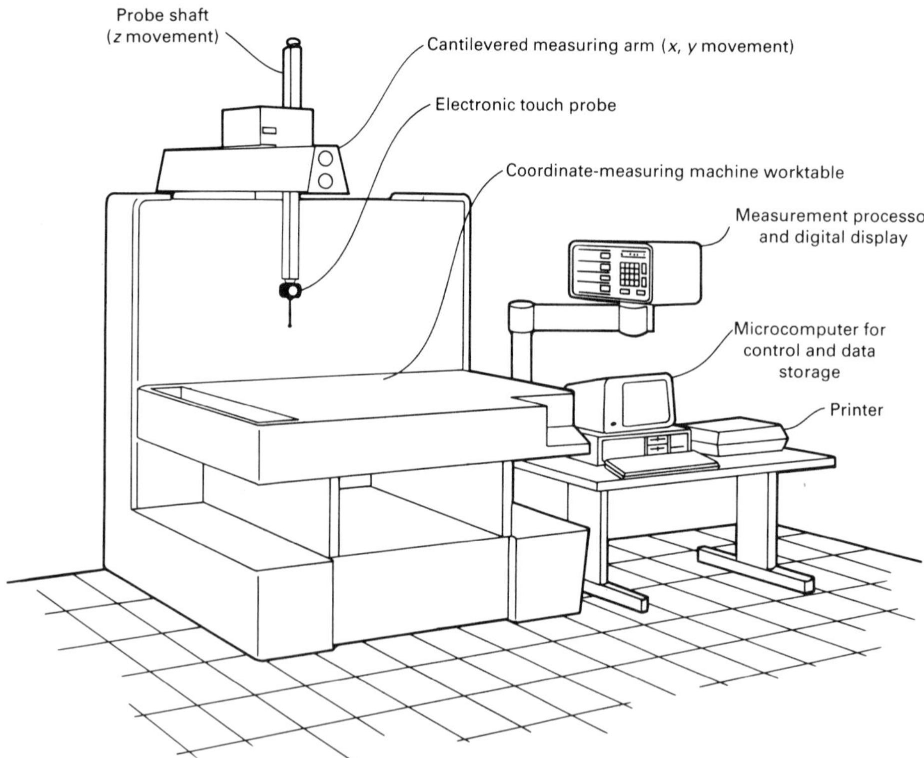

Fig. 1 Equipment used in a typical installation for the computer-aided dimensional inspection of castings showing a coordinate measuring machine and microcomputer

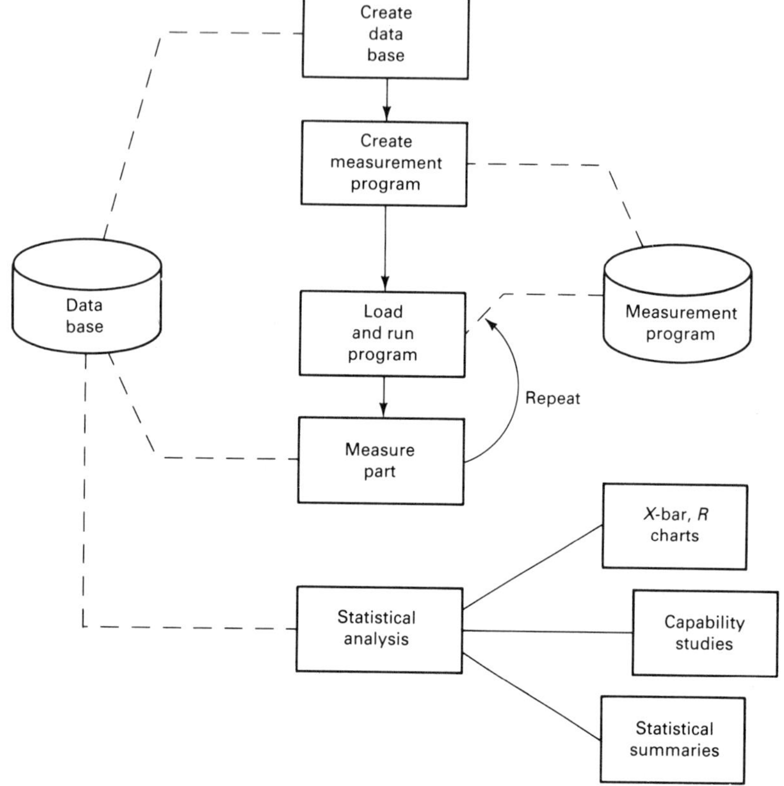

Fig. 2 Flowchart showing typical sequence of operations for computer-aided dimensional inspection

The next step in the setup process is to develop a set of instructions for measuring the part with the coordinate measuring machine. The instructions consist of commands that the coordinate measuring machine uses to establish reference planes and to measure such features as the center points of circular holes.

This measurement program can be entered in either of two ways. In the first method, the operator simply types in a list of commands that he wants the coordinate measuring machine to execute and that give the required dimensions as defined in the part data base. The second method uses a teach mode. The operator actually places a part on the worktable of the coordinate measuring machine and checks it in the proper sequence, while the computer monitors the process and stores the sequence of commands used. In either case, the result is a measurement program stored on the microcomputer that defines in precise detail how the part is to be measured. Special commands can also be included in the measurement program to display operator instructions on the computer screen while the part is being measured.

In developing the measurement program, consideration must be given to the particular requirements of the part being measured. Customer prints will normally show datum planes from which measurements are to be made. When using a cast surface to establish a datum plane, it is good practice to probe a number of points on the surface and to allow the computer to establish a best-fit plane through the points. Similarly, the center points of cast holes can best be found by probing multiple points around the circumference of the hole. Machined features can generally be measured with fewer probe contacts. When measuring complex castings, maximum use should be made of the ability of the coordinate measuring machine to electronically rotate measurement planes without physically moving the part; unclamping and turning a part will lower the accuracy of the overall layout.

Once the setup process is complete, the dimensional inspection of parts from the foundry begins. Based on statistical considerations, a sampling procedure and frequency must be developed. Parts are then selected at random from the process according to the agreed-upon frequency. The parts are brought to the coordinate measuring machine, and the operator calls up the measurement program for that part and executes it. As the part is measured, the dimensions are sent from the coordinate measuring machine to the data base on the microcomputer. Once the measurement process is completed, information such as mold number, shift, date, and serial number should be entered by the operator so that this particular set of dimensions can be identified later. A layout report can then be

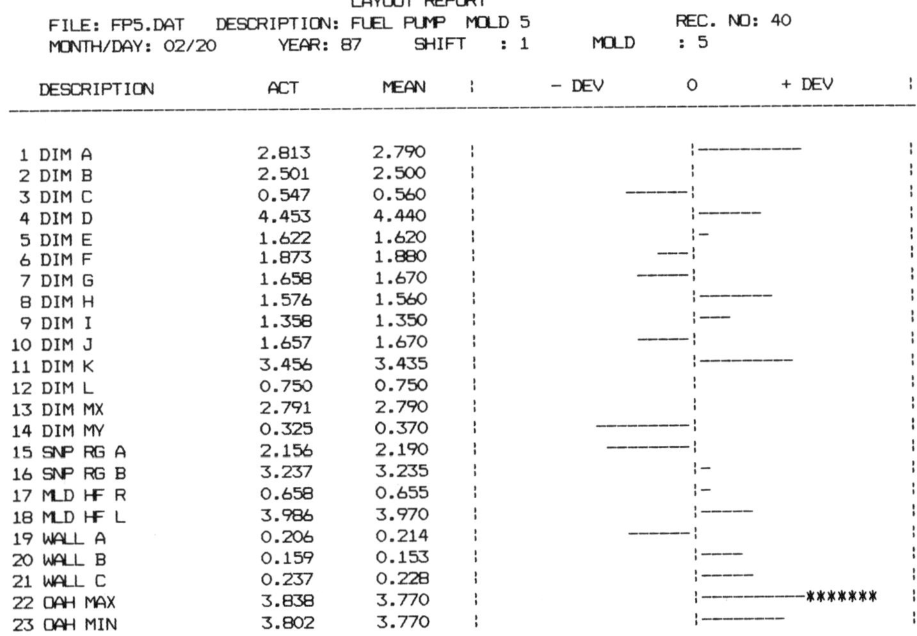

LAYOUT REPORT

FILE: FP5.DAT DESCRIPTION: FUEL PUMP MOLD 5 REC. NO: 40
MONTH/DAY: 02/20 YEAR: 87 SHIFT : 1 MOLD : 5

DESCRIPTION	ACT	MEAN	- DEV 0 + DEV
1 DIM A	2.813	2.790	
2 DIM B	2.501	2.500	
3 DIM C	0.547	0.560	
4 DIM D	4.453	4.440	
5 DIM E	1.622	1.620	
6 DIM F	1.873	1.880	
7 DIM G	1.658	1.670	
8 DIM H	1.576	1.560	
9 DIM I	1.358	1.350	
10 DIM J	1.657	1.670	
11 DIM K	3.456	3.435	
12 DIM L	0.750	0.750	
13 DIM MX	2.791	2.790	
14 DIM MY	0.325	0.370	
15 SNP RG A	2.156	2.190	
16 SNP RG B	3.237	3.235	
17 MLD HF R	0.658	0.655	
18 MLD HF L	3.986	3.970	
19 WALL A	0.206	0.214	
20 WALL B	0.159	0.153	
21 WALL C	0.237	0.228	
22 OAH MAX	3.838	3.770	*******
23 OAH MIN	3.802	3.770	

Fig. 3 Example layout report showing all dimensions measured on a single casting, with visual indication of deviations from print mean. Note out-of-tolerance condition indicated by asterisks.

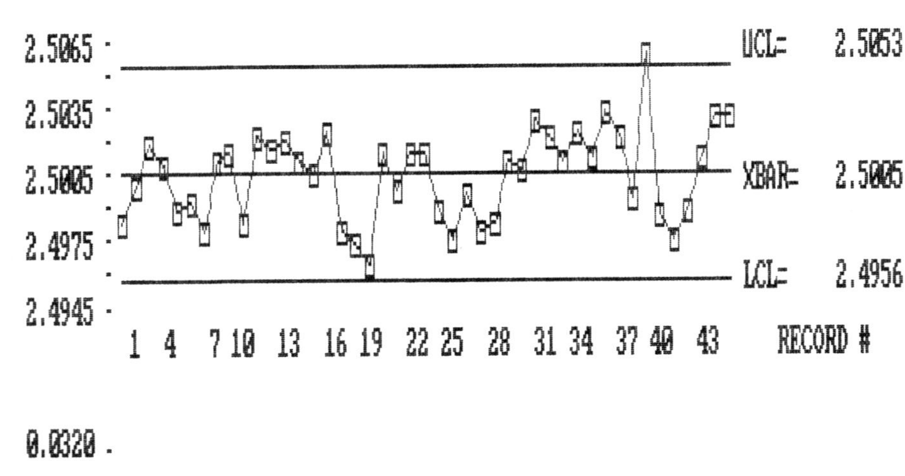

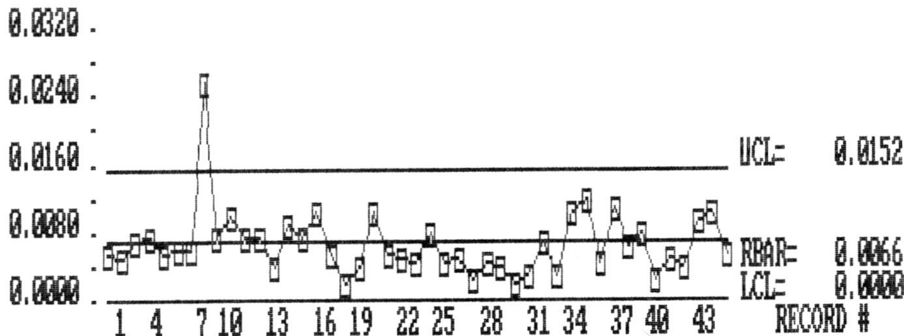

Fig. 4 Control chart with average of groups of measurements (X values) plotted above and ranges within the groups (R) plotted below. Control limits have been calculated and placed on the chart by the computer.

generated to show how well the measured part checked out relative to the specified dimensions and tolerances. Figure 3 shows a sample report in the form of a bar graph, in which any deviation from print tolerance appears as a line of dashes to the left or right of center. A deviation outside of tolerance limits displays asterisks to flag its condition. Such a report is useful in that it gives a quick visual indication of the measurement of one casting.

Statistical analysis permits the mathematical prediction of the characteristics of all the parts produced by measuring only a sample of those parts (the principles and applications of statistics in industrial environments are discussed in the article "Statistical Quality Design and Control" in this Volume). All processes are subject to some amount of natural variation; in most processes, this variation follows a normal distribution (the familiar bell-shaped curve) when the probability of occurrence is plotted against the range of possible values. Standard deviation, a measure of the distance from center on the probability curve, is the principal means of expressing the range of measured values. For example, a spread of six standard deviations (plus or minus three standard deviations on either side of the measured mean) represents the range within which one would expect to find 99.73% of observed measurements for a normal process. This allows the natural variation inherent in the process to be quantified.

Control Charts. With statistical software incorporated into the microcomputer system, the results of numerous measurements of the same part can be analyzed to determine, first, how well the process is staying in control, that is, whether the natural variations occurring in a given measurement are within control limits and whether any identifiable trends are occurring. This is done by using a control chart (Fig. 4), which displays the average values and ranges of groups of measurements plotted against time. Single-value charts with a moving range can also be helpful. The control limits can also be calculated and displayed. With the computer, this type of graph can be generated within seconds. Analysis of the graph may show a developing trend that can be corrected by adjusting the tooling before out-of-tolerance parts are made.

Statistical Summary Report. The second type of analysis shows the capability of the process, that is, how the range of natural variation (as measured by a specified multiple of the standard deviation) compares with the tolerance range specified for a given dimension. An example of a useful report of this type is shown in Fig. 5. This information is of great interest both to the customer and the process engineer because it indicates whether or not the process being used to produce the part can hold the dimensions within the required tolerance limits. The user must be aware that different methods of capability analysis are used by different casting buyers, so the software should be flexible enough to accommodate the various methods of calculation that might be required.

Histograms. An alternative method of assessing capability involves the use of a histogram, or frequency plot. This is a graph that plots the number of occurrences within successive, equally spaced ranges of a given measured dimension. Figure 6

```
            Statistical Summary Report                                   Page 1
            FILE: FP5.DAT              DESCRIPTION: FUEL PUMP  MOLD 5
     For  SHIFT    : ALL       MOLD      : ALL       Start Rec: 1       End Rec: 68
```

	Actual Mean	Spec.	Spec.− Mean	Tol.	Sigma	6 Sigma	Process Capab
1 DIM A	2.8025	2.7900	−0.0125	0.0460	0.0070	0.0419	1.0976
2 DIM B	2.5017	2.5000	−0.0017	0.0460	0.0059	0.0352	1.3072
3 DIM C	0.5537	0.5600	0.0063	0.0460	0.0051	0.0304	1.5152
4 DIM D	4.4517	4.4400	−0.0117	0.0460	0.0083	0.0496	0.9281
5 DIM E	1.6182	1.6200	0.0018	0.0460	0.0074	0.0447	1.0300
6 DIM F	1.8665	1.8800	0.0135	0.0460	0.0218	0.1305	0.3524
7 DIM G	1.6607	1.6700	0.0093	0.0460	0.0088	0.0526	0.8740
8 DIM H	1.5671	1.5600	−0.0071	0.0460	0.0097	0.0580	0.7931
9 DIM I	1.3621	1.3500	−0.0121	0.0460	0.0044	0.0265	1.7354
10 DIM J	1.6597	1.6700	0.0103	0.0460	0.0106	0.0639	0.7201
11 DIM K	3.4469	3.4350	−0.0119	0.0460	0.0093	0.0560	0.8212
12 DIM L	0.7479	0.7500	0.0021	0.0460	0.0047	0.0282	1.6326
13 DIM MX	2.7821	2.7900	0.0079	0.0900	0.0102	0.0609	1.4771
14 DIM MY	0.3338	0.3700	0.0362	0.1000	0.0108	0.0648	1.5433
15 SNP RG A	2.1586	2.1900	0.0314	0.0900	0.0108	0.0648	1.3899
16 SNP RG B	3.2267	3.2350	0.0083	0.0900	0.0114	0.0683	1.3170
17 MLD HF R	0.6615	0.6550	−0.0065	0.0600	0.0036	0.0216	2.7834
18 MLD HF L	3.9807	3.9700	−0.0107	0.0600	0.0083	0.0498	1.2055
19 WALL A	0.2133	0.2140	0.0007	0.0280	0.0084	0.0502	0.5573
20 WALL B	0.1541	0.1530	−0.0011	0.0300	0.0097	0.0581	0.5161
21 WALL C	0.2286	0.2280	−0.0006	0.0400	0.0073	0.0440	0.9097
22 OAH MAX	3.7892	3.7700	−0.0192	0.0800	0.0352	0.2110	0.3791
23 OAH MIN	3.7589	3.7700	0.0111	0.0800	0.0244	0.1464	0.5465

```
 68 Parts included in analysis
Actual Sigma calculated

Process Capability  =        Blueprint Tolerance
                          -------------------------
                                  6 X Sigma
```

Fig. 5 Statistical summary report showing the mean of measured observations, the blueprint specification for the mean, the difference between specified and measured means, the tolerance, the standard deviation of the measured dimensions, and the capability of the process. These calculations are performed for all measured dimensions on the part.

shows an example output report of this type. A graph such as this, which has superimposed upon it the tolerance limits for the dimension being analyzed, allows a quick, qualitative evaluation of the variation and capability of the process. It also allows the normality of the process to be judged through comparison with the expected bell-shaped curve of a normal process.

Other Applications for Computer-Aided Inspection. The general sequence described for semiautomatic dimensional inspection can be applied to a number of other inspection criteria. Examples would include pressure testing or defect detection by electronic vision systems. The statistical analysis of scrap by defect types is also very helpful in identifying problem areas. In some cases, direct data input to a computer may not be feasible, but the benefits of entering data manually into a statistical analysis program should not be overlooked. The computer allows rapid analysis of large amounts of data so that statistically significant trends can be detected and proper attention paid to appropriate areas for improvement. The benefits and costs of each anticipated application of automation to a particular situation, as well as the feasibility of applying state-of-the-art equipment, need to be studied as thoroughly as possible prior to implementation.

Liquid Penetrant Inspection

Liquid penetrant inspection essentially involves a liquid wetting the surface of a workpiece, flowing over that surface to form a continuous and uniform coating, and migrating into cracks or cavities that are open to the surface. After a few minutes, the liquid coating is washed off the surface of the casting, and a developer is placed on the surface. The developer is stained by the liquid penetrant as it is drawn out of the cracks and cavities. Liquid penetrants will highlight surface defects so that detection is more certain. Details of this method can be found in the article "Liquid Penetrant Inspection" in this Volume.

Liquid penetrant inspection should not be confined to as-cast surfaces. For example, it is not unusual for castings of various alloys to exhibit cracks (frequently intergranular) on machined surfaces. A pattern of cracks of this type may be the result of intergranular cracking throughout the material because of an error in composition or heat treatment, or the cracks may be on the surface only as a result of machining or grinding. Surface cracking may result from insufficient machining allowance, which does not allow for complete removal of imperfections produced on the as-cast surface, or it may result from faulty machining techniques. If imperfections of this type are detected by visual inspection, liquid penetrant inspection will show the full extent of such imperfections, will give some indication of the depth and size of the defect below the surface by the amount of penetrant absorbed, and will indicate whether cracking is present throughout the section.

Liquid penetrant inspection is sometimes used in conjunction with another nondestructive method. Such is the case in the following example.

Example 1: Tail Rotor Gearbox Failure on the H-53 Helicopter. The H-53 tail rotor gearbox mounting lugs experienced failures that resulted in at least two aircraft accidents. The gearbox is a magnesium casting, either AZ-91 or ZE-41. Initially, only visual inspection using a 10× magnifying glass was required. Initiation sites for the cracking were found to be localized in a few areas adjacent to the attached lugs (Fig. 7). A manual eddy current inspection was developed that concentrated on the potential initiation sites. If an indication was found, then a fluorescent penetrant inspection was performed as a backup inspection.

Inspection Procedure. An eddy current tester with a flawgate and shielded probes was used to perform the inspection. The instrument is calibrated using standard magnesium alloy (either AZ-91 or ZE-41) calibration blocks. The tester is calibrated on a 0.50 mm (0.020 in.) slot to achieve a minimum 150 μA deflection. The flawgate is calibrated to activate for meter deflection greater than 150 μA. The part is scanned very slowly over all critical areas. As an aid for the inspection of the lug edges, a wooden toothpick is taped to the probe to maintain a constant probe-to-edge distance. A positive indication—meter deflection greater than 150 μA—requires a fluorescent penetrant inspection.

In preparation for the fluorescent penetrant inspection, the paint in the area to be inspected is removed. The required penetrant material sensitivity is level II or III and is used with a halogenated solvent remover and nonaqueous developer. Clean cloths moistened with solvent cleaner are required for the cleaning step. The inspection area is never flushed with solvent. An initial inspection is required before the developer is applied. If no indications are noted, developer is applied. When an indication is noted, it is wiped with a cotton swab moistened with solvent, and the area is inspected for the indication bleedback. A 30-min penetrant dwell time is required. The minimum required dwell time between cleaning and application of the developer is 10 min before inspection. The minimum developer dwell time is half the pene-

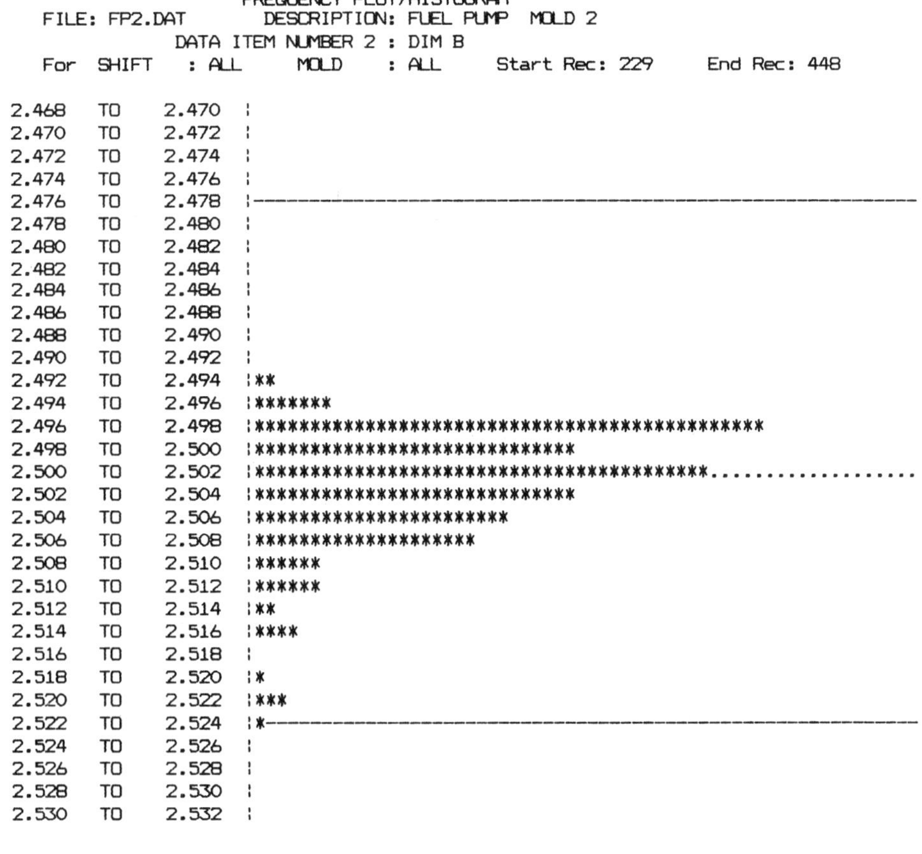

Fig. 6 Frequency plot for one measured dimension showing the distribution of the measurements

trant dwell time, and a maximum developer dwell time is not to exceed the penetrant dwell time.

After the inspection, all developer residue is removed from the mounting lug. Areas of greatest concern include the mount pad radius areas, lug flanges along edges, and the mounting pad. After completion of penetrant inspection, the finishes on the mounting lug are replaced in such a way that the eddy current inspection is not hindered and the visual inspection is enhanced. The visual inspection is enhanced by painting the lugs light gray or white.

Conclusions. These inspections were performed during scheduled inspection intervals until magnesium alloy castings were replaced with A356.0 aluminum alloy castings.

Magnetic Particle Inspection

Magnetic particle inspection is a highly effective and sensitive technique for revealing cracks and similar defects at or just beneath the surface of castings made of ferromagnetic metals. The capability of detecting discontinuities just beneath the surface is important because such cleaning methods as shot or abrasive blasting tend to close a surface break that might go undetected in visual or liquid penetrant inspection.

When a magnetic field is generated in and around a casting made of a ferromagnetic metal and the lines of magnetic flux are intersected by a defect such as a crack, magnetic poles are induced on either side of the defect. The resulting local flux disturbance can be detected by its effect on the particles of a ferromagnetic material, which become attracted to the region of the defect as they are dusted on the casting. Maximum sensitivity of indication is obtained when a defect is oriented in a direction perpendicular to the applied magnetic field and when the strength of this field is sufficient to saturate the casting being inspected.

Equipment for magnetic particle inspection uses direct or alternating current to generate the necessary magnetic fields. The current can be applied in a variety of ways to control the direction and magnitude of the magnetic field.

In one method of magnetization, a heavy current is passed directly through the casting placed between two solid contacts. The induced magnetic field then runs in the transverse or circumferential direction, producing conditions favorable to the detection of longitudinally oriented defects. A coil encircling the casting will induce a magnetic field that runs in the longitudinal direction, producing conditions favorable to the detection of circumferentially (or transversely) oriented defects. Alternatively, a longitudinal magnetic field can be conveniently generated by passing current through a flexible cable conductor, which can be coiled around any metal section. This method is particularly adaptable to castings of irregular shape. Circumferential magnetic fields can be induced in hollow cylindrical castings by using an axially disposed central conductor threaded through the casting.

Small castings can be magnetic particle inspected directly on bench-type equipment that incorporates both coils and solid contacts. Critical regions of larger castings can be inspected by the use of yokes, coils, or contact probes carried on flexible cables connected to the source of current; this setup enables most regions of castings to be inspected. Equipment and techniques associated with this method can be found in the article "Magnetic Particle Inspection" in this Volume.

Eddy Current Inspection

Eddy current inspection consists of observing the interaction between electromagnetic fields and metals. In a basic system, currents are induced to flow in the testpiece by a coil of wire that carries an alternating current. As the part enters the coil, or as the coil in the form of a probe or yoke is placed on the testpiece, electromagnetic energy produced by the coils is partly absorbed and converted into heat by the effects of resistivity and hysteresis. Part of the remaining energy is reflected back to the test coil, its electrical characteristics having been changed in a manner determined by the properties of the testpiece. Consequently, the currents flowing through the probe coil are the source of information describing the characteristics of the testpiece. These currents can be analyzed and compared with currents flowing through a reference specimen.

Eddy current methods of inspection are effective with both ferromagnetic and nonferromagnetic metals. Eddy current methods are not as sensitive to small, open defects as liquid penetrant or magnetic particle methods are. Because of the skin effect, eddy current inspection is generally restricted to depths less than 6 mm (¼ in.). The results of inspecting ferromagnetic materials can be obscured by changes in the magnetic permeability of the testpiece. Changes in temperature must be avoided to prevent erroneous results if electrical con-

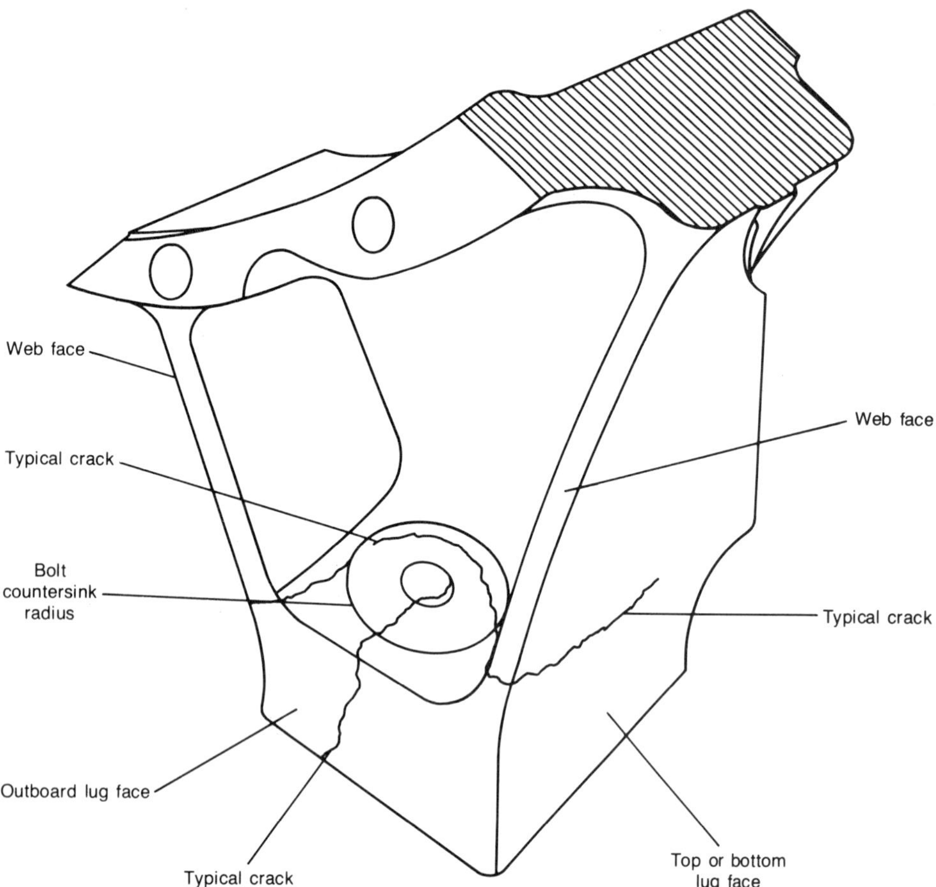

Fig. 7 Helicopter transmission mounting lug. Note: bolt is shown removed for clarity. Bolt is not removed for eddy current inspection.

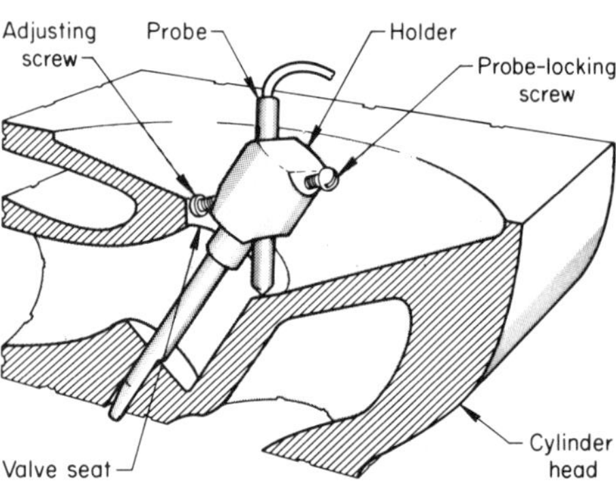

Fig. 8 Eddy current inspection using an adjustable probe for determining case depth of an induction-hardened gray iron valve seat

ductivity or other properties, including metallurgical properties, are being determined.

Applications of eddy current and electromagnetic methods of inspection to castings can be divided into the following three categories:

- Detecting near-surface flaws such as cracks, voids, inclusions, blowholes, and pinholes (eddy current inspection)
- Sorting according to alloy, temper, electrical conductivity, hardness, and other metallurgical factors (primarily electromagnetic inspection)
- Gaging according to size, shape, plating thickness, or insulation thickness (eddy current or electromagnetic inspection)

The following example demonstrates the use of eddy current inspection for the evaluation of an induction-hardened surface on a gray iron casting.

Example 2: Eddy Current Inspection of Hardened Valve Seats on Gray Iron Automotive Cylinder Heads. Exhaust-valve seats on gray iron automotive engine cylinder heads were induction hardened to withstand use with lead-free fuels. The seats were specified to be hardened to 50 HRC to a depth of 1.27 to 2.03 mm (0.050 to 0.080 in.). Failure to achieve these depths could lead to excessive recession of the seat, causing premature failure.

To establish the feasibility of, and the conditions for, eddy current inspection of cylinder heads having valve seats of different diameters, an adjustable probe was made using a 10 mm (0.400 in.) diam standard probe. The end was machined to a 120° conical point, and the probe was mounted in a plastic holder. A scrap valve stem mounted in the holder was used to position the probe on the valve seat. The probe was centered on the valve seat by moving the probe in its holder and by the adjusting screw. After the inspection conditions were established, rugged units for use by production inspectors were made of stainless steel with the protected probe in a fixed position to suit the cylinder-head seat.

In operation, the probe was brought into intimate contact with the hardened surface, and a meter reading was taken. Positioning of the probe on the valve seat is shown in Fig. 8. The meter reading, taken from an instrument calibrated from −100 to +100, was compared with a developed chart for case depth versus meter reading (a linear relationship). A meter reading of 0 was established for 50 HRC at a depth of 1.27 mm (0.050 in.) and −40 for 2.03 mm (0.080 in.). The instrumentation was set up to respond to A_r (resistive component). Inspection was performed at a frequency of 2 kHz. Extensive laboratory and production inspection has shown a standard deviation, σ of ±0.8 mm (±0.03 in.) and a confidence limit of 3σ for the total data generated.

Radiographic Inspection

Radiographic inspection is a process of testing materials using penetrating radiation from an x-ray generator or a radioactive source and an imaging medium, such as x-ray film or an electronic device. In passing through the material, some of the radiation is attenuated, depending on the thickness and the radiographic density of the material, while the radiation that passes through the material forms an image. The radiographic image is generated by variations in the intensity of the emerging beam.

Internal flaws, such as gas entrapment or nonmetallic inclusions, have a direct effect on the attenuation. These flaws create variations in material thickness, resulting in localized dark or light spots on the image.

The term radiography usually implies a radiographic process that produces a permanent image on film (conventional radiography) or paper (paper radiography or xero radiography), although in a broad sense it refers to all forms of radiographic inspection. When inspection involves viewing an image on a fluorescent screen or image intensifier, the radiographic process is termed filmless or real-time inspection. When electronic nonimaging instruments are used to measure the intensity of radiation, the process is termed radiation gaging. Tomography, a radiation inspection method adapted from the medical computerized axial tomography scanner, provides a cross-sectional view of a testpiece. All of the above terms are primarily used in connection with inspection that involves penetrating electromagnetic radiation in the form of x-rays or γ-rays. Neutron radiography refers to radiographic inspection using neutrons rather than electromagnetic radiation.

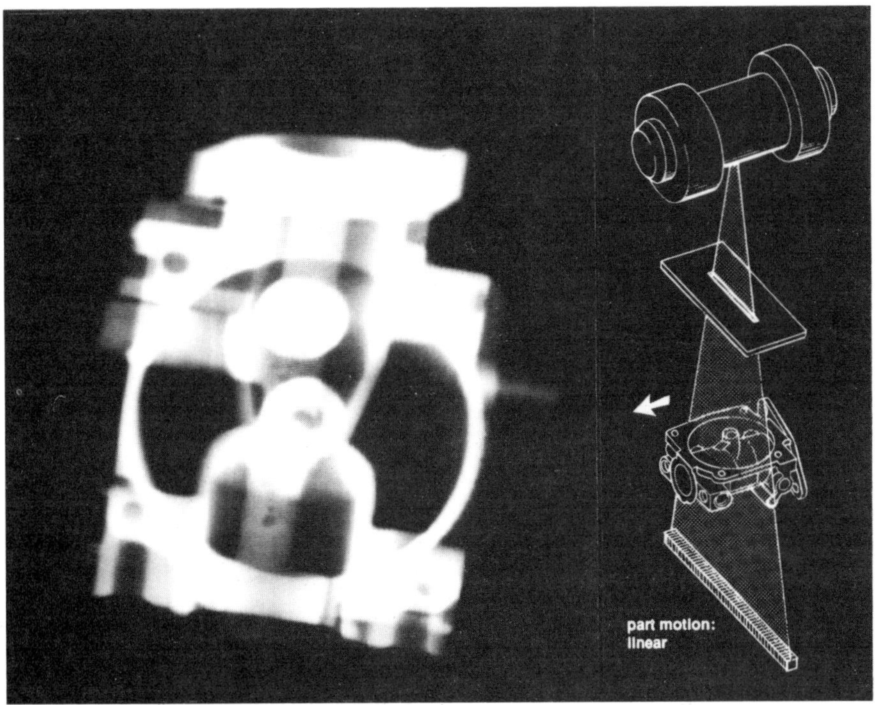

Fig. 9 Digital radiography image of a die cast aluminum carburetor. Porosity appears as dark spots in the area of the center bore, through the vertical center of the image. Courtesy of B.G. Isaacson, Bio-Imaging Research, Inc.

The sensitivity, or the ability to detect flaws, of radiographic inspection depends on close control of the inspection technique, including the geometric relationships among the point of x-ray emission, the casting, and the x-ray imaging plane. The smallest detectable variation in metal thickness lies between 0.5 and 2.0% of the total section thickness. Narrow flaws, such as cracks, must lie in a plane approximately parallel to the emergent x-ray beam to be imaged; this requires multiple exposures for x-ray film techniques and a remote-control parts manipulator for a real-time system.

Real-time systems have eliminated the need for multiple exposures of the same casting by dynamically inspecting parts on a manipulator, with the capability of changing the x-ray energy for changes in total material thickness. These capabilities have significantly improved productivity and have reduced costs, thus enabling higher percentages of castings to be inspected and providing instant feedback after repair procedures. Figures 9 and 10 show real-time digital radiography images of automotive components.

Advances. Several advances have been made to assist the industrial radiographer. These include the computerization of the radiographic standard shooting sketch, which graphically shows areas to be x-rayed and the viewing direction or angle at which the shot is to be taken, and the development of microprocessor-controlled x-ray systems capable of storing different x-ray exposure parameters for rapid retrieval and automatic warm-up of the system prior to use. The advent of digital image-processing systems and microfocus x-ray sources (near point source), producing energies capable of penetrating thick material sections, have made real-time inspection capable of producing images equal to, and in some cases superior to, x-ray film images by employing geometric relations previously unattainable with macrofocus x-ray systems. The near point source of the microfocus x-ray system virtually eliminates the edge unsharpness associated with larger focus devices.

Digital image processing can be used to enhance imagery by multiple video frame integration and averaging techniques that improve the signal-to-noise ratio of the image. This enables the radiographer to digitally adjust the contrast of the image and to perform various edge enhancements to increase the conspicuity of many linear indications (Fig. 11).

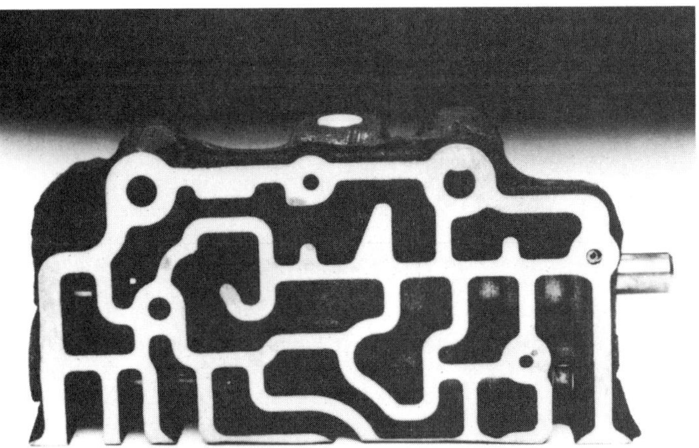

(a)

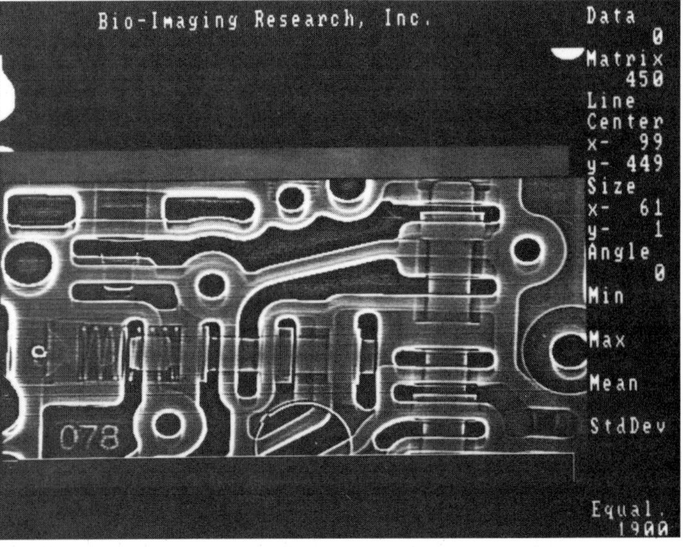

(b)

Fig. 10 Evaluation of cast transmission housing assembly. (a) Photograph of cast part. (b) Digital radiography image used to verify the steel spring pin and shuttle valve assembly through material thicknesses ranging from 3 mm (1/8 in.) in the channels to 25 mm (1 in.) in the rib sections of the casting. Courtesy of B.G. Isaacson, Bio-Imaging Research, Inc.

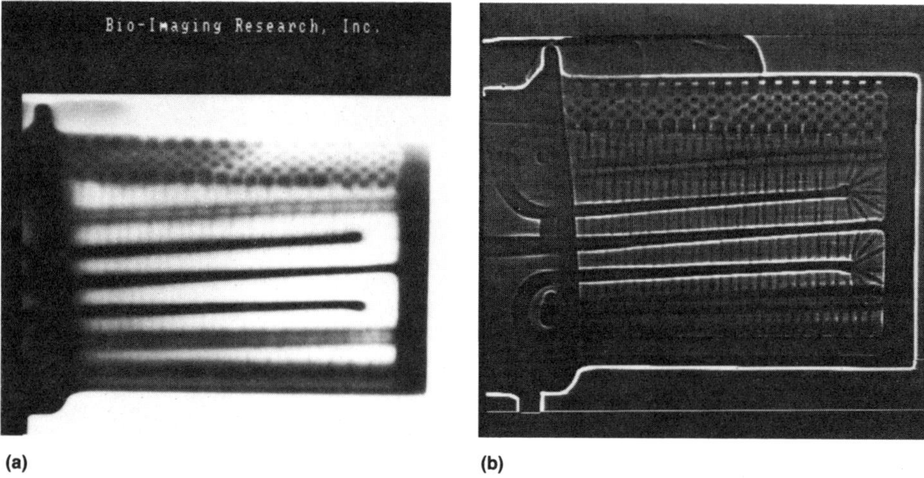

(a) (b)

Fig. 11 Digital radiography images of an investment cast jet engine turbine blade showing detail through a wide range in material thickness. The trailing edge of the blade (along the top of the image) is 2 mm (0.080 in.) thick, the root section of the blade (to the far left in the image) is 19 mm (0.75 in.) thick, and the shelf area (to the right of the root section) is 25 mm (1 in.) thick. The image shown in (a) is unprocessed; the image in (b) is processed to subdue the background and to enhance edges and internal features. Courtesy of B.G. Isaacson, Bio-Imaging Research, Inc.

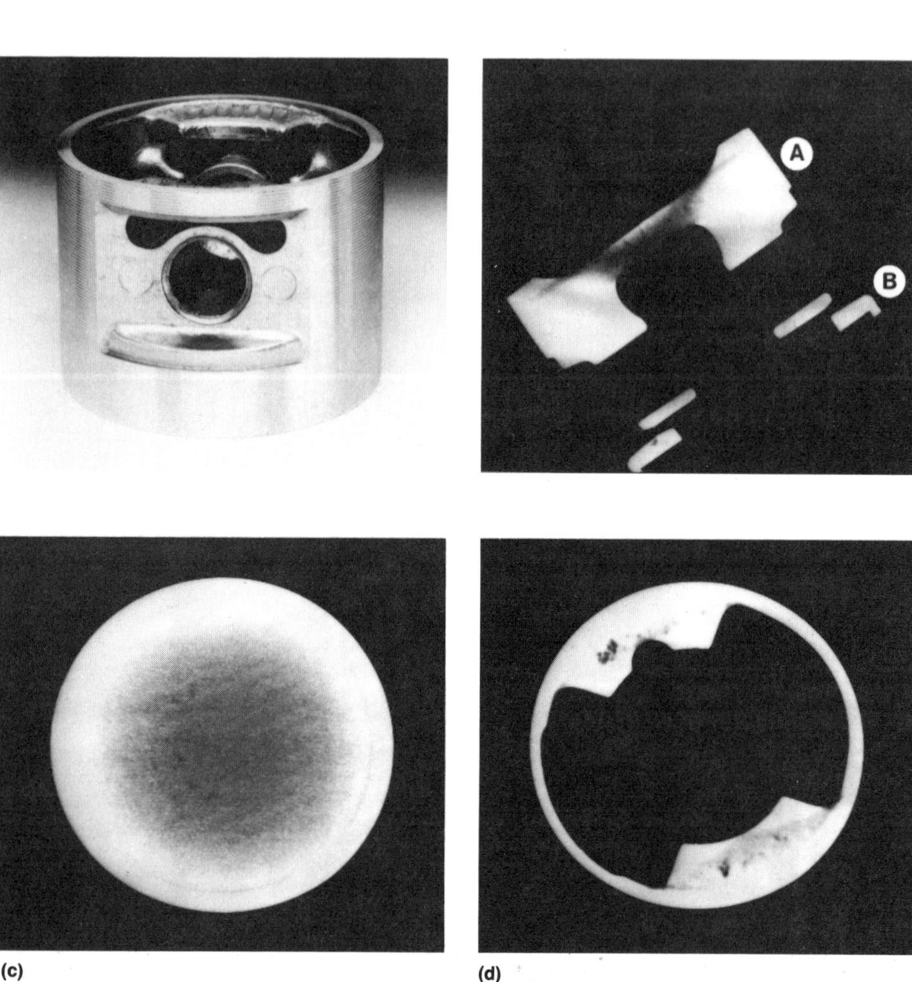

(c) (d)

Fig. 12 Computed tomographic images of a die cast aluminum automotive piston. (a) Photograph of cast part. (b) Vertical slice through the piston shows porosity as dark spots in the crown area (point A) and counterbalance area (point B). (c) Transverse slice through the crown of the piston verifies the porosity; the smallest void that is visible is 0.4 mm (0.016 in.) in diameter. (d) Transverse slice through the counterbalance area also verifies porosity. Dimensional analysis of the piston walls is possible to an accuracy of ±50 μm (±0.002 in.). Courtesy of B.G. Isaacson, Bio-Imaging Research, Inc.

Interpretation of the radiographic image requires a skilled specialist who can establish the correct method of exposing the castings with regard to x-ray energies, geometric relationships, and casting orientation and can take all of these factors into account to achieve an acceptable, interpretable image. Interpretation of the image must be performed to establish standards in the form of written or photographic instructions. The inspector must also be capable of determining if the localized indication is a spurious indication, a film artifact, a video aberration, or a surface irregularity. The article "Radiographic Inspection" in this Volume provides additional information on real-time digital systems.

Computed tomography, also known as computerized axial tomography (or CAT scanning), is a more sophisticated x-ray imaging technique originally developed for medical diagnostic use (Ref 2). It is the complete reconstruction by computer of a tomographic plane, or slice of an object. A collimated (fan-shaped) x-ray beam is passed through a section of the part and is intercepted by a detector on the other side. The part is rotated slightly, and a new set of measurements is made; this process is repeated until the part has been rotated 180°.

The resulting image of the slice (tomogram) is formed by computer calculations based on electronic measurement (digital sampling) of the radiation transmitted through the object along different paths during the rotating scan. The data thus accumulated are used to compute the densities of each point in the cross section, enabling the computer to reconstruct a two-dimensional visual image of the slice.

The shapes of internal features are determined by their computed densities. After one slice is produced through a complete rotation, either the part or the radiation source and detector can be moved and a three dimensional image built up through the scanning of successive slices.

Compared to electronic radiography, computed tomography provides increased sensitivity and detection capabilities. The contrast resolution of a good-quality tomographic image is 0.1 to 0.2%, which is approximately two orders of magnitude better than with x-ray film.

In addition, images are produced in a quantitative, ready-to-use digital format (Ref 3). They provide detailed physical information, such as size, density, and composition, to aid in evaluating defects. Methods are being developed to use this information to predict failure modes or system performance under operating loads. The data can also be easily manipulated to obtain various types of images, to develop automated flaw detection techniques, and to promote efficient archiving. Figures 12 and 13 illustrate the uses of computed tomography for examining castings. More detailed information can be found in the

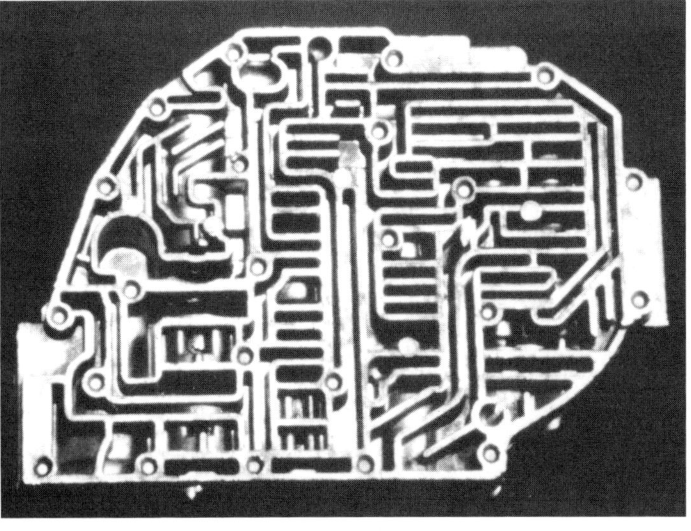

(a)

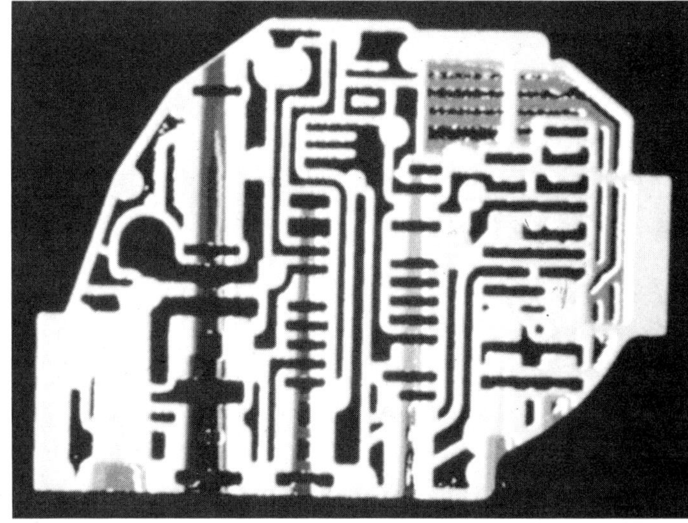

(b)

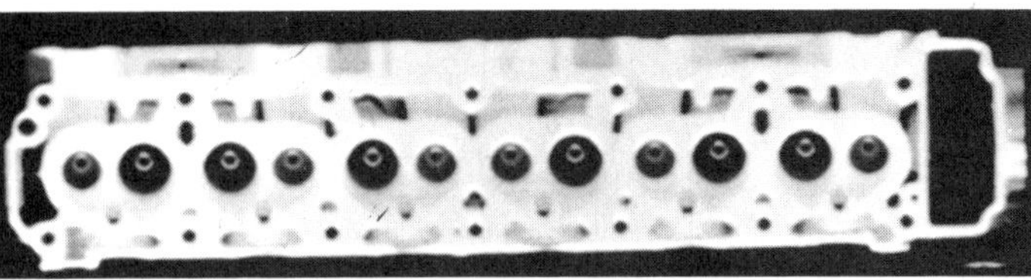

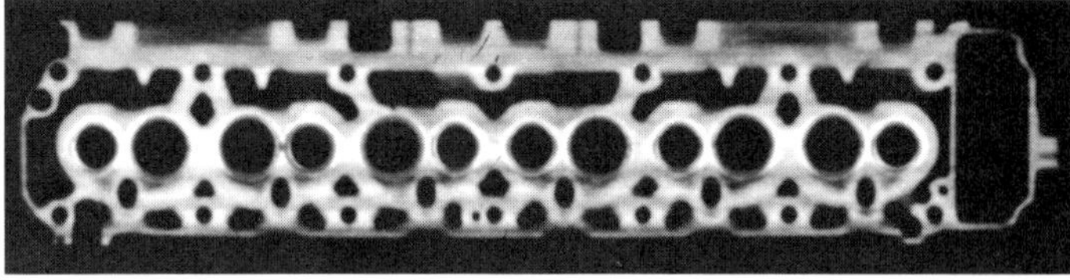

(c)

Fig. 13 Use of computed tomography for examining automotive components. (a) Photograph of a cast aluminum transmission case with (b) corresponding tomographic image. (c) Two three-dimensional images of a cast aluminum cylinder head generated from a set of continuous tomographic scans used to view water-cooling chambers where a leak had been detected. Courtesy of R.A. Armistead, Advanced Research and Applications Corporation

article "Industrial Computed Tomography" in this Volume.

Ultrasonic Inspection

Ultrasonic inspection is a nondestructive method in which beams of high-frequency acoustic energy are introduced into the material under evaluation to detect surface and subsurface flaws and to measure the thickness of the material or the distance to a flaw (see the article "Ultrasonic Inspection" in this Volume for details). An ultrasonic beam will travel through a material until it strikes an interface or defect. Interfaces and defects interrupt the beam and reflect a portion of the incident acoustic energy. The amount of energy reflected is a function of the nature and orientation of the interface or flaw as well as the acoustic impedance of such a reflector. Energy reflected from various interfaces or defects can be used to define the presence and locations of defects, the thickness of the material, or the depth of the defect beneath a surface.

The advantages of ultrasonic tests are as follows:

- High sensitivity, which permits the detection of minute cracks
- Great penetrating power, which allows the examination of extremely thick sections
- Accuracy in measuring of flaw position and estimating defect size

Ultrasonic tests have the following limitations:

- Size-contour complexity and unfavorable discontinuity orientation can pose problems in interpreting the echo pattern
- Undesirable internal structure—for example, grain size, structure, porosity, inclusion content, or fine dispersed precipitates—can similarly hinder interpretation
- Reference standards are required

Effect of Casting Shape. Because castings are rarely simple flat shapes, they are not as easy to inspect as such products as rolled rectangular bars. The reflections of a sound beam from the back surface of a parallel-sided casting and a discontinuity are shown schematically in Fig. 14(a), together with the relative heights and positions of the reflections of the two surfaces on an oscilloscope screen. A decrease in the back reflection at the same time as the appearance of a discontinuity echo is a secondary indication of the presence of a discontinuity. However, if the back surface of the casting at a particular location for inspection is not approximately at a right angle to the incident sound beam, the beam will be reflected to remote parts of the casting and not directly returned to the detector. In this case, as shown in Fig. 14(b), there is no back reflection to monitor as a secondary indication.

Many castings contain cored holes and changes in section, and echoes from holes and changes in section can interfere with echoes from discontinuities. As shown in Fig. 14(c), the echo from the cored hole overlaps the echo from the discontinuity on the oscilloscope screen. The same effect is shown in Fig. 14(d), in which echoes from the discontinuity and the casting fillets at a change in section are shown overlapping on the oscilloscope.

Curved surfaces do not permit adequate or easy coupling of the flat search units to the casting surface, especially with contact double search units. This can be overcome to some extent by using a suitable viscous couplant, but misleading results may be produced because multiple reflections in the wedge of fluid between the search unit and the surface can result in echoes on the screen in those positions where discontinuity echoes may be expected to appear. Because the reflections inside the couplant use energy that would otherwise pass into the casting, the back echo decreases, and this decrease might be interpreted as confirmation of the presence of a discontinuity. On cylindrical surfaces, the indication will change as a double search unit is rotated.

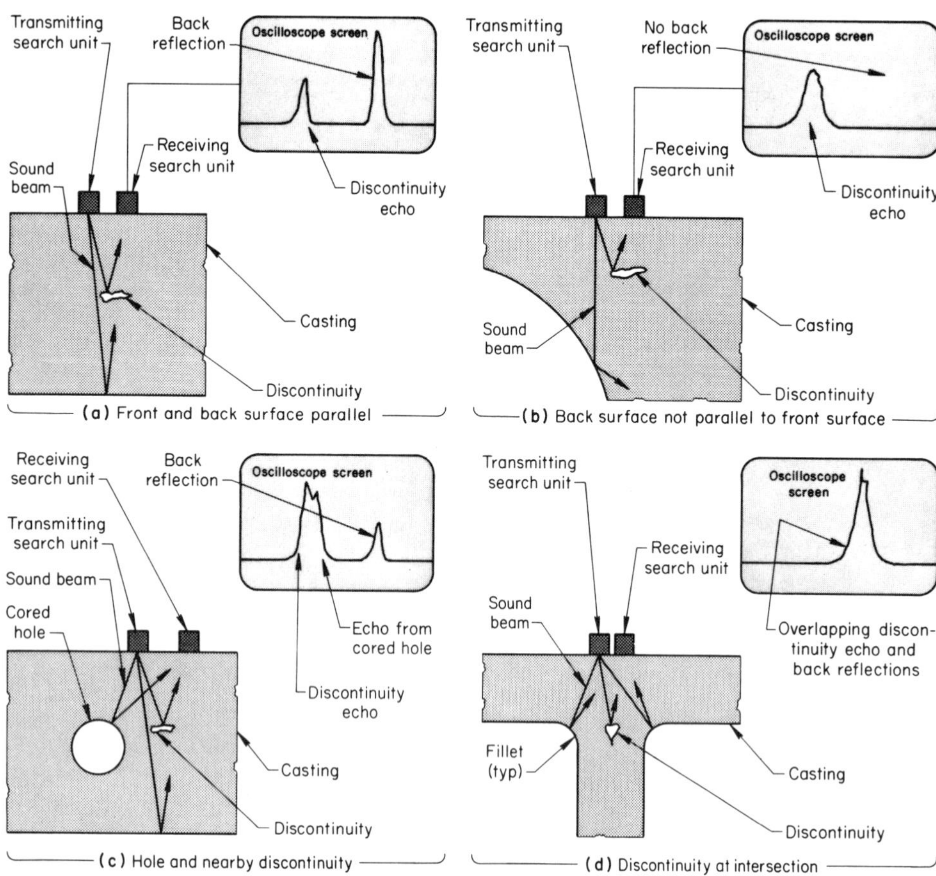

Fig. 14 Schematic of the effect of casting shapes on reflection and oscilloscope screen display of sound beams. See text for discussion.

The wedge effect is least when the division between the transmitting and receiving transducers is parallel to the axis of the cylinder. Wedge effects in the couplant are a particular problem on castings curved in two directions. One solution in this case is to use a small search unit so that the wedge is short, although the resolution and sensitivity may be reduced.

If the surface of the casting to be inspected is of regular shape, such as the bore of a cylinder in an engine block, the front of the search unit can be shaped to fit the curvature of the surface. These curved shapes form an acoustic lens that will alter the shape of the sound beam, but unless the curvature is severe, this will not prevent adequate accuracy in the inspection. Cast-on flat metal pads for application of the ultrasonic search unit are very effective and allow particular areas of the casting to be inspected.

Subsurface Defects. Defects such as small blowholes, pinholes, or inclusions that occur within depths of 3 or 4 mm (0.1 or 0.15 in.) of a cast surface are among the most difficult to detect (Ref 1). They are beyond the limits of sensitivity of conventional magnetic particle methods and are not easily identified by eddy current techniques. They usually fall within the dead zone (the surface layer that cannot be inspected) of conventional single-crystal ultrasonic probes applied directly to a cast surface, although some improvement can be obtained by using twin crystal probes focused to depths not too far below the surface. The other alternative using contact methods of ultrasonic testing is to employ angle probes, but this complicates the procedures and interpretation methods to the point at which they can only be applied satisfactorily under the close control of skilled operators.

Freedom from such surface defects is, however, a very important aspect of the quality of castings. Apart from their effect in reducing bending fatigue properties, such defects are frequently revealed at late stages in the machining of a component, leading to its rejection.

Ultrasonic methods for detecting subsurface defects are much more successful when the dead zone beneath the as-cast surface is virtually eliminated by using immersion methods in which the probe is held away from the cast surface at a known controlled distance, with coupling being obtained through a liquid bath. To make such methods consistent and reliable, the test itself must be automated. Semiautomatic equipment has been developed for examining castings such as cylinder heads by this method (Ref 1). With this equipment, the casting is loaded into a cradle from a roller track and is then transferred using a hoist into the immersion tank until the surface of the casting to be inspected is just submerged in the liquid. Depth of immersion is closely controlled because the customer will not permit liquid to be left in the internal passageways of the cylinder head. The immersed surface of the casting is then scanned manually using an ultrasonic probe held at a fixed distance from the casting surface. This equipment is suitable for testing any casting requiring examination over a flat surface.

Internal Defects. Ultrasonic inspection is a well-established method for the detection of internal defects in castings. Test equipment developments, automated testing procedures, and improvements in determining the size and position of defects, which is essential to assessing whether or not their presence will likely affect the service performance of the casting, have contributed to the increasing use of ultrasonic test equipment.

For determining the position and size of defects, the usual method of presentation of ultrasonic data is an A-scan, in which the amplitude of the echoes from defects is shown on a time base and has well-known limitations (Ref 1). Sizing relies on measuring the drop in amplitude of the echo as the probe is passed over the boundary of a defect or measuring the reduction in the amplitude of the back wall echo due to the scattering of sound by the defect. In most cases, sizing is approximate and is restricted to one or two dimensions. Improvements in data presentation in the form of B-scans and C-scans that present a plan view through the section of the component provide a marked improvement in defining defect positions and size in two or three dimensions. Such displays have been used for automated defect characterization systems in which porosity, cracks, and dross have been distinguished. Because of the requirement to scan the probe over the surface, the application of B-scan and C-scan methods has generally been limited to simple geometric shapes having good surface finish, such as welded plate structures. Application to castings is currently restricted, but greater use of B-scan and C-scan methods is likely with either improved scanning systems or arrays of ultrasonic probes.

Structure evaluation is an area of growing importance for foundry engineers. Ultrasonic velocity measurements are widely used as a means of guaranteeing the nodularity of the graphite structure and, if the matrix structure is known to be consistent, guaranteeing the principal material properties of ductile irons (Ref 1). Velocity measurements have been used to evaluate compacted graphite iron structures to ensure that the desired properties have been consistently obtained. The use of ultrasonic velocity measurements for structure evaluation is discussed in the section "Ductile Iron Castings" in this article.

Leak Testing

Castings that are intended to withstand pressures can be leak tested in the foundry. Various methods are used, according to the type of metal being tested (see the article "Leak Testing" in this Volume). One method consists of pumping air at a specific pressure into the inside of the casting in water at a given temperature. Any leaks through the casting become apparent by the release of bubbles of air through the faulty portions. An alternative method is to fill the cavities of a casting with paraffin at a specified pressure. Paraffin, which penetrates the smallest of crevices, will rapidly find any defect, such as porosity, and will show quickly as an oily or moist patch at the position of the fault. Liquid penetrants can be poured into areas of apparent porosity and time allowed for the liquid to seep through the casting wall.

The pressure testing of rough (unmachined) castings at the foundry may not reveal any leaks, but it must be recognized that subsequent machining operations on the casting may cut into porous areas and cause the casting to leak after machining. Minor seepage leaks can be sealed by impregnation of the casting with liquid or by filling with sodium silicate, a synthetic resin, or other suitable substance. As-cast parts can be impregnated at the foundry to seal leaks if there is to be little machining or if experience has shown that machining does not affect the pressure tightness. However, it is usually preferable to impregnate after final machining of the casting.

Inspection of Ferrous Castings

Ferrous castings can be inspected by most of the nondestructive inspection methods. Magnetic particle inspection can be applied to ferrous metals with excellent sensitivity, although a crack in a ferrous casting can often be seen by visual inspection. Magnetic particle inspection provides good crack delineation, but the method should not be used to detect other defects. Nonrelevant magnetic particle indications occasionally occur on ferrous castings, especially with a strong magnetic field. For example, a properly fused-in steel chaplet can be indicated as a defect because of the difference in magnetic response between low-carbon steel and cast iron. Even the graphite in cast iron, which is nonmagnetic, can cause a nonrelevant indication. Standard x-ray and radioactive-source techniques can be used to make radiographs of ferrous castings, but the typical complexity of shape and varying section thicknesses of the castings may require digital radiography or computed tomography.

Ultrasonic inspection for both thickness and defects is practical with most ferrous castings except for the high-carbon gray iron castings, which have a high damping capacity and absorb much of the input energy. The measurement of resonant frequency is a good method of inspecting some ductile iron castings for soundness and graphite shape. Electromagnetic testing can be used to distinguish metallurgical differences between castings. The criteria for separating acceptable from unacceptable castings must be established empirically for each casting lot.

Gray Iron Castings

Gray iron castings are susceptible to most of the imperfections generally associated with castings, with additional problems resulting from the relatively high pouring temperatures. These additional problems result in a higher incidence of gas entrapment, inclusions, poor metal structure, interrupted metal walls, and mold wall deficiencies.

Gas entrapment is a direct result of gas being trapped in the casting wall during solidification. This gas may be in the metal prior to pouring, may be generated from aspiration during pouring, or may be generated from core and mold materials. Internal defects of this type are best detected by radiography, but ultrasonic and eddy current inspection methods are useful when the defect is large enough to be detected by these methods.

Inclusions are casting defects in which solid foreign materials are trapped in the casting wall. The inclusion material can be slag generated in the melting process, or it can be fragments of refractory, mold sand, core aggregate, or other materials used in the casting process. Inclusions appear most often on the casting surface and are usually detected by visual inspection, but in many cases the internal walls of castings contain inclusions that cannot be visually detected. Internal inclusions can be detected by eddy current, radiographic, or ultrasonic inspection; radiography is usually the most reliable method.

Poor Metal Structure. Many casting defects resulting from metal structure are related to shrinkage, which is either a cavity or a spongy area linked with dendrites or a depression in the casting surface. This type of defect arises from varying rates of contraction while the metal is changing from a liquid to a solid. Other casting defects resulting from varying rates of contraction during solidification include carbide formation, hardness variations, and microporosity.

Internal shrinkage defects are best detected by radiography, although eddy current or ultrasonic inspection can be used. Soft or hard gray iron castings are usually detected by Brinell hardness testing; electromagnetic methods have proved useful on some castings.

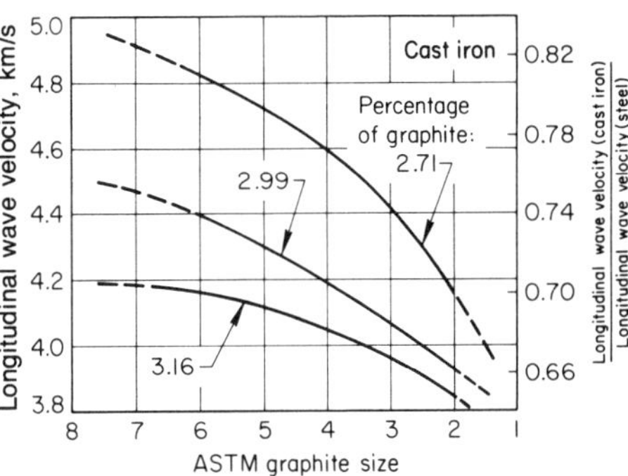

Fig. 15 Variation of longitudinal wave sonic velocity with graphite size for lamellar cast iron containing different percentages of free graphite. Scale at right indicates ratio of sonic velocity in cast iron to corresponding sonic velocity in steel.

Interrupted Metal Walls. Included in this category are such flaws as hot tears, cold shuts, and casting cracks. Cracking of castings is often a major problem in gray iron foundries because of the combination of casting designs and high production rates. Visual inspection or an aided visual method such as liquid penetrant or magnetic particle inspection is used to detect cracks and cracklike flaws in castings.

Mold wall deficiencies are common problems in gray iron castings. They result in surface flaws such as scabs, rattails, cuts, washes, buckles, drops, and excessive metal penetration into the spaces between sand grains. These flaws are generally detected by visual inspection.

Size and Quantity of Graphite. In cast iron, the length of the lamellae (flakes)—that is, the coarseness of the graphite—is expressed by code numbers from 1 to 8, as described in ASTM A 247 and in Volume 15 of the 9th Edition of *Metals Handbook*. These numbers correspond to lamellar lengths ranging from 1.25 to 0.01 mm (0.05 to 0.0004 in.), as viewed in micrographs of the cast iron structure. The dependence of the longitudinal wave sonic velocity on the size and quantity of graphite is shown in Fig. 15; with decreasing coarseness and quantity of graphite, the velocity approaches that in steel. In microlamellar cast iron, the amount of graphite is usually smaller. Therefore, the finer the graphite, the higher the sonic velocity. In both lamellar and spheroidal cast irons, the transverse wave sonic velocity is related to the longitudinal wave velocity, as shown in Fig. 16.

Malleable Iron Castings

Blowholes and spikes are defects often found in malleable iron castings. Spikes are a form of surface shrinkage not normally visible to the naked eye, but they appear as

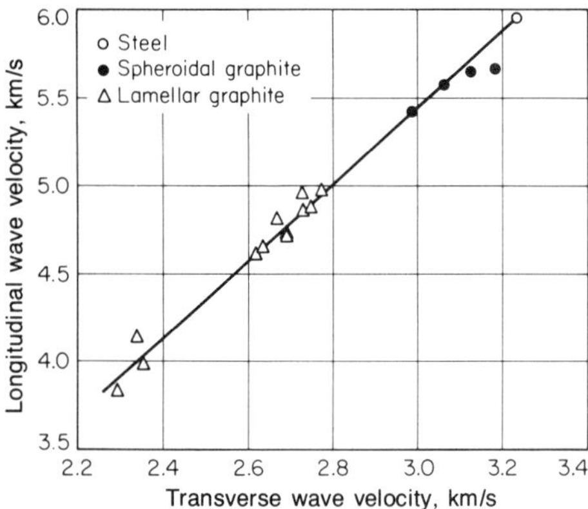

Fig. 16 Relation of longitudinal wave and transverse wave sonic velocities for spheroidal and lamellar cast irons and for steel

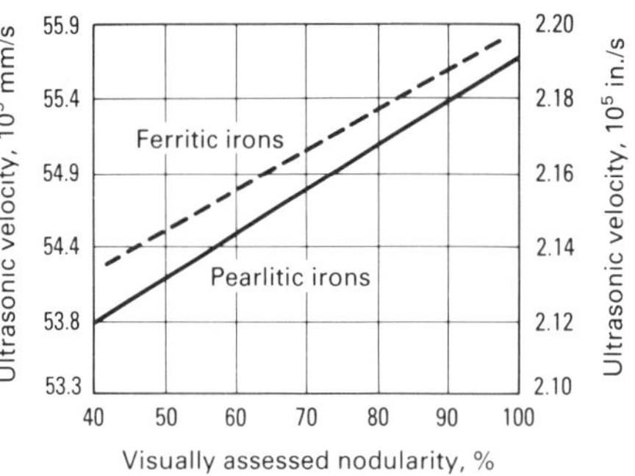

Fig. 17 Ultrasonic velocity versus visually assessed nodularity in ductile iron castings

a multitude of short, discontinuous surface cracks when subjected to fluorescent magnetic particle inspection. Unlike true fractures, spikes do not propagate, but they are not acceptable where cyclic loading could result in fatigue failure. Spikes are usually seen as short indications about 1.6 mm (1/16 in.) long or less and never more than 75 μm (0.003 in.) deep. These defects do not have a preferred orientation, but a random pattern that may or may not follow the direction of solidification. Shrinkage or open structure in the gated area is a defect often found in malleable iron castings that may be overlooked by visual inspection, although it is readily detected by either liquid penetrant or magnetic particle inspection.

Ductile Iron Castings

Ductile iron is cast iron in which the graphite is present in tiny balls or spherulites instead of flakes (as in gray iron) or compacted aggregates (as in malleable iron). The spheroidal graphite structure is produced by the addition of one or more elements to the molten metal.

Soundness and Integrity. Cracks and fine tears that break the surface of the casting but are difficult to detect visually can be revealed with dye penetrants or magnetic particle inspection. Modern techniques of magnetizing the casting, followed by the application of fluorescent magnetic inks, are very effective and widely used.

Methods of sonic testing that involve vibrating the casting and noting electronically the rate of decay of resonant frequency or damping behavior are also used to detect cracked or flawed castings. Internal unsoundness, when not immediately subsurface, can be detected with ultrasonic inspection by the failure to observe a back wall echo when using reflected radiation or by a weakening of the signal in the transmission through the casting. Coupling of the probes and interpretation of the results involve operator skill, but methods are available that consist of partial or total immersion of the casting in a liquid, automatic or semiautomatic handling of the probes, and computer signal processing to ensure more reliable and consistent interpretation of results. Problems arise in detecting very-near-surface defects and when examining thin castings, but the use of angled probes and shear wave techniques has yielded good results.

The soundness of the ductile iron can also be assessed by x-ray or γ-ray examination. The presence of graphite, especially in heavy sections, makes the method more difficult to evaluate than for steels, but the use of image intensification by electronic means offers considerable promise, especially for sections up to 50 mm (2 in.) thick.

Confirmation of Graphite Structure (Ref 4). Both the velocity of ultrasonic transmission and the resonant frequency of a casting can be related to the modulus of elasticity. In cast iron, the change from flake graphite to nodular graphite is related to an increase in both modulus of elasticity and strength; therefore, ultrasonic velocity or resonant frequency measurement can be employed as a guide to nodularity, strength, and other related properties. Because the microscopic estimation of nodularity is a subjective measurement, these other nondestructive examination methods may provide a better guide to some properties provided the matrix remains constant (Ref 5). Figure 17 illustrates how ultrasonic velocity may vary with graphite nodularity.

Ultrasonic transmission measurement is conducted with two probes on either side of the casting. This method provides a guide to the local properties. It must be coupled with a thickness measurement, and automatic equipment is commonly used, often involving immersion of the casting in a tank of fluid (Fig. 18). The calculation does not require calibration of the castings. Simple caliper devices have also been used for examining castings and simultaneously measuring their thickness to provide a calculated value of ultrasonic velocity and a guarantee of nodularity (Ref 6).

Sonic testing involves measurement of the resonant frequency of the casting or rate of decay of resonance of a casting that has first been excited by mechanical or electrical means. This method evaluates the graphite structure of the entire casting and requires calibration of the castings against standard castings of known structure. It is also necessary to maintain casting dimensions within a well-controlled narrow range. Some foundries use sonic testing as a routine method of final inspection and structure guarantee (Ref 7).

The relationship between nodularity, resonant frequency, or ultrasonic transmission velocity and properties has been documented for tensile strength, proof strength, fatigue, and impact strength. Examples are shown in Fig. 19 and 20.

The presence of carbides can also be detected with sonic or ultrasonic measurements provided enough carbides are present to reduce the graphite sufficiently to affect the modulus of elasticity. Discrimination between the effects of graphite variation and carbide amount would require an additional test, such as a hardness test.

Properties Partially Dependent on Graphite Structure (Ref 7). When the matrix structure of ductile iron varies, this variation cannot be detected as easily as variations in graphite structure, and sonic and ultrasonic readings may not be able to reflect variations in mechanical properties. A second measurement, such as a hardness measurement, is then needed to detect matrix variations in the same way as would be necessary to confirm the presence of carbides.

Eddy current or coercive force measurements can be used to detect many changes in casting structure and properties; but the indications from such measurements are difficult to interpret, and the test is difficult to apply to many castings unless they are quite small and can be passed through a coil 100 to 200 mm (4 to 8 in.) in diameter. Eddy current indications are, however, useful for evaluating pearlite and carbide in the iron matrix. Multifrequency eddy current testing uses probes that do not require the casting to pass through a coil, it is less sensitive to casting size, and it allows automatic measurements and calculations to be made; but the results remain difficult to interpret with reliability in all cases. It may, however, be a very good way to detect chill and hard edges on castings of reproducible dimensions.

Inspection of Aluminum Alloy Castings

Effective quality control is needed at every step in the production of an aluminum alloy casting, from selection of the casting method, casting design, and alloy to mold production, foundry technique, machining, finishing, and inspection. Visual methods, such as visual inspection, pressure testing,

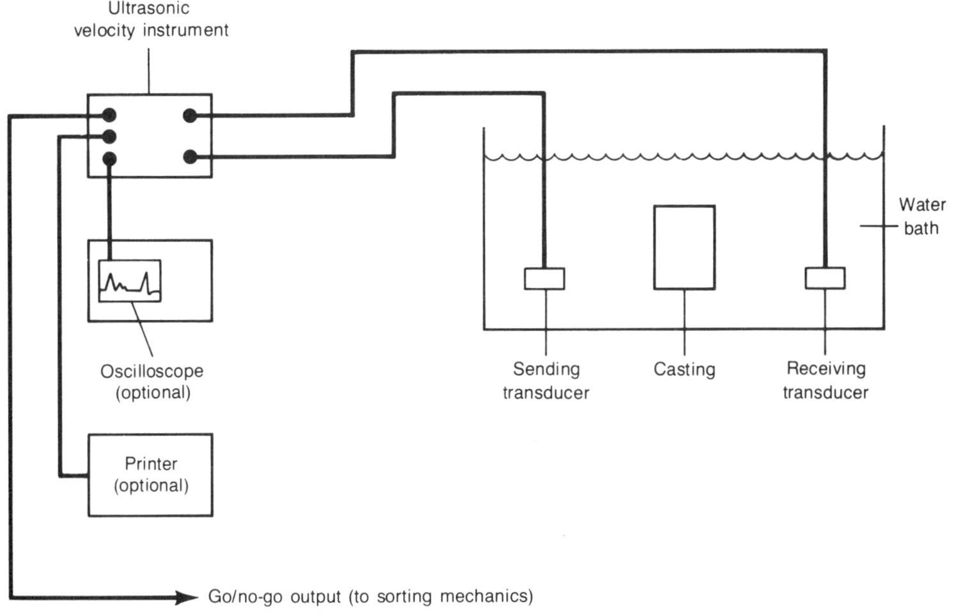

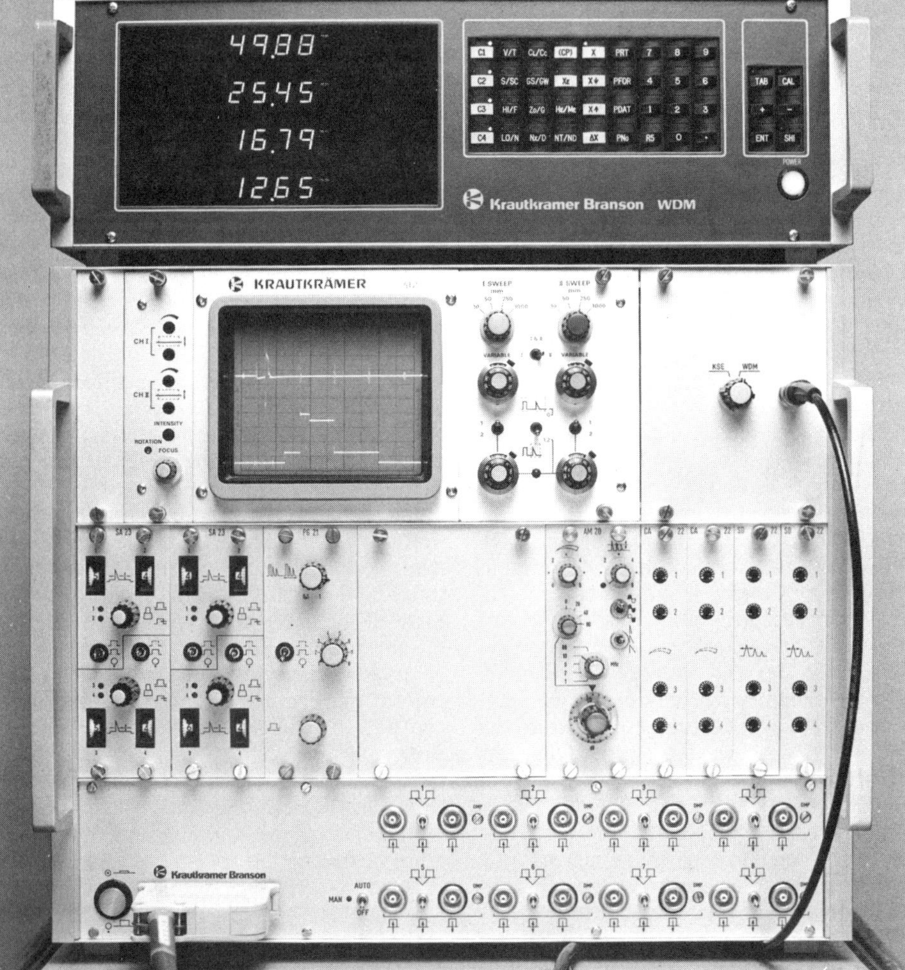

Fig. 18 Ultrasonic test equipment used for determining the thickness, nodularity, and integrity of ductile iron castings. (a) Schematic of setup for the ultrasonic velocity testing for structure evaluation. (b) Photograph of test instrument used for integrity/nodularity studies showing controls and instrumentations. Courtesy of J. Johnston, Krautkramer Branson

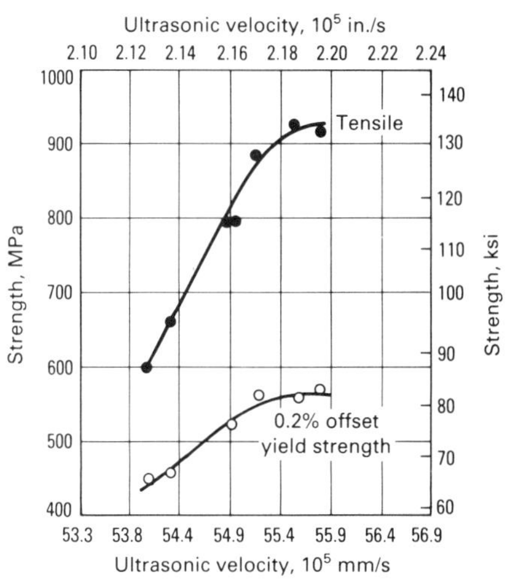

Fig. 19 Ultrasonic velocity versus strength in ductile iron castings

liquid penetrant inspection, ultrasonic inspection, radiographic inspection, and metallographic examination, can be used to inspect for casting quality. The inspection procedure used should be geared toward the specified level of quality. Information on casting processes, solidification, hydrogen content, silicon modification, grain refinement, and other topics related to aluminum alloy castings is provided in the articles "Solidification of Eutectic Alloys," "Nonferrous Molten Metal Processes," and "Aluminum and Aluminum Alloys" in Volume 15 of the 9th Edition of *Metals Handbook*.

Stages of Inspection. Inspection can be divided into three stages: preliminary, intermediate, and final. After tests are conducted on the melt for hydrogen content, for adequacy of silicon modification, and for degree of grain refinement, preliminary inspection may consist of the inspection and testing of test bars cast with the molten alloy at the same time the production castings are poured. These test bars are used to check the quality of the alloy and the effectiveness of the heat treatment. Preliminary inspection also includes chemical or spectrographic analysis of the casting, thus ensuring that the melting and pouring operations have resulted in an alloy of the desired composition.

Intermediate inspection, or hot inspection, is performed on the casting as it is taken from the mold. This step is essential because castings that are obviously defective can be discarded at this stage of production. Castings that are judged unacceptable at this stage can then be considered for salvage by impregnation, welding, or other methods, depending on the type of flaw present and the end use of the casting. More complex castings usually undergo visual and dimensional inspection after the removal of gates and risers.

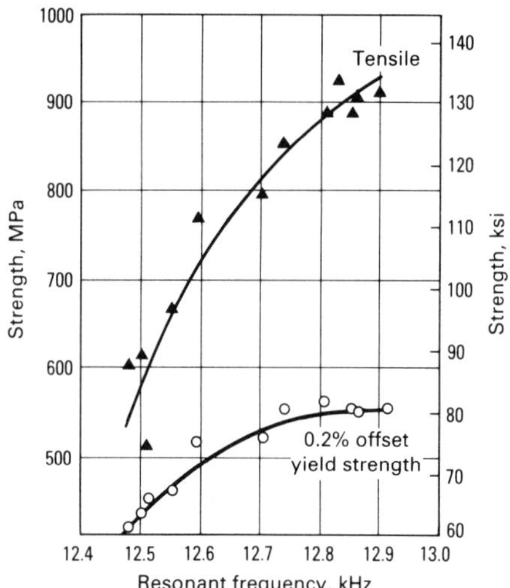

Fig. 20 Strength versus resonant frequency in the nondestructive evaluation of ductile iron test bars

Final inspection establishes the quality of the finished casting through the use of any of the methods previously mentioned. Visual inspection also includes the final measurement and comparison of specified and actual dimensions. Dimensions of castings from a large production run can be checked with gages, jigs, fixtures, or coordinate measuring systems.

Liquid penetrant inspection is extensively used as a visual aid for detecting surface flaws in aluminum alloy castings. Liquid penetrant inspection is applicable to castings made from all the aluminum casting alloys as well as castings produced by all methods. One of the most useful applications, however, is the inspection of small castings produced in permanent molds from alloys such as 296.0, which are characteristically susceptible to hot cracking. For example, in cast connecting rods, hot shortness may result in fine cracks in the shank sections. Such cracks are virtually undetectable by unaided visual inspection, but are readily detectable by liquid penetrant inspection.

All the well-known liquid penetrant systems (that is, water-washable, postemulsifiable, and solvent-removable) are applicable to the inspection of aluminum alloy castings. In some cases, especially for certain high-integrity castings, more than one system can be used. Selection of the system is primarily based on the size and shape of the castings, surface roughness, production quantities, sensitivity level desired, and available inspection facilities.

Pressure testing is used for castings that must be leaktight. Cored-out passages and internal cavities are first sealed off with special fixtures having air inlets. These inlets are used to build up the air pressure on the inside of the casting. The entire casting is then immersed in a tank of water, or it is covered by a soap solution. Bubbles will mark any point of air leakage.

Radiographic inspection is a very effective means of detecting such conditions as cold shuts, internal shrinkage, porosity, core shifts, and inclusions in aluminum alloy castings. Radiography can also be used to measure the thickness of specific sections. Aluminum alloy castings are ideally suited to examination by radiography because of their relatively low density; a given thickness of aluminum alloy can be penetrated with about one-third the power required for penetrating the same thickness of steel.

Aluminum alloy castings are most often radiographed with an x-ray machine, using film to record the results. Real-time (digital) radiography and computed tomography are also widely used and are best suited to detecting shrinkage, porosity, and core shift (Fig. 12 and 13). Gamma-ray radiography is also satisfactory for detecting specific conditions in aluminum castings. Although the γ-ray method is used to a lesser extent than the x-ray method, it is about equally as effective for detecting flaws or measuring specific conditions. Aluminum alloy castings are most often radiographed to detect approximately the same types of flaws that may exist in other types of castings, that is, conditions such as porosity or shrinkage, which register as low-density spots or areas and appear blacker on the film or real-time image screen than the areas of sound metal.

Aluminum ingots may contain hidden internal cracks of varying dimensions. Depending on size and location, these cracks may cause an ingot to split during mechanical working and thermal treatment, or they may appear as a discontinuity in the final wrought product. Once the size and location of such cracks are determined, an ingot can be scrapped, or sections free from cracks can be sawed out and processed further. Because the major dimensions of the cracks are along the casting direction, they present good reflecting surfaces for sound waves traveling perpendicular to the casting direction. Therefore, ultrasonic methods using a wave frequency that gives adequate penetration into the ingot provide excellent sensitivity for 100% inspection of that part of the ingot containing critical cracks. Because of ingot thickness (up to 400 mm, or 16 in.) and the small metal separation across the crack, radiographic methods are impractical for inspection.

Ultrasonic Inspection. Aluminum alloy castings are sometimes inspected by ultrasonic methods to evaluate internal soundness or wall thickness. The principal uses of ultrasonic inspection for aluminum alloy castings include the detection of porosity in castings and internal cracks in ingots.

Inspection of Copper and Copper Alloy Castings

The inspection of copper and copper alloy castings is generally limited to visual and liquid penetrant inspection of the surface, along with radiographic inspection for internal discontinuities. In specific cases, electrical conductivity tests and ultrasonic inspection can be applied, although the usual relatively large cast grain size could prevent a successful ultrasonic inspection.

Visual inspection is simple yet informative. A visual inspection would include significant dimensional measurements as well as general appearance. Surface discontinuities often indicate the presence of internal discontinuities.

For small castings produced in reasonable volume, a destructive metallographic inspection on randomly selected samples is practical and economical. This is especially true on a new casting for which foundry practice has not been optimized and a satisfactory repeatability level has not been achieved.

For castings of some of the harder and stronger alloys, a hardness test is a good means of estimating the level of mechanical properties. Hardness tests are of less value for the softer tin bronze alloys because hardness tests do not reflect casting soundness and integrity.

Because copper alloys are nonmagnetic, magnetic particle inspection cannot be used to detect surface cracks. Instead, liquid penetrant inspection is recommended. Ordinarily, liquid penetrant inspection requires some prior cleaning of the casting to highlight the full detail.

For the detection of internal defects, radiographic inspection is recommended. Radiographic methods and standards are well established for some copper alloy castings (for example, ASTM E 272 and E 310).

As a general rule, the method of inspection applied to some of the first castings made from a new pattern should include all those methods that provide a basis for judgment of the acceptability of the casting for the intended application. Any deficiencies or defects should be reviewed and the degree of perfection defined. This procedure can be repeated on successive production runs until repeatability has been ensured.

Gas Porosity. Copper and many copper alloys have a high affinity for hydrogen, with an increasing solubility as the temperature of the molten bath is increased. Conversely, as the metal cools in the mold, most of this hydrogen is rejected from the metal. Because all the gas does not necessarily escape to the atmosphere and may become entrapped by the solidifying process, gas porosity may be found in the casting.

In most alloys, gas porosity is identified by the presence of voids that are relatively spherical and are bright and shiny inside. Visible upon sectioning or by radiography, they may either be small, numerous, and rather widely dispersed or fewer in number and relatively large. Regardless of size, they are seldom interconnected except in some of the tin bronze alloys, which solidify in a very dendritic mode. In these alloys, the gas porosity tends to be distributed in the interstices between the dendrites.

Shrinkage voids caused by the change in volume from liquid to solid in copper alloys are different only in degree and possibly shape from those found in other metals and alloys. All nonferrous metals exhibit this volume shrinkage when solidifying from the molten condition.

Shrinkage voids may be open to the air when near or exposed to the surface, or they may be deep inside the thicker sections of the casting. They are usually irregular in shape, compared to gas-generated defects, in that their shape frequently reflects the internal temperature gradients induced by the external shape of the casting.

Hot Tearing. The tin bronzes as a class, as well as a few of the leaded yellow brasses, are susceptible to hot shortness; that is, they lack ductility and strength at elevated temperature. This is significant in that tearing and cracking can take place during cooling in the mold because of mold or core restraint. In aggravated instances, the resulting hot tears in the part appear as readily visible cracks. Sometimes, however, the cracks are not visible externally and are not detectable until after machining. In extreme cases, the cracks become evident only through field failure because the tearing was deep inside the casting.

Nonmetallic inclusions in copper alloys, as with all molten alloys, are normally the result of improper melting and/or pouring conditions. In the melting operation, the use of dirty remelt or dirty crucibles, poor furnace linings, or dirty stirring rods can introduce nonmetallic inclusions into the melt. Similarly, poor gating design and pouring practice can produce turbulence and can generate nonmetallic inclusions. Sand inclusions may also be evident as the result of improper sand and core practice. All commercial metals, by the nature of available commercial melting and molding processes, usually contain very minor amounts of small nonmetallic inclusions. These have little or no effect on the casting. Inclusions of significant size or number are considered detrimental. A thorough review of copper alloy melting, refining, and casting practices is available in Volume 15 of the 9th Edition of *Metals Handbook*.

REFERENCES

1. P.J. Rickards, Progress in Guaranteeing Quality Through Nondestructive Methods of Evaluation, *Foundryman Int.*, April 1988, p 196-209
2. P.M. Bralower, Nondestructive Testing. Part I. The New Generation in Radiography, *Mod. Cast.*, Vol 76 (No. 7), July 1986, p 21-23
3. R.A. Armistead, CT: Quantitative 3-D Inspection, *Adv. Mater. Process.*, Vol 133 (No. 3), March 1988, p 42-48
4. A.G. Fuller, P.J. Emerson, and G.F. Sergeant, A Report on the Effect Upon Mechanical Properties of Variation in Graphite Form in Irons Having Varying Amounts of Ferrite and Pearlite in the Matrix Structure and the Use of Nondestructive Tests in the Assessments of Mechanical Properties of Such Irons, *Trans. AFS*, Vol 88, 1980, p 21-50
5. A.G. Fuller, Evaluation of the Graphite Form in Pearlitic Ductile Iron by Ultrasonic and Sonic Testing and the Effect of Graphite Form on Mechanical Properties, *Trans. AFS*, Vol 85, 1977, p 509-526
6. P.J. Rickards, "Progress in Guaranteeing Quality Through Non-Destructive Methods of Evaluation," Paper 21, presented at the 54th International Foundry Congress, New Delhi, The International Committee of Foundry Technical Associations (CIATF), Nov 1987
7. A.G. Fuller, Nondestructive Assessment of the Properties of Ductile Iron Castings, *Trans. AFS*, Vol 88, 1980, p 751-768

Powder Metallurgy Parts*

R.C. O'Brien and W.B. James, Hoeganaes Corporation

THE PROBLEM of forming defects in green parts during compaction and ejection has become more prevalent as parts producers have begun to use higher compaction pressures in an effort to achieve high-density, high-performance powder metallurgy (P/M) steels. In this article, several nondestructive inspection methods are evaluated, with the aim of identifying those that are practical for detecting defects as early as possible in the production sequence.

The most promising nondestructive testing methods for P/M applications include electrical resistivity testing, eddy current and magnetic bridge testing, magnetic particle inspection, ultrasonic testing, x-ray radiography, gas permeability testing, and γ-ray density determination. The capabilities and limitations of each of the techniques are evaluated in this article.

Current Status of P/M Testing

In the ceramics industry, the fraction of the finished-part cost that arises from scrap due to flaws introduced during processing is estimated to average 50%, and it can be as high as 75% (Ref 1). Although the ceramics industry has been mobilized for the past 15 years toward the use of nondestructive evaluation in processing, the P/M industry has built up only a scattered background of experience (Ref 2).

To remain competitive, P/M parts producers have increasingly turned to simplified processing. It has been shown that the physical properties of P/M parts, especially the fatigue strength, are always improved by increasing the density (Ref 3). The need for densification by double pressing can often be avoided by pressing to high density in a single step. However, the use of higher compaction pressures requires the utmost attention to materials selection, tool design, and press setup (Ref 4). A quick, preferably nondestructive method of crack detection would be of great benefit during press setup and for testing the integrity of parts as early as possible in their production sequence.

The growth of nondestructive testing in the 1980s has been explosive, and the field has benefited greatly from computerized image reconstruction techniques applied to radiography, ultrasonic, and even magnetic particle inspection. Commercial test systems are being marketed as fast as the technology is developed, and the metal powder industry should find solutions to its on-line testing needs by reviewing methods being used by other parts fabrication technologies.

In preparing this article, a number of test methods presented themselves as having potential for crack detection in green (unsintered) P/M compacts, and these are recommended for further investigation. For a detailed overview of P/M technology, the reader is referred to Volume 7 of the 9th Edition of *Metals Handbook*.

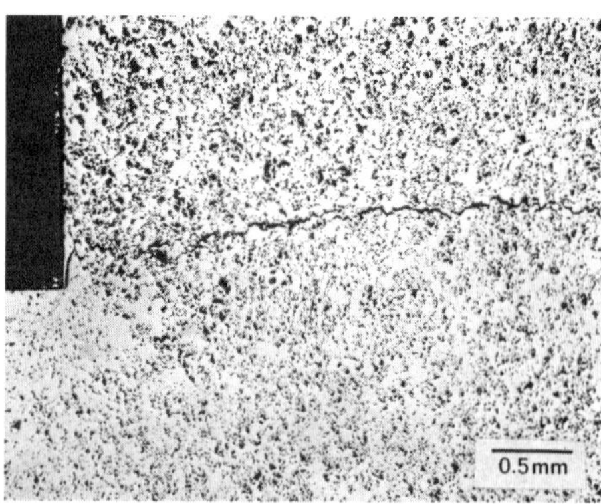

Fig. 1 Ejection crack in sintered P/M steel. Unetched

Summary of Defect Types in P/M Parts

The four most common types of defects in P/M parts are ejection cracks, density variations, microlaminations, and poor sintering.

Ejection Cracks. When a part has been pressed, there is a large residual stress in the part due to the constraint of the die and punches, which is relieved as the part is ejected from the die. The strain associated with this stress relief depends on the compacting pressure, the green expansion of the material being compacted, and the rigidity of the die. Green expansion, also known as spring out, is the difference between the ejected-part size and the die size. A typical value of green expansion for a powder mix based on atomized iron powder pressed at relatively high pressure (600 to 700 MPa, or 45 to 50 tsi) is 0.20%. In a partially ejected compact, for example, the portion that is out of the die expands to relieve the residual stress, while the constrained portion remains die size and a shear stress is imposed on the compact. When the ability of the powder compact to accommodate the shear stress is exceeded, ejection cracks such as the one shown in Fig. 1 are formed.

The radial strain can be alleviated to a degree by increasing the die rigidity and designing some release into the die cavity. However, assuming that the ejection punch motions are properly coordinated, the successful ejection of multilevel parts depends to a large degree on the use of a high-quality powder that combines high green strength with low green expansion and low stripping pressure.

Density Variations. Even in the simplest tool geometry possible—a solid circular cylinder—conventional pressing of a part to an overall relative density of 80% will result in a distribution of density within the part ranging from 72 to 82% (Ref 5). The addition of simple features such as a central hole and gear teeth presents minor problems compared with the introduction of a step or second level in the part. Depending on the severity of the step, a separate, independently actuated punch can be required for each level of the part. During the very early stage of compaction, the powder redistributes itself by flowing between sections of the die cavity. However, when the pressure increases and the powder movement is restricted, shearing of the compact in planes parallel to the punch axis can only be avoided by proper coordination of punch motions. When such shear exists, a density gradient results.

The density gradient is not always severe enough for an associated crack to form upon ejection. However, a low-density area around an internal corner, as shown in Fig. 2, can be a fatal flaw, because this corner is

*Reprinted with permission from *Prevention and Detection of Cracks in Ferrous P/M Parts*, Metal Powder Industries Federation, 1988, p 25-56

Table 1 Comparison of the applicability of various nondestructive evaluation methods to flaw detection in P/M parts

Method	Measured/detected	Applicability to P/M parts(a) Green	Sintered	Advantages	Disadvantages
X-ray radiography	Density variations, cracks, inclusions	C	C	Can be automated	Relatively high initial cost; radiation hazard
Computed tomography	Density variations, cracks, inclusions	C	C	Can be automated; pinpoint defect location	Extremely high initial cost; highly trained operator required; radiation hazard
Gamma-ray density determination	Density variations	A	A	High resolution and accuracy; relatively fast	High initial cost; radiation hazard
Ultrasonic imaging: C-scan	Density variations, cracks	D	B	Sensitive to cracks; fast	Coupling agent required
Ultrasonic imaging: SLAM	Density variations, cracks	D	C	Fast; high resolution	High initial cost; coupling agent required
Resonance testing	Overall density, cracks	D	B	Low cost; fast	Does not give information on defect location
Acoustic emission	Cracking during pressing and ejection	C	D	Low cost	Exploratory
Thermal wave imaging	Subsurface cracks, density variations	D	C	No coupling agent required	Flat or convex surfaces only
Electrical resistivity	Subsurface cracks, density variations, degree of sinter	A	A	Low cost, portable, high potential for use on green compacts	Sensitive to edge effects
Eddy current/magnetic bridge	Cracks, overall density, hardness, chemistry	C	A	Low cost, fast, can be automated; used on P/M valve seat inserts	Under development;
Magnetic particle inspection	Surface and near-surface cracks	C	A	Simple to operate, low cost	Slow; operator sensitive
Liquid dye penetrant inspection	Surface cracks	C	D	Low cost	Very slow; cracks must intersect surface
Pore pressure rupture/gas permeability	Laminations, ejections, cracks, sintered density variations	A	A	Low cost, simple, fast	Gas-tight fixture required; cracks in green parts must intersect surface

(a) A, has been used in the production of commercial P/M parts; B, under development for use in P/M; C, could be developed for use in P/M, but no published trials yet; D, low probability of successful application to P/M

usually a point of stress concentration when the part is loaded in service.

Microlaminations. In photomicrographs of unetched part cross sections, microlaminations such as those shown in Fig. 3 appear as layers of unsintered interparticle boundaries that are oriented in planes normal to the punch axis. They can be the result of fine microcracks associated with shear stresses upon ejection; such microcracks fail to heal during sintering. Because of their orientation parallel to the tensile axis of standard test bars, they have little influence on the measured tensile properties of the bars, but are presumed to be a cause of severe anisotropy of tensile properties.

Poor Sintering. When unsintered particle boundaries result from a cause other than shear stresses, they are usually present because of insufficient sintering time or sintering temperature, a nonreducing atmosphere, poor lubricant burn-off, inhibition of graphite dissolution, or a combination of these. A severe example is shown in Fig. 4. Unlike microlaminations, defects associated with a poor degree of sintering are not oriented in planes.

Nondestructive Tests and Their Applicability to P/M Processing

As described below, applicable inspection methods for P/M parts can be broadly classified into the following categories:

- Radiographic techniques
- Acoustic methods
- Thermal inspection
- Electrical resistivity inspection
- Visual inspection and pressure testing

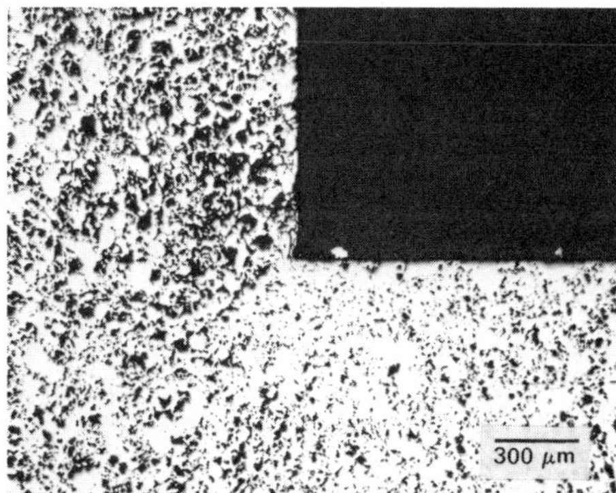

Fig. 2 Density gradient around an internal corner in a part made with a single-piece stepped punch. Unetched

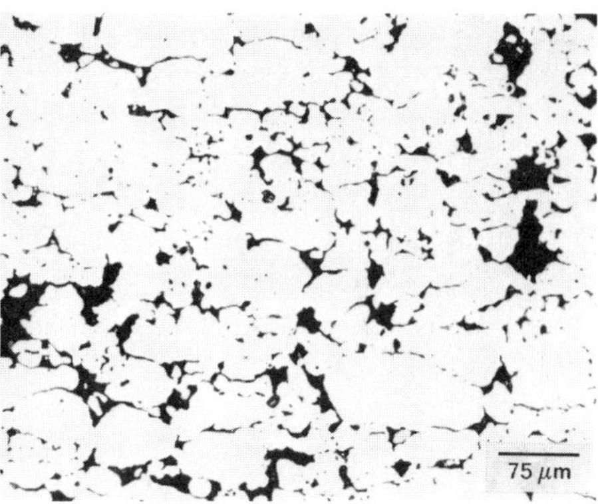

Fig. 3 Microlaminations in sintered P/M steel. Unetched

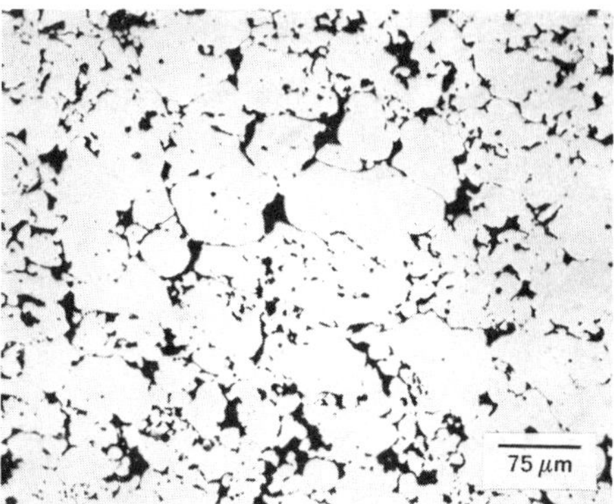

Fig. 4 Poor degree of sintering in P/M compact. Unetched

The techniques covered in this article are summarized in Table 1. Additional information on these procedures can be found in the Section "Methods of Nondestructive Evaluation" in this Volume.

Radiographic Techniques

X-Ray Radiography. Any feature of a part that either reduces or increases x-ray attenuation will be resolvable by x-ray radi-

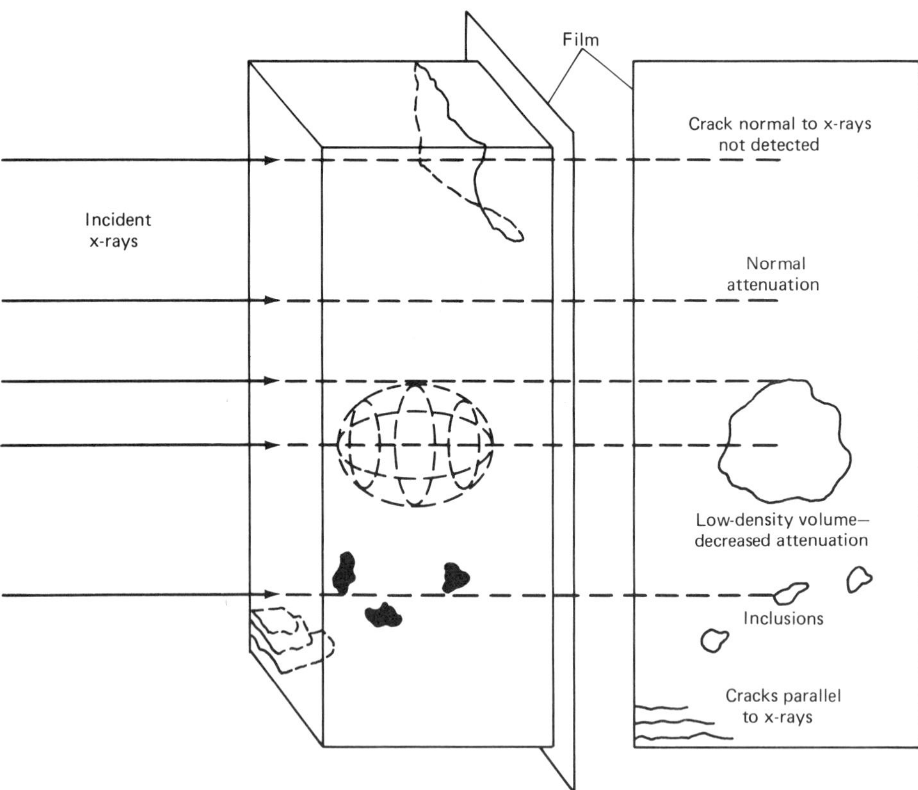

Fig. 5 Schematic of flaws and their x-ray images. Defect types that can be detected by x-ray radiography are those that change the attenuation of the transmitted x-rays. Source: Ref 6

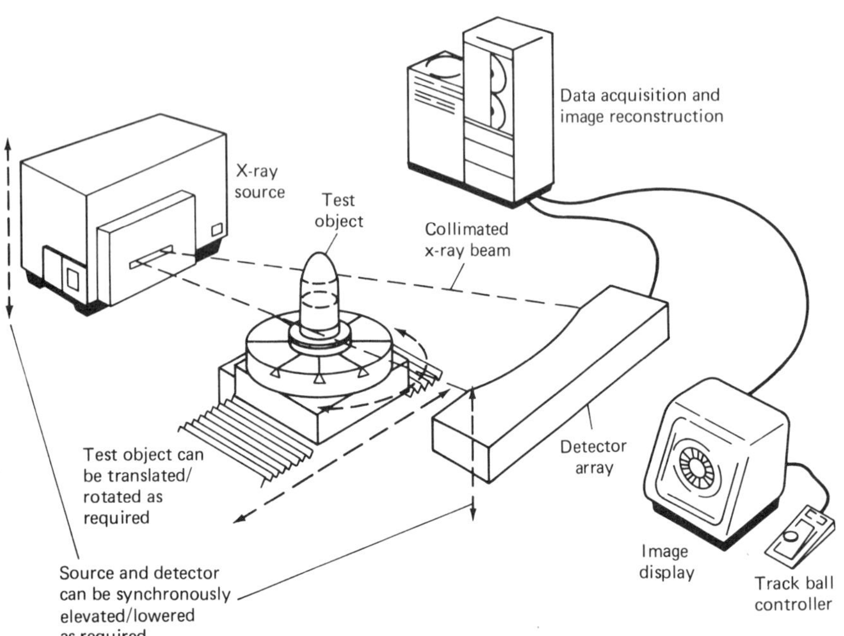

Fig. 6 Schematic of computed tomography, which is the reconstruction by computer of a series of tomographic planes (slices) of an object. The transmitted intensity of the fan-shaped beam is processed by computer and the resulting image is displayed on a terminal. Source: Ref 7

ography. Some types of flaws and their x-ray images are shown in Fig. 5.

The ability to detect defects depends on their orientation to the x-ray source. A crack parallel to the x-rays will result in reduced attenuation of the rays, and the x-ray film will be darker in this region. A thin crack perpendicular to the x-ray will hardly influence attenuation and will not be detected.

Historically, flaw detection by x-ray radiography has been an expensive and cumbersome process suited only to safety-critical and high added value parts. The process has been considerably improved by the development of real-time imaging techniques that replace photographic film. Real-time imaging means that parts can be tested rapidly and accepted or rejected on the spot.

Real-time x-ray systems include image intensifiers or screens that convert x-rays into visible light and discrete detector arrays that convert x-rays into electronic signals (which are reconstructed by computer for video display). The image in all these systems can be recorded and digitized for image enhancement. The ability of the system to detect flaws is, however, still sensitive to defect orientation. Additional information is available in the article "Radiographic Inspection" in this Volume.

Computed tomography is a recently developed version of x-ray radiography that includes highly sophisticated analysis of the detected radiation. A tomographic setup consists of a high-energy photon source, a rotation table for the specimen, a detector array, and the associated data analysis and display equipment, as shown in Fig. 6. The ability to rotate the specimen increases the chance of orienting a defect relative to the x-rays such that it will be detected. The x-ray source and detector array can be raised or lowered to examine different planes through the sample.

In a typical system, the photon source can be a radioisotope such as ^{60}Co, depending on the energy requirements of the individual specimens. The lead aperture around the source acts as a collimator to produce a fan-shaped beam about 5 mm (0.2 in.) thick. The sample is rotated in incremental steps, and the transmitted radiation is detected at each step by computer-controlled detectors situated one every 14 mm (0.55 in.) in a two-dimensional array.

The computer then reconstructs the object using intensity data from a number of scans at different orientations. The output is in the form of a two-dimensional plan in which colors are mapped onto the image according to the intensity of the transmitted radiation. The resolution available depends on the difference in density between the various features of the object. For example, steel pins embedded in polyvinyl chloride plastic are more easily resolvable than aluminum pins of the same diameter (Ref 7). Experiments with P/M samples have shown that density can be measured to better than 1% accuracy, with a spatial resolution of 1 mm (0.04 in.) (Ref 8). The article "Industrial Computed Tomography" in this Volume contains more information on the principles and applications of this technique.

Gamma-Ray Density Determination. Local variations in the density of P/M parts have been detected by measuring the attenuation of γ-rays passing through the part (Ref 9). Depending on the material and the dimensions of the part, density can be measured to an accuracy of ±0.2 to ±0.7%, and

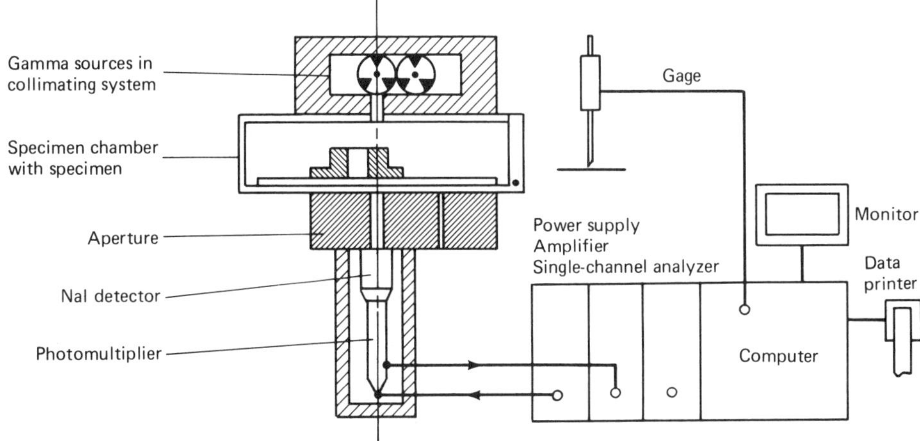

Fig. 7 Schematic of the Gamma Densomat. Source: Ref 9

the technique has been used by P/M parts fabricators in place of immersion density tests as an aid in tool setting.

The apparatus consists of a vertically collimated γ-ray beam originating from a radioisotope. The beam passes through the sample as shown in Fig. 7 and reaches a detector via a 1 mm (0.04 in.) diam aperture, where the transmitted intensity is measured. The detector consists of a sodium iodide scintillation crystal, which in turn excites a photomultiplier. Exposure time is 1 to 2 min; a 4 mm (0.15 in.) aperture can reduce this time to 30 s at the expense of some resolution. The radiation source of the Gamma Densomat is Americium 241 (60 keV). For high-energy beams, Cesium 137 (660 keV) can be substituted.

This method has been shown to be particularly useful in cases where the section of the part to be checked is too small for immersion density measurements (Ref 10). Tool life was extended when the method was used for part density checks in order to avoid overloading.

Acoustic Methods

Ultrasonic Testing. Many characteristics of solids can be determined from the behavior of sound waves propagating in them. Ultrasonic signals impinged on a sample at one surface are transmitted at speeds and attenuated at rates determined by the density, modulus of elasticity, and continuity of the material. The sound waves are reflected from other surfaces of the sample, including cracks as well as free surfaces. They can be picked up and amplified for display on a CRT screen, as described in Ref 6 and the article "Ultrasonic Inspection" in this Volume. The height and position of the flaw/defect peak indicate its size and location. Although much effort has been directed toward relating sound propagation to the physical properties of P/M materials, little has been written on the detection of cracks by the ultrasonic inspection of porous materials.

Ultrasound Transmission in Green Compacts. The characterization of green compacts by ultrasonic techniques appears to be hindered by problems of extreme attenuation of the incident signal. In one case, signals of 1 to 20 MHz were transmitted through an 8 mm (0.3 in.) thick compact of atomized iron with 0.2% graphite added (Ref 11). Although the attenuation did not allow back-wall echo measurement, through-transmission measurements indicated that the transmitted intensity had a maximum at 4 to 5 MHz (Fig. 8). Density was found to influence the transmitted intensity, with specimens at 95% relative density allowing some degree of transmission over the entire range of frequencies tested, while specimens at 87% relative density damped the incident signals entirely.

Figure 9 shows that the velocity of ultrasonic waves in green compacts is about half the velocity in sintered compacts and that it is essentially invariant with density (Ref 13). It has also been shown that the velocity of ultrasound in green parts is highly anisotropic and that the experimental reproducibility is very poor (Fig. 10). It has been proposed that the anisotropy in velocity is due to the orientation of porosity (Ref 15).

The variation in the velocity of ultrasound with applied pressure during the compaction of ceramic powders has been measured *in situ* by fixing transducers to the ends of the punches (Ref 16). Unlike the case of finished green P/M compacts, a clear relationship was found between longitudinal wave velocity and compacting pressure (Fig. 11), probably because the constraint of the punches and die forced the individual particles together, providing an efficient acoustic coupling between particles.

Ultrasound Transmission in Sintered Parts. Early work relating the physical properties of cast iron to the velocity of sound waves suggested the potential for evaluating P/M steels in the same way (Ref 17). As expected, both the velocity of sound in P/M parts and their resonant frequencies have been related to density, yield strength, and tensile strength. Plain carbon steel P/M specimens were used in one series of tests

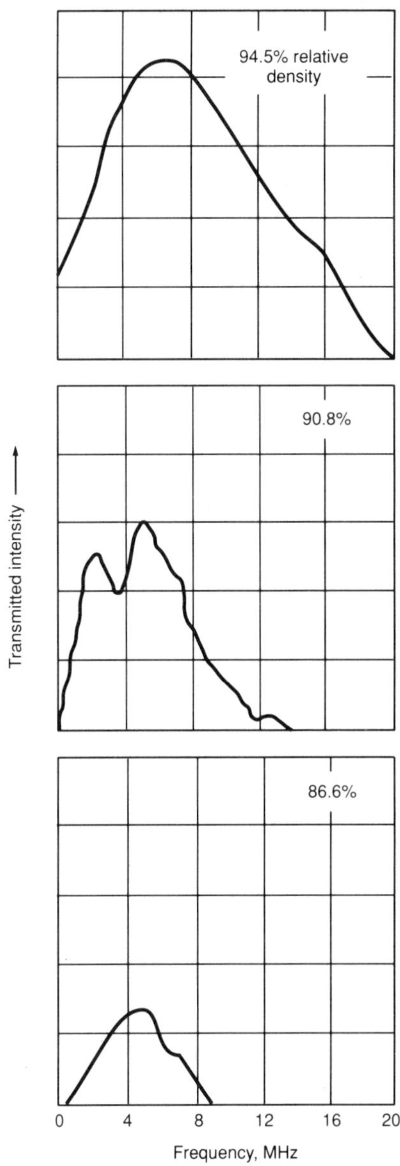

Fig. 8 Ultrasonic spectrum analyzer output showing change in transmitted intensity with density of green compact. Source: Ref 11

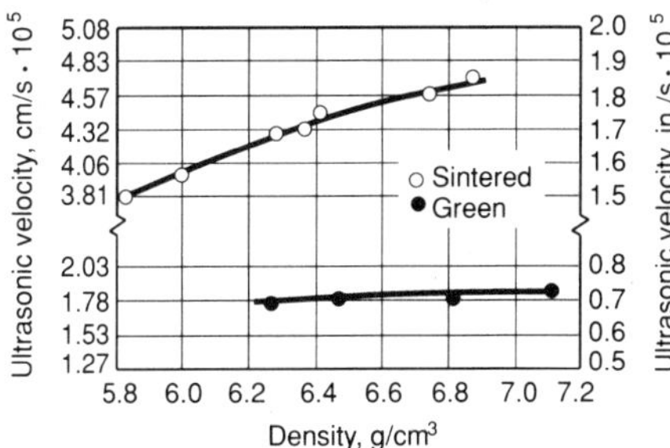

Fig. 9 Effect of density on ultrasonic velocity in green and sintered cylindrical Ancorsteel 1000-B specimens. Source: Ref 12

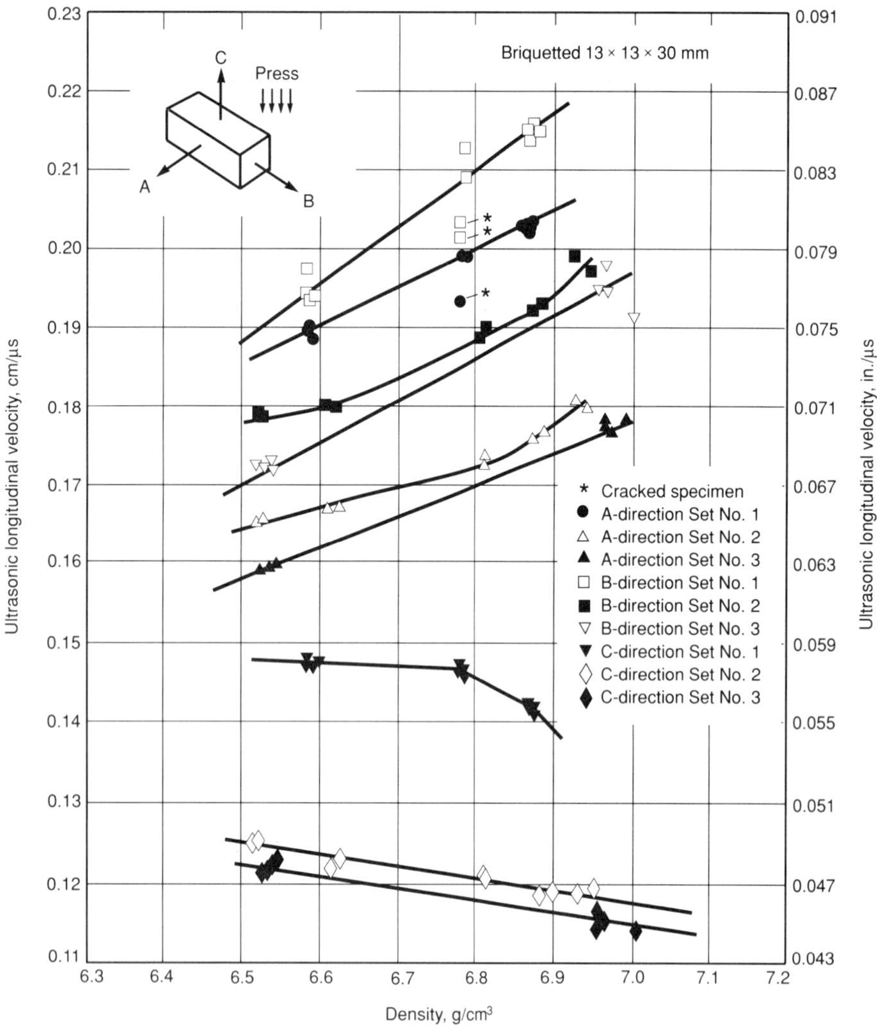

Fig. 10 Anisotropy of ultrasound in green transverse rupture strength bars. Source: Ref 14

and the correlation was found to be close enough for the test to be used as a quick check for the degree of sintering in production P/M parts (Ref 12). Other work has demonstrated the relationship between sound velocity and tensile strength in porous parts (Fig. 12). The same types of relationships have also been documented in powder forgings (Ref 19).

Sintered parts have been found to transmit ultrasound according to the relationships shown in Fig. 13. The highest wave velocities occurred in the pressing direction. An additional distinction was found between the velocities in the longitudinal and lateral axes of an oblong specimen, and these results were shown to be reproducible between different powder lots and specimen groups. The anisotropy of velocity diminished at higher densities and disappeared above 6.85 g/cm^3.

Ultrasonic Imaging: C-Scan. The C-Scan is a form of ultrasonic testing in which the testpiece is traversed by the ultrasound transducer in a computer-controlled scan protocol (Fig. 14). The transmitted intensity is recorded and analyzed by computer, and a gray-mapped image is output.

In one trial, seeded oxide inclusions were detected in porous sintered steels using a C-Scan (Ref 21). The inclusions consisted of admixed particles of chromium oxide and alumina at concentrations of 65 to 120 particles per square centimeter. Inclusions as small as 50 µm in diameter were detected. Additional information on the C-Scan can be found in the articles "Ultrasonic Inspection" and "Adhesive-Bonded Joints" in this Volume.

Ultrasonic Imaging: Scanning Acoustic Microscopy (SAM). Ultrasonic waves can be focused on a point using a transducer and lens assembly, as shown in Fig. 15 and described in the article "Acoustic Microscopy" in this Volume. In this way, the volume of the specimen being examined is highly limited, so that reflections from defects can be closely located at a given depth and position in the specimen. In SAM, the specimen is moved by stepper motors in a raster pattern, and an image of the entire structure can be built up. Scanning acoustic microscopy has been shown to be capable of resolving small surface and subsurface cracks, inclusions, and porosity in sintered, fully dense ceramics (Ref 22).

Ultrasonic Imaging: Scanning Laser Acoustic Microscopy (SLAM). When a continuous plane wave impinges on a sample that is roughly flat in shape, it propagates through and is emitted from the sample with relatively little scattering, retaining its planar nature. When the plane wave is emitted from the sample, it contains information on variations in properties that were encountered in the interior of the sample, which takes the form of variations in intensity with position in the plane. A scanning laser acoustic microscope detects these variations as distortions in a plastic sheet that is placed in the path of the plane wave. The information is gathered by a laser that scans a reflective coating on one side of the sheet, as shown in Fig. 16 and explained in the article "Acoustic Microscopy" in this Volume.

Therefore, although ultrasonic testing is not appropriate for evaluating green P/M parts, it is applicable to the assessment of sintered components. Optimum results dictate careful selection and placement of the transducers because the orientation of the defects influences the ability to detect them. Small defects close to the specimen surface can be masked by surface echoes. Although enhanced image analysis techniques appear beneficial, it is unlikely that the more sophisticated techniques, such as C-Scan and SLAM, will be cost effective for most ferrous P/M parts in the near future.

Resonance Testing. When a structural part is tapped lightly, it responds by vibrating at its natural frequency until the sound is damped. Both the damping characteristics and the natural frequency change with damage to the structure. Changes in the natural frequency can be detected with a spectrum analyzer, as shown in Fig. 17.

Sintered P/M parts behave in a similar manner, and the minimum defect size that can be detected has been determined experimentally by testing the resonant frequency after milling narrow grooves of various depths in the parts (Ref 24). It was found that defects covering 2% of the cross section could always be detected and that smaller defects (down to 0.5%) could be detected under favorable conditions of part geometry. This was later shown to apply to real defects as well as machined grooves (Ref 25). There is no record of the technique having been tried on green parts. However, the extremely high sound-damping capacity of green parts would appear to preclude its use. As with ultrasonic techniques, resonance testing has been used to determine physical properties such as the elastic modulus of materials as well as their defect structure (Ref 26).

Acoustic emissions are sounds generated in a material as stored elastic energy is

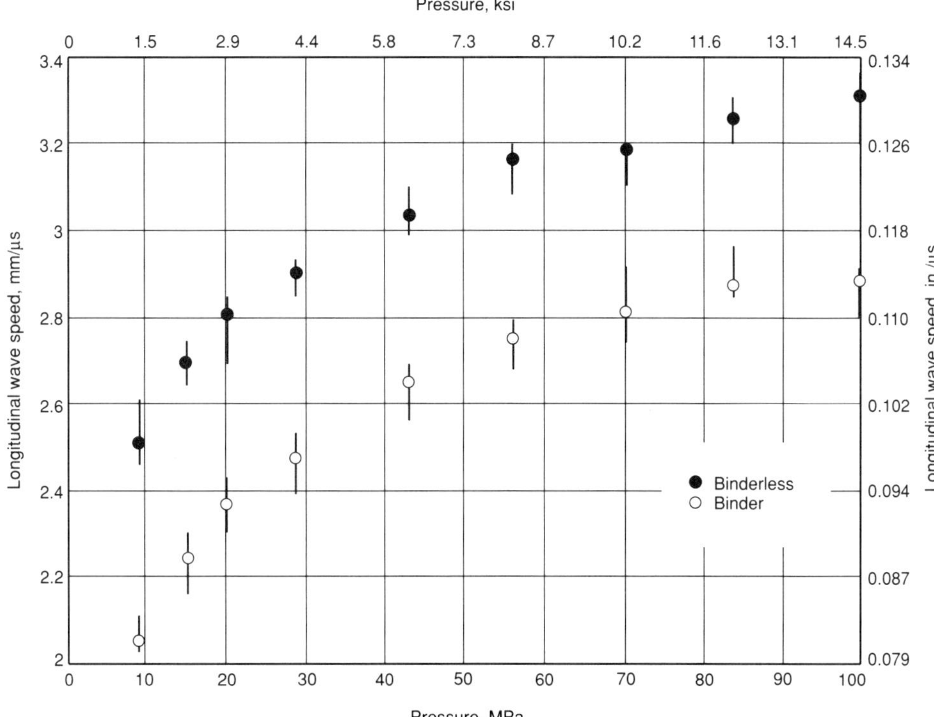

Fig. 11 Ultrasonic wave velocity in ceramic powders, measured during compaction. Source: Ref 16

released in a noncontinuous mode by mechanisms such as transformation and twinning, slip, and fracture (see the article "Acoustic Emission Inspection" in this Volume). The acoustic emission spectra have been characterized for the compressive deformation of powder-forged 4600 steels with carbon contents ranging from 0.3 to 0.9% (Ref 27).

If P/M tooling were monitored for acoustic emissions of the powder during compaction and ejection, it might be possible to distinguish emission peaks due to the release of stored energy as cracks are formed.

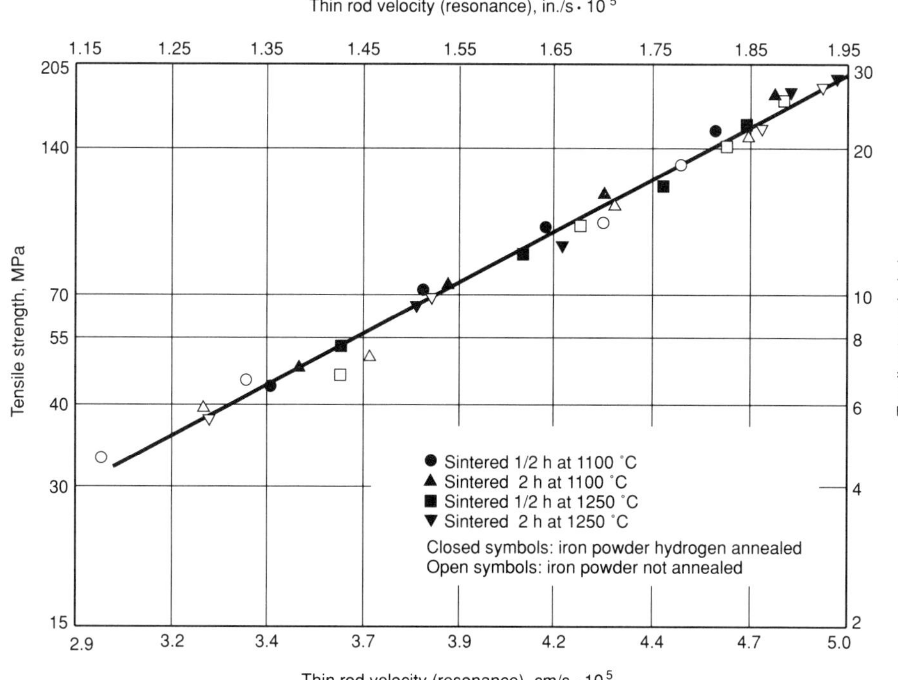

Fig. 12 Correlation of ultrasonic velocity with tensile strength of sintered steel. Source: Ref 18

However, a developmental program would be required to evaluate this concept and practical application is not anticipated.

Thermal Inspection

Thermal Wave Imaging. When a pulsed laser impinges on a surface, the rapidly alternating heating and cooling of the surface is conducted into the body of the specimen, as shown in Fig. 18. These thermal waves have been shown to possess many of the same characteristics as electromagnetic or mechanical waves. They can be reflected and refracted, they can form interference patterns, and they interact with irregularities contained in the transmitting medium. In coincidence with the thermal wave formation, acoustic waves are formed by the alternating expansion and contraction of the area of impingement of the laser on the surface. These photoacoustic waves have the same frequency as the thermal waves (typically 1 MHz) but have a much longer wavelength. They are also affected by scattering and reflection of the thermal waves in the volume immediately surrounding the laser impingement point, and it is this effect that allows detection of flaws. Thermal wave imaging has been used to detect delamination and microcracking in silicon integrated circuits (Ref 28).

Another method of detecting the interactions of thermal waves with defects is optical beam deflection, or the mirage effect (Ref 29). The impingement point of the laser on the surface heats rapidly, and the air around this point is also heated. If there are no irregularities present beneath the surface, this volume of lower-density heated air is roughly hemispherical in shape. A second laser beam that transits this low-density air volume by skimming closely parallel to the specimen surface, as shown in Fig. 19, will be refracted by the density gradient in the same way as it would be by a conventional lens. A four-quadrant detector array gathers the beam deflection data as the specimen surface is scanned by the laser. Subsurface defects are detected as changes in the shape of the density gradient "lens."

Although there is no record of thermal wave imaging having been applied to P/M parts, the damping capacity of green compacts would appear to restrict the potential application of the technique to sintered components only. Full details on the principles and applications of thermal wave imaging can be found in the article "Thermal Inspection" in this Volume.

Electrical Resistivity Testing

Direct Current Resistivity Testing. A voltage field within a conductive solid will create currents that are influenced by structural irregularities, including cracks and porosity. This characteristic has been used to measure carburized case depth in wrought

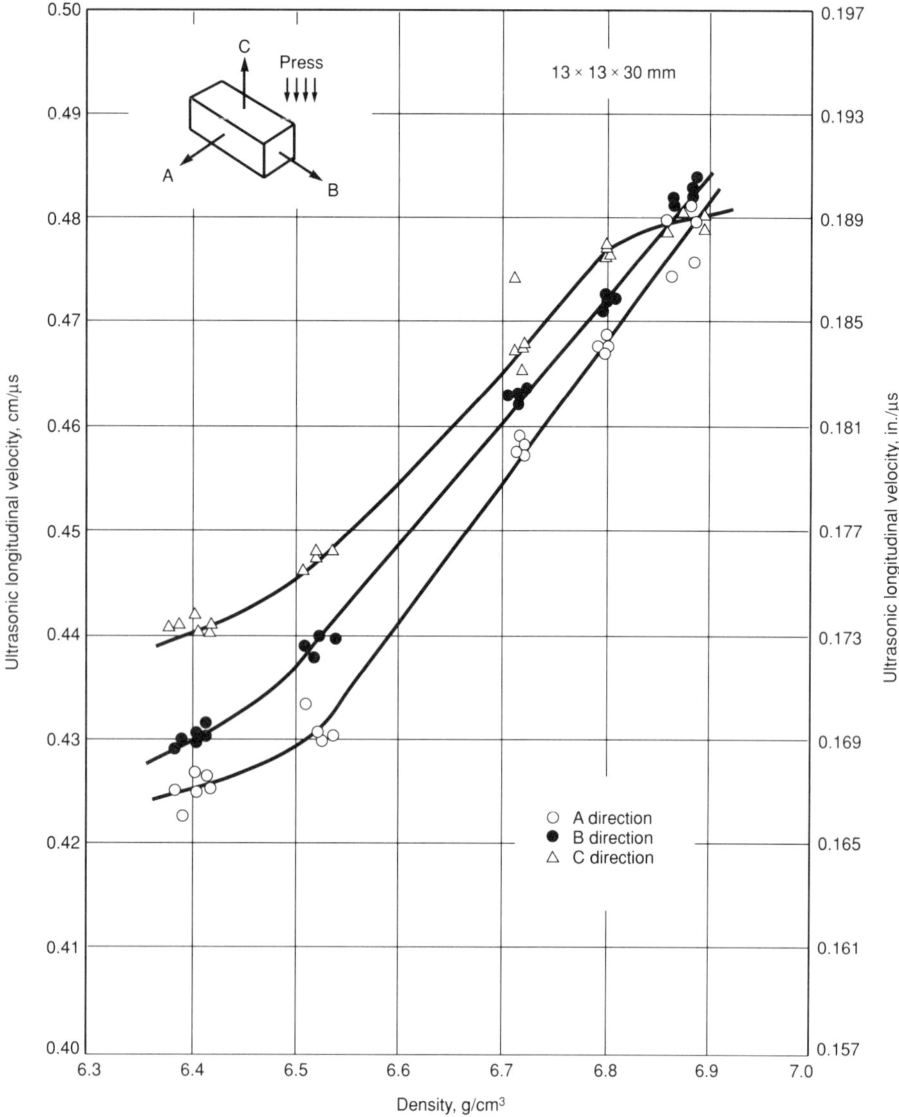

Fig. 13 Anisotropy of ultrasound velocity in sintered transverse rupture strength bars. Source: Ref 14

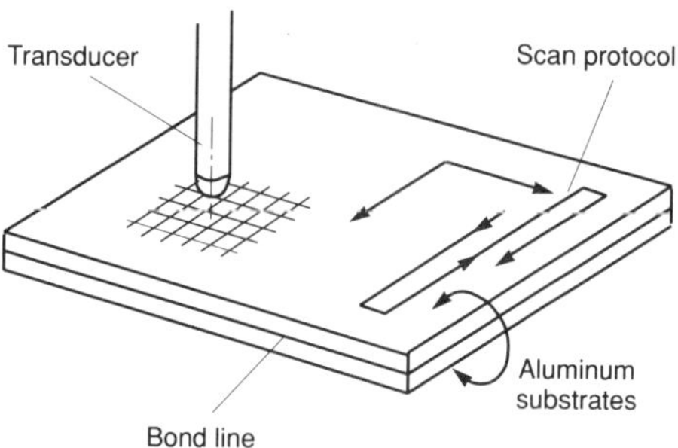

Fig. 14 Schematic of a C-Scan scanning protocol for an adhesive-bonded structure. Source: Ref 20

steels (Ref 30). The arrangement shown in Fig. 20 is used to measure the voltage drop in a current field localized between two electrode probes. This method has been used to detect seeded defects in laboratory specimens. It has also been successfully applied to the production of sintered steel parts (Ref 31), as described in Examples 1 to 3.

Although the resistivity of green compacts is an order of magnitude higher than that after sintering, the same technique has been shown to apply (Ref 30). Green-state specimens with laboratory-simulated cracks of the type shown in Fig. 21 have been subjected to resistivity inspection with encouraging results. If the probe electrodes span the plane containing the defects and if a series of measurements is made along the edge of the plane, the resistivity varies when defects are present, as shown in Fig. 22. Other tests on green parts are described in Ref 30 and 31.

There are two potential contributors to variability in the resistivity test. First, in addition to cracks, the edges and corners of the parts distort the current fields. The internal corners of parts are often the sites of green cracks. Testing the volume of material immediately underlying the corners necessitates the use of specially made electrode probe sets. Another variable influencing the resistivity inspection of green compacts is the nature of the oxide layers on the particles. When the oxide layer is altered with a thermal treatment, the resistivity of the green part decreases (Ref 32).

Another study has yielded the relative density/conductivity relationship shown in Fig. 23, suggesting that resistivity tests could be used as a rapid check for localized density variations. As with ultrasound, the elastic modulus and the toughness of porous steels can also be distinguished by resistivity checks (Ref 34).

The direct current resistivity test can be used on any conductive material; it is not limited to ferromagnetic materials. Although further development is needed, resistivity measurements appear to be one of the most promising techniques for the nondestructive evaluation of both green and sintered P/M parts. In addition to detecting cracks in green parts, as well as part-to-part density variation, studies have shown that changes in resistivity due to poor carbon pickup during sintering were also detectable (Ref 31). Resistivity testing has also been used later in the processing sequence to screen heat-treated parts for incomplete transformation to martensite. Several uses for resistivity testing are given in the following examples (Ref 31).

Example 1: Automotive Air Conditioner Compressor Part. The resistivity-measuring equipment and hand-held probe are shown in Fig. 24. The part, shown in Fig. 25, was tested for green cracking in the locations marked in Fig. 25(b), which were suspect because of prior experience. The parts could then be sorted for cracks by comparing the measured resistivity with limiting resistivity values that had previously been determined using parts with cracks indicated by magnetic particle testing. The prior test method consisted of sintering, sectioning, and magnetic particle inspection, a 2-h process. This part was also the subject of a series of experiments demonstrating that the resistivity test method had high reproducibility and was not operator sensitive.

Example 2: Automotive Transmission Spacer. The resistivity test was used to screen the parts shown in Fig. 26 for incomplete transformation to martensite upon heat treating. The test is based on the lower resistivity of pearlitic microstructures compared with martensitic microstructures of the same chemistry. To determine a resistivity criterion for the screening of these parts, resistivity was correlated with hardness measurements. A resistivity of 60 μΩ · cm was associated with a hardness of 30

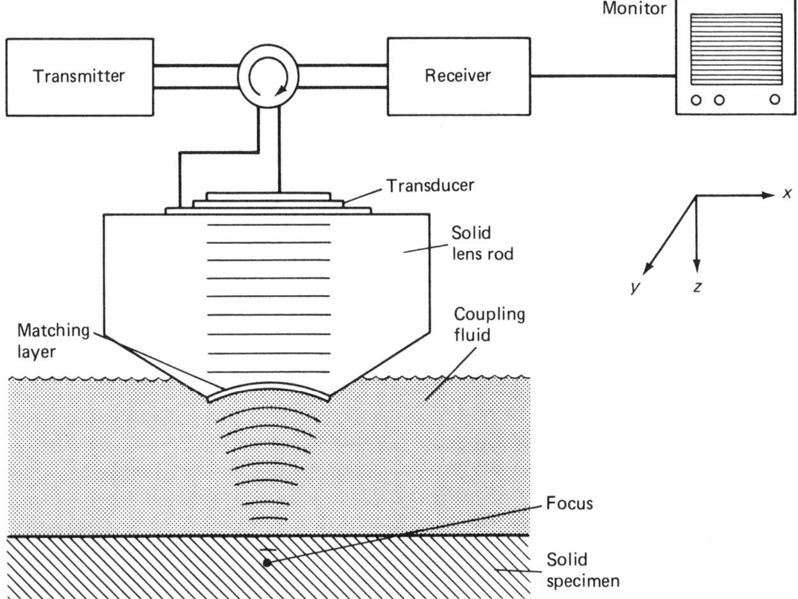

Fig. 15 Schematic of the image-forming process in scanning acoustic reflecting microscopy. Source: Ref 22

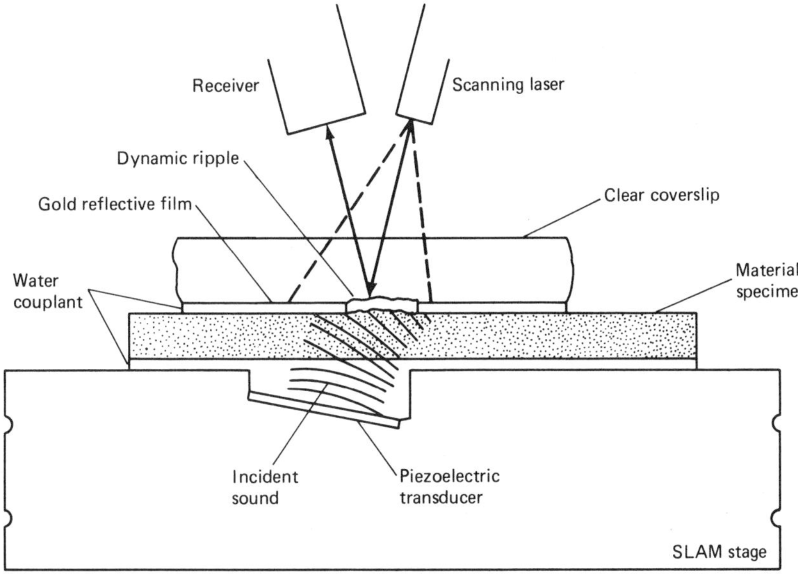

Fig. 16 General configuration used in scanning laser acoustic microscopy. Source: Ref 23

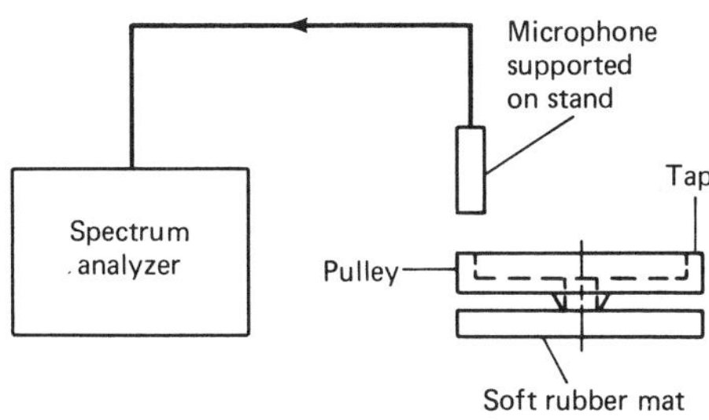

Fig. 17 Schematic of resonance test configuration. Source: Ref 24

HRC, and a go/no-go test strategy was used. The prior test methods for this part were hardness measurements and metallography.

Example 3: Automatic Transmission Clutch Plate. The part, shown in Fig. 27, was pressed, sintered, and sized. The resistivity test was then used to screen for part-to-part density variations to levels below 6.8 g/cm^3, which was shown to be a minimum density level for achieving the radial crush strength specification for the part. Again, a limiting resistivity value was determined for the part; resistivity values below 27.5 $\mu\Omega \cdot$ cm were considered acceptable.

Eddy Current Testing. Another form of resistivity testing is the eddy current test. In this test, instead of producing currents in the part by direct contact with electrodes, eddy currents are induced in the part by an alternating electromagnetic field from an induction coil, as described in the article "Eddy Current Inspection" in this Volume.

Single-Coil Tests. Disruptions in the eddy current path due to any defect that changes the resistivity of the material are detected as extraneous induced voltages in the induction coil. Alternatively, a separate detector

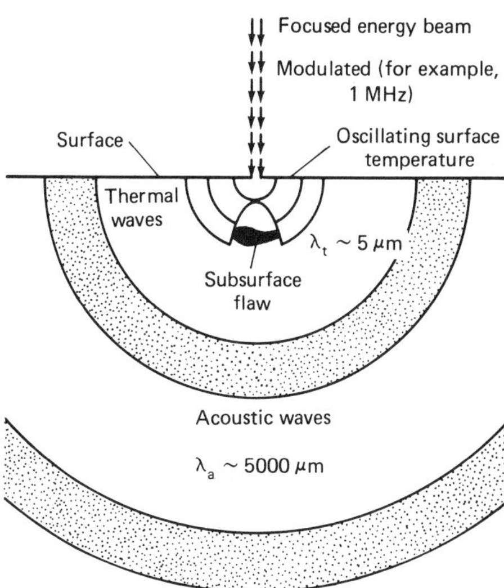

Fig. 18 Transmission of thermal and acoustic waves in thermal wave imaging. Source: Ref 28

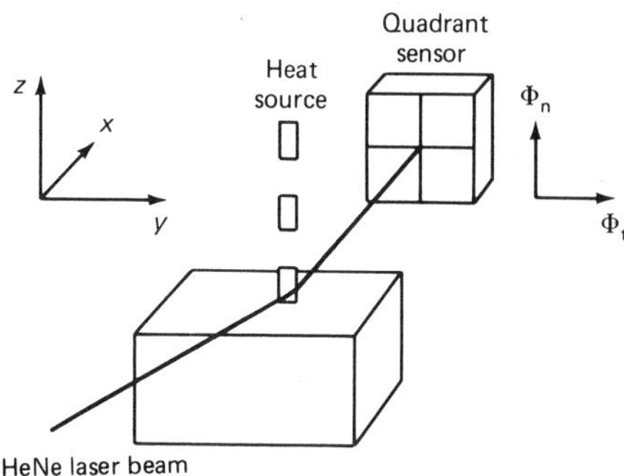

Fig. 19 Detection of interactions between thermal waves and flaws by optical beam deflection (mirage effect). Source: Ref 29

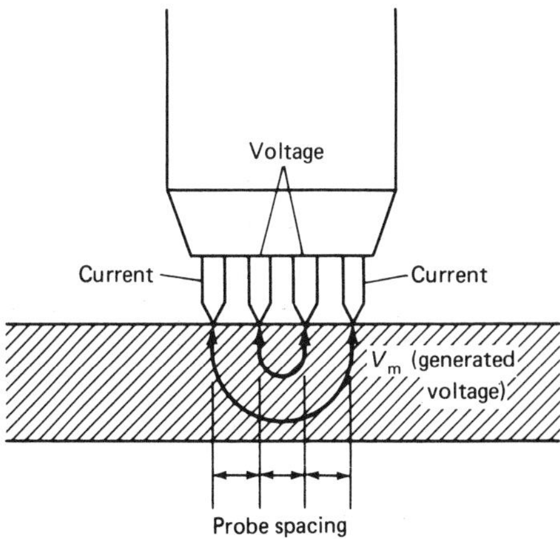

Fig. 20 Four-point probe used in the resistivity test. The outer probe pins are the current leads; the inner pins are the potential leads. Source: Ref 30

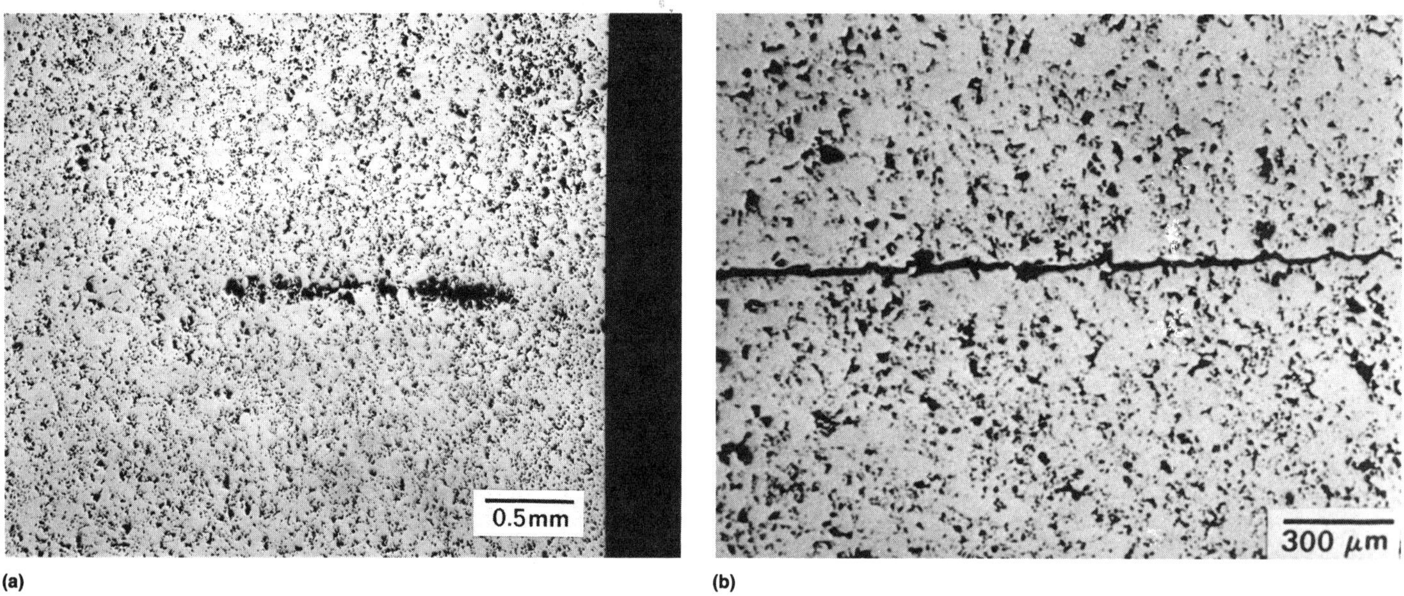

Fig. 21 Defects in green P/M compacts. (a) Artificial defect caused by the inclusion of a fine wax sliver in the die fill. Unetched. (b) Artificial defect produced by compacting a partially filled die at 345 MPa (25 tsi), completing the fill, and carrying out final compaction of the entire part at 620 MPa (45 tsi). Unetched. See also Fig. 22.

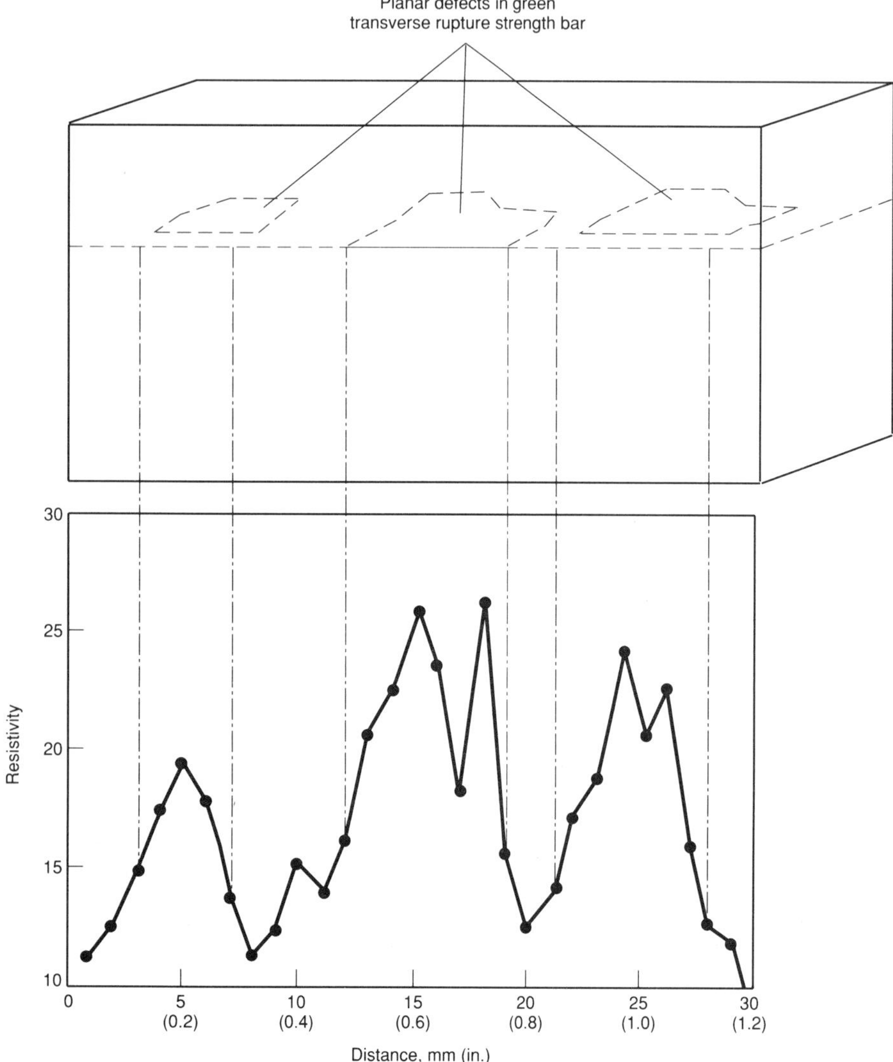

Fig. 22 Variation in resistivity in a green compact was used to locate artificial defects of the type shown in Fig. 21(a) and (b). Source: Ref 30

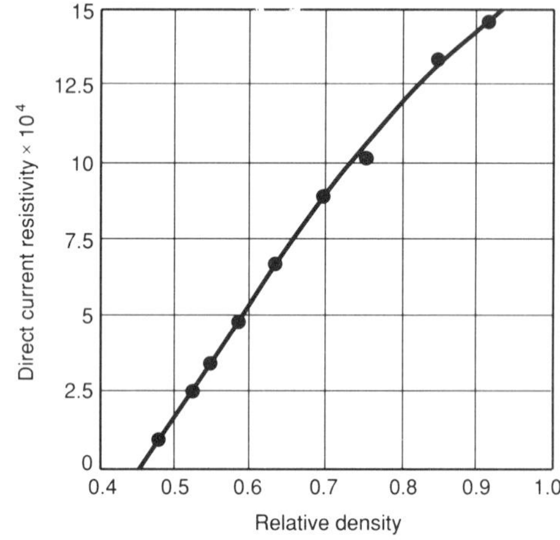

Fig. 23 Variation in resistivity with relative density in sintered iron. Source: Ref 33

coil can be placed in the magnetic field around the testpiece.

The alternating current in the induction coil can vary from 1 to 1000 kHz. The depth of penetration varies with frequency, with the highest frequencies yielding the smallest depths (skin effect). The way in which the eddy current varies as a function of depth is also described in the article "Eddy Current Inspection" in this Volume.

The output of eddy current testing is in the form of an oscilloscope display. An eddy current inspection system can detect changes from point to point in single testpieces (for example, welded tubes) as they move through the coil. For cases where the testpieces consist of a series of discrete parts, a second coil containing a reference can be added to the system; this configuration is called a magnetic bridge comparator.

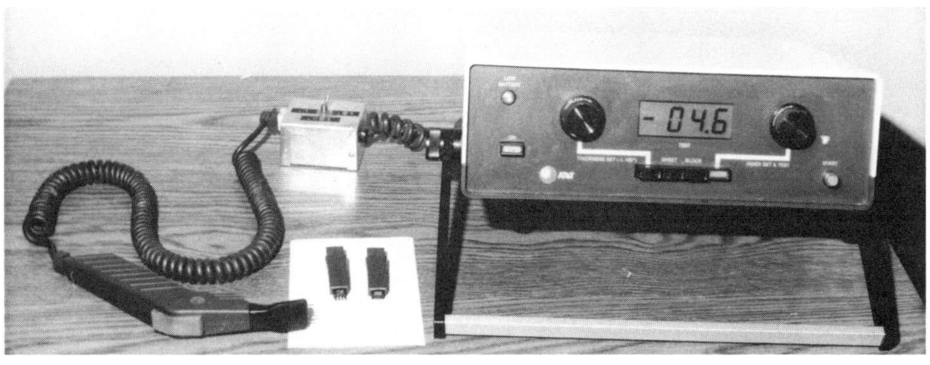

Fig. 24 Resistivity-measurement device for examining P/M parts. Courtesy of R.A. Ketterer and N.F. McQuiddy, Ferraloy

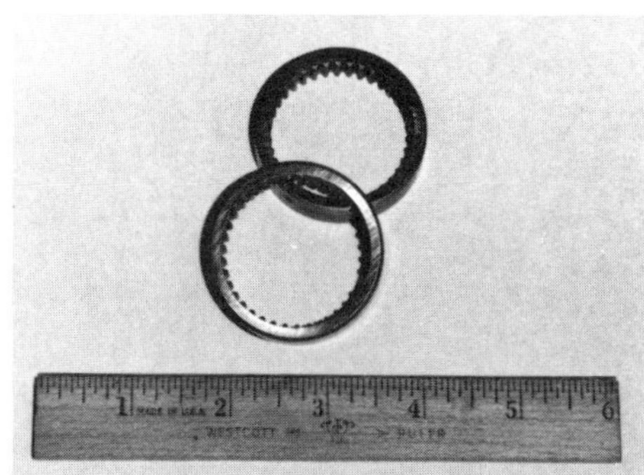

Fig. 26 Automatic transmission spacer examined by resistivity measurement. Courtesy of R.A. Ketterer and N.F. McQuiddy, Ferraloy

(a)

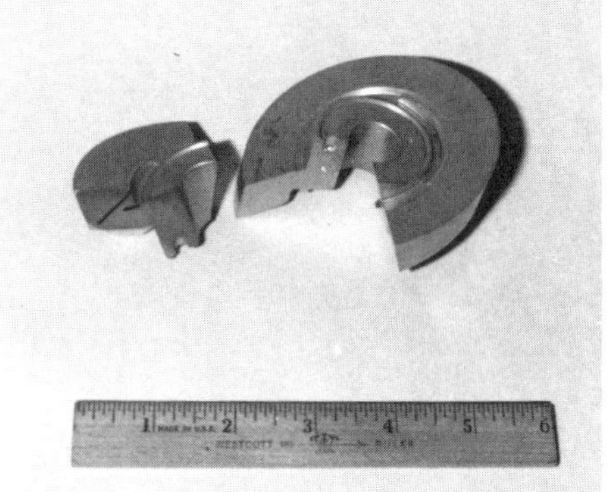

(b)

(c)

Fig. 25 Automotive air conditioner compressor part examined by resistivity measurement. (a) Actual part. (b) Cross section showing flawed areas. (c) Six location test fixture. Courtesy of R.A. Ketterer and N.F. McQuiddy, Ferraloy

Fig. 27 Automatic transmission clutch plate examined by resistivity measurement. Courtesy of R.A. Ketterer and N.F. McQuiddy, Ferraloy

Magnetic Bridge Comparator Testing. When a ferromagnetic part is placed in the core of a coil with an alternating current, a unique set of harmonics characteristic to the part can be detected in the coil. Some of the variables influencing the harmonics are alloy type, core or surface hardness, case depth, and porosity (Ref 35).

In the magnetic bridge comparator arrangement, the harmonic signals from two like coils are compared. The coils are similar and carry the same excitation waveform. One coil contains the part to be inspected and the other a reference part chosen at random from the group to be tested. Differences between the harmonic characteristics of the two parts are displayed as the displacement of a dot from the center of an oscilloscope screen; no displacement means the two parts are alike.

Although the magnetic susceptibility of porous sintered steels is reduced by the pinning of domain-boundary walls by pores, P/M parts are also capable of being analyzed by the magnetic bridge comparator. In one study, 120 P/M production parts were tested in a magnetic bridge comparator. Seventeen of the parts were singled out on the basis of a displacement of the oscilloscope indication, as shown in Fig. 28. These parts were tested for chemistry, hardness, crush strength, and pressed height. For comparison, 25 parts selected at random from the remaining specimens were also tested. Statistically significant differences were found between the groups with regard to carbon content and hardness (Ref 36). The technique has also been successfully applied to powder-forged parts (Ref 37). Although there are no published trials, there is a possibility that the comparator could also be used for testing green compacts.

Visual Inspection and Pressure Testing

Magnetic Particle Inspection. Cracks that exist on or close to the surface of a

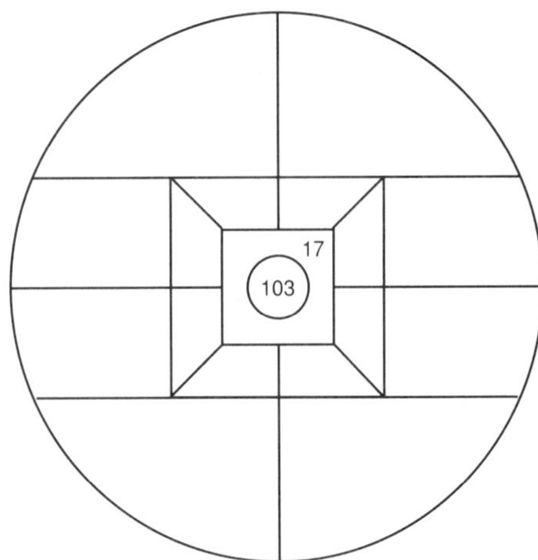

Fig. 28 Magnetic bridge comparator display for a set of 120 sintered parts, in which 17 parts were indicated as differing from the reference part. Source: Ref 36

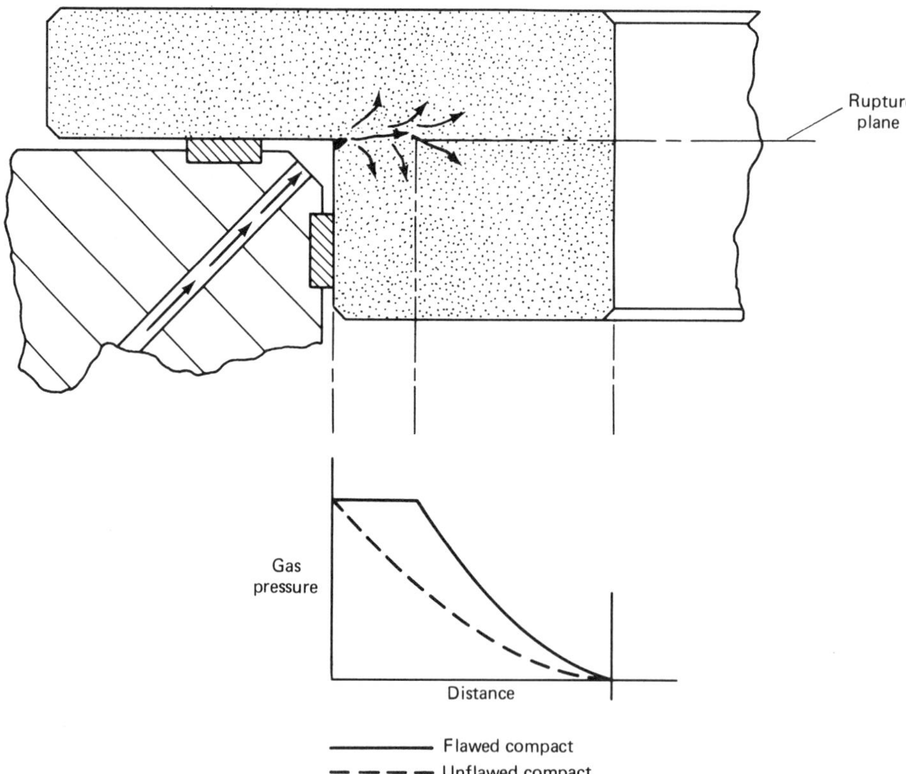

Fig. 29 Pore pressure rupture test for crack detection in green parts. Source: Ref 39

ferromagnetic material in the magnetic field act as magnetic poles, creating localized increases in the field intensity. Iron particles suspended in a fluid at the surface will be preferentially attracted to these high-intensity areas, and these particles can be used to mark the locations of the flaws. The detectability of the particles themselves can in turn be improved by coating with a pigment that contrasts with the part surface or fluoresces under ultraviolet light (see the article "Magnetic Particle Inspection" in this Volume).

This method has been used to inspect finished P/M parts for cracks originating in processing, and it may also be applicable to green compacts. It is also possible to automate the inspection process by using digital image processing (Ref 38).

Liquid Dye Penetrant Inspection. A liquid that wets the surface of the material being inspected will lower its surface energy by residing preferentially in surface cracks and cavities. In the liquid penetrant inspection technique, cracks are detected by removing the dye from the flat surface of the specimen. The dye that is left behind in the cracks is then wicked out onto the surface by a fine particulate layer in which the pore radius is even lower than that of the crack. The penetrant in this particulate developer layer can be detected visually because of its high contrast with the white developer, or it can be mixed with a dye that fluoresces under ultraviolet light. This process is described in the article "Liquid Penetrant Inspection" in this Volume.

The dye penetrant equipment found in P/M shops is generally used only for checking parts of the tooling and machinery for cracks. The dye does not preferentially reside at cracks in P/M parts, because the pore radius and the crack radius are equivalent. However, there might be an application for green parts because the surfaces of

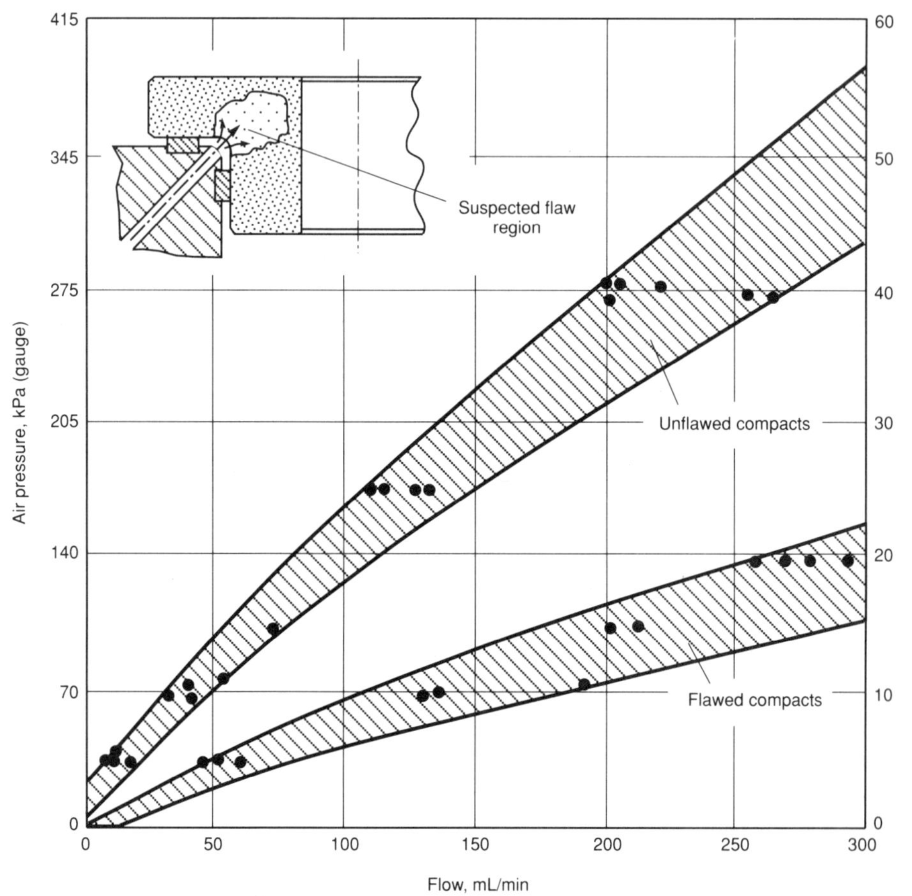

Fig. 30 Detection of flawed compact using the gas permeability technique. Source: Ref 39

green parts are sealed against penetration by liquids through smearing of the metal powder against the die wall and through the formation of a thin coating of dry powder lubricant on the surface. Cracks intersecting the surface may form an opening in this layer that could be detected by the dye penetrant.

Pore Pressure Rupture Testing of Green Compacts. A novel test is available for detecting ejection cracks in green compacts (Ref 39). A pressure seal is formed around a corner or area of a part where experience has shown that cracks are likely to occur. The area is then pressurized to about 3.5 MPa (500 psi) using a fixture such as that shown in Fig. 29. If a crack is present, the gas pressure in the crack will be sufficient to propagate the crack the rest of the way through the part. This would be classed as a proof test rather than a nondestructive test because the part is destroyed if defects are present.

The test can be used in a nondestructive manner on sintered parts. The gas permeability of the pressurized area is measured at reduced pressures, and the presence of cracks or low-density areas is indicated by high permeability, as shown in Fig. 30.

REFERENCES

1. J.W. McCauley, Materials Testing in the 21st Century, in *Nondestructive Testing of High Performance Ceramics*, Conference Proceedings, American Ceramics Society/American Society for Nondestructive Testing, 1987, p 1
2. R.W. McClung and D.R. Johnson, Needs Assessment for NDT and Characterization of Ceramics: Assessment of Inspection Technology for Green State and Sintered Ceramics, in *Nondestructive Testing of High Performance Ceramics*, Conference Proceedings, American Ceramics Society/American Society for Nondestructive Testing, 1987, p 33
3. R.C. O'Brien, "Fatigue Properties of P/M Materials," SAE Technical Paper 880165, Society of Automotive Engineers, March 1988
4. G.F. Bocchini, "High Pressure Compaction, High Pressure Coining, and High Pressure Repressing of P/M Parts," Paper presented at the Prevention and Detection of Cracks in Ferrous P/M Parts Seminar, Metal Powder Industries Federation, 1988
5. F.V. Lenel, *Powder Metallurgy Principles and Application*, Metal Powder Industries Federation, 1980, p 112
6. C. Rain, Uncovering Hidden Flaws, *High Technol.*, Feb 1984
7. B. Chang *et al.*, Spatial Resolution in Industrial Tomography, *IEEE Trans. Nuclear Sci.*, NS30 (No. 2), April 1983
8. H. Heidt *et al.*, Nondestructive Density Evaluation of P/M Objects by Computer Tomography, in *Horizons of Powder Metallurgy*, 1986 International Powder Metallurgy Conference Proceedings, Part II, p 723
9. G. Schlieper, W.J. Huppmann, and A. Kozuch, Nondestructive Determination of Sectional Densities by the Gamma Densomat, *Prog. Powder Metall.*, Vol 43, 1987, p 351
10. C.T. Waldo, Practical Aspects of the Gamma Densomat, in *Horizons in Powder Metallurgy*, 1986 International Powder Metallurgy Conference Proceedings, Part II, p 739
11. J.L. Rose, M.J. Koczak, and J.W. Raisch, Ultrasonic Determination of Density Variations in Green and Sintered Powder Metallurgy Components, *Prog. Powder Metall.*, Vol 30, 1974, p 131
12. B. Patterson, C. Bates, and W. Knopp, Nondestructive Evaluation of P/M Materials, *Prog. Powder Metall.*, Vol 37, 1981, p 67
13. M.F. Termine, "Ultrasonic Velocity Measurements on Green and Sintered P/M Compacts," Unpublished Report, Hoeganaes Corporation, 1985
14. E.P. Papadakis and B.W. Petersen, Ultrasonic Velocity As A Predictor of Density in Sintered Powder Metal Parts, *Mater. Eval.*, April 1979, p 76
15. A. Gallo and V. Sergi, Orientation of Porosity of P/M Materials Evaluated by Ultrasonic Method, in *Horizons in Powder Metallurgy*, 1986 International Powder Metallurgy Conference Proceedings, Part II, p 763
16. M.P. Jones and G.V. Blessing, Ultrasonic Evaluation of Spray-Dried Alumina Powder During and After Compaction, in *NDT of High Performance Ceramics*, Proceedings of the 1987 Conference, American Ceramics Society/American Society for Nondestructive Testing, 1987, p 148
17. B.R. Patterson and C.E. Bates, Nondestructive Property Prediction in Gray Cast Iron Using Ultrasonic Techniques, Paper 65, *Trans. AFS*, 1981, p 369
18. R.H. Brockelman, Dynamic Elastic Determination of the Properties of Sintered Powder Metals, *Perspect. Powder Metall.*, Vol 5, 1970, p 201
19. E.R. Leheup and J.R. Moon, Yield and Fracture Phenomena in Powder Forged Fe-0.2C and Their Prediction by NDT Methods, *Powder Metall.*, Vol 23 (No. 4), 1980, p 177
20. K. Subramanian and J.L. Rose, C-Scan Testing for Complex Parts, *Adv. Mater. Process. inc. Met. Prog.*, Vol 131 (No. 2), 1987, p 40
21. A. Hecht and E. Neumann, Detection of Small Inclusions in P/M Alloys by Means of Nondestructive Ultrasonic Testing, in *Horizons of Powder Metallurgy*, 1986 International Powder Metallurgy Conference Proceedings, Part II, p 783
22. G.Y. Baaklini and P.B. Abel, Flaw Imaging and Ultrasonic Techniques for Characterizing Sintered Silicon Carbide, in *Nondestructive Testing of High Performance Ceramics*, Proceedings of the 1987 Conference, American Ceramics Society/American Society for Nondestructive Testing, 1987, p 304
23. E.R. Generazo and D.J. Roth, Quantitative Flaw Characterization With Scanning Laser Acoustic Microscopy, *Mater. Eval.*, Vol 44 (No. 7), June 1986, p 864
24. P. Cawley, Nondestructive Testing of Mass Produced Components by Natural Frequency Measurements, *Proc. Inst. Mech. Eng.*, Vol 199 (No. B3), 1985, p 161
25. P. Cawley, Rapid Production Quality Control by Vibration Measurements, *Mater. Eval.*, Vol 45 (No. 5), May 1987, p 564
26. R. Phillips and W. Franciscovich, Free-free Resonant Frequency Testing of Powder Metal Alloys to Determine Elastic Moduli, *Prog. Powder Metall.*, Vol 39, 1983, p 369
27. Y. Xu, S.H. Carpenter, and B. Campbell, An Investigation of the Acoustic Emission Generated During the Deformation of Carbon Steel Fabricated by Powder Metallurgy Techniques, *J. Acoust. Emiss.*, Vol 3 (No. 2), 1984, p 81
28. A. Rosencwaig, Thermal Wave Imaging, *Science*, Vol 218 (No. 4569), 1982, p 223
29. L.J. Inglehart, Photothermal Characterization of Ceramics, in *Nondestructive Testing of High Performance Ceramics*, Proceedings of 1987 Conference, American Ceramics Society/American Society of Nondestructive Testing, 1987, p 163
30. A. Lewis, "Nondestructive Inspection of Powder Metallurgy Parts Through the Use of Resistivity Measurements," Paper presented at the Prevention and Detection of Cracks in Ferrous P/M Parts Seminar, Metal Powder Industries Federation, 1988
31. R.A. Ketterer and N. McQuiddy, "Resistivity Measurements on P/M Parts: Case Histories," Paper presented at the Prevention and Detection of Cracks in Ferrous P/M Parts Seminar, Metal Powder Industries Federation, 1988
32. E.R. Leheup and J.R. Moon, Electrical Conductivity and Strength Changes in Green Compacts of Iron Powder When Heated in Range 50-400 °C in Air, *Powder Metall.*, Vol 23 (No. 4), 1980, p 217
33. E.R. Leheup and J.R. Moon, Electrical Conductivity Changes During Compaction of Pure Iron Powder, *Powder Metall.*, Vol 21 (No. 4), 1978, p 195

34. E.R. Leheup and J.R. Moon, Relationships Between Density, Electrical Conductivity, Young's Modulus, and Toughness of Porous Iron Samples, *Powder Metall.*, Vol 21 (No. 4), 1978, p 1
35. P. Neumaier, Computer-Aided Tester for Nondestructive Determination of Material Properties, *Metallurg. Plant Technol.*, No. 3, 1987, p 58
36. R.C. O'Brien, "Analysis of Variance of Sintered Properties of P/M Transmission Parts," Unpublished Report, Hoeganaes Corporation, 1984
37. W.B. James, "Quality Assurance Procedures for Powder Forged Materials," SAE Technical Paper 830364, Society of Automotive Engineers, Feb 1983
38. Y.F. Cheu, Automatic Crack Detection With Computer Vision and Pattern Recognition of Magnetic Particle Indications, *Mater. Eval.*, Vol 42 (No. 11), Nov 1984, p 1506
39. I. Hawkes and C. Spehrley, Point Density Measurement and Flaw Detection in P/M Green Compacts, *Mod. Develop. Powder Metall.*, Vol 5, 1970, p 395

Steel Bar, Wire, and Billets

THE INSPECTION OF STEEL BARS will be the focus of this article, but the principles involved also apply for the most part to steel wire. In many cases, as far as nondestructive inspection is concerned, steel bars and wire are the same.

Steel billets, mainly because their irregularities differ significantly from the irregularities found in bars, require special inspection techniques. Billet inspection is covered in the section "Nondestructive Inspection of Steel Billets" in this article.

The primary objective in the nondestructive inspection of steel bars and wire is generally the same as for the inspection of other products, that is, to detect conditions in the material that may be detrimental to the satisfactory end use of the product. There is, however, an additional objective in attempting to detect undesirable conditions in semifinished products such as bars, namely, to eliminate unacceptable material before spending time, money, and energy in manufacturing products that will later be rejected.

The nondestructive inspection of bars and other semifinished products does not impair the product, provides rapid feedback of information, and can be utilized as either an in-line or off-line system. It makes use of several devices, such as visual, audio, and electromagnetic, for the detection of flaws and of variations in composition, hardness, and grain structure. A wide range of selectivity is provided in each device, permitting acceptance or rejection at various specification levels. The most common function of nondestructive inspection in the steel industry is the detection and evaluation of flaws. It is also used for the detection of variations in composition and physical properties. No amount of nondestructive inspection can ensure an absolutely flawless bar, but it does provide a consistent specified degree of quality during everyday operation.

Types of Flaws Encountered

The terms used for the various types of flaws discussed in this article may not be the same in various geographic areas. In many cases, different terms are applied to the same type of flaw. Therefore, this section contains a description and an illustration of each condition. The term flaw is applied to blemishes, imperfections, faults, or other conditions that may nullify acceptability of the material. The term also encompasses such terms as pipe, porosity, laminations, slivers, scabs, pits, embedded scale, cracks, seams, laps, and chevrons, as well as blisters and slag inclusions in hot-rolled products. For products that are cold drawn, die scratches may be added.

Most flaws in steel bars can be traced back to the pouring of the hot metal into molds. Factors that work against obtaining a perfect homogeneous product include:

- The fast shrinkage of steel as it cools (roughly 5% in volume)
- The gaseous products that are trapped by the solidifying metal as they try to escape from the liquid and semisolid metal
- Small crevices in the mold walls, which cause the metal to tear during the stripping operation
- Spatter during pouring, which produces globs of metal frozen on the mold walls because of the great difference in temperature of the mold surfaces and the liquid metal

Pipe is a condition that develops in the nominal top centerline of the ingot as the result of solidification of the molten metal from the top down and from the mold walls to the center of the ingot (Fig. 1). Because of the metal shrinkage and lack of available liquid metal, a cavity develops from the top down and, if not completely cropped before subsequent rolling, becomes elongated and will be found in the center of the final product, as shown in ingot B in Fig. 1.

Porosity is the result of trapped gaseous bubbles in the solidifying metal causing porous structures in the interior of the ingot (Fig. 1). Upon rolling, these structures are elongated and interspersed throughout the cross section of the bar product, as illustrated in Fig. 1.

Inclusions may be the products of deoxidation in the ingot, or they may occur from additives for improving machinability, such as lead or sulfur. Inclusions and their typical location in a steel bar are shown in Fig. 2(a).

Laminations may occur from spatter (entrapped splashes) during the pouring of the steel into the mold. They are elongated during rolling and are usually subsurface in the bar. Figure 2(b) illustrates a lamellar structure opened up by a chipping tool.

Slivers are most often caused by a rough mold surface, overheating prior to rolling, or abrasion during rolling. Very often, slivers are found with seams. Slivers usually have raised edges, as shown in Fig. 2(c).

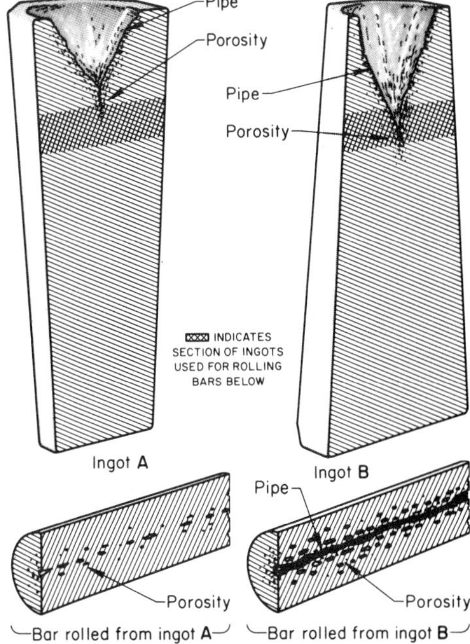

Fig. 1 Longitudinal sections of two types of ingots showing typical pipe and porosity. When the ingots are rolled into bars, these flaws become elongated throughout the center of the bars.

Scabs are caused by splashing liquid metal in the mold. The metal first freezes to the wall of the mold, then becomes attached to the ingot, and finally becomes embedded in the surface of the rolled bar (Fig. 2d). Scabs thus bear some similarity to laminations.

Pits and Blisters. Gaseous pockets in the ingot often become, during subsequent rolling, pits or blisters on the surface or slightly below the surface of bar products. Other pits may be caused by overpickling to remove scale or rust. Pits and blisters are both illustrated in Fig. 2(e).

Embedded scale may result from the rolling or drawing of bars that have become excessively scaled during prior heating operations. The pattern illustrated in Fig. 2(f) is typical.

Cracks and seams are often confused with each other. Cracks with little or no oxide present on their edges may occur when the metal cools in the mold, setting up highly stressed areas. Seams develop from these cracks during rolling as the reheated outer skin of the billet becomes heavily oxidized,

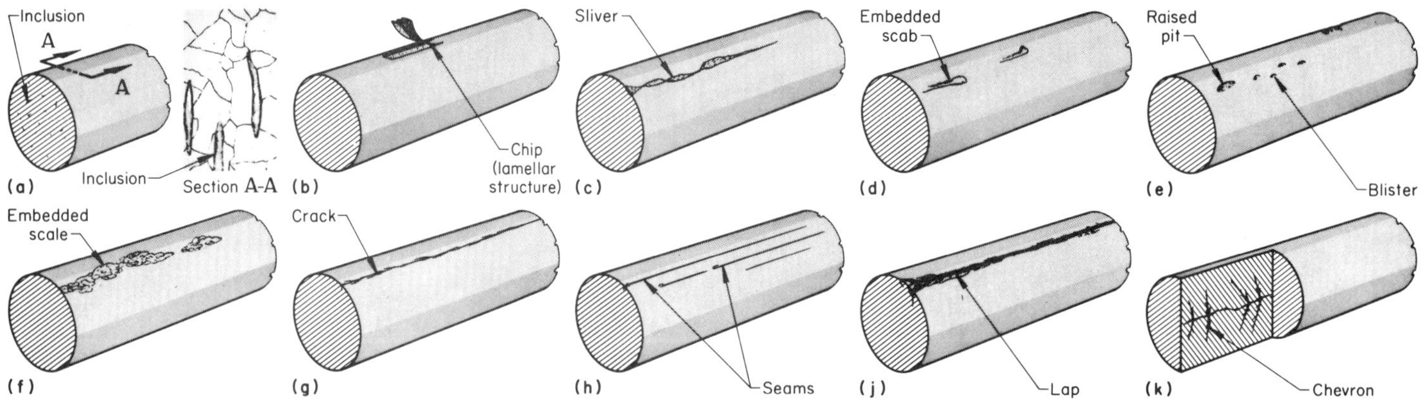

Fig. 2 Ten different types of flaws that may be found in rolled bars. See text for discussion.

transforms into scale, and flakes off the part during further rolling operations. Cracks also result from highly stressed planes in cold-drawn bars or from improper quenching during heat treatment. Cracks created from these latter two causes show no evidence of oxidized surfaces. A typical crack in a bar is shown in Fig. 2(g).

Seams result from elongated trapped-gas pockets or from cracks. The surfaces are generally heavily oxidized and decarburized. Depth varies widely, and surface areas sometimes may be welded together in spots. Seams may be continuous or intermittent, as indicated in Fig. 2(h). A micrograph of a typical seam is shown in Fig. 3.

Laps are most often caused by excessive material in a given hot roll pass being squeezed out into the area of the roll collar. When turned for the following pass, the material is rolled back into the bar and appears as a lap on the surface.

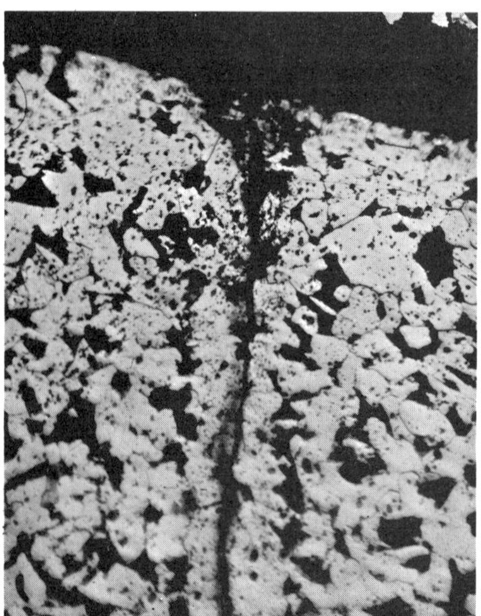

Fig. 3 Micrograph of a seam in a cross section of a 19 mm (3/4 in.) diam medium-carbon steel bar showing oxide and decarburization in the seam. 350×

Chevrons are internal flaws named for their shape (Fig. 2k). They often result from excessively severe cold drawing and are even more likely to occur during extrusion operations. The severe stresses that build up internally cause transverse subsurface cracks.

Methods Used for Inspection of Steel Bars

Almost all inspection of steel bars (other than plain visual inspection) is performed by means of the following four methods, used either singly or in combination:

- Magnetic particle inspection
- Liquid penetrant inspection
- Ultrasonic inspection
- Electromagnetic inspection

Magnetic Particle Inspection

Magnetic particle inspection offers the same visual aid in the nondestructive inspection of bars as it does for castings, forgings, or machined products. The method is used for detecting seams, cracks, and other surface flaws, and, to a limited extent, subsurface flaws. As a rule, the method is not capable of detecting flaws that are more than 2.5 mm (0.1 in.) beneath the surface.

The magnetic particle method utilizes a magnetic field set up in the bar. Flaws cause a leakage of flux if they are at an angle to the flux flow. This flux is due to the lower magnetic permeability of the material in the flaw (air, oxide, or dirt) compared with that of the metal. Because the flux leakage forms magnetic poles, fine iron powder sprinkled on the surface will adhere, indicating the extent of the flaw.

Longitudinal Flaws. Optimum indications are obtained when the magnetic field is perpendicular to the flaws. A similar result is obtained for flaws slightly below the surface, but the surface leakage is less and, consequently, fewer iron particles are attracted to the area, producing a less definite indication.

Various colors of iron powders are commercially available to permit the choice of a color that provides maximum contrast between the powder and the material being inspected. Fluorescent coatings on powders and ultraviolet light can be used to make the indication more vivid. The powders can be applied in dry form, or can be suspended in oil or a distillate and flowed over the workpiece during or after the magnetizing cycle.

Transverse Flaws. To detect flaws transverse to the long axis of the bar being inspected, a solenoid winding or encircling coil is used. For longitudinal-type flaws, circumferential magnetization is utilized; an electric current flowing through the bar sets up a magnetic field at right angles to the long axis of the bar. To protect the bar from arc burns when the current is turned on, electrical contact is usually made by soft metallic pads held firmly against the bar ends.

Power Requirements. The power used can be direct current or alternating current. Direct current may be from batteries or rectified alternating current. Alternating current travels near the surface and should not be used for detecting subsurface flaws. In most cases, the continuous-magnetization system is used for bars because most bars have low retentivity for magnetism; therefore, the residual-magnetism system is not suitable. Finished bars must be demagnetized; otherwise, during manufacturing operations such as machining, steel chips will adhere and possibly cause trouble.

Quantity Requirements. As a rule, the magnetic particle inspection of bars is confined to the inspection of a small quantity of bars, as in a fabricating shop. The method is, in its present state of development, considered too slow and too costly for mass-production inspection, as in a mill. Detailed information on magnetic particle inspection as it is applied to various ferromagnetic products is available in the article "Magnetic Particle Inspection" in this Volume.

Liquid Penetrant Inspection

Liquid penetrant inspection (another visual aid), for several practical reasons, is not extensively used for detecting flaws in

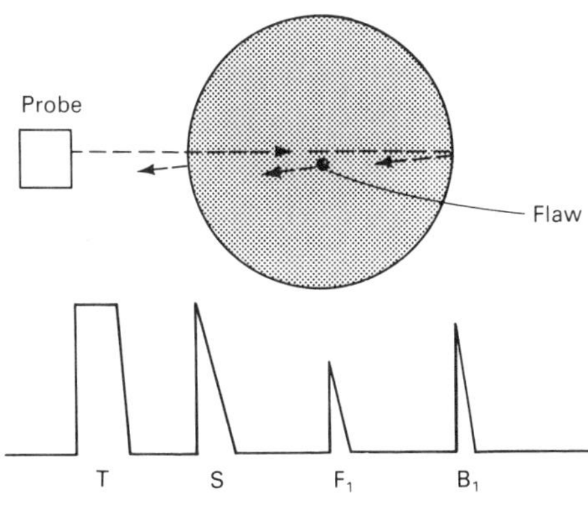

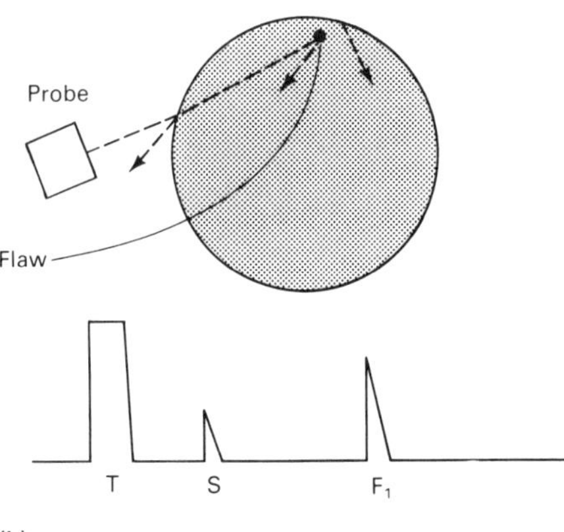

Fig. 4 Schematic showing position of probe relative to flaw inside of bar and resulting wave display obtained for two methods of ultrasonic flaw detection. (a) Normal-beam method. (b) Angle-beam method. Wave display nomenclature: T, transmit pulse; S, surface reflection echo; F_1, flaw echo; B_1, back wall echo. Source: Ref 1

steel bars. These reasons include the following:

- Its use is restricted to the detection of flaws that are open to the bar surface
- Adaptation to automation is limited compared with certain other inspection methods
- Time cycles are too long for the inspection of bars on a mass-production basis

There are exceptions, however, and there are cases where liquid penetrant inspection has been used advantageously for inspecting from one to a few bars, as in a fabricating shop. Specific advantages are:

- Liquid penetrant inspection is extremely sensitive and can sometimes detect surface flaws missed by other methods
- The solvent-removable system (one of the several liquid penetrant systems) in particular is extremely flexible and can be used for inspecting bars or portions of bars in virtually any location, including in the field

Detailed information on liquid penetrant inspection is available in the article so titled in this Volume.

Table 1 Specifications of a rotating-type ultrasonic flaw detection system

Parameter	Specifications
Dimension of material, mm (in.)	15–32 (0.59–1.26)
Testing method	Normal-beam method and angle-beam method
Testing frequency, MHz	10 and 5
Number of rotations of probe, rev/min	1000
Signal transmit	Noncontact rotation transmit
Marker	One each for near-surface flaw and inside flaw

Source: Ref 1

Ultrasonic Inspection

Ultrasonic inspection is done with high-frequency (about 1 to 25 MHz) sound waves and can successfully detect internal flaws in steel bars. Most often, the ultrasonic inspection of steel bars is restricted to large-diameter bars and to applications where high integrity is specified. Also, because of the limitations of ultrasonic inspection for detecting surface flaws, it is ordinarily used in conjunction with some other method that is more suitable for inspecting bar surfaces.

An ultrasonic beam has the valuable property that it will travel for long distances practically unaltered in a homogeneous liquid or solid, but when it reaches an interface with air (for example, at a crack or at the surface of a metal body), it is almost completely reflected. The ultrasonic beam is generated by applying a high-frequency voltage to a piezoelectric crystal, which is thus brought into mechanical oscillation. This energy in turn is fed to the workpiece by a liquid couplant.

The technique most commonly used in the nondestructive inspection of bars or barlike workpieces is the pulse-echo technique. Short pulses of ultrasonic energy are passed through the bar. The sweep voltage of the time base is coordinated with the pulse-repetition frequency so that the reflections are indicated on an oscilloscope screen. A certain amount of energy is reflected at the interface between probe and specimen, giving the first transmission signal. The probe can either have two separate crystals, a transmitter and a receiver, or have only one, which is used alternately as transmitter and receiver.

The ultrasonic method is characterized by high sensitivity and very deep penetration, but in addition to its surface limitations, its production speed is relatively low. A liquid couplant is necessary and can be a source of interference. This method is suitable for testing ingots, billets, plate, and tubes in addition to bars or barlike workpieces.

In certain cases, ultrasonic inspection has been automated. Typical products that are ultrasonically inspected using automated

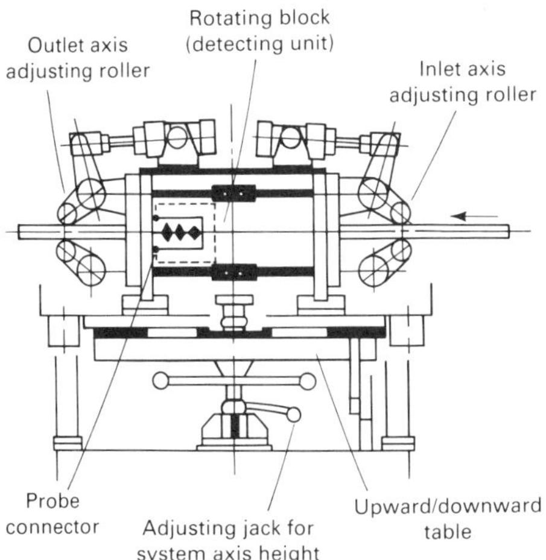

Fig. 5 Schematic of a typical rotating-type ultrasonic flaw detection system. Source: Ref 1

equipment are forged axle shafts (which are, in effect, extruded bars) and rolled bars.

Cold-Drawn Bars (Ref 1). The most effective method for the inside flaw inspection of cold-drawn bars is ultrasonic flaw detection. However, it is becoming more and more necessary to detect the smaller defects in the near-surface area in accordance with changing the production process. The conventional normal beam method (Fig. 4a) is not satisfactory, because of untested area near the surface.

A testing method for detecting smaller flaws immediately under the surface of cold-drawn bars is the angle-beam method (Fig. 4b), which conveys ultrasonic waves into the material with an angle beam. It can detect the flaws immediately under the surface that are in the dead zone for the conventional normal-beam method. Entire cross-sectional area testing becomes possible with the angle-beam method and the conventional normal-beam method in combination. The testing method to feed the material spirally and to make the probes follow the deflection of the material feeding has already been adopted in practical use for as-rolled steel bars. It is difficult to obtain higher testing speed for cold-drawn bars because of smaller dimensions. Therefore, the following method has been developed in which the material is fed straight and the probes are simultaneously rotated at high speed. Table 1 lists the main specifications of the system, and Fig. 5 shows a schematic of the setup. For bars with smaller dimensions, guide sleeves and tripplet rollers are used to prevent the ultrasonic incident angle to the material from changing because of excessive vibration and/or bending of the material. The water circulation system also incorporates a device that stabilizes the coupling water.

For flaws located immediately under the surface, the angle-beam method record can

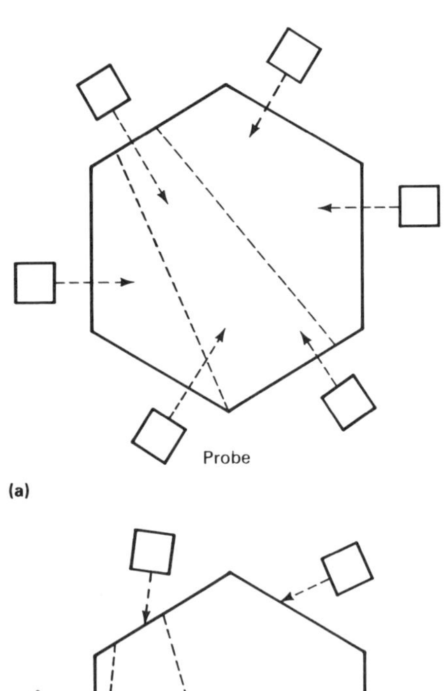

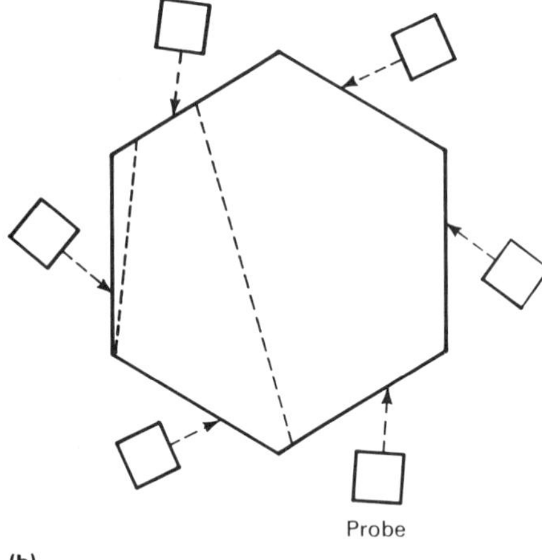

Fig. 6 Dual set of six circumferentially mounted probes used to ultrasonically detect flaws in cold-drawn hexagonal bars. (a) Normal-beam method to detect flaws deep inside bar. (b) Angle-beam method to detect surface and near-surface flaws. Source: Ref 1

detect flaws as small as 0.2 to 0.3 mm (0.008 to 0.012 in.). Flaw echoes this small are not detectable with the normal-beam method.

Cold-Drawn Hexagonal Bars (Ref 1). Requirements for strict quality assurance are increasing for gaging inside flaws to the same level as surface flaws. The conventional testing method is manual detection with the normal-beam method. Because this method requires testing with plural directions, working efficiency is low. Furthermore, an untested zone remains at the area immediately under the surface. Therefore, a testing system using the entire cross section with higher efficiency has been sought.

Higher efficiency has been attained by incorporating an automated ultrasonic flaw detection system with probes for each face of the material to detect separately the flaws located on the inside area and the near-surface area (Fig. 6). Flaws inside the material are detected with the normal-beam method at each face of the material. In this method, the untested zone remains in the near-surface area. Therefore, surface and near-surface area flaws are detected with the angle-beam method at each face of the material. That is, six normal-beam probes and six angle-beam probes are located on the circumference of the materials to be tested, which is conveyed longitudinally. The probe positions are arranged so that the entire cross section can be detected. The probe holder is designed so that all the probes can be adjusted simultaneously by adjusting one when the material size is changed. The coupling medium is a special oil that has low ultrasonic attenuation and causes no rust on the material to be tested. Table 2 lists the specifications of the system.

Table 2 Specifications of an ultrasonic flaw detection system for cold-drawn hexagonal bars

Parameter	Specifications
Dimension of material, mm (in.)	12–32 (0.472–1.260)
Testing method	Normal-beam, 6 channels; angle-beam, 6 channels
Testing frequency, MHz	5
Probe position	Fixed in circumferential direction
Marker	Two for near-surface flaw and inside flaw

Source: Ref 1

Flaws larger than 0.3 mm (0.012 in.) can be detected at the near-surface area. Flaws measuring at least 0.2 mm (0.008 in.) can be detected deep inside the hexagonal bar material.

Ultrasonic Flaw Detection on Cold-Drawn Wires (Ref 1). Surface flaw inspection is important for drawn wires. A rotation-type eddy current flaw detection system is used for quality assurance. However, inside flaw inspection has been urgently needed because on-line inspection has been considered impossible.

In quality assurance for drawn wires, rotating-type eddy current flaw detection has been used in combination with rotating ultrasonic flaw detection to detect surface defects and inside flaws, respectively, in a two-step process. However, the high cost and inefficiency of this method have prompted the development of a system with a rotating-type ultrasonic flaw detection unit that can also detect surface flaws.

An additional die is placed behind the cold-drawing die to stabilize the vibration of the material. A detection unit, which has probes arrayed in a circumferential direction, is placed between these dies. There are three detection modes (Fig. 7):

- Surface wave detection mode for surface defects
- Angle-beam detection mode for near-surface defects
- Normal-beam detection mode for inside defects

The ultrasonic incident angle can be optimized according to material dimensions.

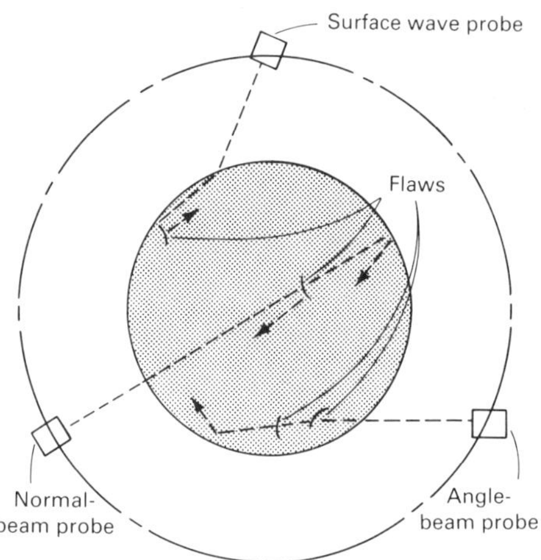

Fig. 7 Principle of ultrasonic flaw detection for cold-drawn wires using three detection mode probe. Source: Ref 1

Table 3 Specifications of an ultrasonic flaw detection system for cold-drawn wires

Parameter	Specifications
Dimension of material, mm (in.)	15–30 (0.590–1.181)
Testing frequency	Normal beam: 10 MHz, 1 channel Angle beam: 5 MHz, 2 channels Surface wave, 5 MHz, 2 channels
Number of rotations of probe, rev/min	1000
Signal transmit	Noncontact rotation transmit
Marker	One each for near-surface flaw and inside flaw

Source: Ref 1

Water, the coupling medium, is always kept in full quantity even in high-speed rotation. Thus, the system has the stable mechanism to provide constant detection.

One advantage of this system is that linear defects can be detected by surface wave detection at the same level as an eddy current method. Another is that the entire cross section can be covered by means of a combination angle-beam/normal-beam method. Table 3 summarizes the specifications of this setup.

Results of experiments with this system showed detectability of 0.1 mm (0.004 in.) minimum flaw depth on surface defects and 0.2 mm (0.008 in.) minimum inside defect size. This system enables the user to inspect the entire cross section of cold-drawn wires to a high degree of accuracy. Detailed information on the fundamentals, equipment, and techniques for ultrasonic inspection is available in the article "Ultrasonic Inspection" in this Volume.

Electromagnetic Inspection Methods

Electromagnetic methods of inspection are used far more extensively for nonde-

Table 4 Specifications of a rotating probe type eddy current flow detection system

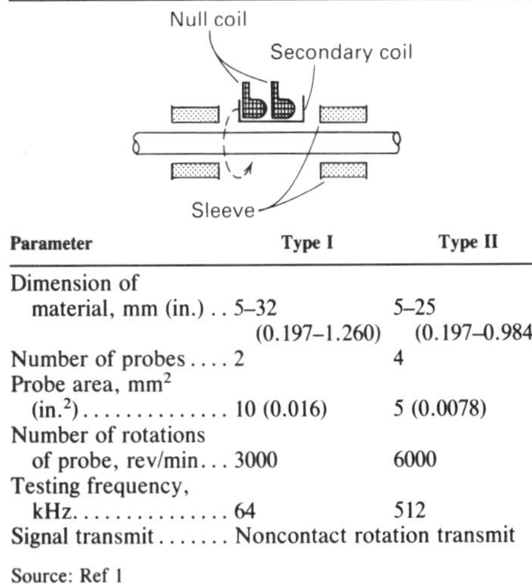

Parameter	Type I	Type II
Dimension of material, mm (in.)	5–32 (0.197–1.260)	5–25 (0.197–0.984)
Number of probes	2	4
Probe area, mm^2 (in.^{2})	10 (0.016)	5 (0.0078)
Number of rotations of probe, rev/min	3000	6000
Testing frequency, kHz	64	512
Signal transmit	Noncontact rotation transmit	

Source: Ref 1

structive inspection of steel bars than any of the methods discussed above. Electromagnetic methods are readily adaptable to automation and can be set up to detect flaws, as well as a number of different compositional and structural variations, in bars on a mass-production basis.

Equipment can be relatively simple, but for mass-production inspection the equipment may be highly sophisticated and costly. Such equipment can not only detect flaws and indicate them on an oscilloscope or other form of readout but can also mark the location of the flaw on the bar before it emerges from the inspection equipment and can automatically sort the bars on the basis of seam depth.

Eddy Current Testing of Cold-Drawn Bars (Ref 1). Surface defects on cold-drawn bars can be inspected by eddy current detection methods using an encircling coil. This method utilizes a rotating probe that detects surface defects with the probe coil rotating at high speed around the circumference of the cold-drawn bars.

The encircling coil method exhibits lower detectability on linear flaws because flaw detection depends on the difference between two test coils in which the material to be tested is encircled. On the other hand, the method of rotating the probe coil at high speed along the circumference of the material to be tested can detect linear defects because it detects bars in spiral scanning.

Table 4 lists the specifications of the detection system. One of the main features is signal transmission in the probe rotation unit by the noncontact rotating transmit method, which requires no maintenance work. Guide sleeves are placed in front of and behind the probe to maintain a constant distance between the probe and the material to be tested, which is important for acceptable performance of the system (Fig. 8).

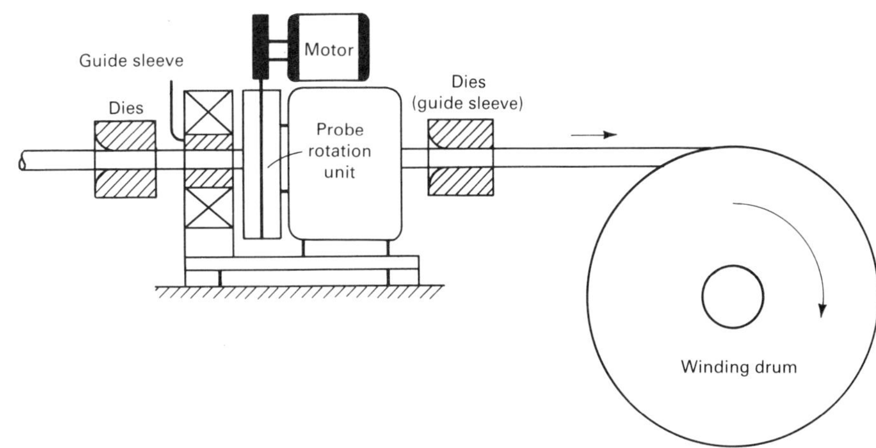

Fig. 8 Schematic of a rotating probe type eddy current flaw detector. Source: Ref 1

Furthermore, the rotation axis of the probe and the axis of the material to be tested are kept in a line by pinch rollers placed in front of and behind the detector. On the probe, a distance sensor is used for the automatic gain control function to provide electric compensation against distance variation.

Figure 9 shows the relation between flaw depth and signal output. Natural flaws produce a larger deviation in signal output than artificially introduced flaws because of the complicated cross-sectional configuration of the flaw, but the minimum detectable flaw depth is 0.1 mm (0.004 in.). Detectable flaw length depends on the feeding speed of the material, the number of probes, and the number of rotations. For example, at a speed of 60 m/min (200 sfm), the full surface is converted, and the minimum detectable flaw length is as long as the length of the probe coil.

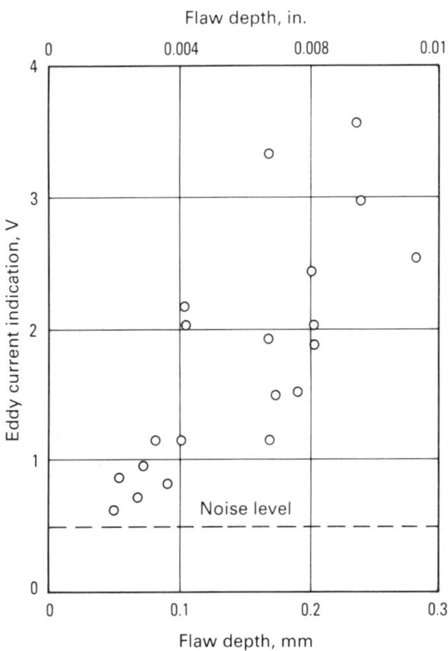

Fig. 9 Plot of eddy current signal output versus flaw depth to gage detectability of flaws in cold-drawn bars. Source: Ref 1

Eddy Current Flaw Detection on Cold-Drawn Hexagonal Bars (Ref 1). Cold-finished steel profiles (hexagonal bars) are mainly used as the raw material for couplers in oil pressure piping, an application for which quality assurance is important. Surface defects on cold-drawn hexagonal bars include cracks derived from the cold-working process as well as material flaws, both of which are long, longitudinal defects. It is impossible to detect these defects by the differential method using encircling coils. The rotating probe

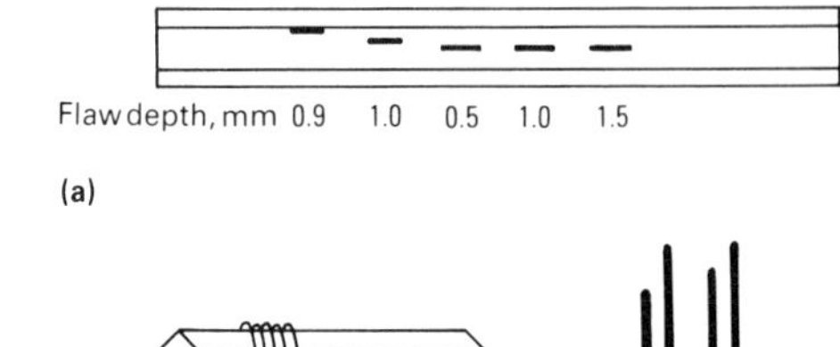

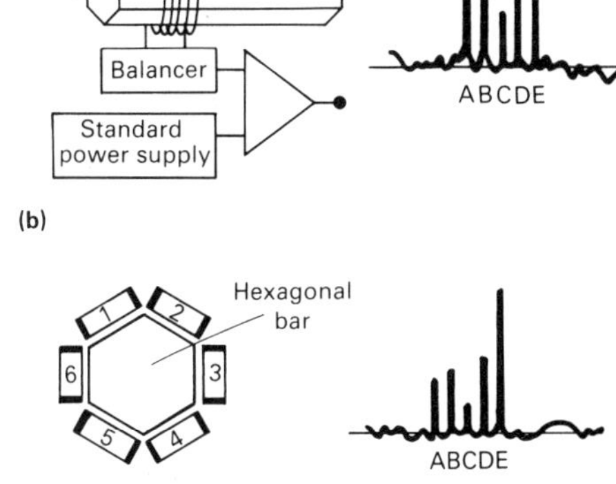

Fig. 10 Eddy current flaw detection method for cold-drawn hexagonal bars. (a) Location of artificial flaws ranging from 0.5 to 19 mm (0.020 to ¾ in.) below probe position. (b) Schematic of setup for standard voltage comparison (encircling coil) method (left) and plot of signals obtained for the designated flaw depths (right). (c) Schematic of setup for differential (six probe coil assembly) method (left) and plot of signals obtained for the designated flaw depths (right). Source: Ref 1

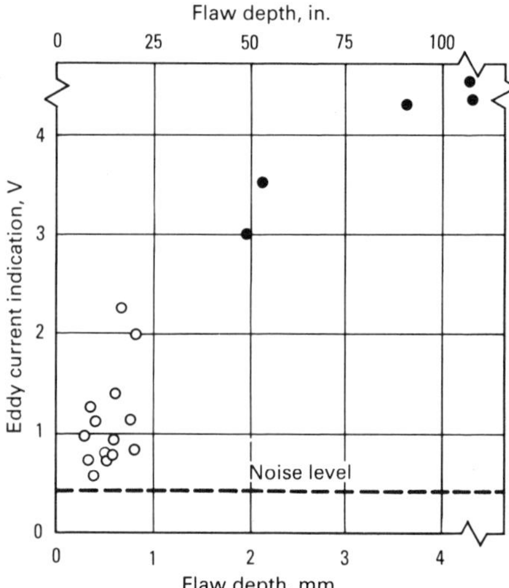

Fig. 11 Plot of eddy current signal output versus flaw depth to measure detectability of flaws—specifically material flaws (open circles) and process-induced cracks (closed circles)—in cold-drawn hexagonal bars. Source: Ref 1

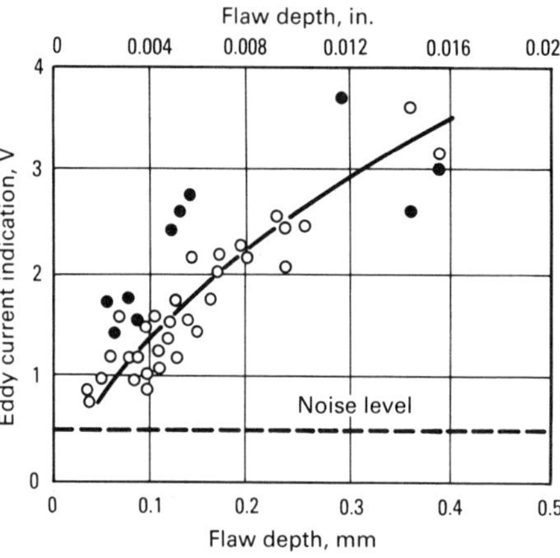

Fig. 12 Plot of eddy current signal output versus flaw depth to measure detectability of flaws, specifically cracks (open circles) and scabs (closed circles), in cold-drawn wires. Source: Ref 1

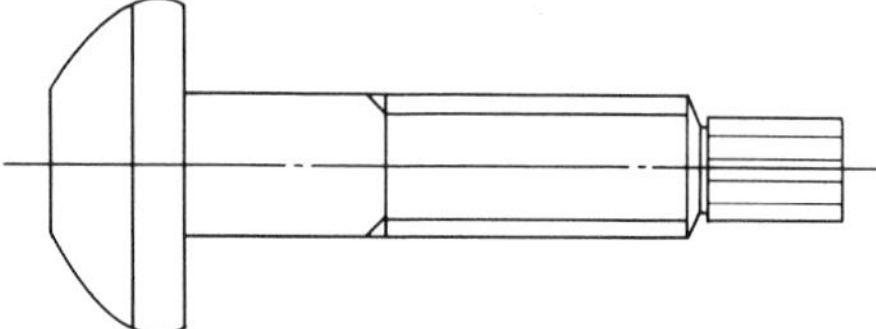

Fig. 13 Schematic of a high-tension sheared bolt

method is also not applicable, because of the hexagonal form. An automated flaw detection system for cracks initiated by the working process was developed using the eddy current flaw detection system by a standard voltage comparison method.

There are two methods for testing cold-finished steel hexagonal bars: the standard voltage comparison method with encircling probes (Fig. 10b) and the differential method with probe assembly (Fig. 10c). There is no effective difference in detectability between these two methods. For the probe assembly method, it is necessary to consider the differences in detectability of each individual probe, which is not necessary for the standard voltage comparison method.

The standard voltage comparison method is inferior in detectability to the rotating probe method, but is less expensive and can efficiently detect cracks resulting from the cold-working process. This method, which can detect material flaws more than 0.6 mm (0.024 in.) deep, is illustrated in Fig. 11.

Eddy Current Flaw Detection of Cold-Drawn Wires (Ref 1). Surface flaw detection on wire drawing line has been conducted by the encircling-type eddy current method. However, this method has difficulty in detecting linear flaws. A rotating probe type eddy current detection method can be effective, as illustrated in Fig. 8 for use on cold-drawn bars. It is important in the rotating probe method to maintain a constant distance between the probe and the material to be tested. The rotating unit is positioned between dies where the smaller vibration of the material is expected. Guide sleeves are used to adjust the rotating axis and the axis of the material to be tested.

Detectability is illustrated in Fig. 12. Flaws having a 0.1 mm (0.004 in.) minimum depth are detectable.

Eddy Current Flaw Detection for a Cold-Forged, High-Tensile Sheared Bolt (Ref 1). Figure 13 shows a general view of a high-tension sheared bolt. This type of bolt has a head with a round cross section and is mainly used for general construction and bridge applications. This bolt is produced by cold forging from cold-drawn wires in the diameter similar to the outside diameter of a threaded part of the bolt. The head is the most severely processed part of the bolt. The circumferential part of the bolt head is formed between punch and die during cold forging. Therefore, cracks tend to occur on the head. Eddy current testing can detect flaws in the bolt head at high speed with the probe rotating method.

Figure 14 shows a general view of the inspection system used. Table 5 lists the main specifications. Bolts are conveyed from hopper to line-up unit. Lined-up bolts are conveyed to the index table by straight feeder and then conveyed intermittently to the rotating detection head and further to the separator.

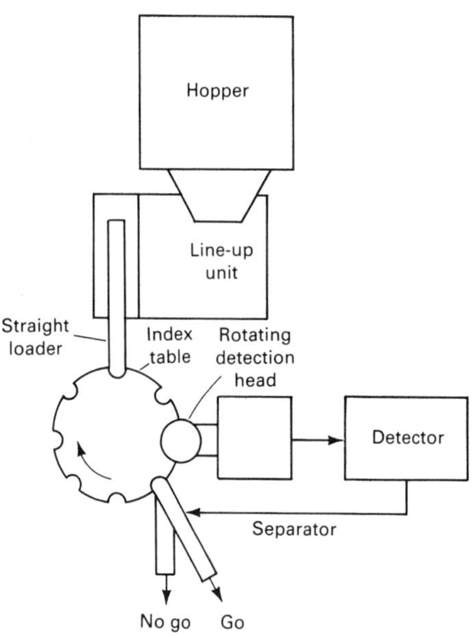

Fig. 14 Schematic of eddy current flaw detection system used to inspect sheared bolt illustrated in Fig. 13. Source: Ref 1

After the bolt heads are detected with the rotating detection head, the bolts are classified as good/no-good and separated according to detection result. Figure 15 shows the operation of the rotating detection head. The rotating detection head repeats the following operations while rotating continuously regardless of the position of the bolt head to be tested:

- A bolt stops immediately under the detection head (Fig. 15a)
- The detection head descends while maintaining rotation (Fig. 15b)
- The detection head approaches the bolt head, scans around the bolt head for two revolutions, and detects the flaws (Fig. 15c)
- Pincerlike probe holders release from the bolt head, and the detection head ascends
- Bolt is conveyed to separator while next bolt is conveyed to the position immediately under detection head

A detection rate of 60 pieces/min was maintained by the mechanism to keep the detection head rotating continuously. Figure 16 shows the relation between flaw depth and signal output. Noise level is high at the circumferential surface of the bolt head because of surface roughness, but the minimum detectable flaw depth is 0.3 mm (0.012 in.). Detailed information on methods of electromagnetic inspection is available in the articles "Eddy Current Inspection" and "Remote-Field Eddy Current Inspection" in this Volume.

Table 5 Specifications of an eddy current detection system for a high-tension sheared bolt

Parameter	Specification
Material	M20
Testing speed, pieces/min	60
Number of rotations of detecting head, rev/min	300
Testing frequency, kHz	125
Probe type	Self induction, self comparison

Source: Ref 1

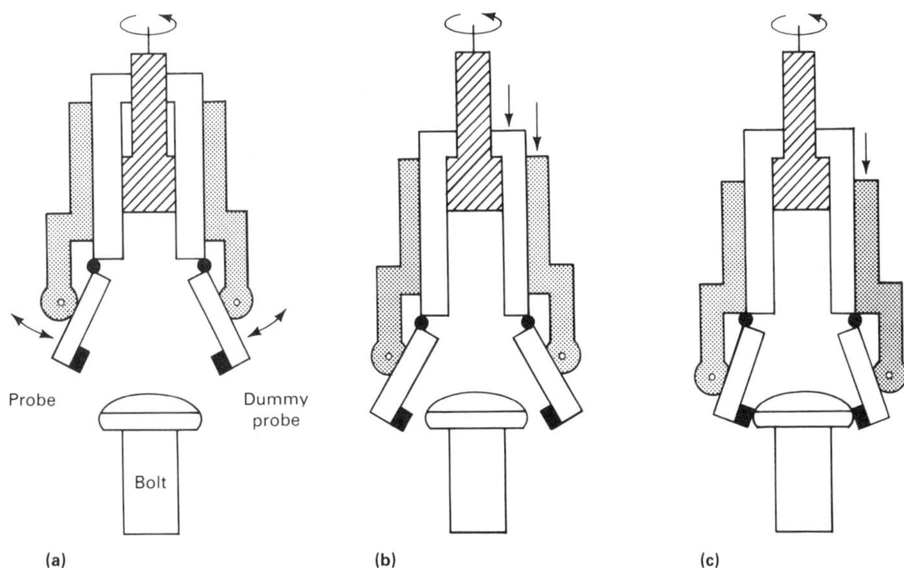

Fig. 15 Operation of rotating eddy current detection head. (a) Shear bolt positioned under rotating detection head. (b) Rotating detection head descends to lower probes into position to inspect bolt head. (c) Probe scans bolt head as bolt undergoes two complete revolutions to detect flaws. Source: Ref 1

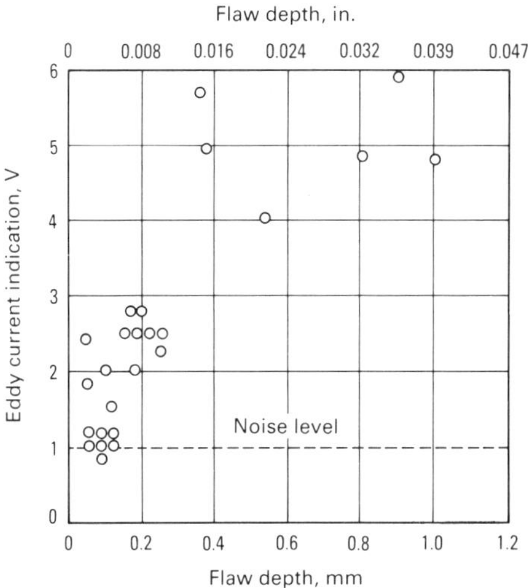

Fig. 16 Plot of eddy current signal output versus flaw depth to measure detectability of flaws in high-tensile sheared bolts. Source: Ref 1

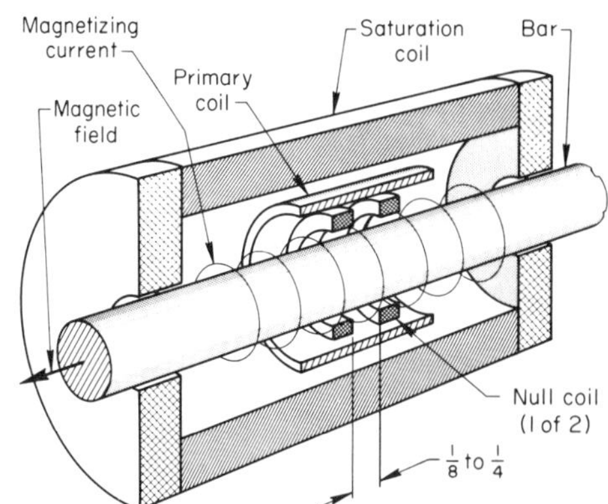

Fig. 17 Coil assembly for the inspection of steel bars by the eddy current system. Dimension given in inches

Electromagnetic Systems

Electromagnetic systems include the systems that use magnetic fields generated by alternating current flowing in a solenoid. A wide range of frequencies is used. As the alternating current flows through the solenoid, the magnetic field generated induces eddy currents within the metal workpiece. These currents are affected by the electrical resistivity (more commonly expressed as electrical conductivity—the reciprocal of resistivity), magnetic permeability, configuration, homogeneity, surface irregularities, and flaws of the metal. The resistivity of the workpiece can vary because of the chemical composition, crystal orientation, structure, and history of mechanical working. Permeability will vary over a broad range, depending on the amount of stress present in the work metal. It increases slightly in the vicinity of a flaw when the bar is subjected to a stress-producing operation.

Electromagnetic systems of flaw detection are broadly classified as:

- Those depending primarily on variations in electrical conductivity
- Those depending primarily on variations in magnetic permeability

Both systems are capable of detecting flaws in ferromagnetic bars. The conductivity-dependent systems can also be used to detect flaws in nonferromagnetic bars.

Eddy Current Systems

When electrical conductivity (resistivity) is the major variable relied upon, the test procedure is known as the eddy current system. The alternating-field intensity is low, permitting the use of a correspondingly small inductor. Most eddy current systems use a constant-voltage input derived from an electronic oscillator with a means of varying the output frequency through a wide range, such as from 0.5 to 1000 kHz, in discrete steps.

For general flaw detection, the range of 1 to 50 kHz is widely used. For ferromagnetic bars, a means must be provided to eliminate or minimize the effects of permeability variation. This is usually accomplished by magnetically saturating the bar being tested. The means for doing this is either a dc solenoid or a strong permanent magnet. A longitudinal section of one type of eddy current coil assembly is shown in Fig. 17, a more detailed drawing of the rotating coil setup shown in Table 4.

Eddy current inspection is especially useful for detecting and evaluating seams in steel bars. With this system, depending on the circuitry used, a difference of as little as 0.025 mm (0.001 in.) in seam depth can be detected. Because of the skin effect, the ability of eddy currents to penetrate into the test metal decreases in proportion to the increase of the frequency.

Eddy current inspection can be used without magnetic saturation for inspecting hot bars in the mill when the metal is above the Curie temperature, because the metal is nonmagnetic at this temperature. Therefore, it follows that the magnetic permeability system cannot be used to inspect hot bars.

Magnetic Permeability Systems

Systems that depend on variations in magnetic permeability can be used for detecting flaws and for detecting differences in composition, hardness, or structure. With appropriate instrumentation, both functions can be acccomplished simultaneously.

Flaw Detection. Magnetic permeability systems usually employ a solenoid (primary coil), which is excited by the standard line frequency of 60 Hz with an adjustable current control to produce magnetic fields from 1000 to 30 000 ampere-turns; however, the solenoid is usually operated in the range of 10 000 to 15 000 ampere-turns. A typical coil arrangement used for permeability systems is shown in Fig. 18.

As seen in Fig. 18, the coil arrangement consists of a primary coil (60 Hz), two null coils (zero-voltage-output coils), and two standard coils. The secondary or pickup coils (null coils) are concentric with the primary coil, connected electrically in opposition, and adjusted to a null or zero-voltage output. The null coils are usually spaced 75 to 102 mm (3 to 4 in.) apart. The reason for this spacing is that a normal seam in a bar tapers into the bar to sound material. The variation in stress level producing

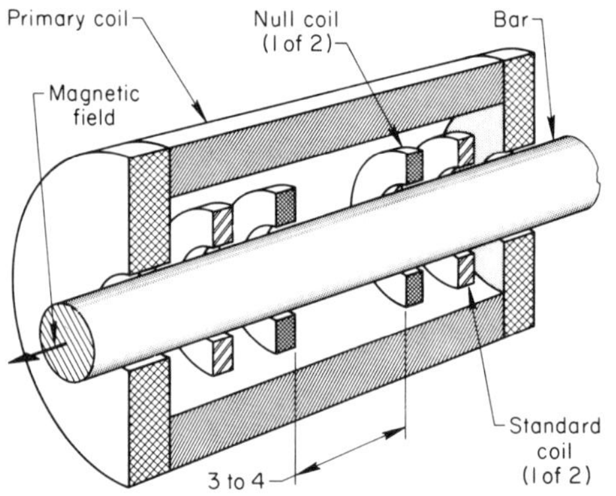

Fig. 18 Coil assembly used for the simultaneous detection of flaws and of variation in composition, structure, and hardness in steel bars. Dimension given in inches

a measurable change in magnetic permeability is related to the change in seam depth found usually within 75 mm (3 in.) of seam length.

Limitations for Flaw Detection. The detection of flaws by permeability systems depends on permeability variations resulting from changes in stress, due to cold work or heat treatment, in the adjacent area of the flaw. These changes are more or less directly proportional to the change in stress up to the elastic limit of the ferrous product.

These systems cannot be used to inspect hot-rolled or annealed bars unless they have been subjected to some uniform cold work, such as rotary straightening for round material or planar-type straightening for square, hexagonal, or flat sections. Gag-straightened bars are not suited to inspection by permeability systems, because of nonuniform high-stress concentrations wherever the ram meets the work metal. Such stresses are far in excess of those for flaws in uniformly stressed material.

The efficiency of flaw detection is a function of uniform residual stress levels within the bar. The five conditions that follow are listed in order of decreasing efficiency for detection of flaws by permeability systems:

- Heat treated, quenched, drawn, and machine straightened
- Cold drawn and machine straightened
- Cold drawn, annealed, and machine straightened
- Hot rolled and machine straightened, centerless ground
- Hot rolled and machine straightened

After straightening, the bars should be aged 24 to 48 h at near room temperature for optimum sensitivity of flaw detection. Aging can be hastened by stress relieving at low temperature in a furnace (up to 260 °C, or 500 °F).

The minimum seam depth that can be detected in cold-drawn, straightened round bars is approximately 0.025 mm (0.001 in.) for each 1.6 mm (1/16 in.) of bar diameter; hexagonal and square bars with

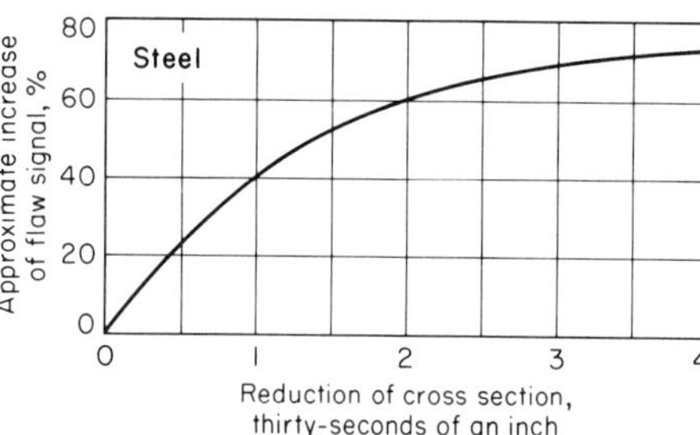

Fig. 19 Relationship between increase of flaw signal and increasing reduction of cross section (increasing draft) for cold-drawn steel bars. Base reference is a hot-rolled bar.

the same processing will be more sensitive. For example, in a 25 mm (1 in.) diam round bar, a 0.41 mm (0.016 in.) seam is readily detected, while a 0.30 to 0.33 mm (0.012 to 0.013 in.) seam can be detected in hexagonal or square bars. The reason for this difference lies in the residual stress levels imparted by the rotary and planar straighteners.

Other flaws, such as laps, slivers, cracks, hard or soft spots, dimensional changes, cupping, chevrons, and pipe, are readily indicated. For subsurface-type flaws, detection is possible only if they lie within the normal penetration range and are of sufficient size to affect the inherent stress level. The penetration is approximately 6.4 mm (1/4 in.) for low-carbon steels, 7.9 mm (5/16 in.) for medium-carbon steels, and up to 13 mm (1/2 in.) for many alloy steels.

One other factor not to be overlooked is the end effect, which prevents end-to-end inspection of the bar. As the front and rear ends of the bar enter and leave the magnetic field, the field is grossly distorted, preventing inspection of the end portions of the bar. For the average inspection speed of 37 to 46 m/min (120 to 150 sfm), the noninspected lengths will be as follows:

Bar diameter		Noninspected length at each end	
mm	in.	mm	in.
6.4–13	1/4–1/2	102–152	4–6
13–25	1/2–1	152–203	6–8
25–50	1–2	203–305	8–12
50–75	2–3	305–406	12–16

The signal obtained for a flaw of given size, as well as the amount of end effect, will vary somewhat with the amount of draft used in drawing the bar. Using the normal 0.8 mm (1/32 in.) draft as the basis for comparison, a 1.6 mm (1/16 in.) draft will increase the signal size by 50%, while a 3.2 mm (1/8 in.) draft will produce an increase of about 90% (Fig. 19).

All the above values hold true only when the secondary test coil is of the proper size; that is, the inside diameter of the coil should be 3.2 to 6.4 mm (1/8 to 1/4 in.) greater than the bar diameter. The diameters of bar stock inspected by these systems generally range from 4.8 to 140 mm (3/16 to 5 1/2 in.).

Equipment for Detecting Flaws. The circuitry may include three types of electronic systems: the null system for the detection of flaws (as explained above and shown in Fig. 18) and two identical standard systems, one of which is used for detecting mixed grades in a given lot of steel and the other for indicating variations of hardness or structure. All systems are independent and provide simultaneous indications with a single pass of the bar through the coil.

The null system utilizes a pair of matched windings that provides for the comparison of a section of the bar with another section spaced some distance from the first. The matched windings are connected in opposition, and the resultant voltage is therefore theoretically zero, making the wave displayed on the oscilloscope a straight line. In practice, however, such a balance is seldom obtained. A small voltage with the waveshape showing two peaks phase displaced 180° can normally be seen on the oscilloscope screen (bar out, Fig. 20). When a bar is placed within the coils, the wave pattern changes (bar in, Fig. 20). Should a flaw of minimum depth be present, the change in the waveshape is too small for measurement, even though there is a differential voltage between the null coils. Therefore, other relationships must be used to provide the desired information.

The use of an electronic gate of any desired width permits these measurements to be made in any section of the wave. For example, the test gate shown in Fig. 20 is adjustable to any position of the 360° cycle. It is normally positioned 8 to 20° on either side of the stress peaks, where experience has revealed the wild stress effects are minimal and waveform changes for flaws are readily detectable. Most systems provide a second electronic gate that can monitor the section of waveshape where flaws cause a change in the saturation level, if this can be reached for the size and grade of material under test. Deflections greater than a predetermined amount will energize a signal that indicates rejection.

Use for Sorting. The two standard systems differ from the null in that only one coil winding for each is utilized on the bar being tested (Fig. 18). The voltage derived from this coil is balanced by a voltage in the instrument that is fully adjustable to the degree that the zero-center meters can be adjusted to their midpoint while the oscilloscope presentation continually shows the distorted waveshape. Should any undesired bars appear within the lot being tested, the meter deflection will then provide power for activation of suitable alarm devices. The selectivity of the section of waveshape to be monitored is provided by an electronic gate, adjustable through 180°. Only half of the full 360° wave is required, the remainder being

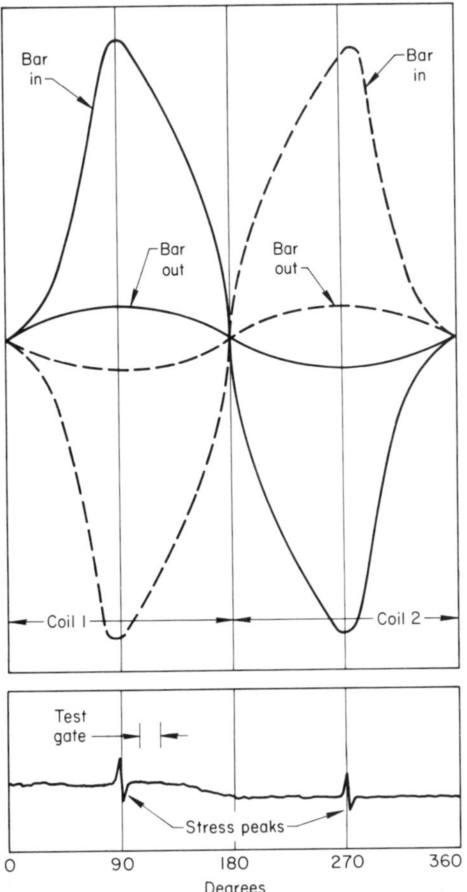

Fig. 20 Waveshape for oscilloscope pattern of a full electrical cycle for empty coils (bar out) and loaded coils (bar in). The position of an electronic gate for viewing an established portion of the cycle is shown.

the negative duplicate of the positive and not shown on the oscilloscope. Both systems (standard coils, Fig. 18) are fully independent and should be operated at different positions of the waveform to obtain as much information as possible during the test. The standard system is used to monitor each bar in a lot for composition, hardness, structure, and size and to indicate the presence of uniform-depth seams, cracks, and laps, which generally escape detection by the null system.

Equipment Requirements

In addition to coil arrangements such as those illustrated in Fig. 17 and 18, a fairly elaborate set of electronic gear is required for inspecting steel bars. Some type of equipment for handling the bars and conveying them through the coils at the desired rate is also required. The degree of sophistication designed into the equipment depends mainly on the number of similar bars to be inspected.

Typical control units are adaptable to either the eddy current or the magnetic permeability systems of inspection. Many variations are commercially available.

Flaw Detection Procedure Using a Permeability System

The following description is presented purely as an example of setting up a procedure for detecting flaws by permeability change using the null coils with one specific type of instrumentation. Procedures vary widely for various types of equipment.

Select the proper size of secondary coil for the bars to be tested, and insert it into the primary-coil unit, making sure all electrical connections are secure. Turn on the primary power, and adjust the coil current to about 8 A as shown on the electrical meter on the power-supply panel. Adjust the combined coil unit so that the material will be concentric with the inside diameter of the secondary coil. Select the appropriate feed-through speed of the conveyor system, and insert a bar into the test coil. Turn the sensitivity on the null-equipment panel clockwise to about midposition, and while the bar is moving in the forward direction, alternately adjust the balance x and y controls to bring the null waveshape, as seen on the oscilloscope, to as near a straight line as possible. Do not stop the bar while making these adjustments, because it heats rapidly and its permeability changes with increasing temperature, resulting in a flaw indication when passed through the test coil on a repeat run. Furthermore, do not adjust the controls when the feed is reversed through the test coil, because the magnetic field is dragged by the bar in the direction of travel and the balance obtained would not be correct for the normal feed direction.

Assuming the oscilloscope controls are adjusted properly, the two stress peaks of the null waveshape will be located equidistant from the edge of the tube face and 180° apart. Locate the index or gate about 10 to 15° to the right of the left stress peak by use of its control knob, and adjust its height by the gain control so that it is easily recognizable. The choice of using a 60-Hz filter, either in or out, is based on past experience when testing the same type of material with a similar setup. Finally, readjust the sensitivity control so that the small fluctuations within the gate do not reach the preset trip level for the various signaling devices.

Sorting Procedure Using a Permeability System

The following example of a procedure for sorting involves the use of the standard coils (Fig. 18) for detecting variations in hardness, grade, and structure, employing three different electronic gates within 180° (one-half cycle). In this case, the same package of electronic gear is used for pickup from both the null and the standard coils.

Position the waveshape switch to the respective A or B standard circuit, and its

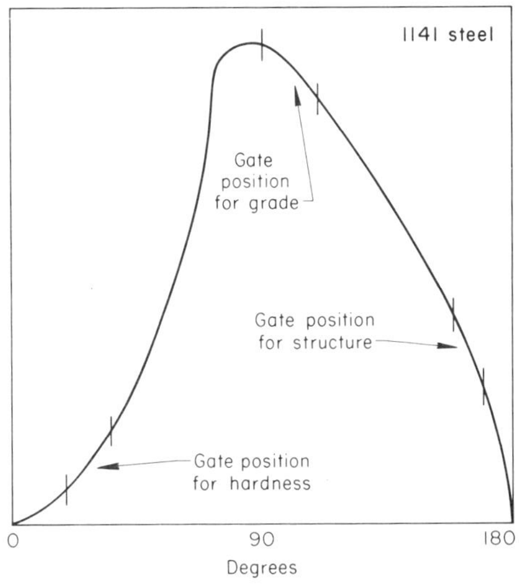

Fig. 21 Waveshape for one-half of an electrical cycle as seen on the screen of an oscilloscope with a bar of 1141 steel in the coil. Three electronic gate positions are indicated for inspecting for hardness, grade, and structure.

oscilloscope trace will appear on the face of the tube in place of the former null trace. Adjust the sensitivity control to the third or fourth step, and with a bar feeding through the test coil, position the gate to its desired location by means of the coarse-gate and fine-gate set controls. The associated zero-center meter will deflect either to the positive or negative side, and by turning the compensator knob in the appropriate direction to deflect it toward its midpoint, the meter can be balanced to zero. A rough adjustment can be made with the bar stationary in the test coil. When the bar is heated, the oscilloscope pattern continually changes. By moving the bar forward two or three times while making the above adjustments, the final balance can be obtained. The same procedure is used for the other circuit, but the gate should not be at the same position on the electrical cycle (Fig. 21). Figure 21 shows the waveshape for one-half of an electrical cycle for a bar of 1141 steel. The three gate positions for detecting variations in hardness, grade, and structure are indicated on the waveshape.

The bar that was used for setup purposes should be set aside until testing of the whole lot of material is completed, because every bar is being compared to the setup bar. Should the meters exhibit a deviation for the whole length of the bar, the bar could be a different grade or contain a full-length seam or crack. A varying meter reading may indicate a change of hardness or structure as well as a deep flaw. The latter will be simultaneously indicated by the null circuit. If an unwanted grade of steel bars has been separated, it is advisable to retest this group, choosing one bar from the unwanted lot for setup, because it is not unusual to find a third grade within a mixed lot of steel.

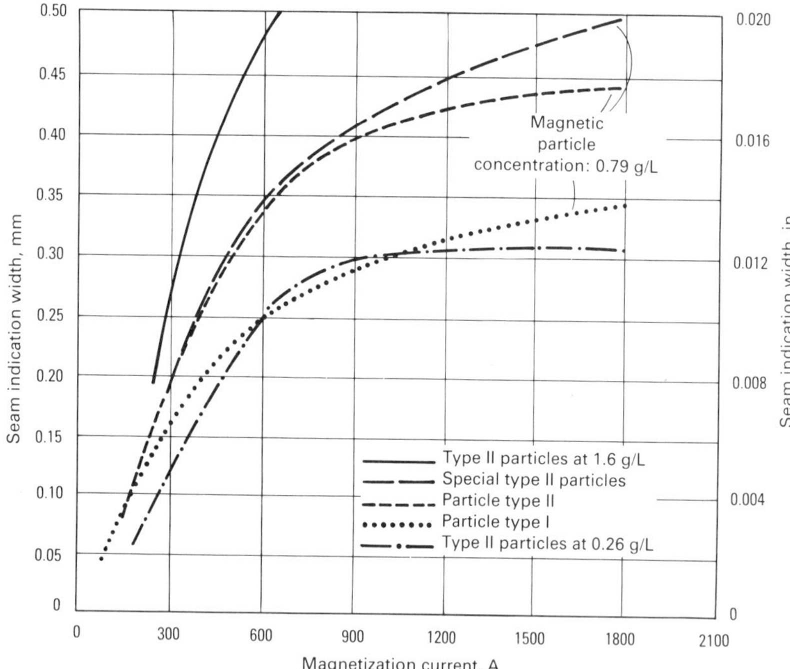

Fig. 22 Seam indication width versus magnetization current for a 105 × 105 mm (4⅛ × 4⅛ in.) 1021-1026 grade steel billet. Seams tested: center of billet face perpendicular to billet surface; seam or portion of seam with width ≧ 0.025 mm (0.001 in.) for a total depth of 0.76 mm (0.030 in.). Magnetic particle sizes: type I, 3 to 28 μm (120 μin. to 0.0011 in.) (major % = 8 to 18 μm, or 320 to 720 μin.); type II, 15 to 60 μm (600 μin. to 0.0024 in.) (major % = 28 to 48 μm, or 0.0011 to 0.0019 in.); special type II, 40 to 74 μm (0.0016 to 0.0030 in.) (major % = 44 to 62 μm, or 0.0017 to 0.0025 in.). Bath application: 76 L (20 gal.) pressurized mixing tank with hose and hand-operated applicator. Source: Ref 2

Nondestructive Inspection of Steel Billets

Steel billets are generally less uniform in section and straightness than steel bars. Furthermore, billet surfaces are usually less refined and therefore not as smooth as bar surfaces. These characteristics of billets make it more difficult to establish methods and procedures for nondestructive inspection. The methods described in this section are applicable to the common rolled round-cornered-square billets.

Surface Preparation

Regardless of the method used for billets, inspection results are greatly improved when they are free from excessive scale and blisters. Tightly adherent scale usually does not interfere with inspection unless it is thick enough to affect the sensitivity of the inspection method. Two common methods for preparing the billet surfaces for inspection are pickling in hot acid and gritblasting.

Magnetic Particle Inspection

The visual inspection of billets has been upgraded by the addition of the magnetic particle method. The principles involved in the magnetic particle inspection of billets are essentially the same as those for the magnetic particle inspection of other ferrous products or product forms (see the article "Magnetic Particle Inspection" in this Volume). As a rule, the wet fluorescent particle system is chosen for inspecting billets.

Magnetic particle inspection has improved the ability to detect obscure flaws, and through the use of mechanical handling equipment, inspection can be accomplished quite rapidly. However, this method is subject to some of the inherent disadvantages of visual inspection. Test results depend on the alertness and eyesight of the operator and on his ability to judge the severity of the flaw. Other variables, such as magnetizing current, particle size, and contamination of the particle bath by foreign substances, can vary the intensity of the flaw indication and thus increase the difficulty of appraising its severity (Fig. 22). Another factor is that, even when the above conditions are controlled, the magnetic leakage field attracting the particles is not uniform across the flat face of a square billet. The field density is maximum at the center of the flat face and almost nonexistent at the corners of magnetized billets. Consequently, the particle concentration and the flaw indication decrease with distance from the center for any one of the four flat faces.

Another magnetic test for the inspection of billets also requires the billet to be magnetized. Instead of using particles, a magnetic tape placed close to the billet surface is used to detect the presence of flaws. The flux leakage resulting from a surface discontinuity is recorded on the tape. The tape is then scanned by a tiny probe coil, which detects the flux leakage recorded on the belt. The probe transforms the magnetic leakage into an electrical signal. The signal transmitted to the control cabinet by means of contactless transformers will, if it is larger than a preset value, operate an alarm or marker. The billet face is usually divided into five to ten tracks, each about 19 mm (¾ in.) wide. The signal operates an electronic trigger circuit, which trips the appropriate indicator for the track in which the flaw lies.

This method eliminates a deficiency of the magnetic particle test in that the interpretation of results is not dependent on an operator. Also, the use of compensating circuitry reduces the differential in sensitivity to flaws from the middle to the corner of the billet.

Rolled Versus Continuous-Cast Billets (Ref 2). The major flaw types for rolled-steel billets are surface or near-surface seams, which are primarily oriented along the length of the billet. For continuous-cast products, in the as-cast state, flaws include surface cracks oriented in the transverse direction and round near-surface flaws such as slag pockets.

There are two basic methods of magnetic field application. For detecting longitudinal seams or cracks, the circular magnetization method is used. For detecting transverse seams or cracks, longitudinal (or coil) magnetization is used. For continuous-cast billets, circular and coil magnetization have been applied simultaneously. To detect multidirectional near-surface flaws and surface flaws, it is convenient to use simultaneously a dc field for circular magnetization and an ac field for longitudinal magnetization.

Generally, the billets are 3 to 14 m (10 to 45 ft) in length. The large billets are often referred to as blooms. To apply the circular field, the current is passed directly through the billet longitudinally. To apply the longitudinal field, the billet is placed along the axis of the coil, and either the billet or the coil is moved relative to the other.

In the wet magnetic particle method, for economy and handling convenience, a common carrier of the particles is water. The particles are formulated so that they will not easily deteriorate in prolonged mixture with water. Other ingredients can be added to facilitate the use of such a mixture, such as a wetting agent to disperse the particles and properly wet the part surface, antifoaming chemicals to reduce the suds due to agitation, and antifreeze to keep the bath from freezing in winter. If needed, fixer or binder can also be added to the bath to make the dried indications durable enough to withstand the normal handling in the mill.

Inspection of Rolled Billets. Typically, particle types with sizes ranging from less than 10 to 70 μm (400 μin. to 0.003 in.) are

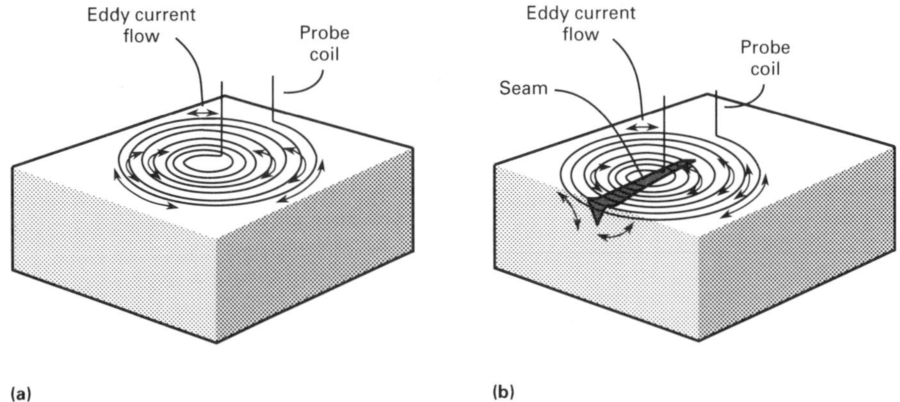

Fig. 23 Schematics of eddy current flow. (a) Eddy current flow around a probe coil for a sound billet. (b) Eddy currents flowing around the seam in a defective billet, thus altering the electrical loading on the probe coil

often adequate to find the significant seams in rolled-billet inspection. The finest-size particle, 6 μm (240 μin.) or less, is good for finding extremely minute flaws—for example, inclusion flaws as small as 25 μm (0.001 in.) wide in 0.25 mm (0.010 in.) thick steel sheets. For these flaws, larger-size particles may be ineffective; significant leakage fields are confined to very small areas. The total forces exerted on the individual particles are usually weak a short distance away, and the relatively bulky size and the momentum of large particles would prevent most of them from getting close enough to the flaws to be strongly attracted and stay there. As a result, either too few magnetic particles are retained at the flaw sites to contrast with the background, or the indications can be easily disturbed by their own weight and by the bath flow.

However, very fine and shallow flaws are not important in practical billet inspection, in which flaw depths greater than 0.64 mm (0.025 in.) are often the only ones of interest. (When reheated, the outer skin of the billet with shallow flaws will be oxidized to become scale and will fall off the part during further rolling operations.) Therefore, particles with nominal diameters of 10 to 50 μm (400 μin. to 0.002 in.) are often used to give large indications for the significant seams and to ignore very minor ones.

The ultimate choice of particle type is not determined by particle size alone. Other factors, such as the magnetic properties, color, and brightness of the particles, must also be considered.

For the proper magnetization current level, a rule of thumb has been developed: Roughly 1000 Adc per 25 mm (1.0 in.) of material diameter is satisfactory for most critical inspections. The equivalent surface field is about 1.2×10^4 A · m^{-1} (157 Oe). For most practical large-billet inspections, less than half this value is used.

Inspection of Continuous-Cast Products. Slag pockets, transverse cracks, and longitudinal cracks are the typical flaws found in continuous as-cast products (billets, blooms, and slabs). The slag pockets are near-surface inclusions. In a few cases, they can be partially exposed to the surface. The inclusions are mainly slags. There may also be voids with coats of oxidelike chemicals on the walls. They are often approximately round, but can be elongated in depth or longitudinal direction. Depending on product type, the largest slag pockets uncovered in the laboratory can be 3.2 to 6.4 mm (1/8 to 1/4 in.) in diameter. The smallest ones may have diameters less than 0.8 mm (1/32 in.). The pockets are generally found within a 2.5 mm (0.1 in.) deep layer under the product surface. Small slag pockets are often found near the surface. Cracks in the continuous-cast products have been observed to be deep and tight, with inclusions. The cracks can occur in the direction of the oscillating mold marks (transverse corner cracks) or perpendicular to that direction. Depending on the casting technique, they can occur in groups and branch out in all directions like a web, covering significant areas.

In general, the discussions about seam indication formation in rolled billets can be applied to crack indications in the continuous-cast product. For example, in the case of slag-pocket indications, the parameter equivalent to the seam width is the indication diameter. However, there are still a few differences between rolled-billet inspection and continuous-cast bloom inspection.

Differences in Magnetization Levels. The magnetization level used in continuous-cast product inspection is usually stronger than that required for rolled billets. Because of the subsurface nature of the flaws and the wider flaw dimensions, the attraction of the magnetic particles to the leakage fields for the flaws is not as intense as the attraction normally found in rolled billets. Therefore, the magnetization levels often must be higher.

The magnetization technique for the continuous-cast product is more complicated because magnetization is needed in both circular and longitudinal directions. When simultaneously applied multidirectional magnetization is used, the combined field effect equals the oscillating field (swinging vector) on the part surface.

Eddy Current Inspection

One advantage of the eddy current inspection of billets over magnetic particle inspection is that uniform results can be obtained without significant involvement of operator judgment. Another advantage is that the corners of a billet can be eddy current inspected to the same degree of thoroughness as the flat surfaces.

There are, however, certain problems involved in the eddy current inspection of billets. Billets are not sufficiently uniform in cross section or straightness to permit their being fed through encircling coils. Also, this type of coil is not very sensitive in detecting the longitudinal-seam type of flaw. Therefore, the eddy current inspection of billets is usually performed by using probe coils. Figure 23(a) illustrates how a probe coil induces eddy currents in the billets, and Fig. 23(b) illustrates how the flow of eddy currents is changed by the presence of a seam in a billet.

It is not feasible, however, because of the inherent cross-sectional shape of square billets, to inspect by merely rotating them around the surface. The probe would bounce and this would excessively vary the distance between the billet surface and the probe coil, thus degrading the accuracy of the results. Also, the rapid bounce could generate false signals. Therefore, another approach must be taken.

Use of Rotary Probe Unit. One approach is the use of a rotary probe unit such as that illustrated in Fig. 24. The significant components of this unit are search probes mounted in a housing to which is attached a tungsten carbide wear shoe and an arrangement of hardened steel rollers. This entire assembly is spring actuated. This unit and usually a second one like it are mounted 180° apart on a drum-shaped assembly. Two probes are mounted in each unit to increase the amount of inspection coverage. Electrical energy reaches the search probes through slip rings. The units rotate around the billet as it progresses through the drum assembly. As can be seen in Fig. 24, the coil in the search probe is, at all times, the same distance from the billet surface, separated only by the tungsten carbide wear shoe. The spring-actuated rollers move as required to aid in holding the search probe on the billet surface as the entire unit revolves around the billet.

One requirement for accurate results is that the speed of the probe relative to the test surface must be maintained within certain limits. Experimental studies have established that the optimum range for this eddy current test is 25 to 43 m/min (80 to 140 sfm). A probe speed below this range results in an inconsistent measurement of flaw

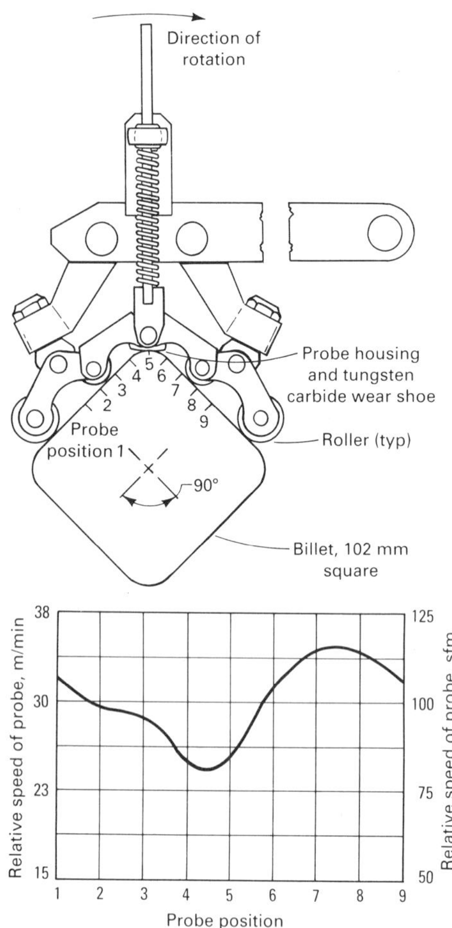

Fig. 24 Rotary probe unit used for the eddy current inspection of steel billets, and graph showing effect of position on speed as the probe unit traverses radially over one quadrant of a 102 mm (4 in.) square billet

depths. Speeds above this range cause objectionable electronic noise.

The graph in Fig. 24 shows the relative speed of the probe at different positions on one quadrant of a 102 mm (4 in.) billet section. Many of the dimensions of the probe assembly components affect the instantaneous velocity of the probe relative to the billet surface. Therefore, the dimensions of the elements must be selected so as to impart to the probe a velocity that is maximum on the flat, where bounce is least likely to occur, and minimum near and around the corner, where the coil has the greatest tendency to leave the surface. The speed fluctuates smoothly within the allowable testing range as the probe assembly is rotated around the billet.

The rotary probe unit shown in Fig. 24 can be used to test billets ranging from 75 to 152 mm (3 to 6 in.) square. For other sizes, the dimensions of the components of the unit must be changed accordingly.

The forward speed of the billet conveyor is between 7.6 and 23 m/min (25 and 75 sfm). The forward speed that can be used depends on the size of the billet and the length of the flaws that must be detected. The inspection is performed on a spiral band on the surface of the billet; the pitch of the spiral (and therefore production rate) is determined by the minimum length of flaw to be detected. Production use in one mill calls for the detection of all flaws 25 mm (1 in.) and longer, resulting in a speed of 12 m/min (40 sfm) for 102 mm (4 in.) billets and 10 m/min (34 sfm) for 127 mm (5 in.) billets. At these speeds and with no delays, 41 to 54 Mg/h (45 to 60 ton/h) can be inspected in one machine.

Marking the flaws is done with spray markers that are mounted near and rotate with the search-probe assemblies. One marker is used per search-probe assembly, and either of the two search probes will activate the marker. With this arrangement, the marks always lead, by a short distance, the actual longitudinal location where the flaw was detected. The circumferential position of the flaw is marked exactly, because the spray markers are displaced from the probe centerline at a distance that is equal to the distance that the faceplate rotates during the reaction time of the marking system.

Calibration of the electronic circuitry is accomplished by moving the entire rotating portion of the machine out of the line of billet travel. A short test billet is placed into the test position. The probes are rotated while contacting the test billet, and the signals are recorded. The electronic circuit is then calibrated with simple dial adjustments.

The test billet contains four simulated seams produced by milling slots 0.76 mm (0.030 in.) deep with a 0.15 mm (0.006 in.) wide cutter. The locations of these milled-slot seams are on the flat, before the corner, on the corner, and after the corner.

REFERENCES

1. N. Matsubara, H. Yamaguchi, T. Hiroshima, T. Sakamoto, and S. Matsumoto, Nondestructive Testing of Cold Drawn Wires and Cold Forged Products, *Wire J. Int.*, March 1986
2. L.C. Wong, Test Parameters of Wet Magnetic Particle Inspection of Steel Billets, *Mater. Eval.*, Nov 1988

Tubular Products

WROUGHT TUBULAR PRODUCTS are nondestructively inspected chiefly by eddy current techniques (including the magnetic flux leakage technique) and by ultrasonic techniques. In general, the eddy current and magnetic flux leakage techniques are applied to products not exceeding 1020 mm (40 in.) in diameter or 19 mm (¾ in.) in wall thickness. On the other hand, ultrasonic inspection is used on tubes ranging from 3.2 to 2030 mm (⅛ to 80 in.) in diameter and from 0.25 to 64 mm (0.01 to 2½ in.) in wall thickness. However, there are many exceptions, and the range of special techniques and applications associated with each inspection method is large. Most welded and seamless tubular products are nondestructively inspected by the manufacturer at the mill.

The many uses to which steel tubular products have been applied form a basis for classifying steel tubular products; for example, the terms casing, tube, and pipe are assigned on the basis of usage, as in water-well casing, oil-well tubing, and drill pipe. A second classification is based on methods of manufacture. Accordingly, all steel tubular products can be classified as either welded or seamless. A third classification applicable to special shapes can be considered subordinate to both of the general classifications above.

The major applications of the nondestructive inspection of tubular products are:

- Detection and evaluation of flaws
- Sorting of mixed stock
- Measurement of dimensions
- Comparative measurement of specific physical and mechanical properties

Of these, the primary application is the detection and evaluation of flaws. Sorting is often an auxiliary application employed for grade or size verification and can be based on chemical composition, dimensions, physical and mechanical properties, or other significant variables. A difficulty encountered in sorting arises when variables of little or no interest affect instrument indications to a greater degree than do the variables of interest. Several applications involving the measurement of dimensions and physical properties are described in this article. Others, such as the measurement of coating thickness and the noncontact measurement of wall thickness and variations in diameter, require highly specialized instrumentation and are discussed elsewhere in this Volume. Additional information on the inspection of pipe and pipelines, including the examination of field-welded girth welds, is available in the articles "Weldments, Brazed Assemblies, and Soldered Joints" and "Boilers and Pressure Vessels" in this Volume.

Selection of Inspection Method

The fundamental factors that should be considered in selecting a nondestructive inspection method, and in selecting from among the commercially available inspection equipment, are the product characteristics, the nature of the flaws, extraneous variables, rate of inspection, the end effect, mill versus laboratory inspection, specification requirements, equipment costs, and the operating costs.

Product Characteristics. Among the product characteristics that may affect the choice of inspection method and equipment are tube or pipe diameter, wall thickness, surface condition, method of fabrication, electrical conductivity, metallurgical condition, magnetic properties (notably permeability), and degree of magnetization.

Nature of Flaws. Both the nature of flaws and of potential but unallowable deviations from certain specified dimensions or properties have a bearing on the selection of inspection methods and equipment. The nature of flaws is often markedly influenced by the method of manufacture. For example, flaws in welded pipe are usually confined to the vicinity of the weld; therefore, an inspection procedure that is confined to the weld area may be adequate. If the welds are resistance welds, the most usual flaws are located in the weld plane and are in effect two-dimensional, having length and width but negligible thickness. On the other hand, if the welds are arc welds, porosity is the most usual flaw. In all welded tubular products, cracks are the most damaging flaws. In seamless tube, the location of flaws is not restricted, but may occur anywhere in or on the tube section.

Extraneous Variables. Many of the measurable variables in tubing and pipe are normal to the product and are not cause for rejection. These extraneous or harmless factors sometimes exert a greater effect on the inspection instrument than do the flaws that must be detected. For example, variations in magnetic permeability are common in steel and generate large signals in instruments that are permeability sensitive. However, the signals often are not pertinent to the test, nor are they cause for rejection. Surface scratches may be cause for rejection in some products and yet may be acceptable in others. Consequently, the inspection method and instrument selected must ignore or minimize variables that will not affect the utility of the part in its intended application.

The rate of inspection required is a major factor in the selection of an inspection procedure. When the value of the part or the hazard associated with its application justifies slow and thorough inspection, the procedure chosen is likely to be radically different from that selected for a mass-produced, low-cost product used in a noncritical application.

End Effect. In some applications, the only portions of the tube that are genuinely critical in its ultimate application are the ends. Unfortunately, with many nondestructive testing instruments, specific problems arise when inspection of the ends is required. End effect is encountered with the eddy current, ultrasonic, and radiographic methods. Consequently, inspection of the entire tube and the ends may require two different procedures; as a result, production speed is reduced and the total cost of inspection is correspondingly increased.

Mill Versus Laboratory Inspection. Although laboratory demonstrations of nondestructive inspection techniques may yield excellent results, subsequent mill performance may be entirely unsatisfactory because of conditions present in the mill that were not present in the laboratory.

Specification requirements may also affect the choice of inspection method and equipment. When the tubular product is covered by a flaw-size specification, all tubes with flaws larger than those specified must be rejected. However, tubes with flaws smaller than the specified rejection level should be accepted. Because many nondestructive inspection systems do not provide for linear adjustment or are incapable of making the required differentiation, this aspect of instrument performance must be carefully investigated.

Equipment cost is usually a major factor in the selection of inspection method and equipment. The initial cost of equipment may occasionally be minor, but in some

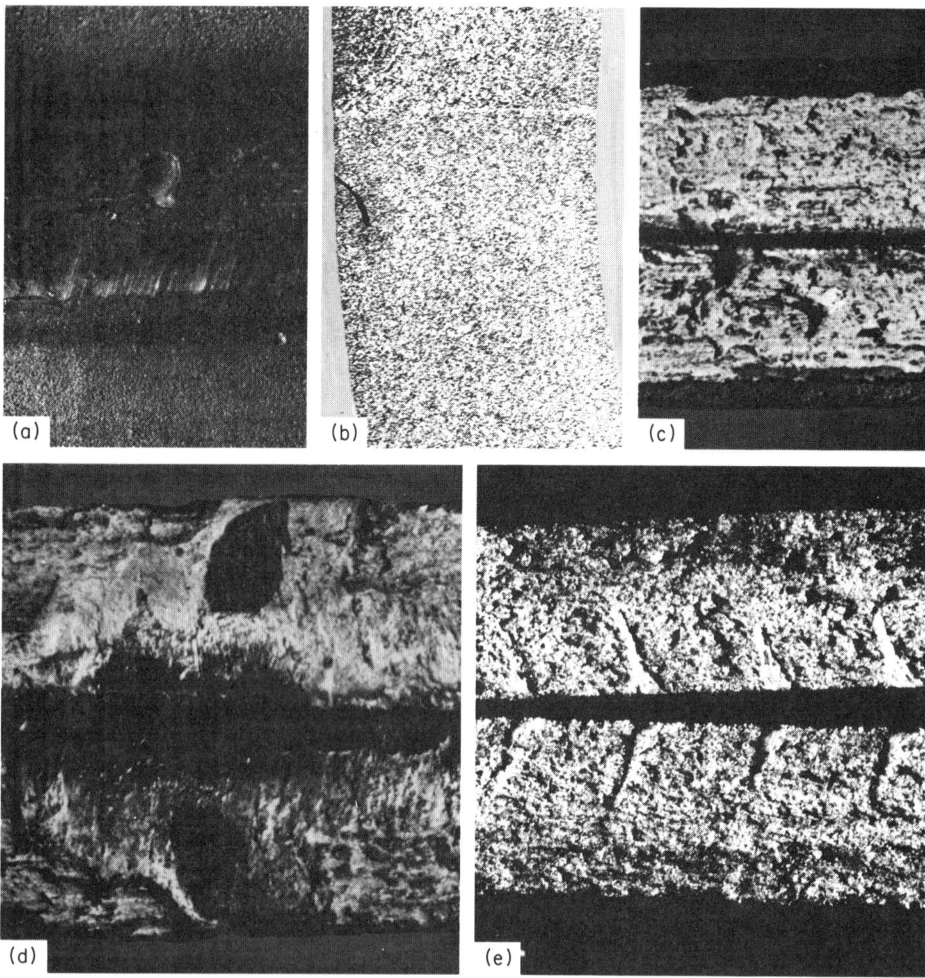

Fig. 1 Typical flaws in resistance-welded steel tubing. (a) Contact marks (electrode burns). (b) Hook cracks (upturned-fiber flaws). (c) Weld-area crack. (d) Pinhole. (e) Stitching. Views (c), (d), and (e) are mating fracture surfaces of welds.

cases installation may cost over $1 million in basic and related equipment. The lowest-cost equipment may be for magnetic particle or liquid penetrant inspection; high-cost installations involve automatic flaw marking, classification of product based on flaw magnitude, computer analysis of results, multiple sorting levels, and many other convenience factors.

The operating cost of inspection procedures and equipment varies widely. In general, it is inversely proportional to the cost of the installation. The more expensive installations are usually completely automatic and are incorporated in a production line whose primary function is something other than inspection. Consequently, inspection adds little to total operating cost. In contrast, the lower-cost installations usually involve a separate operation and require the services of a highly trained, skilled operator.

Resistance-Welded Steel Tubing

The diameters of resistance (longitudinal) welded steel tubing range from about 13 to 914 mm (½ to 36 in.); wall thicknesses, from 0.38 to 13 mm (0.015 to 0.5 in.). Tubing of intermediate and smaller diameters is produced on a draw bench.

Flaws that occur in resistance-welded steel tubing include cold welds, contact marks, cracks, pinholes, and stitching; these are described below. The terminology used to designate such flaws varies; the terms used in this article are those adopted by the American Petroleum Institute (API) (Ref 1).

Cold weld is the term widely used to indicate inadequate or brittle bonding with no apparent discontinuity in the fracture. Cold weld cannot be detected reliably by any nondestructive inspection method currently available.

Contact marks (electrode burns) (Fig. 1a) are intermittent imperfections near the weld line that result from miniature arcs between the welding electrode and the surface of the tube.

Hook cracks (upturned-fiber flaws) (Fig. 1b) are separations within the base metal due to imperfections in the strip edge, which are parallel to the surface and turn toward the outside or inside surface when the edges are upset during welding.

Weld-area cracks (Fig. 1c) are any cracks in the weld area not due to upturned fibers.

Pinholes (Fig. 1d) are minute holes located in the weld line.

Stitching (Fig. 1e) comprises a regular pattern of light and dark areas that are visible when the weld is broken in the weld line. The frequency of variation usually corresponds to weld-current variation. Increased use of ultrahigh-frequency current for welding has minimized the occurrence of stitching.

The nondestructive inspection of resistance-welded tube can be performed continuously on a welding machine or on individual lengths at any stage of processing. When performed on a welding machine, test indications can be used to guide the welding-machine operator in making machine adjustments.

Eddy Current Inspection

In 1929, H.C. Knerr initiated the development of eddy current inspection for Republic Steel Corporation. In 1930, in conjunction with Knerr, C. Farrow demonstrated the detection of flaws in a brass tube by the phase shift in eddy currents generated in the tube. In 1931, he demonstrated that the phase-shift concept could be applied to magnetic steel tube, provided the area being inspected was magnetically saturated (Ref 2). The eddy current inspection method replaced both hydrostatic and pneumatic pressure testing of resistance-welded tubing because of its superiority in detecting weld imperfections. Several of these typical weld imperfections are shown in Fig. 2.

At present, the eddy current methods are probably the most widely used for the inspection of welded steel tubing in diameters up to 75 mm (3 in.), although these methods are not limited to the smaller diameters. In welded tubing, most flaws occur in or near the longitudinal-welded seam, and in most cases a test of a narrow band including the seam is adequate. This makes possible the use of small eddy current probe coils tangent to the seam area, eliminating the diameter limitation. Tubes with diameters up to 406 mm (16 in.) are currently being inspected in production by this method.

Eddy current inspection is usually on tubing having wall thicknesses less than 3.2 mm (⅛ in.), but successful production testing has been reported on tubing having wall thicknesses to 13 mm (½ in.) (Ref 3). Most eddy current tests use differential systems and therefore are most sensitive to flaws that involve a marked change in normal electrical characteristics. If the flaw is of considerable length and of uniform characteristics, it is sometimes necessary to use special arrangements for its detection. Small probe coils continuously compare the

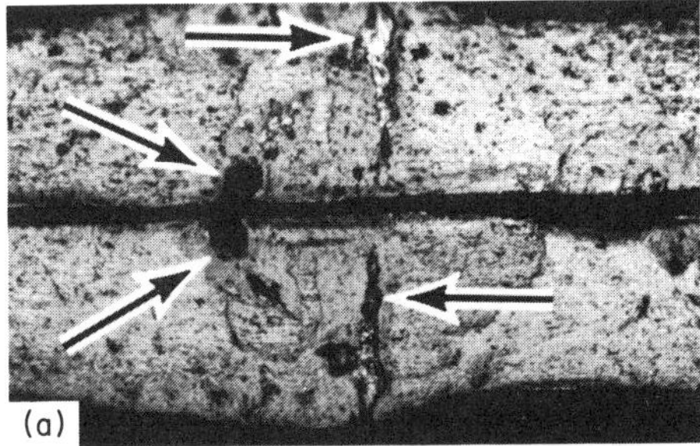
(a)

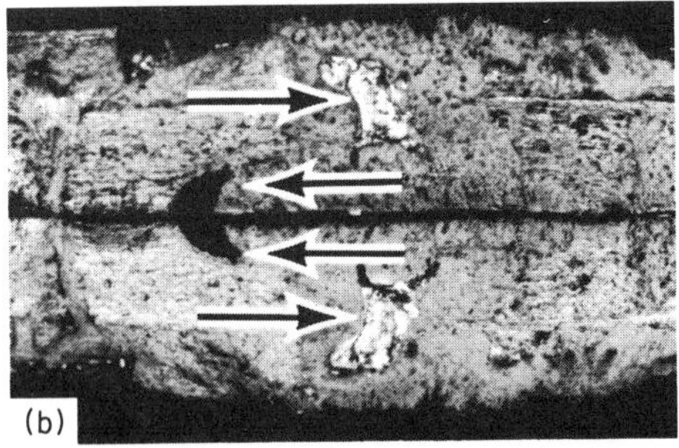
(b)

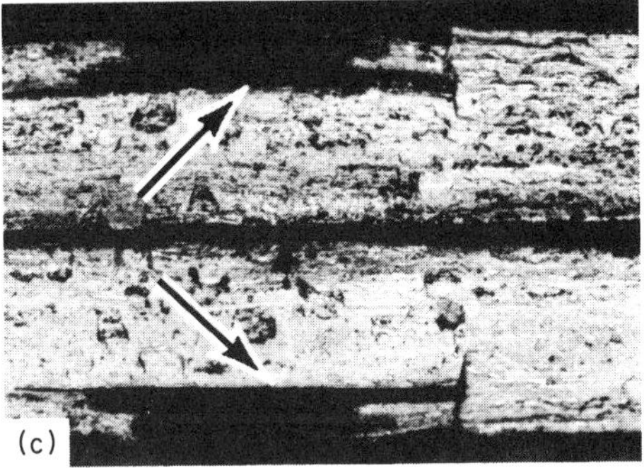
(c)

Fig. 2 Mating fracture surfaces of pipe or tube welds showing imperfections detectable by eddy current inspection but not by pressure testing. (a) Unwelded spot (diagonal arrows) and a nonpenetrating pinhole (horizontal arrows). (b) Unwelded spots, probably caused by entrapped foreign matter. (c) Surface crack in weld

weld zone with the base metal, thus revealing the existence of the elongated or long flaw.

When inspecting for shallow-cracklike surface flaws 3.2 mm (⅛ in.) or less in depth, relatively close correlation between crack depth and signal magnitude has been obtained with a single coil arrangement without magnetic saturation (Ref 4). However, the limited penetration of this arrangement and its need for a surface opening for depth evaluation limit its usefulness.

The speed of inspection by eddy current methods depends in part on many factors, including the size of the flaw that must be detected, the discriminating ability of the circuit used, end-inspection requirements, and the speed of response of the signal circuit. The mathematical relationship of the test-current frequency and linear speed may automatically limit the size of flaw that can be detected. Speeds of 305 m/min (1000 sfm) have been recorded, but the usual speed is 45 to 90 m/min (150 to 300 sfm).

Weld Twist. When using probe coils in eddy current inspection, the twist in the weld sometimes causes a special problem. When the weld twists out of the zone of high sensitivity, the effectiveness of flaw detection is markedly reduced. The problem cannot be solved by increasing the size of the arc segment covered by the probe coil, because this arrangement also reduces sensitivity. One solution involves the use of a series of small probe coils, staggered with respect to the weld line, to ensure continuous coverage (Ref 5, 6). The problem has also been solved in some installations by taking advantage of the electromagnetic difference that exists between the weld zone and the base metal. Special probe coils respond to this difference and automatically rotate the test head or the tube until the weld zone is properly located with respect to the probe coil.

End effect, caused by abrupt changes in the magnetic field, becomes a problem whenever cut lengths are inspected. Various auxiliary circuits, ranging widely in effectiveness, have been developed for suppressing end effect to permit satisfactory inspection closer to the end of the tube. End effect can be minimized by keeping the tube ends in contact as they move through the test coils.

Mechanical variables that may affect inspection results include transverse movement of the tube in the test coil and changes in temperature or linear speed. The contribution of these factors to test results is sometimes difficult to determine in the laboratory, but they may create serious problems in production testing.

Equipment costs for eddy current inspection can vary widely, depending on the extent of refined circuitry, automatic handling and sorting equipment, computer analyzers, or special auxiliary equipment that may be needed.

The operating costs of a well-designed eddy current system are among the lowest of any nondestructive inspection method. After the system has been properly adjusted, it can be operated by unskilled workers. When automatic marking is provided, the inspection can frequently be combined with another operation without appreciably increasing the cost of the latter operation.

Advantages and Limitations. All flaws in resistance welds except cold weld, are readily detected by eddy current methods. Cold weld is by far the most difficult of all flaws to detect by any of the nondestructive inspection methods. Although the other types of flaws listed above can be detected by eddy current methods, it should not be inferred that all eddy current instruments will detect all of these flaws. The range of capabilities of commercial eddy current instruments is extensive, and conclusions regarding their capabilities often require actual tests. Because eddy current test coils may either surround or be adjacent to the tube being tested, the variety of coil designs, arrangements, and combinations constitutes another major group of variables affecting equipment capabilities. In general, eddy current instruments have the advantages of speed in testing and convenience in operating, marking, and sorting. Perhaps their most universal disadvantage is their inability to inspect completely to the ends of tubes. Additional information is available in the articles "Eddy Current Inspection" and "Remote-Field Eddy Current Inspection" in this Volume.

Flux Leakage Inspection

Flux leakage (or magnetic field perturbation) inspection is similar to eddy current inspection but requires magnetization of the tube and is limited to the inspection of ferromagnetic materials. When the tube is magnetized to near saturation, the magnetic flux passing through the flaw zone is diverted by the flaws. Detectors of various types detect the diverted flux when either the detector or the tube is moved in a direction that causes the detector to cut through the diverted flux. This in turn produces a signal to reveal the presence of the flaw.

Various means are used to magnetize the tube. A current-carrying conductor inside the tube produces a circular magnetic field, magnetizing the tube in a circumferential direction. The magnetic flux is diverted by the longitudinal component of any flaws in its path. The probe, moving through the diverted flux, generates a signal roughly proportional to the size of the flaw. On a longitudinal-welded seam, an electromagnet with pole pieces on each side of the weld can be used to magnetize the weld area, with flux passing transversely across the seam. The magnetic flux is diverted by the longitudinal component of any flaw in the weld, and the flaw can be detected electronically. To detect transverse flaws, the tube may be magnetized longitudinally by an encircling conductor. The flux is then diverted by the transverse component of any flaw present, and the probe moving through the diverted or leakage flux reveals the presence of the flaw.

Hall probes are the detectors ordinarily used. In all applications, there must be relative movement between the probes and the diverted flux so as to generate a signal and to indicate the presence of a flaw. The relative motion can be achieved by rotating or oscillating either the tube or the probes. As in eddy current inspection, various types of instrumentation have been developed and are available commercially (Ref 7-10).

Limitations. Because of the nature of the flux leakage test, tube diameter is not a limitation, but the wall thickness that can be tested is limited by the ability of the magnetic flux to penetrate the wall and the ability of the sensor to sense flaws at a distance from the wall. Production applications have been used on tubing having wall thicknesses up to 25 mm (1 in.), but 7.6 mm (0.3 in.) is the usual limit. At wall thicknesses in excess of 7.6 mm (0.3 in.), sensitivity becomes a serious problem. However, current developments are improving capabilities in testing thick walls (Ref 8-10).

Although the flux leakage method usually detects flaws that are longitudinally oriented, the principle of the flux leakage method can be used in the design of equipment for detecting transverse flaws. Pinholes, with minimal longitudinal dimensions, and subsurface flaws are difficult to detect. For reliable detection of isolated pinholes, the pitch of the helical inspection path must be small, and the production rate is correspondingly limited. Sensitivity to subsurface flaws drops rapidly as the flaws are located farther from the surface. To detect inside-surface flaws such as cracks and gouges, flux leakage equipment requires special design features for reliable quantitative evaluation.

Speed of inspection is a function of the dimensions of the elements involved and the maximum tolerable length of flaw. Because the tube or the probe must be rotated or oscillated, only a helical or zigzag path is inspected, and the pitch of the helix or the distance between reversals must be less than the maximum tolerable length of flaw. When the tube must be fed over a central conductor for magnetization and then removed for inspection, high-speed production is hindered. The use of multiple probes reduces the actual testing time in proportion to the number of probes, but the time required to feed the tube over the conductor remains constant. Such installations are operated in production at speeds as high as 15 m/min (50 sfm). Installations using external magnetization are reported to operate at speeds to 60 m/min (200 sfm).

Weld twist presents a problem in any installation in which only the weld is inspected. In flux leakage inspection, the problem is solved by increasing the magnitude of the arc covered by the oscillating probe.

End Effect. As in eddy current inspection, the error caused by end effect can be minimized by butting the ends of the tubes together during the test.

Mechanical conditions, such as tube ovality, variations in linear speed, and transverse movement of the tube, have adverse effects on the test results and must be controlled.

Equipment Costs. As with eddy current equipment, the cost of equipment for flux leakage inspection varies. An elementary unit costs slightly more than a comparable eddy current unit because of the need for rotating devices. The addition of auxiliary equipment, such as automatic markers, recorders, computer analyzers, and special handling devices, markedly increases cost.

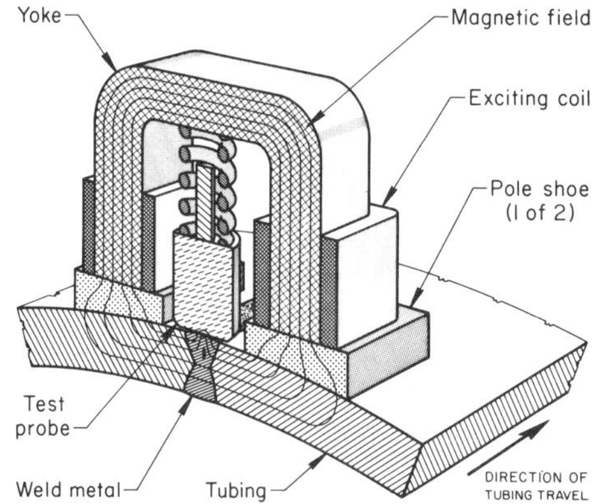

Fig. 3 Setup for the flux leakage inspection of welded steel tubing

Operating costs, which are relatively low, depend on the degree of automation and the degree to which the inspection can be combined with other operations. Flux leakage tests can sometimes be combined with another operation—for example, welding. As the tube emerges from the welding operation, it enters the field of the electromagnetic yoke (Fig. 3), which generates a flux in the weld area. The oscillating probe detects any flux diverted by a flaw in the weld. In most cases, however, the need for movement of the probe through the diverted flux makes the combination less desirable than systems with no moving parts.

Ultrasonic Inspection

Ultrasonic inspection is one of the most widely used methods for inspecting tubular products (see the article "Ultrasonic Inspection" in this Volume). Widespread use of the ultrasonic method on tubular products was made practical by the development of angle-beam shear-wave testing, immersion testing, and focused transducers. As with the eddy current and flux leakage methods, ultrasonic inspection can be applied either to the entire tube or to the weld only.

Ultrasonic inspection of the entire welded tube is usually limited to small-diameter, drawn products, in which the weld cannot easily be distinguished from the remainder of the tube. The tubing may have a diameter as small as 3.2 mm (1/8 in.) and a wall thickness of only 0.25 mm (0.01 in.). These small products are usually inspected while immersed in water (immersion inspection). They are rotated as they pass longitudinally through a glanded immersion tank. The immersed transducers must be carefully selected for tube diameter, wall thickness, and type of imperfection to be located. The transducer, focal length, response to outside diameter and inside diameter calibration notches, instrumentation pulse rate, gate adjustment for flaw alarm, and speed of tube travel are all variables to be taken into consideration. Inspection is usually performed slowly (~0.9 m/min, or 3 sfm). Tubes must be clean, straight, round, and of uniform dimensions. All types of flaws that commonly occur in resistance welds, except cold weld, can be detected by ultrasonic inspection.

Most ultrasonic inspection of resistance-welded tubing is restricted to the weld zone and is performed immediately after the welding operation. Components and adjustments for inspecting the weld must be carefully selected and accurately controlled. The transducer must be appropriate for the size and type of flaws to be detected. Focused transducers are generally preferred. The shear-wave angle must be selected for the best evaluation of imperfections. The angle often used is 45°, but tests have revealed that angles between 50 and 70° yield signals more directly proportional to the area of flaws in the weld plane (Ref 3, 11).

In the inspection of pipe, provision must be made to maintain the spacing between the transducer and the pipe constant within close tolerances as the pipe moves past the transducer. The couplant should preferably be continuously delivered to the surface of the pipe through openings in the transducer mounting. Coupling through a water jet is also used. Particular attention should be given to the detection of short flaws. Some ultrasonic pipe-inspection equipment will not detect flaws shorter than 6.4 mm (1/4 in.), which will not satisfy the inspection requirements for most resistance-welded pipe.

A disadvantage of the ultrasonic method in tube inspection is its high sensitivity to minor scratches and to elongated dimensional changes, such as the ridge left when the weld flash is not completely removed or rolled down. However, proper selection of inspection equipment can minimize this problem (Ref 11). An important development is the wheel-type search unit. The transducer of the wheel-type search unit is mounted on the axle of a liquid-filled wheel and is held in a fixed position as the wheel rotates. The surface of the wheel is flexible and adapts itself to the surface condition of the tube as it rolls over it. A small amount of liquid couplant, usually water, is required between the surface of the wheel tire and the surface of the tube. This arrangement provides most of the advantages of immersion testing without the necessity of immersing the tube.

Speed of inspection is limited by the pulse rate of the ultrasonic equipment and by the maximum length of a tolerable imperfection. Speeds as high as 69 m/min (225 sfm) have been reported, but unless multiple inspection heads are used, speed is ultimately dependent on the rejectable flaw size.

Weld twist can present a problem; as the weld twists away from the critical location, transducer sensitivity drops sharply. To maintain the weld and the transducer in the correct mechanical relationship, the weld can be positioned automatically by the use of an electromagnetic control.

End effect, although less of a problem than in eddy current inspection, is a factor in ultrasonic inspection, and supplementary testing may be necessary if inspection of the tube ends is critical. The supplementary test can be made with ultrasonic equipment of special design.

Mechanical variables are critical in contact ultrasonic testing. Spacing between the transducer and the surface of the tube, angle of transducer, and sidewise movement of tube must be accurately controlled. These variables can sometimes be better controlled in immersion testing.

The equipment costs of ultrasonic inspection equipment are highly dependent on the amount of auxiliary equipment included. Accessories such as automatic marking devices, computer analyzers, and material-handling equipment can markedly increase equipment costs, especially for the inspection of heavy pipe.

The operating costs of ultrasonic inspection, in accord with other inspection methods, depend on whether inspection is operated separately or combined with another operation. For example, if inspection is incorporated into the welding line, an inspector usually is not required, and the operating costs are minimal.

Example of Practice. Details of an ultrasonic inspection procedure applied to resistance-welded stainless steel tubing that was used to clad nuclear fuel elements are given in the following example.

Example 1: Ultrasonic Inspection of Welded Type 304 Stainless Steel Tubing Used To Clad Nuclear Fuel Elements. The ultrasonic method, employing the immersion pulse-echo shear-wave technique, was used in the inspection of resistance-welded type 304 stainless steel tubing that was used as cladding for nuclear fuel elements. A typical tube size was 9.88 ± 0.013 mm (0.389 ± 0.0005 in.) ID and 0.419 mm (0.0165 in.) ± 5% wall thickness. A standard length was 4.6 m (15 ft), and the tubing was inspected in the 10 to 15% cold-worked condition.

The reference standard, which was a tube selected from the same production lot as that inspected, contained longitudinal and transverse notches, 1.59 mm (0.0625 in.) long by 0.041 mm (0.0016 in.) deep by 0.10 mm (0.004 in.) wide, prepared by electrical discharge machining, in outside and inside walls.

Test Conditions. The tubing, in a clean and dry condition, was inspected before being cut to final length. The tubing was propelled through an immersion water tank on a preselected helix by a variable-speed drive. Throughput speed was 2.4 m (8 sfm). The tubing was rigidly supported to maintain accurate alignment between tubing and search unit.

Direction of Testing. Normally, testing was conducted in two directions only—one a circumferential mode, searching for longitudinal outside or inside notches, and the other a longitudinal mode, searching for circumferential or transverse notches. When required, two additional searches and search units were employed to search in directions opposite to those mentioned. Proper balance between outside and inside notches was obtained by careful alignment of search units and gate adjustments on the CRT.

Search Units. Focused search units, using either spherical or cylindrical transducers, were used. Frequencies ranged from 3.8 to 4.8 MHz, and the active beam profile was equal to, or shorter than, the notch length, thus ensuring the most reproducible test. The search unit in the longitudinal direction operated at a frequency of 3.8 MHz with a cylindrical line focus of 1.1 × 0.51 mm (0.045 × 0.020 in.). The search unit in the transverse direction operated at 4.8 MHz with a focus of 0.89 × 0.51 mm (0.035 × 0.020 in.). Commercially available electronic equipment with a pulse rate of at least 5 kHz, a gated alarm, and a pulse-stretching circuit was used. Readout was by a chart recorder.

Magnetic Particle Inspection

The principal use of the magnetic particle method in the inspection of resistance-welded pipe is largely limited to the inspection of pipe ends. In some pipe applications, the ends of the pipe are the sections most critically loaded, and magnetic particle inspection of the ends supplements inspection of the remainder of the pipe by other methods. In the past, the method was widely used to inspect the entire area. However, its inability to detect significant subsurface flaws, even when the magnetic particles are coated with a fluorescent, and its dependence on human vision and judgment led to its replacement by eddy current and ultrasonic methods. The magnetic particle method is still used in the mill to help establish the precise location of flaws previously detected by other inspection methods. Additional information is available in the article "Magnetic Particle Inspection" in this Volume.

Liquid Penetrant Inspection

Liquid penetrants (visible-dye and fluorescent) are ordinarily used on nonferromagnetic materials, which constitute only a small fraction of resistance-welded tubular products. Testing speeds are extremely slow, and use of these methods can be justified only when the hazard involved in end use justifies extreme inspection precautions. In such cases, the penetrant methods usually supplement other methods. Additional information is available in the article "Liquid Penetrant Inspection" in this Volume.

Radiographic Inspection

Radiographic methods of inspection cannot be used successfully on the longitudinal seam of resistance-welded pipe, because the predominant flaws are essentially two dimensional and have little or no effect on the radiographic film. However, when the ends of resistance-welded pipe are butt welded together, arc welding is frequently used, and the method normally used to inspect arc-welded joints is radiography. Additional information is available in the articles "Radiographic Inspection," "Industrial Computed Tomography," and "Neutron Radiography" in this Volume.

Double Submerged Arc Welded Steel Pipe

Most arc-welded tubular products are produced by the double submerged arc process, in which the seam is welded in two passes, one on the outside and the other on the inside. Tube and pipe diameters range from 457 to 2030 mm (18 to 80 in.), and wall thicknesses from 6.4 to 19 mm (¼ to ¾ in.). Nondestructive inspection is usually confined to the weld area.

Flaws. Some of the flaws usually encountered in double submerged arc welds are incomplete fusion, incomplete penetration, offset of plate edges, out-of-line weld beads, porosity, slag inclusions, and weld-area cracks (Ref 1). These flaws are illustrated in Fig. 4 and described below.

Incomplete fusion (Fig. 4a) is a lack of complete coalescence of some portion of the metal in a weld joint.

Incomplete penetration (Fig. 4b) is a condition in which the weld metal does not continue through the full thickness of the joint.

Offset of plate edges (Fig. 4c) is the radial offset of plate edges in the weld seams.

Out-of-line weld beads (off-seam) (Fig. 4d) is a condition in which the inner or outer weld beads, or both, are sufficiently out of radial alignment with the abutting edges of the joint to cause incomplete penetration.

Porosity (gas pocket) (Fig. 4e) consists of cavities in a weld caused by gas entrapped during solidification. Porosity may occur as subsurface or surface cavities.

Slag inclusions (Fig. 4f) are nonmetallic solid material trapped in the weld deposit or between weld metal and base metal.

Weld-area cracks (Fig. 4g) are cracks that occur in the weld deposit, the fusion line, or the heat-affected zone.

Radiographic Inspection

Film x-ray was the first nondestructive inspection method applied to the quality control of double submerged arc welded

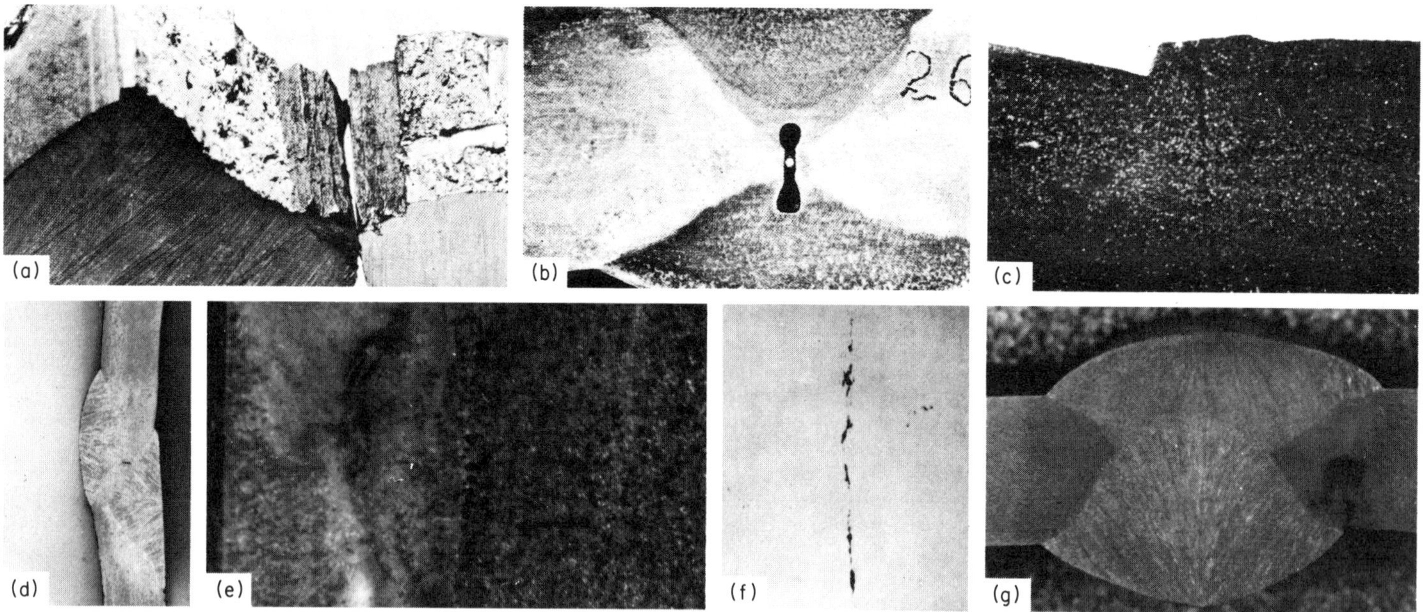

Fig. 4 Typical flaws in double submerged arc welded steel pipe. (a) Incomplete fusion. (b) Incomplete penetration. (c) Offset of plate edges. (d) Out-of-line weld beads (off seam). (e) Porosity (gas pocket). (f) Slag inclusions. (g) Weld-area crack

steel tubular products. The film technique was expensive and was subsequently replaced by the image intensifier with fluorescent screen, an arrangement that requires the inspector, stationed in a darkened room, to observe the fluorescent screen continuously while the pipe travels over the x-ray tube. Rejection of the pipe depends on an appraisal of the degree of darkness registered on the screen. This method is still in use but is being supplemented, and sometimes replaced, by ultrasonic inspection.

Limitations. In addition to high cost and dependence on the human factor, a lesser capacity for discrimination is another important reason for the replacement of x-ray radiographic techniques by ultrasonic inspection methods. Radiography is sensitive to flaws only when the flaws significantly alter the ability of the material to absorb radiation; this usually occurs when a flaw changes the effective metal thickness by 2% or more or when the flaw causes a change in density equivalent to at least a 2% change in metal thickness. The technique is sensitive to small gas pockets but is relatively insensitive to tightly closed cracks (Ref 12). Because cracks, especially those on the surface, are much more damaging than small totally embedded gas pockets, it is often essential to supplement radiographic inspection with another method that is more sensitive to cracks (Ref 13).

Speed of inspection by the continuous radiographic method depends on the eyesight, alertness, and judgment of the inspector and is usually in the range of 3 to 9 m (10 to 30 sfm).

Weld Twist. Because of the thick cross section and the width of the weld bead characteristic of double submerged arc welded steel pipe, weld twist is usually not a problem.

End effect is an important consideration. Consequently, most specifications require that the continuous radiographic inspection of pipe ends be supplemented by magnetic particle inspection, film x-ray inspection, or both.

Equipment Costs. Most of the expenditure is directed toward the mechanical arrangements needed for handling the large-diameter, thick-wall product. Precise positioning, uniform linear speed, and convenient controls for slowing speed to inspect flaw indications are essential for satisfactory inspection.

The operating costs of continuous radiographic inspection are comparatively high because of the slow speeds and the need for two operators in most installations.

Ultrasonic Inspection

In the ultrasonic inspection of double submerged arc welded pipe, the excess bead at the welded seam presents a problem in that the edges of the bead generate false indications. On the other hand, the excess bead provides a guide for maintaining the inspection head in the correct position as it tracks the weld. One solution to the problem of extraneous reflections is the use of two search units, each with two or more focused transducers and appropriate electronic accessories (Ref 14). In this method, the pipe moves into the testing station and is brushed and prewetted. An optical system locates the weld precisely. As the pipe moves forward, the two search units are lowered into the test position. As the end of a pipe leaves the tester, the search units are automatically raised from the pipe. The transducers are positioned to produce shear waves at an angle of approximately 70° in the pipe. The electronic controls are adjusted so that part of the weld zone and a 25 mm (1 in.) wide band of adjacent strip are inspected by each search unit. The inspected portions of the weld zone overlap to ensure thoroughness of inspection. Flaws exceeding a predetermined standard automatically operate the appropriate marker.

The ultrasonic method is not as sensitive to gas pockets as the radiographic method but is much more sensitive to tightly closed cracks. Such cracks can be detected even when they are located at the junction of the weld bead and the base metal.

Arc-Welded Nonmagnetic Ferrous Tubular Products

Austenitic stainless steel and other nonmagnetic ferrous tubular products are, except for seamless tubing, usually fabricated from plate, sheet, and strip by forming and arc welding, frequently by the gas tungsten-arc process. The weld flaws encountered are similar to those found in double submerged arc welded products, except for those resulting from deposition of the second bead. Although tube diameters range from 3.2 to 762 mm (⅛ to 30 in.) and wall thicknesses from 0.10 to 9.53 mm (0.004 to 0.375 in.), the small-diameter, thin-wall products predominate. Tubes having diameters from about 3.2 to 102 mm (⅛ to 4 in.) are produced on a draw bench.

Eddy Current Inspection. Although the austenitic stainless steels are nominally nonmagnetic, some of these alloys will develop magnetic constituents (notably ferrite) in the weld fusion zone and in cold-worked areas, thus causing variations in permeability. Consequently, magnetic saturation is required for the reliable eddy current detection of small flaws in the weld zones of these alloys. In addition, because of their lower electrical conductivity, stainless steels require higher frequencies than those used for the eddy current inspection of carbon steels. In other respects, the practices employed in eddy current inspection are the same as those applied to carbon steel products.

Because diameters in the range of 3.2 to 75 mm (⅛ to 3 in.) account for most of welded stainless steel tube production, encircling inductor and detector coils can be used in inspection. Usually, the coils are close fitting, with an allowable maximum clearance of only 1.6 mm (1/16 in.). The entire tube is inspected. For calibration purposes, a hole drilled through the tube wall is ordinarily used. Allowable flaws are described in ASTM A 450. In general, inspection speeds range from 7.6 to 46 m/min (25 to 150 sfm), although speeds up to 140 m/min (450 sfm) have been attained.

Ultrasonic inspection is performed on tubes ranging from 3.2 to 762 mm (⅛ to 30 in.) in diameter with wall thicknesses ranging from 0.10 to 9.52 mm (0.004 to 0.375 in.). Barium titanate, lithium sulfate, quartz, or lead zirconate can be used as the transducer element; operating frequencies range from 2.25 to 15 MHz. In general, the pulse-echo shear-wave technique is used with immersed, focused transducers. For larger-diameter tubes, where the size can cause practical difficulties in immersion and where the curvature of the tube is slight, flat crystals with conventional contact coupling can be used. In immersion testing, the shear-wave angle in the testpiece ranges from 45 to 70°.

When the application is critical, ultrasonic inspection must be supplemented by other methods, such as eddy current and liquid penetrant inspection. On larger-diameter tubing, only the weld is tested. Usually, the transducer is stationary and the tube is moved axially. With smaller-diameter tubing, the entire tube is tested. The transducer is usually stationary while the rotating tube is moved axially. The pitch of the helix inspected is a function of the maximum length of the flaw that can be tolerated. The calibration tube contains one or more milled slots, as required by the applicable specification.

Production rate depends largely on the amount of metal being inspected—that is, the weld zone only or the entire tube—and can range from a few inches to about 1.5 m/min (5 sfm). Speeds as high as 6 m/min (20 sfm) have been attained.

Liquid Penetrant Inspection. Penetrants used for the inspection of tubing may be either the fluorescent or nonfluorescent (visible-dye) type. The fluorescent penetrants may be water soluble or an emulsion. For a reliable test, the tube must always be cleaned before testing. Because the appraisal of flaws is based on visual judgment, inspection is usually restricted to the outside surface unless the tube is large enough and short enough to permit inspection of the inside surface. Production rate is highly dependent on the mechanical handling facilities available and ranges from a few inches per minute to as much as 3 m/min (10 sfm).

Radiographic Inspection. Some users specify that film x-ray must be used (Ref 15). Radiography is used to inspect tubes ranging in diameter from 6.4 to 762 mm (¼ to 30 in.) and in wall thickness from 0.64 to 9.53 mm (0.025 to 0.375 in.). Calibration is by penetrameter. Either the tube or the tester may move with respect to the other. The usual flaw detection difficulties associated with radiography are compounded with the smaller diameters because the x-ray beam must pass through two wall thicknesses. Production speed ranges from the average of a few inches per minute to as much as 3.7 m/min (12 sfm).

Continuous Butt-Welded Steel Pipe

Most carbon steel pipe is produced by a process that continuously forms hot strip (skelp) into tubular shape, then passes it through a welding horn and welding stand, where the edges are pressed firmly together and pressure welded to provide the longitudinal seam. The strip ends are flash butt welded together without stopping the line, thus making the process truly continuous. A multiple-stand stretch-reducing mill brings the welded pipe to its finished size. As the name implies, this mill is used for reducing the diameter of the pipe being produced by simultaneously applying pressure and tension. Pipe diameters range from 22 to 114 mm (⅞ to 4½ in.) and wall thicknesses from 1.9 to 12.7 mm (0.076 to 0.500 in.).

The inspection of continuous butt-welded pipe is directed primarily toward in-plant quality control, although it is required by some specifications. The eddy current method is employed almost exclusively and is applied not only to cold pipe but also to hot pipe at a temperature above the Curie point. Encircling inductor coils and differentially wound detector coils are used with a coil-to-pipe spacing of 6.4 mm (¼ in.) or less and an operating frequency of 2.5 kHz. A saturating field of about 70 000 ampere-turns is required for cold pipe, which is nonmagnetic at temperatures above the Curie point. The coil assembly for hot pipe must be water cooled and encased in heat-resistant material (Ref 16). Marking the hot pipe requires a special high-temperature paint.

The eddy current inspection setup is usually calibrated to reject all flaws, internal and external, that extend more than 12½% through the wall. The calibration standard is usually a drilled hole.

The speed of inspection corresponds with the speed of the mill and, for cold pipe, ranges from 53 to 150 m/min (175 to 500 sfm). Higher speeds are employed for inspecting hot pipe.

Spiral-Weld Steel Pipe

Spiral-weld steel pipe is made principally in large diameters and is a relatively low-production product. The pipe is either resistance or submerged arc welded.

Resistance-welded pipe is inspected by either the eddy current or the ultrasonic method. The ultrasonic or radiographic method is used to inspect submerged arc welded pipe. Usually, the ends of pipe welded by either process are also inspected by a supplementary technique, such as film x-ray or magnetic particle inspection.

Seamless Steel Tubular Products

Steels melted by various processes can be successfully converted into seamless tubes. In general, killed steels made by open-hearth, electric-furnace, and basic-oxygen processes are used. Because of the severity of the forging operation involved in piercing, the steels used for seamless tubes must have good characteristics with respect to both surface and internal soundness. A sound, dense cross section, free from center porosity or ingot pattern, is the most satisfactory for seamless tubes. Metallurgical developments have contributed greatly to the improvement of steels for seamless tubes; as a result, the seamless process has been extended to include practically all of the carbon and alloy grades of steel.

Flaws in seamless tubular products may occur at any point on the outside and inside surfaces or within the tube wall (Ref 1). The flaws usually encountered are blisters, gouges, laminations, laps, pits, plug scores, rolled-in slugs, scabs, and seams. These are illustrated in Fig. 5 and described below.

Blisters (Fig. 5a) are raised spots on the surface of the pipe caused by the expansion of gas in a cavity within the wall.

Gouges (Fig. 5b) are elongated grooves or cavities caused by the mechanical removal of metal.

Laminations (Fig. 5c) are internal metal separations creating layers generally parallel to the surface.

Laps (Fig. 5d) are folds of metal that have been rolled or otherwise worked against the surface but that have not been fused into sound metal.

Pits (Fig. 5e) are depressions resulting from the removal of foreign material rolled into the surface during manufacture.

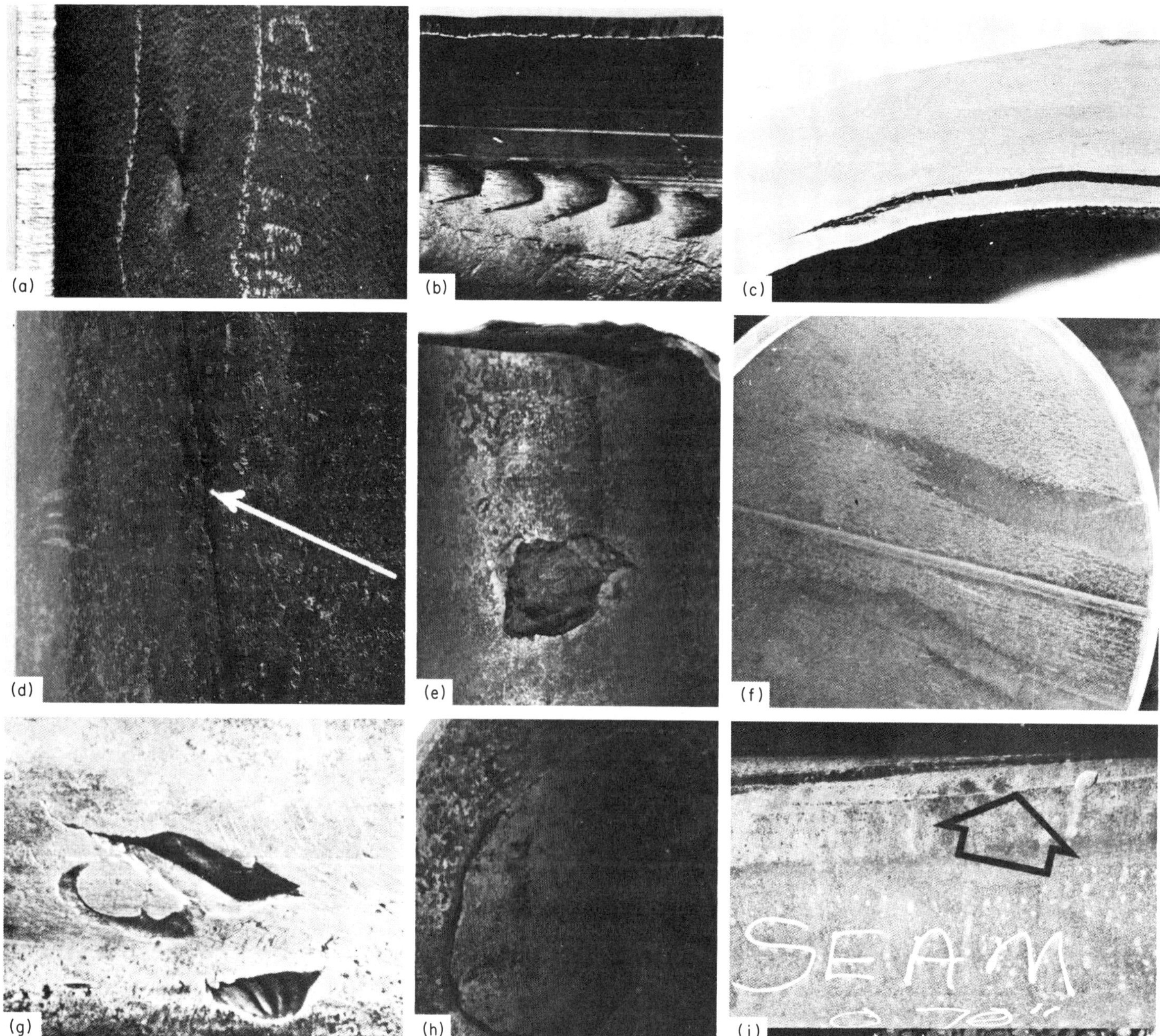

Fig. 5 Typical flaws in seamless tubing. (a) Blister. (b) Gouge. (c) Lamination. (d) Lap (arrow). (e) Pit. (f) Plug scores. (g) Rolled-in slugs. (h) Scab. (j) Seam (arrow)

Plug scores (Fig. 5f) are internal longitudinal grooves, usually caused by hard pieces of metal adhering to the mandrel, or plug, during plug rolling.

Rolled-in slugs (Fig. 5g) are foreign metallic bodies rolled into the metal surface, usually not fused.

Scabs (Fig. 5h) are flaws in the form of a shell or veneer, generally attached to the surface by sound metal. Usually, scabs originate as ingot flaws.

Seams (Fig. 5j) are crevices in rolled metal that have been closed by rolling or other work but have not been fused into sound metal.

Ultrasonic inspection is probably the method most commonly used on seamless tubular products, which range from 3.2 to 660 mm (⅛ to 26 in.) in diameter and from 0.25 to 64 mm (0.01 to 2½ in.) in wall thickness. The tube, while rotating, is usually moved longitudinally past the transducers, thus providing inspection along a helical path. In a typical installation, six transducers inspect the rotating tube as it progresses through the machine. Four transducers are below the tube and are coupled to it by water columns; two transducers are above the tube and make contact through fluid-filled plastic wheels. This machine is capable of handling tubes ranging from 50 to 305 mm (2 to 12 in.) in diameter.

As with welded tubing, the smaller diameters of seamless products are inspected while immersed, and the larger diameters are inspected by direct contact. Transducer crystals may be quartz, lithium sulfate, barium titanate, or lead zirconate. The frequency ordinarily used is 2.25 MHz. The transducer may or may not be focused, depending on the tube diameter and the nature of the flaws anticipated. In most cases, the ultrasonic shear-wave angle is 45° but may be as large as 70°. The usual couplant is water, but oil is sometimes used.

Because all installations involve rotation of either the tube or the transducers, the inspection invariably follows along a helical path. The pitch and width of the helical path vary widely, depending on the characteristics of the equipment and the specifications

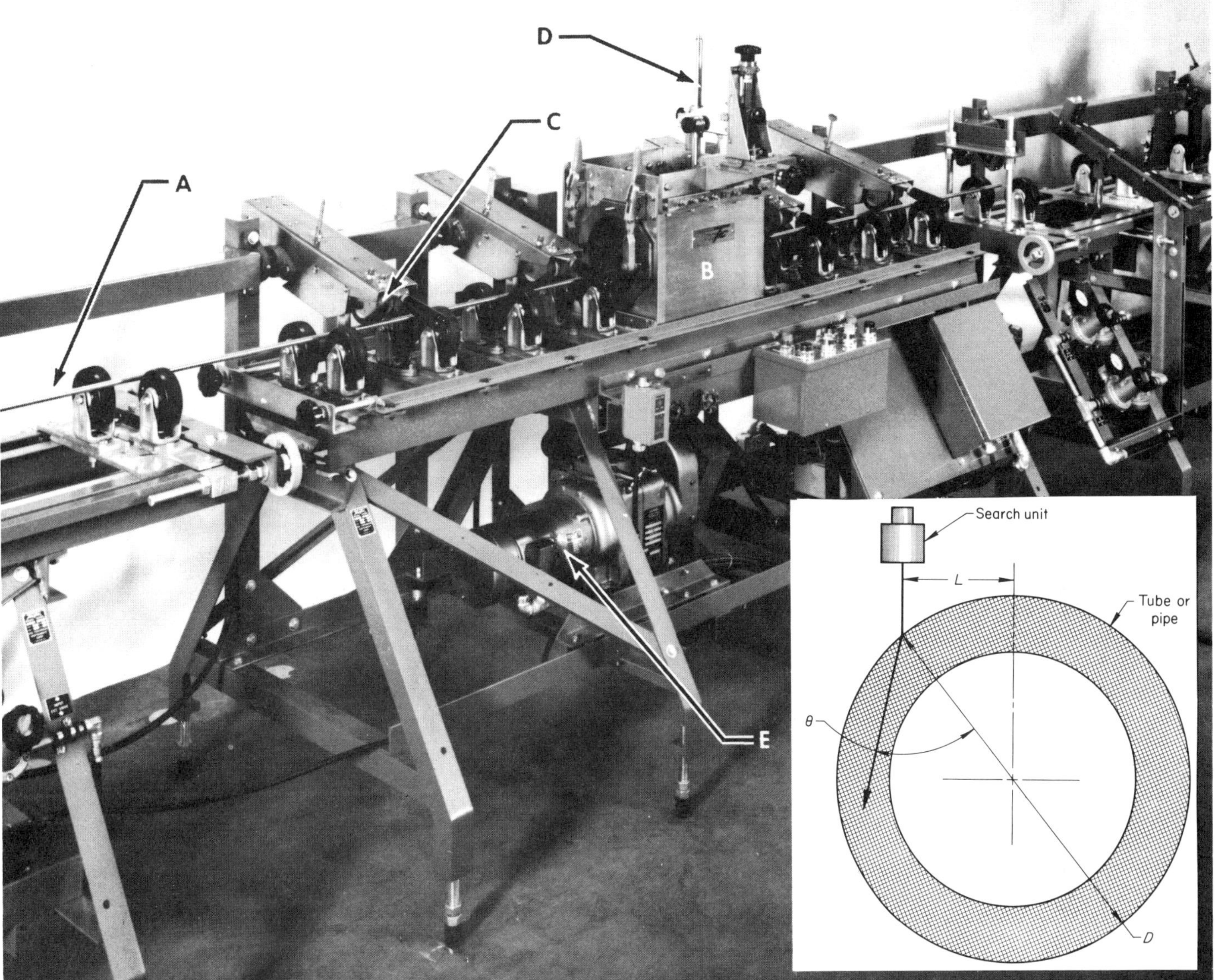

Fig. 6 Unit used for the ultrasonic inspection of seamless and welded stainless austenitic steel tubular products. A, tube being inspected; B, immersion tank; C, drive wheels; D, search-unit tube; E, drive mechanism. Inset shows lateral displacement of the search unit for circumferential inspection to detect longitudinal flaws.

to be met. The pitch is usually between 9.5 and 13 mm (3/8 and 1/2 in.), which translates to two to three revolutions per each 25 mm (1 in.) of longitudinal travel. Almost all forms of visible and audible alarms, as well as automatic recorders, are used with the ultrasonic equipment.

All types of flaws in seamless tubes can be detected by ultrasonic methods, but the minimum flaw dimensions, the degree of sensitivity, the flexibility of adjustment, and the accuracy of calibration all vary widely with the basic instrumentation and the supplementary components chosen. The flaw used most frequently for calibration is a longitudinal slot. The depth of slot may vary from 3 to 12½% of the wall thickness, depending on the end use of the product and the specification involved. The length of the slot may be as much as 38 mm (1½ in.) but is usually specified as twice the width of the transducer. Widths of slots should be kept at a minimum and should never exceed twice the depth. The frequency of calibration checks depends on the criticality of the tube application. In a few cases, a calibration check is required after every tube, but once every 4 h of production is usually considered adequate.

Speed of inspection also varies widely, depending on many variables, especially the maximum tolerable length of flaw and the number of transducers used. The current range of inspection speed is 0.6 to 46 m/min (2 to 150 sfm); the upper limit can be increased by increasing the number of transducers. The following example describes the setup used for the ultrasonic inspection of stainless steel tubular products, with special emphasis on the calibration procedures used for flaw detection.

Example 2: Ultrasonic Inspection of Seamless and Welded Austenitic Stainless Steel Tubular Products. Seamless and welded austenitic stainless steel tubular products were inspected by ultrasonics, using a system that rotated and moved the pipe or tube past a stationary ultrasonic search unit. The system was equipped to inspect by either contact or immersion. The inspection unit contained a rotational and longitudinal drive mechanism and a relatively small open-ended and glanded immersion tank through which the tube or pipe passes (Fig. 6). The tank also contained the search units.

To inspect, a reference tube or pipe was usually first placed in the unit. The reference tube or pipe was of a convenient length (usually 1.2 to 1.8 m, or 4 to 6 ft) and similar in type, size, and wall thickness to the product to be inspected. The reference tube contained notches in both outside and inside walls. Generally, the depth of the notches varied from 3 to 12½% of the wall thickness, with the shallower notch being used for seamless tube and the deeper notch for welded tube. Most notches ran in the longitudinal direction and were approximately 25 mm (1 in.) long. Typically, the width of a notch was not more than twice its depth. In some cases, transverse notches were used as well as longitudinal notches. The dimensions of the transverse notch were the same as those of the longitudinal notch. The depth of notches was never less than 0.10 mm (0.004 in.).

With the reference standard placed in the inspection unit, suitable search units were selected. For the detection of longitudinal flaws, a line-focused search unit was generally used. The focal length varied from approximately 25 to 114 mm (1 to 4½ in.). The short focal length was used for small-diameter tubing, and the longer focal tubing length was used for larger-diameter tubing. The length of the beam (lined up with respect to the longitudinal axis of the tube or pipe) varied from approximately 6.4 to 25 mm (¼ to 1 in.). The length of the beam used depended on the size of the flaw to be detected. Shorter beam lengths were used to detect smaller (shorter) flaws. A beam 13 mm (½ in.) long was normally used. To detect transverse flaws, a spherically focused search unit could be used in addition to a line-focused unit.

The search unit was centered with respect to the reference tube, ensuring that a water gap between the search unit and the tube was equal to the focal length of the search unit minus the radius of the tube. Angular adjustment and positioning were performed to obtain a maximum response from the tube wall indicative of proper centering. For circumferential inspection to detect longitudinal flaws, the search unit was displaced laterally or set off from the center of the tube to obtain the proper angle at which the sound would travel through the tube wall (inset, Fig. 6). Generally, inspection was performed using a shear wave traveling at an angle of 45° around the tube. The offset distance required was calculated using the formula:

$$L = (D/2)(v_w v_m) \sin \theta \qquad \text{(Eq 1)}$$

where L is the offset distance, D is the diameter of the tube or pipe, v_w is the velocity of sound in water, v_m is the velocity of sound in the tube, and θ is the desired refracted angle in the tube or pipe. The value for the velocity of sound in the tube depended on whether a longitudinal or shear wave was to be used. The offset value obtained from Eq 1 was an approximation. The actual amount of offset was adjusted from that calculated to obtain best presentation and equalization of responses from outside and inside notches.

When the search unit and reference tube or pipe had been properly located with respect to each other, the reference tube was rotated and driven longitudinally past the search unit. Controls of the ultrasonic instrument were adjusted to display a clear response from both inside and outside notches on an oscilloscope screen. The controls of the flaw-alarm module of the instrument were adjusted to position the gate properly to include the signals from the notches and to activate the alarm when they were detected. The pulse-repetition rate of the instrument was adjusted high enough to ensure the detection of all notable flaws at the speed of inspection.

The inspection speed was controlled by the rate of rotation and longitudinal movement per revolution (pitch) of the tube. The allowable pitch was a function of the length of the line-focused ultrasonic beam and the size of the flaw to be detected. Normally, when the reference standard was passed through the unit, the controls had been adjusted to provide more than one signal from each of the notches. When this had been established, the system was properly calibrated for production inspection.

Ultrasonic Inspection Precautions. Generally, a chart recorder is employed to provide a permanent record of the inspection described in Example 2. Multiple search units can be used in the immersion tank to provide several simultaneous inspections during one pass of the pipe or tube through the unit. Specifications may call for circumferential inspection from two directions because the reflection from a flaw may vary, depending on the direction in which the ultrasonic beam strikes it. In addition to the circumferential inspection to detect longitudinal flaws, a longitudinal (axial) inspection may be required to detect transverse flaws. Also, it may be desirable to ultrasonically measure wall thickness, eccentricity, or both. All these tests can be performed at the same time by utilizing search units that are designed for the tests and that are properly positioned in the tank.

Normally, rejection is based on the presence of flaw indications exceeding those from the reference notch. Reworking and reinspection are generally permitted if other requirements, such as minimum wall thickness, are satisfied.

Other refinements are included in, or can be added to, the inspection system. For example, the feeding of tubes to the unit and their withdrawal can be automated. Various audible and visible alarm systems and marking devices can be added.

Normally, the water used in the system is filtered and deaerated. Air entrapped in the water can produce false indications. Similarly, water on the inside surfaces of tubes will produce false signals, and these surfaces must be kept dry. Tubes are connected to each other by stoppers or by taping the ends together. The glands at each end of the water tank must be cut in a manner that will allow passage of the tubes without undue loss of water couplant. Finally, air must be prevented from being drawn into the entry gland along with the tube. This is usually accomplished by directing a stream of water over the outside of the tube just before it enters the gland.

In eddy current inspection, use of an encircling detector coil is limited to a maximum tube diameter of about 75 mm (3 in.). As tube diameter increases, the ratio of flaw size to tube diameter decreases; consequently, the flaw is increasingly more difficult to detect. This problem is overcome by using several small probe coils (Ref 6, 7, and 17) and with spinning probes (Ref 8). When probe coils are used, the flaw becomes a significantly high percentage of the zone surveyed. Because independently mounted probes ride over the tube surface, good magnetic coupling is ensured. Magnetic saturation is used to obtain maximum sensitivity to flaws close to, or on, the inside surface of the tube, and the frequency of the test current is kept relatively low, sometimes as low as 1 kHz. Internal spinning probes can also be used if a lower production rate can be tolerated (Ref 8).

In some installations, the eddy current test with magnetic saturation is supplemented by a probe-type eddy current test, which in effect provides high-sensitivity inspection of the surface (Ref 18). Four probe coils, as shown in Fig. 7, each serving as both inductor and detector, rotate about the tube as it moves longitudinally through the rotating assembly. Magnetic saturation is not required. The segment of the tube in which the flaw occurs is identified by markers. In Fig. 8, the test head is shown ready for use, with eight paint-spray guns in position for marking the proper zone.

Eddy current inspection is used on seamless tubular products ranging from 3.2 to 244 mm (⅛ to 9⅝ in.) in diameter and from 0.25 to 14.0 mm (0.01 to 0.55 in.) in wall thickness. In most cases, magnetic saturation is used, but when the primary concern is the detection of surface imperfections, small probe coils without magnetic saturation are used. If the steel is entirely nonmagnetic, no saturation is required in any system. The frequency used ranges from 1 to 400 kHz and depends on such variables as the wall thickness of the pipe or tube, the coil design and arrangement, and the use of saturation. The spacing between the pipe or tube and the test coil(s) varies widely. However, for high sensitivity and accuracy, it should be kept to a minimum and is occasionally kept to as little as 0.25 mm (0.01 in.). This clearance is insufficient for practical use, and the usual clearance for production testing ranges from 1.6 to 19 mm (1⁄16 to ¾ in.).

Although all flaws that usually occur in seamless pipe or tube can be detected by

Fig. 7 Unit used for the probe-type eddy current inspection of seamless steel tubing. A, outer cover, containing test head (Fig. 8), in open position; B, one of four rotating eddy current probe coils; C, reference-standard testpiece in position for calibration; D, one of eight paint-spray guns for marking

eddy current methods, external flaws are more easily detected than internal flaws. Laminations are the most difficult flaws to detect. Some installations are intended to detect surface flaws only.

Flux leakage techniques are used for the inspection of seamless tubular products ranging from 32 to 914 mm (1¼ to 36 in.) in diameter and from 3.2 to 19 mm (⅛ to ¾ in.) in wall thickness. Because the flaws sought usually have a significant longitudinal dimension, transverse magnetic fields are usually used. Longitudinal fields can be used but are rarely considered necessary. As a rule, the transverse magnetic field is produced by a current passing through a conductor located in the center of the pipe or tube. In some cases, the field is produced between the poles of an electromagnet or a permanent magnet whose pole pieces are shaped to fit the pipe or tube diameters. The signal-generating movement of either the tube or the probe with respect to the other is accomplished by rotating either tube or probe.

Sensitivity to inside surface flaws is a problem when using these methods; the problem becomes more serious as the wall thickness increases. In some cases, the solution to the problem is a rotating internal probe moving through the tube or kept stationary while the tube moves (Ref 8). Other installations depend on electronic filters and the difference in frequency between the signals for internal and surface flaws (Ref 10).

The production rate of flux leakage inspection depends on many factors. The maximum permissible speed of probe or tube movement with respect to the other is about 90 m/min (300 sfm). The circuits will respond almost instantaneously when inspection speeds are kept below this speed limit. The principal limiting factor in production-output speed then becomes the maximum tolerable length of the flaw, which in turn governs the pitch of the helix inspected. However, the production rate can be increased to any desired level by the simultaneous inspection of several segments of the same pipe or tube. Actual inspection speeds range from 0.9 to 60 m/min (3 to 200 sfm), depending on the diameter, the system used, sensitivity required, and other variables. The methods that use external magnets can inspect at a much higher overall rate than those that depend on an internal conductor for magnetization.

Magnetic particle and liquid penetrant inspection methods are simple and economical and are most useful for surface inspection in specialized, small-scale applications. When applied to welded tubing, their use can be restricted to the weld zone. However, when applied to seamless tubing, there are no surface restrictions. The inability of these methods to locate small flaws beneath the surface and their dependence on the vision, alertness, and judgment of the inspector limit their usefulness in meeting modern specifications for seamless steel tubing.

Radiographic Inspection. The principal application of radiography to seamless tubing, as with welded tubing, is the inspection of girth welds joining the ends of tubes. Even in this application, it is apparent that it should be supplemented by magnetic particle inspection (Ref 12).

Finned Tubing

The production of finned tubing for use in heat exchangers, notably, the heat exchangers used in nuclear reactor installations, has been increasing. These tubes are normally designed with three or six fins, and the presence of these fins precludes the use of several of the conventional inspection techniques. Among the methods considered to be feasible are the magnetic particle, liquid penetrant, and eddy current methods. The tube wall just below the outer surface can

Fig. 8 Test head of the eddy current inspection unit shown in Fig. 7. A, orifice for test pipe or tube; B, one of eight paint-spray guns for marking; C, reference-standard testpiece

be inspected with a high degree of efficiency and speed using special eddy current techniques (Ref 19). For this application, external coils are designed to fit the external tube contours precisely. The transmission method, in which the exiting and receiving coils are placed on different sides of the tube wall, is used. Although ultrasonic inspection can be used, it is prohibitively slow and expensive because of the finned-tube contours.

Duplex Tubing

Exploratory tests have demonstrated the feasibility of applying several nondestructive inspection methods to the inspection of duplex tubing used in the atomic energy industry. The tubes, which are made by metallurgically bonding two concentric pieces of 2.25Cr-1Mo steel tubing to make a single piece, have a composite wall thickness of 3.86 mm (0.152 in.) and an outside diameter of 22 mm (⅞ in.). The outer 22 mm (⅞ in.) diam tube is fabricated with four equally spaced axial grooves on the inside surface. These grooves complicate testing, but laboratory experience indicates that a useful inspection of the assembled tubes can be accomplished by the ultrasonic method (Ref 20). Complete inspection requires scanning of the inside 14.5 mm (9/16 in.) diam tube as well as the outer tube surfaces. Eddy current techniques can also be used for inspecting duplex tubing, as will be shown in the next section, which includes an example of the eddy current inspection of aluminum tubing (Example 3).

Nonferrous Tubing*

A wide variety of nonferrous alloy tubing, such as tubing made of brass, copper, aluminum, nickel, and zirconium, can be inspected for cracks, seams, splits, and other flaws. Eddy current inspection is the method most widely used, followed by the ultrasonic and liquid penetrant methods.

Eddy Current Inspection

When eddy current inspection is employed for nonferrous tubing, the range of tube diameters normally permits the use of an encircling coil. The typical flaws respond well to differential-type coils. The frequencies employed usually range from 1 to 25 kHz. The choice of frequency is generally dependent on the electrical conductivity and wall thickness of the tubing. Because magnetic saturation is not required, the inspection equipment is simpler and more compact than that used on ferromagnetic tubing. Tubes range from 3.2 to 89 mm (⅛ to 3½ in.) in diameter, with wall thicknesses from 0.25 to 14.2 mm (0.01 to 0.56 in.). Testing speeds to 370 m/min (1200 sfm) have been reported. On a limited basis, eddy current inspection is also being applied to finned copper tubing.

*Examples 3 to 6 were prepared by J.P. Crosson, Lucius Pitkin, Inc.

Example 3: Corrosion of Duplex 3003-H14 Aluminum Heat Exchanger Tubes Clad With 7072 Aluminum on Inner Surface. An eddy current inspection was performed at a petrochemical complex on a heat exchanger containing 1319 tubes. It was advised that the tube material was aluminum alloy 3003 in the H14 condition. The tube side environment was salt water at a pressure of 240 kPa (35 psig) at an inlet temperature of 30 °C (85 °F) and an outlet temperature of 40 °C (100 °F). The shell side environment contained hydrocarbons, hydrogen sulfide, water, hydrogen, and ammonia. The shell side stream entered at 125 °C (260 °F) and exited at 45 °C (110 °F) at 380 kPa (55 psig).

Eddy current inspection of the heat exchanger revealed that 300 out of 1319 tubes exhibited indications characteristic of broad, inner-surface pitting corrosion. The suspect tubes were randomly scattered about the tube bundle.

A tube that exhibited strong indications of inner-surface pitting was pulled, split, and visually examined. The entire outer surface of the tube was covered with a thin, uniform black scale. The inner surface of the tube was coated with light-brown silt deposits. Broad, shallow corrosion pits were randomly scattered over the entire inner surface. Without exception, at each of the corroded areas, the depth of the corrosion attack measured 0.28 mm (0.011 in.). Figure 9(a) shows the outer and inner surfaces of the subject tube.

Metallographic examination revealed the tube to be of duplex design; that is, it was clad on the inner surface. The corrosive attack was limited to the cladding thickness of 0.28 mm (0.011 in.) as shown in Fig. 9(b). Qualitative spectrographic analysis revealed the aluminum alloy 3003 tube material to be clad on the inner surface with aluminum alloy 7072.

The examination revealed that the subject tubes had undergone galvanic corrosion. The 7072 cladding sacrificially corroded as it electrochemically protected the substrate alloy; 7072 cladding alloy is sufficiently anodic in the tube side environment to provide cathodic protection. Consequently, the corrosive attack penetrated only to the interface, then proceeded laterally along the substrate surface, confined to the cladding. To the extent that the cladding remained over a sufficient area, the substrate material was protected. Therefore, although the tubes exhibited corrosive attack, it was apparent that considerable service life re-

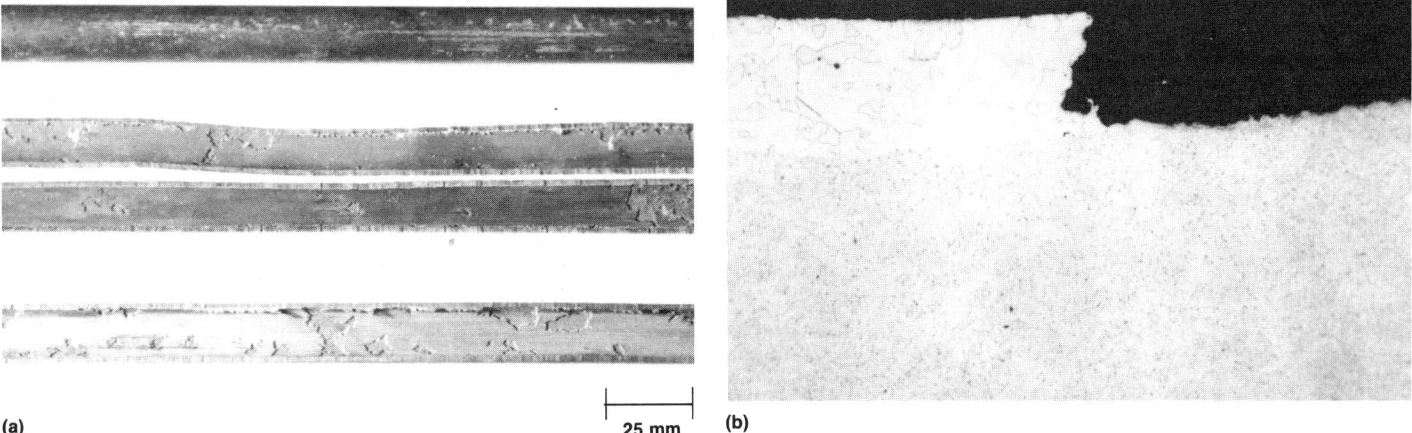

Fig. 9 Corrosion in duplex 3003-H14 aluminum heat exchanger tubes clad with 7072 aluminum on inner surface. (a) Macrograph showing aluminum tube samples removed from the subject heat exchanger unit after eddy current inspection. The outer surface of the tube is shown at the top. The center two sections illustrate the condition of the inner surface before cleaning, and the bottom section shows the inner surface after removal of the light silt deposits. Broad patches of corrosion attack are evident on the inner surface. (b) Micrograph of the inner surface of the duplex tube showing corrosion of the 7072 cladding (top section of photograph). The corrosive attack was limited to the clad thickness (0.28 mm, or 0.011 in.), and the substrate 3003 alloy was not affected. 50×. Courtesy of J.P. Crosson, Lucius Pitkin, Inc.

mained, and tube replacement was not recommended.

Example 4: Stress-Corrosion Cracking of Copper Absorber Tubes in an Air-Conditioning Unit. Eddy current inspection was performed on a leaking absorber bundle in an absorption air-conditioning unit. The inspection revealed cracklike indications in approximately 50% of the tubes. The tube material was phosphorus-deoxidized copper.

Two tubes with indications were pulled and examined visually and metallographically to determine the cause of cracking. The outer surfaces of the tubes were irregularly stained with a green-blue-black film, apparently the result of reaction with the shell side lithium bromide solution. The inner surfaces were covered with a thin, crusty green-black deposit that easily flaked off from the surface. No significant corrosive attack was observed on the inner surfaces.

One of the pulled tubes exhibited a large, irregular longitudinal crack, as shown in Fig. 10(a). The other tube exhibited a very fine longitudinal crack, as shown in Fig. 10(b).

Metallographic examination revealed that the cracks originated at the outer surface, were intergranular in nature, and were free of any localized grain deformation. Such features are characteristic of stress-corrosion cracking in a copper heat exchanger tube. The crack path is shown in Fig. 10(c).

The service contractor responsible for maintaining the absorption unit suspected that mercury contamination from a manometer was the cause of the stress-corrosion cracking. However, an electron probe microanalysis performed on a microspecimen did not reveal any mercury at or near the cracks.

Chemical analysis of the lithium bromide solution revealed significant quantities of nitrates in the solution. Such nitrates are normally added to the lithium bromide solution to act as corrosion inhibitors.

The results of the examination indicated that the absorber tubes failed by stress-corrosion cracking initiated by ammonia contamination in the lithium bromide solution. Cracking was from the shell side or outer surface of the tubes, where the tubes were in contact with nitrate-inhibited lithium bromide. The source of ammonia was apparently the reduction of nitrates by hydrogen evolved during corrosion of the steel shell and/or tubes.

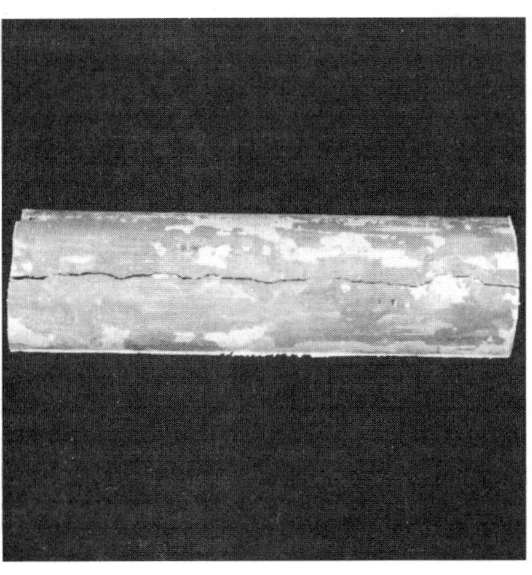

Fig. 10 Longitudinal crack and intergranular stress-corrosion cracks in copper air-conditioning absorber tubes. (a) Longitudinal crack in one of the subject absorber tubes. 0.75×. (b) Macrograph of fine, irregular crack observed on the outer surface of the second absorber tube after light acid cleaning to remove the corrosion product. 2×. (c) Micrograph showing profiles of the primary crack and two fine secondary cracks at the outer surface of the subject absorber tube. The crack profiles are typical of stress-corrosion cracking, that is, intergranular and free of any localized grain deformation. 75×. Courtesy of J.P. Crosson, Lucius Pitkin, Inc.

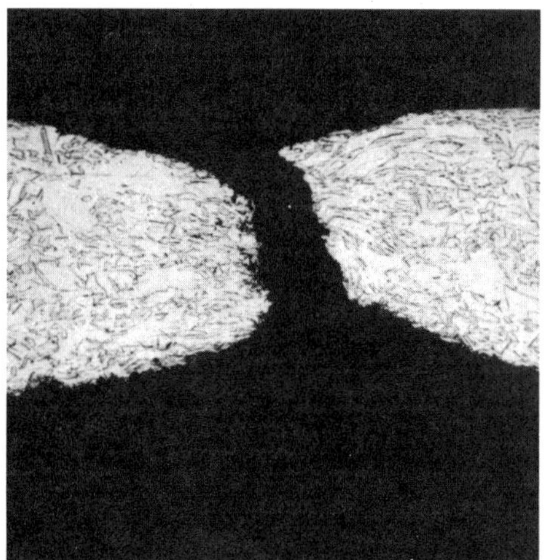

Fig. 11 Micrograph of a transverse section of a burst copper evaporator tube showing the longitudinal rupture present in one of the failed tubes. At the fracture, grain deformation and necking down of the tube wall are evident. Such features are characteristic of overload failure in a ductile material. 55×. Courtesy of J.P. Crosson, Lucius Pitkin, Inc.

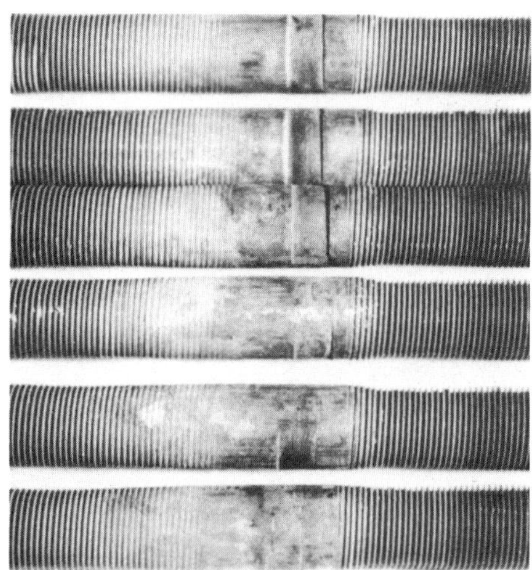

Fig. 12 Grooves at the tube support saddles formed by fretting corrosion attack due to vibrational contact between the copper tubes and the steel tube support plates. Courtesy of J.P. Crosson, Lucius Pitkin, Inc.

Example 5: Burst Copper Evaporator Tubes in an Absorption Air-Conditioning Unit. An eddy current survey of the copper evaporator (chiller) tubes in an absorption air-conditioning unit revealed two tubes in the evaporator bundle with indications typical of longitudinal cracks. The two tubes were pulled for visual and metallographic examination.

One tube exhibited a 27 mm (1 1/16 in.) long longitudinal crack adjacent to the location of the first tube support. The second tube exhibited a 14.5 mm (9/16 in.) long longitudinal crack 1735 mm (68 1/4 in.) from the front end. The tube surface adjacent to the cracks appeared to be bulged and smeared. The smearing was apparently a result of removing the tubes from the unit.

Splitting the tubes revealed the inner surfaces to exhibit a thin, normal, uniform oxide scale. The tube wall beneath the scale was free of any significant corrosion.

Metallographic examination revealed significant necking down and grain distortion at the fracture surfaces, as shown in Fig. 11. The fracture features were characteristic of an overload failure in a ductile material. The general microstructure away from the fracture consisted of twinned equiaxed grains typical of annealed copper and was considered satisfactory.

The results of the examination indicated that the ruptured tubes failed as a result of excessive internal pressure, as evidenced by the necking down of the tube wall at the fracture. The normal operating pressure of approximately 690 kPa (100 psi) produces a hoop stress in the tube wall of approximately 7.6 MPa (1100 psi), which was well below the yield strength of the material. Therefore, the failure could not be related to minor fluctuations in operation pressure. The source of the excessive internal pressure was most probably a freeze-up of the tube side water that occurred during interruption of the tube side flow or misoperation of the unit.

Example 6: Worn Copper Condenser and Chiller Tubes in a Centrifugal Air-Conditioning Unit. Eddy current inspection was performed on the copper chiller and condenser tubes of a centrifugal air-conditioning unit. The results of this inspection revealed the presence of seven chiller tubes exhibiting indications characteristic of a decrease in wall thickness at a location corresponding to the forward tube support plate. Subsequent removal of these tubes confirmed the presence of tube-wall wear at the saddle (smooth portion) of the tube, which sits in the tube support. The tube-wall wear at the support saddles was measured between 10 and 60% of the original wall thickness.

Two condenser tubes were also observed with wear at the support saddles. The decrease in wall thickness was measured to be 20% and 50% for the two tubes. Figure 12 shows several tubes with tube-wall wear at the tube support saddles. Results of the inspection attributed the wear to fretting corrosion attack caused by vibrational contact between the tubes and the steel tube support plate.

Immersion ultrasonic inspection is used on tubes ranging in diameter from 6.4 to 254 mm (1/4 to 10 in.), with wall thicknesses as small as 0.25 mm (0.01 in.). In some installations, four channels are used, two for the detection of transverse flaws and two for longitudinal flaws. Because these tests are usually critical, they are performed at low speeds, usually less than 3 m/min (10 sfm).

The liquid penetrant inspection of nonferrous tubular products is performed in the conventional manner, as described in the article "Liquid Penetrant Inspection" in this Volume.

In-Service Inspection*

The demand for the in-service inspection of tubular products has been increasing in many industries, including the petroleum, chemical, nuclear, and steelmaking industries. When there is access to the outer surface of the tubing, several of the conventional inspection methods can be used, but when the outside surface is inaccessible, the problem is more complex because the test equipment must pass through the tube. This requirement sharply limits the inspection methods and equipment that can be used.

Tubular Products in Commercial Applications

Equipment used to inspect from the inside surface must use transducers that move with respect to the shape of the testpiece being inspected. Even with this limitation, the variety of eddy current, flux leakage, and ultrasonic devices available is large enough that commercial equipment can usually be found that is capable of meeting all requirements.

Both eddy current equipment and flux leakage equipment have been successful in applications that require passage through the tube. In one application employing eddy current equipment, a combination exciting and detector probe is rotated as it passes through the tube of a reactor heat exchanger. The equipment has high sensitivity and operates at high speed (Ref 19). If the tubes are made of a nominally nonmagnetic metal that is slightly and variably magnetic, such as a Monel alloy or an austenitic stainless steel, it is necessary to magnetically saturate the tubes, using either a direct current field or a permanent magnet. A combination of the three units—inductor, detector, and saturator—built into a small probe has been successfully used to detect small flaws in the 13 mm (1/2 in.) diam by 1.24 mm (0.049 in.) wall tubes of a nuclear power plant steam generator (Ref 21).

Acoustic emission is a promising technique for the nondestructive inspection of buried pipe. Holes are drilled through the soil to the pipe at intervals of perhaps 120 m (400 ft) to allow sensors to be placed on the pipe. A heavy truck, driven over the area in which the pipe is buried, produces simultaneous indications on the instrumentation connected to each sensor. Analysis of the indication determines the presence or absence of cracks. Examples 7 through 10 show typical uses of the ultrasonic inspec-

*Example 7 was prepared by L.D. Cox, General Dynamics Corporation. Examples 8 to 10 were prepared by J.P. Crosson, Lucius Pitkin, Inc.

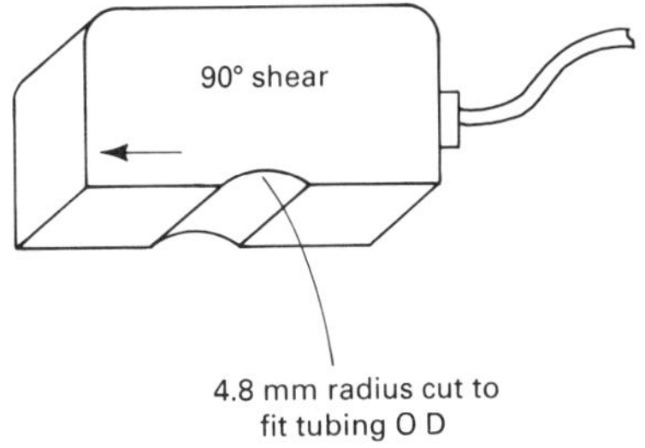

(a)

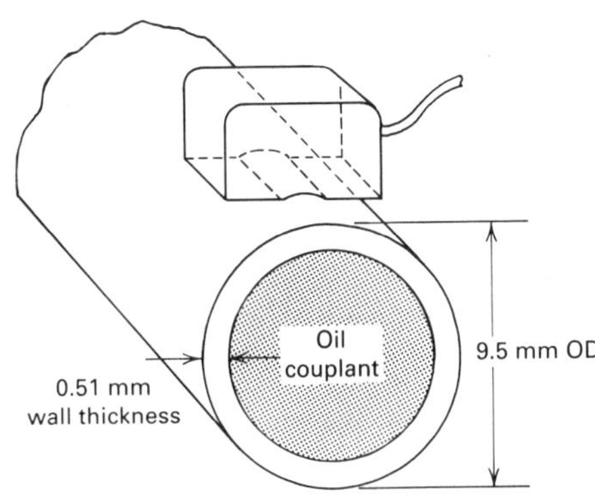

(b)

Fig. 13 15-MHz Rayleigh surface wave transducer (90° shear) used for detecting intergranular attack and root weld cracks. (a) Transducer with radius machined on transducer shoe to allow device to conform to tubing outside diameter. (b) Transducer positioned on tube outside diameter to couple to tube using a lightweight oil couplant. Source: L.D. Cox, General Dynamics Corporation

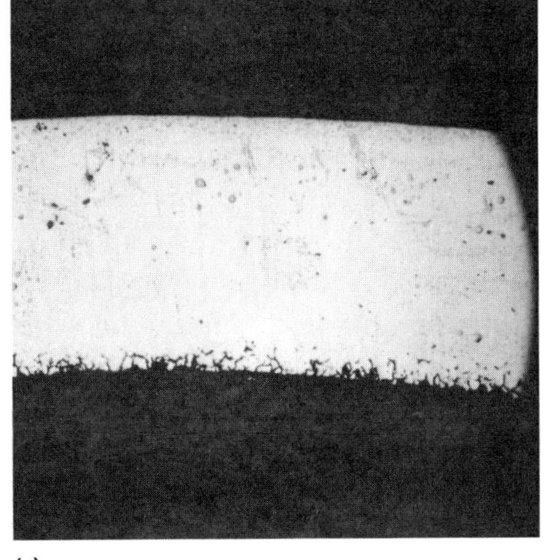

(a)

(b)

Fig. 14 Intergranular attack of 0.51 mm (0.020 in.) wall thickness, Fe-21Cr-6Ni-9Mn stainless steel tubing inside diameter. (a) 60×. (b) 85×. Courtesy of L.D. Cox, General Dynamics Corporation

tion and eddy current inspection of tubular products.

Example 7: Intergranular Attack and Root Cracks in Welded Austenitic Stainless Steel Tubing Detected Using Contact Shear-Wave Ultrasound Techniques. Contact shear-wave inspection techniques are very effective for detecting intergranular attack and/or root weld cracking in stainless steel tubing. These techniques were developed to overcome the limitations of inspecting formed and installed tubing in aircraft or other areas having limited accessibility.

This example demonstrates techniques applied to 9.5 mm (3⁄8 in.) outside diameter stainless steel tubing with a 0.51 mm (0.020 in.) wall thickness. However, these techniques can be modified to enable detection of intergranular attack and/or root weld cracking in various materials and sizes.

A 15-MHz Rayleigh (surface) wave transducer is machined with a 4.8 mm (3⁄16 in.) radius to fit the 9.5 mm (3⁄8 in.) outside diameter of the 0.51 mm (0.020 in.) wall thickness tubing to be inspected (Fig. 13a). The transducer is then positioned on the tubing, as shown in Fig. 13(b). No prior preparation of the sample was required. Mixed-mode shear wave is induced in the tubing to detect intergranular attack (Fig. 14) using a ring-pattern CRT display (Fig. 15). This transducer is also very sensitive and can be used for detecting root weld cracks (Fig. 16). The ultrasound instrument will be set up for monitoring discrete echoes from the root crack. The display produced is shown in Fig. 17.

Example 8: Eddy Current Inspection of Pitting and Stress-Corrosion Cracking of Type 316 Stainless Steel Evaporator Tubes in a Chemical Processing Operation. Eddy current inspection was performed on a vertical evaporator unit used in a chemical processing plant. The evaporator contained 180 tubes 25 mm (1 in.) in diameter.

It was advised that the tube material was type 316 stainless steel. The shell side fluid was condensate and gaseous methylene chloride, while the tube side fluid was contaminated liquid methylene chloride.

Eddy current inspection revealed 101 tubes that exhibited severe outer surface pitting and cracklike indications near each tube sheet. Several tubes exhibiting strong indications were pulled and examined visually and metallurgically.

It was observed that the indications correlated with rust-stained, pitted, and cracked areas on the outer surfaces. The observed condition was most severe along the portions of the tubes located between the upper tube support and top tube sheet. Figures 18(a) and (b) show a pitted and cracked area before and after dye-penetration application.

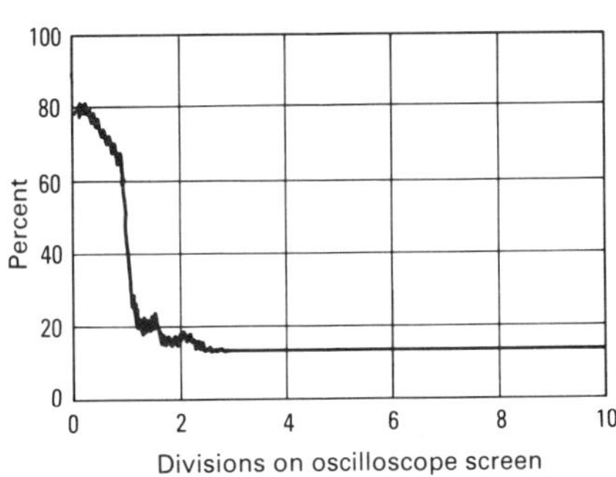

(a)

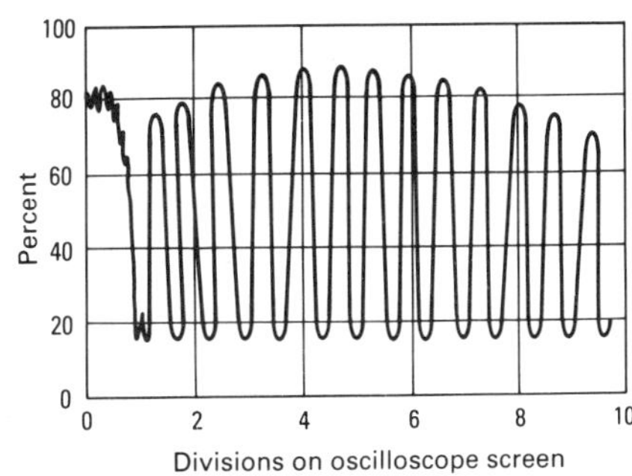

(b)

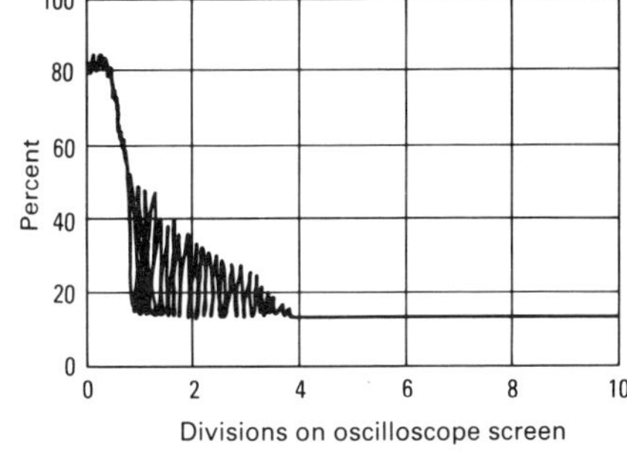

(c)

Fig. 15 Mixed-mode shear wave used to detect intergranular attack showing oscilloscope screen display for (a) transducer in air, (b) transducer coupled to an acceptable tube having no defects due to intergranular attack, and (c) transducer coupled to tube rejected because of intergranular attack. Significant attenuation of the ultrasonic signal in (c) is due to scatter. Source: L.D. Cox, General Dynamics Corporation

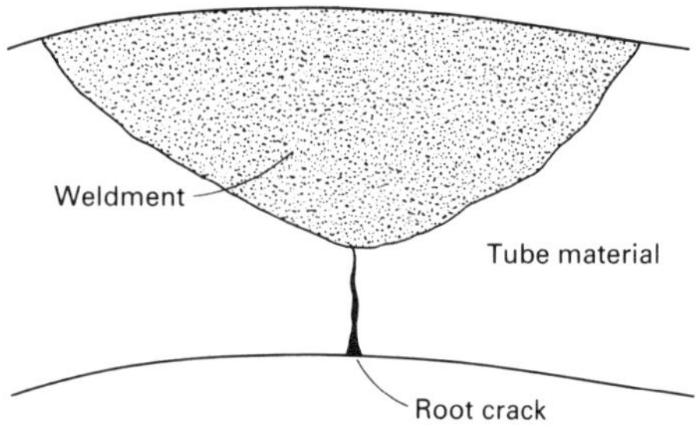

Fig. 16 Cross section of a tube having a crack at the root of the weld seam. Source: L.D. Cox, General Dynamics Corporation

Metallographic examination revealed that the cracking initiated from the outer surface, frequently at pits, and penetrated the tube wall in a transgranular, branching fashion. The crack features were characteristic of chloride stress-corrosion cracking. In many cases, the cracking, rather than penetrating straight through the tube wall, veered off in a tangential direction at or about mid-wall, suggesting the possibility of a change in the residual stress-field from tube drawing. Figure 18(c) shows stress-corrosion cracking originating from pits on the outer surface of the tube.

The results of the examination indicated that the subject tube failures occurred by way of stress-corrosion cracking as a result of exposure to a wet-chloride-containing environment. Therefore, a change in tube material was recommended to avoid future failures and loss of service.

Example 9: Eddy Current Inspection of a Pitted Type 316 Stainless Steel Condenser Tube. Eddy current inspection was performed on approximately 200 stainless steel tubes in a main condenser unit aboard a container ship. The stainless steel tubes comprised the upper two tube rows in the condenser. The tube material was reported to be type 316 stainless steel; this was confirmed by subsequent chemical analysis. The remaining tubes were 90Cu-10Ni. Recurring leaks had occurred in the stainless steel tubes, but no leaks had occurred in the copper-nickel tubes.

Eddy current indications typical of inner surface pitting were observed in 75% of the stainless steel tubes inspected. A tube exhibiting a strong indication was pulled from the condenser and examined visually and metallographically.

Visual examination of the outer surface revealed occasional patches of rust-colored deposit at the locations of the eddy current indications. No apparent defects of any type were observed on the outer surface.

Subsequent splitting of the tube revealed several areas of severe pitting corrosion attack on the inner surface at locations corresponding to the eddy current indications. The corrosion progressed in such a way as to hollow out the wall thickness, and at several locations the pits had completely penetrated the wall thickness. The pitting corrosion attack tended to be close to the bottom of the tube and essentially in line along the tube sample length.

Figure 19(a) shows a severely pitted location. Metallographic examination revealed the attack to be broad and transgranular in

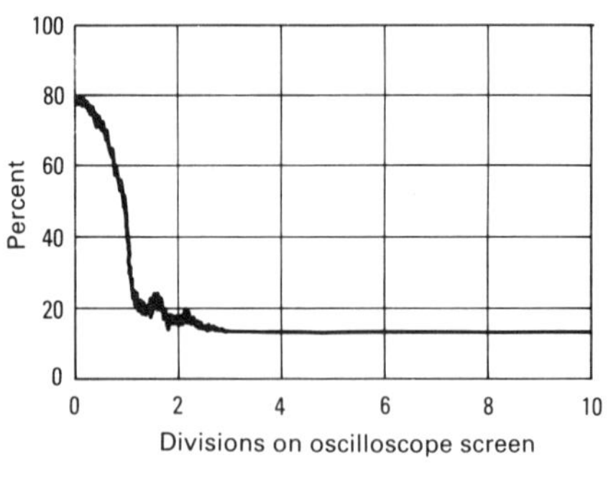

(a)

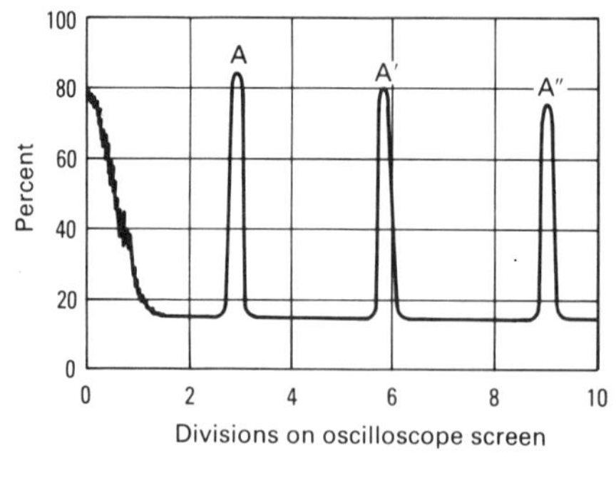

(b)

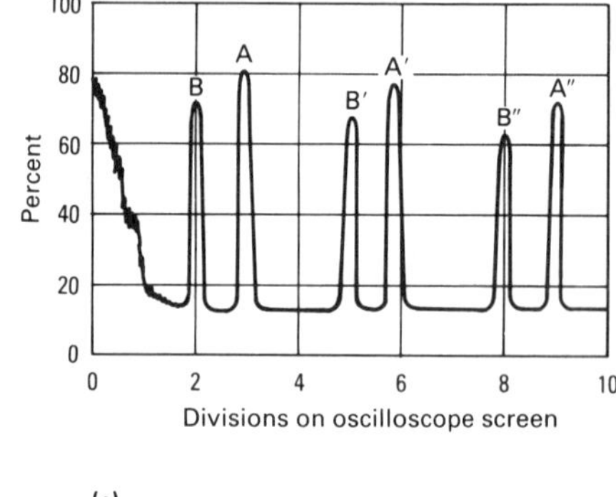

(c)

Fig. 17 Plots obtained on oscilloscope screen with ultrasonic device set up to monitor discrete echoes from a root crack: (a) transducer in air; (b) transducer coupled to tube devoid of root crack defects; and (c) transducer coupled to tube rejected due to presence of a root crack. Signal A in (b) and (c) is due to reflection from the transducer/tube OD contact point. Signal B in (c) is the root crack signal [when the transducer is indexed circumferentially, the A signal will be stationary (no change in time-of-flight) while the B signal will shift]. Source: L.D. Cox, General Dynamics Corporation

(a)

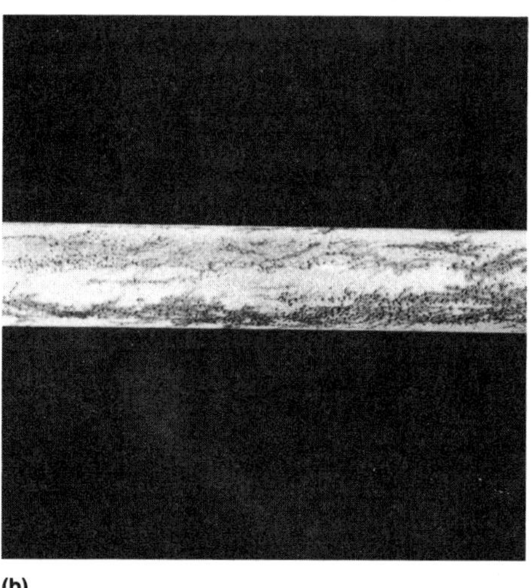

(b)

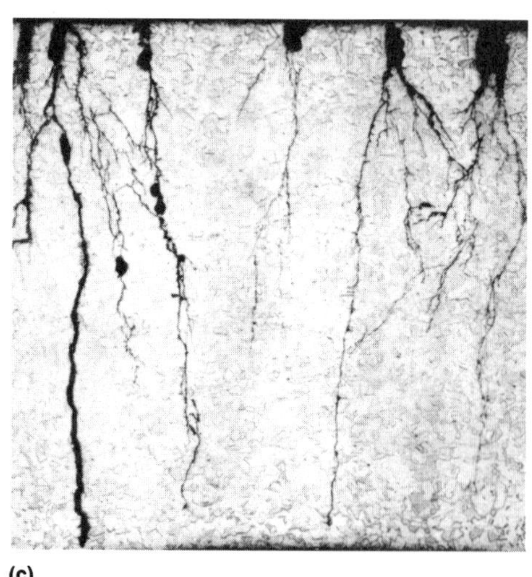

(c)

Fig. 18 Pitting and stress corrosion in type 316 stainless steel evaporator tubes. (a) Rust-stained and pitted area near the top of the evaporator tube. Not clear in the photograph, but visually discernible, are myriads of fine, irregular cracks. (b) Same area shown in (a) but after dye-penetrant application to delineate the extensive fine cracks associated with the rust-stained, pitted surface. (c) Numerous multibranched, transgranular stress-corrosion cracks initiating from the outer surface pits. 35×. Courtesy of J.P. Crosson, Lucius Pitkin, Inc.

(a)

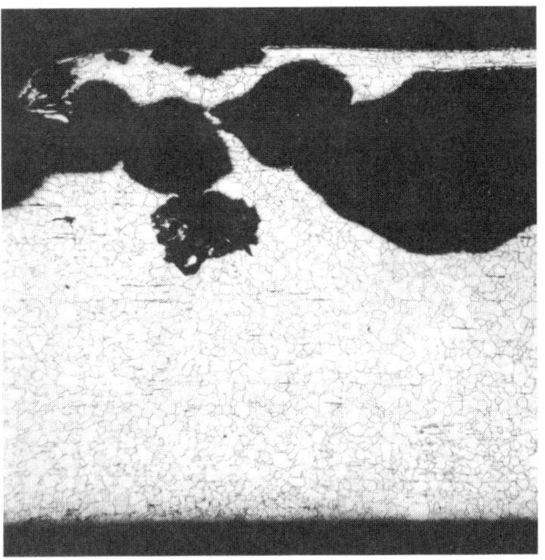

(b)

Fig. 19 Pitted type 316 stainless steel condenser tube. (a) Inner surface of main condenser tube showing extensive but localized pitting corrosion attack. 1×. (b) Longitudinal section passing through a pitted area showing extensive pitting that had progressed beneath the inner surface of the main condenser tube. 55×. Courtesy of J.P. Crosson, Lucius Pitkin, Inc.

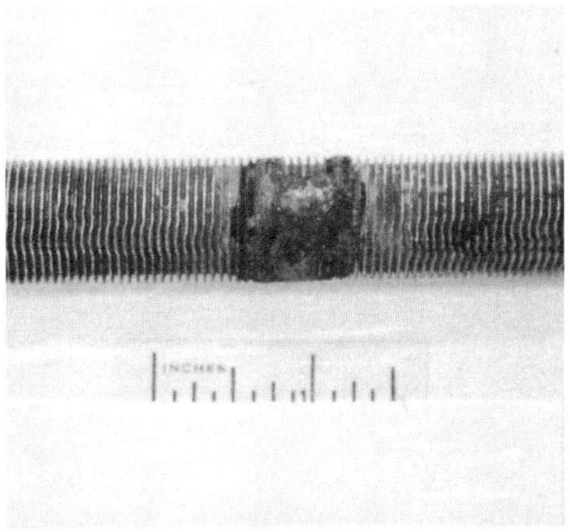

Fig. 20 Magnetic corrosion product embedded in the tube fins at the tube support of a steel tube. The corrosion product caused by eddy current indication characteristic of tube-wall wear at the support. Courtesy of J.P. Crosson, Lucius Pitkin, Inc.

nature without any corrosion product buildup at or around the pits. Figure 19(b) shows the manner in which the pitting had penetrated into and beneath the inner surface.

The results of the examination revealed that the subject stainless steel condenser tube had failed as a result of pitting corrosion attack, which initiated at the inner surface and progressed through the tube wall. That the pitting was essentially on the bottom of the tubes was strong evidence of deposit-type pitting corrosion attack.

Deposit attack occurs when foreign material carried by the tube side fluid settles or deposits on the inner surface, generally at the bottom of the tube. The deposit shields the tube surface, creating a stagnant condition in which the fluid beneath the deposit becomes deficient in oxygen compared to the free-flowing fluid around the deposit. The difference in oxygen content results in the formation of an oxygen concentration cell in which the smaller, oxygen-deficient sites become anodic with respect to the larger oxygenated cathodic sites. As a result, pitting corrosion attack occurs at the anodic sites.

In stainless steel, the condition is further aggravated by the fact that type 316 stainless steel performs best in a service where the fluid is oxidizing and forms a passive film on the surface of the tube. If there is an interruption in the film, as may be caused by chemical breakdown through decomposition of organic materials or mechanically by abrasion, and if the damage film is not reformed, pitting corrosion will initiate and grow at the damaged site. In main condenser service, certain deposits, such as shells, sand, or decomposing sea life, can initiate breakdown of the passive film.

Example 10: Eddy Current Inspection of a Magnetic Deposit Located on a Steel Tube at Tube Sheet Joint in a Centrifugal Air-Conditioning Unit. In this case, defective tubes were not detected. However, the results of the eddy current inspection were directly influenced by a previous tube failure in the unit.

Eddy current inspection of the condenser bundle of a centrifugal air-conditioning unit revealed several tubes with indications typical of tube-wall wear at locations corresponding to the tube supports. One of the tubes exhibiting indications was pulled and visually examined. A tightly adherent magnetic deposit was observed at the area of the tube in contact with the first tube support. Splitting the tube revealed the deposit to be tightly packed in the fins, as shown in Fig. 20. This tube was of tru-finned rather than skip-finned design; that is, the tube did not have smooth support saddles where it was in contact with the tube support plate. Instead, the tube was finned from end to end. Therefore, although the test instrument parameters were selected to phase out the magnetically induced indications from the steel tube supports, the magnetic deposit, which was tightly embedded between the fins, was closer to the internal test probe and caused an indication that was interpreted as tube-wall wear.

Through further investigation it was determined that a previous tube failure had caused the Freon on the shell side to become contaminated with water. This condition proved corrosive to the steel supports and shell and subsequently caused the magnetic corrosion deposit observed at the tube support.

Oil-Country Tubular Products

The application of nondestructive inspection to the tubular products of the oil and gas distribution industry is extensive and is vital to successful operation. The American Petroleum Institute has, with international cooperation and international acceptance, developed tubing and pipe specifications that include many rigorous requirements for nondestructive inspection (Ref 22). Inspection installations range from simple magnetic particle installations to complex assemblies of machinery whose continuous productivity is completely dependent on the reliability and accuracy of its nondestructive inspection equipment. The larger installations may use ultrasonic, eddy current, flux leakage, or radiographic equipment, singly or in combination, and can be supplemented by magnetic particle inspection.

The inspections of pipe or casing can be performed during manufacture, when it is received on site, while it is in service, or when it must be inspected for reuse or resale. When inspection is included in the manufacturing operation, tests are usually performed immediately after the pipe is produced and again after processing has been completed.

Industry has promoted the development and use of highly sophisticated equipment for the in-service inspection of pipe in diameters of 75 mm (3 in.) and larger. In one of several different pipe crawlers available commercially, the probe travels through the gas lines and, by means of flux leakage measurements, reports on the condition of the pipe. Another type of in-service inspection unit, which is shown in Fig. 21, includes tight-fitting seals so that it can be propelled through the pipelines by the oil or gas being carried. The traveling unit includes not only the test instrumentation and a tape recorder but also a power supply so that it is completely self-sufficient, requiring no connection outside the pipe. The sections of this unit are connected by universal joints to permit passage around bends. When the unit completes its cycle, total information on the condition of the pipe is immediately available.

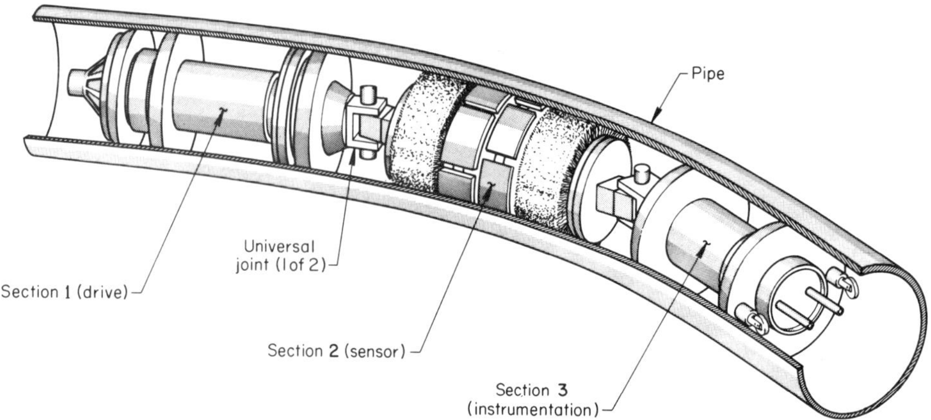

Fig. 21 Self-contained flux leakage inspection unit used in oil and gas pipeline for in-service inspection

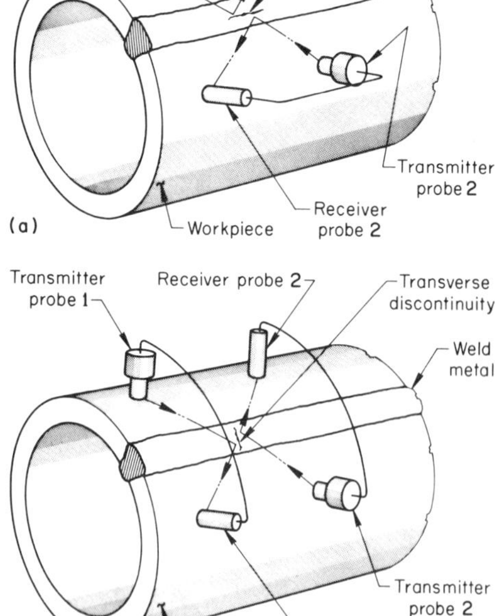

Fig. 22 Diagram of arrangements of probes in the ultrasonic inspection of submerged arc welded pipe for the detection of (a) longitudinally oriented and (b) transversely oriented discontinuities

One of the most important inspection procedures in this industry involves the inspection of girth welds joining the ends of pipes to each other or to fittings and bends. Although radiographic tests are widely favored for this application (Ref 23), supplementary tests are needed to detect the tightly closed flaws not detected by radiography (Ref 12). This industry also uses automated inspection of small tubular pipe couplings. One machine separates acceptable couplings from rejectable couplings automatically and requires no operator. The couplings are fed into the machine from a cutoff lathe. After automatic inspection to API specifications, rejectable couplings are diverted to a reject receptacle.

Nondestructive Inspection of Steel Pipelines (Ref 24)

The nondestructive inspection of welds in steel pipe is used to eliminate discontinuities that could cause failure or leakage. Most steel pipes for gas transmission are made by the hot rotary forging of pierced billets or by forming plate or strip and then welding by either the submerged arc or the resistance process. Pipes are usually made to one of the API specifications, with supplementary requirements if necessary.

Submerged Arc Welded Pipe. The shrinkage of liquid metal upon solidification results in primary piping in the ingot, which can cause laminations oriented in the plane of the plate or strip rolled from the ingot. Laminations can also result from secondary piping and from large inclusions. Laminations can nucleate discontinuities during welding or propagate to form a split in the weld. They cannot be detected by radiography, because of their orientation, but they can be detected by ultrasonics or can be seen when they occur as skin laminations.

The API specifications do not require nondestructive inspection of the plate or strip before welding, but ultrasonic inspection is mandatory in some customer requirements. Generally, the periphery of the plates must be examined to ensure that edges to be welded are free of laminations.

Pulse-echo ultrasonic inspection has been used for most plate-inspection specifications. This method cannot distinguish laminations near the surface remote from the probe, because the echoes from the laminations cannot be inspected from the back echo. Some specifications require that the plate be inspected from both surfaces or that transmission methods be used.

Manual scanning, although time consuming, is feasible because the echo pattern from laminations persists on the oscilloscope screen and is easy to interpret. However, if laminations are fragmented or at an oblique angle to the surface of the plate, there is no distinct flaw echo, but merely a loss of back echo. In most pulse-echo equipment, no account is taken of this loss; therefore, transmission methods are preferable for plate inspection. Such methods normally require mechanization with automatic recording of the results, and inspection systems based on these methods have been installed in some plate mills.

In both transmission and pulse-echo inspection, the probe area represents the area of the plate under inspection at any instant. Shear-wave angle probes are used to detect lamination, but the method is not reliable, particularly for the thin plates used in pipe manufacture. Laminations can also be detected by ultrasonic Lamb waves, which can inspect a zone extending across some or all of the plate. Lamb waves have been used on steel sheet, but they cannot be excited in plates more than 6.4 mm (¼ in.) thick using standard equipment. Special equipment is now available for plate up to 13 mm (½ in.) thick.

The main flaws that occur in submerged arc welds are incomplete fusion and incomplete penetration between the inside and outside weld beads or between the base metal and the filler metal, cracks, undercut or underfill, and overflow. The API standards require full-length inspection of welds by radiography or ultrasonics. Fluorescent screens or television screens are permitted for radiography, and they are often used because they are less expensive than radiographic film, although less discriminating. Fluoroscopy is inherently less sensitive to the more critical flaws, cracks, incomplete sidewall fusion, and incomplete penetration. Ultrasonic inspection is more sensitive to serious flaws and can be automated.

The arrangements of transmitter and receiver probes in ultrasonic inspection of submerged arc welded pipe for detection of longitudinally and transversely oriented discontinuities are shown in Fig. 22. The region of the oscilloscope time base corresponding to the weld region is analyzed electronically, and echoes above the amplitude of the reference derived from the calibration block actuate a relay that can operate visible or audible warnings, paint sprays, or pen recorders. In some installations, only one probe is used on each side of the weld, and detection of discontinuities that are oriented transversely to the weld is not possible. To ensure correct lateral positioning of the probes on the weld, they are mounted on a frame, which is then moved along the weld; alternatively, the pipe can be moved past stationary probes.

Accurate positioning of the probes over the welds is difficult because the width, shape, and straightness of the weld bead vary. The inspection area is limited in order to reduce confusion between echoes from flaws within the weld and the boundaries of the weld reinforcement. Acoustic coupling can be reduced by the probes riding up on weld spatter, by drifting of the scanning frame, by loss of coupling water, or by loose mill scale.

General practice is to use automatic ultrasonic inspection methods and to radiograph those regions of the pipe suspected of containing discontinuities. If radiography does not reveal an objectionable flaw, the ultrasonic indication is ignored and the pipe is accepted. This procedure would accept cracks or laminations parallel to the plate surface that, because of their orientation, cannot be detected by radiography. As an alternative approach, regions that give an ultrasonic flaw indication should be inspected radiographically and by manual ultrasonics. If the original ultrasonic indication was from a discontinuity shown by the radiograph to be acceptable within the specification or if the manual ultrasonic inspection revealed that the indication was a spurious echo arising from a surface wave or from local weld shape, then the pipe was accepted; if the radiograph showed an objectionable flaw, then the pipe was rejected. If there was no explanation for the echo, it was assumed to have arisen from a discontinuity adversely oriented for radiography.

Seamless Pipe. There are two sources of flaws in roll forged seamless pipe: inhomogeneities and the manufacturing process. Inhomogeneities in the ingot such as primary and secondary ingot pipe can be carried into the roll-forged product and can cause flaws in a similar manner to the formation of laminations in steel plate. Such flaws are likely to have a major dimension oriented in the plane of the pipe wall. In manufacturing, the rolls and the mandrel can cause surface discontinuities such as tears and laps, and such discontinuities will have substantial orientation normal to the pipe wall. In addition, pipes and tubes produced by working pierced billets are prone to eccentric wall thickness, with the eccentricity varying along the length of the pipe.

The API specifications that cover seamless line pipe require neither nondestructive inspection nor wall thickness measurement away from the pipe ends. In some mills, destructive inspections are carried out on samples cut from each pipe end to determine the presence of primary pipe flaws. In some API standards (casing, tubing, and drill pipe), nondestructive inspection is optional, but in other standards (high-strength casing and tubing) nondestructive inspection of the full pipe length is mandatory. Magnetic particle, ultrasonic, or eddy current inspection methods are permitted.

Magnetic particle inspection methods have little or no sensitivity to discontinuities that do not show on the surface and are likely to detect laminar discontinuities resulting from ingot piping. Although surface laps are amenable to magnetic crack detection, it would be difficult to apply the inspection method to internal-surface discontinuities.

Eddy current inspection methods can be used to inspect seamless tubing. Very rapid inspection rates are possible with the encircling-coil system. When pipe is passed through a coil fed with alternating current, the resistive and reactive components of the coil are modified; the modification depends on dimensions (and therefore indirectly on discontinuities), electrical conductivity and magnetic permeability, and the annulus between the pipe and the coil (and therefore the outside diameter of the pipe). The analysis to determine which effect is causing any modification is complex.

Eddy current methods are extensively used for the inspection of small, nonferrous tubes, but ferrous material causes complications from magnetic permeability. The initial permeability is affected by residual-stress level. Roll-forged pipe may have varying amounts of residual cold work, depending on the original soaking conditions and the time taken to complete forging. The effect can be alleviated by applying a magnetically saturated field; equipment that can produce a magnetically saturated field has been installed in steel tube mills. However, saturation becomes more difficult as pipe diameter increases.

Radiographic inspection methods, employing either x-ray or γ-ray transmission, can be used with a scintillation counter to estimate the wall thickness of pipe. The accuracy of scintillation counters depends on the size of the count for a given increment of thickness; the count increases with the time the increment is in the beam. As a result, the count, and therefore the accuracy, increases with decreasing scanning rate. When large-diameter pipes are scanned at realistic rates, eccentricity is usually averaged out.

Ultrasonic inspection methods can detect discontinuities oriented both in the plane of, and normal to, the pipe wall. Discontinuities in the plane of the wall can be detected by using a compression-wave probe scanning at normal incidence. For discontinuities normal to the wall, the beam is converted to shear wave and propagated around or along the tube. The pipe is rotated and moved longitudinally relative to the probes, thus giving a helical scan.

The reliability of mechanized scanning is a function of acoustic coupling, and optimum results are achieved with immersion coupling. The efficiency of acoustic coupling through large columns of water is lower but much more consistent than that through the thin liquid films used in contact scanning. Immersion methods also eliminate probe wear and the requirement for specially contoured probes to accommodate each pipe size.

Alternatively, immersion coupling by a column of water flowing between the probe and the pipe can be used. With this method, probe-rotation scanning is possible. Advantage can be taken of the smaller inertia of the probes to increase the scanning rate, and therefore the speed of inspection, by about an order of magnitude.

When an ultrasonic beam propagates radially through the pipe wall, the time interval between successive back echoes reflected from the bore surface is directly proportional to the wall thickness. If the first back echo is used to trigger a high-speed electronic counter whose frequency is such that it will produce a count of 100 during the time taken to receive four echoes in 25 mm (1 in.) thick plate and if a subsequent back echo is used to stop the counter, a count proportional to the wall thickness is produced. By changing the frequency of the counter oscillator, it is possible to change the thickness range inspected or to accommodate different materials. Information from the counter can be fed to a chart recorder, thus continuously recording the wall thickness. Lamination would be recorded as an abrupt localized reduction in wall thickness.

For the detection of cracks and laplike discontinuities, the display on the flaw detection oscilloscope is gated. The discontinuity can then be recorded in its position around the circumference. Chart length can be made proportional to pipe length, thus facilitating discontinuity location and extent in relation to pipe length and variation in wall thickness around and along the pipe. Alternatively, information can be monitored in a go/no-go method to provide a paint spray that identifies the locations of significant discontinuities.

Resistance-Welded Pipe. The type of discontinuity usually responsible for the failure of resistance-welded pipe is incomplete fusion, with associated oxide film. The nondestructive inspection of larger-diameter resistance-welded pipes is normally restricted to inspection of the weld region. Systems similar to those used for the inspection of submerged arc welded seams can be employed, although the arrangements for tracking the probes with respect to the weld reinforcements are not applicable. Because of problems associated with accurate weld tracking, it is necessary that small variations in weld-probe separation should only cause acceptably small variations in discontinuity detection sensitivity.

Probe angles of 60 to 65° give satisfactory results, and coverage of the weld depth is achieved by using two probes on each side of the weld. Such a system is also relevant to the ultrasonic inspection of submerged arc welded pipes.

Nondestructive Inspection of Pipeline Girth Welds

The API specifications do not require that all girth welds be inspected; the use of radiography and the extent of coverage are optional. Generally, it has been the practice to inspect 10% of the weld length. Where the integrity of a pipeline is vital, as in high-pressure gas-transmission systems, it is advisable to consider a 100% inspection, especially when inspection is not a large proportion of the total cost of the pipeline. Where operating conditions are less critical, however, it may be possible to be less critical regarding the size and type of discontinuity permissible.

Radiographic inspection. Characteristic discontinuities in pipeline welds are slag, elongated piping in root, scattered piping and porosity, burn-throughs in the root, incomplete root penetration, incomplete sidewall fusion, and cracks, which often break the inner surface in the heat-affected zone. Except for cracks and incomplete sidewall fusion, these discontinuities are amenable to detection by radiography. Open cracks can be detected, but tighter cracks, even though favorably oriented, are detectable only by optimum practice. Some cracks may not be revealed at all.

Assuming good radiographic techniques, radiographic quality depends on the choice of conditions that control the contrast and definition of the radiograph; the detection of discontinuities improves with increasing contrast and fine definition. Contrast can be assessed in terms of the thickness sensitivity, which can be conveniently estimated by image-quality indicators.

Many factors other than good radiographic techniques influence radiographic contrast. For pipeline radiography, radiation energy is probably the most important. The absorption of radiation by steel decreases with increasing energy; the absorption coefficient at 150 kV is about three times that at 700 kV. For optimum detection of small changes in thickness, the absorption should be as high as possible so that large differences in exposure, consistent with a reasonable amount of energy being transmitted to provide a realistic overall exposure, result at the film. Gamma radiation from a ^{192}Ir source is approximately equivalent to x-rays generated at 700 kV and therefore will not be absorbed sufficiently to give good contrast sensitivity. For wall thicknesses typical of pipelines, x-rays generated at about 150 to 175 kV have reasonable absorption.

For the detection and correct identification of discontinuities from the radiographic image, the delineation of the shadow must be sharp. The principal sources of unsharpness in radiographic images are geometric unsharpness, resulting from the finite size of radiation sources; unsharpness in the film resulting from the kinetic energy of the radiation, grain size of the emulsion and degree of development; and unsharpness resulting from the intensifying screen. Radiographs on pipelines are generally made under less-than-satisfactory conditions; nevertheless, it should be possible to avoid vibration and other forms of relative movement of the source and film during exposure so that the total geometric unsharpness is the penumbra effect. With piping, geometric unsharpness is not large.

An important side effect of unsharpness is that when the unsharpness is greater than the width of a flaw, the contrast resulting from the flaw is reduced from its theoretical value; the greater the unsharpness, the greater the contrast reduction. For tight cracks, the contrast may be so reduced that the change in tone is below the threshold for detection.

On-site conditions may reduce the capability of radiography to detect flaws. Under some conditions, it is difficult to maintain correct developer temperature. The operator is often pressured to keep pace with the welding crews; also, weather and working conditions may be adverse. Suitable equipment and adequate planning should overcome these problems. The more difficult problem is the repetitiveness of the procedure, which causes the operators to lose concentration and gradually to devote less attention to detail.

For most pipe sizes it is necessary to make three exposures to cover the circumference of the weld because the length of weld that can be covered in one exposure (the diagnostic film length) is limited by fade at each end of the film. Panoramic techniques have been used on some larger-diameter pipes. The source is held in a spider arrangement and positioned on the pipe axis; the films are placed around the outer surface of the pipe at the weld. In this manner, the entire weld can be radiographed in one exposure. This exposure is shorter than one of the exposures required in the double-wall, single-image technique because the radiation has to propagate through only one wall of the pipe. In practice, the radiation source can be manually positioned only inside pipes having a diameter of 762 mm (30 in.) or more; even then, conditions must be good. Crawler devices are available that are mechanically propelled through the pipe, with the exposure being operated from an external control (see the section "In-Motion Radiography" of the article "Radiographic Inspection" in this Volume).

On pipelines where the rate of welding is low and the investment on crawlers is not justified, x-ray sets can be clamped onto the outside of the pipe and radiography implemented by the double-wall, single-image technique. Gamma radiography has been favored for pipeline radiography because of its convenience and lower cost. Source containers are more compact and portable than x-ray generators and do not require a power supply.

Panoramic x-ray radiography is barely feasible without crawler devices because of the difficulty of manually maneuvering the cumbersome x-ray sets and control units inside a pipe. Because of the potential for increased use of x-ray radiography on pipelines, there has been considerable effort applied to development of x-ray crawlers.

Ultrasonic Inspection. Welds are usually ultrasonically inspected by a pulse-echo reflection technique. Before the inspection of a weld, the pipe should be checked for laminations that may divert the beam from its theoretical path.

Discontinuities can be identified most reliably by accurate positioning of the source of the discontinuity echo, preferably during scanning from more than one direction. Skilled operators may be able to gain additional information on type of discontinuity from the shape of the echo on the oscilloscope screen, but the display on battery-operated flaw detectors used in daylight is not sufficiently distinct for the technique to be employed on pipelines.

The more significant discontinuities occur in the root of the weld, where discrimination between sources of echo reflection is more difficult. Acceptable features such as full root penetration cause echoes comparable in magnitude to those from root underbead cracks or incomplete penetration. Accurate positioning of the probe with respect to the weld centerline is necessary, and it has been suggested that the required accuracy can be achieved only by marking and machining the pipe ends before welding. Positioning from the center of the weld cap is only approximate because the weld is not necessarily symmetrical about the centerline through the root. Even with the premarking, it is difficult for an operator to locate the ultrasonic probe accurately and still be in a position to view the instrument screen. Thinner-wall pipes reduce the differences in beam path and probe position for discrimination between the various features of the root. Also, the weld must be examined with the probe-to-weld distance increased to avoid confusion between echoes from the weld and those from probe noise. This increases the effect of beam spread and may lead to extraneous echoes from the cap reinforcement.

The skip distance and beam-path length vary as the wall thickness varies. Variations in wall thickness between nominally the same classes of submerged arc welded pipe range from 10 to 15%, but in seamless pipe a ±10% variation along the length or around the circumference at a given position along the length is common. Although it is possible to measure wall thickness accurately by ultrasonics, it is not feasible to measure wall thickness concurrently with scanning the weld. Surface roughness can cause considerable variations in beam angle. Weld spatter can reduce the effectiveness of the coupling, and also alter beam angle by lifting part of the probe off the pipe.

Ultrasonic inspection on girth welds was originally used to determine which welds to radiograph. If the radiograph did not detect anything, it was the practice on most pipelines to accept the radiographic evidence and not that from ultrasonics. Now that pipelines are being examined 100% by radiography, the role of ultrasonics has changed to that of detecting root underbead cracks that may escape detection by radiography and of providing supplementary evidence to aid in the interpretation of radiographic images of weld-root regions.

Surface Crack Detection. Root underbead cracks break the surface of the pipe in the bore and can be detected with liquid penetrant and magnetic particle inspection. The weld area can be magnetized using a yoke powered by permanent magnets. Both methods are sensitive under ideal conditions, but liquid penetrants require very clean surfaces. Magnetic particle crack detection is therefore preferred for pipeline applications. Interpretation of the indications is not a problem, except for the confusion that may arise from the tendency of sharp changes in root profile to give a slight crack indication.

REFERENCES

1. "Nondestructive Testing Terminology," Bulletin 5T1, American Petroleum Institute, 1974
2. H.C. Knerr and C. Farrow, Method and Apparatus for Testing Metal Articles, U.S. Patent 2,065,379, 1932
3. W.C. Harmon, "Automatic Production Testing of Electric Resistance Welded Steel Pipe," Paper presented at the ASNT Convention, New York, American Society for Nondestructive Testing, Nov 1962
4. W.C. Harmon and I.G. Orellana, Seam Depth Indicator, U.S. Patent 2,660,704, 1949
5. J.P. Vild, "A Quadraprobe Eddy Current Tester for Tubing and Pipe," Paper presented at the ASNT Convention, Cleveland, American Society for Nondestructive Testing, Oct 1970
6. H. Luz, Die Segmentspule—ein neuer Geber für die Wirbelstromprüfung von Rohren, *BänderBlecheRohre*, Vol 12 (No. 1), Jan 1971
7. W. Stumm, Tube-Testing by Electromagnetic NDT (Non-Destructive Testing) Methods: I, *Non-Destr. Test.*, Vol 7 (No. 5), Oct 1974, p 251-258
8. F. Förster, The Nondestructive Inspection of Tubings for Discontinuities and Wall Thickness Using Electromagnetic Test Methods: I, *Mater. Eval.*, Vol 28 (No. 4), April 1970, p 21A-25A, 28A-31A
9. F. Förster, The Nondestructive Inspection of Tubings for Discontinuities and Wall Thickness Using Electromagnetic Test Methods: II, *Mater. Eval.*, Vol 28 (No. 5), May 1970, p 19A-23A, 26A-28A
10. P.J. Bebick, "Locating Internal and Inside Diameter Defects in Heavy Wall Ferromagnetic Tubing by the Leakage Flux Inspection Method," Paper presented at the ASNT Convention, Cleveland, American Society for Nondestructive Testing, Oct 1974
11. H.J. Ridder, "New Nondestructive Technology Applied to the Testing of Pipe Welds," Paper presented at the ASME Petroleum Conference, New Orleans, American Society of Mechanical Engineers, Sept 1972
12. R.F. Lumb and G.D. Fearnebaugh, Toward Better Standards for Field Welding of Gas Pipelines, *Weld. J.*, Vol 54 (No. 2), Feb 1975, p 63-s to 71-s
13. M.J. May, J.A. Dick, and E.F. Walker, "The Significance and Assessment of Defects in Pipeline Steels," British Steel Corporation, June 1972
14. W.C. Harmon and T.W. Judd, Ultrasonic Test System for Longitudinal Fusion Welds in Pipe, *Mater. Eval.*, March 1974, p 45-49
15. "Inspection, Radiographic," Military Standard 453A, May 1962
16. W. Stumm, New Developments in the Eddy Current Testing of Hot Wires and Hot Tubes, *Mater. Eval.*, Vol 29 (No. 7), July 1971, p 141-147
17. F.J. Barchfeld, R.S. Spinetti, and J.F. Winston, "Automatic In-Line Inspection of Seamless Pipe," Paper presented at the ASNT Convention, Detroit, American Society for Nondestructive Testing, Oct 1974
18. T.W. Judd, Orbitest for Round Tubes, *Mater. Eval.*, Vol 28 (No. 1), Jan 1970, p 8-12
19. F. Förster, Sensitive Eddy-Current Testing of Tubes for Defects on the Inner and Outer Surfaces, *Non-Destr. Test.*, Vol 7 (No. 1), Feb 1974, p 25-35
20. K.J. Reimann, T.H. Busse, R.B. Massow, and A. Sather, Inspection Feasibility of Duplex Tubes, *Mater. Eval.*, Vol 33 (No. 4), April 1975, p 89-95
21. V.S. Cecco and C.R. Bax, Eddy Current In-Situ Inspection of Ferromagnetic Monel Tubes, *Mater. Eval.*, Vol 33 (No. 1), Jan 1975, p 1-4
22. "Specification for Line Pipe," API 5L, American Petroleum Institute, 1973
23. "Standard for Welding Pipe Lines and Related Facilities," API 1104, American Petroleum Institute, 1968
24. R.F. Lumb, Non-Destructive Testing of High-Pressure Gas Pipelines, *Non-Destr. Test.*, Vol 2 (No. 4), Nov 1969, p 259-268

Weldments, Brazed Assemblies, and Soldered Joints

THE SELECTION of a method for inspecting weldments, brazed assemblies, and soldered joints for flaws (referred to as discontinuities in welding terminology) depends on a number of variables, including the nature of the discontinuity, the accessibility of the joint, the type of materials joined, the number of joints to be inspected, the detection capabilities of the inspection method, the level of joint quality required, and economic considerations. Regardless of the method selected, established standards must be followed to obtain valid inspection results.

In general, nondestructive inspection methods (NDI) are preferred over destructive inspection methods. Sections can be trepanned from a joint to determine its integrity; however, the joint must be refilled, and there is no certainty that discontinuities would not be introduced during repair. Destructive inspection is usually impractical, because of the high cost and the inability of such methods to accurately predict the quality of those joints that were not inspected.

This article will review nondestructive methods of inspection for weldments (including diffusion-bonded joints) and brazed and soldered joints. More detailed information on the techniques discussed can be found in the Sections "Inspection Equipment and Techniques," and "Methods of Nondestructive Evaluation" in this Volume.

Weldments

Weldments made by the various welding processes may contain discontinuities that are characteristic of that process. Therefore, each process, as well as the discontinuities typical of that process, are discussed below. Explanations of welding processes, equipment and filler metals, and welding parameters for specific metals and alloys are available in Volume 6 of the 9th Edition of *Metals Handbook*.

Discontinuities in Arc Welds

Discontinuities may be divided into three broad classifications: design related, welding process related, and metallurgical.

Design-related discontinuities include problems with design or structural details, choice of the wrong type of weld joint for a given application, or undesirable changes in cross section. These discontinuities, which are beyond the scope of this article, are discussed in the articles "Joint Design and Preparation" and "Residual Stresses and Distortion" in Volume 6 of the 9th Edition of *Metals Handbook*.

Discontinuities resulting from the welding process include:

- *Undercut*: A groove melted into the base metal adjacent to the toe or root of a weld and left unfilled by weld metal
- *Slag inclusions*: Nonmetallic solid material entrapped in weld metal or between weld metal and base metal
- *Porosity*: Cavity-type discontinuities formed by gas entrapment during solidification
- *Overlap*: The protrusion of weld metal beyond the toe, face, or root of the weld
- *Tungsten inclusions*: Particles from tungsten electrodes that result from improper gas tungsten arc welding procedures
- *Backing piece left on*: Failure to remove material placed at the root of a weld joint to support molten weld metal
- *Shrinkage voids*: Cavity-type discontinuities normally formed by shrinkage during solidification
- *Oxide inclusions*: Particles of surface oxides that have not melted and are mixed into the weld metal
- *Lack of fusion (LOF)*: A condition in which fusion is less than complete
- *Lack of penetration (LOP)*: A condition in which joint penetration is less than that specified
- *Craters*: Depressions at the termination of a weld bead or in the molten weld pool
- *Melt-through*: A condition resulting when the arc melts through the bottom of a joint welded from one side
- *Spatter*: Metal particles expelled during welding that do not form a part of the weld
- *Arc strikes (arc burns)*: Discontinuities consisting of any localized remelted metal, heat-affected metal, or change in the surface profile of any part of a weld or base metal resulting from an arc
- *Underfill*: A depression on the face of the weld or root surface extending below the surface of the adjacent base metal

Metallurgical discontinuities include:

- *Cracks*: Fracture-type discontinuities characterized by a sharp tip and high ratio of length and width to opening displacement
- *Fissures*: Small cracklike discontinuities with only a slight separation (opening displacement) of the fracture surfaces
- *Fisheye*: A discontinuity found on the fracture surface of a weld in steel that consists of a small pore or inclusion surrounded by a bright, round area
- *Segregation*: The nonuniform distribution or concentration of impurities or alloying elements that arises during the solidification of the weld
- *Lamellar tearing*: A type of cracking that occurs in the base metal or heat-affected zone (HAZ) of restrained weld joints that is the result of inadequate ductility in the through-thickness direction of steel plate

The observed occurrence of discontinuities and their relative amounts depend largely on the welding process used, the inspection method applied, the type of weld made, the joint design and fit-up obtained, the material utilized, and the working and environmental conditions. The most frequent weld discontinuities found during manufacture, ranked in order of decreasing occurrence on the basis of arc-welding processes, are:

Shielded metal arc welding (SMAW)
- Slag inclusions
- Porosity
- LOF/LOP
- Undercut

Submerged arc welding (SAW)
- LOF/LOP
- Slag inclusions
- Porosity

Flux cored arc welding (FCAW)
- Slag inclusions
- Porosity
- LOF/LOP

Gas metal arc welding (GMAW)
- Porosity
- LOF/LOP

Gas tungsten arc welding (GTAW)
- Porosity
- Tungsten inclusions

The commonly encountered inclusions—as well as cracking, the most serious of weld defects—will be discussed in this section.

Gas porosity can occur on or just below the surface of a weld. Pores are characterized by a rounded or elongated teardrop shape with or without a sharp point. Pores can be uniformly distributed throughout the weld or isolated in small groups; they can also be concentrated at the root or toe of the weld. Porosity in welds is caused by gas entrapment in the molten metal, by too much moisture on the base or filler metal, or by improper cleaning of the joint during preparation for welding.

The type of porosity within a weld is usually designated by the amount and distribution of the pores. Some of the types are classified as follows:

- *Uniformly scattered porosity*: Characterized by pores scattered uniformly throughout the weld (Fig. 1a)
- *Cluster porosity*: Characterized by clusters of pores separated by porosity-free areas (Fig. 1b)
- *Linear porosity*: Characterized by pores that are linearly distributed (Fig. 1c). Linear porosity generally occurs in the root pass and is associated with incomplete joint penetration
- *Elongated porosity*: Characterized by highly elongated pores inclined to the direction of welding. Elongated porosity occurs in a herringbone pattern (Fig. 1a)
- *Wormhole porosity*: Characterized by elongated voids with a definite worm-type shape and texture (Fig. 2)

Radiography is the most widely used nondestructive method for detecting subsurface gas porosity in weldments. The radiographic image of round porosity appears as round or oval spots with smooth edges, and elongated porosity appears as oval spots with the major axis sometimes several times longer than the minor axis. The radiographic image of wormhole porosity depends largely on the orientation of the elongated cavity with respect to the incident x-ray beam. The presence of top-surface or root reinforcement affects the sensitivity of inspection, and the presence of foreign material, such as loose scale, flux, or weld spatter, may interfere with the interpretation of results.

Ultrasonic inspection is capable of detecting subsurface porosity. However, it is not extensively used for this purpose except to inspect thick sections or inaccessible areas where radiographic sensitivity is limited. Surface finish and grain size affect the validity of the inspection results.

Eddy current inspection, like ultrasonic inspection, can be used for detecting subsurface porosity. Normally, eddy current inspection is confined to use on thin-wall welded pipe and tubing because eddy currents are relatively insensitive to flaws that do not extend to the surface or into the near-surface layer.

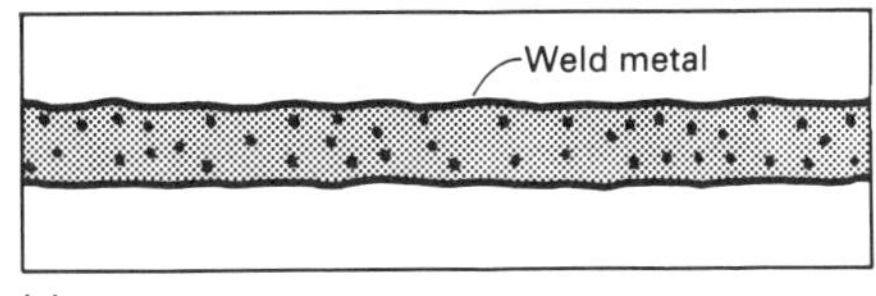

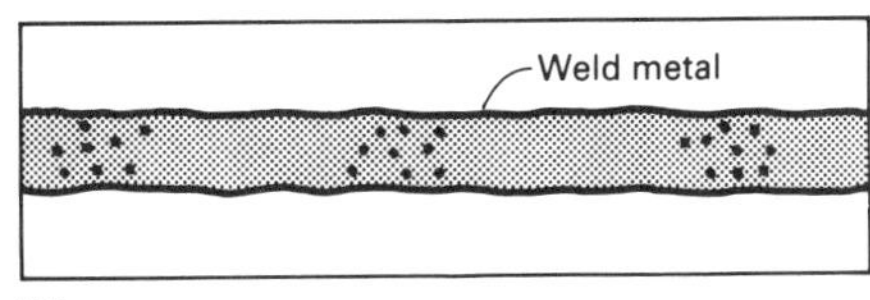

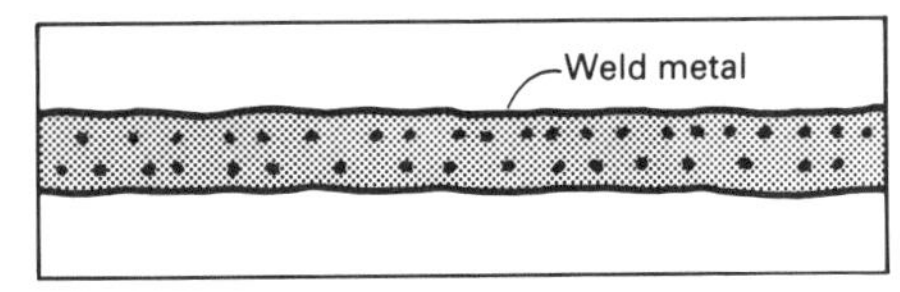

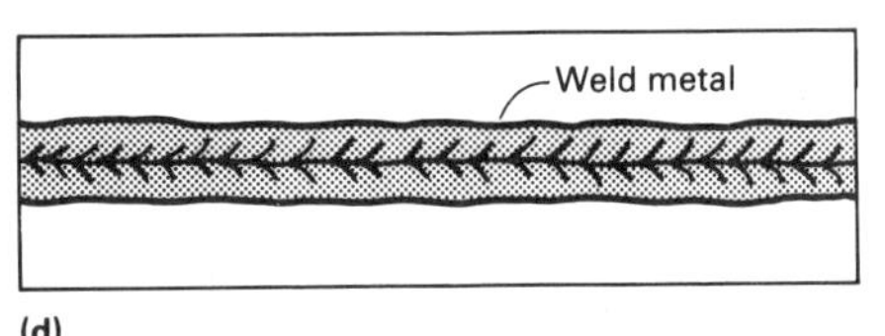

Fig. 1 Types of gas porosity commonly found in weld metal. (a) Uniformly scattered porosity. (b) Cluster porosity. (c) Linear porosity. (d) Elongated porosity

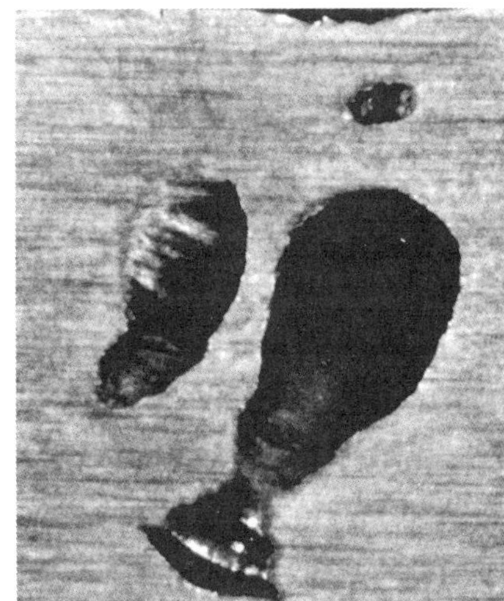

Fig. 2 Wormhole porosity in a weld bead. Longitudinal cut. 20×

Magnetic particle inspection and liquid penetrant inspection are not suitable for detecting subsurface gas porosity. These methods are restricted to the detection of only those pores that are open to the surface.

Slag inclusions may occur when using welding processes that employ a slag covering for shielding purposes. (With other processes, the oxide present on the metal surface before welding may also become entrapped.) Slag inclusions can be found near the surface and in the root of a weld (Fig. 3a), between weld beads in multiple-pass welds (Fig. 3b), and at the side of a weld near the root (Fig. 3c).

During welding, slag may spill ahead of the arc and subsequently be covered by the weld pool because of poor joint fit-up, incorrect electrode manipulation, or forward arc blow. Slag trapped in this manner is generally located near the root. Radical motions of the electrode, such as wide weaving, may also cause slag entrapment on the sides or near the top of the weld after the slag spills into a portion of the joint that has not been filled by the molten pool. Incomplete removal of the slag from the previous pass in multiple-pass welding is another common cause of entrapment. In multiple-pass welds, slag may be entrapped any number of places in the weld between passes. Slag inclusions are generally oriented along the direction of welding.

Three methods used for the detection of slag below the surface of single-pass or multiple-pass welds are magnetic particle, radiographic, and ultrasonic inspection. Depending on their size, shape, orientation, and proximity to the surface, slag inclusions can be detected by magnetic particle inspection with a dc power source, provided the material is ferromagnetic. Radiography can be used for any material, but is the most expensive of the three methods. Ultrasonic inspection can also be used for any material and is the most reliable and least expensive method. If the weld is machined to a flush

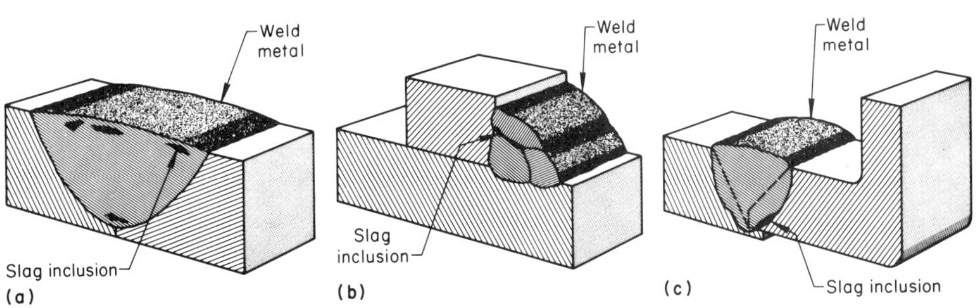

Fig. 3 Sections showing locations of slag inclusions in weld metal. (a) Near the surface and in the root of a single-pass weld. (b) Between weld beads in a multiple-pass weld. (c) At the side of a weld near the root

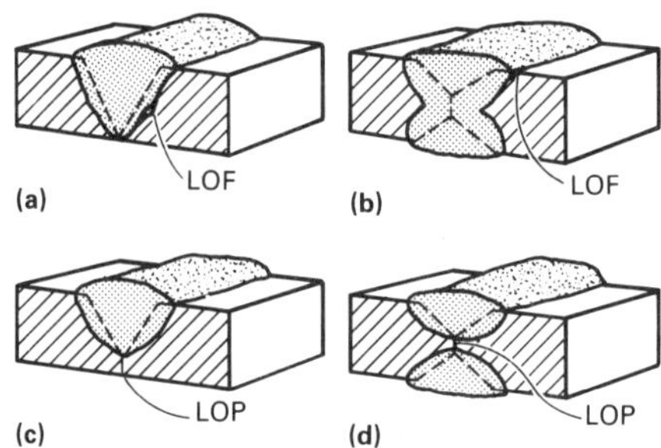

Fig. 4 Lack of fusion in (a) a single-V-groove weld and (b) double-V-groove weld. Lack of penetration in (c) a single-V-groove and (d) a double-V-groove weld

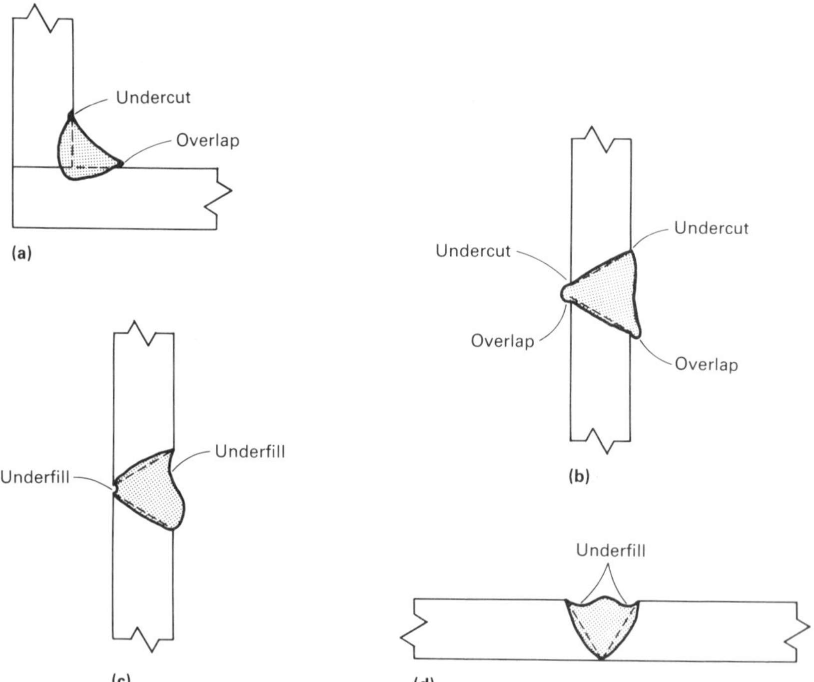

Fig. 5 Weld discontinuities affecting weld shape and contour. (a) Undercut and overlapping in a fillet weld. (b) Undercut and overlapping in a groove weld. (c) and (d) Underfill in groove welds

contour, flaws as close as 0.8 mm (1/32 in.) to the surface can be detected with the straight-beam technique of ultrasonic inspection, provided the instrument has sufficient sensitivity and resolution. A 5- or 10-MHz dual-element transducer is normally used in this application. If the weld cannot be machined, near-surface sensitivity will be low because the initial pulse is excessively broadened by the rough, as-welded surface. Unmachined welds can be readily inspected by direct-beam and reflected-beam techniques, using an angle-beam (shear-wave) transducer.

Tungsten inclusions are particles found in the weld metal from the nonconsumable tungsten electrode used in GTAW. These inclusions are the result of:

- Exceeding the maximum current for a given electrode size or type
- Letting the tip of the electrode make contact with the molten weld pool
- Letting the filler metal come in contact with the hot tip of the electrode
- Using an excessive electrode extension
- Inadequate gas shielding or excessive wind drafts, which result in oxidation
- Using improper shielding gases such as argon-oxygen or argon-CO_2 mixtures, which are used for GMAW

Tungsten inclusions, which are not acceptable for high-quality work, can only be found by internal inspection techniques, particularly radiographic testing.

Lack of fusion and lack of penetration result from improper electrode manipulation and the use of incorrect welding conditions. Fusion refers to the degree to which the original base metal surfaces to be welded have been fused to the filler metal. On the other hand, penetration refers to the degree to which the base metal has been melted and resolidified to result in a deeper throat than was present in the joint before welding. In effect, a joint can be completely fused but have incomplete root penetration to obtain the throat size specified. Based on these definitions, LOF discontinuities are located on the sidewalls of a joint, and LOP discontinuities are located near the root (Fig. 4). With some joint configurations, such as butt joints, the two terms can be used interchangeably. The causes of LOF include excessive travel speed, bridging, excessive electrode size, insufficient current, poor joint preparation, overly acute joint angle, improper electrode manipulation, and excessive arc blow. Lack of penetration may be the result of low welding current, excessive travel speed, improper electrode manipulation, or surface contaminants such as oxide, oil, or dirt that prevent full melting of the underlying metal.

Radiographic methods may be unable to detect these discontinuities in certain cases, because of the small effect they have on x-ray absorption. As will be described later, however, lack of sidewall fusion is readily detected by radiography. Ultrasonically, both types of discontinuities often appear as severe, almost continuous, linear porosity because of the nature of the unbonded areas of the joint. Except in thin sheet or plate, these discontinuities may be too deep-lying to be detected by magnetic particle inspection.

Geometric weld discontinuities are those associated with imperfect shape or unacceptable weld contour. Undercut, underfill, overlap, excessive reinforcement, fillet shape, and melt-through, all of which were defined earlier, are included in this grouping. Geometric discontinuities are shown schematically in Fig. 5. Radiography is used most often to detect these flaws.

Cracks can occur in a wide variety of shapes and types and can be located in numerous positions in and around a welded joint (Fig. 6). Cracks associated with welding can be categorized according to whether they originate in the weld itself or in the base metal. Four types commonly occur in the weld metal: transverse, longitudinal, crater, and hat cracks. Base metal cracks can be divided into seven categories: transverse cracks, underbead cracks, toe cracks, root cracks, lamellar tearing, delaminations, and fusion-line cracks.

Weld metal cracks and base metal cracks that extend to the surface can be detected by liquid penetrant and magnetic particle inspection. Magnetic particle inspection can also detect subsurface cracks, depending on their size, shape, and proximity to the surface. Although the orientation of a crack with respect to the direction of the radiation

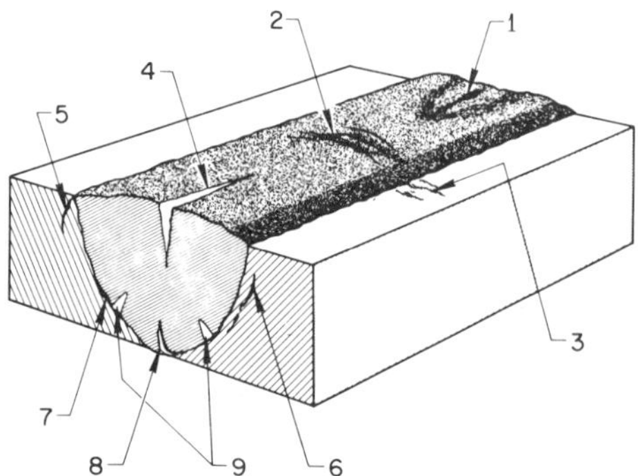

Fig. 6 Identification of cracks according to location in weld and base metal. 1, crater crack in weld metal; 2, transverse crack in weld metal; 3, transverse crack in HAZ; 4, longitudinal crack in weld metal; 5, toe crack in base metal; 6, underbead crack in base metal; 7, fusion-line crack; 8, root crack in weld metal; 9, hat cracks in weld metal

beam is the dominant factor in determining the ability of radiography to detect the crack, differences in composition between the base metal and the weld metal may create shadows to hide a crack that otherwise might be visible. Ultrasonic inspection is generally effective in detecting most cracks in the weld zone.

Transverse cracks in weld metal (No. 2, Fig. 6) are formed when the predominant contraction stresses are in the direction of the weld axis. They can be hot cracks, which separate intergranularly as the result of hot shortness or localized planar shrinkage, or they can be transgranular separations produced by stresses exceeding the strength of the material. Transverse cracks lie in a plane normal to the axis of the weld and are usually open to the surface. They usually extend across the entire face of the weld and sometimes propagate into the base metal.

Transverse cracks in base metal (No. 3, Fig. 6) occur on the surface in or near the HAZ. They are the result of the high residual stresses induced by thermal cycling during welding. High hardness, excessive restraint, and the presence of hydrogen promote their formation. Such cracks propagate into the weld or beyond the HAZ into the base metal as far as is needed to relieve the residual stresses.

Underbead cracks (No. 6, Fig. 6) are similar to transverse cracks in that they form in the HAZ because of high hardness, excessive restraint, and the presence of hydrogen. Their orientation follows the contour of the HAZ.

Longitudinal cracks (No. 4, Fig. 6) may exist in three forms, depending on their positions in the weld. Check cracks are open to the surface and extend only partway through the weld. Root cracks extend from the root to some point within the weld. Full centerline cracks may extend from the root to the face of the weld metal.

Check cracks are caused either by high contraction stresses in the final passes applied to a weld joint or by a hot-cracking mechanism.

Root cracks are the most common form of longitudinal weld metal crack because of the relatively small size of the root pass. If such cracks are not removed, they can propagate through the weld as subsequent passes are applied. This is the usual mechanism by which full centerline cracks are formed.

Centerline cracks may occur at either high or low temperatures. At low temperatures, cracking is generally the result of poor fit-up, overly rigid fit-up, or a small ratio of weld metal to base metal.

All three types of longitudinal cracks are usually oriented perpendicular to the weld face and run along the plane that bisects the welded joint. Seldom are they open at the edge of the joint face, because this requires a fillet weld with an extremely convex bead.

Crater cracks (No. 1, Fig. 6) are related to centerline cracks. As the name implies, crater cracks occur in the weld crater formed at the end of a welding pass. Generally, this type of crack is caused by failure to fill the crater before breaking the arc. When this happens, the outer edges of the crater cool rapidly, producing stresses sufficient to crack the interior of the crater. This type of crack may be oriented longitudinally or transversely or may occur as a number of intersecting cracks forming the shape of a star. Longitudinal crater cracks can propagate along the axis of the weld to form a centerline crack. In addition, such cracks may propagate upward through the weld if they are not removed before subsequent passes are applied.

Hat cracks (No. 9, Fig. 6) derive their name from the shape of the weld cross section with which they are usually associated. This type of weld flares out near the weld face, resembling an inverted top hat. Hat cracks are the result of excessive voltage or welding speed. The cracks are located about halfway up through the weld and extend into the weld metal from the fusion line of the joint.

Toe and root cracks (No. 5 and 8, Fig. 6) can occur at the notches present at notch locations in the weld when high residual stresses are present. Both toe and root cracks propagate through the brittle HAZ before they are arrested in more ductile regions of the base metal. Characteristically, they are oriented almost perpendicular to the base metal surface and run parallel to the weld axis.

Lamellar tearing is the phenomenon that occurs in T-joints that are fillet welded on both sides. This condition, which occurs in the base metal or HAZ of restrained weld joints, is characterized by a steplike crack parallel to the rolling plane. The crack originates internally because of tensile strains produced by the contraction of the weld metal and the surrounding HAZ during cooling. Figure 7 shows a typical condition.

Fusion-line cracks (No. 7, Fig. 6) can be classified as either weld metal cracks or base metal cracks because they occur along the fusion line between the two. There are no limitations as to where along the fusion line these cracks can occur or how far around the weld they can extend.

Discontinuities Associated With Specialized Welding Processes

The preceding section has dealt mainly with the discontinuities common to conventional arc-welding processes. In addition, there are certain more specialized welding methods that may have discontinuities unique to them. These methods include electron beam, plasma arc, electroslag, friction, and resistance welding. In general, the types of discontinuities associated with these processes are the same as those associated with conventional arc welding; however, because of the nature of the processes and the joint configurations involved, such discontinuities may be oriented differently from those previously described, or they may present particular problems of location and evaluation.

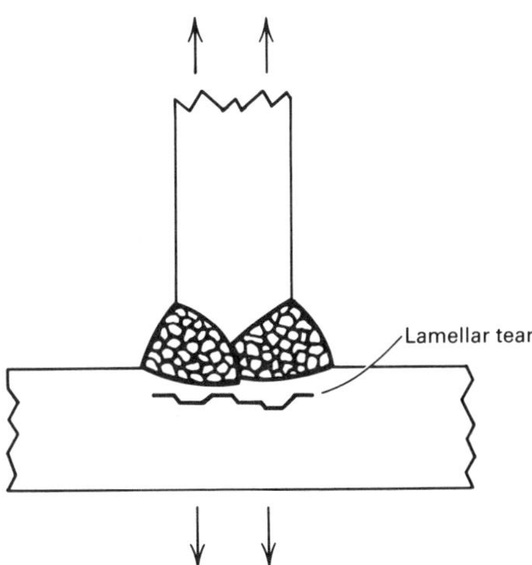

Fig. 7 Lamellar tear caused by thermal contraction strain

Electron Beam Welding

In electron beam welding, as in all other welding processes, weld discontinuities can be divided into two major categories:

- Those that occur at, or are open to, the surface
- Those that occur below the surface

Surface flaws include undercut, mismatch, underfill, reinforcement, cracks, missed seams, and LOP. Subsurface flaws include porosity, massive voids, bursts, cracks, missed seams, and LOP. Figure 8 shows poor welds containing these flaws, and a good weld with none of them.

Surface discontinuities such as undercut, mismatch, reinforcement, and underfill are macroscopic discontinuities related to the contour of the weld bead or the joint. As such, they are readily detected visually or dimensionally. Surface discontinuities such as cracks are usually detected visually using liquid penetrant inspection or using magnetic particle inspection if the material is ferromagnetic.

When liquid penetrants are used to inspect a weld for surface discontinuities such as cracks, missed seams, and LOP, the surface to be inspected must be clean and the layers of metal smeared from machining or peened from grit- or sandblasting must be removed. Generally, some type of etching or pickling treatment works well, but the possibility of hydrogen pickup from the treatment must be considered.

Occasionally, special inspection procedures must be employed to detect some types of surface discontinuities. Missed

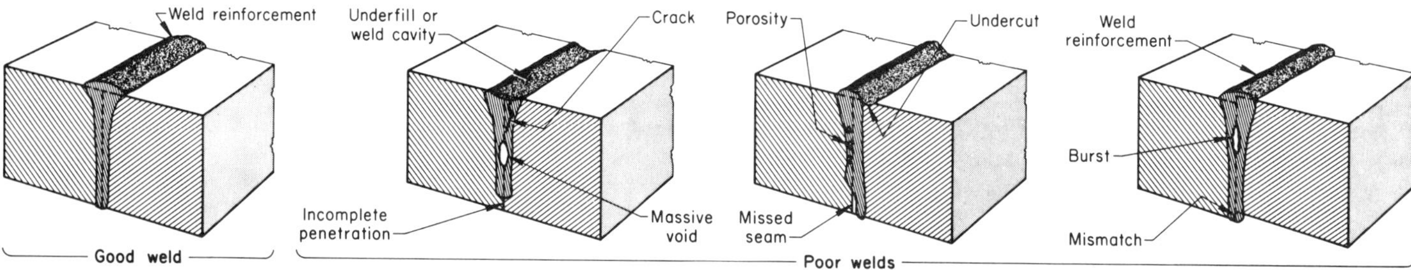

Fig. 8 Electron beam welds showing flaws that can occur in poor welds and the absence of flaws in a good weld with reinforcement

seams and LOP are often difficult to detect because they are frequently associated with complex weld joints that prevent direct viewing of the affected surface.

Because of this difficulty, missed seams are often detected using a visual witness-line procedure, in which equally spaced parallel lines are scribed on both sides of the unwelded joint at the crown and root surfaces. Missed seams, which result from misalignment of the electron beam with the joint such that the fusion zone fails to encompass the entire joint, are detected by observing the number of witness lines remaining on either side of the weld bead. By establishing the relationship between the width of the fusion zone and the spacing of the witness lines, reasonably accurate criteria for determining whether the joint has been contained within the weld path (and therefore whether missed seams are present) can be developed.

Lack of penetration discontinuities occur when the fusion zone fails to penetrate through the entire joint thickness, resulting in an unbonded area near the root of the joint. These discontinuities are best detected by etching the root surface and observing the macroscopic shape and width of the fusion zone for full and even penetration.

An alternative method for inspecting complete weld penetration is that of immersion pulse-echo ultrasonic testing. The planet gear carrier assembly shown in Fig. 9(a) consists of three decks (plates) and eight curved spacer sections. The assembly, which is made from SAE 15B22M boron-treated structural steel, is held together by 16 welds. Weld integrity is monitored by ultrasonic methods.

The parts are mounted on a turntable on a locating fixture so that the welds along the outside diameter are accessible. Two transducers, located above and below the assembly, are used. Figure 9(b) shows the upper transducer in position to test the welds. An overall view of the tank, turntable, controls, and reject/accept light panel is shown in Fig. 10. The use of such a system involves little downtime and enables a high quality level to be maintained. More information on ultrasonic test methods is presented later in this article.

Subsurface discontinuities are generally considerably more difficult to detect than surface discontinuities because observation is indirect. The two most reliable and widely used NDI methods are radiography and ultrasonics.

Volume-type discontinuities such as porosity, voids, and bursts are detected by radiographic inspection, provided their cross sections presented to the radiating beam exceed 1 to 2% of the beam path in the metal. Discontinuities that present extremely thin cross sections to the beam path, such as cracks, missed seams, and LOP, are detectable with x-rays only if they are viewed from the end along their planar dimensions.

Ultrasonic inspection can detect most volume discontinuities as well as planar discontinuities. Planar discontinuities are best detected normal to the plane of the discontinuities, but missed seams and LOP often appear as continuous porosity when viewed looking down from the crown to the root of the weld in the plane of the discontinuity.

Because of the inherent dependence of both radiographic and ultrasonic inspection

(a)

(b)

Fig. 9 Planet gear carrier assembly (a) showing the four welds in one stack that connect the three decks. The top and top-center welds are tested by the top ultrasonic transducer, and the bottom and bottom-center welds are inspected by the bottom transducer. (b) Close-up of upper transducer in position to test the welds. Courtesy of John M. St. John, Caterpillar, Inc.

Fig. 10 Overall view of the ultrasonic unit used to test the electron beam welded assembly shown in Fig. 9. Courtesy of John M. St. John, Caterpillar, Inc.

on the shape and orientation of flaws and because each of the two methods can generally detect those flaws that the other misses, it is most advisable to complement one method with the other. Furthermore, to increase the likelihood of properly viewing a flaw, one of the methods should be employed in at least two (preferably perpendicular) directions.

Plasma Arc Welding

Discontinuities that occur in plasma arc welds include both surface and subsurface types, as shown in Fig. 11.

Surface discontinuities such as irregular reinforcement, underfill, undercut, and mismatch that are associated with weld bead contour and joint alignment are easily detected visually or dimensionally. Lack of penetration is also detected visually through the absence of a root bead. Weld cracks that are open to the surface are detected with liquid penetrants. Surface contamination, which results from insufficient shielding-gas coverage, is detected by the severe discoloration of the weld bead or adjacent HAZ.

Subsurface discontinuities are generally more prevalent in manual than in automatic plasma arc welding and are detected primarily by radiographic or ultrasonic inspection.

Porosity is by far the most commonly encountered discontinuity. Radiographic inspection is limited to detecting pores greater than approximately 1 to 2% of the joint thickness. Visibility is greater if both the crown and root beads are machined flush. Ultrasonic inspection can detect porosity if the joint is machined flush and joint thickness exceeds approximately 1.3 mm (0.050 in.).

Tunneling, as shown in Fig. 11, is a severe void along the boundary of the fusion zone and the HAZ. This discontinuity results from a combination of torch alignment and welding conditions (particularly travel speed). Tunneling is readily detectable by radiographic inspection.

Lack of fusion discontinuities occur in either single-pass or multiple-pass repair welds (Fig. 11). These discontinuities result from insufficient heat input to permit complete fusion of a particular weld bead to the part. Incomplete fusion can be detected by radiographic or ultrasonic inspection. Depending on the orientation of the discontinuity, one method may have an advantage over the other, so both should be used for optimum inspection.

Subsurface weld cracks, regardless of their cause, are detectable by radiographic and ultrasonic inspection.

Subsurface contamination in plasma arc welding results when copper from the torch nozzle is expelled into the weld. This is caused by excessive heat, usually produced in manual repair welding when the torch nozzle is placed too close to the weld, particularly in a groove. The resulting contamination, which may be detrimental, is undetectable by conventional NDI methods. The only way of detecting copper contamination is by alerting the operator to watch for copper expulsion, which then must be machined out.

Electroslag Welding

Electroslag welding involves the use of copper dams over the open surfaces of a butt joint to hold the molten metal and the slag layer as the joint is built up vertically. Wire is fed into the slag layer continuously and is melted by the heat generated as current passes through the highly resistant slag layer.

Generally, electroslag welds are inspected with the same nondestructive examination (NDE) methods as other heavy-section welds. With the exception of procedure qualification, all testing is nondestructive because of the sizes used. Techniques such as radiography and ultrasonic inspection are most often used, while visual, magnetic particle, and liquid penetrant testing are used also. Internal defects are generally more serious. Radiography and ultrasonic tests are the best methods for locating internal discontinuities.

Because of the nature of the process, LOF is rare. If fusion is achieved on external material edges, then fusion is generally

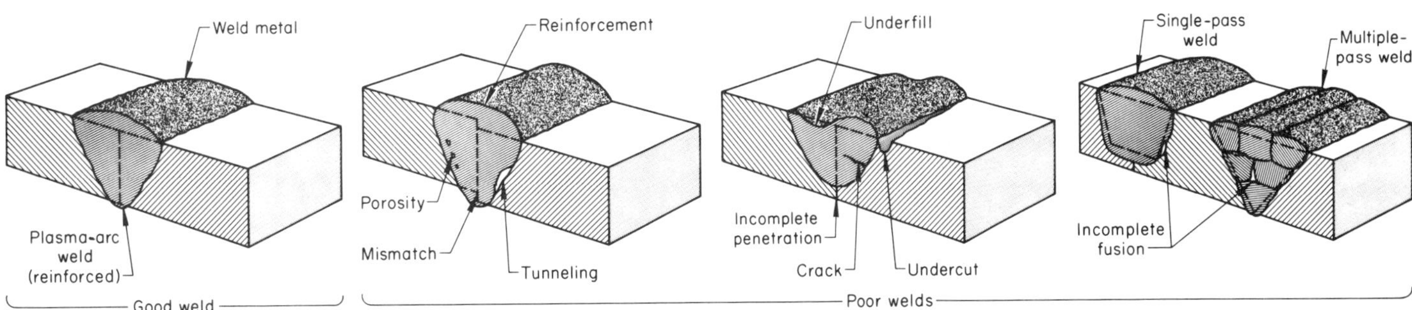

Fig. 11 Plasma arc welds showing flaws that can occur in poor welds and the absence of flaws in good reinforced weld

complete throughout. Cracking may occur either in the weld or the HAZ. Porosity may either take the form of a rounded or a piped shape; the latter is often called wormhole porosity. Ultrasonic inspection is probably the quickest single method for inspecting any large weldment. If defects should occur, they appear as porosity or centerline cracking. Ultrasonic inspection is effective for locating either type of defect; however, only well-qualified personnel should set up the equipment and interpret the test results.

Electroslag welding results in large dendritic grain sizes because of the slow cooling rate. Inexperienced personnel often use high sensitivity and actually pick up the large coarse grains; when such welds are sectioned, usually no defects are present. Inspectors must learn to use low sensitivity to obtain good results when inspecting electroslag welds. Magnetic particle inspection is not a particularly good inspection method, because the areas examined by this technique are primarily surface or near surface. This is only a small percentage of the total weld; the only useful information is either checking the ends for craters, cracks, or centerline cracking or possibly for lack of edge fusion on the weld faces. Usually, a visual examination gives the same result unless the defect is subsurface. Visual examinations are only effective for surface defects, which are not common in this process.

Friction Welding

If impurities are properly dispelled during upsetting, friction or inertia welds are generally free of voids and inclusions. Incomplete center fusion can occur when flywheel speed is too low, when the amount of upset is insufficient, or when mating surfaces are concave. Tearing in the HAZ can be caused by low flywheel speed or excessive flywheel size. Cracks can occur when materials that are prone to hot shortness are joined. The penetration of a split between the extruded flash into the workpiece cross section is most prevalent during the welding of thin-wall tubing using improper conditions that do not allow for sufficient material upset.

The area where LOF generally occurs is at or near the center of the weld cross section. Because this is a subsurface discontinuity, detection is limited to radiographic or ultrasonic inspection; ultrasonic inspection is more practical. The longitudinal wave test (either manual-contact or immersion method) with beam propagation perpendicular to the area of LOF gives the most reliable results. This test can be performed as long as one end of the workpiece is accessible to the transducer.

Penetration of the split between the extruded flash on the outer surface of a tube is readily detected by liquid penetrant or magnetic particle inspection after the flash has been removed by machining. A split between the weld flash on the inner surface of the tube can be detected by ultrasonic inspection using the angle-beam technique with manual contact of the transducer to the outside surface of the tube. The transducer contacts the tube so that the sound propagates along the longitudinal axis of the tube through the weldment.

Resistance Welding

Resistance welding encompasses spot, seam, and projection welding, each of which involves the joining of metals by passing current from one side of the joint to the other. The types of discontinuities found in resistance welds include porosity, LOF, and cracks. Porosity will generally be found on the centerline of the weld nugget. Lack of fusion may also be manifested as a centerline cavity. Either of these can be caused by overheating, inadequate pressure, premature release of pressure, or late application of pressure. Cracks may be induced by overheating, removal of pressure before weld quenching is completed, improper loading, poor joint fit-up, or expulsion of excess metal from the weld.

Weld Appearance. On the surface of a resistance spot welded assembly, the weld spot should be uniform in shape and relatively smooth, and it should be free of surface fusion, deep electrode indentations, electrode deposits, pits, cracks, sheet separation, abnormal discoloration around the weld, or other conditions indicating improper maintenance of electrodes or functioning of equipment. However, surface appearance is not always a good indicator of spot weld quality, because shunting and other causes of insufficient heating or incomplete penetration usually leave no visible effects on the workpiece.

The common practice for monitoring spot weld quality in manufacturing operations is the teardown method augmented with pry testing and visual inspection (Ref 1). In visual inspection, the operator uses the physical features of the weld surface, such as coloration, indentation, and smoothness, for assessing the quality of the weld. In pry testing, a wedge-shaped tool is inserted between the metal sheets next to the accessible welds, and a prying action is performed to see if the sheets will separate in the weld zone. The teardown method consists of physically tearing apart the welded members with hammers and chisels to determine the presence and size adequacy of fused metal nuggets at the spot weld site. The specifications require that the parent metal be torn and that the weld nugget remain intact. The destructiveness and/or inadequacy of these common inspection methods has long been recognized, and as a result, nondestructive methods have been extensively studied. The pulse-echo ultrasonic inspection of spot welds is now feasible and is discussed in the section "Ultrasonic Inspection" in this article. The use of acoustic emission is also discussed below.

Diffusion Bonding (Ref 2-5)

Diffusion bonding is a metal joining process that requires the application of controlled pressures at elevated temperatures and usually a protective atmosphere to prevent oxidation. No melting and only limited macroscopic deformation or relative motion between the faying surfaces of the parts occur during bonding. As such, the principal mechanism for joint formation is solid-state diffusion. A diffusion aid (filler metal) may or may not be used. Terms that are also frequently used to describe the process include diffusion welding, solid-state bonding, pressure bonding, isostatic bonding, and hot press bonding. Diffusion bonding has the advantage of producing a product finished to size, with joint efficiencies approaching 100%. Details of the process are given in the article "Solid-State Welding" in Volume 6 of the 9th Edition of *Metals Handbook*.

Discontinuities in Diffusion Bonds. In the case of fusion welds, the detection of discontinuities less than 1 mm (0.04 in.) in size is not generally expected. In diffusion bonding, in which no major lack of bonding occurs, individual discontinuities may be only micrometers in size. To understand how discontinuities form in diffusion-bonded structures, it is first necessary to consider the principles of the process.

As illustrated in Fig. 12, metal surfaces have several general characteristics:

- Roughness
- An oxidized or otherwise chemically reacted and adherent layer
- Other randomly distributed solid or liquid products such as oil, grease, and dirt
- Adsorbed gas, moisture, or both

Because of these characteristics, two necessary conditions that must be met before a satisfactory diffusion bond can be made are:

- Mechanical intimacy of metal-to-metal contact must be achieved
- Interfering surface contaminants must be disrupted and dispersed to permit metallic bonding to occur (solvent cleaning and inert gas atmospheres can reduce or eliminate problems associated with surface contamination and oxide formation, respectively)

For a given set of processing parameters, surface roughness is probably the most important variable influencing the quality of diffusion-bonded joints. The size of the discontinuities (voids) is principally determined by the scale of roughness of the surfaces being bonded. The degree of surface roughness is dependent on the material and fabrication/machining technique used. It has been shown that bonding becomes easier with finer surface roughness prior to

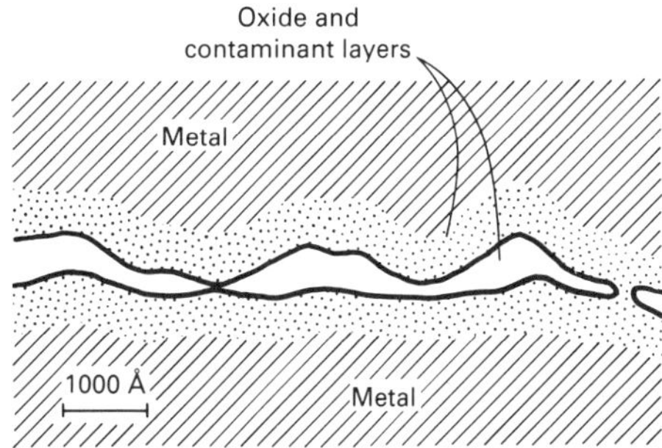

Fig. 12 Characteristics of a metal surface showing roughness and contaminants present. Source: Ref 5

bonding. Figure 13 compares the surface roughness values produced by a variety of fabrication methods. More detailed information on surface roughness can be found in the article "Surface Finish and Surface Integrity" in Volume 16 of the 9th Edition of *Metals Handbook*.

The mechanism of bond formation in diffusion bonding is believed to be the deformation of the surface roughness in order to cause metal-to-metal contact at asperities, followed by the removal of interfacial voids and cracks by diffusional and creep processes. For conventional diffusion bonding without a diffusion aid, the three-stage mechanistic model shown in Fig. 14 describes bond formation. In the first stage, deformation of the contacting asperities occurs primarily by yielding and by creep deformation mechanisms to produce intimate contact over a large fraction of the interfacial area. At the end of this stage, the joint is essentially a grain boundary at the areas of contact with voids between these areas. During the second stage, diffusion becomes more important than deformation, and many of the voids disappear as the grain-boundary diffusion of atoms continues. Simultaneously, the interfacial grain boundary migrates to an equilibrium configuration away from the original plane of the joint, leaving many of the remaining voids within the grains. In the third stage, the remaining voids are eliminated by the volume diffusion of atoms to the void surface (equivalent to diffusion of vacancies away from the void). Successful completion of stage three is dependent on proper surface processing and joint processing.

Successful nondestructive evaluation of diffusion-bonded joints requires that the maximum size and distribution of discontinuities be determined. However, many conventional NDE methods and equipment are not adequate for discontinuity determination.

Fluorescent penetrants provide excellent detection capability for lack of bonding as long as the interfacial crack breaks the surface. They are completely ineffective for defects that have no path to the surface.

Conventional film radiography is not suitable for detecting the extremely small defects

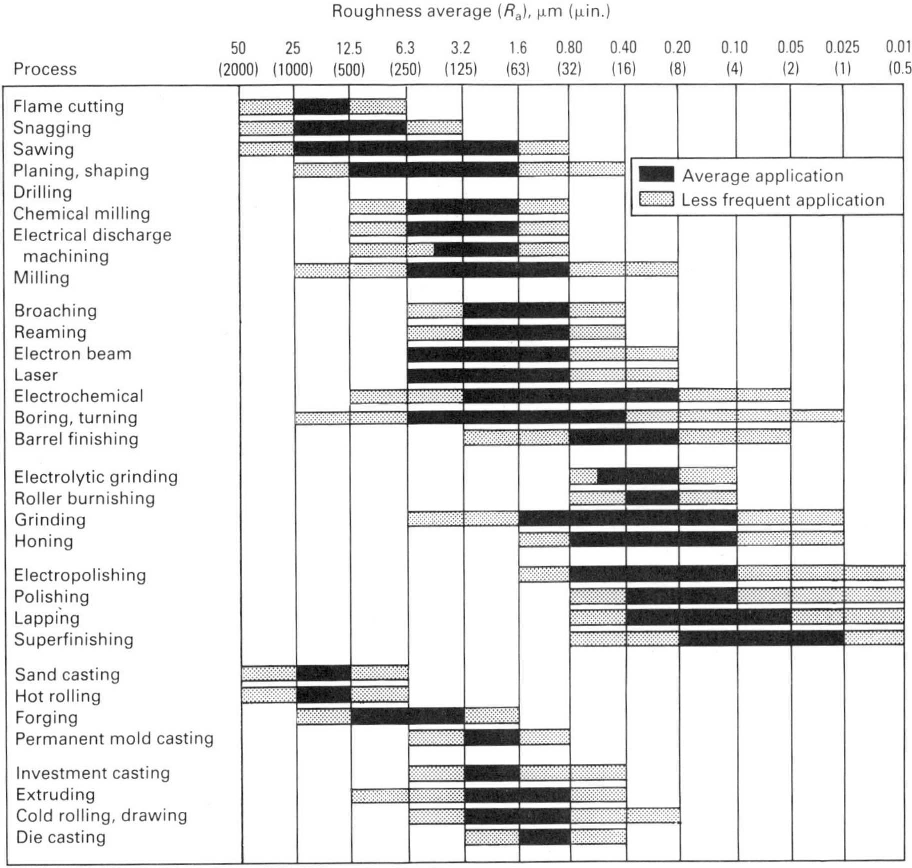

Fig. 13 Surface roughness produced by common production methods. The ranges shown are typical of the processes listed. Higher or lower values can be obtained under special conditions.

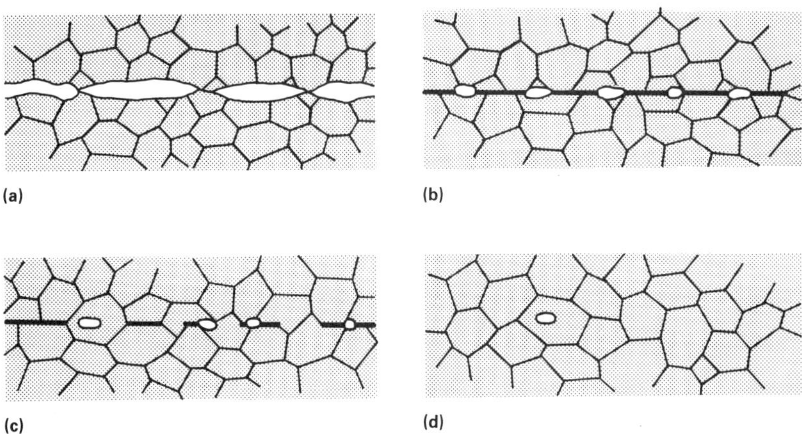

Fig. 14 Three-stage mechanistic model of diffusion welding. (a) Initial asperity contact. (b) First-stage deformation and interfacial boundary formation. (c) Second-stage grain-boundary migration and pore elimination. (d) Third-stage volume diffusion and pore elimination. Source: Ref 5

involved in diffusion bonding, but the use of x-ray microfocus techniques coupled with digital image enhancement offers an improvement in resolution. Discontinuities as small as 50 μm (0.002 in.) have been detected.

Conventional ultrasonic testing has some applications, although only significant lack of bonding or clusters of smaller defects can be reliably detected. High-resolution flaw detection involving frequencies approaching 100 MHz appears to have distinct advantages over conventional testing. Scanning acoustic microscopy also appears to offer excellent possibilities in diffusion bond inspection. Eddy current and thermal methods are relatively unsatisfactory for most applications.

Methods of Nondestructive Inspection

The nondestructive inspection of weldments has two functions:

- Quality control, which is the monitoring of welder and equipment performance and of the quality of the consumables and the base materials used
- Acceptance or rejection of a weld on the basis of its fitness for purpose under the service conditions imposed on the structure

The appropriate method of inspection is different for each function. If evaluation is a viable option, discontinuities must be detected, identified, located exactly, sized, and their orientation established, which limits its inspection to a volumetric technique.

Weld discontinuities constitute the center of activity with the inspection of welded constructions. The most widely used inspection techniques used in the welding industry are visual, liquid penetrant, magnetic particle, radiographic, ultrasonic, acoustic emission, eddy current, and electric current perturbation methods. Each of these techniques has specific advantages and limitations. Existing codes and standards that provide guidelines for these various techniques are based on the capabilities and/or limitations of these nondestructive methods.

Selection of Technique. A number of factors influence selection of the appropriate nondestructive test technique for inspecting a welded structure, including discontinuity characteristics, fracture mechanics requirements, part size, portability of equipment, and other application constraints. These categories, although perhaps unique to a specific inspection problem, may not clearly point the way to the most appropriate technique. It is generally necessary to exercise engineering judgment in ranking the importance of these criteria and thus determining the optimum inspection technique.

Characteristics of the Discontinuity. Because nondestructive techniques are based on physical phenomena, it is useful to describe the properties of the discontinuity of interest, such as composition and electrical, magnetic, mechanical, and thermal properties. Most significant are those properties that are most different from those of the weld or base metal. It is also necessary to identify a means of discriminating between discontinuities with similar properties.

Fracture mechanics requirements, solely from a discontinuity viewpoint, typically include detection, identification, location, sizing, and orientation. In addition, complicated configurations may necessitate a nondestructive assessment of the state of the stress of the region containing the discontinuity. In the selection process, it is important to establish these requirements correctly. This may involve consultation with stress analysts, materials engineers, and statisticians.

Often, the criteria may strongly suggest a particular technique. Under ideal conditions, such as in a laboratory, the application of such a technique might be routine. In the field, however, other factors may force a different choice of technique.

Constraints tend to be unique to a given application and may be completely different even when the welding process and metals are the same. Some of these constraints include:

- Access to the region under inspection
- Geometry of the structure (flat, curved, thick, thin)
- Condition of the surface (smooth, irregular)
- Mode of inspection (preservice, in-service, continuous, periodic, spot)
- Environment (hostile, underwater, and so on)
- Time available for inspection (high speed, time intensive)
- Reliability
- Application of multiple techniques
- Cost

Failure to consider adequately the constraints imposed by a specific application can render the most sophisticated equipment and theory useless. Moreover, for the simple or less important cases of failure, it may be unnecessary. Once criteria have been established, an optimum inspection technique can be selected, or designed and constructed.

The terms accuracy, sensitivity, and reliability are used loosely in NDE. Often, they are discussed as one term to avoid distinguishing among the specific aspects of these terminologies.

Accuracy is the attribute of an inspection method that describes the correctness of the technique within the limits of its precision. In other words, the technique is highly accurate if the indications resulting from the technique are correct. This does not mean that the technique was able to detect all discontinuities present, but rather that those indicated actually exist.

Sensitivity, on the other hand, refers to the capability of a technique to detect discontinuities that are small or that have properties only slightly different from the material in which they reside. Figure 15 schematically illustrates the concepts of accuracy and sensitivity in the context of detection probability. In general, sensitivity is gained at the expense of accuracy because high sensitivity increases the probability of false alarms.

Reliability is a combination of both accuracy and sensitivity. Three factors influence reliability: inspection procedure, including the instrumentation; human factors (inspector motivation, experience, training, education, and so on); and data analysis. Uncalibrated equipment, improper application of technique, and inconsistent quality of accessory equipment (transducers, couplant, film, chemicals, and so on) may affect accuracy and, in some cases, sensitivity. Poor inspector technique, unfamiliar response, lack of concentration, and other human factors can combine to reduce reliability. Data analysis, or the lack of it, can influence reliability as well; generally, inspection is performed under conditions in which detection probability is less than 100% and is not constant with discontinuity severity. Consequently, statistics must be employed to establish the level of confidence that may be attached to the inspection results.

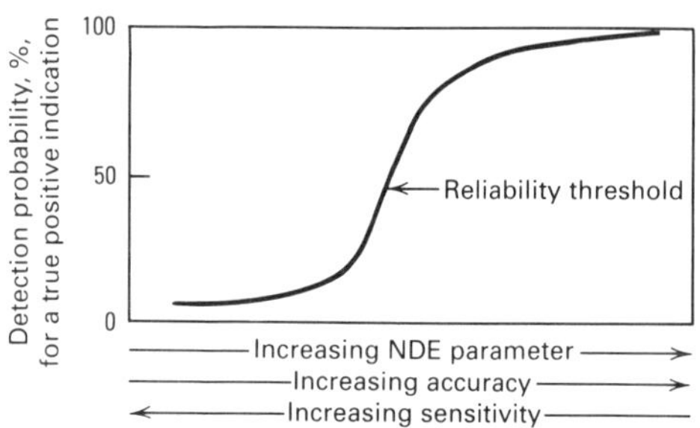

Fig. 15 Detection probability for a true positive indication. High sensitivity increases the likelihood of false indications. The minimum NDE parameter size required to establish fitness-for-purpose must lie to the right of the transition region, or reliability threshold, to achieve satisfactory reliability.

High sensitivity with low accuracy may be far worse, from the viewpoint of reliability, than low sensitivity with high accuracy, especially if the sensitivity level is adequate for detecting the weld discontinuities in question. As a general rule, the transition region of the detection probability curve indicates the degree of reliability. If this region occurs with the limits encompassed by inspection capabilities, which are smaller than the values required for evaluating the fitness-for-purpose of the welds being inspected, reliability is satisfactory. If, on the other hand, the region occurs at values higher than those required, reliability is unsatisfactory. The transition region can be viewed as the reliability threshold. More detailed information on the reliability of nondestructive test data can be found in the Section "Quantitative Nondestructive Evaluation" in this Volume.

Visual Inspection

For many noncritical welds, integrity is verified principally by visual inspection. Even when other nondestructive methods are used, visual inspection still constitutes an important part of practical quality control. Widely used to detect discontinuities, visual inspection is simple, quick, and relatively inexpensive. The only aids that might be used to determine the conformity of a weld are a low-power magnifier, a borescope, a dental mirror, or a gage. Visual inspection can and should be done before, during, and after welding. Although visual inspection is the simplest inspection method to use, a definite procedure should be established to ensure that it is carried out accurately and uniformly.

Visual inspection is useful for checking the following:

- Dimensional accuracy of weldments
- Conformity of welds to size and contour requirements
- Acceptability of weld appearance with regard to surface roughness, weld spatter, and cleanness
- Presence of surface flaws such as unfilled craters, pockmarks, undercuts, overlaps, and cracks.

Although visual inspection is an invaluable method, it is unreliable for detecting subsurface flaws. Therefore, judgment of weld quality must be based on information in addition to that afforded by surface indications.

Additional information can be gained by observations before and during welding. For example, if the plate is free of laminations and properly cleaned and if the welding procedure is followed carefully, the completed weld can be judged on the basis of visual inspection. Additional information can also be gained by using other NDI methods that detect subsurface and surface flaws.

Dimensional Accuracy and Conformity. All weldments are fabricated to meet certain specified dimensions. The fabricator must be aware of the amount of shrinkage that can be expected at each welded joint, the effect of welding sequence on warpage or distortion, and the effect of subsequent heat treatment used to provide dimensional stability of the weldment in service. Weldments that require rigid control of final dimensions usually must be machined after welding. Dimensional tolerances for as-welded components depend on the thickness of the material, the alloy being welded, the overall size of the product, and the particular welding process used.

The dimensional accuracy of weldments is determined by conventional measuring methods, such as rules, scales, calipers, micrometers, and gages. The conformity of welds with regard to size and contour can be determined by a weld gage. The weld gage shown in Fig. 16 is used when visually inspecting fillet welds at 90° intersections. The size of the fillet weld, which is defined by the length of the leg, is stamped on the gage. The weld gage determines whether or not the size of the fillet weld is within allowable limits and whether there is excessive concavity or convexity. This gage is designed for use on joints between surfaces that are perpendicular. Special weld gages are used when the surfaces are at angles other than 90°. For groove welds, the width of the finished welds must be in accordance with the required groove angle, root face, and root opening. The height of reinforcement of the face and root must be consistent

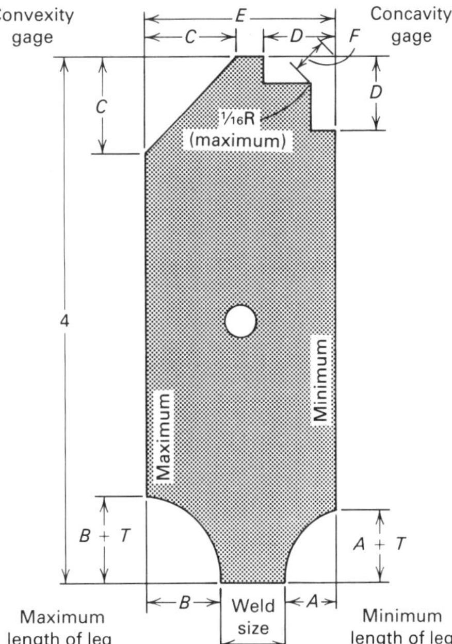

A .... Minimum allowable length of leg
B .... Maximum allowable length of leg
C .... 1.414 times maximum allowable throat size (specifies maximum allowable convexity)
D .... Maximum allowable length of leg when maximum allowable concavity is present
E .... *A* plus *B* plus nominal weld size (or nominal length of leg)
F .... Minimum allowable throat size (specifies maximum allowable concavity)
T .... Additional tolerance for clearance of gage when placed in the fillet

Fig. 16 Gage for visual inspection of a fillet weld at a 90° intersection. Similar gages can be made for other angles. Dimension given in inches

with specified requirements and can be measured by a weld gage.

Appearance Standards. The acceptance of welds with regard to appearance implies the use of a visual standard, such as a sample weldment or a workmanship standard. Requirements as to surface appearance differ widely, depending on the application. For example, when aesthetics are important, a smooth weld that is uniform in size and contour may be required.

The inspection of multiple-pass welds is often based on a workmanship standard. Figure 17 indicates how such standards are prepared for use in the visual inspection of groove and fillet welds. The workmanship standard is a section of a joint similar to the one in manufacture, except that portions of each weld pass are shown. Each pass of the production weld is compared with corresponding passes of the workmanship standard.

Discontinuities. Before a weld is visually inspected for discontinuities such as unfilled craters, surface holes, undercuts, overlaps, surface cracks, and incomplete joint penetration, the surface of the weld should be cleaned of oxides and slag. Cleaning must be done carefully. For example, a chipping hammer used to remove slag could leave hammer marks that can hide fine cracks. Shotblasting can peen the surface of relatively soft metals and hide flaws. A stiff wire brush and sandblasting have been found to be satisfactory for cleaning surfaces of slag and oxides without marring. Additional information on the uses and equipment associated with the visual examination of parts and assemblies is available in the article "Visual Inspection" in this Volume.

Magnetic Particle Inspection

Magnetic particle inspection is a nondestructive method of detecting surface and near-surface flaws in ferromagnetic materials. It consists of three basic operations:

- Establishing a suitable magnetic field in the material being inspected
- Applying magnetic particles to the surface of the material
- Examining the surface of the material for accumulations of the particles (indications) and evaluating the serviceability of the material

Capabilities and Limitations. Magnetic particle inspection is particularly suitable for the detection of surface flaws in highly ferromagnetic metals. Under favorable con-

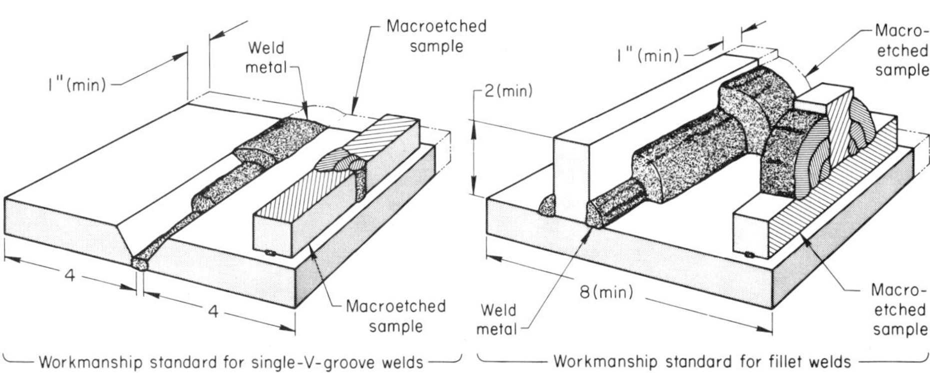

Fig. 17 Workmanship standard for visual comparison during inspection of single-V-groove welds and fillet welds. Dimensions given in inches

ditions, those discontinuities that lie immediately under the surface are also detectable. Nonferromagnetic and weakly ferromagnetic metals, which cannot be strongly magnetized, cannot be inspected by this method. With suitable ferromagnetic metals, magnetic particle inspection is highly sensitive and produces readily discernible indications at flaws in the surface of the material being inspected. Details on the application of this method are available in the article "Magnetic Particle Inspection" in this Volume.

The types of weld discontinuities normally detected by magnetic particle inspection include cracks, LOP, LOF, and porosity open to the surface. Linear porosity, slag inclusions, and gas pockets can be detected if large or extensive or if smaller and near the surface. The recognition of patterns that indicate deep-lying flaws requires more experience than that required to detect surface flaws.

Nonrelevant indications that have no bearing on the quality of the weldment may be produced. These indications are magnetic particle patterns held by conditions caused by leakage fields. Some of these conditions are:

- Particles held mechanically or by gravity in surface irregularities
- Adherent scale or slag
- Indications at a sharp change in material direction, such as sharp fillets and threads
- Grain boundaries. Large grain size in the weld metal or the base metal may produce indications
- Boundary zones in welds, such as indications produced at the junction of the weld metal and the base metal. This condition occurs in fillet welds at T-joints, or in double-V-groove joints, where 100% penetration is not required
- Flow lines in forgings and formed parts
- Brazed joints. Two parts made of a ferromagnetic material joined by a nonferromagnetic material will produce an indication
- Different degrees of hardness in a material, which will usually have different permeabilities that may create a leakage field, forming indications

Operational Requirements. The magnetic particle inspection of weldments requires that the weld bead be free of scale, slag, and moisture. For maximum sensitivity, the weld bead should be machined flush with the surface; however, wire brushing, sandblasting, or gritblasting usually produces a satisfactory bead surface. If the weld bead is rough, grinding will remove the high spots.

Weldments are often inspected using the dry particle method. A powder or paste of a color that gives the best possible contrast to the surface being inspected should be used.

The type of magnetizing current used depends on whether there are surface or subsurface discontinuities. Alternating current is satisfactory for surface cracks, but if the deepest possible penetration is essential, direct current, direct current with surge, or half-wave rectified alternating current is used.

The voltage should be as low as practical to reduce the possibility of damage to the surface of the part from overheating or arcing at contacts. Another advantage of low voltage is freedom from arc flashes if a prod slips or is withdrawn before the current is turned off.

The field strength and flux density used must be determined for each type of weldment. An overly strong field will cause the magnetic particles to adhere too tightly to the surface and hinder their mobility, preventing them from moving to the sites of the flaws. Low field strengths result in nondiscernible patterns and failure to detect indications.

Inspections can be made using the continuous-field and residual-field methods. In the continuous-field method, magnetic particles are placed on the weldment while the current is flowing. In the residual-field method, the particles are placed on the weldment after the magnetizing current is turned off. Residual magnetic fields are weaker than continuous fields. Consequently, inspections using the residual-field method are less sensitive.

The need for the demagnetization of weldments after magnetic particle inspection must be given serious consideration. Where subsequent welding or machining operations are required, it is good practice to demagnetize. Residual magnetism may also hinder cleaning operations and interfere with the performance of instruments used near the weldment.

Liquid Penetrant Inspection

Liquid penetrant inspection is capable of detecting discontinuities open to the surface in weldments made of either ferromagnetic or nonferromagnetic alloys, even when the flaws are generally not visible to the unaided eye. Liquid penetrant is applied to the surface of the part, where it remains for a period of time and penetrates into the flaws. For the correct usage of liquid penetrant inspection, it is essential that the surface of the part be thoroughly clean, leaving the openings free to receive the penetrant. Operating temperatures of 20 to 30° C (70 to 90° F) produce optimum results. If the part is cold, the penetrant may become chilled and thickened so that it cannot enter very fine openings. If the part or the penetrant is too hot, the volatile components of the penetrant may evaporate, reducing the sensitivity.

After the penetrating period, the excess penetrant remaining on the surface is removed. An absorbent, light-colored developer is then applied to the surface. This developer acts as the blotter, drawing out a portion of the penetrant that had previously seeped into the surface openings. As the penetrant is drawn out, it diffuses into the developer, forming indications that are wider than the surface openings. The inspector looks for these colored or fluorescent indications against the background of the developer. Details of the mechanics of this method are given in the article "Liquid Penetrant Inspection" in this Volume.

Radiographic Inspection*

Radiography is a nondestructive method that uses a beam of penetrating radiation such as x-rays and γ-rays (see the article "Radiographic Inspection" in this Volume). When the beam passes through a weldment, some of the radiation energy is absorbed, and the intensity of the beam is reduced. Variations in beam intensity are recorded on film or on a screen when a fluoroscope or an image intensifier is used. The variations are seen as differences in shading that are typical of the types and sizes of any discontinuities present. Radiography is used to detect subsurface discontinuities as well as those that are open to the surface.

Penetrameters. Most U.S. codes and specifications for radiographic inspection require that the sensitivity to a specific flaw size be indicated on each radiograph. Sensitivity is determined by placing a penetrameter (image-quality indicator) that is made of substantially the same material as the specimen directly on the specimen, or on a block of identical material of the same thickness as the specimen, prior to each exposure.

The penetrameters are usually placed on the side of the specimen nearest the source, parallel and adjacent to the weld at one end of the exposed length, with the small holes in the penetrameter toward the outer end. Penetrameter thickness is usually specified to be 2% of the specimen thickness (through the weld zone), although 1 and 4% penetrameter thicknesses are also common. American Society for Testing and Materials and American Society of Mechanical Engineers plaque-type penetrameters have three small drilled holes, whose diameters equal one, two, and four times the penetrameter thickness. Image quality is determined by the ability to distinguish both the outline of the penetrameter and one or more of the drilled holes on the processed radiograph; for example, 2-2T sensitivity is achieved when a 2% penetrameter and the hole that

*The reference radiographs discussed in this section and shown in Fig. 18 to 37 were generated from single-V-groove weld samples fabricated from 9.5 mm (0.375 in.) thick carbon steel plates using gas metal arc welding, shielded metal arc welding, and gas tungsten arc welding.

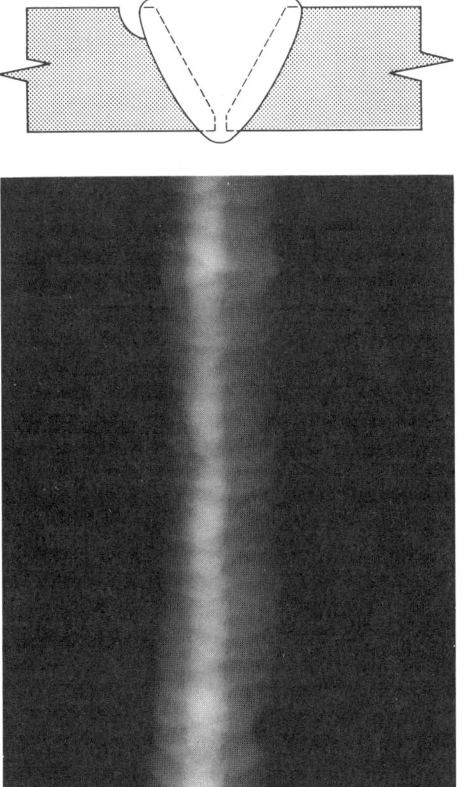

Fig. 18 External undercut, which is a gouging out of the piece to be welded, alongside the edge of the top or external surface of the weld. Radiographic image: An irregular darker density along the edge of the weld image. The density will always be darker than the density of the pieces being welded. Welding process: SMAW. Source: E.I. Du Pont de Nemours & Company, Inc.

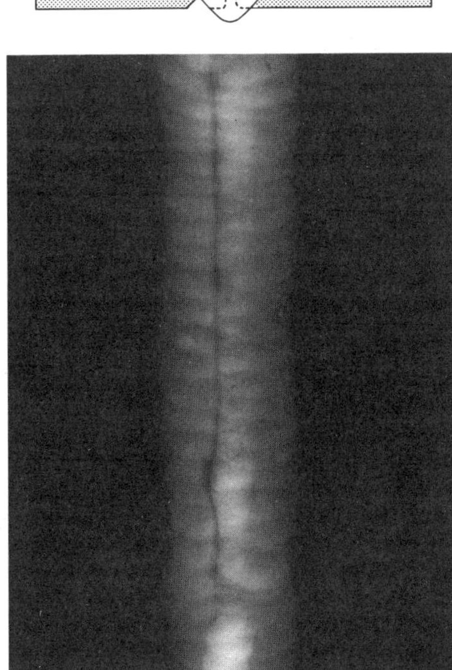

Fig. 19 Internal (root) undercut, which is a gouging out of the parent metal, alongside the edge of the bottom or internal surface of the weld. Radiographic image: An irregular darker density near the center of the width of the weld image and along the edge of the root pass image. Welding process: SMAW. Source: E.I. Du Pont de Nemours & Company, Inc.

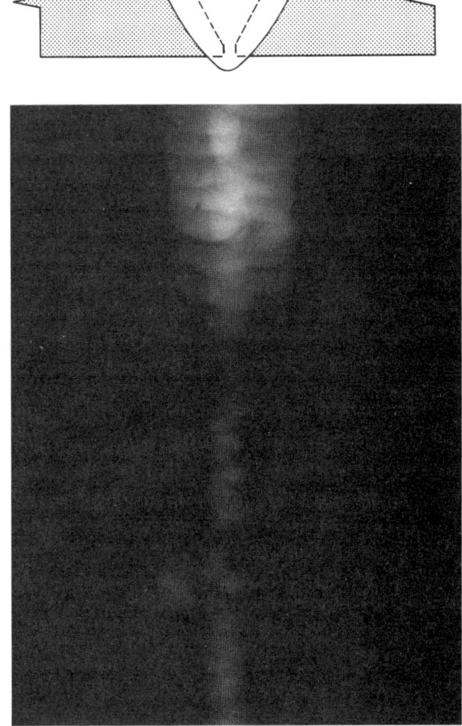

Fig. 20 External concavity or insufficient fill, which is a depression in the top of the weld, or cover pass, indicating a thinner-than-normal section thickness. Radiographic image: A weld density darker than the density of the pieces being welded and extending across the full width of the weld image. Welding process: SMAW. Source: E.I. Du Pont de Nemours & Company, Inc.

has a diameter of twice the penetrameter thickness are visible. Also, the image of raised numbers that identify the actual thickness (not the percentage thickness) of a penetrameter should appear clearly, superimposed on the image of the penetrameter.

Surface discontinuities that are detectable by radiography include undercuts (Fig. 18 and 19), longitudinal grooves, concavity at the weld root (Fig. 20 and 21), incomplete filling of grooves, excessive penetration (Fig. 22), offset or mismatch (Fig. 23 and 24), burn-through (Fig. 25), irregularities at electrode-change points, grinding marks, and electrode spatter. Surface irregularities may cause density variations on a radiograph. When possible, they should be removed before a weld is radiographed. When impossible to remove, they must be considered during interpretation.

Undercuts result in a radiographic image of a dark line of varying width and density. The darkness or density of the line indicates the depth of the undercut.

Longitudinal grooves in the surface of weld metal produce dark lines on a radiograph that are roughly parallel to the weld seam but are seldom straight. These dark lines have diffused edges and should not be mistaken for slag lines, which are narrow and more sharply defined.

Concavity at the weld root occurs only in joints that are welded from one side, such as pipe joints. It appears on the radiograph as a darker region than the base metal.

If weld reinforcement is too high, the radiograph shows a lighter line down the weld seam. There is a sharp change in image density where the reinforcement meets the base metal. Weld reinforcements not ground completely smooth show irregular densities, often with sharp borders. When excess metal is deposited on a final pass, it may overlap the base metal, causing LOF at the edge of the reinforcement. Although there is a sharp change in image density between reinforcement and base metal, the edge of the reinforcement image is usually irregular.

Irregularities at electrode-change points may be either darker or lighter than the adjacent areas.

Grinding marks appear as darker areas or lines in relation to the adjacent areas in the radiograph. Electrode spatter will appear as globular and lighter on the radiograph and should be removed before radiographic inspection.

As material thickness increases, radiography becomes less sensitive as an inspection method. Thus, for thick material, other NDI methods are used before, during, and after welding on both the base metal and weld metal.

Subsurface discontinuities detectable by radiography include gas porosity (Fig. 26-28), slag inclusions (Fig. 29 and 30), cracks (Fig. 31-33), LOP (Fig. 34), LOF (Fig. 35 and 36), and tungsten inclusions (Fig. 37). On a radiograph, a pore appears as a round or oval dark spot with or without a rather sharp tail. The spots caused by porosity are often of varying size and distribution. A wormhole appears as a dark rectangle if its long axis is perpendicular to the radiation beam, and it appears as two concentric circles, one darker than the other, if the long axis is parallel to the beam. Linear porosity is recorded on radiographs as a series of round dark spots along a line parallel to the direction of welding.

Slag inclusions appear along the edge of a weld as irregular or continuous dark lines

Fig. 21 Internal concavity (suck back), which is a depression in the center of the surface of the root pass. Radiographic image: An elongated irregular darker density with fuzzy edges, in the center of the width of the weld image. Welding process: GTAW-SMAW. Source: E.I. Du Pont de Nemours & Company, Inc.

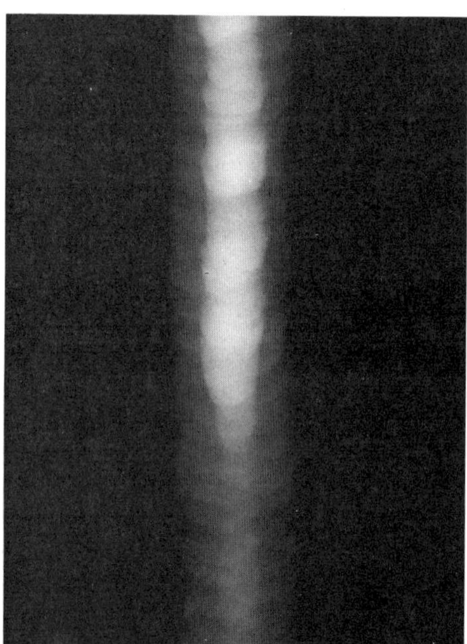

Fig. 22 Excessive penetration (icicles, drop-through), which is extra metal at the bottom (root) of the weld. Radiographic image: A lighter density in the center of the width of the weld image, either extended along the weld or in isolated circular drops. Welding process: SMAW. Source: E.I. Du Pont de Nemours & Company, Inc.

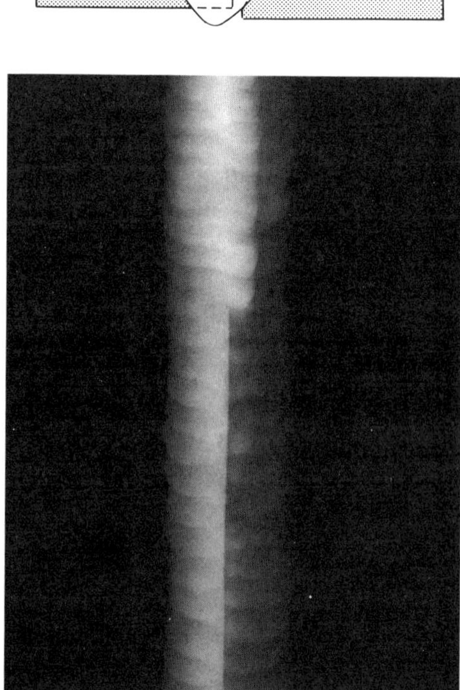

Fig. 23 Offset or mismatch with LOP, which is a misalignment of the pieces to be welded and insufficient filling of the bottom of the weld or root area. Radiographic image: An abrupt density change across the width of the weld image with a straight longitudinal darker-density line at the center of the width of the weld image along the edge of the density change. Welding process: SMAW. Source: E.I. Du Pont de Nemours & Company, Inc.

on the radiograph. Voids are sometimes present between weld beads because of irregular deposition of metal during multiple-pass welding. These voids have a radiographic appearance that resembles slag lines.

The radiographic image of a crack is a dark narrow line that is generally irregular. If the plane of the crack is in line with the radiation beam, its image is a fairly distinct line. If the plane is not exactly in line with the radiation beam, a faint dark linear shadow results. In this case, additional radiographs should be taken at other angles.

Lack of penetration shows on a radiograph as a very narrow dark line near the center of the weld. The narrowness can be caused by drawing together of the plates being welded, and the LOP may be very severe. Slag inclusions and gas holes are sometimes found in connection with LOP and cause the line to appear broad and irregular.

The radiographic image of incomplete fusion shows a very thin, straight dark line parallel to and on one side of the weld image. Where there is doubt, additional radiographs should be made with the radiation beam parallel to the bevel face. This will increase the possibility of the LOF appearing on the radiograph.

Tungsten inclusions appear either as single light spots or as clusters of small light spots. The spots are usually irregular in shape, but sometimes a rectangular light spot will appear.

Real-time radiography, which involves the display of radiographic images on television monitors through the use of an image converter and a television camera, is a rapidly developing method for weld inspection (Ref 6). One of the main advantages of real-time radiography for weld inspection is the cost savings that results from reducing the use of x-ray films. However, the possibility of expanding such an inspection system to include automatic defect evaluation by the image-processing system can yield significantly greater advantages. Automatic defect evaluation systems will result in objective and reproducible x-ray inspection, independent of human factors. Until now, the human brain has been much faster in analyzing and classifying the large range of flaw types found in welded joints. Computer programs for the efficient automated evaluation of weld radiographs are currently being developed and refined. More detailed information on the application of real-time systems for weld inspection can be found in the article ''Radiographic Inspection'' in this Volume.

Ultrasonic Inspection

In ultrasonic inspection, a beam of ultrasonic energy is directed into a specimen, and either the energy transmitted through the specimen is measured or the energy reflected from interfaces is indicated. Normally, only the front (entry) and back surfaces plus discontinuities within the metal produce detectable reflections, but in rare cases, the HAZs or the weld itself may act as reflecting interfaces.

Scanning Techniques. Figure 38 shows how a shear wave from an angle-beam transducer progresses through a flat testpiece—by reflecting from the surfaces at points called nodes. The linear distance between two successive nodes on the same surface is called the skip distance and is important in defining the path over which the transducer should be moved for reliable and efficient scanning of a weld. The skip distance can be easily measured by using a

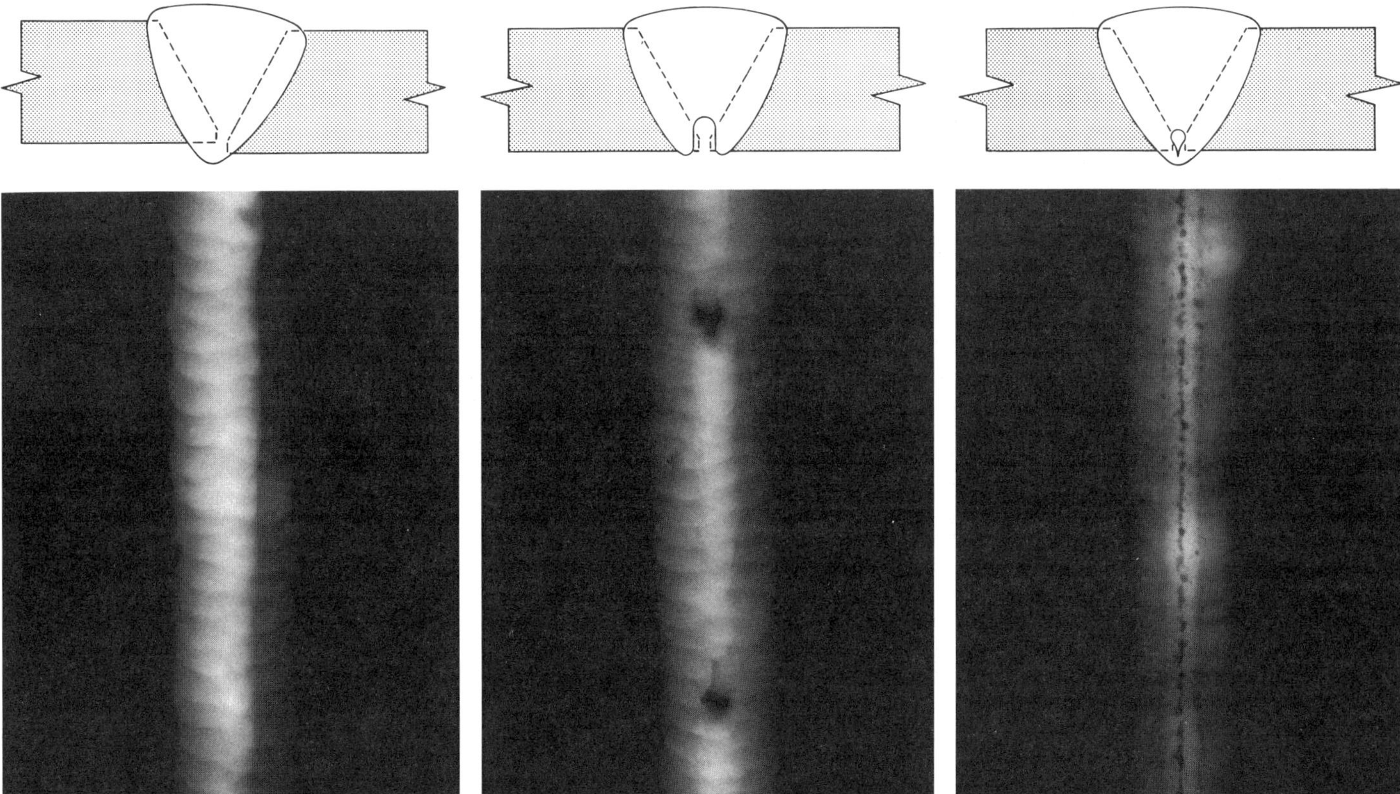

Fig. 24 Offset or mismatch (high-low), which is a misalignment of the pieces to be welded. Radiographic image: An abrupt change in film density across the width of the weld image. Welding process: SMAW. Source: E.I. Du Pont de Nemours & Company, Inc.

Fig. 25 Burn-through, which is a severe depression or a crater-type hole at the bottom of the weld but usually not elongated. Radiographic image: A localized darker density with fuzzy edges in the center of the width of the weld image. It may be wider than the width of the root pass image. Welding process: SMAW. Source: E.I. Du Pont de Nemours & Company, Inc.

Fig. 26 Root pass aligned porosity, which involves rounded and elongated voids in the bottom of the weld aligned along the weld centerline. Radiographic image: Rounded and elongated darker-density spots, which may be connected, in a straight line in the center of the width of the weld image. Welding process: GMAW. Source: E.I. Du Pont de Nemours & Company, Inc.

separate receiving transducer to detect the nodes or by using an angle-beam test block, or it can be calculated. Once the skip distance is known, the region over which the transducer should be moved to scan the weld can be determined. This region should extend the entire length of the weld at a distance from the weld line of approximately ½ to 1 skip distance, as shown in Fig. 39. A zigzag scanning path is used, either with sharp changes in direction (Fig. 39) or with squared changes (Fig. 40).

To detect longitudinal discontinuities in full-penetration butt and corner welds that are not ground flush, the transducer is oscillated to the left and right in a radial motion, with an included angle of approximately 30°, while scanning perpendicularly toward the weld, as shown in Fig. 40(a). The longitudinal movement necessary to advance the transducer parallel to the weld should not exceed 75% of the active width of the transducer per transverse scan. The weld should be scanned from both sides on one surface or from one side on both surfaces to ensure that nonvertically oriented flat discontinuities are detected. This type of discontinuity can be distinguished from vertically oriented flat discontinuities because the signal amplitudes from the two sides are different.

To detect transverse discontinuities in welds that are not ground flush, the transducer is placed on the base metal surface at the edge of the weld. The sound beam is directed by angling the transducer approximately 15% toward the weld from the longitudinal-weld axis, as shown in Fig. 40(a). Scanning is performed by moving the transducer along the edge of the weld either in one direction along both sides of the weld or in opposite directions along one side of the weld.

To detect longitudinal discontinuities in welds that are ground flush, the transducer is oscillated to the left and right in a radial motion, with an included angle of approximately 30°, while scanning across the weld as shown in Fig. 40(b). The longitudinal movement necessary to advance the transducer parallel to the weld must not exceed 75% of the active width of the transducer per transverse scan. When possible, the weld is scanned from one surface on two sides of the weld. When this is not possible, the weld can be scanned from one side on two surfaces or from one side on one surface using at least one full skip distance.

To detect transverse discontinuities in welds that are ground flush, the transducer is oscillated to the left and right in a radial motion, with an included angle of approximately 30°, as shown in Fig. 40(b), while scanning along the top of the weld from two opposing directions. If the width of the weld exceeds the width of the transducer, parallel scans should be performed, with each succeeding scan overlapping the previous one by a minimum of 25% of the active width of the transducer.

The entire volume of full-penetration welds in corner joints should be scanned with shear waves by directing the sound beam toward, or across and along, the axis of the weld, as shown in Fig. 41. If longitudinal wave testing is utilized, the weld is scanned by moving the transducer over the weld with overlapping paths. Each succeeding scan should overlap the previous scan by at least 25% of the active width of the transducer.

For the detection of discontinuities in the root area in T-joints (such as lack of root fusion), the width of the inspection zone should be limited to the thickness of the attachment member. The width of the in-

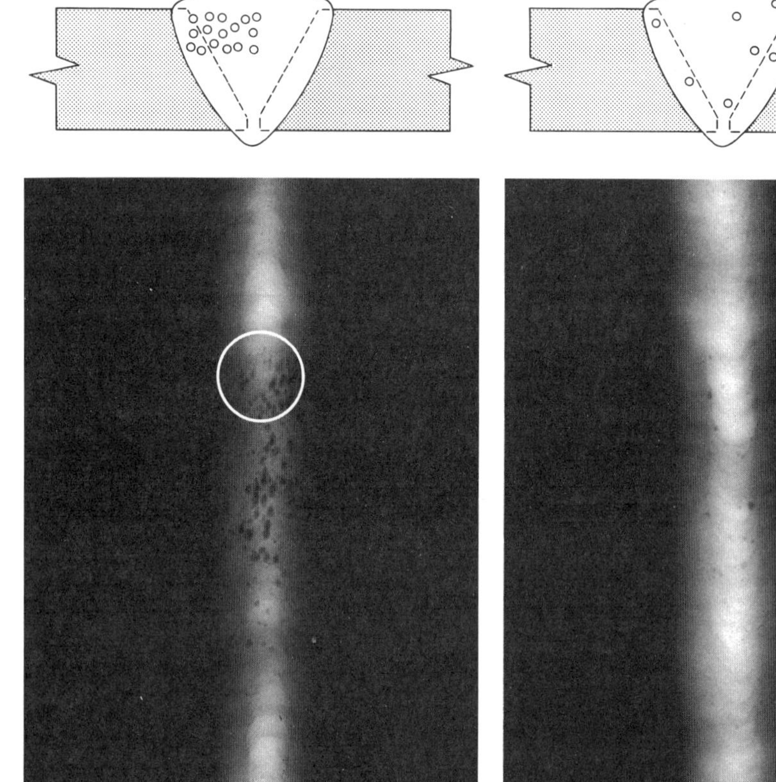

Fig. 27 Cluster porosity, which involves rounded or slightly elongated voids grouped together. Radiographic image: Rounded or slightly elongated darker-density spots in clusters with the clusters randomly spaced. Welding process: SMAW. Source: E.I. Du Pont de Nemours & Company, Inc.

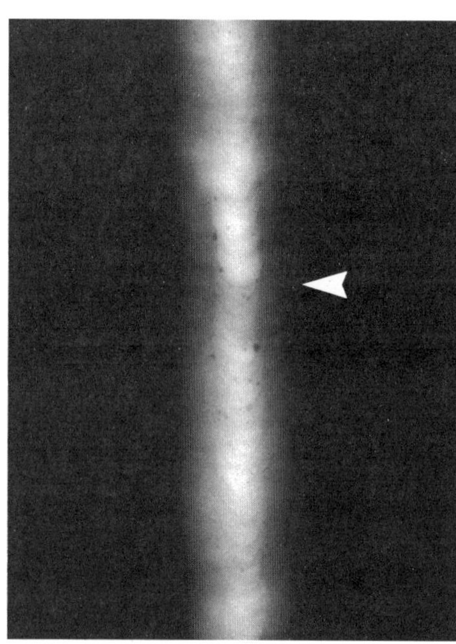

Fig. 28 Scattered porosity, which involves rounded voids random in size and location. Radiographic image: Rounded spots of darker densities random in size and location. Welding process: SMAW. Source: E.I. Du Pont de Nemours & Company, Inc.

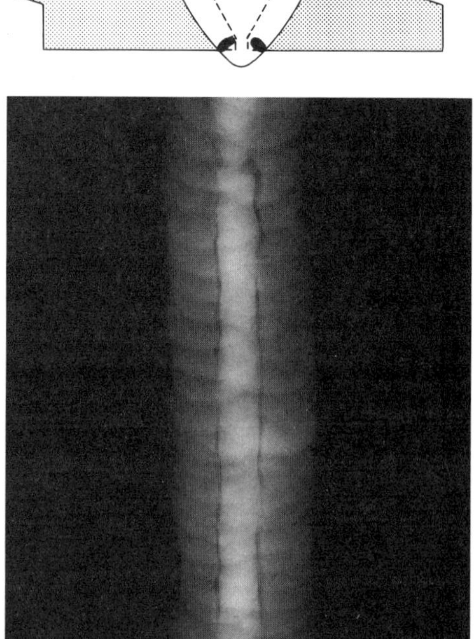

Fig. 29 Elongated slag lines (wagon tracks), which are impurities that solidify on the surface after welding and were not removed between passes. Radiographic image: Elongated, parallel, or single darker-density lines, irregular in width and slightly winding in the lengthwise direction. Welding process: SMAW. Source: E.I. Du Pont de Nemours & Company, Inc.

spection zone is located using ultrasonics or mechanical means and marked on the test surface. Shear-wave scanning for discontinuities in the base metal of any T-joint configuration should be performed whenever the surface opposite the attachment member is accessible. This scanning procedure can also be applied to partial-penetration welds in T-joints.

Coverage in each direction begins from the nearest section of the joint to beyond the centerline of the weld. The angle-beam transducer is directed at the particular area of interest and oscillated to the left and right in a radial motion, with an included angle of approximately 30°, while scanning perpendicularly toward the inspection zone. The inspection zone depth should be limited to the through-member-plate thickness minus 6 mm (¼ in.). The movement necessary to advance the transducer parallel to the inspection zone should not exceed 75% of the active width of the transducer per perpendicular scan.

Discontinuity Signals. Cracks and LOF discontinuities present essentially flat reflectors to the ultrasonic beam. If the beam is perpendicular to the place of the discontinuity, the amplitude of the signal is high; but if the beam strikes the discontinuity at an angle, most of the ultrasonic energy is reflected away from the transducer, and the reflected signal has a small amplitude that will vary with the angle. Because both cracks and sidewall LOF discontinuities produce similar reflected signals, they cannot be distinguished from one another by the signal amplitude or signal shape on the viewing screen when scanning is done from only one side. Therefore, the weld should be inspected from two sides, as shown in Fig. 42. If the discontinuity is vertically oriented, such as a centerline crack would be, the reflected signals received during a scan of each side should have approximately the same amplitude. If the discontinuity is in an inclined position, such as a sidewall LOF discontinuity would be in many joint designs, there will be an appreciable difference between the signal amplitudes.

A slag inclusion in a butt weld may produce a reflected signal with the same amplitude as that received from a crack or LOF discontinuity. However, scattered ultrasonic energy produces a relatively wide and high signal; as the transducer is manipulated around the slag inclusion, the signal height does not decrease significantly, but the edges of the signal vary. The same shape of reflected signal should be displayed when the weld is scanned from the opposite side of the weldment.

The signals that are reflected from porosity (gas pockets) are usually small and narrow. The signal amplitude will vary if the transducer is manipulated around the gas pocket or if the gas pocket is scanned from the opposite side of the weld. Cluster porosity (groups of gas pockets, as shown in Fig. 27) usually produces displays with a number of small signals. Depending on the number of gas pockets and their orientation to the ultrasonic beam, the displayed signals will be stationary or will be connected with one another.

Lack of fusion, weld root cracks, and LOP give essentially the same type of signal on an oscilloscope screen; the reflected signals are narrow and appear at the same location. The best way to differentiate among these flaws is to determine the extent of the flaw in the transverse direction.

Weld undercutting is distinguishable from sidewall LOF. The signals reflected from undercutting are approximately equal in amplitude when scanned from both sides. The signals produced by a sidewall LOF discontinuity vary considerably in amplitude when scanned from both sides.

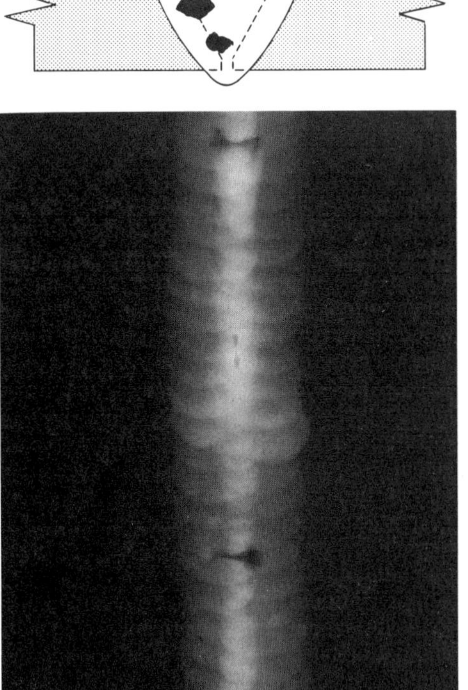

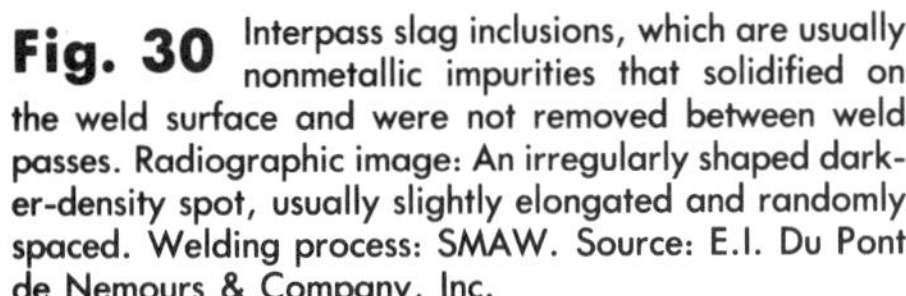

Fig. 30 Interpass slag inclusions, which are usually nonmetallic impurities that solidified on the weld surface and were not removed between weld passes. Radiographic image: An irregularly shaped darker-density spot, usually slightly elongated and randomly spaced. Welding process: SMAW. Source: E.I. Du Pont de Nemours & Company, Inc.

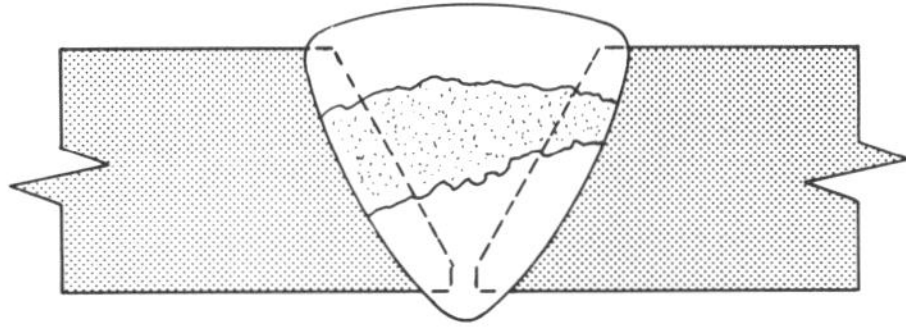

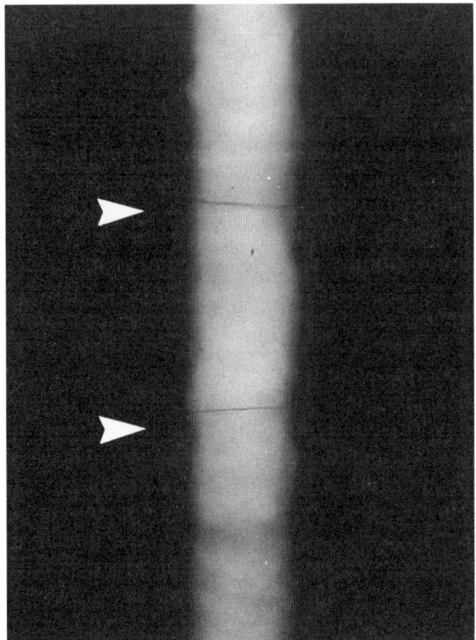

Fig. 31 Transverse crack, which is a fracture in the weld metal running across the weld. Radiographic image: Feathery, twisting line of darker density running across the width of the weld image. Welding process: GMAW-GTAW. Source: E.I. Du Pont de Nemours & Company, Inc.

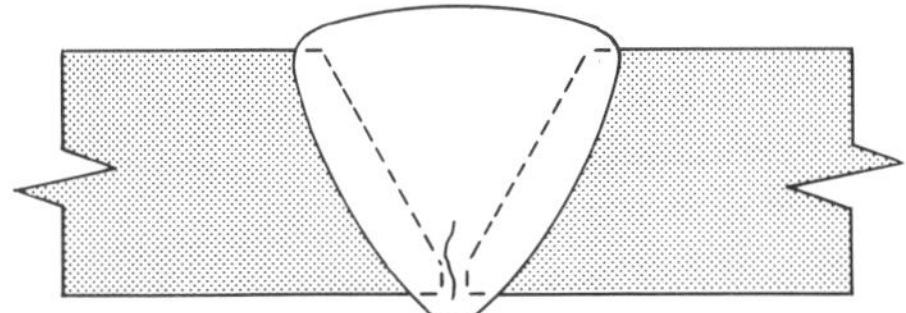

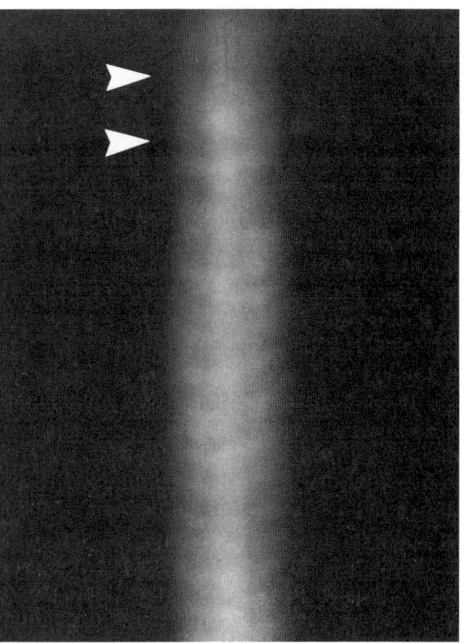

Fig. 32 Longitudinal crack, which is a fracture in the weld metal running lengthwise in the welding direction. Radiographic image: Feathery, twisting lines of darker density running lengthwise along the weld at any location in the width of the weld image. Welding process: GMAW-SMAW. Source: E.I. Du Pont de Nemours & Company, Inc.

In many cases, a weld is made when two misaligned parts must be joined; this condition is termed weld mismatch (Fig. 23 and 24). The inspector must not confuse a signal reflected from a root crack with one reflected from the misaligned edge. A narrow signal is usually produced when the ultrasonic beam strikes the misaligned edge. In most cases, no reflected signal will be received if the misaligned edge is scanned from the opposite side.

Ultrasonic Inspection of Spot Welds in Thin-Gage Steel. With the development of high-frequency transducers (12 to 20 MHz), pulse-echo ultrasonic inspection of spot welds in very thin gage sheet metal (0.58 mm, or 0.023 in.) is now possible (Ref 1). The ultrasonic test for spot weld nugget integrity relies on an ultrasonic wave to measure the size of the nugget. The size is in three dimensions, including thickness as well as length and breadth (or diameter for a circular spot). The successful measurement of nugget size places several requirements on the ultrasonic wave path, wave velocity, and wave attenuation.

Wave Path. The first requirement is that the ultrasonic wave be in the form of a beam directed perpendicular to the faces of the metal sheets and through the center of the nugget (Fig. 43). Two diameters of nuggets are shown: larger than the beam and smaller than the beam.

In general, an ultrasonic wave will be reflected when it impinges on an interface where the density and/or the ultrasonic velocity change. Examples are water-to-metal and metal-to-air. In Fig. 43, reflections will occur at the outer surfaces of the two sheets and at the interface (air) between the two sheets if the nugget is small, as in Fig. 43(c). The nugget-to-parent-metal boundary will not produce perceptible reflections, refraction, or scattering, because the changes in density and velocity are a tenth of a percent or less, while the air-to-steel difference is more than 99.9%.

Typical oscilloscope displays showing the pulse-echo patterns for these two nugget-to-beam-diameter ratios are shown in Fig. 43(b) and (d). The difference in the echo patterns permits the distinction to be made between adequate and undersize welds.

Velocity/Thickness Gaging. The beam path shown in Fig. 43(a) illustrates the situation in which the ultrasonic beam should indicate an acceptable nugget. The beam will be reflected only at the outer surfaces (1 and 3) of the pair of sheets as joined. To make this reflection sequence visible, the ultrasonic beam must consist of a short pulse that can reverberate back and forth between the outer faces and produce separate echoes when viewed on an oscilloscope. The picture observed is shown in Fig. 43(b). The pulse must be short enough to resolve the double thickness of the two joined sheets.

Similarly, the beam path shown in Fig. 43(c) illustrates the situation in which the ultrasonic beam should indicate an undersize nugget. The beam will be reflected in the single thickness of the upper sheet around the perimeter of the nugget. Therefore, on the oscilloscope, echoes will appear between the principal echoes arising from the portion of the beam traversing the nugget (Fig. 43d). In terms of thickness gaging, the ultrasonic pulse in the beam must be short enough to resolve the thickness of one layer of sheet metal.

Attenuation. The thickness of the nugget can be measured only indirectly because the thickness gaging function (described above) can measure only the thickness between outer faces in the nugget area. The nugget itself is measured by the effect of its grain structure on the attenuation of the ultrason-

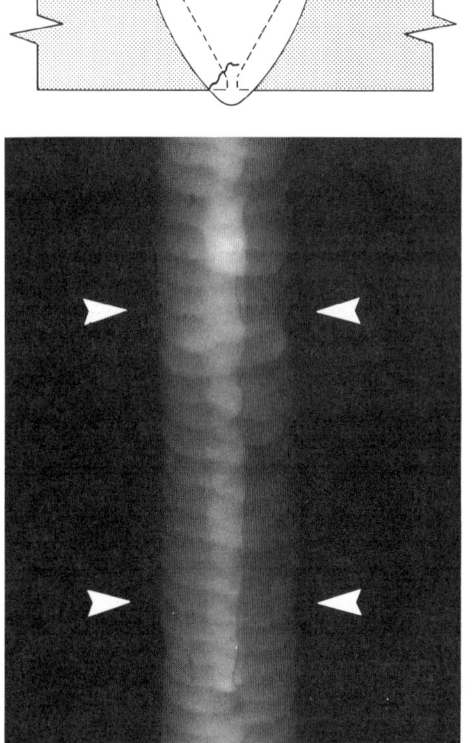

Fig. 33 Longitudinal root crack, which is a fracture in the weld metal at the edge of the root pass. Radiographic image: Feathery, twisting lines of darker density along the edge of the image of the root pass. The twisting feature helps to distinguish the root crack from incomplete root penetration. Welding process: SMAW. Source: E.I. Du Pont de Nemours & Company, Inc.

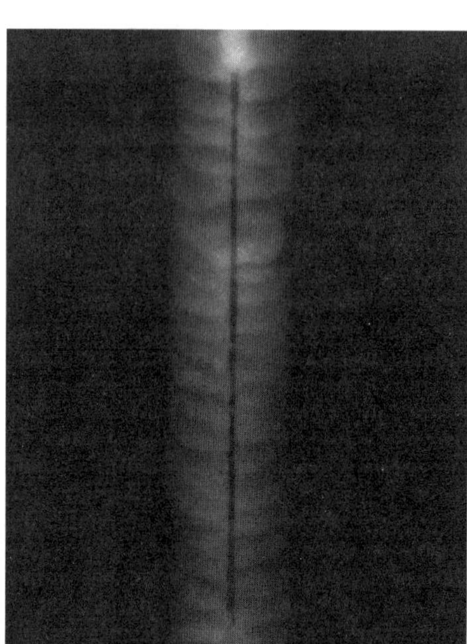

Fig. 34 Lack of penetration, which occurs when the edges of the pieces have not been welded together, usually at the bottom of single-V-groove welds. Radiographic image: A darker-density band, with very straight parallel edges, in the center of the width of the weld image. Welding process: SMAW. Source: E.I. Du Pont de Nemours & Company, Inc.

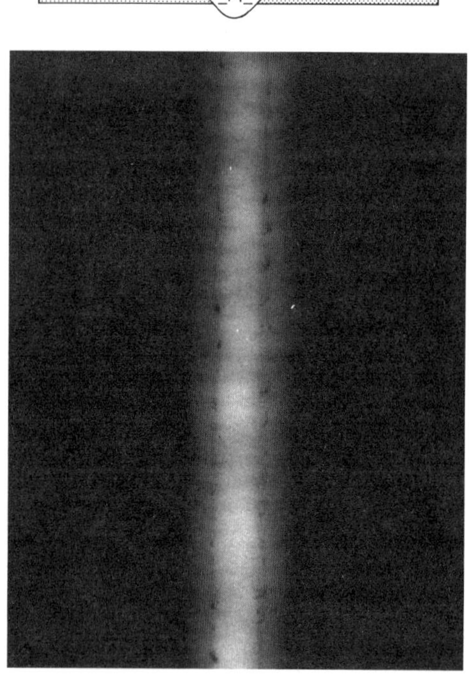

Fig. 35 Lack of sidewall fusion, which involves elongated voids between the weld beads and the joint surfaces to be welded. Radiographic image: Elongated parallel, or single, darker-density lines, sometimes with darker-density spots dispersed along the LOF lines, which are very straight in the lengthwise direction and not winding like elongated slag lines. Although one edge of the LOF lines may be very straight as with LOP, lack of sidewall fusion images will not be in the center of the width of the weld image. Welding process: GMAW. Source: E.I. Du Pont de Nemours & Company, Inc.

ic wave in the beam. As the wave reflects back and forth between the outer faces of the welded sheets, its amplitude is attenuated (dies out). The attenuation (rate of decay) of the ultrasonic wave depends on the microstructure of the metal in the beam. In the spot welds under consideration, the attenuation is caused principally by grain scattering. The grains scatter the ultrasonic energy out of the coherent beam, causing the echoes to die away. In most metals, coarse grains scatter more strongly than fine grains.

Because a nugget is a melted and subsequently refrozen cast microstructure with coarser grains than the adjacent cold-rolled parent metal, the nugget will scatter more strongly than the remaining parent metal. It follows that a nugget will produce higher attenuation than the parent metal and that a thick nugget will result in higher attenuation than will a thin nugget. Therefore, a thin nugget can be distinguished from a thick nugget by the rate of decay of the echoes in the case in which the diameters of both nuggets are equal. Typical echoes from a thick nugget area and from a thin nugget are shown in Fig. 44. It is clear that a trained observer could differentiate between the two welds on the basis of the decay patterns. Given this observation, it is obvious that the pulse-echo ultrasonic method at normal incidence could perform the required measurements on spot welds in metals with coarse-grain nuggets and fine-grain parent sheet metal.

Acoustic Emission Monitoring

Acoustic emissions are impulsively generated small-amplitude elastic stress waves created by deformations in a material. The rapid release of kinetic energy from the deformation mechanism propagates elastic waves from the source, and these are deteced as small displacements on the surface of the specimen. The emissions indicate the onset and continuation of deformation and can be used to locate the source of deformation through triangulation techniques. Details of the process are available in the article "Acoustic Emission Inspection" in this Volume.

Acoustic emissions can be used to assess weld quality by monitoring during or after welding. In weldments, regions having LOP, cracking, porosity, inclusions, or other discontinuities can be identified by detecting the acoustic emissions originating at these regions. During welding processes, acoustic emissions are caused by many factors, including plastic deformation, melting, friction, solidification, solid-solid phase transformations, and cracking. In some cases, the monitoring of acoustic emissions during welding can include automatic feedback control of the welding process. In large-scale automatic welding, the readout equipment can be conveniently located near the welder controls or in a quality-monitoring area.

The locations of acoustic sources on a weld line can be presented in a variety of ways. One technique displays the number of events versus distance along the weld on an oscilloscope screen or an *x-y* plotter. Another technique uses a digital-line printer that gives the time of the event, its location, and its intensity. This information facilitates appraisal of the severity of each source. Once the acoustic emission sources are graded, other NDI methods can be used to study the discontinuity in detail.

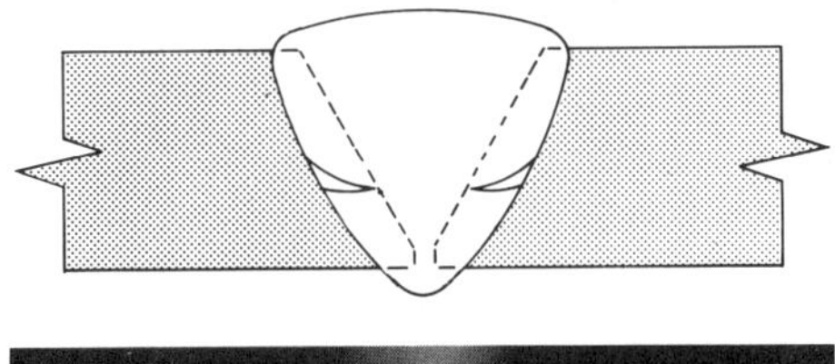

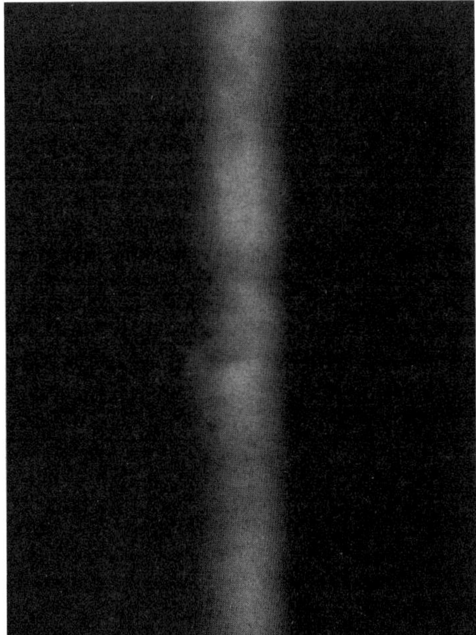

Fig. 36 Interpass cold lap, which involves LOF areas along the top surface and edge of lower passes. Radiographic image: Small spots of darker densities, some with slightly elongated tails, aligned in the welding direction and not in the center of the width of the weld image. Welding process: GMAW. Source: E.I. Du Pont de Nemours & Company, Inc.

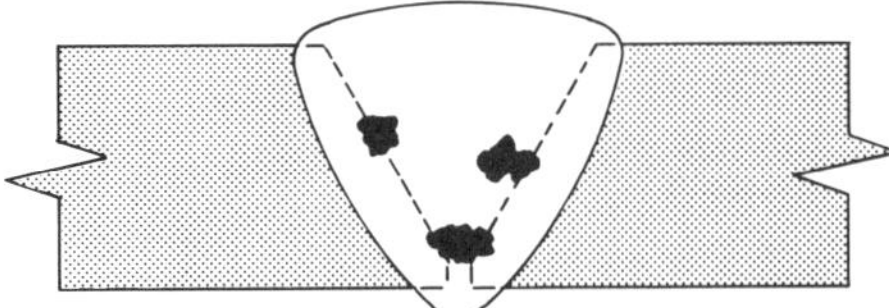

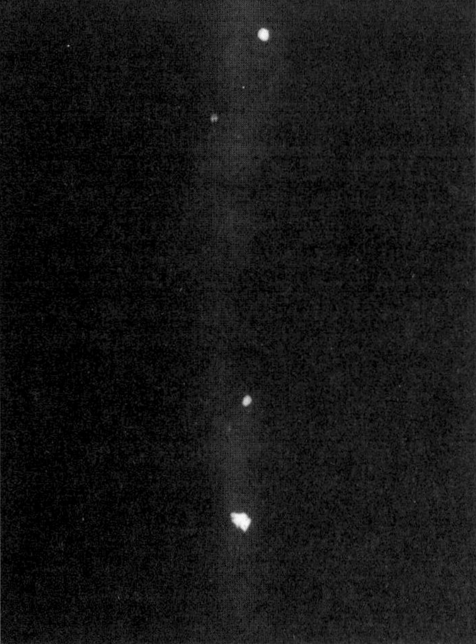

Fig. 37 Tungsten inclusions, which are random bits of tungsten fused into but not melted into the weld metal. Radiographic image: Irregularly shaped, lower-density spots randomly located in the weld image. Welding process: GTAW. Source: E.I. Du Pont de Nemours & Company, Inc.

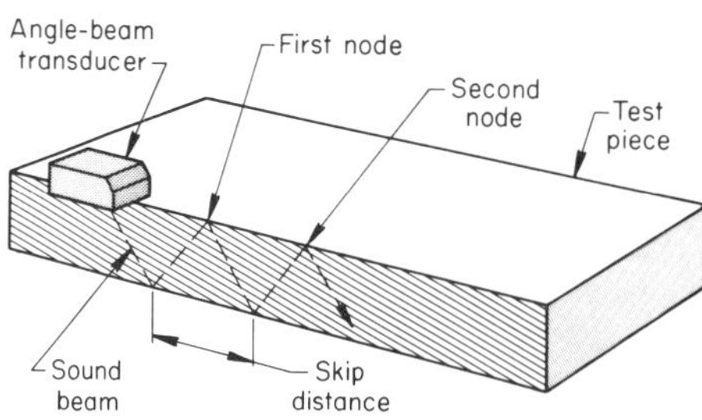

Fig. 38 Sound beam path in a flat testpiece being ultrasonically inspected with a shear wave from an angle-beam transducer, showing the skip distance between the nodes where the beam reflects from the surfaces

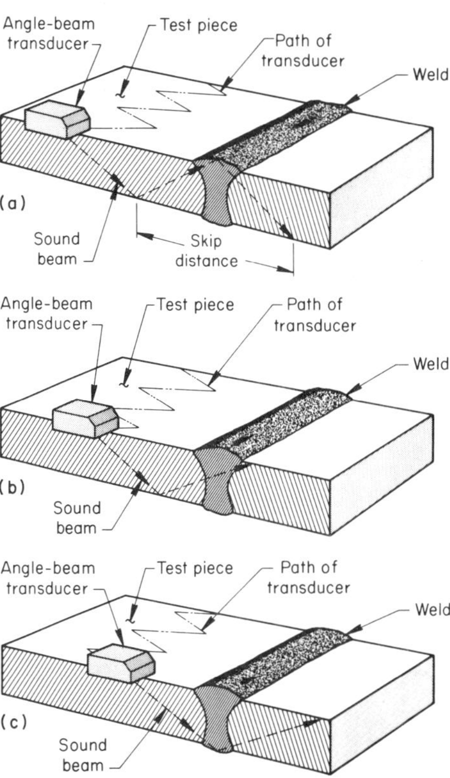

Fig. 39 Three positions of the contact type of transducer along the zigzag scanning path used during ultrasonic inspection of welded joints. The movement of the sound beam path across the weld is shown on a section taken along the centerline of the transducer as it is moved from the far left position in the scanning path (a), through an intermediate position (b), to the far right position (c).

It is sometimes difficult to achieve a good acoustic coupling between the sensor (or sensors) and the part. This is especially true for welding processes that have a fixed arc and a moving part. For postweld acoustic emission monitoring, a stimulus must be applied after complete cooling of the weld. This involves the application of a mechanical load or sometimes a thermal stress to the part or structure. For the field inspection of in-service welds for cracks or other flaws, welds are commonly subjected to stresses that just exceed the maximum stress previously experienced by the weld. The excess stress is necessary to produce acoustic emissions due to plastic deformation at the crack tip. The application of this stress, however, is often difficult to do and in some cases is undesirable. An alternative method that can often clearly detect existing cracks in a weld is to cyclically load the weld at low-stress levels. The relative motions of the crack surfaces produce frictional excitation of stress waves.

Acoustic emission monitoring for evaluation of quality and control of welding processes requires preliminary studies for each application to establish such operating conditions as the number, location, and mounting of sensors; gain settings; filtering; data presentation; and data interpretation. These studies normally include correlation with other nondestructive and destructive methods of inspection. For example, the detection of tungsten inclusions from GTAW joints by acoustic emission can also involve inclusion classification by radiography.

The postweld monitoring of weldments includes both quality control inspection during the period between the completion of the weld and additional fabrication of the part and nondestructive inspection of in-service weldments. An example of the former is the inspection of butt-welded plates, such as those fabricated in the building of a ship. The following example demonstrates the feasibility of immediate postweld acoustic emission monitoring of butt-welded plates.

Example 1: Acoustic Emission Monitoring of Butt Welds in Low-Carbon Steel Plates. Using a dc arc welder, two low-carbon steel plates 460 mm (18 in.) wide by 610 mm (24 in.) long and 3.2 mm (0.125 in.) thick were butt welded to form a test plate approximately 1.2 m (48 in.) long, as shown in Fig. 45. Extra precautions were taken to produce a sound weld. After the weld metal was cool, the plate was placed in a fixture so that acoustic emission from the weld metal could be monitored.

The plate was supported horizontally by a system of rubber rollers. Weights were placed on each end of the plate to hold it firmly on the rollers. Mild bending was induced in the butt joint by raising and lowering the middle of the plate with a hydraulic jack. Acoustic emissions were monitored during loading and unloading oscillation and while the plate was held in a stress state, using sensors attached to the top side of the plate at each end of the weld. These emissions were used to calibrate the oscilloscope screen and were the basis for

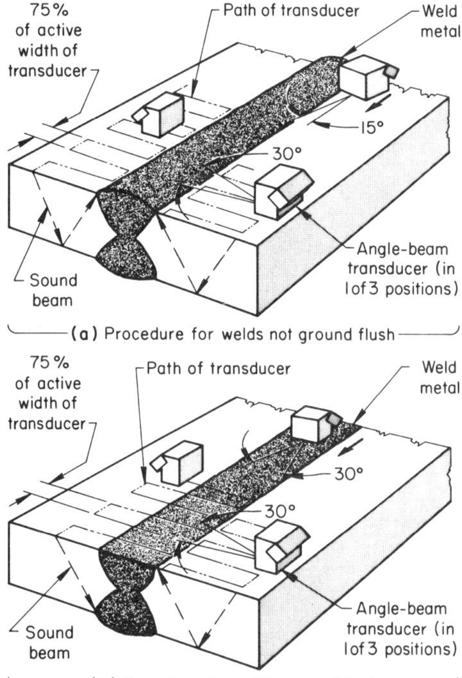

Fig. 40 Ultrasonic scanning procedures to detect longitudinal and transverse discontinuities in welds that (a) are not ground flush and (b) are ground flush

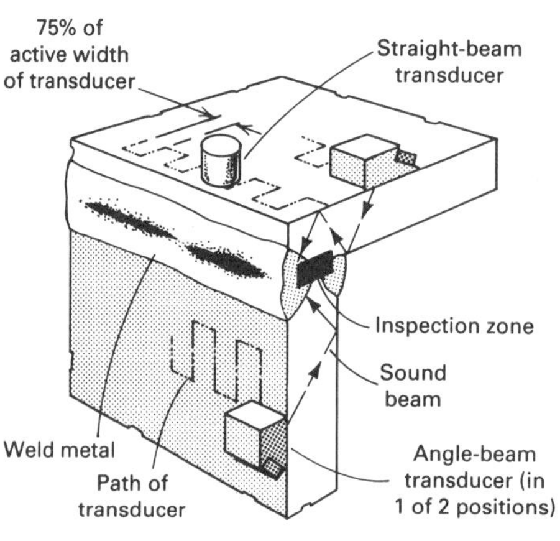

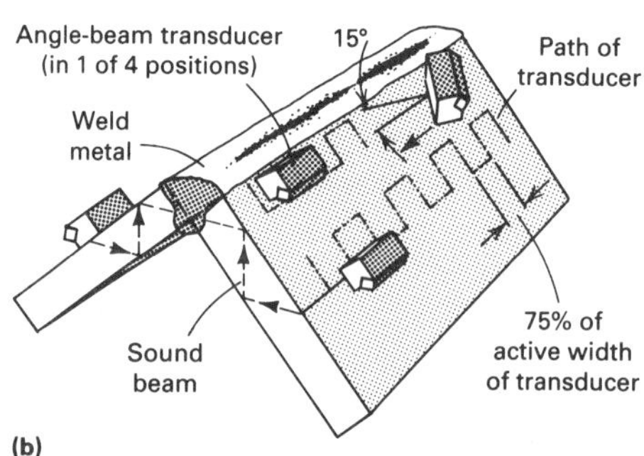

Fig. 41 Ultrasonic scanning procedure for full-penetration groove weld (a) and double-fillet welds (b) in corner joints

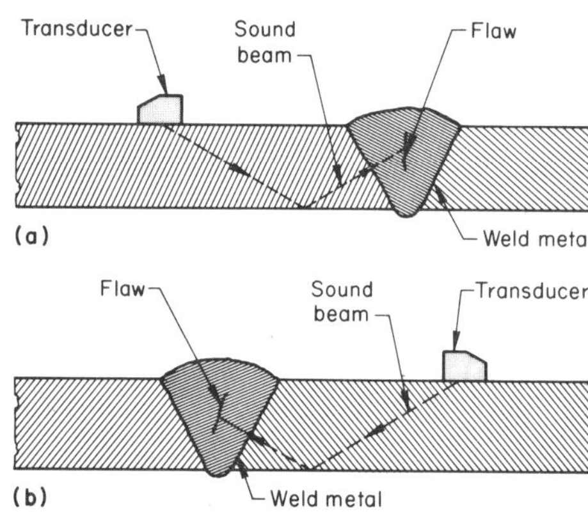

Fig. 42 Transducer scanning positions for distinguishing between weld metal flaws that are (a) vertically oriented and (b) in an inclined position

subsequent inspections of plates having known or suspected discontinuities in the weld metal.

The output signal of each sensor was amplified (75-dB gain), filtered (50 kHz to 1 MHz), and displayed on a dual-beam oscilloscope. Single-sweep traces of both signals were triggered at the arrival of the first signal. These were photographed from the dual-beam display, and successive photographs permitted documentation of the difference between the arrival times at the two sensors for several sources. The locations of the predominant sources were then inferred.

After the preliminary experiments, a second test plate was prepared and butt welded across its entire width. The weld was made with intentional poor penetration and slag inclusions in a region centered about 100 mm (4 in.) on one side of the centerline of the plate, as indicated by the shaded area in Fig. 45(b). Acoustic emissions were recorded during bending and oscillation. The location and number of acoustic emissions are given graphically in Fig. 45(b). A good correlation existed between the tabulated source locations and the locations of known flaws.

A third test plate was prepared by making a saw cut 180 mm (7 in.) long, from one edge toward the center, along the transverse centerline. This saw cut was repaired by welding, then monitored for acoustic emissions in the same manner as the second test plate. Radiographic inspection of the plate revealed two regions of discontinuities in the weld, as indicated by the shaded areas in Fig. 45(c). A region of very poor penetration between 130 and 150 mm (5 and 6 in.) from the longitudinal centerline of the plate is shown by the darker shaded area in Fig. 45(c). This corresponds to the large number of acoustic emissions occurring in the region. Also shown in Fig. 45(c) are other regions from which acoustic emissions originated, indicating discontinuities in the weld metal. From the results obtained on these test plates, it was concluded that it is feasible to use acoustic emission monitoring as a method of assessing the structural integrity of butt-welded joints.

In-Service Monitoring. One application of in-service acoustic emission monitoring of welds involved the locating of defective or deteriorated welds in buried pipelines. Gas distribution pipelines that had been in service for many years needed inspections for structural integrity, especially on oxy-acetylene welds. Although the location of the buried pipe was known, the locations of the welds were not. In preliminary studies, it was found that acoustic emission signals from weld discontinuities would propagate several hundred feet down the pipe. Therefore, a very efficient method of locating the defective welds was devised. The loading stimulus, an extra heavy vehicle moving slowly along the surface above the pipe,

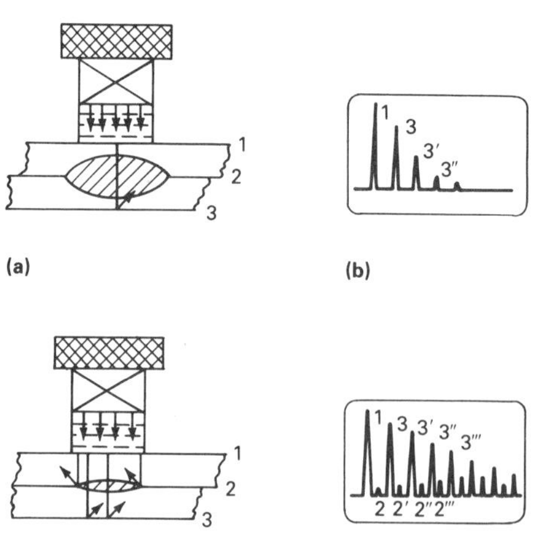

Fig. 43 Schematic illustrating setup for the pulse-echo ultrasonic inspection of resistance-welded spot welds. (a) Wave paths in satisfactory weld. (b) Resulting echoes. (c) Wave paths in an unsatisfactory weld. (d) Resulting echoes. Source: Ref 1

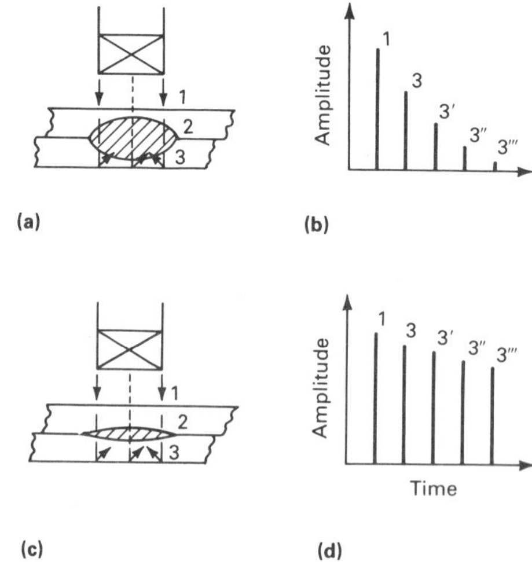

Fig. 44 Ultrasonic thickness measurements of resistance spot weld nuggets. (a) Satisfactory weld. (b) Resulting attenuation of the ultrasonic wave. (c) Unsatisfactory weld. (d) Resulting wave attenuation. Source: Ref 1

induced bending stresses that were sufficient to cause weld discontinuities to emit.

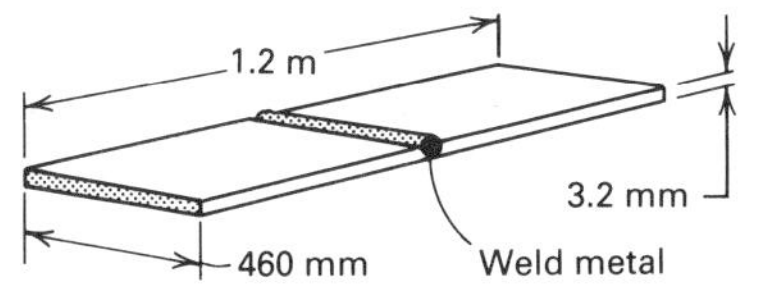

(a)

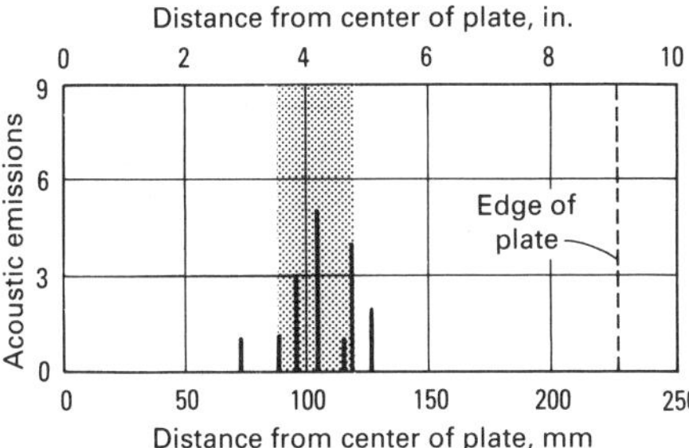

(b)

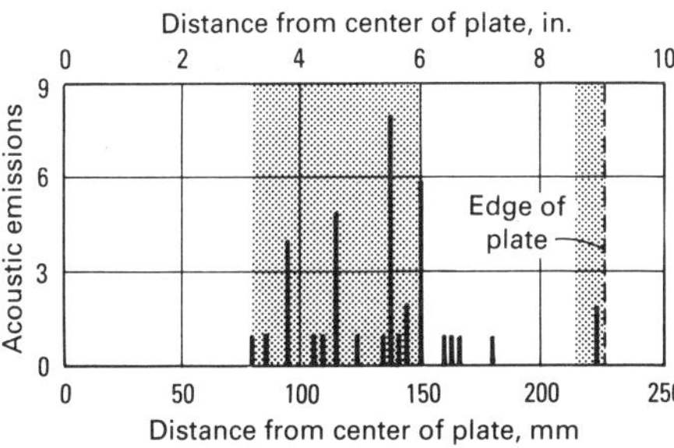

(c)

Fig. 45 Acoustic emission monitoring of butt welds in low-carbon steel test plates. (a) Test plate. Three were prepared; the first, with a sound weld, was used to establish conditions for acoustic emission monitoring of the others. (b) Location and number of acoustic emissions in the second test plate, which had a region of intentional poor penetration and slag inclusions (shaded area). (c) Location and number of acoustic emissions in the third test plate, in which flaw indications were revealed by radiographic inspection (shaded areas).

Sensors were placed on either side of the suspect weld, and the signals reaching the sensors were monitored with portable equipment. By comparing the time required for the emissions to reach each pair of sensors and by comparing the intensities of the signals, the locations of the defective weld were determined. The welded joints were then excavated, and the welds were further inspected or repaired.

Monitoring During Welding. Arc-welding processes are inherently ultrasonically noisy—particularly so in continuous high-frequency welding. However, the acoustic emissions detected during proper welding without discontinuity creation have steady characteristics. When cracking, excessive slag inclusion, or a significant change in the weld conditions occurs, the acoustic emission levels change correspondingly. Therefore, online monitoring during welding gives immediate indication of variations in the quality of the weld. Cold cracking can be detected by monitoring the welded structure for minutes, or even hours, after welding.

Acoustic emissions result from multiple causes during resistance spot welding. The making of a resistance spot weld consists of setdown of the electrodes, squeeze, current flow, hold time, and lift off. Many acoustic emissions are produced during these various steps. The most commonly observed signals are shown schematically in Fig. 46. The ultrasonic noise during setdown and squeeze can be related to the conditions of the electrodes and the surface of the parts. The large, but brief, signal at current initiation can be related to the initial resistance and the cleanliness of the parts. During current flow, acoustic emission results from plastic deformation, friction, melting, and expulsions. The signals associated with expulsion (spitting and flashing) are generally large in amplitude and can be easily distinguished from the rest of the emissions associated with nugget formation. When current flow ceases, some materials exhibit appreciable solidification noise that can be related to nugget size and inclusions. As the nugget cools in the hold period, acoustic emissions can result from solid-solid phase transformations and cracking. Finally, as the electrode is lifted, noise is produced by the separation of the electrode from the part. This noise, or signal, can be related to the size of the nugget as well as to the visual appearance of the weld.

Any measure of the cumulative acoustic emissions during resistance spot welding cannot be expected to relate clearly to weld quality. On the other hand, by using both time discrimination and multiple detection levels, the various segments of acoustic emissions can be separately measured and related to various indicators of quality. Commercial instrumentation is available that is capable of separately monitoring several of the acoustic emission segments. For example, the expulsion count, phase-transformation count, and cracking count can be monitored and recorded for each resistance spot weld, giving a permanent record of quality. In the following example, a number of acoustic emissions during martensitic phase transformation were found to relate to the strength of the resistance spot weld.

Example 2: Determination of Strength of Resistance Spot Welds by Acoustic Emission Monitoring. Twenty carbon steel coupons were spot welded at identical settings of heat and weld cycles in a 75-kVA ac spot welder. An acoustic emission sensor was attached to the lower electrode. The acoustic emission monitor was gated to begin 120 ms after the current ceased and to stop 240 ms later. Within that time interval, martensitic phase transformation occurred as the nugget cooled. Because the total volume of material experiencing the transformation is related to both the nugget size and the area of diffusion bonding, the acoustic emissions during the phase transformation are related to the strength of the weld. The specimens were pulled to failure in tension shear, and the ultimate strength was recorded. The results are shown in Fig. 47. The tests indicated that a 10% variation in weld strength corresponded to about a 30% change in acoustic emissions.

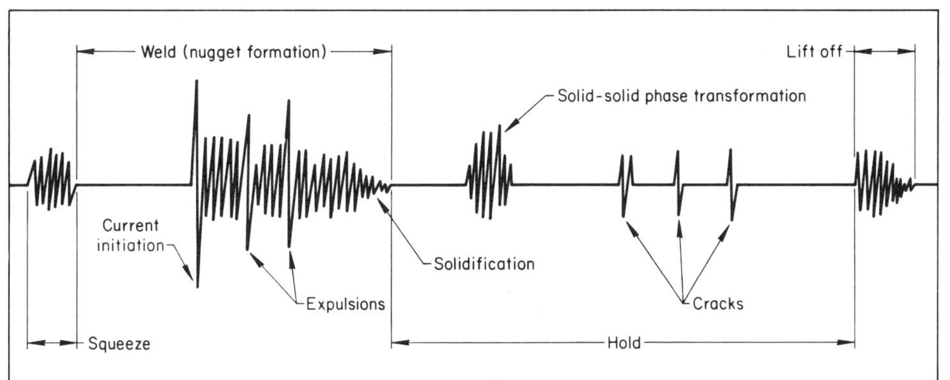

Fig. 46 Schematic showing typical acoustic emission signals obtained during various stages of resistance spot welding

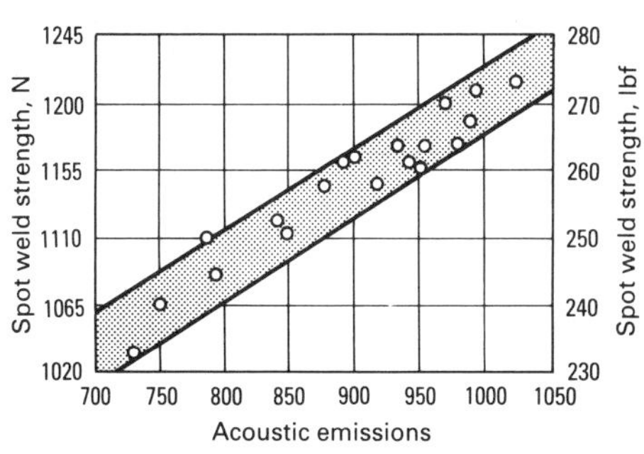

Fig. 47 Relation of strength of resistance spot welds in carbon steel coupons to number of acoustic emissions during the martensitic phase transformation

Leak Testing

Welded structures are leak tested to measure the integrity of the structure for containing gases, fluids, semisolids, and solids and for maintaining pressures and vacuums. The more common leak-testing methods used are (in order of increasing sensitivity):

- Odor from tracer gas
- Pressure change
- Pressurized liquid (generally water) and visual observation
- Pressurized gas using a leak detection solution
- Tracer gas using thermal leak detectors
- Helium using a mass spectrometer during pressure and vacuum tests

Other methods less frequently used are acoustical detection of gas flow through a leak and use of radioactive tracer gas. Detailed descriptions of the various methods used are presented in the article "Leak Testing" in this Volume.

Weld flaws that contribute to leakage of a structure are porosity, LOF or LOP, and cracks. Cracks are of particular concern because they may propagate when the structure is proof tested or otherwise tested for structural integrity. Therefore, it is preferred that leak testing be done after completion of the structural tests.

Selection of a leak testing method depends on the environment in which the structure is used and the potential danger and economic impact involved in the event of a service failure. The acceptance criteria should include a numerical expression of the allowable leak rate; the frequently used expression "shall be free from leaks" is meaningless.

Eddy Current and Electric Current Perturbation Inspection

Eddy current inspection is based on the principles of electromagnetic induction and is used to identify or to differentiate between a wide variety of physical, structural, amd metallurgical conditions in electrically conductive ferromagnetic and nonferromagnetic metals. Normally, eddy current inspection is used only on thin-wall welded pipe and tubing for the detection of longitudinal-weld discontinuities, such as open welds, weld cracks, and porosity. The application of this method to tubular goods is discussed in the articles "Eddy Current Inspection", "Tubular Products", and "Boilers and Pressure Vessels" in this Volume.

The electric current perturbation method consists of establishing an electric current flow in the part to be inspected (usually by means of an induction coil) and detecting the magnetic field associated with perturbations in the current flow around defects by using a separate magnetic field sensor (Ref 7). This technique is applicable to the detection of both very small surface cracks as well as subsurface cracks in both high- and low-conductivity, nonferromagnetic materials, such as titanium and aluminum alloys. The principles, equipment, and applications associated with this method are outlined in the article "Electric Current Perturbation NDE" in this Volume.

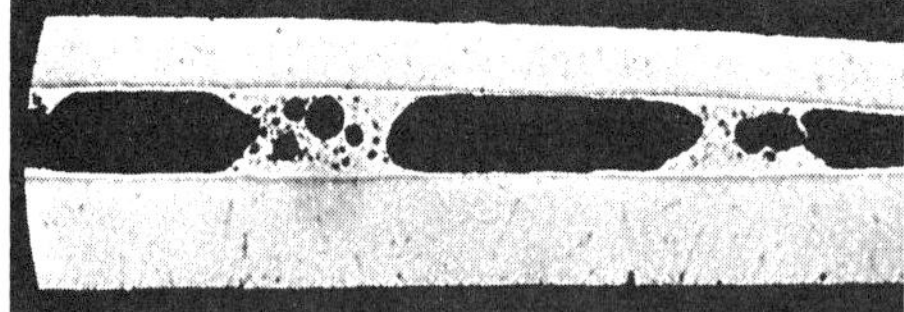

Fig. 48 Voids resulting from lack of fill between the faying surfaces of a lap joint between two sheets of Hastelloy X brazed with BNi-1 filler metal. Unetched. 16½×

Brazed Assemblies

Brazing is defined by the American Welding Society as a group of welding processes that produce coalescence of materials by heating them to a suitable temperature and by using a filler metal having a liquidus above 450 °C (840 °F) and below the solidus of the base metal. The filler metal is distributed between the closely fitted faying surfaces of the joint by capillary action.

The temperature limitation of 450 °C (840 °F) differentiates brazing from soft soldering, which involves the use of filler metals having a liquidus below 450 °C (840 °F). To clarify the difference between brazing and conventional welding, it should be pointed out that in brazing the base materials being joined are never melted, while in most welding processes the base metals are melted (exceptions are those welding processes that utilize pressure in conjunction with heat).

There are six brazing processes included under the group heading of brazing. These processes are torch brazing, furnace brazing, induction brazing, dip brazing, resistance brazing, and infrared brazing.

Brazing can also be classified according to the major constituents of the more common types of filler metals used:

- Aluminum brazing
- Silver brazing
- Copper brazing
- Nickel brazing
- Precious metal brazing

There are five essential properties for brazing filler metal:

- Ability to wet and make a strong, sound bond on the base metal
- Suitable melting temperature and flow properties to permit distribution by capillary attraction in properly prepared joints
- A composition of sufficient homogeneity and stability to minimize separation by liquation under the brazing conditions to be encountered. Excessively volatile constituents in filler metals may be objectionable
- Capability of producing a brazed joint that will meet service requirements, such as required strength and corrosion resistance
- Depending on the requirements, ability to produce or avoid interactions between base metal and filler metal

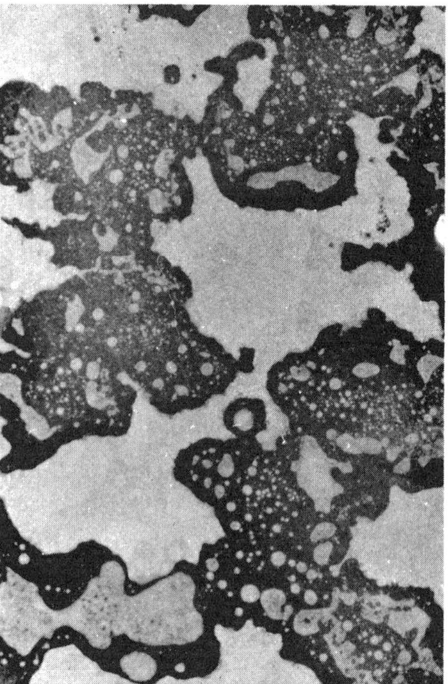

Fig. 49 Radiograph showing entrapped flux (dark areas) in a low-carbon steel joint torch brazed with BAg-1 filler metal (light areas). ×1

Flaws Commonly Found in Brazed Joints

Flaws exhibited by brazed joints are usually of the following types:

- Lack of fill
- Flux entrapment
- Noncontinuous fillet
- Base metal erosion

Lack of Fill. Voids resulting from lack of fill can be the result of improper cleaning of the faying surfaces, improper clearances, insufficient brazing temperatures, or insufficient brazing filler metal (Fig. 48).

Flux entrapment normally occurs during torch brazing, induction brazing, or furnace brazing, when reducing atmospheres are not employed. As the term implies, flux becomes trapped within the joint by the braze metal and prevents coverage. Figure 49 is a radiograph of a torch-brazed joint in which flux entrapment was a serious problem.

Noncontinuous Fillet. A brazed joint in which a large void in the fillet is evident is shown in Fig. 50. Such a void is discernible by visual examination and may or may not be acceptable, depending on requirements. For example, if the void in the fillet did not extend through the entire braze width, the joint would still be leaktight, which was the major requirement of the brazement. On the other hand, if 100% braze fillet was needed because of stress requirements, the assembly would be unacceptable.

Base Metal Erosion. Certain brazing filler metals will readily alloy with the base metals being brazed, causing the constituents of the base metal to melt and, in some cases, creating an undercut condition or the actual disappearance of the faying surfaces. This is called base metal erosion. Extreme erosion in type 304 stainless steel brazed with a nickel-chromium-boron filler metal is shown in Fig. 51; a similar joint without erosion is shown in Fig. 52. Erosion may not be serious where thick sections are to be joined, but it cannot be permitted where relatively thin sections are used.

Three factors influencing base metal erosion are brazing temperature, time at temperature, and the amount of brazing filler metal available or used in making the joint. As the brazing temperature exceeds the melting point of the filler metal, interaction between the molten filler metal and the base metal accelerates. The brazing temperature should therefore be kept low, provided, of course, that it is sufficient for proper flow of the filler metal to fill the joint. Similarly, time at temperature should be kept to a minimum to prevent excessive interaction between the molten filler metal and the base metal. Finally, the amount of filler metal required to fill the joint and provide the necessary fillet size should be closely controlled. Filler metal present in excess of the amount required is likely to react with the base metal, creating severe or excessive erosion in proportion to the amount of excess filler metal.

Joint Integrity

Some form of discontinuity usually occurs in all types of brazed joints. The degree and severity vary from a minor pinhole in the filler metal to gross discontinuities. Lack of fill or flux entrapment can vary from slight to nearly 100%. Erosion of the base metal can be nonexistent or can cause complete destruction of the joint.

Requirements for brazed joints are many and varied. As with other accepted joining processes, it is important that brazed joints be properly designed and engineered for the use intended. Significant factors involved are selection of proper base metals and brazing filler metal for compatibility and strength, proper fits and clearances, proper brazing process, and cleanliness of the surface to be brazed. Furthermore, it must be

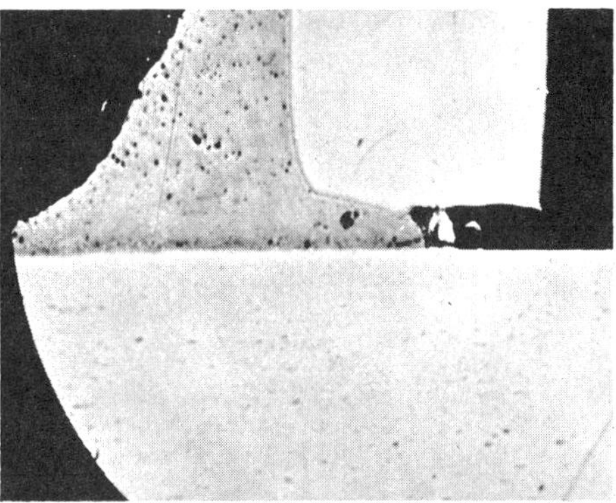

Fig. 50 Incomplete penetration of filler metal (BAg-1) in a brazed joint between copper components. 20×

determined what requirements are necessary for withstanding the service conditions to which the finished brazement will be exposed.

Primarily, brazed joints are designed for mechanical performance, electrical conductivity, or pressure-tightness. The braze quality requirements, therefore, should reflect the end use for which the joint was designed.

Methods of Inspection

Inspection of the completed assembly or subassembly is the last step in the brazing operation and is essential for ensuring satisfactory and uniform quality of the brazed unit. This inspection also provides a means for evaluating the adequacy of the design and the brazing operation with regard to ultimate integrity of the brazed unit.

Destructive methods such as peel tests, impact tests, torsion tests, and metallographic examination are initially used to determine whether the braze design meets the specified requirements. In production, these methods are employed only with random selection or lot testing of brazed joints. In lot testing, samples representing a small specified percentage of all production are tested to destruction. The results of these tests are assumed to apply to the entire production, and the joints in the various lots or batches are accepted or rejected accordingly. When used as a check on an NDI method, such as visual examination, a production part can be selected at regular intervals and the joint tested to destruction so that rigid control of brazing procedures is maintained.

The inspection method chosen to evaluate the final brazed component should depend on service and reliability requirements. In many cases, the inspection methods are specified by the ultimate user or by regulatory codes. In establishing codes or specifications for brazed joints, the same approach should be used as in the setting up of standards for any other phase of manufacturing. The standards should be based, if possible, on requirements that

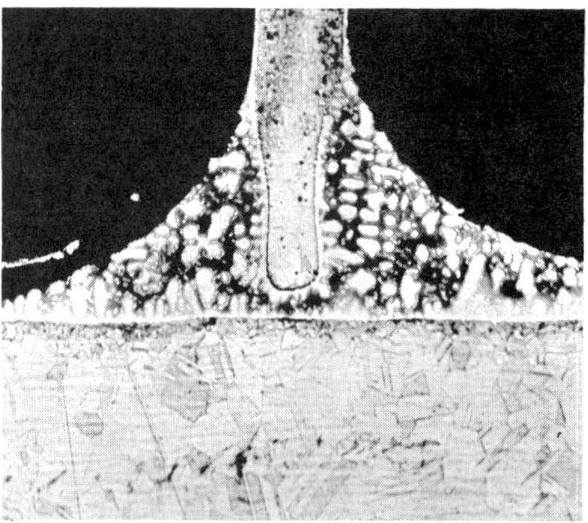

Fig. 51 Excessive erosion of type 304 stainless steel base metal by BNi-1 filler metal. Compare with the noneroded joint shown in Fig. 52. 20×

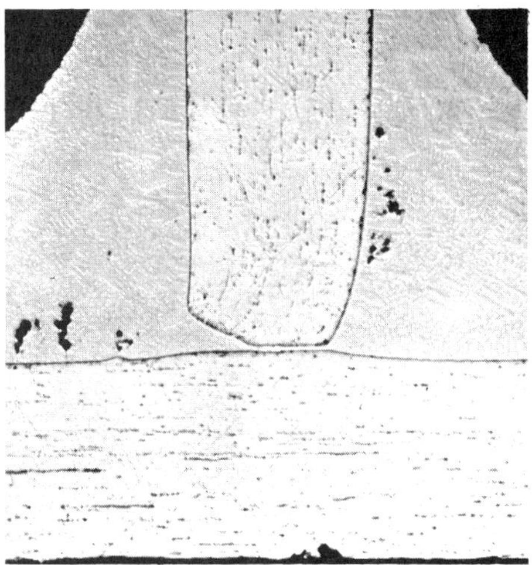

Fig. 52 Joint between type 304 stainless steel components brazed with BNi-1 filler metal, in which no base metal erosion occurred. Note characteristic sheared edge on one component and small voids in the filler metal. 50×

have been established by prior service or history.

Visual Inspection

Visual inspection is the most widely used of the nondestructive methods for evaluating brazed joints. However, as with all other methods of inspection, visual inspection will not be effective if the joint cannot be readily examined. Visual inspection is also a convenient preliminary test where other inspection methods are used.

When brazing filler metal is fed from one side of the joint or replaced within the joint at or near one side so that visual examination of the opposite side of the joint after brazing shows a continuous fillet of filler metal, it can usually be assumed that the filler metal has flowed through the joint by capillary attraction and that a sound joint has been obtained. On the other hand, if the joint can be inspected only on the side where the filler metal is applied, it is quite possible that an unsatisfactory joint has been produced, even though a satisfactory fillet is in evidence to the inspector.

Visual inspection cannot reveal internal discontinuities in a brazed joint that result from flux entrapment or lack of fill. Occasionally, gross erosion can be detected.

Proof Testing

Proof testing is a method of inspection that subjects the completed joints to loads slightly in excess of the loads to be applied during their subsequent service life. These loads can be applied by hydrostatic methods, tensile loading, spin testing, or numerous other methods. Occasionally, it is not possible to ensure a serviceable part by any of the other nondestructive methods of inspection, and proof testing then becomes the most satisfactory method.

Pressure Testing

Pressure testing of brazed assemblies is a method of leak testing and is usually confined to vessels and heat exchangers where liquid, gas, or air tightness is required. Several methods of pressure testing can be employed. Most use either air or gas, depending on the application of the vessel or heat exchanger. In most cases, the test pressures are greater than those to which the assembly will be subjected in service and are specified by the user.

One or more of the following three procedures are generally employed for pressure testing:

- All openings are closed. Air or gas is injected into the assembly until the specified pressure is reached. The inlet sources are closed off, the assembly is allowed to sit for a period of time, and pressure decreases are then measured on a gage
- All openings in the assembly are closed except one, which is fitted with an inlet-pressure line. With the assembly submerged in a tank of water, air or gas is admitted through the inlet line until a specified pressure is reached. The inspector then looks for bubbles rising through the water
- All openings are closed, and the assembly is pressurized to the specified pressure. Then a leak-detecting solution, of which there are several commercially available, is brushed on the joints to be inspected. If any of the joints leak, bubbling will occur

Vacuum-and-helium testing is generally used in inspecting assemblies where it is imperative that the most minute leak be detected. This method of inspection is often employed on nuclear reactor hardware. It is also extensively used in the inspection of refrigeration equipment.

The assembly to be inspected is connected to a vacuum system, and the vacuum is monitored by a mass spectrometer. Helium gas is flushed around the brazed joint; if any minute leak is present, the helium, because of its small molecule, will be pulled in by vacuum and register on the mass spectrometer, thus indicating the leak.

A more sensitive technique is pressurizing the assembly with helium while the assembly is contained in a sealed plastic bag. After pressurizing for a period of time (for example, 24 h), the atmosphere in the bag is analyzed for the presence of helium. Additional information on pressure testing techniques can be found in the article "Leak Testing" in this Volume.

Ultrasonic Inspection

Ultrasonic inspection, although not extensively used in the evaluation of brazed joints, can be the only method applicable in certain cases. The use of ultrasonic inspection depends largely on the design of the joint and the configuration of the adjacent areas of the brazed assembly. Advancements in ultrasonic inspection may increase the utility of this process so that brazed joints can be evaluated with reliability.

Radiographic Inspection

Radiographic inspection is commonly used for the nondestructive evaluation of brazed joints following visual examination. In almost all cases, the radiation beam is directed at about 90° to the plane of the joint, and the radiograph is taken through the thickness of the braze metal.

X-rays readily discern the differences in density between the brazing filler metal and the base metal. Care must be exercised, however, because joints between sections of varying thicknesses can produce radiographs that are misleading and difficult to interpret. Also, it is often difficult to determine whether a joint has been penetrated fully or not at all; both situations yield radiographs in which there is a full fillet visible around the joint, and the gap in the joint itself has uniform radiographic density. By contrast, partly filled joints, voids in the braze metal, and inclusions are relatively easy to find with radiography.

The filler metal in brazed joints is very thin—from 0.013 to 0.25 mm (0.0005 to 0.010 in.) in thickness. When radiographs are made of brazed joints between thick components, the process may be unable to record the braze metal as a difference in density; at least 2% difference is usually needed for good sensitivity.

Liquid Penetrant Inspection

Liquid penetrant inspection is another nondestructive method for determining the reliability of brazed joints and assemblies. This inspection method produces a visual image of a discontinuity in the surface of the braze and reveals the nature of a discontinuity without impairing the parent metal. Acceptable and unacceptable components or assemblies can be separated in accordance with predetermined standards.

There are certain advantages obtained from the liquid penetrant inspection of brazed assemblies. However, a brazed joint or component should be visually inspected first, then inspected by a liquid penetrant method to resolve any doubt concerning joint integrity.

Visual examination is restricted to those discontinuities that can be detected by the unaided eye. Liquid penetrant carries visual inspection a step further by increasing the detectability of fine cracks or openings. Discontinuities such as LOF, cracks that x-rays cannot show because of orientation, and porosity and laps become visible with this technique. Liquid penetrants do not disclose subsurface discontinuities such as voids, cracks, or flux entrapment; radiography is best used to discover such discontinuities.

Selection of the specific liquid penetrant system for the inspection of brazed assemblies depends on the same factors as those that affect system selection for other workpieces. The water-washable, postemulsifiable, and solvent-removable systems have been successfully used for inspecting brazed assemblies.

Inspection using liquid penetrants should not be performed prior to brazing unless adequate cleaning steps, such as vapor degreasing, are taken to remove entrapped penetrant fluid. If permitted to remain during the brazing cycle, this fluid can contaminate the furnace atmosphere and braze metal, producing flaws.

Inspection With Thermally Quenched Phosphors

Inspection with thermally quenched phosphors is a means of nondestructively detecting flaws such as voids and unbonded areas of laminated honeycomb brazed structures where thin (<1.6 mm, or 1/16 in., thick) sheets are used. Thermally quenched phosphor powder mixed with a carrier of polyvinyl alcohol binder is sprayed onto the surface of the part to be inspected. After the surface has dried at room temperature, the part is placed in a darkened booth under a filtered ultraviolet light source. Heat is applied uniformly over the surface of the part, using a heat gun having a blower. Generally, the heat should be applied quickly. The surface of the part that was heated will initially appear uniformly dark; almost immediately, any unbonded areas of the underlying bonded structure will show up very dark on the surface. Voids or unbonded areas that appear much darker than the rest of the area should be promptly marked for later reference because as the heated surface cools, the color of the surface will become uniform.

The part can be reinspected after the surface has cooled to room temperature by reheating the surface. Phosphor coating can be removed quickly by washing in warm water, and no residue or contaminants will remain. Additional information on infrared-sensitive imaging is available in the article "Thermal Inspection" in this Volume.

Soldered Joints

Soldering is a process of ensuring a permanent joint or of building up a surface on metallic parts by means of a metallic filler (solder), which is either introduced in a liquid state between the surfaces to be joined or becomes liquefied by heating below 450 °C (840 °F) and below the melting point of the base metal. The action of capillary forces causes the liquid solder to fill the gap between the faces of the parts being joined. A temperature of about 450 °C (840 °F) is usually accepted as the upper limit of the melting temperature of solder. Detailed information on soldering alloys, solder joint design, and soldering processes can be found in the article "Soldering" in Volume 6 of the 9th Edition of *Metals Handbook*. In addition, the inspection of soldered joints used in electronic components and assemblies will be covered in Volume 1 of *Electronic Materials Handbook*, which will also be published by ASM INTERNATIONAL.

There are three basic steps to achieving reliability of soldered joints, and each step requires some degree of inspection and quality control. These steps are material selection, joint design, and the soldering process.

Matching the flux to the base metal is an important consideration. The flux is a substance such as rosin or borax, applied to the surfaces to be joined just before or during soldering, that cleans the surfaces of oxides and facilitates their union. If an inadequate flux is used, the solder will not wet the metal properly, regardless of the time spent in making the connection. Correct solder alloy is another consideration. Not all solder alloys have the same properties, and it is important that the solder alloy match the application. A simple solderability test can determine whether the base metal, flux, and solder are compatible.

The joint design must be such that the joint has adequate current-carrying capacity (if an electrical connection) and strength. Without these properties, the joint may overheat or fail mechanically under stress. Joint designs that do not allow for visual inspection should be rejected at the design stage, and dimensional requirements should be established so that the inspector has fixed guides to follow. Adequate control must be maintained over the soldering procedures to eliminate variations in such conditions as soldering time, temperature, density of flux, and cleaning efficiency.

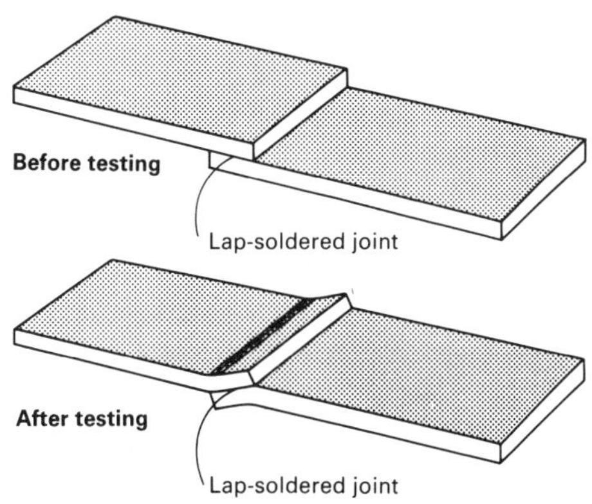

Fig. 53 Lap-soldered joints showing flat seam area before tensile testing, and bending that may occur during tensile testing

The inspection and testing of soldered joints should be considered in the final steps of the soldering operation. If not dictated by codes or specifications prepared by the user, the inspection and testing procedures should be selected by the manufacturer of the soldered joint, with due consideration given to the service conditions.

Workmanship samples are often prepared and used as guides to minimum levels of quality. Inspection and testing procedures are of two types: destructive and nondestructive.

Destructive Inspection

Destructive tests on soldered joints are usually performed on preproduction samples or test specimens to determine the suitability of a given solder, flux, or soldering procedure for a soldered joint of given mechanical properties. Preproduction samples or test specimens are made using the work metal, joint design, solder composition, flux, and soldering procedures specified for the proposed design. Short-time or long-time testing of soldered joints can be performed at elevated temperatures and under vibrating and static tensile loads as deemed necessary.

Difficulties may arise in testing the tensile strength of test specimens cut from thin sheets that are lap soldered. Sometimes during testing, because of the noncoincidence of the axes of symmetry in the lap-soldered halves of the test specimen, the seam area may assume an angle with the direction of the tensile stresses (Fig. 53). This leads to bending stresses that are difficult to determine. Shear strength values, determined according to the load divided by the lap area of the seam, do not reflect the true and intricate process of destruction of the soldered joint. Bending stresses do not occur in such tests if test specimens that are double lap soldered are used.

Soldered parts that require tightness in operation should be pressure tested. Low-pressure tests can be carried out with air, but high-pressure tests require the use of fluids. Because of possible damage, parts that have been pressure tested at or above the working pressure should be discarded.

Visual Inspection

Nearly all soldered joints are visually inspected for defects. The hand solderer visually inspects the joint as it is made and cleaned. The operator of a mass soldering machine usually gives at least a quick glance to each printed circuit board to ensure that the machine is still working properly. The only joints that need not be visually inspected are those for which a convenient nondestructive test is available.

A soldered joint surface should be shiny, smooth, and free of cracks, porosity, or holes. Ordinarily, no flux residues should remain. The transition from the edge of the solder surface to the exposed base metal should show a smooth profile. There should be no extraneous solder beyond the intended amount. Any area of base metal that has contacted the solder should show evidence of wetting.

The reasons for conditions different from those above include the heat sink effect of a nearby large conductive mass or a restraint of solder flow caused by the design of the joint. The inspector must question any condition that cannot be explained because it may indicate deeper problems. For example, slight dewetting around the pad of a printed circuit board is sometimes considered acceptable, but if the dewetting were to extend over the solder/land interface, the strength and reliability of the joint could be seriously impaired.

The degree to which any flaw is cause for rejection must be determined for each application. It is not possible to generalize. For example, some military specifications require that all holes on a soldered printed circuit board be filled with solder, while a commercial application, such as a small radio, would not require such perfection. The differences are often small, and good-quality joints are found in all types of applications.

Normally, inspection is facilitated by the use of a magnifying device that should provide no more than a 10× magnification factor, because too much magnification may make very minor flaws seem more important than they actually are. However, very small joints, such as in microelectronics, require additional magnification to bring the image up to "normal" size. Visual inspection of soldered joints requires training and experience for the inspectors. Training assistance, ranging from simple training aids up to full courses of instruction, is commercially available. Despite this requirement, no other method of testing is currently competitive with visual inspection for a wide range of soldered joints. Common discontinuities in soldered assemblies that are detectable by visual inspection are covered in the following sections.

Dewetting is a phenomenon in which solder flows over a surface and wets it, but before the solder solidifies, a change in the relative surface tension forces takes place so that the solder withdraws into ridges and globules similar to water on a greasy surface. The areas between solder globules retain the color of solder, but they have poor solderability, and this surface can be made to wet properly only with great effort, if at all. The exact mechanism of dewetting is not well understood, but dry abrasive cleaning that leaves particles imbedded in copper is one cause.

Nonwetting is more obvious than dewetting because, in most cases, the base metal retains its original color. Like dewetting, nonwetting is a major flaw because the solder joint is not continuous and is therefore weak. In most cases, nonwetting in very small amounts scattered across the surface can be tolerated. Typically, nonwetting can be traced to insufficient flux activity or inadequate time and/or temperature during soldering.

A flaw related to dewetting or nonwetting is the poor filling of capillary spaces, such as the holes on printed circuit boards. Nonwetting or dewetting may cause the poor fill, or excessive joint clearances may prevent joint completion. If the latter is the cause, the joint may still be acceptable, depending on the required soldering quality level.

Dull or rough solder surfaces can result from overheating or underheating. Overheating is not so damaging to the solder itself, but may damage surrounding heat-sensitive materials. Underheating produces poor contact of the solder with the base metal, but can probably be repaired. A dull, not necessarily rough, surface may also result from vibration or movement during solidification or from certain solder contaminants. Contamination is often inconsequential, but vibration could mean that a degree of weakness has been introduced into the joint structure.

Bridging. Bridges of solder spanning the space between solder joints cannot be tolerated, especially in electrical connections. Assuming that the spacing of the connections is adequate (a design parameter that depends on many factors), the most likely cause of bridging is contamination of the solder by an element that promotes oxides in the solder, such as excess cadmium or zinc.

A related phenomenon is icicling, which resembles incomplete bridges. These spikes of material attached to the joints are normally associated with wave or drag soldering. The same impurity that causes bridging can also cause icicles by interfering with the drainage of the solder as the printed circuit board is removed. Another possible cause is improper alignment of the path of the board when exiting the bath.

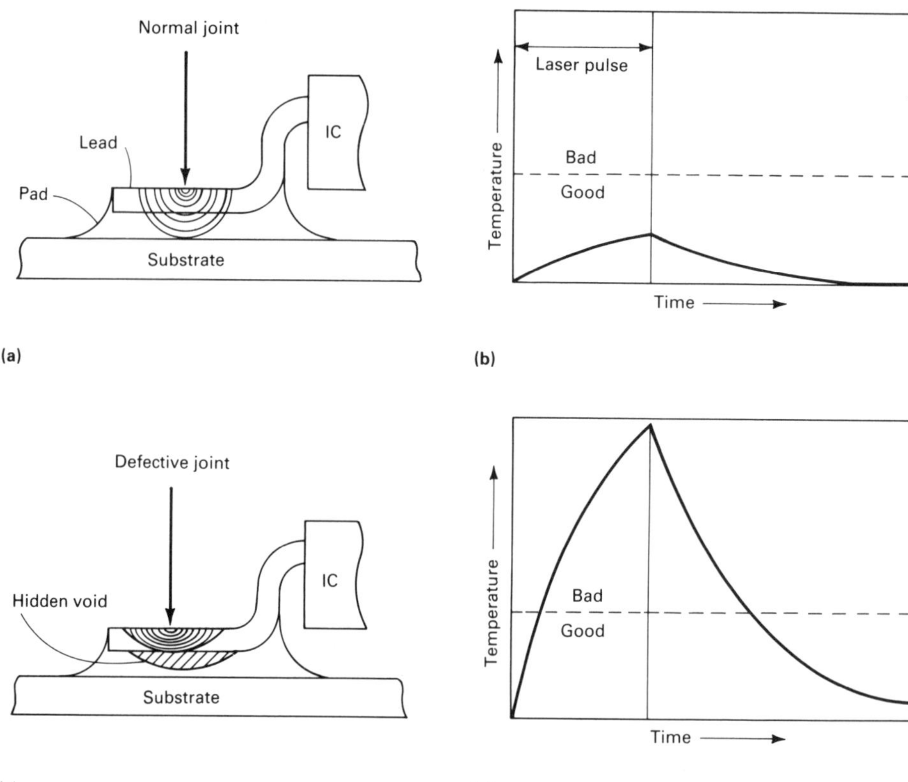

Fig. 54 Schematic of laser/thermal inspection of soldered joints. (a) Cross section of an acceptable lap joint during laser beam heating. (b) Heating curve taken during the laser pulse of acceptable joint. (c) Defective lap joint containing an air pocket. (d) Thermal signal of the defective joint. IC, integrated circuit. Source: Ref 9

Porosity in the solder joint, if small and scattered, poses no threat to reliability. Large pores can significantly decrease the volume of solder in the joint area and therefore the strength of the joint. Pores that break the solder surface can trap corrosive materials or can extend down to the base metal, allowing local corrosion. Such gross porosity is formed when a bubble of air, flux vapor, or water vapor is trapped in the solidifying solder. In some cases, such defects can only be found through sectioning of the soldered joint, but usually there is some other evidence of a problem to alert the inspector.

Excessive Solder. Another discontinuity is excess solder in the joint areas, which may mask other discontinuities. Obviously, where visual inspection is the only possibility, any condition that interferes with the inspection cannot be tolerated.

Nondestructive Inspection*

For a few soldering applications, especially those for which the solder forms a fluid seal, pressure or vacuum testing can be used to check for leakage. The choice is usually determined by the intended application. For example, food cans that will be vacuum packed are tested by sealing and evacuating the can. The special test machine then monitors for any loss of vacuum. Vehicle radiators and other heat exchangers that operate under pressure are leak tested by creating an air pressure inside the part being tested while it is immersed in a water tank. The test pressure is normally selected to be at least as great as the pressure level to be encountered in use. Plumbing joints are usually checked visually for leaks by test operation of the system.

For large, flat areas, such as the mounting of photovoltaic cells for solar arrays, x-radiography has been used. Unfortunately, the design of the soldered joint rarely allows this test method.

New developments include acoustic emission, laser inspection techniques, and acoustic microscopy. Because acoustic emission requires slight deformation of the solder to generate a signal, there is some risk of damaging the joint during testing.

The laser inspection technique is most easily applied to cases in which many joints of the same geometry are to be tested. The basic concept of the technique is a burst of heat generated in the joint by the laser. The amount of heat absorbed is monitored and related to joint quality (Ref 8, 9). Figure 54(a) shows a cross section of a normal lap joint during laser beam heating. The sur-

*Example 3 was prepared by Dr. Robert S. Gilmore, General Electric Research and Development Center.

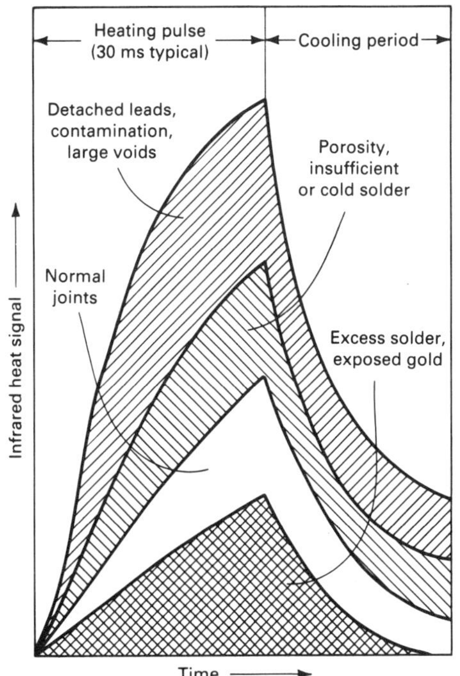

Fig. 55 Typical thermal signatures for acceptable and defective soldered joints inspected by the laser beam heating technique. Source: Ref 9

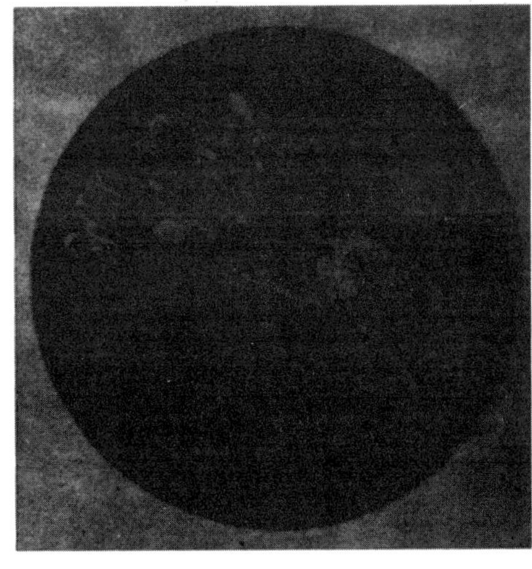

(a)

(b)

Fig. 56 X-ray images of silicon-solder sample. (a) X-ray shadowgraph after initial solder bond in vacuum. Light-gray areas are voids in the solder. (b) X-ray shadowgraph of sample after reheating at atmospheric pressure. Note that most of the major voids have been closed. Courtesy of Robert S. Gilmore, General Electric Research and Development Center

face-generated heat flows normally through the lead and into the underlying solder because of the intimate thermal contact. Figure 54(b) shows a heating curve taken during the laser pulse, which represents a normal amount of observed surface warming.

A defective joint containing an air pocket beneath the lead is shown in Fig. 54(c). The trapping of heat in the lead is depicted by the compressed family of isothermal contours, resulting in a higher thermal signal, as shown in Fig. 54(d).

A composite of thermal signatures for a variety of acceptable and unacceptable solder joints is shown in Fig. 55. The unshaded area denotes a reference region in which acceptable thermal signals might fall, taking account of normal variation in solder mass, heat sinking, and surface quality. The signatures of defects would fall into the shaded areas, each having a finite extent because each defect type has varying degrees of severity.

In acoustic microscopy, magnified acoustic images of the elastic structure of the surface or interior of a solid are produced by passing high-frequency focused acoustic pulses through the material and displaying the received signals in image form as shades of gray (Ref 10). Usually, much higher frequencies are used than in other kinds of acoustic imaging. The acoustic pulses are focused directly, or the energy used to produce the pulses is focused into a very small region. When a small area of a specimen is scanned with a series of focused pulses and the transmitted or reflected signals are viewed on an imaging system (a television picture tube, a gray-scale recorder, or some other scanned display), the magnification of the resulting image is equal to the ratio at the diameter of the display to the diameter of the area scanned. At present, acoustic microscopy includes several methods for the formation of magnified acoustic images. The three discussed in the article "Acoustic Microscopy" in this Volume are scanning acoustic microscopy, C-mode scanning acoustic microscopy, and scanning laser acoustic microscopy. Acoustic images can also be produced by C-scan ultrasonic imaging as described in the article "Ultrasonic Inspection" in this Volume. The following example demonstrates the use of combined x-ray inspection and acoustic microscopy for determining soldered joint integrity.

Example 3: Determination of Silicon-Solder Attachment Integrity. Two [111] cut silicon wafers, 40 mm (1.5 in.) in diameter and 0.32 mm (0.0125 in.) thick, were vacuum soldered together. The solder foil was initially 25 μm (0.001 in.) thick before melting. The sample was fabricated to study silicon-solder integrity through the use of acoustic microscopy and x-ray shadowgraphs.

Sample Preparation. The sample was prepared in the following manner:

- A solder foil was placed between the two prepared silicon wafer surfaces. The assembly was placed in a vacuum apparatus, on a surface that could be heated electrically, with an axial load applied perpendicular to the plane of the wafers
- A vacuum was applied (~135 Pa, or 1 torr), the sample was heated, the solder melted, and then the assembly was cooled under vacuum. The x-ray image shown in Fig. 56(a) was taken at this step in the sample preparation. Figure 56(a) clearly shows a number of voids in the solder layer
- Because of the voids in the solder layer, the sample was reheated, this time in a dry nitrogen glove box, at atmospheric pressure. An additional x-ray image (Fig. 56b) indicated that the large voids had now been closed. This was possible at the modest pressure applied by the atmosphere because the voids contained only vacuum. However, an acoustic microscope image (Fig. 57), in addition to displaying grain structure and small voids, indicated that the solder still was not well attached and had probably never achieved good attachment over most of the initial solder foil area

Inspection Procedure/Defect Analysis. The microfocus x-ray images were made at 100 kV with Polaroid film. The wafer and film were placed at a source-to-target distance of approximately 750 mm (30 in.).

The acoustic micrographs were made with an ultrasonic microscope using a 6 mm (0.25 in.) diam 50-MHz transducer with a focal length of 13 mm (0.5 in.) in water. Figure 57 was made by focusing the ultrasonic beam on the solder layer and scanning it at a pulse-to-pulse and line-to-line spacing of approximately 25 μm (0.001 in.). The gray-scale display is of the amplitude of the ultrasonic echo from the solder layer. The solder grains are imaged because their elastic anisotropy results in a different echo amplitude at the silicon/solder interface.

Conclusions. X-ray images clearly indicated voids where the pores contained substantial volume. However, because the x-rays produced no difference in density, they could not detect a lack of attachment between a solder layer of uniform thickness and the two wafers, where the lack of attachment was not

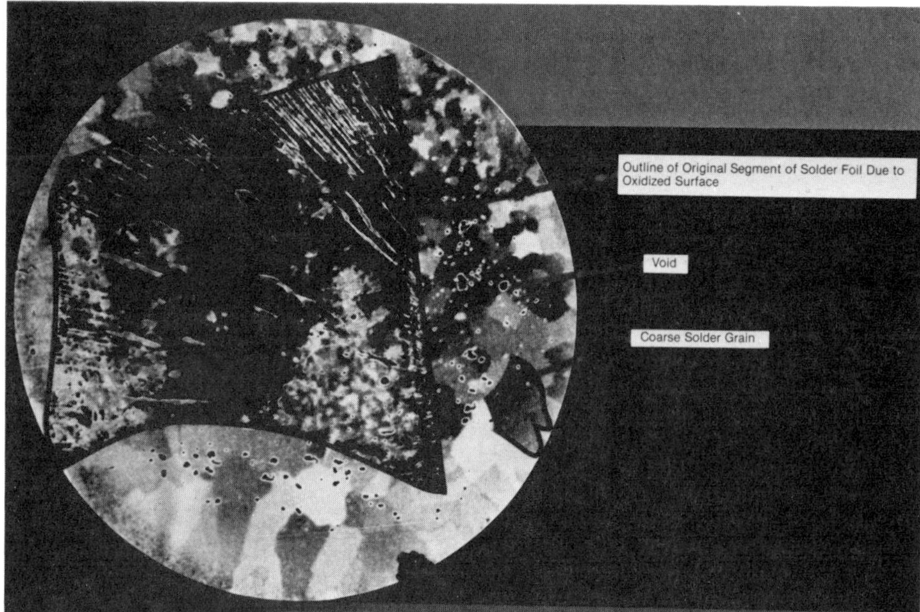

Fig. 57 Acoustic micrograph made with a 50-MHz transducer focused through the silicon on the solder bond. The focused beam diameter is approximately 75 μm (0.003 in.). Courtesy of Robert S. Gilmore, General Electric Research and Development Center

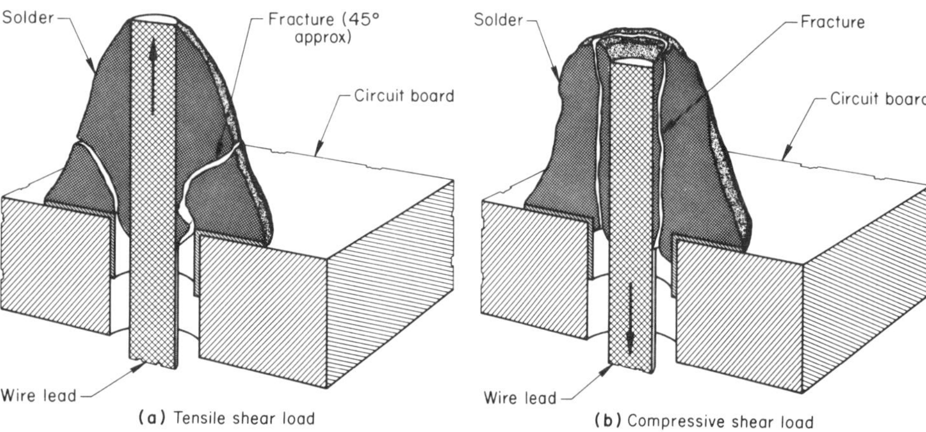

Fig. 58 Fracture mode (a) in excessive tensile shear load of solder; cup-and-cone fracture line is at 45° to the tensile axis. (b) Fracture mode in excessive compressive shear load of solder; fracture surface is cylindrical and is parallel to the load axis.

accompanied by a pore volume. Small voids also showed poor detectability when they resulted in an intensity change of less than a shade of gray (−3 dB) in the x-ray transmission shadowgraph image.

The acoustic image (Fig. 57) clearly displays the attachment integrity of the final assembly as well as evidence of the stages in its preparation. The grain size in the well-attached portion of the solder layer is clearly displayed, as are the voids in this layer. The outlines of the original x-ray detected voids are indicated as well as the grain structure of the solder now filling these areas and the location of a few small voids still remaining. The lack of attachment over large regions of the wafer is indicated, and the fabricators of the sample verified that these regions clearly outline the original solder foil area. It seems reasonable to conclude that a combination of x-ray and acoustic evaluations could supply the most information on silicon-solder attachment, but for this specific geometry the acoustic images provided the most information.

Cracked Soldered Joints

Cracks occur in soldered joints for a number of reasons, and each type of crack may have a distinctive appearance. Among the causes of cracks in soldered joints are excessive mechanical loads, differential thermal expansion, foreign material, improper joint clearances, and intermetallic compound layers.

Excessive mechanical loads produce fractures that are characterized by the nature of the applied stress. As the load increases, the cracks become wider and deeper until fracture occurs. If the load on the base metal creates excessive tensile shear loading on the solder, the specimen will exhibit a typical cup-and-cone fracture with the angle of the cone approximately 45° to the applied load, as shown in Fig. 58(a). If the solder is overloaded in compressive shear, the fracture surface will be parallel to the applied load, as shown in Fig. 58(b). If the joint was of questionable strength before soldering, the inspection process itself may overload the solder and produce cracks.

Differential thermal expansion can cause solder cracks; these generally consist of a series of closely spaced hairline cracks. To eliminate this type of crack, design specifications concerning the choice of base metals, solder alloys, and amount of thermal cycling permitted should be revised.

Foreign material such as flux inclusions and gas bubbles can act as nucleation sites for cracks. Small, random cracks frequently radiate from the inclusion or bubble to the outer surface. The remedy lies in more care and skill in the soldering operation.

Improper joint clearances may prevent proper flow of molten solder and thus create a joint of subnormal strength, which is vulnerable to cracks. A clearance of less than 75 μm (0.003 in.) may result in flux entrapment, poor solder penetration, and voids. Small, random cracks may radiate from the voids to the outer surface. On the other hand, a clearance greater than 125 μm (0.005 in.) may impair the capillary flow of molten solder and also form voids. In this situation, cracks may form parallel to the joint at the point of incomplete fill. The most satisfactory capillary flow of solder is achieved with joint clearances between 75 and 125 μm (0.003 and 0.005 in.).

Intermetallic compound layers formed in the solder may cause embrittlement and subsequent cracking. Generally, the cracks are parallel to the surface of the part and open to the surface for a comparatively long distance. They may result from high temperature, or excessive time at temperature, during the soldering operation, or from the alloy combinations chosen for the joint. In high-reliability systems, for example, the connections are frequently gold plated before soldering to ensure good corrosion resistance and adequate solderability. However, if the gold finish is only partly dissolved in the solder, a layered structure may form that is very susceptible to cracking. This structure consists of layers of gold plate, of the $AuSn_4$ intermetallic compound, and of solder. It is the interface between the soft gold and the hard $AuSn_4$ that initiates the cracks. A printed circuit board connection with a crack between the gold plate and the $AuSn_4$ is shown in Fig. 59.

Corrosion of Soldered Joints

Either galvanic or chemical corrosion can attack a soldered joint to a degree that may

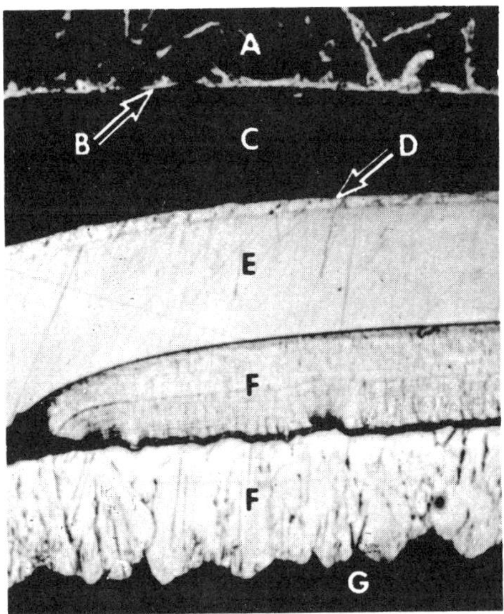

Fig. 59 Section through a cracked soldered connection on a printed circuit board. At A is the tin-lead solder with a thin layer of the intermetallic compound $AuSn_4$ (B) on its underface. C is the crack that opened at the interface between the $AuSn_4$ and a layer of gold plate at D. E is a layer of nickel plating over two layers of copper plate (F) on the circuit board (G). 500×

severely weaken its structural integrity or completely destroy it. In electrical connections, the occurrence of corrosion in a soldered joint may either increase the circuit resistance by impairment of contact or cause current leakage through surface deposits of conductive corrosion products. Problems in relay contacts can arise because of deposits of nonconductive corrosion products on conductive surfaces. Although the corrosion products due to many causes of corrosion are visible as white or green powder or dust around the soldered joint, corrosion due to flux residues will not necessarily result in discernible corrosion products.

Galvanic corrosion will occur when a sufficiently large difference in electromotive force exists between base metals or between solder and base metal, provided that an electrolyte is present on the soldered joint.

Chemical corrosion of soldered joints is frequently caused by inadequate removal of corrosive flux residues left from the soldering operation. Other causes of chemical corrosion include residual plating or etching solutions and contamination from packaging materials, handling procedures (including human perspiration), and environmental conditions. Additional information on the corrosion characteristics of soldering alloys and soldered assemblies is available in Volume 13 of the 9th Edition of *Metals Handbook*.

Cleanliness in Soldered Joints

It is important that traces of flux or flux residues be removed during postsoldering cleaning procedures. In addition, many solder operations are by nature contributors of small solder droplets. These small metallic particles can eventually be dislodged in the assembly and can cause short circuits and other undesirable side effects. When an assembly is improperly cleaned after soldering or fluxes are not removed, visual inspection is physically hampered. Cleanliness is therefore one of the items that should be specified in any proper procedure for the inspection of soldered joints.

Developments in high-reliability low-current circuitry (especially printed circuits) have emphasized the importance of cleanliness of electronic assemblies because of the danger of corrosion. Corrosion damage to conductors can increase circuit resistance, and high resistance is undesirable. Corrosion can also cause physical failure of the conductors by weakening and embrittlement. In addition, corrosion products themselves can cause current leakage. Current leakages are particularly bad because they are not consistent; humidity changes in the atmosphere cause variations in and sometimes intermittent occurrences of current leakage. Corrosion products also can cause contamination throughout the whole system in the form of nonconductive deposits on mechanical contacts and relay surfaces. Corrosion is definitely a problem, but corrosion is not always caused by fluxes. Additional information on the sources of corrosion in electronic components is available in the articles "Corrosion in the Electronics Industry" and "Case Histories and Failures of Electronics and Communications Equipment" in Volume 13 of the 9th Edition of *Metals Handbook*.

Cleanliness is easily checked after the proper cleaning procedures are followed. The danger to the assembly stems from the presence of ionizable materials, mostly chlorides. The easiest and most thorough check is done with a conductivity cell and distilled and demineralized water, either leaching the ionizable materials off the surfaces in premeasured amounts of water or submerging the whole assembly into a container of the water and measuring the resistivity of the water. This indicates the presence of ionizable materials. However, because chlorides are the major contributors in the corrosion mechanism and are the most abundant form of ionic contamination on the assembly, standardized, simple silver nitrate tests can be used to establish the presence of chlorides.

Cleanliness after soldering is important because soldering is usually the last step in a long process of assembly. However, precautions must be taken to prevent recontamination of the surfaces; otherwise, the purpose of cleanliness and of the check for corrosion products in the final inspection is defeated.

ACKNOWLEDGMENT

The contributions of the following individuals were critical in the preparation of this article: W.H. Kennedy, Canadian Welding Bureau; Robert S. Gilmore, General Electric Research and Development Center; and John M. St. John, Caterpillar, Inc. Special thanks are also due to Michael Jenemann, Product Manager, NDT Systems, E.I. Du Pont de Nemours & Company, Inc., for supplying the reference radiographs of welds shown in Fig. 18 to 37. Finally, the efforts of the ASM Committee on Weld Discontinuities and the ASM Committee on Soldering from Volume 6 of the 9th Edition of *Metals Handbook* are gratefully acknowledged; material from the aforementioned Volume was used in this article.

REFERENCES

1. T.M. Mansour, Ultrasonic Inspection of Spot Welds in Thin-Gage Steel, *Mater. Eval.*, Vol 46 (No. 5), April 1988, p 650-658
2. P. Kapranos and R. Priestner, NDE of Diffusion Bonds, *Met. Mater.*, Vol 3 (No. 4), April 1987, p 194-198
3. P.G. Partridge, "Diffusion Bonding of Metals," Advisory Group for Aerospace Research and Development (NATO), Aug 1987, p 5.1-5.23
4. G. Tober and S. Elze, "Ultrasonic Testing Techniques for Diffusion-Bonded Titanium Components," Advisory Group for Aerospace Research and Development (NATO), July 1986, p 11.1-11.10
5. *Diffusion Welding and Brazing*, Vol 3, 7th ed., *Welding Handbook*, American Welding Society, 1980, p 311-335
6. J.J. Munro, R.E. McNulty, and W. Nuding, Weld Inspection by Real-Time Radiography, *Mater. Eval.*, Vol 45 (No. 11), Nov 1987, p 1303-1309
7. G.L. Burkhardt and B.N. Ranganathan, Flaw Detection in Aluminum Welds by the Electric Current Perturbation Method, in *Review of Progress in Quantitative Nondestructive Evaluation*, Vol 4A, Plenum Press, 1985, p 483-490
8. R. Vanzetti, "Intelligent Laser Soldering Inspection and Process Control, Electronics Reliability and Measurement Technology," NASA CP-2472, National Aeronautics and Space Administration, June 1986, p 85-94
9. A.C. Traub, Parts Inspection by Laser Beam Heat Injection, *NDT Int.*, Vol 21 (No. 2), April 1988, p 63-69
10. R.S. Gilmore, Acoustic Microscopy, in *Encyclopedia of Materials Science and Engineering*, Pergamon Press, 1986, p 38-43

Adhesive-Bonded Joints*

Donald J. Hagemaier, Douglas Aircraft Company, McDonnell Douglas Corporation

ADHESIVE-BONDED JOINTS in which adhesives are used to join and reinforce materials (Fig. 1), are extensively used in aircraft components and assemblies where structural integrity is critical. Figures 2 and 3 show typical adhesive-bonded joints used in aircraft structural assemblies. However, the structural components are not limited to aircraft applications, but can be translated to commercial and consumer product applications as well. In terms of production cost, ability to accommodate manufacturing tolerances and component complexity, facility and tooling requirements, reliability, and repairability, adhesive bonding is very competitive when compared to both co-cure and mechanical fastener joining methods.

Variables such as voids (lack of bond), inclusions, or variations in glue line thickness are present in adhesive-bonded joints and will affect joint strength. This article addresses the problem of how to inspect bonded assemblies so that all discrepancies are identified. Once the flaws are located, engineering judgment and analysis can determine if adequate strength exists or if the parts need to be reworked or scrapped. When very large and costly assemblies have been bonded, the importance of good nondestructive inspections is fully appreciated.

This article describes several techniques and presents drawbacks and limitations where they exist. No single method or procedure can locate every flaw, so it is important for personnel to be aware of the available techniques and their limitations. Bonded assemblies are inspected immediately after the adhesive cure cycle, after the parts have been cleaned and removed from the manufacturing jigs or fixtures. The next important time for inspection is after the assemblies have been in use and the condition of the parts is known. It is necessary to determine if new delaminations have occurred or if any metal cracks have developed. Inspection in the field requires the instruments to be easily transported and to be used in confined places.

The commonly used terms describe what is to be done. For example, nondestructive testing (NDT) comprises the testing principles methodology, nondestructive inspection (NDI) involves inspection to meet an established specification or procedure, and nondestructive evaluation (NDE) is the actual examination of materials, components, and assemblies to define and classify anomalies or discontinuities in terms of size, shape, type, and location. This article specifically addresses the three terms as they apply to adhesive-bonded joints within structures.

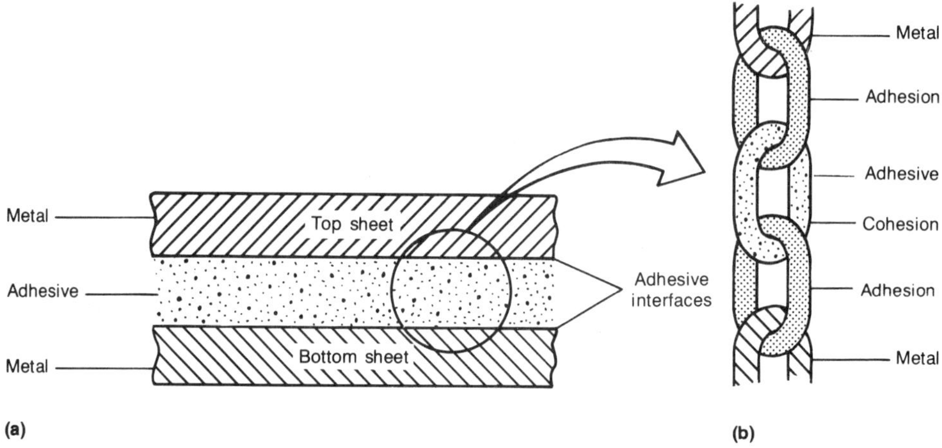

Fig. 1 Schematic of a metal-to-metal adhesive-bonded joint. (a) Adhesive sandwiched between two metal sheets. (b) Analogy of adhesive-bonded joint components to individual links in a chain

Inspection reference standards must be developed for each instrument, and the standards must be made available to and used by inspectors to standardize their equipment and to permit them to detect anomalies that will indicate a flaw. Inspectors must also be familiar with the internal details of the assembly being inspected so that they can distinguish between flaws and legitimate structural details. For example, how many layers of metal were bonded at any one spot?

Description of Defects

There are a wide variety of flaws or discontinuities that can occur in adhesive-bonded structures. Table 1 lists possible generic flaw types and their producing mechanisms. Metal-to-metal voids or unbonds are the most frequently occurring rejectable flaws, as indicated in Table 2.

Interface defects are the result of errors made during the pretreatment cycle of the adherends prior to the actual bonding process. In practice, pretreatment flaws are reduced by careful process control and by adherence to specification requirements and inspection before proceeding with the bond cycle. Controls generally include the waterbreak test and measurement of the anodic layer and primer thickness. Interface defects can be caused by improper or inadequate degreasing, deoxidizing, anodizing, drying, damage to the anodizing layer, or excessive primer thickness.

Interface defects are generally not detectable by state-of-the-art NDT methods. Therefore, test specimens are processed along with production parts and sent to the laboratory for evaluation. Applicable wedge crack specimens, lap shear specimens, or honeycomb flatwise tension specimens are fabricated and tested to determine if the process meets specification requirements before the bonding cycle starts.

Considerable effort at Fokker in the Netherlands led to the important discovery that the ideal oxide configuration for adhesion on aluminum alloys can be detected by inspection with an electron microscope at suitable magnification (Ref 2-4). To inspect with the electron microscope, a piece of the

*Reprinted from *Adhesive Bonding of Aluminum Alloys*, Marcel Dekker, 1985, p 337-423. With permission

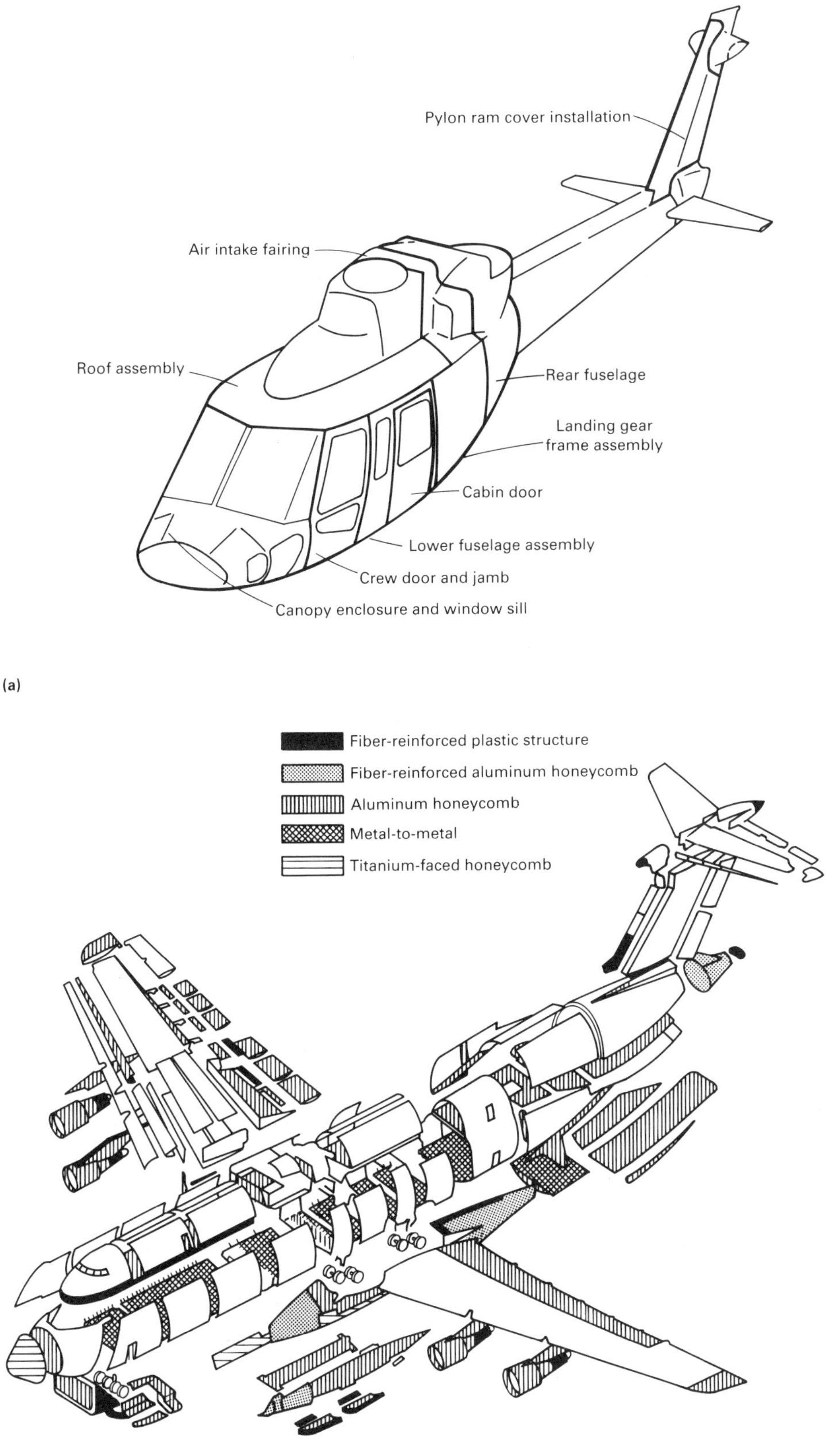

Fig. 2 Typical applications of adhesive-bonded joints in aircraft. (a) Helicopter components. (b) Lockheed C-5A transport plane, with various types of honeycomb sandwich structures totaling 2230 m^2 (24 000 ft^2) in area

structure must be removed. As a consequence, the electron microscope became a useful tool for adhesion quality control.

Another physical parameter that was used as a basis for the NDT of surfaces for the ability to bond was contact potential, which is measured by a proprietary method developed by Fokker known as a contamination tester (Ref 3, 5). This instrument is based on Kelvin's dynamic-condenser method but avoids the disturbances usually associated with it. There is sufficient evidence that the contamination tester is able to detect nondestructively the absence of the optimum oxide configuration arising from incomplete anodizing and/or subsequent contamination (Ref 4).

More recently, Couchman *et al.* (Ref 6) at General Dynamics developed an adhesive bond strength classifier algorithm that can be used to build an adhesive bond strength tester. Lap shear specimens were fabricated using Reliabond 398 adhesive. The test specimens include:

- A control set with optimum bond strength
- An undercured set
- A weak bond produced by an unetched surface
- A thin-bond adhesive that was cured without a carrier

The weakest bond was observed to fail at 725 kPa (105 psi), while the strongest held to 15.7 MPa (2.27 ksi). Tabulated results showed the following:

- *Undercured set*: 725 to 6410 kPa (105 to 930 psi)
- *Unetched surface set*: 5.79 to 7.58 MPa (0.840 to 1.10 ksi)
- *Control set*: 13.4 to 14.8 MPa (1.94 to 2.15 ksi)
- *Thin adhesive set*: 14.1 to 15.6 MPa (2.05 to 2.27 ksi)

The accept/reject value was set at 13.1 MPa (1.90 ksi), and all specimens were classified correctly. The important factor is that an interface defect (unetched surface), which results in poor adhesion, was detected.

Defects within the cured adhesive layer can be one or more of the following:

- Undercured or overcured adhesive
- Thick adhesive resulting in porosity or voids due to improper bonding pressure or part fit-up
- Frothy fillets and porous adhesive caused by too fast a heat-up rate
- Loss of long-term durability due to excessive moisture in the adhesive prior to curing

In normal cases, the curing time is very easily controlled. The curing temperature and temperature rate are controlled by proper positioning of the thermocouples on the panel and by regulating the heat-up rate.

Thick glue lines occur in a bonded assembly due to inadequate mating of the facing sheets or blocked fixing rivets, and they result mostly in porosity and voids. However, a thick glue line made with added layers of adhesive is usually free of porosity. Porosity has a significant effect on the strength of the adhesive, with higher porosity related to a greater loss in strength and a

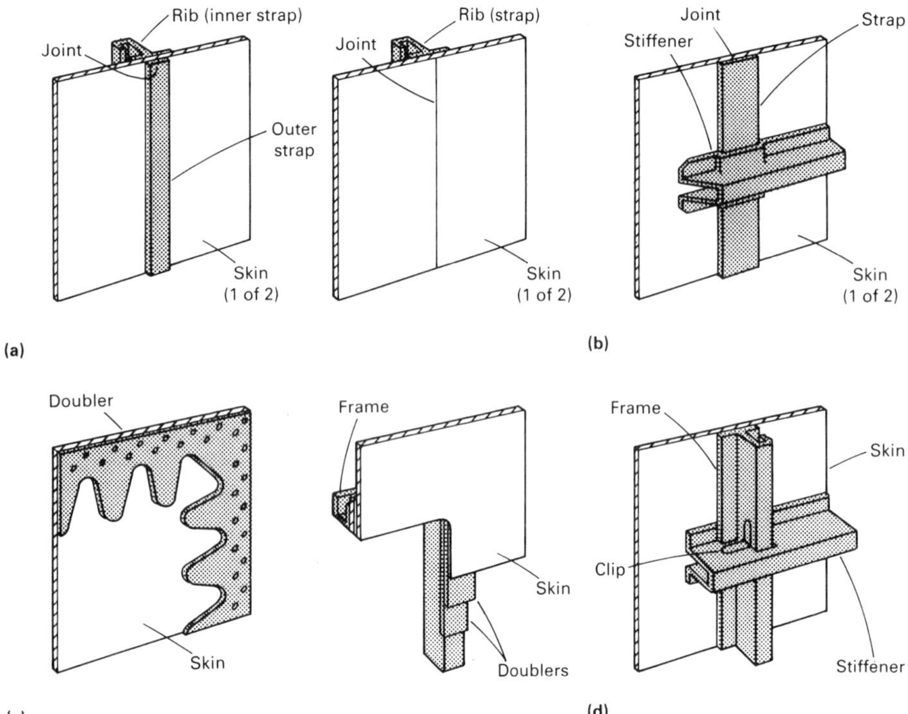

Fig. 3 Some typical adhesive-bonded joints used to join components in structural assemblies. (a) Skin splices. (b) Stiffener runout. (c) Bonded doublers. (d) Shear clip

void condition resulting in no strength. These defects occur quite frequently (Table 2). Porosity can also be caused by the inability of volatiles to escape from the joint, especially in large-area bond lines. Excessive moisture in the adhesive prior to curing can be prevented by controlling the humidity of the lay-up room. The entrapped moisture, after curing, cannot be detected by NDT methods unless it results in porosity.

Other defects that occur during fabrication can include:

- No adhesive film
- Protective film left on adhesive
- Foreign objects (inclusions)

In practice, these conditions must be prevented by process control and training of the personnel engaged in the bonding operation. The first two conditions occur infrequently. Shavings, chips, wires, and so on, can result in porosity or voids. Honeycomb core assemblies have been found with all types of foreign material.

Metal-To-Metal Defects

Voids. A void is any area that should contain, but does not contain, adhesive. Voids are found in a variety of shapes and sizes and are usually at random locations within the bond line. Voids are generally surrounded by porosity if caused by a thick bond line and may be surrounded by solid adhesive if caused by entrapped gas from volatiles.

Unbonds or Disbonds. Areas where the adhesive attaches to only one adherend are termed unbonds. Unbonds can be caused by inadequate surface preparation, contamination, or improperly applied pressure. Because both adherends are not bonded, the condition is similar to a void and has no strength. Unbonds or disbonds are generally detectable by ultrasonic or sonic methods.

Porosity. Many adhesive bond lines have some degree of porosity, which may be either dispersed or localized. The frequency and/or severity of porosity is random from one assembly to the next. Porosity is defined as a group of small voids clustered together or in lines. The neutron radiograph shown in Fig. 4(a) confirms the presence of porosity in the bond lines visible to the eye in Fig. 4(b). Porosity is detected equally well using conventional radiography or neutron radiography (Fig. 5). Scattered linear and dendritic porosity is usually found in adhesives supported with a matte carrier. Linear porosity generally occurs near the outer edge of a bonded assembly and in many cases forms a porous frame around a bonded laminate. Porosity is usually caused by trapped volatiles and is also associated with thick (single-layer adhesive) bond lines that did not have sufficient pressure applied during the cure cycle. The reduced bond strength in these porous areas is directly related to its density frequency and/or severity.

Porous or Frothy Fillets. This condition results from too high a heat-up rate during curing. The volatiles are driven out of the adhesive too rapidly, causing bubbling and a porous bond line, which is distinguished by the frothy fillets. This defect is visually detectable and should also be seen in the test specimens processed within the production parts.

Lack of Fillets. Visual inspection of a bonded laminate can reveal areas where the adhesive did not form a fillet along the edge of the bonded adherends or sheets. In long, narrow joints, a lack of fillet on both sides generally indicates a complete void. This defect is considered serious because the high stresses near the edges of a bond joint can cause a cracked adhesive layer due to shear or peel forces. A feeler gage can be used to determine the depth of the defect into the joint. If the gap is too tight for a feeler gage, ultrasonic or radiographic techniques can be used to determine the depth of the edge void.

Fractured or Gouged Fillets. These defects are detected visually. Cracked fillets are usually caused by dropping or flexing the bond assembly. Gouges are usually made with tools such as drills or by impact with a sharp object. Fractured and gouged bond lines are considered serious for the reasons stated earlier for lack of fillets.

Adhesive Flash. Unless precautions are taken, adhesive will flow out of the joint and form fillets plus additional adhesive flow on mating surfaces. Although the condition is not classified as a defect, it is considered unacceptable if it interferes with ultrasonic inspection at the edges of the bonded joint where stresses are highest.

Burned Adhesive. The adhesive may be burned during drilling operations or when bonded assemblies are cut with a band saw. The burned adhesive is essentially overcured, causing it to become brittle and to separate from the adherend. Also, the cohesive strength of the burned adhesive is drastically reduced. Figure 6 shows burned adhesive around hole 9 caused by improper drilling, as well as bond delamination along the edge of the panel adjacent to holes 16 through 20 caused by band sawing. Improper drill speed or feed coupled with improper cooling can cause these types of defects. The burned adhesive around hole 9 in Fig. 6 is detectable by ultrasonic C-scan recording methods (Fig. 7).

Adherend Defects

Adherend defects can be detected visually and do not include the processing procedures.

Fractures (Cracks). Cracks in the adherend, whatever the cause, are not acceptable.

Double-Drilled or Irregular Holes. Some bonded assemblies may contain fastener holes. When holes are drilled more than once, have irregular shapes, or are formed with improper tools, they are considered defects. The load-carrying capacity of the fasteners is unevenly distributed to the ad-

Table 1 Generic flaw types and flaw-producing mechanisms

Flaw-producing mechanism	Generic flaw type: Metal-to-metal	Metal-to-core	Core	Surface	Adhesive
Disbonds, internal	X				
Disbonds, part edge	X	X			
Disbonds, high core	X				
Porosity	X	X	X		X
Unremoved protective release film from adhesive	X	X			
Foam adhesive in film adhesive bond line	X	X			
Cut adhesive	X	X			X
Adhesive gaps	X	X			
Missing adhesive	X	X	X		
Weak bonds	X	X	X		
Extra layers of film adhesive	X	X			
Foreign objects	X	X	X		X
Double drilled or irregular holes	X			X	
Disbonds, low core		X			
Void or gap, chemical milled land		X			
Void or gap, doublers		X			
Missing fillets		X			
Void, closure-to-core		X			
High-density inclusions (chips, etc.)	X	X			X
Voids, foam joint		X	X		
Disbond, shear ties	X	X			
Lack of sealant at fasteners		X			
Thick foam adhesive		X			
Broken fasteners	X	X			
Crushed core			X		
Wrinkled core			X		
Condensed core			X		
Distorted core			X		
Blown core			X		
Node bond separation			X		
Missing core (short core)		X	X		
Cut core		X	X		
Water in core			X		
Cracks				X	
Scratches				X	
Blisters				X	
Protrusions				X	
Indentations (dents/dings)				X	
Wrinkles				X	
Pits				X	

Source: Ref 1

Table 2 Frequency of rejectable flaws in adhesive-bonded assemblies

Defect	Number of defects	Percentage of total
Metal-to-metal voids and disbonds	378	74
Skin-to-core voids and disbonds	19	3
Gap in core-to-closure bond	9	2
Lack of foaming adhesive or voids in foaming adhesive	22	4
Difference in core density	6	2
Lack of fillets	1	1
Crushed or missing core	32	6
Short core	40	8
Total	**507**	**100**

herends, resulting in high local stresses that may cause fracture during service.

Dents, Dings, and Wrinkles. These defects are serious only when extensive in nature, as defined by applicable acceptance criteria or specifications. They are most detrimental close to, or at, a bond joint. Dents are usually caused by impact with blunt tools or other objects and are usually rounded depressions. Dings result from impact with sharp objects or when an assembly is bumped at the edge. Dents or dings may cause bond line or adherend fractures. Wrinkles are bands of distorted adherends and are usually unimportant.

Scratches and Gouges. A scratch is a long, narrow mark in the adherend caused by a sharp object. Deep scratches are usually unacceptable, because they can create a stress raiser which may generate a metal crack during service. On the other hand, gouges are blunt linear indentations in the adherend surface. Deep gouges, like scratches, are generally unacceptable.

Honeycomb Sandwich Defects

The most prominent defects found or generated in honeycomb sandwich assemblies are summarized in Table 1. Adhesion and/or cohesion defects may also occur in bonded honeycomb sandwich assemblies. The metal-to-metal closure areas for honeycomb panels may exhibit the types of defects discussed in the preceding section. In addition, sandwich assemblies can have defects in the honeycomb core, between the core and skins, between core and closure, at chemically milled steps, and in core splices. These bond areas are shown in Fig. 8 for a typical honeycomb assembly.

Water in Core Cells. Upon completion of the bonded assembly, some manufacturers perform a hot-water leak test to determine if the assembly is leakproof. If the assembly emits bubbles during the leak test, the area is marked and subsequently repaired. To ensure that all bonded areas are inspected and that no water remains trapped in the assembly, it is then radiographed. This is important because water can turn into ice during operational service and rupture the cells, or it can initiate corrosion on the skin or core. Water in the core can be detected radiographically when the cells are filled to at least 10% of the core height. Also, x-ray detection sensitivity is dependent on the sandwich skin thickness and radiographic technique. An additional problem is the ability to determine whether the suspect area has excessive adhesive, filler, or water. Water images usually have the same film density from cell to cell or for a group of cells, while adhesive or filler images may vary in film density within the cells or show indications of porosity. A radiographic positive print of moisture in honeycomb is shown in Fig. 9.

Crushed Core. A crushed honeycomb core may be associated with a dent in the skin or may be caused by excessive bonding pressure on thick core sections. Crushing of the core greatly diminishes its ability to support the facing sheets. Figure 10 shows an x-ray positive print of crushed core. Generally, crushed core is most easily detected with angled x-ray exposures. Crushed core is defined as localized buckling of the cell walls at either face sheet, when associated with the halo effect on a radiograph. On the other hand, for wrinkled core, the cell walls are slightly buckled or corrugated. Radiographically, the condition appears as parallel lines in the cell walls. A wrinkled core is generally acceptable.

Condensed core occurs when the edge of the core is compressed laterally. Lateral compression may result from bumping the edge of the core during handling or lay-up, or slippage of detailed parts during bonding. The condition occurs most often near honeycomb edge closures. Figure 11 shows a positive radiograph of various degrees of condensed core.

Node separation results when the foil ribbons are separated at their connecting points or nodes, as shown schematically in Fig. 12 and in the photograph in Fig. 13. Node separation usually occurs during core fabrication. It may also result from pressure buildup in cells as a result of vacuum bag leaks or failure, which allows the pressurizing gas to enter the assembly and core cells.

Blown core occurs as a result of a vacuum bag leak or because of a sudden change in pressure during the bonding cycle. The pressure change produces a side loading on the cell walls that can either distort the

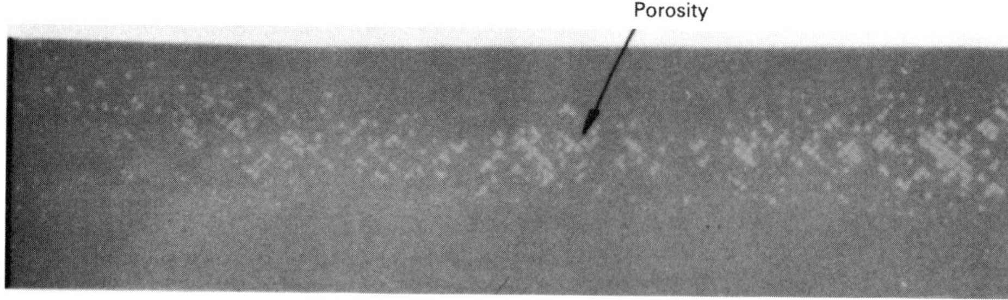

(a)

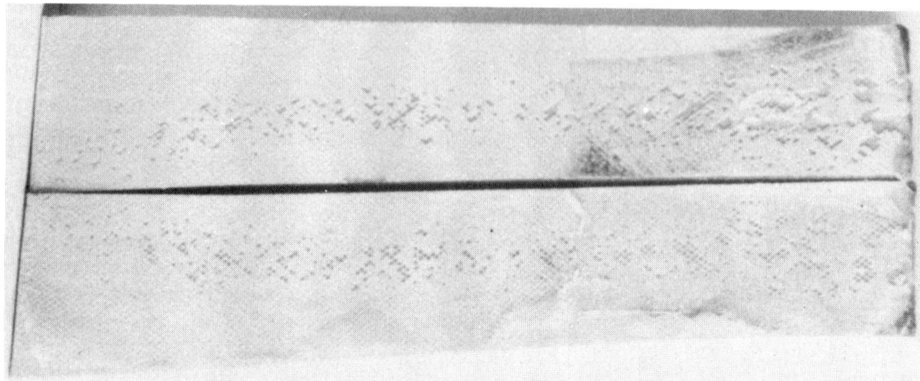

(b)

Fig. 4 Neutron radiograph (a) and visual confirmation (b) of porosity in an adhesive-bonded joint

cell walls or break the node bonds. Radiographically, this is indicated as:

- Single-cell damage, usually appearing as round or elliptical cell walls with partial node separation
- Multicell damage, usually appearing as a curved wave front of core ribbons that are compressed together

The blown core condition is most likely to occur at the edge of the assembly in an area close to the external surface where the greatest effect of sudden change in pressure occurs. This condition is most prevalent whenever there are leak paths, such as gaps in the closure ribs to accommodate fasteners, or chemically milled steps in the skin where the core may not fit properly. When associated with skin-to-core unbonds, the condition is detectable by pulse-echo and through transmission ultrasonic techniques. The condition is readily detectable by radiography when the x-ray beam centerline is parallel to the core cell walls, as shown in Fig. 14.

Voids in Foam Adhesive Joints. Defects in core-to-closure, core-to-core splice, core-to-trailing edge fitting, and chemically milled steps (Fig. 8) in foam adhesive joints can result from the following conditions:

- The foam adhesive can slump or fall, leaving a void between the core and the face skins. This particular condition is most readily detected by ultrasonics
- The core edge dimension may be cut undersize and the foam does not expand uniformly to fill the gap between the closure and the core
- The foaming adhesive can fail to expand and surround the core tangs, as illustrated in Fig. 15

Protective Film Left on Adhesive. Protective films are usually given a bright color so that they can be seen and removed from the adhesive before bonding. If they remain on the adhesive, the adhesive is prevented from contacting one of the adherends. This condition is very difficult to detect by ultrasonic techniques and x-ray radiography, with which it would appear as a mottled condition. It generally produces porosity, especially at the perimeter, which aids in its detection.

Metal-to-Metal Defects. The most prominent defects are voids and porosity. Disbonds can occur under the following conditions:

- When the core is slightly higher than a closure member
- Lack of applied pressure from tooling
- Entrapped volatiles in the bond joint prior to cure
- Excessive moisture in the adhesive prior to cure
- Contaminated honeycomb core

Voids and porosity are detectable by low-kilovoltage (15 to 50 kV) x-ray techniques using a beryllium-window x-ray tube, by thermal neutron radiography, and by ultrasonic C-scan techniques employing small-diameter or focused search units operating at 5 to 10 MHz. If the flaw is the result of insufficient pressure, the adhesive will be porous, as shown in Fig. 4 and 5. The lower the kilovoltage and/or the thicker or denser the adhesive, the higher the resolution of the flaw image. In general, the flaw size detectable by radiography is smaller than that detected by ultrasonics. Also, some adhesives, such as AF-55 and FM-400, are x-ray opaque, which yields a much higher contrast among voids, porosity, and solid adhesive (Fig. 16). Disbonds (separation between adhesive and adherend) are best detected by ultrasonic techniques.

One of the most important types of voids in honeycomb assemblies is the leakage-type void, which is oriented normal to the metal-to-metal bond line and penetrates to the core. Moisture can penetrate such a void during operational service and cause corrosion or ice damage to the core. This type of void is illustrated in Fig. 17(a) and (b).

Foam Intrusion. Another type of defect that can cause problems in radiography is foam intrusion into a metal-to-metal joint near a closure. This type of flaw is illustrated in Fig. 17(a) and (c). Ultrasonically, it will appear bonded; therefore, ultrasonics cannot confirm the radiographic findings. Because this may or may not be a defect, it must be determined by engineering analysis.

Skin-to-Core Voids at Edges of Chemically Milled Steps or Doublers. This condition occurs when the adhesive fails to bridge the gap at the edges of chemically milled or laminated steps or doublers (Fig. 8). This is detected radiographically as a dark line or an elliptically shaped dark image. Ultrasonically, it will appear as a linear void along the chemically milled step position.

Missing Fillets. As pressure is applied during the bonding cycle, adhesive fillets are formed at the edges of each honeycomb cell. Fillets will not be formed if pressure is not maintained. If the adhesive is x-ray opaque, this condition is readily detected by directing the radiation at an angle of approximately 30° with respect to the centerline of the core or closure web. If the adhesive is not x-ray opaque, then ultrasonic, eddy sonic, and tap tests can be used to locate the area having unbonded cells. Ultrasonic C-scan can be used to record skin-to-adhesive and adhesive-to-core voids. Adhesive-to-core voids are more difficult to detect than skin-to-adhesive voids.

Short core can exist if the core edges are cut shorter than the assembly into which it will be placed and bonded. It is detectable by x-ray radiography and is evident as core edges that do not tie into the closure via the foaming adhesive.

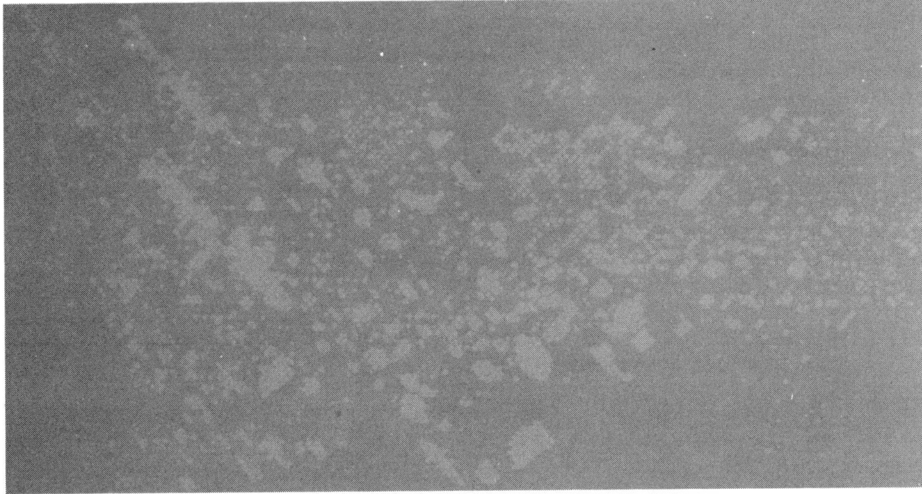

(a)

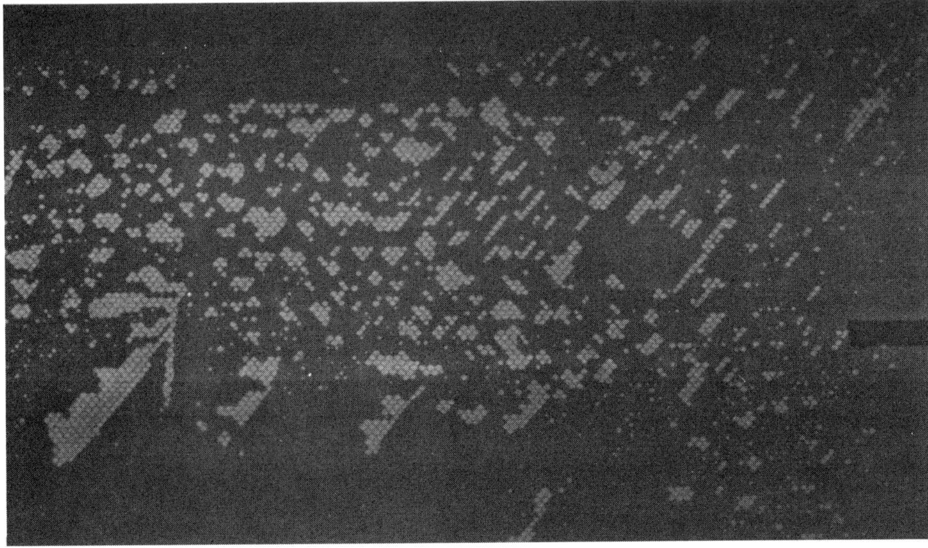

(b)

Fig. 5 Porosity in an adhesive-bonded joint. (a) X-ray radiograph. (b) Neutron radiograph

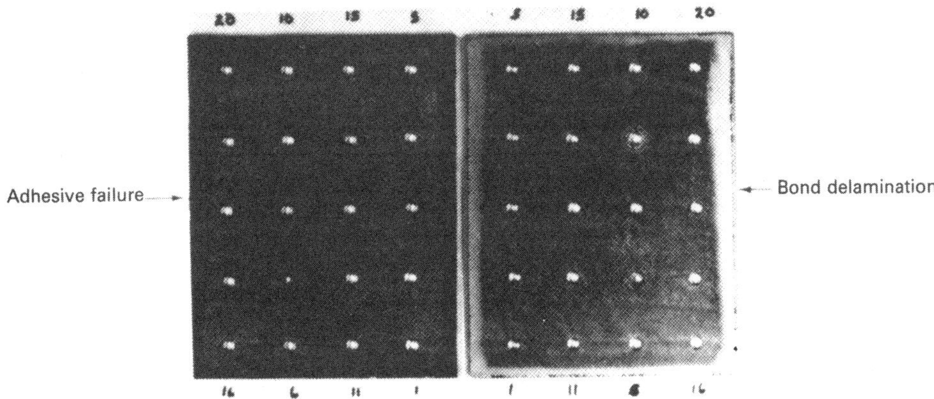

Fig. 6 Examples of burned adhesive

Foreign Objects. Honeycomb assemblies may contain foreign objects as a result of poor fabrication practices. Nuts, rivets, small fasteners, metal chips, and similar items can be left in the honeycomb cells before bonding or as a result of drilling operations for fastener installations near hinge points. These objects are easily detected by x-ray radiography. They are usually not detrimental if they can be potted in place with a room-temperature adhesive.

Repair Defects

All the flaws or defects defined previously for metal-to-metal or honeycomb sandwich assemblies can occur during salvage or repair of these components. Repairs must be of good quality in order to maintain the reliability of the bonded structure during continued service.

In-Service Defects

Most defects or flaws caused during service originate from impact damage, corrosion, and poor fabrication.

Impact Damage. Many bonded assemblies are made from thin materials and are susceptible to damage by impact. Damage can be caused by small arms projectiles, work stands, dropped tools, personnel walking on no-step assemblies, stones or other debris thrown by aircraft wheels, and similar damage. Impact imposes strain on the adhesive, causing it to crack or separate from the adherends. Impact can cause crushed honeycomb core, resulting in loss of strength. The crushed core can resonate during service and slowly degrade the adhesive by fatigue until it debonds from the adherend or until cracks occur adjacent to the skin-to-core fillets. Fortunately, impact damage will generally leave a mark in the surface of the part. These surface marks pinpoint or indicate possible subsurface damage, which can be evaluated by NDT inspections.

Corrosion can be found in all structural concepts. However, a good example of bonding excellence is the honeycomb acoustic panels for the DC-10 and 747 engine inlets. These applications are quite demanding because of the combination of sonic and ambient environment being introduced into the perforated sandwich structure. Stable oxide surface preparation is an essential part of the bond foundation. Improper surface preparation can result in an unstable oxide layer, which may allow the entry of moisture, delamination, and/or crevice corrosion (Fig. 18).

Moisture entry to the core of a honeycomb assembly eventually leads to corrosion of the core. This occurs when moisture moves along the bond line to the individual cells. Moisture moves more rapidly in an assembly if the core is perforated. Fortunately, this moisture problem is detectable by NDT inspection methods. Water is de-

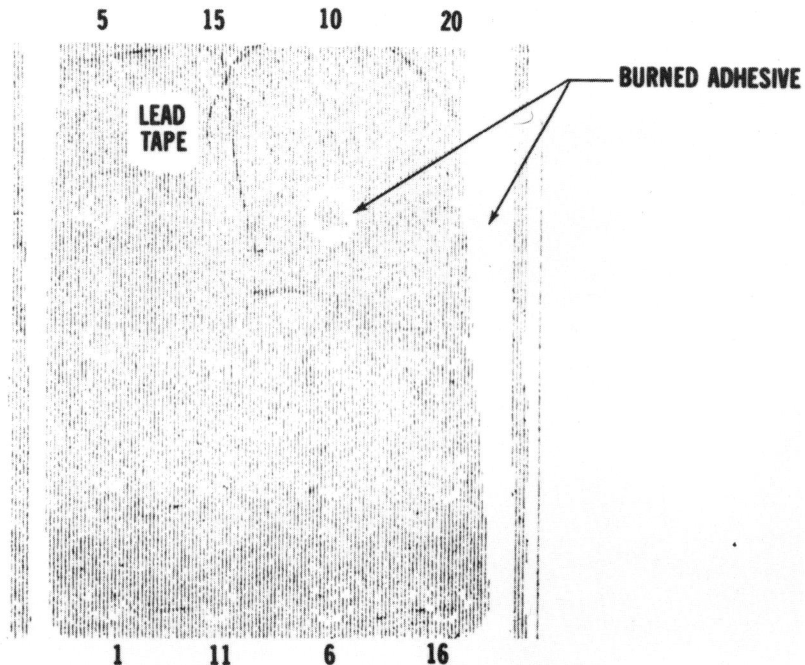

Fig. 7 Ultrasonic C-scan recording of right side of Fig. 6

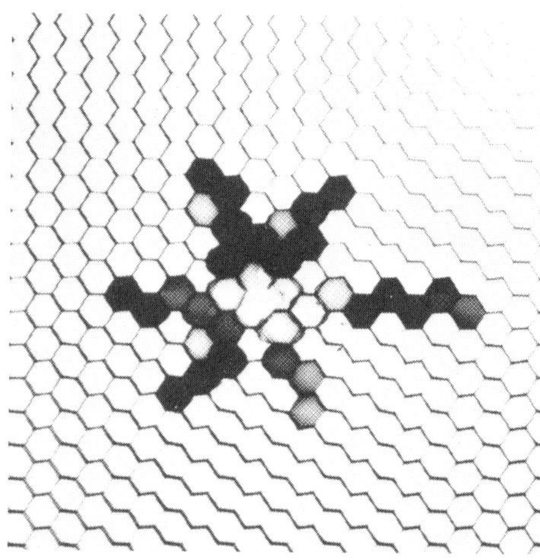

Fig. 9 Positive print from x-ray negative showing water intrusion into honeycomb cells

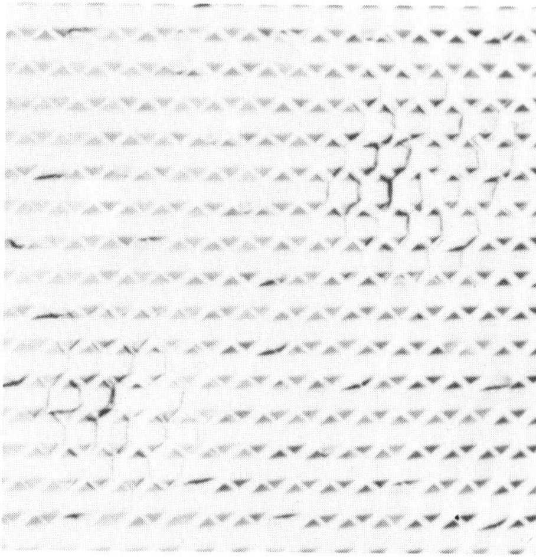

Fig. 10 Positive print from x-ray negative showing crushed honeycomb core

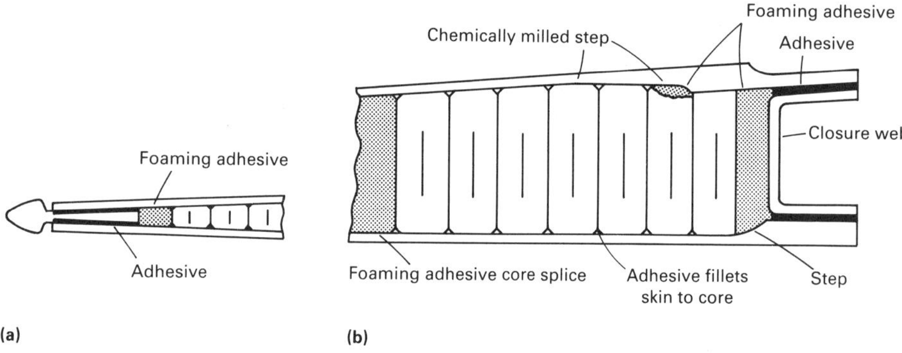

Fig. 8 Typical configuration of bonded honeycomb assembly. (a) Trailing edge. (b) Leading edge

tectable by x-ray radiography or by acoustic emission testing with a hot-air gun or heat lamp to cause boiling or cavitation of the water (Ref 7-9). Core corrosion and subsequent core-to-skin delamination is detectable by a variety of NDT methods, such as x-ray, contact ultrasonic ringing, sonic bond testers, eddy sonic, tap test, and acoustic emission with a heat source.

Figure 19 shows a section from a commercial aircraft wing trailing edge that had moisture entry at the leading and trailing edges, resulting in bond line corrosion. The corrosion in this panel was detected by x-ray radiography, bond testers, tap test, neutron radiography (which looked much like Fig. 19), and acoustic emission using a heat source.

Poor Fabrication (Flaws or Weak Bonds). In-service failures can occur from weak bonds (adhesion failures) caused by poor surface preparation, unstable oxide failure, and corrosion of alcladding and base aluminum alloy (Fig. 18). Many tests show that the alclad on 7000-series aluminum sheet is very susceptible to corrosion attacks in the bond line and must be thoroughly tested before consideration for bonding operations. The 2000-series alclad aluminum alloys are not as susceptible to this condition. Both series of alloys with no alclad surface but with good surface preparation, adhesive, and corrosion-inhibiting primer will have no corrosion problems.

Another manufacturing condition that leads to adhesion failures is caused by improper cleaning of the honeycomb core prior to bonding. This condition results if glycol (which is often used to support the core when it is machined) is not completely removed prior to bonding. Incomplete removal of glycol will cause a weak bond to exist, and in-service stresses may cause a skin-to-core delamination. This condition is controlled by adding a fluorescent tracer to the glycol. After the core is machined and cleaned, it is inspected, using an ultraviolet (black) light, for any residual glycol prior to bonding.

Applications and Limitations of NDT to Bonded Joints

A variety of NDT methods are available for inspection. Only the methods applicable to the inspection of bonded structures will be discussed in this section. All the methods or techniques presented can be used in fabrication inspections, while only a limited number are applicable to on-aircraft or in-service inspection. The following methods have proved to be the most successful in detecting flaws in bonded laminates and honeycomb assemblies.

Visual Inspection

All metal details must be inspected to ensure conformity with design. Before large assemblies are ready for bonding, the details are assembled in the bonding jig as though bonding were to occur. In place of

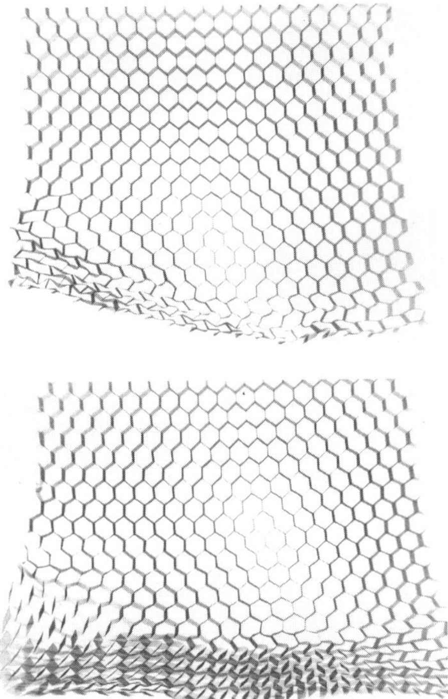

Fig. 11 Two x-ray positive radiographs showing various degrees of condensed core

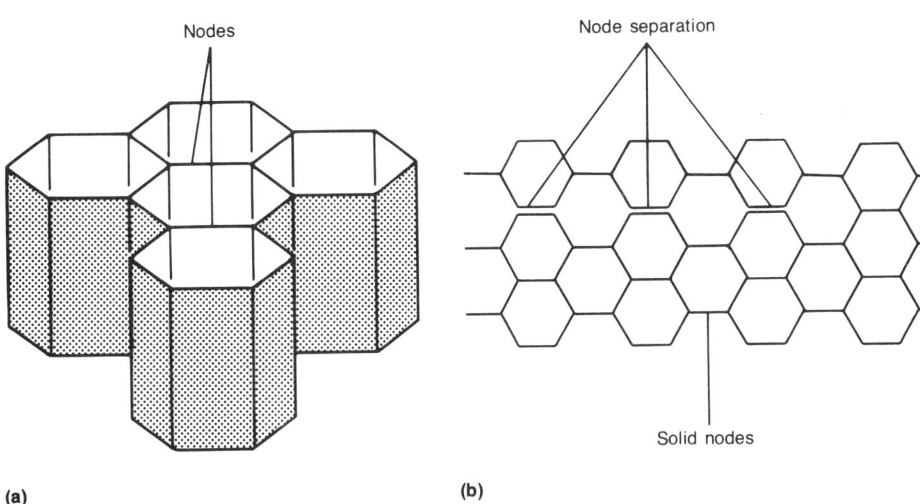

Fig. 12 Examples of honeycomb core separation. (a) Joined and solid nodes. (b) Node separation

the adhesive, a sheet of Verifilm is used. This material has nearly the same flow characteristics as the adhesive, but it is prevented from bonding by release film so that it will not stick to the details. The whole assembly is placed in an autoclave or press just as if it were being bonded. After the heat-pressure cycle, the parts are disassembled, and the Verifilm is inspected visually to determine if the pressure marks are uniform throughout. A uniform marking gives good assurance of proper pressure at the bond lines. All areas showing no pressure must be inspected and parts modified to obtain proper fit and pressure during the cure cycle. A poor showing on the Verifilm is cause to rerun the check.

When the details go through an anodize cycle, a visual check can be made for phosphoric acid anodize but not for other types of anodize. For visual verification of phosphoric anodize, the inspector looks through a polarizing filter at an angle of approximately 5 to 10° from parallel to the surface of the anodized part. The surface of the panel is well lighted by a fluorescent tube, and the inspector rotates the polarizing filter while observing the anodized surface. If the panel has been properly anodized, the inspector, viewing at an angle of 50 to 100° from the panel surface, will see a change in hues equivalent to the colors of the rainbow.

From the opened wedge crack specimen, the inspector looks at the failed surface to determine the type of failure. Areas of poor adhesion occur where the adhesive has separated from the substrate. This condition is manifested by variations in color and texture. A cohesive failure occurs through the adhesive and is uniform in color and texture.

After an assembly is bonded, it is inspected visually for scratches, gouges, dings, dents, or buckles. The adhesive flash and fillets can be inspected for cracks, voids, or unbonds at the edge. Feeler gages can then be used to determine the depth of the unbond at the edge. The in-service visual inspection of bonded joints can reveal cracked metal or adhesive fillets, delamination or debonding due to water intrusion or corrosion, impact or foreign object damage, and blisters, dents, or other mechanical damage.

Ultrasonic Inspection

A number of different types of ultrasonic inspection using pulsed ultrasound waves at 2.25 to 10 MHz can be applied to bonded structures. Following is a brief description of the various ultrasonic techniques being used to inspect bonded structures. Additional information is available in the article "Ultrasonic Inspection" in this Volume.

Contact Pulse Echo. In this technique, the ultrasonic beam is transmitted and received by a single search unit placed on one surface of the part (Fig. 20a). The sound is transmitted through the part, and reflections are obtained from voids at the bond line. If the bond joint is of good quality, the sound will pass through the joint and will be reflected from the opposite face, or back side. Bonded aircraft structure is usually composed of thin skins, which result in multiple back reflections on the CRT screen of the pulser/receiver. The appearance of multiple reflections on the CRT has prompted some inspectors to term this the ringing technique. When a void is present, the reflection pattern changes on the CRT and no ringing is seen. For bonded parts having skins of different thickness, inspection should be conducted from the thin skin side.

Contact through transmission (Fig. 20b) is useful for inspecting flat honeycomb panels and metal-to-metal joints. Special search-unit holding devices have been fabricated so that the test can be performed by one inspector. Such a device is used for inspecting the metal-to-metal closures of a bonded honeycomb panel. Longer and wider-spaced holders have been fabricated from tubing. The holding tool should be custom designed for the assembly being inspected. To perform the test, liquid couplant must be applied to both sides of the assembly.

Immersion Method. For this method, the assembly must be immersed in a tank of water, or water squirters must be used to

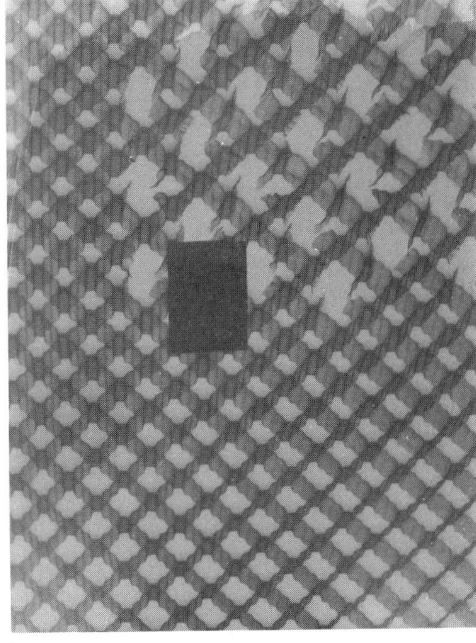

Fig. 13 Radiograph illustration of node separation

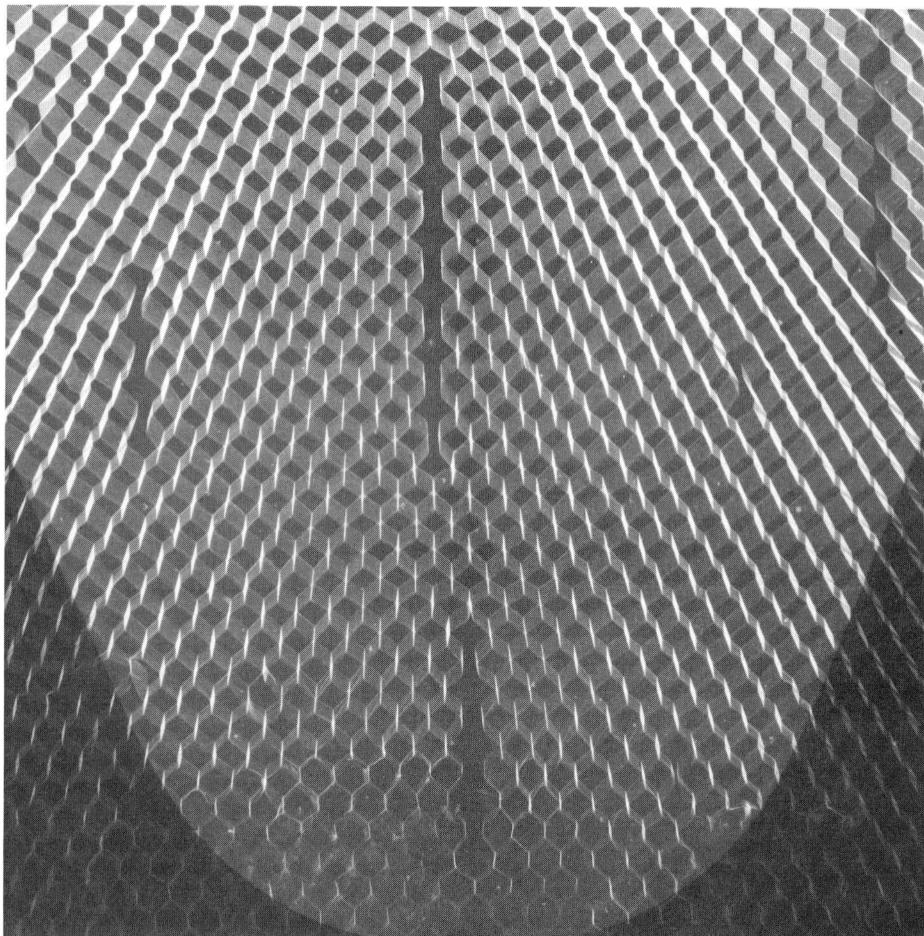

Fig. 14 Radiographic illustration of a blown core

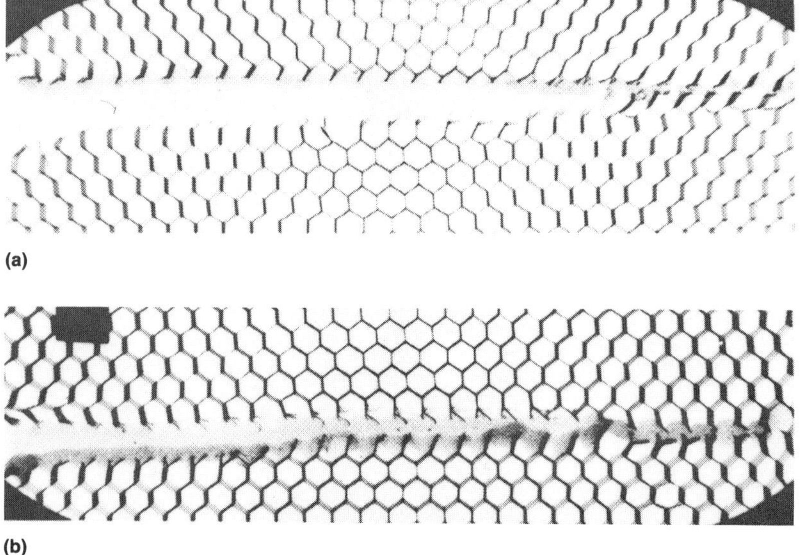

Fig. 15 Positive print from x-ray radiograph of foaming adhesive at core splice. (a) Lack of or unacceptable foaming at core splice. (b) Acceptable foaming at core splice

act as a couplant for the ultrasonic beam. There are three fundamental techniques: pulse echo (Fig. 20c), through transmission (Fig. 20d), and reflector plate (Fig. 20e). Typical CRT displays for bonded and unbonded samples using these three immersion techniques are shown in Fig. 21. These techniques are applicable to bonded laminates and honeycomb structures. The choice of technique is partially based on the thickness and configuration of the bonded assembly as well as physical size.

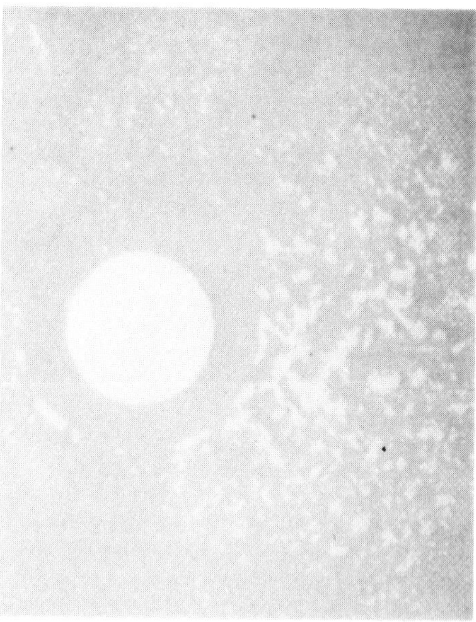

Fig. 16 Positive print from neutron radiograph showing void and porosity in an adhesive bond line. Two 13 mm (½ in.) thick aluminum adherends bonded with EA 9628 adhesive are shown; AF-55 adhesive bond would have yielded similar results with x-ray radiography.

Bonded structures are generally inspected by the immersion method using a C-scan recorder. A C-scan recorder is an electrical device that accepts signals from the pulser-receiver and prints out a plan-view record of the part on a wet or dry facsimile paper recording (Fig. 22). To obtain the recording, the bonded panel is placed under water in a tank, and the ultrasonic search unit is automatically scanned over the part. The ultrasonic signals for bond-unbond conditions are determined from built-in defect reference standards. The signals are displayed on the pulser-receiver CRT screen, and the signal amplitude is used to operate the recording alarm after setting the electronic gate around the signal of interest (Fig. 23). Generally, high-amplitude signals will record, but low-amplitude signals will not record. The recording level is adjustable to select the desired signal amplitude. Figure 23 illustrates a typical ultrasonic C-scan recording system.

Ultrasonic immersion techniques, employing a C-scan recorder, are extensively used by aircraft manufacturers to inspect adhesive-bonded assemblies. The C-scan recordings provide a permanent record with information on the size, orientation, and location of defects in bonded assemblies. The C-scan systems are designed to inspect particular parts and therefore vary considerably in size and configuration. Computer-operated controls have been incorporated into some systems to control the scanning motions of the search units and to change instrument gain at changes of thickness in the assembly.

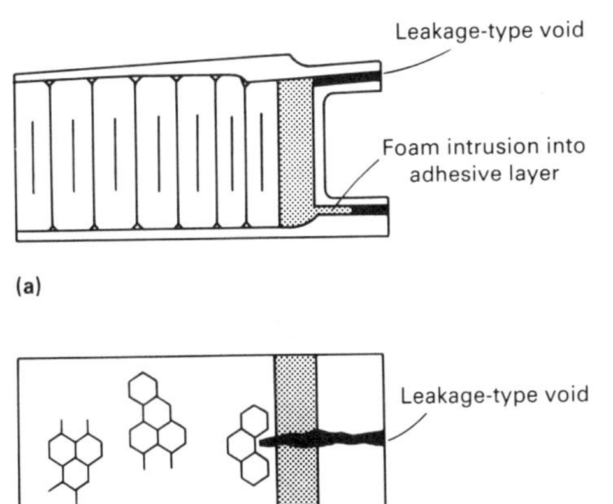

(a)

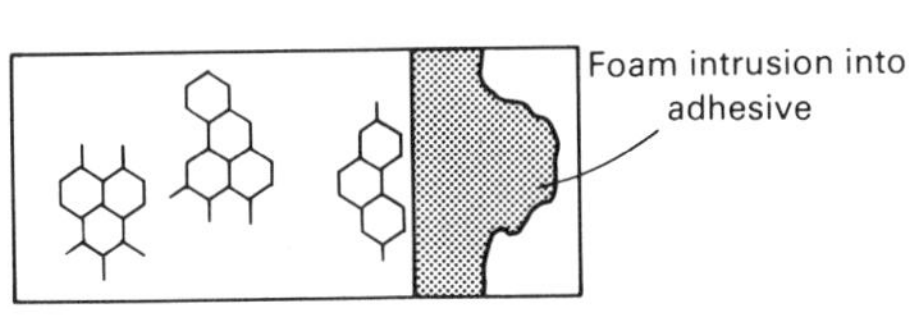

(b)

Foam intrusion into adhesive

(c)

Fig. 17 Schematics of leakage-type void and foam intrusion in metal-to-metal joints. (a) Leakage-type void, and foam intrusion into adhesive layer caused by excessive gap between extrusion and skins. (b) Leakage-type void. (c) Foam intrusion into adhesive

The through transmission and reflector plate techniques are easier to perform than the pulse-echo technique and are useful for producing C-scan recordings of test specimens and flat laminates. Special equipment is required for large panels, squirters for honeycomb panels that cannot be immersed, and contour followers for contoured parts.

Ultrasonic Inspection Limitations. The ultrasonic method suffers from destructive wave interference thickness. Adhesive thickness interference effects are most notable at the pinch-off zone near the edges of

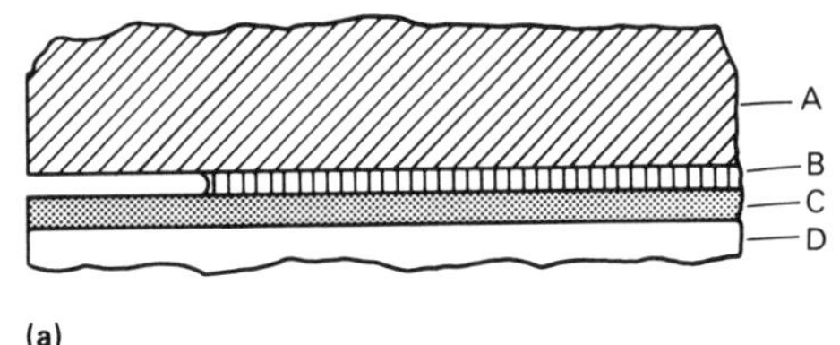

(a)

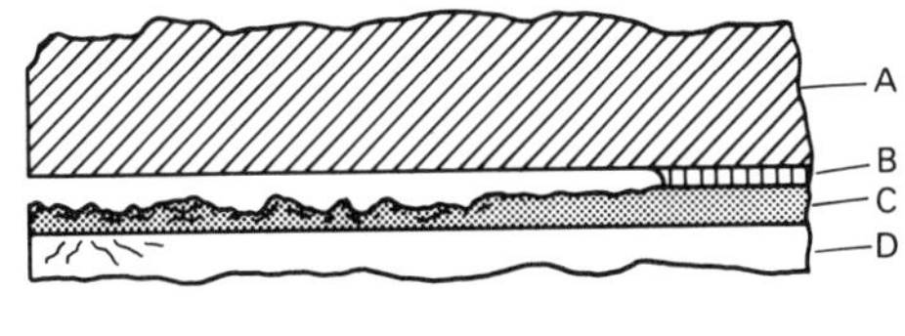

(b)

Fig. 18 Schematics illustrating the causes of adhesive delamination for a metal adherend. (a) Results of moisture entry in the unstable oxide. (b) Corrosion of cladding and base aluminum. A, adhesive primer system; B, oxide; C, alcladding; D, base aluminum

laminates bonded with adhesive using a mat (nonwoven) carrier. Metal thickness interference effects (Fig. 24) are usually obtained from any laminate having a bonded tapered doubler. The problem with the interference effects is that they make the parts appear unbonded. The phenomenon is complex and will require further study before its effects can be eliminated.

Ultrasonic bond test methods or neutron radiography is used to determine the acceptability of bonded assemblies yielding C-scan results showing interference effects (see the section "Ultrasonic Bond Testers" in this article). Figure 25 shows a through transmission C-scan of a bonded honeycomb assembly. The lead tape (generally used for radiography) attenuates the ultrasonic beam and is therefore used to simulate voids and also as an index or orientation marker to relate the location of flaws in the part to the C-scan recording. In the contact pulse-echo ultrasonic inspection of bonded laminates, inspection from the thin adherend side produces the best results.

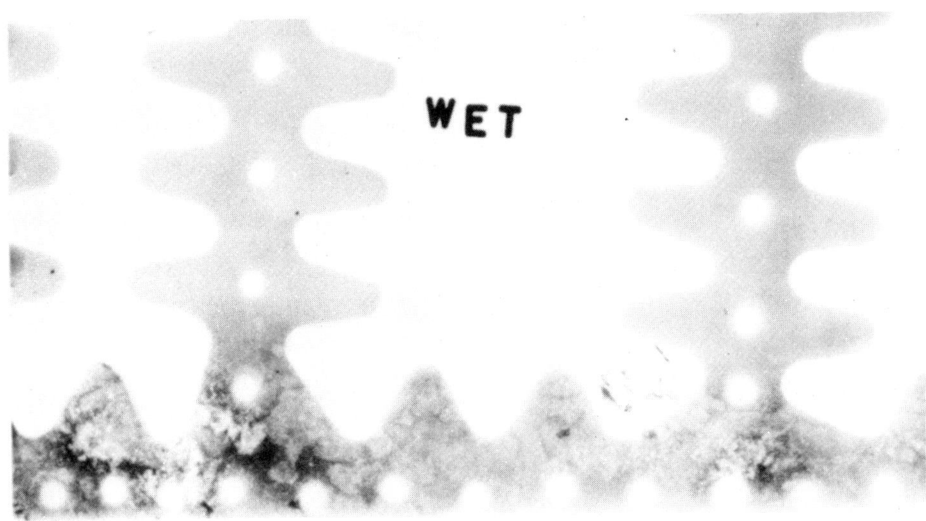

Fig. 19 Positive print from x-ray radiograph showing interface corrosion of an adhesive-bonded aluminum laminate

Ultrasonic Bond Testers

A wide variety of ultrasonic bond testers have been developed over the past 20 years. As a consequence, this mode of bond inspection probably requires the most study because of the number of instruments available and the claims of the manufacturers. Various independent studies have been conducted in an attempt to clarify the situation (Ref 2, 3, 10–20). The results of these studies revealed that all the instruments are capable, in varying degrees, of determining the quality of the bond. None of the instruments is capable of establishing the adhesive quality of a bond that is defective as a result of:

- Poor surface preparation of the substrate
- Insufficient cure temperature
- Contamination of the adhesive or substrate prior to bonding

The conclusions indicate that ultrasonic bond test instruments are the most reliable and sensitive for detecting voids, porosity, thick adhesive, and corrosion at the bond line, and that they can be used to inspect metal-to-metal and honeycomb structures.

Instruments such as the Coinda Scope, Stub Meter (Ref 13, 14), Sonic Resonator (Ref 11), and Arvin Acoustic-Impact Tester (Ref 15) have not found general acceptance and are no longer on the market. Following is a short description of the ultrasonic bond testers currently being used to inspect adhesive-bonded assemblies.

The Fokker Bond Tester (Fig. 26) is based on the analysis of the dynamic characteristics of the mass-spring and dashpot system formed by the combination of the bonded assembly with a piezoelectric transducer having known mass and resonance characteristics. Changes in the viscoelastic properties of the adhesive layer are detected as variations in the resonance frequency and impedance of the system. The calibrated body acts as a transducer that can be driven at different frequencies. The dimensions of the transducer are chosen in relation to the total thickness of the metal sheet to be tested and to the required mode of resonance. The response of the total system, as shown by the impedance curve over the swept frequency band, is displayed on a CRT display (A scale) (Fig. 27), and the peak-to-peak amplitude of the curve is shown by a microammeter (B scale).

For inspecting metal-to-metal bonded joints, the probe containing the calibrated body is placed on top of a piece of sheet material with the same thickness as the upper sheet material of the bonded laminate. The central frequency of the oscillator is selected such that the lowest point of the

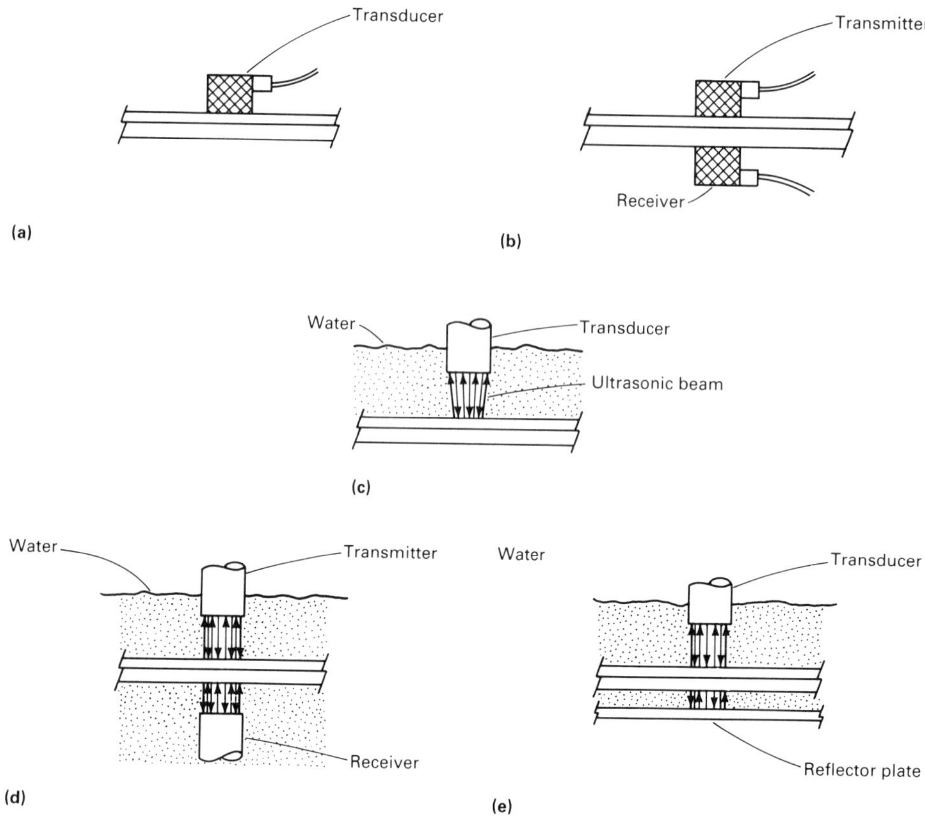

Fig. 20 Ultrasonic inspection techniques. (a) Contact pulse echo with a search unit combining a transmitter and receiver. (b) Contact through transmission. Transmitting search unit on top and receiving search unit on bottom. (c) Immersion pulse echo with search unit (transmitter/receiver) and part being inspected under water. (d) Immersion through transmission with both search units (transmitter and receiver) and part under water. (e) Immersion reflector plate. Same as (c) but search unit requires a reflector plate below the part being inspected

impedance curve is in the center of the A scale. Simultaneously, the B scale is adjusted to 100 (Fig. 27). Calibration of the instrument on a nonbonded sheet ensures that, in all cases of a complete void, the peak position will return to the center of the A scale and that the B scale reading will be 100 (Fig. 28). The next calibration places the probe on a piece of metal sheet equivalent in thickness to all the metal sheets in the subject bonded laminate. The peak obtained in this test is the resonance frequency to be expected of an ideally bonded laminate (Fig. 27 and 28).

The quality of cohesion from any test can be accurately determined by comparing the instrument reading with established correlation curves. In practice, the acceptance limits are based on the load or stress requirements of the adhesive for each joint. The accuracy of the prediction of quality depends primarily on knowing the manufacturing variables and the accuracy of the nondestructive and destructive tests conducted in accordance with MIL-STD-860 (Ref 21). The destructive test for metal-to-metal joints uses the lap shear specimen and, for honeycomb panels, uses the tension specimen for bond strength correlations. Comparative tests by several investigators indicate the Fokker bond tester to be more reliable in quantitatively measuring bond strength as related to voids and porosity in the joint (Ref 2, 3, 5, 10, 16, 18, 22, 23). For the inspection of honeycomb, calibration is accomplished in the same manner except that the micrometer (B scale) on the instrument is used. The degree of quality is reflected on the B scale. Low-quality bonds will give a high B reading, and good-quality bonds will give a low B reading.

The Fokker bond tester has been successfully applied to a wide variety of bonded sandwich assemblies and overlap-type joints of various adherends, adhesive materials, and configurations. The method has proved suitable for joints having reasonably rigid adherends, including metallic and nonmetallic materials. Highly elastomeric or porous adherends attenuate ultrasonic response. The method is most sensitive to the properties of the adhesive and is particularly sensitive to voids, porosity, and incomplete wetting (unbond). Fokker tests readily detect:

- Voids in either adhesive or nonmetallic adherend materials
- Cracks and delaminations in adherends
- Unbonded, flawed, ruptured, unspliced, or crushed honeycomb core

Data indicate the test to be capable of measuring bond degradation caused by such factors as moisture, salt spray, corrosion, heat aging, weathering, and fatigue.

NDT-210 Bond Tester (Fig. 29). Sound waves from a resonant transducer are transmitted into, and received from, the bonded structure. Unbonds or voids in the structure alter the sound beam characteristics, which in turn affect the electroacoustic behavior (impedance) of the transducer. A frequency range extending from less than 100 Hz to more than 6 MHz is provided for maximum performance. The unit automatically adjusts to the resonant frequency of the probe. A precision automatic gain control oscillator maintains a constant voltage at all frequencies, eliminating the need for manual adjustment. Test response is displayed on a 114 mm (4½ in.) wide meter. The operator can set an adjustable alarm to trigger when the response exceeds a preset level.

Metallic, nonmetallic, or a combination of joints can be inspected for voids and unbonds. The joints can be adhesively bonded, brazed, or diffusion bonded.

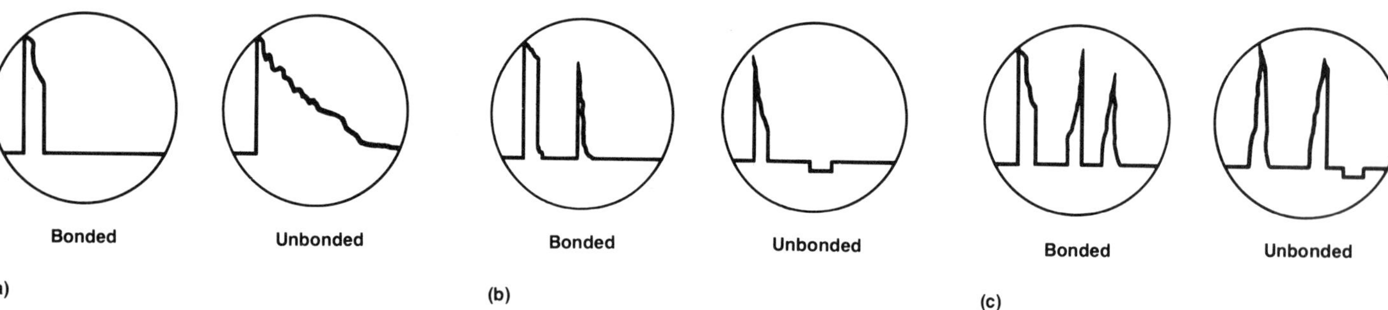

Fig. 21 Typical CRT C-scan displays obtained for both bonded and unbonded structures using three pulsed ultrasound techniques. (a) Pulse echo. (b) Through transmission. (c) Reflector plate

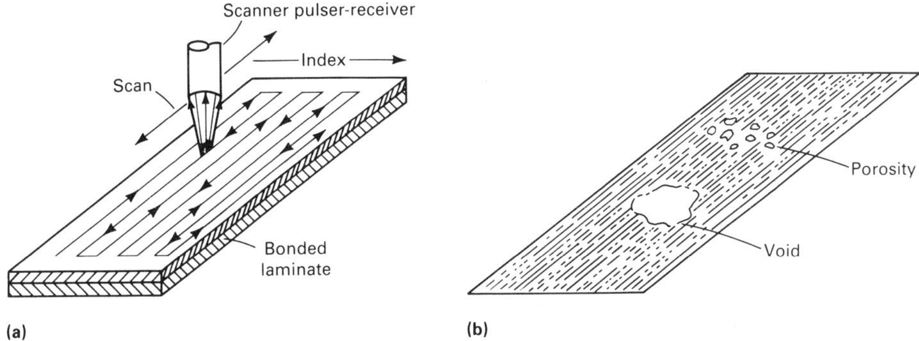

Fig. 22 Schematics of immersion ultrasonic C-scan (a) scanner motion and (b) plan-view record of C-scan recording on facsimile paper

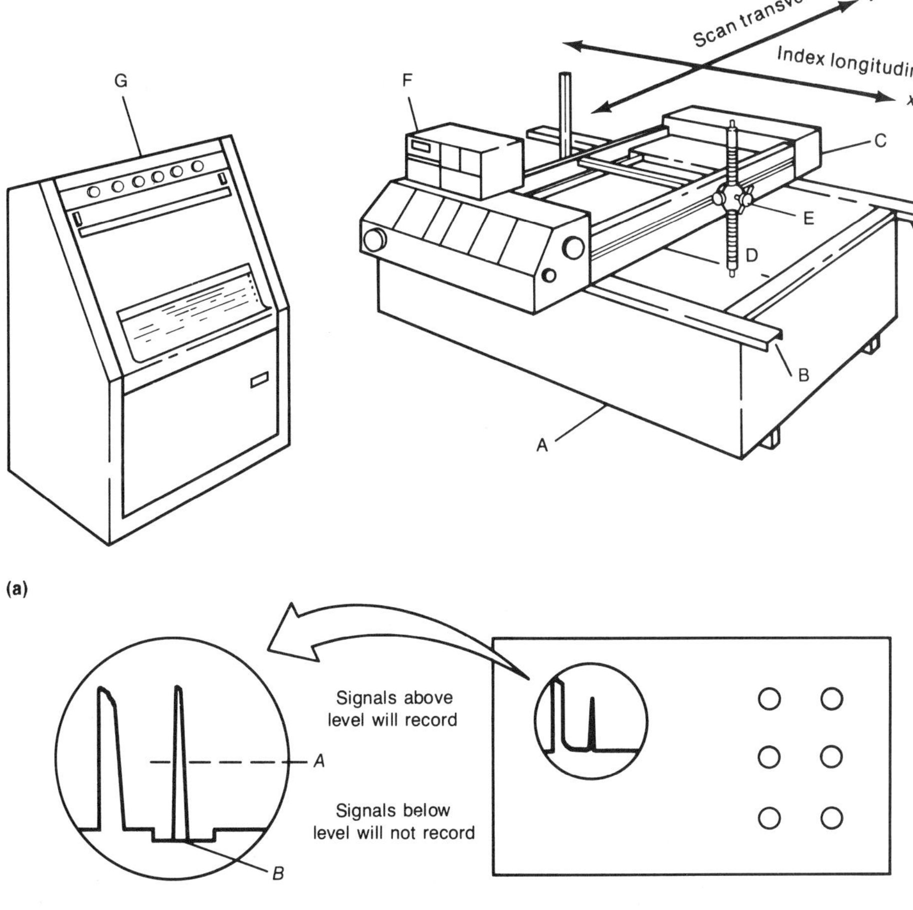

Fig. 23 Ultrasonic C-scan immersion system. (a) The system consists of a large water tank (A) with side rails (B), which support the movable bridge (C). The search unit (D) is held at a proper distance above the part by the scanner arm (E), moves across the part, and is then indexed along the longitudinal axis. The visual pulser-receiver unit (F) is mounted on the bridge. A permanent record is made by the C-scan recorder (G). (b) Close-up of pulser-receiver on the right and a close-up of the CRT presentation on the left. A recording level (*A*) is set on the instrument. All signals above will record on the C-scan machine, and signals below will not show. Item *B* is an automatic recording gate associated with the mechanics of record making.

The Shurtronics Mark I harmonic bond tester (Fig. 30), now identified as the Advanced Bond Evaluator, is a portable, low-frequency, eddy sonic instrument that uses a single transducer for the inspection of thin metal laminations and metal-to-honeycomb bonded structures. The instrument utilizes a coil that electromagnetically vibrates the metal face sheet. This coil induces a pulsed eddy current flow in a conductive material by an oscillating electromagnetic field in the probe. The alternating eddy current flow produces an accompanying alternating magnetic field in the part. The attraction of the magnetic fields causes acoustomechanical vibrations in the part. When a structural variation is encountered, the change in acoustic response is detected by a broadband receiving microphone located inside the eddy current probe.

An ultrasonic oscillator generates a 14- to 15-kHz electric signal in the probe coil, creating an oscillating electromagnetic field. The resulting acoustomechanical vibration in the testpiece is detected by a microphone with a bandwidth of 28 to 30 kHz. The microphone signal is filtered, amplified, and displayed on a microammeter. The instrument is calibrated to read just above zero for a condition of good bond and to read full scale for unbonds over 13 mm (½ in.) in diameter or larger. This method does not require a liquid couplant on the part surface. It is useful for the quick detection of unbonds in large-area bonded metal laminates, metal-to-nonmetal laminates, or honeycomb structures. It is highly effective for detecting skin-to-core unbonds or voids in acoustic-honeycomb panels utilizing perforated facing sheets. The sensitivity of the instrument decreases rapidly for skin thickness over 2.0 mm (0.080 in.) and near the edge of the part.

The Advanced Bond Evaluator (ABE) bond tester (Fig. 31) has the following improvements:

- *1% alarm accuracy*: Vernier control ensures test repeatibility with zero hysteresis
- *Phase comparator*: Detects defects in nonmetallic composite materials and provides additional information on metallic structures
- *Portable operation*: Self-contained batteries and built-in recharger
- *Modular construction*: Plug-in circuit boards for fast field servicing and maintenance

The Sondicator is a pulsed transmit/receive ultrasonic portable instrument that is capable of operating in a very low (25 to 50 kHz) acoustic frequency range. The instrument operates within this range at a selected single frequency obtained by manual tuning for best instrument performance. The model S-2 contains two meters and associated electronic circuitry, along with a test probe containing two Teflon-tipped transducers mounted approximately 19 mm (¾ in.) apart. One transducer imparts vibration at 25 kHz into the surface. Vibrations travel laterally through the material to the other transducer. The second transducer detects the amplitude and phase relationship of the vibrating surface, directing the associated signals to the two meters of the test equipment. If the probe encounters a delaminated or unbounded area, it will in effect be introducing vibration into an area that is thinner than the bonded area. The amplitude of vibration will increase and the phase of vibration will shift as the thinner section vibrates more vigorously. The needles on

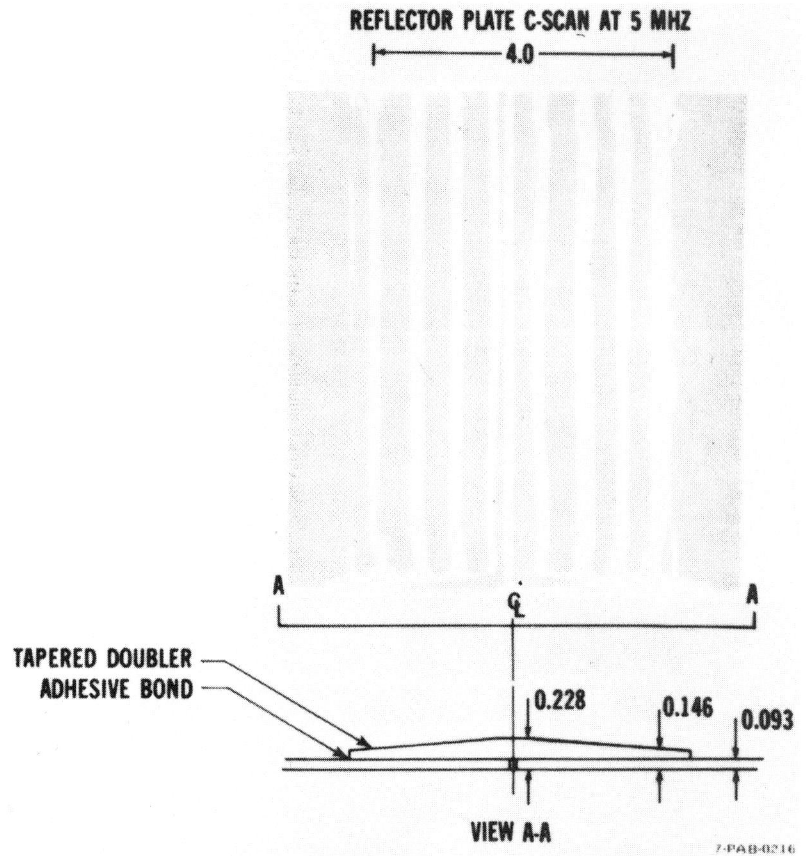

Fig. 24 Reflector plate C-scan at 5 MHz of ultrasonic destructive interference effects due to varying adherend thickness

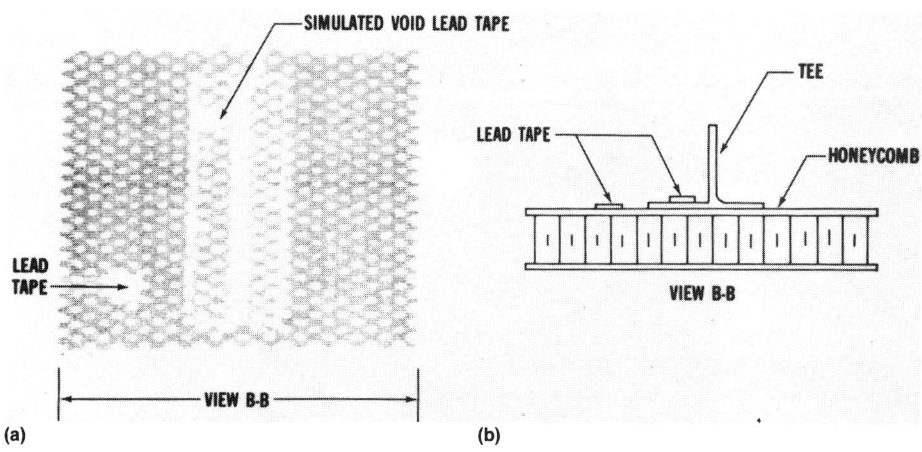

Fig. 25 Through transmission C-scan of adhesive-bonded honeycomb. (a) C-scan of section B–B at 5 MHz using lead tape to simulate void. (b) Cross-sectional view B–B of bonded honeycomb showing position of lead tape

the two meters will move toward each other.

The instrument is primarily used to detect delaminations of the bond or laminar-type voids in metal and nonmetal structures. The inspection can be performed from one face of the structure or by through transmission. It requires no liquid couplant. Multiple bond lines and part edges will reduce the sensitivity of the instrument. Because of the directionality of the sound from the transmitter to the receiver, the part must be scanned uniformly. The unit is capable of detecting defects 25 mm (1 in.) in diameter and larger in most materials. Metallic and nonmetallic materials can be inspected without changing probes. Audible and visible alarms are activated by received signals. The instrument can detect internal delaminations and voids in bonded wood, metal, plastic, hard rubber, honeycomb structures, Styrofoam, and composites. It can measure

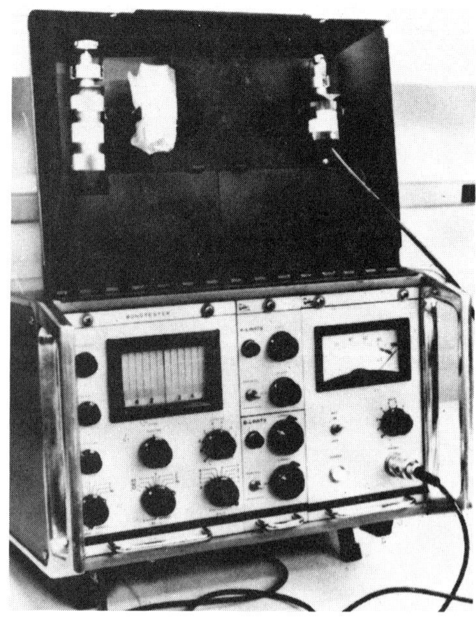

Fig. 26 Fokker bond tester

skin or face sheet thickness in composite structures as a change in amplitude or phase (Fig. 32).

Bondascope 2100. The Bondascope (Fig. 33) is an advanced microprocessor-based device that operates on the principle of ultrasonic impedance analysis. This technique allows the total ultrasonic impedance vector for the material to be monitored as a flying dot on a scope display. With the center of the scope screen as the origin (reference), impedance phase changes are displayed circumferentially, while impedance amplitude changes are displayed radially. Thus, the position of the dot on the scope immediately reveals the phase and amplitude of the impedance of the material (Fig. 33a). Therefore, defects at different bond lines each possess a characteristic dot position on the scope display.

With the addition of a 640-KB random access memory (RAM) portable computer having a 229 mm (9 in.) built-in color monitor, a high-speed plug-in data acquisition card (DAC), a software package, and a 1.4 to 1.8 kg (3 to 4 lb) scanner incorporating a ball-bearing gimbal probe holder, the Bondascope 2100 becomes an integral part of the PortaScan (portable ultrasonic color scan imaging) system that can be used for both C-scan imaging and set-up flying dot (impedance plane) imaging (Fig. 33b).

Figure 34 illustrates the typical Bondascope response to unbonds at different bond lines (depths) in a multilayered adhesive-bonded laminate. The instrument was calibrated so that the dot was in the center of the screen (origin) when the probe was placed on a well-bonded section of the laminate. The numbered dots represent the signals obtained when the probe was placed over regions having unbonds located at dif-

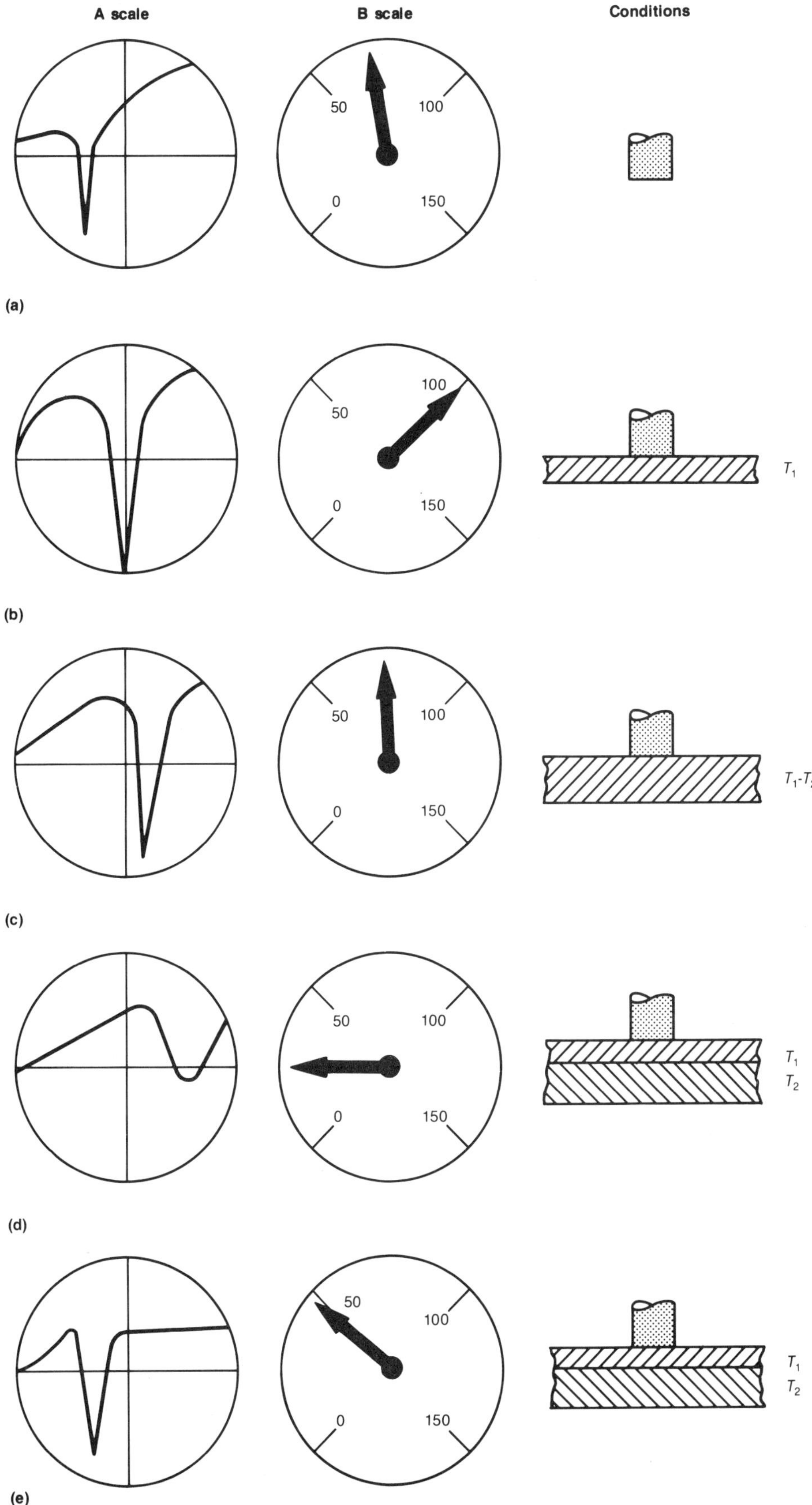

Fig. 27 Several typical Fokker bond tester readings for specific bond conditions. (a) Probe held in air. (b) Probe is on top of the upper adherend (no bonds) and is a calibration for unbond condition. (c) Probe placed on a single piece of metal as thick as combined parts being bonded and calibrated for ultimate quality. (d) Probe placed on bonded joint with good-quality bond (no voids); high-strength bond. (e) Probe placed on bonded joint with porosity; low-strength bond

ferent respective bond line depths in the laminate. The dot labeled "air" is the signature obtained when the probe was off the sample. Thus, improper ultrasonic coupling into the material is quickly recognized.

A meter on the Bondascope, by means of selectable push buttons, can display the signal amplitude, phase, or the vertical resolved component of the amplitude. Through the use of a phase-rotator control, the signal response can be positioned so that nonrelevant signals are suppressed from the meter/alarm readings.

A keyboard push-button matrix allows the operator to digitally program calibration-sample reference dots on the scope display to aid in interpreting signals obtained during actual testing. This type of storage display not only facilitates operation and interpretation but also eliminates the confusing permanent streaks that would occur if a conventional storage scope were used to display the flying dot.

The Bondascope is suitable for inspecting metallic and nonmetallic bonded structures (multilayered laminates or honeycomb). Graphite-epoxy and other fiber-matrix composites can also be tested with this device.

The NovaScope is a sophisticated high-resolution, pulsed ultrasonic thickness gage with a digital readout and scope display for qualifying the echo pattern. It is primarily intended for either manual contact probe or focused immersion (squiter) applications for which the conventional digital-only readout gages are not suitable. This is the case for structures having complex, difficult-to-test shapes or whenever the product is in motion. The instrument has a variety of controls for optimizing the test response to the desired ultrasonic echo periods.

Figure 35 shows the resolving power of the instrument for gaging thin materials. In this case, the material is a steel sheet only 0.15 mm (0.006 in.) thick. The upper trace is the A-scan echo pattern, while the lower trace shows the thickness gate adjusted between two successive echoes from the thickness of 0.15 mm (0.006 in.). This instrument is useful for testing thin composite laminates for delaminations, for detecting corrosion in aircraft skins, and for bond line thickness gaging.

Ultrasonic and Bond Tester Method Sensitivity. The defect detection sensitivity of ultrasonic techniques with respect to defect size and location is highly dependent on the changing conditions of part complexity, number of bond lines, operator experience, and so on. Most instruments detect major defects most of the time, but in some cases special techniques and skills are needed to conduct a reliable inspection. Whenever possible, bonded structures should be inspected from opposite surfaces to detect small defects. In the ultrasonic bond test methods, different-size search units or probes are available. The smallest flaw that

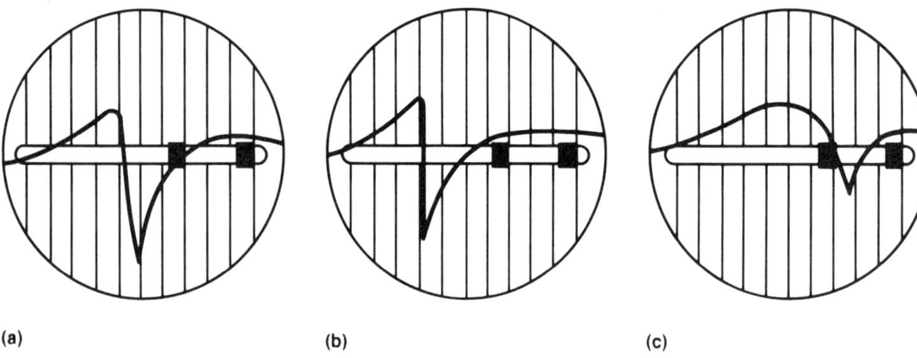

Fig. 28 Typical displays of Fokker bond tester A scale for various qualities of adhesive-bonded joints. (a) Central frequency, no strength. (b) Higher frequency, low strength. (c) Lower frequency, high strength

Fig. 29 NDT-210 bond tester

can be readily detected is of the order of one-half the diameter of the search unit or probe element.

Test Frequency. The higher the test frequency, the smaller the flaw that can be detected. Ultrasonic testing at 5 or 2.25 MHz will detect smaller flaws than can be detected with the Sondicator (25 to 50 kHz) and the harmonic bond tester (15 kHz). With multiple bond lines in a structure, smaller flaws become progressively more difficult to detect from the surface through succeeding bond lines. This does not apply to ultrasonic through transmission or reflector plate testing, in which the small-diameter sound beam passes completely through the part. Figure 36 shows the relative defect sensitivity of ultrasonic and bond tester techniques.

X-Ray Radiography

Radiography is a very effective method of inspection that allows a view into the interior of bonded honeycomb structures. The radiographic technique provides the advantage of a permanent film record. On the other hand, it is relatively expensive, and special precautions must be taken because of the potential radiation hazard. With the radiographic method, inspection must be conducted by trained personnel. This method utilizes a source of x-rays to detect discontinuities or defects through differential densities or x-ray absorption in the material. Variations in density over the part are recorded as various degrees of exposure on the film. Because the method records changes in total density through the thickness, it is not preferred for detecting defects (such as delaminations) that are in a plane normal to the x-ray beam.

Some adhesives (such as AF-55 and FM-400) are x-ray opaque, allowing voids and porosity to be detected in metal-to-metal bond joints (Fig. 5a). This is extremely advantageous, especially for complex-geometry joints, which are difficult to inspect during fabrication. The x-ray inspection should be performed at low kilovoltage (25 to 75 kV) for maximum contrast. A beryllium-window x-ray tube should always be used when radiographing adhesive-bonded structures. To hasten exposures, medium-speed, fine-grain film should be used. Selection of the film cassette should be given special consideration because some cassette materials produce an image on the film at low kilovoltages.

Neutron Radiography

Neutron radiography is very similar to x-ray or γ-ray radiography in that both depend on variations in attenuation to

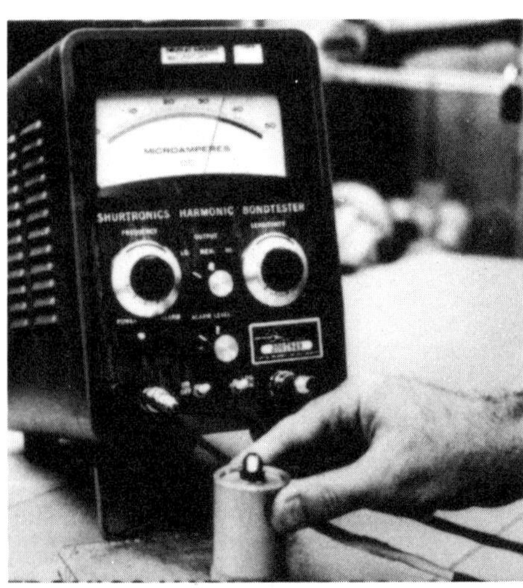

Fig. 30 Shurtronics harmonic bond tester

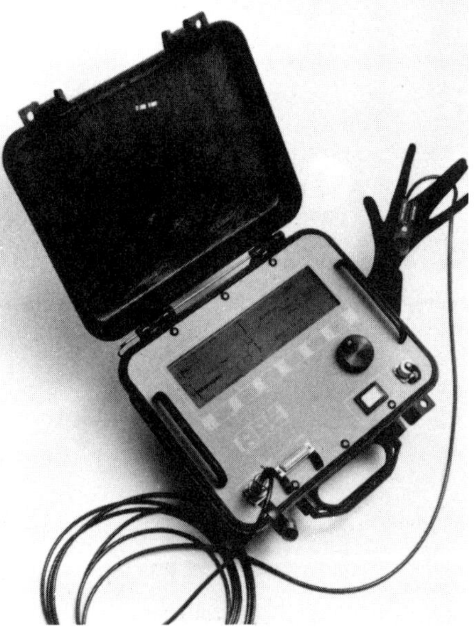

Fig. 31 Advanced Bond Evaluator bond tester. Courtesy of M. Gehlen, UniWest/Shurtronics Corporation

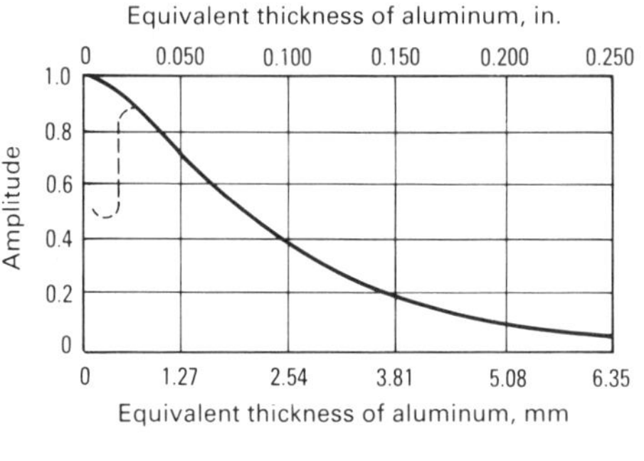

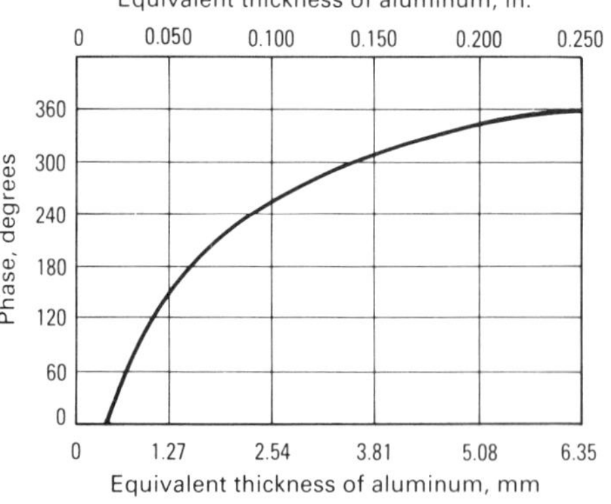

Fig. 32 Variation of amplitude (a) and phase (b) as detected with a Sondicator

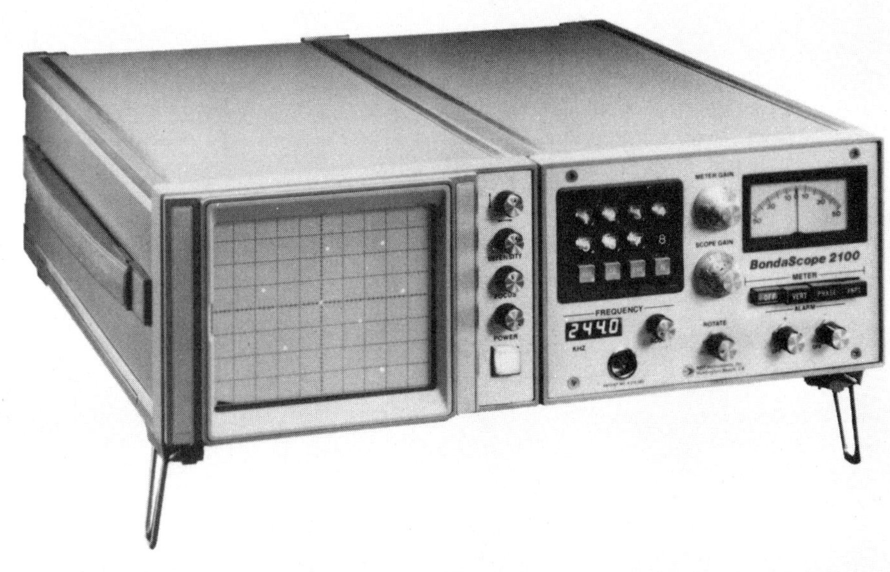

(a)

(b)

Fig. 33 Bondascope 2100. (a) Front view of unit showing CRT readout of bond line depths. (b) Unit used as a component in a PortaScan bond testing system. Courtesy of R.J. Botsco, NDT Instruments

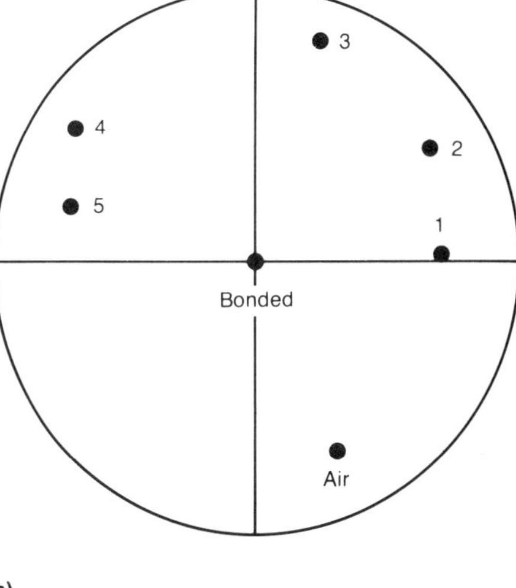

(a)

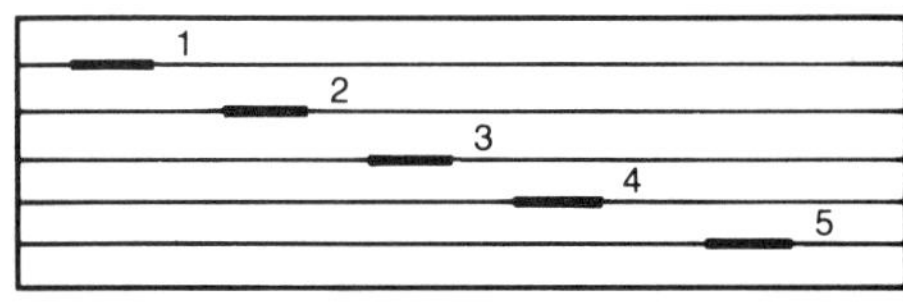

(b)

Fig. 34 Bondascope ultrasonic impedance plane presentation for multilayer laminar. (a) Orientation of Bondascope display of the unbonds shown in (b). (b) Multilayer bonded laminate with unbonds located at 1, 2, and so on

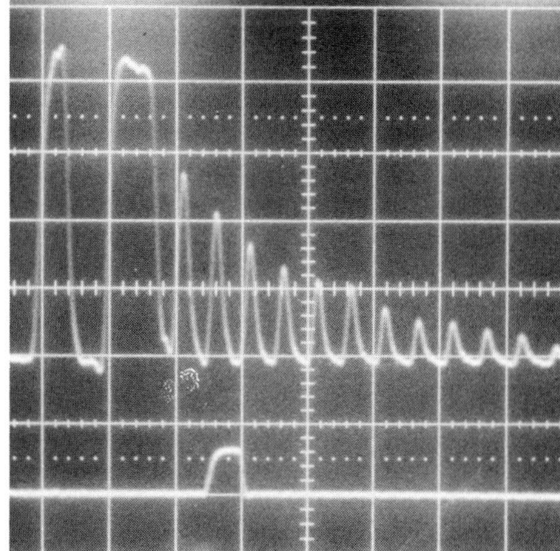

Fig. 35 Novascope CRT display of a 0.15 mm (0.006 in.) steel sheet. Upper trace: A-scan echo pattern. Lower trace: Thickness gate reading obtained from two successive echoes. Courtesy of R.T. Anderson, NDT Instruments

achieve object contrast on the resultant radiograph. However, significant differences exist in the effectiveness of the two methods, especially when certain combinations of elements are examined. The mass absorption coefficients of the different elements for x-rays assume a near-linear increase with atomic number, while the coefficients for thermal neutrons show no such proportionality. The attenuation of x-rays is largely determined by the density of the material being examined. Thus, thicker and/or denser materials appear more opaque. The absorption for thermal neutrons is a function of both the scattering and capture probabilities for each element. The density or thickness of a material or component is less important in determining whether it will be transparent or opaque to the passage of neutrons. For example, x-rays will not pass through lead easily, yet they will readily penetrate hydrocarbons. In contrast, neutrons will penetrate lead and are readily absorbed by the hydrogen atoms in an adhesive or hydrocarbon material.

Neutron Sources. Neutrons are produced from accelerator, radioisotope, or reactor sources. Neutrons, like x-rays, can be produced over an enormous energy range with large differences in attenuation at the various energy levels. The major efforts in neutron radiography have been performed using thermal neutrons because the best detectors exist in this energy regime. Neutron sources generally produce γ-rays of moderate intensity, so that a neutron detector sensitive to γ-ray radiation has a γ-ray image superimposed on the neutron image.

Direct Versus Transfer Method. The most widely used imaging method is conventional x-ray film with converter screens. The rate of radioactive emission of the converter screen divides the photographic imaging into prompt emission or delayed emission. The prompt-emission converters (gadolinium or rhodium) require that the film be present during neutron exposure. This process is referred to as the direct

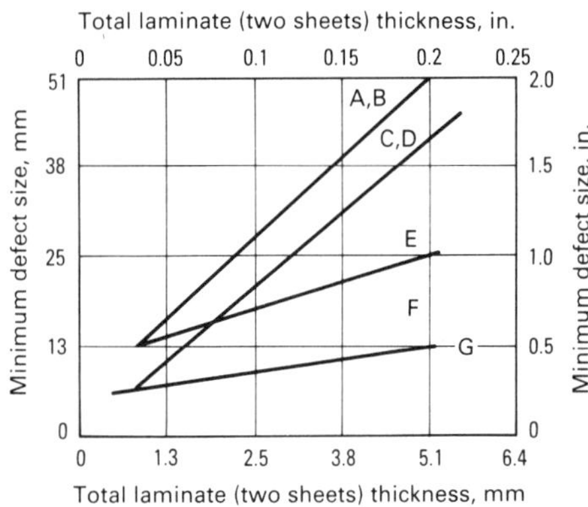

Fig. 36 Relative sensitivity of instruments used to inspect bonded laminates of increasing total thicknesses. A, Sondicator, contact method; B, Shurtronics harmonic bond tester; C, Fokker bond tester; D, NDT-210 bond tester; E, Sondicator, through transmission method; F, ultrasonic pulse-echo; G, ultrasonic through transmission

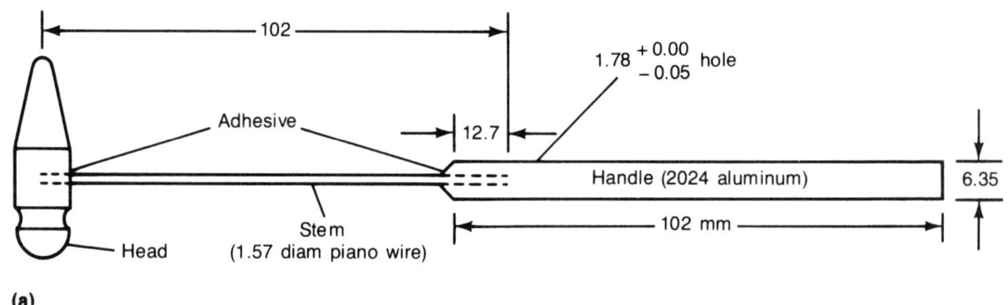

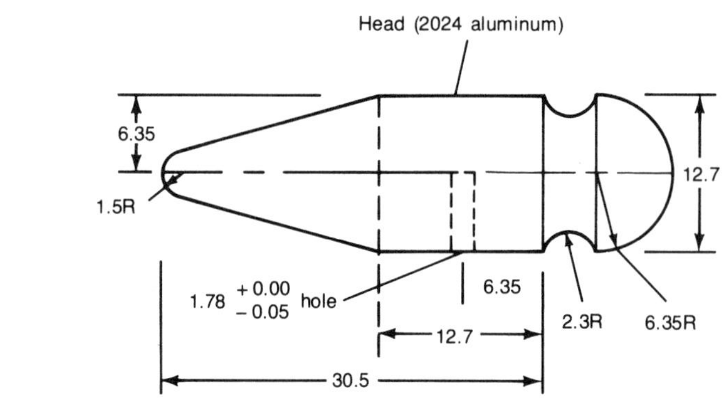

Fig. 37 Construction details for an inspection tap hammer. (a) Complete assembly. (b) Tap hammer head. Liquid/paste adhesive can be used if desired. The hole in the handle/head can be reduced to provide an interference fit and to preclude the need for the adhesive. Dimensions given in millimeters

method. The delayed-emission converters (indium or dysprosium) allow for activation of the converter and transfer of the induced image to the film after neutron exposure. This technique is referred to as the transfer method. In the original film, the neutron opaque areas appear lighter than the surrounding material. Image contrast can be increased by making a contact positive print from the film negative. When this is done, the neutron opaque area will be darker than the surrounding material.

Evaluation of Adhesive-Bonded Structures. Hydrogenous (adhesive) materials inspected by neutron radiography can be delineated from other elements in many cases where x-ray radiography is inadequate. However, the neutron radiography inspection method does not appear to be cost effective for the routine inspection of adhesive-bonded structures. It is extremely useful for evaluating the quality of built-in defect reference standards or for failure analysis. If the adhesive is not x-ray opaque, neutron radiography can be used to detect voids and porosity. The hydrogen atoms in the adhesive absorb thermal neutrons, rendering it opaque. Detailed information on this inspection method is available in the article "Neutron Radiography" in this Volume.

Tap Test

Perhaps the simplest inspection method for ensuring that a bond exists between the honeycomb and the facing sheet is that of coin tapping. An unbond is readily apparent by a change in the tone or frequency of sound produced when an adhesive-bonded structure is tapped with a coin or rod as compared to the sound produced for a bonded area. Coins such as silver dollars or half-dollars are used for this test. For standardized production testing, a 13 mm (½ in.) diam solid nylon or aluminum rod, 102 mm (4 in.) in length, with the testing end smoothly rounded to a 6.4 mm (¼ in.) radius, is used. Another version of a tap tester is the aluminum hammer shown in Fig. 37.

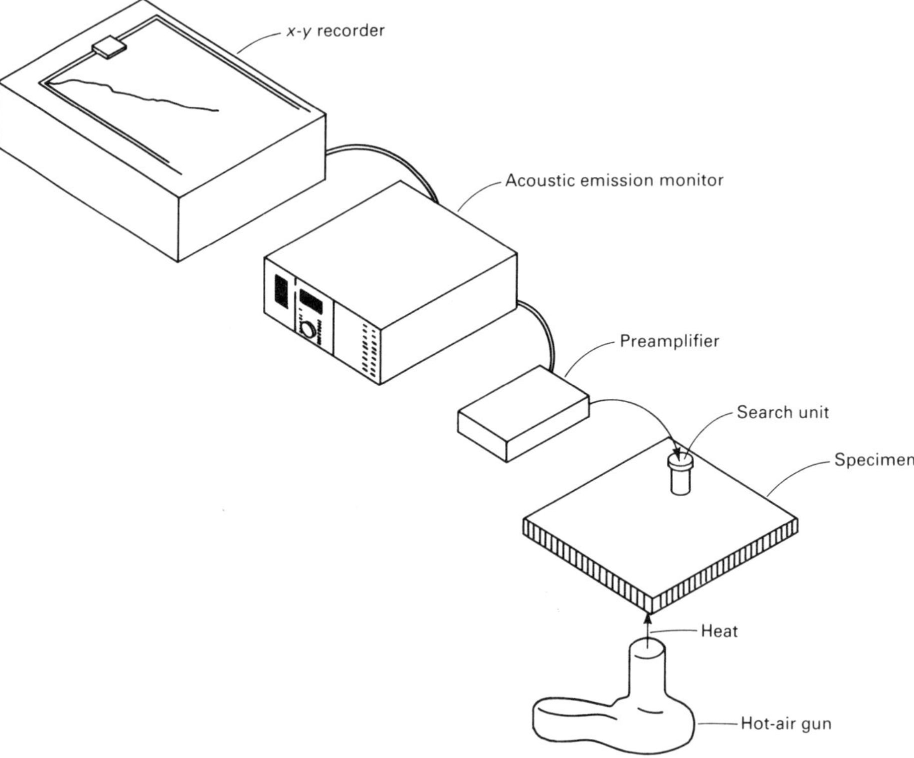

Fig. 38 Acoustic emission detection of active corrosion in adhesive-bonded structures

Tap testing with a coin or a small aluminum rod or hammer is useful for locating large voids or disbonds of the order of 38 mm (1½ in.) in diameter or larger in metal-to-metal or thin facing-sheet honeycomb assemblies. Tap testing is limited to the detection of upper facing sheet to adhesive

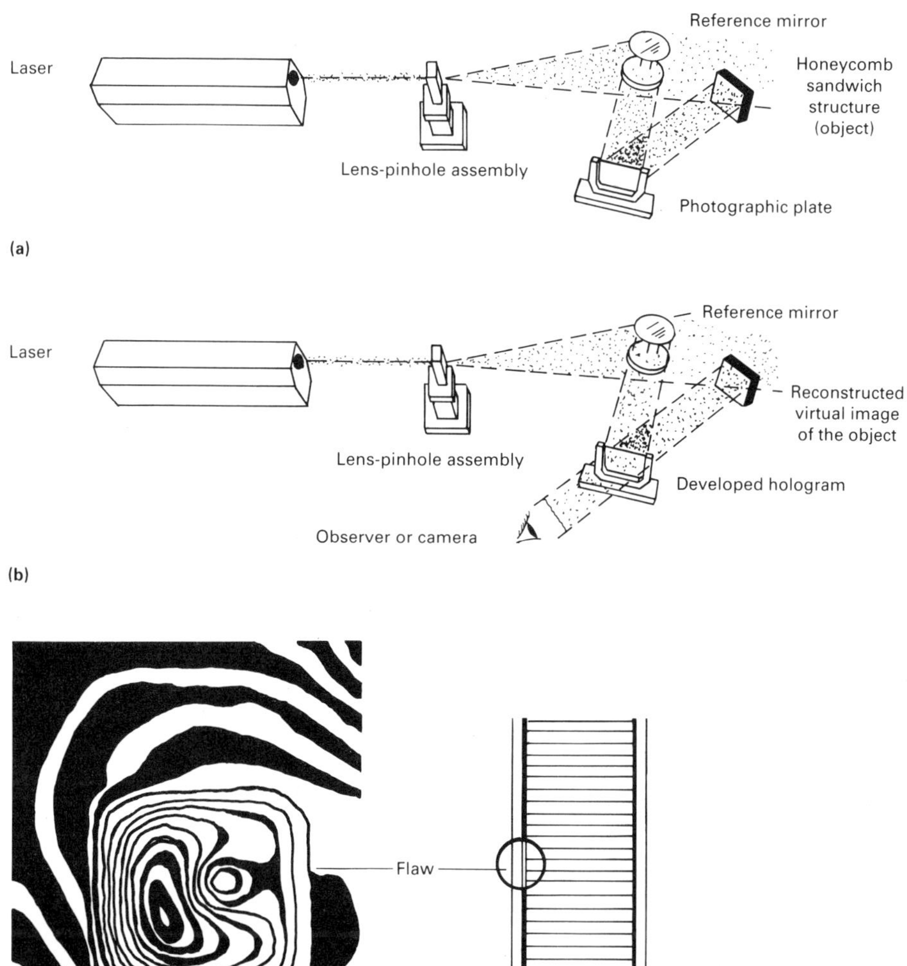

Fig. 39 Holographic recording and viewing. (a) Setup for recording hologram. (b) Setup for real-time viewing of hologram. (c) Example of real-time hologram of a flaw (void in adhesive) in a honeycomb core assembly. The part was heated to give thermal stressing.

disbonds or voids. It will not detect voids or disbonds at second, third, or deeper adherends. The test method is subjective and may yield wide variations in test results. On thin face sheets, the coin tap will produce undesirable small indentations. In this case, an instrument such as the Shurtronics harmonic bond tester can be used effectively.

Acoustic Emission

In certain cases, acoustic emission techniques are more effective than x-ray or conventional ultrasonic methods in detecting internal metal corrosion and moisture-degraded adhesive in bonded panels. The principle is based on the detection of sound or stress wave signals created by material undergoing some physical or mechanical transformation. Regarding the detection of bond line corrosion, the acoustic signals apparently arise from cavitation or the boiling of moisture within the joint. If the corroded joint is dried out before the acoustic emission test is performed, no acoustic response is obtained.

The equipment consists basically of an amplifier and a piezoelectric sensor with a resonant frequency of about 200 kHz. The emission level is recorded on a chart or counter. Other equipment includes a signal processor, a search unit, a 50-dB preamplifier, an *x-y* chart recorder or counter, and a hot-air gun (Fig. 38). Simple heating methods employing a hot-air gun or heat lamp are used to increase emissions from active corrosion sights or to boil moisture. Heating can be done from the search-unit side.

The panel to be tested is heated to about 65 °C (150 °F) by holding the hot-air gun within 50 to 75 mm (2 to 3 in.) of the surface of the panel for about 15 to 30 s. Immediately after heating, the transducer or search unit is placed a short distance from the heated spot. The transducer is held in position for 15 to 30 s to obtain a complete record of any emission in the heated area. The inspection is conducted on a 152 mm (6 in.) grid. An important consideration during the test is the manner in which the transducer is held against the part surface. Because movement of the transducer can produce appreciable noise, care must be taken in its placement and holding.

Corrosion has been detected in a number of adhesive-bonded honeycomb structures (Ref 7-9). A direct inspection cost savings of more than 75% over comparative x-ray inspection has been achieved. Additional information on this inspection method is available in the article "Acoustic Emission Inspection" in this Volume.

Special NDT Methods

Holographic Interferometry. Defects such as core-to-skin and core-splice voids, delaminations, crushed core, and bond line corrosion can usually be detected by holographic nondestructive test techniques (Ref 3, 24). A hologram of the test specimen is first recorded by means of laser light reflected from its surface and superimposed on a mutually coherent reference beam in the plane of a high-resolution photographic emulsion. The hologram provides a complete record of all visual information about the entire illuminated surface of the specimen, including the phase and the amplitude of the reflected wave front.

The specimen is then stressed in one of several possible ways, including heat, pressure change, evacuation, and acoustic vibration. A surface displacement differential of only a few microinches between a defective and a good-quality bond is adequate to produce a holographic image. Differential surface displacements, caused by subsurface anomalies, are observed or recorded with one or more of the following three holographic techniques:

- *Real time*: The stress-illuminated specimen is viewed through its developed hologram (made in a different state of stress)
- *Time lapse*: Two holographic exposures are made on the same plate, with the specimen in two states of stress, and then reconstructed
- *Time average*: The hologram is recorded during many cycles of sinusoidal vibration of the specimen

A fluid-gate photographic plate holder provides for the in-place development of holograms and enhances the speed of the operation (Ref 24). Because of the acute sensitivity of holographic interferometry to small disturbances, it is necessary to reduce spurious and unwanted acoustic noise and temperature gradients within the environment of the testing system. Floor vibrations arising from heavy traffic, the loading/unloading of vehicles, and heavy-duty machinery generate noise that must be minimized.

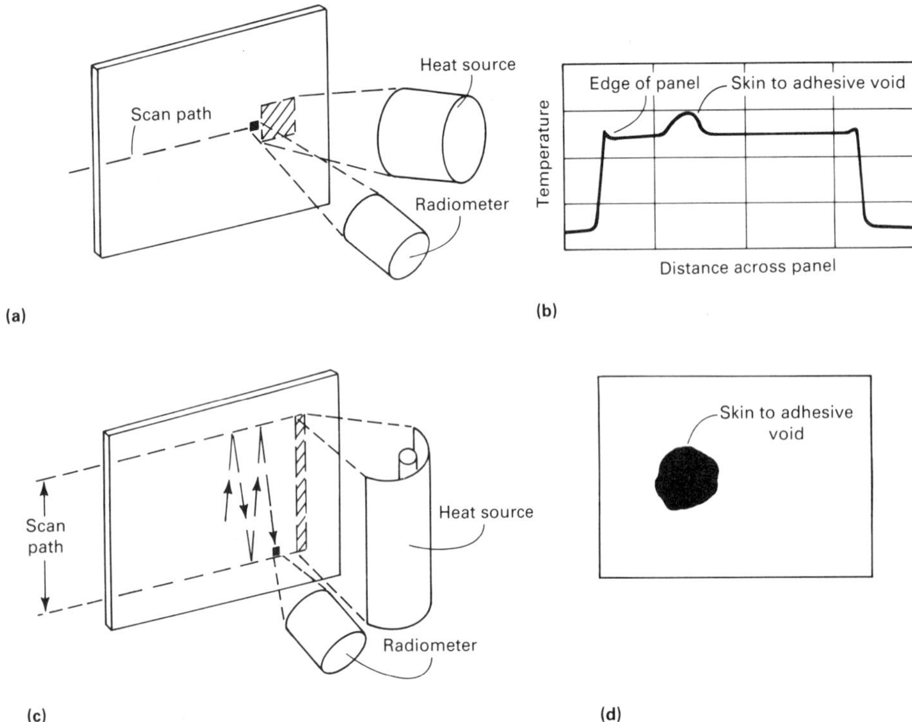

Fig. 40 Schematics and readouts illustrating infrared radiometer tests. (a) Arrangement for line scan test. (b) Example of readout and method of void detection. (c) Arrangement for area scan test. (d) Example of readout and method of void detection

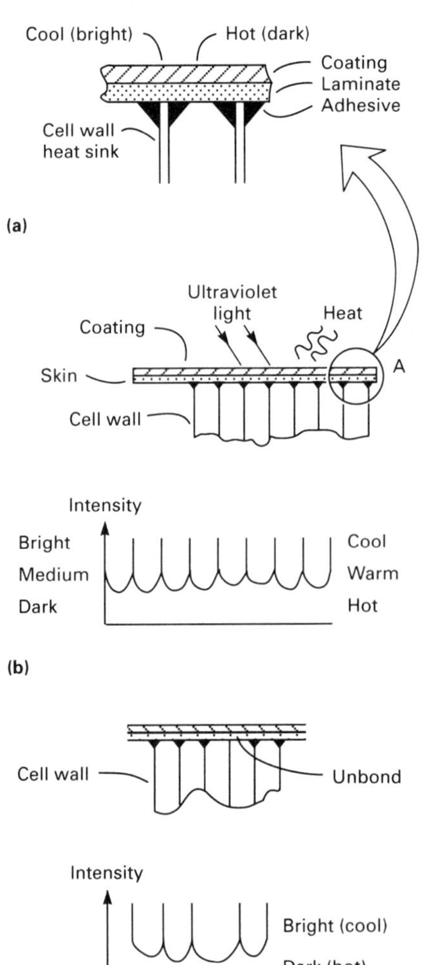

Fig. 41 Thermoluminescent coating technique on boron composite aluminum honeycomb flap assembly. (a) Close-up of honeycomb structure. (b) Plot of intensity versus location for bonded joint at locations shown. (c) Plot of intensity versus location for unbonded joint at location shown

Systems for testing sandwich structures generally use helium-neon gas lasers, which normally deliver between 60 and 80 mW of power at 632.8 nm (6328 Å). This power is sufficient to examine at least 0.19 m^2 (2 ft^2) of surface area, and the deep red color of the light is close enough to the sensitivity peak of the eye for good fringe-to-background contrast.

A typical real-time hologram showing a skin-to-core void revealed by thermal stressing is illustrated in Fig. 39. In large measure, the basic limitation of holography is related to the stressing techniques utilized. Holography cannot be used without a surface manifestation of a defect during stressing. If the material thickness precludes detection of a resolvable fraction of the fringe spacing, then the technique is ineffective. It is not useful for inspecting complex laminates, because of the inability to stress the void areas. Thermal stressing of aluminum is unsatisfactory, because of its high thermal conductivity (heat is transferred laterally rather than vertically through the joint). Holography is satisfactory for inspecting honeycomb structures, but few voids exist at the skin-to-core interface. It is limited in locating voids at honeycomb closures or multilaminate metal-to-metal areas. Numerous flaws in the closure areas of bonded honeycomb assemblies (voids and porosity) detected by ultrasonics and x-ray methods were not detected by holographic interferometry.

At Fokker-VFW, a holographic installation for testing bonds was developed and put into use for production inspection (Ref 3). The installation uses a single hologram for components up to 6 m (20 ft) in length. It has adjustable optics for optimum sensitivity and component scanning. The defect area can be magnified and the results presented on a highly sensitive video monitor which can record the information. The bonded components can be deformed by either vibrations or thermal effects.

Interference holography shows much more than just bond defects, and training is needed for personnel to learn to distinguish areas that are defective from those that are merely not ideal (Ref 2, 3, 21). Core-splice location and quality can be inspected, and core machining anomalies are visible. Also, the thermal method is the most practical deformation method for sandwich structures, especially for graphite-epoxy skins bonded to aluminum core. Extensive studies have shown that thermal deformation is difficult to use for metal-to-metal structures (Ref 21). These difficulties led workers at Fokker to concentrate on vibration testing because of the relation between resonance properties and cohesion quality. Much time will be needed to develop a universal system of quality testing based on interference holography. However, a specific interference holography system has been developed for the inspection of truck and aircraft tires (Ref 25). A wide variety of tire defects have been detected and categorized for both new and retread tires and have been related to durability. Detailed information on this inspection method is available in the article "Optical Holography" in this Volume.

Infrared or Thermal Inspection. The thermal NDT of adhesive-bonded structures has been performed by a variety of techniques, including the following:

- Infrared radiometer testing (Ref 22, 26)
- Thermochromic or thermoluminescent coatings (Ref 22, 27)
- Liquid crystals (cholesteric) (Ref 28, 29)

For infrared radiometer testing, the detector employs a moving heat source and records variations in heat absorption or emission while scanning the part surface. The scanners, or detectors, used in infrared testing are called radiometers. The radiometer generates an electrical signal exactly proportional to the incident radiant flux. Because scanning and temperature sensing

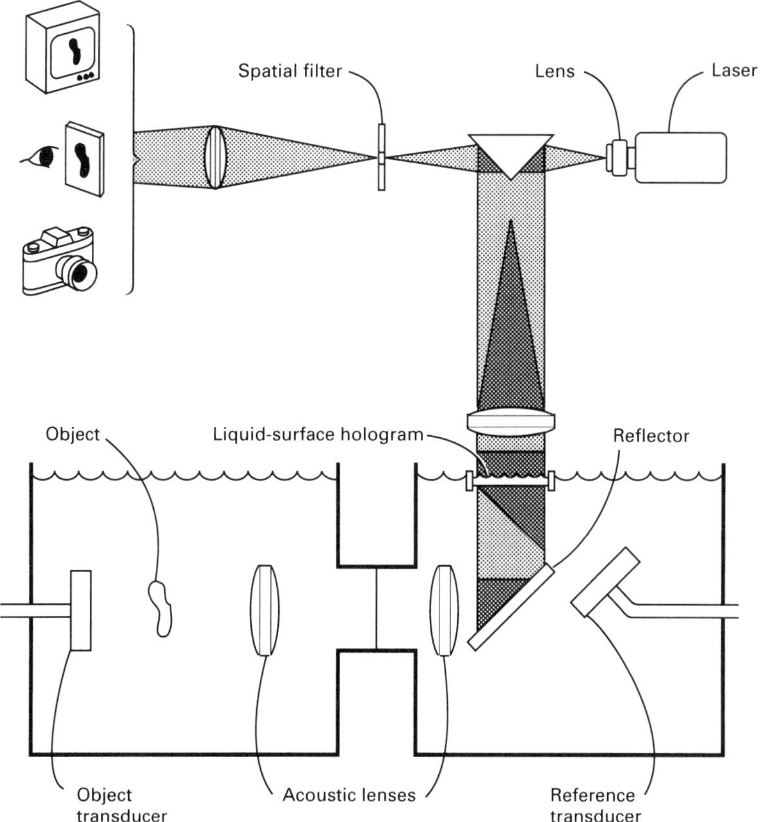

Fig. 42 Schematic of a liquid-surface imaging system

are performed without contact, the observed surface is not disturbed or modified in any way. The thermal pattern from the part under test can be observed on a CRT, storage (memory) tube, or *x-y* recorder.

Tests are performed by heating the sample with a visible-light heat source (quartz lamp or hot-air gun), then observing the surface heating effects with the radiometer. Heat is applied to both bonded and unbonded areas. The bonded areas will conduct more heat than the unbonded areas because of good thermal conduction to the honeycomb structure. The test panel is first coated with flat-black paint to ensure uniform surface emissivity. The sample is then scanned as shown in Fig. 40. Graphs are obtained by connecting the horizontal movement of the sample to one axis of an *x-y* recorder and the radiometer output to the other axis. Typical line scan and area scan test results are shown in Fig. 40(b) and 40(d), respectively. Similar results are obtained with the AGA Thermovision (in the pitch or catch or through transmission modes) (Ref 22). Good results are generally obtained when testing graphite-fiber or boron-fiber composite face sheets bonded to aluminum core. Honeycomb structures fabricated exclusively from aluminum (skin and core), or aluminum skin bonded to plastic core, are difficult to inspect by infrared methods because of the lateral heat flow in the aluminum face sheets.

Thermochromic or Thermoluminescent Coatings. The use of an ultraviolet-sensitive coating containing a thermoluminescent phosphor that emits light under excitation by ultraviolet radiation (black light) permits the direct visual detection of disbonds as dark regions in an otherwise bright (fluorescent) surface (Ref 22, 27). The coating is sprayed on and dried to a 75 to 130 μm (3 to 5 mil) thick plastic film. Defects appear as darkened areas when the panel is heated to 60 °C (140 °F) and viewed under ultraviolet light (Fig. 41). The defective regions can then be marked on the plastic film with a felt-tip pen. The rate of fluorescent reversal and retention depends on the thermal conductivity of the underlying structure.

Thermochromic paints consist of a mixture of temperature-indicating materials that change color when certain temperatures are reached. Thirty-six materials covering the temperature range of 40 to 1628 °C (104 to 2962 °F) are available, with the low-temperature materials being used for bond inspection. The low-temperature paint changes from light green to vivid blue upon reaching 40 °C (104 °F). The color change will last (in the defect area) for about 15 min based on the relative humidity. The dark areas can be made to revert back to the original color by applying moisture in the form of steam. The tests can be repeated a number of times without destroying the properties of the paint. The paints are easy to apply and remove.

Liquid crystals are a mixture of cholestric compounds that change color when their temperature changes as little as 0.84 °C (1.5 °F) and they always attain the same color at a given temperature for a specific crystal composition. After suitable surface preparation or cleaning, a thin coating of liquid crystals is applied by spray or brush to the

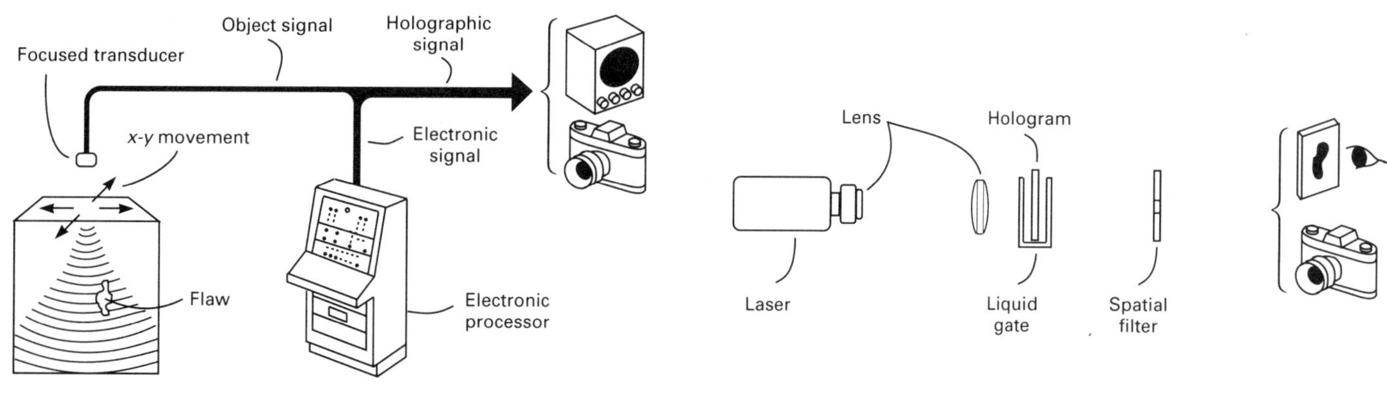

Fig. 43 Schematic of (a) pulse echo acoustical holography system and (b) holographic reconstruction unit

Table 3 Correlation of NDT results for built-in defects in laminate panels

	NDE method(a)												
	Radiography		Ultrasonic bond testers				Ultrasonic inspection(f)						
Laminate defects	Low-kilovolt x-ray	Neutron radiography	Fokker bond tester	Sondicator	Harmonic bond tester	210 sonic bond tester	Contact pulse echo	Contact through transmission	Immersion C-scan pulse echo	Immersion C-scan through transmission	Immersion C-scan reflector plate	Coin tap test	Remarks
1. Void	(b)	D	D	D	D	D	D	D	D(c)	D	D	PD(e)	b,c,e,g
2. Void (C-14 repair)	(b)	D	D	D	D	D	D	D	D(c)	D	D	PD(e)	c,e
3. Void (9309 repair)	(b)	D	D	D	D	D	D	D	D(c)	D	D	PD(e)	c,e
4. Lack of bond (skin to adhesive)	(b)	ND	D	D	D	D	D	D	D(c)	D	D	PD(e)	c,e
5. Manufacturer's sheet (FM123-41)	(b)	PD(d)	D	D	ND	PD	ND	PD	D(c)	D(d)	D(d)	ND	c,d
6. Thick adhesive (1, 2, 3 ply)	(b)	ND	D	ND	ND	D	ND	ND	ND(e)	PD(c)	D(c)	ND	c,e
7. Porous adhesive	(b)	D	D	ND	ND	D	D	D	D	D	D	ND	b
8. Burned adhesive	(b)	ND	D	ND	ND	D	D	D	PD	D	D	ND	h
9. Corroded joint	D	D	D	PD	PD	D	D	D	PD	D	D	PD	i

(a) ND, not detected; PD, partial detection; D, detected. (b) Panels were made using FM-73, which is not x-ray opaque. With x-ray opaque adhesive, defects 1, 2, 3, 5, 6, 7 are detected. (c) Method suffers from ultrasonic wave interference effects caused by tapered metal doubles or variations in adhesive thickness. (d) Manufacturer's separator sheet not detectable but developed porosity and an edge unbond during cure cycle, which was detectable. (e) MIL-C-88286 (white) external topcoat and PR1432G (green) plus MIL-C-83019 (clear) bilge topcoat dampened the pulse-echo response. (f) Contact surface wave was tried but did not detect any built-in defects. (g) Minimum detectable size approximately equal to size of probe being used. (h) Caused by drilling holes or band sawing bonded joints. (i) Moisture in bond joint (Armco 252 adhesive, Forest Products Laboratory etch)

test object surface. When the object is correctly heated (relatively low temperatures) with a heat lamp or hot-air gun, the defects are indicated by differences in color (Ref 28, 29). Unfortunately, the color continues to change through a specific color band as it is heated and cooled. Therefore, the defects must be marked on the surface of the part as they appear and disappear. Photographs can be taken as a record of the test results after specific defects are located. The test can be repeated a number of times without destroying the liquid crystals. In some cases, black paint is required under the liquid crystal coating to obtain uniform emissivity and color contrast in the defect areas. The theory and results obtained using liquid crystals are virtually the same as for thermoluminescent and thermochromic coatings. Detailed information on thermal NDT is available in the article "Thermal Inspection" in this Volume.

Leak Test. A hot-water leak test is generally conducted on all bonded honeycomb assemblies immediately after fabrication or repair. The test is performed by immersing the part in a shallow tank of water heated to about 65 °C (150 °F). The heat causes the entrapped air to expand, and if there are any leakage paths, bubbles will be generated at the leakage site. The panel is monitored visually for escaping bubbles, and the leakage site is marked on the panel for subsequent sealing. A part that leaks after fabrication will generally develop problems in service. After the leak test, the assembly is radiographed for water that may have become entrapped during the leak test. Additional information is available in the article "Leak Testing" in this Volume.

Acoustical holography provides a way to observe the interior properties of composite laminates or adhesive-bonded joints. This acoustical technique employs ultrasound to obtain three-dimensional information on the

Table 4 Correlation of NDT results for built-in defects in honeycomb structures

	NDE method(a)												
	Radiography		Ultrasonic bond testers				Ultrasonic inspection						
Honeycomb defects	Low-kilovolt x-ray	Neutron radiography	Fokker bond tester	Sondicator	Harmonic bond tester	210 bond tester	Contact pulse echo	Contact through transmission	Contact shear wave	Immersion C-scan pulse echo	Immersion C-scan through transmission	Coin tap test	Remarks
1. Void (adhesive to skin)	ND(b)	D	D	D	D	D	D	D	ND	D	D	D	Replacement(b) standard
2. Void (adhesive to skin) repair with C-14	...	...	...	...	...	...	...	...	...	...	...	...	No void improperly made
3. Void (adhesive to skin) repair with 9309	...	...	...	...	...	...	...	...	...	...	...	...	No void improperly made
4. Void (adhesive to core)	ND	D	D	...	D	D	ND	D	ND	D	D	D	Replacement standard
5. Water intrusion	D	D	ND	ND	ND	ND	ND	D	ND	ND	D	D	...
6. Crushed core (after bonding)	D(c)	PD	PD(e)	ND	PD(e)	ND	ND	PD(e)	ND	PD(e)	PD(e)	D	(c, e)
7. Manufacturer's separator sheet (skin to adhesive)	ND(d)	D	ND	ND	PD	ND	PD	D	ND	D	D	PD	(d)
8. Manufacturer's separator sheet (adhesive to core)	ND(d)	D	ND	ND	ND	ND	D	ND	ND	D	PD	D	(d)
9. Void (foam to closure)	D	D	D	D	D	D	D	D	ND	D	D	D	...
10. Inadequate tie-in of foam to core	D	D	D	D	D	D	ND	D	ND	D	ND	D	...
11. Inadequate depth of foam at closure	D	D	D	D	ND	PD(f)	ND	ND	ND	ND	ND	PD	...
12. Chem-mill step void	ND(d)	D	ND	ND	ND	ND	ND	ND	ND	ND	ND	ND	(d)

(a) ND, not detected; PD, partial detection; D, detected. (b) Panels were made using FM-73, which is not x-ray opaque. (c) The 0.13 and 0.25 mm (0.005 and 0.010 in.) crushed core detected by straight and better by angle shot. (d) Has been detected by x-ray when adhesive was x-ray opaque (FM-400). (e) Detects 0.25 mm (0.010 in.) crush core. (f) Discloses defect at a very high sensitivity

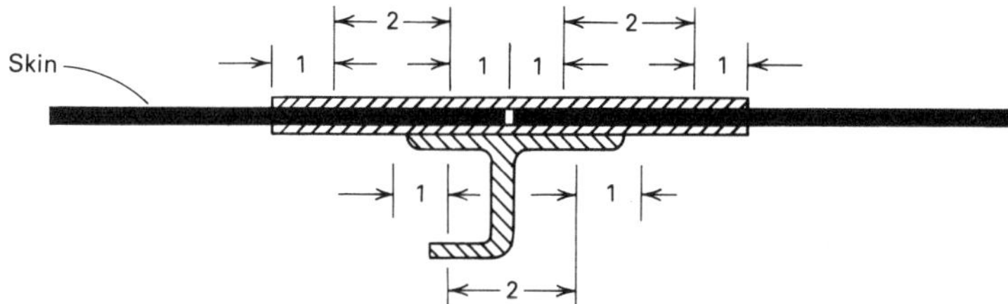

Fig. 44 Schematic of bonded joint divided into quality zones based on design shear stresses. The example shown is a longitudinal skin splice with an internal longeron. Quality grade 1 is within 13 mm (½ in.) of the edge of any bonded member, and grade 2 is the bond between any two grade 1 bonds.

internal structure of the test specimen. The acoustical hologram is the converter, or recorder, which allows acoustic information to be visualized much as film is the converter or recorder for light. The acoustic imagers employ different holographic recording techniques, depending on the specific application of the instrument (Ref 30).

Liquid-Surface Acoustical Holography. In this method, the liquid surface acts as a dynamic film for momentary storage of the hologram while it is converted to a visual image through the use of coherent laser light. This technique (Fig. 42) allows real-time acoustical imaging, supplying the operator with an instantaneous view of all internal structures. Because the part must be moved through the fixed acoustic beam, part size is a limiting factor in this technique.

Scanning acoustical holography employs a scanning technique to construct the hologram. The hologram is then recorded on either transparency film or a storage oscilloscope. This technique provides a permanent record of the holographic information and allows the operator to observe the reconstructed image at a later time. By illuminating either the liquid-surface hologram or the transparency film with laser light, the operator can immediately observe all interior properties of the test sample. Figure 43 schematically illustrates a scanning holographic system. The capability of acoustical holography to reveal flaws in bonded structure is covered in Ref 31. One drawback to the use of scanning acoustical holography is the high cost of the equipment. In addition, considerable time is needed to reconstruct the various holograms of the entire part.

Selection of NDT Method

A universally accepted single test method for evaluating bonded structures has not yet been developed. Therefore, a combination of test methods is required for complete and reliable inspection. Generally, selection of the test method is based on:

- Part configuration and materials of construction
- Types and sizes of flaws to be detected
- Accessibility to the inspection area
- Availability of equipment/personnel

The application of NDT methods in the production cycle for honeycomb sandwich panels generally includes the following:

- Material property tests
- Surface preparation checks (wedge crack, T-peel, and so on)
- Verifilm (prebond) tooling check
- Visual inspection
- Hot-water leak check
- Radiographic check for water and other internal discontinuities
- Ultrasonic C-scan for high-resolution inspection of skin-to-core bond
- Manual inspection of doublers, close-out members, and so on, using acoustic methods such as pulse-echo ultrasonic,

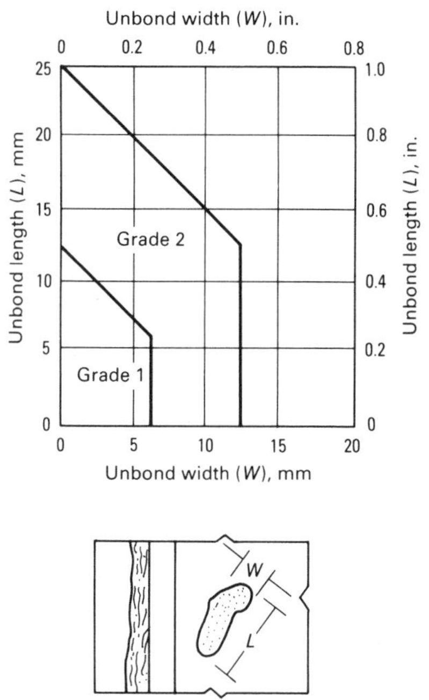

Fig. 45 Definition of acceptance grades for voids or unbonds

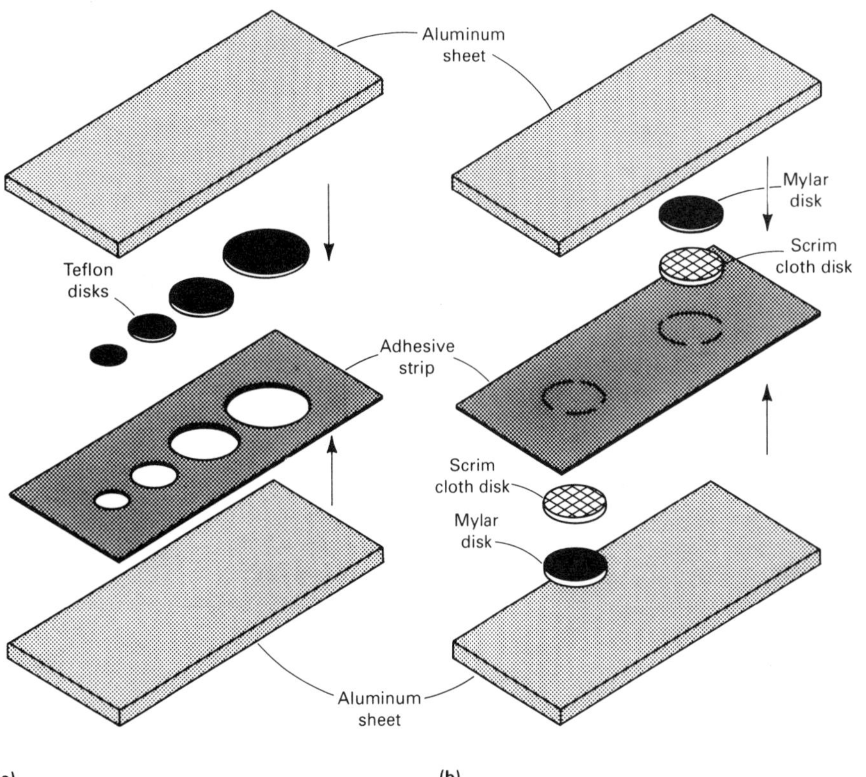

Fig. 46 Two methods of fabricating unbond reference standards. (a) Method 1 uses two pieces of aluminum sheet that have been anodized and primed and then bonded together with a piece of adhesive that has various-size cutouts filled with 0.13 mm (0.005 in.) thick Teflon disks filling the holes. (b) Method 2 uses metal as in method 1, but the voids are produced by placing a disk sandwich of Mylar and scrim cloth of desired size on the uncured adhesive. The assembly is then bonded.

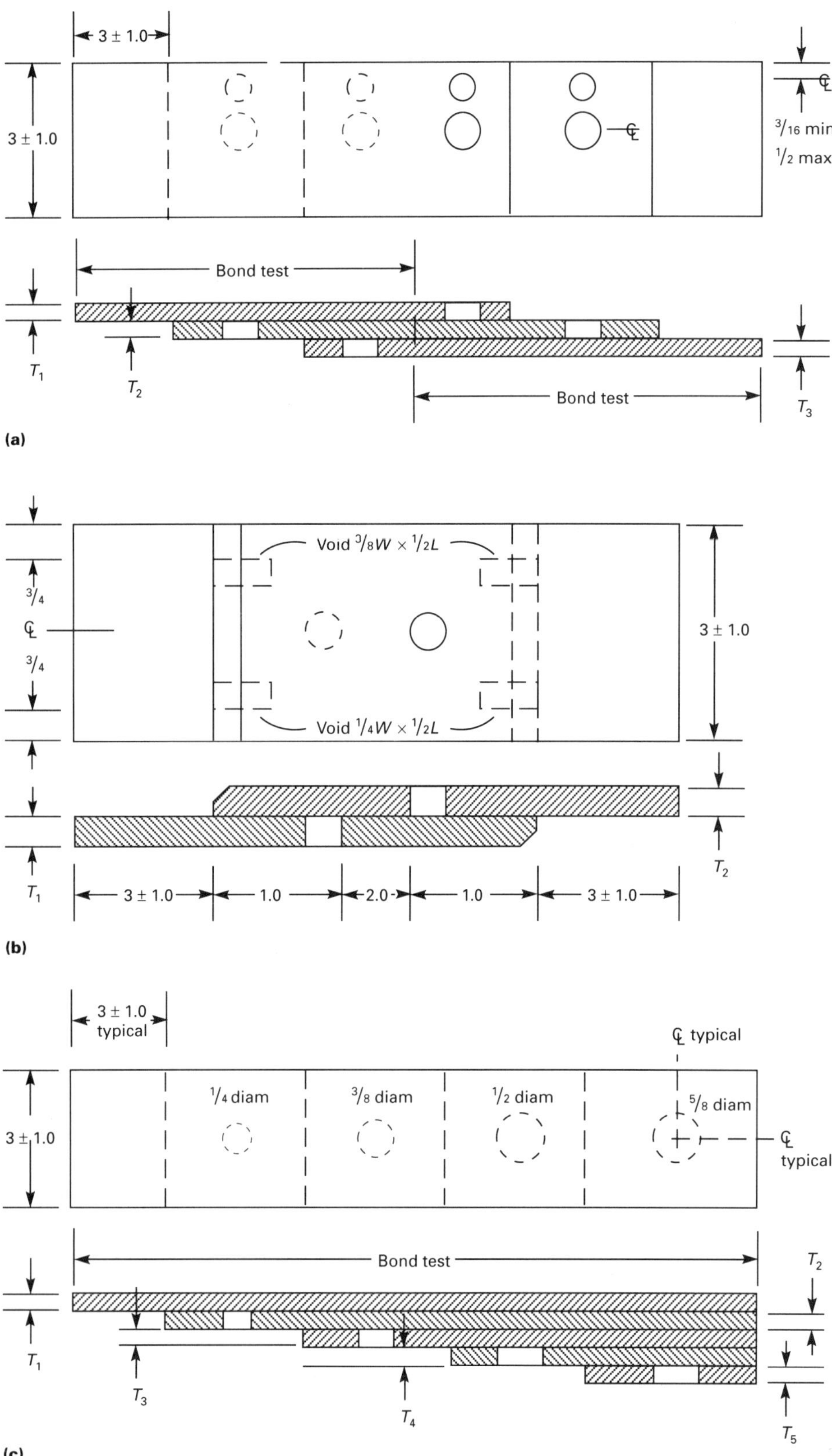

Fig. 47 Additional methods for fabricating unbond reference standards. (a) Method 3 uses three flat, unchamfered aluminum sheets of different thicknesses. This standard can be inspected from either side. Hole diameters can vary from 6.4 to 13 mm (¼ to ½ in.) and are free of adhesive. (b) Method 4 uses two chamfered aluminum sheets of different thicknesses with drilled holes 9.5 or 13 mm (⅜ or ½ in.) in diameter (free of adhesive) and edge voids made by inserting Teflon shims of the size shown. Shims must be removed after bonding. This standard can be inspected from either side. (c) Method 5 is a flat step standard using a buildup of various sheet thicknesses that matches the structure to be inspected. There should be no adhesive at the base of the holes. This standard should be inspected from the top side only. Dimensions given in inches

Shurtronic harmonic bond tester, or Fokker bond tester

Built-in defects were produced in typical laminate and honeycomb specimens. These specimens were evaluated by state-of-the-art NDT methods to determine which methods reliably detected the different defects. The results of this investigation are given in Table 3 for adhesive-bonded laminates and in Table 4 for adhesive-bonded honeycomb structures (Ref 6). The NDT methods described in this section are used to test sandwich structures having aluminum or titanium skins and aluminum core. Most of the methods can be effectively applied to structures with metal facing sheets and nonmetallic core and structures with graphite fiber, boron fiber, and fiberglass face sheets bonded to aluminum core. Only a limited number of the methods are applicable for structures having both core and facing sheets made from nonmetallic materials.

NDT in the Product Cycle

Nondestructive testing serves as a basis for process control and serves to establish quality control standards. The acceptance or rejection of a bonded assembly is directly related to the quality level desired by the designer. The quality level, in turn, is based on the importance of the part or component in terms of performance and safety. The frequency/severity flaw criteria and sensitivity of the test should be based on the desired quality level. To avoid unwarranted inspection costs, it is advisable to divide the bonded joint into zones based on stress levels or criticality of part function. Bonded joints are generally designed to withstand shear loads. Stress analysis of bonded joints reveals that the outer edges of a bonded joint will be subjected to higher stress levels than the center of the joint. Therefore, higher quality may be desired at the edges, and parts should be zoned accordingly. The zoned area should be dimensioned so that the inspector can define the two different zones prior to inspection. Figure 44 shows an example of quality zoning for bonded laminate skin splice. A definition of inspection grade numbers versus allowable void sizes is illustrated in Fig. 45.

For multiple bond line joints, it may not be possible to inspect to a high quality at the edge of all adherends, because of the loss of test sensitivity at each successive bond line below the surface. If the edges of a bond joint are stepped or staggered, high-quality inspection is possible at the edges of all bond lines. The width of the step should be large enough to accommodate the test instrument probe. Joint edges that are not stepped should be inspected from opposite sides.

Acceptance/Rejection Criteria

Before inspection can be performed on a bonded structure, accept/reject criteria must be established by engineering. These criteria are usually in the form of frequency (number of flaws per unit area) and/or severity (maximum allowable size flaw data). To avoid unnecessary rejections, engineering must not specify unrealistic or overly conservative criteria. The frequency and severity criteria should be based on sound engineering judgement as related to calculated levels. It should be noted that flaws do not grow under cycle loads if good, durable adhesives are used and if generous overlaps exist.

Military Guidelines. For military contracts, the guidelines or requirements for preparing such criteria are specified in MIL-A-83377 (Ref 32). Specific acceptance criteria for adhesive-bonded aluminum honeycomb structures are specified in MIL-A-83376 (Ref 33). Specification MIL-A-83377 states that the contractor shall prescribe nondestructive tests and a complete NDT process specification when required by MIL-I-6780 in evaluating the quality of adhesive-bonded structures (Ref 34). The NDT process specification must be submitted for approval by the government prior to bonded assembly production. The contractor is required to specify the type and size of the defects that are acceptable. The size of the defect must be consistent with the capability of the NDT method to be used, realizing that NDT methods and instrumentation may have different capabilities. Application of the various NDT methods is limited by material types and design configuration. For aluminum honeycomb sandwich, the allowables of MIL-A-83376 can be used, if determined by the contractor to be acceptable minimums for the particular application.

Specification MIL-A-83377 also requires that the NDT of adhesive-bonded structures be performed only by qualified NDT personnel. Personnel qualification requirements for NDT are specified in MIL-STD-410 (Ref 35).

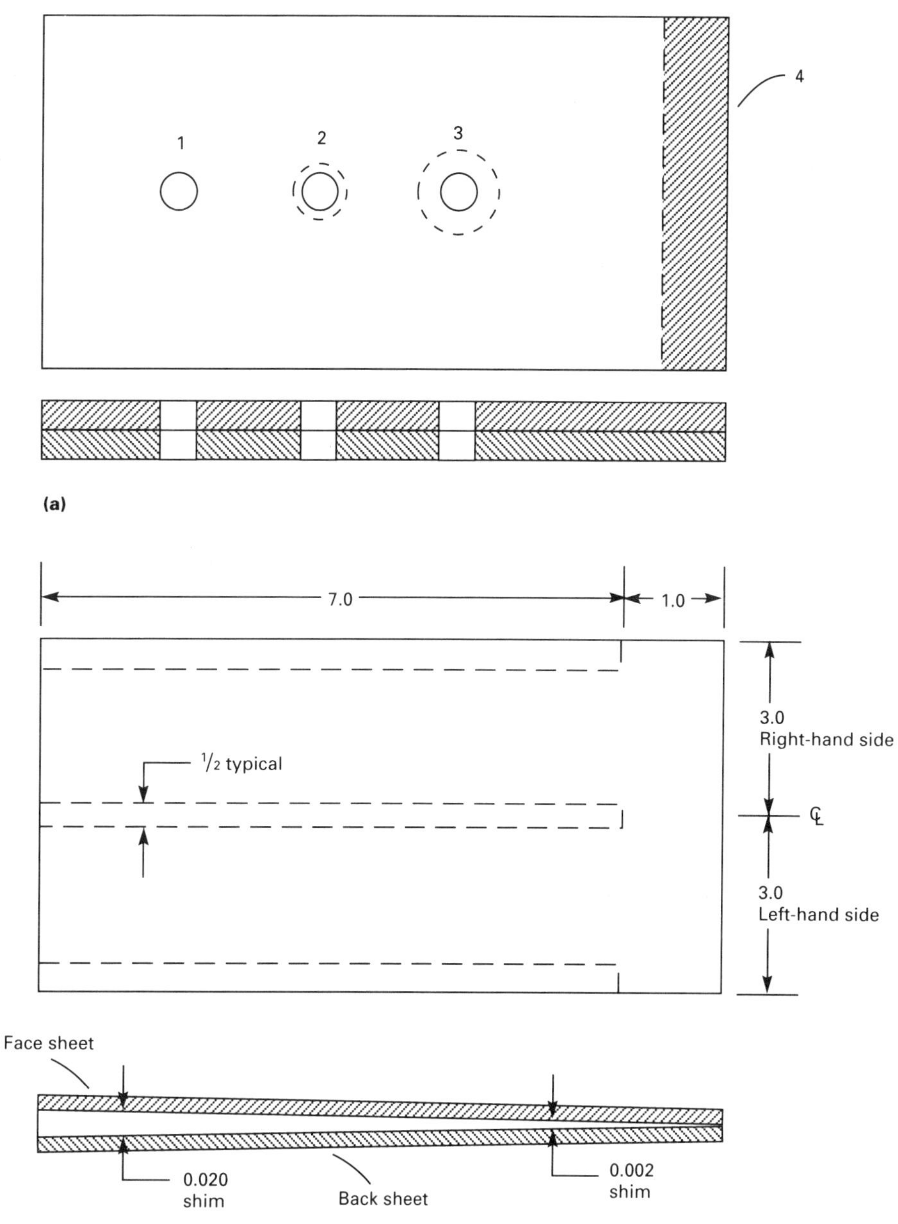

Fig. 48 Additional methods for fabrication unbond reference standards. (a) Method 6 presents a burned-adhesive standard in which (1) is a good-quality hole, (2) is a slightly burned adhesive, (3) is a grossly burned adhesive, and (4) is a grossly burned edge caused by abusive band sawing. (b) Method 7 becomes a variable-quality standard by inserting tapered shims between face and back sheet before bonding. Three 13 mm (½ in.) wide shims are shown in place. Tapered shims cause increasing porosity going from right to left. Half of the face sheet can be removed after inspection to reveal the condition of the adhesive. Dimensions given in inches

Records and Specifications

It is recommended that four basic types of inspection records be prepared and maintained. They are specifications, inspection log, detailed written test procedure, and rejection or nonconformance record.

Specifications. Test method specifications should be obtained for each production inspection instrument to be used—for example, Fokker bond tester (Ref 4), 210 bond tester, harmonic bond tester, Sondicator, and so on). Each specification should specify controls for the particular method and should define reference standard configurations. In addition, an acceptance criteria specification should be prepared.

Inspection Log. The production inspection personnel should maintain a daily inspection log. This log should contain the part number, part dash number, part serial number, fabrication outline serial number, inspector name or stamp, date and method of inspection, and a remarks column to indicate pass or reject. The rejection tag number can be entered, along with a description of the flaw, noting its exact location.

Detailed Written Test Procedure. Nondestructive testing inspections should follow a detailed written procedure for each component tested. The written procedure must comply with the test method specification requirements and, as a minumum, must include:

- A sketch of the part or configuration showing in detail the part thickness, alloy material and temper, and the adhesive system used
- Manufacturer and model number of instrumentation to be used
- Model or designation number of probe to be used
- Any recording, alarm, or monitoring equipment and test fixtures or inspection aids that are to be used
- Applicable reference standard(s) serial number and design description

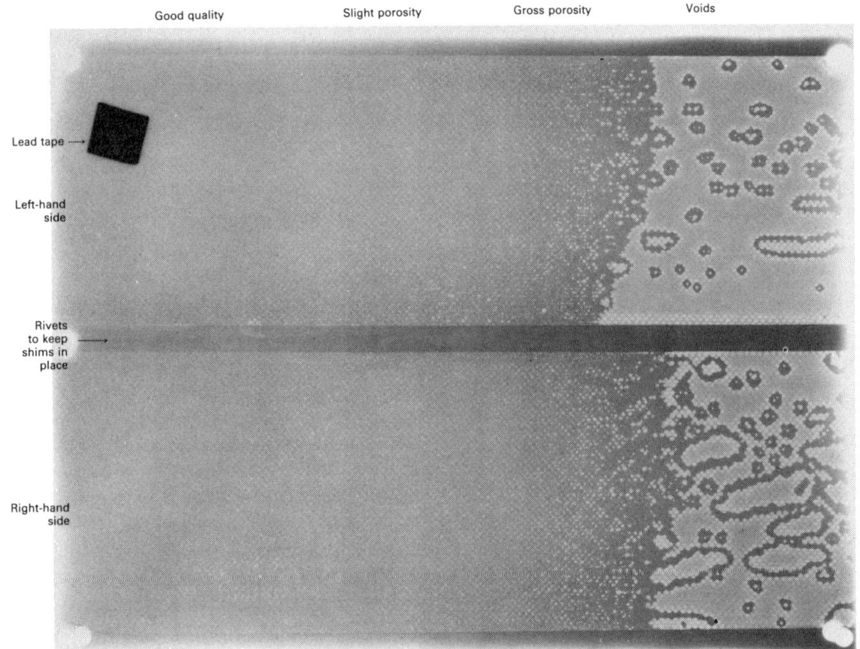

Fig. 49 Positive print from x-ray negative of tapered-shim variable-quality standard shown as method 7 in Fig. 48

- Type of fluid couplant to be used
- Instrument calibration procedure, if required
- Testing plan. The part surface from which the test will be performed, the testing technique(s) for each joint, and the maximum probe index or shift per scan should be described
- Acceptance limits for each joint of the assembly
- Reference point or identifying mark, which is placed on each part prior to inspection; this mark is used for orientation when reviewing inspection results
- Outline of the areas where multiple layers of adhesive have been used
- Identification of section area changes not visible to the inspector
- Verification of the test procedure on a production part prior to approval

Rejection or Nonconformance Record. All nonconforming flaws should be described on a rejection tag. The inspection agency should outline on the part surface the location and shape of unacceptable bond line flaws. If it is necessary to mark the part permanently, the inspector must remove the couplant and wipe with solvent the area to be marked, and then reapply the marks. Next, the marks are covered with a very light coat of clear epoxy enamel. Marking the problem areas in this way facilitates their location after the part has been in service.

Reference Standards

Reference test standards for calibrating and standardizing the bond inspection instruments are essential. Even with adequate reference standards, inspection often produces conflicting information or a deviation of results. In these cases, more than one inspection method should be used in order to improve the accuracy of the data and to define the deviating condition. Under ideal conditions, the test standard with built-in defects of selected sizes should closely duplicate the structure adherends to be inspected. The laminate skin should be of the same thickness, and the honeycomb core should be of the same size and density. Other variations to be encountered in the test specimen, such as tapered core, chemically milled skins and doublers, and so on, should be incorporated into the test standard.

Except for built-in defects, the test standard for adhesive-bonded structures should be fabricated in the same manner as the production assembly. The void or defect should be introduced in the same bond line as that to be inspected in the structure. The standard can be a series of simple test specimens composed of details identical to the several areas of the assembly to be inspected and having a bond of good quality but with controlled or known defect locations and size.

Attempts to produce voids or controlled understrength bonds by the local application of grease or other foreign materials have been found to be ineffective. All adhesives and primers vary and may not respond to the following suggested methods of standard preparation. Therefore, the finished standards should be evaluated by various NDT methods prior to validation of their use as an inspection standard. Neutron radiography is an excellent method for verification if voids, porosity, or both have been generated in the standard. The standard must then be inspected by the test instrument chosen for the particular part to confirm its ability to locate the flaw.

Metal-To-Metal Reference Standards. Various methods have been developed for producing simulated voids, porosity, or unbonds in metal-to-metal bonded laminates. Numerous ways to produce reference standards for metal-to-metal bonded joints or assemblies are described in Ref 10. Because of the variety of such standards, it is not possible to describe all of them in this article, and it is recommended that the reader consult the references for additional information.

Figure 46 shows two methods for producing standards for simulated skin-to-adhesive unbonds. The method 1 or 2 standards are useful for all NDT application methods provided the Teflon or Mylar inserts do not bond to the adherends.

Void, Porosity, and Unbond Standards. Voids or unbonds of specified sizes are difficult to produce because free-flowing adhesive may fill the anticipated void or unbond area. One researcher was successful in producing natural flaws in the adhesive layer of metal-to-metal bond joints (Ref 1). By trial-and-error methods, defect conditions such as a void produced by removing a section of adhesive, a void produced by inserting a 0.64 mm (0.025 in.) diam wire between adherends, and porosity caused by inserting a 0.38 mm (0.015 in.) diam wire between adherends prior to bonding can be produced at specific locations in a bonded joint and with the desired sizes to be detected in an x-ray radiograph.

The methods 3, 4, and 5 standards (Fig. 47) are useful for manual bond test methods (such as Fokker, Harmonic, and 210 bond test applications) and for pulse-echo ultrasonics. The holes at the edge and center of the method 3 standard are required if pinch-off of adhesive occurs near the edge with thicker adhesive near the center of the standard. For maximum sensitivity, a bond joint should be inspected from both sides, and method 3 and 4 standards are used. If the joint can be inspected from only one side, the method 5 standard is used. Holes are drilled after the part is bonded.

Standards containing porosity can be made by the wire insertion method or by bonding a large-area panel 0.6 × 0.6 m (2 × 2 ft, minimum) at low pressure (275 kPa, or 40 psi, maximum). In bonding with a mat (nonwoven) carrier, the adhesive pinches off at the outer edge, producing a picture frame of gross porosity close to the edge. Scattered porosity and some small voids are produced throughout the rest of the panel. After ultrasonic C-scan or radiographic inspection, representative specimens or small

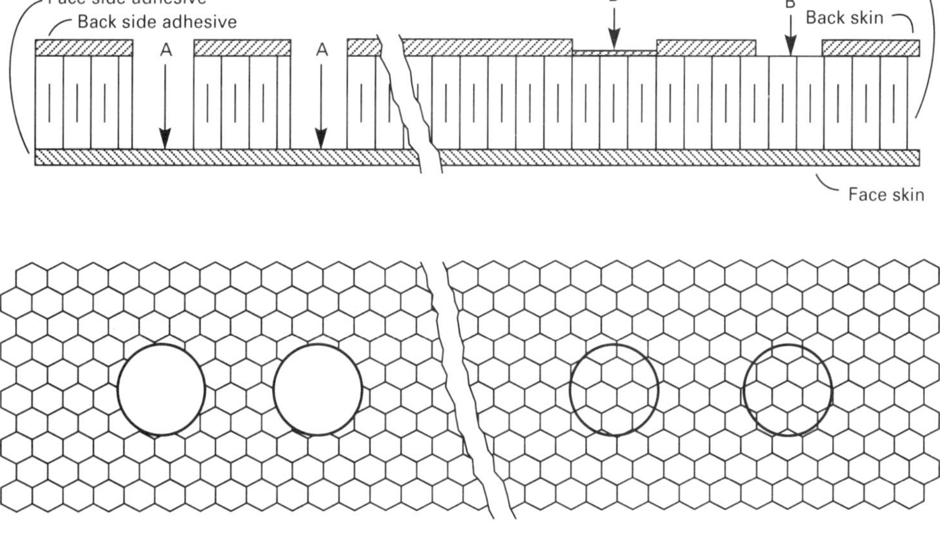

Fig. 50 Construction details for making a honeycomb standard for voids and unbonds. (a) Back skin and core cut out to desired diameter. Cutting through face sheet adhesive shows skin-to-adhesive unbond; cutting to the adhesive shows adhesive-to-core unbond. (b) Cutting through back skin first to the adhesive to show skin-to-adhesive unbond and then through the adhesive to show adhesive-to-core unbond. Note that Teflon disks can be used to help separate adhesive from skins at cutouts. All tests are to be made from the face skin side.

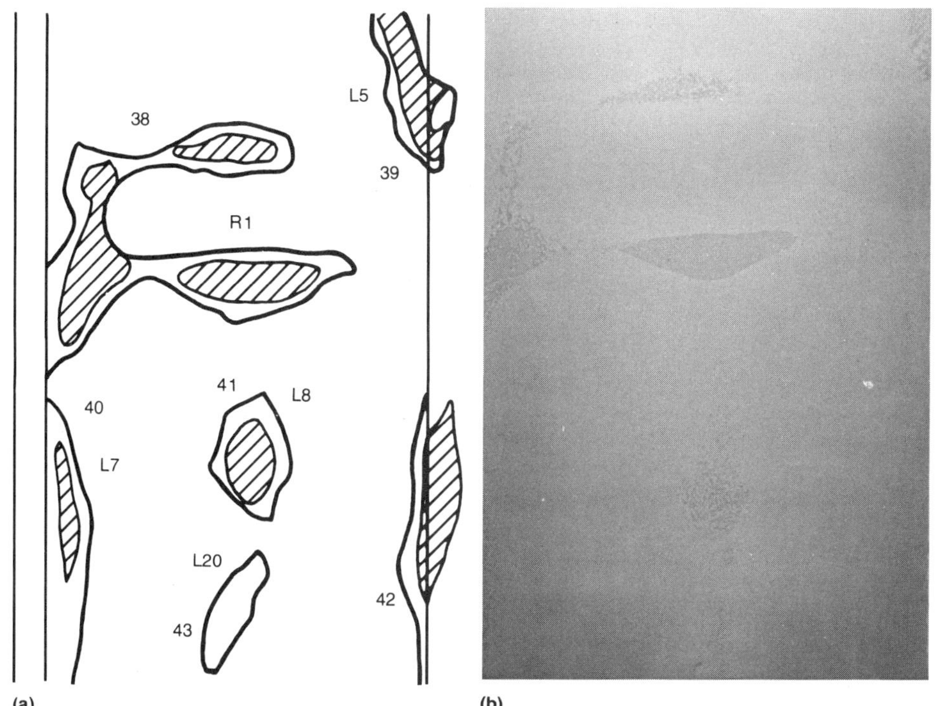

Fig. 51 Correlation of NDI results. The same specimen was inspected by (a) Fokker bond tester and (b) neutron radiography. See text for discussion.

standards can be removed from the panel. The methods for producing voids and porosity are useful for creating defects at specific locations in slow-cycle fatigue test specimens. This type of specimen is used to study the effects of defects under load, cycle rate, or environment.

Burned-Adhesive Standards. Method 6 (Fig. 48) is used to fabricate a standard representing burned adhesive caused by improper drilling or sawing of bonded panels. If improper drills, excessive drill speed and feed, or no coolant is used, burning of the adhesive adjacent to the hole may occur.

Variable-Quality Standards. Method 7 (Fig. 48) is a way to fabricate a variable-quality standard. Through the use of tapered shims, good quality is obtained at the thin end and poor quality at the thick end. Figure 49 shows a positive reproduction made from an x-ray negative of AF-55 adhesive in a tapered standard. For visual correlation, only the right-hand upper skin need be removed from the panel, thus revealing the variable adhesive quality. The left-hand side is kept intact to determine variations in NDT instrument response to different adhesive quality conditions.

Calibration of Bond Testers. When the size of the unbond or void is not of primary concern or to calibrate instruments to a bond/no-bond response, the following standards prove adequate. These standards are especially useful for resonant-type instruments such as the Fokker, Harmonic, Sondicator, and 210 bond testers. They relate well to the bonded part when a 120 to 175 °C (250 to 350 °F) curing epoxy or epoxy phenolic resin system, which does not attenuate the ultrasonic energy, is used. Basically, the standards are made by cutting 38 mm (1.5 in.) diam holes in a 13 mm (½ in.) thick plywood panel and then bonding 75 × 75 mm (3 × 3 in.) aluminum sheets, using room-temperature curing adhesives, so that the center of the sheet is centered over the holes in the plywood. The thickness of the aluminum sheets is chosen to match the thickness of the aluminum adherends of the test part. The number of sheets is governed by the number of adherends in the test part. When the aluminum sheet is bonded to a wooden base, it does not resonate freely and reacts similarly to unbonds in the part being inspected.

Honeycomb Reference Standards. Method 1 or 2 (Fig. 46) is used to create a void or unbond reference standard for honeycomb. For honeycomb standards, the unbonds are created between the adhesive-to-skin and adhesive-to-core interfaces. Skin-to-adhesive voids or unbonds are usually easier to detect than adhesive-to-core unbonds. Another way to prepare a void or unbond standard for a honeycomb panel is illustrated in Fig. 50. This standard is useful when attempting to detect unbonds on both sides of the panel if inspection can be performed from only one side.

Honeycomb crushed core standards are made by inserting 25 mm (1 in.) diam, 0.13 and 0.25 mm (0.005 and 0.010 in.) thick metal disks between the skin and core and then applying pressure. The disks are removed and the panel is bonded. The 0.13 mm (0.005 in.) crushing is very subtle and is difficult to detect radiographically. The 0.25 mm (0.010 in.) crushing is easily detectable.

Water intrusion standards are produced by hot-bonding one skin to the core, adding various quantities of water into the cells,

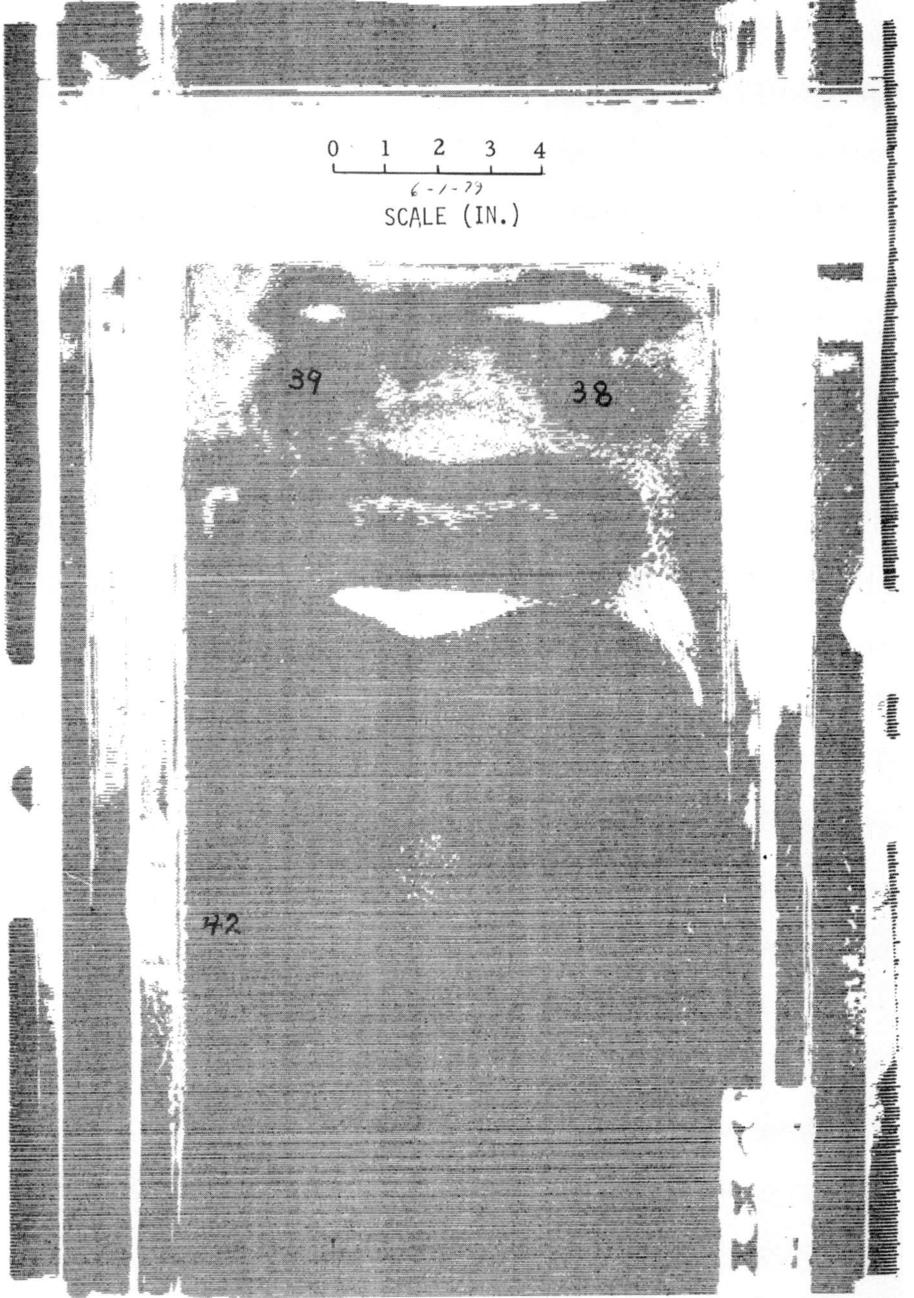

Fig. 52 Ultrasonic C-scan recording of panel in Fig. 51 showing voids and porosity. Through transmission method was used with low-gain (31 dB).

and then cold-bonding the second facing sheet in place.

Core-splice standards are made by cutting the edges of the core with a taper and then applying foaming adhesive to one side of the splice before bonding. Voids in the foaming adhesive are detectable by low-kilovoltage radiography. Core-to-closure voids are made by eliminating foaming adhesive at small areas along the joint.

Separator sheet standards are produced by bonding-in a desired size of the material at the skin-to-adhesive and adhesive-to core interfaces.

A blown core standard can be made by mechanically damaging the core prior to bonding.

Corroded core or core-to-adhesive corrosion standards can be made by bonding only one skin to the core (open-face honeycomb) and then alternately spraying the honeycomb side in a salt spray system. The degree of corrosion can be monitored visually. If it is desired to close up the panel, the second facing sheet can be cold-bonded to the other side of the core.

Substitute Standards. When the ideal standards are not available, other standards of construction similar to that of the test part can be used. However, honeycomb standards should not be used for metal-to-metal laminate structure, and vice versa.

Inspection Without Standards. When standards are not available, a known undamaged area can be used as the standard. Repaired areas are inspected by comparing them to an unrepaired area. The instrument reading may change because of structural changes resulting from the repair. Repeated inspection scans must be conducted under various instrument settings, and the inspection should be verified by other methods and/or instruments. A knowledge of the part configuration (for example, number of bond lines, adhesive thickness, skin thickness, and so on) is essential for interpreting the test results.

Evaluation and Correlation of Inspection Results

The experience, confidence, and effectiveness of NDI technicians and engineering personnel are gained by correlating NDI results with destructive testing. Comparisons can be made between the NDI results and the actual size, shape, location, and type of defect. Mistakes will occur in the early stages of the bonding program, but if accurate records, photographs, and sketches are kept, they can serve as effective training devices. Sections of defective parts should be kept for training purposes. All these items add to the confidence of the inspector in evaluating and accurately reporting defects. They also add to the engineer's confidence in the inspector and in the data reported by the inspector.

Procedure

Before new adhesive-bonded assemblies are fabricated on a mass production basis, the first assembly is evaluated by numerous methods. Inspections are performed and NDT results are evaluated to detect any variations in the part that are indicative of discontinuities. The part is then cut up, and special precautions are taken to section through the discontinuities or to separate the joint at the discontinuity to identify and verify the NDI results. To determine which NDT method(s) detected the discontinuity, the NDI technicians use different color markings—for example, blue for x-ray, green for ultrasonic, red for a bond tester, and so on. These markings can be affixed to the surface of the part or placed on a transparent overlay of the part. When the part is sectioned, the results will identify which NDT method(s) correctly identified the location, size, shape, and type of discontinuity. Although it is expensive, cutting the adhesive-bonded assemblies is sometimes the only way to determine the nature of the discontinuity or to verify its exis-

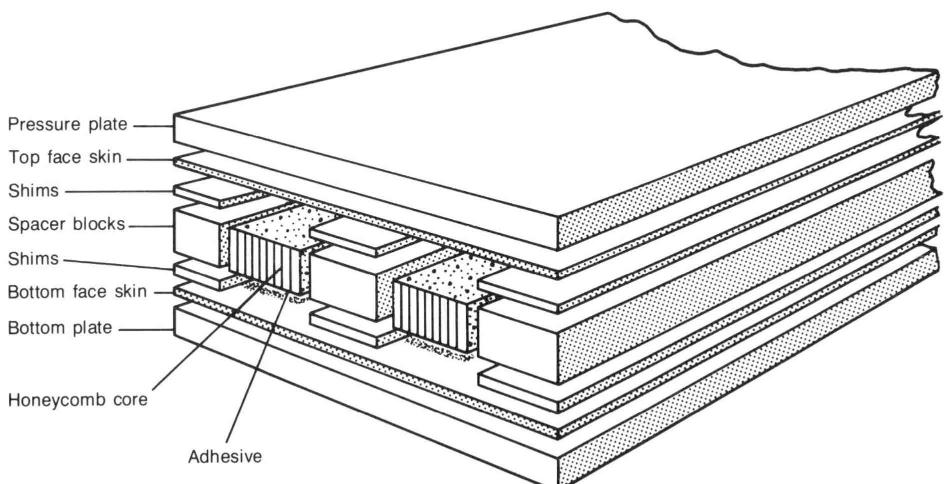

Fig. 53 Schematic of honeycomb sandwich test panel

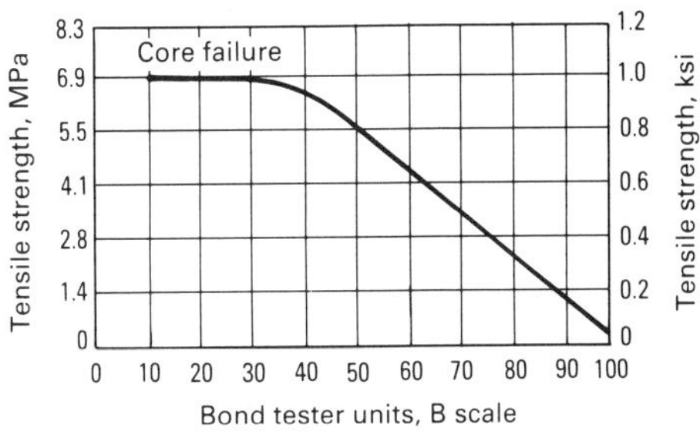

Fig. 54 Typical Fokker B-scale correlation curve versus honeycomb flatwise tensile strength. Cell structure: 0.64 mm (0.025 in.) face skin, 3.18 mm (0.125 in.) cell size, and 0.025 mm (0.001 in.) foil thickness. Probe used was No. 3414.

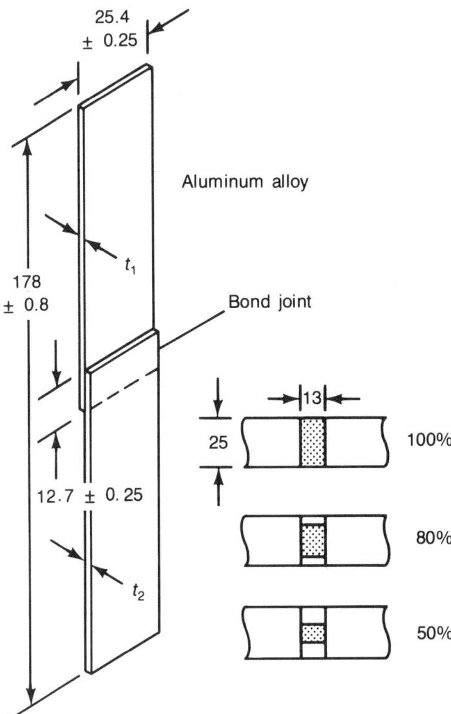

Fig. 55 Lap shear test coupon. Dimensions given in millimeters

tence. All inspection results should be carefully documented for future reference.

Once a highly confident inspection technique is established, the remaining defective assemblies can be used for developing repair procedures. The original NDI techniques can then be used for determining if the repairs are effective. In some cases, defective parts are consistently produced, and the material, production process, or design must be changed to correct the condition. This condition is usually evident when the same type of defect occurs in the same location in all manufactured parts.

Reliable feedback from manufacturing personnel is beneficial to the NDI technician. This information can include such items as too high or too low an oven temperature, incorrect heat-up rate, leaks in the vacuum bag, additional layers of adhesive added at a particular spot in the assembly, and so on. With this information, the NDI technician is better prepared to inspect a particular group of parts.

Destructive Correlations of NDI Results

Following is a typical example of correlating NDI results by destructive testing. In this example, a metal-to-metal joint containing a bonded doubler and longerons and frames was rejected for numerous voids and porosity. The initial inspection was performed using the Fokker bond tester, and the defective areas were marked on the surface of the part, as shown in Fig. 51(a). The inspection results were verified using the 210 bond tester. The part was then sent out for thermal neutron radiography. There was excellent correlation between the Fokker bond test results and the neutrograph (Fig. 51b).

The panel was then ultrasonically C-scanned using the through transmission technique. Again, the correlation was excellent, as shown in Fig. 52. A plastic overlay was made to record the Fokker bond tester results.

The part was then subjected to chemical milling to dissolve the aluminum adherends, leaving the adhesive layer intact. The adhesive thickness was measured and recorded. The destructive to nondestructive correlation was again excellent and revealed that the adhesive thickness in and around the defects was greater (0.33 to 0.46 mm, or 0.013 to 0.018 in.) than in the nondefective areas (0.18 to 0.28 mm, or 0.007 to 0.011 in.). In addition to the correlations noted above, lap shear specimens could have been made through the different defect conditions, and variations in bond strength could have been determined.

Because of the variety of defects that can occur in honeycomb assemblies, defects detected by one test method should be verified by an additional, but similar test method. Time and facilities permitting, the assembly should also be x-rayed in the defective area.

Ultrasonic Wave Interference Effects

In the ultrasonic inspection of adhesive-bonded joints, the following two conditions may cause ultrasonic wave interference (Fig. 24) effects:

- When inspecting bond lines under tapered metal doublers
- When inspecting close to the edge of a bonded laminate and the adhesive is thin due to pinch-off (Ref 10)

Ultrasonic theory predicts that constructive interference (increased amplitude) occurs when the transmitted and reflective waves are in phase with one another (Ref 36, 37). Conversely, destructive interference (decreased amplitude) occurs when the transmitted and reflected waves are 180° out of phase with one another. Destructive wave interference causes a drop in signal amplitude, and the C-scan recorder will not record in the area of interference. An inexperienced operator might misinterpret these areas as voids or unbonds. However, interference effects (caused by tapered adherends) are uniform across the part, resulting in bands of light and dark on the C-scan recording (Ref 10). When destructive interference effects are shown in C-scan recordings of bonded splices having tapered adherends, the bond joints are given another inspection by a different method, such as the Fokker bond tester.

Destructive wave interference effects can also be seen in C-scan recordings of adhesive-bonded laminates made with an adhesive having a mate carrier. It is less evident in laminates bonded with adhesive having a woven carrier. The destructive interference

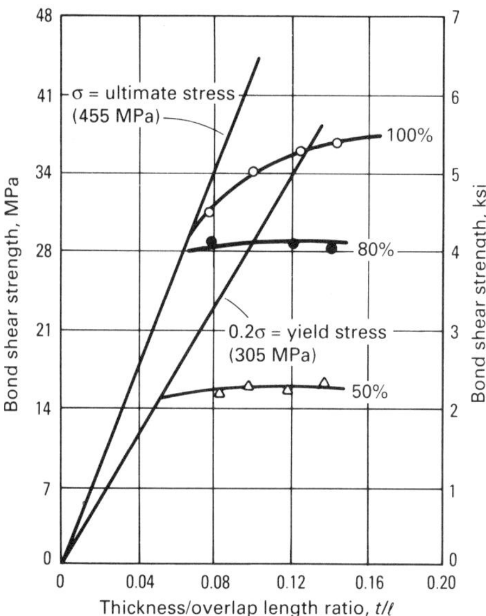

Fig. 56 Variation in lap shear strength with t/ℓ ratio and bond length using aluminum alloy 2024-T3 and adhesive FM-73. t_1/t_2 (mm): 1.0/1.0, 1.3/1.3, 1.6/1.6, and 1.8/1.8.

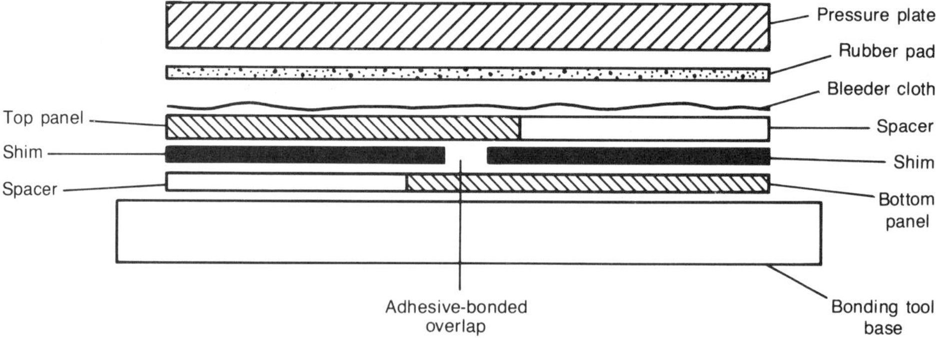

Fig. 57 Metal-to-metal lap shear test panel

effects are most evident near the outer edges of the assembly where the adhesive is pinched off to thicknesses of 0.08 mm (0.003 in.) or less. The interference effects are frequently dependent and therefore will change position if the test frequency is altered (Ref 36, 37). Less wave interference is experienced with through transmission testing than with pulse-echo testing at similar frequencies. Bonded laminates exhibiting destructive wave interference effects require inspection by another test method in the area in question to verify bond quality.

Correlation of Bond Strength and Fokker Bond Tester Readings

The ability of the Fokker bond tester to measure bond strength is completely dependent on the accuracy of the correlation curves, which plot Fokker readings versus either lap shear strength (for metal-to-metal joints) or flatwise tensile strength (for honeycomb joints). For metal-to-metal joints, the correlation is made between the Fokker A-scale readings and tensile lap shear strength. For honeycomb structure, the correlation is made between Fokker B-scale readings and flatwise tensile strength.

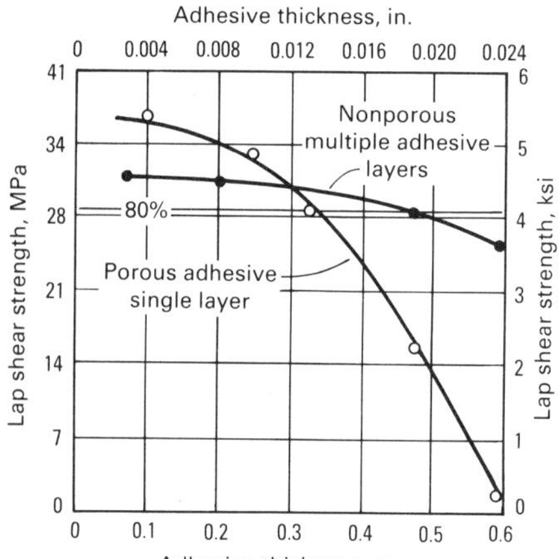

Fig. 58 Lap shear strength versus adhesive thickness for porous and nonporous adhesives. Adherends, 1.6 mm/1.6 mm; adhesive, FM-73 with mat carrier; primer, BR127

Flatwise tensile specimens are prepared in accordance with MIL-STD-401 (Ref 38), except that the adherends and core should simulate those of the bonded test assembly being tested. Fabrication procedures should be identical to those used in preparing the assembly to be tested. Specimens are prepared with various degrees of bond quality by the introduction of porosity in the cured adhesive. The degree of porosity increases with the thickness of the adhesive layer. Figure 53 illustrates the procedure for honeycomb-to-skin joints.

Nondestructive testing of the panels is performed in accordance with procedures outlined in MIL-STD-860 (Ref 23) or as specified in the Fokker bond tester training manual. Instrument readings should be taken within predetermined areas of the test specimen. Each reading should be recorded and correlated with the probe location on the specimen. The specimens are then destructively tested, and the tensile strength values are recorded. Flatwise tensile strength should be plotted versus average ultrasonic readings for each specimen and a representative curve drawn through the plotted points. Two correlation curves should be drawn for specimens with dissimilar thickness adherends. A typical correlation curve for Fokker B-scale readings versus flatwise tensile strength is shown in Fig. 54.

Effect of Porosity and Voids on Bond Testing. A relationship can be established between Fokker bond tester A-scale frequency shift and lap shear strength as related to porosity. Tests have shown that the percentage of porosity increases with adhesive thickness (single layer) (Ref 5, 10). Porosity first appears at an adhesive thickness of 0.25 mm (0.010 in.) and becomes almost completely void at 0.51 mm (0.020 in.) in thickness.

The first step is to produce high-quality lap joints (25.4 mm, or 1.0 in., wide by 13 mm, or ½ in., overlap) containing 100, 80, and 50% width bond lines, as shown in Fig. 55. The test specimens are approximately 178 to 203 mm (7 to 8 in.) long, and a minimum of six specimens should be made for each percentage bond and thickness, that is, 1.0 mm/1.0 mm (0.040 in./0.040 in.) (in general, the t_1 thickness is equal to the t_2 thickness) and so on. Typical average lap shear values for 2024-T3 and FM-73 adhesive are shown in Fig. 56.

The second step is to make specimens with increasing thickness of the single adhesive layer. This will produce specimens of variable quality resulting from porosity. To obtain the quality variation, spacer shims are used, as shown in Fig. 57. The thickness of the spacers is increased in steps (0.10 mm, or 0.004 in., per step) from 0.10 to 0.51 mm (0.004 in. to 0.020 in.), resulting in a highly porous bond joint. The lap shear specimens are numbered and neutron radiographed. The specimens are then Fokker bond tested, and the A-scale readings are recorded for each specimen. The lap shear coupons are cut from each panel, and the adhesive thickness (for each specimen) is accurately measured. The adhesive will generally be found to be slightly thicker than the associated shim. The lap shear test is performed in accordance with Federal Test Method Standard No. 175, method 1033, or Federal Specification MMM-A-132. The neutron radiographs indicate areas of nonuniformity of porosity, and results from these nonuniform areas should be discarded.

The third step is to prepare shimmed specimens using additional layers of adhesive. This will produce good-quality specimens of thick adhesive. The panels are neutron radiographed, Fokker bond tested, cut into specimens, and lap shear tested, with the data subsequently evaluated. Plots of lap shear strength versus adhesive thickness for the single porous layer and multiple

Table 5 Fokker cohesive bond strength acceptance limits for single bond line using FM-73 adhesive

Lower sheet thickness mm	in.	Model 70 probe	Upper limit(b)	Lower limits(a) Class A(c) 80%	Class A/B(c) 65%	Class B(c) 50%	Class C(c) 25%
1.0	0.040	3814	>R8	>L54	>L40	>L25	>L10
1.3	0.050	3814	>R10	>L48	>L36	>L20	>L9
1.6	0.063	3814	>R12	>L42	>L30	>L18	>L8
1.8	0.071	3814	>R16	>L40	>L28	>L18	>L7
2.1	0.081	3814	>R18	>L38	>L25	>L15	>L6

(a) Skin/stringer and skin/frame bonds shall be Class minimum; doubler/skin bonds shall be Class A/B minimum. Additional adhesive layers shall not yield readings less than the lower limits by the following factors: 2 layers (−15); 3 layers (−25); 4 layers (−30). (b) R, right-hand side. (c) L, left-hand side

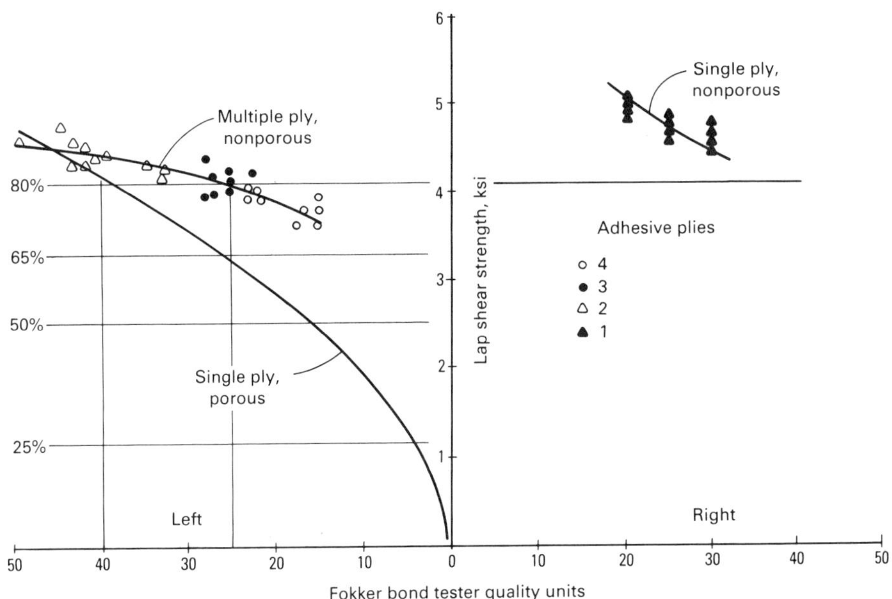

Fig. 59 Lap shear strength versus Fokker bond tester quality units. Bond tester, model 70; probe, 3814; adherends, 1.6 mm/1.6 mm; adhesive, FM-73; primer, BR127

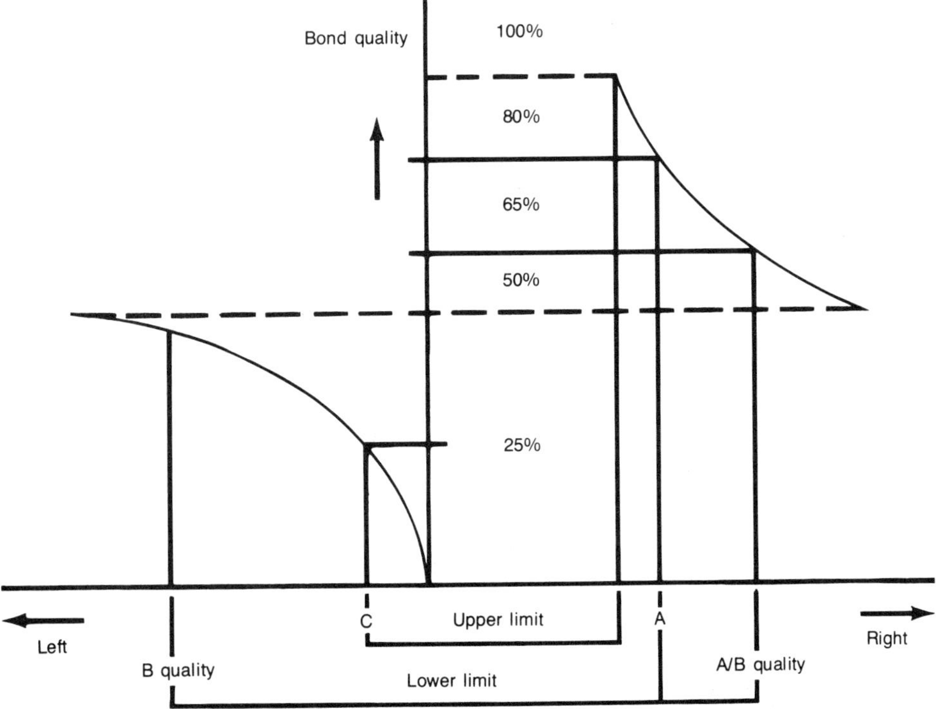

Fig. 60 Establishment of acceptance limits (sheet to sheet)

nonporous layers of FM-73 are shown in Fig. 58.

The fourth step is to plot lap shear strength versus Fokker A-scale readings for each adherend thickness (Fig. 59). If the thickness of the adherends is not equal, the back (unprobed) thickness is plotted. Unequal-thickness adherend specimens are tested from both sides and two quality curves are established. With reference to Fig. 59 and 60, the bond strength can be determined for 100, 80, and 50% specimens having various thickness or t/ℓ ratios. By evaluating all the data, the following classes for single-layer (porous) and multiple-layer (nonporous) joints can be specified:

- *Class A*: Greater than 80%
- *Class A/B*: Greater than 65%
- *Class B*: Greater than 50%
- *Class C*: Greater than 25%

A-Scale Frequency Shift Versus Bond Quality. The large scatter in results for the 25% quality level specimens can be attributed to the extensive porosity in these specimens. This does not pose a problem to the inspector, because quality levels below 50% are usually not acceptable. The actual relationship between the A-scale frequency shift (as seen by the inspector) and the bond quality (class) is illustrated in Fig. 60.

Cohesive Bond Strength Acceptance Limits. Finally, the acceptance limits are incorporated into a table for use by the inspector. Table 5 lists the acceptance limits for single bond line joints with FM-73 adhesive and specific thickness adherends. Establishing Fokker quality diagrams for multiple bond line joints is possible but complicated. Fokker-VFW should be contacted to obtain the recommended procedures for developing such quality diagrams.

REFERENCES

1. M.T. Clark, "Definition and Non-Destructive Detection of Critical Adhesive Bond-Line Flaws," AFML-TR-78-108, U.S. Air Force Materials Laboratory, 1978
2. R.J. Schliekelmann, Non-Destructive Testing of Adhesive Bonded Metal-to-Metal Joints, *Non-Destr. Test.*, April 1972
3. R.J. Schliekelmann, Non-Destructive Testing of Bonded Joints—Recent Developments in Testing Systems, *Non-Destr. Test.*, April 1975
4. P. Bijlmer and R.J. Schliekelmann, The Relation of Surface Condition After Pretreatment to Bondability of Aluminum Alloys, *SAMPE Q.*, Oct 1973
5. K.J. Rienks, "The Resonance/Impedance and the Volta Potential Methods for the Nondestructive Testing of Bonded Joints," Paper presented at the

Eighth World Conference on Nondestructive Testing, Cannes, France, 1976

6. J.C. Couchman, B.G.W. Yee, and F.M. Chang, Adhesive Bond Strength Classifier, *Mater. Eval.*, Vol 37 (No. 5), April 1979
7. J. Rodgers and S. Moore, *Applications of Acoustic Emission to Sandwich Structures*, Acoustic Emission Technology Corporation, 1980
8. J. Rodgers and S. Moore, "The Use of Acoustic Emission for Detection of Active Corrosion and Degraded Adhesive Bonding in Aircraft Structures," Sacramento Air Logistics Center (SM/ALC/MMET), McClellan Air Force Base, 1980
9. The Sign of a Good Panel Is Silence, *Aviat. Eng. Maint.*, Vol 3 (No. 4), April 1979
10. D. Hagemaier and R. Fassbender, Nondestructive Testing of Adhesive Bonded Structures, *SAMPE Q.*, Vol 9 (No. 4), 1978
11. R. Botsco, The Eddy-Sonic Test Method, *Mater. Eval.*, Vol 26 (No. 2), 1968
12. J.R. Kraska and H.W. Kamm, "Evaluation of Sonic Methods for Inspecting Adhesive Bonded Honeycomb Structures," AFML-TR-69-283, U.S. Air Force Materials Laboratory, 1970
13. N.B. Miller and V.H. Boruff, "Evaluation of Ultrasonic Test Devices for Inspection of Adhesive Bonds," Final Report ER-1911-12, Martin Marietta Corporation, 1962
14. J. Arnold, "Development of Non-Destructive Tests for Structural Adhesive Bonds," WADC-TR-54-231, Wright Air Development Center, 1957
15. R. Schroeer, *et al.*, The Acoustic Impact Technique—A Versatile Tool for Non-Destructive Evaluation of Aerospace Structures and Components, *Mater. Eval.*, Vol 28 (No. 2), 1970
16. H.M. Gonzales and C.V. Cagle, Nondestructive Testing of Adhesive Bonded Joints, in *Adhesion*, STP 360, American Society for Testing and Materials, 1964
17. D.F. Smith and C.V. Cagle, Ultrasonic Testing of Adhesive Bonds Using the Fokker Bond Tester, *Mater. Eval.*, Vol 24, July 1966
18. H.M. Gonzales and R.P. Merschell, Ultrasonic Inspection of Saturn S-II Tank Insulation Bonds, *Mater. Eval.*, Vol 24, July 1966
19. R.J. Botsco, Nondestructive Testing of Composite Structures With the Sonic Resonator, *Mater. Eval.*, Vol 24, Nov 1966
20. R. Newschafer, Assuring Saturn Quality Through Non-Destructive Testing, *Mater. Eval.*, Vol 27 (No. 7), 1969
21. "Fokker Ultrasonic Adhesive Bond Test," MIL-STD-860, 1978
22. R.J. Schliekelmann, Non-Destructive Testing of Adhesive Bonded Joints, in *Bonded Joints and Preparation for Bonding*, AGARD-NATO Lecture Series 102, Advisory Group for Aerospace Research and Development, 1979
23. D.J. Hagemaier, Bonded Joints and Non-Destructive Testing: Bonded Honeycomb Structures, 2, *Non-Destr. Test.*, Feb 1972
24. D. Wells, NDT of Sandwich Structures by Holographic Interferometry, *Mater. Eval.*, Vol 27 (No. 11), 1969
25. R.M. Grant, "Conventional and Bead to Bead Holographic Non-Destructive Testing of Aircraft Tires," Paper presented at the 1979 ATA NDT Forum, Seattle, WA, Sept 1979
26. P.R. Vettito, A Thermal I.R. Inspection Technique for Bond Flaw Inspection, *J. Appl. Polym. Sci. Appl. Polym. Symp. No. 3*, 1966
27. C. Searles, Thermal Image Inspection of Adhesive Bonded Structures, in *Proceedings of the Symposium on the NDT of Welds and Materials Joining*, 1968
28. W. Woodmansee and H. Southworth, Detection of Material Discontinuities With Liquid Crystals, *Mater. Eval.*, Vol 26 (No. 8), 1968
29. S. Brown, Cholestric Crystals for Non-Destructive Testing, *Mater. Eval.*, Vol 26 (No. 8), 1968
30. D.H. Collins, Acoustical Holographic Scanning Techniques for Imaging Flaws in Reactor Pressure Vessels, in *Proceedings of the Ninth Symposium on NDE* (San Antonio), April 1973
31. *A Sample of Acoustical Holographic Imaging Tests*, Holosonics Inc.
32. "Adhesive Bonding (Structural) for Aerospace Systems, Requirements for," MIL-A-83377
33. "Adhesive Bonded Aluminum Honeycomb Sandwich Structure, Acceptance Criteria," MIL-A-83376
34. "Inspection Requirements, Nondestructive: For Aircraft Materials and Parts," MIL-I-6870 (ASG)
35. "Nondestructive Testing Personnel Qualification and Certification," MIL-STD-410
36. B.G. Martin, *et al.*, Interference Effects in Using the Ultrasonic Pulse-Echo Technique on Adhesive Bonded Metal Panels, *Mater. Eval.*, Vol 37 (No. 5), 1979
37. P.L. Flynn, Cohesive Bond Strength Prediction for Adhesive Joints, *J. Test. Eval.*, Vol 7 (No. 3), 1979
38. "Sandwich Construction and Core Materials; General Test Methods," MIL-STD-401

Boilers and Pressure Vessels

DURING THE FABRICATION of a boiler, pressure vessel, and such related components as boiling water reactor piping or steam generator tubes, various types of nondestructive inspection (NDI) are performed at several stages of processing, mainly for the purpose of controlling the quality of fabrication. In-service inspection is used to detect the growth of existing flaws or the formation of new flaws. This can be done while the vessel is in operation or down for servicing. The inspection methods used include visual, radiographic, ultrasonic, liquid penetrant, magnetic particle, eddy current, and acoustic emission inspection, as well as replication microscopy and leak testing. The assurance of component quality depends largely on the adequacy of NDI equipment and procedures and on the qualification of personnel conducting the inspection. In many cases, nondestructive inspection, both prior to and during fabrication, must be done to sensitivities more stringent than those required by specifications. The use of timely inspection and rigid construction standards results in the reduction of both the costs and delays due to rework.

Quality planning starts during the design stage. For inspections to be meaningful, consideration must be given to the condition of the material, the location and shape of welded joints, and the stages of production at which the inspection is to be conducted. During fabrication, quality plans must be integrated with the manufacturing sequence to ensure that the inspections are performed at the proper time and to the requirements of the applicable standard. In the newest nuclear plants, quality design planning includes:

- Avoidance of complex weld geometries to facilitate attachment of ultrasonic transducers to the surface at the best positions
- The increased use of ring forgings for pressure vessel components; this means that there are no longitudinal welds that have to be inspected in service. The result is a reduction in the amount of in-service inspection and man-rem exposures
- Incorporating large numbers of access points for introducing mechanized inspection equipment, which can be operated remotely, thus avoiding exposures to operators and enabling more accurate processing than is possible with handheld inspection equipment
- The elimination of welds between cast austenitic components; inspection of welds through cast welds is difficult because they are opaque to ultrasonic inspection to a large degree (Ref 1)

Boiler and Pressure Vessel Code and Inspection Methods*

Pressure vessels—both fossil fuel and nuclear—are manufactured in accordance with the rules of the applicable American Society of Mechanical Engineers (ASME) Boiler and Pressure Vessel Code (Ref 2). For nuclear vessels, section XI of the ASME code establishes rules for continued nondestructive inspections at periodic intervals during the life of the vessel. One feature of the rules in section XI is the mandatory requirement that the vessel be designed so as to allow for adequate inspection of material and welds in difficult-to-reach areas. Section III of the code describes the material permitted and gives rules for design of the vessel, allowable stresses, fabrication procedure, inspection procedure, and acceptance standards for the inspections.

Pressure vessels are constructed in various sizes and shapes, and some of the largest are those manufactured for the nuclear power industry. Some pressure vessels are more than 6 m (20 ft) in diameter and 20 m (70 ft) in length and weigh almost 900 Mg (1000 tons). Thickness of the steel in the walls of these vessels ranges from about 150 mm (6 in.) to more than 400 mm (15 in.), although many pressure vessels and components are fabricated from much thinner material. Joining of the many vessel sections is accomplished by welding. Welders of pressure vessels are qualified according to section IX of the ASME Boiler and Pressure Vessel Code, and welding is done in accordance with qualified welding procedures. Nondestructive inspection of welds is only a part of the inspection requirements; the materials themselves must be inspected prior to welding. For pressure vessels that are not constructed according to the ASME code, it is a matter of agreement between the manufacturer and the user as to whether NDI methods are to be employed and which method or methods are to be used.

*The material on replication microscopy in this section was prepared by H.I. Newton, Babcock & Wilcox.

Nondestructive Inspection Methods. An appendix to each section of the ASME code establishes the methods for performing nondestructive inspection to detect surface and internal discontinuities in materials. Four inspection methods are acceptable: radiographic, magnetic particle, liquid penetrant, and ultrasonic inspection. All these methods are mandatory for nuclear vessels, for section III, and for division 2 of section VIII of the code. Ultrasonic inspection is listed in division 1 of section VIII of the code as nonmandatory. Leak testing, eddy current inspection, acoustic emission inspection, and visual inspection are included in section V. Details as to which method is to be used and the required acceptance standards are specified in the appropriate articles on materials and fabrication. All NDI personnel must be qualified and certified to SNT-TC-1A procedures (Ref 3).

Radiographic Inspection. Methods of radiographic inspection are extensively detailed in the ASME codes; radiography using either x-rays or radioisotopes as the radiation source is permitted. Radiography is the oldest inspection method detailed in the codes and is probably the most understood and the most widely accepted. A principal reason for its wide use is that radiography provides a permanent record of the results of the inspection. This record is important because the inspector can review the radiographs at any time to ensure that federal, state, or insurance requirements have been met.

Acceptance standards were developed according to the limits of radiography (what can or cannot be detected by the method) and by the quality level obtainable by the manufacturing practices used to produce the vessels. Essentially, the acceptance standards do not permit the existence of indications of the following types of flaws: cracks, incomplete fusion, incomplete penetration, slag inclusions exceeding a given size that is not related to the thickness of the part, and porosity that exceeds that presented in illustrated charts provided in the codes. These standards result from the ability to distinguish among porosity, slag, and incomplete fusion in the radiograph; more important, they also mean that no indications of cracks or of incomplete fusion are permitted.

Magnetic Particle Inspection. The procedures for magnetic particle inspection reference ASTM E 709 (Ref 4) or section V of the ASME code for the method. Acceptance standards permit no cracks, but rounded indications of discontinuities are permitted provided they do not exceed a certain size or number in a specified area. Magnetic particle inspection is universally used on ferromagnetic parts, on weld preparation edges of ferromagnetic materials, and on the final welds after the vessel has been subjected to the hydrostatic test. A magnetic particle inspection must be conducted twice on each area, with the lines of magnetic flux during the second application at approximately 90° to the lines of magnetic flux in the first application. Depending on the shape of the part and its location at the time of inspection, magnetization can be done by passing a current through the part or by an encircling coil and sometimes by a magnetic yoke. The acceptance level is judged by a qualified operator and is subject to review by an authorized code inspector.

Liquid penetrant inspection is usually employed on nonferromagnetic alloys, such as some stainless steels and high-nickel alloys. The acceptance standards are the same as for magnetic particle inspection and are also judged by an operator, subject to review by a code inspector. The methods are specified to those contained in ASTM E 165 (Ref 5) or section V of the ASME code. Water-washable, postemulsifiable, or solvent-removable penetrants can be used. A water-washable color-contrast penetrant is usually employed because it is easy to handle, requires no special ventilation, and is nontoxic. Sometimes, special requirements dictate the use of either a solvent-removable color-contrast penetrant or a fluorescent penetrant.

Ultrasonic inspection is used to inspect piping, pressure vessels, turbine rotors, and reactor coolant pump shafts. Straight-beam ultrasonic inspection is specified to detect laminations in plates and to detect discontinuities in welds and forgings. This technique is described in general and specific terms in section XI of the ASME code, in the United States Nuclear Regulatory Commission Regulatory Guide 1.150 (Ultrasonic Testing of Reactor Vessel Welds During Preservice and Inservice Examinations), and in companion reports written by utility *ad hoc* committees (Ref 1). Angle-beam inspection is specified for welds, and a more detailed procedure is presented, including reporting requirements. It is mandatory, however, that ultrasonic inspection, either by straight beam or angle beam, be conducted to a detailed written procedure. These procedures are usually developed by the manufacturer. Specifications and standards for steel pressure vessels are given in ASTM A 577 (Ref 6), A 578 (Ref 7), and A 435 (Ref 8). Acceptance standards for the inspection of welds by ultrasonics closely parallel the acceptance standards for radiography. Cracks, incomplete fusion, and incomplete penetration are not permitted. The size permitted for other linear indications is the same for the slag permitted by radiography. However, ultrasonic inspection can detect cracks better than radiography, but it is sometimes difficult to separate cracks from other linear indications by ultrasonics. Furthermore, ultrasonic inspection procedures refer to the amplitude of the signal obtained from a calibration notch, hole, or reflector placed in a standard reference block, but not all slag inclusions or cracks in an actual workpiece present a similar response to that obtained from the artificial calibrator.

Advanced ultrasonic systems (see the section "In-Service Quantitative Evaluation" in this article) and the improvements in codes and regulations have combined to make ultrasonic inspection one of the most commonly used nondestructive methods in the power industry. Advanced ultrasonic methods are intended to ensure that the vessel remains fit for continued service by detecting and sizing defects that could degrade structural integrity.

Acoustic emission (AE) inspection has been used for the following applications (Ref 9):

- Inspection of chemical and petrochemical vessels
- Monitoring nuclear plant components or systems during hydrotests, plant operation, or preservice pressure testing of the primary system
- Monitoring during pressure testing of intentionally flawed vessels
- Monitoring fiber-reinforced plastic tanks, with the major problems being associated with poor manufacturing techniques that allow dust and other foreign objects to be mixed in with the resin
- Monitoring liquefied petroleum gas storage tanks. The main problems associated with this type of test program include the different propagation paths and attenuation coefficients caused by the geometry of the vessel and the correct transducer locations and spacing

Rupture tests on experimental vessels with wall thicknesses up to 150 mm (6 in.) have been monitored, and such tests help define the acoustic emission response patterns that can be used to recognize incipient vessel failures. However, many of the vessel rupture tests monitored by acoustic emission have been conducted mainly to provide fracture mechanics data, and the acoustic emission monitoring was an add-on feature (Ref 9).

Acoustic emission tests are often conducted during preshutdown operations in an effort to identify areas requiring special maintenance or during special tests while under slightly varied operational conditions. Data on wave propagation and failure mechanisms have been recorded and used to develop reliable acoustic emission evaluation techniques (Ref 10).

The advancement of AE inspection techniques includes the introduction of an AE methodology into Section V, Article 12, of the ASME Boiler and Pressure Vessel Code as a December 1988 Addendum. Other AE methodologies in the inspection of fiberglass and metal pressure vessels are described in the article "Acoustic Emission Inspection" in this Volume.

Eddy Current Inspection. One of the rapidly increasing applications for this method is the inspection of thin-wall (0.9 to 1.5 mm, or 0.035 to 0.060 in.) Inconel alloy steam generator tubing (Ref 11, 12) and heat exchanger tubing (Ref 13). The focus of steam generator tube inspection by eddy current has shifted from concentrating solely on the detection and characterization of tube wall denting and wastage to more complex tube wall degradation mechanisms. The newer degradation mechanisms consist of intergranular attack (IGA), stress-corrosion cracking (SCC), mechanical wear, and pitting in the presence of copper. A summary of steam generator tubing defects is shown in Fig. 1. In response to the complexity of these newer problems, eddy current instrumentation has also evolved from a single-frequency to a multiple-frequency configuration. The analog instrumentation has been replaced by digital multiple-frequency instrumentation, offering more consistent data acquisition. The wider dynamic range offered by the digital instrumentation allows analysis of eddy current data obtained from traditionally difficult areas, such as dented tube support/tube sheet interfaces and roll transition/roll expansion areas within the tube sheet. Additional information, including typical data produced by multiple-frequency instrumentation, can be found in the article "Eddy Current Inspection" in this Volume.

Another advancement is that of remote-field eddy current testing, which has been used to examine nuclear fuel rods and other tubular products. The article "Remote-Field Eddy Current Inspection" in this Volume can be consulted for details and results of analyses.

Replication microscopy, or field metallography, is an effective and economical means of obtaining an image of a component surface, permitting both on-site and laboratory examination and evaluation of the metallurgical condition of the material (see the article "Replication Microscopy Techniques for NDE" in this Volume). Material sample removal from critical components is a costly and time-consuming process that often necessitates part re-

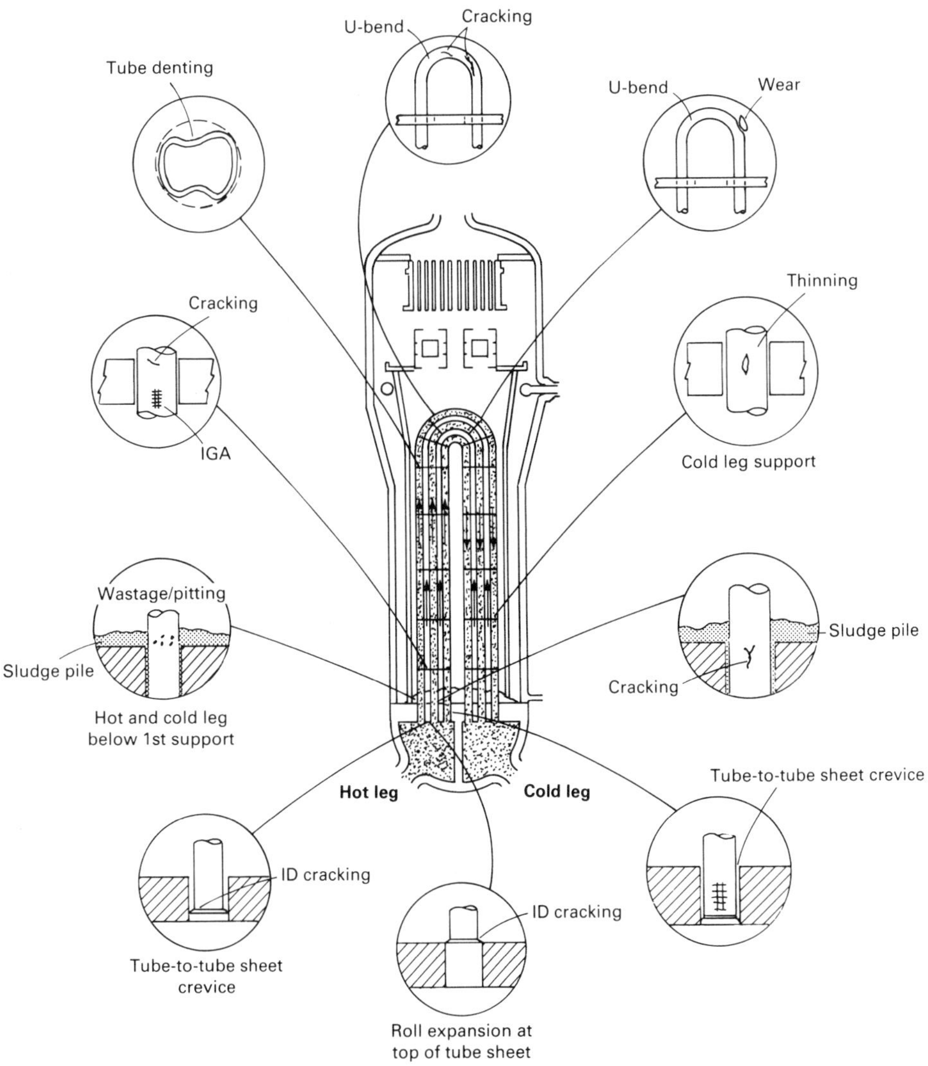

Fig. 1 Locations of known tube wall degradations in recirculating steam generators. Source: Ref 11

placement, weld repair, and stress relief. The replication examination is completed without having to cut out a portion of the component, and the metallurgical data can be determined in fine detail. Therefore, material evaluation can be performed with a nondestructive and cost-effective method that does not require the removal of material samples.

Specimen (Surface) Preparation. Critical surface locations for inspection are identified, and external insulation is removed. The area to be replicated is polished with progressively finer grits and compounds. The final compound is a 0.25 μm diamond polishing paste, which ensures a surface finish of 1 μm or better.

The surface is then etched to reveal the microstructural detail (Fig. 2). A plastic film is applied, cured, and carefully removed. The surface of the film in contact with the polished area retains a precise, reverse image of the etched component surface.

Procedure/Defect Analysis. In the laboratory, the replica is examined using standard light (optical) microscopy, or it can be viewed at high magnification with an electron microscope. To improve resolution, the replica is enhanced by shadowing with sputtering or vacuum evaporation techniques.

An actual replica magnified to 400× is shown in Fig. 3. The microstructure indicates that linked creep voids exist along the grain boundaries in this component, suggesting that creep is active and that cracking is imminent. The progression of damage from microstructural voiding to linked voids to cracking is schematically illustrated in Fig. 4. Replicas can reveal all the microvoids, microcracks, and microstructural features in exact detail, just as if the metal surface itself were being viewed.

The principal application of the replication technique is in revealing metallurgical anomalies and incipient damage (cracking). In the absence of fatigue and/or preexisting defects, creep cracks are theorized to initiate by the formation, growth, and linkup of voids into microcracks, which in turn consolidate to form macrocracks (Fig. 5). The propagation of the macrocracks can then lead to final failure.

In actuality, a creep-fatigue interaction better describes the type of damage mechanism most often encountered. It is theorized that the creep mechanism described previously is accomplished by a fatigue mechanism. In a creep-fatigue interaction, one may observe a mixture of intergranular and transgranular crack growth surfaces

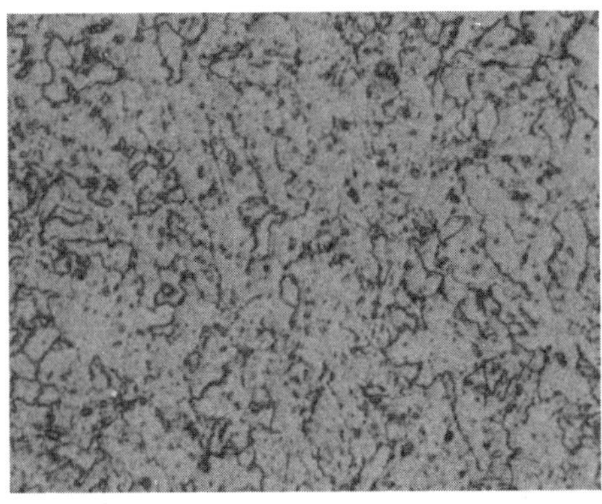

(a)

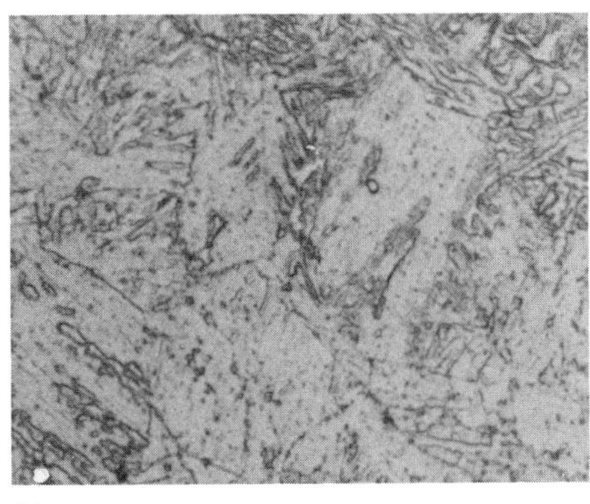

(b)

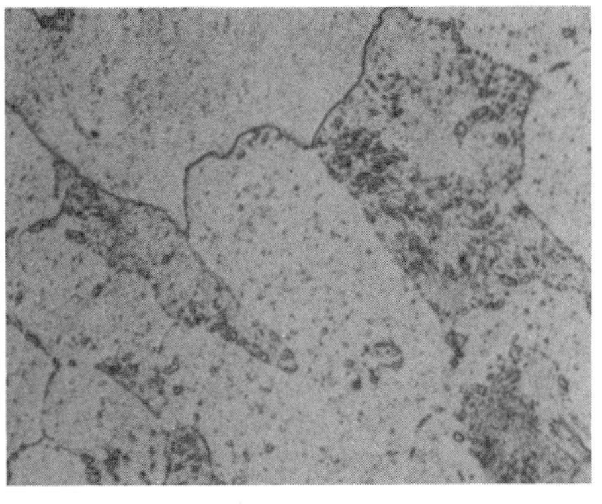

(c)

Fig. 2 Weld microstructures showing no evidence of microvoids in a secondary superheater outlet header made from 1.25Cr-0.5Mo steel. (a) Weld metal. (b) Heat-affected zone. (c) Base metal. Weld location: nozzle-to-header weld (header side of weld). All 400×. Courtesy of H.I. Newton, Babcock & Wilcox

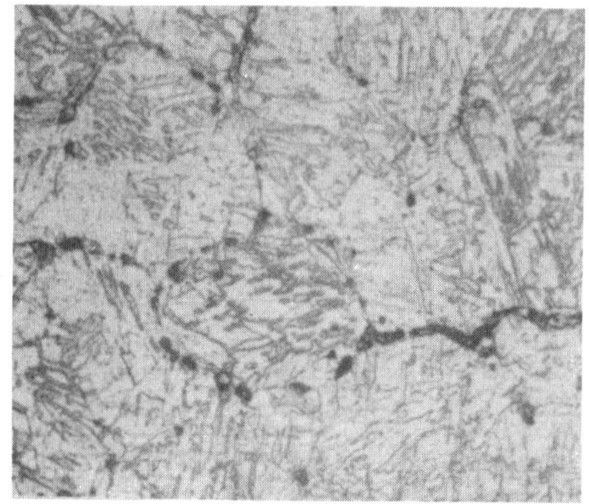

Fig. 3 Replica showing random and linked microvoids in the heat-affected zone of a secondary superheater outlet header weld. 400×. Courtesy of H.I. Newton, Babcock & Wilcox

with the absence of voiding. The presence of preexisting flaws affords an initiation site for cracking and subsequent material degradation. Actively growing cracks are evidence that the condition of a component is deteriorating under service loads; this warrants monitoring, repair, or replacement. Creep-fatigue interaction is discussed in detail in the article so titled in Volume 8 of the 9th Edition of *Metals Handbook*.

The advantages of the replication techniques are many. It is nondestructive. An accurate image of the component surface is also obtained. In addition, early detection of damage and the evaluation of existing failure mechanisms are possible. Replicas also reinforce decision making with regard to the need to replace or repair critical degraded components.

Limitations. There are also disadvantages with the replication technique. The analysis is limited to the accessible surface of the component. The replication technique is also site sensitive. In addition, no compositional analysis of the microstructural constituents is possible with surface replicas, although extraction replicas have been used in chemical analysis (see the section "Precipitate Analysis" in the article "Replication Microscopy Techniques for NDE" in this Volume). With these limitations in mind, it is recommended that replication be used in conjunction with other nondestructive techniques to provide a complete overall evaluation of a component.

Inspection of Plates, Forgings, and Tubes

Plates up to 300 mm (12 in.) thick, such as those conforming to ASME SA-533, grade B (Fe-Mn-Ni-Mo) are used for pressure vessel shell sections. All plates are ultrasonically inspected 100%. Pulse-echo search units have been adapted for this inspection. The search unit consists of an immersion-type longitudinal wave or shear wave transducer encased in a container holding a liquid couplant. Because the container can conform to minor surface irregularities on the material being inspected, much of the surface conditioning previously required for contact inspection is eliminated. Good, sound penetration and a stable back reflection are obtained with this type of unit.

Straight-beam inspection is conducted over the entire surface of the plate. Calibration for straight-beam inspection is accomplished by adjusting the back-reflection amplitude from the opposite surface to approximately 75% of the height of the CRT. Discontinuities causing a loss of back reflection that cannot be contained in a 75 mm (3 in.) diam circle are unacceptable. Laminar-type discontinuities, which are the most common type of imperfections encountered in plate materials, are readily detected by straight-beam inspection.

The angle-beam inspection of plates is included in many specifications. Scanning is conducted in overlapping passes over the entire surface in two directions at 90° to

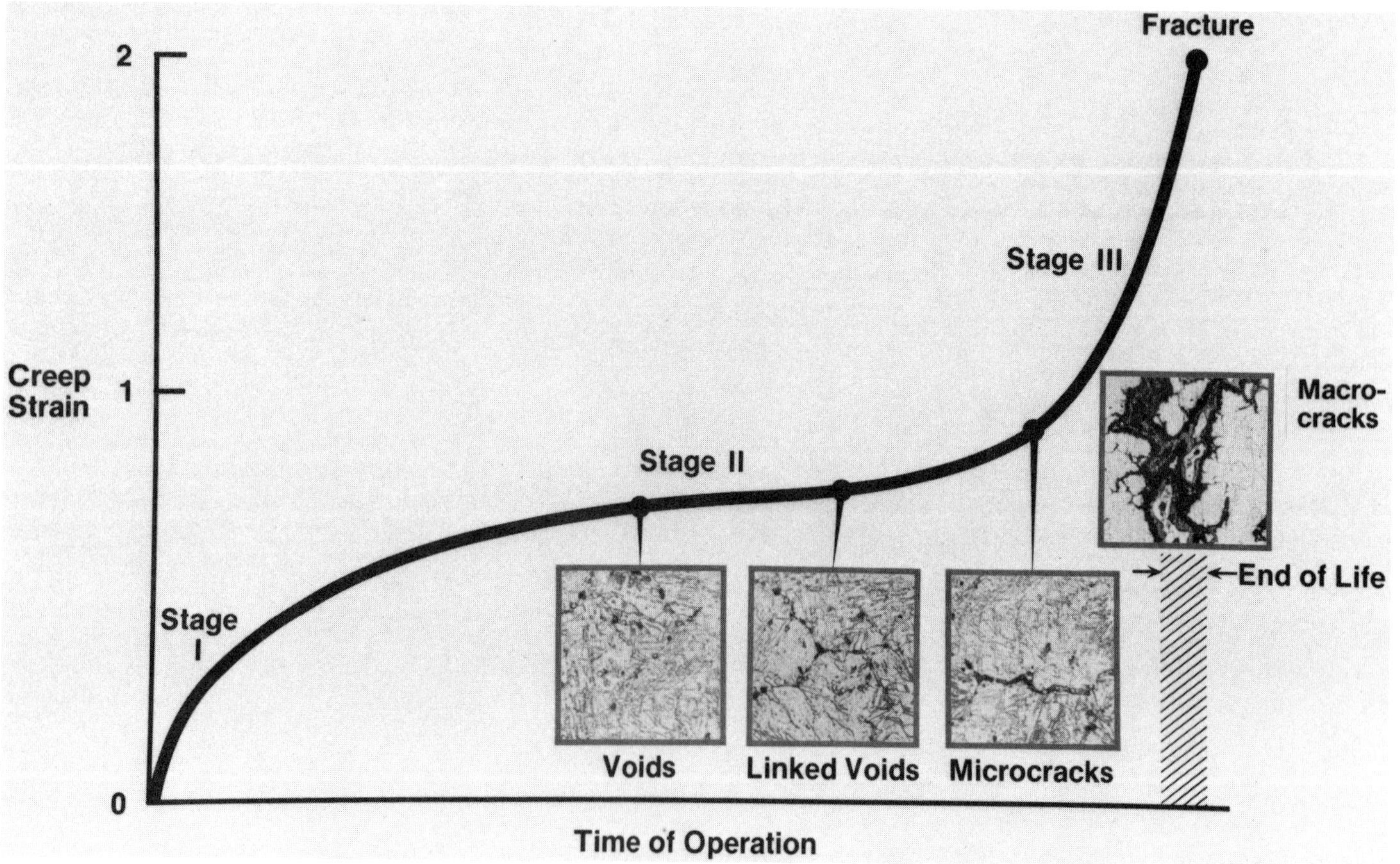

Fig. 4 Progression of creep damage over time as depicted by replica photomicrographs. Courtesy of H.I. Newton, Babcock & Wilcox

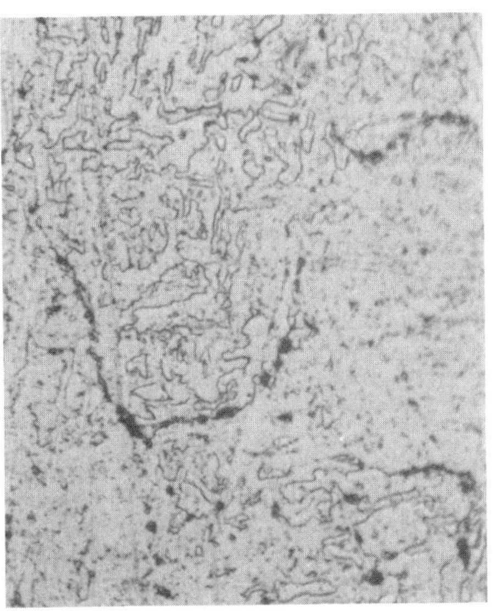

(a)

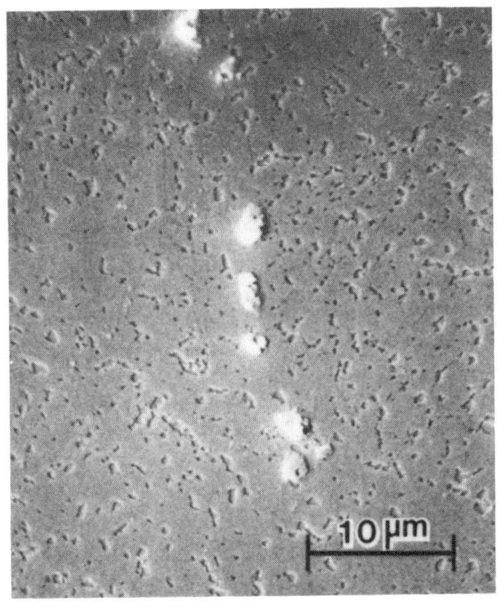

(b)

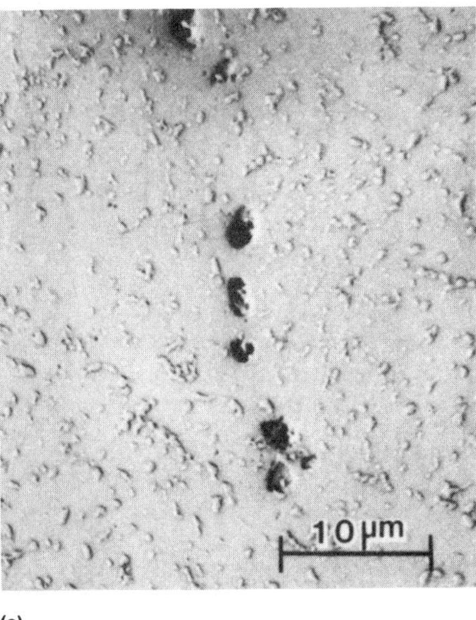

(c)

Fig. 5 Replicas of creep damage in pressure vessel components. (a) Random and linked microvoids in a 1.25Cr-0.5Mo main steam line weld. 425×. Courtesy of H.I. Newton, Babcock & Wilcox. (b) and (c) A comparison of scanning electron micrographs of a replica from 1.25Cr-0.5Mo material with aligned creep cavities. (b) Normal contrast. (c) Reversed image contrast. Etched with 4% picral. Courtesy of E.V. Sullivan, Aptech Engineering Services, Inc.

each other. Calibration is accomplished by adjusting the signal to a square notch milled to a depth of 3% of the plate thickness. Discontinuity indications equal to or exceeding the signal amplitude from the notch are cause for rejection.

During the ultrasonic inspection of plates, close attention must be given to signal patterns that indicate the presence of discontinuities, such as secondary piping, which do not normally produce significant response indications on the oscilloscope. Discontinuities of this type can result in untimely rejection or high rework costs if not detected prior to the start of fabrication.

Forgings used for pressure vessel flanges, head flanges, and nozzles are commonly purchased to the requirements of ASME specifications, such as those of ASME SA-508, class 2 (Fe-Mn-Ni-Cr-Mo-V), and are heat treated and nondestructively inspected by the supplier. Ultrasonic inspection is required on two surfaces of the forging using the straight-beam inspection method; for ring or hollow round shapes, angle-beam inspection is specified. These inspections must be carried out prior to contour machining to ensure complete volumetric coverage. Inspection is repeated to the maximum extent practical after machining and heat treatment are completed. Discontinuity indications observed by straight-beam inspection that cause a loss of back reflection, and by angle-beam inspection that are equal to a 3%-of-plate-thickness notch, will result in the rejection of the component.

Ferromagnetic forgings are inspected by magnetic particle inspection methods after final heat treatment and contour machining. Linear discontinuities such as laps, cracks, or inclusions are cause for rejection.

Tubular products can receive ultrasonic and liquid penetrant inspections prior to release for fabrication. As stated above, eddy current inspection is common during the in-service inspection of tubular products.

Nondestructive Inspection During Fabrication

Plates are formed into shell cylinder segments or into hemispherical segments for the top and bottom heads. Shell plates of thick material are hot formed by pressing to the radius required. Plates for head segments are hot formed over dies machined to the proper contour. After the hot forming is completed, the plates are allowed to cool and are then inspected to confirm that the proper radius has been obtained. The formed section is then reheated, water quenched, and tempered to obtain the required mechanical properties. Samples are cut from the heat-treated material and tested to determine the tensile and impact properties of the material before further fabrication is permitted.

Fit-Up and Welding. Edges of plates that are machined in preparation for welding are magnetic particle inspected. Surface discontinuities, other than laminar types less than 25 mm (1 in.) long, are removed. Discontinuities with any significant thickness are further investigated by radiographic or ultrasonic inspection to determine their nature and extent.

Welding of the longitudinal seams in the shell cylinders is commonly performed by the multilayer submerged arc process. Single arc and tandem arc procedures are used, with preheating and postheating when specified. Welding of head-torus sections can be done using the horizontal submerged arc process, in which the weld beads are deposited through the thickness of the material. With the horizontal submerged arc process, the axis of the weld is in a vertical position, and each weld bead is deposited from the inside to the outside of the seam progressively upward until the complete length is welded. A manual welding procedure (usually, shielded metal arc) in which the axis of the weld is in the vertical position is sometimes used instead of the submerged arc process. These processes require NDI techniques suitable for adequate interpretation of discontinuity size. After welding and interstage stress relieving of the longitudinal seams are completed, the weld is magnetic particle inspected and radiographed.

The design of the weld joint and the welding procedure used are important considerations in determining the extent of nondestructive inspection and the adequacy of the results. The narrow-groove weld joint illustrated in Fig. 6 provides for good radiographic inspection, including crack detection. Those cracks that are more likely to occur along the sidewall of a weld are essentially parallel to the radiographic beam in the joint design. When the radiographs are made with equipment and techniques that obtain good sensitivity in these thick-section welds, cracks of this type can be readily detected. Welds having wide-angle sidewalls require additional angle radiography or ultrasonic inspection.

The depth at which slag inclusions are beneath the surface and the thickness of the inclusions usually cannot be determined by radiographs taken normal to the weld surface. Discontinuity indications observed on

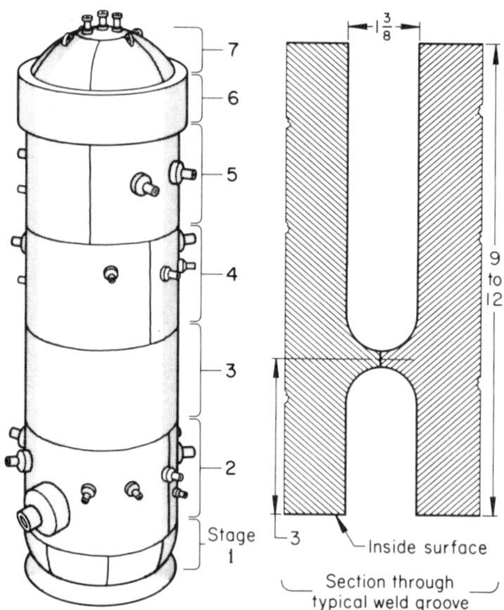

Fig. 6 Sequence of assembly and typical narrow-groove joint used for the welding of thick plates in pressure vessels

the film usually require additional angle radiography or ultrasonic inspection to determine the depth and thickness of the slag inclusions. Acceptance standards for welds made in the vertical position are more rigid than acceptance standards applied to other welding procedures.

Clad Vessels. Certain types of pressure vessels, such as some nuclear vessels or chemical-processing vessels, are fabricated from carbon or alloy steels whose inside surfaces are clad with a layer of stainless steel. With these types of vessels, the inner surfaces are machined smooth to provide a suitable surface for cladding. Machining is done after the shell and head-torus subassemblies have been completed. After machining, the smoothed surfaces are magnetic particle inspected; no indications of linear discontinuities are permitted.

Also at this manufacturing stage, the formed and heat-treated material is again ultrasonically inspected. This inspection is performed using the straight-beam method over the entire surface of the material. Changes in the acoustic quality of the material as a result of heat treatment may cause minor differences in the signal amplitude from those indications previously recorded.

The inner surfaces of large pressure vessels are clad, oscillating multiple-wire systems or strip electrodes, as described in the article "Weld Overlays" in Volume 6 of the 9th Edition of *Metals Handbook*. Three-wire and six-wire self-shielded flux-cored wire, plasma arc processes, and electroslag overlays are also used for pressure vessel applications.

Liquid penetrant inspection is performed on all clad surfaces. When properly cleaned, the as-deposited clad surface is adequate for liquid penetrant inspection. Solvent-removable red-dye penetrants are often used for this inspection. Acceptance standards do not allow linear indications of any type. Rounded indications with less than 1.5 mm (1/16 in.) maximum dimension are acceptable. Mating surfaces of machined flanges and surfaces of O-ring seals are inspected to more rigid acceptance standards. Sealing surfaces must be free of any indications.

Calibration for inspection of the cladding is accomplished by setting up the inspection equipment on a reference block clad by the same process as the material being inspected. A 10 mm (1/3 in.) wide slot is machined through the backing material to the cladding interface. The ultrasonic pattern observed from the slot in the reference block provides the acceptance standard for the cladding.

Inspections are conducted from the clad side of the material, except for small-diameter nozzles. When inspecting from the side opposite the cladding, close observation must be given to the presence of laminar discontinuities in the base material that could mask the area of inspection. Cladding on mating flange surfaces is inspected to a more rigid standard by pitch-catch echo testing. The two-transducer system is calibrated on a 1.5 mm (1/16 in.) diam, side-drilled hole at the interface of the cladding and base material. Discontinuity indications exceeding the amplitude of the signal from the calibration hole are unacceptable.

Assembly. Shell cylinders and flanges are joined together by welding the girth seams using the submerged arc process. Back grooves, final surfaces, and surfaces prepared for welding are magnetic particle inspected. Radiography is performed after cladding the inside of the welded girth seam.

Nozzle openings are cut in the shell cylinder and are then machined to the size and shape required for fitting the nozzles. The narrow-groove weld joint, through the shell thickness, is used for the nozzle-to-shell weld.

Nozzle welds are made by either the manual or the automatic submerged arc process. In addition to magnetic particle and radiographic inspection, all nozzle welds are subjected to ultrasonic inspection. The techniques for the ultrasonic inspection of welds contained in the ASME code are used as the basic method of inspection. Angle-beam inspection is performed on the outer surface of the shell adjacent to the nozzle weld. The sound beam is aimed about 90° to the weld and manipulated laterally and longitudinally so that maximum coverage is obtained.

Sound-beam dispersion from the effect of the clad interface prevents full-node sound transmission. Only the discontinuities encountered in the initial sound path through the material can be consistently detected by the angle beam. Straight-beam inspection is performed by directing the sound into the weld from the inside of the nozzle. The combination of straight-beam and angle-beam inspection provides full inspection coverage of the weld.

For the detection of cracks in welds, ultrasonics and radiography have complementary advantages and disadvantages. Vertical, subsurface cracks in a weld that are readily detectable by radiographic methods are extremely difficult to detect by ultrasonics. Discontinuities oriented in a plane other than normal to the weld surface can be best detected ultrasonically.

Various beam angles and frequencies are used to establish the best ultrasonic procedure that could be applied to production welds. The standard 45° transducer at a frequency of 2.25 MHz is commonly used for the detection of cracks. However, to provide additional assurance, scanning should be performed at a gain setting of at least ten times the calibration sensitivity.

Final Inspection. After assembly is completed and all inspections are verified, the vessel is subjected to a final stress-relieving treatment. Magnetic particle inspection of all exposed ferritic material is again performed, using the yoke method. Arc strikes that are likely to occur with the prod method are thus avoided on the final stress-relieved surface.

Final machining of the vessel and head is done after stress relieving. All machined surfaces are inspected by liquid penetrants after the vessel is dimensionally inspected.

Vessel and head inside surfaces are thoroughly cleaned, and the vessel is positioned vertically for hydrostatic testing. After hydrostatic testing, all exposed ferritic welds are magnetic particle inspected. Instrumentation and attachment welds are liquid penetrant inspected.

A baseline ultrasonic inspection can be performed on the completed vessel. This inspection provides a record of ultrasonic indications that can be used as a reference for future in-service inspections of the vessel. The data generated from this inspection are useful only if the method of inspection during the in-service inspection is compatible with the baseline inspection method. The effect of the cladding interface on the ultrasonic sound beam requires increased sensitivity settings on the equipment if inspection is conducted from the inside of the vessel. Beam-angle changes due to the irregular cladding interface require exact duplication of the position of the transducer and the entrance angle of the sound beam if reproducible results are to be obtained. Equipment capable of remotely controlling these variables is essential if these inspections are to be conducted.

Visual Inspection of Pressure Vessels

Visual inspection can readily detect misalignment, movement of mating parts, surface contamination, cracks, and other sur-

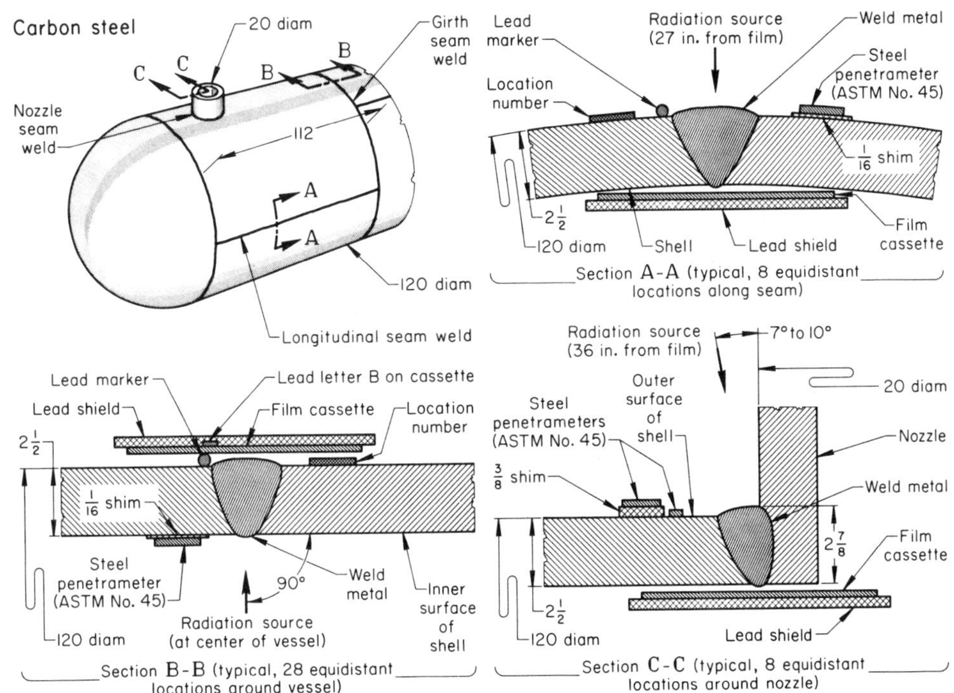

Fig. 7 Nuclear pressure vessel, of 65 mm (2½ in.) thick carbon steel, in which longitudinal, girth, and nozzle welds were radiographically inspected using 50-Ci cobalt-60 as the radiation source, Eastman Kodak type AA radiographic film, and the inspection setups shown. Dimensions given in inches

face discontinuities. Direct examination is possible when the eye or an equivalent detector can be placed within 600 mm (24 in.) of the surface to be inspected. Where radiation levels are too high or access is limited, remote inspections can be used; telescopes, periscopes, borescopes, fiber optics, or television monitoring systems are available (see the article "Visual Inspection" in this Volume).

Visual inspection requirements are extensive in section XI of the ASME Boiler and Pressure Vessel Code. In general, all weld areas not receiving volumetric inspections must be visually inspected after hydrostatic testing. Also, comprehensive inspections are applied to bolting, piping, pumps, and valves.

Radiographic Inspection of Pressure Vessels

A radiograph is basically a two-dimensional picture of the intensity distribution of a form of radiation that is projected from a source (ideally, a point source) and that has passed through a material that attenuates the intensity of the radiation. Voids, changes in thickness, or regions of different composition will, under favorable circumstances, attenuate the radiation by different amounts, producing a projected shadow of themselves. Three forms of penetrating radiation are used in radiography: x-rays, γ-rays, and neutrons. X-rays were the earliest to be used. Detailed information on radiography is available in the article "Radiographic Inspection" in this Volume.

Radiography is normally used to examine welded seams in fabricated pressure vessels. The main reasons for the use of radiography are the following:

- Radiography permits internal inspection of a component
- Radiography supplies a permanent visual record
- Radiography is generally sensitive to discontinuities commonly present in welds

The use of radiography is generally restricted to the inspection of welds for the following reasons:

- The cost of performing radiographic inspections is high when compared to other available methods
- Radiographic inspection of large areas is relatively slow
- Radiography is insensitive to laminar-type discontinuities

Radiography can be used, if the proper equipment is available, to inspect welds in steel pressure vessels with walls up to 400 mm (16 in.) thick. The only conditions that could limit the use of radiography for the inspection of welds in pressure vessels are, first, a lack of accessibility to both sides of the weld for placement of the film and the radiation source and, second, a high degree of variability in shape and thickness in the area being inspected. Both single-wall and double-wall techniques are commonly employed for the radiographic inspection of pipe. Detailed discussions of these techniques are available in the article "Radiographic Inspection" in this Volume. The following examples describe standard radiographic techniques used to inspect common welded seams in pressure vessels.

Example 1: Radiographic Inspection of Longitudinal, Girth, and Nozzle Welds in a Carbon Steel Pressure Vessel. The pressure vessel shown in Fig. 7 was used in a commercial nuclear application and was inspected in accordance with section III of the ASME Boiler and Pressure Vessel Code. The material was 65 mm (2½ in.) thick carbon steel. The weld joints were single-V grooves and were welded using the submerged arc process for both the longitudinal and girth seams. Shielded metal arc welding was used to join the nozzle to the vessel.

For radiographic inspection, cobalt-60 was selected as the radiation inspection because of the thickness of the material, and Eastman Kodak type AA film was selected for its relative speed to give the shortest exposure time. The combination of Co-60 and type AA film can sometimes result in marginal sensitivity that occasionally requires the reshooting of a specific region.

Each longitudinal seam was radiographically inspected with eight shots spaced 350 mm (14 in.) apart using a 180 × 430 mm (7 × 17 in.) film. The penetrameters were steel ASTM No. 45 with a 1.5 mm (1/16 in.) thick shim. The Co-60 radiation source was placed 690 mm (27 in.) from the film. The cassette was loaded with a 0.25 mm (0.010 in.) thick lead screen, two pieces of film, a 0.25 mm (0.010 in.) thick lead screen, and a 3 mm (1/8 in.) thick lead back shield. Section A–A in Fig. 7 shows the setup for radiographic inspection of the longitudinal seam.

The girth weld was inspected using the panoramic technique, in which the Co-60, 50-Ci radiation source was placed at the center of the vessel and 28 film cassettes were placed around the outer surface (section B–B, Fig. 7). The technique involves only one exposure of 25 min to radiograph the entire seam, but its success depends on the placement of the radiation source exactly in the center of the vessel. The disadvantages of this technique include a lengthy setup time, a fairly complicated setup, and additional film costs if an error is made.

An alternative technique is to place the radiation source outside the vessel and inspect the seam with 28 separate exposures. This method would result in higher-quality radiographs, but a much longer total exposure time would be required.

Three equally spaced steel ASTM No. 45 penetrameters with 1.5 mm (1/16 in.) thick shims were used for the 28 radiographs. The film size and the types and thicknesses of screens and filters were the same as those used for the longitudinal seam welds.

The nozzle seam weld was radiographed using eight 13-min exposures. As shown in section C–C in Fig. 7, the radiation source

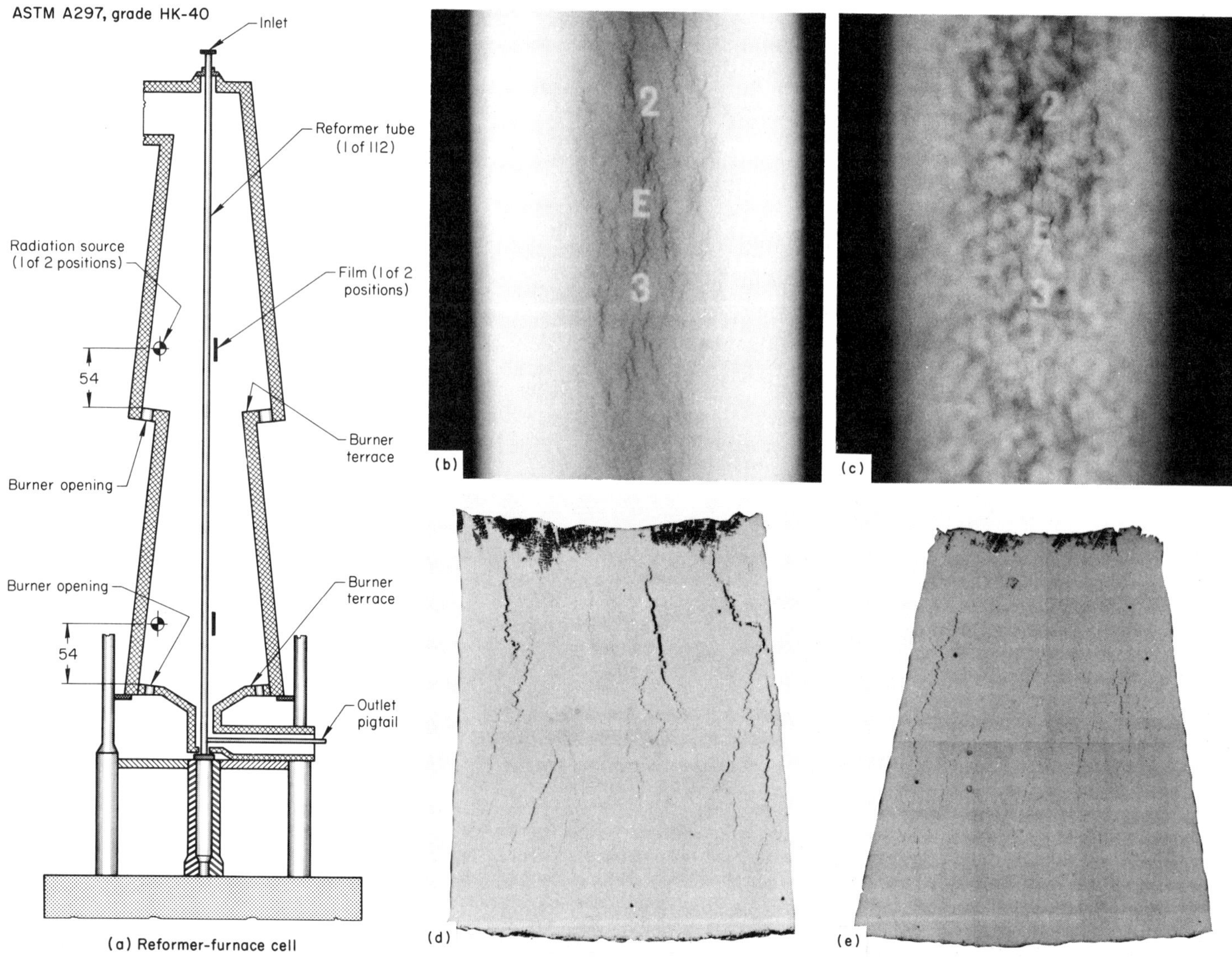

Fig. 8 Reformer-furnace cell from which cast tubes of ASTM A297, HK-40, heat-resistant alloy were radiographically inspected for the detection of creep fissuring. (a) Schematic of furnace cell showing positions of radiographic sources and films. Dimensions given in inches. (b) Radiograph of a section removed from a failed tube that contained no catalyst showing fissures near the ruptured area. (c) Same section as (b) but containing a catalyst. Fissures are visible but less apparent. (d) Macrograph showing fissures in a tube that were detected by radiography. 6×. (e) Macrograph showing fine fissuring that was not indicated by radiography. 6×

was placed outside the vessel and was offset at an angle of 7 to 10° away from the nozzle. The source-to-film distance was 900 mm (36 in.). The variation in weld thickness at the nozzle required that two penetrameters be used. One was flush to the sidewall, and the other had a 9 mm (3⁄8 in.) thick shim, which compensated for the difference in weld thickness. This two-penetrameter setup qualified the sensitivity of the radiograph because of the differential thickness of the weld being inspected. A sensitivity of 2-2T and a density of 2.0 were specified for the radiographs of all three types of welds.

Example 2: Radiographic Inspection for Creep Fissures in Reformer-Furnace Tubes (Ref 14). About 1 year after start-up, two steam-methane reformer furnaces were subjected to short-time heat excursions because of a power outage, which resulted in creep bulging in the Incoloy 800 outlet pigtails, requiring complete replacement. It was thought that during this heat excursion some of the reformer tubes experienced slight bulging, about 1%, by plastic deformation. However, this type of bulging does not necessarily shorten the life of a tube in terms of creep.

Each furnace had three cells, consisting of 112 vertical tubes per cell, each filled with a nickel catalyst. The tubes were centrifugally cast from ASTM A297, grade HK-40 (Fe-25Cr-20Ni-0.40C), heat-resistant alloy. The tubes had an outside diameter of 150 mm (5.85 in.) and a minimum wall thickness as-cast of 20 mm (0.802 in.), and they were more than 13.7 m (45 ft) in length. The design limits for the tubes were 2200 kPa (320 psi), with a surface temperature of 958 °C (1757 °F). Operating limits were 2140 kPa (310 psi) at a surface temperature of 943 °C (1730 °F). The furnace cell is illustrated in Fig. 8(a).

The first tube failure occurred after 33 000 h of operation. The unit was shut down, and the failed tube was removed for metallurgical inspection. The results indicated that the tube failed from creep rupture (stress rupture). The tube failure instigated a project for detecting midwall creep fissuring. Radiography had been reported to be limited to detecting only severe third-stage creep and not the early stages.

Laboratory Radiography. Preliminary studies were made to determine type of radiation source and strength, radiation source-to-film distance, exposure time, and effect of catalyst in tube. Figure 8(b) shows a laboratory radiograph of a section removed from the failed tube without catalyst. Fissures were clearly visible near the ruptured area and diminished to nondetectable fissures at a point 400 mm (16 in.) below the rupture. The same tube section radiographed under similar conditions but with catalyst in the tube is shown in Fig. 8(c). Fissures are visible, but it is apparent that the catalyst reduced the sensitivity by masking the smaller fissures.

Liquid penetrant inspection and macroexamination of the specimen tube revealed the gross fissuring that was easily detectable and those fissures that were undetectable by radiography. The largest fissures undetectable by radiography were approximately 3 mm (0.125 in.) long and 0.13 mm (0.005 in.) wide. (Fissures have been detected that were between 5 and 10 mm, or 0.200 and 0.400 in., long.)

Based on the radiography and macroexamination results, it was decided that future tube replacements would be inspected by radiography, preferably with the catalyst removed. If the fissure were large enough to show on a radiograph, either with or without the catalyst, the tube could be expected to fail within 1 year.

In-Service Radiographic Inspection. During shutdown of one furnace, the first full test of the radiographic inspection method was made. All the tubes were measured for evidence of outside diameter growth, and the suspect areas were strapped to identify the hottest zones. Growth to failure in HK-40 material having a rough exterior surface is difficult to measure, because total creep of only about 1% has occurred by the time of failure. Gaging did not show much, but strapping on some tubes clearly identified bulging in the hottest areas about 1.4 m (54 in.) above the burner terrace. This was used as the basis for radiographing all other tubes.

The catalyst was removed from the tubes in preparation for radiography. Two IR-192 radiation sources having a strength of 3300 GBq (90 Ci) and one 3700 GBq (100 Ci) source were used, mounted in jigs at distances of 750 and 900 mm (30 and 36 in.), respectively, from the film. A tungsten collimator was also used to limit the emission to a single tube and permit the technicians to remain in the firebox during exposure. Exposure time was 7 to 10 min per shot, and approximately 700 shots were taken.

Shortly after the first furnace was back on stream, the second furnace was shut down and the tubes radiographed. Using the same techniques but with a 3900 GBq (105 Ci) source, time to radiograph the second furnace was reduced about 25%.

Twenty-four tubes in the first furnace and 53 in the second furnace showed significant fissuring. One of these fissured tubes was left in the first furnace and 15 in the second; when these tubes eventually failed, they would provide an indication of remaining life after a known radiographic examination.

One of the tubes that had been removed from the furnace was sectioned at two places—through a radiographic indication of a fissure and through an area containing no indications. Macrographs of these sections are shown in Fig. 8(d) and (e). Figure 8(d) shows fissuring that was detected by radiography. Fine fissuring that was not seen on the radiograph is shown in Fig. 8(e).

Conclusion. Radiography was a practical, economical method of detecting the creep fissuring, and it provided advance information for purchase of replacement tubes. However, because radiography was limited to detecting fissures caused by third-stage creep in tubes from which the nickel catalyst had been removed and because of the cost of removing the catalyst, ultrasonic techniques were developed for inspecting the tubes. These techniques are described in Example 4 in this article.

Ultrasonic Inspection of Pressure Vessels

Volumetric inspections are used to inspect the volume of material bounded by the surfaces of components and of piping. Theoretically, any source of energy that penetrates the volume of a material can be used. In practice, however, only x-rays, γ-rays, and ultrasonic waves are used. In nuclear vessels, which are housed in a containment building, radiographic inspection methods for in-service inspection currently have limited use because of the need for access to both surfaces and because of the high γ-ray background in most areas of the containment building. When radiography is used, other inspection methods such as acoustic emission can also be used for additional monitoring. However, a large number of inspections are performed using various ultrasonic techniques.

As described in the article "Ultrasonic Inspection" in this Volume, ultrasonic waves are generated by piezoelectric transducers that convert high-frequency electrical signals into mechanical vibrations. These mechanical vibrations form a wave front, which is coupled to the vessel being inspected through the use of a suitable medium. Several wave modes can be used for inspection, depending on the orientation and location of the discontinuities that exist. Longitudinal, shear, and surface waves are used separately in different techniques to reveal discontinuities that are respectively parallel to, at an angle to, and on or near the surface from which the inspection is performed. These inspections are made with longitudinal wave transducers, pulse-echo, through transmission, pitch-catch, or delta techniques. Most of the inspections are performed using pulse-echo straight longitudinal wave beams and angled shear wave beams from a single transducer.

Manual ultrasonic inspection is performed with single-transducer pulse-echo techniques. The ultrasonic wave is coupled to the component being inspected through a couplant (usually a glycerin or light oil) that transports the ultrasound between the face of the transducer and the surface of the component. In longitudinal wave straight-beam inspection, ultrasonic waves enter normal to the surface and will detect discontinuities perpendicular to the direction of wave propagation. In angle-beam inspection, the longitudinal wave is coupled through a plastic wedge at an angle to the surface. The wave is again coupled to the component (still a longitudinal wave) and undergoes a mode conversion to a shear wave. The angle of the resulting shear wave is dependent on the ratio of the combination of ultrasonic velocities in the plastic wedge and the metal component with the angle of the incident longitudinal wave in the plastic wedge. Shear wave angles from 40 to 75° are used for inspection. Indications are recorded and plotted from reference locations on the weld metal.

Automated Ultrasonic Inspection. For accurate inspections, immersion ultrasonic inspection provides the highest level of inspection speed, accuracy, and repeatability. The coupling of ultrasonic waves is done with much less variability in couplant thickness and transducer pressure than occurs in manual techniques. The immersion method permits automated scanning and digital recording of data with a high degree of precision and repeatability. Additional advanced automated methods are described below in the section "In-Service Quantitative Evaluation" in this article.

Thickness Measurements. Inspection should properly begin with thickness measurements of the shell, heads, nozzles, and piping. Well-documented inspection records are necessary to complete a satisfactory inspection. Records can indicate where to expect metal loss or corrosion on the component and therefore enable the suspect areas to be thoroughly inspected. A comparison of measurements obtained during previous inspections will determine the amount of loss and corrosion rates. Original or nominal thicknesses taken from specifications usually have tolerances too great to make them reliable in the determination of corrosion rates. Vessels having a history of minimal corrosion require only a moderate amount of inspection, while those having a high corrosion rate or history of attack must be more thoroughly inspected. Inspection experience and records provide the key to how much coverage should be given. If the vessel is open, the thickness measurements can be made from the internal surfaces at any location the inspector desires—an approach much more flexible than dependence on fixed corrosion-gaging points.

The distance between thickness measurements can vary depending on the coverage desired, but a sufficient number of readings should be obtained to ensure a correct determination of vessel condition (distances

Table 1 Correction factors for ultrasonic thickness measurements in carbon steels at elevated temperatures

Temperature		Measurement				
		Ultrasonic		Micrometer		
°C	°F	mm	in.	mm	in.	Correction factor
40	100	13.11	0.516	13.11	0.516	0
95	200	13.15	0.518	13.13	0.517	0.998
150	300	13.26	0.522	13.14	0.5175	0.991
205	400	13.33	0.525	13.15	0.518	0.987
260	500	13.36	0.526	13.16	0.5185	0.986
315	600	13.41	0.528	13.18	0.519	0.983
370	700	13.46	0.520	13.20	0.520	0.981
400	750	13.51	0.532	13.26	0.522	0.981

Source: Ref 15

generally range from ⅓ to 1 m). In determining repair areas or the size and location of a patch, a grid pattern laid out on the vessel can be useful.

Rough or badly pitted surfaces should have a small area ground smooth to permit the transducer to make good contact. Transducers having a rubber membrane as a protective facing contact better on mildly pitted surfaces because the rubber will tend to conform to the surface. Removal of loose scale or dirt by scraping or filing is often sufficient surface preparation.

It is not necessary to drill gage holes or use other destructive methods to determine wall thickness. Insulation, if present, must be removed at gage locations for external measurements, but the removal of insulation is not required if the measurements are to be taken from the inside.

Special couplants enable many external measurements to be taken while the vessel is operating at temperatures of 370 °C (700 °F) or higher. Corrections should be made for errors introduced by the high temperature. Table 1 lists several correction factors for ultrasonic thickness measurements in carbon steels at elevated temperatures. Additional information can be found in Ref 15.

Flaw detection prior to fabrication is usually concerned with the inspection of steel or alloy plates for internal laminations, segregates, and similar discontinuities because they can be quite harmful under certain operating conditions. For example, hydrogen blistering can occur at a lamination, and stress-corrosion cracking, fatigue, intergranular corrosion, and hydrogen embrittlement are often accelerated by such discontinuities.

For flaw detection, a continuous-scan method is often used, in which the transducer is carried on a rigid carriage and the ultrasound waves are coupled into the plate by a stream of water. In some applications, the plate can be immersed, which simplifies the operation. To eliminate some of the human errors, a flaw-alarm system is set up so that a flaw of predetermined size will trigger a warning of some kind.

For the inspection of a weld, the weld shape may be such that the use of longitudinal waves is impossible, and a flaw such as a crack at the fusion zone probably will not present enough reflective area to be detected from directly above it. Nozzles located in hemispherical heads present special problems, and special techniques must be used to ensure a thorough inspection. Using shear waves, the sound can be directed into a suspect area from almost any angle. Large-scale drawings of the suspect area are needed to lay out correct angles and distances for the transducers (Fig. 9). For components having a complicated design, a plastic model can be an aid to better visualization of the sound path and angles of reflection. Thought given to the best approach and techniques can save hours in inspection and can ensure proper coverage. Angle transducers of 45 and 60° will detect almost any weld discontinuity, particularly if both internal and external surfaces are available for inspection. If the internal surface is not available, the nozzle can be filled with water and a search unit lowered to the correct depth to inspect the weld. The search unit can be adjusted to any angle and turned from side to side while in use. A plastic wedge on the search unit can be used to direct the sound beam through the nozzle body. Correctly angled, the beam will be intercepted by any radial flaw in the nozzle body and can also reach the attachment weld on the far side. The wedge can be shaped on a test block duplicating a nozzle section, and the sound path can be determined by through transmission (one transducer transmitting the signal and another, placed opposite, receiving the signal).

Shear wave search units intended for inspection at high temperatures may have the plastic wedge made of a special heat-resistant material. Such materials usually increase attenuation, and calibration must be made at temperatures close to that of the surface of the part being inspected if detection of a flaw of given size is to be ensured.

Examples of Application. The use of ultrasonics in the detection of discontinuities in a pressure vessel is discussed in the following examples. Additional information on in-service inspection is presented in the following section of this article.

Example 3: Use of Ultrasonic Inspection to Detect Creep Rupture in Stainless Steel Headers of an Ammonia-Plant Reformer Furnace (Ref 16). A leak was detected in one of the coils in the radiant section of a primary reformer furnace used in an ammonia plant. Furnace temperatures were reduced immediately, but it was necessary to continue firing of the furnace at about 540 to 595 °C (1000 to 1100 °F) for 16 h to reduce the catalyst. Subsequent shutdown inspection revealed that the bottom of one of seven outlet headers had ruptured, causing a section about 100 mm (4 in.) wide by 460 mm (18 in.) long to fall to the furnace floor. The outlet headers were 100 mm (4 in.) nominal diameter schedule 120 (11 mm, or 0.438 in., wall) pipe about 10 m (34 ft) long and were made of ASME SA-452, grade TP316H, stainless steel. In the absence of hot spots, the surface of the outer tube ranged from 845 to 855 °C (1550 to 1575 °F). The unit had been on stream about 29 000 h prior to failure.

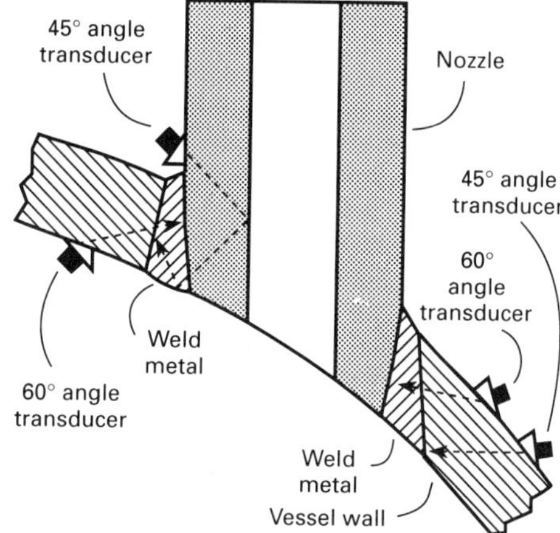

Fig. 9 Typical layout used on large-scale drawings to determine transducer angles and distances for inspection of difficult-to-reach weld areas

To get the unit back in service quickly, a section about 1.4 m (4½ ft) in length was replaced. The amount of header metal replaced was based on results from both visual and liquid penetrant inspections of the outer surface of the header. The inner surface, where accessible, was also checked visually and by liquid penetrant inspection. The six other headers were inspected at selected locations, and no surface cracks were detected.

Investigation of the Removed Section. Metallographic examination of the 1.4 m (4½ ft) section of the header that had been removed revealed that it had failed as a result of intergranular fissuring and oxidation, commonly termed creep rupture. Severe intergranular fissuring was found throughout the cross section of the header sample. Primary cracking occurred intergranularly through the grain boundaries, typical of high-temperature fissuring. Close observation of the crack paths linking some of the larger fissures, however, did reveal local areas having grains that were much

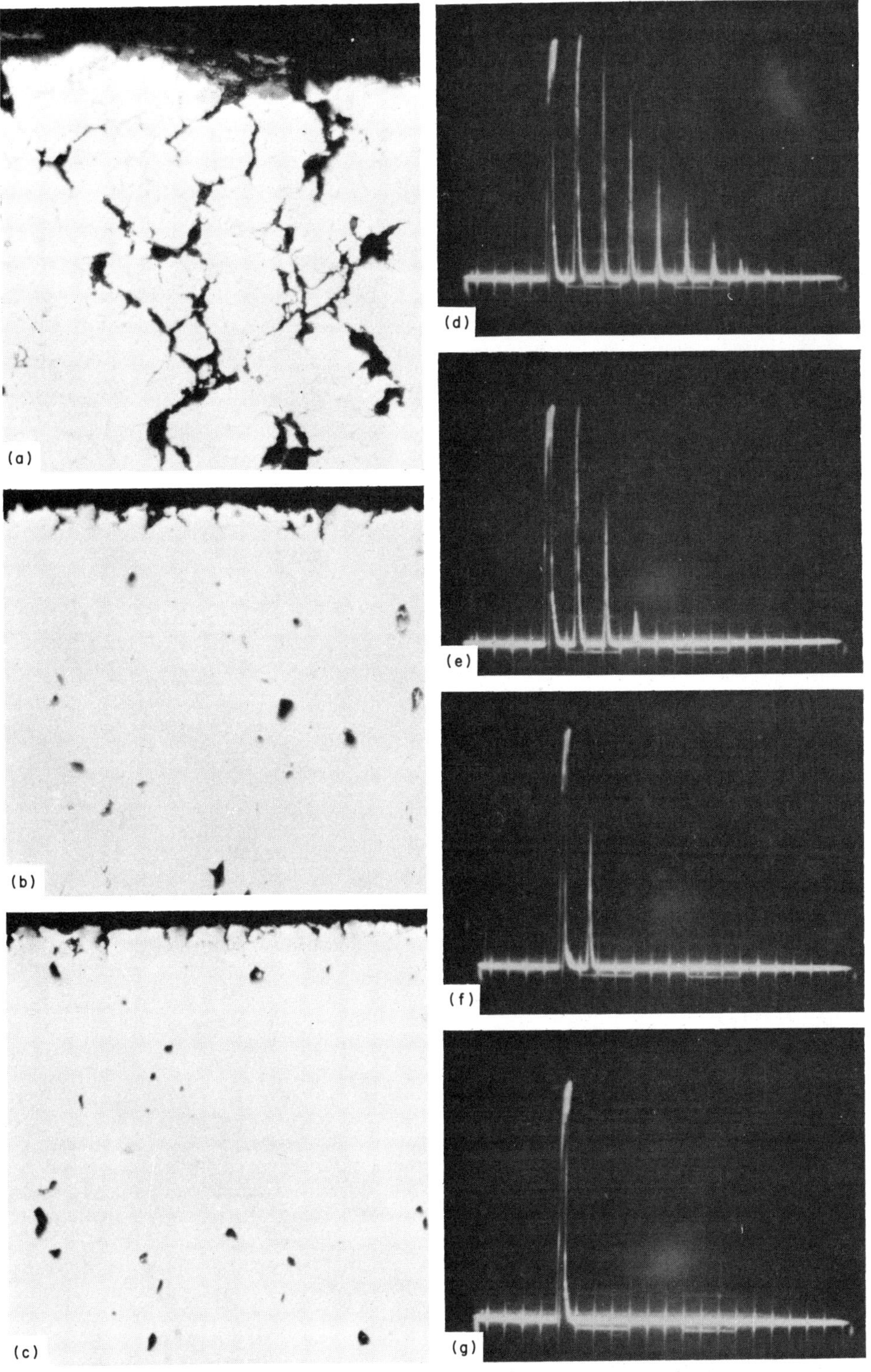

Fig. 10 Unetched microscopic appearance of three samples of failed pipe from a reformer-furnace header of TP316H stainless steel pipe and corresponding ultrasound responses of the three samples compared to the response of unused pipe. Micrographs of (a) sample A, (b) sample B, and (c) sample C. All 100×. Black voids are intergranular fissures. Ultrasound responses for (d) unused pipe, (e) sample C, (f) sample B, and (g) sample A. Strong reflections were obtained from the unused pipe, and attenuation increased as the number of fissures increased in damaged pipes.

smaller than those of the header metal. The presence of these very small grains could only be the result of cold deformation and subsequent recrystallization at elevated temperatures. Therefore, some plastic strain and cracking had occurred at temperatures well below the operating temperature, apparently as a result of thermal expansion and contraction stresses during startups and shutdowns, and then the cold-strained areas had recrystallized subsequently when they were reheated to the service temperature. Therefore, it follows that all of the fissuring did not take place during any one operating period. Also, the microstructure in the ruptured area gave the following evidence that the header metal had not been subjected to gross overheating in service:

- The header metal still had a general grain structure that would be classified as being of fine size, while severe grain coarsening would have occurred if the header metal had been heated at temperatures between 1095 and 1315 °C (2000 and 2400 °F)
- The microconstituents present in the grains would have been dissolved at grossly elevated temperatures and would have been of much finer size if they had subsequently precipitated out of solid solution during the final 16 h of firing at 595 °C (1100 °F) to reduce catalyst

It was concluded from this investigation of the ruptured area that the header had failed by conventional long-time creep rupture as a result of exposure to operating temperatures probably between 900 and 955 °C (1650 and 1750 °F), rather than by short-time exposure at grossly elevated temperatures. It was suspected that the furnace headers might not be as sound as the visual and liquid penetrant inspections of the surface metal had indicated; therefore, three ring sections from the removed 1.4 m (4½ ft) long header section were selected for further study. The sections were taken from the ruptured area (sample A), from a slightly bulged but nonruptured area (sample B), and from visually sound metal about 0.6 m (2 ft) from the rupture (sample C).

Inspection of the three samples revealed the presence of pinhead-size intergranular fissures throughout the cross sections of samples B and C (Fig. 10). This indicated that the header metal at these locations had been in an advanced stage of creep rupture even though the fissures or voids were of small size. Sample C had fewer pinhead voids than sample B. With continued service, the pinhead voids would have increased in number and grown in size and eventually would have linked up and caused failure. Generally, the microstructure of the inner surface of the header in the area of rupture (sample A, Fig. 10a) contained intergranular fissures (black areas), a relatively fine grain structure, precipitated microconstituents in the grains, and considerable amounts of carbides in the nonfissured grain boundaries. The presence of the carbides was evidence of carbon pickup in service.

Sample C had been removed from a visually undamaged header area located about 0.6 m (2 ft) from the rupture and about 250 mm (10 in.) from the sound end of the header section. It was considered possible that the header metal that was in service and only 250 mm (10 in.) away from the area of sample C was also in an advanced stage of creep rupture. It appeared probable that localized areas of the other six headers were in critical condition or that the problem was of a general nature. Therefore, it was decided that the remaining headers in service should be inspected as soon as possible because obtaining replacement parts might require several months of lead time. An ultrasonic attenuation method was considered to be the only possible nondestructive method for this evaluation, and development of a suitable system for use in the field was undertaken.

Ultrasonic Inspection Equipment and Standards. It was determined that optimum results that permitted the determination of changes in attenuation as a function of the number of cracks present required the use of a 22-MHz search unit in conjunction with a pulse-echo instrument. To minimize surface effects and curvature, a short water column was used to couple the sonic energy to the headers. This water column was enclosed in a Lucite cylinder 32 mm (1.25 in.) long with a 6 mm (0.25 in.) thick polyurethane-foam gasket at the bottom. The pliable gasket conformed to the curved surfaces of the header. The search unit was held in position by a setscrew. Water flowed through the unit by gravity and escaped through a small hole in the Lucite cylinder at a point just above the front surface of the transducer.

A section of unused type 316H stainless steel pipe and the three sections of header metal from the rupture area were used for reference. Strong reflections were obtained from the undamaged pipe (Fig. 10); increased attenuation was obtained from damaged piping as the number of cracks increased, as indicated by the ultrasonic responses in Fig. 10(e) to (g).

The results of the test correlated well with the characterization of the number of cracks observed by optical microscopy (Fig. 10a to c). Approximately 1 year later, with 37 336 h on stream, the plant was shut down, and in-service inspection was conducted using the ultrasonic technique developed.

In-Service Ultrasonic Inspection. The standards and equipment used to develop the ultrasonic technique were used for the field inspection of the reformer furnace. Ultrasonic readings were taken on the horizontal header metal between each vertical tube and on both sides of any header welds. Over 350 readings were made. In addition, at some locations, readings were taken on the top, bottom, and sides. It was found that the headers all contained internal voids throughout their lengths, of varying numbers between those of samples B and C. The inspection results indicated that all headers were in an advanced stage of creep rupture (stress rupture) but that no areas had fissured to a degree that they needed immediate replacement. The replacement piping that had been installed during the previous shutdown 13 months earlier gave a "good" signal similar to unused material. The technique developed for the field survey was not adaptable for determining the conditions of the welded joints, so none of the readings represents welds.

On the basis of results of the in-service inspection, two conclusions were reached. First, the furnace was deemed serviceable, and second, in the absence of local hot spots, the headers would survive for a reasonable period of time.

Example 4: Ultrasonic Inspection for Creep Fissures in Reformer-Furnace Tubes (Ref 17). Preliminary studies were made to establish whether sound would transmit through cast heat-resistant alloy HK-40 tubes from a reformer furnace. Several tests were conducted, and the most likely method was found to be the measurement of attenuation losses in a dual sensor using the through transmission method in an immersion tank. The tests showed that sound would transmit in HK-40 tubes with the use of a low-frequency transducer. Also, the rough outer surface of a centrifugal casting was the significant factor, rather than poor transmission in the large-grain cast microstructure. Immersion methods minimized the surface roughness condition and provided sufficient acoustic energy in the tubes. A dual search unit, with a sending transducer and a receiving transducer to produce a refracted-angle sound beam, was the best means for passing sound through the tube and across the plane of fissure formation. The attenuation of this beam, measured by the receiver, was a measure of fissure density.

Figure 11 shows the results of ultrasonic C-scan amplitude recordings on three sample tubes in an immersion tank. By moving the dual search unit at a constant speed longitudinally along the tube and indexing at the end of each traverse, and continuously recording the received ultrasonic signal, attenuation patterns in fissured tubes were recorded. Sample tubes with known fissures showed that some or all of the sound transmission was interrupted, depending on the density of the fissures (compare Fig. 11a and b). Sample tubes with no fissures showed full sound transmission (Fig. 11c).

Strip chart recordings of sound attenuation were made on sections of sample tubes. By evaluation of the strip charts, the tubes were categorized as to sound attenuation and a grading scale of 1 to 5 was arbitrarily established. A rating of 1 represented good sound transmission through unfissured tubes, and a rating of 5 represented little or no sound transmission through severely fissured tubes. Intermediate values represented various degrees of fissuring.

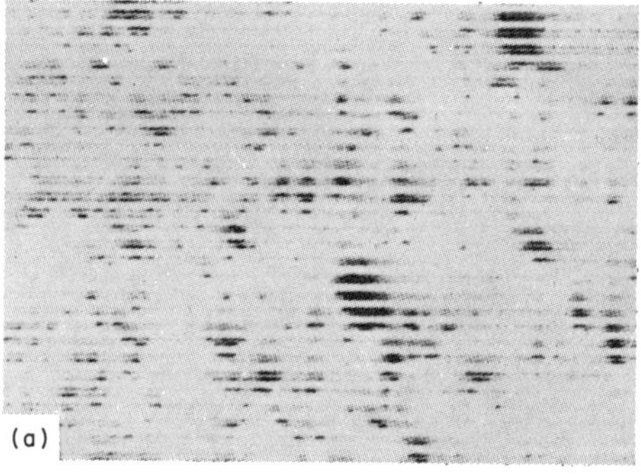

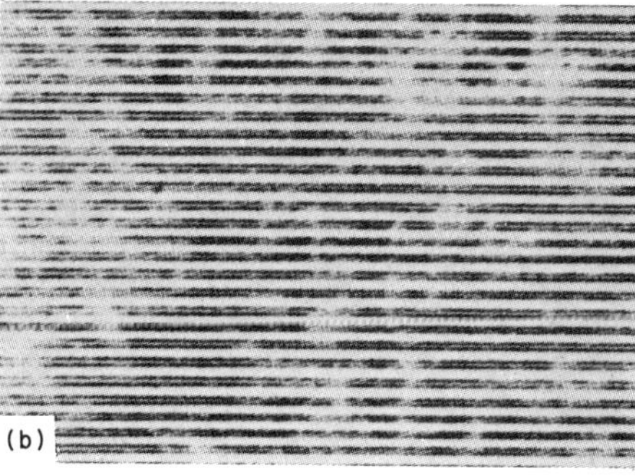

Fig. 11 Ultrasonic C-scan recording of three sample alloy HK-40 tubes from a steam-methane reformer furnace. (a) Severely fissured tube. (b) Tube with small fissures. (c) Tube with no fissures. Light areas represent sound attenuation (no through signal); black areas represent no sound attenuation (through signal).

Field Tests. On the basis of the preliminary tests, a field unit was built for subsequent furnace inspections that could be clamped on a tube and mechanically operated. The basic principle of the field unit is shown in Fig. 12.

The first full plant test with the unit was conducted with catalyst in the tubes. At that time, the furnace had operated over 50 000 h. The unit was calibrated on sample tubes with known conditions. Two men were in the furnace with voice communications to a third man outside the furnace operating the supporting ultrasonic equipment. With the unit clamped to the tube, a single scan was made at the critical hot area at each level. Ultrasonic scanning time at each location was less than 30 s, during which the attenuation information was recorded on a strip chart for later interpretation. About 32 h was required to inspect the entire furnace.

The tubes were graded on the scale of 1 to 5 established during testing of the sample

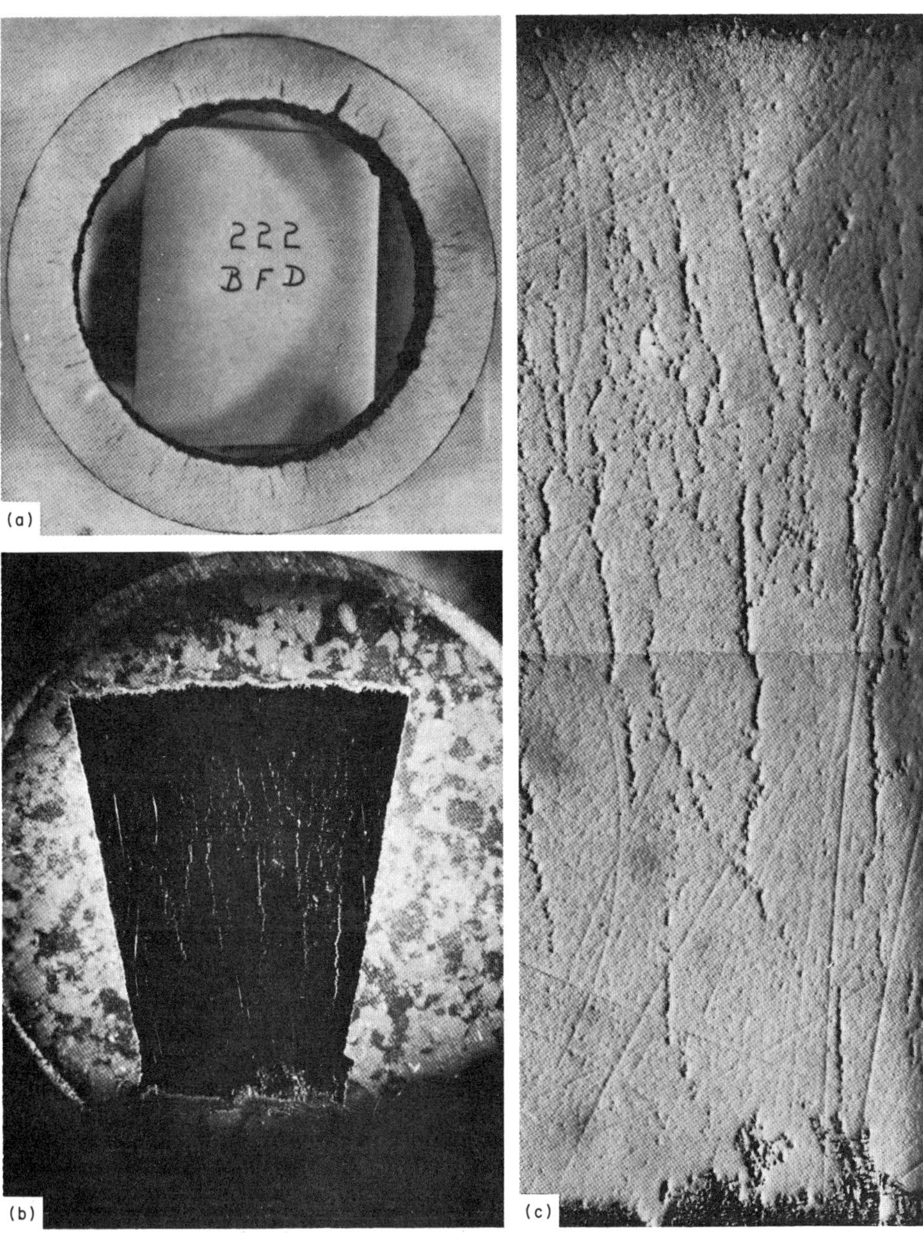

Fig. 12 Creep fissures in a centrifugally cast HK-40 reformer-furnace tube that are detectable by ultrasonic inspection and by radiography with nickel catalyst in tube. (a) Tube cross section. 0.45×. (b) Tube wall. 2.5×. (c) Enlargement of inside diameter portion of wall shown at bottom in (b). 7.5×

tubes. Of the 295 tubes tested, about 15% had ratings of 5. About 30% of these tubes were replaced; the remainder were left in the furnace for a time-to-failure test. Of the 15 known fissured tubes left in the furnace at the previous shutdown (see Example 2), only 3 showed severe creep fissures by ultrasonic inspection and were replaced. The remaining 12 tubes had ratings of 4.

Metallographic examination was made of specimens taken from the removed tubes to determine the reliability of the ultrasonic unit. Results indicated that the unit was so sensitive that it could detect mild third-stage creep (not detectable by radiography) as well as severe fissures (detectable by radiography). Because of the sensitivity of the unit, there was some difficulty in differentiating between mild and severe fissuring. Radiography with catalyst in the tubes can detect only the most severe creep fissures, but it was used as the basis on which tubes were replaced.

The tubes with indications of severe creep fissures (ratings of 5) were radiographed. Seven tubes showed definite fissures on the radiographs, even with catalyst in place. These seven tubes and seven additional tubes that showed questionable fissures were replaced. One of the known fissured tubes left in the furnace during the previous shutdown showed questionable fissures on the radiograph taken with catalyst and had an ultrasonic rating of 5. This tube was again allowed to remain in the furnace for a time-to-failure test. Metallographic features of a tube with severe creep fissures that had a rating of 5 with the ultrasonic unit, subsequently confirmed with radiography (with catalyst in the tube), are shown in Fig. 12.

Conclusions. The ultrasonic unit is a good field inspection tool for centrifugally cast alloy HK-40 tubes. Further refinement of the device is necessary to discriminate between mild and severe fissures and thus eliminate the need for radiography to determine the most severe fissures.

An ultrasonic inspection system provides the following advantages:

- Higher speed
- Increased coverage at lower cost
- Increased sensitivity
- Need for fewer radiographs
- Elimination of the need for removal of catalyst to effect inspection

In addition, maintenance work need not be interrupted while inspection is in progress, as it must be for radiography.

In-Service Quantitative Evaluation (Ref 18-20)

The structural integrity of the reactor pressure vessel receives considerable attention because the vessel is the primary containment for the reactor coolant. In the United States, periodic in-service examination of the vessel is performed according to section XI of the ASME Boiler and Pressure Vessel Code. Ultrasonic methods of quantitative nondestructive evaluation (NDE) are those most commonly used to accomplish in-service examinations. Nearly all of the examinations are performed with remote-controlled equipment. Many innovative devices and specialized ultrasonic techniques have been developed to examine components with complex geometry, which are often extremely difficult to access. Some of the more difficult areas to examine are the under-clad region nozzle inner radii, nozzle-to-shell welds, dissimilar-metal welds in the safe-ends, and seam welds in areas of complex shape (Ref 18).

Fracture mechanics is used to evaluate indications detected during in-service examinations. Accurate measurements of the sizes and locations of all defects are required. Furthermore, the probabilistic failure prediction methodology now being used requires additional information on the probability of detecting flaws with each NDE technique. These requirements are the driving force in the current trend toward additional regulatory requirements for the quantitative demonstration of NDE performance (Ref 18).

The probability of detection as a function of flaw size and the accuracy of flaw size measurement are the important characteristics of each NDE technique that are to be measured in performance demonstration (Ref 18). Much of the work to date in the area of performance demonstrations and quantitative NDE of pressure vessels has been carried out at the Electric Power Research Institute (EPRI). Some of the published EPRI work is contained in the extensive list of Selected References found at the end of this article. Additional information on the principles of quantitative analysis can be found in the Section "Quantitative Nondestructive Evaluation" in this Volume.

In designing NDE performance demonstrations, both the examination sensitivity and the mechanical precision of the scanning device must be addressed to determine if they are adequate to detect and size the defects of concern (Ref 18). Accordingly, the scope of a demonstration can be broken into three conveniently separate parts:

- The mechanical handling system
- The ultrasonic system
- The data-recording system

Mechanical Handling (Ref 18). The intent of testing the mechanical system is to measure and document the accuracy, backlash, and repeatability of the complete remote positioning system over its full range of operation. The tolerances of the mechanical system, when combined with those of the ultrasonic system, must give adequate flaw sizing and location capability for any required fracture mechanics analysis.

Tests of Ultrasonic and Data-Recording Systems (Ref 18). The intent of these tests is to demonstrate that the intended inspection procedures are capable of detecting and sizing all flaws of potential concern to the safety of the reactor pressure vessel. The vessel regions to be addressed can be categorized as follows:

- The region of the vessel in the vicinity of the cladding/base metal interface
- The nozzle regions, including nozzle inner radius, nozzle-to-vessel welds, and safe-end welds (up to and including the pipe-to-safe-end weld)
- The remaining welds that can be further categorized as circumferential or longitudinal welds. These are separately identified because of the different relative directions of weld, cladding, and curvature

Full-sized specimens representing the appropriate component are required. Intentional flaws are introduced of the size and type of concern. For example, these may be thermal fatigue cracks in the nozzle inner radius or intergranular stress-corrosion cracks in the nozzle-to-safe-end weld. The number of defects required in the demonstration is dependent on the purpose of that demonstration. Tests can be classified as either performance demonstration or validation. Additional information can be found in Ref 18 and in the Selected References that follow this article.

NDE of Clad Vessels (Ref 19). In the early 1980s, there was much concern in the United States about pressurized thermal shock, a series of events that started with a considerable primary fluid loss, the addition of cold make-up water, and the subsequent repressurization of the system. Under these conditions, small cracks (6 mm, or 0.25 in., in depth), if present immediately under the stainless steel vessel clad, could act as initiation sites for crack growth. This concern caused many reactor pressure vessel inspection development efforts to be focused on the detection of small cracks in the innermost region of the vessel inner surface. The following sections discuss several advanced systems developed at EPRI for the detection and sizing of flaws, the automatic discrimination of flaw signals, and computer-aided sizing from signals using crack tip diffraction methods.

The ultrasonic data recording and processing system (UDRPS) is a high-speed, general-purpose device that consists of a large minicomputer, high-speed data channel processor, color video display, and disk/tape storage devices. The UDRPS uses a detection criterion that is based on the signal-to-local-noise ratio threshold and the apparent motion of the target within the field of view of the moving transducer. Resulting patterns of indications are color coded for signal-to-noise ratio and are viewed by an analyst for the presence of formations suggestive of defects. Crack length and depth are also estimated from several image display modes that are available to the operator. Figure 13 shows the UDRPS results on a test block. Crack length and depth measurement capabilities are shown in Fig. 14.

Flaw discriminators provide the ability to distinguish among ultrasonic signal types. Feasibility studies have been conducted to demonstrate the use of integrated ultrasonic inspection and pattern recognition systems for distinguishing among slag inclusions, cracks, and spurious clad-noise signals.

Pressure Vessel Imaging Systems. The present inspection method for weld zones of nuclear reactor pressure vessels uses a pulse-echo ultrasonic technique for both preservice and in-service inspections. Pulse-echo inspection data are not sufficiently accurate to satisfy the demands of structural analysis by fracture mechanics methods. This has resulted in the development of imaging systems that combine conventional ultrasonic inspection techniques (B-scan, C-scan, pulse echo) with acoustic holography, which provides real-time three-dimensional estimations of flaw size and depth and more accurate information for fracture mechanics analysis.

Computer-aided sizing through crack tip diffraction involves the use of advanced digital processing methodologies that utilize spectral features of signals diffracted from the under-clad crack tips to distinguish more accurate depth estimates from less reliable ones. This technology is based on the development of sizing algorithms that are available as computer codes.

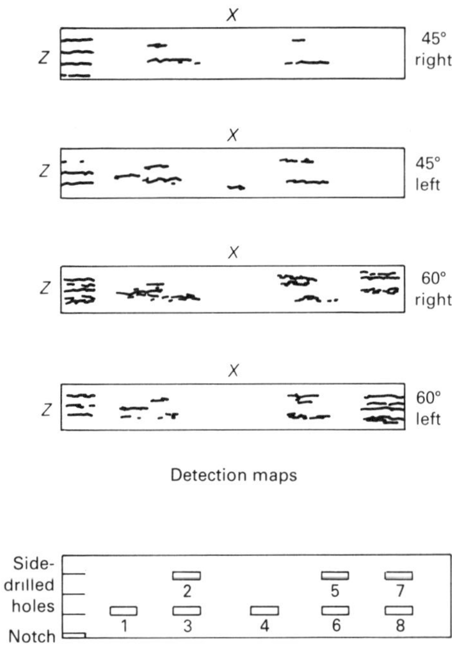

Fig. 13 UDRPS flaw detection result on a heavy-section test block. Source: Ref 19

Acoustic Emission Monitoring of Pressure Vessels

When discontinuities exist in a metal, a stress concentration occurs at the tips of discontinuities; under increasing applied stress, deformation occurs first at these discontinuities. This deformation, which may be plastic flow, microcracking, or even large-scale cracking, produces signals that, by means of suitable amplification, can be recorded as acoustic emission.

The emission is converted to an electrical signal by means of a piezoelectric transducer, which is mounted on the pressure vessel. The transducer is contained in a simple housing, which can be bonded to the vessel with glue or with a film of grease. The signal is then amplified (often using a preamplifier close to the transducer), filtered to remove low-frequency extraneous mechanical and electrical noise, and then recorded. Different types of analysis are used, and the signals can be shaped as pulses to aid quantitative interpretation of data.

Stressing of a discontinuity can produce a continuous emission and bursts of high amplitude. Analysis of these signals can be made in any of several ways; for example,

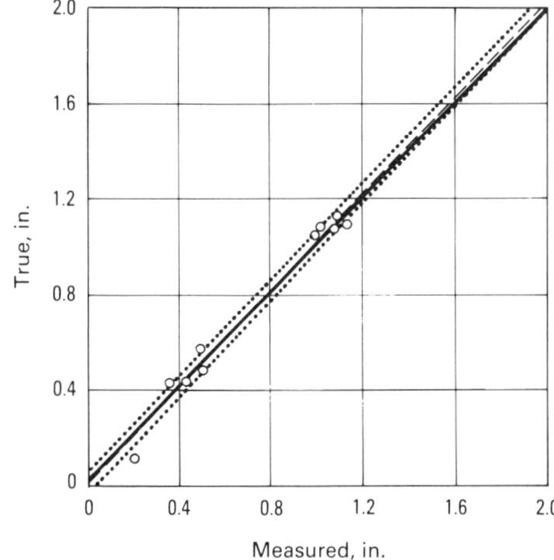

Statistical summary	
Slope	1.010
Intercept	0.012
Sample correlation coefficient	0.990
Standard error of estimate	0.045
Mean error	−0.019
Standard deviation of error	0.002
Root mean square	0.049
Population	16

(a)

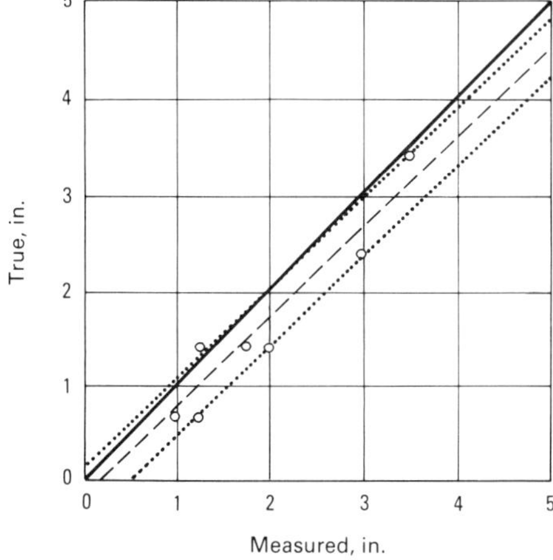

Statistical summary	
Slope	0.936
Intercept	−0.148
Sample correlation coefficient	0.939
Standard error of estimate	0.295
Mean error	−0.268
Standard deviation of error	0.090
Root mean square	0.402
Population	6

(b)

Fig. 14 UDRPS estimates of the flaw depths (a) and flaw lengths (b) in a heavy-section test block. Inspection technique: 45° shear and 60° shear. Source: Ref 19

each ring of the transducer above a set threshold can be counted or, alternatively, each high-amplitude burst can be counted as a discrete pulse. The number of counts per second or the integrated count is then compared with vessel pressure for strain.

Application. A series of search units is used for monitoring the emission of sound energy from stressed pressure vessels. All the outputs can be digitally stored for subsequent analysis. Additionally, some of the circuits can be continuously monitored to give an immediate indication of high acoustic activity and therefore the existence of a severe discontinuity that could cause fracture. Any source of acoustic emission can be located by measurement of the time taken for the signal to reach different search units at known locations. On-line or subsequent location techniques can be used, preferably in conjunction with a small computer. The number of search units necessary depends on the degree of accuracy required in defining the sources of the emissions.

Most of the information in any signal generated from a source within a metal occurs in the 1 to 2 MHz frequency range; however, at these frequencies, the signal attenuation is high, so that the higher frequencies in the stress wave are soon attenuated to near the background noise over relatively short distances (a few meters). The lower-frequency components of the wave in the 50 to 500 kHz frequency range can transmit over larger distances, because of less attenuation. The lower-frequency signals in the 50-kHz range can be detected over relatively large distances on an uncoated surface. This means that some general source location can be performed using the lower-frequency components of the acoustic emission signal for large transducer separations, provided the signal attenuation (which increases when a vessel is coated and/or buried) does not reduce the surface wave to the extent that the true signal is lost in the background noise (Ref 9).

Source identification, however, requires the use of wide-band transducers operating at high frequencies and located as close as possible to the source of the emission. Only then is it possible to obtain some indication of the original waveform, and even then it will have been modified by its passage through the material to the surface so that care must be used in interpretation.

Other factors influencing the interpretation of results include:

- Defect type, size, shape, orientation, and location
- Chemical composition of the vessel material
- Vessel geometry and its associated pipework
- Epoxy-type coatings

Details on these factors can be found in Ref 9 and in the article "Acoustic Emission Inspection" in this Volume.

Proof Testing. Because the acoustic emission technique depends on a changing state of stress, especially around a discontinuity, the most convenient time for application to most pressure vessels is at the first proof test. With a sufficient number of search units, it is possible to monitor the entire vessel and locate the areas where discontinuities exist. Subsequent ultrasonic inspection can confirm the existence of very small acceptable discontinuities in the position indicated by the acoustic emission technique.

Test to Failure (Ref 9). In an attempt to improve the interpretation of acoustic emission data, tests have been conducted on operating vessels, static vessels, and vessels deliberately tested to failure. Data on wave propagation and failure mechanisms have been recorded and used to develop a reliable acoustic emission integrity evaluation technique. The following example illustrates the use of these tests.

Example 5: Test to Failure of a 10-Year-Old Pressure Vessel (Ref 9). A vessel that had been in service for over 10 years and was operated at a normal pressure of 9.6 MPa (1400 psi) at 550 °C (1020 °F) was tested by acoustic emission and analyzed by subsequent fractographic examination. The defects causing its removal from service were extensive cracks in the nozzle reinforcement. These cracks ran circumferentially around the nozzle penetration, and those on the inside surface of the vessel were up to 25 mm (1 in.) deep. The test program was designed to pressurize the vessel to failure after many pressure cycles designed to accelerate the growth of flaws in the nozzle by cyclic fatigue. Numerous pressurizations were carried out at progressively increasing pressures, including a group of 995 cycles in the range of 7 to 24 MPa (1000 to 3500 psi).

The last four cycles were monitored with acoustic emission. The last three cycles were up to 58 MPa (8400 psi), and in the final pressurization, the failure of the vessel occurred at 62 MPa (9000 psi). There was considerable acoustic emission activity detected in the early stages of the test program, but this is thought to have resulted from movement of the vessel supported by observations using television monitors. After the pressure had exceeded 56 MPa (8100 psi), this acoustic emission activity diminished concurrently with the vessel becoming more settled on its supports. Two active source areas were localized at the bottom of adjoining nozzles. It was noted in this test that some of the location patterns were distorted due to some propagation paths being around nozzle holes in the vessel. During the final stages of the test, the acoustic emission event rate further diminished, which was partly due to the reduced pumping rate caused by the vessel expansion. More acoustic emission activity was detected from the base of the central nozzle (No. 3, Fig. 15) than most of the other

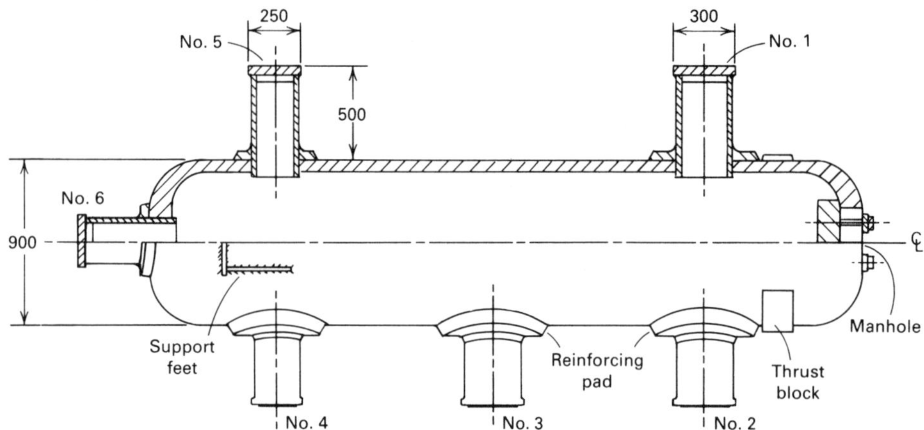

Fig. 15 Schematic of pressure vessel tested to failure and monitored by acoustic emission. Dimensions given in millimeters. Source: Ref 9

locations, and this primary initiation point for failure was later confirmed by independent fractography examination.

The failure was almost entirely brittle in character and appeared to be from localized exhaustion of ductility, with no apparent involvement of any significant preexisting defect. It appeared to propagate axially along the vessel from the initiation point and then to branch circumferentially between nozzle Nos. 2, 3, and 4, as shown in Fig. 15. Half of the vessel then lifted and separated, causing an axial fracture diametrically opposite the underside of these nozzles. This fracture acted as a hinge, which allowed this section to lift. The only evidence of ductile fracture was in a narrow shear lip adjacent to the external surface of the vessel.

The absence of a flaw at the point of initiation was somewhat surprising; however, the highly localized acoustic emission activity and subsequent failure may have resulted from a steep strain gradient adjacent to the reinforced and highly restrained nozzle penetration. The failure of the preexisting cracks to propagate confirms predictions of stress analysis that they were not located in highly stressed regions during simple pressurization. Their formation during service resulted from loadings imposed by external supports and pipework, rather than from internal pressure.

The conditions encountered in this test are clearly different from those that would exist in a nondestructive proof test. The vessel suffered extensive plastic deformation at pressure considerably higher than any conceivable proof test pressure. Also, the failure occurred by the local exhaustion of ductility in a highly strained region, rather than by the initiation of yielding in a highly stressed region. However, the ability of the equipment to identify localized acoustic emission sources was clearly demonstrated.

Inspection During Fabrication. During fabrication, acoustic monitoring can be used to detect cracking during or after the welding process. Stress-relief cracking can be identified as it occurs, although provision must be made for keeping the transducers at a low temperature. This is normally done by the use of waveguides, the extremities of which hold the transducers outside the stress-relieving furnace.

In-service inspection may consist of monitoring during periodic proof testing, during normal pressure cycles, or continuously during normal operation. When the vessel is pressurized to a level less than that to which it has been previously subjected (during, for example, the proof tests), little or no acoustic emission occurs. Therefore, on subsequent pressurizations a quiet vessel will be obtained unless a crack has extended in service because of corrosion or fatigue. On pressurizing after crack growth, the stress system at the enlarged crack will be changed from that in the proof test, and further emission will be obtained.

Flaw Location. In many cases, especially on large pressure vessels, it becomes necessary not only to detect acoustic emissions but also to locate the source of the signals. This can be accomplished by uniformly spacing multiple search units over the surface area of a pressure vessel and monitoring the time of arrival of the signals to the various search-unit locations. Because of the high velocity of sound and the relatively close spacing of search units on a steel vessel, time resolutions must be made in microseconds to locate the source to within a centimeter. In most cases, inspection requirements are such that data must be available in a short period of time. Therefore, most systems of this type utilize a computer for handling and displaying the data.

REFERENCES

1. Outlook on Nondestructive Examination, *Nucleonics Week*, 30 June 1988
2. ASME Boiler and Pressure Vessel Code: Section II—Material Specifications, Part A—Ferrous Materials; Section III, Division 1—Nuclear Power Plant Components; Section V—Nondestructive Examination; Section VIII—Division 1—Pressure Vessels, Division 2—Alternative Rules for Pressure Vessels; Section IX—Welding and Brazing Qualifications; Section XI—Rules for Inservice Inspection of Nuclear Power Plant Components, American Society of Mechanical Engineers
3. "Recommended Practice for Nondestructive Testing Personnel Certification," SNT-TC-1A, American Society for Nondestructive Testing, 1988
4. "Standard Recommended Practice for Magnetic Particle Examination," E 709, *Annual Book of ASTM Standards*, American Society for Testing and Materials
5. "Standard Practice for Liquid Penetrant Inspection Method," E 165, *Annual Book of ASTM Standards*, American Society for Testing and Materials
6. "Standard Specification for Ultrasonic Angle-Beam Examination of Steel Plates," A 577, *Annual Book of ASTM Standards*, American Society for Testing and Materials
7. "Standard Specification for Straight-Beam Ultrasonic Examination of Plain and Clad Steel Plates for Special Applications," A 578, *Annual Book of ASTM Standards*, American Society for Testing and Materials
8. "Standard Specification for Straight-Beam Ultrasonic Examination of Steel Plates," A 435, *Annual Book of ASTM Standards*, American Society for Testing and Materials
9. B.R.A. Wood, Acoustic Emission Applied to Pressure Vessels, *J. Acoust. Emiss.*, Vol 6 (No. 2), 1989, p 125-132
10. J.C. Spanner, Acoustic Emission in Pressure Vessels, in *Pressure Vessel and Piping Technology—1985: A Decade of Progress*, American Society of Mechanical Engineers, 1985, p 613-632
11. K. Krzywosz, Recent NDE Experiences With PWR Steam Generator Tubing Inspection, in *NDE in the Nuclear Industry*, ASM INTERNATIONAL, 1987, p 157-167
12. R.H. Ferris, A.S. Birks, and P.G. Doctor, Qualification of Eddy Current Steam Generator Tube Examination, in *NDE in the Nuclear Industry*, ASM INTERNATIONAL, 1987, p 71-73
13. V.S. Cecco and F.L. Sharp, Special Eddy Current Probes for Heat Exchanger Tubing, in *NDE in the Nuclear Industry*, ASM INTERNATIONAL, 1987, p 109-174
14. R.R. Dalton, Radiographic Inspection of Cast HK-40 Tubes for Creep Fissures, *Mater. Eval.*, Vol 30 (No. 12), Dec 1972, p 249-253
15. D.J. Evans, Field Application of Nonde-

structive Testing in the Petroleum and Petrochemical Industries, in *Materials Engineering and Sciences Division Biennial Conference*, American Institute of Chemical Engineers, 1970, p 484-487
16. B. Ostrofsky and N.B. Heckler, Detection of Creep Rupture in Ammonia Plant Reformer Headers, in *Materials Engineering and Sciences Division Biennial Conference*, American Institute of Chemical Engineers, 1970, p 472-476
17. R.R. Dalton, Ultrasonic Inspection of Cast HK-40 Tubes for Creep Fissures, *Mater. Eval.*, Vol 32 (No. 12), Dec 1974, p 264-268
18. A.J. Willets, F.V. Ammirato, and J.A. Jones, Objectives and Techniques for Performance of In-Service Examination of Reactor Pressure Vessels, in *Performance and Evaluation of Light Water Reactor Pressure Vessels*, American Society of Mechanical Engineers, 1987, p 79-86
19. G.J. Dav and M.M. Behravesh, U.S. Developments in the Ultrasonic Examination of Pressure Vessels, *Int. J. Pressure Vessels Piping*, Vol 28, 1987, p 3-17
20. P.C. Riccardella, J.F. Copeland, and J. Gilman, Evaluation of Flaws or Service Induced Cracks in Pressure Vessels, in *Performance and Evaluation at Light Water Reactor Pressure Vessels*, American Society of Mechanical Engineers, 1987, p 87-94

SELECTED REFERENCES

Replication Microscopy

- J.J. Balaschak and B.M. Strauss, Field Metallography in Assessment of Steam Piping in Older Fossil Power Plants, in *Microstructural Science*, Vol 15, ASM INTERNATIONAL, 1987, p 27-36
- C.J. Bolton, B.F. Dyson, and K.R. Williams, Metallographic Methods of Determining Residual Creep Life, *Mater. Sci. Eng.*, Vol 46, 1980, p 231-239
- P.B. Ludwigsen, The Replica Method for Inspection of Material Structures and Crack Detection, *Structure*, No. 15, Sept 1987, p 3-5
- B. Neubauer and U. Wedel, NDT: Replication Avoids Unnecessary Replacement of Power Plant Components, *Power Eng.*, Vol 88 (No. 5), May 1984, p 5
- N. Nilsvang and G. Eggeler, A Quantitative Metallographic Study of Creep Cavitation in a 12% Chromium Ferritic Steel (X20 CrMoV 12 1), *Pract. Metallogr.*, Vol 24, 1987, p 323-335
- E.V. Sullivan, Field Metallography Equipment and Techniques, in *Microstructural Science*, Vol 15, ASM INTERNATIONAL, 1987, p 3-11

Quantitative NDE (1980)

- A.J. Boland *et al.*, "Development of Ultrasonic Tomography for Residual Stress Mapping," EPRI NP-1407, Electric Power Research Institute, May 1980
- A.E. Holt, "Defect Characterization by Acoustic Holography—Volume 1: Imaging in Field Environments," EPRI NP-1534, Electric Power Research Institute, Sept 1980
- P.H. Hutton, "Development of an Acoustic Emission Zone Monitor and Recorder for BWR Pipe-Cracking Detection," EPRI NP-1408, Electric Power Research Institute, June 1980
- W. Lord, "Magnetic Flux Leakage for Measurement of Crevice Gap Clearance and Tube Support Plate Inspection," EPRI NP-1427, Electric Power Research Institute, June 1980
- W. Lord and R. Palanisamy, Magnetic Probe Inspection of Steam Generator Tubing, *Mater. Eval.*, Vol 38 (No. 5), May 1980
- V.I. Neeley *et al.*, "Technology Transfer Phase of Advanced Ultrasonic Nuclear Reactor Pressure Vessel Inspection System," EPRI NP-1535, Electric Power Research Institute, Sept 1980
- G.P. Singh and J.L. Rose, "Ultrasonic Field Analysis Program for Transducer Design in the Nuclear Industry," EPRI NP-1335, Electric Power Research Institute, Feb 1980
- S. Wenk *et al.*, "NDE Characteristics of Pipe Weld Defects," EPRI NP-1590-SR, Electric Power Research Institute, Sept 1980

Quantitative NDE (1981)

- D.G. Eitzen *et al.*, "Fundamental Development for Quantitative Acoustic Emission Measurements," EPRI NP-2089, Electric Power Research Institute, Oct 1981
- D.G. Eitzen *et al.*, "Summary of Fundamental Development for Quantitative Acoustic Emission Measurements," EPRI NP-1877, Electric Power Research Institute, June 1981
- T.D. Jamison *et al.*, "Studies of Section XI Ultrasonic Repeatability," EPRI NP-1858, Electric Power Research Institute, May 1981
- W. Lord, "Development of a Finite Element Model for Eddy-Current NDT Phenomena," EPRI NP-2026, Electric Power Research Institute, Sept 1981
- W.R. McDearman *et al.*, "Steam Generator Support Plate Radiography," EPRI NP-2042, Electric Power Research Institute, Sept 1981
- E.J. Parent *et al.*, "Profilometry for Steam Generator Tube Dents," EPRI NP-2141, Electric Power Research Institute, Nov 1981
- C.O. Ruud, "Review and Evaluation of Nondestructive Methods for Residual Stress Measurements," EPRI NP-1971, Electric Power Research Institute, Sept 1981

Quantitative NDE (1982)

- S. Brown, "Field Experiences With Multifrequency-Multiparameter Eddy Current Technology," EPRI NP-2299, Electric Power Research Institute, March 1982
- W.E. Cramer *et al.*, "Application of an Eddy Current Technique to Steam Generator U-Bend Characterization," EPRI NP-2339, Electric Power Research Institute, April 1982
- G.J. Dau *et al.*, Automatic Analysis of Eddy Current Signals, in *Proceedings of the 5th International Conference on Inspection of Pressurized Components*, Institute of Mechanical Engineers, 1982
- E.S. Furgason and V.L. Newhouse, "Evaluation of Pulse-Echo Ultrasound for Steam Generator Tube-to-Support Plate Gap Measurement," EPRI NP-2285, Electric Power Research Institute, June 1982
- M.E. Lapides, Factors Influencing Detection, Location, and Sizing of Flaws: Intergranular Stress Corrosion Cracking (IGSCC), in *Proceedings of the 5th International Conference on Inspection of Pressurized Components*, Institute of Mechanical Engineers, 1982
- M.E. Lapides and T.U. Marston, In-Service Inspection of Heavy Section Castings: Techniques, Results, Implication, in *Proceedings of the Seminar on Improvements in Power Plant Casting Quality* (St. Charles, IL), American Society for Metals, 1982
- J.L. Rose *et al.*, "A Physically Modelled Feature Based Ultrasonic System for IGSCC Classification," Paper presented at the ASNT Spring Conference (Boston), American Society for Nondestructive Testing, March 1982

Quantitative NDE (1983)

- C. Bradshaw, Benefits of Automatic Steam Generator Tube Inspection Data Acquisition, in *Proceedings of the SMIRT-7 Post Conference* (Monterey, CA), 1983
- S.D. Brown, "Eddy Current NDE for Intergranular Attack," EPRI NP-2862, Electric Power Research Institute, Feb 1983
- S.D. Brown, "Steam Generator U-Bend Eddy Current NDE," EPRI NP-3010, Electric Power Research Institute, April 1983
- M.E. Lapides, "Radiographic Detection of Intergranular Stress Corrosion Cracking: Analysis, Qualification, and Field Testing," EPRI NP-3164-SR, Electric Power Research Institute, Oct 1983
- W. Lord, "Magnetic Flux Leakage for Measurement of Crevice Gap Clearance and Tube Support Plate Inspection," EPRI NP-2857, Electric Power Research Institute, Feb 1983

- W.R. McDearman *et al.*, "Steam Generator Support Plate Radiographic Evaluation System," EPRI NP-3253, Electric Power Research Institute, Jan 1983
- V.I. Neeley, "Development of a Production Prototype Pressure Vessel Imaging System," EPRI NP-3253, Electric Power Research Institute, Oct 1983
- R.B. Thompson *et al.*, "A Prototype EMAT System for Inspection of Steam Generator Tubes," EPRI NP-2836, Electric Power Research Institute, Jan 1983

Quantitative NDE (1984)

- F.L. Becker, Effective Demonstrations for Under Clad Crack Detection, in *Non-Destructive Examination for Pressurised Components*, Elsevier, 1984
- R.L. Beverly and R.A. Baker, "Evaluation of Nondestructive Examinations of Intergranular Stress Corrosion Cracking Countermeasures," EPRI NP-3324-LD, Electric Power Research Institute, March 1984
- C.L. Bradshaw, Automatic Denting Analysis of Steam Generator Tubing, in *Non-Destructive Examination for Pressurised Components*, Elsevier, 1984
- G.J. Dau and M.M. Behravesh, Status of Intergranular Stress Corrosion Crack Depth Sizing, in *Non-Destructive Examination for Pressurised Components*, Elsevier, 1984
- M.E. Lapides, The Challenge of Continuous Flaw Monitoring in Electric Utilities Practice, *J. Pressure Vessel Technol.*, Vol 106, Aug 1984
- M.E. Lapides, MINAC: Portable, High Energy Radiographic Source; Experience and Applications, in *Non-Destructive Examination for Pressurised Components*, Elsevier, 1984
- M.E. Lapides, Radiographic Detection of Crack-Like Defects in Thick Sections, *Mater. Eval.*, Vol 42 (No. 6), May 1984
- C.R. Mikesell and S.N. Liu *et al.*, Detection of Intergranular Stress Corrosion Cracking Using Automated Ultrasonic Techniques, in *Proceedings of the Nondestructive Testing and Electrochemical Methods of Monitoring Corrosion in Industrial Plants*, ATME E-7/G-1 Symposium, Montreal, Canada, May 1984

Quantitative NDE (1985)

- M.L. Fleming, Inspection of Pipe, Tubing and Plate, in *Proceedings of the 1985 ASNT Spring Conference* (Washington, D.C.), American Society for Nondestructive Testing, 1985
- M.L. Fleming, Field Experience in Automated Ultrasonic Inspection of Stainless Steel, in *Proceedings of the 1985 Pressure Vessels and Piping Conference* (New Orleans, LA), American Society of Mechanical Engineers, 1985
- V.I. Neeley, "Application of Medical Ultrasonic Testing Technology to the Utility Industry," EPRI NP-4034, Electric Power Research Institute, May 1985
- R.B. Thompson *et al.*, "Ultrasonic Scattering From Intergranular Stress Corrosion Cracks: Derivation and Application of Theory," EPRI NP-3822, Electric Power Research Institute, Jan 1985

Quantitative NDE (1986)

- F.V. Ammirato *et al.*, "Development of Improved Procedure for Examination of Dissimilar-Metal Welds in BWR Nozzle-to-Safe-End Welds," EPRI NP-4606-LD, Electric Power Research Institute, May 1986
- M.J. Avioli, Jr., Modeling Ultrasonic Flaw Detection, *EPRI J.*, Sept 1986
- F.L. Becker *et al.*, NDT of Steam Piping and Headers, in *Proceedings of the Fossil Plant Inspections Workshop*, Electric Power Research Institute, 1986
- E.S. Furgason *et al.*, "Digital Techniques to Improve Flaw Detection by Ultrasound Systems," EPRI NP-4878, Electric Power Research Institute, Oct 1986
- B.P. Hildebrand, "Investigation of Advanced Acoustic and Optical Nondestructive Evaluation Techniques," EPRI NP-4897, Electric Power Research Institute, Oct 1986
- K. Krzywosz, Trends and Recent Developments in NDE of Steam Generator Tubes, in *Proceedings of the Fifth Annual Steam Generator NDE Workshop* (Myrtle Beach, SC), Steam Generator Owners Group II, 1986
- S.M. Walker, Ultrasonic Examination of Corrosion-Resistant Clad Weldments, in *Proceedings of the Southwest Research Institute 14th Nuclear Power Educational Seminar*, Southwest Research Institute, 1986
- A.J. Willetts *et al.*, "Evaluation of the Ultrasonic Data Recording and Processing System (UDRPS)," EPRI NP-4397, Electric Power Research Institute, Jan 1986

Quantitative NDE (1987)

- F. Ammirato and S. Walker *et al.*, Examination of Dissimilar Metal Welds in BWR Nozzle-to-Safe-End Joints, in *Proceedings of the 8th International Conference on NDE in the Nuclear Industry* (Orlando, FL), ASM INTERNATIONAL, 1987
- L. Becker and S. Walker *et al.*, NDE of Fossil Plant Steam Piping, in *Proceedings of the 8th International Conference on NDE in the Nuclear Industry* (Orlando, FL), ASM INTERNATIONAL, 1987
- M.M. Behravesh *et al.*, Status of Advanced UT Systems for the Nuclear Industry, *Nucl. Eng. Des.*, No. 102, 1987, p 265-273
- D. Kedem, "Computed Tomography for Thick Steel Pipe and Castings," EPRI NP-5107-LD, Electric Power Research Institute, March 1987
- K. Krzywosz, Recent NDE Experience With PWR Steam Generator Tubing Inspection, in *Proceedings of the 8th International Conference on NDE in the Nuclear Industry* (Orlando, FL), ASM INTERNATIONAL, 1987
- D. MacDonald and E.K. Kietzman, Comparative Evaluation of Acoustic Holography Systems, in *Proceedings of the 8th International Conference on NDE in the Nuclear Industry* (Orlando, FL), ASM INTERNATIONAL, 1987
- D. MacDonald and S.M. Walker, "Effects of Ultrasonic Equipment Variations on Crack Length Measurements," EPRI NP-5485, Electric Power Research Institute, Oct 1987
- G. Selby, "Ultrasonic Examination of BWR Replacement Pipe Joint Mock-Ups," EPRI NP-5438-LD, Electric Power Research Institute, Aug 1987
- R.B. Thompson *et al.*, "Modeling Ultrasonic Inspection of Nuclear Components—Beam Models and Applications," EPRI NP-5330, Electric Power Research Institute, Aug 1987
- A.J. Willetts and E.K. Kietzman, UT Techniques for Detection and Sizing of Under-Clad Cracks in Reactor Pressure Vessels, in *Proceedings of the 8th International Conference on NDE in the Nuclear Industry* (Orlando, FL), ASM INTERNATIONAL, 1987

Quantitative NDE (1988)

- F. Ammirato, "NDE of Cast Piping in the Nuclear Industry," Paper presented at the IAEA Specialists Meeting on the Inspection of Austenitic and Dissimilar Metal Welds, Espoo, Finland, International Atomic Energy Agency, June 1988
- F. Ammirato *et al.*, Ultrasonic Examination of Dissimilar-Metal Welds in PWR and BWR Plants, in *Non-Destructive Examination in Relation to Structural Integrity*, Proceedings of the Post-SMiRT Seminar #3, Elsevier, 1988
- L. Goldberg, "Reliability of Magnetic Particle Inspection Performed Through Coatings," EPRI NP-5919, Electric Power Research Institute, July 1988
- J.D. Heald, "Evaluation of a New Gamma-Scanning Method for Detection and Sizing of Intergranular Stress Corrosion Cracking," EPRI NP-5759-LD, Electric Power Research Institute, May 1988
- B.P. Hildebrand, "Stepped Frequency Imaging for Flaw Monitoring," EPRI NP-6033, Electric Power Research Institute, Sept 1988
- B.P. Newberry and R.B. Thompson, Prediction of Surface Induced Ultrasonic Beam Distortions, in *Review of Progress in Quantitative NDE-8*, Plenum Press, 1988
- J.L. Rose, "Wave-Propagation Studies for Improved Ultrasonic Testing of Cen-

trifugally Cast Stainless Steel," EPRI NP-5979, Electric Power Research Institute, Aug 1988

- J. Saniee, T. Wong, and N.M. Bilgutay, Optimal Ultrasonic Flaw Detection Using a Frequency Diversity Technique, in *Review of Progress in Quantitative NDE-8*, Plenum Press, 1988
- T. Sasahara and F. Ammirato, "Automated Ultrasonic Pipe Examination and Interpretation," EPRI NP-5760, Electric Power Research Institute, April 1988
- R. Shankar, Field Application of Integrated Ultrasonic Feature-Based and Imaging-Based Analysis, in *Proceedings of the 9th International Conference on Non-Destructive Evaluation in the Nuclear Industry*, ASM INTERNATIONAL, 1988
- R. Shankar *et al.*, Feature-Enhanced Ultrasonic Imaging—Application of Signal Processing and Analysis, in *Non-Destructive Examination in Relation to Structural Integrity*, Proceedings of the Post-SMiRT Seminar #3, Elsevier, 1988

Quantitative Nondestructive Evaluation

Introduction

Vicki E. Panhuise, Allied-Signal Aerospace Company, Garrett Engine Division

THE RELIABILITY of a nondestructive inspection (NDI) procedure was defined in the article "Reliability of Flaw Detection by Nondestructive Inspection" in Volume 11 of the 8th Edition of *Metals Handbook* as a quantitative measure of the efficiency of that procedure in finding flaws of specific type and size. During the years since that article was published, many inspection reliability programs have been conducted, and various quantitative measurements have been cited to express the procedure capabilities. To establish the basis for NDI method reliability, it is necessary to review the history of NDI and its relation to reliability methods.

Historical Development of Quantitative Measurement Techniques

Nondestructive inspection methods are specified for material and/or component inspection requirements to maintain the necessary quality for the final service life of the material/component. In most industries, the inspection requirements are defined in a specification that describes the sensitivity level of the inspection method as well as the rejectable flaw size.

An example of a specification requirement is given in Table 1; the specification is based on longitudinal wave inspection using flat-bottom holes (FBH). It defines the ultrasonic inspection requirements for product over 13 mm (0.5 in.) thick. Table 1 indicates the defect detection/rejection limits for each quality class material. For example, quality class AA requires that single discontinuities be detected at a sensitivity level equivalent to a No. 3 (1.2 mm, or 3/64 in., diam) FBH. In addition, multiple discontinuities* shall be detected at a sensitivity level equivalent to a No. 1 (0.4 mm, or 1/64 in., diam) FBH.

Each specification also defines a procedure for demonstrating the capability to detect defects at the required sensitivity levels. In AMS 2630A, this defined procedure is a calibration technique used to set up the ultrasonic instrumentation. Ultrasonic standards have been designed to establish the performance of the inspection system. Procedures have been defined by the American Society for Testing and Materials for the manufacture of these ultrasonic reference blocks for longitudinal wave testing (Ref 2, 3). Finally, the specifications require specific training for the inspectors. For conformance with AMS 2630A, personnel must be certified to and function within the limits of their levels of certification as specified by the American Society for Nondestructive Testing (Ref 4). Further details of ultrasonic inspection procedures are described elsewhere in this Volume.

*Multiple discontinuities are defined in AMS 2630A as two or more indications above the level established for the class that occur within 16 cm^3 (1.0 $in.^3$) of the inspected surface.

These specification requirements were designed to control the inspection processes and the quality of the inspection results. However, several catastrophic failures of major engineering systems (such as the F-111, space shuttle, nuclear reactors, and the Alaskan pipeline) and the development of advanced materials such as composites were major forces in the development and application of NDI technology (Ref 5). In concert with these events, a new design method using linear elastic fracture mechanics (LEFM) required inspection methodology for detecting defects in production and in service (for a description of LEFM, see Volume 8 of the 9th Edition of *Metals Handbook*). Linear elastic fracture mechanics design assumes the presence of structural defects and then allows the designer to answer the following questions:

- What is the critical flaw size that will cause failure for a given component subject to service stress and temperature conditions?
- How long can a precracked structure be safely operated in service?
- How can a structure be designed to prevent catastrophic failure from preexisting cracks?
- What inspections must be performed to prevent catastrophic failure?

The ability to answer these questions forms the basis of *nondestructive evaluation* (NDE), which involves damage-tolerant design approaches and is centered on the philosophy of ensuring safe operation in the presence of flaws. The U.S. Air Force has used damage tolerance analysis, as shown in Fig. 1. The initial component in the as-manufactured condition is assumed to have a flaw of length a_0. This flaw length is based on the manufacturing inspection capability or material flaw size distribution. The growth of the flaw is predicted for service usage and will reach a critical flaw size, a_f, after t_0 flight hours. The current Air Force philosophy requires an inspection at half the time required for the potential crack to grow to critical size. This inspection is assumed to detect and remove any flaw of size larger than a_{NDE}. The assumption creates the requirement to determine NDE probability of detection (POD) for this current design practice, as described in the articles that follow in this Section.

The driving functions described above caused the NDE/NDI community to evaluate the inspection capabilities. The first evaluation in the aerospace industry was conducted by Lockheed in the 1970s and

Table 1 Typical NDI acceptance criteria

Five classes of ultrasonic quality are established for longitudinal wave inspection. These classes are defined for inspection involving flat-bottom reflectors in ultrasonic references.

Quality class	Single discontinuity(a), FBH No.	Multiple discontinuities(a), FBH No.	Maximum linear discontinuity, in.	Maximum loss of back reflection, %
AA	3	1(b)	No. 1 response for 0.12	50
A1	3	2(c)	No. 2 response for 1.00	50
A	5	3	No. 3 response for 1.00	50
B	8	5	No. 5 response for 1.00	50
C	As established by purchaser and vendor for specific part			

(a) FBH numbers indicate diameter in multiples of 0.4 mm (1/64 in.) of FBH in ultrasonic reference. (b) 11% of a No. 3 FBH is equivalent to a No. 1 FBH and can be used in place of the response from the No. 1 FBH. (c) 44% of a No. 3 FBH is equivalent to a No. 2 FBH and can be used in place of the response from the No. 2 FBH. Source: Ref 1

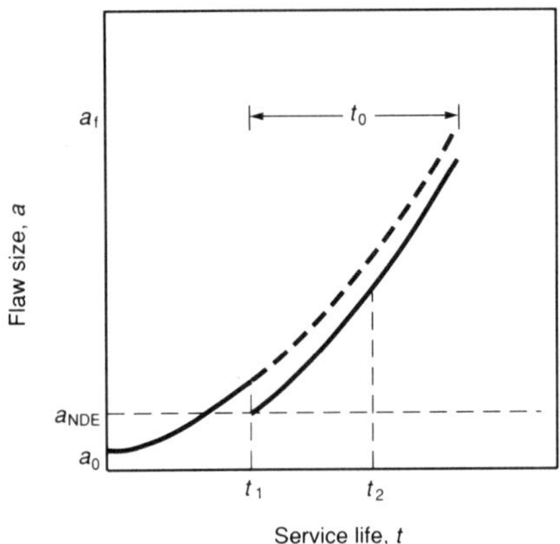

Fig. 1 Crack growth life curve for damage-tolerant design

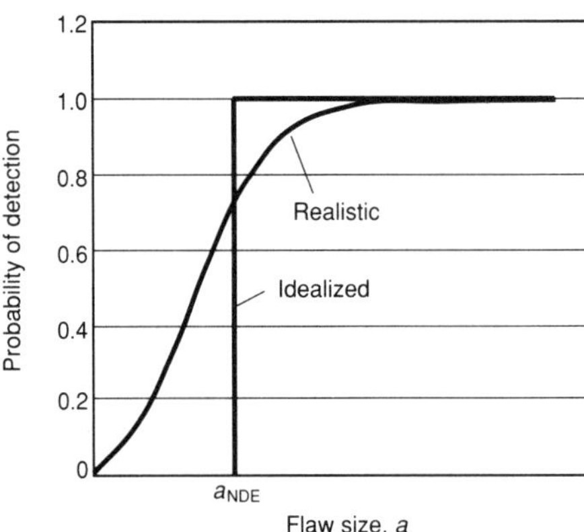

Fig. 2 Schematic of probability of detection curves

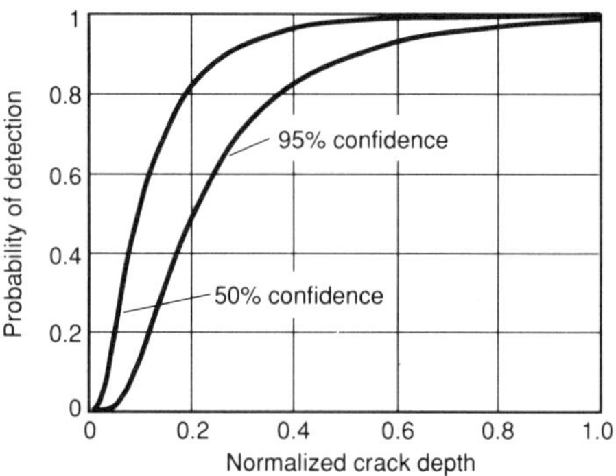

Fig. 3 POD curve for the automated eddy current inspection of titanium bolt holes

was called the "Have Cracks, Will Travel" program (Ref 6). This program was established to determine if airframe inspection would repeatedly detect cracks so that engineers could use LEFM philosophy for design. The study concluded that the overall reliability of NDI performed by the Air Force and evaluated in the program falls below that which has been previously assumed by established guidelines (Ref 6). The most significant result of the program was that the 90/95 reliability criteria could not be attained for any flaw size with typical inspection techniques applied by the average technician. Many advancements have been made in NDE/NDI technology to improve the inspection capability since this study. These technologies are described throughout this Volume.

NDE Reliability

The following article in this Section—"Fracture Control Philosophy"—describes the U.S. Air Force philosophy for current design practices. Damage-tolerant design requires knowledge of the reliability of the inspection technique used to detect flaws or damage. In the initial stages of using this design approach, a one-number characterization of NDE capability was in use. The one-number characterization was the minimum crack (flaw) length for which there is a fixed degree of confidence that at least a fixed probability of cracks will be detected. Typically, the minimum crack length was chosen such that, at the 95% confidence level (CL), at least 90% of all cracks greater than this length will be detected. The number is referred to as the 90/95 (POD/CL) crack length.

Probability of detection functions for describing the reliability of an NDE technique have been the subject of many studies. The ideal inspection system POD function is shown schematically in Fig. 2. All flaws larger than a_{NDE} would be detected all of the time, while all flaws smaller than a_{NDE} would not. Obviously, no ideal system exists, and the POD functions used produce a continuous curve. Figure 3 shows a POD curve for an automated eddy current inspection of titanium bolt holes. Two confidence level POD curves are shown, 50 and 95%. Berens and Hovey completed a study that compared POD analysis techniques to determine the optimum calculation method (Ref 7). Additional information concerning the analysis of NDE data for the determination of reliability is available in the article "NDE Reliability Data Analysis" in this Section.

The accuracy of the POD function is dependent on the demonstration program design. The data acquired during the demonstration program testing can influence the resulting prediction of reliability. For example, consider the cases presented in Fig. 4, which graphically presents the specimen/crack depth histogram for two case studies. If the design requirement is such that a_{NDE} (that is, 90/95 crack depth) must be 250 μm (10 mils) or less, only the specimens in case study No. 2 could demonstrate this capability. That is, to demonstrate 90/95 crack size, the specimens fabricated for the demonstration program must have cracks less than a_{NDE} and some larger. If it is assumed that all the flaws were detected in both cases, the 90/95 crack sizes predicted are as follows:

Case study	90/95 crack size μm	mils
1	417	16.40
2	119	4.70

This case study, which demonstrates the criticality of planning the reliability demonstration program, showed only the importance of specimen flaw size to the final outcome of the reliability study.

In the following articles in this Section, fracture control philosophy is discussed, reliability demonstrations that have been completed are reviewed, and the analysis of NDE data is examined. The final article in this Section provides information on advanced modeling studies for predicting reli-

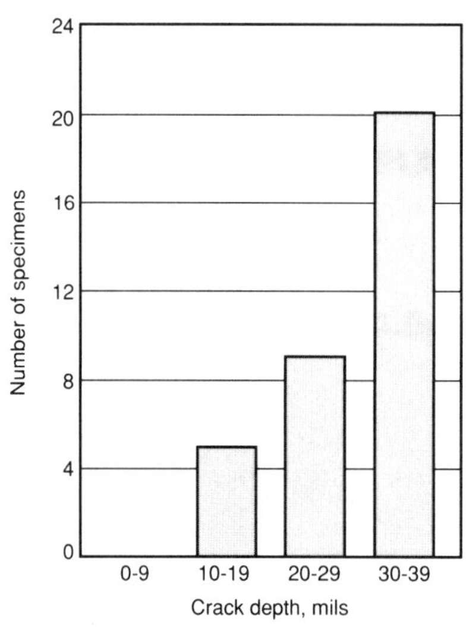

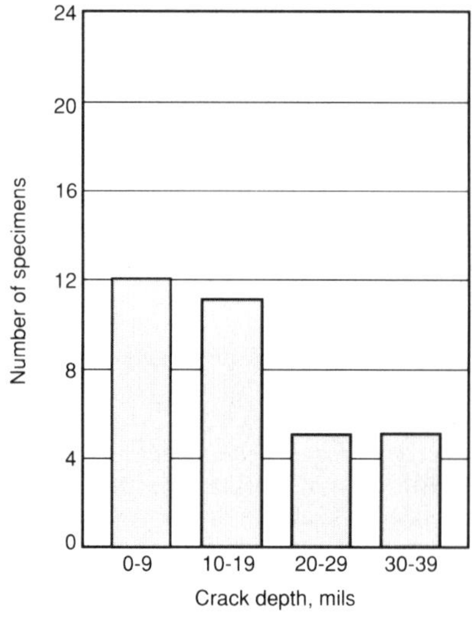

Fig. 4 Specimen crack depth distribution for two case studies. (a) Case study No. 1. (b) Case study No. 2

ability for the inspection of a particular component. The objective of this Section is to provide the necessary information and resources to conduct an effective NDE reliability program.

REFERENCES

1. "Ultrasonic Inspection of Product over 0.5 in. (13 mm) Thick," Aerospace Material Specification 2630A, Society of Automotive Engineers, 1980 (original release 1960)
2. "Standard Practice for Fabricating and Checking Aluminum Alloy Ultrasonic Standard Reference Blocks," E 127, *Annual Book of ASTM Standards*, American Society for Testing and Materials
3. "Standard Recommended Practice for Fabrication and Control of Steel Reference Blocks Used in Ultrasonic Inspection," E 428, *Annual Book of ASTM Standards*, American Society for Testing and Materials
4. "Recommended Practice, Personnel Qualification and Certification in NDT," SNT-TC-IA, American Society for Nondestructive Testing
5. W. Rummel, Recommended Practice for a Demonstration of Nondestructive Evaluation (NDE) Reliability on Aircraft Production Parts, *Mater. Eval.*, Vol 40 (No. 9), Aug 1982, p 922
6. W.H. Lewis, *et al.*, "Reliability of Nondestructive Inspections—Final Report," Report SA-ALC/MME 76-6-38-1, United States Air Force, Air Logistics Center, Kelly Air Force Base, 1978
7. A.P. Berens and P.W. Hovey, "Flaw Detection Reliability Criteria Volume I—Methods and Results, Final Report," Report AFWAL-TR-84-4022 Volume I, United States Air Force Material Laboratory, Wright-Patterson Air Force Base, 1984

Fracture Control Philosophy

William D. Cowie, United States Air Force, Aeronautical System Division, Propulsion Directorate

FRACTURE CONTROL PHILOSOPHIES are being used in the design, development, and life management of United States Air Force (USAF) turbine engine and airframe components. This article describes the fracture control program for turbine engine components. The section "Applications (Case Studies)" of the article "Applications of NDE Reliability to Systems" in this Volume provides an overview of fracture control programs for both airframe and turbine engine components.

The establishment of a fracture control philosophy and the implementation of a fracture control program have been integral components of the USAF turbine engine development process since 1978. They have been applied to new engine programs as part of the USAF Engine Structural Integrity Program (ENSIP) described in military standard MIL-STD-1783 (issued formally in 1984). This military standard was reviewed and approved by the Aerospace Industries Association of America in 1982.

Fracture control philosophy has also been applied to existing inventory USAF engines through structural durability and damage tolerance assessments. In all, fracture control programs have been applied or are being applied to the F-100, TF-34, F100-PW-220, F100-PW-229, F110-GE-100, F110-GE-129, F101-GE-102, F109-GA-100, F-119, F120, and T406 engines and have resulted in the implementation of enhanced nondestructive evaluation (NDE) methods (for example, eddy current inspection) at manufacturing and at field/depot. These inspections have been successful in detecting early cracking and in accelerating corrective actions. Several developmental efforts in the last 5 years have identified fluorescent penetrant inspection process improvements that must be implemented within industry and Air Force depots to improve flaw detection reliability. The need to quantify detection reliability for imbedded defects is also identified.

The engine development process has been evolutionary in terms of the application of upgraded requirements. The new process of fracture control, sometimes referred to as damage tolerance, is contained in ENSIP, and it involves material selection as well as design and life management. Recent experience clearly demonstrates that the damage tolerance requirement is cost effective when assessed on a life cycle basis.

Overview of ENSIP

In the past 16 years, a large number of structural problems have occurred in USAF gas turbine engines. Many of these were safety problems that resulted in loss of aircraft, and an even greater number affected durability, causing a high level of maintenance and modification costs. All of these problems have adversely affected fleet readiness. The Engine Stuctural Integrity Program was intended to reduce these problems substantially and was developed based on the following specific lessons:

- It is unrealistic (and can be dangerous) to assume defect-free structure in safety-of-flight components
- Critical parts (and part details) and potential failure modes must be identified early and appropriate control measures implemented
- Internal thermal and vibratory environments must be identified early in the engine development
- Predicted analytical stresses must be verified by test for complex components
- Materials and processes must be adequately characterized (particularly, the fracture properties)
- Design stress spectra, component test spectra, and full-scale engine test spectra must be based on the anticipated service usage of the engine, that is, accelerated mission-related testing
- Potential engine/airframe structural interactions must be defined and accounted for
- Management procedures (such as individual engine tracking procedures and realistic inspection and maintenance requirements) must be defined and enforced

The Engine Structural Integrity Program was established by the Air Force to provide an organized and disciplined approach to the structural design, analysis, development, production, and life management of gas turbine engines, with the goal of ensuring engine structural safety, increased service readiness, and reduced life cycle costs. The five major tasks associated with ENSIP are the development of design information; design analysis and component and material characterization; component and core engine testing; ground and flight engine testing; and production quality control and engine life management. Each major task is subdivided into a number of subtasks (Table 1) that guide the development process.

The Engine Structural Maintenance Plan represents the output of the ENSIP program. This plan identifies and defines individual part life limits, the necessary inspection periods for each fracture-critical part, and the inspection procedure. The basic components of the Engine Structural Maintenance Plan are as follows:

- Structural safety is obtained in ENSIP by requiring a structure with a damage tolerance that is capable of accommodating flaws induced either in manufacture or service
- Durability design requirements stipulate that the economic life of the engine must exceed the specified design service life of the aircraft when flown to the design usage spectra
- Maintainability criteria require that old parts fit and function with new parts, that repair life be defined, and that inspectability and structural diagnostics be designed into the engine and its components
- A materials and process characterization plan controls materials development through key engine development points
- Environmental definition requirements specify the thermal, dynamic, and steady-state stress; the stress spectra; and the component sensitivities
- A comprehensive ground test policy is utilized to ensure compliance with safety, durability, and maintainability requirements
- A usage and tracking policy is used to form the basis of an engine life management program

Conventional Life Management

The conventional fatigue design approach used in the past is illustrated in Fig. 1. Components were designed so that the low-

Table 1 Tasks of the engine structural integrity program

Task I: Design information	Task II: Design analysis and material characterization and development tests	Task III: Component and core engine tests	Task IV: Ground and flight engine tests	Task V: Engine life management
Development plans	Design duty cycle	Component tests	Ground engine tests	Updated analyses
ENSIP master	Material characterization	Strength	Strength	Structural maintenance plan
Durability and damage control	Design development tests	Vibration	Damage tolerance	Operational usage survey
Material and process characterization	Structural/thermal analysis	Damage tolerance	Accelerated mission test	Individual engine tracking
Corrosion prevention and control	Installed engine inspectability	Durability	Thermal survey	Durability and damage tolerance control actions (production)
Inspection and diagnostics	Manufacturing and quality control	Core engine tests	Vibration strain and flutter boundary survey	
Operational requirements		Thermal survey	Flight engine tests	
Design service life and usage requirements		Vibration strain and flutter boundary survey	Fan strain survey	
Design criteria			Thermal survey	
			Installed vibration	
			Deterioration	

cycle fatigue (LCF) limit exceeded the required usage interval in terms of engine flight hours or cycles. The LCF limit was based on a lower bound of three standard deviations (-3σ or 1/1000 probability) of the distribution of crack initiation times. The lower-bound limit was chosen to minimize the occurrence of cracking and resultant failure and the need for repair (economics). The main concern with the conventional approach is that insufficient recognition or provision exists regarding the impact that initial defects can have on total component life (component failure can and has occurred prior to reaching the LCF limit). Implementation of the conventional life philosophy requires that all like components be removed from service upon expiration of the lower-bound life. Because there is wide scatter in fatigue data for materials, the analysis is usually based on the material properties at -3σ, which implies that all components are treated as the worst-case component in the population.

Further, the fatigue design philosophy is strongly based on the premise that the material is free of initial defects. This philosophy makes no allowance for material, manufacturing, producibility, and handling anomalies or defects. In MIL-STD-1783, the fracture control philosophy and policy requirement have been incorporated to prevent safety or flight structural failures caused by material defects, manufacturing defects, or fatigue-induced cracks. In numerous engine failures in the past, initial defects grew in size because of the use of high-strength, low fracture toughness material. In addition, improper detail design resulted in high stress levels and structural discontinuities, and adequate quality control was lacking.

The primary need for damage tolerance requirements or fracture control policy has occurred because of the ever-present drive to minimize engine weight through the development of materials with increased strength and resistance to crack initiation. An undesirable but attendant feature of these new materials has been a decreased resistance to crack progagation (Fig. 2 and 3). As a result of these trends, the F100 and

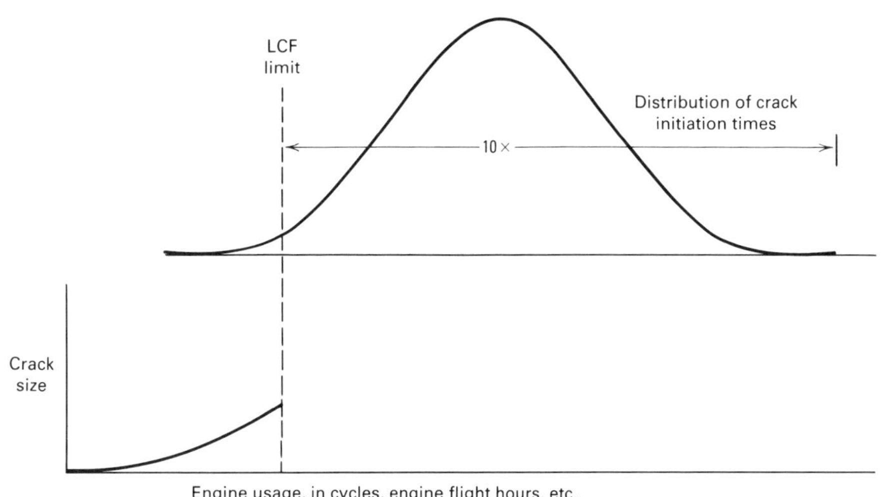

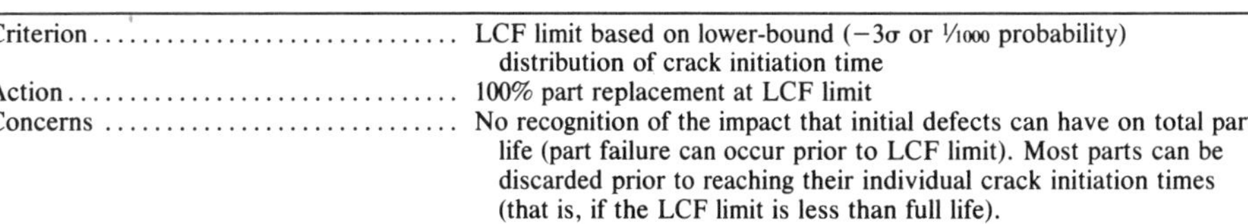

Criterion	LCF limit based on lower-bound (-3σ or 1/1000 probability) distribution of crack initiation time
Action	100% part replacement at LCF limit
Concerns	No recognition of the impact that initial defects can have on total part life (part failure can occur prior to LCF limit). Most parts can be discarded prior to reaching their individual crack initiation times (that is, if the LCF limit is less than full life).

Fig. 1 Conventional approach to the life management of cyclic-limited engine components. LCF, low-cycle fatigue

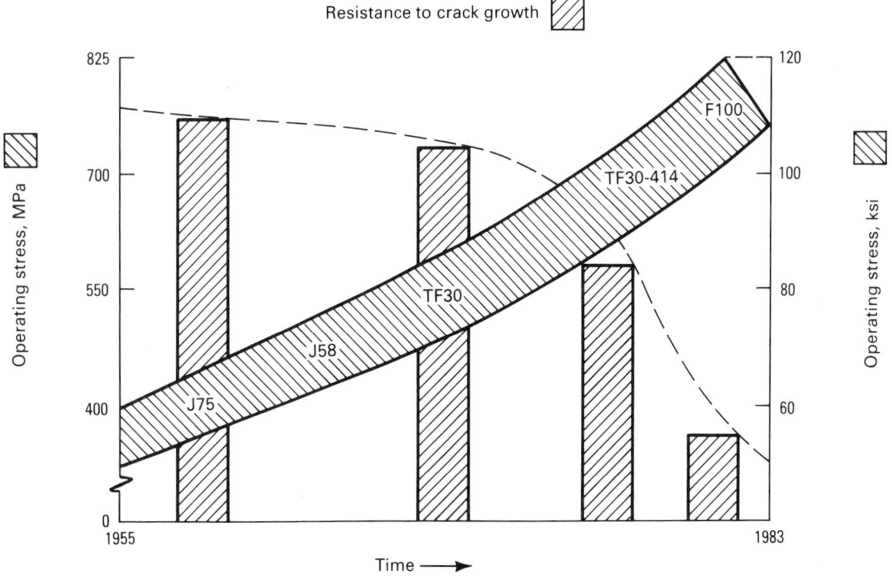

Fig. 2 Trends in the operating stress and material characteristics (resistance to crack growth) of aircraft engine components

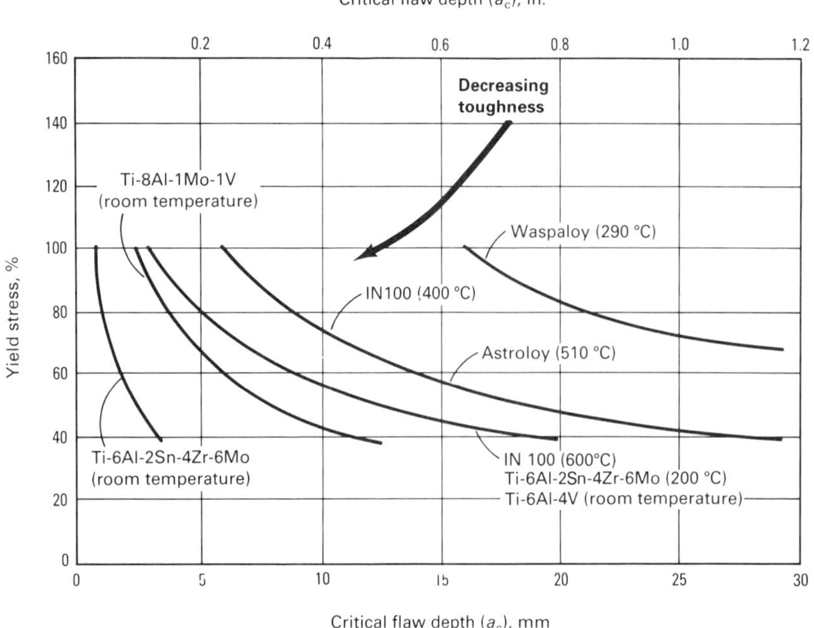

Fig. 3 Yield stress versus critical flaw depth of some materials used in aircraft engine components

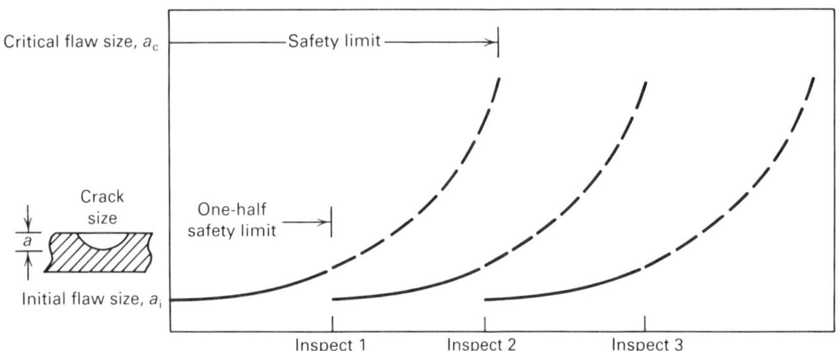

Fig. 4 Damage tolerance approach to life management of cyclic-limited engine components. The safety limit or residual life is the time for the initial flaw to grow and cause failure. The size of the initial flaw, a_i, is based on the inspection method or material defect distribution (for imbedded defects).

TF34 engine durability and damage tolerance assessments were performed. The assessment of engine durability and damage tolerance recommended:

- Updated replacement limits
- Upgraded inspection requirements
- Preferred structural improvements
- Candidate life extension components

ENSIP and Fracture Control Philosophy Policy

Damage tolerance is defined as the ability of the engine to resist failure due to the presence of flaws, cracks, or other damage for a specified period of usage. The damage tolerance or fracture control philosophy used in ENSIP is shown in Fig. 4. Components are designed for crack growth so that the safety limit exceeds two times the required inspection interval. The safety limit or residual life is the time for assumed initial flaws to grow and cause failure. Because the requirement is to inspect at one-half the safety limit, the design goal for the safety limit is two times the required design life (that is, no inspections). The minimum design requirement for the safety limit is two times the planned depot visit interval. An important aspect of the damage tolerance requirement is that it applies only to fracture-critical components.

Fracture-critical components are defined as those components whose failure will result in probable loss of the aircraft due to noncontainment or, for single-engine aircraft, power loss that presents sustained flight because of direct part failure or by causing other progressive part failures. Damage tolerance requirements are applied only to fracture-critical components (that is, components that must maintain their integrity during flight) and not, in general, to durability-critical components (that is, components that affect maintenance schedules). As expected, component classification is affected by aircraft engine configuration (single engine or multiengine). Component classification is established early and is identified in the contract.

Initial Flaw Size. Initial flaws are assumed to exist in fracture-critical components. Experience has shown that premature cracking (that is, crack initiation prior to the LCF limit) occurs at high-stress areas and where components initially contained both material- and manufacturing-related quality variations (voids, inclusions, machining marks, scratches, sharp cracks, and so on). The fracture control or damage tolerance requirement assumes a sharp crack as the initial flaw when characterizing these abnormal initial conditions. Thc assumed initial imbedded flaw sizes are based on the intrinsic material defect distribution or the NDE methods to be used during manufacture. The assumed surface flaw size also depends on the NDE capability. An inspection reliability of 90% probability of detection (POD) at the lower-bound 95% confidence level (CL) is required for the assumed initial flaw sizes.

The assumed initial flaw size to account for intrinsic material defect distribution should encompass 99.99% of the defect population if a scatter factor of two is used to establish the inspection interval, or 99.9% if a scatter factor of one is used. If embedded defects cannot be inspected in service, the 99.99 percentile (or the 99.9 percentile) is used to satisfy the design life requirement.

An initial flaw size not less than 0.75 mm (0.030 in.) in length (for surface flaws) or 0.4 × 0.4 mm (0.015 × 0.015 in.) in size (corner cracks) for nonconcentrated stress areas (bores, webs, and so on) is required. Initial flaw sizes for other surface locations (holes, fillets, scallops, and so on) will be consistent with the demonstrated capability (90% POD/95% CL) of the inspection systems proposed for use. It is recommended that the initial design and sizing of components be based on 0.75 mm (0.030 in.) long surface flaws or 0.4 × 0.4 mm (0.015 × 0.015 in.) corner cracks at all locations. This design recommendation is based on the initial flaw size that can be detected with fluorescent penetrant inspection. This includes fully automated fluorescent penetrant inspection systems that are being developed to meet the 0.75 mm (0.030 in.) and 0.4 × 0.4 mm (0.015 × 0.015 in.) inspection criteria.

These flaw sizes are intended to represent the maximum size of the damage that can be present in a critical location after manufacture and/or inspection. The specification of these flaw sizes is based on the demonstrated flaw detection capability of the nondestructive inspection (NDI) method. During

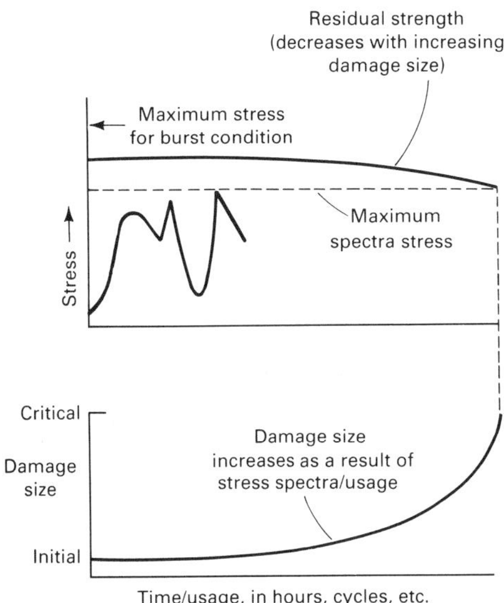

Fig. 5 Diagram of the residual strength requirement

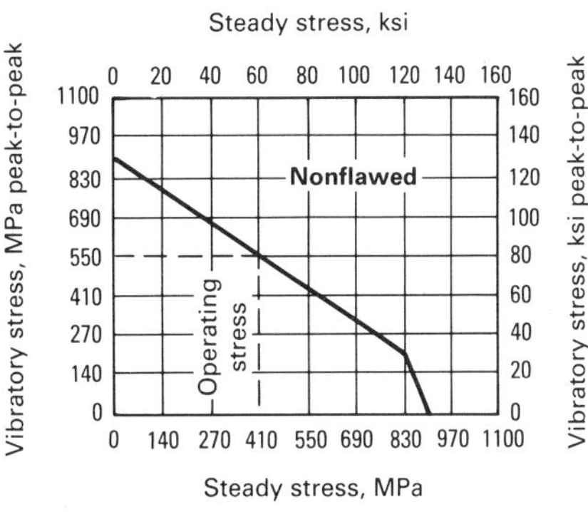

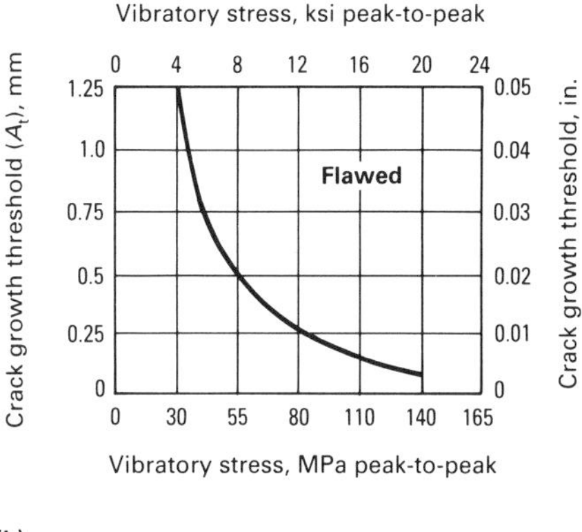

Fig. 6 Interaction of vibratory stress and initial flaws. (a) Large vibratory stress required to initiate crack. (b) Low vibratory stress will propagate cracks. The crack growth threshold, A_t, represents the threshold of vibratory motion that will cause the growth of a given crack size.

design of the components, the assumed initial flaw size that is appropriate for various NDI methods is:

- 0.75 mm (0.030 in.) surface length where the NDI method is fluorescent penetrant inspection
- 0.25 mm (0.010 in.) surface length where the NDI method is eddy current or ultrasonic inspection
- 1.3 mm^2 (0.002 $in.^2$) area for imbedded defects utilizing ultrasonic inspection
- 5 mm (0.200 in.) surface length and imbedded sphere = 0.2 x thickness for weldments
- When initial flaw sizes are based on material defect distribution, selected size shall encompass 99.99% of the distribution
- Demonstration that assumed flaw sizes can be reliably detected with a 90% POD and a 95% CL

The capabilities of the NDI method must be demonstrated by the contractor. The design of NDE reliability experiments is discussed in the article "NDE Reliability Data Analysis" in this Volume.

Residual strength is defined as the load-carrying capability of a component at any time during the service exposure period, considering the presence of damage and accounting for the growth of damage as a function of exposure time. The requirement is to provide limit load residual strength capability throughout the service life of the component. In other words, the minimum residual strength for each component (and location) must be equal to the maximum stress that occurs within the applicable stress spectra based on the design duty cycle. Normal or expected overspeed due to control system tolerance and engine deterioration is included in the residual strength requirement, but fail-safe conditions, such as burst margin, are excluded. The residual strength requirement is illustrated in Fig. 5.

Inspection Intervals. It is highly desirable to have no damage tolerance inspections required during the design lifetime of the engine. This in-service noninspectable classification requires that components be designed such that the residual life or safety limit be twice the design life. Designing components as in-service noninspectable is a requirement for those components or locations that cannot be inspected during the depot maintenance cycle.

However, the weight penalty incurred to achieve a safety limit/residual life/damage growth interval twice the design life may be prohibitive on some components/locations. Therefore, in-service inspections will be allowed on some components subject to justification. The basis for the justification is characterization of the costs as a function of the requirements as established by trade studies. Cost is usually expressed in terms of weight or life cycle cost, and the requirement in terms of safety limit/residual life/damage growth interval.

The depot or base-level inspection interval for damage tolerance considerations should be compatible with the overall engine maintenance plan. Once again, it is highly desirable that the inspection interval be equal to the design service life of the parts in the hot gas path (that is, the hot-part design service life, which is equal to one-half the design lifetime of the engine) because this is the expected minimum depot or maintenance interval for the engine or module. It is required that the minimum damage tolerance inspection interval be contained in the contract specification.

Flaw Growth. It is required that the assumed initial flaw sizes will not grow to critical size and cause failure of a component due to the application of the required residual strength load in two times the inspection interval. The flaw growth interval is set equal to two times the inspection interval to provide a margin for a variability that exists in the total process (that is, inspection reliability, material properties, usage, stress predictions, and so on). Factors other than two should be used when individual assessments of the variables that affect crack growth can be made (for example, to account for observed scatter in crack growth during testing).

It is important that the effects of vibratory stress on unstable crack growth be accounted for in establishing the safety limit. Experience shows that the threshold crack size can be significantly less than the critical crack size associated with the material fracture toughness, depending on the material, the major stress cycle, and the vibration stress. As shown in Fig. 6, the conventional Goodman diagram may not disclose the true sensitivity of initial defects to vibratory stresses. The threshold crack size must be established at each individual sustained-power condition (idle, cruise, intermediate) using the appropriate values of steady stress and vibratory stress. The smallest threshold crack size will be used as a limiting value in calculating the safety limit if it is less than the critical crack size associated with the material fracture toughness.

Fracture Control Verification

Verification that the fracture control policy is met is accomplished by the development and implementation of a Damage Tolerance Control Plan, by analysis and test, and by the implementation of reliable inspection methods during manufacture and field/depot maintenance.

A Damage Tolerance Control Plan is prepared that identifies and schedules each of the tasks and interfaces in the functional areas of design, materials selection, tests, manufacturing control, and inspection. Specific tasks that are addressed in the Damage Tolerance Control Plan are:

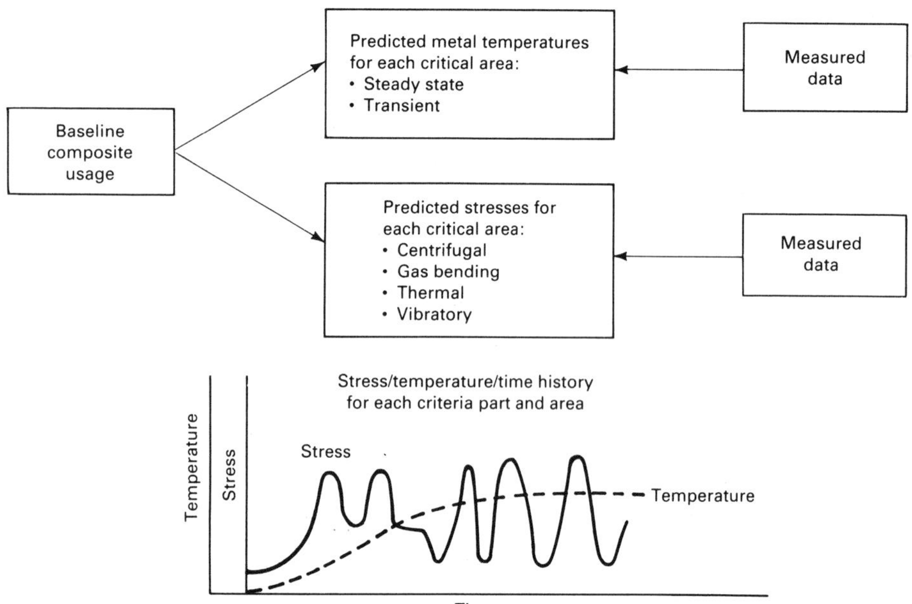

Fig. 7 Development of stress spectra

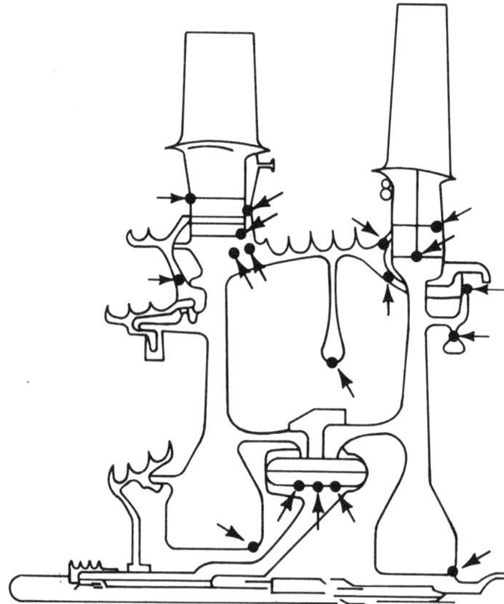

Fig. 9 Typical setup for a cyclic spin test (heated spin pit) of turbine rotor. Arrows indicate critical locations that were preflawed to simulate the worst expected damage.

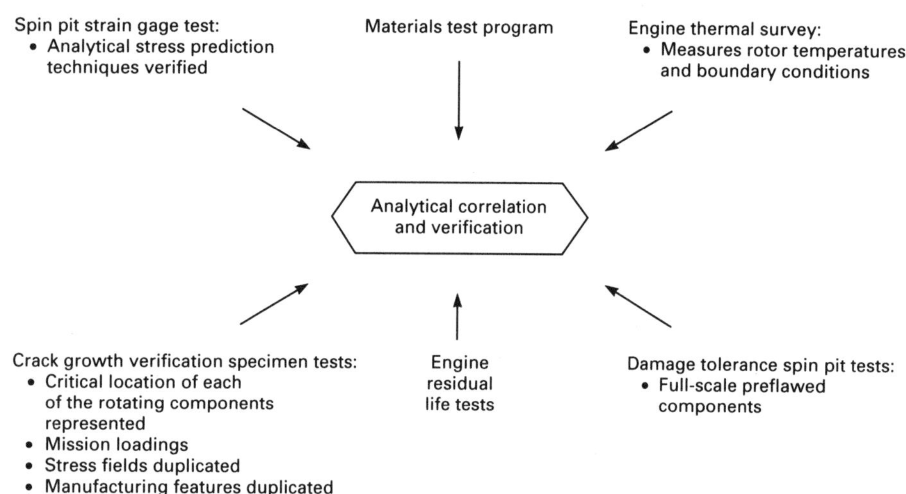

Fig. 8 Residual life analysis and test procedure

- Trade studies for design concepts/material/weight/performance/cost
- Analysis
- Development and qualification tests
- Fracture-critical parts list
- Zoning of drawings
- Basic materials fracture data
- Material properties controls
- Traceability
- NDI requirements

Most of the tasks to be contained in the Damage Tolerance Control Plan have been accomplished by engine manufacturers in past development and production programs. However, the damage tolerance requirement now established by the Air Force imposes the need for new tasks as well as tighter controls and more involvement among the functional areas. Experience indicates that the development and implementation of a Damage Tolerance Control Plan are difficult, but experience also shows very strongly that the development of a plan results in an improved understanding of what must be done. The importance of having a plan rests on the involvement of multiple functional areas and the criticality of having assigned responsibilities for each task.

A particularly important part of the Damage Tolerance Control Plan is the requirement for early trade studies for design concepts/materials/weight/performance/cost. These trade studies are critical to defining cost versus requirement (for example, weight impact versus inspection interval).

Analysis and Test Correlation Requirement. The procedure for the analysis of damage tolerance includes the development of stress spectra (Fig. 7) and the analysis and testing of residual life (Fig. 8). Particular emphasis is placed on establishing a correlation between analytical predictions and test measurements for the growth of cracks in critical areas. Refined analysis models that predict the stress state at and away from the surface, as well as multiple cyclic tests of coupons, subcomponents, and full-scale components in the presence of initial damage, are required. As with the overall ENSIP development philosophy, damage tolerance analysis and tests are conducted early with minimum impact. Test requirements include:

- Specimen tests to define basic material fracture data, such as the plane-strain fracture toughness (K_{Ic}), the plane-stress fracture toughness (K_c), the threshold stress intensity for stress-corrosion cracking (K_{Iscc}), and the crack growth curve, which defines the crack growth rate da/dN (where a is the crack length and N is the number of cycles)
- Subcomponent tests to evaluate crack growth at typical critical features, such as bolt holes and fillet radii
- Spin pit test of full-scale components, such as disks, spacers, or rotors
- Engine testing of preflawed components

A typical test configuration for a cyclic spin test is shown in Fig. 9, and typical empirical data are shown in Fig. 10 to 12. Verification requirements may also include engine test with components that are preflawed or cracked in critical locations to determine the effects of the real environment (temperature and gradient, vibration, and so on). Such tests must be closely controlled and monitored using the inspec-

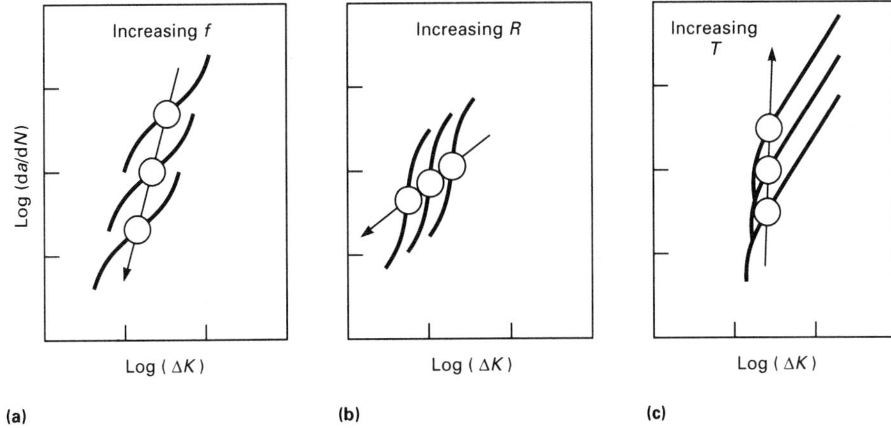

Fig. 10 Crack progagation as influenced by increases in three parameters. (a) Frequency, *f*. (b) Stress ratio, *R*. (c) Temperature, *T*. The variable *K* is the calculated stress at the crack tip.

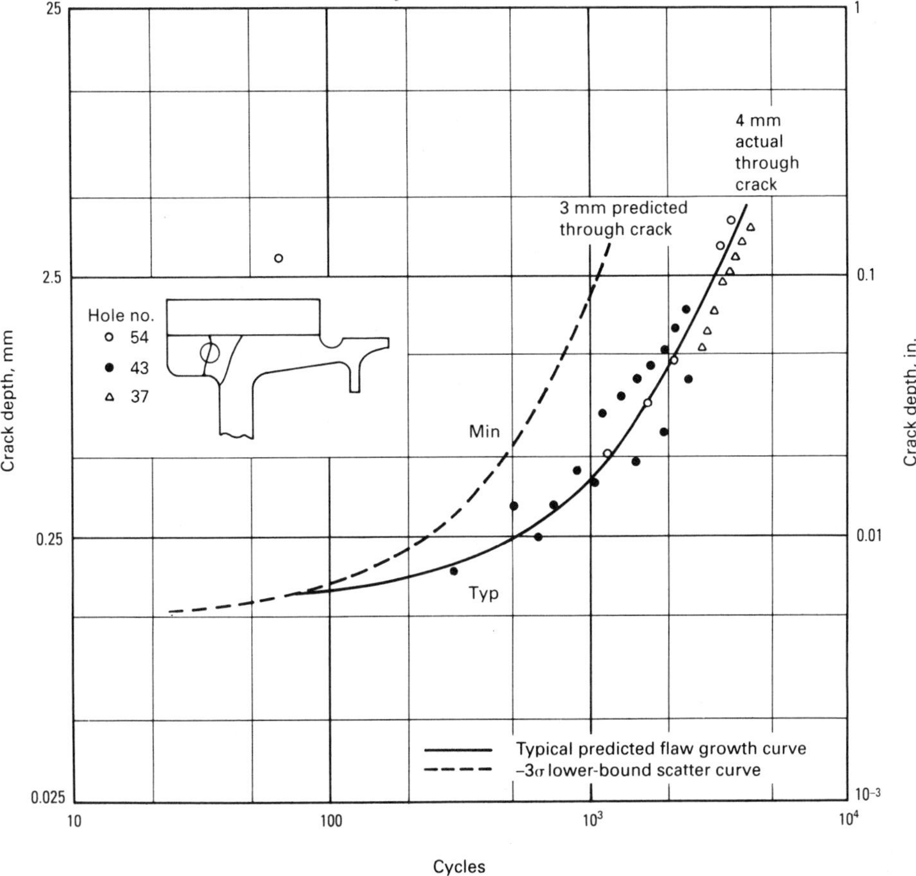

Fig. 11 Actual and predicted flaw growth for an engine test on a second-stage high-pressure turbine disk forward cooling air hole

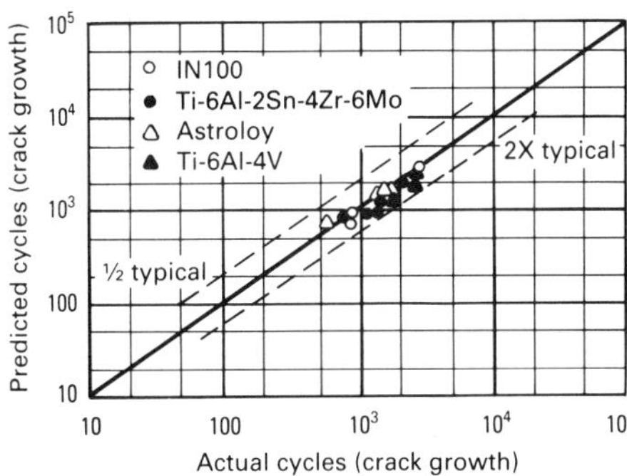

Fig. 12 Methodology correlation (specimens and components)

tion requirements planned for service engines to ensure safety of the test engine.

Nondestructive evaluation requirements are implemented on fracture-critical parts during manufacture and during field/depot inspection to ensure safety. Specific inspection requirements are derived through design analysis trade-offs among initial flaw size assumption, stress level, and material properties for a given usage (stress environment spectrum). As discussed in the section "Initial Flaw Size" in this article, a flaw size assumption (for surface flaws) of less than 0.75 mm (0.030 in.) requires implementation of enhanced NDE (that is, eddy current inspection). The ability of eddy current inspection to reliably detect surface flaws having depths of 0.13 mm (0.005 in.) has been demonstrated in several applications. Primary emphasis on the use of eddy current inspection is for stress concentration areas in which a small flaw size assumption is required to achieve the necessary residual life without excessive weight penalty. Typical probability of detection data for eddy current inspection are shown in Fig. 13.

In general, fluorescent penetrant inspection is specified for areas in which the detection of a flaw with a surface length of 0.75 mm (0.030 in.) or larger is required to achieve the specified residual life. However, the ability of current fluorescent penetrant inspection processes to reliably detect 0.75 mm (0.030 in.) flaws is not clear. Therefore, in some cases, eddy current inspection may be specified for large areas if susceptibility data (that is, probabilistic data on the capability of fluorescent penetrant inspection) indicate the need. Data generated on numerous demonstration programs clearly indicate that the fluorescent penetrant inspection process can be significantly improved through upgraded training, equipment, and procedures (proper cleaning, including etch, hydrophilic emulsifier, and wet developer). These demonstration programs have been conducted on several engine development programs (F100, F101, F110, F110-GE-129, F100-PW-229) and laboratory technology programs (Air Force Wright Research and Development Center). Some typical detection improvements that have been demonstrated for upgraded fluorescent penetrant inspection processes are shown in Fig. 14. The critical need is to implement the best fluorescent penetrant inspection process within industry and within the Air Force logistics centers because this method will likely remain the most widely used for inspecting large areas for cracklike damage.

Another critical NDE need is to quantify the POD of ultrasonics to detect imbedded defects in bulk volumes and to develop inspection methods for finished shapes. Very limited data indicate that reliable detection limits may be as large as 1.3 mm^2 (0.002 in.^{2}) (approximately equal to a planar disk of 1.2 mm, or 3⁄64 in., diameter). The goal is to develop and implement ultrasonic inspection methods such that a residual life

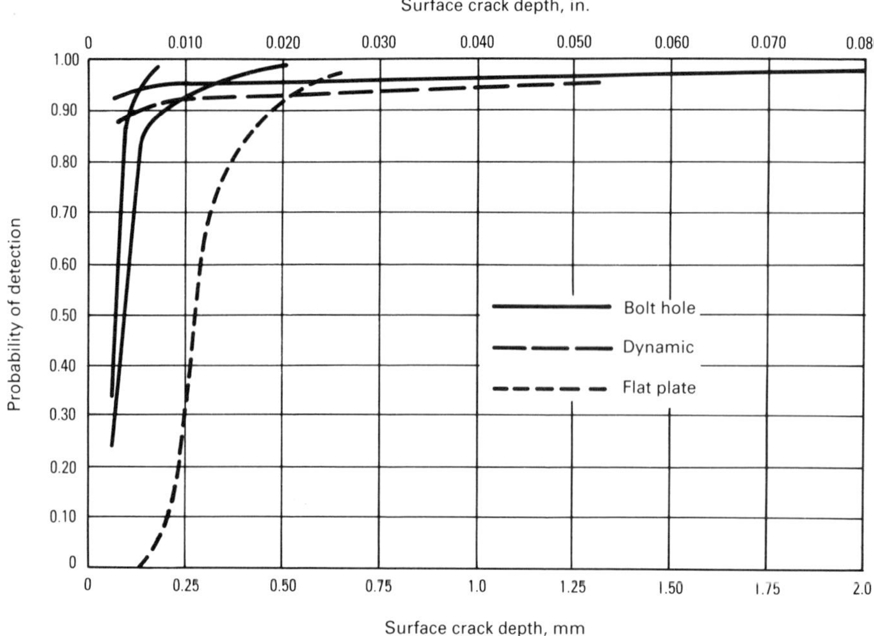

Fig. 13 POD curves with eddy current inspection. All curves are for lower 95% confidence limit.

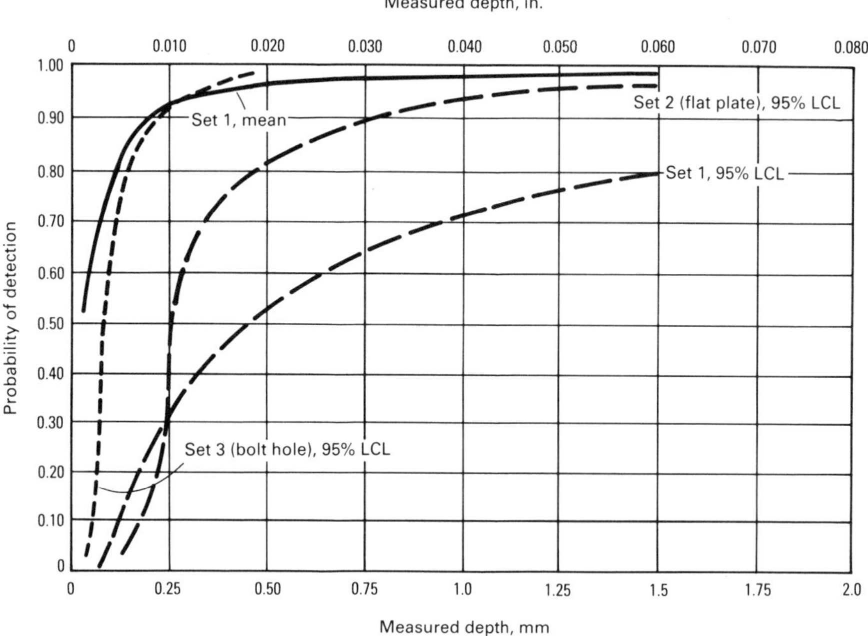

Fig. 14 Probability of detection with fluorescent penetrant inspection. LCL, lower confidence limit

equal to two times the required life or the inspection interval can be achieved, assuming the largest undetectable flaw size without excessive impact on weight.

Retirement-for-Cause (RFC). Traditionally, components whose dominant failure mode is low-cycle fatigue have been designed to a crack initiation criterion. With this approach, only 1 component in a population of 1000 would have actually initiated a crack, and the remaining 999 components would be discarded with substantial undefined useful life to crack initiation remaining. Figure 15 shows that the difference between the number of cycles to reach the design-allowable curve and the population average curve for an average component would have consumed only 10% or less of its potential useful life-to-crack initiation. Under the initiation criterion, there is no way to utilize this potential life without accepting a higher probability of failure of the remaining components.

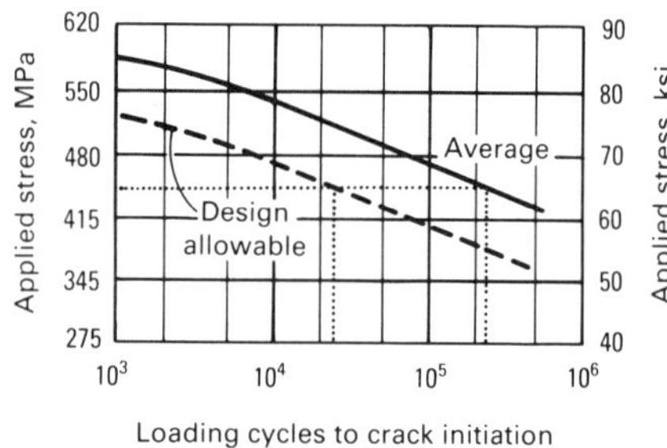

Fig. 15 Stress versus loading cycles to crack initiation for Inconel 718. Temperature is 540 °C (1000 °F), and the ratio of alternating stress to mean stress (A ratio) = 1.

Under the RFC concept, this additional useful life can be utilized by adopting a rejection criterion that uses each component in a population until it specifically initiates a crack, rather than rejecting the entire population on the behavior of the statistical minimum. The development of fracture mechanics concepts over the last several years has permitted the degree of predictability for crack progagation rates necessary to implement such an approach on a safe basis.

The RFC concept would apply the fracture control philosophy (Fig. 4) to life management. In using RFC as an operating system, all components would be inspected first at the end of a safety limit period divided by an appropriate safety margin, and only those components containing detectable cracks equal to or greater than a_i would be retired or repaired. All others would be returned for additional service (with the assumption that if a flaw existed, it would be smaller than a_i for another inspection interval). In this way, the crack propagation residual life is continually reset to a safe value. By following this approach, components are rejected only for cause (cracks), and each component is allowed to operate for its own specific crack initiation life. It should be noted that if a crack is missed at the first inspection interval, another chance should exist to find a larger crack.

It is clear that not all fatigue-limited components can be handled in this way and that each component must be evaluated individually to determine the technical feasibility of RFC. Low-cycle fatigue is a real physical phenomenon that is not directly associated with the presence of defects. Any criterion, initiation or otherwise, that allows components to run beyond a design-allowable life (-3σ, for example) or beyond the average life will inherently result in a significant increase in the probability of a large number of cracks and possible failures. Such an increase in risk may be acceptable, but must be understood and evaluated. Any inspection process has associated with it a finite probability of detection and therefore a finite probability of missing real cracks (not just defects assumed to be there and assumed to be in the form of sharp cracks). Missing real cracks and presuming that crack growth knowledge is sufficient to detect cracks at the next inspection has significantly more risk associated with it than concern over possible defects.

The economic feasibility of RFC must also be evaluated. The inspection interval must be such that it does not place undue constraints on the operation of the component or that the cost of the necessary teardown and inspection does not negate the advantage of the life extension. It seems unlikely that RFC can be applied to components limited by high-cycle fatigue considerations, but for many high-cost components limited by low-cycle fatigue, such as engine disks, this approach does offer significant economic advantages.

It is also clear that in applying RFC, nondestructive evaluation becomes a critical factor. The crack length determines the residual life of the component, and its detection is limited by the resolution and reliability of the inspection system employed. In many cases, the decision as to whether or not RFC can be applied to a component will be predicated upon the ability of available NDE approaches to detect the initial flaw, a_i, with sufficient sensitivity and reliability. Because RFC procedure is based on fracture control concepts, the NDE techniques can be selected, refined, and focused on a particular local area, rather than attempting to critically inspect large areas.

SELECTED REFERENCES

- W.D. Cowie and T.A. Stein, "Damage Tolerant Design and Test Considerations in the Engine Structural Integrity Program," Paper presented at the propulsion session of the 21st Structures, Structural Dynamics, and Materials Conference, Seattle, WA, American Institute of Aeronautics and Astronautics, May 1980
- T.T. King (ASD/EN), W.D. Cowie (ASD/YZEE), and W.H. Reimann (AFWL/MLLN), "Damage Tolerance Design Concepts for Military Engines," Paper presented at AGARD Conference 393, San Antonio, TX, Advisory Group for Aerospace Research and Development, April 1985
- C.F. Tiffany and W.D. Cowie, "Progress on the ENSIP Approach to Improved Structural Integrity in Gas Turbine Engine/An Overview," Paper 78-WA/GT-13, presented at the Winter Annual Meeting (San Francisco), American Society of Mechanical Engineers, Dec 1978

Applications of NDE Reliability to Systems

Ward D. Rummel, Martin Marietta Astronautics Group, and Grover L. Hardy and Thomas D. Cooper, Wright Research & Development Center, Wright-Patterson Air Force Base

THE SUCCESS of a reliable NDE application depends greatly on the expertise and thoroughness of the NDE engineering that is performed. This involves comprehensive analyses to define the relationships between the NDE measurements to be made and the impact on the system functions being assessed and on the capability to implement the NDE measurements to discriminate to the expected level of acceptance applied. Most failures in NDE systems applications and in the automation of an NDE system can be attributed to failures in NDE engineering and to unrealistic NDE performance expectations. All modern engineering is based on comprehensive applications of principles that can be implemented by qualitative measurements and predictive modeling and on comprehensive procedural applications of principles based on prior art and systems test data. Modern engineering methods are equally applicable to the use of NDE on a material, component, structure, or system. The principles and data available are, however, not as well defined, generally recognized, or understood as those of other engineering disciplines.

General Considerations

Prior Art. Particular concern must be given to NDE assessment and analyses based on prior art. Although NDE methods and principles have been applied since the beginning of time and have been applied specifically to quantitative materials evaluation during the last decade, the perceived performance level is often considerably different from the actual performance level. Differences can be attributed to the primitive level of understanding of materials, engineering, and NDE engineering sciences and principles; excessive optimism in effecting early application of an NDE procedure; economic and social pressure to solve problems with troublesome engineering systems; and the attitudes and practices of our legal system.

Progressive developments in the evaluation of NDE reliability have established a new dimension for the assessment of NDE performance. Applications of prior NDE art must be judiciously examined to determine suitability for new applications. Quantification of the performance level, calibration (or process control) methods, and acceptance criteria are particularly important in extrapolating the applicability of prior NDE data to a new NDE engineering problem. Conversely, the lessons learned in applications of prior art can be very useful in design and calibration and process control procedures, in establishing the characteristics and performance of materials and equipment used, in anticipating the problems and controls necessary to effect application in a production environment, and in assessing human factors relevant to the application.

Incorporation and integration of the qualitative factors and considerations in the application of prior art are essential for making the transition from laboratory test data to production line use. However, careful analysis and criteria must be applied to quantitative data from application of prior art. Quantitative assessment of ongoing NDE process applications has shown that performance levels may vary considerably in NDE applications to established specification/process requirements. Performance variations are rarely integrated into overall system reliability estimates and management is rarely accurate in recognizing and identifying superior performance by human operators. Quantitative data must be supported by actual NDE system measurements and accurate descriptions of NDE materials, equipment, procedures, and human operator qualifications to be seriously considered in NDE system design or qualification by similarity. It must be emphasized that quantitative data are not necessary for the multitude of NDE applications used to add confidence to routine process control and other applications in which the NDE procedure does not constitute final acceptance of performance characteristics. Quantitative data are required when the NDE measurement/acceptance is integral to design acceptance and/or performance acceptance.

NDE Response. The response from an NDE system or process may take the form of a signal output (or outputs) or a direct or indirect image. Acceptable conditions can be differentiated from unacceptable conditions by threshold discrimination from the electronic output or by pattern recognition and threshold discrimination by image analyses. Discrimination can be automated or performed by a human operator. Discrimination of threshold electronic signals can be automated or gated to alert the human operator. The consistency and reliability of electronic signal discrimination can often be improved by automating the discrimination process. Superior consistency and reliability of pattern recognition and discrimination level for images are usually achieved by the human operator. The feasibility of application of NDE to a system is dependent on the establishment and characterization of a relationship between the response from an NDE output and a desired engineering system performance parameter.

A direct or indirect relationship between an NDE response and a system performance characteristic may be functional under laboratory conditions, but may be impractical in applications under production or service conditions. Factors such as calibration, acceptance criteria, component accessibility, surface condition, inspection material compatibility, and inspection environment must be assessed to determine that a positive relationship between NDE response and system performance can be maintained.

NDE System Management and Schedule. The implementation of a reliable NDE procedure is dependent on allowing time to collect data, perform the critical analyses, apply required resources, and develop operator (personnel) skills. Many critical NDE procedures have been implemented as a result of unanticipated engineering system failures. The economic and social pressures resulting from an engineering system failure must be judiciously balanced against the required time and resources necessary to perform disciplined and thorough NDE en-

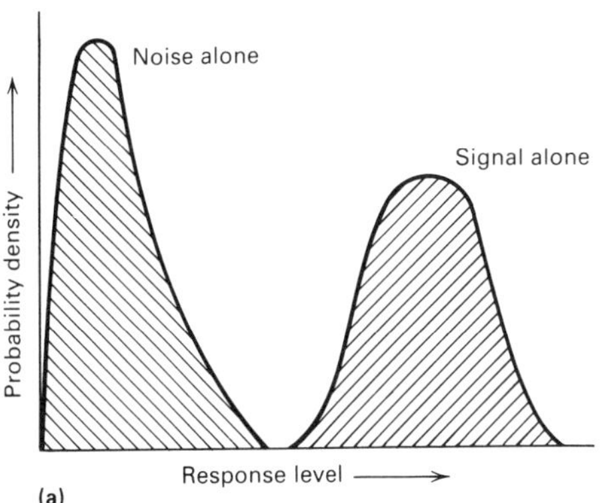

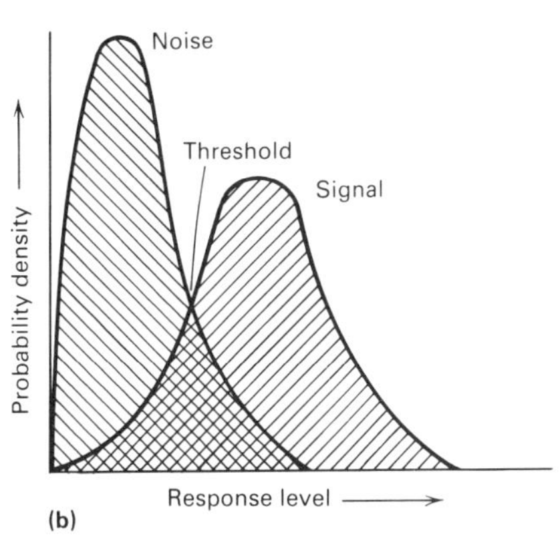

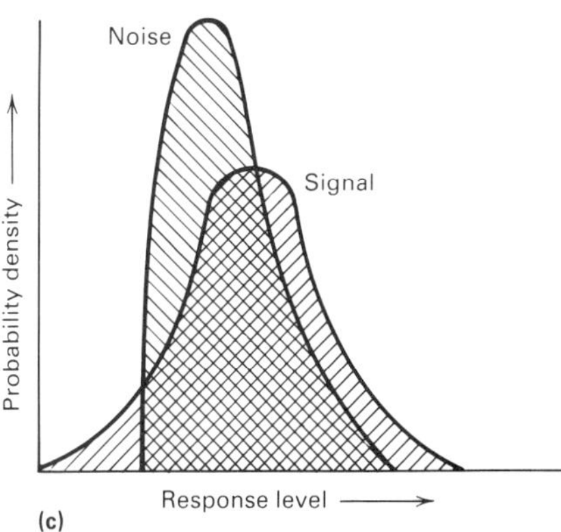

Fig. 1 Signal/noise density distribution for a large flaw (a), a medium flaw (b), and a small flaw (c)

gineering analyses, procedure development, and procedure validations. After the required procedures have been implemented, NDE system/process control must be maintained to ensure a consistent level of discrimination. Shortcuts in NDE engineering, NDE procedure development, and NDE system/process control increase risks in system performance, may not reduce the risk of engineering system failure, and may contribute to a false confidence level in system performance.

NDE Engineering

The difference between NDE engineering and the classical engineering disciplines result from the variety of problems and the indirect nature of NDE measurements on engineering system performance. The functional performance of most NDE methods can be measured, controlling parameters can be documented, and performance output can be modeled; however, the interaction of the NDE method with the test object necessitates the generation of new response parameters and characteristics for many new applications. In addition, variations in material properties, geometry, surface condition, access, or environmental conditions may modify the NDE responses. Therefore, NDE engineering is an essential element of critical engineering system design, qualification, acceptance, and life cycle management. A critical design is not complete until the NDE engineering has been performed and NDE system/process performance validated to functional design requirements and acceptance criteria levels.

Procedure Selection/Development. Trade studies to identify and select candidate NDE procedures are needed for establishing the most economical and reliable procedure that meets acceptance requirements. The process may be satisfied by the assessment of prior art and applications to similar problems or may require research and development of totally new procedures. For demanding applications, a combination of complementary NDE procedures may be required to meet the acceptance criteria objectives. One or more methods can be further characterized and assessed to ensure that the performance objectives, NDE performance margins, NDE costs, engineering system performance risks, and the risks of NDE system/process false alarms can be balanced in overall engineering system management.

System/Process Performance Characteristics. Although care, discipline, and control measures are applied to ensure a consistent output from an NDE system or process, the output will vary within the established control parameters and as a result of slight variations in engineering hardware materials properties, geometry, surface condition, and so on. If repetitive applications are made, a probability density distribution of signal/image output will be generated. This distribution is similar to that obtained by repetitive measurements of a dimension such as a hole diameter or the length of a bolt.

Nondestructive measurements are usually indirect, and positive signals may be generated from nonrelevant sources, such as surface roughness, grain structure, and geometry variations. Such signals constitute the application noise inherent in a specific NDE process or procedure. Discrimination of NDE signal/image outputs must be derived from those signal levels/amplitudes that exceed the level of the application noise (Fig. 1). Analysis of signal and signal plus noise are common in electronic devices, optics, and other discrimination processes. Similarly, the signal-to-noise margin (ratio) is a primary factor in establishing the level of discrimination of an NDE procedure. Signal/noise reduction procedures can be used to enhance the overall performance of an NDE procedure. However, it is important to recognize that the dominant noise source in an NDE process is not electronic noise that may be reduced by filtering, multiple sampling, and averaging techniques, but is instead the noise due to nonrelevant signals generated in applying the NDE procedure to a specific hardware element.

Conditional Probability in NDE Discrimination. Nondestructive evaluation involves the measurement of complex parameters with inherent variations in both the measurement process and the test object. The output from such a measurement/decision process can be analyzed as a problem in conditional probability. When an NDE assessment is performed for the purpose of crack detection, the outcome is not a simple accept/reject (binary) process, as is frequently envisioned. It is actually the product of conditional acceptance due to the interdependence of the measurement and decision responses. Figure 2 shows the four possible outcomes that result from the application of NDE procedure for crack detection. As shown in Fig. 2, the possible outcomes from an inspection process are:

- *True positive* (TP): A crack exists and is detected, where $M(\mathrm{A,a})$ is the total number of true positives and $P(\mathrm{A,a})$ is the probability of a true positive
- *False positive* (FP): No crack exists but one is identified, where $M(\mathrm{A,n})$ is the total number of false positives and $P(\mathrm{A,n})$ is the probability of a false positive
- *False negative* (FN): A crack exists but is not detected, where $M(\mathrm{N,a})$ is the total number of false negatives and $P(\mathrm{N,a})$ is the probability of a false negative
- *True negative* (TN): No crack exists and none is detected, where $M(\mathrm{N,n})$ is the total number of true negatives and $P(\mathrm{N,n})$ is the probability of a true negative

The interdependence of these matrix quantities can be expressed as:

$M(\mathrm{A,a}) + M(\mathrm{N,a}) =$ (TP) and (FN) outcomes giving the total opportunities for positive calls (Total number of defects)

and

$M(\mathrm{A,n}) + M(\mathrm{N,n}) =$ Total opportunities for false alarms from the possible (FP) and (TN) outcomes

Because of the interdependent relationship, only two independent probabilities need be considered to quantify the inspection/decision task. The probability of detec-

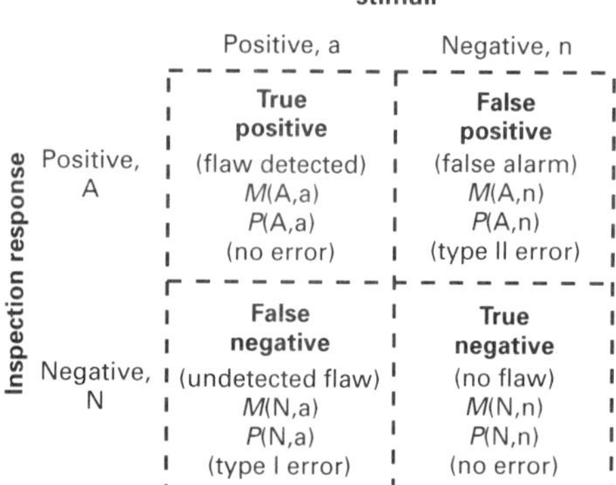

Fig. 2 Matrix of four possible outcomes from an NDE procedure for flaw detection

tion (POD) or probability for a true positive $P(\text{A,a})$ can be expressed as:

$$P(\text{A,a}) = \frac{M(\text{A,a})}{M(\text{A,a}) + M(\text{N,a})} \text{ or } \frac{\text{Total true positive calls}}{\text{Total number of defects}}$$

Similarly, the probability of false alarms (POFA) or the probability for a false positive $P(\text{A,n})$ can be expressed as:

$$P(\text{A,n}) = \frac{M(\text{A,n})}{M(\text{A,n}) + M(\text{N,n})} \text{ or } \frac{\text{Total false alarms}}{\text{Opportunities for false alarms}}$$

Signal/Noise Relationships. The desired results of the application of NDE procedure are crack detection (signal present) or crack nondetection (signal absent). The basis for detection is that of sensing a signal response and determining that the signal response is above a predetermined threshold. Both sensing and interpretation are dependent on the signal (plus noise) and the noise (application background or response to nonrelevant parameters) that are subjected to the discrimination media (programmed machine discriminator or human operator).

If an NDE procedure is repetitively applied to a single flaw of a given size (in a part with a fixed geometry, surface condition, and so on), the output responses can be used to plot probability density distributions of both flaw signal and application noise responses. Under ideal conditions, such as the response from a large flaw, the signal and noise distributions will be well separated, as shown in Fig. 1(a).

The discrimination of flaw responses from application noise responses is a simple process; POD will be high, and the POFA will be low. In practical engineering applications, the flaw size is not fixed (and is rarely large), and the discrimination process is more complex. Indeed, the discrimination process is applied to a continuous range of flaw sizes, where the capability for discrimination is dependent on the inherent performance characteristic of the NDE procedure and on the separation of the signal (plus noise) from the noise response of the process.

If the NDE procedure is applied to a single flaw of intermediate size (in a part with the same fixed surface finish, geometry, and so on), the output responses can be used to generate probability density distributions for signal and noise, as shown in Fig. 1(b). For this flaw size, the distributions overlap (in part), and the capability for discrimination is dependent on the response from a single set of output signals within these distributions. If the single set has output signals that are well separated (that is, signals at the outer extremes of the distributions), the output response will be interpreted as acceptable (no flaw condition) for those cases where the threshold response acceptance level is located between the signal and noise signals. If the single set of outputs lies at the inner extremes of the distributions, the output response may be interpreted as acceptable (no flaw or undetected flaw condition) or may be interpreted as unacceptable (false alarm condition) for the same threshold response acceptance level. For this condition, the POD will be lower and the POFA will be higher than for the case of discrimination with positive signal/noise separation margins. If the process is repeated for a small flaw (under the same operating conditions), the signal and noise response distribution will approach coincidence, as shown in Fig. 1(c). The POD will be low, and the POFA will be high.

It is clear that the performance capability of a given NDE procedure is dependent on the nature and distribution of the signal outputs generated under the conditions of application. It is also clear that the threshold acceptance criterion applied in the discrimination process is an important factor in the successful application of a procedure. Consider the application of an NDE procedure to a large flaw under conditions that produce a significant separation of probability density distributions of signal and noise, as shown in Fig. 3. If the threshold acceptance criterion (represented by the vertical arrow) is placed at too high a level (Fig. 3a), some of the flaws will be missed (reduce POD). If the acceptance criterion is placed at a proper level (Fig. 3b), clear discrimination will result (high POD). If the acceptance criterion is placed too low (Fig. 3c), all of the flaws will be rejected; but some false alarms will result, and good parts will be rejected (high POFA).

The NDE procedure performance characteristics of primary importance are the signal-to-noise ratio (separation margin) and the threshold acceptance criteria applied in the discrimination process. Optimum NDE procedure performance can be obtained by characterizing an NDE procedure and by matching the threshold acceptance criteria

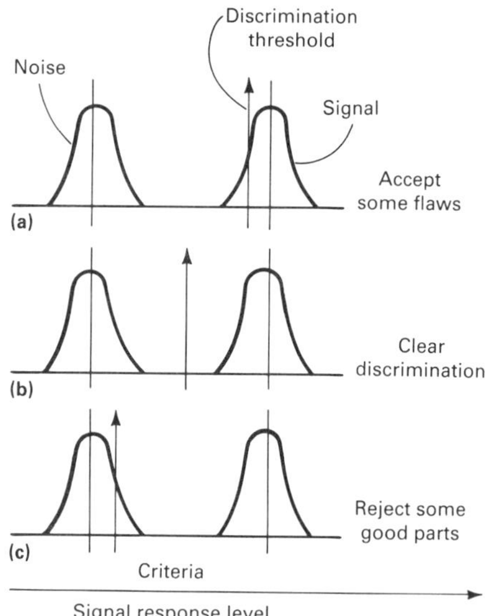

Fig. 3 Influence of acceptance criterion (vertical arrow) on process discrimination. (a) Acceptance criterion too high. (b) Acceptance criterion at proper level. (c) Acceptance criterion too low

to the performance capabilities of the NDE procedure. Such characterization also enables the assessment and quantification of risks that result from changes in acceptance criteria.

Reference Standards. Historically, reference standards for NDE methods of defect detection have been used to ensure the reproducibility of the application of the method(s) and to establish an acceptable quality of the process rather than to establish the dimensions or other applicable parameter of the defect or anomaly. Some methods (such as those used for thickness gaging or electrical conductivity determination) were able to provide an extremely accurate measurement of the appropriate parameter, usually by extrapolation between two known and closely spaced reference standards representative of the condition to be determined. The remaining methods, as typified by radiography, offer at best only a crude estimate of the dimensions, orientation, shape, or other characteristics of the detected defects.

The primary reason for not using reference standards for the quantitative evaluation of defects was, and still is, that the NDE methods respond to most of the parameters of a defect simultaneously; in most cases, there is no way to separate the response from a single parameter, or there is not an accurate response to a single parameter in other cases. Consider, for example, the case of penetrant inspection. The indications formed are usually greater in length, width, or area than the discontinuity present because of the flow of the penetrant material out of the discontinuity during development. When this excess material is removed and the indication is viewed as it starts to appear again, the full length of the discontinuity, if it is linear, will initially not be revealed, because the ends provide little or no penetrant for formation of the indication. As the indication continues to form, it will eventually reach the same length as the discontinuity and will then continue to grow as additional penetrant flows to the surface. This lack of response to the extremities of a discontinuity is common to all NDE methods and illustrates the difficulties associated with using NDE methods for sizing defects.

Quantitative NDE, however, requires that a good estimate be made of the defect size that is detected or, more important, the size of the largest defect that might be left in the part. Because the NDE methods, by the laws of physics, are inherently inaccurate in sizing, the only available approach is to make a conservative estimate of the size of the defects that can remain. The approach requires a second type of standard used along with the conventional reference standards. This second type, called a qualification standard, contains defects that represent the worst case for both flaw detection and crack growth (generally a surface fatigue crack). Fatigue cracks have the advantages that they can be grown in the laboratory and, when produced under well-controlled conditions, have predictable geometries. The qualification standards are then used in sets to define the lower limit of the flaw size that a given NDE process can reliably detect. The conventional reference standards are used to control the NDE process; therefore, once the qualification standards establish the process sensitivity, the reference standards can be used to ensure that the sensitivity is maintained. Consequently, there are certain requirements that reference standards must meet.

First, reference standards must produce a response comparable to that produced by the smallest qualification standard flaw that is considered reliably detectable. This comparability includes not only response to the flaw itselt but also the geometry in which the flaw is contained. In the ultrasonic method, for example, the response to a fatigue crack located in the center of a flat plate can be far different from the response to the same size crack that is located in the bore of a large-diameter hole. Consequently, application of the ultrasonic method for inspection of the two geometries requires qualification standards as well as reference standards for both geometries.

Second, reference standards for a specific inspection must be relatable to other reference standards used for the same inspection. That is, when several reference standards are available, the responses for each one must be known, and more important, the differences in responses for the standards must be known so that adjustments can be made in the inspection to ensure that a uniform process sensitivity can be maintained. Ideally, the responses of all reference standards for a given inspection should be identical; however, from a practical standpoint this is impossible to achieve. As stated previously, the response of the NDE methods is from a multitude of parameters associated with a given discontinuity.

For example, if an electrically discharge machined slot is selected as a defect for a reference standard for an ultrasonic inspection, exact control over the size of the slot in each standard is not sufficient to guarantee identical responses. Slight variations in the orientation of the slot with respect to the surface of the standard and in the surface finish of the slot itself can cause noticeable differences in the ultrasonic response. This is the worst case; other NDE methods vary in their response to subtle geometric parameters, with the magnetic particle and penetrant methods probably being the most tolerant. However, even these methods are highly sensitive to the width of the flaw used in a reference standard.

Another important property of reference standards is durability. Both the material and the type of flaw in a reference standard must be selected so that the standard will not readily deteriorate or change in the environment in which it will be used. These selections are affected by the NDE method for which the standard is intended. Both ultrasonic and penetrant methods are very sensitive to the presence of foreign material inside the flaw. For ultrasonics, this can affect the amount of energy that is reflected from the flaw. For penetrant inspection, the quantity of penetrant that can enter the defect, and consequently the brightness of the indication, will be reduced. The foreign material may be fluids, soils, or corrosion products. For these reasons, magnesium, ferritic steels, and aluminum are particularly poor choices for reference standards for penetrant inspection and require some type of protection if used as ultrasonic standards.

Eddy current methods are not affected by foreign material in the flaw, but are very sensitive to such surface conditions as scratches, pitting, and corrosion. Magnetic particle methods are sensitive to the width of the defect and to anything, such as cold working, that may change the magnetic permeability of the standard. Radiographic methods are sensitive to the thickness of the standard and to changes in the shape of the flaw. In general, a good choice of material for any method is one that is reasonably hard and forms an adherent, tough, and stable surface oxide layer (such as a titanium or nickel-base alloy), thus providing protection against mechanical damage and the gradual buildup of corrosion products.

Personnel. Unless the inspection process is fully automated, the proficiency of the inspection personnel is the largest variable affecting inspection reliability. This proficiency varies widely not only from inspector to inspector but also with the same inspector, depending on his working environment and his mental condition. For fully automated inspections, the proficiency of the inspector in operating the equipment is important but has little or no impact on inspection reliability.

The first task in addressing the contribution of the inspector to inspection reliability is to ensure that he is knowledgeable of the specific techniques to be used and has the basic proficiency to perform the inspection to the required reliability. Experience has demonstrated that the previous qualifications of the inspector—for example, certification to MIL-STD-410D (Ref 1)—are not sufficient to ensure the desired performance with a new inspection that must be performed with high reliability. Therefore, the most straightforward way to assess proficiency is to require inspector participation in the demonstration of inspection reliability. All inspectors that will be required to perform the inspection should also participate. This not only establishes the reliability of the proposed inspection but also identifies those personnel requiring additional training or experience before they can be expected to perform adequately. Careful observation of the inspector during the demonstration and of the results obtained is necessary to identify the additional training or experience needed. After training and/or additional experience is acquired, the demonstration effort can be repeated to indicate if the inspector has become sufficiently proficient in the inspection technique.

After basic inspection proficiency has been demonstrated, it becomes a supervisory task to ensure that this proficiency is maintained. Control of the work environment of the inspector is important. Distractions such as noise, extremes in temperature, and other irritants should be eliminated to the extent possible. Break periods should be frequent enough to reduce fatigue. Personnel who are ill or otherwise physically impaired should be temporarily assigned other tasks to the extent possible. Other efforts that improve or maintain a good mental attitude are excellent ways to ensure sustained inspection reliability. These include providing acceptable materials and equipment with which to conduct the inspection. Finally, when it is not possible to provide a consistently conducive environment for a highly reliable inspection, two inspectors can perform the same inspection independently to achieve higher reliability than can be obtained with a single inspector. Two inspectors generally will not make identical mistakes.

NDE Process Control

Once a given NDE process has been demonstrated to possess the required reliability, it then becomes necessary to put into effect the controls needed to maintain this reliability. This requires the proper controls over the materials, equipment, personnel, and process documentation. Establishing effective controls requires the identification of the features of these items that are significant in terms of their impact on inspection reliability. These controls are necessary for accommodating the changes that occur with time and for ensuring that replacements for any of the items are adequate.

Control of Inspection Materials. Inspection materials will change with use and will require replacement. Fortunately, it is relatively easy to identify the significant features of the inspection materials and to ensure that in-use as well as replacement items function properly.

For penetrant and magnetic particle materials, the concept of signal-to-noise ratio is useful in determining adequacy. With both processes, defects are detected by the visual indications produced. Under ideal conditions, indications are produced by defects only when the rest of the component or part under inspection is clean. In reality, indications are produced not only by defects but also as a result of processing errors, nonrelevant surface or specimen conditions, and contamination or breakdown of the materials themselves. Also present is a condition called background, which is best considered as noise that tends to be uniformly distributed over the surface of the part. This background is related to the surface finish and serves to limit the inspection sensitivity of the process. In controlling these materials, the important factors are the brightness or relative visibility of the defect indications, the presence of nonrelevant indications from material contamination or breakdown, and the brightness or visibility of the background. The indication brightness is the signal, and the nonrelevant indications and background constitute the noise.

Measuring the brightness of indications produced by either penetrant or magnetic particle materials requires the use of a reference standard with well-characterized defects. The indications produced by either the in-use or replacement materials are compared to indications produced by materials used in the qualification of the inspection process or other materials known to possess the required performance. This can be either a qualitative comparison to photographs or indications produced on a nominally identical reference standard with serviceable materials, or a quantitative measurement of the brightness or visibility. Suitable reference standards for this can be fatigue cracks, small elox slots, or stress cracks in a chromium-plated panel. Similarly, the noise is determined by processing a specimen (such as a gritblasted panel) and by comparing the resultant background and/or presence of nonrelevant indications to those of serviceable materials. The comparison can either be quantitative or qualitative. It is important to note that the measurement of the brightness of the bulk materials is not a good indicator of the brightness of the indications that the materials will produce. This is because the brightness of the indication depends not only on the material brightness but also on the quantity of the material present at the defect site.

X-ray film and the chemicals required for developing can be controlled in a similar manner. The defect indications are regions of different density in the film; this difference is the signal. The difference is a result of the sensitivity of the film to slight differences in radiation intensity and its spatial resolution. Both can be quantitatively measured. The noise derives from the film graininess, which is the inverse of the spatial resolution, and from the background fog or the density that results when unexposed film is developed. Another source of noise is mottling, which can be the result of using unserviceable developing chemicals. Control of the film and chemicals requires the use of reference standards that will produce known images, and the images produced using the in-use or replacement film or chemicals are then compared with those known to be serviceable.

For eddy current and ultrasonic methods, the materials of concern are the probes (eddy current) and the transducer (ultrasonic). Again, the signal-to-noise concept is useful for the maintenance of these items. Both methods rely on these items as the interface between the specimen and the instrumentation. Reference standards with defects that are both well characterized and produce responses that are comparable in magnitude and spatial extent to those used to qualify the inspection process are used to evaluate the signals produced by either the in-use or replacement items. The noise consists of the signals produced when the probes or transducers are passed over nondefective regions of the reference standards, and these can be compared to those of serviceable items. These comparisons are inherently quantitative because the instrumentation produces measurable signals.

Other miscellaneous materials used in quantitative NDE, such as ultrasonic couplant and leak test fluids, are maintained in a similar manner. The performance of the in-use or replacement materials is compared to serviceable materials in laboratory tests to determine adequacy.

Qualification of Inspection Processes. Generally, it is not practical to demonstrate a quantitative NDE capability with production parts and the exact process that is to be used on these parts, because of the immense difficulty in producing production parts with the required range of well-characterized flaws. Consequently, it is necessary to qualify the production process, that is, to ensure that the production process has the same flaw detection capability and reliability as the process demonstrated. This task requires the identification of those portions of the demonstrated process that affect the quantitative capability and the corresponding portions of the production process.

For ultrasonic, eddy current, and radiographic methods, the most straightforward manner of qualifying a production inspection process is to utilize the same type of flaws used in the reference standards and to place these flaws in one or more production parts in critical locations. The production process is then performed in accordance with the process documentation, and if the responses from the flaws are of the correct amplitude and if the noise is at or below the levels encountered during the demonstration, the production process can be considered qualified. During the qualification, the process documentation must be followed exactly as written because it is an important part of the process being qualified. It is necessary to perform this procedure on each part number to be inspected because both geometry and material significantly affect NDE reliability.

For magnetic particle methods, the procedure is more difficult and less certain. It is possible to use devices to show that an adequate magnetic field is present in most areas of most components; such devices can be either thin ferromagnetic shims with scribed figures or gaussmeters. However, the use of such methods is limited to relatively accessible areas and in regions where there is sufficient space to position a shim or a probe. Consequently, one must first choose the geometry of the demonstration specimen carefully such that, given the limitations of the field-measuring devices, a reasonable estimate of the field strength in critical locations on production parts can be made. The process documentation is also very critical and must be followed exactly to ensure that the documentation is capable of producing a reliable inspection.

Penetrant inspection is probably the most difficult process to qualify for production. There are no suitable devices that can be located in critical parts of production parts to ensure that the penetrant system is producing adequate flaw indications. Furthermore, and this also holds true for the magnetic particle process, there is no satisfactory manner of ensuring that the production inspection personnel will observe the defect indications if they are produced. For this reason, qualification of the penetrant process requires the expertise of an experienced NDT engineer and is subjective at best. As with other methods, the process documentation is also an important portion of the process to be qualified, and again the judgment of the NDT engineer is required.

The role of process documentation in ensuring NDE reliability cannot be underestimated. The documentation is the method by which the process instructions are recorded and repeated by inspection personnel, both at the site of the original qualification demonstration and at other physical locations. Small errors in the documentation may go unrecognized by personnel not involved in its preparation and can significantly affect the way in which the process is performed, with an accompanying reduction in inspection reliability.

Qualification of Inspection Equipment. Inspection equipment can be characterized to the extent that, in general, those parameters necessary for ensuring a given level of performance can be identified, quantified, and therefore controlled. In those few cases in which quantification is impossible or impractical, other avenues exist to ensure that either in-use or replacement equipment is adequate for the inspection being performed.

Equipment for penetrant inspection consists of the apparatus used to apply the penetrant materials; to apply and control rinse water; and ovens, timers, and light sources. Adequate application of materials is confirmed by the observation of parts during processing. Adequate rinsing is also confirmed by observation, along with periodic checks of pressure and temperature gages. Oven operation is monitored by periodic checks of thermometers, air flow gages, and the condition of the parts after drying. Light sources, for either fluorescent or visible-dye processes, are periodically measured for intensity with the appropriate instrumentation. Timers can also be readily checked for proper operation.

Ensuring the adequacy of magnetic particle inspection equipment is also straightforward. The equipment provides and controls the magnetizing current and also provides lighting. The amplitude and duration of the magnetizing current is checked periodically with the appropriate electronic equipment. In addition, the induced magnetic fields, either from ac or dc sources, can be checked with gaussmeters. As with penetrant inspection equipment, the intensity of the lighting can be checked with the appropriate instrumentation.

X-ray equipment provides and controls the magnitude and duration of the radiation used in the process. Associated equipment is responsible for the development and viewing of the film or, where nonfilm imaging techniques are employed, actually converts the radiation patterns to visual images and controls the quality of such images. A portion of the control of x-ray equipment consists of periodic checks of voltages, currents, and timers. However, the control of the quality of the radiation, that is, the intensity and spectral distribution of the radiation, is usually not done by direct measurement. Rather, images, either on film or other imaging devices, are produced with well-characterized specimens and then checked for the required contrast sensitivity and spatial resolution. This will also serve as a check of nonfilm imaging equipment. Film-developing equipment is checked by processing specially exposed film and checking the quality of the resultant images. Film-viewing equipment is checked either directly for light output intensity or with exposed film containing images that are visible only with sufficient light output.

Eddy current equipment and associated scanning devices are controlled by both direct measurements and comparisons to equipment known to be operating properly. The frequency and amplitude of the eddy current generating output can be directly measured. The adequacy of the scanning devices is most easily checked by scanning production parts with flaws located in critical areas, as discussed in the section "Qualification of Inspection Processes" in this article. In addition, gates, recorders, alarms, and other auxiliary equipment are readily checked with such parts by ensuring that the proper performance is obtained when the equipment is operating.

Ultrasonic equipment and associated scanning devices are also checked by both direct measurements and comparison to known serviceable equipment. The shape and magnitude of the output pulse, the frequency response and gain of the amplifier and deflection circuits, and the attenuators can be checked with the appropriate electronic equipment. Other devices are best checked as described in the section "Qualification of Inspection Processes" in this article by inspecting production parts with known and characterized flaws in critical locations.

Human Factors. The education, training, and proficiency of human operators involved in the application of an NDE procedure are critical factors in assessing and ensuring the consistency of those processes in which human dexterity, vigilance, and discrimination capability are required. In assessing the influence of human factors on NDE output, all steps of the application of the NDE procedure must be considered—for example, dexterity in scanning, vigilance in reading a scope, and proficiency in pattern recognition and discrimination for those outputs that require image interpretation. Variations in human factors are often considered to be the primary cause of failures in discrimination by application of NDE procedures. Although human factors may have a significant influence on NDE output, process control factors have been purposely listed in the order presented in order to emphasize their importance on NDE output. If inspection equipment, inspection materials, or inspection processes are allowed to vary, the human operator does not have a chance of achieving correct discrimination and interpretation at the final stage in the process.

Education and training are well recognized as important factors in the completion of any critical process. Similarly, working environment, operator temperament, operator health, supportive supervision, and so on, are recognized as contributing factors in the consistent output of a process. These elements must be addressed as part of every good industrial management operation and are assumed (often incorrectly) to be constant for purposes of this discussion. All operators are not, however, created equal. The capability and performance level of an individual operator can be measured, quantified, and audited by the assessment of performance proficiency using test hardware that is representative of the production hardware being assessed. (A level of consistency can be established by performance on similar hardware or on test specimens that bracket the performance limits of the NDE procedure.)

Probability of Detection Curves. The characteristic performance level of a given NDE procedure can be established by subjecting a number of test articles containing cracks in a range of sizes to the procedure. A POD curve can be generated from the results. The procedure can then be repeated by several different operators, and the resulting curves will reflect the variations in performance proficiency within the constraints of the factors that are operator controlled. Superior performance proficiency can be identified and quantified by the POD curves, and superior performers can be grouped and selected to perform critical NDE procedures. The POFA or false call rate is not reflected by the POD curve and should be recorded and used as a secondary characteristic in proficiency assessment. Performance to a minimum level can be used as a condition for operator qualification (and certification to perform on critical NDE procedures). After initial procedure/personnel qualification has been established, qualification of additional operators and requalification of operators must be performed to ensure continuing performance. Such qualification can be performed by repetition of the original qualification process using a full complement of test articles or can be performed by subset sampling procedures.

Relative Operating Characteristic (ROC) Curves. Performance qualification using subset sampling procedures can be conveniently accomplished through the use of ROC curve methods. The ROC curve methods were developed during World War II to qualify the proficiency of radar operators in response to signals on the radar

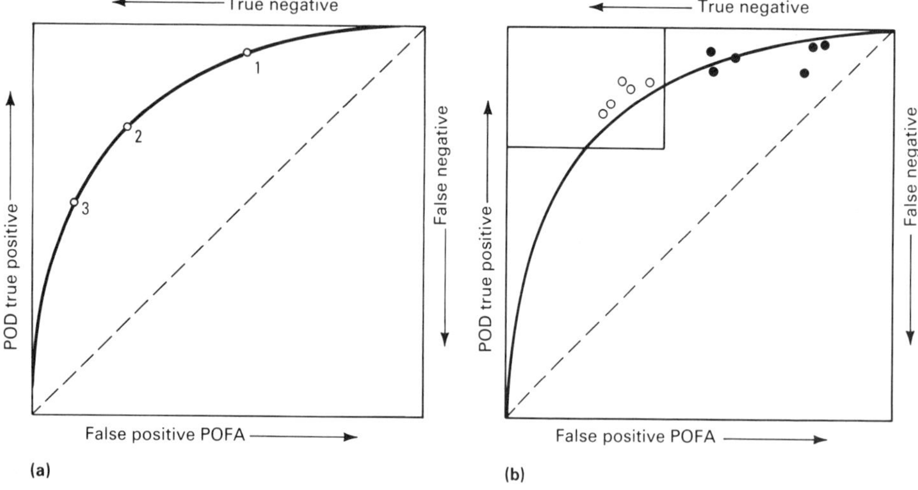

Fig. 4 ROC curve (a) showing performance at three acceptance criteria. (b) Modified ROC analysis showing relative operator proficiency

scopes. The method assumes conditional probability in detection/discrimination and utilizes both POD and POFA performance as the process performance level and discrimination proficiency level of the operator are varied.

Data required in the use of the ROC curve method for subset data analysis are obtained from the original qualification data using the full complement of test articles. If a set of test articles, containing a group of flaws of similar size, is repetitively assessed, performance data on POD and POFA can be calculated. If these data are available for flaws that are at the critical threshold detection limit of the NDE procedure, superior performance will result in a high POD and a low POFA. The performance level for a group of flaws that are near the threshold detection can be documented on an ROC plot, as shown in Fig. 4(a). The solid line generated by the locus of points (shown as 1, 2, and 3) constitutes a desired level of discrimination at the performance level qualified. Performance at lower discrimination levels results in a family of curves that approach the dashed line diagonal, which constitutes discrimination that is independent of the results of the process applied, that is, the random guess line.

The ROC method is implemented by having a subset number of test articles containing flaws of similar size evaluated by an operator using a specific NDE procedure. The resulting data are used to generate a single point on the ROC curve, as shown in Fig. 4(b). The procedure is repeated by several operators of varying levels of proficiency, and the data are plotted on the same curve. The performance of the most qualified operators should fall near the upper left-hand corner of the plot, that is, the more discriminating results. A zone of acceptable performance can be selected from the results obtained by operators who qualified on the full set of test articles. Operators whose performance falls outside of that zone can be identified as being less proficient and subject to further training or reassignment. The ROC method combines POD and POFA performance characteristics and can be supported by statistical sampling analyses to quantify a confidence level in the proficiency level measured.

Alternatives to the ROC qualification method include the use of parallel processing of components by qualified and untested operators to determine a comparative level of performance on specific test articles that contain known and validated flaws. The double-inspection method is often proposed as a method of increasing the confidence level of the results of an NDE process.

Management, by definition, deals with organization and the use of human resources. Superior knowledge, capabilities, environment, job stress, and so on, are recognized elements of job performance by human operators. Human proficiency is not, however, a subject that can be addressed exclusively by social sciences. Hard decisions are required to maintain a level of performance. For example, if operator proficiency is less than required for a given task, provisions for continuing employment or employment on other tasks must be made. Similarly, if an operator qualifies to a given proficiency level, an adjustment in rate of pay may be in order to identify and recognize special qualification. The adjustment would then be lost if the operator fails to requalify on a periodic assessment. Although such adjustments are not common in industrial tasks, special skill qualification, such as flight checks for pilots, are routinely applied.

Although all elements have not been considered, assessments of human factors have shown a number of interesting results as measured by proficiency in POD. Results include the following:

- Management judgment may be effective in determining extremely poor proficiency, but is rarely accurate in judging superior proficiency. Proficiency must be demonstrated and measured
- Operator education and classroom training (beyond a minimum level) have little effect on demonstrated proficiency. On-the-job training is required
- Recentness of performance experience is a significant factor in operator proficiency level. Part-time operators are generally less proficient than full-time operators

The implementation of reliable NDE procedures requires attention to all elements of NDE engineering, procedure characterization and validation, and process control. Implementation elements are serial in nature, and failure in any task in the sequence will result in a lower capability of the overall NDE process. Overall capability can be assessed and quantified by the POD method as an aid to process control or as a method of validating conformance to design requirements.

Applications (Case Studies)

Airframes. In the aerospace industry, the first application of quantitative NDE in a production facility was on airframe structures. Loss of an F-111 aircraft by the propagation of an undetected manufacturing defect led to the incorporation of fracture control in the design and qualification of aircraft structure by the United States Air Force. This was first applied rigorously to the B-1 aircraft and consisted of:

- The identification of critical structural components whose failure would cause the loss of aircraft
- The identification of those areas in the critical structural components experiencing the highest stresses defined
- The estimation of the maximum size of a rogue defect in these areas that could exist without growing to failure in twice the estimated design life

The maximum acceptable size of a rogue defect was established as 6.4 mm (0.250 in.) long by 3.2 mm (0.125 in.) deep for a surface flaw, or 1.3 × 1.3 mm (0.050 × 0.050 in.) for a corner crack in a bolt hole or a structural edge. For an embedded flaw, the size was a 6.4 mm (0.250 in.) diam circle. These sizes were selected because experience had shown that they were large enough to be readily detected during production inspection. If the crack sizes that the designer selected for his structure were to be smaller, the capability of the production NDE facility to find the smaller cracks reliably would have to be demonstrated. For the

B-1, the crack sizes for many of the critical structural members was smaller than those suggested. Redesign of the components to tolerate the larger, suggested flaws would have resulted in unacceptable weight penalties; consequently, the B-1 production NDE facility was subjected to the first quantitative NDE capability evaluation.

A conservative approach was taken in the design of the NDE reliability demonstration program. Fatigue cracks were selected to represent manufacturing defects for surface and corner flaws, and voids created by diffusion bonding were to serve as embedded defects. Flat panels were the geometries selected, either single panels for surface defect areas or panel stackups for fastener hole areas, and rectangular blocks housed the embedded defects. The materials were aluminum, titanium, and steel. The surface finish for the specimens was the same as for production parts. The methods to be demonstrated were penetrant inspection for all materials with surface flaws, magnetic particle for steel panels with surface flaws, eddy current for all materials with bolt holes, and ultrasonic for titanium and steel embedded flaws. The surface and corner flaws were divided into four groups based on size (Table 1). Because of the difficulty of manufacturing embedded flaws, only two sizes were selected: 1.3 mm (0.050 in.) diam circle and a 1.3 × 2.5 mm (0.050 × 0.100 in.) ellipse.

For every flawed specimen presented to an inspector, at least two unflawed specimens would also be included. Sizing of the defects was not required, and no penalties were assessed if the inspector identified a flaw that did not exist. In addition, if the desired reliability was achieved for a given size range, it also had to be met at the next higher range. The reliability criterion was established to be a 90% probability of detection at the 95% confidence level (CL). This would be considered to be met for the largest flaw size in the given ranges if the following detection requirements were met:

Number of observations	Misses
29	0
46	1
61	2
75	3
89	4
103	5

A variety of lessons were learned during the demonstration program. The production inspection processes used at the time were unable to meet the design requirements of 1.3 mm (0.050 in.) depth for corner cracks. The processes were refined and the documentation was completely rewritten before the requirements were met. Second, not all inspection personnel were capable of achieving the required reliability, even with the refined procedures and documentation. As a result, the demonstration program also became a qualification requirement for personnel that would inspect fracture-critical components. Etching was also required before the penetrant inspection could meet the requirements for any of the materials. Embedded flaws could only be produced for titanium. Table 2 summarizes the results demonstrated for the B-1 production NDE facilities at the reliability criterion of a 90% POD at 95% CL.

Fracture control, with the resultant quantitative NDE requirement, has been incorporated into the design of every Air Force aircraft since the B-1. In addition, designs already in existence at the time have been analyzed to determine which inspections would be required in service to ensure attainment of the design life of a particular system. Generally, where the suggested flaw sizes mentioned above could not be tolerated, the assumed design flaw for new designs has been a 0.75 mm (0.030 in.) corner crack and a surface flaw with a depth of 1.25 mm (0.050 in.) and a length of 2.5 mm (0.100 in.). The analysis of existing designs, however, has required the assumption of even smaller design flaws. These instances of having to detect smaller flaws in service have been few, and fortunately, identification of the requirement has occurred in sufficient time to allow adequate NDE engineering to address the problems and explore various noninspection options.

Gas Turbine Engines. Following the successful application of damage tolerance concepts and designs to Air Force aircraft structures, attention began to focus on aircraft gas turbine engines. The reason is that the failure of a high-energy (rotating) component in an aircraft gas turbine engine usually results in catastrophic consequences for that engine. Although most engines are designed to contain failure of the blades, the fracture of a disk or spacer will result in destruction of the engine and can cause significant damage to adjacent structures, such as fuel tanks, major structural members, or other engines. Even the contained failure of a blade can cause immediate engine shutdown, which can also have catastrophic consequences for a high-performance, single-engine fighter. Therefore, the integrity of many high-performance components in gas turbine engines is critical to aircraft safety.

Table 1 Groups of manufactured flaws used to demonstrate the reliability of NDE methods in the B-1 program

Group	Length range, mm	Length range, in.	Depth range, mm	Depth range, in.
Surface flaws				
1	0.75–1.91	0.03–0.075	0.25–0.90	0.01–0.035
2	1.92–2.55	0.076–0.100	0.50–1.25	0.020–0.050
3	2.56–3.81	0.101–0.150	0.50–1.80	0.020–0.070
4	3.83–6.35	0.151–0.250	0.50–3.0	0.020–0.12
Group	**Bore length range, mm**	**Bore length range, in.**	**Radial depth range, mm**	**Radial depth range, in.**
Corner flaws				
1	0.75–1.53	0.03–0.060	0.25–0.76	0.01–0.030
2	1.54–2.55	0.061–0.100	0.78–1.27	0.031–0.050
3	2.56–3.56	0.101–0.140	1.29–1.78	0.051–0.070
4	3.58–5.0	0.141–0.20	1.80–2.5	0.071–0.10

As part of an effort to increase the reliability and reduce the costs of operating and maintaining gas turbine engines for the U.S. Air Force, a program known as the Engine Structural Integrity Program (ENSIP) has evolved over the last few years. Its objective is to establish an approach to defining the structural performance, design, development, verification, and life management requirements for new engines for Air Force aircraft. Military standard MIL-STD-1783 defines ENSIP and is currently being applied to the development of all new engines for the Air Force (Ref 2). The document is written in a generic format so that it can be tailored for use by specific System Program Offices to define an engine that will satisfy their own needs. An accompanying handbook is attached as an appendix to provide specific guidance on the rationale, background criteria, lessons learned, and instructions necessary to tailor specific sections of the standard for application. The technical approach is similar to the Aircraft Structural Integrity Program (ASIP), which is defined in MIL-STD-1530 and which has been successfully used for several years in the design of airframes for Air Force systems (Ref 3).

One of the most significant differences between ENSIP and the traditional approaches used in designing engine struc-

Table 2 Results of production NDE facilities in the B-1 program

Method	Material	Flaw type	Flaw depth, mm	Flaw depth, in.
Penetrant	Aluminum	Surface	0.90	0.035
	Titanium	Surface	0.90	0.035
	Steel	Surface	1.0	0.040
Magnetic particle	Steel	Surface	1.25	0.050
Ultrasonic, shear	Titanium	Embedded	0.9	0.035
	Steel	Embedded	1.25	0.050
Ultrasonic, longitudinal	Titanium	Embedded	1.17	0.046
Eddy current	All	Corner	0.75	0.030

tures is the requirement to apply damage tolerance and durability criteria to critical components. This requires the designer to assume that flaws exist in the engine structure as manufactured and then to design the critical parts so that the flaws cannot grow to the size that will cause failure in the lifetime of the part or at least within some predetermined inspection interval. It also establishes life management requirements and procedures to ensure that the necessary inspections capable of finding flaws in the size range used in design are conducted and that the engine parts are sufficiently durable so that the economic life of the engine is acceptable.

The impact of this approach on the inspection community is very significant. Based on the estimated capability of state-of-the-art inspection methods and procedures, many improvements have had to be made both in manufacturing and depot practice to satisfy the intent of ENSIP. Furthermore, to allow the full implementation of this approach, the problem of defining available inspection capability in quantitative terms must continue to receive attention. Acceptable procedures are being established so that it is possible to define exactly how sensitive the inspection methods are on a statistical basis.

The first application of this technology by the Air Force was made to an already designed and operational aircraft gas turbine engine. The F-100 engine, designed and built by Pratt & Whitney, was already widely used in the dual-engine F-15 and the single-engine F-16 aircraft when, in 1978, a durability and damage tolerance assessment effort was initiated. This engine was selected because it was to be purchased in large numbers for Air Force applications for many years to come and because it was a high-performance machine that was very demanding of materials and designs. The costs of owning and operating this system could quickly become untenable if problems developed that significantly limited the life of critical parts or caused significant downtime for repairs. Following the pattern that had been established for applying ASIP to airframes, a joint Air Force/Pratt & Whitney team was formed to work on-site at the facilities of the contractor to complete the analysis and to provide a viable Force Structural Maintenance Plan that could be implemented at the San Antonio Air Logistics Center, where maintenance responsibility for the engines resided.

After the team was in place and the necessary analytical studies were started, it was recognized early in the program that a quantitative understanding of the inspection procedures used in the manufacture of the engine was lacking and that a reliable definition of the largest flaw that could escape detection during manufacture had never been determined. Qualitative statements were made ex-

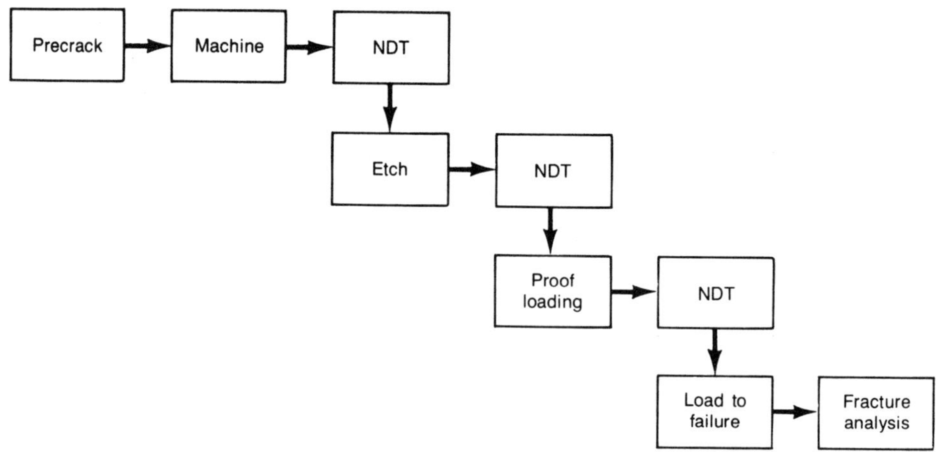

Fig. 5 Experimental test analysis sequence

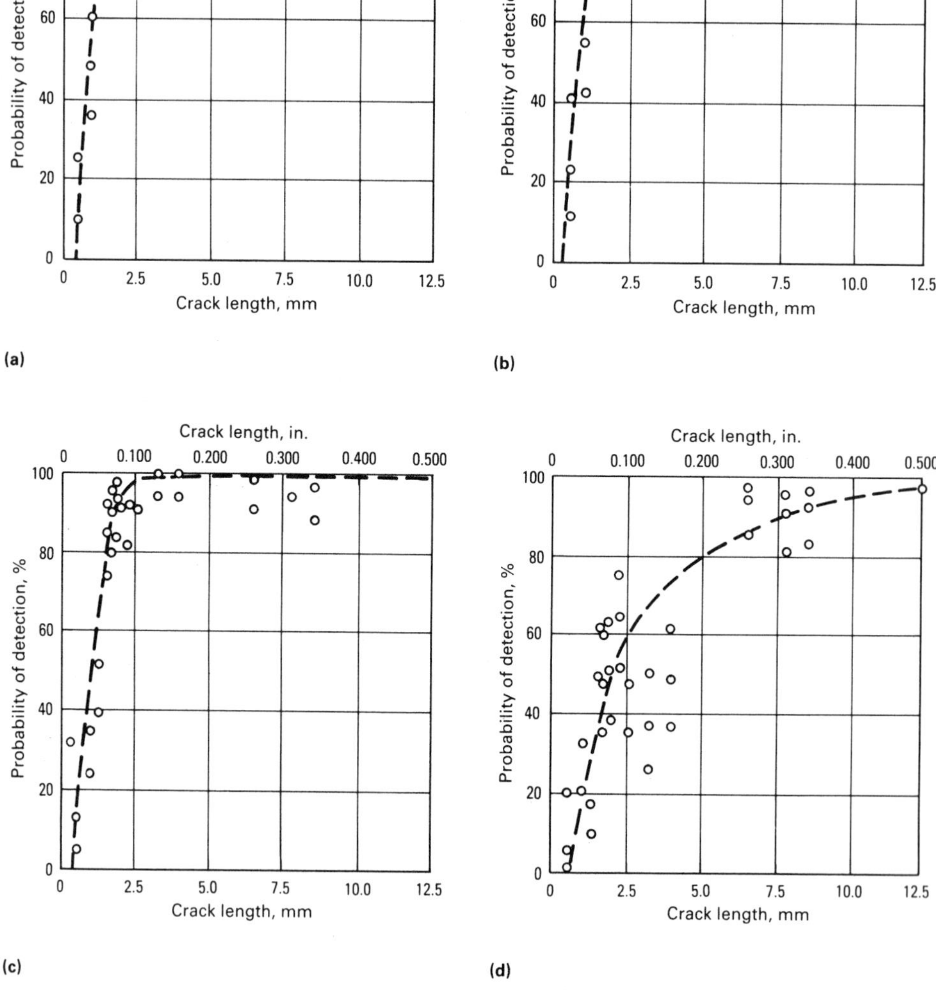

Fig. 6 POD plots for four different NDE methods on the same set of specimens. (a) Penetrant inspection. (b) Ultrasonic inspection. (c) Eddy current inspection. (d) X-ray inspection

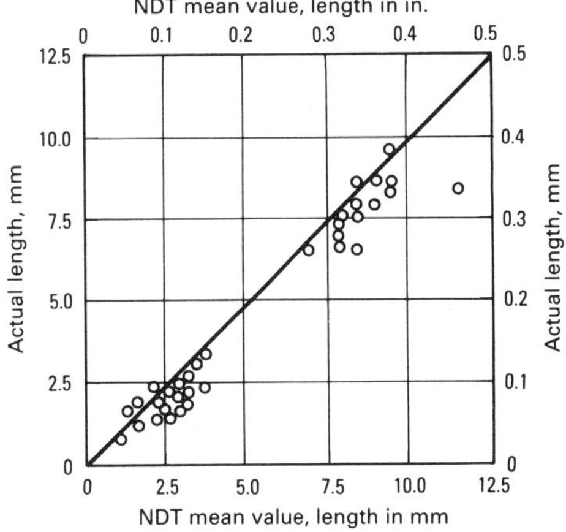

(a)

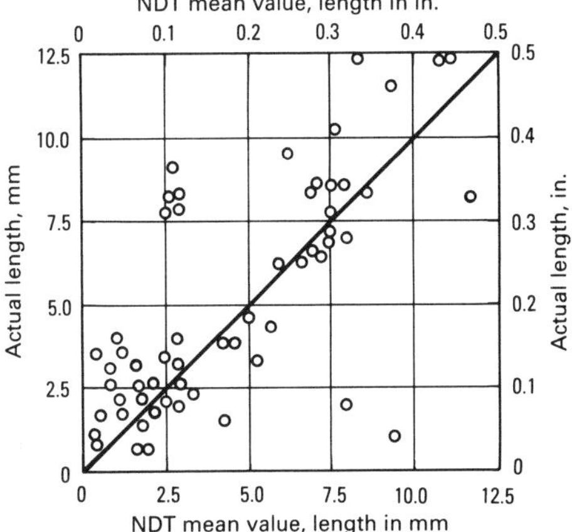

(b)

Fig. 7 Actual versus NDE estimated crack length for etched specimens. (a) Penetrant inspection. (b) X-ray inspection

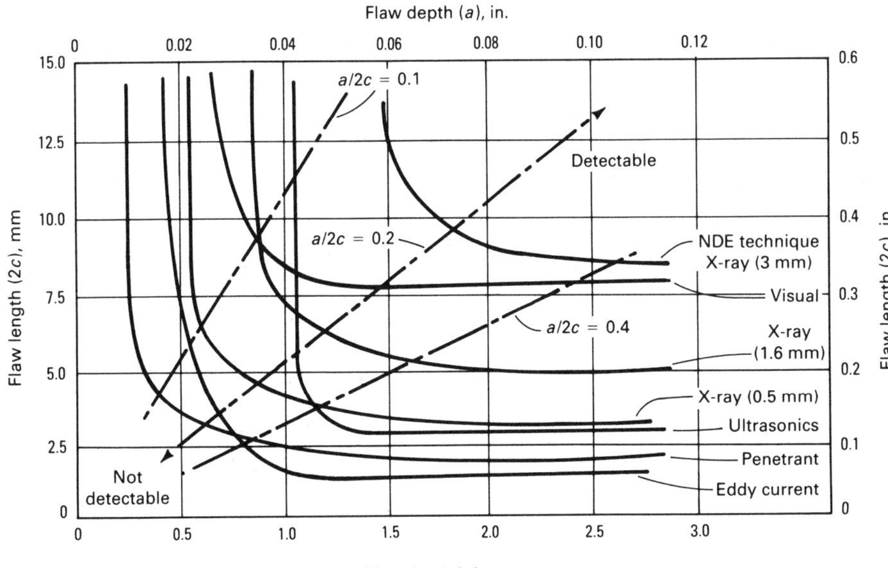

Fig. 8 Combined threshold detection results

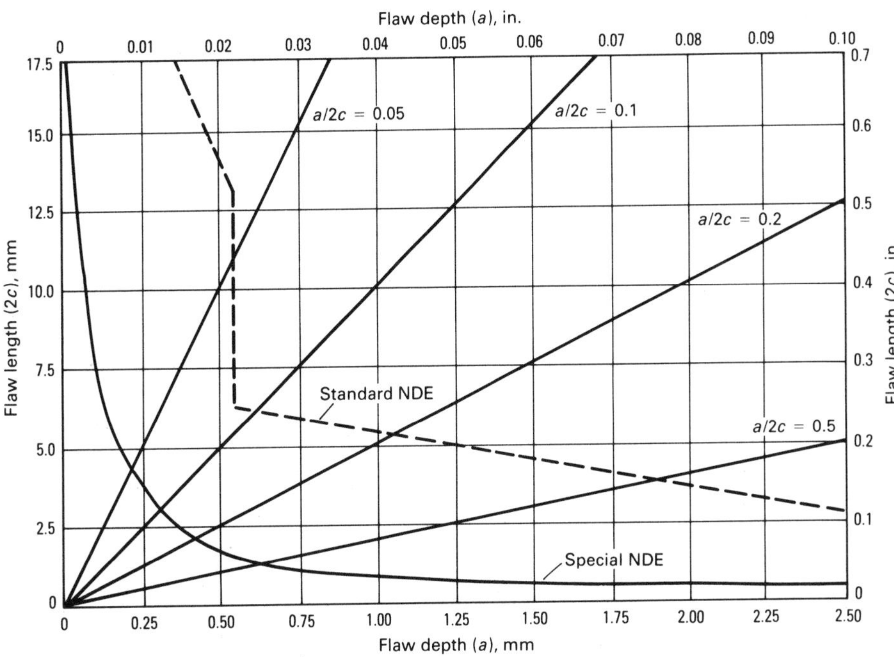

Fig. 9 Plot of design limits

pressing confidence in the inspection methods used, based on the good performance of the engines to that time. It was acknowledged, however, that the fleet was still young and that a quantitative definition of the inspection process was urgently needed.

To fill that need, a joint Air Force Materials Laboratory/Pratt & Whitney effort was established to prepare specimens with known flaws in selected size ranges that would allow quantitative determination of the capability of the inspection methods to be established. The specimens were designed to contain flawed areas in geometrical features that simulated the real areas in the actual hardware. These included holes, the edges of holes, radii, and flat surfaces. Small flaws were generated by damaging the surface, initiating and growing a fatigue crack, and then removing the damaged area until only the desired depth of the flaw that remained was used to produce the specimens. During 1979, some 39 sets of specimens with the desired geometries were fabricated from nickel- and titanium-base alloys for the program. Target crack depths of 0.13, 0.25, and 0.50 mm (0.005, 0.010, and 0.020 in.) with a nominal 3:1 aspect ratio were prepared. Similar specimens with no flaws were also included in each set. The specimens were to be used not only to determine the capability of the manufacturing inspection methods but also to provide guidance concerning the establishment of inspection methods to be used in the depot during maintenance. Fluorescent penetrant inspection and eddy current procedures were evaluated using these specimens. The specimen sets were evaluated by both laboratory and production inspectors.

Because this was the first documented attempt to fabricate flawed specimens of this complexity, a great deal of experience was gained, not only in specimen preparation but also in evaluating the effectiveness of the inspection methods. Fluorescent penetrant inspection was included because, at the time, it was the most extensively applied inspection method used in both manufacturing and depot maintenance. Eddy current methods were included because they had the best potential for finding the very small surface-connected flaws of concern. Based on the work done in this pro-

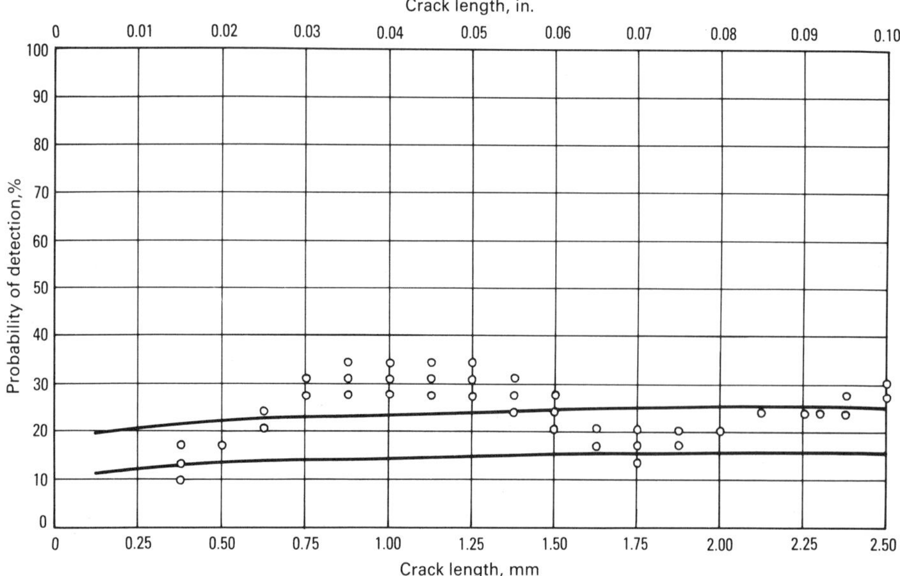

(a)

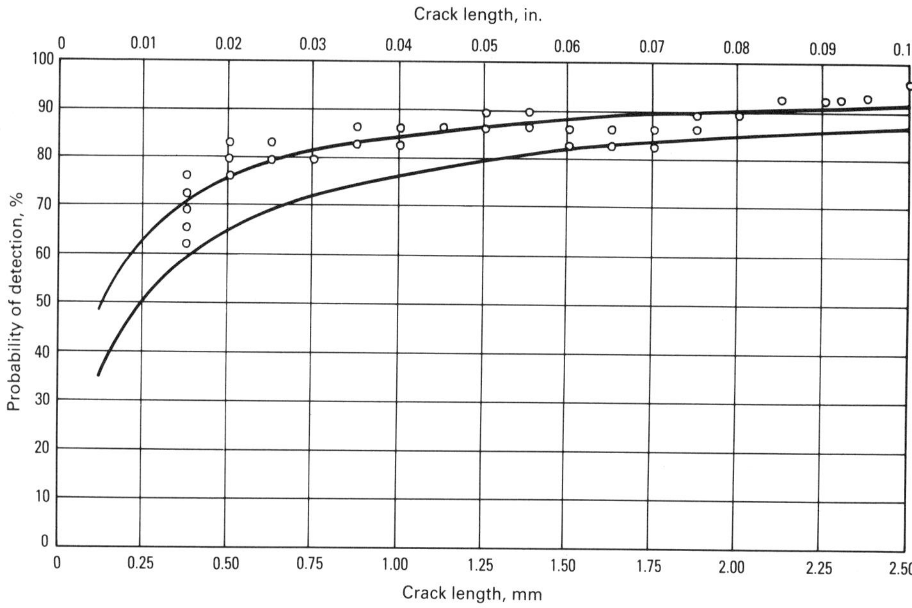

(b)

Fig. 10 POD plotted (a) at a threshold of 3.0 mV and (b) at a threshold of 0.5 mV

gram, the following conclusions were reached:

- The capability of fluorescent penetrant inspection to find small flaws with confidence was affected by many variables, including surface condition, nature of the flaw, the process used and the extent to which it was controlled, and the skills and abilities of the inspectors. The process did not have the necessary reliability to detect very small flaws in many of the critical areas that had been defined
- Eddy current methods appeared to have the best potential for detecting small flaws with the confidence level required
- Eddy current technology and procedures developed in the program could also be adequately automated to make this method viable for use in the depot inspection environment. They could also be adapted for manufacturing

As a result of this study, the decision was made to implement semiautomated eddy current inspection methods at the San Antonio Air Logistics Center to provide reliable inspection of defined areas in selected critical parts. A facility was established at the depot to allow the inspection of critical components as they were cycled through the depot. The eddy current equipment designed by Pratt & Whitney and incorporated into the facility has been demonstrated to have the capability of finding 0.13 mm (0.005 in.) deep flaws with a reliability of 90% POD at 95% CL.

The second application of damage tolerance analysis involved the TF-34 engine, which powers the A-10 ground-support attack aircraft and the S-3A antisubmarine aircraft. In this case, an Air Force/General Electric team was formed at the General Electric Aircraft Engine Business Group facility in May 1981. Once again, a critical part of the assessment activity was to establish the level of quality built into the parts during the years the engines were manufactured. This was needed to provide an indication of the largest flaw that could have been missed by the inspection methods being used at the time. Because production of the engine was essentially completed by the time the assessment started, the only impact that establishing improved NDE procedure could have was on the methods being used at the maintenance and overhaul depot.

In the case of the TF-34 engine, depot responsibility had been assigned to the Navy and was conducted at the Naval Air Newark Facility in Alameda, CA. Once again, the assessment determined that to achieve acceptable inspection intervals for certain critical rotating components, inspection methods more sensitive and reliable than fluorescent penetrant inspection would be required. A special clean room containing eddy current equipment was established at Alameda to allow inspection of the engine disks and spacers that were considered to be the most critical. The equipment selected for this room was the Eddy Current II system developed under an Air Force Materials Laboratory manufacturing technology contract by the General Electric Company. In addition, the fluorescent penetrant inspection facility at Alameda was upgraded to make it more reproducible and reliable. Inspections have been conducted with these systems since 1984.

Other applications of this technology by the U.S. Air Force have been made to all new gas turbine engines now being used. This includes the F-101 engine in the B-1B and the F-110 engine in the F-16. New engines now being designed for Air Force applications will also incorporate ENSIP technology and will therefore have quantitative inspection capabilities as an integral part of their development. Additional information on ENSIP is available in the article "Fracture Control Philosophy" in this Volume.

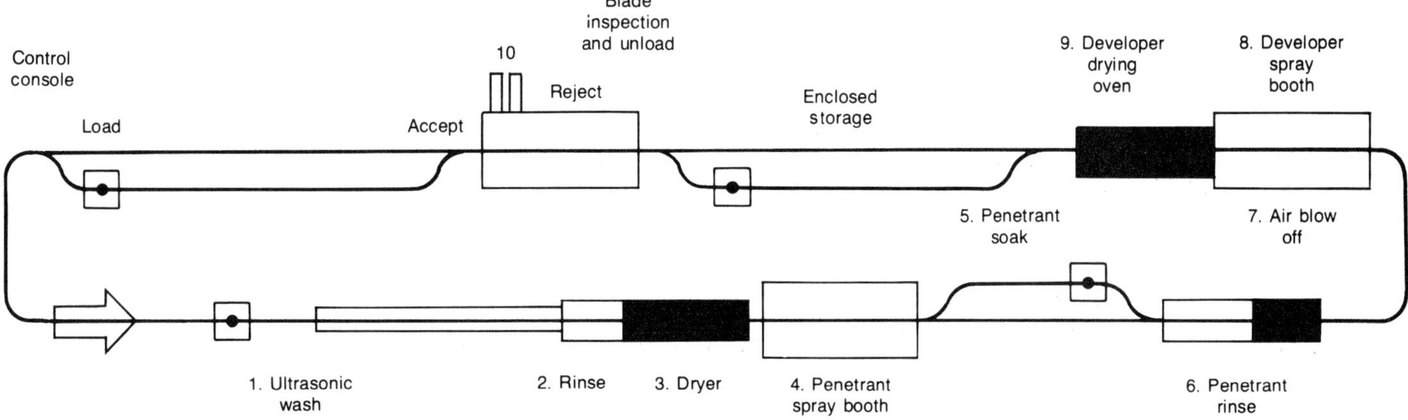

Fig. 11 Flow diagram of an automated fluorescent penetrant inspection system

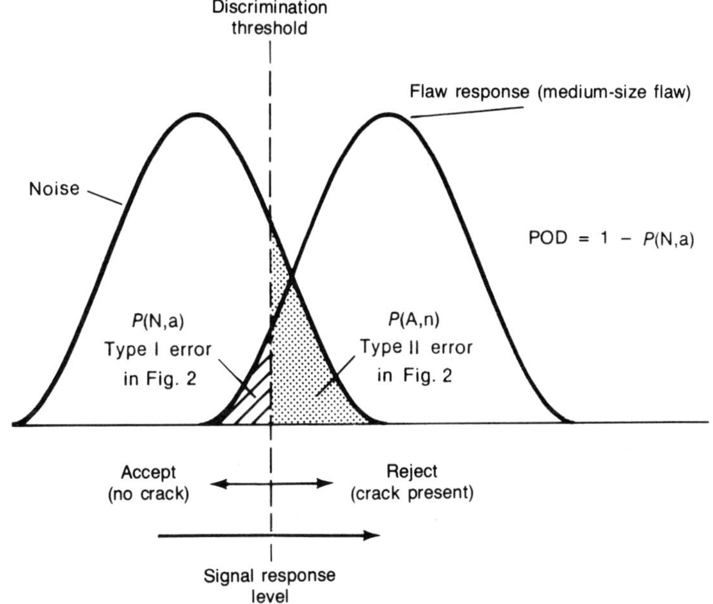

Fig. 12 Probability distribution of a noise signal and flaw indication

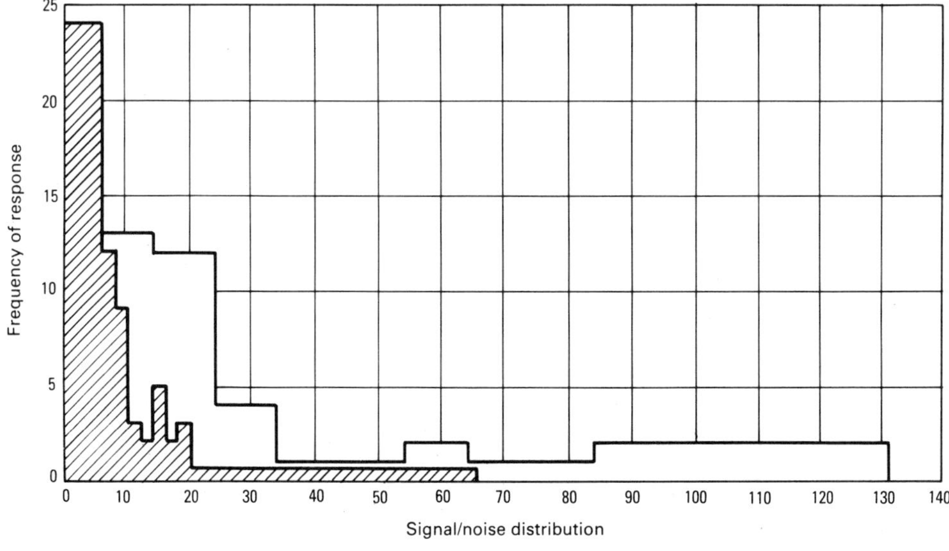

Fig. 13 A sample of the signal/noise distribution for an integrated blade inspection system

Space Shuttle Program. Design requirements for the space shuttle program of the National Aeronautics and Space Administration included the use of fatigue and fracture mechanics principles in all systems designs. The implementation of fatigue and fracture mechanics requires knowledge of the quantitative performance capabilities of the materials, components, and systems acceptance methods. For this application, Rummel *et al.* introduced the concepts of statistical assessments of NDE process performance capabilities (Ref 4). Fatigue cracks in 2219-T87 aluminum alloy were selected as the test specimens for the assessment of various NDE methods. Test specimens were prepared by inducing fatigue cracks of varying size in aluminum alloy sheet specimens (fatigue crack growth from electrodischarge machined starter notches), by machining the specimen surfaces to produce a surface that was representative of production conditions, and by passing the specimens through various NDE procedures and measuring the success in crack detection for the various procedures and operating conditions. The concept of probability of detection as a function of flaw size was introduced, and the performance level of the various NDE procedures was quantified as summarized below.

The experimental test sequence for data gathering, including nondestructive testing (NDT), is shown in Fig. 5. The test specimens were given a precrack (starter notches) that enabled the growth of fatigue cracks of varying size and aspect ratio. A total of 328 cracks were grown in 118 specimens, with flaw length ranging from 0.3 to 18 mm (0.012 to 0.700 in.). The specimens were subjected to x-ray radiographic, liquid penetrant, ultrasonic, and eddy current procedures in the as-machined, after-etch, and after-proof-test conditions. The resulting POD curves for machined and etched surfaces are also shown in Fig. 6.

The precision in crack sizing was also measured, and the results are shown in Fig. 7. The composite threshold detection re-

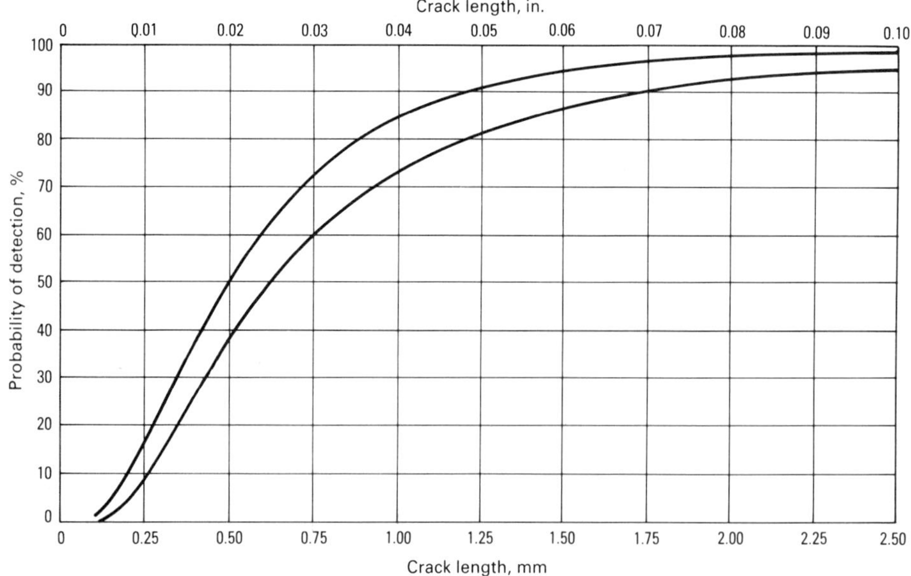

Fig. 14 POD curve for an automated fluorescent penetrant inspection system

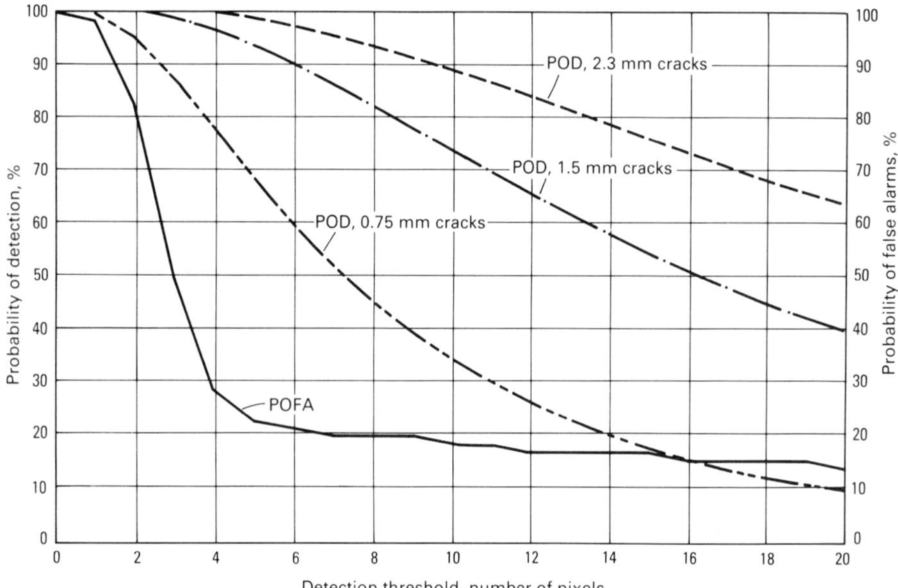

Fig. 15 Performance curves for various discrimination levels

sults were reduced to a plot, as shown in Fig. 8. These data were further simplified to produce the design limits, as shown in Fig. 9. Figure 9 contains design limits for standard NDE and special NDE. The use of special NDE to meet difficult design constraints requires actual demonstration of the performance capabilities of the proposed NDE procedures and a system of controls to ensure that the performance conditions are maintained. These design limits, set by actual NDE performance demonstration, were used as the basis for design and risk management for all space shuttle system components. Assessments and NDE performance capability demonstrations have been continued with space shuttle contractors for all systems production and revalidation.

Special Inspection Systems. As a result of the evolution of damage tolerance requirements, special equipment has been developed in the last few years for increasing the reliability of inspection operations by automating the process and by incorporating computerized data generation and control. This has significantly reduced the dependence on human operators, thus eliminating one of the major sources of error in the process. Most of these automation efforts have been directed toward the inspection of critical aircraft gas turbine engine hardware, primarily in the maintenance depot environment. The systems could be used equally effectively in manufacturing operations. The following are examples of the systems that have been developed over recent years.

Structural Assessment Testing Applications. The NDE performance requirements for overhaul and revalidation of the structural integrity of aircraft engine components (United States Air Force facilities) required the use of special, controlled methods to meet meantime between overhaul and life cycle performance requirements. Special equipment, fixturing, procedures, and personnel training are implemented at the San Antonio Air Logistics Center at Kelly Air Force Base to approach the imposed design constraints. Assessment, demonstration, and validation of the system performance capabilities and reliabilities were necessary to ensure that design requirements were being met.

The special NDE methods were successful in meeting design requirements, and new levels of understanding and performance were gained by the implementation of advanced NDE methods and controls in this special facility. The most significant output from the assessment of these processes was the demonstration of the need for excellence in NDE engineering in validating NDE procedures and in setting acceptance limits. The performance capability of an eddy current method with an acceptance threshold limit set at 3.0 mV is shown in Fig. 10(a). The performance level of the same procedure (same data) with an acceptance limit set at 0.5 mV is shown in Fig. 10(b). This example clearly illustrates the importance of fully characterizing the NDE procedure and managing the NDE procedure within achievable acceptance limits.

Integrated Blade Inspection System. The quantitative assessment of process performance capabilities and process characterization is absolutely necessary in implementing automated NDE systems. At a gas turbine overhaul facility, for example, quantitative assessment methods were applied to the implementation of an integrated blade inspection system with an automated fluorescent penetrant inspection module (Fig. 11). The processed blades are introduced into a robotic handling system that manipulates the blade in a high-gain optical-laser scan readout system to produce a digitized image of the fluorescent penetrant indications. A computerized data processing and image analysis system provides the readout and decision processing to accept or reject the blades.

The optical performance capabilities of the readout system enable the detection of very small indications of varying brightness level. The performance of the system is therefore limited by the fluorescent process capabilities and by the pattern recognition capabilities of the system. Signal/noise distributions must be addressed by the deci-

Fig. 16 Retirement-for-cause inspection facility

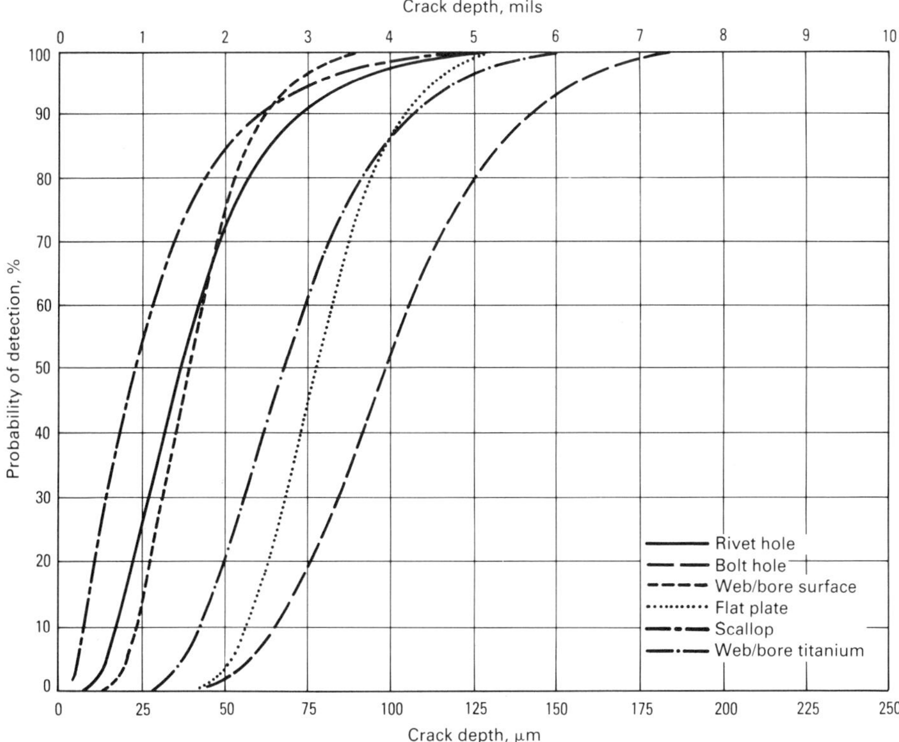

Fig. 17 RFC system POD curves for various geometrical features

sion discrimination process, as shown in Fig. 12. Actual signal/noise distributions measured for the system are shown in Fig. 13. If the discrimination threshold is set too low, a high false call rate will reduce the effectiveness of the automated process. A balance of discrimination at an acceptable false call rate is used to achieve the figure of merit of performance level (Fig. 14) for the system. Signal/noise distribution data can be used to manage the overall process at any desired discrimination/false call level. The process can be modeled to establish performance at varying discrimination levels, as shown in Fig. 15. Quantification and characterization of process parameters are essential for the design, implementation, and management of automated NDE processes.

Retirement-For-Cause (RFC) Inspection Equipment. As described previously, the concept of removing critical rotating components from gas turbine engines because of the initiation of an actual flaw rather than on the basis of an arbitrarily selected time period has already saved the Air Force significant sums of money. One of the key technologies that has allowed the RFC concept to be accepted has been the development of an accurate, repeatable, reliable inspection system capable of finding very small flaws in the parts. The RFC system, developed under an Air Force Manufacturing Technology contract, has been in operation at the San Antonio Air Logistics Center at Kelly Air Force Base since late 1986. It was developed under a contract with Systems Research Laboratories, and although applied specifically to parts for the F-100 engine (used in the F-15 and F-16 aircraft), it was designed to be sufficiently flexible to inspect parts from any gas turbine engine. This capability was achieved through the use of a team of subcontractors on the program that included Pratt & Whitney, General Electric, Garrett, and Allison.

Eddy current methods for surface flaw detection and ultrasonic methods for detecting embedded flaws were the two inspection techniques selected for the RFC system. Significant advances had been made in automating eddy current inspection through the development of the Eddy Current II system by the General Electric Company on earlier Air Force Manufacturing Technology contracts. These stand-alone automated eddy current devices are in use at the Alameda Naval Air Rework Facility, as previously discussed; the Oklahoma City Air Logistics Center; the General Electric Manufacturing facility in Evandale, OH; and other commercial overhaul facilities. The technology developed and proved in these systems provided an excellent base for the evolution of the more complex RFC system.

The RFC inspection system, which is housed in the special facility shown in Fig. 16, consists of an operator console, a system computer, and eddy current and ultrasonic inspection stations. The operator console is used to monitor the operational status of the system, to track inspection status at each NDE station, and to generate inspection data reports. The system computer performs advanced data processing, system-wide communication, and sophisticated high-speed mathematical and scientific data analyses critical to the inspection process. The NDE inspection stations perform the automated part inspections, flaw detection, and signal-preprocessing activities. The system functions essentially independent of any human operator input, with the exception of loading and unloading the parts to be inspected onto the stations.

When the system was installed, a series of critical tests was conducted to establish its flaw size detection capability and reliability. Tests included automatic scans of engine disks and a statistically significant number of representative fatigue-cracked test specimens. Rivet hole inspection data showed a 90% POD with a 95% CL at the 100 μm (4 mil) crack depth range. Bolt hole and flat surface data indicated reliable detection in the desired 125 to 250 μm (5 to 10 mil) depth range. A strong correlation between apparent versus actual flaw depth data was seen in all test data. The ultrasonic inspection data were similarly encouraging. Examples of POD at 95% CL generated for various geometrical configurations are shown in Fig. 17. A similar facility is to be installed at the Air Force engine maintenance depot at the Oklahoma City Air Logistics Center at Tinker Air Force Base.

REFERENCES

1. "Nondestructive Testing Personnel Qualification and Certification," MIL-STD-410D, 25 June 1974
2. "Engine Structural Integrity Program (ENSIP)," MIL-STD-1783 (USAF), 30 Nov 1984
3. "Aircraft Structural Integrity Program, Airplane Requirements," MIL-STD-1530A (USAF), 11 Dec 1975
4. W.D. Rummel, P.H. Todd, Jr., S.A. Frecska, and R.A. Rathke, "The Detection of Fatigue Cracks by Nondestructive Testing Methods," CR-2369, National Aeronautics and Space Administration, Feb 1974

NDE Reliability Data Analysis

Alan P. Berens, University of Dayton Research Institute

INSPECTION SYSTEMS are inevitably driven to their extreme capability for finding small flaws. When applied at this extreme, not all flaws of the same size will be detected. In fact, repeat inspections of the same flaw will not necessarily produce consistent hit or miss indications, and different flaws of the same size may have different detection probabilities. Because of this uncertainty in the inspection process, capability is characterized in terms of the probability of detection (POD) as a function of flaw size, a. At present, the function POD(a) can be estimated only through inspection reliability experiments on specimens containing flaws of known size. Further, statistical methods must be used to estimate the parameters of the POD(a) function and to quantify the experimental error in the estimated capability.

The methods for analyzing NDE reliability data have undergone a considerable evolution since the middle of the 1970s. Formerly, a constant probability of detection of all flaws of a given size was postulated, and binomial distribution methods were used to estimate this probability and its lower confidence bound (Ref 1, 2). This nonparametric method of analysis produced valid statistical estimates for a single flaw size, but required very large sample sizes to obtain reasonable lower confidence bounds on the probability of detection. In the absence of large numbers of representative specimens with equal flaw sizes, various methods were devised for analyzing data based on grouping schemes. Although the resulting POD appeared more acceptable using these schemes, the lower confidence bounds were no longer valid.

In recent years, an approach based on the assumption of a model for the POD(a) function was devised (Ref 3-7). Analyses of data from reliability experiments on nondestructive inspection (NDI) methods indicated that the POD(a) function can be reasonably modeled by the cumulative log normal distribution function or, equivalently, the log-logistics (log odds) function. The parameters of these functions can be estimated using maximum likelihood methods. The statistical uncertainty in the estimate of NDI reliability has traditionally been reflected by a lower (conservative) confidence bound on the POD(a) function. The asymptotic statistical properties of the maximum likelihood estimates can be used to calculate this confidence bound. Details of the mathematics for these maximum likelihood calculations are presented in this article.

Statistical Nature of the NDE Process

In the application of an NDE method there are many factors that can influence whether or not the inspection will result in the correct decision as to the absence or presence of a flaw. In general, NDE comprises the application of a stimulus to a structure and the interpretation of the response to the stimulus. Repeated inspections of a specific flaw can produce different magnitudes of stimulus response because of minute variations in setup and calibration. This variability is inherent in the process. Different flaws of the same size can produce different response magnitudes because of differences in material properties, flaw geometry, and flaw orientation. The interpretation of the response can be influenced by the capability of the interpreter (manual or automated), the mental acuity of the inspector as influenced by fatigue or emotional outlook, and the ease of access and the environment at the inspection site. All these factors contribute to inspection uncertainty and lead to a probabilistic characterization of inspection capability.

There are two related approaches to a probabilistic framework for analyzing inspection reliability data. Originally, inspection results were recorded only in terms of whether or not a flaw was found. Data of this nature are called hit/miss data, and an analysis method for this data type evolved from the original binomial characterization (Ref 3-5). It was later observed that there is more information in the NDE signal response from which the hit/miss decision is made (Ref 6). Because the NDE signal response can be considered to be the perceived flaw size, data of this nature are called $\hat{a}$ data. A second analysis method was developed based on $\hat{a}$ data (Ref 7). Although the analysis frameworks are based on data of different natures, the hit/miss data can be obtained from $\hat{a}$ data. Both methods are based on the same model for the POD(a) function, but different results will be obtained if the two analysis methods are applied to the same data set. The following sections present the two approaches to formulating the POD(a) function.

POD(a) From Hit/Miss Data. In typical NDE reliability studies, relatively few inspections are performed on each flaw in the specimen set. Table 1 presents an example of hit/miss results from fluorescent penetrant inspections by 3 inspectors on 35 cracks in flat plate specimens. Because there are only three inspections, it is impossible to obtain more than a general impression of any change in the chances of crack detection as the cracks get larger. In a study for the Air Force (Ref 3), inspections of cracked specimens were independently conducted by large numbers of inspectors around the country. The data from this program provided considerable insight into the nature of the probability of crack detection.

Figure 1 shows the results of eddy current inspections by 60 different inspections of a set of 41 cracks around countersunk fastener holes in a 1.5 m (5 ft) segment of a C-130 center wing box (Ref 3). Each data point represents the proportion of times that the crack was found. This data set, which is representative of such available data sets, clearly indicates that:

- The chances of detection are correlated with crack size
- Different cracks of the same size can have significantly different crack detection probabilities
- Factors other than size are affecting the chances of detection

Data of this nature also provide the analysis framework for characterizing NDE reliability for the general hit/miss data set.

The POD(a) function is defined as the proportion of all cracks of size a that will be detected in a particular application of an NDE system. Assume that each crack of size a in the potential population of cracks has its own distinct crack detection probability, p, and that the probability density function of the detection probabilities is given by $f_a(p)$. Figure 2 shows a schematic

Table 1 Example of summary data sheet of hit/miss results

The example is based on the fluorescent penetrant inspection of flat plates by three inspectors.

Crack identification	Crack size, mm	Crack size, in.	Inspector(a) A	Inspector(a) B	Inspector(a) C
1	2.21	0.087	0	1	1
2	1.63	0.064	1	0	0
3	0.38	0.015	0	0	0
4	2.84	0.112	1	1	0
5	0.99	0.039	0	0	0
6	4.42	0.174	1	1	1
7	2.97	0.117	1	1	1
8	2.06	0.081	0	0	0
9	0.46	0.018	0	0	0
10	4.22	0.166	1	0	1
11	2.54	0.100	1	0	0
12	1.98	0.078	1	1	0
13	0.64	0.025	0	1	0
14	2.18	0.086	1	0	0
15	2.64	0.104	0	0	1
16	2.49	0.098	0	1	1
17	2.41	0.095	1	0	0
18	1.42	0.056	0	0	0
19	0.20	0.008	0	0	0
20	0.58	0.023	0	0	0
21	6.99	0.275	1	1	1
22	3.30	0.130	0	0	1
23	6.20	0.244	1	1	1
24	2.03	0.080	0	0	0
25	1.85	0.073	0	0	0
26	4.95	0.195	1	1	1
27	0.23	0.009	0	0	0
28	2.13	0.084	0	1	1
29	5.59	0.220	1	1	1
30	1.02	0.040	0	0	0
31	0.99	0.039	0	0	0
32	2.18	0.086	0	0	1
33	0.25	0.010	0	0	1
34	4.09	0.161	0	0	1
35	0.51	0.020	0	0	0

(a) 1 indicates crack was found; 0 indicates crack was not found.

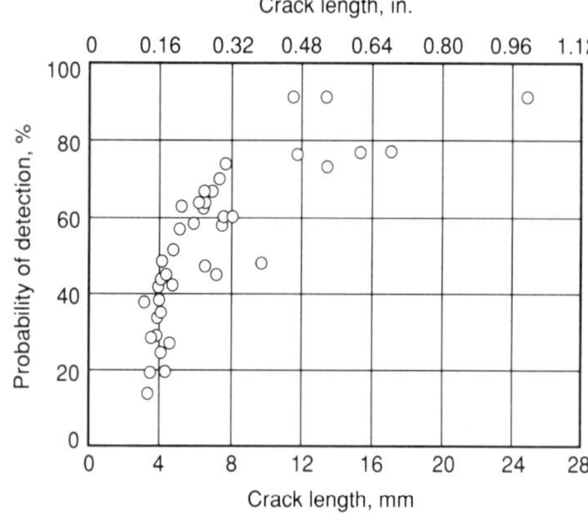

Fig. 1 Example crack detection probabilities from 60 eddy current inspections of each crack

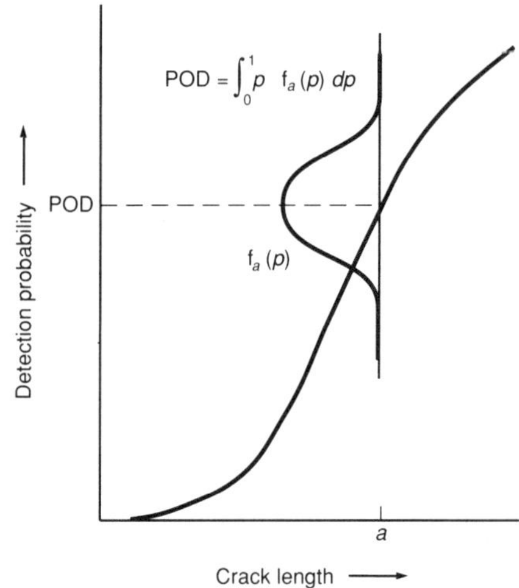

Fig. 2 Schematic of distribution of detection probabilities for cracks of fixed length

representation of this density. The conditional probability of a randomly selected crack from the population having detection probability of p and being detected at the inspection is given by $p\ f_a(p)\ dp$. The unconditional probability of a randomly selected crack from the population being detected is the sum of the conditional probabilities over the range of p, that is:

$$\mathrm{POD}(a) = \int_0^1 p\ f_a(p)\ dp \quad \text{(Eq 1)}$$

Therefore, POD(a) is the average of the detection probabilities for cracks of size a.

Equation 1 implies that the POD(a) function is the curve through the averages of the individual density functions of the detection probabilities. This curve is the regression equation and provides the basis for testing assumptions about the applicability of various POD(a) models. In Ref 4, seven different functional forms were tested for applicability to available POD data, and it was concluded that the log-logistics (log odds) function best modeled the data and provided an acceptable model for the data sets of the study. Note that the log odds model is commonly used in the analysis of binary (hit/miss) data because of its analytical tractability and its close agreement with the cumulative log normal distribution (Ref 8).

Two mathematically equivalent forms of the log odds model have subsequently been used. The earliest form is given by:

$$\mathrm{POD}(a) = \frac{\exp(\alpha + \beta \ln a)}{1 + \exp(\alpha + \beta \ln a)} \quad \text{(Eq 2)}$$

This parametrization can also be expressed as:

$$\ln\left[\frac{\mathrm{POD}(a)}{1 - \mathrm{POD}(a)}\right] = \alpha + \beta \ln(a) \quad \text{(Eq 3)}$$

In the Eq 3 form, the log of the odds of the probability of detection (the left-hand side of Eq 3) is expressed as a linear function of ln (a) and is the source of the name of the log odds model. Note that given the results of a large number of independent inspections of a large number of cracks, the parameters of the model can be fit with a regression analysis. As an example, Fig. 3 shows Eq 3 fit to the data of Fig. 1. This regression approach will not be discussed further, because the maximum likelihood estimates (see the section "Analysis of Hit/Miss Data" in this article) can be applied to much smaller samples of inspection results and can give equivalent answers for large sample sizes.

Although the parametrizations of Eq 2 and 3 are sensible in terms of estimation through regression analyses, α and β are not easily interpretable in physical terms. A mathematically equivalent form of the log odds POD(a) model is given by (Ref 8):

$$\mathrm{POD}(a) = \left\{1 + \exp - \left[\frac{\pi}{\sqrt{3}}\left(\frac{\ln a - \mu}{\sigma}\right)\right]\right\}^{-1} \quad \text{(Eq 4)}$$

In this form, $\mu = \ln a_{0.5}$, where $a_{0.5}$ is the flaw size that is detected 50% of the time, that is, the median detectable crack size. The steepness of the POD(a) function is inversely proportional to σ; that is, the smaller the value of σ, the steeper the POD(a) function. The parameters of Eq 2 and 4 are related by:

$$\mu = \frac{-\alpha}{\beta} \quad \text{(Eq 5)}$$

$$\sigma = \frac{\pi}{(\beta\sqrt{3})} \quad \text{(Eq 6)}$$

The log odds POD(a) function is practically equivalent to a cumulative log normal distribution with the same parameters, μ

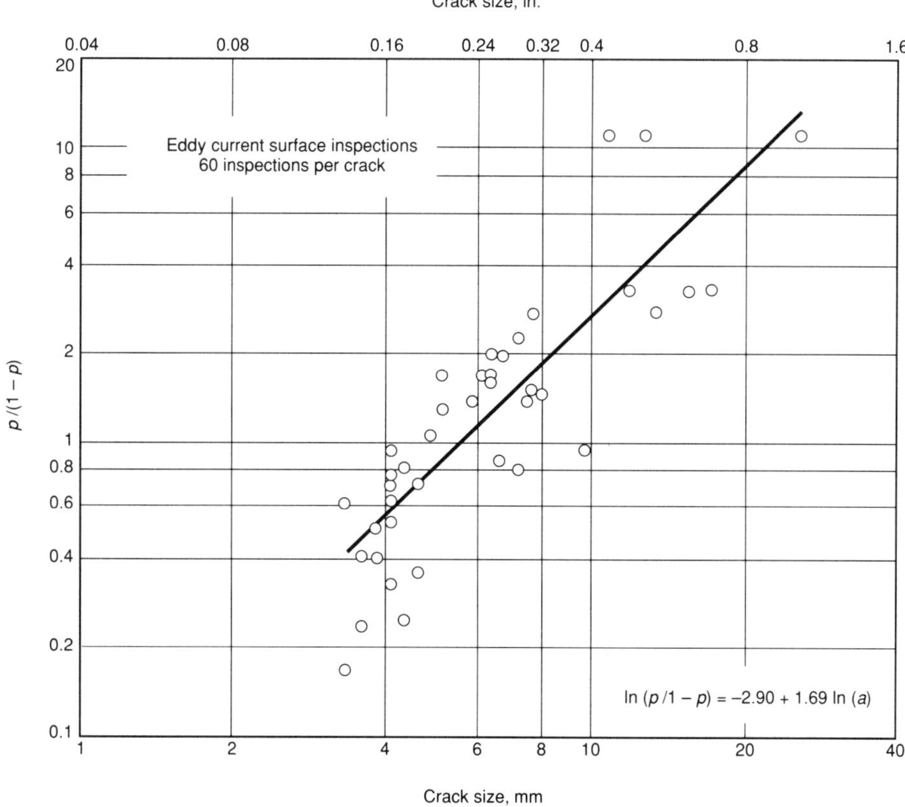

Fig. 3 Example linear relation between log odds of crack detection and log crack size

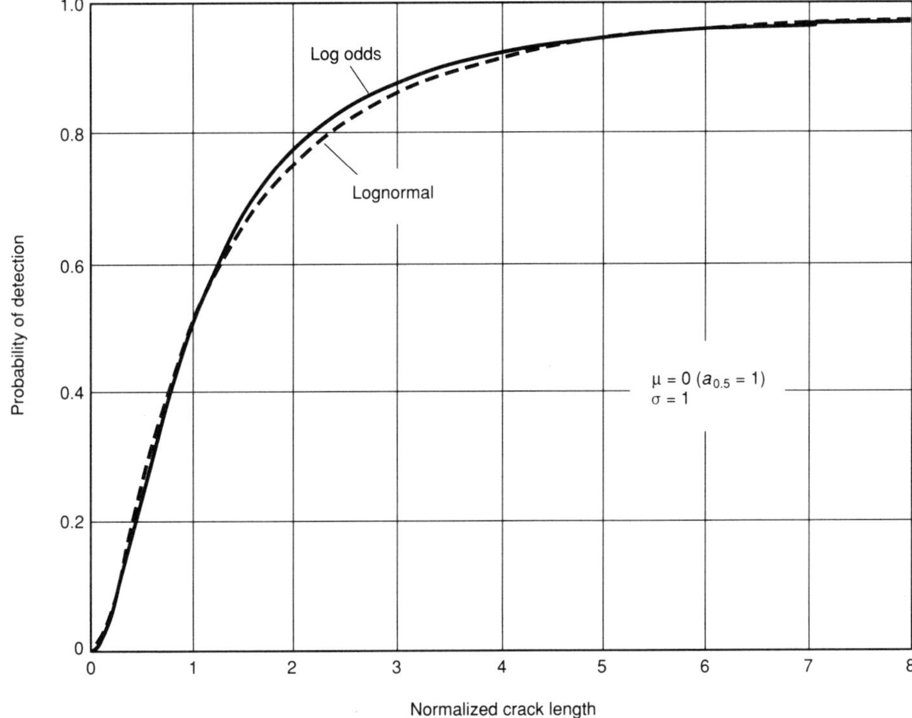

Fig. 4 Comparison of log odds and cumulative log normal models

and σ of Eq 4. Figure 4 compares the log odds and cumulative log normal distribution functions for $\mu = 0$ and $\sigma = 1$. Equation 4 is the form of the log odds model that will be used in the section "Analysis of Hit/Miss Data" in this article.

POD(*a*) From Signal Response Data. The NDE flaw indications are based on interpreting the response to a stimulus. In eddy current or ultrasonic systems, the response might be a peak voltage referenced to a calibration. In fluorescent penetrant inspections, the response would be a combination of brightness and size of the indication. Assume the response can be quantified and recorded in terms of a parameter, $\hat{a}$, that is correlated with flaw size. Then $\hat{a}$ summarizes the information for determining if a positive flaw indication will be given. Only if $\hat{a}$ exceeds a defined decision threshold, $\hat{a}_{dec}$, will a positive indication by given.

As an example of the concept, Table 2 summarizes the results of highly automated eddy current inspections of 28 cracks in flat plate specimens. The three data sets resulted from the use of three probes, with all other factors held constant. The values in Table 2 are the depth of each crack and the peak voltage in counts recorded by the system. Figure 5 shows a plot of the $\hat{a}$ versus a data for probe A. No signal was recorded for 2 of the cracks, because their $\hat{a}$ values were below the recording signal threshold, $\hat{a}_{th}$. These points are indicated by a down arrow at $\hat{a}_{th}$, indicating that the response was at an indeterminable value below the recording signal threshold. Similarly, $\hat{a}$ for 5 of the cracks exceeded the saturation limit, $\hat{a}_{sat}$, of the recording system. These points are indicated by an up arrow at $\hat{a}_{sat}$, indicating that the response was at an indeterminable value above the recording saturation limit. In Fig. 5, the decision threshold is set at 250 counts. Only those cracks whose $\hat{a}$ value is above 250 would have been flagged (detected).

The POD(a) function can be obtained from the relation between $\hat{a}$ and a. If $g_a(\hat{a})$ represents the probability density of the $\hat{a}$ values for fixed crack size a, then:

$$\text{POD}(a) = \int_{\hat{a}_{dec}}^{\infty} g_a(\hat{a})\, d\hat{a} \quad \text{(Eq 7)}$$

This calculation is illustrated in Fig. 6, in which the shaded area under the density functions represents the probability of detection.

In general, the correlating function between $\hat{a}$ and a defines the mean of $g_a(\hat{a})$, that is:

$$\hat{a} = \mu(a) + \delta \quad \text{(Eq 8)}$$

where $\mu(a)$ is the mean of $g_a(\hat{a})$ and δ is a random error term accounting for the differences between $\hat{a}$ and $\mu(a)$. The distributional properties of δ determine the probability density $g_a(\hat{a})$ about $\mu(a)$, as will be shown.

In the $\hat{a}$ data analyzed to date, a linear relation between ln ($\hat{a}$) and ln (a) with normally distributed deviations has proved satisfactory (for example, Fig. 5). This model is expressed by:

$$\ln(\hat{a}) = \beta_0 + \beta_1 \ln(a) + \delta \quad \text{(Eq 9)}$$

where δ is normally distributed with zero mean and constant standard deviation, σ_δ.

Table 2 Example of a summary data sheet of $\hat{a}$ versus a data

The example is based on eddy current inspections of flat plates.

Crack identification	Crack depth, mm	Crack depth, in.	Peak voltage in counts, Probe A	Peak voltage in counts, Probe B	Peak voltage in counts, Probe C
11	0.33	0.013	1052	884	1282
30	1.40	0.055	4095	4095	3831
42	0.38	0.015	1480	1182	1699
2	0.25	0.010	723	624	840
21	0.74	0.029	4095	4095	2249
13	0.48	0.019	2621	2401	1101
19	0.30	0.012	377	809	350
26	0.23	0.009	223	205	277
15	0.56	0.022	1654	3319	1289
29	1.65	0.065	4095	4095	2648
33	0.08	0.003	(a)	(a)	(a)
25	0.25	0.010	669	565	824
32	0.18	0.007	374	379	407
34	0.03	0.001	(a)	(a)	(a)
39	0.18	0.007	409	387	586
12	0.28	0.011	895	690	677
38	0.20	0.008	374	301	549
20	0.79	0.031	4095	4095	1778
28	0.23	0.009	638	454	782
27	0.15	0.006	533	385	631
1	0.08	0.003	150	136	135
35	0.28	0.011	749	660	989
40	0.20	0.008	433	378	591
31	0.36	0.014	879	888	1402
3	0.23	0.009	286	211	352
7	0.23	0.009	298	163	215
16	0.41	0.016	1171	1110	1628
37	2.54	0.100	4095	4095	4095

(a) Peak voltage below the recording level threshold

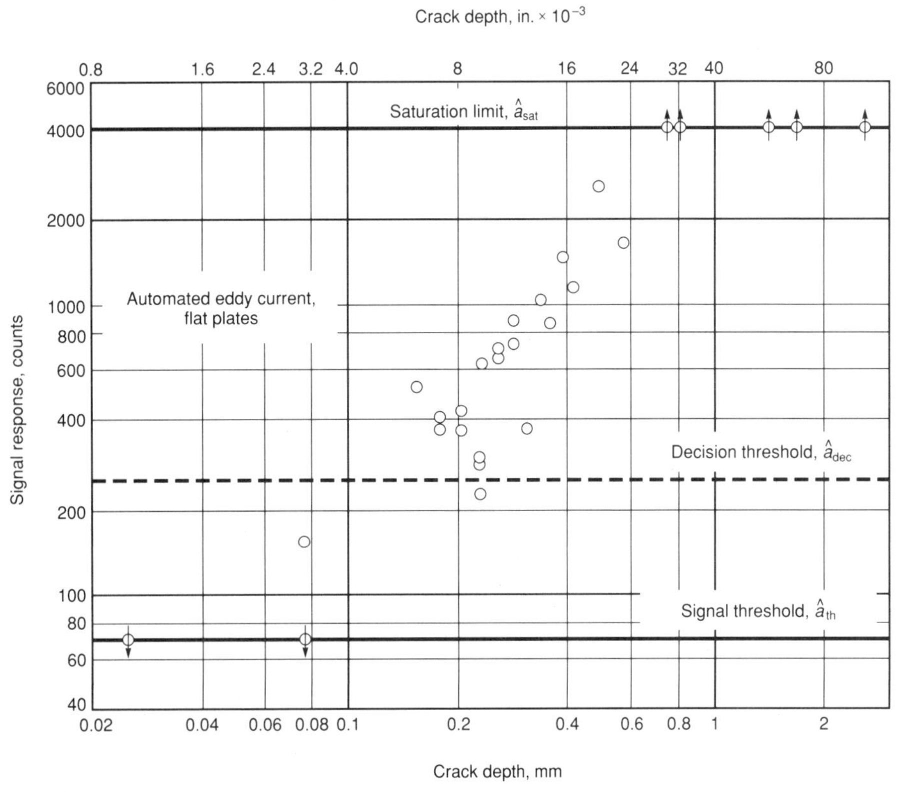

Fig. 5 Example inspection signal response as a function of crack depth

Data have been observed that flatten at the large crack sizes. However, because the decision threshold was far below the nonlinear range, restricting the range of cracks to smaller sizes permitted the application of Eq 9. The normality of δ has proved to be an acceptable assumption.

Assuming that the $\hat{a}$ versus a relation is modeled by Eq 9 and that δ is normally distributed with zero mean and standard deviation of σ_δ, the POD(a) function is calculated as:

$$\text{POD}(a) = \text{Probability } [\hat{a} > \hat{a}_{dec}]$$

$$\text{POD}(a) = \text{Probability } [\ln(\hat{a}) > \ln(\hat{a}_{dec})]$$

$$\text{POD}(a) = 1 - \Phi\left[\frac{\ln(\hat{a}_{dec}) - \beta_0 + \beta_1 \ln(a)}{\sigma_\delta}\right] \quad \text{(Eq 10)}$$

where Φ is the standard normal distribution function. Using the symmetry properties of Φ, Eq 10 can be reduced to:

$$\text{POD}(a) = \Phi\left\{\frac{\ln(a) - [\ln(\hat{a}_{dec}) - \beta_0]/\beta_1}{\sigma_\delta/\beta_1}\right\} \quad \text{(Eq 11)}$$

Equation 11 is a cumulative log normal distribution function with mean and standard deviation of log crack length given by:

$$\mu = \frac{\ln(\hat{a}_{dec}) - \beta_0}{\beta_1} \quad \text{(Eq 12)}$$

$$\sigma = \frac{\sigma_\delta}{\beta_1} \quad \text{(Eq 13)}$$

In the section "Signal Response Analysis" in this article, maximum likelihood methods for estimating β_0, β_1, and σ_δ from $\hat{a}$ versus a data will be presented. Note that the $\hat{a}$ values below the recording threshold and above the saturation limit must be properly accounted for in these analyses. Note also that data from multiple inspections of the same cracks require analysis methods that are dependent on the design of the reliability experiment. Methods for placing lower confidence bounds on the estimated POD(a) function using the sampling distributions of the maximum likelihood estimates of β_0, β_1, and σ_δ are also included in the section "Signal Response Analysis."

Design of NDE Reliability Experiments

An NDE reliability experiment comprises a test matrix of inspections on a set of specimens with known flaw locations and sizes. The specimens are inspected under conditions that simulate as closely as practical the actual application conditions. The experimental design determines the test matrix, and there are four major analysis concerns to be addressed in the experimental design. These are:

- The method of controlling the factors to be evaluated in the experiment
- The method of accounting for the uncontrolled factors in the experiment
- The number of flawed and unflawed inspection sites
- The sizes of the flaws in the specimens

These topics are addressed in the following sections.

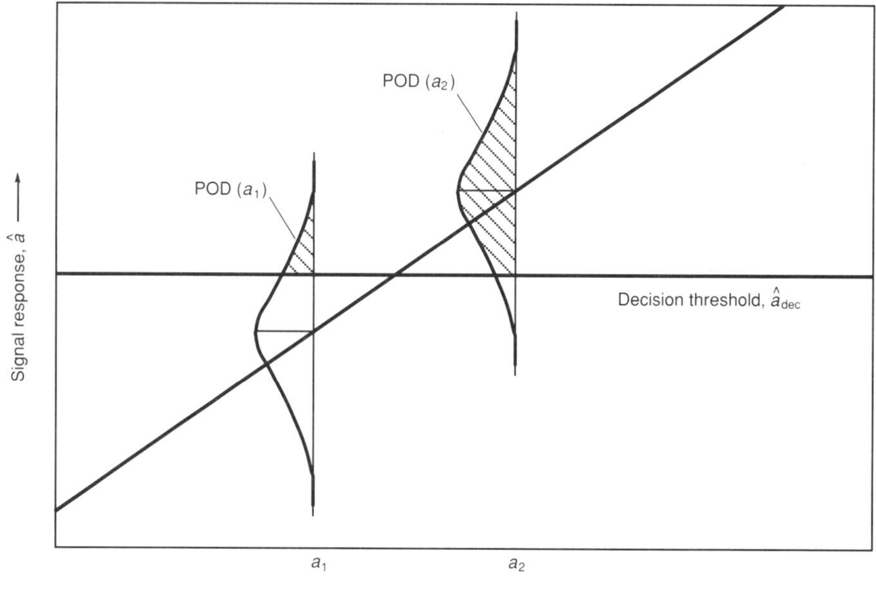

Fig. 6 Schematic of POD(a) calculation from $\hat{a}$ versus a relation

Controlled and Uncontrolled Factors

The primary objective of NDE reliability experiments has been to demonstrate efficacy for a particular application by estimating the POD(a) function and its lower 95% confidence bound. (Although NDE reliability experiments can also be conducted to optimize a system, analyses to meet this objective are beyond the scope of this article.) To demonstrate capability, it is assumed that the protocol for conducting the inspections is well defined for the application, that the inspection process is under control (hit/miss decisions are stable over time), and that all other factors introducing variability into the inspection decision will be representative of the application. The representativeness of these other factors can be ensured either by controlling the factors during the inspection or randomly sampling the factors to be used in the experiment. The methods of accounting for these factors are important aspects of the statistical design of the experiment and significantly influence the statistical properties of the estimates of the POD(a) function parameters. Of particular note in this regard is that k inspections on n flaws is not equivalent to inspections on $n \cdot k$ different flaws, even if the inspections are totally independent.

The most important of the factors introducing variation are:

- Differences in physical properties of cracks of nominally identical sizes
- The basic repeatability of the magnitude of the NDE signal response when a specific crack is independently inspected by a single inspector using the same equipment
- The summation of all the human factors associated with the particular inspectors in the population of interest
- Differences introduced by changes in inspection hardware

These factors must be addressed explicitly or implicitly in every NDE reliability experiment.

In general, the specimens used in NDE reliability experiments are very expensive to obtain and characterize in terms of the sizes of the flaws in the specimens. Therefore, each experiment is based on one set of specimens containing flawed and unflawed inspection sites. Because the results are significantly influenced by the specimens, it must be assumed that the flaws are representative of those that will be present in the structural application. If other factors are to be included in the experiment, they will be based on repeated inspections of the same flaws. From a statistical viewpoint, this restriction on the experimental design limits the sample size to the number of flaws in the specimen set. Because different cracks of the same size can have significantly different crack detection probabilities, multiple inspections of the same crack provide information about the detection probability of only that crack.

The generality of the capability characterization is limited to the application for which the experiment is representative. Either important factors must be typical of the application or random samples must be chosen from the population of interest and repeat inspections performed for these factors. For example, if a single inspector is used to characterize a fluorescent penetrant inspection, it must be assumed that this inspector is typical of all the inspectors in the shop. An alternative might be to choose a random sample of inspectors from the total pool and have each of the selected inspectors perform the experiment.

Depending on the application of the results of the experiment, stratified sampling may be required to obtain a representative sample. For instance, if the capability will apply to two facilities and one of them inspects twice as many components as the other, then that facility should have twice as many inspectors in the experiment. An alternative method is to characterize each facility independently. Care is then required in combining the results for the joint characterization.

Factorial Experiments for Hit/Miss Data. The analysis for the hit/miss data requires that all factors be balanced in any one analysis. When practical, this can be most easily achieved by performing complete factorial experiments. For example, Table 1 contains the results of a two-factor experiment, with the factors being cracks and inspectors. These data can be analyzed as one data set with three inspections per crack. The resulting POD(a) function and its confidence bound would be representative of the population of inspectors from which the sample was drawn.

If the effect of a third factor, for example, different lots of penetrants, were to be included, the entire experiment would be repeated for each of the lots chosen at random from the population of all lots. If three lots were sampled, a total of nine inspections would be performed on each of the flawed specimens, and the resulting POD(a) would apply to the entire inspection process. Suppose, however, that the second and third samples of penetrant were used only by Inspector A. In this case, the two additional sets of inspection data cannot be combined with the other three in a single analysis, because the triple representation of Inspector A would bias the resulting POD(a) function toward his specific capability. The three sets of inspection results for Inspector A can be combined, but the range of applicability of the answer is limited to Inspector A (unless it can be shown or assumed that Inspector A is representative of the entire population).

When many factors must be considered, the number of possible combinations in a factorial experiment can easily become prohibitive. More sophisticated experimental designs (fractional replications, for example) may then be required. In such cases, the assistance of a professional statistician is recommended.

Experimental Design for $\hat{a}$ Data. Inspection-result data in the $\hat{a}$ form contain considerably more information than hit/miss data and, as a consequence, permit more flexibility in the design of the experiment. In $\hat{a}$ analysis, the parameters of the POD(a)

function are estimated from the slope, intercept, and standard deviation of residuals of the ln ($\hat{a}$) versus ln (a) relation, as given by Eq 9, 12, and 13. In Eq 9, δ can be considered to be the sum of random effects, and experiments can be designed to estimate the components of the total variation in δ. For example, operators, probes, and repeatability can be jointly evaluated in a factorial experiment and their effects accounted for in the estimate of POD(a). The statistical model for this experiment would be:

$$\ln(\hat{a}) = \beta_0 + \beta_1 \ln(a) + C_i + O_j + P_k + R_l + \text{(interaction terms)} \qquad \text{(Eq 14)}$$

where C_i, O_j, P_k, and R_l are the random effects due to cracks, operators, probes, and repeats, respectively. The random term, δ of Eq 9, is the sum of all random effects. It can be assumed that the mean and variance of random effect X are zero and σ_x^2, respectively. Then:

$$\sigma_\delta^2 = \sigma_C^2 + \sigma_O^2 + \sigma_P^2 + \sigma_R^2 + \ldots \qquad \text{(Eq 15)}$$

Therefore, β_0 and β_1 can be estimated from a regression analysis, and σ_δ can be estimated from the components of variance using the expected mean squares for the random effects.

In principle, any statistical design from which the components of variance can be estimated can be used in an $\hat{a}$ NDE reliability experiment. However, the analysis methods would be specific to the particular design, and it is beyond the scope of this article to address the general problem. In the section "Signal Response Analysis" in this article, it will be assumed that only the variation due to cracks and one other factor is being investigated. It is recommended that the assistance of a qualified statistician be obtained for more sophisticated experimental designs.

Sample Sizes and Flaw Sizes

Sample sizes in NDE reliability experiments are driven more by the economics of specimen fabrication and characterization than by the desired degree of precision in the estimate of the POD(a) function. Although apparently reasonable POD(a) functions can often be obtained from applying the maximum likelihood analysis to relatively few test results, the confidence bound calculation is based on asymptotic (large sample) properties of the estimates. It should be emphasized that the calculations can also produce totally unacceptable results from the relatively few test results or from data that are not reasonably represented by the assumptions of the models. Therefore, there are minimal sample size requirements that must be met to provide a degree of reasonable assurance in the characterization of the capability of the system.

Larger sample sizes in NDE reliability experiments will, in general, provide greater precision in the estimate of the POD(a) function. However, the sample size is determined from the number of cracks in the experiment, and there is a coupling with the flaw sizes that must also be considered. The effect of this coupling manifests itself differently for the hit/miss and $\hat{a}$ analyses.

Sample Size Requirements for Hit/Miss Analysis. Data from hit/miss experiments are generally not amenable to testing assumptions regarding the form of the POD(a) model. These tests require either large numbers of independent inspections on each flaw of a specimen set or inspection results from an extremely large number of compatible specimens (Ref 3, 4). Number and size considerations in hit/miss experiments are directed at their effect on the sampling properties of the parameters of the POD(a) function (Ref 9).

In the hit/miss analysis, the output of an inspection states only whether or not a crack of known length was found in the inspection (Table 1). There are probabilities associated with the outcomes, and the analysis assumes that this probability increases with flaw size. Because it has been assumed that the inspection process is in a state of control, there is a range of flaw sizes over which the POD(a) function is rising. In this flaw size range of uncertainty, the inspection system has limited discriminating power in the sense that detecting or failing to detect would not be unusual. Such a range might be defined by the interval ($a_{0.10}$, $a_{0.90}$), where a_p denotes the flaw size that has probability of detection equal to p; that is:

$$\text{POD}(a_p) = p \qquad \text{(Eq 16)}$$

Flaws smaller than $a_{0.10}$ would then be expected to be missed, and flaws greater than $a_{0.90}$ would be expected to be detected.

In a hit/miss reliability experiment, flaws outside the range of uncertainty do not provide as much information concerning the POD(a) function as cracks within this range. Cracks in the almost-certain detection range and almost-certain miss range provide very little information concerning probability of detection. Therefore, in the hit/miss experiment, not all flaws convey the same amount of information, and the effective sample size is not necessarily the total number of flaws in the experiment. Adding a large number of very large flaws does not increase the precision in the estimate of the parameters of the POD(a) function.

In a reliability experiment, the location of the increasing range of the POD(a) function is not known. Further, the same sets of specimens are often used in many different experiments. Therefore, it is not possible to fabricate a set of specimens with optimal flaw sizes for a particular experiment. To minimize the chances of completely missing the crack size range of maximum information and to accommodate the multiple uses of specimens, flaw sizes should be uniformly distributed between the minimum and maximum of the sizes of potential interest. A minimum of 60 flaws should be distributed in this range, but as many as economically possible should be used.

Sample Size Requirements for $\hat{a}$ Analysis. The recorded signal response, $\hat{a}$, provides significantly more information for analysis. In particular, the POD(a) model is derived from the correlation of the $\hat{a}$ versus a data, and the assumptions concerning the POD(a) model can be tested using the signal response data. Further, the pattern of $\hat{a}$ responses can indicate an acceptable range of extrapolation. Therefore, the range of crack sizes in the experiment is not as critical in an $\hat{a}$ analysis as in a hit/miss analysis. For example, if the decision threshold in Fig. 5 were set at 250 counts, all but four of the cracks would have been detected. The larger cracks would have provided little information about the POD(a) function in a hit/miss analysis. In an $\hat{a}$ analysis, however, all of the recorded $\hat{a}$ values provided full information concerning the relation between signal response and crack size, and the values at the signal threshold and saturation limit provided partial information. The linearity of the fit, the normality of the deviations, and the constancy of the residual variation can all be easily evaluated from the $\hat{a}$ versus a plot.

Because of the added information in the $\hat{a}$ data, it is recommended that at least 30 flaws be present in experiments whose results can be recorded in this form. Increasing the number of flaws increases the precision of the estimates, so the test set should contain as many flawed specimens as economically feasible.

Unflawed Inspection Sites. In the context of the analyses presented in this section, sample size refers to the number of known flaws in the specimens to be inspected. The total specimen set should also contain at least twice this number of unflawed inspection sites. The unflawed sites are necessary to ensure that the NDE procedure is discriminating between flawed and unflawed sites and to provide an estimate of the false call rate.

Although the false call rate can have important economic consequences, the NDE reliability analyses in this section were dictated by the requirements of damage tolerance analyses. The primary objective was to estimate the chances of missing flaws that might lead to structural failures. The concepts of these NDE reliability analyses can be generalized to include a nonzero probability of a flaw indication when no flaw is present at an inspection site, that is, POD(a = 0) > 0.

Maximum Likelihood Analysis

Parameter estimation based on the principle of maximizing the likelihood of an observed sample of data is a standard statistical technique and is amply described in the literature (Ref 10, 11). The purpose of this section is to summarize the method and its asymptotic sampling distribution properties in the context of analyzing NDE reliability data. Further, a method for using this information to calculate lower confidence bounds on the POD(a) function is also presented.

Parameter Estimation. Let X_i represent the outcome of the ith inspection and $f(X_i;\boldsymbol{\theta})$ represent the probability of obtaining X_i, where $\boldsymbol{\theta} = (\theta_1,\theta_2,\ldots\theta_k)'$ is the vector of the k parameters in the probability model. For example, in a hit/miss experiment, X_i would be 0 or 1 with probability defined by Eq 4, where a is the size of flaw i and $\boldsymbol{\theta} = (\mu,\sigma)'$. In an $\hat{a}$ versus a experiment, X_i is the log of the signal response, and $f(X_i;\boldsymbol{\theta})$ is a normal density function with mean and standard deviation given by $\boldsymbol{\theta} = (\beta_0 + \beta_1 \ln a, \sigma_\delta)'$, as defined in Eq 9. Let $X_1,\ldots,X_n$ represent the results of independent inspections of n flaws. The likelihood, L, of a specific result is given by the likelihood function:

$$L(\boldsymbol{\theta}) = \prod_{i=1}^{i=n} f(X_i;\boldsymbol{\theta}) \qquad \text{(Eq 17)}$$

For a given outcome of the experiment, X_i is known and Eq 17 is a function of $\boldsymbol{\theta}$. The maximum likelihood estimate is the value, $\hat{\boldsymbol{\theta}}$, which maximizes $L(\boldsymbol{\theta})$. For the models considered here, it is more convenient to work with the log $L(\boldsymbol{\theta})$:

$$\log L(\boldsymbol{\theta}) = \Sigma \log f(X_i;\boldsymbol{\theta}) \qquad \text{(Eq 18)}$$

which is also maximized at $\hat{\boldsymbol{\theta}}$. The maximum likelihood estimates are given by the solution of the k simultaneous equations:

$$\frac{\partial \log L(\boldsymbol{\theta})}{\partial \theta_i} = 0,\ i = 1, \ldots, k \qquad \text{(Eq 19)}$$

Asymptotic Sampling Distribution Properties. For the models being used in NDE reliability studies, the maximum likelihood estimates are invariant, consistent, and efficient. Further, they are asymptotically joint normally distributed with means given by the true parameter values, θ_i, and the variance-covariance matrix defined by:

$$V = I^{-1} \qquad \text{(Eq 20)}$$

where I is the information matrix whose elements I_{ij} are the expected (E) values:

$$I_{ij} = -\mathrm{E}\left[\frac{\partial^2}{\partial\theta_i\partial\theta_j}\log f(X_i;\boldsymbol{\theta})\right],\ i, j = 1, \ldots, k \qquad \text{(Eq 21)}$$

In application, the maximum likelihood estimate, $\hat{\boldsymbol{\theta}}$, is substituted for $\boldsymbol{\theta}$ in Eq 21. Therefore, given the results of inspecting a large number of flaws and a specific function for the POD(a) model, the parameters of the model can be estimated, and the

Table 3 Values of γ for lower confidence bounds on the POD(a) function

Sample size	Confidence level, % 90	95	99
20	3.903	5.243	8.401
25	3.884	5.222	8.376
30	3.871	5.208	8.359
40	3.855	5.191	8.338
50	3.846	5.180	8.325
60	3.839	5.173	8.317
80	3.831	5.165	8.306
100	3.827	5.159	8.300
∞	3.808	5.138	8.273

Source: Ref 13

sampling distribution of the parameters will be joint normal with the known variance-covariance matrix. Examples of these equations for the hit/miss and response signal models are given in the sections "Analysis of Hit/Miss Data" and "Signal Response Data," respectively, in this article. In these applications, the assumed models will be the log odds and cumulative log normal distribution functions. However, other models can also be used if evidence is available to support their selection.

Confidence Bounds on the POD(a) Function. Because the POD(a) function is equivalent to a cumulative distribution function and the parameters are being estimated by maximum likelihood, a procedure developed by Cheng and Iles (Ref 12 and 13) can be used to place lower confidence bounds on the POD(a) function. Such bounds are calculated from the variance-covariance matrix of the estimates and reflect the sensitivity of the experiment to both the number and sizes of flaws in the specimens of the experiment.

The assumed POD(a) model is a cumulative log normal distribution function with parameters $\boldsymbol{\theta} = (\mu,\sigma)'$. For distribution functions defined by location and scale parameters (as is the case of the log normal distribution), the information matrix can be written in the form:

$$I(\mu,\sigma) = \frac{n}{\sigma^2}\begin{bmatrix} k_0 & -k_1 \\ -k_1 & k_2 \end{bmatrix} \qquad \text{(Eq 22)}$$

where n is the number of cracks in the experiment. The lower one-sided confidence bound of the POD(a) function is given by:

$$\mathrm{POD}_\alpha(a) = \Phi(\hat{z} - h) \qquad \text{(Eq 23)}$$

where $\Phi(z)$ is the standard cumulative normal distribution, and:

$$\hat{z} = \frac{\ln(a) - \hat{\mu}}{\hat{\sigma}} \qquad \text{(Eq 24)}$$

$$h = \left\{\frac{\gamma}{n\,k_0}\left[1 + \frac{(k_0\hat{z} + k_1)^2}{(k_0k_2 - k_1^2)}\right]\right\}^{0.5} \qquad \text{(Eq 25)}$$

where γ is obtained from Table 3 for the number of cracks in the experiment and the desired confidence level.

Analysis of Hit/Miss Data

Estimation of the parameters of the log odds model for hit/miss data is based directly on the probability of each 0 or 1 result of an inspection. Assume that a balanced experiment has produced k inspections on each of n cracks. For this application, the likelihood function is given by:

$$L(\mu,\sigma) = \prod_{i=1}^{n}\prod_{j=1}^{k} p_i^{Z_{ij}}(1 - p_i)^{1-Z_{ij}} \qquad \text{(Eq 26)}$$

where $Z_{ij} = 0$ or 1 for the jth inspection of the ith flaw producing a miss or a find, respectively, and the probability of detecting a flaw of size a_i is given by:

$$p_i = \frac{h(a)}{1 + h(a)} \qquad \text{(Eq 27)}$$

$$h(a) = \exp\left[\frac{\pi}{\sqrt{3}}\left(\frac{\ln(a_i) - \mu}{\sigma}\right)\right] \qquad \text{(Eq 28)}$$

This form of the POD(a) function is simply a more convenient algebraic form of Eq 4. The vector of parameters to be estimated is defined by $\boldsymbol{\theta} = (\mu,\sigma)'$.

The log likelihood equation is:

$$\ln L(\mu,\sigma) = \sum_i\sum_j [Z_{ij}\ln(p_i) + (1 - Z_{ij})\ln(1 - p_i)] \qquad \text{(Eq 29)}$$

Parameter Estimation in Hit/Miss Analysis. The maximum likelihood estimates are given by the solution to:

$$0 = \frac{\partial \ln L}{\partial\mu} = \sum_i\sum_j\left[\frac{Z_{ij}}{p_i}\frac{\partial p_i}{\partial\mu} + \frac{(1 - Z_{ij})}{(1 - p_i)}\left(\frac{\partial p_i}{\partial\mu}\right)\right] \qquad \text{(Eq 30)}$$

$$0 = \frac{\partial \ln L}{\partial\sigma} = \sum_i\sum_j\left[\frac{Z_{ij}}{p_i}\frac{\partial p_i}{\partial\sigma} + \frac{(1 - Z_{ij})}{(1 - p_i)}\left(\frac{\partial p_i}{\partial\sigma}\right)\right] \qquad \text{(Eq 31)}$$

Taking the derivatives and simplifying yields:

$$0 = \sum_i\sum_j\frac{Z_{ij}}{k} - \sum_i\frac{h(a_i)}{1 + h(a_i)} \qquad \text{(Eq 32)}$$

$$0 = \sum_i\sum_j\frac{Z_{ij}}{k}\ln(a_i) - \sum_i\frac{\ln(a_i)\,h(a_i)}{1 + h(a_i)} \qquad \text{(Eq 33)}$$

Any standard computational method, such as the Newton-Rhapson iterative procedure (Ref 14), can be used to find the solutions to Eq 32 and 33.

Because iterative techniques converge to local maxima, the solution to Eq 32 and 33 may be sensitive to the initial values. A set of initial values based on the method of moments has been found to be useful (Ref 7). These are given by:

$$\mu_1 = X_n - \frac{X_1 p_1}{2} - (0.5)\sum_{i=2}^{n}(X_i - X_{i-1})(p_i - p_{i-1}) \qquad \text{(Eq 34)}$$

$$\sigma = \{X_n^2 - \frac{p_1}{2} X_1^2 - \mu_1^2 - (0.5) \sum_{i=2}^{n} (X_i^2 - X_{i-1}^2)(p_i - p_{i-1})\}^{0.5} \quad \text{(Eq 35)}$$

where $X_1,...,X_n$ are the ordered values of the natural logs of the flaw sizes and p_i is the observed percentage of detections of the ith ordered flaw size. If convergence is not obtained, increasing the initial estimate of σ has often provided convergence. However, Eq 32 and 33 are not always solvable. This will be discussed further in the section "Comments on Hit/Miss Analysis" in this article.

Confidence Bound Calculation in Hit/Miss Analysis. The information matrix is estimated from Eq 21, using $\hat{\mu}$ and $\hat{\sigma}$ for μ and σ. For this POD(a) model, the elements of the information matrix are given by:

$$I_{11} = \frac{1}{\sigma^2}\left[\frac{\pi^2}{3}\right] \sum_i \frac{h(a_i)}{[1 + h(a_i)]^2} - \frac{n\, k_0}{\sigma^2} \quad \text{(Eq 36)}$$

$$I_{12} = I_{21} = \frac{1}{\sigma^2}\left[\frac{\pi^2}{3\sigma}\right] \sum_i \frac{\ln(a_i)\, h(a_i)}{[1 + h(a_i)]^2} = \frac{n\, k_1}{\sigma^2} \quad \text{(Eq 37)}$$

$$I_{22} = \frac{1}{\sigma^2}\left[\frac{\pi^2}{3\sigma^2}\right] \sum_i \frac{(\ln a_i)^2 h(a_i)}{[1 + h(a_i)]^2} = \frac{-n\, k_2}{\sigma^2} \quad \text{(Eq 38)}$$

Note that k_0, k_1, and k_2, the parameters required in the calculation of the lower confidence bound on the POD(a) function, are also defined by Eq 36 to 38. All of the parameters required by Eq 23 to 25 to calculate the lower confidence bound on the POD(a) function are available.

Hit/Miss Analysis Examples. As examples of the application of the method to real data, the parameters of the log odds POD(a) function were obtained for the data in Table 1. Table 4 presents a summary of the parameters of the POD(a) function for each data set of Table 1 and the combination of the three data sets in a single analysis. Figure 7 shows the POD(a) function and the lower 95% confidence bound for Inspector A and the same information when the data from the three inspectors are combined. Adding inspections of the same cracks did not increase the precision of the estimate of the POD(a) function. Figure 8 compares the POD(a) functions for the three inspectors and the composite.

Comments on Hit/Miss Analysis. In a well-designed experiment of sufficient sample size for which the log odds model is a reasonable representation of the POD(a) function, the maximum likelihood hit/miss analysis will provide a valid solution. Conversely, lacking any of these elements, it is possible that either no solution or an unacceptable solution can result. If there is no overlap in the flaw size ranges of the detections and misses, Eq 32 and 33 will not yield a solution. More flaws are needed in the region of increase of the POD(a) function. It is also possible to obtain an estimate of a POD(a) function that decreases with flaw size if the inspection system is poorly designed or not in control and if large flaws tend to be missed more often than small flaws. Both of these types of results are readily apparent, albeit disconcerting.

A third type of unacceptable result is an apparently acceptable POD(a) function but a confidence bound that eventually decreases with flaw size. This situation is most easily understood in terms of the log odds versus log flaw size plot. If the slope is positive, the POD(a) function will appear reasonable, but if it is not significantly greater than zero, the lower confidence bound will eventually decrease with flaw size. Therefore, a decreasing confidence bound is evidence of lack of fit of the log odds model.

Finally, lack of fit of the model is often manifest in large values of σ coupled with small values of μ or extremely wide confidence intervals. Although there are, in general, insufficient data in hit/miss experiments to test hypotheses about the POD(a) model, as a minimum each fit should be subjectively judged. For example, in Fig. 9, the observed detection proportions of each crack in the data of Table 1 are superimposed on the composite POD(a) function and confidence limit from Fig. 7. The uncertainty in the POD(a) function as indicated by the width of the confidence bound seems justified by the plot of the raw data. In this example, if greater precision (narrower confidence bounds) were desired, more cracks in the 2 to 8 mm (0.08 to 0.3 in.) range would be needed in the experiment. Such plots provide an indication of the fit of the model to the data as well as the range of flaw sizes that are contributing to the information from which the POD(a) function is

Table 4 POD(a) parameters for the hit/miss data in Table 1

Parameter	Inspector A	Inspector B	Inspector C	Composite
$\hat{\mu}$	0.96	1.11	0.82	0.96
$\hat{\sigma}$	0.59	1.04	0.87	0.88
a_{50}, mm (in.)(a)	2.62 (0.103)	3.03 (0.119)	2.27 (0.089)	2.61 (0.103)
a_{90}, mm (in.)(b)	5.34 (0.210)	10.6 (0.417)	6.54 (0.257)	7.18 (0.283)
$a_{90/95}$, mm (in.)(c)	21.6 (0.850)	232 (9.13)	38.8 (1.53)	51.0 (2.01)

(a) $a_{50} = \exp(\hat{\mu})$ = estimate of crack size at 50% POD. (b) $a_{90} = \exp(\hat{\mu} + 1.282\,\hat{\sigma})$ = estimate of crack size at 90% POD. (c) $a_{90/95}$ = upper 95% confidence bound on the estimate of a_{90}.

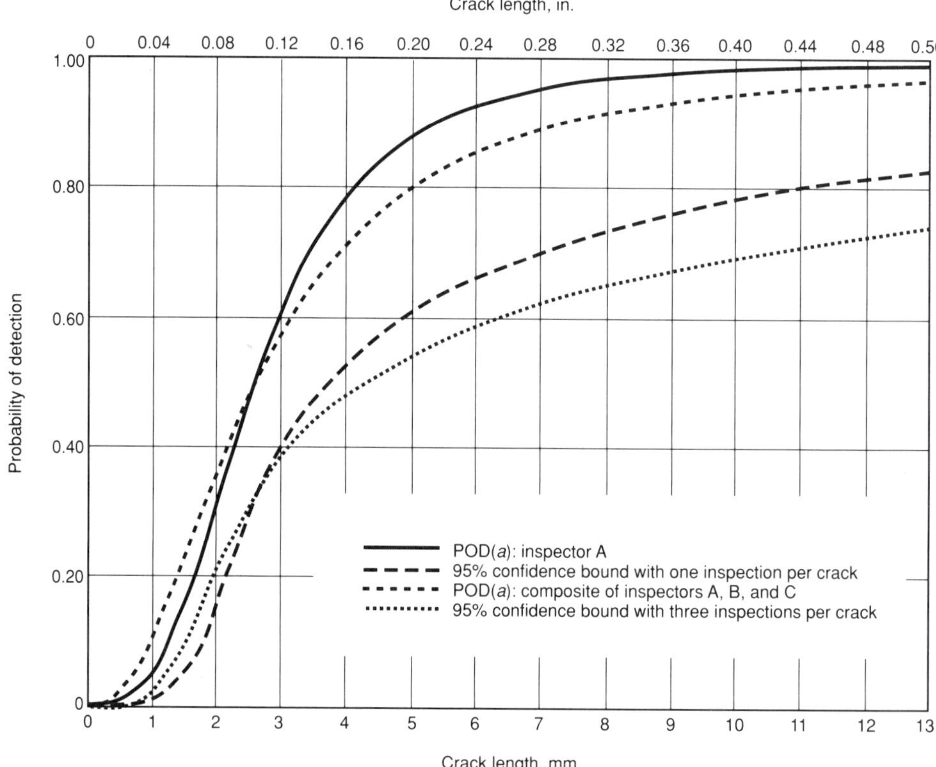

Fig. 7 POD(a) function and lower 95% confidence bound from hit/miss analysis of the data in Table 1 for one inspection per crack (from Inspector A) and for three inspections per crack (from the composite result of Inspectors A, B, and C)

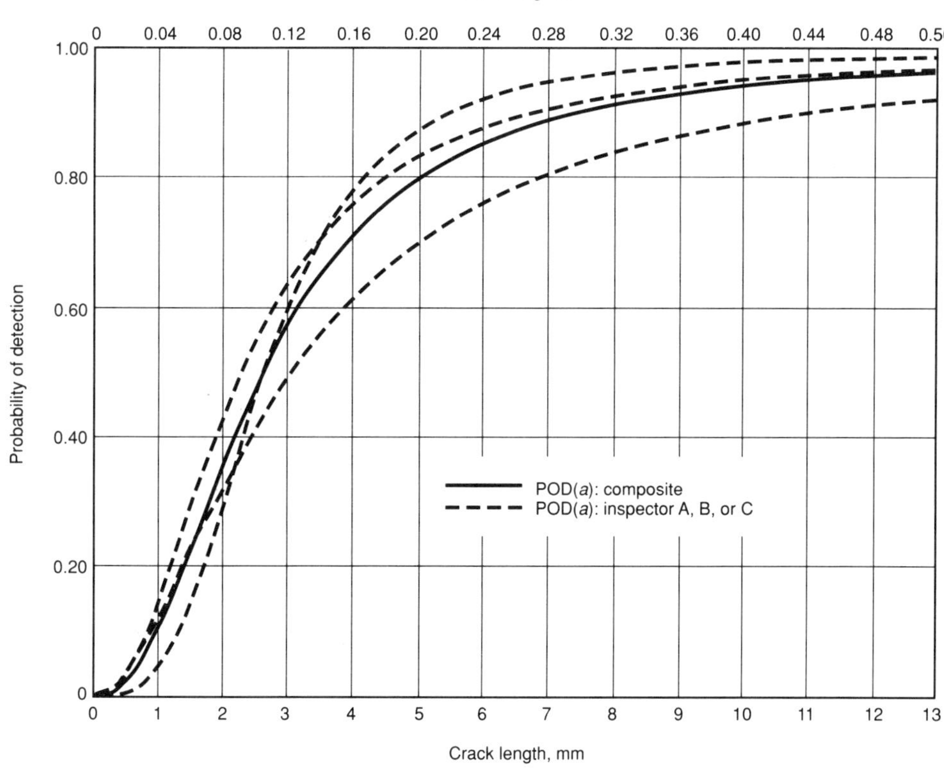

Fig. 8 POD(*a*) functions from hit/miss analysis of the data in Table 1

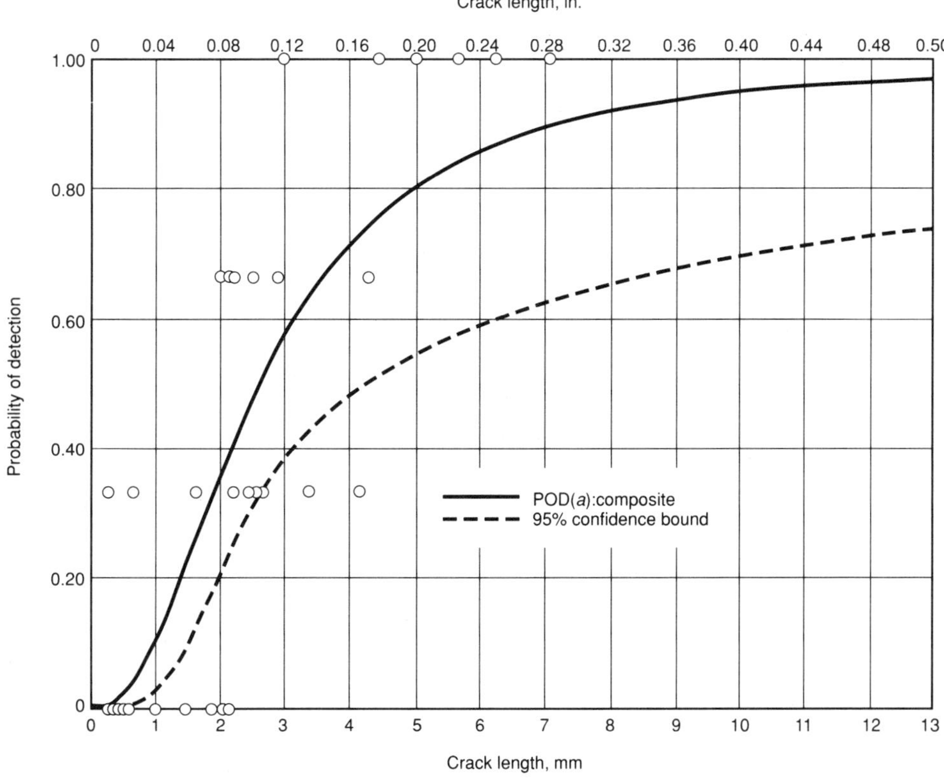

Fig. 9 Example fit of hit/miss POD(*a*) function and lower 95% confidence bound to observed detection probabilities (three inspections per crack)

being estimated. This is true even for experiments in which there is only one inspection per crack and all detection probabilities are plotted at 0 or 1.

Signal Response Analysis

In signal response data analysis, the parameters of the POD(a) function are calculated from parameters of the $\hat{a}$ versus a relation. If all the $\hat{a}$ values are between the signal recording threshold and the saturation limit, a simple regression analysis of ln ($\hat{a}$) versus ln (a) will yield the necessary information to estimate the POD(a) function and its lower confidence bound. In fact, the least squares estimates from the regression analysis also happen to be the maximum likelihood estimates. The analysis presented in this section is applicable to the more general case in which some of the $\hat{a}$ values are censored at the recording threshold or the saturation limit. This more general analysis will give answers identical to those of the regression analysis if all $\hat{a}$ values are available for all of the flaws (Ref 15).

In the response signal analysis, it is assumed that the $\hat{a}$ values for a flaw of size a have a normal distribution, with mean and standard deviation given by:

$$\mu_{\ln(\hat{a})} = \beta_0 + \beta_1 \ln(a) \quad \text{(Eq 39)}$$

$$\sigma_{\ln(\hat{a})} = \sigma_\delta \quad \text{(Eq 40)}$$

where σ_δ does not depend on the crack size. To simplify the notation, let $Y_i = \ln(\hat{a}_i)$ and $X_i = \ln(a_i)$. The random variable:

$$Z = \frac{Y - (\beta_0 + \beta_1 X)}{\sigma_\delta} \quad \text{(Eq 41)}$$

has a standard normal distribution. Let $\phi(z)$ represent the density function of the standard normal distribution:

$$\phi(z) = \frac{1}{\sqrt{2\pi}} \exp\left[-\left(\frac{z^2}{2}\right)\right] \quad \text{(Eq 42)}$$

and $\Phi(z)$ represent the cumulative normal distribution:

$$\Phi(z) = \int_{-\infty}^{z} \phi(\xi)\, d\xi \quad \text{(Eq 43)}$$

The likelihood function is partitioned into three regions:

- Region R, for which $\hat{a}$ values were recorded
- Region T, for which only a maximum value is known (the $\hat{a}$ values fall below the recording signal threshold and cannot be recorded)
- Region S, for which only a minimum $\hat{a}$ value is known (the $\hat{a}$ values fall above the saturation limit and cannot be recorded)

These regions are identified by the open circles, the down arrows, and the up arrows, respectively, in Fig. 5. The likelihood function for the entire sample is the product of the likelihood functions for the three regions:

$$L(\beta_0, \beta_1, \sigma_\delta) = \prod_R L_R \prod_T L_T \prod_S L_S \quad \text{(Eq 44)}$$

But (suppressing the dependency of L on β_0, β_1, and σ_δ):

$$L_R = \prod_{i=1}^{r} \frac{1}{\sigma} \phi(Z_i) \quad \text{(Eq 45)}$$

$$L_T = \prod_{i=1}^{t} \Phi_i(a_{th}) \quad \text{(Eq 46)}$$

$$L_S = \prod_{i=1}^{s} [1 - \Phi_i(a_{sat})] \quad \text{(Eq 47)}$$

because $1/\sigma\ \phi(Z_i)\ dz$ is the probability of observing $\hat{a}_i$ for the ith flaw in R, $\Phi_i(a_{th})$ is the probability of obtaining an ln $\hat{a}_i$ value below the recording threshold for the ith flaw in S, and $1 - \Phi_i(a_{sat})$ is the probability of obtaining an $\hat{a}$ value above the saturation limit for the ith flaw in T. The log of the likelihood function is:

$$\ln [L(\beta_0, \beta_1, \sigma_\delta)] = -r \ln(\sigma) - \frac{1}{2\sigma^2} \sum_R [Y_i - (\beta_0 + \beta_1 X_i)]^2 + \sum_T \ln \Phi_i(a_{th}) + \sum_S \ln [1 - \Phi_i(a_{sat})] \quad \text{(Eq 48)}$$

where r is the number of cracks in R, that is, the number of cracks for which $\hat{a}$ values were recorded.

Parameter Estimation in $\hat{a}$ Analysis. The maximum likelihood estimates are given by the solutions to:

$$0 = \frac{\partial \ln(L)}{\partial \beta_0} = \frac{1}{\sigma}\left[\sum_R Z_i + \sum_S V(Z_i) - \sum_T W(Z_i)\right] \quad \text{(Eq 49)}$$

$$0 = \frac{\partial \ln(L)}{\partial \beta_1} = \frac{1}{\sigma}\left[\sum_R Z_i X_i + \sum_S X_i V(Z_i) - \sum_T X_i W(Z_i)\right] \quad \text{(Eq 50)}$$

$$0 = \frac{\partial \ln(L)}{\partial \sigma} = \frac{1}{\sigma}\left[-r + \sum_R Z_i^2 + \sum_S Z_i V(Z_i) - \sum_T Z_i W(Z_i)\right] \quad \text{(Eq 51)}$$

where:

$$V(Z_i) = \frac{\phi(Z_i)}{1 - \Phi(Z_i)} \quad \text{(Eq 52)}$$

$$W(Z_i) = \frac{\phi(Z_i)}{\Phi(Z_i)} \quad \text{(Eq 53)}$$

Standard numerical methods, such as the Newton-Rhapson iterative procedure (Ref 14), can be used to find the solutions to Eq 49 to 51. Excellent choices for the initial estimates of iterative procedures are the intercept, slope, and standard deviation of residuals obtained from a standard regression analysis of only those $\hat{a}$ values for which a valid response was recorded.

Confidence Bound Calculation in $\hat{a}$ Analysis. Because the POD(a) parameters are calculated from the estimates of the $\hat{a}$ versus a data, the calculation of the lower confidence bound is a five-step process:

- The information matrix for the estimates of β_0, β_1, and σ_δ is obtained using Eq 21
- The variance-covariance matrix of $\hat{\beta}_0$, $\hat{\beta}_1$, and $\hat{\sigma}_\delta$ is obtained by inverting the information matrix (Eq 20)
- The variance-covariance matrix of the estimates of μ and σ are calculated based on a first-order Taylor series expansion of the equations relating β_0, β_1, and σ_δ to μ and σ (Eq 12 and 13)
- The information matrix for $\hat{\mu}$ and $\hat{\sigma}$ is obtained by inverting the variance-covariance matrix to obtain Eq 22
- The calculated values are substituted into Eq 23 to 25 to obtain the lower confidence bound

The elements of the information matrix for $\hat{\theta} = (\beta_0, \beta_1, \sigma_\delta)$ are given by (dropping the subscripts):

$$I_{11} = \frac{-\partial^2 \log(L)}{\partial \beta_0^2} = \frac{1}{\sigma^2}\left[r + \sum_S \lambda(Z) - \sum_T \psi(Z)\right] \quad \text{(Eq 54)}$$

$$I_{12} = I_{21} = \frac{-\partial^2 \log(L)}{\partial \beta_0 \partial \beta_1} = \frac{1}{\sigma^2}\left[\sum_R X + \sum_S X\lambda(Z) - \sum_T X\psi(Z)\right] \quad \text{(Eq 55)}$$

$$I_{13} = I_{31} = \frac{-\partial^2 \log(L)}{\partial \beta_0 \partial \sigma_\delta} = \frac{1}{\sigma^2}\left[2\sum_R Z + \sum_S V(Z) + \sum_S Z\lambda(Z) - \sum_T W(Z) - \sum_T Z\psi(Z)\right] \quad \text{(Eq 56)}$$

$$I_{22} = \frac{-\partial^2 \log(L)}{\partial \beta_1^2} = \frac{1}{\sigma^2}\left[\sum_R X^2 + \sum_S X^2\lambda(Z) - \sum_T X^2\psi(Z)\right] \quad \text{(Eq 57)}$$

$$I_{23} = I_{32} = \frac{-\partial^2 \log(L)}{\partial \beta_1 \partial \sigma_\delta} = \frac{1}{\sigma^2}\left[2\sum_R XZ + \sum_S XV(Z) + \sum_S XZ\lambda(Z) - \sum_T XW(Z) - \sum_T XZ\psi(Z)\right] \quad \text{(Eq 58)}$$

$$I_{33} = \frac{-\partial^2 \log(L)}{\partial \sigma_\delta^2} = \frac{1}{\sigma^2}\left[-r + 3\sum_R Z^2 + 2\sum_S ZV(Z) + \sum_S Z^2\lambda(Z) - 2\sum_T ZW(Z) - \sum_T Z^2\psi(Z)\right] \quad \text{(Eq 59)}$$

$$\lambda(Z) = V(Z)\,[V(Z) - Z] \quad \text{(Eq 60)}$$

$$\psi(Z) = -W(Z)\,[W(Z) + Z] \quad \text{(Eq 61)}$$

Let $V(\hat{\beta}_0, \hat{\beta}_1, \hat{\sigma}_\delta)$ represent the variance-covariance matrix of the maximum likelihood estimates of the ln ($\hat{a}$) versus ln (a) analysis. The value $V(\hat{\beta}_0, \hat{\beta}_1, \hat{\sigma}_\delta)$ is obtained from the inverse of the information matrix. Let the elements of $V(\hat{\beta}_0, \hat{\beta}_1, \hat{\sigma}_\delta)$ be defined by:

$$V(\hat{\beta}_0,\hat{\beta}_1,\hat{\sigma}_\delta) = \begin{bmatrix} \sigma_0^2 & \sigma_{01} & \sigma_{0\delta} \\ \sigma_{01} & \sigma_1^2 & \sigma_{1\delta} \\ \sigma_{0\delta} & \sigma_{1\delta} & \sigma_\delta^2 \end{bmatrix} \quad \text{(Eq 62)}$$

Using a Taylor series expansion about the true values of μ and σ to linearize the relations expressed by Eq 12 and 13, the variance-covariance matrix of $\hat{\mu}$ and $\hat{\sigma}$ is given by (Ref 16):

$$V(\mu,\sigma) = \frac{1}{\beta_1^2} T\, V(\hat{\beta}_0,\hat{\beta}_1,\hat{\sigma}_\delta)\, T \quad \text{(Eq 63)}$$

and the transformation matrix T is defined by:

$$T = \begin{bmatrix} 1 & \mu & 0 \\ 0 & \sigma & -1 \end{bmatrix} \quad \text{(Eq 64)}$$

Multiplying the matrices yields the variances and covariance of $\hat{\mu}$ and $\hat{\sigma}$ as:

$$\sigma_{\hat{\mu}}^2 = \frac{1}{\beta_1^2}[\sigma_0^2 + 2\hat{\mu}\sigma_{01} + \hat{\mu}^2\sigma_1^2] \quad \text{(Eq 65)}$$

$$\sigma_{\hat{\sigma}\hat{\mu}} = \sigma_{\hat{\mu}\hat{\sigma}} = \frac{1}{\beta_1^2}[\hat{\sigma}\sigma_{01} - \sigma_{0\delta} - \hat{\mu}\sigma_{1\delta} + \hat{\mu}\hat{\sigma}\sigma_1^2] \quad \text{(Eq 66)}$$

$$\sigma_{\hat{\sigma}}^2 = \frac{1}{\beta_1^2}[\sigma_\delta^2 - 2\hat{\sigma}\sigma_{1\delta} + \hat{\sigma}^2\sigma_1^2] \quad \text{(Eq 67)}$$

Inverting this variance-covariance yields the values of k_0, k_1, and k_2 required in Eq 25 to calculate the lower confidence bound on the POD(a) function.

Multiple Inspections Per Flaw. Repeat $\hat{a}$ values for the same flaw can be analyzed to estimate the magnitude of the total variability being introduced by factors other than the flaws in the experiment. In essence, the random term, δ, of Eq 9, can be partitioned into components that can be estimated if the experiment is properly designed. The relative magnitude of the components of variance indicates potential areas for improving the system. However, the methods for using the $\hat{a}$ values to generate POD(a) functions and confidence bounds from complex experiments with censored values are still under development.

If there are no censored $\hat{a}$ values, the following analysis provides valid estimates of both the POD(a) function and its confidence bound. If there are censored $\hat{a}$ values, the POD(a) parameter estimates are valid, but the confidence bound is approximate. Much of the data recorded in the $\hat{a}$ format is from automated systems. In these systems, the variability in $\hat{a}$ values has been dominated by that associated with different flaws. This variability is correctly analyzed in the following analysis. In applications to

Table 5 Example of POD(a) parameters determined from the data in Table 2

The decision threshold is 250 counts.

Parameter	Probe A	Probe B	Probe C	Composite
$\hat{\mu}$	−1.89	−1.80	−2.12	−1.96
$\hat{\sigma}$	0.24	0.23	0.40	0.33
a_{50}, mm (in.)(a)	0.15 (0.006)	0.17 (0.007)	0.12 (0.005)	0.14 (0.0055)
a_{90}, mm (in.)(b)	0.21 (0.0083)	0.22 (0.009)	0.20 (0.008)	0.21 (0.0083)
$a_{90/95}$, mm (in.)(c)	0.25 (0.010)	0.27 (0.011)	0.27 (0.011)	0.26 (0.0102)

(a) $a_{50} = \exp(\hat{\mu})$ = estimate of crack size at 50% POD. (b) $a_{90} = \exp(\hat{\mu} + 1.282\hat{\sigma})$ = estimate of crack size at 90% POD. (c) $a_{90/95}$ = upper 95% confidence bound on the estimate of a_{90}.

date, the approximation to account for the secondary sources of variability has been judged to be negligible.

To model an additional source of variability, Eq 9 is rewritten as:

$$Y_{ij} = \ln(\hat{a}_{ij}) = \beta_0 + \beta_1 \ln(a_i) + c_i + r_{ij} \quad i = 1, \ldots, n \quad j = 1, \ldots, k \qquad \text{(Eq 68)}$$

where $\hat{a}_{ij}$ represents the jth observation on the ith flaw, c_i is the random deviation of flaw i from the fit, and r_{ij} is the random deviation about c_i introduced by replicate j of the source of the repeated observations. For this model of ln ($\hat{a}$) versus ln (a):

$$\delta = c_i + r_{ij} \qquad \text{(Eq 69)}$$

Because it is reasonable to assume that the variability introduced by flaws is independent of that introduced by other factors:

$$\sigma_\delta^2 = \sigma_c^2 + \sigma_r^2 \qquad \text{(Eq 70)}$$

where σ_r^2 can be estimated as the pooled-within variance of $\hat{a}_i$ values for each flaw using those flaws for which $\hat{a}$ values were recorded (Ref 17):

$$\hat{\sigma}_r^2 = \frac{\sum_{i=1}^{n^*} \sum_{j=1}^{k_i} (Y_{ij} - \bar{Y}_i)^2}{\Sigma k_i - n^*} \qquad \text{(Eq 71)}$$

where n^* is the number of flaws with uncensored $\hat{a}$ values and k_i is the number of uncensored $\hat{a}$ values for flaw i.

Between-crack variability, σ_c^2, cannot be estimated directly, but can be estimated indirectly from a censored regression analysis. First, the mean log response for each flaw is calculated. This mean response may be a simple average if all $\hat{a}$ values are available, but a mean based on an analysis of censored data (as previously discussed) will be required for the flaws for which $\hat{a}$ values were censored at the decision threshold or saturation limit. The analysis in the section "Parameter Estimation in $\hat{a}$ Analysis" in this article is then used on the model:

$$Y_i = \beta_0 + \beta_1 X_i + c_i + \bar{r} \qquad \text{(Eq 72)}$$

to obtain estimates $\hat{\beta}_0$, $\hat{\beta}_1$, and $\hat{\sigma}_\delta^*$. However:

$$\hat{\sigma}_\delta^{*2} = \hat{\sigma}_c^2 + \hat{\sigma}_{\bar{r}}^2 = \hat{\sigma}_c^2 + \frac{\hat{\sigma}_r^2}{k} \qquad \text{(Eq 73)}$$

Solving Eq 73 for $\hat{\sigma}_c^2$ and substituting into Eq 70 yields:

$$\hat{\sigma}_\delta^2 = \hat{\sigma}_\delta^{*2} + \frac{(k-1)}{k}\hat{\sigma}_r^2 \qquad \text{(Eq 74)}$$

The value $(\hat{\sigma}_\delta^*)^2$ is obtained from the censored regression analysis, and $\hat{\sigma}_r^2$ is obtained directly from the ln ($\hat{a}$) values. Equation 74 provides the estimate of σ_δ to be used in the POD(a) function.

To date, the lower confidence bound has been placed on the POD(a) function using the information matrix derived from the censored regression through the average of the ln ($\hat{a}$) values for each crack. This procedure does not account for the added uncertainty resulting from the estimate of $\hat{\sigma}_r^2$. When $\hat{\sigma}_c^2$ is significantly greater than $\hat{\sigma}_r^2$, the error introduced by neglecting this variation is judged to be negligible. (Most of the applications of the $\hat{a}$ analysis have been on highly automated eddy current systems. In these applications, the variability of ln ($\hat{a}$) values within cracks has been significantly less than that between cracks.) The method of constructing exact confidence bounds is under development. This process can be extended to account for more than one component of the variability of ln ($\hat{a}$) values.

Examples of $\hat{a}$ Analysis. The data in Table 2 resulted from three eddy current inspections of 28 fatigue cracks in flat plates simulating the web/bore of an aircraft engine disk. The three inspections are from the use of three different probes, with all other factors being held fixed. Table 5 presents a summary of the parameters of the POD(a) function for each data set in Table 2 and the combination of the three data sets in a single analysis. The ln ($\hat{a}$) versus ln (a) data from probe A are presented in Fig. 5. This data set had two inspection results below the recording threshold and five results above the saturation limit. The POD(a) function for probe A and its lower 95% confidence bound are shown in Fig. 10. The combined results from the three probes yield the composite POD(a) function and its lower 95% confidence bound, which are also shown in Fig. 10. If the probes were in some sense selected at random from the population of all probes applicable to this equipment and if other factors were assumed to be representative of the application, this composite POD(a) function would be representative of the inspection system. Figure 11 shows the individual POD(a) functions for all three probes and their composite.

Comments on $\hat{a}$ Analysis. There are several advantages of the $\hat{a}$ analysis over the hit/miss analysis in estimating the POD(a) function. These accrue primarily because of the added information contained in the $\hat{a}$ values.

It is not as critical in the experiment to have flaws with sizes in the range of increase of the POD(a) function. Because the analysis does not depend on whether or not flaws were detected, the decision threshold can be arbitrarily set after the experiment. Even if the recording threshold is close to the decision threshold, the method permits extrapolation of the results from larger flaw sizes to the smaller sizes of interest. The extrapolation is reasonable over the flaw size range for which the ln ($\hat{a}$) versus ln (a) relation is linear, and deviations from the fit are normally distributed with constant variance.

A principal advantage of the $\hat{a}$ analysis is that statistical tests of the underlying assumptions are readily available (Ref 17). In most of the experiments that have been analyzed by this method, the assumptions could not be rejected. However, for very large flaws, a tendency for the ln ($\hat{a}$) versus ln (a) relation to bend down has been observed on occasion. In this region, the $\hat{a}$ values are very large, and all of the flaws are easily detected. One method of linearizing the ln ($\hat{a}$) response is to restrict the analysis to a range of smaller flaw sizes by censoring $\hat{a}$ values at a lower saturation limit. This is equivalent to deleting data points and may reduce the sample size to an unacceptable number. In this case, the flaw sizes in the experiment were not appropriate for the application. It should be noted that ignoring this type of nonlinearity tends to produce a nonconservative POD(a) function, because the effect is to produce a smaller value of the median detectable crack size, μ.

When the data do not fit the model, it is possible to obtain results that have no relation to reality. This can occur if ln ($\hat{a}$) is not an increasing linear function of ln (a) or if there are outliers in the data set that have a significant influence on the analysis. The effects of these anomalies are sometimes manifested in unreasonable values of μ or σ or by a lower confidence bound that eventually decreases for large crack sizes. As a minimum, it is recommended that a plot of ln ($\hat{a}$) versus ln (a) be obtained for all experiments. This will permit at least a subjective judgment concerning the assumptions of the analysis.

If the data do not fit the model, it is necessary to ensure that the process is in control and that the experiment was prop-

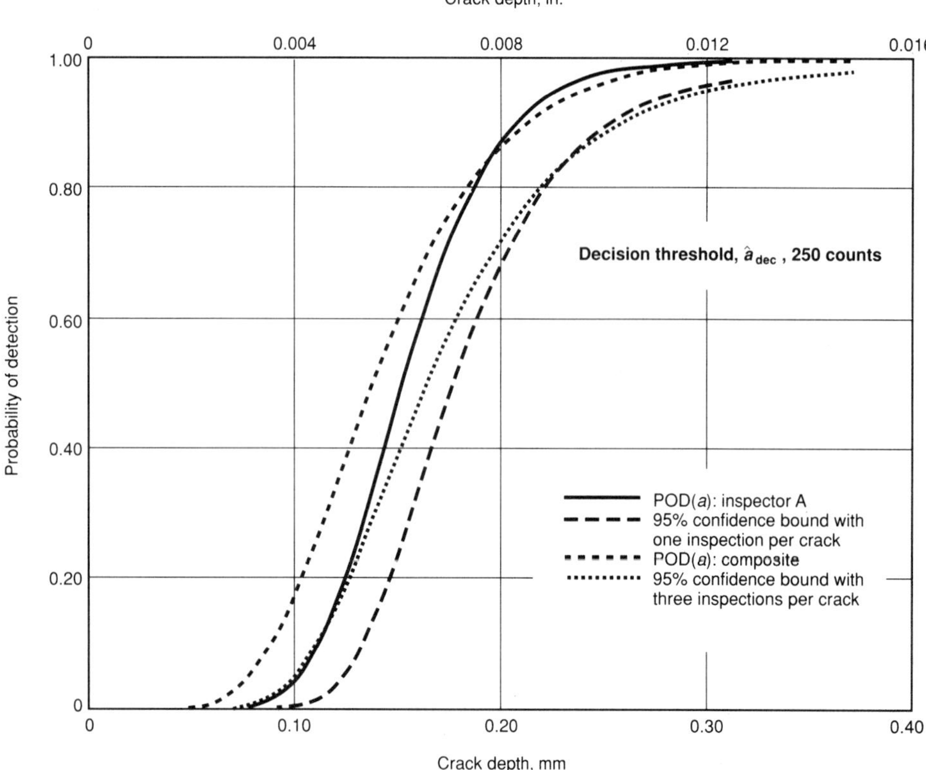

Fig. 10 POD(a) function and lower 95% confidence bound from signal response analysis of the data in Table 2 with one inspection per crack (with probe A) and with three inspections per crack (from the composite result of probes A, B, and C)

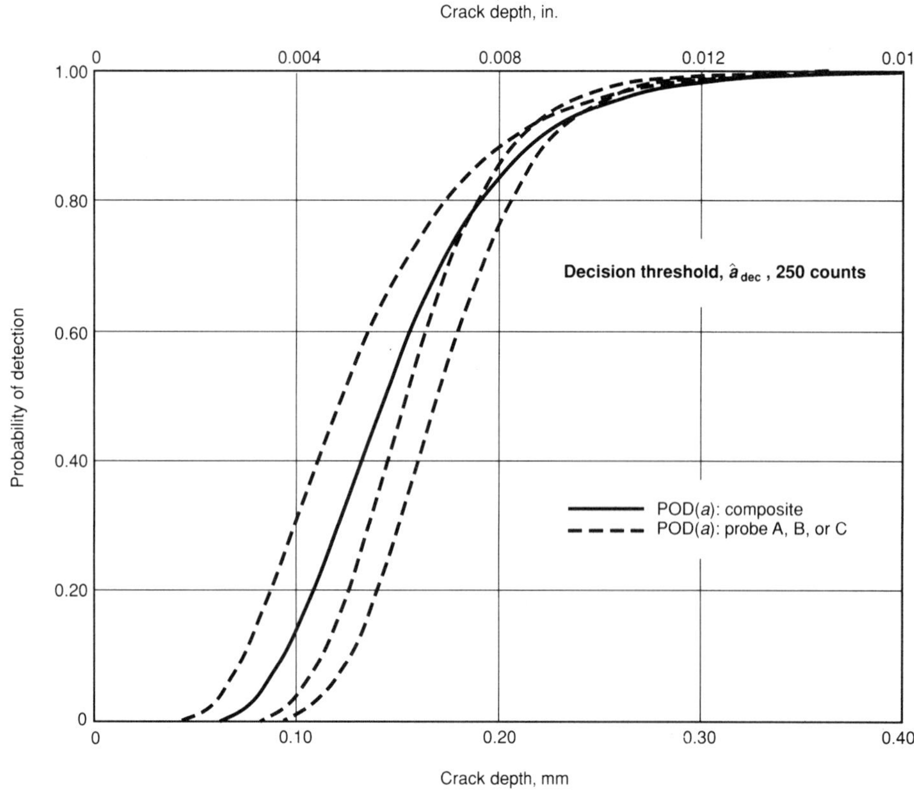

Fig. 11 POD(a) functions from signal response analysis of the data in Table 2

erly designed and executed. It may also be possible to use other relations between $\hat{a}$ and a as the basis of analysis, but these have not yet been explored.

REFERENCES

1. B.G.W. Yee, F.H. Chang, J.C. Couchman, G.H. Lemon, and P.F. Packman, "Assessment of NDE Reliability Data," NASA CR-134991, National Aeronautics and Space Administration, Oct 1976
2. W.D. Rummel, Recommended Practice for Demonstration of Nondestructive Evaluation (NDE) Reliability on Aircraft Production Parts, *Mater. Eng.*, Vol 40, Aug 1982, p 922-932
3. W.H. Lewis, W.H. Sproat, B.D. Dodd, and J.M. Hamilton, "Reliability of Nondestructive Inspections—Final Report," SA-ALC/MME 76-6-38-1, San Antonio Air Logistics Center, Kelly Air Force Base, Dec 1978
4. A.P. Berens and P.W. Hovey, "Evaluation of NDE Reliability Characterization," AFWAL-TR-81-4160, Vol 1, Air Force Wright-Aeronautical Laboratories, Wright-Patterson Air Force Base, Dec 1981
5. A.P. Berens and P.W. Hovey, Statistical Methods for Estimating Crack Detection Probabilities, in *Probabilistic Fracture Mechanics and Fatigue Methods: Applications for Structural Design and Maintenance*, STP 798, J.M. Bloom and J.C. Ekvall, Ed., American Society for Testing and Materials, 1983, p 79-94
6. D.E. Allison *et al.*, "Cost/Risk Analysis for Disk Retirement—Volume I," AFWAL-TR-83-4089, Air Force Wright-Aeronautical Laboratories, Wright-Patterson Air Force Base, Feb 1984
7. A.P. Berens and P.W. Hovey, "Flaw Detection Reliability Criteria, Volume I—Methods and Results," AFWAL-TR-84-4022, Air Force Wright-Aeronautical Laboratories, Wright-Patterson Air Force Base, April 1984
8. D.R. Cox, *The Analysis of Binary Data*, Methuen and Co., 1970
9. A.P. Berens and P.W. Hovey, The Sample Size and Flaw Size Effects in NDI Reliability Experiments, in *Review of Progress in Quantitative Nondestructive Evaluation 4B*, D.O. Thompson and D.E. Chimenti, Ed., Plenum Press, 1985
10. H. Cramer, *Mathematical Methods of Statistics*, Princeton University Press, 1946
11. J.F. Lawless, *Statistical Models and Methods for Lifetime Data*, John Wiley & Sons, 1982
12. R.C.H. Cheng and T.C. Iles, Confidence Bands for Cumulative Distribution Functions of Continuous Random Variables, *Technometrics*, Vol 25 (No.

1), Feb 1983, p 77-86
13. R.C.H. Cheng and T.C. Iles, One Sided Confidence Bands for Cumulative Distribution Functions, *Technometrics*, Vol 32 (No. 2), May 1988, p 155-159
14. A. Ralston, *A First Course in Numerical Analysis*, McGraw-Hill, 1965
15. M. Glaser, Regression Analysis With Dependent Variable Censored, *Biometrics*, Vol 21, June 1965, p 300-307
16. S.R. Searle, *Linear Models*, John Wiley & Sons, 1971
17. W.J. Dixon and F.J. Massey, Jr., *Introduction to Statistical Analysis*, McGraw-Hill, 1957

Models for Predicting NDE Reliability

J.N. Gray, T.A. Gray, N. Nakagawa, and R.B. Thompson, Center for NDE, Iowa State University

FRACTURE CONTROL PHILOSOPHIES, as discussed in previous articles in this Section, depend on a damage-tolerant design. The essential feature of such an approach is the incorporation of redundant load paths so that, even if local failures do occur, the structure will be safe for a period of time and can either be removed from service or repaired. Parts containing readily detectable damage can be used if that damage will remain in stable condition until the next inspection opportunity.

The implementation of such an approach requires the knowledge and integration of stress, flaw size, and failure mechanisms. As shown in Fig. 1 for the case in which fatigue can be modeled by linear-elastic fracture mechanics, cyclic stress excursions, $\Delta\sigma$, and flaw size, a, are required as inputs to fracture mechanics to predict a stress intensity range, ΔK. If the stress intensity exceeds a critical value known as the plane-strain fracture toughness, K_{Ic}, catastrophic failure is imminent. Otherwise, crack growth laws of the form $da/dN = A(\Delta K)^m$ (where N is the number of stress excursions and A and m are constants) can be used to estimate the safe life available before catastrophic failure.

Implementation of this damage-tolerant design approach rests on three methodologies: stress analysis, nondestructive evaluation (NDE), and failure modeling. Incorporation of these methodologies in the design of damage-tolerant components would ideally take advantage of analytical or numerical computations to model the expected lifetime performance of a component. At present, extensive capabilities are in place for modeling stresses and failures, and these are widely used in the design process. However, the modeling of nondestructive evaluation is not nearly as widely accepted. Instead, frequent use is made of empirical rules based on extensive demonstration programs. For both economic and time reasons, there is a significant need to develop a model base for estimating NDE reliability (which is often measured in terms of the probability of flaw detection at given confidence level). It is the purpose of this article to present the current status and future directions of efforts to develop such a capability. This is not intended as a review of international efforts in NDE reliability modeling but rather as a summary of the authors' experience in modeling the inspectability of aerospace components, with emphasis on engine components. Therefore, attention is given to ultrasonic, eddy current, and radiographic inspection. Broader sets of references for the case of ultrasonics can be found in recent review articles (Ref 2, 3). Of particular note is the work performed by the Central Electricity Generating Board in modeling the inspectability of nuclear power generating components (Ref 4).

The details of models for NDE reliability are partially dictated by their envisioned uses (Ref 5, 6), which are conceptually illustrated in Fig. 2. One would like to have a model that predicts the probability of detecting flaws of various sizes. This would clearly require as inputs the design of the component, its history of processing and service, and a specification of the inspecting methodology to be used. Given the specifications of these input parameters, it could first be asked whether the predicted probability of flaw detection of the NDE system is adequate to meet the demands imposed by the required performance of the component. Should the expected probability of detection (POD) be inadequate, the model could be exercised to modify the specifications of the inspection, the design of the component, or the processing or service profiles. The NDE models thus become an integral part of a broader concept known as unified life cycle engineering or simultaneous engineering (Ref 7). The essential feature is that one should consider all aspects of the life of a component in the design process, including the ability to inspect and maintain the component, rather than just the initial costs. This will ultimately lead to more sophisticated networking of models, as shown in Fig. 3. Figure 3 illustrates a number of factors that must be added to the traditional computer-aided design and manufacturing (CAD/CAM) methodologies to produce a design optimized for

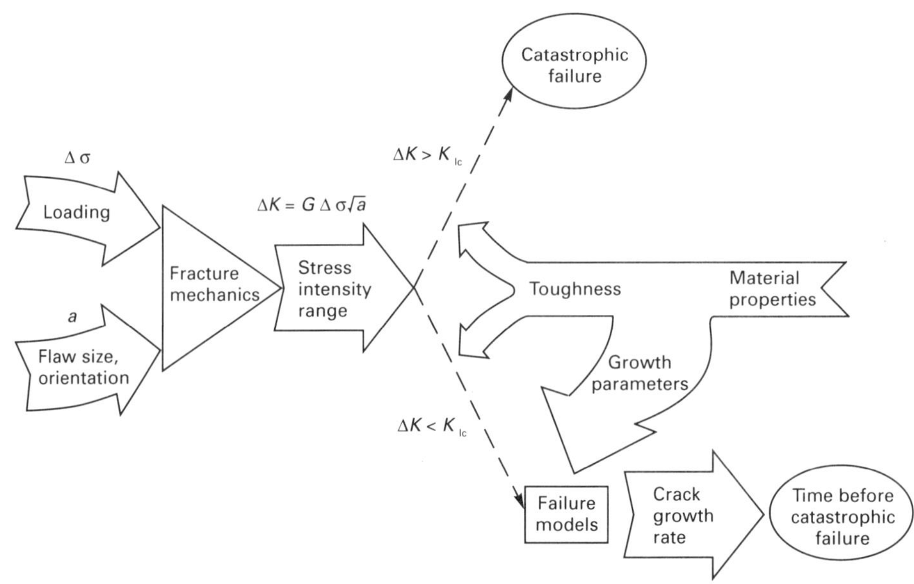

Fig. 1 Methodology of lifetime prediction for metal parts undergoing cyclic fatigue. Source: Ref 1

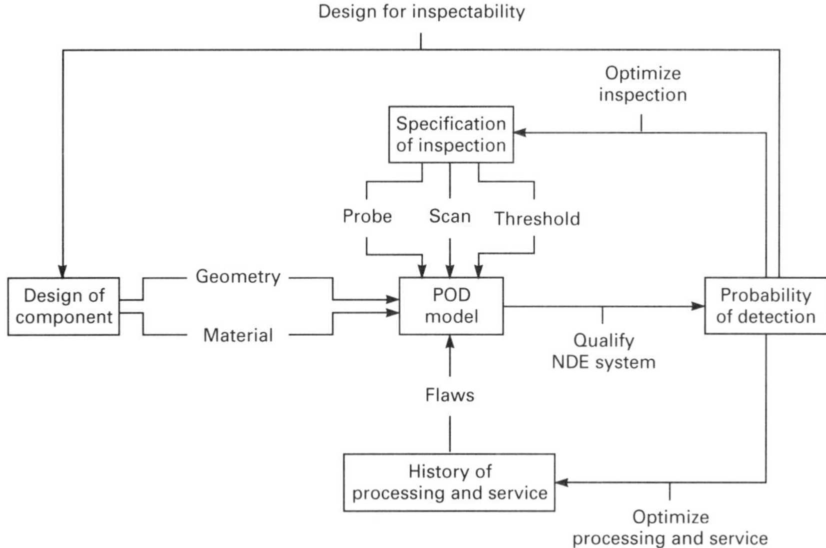

Fig. 2 Diagram of probability of detection model and its application to NDE system qualification and optimization and to computer-aided design for inspectability. POD, probability of detection

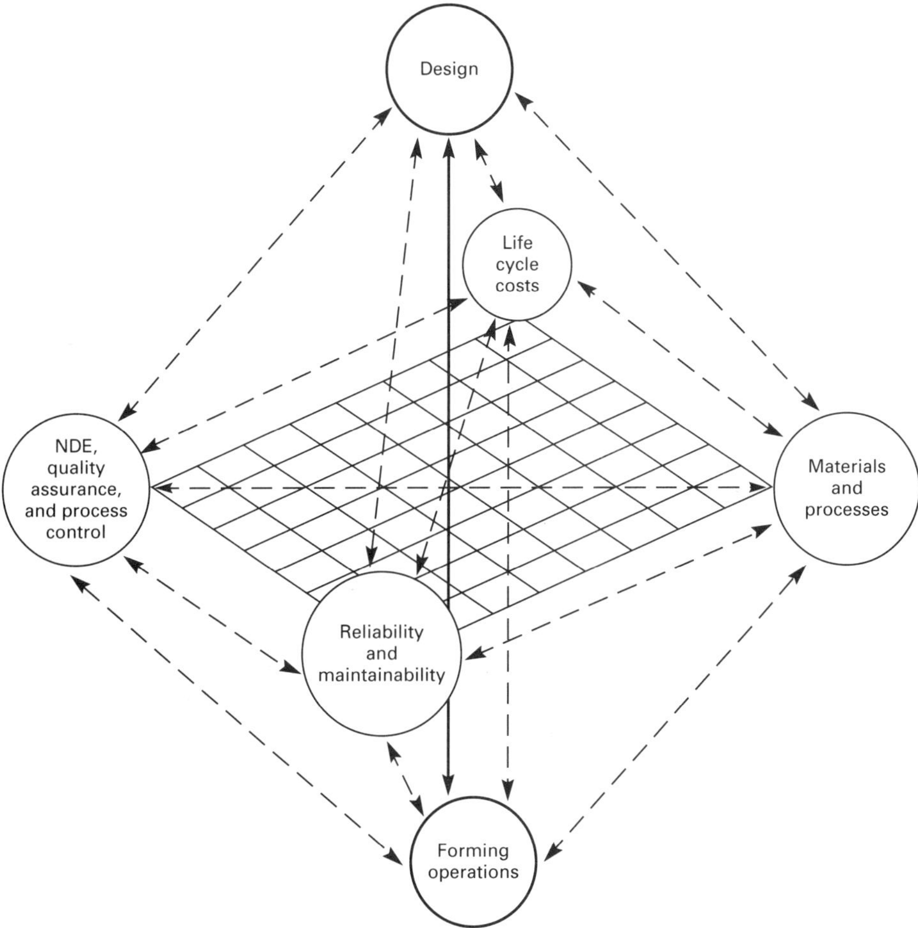

Fig. 3 Schematic of possible linkages needed for unified life cycle engineering

life cycle performance. Multiple interactions among the various factors must be considered to allow the design to address simultaneously all the issues associated with damage tolerance (Ref 8).

Given these objectives, it is obvious that an NDE model must exhibit certain characteristics. First, it must predict the response of a real measurement system, as influenced by the specific characteristics of commercially available probes and instruments, rather than an idealized response based on assumptions such as plane wave illumination. Second, the models should give as outputs the information obtained by real inspection protocols. For example, if a signal strength is compared to a threshold as a criterion for detection, this operation should be simulated by the model. If separate protocols are followed in detection and sizing, these should be described by separate models. Third, the models should be used to develop more reliable standardization approaches. This is necessary to ensure that inspections specified in the design process are uniformly implemented by NDE units at the various manufacturing and maintenance departments encountered by the component in its lifetime. Fourth, it is desirable in certain industries for the models to be configured such that they can be integrated with standard CAD packages. This article discusses some ultrasonic, eddy current, and x-ray radiography models that have been developed to exhibit the characteristics mentioned above. As noted previously, primary emphasis is placed on formulations that have been developed in response to the particular needs of the aerospace industry. This article also presents a broader discussion of possible future applications of a reliability modeling capability.

Ultrasonic Inspection Model

Empirical determinations of ultrasonic inspectability based on demonstration experiments are of limited utility because their predictions cannot, in general, be extrapolated to new situations beyond the bounds of the data set upon which they are based. Additional data are needed in order to apply them to other cases, and the costs and time required to develop such results for all (or many) possible component designs and scan plans are prohibitive. However, the physical principles upon which many different ultrasonic inspection techniques are based, as applied to a variety of components, are quite similar. Therefore, a mathematical model, whose validity has been proved against a relatively small amount of empirical data, can be used to accurately predict inspectability beyond the bounds of the experimental evidence.

The foundation of a physically based mathematical or computer model of ultrasonic inspectability is an analytical formalism, a numerical algorithm, or a combination of the two incorporating the physical principles of the measurement. Such a model consists of descriptions of the waves radiated by the probes, their modification by the geometry of the testpiece, the wave propagation and scattering from defects, and the effects of signal processing and display. The first four sections below review the technical details of each of these elements as developed to describe measurements made in aircraft engine components.

In the fifth section, the use of the models to predict the POD of flaws is discussed. The prediction of the POD for flaws is emphasized because of the importance of that parameter in damage-tolerant design. Other uses of models, such as assisting in the interpretation of data during flaw characterization and sizing, are also important but are not explicitly discussed. Some early applications of POD models are summarized in the sixth section.

Reciprocity Relation. The ultrasonic NDE simulation models described in this section are based on the formalism of the electromechanical reciprocity relationship of Auld (Ref 9). This relationship, when specialized to the case of elastic wave scattering, can be expressed as follows. Assume that two identical ultrasonic transducers, a and b, are placed in a fluid to be used in an immersion, pitch-catch measurement (a single probe, or pulse-echo, configuration is a special case) of a component containing a flaw, F. Let Γ be the ratio of the electrical signal radiated into coaxial line b by the receiving transducer to the electrical signal incident on the transmitting probe from coaxial line a. Then the change, $\delta\Gamma_F$, in this signal induced by the presence of the flaw in the insonified region of the component is given by:

$$\delta\Gamma_F = \frac{1}{4P}\int_{S_F}\left(\dot{\mathbf{u}}_b \cdot \boldsymbol{\sigma}_a - \dot{\mathbf{u}}_a \cdot \boldsymbol{\sigma}_b\right) \cdot \mathbf{n} d^2A \qquad \text{(Eq 1)}$$

where $\dot{\mathbf{u}}_a$ and $\boldsymbol{\sigma}_a$ are the displacement vector field and the stress tensor, respectively, that would be produced in the presence of the flaw when probe a is excited by an electrical power P, $\dot{\mathbf{u}}_b$ and $\boldsymbol{\sigma}_b$ are the fields that would have been established in the absence of the flaw if probe b had been excited, the overhead dot denotes time differentiation, S_F is an arbitrary closed surface enclosing the flaw, and $\mathbf{n}$ is the inward normal to that surface. Equation 1 is an exact result known as the electromechanical reciprocity relationship. Its simplicity, however, belies its intractability, except in a very few special cases of limited practical use.

Measurement Model. A number of approximations and simplifications are necessary in deriving an accurate yet computationally efficient inspection measurement model from Eq 1. One situation for which a useful and accurate model can be extracted from Eq 1 through rather remarkable simplifications is the pulse-echo inspection of isotropic, homogeneous elastic materials containing small flaws of fairly simple shape. This is applicable, for example, to typical ultrasonic inspections of gas turbine aircraft engine components, in which fracture-critical flaws are quite small because of the high stresses created during engine operation.

The assumption of small in this case means that the dimensions of the defect are small relative to the variations in the transverse profile of the ultrasonic beam. The ultrasonic fields can be approximated locally as plane waves whose displacement and stress fields are the same as those of the true fields (Ref 10). Scattered fields can also be simply modeled, provided their variation is not significant over the face of the receiving transducer. These fields can be represented by the product of a spherically spreading wave times the far-field, unbounded medium scattering amplitude of the flaw (Ref 10). This scattering amplitude A is formally defined as:

$$u_s = u_0 A \frac{\exp(-ik_s r)}{r} \qquad \text{(Eq 2)}$$

where u_s is the displacement amplitude of the fields at a distance r from the center of the flaw, u_0 is the amplitude of an illuminating plane wave, and k_s is the wavenumber of the scattered wave mode. A time-harmonic term of the form exp ($i\omega t$) is implicitly assumed in this discussion. However, the model has the capacity to deal with pulses, as described later.

Applying these approximations to Eq 1, the a fields (those produced when probe a is fired and the flaw is present) are expressed as the sum of a plane wave illuminating field and a scattered field due to that plane wave, using the scattering amplitude as in Eq 2. The b field is simply a plane wave term representing illumination by probe b in the absence of the flaw. After some manipulation of the resulting integral and neglecting higher-order terms, a measurement model is derived that represents the measured signal as a product of factors describing the effects of transducer efficiency, transmission through interfaces, attenuation and beam spread, and scattering (Ref 10). Specifically, one finds:

$$\delta\Gamma_F = \beta(T_a C_a)(T_b C_b)\frac{i2A\rho_1 V_b}{k_b a_T \rho_0 V_0}$$
$$\times\exp\{-i[k_0(z_{oa} + z_{ob}) + k_a z_{1a} + k_b z_{1b}]\}$$
$$\times\exp\{-[\alpha_0(z_{oa} + z_{ob}) + \alpha_a z_{1a} + \alpha_b z_{1b}]\} \qquad \text{(Eq 3)}$$

where β is an efficiency factor; $T_a C_a$ and $T_b C_b$ represent local plane wave amplitudes at the flaw depth for the a and b fields (these are the product of interface transmission, T, and beam diffraction, C, factors); A is the scattering amplitude of the flaw; $\rho_1 V_b$ and $\rho_0 V_0$ are the acoustic impedances of the solid and fluid media, respectively; k_b is the wavenumber for the received wave mode; a_T is the transducer radius (which is assumed to be the same for both a and b); and the two exponential terms represent the ultrasonic phase change and attenuation, respectively. Equation 3 represents only one frequency. A frequency spectrum can be obtained via superposition of individual frequency terms. Thus, time-domain waveforms, such as would be observed on an oscilloscope screen, can be obtained by an inverse Fourier transform of this measurement model spectrum. (Further details of this derivation and the determination of the efficiency factor β in Eq 3 can be found in Ref 10.) Work at a number of institutions has led to formulas, algorithms, and data bases for the ultrasonic beam and scattering amplitude factors in Eq 3, so that a variety of useful simulations can be made, including the use of planar or focused probes (Ref 11-14), inspection through planar or curved liquid/solid interfaces (Ref 11, 12, 14, 15), and scattering from both volumetric (Ref 16-18) and cracklike (Ref 17, 19) defects. These aspects of the measurement model will be described in subsequent sections in this article. An alternative formulation, appropriate when the flaw is not small, is found in Ref 4.

As an example, Fig. 4 shows a comparison of an experimental pulse-echo radio-frequency (RF) waveform obtained from a circular flat crack versus two model-based simulations. The magnitude of the Fourier spectrum of each RF signal is also shown. The simulated crack was a circular, disk-shaped cavity 0.8 mm (0.032 in.) in diameter and 0.08 mm (0.003 in.) thick located in a diffusion bond plane of a specimen of a nickel-base powder metal alloy (IN 100). The specimen was a 25 mm (1.0 in.) thick plate. The face of the crack was parallel to the surface of the sample, and a 10-MHz, 6.35 mm (0.25 in.) diam unfocused transducer was tilted approximately 7° from normal to the sample surface to generate a 30° refracted longitudinal wave in the sample.

The scattering amplitudes used in Fig. 4 were either results of the method of optimal truncation (MOOT) (Ref 17), which is a computationally intensive algorithm, or the elastodynamic Kirchhoff approximation (Ref 19). The experimental and MOOT results are very nearly identical in both the time and frequency domains. Because the MOOT results are in quasi-exact agreement with the measured scattering amplitudes in this case, this comparison highlights the accuracy of the measurement model. The Kirchhoff model result fails to reproduce some of the detailed wiggles of the experimental and MOOT results, but quite accurately reproduces the overall signal amplitude (voltage). This amplitude is the measured quantity that is routinely used in ultrasonic flaw detectors. Therefore, the approximation, which in this case yields a simple and computationally efficient model, can simulate a practical inspection problem. Note, however, that the error of the Kirchhoff model will depend on the flaw size and the orientation and polarization of the ultrasonic wave. Care must be taken to ensure that sufficient accuracy is obtained in particular applications by validating the model through comparison with controlled exper-

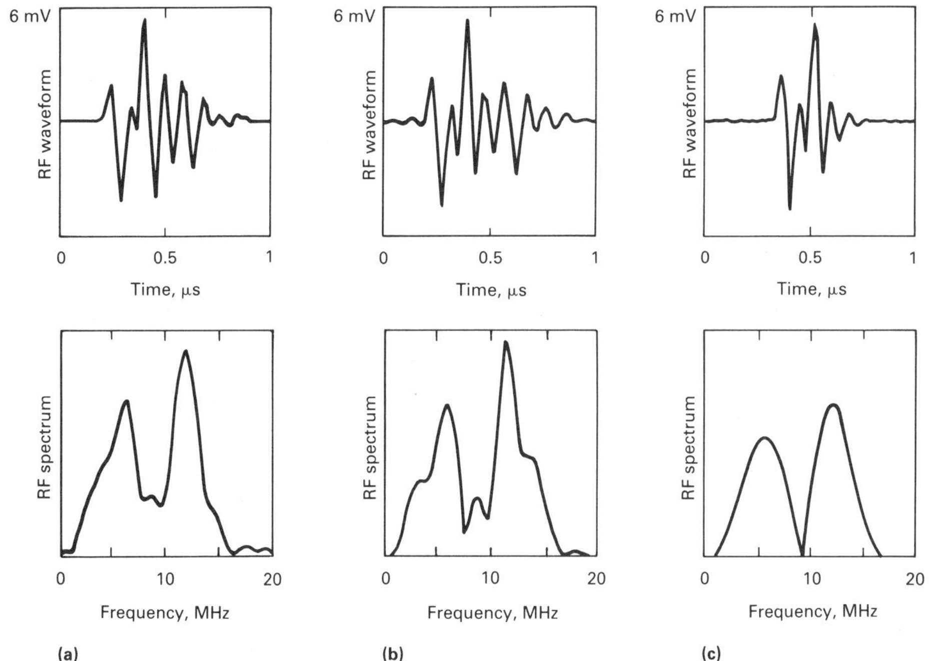

Fig. 4 Comparisons between experimental and model-predicted RF waveforms (top) and their Fourier spectra (bottom) for 30° longitudinal wave backscatter from a 0.8 mm (0.03 in.) diam circular crack in IN100. (a) Experimental measurements of scattering amplitudes (top) and their Fourier spectra (bottom). (b) Model of scattering with method of optimal truncation. (c) Model of scattering with the Kirchhoff approximation

iments and/or more exact theories for special cases.

Beam Models. To perform the preceding comparison, one essential element in the simulation was the representation of the ultrasonic fields in the vicinity of the flaw. Because of the finite size of any realistic transducer, these fields will be quite complex, exhibiting peaks and valleys along the axis of the beam and side-lobes away from the axis. The simulations shown in Fig. 4 represent the case of a scatterer on the axis of the beam, for which a number of approximate beam models have been generated (Ref 11, 13, 14). In a typical automated ultrasonic scan of a component, however, a defect in that component will not generally lie along the axis of the ultrasonic beam. The degree of misalignment will depend, for example, on the coarseness of the scan mesh used to inspect the part. To simulate such an inspection situation, it is necessary to incorporate the full fields of an ultrasonic transducer into the model. This is a formidable task because of the elastic (tensor) nature of wave propagation in a solid and because of the need to consider the interaction of the probing fields with a possibly curved liquid/solid interface at the component surface. However, two approximations have emerged as useful models of transducer radiation patterns: the Gaussian model (Ref 12) and the Gaussian-Hermite model (Ref 13, 14, 20).

The former, and simpler, of these models assumes that the transverse profile of the radiation profile is Gaussian in shape at all distances from the probe. This Gaussian beam model provides a set of simple algebraic formulas that predict diffraction effects (beam spread only), effects of lenses, and refraction/focusing due to transmission through curved liquid/solid interfaces (Ref 12). However, typical ultrasonic transducers do not generate Gaussian radiation patterns. For example, typical piston-type radiators exhibit side-lobes and peaks and nulls along the axis of the probe in the near field. However, in the far field (that is, several times the near-field distance), the Gaussian model, if suitably normalized, does accurately predict the amplitude and width of the main lobe in the radiation pattern of a typical piston-type transducer (Ref 12). One application of this approximation, therefore, is the simulation of the fields near a focal region, which can occur either as a result of an acoustic lens on the probe or the focusing effect of a curved component surface.

The Gaussian-Hermite model is based on a series expansion of the radiated fields of a transducer in terms of a complete set of orthogonal solutions to a reduced wave equation (Ref 13, 14, 20). These functions are products of a Gaussian factor and a Hermite polynomial. The coefficients in the series expansion are obtained by integrating the product of the Gaussian-Hermite functions and the velocity distribution on the face of a probe over its area. This distribution and the shape of the probe face are arbitrary, so that virtually any probe shape, lens type, and so on, can be modeled. The laws for transmission through curved liquid/solid interfaces and propagation in elastic isotropic media are implemented as simple algebraic operations. The primary disadvantage of the Gaussian-Hermite model is that it is a series solution and therefore can require significantly longer computation times than the Gaussian approximation because of the need for a large number of terms, especially in the near field. However, this becomes less of a disadvantage as computational speeds continue to increase as a result of advances in computer hardware.

Scattering Approximations. Another key element in the simulation of ultrasonic inspection of structures is the model, or models, for representing the interaction of the probing ultrasonic fields with defects. In the most general case, this is represented by a complicated and computationally intractable integral, such as Eq 1. In some cases, however, the effects of the probing ultrasonic fields can be separated from the scattering effects. Specifically, under the assumptions that led to the measurement model (Eq 3), the ultrasonic beam can be described by one of the models just mentioned. For example, elastic wave scattering can be modeled through the use of a far field, unbounded medium scattering amplitude, whose definition was given in Eq 2. Fortunately, considerable research effort has been directed over the past several years toward the development of various models, approximations, and solutions for scattering amplitudes of both volumetric and cracklike defects (Ref 3).

For application to the ultrasonic inspection of jet aircraft engine components, a reasonable inventory of scattering models includes formalisms for both volumetric and cracklike flaws. For volumetric flaws, of ellipsoidal shape and arbitrary orientation, both voids and inclusions can be represented by an elastodynamic Kirchhoff approximation (Ref 18). This approximation is exact in its treatment of the strength of the front surface reflection (δ function). It is valid for both longitudinal and shear wave backscatter, such as would be used to simulate pulse-echo inspections. The limitation of this model, however, is that it is accurate only for early-time events in the scattering. Therefore, it does not predict the amplitude of scattered fields that have reverberated within an inclusion; in some cases, these scattered fields can be of higher amplitude than the initial front surface reflection. The Kirchhoff approximation is therefore a conservative model for scattering from volumetric flaws. It does have the benefit of simplicity and computational efficiency.

For cracklike flaws, an elastodynamic Kirchhoff approximation to scattering from internal flat cracks of elliptical shape has been implemented for both pulse-echo and pitch-catch techniques and for longitudinal

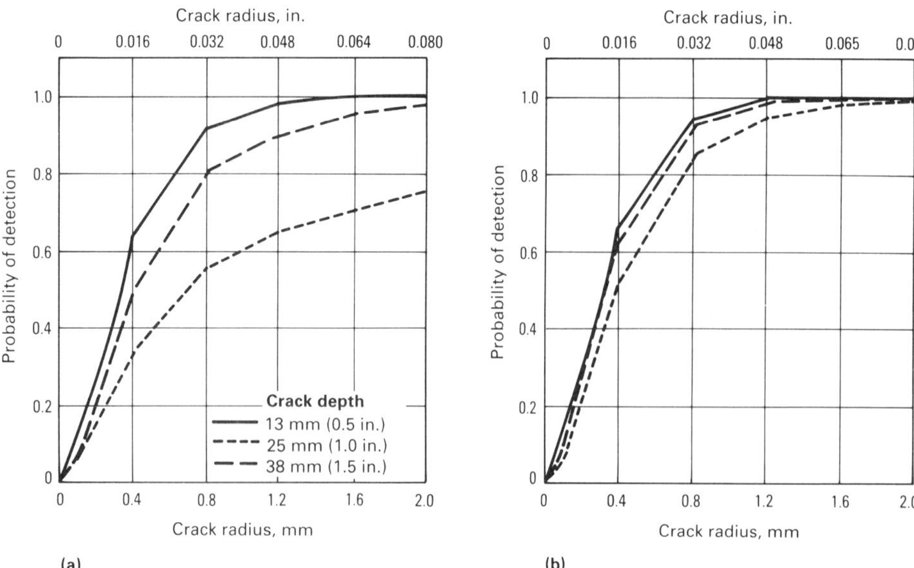

Fig. 5 Predicted influence of scan plan on POD. (a) POD at three depths for axial and circumferential scan increments of 2.5 mm (0.1 in). (b) POD at three depths for axial and circumferential scan increments of 5 and 1.3 mm (0.2 and 0.05 in.), respectively

and shear wave modes (Ref 19). This Kirchhoff approximation accurately predicts the specular content (mirror reflection) of scattering, but does not properly include edge diffraction contributions or surface wave modes. It also does not contain any provision for surface roughness or partial closure of the crack faces. It does, however, yield reasonably accurate predictions of signal amplitudes in the near-specular regime, as can be inferred from Fig. 4. One fairly significant limitation of the model is its inaccuracy in the nonspecular regime. It predicts, for example, that the scattering amplitude is identically zero for edge-on incidence from a crack, which is inconsistent with established crack scattering results (Ref 17, 21). However, the model is of significant utility because an inspection system for detecting cracks would be set up to take advantage of specular orientations, if possible. Furthermore, in this case, the Kirchhoff approximation is a simple and computationally fast algorithm.

Probability of Detection Models. One important application of ultrasonic measurement modeling is the simulation of probability of detection. Detection is appropriately described in terms of a probability for several reasons. A given size and type of defect may occur at random positions and with a range of orientations within each of a set of nominally identical components. The detailed shape of the defect may vary in a way that influences its ultrasonic response differently from its fracture response. Variations in grains, porosity, surface roughness, and so on, as well as the electronic equipment in a detection system, will cause noise, which will interfere with the signals from a flaw. Therefore, a given size and type of defect will exhibit a distribution of signal amplitudes if measured in a population of components containing such defects. Because these signal amplitudes are compared to a preset amplitude threshold in typical flaw detectors, some of the flaws will be missed, while others will be detected. The POD is the ratio of the number of flaws that are detected to the total number of flaws in the inspected components.

A formalism to predict the probability, $p(S, N, T)$, that a given signal, S (video envelope), in the presence of noise with total power N^2 will be detected by exceeding a threshold, T, is given by:

$$p(S, N, T) = \int_T^\infty \frac{r}{N} \exp\left[-\frac{(r-S)^2}{2N^2}\right] i_0\left(\frac{rS}{N^2}\right)\frac{dr}{N} \qquad \text{(Eq 4)}$$

where $i_0(z) = \exp(-z)I_0(z)$, with $I_0(z)$ being the modified Bessel function of the first kind and order zero. This approach, based on work performed by Rice (Ref 22), was developed for a narrow-band signal, as is typical in radar analysis. Equation 4 represents only the probability of detecting a single signal level, S, and does not take into account the distribution of signal amplitudes for variations in the size, shape, and type of flaw.

A POD model for ultrasonics has been developed by using the measurement model to calculate the variability of signal levels as influenced by the position and orientation of the flaws. Then, for a given root-mean-square noise level, N, Eq 4 can be used to represent the probability that the video signal from a specific defect (a given size, shape, type, location, orientation, and so on) in a given component (material, geometry, and so on) and using a specific inspection system will exceed a threshold amplitude. As an example, let p_x and p_θ represent the probability distribution functions for the location, $\mathbf{x}$, and orientation, $\boldsymbol{\theta}$, of a given size, shape, and type of defect, and let $S(\mathbf{x}, \boldsymbol{\theta})$ represent its signal amplitude, as simulated by the measurement model. An expression for POD, assuming a noise level N and a detection threshold amplitude T, can then be written formally as:

$$\text{POD} = \int_{\boldsymbol{\theta}}\int_{\mathbf{x}} p(S(\mathbf{x}, \boldsymbol{\theta}), N, T) dp_{\mathbf{x}} dp_{\boldsymbol{\theta}} \qquad \text{(Eq 5)}$$

where the integrals are taken over the range of orientations and positions of possible flaws (Ref 15). Obvious generalizations of Eq 5 can treat the effects of flaw shape, type, and so on.

Equation 5 does not contain an explicit factor to represent the probability distribution of the presence of flaws; therefore, it predicts the probability that an ultrasonic indication (signal plus noise) will be detected assuming that a flaw is, in fact, present. The use of the POD model to analyze reliability issues, such as the probabilities of falsely accepting flawed components or of falsely rejecting good ones, would require the incorporation into the model of the probability distribution function for flaws as a function of defect size, location within the part, and so on. The result of Eq 5 would then need to be further integrated with respect to that probability distribution.

Applications. Figure 5 shows the results of simulating the detectability (POD) of circular cracks at three different depths below a cylindrical component surface and for two different scan plans (Ref 6). Figure 5(a) illustrates the use of the POD model to quantify the detection capability of an NDE system. For the specific parameters in that simulation, cracks that are otherwise identical have significantly different detectability levels, depending on their depth below the surface of the part. In this example, the curved surface of the component behaves like an acoustic lens that happens to focus the beam at the middle depth (25 mm, or 1 in.). Because of the relatively coarse scan mesh and the reduced beam width in this focal region, there is a significant likelihood that a flaw will be located in a low-amplitude portion of the beam profile and, therefore not be detected. For the other two depths, the beam width is greater than the scan mesh distance. The plot in Fig. 5(b) shows the result of sufficiently refining the scan mesh so that the beam width at the focal region is greater than the distance between scan points. In this case, the POD curves are nearly the same for the three different depths. This example illustrates the use of the POD model for quantifying the capability of a flaw detection system and for suggesting improvements in this area in the system or its operation.

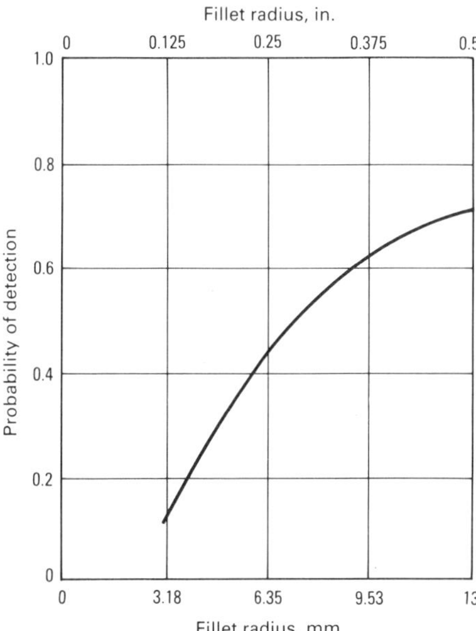

Fig. 6 Simulated POD for a 0.8 mm (0.03 in.) diam circular crack below a bicylindrical fillet as a function of fillet radius of curvature

Another example of the use of the POD model is shown in Fig. 6 (Ref 6). The POD curve is expressed in a rather nonstandard manner, because it does not represent the typical POD versus flaw size plot. Instead, Fig. 6 shows the variation in POD due to modification of a geometrical parameter of a component. Specifically, the component is a simulated turbine disk assumed to contain radially oriented circular cracks below a bicylindrical fillet. The abscissa of Fig. 6 is the in-plane radius of curvature of that fillet. The flaw size is assumed to be constant, representing, for example, the critical flaw size as predicted by fracture mechanics. Most important, Fig. 6 shows that flaw detectability can be improved by modifying the geometry of the component. This information is important for the definition of sonic near-net shapes of components in production, for example, and ensuring in-service inspectability during maintenance. Moreover, because the characteristics of the inspection system are contained explicitly in the POD model, the scan plan required to achieve the necessary detectability levels are easily determined. This concept of predicting component inspectability at the design stage, of determining the component design parameters that favorably influence flaw detectability, and of incorporating the requisite scan procedure into the design data base is perhaps the ultimate application of inspectability modeling.

Eddy Current Inspection Model

The eddy current NDE method has a long history of use (Ref 23-34). Because of its conceptually straightforward design principle, the method allows measurements to be made in a noncontacting, single-sided inspection geometry. Eddy current is therefore a cost-effective and truly nondestructive method of inspection. However, as a fully quantitative method, eddy current inspection has not achieved the level of sophistication found in other methods, such as ultrasonics and x-ray radiography. The reason for the delay in its development is associated with the fundamental nature of the measurement. In essence, the measurement response is a complex function of the probe fields and their interaction with the flaw, and it is difficult to isolate these two contributions. Therefore, one cannot develop simple models in which various effects can be described by separate factors, as in the case of ultrasonics. A significant problem has also been obtaining material property data (such as permeability and conductivity) for complex engineering materials. Welds are especially difficult.

From a modeling perspective, it is necessary to obtain electromagnetic field solutions for a given probe/flaw system by solving Maxwell's equations. When written in the form suitable to eddy current problems, the equations show that the basic dynamics are of a dissipative, nonscattering nature. One well-known consequence is that only near-surface flaws, confined within a region of a finite skin depth, are detectable. Putting this obvious limitation aside, there is a subtler problem posed by the basic principle. Namely, electromagnetic fields spread over a wide region outside the specimen, permitting no simple way of focusing the probe sensitivity to flaw regions. (In the other inspection methods, the beam focusing, for example, can be used for this purpose). The impedance signal, being an integrated quantity, contains not only flaw information but also redundant environment information (such as probe lift-off), which is uninteresting in terms of flaw detection and characterization and may contribute significant noise. Extracting flaw information from impedance signals therefore becomes highly dependent on component geometry and is a computationally intensive task.

Many efforts have been devoted to overcoming this difficulty, and the results have been promising (Ref 25–32). Rapid progress has occurred recently in computational methods with regard to both hardware and software. It appears that the new-generation desktop workstations are sufficiently capable of handling the computational requirements of eddy current data analyses. One may find an even better situation when using eddy current models to assist in design processes, for which state-of-the-art supercomputers may be available. In addition to progress in hardware, researchers are applying certain new software techniques to eddy current problems. With new software running on powerful computers, the difficulty in numerical analyses is becoming obsolete. Recent experimental advances have also proved beneficial. Emphasis is placed, more than before, on making characterizable probes and on operating them in broad frequency ranges. In fact, no fully quantitative eddy current methods can be established without these two capabilities. Incidentally, all existing eddy current POD models discussed in the literature rely on some type of experimental input for noise characterizations (Ref 30, 34).

Reciprocity Relation. On the theoretical side, the appearance of a reciprocity formula, proposed by B.A. Auld (Ref 31), has been regarded as a breakthrough in that it provides a formalism for interpreting measurements in realistic experimental geometries. The formula gives the impedance change ΔZ due to the presence of a flaw in the form:

$$\Delta Z = \frac{1}{I^2} \oint (\mathbf{E}' \times \mathbf{H} - \mathbf{E} \times \mathbf{H}') \cdot \mathbf{n} dS \qquad \text{(Eq 6)}$$

where **E** and **H** denote electric and magnetic fields in the absence of the flaw, and **E**′ and **H**′ denote the fields in the presence of the flaw. The integration in Eq 6 is carried out over a surface enclosing the flaw, with **n** being the inner-directed surface normal. The attractive feature of Eq 6 is that flaw signals can be evaluated as soon as electromagnetic fields are obtained in the vicinity of a flaw region. Equation 6 has been proved effective in impedance evaluations and has been used in many applications. Auld's theory, which was built around a combination of the reciprocity relation and the Born approximation, has been playing a central role in the existing eddy current POD studies.

A word of caution is necessary concerning the reciprocity formula. Although its use is highly recommended, the reciprocity formula is not the ultimate method for conducting impedance evaluations. In fact, the most versatile method is the direct use of Faraday's law, which works under any circumstances. However, this method usually results in the most computationally intensive efforts. A third possibility is to use the energy conservation law. This method, however, is the least versatile because it is not always possible to estimate an amount of energy escaping from the system. There are no general criteria for method selection. A case-by-case comparison is required, taking cost-effectiveness and the required degree of accuracy into account.

Probe-Flaw Interaction Models. To make the eddy current method quantitative, one must be able to predict flaw signals for any given system. Earlier, equivalent circuit methods were attempted (Ref 23). The present consensus is that detailed evaluations of electromagnetic fields are needed to make sufficiently precise predictions.

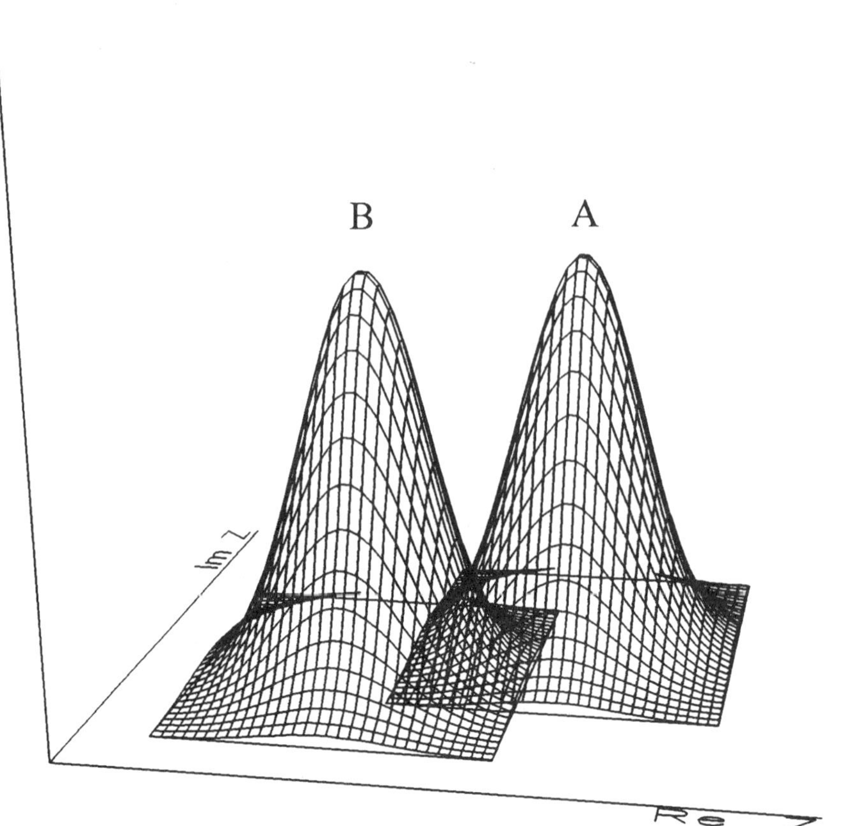

Fig. 7 Illustration of signal distributions. Measured impedance values will follow a certain distribution. Distribution A illustrates an on-flaw signal distribution (signals measured on a flaw), while B illustrates an off-flaw signal distribution.

One therefore needs to solve Maxwell's equations, assuming that all the system parameters (that is, a probe, a part, a flaw, and the inspection geometry) are given. Such mathematical techniques as exact methods, integral equation methods, perturbation theory, and variational methods are frequently used. Various approximation schemes can also be employed. In the literature, the integral equation methods are used most frequently, where the basic differential equations are cast into integral equations using appropriate boundary conditions. Particularly popular are the three-dimensional finite-element method and the boundary integral equation (BIE) method.

Again, selection of the most suitable method for a given situation should be done on a case-by-case basis. The three-dimensional finite-element method is the most versatile, but is also the most expensive. For crack problems, the BIE method may be favored in several aspects. Typically, an eddy current system is piece-wise homogeneous, consisting of several spatial regions separated by sharp boundaries. Moreover, the boundaries are rigid surfaces that are not subject to change over time. Clearly, this is an ideal situation for applying the BIE method. Another favorable aspect is that the aforementioned reciprocity formula uses only field values on boundaries. As long as it is applicable, the BIE approach gives a faster algorithm than the three-dimensional finite-element method. There are, however, situations in which the BIE method fails to work well. For example, it is not effective when dealing with flaws caused by corrosion where conductivity may vary continuously in space. Presumably, this type of flaw is still confined in a local region. If so, it is possible to solve the problem with a combination of the BIE and three-dimensional finite-element methods. To the best of the authors' knowledge, such a combined approach has never appeared in the literature. Perturbation theory may be useful if the variation is small. The perturbation method should certainly be used more frequently in dealing, for example, with a conductivity fluctuation, which is an important noise source to consider in POD models.

Probability of Detection Models. In this section, definitions of probability of detection and probability of false alarm (POFA) will be given in a form suitable for the eddy current inspection method. Various, inequivalent definitions are seen in the literature, but the intention here is to illustrate the basic idea.

Suppose that an eddy current inspection system is specified and that a series of impedance measurements is carried out over a flaw at a given frequency. By nature, the measured signals will scatter around a mean value in a complex impedance plane. Distribution A shown in Fig. 7 illustrates a typical data distribution from a flawed area, while distribution B represents a data set taken under the same controlled conditions but over an unflawed part. Strictly speaking, a flaw signal is given by the difference between the on-flaw and off-flaw signals; the flaw signal corresponds graphically to the shift of the two peak positions, A and B. The fluctuations around the peaks reflect the noise characteristics of the given inspection system.

Once signal/noise distributions are given, one can begin to study the inspectability of the system, as follows. Consider a window in the impedance plane (for example, C in Fig. 8), and make a hypothesis that any signal falling in this window is actually a flaw signal. Then, POD is defined by the probability of this hypothesis being true, which is equal to the integral of distribution A over window C (provided the total integral of A is normalized to unity). Similarly, POFA is given by the integral of B over C, that is, the probability of off-flaw signals being mistaken as flaw signals. The POD and POFA values thus evaluated are functions of C, and when evaluated for various windows, they will form a domain in a POD-POFA plane. Such POD-POFA plots can be used as a measure of inspectability. Alternatively, representatives of POD and POFA values can be used for specification, such as those in which the maximum POD/POFA ratio is achieved.

Therefore, the fundamental task is to determine the on-flaw and off-flaw signal distributions for a given system. One important theoretical task, for example, is to predict the peak positions A and B with model calculations. Noise distributions are more difficult to predict because, by nature, any measurement system will contain certain unpredictable noise sources. Nonetheless, to make eddy current POD packages an effective design aid, as much effort as possible should be devoted to the complete identification and characterization of conceivable noise sources. References 30 and 34 contain specific examples of POD calculations, and Fig. 9 illustrates an example of a POD-POFA plot.

Discussions and Future Directions. Comparing the goal just stated with the current developmental status in the literature (Ref 30, 34) will lead one to expect improvements in eddy current POD models in the following areas. First, theoretical models of limited capabilities have been used for impedance calculations. These limitations inevitably restrict the applicability of the existing POD models. For example, the models have been probe specific, limited to work only for certain measurement geometries, and so on. Any progress in forward calculations will provide more flexible POD models.

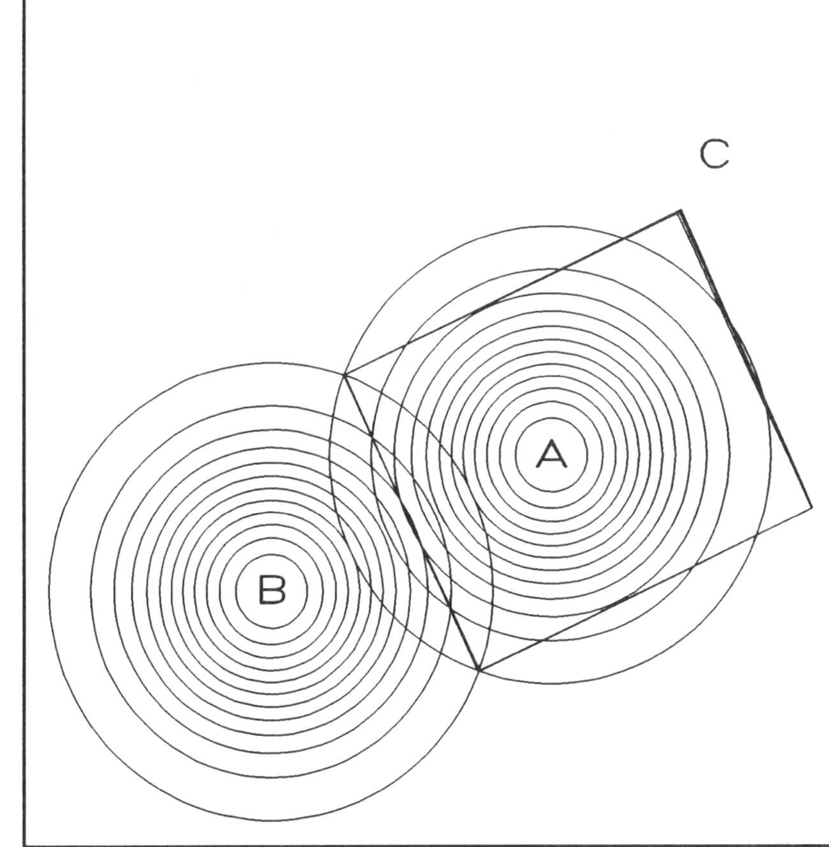

Fig. 8 The same distributions as in Fig. 7 viewed from above. Window C will be used to define POD and POFA as integrals of the distributions over C.

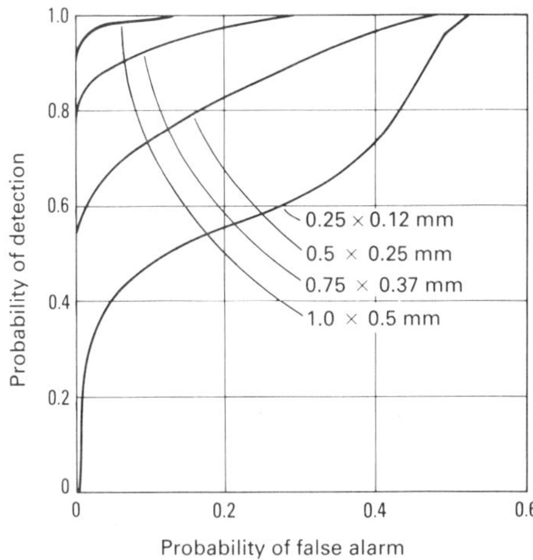

Fig. 9 Sample plot of POD-POFA curves for different flaw sizes. Source: Ref 12

Second, noise characterizations have been accomplished almost exclusively by relying on experimental input. In view of using POD models as a design aid, this is not an ideal situation, because one needs to build a prototype system before calculating POD. Unfortunately, this trend will continue until sufficient data have been accumulated so that reliable general formulas can be extracted empirically. Very little has been done to find purely theoretical noise simulations.

Finally, as a part of software development, a program shell of a POD package should be established, following the general guideline of POD evaluations. Ideally, the package will be written based on a modular programming concept. It may consist of three basic modules: impedance calculation, noise calibration, and POD calculation. Once the shell is written, any progress in either forward calculation or noise characterization can be implemented quickly into the package by replacing the appropriate modules.

There are also two technical recommendations. First, although POD itself can be calculated for a fixed frequency, it is of crucial importance to enable the system to handle a broad range of frequencies. Without this capability, no precise determinations of flaw sizes and shapes can be expected. This applies equally to hardware and software developments. Second, few of the previous measurement techniques and associated POD models fully exploit the complex nature of impedance. Both the real and imaginary parts should be treated on an equal footing as described above in order to make full use of valuable information.

It should be emphasized that, so far, the concept of inspectability has been defined for systems as a whole, so that one can discuss the POD of a system but not the POD of a single component. To validate the notion of the POD of a given probe, it is necessary to integrate the probe into an inspection system in which all the other system components are strictly controlled to meet certain specifications. In the future, such a standardization may become an important issue for comparison purposes.

Radiographic Inspection Model

In radiographic modeling, one is essentially trying to predict how the two-dimensional projection of an irregularly shaped object changes as the position of the object changes with respect to the x-ray beam and detector and, further, to study the degree to which various parameters control image quality (Ref 35, 36). The requirements for a quantitative model of the radiographic process consist of the ability to predict the output of an x-ray generator, the interaction of the x-ray beam with the sample, and the detector characteristics. The capability to handle complex part geometries and nontrivial flaw morphologies is required if the simulation is to have wide application.

Gray *et al.* have developed a full three-dimensional radiography model that generates a two-dimensional (x,y) film image (Ref 37, 38) rather than the one-dimensional source-sample-detector dimension previously modeled (Ref 39). The full shadowing effects and the radial divergence of the x-ray beam in the projections are included, allowing a realistic treatment of complex part geometries. The model consists of:

$$I(x,y,E) = I_0(E) \int_{\text{source}} \frac{e^{-\mu(x,y,E)\rho(x,y)}}{r^2(x,y)} dA \qquad \text{(Eq 7)}$$

$$D(x,y,E) = D_0(1 - e^{-\sigma[(1+\eta)I(x,y,E)t+\delta]}) \qquad \text{(Eq 8)}$$

In Eq 7, E is the energy, I is the intensity immediately above an xy point on the detector, I_0 is the initial intensity produced from the x-ray generator, μ is the energy-dependent linear absorption coefficient, ρ is the x-ray path length through the sample, and r is the distance from the source to the detector.

In Eq 8, D is the film density, σ is the interaction cross section of an x-ray with a film grain, η is the coefficient for the x-ray scattering, t is the time of the film exposure, δ is the natural film fog density, and D_0 is the maximum film density. The integration goes over the area of the source, thus including the affect of source properties on image quality. This model of the x-ray radiographic process has the capability of modeling a very large number of parameters that affect the image quality of an x-ray radiograph. As noted in Eq 7 and 8, the major elements of the model include an x-ray beam model (I_0), the interaction of the beam with a nontrivial sample, and a detector model. These must be combined with a detectability criterion in order to complete a POD prediction.

Beam Model. The model of the initial x-ray beam involves a first-principles prediction of the energy-dependent x-ray intensity spectrum calculated from the bremsstrahlung and characteristic interaction cross sections (Ref 40). These cross sections are calculated from the interaction of the relativistic electron beam with the bound atomic electrons of the target atom and are based on a one-photon production process. The angular dependence of the bremsstrahlung cross section is integrated over all angles. This simplification implies that the calculated intensities will require a scaling factor to match the experimentally measured values. The energy dependence of the spectrum is calculated assuming a thick target attenuation of the electron beam. This beam model has features that allow variation in the beam focal-spot size, shape, and uniformity. Further, the target material and angle of incidence to the electron beam are variables. The model accounts for the inherent filtration of the generated x-ray beam by the target material and the vacuum window of the exit port. Features that the beam model does not address are the generation of off-axis x-rays, the interaction cross sections that involve the generation of more than one photon, the physical mechanics of electron beam focusing, and the variation of the voltage from power supply. These features influence the intensity of the beam and at present are handled in the beam model with an intensity scaling factor.

Sample Interaction Model. The interaction of the x-ray beam with the sample monitors the energy-dependent attenuation through the sample (Ref 41). This portion of the model includes the effects of the sample shape, the flaw morphology, and the configuration of the experimental setup. The sample shape currently used in the simulation is a flat plate with surface roughness. The simple part shape allows the effects of many parameters to be isolated from those involving the complex part shapes.

Two types of flaws are available for study in the simulation. The first is an ellipsoid with nine independent parameters characterizing its shape, orientation, and position. The position of the origin, the three major axes, and three Euler angles give complete, arbitrary control of the position and shape of the ellipsoid. The ellipsoid can be a void or an inclusion. Further, the sample position relative to the source or detector, the thickness, and the material can be chosen as desired, thus providing a wide range of setup geometries. The second flaw is a truncated cone whose apex angle, truncation distance, composition material, and *xyz* position in the sample can all be varied. Additional features are the ability to model density variations in the material and a wide choice of sample materials ranging from low atomic number carbon-base materials to high atomic number materials such as tantalum. As noted above, the sample shape is limited. However, future integration of the simulation into a CAD package will remedy the lack of a wide variety of shapes because any part geometry that the CAD package can generate can be used in the simulation. A note on the introduction of complex CAD-generated part shapes is that the calculation of the POD can take significant amounts of time, at least with present computer capabilities.

Detector Model. If film is used as the detector, a model of the film simulates a number of film speeds, maximum densities, and the energy-dependent effects of the x-ray interaction with the silver halide grains. The film model is derived by considering the number of grains that have absorbed x-ray photons, which absorb an additional photon as a result of the flux incident on the film. The number of photons required for a silver halide grain to be developed is energy dependent. For very low energy (10 keV), several photons are required, while at high energy (1 MeV), one photon can activate several grains. Unfortunately, the energy dependence of grain activation is also dependent on the likelihood of x-ray absorption, a feature that is dependent on the size of the grain and the energy of the x-ray photon. These energy-dependent features are reflected in the x-ray interaction cross section, σ. This cross section, together with the grain size, gives the film model the capability to model the speed of the film (Ref 42). The total amount of silver halide grain present, as represented by the weight of the emulsion, controls the maximum film density, D_0. The fog density, δ, of the unexposed film can be accounted for in the final density.

The film model tracks the random counting statistics noise and in a limited set of sample geometries tracks the noise due to Compton scattering. For simple geometries, the Compton scattering is a uniform intensity at the film and is modeled through the scattering coefficient, η. This is a simplistic approximation to a complex problem. Indeed, the Compton scattering for energies above several hundred kiloelectron volts is the dominant scattering mechanism and can contribute more to the film density than the information-carrying portion of the beam. The additional complexities of part geometry nullify the uniformity of the Compton scattering at the detector plane, making a detailed tracking of the Compton component necessary. The addition of this important effect is the subject of ongoing research.

Detectability. The detectability criterion accounts for the physiological limitation of the human eye to detect, under normal conditions, variations in gray film scale levels. The determination of the limits of gray-scale variation that the human eye can perceive is based on studies of uniform gray background with a circle that has a different gray-scale level (Ref 35). Although this is not the general condition under which radiographs are viewed, this is the criterion currently incorporated into the model. Further modification involves a mechanism to account for degradation of the contrast due to a poor signal-to-noise ratio. One of the features of the simulation is the capability of turning off the noise, thus allowing examination of the noise degradation of the signal. The noise in the film image is, in most cases, dominated by the random fluctuations of the number of photons impinging on a unit area. The result is that the noise magnitude is related to the density of the film; higher noise is associated with low densities, while low noise corresponds to high density. This has led to the incorporation of contrast degradation due to noise and therefore to the inclusion of its influence on flaw signal detectability. This detectability model falls into the amplitude model class of detectability, and although a simple model, it has the major features for adequately describing detectability (Ref 43).

Limitations. At present, there are a number of limitations to the model. They include a limited accounting for the effect of Compton scattering, the need for an intensity scaling factor for the initial beam, and a simple detectability model. Experimental verification of the major components is under way, and the preliminary results are very good. The initial beam model gives results that only need scaling to accurately model the observed x-ray spectrum. The film model can characterize the film response to a photon flux for a wide range of film types. For the full simulation, with only the scaling for the initial beam intensity, the results, when compared to an actual radiograph of a machined cone-shaped void, are in the worst case within a factor or two of the actual film densities. In the best cases, the model results are within 20% of the observed results. The refinement of the model is work that is ongoing.

Applications. With this highly flexible model of the radiographic process, several factors affecting flaw detectability can be studied in a quantitative manner. A sample of the types of effects studied can be seen in Fig. 10 and 11, in which the loss of detectability was simulated for changes in beam hardness and flaw orientation, respectively. The simulation predicts a range of kilovoltages for which adequate contrast and noise levels are present. To perform adequate

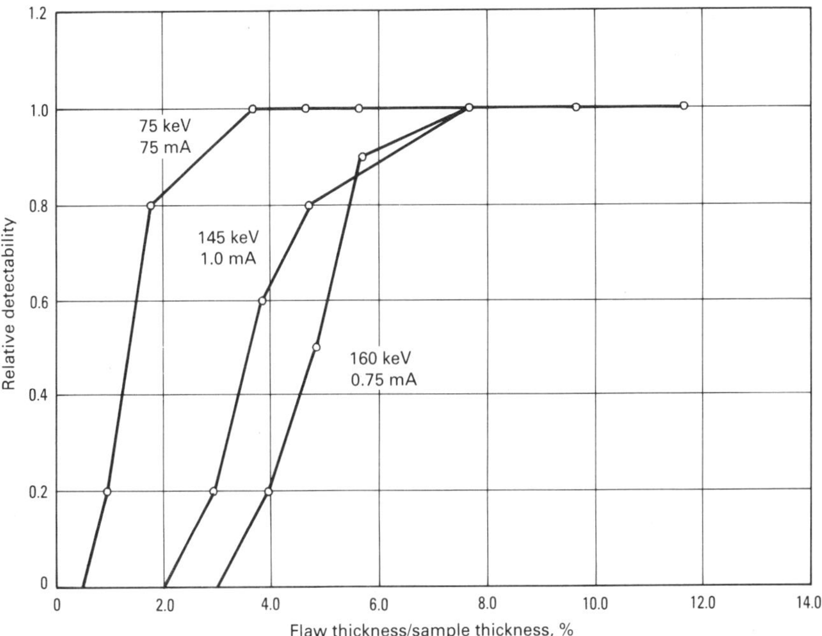

Fig. 10 Simulation results showing the effects on thickness sensitivity of the x-ray beam hardness and the relative size of the flaw to sample thickness

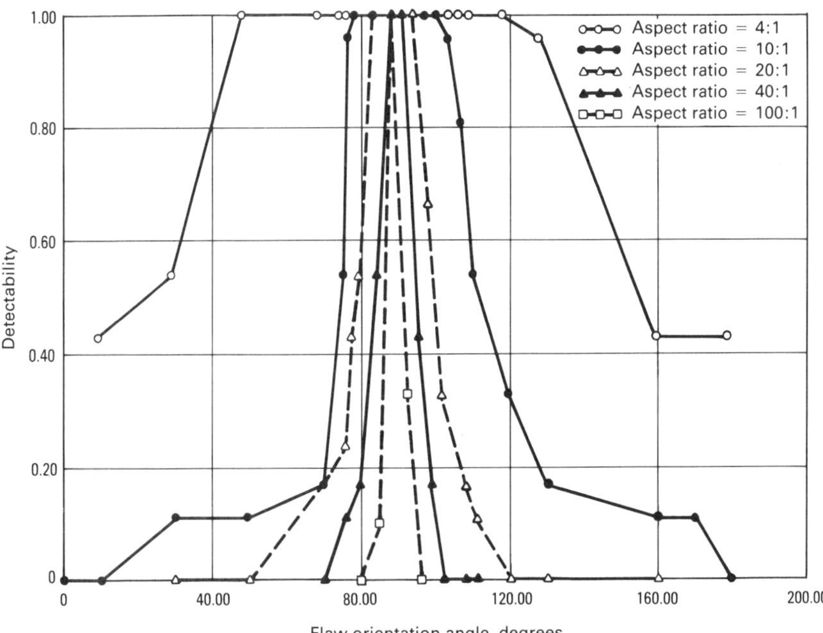

Fig. 11 Detectability of different ellipsoidal flaws as a function of their orientation and shape. The shape of the ellipsoid is changed by varying the aspect ratio, where two of the major axes are kept equal and the third is decreased. The ratio of the constant major axes to the smaller is the aspect ratio. As the aspect ratio increases, the ellipsoid flattens and becomes cracklike. The scale for angular orientation is chosen such that a crack is optimally oriented at 90°.

x-ray inspections, the data must be collected in these plateau regions. It should be noted that the sizes of these optimal regions are also strongly controlled by the flaw shape and size. The loss of detectability was calculated quantitatively based on the physiological limitation of human vision to discern variations in gray shades, the noise level present in the radiograph, and the contrast available.

Figure 10 shows a plot of the effect of beam hardness (the white spectrums of a harder beam have a higher average energy) on the detectability of a flaw where the film type, the sample composition, the x-ray beam shape, and the physical geometric setup were kept the same. The time of the exposure was increased to attain the same film densities for each simulated exposure. As can be seen, the harder beam has less sensitivity but a shorter exposure time, while the softer beam yields greater sensitivity at the cost of longer exposure times. Thus, the trade-off between sensitivity and speed of inspection can be studied.

Figure 11 shows the effect of flaw morphology on detectability. The flaws represented in Fig. 11 have widely varying aspect ratios, thus simulating varying degrees of crack tightness. As the orientation of these cracks with respect to the incident x-ray beam changes, their detectable signal drops abruptly.

There are a number of applications of the forward model of the x-ray radiographic process in the field of production inspection. The first is the collapsing of the full three-dimensional aspect of the model to a one-dimensional line path. This allows quick calculation of the kilovoltage, current, and time required for the exposure of a type of film to the density required. Another is in the area of inspection validation studies. The use of seeded defects and image quality indicators and the sensitivity testing of various parameters are time consuming and expensive, so much so that a full study in which all parameters are varied is rarely done. The use of a simulation program coupled with a set of benchmark trials can greatly expand the scope of this type of certification of an inspection procedure. Modest extensions of the model would facilitate an even wider range of applications. The prediction of a full POD requires only the addition of a probability distribution function describing the flaw distribution, as described in the section "Ultrasonic Inspection Model" in this article. As can be seen from Fig. 10 and 11, the current model is very close to the generation of POD curves for different system parameters. A second area for extension is the simulation of tomographic scans. The addition of a detector model typical of those used in tomographic scans will allow the generation of the projection data required for the reconstruction algorithms.

Future Impact of Models

The three preceding sections describe computer models that can be used to generate quantitative predictions of inspectability for ultrasonic, eddy current, and x-ray film methods. These are all physically based models that explicitly consider the characteristics of the realistic flaws, the typical component geometries and materials, and the practical inspection methods and practices. These or similar models may:

- Replace costly and time-consuming experimental demonstration programs in the prediction of NDE reliability
- Improve the validation and optimization of inspection procedures

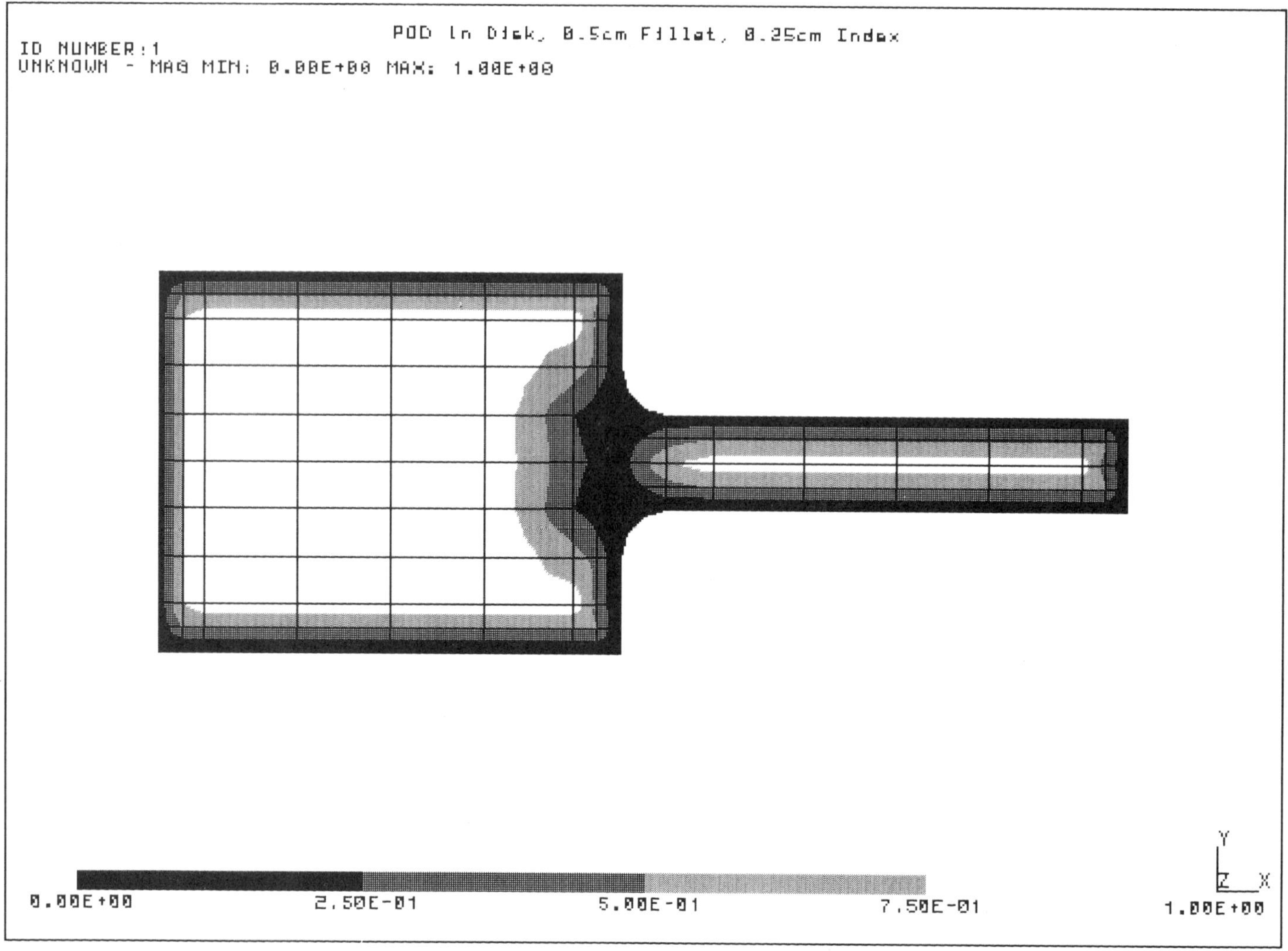

Fig. 12 Example of a CAD-generated display of inspectability (POD) as a function of position within the cross section of a simulated disk. The POD scale ranges from black for the lowest to white for the highest POD values. The low POD in the fillet region is due to the combination of a coarse scan mesh and a tight fillet radius.

- Improve component design and the definition of the life cycles

Another potential application of NDE reliability models may involve integration with CAD systems. This approach could allow the assessment of inspectability during the design stage just as form, fit, and function are analyzed during design. The integration of NDE reliability models with a CAD system may also allow quantitative estimates of inspection reliability as a function of design changes. The following example, illustrated in Fig. 12 and 13, describes this application of NDE/CAD integration for ultrasonic inspectability (Ref 8). The inspectability (POD) model for ultrasonic NDE has achieved the highest state of technical maturity, and initial strides have been made toward its incorporation into CAD systems. However, the general idea is applicable to NDE reliability models of inspection methods other than ultrasonics.

Example 1: Integration of an Ultrasonic Reliability Model with a CAD System. Figures 12 and 13 show the result of a computer simulation of ultrasonic POD for oblate spheroidal inclusions in a rotationally symmetric component, such as a jet engine turbine disk. The display shows the results obtained from a cross section of the component, and the gray shades refer to the POD value, in the range from zero to unity, with black being the lowest and white the highest POD values. In the POD simulation, the inclusions were assumed to have a preferential orientation with their largest face parallel to the component surface, and the component was assumed to be scanned in pulse-echo mode with a focused probe normal to the surface of the part and focused at the surface. The interior points of the component, at which the POD values were calculated, were defined to be the nodes of an automatically generated finite-element mesh that was imposed on the solid model of the part. This artifice allowed the built-in postprocessing graphical interface of the CAD package to be used to display the inspectability results in a format that would be very familiar to a designer who is experienced in the use of the package.

Figure 12 shows the result of a nominal scan plan and component design. In particular, the scan mesh was quite coarse, causing poor detectability in the web section to the right of the cross section and near the surface of the entire component. The fillet region, at the juncture between the web and the thicker bore area to the left of the cross section, was also assumed to have a small radius of curvature. Below this fillet region, the POD values are quite low, which is a result of the combination of the ultrasonic beam being focused on the surface of the part and of the focusing effect of the surface curvature. This caused the fields in the region below the fillet to be rapidly diverging and therefore of low amplitude.

To improve the POD levels throughout the component, two questions must be considered. The first is whether the inspection technique was optimized with respect to the design of the component, and the second is whether the design itself is inspection limit-

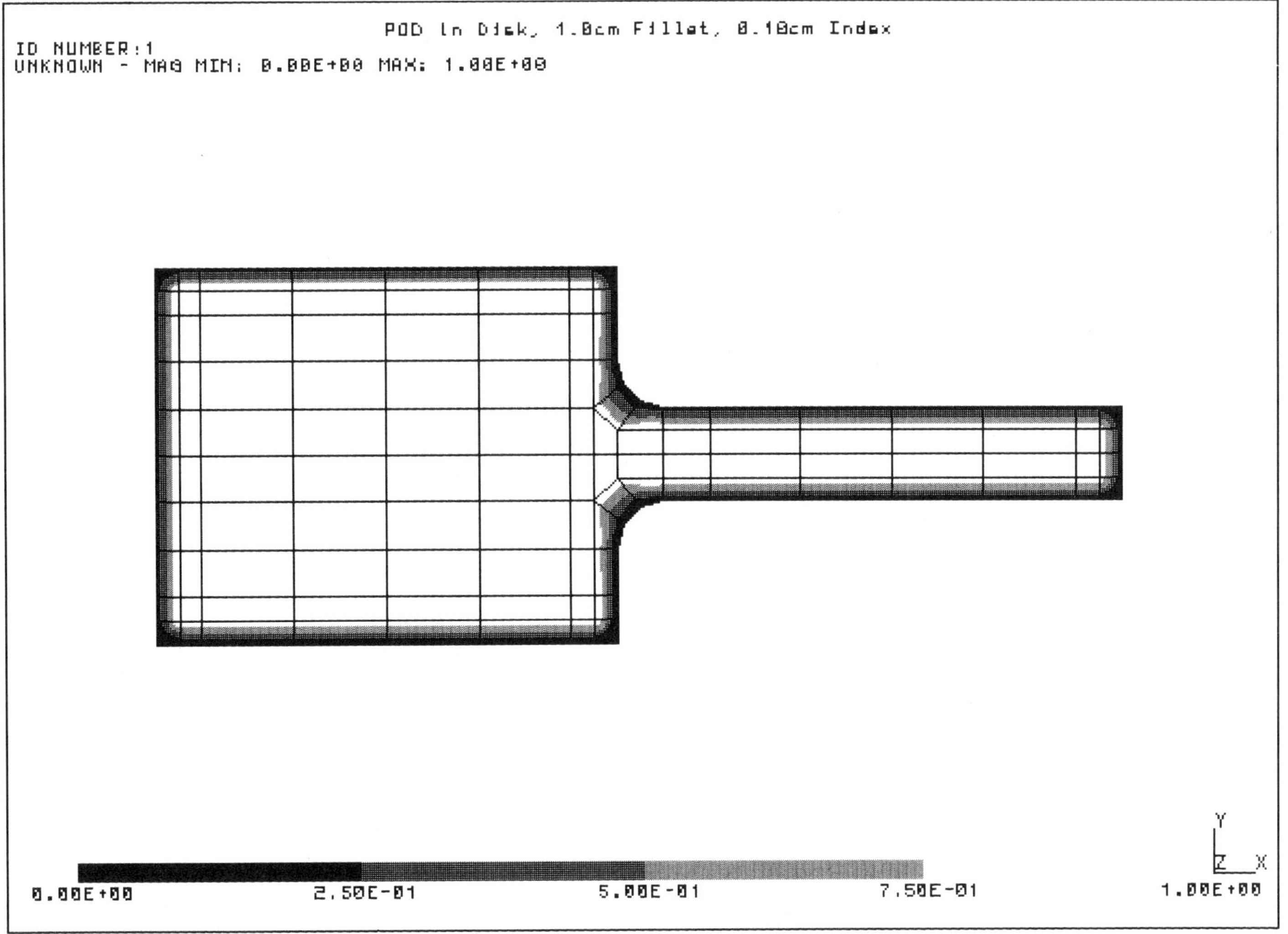

Fig. 13 A display similar to the one shown in Fig. 12, except that the scan mesh has been refined and the fillet has a larger radius of curvature. The combination of scan plan and component design modification has significantly improved inspectability throughout the disk.

ed. The first question is crucial because a simple modification of the scan plan will almost certainly be a better solution than the change in operational characteristics of the component created by a modification of its design. In some cases, however, both the scan plan and the design will require optimization. Figure 13 shows the result of both options; the scan mesh was refined, which caused the enhanced POD in the near-surface and web regions, and the fillet radius of curvature was increased, which improved the POD in the fillet region.

Example 2: Implications on Standardization. As the design applications of reliability models become more widespread, the questions of quantitative standards will become more important. Because the manufacturer will usually be at a remote location from the designer and may be affiliated with a different organization, it will be necessary to communicate the inspection procedure at the time that manufacturing procedures are specified. Absolute calibration standards will then be essential to ensure that the inspection, including specified thresholds, is properly set up and implemented. This may require a move away from the use of reference reflectors, such as flat-bottom holes (Ref 44-45). Recent work in ultrasonics has shown how the measurement model, Eq 3, can be used as the basis for such a procedure (Ref 46).

Example 3: Uses of Artificial Intelligence in Optimization. An important part of the consideration of inspectability in CAD procedures is optimization of the inspection and/or design. This was done intuitively by a highly trained operator in Example 1. However, in the future, it may be desirable to eliminate the operator through the use of artificial intelligence techniques, such as expert systems. Although this has not yet been accomplished, a step in that direction has been the use of the ultrasonic measurement model in the development of an expert system to differentiate cracklike flaws from volumetric flaws (Ref 47).

REFERENCES

1. S.T. Rolfe and J.M. Barson, *Fracture and Fatigue Control in Structures: Application of Fracture Mechanics*, Prentice-Hall, 1977
2. R.B. Thompson and T.A. Gray, Use of Ultrasonic Models in the Design and Validation of New NDE Techniques, *Philos. Trans. R. Soc. (London) A*, Vol 320, 1986, p 329-340
3. R.B. Thompson and H.N.G. Wadley, The Use of Elastic Wave-Material Structure Interaction Theories in NDE Modeling, *CRC Crit. Rev. Solid State Mater. Sci.*, in press
4. J.M. Coffey and R.K. Chapman, Application of Elastic Scattering Theory for Smooth Flat Cracks to the Quantitative Prediction of Ultrasonic Defect Detection and Sizing, *Nucl. Energy*, Vol 22, 1983, p 319-333
5. R.B. Thompson, D.O. Thompson, H.M. Burte, and D.E. Chimenti, Use of

Field-Flaw Interaction Theories to Quantify and Improve Inspection Reliability, in *Review of Progress in Quantitative Nondestructive Evaluation*, Vol 3A, D.O. Thompson and D.E. Chimenti, Ed., Plenum Press, 1984, p 13-29
6. T.A. Gray and R.B. Thompson, Use of Models to Predict Ultrasonic NDE Reliability, in *Review of Progress in Quantitative Nondestructive Evaluation*, Vol 5, D.O. Thompson and D.E. Chimenti, Ed., Plenum Press, 1986, p 911
7. H.M. Burte and D.E. Chimenti, Unified Life Cycle Engineering: An Emerging Design Concept, in *Review of Progress in Quantitative Nondestructive Evaluation*, Vol 6B, D.O. Thompson and D.E. Chimenti, Ed., Plenum Press, 1987, p 1797-1809
8. D.O. Thompson and T.A. Gray, The Role of NDE in Global Strategies for Materials Synthesis and Manufacturing, *Proceedings of the 1988 Fall Meeting*, Materials Research Society, in press
9. B.A. Auld, General Electromechanical Reciprocity Relations Applied to the Calculation of Elastic Wave Scattering Coefficients, *Wave Motion*, Vol 1, 1979, p 3
10. R.B. Thompson and T.A. Gray, A Model Relating Ultrasonic Scattering Measurements Through Liquid-Solid Interfaces to Unbounded Medium Scattering Amplitudes, *J. Acoust. Soc. Am.*, Vol 74, 1983, p 1279
11. R.B. Thompson and T.A. Gray, Analytical Diffraction Corrections to Ultrasonic Scattering Measurements, in *Review of Progress in Quantitative Nondestructive Evaluation*, Vol 2, D.O. Thompson and D.E. Chimenti, Ed., Plenum Press, 1983, p 567
12. R.B. Thompson and E.F. Lopes, The Effects of Focusing and Refraction on Gaussian Ultrasonic Beams, *J. Nondestr. Eval.*, Vol 4, 1984, p 107
13. R.B. Thompson, T.A. Gray, J.H. Rose, V.G. Kogan, and E.F. Lopes, The Radiation of Elliptical and Bicylindrically Focused Piston Transducers, *J. Acoust. Soc. Am.*, Vol 82, 1987, p 1818
14. B.P. Newberry and R.B. Thompson, A Paraxial Theory for the Propagation of Ultrasonic Beams in Anisotropic Solids, *J. Acoust. Soc. Am.*, to be published
15. B.P. Newberry, R.B. Thompson and E.F. Lopes, Development and Comparison of Beam Models for Two-Media Ultrasonic Inspection, in *Review of Progress in Quantitative Nondestructive Evaluation*, Vol 6, D.O. Thompson and D.E. Chimenti, Ed., Plenum Press, 1987, p 639
16. C.F. Ying and R. Truell, Scattering of a Plane Longitudinal Wave by a Spherical Obstacle in an Isotropically Elastic Solid, *J. Appl. Phys.*, Vol 27, 1956, p 1086
17. J.L. Opsal and W.M. Visscher, Theory of Elastic Wave Scattering: Applications of the Method of Optimal Truncation, *J. Appl. Phys.*, Vol 58, 1985, p 1102
18. J.-S. Chen and L.W. Schmerr, Jr., The Scattering Response of Voids—A Second Order Asymptotic Theory, in *Review of Progress in Quantitative NDE*, Vol 7, D.O. Thompson and D.E. Chimenti, Ed., Plenum Press, 1988, p 139
19. L. Adler and J.D. Achenbach, Elastic Wave Diffraction by Elliptical Cracks: Theory and Experiment, *J. Nondestr. Eval.*, Vol 1, 1980, p 87
20. B.D. Cook and W.J. Arnoult III, Gaussian-Laguerre/Hermite Formulation for the Nearfield of an Ultrasonic Transducer, *J. Acoust. Soc. Am.*, Vol 59, 1976, p 9
21. J.D. Achenbach, A.K. Gautesen, and H. McMaken, *Ray Methods for Waves in Elastic Solids*, Pittman Publishing, 1982
22. S.O. Rice, Mathematical Analysis of Random Noise, *Bell Syst. Tech. J.*, Vol 23, 1944, p 282; Vol 24, 1945, p 96
23. H.L. Libby, *Introduction to Electromagnetic Nondestructive Test Methods*, Wiley-Interscience, 1971
24. M. Burrows, "A Theory of Eddy Current Flaw Detection," Ph.D. thesis, University of Michigan, 1964
25. C. Dodd and W.E. Deeds, Analytical Solution to Eddy-Current Probe-Coil Problem, *J. Appl. Phys.*, Vol 39, 1968, p 2829
26. A.H. Kahn, R. Spal, and A. Feldman, Eddy-Current Losses Due to a Surface Crack in a Conducting Material, *J. Appl. Phys.*, Vol 48, 1977, p 4454
27. W.D. Dover, F.D.W. Charlesworth, K.A. Taylor, R. Collins, and D.H. Michael, in *The Measurement of Crack Length and Shape during Fatigue and Fracture*, C.J. Beevers, Ed., EMAS, 1980
28. B.A. Auld, Theoretical Characterization and Comparison of Resonant-Probe Microwave Eddy-Current Testing With Conventional Low-Frequency Eddy-Current Methods, in *Eddy Current Characterization of Materials and Structures*, STP 722, G. Birnbaum and G. Free, Ed., American Society for Testing and Materials, 1981
29. T.G. Kincaid, A Theory of Eddy Current NDE for Cracks in Nonmagnetic Materials, in *Review of Progress in Quantitative NDE*, Vol 1, D.O. Thompson and D.E. Chimenti, Ed., Plenum Press, 1982
30. A.J. Bahr, System Analysis of Eddy-Current Measurements, in *Review of Progress in Quantitative NDE*, Vol 1, D.O. Thompson and D.E. Chimenti, Ed., Plenum Press, 1982; A.J. Bahr and D.W. Cooley, Analysis and Design of Eddy-Current Systems, in *Review of Progress in Quantitative NDE*, Vol 2A, D.O. Thompson and D.E. Chimenti, Ed., Plenum Press, 1983; J.R. Martinez and A.J. Bahr, Statistical Detection Model, in *Review of Progress in Quantitative NDE*, Vol 3A, D.O. Thompson and D.E. Chimenti, Ed., Plenum Press, 1984
31. B.A. Auld, F.G. Muennemann, and D.K. Winslow, *J. Nondestr. Eval.*, Vol 2, 1982, p 1; B.A. Auld, F.G. Muennemann, and M. Riaziat, Quantitative Modeling of Flaw Responses in Eddy Current Testing, in *Research Techniques in Nondestructive Testing*, Vol 7, R.S. Sharpe, Ed., Academic Press, 1984
32. R.E. Beissner, Boundary Element Model of Eddy Current Flaw Detection in Three Dimensions, *J. Appl. Phys.*, Vol 60, 1986, p 352-360; Analytical Green's Dyads for an Electrically Conducting Half-Space, *J. Appl. Phys.*, Vol 60, 1986, p 855-858
33. J.C. Moulder, P.J. Shull, and T.E. Capobianco, Uniform Field Probe: Experiments and Inversion for Realistic Flaws, in *Review of Progress in Quantitative NDE*, Vol 7B, D.O. Thompson and D.E. Chimenti, Ed., Plenum Press, 1987
34. R.E. Beissner, K.A. Bartels, and J.L. Fisher, Prediction of the Probability of Eddy Current Flaw Detection, in *Review of Progress in Quantitative NDE*, Vol 7B, D.O. Thompson and D.E. Chimenti, Ed., Plenum Press, 1988
35. R. Halmshaw, *Physics of Industrial Radiography*, Applied Science, 1966
36. P. DeMeester and W. Aerts, in *Research Techniques in Nondestructive Testing*, Vol VI, R.S. Sharpe, Ed., Academic Press, 1982
37. J.N. Gray, Three Dimensional Modeling of Projection Radiography, in *Review of Progress in Quantitative Nondestructive Evaluation*, Vol 7A, D.O. Thompson and D.E. Chimenti, Ed., Plenum Press, 1988, p 343-348
38. J.N. Gray, F. Inanc, and B.E. Shull, Three Dimensional Modeling of Projection Radiography, in *Review of Progress in Quantitative Nondestructive Evaluation*, Vol 8A, D.O. Thompson and D.E. Chimenti, Ed., Plenum Press, 1989
39. D.E. Rundquist, in *Research Techniques in Nondestructive Testing*, Vol IV, R.S. Sharpe, Ed., Academic Press, 1982
40. C.M. Lee, L. Kissell, R.H. Pratt, and H.K. Tseng, *Phys, Rev.*, Vol A13 (No. 5), 1976, p 1714-1727
41. W.H. McMaster, N. Kerr Del Grande,

J.H. Mallett, and J.H. Hubbell, "Compilation of X-ray Cross Sections," UCRL-50174 Section II Rev. 1, Los Alamos National Laboratory, 1974

42. H.H. Barrett and W. Swindell, in *Radiological Imaging: The Theory of Image Formation, Detection, and Processing*, Vol 1, Academic Press, 1981, p 206
43. L.D. Loo, K. Doi, and C.E. Metz, *Phys. Med. Biol.*, Vol 29 (No. 7), 1984, p 837-856
44. B.R. Tittmann and D.O. Thompson, Approach to a Self-Consistent Calibration Procedure of an Ultrasonic System, *Mater. Eval.*, Vol 35, 1977, p 96-102
45. B.R. Tittmann, D.O. Thompson, and R.B. Thompson, Standards for Quantitative Nondestructive Evaluation, in *Nondestructive Testing Standards—A Review*, M. Berger, Ed., STP 624, American Society for Testing and Materials, 1977, p 295-311
46. D.D. Bennink and A.L. Pate, Investigation of Scatter in Ultrasonic Responses Caused by Variability in Transducer and Material Properties, in *Review of Progress in Quantitative Nondestructive Evaluation*, Vol 7A, D.O. Thompson and D.E. Chimenti, Ed., Plenum Press, 1988, p 621-628
47. S.W. Nugen, K.E. Christensen, L.-S. Koo, and L.W. Schmerr, FLEX—An Expert System for Flaw Classification and Sizing, in *Review of Progress in Quantitative Nondestructive Evaluation*, Vol 7A, D.O. Thompson and D.E. Chimenti, Ed., Plenum Press, 1988, p 445-451

Statistical Methods

Statistical Quality Design and Control*

Richard E. DeVor, University of Illinois, Urbana-Champaign
and Tsong-how Chang, University of Wisconsin, Milwaukee

A MAJOR REVOLUTION in the industrial sector has been taking place in America during the 1980s as manufacturers strive to regain the competitive position they once held in the world marketplace. One element of this revolution has centered around a renewed emphasis on quality, with an approach aimed at preventing defective materials from being manufactured through improved process monitoring and diagnosis and at designing quality into the product from the very beginning. The concepts and methods of Deming and others have had a profound impact on the way quality is viewed from the manufacturing/process perspective. The simple but powerful statistical methods for process control developed by Shewhart some time ago have been successfully revived and applied on a very broad basis. In the engineering design arena, the methods of Taguchi, referred to as off-line quality control, have been successfully used for more than 30 years to provide a sound basis for improved product/process design. From the total system point of view, the concept of company-wide quality control, which has been practiced in Japan for some time, is now receiving considerable attention. In particular, recent emphasis has been placed on quality function deployment as a means to transmit customer needs through the organization both vertically and horizontally.

The work of Shewhart in the 1920s led to a sound approach to the scrutiny of process variation and the diagnosis and removal of process faults. However, the statistical approach to the sampling of process output prior to shipping to determine the extent to which it conformed to specifications dominated the quality field from the 1930s through the 1970s. Unfortunately, it has only recently been recognized that this product control approach to quality control contributes little to the enhancement of competitive position. The recognition that quality and productivity can move together in the right direction only when the process is evaluated, finding the root cause of process faults and taking action to remove them, is today reshaping the meaning and intent of quality control.

The dramatic change in the meaning and application of quality control as a discipline has both a philosophical and an analytical side. These two aspects of the problem must be understood together. This article will present some fundamental elements of statistical thinking and methods for quality design and control. Commonly used techniques employing control charts and the design of experiments are discussed with illustrations.

Quality and Productivity Fundamentals

Dr. W. Edwards Deming is the man largely responsible for bringing statistical thinking and methods for quality improvement to Japan after World War II. An eminent statistician in America during the 1930s and 1940s, Deming was one of the men in the forefront of the statistical quality control scene during World War II. After the war, Deming's philosophies and teachings fell on the deaf ears of American management, and the quality effort fires that burned so brightly during the war years slowly went out. However, the Japanese became aware of Deming's work and invited him to Japan, where he met with and caught the attention of top management.

When Deming first traveled to Japan in the late 1940s, he found a climate that was quite conducive to the promotion of his concepts and methods. On the one hand, Japan appeared to have a solid base of statistical expertise, although its energies had been directed primarily toward mathematical theory and the application of that theory in nonmanufacturing environments such as agriculture. On the other hand, Deming found an industrial leadership base in Japan very eager to listen to what he had to say. With the lessons learned from his experience in attempting to promote quality improvement in America, Deming's first order of business when he arrived in Japan was to conduct a series of top management training seminars in which he laid out what needed to be done to place quality improvement on an institutional basis within any organization. The obligations and responsibilities of management that he spelled out in these seminars came to be known as his Fourteen Points for management.

Perhaps it is the combination of upper management perspective and a firmly rooted background in mathematical statistics that has enabled Deming to sustain his efforts in a leadership position over more than 50 years. He has been able to tell management what they ought to do and then to provide the rigorous analytical tools and methods needed to carry out his directions. His teachings are heavily oriented toward the use of statistical thinking and methods to identify opportunities for quality and productivity improvement. Deming has developed a road map for management to follow to enhance competitive position over the long run. This road map, referred to as his Fourteen Points (management's obligations), is summarized below. It should be noted that in the spirit of his Fourteen Points, Deming continues to this day to refine and improve these tenets. As a result they themselves are continually changing. An excellent discussion of Deming's Fourteen Points is provided in Ref 1. Deming's Fourteen Points are listed below:

- Create constancy of purpose for the improvement of product or service
- Adopt the new philosophy of process control and variation reduction
- Cease dependence on mass inspection for quality control
- End the practice of awarding business on the basis of price
- Improve constantly and forever the system of production and service in order to

*Adapted from R.E. DeVor and T-H. Chang, *Statistical Quality Design and Control: Contemporary Concepts and Methods*, Macmillan, to be published

improve quality and productivity and thus continuously decrease costs
- Institute thorough and better job-related trainings
- Institutionalize leaderships
- Drive out fears, so that everyone may work effectively for the company
- Break down barriers between departments
- Eliminate slogans, exhortations, and targets for the workforce that ask for zero defects and new levels of productivity
- Eliminate work standards on the factory floor
- Remove the barriers that rob employees at all levels in the company of their right to pride of workmanship
- Institute a vigorous program of education and self-improvement
- Put everybody in the organization to work to accomplish the transformation

The Fourteen Points are clearly the responsibility of management. They define the essential elements of the institutionalization of quality and productivity improvement through statistical thinking and methods. Other pioneers in quality planning and management include Dr. J.M. Juran (Ref 2) and Dr. A.V. Feigenbaum (Ref 3).

Quality and Engineering Specifications

In today's economic age, it is simply impractical to use concepts and measures of quality that do not relate the achievement of function in the field to the engineering design process. Definitions of quality that promote improvement to some acceptable plateau of performance will inhibit the continual pursuit of never-ending improvement and will therefore have a weak and perhaps opposing relationship to process performance in terms of efficiency/productivity. Instead, to improve competitive position, the definition of quality should emphasize:

- A design criterion to provide a quantitative basis to move the quality issue upstream to engineering design
- Prevention rather than containment to promote focus on the process, not the product, in a manufacturing sense
- Continual pursuit of never-ending improvement to be strongly tied to, and therefore promote, the issue of productivity
- Consumer versus producer orientation to quantify/measure the loss imparted to the customer as a result of poor quality rather than the loss imparted to the producer

The traditional definitions of quality often fall short in terms of their ability to articulate quality in a way that can foster improvement in competitive position. In particular, the association of quality with conformance to the engineering specification puts the measurement of quality on an attribute basis and provides little more than a shipping criterion when it is essential that quality be articulated in a manner that enables it to be used as a design criterion. The view of quality as conformance to specifications promotes the product control approach to quality control and therefore stands as a significant inhibitor to the adoption of a process control approach to manufacturing and the integration of quality and the design process.

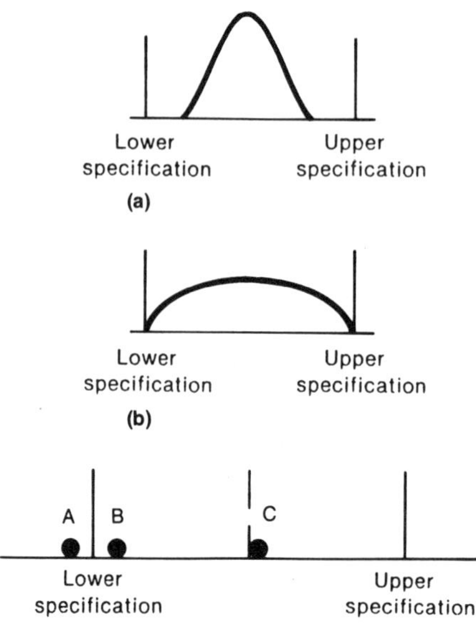

Fig. 1 Graphical depiction of three traditional interpretations of engineering specifications. (a) Desirable. (b) Acceptable. (c) Nominal

Figure 1 illustrates the traditional interpretation of the engineering specification. Figure 1(a) and (b), two representations of a process as a statistical distribution of measurements are shown relative to a certain bilateral specification. Although the bell-shaped process distribution in Fig. 1(a) might be preferred to the more loaf-shaped process distribution in Fig. 1(b), there generally would not be much distinction between the two cases in terms of quality; that is, in both cases, virtually all the processes/parts are conforming to the specifications.

When one begins to consider the dots labeled A, B, and C as representing three different manufactured parts in Fig. 1(c), part A is generally interpreted as unacceptable because it is outside the specifications, while parts B and C are considered acceptable because they are within the specifications. The crucial point in this interpretation is the fact that a distinction is generally not made between parts B and C so far as quality is concerned. In fact, it is unlikely that there is very much difference between parts A and B in terms of functional performance, while part C will likely perform considerably better than either A or B.

Loss Function Concept

The methods of Taguchi have provided a whole new perspective on the way the engineering design process is viewed and have provided further support to the methods of design of experiments for quality design and improvement (Ref 4, 5). The design of experiments as a discipline has a rich and extensive history that finds its origin in the work of Sir Ronald Fisher during the first quarter of the 20th century (Ref 6). From an industrial point of view, the work of Dr. G.E.P. Box and his coworkers has left an indelible mark on the design of experiments in particular and the theory and practice of industrial statistics in general (Ref 7, 8).

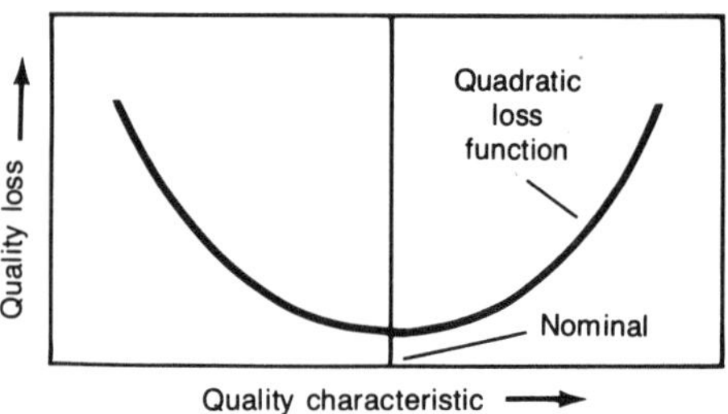

Fig. 2 Plot of quality loss versus quality characteristic to illustrate concept of loss function of quality

The Taguchi approach to quality engineering has a number of significant strengths. In particular, Taguchi has placed a great deal of emphasis on the importance of minimizing variation as the primary means of improving quality. The concepts of designing products such that the performance of each is insensitive to environmental conditions and making this happen at the design stage through the use of design of experiments have been cornerstones of the Taguchi methodology for quality engineering.

Taguchi suggests that it is important to think of quality in terms of the loss imparted to society during product use as a result of functional variation and harmful effects. Harmful effects refer to the side effects that are realized during the use of the product and are unrelated to product function. Functional variation refers specifically to the deviation of product performance from that intended by design, that is, from the nominal. Specific sources of functional variation will be discussed later in this article. Taguchi uses a loss function concept to quantify quality as loss due to functional variation. Figure 2 illustrates the idea of the loss function.

The loss function concept as a way to measure quality clearly suggests that the goal of design and manufacturing is to develop products and processes that perform on target with smallest variation. It places tremendous importance on the reduction of variation to achieve the most consistently performing products and processes. To drive performance toward this overall objective, more must be learned about the way in which the important parameters of the product/process influence performance. Over the years, the methods of design of

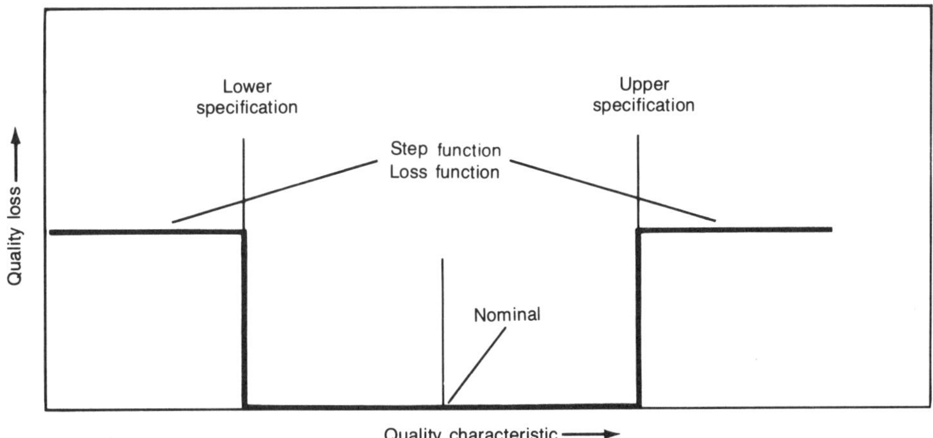

Fig. 3 Plot of quality loss versus quality characteristic to illustrate loss function of engineering specification

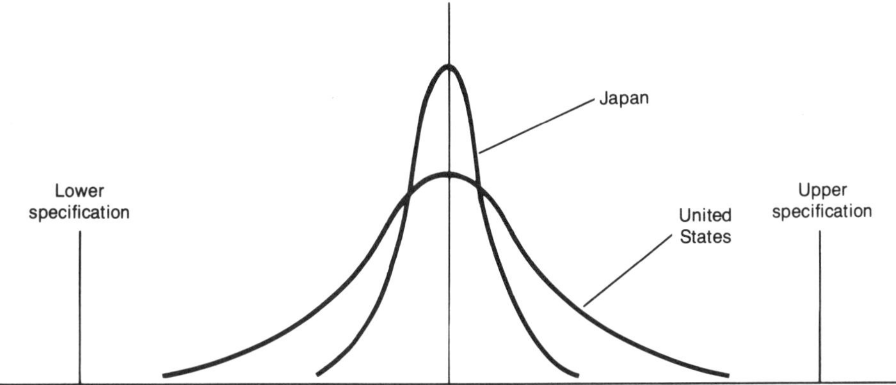

Fig. 4 Comparison of critical dimensions for transmission components manufactured in the United States versus those manufactured in Japan

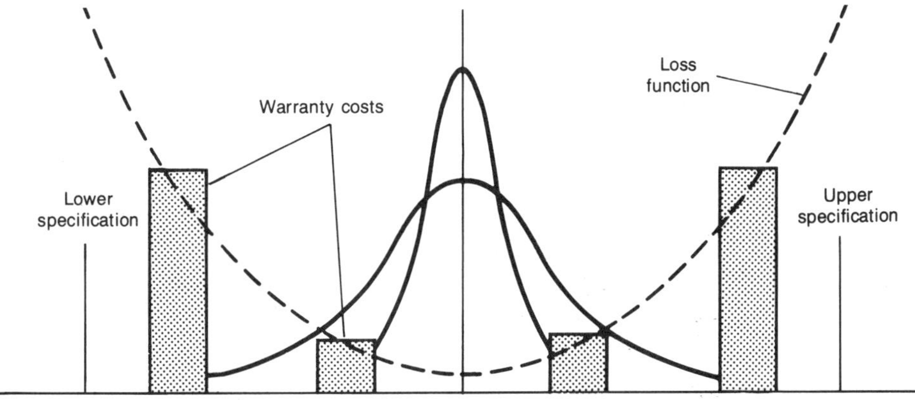

Fig. 5 Loss function interpretation of the engineering specifications in Fig. 4 in terms of warranty costs

experiments have been proved to be an effective way to do this.

Figure 3 provides an interpretation of the loss function concept when overlaid on the traditional engineering specification definition of quality. In this way, no loss would be considered as long as the quality characteristic in question lies within the lower and upper specification. Outside of the specifications, the product is considered to be unacceptable, and a constant loss would therefore be realized. This loss would likely be measured in terms of scrap cost or rework cost. The following case study suggests that this interpretation is illogical.

Variation Reduction and the Loss Function. The loss function view of quality suggests that there exists a clear economic advantage to reducing variation in the performance of a product. The case study described below clearly indicates how such reduced manufacturing imperfection can lead to a reduction in costs and therefore an improvement in competitive position.

Several years ago an automobile company performed a study to compare the manufacturing variations evident in certain transmission components for comparable transmissions made in the United States and in Japan. Random samples of transmissions were selected in each case. The transmissions were disassembled, and a number of critical dimensions were measured and recorded.

Figure 4 illustrates the general findings of the study. In particular, in the case of the U.S. transmissions, the critical dimensions generally consumed about the middle 75% of the tolerance range specified. Normally, one would conclude that the capabilities associated with the manufacturing processes are therefore well within normal expectations. However, the same critical dimensions for the Japanese counterpart of this transmission consumed only about the middle 25% of the same tolerance range. The question that begs to be asked is, Why would the Japanese strive to make the parts to such tight tolerances?

The warranty cost bars shown in Fig. 5 have been plotted at the extremes of the two distributions of the critical dimensions to depict the relative costs associated with variability. It is clear that for the U.S. transmissions the cost associated with repair and replacement is significantly larger than that for their Japanese counterparts. This comparison of the economic data associated with the two transmissions would suggest that there exists a definite relationship between variability and cost in terms of loss incurred due to functional variation. One could think of the warranty costs as literally mapping out the loss function.

Robust Design Approach to Quality Design

The design process in general and the design/manufacturing interface in particular have received considerable attention over the past 20 years. Computer-aided design and manufacture has dealt primarily with the computerization of the design and manufacturing processes and in particular with the translation of design specifications into manufacturing procedures and activities. The computer has greatly facilitated this translation, and as a result, lead times have been reduced. Design modeling and analysis have been strengthened and extended. Standardization and rationalization of both design characteristics and manufacturing process characteristics have been advanced, and the efficiency of the interactive process at the design/manufacturing interface has been improved.

The concepts of design for manufacturability and design for assembly have been the subject of considerable research and development. Numerous specific models have been proposed and refined and will not be discussed here. This work is aimed at over-

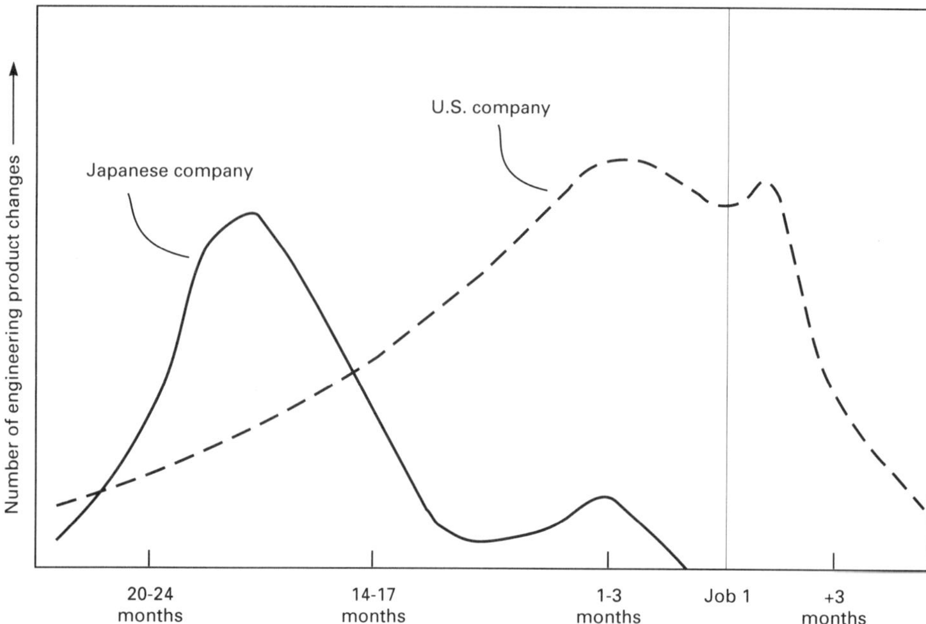

Fig. 6 Comparison of Japanese and American product design life cycles

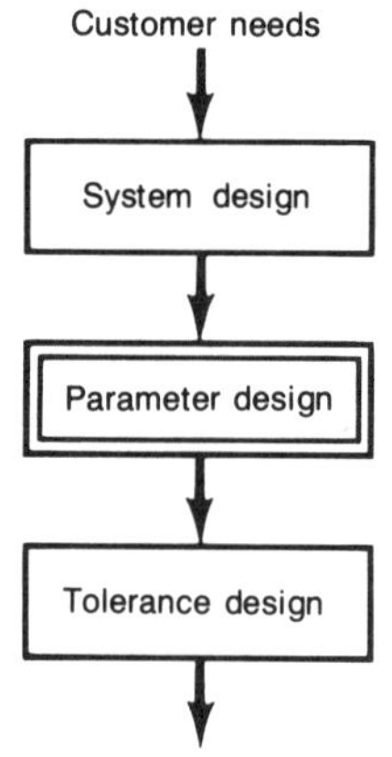

Fig. 7 Block diagram illustrating Taguchi's three-stage design process model

coming the difficulties precipitated by the traditional over-the-wall design philosophy.

The magnitude of the problem with the over-the-wall approach to design and manufacturing can perhaps be measured in terms of the number of times the part drawing is thrown back and forth over the wall. As the initial design concept progresses from prototype testing and development to final design detailing and ultimately to the initiation of production and beyond, the number of design interactions has in the past been too high. In Ref 9, typical Japanese and U.S. companies are contrasted in this regard with an illustration similar to Fig. 6.

Taguchi methods have strong engineering orientation and focus primary attention on the engineering design process, particularly the projection of a three-stage design process model of system design, parameter design, and tolerance design. Figure 7 illustrates the Taguchi design process model (Ref 4, 5).

Systems Design Stage. In the initial stage, systems design (the available science, technology, and experience bases) is used to develop and select the basic design alternative to meet customer needs. A variety of techniques may be useful in specifically mapping the relationship between customer needs and the selection of design configuration and parameters that will effectively meet those needs. The methods of quality function deployment may be most useful at this stage of the design process.

At the parameter design stage, interest focuses on the selection of the specific nominal values for the important design parameters. The overarching selection criterion is to identify those nominal values that minimize the transmission of functional variation to the output performance as a result of the presence of noise factors. It is at the parameter design stage that Taguchi strongly advocates the use of design of experiments methods.

The tolerance design stage of the design process concentrates on the selection of allowable tolerances for the important design parameters. The loss function concept of quality is used to provide a basis for striking the proper economic balance in the selection of design.

Product/Process Performance Factors. Taguchi suggests an engineering interpretation of the varying roles that the important system factors/variables play in influencing product/process performance. Taguchi emphasizes the importance of evaluating quality performance as part of the design process. Taguchi classifies the factors that can influence product/process performance into four categories:

- *Signal factors*: Factors that can be adjusted by the user to attain the target performance. Steering angle, for example, is a signal factor for the steering mechanism of an automobile
- *Control factors*: Product/process design parameters whose values are to be determined during the design process. One design activity is to select the optimum levels of the control factors according to an appropriate design criterion
- *Noise factors*: Factors that are either inherently uncontrollable or impractical to control because of technological/economic reasons. Taguchi further classifies these into outer and inner noises
- *Scaling/leveling factors*: Special cases of control factors. They are factors that can be easily adjusted to achieve a desired functional relationship between a signal factor and the output response. For example, the gearing ratio in a steering mechanism can be easily adjusted to achieve a desired relationship between the turning radius and the steering angle

Signal factors are those variables adjusted to attain the target/nominal performance. The control factors are those variables under the control of the designer. Selection of the nominal values for the control factors is the primary role of parameter design. Noise factors describe those variables that are difficult or impossible to control but whose variation is analyzed to understand the way it is transmitted to the output quality. To minimize the effects of outer and inner noises on product quality, certain countermeasures can be taken. The most important of these countermeasures is that of design. A product is said to be robust if its design is such that its performance is minimally influenced by uncontrollable noise factors.

An automobile design can be said to be robust if, for example, the fuel economy remains fairly constant over a wide range of speeds, road conditions, and wind velocities. This is shown graphically in Fig. 8.

Sources of Variation and Their Countermeasures

To reduce functional variation, that is, to increase the consistency of product/process performance, it is essential to identify the basic sources of functional variation so that appropriate countermeasures can be formulated and implemented. Taguchi suggests that variations in product or process function (also known as functional variations) arise from three basic sources (Ref 4):

- *Outer noise*: External sources or factors that are operating in the environment in which the product is functioning and whose variation is transmitted through the design to the output performance of the product. Examples of outer noise factors are temperature, humidity, contaminants, voltage fluctuations, vibrations from nearby equipment, and variations in human performance

- *Inner noise*: Internal change in product characteristics such as drift from the nominal over time due to deterioration. Inner noise can be precipitated by such factors as mechanical wear and aging
- *Variational noise*: Variation in the product parameters from one unit to another as a result of the manufacturing process. For example, the design nominal for a resistor may be 200 Ω; however, one manufactured resistor may have a resistance of 202 Ω, while another may have a resistance of 197 Ω

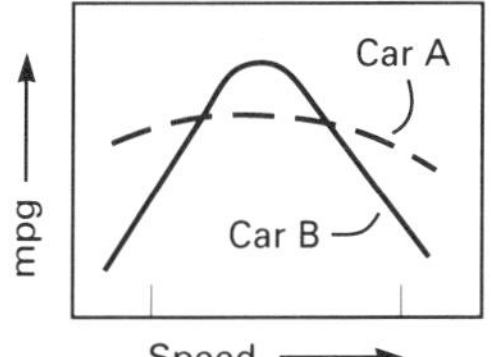

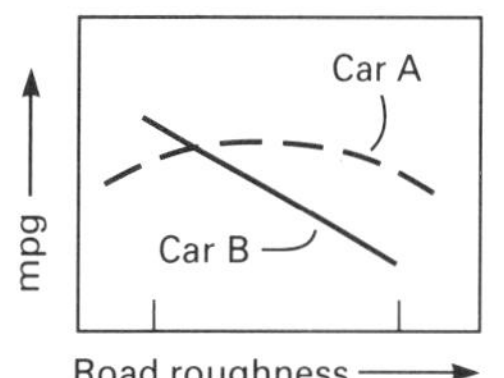

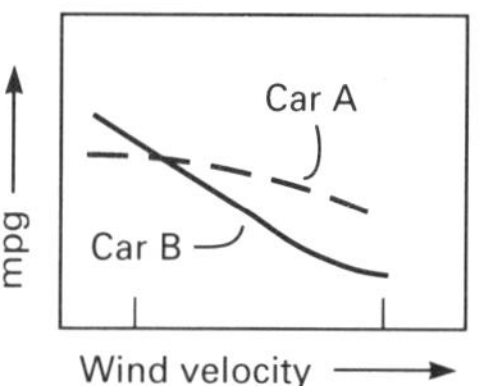

Fig. 8 Effect of three noise factors on automotive fuel efficiency (mpg) to demonstrate a robust product (Car A). Car A offers peak performance over a wide range of conditions for all three factors, while Car B offers peak performance over a limited range of conditions for each individual factor.

The significance of the recognition of the above sources of variation becomes evident as one begins to think in terms of the fundamental countermeasures one might invoke to mitigate the forces of these sources of variation. It quickly becomes clear that the forces of outer noise and inner noise can only be effectively dealt with upstream by the engineering design process. Variational noise is a matter of manufacturing imperfection and can therefore be dealt with, in part, at the process with such techniques as statistical process control. However, mitigating the forces of variational noise should also be considered to be a product and process design issue. In fact, it is likely that variational noise can be dealt with in a more significant and fundamental way if it is thought of as a process and product design problem. Certainly the concept of design for manufacturability and the current emphasis on the simultaneous engineering of products and processes have bearing on this issue.

Figure 9 illustrates how the sources of variation defined above are transmitted to the quality response of the product. At the product design stage, the nominal values of the critical design parameters are selected to produce a prespecified level of performance in the product in terms of one or more quality responses. Sources of variation active in the manufacturing processes for the product cause the design parameters to vary from those values intended by the design process, and this introduces variation into the quality response. Once the product is put into field use, outer noise sources of variation active in the environment in which the product is being asked to function are transmitted through the design and introduce further variation in the quality response. For example, variation in temperature may cause product performance to vary. As time goes by and the use of the product continues, the forces of inner noise are transmitted through the design and introduce still further variation into the quality response. For example, as a result of wear, a critical design parameter (size, resistance, viscosity, and so on) will actually change, thus causing a change in quality performance. From Fig. 9 it is clear that sources of variation enter the picture and affect the ultimate performance of the product during manufacturing and field use.

Shewhart (Ref 10), Deming (Ref 11), and Juran (Ref 12) all clearly point out that quality and productivity problems at the process fall into two basic categories. Shewhart described the variation in the process as arising from either chance causes or assignable causes. Deming refers to systems faults and local faults, while Juran refers to chronic problems and sporadic problems. Each provides a lucid description of the overall situation. Assignable causes, local faults, or sporadic problems arise in a somewhat unpredictable fashion and can usually be dealt with effectively at the workstation by the immediate level of supervision or the operator. On the other hand, system faults/chance causes/chronic problems are ever present, influencing all of the production until the system is changed. It is the attack on the system problems that Deming stresses, because about 80 to 85% of all the problems encountered are of this nature, while only 15 to 20% are of a local nature.

All processes are subject to local faults and system faults:

Local faults Special causes Sporadic problems	System faults Common causes Chronic problems
Broken tools Jammed machine Material contamination Machine setting drift	Poor supervision Poor training Inappropriate methods Poor workstation design
Correctable locally, at the machine, by the operator or first level of supervision	Requires a change in the system; only management can implement changes

In the case of common causes, it is clear by their very nature that management must take the responsibility for their removal. Only management can take the actions necessary to improve the training and supervision of the workers, only management can take the responsibility for the redesign of a poor workstation layout, and only management can establish new methods or procedures. Clearly, it is essential that these two types of problems be properly differentiated so that responsibility can be assigned for solving them.

Process Behavior Over Time

A succession of parts emanating from a process under statistical control will exhibit variability in their measurements because of a constant set of common causes. These variable measurements tend to collect into a predictable pattern of variation that can be easily described by a few simple statistical measures; namely, a mean, a standard deviation, and a frequency distribution. These measures stand as a model that predicts how the process will behave if subject only to a constant set of common causes. Figure 10 illustrates how such a statistical model may emerge in the manufacture of automobile head gaskets. The gasket thickness (Fig. 10a) is the quality characteristic of interest.

Measurements of thickness can be recorded for a succession of gaskets produced by the process (Fig. 10b), and these data can be summarized by statistical measures, such as an average (Fig. 10c), a standard deviation (Fig. 10d), and a distribution (Fig. 10e). If a model can be developed for the process measurements when subject only to a constant system of common causes, then when a major disturbance affects the process measurement it will be seen not to conform to this model. It will clearly stand out in the common cause variability pattern.

Time Order of Production. Although the above statistical description and the interpretation of data are useful, they ignore a crucial characteristic of the data: the time order of production. To properly indicate statistical control or lack of it, the model must be considered as it evolves over time of production. Figure 11 shows the time order evolution of the process. In Fig. 11(a), the process mean and variation stay constant over time, and the process is said to be in statistical control. In Fig. 11(b), a process mean shift occurs at 11:30 a.m. The process is said to have gone out of control. A process can be subject to several types of disturbances (special causes) that can produce a variety of unstable behaviors with respect to either its mean level or the level of variability or both.

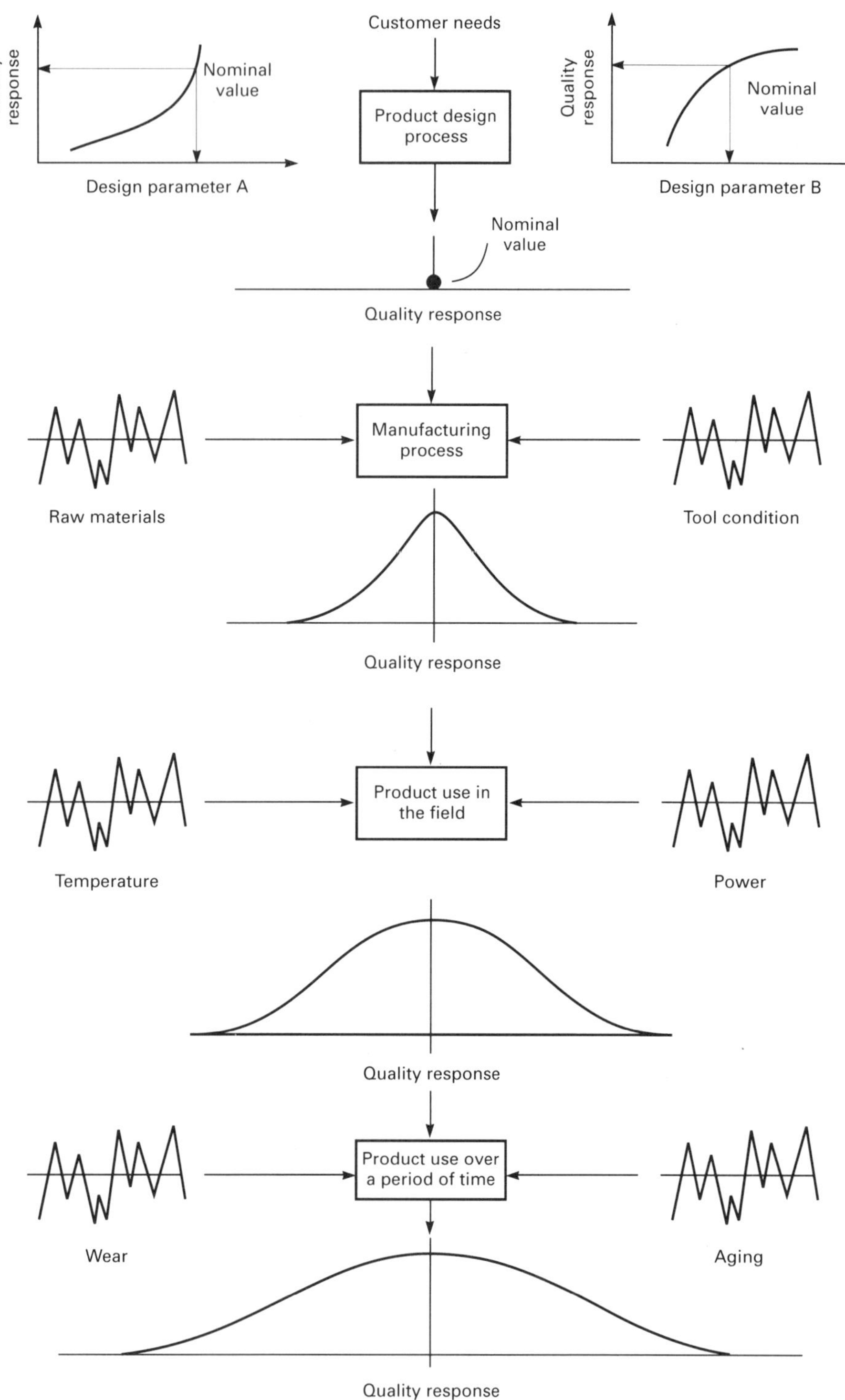

Fig. 9 Block diagram illustrating the impact of numerous sources of variation on the quality performance of a product from conception to eventual discard

Process Control and Process Improvement

Many years ago, Dr. Walter Shewhart showed how such data from a manufacturing process could be developed and interpreted through the use of very simple but profound statistical methods. In his benchmark book on the subject, *Economic Control of Quality of Manufactured Product*, Shewhart establishes from the very beginning the overarching philosophy that drives the control chart concept (Ref 10). The first paragraph of the Preface of this work clearly sets the foundation for what is known today as statistical control:

''Broadly speaking, the object of industry is to set up economic ways and means of satisfying human wants and in so doing to reduce everything possible to routines requiring a minimum amount of human effort. Through the use of the scientific method, extended to take account of modern statistical concepts, it has been found possible to set up limits within which the results of routine efforts must lie if they are to be economical. Deviations in the results of a routine process outside such limits indicate that the routine has broken down and will no longer be economical until the cause of trouble is removed'' (Ref 10).

From the above statement, several factors become immediately obvious:

- The fundamental focus is on the process—''ways and means of satisfying human wants''
- The overarching objective is economic operation of the process—''reduce everything possible to routines requiring a minimum amount of human effort''
- During normal operation, process behavior falls within predictable limits of variation—''It has been found possible to set up limits within which the results of routine efforts must lie if they are to be economical''
- Deviations in performance outside these limits signal the presence of problems that are jeopardizing the economic success of the process—''Deviations in the results of a routine process outside such limits indicate that the routine has broken down and will no longer be economical
- Improvement in quality and productivity requires that attention be directed at the process to find the root cause of the trouble and remove it—''the routine has broken down and will no longer be economical until the cause of trouble is removed''

In addition, there is no mention of the product and the conformance of the product to specifications in the above statement from Shewhart. The total emphasis is on the process and its economic operation.

Bringing a process into a state of statistical control does not necessarily mean that a fundamental improvement has been achieved. Clearly, a bad situation has been rectified by bringing the process into control, and quality and productivity are enhanced. However, bringing a process into control simply means that the process is back to where it should have been to begin with. At this point, it is then possible to begin to assess the present ability of the process to realize the potential it was initially intended to have. It may be failing to realize this potential because the implementation of the process is flawed or because the design of the process itself is flawed. In either case, the root cause(s) of the chronic common cause problem must be identified

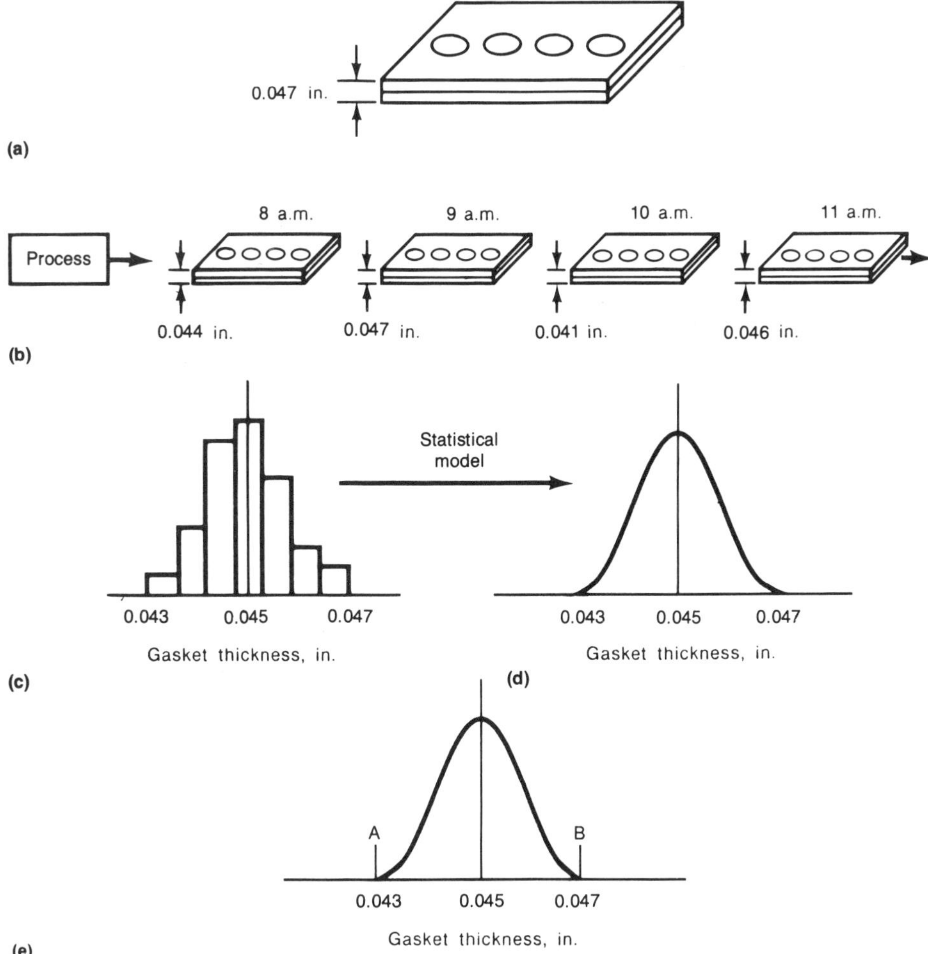

Fig. 10 Schematic and statistical analysis of an automobile head gasket to demonstrate variability of the manufacturing process. (a) Gasket thickness specification. (b) Gasket thickness measurements taken of samples produced at 1-h intervals. (c) Histogram of mean gasket thickness. (d) Plot showing standard deviation of gasket thickness. (e) Plot showing distribution of gasket thickness measurements

and removed at the system level. This constitutes a breakthrough in performance; that is, an improvement in the process has taken place. The results of the essential steps leading to such a breakthrough are shown in Fig. 12.

Fundamental Purposes of Shewhart Control Charts

There are two fundamental purposes that Shewhart control charts can serve:

- As a tool to provide a sound economic basis for making a decision at the machine regarding whether to look for trouble, adjust the process, or leave the process alone
- As a means of assisting in the identification of both improvement opportunities through the detection of sporadic and chronic faults in the process and to help provide the basis to formulate corrective and improvement actions

Although these two purposes may appear to be essentially the same, they are not. The former is strictly an on-line activity, while the latter is an off-line activity. On line, the operator can only recognize and take action against the presence of special causes of variation. If the process, as monitored on line, is free of special causes, the operator is doing the best that he can. In short, if the process is in statistical control, the operator should do nothing at all to disturb the routine behavior of the process.

On the other hand, the fact that the process is in statistical control in no way guarantees that it is a capable process (able to meet customer expectations as they may be communicated through a specification). However, the operator cannot be expected to be responsible for this, though he may have some valuable insight into the problem. A lack of process capability is basically a product/process design and planning problem.

For the techniques of statistical process control (specifically Shewhart control charts) to be successfully employed as an off-line problem identification and problem-solving tool, it is essential to keep in mind that it is a three-step process, as follows:

- Use statistical signals to find improvement opportunities through the identification of process faults
- Use experience, technical expertise, and fault diagnosis methods to find the root cause of the fault that has been identified
- Develop an action plan to correct the fault in a manner that will enable any gains that are realized to be held

This three-step process can be explained by using the classical feedback control system perspective, as shown in Fig. 13. There are five distinct stages in the generic control loop shown in Fig. 13. These five stages facilitate the three-step process in the following way:

- *Step 1*: Use of statistical signals (observation and evaluation)
- *Step 2*: Fault diagnosis (diagnosis)
- *Step 3*: Action plan (decision and implementation)

Construction and Interpretation of $\bar{x}$ and *R* Control Charts

This section focuses on the construction and interpretation of Shewhart control charts for variable data, in particular, $\bar{x}$ and *R* control charts. All the necessary equations and procedures for the calculation and graphical representation of the basic elements of the charts are presented. In addition, this section presents some probabilistic rules to aid in the interpretation of these charts.

Shewhart Control Chart Model

The control chart stands much as a jury in a court of law. The information from each sample is judged in order to determine whether or not it indicates the presence of a special cause disturbance. Unless the evidence is overwhelmingly in favor of the occurrence of a special cause disturbance, a verdict of not guilty is entered. In other words, there is no strong reason to believe that forces other than those of common causes are at work. Such an interpretation is, in fact, based on and consistent with the hypothesis-testing approach to making statistical inferences about the population/process based on information contained in the sample.

If a process under surveillance by periodic sampling maintains a state of good statistical control, this means that its mean level and level of variability remain constant over time. If such is the case, the sample means would have a distribution that follows the normal distribution, and nearly all the sample means would fall within the band of plus or minus three standard deviations of

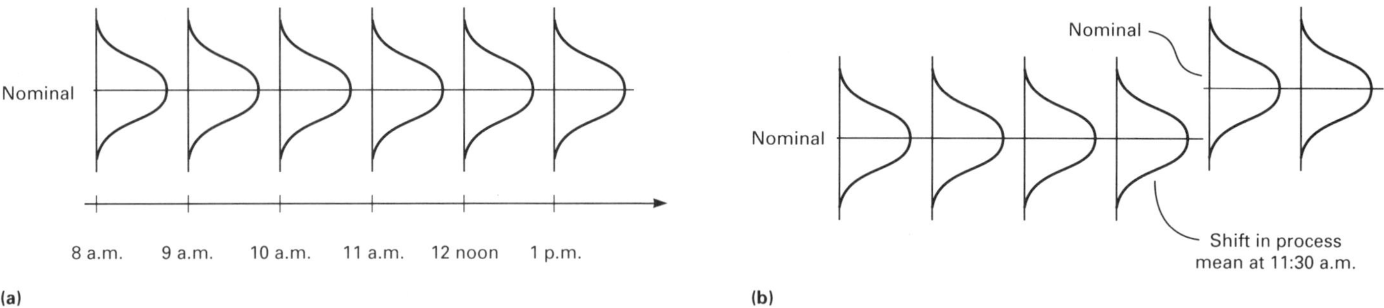

Fig. 11 Changing process behavior over a 5-h time interval. (a) Process in statistical control. (b) Stable variability but out-of-control process due to sustained shift in process mean at 11:30 a.m.

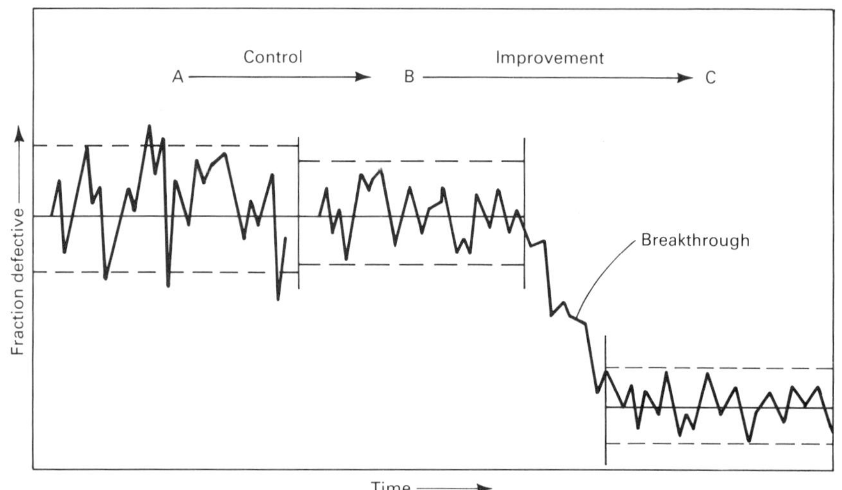

Fig. 12 Impact of having process initially in a state of statistical control versus improvement resulting from a breakthrough in performance

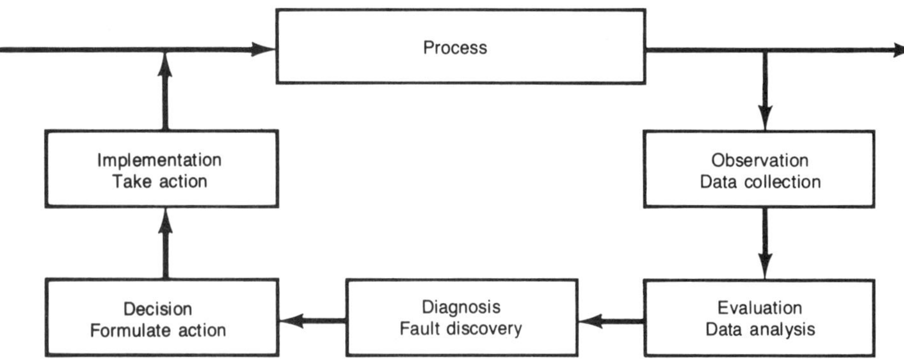

Fig. 13 Classical feedback control system view of statistical process control implementation

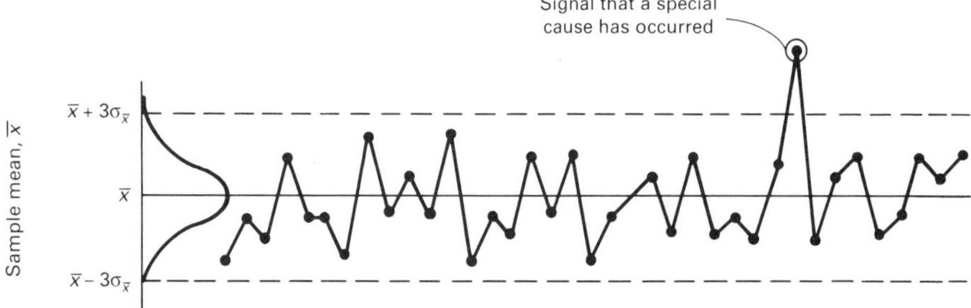

Fig. 14 Use of probabilistic limits to identify unusual process behavior over a period of time

$\bar{x}$, or $\pm 3\sigma_{\bar{x}}$, where $\sigma_{\bar{x}}$ is the standard deviation of $\bar{x}$ about the established process average. Because the chance of realizing a point outside the $3\sigma_{\bar{x}}$ limits is so small if good statistical control is evident, the occurrence of such a sample result must be interpreted as signaling some special/sporadic departure from expected behavior; that is, something other than common cause variation must be present. Figure 14 illustrates the presence of such a special cause. The $3\sigma_{\bar{x}}$ limits about the process average/centerline are referred to as the upper control limit (UCL) and the lower control limit (LCL).

Suppose a sample mean, $\bar{x}$, falls above the upper control limit. Because the chance of this occurring is so small (about ⅛ of 1%) if the process mean is truly at the centerline, it must be assumed that some special cause is present that shifted the process mean to a different level. Unfortunately, the fact that all points fall within the $3\sigma_{\bar{x}}$ limits does not mean that special causes are not present. Departures from expected process behavior may not always manifest themselves on the control chart immediately or for that matter at all. Although narrower limits (for example, $2\sigma_{\bar{x}}$ limits) would allow easier and faster detection of special causes, such limits would also increase the chances of false alarms, that is, times when the process is actually in good statistical control but a sample mean, $\bar{x}$, falls outside the control limits.

Appropriate placement of the upper and lower control limits is an economic issue. The intent would be to fix the limits in such a way as to balance the economic consequences of failing to detect a special cause when it does occur and wrongly identifying the presence of a special cause when it has really not occurred. Experience has led to the use of $3\sigma_{\bar{x}}$ limits as a good balance of these risks.

Generally, chaotic disturbances manifest themselves in two possible ways:

- Shifts or changes in the mean level of the process
- Shifts or changes in the amount of process variability

Figure 14 represents only one side of the coin; thus, a fundamental rule has been violated—that an $\bar{x}$-chart should not be

Table 1 Factors for $\bar{x}$ and R control chart limits

Sample size, n	Factors for control limits: $\bar{x}$-chart A_2	Factors for control limits: R-chart D_3	Factors for control limits: R-chart D_4	Factor for calculating σ_x from range, R d_2
2	1.880	0	3.267	1.128
3	1.023	0	2.573	1.693
4	0.729	0	2.282	2.059
5	0.577	0	2.114	2.326
6	0.483	0	2.004	2.534
7	0.419	0.076	1.924	2.704
8	0.373	0.136	1.864	2.847
9	0.337	0.184	1.816	2.970
10	0.308	0.223	1.777	3.078
11	0.285	0.256	1.744	3.173
12	0.266	0.283	1.717	3.258
13	0.249	0.307	1.693	3.336
14	0.235	0.328	1.672	3.407
15	3.472	0.223	0.347	1.653
16	0.212	0.363	1.637	3.532
17	0.203	0.378	1.622	3.588
18	0.194	0.391	1.608	3.640
19	0.187	0.403	1.597	3.699
20	0.180	0.415	1.585	3.735

Table 2 $\bar{x}$ and R control chart engine block cylinder boring process data of Example 1

Sample number, k	Individual measurements, x(a)(b): 1	2	3	4	5	$\bar{x}$(b)	R(b)
1	205	202	204	209	205	205	7
2	202	196	201	198	202	199.8	6
3	201	202	199	197	196	199	6
4	205	203	196	201	197	200.4	9
5	199	196	201	200	195	198.2	6
6	203	198	192	217	196	201.2	25
7	202	202	198	203	202	201.4	5
8	197	196	196	200	204	198.6	8
9	199	200	204	196	202	200.2	8
10	202	196	204	195	197	198.8	9
11	206	204	202	210	205	205.4	8
12	200	201	199	200	201	200.2	2
13	205	196	201	197	198	199.4	9
14	202	199	200	198	200	199.8	4
15	200	200	201	205	201	201.4	5
16	201	187	209	202	200	199.8	22
17	202	202	204	198	203	201.8	6
18	201	198	204	201	201	201	6
19	207	206	194	197	201	201	13
20	200	204	198	199	199	200	6

(a) Sample/subgroup size, n = 5. (b) Although actual cylinder bore data ranged from 3.5187 to 3.5217 in., only the last three digits in the measurement for x, $\bar{x}$, and R are used.

shown without the associated range chart. By monitoring the process through sampling, the sample mean, $\bar{x}$, can be used to aid in determining when a change in the mean level has occurred. By determining the range, R, for each sample and tracking its behavior over time, a basis is provided for identifying the presence of changes in process variability.

Establishing a control chart for ranges R comparable to the $\bar{x}$-chart requires more specific information about the distribution of the process. In particular, it will be assumed that the individual measurements x are normally distributed. Doing so establishes the relationship between the standard deviation of ranges R and the standard deviation of individual process observations or measurements.

For simplicity, $\pm 3\sigma_R$ limits will be placed about the average range $\bar{R}$ to define the upper and lower control limits for R even though the frequency distribution of ranges is not a symmetric distribution. Obtaining σ_R will be discussed later in this article.

Setting Up $\bar{x}$ and R Control Charts. Once the statistical basis for Shewhart control charts has been established, it remains to define the necessary elements of the charts mathematically and to establish a standard graphical representation. This section deals with the construction of $\bar{x}$ and R control charts. All the necessary equations and general procedures for the calculation and graphical representation of the basic elements of the charts are presented in this section.

The first step in setting up $\bar{x}$ and R control charts is the selection of the samples. It is important that all samples be rational samples. Rational samples are groups of measurements whose variation is attributable only to one constant system of common cause. Sampling from different machines, sampling over extended periods of time, and sampling from product combined from several sources are all nonrational sampling methods and must be avoided. Rational samples will be discussed in more detail later in this article.

As a rule of thumb, 25 to 50 samples should be selected to provide a solid basis for the initiation of the control charts. This will help to ensure more precise estimation of the process mean and standard deviation. The sample/subgroup size should be relatively small (between n = 3 and n = 6).

With k rational samples of n each, the following steps can be used as a guide when constructing $\bar{x}$, R control charts with 3σ limits:

- Calculate the sample mean and sample range for each sample using $\bar{x} = \Sigma x/n$ and $R = x_{max} - x_{min}$
- Calculate the grand mean of the n sample means and the average range using $\bar{\bar{x}} = \Sigma\bar{x}/k$ and $\bar{R} = \Sigma R/k$
- Calculate the control limits for the R-chart. Although the true distribution of sample ranges is not normal and not symmetric, the symmetric limits are conventionally used for the R-charts. With assumed normal distribution for the individual measurements, the following formulas can be used for the calculation of the control limits: $UCL_R = D_4\bar{R}$ and $LCL_R = D_3\bar{R}$
- Calculate the control limits for the $\bar{x}$-chart. Although the required standard deviation (the standard deviation of the sample mean, $\bar{x}$) for setting the limits is σ_x, this value is conveniently estimated by $\bar{R}/(d_2\sqrt{n})$, where d_2 is a function of n. For $3\sigma_{\bar{x}}$ limits, one uses a factor called A_2, which is equal to $3/(d_2\sqrt{n})$ and can be found in Table 1. Thus, the control limits are calculated by: $UCL_x = \bar{\bar{x}} + A_2\bar{R}$ and $LCL_x = \bar{\bar{x}} - A_2\bar{R}$

The appropriate values for d_2, A_2, D_3, and D_4 are obtained from Table 1.

Example 1: Construction of $\bar{x}$ and R Control Charts of an Engine Block Cylinder Bore. The inside diameter of the cylinder bore in an engine block was measured after a boring operation. Measurements were made to 0.0001 in. Samples of size n = 5 were taken to obtain some data to initiate $\bar{x}$ and R control charts for this process. The samples were taken roughly every ½ h. The sample measurements were all taken on the same number cylinder in the block. The results of the first 20 samples are given in Table 2. The actual measurements are of the form 3.5205 in., 3.5202 in., and so on. Table 2 provides only the last three digits in the measurement.

Determination of Trial Control Limits. Based on these first 20 samples, each sample mean $\bar{x}$ and R was calculated as listed in the last two columns of Table 2. The grand average, $\bar{\bar{x}}$, and average range, $\bar{R}$, are $\bar{\bar{x}}$ = 200.62 and $\bar{R}$ = 8.5. For $\bar{x}$-chart:

$$UCL = \bar{\bar{x}} + A_2\bar{R} = 200.62 + (0.58)(8.5) = 205.55$$
$$LCL = \bar{\bar{x}} - A_2\bar{R} = 200.62 - (0.58)(8.5) = 195.69$$

For R-chart:

$$UCL = \bar{R}\,D_4 = 8.5\,(2.11) = 17.935$$
$$LCL = \bar{R}\,D_3 = 8.5\,(0) = 0.0$$

With the above calculations completed, the $\bar{x}$ and R control charts can be constructed. It will be important to start with the R-chart and to get it under statistical control first. This is necessary because the limits of the $\bar{x}$-chart depend on the magnitude of the

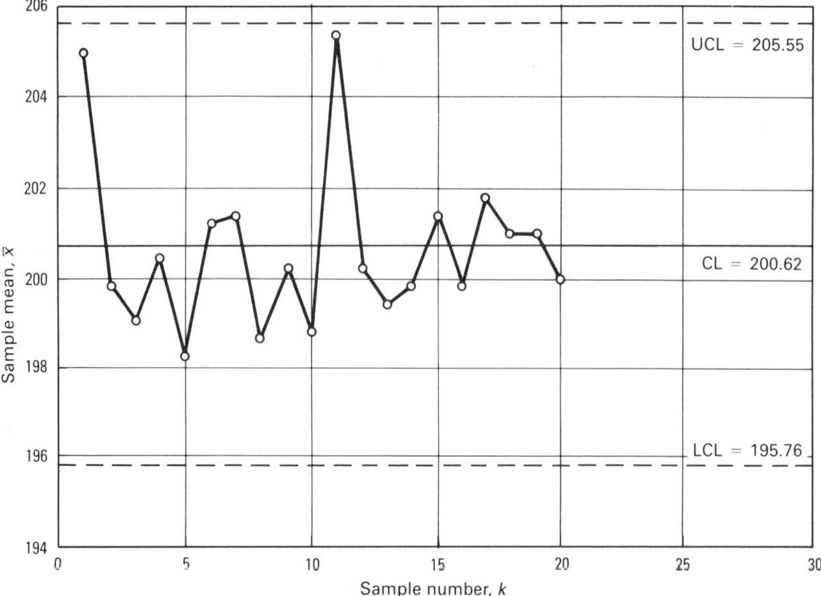

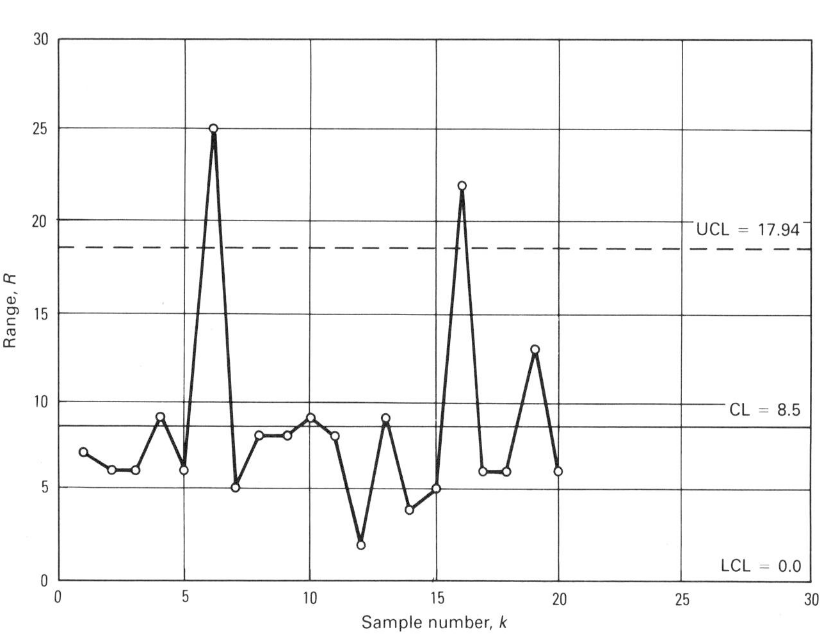

Fig. 15 Control charts for inside diameter measurements of engine block cylinder bores in Table 2. (a) $\bar{x}$-chart. (b) R-chart. Data are for $k = 20$, $n = 5$.

common cause variation of the process measured by $\bar{R}$. If initially some points on the $\bar{R}$-chart exceed the upper control limit (special causes present), the limits in the $\bar{x}$-chart will be inflated and will need to be recalculated after such special cause data are removed.

Interpretation of Initial Charts. Figure 15 shows the $\bar{x}$ and R control charts based on the calculations given above. The R-chart is examined first, and it is found that two points exceed the 3σ limit. This indicates that there are special causes (excess process variability) at least at these points. These points—samples 6 and 16—are then examined to identify the reasons for these special causes. The records show that at these points the regular operator was absent and a less-experienced relief operator ran the production for a short time. As a result, samples 6 and 16, taken over the time the relief operator ran the process, exhibited greater variability, perhaps because of inexperience. Because special causes can be identified for samples 6 and 16, these sample values (both $\bar{x}$ and R) are removed and new limits are calculated. The revised centerlines and limits for $\bar{x}$ and R control charts based on the data after samples 6 and 16 have been deleted are shown in Fig. 16.

Interpretation of Revised Charts. Again, the R-chart is examined first. No points are now outside the control limits, and no other unusual patterns of variability appear to be present. The R-chart seems to be in good statistical control. However, examination of the $\bar{x}$-chart reveals that there are two points above the upper control limit. The investigation of these points reveals the fact that these two samples (1 and 11) occurred at 8:00 a.m. and 1:00 p.m., roughly corresponding to the start-up of the boring machine in the morning and directly after the lunch hour. It was found that the samples were taken from the first few parts made in each case. Once the machine warmed up (approximately 10 min), the problem disappeared. It was decided as a standard policy not to initiate production until 10 min after start-up of the machine.

With this special cause identified and the corresponding values removed, the new limits are calculated. The control charts in Fig. 17 now both show good statistical control. The limits can now be extended, and monitoring of the process can be continued.

Importance of Using Both $\bar{x}$ and R Control Charts. This example points to the importance of maintaining both $\bar{x}$ and R control charts and the significance of first focusing attention on the R-chart and establishing its stability. Initially, no points fell outside the $\bar{x}$-chart control limits and one could be led to believe that this indicates that the process mean exhibits good statistical control. However, the fact that the R-chart was initially not in control caused the limits on the $\bar{x}$-chart to be somewhat wider because of two inordinately large R values. Once these special causes of variability were removed, the limits on the $\bar{x}$-chart became narrower, and two $\bar{x}$ values now fall outside these new limits. Special causes were present in the $\bar{x}$ data, but initially were not recognizable because of the excess variability as seen in the R-chart.

This example also points strongly to the need to have 25 or more samples before initiating control charts. In this case, once special causes were removed, only 16 subgroups remained to construct the charts. This is simply not enough data.

Importance of Rational Sampling

Perhaps the most crucial issue to the successful use of the Shewhart control chart concept is the definition and collection of the samples or subgroups. This section will discuss the concept of rational sampling, sample size, sampling frequency, and sample collection methods and will review some classic misapplications of rational sampling. Also, a number of practical examples of

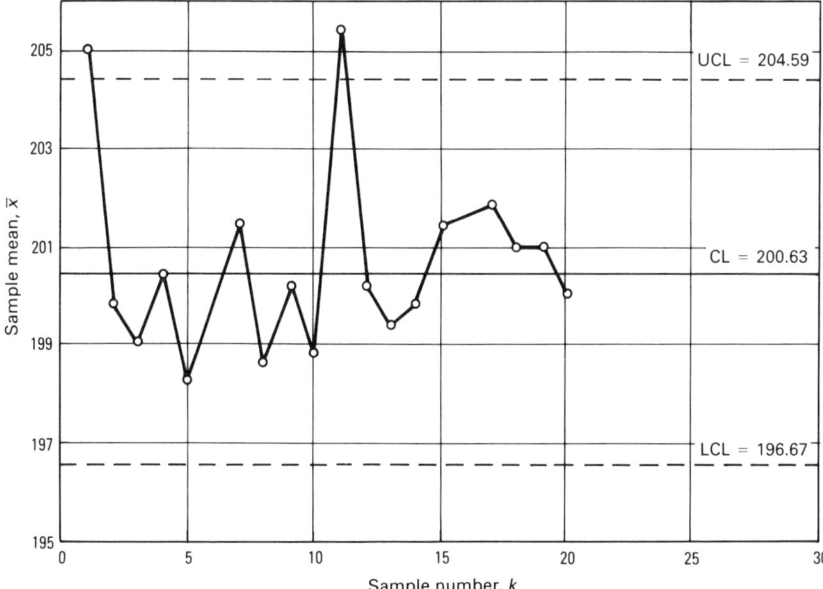

(a)

(b)

Fig. 16 Revised control charts for Table 2 data with sample 6 and sample 16 data eliminated (both have *R* values above UCL in Fig. 15) because they are sources of excess process variability. (a) $\bar{x}$-chart. (b) *R*-chart. Data now have $k = 18$, $n = 5$.

subgroup definition and selection will be presented to aid the reader in understanding and implementing this central aspect of the control chart concept.

Concept of Rational Sampling. Rational subgroups or samples are collections of individual measurements whose variation is attributable only to one unique constant system of common causes. In the development and continuing use of control charts, subgroups or samples should be chosen in a way that provides the maximum opportunity for the measurements within each subgroup to be alike and the maximum chance for the subgroups to differ from one another if special causes arise between subgroups. Figure 18 illustrates the notion of a rational sample. Within the sample or subgroup, only common cause variation should be present. Special causes/sporadic problems should arise between the selection of one rational sample and another.

Sample Size and Sampling Frequency Considerations. The size of the rational sample is governed by the following considerations:

- Subgroups should be subject to common cause variation. The sample size should be small to minimize the chance of mixing data within one sample from a controlled process and one that is out of control. This generally means that consecutive sample selection should be used rather than distributing the sample selection over a period of time. There are, however, certain situations where distributed sampling may be preferred
- Subgroups should ensure the presence of a normal distribution for the sample means. In general, the larger the sample size, the better the $\bar{x}$ distribution is represented by the normal curve. In practice, sample sizes of four or more ensure a good approximation to normality
- Subgroups should ensure good sensitivity to the detection of assignable causes. The larger the sample size, the more likely that a shift of a given magnitude will be detected

When the above factors are taken into consideration, a sample/subgroup size of four to six is likely to emerge. Five is the most commonly used number because of the relative ease of further computation.

Sampling Frequency. The question of how frequently samples should be collected is one that requires careful thought. In many applications of $\bar{x}$ and *R* control charts, samples are selected too infrequently to be of much use in identifying and solving problems. Some considerations in sample frequency determination are the following:

- If the process under study has not been charted before and appears to exhibit somewhat erratic behavior, samples should be taken quite frequently to increase the opportunity to quickly identify improvement opportunities. As the process exhibits less and less erratic behavior, the sample interval can be lengthened
- It is important to identify and consider the frequency with which occurrences are taking place in the process. This might include, for example, ambient condition fluctuations, raw material changes, and process adjustments such as tool changes or wheel dressings. If the opportunity for special causes to occur over a 15-min period is good, sampling twice a shift is likely to be of little value
- Although it is dangerous to overemphasize the cost of sampling in the short term, clearly it cannot be neglected

Common Pitfalls in Subgroup Selection. In many situations, it is inviting to combine the output of several parallel and

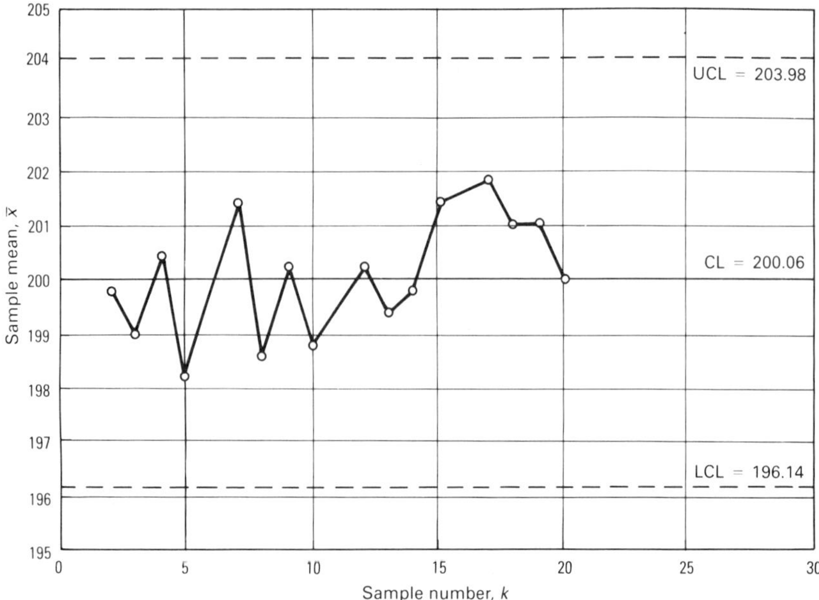

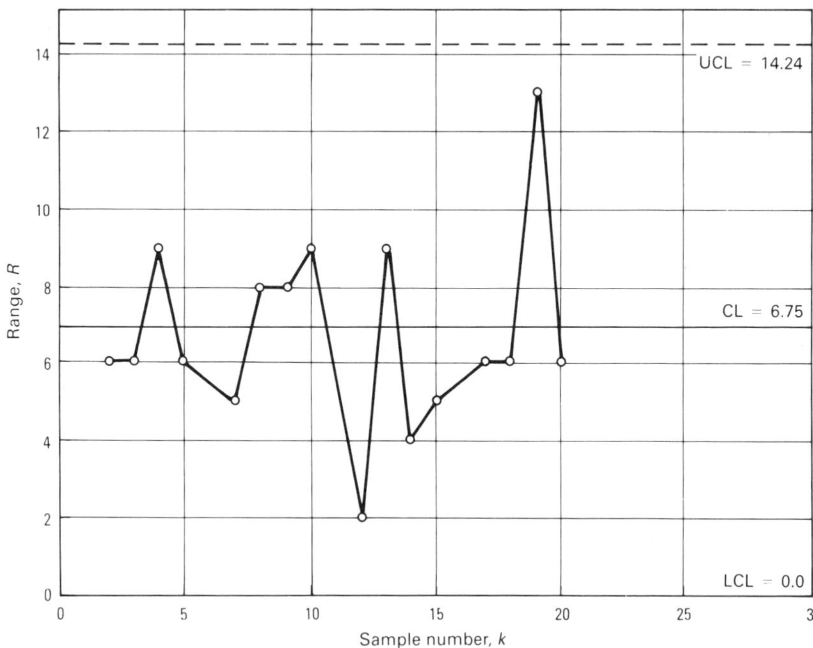

Fig. 17 Second revision control charts for Table 2 data with two more sample deletions (samples 1 and 11, both of which exceed UCL in Fig. 16) resulting from workpieces produced prior to machine being properly warmed up. (a) $\bar{x}$-chart. (b) R-chart. Data now have $k = 16$, $n = 5$.

assumed-to-be-identical machines into a single sample to be used in maintaining a single control chart for the process. Two variations of this approach can be particularly troublesome: stratification and mixing.

Stratification of the Sample. Here each machine contributes equally to the composition of the sample. For example, one measurement each from four parallel machines yields a sample/subgroup of $n = 4$, as seen in Fig. 19. In this case, there will be a tremendous opportunity for special causes (true differences among the machine) to occur within subgroups.

When serious problems do arise, for example, for one or more of the machines, they will be very difficult to detect because of the use of stratified samples. This problem can be detected, however, because of the unusual nature of the $\bar{x}$-chart pattern (recall the previous pattern analysis) and can be rectified provided the concepts of rational sampling are understood.

The R-charts developed from such data will usually show very good control. The corresponding $\bar{x}$ control chart will show very wide limits relative to the plotted $\bar{x}$ values, and their control will therefore appear almost too good. The wide limits result from the fact that the variability within subgroups is likely to be subject to more than merely common causes (Fig. 20).

Mixing Production From Several Machines. Often it is inviting to combine the output of several parallel machines/lines into a single stream of well-mixed product that is then sampled for the purposes of maintaining control charts. This is illustrated in Fig. 21.

If every sample has exactly one data point from each machine, the result would be the same as that of stratified sampling. If the sample size is smaller than the number of machines with different means or if most samples do not include data from all machines, the within-sample variability will be too low, and the between-sample differences in the means tend to be large. Thus, the $\bar{x}$-chart would give an appearance that the $\bar{x}$ values are too far away from the centerline.

Zone Rules for Control Chart Analysis

Special causes often produce unnatural patterns that are not as clear cut as points beyond the control limits or obvious regular patterns. Therefore, a more rigorous pattern analysis should be conducted. Several useful tests for the presence of unnatural patterns (special causes) can be performed by dividing the distance between the upper and lower control limits into zones defined by σ, 2σ, and 3σ boundaries, as shown in Fig. 22. Such zones are useful because the statistical distribution of $\bar{x}$ follows a very predictable pattern—the normal distribution; therefore, certain proportions of the points are expected to fall within the $\pm\sigma$ boundary, between σ and 2σ, and so on.

The following sections discuss eight tests that can be applied to the interpretation of $\bar{x}$ and R control charts. Not all of these tests follow/use the zones just described, but it is useful to discuss all of these rules/tests together. These tests provide the basis for the statistical signals that indicate that the process has undergone a change in its mean level, variability level, or both. Some of the tests are based specifically on the zones defined in Fig. 22 and apply only to the interpretation of the $\bar{x}$-chart patterns. Some of the tests apply to both charts. Unless specifically identified to the contrary, the

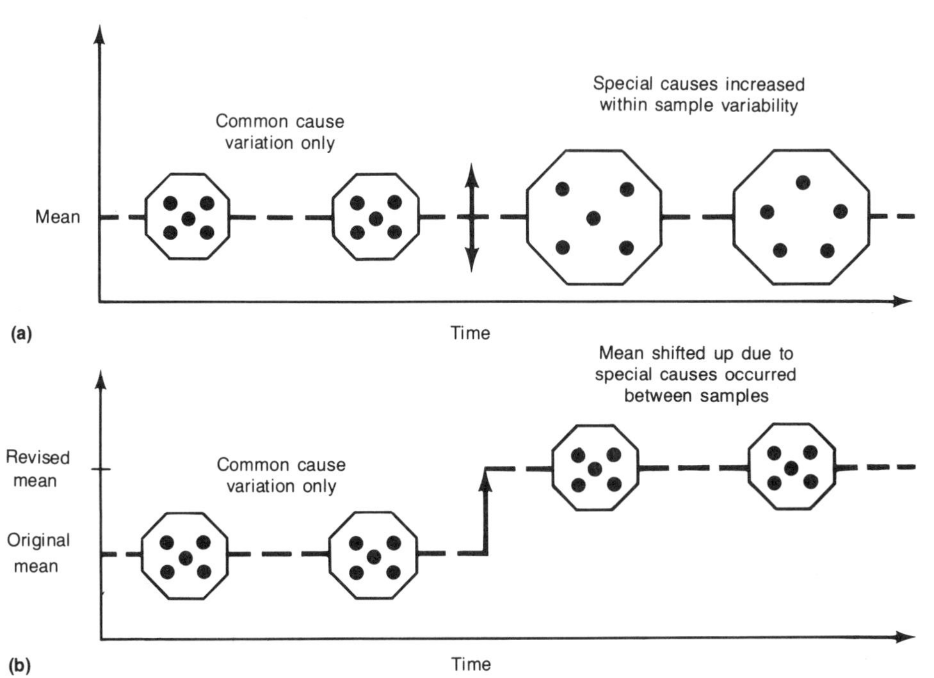

Fig. 18 Graphical depiction of a rational subgroup illustrating effect of special causes on mean. (a) Unshifted. (b) Shifted

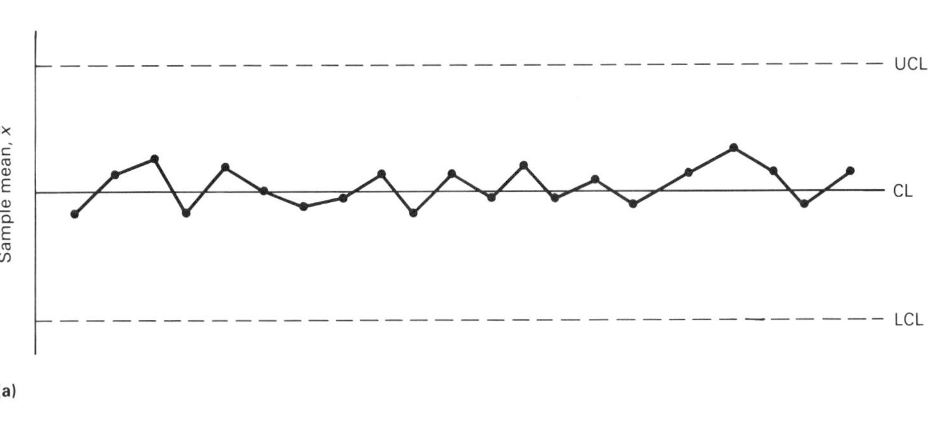

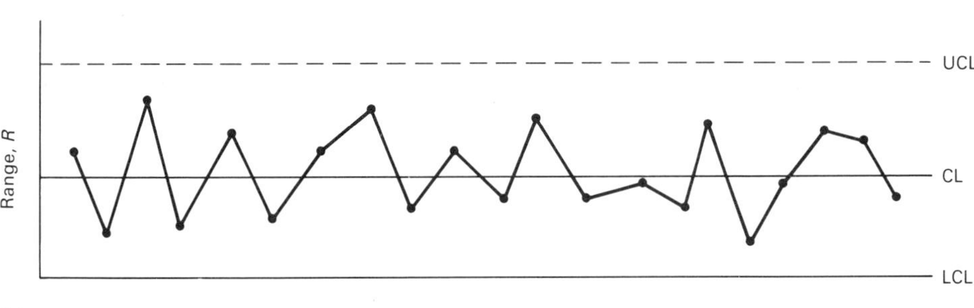

Fig. 20 Typical control charts obtained for a stratified sample selection. (a) $\bar{x}$-chart. (b) R-chart

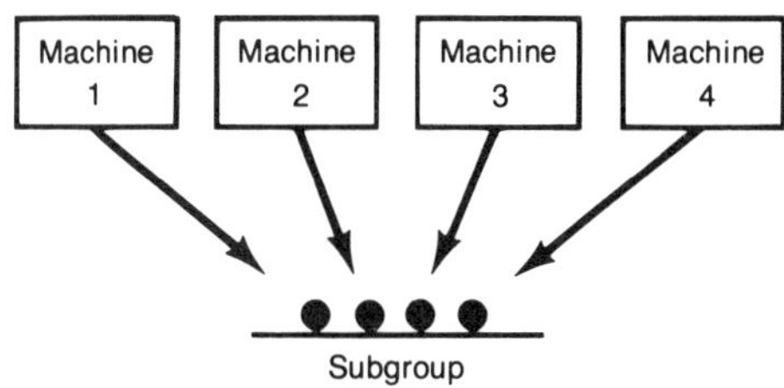

Fig. 19 Block diagram depicting a stratified sample selection

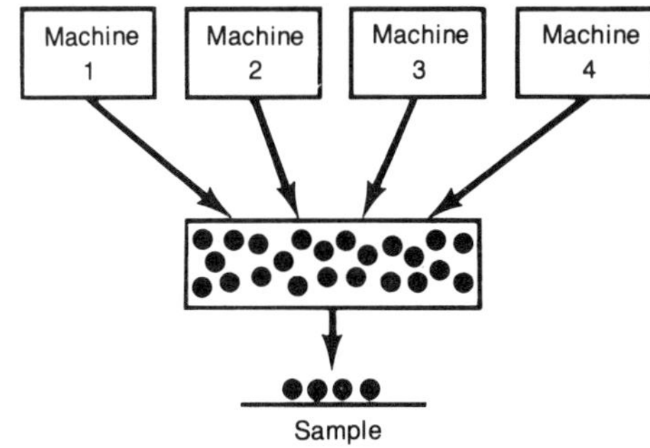

Fig. 21 Block diagram of sampling from a mixture

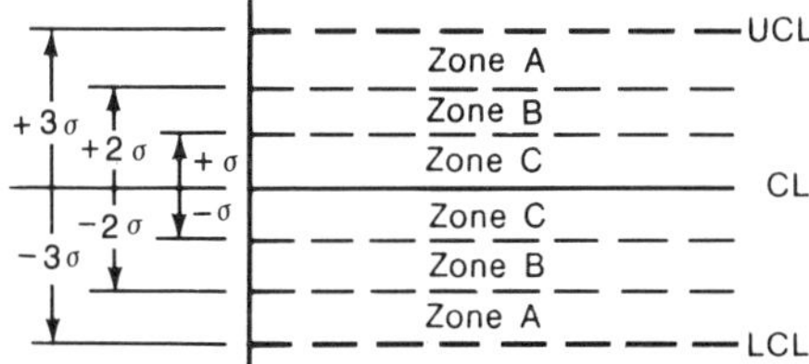

Fig. 22 Control chart zones to aid chart interpretation

tests/rules apply to the consideration of data to one side of the centerline only.

When a sequence of points on the chart violates one of the rules, the last point in the sequence is circled. This signifies that the evidence is now sufficient to suggest that a special cause has occurred. The issue of when that special cause actually occurred is another matter. A logical estimation of the time of occurrence may be the beginning of the sequence in question. This is the interpretation that will be used here. It should be noted that some judgment and latitude should be given. Figure 23 illustrates the following patterns:

- *Test 1 (extreme points)*: The existence of a single point beyond zone A signals the presence of an out-of-control condition (Fig. 23a)
- *Test 2 (2 out of 3 points in zone A or beyond)*: The existence of 2 out of any 3 successive points in zone A or beyond signals the presence of an out-of-control condition (Fig. 23b)
- *Test 3 (4 out of 5 points in zone B or beyond)*: A situation in which there are 4 out of 5 successive points in zone B or beyond signals the presence of an out-of-control condition (Fig. 23c)
- *Test 4 (runs above or below the centerline)*: Long runs (7 or more successive points) either strictly above or strictly below the centerline; this rule applies to both the $\bar{x}$ and R control charts (Fig. 23d)
- *Test 5 (trend identification)*: When 6 successive points on either the $\bar{x}$ or the R control chart show a continuing increase or decrease, a systematic trend in the process is signaled (Fig. 23e)

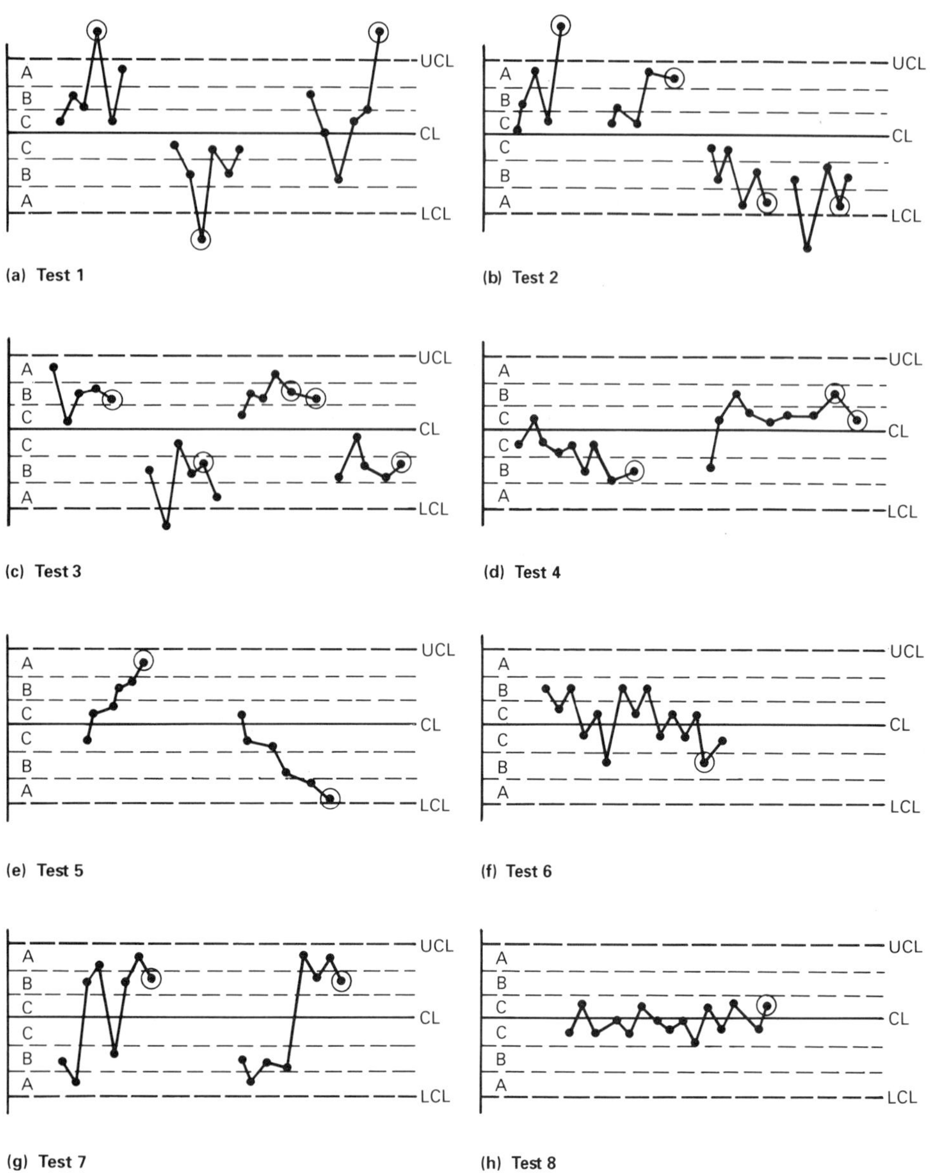

Fig. 23 Pattern analysis of $\bar{x}$-charts. Circled points indicate last point in a sequence of points on a chart that violates a specific rule.

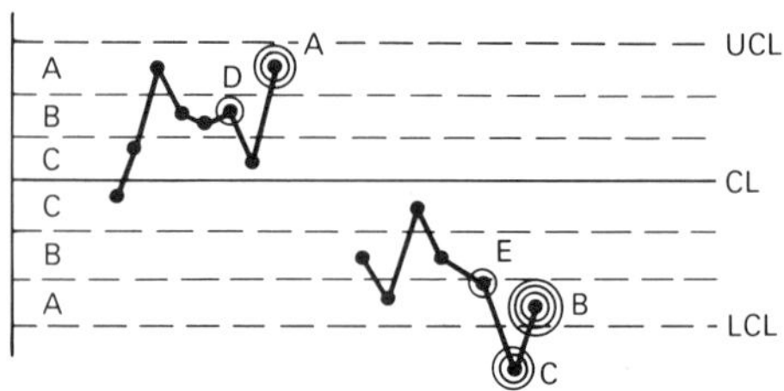

Fig. 24 Example of simultaneous application of more than one test for out-of-control conditions. Point A is a violation of tests 3 and 4; point B is a violation of tests 2, 3, and 4; and point C is a violation of tests 1 and 3. See text for discussion.

- *Test 6 (trend identification)*: When 14 successive points oscillate up and down on either the $\bar{x}$ or R control chart, a systematic trend in the process is signaled (Fig. 23f)
- *Test 7 (avoidance of zone C test)*: When 8 successive points, occurring on either side of the centerline, avoid zone C, an out-of-control condition is signaled. This could also be the pattern due to mixed sampling (discussed earlier), or it could also be signaling the presence of an over-control situation at the process (Fig. 23g)
- *Test 8 (run in zone C test)*: When 15 successive points on the $\bar{x}$-chart fall in zone C only, to either side of the centerline, an out-of-control condition is signaled; such a condition can arise from stratified sampling or from a change (decrease) in process variability (Fig. 23h)

The above tests are to be applied jointly in interpreting the charts. Several rules may be simultaneously broken for a given data point, and that point may therefore be circled more than once, as shown in Fig. 24.

In Fig. 24, point A is circled twice because it is the end point of a run of 7 successive points above the centerline and the end point of 4 of 5 successive points in zone B or beyond. In the second grouping in Fig. 24, point B is circled three times because it is the end point of:

- A run of 7 successive points below the centerline
- 2 of 3 successive points in zone A or beyond
- 4 of 5 successive points in zone B or beyond

Point C in Fig. 24 is circled twice because it is an extreme point and the end point of a group of five successive points, four of which are in zone B or beyond. Two other points (D, E) in these groupings are circled only once because they violate only one rule.

Control Charts for Individual Measurements

In certain situations, the notion of taking several measurements to be formed into a rational sample of size greater than one simply does not make sense, because only a single measurement is available or meaningful at each sampling. For example, process characteristics such as oven temperature, suspended air particulates, and machine downtime may vary during a short period at sampling. Even for those processes in which multiple measurements could be taken, they would not provide valid within-sample variation for control chart construction. This is so because the variation among several such measurements would be primarily attributed to variability in the measurement system. In such a case, special control charts can be used. Commonly used control charts for individual measurements include:

- x, R_m control charts
- Exponentially weighted moving average (EWMA) charts
- Cumulative sum charts (CuSum charts)

Both the EWMA (Ref 13-16) and the CuSum (Ref 17-21) control charts can be used for charting sample means and other statistics in addition to their use for charting individual measurements.

x and R_m (Moving-Range) Control Charts. This is perhaps the simplest type of control chart that can be used for the study of individual measurements. The construction of x and R_m control charts is similar to that of $\bar{x}$ and R control charts except that x stands for the value of the individual measurements and R_m for the moving range, which is the range of a group of n consecutive individual measurements artificially combined to form a subgroup of size n (Fig. 25). The moving range is usually comprised of the largest difference in two or three successive individual measurements. The moving ranges are calculated as shown in Fig. 25 for the case of three consecutive measurements used to form the artificial samples of size $n = 3$.

Because the moving range, R_m, is calculated primarily for the purpose of estimating common cause variability of the process, the artificial samples that are formed from successive measurements must be of very small size to minimize the chance of mixing data from out-of-control conditions. It is noted that x and R_m are not independent of each other and that successive sample R_m values are overlapping.

The following example illustrates the construction of x and R_m control charts, assuming that x follows at least approximately a normal distribution. Here, R_m is based on two consecutive measurements; that is, the artificial sample size is $n = 2$.

Example 2: x and R_m Control Chart Construction for the Batch Processing of White Millbase Component of a Topcoat. The operators of a paint plant were studying the batch processing of white millbase used in the manufacture of topcoats. The basic process begins by charging a sandgrinder premix tank with resin and pigment. The premix is agitated until a homogeneous slurry is obtained and then pumped through the sandgrinder. The grinder output is sampled to check for fineness and gloss. A batch may require adjustments by adding pigment or resin to achieve acceptable gloss. Through statistical modeling of the results of some ash tests, a quantitative method was developed for determining the amount of pigment or resin to be added when necessary, all based on the weight per unit volume (lb/gal.) of the batch. Therefore, it became important to monitor the weight per

$R_1 = 4 - 1 = 3$ $R_3 = 3 - 1 = 2$ $R_5 = 7 - 2 = 5$ $R_7 = 5 - 2 = 3$

1 4 2 1 3 7 2 5 4 1

$R_2 = 4 - 1 = 3$ $R_4 = 7 - 1 = 6$ $R_6 = 7 - 2 = 5$ $R_8 = 5 - 1 = 4$

Fig. 25 Examples of three successive measurements used to determine the moving range

unit volume for each batch to achieve millbase uniformity. Table 3 lists weight per unit volume data for 27 consecutive batches.

In the calculation of averages in Table 3, $\bar{x}$ is an average of all 27 individual measurements, while $\bar{R}_m$ is an average of $27 - 1 = 26$ R_m values because there are only 26 moving ranges for $n = 2$. If the artificial samples were of size $n = 3$, there would be only $27 - 2 = 25$ moving averages.

Once the $\bar{x}$ and $\bar{R}_m$ values are calculated, they are used as centerline values of x and R_m control charts, respectively. The calculation of upper and lower control limits for the R_m control chart is also the same as in $\bar{x}$, R control charts, using the artificial sample size n to determine D_3 and D_4 values. However, the upper and lower control limits for the x chart should always be based on a sample size of one, using $\bar{R}_m$ the same way as $\bar{R}$ in $\bar{x}$ control charts. These calculations are shown below for the example data. For the R_m-chart, for $n = 2$, $D_3 = 0$, and $D_4 = 3.27$:

$$CL_{R_m} = \bar{R}_m = 4.99/26 = 0.192$$
$$UCL_{R_m} = D_4\,\bar{R}_m = (3.27)(0.19) = 0.62$$
$$LCL_{R_m} = D_3\,\bar{R}_m = 0$$

Table 3 x and R_m control chart data for the batch processing of white millbase topcoat component of Example 2

Batch	x, lb/gal.	R_m (a)
1	14.04	
2	13.94	0.10 (14.04 − 13.94 = 0.10)
3	13.82	0.12 (13.94 − 13.82 = 0.12)
4	14.11	0.29 (14.11 − 13.82 = 0.29)
5	13.86	0.25
6	13.62	0.24
7	13.66	0.04
8	13.85	0.19
9	13.67	0.18
10	13.80	0.13
11	13.84	0.04
12	13.98	0.14
13	13.40	0.58
14	13.60	0.20
15	13.80	0.20
16	13.66	0.14
17	13.93	0.27
18	13.45	0.48
19	13.90	0.45
20	13.83	0.07
21	13.64	0.19
22	13.62	0.02
23	13.97	0.35
24	13.80	0.17
25	13.70	0.10
26	13.71	0.01
27	13.67	0.04
	$\bar{x} = 13.77$	$\bar{R}_m = 0.19$

(a) Calculated, $n = 2$

For the x-chart, an estimate of the standard deviation of x is equal to $\bar{R}_m/d_2$, where $d_2 = 1.128$ from Table 1 using $n = 2$. Thus, $3\sigma = (3/d_2)\,\bar{R}_m = (3/1.128)\,\bar{R}_m = 2.66\,\bar{R}_m$:

$$CL_x = 13.77$$
$$UCL_x = \bar{x} + 2.66\,\bar{R}_m = 13.77 + (2.66)(0.19) = 14.28$$
$$LCL_x = \bar{x} - 2.66\,\bar{R}_m = 13.77 - (2.66)(0.19) = 13.26$$

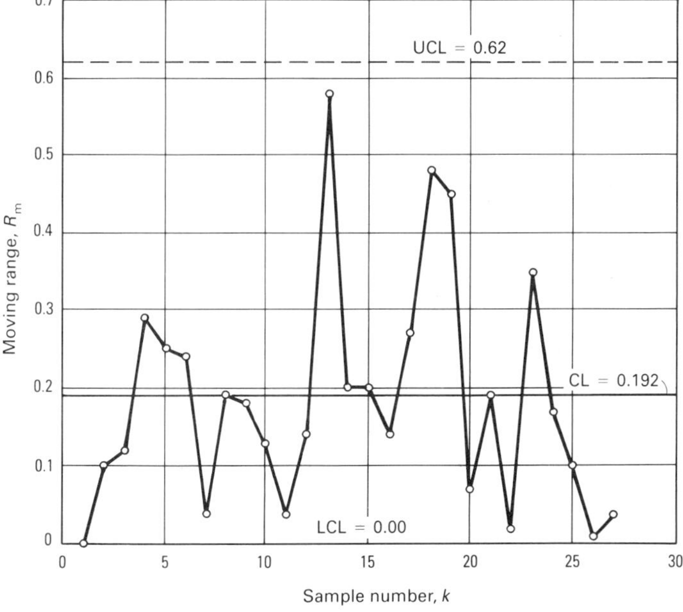

Fig. 26 R_m control chart obtained for white millbase data in Table 3. Data are for $k = 27$, $n = 2$.

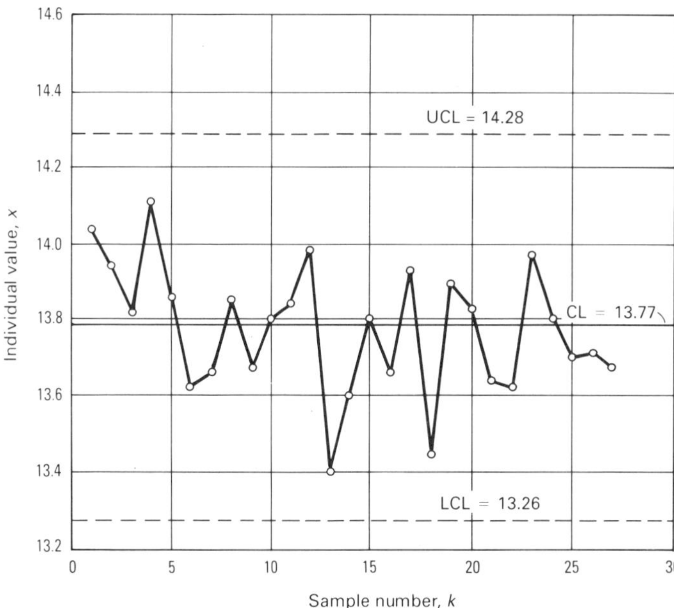

Fig. 27 x control chart obtained for white millbase data in Table 3. Data are for $k = 27$, $n = 2$.

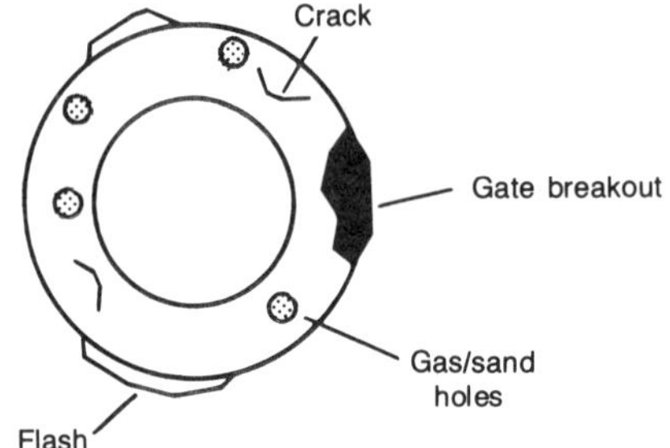

Fig. 28 Typical multiple defects present on an engine valve seat blank to illustrate defect identification in an attribute quality characterization situation

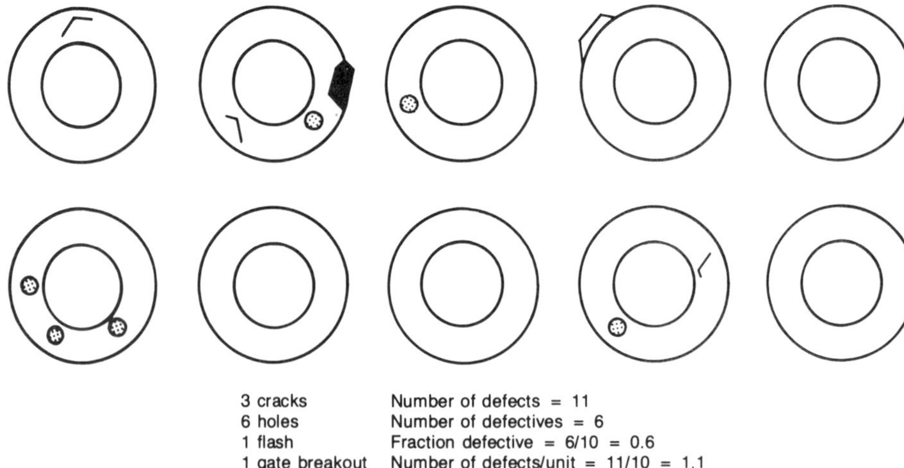

Fig. 29 Analysis of four basic measures of attribute quality characterization used to illustrate the typical defects present in the engine valve seat blank shown in Fig. 28. Out of ten samples tested, four had no defects, three had single defects, and three had multiple defects.

When the individual x values follow a normal distribution, the patterns on the x-chart are analyzed in the same manner as the Shewhart $\bar{x}$-charts. A sample x is circled as a signal of out-of-control values if it falls outside a control limit, if it is the end point of a sequence that violates any of the zone rules, or if it simply indicates a nonrandom sequence. However, tests for unnatural patterns should be used with more caution on an x-chart than on an $\bar{x}$-chart because the individual chart is sensitive to the actual shape of the distribution of the individuals, which may depart considerably from a true normal distribution. Figures 26 and 27 show the x and R_m control charts for the example data.

Shewhart Control Charts for Attribute Data

Many quality assessment criteria for manufactured goods are not of the variable measurement type. Rather, some quality characteristics are more logically defined in a presence-of or absence-of sense. Such situations might include surface flaws on a sheet metal panel; cracks in drawn wire; color inconsistencies on a painted surface; voids, flash, or spray on an injection-molded part; or wrinkles on a sheet of vinyl.

Such nonconformities or defects are often observed visually or according to some sensory criteria and cause a part to be defined simply as a defective part. In these cases, quality assessment is referred to as being made by attributes.

Many quality characteristics that could be made by measurements (variables) are often not done as such in the interest of economy. A go/no-go gage can be used to determine whether or not a variable characteristic falls within the part specification. Parts that fail such a test are simply labeled defective. Attribute measurements can be used to identify the presence of problems, which can then be attacked by the use of $\bar{x}$ and R control charts. The following definitions are required in working with attribute data:

- *Defect*: A fault that causes an article or an item to fail to meet specification requirements. Each instance of the lack of conformity of an article to specification is a defect or nonconformity
- *Defective*: An item or article with one or more defects is a defective item
- *Number of defects*: In a sample of n items, c is the number of defects in the sample. An item may be subject to many different types of defects, each of which may occur several times
- *Number of defectives*: In a sample of n items, d is the number of defective items in the sample
- *Fractional defective*: The fractional defective, p, of a sample is the ratio of the number of defectives in a sample to the total number of items in the sample. Therefore, $p = d/n$

Operational Definitions

The most difficult aspect of quality characterization by attributes is the precise determination of what constitutes the presence of a particular defect. This is so because many attribute defects are visual in nature and therefore require a certain degree of judgment and because of the failure to discard the product control mentality. For example, a scratch that is barely observable by the naked eye may not be considered a defect, but one that is readily seen is. Furthermore, human variation is generally considerably larger in attribute characterization (for example, three different caliper readings of a workpiece dimension by three inspectors and visual inspection of a part by these same individuals yield anywhere from zero to ten defects). It is therefore important that precise and quantitative operational definitions be laid down for all to observe uniformly when attribute quality characterization is being used. The length or depth of a scratch, the diameter of a surface blemish, or the length of a flow line on a molded part can be specified.

The issue of the product control versus process control way of thinking about defects is a crucial one. From a product control point of view, scratches on an automobile grille should be counted as defects only if they appear on visual surfaces, which would directly influence part function. From a process control point of view, however, scratches on an automobile grille should be counted as defects regardless of where they appear because the mechanism creating these scratches does not differentiate between visual and concealed surfaces. By counting all scratches, the sensitivity of the statistical charting instrument used to identify the presence of defects and to lead to their diagnosis will be considerably increased.

A major problem with the product control way of thinking about part inspection is that when attribute quality characterization is being used not all defects are observed and noted. The first occurrence of a defect that is detected immediately causes the part to be scrapped. Often, such data are recorded in scrap logs, which then present a biased view of what the problem may really be. One inspector may concentrate on scratch defects on a molded part and will therefore tend to see these first. Another may think splay is more critical, so his data tend to reflect this type of defect more frequently. The net result is that often such data may then mislead those who may be using it for process control purposes. Figure 28 shows an example of the occurrence of multiple defects on a part. It is essential from a process control standpoint to carefully observe and note each occurrence of each type of defect. Figure 29 shows a typical sample result and the careful observation of each occurrence of each type of defect. In Fig. 29, the four basic measures used in attribute quality characterization are defined for the sample in question.

p-Chart: A Control Chart for Fraction Defective

Consider an injection-molding machine producing a molded part at a steady pace. Suppose the measure of quality conformance of interest is the occurrence of flash and splay on the molded part. If a part has so much as one occurrence of either flash or splay, it is considered to be nonconforming, that is, a defective part.

To establish the control chart, rational samples of size $n = 50$ parts are drawn from production periodically (perhaps, each shift), and the sampled parts are inspected and classified as either defective (from either or both possible defects) or nondefective. The number of defectives, d, is recorded for each sample. The process characteristic of interest is the true process fraction defective p'. Each sample result is converted to a fraction defective:

$$p = \frac{d}{n} \qquad \text{(Eq 1)}$$

The data (fraction defective p) are plotted for at least 25 successive samples of size $n = 50$. The individual values for the sample fraction defective, p, vary considerably, and it is difficult to determine from the plot at this point if the variation about the average fraction defective, $\bar{p}$, is solely due to the forces of common causes or special causes.

Control Limits for the *p*-Chart. It can be shown that for random sampling, under certain assumptions, the occurrence of the number of defectives, d, in the sample of size n is explained probabilistically by the binominal distribution. Because the sample fraction defective, p, is simply the number of defectives, d, divided by the sample size, n, the occurrence of values for p also follows the binominal distribution. Given k rational samples of size n, the true fraction defective, p', can be estimated by:

$$\bar{p} = \sum_{i=1}^{k} p_i \Big/ k \qquad \text{(Eq 2)}$$

or

$$\bar{p} = \sum_{i=1}^{k} d_i \Big/ \sum_{i=1}^{k} n_i \qquad \text{(Eq 3)}$$

Equation 3 is more general because it is valid whether or not the sample size is the same for all samples. Equation 2 should be used only if the sample size, n, is the same for all k samples.

Therefore, given $\bar{p}$, the control limits for the p-chart are then given by:

$$\text{UCL}_p = \bar{p} + 3\sqrt{\bar{p}(1 - \bar{p})/n} \qquad \text{(Eq 4a)}$$

$$\text{LCP}_p = \bar{p} - 3\sqrt{\bar{p}(1 - \bar{p})/n} \qquad \text{(Eq 4b)}$$

Thus, only $\bar{p}$ has to be calculated for at least 25 samples of size n to set up a p-chart. The binomial distribution is generally not symmetric in quality control applications and has a lower bound of $p = 0$. Sometimes the calculation for the lower control limit may yield a value of less than 0. In this case, a lower control limit of 0 is used.

Example 3: A *p*-Chart Applied to Evaluation of a Carburetor Assembly (Ref 22). This example illustrates the construction of a p-chart. The data in Table 4 are inspection results on a type of carburetor at the end of assembly; all types of defects except leaks were noted, and $n = 100$ for all samples. Samples taken numbered $k = 35$.

Using this sample data to establish the p-chart:

$$\bar{p} = \sum_{i=1}^{35} d_i \Big/ \sum_{i=1}^{35} n_i = \frac{73}{3500} = 0.02086$$

$$\hat{\sigma}_p = \sqrt{\frac{\bar{p}(1 - \bar{p})}{n}}$$

$$= \sqrt{\frac{(0.02086)(1 - 0.02086)}{100}} = 0.01429$$

$$3\hat{\sigma}_p = 3\sqrt{\frac{\bar{p}(1 - \bar{p})}{n}}$$

$$= 3(0.01429) = 0.04287$$

Therefore:

$$\text{UCL}_p = 0.02086 + 0.04287 = 0.06373$$
$$\text{LCL}_p = 0.02086 - 0.04287 = -0.02201$$

That is:

$$\text{LCL}_p = 0$$

The plot of the data on the corresponding p-chart is shown in Fig. 30. The process appears to be in statistical control, although eight points lie on the lower control limit. In this case, results of $p = 0$ that fall on the lower control limit should not be interpreted as signaling the presence of a special cause. For a sample size of $n = 100$ and a fraction defective $p' = 0.02$, the binomial distribution gives the probability of $d = 0$ defectives in a sample of 100 to be 0.133. Therefore, a sample with zero defectives would be expected about one out of seven times.

In summary, the p-chart in this example seems to indicate good statistical control, having no extreme points (outside the control limits), no significant trends or cycles, and no runs of sizable length above or below the centerline. At least over this period of data collection, the process appears to be operating only under a common cause system of variation. However, Fig. 30 shows that the process is consistently operating at a 2% defective rate.

Variable Sample Size Considerations for the *p*-Chart. It is often the case that the sample size may vary from one time to another as data for the construction of a p-chart are obtained. This may be the case if the data have been collected for other reasons (for example, acceptance sampling) or if a sample constitutes a day's production (essentially 100% inspection) and production rates vary from day to day. Because the

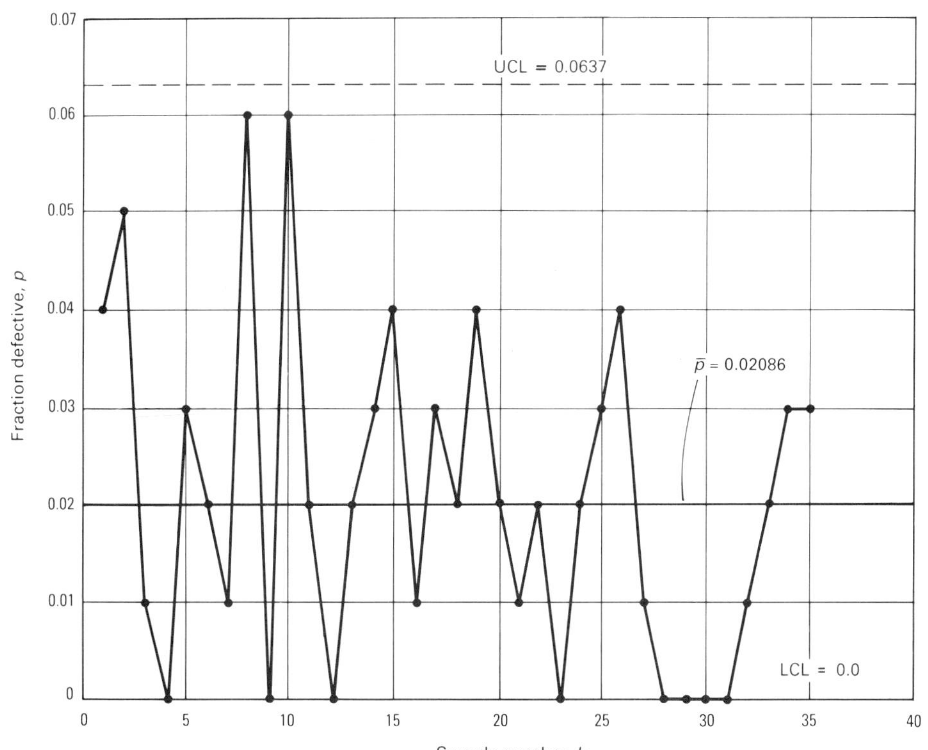

Fig. 30 p control chart obtained for the evaluation of the carburetor assembly data in Table 4. Data are for $k = 35$, $n = 100$.

Table 4 p-chart data for carburetor assembly of Example 3

Sample, k(a)	d	p	Sample, k(a)	d	p
1	4	0.04	19	4	0.04
2	5	0.05	20	2	0.02
3	1	0.01	21	1	0.01
4	0	0.00	22	2	0.02
5	3	0.03	23	0	0.00
6	2	0.02	24	2	0.02
7	1	0.01	25	3	0.03
8	6	0.06	26	4	0.04
9	0	0.00	27	1	0.01
10	6	0.06	28	0	0.00
11	2	0.02	29	0	0.00
12	0	0.00	30	0	0.00
13	2	0.02	31	0	0.00
14	3	0.03	32	1	0.01
15	4	0.04	33	2	0.02
16	1	0.01	34	3	0.03
17	3	0.03	35	3	0.03
18	2	0.02			

(a) $n = 100$

Table 5 c-chart data for plastic-insulated wire of Example 4

Sample number, k	Number of breakdowns	Sample number, k	Number of breakdowns
1	1	16	20
2	1	17	1
3	3	18	6
4	7	19	12
5	8	20	4
6	1	21	5
7	2	22	1
8	6	23	8
9	1	24	7
10	1	25	9
11	10	26	2
12	5	27	3
13	0	28	14
14	19	29	6
15	16	30	8

limits on a p-chart depend on the sample size n, some adjustments must be made to ensure that the chart is properly interpreted. There are several ways in which the variable sample size problem can be handled. Some of the more common approaches are the following:

- Compute separate limits for each individual subgroup. This approach certainly leads to a correct set of limits for each sample, but requires continual calculation of the control limits and a somewhat messy-looking control chart
- Determine an average sample size, $\bar{n}$, and a single set of control limits based on $\bar{n}$. This method may be appropriate if the sample sizes do not vary greatly, perhaps no more than about 20%. However, if the actual n is less than $\bar{n}$, a point above the control limit based on $\bar{n}$ may not be above its own true upper control limit. Conversely, if the actual n is greater than $\bar{n}$, a point may not show out of control when in reality it is
- A third procedure for varying sampling size is to express the fraction defective $\bar{p}$ in standard deviation units, that is, plot $(p - \bar{p})/\sigma_p$ on a control chart where the centerline is zero and the control limits are set at ±3.0. This stabilizes the plotted value even though n may be varying. Note that $(p - \bar{p})/\sigma_p$ is a familiar form; recall the standard normal (Z) distribution. For this method, the continued calculation of the stabilized variable is somewhat tedious, but the chart has a clean appearance, with constant limits of always ±3.0 and constant centerline at 0.0

c-Chart: A Control Chart for Number of Defects

The p-chart deals with the notion of a defective part or item where defective means that the part has at least one nonconformity or disqualifying defect. It must be recognized, however, that the incidence of any one of several possible nonconformities would qualify a part for defective status. A part with ten defects, any one of which makes it a defective, is on equal footing with a part with only one defect in terms of being a defective.

Often it is of interest to note every occurrence of every type of defect on a part and to chart the number of defects per sample. A sample may only be one part, particularly if interest is focusing on final inspection of an assembled product, such as an automobile, a lift truck, or perhaps a washing machine. Inspection may focus on one type of defect (such as nonconforming rivets on an aircraft wing) or multiple defects (such as flash, splay, voids, and knit lines on an injection-molded truck grille).

Considering an assembled product such as a lift truck, the opportunity for the occurrence of a defect is quite large, perhaps to be considered infinite. However, the probability occurrence of a defect in any one spot arbitrarily chosen is probably very, very small. In this case, the probability law that governs the incidence of defects is known as the Poisson law or Poisson probability distribution, where c is the number of defects per sample. It is important that the opportunity space for defects to occur be constant from sample to sample. The Poisson distribution defines the probability of observing c defects in a sample where c' is the average rate of occurrence of defects per sample.

Construction of c-Charts From Sample Data. The number of defects, c, arises probabilistically according to the Poisson distribution. One important property of the Poisson distribution is that the mean and variance are the same value. Then given c', the true average number of defects per sample, the 3σ limits for the c-chart are given by:

$$CL_c = c' \pm 3\sqrt{c'} \quad \text{(Eq 5)}$$

Note that the standard deviation of the observed quantity c is the square root of c'. The Poisson distribution is a very simple probability model, being completely described by a single parameter c'.

When c' is unknown, it must be estimated from the data. For a collection of k samples, each with an observed number of defects c, the estimate of c' is:

$$\bar{c} = \sum_{i=1}^{k} c_i \Big/ k \quad \text{(Eq 6)}$$

Therefore, trial control limits for the c-chart can be established, with possible truncation of the lower control limit at zero, from:

$$UCL_c = \bar{c} + 3\sqrt{\bar{c}}$$
$$LCL_c = \bar{c} - 3\sqrt{\bar{c}}$$

Example 4: c-Chart Construction for Continuous Testing of Plastic-Insulated Wire at a Specified Test Voltage. Table 5 lists the results of continuous testing of a certain type of plastic-covered wire at a specified test voltage. This test causes breakdowns at weak spots in the insulation, which are cut out before shipment.

The original data consisted of the number of breakdowns in successive lengths of 1000 ft each. There may be 0, 1, 2, 3, ..., breakdowns per length, depending on the number of weak spots in the insulation. However, so few defects were obtained during a short period of production by using the 1000 ft length as a unit and the expectancy in terms of the number of breakdowns per length was so small that a longer length of 10 000 ft was used for the unit size for the corresponding c-chart. In general, it is desirable to select the sample size for the c-chart application such that on average ($\bar{n}$) at least one or two defects are occurring per sample.

In most applications, the centerline of the c-chart is based on the estimate of the average number of defects per sample. This estimate can be calculated by:

$$\bar{c} = \frac{\text{Total number of breakdowns}}{\text{Total number of units}} = \frac{187}{30}$$

$$= 6.23$$

Control limits: $\bar{c} \pm 3\sqrt{\bar{c}} = 6.23 \pm 3\sqrt{6.23}$
UCL = 13.72
LCL = 0

The resulting c-chart in Fig. 31 shows the presence of special causes of variation.

u-Chart: A Control Chart for the Number of Defects per Unit

Although in c-chart applications it is common for a sample to consist of only a single unit or item, the sample or subgroup can be comprised of several units. Further, from subgroup to subgroup, the number of units per subgroup may vary, particularly if a subgroup is an amount of production for the shift or day, for example.

In such cases, the opportunity space for the occurrence of defects per subgroup changes from subgroup to subgroup, violating the equal opportunity space assumption on which the standard c-chart is based. Therefore, it is necessary to create some standardized statistic, and such a statistic may be the average number of defects per unit or item where n is the number of items per subgroup. The symbol u is often used to denote average number of defects per unit, that is:

$$u = \frac{c}{n} \quad \text{(Eq 7)}$$

where c is the total number of defects per subgroup of n units. For k such subgroups gathered, the centerline on the u-chart is:

$$\bar{u} = \frac{\sum_{i=1}^{k} c_i}{\sum_{i=1}^{k} n_i} = \frac{\text{Total defects in } k \text{ subgroups}}{\text{Total number of items in } k \text{ subgroups}} \quad \text{(Eq 8)}$$

The trial control limits for the u-chart are then given by:

$$\text{CL}_u = \bar{u} \pm 3\sqrt{\bar{u}/n} \quad \text{(Eq 9)}$$

Example 5: Use of the u-Chart to Evaluate Leather Handbag Lots. Table 6 lists inspection results in terms of defects observed in the inspection of 25 consecutive lots of leather handbags. Because the number of handbags in each lot was different, a constant sample size of $n = 10$ was used. All defects were counted even though two or more defects of the same or different type occurred on the same bag. The u-chart data are as follows (Fig. 32):

(a) Centerline $\bar{u} = \frac{382}{250} = 1.53$

(b) Control Limits for $n = 10$:

$\bar{u} \pm 3\sqrt{\bar{u}/n} = 1.53 \pm 3\sqrt{0.153}$
$= 1.53 \pm 1.17$

$\text{UCL}_u = 2.70$
$\text{LCL}_u = 0.36$

Table 6 u-chart data for leather handbag lot production of Example 5

Sample number, k(a)	Total number of defects	Defects per unit
1	17	1.7
2	14	1.4
3	6	0.6
4	23	2.3
5	5	0.5
6	7	0.7
7	10	1.0
8	19	1.9
9	29	2.9
10	18	1.8
11	18	1.8
12	5	0.5
13	8	0.8
14	11	1.1
15	18	1.8
16	13	1.3
17	22	2.2
18	6	0.6
19	23	2.3
20	22	2.2
21	9	0.9
22	15	1.5
23	20	2.0
23	20	2.0
25	24	2.4
Total	**382**	**38.2**

(a) $n = 10$

Process Capability Assessment

This section presents both traditional and more modern views of process capability and how it is assessed. The presentation stresses the relationship between the control/stability of a process and its capability. The clear distinction between the engineering specification and the statistical control limits in terms of their use and interpretation is emphasized. Using the traditional conformance to the specifications view of process capability, this section illustrates

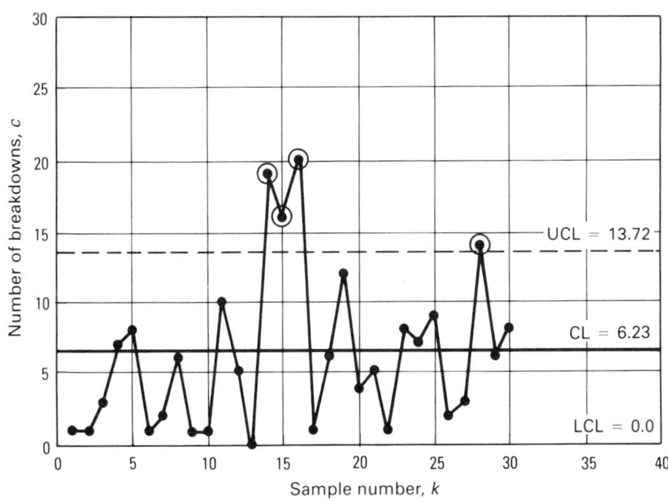

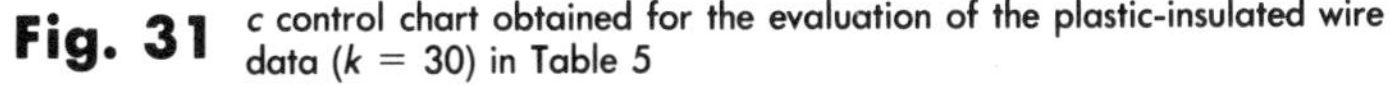

Fig. 31 c control chart obtained for the evaluation of the plastic-insulated wire data ($k = 30$) in Table 5

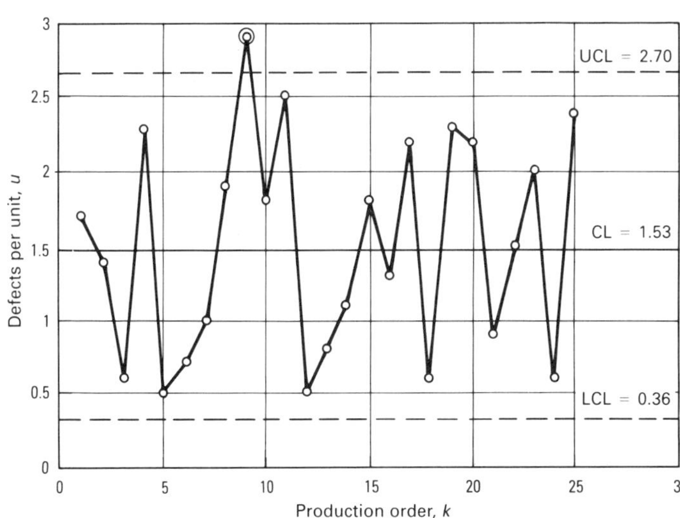

Fig. 32 u control chart obtained for the evaluation of leather handbag lot data in Table 6. Data are for $k = 25$, $n = 10$. Datum for sample 9 is an extreme point because it exceeds value of UCL.

the consequences of a lack of statistical control in terms of the manner and extent to which a process produces parts that meet design intent. This section also presents the loss function approach to the articulation of process capability.

Process Capability Versus Process Control

There are two separate but vitally important issues that must be addressed when considering the statistical representation of process data. These are:

- The ability of the process to produce parts that conform to specifications
- The ability of the process to maintain a state of good statistical control

These two process characteristics are linked in the sense that it will be difficult to assess process capability with respect to conformance to specifications without being reasonably assured of having good statistical control. Although control certainly does not imply conformance, it is a necessary prerequisite to the proper assessment of conformance.

In a statistical sense, conformance to specifications involves the process as a whole; therefore, attention will be focused on the distribution of individual measurements. In dealing with statistical control, summary statistics from each sample, mainly $\bar{x}$ and R, are used; as a result, this involves the distribution of these statistics, not individual measurements.

Because of the above distinction between populations and samples, it is crucial not to compare or confuse part specifications and control limits. In fact, tolerance/specification limits should never be placed on a control chart. This is so because the control chart is based on the variation in the sample means $\bar{x}$, while it is the individual measurements in the sample that should be compared to specifications. Placing specifications on the control chart may lead to the mistaken impression that good conformance exists when, in fact, it does not. This is illustrated in Example 6.

A process may produce a large number of pieces that do not meet the specified production standards, even though the process itself is in a state of good statistical control (that is, all the points on $\bar{x}$ and R control charts are within the 3σ limits and vary in a random manner). This may be because the process is not centered properly; in other words, the actual mean value of the parts being produced may be significantly different from the specified nominal value of the part. If this is the case, an adjustment of the machine to move the mean closer to the nominal value may solve the problem. Another possible reason for lack of conformance to specifications is that a statistically stable process may be producing parts with an unacceptably high level of common cause variation. In summary, if a process is in statistical control but not capable of meeting the tolerances, the problem may be one of the following:

- The process is off-center from the nominal
- The process variability is too large relative to the tolerances
- The process is off-center and with large variation

Example 6: Statistical Assessment of Process Capability for a Workpiece. For a certain part, a dimension of interest was specified by the engineering department as 0.140 ± 0.003 in. Many parts were being rejected on 100% inspection using a go/no-go gage because they failed to meet these tolerances.

It was decided to study the capability of the process using $\bar{x}$ and R control charts. These data were taken from the same machine and operator and at the rate of about one sample per hour.

Both the $\bar{x}$ and R control charts seem to indicate good statistical control with no points exceeding the 3σ limits, a reasonably normal distribution of points between the limits, and no discernible trends, cycles, and so on. Therefore, the calculated sample mean $\bar{\bar{x}} = 0.1406$ in. and the sample standard deviation $\sigma_x = 0.0037$ in. are good estimates. The process can then be evaluated with respect to its conformance to specifications.

To obtain a clear picture of the statistical nature of the data, a frequency histogram was plotted that resembled a normal distribution, but the mean appears to be slightly higher than the nominal value of 0.140 set by the engineering department. Figure 33 shows this population distribution curve centered at $\bar{\bar{x}}$ with a spread of $\sigma_x = 0.0037$ in. The specifications are also shown on this plot.

The shaded areas in Fig. 33 represent the probability of obtaining a part that does not meet specifications. To compute the probability of a part falling below the lower specification, the standard normal distribution, Z, is calculated, and a normal curve table is used:

$$Z = (x - \bar{\bar{x}})/\sigma_x$$

where x is the value of either the lower or upper specification, $\bar{\bar{x}}$ is the estimate for the population mean, and σ_x is the estimate of the process standard deviation.

To find the probability of a point below the lower specification limit, $LSL_x = 0.137$, with $\bar{\bar{x}} = 0.1406$, and $\hat{\sigma}_x = 0.0037$:

$$Z = \frac{0.137 - 0.1406}{0.0037} = -0.97$$

Looking this value up in the normal table produces Prob (0.137 or less) = 0.1660. This means that there is a 16.6% chance of an individual part falling below the specified tolerance.

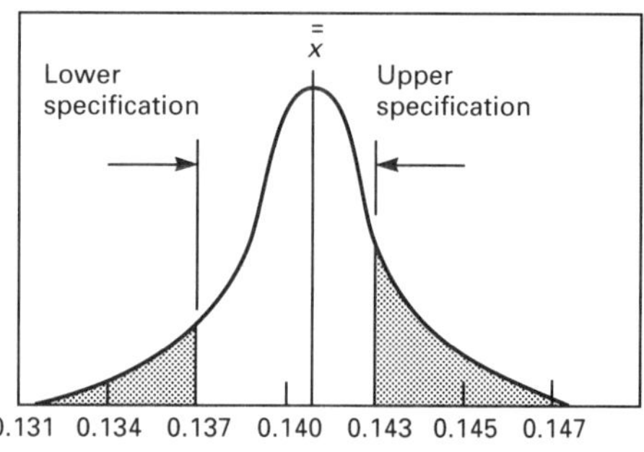

Fig. 33 Normal distribution model for process capability for the data of Example 6. $LSL_x = 0.137$, $USL_x = 0.143$, $\bar{\bar{x}} = 0.1406$, and $\sigma_x = 0.0037$.

To find the probability of an individual part falling above the upper specification limit, $USL_x = 0.143$, $\bar{\bar{x}} = 0.1406$, $\hat{\sigma}_x = 0.0037$, and:

$$Z = \frac{0.143 - 0.1406}{0.0037} = 0.65$$

The probability of a part being above the upper specification limit is equal to 0.2578, based on $Z = 0.65$. The process that does not meet the specifications is therefore 16.6% + 25.78% = 42.37%.

It might be asked whether centering the process at the nominal value of 0.140 would help. To check, a normal curve is constructed, centered at 0.140. The probability of a point below the lower specification is found by computing Z using the nominal value as the population mean:

$$Z = \frac{0.137 - 0.140}{0.0037} = -0.81$$

Looking up the area for this Z value in the normal curve table produces a value of 0.209, which is the probability of getting a value below the lower specification limit. The probability of getting a value above the upper specification is found by:

$$Z = \frac{0.143 - 0.140}{0.0037} = 0.81$$

which from the normal curve table gives an area = 0.209. The total probability of a part not meeting specification is the sum of these, or 0.209 + 0.209 = 0.41, or 41%.

Therefore, recentering the process will not be of much help. The process is in control; no special causes of variability were indicated. However, about 42% of the parts were outside the tolerances. Possible remedies include the following:

- Continue to sort by 100% inspection
- Widen the tolerances, for example, 0.140 ± 0.006
- Use a more precise process; reduce process variation

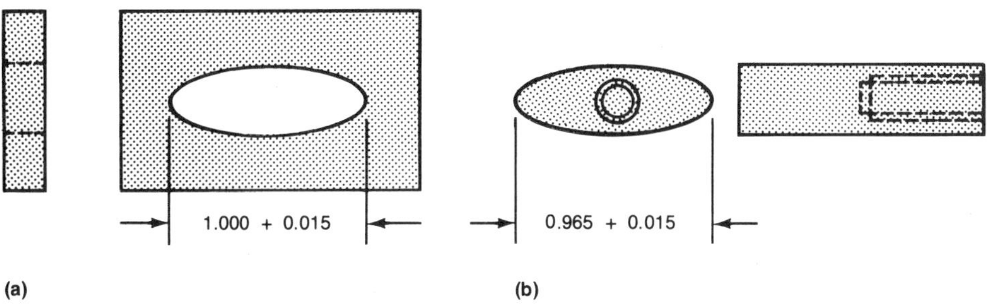

Fig. 34 Machined components statistically analyzed in Example 7. (a) Plate with hole. (b) Pin assembly. Dimensions given in inches

- Use statistical methods to identify variation reduction opportunities for the existing process

Too often, the strategy used (or at least urged) is the second remedy listed above. Clearly, a stronger consideration should be given to the final remedy listed above.

Comparison of Tolerances and Control Limits

It is important to clearly differentiate between specification limits and control limits. The specification limits or tolerances of a part are:

- Characteristic of the part/item in question
- Based on functional considerations
- Related to/compared with an individual part measurement
- Used to establish the conformability of a part

The control limits on a control chart are:

- Characteristic of the process in question
- Based on process variability
- Dependent on sampling parameters, namely, sample size
- Used to identify presence/absence of special cause variation in the process

Control limits and tolerances must never be compared numerically and should not appear together on the same graph. Tolerances are limits on individual measurements and as such can be compared against the process as a whole as represented by many individual measurements collected in the form of a statistical distribution, as was done in Example 6 to assess overall process capability.

Process Capability Indices

It is common to measure process capability in the units of process standard deviations. In particular, it is common to look at the relationship between the process standard deviation and the distance between the upper and lower specification:

$$C_p = \frac{\text{USL}_x - \text{LSL}_x}{\sigma_x} \quad \text{(Eq 10)}$$

The minimum acceptable value for C_p is considered to be 6.

Recently, many companies have begun to use a capability index referred to as C_{pk}. For bilateral specifications, C_{pk} is defined in the following manner. First, the relationship between the process mean $\bar{\bar{x}}$ and the specification limits in the units of standard deviations is determined:

$$Z_{\text{USL}} = \frac{\text{USL} - \bar{x}}{\sigma_x} \quad \text{(Eq 11a)}$$

$$Z_{\text{LSL}} = \frac{\text{LSL} - \bar{x}}{\sigma_x} \quad \text{(Eq 11b)}$$

Then the minimum of these two values is selected:

$$Z_{\min} = \min[Z_{\text{USL}}, -Z_{\text{LSL}}] \quad \text{(Eq 12)}$$

The C_{pk} index is then defined by dividing this minimum value by 3:

$$C_{pk} = \frac{Z_{\min}}{3} \quad \text{(Eq 13)}$$

Commonly, C_{pk} must be $\geqq 1.00$.

Statistical Process Control and the Statistical Tolerance Model

The issue of part tolerancing and, in particular, the statistical assignment and assessment of tolerances are excellent examples of the need for design and manufacturing to understand what each other is doing and why. The best intentions of the design process can go unmet if the manufacturing process is not operated in a manner totally consistent with design intent. To more clearly appreciate the marriage of thinking that must exist between the design and manufacturing worlds, some of the basic assumptions of the tolerancing activity and their relationship to the manufacturing process will be examined. The following sections clearly point to the importance of statistical process control relative to the issue of process capability.

The key concepts in statistical tolerancing are:

- The use of a statistical distribution to represent the design characteristic and therefore the process output for the product/part in question relative to the design specifications
- The notion of random assembly, that is, random part selection from these part process distributions when more than one part is being considered in an assembly
- The additive law of variances as a means to determine the relationship between the variability in individual parts and that for the assembly

To assume that the parts can be represented by a statistical distribution of measurements (and for the assumption to hold in reality), the part processes must be in a state of statistical control. The following example illustrates the importance of statistical process control in achieving design intent in a tolerancing problem.

Example 7: Statistical Tolerance Model for Optimum Fit of a Pin Assembly in a Hole Machined in a Plate. Figure 34 shows two simple parts: a plate with a hole and a pin that will ultimately be assembled to a third part but must pass through the hole in the plate. For the assembly, it is desired for function that the clearance between the plate hole and the pin be at least 0.015 in. but no more than 0.055 in.

To achieve the design requirement stated above, the nominal values and tolerances for the plate hole and pin were statistically derived and are shown in Fig. 34. To arrive at these tolerances, it was assumed that:

- The parts would be manufactured by processes that behave according to the normal distribution
- The process capabilities would be at least 6σ, the processes would be centered at the nominal values given in Fig. 34, and the processes would be maintained in a state of statistical control
- Random assembly would prevail

If these assumptions are met, the processes for the two parts, and therefore the clearance associated with assembled parts, would be as shown in Fig. 35, and the design intent would be met.

Suppose that despite the assumptions made and the tolerances derived, the processes manufacturing the pin and plate hole were not maintained in good statistical control. As a result, the parts actually more nearly follow a uniform/rectangular distribution within the specifications, as shown in Fig. 36. Such could have arisen as a result of sorting or rework of a more variable process(es), in which case the results are doubly distressing, that is, poorly fitting assemblies and increased cost to the system.

Figure 36 shows the distribution of the clearance if the hole and pin dimensions follow the uniform distribution within the specifications. The additive law of variances has been used to derive the variation in the clearance distributions but assuming the uniform distribution for the individual part processes. Some assemblies may not go together at all, some will fit quite tightly and may later bind if foreign matter gets into the gap, and others will fit together with a much larger clearance than desired.

The problem here is not a design problem. The plate hole and pin tolerances have been derived using sound statistical methods. However, if the processes are not in good statistical control and therefore not capable of meeting the assumptions made during design, poor-quality assemblies will follow. It should be noted that the altogether too common process appearance of a uniform distribution of measurements within the specifications can arise in several different ways:

- From processes that have good potential with regard to variation, but are not kept in good statistical control
- From unstable and/or large variation processes that require sorting/rework to meet the specifications
- From processes that are intentionally allowed to vary over the full range of the specifications to take advantage locally of wide specifications relative to the process variation

In all of the three cases mentioned above, additional costs will be incurred and product quality will be eroded. Clearly, statistical process control is crucial to the tolerancing issue in engineering design. Taguchi's loss function model, which is an essential element in tolerance design, assumes similarly that quality characteristics can be represented by a statistical distribution of measurements. Again, for this assumption to be met at the process and therefore in the ultimate product in the field, the manufacturing processes must be maintained in a state of statistical control.

Design of Experiments: Factorial Designs

The process of product design and its associated manufacturing processes and tolerance designs often involve many experiments to better understand the various cause-effect relationships for quality performance of the product and for ease in process control. The sections that follow will present some of the basic concepts and methods for the planning, design, and analysis of experiments.

The purpose of most experimental work is to discover the direction(s) of change that may lead to improvements in both the quality and the productiveness of a product or process. Such endeavors can be referred to as process improvement because product improvement can only be meaningfully measured through its use and that is of course a process. Historically, design of experiments methods have tended to focus more attention on process improvement as contrasted with product design. In this regard, the different view that might be taken toward design of experiments in product design versus processing is probably overstated. In this section, the role of design of experiments in product design is emphasized, as is its use in the simultaneous engineering of products and processes.

In investigating the variation in performance of a given product or process, attention focuses on the identification of those factors that, when allowed to vary, cause performance to vary in some way. Some such factors are qualitative in nature (categorical variables), while others are quantitative (possessing an inherent continuity of change). The situations examined below may consider both qualitative and quantitative variables simultaneously. In fact, an important advantage of the two-level factorial designs that will be discussed is their ability to consider both types of variables within the same test plan.

Mathematical Model

A fundamental problem of design of experiments is that of selecting the appropriate arrangement of test points within the space defined by the design/control and noise variables. Although many different considerations must come into play in selecting a test plan, none can be more fundamental than the notion of the mathematical model. Whether or not explicitly recognized as such, most experimental studies are aimed either directly or indirectly at discovering the relationship between some performance response and a set of candidate variables influencing that response. In general, this relationship can be written as:

$$Y = f(X_1, X_2, \ldots, X_k) + e \qquad \text{(Eq 14)}$$

where Y is the response of interest, f is some unknown functional relationship, $X_1, \ldots, X_k$ are the independent variables, and e is a random error. The functional form f can be thought of as a transfer function. In Taguchi's framework, the variables X_1, $X_2, \ldots, X_k$ are generally partitioned into signal, control, and noise variables.

Sometimes enough is known about the phenomenon under study to use theoretical considerations to identify the form of f. For example, a chemical reaction can be described by a differential equation, which

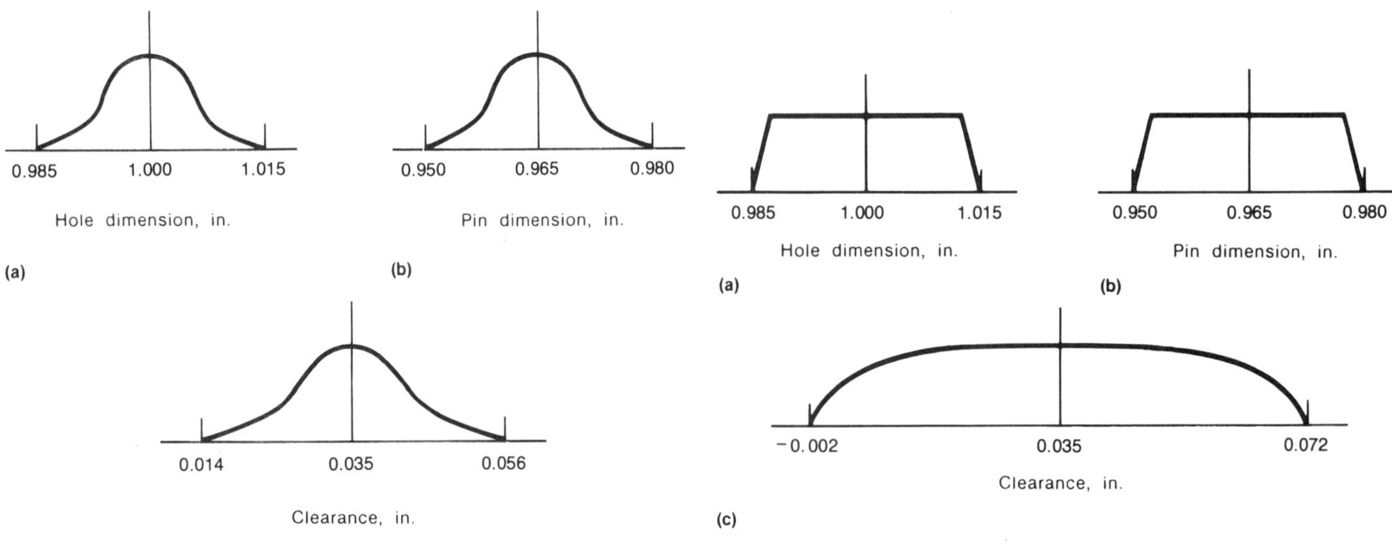

Fig. 35 Statistical basis for satisfying design intent for the hole/pin assembly clearance in Fig. 34. (a) Distribution of hole. (b) Distribution of pin. (c) Distribution of clearance

Fig. 36 Clearance implications of poor process control of plate hole and pin dimensions for components of Fig. 34. (a) Distribution of hole. (b) Distribution of pin. (c) Distribution of clearance

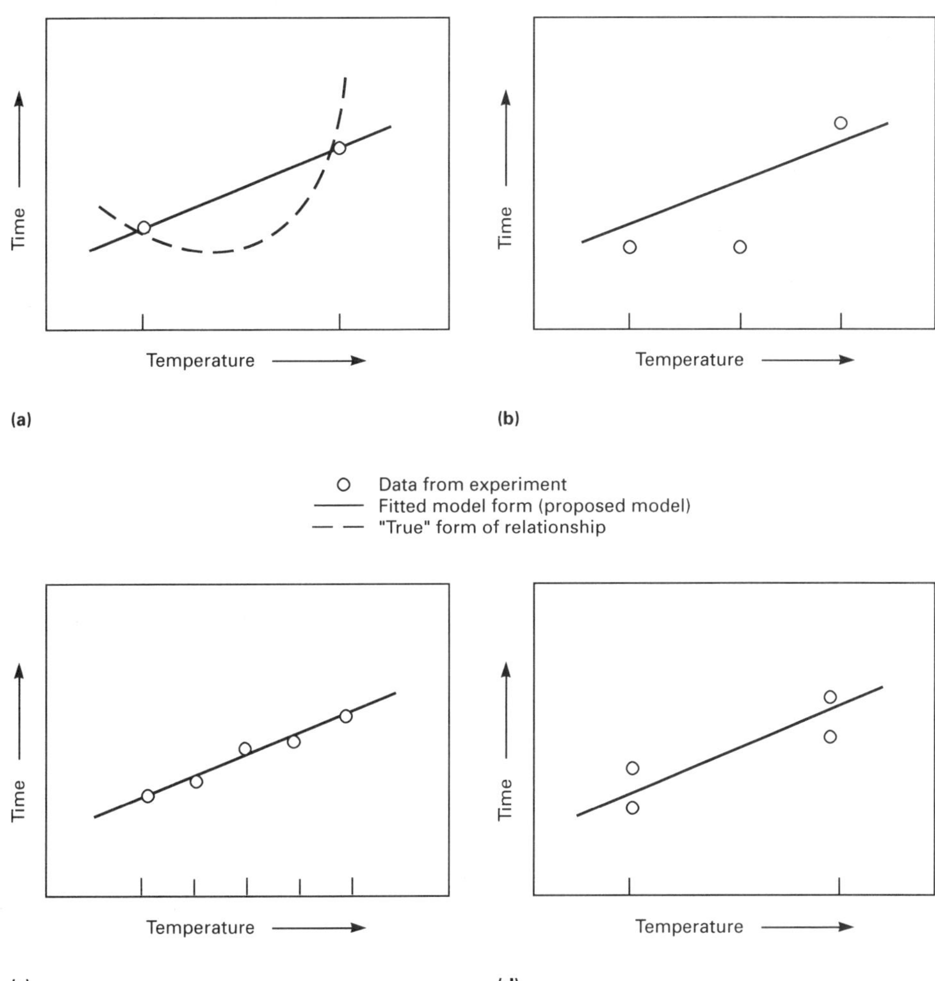

Fig. 37 Comparison of typical time-temperature relationships for true relationship (a) compared to experimental and fitted models (a through d. See text for more details.)

when solved produces a theoretical relationship between the dependent and independent variables. More often than not, however, the knowledge is more sparse, and empirical models must be relied upon that act as mathematical french curves describing relationships through the data; for example:

$$Y = b_0 + b_1X_1 + b_2X_2 + e \quad \text{(Eq 15a)}$$

$$Y = b_0 + b_1X_1 + b_{11}X_1^2 + e \quad \text{(Eq 15b)}$$

Model Building. In most studies, the experimenter begins with a tentative hypothesis concerning the plausible model forms that are to be initially entertained. He must then select an experimental design having the ability to produce data that will:

- Be capable of fitting proposed model(s)
- Be capable of placing the model in jeopardy in the sense that inadequacies in the model can be detected through analysis

The second consideration above is of particular importance to ensure that through a series of iterations the most appropriate model can be determined, while others may be proved less plausible through the data.

If, for example, a curvilinear relationship between temperature and reaction time in a chemical process is suspected, an experiment that studies the process at only two temperatures will be inadequate to reveal this possibility. An experiment with three levels of temperature would, however, allow this possibility to be considered. Figure 37 illustrates several scenarios that emphasize the importance of the relationship between the math model and the associated design of experiment. In Fig. 37, the following points should be noted:

(a) The relationship is actually curvilinear, but such will never be detected by the data
(b) A poor model (straight line) has been hypothesized, but model checking can reveal this and help propose a better model
(c) If the relationship is known to be a straight line, many levels of temperature in the experiment are unnecessary
(d) If it is known *a priori* that the relationship is a straight line, the best test plan would be to study only two relatively extreme levels of temperature and to use additional tests for replication to observe the amount of experimental error

Sequential and Iterative Experimentation. There is always the temptation to carefully design one large experiment that will consider all aspects of the problem at hand. Such a step is dangerous for the following reasons:

- If erroneous decisions and hypotheses about the state of affairs are made, considerable time and experimental resources may be wasted, and the end product may provide little useful information or direction in terms of what to do next
- If knowledge of the underlying situation is limited *a priori*, many factors may be suspected as being important, requiring a very large experiment in terms of number of tests. Ultimately, only a small subset of variables will be found to be of major significance
- In the early stages of experimentation, knowledge of the specific ranges of variables that ought to be studied is not always available. Furthermore, the metrics to be employed to define the variables, responses, or even what responses to observe may not always be clear in the early stages of experimental work
- One large experiment will necessarily cause the testing period to be protracted in time, making it more difficult to control the forces of nuisance variation

For these reasons, it is much more desirable to conduct an experimental program through a series of smaller, often interconnected experiments. This provides the opportunity to modify hypotheses about the state of affairs concerning the situation at hand, to discard variables that are shown to be unimportant, to change the region of study of some or all of the variables, and/or to define and include other measures of process performance. Experimental designs that can be combined sequentially are very useful in this regard. This is often referred to as the sequential assembly of experimental designs.

Revelation of Variable Effects. Often, the variables of importance are not clearly known *a priori*. It is desirable to be able to study several variables together but to independently observe the effect of a change in each one of the variables. Furthermore, it may be deemed important to know if such a variable effect varies with the conditions of the process, that is, when other variables take on varying levels. An arrangement of the tests is called a design, which provides for the opportunity to learn much about the relationships between the variables and the response. In particular:

- The effect of changing any of the variables alone can be observed

- The possibility that the effects measured above can vary as conditions of other variables vary can be observed, that is, the existence of variable interactions

System Noise/Variation

The experimental study of any phenomenon is made difficult by the presence of noise or experimental error. Many factors, not directly under study, are varying over the course of the experiment. These are often referred to as the forces of common cause system variation. Such variation may cloud or mask the effect of change of the factors under study in an experiment. The forces of noise or variation can be better understood or mitigated by several approaches, some of which are strictly experimental design issues.

Statistical Control/Stability Analysis. If the phenomenon under study is already a viable and ongoing process, the pursuit of improvement opportunities through experimentation can be considerably enhanced by employing the techniques of statistical process control. In this way, spurious or sporadic sources of variation can be identified and, through remedial action, removed. Achievement of a stable process will greatly contribute to the ability to more readily observe the effects of purposeful process change. Thus, continued study will further enhance the ability to observe the persistence of changes that might be introduced. Once a process is stabilized, continued attack on the common cause system will lead to a progressively quieter process, further heightening the ability to observe the forces of purposeful process change through experimentation.

Experimental Design Strategies. In many situations, the notion of a stable, ongoing process has little meaning. In the early stages of product or process design or prototype or pilot-plant testing, a stable, consistent process is not present. It is perhaps for this reason (among others) that the body of knowledge known as experimental design was cultivated. Under such situations, the following factors are significant:

- Attempt to identify major sources of variation and take action to ensure that their presence is blocked out from the comparisons made within an experiment. The technique of blocking is useful for this purpose
- Counteract the forces of unknown systematic variation over the period of the experiment by randomization of the tests so that such variation is uniformly and randomly distributed across the trials conducted
- Include replication in the experimental test plan. Multiple tests at the same conditions will provide comparisons that directly measure the amount of variation/experimental error
- Include confirmatory testing as part of the experimental strategy. It will be important that additional trials are run under specific conditions determined from the analysis to verify the improvement opportunities revealed from the experiment

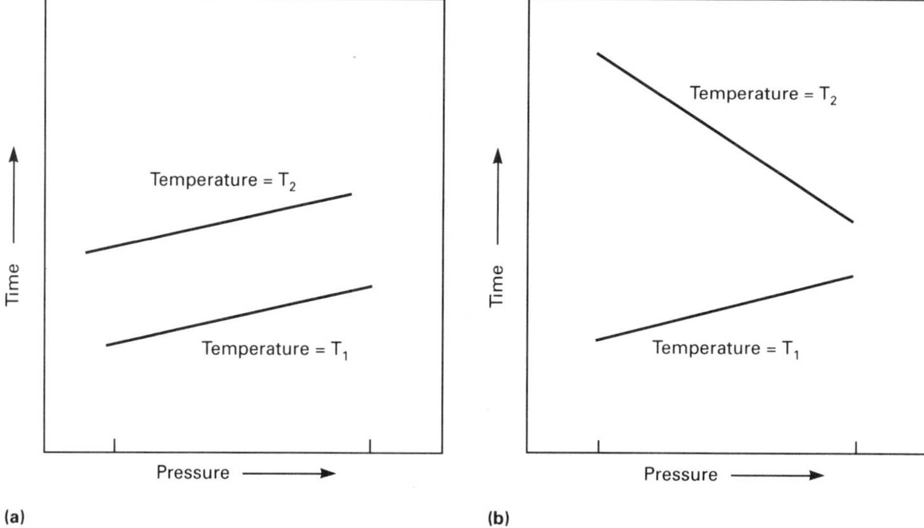

Fig. 38 Graphical depiction of the absence (a) and presence (b) of a two-factor interaction effect

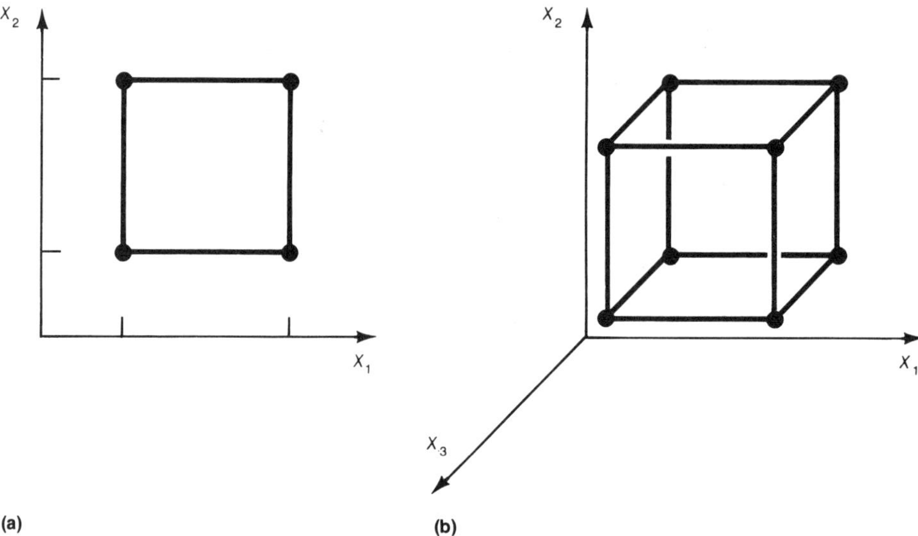

Fig. 39 Two-level factorial design for two (four tests required) (a) and three (eight tests required) (b) variables

The parameter design method is specifically directed at mitigating the forces of noise variation as it may be transmitted through the product/process design to the output performance.

Nature of Variable Interactions

For many products and/or processes, the effects that the important design/control factors have on the system performance responses of interest do not act independently of each other. That is, the effect a certain factor has on the response may be different for different levels of a second factor. When this occurs, the two factors are said to interact or to have an interdependency relationship; that is, a two-factor interaction effect is present. Figure 38 summarizes the nature of the two-factor interaction effect. Figure 38(a) shows that the effect of pressure on time (the slope of the line) is the same regardless of the level of temperature. Therefore, no interaction is present. However, in Fig. 38(b), the effect of pressure on time is clearly seen to vary with temperature. Therefore, a two-factor interaction is present.

Simple Yet Powerful Experimental Design. Many of the problems created by *ad hoc* testing methods and/or methods such as the one-variable-at-a-time approach can be overcome by using an experimental design structure referred to as the two-level factorial design. For such test plans, each factor/variable is studied over only two levels or settings, and all possible combinations are examined. Therefore, the total number of

Table 7(a) Test matrix for simple two-variable, two-level factorial design

Test	X_1	X_2
1	−	−
2	+	−
3	−	+
4	+	+

Table 7(b) Test matrix for simple three-variable, two-level factorial design

Test	X_1	X_2	X_3
1	−	−	−
2	+	−	−
3	−	+	−
4	+	+	−
5	−	−	+
6	+	−	+
7	−	+	+
8	+	+	+

unique tests required for such a test plan is 2^k, where k is the number of variables; for example, for two variables, $2^2 = 4$ test conditions define the test matrix.

Figure 39 shows a graphical representation of the two-level factorial design when two and three variables are under study. The geometric representation is useful from the standpoints of interpreting the variable effects and communicating the purpose and results of the test plan to others. The corners of the square and the cube represent geometrically the conditions of each unique combination of the variable settings.

Tables 7(a) and (b) provide for a more algebraic way to represent the test conditions for a two-level factorial design. The two levels for each factor are often simply referred to as the high and low levels and are represented in coded form as + or +1 and − or −1. This facilitates the determination of the variable effects, given the data. Each row in Tables 7(a) and (b) represents the recipe for a particular test. For example, in the 2^3 factorial in Table 7(b), test 3 is run with variable $1(X_1)$ at its low level, variable $2(X_2)$ at its high level, and variable $3(X_3)$ at its low level.

Experimental Study Using a 2^3 Factorial Design

Example 8: Use of 2^3 Factorial Design to Determine Need for Preheating or Postheating of Welded High-Carbon Steel Rail Bars (Ref 23). High-carbon steel, because of its high strength and low cost, has been extensively used for railway track. The Rail Steel Bar Association sponsored a research project at the Welding Research Laboratory of the University of Wisconsin around 1960 to study whether or not preheating and postheating were necessary for the high-carbon steel to have good-quality welds. After a preliminary investigation by manual arc-welding tests, it appeared that there were three variables affecting the ultimate tensile stress of a weld. Based on the needs of the study and the available funds, it was suggested to run 16 tests that were defined according to a 2^3 factorial design. The three selected variables and their high and low levels are shown below:

Variable	Low level	High level
Ambient temperature (T), °F	0	70
Wind velocity (V), mph	0	20
Bar size (B), ⅛ in.	4	11

First, the variables are coded such that the high level will be denoted by +1 and the low level by −1. By so doing, regardless of the physical conditions represented by the two levels, the basic design of any two-level factorial design becomes a simple arrangement of plus or minus ones. Writing down three columns of the plus and minus ones next to one another produces the desired 2^3 factorial design, which consists of the 8 distinct sets of coded test conditions, as indicated in Table 8. The 8 sets of coded test conditions are given by the 8 rows corresponding to test numbers 1 to 8. These constitute the recipes for running the tests. For example, the actual welding conditions for test number 1 (which is denoted by −1, −1, −1) were a temperature of 0 °F, a wind velocity of 0 mph, and a bar size of ½ in.

Geometric Representation of 2^3 Factorial Design. If the three variables are considered to be three mutually perpendicular coordinate axes X_1, X_2, and X_3, the 2^3 factorial design can be represented geometrically as a cube, as shown in Fig. 40. The circled numbers at the 8 corner points of the cube represent the corresponding test numbers in standard order. The 8 actual test conditions are given in brackets. This geometric configuration will be useful later when the average and interaction effects of the variables are calculated and when the average/main effect and two- and three-factor interaction effects are interpreted.

Test Conduct: Random Order and Replication. In performing experiments, randomization of the test order should be exercised wherever possible. This is important because the standard order discussed previously is a very systematic ordering of the tests in terms of the patterns of the levels of the variables from test to test. In this example, there are 8 test conditions, each of which was replicated once for a total of 16 tests. The main purpose of conducting replicated tests is to provide an estimate of the experimental error of the test method. The observed responses of these welding experiments were the ultimate tensile stresses of the welds. Because of replication, 2 responses (y_{ai}, y_{bi}) were observed at each of the 8 distinct sets of test conditions. Therefore, there were a total of 16 responses (Table 9).

The tests were performed in a random fashion, as indicated by the test order. The average of the 2 responses observed at each set of test conditions (that is, $y_{ai} + y_{bi}/2$, denoted by $\bar{y}_i$) was determined, and the 8 average responses ($\bar{y}_i$, where $i = 1, 2, ..., 8$) are given in the far right-hand column of Table 9. These average responses will be used to calculate the average main and average interaction effects for the variables under study.

Table 8 Coded test conditions based on actual test conditions for the steel rail bars of Example 8

	Actual test conditions			Coded test conditions		
Test number	Temperature, °F	Wind velocity, mph	Bar size, ⅛ in.	X_1	X_2	X_3
1	0	0	4	−1	−1	−1
2	70	0	4	+1	−1	−1
3	0	20	4	−1	+1	−1
4	70	20	4	+1	+1	−1
5	0	0	11	−1	−1	+1
6	70	0	11	+1	−1	+1
7	0	20	11	−1	+1	+1
8	70	20	11	+1	+1	+1

Table 9 Results of the 16 welding experiments for the steel rail bars of Example 8

Because of replication, two responses (y_{ai}, y_{bi}) were observed for each of the eight distinct sets of test conditions.

				Replicate 1		Replicate 2		
				Order of		Order of		
i	X_1	X_2	X_3	y_{ai}	y_{ai}	y_{bi}	y_{bi}	Average, $\bar{y}$
1	−1	−1	−1	6	84.0	3	91.0	87.5
2	+1	−1	−1	8	90.6	7	84.0	87.3
3	−1	+1	−1	1	69.6	5	86.0	77.8
4	+1	+1	−1	2	76.0	4	98.0	87.0
5	−1	−1	+1	5	77.7	8	80.5	79.1
6	+1	−1	+1	3	99.7	1	95.5	97.6
7	−1	+1	+1	4	82.7	2	74.5	78.6
8	+1	+1	+1	7	93.7	6	81.7	87.7

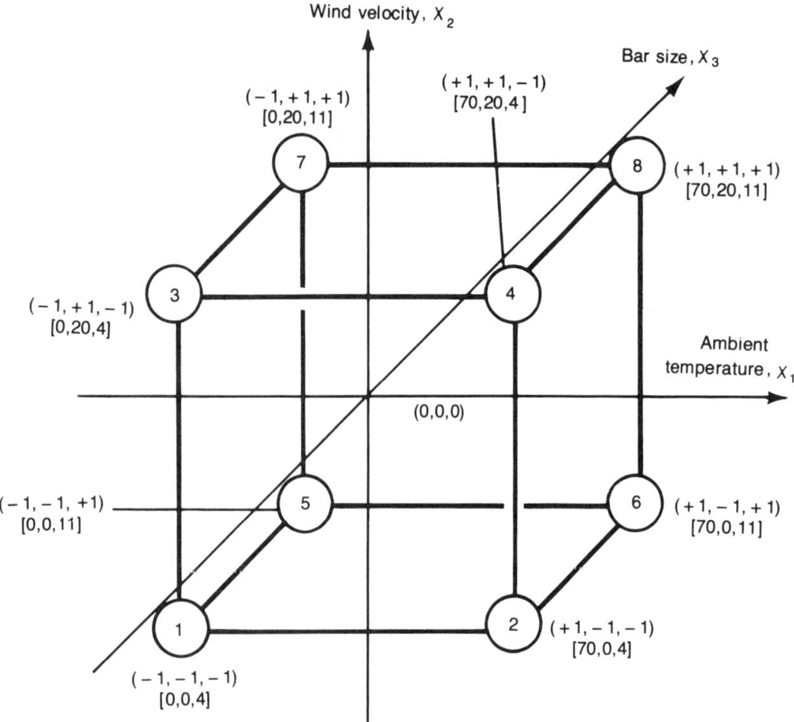

Fig. 40 Geometric representation of the 2^3 factorial design

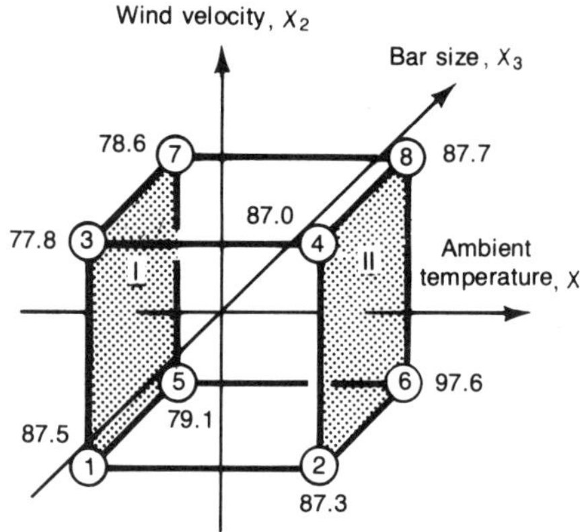

Fig. 41 Geometric representation of the effect on ultimate tensile stress, given at the corners of the cube, of changes in ambient temperature when wind velocity and bar size are held constant

Calculation of Variable Effects on Weld Strength. Given the results in Table 9, it is possible to evaluate the individual and joint influences of ambient temperature, wind velocity, and bar size on the ultimate tensile stress of the weld. To help determine what these variable effects are, what they mean, and how they can be calculated, the geometrical representation of the experimental design is used. The cube shown in Fig. 41 depicts the 2^3 design geometrically. The average response (ultimate tensile strength) for each test is given at the corners of the cube.

Calculation of Average Main Effects. As Fig. 41 shows, there are four comparisons of test results or contrasts that indicate how ultimate tensile stress changes when ambient temperature changes, with wind velocity and bar size being held constant. The effect of ambient temperature alone can be determined by looking across the cube from left to right. The differences in the results within each of the 4 pairs of tests reflect the effect of ambient temperature alone on ultimate tensile stress. The differences in units of ksi are:

- *Test No. 1 and 2*: $\bar{y}_2 - \bar{y}_1 = 87.3 - 87.5 = -0.2$
- *Test No. 3 and 4*: $\bar{y}_4 - \bar{y}_3 = 87.0 - 77.8 = 9.2$
- *Test No. 5 and 6*: $\bar{y}_6 - \bar{y}_5 = 97.6 - 79.1 = 18.5$
- *Test No. 7 and 8*: $\bar{y}_8 - \bar{y}_7 = 87.7 - 78.6 = 9.1$

The average effect of ambient temperature, designated by E_1, is defined as the average of the above four differences. Note that this average effect is also commonly referred to as a main effect. That is:

$$E_1 = \tfrac{1}{4}[(\bar{y}_2 - \bar{y}_1) + (\bar{y}_4 - \bar{y}_3) + (\bar{y}_6 - \bar{y}_5) + (\bar{y}_8 - \bar{y}_7)]$$
$$E_1 = \tfrac{1}{4}[(-0.2) + (9.2) + (18.5) + (9.1)]$$
$$E_1 = 9.15 \text{ ksi}$$

Geometrically, the average (main) effect of ambient temperature, E_1, is the difference between the average test result on plane II (high level of ambient temperature) and the average test result on plane I (low level of ambient temperature), as shown in Fig. 41.

The average effect of ambient temperature indicated that, on the average, over the ranges of the variables studied in this investigation, the effect of changing the ambient temperature from its low level to its high level is to increase the ultimate tensile stress by 9.150 ksi. However, the individual differences (0.2, 9.2, 18.5, and 9.1 ksi) are actually quite erratic. The average effect, therefore, must be interpreted with considerable caution because this effect is not particularly consistent over the four unique combinations of wind velocity and bar size.

Again, referring to Fig. 41, the following pairs of tests can be compared or contrasted to determine the effect of wind velocity. The average effect of wind velocity, E_2, can be obtained by taking the average of the four individual differences, which are in units of ksi:

- *Test No. 1 and 3*: $\bar{y}_3 - \bar{y}_1 = 77.8 - 87.5 = -9.7$
- *Test No. 2 and 4*: $\bar{y}_4 - \bar{y}_2 = 87.3 - 87.0 = -0.3$
- *Test No. 5 and 7*: $\bar{y}_7 - \bar{y}_5 = 78.6 - 79.1 = -0.5$
- *Test No. 6 and 8*: $\bar{y}_8 - \bar{y}_6 = 87.7 - 97.6 = -9.9$

The value E_2 for tests 1 to 8 is calculated as follows:

$$E_2 = \tfrac{1}{4}[(\bar{y}_3 - \bar{y}_1) + (\bar{y}_4 - \bar{y}_2) + (\bar{y}_7 - \bar{y}_5) + (\bar{y}_8 - \bar{y}_6)]$$
$$E_2 = \tfrac{1}{4}[(-9.7) + (-0.3) + (-0.5) + (-9.9)]$$
$$= -5.1 \text{ ksi}$$

Similarly, the average effect of bar size, E_3, is 0.85 ksi.

Interpretation of Average/Main Effects. The average or main effect of a variable has been defined as the amount of change observed in the response, on the average, when only that variable changes from its low to high value or level. It is an average effect because within the experimental design there are generally several comparisons or contrasts that can be made that measure how much the response changes when only a certain variable changes. The sign and the magnitude of an average/main effect have the following meaning:

- The sign indicates the direction of the effect
- The magnitude indicates the strength of the effect

If the numbers comprising this average are quite similar in magnitude (and of course, sign), the effect of that particular variable would appear to be independent of the particular level(s) of other variable(s). However, if the numbers comprising this

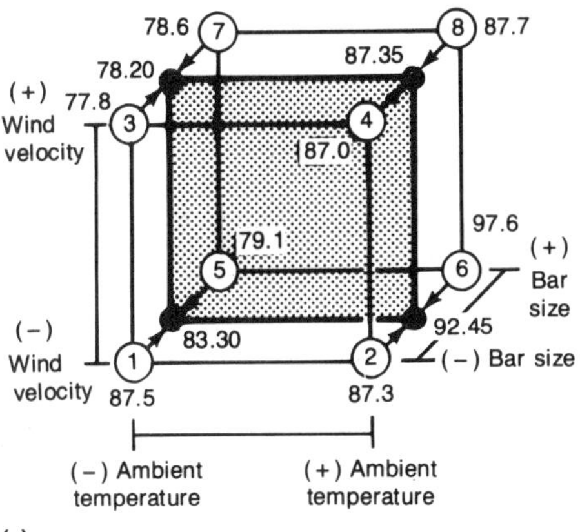

(a)

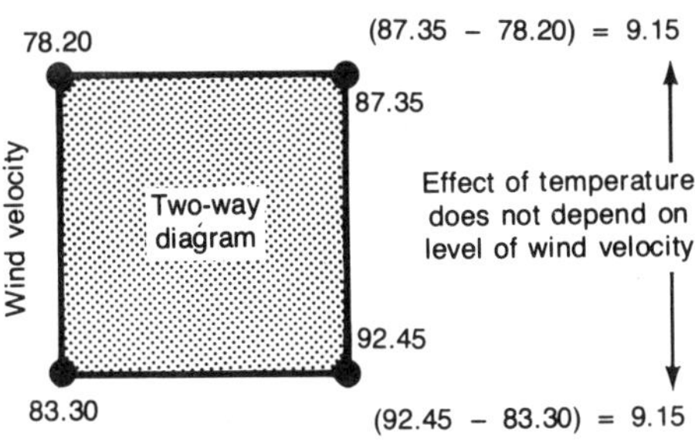

Ambient temperature

E_{12} = [(87.35 − 78.20) − (92.45 − 83.30)]/2
E_{12} = 0

(b)

Fig. 42 Two-factor interaction between temperature and wind velocity obtained by compressing three-dimensional cube (a) in the bar size direction to obtain two-dimensional square plane (b)

average are quite different in magnitude (or even sign), the effect of that particular variable depends on the level(s) that the other variable(s) assumes. In this case, the average/main effect does not hold much significance, and the response of interest must be examined when two or more variables change simultaneously.

Meaning of Variable Interactions. In calculating the average/main effect of ambient temperature, the amount (and even direction) of change in weld strength with change in temperature seemed to depend quite heavily on the particular levels of wind velocity and bar size:

Wind velocity, mph	Bar size, in.	Effect of temperature on weld strength, ksi
0	4/8	−0.2
20	4/8	+9.2
0	11/8	+18.5
20	11/8	+9.1

What is the particular nature of the dependency of the effect of temperature on wind velocity and bar size? The degree of dependency of the temperature effect on the particular levels of wind velocity and bar size can be appreciated via the following graphical summaries. Referring to the geometrical representation of the 2^3 factorial in Fig. 41, the interaction between temperature and wind velocity can be examined by compressing the cube in the bar size direction as shown in Fig. 42(a). Compressing the cube means that the response values for a given temperature and wind velocity combination—for example, (−), (−)—are averaged across the high and low levels of bar size. The result is that the cube (Fig. 42a) becomes a square (Fig. 42b). Similarly, the other two interactions are found to be:

$$E_{13} = [(92.65 - 78.85) - (87.15 - 82.65)]/2 = 4.65 \text{ ksi}$$

$$E_{23} = [(83.15 - 82.40) - (88.35 - 87.40)]/2 = -0.10 \text{ ksi}$$

A Simplified Method for Main and Interaction Effects. Although the geometrical representation of the experimental design has been quite useful, a simpler and more general method will be needed for calculation of the variable main and interaction effects. A simplified calculation procedure, which is easily extended for analyzing two-level factorial designs in any number of variables, is described below.

The mathematical model form associated with a 2^2 factorial is:

$$Y = b_0 + b_1X_1 + b_2X_2 + b_{12}X_1X_2 + e$$

where the coefficients b_1 and b_2 correspond to the average/main effects of variables 1 and 2, respectively, and the coefficient b_{12} corresponds to the interaction effect between variables 1 and 2.

Referring to the design matrix for Example 8 (Table 9), estimates of the interaction effects E_{12}, E_{13}, E_{23}, and E_{123} can be obtained by forming the cross-product columns X_1X_2, X_1X_3, X_2X_3, and $X_1X_2X_3$, as indicated in Table 10. This complete seven-column matrix will be referred to as the calculation matrix. The cross-product columns of ± signs are simply the inner-products of the individual columns, for example, $X_1X_2 = (X_1)(X_2)$.

To calculate any one of the average effects or interactions, merely multiply, element by element, the appropriate column by the column of average responses, sum algebraically, and divide the sum by 4 (that is, $2^3/2$; in general, for a 2^k factorial design, the sum should be divided by $2^k/2$). For example, to calculate the average effect of ambient temperature, E_1, it is necessary to proceed as indicated in Table 11. The sum is 36.6 × 10^3 psi or 36.6 ksi, which, when divided by 4, yields 9.150 ksi for the average effect of temperature, the same answer obtained previously.

The two-variable and three-variable interactions can be calculated by the same method (Table 12). For example, to calculate the interaction between ambient temperature and wind velocity, E_{12}, multiply the column X_1X_2 by the column of $\bar{y}_i$, sum algebraically, and then divide the sum by $2^3/2 = 4$. Dividing the sum by 4 yields the answer obtained previously (namely, 0) for this two-variable interaction.

Judging the Importance of Average Effects. To judge the relative importance of a variable effect based on the calculated average effects and interactions, it is necessary to make use of the test replications to

Table 10 Calculation matrix for the data obtained for the steel rail bars of Example 8

Test	Main effects X_1	X_2	X_3	Interactions X_1X_2	X_1X_3	X_2X_3	$X_1X_2X_3$	$\bar{y}$
1	−1	−1	−1	+1	+1	+1	−1	87.5
2	+1	−1	−1	−1	−1	+1	+1	87.3
3	−1	+1	−1	−1	+1	−1	+1	77.8
4	+1	+1	−1	+1	−1	−1	−1	87.0
5	−1	−1	+1	+1	−1	−1	+1	79.1
6	+1	−1	+1	−1	+1	−1	−1	97.6
7	−1	+1	+1	−1	−1	+1	−1	78.6
8	+1	+1	+1	+1	+1	+1	+1	87.7

Table 11 2^k factorial design calculation of the average effect of ambient temperature on steel rail bar ultimate tensile strength in Example 8

X_1	$\bar{y}_i$		$X_1\bar{y}_i$			$X_1\bar{y}_i$
−1	87.5		(−1)	(87.5)		−87.5
+1	87.3		(+1)	(87.3)		+87.3
−1	77.8		(−1)	(77.8)		−77.8
+1	87.0	=	(+1)	(87.0)	=	+87.0
−1	79.1		(−1)	(79.1)		−79.1
+1	97.6		(+1)	(97.6)		+97.6
−1	78.6		(−1)	(78.6)		−78.6
+1	87.7		(+1)	(87.7)		+87.7
				Sum	=	**+36.6**

Table 12 2^k factorial design calculation of interaction between ambient temperature and wind velocity on steel rail bar ultimate tensile strength in Example 8

X_1X_2	$\bar{y}_i$		X_1X_2	$\bar{y}_i$		$X_1X_2\bar{y}_i$
+1	87.5		(+1)	(87.5)		+87.5
−1	87.3		(−1)	(87.3)		−87.3
−1	77.8		(−1)	(77.8)		−77.8
+1	87.0	=	(+1)	(87.0)	=	+87.0
+1	79.1		(+1)	(79.1)		+79.1
−1	97.6		(−1)	(97.6)		−97.6
−1	78.6		(−1)	(78.6)		−78.6
+1	87.7		(+1)	(87.7)		+87.7
				Sum	=	**0.0**

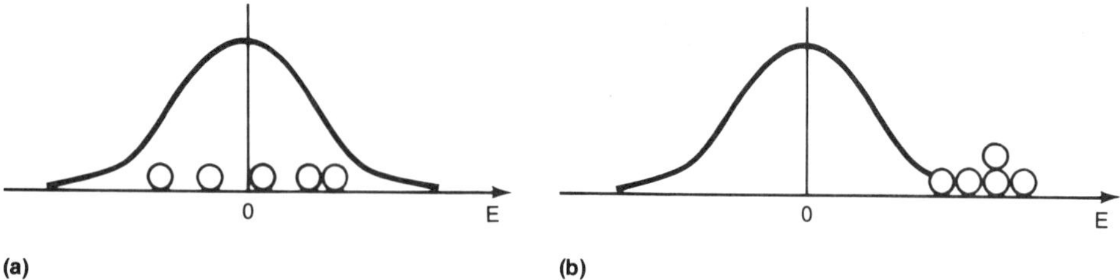

Fig. 43 Judging the relative importance of a variable effect. (a) Effect not important. (b) Effect important. Open circles represent effect estimates based on multiple experiment results.

estimate the error expected in an effect estimate. In particular, it is necessary to:

- Estimate the experimental error via the variance of an individual observation
- Estimate the error/variance associated with an average effect and/or interaction effect, either of which is simply a linear combination (sum) of several individual observations
- Construct the statistical distribution of effect estimates and examine the range of estimates that could arise

Figure 43 shows a graphical depiction of what could arise from the above-mentioned test replications. In case 1 (Fig. 43a), the effect estimates could easily arise from a distribution with a mean of zero. In case 2 (Fig. 43b), the implication is that they could not. The above procedures come under the heading of what is commonly referred to as a statistical test of significance. The two most common approaches in this regard are the hypothesis-testing method and the confidence interval method. Additional information is available in Ref 8 and 24.

Confidence Interval Method. Each of the 8 tests in Example 8 was replicated once (Table 9), so that there were actually 16 individual observations on the ultimate tensile stress. The error or variance of each of these 16 observations will now be estimated. The amount of error in an observation is assumed to be about the same for all observations; that is, the true variance is the same for all 16 observations, and the observations are independent.

For test 1, the two observations are 84.0 ksi and 91.0 ksi. A sample variance for this test condition, designated as s_1^2, can be calculated in the usual way:

$$s_1^2 = \frac{(y_{a1} - \bar{y}_1)^2 + (y_{b1} - \bar{y}_1)^2}{2 - 1} = 24.50$$

In this example, 8 different and completely independent sample variances ($s_1^2, s_2^2, ..., s_8^2$) are calculated, 1 for each unique test condition. The 8 sample variances are calculated to be:

$s_1^2 = 24.50$	$s_5^2 = 3.92$
$s_2^2 = 21.78$	$s_6^2 = 8.82$
$s_3^2 = 134.48$	$s_7^2 = 33.62$
$s_4^2 = 242.00$	$s_8^2 = 72.00$

Because there is assumed to be a common variance for all 16 observations, an estimate for the variance is the pooled sample variance s_p^2 of the 8 estimated variances $s_1^2, s_2^2, ..., s_8^2$. In this case:

$$s_p^2 = \frac{[(y_{a1}-\bar{y}_1)^2+(y_{b1}-\bar{y}_1)^2]+ . . . +[(y_{a8}-\bar{y}_8)^2 + (y_{b8}-\bar{y}_8)^2]}{(2 - 1) + . . . + (2 - 1)}$$

$$s_p^2 = \frac{24.50 + 21.78 + . . . + 72.00}{8} = 67.64 \quad \text{(Eq 16)}$$

In general, an effect is determined from:

$$E = \frac{2}{N}[y_1 + y_2 + . . . + y_N] \quad \text{(Eq 17)}$$

where N is the total number of unique test/trial results.

In this general case, the variance of an effect is given by:

$$\sigma^2_{\text{effect}} = \frac{4\sigma_x^2}{N} \quad \text{(Eq 18)}$$

An estimate of the variance of an effect (s^2_{effect}) is obtained by substituting the pooled sample variance, s_p^2, for σ_x^2.

A confidence interval for a certain parameter can be calculated on the basis of the sample statistic. Here the statistics of interest are the average effects, E_1, E_2, E_3, and the interactions, E_{12}, E_{13}, E_{23}, and E_{123}. Because the sample variances of the average effects and interactions are all estimated by $s_p^2/4$, the result, for $100(1 - a)\%$ confidence intervals using the Student's t-statistic, is:

$$E \pm t_{(v,\, 1 - a/2)} \sqrt{\frac{s_p^2}{4}} \quad \text{(Eq 19)}$$

where $v = 8$ is the degree of freedom and $a = 0.05$ is the level of significance. Therefore, the 95% confidence intervals for each of the sample average effects are therefore:

$$E \pm 2.306 \sqrt{\frac{67.64}{4}} = E \pm 9.48$$

Table 13 gives the seven statistics (E_1, E_2, E_3, E_{12}, E_{13}, E_{23}, and E_{123}) together with their corresponding 95% confidence intervals. From Table 13, it would appear that none of the estimated effects is important/significant.

Unreplicated Factorial Designs. There are times when it is either not feasible or not desirable to include replication in a two-level factorial design. In such cases, it is therefore not possible to obtain a direct estimate of the experimental error; as a result, it becomes more difficult to assess the relative importance of the variable effects. There are, however, at least two methods that can be employed to aid in the assessment of the relative importance of variable effects estimated from the results of an unreplicated two-level factorial design. These two methods are:

- The use of normal probability plots of the effect estimates
- The use of higher-order interaction effect estimates as a means to obtain an estimate of the experimental error

These methods are discussed in Ref 7 and 25.

Table 13 Calculation of 95% confidence levels for average effect and interaction statistics of steel rail bars of Example 8

Average effects	95% confidence interval, psi
Ambient temperature (E_1)	9150 ± 9480
Wind velocity (E_2)	−5100 ± 9480
Bar size (E_3)	850 ± 9480
Ambient temperature × wind velocity (E_{12})	0 ± 9480
Ambient temperature × bar size (E_{13})	−4650 ± 9480
Wind velocity × bar size (E_{23})	−100 ± 9480
Ambient temperature × wind velocity × bar size (E_{123})	4700 ± 9480

Two-Level Fractional Factorial Designs

Although the class of two-level factorial designs appears to be an efficient way to deal simultaneously with several factors, this efficiency quickly disappears as the number of variables to be studied grows. Because the two-level factorial requires the consideration of all possible combinations of k variables at two levels each, a ten-variable experiment would require 2^{10} = 1024 tests.

In dealing with phenomena involving continuous variables, the relationship that describes the influence of the variables on a response of interest often constitutes a relatively smooth response surface. When dealing with qualitative (discrete) variables, it is usually the case that the responses are similar at different levels of such variables. Therefore, the higher-order variable interactions are often very small or negligible in magnitude. In fact, interaction effects involving three factors or more can be ignored for the purpose of screening out the much more important main effects and two-factor interaction effects from a limited number of tests.

Experimentation with large numbers of variables generally arises out of uncertainty

Table 14 Calculation matrix obtained by expanding the design matrix

Test number	Average	1	2	3	4	12	13	14	23	24	34	123	124	134	234	1234
1	+	−	−	−	−	+	+	+	+	+	+	−	−	−	−	+
2	+	+	−	−	+	−	−	+	+	−	−	+	−	−	+	+
3	+	−	+	−	+	−	+	−	−	+	−	+	−	+	−	+
4	+	+	+	−	−	+	−	−	−	−	+	−	−	+	+	+
5	+	−	−	+	+	+	−	−	−	−	+	+	+	−	−	+
6	+	+	−	+	−	−	+	−	−	+	−	−	+	−	+	+
7	+	−	+	+	−	−	−	+	+	−	−	−	+	+	−	+
8	+	+	+	+	+	+	+	+	+	+	+	+	+	+	+	+

as to which variables have the dominant influence on the response of interest. In the end, however, only a subset of all these variables will be proved to be important. This phenomenon is referred to as the sparsity of variable effects (Ref 8). If this is the case, one should be able to conduct fewer tests than required for the full factorial without much loss of relevant information.

Consequences of Fractionation

In the rail steel bar problem, three variables (ambient temperature, wind velocity, and bar size) were studied to determine their possible effect on the ultimate tensile strength of the welded bars. A 2^3 full factorial design was run, and the three main effects, three two-factor interaction effects, and the one three-factor interaction effect could all be estimated separately. Suppose now that the investigator had wished to consider a fourth variable, type of welding flux, but that the full factorial, $2^4 = 16$ tests, could not be considered. Rather, only 8 tests can be run. Is it possible to conduct an experiment on 4 factors with only 8 tests, and if so, what useful information can be obtained from the results?

Based on the assumption about the negligible third and higher-order interaction effects, one could consider assigning the column of plus and minus signs associated with the **123** interaction to be the fourth variable and estimate the average main effect of variable **4** using this column. In using the **123** column to introduce a fourth variable into the experiment, the new design matrix that defines the 8 tests to be conducted becomes:

Test	Variables 1	2	3	4 (123)(a)
1	−	−	−	−
2	+	−	−	+
3	−	+	−	+
4	+	+	−	−
5	−	−	+	+
6	+	−	+	−
7	−	+	+	−
8	+	+	+	+

(a) **123** column replaced by variable **4**

Once the tests are conducted in accordance with the test recipes defined by the design matrix, the calculation matrix is determined to provide for the estimation of the interaction effects. By expanding the design matrix above, Table 14 provides the calculation matrix obtained by forming all possible products of columns 1 through 4.

Examination of the calculation matrix in Table 14 reveals that many of the columns are identical. In particular, of the 16 columns, only 8 are unique; each unique column appears twice. The following pairs of variable effects are represented in the calculation matrix by the same column of plus and minus signs:

1 and 234	12 and 34
2 and 134	13 and 24
3 and 124	23 and 14
4 and 123	Ave and 1234

What does all this mean? For example, when the **12** column is multiplied by the data, summed, and divided by 4, is the result an estimate of the two-factor interaction **12**? Or the two-factor interaction **34**? Or both? The interactions **12** and **34** are said to be confounded or confused. They are aliases of the unique column of plus and minus signs defined by (+ − − + + − − +). Use of this column for effect estimation produces a number (estimate) that is actually the combined total of the two-factor interaction effects **12** and **34**. Similarly, **1** and **234** are confounded effects, **2** and **134** are confounded effects, and so on. It seems that the innocent act of using the **123** column to introduce a fourth variable into a 2^3 full factorial scheme has created a lot of confounding among the variable effects, but if all the three-factor interactions are negligible, one obtains clear estimates of all 4 main effects from only 8 tests.

The 2^{k-p} Fractional Factorial Designs. The four-variable, eight-test, two-level experiment discussed thus far is referred to as a two-level fractional factorial design because it considers only a fraction of the tests defined by the full factorial. In this case, a one-half fraction design has been created. It is commonly referred to as a 2^{4-1} fractional factorial design. It is a member of the general class of 2^{k-p} fractional factorial designs. In these designs, the following factors must be considered:

- k variables are examined in 2^{k-p} tests
- Require that p of the variables be introduced into the full factorial in k-p variables
- Assign them to interaction effects in the first k-p variables. This assignment is done through the use of relationships known as generators (Ref 7, 8)
- These generators can then be used to establish the defining relationship (Ref 7, 8), which completely reveals the confounding/alias structure of the experimental design

Resolution of Two-Level Fractional Factorial Designs. As discussed previously, the introduction of additional variables into full two-level factorials gives rise to confounding or aliasing of variable effects. It would be desirable to make this introduction in such a way as to confound low-order effects (main effects and two-factor interactions)—not with each other but with higher-order interactions. Then, under the assumption that third and higher-order interactions can be neglected, the low-order effects become, in a sense, unconfounded by this assumption.

To illustrate, consider the study of 5 variables in just 16 tests (the full factorial would require $2^5 = 32$ tests). One additional variable—the fifth variable—must be introduced into a $2^4 = 16$ run base design. Any of the interactions in the first 4 variables could be used for this purpose: **12**, **13**, **14**, **23**, **24**, **34**, **123**, **124**, **134**, **234**, **1234**. If any one of the two-factor interactions are used, for example, **5** = **12**, then the design generator becomes **I** = **125**, which is also the defining relationship. Therefore, at least some of the main effects will be confounded with two-factor interactions, namely, **1** = **25**, **2** = **15**, **5** = **12**.

Selecting Preferred Generators. If any one of the three-factor interactions is used to introduce the fifth variable, the situation is greatly improved, at least for the estimation of main effects. For example, if **5** = **123**, then **I** = **1235** is the generator and defining relationship; therefore, some main effects are confounded with, at worst, three-factor interactions, while two-factor interactions are confounded with each other, for example:

1 = 235	12 = 35
2 = 135	13 = 25
3 = 125	23 = 15
5 = 123	

If the four-factor interaction in the first 4 variables is used to introduce the fifth variable, an even more desirable result is obtained (the best under these circumstances): **5** = **1234**. The generator and defining relationship is **I** = **12345**. Therefore:

1 = 2345	15 = 234
2 = 1345	23 = 145
3 = 1245	24 = 135
4 = 1235	25 = 134
5 = 1234	34 = 125
12 = 345	35 = 124
13 = 245	45 = 123
14 = 235	

In this last case, all main effects are confounded with four-factor interactions, and all two-factor interactions are confounded with three-factor interactions.

Concept of Resolution. The varying confounding structures produced by using different orders of variable interactions to introduce the fifth variable in the example above are described by the concept of the resolution of fractional factorial designs. The resolution of a two-level fractional factorial design is defined to be equal to the number of letters (numbers) in the shortest-length word (term) in the defining relationship, excluding **I**. If the defining relationship of a certain design is **I** = **124** = **135** = **2345**, then the design is of Resolution III. If the defining relation of a certain design is **I** = **1235** = **2346** = **1456**, then the design is of Resolution IV. The last design examined above, which had the defining relationship **I** = **12345**, is a Resolution V design. These concepts can be summarized as follows:

- If a design is of Resolution III, this means that at least some main effects are confounded with two-factor interactions
- If a design is of Resolution IV, this means that at least some main effects are confounded with three-factor interactions, while at least some two-factor interactions are confounded with other two-factor interactions
- If a design is of Resolution V, this means that at least some main effects are confounded with four-factor interactions, and some two-factor interactions are confounded with three-factor interactions

Example 9: Design Resolution/Selection of Generators. A 2^{6-2} fractional factorial design is set up by introducing variables **5** and **6** via **5** = **123**, **6** = **1234**. What is the resolution of this design?

The design generators are **I** = **1235** = **12346**, with the defining relationship **I** = **1235** = **12346** = **456**. Therefore, the design is of Resolution III.

What would the resolution be if the generators were **5** = **123** and **6** = **124**? Because the defining relationship is **I** = **1235** = **1246** = **3456**, the design is of Resolution IV. It is clear that the selection of the proper design generators is very important.

Additional Observations on Design Resolutions. From the above discussion on design resolution, several observations can be made:

- Higher-resolution designs seem more desirable because they provide the opportunity for low-order effect estimates to be determined in an unconfounded state, assuming higher-order interaction effects can be neglected
- The more variables considered in a fixed number of tests, the lower the resolution of the design becomes
- There is a limit to the number of variables that can be considered in a fixed number of tests while maintaining a prespecified resolution requirement
- No more than (n − 1) variables can be examined in n tests (n is a power of 2, for example, 4, 8, 16, 32, ...) to maintain a design resolution of at least III. Such designs are commonly referred to as saturated designs. Examples are 2^{3-1}, 2^{7-4}, 2^{15-11}, 2^{31-26}. For saturated designs, all interactions in the base design variables are used to introduce additional variables

Table 15 2_{III}^{7-4} family of fractional factorials

Fraction	Generators	When combined with principal fraction gives:
Principal	I = 124 I = 135 I = 236 I = 1237	. . .
A1	I = −124 I = −135 I = −236 I = −1237	All main effects
A2	I = −124 I = −135 I = 236 I = −1237	1, 12, 13, 14, 15, 16, 17
A3	I = −124 I = 135 I = −236 I = −1237	2, 12, 23, 24, 25, 26, 27
A4	I = 124 I = −135 I = −236 I = −1237	3, 13, 23, 34, 35, 36, 37
A5	I = −124 I = 135 I = 236 I = 1237	4, 14, 24, 34, 45, 46, 47
A6	I = 124 I = −135 I = 236 I = 1237	5, 15, 25, 35, 45, 56, 57
A7	I = 124 I = 135 I = −236 I = 1237	6, 16, 26, 36, 46, 56, 67
A8	I = 124 I = 135 I = 236 I = −1237	7, 17, 27, 37, 47, 57, 67

Importance of Sequential Experimentation

In the early stages of an investigation, it often seems that many variables are of potential importance. A project group or task force can draw up a list of 5 to 15 or more variables. In the final analysis, perhaps only 2 or 3 of these variables will prove to be important. The problem is, Which ones? The first task at hand is to conduct some experiments that will quickly reduce the number of variables under study to the few seemingly important ones that will then be the focus of further experimentation. For this screening task, two-level fractional factorial designs constitute a powerful and efficient tool. If the investigator attempts to get his arms around the entire problem by designing one comprehensive experiment, the resource requirements will probably be extensive, and the final results of the experiment may be inconclusive because of poor selection of variable levels and/or a poorly controlled experimental environment precipitated by a large experiment.

It would be wise to take a sequential approach, building up knowledge more gradually through a series of related experiments. In this regard, two-level fractional factorial designs serve as useful building blocks in a course of sequential experimentation. This notion is important to ensure that a lack of knowledge early in an investigation will not lead to the waste of experiment resources—for example, inappropriate variables or variable levels chosen in the context of one large experiment. It is possible to identify families of experiments that combine or piece together well. A key point here is that these related experiments can provide several different alternatives for a second experiment, depending on the results and inferences drawn from the first experiment. The sequential assembly aspect of related experiments is discussed below.

Example 10: Notion of Families of Fractional Factorials. All 16 one-sixteenth fractions of the 2^7 factorial are related through the generators **I** = **±124**, **I** = **±135**, **I** = **±236**, **I** = **±1237**. The terms remain the same, only the signs are changed for different fractions. The original design (all + signs on the generators) is called the principal fraction. The remaining 15 are the alternate fractions.

The key point is that, depending on the interpretation of the results of the principal fraction, any one of several other alternate fractions can be chosen to achieve a particular result when the two fractions are combined. A general strategy used in running successive experimental designs is to choose judiciously among the members of a given family of fractional factorials.

Clearly, given the principal fraction, several alternatives are available among the remaining alternate fractions, each providing a different set of information/effect estimates from the combined design. Table 15 provides some relevant information concerning the family of fractional factorials associated with a seven-variable experiment using eight-test designs.

Orthogonal Arrays

The theory of fractional factorial designs was first worked out by Finney (Ref 26) and Rao (Ref 27). Many highly fractionated designs were introduced by Tippett (Ref 28), Plackett and Burman (Ref 29), and others. Some of these were referred to as magic squares and orthogonal arrays. Two-level and three-level fractional factorial designs gained widespread attention and industrial application, beginning in the 1950s. Box and Hunter (Ref 7) provided much useful guidance to the practitioner in the adroit use of these experimental design structures.

In the late 1970s, the use of orthogonal arrays for quality design and improvement by Taguchi (Ref 4) and others gained widespread acceptance in industry. This has led to considerable discussion concerning the

Table 16 An L_{16} (2^{15}) orthogonal array

Test	*F* 1	*A* 2	e(a) 3	*B* 4	e(a) 5	*A* × *B* 6	*E* 7	C 8	*H* 9	e(a) 10	*B* × *D* 11	e(a) 12	*A* × *D* 13	*G* 14	*D* 15
1	1	1	1	1	1	1	1	1	1	1	1	1	1		
2	1	1	1	1	1	1	1	2	2	2	2	2	2	2	2
3	1	1	1	2	2	2	2	1	1	1	1	2	2	2	2
4	1	1	1	2	2	2	2	2	2	2	2	1	1	1	1
5	1	2	2	1	1	2	2	1	1	2	2	1	1	2	2
6	1	2	2	1	1	2	2	2	2	1	1	2	2	1	1
7	1	2	2	2	2	1	1	1	1	2	2	2	2	1	1
8	1	2	2	2	2	1	1	2	2	1	1	1	1	2	2
9	2	1	2	1	2	1	2	1	2	1	2	1	2	1	2
10	2	1	2	1	2	1	2	2	1	2	1	2	1	2	1
11	2	1	2	2	1	2	1	1	2	1	2	2	1	2	1
12	2	1	2	2	1	2	1	2	1	2	1	1	2	1	2
13	2	2	1	1	2	2	1	1	2	2	1	1	2	2	1
14	2	2	1	1	2	2	1	2	1	1	2	2	1	1	2
15	2	2	1	2	1	1	2	1	2	2	1	2	1	1	2
16	2	2	1	2	1	1	2	2	1	1	2	1	2	2	1

(a) e = estimate of error

relative merits of these designs and the methods of design selection *vis-à-vis* the methodology of the more general class of fractional factorial designs. In particular, it is important to examine the philosophical framework and interpretation of these similar (often identical) design structures.

Issue of Confounding. To many, the most significant property of the designs is orthogonality, that is, the ability to separate out the individual effects of several variables on a response of interest. The term orthogonal arrays as used by Taguchi implies having this property of orthogonality or producing effects that are not confounded. In fact, whenever anything less than a full factorial is under consideration, confounding or mixing of effects is present by definition. Such confounding can only be removed by assumptions made about the physical system and confirmed through some analyses of the results.

Similarly, when two-level factorials are fractionated, the ability to determine all the interactions among a set of factors is lost. However, judicious fractionation leads to the ability to obtain knowledge on low-order interactions under the assumption that higher-order interactions are of negligible importance. Because the orthogonal arrays, as applied to Taguchi methods, are highly fractionated factorial designs, the information these arrays produce is a function of two elements: the nature of their confounding patterns and the assumptions made about the physical system they are applied to. Often, Taguchi's use of linear graphs for design selection fails to produce the highest possible degree of resolution, as the following example will illustrate.

Example 11: Examination of the Use of an $L_{16}(2^{15})$ Orthogonal Array (Ref 4). The notation for various orthogonal arrays by Taguchi and Wu is generally of the following form for designs without mixing levels $L_n(r^m)$, where n is number of tests, r is number of levels of each factor, and m is the maximum number of factors, including selected interactions, that the design can study. For example, L_{16} (2^{15}) refers to a design with 16 tests to study up to 15 two-level factors and/or interactions. An $L_{16}(2^{15})$ orthogonal array is given in Table 16.

In this application of the $L_8(2^{15})$ orthogonal array, the following assignments have been made:

- Eight columns have been assigned to variable main effects *A*, *B*, *C*, *D*, *E*, *F*, *G*, *H*
- Three columns have been assigned to two-factor interactions *A* × *B*, *B* × *D*, *A* × *D*
- Four columns have been assigned to the estimation of error

Close examination of the $L_{16}(2^{15})$ orthogonal array shows that the columns assigned to *E*, *G*, *H*, and *C* are the following products of the columns assigned to *D*, *B*, *A*, and *F*:

5 = 234	6 = −14
($E = B \times A \times F$)	$H = -(C \times F)$
7 = 123	8 = −1234
($G = C \times B \times A$)	$D = -(C \times B \times A \times F)$

It follows then from the associated defining relationship that this $L_{16}(2^{15})$ orthogonal array is not only a 2^{8-4} fractional factorial but also a Resolution III 2^{8-4} design. This is surprising because it is possible to design a 16-run, two-level fractional factorial to examine eight variables (Resolution IV) using generators such as:

5 = 234	7 = 123
(I = 2345)	(I = 1237)
6 = 134	8 = 124
(I = 1346)	(I = 1248)

The $L_{16}(2^{15})$ orthogonal array under examination, when viewed as a 2^{8-4} fractional factorial design, produces the following 16 linear combinations of effects, which can be estimated from the data. It is assumed that three-factor interactions and higher can be neglected:

l_0 estimates	*I* (mean)
l_1 estimates	1 − 46 − 58
l_2 estimates	2
l_3 estimates	3
l_4 estimates	4 − 16 − 78
l_5 estimates	12 + 37
l_6 estimates	13 + 27
l_7 estimates	6 − 14 − 57
l_8 estimates	23 + 45 + 17 + 68
l_9 estimates	24 + 35
l_{10} estimates	34 + 25
l_{11} estimates	7 − 48 − 56
l_{12} estimates	26 + 38
l_{13} estimates	36 + 28
l_{14} estimates	5 − 18 − 67
l_{15} estimates	8 − 15 − 47

It is interesting to note the following for this particular 2^{8-4} fractional factorial:

- Only two of the eight main effects are not confounded with two-factor interactions. The other six main effects are each confounded with two two-factor interactions
- One of the desired two-factor interactions (23) is confounded with three other two-factor interactions
- The four columns assigned to the estimation of error each represent pairs of two-factor interactions confounded with each other

An Alternate $L_{16}(2^{15})$ Orthogonal Approach. It would appear that the $L_{16}(2^{15})$ orthogonal array using the generators given below is more suitable for the purposes of Taguchi methods (or any methods for that matter), particularly given the assumptions generally made about interaction effects:

5 = 234	7 = 123
6 = 134	8 = 124

From this 2^{8-4} fractional factorial design, the linear combination of effects (ignoring interactions of three or more factors) given in Table 17 would result.

The following conclusions can be drawn from Table 17:

Table 17 Linear combination of effects obtained from 2^{8-4} fractional factorial design

		Effect name in Taguchi experiment
l_0 estimates	*I* (mean)(a)	
l_1	$\underline{1}$	*D*
l_2	$\underline{2}$	*B*
l_3	$\underline{3}$	*A*
l_4	$\underline{4}$	*F*
l_5	$\underline{12}$ + 37 + 48 + 56	*B* × *D*
l_6	$\underline{13}$ + 46 + 37 + 58	*A* × *D*
l_7	14 + 36 + 28 + 57	*e*(b)
l_8	$\underline{23}$ + 45 + 17 + 68	A × B
l_9	24 + 35 + 18 + 67	*e*(b)
l_{10}	34 + 25 + 16 + 78	*e*(b)
l_{11}	$\underline{7}$	*H*
l_{12}	$\underline{8}$	*C*
l_{13}	$\underline{6}$	*G*
l_{14}	$\underline{5}$	*E*
l_{15}	15 + 25 + 47 + 38	*e*(b)

(a) Underscored effects identify corresponding effects. For example, $\underline{1}$ is *D*, $\underline{2}$ is *B*, and so on, in Taguchi's experiment. (b) e = estimate of error

- All eight main effects are confounded with three-factor and higher—only two previously
- All three desired two-factor interactions are confounded with three other two-factor interactions—only one previously
- All four error estimates are a set of four confounded two-factor interactions; previously, they were sets of two confounded two-factor interactions. This is a trade-off for the estimation of two of the desired two-factor interactions: $B \times D$ and $A \times D$

Choice of Variables in Calculation Matrix. When fewer than $k = n - 1$ variables are studied in $2^{k-p} = n$ tests, some choice exists in terms of how variables (main effects) are assigned to columns in the $L_n(2^k)$ array (the calculation matrix). Some choices are often better than others in terms of the resulting design resolution. Under the Taguchi approach to design of experiments, the philosophy toward interaction effects is quite different. As a result, the bottom line is that unless they are expressly identified and assigned to the orthogonal array, they are assumed to be neglectable. If such an attitude is adhered to, it is perhaps of even greater importance to place emphasis on design resolution and the careful examination of the alias structure of the experimental design. When higher-resolution designs (arrays) are available, it is of utmost importance to seek them out in an effort to protect the assumptions used. Again, Example 11 demonstrates the importance of design resolution and the judicious choice of design generators. In the context of Taguchi's approach to design of experiments, this is even more important because the sequential assembly concept is not used; that is, single experiments tend to be employed on any given problem.

Implementing Robust Design

Recently, a number of different approaches have been proposed as possible ways to implement the robust design concept of Taguchi. These vary from purely analytical approaches, to computer simulation using product/process mathematical models and/or Monte Carlo methods, to the use of physical experimentation. In most of these approaches, the use of experimental design strategies, including two-level and multilevel factorial and fractional factorial designs and orthogonal arrays, has been extensive.

Taguchi views the design process as evolving in three distinct phases or steps:

- System design
- Parameter design
- Tolerance design

It is perhaps this broad umbrella that he places over his concepts and methods for quality design and improvement that makes his approach so widely accepted by the engineering community. As discussed previously, Taguchi considers engineering design as the central issue and statistical methods as just one of several tools to accomplish his objectives in engineering.

In addition to the different approaches to generating data on product/process performance, the issue of the specific measures of performance to use in the facilitation of parameter design needs to be considered. Taguchi and his colleagues make extensive use of the signal-to-noise ratio as a measure of performance. Others evaluate the mean and the variance or standard deviation of performance separately. The relative merits of these varying approaches with regard to performance evaluation will be discussed later in this article. The separate analysis of mean response and variation in response is discussed below.

Degree of Control

The issue of the control of a certain factor is actually one of economics; therefore, any factor might be thought of in terms of its degree of control rather than as controllable or not controllable. Suppose that an injection-molding process is under consideration because of a part shrinkage problem. A brainstorming session might lead to the following list of factors as potentially having an influence on part shrinkage:

- Cycle time
- Percentage of regrind versus virgin raw material
- Gate configuration
- Holding pressure
- Screw speed
- Mold temperature
- Cavity surface finish
- Raw material moisture content
- Raw material melt flow index
- Mold cooling water temperature

Although any or all of these factors can influence part shrinkage, they are not all controlled or manipulated to the same degree or with the same ease, economically. Some of these factors can be controlled at the engineering design process—for example, the gate configuration or the cavity surface finish of the mold. Others are controllable at the molding machine—for example, screw speed or cycle time. Still others are properties of the incoming raw material and are thus under the control of the supplier—for example, melt flow index. Finally, certain factors can be best categorized as defining part of the environment in which the process must function—for example, mold cooling water temperature and percentage of regrind versus virgin raw material. Two case studies will be discussed that exemplify some of the different ways in which the concept of parameter design can be implemented to develop robust product and process designs.

Analytical Approaches to Parameter Design

It is important to keep a broad perspective on the contributions of Taguchi and the wide variety of ways in which they can be utilized. Perhaps the single overarching concept that drives the philosophies of Taguchi is the importance placed on the consistency of performance of products and processes, that is, the importance of variation reduction. In essence, as designs are formulated and output performances evaluated, either analytically or experimentally, as much or more emphasis should be placed on variation in performance as on performance on average. There are many ways to invoke this overarching concept.

Although much of the knowledge used in a given design situation is derived from experimentation, an ever-expanding baseline of knowledge exists that is being derived from first principles/physical laws and the utilization of sophisticated modeling methods such as finite-element modeling. As closed form solutions become available, it becomes possible to evaluate them analytically or numerically using classical optimization methods. The performance function should be examined in terms of both its expected value and its variance function. Searching the variance function to find those values for the design/control variables that minimize the variance in performance then amounts to invoking the parameter design method.

Example 12: Parameter Design Methods Applied to Analysis of Effort Required to Close an Automobile Door (Ref 30). An explicit model is used for the mean square error of design performance in an automobile door-closing-effort problem to identify the values of the design parameters that provide the most consistent design configuration (minimum variance) while maintaining an acceptable level of average performance. In this case study, the following logic/methodology is employed.

Problem Formulation. The first step in any design optimization problem is to formulate the design performance measure of interest. In this case study, automobile door fit is the issue, particularly the sensitivity of the closing effort to variations in door positioning as it is hung in place. Figure 44 illustrates a car door and the locations of the three positioning points used to hang the door. The closing effort is a function of the compression of the weatherstripping around the door periphery. The closing effort can be expressed analytically as a function of the weatherstripping diametral stiffness, the weatherstripping diameter and length, and the nominal values of the door position points that define the door/frame gap. The basic design problem is then to find the nominal values for the door position points that minimize the variation in closing effort as a result of the tolerance range assigned to

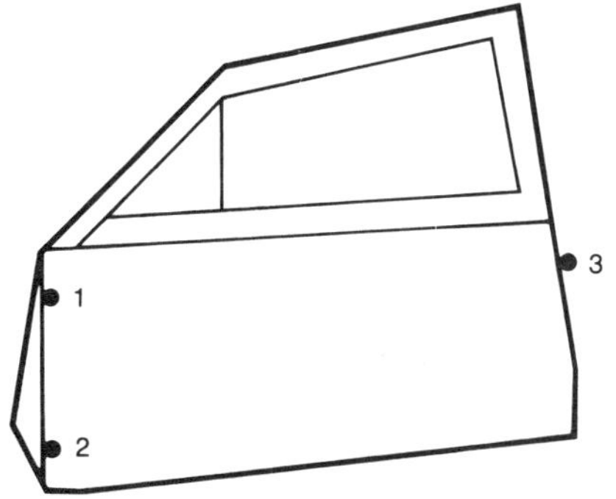

Fig. 44 Schematic of an automobile door showing locations (solid circles) of three positioning points used to hang the door

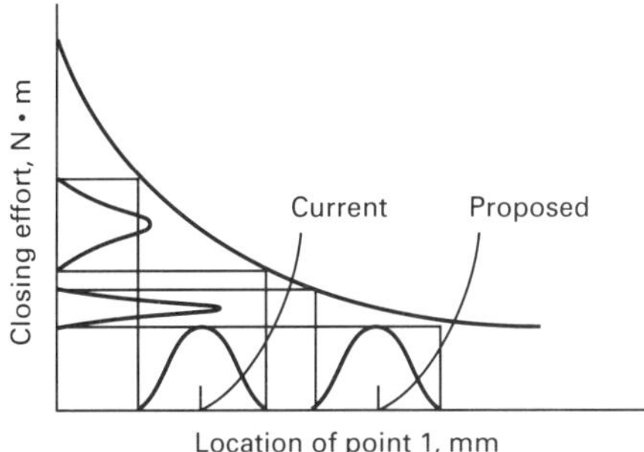

Fig. 45 Plot of closing effort versus point 1 location for positioning the door shown in Fig. 44 with fixed values of location points 2 and 3

the door position points. This problem was formulated in an attempt to exploit the underlying nonlinearities in the system to improve performance consistency.

Model Derivation. Under the assumption that the door gap is linearly proportional to the door position point locations, it is possible to derive the door-closing effort (in stored energy) as a quadratic function of the position point locations. The position location points are the control factors in this problem. The quadratic function for closing effort as a function of the three hanging positions has several unknown parameter values. Although the closing effort will clearly vary as a function of variation from car to car in the weatherstripping thickness, car body, size, and so on, only the variation in the door hanging position from car to car was examined in this case study.

Model Fitting. The parameters of the quadratic function for closing effort were estimated by first calculating the door gap at 15 points around the door as a function of variations in the door hanging positions and then using the calculated gap values to determine the closing-effort values. The variations in the hanging position control factors were introduced using a three-level factorial design, which required a total of 27 evaluations of the gap and the stored energy at each of the 15 locations around the door. These data were then used to estimate the coefficients of the quadratic closing-effort function.

Design Optimization. The variance of the closing-effort quadratic function can now be derived analytically, and this variance function can be subjected to a classical optimization routine. Such an optimization algorithm will seek the values of the three door position locations that minimize the variance of the closing effort. This optimization is subject to the constraint that the mean closing effort is less than or equal to a prespecified target value. The result is the determination of the nominal values for the design/control factors, that is, the door positioning locations that minimize the car-to-car closing-effort variation because of variation in these locations. In this design optimization problem, the control factors are the nominal values for the door position locations, and the noise factors are the tolerance/variations in the door position locations.

Figure 45 shows a plot of the derived functional relationship between closing effort and the location of point 1 for positioning the door for fixed values of location points 2 and 3. Figure 45 clearly shows how the parameter design method exploits a basic system nonlinearity by seeking a new nominal value for the location of point 1 that reduces the transmitted variation to the closing effort due to the variation in the location of point 1 about its nominal value.

Table 18 Design matrix for fuel gage experiment of Example 13

Test number	x_1	x_2	x_3	x_4	Y, %
1	Loose	Visual	Vendor A	90	14
2	Tight	Visual	Vendor A	90	10
3	Loose	Other	Vendor A	90	7
4	Tight	Other	Vendor A	90	8
5	Loose	Visual	Vendor B	90	18
6	Tight	Visual	Vendor B	90	9
7	Loose	Other	Vendor B	90	7
8	Tight	Other	Vendor B	90	10
9	Loose	Visual	Vendor A	93	15
10	Tight	Visual	Vendor A	93	9
11	Loose	Other	Vendor A	93	14
12	Tight	Other	Vendor A	93	7
13	Loose	Visual	Vendor B	93	7
14	Tight	Visual	Vendor B	93	12
15	Tight	Other	Vendor B	93	9
16	Tight	Other	Vendor B	93	12

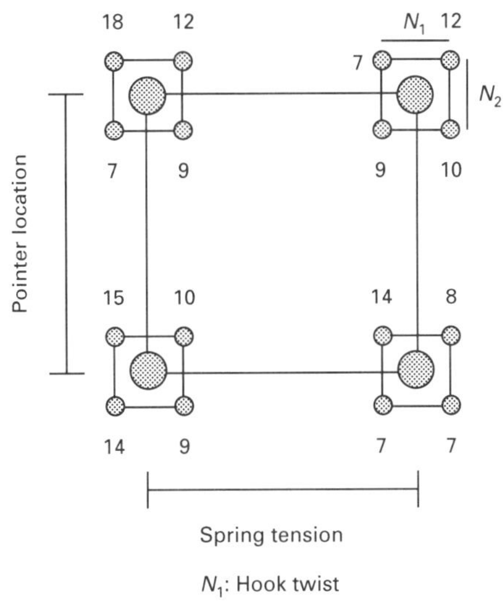

Fig. 46 Inner/outer experimental design structure for the fuel-gage study data in Table 18. Numbers at corners of outer design squares are fallback (% deflection) values, *Y*.

Experimental Approaches to Parameter Design

The parameter design concept can be implemented through physical experimentation in many different ways. The following example illustrates the explicit use of an inner/outer design for robust design. The example was developed from the results of a larger experiment and is presented to demonstrate how simple experiments such as two-level factorial designs can be used to study and improve the performance of an assembly process using the concept of robust design.

Example 13: Optimization of Automotive Fuel Gage Reading to Actual Fuel Tank Quantity. Customer complaints regarding the difference between the fuel gage reading and the actual amount of fuel in the tank led to an investigation of the process of gage calibration. Seven variables were tentatively identified as being of possible importance to the position (% deflection or fallback) of the indicator in the fuel gage.

Initial Design of Experiments. A 2^{7-3} fractional factorial design experiment was constructed and carried out. A portion of the results will be examined that involve four of the seven variables determined to be the most important. These four variables are:

Variable No.	Variable	Low level	High level
1	Spring tension	Loose	Tight
2	Method of pointer location	Visual	Other (standard)
3	Bimetal hook twist	Vendor A	Vendor B
4	Bimetal bracket bend	90°	93°

Variables 1 and 2 describe adjustments that can be made during the calibration process on the assembly line. Variables 3 and 4 describe the condition (considered undesirable) of certain purchased parts that are assembled into the gage and may affect the calibration process. The 2^{7-3} fractional factorial design was collapsed into a 2^4 full factorial design to study the effects of these variables, as they are purposely changed, on the response in question.

Table 18, the design matrix, lists the settings for the 16 tests conducted in terms of the four variables defined above. The tests are listed in standard order, although they were actually run in a randomized test sequence.

The main effects of each of the four factors, as well as the two-, three-, and four-factor interactions, were estimated from the data and are listed below:

E_1 = −2.50	$*E_{24}$ = 0.50
$*E_2$ = 0.00	E_{34} = −1.25
E_3 = −1.75	
E_4 = 3.00	$*E_{123}$ = 0.75
	E_{124} = −2.50
$*E_{12}$ = 0.50	$*E_{134}$ = 0.75
E_{13} = 1.75	$*E_{234}$ = 0.25
E_{14} = −1.00	
E_{23} = 2.25	E_{1234} = −2.25

The results of the larger experiment, in fact, showed that these four variables tended to exhibit the stronger effect estimates, both main effects and interactions. A normal probability plot of the estimated effects could be interpreted as suggesting that only the effect estimates indicated by the asterisk might be considered insignificant. This interpretation leaves the investigator somewhat perplexed as to what exactly to do next. Many interactions seem important, yet it is not at all clear how the process can best be adjusted/altered.

It would appear that the main effects of variables 1 and 4 might be two of the more important effects:

- $E_1 = -2.50$, implies that, on the average, tighter spring tension reduces fallback (deflection)
- $E_4 = +3.00$, implies that larger bimetal bracket bends (93°) increase deflection, which is undesirable

The latter inference suggests that closer adherence to the 90° bend, which is the desired nominal value, may have to be requested from the bimetal bracket supplier. This essentially means that the supplier would be required to tighten the tolerance on this angle. Rather than taking this more costly approach, it might be useful to consider the variation in this part dimension as a noise variable and then determine how the transmission of the variation in the bimetal bracket bend could be reduced through the manipulation of control factors during the calibration process.

Alternative Experimental Design and Analysis. A more thoughtful examination of the four variables under study seems to indicate that they fall into two basic categories. The first category consists of:

- *Variable 1*: Spring tension
- *Variable 2*: Method of pointer location

The second category consists of:

- *Variable 3*: Bimetal hook twist
- *Variable 4*: Bimetal bracket bend

Variables 1 and 2 can be easily adjusted at the discretion of the operator/setup person. Variables 3 and 4 describe certain aspects of the condition of two purchased parts. Although the vendors have been given certain requirements, strong control is not exercised over these factors (which determine the condition of the parts coming in) without tightening the specification requirements—a costly proposition. Therefore, these two variables might be considered to be noise variables, and perhaps the robust design/parameter design concept can be used to see how adjustments in variables 1 and 2 could be made to reduce the transmitted variability due to variables 3 and 4.

As a result, instead of thinking in terms of a 16-run experiment design that considers four variables at two levels each, one should think in terms of two 2^2 factorial experiments:

- The inner design is a 2^2 factorial in the two controllable variables (spring tension and method of pointer location)
- The outer design is a 2^2 factorial in the two noise variables (bimetal hook twist and bimetal bracket bend)

The outer design is conducted at each of the four unique variable settings of the inner design. A graphical representation of this experiment design is shown in Fig. 46. The numbers in the corners of each of the four outer design squares represent the fallback results for each of the 16 unique trails. Referring to the data in Fig. 46, $\bar{Y}$ and σ_Y, the mean and standard deviation of the % deflection, are determined at each of the four test combinations of the inner design:

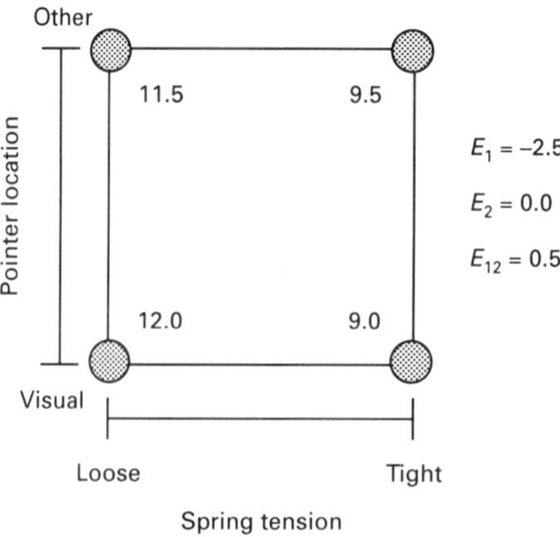

Fig. 47 Effect of control factors on average performance. Numbers at corners of square are the means of the % deflection, $\bar{Y}$.

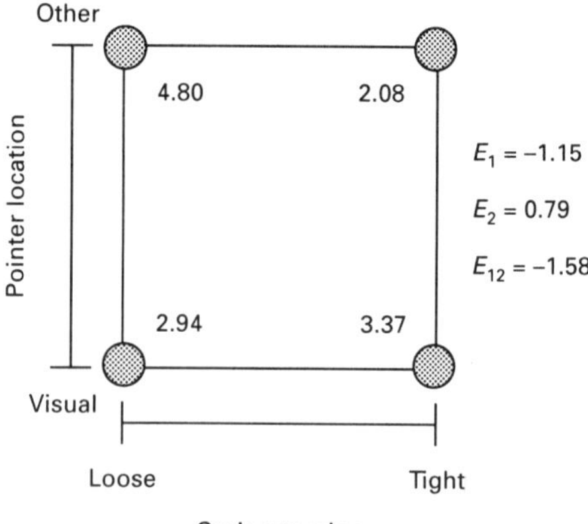

Fig. 48 Effect of control factors on variation in performance. Numbers at corners of square are the standard deviations of the % deflection, σ_Y.

Test no.	X_1	X_2	$\bar{Y}$	σ_Y
1	−	−	12.0	2.94
2	+	−	9.0	3.37
3	−	+	11.5	4.8
4	+	+	9.5	2.08

By analyzing the above responses (average fallback and standard deviation of fallback) separately, the settings/values can be determined for the control variables that minimize average fallback and give rise to minimum variation in fallback, that is, the most robust process design.

Results. Figures 47 and 48 illustrate the results of the analyses. The interpretation of the results is as follows:

- In terms of average response, % deflection is reduced by increasing spring tension (Fig. 47). This reduction is about the same for both methods of pointer location. On the average, the pointer location method has little effect on % deflection
- In terms of standard deviation, it is best to use the other (standard) method of pointer location along with tighter spring tension (Fig. 48)

It appears that the nominal response can be controlled by the manipulation of spring tension, while the variation in response is controlled by manipulation of the method of pointer location.

REFERENCES

1. W.W. Scherkenbach, *The Deming Route to Quality and Productivity: Road Maps and Roadblocks*, Mercury Press/Fairchild Publications, 1987
2. J.M. Juran, *Juran on Planning for Quality*, The Macmillan Free Press, 1986
3. A.V. Feigenbaum, *Total Quality Control*, McGraw-Hill, 1983
4. G. Taguchi and Y. Wu, *Introduction to Off-Line Quality Control*, Central Japan

Quality Control Association, 1979

5. G. Taguchi, *On-Line Quality Control During Production*, Japanese Standards Association, 1981
6 R.A. Fisher, *Design of Experiments*, 8th ed., Hafner Publishing, 1966
7. G.E.P. Box and J.S. Hunter, The 2k-p Fractional Factorial Designs, Part I and Part II, *Technometrics*, Vol 3, 1961
8. G.E.P. Box, W.G. Hunter, and J.S. Hunter, *Statistics for Experimenters*, John Wiley & Sons, 1978
9. L.P. Sullivan, Reducing Variability: A New Approach to Quality, *Qual. Prog.*, Vol 17 (No. 7), 1985, p 15–21
10. W.A. Shewhart, *The Economic Control of Quality of Manufactured Product*, Van Nostrand, 1931 (ASQC, 1980)
11. W.E. Deming, *Quality, Productivity, and Competitive Position*, MIT Center for Advanced Engineering, 1982
12. J.M. Juran, Quality Problems, Remedies, and Nostrums, *Ind. Qual. Con.*, Vol 22, 1966, p 647–653
13. S.W. Roberts, Control Charts Based on Geometric Moving Averages, *Technometrics*, Vol 1, 1959, p 234-250
14. A.L. Sweet, Control Charts Using Coupled Exponentially Weighted Moving Averages, *Trans. IIE*, Vol 18 (No. 1), 1986, p 26-33
15. A.W. Wortham and G.F. Heinrich, Control Charts Using Exponential Smoothing Techniques, *Trans. ASQC*, Vol 26, 1972, p 451-458
16. A.W. Wortham, The Use of Exponentially Smoothed Data in Continuous Process Control, *Int. J. Prod. Res.*, Vol 10 (No. 4), 1972, p 393-400
17. A.F. Bissell, *An Introduction to CuSum Charts*, The Institute of Statisticians, 1984
18. "Guide To Data Analysis and Quality Control Using CuSum Techniques," BS5703 (4 parts), British Standards Institution, 1980-1982
19. J.M. Lucas, The Design and Use of V-Mask Control Scheme, *J. Qual. Technol.*, Vol 8 (No. 1), 1976, p 1-12
20. J. Murdoch, *Control Charts*, Macmillan, 1979
21. J.S. Oakland, *Statistical Process Control*, William Heinemann, 1986
22. I. Burr, *Statistical Quality Control Methods*, Marcel Dekker, 1976
23. S.M. Wu, Analysis of Rail Steel Bar Welds By Two-Level Factorial Design, *Weld. J. Research Supplement*, April 1964
24. R.E. DeVor and T.H. Chang, Quality and Productivity Design and Improvement: Module 2, in *Statistical Process Control*, Ford Motor Company, Plastic Products Division, 1988
25. R.E. DeVor and T.H. Chang, Statistical Methods for Quality and Productivity Design and Improvement, in *The Tool and Manufacturing Engineer's Handbook*, Vol 4, Society of Manufacturing Engineers, 1987
26. D.J. Finney, The Fractional Replication of Factorial Arrangements, *Ann. Eugen.*, Vol 12, 1945, p 291-301
27. C.R. Rao, Factorial Experiments Derivable From Combinatorial Arrangements of Arrays, *J. Royal Stat. Soc.*, Vol B9, 1947, p 128-140
28. L.H.C. Tippett, *Applications of Statistical Methods to the Control of Quality in Industrial Production*, Manchester Statistical Society, 1934
29. R.L. Plackett and J.P. Burman, Design of Optimal Multifactorial Experiments, *Biometrika*, Vol 23, 1946, p 305-325
30. H.L. Oh, Variation Tolerant Design, in *ASME Proceedings on Quality, Design, Planning, and Control*, American Society of Mechanical Engineers, 1987, p 137-145

Metric Conversion Guide

This Section is intended as a guide for expressing weights and measures in the Système International d'Unités (SI). The purpose of SI units, developed and maintained by the General Conference of Weights and Measures, is to provide a basis for worldwide standardization of units and measure. For more information on metric conversions, the reader should consult the following references:

- "Standard for Metric Practice," E 380, *Annual Book of ASTM Standards*, American Society for Testing and Materials, 1916 Race Street, Philadelphia, PA 19103
- "Metric Practice," ANSI/IEEE 268–1982, American National Standards Institute, 1430 Broadway, New York, NY 10018
- *The International System of Units*, SP 330, 1986, National Institute of Standards and Technology. Order from Superintendent of Documents, U.S. Government Printing Office, Washington, DC 20402-9325
- *Metric Editorial Guide*, 4th ed. (revised), 1985, American National Metric Council, 1010 Vermont Avenue NW, Suite 1000, Washington, DC 20005–4960
- *ASME Orientation and Guide for Use of SI (Metric) Units*, ASME Guide SI 1, 9th ed., 1982, The American Society of Mechanical Engineers, 345 East 47th Street, New York, NY 10017

Base, supplementary, and derived SI units

Measure	Unit	Symbol
Base units		
Amount of substance	mole	mol
Electric current	ampere	A
Length	meter	m
Luminous intensity	candela	cd
Mass	kilogram	kg
Thermodynamic temperature	kelvin	K
Time	second	s
Supplementary units		
Plane angle	radian	rad
Solid angle	steradian	sr
Derived units		
Absorbed dose	gray	Gy
Acceleration	meter per second squared	m/s^2
Activity (of radionuclides)	becquerel	Bq
Angular acceleration	radian per second squared	rad/s^2
Angular velocity	radian per second	rad/s
Area	square meter	m^2
Capacitance	farad	F
Concentration (of amount of substance)	mole per cubic meter	mol/m^3
Conductance	siemens	S
Current density	ampere per square meter	A/m^2
Density, mass	kilogram per cubic meter	kg/m^3
Electric charge density	coulomb per cubic meter	C/m^3
Electric field strength	volt per meter	V/m
Electric flux density	coulomb per square meter	C/m^2
Electric potential, potential difference, electromotive force	volt	V
Electric resistance	ohm	Ω
Energy, work, quantity of heat	joule	J
Energy density	joule per cubic meter	J/m^3
Entropy	joule per kelvin	J/K
Force	newton	N
Frequency	hertz	Hz
Heat capacity	joule per kelvin	J/K
Heat flux density	watt per square meter	W/m^2
Illuminance	lux	lx
Inductance	henry	H
Irradiance	watt per square meter	W/m^2
Luminance	candela per square meter	cd/m^2
Luminous flux	lumen	lm
Magnetic field strength	ampere per meter	A/m
Magnetic flux	weber	Wb
Magnetic flux density	tesla	T
Molar energy	joule per mole	J/mol
Molar entropy	joule per mole kelvin	$J/mol \cdot K$
Molar heat capacity	joule per mole kelvin	$J/mol \cdot K$
Moment of force	newton meter	$N \cdot m$
Permeability	henry per meter	H/m
Permittivity	farad per meter	F/m
Power, radiant flux	watt	W
Pressure, stress	pascal	Pa
Quantity of electricity, electric charge	coulomb	C
Radiance	watt per square meter steradian	$W/m^2 \cdot sr$
Radiant intensity	watt per steradian	W/sr
Specific heat capacity	joule per kilogram kelvin	$J/kg \cdot K$
Specific energy	joule per kilogram	J/kg
Specific entropy	joule per kilogram kelvin	$J/kg \cdot K$
Specific volume	cubic meter per kilogram	m^3/kg
Surface tension	newton per meter	N/m
Thermal conductivity	watt per meter kelvin	$W/m \cdot K$
Velocity	meter per second	m/s
Viscosity, dynamic	pascal second	$Pa \cdot s$
Viscosity, kinematic	square meter per second	m^2/s
Volume	cubic meter	m^3
Wavenumber	1 per meter	1/m

Conversion factors

To convert from	to	multiply by
Angle		
degree	rad	1.745 329 E − 02
Area		
in.^{2}	mm^2	6.451 600 E + 02
in.^{2}	cm^2	6.451 600 E + 00
in.^{2}	m^2	6.451 600 E − 04
ft^2	m^2	9.290 304 E − 02
Bending moment or torque		
lbf · in.	N · m	1.129 848 E − 01
lbf · ft	N · m	1.355 818 E + 00
kgf · m	N · m	9.806 650 E + 00
ozf · in.	N · m	7.061 552 E − 03
Bending moment or torque per unit length		
lbf · in./in.	N · m/m	4.448 222 E + 00
lbf · ft/in.	N · m/m	5.337 866 E + 01
Current density		
A/in.^{2}	A/cm^2	1.550 003 E − 01
A/in.^{2}	A/mm^2	1.550 003 E − 03
A/ft^2	A/m^2	1.076 400 E + 01
Electricity and magnetism		
gauss	T	1.000 000 E − 04
maxwell	μWb	1.000 000 E − 02
mho	S	1.000 000 E + 00
Oersted	A/m	7.957 700 E + 01
Ω · cm	Ω · m	1.000 000 E − 02
Ω circular-mil/ft	μΩ · m	1.662 426 E − 03
Energy (impact, other)		
ft · lbf	J	1.355 818 E + 00
Btu (thermochemical)	J	1.054 350 E + 03
cal (thermochemical)	J	4.184 000 E + 00
kW · h	J	3.600 000 E + 06
W · h	J	3.600 000 E + 03
Flow rate		
ft^3/h	L/min	4.719 475 E − 01
ft^3/min	L/min	2.831 000 E + 01
gal./h	L/min	6.309 020 E − 02
gal./min	L/min	3.785 412 E + 00
Force		
lbf	N	4.448 222 E + 00
kip (1000 lbf)	N	4.448 222 E + 03
tonf	kN	8.896 443 E + 00
kgf	N	9.806 650 E + 00
Force per unit length		
lbf/ft	N/m	1.459 390 E + 01
lbf/in.	N/m	1.751 268 E + 02
Fracture toughness		
ksi $\sqrt{\text{in.}}$	MPa $\sqrt{\text{m}}$	1.098 800 E + 00
Heat content		
Btu/lb	kJ/kg	2.326 000 E + 00
cal/g	kJ/kg	4.186 800 E + 00
Heat input		
J/in.	J/m	3.937 008 E + 01
kJ/in.	kJ/m	3.937 008 E + 01
Length		
Å	nm	1.000 000 E − 01
μin.	μm	2.540 000 E − 02
mil	μm	2.540 000 E + 01
in.	mm	2.540 000 E + 01
in.	cm	2.540 000 E + 00
ft	m	3.048 000 E − 01
yd	m	9.144 000 E − 01
mile	km	1.609 300 E + 00
Mass		
oz	kg	2.834 952 E − 02
lb	kg	4.535 924 E − 01
ton (short, 2000 lb)	kg	9.071 847 E + 02
ton (short, 2000 lb)	kg × 10^3(a)	9.071 847 E − 01
ton (long, 2240 lb)	kg	1.016 047 E + 03
Mass per unit area		
oz/in.^{2}	kg/m^2	4.395 000 E + 01
oz/ft^2	kg/m^2	3.051 517 E − 01
oz/yd^2	kg/m^2	3.390 575 E − 02
lb/ft^2	kg/m^2	4.882 428 E + 00
Mass per unit length		
lb/ft	kg/m	1.488 164 E + 00
lb/in.	kg/m	1.785 797 E + 01
Mass per unit time		
lb/h	kg/s	1.259 979 E − 04
lb/min	kg/s	7.559 873 E − 03
lb/s	kg/s	4.535 924 E − 01
Mass per unit volume (includes density)		
g/cm^3	kg/m^3	1.000 000 E + 03
lb/ft^3	g/cm^3	1.601 846 E − 02
lb/ft^3	kg/m^3	1.601 846 E + 01
lb/in.^{3}	g/cm^3	2.767 990 E + 01
lb/in.^{3}	kg/m^3	2.767 990 E + 04
Power		
Btu/s	kW	1.055 056 E + 00
Btu/min	kW	1.758 426 E − 02
Btu/h	W	2.928 751 E − 01
erg/s	W	1.000 000 E − 07
ft · lbf/s	W	1.355 818 E + 00
ft · lbf/min	W	2.259 697 E − 02
ft · lbf/h	W	3.766 161 E − 04
hp (550 ft · lbf/s)	kW	7.456 999 E − 01
hp (electric)	kW	7.460 000 E − 01
Power density		
W/in.^{2}	W/m^2	1.550 003 E + 03
Press capacity		
See Force		
Pressure (fluid)		
atm (standard)	Pa	1.013 250 E + 05
bar	Pa	1.000 000 E + 05
in. Hg (32 °F)	Pa	3.386 380 E + 03
in. Hg (60 °F)	Pa	3.376 850 E + 03
lbf/in.^{2} (psi)	Pa	6.894 757 E + 03
torr (mm Hg, 0 °C)	Pa	1.333 220 E + 02
Specific heat		
Btu/lb · °F	J/kg · K	4.186 800 E + 03
cal/g · °C	J/kg · K	4.186 800 E + 03
Stress (force per unit area)		
tonf/in.^{2} (tsi)	MPa	1.378 951 E + 01
kgf/mm^2	MPa	9.806 650 E + 00
ksi	MPa	6.894 757 E + 00
lbf/in.^{2} (psi)	MPa	6.894 757 E − 03
MN/m^2	MPa	1.000 000 E + 00
Temperature		
°F	°C	5/9 · (°F − 32)
°R	°K	5/9
Temperature interval		
°F	°C	5/9
Thermal conductivity		
Btu · in./s · ft^2 · °F	W/m · K	5.192 204 E + 02
Btu/ft · h · °F	W/m · K	1.730 735 E + 00
Btu · in./h · ft^2 · °F	W/m · K	1.442 279 E − 01
cal/cm · s · °C	W/m · K	4.184 000 E + 02
Thermal expansion		
in./in. · °C	m/m · K	1.000 000 E + 00
in./in. · °F	m/m · K	1.800 000 E + 00
Velocity		
ft/h	m/s	8.466 667 E − 05
ft/min	m/s	5.080 000 E − 03
ft/s	m/s	3.048 000 E − 01
in./s	m/s	2.540 000 E − 02
km/h	m/s	2.777 778 E − 01
mph	km/h	1.609 344 E + 00
Velocity of rotation		
rev/min (rpm)	rad/s	1.047 164 E − 01
rev/s	rad/s	6.283 185 E + 00
Viscosity		
poise	Pa · s	1.000 000 E − 01
stokes	m^2/s	1.000 000 E − 04
ft^2/s	m^2/s	9.290 304 E − 02
in.^{2}/s	mm^2/s	6.451 600 E + 02
Volume		
in.^{3}	m^3	1.638 706 E − 05
ft^3	m^3	2.831 685 E − 02
fluid oz	m^3	2.957 353 E − 05
gal. (U.S. liquid)	m^3	3.785 412 E − 03
Volume per unit time		
ft^3/min	m^3/s	4.719 474 E − 04
ft^3/s	m^3/s	2.831 685 E − 02
in.^{3}/min	m^3/s	2.731 177 E − 07
Wavelength		
Å	nm	1.000 000 E − 01

(a) kg × 10^3 = 1 metric ton or 1 megagram (Mg)

SI prefixes—names and symbols

Exponential expression	Multiplication factor	Prefix	Symbol
10^{18}	1 000 000 000 000 000 000	exa	E
10^{15}	1 000 000 000 000 000	peta	P
10^{12}	1 000 000 000 000	tera	T
10^{9}	1 000 000 000	giga	G
10^{6}	1 000 000	mega	M
10^{3}	1 000	kilo	k
10^{2}	100	hecto(a)	h
10^{1}	10	deka(a)	da
10^{0}	1	BASE UNIT	
10^{-1}	0.1	deci(a)	d
10^{-2}	0.01	centi(a)	c
10^{-3}	0.001	milli	m
10^{-6}	0.000 001	micro	μ
10^{-9}	0.000 000 001	nano	n
10^{-12}	0.000 000 000 001	pico	p
10^{-15}	0.000 000 000 000 001	femto	f
10^{-18}	0.000 000 000 000 000 001	atto	a

(a) Nonpreferred. Prefixes should be selected in steps of 10^3 so that the resultant number before the prefix is between 0.1 and 1000. These prefixes should not be used for units of linear measurement, but may be used for higher order units. For example, the linear measurement, decimeter, is nonpreferred, but square decimeter is acceptable.

Abbreviations and Symbols

a crack length; flaw size; depth; level of significance
A ampere
A amplitude; area
Å angstrom
ac alternating current
A/D analog-to-digital converter
AE acoustic emission
AFS American Foundrymen's Society
AGC automatic gain control
AIA Aerospace Industries Association of America
AIPD acoustic impedance polarity detector
AISI American Iron and Steel Institute
AMS Aerospace Material Specifications
API American Petroleum Institute
ASIP Aircraft Structural Integrity Program
ASTM American Society for Testing and Materials
atm standard atmosphere
B bar size; magnetic flux density
B magnetic flux vector
bcc body-centered cubic
bcd binary coded decimal
BIE boundary integral equation
Bq becquerel
BWO backward wave oscillator
c speed of light; specific heat
C coulomb
CAD/CAM computer-aided design/computer-aided manufacturing
CARP Committee for Acoustic Emission in Reinforced Plastics
CCD charge-coupled device
cd candela
CFA crossed-field amplifier
CFRP carbon fiber-reinforced plastic
Ci curie
CID charge-injected device
CIE Commission Internationale de l'Eclairage (International Commission on Illumination)
CL control limit; confidence level
CMM coordinate measuring machine
CMS coordinate measuring system
COD crack opening displacement
CRT cathode ray tube
C-SAM C-mode scanning acoustic microscopy
cSt centiStokes
CT computed tomography
CuSum cumulative sum
CW continuous wave
d deuteron
d used in mathematical expressions involving a derivative (denotes rate of change); depth; diameter
D diameter; duration; dislocation density
da/dN fatigue crack growth rate
DAC digital-to-analog converter
dB decibel
dBae decibel of acoustic emission amplitudes
dc direct current
DCC direct computer control
diam diameter
DIN Deutsche Industrie-Normen (German Industrial Standards)
DIP dual-in-line package
DLM digital light meter
dpi dots per inch
DR digital radiography; digital radiograph
e natural logarithm base, 2.71828; random error
E modulus of elasticity; electric field strength; emissive power; electromotive force; energy counts (acoustic emission); applied voltage
EC eddy current
ECP electric current perturbation
EDM electrical discharge machining
EMA electromagnetic acoustic (transducer)
ENSIP Engine Structural Integrity Program
EPA Environmental Protection Agency
Eq equation
et al. and others
ET eddy current testing
eV electron volt
EWMA exponentially weighted moving average
f frequency; transfer function
F eddy current inspection test frequency
FERPIC ferroelectric photoconductor image camera
FET field-effect transistor
FFT fast Fourier transform
FID flame ionization detector
Fig. figure
fm femtometer
FM frequency modulation
FMR ferromagnetic resonance
FR felicity ratio
FRP fiber-reinforced plastic
ft foot
ftc footcandle
ft-L footlambert
g gram
G gauss
G_D film gradient
GBq gigabecquerel
GPa gigapascal
GTO gate turnoff
Gy gray (unit of absorbed radiation)
h convection heat transfer coefficient
h hour
H magnetic field strength
H magnetic field vector
HAZ heat-affected zone
HB Brinell hardness
HIP hot isostatic pressing
HNDT holographic nondestructive test
HR Rockwell hardness (requires scale designation, such as HRC for Rockwell C hardness)
HSL hue, saturation, luminance
i $\sqrt{-1}$ for a complex variable; a summation index
i unit vector along the *x*-axis
I intensity; electrical current
$\langle I \rangle$ average intensity
IACS International Annealed Copper Standard
IBIS integrated blade inspection system
IC integrated circuit
ID inside diameter
IF intermediate frequency
IGA intergranular attack
IMPATT impact avalanche transit time

in. inch
IQI image quality indicator
IR infrared
j $\sqrt{-1}$ for a complex variable; a summation index
j unit vector along the y-axis
J joule
k thermal conductivity; wave number
k unit vector along the z-axis
K stress intensity factor; kernel
K_c plane-stress fracture toughness
K_{Ic} plane-strain fracture toughness
K_{Iscc} threshold stress intensity for stress-corrosion cracking
kg kilogram
km kilometer
kPa kilopascal
ksi 1000 lb (1 kip) per square inch
kV kilovolt
kVp kilovolt peak
l average rate of occurrence of defects per sample; length
L length; likelihood (function)
lb pound
LCC life cycle cost
LCF low cycle fatigue
LCL lower control limit; lower confidence limit
LED light-emitting diode
LEL lower explosive limit
LINAC linear accelerator
LLW leaky Lamb wave
lm lumen
ln natural logarithm (base e)
LNG liquified natural gas
LO local oscillator
LPF low-pass filter
lpm line pairs per millimeter
LSP laser speckle photography
LUT look-up table
lx lux
m meter
m maximum number of factors, including selected interactions, that the design can study
M_a molecular weight of air
M_t molecular weight of tracer gas
MARSE measured area under the rectified signal envelope
MDL minimum detectable leakage
MFD magnetic field disturbance
mg milligram
Mg megagram (metric tonne)
MIG metal inert gas (welding)
MIL-STD military standard
MIVC magnetically induced velocity change
min minimum; minute
mL milliliter
MLCD microwave liquid crystal display
mm millimeter
MOOT method of optimal truncation
MPa megapascal
mpg miles per gallon
mrem millirem
mSv millisievert
MTF modulation transfer function
mV millivolt
MVp megavolt peak
n neutrons
N newton
n index of refraction; fringe number; number of tests
$\bar{n}$ average sample size
n unit vector normal to the surface
N number of cycles; summation index; fringe order; length of the near field
N_K Knudsen number
N_{Re} Reynolds number
NBS National Bureau of Standards
NDE nondestructive evaluation
NDI nondestructive inspection
NDT nondestructive testing
Nd: YAG neodymium: yttrium-aluminum-garnet (laser)
nm nanometer
NMR nuclear magnetic resonance
No. number
NTSC National Television Systems Committee
OD outside diameter
Oe oersted
OSHA Occupational Safety and Health Administration
oz ounce
p probability
P pressure
Pa pascal
PC printed circuit
PCB printed circuit board
pH negative logarithm of hydrogen ion activity
PH precipitation hardening
PIND particle impact noise detection
PLC programmable logic control
P/M powder metallurgy
PMT photomultiplier tube
POD probability of detection
POFA probability of false alarm
ppm parts per million
ppt parts per trillion
psi pounds per square inch
psig gage pressure (pressure relative to ambient pressure) in pounds per square inch
Q calibrated leakage flow
r number of levels of each factor; original pixel value; radius; range
R roentgen
R resistance; reflection coefficient; rise time; gas constant; radius
R_a arithmetic average roughness
R_s effective shunt resistance
rad radiation absorbed dose
radac rapid digital automatic computing
rbe relative biological effectiveness
ref reference
rem roentgen equivalent man; remainder
RF radio frequency
RFEC remote-field eddy current
RGB red, green, blue
RHM roentgen per hour at one meter
RMM roentgen per minute at one meter
rms root mean square
ROC relative operating characteristic
RSSS radiographic standard shooting sketch
RTR real-time radiography
s second
S siemens
S sensitivity; speed; spacing
S signal input
SAE Society of Automotive Engineers
SAM scanning acoustic microscopy
SAT structural assessment testing
SDD source-detector distance
SDP standard depth of penetration
SEC secondary electron-coupled (vidicon)
SEM scanning electron microscopy
sfm surface feet per minute
SiIMPATT silicon impact avalanche transit time
SIT silicon intensified target
SLAM scanning laser acoustic microscopy
SLR single lens reflex
S/N signal-to-noise (ratio)
SOD source-object distance
SOF shape and orientation factor of the flaw
SPC statistical process control
SPDT single pole double throw
Sv sievert
std standard
t time; thickness
T tesla
T temperature; transmission coefficient
TAB tape-automated bonding
TE transverse electric (mode)
TEM transmission electron microscopy

TEM transverse electromagnetic (wave)
TM transverse magnetic (mode)
TIG tungsten inert gas (welding)
TOF time of flight
tsi tons per square inch
U_g geometric unsharpness
UCL upper control limit
ULCE unified life cycle engineering
USAF United States Air Force
UST underground storage tank
UT ultrasonic testing
ν degree of freedom; velocity
V volume; velocity; variance
VHS very high speed
VHSIC very high speed integrated circuit
VLSI very large scale integration
vol volume
W watt
WORM write once read many
$\bar{x}$ sample mean
YIG yttrium iron garnet
yr year
Z impedance; atomic number; standard normal distribution
° angular measure; degree
°C degree Celsius (centigrade)
°F degree Fahrenheit
$\leftrightarrow$ direction of reaction
$\div$ divided by
$=$ equals
$\cong$ approximately equals
$\neq$ not equal to
$\equiv$ identical with
$>$ greater than
$\gg$ much greater than
$\geq$ greater than or equal to
∞ infinity
$\propto$ is proportional to; varies as
$\int$ integral of
$<$ less than
$\ll$ much less than
$\leq$ less than or equal to
$\pm$ maximum deviation
$-$ minus; negative ion charge
$\times$ diameters (magnification); multiplied by
$\cdot$ multiplied by
/ per
% percent
$+$ plus; positive ion charge
$\sqrt{}$ square root of
$\sim$ approximately; similar to
∂ partial derivative
α thermal diffusivity; angle of incidence
β angle of refraction
γ shear strain
δ skin depth
Δ change in quantity; increment; range; phase shift
ϵ normal strain; emissivity; dielectric coefficient
η viscosity
θ angle
λ wavelength; mean free path
μ linear attenuation coefficient; magnetic permeability; the mean (or average) of a distribution
μin. microinch
μm micron (micrometer)
ν Poisson's ratio; frequency
π pi (3.141592)
ρ density; resistivity
σ stress; standard deviation; electrical conductivity
Σ summation of
τ shear stress
ϕ angle of refraction
ω circular frequency (angular velocity)
Ω ohm

Greek Alphabet

Α, α alpha	**Ι, ι** iota	**Ρ, ρ** rho
Β, β beta	**Κ, κ** kappa	**Σ, σ** sigma
Γ, γ gamma	**Λ, λ** lambda	**Τ, τ** tau
Δ, δ delta	**Μ, μ** mu	**Υ, υ** upsilon
Ε, ε epsilon	**Ν, ν** nu	**Φ, φ** phi
Ζ, ζ zeta	**Ξ, ξ** xi	**Χ, χ** chi
Η, η eta	**Ο, ο** omicron	**Ψ, ψ** psi
Θ, θ theta	**Π, π** pi	**Ω, ω** omega

Index

A

B

E

F

J

K

M

N

Q

R

S

U

V